Periodic Table of the Elements

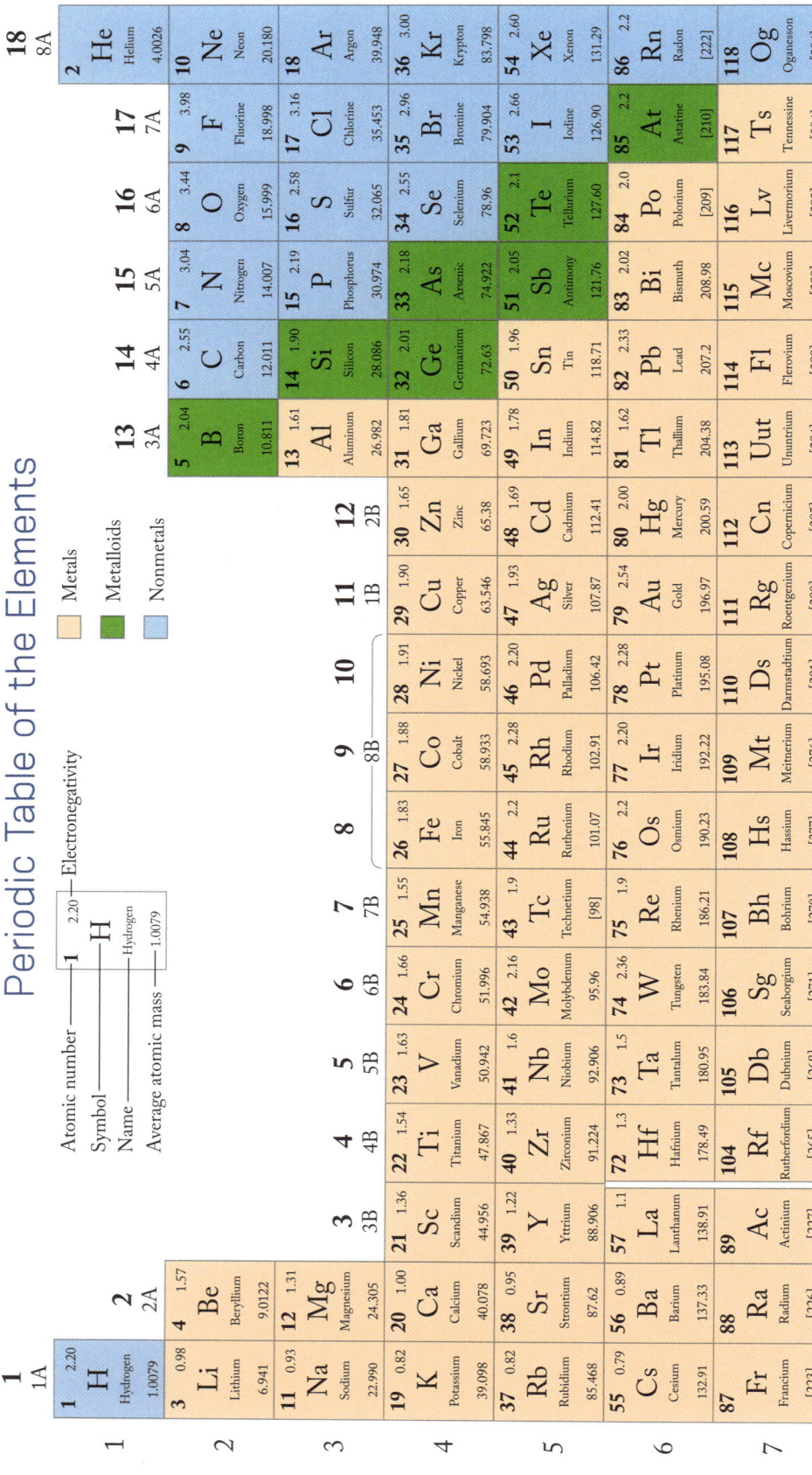

Metals

Metalloids

Nonmetals

Atomic number — **1** 2.20 — Electronegativity
Symbol — **H**
Name — Hydrogen
Average atomic mass — 1.0079

		18 / 8A
		2 He Helium 4.0026

| 1 / 1A | 2 / 2A | 3 / 3B | 4 / 4B | 5 / 5B | 6 / 6B | 7 / 7B | 8 / 8B | 9 / 8B | 10 / 8B | 11 / 1B | 12 / 2B | 13 / 3A | 14 / 4A | 15 / 5A | 16 / 6A | 17 / 7A | 18 / 8A |

1 2.20 H Hydrogen 1.0079

Period 1
Period 2: **3** 0.98 Li Lithium 6.941; **4** 1.57 Be Beryllium 9.0122; **5** 2.04 B Boron 10.811; **6** 2.55 C Carbon 12.011; **7** 3.04 N Nitrogen 14.007; **8** 3.44 O Oxygen 15.999; **9** 3.98 F Fluorine 18.998; **10** Ne Neon 20.180

Period 3: **11** 0.93 Na Sodium 22.990; **12** 1.31 Mg Magnesium 24.305; **13** 1.61 Al Aluminum 26.982; **14** 1.90 Si Silicon 28.086; **15** 2.19 P Phosphorus 30.974; **16** 2.58 S Sulfur 32.065; **17** 3.16 Cl Chlorine 35.453; **18** Ar Argon 39.948

Period 4: **19** 0.82 K Potassium 39.098; **20** 1.00 Ca Calcium 40.078; **21** 1.36 Sc Scandium 44.956; **22** 1.54 Ti Titanium 47.867; **23** 1.63 V Vanadium 50.942; **24** 1.66 Cr Chromium 51.996; **25** 1.55 Mn Manganese 54.938; **26** 1.83 Fe Iron 55.845; **27** 1.88 Co Cobalt 58.933; **28** 1.91 Ni Nickel 58.693; **29** 1.90 Cu Copper 63.546; **30** 1.65 Zn Zinc 65.38; **31** 1.81 Ga Gallium 69.723; **32** 2.01 Ge Germanium 72.63; **33** 2.18 As Arsenic 74.922; **34** 2.55 Se Selenium 78.96; **35** 2.96 Br Bromine 79.904; **36** 3.00 Kr Krypton 83.798

Period 5: **37** 0.82 Rb Rubidium 85.468; **38** 0.95 Sr Strontium 87.62; **39** 1.22 Y Yttrium 88.906; **40** 1.33 Zr Zirconium 91.224; **41** 1.6 Nb Niobium 92.906; **42** 2.16 Mo Molybdenum 95.96; **43** 1.9 Tc Technetium [98]; **44** 2.2 Ru Ruthenium 101.07; **45** 2.28 Rh Rhodium 102.91; **46** 2.20 Pd Palladium 106.42; **47** 1.93 Ag Silver 107.87; **48** 1.69 Cd Cadmium 112.41; **49** 1.78 In Indium 114.82; **50** 1.96 Sn Tin 118.71; **51** 2.05 Sb Antimony 121.76; **52** 2.1 Te Tellurium 127.60; **53** 2.66 I Iodine 126.90; **54** 2.60 Xe Xenon 131.29

Period 6: **55** 0.79 Cs Cesium 132.91; **56** 0.89 Ba Barium 137.33; **57** 1.1 La Lanthanum 138.91; **72** 1.3 Hf Hafnium 178.49; **73** 1.5 Ta Tantalum 180.95; **74** 2.36 W Tungsten 183.84; **75** 1.9 Re Rhenium 186.21; **76** 2.2 Os Osmium 190.23; **77** 2.20 Ir Iridium 192.22; **78** 2.28 Pt Platinum 195.08; **79** 2.54 Au Gold 196.97; **80** 2.00 Hg Mercury 200.59; **81** 1.62 Tl Thallium 204.38; **82** 2.33 Pb Lead 207.2; **83** 2.02 Bi Bismuth 208.98; **84** 2.0 Po Polonium [209]; **85** 2.2 At Astatine [210]; **86** 2.2 Rn Radon [222]

Period 7: **87** 0.7 Fr Francium [223]; **88** 0.9 Ra Radium [226]; **89** 1.1 Ac Actinium [227]; **104** Rf Rutherfordium [265]; **105** Db Dubnium [268]; **106** Sg Seaborgium [271]; **107** Bh Bohrium [270]; **108** Hs Hassium [277]; **109** Mt Meitnerium [276]; **110** Ds Darmstadtium [281]; **111** Rg Roentgenium [280]; **112** Cn Copernicium [285]; **113** Uut Ununtrium [284]; **114** Fl Flerovium [289]; **115** Mc Moscovium [288]; **116** Lv Livermorium [293]; **117** Ts Tennessine [294]; **118** Og Oganesson [294]

6 Lanthanides
58 Ce Cerium 140.12; **59** Pr Praseodymium 140.91; **60** Nd Neodymium 144.24; **61** Pm Promethium [145]; **62** Sm Samarium 150.36; **63** Eu Europium 151.96; **64** Gd Gadolinium 157.25; **65** Tb Terbium 158.93; **66** Dy Dysprosium 162.50; **67** Ho Holmium 164.93; **68** Er Erbium 167.26; **69** Tm Thulium 168.93; **70** Yb Ytterbium 173.05; **71** Lu Lutetium 174.97

7 Actinides
90 Th Thorium 232.04; **91** Pa Protactinium 231.04; **92** U Uranium 238.03; **93** Np Neptunium [237]; **94** Pu Plutonium [244]; **95** Am Americium [243]; **96** Cm Curium [247]; **97** Bk Berkelium [247]; **98** Cf Californium [251]; **99** Es Einsteinium [252]; **100** Fm Fermium [257]; **101** Md Mendelevium [258]; **102** No Nobelium [259]; **103** Lr Lawrencium [262]

The U.S. system as well as the system recommended by the International Union of Pure and Applied Chemistry (IUPAC) has been used to label the groups in this periodic table. The system used in the United States includes a letter and a number (1A, 2A, 3B, 4B, etc.), which is close to the system developed by Mendeleev. The IUPAC system uses numbers 1–18 and has been recommended by the American Chemical Society (ACS). Although both numbering systems are shown here, the U.S. system is used predominantly in the book.

Organic Chemistry
Principles and Mechanisms

SECOND EDITION

Joel M. Karty
Elon University

W. W. NORTON
NEW YORK · LONDON

To Pnut, Fafa, and Jakers

W. W. Norton & Company has been independent since its founding in 1923, when William Warder Norton and Mary D. Herter Norton first published lectures delivered at the People's Institute, the adult education division of New York City's Cooper Union. The firm soon expanded its program beyond the Institute, publishing books by celebrated academics from America and abroad. By midcentury, the two major pillars of Norton's publishing program—trade books and college texts—were firmly established. In the 1950s, the Norton family transferred control of the company to its employees, and today—with a staff of four hundred and a comparable number of trade, college, and professional titles published each year—W. W. Norton & Company stands as the largest and oldest publishing house owned wholly by its employees.

Editor: Erik Fahlgren
Associate Managing Editor, College: Carla L. Talmadge
Editorial Assistant: Sara Bonacum
Managing Editor, College: Marian Johnson
Managing Editor, College Digital Media: Kim Yi
Production Manager: Eric Pier-Hocking
Media Editor: Chris Rapp
Associate Media Editor: Arielle Holstein
Media Project Editor: Jesse Newkirk
Assistant Media Editor: Doris Chiu
Ebook Production Manager: Mateus Manço Teixeira
Ebook Production Coordinator: Lizz Thabet
Marketing Manager, Chemistry: Stacy Loyal
Design Director: Jillian Burr
Photo Editor: Travis Carr
Permissions Manager: Megan Schindel
Composition: Graphic World
Illustrations: Imagineering
Manufacturing: Transcontinental

Permission to use copyrighted material is included at the back of the book.

Library of Congress Cataloging-in-Publication Data

Names: Karty, Joel, author.
Title: Organic chemistry : principles and mechanisms / Joel M. Karty, Elon
   University.
Description: Second edition. | New York : W.W. Norton & Company, [2018] |
   Includes index.
Identifiers: LCCN 2017042262 | ISBN 9780393630756 (hardcover)
Subjects: LCSH: Chemistry, Organic—Textbooks.
Classification: LCC QD253.2 .K375 2018 | DDC 547—c23 LC record available at https://lccn.loc.gov/2017042

W. W. Norton & Company, Inc., 500 Fifth Avenue, New York, NY 10110

wwnorton.com

W. W. Norton & Company Ltd., 15 Carlisle Street, London W1D 3BS

1 2 3 4 5 6 7 8 9 0

About the Author

JOEL KARTY earned his B.S. in chemistry at the University of Puget Sound and his Ph.D. at Stanford University. He joined the faculty of Elon University in 2001, where he currently holds the rank of full professor. He teaches primarily the organic chemistry sequence and also teaches general chemistry. In the summer, Joel teaches at the Summer Biomedical Sciences Institute through the Duke University Medical Center. His research interests include investigating the roles of resonance and inductive effects in fundamental chemical systems and studying the mechanism of pattern formation in Liesegang reactions. He has written a very successful student supplement, *Get Ready for Organic Chemistry*, Second Edition (formerly called *The Nuts and Bolts of Organic Chemistry*).

Brief Contents

Contents

1 Atomic and Molecular Structure   1

> **THE ORGANIC CHEMISTRY OF BIOMOLECULES**
>

INTERCHAPTER

A Nomenclature: The Basic System for Naming Organic Compounds

Alkanes, Haloalkanes, Nitroalkanes, Cycloalkanes, and Ethers   52

INTERCHAPTER

Naming Alkenes, Alkynes, and Benzene Derivatives 152

4 Isomerism 1
Conformers and Constitutional Isomers 165

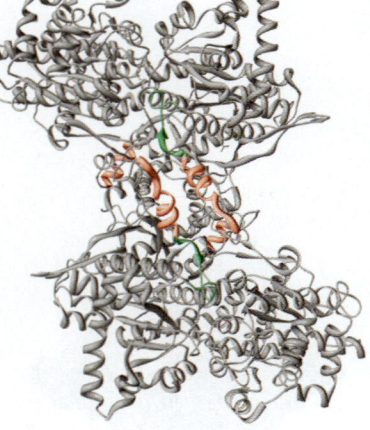

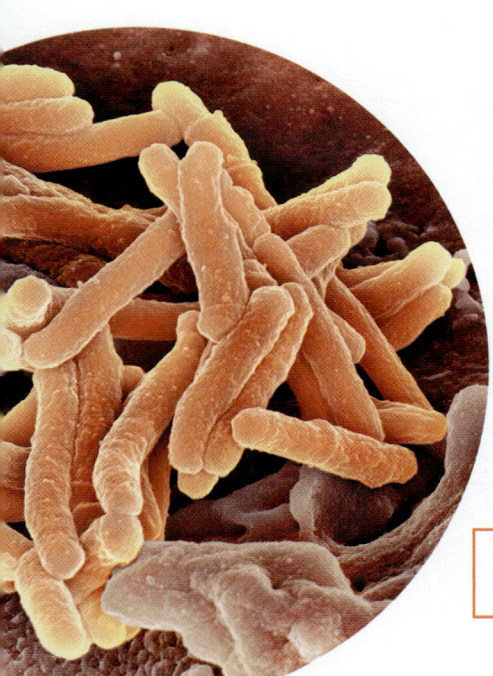

11 Electrophilic Addition to Nonpolar π Bonds 1
Addition of a Brønsted Acid   563

12 Electrophilic Addition to Nonpolar π Bonds 2
Reactions Involving Cyclic Transition States   601

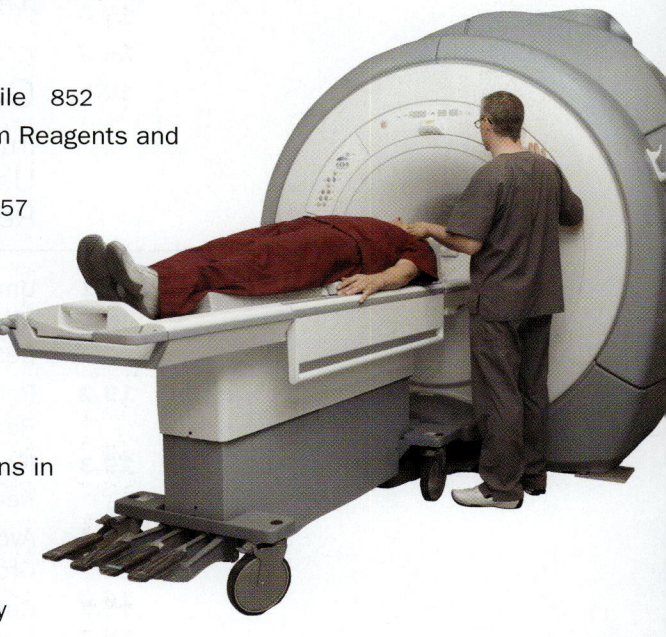

19   Organic Synthesis 2

Intermediate Topics in Synthesis Design, and Useful Redox and Carbon–Carbon Bond-Forming Reactions   946

20   Nucleophilic Addition–Elimination Reactions 1

The General Mechanism Involving Strong Nucleophiles   1000

25 Reactions Involving Free Radicals 1247

INTERCHAPTER

 # G Fragmentation Pathways in Mass Spectrometry 1295

26 Polymers 1307

Biochemistry Topics

Interest Boxes

Connections Boxes

Green Chemistry Boxes

Mechanisms

Preface

Focused on the Student, Organized by Mechanism

When an organic reaction is presented to a novice, only the structural differences between the reactants and products are immediately apparent. Students tend to see only *what* happens, such as the transformation of one functional group into another, changes in connectivity, and aspects of stereochemistry. It should therefore not be surprising that students, when presented reactions, are tempted to commit the reactions to memory. But there are far too many reactions and accompanying details for memorization to work in organic chemistry.

This is where mechanisms come into play. Mechanisms allow us to understand the sequences of elementary steps—the step-by-step pathways—that convert the reactants to products, so we can see *how* and *why* reactions take place as they do. Moreover, the mechanisms that describe the large number of reactions in the course are constructed from just a handful of elementary steps, so mechanisms allow us to see *similarities* among reactions that are not otherwise apparent. In other words, mechanisms actually *simplify* organic chemistry. Thus, teaching students mechanisms—enabling students to understand and simplify organic chemistry—is an enormous key to success in the course.

At the outset of my teaching career, I fully appreciated the importance of mechanisms, so during my first couple years of teaching, I emphasized mechanisms *very* heavily. I did so under a functional group organization where reactions are pulled together according to the functional groups that react. That is the organization under which I learned organic chemistry, and it is also the way that most organic chemistry textbooks are organized. Despite my best efforts, the majority of my students struggled with even the basics of mechanisms and, consequently, turned to flash cards as their primary study tool. They tried to memorize their way through the course, which made matters worse.

I began to wonder what impact the *organization*—an organization according to functional group—had on deterring my students from mechanisms. I had good reason to be concerned because, as I alluded to earlier, functional groups tend to convey *what*, whereas mechanisms convey *how* and *why*. What kinds of mixed messages were my students receiving when I was heavily emphasizing mechanisms, while the organization of the material was giving priority to functional groups? To probe that question, I made a big change to my teaching.

The third year I taught organic chemistry, I rearranged the material to pull together reactions that had the same or similar mechanisms—that is, I taught under a *mechanistic organization*. I made no other changes that year; the course content, course structure, and my teaching style all remained the same. I even taught out of the same textbook. But that year I saw dramatic improvements in my students' mastery of mechanisms.[1] Students had *control* over the material, which proved to be a tremendous motivator. They were better able to solve different kinds of problems with confidence. Ultimately, I saw significant

---

[1]Bowman, B. G.; Karty, J. M.; Gooch, G. Teaching a Modified Hendrickson, Cram and Hammond Curriculum in Organic Chemistry. *J. Chem. Educ.* **2007**, *84*, 1209.

improvements in student performance, morale, and retention. I was convinced that students benefit remarkably from learning under a mechanistic organization.

My goal in writing this book is to support instructors who are seeking what I was seeking: getting students to use mechanisms to learn organic chemistry in order to achieve better performances and to have better experiences in their organic courses. Using a functional group organization to achieve these outcomes can be an uphill battle because of the high priority that it inherently places on functional groups. This textbook, on the other hand, allows students to receive the same message from both their instructor and their textbook—a clear and consistent message that mechanisms are vital to success in the course.

A Closer Look: Why Is a Mechanistic Organization Better?

Consider what the novice sees when they begin a new *functional group chapter*. In an alcohols chapter, for example, students first learn how to recognize and name alcohols, then they study the physical properties of alcohols. Next, students might spend time on special spectroscopic characteristics of alcohols, after which they learn various routes that can be used to synthesize alcohols from other species. Finally, students move into the heart of the chapter: new reactions that alcohols undergo and the mechanisms that describe them. Within a particular functional group chapter, students find themselves bouncing among *several themes*.

Even within the discussion of new reactions and mechanisms that a particular functional group can undergo, students are typically faced with widely varying reaction types and mechanisms. Take again the example of alcohols. Students learn that alcohols can act as an acid or as a base; alcohols can act as nucleophiles to attack a saturated carbon in a substitution reaction, or to attack the carbon atom of a polar π bond in a nucleophilic addition reaction; protonated alcohols can act as electrophiles in an elimination reaction; and alcohols can undergo oxidation, too.

With the substantial jumping around that takes place within a particular functional group chapter, it is easy to see how students can become overwhelmed. Under a functional group organization, students don't receive intrinsic and clear guidance as to what they should focus on, not only within a particular functional group chapter, but also from one chapter to the next. Without clear guidance, and without substantial time for focus, students often see no choice but to memorize. And they will memorize what they perceive to be most important—predicting products of reactions, typically ignoring, or giving short shrift to, fundamental concepts and mechanisms.

Under the mechanistic organization in this book, students experience a *coherent story* of chemical reactivity. The story begins with molecular structure and energetics, and then guides students into reaction mechanisms through a few transitional chapters. Thereafter, students study how and why reactions take place as they do, focusing on one type of mechanism at a time. Ultimately, students learn how to intuitively use reactions in synthesis. In this manner, students have clear and consistent guidance as to what their focus should be on, both within a single chapter and throughout the entire book.

The *patterns* we, as experts, see become clear to students when they learn under this mechanistic organization. Consider the following four mechanisms:

(P-1)

(P-2)

(P-3)

(P-4)

The mechanism in Equation P-1 is for a Williamson synthesis of an ether; the one in Equation P-2 is for an alkylation of a terminal alkyne; the one in Equation P-3 is for an alkylation of a ketone; and the one in Equation P-4 is for the conversion of a carboxylic acid to a methyl ester. In these four reactions, the reactants are an alcohol, an alkyne, a ketone, and a carboxylic acid. In a functional group organization, these reactions will be taught in *four separate chapters*. Because all four reaction mechanisms are identical—a deprotonation followed by an S_N2 step—all four reactions are taught in the *same* chapter in this book: Chapter 10.

Seeing these patterns early, students more naturally embrace mechanisms and use them when solving problems. Moreover, as students begin to see such patterns unfold in one chapter, they develop a better toolbox of mechanisms to draw on in subsequent chapters. Ultimately, students gain *confidence* in using mechanisms to predict what will happen and why. I believe this is vital to their success throughout the course and later on admission exams such as the MCAT.

Details about the Organization

Continuing with the success of the first edition, the book remains divided into three major parts:

Part I: Atomic and molecular structure

- Chapter 1: Atomic structure, Lewis structures and the covalent bond, and resonance theory, culminating in an introduction to functional groups
- Chapter 2: Aspects of three-dimensional geometry and its impacts on intermolecular forces
- Chapter 3: Structure in terms of hybridization and molecular orbital (MO) theory
- Chapters 4 and 5: Isomerism in its entirety, including constitutional isomerism, conformational isomerism, and stereoisomerism

Much of the material in Chapters 1–5 will be new to students, such as organic functional groups, protic and aprotic solvents, effective electronegativity, conformers and cyclohexane chair structures, and stereoisomers. Chapters 1–5 also contain a significant amount of material that students will recognize from general chemistry, such as electronic configurations, Lewis structures and resonance, intermolecular forces, VSEPR theory and hybridization, and constitutional isomers. Because most students do not retain everything they should from general chemistry, I have made the general chemistry topics in this textbook more extensive than in other textbooks. Knowing that this extended coverage is in the book, instructors should feel comfortable covering as much or as little of it as they see fit for their students.

Part II: Developing a toolbox for working with mechanisms

- Chapters 6 and 7: Ten common elementary steps of mechanisms
- Chapter 8: Beginnings of multistep mechanisms using S_N1 and E1 reactions as examples

Mechanisms are vital to succeeding in organic chemistry, but before tackling mechanisms, students must have the proper tools. Chapters 6–8 give students those tools, dealing with aspects of elementary steps in Chapters 6 and 7 before dealing with aspects of multistep mechanisms in Chapter 8. Therefore, the chapters in Part II act a transition from Part I to Part III, which deals more intently with reactions.

Chapter 7 is a particularly important part of this transition. Students learn how to work with elementary steps in Chapter 7 in a low-risk environment, where there are no demands to predict products. Thus, there is no pressure to memorize overall reactions. Furthermore, the fact that Chapter 7 brings together the 10 most common elementary steps—making up the mechanisms of the many hundreds of reactions students will encounter through Chapter 23—sends a strong message to students that mechanisms *simplify* organic chemistry. In turn, students take to heart from the outset that mechanisms are worthwhile to learn.

Part III: Major reaction types

- Chapters 9 and 10: Nucleophilic substitution and elimination
- Chapters 11 and 12: Electrophilic addition
- Chapters 17 and 18: Nucleophilic addition
- Chapters 20 and 21: Nucleophilic addition–elimination
- Chapters 22 and 23: Aromatic substitution
- Chapter 24: Diels–Alder reactions and other pericyclic reactions
- Chapter 25: Radical reactions
- Chapter 26: Polymerization

Several of these chapters come in pairs, where the first chapter is used to introduce key ideas about the reaction or mechanism and the second chapter explores the reaction or mechanism to greater depth and breadth.

Pairing the chapters this way provides flexibility. An instructor could teach all of the chapters in order. Alternatively, following the guidelines set by the American Chemical Society, an instructor could teach the first of each paired chapter in the first semester as part of "foundational" coursework. Then, the remaining chapters would represent "in-depth" coursework for the second semester. Teaching the chapters in this order would also allow an instructor to teach carbonyl chemistry in the first semester.

Interspersed in Part III are chapters dealing with multistep synthesis (Chapters 13 and 19), conjugation and aromaticity (Chapter 14), and spectroscopy (Chapters 15 and 16). The spectroscopy chapters are self-contained and can be taught earlier, at the instructor's discretion. They can even be taught separately in the laboratory. The spectroscopy chapters are movable like this because, with the mechanistic organization of the book, important aspects of spectroscopy are not integrated in reaction chapters like they typically are in a functional group text.

The two chapters devoted to multistep synthesis (Chapters 13 and 19), on the other hand, are strategically located. Chapter 13 appears after students have spent several chapters working with reactions. Having quite a few reactions under their belts, students can appreciate retrosynthetic analysis, as well as cataloging reactions as functional group transformations or reactions that alter the carbon skeleton. Moreover, Chapter 13 appears early enough so students can practice their skills devising multistep syntheses throughout the entire second half of the book; each subsequent chapter has multiple synthesis problems. Additionally, Chapter 13 is an excellent review of reactions students learned to that point in the book, so it could be taught at the end of the first semester as a capstone, or it could be taught at the beginning of the second semester to help jog students' memories in preparation for second semester.

Chapter 19 is delayed a few more chapters because it deals with content related to reactions from Chapter 18, including protecting groups and choosing carbon–carbon bond-forming reactions that result in the desired relative positioning of functional groups. The multistep synthesis topics in Chapter 19 are somewhat more challenging than the ones in Chapter 13, so whereas Chapter 13 should be covered in most mainstream courses, instructors can choose to cover only certain sections of Chapter 19.

I have found that treating multisynthesis in dedicated chapters makes it more meaningful to students. When I taught synthesis under a functional group organization, it became a distraction to the reactions that students were simultaneously learning. I also found that students often associated a synthetic strategy only with the functional group for which it was introduced. For example, when the idea of protecting groups is introduced in the ketones/aldehydes chapter of a textbook organized by functional group, students tended to associate protecting groups with ketones and aldehydes *only*. My dedicated synthesis chapters help students focus on synthesis without compromising their focus on reactions. Furthermore, synthesis strategies are discussed more holistically, so students can appreciate them in a much broader context rather than being applicable to just a single functional group.

Another major organizational feature of the book pertains to nomenclature. Nomenclature is separated out from the main chapters, in five relatively short interchapters— Interchapters A, B, C, E, and F. Separating nomenclature from the main chapters in this way removes distractions. It also allows students to focus on specific rules of nomenclature instead of specific compound classes. With each new nomenclature interchapter, the complexity of the material increases by applying the new rules to the ones introduced earlier.

The instructor has flexibility as to how to work with these nomenclature interchapters. They can be covered in lecture or easily assigned for self-study. They can be split over two semesters or could all be covered in the first semester. The locations of the interchapters in the book (i.e., immediately after Chapters 1, 3, 5, 7, and 9), however, should be taken as indicators as to the earliest that each interchapter should be assigned or taught. Covering a nomenclature interchapter substantially earlier than it appears in the book would expose students to compound classes well before those types of compounds are dealt with in the main chapters.

Finally, the application of MOs toward chemical reactions is separated from the main reaction chapters, and is presented, instead, as an optional, self-contained unit— Interchapter D. This interchapter appears just after Chapter 7, the overview of the 10 most common elementary steps. Each elementary step from Chapter 7 is revisited from the perspective of frontier MO theory. Because this interchapter is optional, chapters later in the book do not rely on coverage of this material.

Presenting this frontier MO theory material together in an optional unit, as I have done in Interchapter D in this book, offers two main advantages to students. First, it removes a potential distraction from the main reaction chapters and, being optional, instructors have the choice of not covering it at all. Another advantage comes from the fact that the MO pictures of all 10 common elementary steps appear together in the interchapter. Therefore, instructors who wish to cover this interchapter can expect their students to come away with a better understanding of the bigger picture of MO theory as it pertains to chemical reactions.

Focused on the Student

While the organization provides a coherent story, I've included pedagogy that promotes active learning and makes this book a better tool for students.

Strategies for Success. I wrote these sections to help students build specialized skills they need in this course. For example, Chapter 1 provides strategies for drawing all resonance structures of a given species, and sections in Chapters 2 and 3 are devoted to the importance of molecular modeling kits in working with the three-dimensional aspects of molecules and also with the different rotational characteristics of single and double bonds. In Chapter 4, students are shown step by step how to draw chair conformations of cyclohexane and how to draw all constitutional isomers of a given formula. Chapter 5 provides help with drawing mirror images of molecules. One Strategies for Success section in Chapter 6 helps students estimate pK_a values and another helps students rank acid and base strengths based only on their Lewis structures. In Chapter 14, I include a section that shows students how to use the Lewis structure to assess conjugation and aromaticity, and Chapter 16 has a section that teaches students the chemical distinction test for nuclear magnetic resonance.

> ## 4.7 Strategies for Success: Drawing Chair Conformations of Cyclohexane
>
> Given the abundance of cyclohexane rings, it would soon become cumbersome if we always had to represent chair conformations three-dimensionally as ball-and-stick models (Fig. 4-23a) or in dash–wedge notation (Fig. 4-23b). Chemists, therefore, have devised the shorthand notation for drawing chair conformations shown in Figure 4-23c.

Your Turn exercises. Getting students to read *actively* can be challenging, so I wrote the Your Turns in each chapter to motivate this type of behavior. Your Turns are basic exercises that ask students to either answer a question, look something up in a table, construct a molecule using a model kit, or interact with art in a figure or data in a plot. These exercises are also intended to be "reality checks" for students as they read. If a student cannot solve or answer a Your Turn exercise easily, then that student should interpret this as a signal to either reread the previous section(s) or seek help. Short answers to all Your Turns are provided in the back of the book and complete solutions to these exercises are provided in the *Study Guide and Solutions Manual*.

YOUR TURN 6.5

Verify the preceding statement that diethyl ether would be a suitable solvent for $(CH_3)_2N^-$. To do so, use Table 6-1 to fill in the boxes below with the appropriate pK_a values and label which one is the stronger acid. Indicate which side of the reaction is favored at equilibrium. Is this the same side that is favored in Equation 6-11?

YOUR TURN 13.2

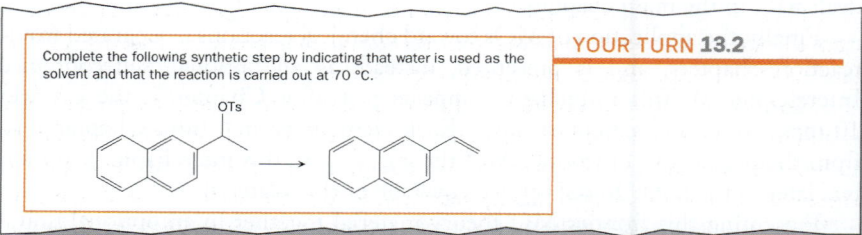

Complete the following synthetic step by indicating that water is used as the solvent and that the reaction is carried out at 70 °C.

Consistent and effective problem-solving approach. Helping students become expert problem solvers, in this course and beyond, is one of my major goals. I have developed the Solved Problems in the book to train students how to approach a problem. Each Solved Problem is broken down into two parts: *Think* and *Solve*. In the Think part, students are provided a handful of guiding *questions* that I want them to be asking as they approach the problem. In the Solve part, those questions are answered and the problem is solved. This mirrors the strategy I use to help students during office hours, and we have used these same steps for *every* problem in the *Study Guide and Solutions Manual* that accompanies the book.

Biochemistry and the MCAT. Most students taking organic chemistry are biology majors or are seeking a career in a health profession. They appreciate seeing how organic chemistry relates to their interests and look for ways in which this course will prepare them for the admissions exams (such as the MCAT) that may have a large impact on their future.

Rather than relegating biochemistry to the end of the book, I have placed self-contained Organic Chemistry of Biomolecules sections at the ends of several chapters, beginning with Chapter 1. The topics chosen for these sections cover many of the topics dealt with on the MCAT, which means that the Organic Chemistry of Biomolecules sections are not *in addition to* what students are expected to know for the MCAT; they are topics that students *should know* for the test. In even the earliest of chapters, students have the tools to start learning aspects of this traditional biochemistry coverage. More importantly, these sections provide reinforcement of topics. In each biomolecules section, the material is linked directly back to concepts encountered earlier in the chapter.

These Organic Chemistry of Biomolecules sections are both optional and flexible. Instructors can decide to cover only a few of these topics or none at all, and can do so either as they appear in the book or as special topics at the end of the second semester.

A range of interesting applications. In addition to the Organic Chemistry of Biomolecules sections, most chapters have two special interest boxes. These boxes apply a concept in the chapter to some depth toward a discovery or process that can have significant appeal to students, perhaps delving into a biochemical process or examining new and novel materials. In addition to reinforcing concepts from the chapter, these boxes are intended to provide *meaning* to what students are learning, and to motivate students to dig deeper.

In addition to these special interest boxes, several Connections boxes in each chapter provide glimpses into the everyday utility of molecules that students have just seen.

New to the Second Edition

Organization of end-of-chapter problems. At the end of each chapter, problems are grouped by concept or section so students can easily identify the types of problems they need to work on. A set of Integrated Problems follows those sets of focused problems. These Integrated Problems require students to bring together major concepts from multiple sections within the chapter, or from multiple chapters, as they would on an exam. These problems also help students stay familiar with material from earlier in the book, thus reducing the time that students would need to spend separately for review. In addition to organizing problems this way, problems that relate to aspects of synthesis are labeled (SYN), so students and instructors can find those types of problem quickly.

More than 300 new problems. Based on user and reviewer feedback, several new problems have been added to each chapter to provide students even more opportunities to hone their problem-solving skills and to assess their mastery of the material. Some of these new problems are specifically geared toward material from the Organic Chemistry of Biomolecules sections from within the chapter, and are grouped together among the end-of-chapter problems to make them easily identifiable.

More Solved Problems. The first edition provided students with about seven Solved Problems per chapter on average. Several new Solved Problems have been added, bringing the average to about eight per chapter. This gives students more opportunities to receive guidance on the strategies they should use when solving a problem. In addition, Solved Problems have been added to each nomenclature interchapter. Nomenclature builds in complexity as new rules are introduced, and each Solved Problem is designed to help students navigate those new rules.

SOLVED PROBLEM 11.3

Predict the major product when indene is treated with HCl.

Think Which C=C double bond will undergo electrophilic addition? What are the *possible* products and the corresponding carbocation intermediates from which they are produced? Which carbocation intermediate is more stable?

Solve The rightmost C=C double bond is the one that will undergo electrophilic addition. The others make up a benzene ring and are much too stable to react under these conditions. The two possible products of HCl addition differ by which C atom gains the H⁺ and which gains the Cl⁻:

THE ORGANIC CHEMISTRY OF BIOMOLECULES

1.14 An Introduction to Proteins, Carbohydrates, and Nucleic Acids: Fundamental Building Blocks and Functional Groups

---

CONNECTIONS

4-Methylphenol, also called *para*-cresol, is one of the compounds responsible for the odor of pigs and is also found in human sweat. One of the main uses of 4-methylphenol is in the production of antioxidants.

Nomenclature presented in five interchapters rather than four. In the first edition, nomenclature was presented in four interchapters. The fourth nomenclature interchapter dealt with all compound classes that call for the addition of a suffix, including amines, alcohols, ketones, aldehydes, and carboxylic acids and their derivatives. Users found this to be too much material for one chapter, so in the second edition, that interchapter has been split into two: Interchapters E and F. Interchapter E deals with alcohols, amines, ketones, and aldehydes; Interchapter F deals with carboxylic acids and their derivatives.

Addition of green chemistry. Based on user feedback, I have added a new section on green chemistry to Chapter 13, the first devoted chapter on multistep synthesis. Section 13.8b provides an overview of green chemistry and its importance, and then delves into three of the 12 main principles of green chemistry outlined by the American Chemical Society: less toxic reagents and solvents; safer synthesis routes; and minimizing by-products and other waste. In subsequent reaction chapters, students will find Green Chemistry boxes in the margin notes, which highlight green aspects of some reactions and provide green alternatives to others. For students planning on a career in chemistry, the goal is to instill in them the importance of considering green chemistry when designing and carrying out a synthesis. All students should know what green chemistry is, and should come to appreciate the fact that chemists in the 21st century are increasingly prioritizing the well-being of our planet.

New strategies to help students analyze IR, NMR, and mass spectra. Even with a strong foundation in the principles that underlie IR and NMR spectroscopy and mass spectrometry, it can still be quite a challenge for students to analyze a spectrum in a way that brings the individual pieces of information together. To help students along these lines in the first edition, I presented spectra of unknowns and then brought students through the analysis methodically, although somewhat passively. New to the second edition, I now present separate strategies up front to analyze IR, NMR, and mass spectra, with sequential steps that students can follow. Then I show students how to apply these strategies toward the analysis of spectra of unknowns. Students are encouraged to develop other strategies that might work better for them, but until then, students have an effective strategy that they can use and rely on.

Oxidation states moved to Chapter 17. In the first edition, calculating oxidation states of atoms was presented in Chapter 1 alongside the calculation of formal charges. Although grouping those two topics together makes sense because of the similarities between the two methods, users reported that students weren't sufficiently applying the ideas of oxidation states toward redox reactions until Chapter 17. Therefore, in the second edition, I moved the calculation of oxidation states to Section 17.3b, where hydride reductions are discussed.

Nobel Prize–winning coupling and metathesis reactions. Because of their importance to organic chemistry, transition metal coupling reactions and alkene metathesis reactions have been added to the second edition. These include: coupling reactions involving dialkylcurprates; the Suzuki reaction; the Heck reaction; and the Grubbs reaction. The utility of these reactions is primarily in organic synthesis, specifically in the formation of new carbon–carbon bonds, so these reactions have been added to Chapter 19, the second chapter devoted to organic synthesis.

Azo coupling and azo dyes. The presentation of azo coupling and a short discussion on azo dyes have been added to Chapter 23, the second chapter on aromatic substitution reactions. The benefits of this section are twofold. First, it is an application of diazotization (Chapter 22) and substituent effects in aromatic substitution (Chapter 23), so it provides reinforcement of newly learned concepts. Second, students can easily relate to dyes, so it is an excellent example of the daily impacts organic chemistry has on students' lives.

Connections boxes. Students often ask, "How does organic chemistry apply to me?" or, "Why should I care about organic chemistry?" For the chemistry major or the student going on to medical school or another health profession, the long-term answer might be apparent. Connections boxes, which are new to the second edition, are designed to help answer that question as it relates to the immediate. In the margins of each chapter, students will find several Connections boxes that highlight the importance or application of a molecule that was just encountered. Students might see that the molecule is integral in the synthesis of a pharmaceutical drug, or that the molecule is important in the manufacture of a material that students use daily. More than just helping provide an answer to the above questions, these Connections boxes also help keep students *interested* in the material, and an interested student is a more successful student.

Acknowledgments

Special thanks to my wife Valerie and my boys Joshua and Jacob for being my biggest fans. Their love and immense support throughout my work on the second edition not only helped push me to the finish line, but they continue to make my achievements worthwhile.

Many thanks to my colleagues in the chemistry department at Elon for your understanding, especially Dan Wright and Karl Sienerth, who served as my department chairs during this endeavor. And a tremendous thank-you to Kathy Matera for the real-time feedback you have given me over the years, and for all the times I barged into your office to pick your brain when I was in the midst of working through a quandary with the book.

I remain indebted to my students. Thank you for bringing such great energy to learning organic chemistry year in and year out, and thank you for allowing me to learn from you.

I continue to be amazed with the members of the Norton team. Erik Fahlgren, thank you for your continued belief in me and in the potential this book has to help teachers teach and to help students learn. John Murdzek, your insights in the developmental process have truly enabled me to reach students more effectively and meaningfully. Arielle Holstein and Sara Bonacum, thank you for being the glue that has held this entire project together. And a further congratulations to Arielle for your new position as Associate Media Editor; thank you for the great work you have done on the book's ancillaries. To Carla Talmadge and Connie Parks, many thanks for holding me to a high standard in the copyediting and page proofing stages. Travis Carr and Elyse Rieder, I admire your patience and persistence when I need just the right photo. Christopher Rapp and Christine Pruis, thank you for your work on the online resources for the book, which not only add value to the book but also make the book more effective. Lisa Buckley, what a fantastic job on the interior design, giving the book a warm and inviting feel. And Stacy Loyal, you continue to amaze me with your vision and the creativity you bring to marketing the book.

A special thanks, once again, to Marie Melzer. With the energy and the insight that you have continued to bring, I could not imagine a better coauthor on the *Study Guide and Solutions Manual*. And to Steve Pruett, I truly value your work on the polymers chapter in the first edition.

Finally, I am indebted to the many reviewers, whose feedback has been instrumental in making several significant improvements over the first edition. I am especially grateful to Joachim Schantl, who accuracy-checked nearly the *entire* book! I am in awe of your breadth and depth of knowledge, as well as your attention to detail. Many, many thanks.

Reviewers of Second Edition

Aron Anderson, Gustavus Adolphus College

Niels Andersen, University of Washington

Amelia Anderson-Wile, Ohio Northern University

Christina Bagwill, Saint Louis University

Joshua Beaver, University of North Carolina, Chapel Hill

David Bergbreiter, Texas A & M University

Shannon Biros, Grand Valley State University

Dan Blanchard, Kutztown University

Elizabeth Blue, Campbell University

Luc Boisvert, University of Puget Sound

Michelle Boucher, Utica College

Rick Bunt, Middlebury College

Nancy Carpenter, University of Minnesota, Morris

Timothy Clark, University of San Diego

Kimberly Cousins, California State University, San Bernardino

Ashton Cropp, Virginia Commonwealth University

Anna Drotor, Metropolitan State University of Denver

Nathan Duncan, Maryville College

Brendan Dutmer, Highland Community College

Todd Eckroat, Penn State, Behrend

Daniel Esterline, Thomas More College

Amanda Evans, California State University, Fullerton

Christoph Fahrni, Georgia State Institute of Technology

Suzanne Fernandez, Lehigh University

Michael Findlater, Texas Tech University

Abbey Fischer, University of Wisconsin–Barron County

Stephen Foley, University of Saskatchewan

Malcolm Forbes, Bowling Green State University

Denis Fourches, North Carolina State University

Andrew Frazer, University of Central Florida

Gregory Friestad, University of Iowa

Brian Frink, Lakeland College

Brian Ganley, University of Missouri

Kevin Glaeske, Wisconsin Lutheran College

Sarah Goforth, Campbell University

Harold Goldston Jr., Des Moines Area Community College

Anne Gorden, Auburn University

Dustin Gross, Sam Houston State University

Matthew Hart, Grand Valley State University

Allan Headley, Texas A & M University

Ian Hill, Gustavus Adolphus College

Daniel Holley, Columbus State University

Robert Hughes, East Carolina State University

Philip Hultin, University of Manitoba

William Jenks, Iowa State University

Bob Kane, Baylor University

Kristopher Keuseman, Mount Mercy University

Brett Kite, Shenandoah University

Jeremy Klosterman, University of California, San Diego

Kazunori Koide, University of Pittsburgh

Shane Lamos, St. Michael's College

Nicholas Leadbeater, University of Connecticut

Carl Lecher, Marian University

Larry Lee, Camosun College

Diana Leung, University of Alabama, Tuscaloosa

Nicholas Llewellyn, Emory University

Carl Lovely, University of Texas, Arlington

Breeyawn Lybbert, University of Wisconsin

Helena Malinakova, University of Kansas

Richard Manderville, University of Guelph

Kristen Mascall, Brandeis University

Eugene Mash, University of Arizona

Daniell Mattern, University of Mississippi

Jimmy Mays, University of Tennessee, Knoxville

Vanessa McCaffrey, Albion College

Justin Mohr, University of Illinois, Chicago

Suazette Mooring, Georgia State University

Jesse More, Loyola University

Andrew Morehead, East Carolina University

Cheryl Moy, University of North Carolina, Chapel Hill

R. Scott Murphy, Regina University

Joan Mutanyatta-Comar, Georgia State University

David Nagib, Ohio State University

Felix Ngassa, Grand Valley State University

Taeboem Oh, California State University, Northridge

Joshua Osbourn, West Virginia University

Keith Pascoe, Georgia State University

Gitendra Paul, Malcolm X College

Michael Pelter, Purdue University Northwest

Angela Perkins, University of Minnesota

Joanna Petridou-Fischer, Spokane Falls Community College

Tarakeshwar Pilarsetty, Arizona State University

Smitha Pillai, Arizona State University

Kyle Plunkett, Southern Illinois University

Pamela Pollet, Georgia Institute of Technology

Brian Popp, West Virginia University

Walda Powell, Meredith College

Stephen Pruett, Jefferson Community and Technical College

Frank Rossi, State University of New York, Cortland

Nicholas Salzameda, California State University, Fullerton

Robert Sammelson, Ball State University

Joachim Schantl, University of Florida

Jacob Schroeder, Clemson University

Reza Sedaghat-Herati, Missouri State University

Jia Sheng, University of Albany

Abbas Shilabin, East Tennessee State University

Matthew Siebert, Missouri State University

Chatu Sirimanne, California State University, Los Angeles

Heather Sklenicka, Rochester Community and Technical College

Mike Slade, University of Evansville

Greg Slough, Kalamazoo College

Gary Spessard, University of Arizona

Nicholas Stephanopoulos, Arizona State University

Robert Ternansky, University of California, San Diego

Sadanandan Velu, University of Alabama, Birmingham

Martin Walker, The State University of New York, Potsdam

Don Warner, Boise State University

Michael Weaver, University of Florida

Lyndon West, Florida Atlantic University

Anne Wilson, Butler University

Kai Ylijoki, Saint Mary's University

Yimin Zhu, Pennsylvania State University, Altoona

Reviewers of the First Edition

Robert Allen, Arkansas Tech University

Herman Ammon, University of Maryland

Carolyn Anderson, Calvin College

Aaron Aponick, University of Florida

Phyllis Arthasery, Ohio University

Jared Ashcroft, Pasadena City College

Athar Ata, University of Winnipeg

Jovica Badjic, Ohio State University

John Bellizzi, University of Toledo

Daniel Berger, Bluffton University

Anthony Bishop, Amherst College

Rebecca Broyer, University of Southern California

Larry Calhoun, University of New Brunswick

Shawn Campagna, University of Tennessee, Knoxville

Nancy Carpenter, University of Minnesota, Morris

Brad Chamberlain, Luther College

Robert Coleman, Ohio State University

Tammy Davidson, University of Florida

Lorraine Deck, University of New Mexico

Sergei Dzyuba, Texas Christian University

Jeff Elbert, University of Northern Iowa

Seth Elsheimer, University of Central Florida

Eric Finney, University of Washington

Andrew Frazer, University of Central Florida

Larry French, St. Lawrence University

Gregory Friestad, University of Iowa

Brian Frink, Lakeland University

Anne Gorden, Auburn University

Christopher Gorman, North Carolina State University

Oliver Graudejus, Arizona State University

Robert Grossman, University of Kentucky

Daniel Gurnon, DePauw University

Jeffrey Hansen, DePauw University

Bryan Hanson, DePauw University

Andrew Harned, University of Minnesota

Stewart Hart, Arkansas Tech University

John Hershberger, Arkansas State University

Gail Horowitz, Brooklyn College

Roger House, Auburn University

Philip Hultin, University of Manitoba

Kevin Jantzi, Valparaiso University

Amanda Jones, Wake Forest University

Jeff Jones, Washington State University

Paul Jones, Wake Forest University

Robert Kane, Baylor University

Arif Karim, Austin Community College

Steven Kass, University of Minnesota

Stephen Kawai, Concordia University

Valerie Keller, University of Chicago

Mark Keranen, University of Tennessee at Martin

Kristopher Keuseman, Mount Mercy College

Angela King, Wake Forest University

Jesudoss Kingston, Iowa State University

Francis Klein, Creighton University

Jeremy Klosterman, Bowling Green State University

Dalila Kovacs, Grand Valley State University

Jason Locklin, University of Georgia

Brian Long, University of Tennessee, Knoxville

Claudia Lucero, California State University, Sacramento

David Madar, Arizona State University Polytechnic

Kirk Manfredi, University of Northern Iowa

Eric Masson, Ohio University

Anita Mattson, Ohio State University

Gerald Mattson, University of Central Florida

Jimmy Mays, University of Tennessee, Knoxville

Alison McCurdy, California State University, Los Angeles

Dominic McGrath, University of Arizona

Mark McMills, Ohio University

Marie Melzer, Old Dominion University

Ognjen Miljanic, University of Houston

Justin Miller, Hobart and William Smith Colleges

Stephen Miller, University of Florida

Barbora Morra, University of Toronto

Joseph O'Connor, University of California, San Diego

James Parise, University of Notre Dame

Gitendra Paul, Malcolm X Community College

Noel Paul, Ohio State University

James Poole, Ball State University

Christine Pruis, Arizona State University

Harold Rogers, California State University, Fullerton

Sheryl Rummel, Pennsylvania State University

Nicholas Salzameda, California State University, Fullerton

Adrian Schwan, University of Guelph

Colleen Scott, Southern Illinois University, Carbondale

Alan Shusterman, Reed College

Joseph Simard, University of New England

Chad Snyder, Western Kentucky University

John Sorensen, University of Manitoba

Levi Stanley, Iowa State University

Laurie Starkey, California State University, Pomona

Tracy Thompson, Alverno College

Nathan Tice, Butler University

John Tomlinson, Wake Forest University

Melissa VanAlstine-Parris, Adelphi University

Nanine Van Draanen, California Polytechnic State University

Qian Wang, University of South Carolina

Don Warner, Boise State University

Haim Weizman, University of California, San Diego

Lisa Whalen, University of New Mexico

James Wilson, University of Miami

Laurie Witucki, Grand Valley State University

James Wollack, St. Catherine University

Andrei Yudin, University of Toronto

Michael Zagorski, Case Western Reserve University

Rui Zhang, Western Kentucky University

Regina Zibuck, Wayne State University

Eugene Zubarev, Rice University

James Zubricky, University of Toledo

Additional Resources

For Students

Study Guide and Solutions Manual

by Joel Karty, Elon University, and Marie Melzer

Written by two dedicated teachers, this guide provides students with fully worked solutions to all unworked problems in the text. Every solution follows the Think and Solve format used in the textbook, so the approach to problem solving is modeled consistently.

Smartwork5 (digital.wwnorton.com/karty2)

Smartwork5 is the most intuitive online tutorial and homework system available for organic chemistry. A powerful engine supports and grades a wide variety of problems written for the text, including numerous arrow-pushing problems. *Every* problem in Smartwork5 has hints and answer-specific feedback to coach students and provide the help they need, when they need it. Problems in Smartwork5 link directly to the appropriate page in the ebook so students have an instant reference and are prompted to read.

Assigning, editing, and administering homework within Smartwork5 is easy. Instructors can select from Norton's bank of more than 3200 high-quality, class-tested problems. Using the sort and search features, instructors can identify problems by chapter section, learning objective, question type, and more. Instructors can use premade assignments provided by Norton authors, modify those assignments, or create their own. Instructors also have access to intuitive question authoring tools—the same ones Norton authors use. These tools make it easy to customize the question content to fit the course needs. Smartwork5 integrates seamlessly with most campus learning management systems and can be used on computers and tablets.

The Smartwork5 course features:

- **An expert author team.** The Smartwork5 course was authored by instructors who teach at a diverse group of schools: Arizona State University, Florida State University, Brigham Young University, Butler University, and Mesa Community College. The authors have translated their experience in teaching a diverse student population by creating a library of problems that will appeal to instructors at all schools.
- **An upgraded drawing tool.** Smartwork5 contains an upgraded 2-D drawing tool that mimics drawing on paper, reduces frustration, and helps students focus on the problem at hand. This intuitive drawing tool supports multistep mechanism and multistep synthesis problems and provides students with answer-specific feedback for every problem.
- **Ease of use for students.** The 2-D drawing tool has a variety of features that make drawing easy and efficient. Students are provided with templates including a variety of common rings and a carbon chain drawing tool. In addition, Smartwork5 presents students with commonly used elements, a simple click to add lone pairs option, and ease-of-use features such as undo, redo, simple-click erase, and zoom-in/zoom-out.
- **Question variety.** The Smartwork5 course offers a diverse set of problems including:
  - Nomenclature problems
  - Multistep Mechanism problems
  - Multistep Synthesis problems
  - Reaction problems
  - Spectroscopy problems
- Conceptual question types include:
  - Multiple-choice/multiple select
  - Ranking
  - Sorting
  - Labeling
  - Numeric entry
  - Short answer
- **Pooled problems.** Smartwork5 features sets of pooled problems for multistep mechanisms and nomenclature to promote independent work. Groups of similar problems are "pooled" into one problem so different students receive different problems from the pool. Instructors can choose our preset pools or create their own.

Ebook (digital.wwnorton.com/karty2)

An affordable and convenient alternative to the print text, the Norton Ebook lets students access the entire book and much more: They can search, highlight, and take notes with ease. The Norton Ebook allows instructors to share their notes with students. The ebook can be viewed on computers and tablets and will stay synced between devices. The online ebook is available at no extra cost with the purchase of a new print text or it may be purchased stand-alone with Smartwork5.

Molecular Model Kits

Norton partners with two model kits and can package either with the textbook for an additional cost.

Darling Molecular Model Kit. Atoms with their valences already attached are constructed by snapping together V-shaped pieces in a jigsaw style, emphasizing bond angles and symmetry elements of the atoms. Double bonds are independent, rectangular units to emphasize the planarity of sp^2-hybridized atoms. Large substituents can be represented by various colored marker balls.

This kit includes 120 pieces:

- 57 sp^3 pieces (black, red, blue, silver black, turquoise, gray)
- 16 sp^2 pieces (gray)
- 18 marker balls (white, red, green, blue)
- 7 double bonds (gray)
- 6 half double bonds (gray, red)
- 2 trigonal atoms (gray)
- 2 linear bonds (gray)
- 2 linear triple bonds (gray)
- 4 bond extenders
- 4 octahedral pieces (pink)
- 2 Atom Visions™ balls

HGS Molecular Structure Model Kit. The HGS kit reflects the traditional ball-and-stick model for constructing molecules. Conjugation can be illustrated using trigonal planar atoms that have five holes to accommodate three bonds and the two lobes of a p orbital. Double bonds can be constructed using curved sticks to occupy two valences of a tetrahedral atom.

This kit includes 210 pieces:

- 30 tetrahedral carbon atoms (black)
- 14 trigonal planar carbon atoms (black)
- 30 hydrogen atoms (light blue)
- 4 oxygen atoms (red)
- 6 nitrogen atoms (blue)
- 4 chlorine atoms (green)
- 2 metal atoms (grey)
- 12 orbital plates (green, blue)
- 108 bond pieces, 5 types (light blue, orange, green, yellow, white)

Please contact your Norton representative about ordering and pricing options for packaging model kits.

For Instructors

Instructor's Guide

by Michelle Boucher, Utica College, and Cliff Coss, Northern Arizona University

Written by users of the first edition, the *Instructor's Guide* is an invaluable resource for instructors organizing their course by mechanism for the first time. Based on their experience, Michelle and Cliff provide a brief overview of every chapter followed by a section-by-section summary that illustrates how easy and rewarding it is to teach a mechanistically organized course. In addition to providing an easy transition, the authors offer other resources, such as class-tested clicker questions that instructors may choose to incorporate into their course. While this guide is an excellent resource for adopters, it may also answer questions for instructors who are interested in a mechanistic organization but are concerned about the transition. The *Instructor's Guide* includes a chapter for each of the 26 chapters in the textbook, plus a chapter for the molecular orbital theory interchapter and a chapter for each of the nomenclature interchapters.

Clickers in Action: Active Learning in Organic Chemistry

by Suzanne M. Ruder, Virginia Commonwealth University

This instructor-oriented resource provides information on implementing clickers in organic chemistry courses. Part I gives instructors information on how to choose and manage a classroom response system, develop effective questions, and integrate the questions into their courses. Part II contains 140 class-tested, lecture-ready questions. Most questions include histograms that show actual student response, generated in large classes with 200–300 students over multiple semesters. Each question also includes insights and suggestions for implementation. The 140 questions from the book are sorted to correspond to the chapters in the textbook.

Test Bank

by James Wollack, St. Catherine University, Jennifer Griffith, Western Washington University, and Chris Markworth, Western Washington University

After teaching with the first edition, our authors have written problems that will make assessing your students easy. Whether your exams are multiple choice, short answer and require drawing, or both, the variety and quality of the problems in the test bank will exceed your needs. The test bank contains approximately 1600 multiple-choice and short-answer questions classified by section and difficulty level. It is available with Exam-View Test Generator software, allowing instructors to effortlessly create, administer, and manage assessments. The convenient and intuitive test-making wizard makes it easy to create customized exams with no software learning curve. Other key features include the ability to create paper exams with algorithmically generated variables and export files directly to Blackboard, Canvas, Desire2Learn, and Moodle.

Instructor's Resources: Flash Drive

This helpful classroom presentation tool features:

- Select photographs and every piece of line art in JPEG format
- Select photographs and every piece of line art in PowerPoint
- Lecture PowerPoint slides with integrated figures from the book
- *Instructor's Guide* in PDF format
- Test bank in PDF, Word, and ExamView formats
- Approximately 500 lecture-ready questions, in PowerPoint, from *Clickers in Action* as well as Joel Karty's course

Downloadable Instructor's Resources (wwnorton.com/instructors)

This instructor-only, password-protected site features instructional content for use in lecture and distance education, including test-item files, PowerPoint lecture slides, images, figures, and more. The instructor's website includes:

- Select photographs and every piece of line art in JPEG format
- Select photographs and every piece of line art in PowerPoint
- Lecture PowerPoint slides with integrated figures from the book
- *Instructor's Guide* in PDF format
- Test bank in PDF, Word, and ExamView formats
- Approximately 500 lecture-ready questions, in PowerPoint, from *Clickers in Action* as well as Joel Karty's course

Author Blog: www.teachthemechanism.com

In July 2012, Joel Karty started a blog about his approach and his experience teaching a course organized by mechanism. Now there are more than 120 guest blog posts written by professors who use Joel's book, garnering nearly 60,000 views and 20+ active conversations. What once was an informational blog has now grown into a platform for a community of instructors to share their experiences and insights, have open-forum discussions, view sample materials, and watch videos of Joel as he discusses a number of topics, including how he believes a mechanistic organization allows users of his book to have increased expectations about student understanding. You are encouraged to visit the blog and join the community.

Preface for the Student
Organic Chemistry and You

You are taking organic chemistry for a reason—you might be pursuing a career in which an understanding of organic chemistry is crucial, or the course might be required for your particular field of study, or both. You might even be taking the course simply out of interest. Regardless of the reason, organic chemistry impacts your life in significant ways.

Consider, for example, the growing concern about the increasing resistance of bacteria to antibiotics over the past several decades. Perhaps no germ has caused more alarm than methicillin-resistant *Staphylococcus aureus* (MRSA), a type of bacteria responsible for staph infections. Methicillin is a member of the penicillin family of antibiotics, and resistance to methicillin in these bacteria was first observed in 1961. Today MRSA, which has been called a *superbug*, is resistant to most antibiotics, including *all* penicillin-derived antibiotics.

A breakthrough in the fight against MRSA occurred in 2006 with the discovery of a compound called platensimycin, isolated from *Streptomyces* spores. The way that platensimycin targets bacteria is different from that of any other antibiotic in use and, therefore, it is not currently susceptible to bacterial resistance.

Streptomyces spores   **Platensimycin**

Platensimycin is found in a type of South African mushroom, *Streptomyces platensis*, and was discovered by screening 250,000 natural product extracts for antibacterial activity. Sheo B. Singh (Merck Research Laboratories) and coworkers determined the structure of platensimycin using a technique called nuclear magnetic resonance (NMR) spectroscopy, which we discuss in Chapter 16. Not long after, K. C. Nicolaou and coworkers from the Scripps Research Institute (La Jolla, California) and the University of California, San Diego, were the first to devise a synthesis of platensimycin from other readily available chemicals.

The story of platensimycin, from discovery to synthesis, involves several of the subdisciplines that make up the field of organic chemistry.

- **Biological chemistry (biochemistry):** The study of the behavior of biomolecules and the nature of chemical reactions that occur in living systems.
- **Structure determination:** The use of established experimental techniques to determine the structure of newly discovered compounds.
- **Organic synthesis:** The design of pathways for making new compounds from existing, readily available compounds by means of known organic reactions.

Because each of these areas typically focuses on solving existing and practical problems, they are considered to be *applied* areas of organic chemistry. However, other areas of organic chemistry, considered to be *theoretical* in nature, provide the foundations on which such applications rest. They focus on answering questions about the *how* and *why* of chemical processes. For example, an understanding of the basic principles of NMR spectroscopy (an analytical technique discussed in Chapter 16) underlies our ability to determine molecular structure. Understanding the principles that govern organic reactions (such as those involved in the synthesis of platensimycin) may allow us to enhance yields, not only by altering reaction conditions, but also perhaps by devising

(a)

(b)

(c)

(d)

FIGURE P.1 **Some uses of plastics** Plastics, which are designed and created in the laboratories of organic chemists, are found in a wide range of products, such as (a) food packaging, (b) an artificial heart, (c) body armor made from Kevlar, and (d) a Boeing 787, a commercial jet whose body consists largely of composite materials made from plastics and carbon fibers.

(a)

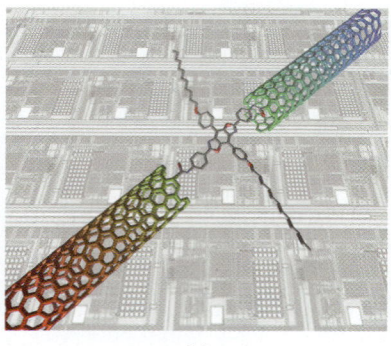
(b)

FIGURE P.2 **Organic chemistry in the electronics industry** (a) A smartphone whose display is made from organic light-emitting diodes. (b) A molecular switch in which an organic molecule joins together two carbon nanotubes—sheets of carbon in the form of cylinders with a diameter on the order of 10^{-9} meter.

entirely new synthesis schemes. And understanding platensimycin's specific mode of attack on bacteria will likely guide us in modifying its chemical structure to make it even more effective.

The story of platensimycin showcases the importance of organic chemistry in the pharmaceutical industry, but organic chemistry is at the center of other high-profile areas as well, including the fabrication of new materials such as plastics (the topic of Chapter 26). The durability and chemical stability of plastics have made them excellent choices for use in food packaging (Fig. P.1a) and the fabrication of the artificial heart (Fig. P.1b). Plastics are the source of synthetic fibers such as nylon and polyester, which are often used in the clothing industry, as well as Kevlar, which is used to make body armor (Fig. P.1c). Composite materials made from plastic and carbon fibers are so strong that some commercial jets are now constructed with a body made largely from plastics (Fig. P.1d).

Organic chemistry has also been at the forefront of generating new materials for electronic devices. Organic light-emitting diodes (OLEDs) are the main components of electronic displays for many high-end smartphones (Fig. P.2a), and single organic molecules can be used to make electronic switches tens of thousands of times smaller than those used in today's integrated circuits (Fig. P.2b).

Perhaps even more important to our lives is the impact that organic chemistry can have on our ability to understand, and solve, environmental problems, such as overflowing landfills (Fig. P.3a), the destruction of the stratospheric ozone layer (Fig. P.3b), and global warming (Fig. P.3c). Organic chemistry, for example, is helping provide new ways to recycle waste materials. Additionally, organic chemistry has been used to engineer new coolants that are safer for the environment than the chlorofluorocarbons (CFCs) used in the late 20th century in refrigerators and air conditioners. Finally, organic chemistry may lead us to economically feasible processes by which we can synthesize hydrogen gas, a fuel whose combustion product is only water. This could be a welcome alternative to coal and oil, whose combustion products not only cause air and water pollution, but also generate carbon dioxide, one of several greenhouse gases responsible for global warming.

Because organic chemistry is important in so many ways, you will find two special interest boxes in the main part of each chapter, which show how the material in the chapter directly connects to issues that you might find more relevant or more interesting. Take the time to read those boxes, and consider researching them even further. In addition to those special interest boxes, you will find several Connections boxes in the margins of each chapter, each of which provides a glimpse into how a molecule you just encountered relates to an aspect of everyday life.

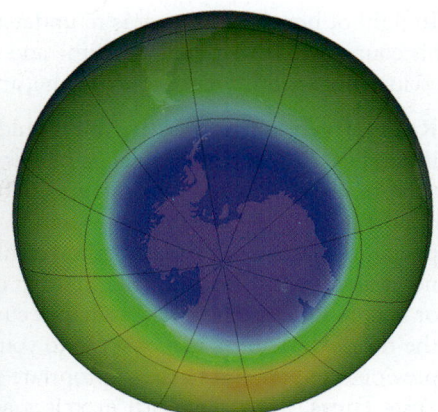

(b) September 2016

(a)

(c)

FIGURE P.3 Organic chemistry and the environment Organic chemistry continues to play a significant role in solving environmental problems, such as (a) overflowing landfills, (b) ozone depletion (the area in blue represents the ozone hole over Antarctica), and (c) global warming (the ice sheets in Montana's Glacier National Park have been melting at a dramatically accelerating rate over the past 90 years).

Some Suggestions for Studying

Perhaps you have heard that organic chemistry is difficult. Perhaps you have heard that it requires an enormous amount of memorization. Are these statements true? It depends on how you approach the course. What is true is that this book contains a lot of information—*much more than you can memorize.* There is a better way.

Organic chemistry can be *understood* through models and theories that are built on *fundamental concepts.* Consider, for example, that when two compounds react under a given set of conditions, the outcome of that reaction is precisely the same each and every time. Is this because the reactant molecules have memorized what products they are supposed to make? No—they are obeying certain chemical laws, and those laws can be learned.

You will spend considerable effort throughout this course developing those models and theories. *Reaction mechanisms*—detailed steps that show how reactions take place—are among the most important ideas to develop. If you devote your time and energy to understanding them and learning how they are applied toward solving problems, you will find that much of organic chemistry can be conquered without rote memorization, and you will find the course to be quite rewarding and enjoyable. Moreover, the skills you develop in organic chemistry will apply to complex situations you will face beyond this course.

If you are planning on a career in a health profession, it is particularly important for you to focus on understanding and applying concepts as opposed to memorizing. On standardized exams like the MCAT, you will often need to choose between answers that look equally good to students who have memorized the material. To a student who is well versed in applying concepts and mechanisms toward solving problems, on the other hand, those choices are more easily discernible.

In light of how important it is to understand concepts and mechanisms, your success in this course will demand a lot of time and devotion. Therefore, you should consider the following suggestions for using that time, and this book, most efficiently:

- **Read actively and diligently.** You should try to read the assigned sections before class if possible. Reading prior to class means that you will see the material for the second time in class. This will allow you to better process information and give you ample opportunity to ask pertinent questions. When you read, you should have a pen or pencil in hand so you can underline or highlight what you feel is important, and take notes about what you find enlightening or confusing. When the text refers to a figure or reaction mechanism, take that as a cue to study that figure now. Be sure that what the text is describing makes sense to you before you move on. If you are referred to a previous chapter, flip to the appropriate page to refresh your memory.
- **Your Turns.** The Your Turn exercises are relatively short activities that ask you to complete a task based on what you have just read. These exercises were developed to help you remain *actively engaged* while you read. They should also help you quickly evaluate whether you understand the topic at hand. I encourage you to *work through all Your Turn exercises in each chapter* and quickly check the answers in the back of the book. Feedback from students who have used this book supports this advice.
- **Problems.** As with anything new you attempt, mastery requires practice. Most of your practice should come from solving problems. I have included more than 2000 problems throughout this book. Many are integrated into the chapters, but most are gathered at the end of each chapter. Take the time to work through as many problems as possible, and use them to assess areas of strength and weakness.

That said, it's time to get started. Keep your focus on concepts and mechanisms, work hard, and ask questions!

Organic chemistry is often referred to as the chemistry of life because biological compounds such as DNA, proteins, and carbohydrates are themselves organic molecules. In this chapter, we examine some of the bonding characteristics of these and other organic molecules, which are constructed primarily from carbon, hydrogen, nitrogen, and oxygen.

1

Atomic and Molecular Structure

Organic chemistry is often called "the chemistry of life" because certain types of compounds, and the reactions they undergo, are suitable to sustain life, while others are not. What are the characteristics of such compounds and what advantages do those compounds afford living organisms? Here in Chapter 1 we begin to answer these questions.

We review several aspects of atomic and molecular structure typically covered in a general chemistry course, including ionic and covalent bonding, the basics of Lewis dot structures, and resonance theory. We then begin to tighten our focus on organic molecules, presenting various types of shorthand notation that organic chemists often use and introducing you to functional groups commonly encountered in organic chemistry.

Toward the end of this chapter, we shift our focus to examining specific classes of biomolecules: amino acids, monosaccharides, and nucleotides. Not only does such a discussion provide insight into the relevance of organic chemistry to biological systems, but it also reinforces specific topics covered in the chapter, such as functional groups.

1.1 What Is Organic Chemistry?

Organic chemistry is the branch of chemistry involving *organic compounds*. What, then, is an organic compound?

In the late 1700s, scientists defined an **organic compound** as one that could be obtained from a *living* organism, whereas **inorganic compounds** encompassed

Chapter Objectives

On completing Chapter 1 you should be able to:

1. Distinguish organic compounds from inorganic ones.
2. Explain the advantages that come from carbon being the basis of organic molecules.
3. Describe the basic structure of an atom and understand that the vast majority of its volume is taken up by electrons.
4. Determine the ground state electron configuration of any atom in the first three rows of the periodic table and distinguish valence electrons from core electrons.
5. Define bond length and bond energy and understand how these two quantities depend on the number of bonds between a given pair of atoms.
6. Draw the Lewis structure of a species, given only its connectivity and total charge.
7. Differentiate between a nonpolar covalent bond, a polar covalent bond, and an ionic bond, and distinguish a covalent compound from an ionic compound.
8. Assign the formal charge to any atom in a molecular species, given only its Lewis structure.
9. Describe what a resonance structure is and explain the effect that resonance has on a species' stability.
10. Draw all resonance structures of a given species, as well as its resonance hybrid, and determine the relative stabilities of resonance structures.
11. Draw and interpret Lewis structures, condensed formulas, and line structures.
12. Explain why functional groups are important and identify functional groups that are common in organic chemistry.

everything else. It was believed that organic compounds could *not* be made in the laboratory; instead, only living systems could summon up a mysterious "vital force" needed to synthesize them. This belief was called **vitalism**. By this definition, many familiar compounds, such as glucose (a sugar), testosterone (a hormone), and deoxyribonucleic acid (DNA), are *organic* (Fig. 1-1).

This definition of organic compounds broke down in 1828, when Friedrich Wöhler (1800–1882), a German physician and chemist, synthesized urea (an organic compound known to be a major component of mammalian urine) by heating a solution of ammonium cyanate (an inorganic compound; Equation 1-1).

An inorganic compound

$(NH_4)^+(NCO)^-$

Ammonium cyanate

$\xrightarrow{\text{Heat}}$

An organic compound

$$\underset{\text{Urea}}{H_2N\overset{\overset{\displaystyle O}{\|}}{C}NH_2}$$

(1-1)

If vitalism couldn't account for the distinction between organic and inorganic compounds, what could? Gradually, chemists arrived at our modern definition:

An **organic compound** contains a substantial amount of carbon and hydrogen.

This definition, however, is still imperfect, because it leaves considerable room for interpretation. For example, many chemists would classify carbon dioxide (CO_2) as *inorganic* because it does not contain any hydrogen atoms, whereas others would argue that it is *organic* because it contains carbon and is critical in living systems. In plants, it is a starting material in photosynthesis, and in animals, it is a by-product of respiration. Similarly, tetrachloromethane (carbon tetrachloride, CCl_4) contains no hydrogen, but many would classify it as an organic compound. Butyllithium (C_4H_9Li), on the other hand, is considered by many to be inorganic, despite the fact that 13 of its 14 atoms are carbon or hydrogen. Although this definition of an organic compound has its inadequacies, it does allow chemists to classify most molecules.

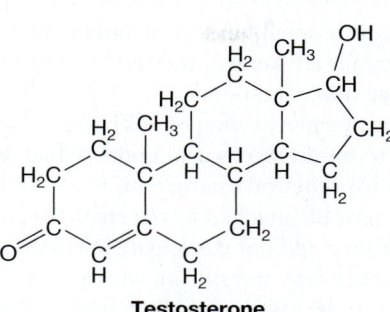

Glucose   **Testosterone**   **DNA**

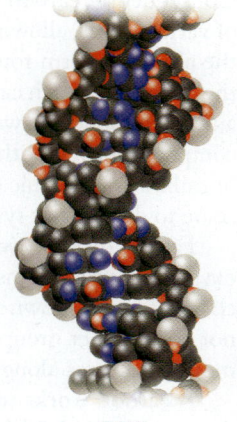

FIGURE 1-1 Some familiar organic compounds Glucose, testosterone, and DNA are organic compounds produced by living organisms.

The birth of organic chemistry as a distinct field occurred around the time that vitalism was dismissed, making the discipline less than 200 years old. However, humans have taken advantage of organic reactions and the properties of organic compounds for thousands of years! Since about 6000 BC, for example, civilizations have fermented grapes to make wine. Some evidence suggests that Babylonians, as early as 2800 BC, could convert oils into soaps.

Many clothing dyes are organic compounds. Among the most notable of these dyes is **royal purple**, also called **Tyrian purple**, which was obtained by ancient Phoenicians from a type of aquatic snail called *Bolinus brandaris* (Fig. 1-2). These organisms produced the compound in such small amounts, however, that an estimated 10,000 of them had to be processed to obtain a single gram of dye. Therefore, the dye was available almost exclusively to those who had substantial wealth and resources—royalty.

Organic chemistry has matured tremendously since its inception. Today, we can not only use organic reactions to reproduce complex molecules found in nature, but also engineer new molecules never before seen.

Bolinus brandaris

Royal purple

FIGURE 1-2 Royal purple Ancient Phoenicians processed about 10,000 aquatic snails, *Bolinus brandaris* (*top*), to yield 1 g of royal purple dye. The structure of the molecule responsible for the dye's color is shown (*bottom*).

1.2 Why Carbon?

Why does the carbon atom play such a central role in the chemistry of life and what is so special about it? First of all, the compounds possible when carbon is their chief structural component are incredibly *diverse*. As we see in Section 1.6, the carbon atom can form four covalent bonds to other atoms—especially other carbon atoms.

A chain of carbon atoms with single bonds only

A chain of carbon atoms with a double and triple bond

A *branched* chain of carbon atoms

With a chain of oxygen atoms, no double bonds, triple bonds, or *branching* is possible.

Consequently, carbon atoms can link together in chains of almost any length and rings of various sizes, allowing for an enormous range in molecular size and shape. Moreover, the ability to form four bonds means there is potential for *branching* at each carbon in the chain. And each carbon atom is capable of forming not only single bonds, but double and triple bonds as well. These characteristics make possible a tremendous number of compounds, even with a relatively small number of carbon atoms. Indeed, to date, tens of millions of organic compounds are known, and the list is growing rapidly as we continue to discover or synthesize new compounds.

Far less diversity would be possible in compounds based on another element, such as oxygen. Oxygen atoms tend to form two covalent bonds, which would allow for a linear chain only (as shown in the hypothetical example on p. 3). No branching could occur, nor could other groups or atoms be attached to the chain except at the ends. Furthermore, the atoms along the chain could not participate in either double or triple bonds.

If carbon works so well, then why *not* silicon, which appears just below carbon in the periodic table? Elements in the same group (column) of the periodic table tend to exhibit similar chemical properties, so silicon, too, can form four covalent bonds, giving it the same potential for diversity as carbon.

The answer is *stability*. As we see in Section 1.4, the carbon atom forms rather strong bonds with a variety of atoms, including other carbon atoms. For example, it takes 339 kJ/mol (81 kcal/mol) to break an average C—C single bond, and 418 kJ/mol (100 kcal/mol) to break an average C—H bond. By contrast, it takes only 223 kJ/mol (53 kcal/mol) to break a typical Si—Si bond. The strength of typical bonds involving carbon atoms goes a long way toward keeping biomolecules intact—an essential characteristic for molecules whose job is to store information or provide cellular structure.

Even though organic molecules are based on the carbon atom, what would life be like, hypothetically, if silicon atoms were to replace carbon atoms in biomolecules such as glucose ($C_6H_{12}O_6$)? Glucose is broken down by our bodies through respiration to extract energy, according to the overall reaction in Equation 1-2. One of the by-products is carbon dioxide, a gas, which is exhaled from the lungs. In a world in which life is based on silicon, glucose would be $Si_6H_{12}O_6$, and its by-product would be silicon dioxide (SiO_2), as shown in Equation 1-3. Silicon dioxide, a solid, is the main component of sand; in its crystalline form, it is known as quartz (Fig. 1-3).

$$C_6H_{12}O_6(s) + 6\ O_2(g) \rightarrow 6\ CO_2(g) + 6\ H_2O(\ell) \qquad (1\text{-}2)$$

$$Si_6H_{12}O_6(s) + 6\ O_2(g) \rightarrow 6\ SiO_2(s) + 6\ H_2O(\ell) \qquad (1\text{-}3)$$

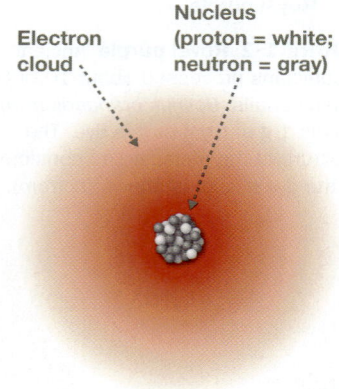

FIGURE 1-3 Quartz crystal Quartz (silicon dioxide) is the silicon analog of carbon dioxide. Whereas carbon dioxide is gaseous, silicon dioxide is a solid.

FIGURE 1-4 Basic structure of the atom Atoms are composed of a nucleus surrounded by a cloud of electrons. Protons (white) and neutrons (gray) make up the nucleus. (This figure is not to scale. If it were, the size of the electron cloud, which is much larger than the size of the nucleus, would have a radius on the order of 500 meters!)

1.3 Atomic Structure and Ground State Electron Configurations

In Section 1.2, we saw that carbon's bonding characteristics are what give rise to the large variety of organic molecules. Those bonding characteristics, and the bonding characteristics of all atoms, are governed by the electrons that the atom has.

This section, then, is devoted to the nature of electrons in atoms. We first review the basic structure of an atom, followed by a discussion of orbitals and shells. Finally, we review electron configurations, distinguishing between *valence electrons*—electrons that can be used for bonding—and *core electrons*.

1.3a The Structure of the Atom

At the center of an atom (Fig. 1-4) is a positively charged nucleus, composed of *protons* and *neutrons*. Surrounding the nucleus is a cloud of negatively charged *electrons*, attracted to the nucleus by simple **electrostatic forces** (the forces by which opposite

Chemistry with Chicken Wire

Even though carbon takes center stage in organic chemistry, organic molecules invariably include other atoms as well, such as hydrogen, nitrogen, oxygen, and halogen atoms. Some of the most exciting chemistry today, however, involves extended frameworks of *only* carbon. A single flat sheet of such a framework is called *graphene*, and resembles molecular chicken wire. Wrapped around to form a cylinder, a graphene sheet forms what is called a *carbon nanotube*. Pure carbon can even take the form of a soccer ball—the so-called *buckminsterfullerene*.

| A sheet of graphene | A carbon nanotube | Buckminsterfullerene |

These structures themselves have quite interesting electronic properties, giving them a bright future in nanoelectronics. Carbon nanotubes and buckminsterfullerenes have high tensile strength, moreover, giving them potential use for structural reinforcement in concrete, sports equipment, and body armor. Chemical modification gives these structures an even wider variety of potential uses. Graphene oxide, for example, has promising antimicrobial activity, and attaching certain molecular groups to the surface of a carbon nanotube or buckminsterfullerene has potential for use as drug carriers for cancer therapeutics.

charges attract one another and like charges repel one another). Individual electrons are incredibly small, even much smaller than the nucleus, but the space that electrons occupy (i.e., the *electron cloud*) is much larger than the nucleus. In other words:

- The size of an atom is essentially defined by the size of its electron cloud.
- The vast majority of an electron cloud (and thus the vast majority of an atom) is empty space.

Table 1-1 lists the mass and charge of each of these elementary particles. Notice that the masses of the proton and neutron are significantly greater than that of the electron, so the mass of an atom is essentially the mass of just the nucleus.

An atom, by definition, has no net charge. Consequently, the number of electrons in an atom must equal the number of protons. The number of protons in the nucleus, called the **atomic number (Z)**, defines the element. For example, a nucleus that has six protons has an atomic number of 6, and can only be a carbon nucleus.

If the number of protons and the number of electrons are unequal, then the entire **species** (that particular combination of protons, neutrons, and electrons) bears a net charge, and is called an **ion**. A negatively charged ion, an **anion** (pronounced AN-eye-on), results from an excess of electrons. A positively charged ion, a **cation** (pronounced CAT-eye-on), results from a deficiency of electrons.

TABLE 1-1 Charges and Masses of Subatomic Particles

Particle	Charge (e)[a]	Mass (u)[b]
Proton	+1	~1
Neutron	0	~1
Electron	−1	~0.0005

[a]e = Elementary charge.
[b]u = Unified atomic mass unit.

SOLVED PROBLEM 1.1

How many protons and electrons does a cation of the carbon atom have if its net charge is +1?

Think How many protons are there in the nucleus of a carbon atom? Does a cation have more protons than electrons, or vice versa? How many more, given the net charge of the species?

Solve A carbon atom's nucleus has six protons. A cation with a +1 charge should have one more proton than it has electrons, so this species must have five electrons.

PROBLEM 1.2 **(a)** How many protons and electrons does an anion of the carbon atom have if its net charge is −1? **(b)** How many protons and electrons does a cation of the oxygen atom have if its net charge is +1? **(c)** How many protons and electrons does an anion of the oxygen atom have if its net charge is −1?

1.3b Atomic Orbitals and Shells

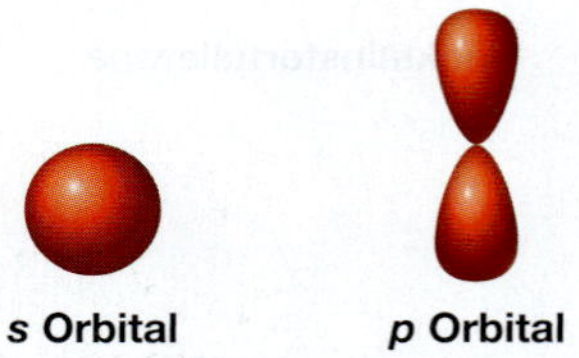

Electrons in an isolated atom reside in **atomic orbitals**. As we shall see, the exact location of an electron can never be pinpointed. An orbital, however, specifies the region of space where the *probability* of finding a given electron is high. More simplistically, we can view orbitals as "rooms" that house electrons. Atomic orbitals are examined in greater detail in Chapter 3; for now, it will suffice to review some of their more basic concepts.

- Atomic orbitals have different shapes. An *s* orbital, for example, is a sphere, whereas a *p* orbital has a dumbbell shape with two lobes (Fig. 1-5). Each orbital is centered on the nucleus of its atom or ion.
- Atomic orbitals are organized in *shells* (also known as *energy levels*). A **shell** is defined by the **principal quantum number**, *n*. There are an infinite number of shells in an atom, given that *n* can assume any integer value from 1 to infinity.
  - The first shell ($n = 1$) contains only an *s* orbital, called 1*s*.
  - The second shell ($n = 2$) contains one *s* orbital and three *p* orbitals, called 2*s*, $2p_x$, $2p_y$, and $2p_z$.
  - The third shell ($n = 3$) contains one *s* orbital, three *p* orbitals, and five *d* orbitals.
- Up to two electrons are allowed in any orbital.
  - Therefore, the first shell can contain up to two electrons (a **duet**).
  - The second shell can contain up to eight electrons (an **octet**).
  - The third shell can contain up to 18 electrons.
- With increasing shell number, the *size* and *energy* of the atomic orbital increase. For example, comparing *s* orbitals in the first three shells, the size and energy increase in the order 1*s* < 2*s* < 3*s*, as shown in Figure 1-6. Similarly, a 2*p* orbital is smaller in size and lower in energy than a 3*p* orbital.
- Within a given shell, an atomic orbital's energy increases in the following order: *s* < *p* < *d*, etc. In the second shell, for example, the 2*s* orbital is lower in energy than the 2*p*.

1.3c Ground State Electron Configurations: Valence Electrons and Core Electrons

The way in which electrons are arranged in atomic orbitals is called the atom's **electron configuration**. The *most stable* (i.e., the lowest energy) electron configuration is called the **ground state** configuration. Knowing an atom's ground state configuration provides insight into the atom's chemical behavior, as we will see.

s Orbital   p Orbital

FIGURE 1-5 Orbitals Orbitals represent regions in space where an electron is likely to be. An s orbital is spherical, and a p orbital is a dumbbell.

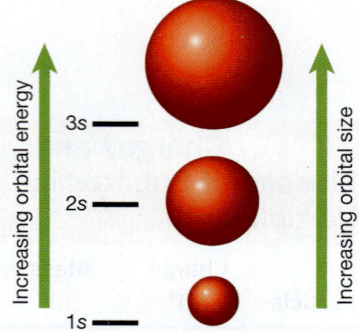

FIGURE 1-6 Relationship between principal quantum number, orbital size, and orbital energy As the shell number of an orbital increases, its size and energy increase, too. The horizontal black lines indicate each orbital's energy.

With the relative energies of atomic orbitals established, an atom's ground state electron configuration can be obtained by applying the following three rules:

1. **Pauli's exclusion principle:** No more than two electrons (i.e., zero, one, or two electrons) can occupy a single orbital; two electrons in the same orbital must have opposite spins.
2. **Aufbau principle:** Each successive electron must fill the lowest energy orbital available.
3. **Hund's rule:** Before a second electron can be paired in the same orbital, all other orbitals *at the same energy* must contain a single electron.

According to these three rules, the first 18 electrons fill orbitals as indicated in Figure 1-7. Each arrow represents an electron, and the direction of the arrow—up or down— represents the electron's spin.

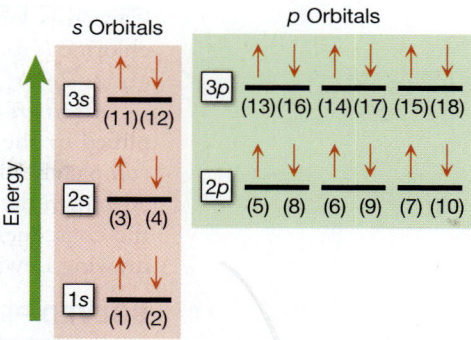

FIGURE 1-7 Energy diagram of atomic orbitals for the first 18 electrons The order of electron filling is indicated in parentheses. Each horizontal black line represents a single orbital. Each successive electron fills the lowest energy orbital available. Notice in the 2*p* and 3*p* sets of orbitals that no electrons are paired up until the addition of the fourth electron.

YOUR TURN **1.1**

In Figure 1-7, place a box around all of the orbitals in the second shell and label them.

Answers to Your Turns are in the back of the book.

In the ground state, the six electrons found in a carbon atom would fill the orbitals as shown in Figure 1-8, with two electrons in the 1*s* orbital, two electrons in the 2*s* orbital, and one electron in each of two different 2*p* orbitals (it doesn't matter which two). The shorthand notation for this electron configuration is $1s^2 2s^2 2p^2$.

Knowing the ground state electron configuration of an atom, we can distinguish *valence* electrons from *core* electrons.

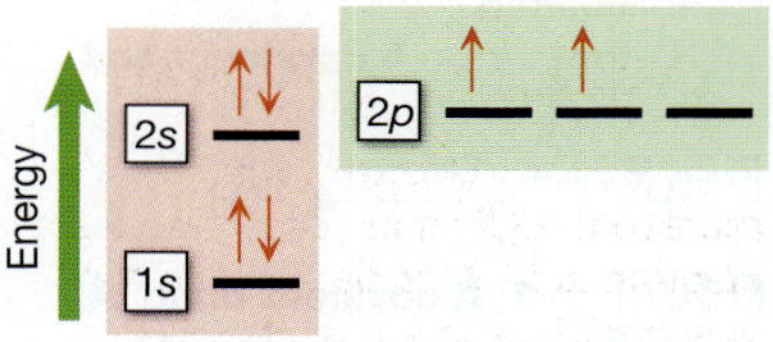

FIGURE 1-8 Energy diagram for the ground state electron configuration of the carbon atom This configuration is abbreviated $1s^2 2s^2 2p^2$.

- **Valence electrons** are those occupying the highest energy (i.e., valence) shell. For the carbon atom, the valence shell is the $n = 2$ shell.
- **Core electrons** occupy the remaining lower energy shells of the atom. For the carbon atom, the core electrons occupy the $n = 1$ shell.

Valence electrons are important because, as we discuss in Section 1.5, they participate in covalent bonds. As we can see in Figure 1-8, for example, carbon has four valence electrons and two core electrons, so bonding involving carbon is governed by those four valence electrons.

YOUR TURN **1.2**

In Figure 1-8, place a circle around the valence electrons and label them. Place a box around all of the core electrons and label them.

We can use the periodic table to quickly determine how many valence electrons an atom has (a copy of the periodic table appears inside the book's front cover).

The number of valence electrons in an atom is the same as the atom's *group number*.

Carbon is located in group 4A, consistent with its four valence electrons, whereas chlorine (group 7A) has seven. According to its ground state electron configuration $(1s^22s^22p^63s^23p^5)$, chlorine's valence electrons occupy the third shell.

Atoms are especially stable when they have completely filled valence shells. This is exemplified by the **noble gases** (group 8A), such as helium and neon, because they have completely filled valence shells and they do *not* form bonds to make compounds. Although the specific origin of this "extra" stability is beyond the scope of this book, the consequences are the basis for the octet and duet rules we routinely use when drawing Lewis structures (Section 1.5).

SOLVED PROBLEM 1.3

Write the ground state electron configuration of the nitrogen atom. How many valence electrons does it have? How many core electrons does it have?

Think How many total electrons are there in a nitrogen atom? What is the order in which the atomic orbitals should be filled (see Fig. 1-7)? What is the valence shell and where do the core electrons reside?

Solve There are seven total electrons ($Z = 7$ for N). The first two are placed in the 1s orbital and the next two in the 2s orbital, leaving one electron for each of the three 2p orbitals. The electron configuration is $1s^22s^22p^3$. The valence shell is the second shell, so there are five valence electrons and two core electrons.

PROBLEM 1.4 Write the ground state electron configuration of the oxygen atom. How many valence electrons and how many core electrons are there?

Each electron belongs to an isolated H atom.   A covalent bond

H·   ·H ⟶ H:H

FIGURE 1-9 A covalent bond A covalent bond is the sharing of two electrons between nuclei.

1.4 The Covalent Bond: Bond Energy and Bond Length

In a compound, nuclei are held together by chemical bonds. Two types of fundamental bonds in chemistry are the *covalent bond* and the *ionic bond* (see Section 1.8). A **covalent bond** is characterized by the *sharing of valence electrons* between two or more atoms, as shown for two H atoms in a molecule of H_2 (hydrogen gas) in Figure 1-9.

In Section 1.5, we will explore how various molecules can be constructed from atoms through the formation of covalent bonds, but first let's examine the nature of covalent bonds more closely. Why do they form at all?

We can begin to answer this question by examining Figure 1-10a, which illustrates how the energy of two H atoms changes as a function of the distance between their nuclei. In particular, when two H atoms separated by a large distance are brought together, their total energy begins to decrease.

Lower energy corresponds to greater stability.

At one particular internuclear distance, the energy of the molecule is at a minimum, while at shorter distances the energy rises dramatically.

The internuclear distance at which energy is the lowest is called the **bond length** of the H—H bond. The energy that is required to remove the H atoms from that internuclear distance to infinity (toward the right in the figure) is the **bond strength**, or **bond energy**, of the H—H bond.

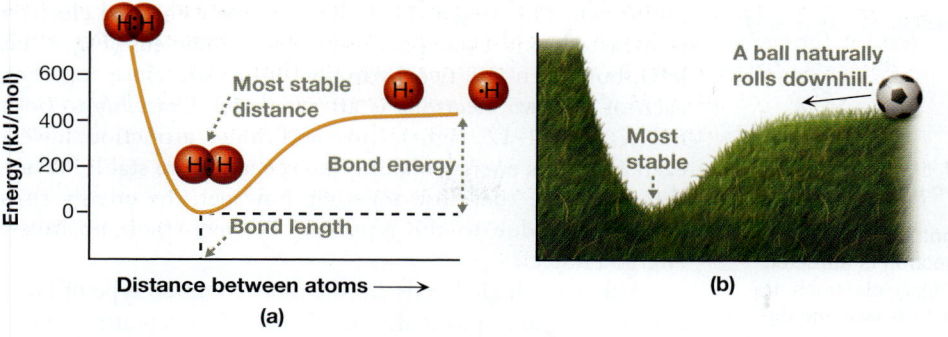

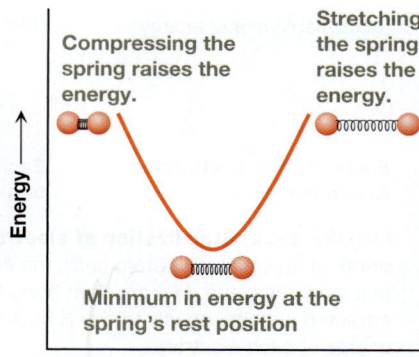

FIGURE 1-10 Formation of a chemical bond (a) Plot of energy as a function of the internuclear distance for two H atoms. The H atoms are most stable at the distance at which energy is a minimum. (b) A ball at the top of a hill becomes more stable at the bottom of the hill, and therefore tends to roll downhill.

FIGURE 1-11 The spring model of a covalent bond The energy curve of a spring connecting two masses resembles that of the covalent bond shown in Figure 1-10a. Both stretching and compressing the spring from its rest position increase the energy in the spring.

This idea is analogous to a ball rolling down a hill (Fig. 1-10b). A ball at the top of a hill has more potential energy than a ball at the bottom, so the ball at the top tends to roll downhill, coming to rest at the bottom. By the same token, it takes energy to roll the ball from the bottom of the hill back to the top.

YOUR TURN **1.3**

Estimate the bond energy of the bond represented by Figure 1-10a.

It is often convenient to *think of a covalent bond as a spring that connects two atoms.* Just as it takes energy to lengthen or shorten a covalent bond from its bond length, it takes energy to stretch or compress a spring from its rest position, as shown in Figure 1-11.

SOLVED PROBLEM 1.5

In the diagram shown here, which curve represents a stronger covalent bond?

Think How can bond breaking be represented for each curve? Which of those processes requires more energy?

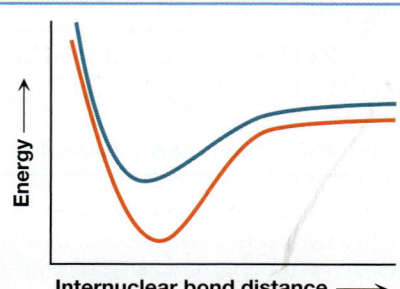

Solve Bond breaking is represented by climbing from the bottom of the curve toward the right (i.e., the internuclear bond distance increases toward the right). For this process, more energy is required for the red curve, so the red curve represents a stronger bond.

CONNECTIONS The behavior of covalent bonds as springs (Fig. 1-11) is what enables greenhouse gases like carbon dioxide (CO_2) and methane (CH_4) to absorb infrared radiation and warm the atmosphere.

PROBLEM 1.6 Which of the two curves in Solved Problem 1.5 represents a longer bond?

Why are two hydrogen atoms connected by a covalent bond lower in energy than two isolated hydrogen atoms? Largely it is because of the additional electrostatic attraction experienced by electrons when they are *shared* between nuclei. In

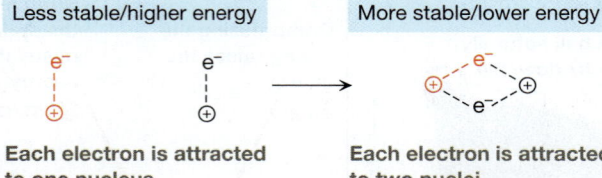

Less stable/higher energy	More stable/lower energy
Each electron is attracted to one nucleus.	Each electron is attracted to two nuclei.

FIGURE 1-12 Stabilization of electrons in a covalent bond In an isolated H atom (*left*), the electron is attracted to a single nucleus. In a covalent bond (*right*), electrons are attracted simultaneously to two H nuclei, thus lowering the energy of each electron.

each *isolated* hydrogen atom, the negatively charged electron is attracted to just one positively charged nucleus (Fig. 1-12, left), but when the two hydrogen atoms are close together, each of the two electrons is attracted *simultaneously* to both nuclei (Fig. 1-12, right). This additional attraction lowers each electron's energy, making the species more stable. When the atoms are too close together, however, the energy rises dramatically due to the repulsion between their positively charged nuclei.

Although single bonds are the most common type of bond found in organic molecules, we frequently encounter double bonds and triple bonds as well. The main difference among single, double, and triple bonds is the number of electrons involved. In a single bond, two electrons are shared between two nuclei; in a double bond, four electrons are shared; and in a triple bond, six electrons are shared.

Tables 1-2 and 1-3 list average bond energies for a variety of common bonding partners found in organic species. Table 1-2 contains only single bonds, whereas

TABLE 1-2 Average Bond Energies of Common Single Bonds

	H kJ/mol	H kcal/mol	C kJ/mol	C kcal/mol	N kJ/mol	N kcal/mol	O kJ/mol	O kcal/mol	F kJ/mol	F kcal/mol	Cl kJ/mol	Cl kcal/mol	Br kJ/mol	Br kcal/mol	I kJ/mol	I kcal/mol	Si kJ/mol	Si kcal/mol
H	436	104	418	100	389	93	460	110	569	136	431	103	368	88	297	71	301	72
C	418	100	339	81	289	69	351	84	439	105	331	79	280	67	238	57	289	69
N	389	93	289	69	159	38	180	43	272	65	201	48	243	58	169	40	355	85
O	460	110	351	84	180	43	138	33	209	50	209	50	222	53	238	57	430	103
F	569	136	439	105	272	65	209	50	159	38	251	60	251	60	280	67	586	140
Cl	431	103	331	79	201	48	209	50	251	60	243	58	222	53	209	50	402	96
Br	368	88	280	67	243	58	222	53	251	60	222	53	192	46	180	43	289	69
I	297	71	238	57	169	40	238	57	280	67	209	50	180	43	151	36	209	50
Si	301	72	289	69	355	85	430	103	586	140	402	96	289	69	209	50	223	53

TABLE 1-3 Average Bond Energies and Bond Lengths of Single and Multiple Bonds

Atoms Bonded Together	SINGLE BOND ENERGY kJ/mol	SINGLE BOND ENERGY kcal/mol	SINGLE BOND LENGTH picometers (pm)	DOUBLE BOND ENERGY kJ/mol	DOUBLE BOND ENERGY kcal/mol	DOUBLE BOND LENGTH pm	TRIPLE BOND ENERGY kJ/mol	TRIPLE BOND ENERGY kcal/mol	TRIPLE BOND LENGTH pm
C—C	339	81	154	619	148	134	812	194	120
C—N	289	69	147	619	148	129	891	213	116
C—O	351	84	143	720	172	120	1072	256	113
N—N	159	38	145	418	100	125	946	226	110
O—O	138	33	148	498	119	121			

Table 1-3 compares bond energies and lengths for some common single, double, and triple bonds.

Notice in Table 1-3 that *as the number of bonds increases between a pair of atoms, the bond energy increases and the bond length decreases.* Thus, double and triple bonds can be viewed as shorter, stronger springs than single bonds (Fig. 1-13).

Refer to Tables 1-2 and 1-3 to answer the following questions, which are designed to acquaint you with the range of strengths of common bonds.

(a) What is the value of the strongest *single* bond listed? _____

(b) What bond does that correspond to? _____

(c) What is the value of the weakest *single* bond? _____

(d) What bond does that correspond to? _____

(e) What is the value of the strongest bond of any type? _____

(f) What bond does that correspond to? _____

YOUR TURN 1.4

Bond strength increases; bond length decreases →

X—Y	X=Y	X≡Y
Single bond	**Double bond**	**Triple bond**

FIGURE 1-13 Bond strength and bond length For a particular pair of atoms that are covalently bonded (X and Y), a triple bond is shorter and stronger than a double bond, which is shorter and stronger than a single bond.

Turning an Inorganic Surface into an Organic Surface

Gold (Au) is a relatively unreactive metal, which is one reason it is widely used in jewelry and high-end electronic components. Gold, however, has a relatively high affinity for sulfur (S); the Au—S bond energy is roughly 120–150 kJ/mol (~30–40 kcal/mol), or nearly half the strength of a typical C—C bond. Sulfur also forms a relatively strong bond to carbon, about 290 kJ/mol (70 kcal/mol). This dual affinity of sulfur has enabled chemists to use sulfur to anchor organic groups to the surface of gold. The reaction is extremely easy to carry out: A sample of gold metal is simply immersed in a solution of an alkanethiol [$CH_3(CH_2)_n$—SH] and the Au—S bond forms spontaneously, yielding a *self-assembled monolayer* (shown below) in which the alkyl chains extend away from the gold surface. Effectively, then, the inorganic gold surface is converted into an organic one.

These self-assembled monolayers have a wide variety of applications, such as protecting the gold surface from substances that would otherwise cause corrosion. More interestingly, by changing the organic portion to which the thiol group (—SH) is attached, the gold surface can be programmed to have specific affinity for other molecules. Gold nanoparticles (GNP), for example, have been capped with thiol-terminated molecules that enable the GNP to form bonds to epidermal growth-factor receptor antibodies. The resulting antibody–GNP conjugates have been used to image cancer cells.

Alkyl chain
Sulfur
GOLD

1.5 Lewis Dot Structures and the Octet Rule

To understand a molecule's chemical behavior, it is necessary to know its **connectivity**—that is, which atoms are bonded together, and by what types of bonds (single, double, or triple). It is also useful to know which valence electrons participate in bonding and which do not. **Lewis dot structures** (or, **Lewis structures**) are a convenient way to convey this information. Let's review some basic conventions of Lewis structures.

CONNECTIONS An isolated chlorine atom is referred to as a chlorine radical. Chlorine radicals in the stratosphere catalyze the breakdown of stratospheric ozone. Unnaturally high concentrations of chlorine radicals in the stratosphere, produced from synthetic coolants called chlorofluorocarbons (CFCs), have, since the middle of the 20th century, led to the development of the ozone hole over Antarctica.

- *Lewis structures take into account only valence electrons.* In a complete Lewis structure, such as the following for isolated C and Cl atoms, *all* valence electrons are shown:

  4 valence electrons   7 valence electrons

  $\cdot\dot{C}\cdot$ $:\dot{\ddot{C}l}\cdot$

- *Bonding and nonbonding electrons are clearly shown.*
  - Single, double, and triple bonds are indicated by one, two, or three lines (i.e. —, =, or ≡), respectively, which represent the sharing of two, four, or six electrons, respectively. Thus, each line represents a shared pair, or **bonding pair**, of electrons.
  - Nonbonding electrons are indicated by dots, and are usually paired (:). These are called **lone pairs** of electrons. In some species, nonbonding electrons are **unpaired**, and are represented by single dots. These species are called **free radicals** and are discussed in greater detail in Chapter 25.
- *Atoms in Lewis structures obey the duet rule and the octet rule.* Atoms are especially stable when they have complete valence shells: two electrons (a duet) for hydrogen and helium, and eight electrons (an octet) for atoms in the second row of the periodic table. These duets and octets can be achieved through the formation of covalent bonds, where valence electrons are *shared* between atoms. Examples are shown in Figure 1-14.

FIGURE 1-14 Covalent bonding: Sharing electrons to produce full valence shells In each of these molecules, all atoms have completely filled valence shells: H has a share of two electrons (a, b, and d), whereas C (b and c), O (c), and N (d) each have an octet of electrons made up of a total of 8 shared and unshared valence electrons.

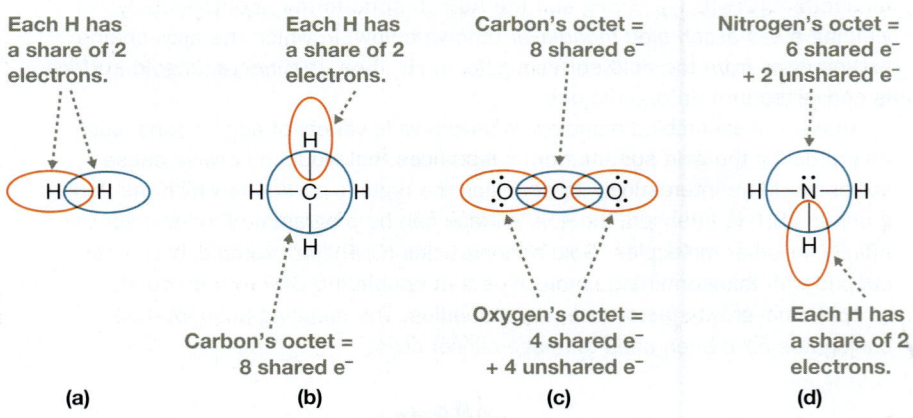

Each H has a share of 2 electrons.

Each H has a share of 2 electrons.

Carbon's octet = 8 shared e⁻

Nitrogen's octet = 6 shared e⁻ + 2 unshared e⁻

Carbon's octet = 8 shared e⁻

Oxygen's octet = 4 shared e⁻ + 4 unshared e⁻

Each H has a share of 2 electrons.

(a)   (b)   (c)   (d)

YOUR TURN 1.5

Using Figure 1-14 as your guide, circle the electrons in the Lewis structure of CH_3OH that represent carbon's octet, oxygen's octet, and each hydrogen's duet.

Because of their widespread use in organic chemistry, you must be able to draw Lewis structures quickly and accurately. The following steps allow you to do so in a systematic way.

Steps for Drawing Lewis Structures

1. Count the total number of *valence* electrons in the molecule.
   a. The number of valence electrons contributed by each atom is the same as its group number (H = 1, C = 4, N = 5, O = 6, F = 7).
   b. Each negative charge increases the number of valence electrons by one; each positive charge decreases the number of electrons by one.

2. Write the skeleton of the molecule, showing only the atoms and the single bonds required to hold them together.
   a. If molecular connectivity is not given to you, the central atom (the one with the greatest number of bonds) is usually the one with the lowest electronegativity. (Electronegativity is reviewed in Section 1.7.)

3. Subtract two electrons from the total in Step 1 for each single covalent bond drawn in Step 2.

4. Distribute the remaining electrons as lone pairs.
   a. Start with the outer atoms and work inward.
   b. Try to achieve an octet on each atom other than hydrogen.

5. If there is an atom with less than an octet, increase the atom's share of electrons by converting lone pairs from *neighboring* atoms into bonding pairs, thereby creating double or triple bonds.

CONNECTIONS Methanol (CH_3OH, Your Turn 1.5), also called methyl alcohol or wood alcohol, is used industrially as a feedstock to produce formaldehyde, which is integral in the production of some plastics, paints, and explosives. Methanol is also the primary fuel used in many types of racing vehicles.

SOLVED PROBLEM 1.7

Draw a Lewis structure of HCO_2^-, where carbon is the central atom.

Think Consider the steps for drawing Lewis structures. Which atoms must be bonded together?

Solve First, count the total number of valence electrons.

1 H atom:	1 × 1 valence electron	= 1 valence electron
1 C atom:	1 × 4 valence electrons	= 4 valence electrons
2 O atoms:	2 × 6 valence electrons	= 12 valence electrons
−1 charge:	1 × 1 valence electron	= 1 valence electron
	Total:	= 18 valence electrons

Carbon is the central atom, so six electrons must be used to connect C to the other three atoms. This leaves 12 more that can be placed as lone pairs around the O atoms to achieve octets, as shown at the right. To give C its octet, one of those lone pairs is converted to a C=O double bond.

Convert a lone pair into a bonding pair to give C its octet.

PROBLEM 1.8 Draw a Lewis structure for C_2H_3N. One carbon is bonded to three hydrogen atoms and to the second carbon. The nitrogen atom is bonded only to a carbon atom.

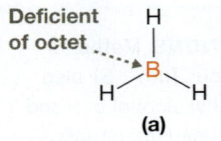

Deficient of octet

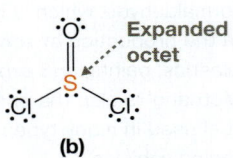

Expanded octet

(b)

FIGURE 1-15 **Exceptions to the octet rule** (a) Boron is in the second row and has less than an octet. (b) Sulfur is in the third row and has an expanded octet.

CONNECTIONS Borane (BH$_3$) and thionyl chloride (SOCl$_2$), shown in Figure 1-15, are important reagents in organic synthesis. BH$_3$ is involved in hydroboration (Section 12.6). SOCl$_2$ is commonly used to replace an –OH group with –Cl in certain types of molecules (Section 21.4). SOCl$_2$ is also used in lithium–thionyl chloride batteries, where it acts as the positive electrode.

In some molecular species, not all atoms have a complete valence shell. In borane (BH$_3$, Fig. 1-15a), for example, the B atom has a share of only six valence electrons. It is two electrons short of an octet because there are not enough valence electrons available to achieve a complete valence shell for all atoms in the molecule. Similarly, in thionyl chloride (SOCl$_2$, Fig. 1-15b), the S atom has a share of 10 electrons but, being in the third row of the periodic table, its valence shell can contain up to 18 electrons. There are even some examples of molecular species in which hydrogen has a share of fewer than two electrons, but these are somewhat rare. (Such examples will be discussed as necessary.)

Because of the emphasis that organic chemistry has on atoms from the second row of the periodic table, we often talk about atoms from the third row and below in terms of an octet. In SOCl$_2$, for example, both Cl atoms have a share of eight valence electrons, so we say that each Cl atom has an octet. The S atom, on the other hand, has what is called an **expanded octet**, given its share of 10 electrons. Be careful, however, when using this terminology, because only atoms in the third row and below can have an expanded octet.

Atoms in the second row are forbidden to exceed the octet!

PROBLEM 1.9 For each structure below, determine whether it is a legitimate Lewis structure. If not, explain why not.

(a) **(b)** **(c)**

(d) **(e)**

1.6 Strategies for Success: Drawing Lewis Dot Structures Quickly

After you've used the systematic steps (p. 13) to construct Lewis structures of several species, you may begin to notice that each type of atom tends to form a specific number of bonds and to have a specific number of lone pairs of electrons. Table 1-4 summarizes these patterns. What you will learn later is that *those are the number of bonds and lone pairs for atoms that bear no formal charge.* Atoms that are charged have combinations of bonds and lone pairs that are different from those in Table 1-4.

We can now use the patterns shown in Table 1-4 to complete the Lewis structure in Solved Problem 1.10.

TABLE 1-4 Common Numbers of Covalent Bonds and Lone Pairs for Selected Uncharged Atoms

Atom	Number of Bonds	Number of Lone Pairs	Examples
H	1	0	—H
C	4	0	—C— =C< ≡C— =C=
N	3	1	—N̈— =N̈· ≡N:
O	2	2	—Ö— =Ö:
X (X = F, Cl, Br, I)	1	3	—F̈: —C̈l: —B̈r: —Ï:
Ne	0	4	:N̈e:

SOLVED PROBLEM 1.10

Complete the Lewis structure for the compound whose skeleton is shown at the right. Assume that all atoms are uncharged.

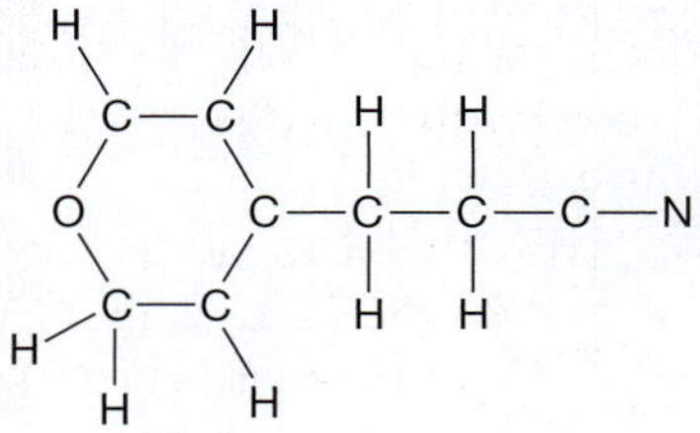

Think Which atoms (other than hydrogen) have an octet and which atoms don't? How many bonds and lone pairs are typical for each element?

Solve The atoms not shown with an octet are highlighted in red below on the left. According to Table 1-4, we need to add two lone pairs to O, convert two C—C single bonds of the ring into double bonds, convert the C—N single bond into a triple bond, and add a lone pair to N.

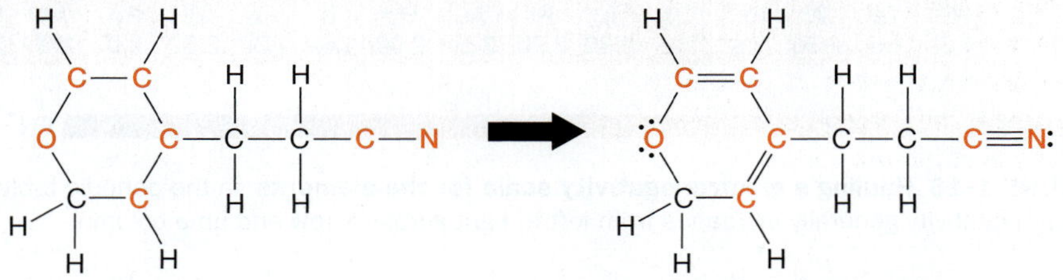

Red atoms = no octet   All atoms have octet.

PROBLEM 1.11 Complete the Lewis structure for the molecule with the connectivity shown at the right. Assume that all atoms have the number of bonds and lone pairs listed in Table 1-4.

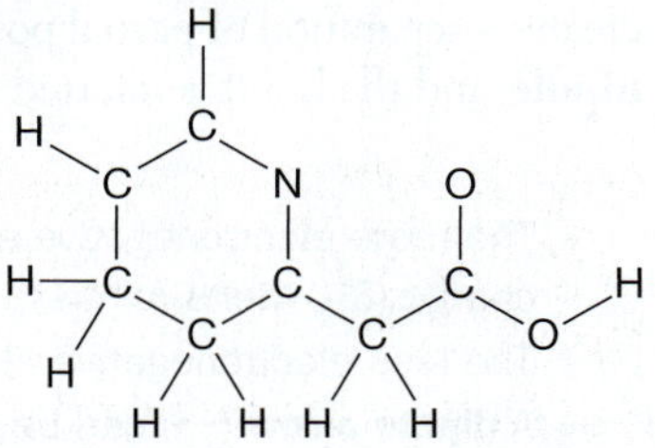

1.7 Electronegativity, Polar Covalent Bonds, and Bond Dipoles

We've seen that covalent bonds are characterized by the sharing of electrons between two atomic nuclei. If the atoms are identical, the electrons are shared equally and the bond is called a **nonpolar covalent bond**. Otherwise, one nucleus will attract electrons more strongly than the other. The ability to attract electrons in a covalent bond is defined as the element's **electronegativity (EN)**.

There are a variety of different electronegativity scales that assign values to each element, but the one devised by Linus Pauling, which ranges from 0 to about 4 (see Figure 1-16), is perhaps the most well known. For main group elements, electronegativity values exhibit the following periodic trends:

- Within the same row, electronegativity values tend to increase from left to right across the periodic table.
- Within the same column, they tend to increase from bottom to top.

As a result, the elements with the highest electronegativities (not counting the noble gases) tend to be in the upper right corner of the periodic table (e.g., N, O, F, Cl, and Br), whereas the elements with the lowest electronegativities tend to be in the lower left corner (e.g., K, Rb, Cs, Sr, Ba).

Electronegativity (EN) using the Pauling scale

Group (vertical)	1	2	3	4	5	6	7	8	9	10	11	12	13	14	15	16	17	18
Period (horizontal)	1A	2A											3A	4A	5A	6A	7A	8A
1	H 2.20																	He
2	Li 0.98	Be 1.57											B 2.04	C 2.55	N 3.04	O 3.44	F 3.98	Ne
3	Na 0.93	Mg 1.31											Al 1.61	Si 1.90	P 2.19	S 2.58	Cl 3.16	Ar
4	K 0.82	Ca 1.00	Sc 1.36	Ti 1.54	V 1.63	Cr 1.66	Mn 1.55	Fe 1.83	Co 1.88	Ni 1.91	Cu 1.90	Zn 1.65	Ga 1.81	Ge 2.01	As 2.18	Se 2.55	Br 2.96	Kr 3.00
5	Rb 0.82	Sr 0.95	Y 1.22	Zr 1.33	Nb 1.6	Mo 2.16	Tc 1.9	Ru 2.2	Rh 2.28	Pd 2.20	Ag 1.93	Cd 1.69	In 1.78	Sn 1.96	Sb 2.05	Te 2.1	I 2.66	Xe 2.60
6	Cs 0.79	Ba 0.89	La 1.1	Hf 1.3	Ta 1.5	W 2.36	Re 1.9	Os 2.2	Ir 2.20	Pt 2.28	Au 2.54	Hg 2.00	Tl 1.62	Pb 2.33	Bi 2.02	Po 2.0	At 2.2	Rn 2.2

EN = 1.0   EN = 4.0

FIGURE 1-16 Pauling's electronegativity scale for the elements In the periodic table, electronegativity generally increases from left to right across a row and up a column.

In a covalent bond, electrons are more likely to be found near the nucleus of the more electronegative atom and less likely near the nucleus of the less electronegative atom. This creates a separation of partial positive and negative charges along the bond, called a **bond dipole**, and the bond is referred to as a **polar covalent bond**. More specifically:

- The more electronegative atom of a covalent bond bears a partial negative charge (δ^-, "delta minus").
- The less electronegative atom bears a partial positive charge (δ^+, "delta plus").
- A **dipole arrow** ($\mapsto$) can be drawn from the less electronegative atom (δ^+) toward the more electronegative atom (δ^-).

These ideas are shown for HF, CH_4, and CO_2 in Figure 1-17.

The magnitude of a bond dipole depends on the *difference* in electronegativity between the atoms involved in the bond. A larger difference in electronegativity results in a larger bond dipole. Relative magnitudes of a bond dipole are often depicted by the lengths of dipole arrows. For example, the difference in electronegativity between hydrogen and fluorine is larger than that between carbon and hydrogen. In Figure 1-17, therefore, the dipole arrow along the H—F bond is longer than the bond dipole arrows along the H—C bonds.

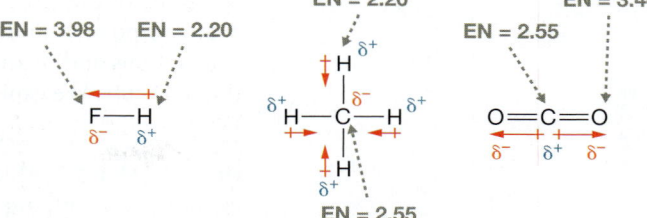

FIGURE 1-17 **Bond dipoles in various molecules** The dipoles are represented by the red arrows. Each arrow points from the less electronegative atom (δ^+) to the more electronegative atom (δ^-). The length of the arrow indicates the relative magnitude of the bond dipole. EN = electronegativity.

YOUR TURN **1.6**

The Lewis structure of BH_3 is shown here. Write the electronegativity next to each atom. Along one of the B—H bonds, draw the corresponding dipole arrow, and add the δ^+ and δ^- symbols.

PROBLEM 1.12 For each *uncharged* molecule at the right, **(a)** complete the Lewis structure by adding multiple bonds and/or lone pairs, and **(b)** draw dipole arrows along each polar covalent bond. Pay attention to the lengths of the arrows.

(i)   (ii)   (iii)

Another useful way to illustrate the distribution of charge along a covalent bond is with an **electrostatic potential map**, examples of which are shown in Figure 1-18. An electrostatic potential map depicts a molecule's electron cloud in colors that indicate its *relative* charge. Red corresponds to a buildup of negative charge, whereas blue represents a buildup of positive charge. Colors in between red and blue in the spectrum, such as green, represent a more neutral charge.

More positive charge   More negative charge

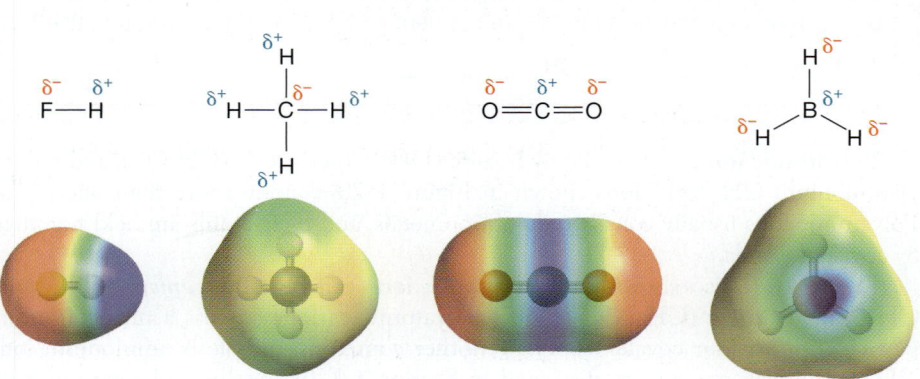

FIGURE 1-18 **Electrostatic potential maps of four molecules** Red indicates a buildup of negative charge; blue indicates a buildup of positive charge.

Some of the structures in Figures 1-17 and 1-18 demonstrate that a single molecule can possess more than one bond dipole. When this occurs, the bond dipoles' orientations and relative magnitudes dictate the overall distribution of charge within the molecule. We explore this concept further in Chapter 2.

PROBLEM 1.13 Which of the compounds below is consistent with the electrostatic potential map shown? Explain.

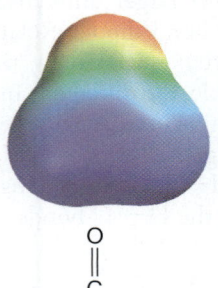

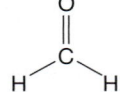 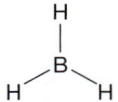 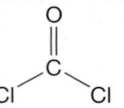

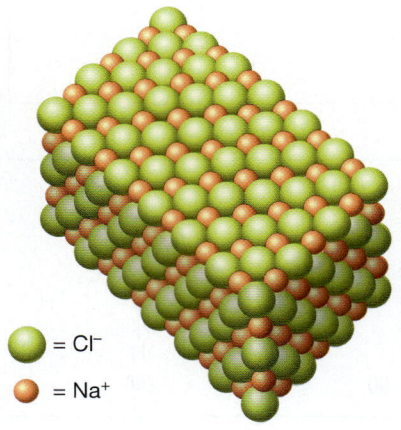

= Cl⁻

= Na⁺

FIGURE 1-19 Representation of the solid crystal structure of NaCl(s) The Na⁺ and Cl⁻ ions are held together by electrostatic forces, called ionic bonds, whereby opposite charges attract. As a solid, the ions form a regular array called a crystal lattice.

1.8 Ionic Bonds

When elements in a compound have large enough differences in electronegativity, **ionic bonding** can occur. Rather than sharing electrons, as in a covalent bond, the more electronegative atom acquires electrons given up by the less electronegative atom, forming oppositely charged ions. The electrostatic attraction between the positively charged cations and the negatively charged anions constitutes the ionic bond. Sodium chloride (NaCl), for example, consists of sodium cations (Na⁺) and chloride anions (Cl⁻). Sodium is a metal from the left side of the periodic table, so it has a low electronegativity value (0.93), whereas chlorine is a nonmetal from the right side of the periodic table, so it has a high electronegativity value (3.16).

Most ionic compounds (e.g., NaCl and MgBr₂) consist of a metal and a nonmetal, whereas most covalent compounds (e.g., CH₄) consist of nonmetals only.

While covalent compounds are generally found as discrete uncharged molecules, the ions in an ionic solid are arranged in a regular array, called a **crystal lattice**, as shown for NaCl in Figure 1-19. As we will discuss in greater detail in Section 2.7, *ionic compounds* such as this tend to dissolve in water as their constituent ions.

YOUR TURN 1.7

Calculate the difference in electronegativity between the elements in NaCl (an ionic compound) and those in CH₄ (a covalent compound). What do you notice?

NaCl _____   CH₄ _____

Polyatomic ions, such as the hydroxide (HO⁻), methoxide (CH₃O⁻), and methylammonium $CH_3NH_3^+$ ions shown in Figure 1-20, contain more than one atom. Polyatomic ions usually consist only of nonmetals, and their atoms are held together by covalent bonds.

Figure 1-20 also shows that polyatomic ions can be either *anions* (HO⁻ and CH₃O⁻) or *cations* ($CH_3NH_3^+$). Most polyatomic cations possess a nitrogen atom participating in four covalent bonds (another common example is ammonium ion, NH_4^+). The reasons why are discussed in Section 1.9. Interestingly, the existence of these kinds of polyatomic cations makes possible the formation of ionic compounds composed entirely of nonmetals (e.g., CH₃NH₃Cl or NH₄Br).

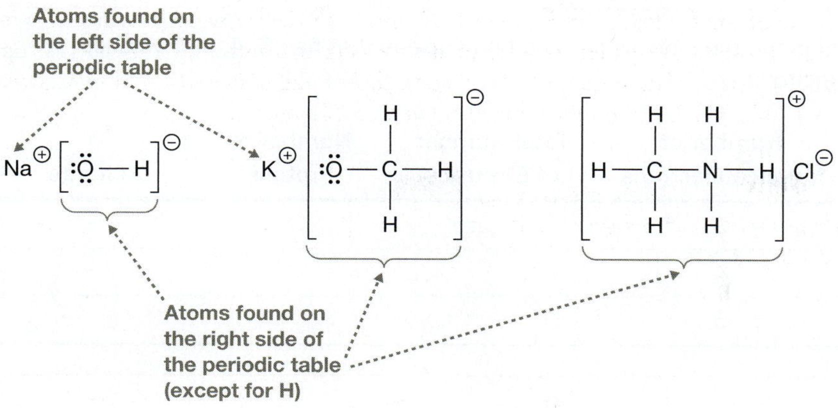

Atoms found on the left side of the periodic table

Atoms found on the right side of the periodic table (except for H)

SOLVED PROBLEM 1.14

Identify each of the following as either an ionic compound (i.e., one containing ionic bonds) or a covalent compound (i.e., one containing only covalent bonds): **(a)** NH_4CHO_2; **(b)** $LiOCH_2CH_3$; **(c)** $CH_3CH_2CH_2OH$.

Think Does the compound contain elements from both the left and right sides of the periodic table? Does the compound contain any recognizable polyatomic cations?

Solve

(a) Ionic compound: NH_4^+ is a common polyatomic cation, leaving CHO_2^- as the anion.

(b) Ionic compound: Li is a metal from the left side of the periodic table, and the remaining elements, which compose the $CH_3CH_2O^-$ anion, are nonmetals from the right side.

(c) Covalent compound: All elements are nonmetals from the right side of the periodic table, and no recognizable polyatomic ions are present.

PROBLEM 1.15 Which of the following are ionic compounds (i.e., ones containing ionic bonds) and which are covalent compounds (i.e., ones containing only covalent bonds)?

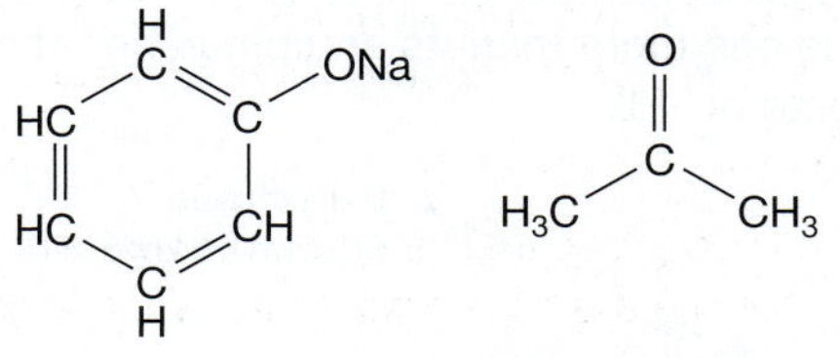

A   B   C

1.9 Assigning Electrons to Atoms in Molecules: Formal Charge

In an *isolated* atom or atomic ion, charge is determined by the difference between the atom's group number and the actual number of valence electrons it possesses. A carbon atom, for example, has zero charge if it possesses four valence electrons—its group number is 4. It carries a charge of −1 if it has five valence electrons and it carries a charge of +1 if it has only three valence electrons.

YOUR TURN 1.8

Fill in the table below for a carbon nucleus that has 3, 4, or 5 valence electrons.

Number of Valence Electrons	Total Number of Electrons	Number of Protons	Charge
3			
4			
5			

In a molecule or polyatomic ion, we can also assign a charge to an individual atom by computing the difference between the atom's group number and the number of valence electrons it possesses. But how do we assign electrons to atoms involved in covalent bonds, where electrons are being *shared*?

One way to do so gives us the atom's **formal charge**, according to these two rules:

- Both electrons of a lone pair are assigned to the atom on which they appear.
- In a given covalent bond, half the electrons are assigned to each atom involved in the bond.

> **CONNECTIONS** The formate anion, HCO_2^-, is produced in significant amounts in the mitochondria of embryonic liver cells and also in cancer cells.

Solved Problem 1.16 shows how to apply these rules to a Lewis structure of HCO_2^-.

SOLVED PROBLEM 1.16

Determine the formal charge on every atom in the methanoate anion (formate anion, HCO_2^-), shown at the right.

Think How are lone pairs assigned? How are bonding pairs of electrons assigned to atoms?

Solve The figure below shows how the formal charge rules are applied to assign valence electrons to each atom. Notice that each pair of electrons in a covalent bond is split evenly. The H atom, the C atom, and the double-bonded O atom are assigned formal charges of 0 because they have the same number of valence electrons as their corresponding group numbers. The single-bonded O is assigned seven valence electrons, which is one more than its group number of 6. It is therefore assigned a formal charge of −1.

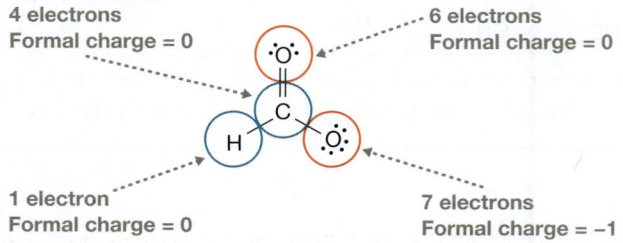

Assigning formal charges to atoms is simply a way of distributing the electrons that are already present within a particular species. Therefore:

The formal charges of all atoms must sum to the total charge of the species.

Sum the formal charges assigned in Solved Problem 1.16. What do you notice?

PROBLEM 1.17 Determine the formal charge of each atom in the molecule at the right, which belongs to a class of compounds called *oximes*. *Note:* Assume that each atom has a filled valence shell.

CONNECTIONS Oximes are important intermediates in organic synthesis, including the industrial production of nylon-6. Some oximes are used as nerve-agent antidotes, and one is an artificial sweetener.

1.10 Resonance Theory

Some species are not described well by Lewis structures. In HCO_2^-, for example, both of the C—O bonds are identical experimentally; they have the same bond length and bond strength, intermediate between those of a single and a double bond (Fig. 1-21a). However, the Lewis structure of HCO_2^- (Fig. 1-21b) shows one C—O single bond and one C=O double bond, suggesting that the two bonds are different. Furthermore, experiments show that both oxygen atoms carry an identical partial negative charge (δ^-), whereas the Lewis structure suggests that one oxygen atom bears a full negative charge and the other is uncharged. In other words, the Lewis structure suggests that the C=O double bond and the −1 charge are both confined to a specific region in space, or **localized**, whereas in the real species, they are spread out, or **delocalized**.

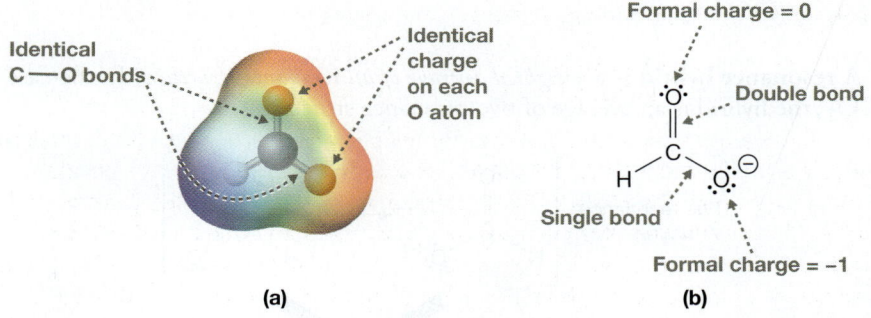

Identical C—O bonds

Identical charge on each O atom

Formal charge = 0

Double bond

Single bond

Formal charge = −1

(a)

(b)

FIGURE 1-21 Limitations of Lewis structures The actual features of HCO_2^- (a) disagree with those suggested by the Lewis structure (b). In the actual structure, both carbon–oxygen bonds are identical and the charge on each O atom is the same. The Lewis structure indicates a C—O single bond and a C=O double bond, as well as different charges on each O atom.

How, then, do we reconcile the differences between the Lewis structures of species like HCO_2^- and their observed characteristics? The answer is through **resonance theory**, the key points of which can be summarized by the following rules:

Rule 1

Resonance exists in species for which there are two or more valid Lewis structures.

For such species, each valid Lewis structure is called a **resonance structure** or a **resonance contributor**. In HCO_2^-, there are two possible resonance contributors. As we can see on the next page, they differ in how we apply Step 5 for drawing Lewis structures— that is, which neighboring lone pair is used to construct the C=O double bond.

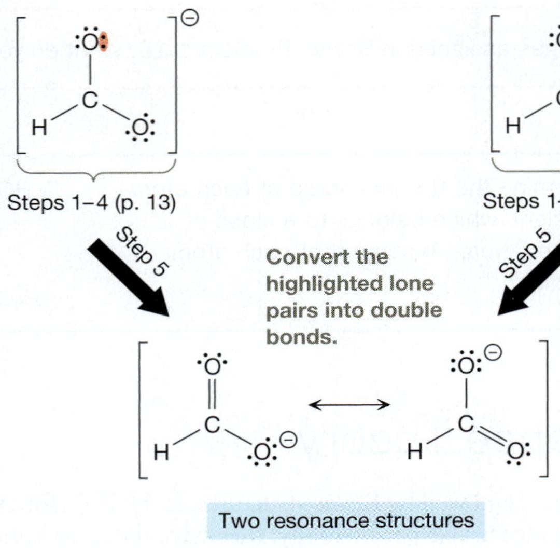

Convert the highlighted lone pairs into double bonds.

Steps 1–4 (p. 13)   Steps 1–4 (p. 13)

Step 5   Step 5

Two resonance structures

As a result:

Resonance structures differ only in the placement of their valence electrons, not their atoms.

Rule 2

Resonance structures are imaginary; the one, true species is represented by the resonance hybrid.

A **resonance hybrid** is a *weighted average of all resonance structures*. In the case of HCO_2^-, the hybrid is an average of two resonance structures:

Two resonance structures

Take the average.

Single bond plus a partial bond

Partial charges

Resonance hybrid

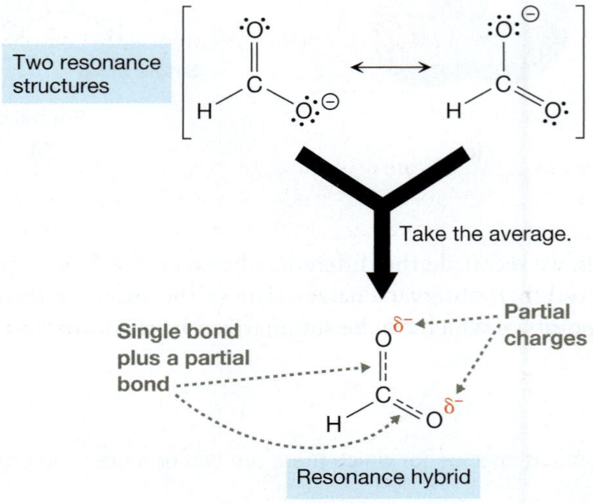

Averaging the two C—O bonds (a single and a double bond) makes each one an identical 1.5 bond—more than a single bond, but less than a double bond. (A partial bond is represented in a resonance hybrid by a dashed line connecting the two atoms.) Averaging the charge on each oxygen atom gives each an identical −0.5 charge. This resonance hybrid is now consistent with experimental results.

Duck

Otter

Duck-billed platypus

FIGURE 1-22 An analogy of resonance A duck-billed platypus has characteristics of a duck and an otter, just as a resonance hybrid has characteristics from its resonance contributors.

To help us remember that resonance structures are imaginary, we draw square brackets, [ ], around the group of resonance structures and we place double-headed single arrows (↔) between them, as shown at the right. These are *not* equilibrium arrows (⇌), which indicate a chemical reaction, because a compound does not rapidly interconvert between its resonance structures. Rather, think of a resonance hybrid in the same way as you might think of a duck-billed platypus (Fig. 1-22). That is, a platypus has characteristics of both a duck and an otter, but it is a unique species that does *not* rapidly interconvert between those two animals!

Rule 3
The resonance hybrid looks most like the lowest energy (most stable) resonance structure.

The two resonance contributors of HCO_2^- are *equivalent*. Each is composed of one H—C bond, one C=O double bond, and one C—O single bond, and each has a −1 formal charge on the singly bonded O. As a result, each structure contributes equally to the resonance hybrid.

In cases in which resonance structures are inequivalent, we must be able to determine their relative energies to determine their relative contributions to the hybrid. In general:

A resonance structure is lower in energy (i.e., more stable) with:
- a greater number of atoms having an octet
- more covalent bonds
- fewer atoms having a nonzero formal charge

The issues surrounding formal charge and resonance structures are covered in greater detail in Chapter 6.

SOLVED PROBLEM 1.18

Which of these resonance structures makes a greater contribution to the resonance hybrid?

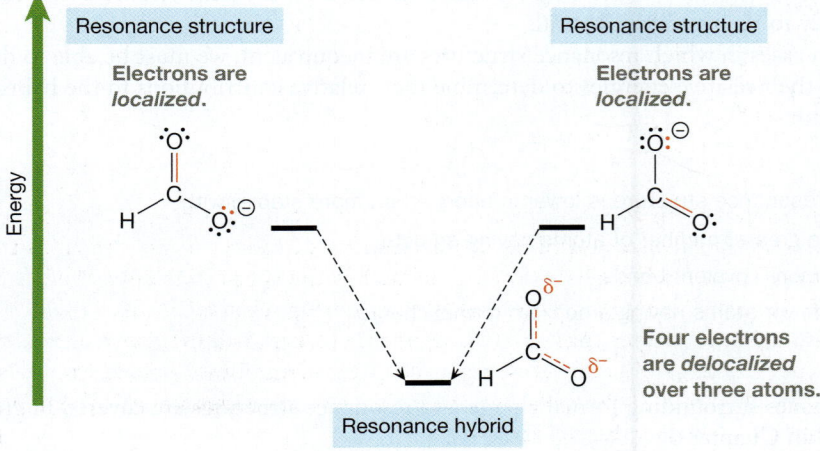

Think Are both resonance structures equivalent? If not, which has more atoms with an octet? Which structure has more bonding pairs of electrons? Which structure has fewer atoms with a nonzero formal charge?

Solve The resonance structures are inequivalent. In the structure on the right, all nonhydrogen atoms have an octet, whereas in the structure on the left, the central carbon atom has less than an octet. Additionally, there is one additional bond in the structure on the right, between the C and O. For both of these reasons, the structure on the right makes a much greater contribution to the overall resonance hybrid.

PROBLEM 1.19 Which of these resonance structures makes a greater contribution to the resonance hybrid? *Note:* Formal charges are not shown.

CONNECTIONS Cationic species like this are reactive intermediates in various organic reactions, including substitution (Chapter 8), elimination (Chapter 8), and addition reactions (Chapter 11).

Rule 4

Resonance provides stabilization.

The stabilization due to resonance results from the **delocalization** of electrons; that is, electrons have lower energy when they are less confined. In HCO_2^-, four electrons are delocalized over three atoms (Fig. 1-23). In a single resonance structure, on the other hand, those electrons are *localized*. The extent by which a species is stabilized in this fashion is called its **resonance energy** or its **delocalization energy**.

Rule 5

Resonance stabilization is usually large when resonance structures are equivalent.

FIGURE 1-23 Resonance stabilization In each resonance structure of HCO_2^-, the four electrons in red are localized. In the resonance hybrid, those four electrons are delocalized over three atoms. Delocalization results in lower energy and greater stability.

FIGURE 1-24 illustration

Six electrons are *delocalized* over six carbon atoms.

CONNECTIONS Benzene (Fig. 1-24) is an aromatic hydrocarbon that is naturally found in crude oil and, because of its high octane number, is an important component of gasoline.

Rule 5 is an outcome of Rules 3 and 4. If resonance structures are equivalent, then they will contribute equally to the hybrid, allowing electrons the greatest possible delocalization. The two resonance structures of HCO_2^- are equivalent, so its resonance energy is quite substantial. Another example is benzene, whose resonance structures and hybrid are shown in Figure 1-24. Benzene's resonance energy is estimated to be about 150 kJ/mol (36 kcal/mol), which is nearly half the energy of a C—C bond! We discuss benzene in much greater depth in Chapter 14, where we explain that much of benzene's resonance energy is due to a phenomenon called *aromaticity*.

Acetic acid (CH_3CO_2H) is *not* stabilized greatly by resonance. Although it has two resonance structures, only the one on the left in Figure 1-25 contributes significantly. The one on the right is much higher in energy because of the presence of the two charges. Therefore, the electrons involved in resonance are not very highly delocalized.

Small contribution to the hybrid

FIGURE 1-25 **Nonequivalent resonance structures** The two resonance structures of acetic acid are nonequivalent, so they contribute unequally to the resonance hybrid. The one on the left is lower in energy, so it has the greater contribution.

Rule 6

All else being equal, the greater the number of resonance structures, the greater is the resonance stabilization.

Rule 6 is an outcome of Rule 4, because additional resonance structures mean electrons are more widely delocalized, which lowers the energy even more.

1.11 Strategies for Success: Drawing All Resonance Structures

CONNECTIONS Acetic acid, when it is diluted to about 5% by volume, is familiar to us as vinegar. In organic chemistry, acetic acid is used as a reagent in a wide variety of organic reactions. It can also be used as a solvent.

It is important to be able to draw *all* resonance structures of a species. That's because all resonance structures contribute to the features of the resonance hybrid, and the total number of resonance structures is related to the species' stability. We therefore devote this section to giving you insights into drawing resonance structures.

To begin, remember that any two resonance structures differ only in where the *electrons* are located. Therefore, one resonance structure can be obtained from another just by moving valence electrons, but *atoms must remain frozen in place!* We emphasize this explicitly by using **curved arrows**.

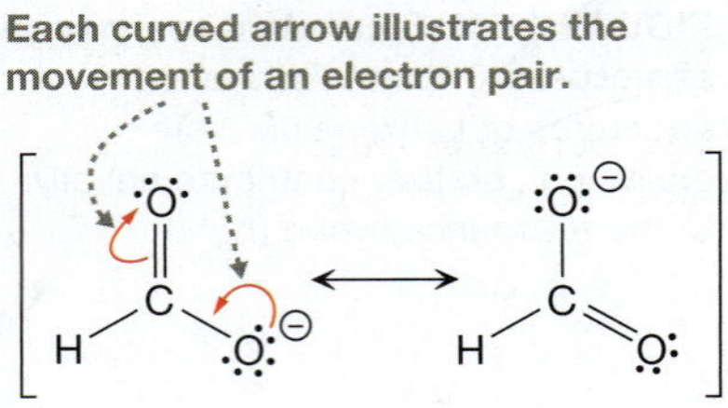

Each curved arrow illustrates the movement of an electron pair.

FIGURE 1-26 Curved arrow notation in resonance Shifting the four electrons indicated by the two curved arrows converts the resonance structure on the left into the one on the right.

A curved arrow, illustrates the movement of a pair of valence electrons.

- A curved arrow can originate from a lone pair of electrons or from the center of a covalent double or triple bond to indicate the specific pair of electrons that is being moved.
- The arrow points to an atom if the electrons being moved become a lone pair. Otherwise, the arrow points to the center of an existing single or double bond to represent the formation of a new double or triple bond there.

This is shown for HCO_2^- in Figure 1-26. The curved arrow on the right represents the conversion of a lone pair into a new bond between C and O. The other curved arrow represents a bond from the C=O double bond being converted into a lone pair on the O atom.

YOUR TURN 1.10

Supply the necessary curved arrows to go from the resonance structure on the left to the resonance structure on the right.

Any of the following four key features in a Lewis structure suggest that another resonance structure exists:

1. A lone pair of electrons on an atom is adjacent to a multiple (i.e., double or triple) bond.
2. An incomplete octet on an atom is adjacent to a multiple bond.
3. A lone pair of electrons on an atom is adjacent to an atom with an incomplete octet.
4. There is a ring of alternating single and multiple bonds.

The first of these features, in fact, appears in HCO_2^-, which is the example we used to introduce resonance (Figure 1-26). This is shown more explicitly in Figure 1-27a. Examples of species with the other features are shown in Figure 1-27b, 1-27c, and 1-27d.

YOUR TURN 1.11

For each pair of resonance structures, supply the necessary curved arrow(s) to convert the resonance structure on the left into the one on the right.

(a)

(b)

You should take the time to study these features. Notice, in particular, that each feature corresponds to a specific number of curved arrows drawn to convert one resonance structure into another. When Feature 1 is present (Fig. 1-27a), two curved arrows are used. One curved arrow accounts for the conversion of the lone pair into another bonding pair of electrons, and the second curved arrow accounts for the conversion of a bond from the original multiple bond into an additional lone pair. When Feature 2 is present (Fig. 1-27b), just one curved arrow is used, which represents the breaking of a bond from the multiple bond and the simultaneous formation of a new bond to the atom lacking the octet. Similarly, just one curved arrow is used when

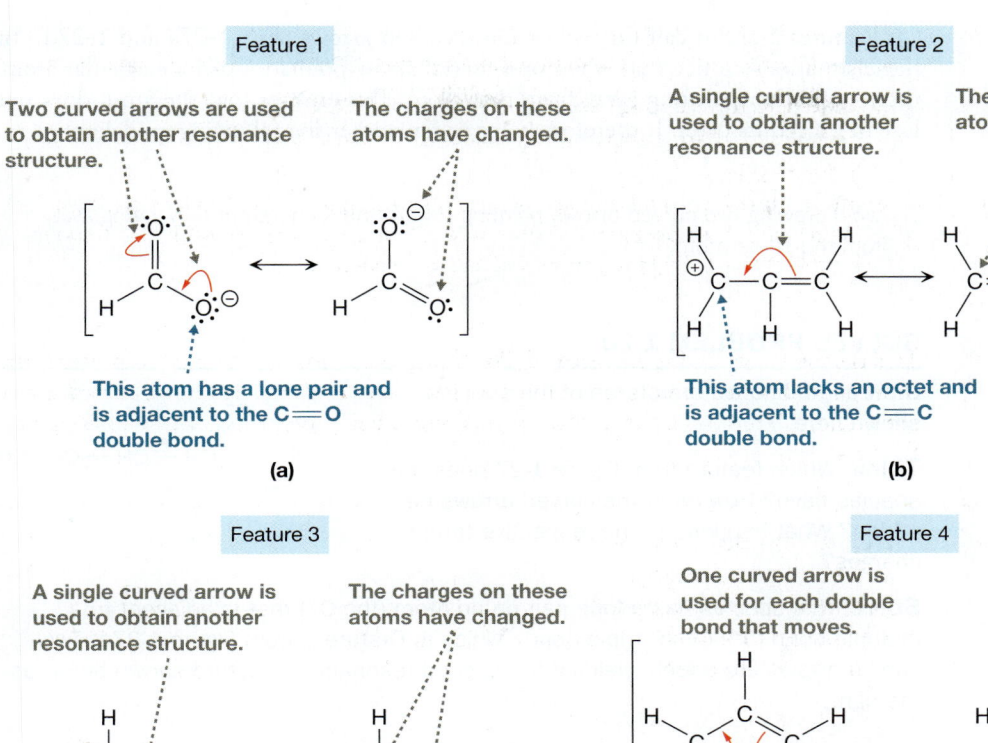

FIGURE 1-27 Features that indicate the existence of another resonance structure (a) Feature 1: a lone pair on an atom adjacent to a multiple bond. (b) Feature 2: an atom lacking an octet adjacent to a multiple bond. (c) Feature 3: an atom with a lone pair of electrons adjacent to an atom lacking an octet. (d) Feature 4: a ring of alternating single and multiple bonds.

Feature 3 is present (Fig. 1-27c); in this case, the curved arrow represents the conversion of a lone pair into an additional covalent bond to the atom lacking the octet. Finally, when Feature 4 is present, one curved arrow is required for each multiple bond that is involved in resonance. For the molecule in Figure 1-27d, three curved arrows are used, but this feature applies to rings of other sizes, too.

For Features 1–3 (Fig. 1-27a, 1-27b, and 1-27c), the formal charges of some atoms will change on going from one resonance structure to another. Although you can recalculate the formal charge on each atom after you draw a new resonance structure, it is helpful to apply the following patterns:

- When a lone pair on an atom is converted into a bonding pair, the formal charge on that atom becomes more positive by 1; when the reverse happens, the formal charge on the atom becomes more negative by 1.
- When an atom initially lacking an octet gains its octet, the formal charge on that atom becomes more negative by 1; when the reverse happens, the formal charge on that atom becomes more positive by 1.

Features 1 and 4 call for two or more curved arrows (Fig. 1-27a and 1-27d). In these situations, notice that when one curved arrow points toward a particular atom, there is another pointing away from that atom. This ensures that the atom does not lose or exceed its octet. It therefore helps to remember the following guideline:

> Avoid drawing two curved arrows pointing toward the same atom or pointing away from the same atom.

SOLVED PROBLEM 1.20

Draw all resonance structures of the species shown here.

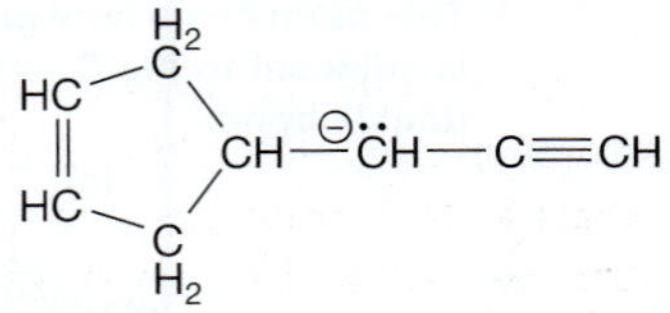

Think Which feature from Figure 1-27 does this species have? How must the curved arrows be added? What happens to the respective formal charges?

Solve This species has a lone pair on an atom (the C⁻) that is adjacent to a multiple bond (the C≡C triple bond), which is Feature 1 from Figure 1-27a. Two curved arrows are added, yielding the second resonance structure shown below at the right.

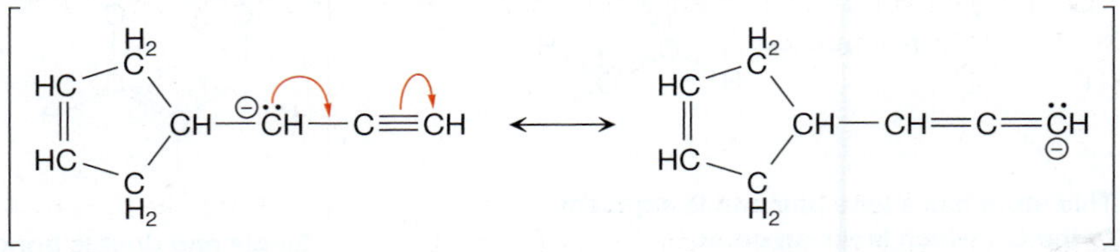

PROBLEM 1.21 Draw a resonance structure for each of the following species.

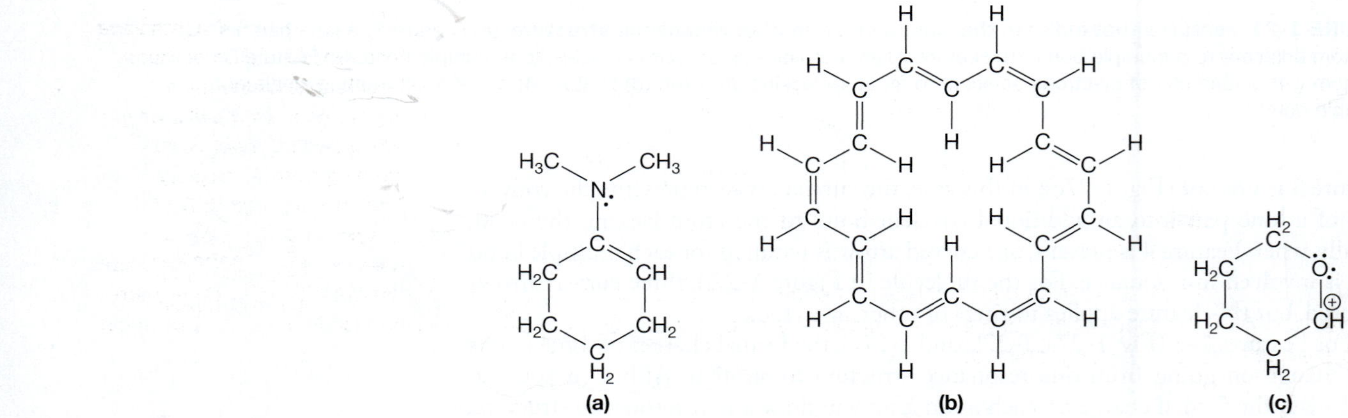

(a) (b) (c)

Frequently, a species will have more than two resonance structures. When drawing all resonance structures of these species, it helps to approach the problem systematically.

1. Identify which feature in Figure 1-27 the species has.
2. Add the curved arrows that the feature calls for and draw the corresponding resonance structure by moving the appropriate valence electrons.
3. Examine the new resonance structure to see if it has a feature from Figure 1-27. If so, repeat Steps 2 and 3 above.
4. Continue this process until you have exhausted all your options for moving electrons.

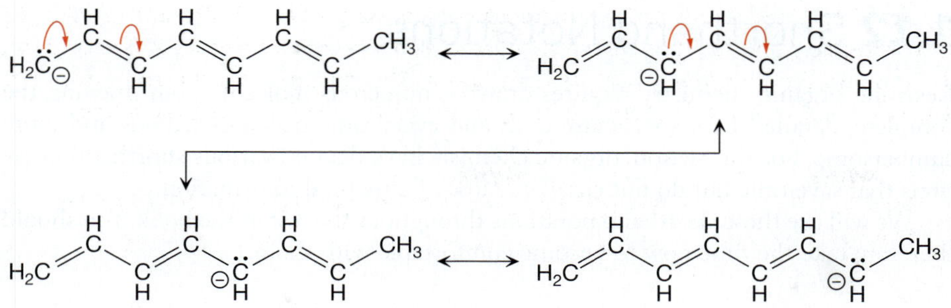

FIGURE 1-28 **Multiple resonance structures involving an atom with a lone pair adjacent to a multiple bond** Two curved arrows are drawn each time to convert one resonance structure into another.

An example of how to apply this strategy is shown in Figure 1-28. The first structure exhibits Feature 1 from Figure 1-27a, a lone pair on an atom (C^-) that is adjacent to a multiple bond (the leftmost C=C bond). To arrive at the second structure, we add two curved arrows and move the valence electrons accordingly to arrive at the second resonance structure. Examining the second resonance structure, we see the same feature appear, so we add two curved arrows again to arrive at the third resonance structure. This can be done one more time to arrive at the final structure.

YOUR TURN **1.12**

Draw in the curved arrows needed to convert the third resonance structure in Figure 1-28 to the fourth.

PROBLEM 1.22 Draw all resonance structures for each of the following species. Make sure to include all appropriate curved arrows.

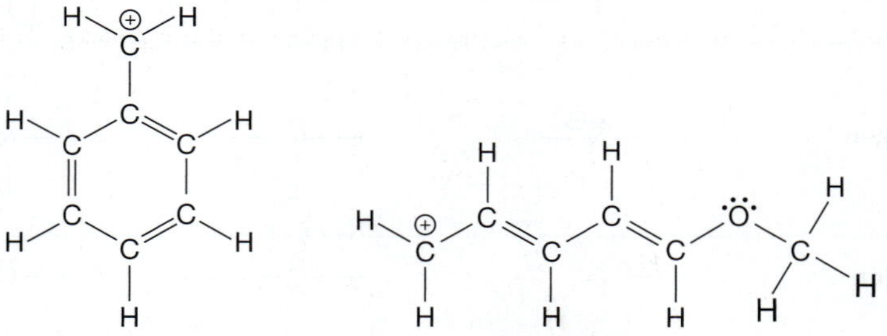

(a)

(b)

CONNECTIONS Naphthalene [Problem 1.23(b)] is perhaps best known as the main ingredient in mothballs, which are used to protect stored clothing from damage by moth larvae. Naphthalene is a possible carcinogen, however, and in 2008 mothballs containing naphthalene were banned in the European Union.

PROBLEM 1.23 Draw all resonance structures of each of the following species. Be sure to include curved arrows to indicate which pairs of electrons are being shifted and how they are being shifted.

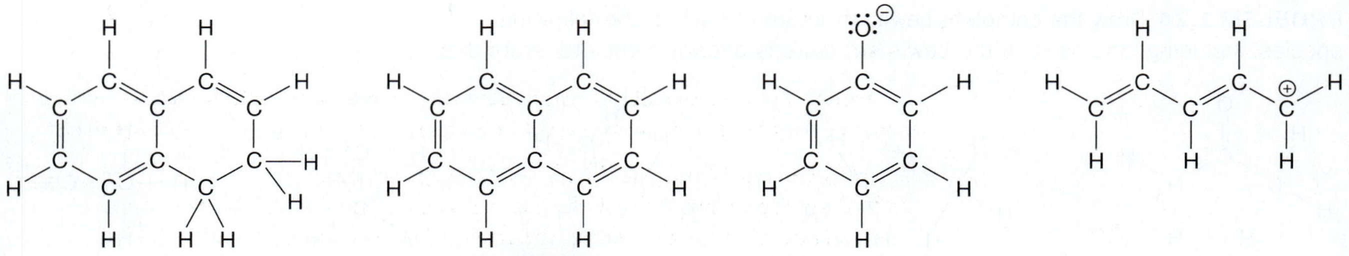

(a)

(b)

Naphthalene

(c)

(d)

1.12 Shorthand Notations

Learning organic chemistry requires drawing numerous molecules, but drawing the complete, detailed Lewis structure each and every time becomes tedious and quite cumbersome. For this reason, organic chemists have devised various shorthand notations that save time but do not entail any loss of structural information.

We will use these shorthand notations throughout the rest of the book. You should therefore take the time now to become comfortable with them.

1.12a Lone Pairs and Charges

Lone pairs of electrons are frequently omitted. They have vital roles in many organic reactions, however, so you must be able to put them back in as necessary. Doing so requires knowledge of how formal charge relates to the numbers of bonds and lone pairs on various atoms, as illustrated in Table 1-5. Pay particular attention to the fact that the atoms have octets in all of the scenarios shown, except in the case of a carbocation (C^+).

TABLE 1-5 Formal Charges on Atoms with Various Bonding Scenarios

	FORMAL CHARGE		
Atom	−1	0	+1
Carbon	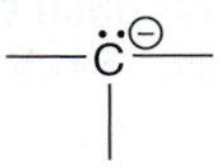	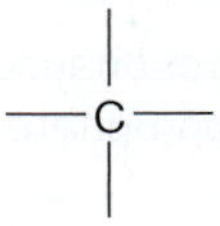	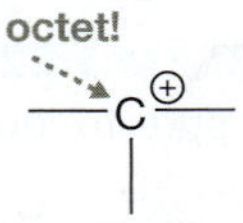 No octet!
Nitrogen	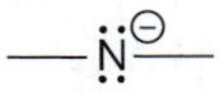	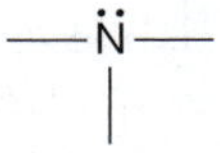	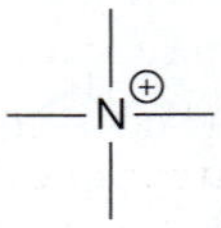
Oxygen	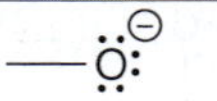	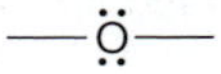	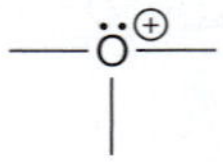
Halogen (X = F, Cl, Br, I)		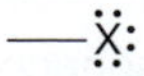	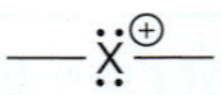

CONNECTIONS Acetamide [Problem 1.24(a)] can be used as a plasticizer, an additive that increases the plasticity of a material. Acetamide also has some use as an industrial solvent.

PROBLEM 1.24 Draw the complete Lewis structure of each of the following species, including lone pairs. If the Lewis structure is already complete, state this.

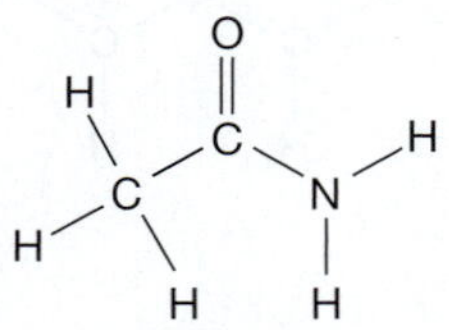

Acetamide

(a)   (b) 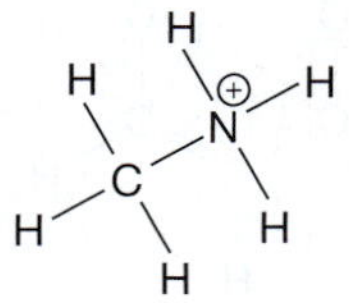   (c) 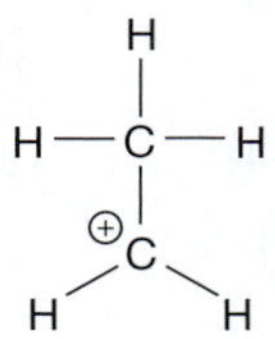   (d) 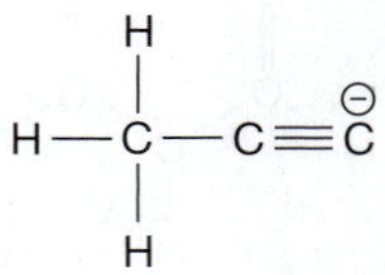   (e)

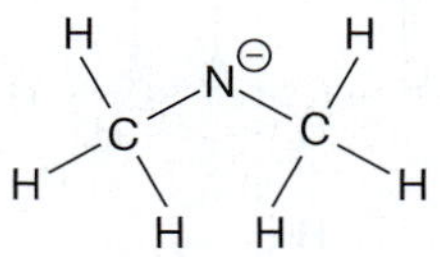

1.12b Condensed Formulas

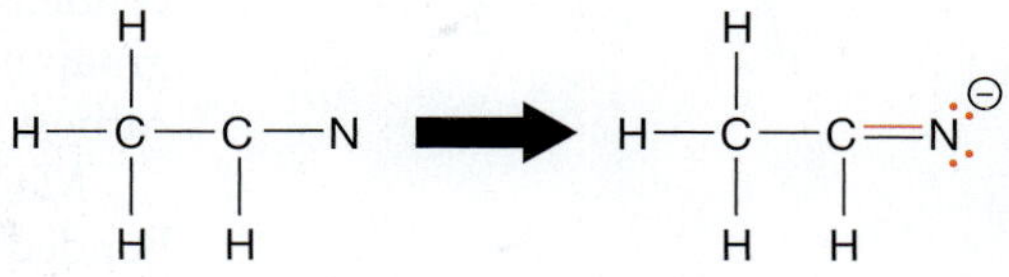

Condensed formulas allow us to include molecules and molecular ions as part of regular text. Each nonhydrogen atom is written explicitly, followed immediately by the number of hydrogen atoms that are bonded to it. Adjacent nonhydrogen atoms in the condensed formula are interpreted as being covalently bonded to each other. For example, the condensed formula CH_3CHN^- indicates that there is a central C atom bonded to another C atom and to a N atom, giving rise to the skeleton that appears on the left in Figure 1-29. The structure is completed by adding the electrons shown in red—a double bond between C and N and two lone pairs on N.

FIGURE 1-29 Lewis structure of CH_3CHN^- To convert from the skeleton on the left to the Lewis structure on the right, an extra C—N bond and two lone pairs (in red) must be added.

SOLVED PROBLEM 1.25

Draw the Lewis structure for crotonaldehyde, $CH_3CHCHCHO$.

Think Which nonhydrogen atoms are bonded together? How can we add bonds and lone pairs to maximize the number of octets and also conform to the total charge of zero? How many total valence electrons must be accounted for in this compound?

Solve The condensed formula indicates the atoms should be connected as shown below on the left. To arrive at a total charge of zero, we give each C atom a total of four bonds and we give the O atom two bonds and two lone pairs. Double-checking the structure, notice that it has 28 valence electrons, as it should.

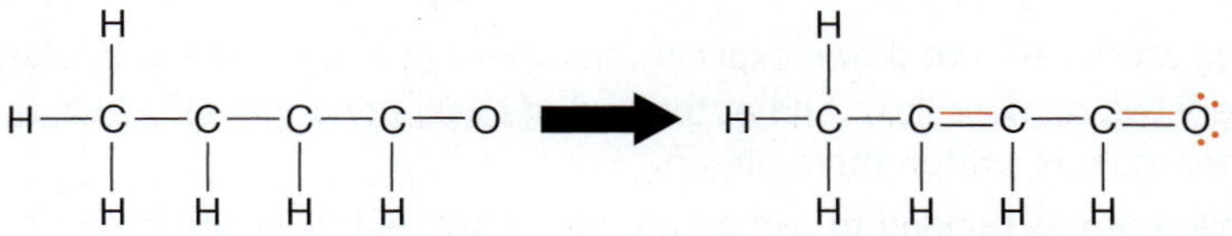

Crotonaldehyde

> **CONNECTIONS**
> Crotonaldehyde is found in some foodstuffs, such as soybean oil. It is also used as a precursor in the industrial synthesis of vitamin E and sorbic acid, a food preservative.

Often a molecule will have multiple CH_2 groups bonded together. In these cases, we can simplify a condensed formula using the notation $(CH_2)_n$. Thus, $CH_3CH_2CH_2CH_2CH_3$ simplifies to $CH_3(CH_2)_3CH_3$.

Parentheses are also used to clarify situations in which three or four groups are attached to the same atom. In 2-methylbutane (Fig. 1-30a), for example, the second C atom from the left is bonded to two CH_3 groups and one CH_2CH_3 group. Its

FIGURE 1-30 Condensed formulas and branching Condensed formulas are shown on the top and Lewis structures are shown on the bottom. (a) Parentheses in a condensed formula denote that the group is attached to the previous C. (b) The CO_2 notation represents a C atom that is doubly bonded to one O atom and singly bonded to another. (c) Rings are generally not shown in their condensed formulas, but they are commonly shown in their partially condensed form.

$CH_3CH(CH_3)CH_2CH_3$

Same as

2-Methylbutane

(a)

CH_3CO_2H

Same as

**Ethanoic acid
(Acetic acid)**

(b)

$-CH_2CH_2CH_2CH_2CH_2CH_2-$ or

Same as

Cyclohexane

(c)

This is a partially condensed form.

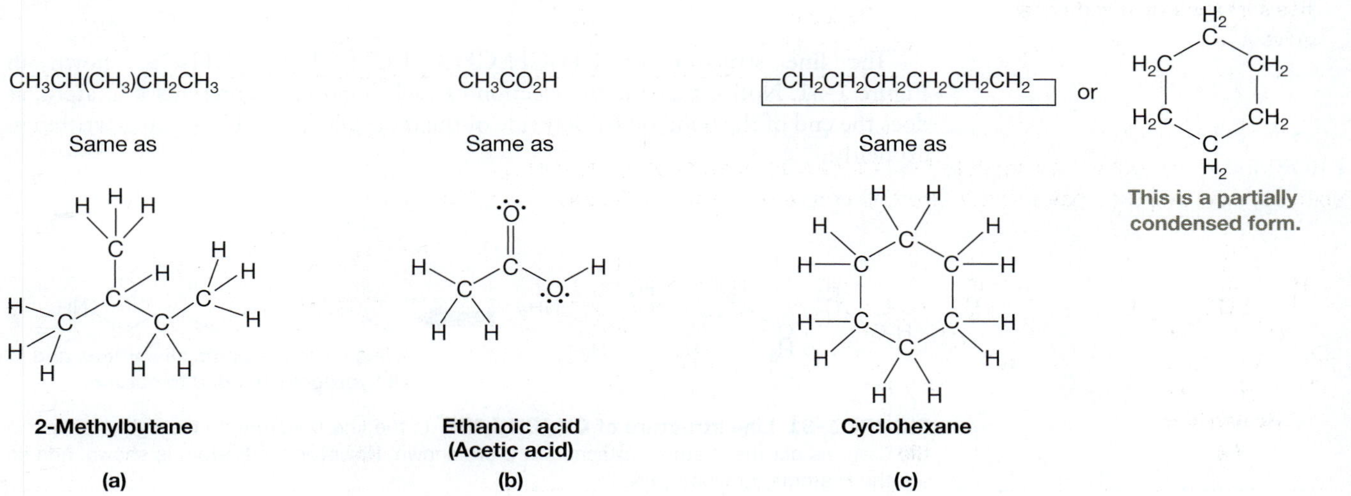

condensed formula is written as $CH_3CH(CH_3)CH_2CH_3$, indicating that the CH_3 group in parentheses and the CH_2CH_3 group are both bonded to the preceding CH carbon.

Many common organic compounds contain structures in which a carbon atom is bonded to one oxygen atom by a double bond and to a second oxygen atom by a single bond. We can write this group in the form $-CO_2-$, as shown in Figure 1-30b for acetic acid (CH_3CO_2H). Another common bonding arrangement is $-CH{=\!=}O$, which is often abbreviated as CHO, as shown previously in Solved Problem 1.25.

Condensed formulas for cyclic structures, like cyclohexane (Fig. 1-30c), are problematic. If the formula is written on a single line of text, the leftmost C atom must be bonded to the rightmost one to complete the ring. This would appear as follows: $\lceil CH_2CH_2CH_2CH_2CH_2CH_2 \rceil$. Because this is so cumbersome, we generally do not represent rings in their fully condensed forms. Instead, rings are often depicted in their *partially* condensed form, as shown at the right in Figure 1-30c.

1.12c Line Structures

Line structures, like condensed formulas, are compact and can be drawn quickly and easily. Unlike condensed formulas, however, they are not intended to be written as part of text. The rules for drawing line structures are as follows:

Rules for Drawing Line Structures

- Carbon atoms are not drawn explicitly, but they are implied at the intersection of every two or more lines and at the end of every bond that is drawn, unless another atom is written there already.
- Hydrogen atoms bonded to carbon are *not* drawn, but hydrogen atoms bonded to all other atoms are.
- All noncarbon and nonhydrogen atoms, called **heteroatoms**, are drawn.
- Bonds to hydrogen are not drawn, but all other bonds are drawn explicitly.
- Several carbon atoms bonded in a single chain are represented by a zigzag structure: $\wedge\!\!\wedge$. 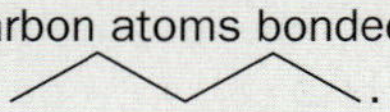
- Enough hydrogen atoms are assumed to be bonded to each carbon atom to fulfill the carbon atom's octet, with close attention paid to the formal charge on carbon (see Table 1-5, p. 30).
- Lone pairs of electrons are generally not shown unless they are necessary to emphasize an important aspect of an atom.

The line structure of $CH_3CH_2CH_2CH_2CH_2CH_2CH_2NH_2$ is shown in Figure 1-31. Notice that the intersection of each bond line represents a C atom, as does the end of the bond on the left side of the molecule. The NH_2 group is written in explicitly.

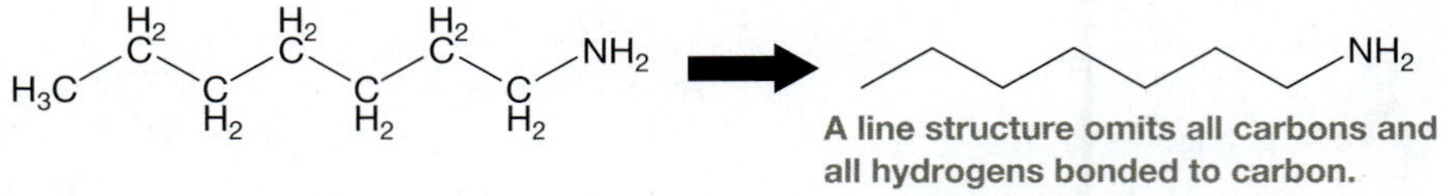

A line structure omits all carbons and all hydrogens bonded to carbon.

FIGURE 1-31 **Line structure of $CH_3(CH_2)_6NH_2$** In the line structure on the right, neither the C atoms nor the H atoms attached to C are shown. However, the N atom is shown, and so are the H atoms attached to N.

PROBLEM 1.26 Redraw each of the following Lewis structures as the corresponding line structure.

Pyrrole
(a)

Benzoic acid
(b)

(c)

CONNECTIONS Pyrrole is a structural component of the heme group of hemoglobin, which is the protein responsible for oxygen transport in blood. Benzoic acid occurs naturally in some plants. It is a component of ointments used to treat fungal skin diseases, and is a precursor to sodium benzoate, a food preservative.

PROBLEM 1.27 For each of the following line structures, draw in all carbon atoms, hydrogen atoms, and lone pairs.

(a)

(b)

(c)

YOUR TURN 1.13

A student mistakenly interprets the line structure at the right as but-2-yne ($CH_3C\equiv CCH_3$). What mistake did the student make?

SOLVED PROBLEM 1.28

Why is it incorrect to draw resonance structures for propene as shown at the right?

Think What would the structures look like as complete Lewis structures (i.e., with all hydrogen atoms drawn in)? What rule is broken in the transformation from the first structure to the second?

Solve If all atoms and bonds are included, then the transformation would appear as follows.

Bond broken   Bond formed

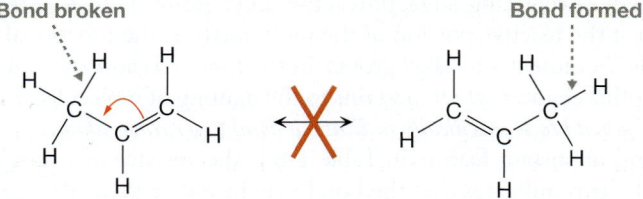

This process entails moving a H atom from the C on the left to the C on the right. When drawing resonance structures, however, only electrons can be moved. All atoms must remain frozen in place.

PROBLEM 1.29 Draw all resonance structures for each of the ions shown here using only line structures.

(a)

(b)

(c)

1.13 An Overview of Organic Compounds: Functional Groups

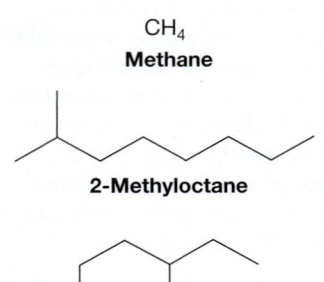

CH$_4$
Methane

2-Methyloctane

Ethylcyclohexane

FIGURE 1-32 Some alkanes
Alkanes consist entirely of carbon and hydrogen and have only single bonds.

Certain types of bonds and arrangements of atoms tend to appear quite frequently in organic chemistry. The most common types of bonds we encounter are C—C and C—H single bonds. Compounds consisting of nothing but C—C and C—H single bonds are called **alkanes**. Examples include methane, 2-methyloctane, and ethylcyclohexane (Fig. 1-32).

Alkanes tend to be among the most *unreactive* organic compounds. For this reason, liquid alkanes (e.g., pentane and hexane) are often used as solvents in which to carry out organic reactions.

Alkanes are relatively inert in part because C—C and C—H bonds are so strong; and, for a reaction to occur, these bonds must break. In fact, the data in Table 1-2 (p. 10) show that these are among the strongest single bonds we will encounter.

Alkanes also tend to be unreactive because C—C and C—H bonds are either nonpolar or are only very slightly polar. In Chapter 7, we explain that a large driving force behind many chemical reactions depends on the existence of substantial bond dipoles.

When other types of bonds and other types of atoms are introduced, then organic molecules tend to be more reactive. In some cases, this may be due to the introduction of weaker bonds, and in other cases, to the presence of large bond dipoles.

As we will learn throughout the remainder of this book, specific arrangements of atoms connected by specific types of bonds tend to react in characteristic ways. In fact, even though such groups may consist of only two or three atoms, they dictate the *function* (i.e., the chemical behavior) of the entire molecule. Hence these structural components are called **functional groups**. Consequently:

> An organic molecule's reactivity is governed by the functional groups it has.

The functional groups that appear most often are listed in Table 1-6, along with the compound classes in which they are found. Take the time to commit these functional groups and compound classes to memory and review them frequently. Pay particular attention to the fact that some bonding arrangements, such as the **hydroxyl group** (O—H) and the **carbonyl group** (C=O), appear in multiple functional groups.

There are two important features in Table 1-6 that you must understand. The first is "R," the symbol used to represent an **alkyl group**, which is an attached group that consists only of carbon and hydrogen. As with alkanes, alkyl groups tend to be unreactive. The benefit of replacing large, unreactive alkyl groups with R is that it allows us to focus on just the reactive portion of the molecule (i.e., the functional group).

We will learn more about alkyl groups in the first *nomenclature* unit (Interchapter A), following this chapter, which pertains to the naming of molecules. For the purpose of Table 1-6, *a bond to R specifically indicates a bond to a carbon atom.*

The second important feature in Table 1-6 is the absence of atoms at the ends of specific bonds. This indicates that the bond may be either to an alkyl group (R) or to a hydrogen atom.

With these in mind, the following conclusions can be drawn from Table 1-6:

1. *Functional groups can differ in type of atom.* R—S—H and R—O—H for example, are different functional groups.
2. *Functional groups can differ in type of bond.* C=C and C≡C are different functional groups.
3. *Some functional groups contain the bonding arrangements of smaller functional groups.* A carboxyl group (–CO$_2$H), for example, contains both the C—O—H and C=O arrangements.

4. *Alkanes are considered to have no functional groups.* They are relatively unreactive, and when they do react, the reaction tends to be very unselective (see Chapter 25).

5. *Rings generally do not constitute new functional groups.* In most cases, the reactivity of a functional group that is part of a ring is very similar to its reactivity in an open chain.

    a. *Arenes* are exceptions, which are discussed in depth in Chapter 14.

    b. Another exception is an *epoxide*, which is a three-membered-ring *ether*. As we discuss in Chapters 2 and 4, this is because small rings are highly strained, which tends to make them more reactive than an analogous open-chain ether group.

TABLE 1-6 **Common Functional Groups**[a]

Functional Group (Red)	Compound Class	Functional Group (Red)	Compound Class	Functional Group (Red)	Compound Class
	Alkene		Thiol		Nitrile
	Alkyne		Ether		Ketone
	Arene or Aromatic compound		Acetal		Aldehyde
(X = F, Cl, Br, I)	Alkyl halide		Hemiacetal		Carboxylic acid
	Alcohol		Epoxide		Ester
	Phenol		Amine		Amide

[a]R indicates an alkyl group, and the absence of an atom at the end of a bond indicates that either R or H may be attached.

PROBLEM 1.30 In Table 1-6, there are seven functional groups that contain the bonding arrangements of simpler functional groups. A carboxyl group ($-CO_2H$) is one of them, as noted previously. What are the other six and what functional groups do they appear to contain?

CONNECTIONS
Cyclohexanone is an important industrial compound because it is a precursor in the synthesis of nylon, the material used to make this parachute.

SOLVED PROBLEM 1.31

Will the chemical reactivity of these compounds be similar or significantly different? Explain.

Cyclohexanone   **Hexan-3-one**

Think Are their functional groups the same or different? What impact will the ring have on the reactivity of cyclohexanone?

Solve Both compounds have a carbonyl group of the form $C_2C{=}O$, characteristic of ketones. Although cyclohexanone's carbonyl group is part of a ring, this should not significantly alter its reactivity relative to that of the open-chain hexan-3-one (see conclusion 5). As a result, both compounds should behave similarly.

CONNECTIONS
δ-Valerolactone (Problem 1.32) is used as a precursor in the industrial synthesis of polyesters, whereas pentanoic acid is used to produce esters that have pleasant odors or pleasant flavors. These esters are used in perfumes and cosmetics or as food additives.

PROBLEM 1.32 Will the chemical reactivity of δ-valerolactone and pentanoic acid be similar or significantly different? Explain.

δ-Valerolactone   **Pentanoic acid**

YOUR TURN 1.14

Circle the functional groups present in cyclohexanone and hexan-3-one in Solved Problem 1.31 and use Table 1-6 to verify that they are both ketones.

SOLVED PROBLEM 1.33

Circle each of the functional groups present in ebalzotan, which was developed as an antidepressant and an antianxiety agent. Also, name the compound class that each functional group characterizes.

Think Are there *bonds* present other than C—C and C—H single bonds? Are there *atoms* present other than carbon and hydrogen? Are there any special rings present?

Ebalzotan

Solve The functional groups are circled at the right and the compound classes they characterize are labeled. There are three C=C double bonds alternating with single bonds in a complete ring, characteristic of an arene. These are *not* separate C=C groups that characterize alkenes. There are two O atoms and two N atoms present, each providing the potential for another functional group. The N atom at the right is part of a N—C bond, characteristic of an amine. The O atom at the top is part of a C—O—C functional group, characteristic of an ether. The O atom and the N atom at the bottom left are part of the O=C—N functional group, characteristic of an amide.

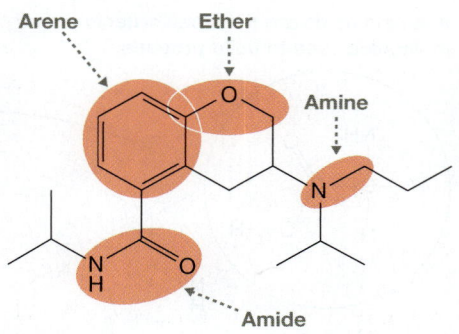

PROBLEM 1.34 Circle each of the functional groups present in ciprofloxacin, a powerful antibiotic commonly sold under the brand name Cipro. Also, name the compound class that each functional group characterizes.

Ciprofloxacin (Cipro)

THE ORGANIC CHEMISTRY OF BIOMOLECULES

1.14 An Introduction to Proteins, Carbohydrates, and Nucleic Acids: Fundamental Building Blocks and Functional Groups

Organic chemistry plays a pivotal role in biology at the molecular level. As we saw in Section 1.1, the original definition of an organic compound was a substance of plant or animal origin, and even today many people still think of "organic" as synonymous with "natural," and distinct from a "synthetic" substance made in a laboratory or an industrial plant. To the chemist, however, organic molecules are those composed chiefly of carbon and hydrogen, and usually other nonmetals such as oxygen, nitrogen, phosphorus, sulfur, and the halogens (fluorine, chlorine, bromine, and iodine). The particular organic molecules found almost exclusively in living organisms are more properly called **biomolecules**.

Because they are generally the products of millions of years of evolution, the structures of biomolecules are highly adapted to serve specific biological functions. Some of these functions are described here in Section 1.14. It is important to realize, however, that these molecules behave basically the same way in living cells as they do in a flask. The chemistry of living cells is thus ultimately the chemistry of organic compounds.

There are four major classes of biomolecules that we will discuss: *proteins, carbohydrates, nucleic acids,* and *lipids.* In this unit on biomolecules, we discuss aspects of proteins, carbohydrates, and nucleic acids, leaving lipids to Section 2.10. We choose this organization because proteins, carbohydrates, and nucleic acids have a common structural attribute that lipids do not.

> Proteins, carbohydrates, and nucleic acids are typically very large structures, with molar masses that can reach the millions, but are constructed from relatively few types of small organic molecules.

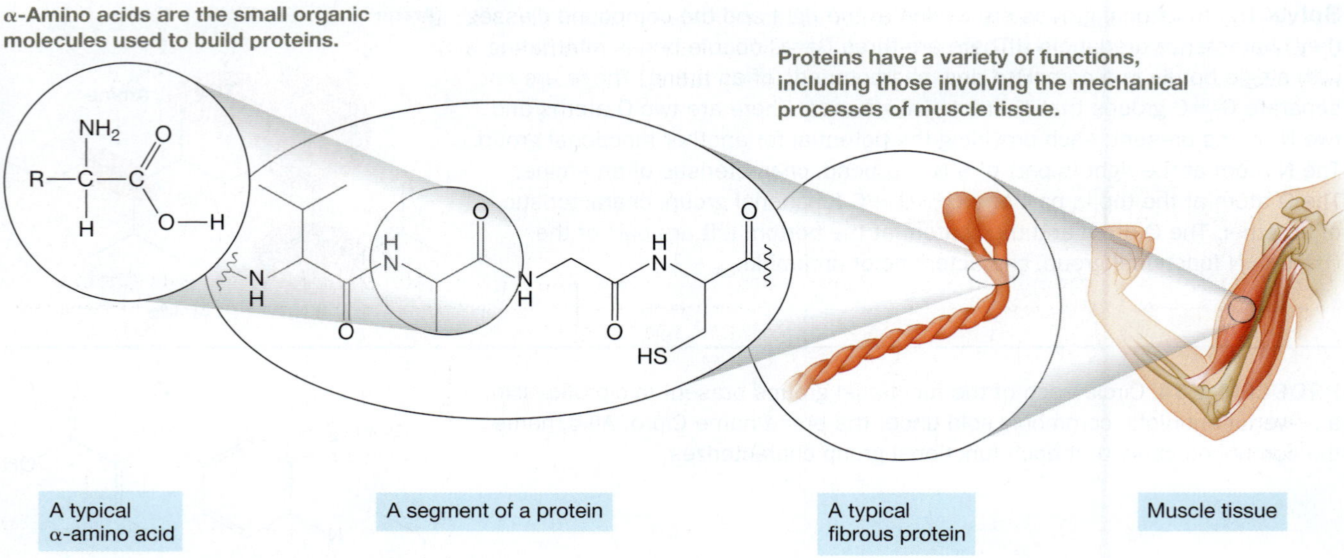

α-Amino acids are the small organic molecules used to build proteins.

Proteins have a variety of functions, including those involving the mechanical processes of muscle tissue.

| A typical α-amino acid | A segment of a protein | A typical fibrous protein | Muscle tissue |

FIGURE 1-33 The size hierarchy of proteins Amino acids (*left*) are covalently bonded together to form proteins. Proteins, in turn, are integrated into more familiar structures, such as muscle tissue.

By contrast, lipids are relatively small- to medium-sized molecules, characterized by their insolubility in water—a topic that is covered in Chapter 2.

Moreover, the small organic molecules that compose large proteins, carbohydrates, and nucleic acids are recognizable because they have distinct structural features containing a variety of the characteristic functional groups listed in Table 1-6 (p. 35). We will examine some of those features in the remainder of this unit.

1.14a Proteins and Amino Acids

Proteins are quite versatile biomolecules. Some, such as actin and myosin, are responsible for the mechanical processes involving muscle tissue. Others, such as cortactin, are responsible for regulating cell shape. There are also many different proteins that are classified as **enzymes**, which act as catalysts for biological reactions—they facilitate biological reactions but are not consumed while doing so. Acetylcholinesterase, for example, catalyzes the breakdown of acetylcholine, a neurotransmitter.

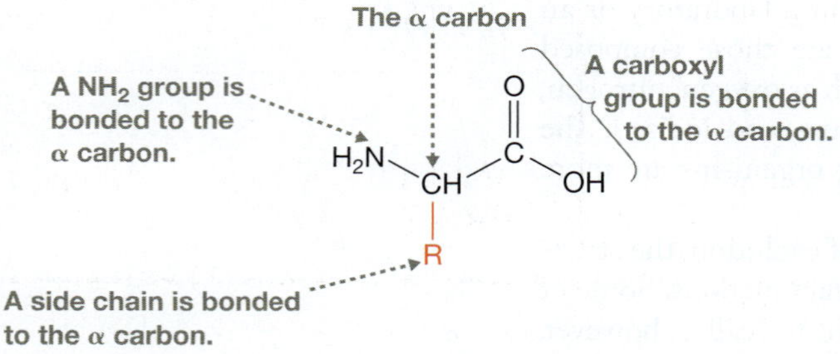

Proteins are constructed from relatively few types of small organic molecules called *α-amino acids*. As a result, we can view proteins on different size scales, as shown in Figure 1-33. The general structure of an amino acid is shown in Figure 1-34.

All α-amino acids have certain functional groups with specific relative locations in common:

The α carbon

A NH₂ group is bonded to the α carbon.

A carboxyl group is bonded to the α carbon.

A side chain is bonded to the α carbon.

FIGURE 1-34 An α-amino acid All α-amino acids have the same basic structure shown in black and differ by the identity of the side chain (R) highlighted in red.

- An **α-amino acid** contains both a NH_2 and a CO_2H group, which are characteristic of amines and carboxylic acids, respectively.
- Both the NH_2 and CO_2H groups are attached to the same carbon atom, called the **α (alpha) carbon**.
- The α carbon is also covalently bonded to a **side chain**, which is sometimes called an **R group**.

The side chain of an amino acid is what distinguishes one amino acid from another. In particular, there are 20 naturally occurring side chains, so there are 20 different naturally occurring amino acids. Their structures are shown in Table 1-7.

TABLE 1-7 The 20 Naturally Occurring Amino Acids

Name (abbreviation)[a]	Side Chain (R)	Name (abbreviation)[a]	Side Chain (R)	Name (abbreviation)[a]	Side Chain (R)
NONPOLAR AMINO ACIDS					
Alanine (Ala or A)	$-CH_3$	Glycine (Gly or G)	$-H$	Isoleucine (Ile or I)	
Leucine (Leu or L)		Methionine (Met or M)	$-(CH_2)_2SCH_3$	Phenylalanine (Phe or F)	
Proline (Pro or P)		Tryptophan (Trp or W)		Valine (Val or V)	
POLAR AMINO ACIDS					
Asparagine (Asn or N)		Cysteine (Cys or C)	$-CH_2SH$	Glutamine (Gln or Q)	
Serine (Ser or S)	$-CH_2OH$	Threonine (Thr or T)		Tyrosine (Tyr or Y)	
ACIDIC AMINO ACIDS					
Aspartic acid (Asp or D)		Glutamic acid (Glu or E)			
BASIC AMINO ACIDS					
Arginine (Arg or R)		Lysine (Lys or K)		Histidine (His or H)	

[a]Each amino acid has a one- and three-letter abbreviation.

Notice in Table 1-7 that amino acid side chains contain a variety of functional groups. The side chain of serine, for example, contains an OH group, characteristic of alcohols, whereas aspartic acid has a CO_2H group, characteristic of carboxylic acids, and lysine has a NH_2 group, characteristic of amines. Some amino acids, such as phenylalanine, contain an aromatic ring.

YOUR TURN 1.15

Use Table 1-7 to help you circle the functional groups just mentioned for serine, aspartic acid, lysine, and phenylalanine.

The identity of the side chain is what distinguishes the properties of one amino acid from another, and these side chains can be polar or nonpolar, and can be acidic or basic. These properties, in turn, are largely what govern an amino acid's function when it forms part of a protein. We explore these kinds of ideas in greater depth as we continue our discussion of biomolecules throughout this book.

PROBLEM 1.35 Identify all of the amino acids whose side chains contain functional groups found in these compound classes: **(a)** alcohol; **(b)** amide; **(c)** carboxylic acid; **(d)** amine; **(e)** arene.

1.14b Carbohydrates and Monosaccharides

Carbohydrates, also called **saccharides**, serve a variety of biological functions. Sugar, starch, and glycogen, for example, are fuels for primary metabolic pathways such as glycolysis, and cellulose is the structural component of the cell walls in plants. Derivatives of carbohydrates are involved in many important processes, too, such as those related to blood clotting and the immune system.

Carbohydrates are characterized by their chemical composition:

- Carbohydrates are composed of only carbon, oxygen, and hydrogen.
- There are two hydrogen atoms for every oxygen atom, regardless of how many carbon atoms are present, giving each carbohydrate the general molecular formula $C_xH_{2y}O_y$.

Thus, hydrogen and oxygen are present in the same ratio as in water, so the general formula of a carbohydrate can be written $C_x(H_2O)_y$. This is why "hydrate" appears in the name. (The bonding scheme in a carbohydrate, however, is quite different from that in water.)

Carbohydrates are frequently very large molecules, called **polysaccharides**. Polysaccharides are constructed from just a few types of smaller molecules, called **monosaccharides** or **simple sugars**. Thus, carbohydrates can be viewed on the different size scales shown in Figure 1-35.

Monosaccharides are themselves carbohydrates, with one additional restriction to their chemical composition:

In a monosaccharide, the number of oxygen atoms is the same as the number of carbon atoms, giving it the general formula $C_xH_{2x}O_x$.

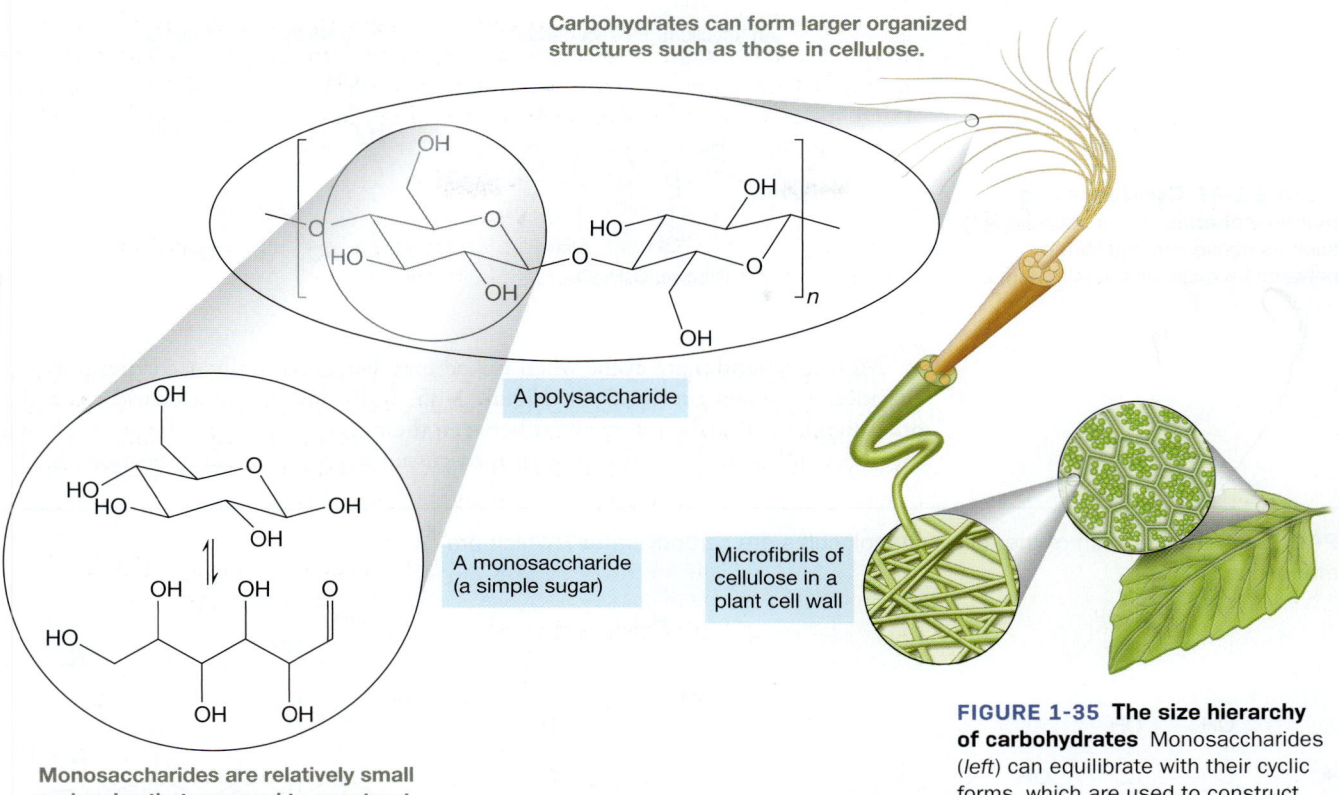

Carbohydrates can form larger organized structures such as those in cellulose.

A polysaccharide

A monosaccharide (a simple sugar)

Microfibrils of cellulose in a plant cell wall

Monosaccharides are relatively small molecules that are used to construct larger carbohydrates.

FIGURE 1-35 The size hierarchy of carbohydrates Monosaccharides (*left*) can equilibrate with their cyclic forms, which are used to construct polysaccharides (*middle*). Cellulose (*right*) is a polysaccharide that makes up the cell walls in plants.

As shown in Figure 1-36, for example, ribose has the molecular formula $C_5H_{10}O_5$, whereas glucose and fructose have the formula $C_6H_{12}O_6$.

Notice in these structures that each C atom is bonded to a single O atom and that one O atom in each monosaccharide is part of a C=O group that is characteristic of an aldehyde or ketone. The remaining O atoms are part of C—OH groups, characteristic of alcohols.

YOUR TURN **1.16**

Circle and label all of the functional groups in the structure of glucose in Figure 1-36. In which compound class is each functional group typically found?

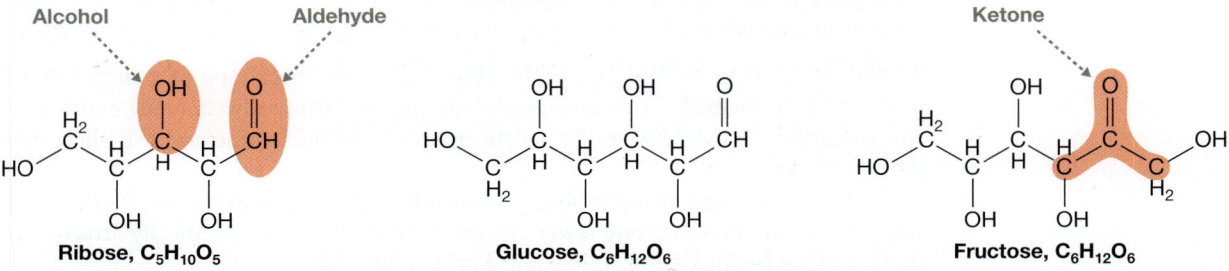

FIGURE 1-36 Some monosaccharides The C—OH group (common to alcohols) and the C=O group (common to aldehydes and ketones) appear in these monosaccharides.

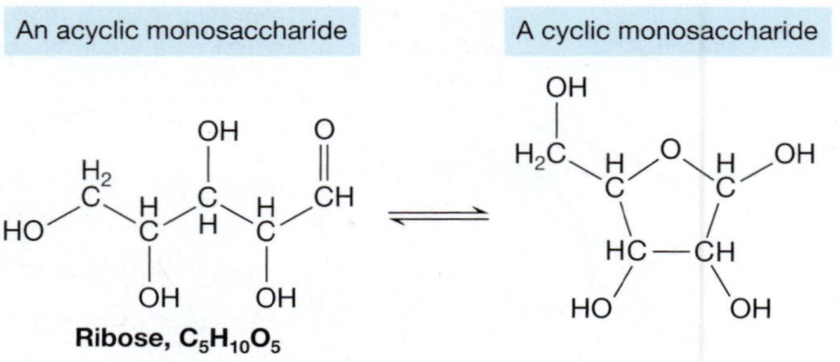

An acyclic monosaccharide   A cyclic monosaccharide

FIGURE 1-37 Cyclization of a monosaccharide A monosaccharide such as ribose can equilibrate between its cyclic and acyclic forms.

Ribose, $C_5H_{10}O_5$

Monosaccharides are cyclic when linked together in naturally occurring polysaccharides, as shown previously in Figure 1-35. Individual monosaccharides, on the other hand, continually interconvert between their cyclic and acyclic forms, as shown for ribose in Figure 1-37. This process is discussed in greater detail in Section 18.13.

PROBLEM 1.36 Which of the following molecules are carbohydrates? Which are monosaccharides?

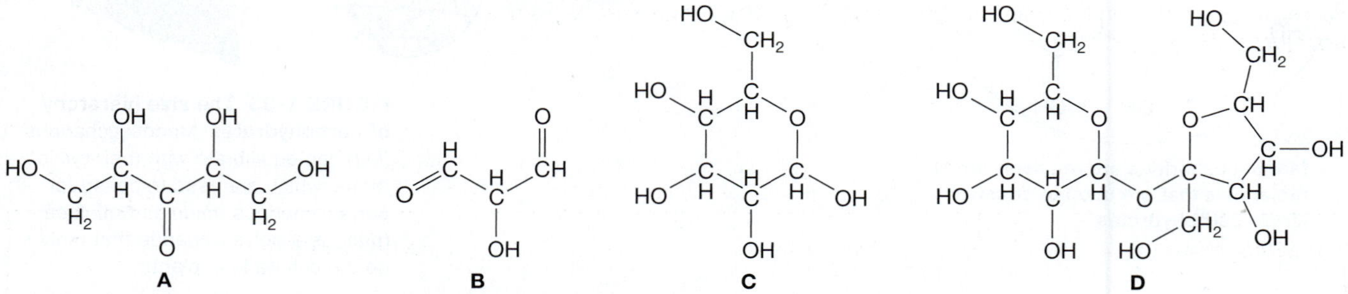

A   B   C   D

1.14c Nucleic Acids and Nucleotides

A **nucleic acid** is a large molecular chain that is primarily associated with the storage and transfer of genetic information. A pair of intertwined nucleic acids form the double-helical **deoxyribonucleic acid (DNA)**, which stores genetic information, and **ribonucleic acid (RNA)**, which participates in protein synthesis. Each strand of the double helix is a nucleic acid. Nucleic acids themselves are constructed from relatively small molecular units called **nucleotides**. Thus, just as we saw with proteins and carbohydrates, nucleic acids can be viewed on the different size scales shown in Figure 1-38.

All nucleotides have three distinct components (Fig. 1-39a):

- an inorganic phosphate ($—OPO_3$) group,
- a cyclic monosaccharide (or sugar), and
- a nitrogenous base.

In both RNA (Fig. 1-39b) and DNA (Fig. 1-39c), the backbone of a single strand consists of alternating sugar and phosphate groups. Thus, adjacent nucleotides are connected by a bond between the sugar group of one nucleotide and the phosphate group of another.

In RNA, the sugar group is ribose, whereas in DNA, it is deoxyribose. As the name suggests, deoxyribose has one fewer oxygen atom than ribose, specifically attached to the 2′ carbon (compare Fig. 1-39b and Fig. 1-39c).

Nucleotides are distinguished from one another by the identity of the nitrogenous base. In both RNA and DNA, it is the specific sequence of the nitrogenous bases that determines the genetic information that is stored or carried. There are four types of

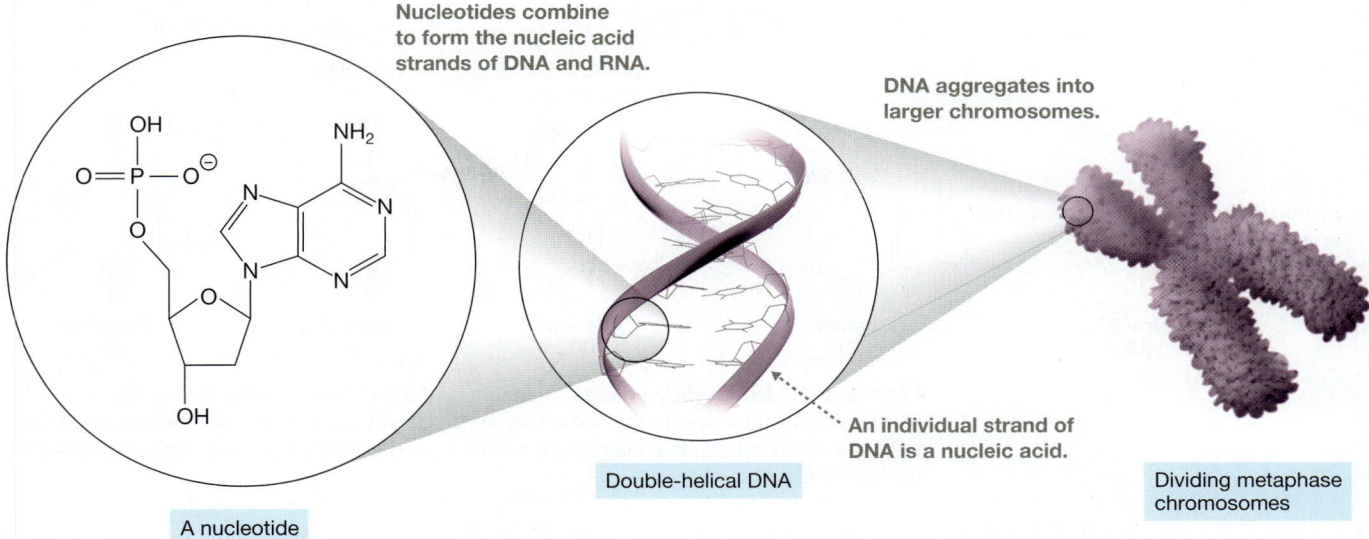

Nucleotides combine to form the nucleic acid strands of DNA and RNA.

A nucleotide

Double-helical DNA

An individual strand of DNA is a nucleic acid.

DNA aggregates into larger chromosomes.

Dividing metaphase chromosomes

FIGURE 1-38 The size hierarchy of nucleic acids Nucleotides (*left*) are used to construct nucleic acids, which form the strands of RNA and DNA (*middle*). Chromosomes (*right*) are aggregates of DNA.

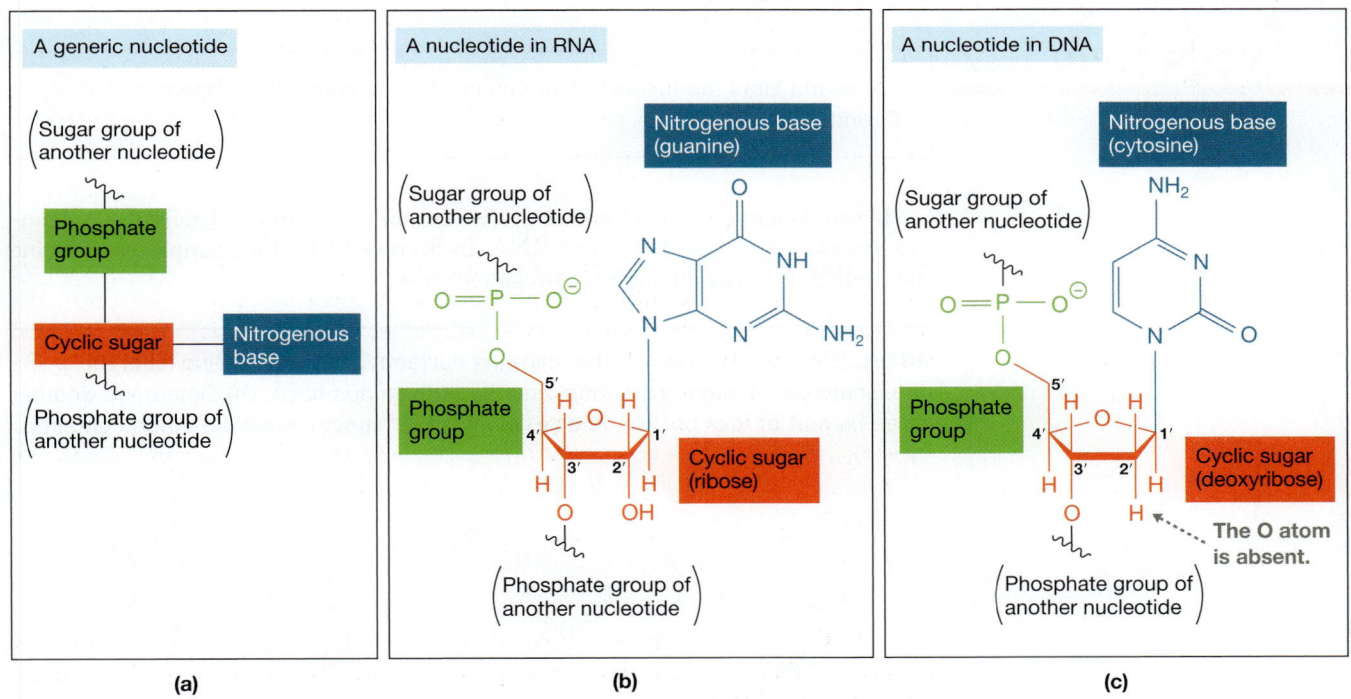

A generic nucleotide

Sugar group of another nucleotide

Phosphate group

Cyclic sugar

Nitrogenous base

Phosphate group of another nucleotide

(a)

A nucleotide in RNA

Nitrogenous base (guanine)

Sugar group of another nucleotide

Phosphate group

Cyclic sugar (ribose)

Phosphate group of another nucleotide

(b)

A nucleotide in DNA

Nitrogenous base (cytosine)

Sugar group of another nucleotide

Phosphate group

Cyclic sugar (deoxyribose)

The O atom is absent.

Phosphate group of another nucleotide

(c)

FIGURE 1-39 Composition of a nucleotide (a) Every nucleotide has three components: a phosphate group (green), a cyclic sugar (red), and a nitrogenous base (blue). The backbone of a nucleic acid consists of alternating sugar and phosphate groups. (b) A nucleotide in a ribonucleic acid. The cyclic sugar component must be ribose. The nitrogenous base shown is guanine, but could also be adenine, cytosine, or uracil. (c) A nucleotide in a deoxyribonucleic acid. The cyclic sugar must be deoxyribose, in which the ribose oxygen indicated is not present. The nitrogenous base shown is cytosine, but could also be adenine, guanine, or thymine.

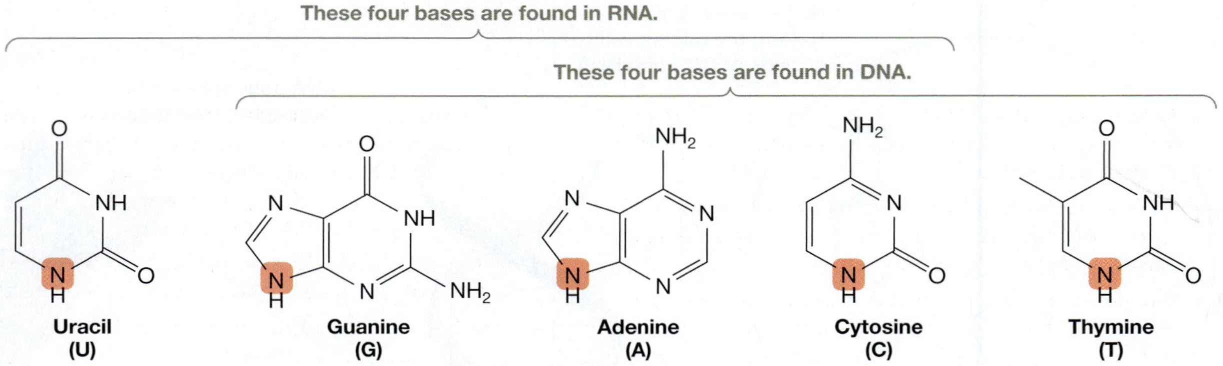

| Uracil (U) | Guanine (G) | Adenine (A) | Cytosine (C) | Thymine (T) |

FIGURE 1-40 **Nitrogenous bases** The identity of a nucleotide is specified by the nitrogenous base bonded to the sugar ring. U, G, A, and C are found in RNA, whereas G, A, C, and T are found in DNA. In a nucleotide, carbon 1′ of the ribose sugar unit connects to the highlighted nitrogen atom.

nitrogenous bases that appear in RNA: uracil, guanine, adenine, and cytosine (abbreviated U, G, A, and C, respectively; Fig. 1-40). There are four types of nitrogenous bases that appear in DNA, too: G, A, C, and thymine (T). Thus, three of the bases in DNA are the same as the bases in RNA. Only T (in DNA) is different from U (in RNA).

Notice the functional groups in these nitrogenous bases. An O=C—N group, for example, which is characteristic of an amide, appears in guanine, and a C=C bond, which characterizes an alkene, appears in uracil and thymine. (In Chapter 14, we learn that this C=C bond is better classified as being part of an aromatic ring rather than an alkene.)

YOUR TURN 1.17

Circle and label the functional groups mentioned above that appear in U, G, A, C, and T in Figure 1-40.

In subsequent units on biomolecules, we examine some of the details of the chemical processes involving DNA and RNA. In Section 14.11, for example, we examine the complementarity among the nitrogenous bases.

PROBLEM 1.37 For each of the following nucleotides, **(a)** circle and label the phosphate group, the sugar group, and the nitrogenous base; **(b)** Determine whether it can be part of RNA or DNA; and **(c)** identify the nitrogenous base that it contains.

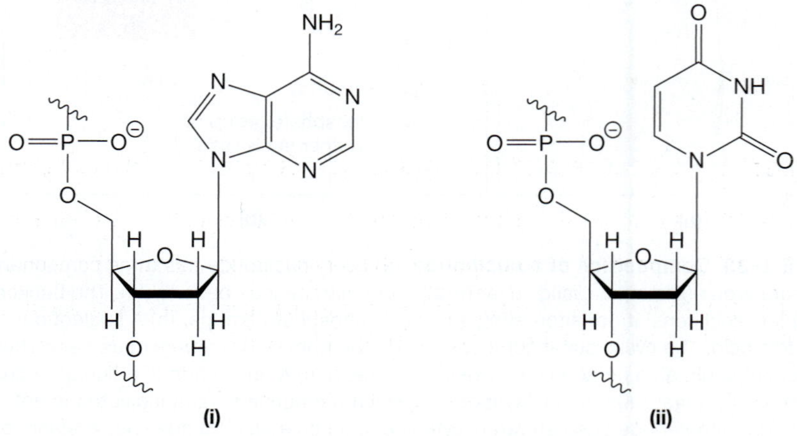

(i)   (ii)

Chapter Summary and Key Terms

- **Organic chemistry** is the subdiscipline of chemistry in which the focus is on compounds containing carbon atoms. **(Section 1.1; Objective 1)**

- Carbon's ability to form four strong covalent bonds is what gives rise to the great variety of organic compounds known. **(Section 1.2; Objective 2)**

- No more than two electrons can occupy an atom's first shell. As many as eight electrons can occupy an atom's second shell. A second or higher level shell containing eight electrons (an **octet**) is especially stable (the octet rule). **(Section 1.3b; Objective 3)**

- An atom's ground state (i.e., lowest energy) electron configuration is derived using the following three rules. **(Section 1.3c; Objective 4)**

  - **Pauli's exclusion principle**: Up to two electrons, opposite in spin, may occupy an orbital.
  - The **aufbau principle:** Electrons occupy the lowest energy orbitals available.
  - **Hund's rule**: If orbitals have the same energy, then each orbital is singly occupied before a second electron fills it.

- Sharing of a pair of electrons by two atoms lowers the energy of the system, creating a **covalent bond**. Some atoms can share two or three pairs of electrons, creating double or triple bonds, respectively. Double bonds are shorter and stronger than single bonds, and triple bonds are shorter and stronger than double bonds. **(Section 1.4; Objective 5)**

- **Lewis structures** illustrate a molecule's **connectivity**. They account for *all* valence electrons, differentiating between bonding pairs and **lone pairs** of electrons. They indicate, moreover, which atoms are bonded together and by what types of bonds. Lewis structures are constructed so that the maximum number of atoms have filled valence shells (i.e., duets or octets). **(Section 1.5; Objective 6)**

- The electrons in a **nonpolar covalent bond** are shared equally. **Polar covalent bonds** arise when atoms with moderate differences in electronegativity are bonded together. **Ionic bonds** arise when there are large differences in electronegativity. **(Sections 1.7 and 1.8; Objective 7)**

- **Formal charge** represents a particular way in which valence electrons are assigned to individual atoms within a molecule. To determine formal charges, each pair of electrons in a covalent bond is split evenly between the two atoms bonded together. **(Section 1.9; Objective 8)**

- Resonance exists when two or more valid Lewis structures can be drawn for a given molecular species. **(Section 1.10)**

  - Each Lewis structure is imaginary and is called a **resonance structure**. **(Objective 9)**
  - The properties of the one, true species—the **resonance hybrid**—represent a weighted average of all the resonance structures. **(Objective 10)**
  - The greater the stability of a given resonance structure, the greater its contribution to the resonance hybrid. **(Objective 9)**

- Resonance structures are related by hypothetically shifting around lone pairs of electrons, and pairs of electrons from double and triple bonds. The atoms in the molecule must remain frozen in place. **(Section 1.11; Objective 10)**

- Shorthand notation is used throughout organic chemistry to draw molecules more quickly and efficiently. Lone pairs are often omitted. **(Section 1.12; Objective 11)**

  - **Condensed formulas** are used primarily to write a molecule in a line of text, with hydrogens written adjacent to the atom to which they are bonded.
  - **Line structures** show all bonds explicitly, except bonds to hydrogen. Hydrogen atoms are omitted if they are bonded to carbon. Carbon atoms are not drawn explicitly, but are assumed to reside at the intersection of two lines and at the end of a line (unless otherwise indicated).

- **Functional groups** are common bonding arrangements of relatively few atoms. Functional groups dictate the behavior of entire molecules—that is, molecules with the same functional groups tend to behave similarly. **(Section 1.13; Objective 12)**

Problems

1.3 Atomic Structure and Ground State Electron Configurations

1.38 In which of the following orbitals does an electron possess the most energy? 2s; 3s; 4s; 3p; 2p

1.39 For each pair, identify the orbital in which an electron possesses more energy. **(a)** 4s or 5s; **(b)** 5p or 5d

1.40 Write the ground state electron configuration of each of the following atoms. For each atom, identify the valence electrons and the core electrons. **(a)** Al; **(b)** S; **(c)** O; **(d)** N; **(e)** F

1.4 The Covalent Bond: Bond Energy and Bond Length

1.41 Consider a molecule of N_2. For each pair of distances between the two N atoms, determine which distance represents a higher energy. Explain. **(a)** 50 pm or 75 pm; **(b)** 75 pm or 110 pm; **(c)** 110 pm or 150 pm; **(d)** 150 pm or 160 pm

1.42 In which of the following molecules is the bond between carbon and nitrogen the shortest? In which molecule is it the strongest? $H-C\equiv N$; $H_2C=NH$; H_3C-NH_2

1.43 According to Table 1-2, which of the following compounds has the strongest single bond? The weakest? HCl; CH_4; H_2O; HF; Cl_2

1.5–1.8 Lewis Dot Structures, Polarity, and Ionic Bonds

1.44 Draw Lewis structures for each of the following molecules: **(a)** CH_5N (contains a bond between C and N); **(b)** CH_3NO_2 (contains a bond between C and N but no bonds between C and O); **(c)** CH_2O; **(d)** CH_2Cl_2; **(e)** BrCN

1.45 For each structure below, determine whether it is a legitimate Lewis structure. If it is not, explain why not.

(a)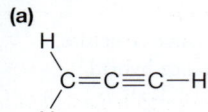

(b)

(c)

(d)

(e)

1.46 Complete the Lewis structure for each of the following molecules using the information provided in Table 1-4. You may assume that all formal charges are zero. All H atoms are shown; add only bonding pairs and lone pairs of electrons.

(a)

(b)

(c)

1.47 Rank the following in order of increasing negative charge on carbon. CH_3-CH_3; CH_3-MgBr; CH_3-Li; CH_3-F; CH_3-OH; CH_3-NH_2

1.48 Which of the following contains an ionic bond? **(a)** H_2; **(b)** NaCl; **(c)** NaOH; **(d)** CH_3ONa; **(e)** CH_4; **(f)** $HOCH_2CH_3$; **(g)** $LiNHCH_3$; **(h)** $CH_3CH_2CO_2K$; **(i)** $C_6H_5NH_3Cl$

1.9 Assigning Electrons to Atoms in Molecules: Formal Charge

1.49 In the methoxide anion (CH_3O^-), is it possible for a double bond to exist between C and O, given that the negative charge resides on O? Explain why or why not.

1.50 Draw Lewis structures for each of the following ions. One atom in each ion has a formal charge that is not zero. Determine which atom it is, and what the formal charge is. **(a)** the C_2H_5 anion; **(b)** the CH_3O cation; **(c)** the CH_6N cation; **(d)** the CH_5O cation; **(e)** the C_3H_3 anion (all three H atoms are on the same carbon)

1.51 Identify the formal charge on each atom in the following species. Assume that all valence electrons are shown.

1.52 The structure at the right is a skeleton of an anion having the overall formula $C_6H_6NO^-$. The hydrogen atoms are not shown.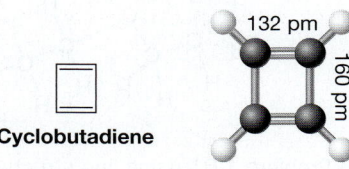

(a) Draw a complete Lewis structure in which the -1 formal charge is on N. Include all H atoms and valence electrons.

(b) Do the same for a Lewis structure with the -1 formal charge on O.

(c) Do the same for a Lewis structure with the -1 formal charge on the C atom that is bonded to three other C atoms.

1.10 and 1.11 Resonance Theory; Drawing All Resonance Structures

1.53 Draw all resonance contributors for each of the following molecules or ions. Be sure to include the curved arrows that indicate which pairs of electrons are shifted in going from one resonance structure to the next.

(a) CH_3NO_2

(b) $CH_3CO_2^-$

(c) $CH_3CHCHCH_2^-$ (the ion has two C—C single bonds)

(d) C_5H_5N (a ring is formed by the C and N atoms, and each H is bonded to C)

(e) C_4H_5N (a ring is formed by the C and N atoms, the N is bonded to one H, and each C is bonded to one H)

1.54 Draw the resonance hybrid of CH_3NO_2 in Problem 1.53(a).

1.55 **(a)** Draw all resonance contributors of sulfuric acid, H_2SO_4 (the S atom is bonded to four O atoms). **(b)** Which resonance structure contributes the most to the resonance hybrid? **(c)** Which resonance structure contributes the least to the resonance hybrid?

1.56 Experiments indicate that the carbon–carbon bonds in cyclobutadiene are of two different lengths. Argue whether or not cyclobutadiene has a resonance structure.

Cyclobutadiene

132 pm

160 pm

1.57 The two species shown are structurally very similar. Draw all resonance structures for each species and determine which is more stable. Explain.

(a)   **(b)**

1.58 The two species shown are structurally very similar. Draw all resonance structures for each species and determine which is more stable. Explain.

(a)   **(b)**

1.12 Shorthand Notations

1.59 Redraw the following structure of glucose as a line structure.

Glucose

1.60 Redraw the following line structure of sucrose as a complete Lewis structure. Include all hydrogen atoms and lone pairs.

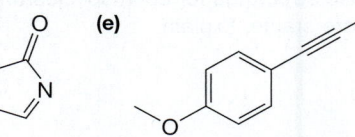

Sucrose

1.61 Draw the following Lewis structures using condensed formulas.

(a)

$$H-\underset{\underset{H}{|}}{\overset{\overset{H}{|}}{C}}-\underset{\underset{H}{|}}{\overset{\overset{H}{|}}{C}}-\underset{\underset{H}{|}}{\overset{\overset{H}{|}}{C}}-\underset{\underset{H}{|}}{\overset{\overset{H}{|}}{C}}-H$$

(b)

$$H-\underset{\underset{H}{|}}{\overset{\overset{H}{|}}{C}}-\underset{H}{\overset{H}{C}}=\underset{H}{\overset{H}{C}}-\underset{\underset{H}{|}}{\overset{\overset{H}{|}}{C}}-H$$

(c)

(d)

(e)

(f)

$$H-\underset{\underset{H}{|}}{\overset{\overset{H}{|}}{C}}-O-\underset{\underset{H}{|}}{\overset{\overset{H}{|}}{C}}-\underset{\underset{H}{|}}{\overset{\overset{H}{|}}{C}}-H$$

1.62 Draw the molecules in Problem 1.61 using line structures.

1.63 Draw Lewis structures for the following molecules. Include all lone pairs and H atoms.

(a)

(b)

(c)

(d)

(e)

(f)

(g)

(h)

(i)

1.64 Draw each of the species in Problem 1.63 as a condensed formula.

1.65 Redraw the given line structure of cholesterol as a condensed formula. What advantages do line structures have?

Cholesterol

1.13 An Overview of Organic Compounds: Functional Groups

1.66 Circle each functional group in glucose (Problem 1.59) and sucrose (Problem 1.60). What compound class is characteristic of each of those functional groups?

1.67 Dimethyl sulfide contains a functional group that is not listed in Table 1-6. Which functional group in Table 1-6 do you think its reactivity might resemble?

$$H_3C—S—CH_3$$

Dimethyl sulfide

1.68 Which one of the following compounds do you think will behave most similarly to ethanol (CH_3CH_2OH)? Explain.

H_2O CH_3OCH_3 CH_3CO_2H

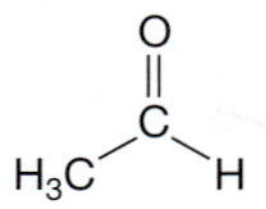

A   **B**   **C**   **D**

1.69 Identify all functional groups that are present in strychnine, a highly toxic alkaloid used as a pesticide to kill rodents, whose line structure is shown here. What compound class is characteristic of each of those functional groups?

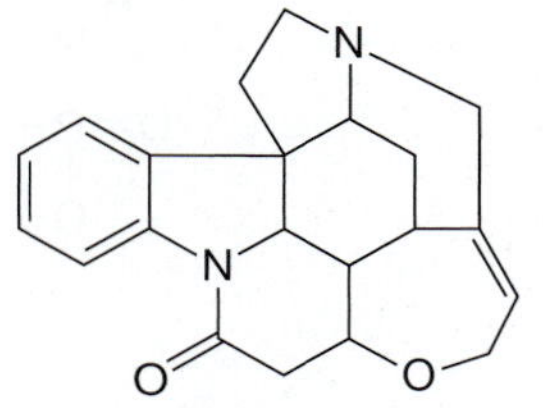

Strychnine

1.70 Identify all functional groups that are present in doxorubicin, a drug used as an antibiotic and cancer therapeutic, whose line structure is shown here. What compound class is characteristic of each of those functional groups?

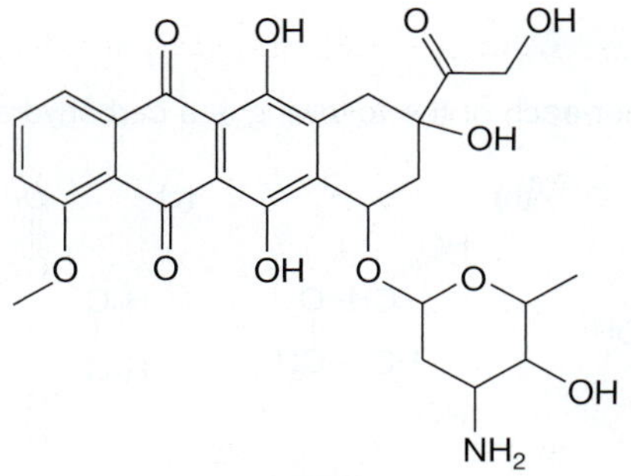

Doxorubicin

1.14 The Organic Chemistry of Biomolecules

1.71 The R group in alanine is $–CH_3$, whereas the R group in aspartic acid is $–CH_2CO_2H$. After consulting Figure 1-34, draw the complete Lewis structure for each of these amino acids.

1.72 Shown here is a tripeptide, which consists of three amino acids linked together in a chain. Circle and name each amino acid.

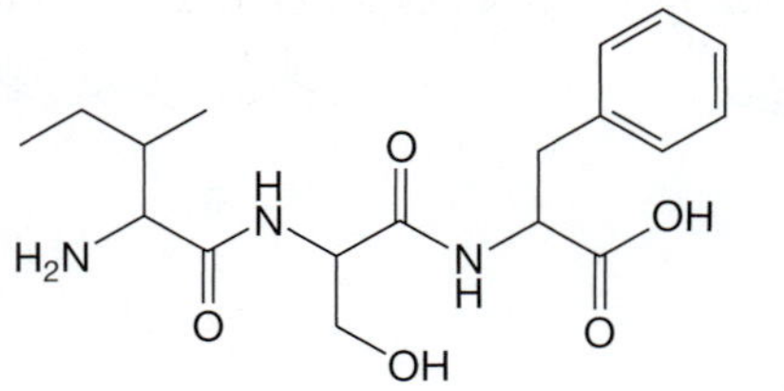

1.73 This is a segment of a nucleic acid. **(a)** Is this a segment of RNA or DNA? **(b)** Circle and name each nucleotide.

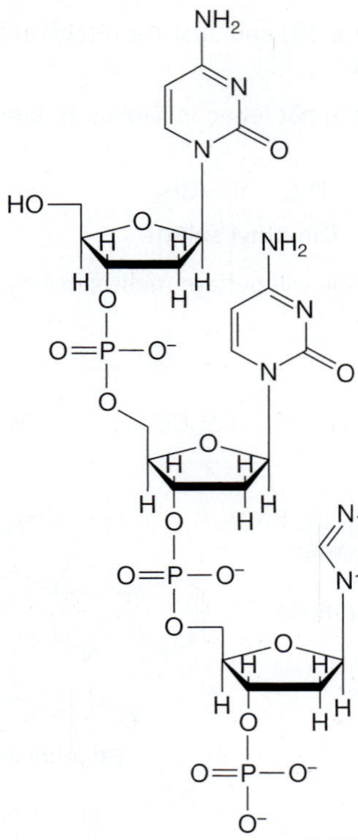

1.74 Determine whether each of the following is a carbohydrate. If so, is it also a monosaccharide?

(a)

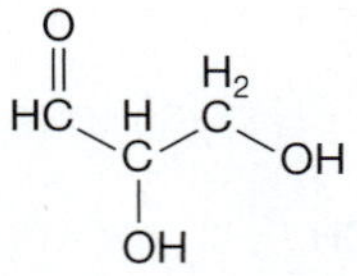

(b)

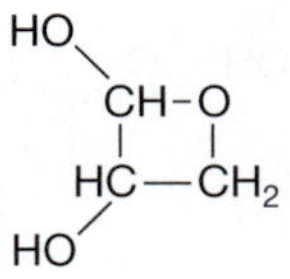

(c)

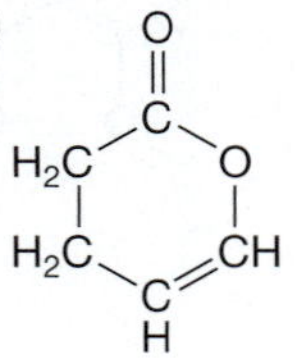

Integrated Problems

1.75 Which of the following pairs are *not* resonance structures of one another? All lone pairs of electrons may or may not be shown. Identify where each lone pair that is not shown belongs.

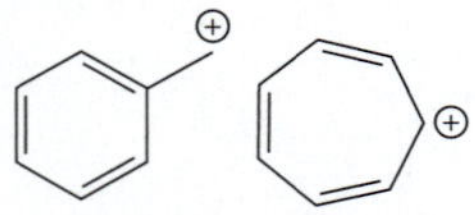

A

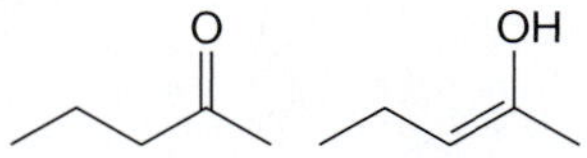

B

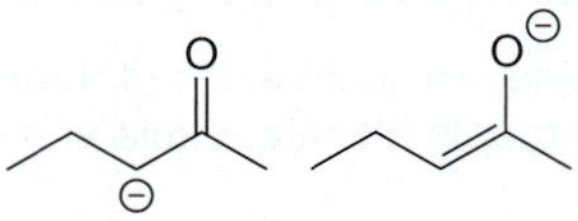

C

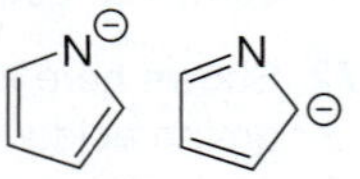

D

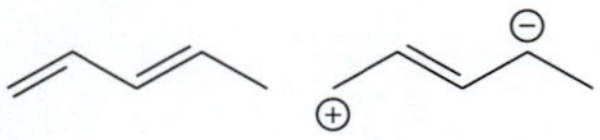

E

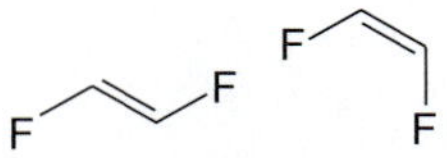

F

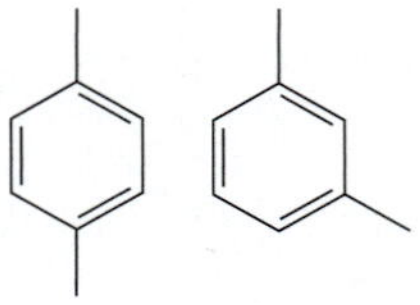

G

H

1.76 Which of the following electrostatic potential maps best represents nitromethane (CH_3NO_2)? Explain.

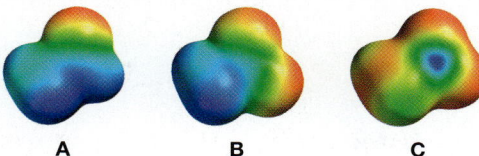

A   B   C

1.77 Which of the following species is responsible for the electrostatic potential map provided? Explain.

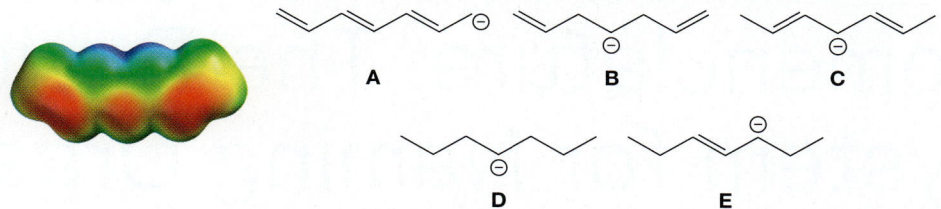

A   B   C

D   E

1.78 Draw the structure of a molecule with formula C_5H_9N that contains functional groups characteristic of **(a)** an alkyne and an amine; **(b)** two alkenes and an amine; **(c)** a nitrile.

1.79 Draw all of the resonance structures for each of the following species. Be sure to include the curved arrows that indicate which pairs of electrons are shifted in going from one resonance structure to the next. Draw the resonance hybrid of each species.

(a)   **(b)**   **(c)**   **(d)**

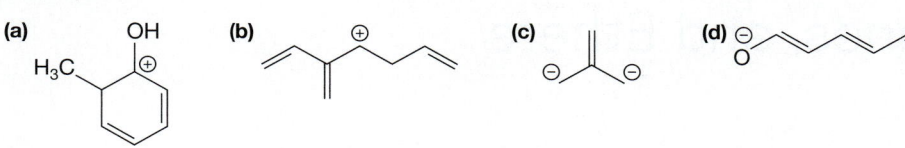

1.80 **(a)** Draw all valid resonance contributors for this ion. Show how the electrons can be moved using curved arrows. **(b)** Draw the resonance hybrid.

1.81 Diazomethane has the formula H_2CN_2. Draw all valid resonance contributors for diazomethane and, using Table 1-3, propose which one contributes more to the resonance hybrid. *Hint:* There are no structures that avoid charged atoms.

1.82 **(a)** Draw all resonance contributors of the following ion. In drawing each additional resonance structure, use curved arrows to indicate which pairs of electrons are being shifted. **(b)** Draw the resonance hybrid. **(c)** Which C—C bond is the longest?

Nomenclature: The Basic System for Naming Organic Compounds

Alkanes, Haloalkanes, Nitroalkanes, Cycloalkanes, and Ethers

A.1 The Need for Systematic Nomenclature: An Introduction to the IUPAC System

Learning how to name specific substances—that is, learning **nomenclature**—is an essential task. Nomenclature is not unique to chemistry; every discipline of study, from accounting and art history to zoology, has its specialized vocabulary. Many terms in these fields can be as technical as anything in the physical sciences.

Biology, for example, includes taxonomy, the classification of organisms and the study of their relationships. The number of species living on our planet is conservatively estimated at 1.5 million (other estimates exceed 30 million). To deal with this great diversity, scientists have adopted, by international agreement, a single language to be used on a worldwide basis. All organisms are given a specific name in Latin, such as *Homo sapiens* (human beings), *Canis familiaris* (the domestic dog), or *Escherichia coli* (a common bacterium of the human digestive system). Like any language, the language of taxonomy evolves with use; some terms commonly used at one time are later declared obsolete and replaced with more precise or more relevant ones.

Organic chemistry is not very different. There are millions of organic compounds, a number comparable to the number of known biological species. In ancient and medieval times, chemists or alchemists sometimes assigned a name to a newly discovered substance that indicated its natural animal or plant source. Vanillin, for example, is the primary component extracted from the vanilla bean, and geraniol is derived from the geranium plant (Fig. A-1a and A-1b). In other cases, the name for a substance was suggested by its physical characteristics or by its chemical or medicinal properties. For example, azulene, which is a dark blue crystalline solid, derives its name from *azul*, the Spanish word for "blue" (Fig. A-1c). And morphine, a powerful sedative isolated from opium, was named for Morpheus, the Greek god of dreams (Fig. A-1d).

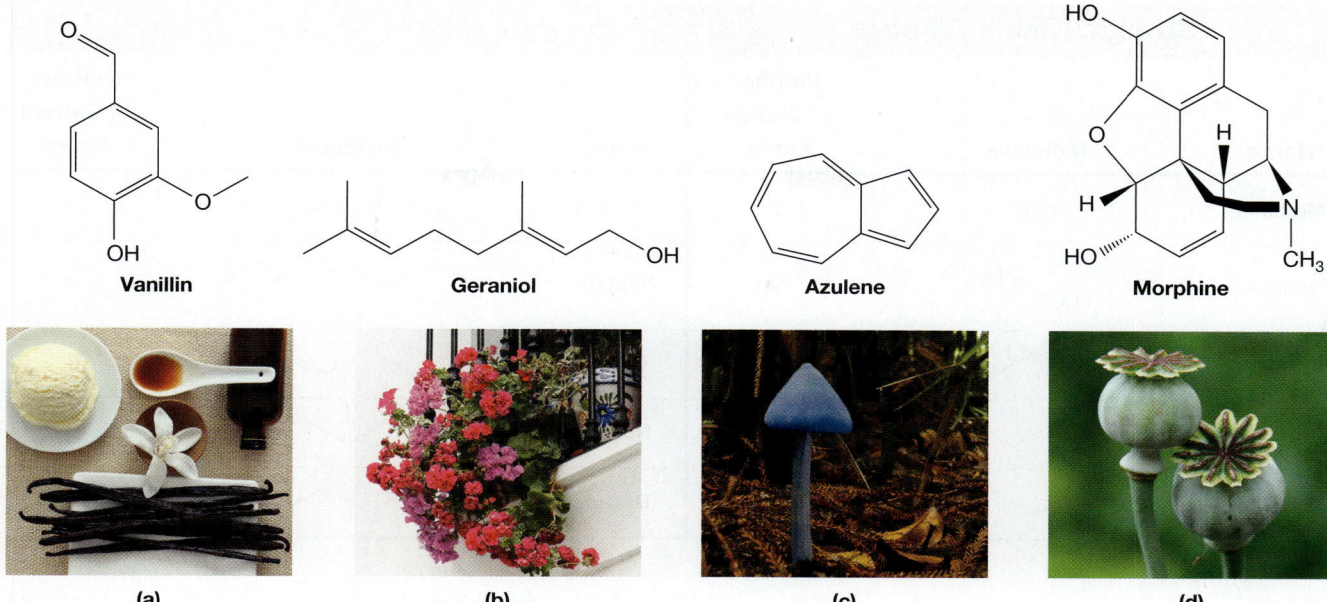

Vanillin   **Geraniol**   **Azulene**   **Morphine**

(a)   (b)   (c)   (d)

FIGURE A-1 Naturally occurring organic compounds (a) Vanillin can be isolated from vanilla extract. (b) Geraniol occurs in geranium oil. (c) Azulene gives this mushroom its blue color. (d) Morphine is derived from opium poppy.

About 100 years ago, chemists realized the necessity for standardizing the names of substances, as well as other chemical terms. Consequently, the **International Union of Pure and Applied Chemistry (IUPAC)** was established in 1919. Immediately, work began on adopting a system of nomenclature for both inorganic and organic substances. Today, while many substances are known by more than one name, there is one *standard* chemical name. More specifically, that name is based on rules that relate to the molecule's structure. Consequently:

- Given a molecular structure, you can derive its IUPAC name, piece by piece.
- Given the IUPAC name, you can draw its structure, piece by piece.

In the sections that follow, we step through the various nomenclature rules and provide several examples, and you will have many opportunities to practice those rules.

A.2 Alkanes and Substituted Alkanes

We begin by learning the names of the **straight-chain alkanes**, or **linear alkanes**, shown in Table A-1, because they form the basis for the entire system of organic nomenclature. Recall from Section 1.13 that *alkanes* comprise the simplest class of organic compounds, because they contain only C—C and C—H single bonds—that is, they contain *no functional groups*. Straight-chain alkanes, furthermore, form *one continuous chain* from one end of the molecule to the other: No carbon atom is bonded to more than two others. The following two rules govern the naming of such compounds:

Naming Straight-Chain Alkanes

- All straight-chain alkanes have the suffix *ane*.
- A prefix (e.g., *meth*, *eth*, *prop*, *but*) is used to identify the number of carbon atoms in the chain.

TABLE A-1 Straight-Chain Alkanes

Name	Molecule	Number of Carbon Atoms	Name	Molecule	Number of Carbon Atoms
Methane	CH_4	1	Hexane		6
Ethane	H_3C—CH_3	2	Heptane		7
Propane		3	Octane		8
Butane		4	Nonane		9
Pentane		5	Decane		10

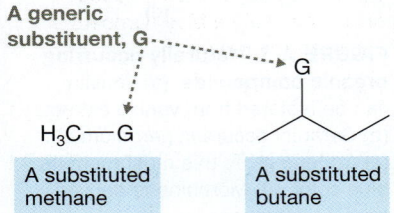

A generic substituent, G

H_3C—G

A substituted methane

A substituted butane

FIGURE A-2 Substituted alkanes
G represents a generic substituent on methane (*left*) and butane (*right*).

Because the names in Table A-1 are fundamental to organic nomenclature, you should take the time to commit them to memory.

If a hydrogen atom of an alkane is replaced by another atom or group of atoms, we say that the alkane is *substituted*, and we call the atom or group of atoms that replaces the hydrogen a **substituent**. In the molecule on the left in Figure A-2, for example, a generic substituent G has replaced one H atom of methane (CH_4), so the molecule is a substituted methane. The molecule on the right is a substituted butane.

PROBLEM A.1 How would you describe each of the molecules shown here, where G is a generic substituent?

(a)

(b)

PROBLEM A.2 Using G as a generic substituent, draw **(a)** a substituted propane and **(b)** a substituted hexane.

There can be multiple substituents in a molecule and the IUPAC name must account for all of them. The rules for doing so depend on which of two types of substituents the molecule has. This chapter deals with the easier case involving *halo*, *nitro*, *alkyl*, and *alkoxy* substituents, all of which require only prefixes to be added to the IUPAC name. The other type of substituents involves considerations of both prefixes and suffixes and is dealt with in Interchapter E.

A.3 Haloalkanes and Nitroalkanes: Roots, Prefixes, and Locator Numbers

To specify a particular substituent in an IUPAC name, you must be able to recognize the substituent in the molecule's structure and you must know the corresponding name for the substituent.

- The *halo* substituents, –F, –Cl, –Br, and –I, are called *fluoro*, *chloro*, *bromo*, and *iodo*, respectively.
- The *nitro* substituent is –NO_2.

With this information, we can name a substituted straight-chain alkane that has a single halo or nitro substituent, so-called **haloalkanes** and **nitroalkanes**, according to the following rules.

Naming a Straight-Chain Alkane with One Substituent

1. Identify the **main chain** or **parent chain** as the *longest continuous chain of carbon atoms*. The name of the corresponding alkane from Table A-1 will be the **root** of the molecule's name.
2. Identify the substituent and add the name of the substituent as a *prefix* to the left of the root.
3. Number each carbon atom of the chain sequentially, beginning with C1 at one end of the chain, so that the carbon atom to which the substituent is bonded receives the *lowest possible number*. The number assigned to the carbon that is bonded to the substituent is called the **locator number** or **locant**.
4. Write the locator number to the left of the substituent name and add a hyphen to separate the locator number from the rest of the name.

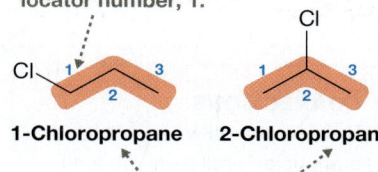

Numbering starts from this end to give chlorine the smaller locator number, 1.

1-Chloropropane   **2-Chloropropane**

The longest continuous carbon chain has three carbon atoms, so the root is propane.

FIGURE A-3 Naming a straight-chain alkane with a single substituent The highlighted carbons indicate the longest continuous carbon chain. The blue numbers are the locator numbers.

Examples of how to apply these rules are shown in Figure A-3 for 1-chloropropane and 2-chloropropane.

SOLVED PROBLEM A.3

What is the IUPAC name of this molecule?

Think How many carbon atoms does the longest continuous carbon chain have? What is the corresponding root? What are the two choices for assigning C1? Which of those choices would give the NO_2 group the lower locator number? How do you add the locator number and the name of the substituent to the IUPAC name?

Solve The longest continuous carbon chain has five carbon atoms, as highlighted at the right, so the root is *pentane*. C1 must be one of the carbon atoms at the end of the chain, so there are two possible numbering systems. In the first numbering system, the locator number for the NO_2 group is 4, whereas in the second numbering system it is 2. Therefore, the second choice is correct. Finally, we add the locator number, a hyphen, and the substituent name as a prefix, so the complete IUPAC name is 2-nitropentane.

Numbering starts from this end to give the nitro group the smaller locator number, 2.

INCORRECT   CORRECT

PROBLEM A.4 Write the IUPAC name for each of the following molecules.

(a)   (b)   (c)   (d)

Sometimes there will be only one locator number possible for a particular substituent, regardless of which end of the chain is assigned C1.

When there is only one number locator possible for a substituent, the number locator is not added to the IUPAC name.

This is the case for bromomethane and iodoethane, shown below. Bromomethane has just a single carbon, so the Br substituent must be attached to C1. Iodoethane has two carbon atoms, so you might think that the I substituent could be bonded to C1 or C2, yielding 1-iodoethane and 2-iodoethane as possible IUPAC names.

CONNECTIONS
Bromomethane was used as a pesticide until being phased out in the early 2000s because it is believed to deplete Earth's ozone layer.

The only locator number possible for Br is 1.

The name 2-iodoethane does not give the smallest locator number for I, which is 1.

Bromomethane

Iodoethane

The name 2-iodoethane, however, violates Rule 3 (p. 55) because the choice for assigning C1 does *not* give the lowest possible number to the carbon atom that is attached to the substituent.

If there are two or more substituents in the molecule, then there are additional considerations, not only to determine the numbering system for the main chain, but also to add prefixes for the substituents.

Establishing the Numbering System for a Main Chain That Has Two or More Substituents

- Choose C1 so that the substituent that is encountered first has the lowest locator number.
- If each choice for C1 results in a tie for the locator number of the first substituent, then C1 is chosen so that the second substituent encountered has the lowest locator number.
- If there is still a tie, keep looking to break the tie by giving the lowest possible locator number to the next substituent encountered.
- If a tie cannot be broken in this way, repeat these steps so that the lowest locator number is assigned to the substituent whose name is alphabetically first.

The examples below show the proper numbering system for molecules containing multiple generic substituents, G.

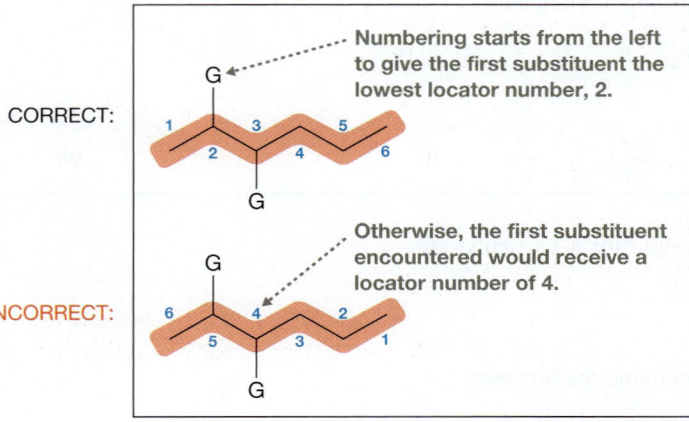

CORRECT:

Numbering starts from the left to give the first substituent the lowest locator number, 2.

INCORRECT:

Otherwise, the first substituent encountered would receive a locator number of 4.

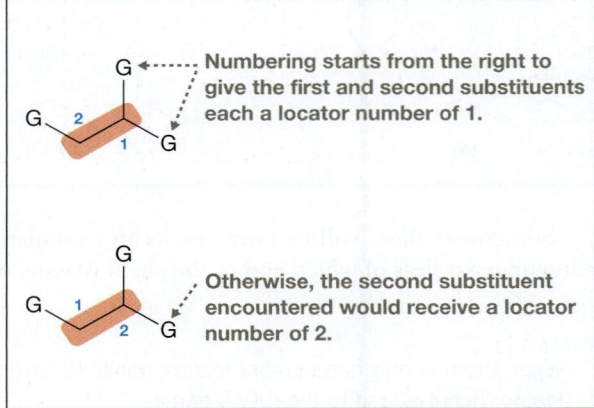

Numbering starts from the right to give the first and second substituents each a locator number of 1.

Otherwise, the second substituent encountered would receive a locator number of 2.

Adding Prefixes When the Main Chain Has Two or More Substituents

- The names of all substituents appear as prefixes before the root, and they are ordered alphabetically according to the substituent's name.
- When there are two or more of the *same* substituent, the appropriate prefix is added just before the name of the substituent:

Number of Substituents	2	3	4	5	6	7	8	9	10
Prefix	di	tri	tetra	penta	hexa	hepta	octa	nona	deca

These multiplying prefixes are not considered part of the substituent name when alphabetizing.

- One locator number is added for *each* substituent.
- If there is just one of a particular substituent, the locator number appears immediately to the left of the substituent it describes.
- If there two or more of the same substituent, the locator numbers appear immediately to the left of the prefix *di*, *tri*, etc. Furthermore, the locator numbers are written in increasing numerical order, separated from each other by commas.
- Hyphens are added to separate a letter in the name from a locator number.
- No spaces should appear in these IUPAC names.

SOLVED PROBLEM A.5

What is the IUPAC name of this molecule?

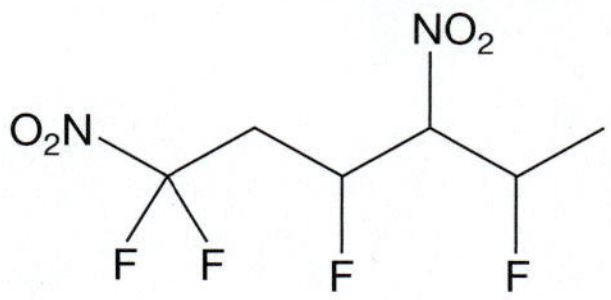

Think How many carbon atoms does the longest continuous carbon chain have and what is the corresponding root? On which end of the chain should numbering begin? What are the names of the substituents and which comes first alphabetically? How do you indicate the number of each kind of substituent and what locator number is assigned to each substituent?

Solve The longest continuous carbon chain has six carbon atoms, so the root is *hexane*. Numbering must begin on the left to give the first substituent encountered the smallest locator number, 1, as shown at the right.

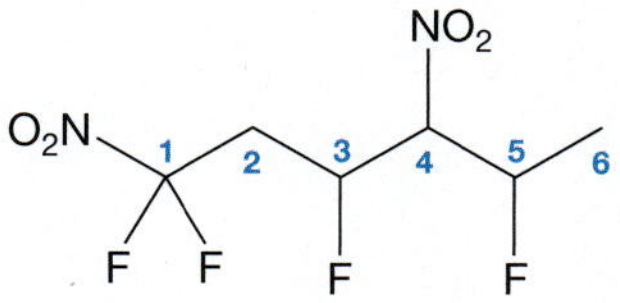

There are two types of substituents, *nitro* and *fluoro*. "Fluoro" comes first alphabetically, so it appears first in the IUPAC name. The prefix tetra is added to indicate four fluoro substituents, and the locator numbers 1,1,3,5 are added to indicate their positions along the chain. The prefix *di* is added to indicate two nitro substituents, and the corresponding locator numbers are 1,4. The IUPAC name, therefore, is 1,1,3,5-tetrafluoro-1,4-dinitrohexane.

PROBLEM A.6 Write the IUPAC name for each of the following molecules.

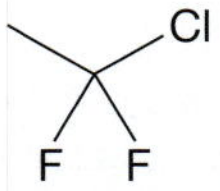

(a) Freon 142b

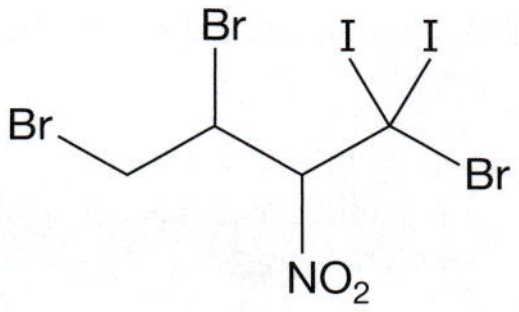

(b)

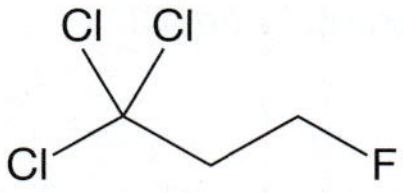

(c)

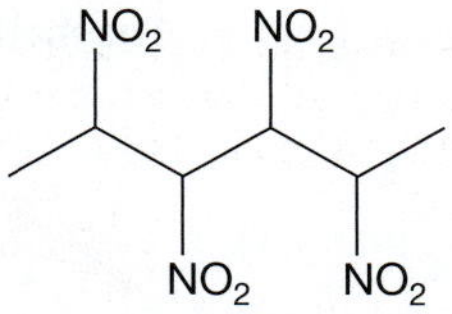

(d)

CONNECTIONS Freon 142b is primarily used as a refrigerant.

As an example where the alphabetical order of the substituent names establishes which end of the carbon chain is C1, consider 1-bromo-3-fluoropropane. Numbering from either end of the chain would result in locator numbers of 1 and 3, but "bromo" comes before "fluoro" alphabetically, so the Br substituent receives the lower locator number.

"Bromo" comes before
"fluoro" alphabetically.

F $\overset{3}{\diagup}$ $\overset{2}{\diagdown}$ $\overset{1}{\diagup}$ Br   (not   F $\overset{1}{\diagup}$ $\overset{2}{\diagdown}$ $\overset{3}{\diagup}$ Br)

1-Bromo-3-fluoropropane   **1-Fluoro-3-bromopropane**

A.4 Alkyl Substituents: Branched Alkanes and Substituted Branched Alkanes

Alkyl groups are substituents that contain only carbon and hydrogen atoms that are all connected by single bonds. You can envision an alkyl group constructed by removing a hydrogen atom from an alkane, leaving one of the carbon atoms available for bonding. The most straightforward alkyl groups to name are *straight-chain alkyl groups*, which are derived by removing a hydrogen atom from the terminal carbon of a straight-chain alkane. Some examples are shown below.

Straight-chain alkyl groups

Bond available to attach to main chain

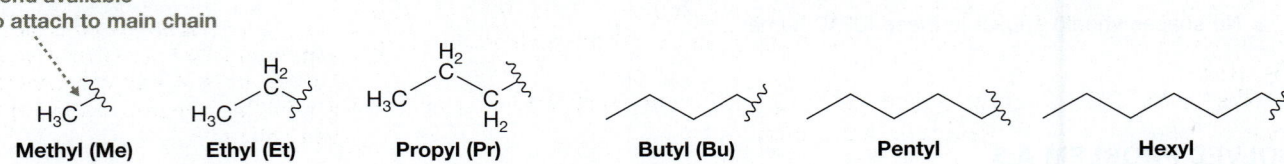

Methyl (Me)   **Ethyl (Et)**   **Propyl (Pr)**   **Butyl (Bu)**   **Pentyl**   **Hexyl**

Naming these alkyl substituents is relatively straightforward.

Naming Straight-Chain Alkyl Groups

- Begin with the name of the straight-chain alkane that has the same number of carbon atoms as the alkyl substituent.
- Replace the *ane* suffix with the new suffix *yl*.

The CH_3- substituent, for instance, is called the **methyl group** because it has a single carbon atom, the same as methane (CH_4). Similarly, the CH_3CH_2- substituent is called the **ethyl group** because it is derived from ethane, CH_3CH_3. Sometimes abbreviations are used for simple alkyl substituents like these, as indicated in parentheses above, and these abbreviations can be used in molecular structures.

When the main chain of a molecule has attached alkyl groups, the molecule is said to be **branched** because not all of the carbon atoms connected together form a single straight chain. The rules for naming these types of molecules are no different from the rules we used to name halo- and nitro-substituted alkanes. Examples are shown for 3-methylhexane and 4,4-dichloro-5-propyloctane.

The methyl group is on C3.   The propyl group is on C5.

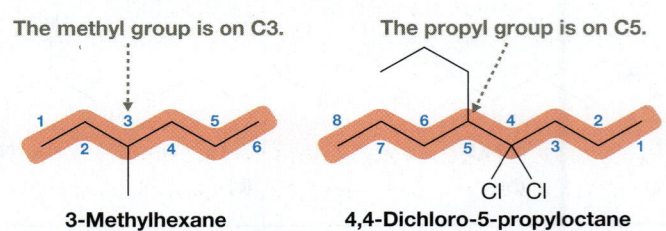

3-Methylhexane   **4,4-Dichloro-5-propyloctane**

You must be careful when dealing with alkyl substituents, however, because they can influence what you might perceive to be the longest continuous chain of carbons. The molecule on the left below, for example, might appear to be 1-methylpentane, but the methyl group is part of the longest continuous chain of six carbons, making the molecule simply hexane. Likewise, on the right, what might appear to be 2-ethylbutane is in fact 3-methylpentane instead.

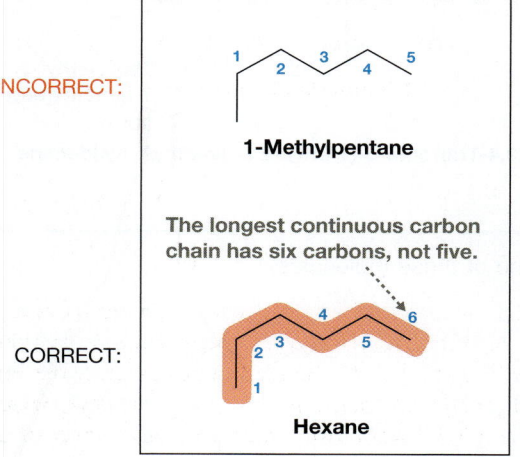
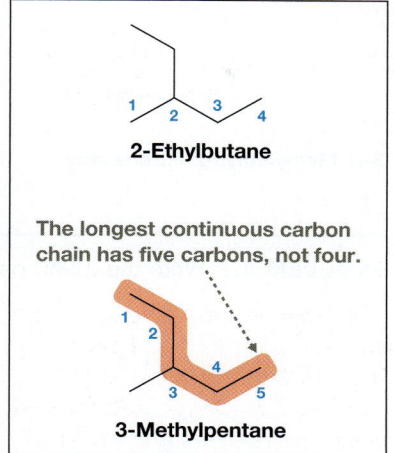

Not all alkyl groups are straight-chain alkyl groups. Rather, an alkyl group can also be branched, having more than one terminal carbon atom. An example is shown in Figure A-4.

The IUPAC rules for naming these branched alkyl groups apply the same logic as some of the rules we have already encountered.

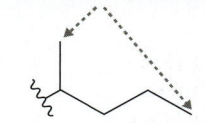

FIGURE A-4 **A branched alkyl substituent** Notice that, beginning from the point of attachment, there is more than one path to arrive at a terminal carbon.

Naming Branched Alkyl Groups

- Identify the longest continuous chain of carbons in the alkyl group, beginning at the substituent's point of attachment. This establishes the *main chain of the branched alkyl group*.
- Number the carbon atoms of the branched alkyl group's main chain, with C1 being the atom at the point of attachment.
- The *root name* of the branched alkyl group is assigned as the name of the straight-chain alkyl group (i.e., *ethyl*, *propyl*, etc.) that has the same number of carbons as the main chain of the branched alkyl group.
- Add prefixes and locator numbers to account for alkyl groups attached along the substituent's main chain.

The first branched alkyl group in Figure A-5, for example, is named 1-methylethyl because the longest carbon chain beginning from the point of attachment has two carbons, and off of that chain a methyl group is attached at C1. The second branched alkyl group is 1-ethyl-2,2-dimethylbutyl because the longest carbon chain beginning from the point of attachment has four carbon atoms, an ethyl group is attached to that chain at C1, and two methyl groups are attached at C2.

Incorporating branched alkyl groups into an IUPAC name follows the same rules as straight-chain alkanes, with one additional rule.

Incorporating Branched Alkyl Groups into an IUPAC Name

Enclose the entire name of a branched alkyl group in parentheses.

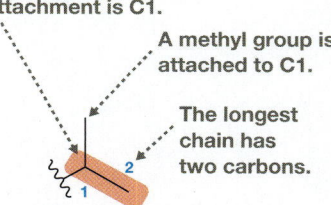

The 1-methylethyl group

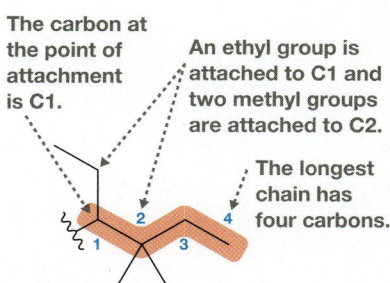

The 1-ethyl-2,2-dimethylbutyl group

FIGURE A-5 **Naming branched alkyl substituents** Numbering begins at the point of attachment. The main chain of the alkyl group is highlighted.

These parentheses help to avoid confusion between the numbering system for the substituent's main chain and that for the main chain of the entire molecule. Examples are shown below.

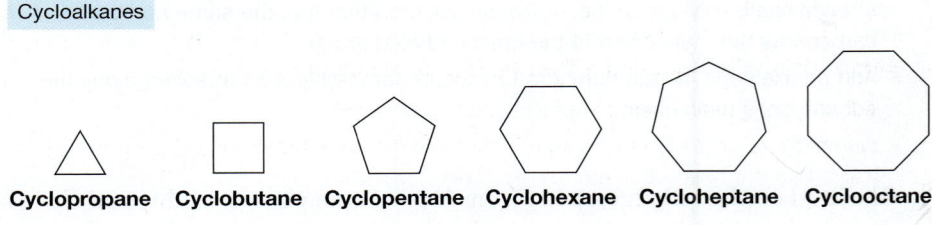

3-(1-Methylethyl)-2-nitrohexane

2,3,4-Tribromo-5-(1-ethyl-2,2-dimethylbutyl)decane

PROBLEM A.7 Write the IUPAC name of these molecules.

(a)

(b)

PROBLEM A.8 Draw the structure that corresponds to each IUPAC name.
(a) 3,3-dibromo-5-(2-methylpropyl)nonane; **(b)** 4,5-di(1,1-dimethylethyl)octane

---

A.5 Cyclic Alkanes and Cyclic Alkyl Groups

If a molecule contains at least one ring made entirely of carbon atoms, then a ring could establish the root as a **cycloalkane**, such as the ones below.

Cycloalkanes

Cyclopropane   Cyclobutane   Cyclopentane   Cyclohexane   Cycloheptane   Cyclooctane

Alternatively, the rings could be treated as substituents, called **cycloalkyl groups**, exemplified by the following:

Cycloalkyl groups

Cyclopropyl   Cyclobutyl   Cyclopentyl   Cyclohexyl

Whether a ring is treated as a root or a substituent depends on the relative number of carbon atoms it has.

Establishing the Root of a Molecule That Contains a Carbon Ring

- If the largest carbon ring has as many or more carbons than the longest continuous straight carbon chain, then the ring establishes the root as a cycloalkane.
- Otherwise, the longest continuous carbon chain establishes the root and the ring is treated as a cycloalkyl substituent.

For example, in the first molecule below, the root is *octane* because the ring has five carbons, not eight or more. In the second molecule, the *cyclooctane* ring establishes the root because the longest continuous straight chain from one terminal carbon to the other has only seven carbons, which is fewer than the eight in the ring.

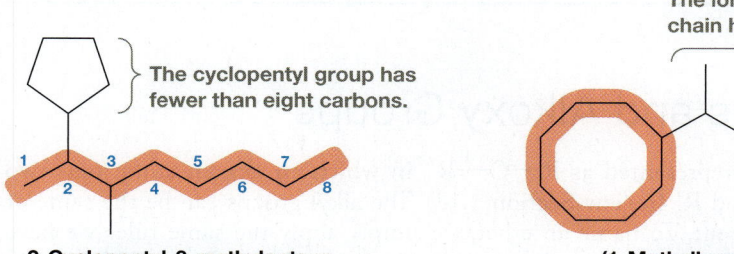

2-Cyclopentyl-3-methyloctane

The cyclopentyl group has fewer than eight carbons.

The longest continuous chain has seven carbons.

(1-Methylhexyl)cyclooctane

If the molecule's root is a cycloalkane, we establish the numbering system just as we did with alkanes. Namely, C1 is chosen so that the first substituent encountered receives the lowest possible locator number. If this results in a tie between two or more carbons of the ring, then we break the tie by choosing C1 so as to give the next substituent encountered the lowest possible locator number, and so on.

Because of the cyclic nature of the ring, you should be aware of the following consequences of the rule for establishing numbering systems.

Establishing the Numbering System of a Cycloalkane

- C1 has the greatest number of attached substituents.
- Numbering increases clockwise or counterclockwise around the ring so that the next substituent is encountered the earliest.

SOLVED PROBLEM A.9

Write the IUPAC name for this molecule.

Think How many carbon atoms are in the longest carbon chain? The largest carbon ring? Which one establishes the root? Which carbon atom gives the lowest locator number to the first substituent? To the second substituent? Should numbering increase clockwise or counterclockwise in order to encounter the next substituent the earliest?

Solve The longest carbon chain has one carbon and the largest carbon ring has six, so the root is *cyclohexane*. The top carbon of the ring is chosen to be C1 because the locator number is 1 for both the first and second substituents (both CH_3 groups in this case). If, on the other hand, C1 were chosen to be one attached to either Br or Cl, then the locator number of the next substituent would be 2. Next, we choose to increase the numbers counterclockwise to give the next substituent, Br, the lowest possible number, 3. Once the numbering system is established, prefixes are added just as before. The name of this molecule is 3-bromo-4-chloro-1,1-dimethylcyclohexane.

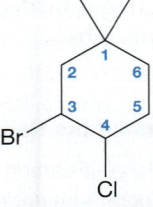

3-Bromo-4-chloro-1,1-dimethylhexane

PROBLEM A.10 Write the IUPAC name for each of the following molecules.

(a)   (b)   (c)   (d)

A.6 Ethers and Alkoxy Groups

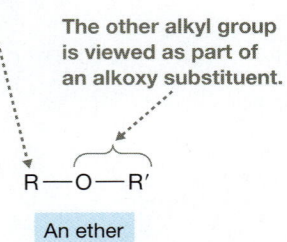

The alkyl group with the longer chain of C atoms establishes the root.

The other alkyl group is viewed as part of an alkoxy substituent.

R—O—R′

An ether

FIGURE A-6 Naming an ether An ether is composed of an alkyl group (R) and an alkoxy group (–OR′).

An ether can be represented as R—O—R′, in which an oxygen atom joins two alkyl groups, R and R′ (review Section 1.13). The alkyl groups can be the same or they can be different. To name an ether, we simply apply the same rules we have learned so far, but we must recognize that the longest carbon chain could belong to either of the alkyl groups. Assuming that R contains the longest carbon chain, then the remaining portion, –OR′ is treated as a substituent called an **alkoxy group**, as shown in Figure A-6.

Once the root and the alkoxy group have been identified, we proceed with constructing the IUPAC name.

> **Naming an Ether**
>
> • The IUPAC name of an ether has the form *alkoxyalkane*.
> • The specific name of the alkoxy substituent (–OR′) is obtained by removing the suffix *yl* from the name of the corresponding alkyl group (R′) and adding the suffix *oxy*.

Thus, –OCH_3 is the **methoxy group** and –OCH_2CH_3 is the **ethoxy group**. The following examples demonstrate this naming system:

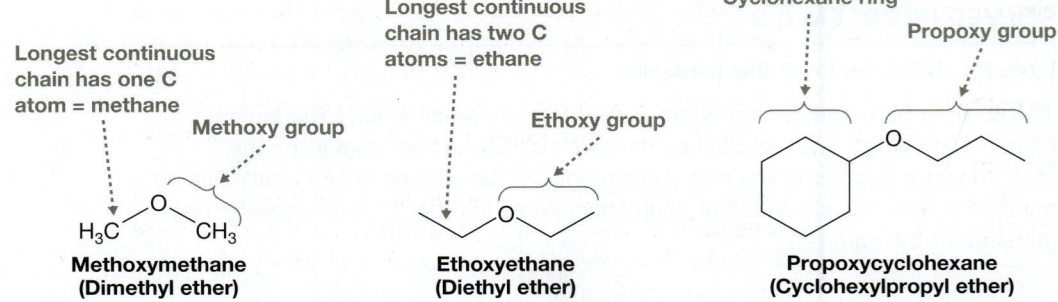

Longest continuous chain has one C atom = methane

Methoxy group

Longest continuous chain has two C atoms = ethane

Ethoxy group

Cyclohexane ring

Propoxy group

Methoxymethane
(Dimethyl ether)

Ethoxyethane
(Diethyl ether)

Propoxycyclohexane
(Cyclohexylpropyl ether)

In methoxymethane, the longest carbon chain has one carbon, so the root is *methane*. The remaining alkoxy substituent is $-OCH_3$, the *methoxy* group. In ethoxyethane, the longest carbon chain has two carbons, so the root is *ethane*, and the remaining $-OCH_2CH_3$ group is the *ethoxy* substituent. In propoxycyclohexane, *cyclohexane* is the root because it has more C atoms (six) than the propyl group (three), and the remaining $-OCH_2CH_2CH_3$ group is the *propoxy* substituent. (The names in parentheses under the molecules are common names, discussed in Section A.7.)

In the preceding compounds, numbers are unnecessary to locate the alkoxy substituents because only one locator number is possible. In the following examples, however, substituent locators are necessary:

1-Methoxypropane
(Methyl propyl ether)

2-Methoxypropane
(Isopropyl methyl ether)

1,4-Diethoxycyclohexane

3,3-Dichloro-2-methoxypentane

2-Methoxy-1,1,3,3-tetramethylcyclopentane

PROBLEM A.11 What is the IUPAC name for each of the following ethers?

(a)   (b)   (c)

PROBLEM A.12 Draw the structure for each of the following compounds.
(a) 3,3-diethoxypentane; **(b)** 1-chloro-5-methoxyhexane; **(c)** 1,2-diethoxy-1-methylcyclopentane

---

A.7 Trivial Names or Common Names

In Section A.1, we mentioned that most organic compounds were originally given names based on their properties or origin. As the language of chemistry grew and evolved, a sizable number of these **trivial names** or **common names** came into regular use, and eventually many of them became integrated into the IUPAC system. Nonscientists sometimes know these trivial names (e.g., formaldehyde, glucose, or adrenaline), even if they don't know the exact structures of these substances.

Although we will minimize the use of trivial names in this book, they are so frequently employed by a substantial number of professionals that you will need to know them to communicate effectively. Thus, where appropriate, we will provide both the IUPAC name and the trivial name together.

In this book, trivial names will generally appear in parentheses, whereas IUPAC names will not.

A.7a Trivial Names of Alkanes and Alkyl Groups

In Section A.2, we learned how to name straight-chain alkanes such as butane and pentane. Historically, to differentiate these from their branched counterparts, straight-chain alkanes were given the prefix *n*–, which stands for *normal*. Common branched alkanes with the same number of carbon atoms were given the same root but a different prefix. As shown in Figure A-7, for example, *n*-butane is the trivial name for butane, whereas isobutane, the only branched alkane with four carbon atoms, is the trivial name for methylpropane. Similarly, *n*-pentane, isopentane, and neopentane all have five carbon atoms and are the trivial names for pentane, methylbutane, and dimethylpropane, respectively.

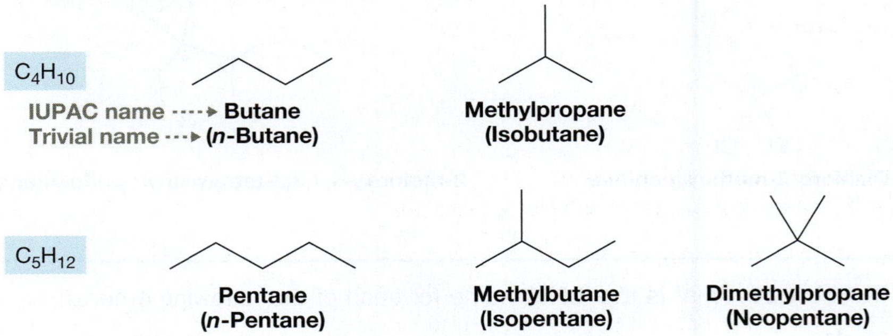

C_4H_{10}

IUPAC name ····▶ **Butane**
Trivial name ···▶ **(*n*-Butane)**

Methylpropane
(Isobutane)

C_5H_{12}

Pentane
(*n*-Pentane)

Methylbutane
(Isopentane)

Dimethylpropane
(Neopentane)

FIGURE A-7 Trivial names of alkanes Trivial names of these molecules appear in parentheses.

A much more common use of trivial names involves branched alkyl substituents. The examples shown in Figure A-8 appear frequently and should be committed to memory. In fact, the branched alkyl groups in Figure A-8 (isopropyl, isobutyl, *sec*-butyl, *tert*-butyl, isopentyl, and neopentyl) appear so frequently that they have become accepted IUPAC names.

Note that we have already learned how to name each of these alkyl substituents under the IUPAC system. The *n*-butyl group, for example, has the IUPAC name butyl, and the neopentyl group has the IUPAC name 2,2-dimethylpropyl.

To help learn the trivial names for the alkyl substituents below, you should know that the prefixes *sec* and *tert* stand for secondary and tertiary, respectively, and reflect the type of carbon atom at the point of attachment. In general, we distinguish carbon

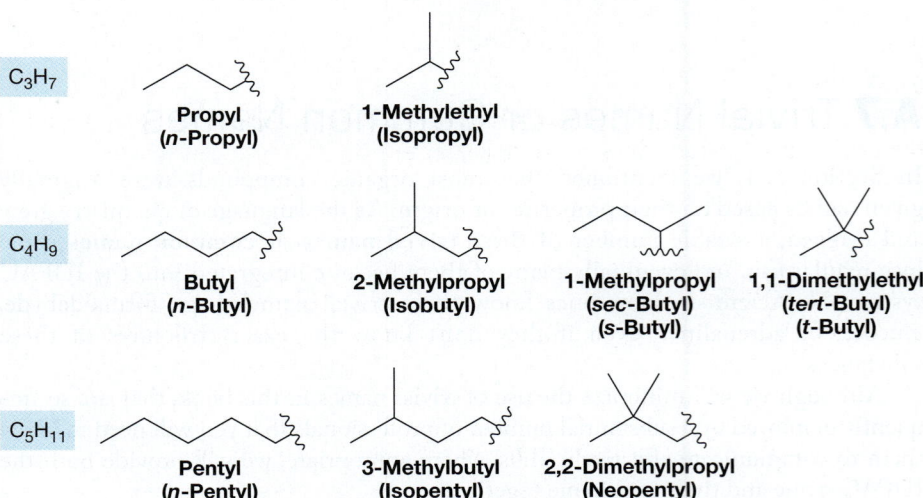

C_3H_7

Propyl
(*n*-Propyl)

1-Methylethyl
(Isopropyl)

C_4H_9

Butyl
(*n*-Butyl)

2-Methylpropyl
(Isobutyl)

1-Methylpropyl
(*sec*-Butyl)
(*s*-Butyl)

1,1-Dimethylethyl
(*tert*-Butyl)
(*t*-Butyl)

C_5H_{11}

Pentyl
(*n*-Pentyl)

3-Methylbutyl
(Isopentyl)

2,2-Dimethylpropyl
(Neopentyl)

FIGURE A-8 Trivial names of common alkyl substituents Trivial names of these substituents appear in parentheses.

atoms based on the number of *other* carbon atoms to which they are directly bonded (Fig. A-9).

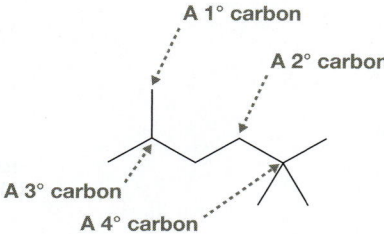

- A *primary carbon* (abbreviated 1°) is directly bonded to one other carbon atom.
- A *secondary carbon* (abbreviated 2°) is directly bonded to two other carbon atoms.
- A *tertiary carbon* (abbreviated 3°) is directly bonded to three other carbon atoms.
- A *quaternary carbon* (abbreviated 4°) is directly bonded to four other carbon atoms.

FIGURE A-9 Types of carbon Carbons in a molecule are classified by the number of other carbon atoms that are directly attached.

Thus, examining the *sec*-butyl group more closely (Fig. A-10), we can see that the C atom at the point of attachment is a secondary (2°) carbon. The *tert*-butyl substituent, on the other hand, is attached by a tertiary (3°) carbon.

PROBLEM A.13 How would you classify the three unlabeled C atoms in the *sec*-butyl and *tert*-butyl groups in Figure A-10?

SOLVED PROBLEM A.14

Name this molecule, using names from Figure A-8 for the various alkyl substituents where appropriate.

Think What is the longest carbon chain or largest carbon ring? What is the corresponding root? How do you establish the numbering system to allow each successive substituent to be encountered the earliest? Are there any recognizable alkyl groups appearing in Figure A-8?

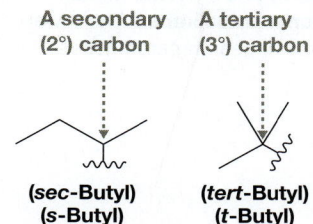

A secondary (2°) carbon   A tertiary (3°) carbon

(*sec*-Butyl)   (*tert*-Butyl)
(*s*-Butyl)   (*t*-Butyl)

FIGURE A-10 Distinguishing C₄H₉ groups A *sec*-butyl group is attached by a secondary C. A *tert*-butyl group is attached by a tertiary C.

Solve The largest carbon ring is *cyclohexane*. To allow each successive substituent to be encountered the earliest, C1 is assigned to the carbon that is attached to the methyl group, and the numbers increase clockwise around the ring. The other two substituents, *isopropyl* and *tert-butyl*, appear in Figure A-8, so we can name the molecule 4-(*tert*-butyl)-2-isopropyl-1-methylcyclohexane. (Notice that the substitutent names appear in alphabetical order.)

Isopropyl group ·······▶ {   } ◀······· *tert*-Butyl group

PROBLEM A.15 Name the following molecules, using names from Figure A-8 for the various alkyl substituents where appropriate.

(a)   (b)   (c)

H_3C-I
Iodomethane
(Methyl iodide)

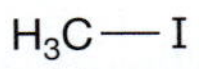

Chloroethane
(Ethyl chloride)

Cl — Cl
Dichloromethane
(Methyl dichloride)
(Methylene chloride)

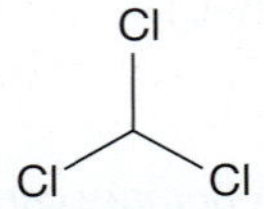

Trichloromethane
(Methyl trichloride)
(Chloroform)

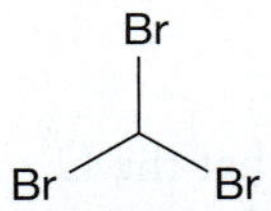

FIGURE A-11 **Trivial names of some haloalkanes** The trivial names appear in parentheses.

Tribromomethane
(Methyl tribromide)
(Bromoform)

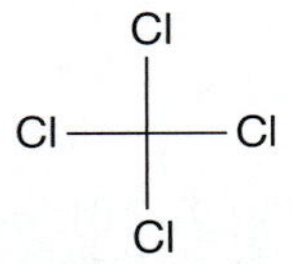

Tetrachloromethane
(Methyl tetrachloride)
(Carbon tetrachloride)

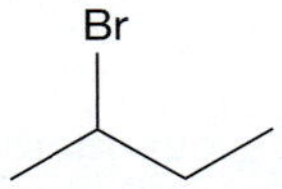

2-Bromobutane
(sec-Butyl bromide)
(s-Butyl bromide)

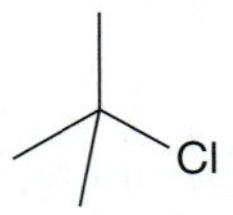

2-Chloro-2-methylpropane
(tert-Butyl chloride)
(t-Butyl chloride)

A.7b Trivial Names of Haloalkanes

Figure A-11 shows the trivial names for a variety of haloalkanes. Some of these names derive from an older system of naming haloalkanes, which mirrors the one for naming *ionic* compounds like NaCl (sodium chloride), where the metal is a cation (positively charged) and the halogen is a halide anion (negatively charged). Under this system, the names of haloalkanes take the form *alkyl halide*.

Trivial Names for Alkyl Halides

- The name of the alkyl group (R) is followed by the name of the halide anion (F^- = fluoride, Cl^- = chloride, Br^- = bromide, and I^- = iodide).
- A space separates the two parts of the trivial name.

Under this system, the alkyl group can be one that has been incorporated into the IUPAC system, as is the case for methyl iodide (CH_3I), *sec*-butyl bromide, and *tert*-butyl chloride.

In addition to the trivial name derived from the above system, some haloalkanes have other trivial names as well. Important ones include methylene chloride (CH_2Cl_2), chloroform ($CHCl_3$), bromoform ($CHBr_3$), and carbon tetrachloride (CCl_4).

PROBLEM A.16 What are the IUPAC names for **(a)** cyclopentyl iodide and **(b)** *n*-hexyl bromide?

Trivial names for ethers

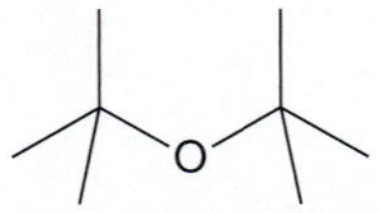

(Di-*tert*-butyl ether)

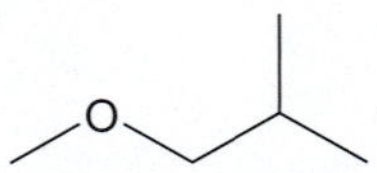

1-Methoxy-2-methylpropane
(Isobutyl methyl ether)

FIGURE A-12 **Trivial names of some ethers** The trivial names appear in parentheses.

A.7c Trivial Names of Ethers

Trivial names for ethers (R—O—R′) primarily come from an older naming system that simply identified the alkyl groups to which the oxygen atom is attached (i.e., R and R′).

Trivial Names for Ethers

- Write the name of each alkyl group attached to O in alphabetical order, followed by "ether." The name has the form *alkyl alkyl ether*, with each word separated by a space.
- If the alkyl groups are identical, the prefix *di* is used.

The trivial name for 1-methoxypropane ($CH_3CH_2CH_2OCH_3$), for example, is methyl propyl ether, and the trivial name for ethoxyethane ($CH_3CH_2OCH_2CH_3$) is diethyl ether. Other examples are shown in Figure A-12.

PROBLEM A.17 What is the IUPAC name for **(a)** *tert*-butyl ethyl ether; **(b)** cyclohexyl methyl ether; **(c)** *sec*-butyl propyl ether?

CONNECTIONS Often simply referred to as *ether*, diethyl ether is a very common organic solvent that was once used to anesthetize patients during medical procedures.

Problems

A.3 Haloalkanes and Nitroalkanes: Roots, Prefixes, and Locator Numbers

A.18 Draw structures for the following haloalkanes. **(a)** 1,2,3-tribromohexane; **(b)** 2,2,3,3,4-pentachlorohexane; **(c)** 1,2-dichloro-4-nitrohexane; **(d)** 1,2-dichloro-3-methoxycyclopentane

A.19 Draw structures for the following haloalkanes. **(a)** 2,2-dichloro-3-cyclopropylbutane; **(b)** 1-bromo-1-chloro-1-iodobutane; **(c)** 2-bromo-1,1-diiodohexane; **(d)** 3-chloro-1,1,2,2-tetrafluoropentane

A.20 Draw structures for the following molecules. **(a)** 3-bromo-2-nitropentane; **(b)** 2,2-dichloro-4,4,5-trinitroheptane; **(c)** 1,2,3,4-tetranitrobutane; **(d)** 6-iodo-1,2-difluorohexane

A.21 Write the IUPAC name for each of the following molecules.

(a) **(b)** **(c)**

A.4 Alkyl Substituents: Branched Alkanes and Substituted Branched Alkanes

A.22 Given each of the IUPAC names provided, draw the corresponding structure. **(a)** 2-methylhexane; **(b)** 3-methylhexane; **(c)** 2,3-dimethylbutane; **(d)** 2,2,3-trimethylbutane

A.23 Given each of the IUPAC names provided, draw the corresponding structure. **(a)** 2,2,4-trimethylpentane; **(b)** 3-ethyl-2,3-dimethylpentane; **(c)** 2,2,3,3-tetramethylhexane

A.24 Given each of the IUPAC names provided, draw the corresponding structure. **(a)** 4-(1-methylethyl)heptane; **(b)** 3-(1,1-dimethylethyl)-4-(1,2-dimethylpropyl)decane

A.25 Given each of the structures provided, write the corresponding IUPAC name.

(a) **(b)** **(c)**

A.26 Write the IUPAC name for each of the following molecules.

(a) **(b)** **(c)** **(d)**

A.5 Cyclic Alkanes and Cyclic Alkyl Groups

A.27 Given each of the IUPAC names provided, draw the corresponding structure. **(a)** 1,1-dimethylcyclohexane; **(b)** 1,2-dimethylcyclohexane; **(c)** 1,2,3-trimethylcyclobutane

A.28 Given each of the IUPAC names provided, draw the corresponding structure. **(a)** 1-cyclopentylhexane; **(b)** cyclohexylcyclohexane; **(c)** 1,2-dicyclopropylnonane

A.29 Given each of the IUPAC names provided, draw the corresponding structure. **(a)** 1-(1,1-dimethylethyl)-2,4-diethylcyclohexane; **(b)** 1,4-dibutyl-2-(1-methylpropyl)cyclooctane; **(c)** 1,1-dicyclopropyl-3-(1,1-dimethylethyl)cycloheptane

A.30 Given each of the structures provided, write the corresponding IUPAC name.

(a)   (b)   (c)   (d) 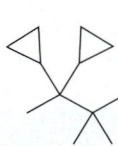   (e)

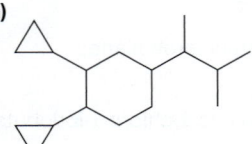   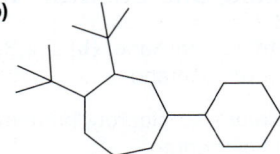   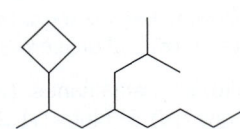

A.31 Given each of the structures provided, write the corresponding IUPAC name.

(a)   (b)   (c)

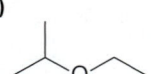

A.6 Ethers and Alkoxy Groups

A.32 Given each of the IUPAC names provided, draw the corresponding structure. **(a)** 1-ethoxypropane; **(b)** 2-ethoxypropane; **(c)** 1,2,3-trimethoxybutane

A.33 Given each of the IUPAC names provided, draw the corresponding structure. **(a)** 1-ethoxy-3-methoxyhexane; **(b)** 1,5-dipropoxypentane; **(c)** 4-butoxy-1,2-dimethoxyheptane

A.34 Given each of the IUPAC names provided, draw the corresponding structure. **(a)** 2-cyclopropoxypentane; **(b)** 1,2-dimethoxy-4-propylcyclohexane; **(c)** 4-(1,1-dimethylethyl)-1,2-dipropoxycyclooctane

A.35 Given each of the structures provided, write the corresponding IUPAC name.

(a)   (b)   (c)   (d)

A.36 Given each of the structures provided, write the corresponding IUPAC name.

(a)   (b)   (c)   (d)

A.7 Trivial Names or Common Names

A.37 Draw the structures of the following compounds. **(a)** 1-isobutyl-4-isopropylcyclohexane; **(b)** *tert*-butylcyclopentane; **(c)** 3,3-diisopropyloctane

A.38 What IUPAC name could be used instead of 3-*tert*-butylhexane?

A.39 In the compound 3-*tert*-butylhexane, how many primary carbons are there? How many secondary carbons are there?

A.40 Draw the structure that corresponds to each of the following names. **(a)** 4-methyl-1-neopentylcyclohexane; **(b)** isobutylcyclobutane; **(c)** 5-sec-butylnonane

A.41 Write the name of each molecule, using names from Figure A-8 (p. 64) for substituents where appropriate.

(a)   (b)   (c)

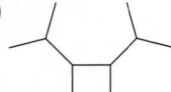   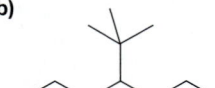

A.42 Draw structures and provide IUPAC names for each of the following. **(a)** *n*-butyl chloride; **(b)** *s*-butyl iodide; **(c)** *t*-butyl fluoride; **(d)** neopentyl bromide; **(e)** diisopropyl ether

A.43 What are the common names for **(a)** 1-propoxybutane and **(b)** 2-ethoxybutane?

Integrated Problems

A.44 Given each of the IUPAC names provided, draw the corresponding structure.
  (a) 2,4-dicyclopropyl-2-ethoxyhexane
  (b) 1,2-dichloro-1-(2-methylpropyl)-4-nitrocyclohexane
  (c) 1,3-dicyclopentyl-1,2,3,4-tetramethoxycyclooctane
  (d) 1-cyclobutyl-4-(1,1-dimethylethyl)-2,4-dinitrononane
  (e) 1-(1,1-dimethylbutyl)-2-ethoxy-1,2,3-trinitrocyclobutane
  (f) 1,2,4-tricyclopropyl-1-(2,2-dichloropentyl)cyclohexane
  (g) 4-(2-chloro-1-methoxyethyl)-1,1-dinitroheptane
  (h) 3,3,4-trichloro-1-cyclohexoxy-4-(1,1-dichloroethyl)decane

A.45 Given each of the structures provided, write the corresponding IUPAC name.

A.46 What is the IUPAC name for each of the following compounds?

2

Geckos can climb effortlessly on almost every surface. Their ability to do so is attributed to ultrafine hairs on their feet, which give rise to a very large contact surface area. This allows for rather strong dispersion forces, one of the intermolecular interactions we examine in this chapter.

Three-Dimensional Geometry, Intermolecular Interactions, and Physical Properties

arbon dioxide (CO_2) and methanoic acid (HCO_2H), also called formic acid, are similar in their chemical makeup, but the boiling point of CO_2 is $-78\ °C$, whereas that of HCO_2H is $101\ °C$. Moreover, CO_2 is only slightly soluble in water, whereas HCO_2H is infinitely soluble in water. Why are their physical properties so vastly different?

$$O = C = O$$
Carbon dioxide

Molar mass = 44 g/mol
Boiling point = −78 °C
Water solubility = slight

**Methanoic acid
(Formic acid)**

Molar mass = 46 g/mol
Boiling point = 101 °C
Water solubility = infinite

As we discuss here in Chapter 2, these compounds behave differently because they experience different *intermolecular interactions*. Those intermolecular interactions are

Chapter Objectives

On completing Chapter 2 you should be able to:

1. Predict both the electron and molecular geometries about an atom, given only a Lewis structure.
2. Recognize molecules that have angle strain.
3. Draw accurate three-dimensional representations of molecules using dash–wedge notation, and be able to interpret the three-dimensional structures of molecules drawn in dash–wedge notation.
4. Determine whether a molecule is polar or nonpolar.
5. Explain how functional groups help determine a species' physical properties.
6. Describe the origin of the various intermolecular interactions discussed and how they govern a species' boiling point, melting point, and solubility.
7. Predict the relative boiling points, melting points, and solubilities of different species, given only their Lewis structures.
8. Distinguish a protic solvent from an aprotic solvent, and explain the role of each type of solvent in the solubility of an ionic compound.
9. Identify the structural features of soaps and detergents and explain how these contribute to their cleansing properties.

governed, in turn, by a variety of factors, including the three-dimensional shapes of the molecules and the functional groups they contain. We begin, therefore, with a review of the factors that determine molecular geometry and then discuss the different types of intermolecular interactions that are important in organic chemistry.

These topics have a broad relevance to many aspects of organic chemistry. Toward the end of this chapter, we explain that intermolecular interactions determine how soaps and detergents function and contribute to the properties of cell membranes. In Chapter 5, we explain how molecular geometry is central to the important concept of *chirality*—that is, whether a molecule is different from its mirror image. And, in Chapter 9, we explain how intermolecular interactions can have a dramatic effect on the outcome of chemical reactions.

2.1 Valence Shell Electron Pair Repulsion (VSEPR) Theory: Three-Dimensional Geometry

To understand many aspects of molecular geometry, chemists routinely work with two models. One, which we discuss here, is **valence shell electron pair repulsion (VSEPR) theory**. The other, which we discuss in Chapter 3, uses the concepts of *hybridization* and *molecular orbital (MO) theory*. Although hybridization and MO theory constitute a more powerful model than VSEPR theory, VSEPR theory remains extremely useful because of its simplicity: Its concepts are easier to grasp and it allows us to arrive at answers much more quickly.

2.1a Basic Principles of VSEPR Theory

The basic ideas of VSEPR theory are as follows:

1. Electrons in a Lewis structure are viewed as groups.
   - A lone pair of electrons, a single bond, a double bond, and a triple bond each constitute one *group* of electrons (Table 2-1).

TABLE 2-1 Various Types of Electron Groups in VSEPR Theory

Type of Group	Total Number of e$^-$	Number of Groups
1 Lone pair	2	1
1 Single bond	2	1
1 Double bond	4	1
1 Triple bond	6	1

TABLE 2-2 Correlations between Electron Geometry and Bond Angle in VSEPR Theory

Number of Electron Groups	Electron Geometry	Approximate Bond Angle
2	Linear	180°
3	Trigonal planar	120°
4	Tetrahedral	109.5°

2. The negatively charged electron groups strongly repel one another, so they tend to arrange themselves as far away from each other as possible.
   - Two electron groups tend to be 180° apart (a linear configuration).
   - Three groups tend to be 120° apart (a trigonal planar configuration).
   - Four groups tend to be 109.5° apart (a tetrahedral configuration).
3. **Electron geometry** describes the orientation of the *electron groups* about a particular atom. These configurations are summarized in Table 2-2.
4. **Molecular geometry** describes the arrangement of *atoms* about a particular atom. Because atoms must be attached by bonding pairs of electrons, an atom's molecular geometry is governed by its electron geometry.

The common molecular geometries, which are summarized in Table 2-3, lead to the following conclusions:

- If all the electron groups are bonds (depicted in gray), then there is an atom attached to each electron group and the molecular geometry is the *same* as the electron geometry.

TABLE 2-3 Molecular Geometries in VSEPR Theory[a]

Number of Bonded Atoms/Groups	ELECTRON GEOMETRY		
	Linear (180°)	Trigonal Planar (120°)	Tetrahedral (109.5°)
2	Linear	Bent	Bent
3		Trigonal planar	Trigonal pyramidal
4			Tetrahedral

[a]Bonding electron groups are depicted with gray sticks; nonbonding electron groups are depicted as yellow sticks terminating in a red lone pair.

• If one or more of the electron groups is a lone pair (depicted in yellow and red), then the molecular geometry is *different* than the electron geometry.

Some examples of molecules containing central atoms without lone pairs are shown in Figure 2-1. For these atoms, the electron and molecular geometries are the same.

H—C—C≡N:
H

180°

Electron geometry = linear
Molecular geometry = linear

(a) Ethanenitrile (Acetonitrile)

:Ö:
∥
H—C—C—C—H
H H H

~120°

Electron geometry = trigonal planar
Molecular geometry = trigonal planar

(b) Propanone (Acetone)

H H
H—C—C—H
H H

~109.5°

Electron geometry = tetrahedral
Molecular geometry = tetrahedral

(c) Ethane

FIGURE 2-1 Compounds in which the central atom lacks lone pairs Because there are no lone pairs about the central atom, the molecular geometries of acetonitrile, acetone, and ethane are identical to their electron geometries.

CONNECTIONS Acetonitrile (Fig. 2-1a) is a polar organic solvent used in the manufacture of DNA oligonucleotides and numerous pharmaceuticals. Acetone (Fig. 2-1b) is another organic solvent and is the active ingredient in nail polish remover. Ethane (Fig. 2-1c) is a gas that is important in the petrochemical industry.

YOUR TURN 2.1

How many electron groups surround **(a)** the triply bonded C in Figure 2-1a, **(b)** the central C in Figure 2-1b, and **(c)** each C atom in Figure 2-1c?

Answers to Your Turns are in the back of the book.

In 2-aminoethanol (ethanolamine, Fig. 2-2, next page), the N and O atoms have molecular geometries that are different than their electron geometries. The electron geometry of the N atom is tetrahedral because it is surrounded by four electron groups: three single bonds and one lone pair. Its molecular geometry, however, which describes only the orientation of the three single bonds, is trigonal pyramidal. Likewise, the O atom of the OH group has a tetrahedral electron geometry (two single bonds and two lone pairs), but its molecular geometry is bent.

SOLVED PROBLEM 2.1

Imines, which are characterized by a C=N double bond, are commonly used as intermediates in organic synthesis. Use VSEPR theory to predict the electron and molecular geometries about the nitrogen atom in the acetone imine molecule shown.

:NH
∥
C
H₃C   CH₃

CONNECTIONS 2-Aminoethanol (Fig. 2-2) is commonly used in the production of a variety of industrial compounds, including shampoos and detergents. It is also used as an injectable treatment for hemorrhoids.

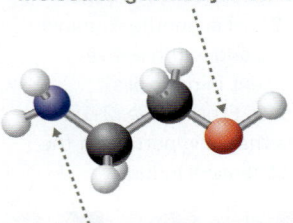

2-Aminoethanol

Electron geometry = tetrahedral
Molecular geometry = bent

Electron geometry = tetrahedral
Molecular geometry = trigonal pyramidal

FIGURE 2-2 Lewis structure and VSEPR geometries about the atoms in 2-aminoethanol The electron geometries about the NH₂ nitrogen and the OH oxygen atoms are both tetrahedral, but they have different molecular geometries because N has one lone pair and O has two.

YOUR TURN 2.2

Electron geometry = trigonal planar
Molecular geometry = trigonal planar

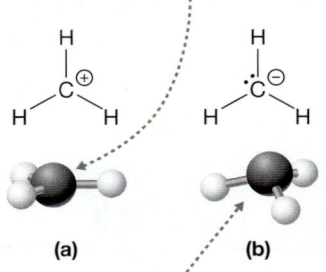

(a)   (b)

Electron geometry = tetrahedral
Molecular geometry = trigonal pyramidal

FIGURE 2-3 Lewis structures and three-dimensional geometries of the methyl cation and methyl anion The central C atom in CH₃⁺ is surrounded by three single bonds only (i.e., no lone pairs), so its electron and molecular geometries are the same. The central C atom in CH₃⁻, on the other hand, is surrounded by three single bonds and one lone pair, so its electron and molecular geometries are different.

Think How many electron groups surround the N atom? Are any of them lone pairs?

Solve There are three groups of electrons around the nitrogen atom: one double bond, one single bond, and one lone pair. According to VSEPR theory, therefore, its electron geometry is trigonal planar, and its molecular geometry is bent.

Electron geometry = trigonal planar
Molecular geometry = bent

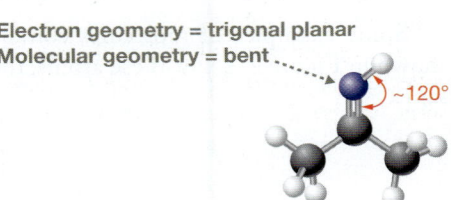
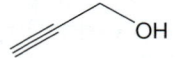

~120°

PROBLEM 2.2 Prop-2-yn-1-ol (propargyl alcohol) is used as an intermediate in organic synthesis and can be polymerized to make poly(propargyl alcohol). Use VSEPR theory to predict the electron and molecular geometries about each nonhydrogen atom in the molecule.

**Prop-2-yn-1-ol
(Propargyl alcohol)**

The rules of VSEPR theory apply equally well to ions. Figure 2-3a shows, for example, that the methyl cation, CH_3^+, has a trigonal planar electron geometry, consistent with a carbon atom that is surrounded by three groups of electrons (i.e., three single bonds). The methyl anion (Fig. 2-3b), on the other hand, is surrounded by four groups of electrons (i.e., three single bonds and a lone pair). Its electron geometry, therefore, is tetrahedral, and its molecular geometry is trigonal pyramidal.

Circle each electron group in the Lewis structures of CH_3^+ and CH_3^- in Figure 2-3.

2.1b Angle Strain

Geometric constraints can force an atom to deviate significantly from its **ideal bond angle**—that is, the bond angle predicted by VSEPR theory. Most commonly this happens in ring structures. For example, the carbon atoms in a molecule of cyclopropane (Fig. 2-4a) should have an ideal bond angle of 109.5°, given that each carbon is surrounded by four groups of electrons (four single bonds). To form the ring, however,

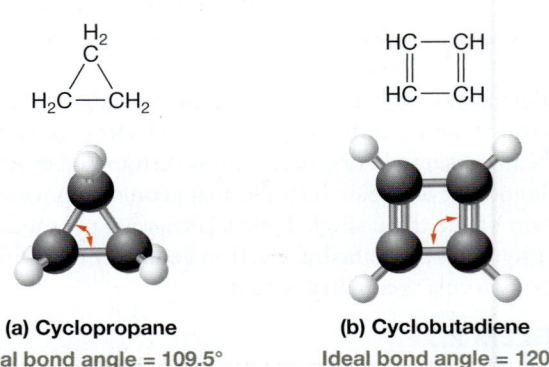

(a) Cyclopropane
Ideal bond angle = 109.5°
Real bond angle = 60°

(b) Cyclobutadiene
Ideal bond angle = 120°
Real bond angle = 90°

FIGURE 2-4 Examples of angle strain In cyclopropane (a), the ideal C—C—C bond angle is 109.5° but the actual angle is 60°. In cyclobutadiene (b), the ideal C—C—C bond angle is 120° but the actual angle is 90°.

the C—C—C bond angles must be 60°. Similarly, each carbon atom of cyclobutadiene (Fig. 2-4b) has an ideal bond angle of 120°, but geometric constraints force the angles in the ring to be 90°.

The deviation of a bond angle from its ideal angle results in an increase in energy, called **angle strain**. Angle strain weakens bonds and makes a species more reactive. In some cases, excessive angle strain can preclude the existence of a molecule altogether.

2.2 Dash–Wedge Notation

Although molecules are three-dimensional, representing them on paper is confined to the two dimensions of the page. To work around this problem, chemists use **dash–wedge notation**, which provides a means to represent atoms both in front of and behind the plane of the paper. Dash–wedge notation has three components:

Rules for Dash–Wedge Notation
- A straight line (—) represents a bond that is in the plane of the paper.
  - Atoms at either end of the bond are also in the plane of the paper.
- A wedge (◄) represents a bond that comes out of the plane of the paper and points toward you.
  - In general, the atom at the thinner end of the wedge is in the plane of the paper, whereas the atom bonded at the thicker end is in front of the page.
- A dash (⫰) represents a bond that is pointed away from you.
  - In general, you may assume in this book that the atom bonded at the thicker end of the dash is behind the plane of the paper. (You may see different conventions in other books.)

Using this dash–wedge notation, Figure 2-5a and 2-5b shows two ways to represent the tetrahedral atom in CH_4. Both illustrations represent the same molecule with the same 109.5° bond angles. The only difference is the vantage point from which you view the molecule. Figure 2-5a is the most common representation, whereas Figure 2-5b is the basis for a shorthand notation of tetrahedral carbon atoms, called the *Fischer projection*, which we introduce in Chapter 5. In Figure 2-5b, only the central C atom lies in the plane of the paper. The atoms at the ends of the wedge bonds are in front of the plane of the paper and the atoms at the ends of the dash bonds are behind.

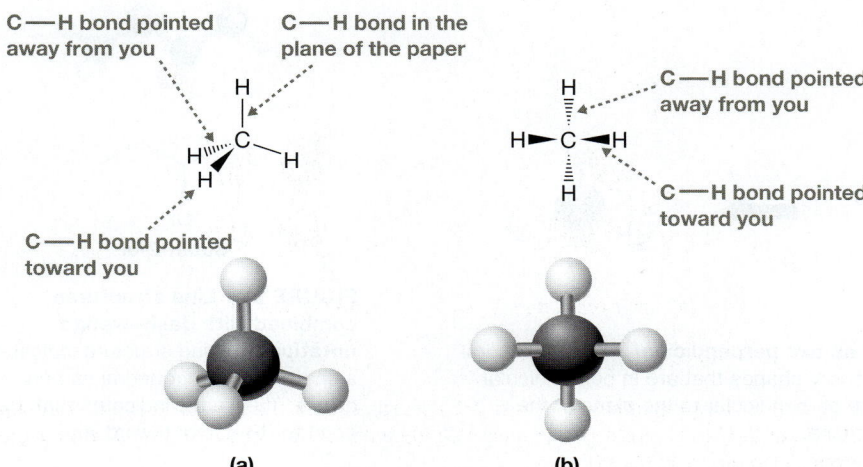

(a)   **(b)**

FIGURE 2-5 Representations of CH_4 using dash–wedge notation
The two different depictions imply views of the molecule from different vantage points.

Ball-and-stick representations of NH_4^+ from two different vantage points are provided. Next to each one, draw the corresponding cation using dash–wedge notation.

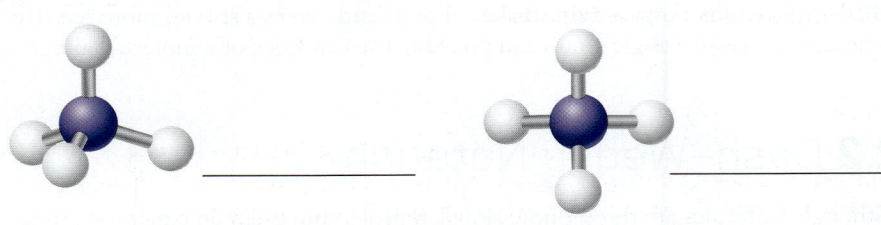

Notice in Figure 2-5 that the four bonds of a tetrahedral atom define two *perpendicular* V shapes (i.e., each H—C—H angle). Keep this in mind whenever you draw the depiction in Figure 2-5a.

- One V is in the plane of the paper, whereas the other is perpendicular to the plane of the paper.
- The Vs must open in *opposite directions*! This is shown explicitly in Figure 2-6.

YOUR TURN **2.4**

In the Lewis structure in the upper right corner of Figure 2-6, trace the V that is in the plane of the page and draw an arrow in the direction in which it opens. Do the same to the V that is perpendicular to the page.

CONNECTIONS Butan-2-ol (Fig. 2-7) is used industrially as a precursor to butan-2-one, which has applications as an industrial solvent and as a welding agent for connecting polystyrene parts of scale models.

Dash–wedge notation can be combined with line structures to illustrate the three-dimensional geometry of more complex molecules, such as butan-2-ol shown in Figure 2-7.

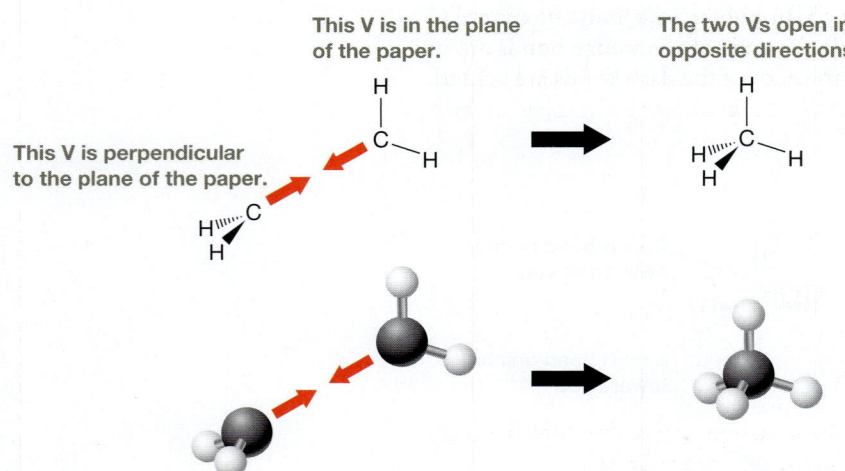

This V is in the plane of the paper.

The two Vs open in opposite directions.

This V is perpendicular to the plane of the paper.

FIGURE 2-6 Tetrahedral geometry viewed as two perpendicular Vs A tetrahedral atom can be viewed as the fusing together of two V shapes that are in perpendicular planes—one in the plane of the paper and one perpendicular to the plane of the paper. The two Vs must open in opposite directions.

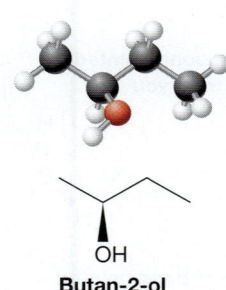

Butan-2-ol

FIGURE 2-7 Line structures combined with dash–wedge notation The line structure indicates a chain of four tetrahedral carbon atoms. The wedge indicates that the bond to OH points toward you.

In the box provided, draw the line structure of butan-2-ol using dash–wedge notation. Note that the C—O bond points away from you.

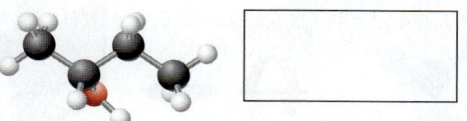

PROBLEM 2.3 Draw line structures of each of these molecules using dash–wedge notation. Assume that no atoms have formal charges. *Note:* Black = carbon, white = hydrogen, green-yellow = chlorine, and blue = nitrogen.

(a)   (b)   (c)

PROBLEM 2.4 This shows a common mistake made with dash–wedge notation. Explain what is incorrect about the structure and then fix it. (It may help to build a model of it.)

Br
Br

2.3 Strategies for Success: The Molecular Modeling Kit

Much of organic chemistry requires us to manipulate molecules in three dimensions. Unfortunately, we are limited to two dimensions when we represent a molecule on paper, even when we use dash–wedge notation. **Molecular modeling kits** can help. Instead of having to rotate a three-dimensional image mentally, you can construct real models and rotate them in your hands. For example, let's use a modeling kit to determine what the following cyclopentane derivative looks like after it has been flipped over vertically.

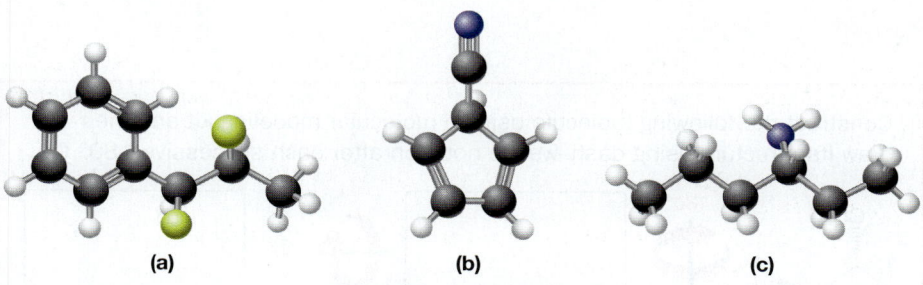

Flip 180°   ?   (2-1)

You may develop your own process for these kinds of manipulations, but for now carry out the following steps, which are depicted in Figure 2-8 (next page):

1. Construct the molecular model exactly as indicated in the accurate dash–wedge notation.
2. Rotate the molecule as indicated.
3. Redraw the molecule in its dash–wedge notation using the rotated model as a guide.

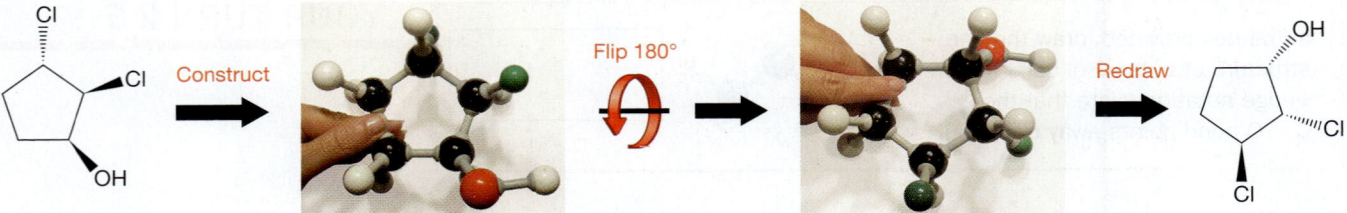

FIGURE 2-8 Model kits and 3-D manipulations To draw the molecule in Equation 2-1 after it is flipped 180°, (1) construct the molecule with a model kit, (2) flip the molecule over 180°, and (3) use the model as a guide to redraw the molecule in dash–wedge notation.

YOUR TURN 2.6

Construct the following molecule using a molecular modeling kit and then draw its structure using dash–wedge notation after each successive 180° flip.

2.4 Net Molecular Dipoles and Dipole Moments

Recall from Section 1.7 that a *polar covalent bond* arises when atoms having different electronegativities are bonded together, and the resulting bond dipole points toward the more electronegative atom. If there is only one polar covalent bond in the entire molecule, as in HF (Fig. 2-9a), then the molecule will have a **net molecular dipole**, or **permanent dipole**. That is, one end of the molecule will bear a partial positive charge and the other a partial negative charge of equal magnitude. In HF, the partial negative charge builds up on the side with the fluorine atom (electronegativity = 3.98; see Fig. 1-16, p. 16), while the partial positive charge is left in the vicinity of the hydrogen atom (electronegativity = 2.20), so the net molecular dipole points toward F and the entire molecule is **polar**.

FIGURE 2-9 Bond dipole moments and net molecular dipole moments Bond dipole moments are shown as thin black arrows and net molecular dipole moments as thick red arrows. (a) HF is a polar molecule, with a net dipole moment of 1.8 D pointing toward the F atom. (b) CO_2 is nonpolar, because the two C=O bond dipoles point in exactly opposite directions and completely cancel by vector addition. (c) Water is a polar molecule, with a net dipole moment of 1.8 D pointing from the point midway between the H atoms toward the O atom.

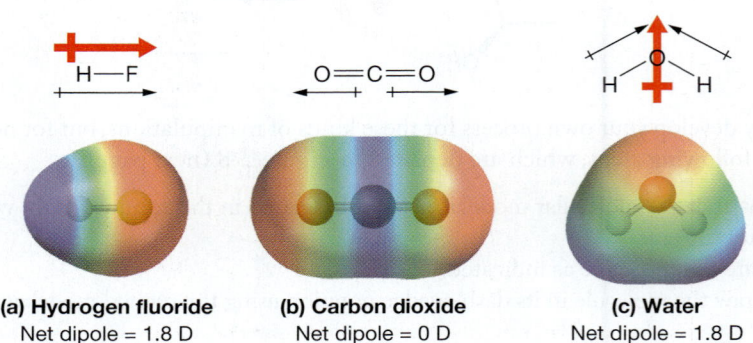

(a) Hydrogen fluoride
Net dipole = 1.8 D

(b) Carbon dioxide
Net dipole = 0 D

(c) Water
Net dipole = 1.8 D

The **dipole moment** of a molecule is a measure of the *magnitude* of its dipole and is reported in units of **debye (D)**. To get a feel for this type of unit, it helps to know that a molecule of HF, which is quite polar, has a dipole moment of 1.8 D. Strictly **nonpolar** molecules, such as H_2, have a dipole moment of 0 D.

A dipole moment is a **vector**, which has both magnitude and direction, so we arrive at the net dipole moment of a molecule by adding the vectors of the bond dipoles together. CO_2 (Fig. 2-9b), for example, is nonpolar, despite having two polar covalent bonds. CO_2 is linear and symmetrical, so the two C=O bond dipoles point in exactly opposite directions, resulting in complete cancellation. Water (Fig. 2-9c), on the other hand, is polar because its two bond dipoles do not point in exactly opposite directions.

YOUR TURN 2.7

Like CO_2, BeH_2 is a linear nonpolar molecule. Using the structure shown, draw dipole arrows above each Be—H bond. (For the necessary electronegativity values, see Fig. 1-16, p. 16.)

H—Be—H

Tetrahedral molecules that have polar covalent bonds can be polar or nonpolar, depending on their symmetry. CCl_4, for example, is nonpolar because the four bond dipoles completely cancel. We can see this more clearly in Figure 2-10a, where we mentally break the four bonds into two perpendicular Vs consisting of CCl_2 groups,

CONNECTIONS Historically, CCl_4 (Fig. 2-10a) was used as a pesticide and was a common organic solvent, and was also used widely in fire extinguishers. We now know, however, that substantial exposure to CCl_4 causes acute liver failure and other severe health problems, including cancer, so these uses have largely been phased out.

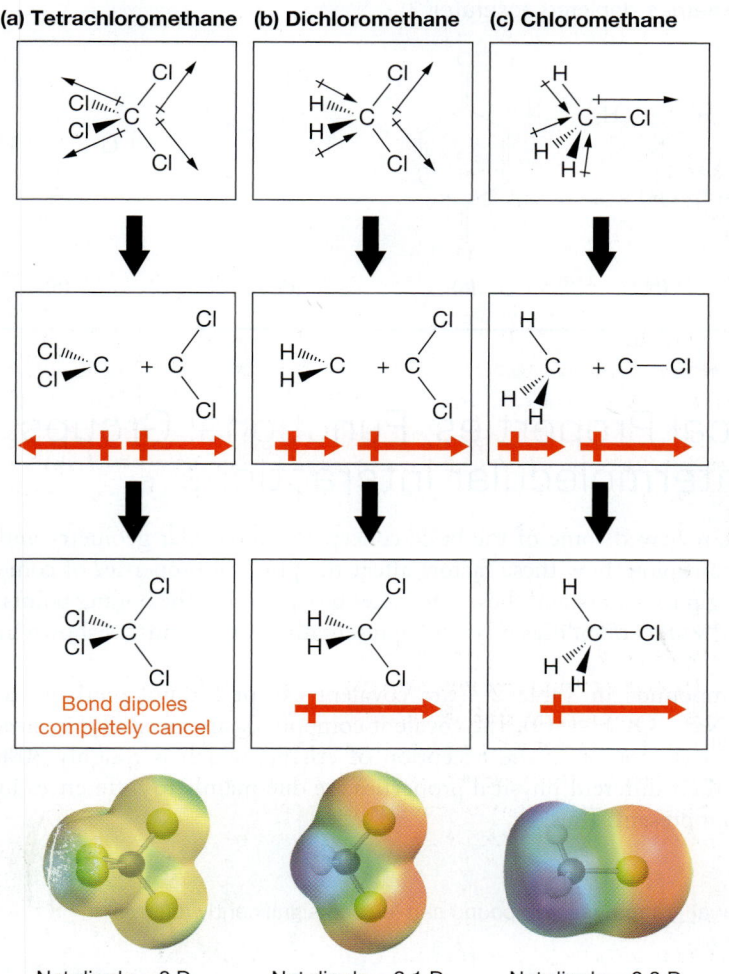

(a) Tetrachloromethane **(b) Dichloromethane** **(c) Chloromethane**

Add together bond dipoles of separate fragments.

Add together net dipoles of each fragment.

Bond dipoles completely cancel

Net dipole = 0 D   Net dipole = 2.1 D   Net dipole = 2.3 D

FIGURE 2-10 Vector addition of bond dipoles in tetrahedral molecules (a) Tetrachloromethane, (b) dichloromethane, and (c) chloromethane all have a tetrahedral molecular geometry. Individual bond dipoles are shown in the top row. The molecule is split into its constituent parts in the second row, and the net dipole of each part is indicated by the thick red arrow. In the third row, the net dipole of the entire molecule is indicated. The bottom row shows each species' electrostatic potential map.

as we discussed in Section 2.2. The resulting net dipoles of the two Vs point in opposite directions and therefore cancel.

YOUR TURN 2.8

A molecule of CF$_4$ is shown decomposed into its two Vs. Draw the bond dipoles along the two C—F bonds in each V and also draw the net dipole resulting from their vector addition.

CONNECTIONS CH$_2$Cl$_2$ (Fig. 2-10b), commonly called methylene chloride, has uses as a paint stripper and degreaser, and has been used in the food industry to decaffeinate coffee. Because of its low boiling point, it is used as the liquid inside the "drinking bird" toy.

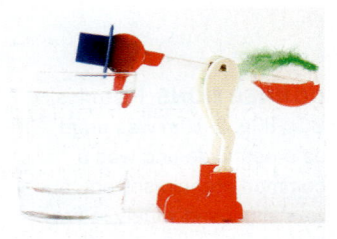

CH$_2$Cl$_2$ and CH$_3$Cl, on the other hand, are both polar. As shown in Figure 2-10b, we can mentally split CH$_2$Cl$_2$ into two Vs, one consisting of CH$_2$ and one consisting of CCl$_2$. Each of these pieces has a net dipole and, instead of canceling, they reinforce each other.

In CH$_3$Cl (Fig. 2-10c), we can treat the net molecular dipole moment as the sum of the net dipoles from the pyramidal CH$_3$ group and the C—Cl bond. These dipoles, too, point in the same direction instead of canceling.

PROBLEM 2.5 Which of the following molecules are polar? For those that are, draw a dipole arrow indicating the direction of the net molecular dipole. *Hint:* Are the molecular geometries depicted accurately?

(a)   (b)   (c)   (d)   (e)

CONNECTIONS CH$_3$Cl (Fig. 2-10c) was once used as a refrigerant, but is no longer due to its toxicity and flammability. Nowadays, it finds use as a local anesthetic and as an herbicide.

2.5 Physical Properties, Functional Groups, and Intermolecular Interactions

Now that we have reviewed some of the basic concepts of molecular geometry and polarity, it's time to explore how these factors affect the physical properties of compounds. We can begin to understand these influences by examining the boiling points, melting points, and water solubilities of some representative compounds, as shown in Table 2-4.

All of the compounds in Table 2-4 are covalent except sodium methanoate (sodium formate, Na$^+$ $^-$OCH=O). The covalent compounds are all similar in size, shape, and molar mass, too, with the exception of ethane, which is roughly 30% lighter. Therefore, their different physical properties are due mainly to differences in the *functional groups* present. Thus:

Different functional groups in a compound can lead to significantly different physical properties.

TABLE 2-4 Physical Properties of Representative Compounds

Compound	Molar Mass (g/mol)	Boiling Point (°C)	Melting Point (°C)	Solubility in Water (g/100 g H_2O)	Dipole Moment (D)	Dominant Intermolecular Interaction
O ‖ HC ⟍O⊖⊕Na **Sodium methanoate (Sodium formate)**	N/A	>253	253	77	N/A	Ion–ion
O ‖ HC ⟍OH **Methanoic acid (Formic acid)**	46	101	8	Infinite	1.4	Hydrogen bonding
OH | H_2C⟍ CH_3 **Ethanol**	46	78	−114	Infinite	1.7	Hydrogen bonding
O ‖ HC⟍ CH_3 **Ethanal (Acetaldehyde)**	44	20	−117	>100	3.0	Dipole–dipole
H_3C⟍O⟋CH_3 **Dimethyl ether**	46	−25	−139	6.9	1.3	Dipole–dipole
CH_2 ‖ HC⟍ CH_3 **Propene**	42	−48	−185	0.00061	0.3	Induced dipole–induced dipole
CH_3 | H_2C⟍ CH_3 **Propane**	44	−45	−188	0.00039	0	Induced dipole–induced dipole
H_3C⟍ CH_3 **Ethane**	30	−89	−183	0.006	0	Induced dipole–induced dipole

CONNECTIONS Sodium methanoate is used as a dye activator in fabric dyeing processes because it helps promote the fixation of a dye to the fabric. It has also been used in the food industry as a preservative and flavor enhancer.

CONNECTIONS Formic acid is found in the venom of ants, and has several uses in industry and agriculture. It is used in the production of leather and in dyeing textiles, and is also used to treat animal feed because of its properties as a preservative and an antibacterial agent.

YOUR TURN 2.9

Circle the functional group that is present in each covalent compound in Table 2-4 and identify the compound class to which each molecule belongs.

Why do functional groups have such a profound effect on the physical properties of organic compounds? Functional groups can differ in the atoms they possess or in the arrangement of those atoms in space, both of which will impact how charge is distributed within a molecule. This will affect the ways in which various species attract (or repel) each other, so-called **intermolecular interactions** (also called **intermolecular forces**). We will examine the following types of intermolecular interactions:

- Ion–ion interactions,
- Dipole–dipole interactions,
- Hydrogen bonding,
- Induced dipole–induced dipole interactions (or London dispersion forces), and
- Ion–dipole interactions.

The first four of these intermolecular interactions are discussed in Section 2.6 in the context of boiling points and melting points. We examine the fifth and final intermolecular interaction in Section 2.7 in the context of a compound's solubility in a given solvent.

Even though these intermolecular interactions are given different names, they all originate from the same fundamental law: *opposite charges attract*. As a result, the strength of each intermolecular interaction depends on the concentrations of charge involved.

All else being equal, the greater the concentrations of charge that are involved in an intermolecular interaction, the stronger is the resulting attraction.

2.6 Melting Points, Boiling Points, and Intermolecular Interactions

Melting points and boiling points can provide a wealth of information about intermolecular interactions. To help you see why, study Figure 2-11 to review what the different phases of a compound look like on the molecular level.

FIGURE 2-11 Microscopic structure of the three phases of matter (a) In a crystalline solid, molecules or ions form a well-ordered structure called a crystal lattice, in which movement is limited to vibration and intermolecular forces are maximized. (b) In a liquid, molecules are free to move around, because intermolecular forces are somewhat less substantial. (c) In the gas phase, molecules are so far apart that intermolecular forces are effectively absent.

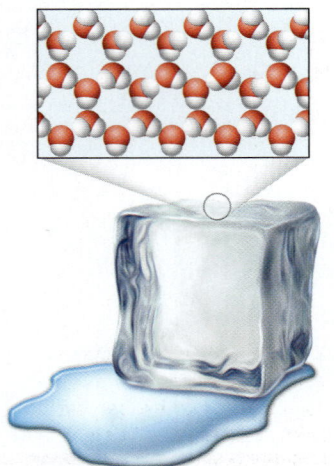
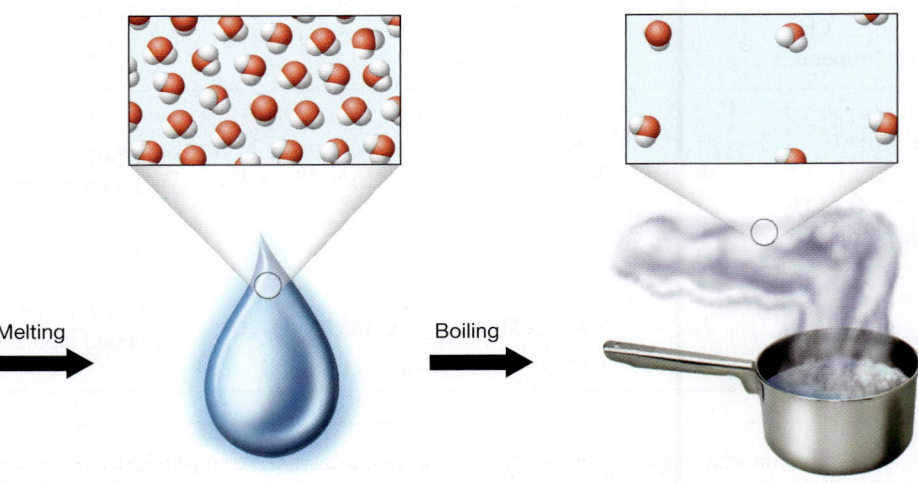

(a) **Solid:** Intermolecular interactions are maximized.

(b) **Liquid:** Intermolecular interactions are less substantial.

(c) **Gas:** Intermolecular interactions are effectively absent.

- Solids consist of atoms, ions, or molecules that are in contact with one another and are essentially immobile. This allows intermolecular interactions to be maximized.
- In a liquid, the species are also in close contact, but they can rotate and slide past one another, so intermolecular interactions are less substantial than in a solid.
- In a gas, the species are far apart and are effectively isolated from one another, so they can move freely. Intermolecular interactions are essentially nonexistent.

Melting, therefore, decreases the intermolecular interactions that exist in the solid phase, and boiling effectively overcomes the remaining intermolecular interactions that exist in the liquid phase. Consequently, as the strength of the intermolecular interactions that exist in a particular substance increases, more energy (in the form of heat) is required for the substance to melt or boil.

- Melting points increase as the intermolecular interactions in a solid increase.
- Boiling points increase as the intermolecular interactions in a liquid increase.

Let's now apply these ideas to interpret the relative strengths of various types of intermolecular interactions.

2.6a Ion–Ion Interactions

Of all the compounds in Table 2-4 on page 81, sodium methanoate (sodium formate, Na^+ $^-OCH{=}O$) has the highest melting point and boiling point, suggesting that it has particularly strong intermolecular attractions in both its solid and liquid phases. Sodium methanoate is an ionic compound, composed of Na^+ and HCO_2^- ions held together (as we saw in Chapter 1) by the electrostatic attraction of oppositely charged ions, called *ionic bonds* or, more generally, **ion–ion interactions**.

Ion–ion interactions are the strongest intermolecular interactions because ions have very high concentrations of positive and negative charge.

2.6b Dipole–Dipole Interactions

Referring again to Table 2-4, notice that the compounds with significant dipole moments—methanoic acid (formic acid), ethanol, ethanal (acetaldehyde), and methoxymethane (dimethyl ether)—have boiling points and melting points that are significantly higher than those of the remaining compounds, which are essentially nonpolar.

Polar molecules are attracted to each other more strongly than similar nonpolar molecules.

The basis for this behavior is dipole–dipole interactions. **Dipole–dipole interactions** arise because the positive end of one molecule's *net dipole* is attracted to the negative end of another's, as shown in Figure 2-12. Therefore:

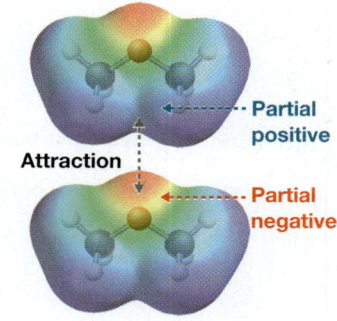

Partial positive

Attraction

Partial negative

FIGURE 2-12 Dipole–dipole interaction The dominant intermolecular force between two molecules of dimethyl ether is a dipole–dipole interaction. The positive end of one ether molecule attracts the negative end of the other.

All other factors being equal, the strength of dipole–dipole interactions increases as the dipole moment increases.

Notice in Table 2-4 (p. 81), for example, that the highly polar ethanal ($CH_3CH{=}O$) has a higher melting point and a higher boiling point than the less polar dimethyl ether (CH_3OCH_3).

YOUR TURN 2.10

The boiling point of CH_3CH_2F is -37.1 °C. How does this compare to the boiling point of $CH_3CH{=}O$? Which compound, therefore, is more polar?

Although dipole–dipole interactions can be quite strong, they involve the attraction of *partial* charges, which are smaller in magnitude than the *full* charges in ion–ion interactions. Therefore:

Dipole–dipole interactions are generally much weaker than ion–ion interactions.

As a result, compounds such as ethanal (acetaldehyde) melt and boil at lower temperatures than ionic compounds.

SOLVED PROBLEM 2.6

Which compound has the higher melting point, $NaOCH_2CH_3$ or $CH_3CH{=}O$?

Think What is the strongest intermolecular interaction in $NaOCH_2CH_3$? In $CH_3CH{=}O$? Which type of interaction is stronger? How does that affect the melting point?

Solve $NaOCH_2CH_3$ is an ionic compound. It consists of $CH_3CH_2O^-$ and Na^+, which are held together by ion–ion interactions. $CH_3CH{=}O$, on the other hand, is a polar covalent compound, so it experiences dipole–dipole interactions. Because ion–ion interactions are stronger than dipole–dipole interactions, the melting point of $NaOCH_2CH_3$ is higher than the melting point of $CH_3CH{=}O$.

PROBLEM 2.7 Which compound has the higher boiling point, $NaOCH_2CH_3$ or $CH_3CH{=}O$?

PROBLEM 2.8 Which compound has the higher boiling point, CH_4 or CH_3F?

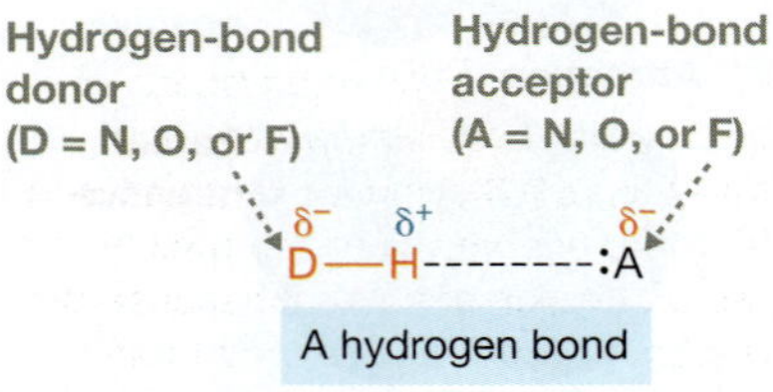

Hydrogen-bond
donor
(D = N, O, or F)

Hydrogen-bond
acceptor
(A = N, O, or F)

A hydrogen bond

FIGURE 2-13 Hydrogen bonds
A hydrogen bond consists of a hydrogen-bond donor (D—H) and a hydrogen-bond acceptor (:A). If the D and A atoms are uncharged, they must be nitrogen, oxygen, or fluorine for the hydrogen bond to be substantial.

2.6c Hydrogen Bonding

Dipole–dipole interactions alone cannot account for some of the data in Table 2-4 (p. 81). Methanoic acid (formic acid, HCO_2H) and ethanol (CH_3CH_2OH), for example, have substantially higher melting points and boiling points than ethanal (acetaldehyde, $CH_3CH{=}O$), despite the fact that ethanal's net dipole moment is the highest of the three. These apparent anomalies arise because methanoic acid and ethanol can form *hydrogen bonds*, whereas acetaldehyde cannot.

Each **hydrogen bond** (H bond) requires a *hydrogen-bond donor* and a *hydrogen-bond acceptor*, as shown in Figure 2-13.

- A **hydrogen-bond donor** is a covalent bond, D—H, where D is a highly electronegative atom, such as N, O, or F.
- A **hydrogen-bond acceptor**, A, can be any atom with a large concentration of negative charge and a lone pair of electrons. For the hydrogen bond to be substantial, an uncharged H-bond acceptor must be F, O, or N.

The high electronegativity of the F, O, or N atom ensures that the D—H bond is highly polar, giving the H atom a large partial positive charge. This sets up a strong attraction between the H atom and the oppositely charged acceptor.

The actual H bond is created when a lone pair of electrons on the H-bond acceptor is shared with the hydrogen atom of the H-bond donor. The H bond is often depicted by a dashed line (different from the dash bond that is part of dash–wedge notation). Figure 2-14 shows how hydrogen bonding might be depicted between two molecules of ethanol (Fig. 2-14a) and between two molecules of methanoic acid (Fig. 2-14b). Notice that either O atom in methanoic acid can serve as a H-bond acceptor.

YOUR TURN 2.11

Circle and label the H-bond donor and H-bond acceptor in *each* H bond shown in Figure 2-14b.

Hydrogen bonding in uncharged species involves only *partial* charges, so like dipole–dipole interactions:

Hydrogen bonding is weaker than ion–ion interactions.

Hydrogen bonds, however, are distinct from dipole–dipole interactions for two main reasons: (1) fluorine, oxygen, and nitrogen are highly electronegative, so the partial charges involved are large, and (2) the hydrogen atom is very small, which allows the partial positive charge on hydrogen to be very close to the partial negative charge on the H-bond acceptor. For these reasons:

Hydrogen bonding is often (but not always) stronger than dipole–dipole interactions.

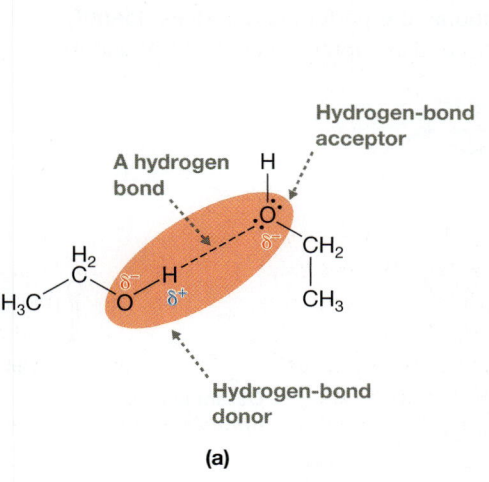

(a)

(b)

FIGURE 2-14 Hydrogen bonding between ethanol molecules and between methanoic acid molecules (a) A hydrogen bond can form between two molecules of ethanol because the H atom is covalently bonded to an O atom in one molecule and is attracted to the second molecule's O atom, which has a lone pair of electrons and a significant partial negative charge. The H bond is indicated by a dashed line. (b) Two different H bonds between two molecules of methanoic acid (formic acid).

PROBLEM 2.9 Which pair of species will give rise to the strongest intermolecular interactions? The weakest?

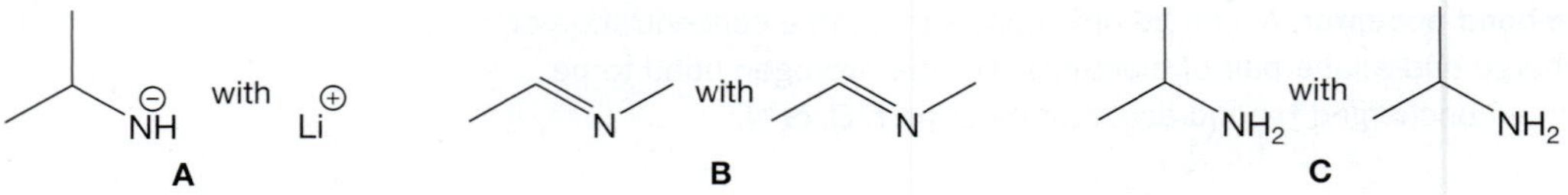

A   with   B   with   C   with

It is important to be able to gauge the *extent* of hydrogen bonding among molecules of a particular compound—that is, the collective strength of all the hydrogen bonds present. As a general rule:

> The extent of hydrogen bonding increases as the total number of *potential* H-bond donors and acceptors increases between the species involved.

With more potential donor–acceptor pairs involving two molecules, there are more ways in which hydrogen bonding can take place.

This explains why methanoic acid (formic acid) has a higher boiling point and melting point than ethanol. As shown in Figure 2-15a, a single molecule of ethanol has one potential H-bond donor and one potential H-bond acceptor. In a pair of ethanol molecules undergoing the intermolecular interaction, therefore, there are two potential donors and two potential acceptors. In a single molecule of methanoic acid (formic acid, Fig. 2-15b), on the other hand, there is one potential donor and *two* potential acceptors. In a pair of methanoic acid molecules, therefore, there are two potential donors and *four* potential acceptors. As a result, there are more ways in which hydrogen bonding can take place in methanoic acid than in ethanol.

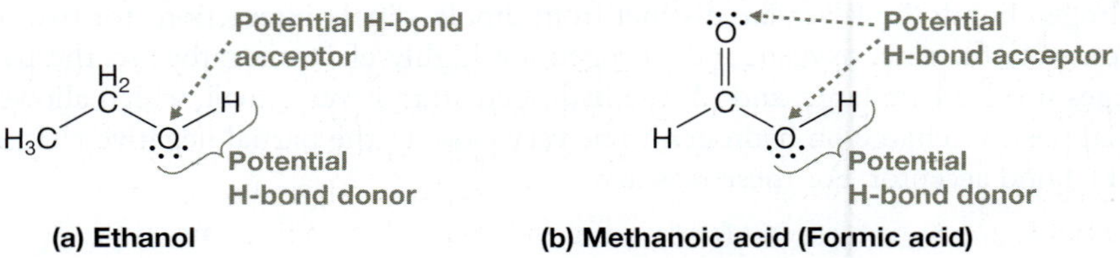

FIGURE 2-15 Potential hydrogen bond donors and acceptors (a) In a molecule of ethanol, the OH bond is a potential H-bond donor and the O atom is a potential H-bond acceptor. (b) In a molecule of formic acid, the OH bond is a potential H-bond donor and each O atom is a potential H-bond acceptor.

(a) Ethanol   (b) Methanoic acid (Formic acid)

YOUR TURN 2.12

High levels of cholesterol, a naturally occurring steroid, have been linked to cardiovascular disease. Octanoic acid (caprylic acid) is a fatty acid found in milk and is used commercially to manufacture perfumes and dyes. Identify the number of *potential* H-bond donors and acceptors in cholesterol and in octanoic acid.

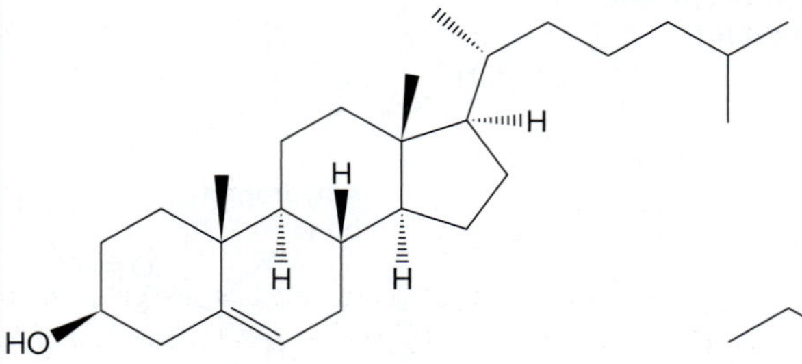

Cholesterol 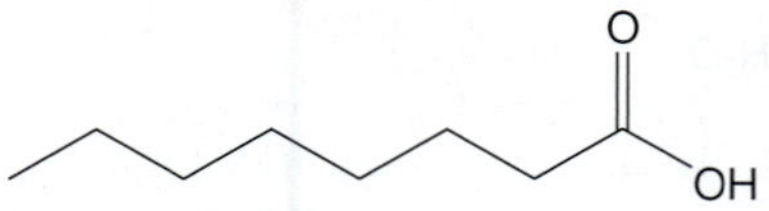   Octanoic acid (Caprylic acid)

PROBLEM 2.10 How many potential H-bond donors and H-bond acceptors are there in each of the following molecules?

(a)　　　　(b)　　　　(c)　　　　(d)　　　　(e)

PROBLEM 2.11 Which functional groups in Table 1-6 (p. 35) possess at least one H-bond acceptor but no H-bond donors? Which functional groups possess at least one H-bond donor and one H-bond acceptor? Which functional groups possess no H-bond donors and no H-bond acceptors?

SOLVED PROBLEM 2.12

1,2-Ethanediol (ethylene glycol, **A**) is used as an automotive antifreeze. Hydroxyacetaldehyde (**B**) is believed to be an intermediate in the metabolism of proteins and carbohydrates. Which of these compounds would you expect to have a higher boiling point? Why?

A

B

Think What is the most important intermolecular interaction that will occur between two molecules of **A**? Between two molecules of **B**? Which interaction is stronger, and how would it affect the boiling points of **A** and **B**?

Solve Both **A** and **B** are polar molecules, so dipole–dipole interactions should be present in a pair of each type of molecule. However, hydrogen bonding, which is often stronger than dipole–dipole interactions, is also possible: Each molecule contains at least one potential H-bond donor (an OH bond) and at least one potential H-bond acceptor (an O atom). To estimate which pair of molecules has greater hydrogen bonding, we count the total number of potential H-bond donors and acceptors. In two molecules of **A**, there are four potential donors and four potential acceptors, because there are two donors and two acceptors from each molecule. In two molecules of **B**, there are two potential donors and four potential acceptors, because there are one donor and two acceptors from each molecule. As a result, we would expect compound **A** to have more hydrogen bonding, and thus a higher boiling point.

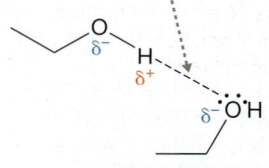

O is more electronegative than N, so this hydrogen bond is stronger.

Ethanol
Boiling point = 78 °C

Ethanamine
Boiling point = 17 °C

FIGURE 2-16 Hydrogen bonding and electronegativity Hydrogen bonding is stronger in ethanol than in ethanamine because O is more electronegative than N, thus giving rise to larger concentrations of positive and negative charges. As a result, the boiling point of ethanol is higher than the boiling point of ethanamine.

PROBLEM 2.13 Which compound, **C** or **D**, would you expect to have a higher boiling point? Why?

C　　　　**D**

The strength of hydrogen bonding also depends on the concentrations of charge in the H-bond donors and H-bond acceptors. Ethanamine ($CH_3CH_2NH_2$), for example, has a lower boiling point than ethanol (CH_3CH_2OH) because N is less electronegative than O, resulting in smaller concentrations of charge in ethanamine than in ethanol (Fig. 2-16).

PROBLEM 2.14 Which H bond would you expect to be stronger, **E** or **F**? Why?

$$\underset{\textbf{E}}{\overset{CH_3}{\underset{CH_3}{:N}}-H\text{-------}\overset{CH_3}{\underset{CH_3}{:N}}-H} \qquad\qquad \underset{\textbf{F}}{:\ddot{F}-H\text{-------}:\ddot{F}-H}$$

2.6d Induced Dipole–Induced Dipole Interactions (London Dispersion Forces)

Nonpolar molecules must attract each other through intermolecular interactions; otherwise, nonpolar compounds could never condense from gas to liquid. The dominant intermolecular interaction between nonpolar molecules is called **induced dipole–induced dipole interactions**, or **London dispersion forces**.

How do induced dipole–induced dipole interactions arise? Although the *average* electron distribution in a molecule such as propane ($CH_3CH_2CH_3$) does not give rise to a significant permanent dipole (Fig. 2-17a), any electron cloud can be distorted—that is, any electron cloud is **polarizable**. Electrons are constantly moving around and at some instant in time, there may be more electrons on one side of the molecule than there are on the other. The extra electrons on that one side give rise to an **instantaneous dipole** (Fig. 2-17b), which can alter the electron distribution on a second molecule by repelling or attracting nearby electrons. The second molecule then develops an **induced dipole** that is attracted to the first molecule (Fig. 2-17c).

> Although they are most pronounced in nonpolar molecules, induced dipole–induced dipole interactions are present when any two species interact.

To gain a sense of the relative strength of induced dipole–induced dipole interactions, notice in Table 2-4 (p. 81) that the nonpolar compounds tend to have the lowest boiling points and melting points. This suggests that:

> Induced dipole–induced dipole interactions are generally the weakest of all intermolecular forces.

FIGURE 2-17 Induced dipole–induced dipole interaction (a) On average, two isolated molecules of propane are nonpolar. (b) Electrons are not static, however, so electron density can build up on one side of a molecule at some instant in time, resulting in a temporary dipole. (c) That temporary dipole, in turn, can alter the electron distribution of a second, adjacent molecule, giving the second molecule an induced dipole. The oppositely charged ends of these induced dipoles attract one another.

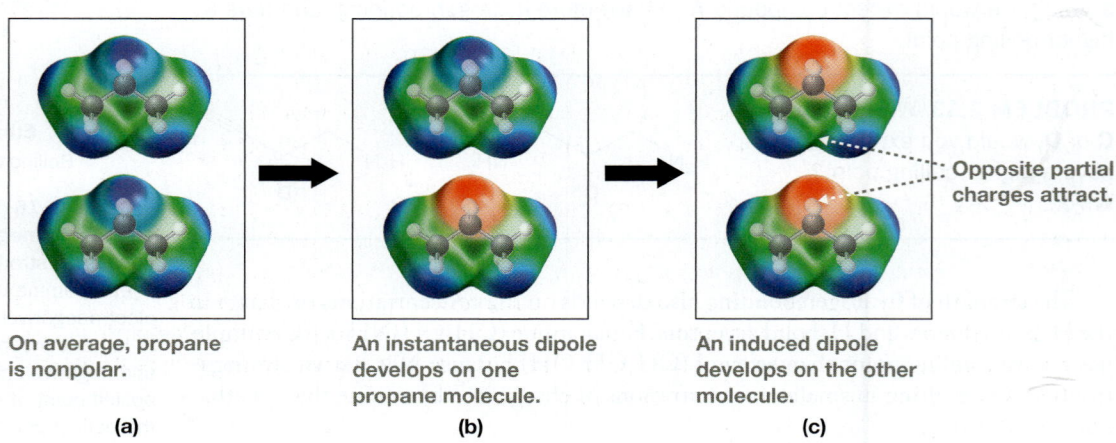

On average, propane is nonpolar.

(a)

An instantaneous dipole develops on one propane molecule.

(b)

An induced dipole develops on the other molecule.

(c)

Opposite partial charges attract.

TABLE 2-5 Melting Points and Boiling Points of Nonpolar Compounds

Molecule	Total Number of Electrons	Melting Point (°C)	Boiling Point (°C)	Molecule	Total Number of Electrons	Melting Point (°C)	Boiling Point (°C)
CH_4 Methane	10	−182	−161	Dimethylpropane	42	−17	10
CH_3CH_3 Ethane	18	−183	−89	Pentane	42	−130	36
Propane	26	−188	−42	Br_2 Bromine	70	−7	59
Cl_2 Chlorine	34	−101	−34	I_2 Iodine	106	114	184
Butane	34	−138	−1				

Their strength is highly variable, however, and can sometimes be quite significant, as shown in Table 2-5. A small molecule like CH_4 is a gas at room temperature, indicating that its induced dipole–induced dipole interactions are extremely weak. A much heavier molecule like I_2, on the other hand, is a solid at room temperature, indicating that its induced dipole–induced dipole interactions are quite strong—even stronger than the hydrogen bonding in water, which is a liquid at room temperature!

The precise strength of induced dipole–induced dipole interactions in a particular species depends on its **polarizability**, which can be defined as the ease with which its electron cloud is distorted.

CONNECTIONS Elemental iodine has use as a disinfectant. Because of its deep violet color and reactivity with compounds such as alkenes, I_2 is used in analytical chemistry to determine end points of titrations.

Climbing Like Geckos

In the opening to this chapter, we highlighted the secret behind the ability of geckos to climb just about any surface effortlessly: dispersion forces. Geckos have a specialized hierarchical structure in their toes, culminating in large numbers of very small hairs less than 200 nm in diameter, called setae. This creates a very large contact surface area for each toe, giving a gecko's foot an adhesive pressure up to about 30 pounds per square inch on glass. This means that the gecko can, in principle, cling to glass by just one toe.

Scientists have known the gecko's secret for many years but were unable to harness it until recently, when Professors Duncan Irschick and Alfred Crosby of the University of Massachusetts Amherst unveiled Geckskin in 2014. Geckskin is made from polydimethylsiloxane, a common polymer, and the adhesive pad mimics the structure of the gecko's toe to maximize its contact surface area. One of the keys to the success of Geckskin is that it is woven into a synthetic tendon to maintain both stiffness and rotational freedom. And successful it is. A piece about the size of an index card has been shown to hold 700 pounds on a smooth wall and is easily removed without leaving any residue. Inspired by this, the Defense Advanced Research Projects Agency (DARPA) developed a system that enables an adult human weighing more than 200 pounds to scale a 25-foot vertical glass wall.

A species with relatively few electrons like CH_4 (melting point = -182 °C, boiling point = -161 °C), which has 10 total electrons, is not very polarizable. This is why its induced dipole–induced dipole interactions are weak. On the other hand, pentane ($CH_3CH_2CH_2CH_2CH_3$), with 42 electrons, is significantly more polarizable, giving it a higher melting point (-130 °C) and boiling point (36 °C). Iodine (I_2) has 106 electrons, so its melting point (114 °C) and boiling point (184 °C) are even higher.

YOUR TURN 2.13

On the graph provided, plot boiling point as a function of total number of electrons for the straight-chain alkanes in Table 2-5 (i.e., methane, ethane, propane, butane, and pentane). What do you notice?

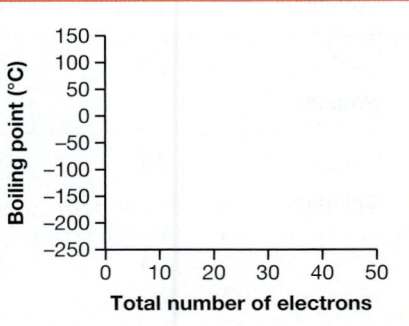

Another important factor governing the strength of induced dipole–induced dipole interactions is **contact surface area**.

For instance, pentane [$CH_3(CH_2)_3CH_3$] and dimethylpropane [neopentane, $C(CH_3)_4$] have the same formula (C_5H_{12}), and thus the same number of electrons, but pentane has a more extended shape than the relatively compact dimethylpropane. As a result, two molecules of pentane have a greater surface area available for interaction than two molecules of dimethylpropane do, as shown in Figure 2-18. Greater contact surface area means that pentane has more effective induced dipole–induced dipole interactions and a higher boiling point than dimethylpropane.

YOUR TURN 2.14

In Figure 2-18, shade in the contact surface area for each pair of molecules. Indicate which pair of molecules has greater contact surface area.

PROBLEM 2.15 Which molecule, in each pair, would you expect to have a higher boiling point? Why?

2,2-Dimethylbutane   2-Methylpentane   1,1-Dimethylcyclopropane   1,2-Dimethylcyclopropane

(a)   (b)

2.7 Solubility

The general rule of solubility is that "like dissolves like." This means that polar compounds tend to be soluble in polar solvents but insoluble in nonpolar solvents, and nonpolar compounds tend to be soluble in nonpolar solvents but insoluble in polar solvents. These tendencies are outcomes of the intermolecular interactions at play in the pure solute, the pure solvent, and the mixture of the two—the solution.

To see how, we begin with the concept of *entropy*. As you may recall from general chemistry, **entropy** is a thermodynamic quantity that increases with the number of equivalent ways the energy in a system can be arranged. Many people like to think of entropy as a *measure of disorder*, and a system with greater entropy (i.e., one that is more disordered) tends to be more likely to occur than one with less entropy. Importantly, two substances have greater entropy when they are mixed. Therefore:

> An increase in entropy provides the driving force for two substances to mix.

If entropy were the only factor, then a given solute would be soluble in any solvent. However, the intermolecular interactions that exist in the mixed and unmixed states must also be considered.

> Substances tend *not* to mix if the intermolecular interactions in the pure solute and pure solvent are sufficiently stronger than those in the solution.

This explains why hexane ($CH_3CH_2CH_2CH_2CH_2CH_3$), a nonpolar compound, is essentially insoluble in water. As shown in Figure 2-19, induced dipole–induced

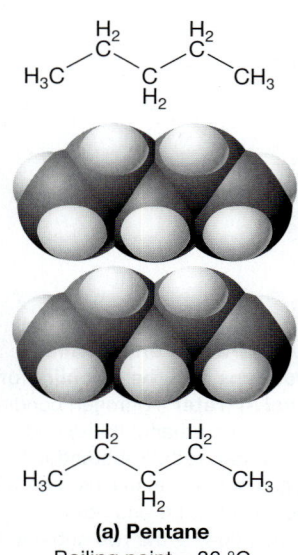

(a) Pentane
Boiling point = 36 °C

(b) Dimethylpropane
Boiling point = 10 °C

FIGURE 2-18 Contact surface area and induced dipole–induced dipole interactions The contact surface area between two molecules of pentane (a) is greater than that between two molecules of dimethylpropane (b). The result is stronger induced dipole–induced dipole interactions in pentane and a correspondingly higher boiling point.

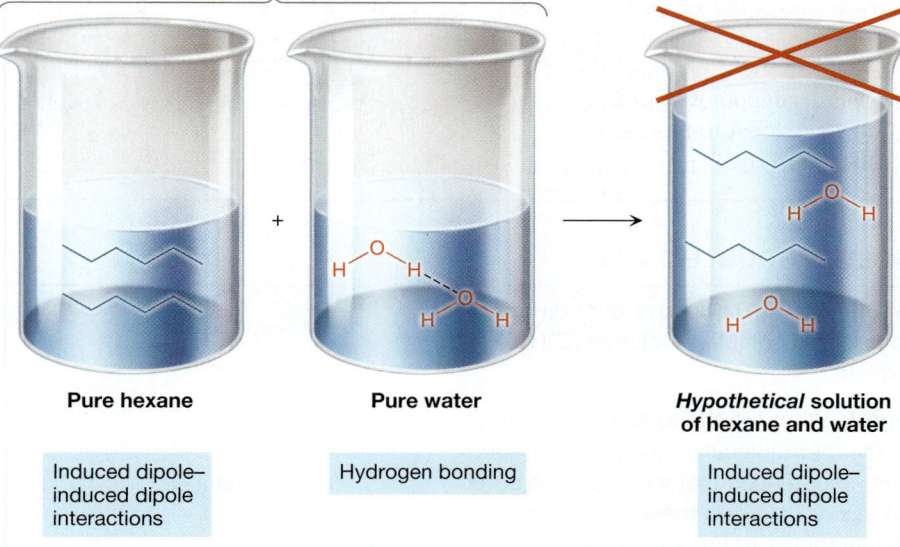

Significantly stronger intermolecular interactions exist in the separated substances, so the compounds are not miscible.

Pure hexane	Pure water	*Hypothetical* solution of hexane and water
Induced dipole–induced dipole interactions	Hydrogen bonding	Induced dipole–induced dipole interactions

FIGURE 2-19 Intermolecular interactions and the insolubility of hexane in water The dominant intermolecular interactions present in pure hexane are induced dipole–induced dipole interactions. The dominant intermolecular interaction in pure water is hydrogen bonding. When we consider the *hypothetical* solution, induced dipole–induced dipole interactions remain but hydrogen bonding is diminished. This favors the pure substances on the left, which explains why hexane is insoluble in water.

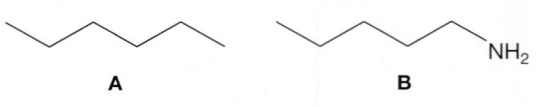

Similar intermolecular interactions exist in the mixed and unmixed states, so the mixed state is favored due to entropy.

FIGURE 2-20 Intermolecular interactions and the solubility of ethanol in water Hydrogen bonding exists in pure ethanol (blue) and in pure water (red). Substantial hydrogen bonding also exists between molecules of water and ethanol in a solution, allowing the two substances to dissolve readily and in any proportion.

Pure ethanol + Pure water → Solution of ethanol and water

Hydrogen bonding   Hydrogen bonding   Hydrogen bonding

dipole interactions dominate in pure hexane, whereas hydrogen bonding dominates in pure water. In the *hypothetical* solution, induced dipole–induced dipole interactions dominate. The much stronger hydrogen bonding that exists in the unmixed state is what prevents the solute and solvent from mixing.

The story changes when the solute and solvent are both polar or both nonpolar. Ethanol (CH_3CH_2OH), for example, is infinitely soluble in water. Once again, the strongest intermolecular interaction that exists in the pure separated substances is hydrogen bonding, both in water and in ethanol (Fig. 2-20). Unlike the hexane–water example, however, substantial hydrogen bonding also exists between the solute and solvent molecules in the ethanol–water solution. In this case, the intermolecular interactions are similar in strength in the mixed and unmixed states, so the increase in entropy causes the substances to mix.

YOUR TURN 2.15

Which compound, **A** or **B**, do you expect to be more soluble in H_2O?

A   B (NH$_2$)

CONNECTIONS Toluene is an important organic solvent, but it is also a precursor in the production of 2,4,6-trinitrotoluene (TNT) and other industrial compounds. In biochemistry, red blood cells can be lysed by toluene to extract hemoglobin.

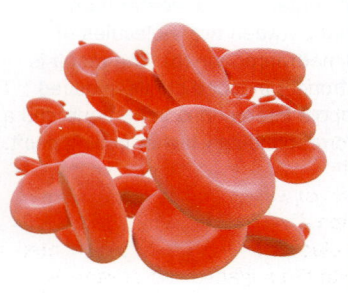

SOLVED PROBLEM 2.16

Would you expect butan-1-ol or diethyl ether to be more soluble in toluene ($C_6H_5CH_3$)? Explain.

CH_3

Toluene

Think What are the relative strengths of the intermolecular interactions in the pure substances that would be disrupted on mixing? How do these compare to the strengths of the intermolecular interactions that would be gained?

Butan-1-ol   OH   Diethyl ether   O

Solve Toluene is nonpolar, so the only interactions that can exist between toluene and each of the given compounds involve a relatively weak induced dipole on toluene. As such, the solute–solvent interactions are roughly the same in both mixtures. The dominant interactions that would be lost from pure diethyl ether are dipole–dipole interactions, whereas hydrogen bonding would be lost from pure butan-1-ol. Because hydrogen bonding is typically stronger than dipole–dipole interactions, mixing is less favorable for butan-1-ol, making diethyl ether more soluble in toluene.

PROBLEM 2.17 Which compound, **C** or **D**, would you expect to be more soluble in toluene ($C_6H_5CH_3$)? Explain.

C   D

2.7a The Solubility of Ionic Compounds: Ion–Dipole Interactions and Solvation

Based on what we've discussed so far, it might seem peculiar that an ionic compound such as sodium methanoate (sodium formate, $Na^+HCO_2^-$) dissolves in a polar solvent like water, because doing so eliminates ion–ion interactions, the strongest of the intermolecular forces. Yet sodium methanoate *does* dissolve in water, and its water solubility (like that of most ionic compounds) is quite high (Table 2-4, p. 81). Why?

When an ionic compound like sodium methanoate dissolves in water, it does so as its individual ions, Na^+ and HCO_2^-, not as uncharged formula units. These free ions can interact with water molecules via **ion–dipole interactions**, a kind of intermolecular interaction different from the ones we have examined thus far. In these particular ion–dipole interactions, the positive end of water's dipole attracts a HCO_2^- anion and the negative end attracts a Na^+ cation (Fig. 2-21).

Notice that an ion–dipole interaction involves a full charge and a partial charge. The concentration of charge involved in an ion–dipole interaction is generally less than in an ion–ion interaction but more than in a dipole–dipole interaction. Consequently:

> The strength of an ion–dipole interaction is intermediate between that of an ion–ion interaction and that of a dipole–dipole interaction.

How, then, are ion–dipole interactions capable of overcoming the stronger ion–ion interactions to dissolve an ionic compound? A major factor is **solvation**, depicted in Figure 2-22 (next page), in which an individual ion participates in *multiple* ion–dipole interactions with the solvent. When this occurs, the ions are said to be **solvated**. The collective stability from all of these ion–dipole interactions can be substantially greater than that of the ion–ion interactions in an ionic compound, thus making the mixture more favored (more stable).

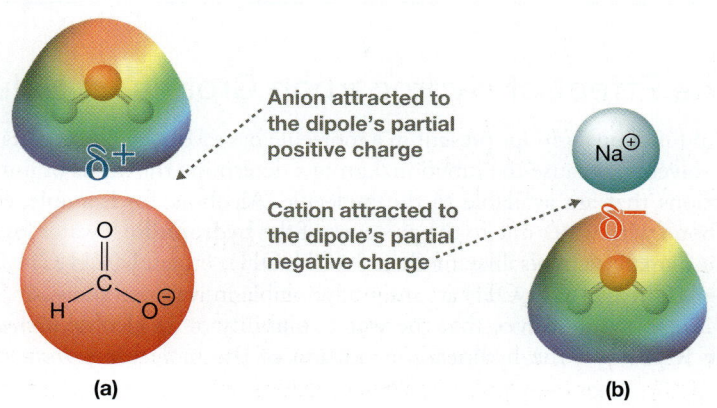

Anion attracted to the dipole's partial positive charge

Cation attracted to the dipole's partial negative charge

(a)   (b)

FIGURE 2-21 Ion–dipole interactions (a) An ion–dipole interaction between HCO_2^- and H_2O. The anion interacts with the positive end of water's dipole. (b) An ion–dipole interaction between Na^+ and a molecule of water. The cation interacts with the negative end of water's dipole.

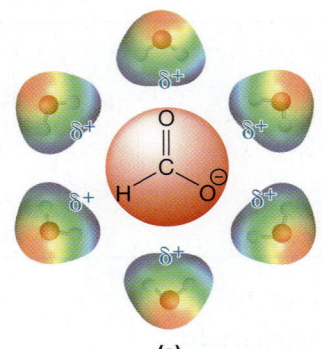

The positive end of water's dipole solvates the methanoate anion via multiple ion–dipole interactions.

(a)

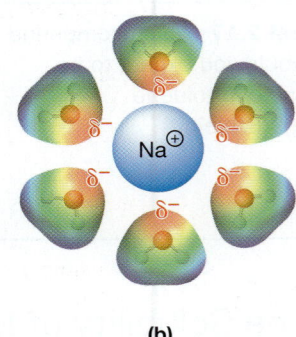

The negative end of water's dipole solvates the sodium cation via multiple ion–dipole interactions.

(b)

FIGURE 2-22 Solvation Ionic compounds can dissolve in water as a result of solvation of the respective ions. (a) The positive end of water's dipole solvates HCO_2^-. (b) The negative end of water's dipole solvates Na^+.

YOUR TURN 2.16

In Figure 2-22, how many *total* ion–dipole interactions are depicted for the two ions?

The strength of an individual ion–dipole interaction depends on the polarity of the solvent molecule, so the collective stability provided by solvation does, too (see Solved Problem 2.18). This is important because, as we discuss in Chapter 9, a solvent's ability to solvate ions can have a dramatic effect on the outcome of reactions.

SOLVED PROBLEM 2.18

In which solvent would you expect NaCl to be more soluble, diethyl ether ($CH_3CH_2OCH_2CH_3$) or propanal ($CH_3CH_2CH{=}O$)?

Think What are the most important intermolecular interactions that would be disrupted on dissolution? What intermolecular interactions would be gained in the solution? In which solvent are those intermolecular interactions stronger?

Solve Very strong ion–ion interactions are disrupted when NaCl dissolves. Both diethyl ether and propanal are polar molecules, so ion–dipole interactions would be present in solution with both solvents. Propanal, however, has a much larger dipole moment (compare the dipole moments of ethanal and dimethyl ether in Table 2-4, p. 81), so it will better solvate the Na^+ and Cl^- ions.

PROBLEM 2.19 NaCl is more soluble in methanol (CH_3OH) than in propan-1-ol ($CH_3CH_2CH_2OH$). Which solvent is better at solvating ions?

2.7b The Effect of Hydrocarbon Groups on Solubility

The types of functional groups present in a molecule have a direct effect on its solubility in various solvents, because the functional groups determine the kinds of intermolecular interactions that are available to the molecule. Alcohols, for example, can always hydrogen bond with water due to the presence of the **hydrophilic** ("water loving") OH group. Thus, small alcohols like methanol (CH_3OH), ethanol (CH_3CH_2OH), and propan-1-ol ($CH_3CH_2CH_2OH$) are infinitely soluble in water, as shown in Table 2-6.

Table 2-6 shows, however, that the water solubility of an alcohol *decreases* as the size of the R group—the hydrocarbon portion of the molecule—*increases*. This is

TABLE 2-6 Water Solubility of Some Simple Alcohols (R—OH)

Alcohol	Water Solubility (g/100 g H_2O)	Alcohol	Water Solubility (g/100 g H_2O)
CH_3OH Methanol	Infinitely soluble	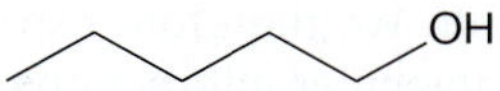 Pentan-1-ol	2.7
CH_3CH_2OH Ethanol	Infinitely soluble	 Hexan-1-ol	0.6
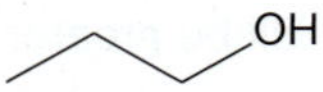 Propan-1-ol	Infinitely soluble	 Heptan-1-ol	0.1
 Butan-1-ol	7.7		

because R groups are highly nonpolar; they are **hydrophobic** ("water fearing"). We can generalize this trend as follows:

> A species behaves more like a nonpolar alkane as the size of its alkyl group increases.

PROBLEM 2.20 Which functional groups in Table 1-6 (p. 35) would be considered hydrophilic? Which would be considered hydrophobic?

Monoalcohols (i.e., alcohols with one OH group) are generally considered insoluble in water if they contain six or more carbons. The effect of the alkyl group, however, can be overcome by increasing the number of hydrophilic functional groups. Typically, molecules are found to be highly water soluble if the ratio of their total number of carbon atoms to the number of hydrophilic functional groups is less than about 3:1. Sucrose (table sugar), for example, is highly soluble in water despite the fact that it has 12 carbon atoms (Fig. 2-23). This is because it contains eight hydrophilic OH groups, so its ratio of carbon atoms to hydrophilic groups is $12:8 = 1.5:1$, which is less than $3:1$.

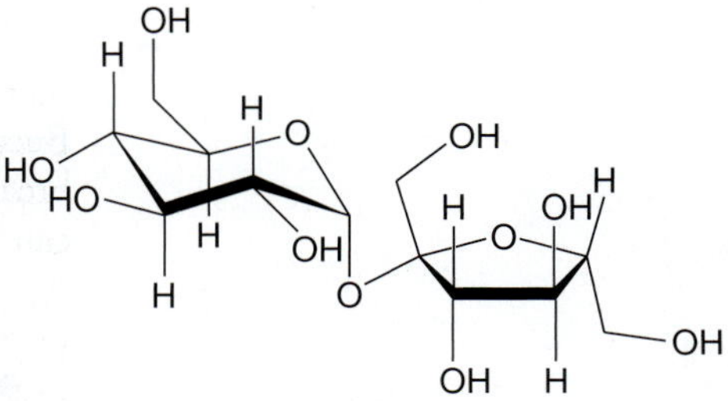

SOLVED PROBLEM 2.21

Would you expect each of the following compounds to be soluble in water? Why or why not?

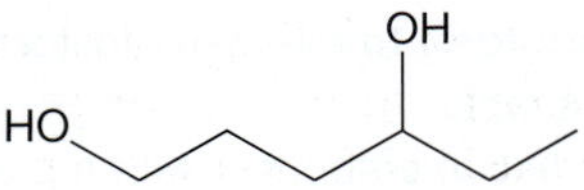

Sucrose, $C_{12}H_{22}O_{12}$

FIGURE 2-23 The water solubility of sucrose The large number of OH groups compared to C atoms makes sucrose soluble in water.

A

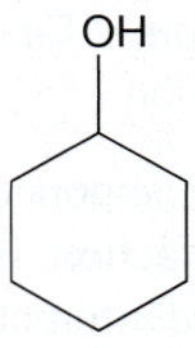

B

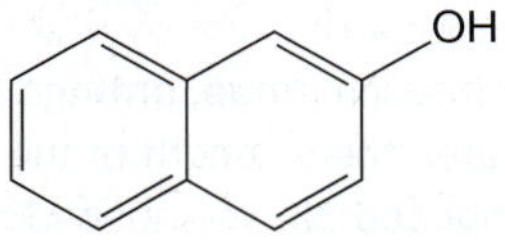

C
2-Naphthol

Sudan I

Think What is the ratio of alcohol groups to the number of carbon atoms of the hydrocarbon group?

Solve Molecule **A** contains six carbons and two hydrophilic OH groups, giving it a ratio of 3:1. We therefore expect **A** to be highly soluble in water. For molecules **B** and **C**, however, the ratios are 6:1 and 10:1, respectively, both of which exceed the 3:1 cutoff. We therefore expect nonpolar groups in **B** and **C** to dominate the intermolecular interactions, causing the molecules to be *insoluble* in water.

PROBLEM 2.22 Like alcohols, aldehydes (R—CH=O) become less soluble in water as the number of carbon atoms in the alkyl group R increases. Do you think the maximum number of carbon atoms for water-soluble aldehydes will be greater than or less than that for water-soluble alcohols? Explain.

2.8 Strategies for Success: Ranking Boiling Points and Solubilities of Structurally Similar Compounds

One type of problem you will face is ranking the boiling points, melting points, or solubilities of a set of compounds, given only their molecular structures. Suppose, for example, that you were asked to rank the boiling points of the following compounds, from lowest to highest:

A   B   C   D   E   F   G

Because there are a number of factors at play simultaneously, you should find ways to break this large problem into smaller, more manageable pieces. The following method can help, but you might also discover another that suits you better.

Ranking Boiling Points

1. Identify the most important intermolecular interactions present in each pure compound and group the compounds accordingly.
2. Arrange those groups of compounds according to the typically observed strength of the intermolecular interactions, which increase from low to high as follows:

    Induced dipole–induced dipole < Dipole–dipole < Hydrogen bonding
    < Ion–dipole < Ion–ion

3. Within each group, arrange the compounds according to the major factor that dictates the strength of the interaction. For example,
    • Induced dipole–induced dipole interactions become stronger with a greater polarizability.

Applying Steps 1 and 2 to compounds **A–G**, we arrive at the following groupings:

	A	**D**	**F**	**G**		**C**		**B**	**E**

Group 1 = nonpolar Induced dipole–induced dipole interactions	**Group 2 = polar** Dipole–dipole interactions	**Group 3 = polar** Hydrogen bonding

Next we can apply Step 3 to each group of molecules. For the nonpolar molecules in Group 1, polarizability dictates the strength of the induced dipole–induced dipole interactions, and polarizability tends to increase with the number of total electrons. Therefore, the strength of the intermolecular interactions increases in the order **D < A < F < G**.

For the Group 3 molecules, which undergo hydrogen bonding, we examine the number of potential donors and acceptors in a pair of each. In a pair of molecules **B**, there are two donors and two acceptors, whereas in a pair of molecules **E**, there are four of each. Therefore, H bonding is more extensive in **E** than in **B**.

Putting it all together, we predict that the strength of intermolecular interactions increases as follows:

	D	**A**	**F**	**G**	**C**	**B**	**E**
Boiling point:	80 °C	111 °C	139 °C	173 °C	181 °C	202 °C	281 °C

We therefore predict that the boiling points of these compounds increase in the same order (actual boiling points are included for comparison).

In solving the ranking problem, we assumed that the molecules possessing only induced dipole–induced dipole interactions (Group 1) had the weakest intermolecular interactions and the lowest boiling points. This was a good assumption in this case because the polarizabilities of all the molecules were sufficiently similar. As Solved Problem 2.23 shows, however, we must take into account significant differences in polarizability.

SOLVED PROBLEM 2.23

The boiling point of 1,4-dibromobenzene, a nonpolar compound, is 219 °C, whereas the boiling point of 1,2-dichlorobenzene, a polar compound, is 181 °C. Explain.

Think What intermolecular interactions are present in each compound? Are they of similar strength in each compound?

Solve 1,4-Dibromobenzene is nonpolar, so it has only induced dipole–induced dipole interactions. 1,2-Dichlorobenzene, on the other hand, is polar, so it has both dipole–dipole interactions and induced dipole–induced dipole interactions. Dipole–dipole interactions are generally assumed to be the more important interaction when comparing compounds with similar polarizabilities. In this case, however, the two molecules should have different polarizabilities. More specifically, 1,4-dibromobenzene should be significantly more polarizable than 1,2-dichlorobenzene because it has 36 more electrons (146 vs. 110), the difference between two Br atoms and two Cl atoms. As a result, induced dipole–induced dipole interactions are stronger in 1,4-dibromobenzene—enough to give 1,4-dibromobenzene the higher boiling point.

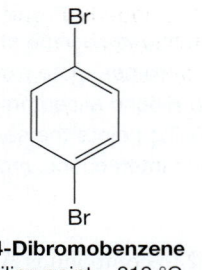

1,4-Dibromobenzene
Boiling point = 219 °C

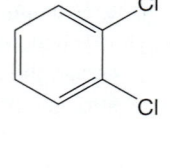

1,2-Dichlorobenzene
Boiling point = 181 °C

PROBLEM 2.24 Which of the compounds shown at the left do you think has the higher boiling point? Explain.

A

B

We can apply a similar strategy to predicting relative solubilities in a given solvent. However, we must consider the disruption of the intermolecular interactions in the pure separated substances as well as the formation of the intermolecular interactions in the mixture. This is illustrated in Solved Problem 2.25.

SOLVED PROBLEM 2.25

Rank compounds **A–E** from lowest solubility in hexane, $CH_3(CH_2)_4CH_3$, to highest solubility in hexane.

A   B   C   D   E

Think What are the most important intermolecular interactions that exist in the isolated substances and what are their relative strengths? What are the most important intermolecular interactions that exist in the solution?

Solve In a solution of each compound **A–E** with hexane, the intermolecular interactions will be quite weak because hexane is nonpolar and cannot participate in hydrogen bonding, so the intermolecular interactions in the pure substances will govern the relative solubilities. As pure substances, the relative strengths of the intermolecular interactions are as follows:

Increasing strength of intermolecular interactions

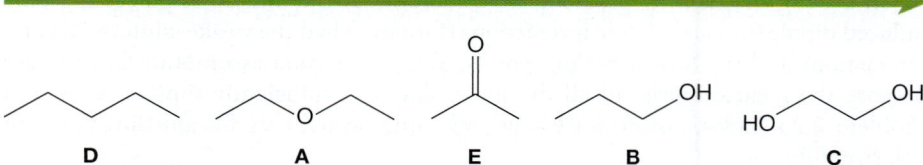

D   A   E   B   C

Compound **D** is nonpolar and therefore possesses only induced dipole–induced dipole interactions. Compounds **A** and **E** are polar and therefore possess dipole–dipole interactions; those interactions are stronger in **E**, moreover, because it has a larger dipole moment. Compounds **B** and **C** both undergo hydrogen bonding, but it is more extensive in **C** because **C** has more H-bond donors and acceptors. To arrive at the relative solubilities, we reverse this order because stronger intermolecular interactions in the unmixed substances will decrease solubility:

Increasing solubility in hexane

HO⏜⏜OH < ⏜⏜⏜OH < (structure E) < ⏜⏜O⏜⏜ < ⏜⏜⏜⏜

    C   **B**   **E**   **A**   **D**

PROBLEM 2.26 Rank the compounds in Solved Problem 2.25 in order from least soluble in water to most soluble.

2.9 Protic and Aprotic Solvents

As we learned in Section 2.7, ionic compounds can have very high solubility in a polar solvent such as water, due to the strong solvation set up by ion–dipole interactions in the resulting solution. For example, about 35 g of sodium chloride (NaCl) dissolves in 100 mL of water (dipole moment = 1.9 D). Dimethyl sulfoxide (DMSO) is even more polar (dipole moment = 4.0 D), but dissolves only about 0.4 g NaCl.

$$35\text{ g }\ce{NaCl}(s) \xrightarrow{\ce{H2O}} \ce{Na}^{\oplus} + \ce{Cl}^{\ominus} \qquad (2\text{-}2a)$$

NaCl is highly soluble in water but not in DMSO.

$$0.4\text{ g }\ce{NaCl}(s) \xrightarrow{\text{DMSO}} \ce{Na}^{\oplus} + \ce{Cl}^{\ominus} \qquad (2\text{-}2b)$$

There must be more to the issue than just the magnitude of the solvent's dipole moment. What also comes into play is how *close* the ions can approach the partial charges of the polar solvent molecule. As shown in Figure 2-24, the partial negative and partial positive charges of water are well exposed to the $\ce{Na+}$ and $\ce{Cl-}$ ions, respectively, so these ions are able to approach the partial charges rather closely. Thus, the solvation of each ion is relatively strong.

The negative end of water's dipole solvates the sodium cation strongly.

The positive end of water's dipole solvates the chloride anion strongly.

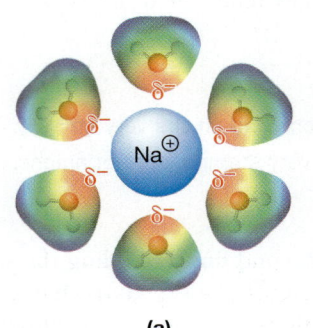

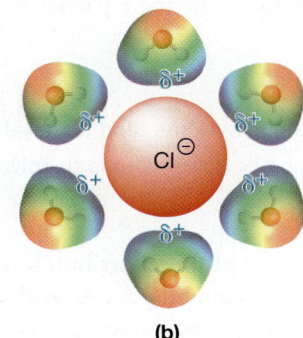

(a)   (b)

FIGURE 2-24 Solvation of NaCl in water (a) The $\ce{Na+}$ ion is strongly solvated because the partial negative charge of water is well exposed. (b) The $\ce{Cl-}$ ion is strongly solvated because the partial positive charge of water is well exposed.

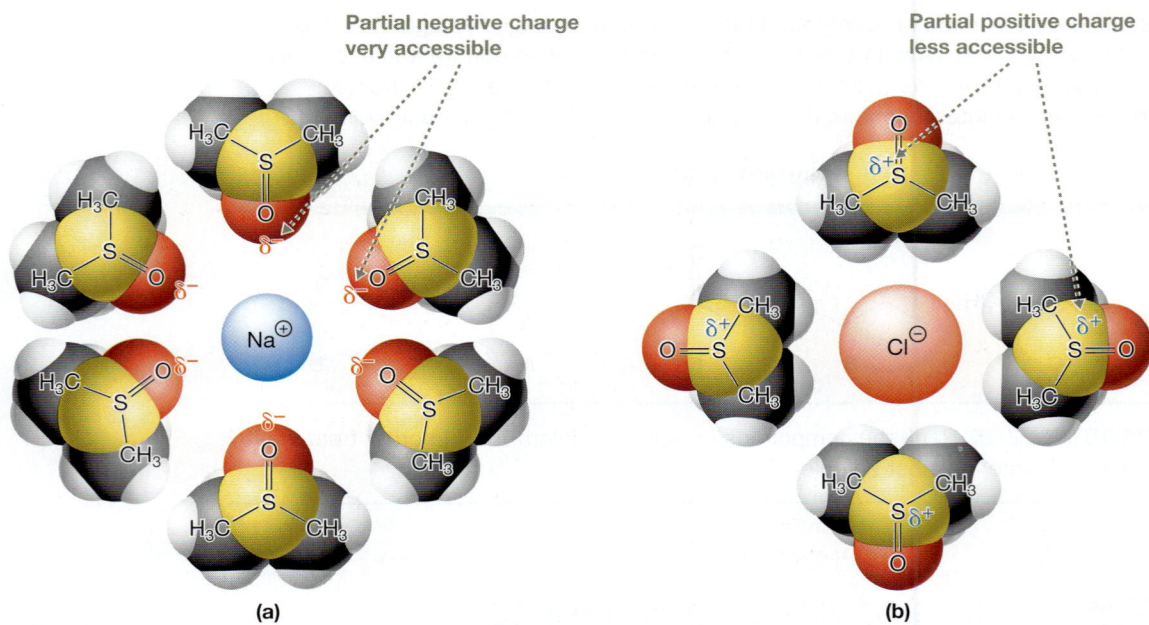

Partial negative charge very accessible

Partial positive charge less accessible

(a)

(b)

FIGURE 2-25 Solvation in DMSO (a) Na$^+$ is solvated strongly in DMSO because the partial negative charge of DMSO is well exposed. (b) Cl$^-$ is not solvated very strongly by DMSO, because DMSO's partial positive charge is buried inside the molecule.

Figure 2-25a shows that the partial negative charge of DMSO is also well exposed, so Na$^+$ is strongly solvated by DMSO, just as it is strongly solvated by water. However, notice that the partial positive charge of DMSO is flanked by two bulky CH$_3$ groups. This makes it difficult for the Cl$^-$ ion to approach DMSO's partial positive charge (Fig. 2-25b). We say that the methyl groups introduce **steric hindrance**. Thus, Cl$^-$ is not solvated very strongly.

These characteristics of water and DMSO, as they pertain to the solubility of ionic compounds, are not unique to just these two solvents. Water is an example of a **protic solvent**, because it possesses a H-bond donor—in this case, an O—H covalent bond. DMSO has no H-bond donors, so it is an **aprotic solvent**. Other examples of polar protic solvents and polar aprotic solvents are shown in Table 2-7.

YOUR TURN 2.17

In Table 2-7, circle and label each potential hydrogen-bond donor.

Like water, all polar protic solvents have easily accessible partial negative and positive charges. Polar aprotic solvents, on the other hand, have well-exposed partial negative charges, but their partial positive charges tend not to be very accessible. Therefore:

- Polar protic solvents tend to solvate both cations and anions very strongly.
- Polar aprotic solvents tend to solvate cations very strongly, but not anions.

The importance of these attributes extends far beyond understanding the solubilities of ionic compounds. As we will see in Chapter 9, the solvation characteristics of protic and aprotic solvents play a major role in governing the outcomes of reactions.

TABLE 2-7 Common Polar Protic Solvents and Polar Aprotic Solvents

POLAR PROTIC SOLVENTS		POLAR APROTIC SOLVENTS	
Structure	Name	Structure	Name
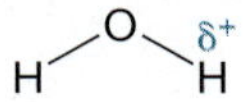	Water	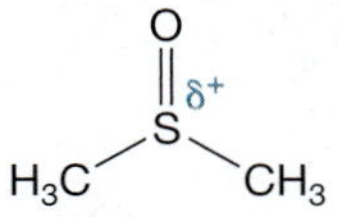	Dimethyl sulfoxide (DMSO)
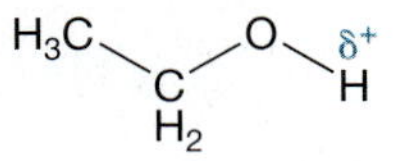	Ethanol	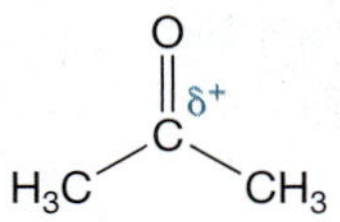	Propanone (Acetone)
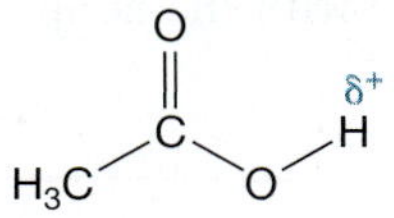	Ethanoic acid (Acetic acid)	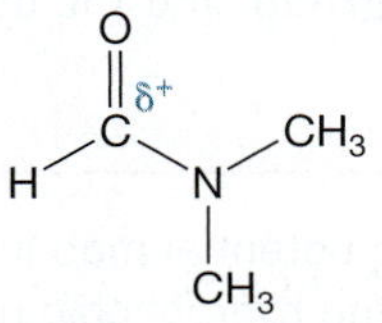	N,N-Dimethyl-formamide (DMF)

PROBLEM 2.27 Will $KSCH_3$ be more soluble in ethanol or acetone? Explain.

2.10 Soaps and Detergents

Soaps and enzyme-free **detergents** belong to a class of compounds called **surfactants**. They remove dirt and oil from our hands, clothes, and dinnerware, all with no chemical reaction occurring in the process (i.e., no covalent bonds are broken or formed). Instead, the cleansing ability of soaps and detergents depends entirely on intermolecular interactions.

The molecules specifically responsible for a soap's cleansing properties are typically *salts of fatty acids*, which are ionic compounds of the form $RCO_2^- \ Na^+$ or $RCO_2^- \ K^+$. In these compounds, R is a long hydrocarbon chain, and the fatty acid salts generally contain from 12 to 18 carbons. Examples include potassium oleate and sodium palmitate:

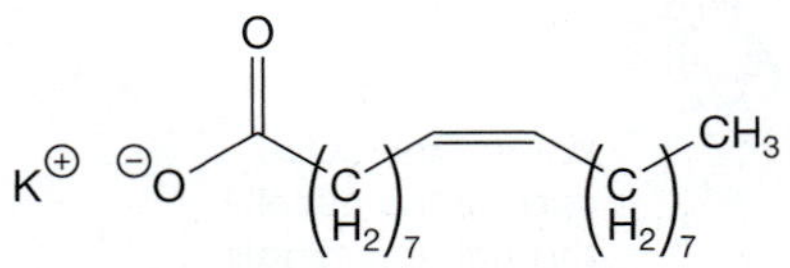

Potassium oleate
$C_{18}H_{33}O_2K$

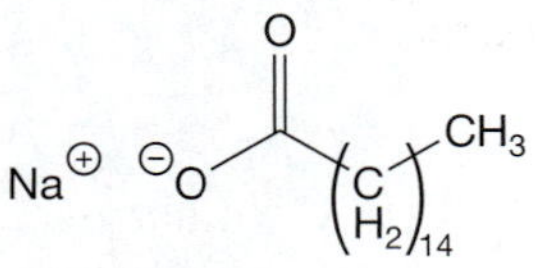

Sodium palmitate
$C_{16}H_{31}O_2Na$

When a soap dissolves in water, it does so as its individual ions: the metal cation (Na^+ or K^+) and the carboxylate anion (RCO_2^-). Of these two species, the carboxylate anion is the one that is directly responsible for the soap's cleansing properties, because it has vastly different characteristics at its two ends (Fig. 2-26). Specifically:

A significant negative charge is located on the portion of the ion containing the $-CO_2^-$ group (the ionic head group), making it hydrophilic. At the same time, the hydrocarbon tail is very nonpolar, making it hydrophobic.

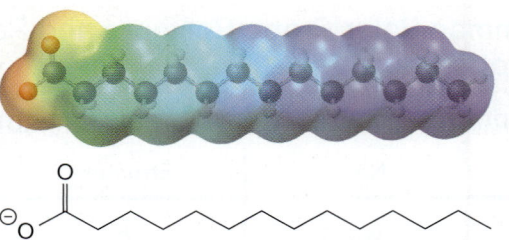

Soaps work because one end of the molecular species is very hydrophilic and the other end is very hydrophobic.

In the case of fatty acid carboxylates, the hydrophilic end is the one with $-CO_2^-$, called the **ionic head group**, and the hydrophobic end is the one with the nonpolar **hydrocarbon tail**.

YOUR TURN 2.18

On the electrostatic potential map in Figure 2-26, circle the hydrophilic region and label it. Circle the hydrophobic region and label it.

In water, the carboxylate anions from the fatty acid salts form spherical aggregates, called **micelles** (Fig. 2-27). In a micelle, the nonpolar tails are on the inside of the sphere, where they can interact with one another via extensive induced dipole–induced dipole interactions, whereas the charged head groups are on the outside, where they can form the greatest number of ion–dipole interactions with the surrounding water molecules. As a result, micelles are highly solvated (Section 2.7a).

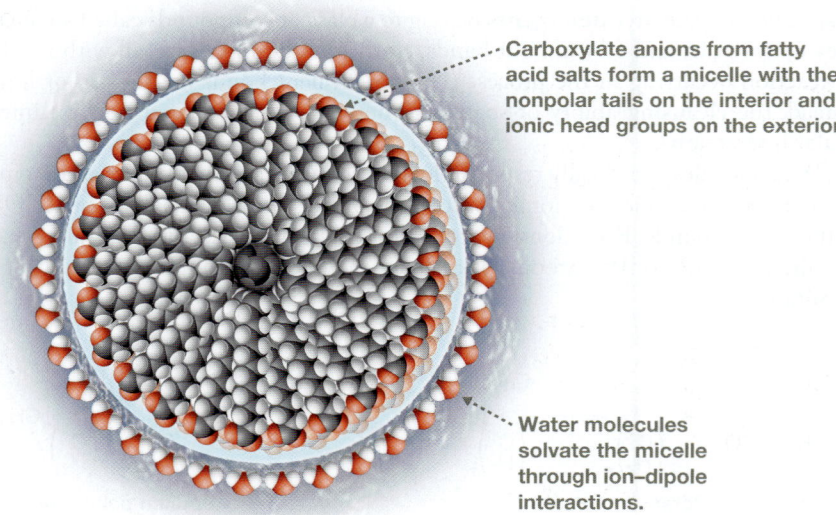

Carboxylate anions from fatty acid salts form a micelle with the nonpolar tails on the interior and ionic head groups on the exterior.

Water molecules solvate the micelle through ion–dipole interactions.

FIGURE 2-27 The structure of a micelle In water, carboxylate anions (RCO_2^-) from fatty acid salts form *micelles*, in which the nonpolar tails are on the inside and the ionic head groups are on the outside. Micelles are soluble in water because they are heavily solvated due to the extensive ion–dipole interactions that take place between the ionic head groups and the water molecules of the surrounding solution.

Enzyme Active Sites: The Lock-and-Key Model

Enzymes are proteins that catalyze biological reactions. In many cases, enzymes enhance reaction rates by several orders of magnitude. What is particularly fascinating is how specific each enzyme's role is, so much so that the name of an enzyme derives from the reaction that it catalyzes. Consider the enzyme fructose 1,6-bisphosphate aldolase, which is responsible for cleaving fructose 1,6-bisphosphate into dihydroxyacetone phosphate and glyceraldehyde-3-phosphate—a reaction that is integral in breaking down sugars.

The general model that accounts for the specificity of enzymes is often referred to as the *lock-and-key model*. An enzyme, which is typically a rather large molecule, assumes a three-dimensional shape that defines an *active site*—a pocket in which the reaction takes place. The *substrate* (i.e., the reactant) docks with the enzyme in the active site to form an enzyme–substrate complex, the reaction takes place, and subsequently the products are released. Under the lock-and-key model, only certain, specific substrates fit into the active site.

Fitting into an active site means more than just spatial fitting; it means that the active site is specially designed to provide optimal intermolecular interactions for the substrate. We can see this by examining a portion of the enzyme–substrate complex in the fructose 1,6-bisphosphate aldolase reaction.

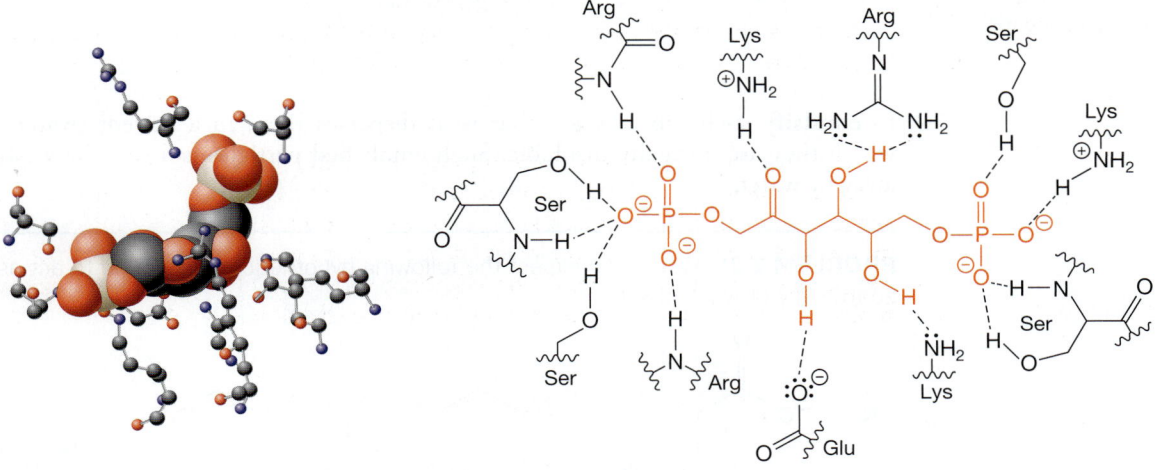

On the left, fructose 1,6-bisphosphate appears as a space-filling model, and the important side groups of some amino acids in the active site appear as ball-and-stick models. On the right, the substrate and side groups appear as line structures. Notice how the locations of the side groups are optimal for numerous specific hydrogen-bonding and ion–dipole interactions.

PROBLEM 2.28 Ethyl ethanoate (ethyl acetate, $CH_3CO_2CH_2CH_3$), a relatively small ester, is insoluble in water. Knowing this, would you expect the following compound to form micelles in water? Explain.

Dirt, grease, and oils are nonpolar substances, so they are insoluble in pure water. They are soluble in a solution of soap and water, however, because the hydrophobic tails of the soap molecules dissolve in droplets of oil or grease, leaving their hydrophilic head groups available for solvation by water (Fig. 2-28). Soap is said

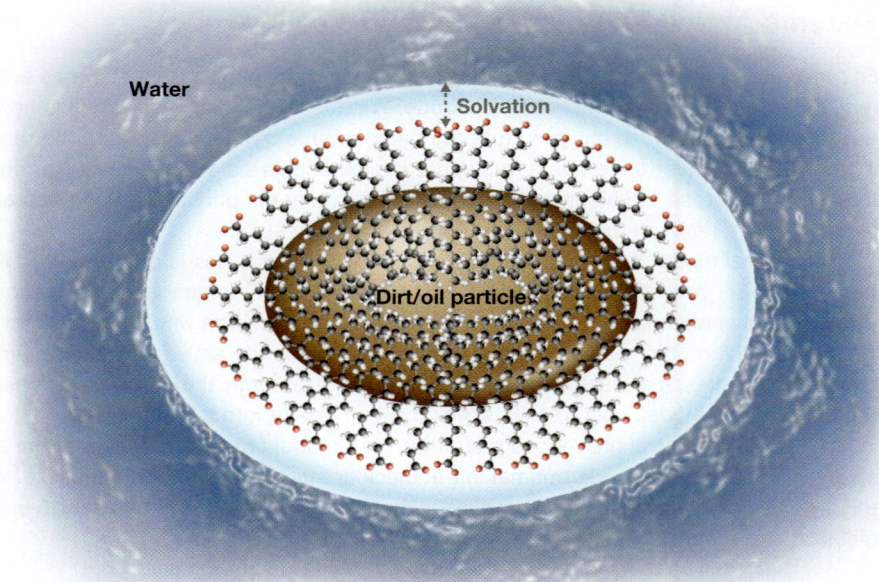

Water

Solvation

Dirt/oil particle

FIGURE 2-28 An oil droplet emulsified in water by soap The nonpolar tails of soap molecules dissolve in a droplet of oil, which can then be washed away by water molecules interacting with the exposed hydrophilic head groups.

to **emulsify** such substances—that is, it disperses them in a solvent (water) in which they are normally insoluble. Such emulsified particles can then be washed away by water.

PROBLEM 2.29 Would you expect the following hypothetical compound to act as a soap? Why or why not?

Hard water poses a problem for soaps. Water is considered *hard* if it contains a significant concentration of Mg^{2+} or Ca^{2+} ions. These ions bind more strongly to the negative charge of the carboxylate than Na^+ or K^+ ions do (see Problem 2.58 at the end of the chapter), causing the long-chain fatty acids to precipitate from water (Equation 2-3).

Ca^{2+} + 2

(2-3)

Ca

Soap scum
(Insoluble in water)

Not only does this decrease the effectiveness of soap, but the precipitate is a nuisance known as **soap scum**. A common way to combat this problem is to first run water through a *water softener*, which contains an ion-exchange resin that removes hard water ions and replaces them with Na^+ ions.

Unlike soaps, detergents are effective cleansers in hard water. Detergents such as sodium dodecyl sulfate are similar in structure to soaps, containing an ionic head group (sulfate, $—OSO_3^-$) and a nonpolar hydrocarbon tail (Fig. 2-29). As a result, they behave in much the same way as soaps, forming micelles in solution that are capable of emulsifying dirt, grease, and oils. The main difference is that metal ions in hard water do not bind to the sulfate portion of detergents as strongly as they do to the carboxylate portion of soaps. (See Problem 2.59 at the end of the chapter.) This allows detergents to remain active even in hard water.

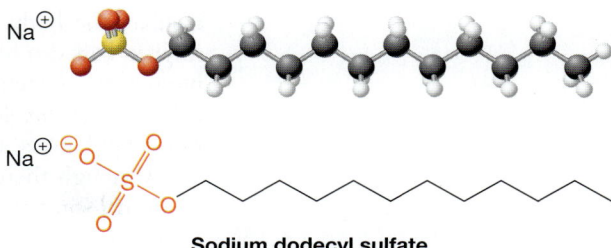

Sodium dodecyl sulfate

FIGURE 2-29 **Detergents** In a detergent such as sodium dodecyl sulfate, the $—OSO_3^-$ head group is very hydrophilic and the long nonpolar tail is hydrophobic.

THE ORGANIC CHEMISTRY OF BIOMOLECULES

2.11 An Introduction to Lipids

Section 1.14 introduced three of the four major classes of biomolecules: proteins, carbohydrates, and nucleic acids. Here in Section 2.11, we introduce the fourth: *lipids*.

Recall that proteins, carbohydrates, and nucleic acids can be very large molecules, but are constructed from just a few types of small organic molecules. Proteins, for instance, are constructed from α-amino acids, polysaccharides are constructed from monosaccharides (simple sugars), and nucleic acids are constructed from nucleotides. Moreover, these smaller molecules can be unambiguously identified by the specific functional groups they possess.

Lipids, on the other hand, are characterized by their solubility:

> **Lipids** are biomolecules that are relatively *insoluble* in water.

Thus, most lipids are highly nonpolar, consisting primarily of carbon and hydrogen, with very little oxygen or nitrogen content. Consequently, they tend to be soluble in nonpolar or weakly polar organic solvents, such as diethyl ether.

Because lipids are characterized in such a broad way, they can have a variety of structures. Nevertheless, lipids can be divided into subclasses based on the structural features they have in common. We examine four of these subclasses in this unit: *fats*, *steroids*, *phospholipids*, and *waxes*. In particular, we describe some of the biological functions that lipids have, along with the structural features shared by lipids within a specific subclass.

2.11a Fats, Oils, and Fatty Acids

Fats can be of animal or plant origin. Animal fats, such as lard, are generally solids at room temperature. Fats from plants, however, are generally liquids at room temperature, and are thus more properly called **oils**. These include corn oil, olive oil, peanut oil, safflower oil, coconut oil, and many more.

Perhaps the most common biological function of fats and oils is to store energy, but they have a variety of other functions as well. For example, fats are needed for the intake of so-called fat-soluble vitamins, such as vitamins A, D, E, and K. Fats can be used to insulate the body against heat loss and to insulate internal organs against physical impact. Fats can help protect organisms from foreign substances—either chemical or biological—by temporarily locking them away in new fat tissue. And, fats contain **fatty acids**, which are long-chain carboxylic acids that help regulate blood pressure and blood lipid levels, and play major roles in the inflammatory response to injury.

Although there are several different kinds of fats and oils (having different plant or animal sources), they all share the same general structure:

$$\text{(2-4)}$$

Notice that a fat or oil contains three adjacent ester groups, each of which can be produced from a *fatty acid* and an OH group from *glycerol*. (The chemical reactions involved are discussed in Chapter 21.) Thus, a fat or oil is often described as a **triacylglycerol** or a **triglyceride**.

YOUR TURN 2.19

Circle and label the two unlabeled ester groups in the fat molecule in Equation 2-4.

Because all fats and oils have the same general structure, the identities of the fatty acids are what distinguish one fat or oil from another. Some examples are shown in Table 2-8.

All natural fatty acids have an even number of carbons because, in nature, fatty acids are synthesized from *acetyl coenzyme A*, a two-carbon-atom source. Most fatty acids have between 14 and 18 carbon atoms, although they can have as few as 4 and as many as 22.

PROBLEM 2.30 **(a)** Draw the triacylglycerol in which all three fatty acids are lauric acid. **(b)** Draw another in which one fatty acid is linoleic acid and two are oleic acid.

2.11b Phospholipids and Cell Membranes

Like a fat or an oil, a **phospholipid** consists of a glycerol backbone attached to fatty acids via ester linkages, as shown in Figure 2-30a.

Whereas a fat or an oil has three fatty acids, making it a *triglyceride*, a phospholipid has only two fatty acids, making it a **diglyceride**. In place of a third fatty acid, the glycerol backbone is bonded to a phosphate ($-OPO_3^-$) group, the same type of group

TABLE 2-8 Some Common Fatty Acids

Fatty Acid	Structure	Sources
Lauric acid $C_{12}H_{24}O_2$		Coconut oil, palm kernel oil
Palmitic acid $C_{16}H_{32}O_2$		Coconut oil, palm oil, palm kernel oil, meats, cheeses
Stearic acid $C_{18}H_{36}O_2$		Animal fats, cocoa butter
Oleic acid $C_{18}H_{34}O_2$		Olive oil, pecan oil, peanut oil
Linoleic acid $C_{18}H_{32}O_2$		Safflower oil, grape seed oil
Linolenic acid $C_{18}H_{30}O_2$		Kiwifruit seeds, flax

that appears in nucleotides in DNA and RNA (Section 1.14). Typically, the phosphate group is bonded to a second molecular fragment, such as choline (Figure 2-30b).

Because of their structural similarities to fats and oils, phospholipids can efficiently store energy, too. More importantly, however, phospholipids are integral in the formation of cell membranes. Such a role is possible because the two ends of a phospholipid have vastly different properties—the phosphate portion is ionic (hydrophilic), whereas the alkyl chains of the fatty acids are nonpolar (hydrophobic). These characteristics are highlighted by the simplified representation of a phospholipid on

FIGURE 2-30 Diglycerides in nature A phospholipid (a) and a phosphatidylcholine (b) are both diglycerides because two fatty acids (blue) are linked to the glycerol backbone (red).

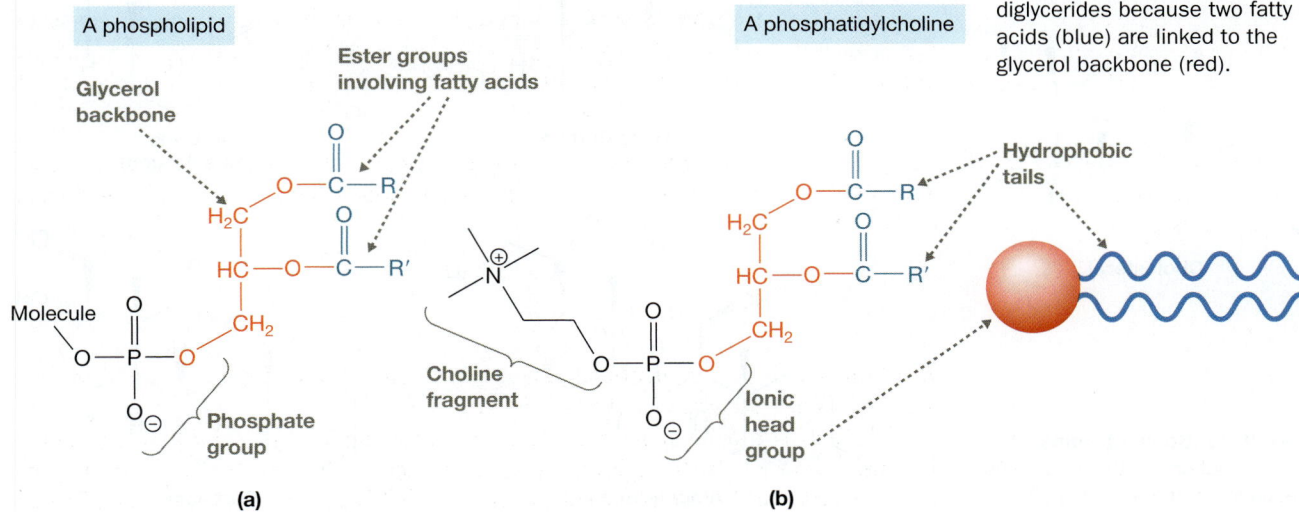

(a)   (b)

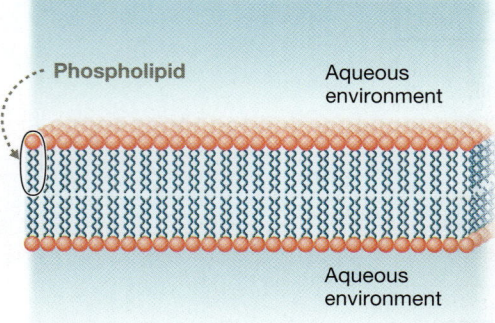

A lipid bilayer

Phospholipid

Aqueous environment

Aqueous environment

(a)

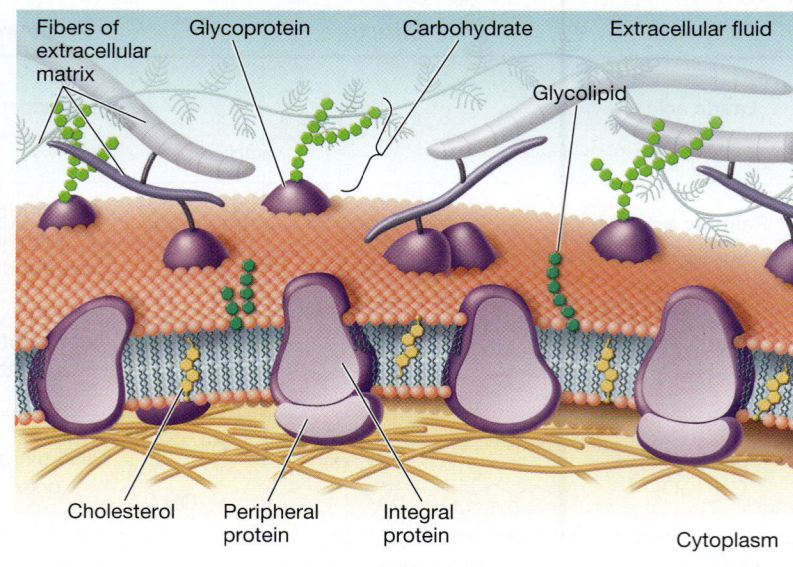

Cell membrane

Fibers of extracellular matrix

Glycoprotein

Carbohydrate

Extracellular fluid

Glycolipid

Cholesterol

Peripheral protein

Integral protein

Cytoplasm

(b)

FIGURE 2-31 The lipid bilayer and cell membrane (a) In an aqueous environment, phospholipids organize into a lipid bilayer. The ionic head groups remain on the exterior of the membrane and are thus stabilized by solvation. The hydrophobic tails aggregate together to escape the aqueous environment. (b) The lipid bilayer is the basis of cell membranes.

the right of Figure 2-30b, in which the red sphere represents the ionic head group and the two blue wavy lines represent the two nonpolar alkyl chains.

Under physiological conditions, the hydrophobic tails from multiple phospholipids associate into a **lipid bilayer** to escape the aqueous environment, as illustrated in Figure 2-31a. This lipid bilayer is the basis of a cell membrane, the basic features of which are shown in Figure 2-31b.

YOUR TURN 2.20

Identify and label the hydrophobic regions in Figure 2-31a and 2-31b.

Cholesterol

Testosterone

Estrone (An estrogen)

Aldosterone

Cortisone

FIGURE 2-32 Some steroids The biological functions of these steroids are described in the text (p. 109).

The cell membrane's hydrophobic interior and hydrophilic exterior are critical, because they restrict the free passage of most molecules from one side of the membrane to the other. Hydrophobic molecules, such as cholesterol, reside in the membrane's interior. Hydrophilic molecules, such as carbohydrates, are located on the outside. As we can see, however, some specialized proteins can exist in both regions simultaneously. Although we will not examine the details, these proteins tend to have separate hydrophilic and hydrophobic regions, and some are capable of shuttling molecules across the membrane.

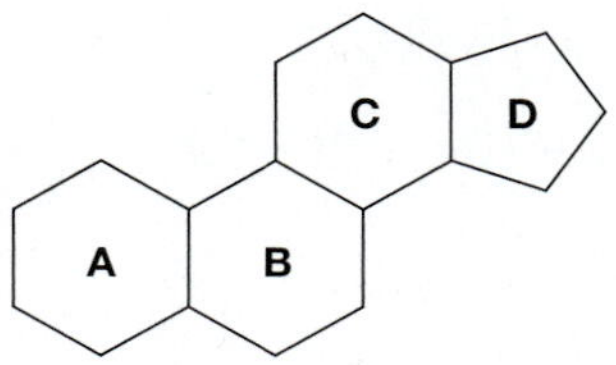

FIGURE 2-33 The steroid ring system All steroids are made of three six-membered rings and one five-membered ring, designated as the A, B, C, and D rings, fused in the arrangement shown.

2.11c Steroids, Terpenes, and Terpenoids

Steroids (Fig. 2-32) have a wide variety of biological functions. *Cholesterol*, one of the most well-known steroids, is responsible for maintaining the permeability of cell membranes in mammals. *Testosterone* and *estrone* (an estrogen), on the other hand, are male and female sex hormones, respectively. *Aldosterone* helps regulate blood pressure, and *cortisone* suppresses the immune system and is used to treat inflammatory conditions.

All steroids have the same basic ring structure (Fig. 2-33), in which three six-membered rings and one five-membered ring are fused together—that is, they have a bond in common. These are designated simply as the A, B, C, and D rings. This ring structure is a common feature among steroids because every steroid is produced via chemical modification of *lanosterol*, as shown in Equation 2-5. Lanosterol, in turn, is produced from *squalene*.

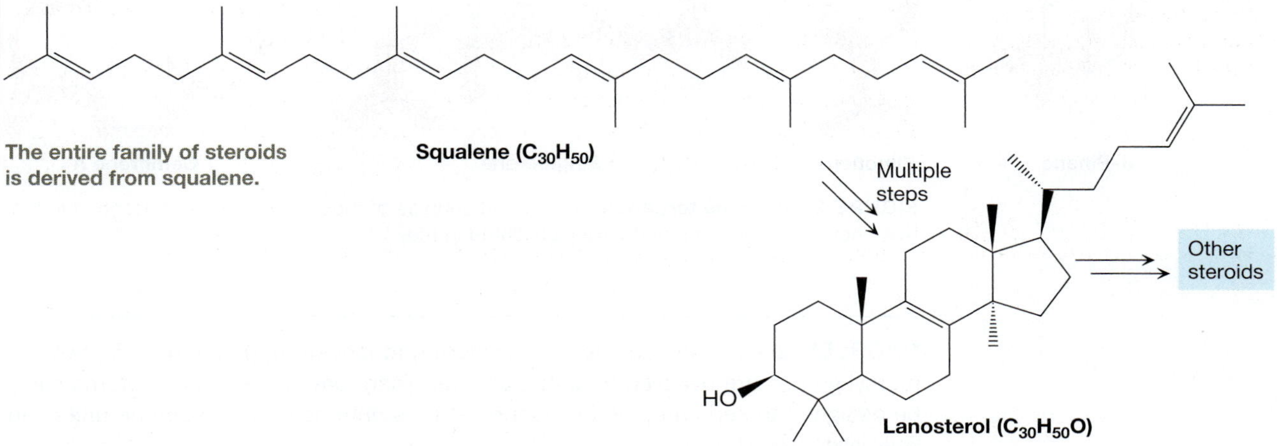

The entire family of steroids is derived from squalene.

Squalene ($C_{30}H_{50}$)

Multiple steps

Other steroids (2-5)

Lanosterol ($C_{30}H_{50}O$)

Squalene itself belongs to a family of compounds called *terpenes*. A **terpene** is a naturally produced hydrocarbon whose carbon backbone can be divided into separate five-carbon units called **isoprene units**:

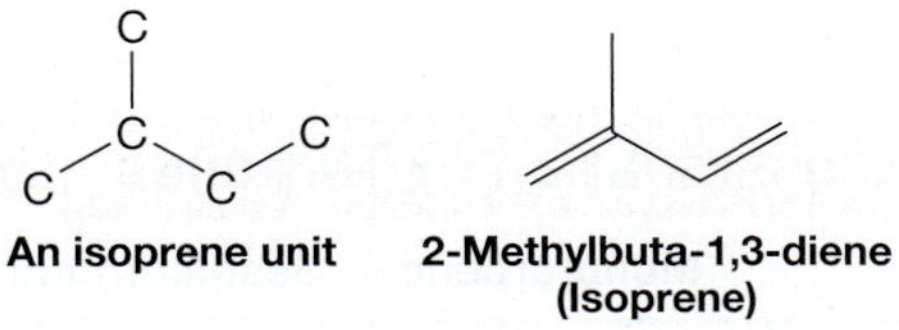

An isoprene unit

2-Methylbuta-1,3-diene (Isoprene)

Isoprene units are highlighted in red.

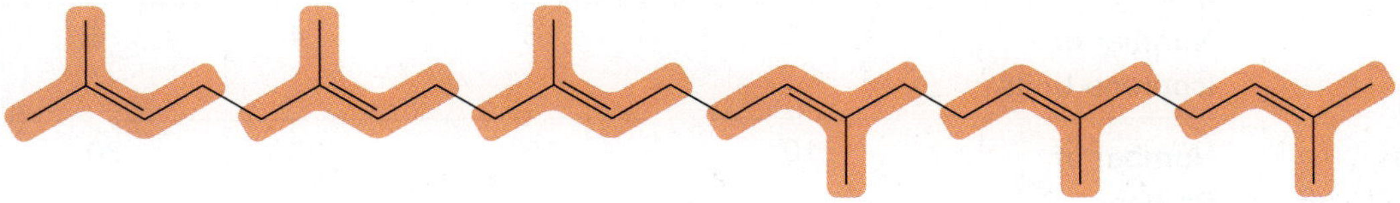

Squalene

They are called isoprene units because they resemble 2-methylbuta-1,3-diene, the compound whose common name is isoprene. Terpenes such as squalene, however, are not actually synthesized from isoprene.

Terpenes (Fig. 2-34) are abundant in nature and are frequently responsible for the aromas of natural products. α-Pinene, for example, is a constituent of pine resin; limonene, which can be isolated from the rinds of lemons, has a strong citrus odor; zingiberene is a component of ginger oil; and cembrene A is isolated from certain corals.

Terpenes are classified according to the number of pairs of isoprene units they contain, as shown in Table 2-9. Based on these definitions, α-pinene is a monoterpene because it contains two isoprene units; zingiberene is a sesquiterpene because it contains three isoprene units; cembrene A is a diterpene because it contains four isoprene units; and squalene is a triterpene because it contains six isoprene units.

| A monoterpene | A monoterpene | A sesquiterpene | A diterpene |

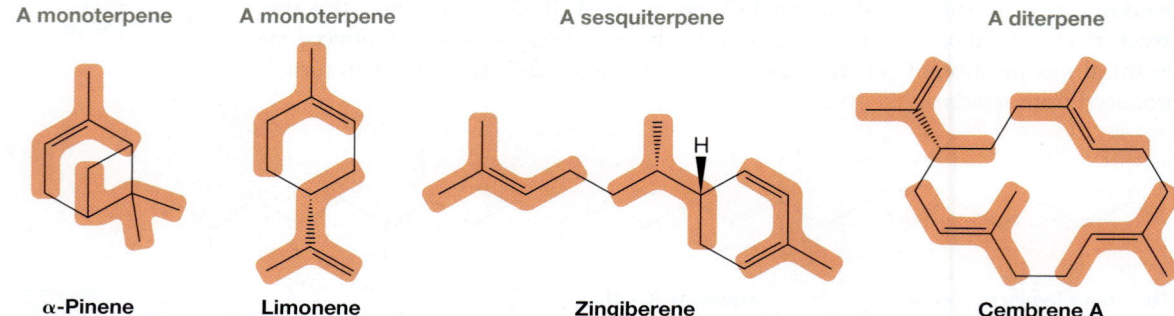

| α-Pinene | Limonene | Zingiberene | Cembrene A |

FIGURE 2-34 **Some terpenes** The natural sources of these terpenes are described in the text above. The isoprene units are highlighted in red.

PROBLEM 2.31 When squalene is converted to lanosterol (Equation 2-5), two methyl groups are believed to shift locations. Therefore, not all carbon atoms can be assigned to separate, distinct isoprene units. Into how many isoprene units can you divide lanosterol?

Many natural products are produced from chemical modifications to terpenes, in which the carbon backbone is altered and/or atoms other than just carbon or

TABLE 2-9 Classification of Terpenes				
	Monoterpene	Sesquiterpene	Diterpene	Triterpene
Number of Pairs of Isoprene Units	1	$1\frac{1}{2}$	2	3
Number of Isoprene Units	2	3	4	6
Number of Carbons	10	15	20	30

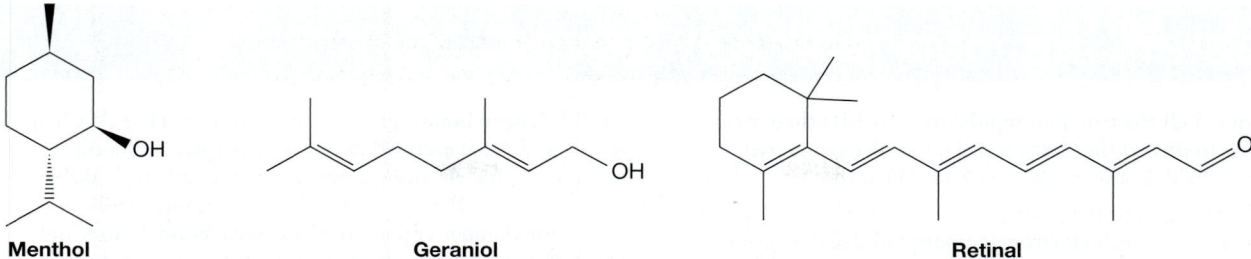

Menthol **Geraniol** **Retinal**

FIGURE 2-35 Some terpenoids Terpenoids are derived from chemical modifications of terpenes.

hydrogen are introduced. All of these compounds are called **terpenoids**. Thus, squalene is a terpene, but the steroids that are derived from it are terpenoids. Other terpenoids are shown in Figure 2-35, including menthol (obtained from mint oil), geraniol (a main constituent of rose oil), and retinal (involved in the chemistry of vision).

2.11d Waxes

Waxes are secretions from plants and animals that are solid at room temperature but melt at relatively low temperatures. Waxes are typically soft and malleable and are very hydrophobic. Bees use beeswax (Fig. 2-36a) to store honey and protect their eggs. More commonly in nature, however, wax is used as a protective coating. Carnauba wax, for example, which is found on the leaves of the carnauba palm tree native to Brazil (Fig. 2-36b), helps prevent excessive loss of water. Lanolin, found on wool, helps sheep shed water from their coats (Fig. 2-36c). Finally, earwax in the human ear canal (Fig. 2-36d) helps protect against bacteria and other foreign substances.

Waxes are generally mixtures of compounds, consisting principally of long-chain esters and alkanes. Tetracosyl hexadecanoate, for example, is a main constituent of beeswax:

Principal component of beeswax

Tetracosyl hexadecanoate

Any polarity that might originate from the ester functional group is overwhelmed by the large, highly nonpolar alkyl groups, making the molecule hydrophobic overall.

(a) (b) (c) (d)

FIGURE 2-36 Sources of various kinds of natural wax (a) Beeswax is used as cells for storing honey and protecting eggs. (b) The carnauba palm has wax deposits on its leaves to protect from excessive water loss. (c) Wool has a coating of lanolin, which helps shed water. (d) Human earwax protects the ear canal from foreign substances.

Chapter Summary and Key Terms

- **Valence shell electron pair repulsion (VSEPR) theory** can be used to predict the electron and molecular geometries about individual atoms. **(Section 2.1a; Objective 1)**
  - Electron groups (i.e., lone pairs or bonds) repel each other to yield an atom's **electron geometry; molecular geometry**, which is derived from the electron geometry, is the molecule's geometry based solely on the orientation of its chemical bonds.
  - Two electron groups yield a *linear* electron geometry, three electron groups yield a *trigonal planar* electron geometry, and four electron groups yield a *tetrahedral* electron geometry.
- Bond angles that deviate from the ideal VSEPR geometry possess **angle strain. (Section 2.1b; Objective 2)**
- **Dash–wedge notation** is used to depict three-dimensional molecular geometry; a wedge (◄) represents a bond pointing toward you, whereas a dash (·····|||) represents a bond pointing away from you. **(Section 2.2; Objective 3)**
- Bond dipole moments are treated as vectors and added together to yield a molecule's **net molecular dipole moment**, or **permanent dipole moment**. If bond dipoles do not perfectly cancel, then the molecule is **polar**. If the bond dipoles perfectly cancel, then the molecule is **nonpolar. (Section 2.4; Objective 4)**
- Differences in the physical properties of compounds similar in their makeup and mass are due to the presence of different functional groups. **(Section 2.5; Objective 5)**
- Boiling points and melting points increase as the strength of the **intermolecular interactions** increases in the liquid and solid phases, respectively. **(Section 2.5; Objectives 6 and 7)**
- Two compounds are generally soluble in each other if the intermolecular forces in the solution are roughly the same strength as or stronger than those that exist in the pure separated compounds. **(Section 2.7; Objective 7)**
- Intermolecular forces differ in strength. The greater the concentration of charges on the molecules interacting, the stronger their attraction. **(Objective 6)** The strength of intermolecular forces generally decreases in the following order:
  - **Ion–ion interactions:** Attraction between two oppositely charged ions. **(Section 2.6a)**
  - **Ion–dipole interactions:** Attraction between a positive or negative ion and a net molecular dipole. **(Section 2.7a)**

- **Hydrogen bonding:** Takes the form D—H·····A, where D and A are each either nitrogen, oxygen, or fluorine (i.e., highly electronegative elements that have available lone pairs). The extent of hydrogen bonding increases as the number of potential **hydrogen-bond donors** and **hydrogen-bond acceptors** increases. **(Section 2.6c)**
- **Dipole–dipole interactions:** Attractions between two net molecular dipoles. The strength of the attraction increases as the magnitudes of the molecular dipoles involved increase. **(Section 2.6b)**
- **Induced dipole–induced dipole interactions (London dispersion forces):** Attraction between two temporary dipoles. These interactions increase as the **polarizabilities** of the species involved increase and as the **contact surface area** increases. Polarizability generally increases with the number of total electrons. **(Section 2.6d)**
- With enough electrons, induced dipole–induced dipole interactions can become the dominant intermolecular force. **(Section 2.6d; Objective 6)**
- Even though ion–dipole interactions are weaker than ion–ion interactions, ionic compounds may dissolve in polar solvents as a result of **solvation** of the ions. **(Section 2.7a; Objectives 6 and 7)**
- As the size of hydrocarbon groups increases, compounds become less polar and less soluble in polar solvents. Mono-alcohols with six or more carbon atoms are insoluble in water. **(Section 2.7b; Objectives 6 and 7)**
- A **protic solvent** contains a H-bond donor, whereas an **aprotic solvent** does not. Ionic compounds tend to be much more soluble in protic solvents due to the strong solvation of anions. **(Section 2.9; Objective 8)**
- **Soaps** and **detergents** are long-chain molecules that are ionic on one end and highly nonpolar on the other. They dissolve in water as **micelles**, in which the **ionic head groups** are arranged on the outside and the **hydrocarbon tails** are grouped on the inside. **(Section 2.10; Objective 9)**
- When molecules of soaps and detergents encounter a particle of dirt, grease, or oil, the nonpolar tails dissolve in the particle, while the ionic head groups remain exposed to water. The **emulsified** particle is water soluble and can be washed away by water. **(Section 2.10; Objective 9)**

Problems

2.1–2.3 Valence Shell Electron Pair Repulsion (VSEPR) Theory; Dash–Wedge Notation; The Molecular Modeling Kit

2.32 Brassinolide, a naturally occurring steroid derivative found in a wide variety of plants, is thought to promote plant growth. Identify the electron geometry for each atom indicated. Where applicable, describe the atom's molecular geometry and estimate the bond angle.

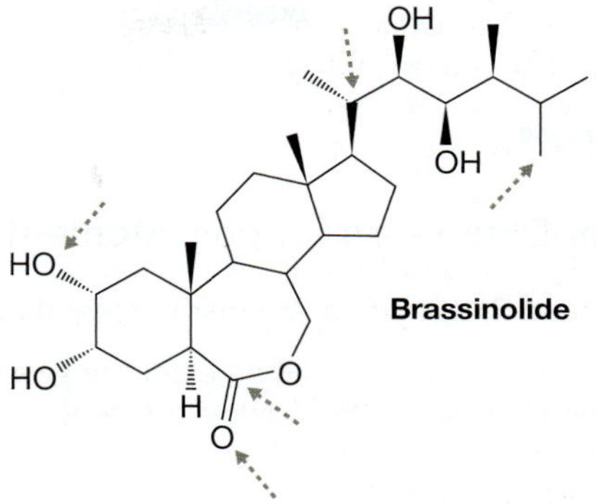

Brassinolide

2.33 Falcarinol, a naturally occurring pesticide found in carrots, is being studied as an anticancer agent. Identify the electron geometry for each atom indicated. Where applicable, describe the atom's molecular geometry and estimate the bond angle.

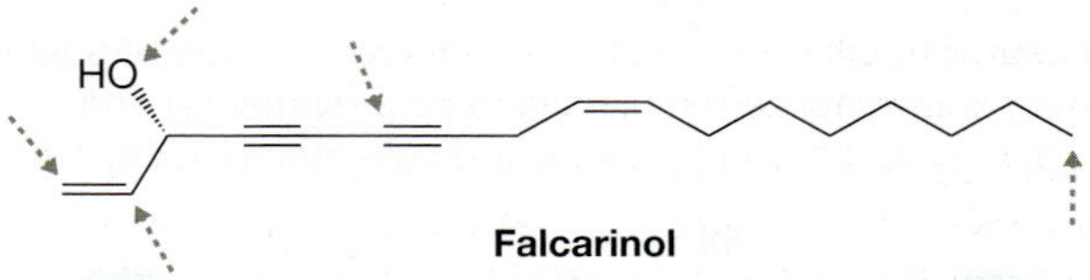

Falcarinol

2.34 Identify the electron geometry about each charged atom. Where appropriate, indicate the molecular geometry and approximate bond angle as well.

(a) 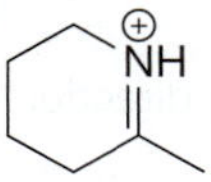 (b) 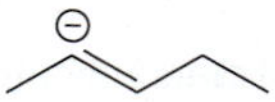 (c) 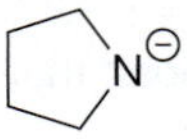 (d) 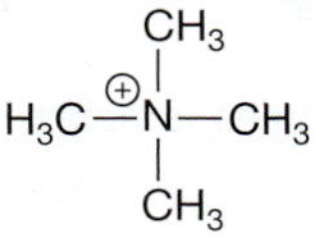 (e) 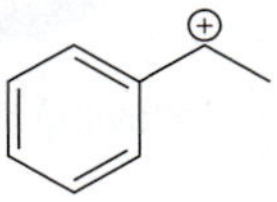 (f)

2.35 Which molecule, **A** or **B**, has greater angle strain? Explain.

A   B

2.36 Add dash–wedge notation to each line structure provided to accurately depict the ball-and-stick model appearing above it.

(a) 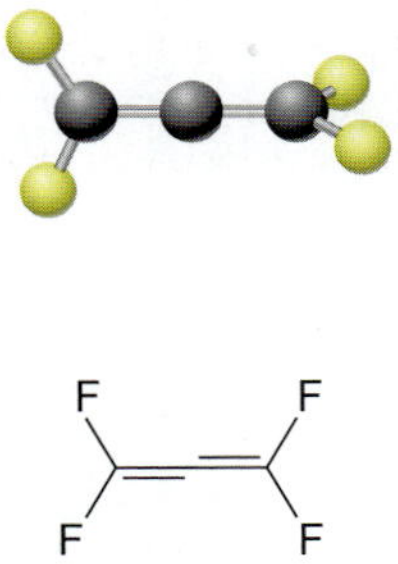 (b) 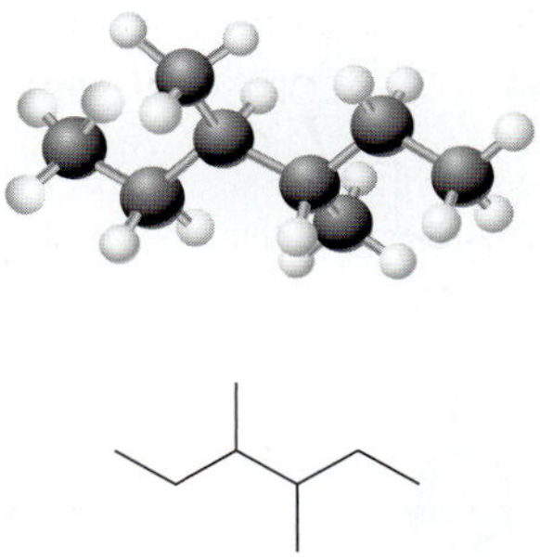 (c) 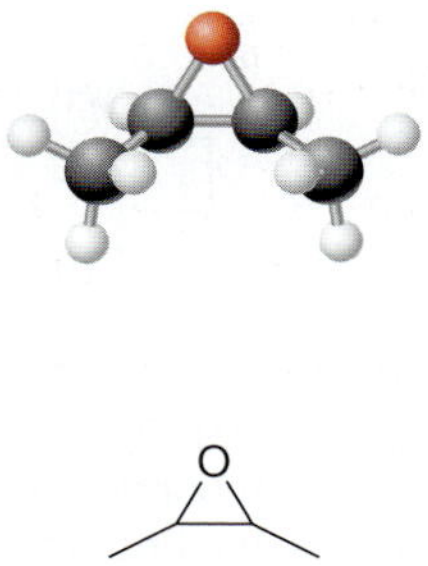 (d)

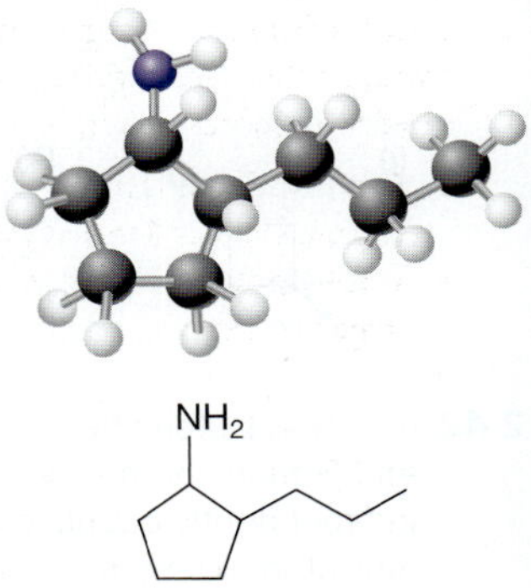

2.37 The structure of D-glucose using dash–wedge notation is shown below. Draw its structure using dash–wedge notation after each of the indicated rotations has taken place.

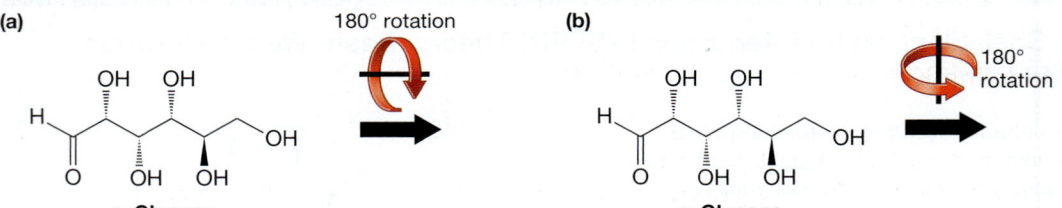

(a) 180° rotation

D-Glucose

(b) 180° rotation

D-Glucose

2.4 Net Molecular Dipoles and Dipole Moments

2.38 Rank BF_3, BF_2H, and BFH_2 from least polar to most polar.

2.39 The dipole moments of a ketone and an ester are given. Why is the magnitude of the ester's dipole moment smaller?

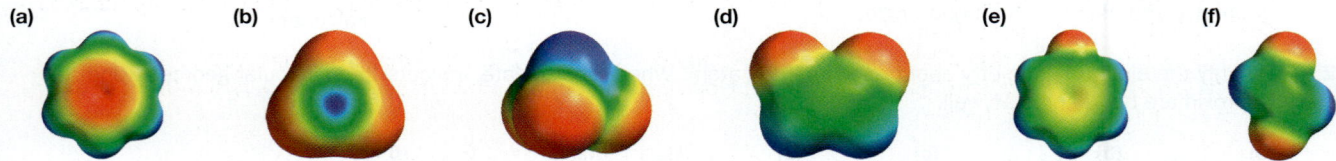

3.0 D   1.8 D

2.40 Use the following electrostatic potential maps of a variety of uncharged molecules to determine which ones are polar (i.e., have nonzero molecular dipole moments) and which ones are nonpolar. For each one that is polar, indicate the direction of the net molecular dipole moment. (You may want to review the color scheme on p. 17.)

(a)   **(b)**   **(c)**   **(d)**   **(e)**   **(f)**

2.41 Determine whether or not each of the following molecules is polar. For those that are polar, indicate the direction of the net dipole moment.

(a) F⌐⌐F   **(b)** F F   **(c)** Cl≡Cl   **(d)** Cl≡CH₃   **(e)** Cl≡Br

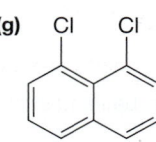   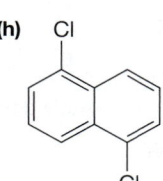

(f)   **(g)** Cl Cl   **(h)** Cl ... Cl   **(i)** Br ... Br Br

(j) Cl ... Cl   **(k)** Cl ... Cl   **(l)** Cl ... Cl   **(m)** Cl Cl Cl Cl   **(n)** Br Br Cl Cl

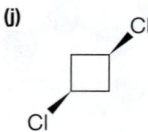   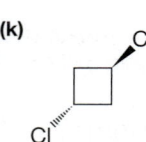   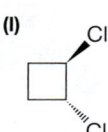   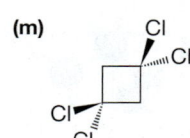   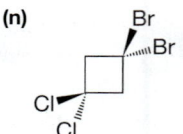

2.42 In 1874, Dutch chemist Jacobus van't Hoff (1852–1911) and French chemist Joseph Le Bel (1847–1930) independently deduced that a carbon atom bonded to four atoms assumes a tetrahedral geometry. Prior to that time, it was believed that tetravalent carbons assumed a square planar geometry. One piece of evidence that can be used to support a tetrahedral geometry is the fact that molecules with the general formula CX_2Y_2 (where X and Y are either a hydrogen or a halogen atom) are always polar. Explain how this supports a tetrahedral geometry and rules out a square planar geometry.

Tetrahedral geometry   **Square planar geometry**

2.6–2.7 Melting Points, Boiling Points, Solubility, and Intermolecular Interactions

2.43 Which ions or molecules would be attracted to CH_3^+? Indicate the type of intermolecular forces involved in the attraction. **(a)** H_2O; **(b)** Na^+; **(c)** Cl^-; **(d)** F^-; **(e)** $H_2C{=}O$

2.44 Which pair of ions will attract each other most strongly? Explain.

$$Mg^{2+}\ O^{2-} \qquad Al^{3+}\ O^{2-} \qquad Na^+\ O^{2-}$$

$$\textbf{A} \qquad\qquad \textbf{B} \qquad\qquad \textbf{C}$$

2.45 Would you expect 1,2-difluorobenzene or 1,3-difluorobenzene to have a higher boiling point? Explain.

1,2-Difluorobenzene   **1,3-Difluorobenzene**

2.46 Draw the complete Lewis structure of $(CH_3)_2CHCH(NH_2)CO_2H$ and identify all H-bond donors and all H-bond acceptors.

2.47 The substance with the lowest known boiling point (4 K) is helium, an atomic element that has two electrons. Hydrogen is a diatomic molecule and also has two electrons, but its boiling point is significantly higher, at 20.28 K.
(a) What is the dominant intermolecular force between a pair of helium atoms and a pair of H_2 molecules?
(b) Why do you think H_2 has a higher boiling point?

2.48 In this chapter, we did not discuss the interaction that is primarily responsible for the attraction that exists between an anion like CH_3O^- and a nonpolar molecule like Br_2.
(a) What name would be ascribed to the strongest interaction that exists between those two species?
(b) For which pair of species would you expect that type of attractive interaction to be stronger: between CH_3O^- and Br_2 or between CH_3O^- and I_2? Why?

2.49 Which organic solvent—**A**, **B**, **C**, or **D**—would be most soluble in water? Explain your choice.

$$CH_3CH_2CH_2CH_2CH_2CH_3 \qquad CH_3OCH_2CH_2CH_2CH_3 \qquad CH_3OCH_2CH_2OCH_3 \qquad CH_3CH_2CH_2CH_3$$

$$\textbf{A} \qquad\qquad\qquad \textbf{B} \qquad\qquad\qquad \textbf{C} \qquad\qquad\qquad \textbf{D}$$

2.50 In which of the following alcohol solvents do you think NaCl would be the most soluble?

$$CH_3OH \qquad CH_3CH_2OH \qquad CH_3CH_2CH_2OH \qquad CH_3CH_2CH_2CH_2OH$$

$$\textbf{A} \qquad\quad \textbf{B} \qquad\qquad \textbf{C} \qquad\qquad\quad \textbf{D}$$

2.51 The following amines have the same molecular formula ($C_5H_{13}N$), but their boiling points are significantly different. Explain why.

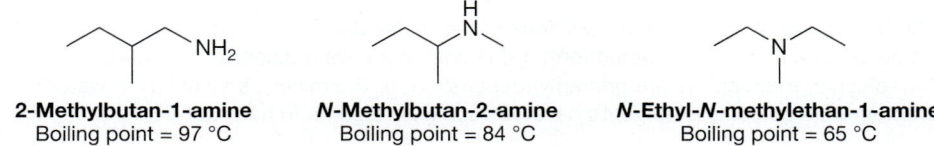

2-Methylbutan-1-amine   **N-Methylbutan-2-amine**   **N-Ethyl-N-methylethan-1-amine**
Boiling point = 97 °C   Boiling point = 84 °C   Boiling point = 65 °C

2.8 Ranking Boiling Points and Solubilities of Structurally Similar Compounds

2.52 Rank molecules **A–H** in order from lowest to highest boiling point.

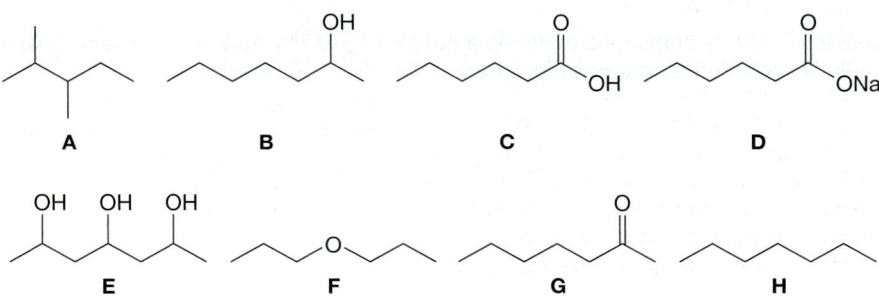

2.53 Rank molecules **A–D** in order from lowest to highest boiling point. Briefly rationalize your choice.

$$CH_3CH_2OCH_2CH_3 \qquad CH_3CH_2CH_2CH_2CH_3 \qquad CH_3CH(OH)CH_2CH_3 \qquad FCH_2CH_2OCH_2CH_3$$

$$\textbf{A} \qquad\qquad\qquad \textbf{B} \qquad\qquad\qquad \textbf{C} \qquad\qquad\qquad \textbf{D}$$

2.54 Rank molecules **A–D** in order from lowest to highest boiling point. Briefly rationalize your choice.

$$CH_3CH_2CH(OH)CH_2OH \qquad CH_3CH_2CH_2CH_2OH \qquad CH_3CH(OH)CH(OH)CH_2OH \qquad FCH_2CH_2CH_2CH_3$$

$$\textbf{A} \qquad\qquad\qquad \textbf{B} \qquad\qquad\qquad \textbf{C} \qquad\qquad\qquad \textbf{D}$$

2.9 Protic and Aprotic Solvents

2.55 Identify each compound below as either a *protic* solvent or an *aprotic* solvent.

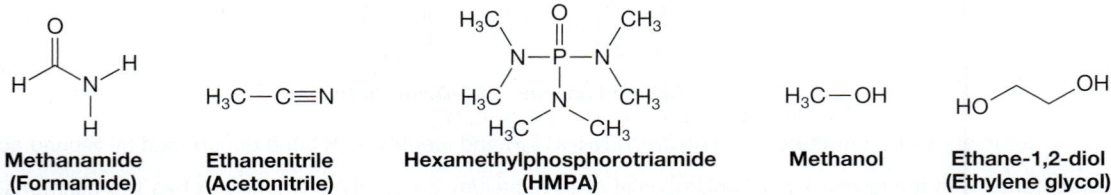

| Methanamide (Formamide) | Ethanenitrile (Acetonitrile) | Hexamethylphosphorotriamide (HMPA) | Methanol | Ethane-1,2-diol (Ethylene glycol) |

2.56 In which solvents in Problem 2.55 do you think NaCl will have a substantially greater solubility than in dimethyl sulfoxide (DMSO)? Explain.

2.57 Do you think NaCl is more soluble in acetone, $(CH_3)_2C{=}O$, or di-*tert*-butyl ketone, $[(CH_3)_3C]_2C{=}O$? Explain.

2.10 Soaps and Detergents

2.58 Recall from Section 2.10 that hard water ions like Ca^{2+} and Mg^{2+} bind more strongly to the carboxylate groups in soap than Na^+ and K^+ ions do. Why do you think this is so?

2.59 Recall from Section 2.10 that metal ions in hard water do not bind to the sulfate portion of detergents as strongly as they do to the carboxylate portion of soaps (alkyl sulfate = $R{-}OSO_3^-$; carboxylate = $R{-}CO_2^-$). Suggest a reason, based on intermolecular forces, for why this is so.

2.60 Hexadecyltrimethylammonium chloride, $CH_3(CH_2)_{15}N(CH_3)_3^+ \ Cl^-$, is one of a class of cationic detergents, commonly used in shampoos and as "clothes rinses."
(a) Identify the hydrophilic head group and the hydrophobic tail.
(b) Draw a depiction of a micelle that would form if this compound were dissolved in water.
(c) What are the intermolecular forces that are primarily responsible for the micelle's solubility in water?

2.61 Detergents need not be ionic. Pentaerythrityl palmitate (shown here) is a nonionic detergent used in dishwashing liquids.
(a) Identify the hydrophilic and hydrophobic portions of the molecule.
(b) Draw a depiction of a micelle that would form if this compound were dissolved in water.
(c) What intermolecular interactions are primarily responsible for the micelle's solubility in water?
(d) What advantages do nonionic detergents have over ionic detergents in hard water?

2.62 Would you expect the following compound to act as a soap? Why or why not? We explain in Chapter 3 that a chain composed of multiple, adjacent, single bonds is flexible. How does this new information affect your answer?

2.11 The Organic Chemistry of Biomolecules

2.63 The compound shown here is a terpenoid. **(a)** Circle the individual isoprene units. **(b)** Determine whether it is derived from a monoterpene, sesquiterpene, diterpene, etc.

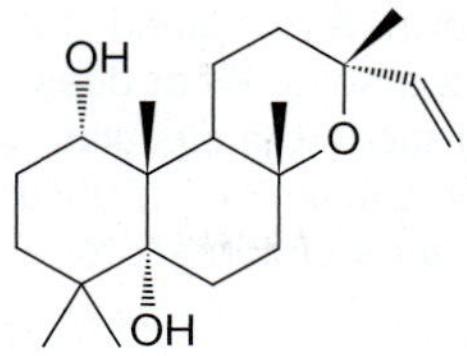

2.64 Draw a fat or oil molecule that is constructed from **(a)** three molecules of stearic acid and **(b)** two molecules of oleic acid and one molecule of linolenic acid.

2.65 Identify each of the following molecules as a steroid, a fatty acid, or a wax.

(a)

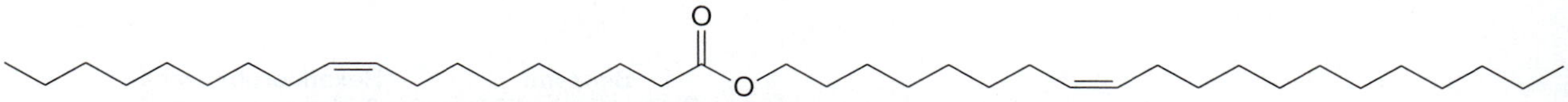

Isolated from the seeds of the jojoba plant

(b)

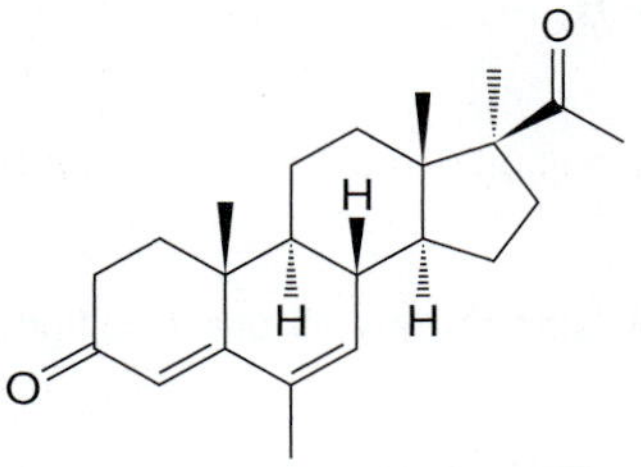

Medrogestone, a synthetic drug

(c)

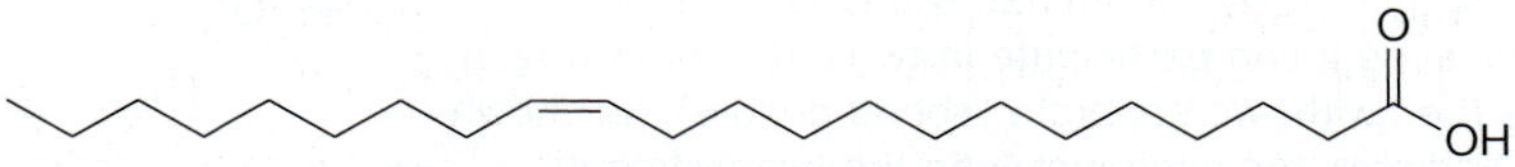

Erucic acid, isolated from mustard seed

Integrated Problems

2.66 Propan-1-ol ($CH_3CH_2CH_2OH$) and propan-2-ol [$(CH_3)_2CHOH$] are both alcohols that have the same formula (C_3H_8O) but significantly different boiling points: 97.5 °C for propan-1-ol and 82 °C for propan-2-ol. Explain.

2.67 The boiling points of several cyclic alkanes and ethers are listed in the table shown here. For small rings, there is a significant difference in boiling points between the cyclic alkane and the cyclic ether (e.g., −76 °C vs. 10.7 °C), but for large rings, the boiling points are nearly the same (e.g., 80.7 °C vs. 88 °C). Explain.

Cyclic Alkane	Boiling Point (°C)	Cyclic Ether	Boiling Point (°C)
	−76		10.7
	−13	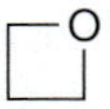	50
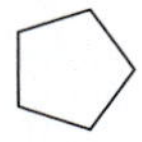	49	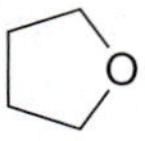	66
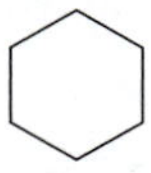	80.7	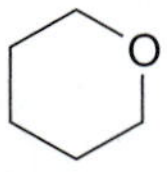	88

2.68 Naturally occurring unsaturated fatty acids, such as **A**, contain only cis double bonds. A corresponding saturated fatty acid (**B**) is also shown. **A** is a liquid at room temperature, whereas **B** is a solid. What does this suggest about the strength of the intermolecular forces in each substance? Can you explain? *Hint:* Build a model and compare the surface area of contact for a pair of each molecule.

Liquid

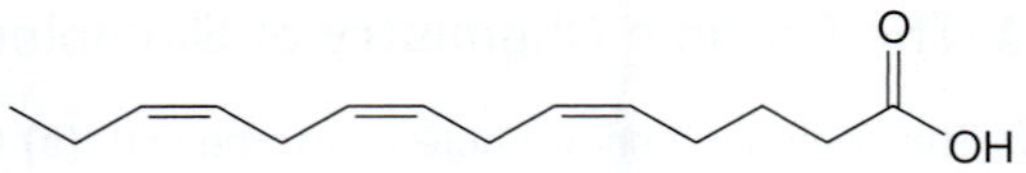

A

Solid

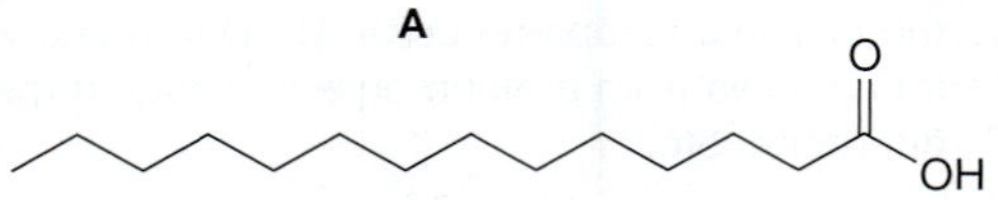

B

2.69 Benzene and hexafluorobenzene have nearly identical boiling points, despite the fact that hexafluorobenzene has significantly more total electrons than benzene. Why do you think this is so?

Benzene
Boiling point = 80 °C

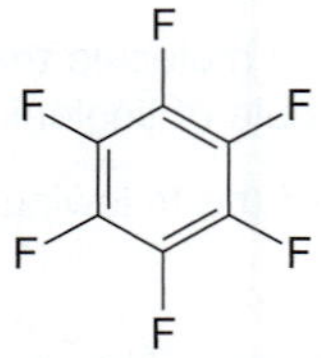

Hexafluorobenzene
Boiling point = 81 °C

2.70 An uncharged oxygen atom has two lone pairs of electrons, so it can participate in two different hydrogen bonds (i.e., with two separate H-bond donors), as shown here. Estimate the angle between the two hydrogen bonds.

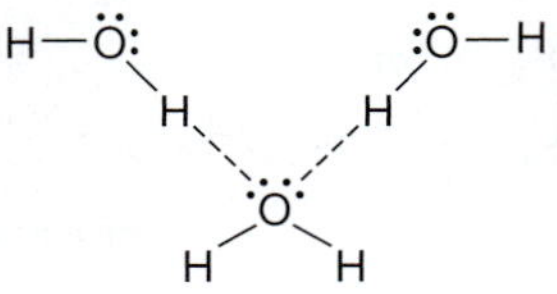

2.71 All of the following alcohols have the same molecular formula ($C_5H_{12}O$), but they have significantly different boiling points. Explain why.

Pentan-1-ol
Boiling point = 136–138 °C

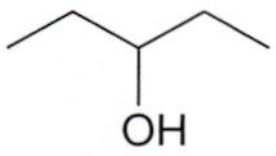

Pentan-3-ol
Boiling point = 114–115 °C

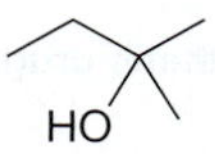

2-Methylbutan-2-ol
Boiling point = 102 °C

2.72 Explain why 1,2-dihydroxybenzene and 1,3-dihydroxybenzene have such different boiling points.

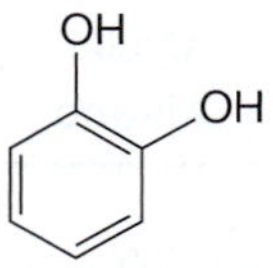

1,2-Dihydroxybenzene
Boiling point = 245 °C

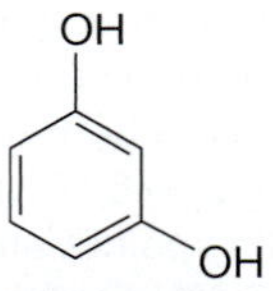

1,3-Dihydroxybenzene
Boiling point = 281 °C

Very small particles such as electrons behave as waves, much like the ones on a water surface. As we see in this chapter, their wavelike behavior is what gives rise to orbitals and bonds.

Orbital Interactions 1
Hybridization and Two-Center Molecular Orbitals

Despite their many uses, Lewis structures and VSEPR theory have shortcomings. We saw in Chapter 1, for example, that Lewis structures cannot account for *electron delocalization* over multiple atoms. Hence the need for resonance theory (Section 1.10), which treats the true structure of a molecule as a weighted average of all resonance contributors.

VSEPR theory, moreover, is incapable of predicting a molecule's *extended* geometry—the relative positions of atoms separated by more than one atom. For example, we know from experiment that a molecule of ethene (Fig. 3-1), $H_2C{=}CH_2$, is entirely planar—all six atoms lie in the same plane—and we also know that there is no rotation of one CH_2 group relative to the other. Although VSEPR theory does successfully predict that the geometry about each individual carbon atom is trigonal planar, it cannot predict the absence of rotation about the $C{=}C$ bond.

In this chapter, we present a more powerful model of bonding that combines the concept of *hybridization* with *molecular orbital theory*. This model describes how *atomic orbitals* (the orbitals that are occupied by electrons in isolated atoms) combine to form *molecular orbitals* (the orbitals occupied by electrons in molecules). Together, hybridization and molecular orbital theory allow us to understand and make predictions about molecular geometry that span several atoms instead of just one. These concepts also provide substantially more information about the *energetics* involved in bonding (which we show in Chapter 7 is essential for accurately describing the dynamics of chemical reactions), as well as various aspects of spectroscopy, which is how molecules interact with light (an important subject with many applications that we discuss in Chapters 15 and 16).

Ethene is a planar molecule; rotations do not occur about the $C{=}C$ bond.

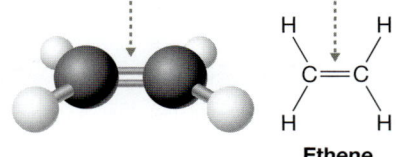

FIGURE 3-1 Ethene The above characteristics of ethene can be understood by orbital interactions.

On completing Chapter 3 you should be able to:

1. Explain the significance of an orbital.
2. Draw s and p atomic orbitals, along with their relative phase and orientation.
3. Explain the importance of molecular orbitals and describe how molecular orbitals can be generated from atomic orbitals via constructive interference and destructive interference; and identify the highest-occupied molecular orbital (HOMO) and the lowest-unoccupied molecular orbital (LUMO) in a molecule.
4. Identify whether a molecular orbital is bonding, nonbonding, or antibonding.
5. Identify whether a molecular orbital has either σ or π symmetry and explain how σ and π symmetry relate to rotations about bonds.
6. Explain the importance of hybridization and describe how hybrid atomic orbitals can be generated from pure

s and p atomic orbitals via constructive interference and destructive interference.

7. Specify the hybridization of an atom in a molecule, given only the molecule's Lewis structure.
8. Determine the total number of σ and π bonds in a molecule, given only its Lewis structure.
9. Construct the atomic orbital contribution picture of a molecule and its molecular orbital energy diagram.
10. Specify when cis and trans configurations exist for a given double bond and identify whether a configuration is cis or trans.
11. Use molecular modeling kits to exhibit rotational characteristics about bonds.
12. Explain how and why an atom's hybridization affects bond lengths and bond strengths, as well as how hybridization relates to the atom's effective electronegativity.

To understand molecular orbital theory, we must first discuss some important results from *quantum mechanics*. Not only does quantum mechanics dictate the characteristics of atomic orbitals, but it also provides the rules for constructing hybrid atomic orbitals and molecular orbitals.

3.1 Atomic Orbitals and the Wave Nature of Electrons

In Chapter 1, we briefly reviewed several aspects of quantum mechanics that you may have encountered in general chemistry, including the concept of an orbital and the meaning of quantum numbers. The central idea of **quantum mechanics** is that very small particles like electrons exhibit *wave–particle duality*:

Electrons have traits that are characteristic of both particles and waves.

An electron has characteristics of a particle because it has mass and can undergo collisions with other forms of matter. It also has characteristics of a wave because it can exhibit phenomena such as *constructive interference* and *destructive interference*. We discuss these phenomena later in this chapter. Here in Section 3.1, we explore other outcomes of an electron's wavelike behavior, including the electron's probability of being found at a particular location in space and its phase.

3.1a The 1s and 2p Orbitals: Probability and Phase

One crucial result of the electron's wave–particle duality is the **Heisenberg uncertainty principle**, which states that the uncertainty in a measurement of an electron's position is inversely proportional to the uncertainty in a measurement of its momentum. In other words:

We cannot know the exact position of an electron in an atom or molecule, but we can know the *probability* of finding the electron at a given location in space.

The probability of finding an electron at a certain position is dictated entirely by the **orbital** to which the electron belongs. For all orbitals, the probability of finding an electron at large distances away from the nucleus is very small but never equals zero, so it is convenient to represent an orbital as a surface that encompasses 90% of the total probability. Thus, we can think of an orbital as a "room" in which to house electrons. For a 1*s* orbital, that surface is a sphere about the nucleus (Fig. 3-2a), whereas for a 2*p* orbital, that surface is *dumbbell shaped*, making two *lobes* on either side of the nucleus (Fig. 3-2b). Although the lobes are actually distorted spheres, as shown on the left of Figure 3-2b, the 2*p* orbital is often simplified to the picture shown on the right of Figure 3-2b.

One of the main differences between the 1*s* and 2*p* orbitals is that the 2*p* orbital has a *node*, whereas the 1*s* orbital does not. In general, a **node** is a surface along which there is no probability of finding an electron. For the 2*p* orbital shown in Figure 3-2b, that surface is the vertical plane that comes out of the page; it is more accurately called a **nodal plane**.

The representations in Figure 3-2 also show that 1*s* and 2*p* orbitals have relative **phase**, which can be either positive or negative. The phase of an orbital is not a measurable quantity, but it does play an important role in how orbitals interact with one another (Section 3.2). Therefore, it is important to recognize whether the phases of two orbitals are the same or opposite in a particular region of space. In this book, we will consistently apply the following convention:

- An electron with positive phase in a particular region of space will be indicated by dark shading of its orbital there.
- Negative phase will be indicated by light shading.

Notice in Figure 3-2a that the entire 1*s* orbital is shaded darkly, indicating that the phase is the same everywhere, in this case positive. (Because phase cannot be measured, it would have been equally correct to show the orbital with negative phase.) In the 2*p* orbital (Fig. 3-2b), the lobe on the right is shaded darkly, indicating positive phase, and the one on the left is shaded lightly, indicating negative phase. (Again, because phase is not measurable, it would have been equally correct to depict the orbital with positive phase on the left and negative phase on the right.)

Unlike the 1*s* orbital, the 2*p* orbital can be assigned an *orientation*. In a given atom there are three mutually perpendicular 2*p* orbitals, called $2p_x$, $2p_y$, and $2p_z$ (Fig. 3-3),

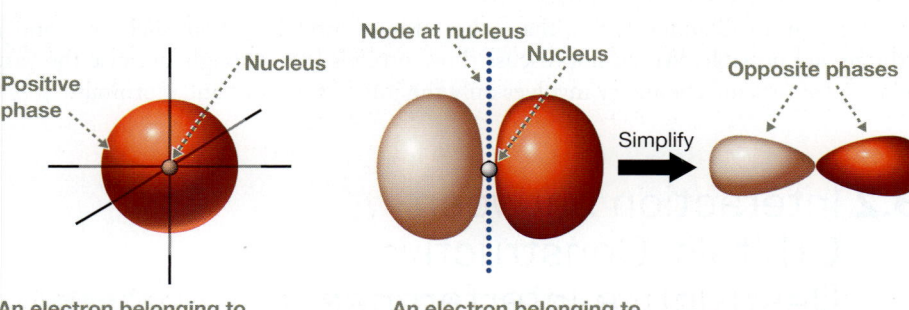

An electron belonging to a 1s orbital has a 90% probability of being found inside this sphere.

(a)

An electron belonging to a 2p orbital has a 90% probability of being found inside these lobes.

(b)

FIGURE 3-2 Representations of the 1s and 2p orbitals (a) The 1s orbital is spherical and has the same phase everywhere. (b) The 2p orbital consists of two spherical lobes of opposite phase, located on opposite sides of the nucleus. A simplified version is shown on the right.

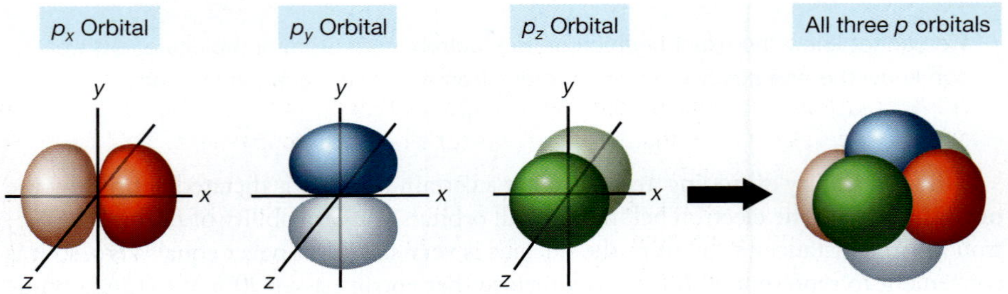

| p_x Orbital | p_y Orbital | p_z Orbital | All three p orbitals |

FIGURE 3-3 Orientations of the three 2p orbitals The p_x, p_y, and p_z orbitals are aligned along the x, y, and z axes, respectively. In the image at the far right, all three p orbitals are shown together.

aligned along the x, y, and z axes, respectively. Aside from their orientations in space, all three orbitals are exactly the same.

YOUR TURN 3.1

In Figure 3-3, draw the dotted line that would represent the nodal plane for the p_x and p_y orbitals. For the p_z orbital, is the nodal plane parallel or perpendicular to the plane of the page? _____

Answers to Your Turns are in the back of the book.

Before moving on, we summarize the important points about $1s$ and $2p$ orbitals:

- A 1s orbital:
    - is spherical, so it has no orientation.
    - has the same phase everywhere.
- A 2p orbital:
    - consists of two lobes with a nodal plane at the nucleus.
    - is opposite in phase on either side of the nucleus.
    - has a specific orientation that defines it as either p_x, p_y, or p_z.

3.1b Other Orbitals

As we saw in Chapter 1, the $1s$ and the $2p$ orbitals are the lowest-energy s and p orbitals, respectively. Higher-energy s and p orbitals (i.e., those belonging to higher shells) are larger, but the shape of an orbital is independent of the shell to which it belongs. That is, the $2s$ and $3s$ orbitals are both spherical, although the size of the orbitals increases in the order $1s < 2s < 3s$. Similarly, the $3p$ orbital consists of two lobes, each larger than the lobes of the $2p$ orbital. As a result, all s orbitals behave the same qualitatively and all p orbitals behave the same qualitatively.

Recall from Chapter 1, too, that other atomic orbitals (AOs), such as d and f orbitals, are possible. We do not discuss those orbitals here, though, because the vast majority of organic chemistry involves only the interactions of s and p orbitals.

3.2 Interaction between Orbitals: Constructive and Destructive Interference

Section 3.1 reviewed the main characteristics of electrons in s and p orbitals of an isolated atom. To account for covalent bonds and the various geometries in *molecules*, chemists often invoke *molecular orbitals* and *hybrid atomic orbitals*. Both are the

Quantum Teleportation

The teleportation of humans from one place to another is a popular plot device in science fiction movies. The basic idea is that a teleporter scans a person to extract all of his or her information, and that information is transmitted to a receiving location. The original person is disintegrated and, using the transmitted information, a perfect replica is constructed at the receiver. Is this possible?

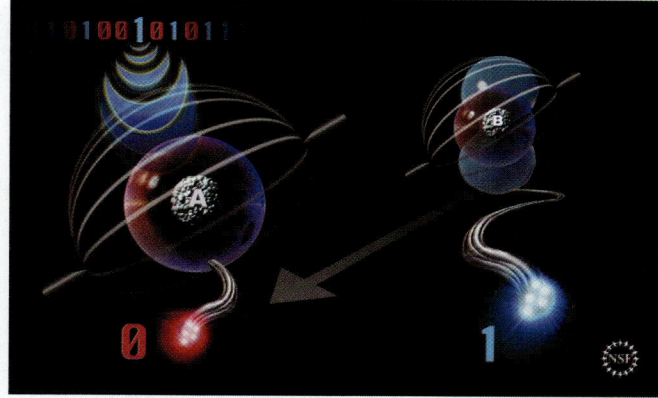

Historically, teleportation has not been taken seriously, because extracting all of an object's information using a measuring or scanning device was thought to be a violation of the Heisenberg uncertainty principle (Section 3.1a). According to the uncertainty principle, the more accurately an object's information is measured, the more the object's state is disrupted. Thus, the object would be altered before all of its information could be extracted.

Although teleporting macroscopic objects is still science fiction, scientists have found a way to teleport information between *entangled* subatomic particles—particles that first interact physically and are then separated. If a third particle—a target particle—comes in contact with one of the entangled particles, that entangled particle can act as a scanner. That entangled particle and the target particle are disrupted, and the second entangled particle assumes the original state of the target. Quite impressively, this quantum teleportation has been achieved over distances as great as 89 miles.

Quantum teleportation is more than just a novelty; it has potential applications in quantum computing, a field that is in its infancy but has the potential to revolutionize computing speed. In quantum computing, quantum particles are used to carry out operations. The resulting information, however, is too sensitive to be extracted by traditional means. Quantum teleportation may provide a mechanism by which to do so, thus helping to propel quantum computing out of its infancy.

product of orbital interactions involving two or more AOs and are outcomes of the wave nature of electrons.

To better understand such orbital interactions, consider two waves on a rope, generated by moving the ends of the rope back and forth rapidly. The waves can be on the same side of the rope (Fig. 3-4, next page), having the *same phase*, or they can be on opposite sides (Fig. 3-5, next page), having *opposite phases*. In both cases, the waves on the rope propagate toward each other and, when the waves overlap, they interact to produce a new wave. If the original waves had the same phase, they undergo **constructive interference** and the new wave is larger. Conversely, if the original waves were on opposite sides, they undergo **destructive interference** and the new wave is diminished in size or canceled entirely.

Similar to waves on a rope, electrons are waves and have phase, too. Unlike waves on a rope, where phase is established by the side on which the wave appears, the phase of an electron is depicted by whether its orbital is shaded darkly or lightly in a particular region of space. And, just like waves on a rope, when orbitals overlap in the same region of space, they undergo constructive and destructive interference according to their relative phases:

- Constructive interference occurs in regions of space where two orbitals have the same phase.
- Destructive interference occurs where the respective orbitals have opposite phases.

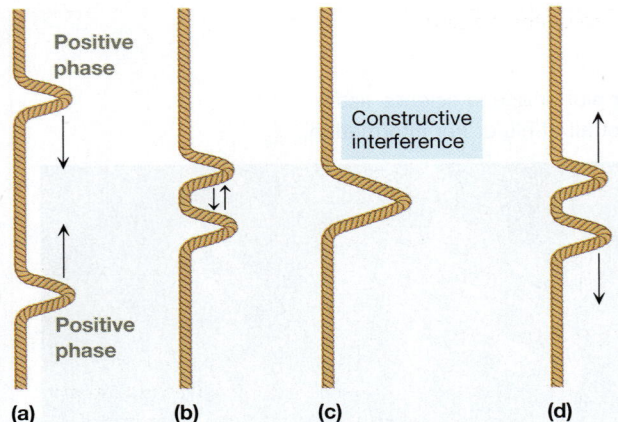

FIGURE 3-4 **Constructive interference between two waves along a rope** (a and b) Waves with the same phase (i.e., waves on the same side of the rope) propagate toward each other. (c) When the waves meet, they undergo *constructive interference*, temporarily creating a new wave with twice the amplitude of the original ones. (d) The waves continue to propagate toward opposite ends of the rope.

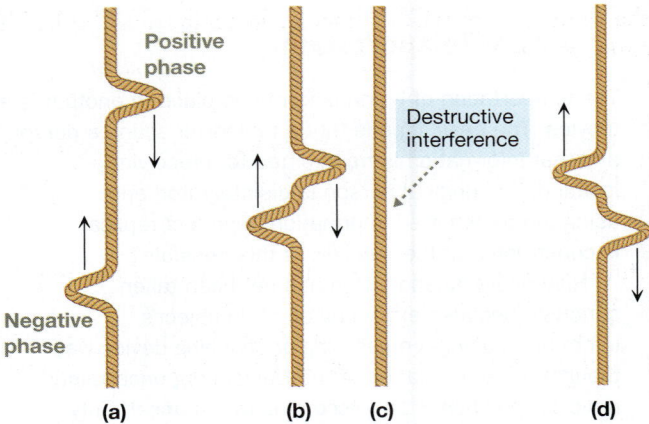

FIGURE 3-5 **Destructive interference between waves generated along a rope** (a and b) Waves with opposite phase (i.e., waves on opposite sides of the rope) propagate toward each other. (c) When the two waves meet, they undergo *destructive interference*, temporarily canceling each other. (d) The waves continue to propagate toward opposite ends of the rope.

As we see in the sections that follow, these types of interactions can generate orbitals different from those that are interacting—namely, *molecular orbitals* and *hybrid atomic orbitals*.

YOUR TURN 3.2

(a) Redraw Figure 3-4 with both initial waves generated on the left side of the rope to represent negative phases. **(b)** Redraw Figure 3-5, but this time generate the top wave on the left side of the rope (negative phase) and generate the bottom wave on the right side of the rope (positive phase).

3.3 An Introduction to Molecular Orbital Theory and σ Bonds: An Example with H_2

In Chapter 1, we saw that a covalent bond results from the sharing of electrons between atoms. AOs alone (i.e., *s*, *p*, etc.), however, cannot account for such electron sharing. To do so, we use *molecular orbital theory*.

The central concept of **molecular orbital theory** is that all electrons in a molecule can be thought of as occupying orbitals called **molecular orbitals (MOs)**. Like AOs, MOs can each accommodate up to two electrons. Therefore, the two electrons that form a single bond occupy one MO, the four electrons of a double bond occupy two MOs, and the six electrons of a triple bond occupy three MOs.

Molecular orbitals are constructed from the AOs of different atoms. When two atoms are brought close enough together (i.e., about a bond length apart), the AOs of one atom significantly overlap the AOs of the other atom, enabling them to undergo constructive and destructive interference, or *mix*, to produce new orbitals. Generating MOs in this way is often referred to as the **linear combination of atomic orbitals (LCAOs)**.

Although several interactions among AOs from two bonded atoms are imaginable, there are relatively few that must be considered because only AOs of roughly the same energy will interact to generate new MOs. This leads to a convenient simplification:

> The important orbital interactions from two atoms are usually limited to those involving just the *valence shell* AOs.

This is consistent with one of the rules for constructing Lewis structures: Covalent bonds involve the sharing of *valence electrons* only.

This principle applies to the MO picture of the H_2 molecule, in which the two hydrogen atoms are held together by a single covalent bond:

H—H

Hydrogen molecule

In each *isolated* hydrogen atom, the valence shell is the $n = 1$ shell, which contains only a $1s$ orbital. Therefore, only the $1s$ orbitals of the hydrogen atoms are considered when constructing the MOs of H_2.

One way in which these $1s$ orbitals can mix is with the same relative phase, as shown in Figure 3-6a. In this case, both $1s$ orbitals are darkly shaded, indicating the same phase, so *constructive interference* will result in an orbital that has been built up in the region of overlap.

A second way in which the two $1s$ orbitals can mix is with opposite phases, as shown in Figure 3-6b, where one AO is darkly shaded and the other is lightly shaded. This will lead to destructive interference, resulting in an orbital that has been diminished in the region of overlap. More specifically, there is a plane between the two nuclei in which the orbitals completely cancel, producing a *nodal plane*.

Can still other orbitals be produced by mixing two $1s$ orbitals, using combinations of phases that are different from those in Figure 3-6? The answer is no. The molecular orbitals shown in Figure 3-6 are the only unique ones produced. This concept is explored in Solved Problem 3.1 and Problem 3.2.

CONNECTIONS H_2 has a wide variety of uses industrially, such as in the processing of fossil fuels and the production of ammonia, NH_3. It is also used as a reducing agent in the synthesis of chemicals and as a fuel source in the hydrogen fuel cell.

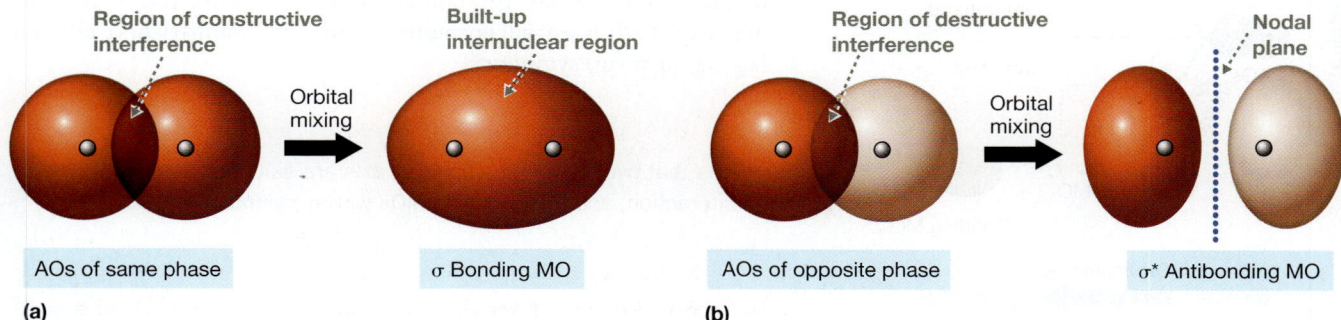

FIGURE 3-6 MO formation from two 1s orbitals (a) The interaction of two pure 1s AOs with the same phase leads to constructive interference between the two nuclei. The result is an elliptical MO, which has been built up in the internuclear region. This MO is designated a σ MO because the AO overlap occurs along the internuclear axis. (b) The interaction of two pure 1s AOs with opposite phase leads to destructive interference between the two nuclei. The result is a MO that has been diminished in the internuclear region, and possesses a nodal plane between the two nuclei. This MO is designated σ*.

SOLVED PROBLEM 3.1

This is a representation of two 1s orbitals overlapping, each with a phase that is opposite the ones in Figure 3-6a. **(a)** Draw the MO that will result from this orbital interaction. **(b)** Is the resulting MO unique compared to the one shown at the right of Figure 3-6a?

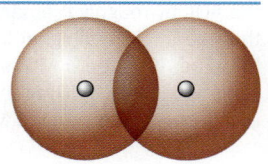

Think In the region in which the orbitals overlap, are the phases the same or opposite? Will that lead to constructive interference or destructive interference?

Solve Both orbitals are shaded lightly, indicating that they have the same relative phase: in this case, negative. Therefore, constructive interference will take place, and the resulting MO is one that has been built up in the overlap region.

The resulting MO is all a single phase (negative) and has precisely the same shape as the MO shown in Figure 3-6a. Because phase is not a measurable quantity, the two MOs are indistinguishable and thus are not unique.

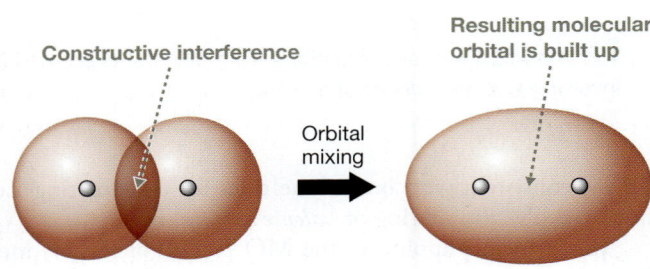

Constructive interference

Orbital mixing

Resulting molecular orbital is built up

PROBLEM 3.2 This is a representation of two 1s orbitals overlapping, each with a phase that is opposite the ones in Figure 3-6b. **(a)** Draw the MO that will result from this orbital interaction. **(b)** Is the resulting MO unique compared to the one shown at the right of Figure 3-6b?

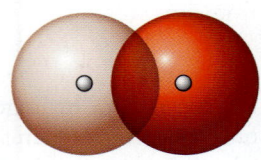

Looking back, we can see that when two 1s orbitals are mixed, two unique new orbitals are produced. This idea can be generalized as follows:

When *n* orbitals are mixed, *n* different orbitals must be produced.

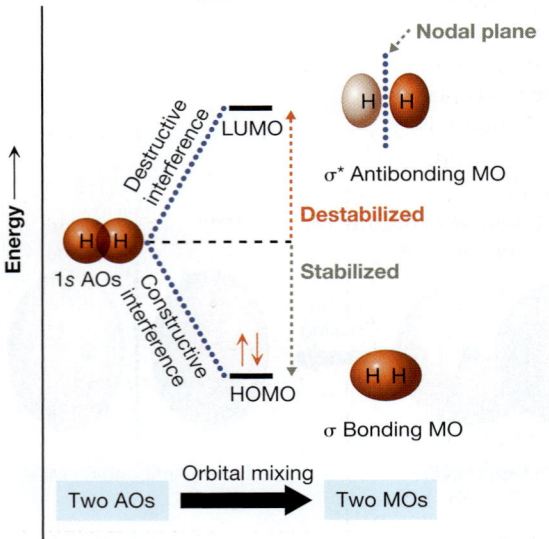

Nodal plane

LUMO

σ* Antibonding MO

Destabilized

Stabilized

Destructive interference

1s AOs

Constructive interference

HOMO

σ Bonding MO

Two AOs → Orbital mixing → Two MOs

Energy →

We can think of this as an effective **conservation of number of orbitals**, a concept that will prove to be quite useful throughout the rest of this chapter and in Chapter 14.

An important property of any MO is its *symmetry*, which is dictated by the locations in which the overlap of AOs takes place relative to the bonding axis—the line connecting the two nuclei. The vast majority of MOs we will encounter in organic chemistry have either σ (sigma) or π (pi) symmetry.

AOs that overlap *along* a bonding axis are said to undergo a σ interaction, and they produce MOs with **σ symmetry**.

In Section 3.6, we will see that this is not the case for MOs of π symmetry. With this in mind, notice that in both Figure 3-6a and 3-6b, the 1s orbitals overlap along the bonding axis, so each of the MOs produced indeed has σ symmetry.

FIGURE 3-7 MO energy diagram of H₂ The overlap of two 1s AOs from separate H atoms results in the formation of one σ MO and one σ* MO. The individual AOs are on the left and the resulting MOs are on the right. The σ MO is lower in energy than the AOs by roughly the same amount that the σ* MO is higher in energy than the AOs. The σ MO is the highest occupied MO (HOMO) and the σ* MO is the lowest unoccupied MO (LUMO).

Not only do the MOs in Figure 3-6 differ in shape, but they also differ in energy, as shown in Figure 3-7. Notice that the σ MO that is produced from 1s AOs having the same phase is *lower* in energy (more stable) than a 1s AO. This is because that MO, having undergone constructive interference, gives an electron a greater probability of being found in the internuclear region, where it is simultaneously attracted to both nuclei. On the other hand, the σ* MO produced from 1s AOs having opposite phase is *higher* in energy (less stable) than a 1s AO because the node largely excludes electrons from the internuclear region. In fact, σ* is destabilized slightly more than σ is stabilized.

Based on the energy changes that accompany the formation of MOs from AOs, we can describe MOs as *bonding*, *antibonding*, or *nonbonding*.

- A **bonding MO** is significantly lower in energy than its contributing AOs.
- An **antibonding MO** is significantly higher in energy than its contributing AOs.
- A **nonbonding MO** has about the same energy as its contributing AOs.

Therefore, the MO that is produced from 1s AOs of the same phase is a bonding MO, whereas the one produced from 1s AOs of opposite phase is an antibonding MO. Frequently, the asterisk or star symbol (*) is used to distinguish an antibonding MO from a bonding MO, so the orbitals that are produced can simply be called σ and σ* (sigma-star), as shown in Figure 3-7.

Once the MOs have been established for a molecule, they can be filled with electrons as usual, beginning with the lowest energy. For H_2, both electrons will occupy the σ MO, leaving the higher-energy σ* MO empty (Fig. 3-7). Thus, the single bond we see in H_2's Lewis structure represents a pair of electrons in the σ MO and can therefore be called a **σ bond** (Fig. 3-8).

The energy diagram in Figure 3-7 explains why the two H atoms in H_2 remain bonded together. In the molecule, the two electrons occupying the σ MO are lower in energy (*more stable*) than they would be in the 1s AOs of the isolated H atoms.

Two MOs, called the *highest occupied MO* (or *HOMO*) and the *lowest unoccupied MO* (or *LUMO*), are of particular interest in any molecule. As their names suggest, the **highest occupied molecular orbital** is the highest-energy MO that contains an electron, whereas the **lowest unoccupied molecular orbital** is the lowest-energy MO that is empty. In H_2, the HOMO is the σ MO and the LUMO is the σ* MO (Fig. 3-7).

Two electrons occupying a σ MO

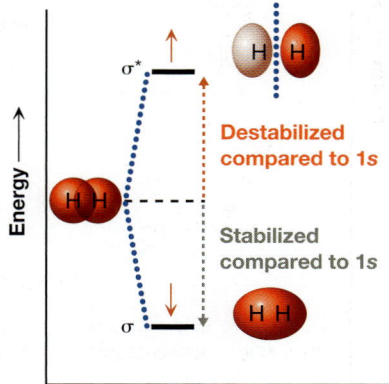

σ Bond

H——H

(a)   (b)

FIGURE 3-8 Comparison between H_2's MO picture and its Lewis structure (a) The MO picture of the H_2 molecule showing just the occupied bonding MO. (b) The Lewis structure of H_2, indicating that the two bonding electrons occupy a σ MO.

SOLVED PROBLEM 3.3

In Chapter 15, we will learn that light can promote an electron to a higher-energy MO. Suppose that light is used to promote one of the two electrons in the H_2 molecule from the σ MO to the σ* MO, as shown at the right. Is this state more stable, less stable, or about the same as the two separated hydrogen atoms?

Think Is an electron in the σ MO stabilized or destabilized compared to an electron in the 1s AO? Is an electron in the σ* MO stabilized or destabilized relative to an electron in the 1s AO?

Solve Compared to electrons in the 1s AO, one electron is stabilized in the σ MO, whereas the other is destabilized by a slightly greater amount in the σ* MO. When the energies of the electrons are added up, the H_2 molecule in the state described has an energy similar to (though slightly higher than) the isolated atoms.

Energy ——→

σ*

Destabilized compared to 1s

Stabilized compared to 1s

σ

YOUR TURN 3.3

In the H_2 molecule described in Solved Problem 3.3, what is the HOMO?

PROBLEM 3.4 Suppose one electron is removed from the H_2 molecule, leaving H_2^+. Draw the energy diagram similar to Figure 3-7 for this species. Is H_2^+ more stable, less stable, or about the same as the isolated H atom and H^+ ion?

3.4 Hybrid Atomic Orbitals and Geometry

As we saw in Chapter 2, the geometries about atoms commonly found in organic compounds are linear, trigonal planar, or tetrahedral, forming bond angles of 180°, ~120°, and ~109.5°, respectively. To account for these geometries, chemists often invoke the concept of *hybridization*, in which *hybrid* AOs are involved in bonding. A **hybrid atomic orbital** is a cross between two or more pure AOs from the valence shell of a single atom. Typically, the 2s and the 2p orbitals are the ones involved in hybridization, resulting in one of three types of hybrid AOs: sp, sp^2, or sp^3. We examine each type of hybridization in turn.

3.4a *sp* Hybrid Orbitals

To account for linear geometries about atoms (180° bond angles), chemists invoke sp hybrid atomic orbitals.

> Two **sp hybrid atomic orbitals** are produced from mixing the single s orbital and one of the three p orbitals—p_x, p_y, or p_z—from the valence shell.

Notice that the number of orbitals is conserved in the mixing process. Moreover, each sp hybrid orbital that is formed has 50% **s character** and 50% **p character**. In other words, an sp hybrid orbital is distinct from either an s orbital or a p orbital, but its characteristics (e.g., shape and energy) resemble both orbitals equally.

Figure 3-9 shows how you can imagine two sp hybrid AOs being produced from the mixing of an s orbital with a p_x orbital (chosen arbitrarily over the p_y and p_z orbitals). In Figure 3-9a and 3-9b, the phase of the s AO is the same, but p AO phases are reversed.

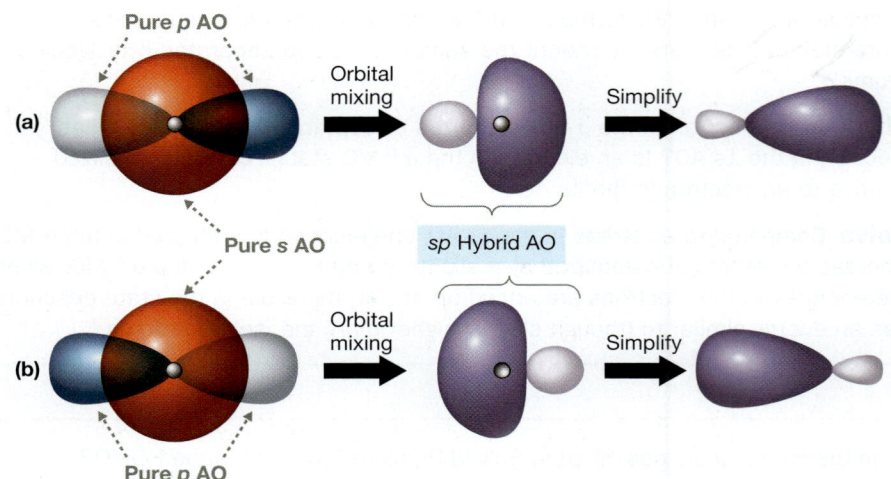

FIGURE 3-9 Generation of two sp hybrid orbitals from pure AOs An s orbital and a p orbital of the same atom can mix with different phase combinations. (a) Constructive interference (same phase of the orbitals) occurs to the right of the nucleus and destructive interference (opposite phases) occurs to the left, so the resulting sp hybrid orbital (middle) becomes larger on the right and smaller on the left. (b) Constructive interference occurs to the left and destructive interference occurs to the right, so the resulting sp hybrid orbital (middle) becomes larger on the left and smaller on the right. Simplified versions of the hybrid orbitals (right) are often shown without the small lobe.

On the left side of Figure 3-9a, notice that the right lobe of the contributing *p* orbital has the same phase as the contributing *s* orbital (both are darkly shaded). Therefore, *constructive interference* takes place to the right of the nucleus. The resulting *sp* hybrid orbital (middle) is built up to the right of the nucleus, denoted by the larger lobe. Notice, too, that the contributing *s* and *p* orbitals are opposite in phase to the left of the nucleus (the *s* orbital is darkly shaded but the lobe of the *p* orbital is lightly shaded), so *destructive interference* takes place there. The resulting *sp* hybrid orbital is diminished to the left of the nucleus.

YOUR TURN 3.4

In Figure 3-9a, circle and label the areas of constructive and destructive interference.

The *sp* hybrid orbital that is generated from this process (Fig. 3-9a, middle) has two lobes that differ in size (one is much larger than the other) and are opposite in phase, and the larger lobe contains the nucleus. For convenience, we usually draw the simplified version at the right of Figure 3-9a, and often the small lobe is omitted entirely for simplicity. (The small lobe, however, can play a vital role in chemical reactions, as discussed in Interchapter D.)

In Figure 3-9b, the left lobe of the contributing *p* orbital has the same phase as the contributing *s* orbital, whereas the right lobe has the opposite phase. Consequently, constructive interference takes place to the left of the nucleus and destructive interference to the right. As a result, the *sp* hybrid orbital has the large lobe on the left and the small lobe on the right. Once again, for convenience, we usually draw the orbital as shown at the right of Figure 3-9b and often omit the small lobe entirely.

YOUR TURN 3.5

In Figure 3-9b, circle and label the areas of constructive and destructive interference.

The two *sp* hybrid orbitals we have presented are the only unique orbitals that can be generated by mixing a 2*s* orbital with a 2*p*$_x$ orbital. We can conceive of other linear combinations of the contributing *s* and *p* orbitals (see Problems 3.5 and 3.6), but the hybrid orbitals that are generated are redundant with the ones in Figure 3-9.

PROBLEM 3.5 Derive the hybrid orbital that would result from the interaction illustrated at the right, in which the phases of the s and p$_x$ orbitals are opposite those in Figure 3-9a. Is the resulting orbital different from the ones in Figure 3-9a and 3-9b? Explain.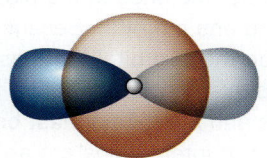

PROBLEM 3.6 Derive the hybrid orbital that would result from the interaction illustrated at the right, in which the phases of the s and p$_x$ orbitals are opposite those in Figure 3-9b. Is the resulting orbital different from the ones in Figure 3-9a and 3-9b? Explain.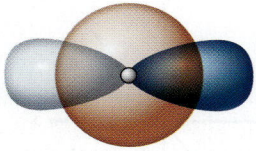

The only difference between the two *sp* hybrid orbitals in Figure 3-9 is the direction the larger lobes point in space—they point 180° apart. When the *p*$_x$ orbital is used for hybridization, the *sp* hybrid orbitals are aligned along the

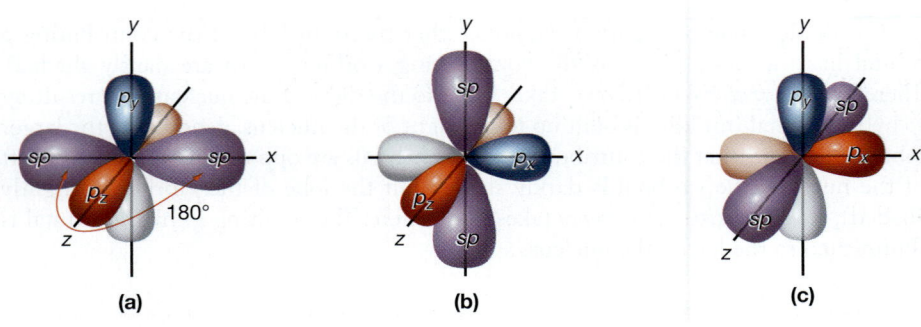

FIGURE 3-10 Three different but equivalent depictions of an sp-hybridized atom (a) An sp-hybridized atom has two sp hybrid AOs directed 180° apart and two unhybridized p orbitals. In this depiction, the sp hybrid orbitals are aligned along the x axis, indicating that the p_x orbital was used for hybridization. (b) In this depiction, the p_y orbital was used for hybridization. (c) In this depiction, the p_z orbital was used for hybridization. In all cases, the small lobes of the hybrid AOs are omitted for clarity.

x axis. The p_y and p_z orbitals that are left alone during the hybridization process remain perpendicular to the x axis, as well as perpendicular to each other, as shown in Figure 3-10a. An atom that exhibits these valence orbitals is said to be *sp-hybridized*.

In the previous treatment, the p_x orbital was *arbitrarily* chosen to be the p orbital used for hybridization. Figure 3-10b and 3-10c show what sp-hybridized atoms would look like if the $2p_y$ or $2p_z$ orbital, respectively, was used for hybridization instead. The only difference among all three cases is the orientation of the orbitals. Therefore, we can make the following generalizations about an sp-hybridized atom:

- Any sp-hybridized atom has two sp hybrid orbitals and two unhybridized p orbitals.
- The two sp hybrid orbitals are aligned along the same axis and point in opposite directions.
- The unhybridized p orbitals are aligned along axes perpendicular to the axis containing the sp hybrid orbitals.

SOLVED PROBLEM 3.7

Consider the pair of sp hybrid orbitals at the right. Draw the orbital interaction that would be necessary to generate each sp hybrid orbital from pure AOs.

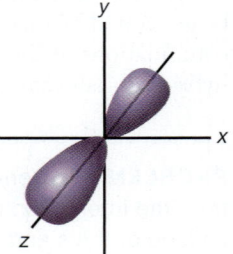

Think Along which axis are these sp hybrid orbitals aligned? What orbitals must therefore interact? What is the phase of the large lobe of each hybrid orbital?

Solve Like any pair of sp hybrid orbitals, the ones in the diagram must result from the interaction between the s and one p orbital. Because the sp hybrid orbitals are aligned along the z axis, it must be the p_z orbital that has been used for hybridization, leaving the p_x and p_y orbitals unhybridized. The large lobe is shaded darkly in both hybrid orbitals, so the s and p_z orbitals must both be shaded darkly where the overlap occurs. The respective interactions are as follows:

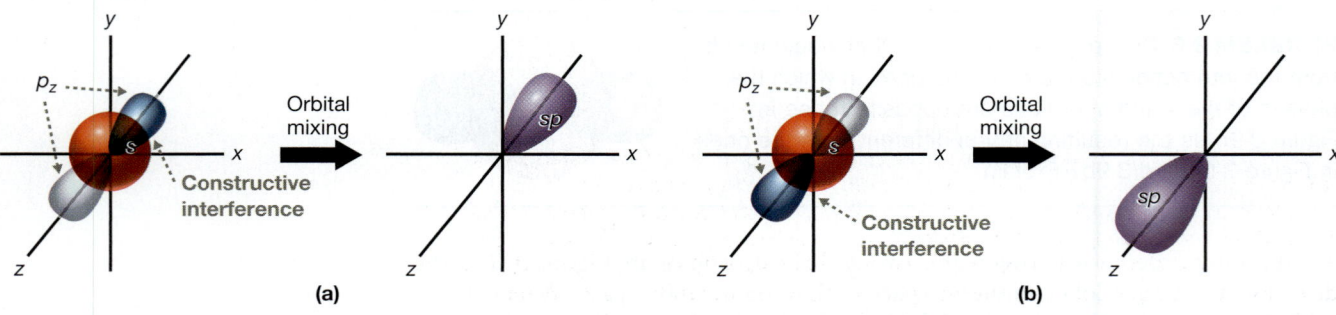

PROBLEM 3.8 Consider this pair of *sp* hybrid orbitals. Draw the orbital interaction that would be necessary to generate each *sp* hybrid orbital from pure AOs.

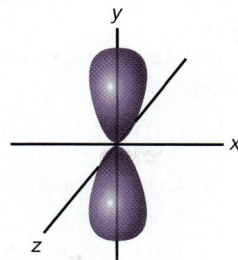

3.4b *sp²* Hybrid Orbitals

Just as *sp* hybrids enable us to conceptualize the AOs of an atom with a linear geometry, *sp²* hybrids do so for an atom with a trigonal planar geometry.

> Three **sp² hybrid atomic orbitals** are produced from mixing the single s orbital and two of the three p orbitals from the valence shell.

Each *sp²* hybrid AO has one-third *s* character and two-thirds *p* character. An *sp²* hybrid orbital, therefore, resembles a *p* orbital more than it does an *s* orbital.

The formation of *sp²* hybrid AOs involves constructive and destructive interference among three orbitals simultaneously. The resulting *sp²* hybrid AOs are similar to *sp* hybrid orbitals in the following ways:

- Each sp² hybrid orbital consists of a large lobe and a small lobe, opposite in phase to each other.
- All three sp² hybrids are identical to one another in shape, but have different orientations in space.

Whereas the two *sp* hybrid orbitals point 180° apart, the three *sp²* hybrid orbitals point 120° apart and are all in the same plane. See Figure 3-11, which shows the three

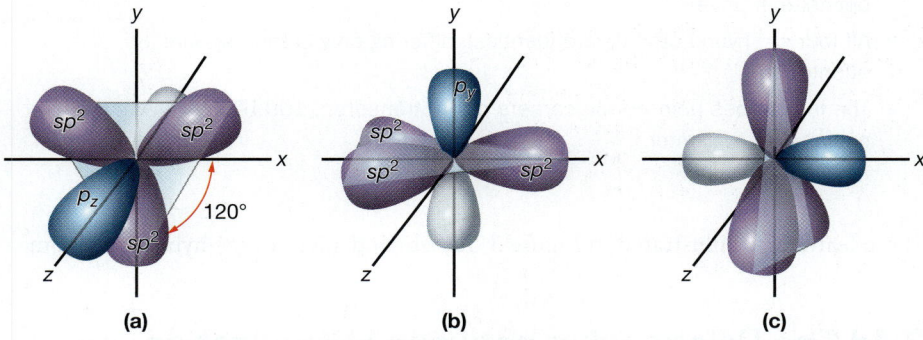

(a)   (b)   (c)

FIGURE 3-11 Three different but equivalent depictions of an *sp²*-hybridized atom
(a) The four valence AOs of an sp²-hybridized atom in which the p_x and p_y orbitals were used for hybridization. The three sp² hybrid AOs are 120° apart and occupy the xy plane, perpendicular to the unhybridized p_z AO. (b) An sp²-hybridized atom in which the p_x and p_z orbitals were used for hybridization, leaving the p_y orbital unhybridized. (c) An sp²-hybridized atom in which the p_y and p_z orbitals were used for hybridization, leaving the p_x orbital unhybridized. In all cases, the small lobes of the hybrid AOs are omitted for clarity.

different possible combinations of s and p orbitals that can form an sp^2-hybridized atom.

Based on Figure 3-11, the following generalizations are true for an sp^2-hybridized atom:

- Any sp^2-hybridized atom has three sp^2 hybrid orbitals and one unhybridized p orbital.
- The three sp^2 hybrid orbitals are all in the same plane, pointing to the corners of a triangle; they are *trigonal planar*.
- The unhybridized p orbital is perpendicular to the *plane* that contains the hybrid orbitals.

YOUR TURN 3.6

In Figure 3-11c, label each orbital as either sp^2, p_x, p_y, or p_z.

3.4c sp^3 Hybrid Orbitals

Whereas sp and sp^2 hybrid orbitals are invoked to explain atoms having linear and trigonal planar geometries, respectively, sp^3 hybrid orbitals are invoked to explain the bonding in atoms with tetrahedral geometries.

Four **sp^3 hybrid atomic orbitals** are produced from mixing the single s orbital and all three p orbitals from the valence shell.

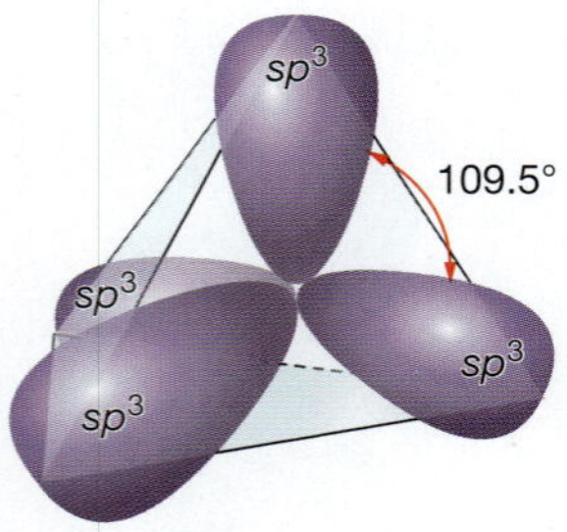

Each sp^3 hybrid AO, therefore, has one-fourth s character and three-fourths p character. Drawing on what we learned from the other types of hybridization, we can state the following key points about sp^3 hybrid orbitals.

- Each sp^3 hybrid orbital consists of a large lobe and a small lobe, which are opposite in phase.
- All four sp^3 hybrid orbitals are identical, differing only in their spatial orientation.
- The large lobes point to the corners of a tetrahedron, 109.5° apart, with the nucleus at the center.

These features are illustrated in Figure 3-12, which depicts an sp^3-hybridized atom.

FIGURE 3-12 The four sp^3 hybrid orbitals of an sp^3-hybridized atom
No unhybridized orbitals remain in the valence shell. The large lobes of the sp^3 hybrid orbitals point to the corners of a tetrahedron, so they are 109.5° apart. The small lobes of the hybrid AOs are omitted for clarity.

3.4d The Relationship between Hybridization and VSEPR Theory

We introduced hybridization as a way to account for the different geometries atoms can have. As Table 3-1 shows, the geometries defined by the hybrid AOs are what specifically correlate with the various VSEPR geometries.

Using VSEPR theory, then, we can deduce an atom's electron geometry from its total number of electron groups, and from its electron geometry we can infer its hybridization. This process is demonstrated in Solved Problem 3.9.

SOLVED PROBLEM 3.9

What is the hybridization of the N atom in this molecule?

Think What is the electron geometry about the N? What is the relationship between electron geometry and hybridization?

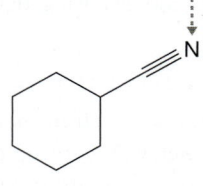

Solve According to VSEPR theory, the N atom is surrounded by two electron groups—a lone pair and a triple bond. The electron geometry, therefore, is linear. According to Table 3-1, the atom must be *sp* hybridized.

TABLE 3-1 The Relationship between Hybridization and VSEPR Theory

Atom's Hybridization	Orientation of Hybrid Orbitals	Bond Angles	Electron Geometry in VSEPR Theory
sp		180°	Linear
sp²		120°	Trigonal planar
sp³		109.5°	Tetrahedral

PROBLEM 3.10 What is the hybridization of the atom indicated in each of the following molecules?

(a) (b) (c) (d) (e)

PROBLEM 3.11 **(a)** Draw the structure of a molecule with the formula C_5H_8 that has two *sp*-hybridized atoms. **(b)** Draw the structure of another molecule with that formula that has four *sp²*-hybridized atoms.

3.5 Valence Bond Theory and Other Orbitals of σ Symmetry: An Example with Ethane (H_3C—CH_3)

According to the following Lewis structure, ethane contains seven single bonds, representing a total of 14 valence electrons:

$$H_3C-CH_3$$

Ethane

In what kinds of orbitals do those electrons reside?

One way to answer this question is by applying MO theory, in which all pure AOs (i.e., *s*, *p*, etc.) from all atoms interact simultaneously. Alternatively, we can simplify

CONNECTIONS The principal use of ethane is in the industrial production of ethene, $H_2C=CH_2$, which has a wide variety of uses. As a liquid, ethane also has use in cryogenics because of its ability to freeze samples more quickly than liquid nitrogen.

the picture by applying **valence bond (VB) theory**, the basis of which can be stated as follows:

> • Atoms that are bonded to two or more other atoms are considered to be hybridized.
> • Hybridized atoms contribute their valence orbitals for mixing, which can include hybrid and pure AOs.
> • New orbitals are produced by mixing just two AOs at a time, one from each of two adjacent atoms.

To apply these ideas to ethane, recall from VSEPR theory that each carbon atom has a tetrahedral electron geometry. According to Table 3-1, then, each carbon atom is sp^3 hybridized. Therefore, each carbon atom contributes four sp^3 hybrid orbitals that are arranged tetrahedrally about the nucleus (Section 3.4c), as shown in Figure 3-13. Notice that there are seven orbital interactions involving AOs from adjacent nuclei. Six interactions involve the overlap between an sp^3 hybrid orbital from C and a $1s$ orbital from H (purple with red), and one interaction involves the overlap between two sp^3 hybrid C orbitals (purple with purple).

All seven of these interactions are σ interactions because the overlap of AOs occurs *along* a bonding axis. Similar to H_2 (Fig. 3-7, p. 126):

> Each σ interaction adds both a σ and a σ* MO to the molecule.

Therefore, seven of the new orbitals are σ and seven are σ*, as illustrated in Figure 3-14.

The aufbau principle (Chapter 1) dictates that the 14 valence electrons fill the seven lowest-energy orbitals available to them—the seven σ MOs (Fig. 3-14, right). The σ* MOs remain unoccupied. Therefore, the HOMO must be the σ MO with the highest energy, whereas the LUMO must be the σ* MO with the lowest energy.

The energy diagram in Figure 3-14 explains why the atoms in ethane remain bonded together. The 14 valence electrons in the σ MOs are lower in energy (more stable) than they would be in the AOs of the isolated atoms.

Notice that there are seven filled σ MOs in ethane, the same as the number of single bonds that appear in its Lewis structure. In general:

> Each single bond in a Lewis structure corresponds to a filled σ MO.

For this reason, it is common to refer to each single bond as a σ bond (Fig. 3-15).

FIGURE 3-13 AO overlap in ethane (a) Dash–wedge notation for a molecule of ethane, H_3C—CH_3, showing that each C atom is sp^3 hybridized. (b) The AOs contributed by each atom and their interaction in the ethane molecule. The seven pairs of interacting orbitals produce 14 MOs of σ symmetry: seven bonding and seven antibonding.

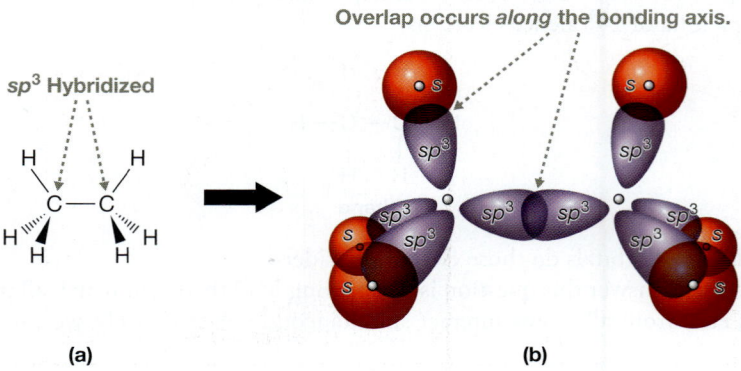

(a)   **(b)**

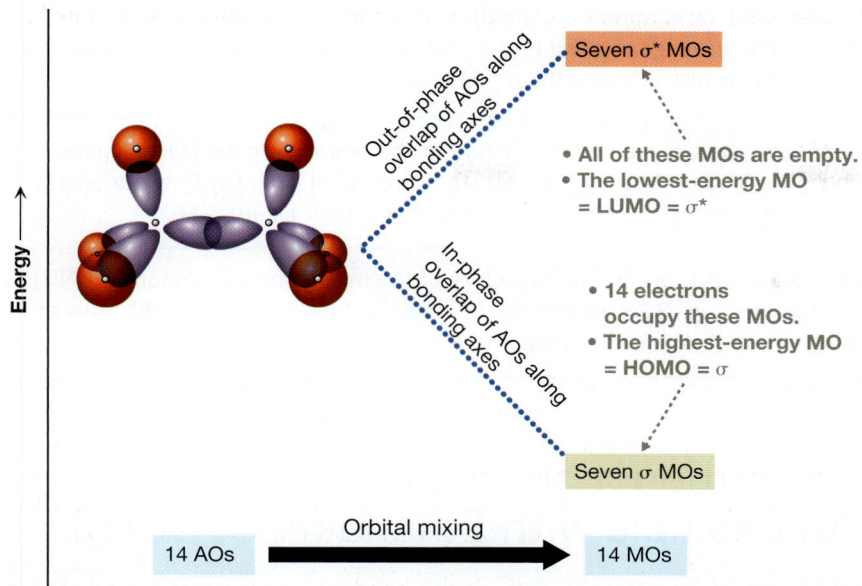

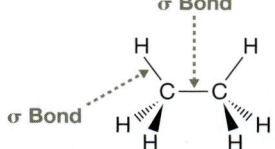

FIGURE 3-14 Energy diagram for the formation of ethane from its constituent atoms The individual AOs are shown on the left. Interaction among those orbitals generates 14 MOs of σ symmetry: seven bonding (σ) MOs and seven antibonding (σ*) MOs (right).

FIGURE 3-15 Bonds and Lewis structures In a Lewis structure, a single bond corresponds to a σ bond. The C—C bond and the six C—H bonds in ethane, therefore, are all σ bonds.

SOLVED PROBLEM 3.12

(a) Draw an orbital picture of methane (CH_4), similar to that in Figure 3-13, indicating the important overlap of AOs. **(b)** Draw an energy diagram for this molecule, similar to the energy diagram in Figure 3-14.

Think What is the hybridization of C? What valence shell orbitals does it contribute? What orbitals do the H atoms contribute? How many total orbitals should be produced from AO mixing?

Solve The C atom is sp^3 hybridized, so it contributes four sp^3 hybrid orbitals in the valence shell. Each H atom contributes a $1s$ orbital. The overlap of AOs appears on the left in the following diagram:

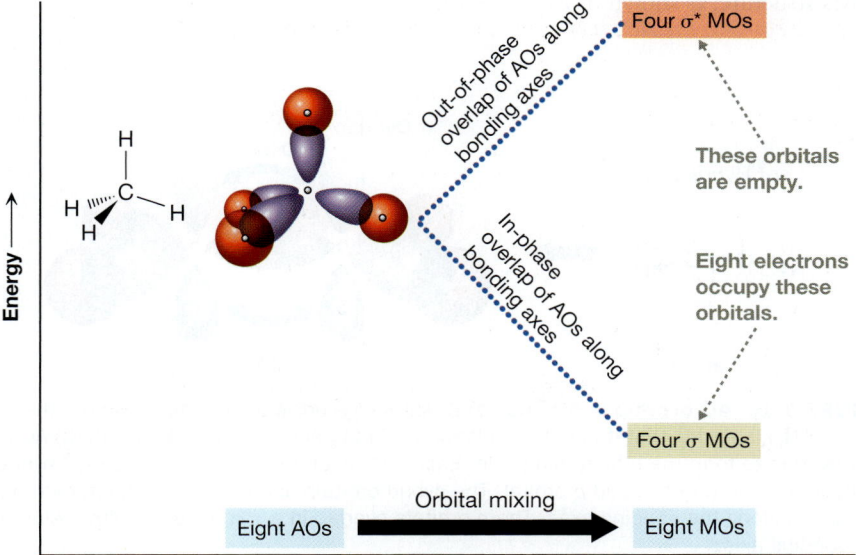

CONNECTIONS Methane is the principal fuel in natural gas, which accounts for roughly one-quarter of all energy usage in the United States. Methane is also the starting material used in industry to produce hydrogen gas and carbon monoxide. Care must be taken to avoid the release of methane into the atmosphere because it is a greenhouse gas at least 30 times more potent than carbon dioxide.

The energies of the MOs appear on the right. The eight AOs that overlap do so along the bonding axes, generating four σ and four σ* MOs. Eight valence electrons completely fill the σ MOs, leaving the σ* MOs empty.

PROBLEM 3.13 **(a)** Draw an orbital picture of the ammonium ion (NH_4^+) similar to that in Figure 3-13, indicating the important overlap of AOs. **(b)** Draw an energy diagram for this species, similar to the energy diagram in Figure 3-14.

PROBLEM 3.14 **(a)** Draw an orbital picture of the methylammonium ion ($CH_3NH_3^+$) similar to that in Figure 3-13, indicating the important overlap of AOs. **(b)** Draw an energy diagram for this species, similar to the energy diagram in Figure 3-14.

3.6 An Introduction to π Bonds: An Example with Ethene (H_2C═CH_2)

Ethene (H_2C═CH_2) contains a C═C double bond, which represents a total of four electrons. What kinds of orbitals do they occupy?

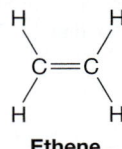

Ethene

We begin by identifying the hybridization of each C atom. According to VSEPR theory, each C atom has a trigonal planar electron geometry, which, according to Table 3-1, makes it sp^2 hybridized. Thus, three hybrid AOs are arranged in a trigonal planar fashion about each C nucleus, as shown in purple in Figure 3-16. A single *p* orbital (blue) on each C atom is left unhybridized. Each is perpendicular to the plane defined by the three hybrid orbitals (recall Fig. 3-11, p. 131).

Based on Figure 3-16b, there are five pairs of adjacent AOs that interact along the bonding axes, 10 orbitals in all (the six hybrid orbitals and the four *s* orbitals), which give rise to 10 total orbitals of σ symmetry: five σ MOs and five σ* MOs. As discussed in Section 3.5, each of these interactions can be represented by a single bond in the Lewis structure, as shown in Figure 3-17.

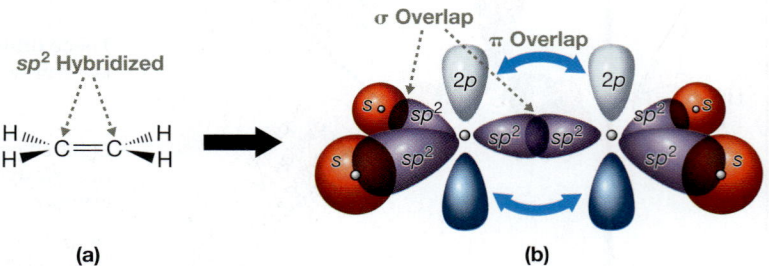

(a) (b)

FIGURE 3-16 AO overlap in ethene (a) Dash–wedge notation of a molecule of ethene (H_2C═CH_2), showing that it is entirely planar and that each C is sp^2 hybridized. (b) AOs that overlap to form the ethene molecule. Each carbon atom contributes three sp^2 hybrid orbitals and one unhybridized *p* orbital. The hybrid orbitals and the s orbitals interact in a σ fashion, along the bonding axes. The *p* orbitals overlap in a π manner on either side of the bonding axis.

This leaves two unhybridized *p* orbitals that are parallel to each other. Oriented as they are in Figure 3-17, these *p* orbitals overlap above and below the bonding axis connecting the two C atoms. As with any pair of AOs that mix, two new orbitals must be generated. One is generated by mixing the orbitals having the same phase (Fig. 3-17a), and the other is generated by reversing the phase of one orbital before mixing (Fig. 3-17b).

In Figure 3-17a, constructive interference takes place above and below the bonding axis because each pair of interacting lobes has the same phase, so the resulting MO is built up in those regions. In Figure 3-17b, the interacting lobes have opposite phases, leading to destructive interference in those regions. In fact, there is complete cancellation — a nodal plane — midway between the two nuclei, so the new orbital that is generated consists of four separate lobes.

These new orbitals do not have σ symmetry, because the overlap of AOs does not take place *along* the bonding axis. Instead, they are described as having *π symmetry*.

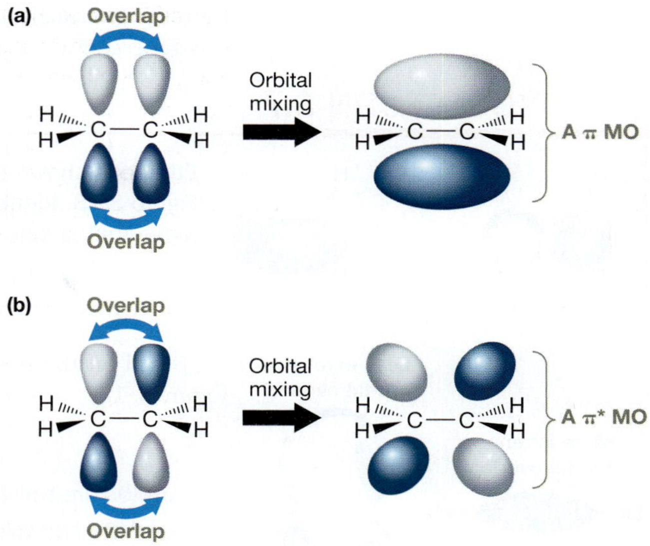

> An orbital of **π symmetry** is generated by AO overlap *on either side of* the bonding axis.

The new orbital in Figure 3-17a is a *π bonding* MO because the buildup of the orbital in the internuclear region lowers its energy relative to that of the *p* AOs from which it is generated (Fig. 3-18). The new orbital in Figure 3-17b is a *π* antibonding* MO. It has an additional nodal plane in the internuclear region that is *perpendicular* to the bonding axis. Just as in a σ* MO, this additional node largely excludes electrons

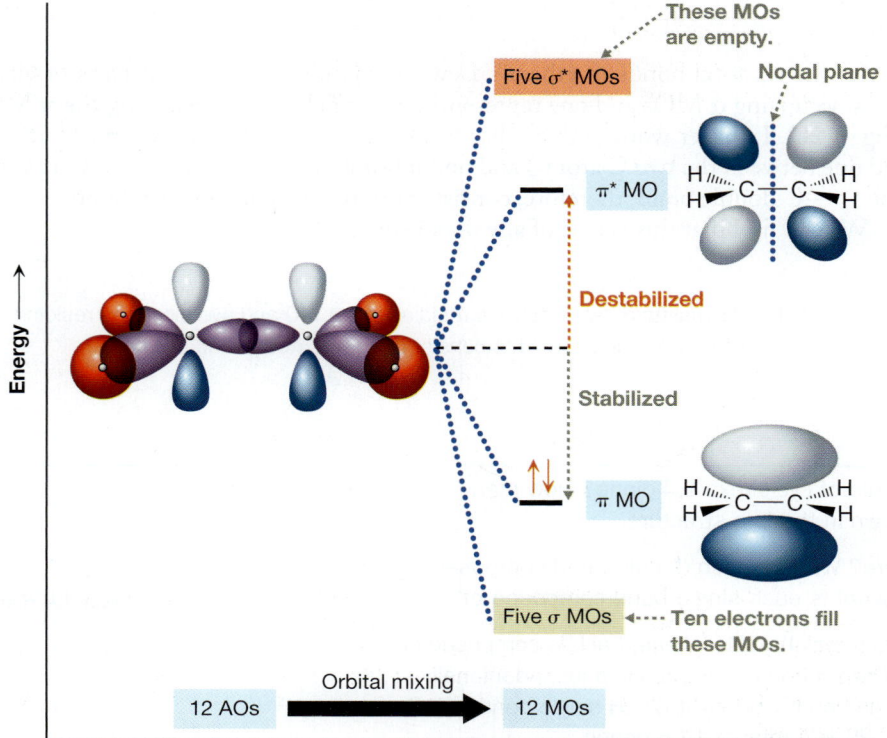

FIGURE 3-17 MOs of π symmetry in ethene (a) Mixing two *p* AOs with the same phase (left) results in constructive interference to generate a bonding MO of π symmetry, which has a lobe on either side of the bonding axis (right). (b) Mixing two *p* AOs with opposite phase (left) results in destructive interference to generate an antibonding MO of π symmetry, which is composed of a total of four lobes (right).

FIGURE 3-18 Energy diagram for the formation of ethene (H_2C=CH_2) from its constituent atoms The AOs are shown on the left and the resulting MOs are shown on the right. The energies of the five σ MOs are lower than that of the π MO, whereas the energies of the five σ* MOs are higher than that of the π* MO.

from the internuclear region, which raises its energy relative to the p AOs from which it is generated.

YOUR TURN 3.7

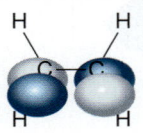 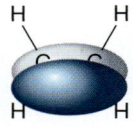 Ethene is shown twice, each with the molecular plane perpendicular to that in Figure 3-18. Identify the π and π^* MOs, and in the π^* MO, draw the nodal plane that is perpendicular to the bonding axis.

Based on the relative energies of the orbitals with σ and π symmetry shown in Figure 3-18:

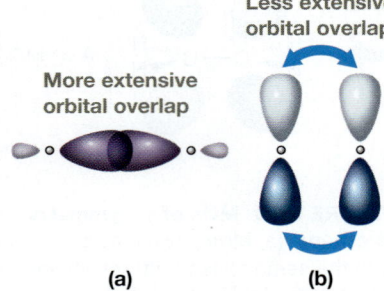

FIGURE 3-19 **Comparison of orbital overlap for orbitals of σ and π symmetry** (a) Overlap of two hybrid AOs in the formation of a MO with σ symmetry. (b) Overlap of two p AOs in the formation of a MO with π symmetry. There is significantly more extensive orbital overlap in (a) than in (b).

- σ MOs are typically lower in energy than π MOs.
- σ^* MOs are typically higher in energy than π^* MOs.

The end-on overlap between orbitals that generates MOs with σ symmetry (Fig. 3-19a) is significantly more extensive than the side-by-side overlap of p orbitals that generates MOs with π symmetry (Fig. 3-19b). Consequently, there is more substantial interaction in σ and σ^* orbitals than there is in π and π^* orbitals.

With the relative energies of the orbitals shown in Figure 3-18, we can fill in the 12 valence electrons to generate ethene's electron configuration. The first 10 valence electrons completely fill the five σ MOs, and the last two fill the π MO. All of the antibonding MOs (i.e., σ^* and π^*) remain empty.

YOUR TURN 3.8

Label the HOMO and LUMO in Figure 3-18.

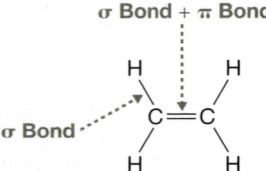

FIGURE 3-20 **Double bonds and Lewis structures** Two electrons of a double bond occupy a σ MO and the other two electrons occupy a π MO. A double bond, therefore, consists of a σ bond and a π bond.

Of the six total bonds in ethene's Lewis structure, five represent pairs of electrons occupying σ MOs and one represents a pair of electrons occupying the π MO (Fig. 3-20). In other words, ethene has five σ bonds (four joining C and H atoms and one between the two C atoms) and one **π bond** (also between the two C atoms). The C=C double bond, therefore, consists of both a σ bond and a π bond.

We will find that this is true of any double bond:

A double bond consists of two electrons residing in a σ MO and two electrons residing in a π MO, so a double bond is said to consist of both a σ bond and a π bond.

SOLVED PROBLEM 3.15

How many π bonds are there in a molecule of cyclohexa-1,4-diene? How many σ bonds? *Hint:* Some bonds are not shown in the line structure.

Cyclohexa-1,4-diene

Think How many double bonds are there? What is each double bond composed of? How many single bonds are there? What is each single bond composed of?

Solve There are two C=C double bonds in cyclohexa-1,4-diene. Each is composed of one π bond and one σ bond, for a total of two π bonds and two σ bonds. Additionally, there are 12 single bonds: four C—C single bonds and eight C—H single bonds. Each C—C and C—H single bond is a σ bond, giving a total of 14 σ bonds.

PROBLEM 3.16 How many π bonds and how many σ bonds are in each of the following species? *Hint:* Some bonds are not shown in line structures.

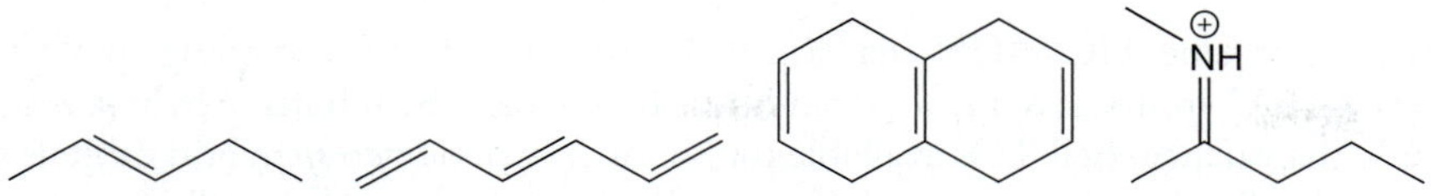

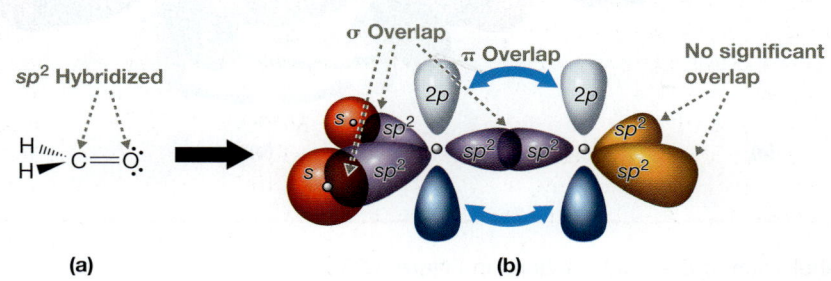

(a)   (b)   (c)   (d)

3.7 Nonbonding Orbitals: An Example with Formaldehyde (H_2C=O)

In the Lewis structure of methanal (formaldehyde, H_2C=O), there are two lone pairs of electrons on O. In what kind of orbitals do they reside? To begin answering this question, notice that both the C and O atoms have a trigonal planar geometry, so they are both sp^2 hybridized (Table 3-1, p. 133). This leads to the orbital contribution picture shown in Figure 3-21.

FIGURE 3-21 Overlap of valence shell AOs in methanal (formaldehyde, H_2C=O) (a) Dash–wedge notation, showing that the molecule is entirely planar and that the O atom has two lone pairs. (b) The AOs in formaldehyde that overlap to produce new orbitals. The carbon and oxygen atoms each contribute three sp^2 hybrid orbitals and one unhybridized p orbital.

There are three pairs of adjacent overlapping AOs (six orbitals in all) that are involved in significant overlap along bonding axes: the four hybrid AOs in purple and the two s orbitals in red. These interactions give rise to six new MOs: three σ MOs and three σ* MOs. Also, similar to ethene, there are parallel adjacent p orbitals (one on C and one on O) that overlap on either side of the bonding axis, giving rise to one π MO and one π* MO.

Unlike the C atoms in ethene, however, the O atom in formaldehyde has two hybrid AOs (shown in orange in Fig. 3-21b) that do not overlap significantly with other AOs. Thus, there are two MOs that are relatively close in energy to the AOs from which they are derived. These are *nonbonding* MOs and can be associated with the lone pairs of electrons in the Lewis structure shown in Figure 3-22.

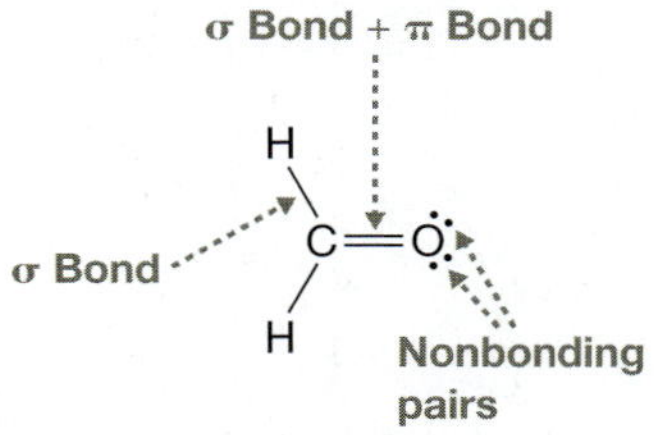

FIGURE 3-22 Lone pairs and Lewis structures Whereas single and double bonds in Lewis structures represent electron pairs occupying bonding MOs, lone pairs represent electrons occupying nonbonding MOs.

PROBLEM 3.17 Draw an orbital picture of H_2C=NH, similar to that in Figure 3-21. How many nonbonding MOs do you expect?

PROBLEM 3.18 For each of the compounds shown here, determine the number of σ bonds, the number of π bonds, and the number of electrons occupying nonbonding MOs.

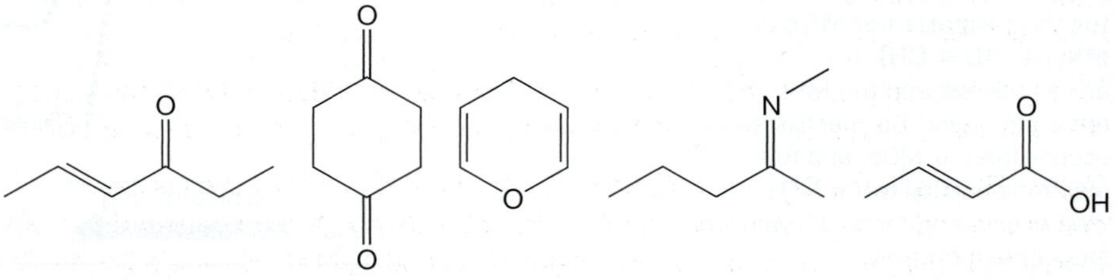

(a)   (b)   (c)   (d)   (e)

3.8 Triple Bonds: An Example with Ethyne (HC≡CH)

Ethyne (acetylene, HC≡CH) has a C≡C triple bond and, according to VSEPR theory, both C atoms have linear geometries. This makes them both sp hybridized. As a result, the valence shell AOs available for bonding include two sp hybrid orbitals and two unhybridized p orbitals on each C atom. The hybrid orbitals form a linear arrangement, as shown in Figure 3-23. Both of the p orbitals on each C (green and blue) are perpendicular to the line containing the hybrid orbitals, and are perpendicular to each other.

There are three σ interactions (Fig. 3-23), giving rise to three σ MOs and three σ* MOs (Fig. 3-24). Additionally, the C atoms contribute two pairs of overlapping p AOs, one shown in blue (p_y) and the other in green (p_z) in Figures 3-23 and 3-24. Each pair gives rise to one π MO and one π* MO.

CONNECTIONS Acetylene is one of the hottest burning fuels, reaching temperatures over 3300 °C (6000 °F), which made it popular in the welding industry until a few decades ago. Arc welding, which uses electrical arcs, is now the technique of choice in most applications.

FIGURE 3-23 **The overlap of valence shell AOs in ethyne (acetylene)** (a) The Lewis structure of ethyne (HC≡CH) shows sp hybridization at each C. (b) The p AOs in blue overlap each other, as do the p AOs in green. Two sp hybrid orbitals (purple) overlap, and each $1s$ orbital (red) from hydrogen overlaps an sp hybrid orbital on carbon.

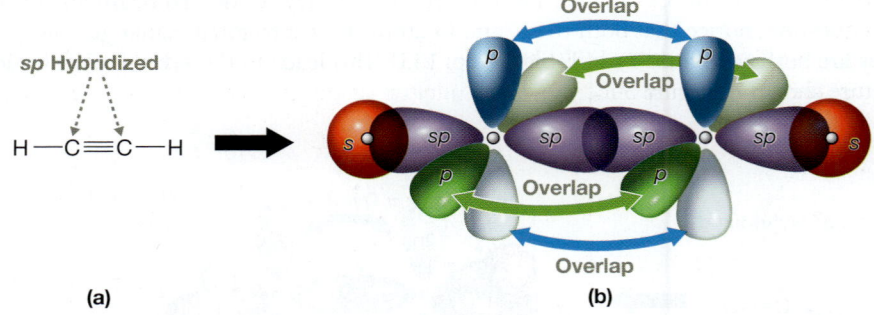

(a)   (b)

YOUR TURN 3.9

Label each of the σ interactions in Figure 3-23b.

The 10 valence electrons completely fill the three σ MOs and the two π MOs. The π* and σ* MOs remain empty.

FIGURE 3-24 **Energy diagram for the formation of MOs in ethyne (HC≡CH)** The AOs are on the left and the MOs are on the right. Six electrons occupy three σ MOs, and four electrons occupy two π MOs, thus accounting for all 10 valence electrons in ethyne.

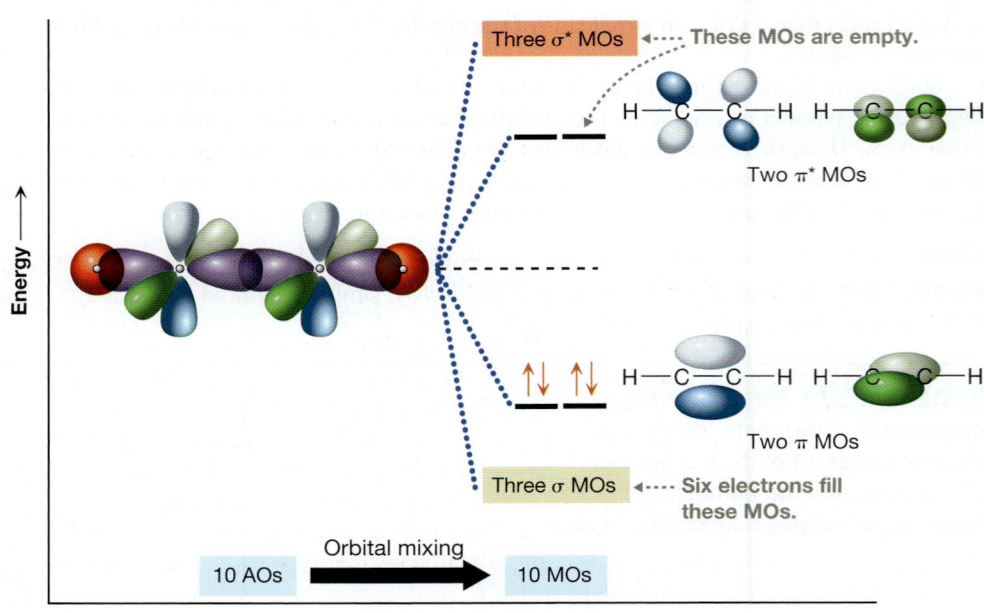

Label the HOMO and the LUMO in Figure 3-24.

Because both π MOs between the two C atoms are filled, two of the three bonds in a C≡C triple bond are π bonds (Fig. 3-25). The remaining bond is a σ bond, representing one of the pairs of electrons occupying a σ MO in Figure 3-24. This can be generalized:

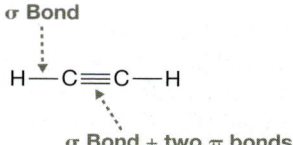

FIGURE 3-25 Triple bonds and Lewis structures Two electrons from a triple bond occupy a σ MO, and the other four electrons occupy two π MOs. A triple bond, therefore, consists of one σ bond and two π bonds.

Any triple bond is composed of one σ bond and two π bonds.

PROBLEM 3.19 Draw the orbital picture of H—C≡N, which should be similar to Figure 3-23.

PROBLEM 3.20 For each of the following species, determine the number of σ bonds, the number of π bonds, and the number of electrons occupying nonbonding orbitals.

(a)   (b)   (c)   (d)

CONNECTIONS HCN (Problem 3.19) is hydrogen cyanide, also called hydrocyanic acid. It is used industrially to make precursors of some polymers such as poly(methyl methacrylate) and nylon-66. These larvae of the eucalyptus leaf beetle release HCN as a defense mechanism.

3.9 Bond Rotation about Single and Double Bonds: Cis and Trans Configurations

There is a significant difference between single and double bonds that goes well beyond the differences in bond strength and bond length that we reviewed in Chapter 1:

Free rotation can occur about single bonds but not about double bonds.

In a molecule of ethane (H_3C—CH_3), for example, the two CH_3 groups freely rotate relative to one another about the single bond that connects them (Fig. 3-26a). In a molecule of ethene (H_2C=CH_2), on the other hand, no rotation occurs. Instead, the entire molecule is planar, with the two CH_2 groups locked in place (Fig. 3-26b).

The CH_3 groups in ethane can rotate freely because, *during rotation, the σ bond that connects the two groups is unaffected* (Fig. 3-27a, next page). The picture is somewhat different with a molecule containing a double bond, which consists of one σ bond and one π bond. Imagine trying to rotate the CH_2 groups in H_2C=CH_2 relative to each other. The σ bond of the double bond would be unaffected, but the π bond would be broken because rotation would destroy the overlap between the *p* orbitals (Fig. 3-27b). The substantial energy cost associated with breaking the π bond is what locks the CH_2 groups of ethene in place.

Free rotation occurs about single bonds.

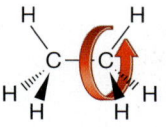

(a) Ethane

CH_2 groups are locked in place.

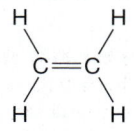

(b) Ethene

FIGURE 3-26 Bond rotation Groups connected by a single bond rotate freely relative to each other, but groups connected by a double bond do not.

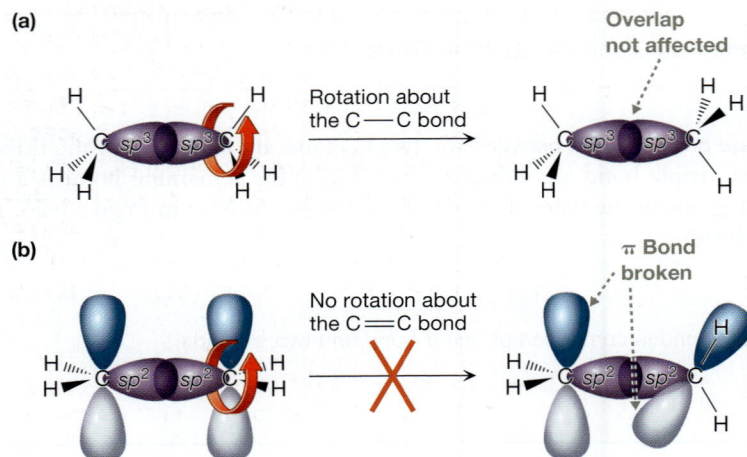

(a)

Rotation about the C—C bond

Overlap not affected

(b)

No rotation about the C=C bond

π Bond broken

FIGURE 3-27 Orbital overlap during rotations about single and double bonds (a) Two different orientations of the CH₃ groups relative to each other during rotation about the C—C bond in H₃C—CH₃. The overlap of the hybrid orbitals is unaffected by the rotation, so this rotation is allowed in the molecule. (b) Two different orientations of the CH₂ groups relative to each other during rotation about the C=C bond in H₂C=CH₂. Although the overlap of the hybrid orbitals is unaffected, overlap between the p orbitals is destroyed, so the π bond must be broken first for rotation to occur. Because breaking the π bond requires a significant input of energy, rotation about a double bond does not normally take place.

For the π bond in H₂C=CH₂ to remain intact, all six atoms must be in the same plane. In general:

> When two atoms are connected by a double bond, those atoms and any atoms to which they are directly bonded prefer to lie in the same plane.

This principle is illustrated in Figure 3-28.

These atoms must all lie in the same plane.

FIGURE 3-28 Planar geometry imposed by a double bond In a molecule of 2-methylhept-2-ene, the five C atoms and the H atom indicated must all lie in the same plane.

PROBLEM 3.21 In each of the molecules shown here, circle all of the atoms that are required to be in the same plane. (Because these are line structures, some H atoms may need to be added back in.)

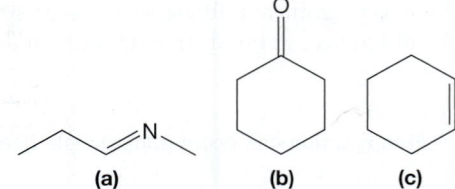

(a) (b) (c)

Because no rotation takes place about double bonds, molecules that differ by the exchange of two groups on one of the doubly bonded atoms are said to have different *configurations*. Molecules like this will not interconvert, so they can typically be isolated from each other and studied separately. This is the case with the two molecules of 1,2-dichloroethene shown in Figure 3-29, which have significantly different boiling points. Further, if one non-H substituent is attached to each atom of the double bond, the configurations can be named either *cis* or *trans*: It is **cis** if the two non-H substituents are on the *same side* of the double bond and it is **trans** if they are on *opposite sides*.

PROBLEM 3.22 List all of the intermolecular interactions present in a sample of *cis*-1,2-dichloroethene and *trans*-1,2-dichloroethene. Using this information, explain the difference in their boiling points.

Not all molecules with double bonds have two distinct configurations. Fluoroethene (CH_2=CHF), for example, does not (see Your Turn 3.11). Exchanging the atoms on either C of the double bond returns the same molecule as before the rotation, in this case because the C on the left is bonded to two H atoms. In general:

> A double bond will *not* have two distinct configurations if one atom of the double bond possesses two identical atoms or groups.

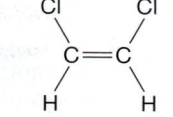

cis-1,2-Dichloroethene
Boiling point = 60.3 °C

trans-1,2-Dichloroethene
Boiling point = 47.5 °C

FIGURE 3-29 Cis and trans configurations These two molecules differ by the exchange of H and Cl on one of the doubly bonded C atoms, and the two configurations do not interconvert. The H atoms are on the same side of the double bond in the cis configuration and are on opposite sides in the trans configuration.

YOUR TURN 3.11

These molecules differ by exchanging the H and F atoms bonded to the C atom on the right. Using a molecular modeling kit, construct the two molecules and try to line one up with the other. Are they the same or different? (Try all orientations of the two molecules.)

CONNECTIONS Fluoroethene, also called vinyl fluoride, is used to make polyvinyl fluoride, a polymer developed by DuPont. Polyvinyl fluoride was given the brand name Tedlar, and it is typically produced in thin films and used as protective coatings. This Goodyear blimp, for example, is protected by Tedlar.

SOLVED PROBLEM 3.23

Does this molecule have two distinct configurations about the double bond?

Think Does the exchange of two groups attached to a doubly bonded C give rise to a different molecule? How can you tell?

Solve One way to answer this question is to compare molecules before and after the exchange.

Exchange Cl and CH₃

These molecules are different because the first one has the Cl atoms on opposite sides of the double bond, whereas the second one has the Cl atoms on the same side. Alternatively, look at the substituents (i.e., the atoms or groups) attached to each C atom joined by the double bond. On the left-hand C atom they are different from each other (H and Cl), and on the right-hand C atom they are also different from each other (Cl and CH_3), so two unique configurations about the double bond must exist.

PROBLEM 3.24 Determine which of molecules **A–D** have a double bond for which two distinct configurations exist.

A

B

C

D

Some molecules have more than one C=C double bond, and each one can potentially exist in cis and trans forms. There are three double bonds, for example, in a molecule of α-linolenic acid, a natural fatty acid. Each double bond in α-linolenic acid is in the cis configuration, indicated in the following line structure by the fact that all of the hydrogen atoms (not shown) bonded to the C=C are on one side of the double bonds and all of the carbon atoms bonded to the C=C are on the other:

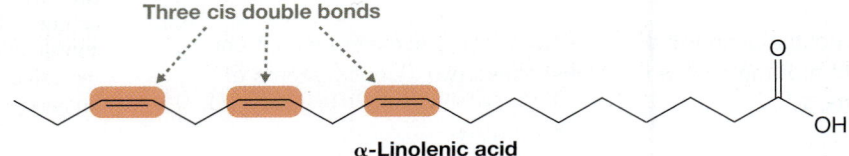

Three cis double bonds

α-Linolenic acid

PROBLEM 3.25 Draw the form of α-linolenic acid in which all three double bonds are trans. How many unique structures can be made by changing the cis/trans configurations about the double bonds in α-linolenic acid?

3.10 Strategies for Success: Molecular Models and Extended Geometry about Single and Double Bonds

In Section 2.3, we saw that a molecular modeling kit can help you visualize three-dimensional molecules. Molecular models are particularly helpful for studying the *rotational* characteristics of the single and double bonds discussed in Section 3.9. Figure 3-30 shows molecular models of ethane (H_3C-CH_3) and ethene ($H_2C=CH_2$) constructed using different modeling kits. Take the time to build these two molecules with your own kit and try to rotate one group relative to the other. You will find that your model kit allows groups to rotate about single bonds but not double bonds, mimicking exactly what takes place in real molecules.

Rotation about single bonds is allowed.

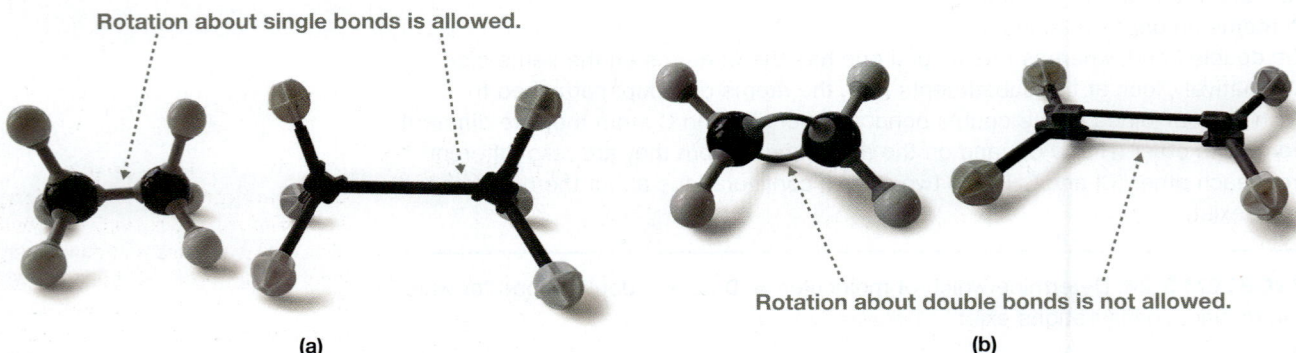

Rotation about double bonds is not allowed.

(a)   (b)

FIGURE 3-30 Model kits and rotational characteristics of molecules (a) Molecular models of ethane (H_3C-CH_3) made with different model kits allow one CH_3 group to rotate freely with respect to the second. (b) Molecular models of ethene ($H_2C=CH_2$) show that rotation of one CH_2 group relative to the other is restricted.

Modeling kits also do a good job showing accurate *extended* geometries about double bonds—that is, the three-dimensional location of the atoms directly attached to each double bond. Because of the planar nature of double bonds, for example, it might be tempting to think that the entire propa-1,2-diene (allene, $H_2C=C=CH_2$) molecule is planar. Experiments show, however, that one CH_2 group is perpendicular to the other. As we can see in Figure 3-31, these are the same results we obtain from molecular modeling kits! (See also Problem 3.41.)

(See also Problem 3.41.)

YOUR TURN 3.12

(a) Using a molecular modeling kit, construct a molecule of $H_2C=C=C=CH_2$. Is the entire molecule planar? **(b)** Next, construct a molecule of $H_2C=C=C=C=CH_2$. Is that molecule entirely planar?

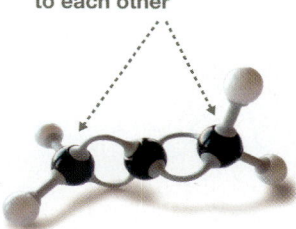

CH₂ groups perpendicular to each other

FIGURE 3-31 Molecular modeling kits and extended geometry A molecular model of propa-1,2-diene (allene, $H_2C=C=CH_2$) constructed with a modeling kit shows that the two CH_2 groups are perpendicular to each other, which agrees with experimental results.

3.11 Hybridization, Bond Characteristics, and Effective Electronegativity

Thus far, we have not treated the various types of hybrid orbitals as being significantly different, but their differences do have some important consequences in chemistry. One is the effect on bond length and bond strength.

> As the hybridization of an atom goes from sp^3 to sp^2 to sp, its bonds become shorter and stronger.

This can be seen in Figure 3-32 (top) with the C—H single bonds in ethane (H_3C—CH_3), ethene ($H_2C=CH_2$), and ethyne ($HC\equiv CH$).

In each of these molecules, the C—H bond is made from a $1s$ AO from H and a hybrid AO from C (Fig. 3-32, bottom). Therefore, the difference is in the hybridization of the AO from C. More specifically, the *s* character of that orbital increases going

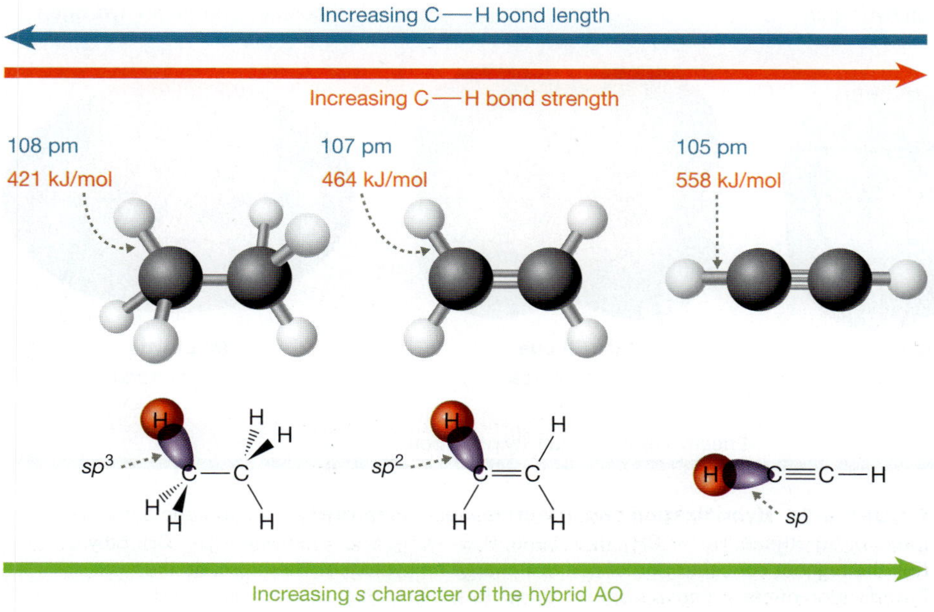

Increasing C—H bond length

Increasing C—H bond strength

| 108 pm | 107 pm | 105 pm |
| 421 kJ/mol | 464 kJ/mol | 558 kJ/mol |

sp^3 ···· H—C—C—H with H's

sp^2 ···· $C=C$ with H's

$H—C\equiv C—H$ ···· sp

Increasing *s* character of the hybrid AO

FIGURE 3-32 Relationship between hybridization, bond length, and bond strength (*top*) The bond distances decrease and bond energies increase for the C—H bonds on going from ethane to ethene to ethyne. (*bottom*) In all cases, the C—H bond involves the overlap of the 1s orbital of a H atom with a hybrid orbital from C. The s character of the C atom's hybrid orbital increases in the order $sp^3 < sp^2 < sp$.

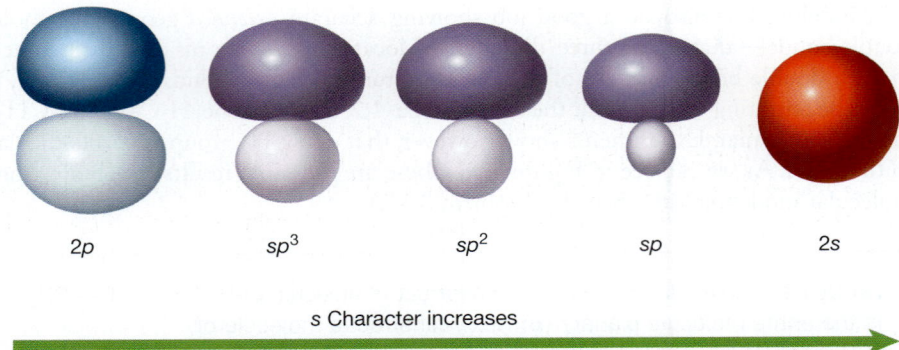

2p sp^3 sp^2 sp 2s

s Character increases

Orbital becomes more compact

FIGURE 3-33 Comparison of pure and hybrid orbitals As s character increases, an orbital becomes more compact and therefore holds electrons closer to the nucleus, on average.

from sp^3 to sp^2 to sp (Fig. 3-32). A 2s AO is more compact than a 2p AO, so the hybrid AO becomes more compact as the s character increases, as illustrated in Figure 3-33. With a shorter hybrid AO, bonds made from it must also be shorter in order to maintain sufficient orbital overlap. And, as we learned in Section 1.4, shorter bonds connecting the same atoms are typically stronger.

YOUR TURN 3.13

Review the percent s character in each of the hybrid AOs in Figure 3-33 and write that percentage above the appropriate AO.

In addition to bond lengths and bond strengths, hybridization can also affect bond polarity. This can be seen in the electrostatic potential maps in Figure 3-34, where the concentration of positive charge (blue) on H increases on going from ethane (sp^3) to ethene (sp^2) to ethyne (sp).

As the hybrid AO on carbon becomes more compact from an increased s character, electrons occupying that orbital are held closer to the nucleus. The C atom therefore

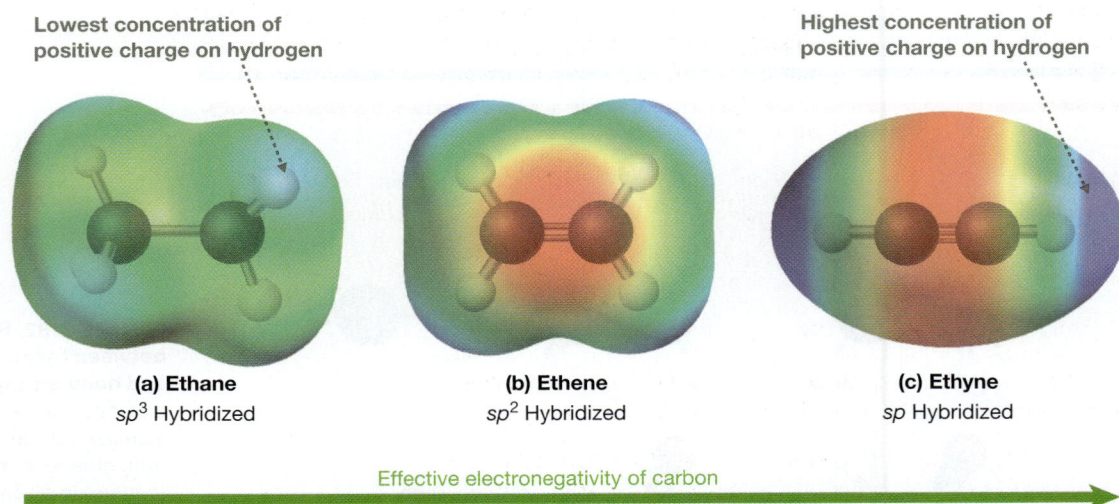

Lowest concentration of positive charge on hydrogen

Highest concentration of positive charge on hydrogen

(a) Ethane
sp^3 Hybridized

(b) Ethene
sp^2 Hybridized

(c) Ethyne
sp Hybridized

Effective electronegativity of carbon

FIGURE 3-34 Hybridization and effective electronegativity Electrostatic potential maps of (a) ethane, $H_3C—CH_3$, (b) ethene, $H_2C=CH_2$, and (c) ethyne, $HC\equiv CH$. Ethyne has the highest concentration of positive charge on hydrogen. Therefore, as the hybridization of carbon goes from sp^3 to sp^2 to sp, the effective electronegativity of carbon increases.

Carbyne: The World's Strongest Material

In Section 3.11, we saw that an atom's bonds tend to get stronger as the atom's hybridization goes from sp^3 to sp^2 to sp. Scientists are looking to take advantage of this property by developing materials composed entirely of sp-hybridized carbon atoms. Because each carbon atom would have a linear geometry, the material, called carbyne, would necessarily consist of one-dimensional chains, a portion of whose structure might be represented as follows:

$$\text{{-}C}\equiv\text{C}-\text{C}\equiv\text{C}-\text{C}\equiv\text{C}-\text{C}\equiv\text{C}-\text{C}\equiv\text{C-}$$

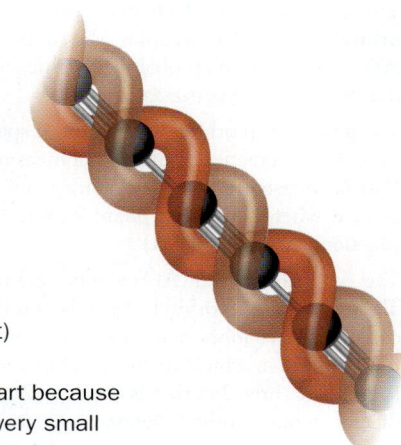

In 2013, Professor Boris Yakobson at Rice University carried out a theoretical study on carbyne. He concluded that carbyne is three times stiffer than diamond and its tensile strength (i.e., its ability to withstand stretching) is twice that of graphene, the two-dimensional sheet of carbon atoms that wraps into cylinders to form carbon nanotubes (see Chemistry with Chicken Wire in Section 1.3). And that's saying something, because carbon nanotubes have found applications in toughening materials in everyday use, including those used in tennis racquets and bicycle frames.

In addition to its mechanical properties, carbyne is theorized to have other properties that are potentially very useful. For example, it has a band gap (a property that determines its electrical conductivity) that varies from 3.2 to 4.4 eV simply by stretching the chain 10%. And molecules can be attached to the ends of the chain that, when twisted 90°, give the chain a helical HOMO (shown at the right) and turn carbyne into a magnetic semiconductor.

Unfortunately, the potential for carbyne is far from being fully realized, in large part because it is difficult to synthesize. To date, researchers have been able to synthesize only very small chains, up to 44 carbon atoms long. If this hurdle can be overcome, however, carbyne has a bright future.

behaves as if its electronegativity is increasing. However, because the nucleus is not changing (it is carbon in all cases), the actual electronegativity is not changing either. Instead, we say that the carbon atom's **effective electronegativity** increases. In general:

As the hybridization of an atom goes from sp^3 to sp^2 to sp, its effective electronegativity increases.

SOLVED PROBLEM 3.26

In which molecule, **A** or **B**, do you think the C=C distances are longer? In which molecule are the C=C bonds stronger?

Think What is the hybridization of each atom involved in the C=C bonds? How does that affect bond length and bond strength?

Solve In **A**, each C=C bond involves two sp^2-hybridized C atoms. In **B**, each C=C bond involves one sp^2-hybridized C atom and one sp-hybridized C atom. With greater s character, the hybrid orbitals are more compact, so the C=C bonds in **B** are shorter and stronger than the ones in **A**.

PROBLEM 3.27 **(a)** In which molecule is the C—C bond shorter? In which molecule is it stronger? Explain. **(b)** Answer the same questions about the carbon–nitrogen bonds.

$$H_3C—C\equiv N$$

$$\underset{H_3C}{}\overset{NH}{\underset{}{C}}\underset{CH_3}{}$$

Chapter Summary and Key Terms

- **Quantum mechanics** treats electrons as waves. The probability of finding an electron in space is governed by the **orbital** the electron occupies. The surface that encompasses 90% of an electron's probability defines the shape of an orbital. **(Section 3.1a; Objective 1)**

- The **phase** of an orbital in a region of space is important in orbital interactions but is not a measurable quantity. Positive phase is indicated by dark shading of the orbital's surface, whereas negative phase is indicated by light shading. **(Section 3.1a; Objective 1)**

- The $1s$ atomic orbital (AO) is spherical and is all one phase. There are three $2p$ orbitals. Each is dumbbell shaped and consists of two lobes that are opposite in phase, separated by a **nodal plane** in which an electron has zero probability of being found. The three $2p$ orbitals (i.e., $2p_x$, $2p_y$, and $2p_z$) are perpendicular to one another. **(Section 3.1a; Objective 2)**

- Atomic s and p orbitals in higher shells are similar in shape and behavior to the $1s$ and $2p$ orbitals, respectively. **(Section 3.1b; Objective 2)**

- Orbitals interact through **constructive interference** and **destructive interference**. **(Section 3.2; Objective 3)**

- Electrons that bond atoms together occupy **molecular orbitals (MOs)**, which arise from overlapping AOs of separate atoms. Each MO can accommodate up to two electrons. **(Section 3.3; Objective 3)**

- A **bonding MO** results from in-phase overlap in the bonding region and is significantly lower in energy than the AOs from which it is constructed. An **antibonding MO** results from out-of-phase overlap in the bonding region and is significantly higher in energy than the contributing AOs. A **nonbonding MO** is similar in energy to the contributing AOs. **(Section 3.3; Objective 4)**

- The highest-energy MO containing an electron is called the **highest occupied molecular orbital (HOMO)** and the lowest-energy empty MO is called the **lowest unoccupied molecular orbital (LUMO)**. **(Section 3.3; Objective 3)**

- When orbitals mix, the total number of orbitals is conserved: The interaction of n orbitals results in n unique new orbitals. **(Section 3.3; Objective 3)**

- MOs of σ symmetry are generated by the interaction of AOs along bonding axes. Two electrons occupying a σ MO establish a **σ bond**. **(Sections 3.3 and 3.5; Objective 5)**

- Two sp **hybrid orbitals** are produced when one s and one p orbital from the valence shell mix together. Each has one large lobe and one small lobe that point in opposite directions. In an

sp-hybridized atom, two p orbitals remain unhybridized; they are aligned perpendicular to each other and perpendicular to the sp orbitals. **(Section 3.4a; Objective 6)**

- Three sp^2 **hybrid orbitals** are produced when one s and two p orbitals from the valence shell mix together. All three hybrid orbitals lie in one plane and point to the corners of a triangle. In an sp^2-hybridized atom, there is one p orbital that remains unhybridized; it is perpendicular to the plane containing the hybrid orbitals. **(Section 3.4b; Objective 6)**

- Four sp^3 **hybrid orbitals** are produced when one s and three p orbitals from the valence shell mix together. The four hybrid orbitals point to the corners of a tetrahedron. **(Section 3.4c; Objective 6)**

- Hybridization is intimately related to VSEPR electron geometry (Table 3-1): sp-Hybridized atoms have a linear geometry, sp^2-hybridized atoms are trigonal planar, and sp^3-hybridized atoms are tetrahedral. **(Section 3.4d; Objective 7)**

- MOs of π **symmetry** are generated by the interaction of adjacent parallel p orbitals, where the orbital overlap takes place on either side of the bonding axis. Two electrons occupying a π MO constitute a **π bond**. **(Section 3.6; Objective 5)**

- A double bond in a Lewis structure consists of one σ bond and one π bond. A triple bond consists of one σ bond and two π bonds. **(Sections 3.6 and 3.8; Objective 8)**

- A lone pair of electrons in a Lewis structure represents a pair of electrons occupying a nonbonding MO. **(Section 3.7; Objectives 4 and 9)**

- **Free rotation** can occur about single bonds but not about double bonds. Rotation about a double bond can only occur by breaking the π bond. **(Section 3.9; Objective 5)**

- For a given double bond, **cis** and **trans** configurations are possible if the exchange of two groups on one of the doubly bonded atoms results in a different molecule. Two groups are cis to each other if they are on the *same* side of a double bond and they are trans to each other if they are on *opposite* sides. **(Section 3.9; Objective 10)**

- Molecular modeling kits display the rotational characteristics about single and double bonds and provide accurate representations of *extended* geometries involving double bonds. **(Section 3.10; Objective 11)**

- Single bonds involving a hybrid orbital become shorter and stronger as the s **character** of the hybrid orbital increases. **Effective electronegativity** also increases as the s character of the atom's hybridization increases. **(Section 3.11; Objective 12)**

Problems

3.3 An Introduction to Molecular Orbital Theory and σ Bonds: An Example with H_2

3.28 Suppose a linear molecule were constructed from three atoms, all of which are found in the second row of the periodic table. How many valence shell AOs would these three atoms contribute toward the production of MOs? How many MOs would be produced by the mixing of these valence shell AOs? *Hint:* The answer is independent of which orbitals overlap.

3.29 One of the orbital interactions we did not consider in this chapter is that between an *s* AO from one atom and a *p* AO from another atom in the fashion shown. These orbitals will not interact while in this orientation. Explain why.

3.30 The bond in H—Cl can be explained by the overlap between an *s* orbital from hydrogen and a *p* orbital from chlorine, as shown in this diagram. **(a)** Draw the bonding and antibonding MOs that would result from such an interaction. **(b)** What is the symmetry of each of these MOs?

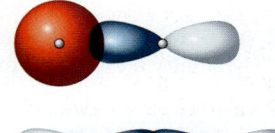

3.31 The bond in Cl_2 can be explained by the end-on overlap between two *p* AOs, as shown in this diagram. **(a)** Draw the bonding and antibonding MOs that would result from such an interaction. **(b)** What is the symmetry of each of these MOs?

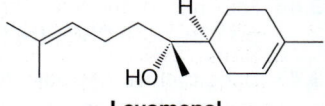

3.4 Hybrid Atomic Orbitals and Geometry

3.32 For each of the following species, determine **(a)** the electron geometry and **(b)** the hybridization for all nonhydrogen atoms.
(i) CH_3NH_2 **(ii)** $CH_3N{=}O$ **(iii)** CH_2Cl_2 **(iv)** BrCN

3.33 For each of the following species, determine **(a)** the electron geometry and **(b)** the hybridization for all nonhydrogen atoms.
(i) $C_2H_5^+$ **(ii)** $C_2H_5^-$ **(iii)** $CH_2{=}OH^+$ **(iv)** CH_6N^+
(v) CH_5O^+ **(vi)** $C_3H_3^-$ (all hydrogens are on the same carbon)

3.34 Levomenol, a naturally occurring sesquiterpene alcohol, has a sweet-smelling aroma and has been used as a component in fragrances. It also is known to have antimicrobial and anti-inflammatory properties. Determine the electron geometry, hybridization, and molecular geometry for each nonhydrogen atom in levomenol.

Levomenol

3.35 Which of the following molecules has a single bond connecting an *sp*-hybridized atom to an sp^2-hybridized atom?

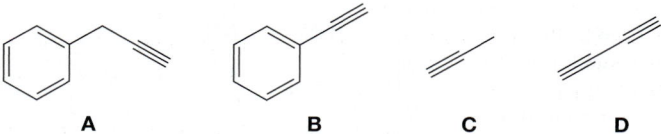

3.5–3.8 An Introduction to π Bonds: An Example with Ethene ($H_2C{=}CH_2$)

3.36 Determine the total number of σ bonds and the total number of π bonds in levomenol (Problem 3.34).

3.37 How many total electrons reside in MOs of π symmetry in this cation?

3.38 Norethynodrel is a synthetic hormone used in Enovid, the first oral contraceptive. **(a)** Determine the hybridization of each nonhydrogen atom. **(b)** How many total σ bonds and π bonds does norethynodrel have?

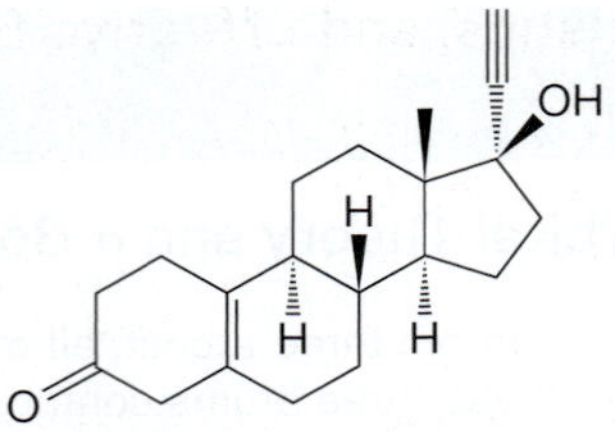

Norethynodrel

3.39 β-Carotene is the compound responsible for the orange color of carrots and is the precursor to vitamin A. Judging from the Lewis structure shown here, how many π bonds does β-carotene have?

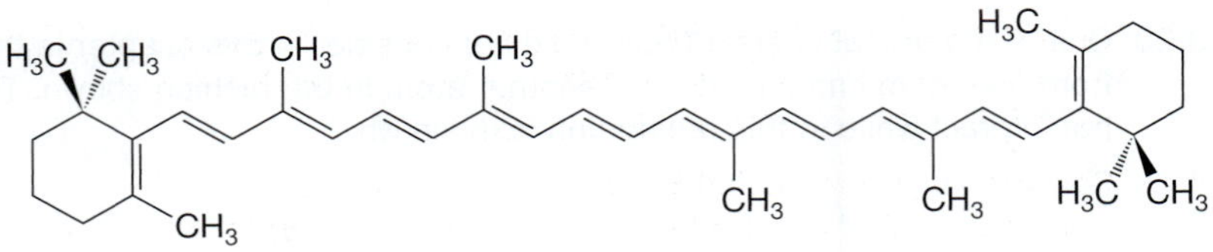

β-Carotene

3.40 **(a)** Draw the orbital picture for :C≡O: showing the explicit overlap of the contributing AOs. **(b)** How many MOs of π symmetry are there in total? **(c)** Draw the orbital energy diagram for CO and identify the HOMO and LUMO.

3.41 **(a)** Draw the molecular orbital picture for propa-1,2-diene, $H_2C=C=CH_2$. *Hint:* The three-dimensional geometry is shown in the chapter. **(b)** Draw the MO energy diagram for propa-1,2-diene. What is the HOMO? What is the LUMO?

3.42 Draw the MO picture for buta-1,2,3-triene, $H_2C=C=C=CH_2$. *Hint:* See Your Turn 3.12.

3.43 Draw the AO contribution picture of $CH_3CH_2^+$ and the MO energy diagram. What is the HOMO? What is the LUMO?

3.44 Suppose that an electron were added to $CH_3CH_2^+$, the species in Problem 3.43, yielding uncharged CH_3CH_2. What is the HOMO? What is the LUMO?

3.9 Bond Rotation about Single and Double Bonds: Cis and Trans Configurations

3.45 Bombykol is a pheromone produced by silkworm moths. Label each C=C double bond's configuration as cis or trans.

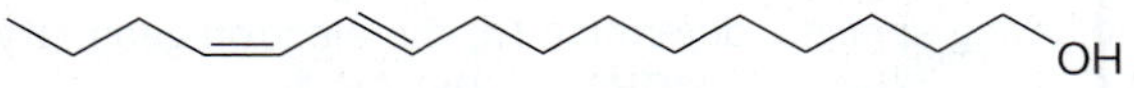

Bombykol

3.46 For which of the following molecules are there two unique configurations about the double bond? Explain.
(a) $(CH_3)_2C=CHCl$; **(b)** $H_2C=CHCH_2CH_3$; **(c)** $ClHC=CHBr$; **(d)** $HC≡CCH=CHCl$

3.47 Does cyclooctene have two distinct configurations about its C=C bond? *Hint:* Consider using a model kit.

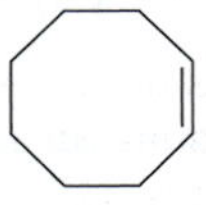

Cyclooctene

3.48 Adenine, cytosine, guanine, and thymine are the four nitrogenous bases found in DNA. For each molecule, identify all of the nonhydrogen atoms that are required to be in the same plane.

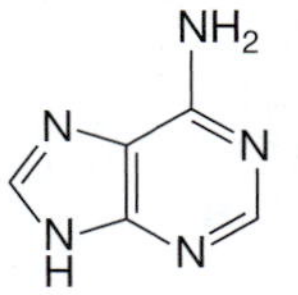

Adenine 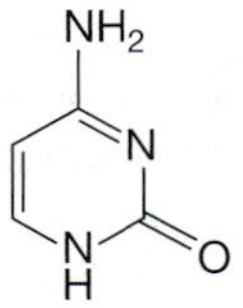 **Cytosine** 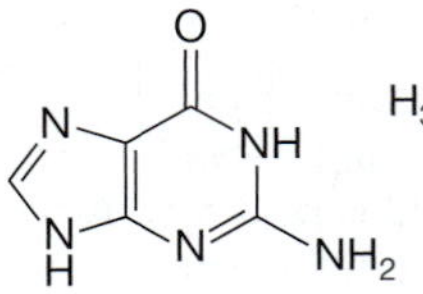 **Guanine** 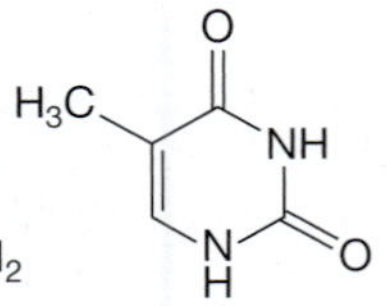 **Thymine**

3.49 Do all of the atoms in buta-1,3-diene have to reside in the same plane? Why or why not?

Buta-1,3-diene

3.11 Hybridization, Bond Characteristics, and Effective Electronegativity

3.50 The boiling point of *cis*-but-2-ene is 3.7 °C, whereas that of *trans*-but-2-ene is 0.9 °C. Explain. *Hint:* Identify which intermolecular interaction is responsible for the difference in boiling points.

cis-**But-2-ene**
Boiling point = 3.7 °C

trans-**But-2-ene**
Boiling point = 0.9 °C

3.51 There are two C—C single bonds in penta-1,3-diyne. **(a)** Which of those bonds would you expect to be stronger? **(b)** Which of those bonds would you expect to be shorter? **(c)** The molecule is moderately polar, with a dipole moment of 1.37 D. In which direction would you expect the dipole moment to point? Explain.

Penta-1,3-diyne

3.52 Consider molecules **A** and **B**. **(a)** In which molecule would you expect the C—Cl bond to be stronger? **(b)** In which molecule would you expect the C—Cl bond to be shorter? **(c)** In which molecule would you expect there to be the greater concentration of negative charge on the Cl atom? Explain.

A   **B**

Integrated Problems

3.53 Suppose a linear molecule were constructed from three atoms, all of which are found in the second row of the periodic table. Suppose that two orbitals from the first atom were to mix with two orbitals from the second atom and that two other orbitals from the second atom were to mix with two orbitals from the third atom. In the resulting molecule, how many bonding MOs would there be in total? How many antibonding MOs? How many nonbonding MOs?

3.54 Octocrylene is an ingredient found in topical sunscreens. It is a water-resistant molecule that helps protect skin against harmful UVA and UVB radiation. **(a)** What is the hybridization of each nonhydrogen atom? **(b)** Circle all atoms bonded to the acyclic C=C double bond that are required to be in the same plane. **(c)** Are there two unique configurations possible about the acyclic C=C double bond? Explain. **(d)** Which of the two C—C single bonds indicated by arrows would you expect to be shorter? Explain.

Octocrylene

3.55 In the chapter, we mentioned that ethene (H_2C=CH_2) does not undergo free rotation about the C=C double bond because the π bond must be broken to achieve the twisted ethene configuration in which the two CH_2 groups are perpendicular to each other (Fig. 3-27b, p. 142). **(a)** Draw the MO energy diagram for the twisted ethene and identify the HOMO and LUMO. **(b)** By comparing the diagram for this molecule to that in Figure 3-18 (p. 137), explain why the molecule is more stable when it is all planar.

3.56 An amide is typically drawn with a single bond connecting the carbonyl C atom to the N atom. If this representation were accurate, we would expect the N atom to be pyramidal, and we would also expect C—N to undergo free rotation. In actuality, however, the N atom is rather planar, and the rotation is quite hindered—properties that give rise to important secondary structures of proteins, such as α helices and β sheets (Chapter 26). Explain these properties of the amide.

An amide

Naming Alkenes, Alkynes, and Benzene Derivatives

Interchapter A discussed how to name basic alkanes and cycloalkanes—hydrocarbons with only single bonds—as well as haloalkanes, nitroalkanes, and ethers. This interchapter on nomenclature extends what we have already learned, enabling us to name *alkenes* (molecules that contain the C=C group) and *alkynes* (molecules that contain the C≡C group). We also discuss how to name simple *benzene derivatives* (compounds that contain a benzene ring), which are related to alkenes because the Lewis structure for benzene (C_6H_6) contains three C=C bonds.

B.1 Alkenes, Alkynes, Cycloalkenes, and Cycloalkynes: Molecules with One C=C or C≡C

In Interchapter A, we learned that our first task in determining an IUPAC name is to establish the root (also called the parent compound) by identifying the longest continuous carbon chain or largest carbon ring—that is, the main chain or ring. Similarly, to name an alkene, alkyne, cycloalkene, or cycloalkyne, we must first determine the root, but the rules for doing so are modified:

Determining the Root of a Compound Containing One C=C or C≡C

- Identify the longest continuous carbon chain or largest carbon ring that contains the *entire* C=C or C≡C bond.
- To assign the root, begin with the name of the analogous alkane or cycloalkane and replace the suffix *ane* by
  - *ene* to account for a C=C functional group.
  - *yne* to account for a C≡C functional group.

These rules are applied to the following molecules.

The root is *pentene* because the longest carbon chain containing C=C has five carbons.

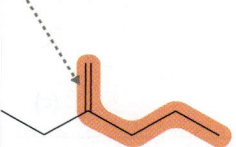

The root is *cyclobutene* because the largest carbon ring containing C=C has four carbons.

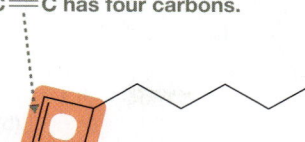

The root is *hexyne* because the longest carbon chain containing C≡C has six carbons.

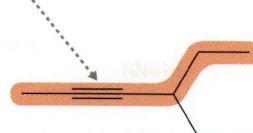

Notice in each of the first two molecules above that the longest carbon chain has more carbon atoms than specified by the root. In the first molecule, the longest carbon chain has six carbon atoms, and in the second molecule, it has five carbons. But in neither of those cases does the longest carbon chain *completely* contain the C=C group.

PROBLEM B.1 Determine the root for each of the following molecules.

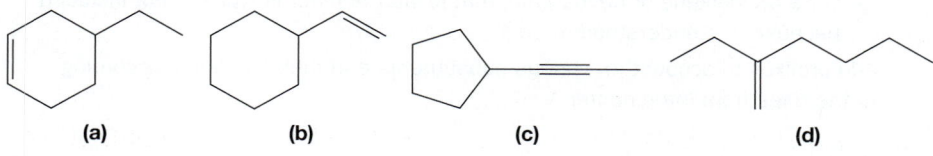

(a) (b) (c) (d)

Once the root is determined, the next step is to establish the numbering system for that main chain or ring. In Interchapter A, which dealt with roots containing only C—C single bonds, the numbering system was dictated only by the locations of the substituents. For an alkene or alkyne, the location of the C=C or C≡C has priority:

> **The Numbering System for a Main Chain or Ring Containing One C=C or C≡C**
>
> - For a chain, C1 begins at the end that allows the C=C or C≡C to be encountered the earliest. If there is a tie beginning from each end, then break the tie using the rules from Interchapter A to minimize the locator numbers (or locants) for the substituents.
> - For a ring, the carbon atoms of the C=C or C≡C must be C1 and C2. Which of those carbon atoms is C1 is determined by the rules from Interchapter A to minimize the locator numbers for the substituents.

The following molecules demonstrate how these rules are applied.

Numbering begins at this end so the C=C is encountered the earliest.

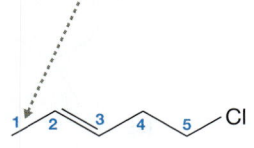

The C=C is the same distance from either end, so numbering begins here to minimize the locator number for the substituent.

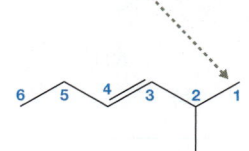

This numbering system allows the C=C atoms to be assigned C1 and C2 and also minimizes the locator numbers for the NO_2 groups.

PROBLEM B.2 Determine the numbering system for each of the molecules shown here.

(a)

(b)

(c)

Finally, the IUPAC name can be constructed according to the following rules:

> **Writing the IUPAC Name of a Molecule Containing One C=C or C≡C**
>
> - Write the root name:
>   - For an acyclic alkene or alkyne, identify the lower of the two locator numbers for the C=C or C≡C atoms and write it immediately before the *ene* or *yne* suffix. As always, use hyphens to separate numbers from letters.
>   - For a cycloalkene or cycloalkyne, that locator number is typically not included because it is understood to be 1.
> - Add prefixes to account for various substituents and their locations, according to the rules from Interchapter A.

Solved Problem B.3 shows how these rules are applied. Make sure you understand that example, and then work through the problems that follow it.

SOLVED PROBLEM B.3

Write the IUPAC name for the molecule at the right.

Think What is the longest carbon chain that entirely contains the C=C? What is the corresponding root name? On which end of the chain should numbering begin to allow the C=C group to be encountered the earliest? Which locator number of the C=C atoms should be included in the IUPAC name?

Solve The longest carbon chain that entirely contains the C=C has six carbon atoms, so the root is *hexene*. Numbering begins at the terminal carbon at the top to encounter the C=C group the earliest, so the locator numbers for those two carbon atoms are C2 and C3. The lower of those two numbers, 2, is added to the root immediately before the *ene* suffix, and prefixes are added to account for the chloro and the 1-methylethyl substituents. The IUPAC name is 4,4-dichloro-3-(1-methylethyl)hex-2-ene.

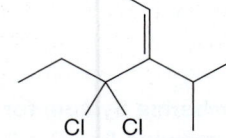

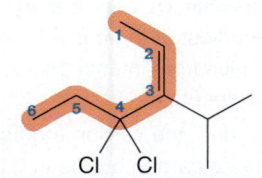

4,4-Dichloro-3-(1-methylethyl)hex-2-ene

PROBLEM B.4 What is the IUPAC name of each of the compounds in Problem B.2?

PROBLEM B.5 Write the IUPAC name for each of the following compounds.

(a)

(b)

(c)

For alkene and alkyne roots containing two or three carbon atoms—namely, ethene, ethyne, propene and propyne—it is not necessary to add a locator number for the alkene or alkyne group. For these cases, the lower of the two locator numbers must be 1, so adding the locator number would be redundant.

Also, placing the locator number immediately before the *ene* or *yne* suffix reflects a relatively recent change to the IUPAC rules. Historically, the locator number was placed immediately before the root, so you will commonly encounter names such as 2-methyl-1-pentene, but the current IUPAC rules call for 2-methylpent-1-ene.

B.2 Molecules with Multiple C=C or C≡C Bonds

For a molecule that has more than one double bond or triple bond, the IUPAC name must indicate the number of double bonds or triple bonds present as well as their locations.

Naming a Molecule with More Than One C=C or C≡C

- Establish the root as the longest carbon chain or largest carbon ring that contains the greatest number of entire C=C or C≡C groups.

- Number the chain or ring so that each successive C=C or C≡C group is encountered the earliest.

- For *each* pair of C=C or C≡C atoms, identify the lower of the two locator numbers.

- Immediately before the *ene* or *yne* suffix, add the letter "a" followed by the above set of locator numbers, and then add a prefix (*di*, *tri*, etc.) to specify how many double or triple bonds are present.

The following examples show how these rules are applied:

The root contains both C=C bonds. Numbering starts at the top C to give the propyl group the lowest locator number.

The root contains both C=C bonds. Numbering starts at the bottom-right C so both C=C groups are encountered the earliest and the locator number of the methoxy group is minimized.

The root contains all three C≡C groups. Numbering starts from the right so the C≡C groups are encountered the earliest.

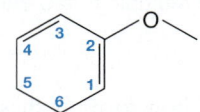

2-Propylpenta-1,4-diene

2-Methoxycyclohexa-1,3-diene

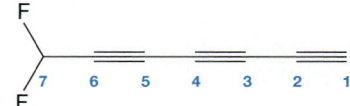

7,7-Difluorohepta-1,3,5-triyne

In the first molecule, the longest carbon chain containing both C=C groups has five carbons. Numbering from either end of the chain results in a tie for the locator numbers of the C=C groups, so the tie is broken by giving the propyl group the lowest number. In the second molecule, the numbering system is chosen to give both C=C groups and the methoxy substituent the lowest numbers. In the third molecule, numbering begins from the right so that the first C≡C group is encountered the earliest.

SOLVED PROBLEM B.6

Write the IUPAC name for this molecule.

Think What is the longest carbon chain that entirely contains all of the C=C groups? On which end of the chain should numbering begin so that the first C=C group is encountered the earliest? How do you determine the set of locator numbers for the C=C groups? When do you add the letter "a" before the suffix?

Solve There are three C=C groups and the longest carbon chain that contains all of them has seven C atoms. Because of the multiple C=C groups, the letter "a" is added before the suffix, so this is a *heptatriene*. Numbering begins at the top-right terminal carbon because that gives the first C=C group the lowest locator number (in this case, 1), so the set of locator numbers assigned to the C=C groups is 1,3,5. The complete IUPAC name is written at the right.

2-Ethyl-6-nitro-3-propylhepta-1,3,5-triene

PROBLEM B.7 Write the IUPAC name for each of the compounds shown here.

(a)

(b)

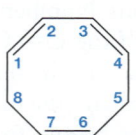

(c)

If a molecule contains both C=C and C≡C functional groups, then it is called an *enyne* and we must apply the following additional rules:

Naming a Molecule That Has Both C=C and C≡C Groups

- Establish the root as the longest carbon chain or largest carbon ring that contains both groups and change the suffix to *enyne*.
- The numbering system must then minimize the locator number for each successive C=C or C≡C group.
  - The C=C and C≡C groups are given equal priority unless there is a tie between two numbering systems.
  - If there is a tie, the tie is broken by giving priority to the C=C group.
- Add the locator number of the C=C immediately before *en* and add the locator number of the C≡C immediately before *yne*.
- Account for multiple C=C and C≡C groups using the same rules as before.

Examples of how to apply these rules are shown below. Numbering starts from the left in the first molecule to give the C=C group the lowest locator number, whereas numbering starts from the right in the second molecule to give the C≡C the lowest number. The third example shows how to account for two C=C groups.

Hex-1-en-4-yne

Hex-4-en-1-yne

Cycloocta-1,3-dien-6-yne

PROBLEM B.8 Draw the structures for pent-3-en-1-yne and 1,2-dimethylcycloocta-1,3-dien-6-yne.

PROBLEM B.9 What is the name of this compound?

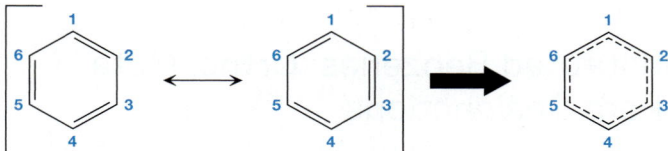

B.3 Benzene and Benzene Derivatives

The Lewis structure of **benzene** (C_6H_6), shown in Figure B-1, consists of three C=C double bonds that alternate with three C—C single bonds in a six-membered ring.

π **Electrons delocalized around the ring**

FIGURE B-1 Benzene The Lewis structures of benzene are shown inside the brackets, and the resonance hybrid is shown on the right.

Based on the rules presented so far for naming alkenes, it might be tempting to rename benzene as "cyclohexa-1,3,5-triene." This name is unacceptable, however, because benzene behaves significantly different than a typical alkene. Benzene is an *aromatic* compound (see Chapter 14 for an in-depth discussion) and, as its resonance hybrid indicates, its π electrons are fully delocalized around the ring. Thus, all six carbon atoms of benzene are identical, and we cannot formally assign any of the double bonds to a single pair of carbon atoms.

In recognition of the unique behavior of the benzene ring, the following rules are applied to naming relatively simple molecules in which benzene has attached substituents, so-called **benzene derivatives**:

Naming Benzene Derivatives

- The IUPAC name receives the root *benzene*.
- Establish the numbering system of the ring to give the lowest locator number to each successive substituent.
- Add prefixes to account for the number, type, and location of each substituent.

These ideas are applied in Solved Problem B.10. Make sure you understand that example and then work through the problems that follow it.

SOLVED PROBLEM B.10

Provide the IUPAC name for this molecule.

Think Is there a six-membered ring of alternating C—C and C=C bonds? What is the root assigned in such cases? Where should C1 be to allow each successive substituent to be encountered the earliest? Should numbering proceed clockwise or counterclockwise?

Solve The six-membered ring of alternating C—C and C=C bonds requires the root *benzene*. The numbering system is then established by the substituents. To encounter each successive substituent the earliest, numbering begins at the C atom attached to Cl and proceeds clockwise. The IUPAC name, therefore, is 1-chloro-4-ethyl-2-nitrobenzene.

1-Chloro-4-ethyl-2-nitrobenzene

PROBLEM B.11 What is the name of each of the compounds shown here?

(a) (b) (c)

CONNECTIONS
1,2,3-Trimethylbenzene (Problem B.12) is mixed with other hydrocarbons to formulate a fuel stabilizer for jet fuel. This stabilizer helps prevent solid particles from forming, which could be harmful to the engine.

PROBLEM B.12 Draw the structures of 1,2,3-trimethylbenzene and 4-bromo-2-chloro-1-nitrobenzene.

B.3a Disubstituted Benzenes: Ortho, Meta, and Para Designations

Although the numbering system outlined in the previous section is always valid, the following nonnumerical system can be used in the special case of **disubstituted benzenes**, in which two substituents are attached to the benzene ring.

Relative Positioning of Substituents in Disubstituted Benzenes

- *ortho* = *o* = 1,2-positioning
- *meta* = *m* = 1,3-positioning
- *para* = *p* = 1,4-positioning

Note: When more than two substituents are attached to the ring, the ortho, meta, and para prefix system is never used.

The examples of disubstituted benzenes below show how to use the ortho, meta, and para notation.

1-Bromo-2-chlorobenzene
ortho-Bromochlorobenzene
o-Bromochlorobenzene

1,3-Dibromobenzene
meta-Dibromobenzene
m-Dibromobenzene

1,4-Dinitrobenzene
para-Dinitrobenzene
p-Dinitrobenzene

Make sure that the ortho/meta/para designation is applied only when benzene is the root. It would be a mistake, for example, to name 1,2-dibromocyclohexane as *ortho*-dibromocyclohexane.

PROBLEM B.13 Draw the structures for *p*-dichlorobenzene and *m*-bromoethoxybenzene. What would the IUPAC names be using locator numbers instead?

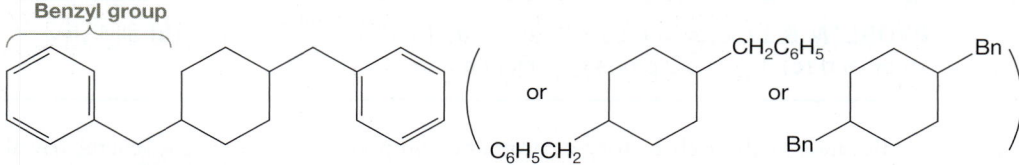

(a) The phenyl (Ph) substituent

(b) The phenylmethyl or benzyl (Bn) substituent

FIGURE B-2 Substituents containing a benzene ring (a) The phenyl substituent. (b) The phenylmethyl or benzyl substituent. Incorporating these substituents into the IUPAC name can often lead to a simpler name than if benzene were treated as the root.

B.3b Phenyl and Benzyl Substituents

If a substituent on a benzene ring is sufficiently complicated, the molecule could be more straightforward to name if we treat the benzene ring as belonging to a substituent instead. One example is C$_6$H$_5$–, the **phenyl (Ph)** substituent (pronounced the same as the spice fennel), as shown in Figure B-2a. Another example is C$_6$H$_5$CH$_2$–, the **phenylmethyl** substituent, also called the **benzyl (Bn)** substituent, as shown in Figure B-2b. Examples of how to incorporate these substituents into a molecule's name and how to abbreviate them in the structure are shown below.

Phenyl group

or C$_6$H$_5$ or   Ph

3-Phenylhept-2-ene

Benzyl group

or C$_6$H$_5$CH$_2$ CH$_2$C$_6$H$_5$ or   Bn   Bn

1,4-Dibenzylcyclohexane

PROBLEM B.14 Draw the structures for **(a)** 2-phenyl-1-hexene and **(b)** 1,5-diphenylpentane.

PROBLEM B.15 What is the name of this compound?

Ph—C—C—Ph with Ph Ph above and Ph Ph below

CONNECTIONS Propylene (Fig. B-3) is used primarily as the starting material to produce polypropylene, a plastic that has many applications, such as food containers, chairs, medical sutures, and diapers.

B.4 Trivial Names Involving Alkenes, Alkynes, and Benzene Derivatives

As with any class of organic compounds, alkenes, alkynes, and benzene derivatives have trivial names that are firmly entrenched in nomenclature. Some of the most common ones for alkenes and alkynes are given in Figure B-3 and should be committed to memory.

H$_2$C=CH$_2$ HC≡CH

Ethene (Ethylene)   **Propene (Propylene)**   **Ethyne (Acetylene)**   **Methylpropene (Isobutylene)**

FIGURE B-3 Some trivial names of alkenes and alkynes IUPAC names are provided first, and trivial names are in parentheses.

CONNECTIONS Isobutylene, (Fig. B-3) is used to manufacture methyl *tertiary*-butyl ether (MTBE), ethyl *tertiary*-butyl ether (ETBE), and isooctane, which are all fuel additives. It is also polymerized to make polyisobutylene, known as butyl rubber.

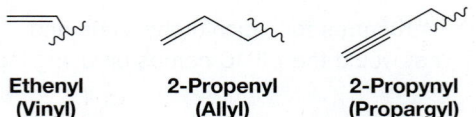

Ethenyl   **2-Propenyl**   **2-Propynyl**
(Vinyl)   **(Allyl)**   **(Propargyl)**

FIGURE B-4 **Some trivial names of substituents containing the C══C or C≡══C bond**
IUPAC names are provided first, and trivial names are in parentheses.

Trivial names are also commonly used for substituents containing C══C or C≡══C bonds. Some common examples are given in Figure B-4—they should be committed to memory, too.

Some examples with these common names are as follows:

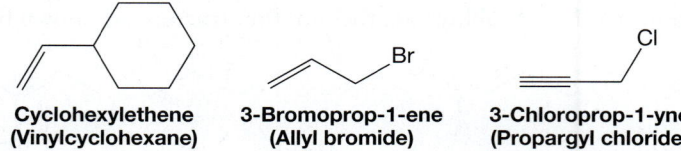

Cyclohexylethene   **3-Bromoprop-1-ene**   **3-Chloroprop-1-yne**
(Vinylcyclohexane)   **(Allyl bromide)**   **(Propargyl chloride)**

PROBLEM B.16 Name the molecule shown here using trivial names where appropriate.

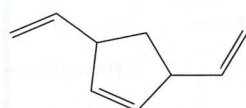

PROBLEM B.17 Draw the structures of **(a)** 1,4-divinylcyclohexane, **(b)** allyl vinyl ether, and **(c)** 1-chloro-3-propargylcyclopentane.

Because of the rich history of aromatic compounds in chemistry, several trivial names are in widespread use. Some of them are given in Figure B-5 and they should be committed to memory.

A handful of these trivial names, such as toluene and anisole, have been adopted by the IUPAC system, which is why those two names are not in parentheses in Figure B-5 (we will encounter other examples in Interchapters E and F). In such cases, the entire substituted benzene molecule establishes the root and C1 is assigned to the carbon that is attached to the substituent on which the molecule's name is based (methyl in the case of toluene, and methoxy in the case of anisole). Examples of how to name substituted toluenes and substituted anisoles are shown on the next page.

CONNECTIONS Anisole (Fig. B-5) is used to manufacture pharmaceuticals, insect pheromones, and perfumes. It is also used to manufacture anethole, a flavoring compound found in anise that is several times sweeter than sugar.

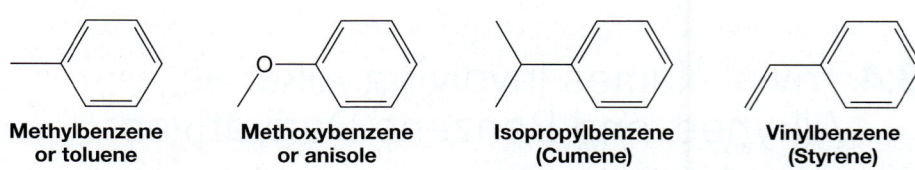

Methylbenzene   **Methoxybenzene**   **Isopropylbenzene**   **Vinylbenzene**
or toluene   **or anisole**   **(Cumene)**   **(Styrene)**

FIGURE B-5 **Some trivial names of molecules containing the phenyl ring (C₆H₅, Ph)**
IUPAC names appear outside of parentheses, whereas trivial names appear inside parentheses.

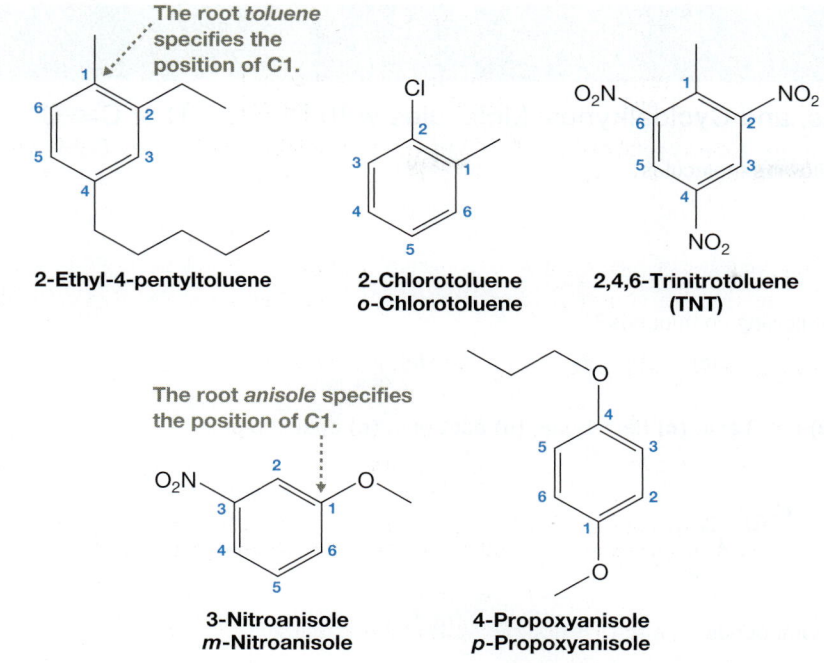

2-Ethyl-4-pentyltoluene

The root *toluene* specifies the position of C1.

2-Chlorotoluene
***o*-Chlorotoluene**

2,4,6-Trinitrotoluene (TNT)

The root *anisole* specifies the position of C1.

3-Nitroanisole
***m*-Nitroanisole**

4-Propoxyanisole
***p*-Propoxyanisole**

PROBLEM B.18 Draw the structures and give alternate IUPAC names for **(a)** *m*-bromotoluene, **(b)** 2,5-dinitrotoluene, and **(c)** *p*-chloroanisole.

Some trivial names describe two substitutents on benzene, not just one. The most common of these is xylene, the trivial name for dimethylbenzene, which can have either an ortho, meta, or para positioning of the methyl substituents, as shown in Figure B-6.

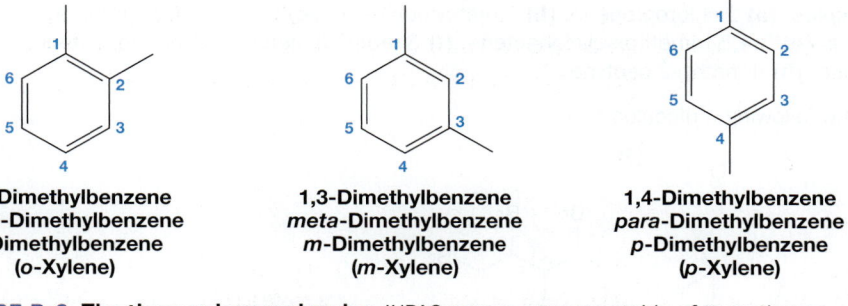

1,2-Dimethylbenzene
***ortho*-Dimethylbenzene**
***o*-Dimethylbenzene**
(*o*-Xylene)

1,3-Dimethylbenzene
***meta*-Dimethylbenzene**
***m*-Dimethylbenzene**
(*m*-Xylene)

1,4-Dimethylbenzene
***para*-Dimethylbenzene**
***p*-Dimethylbenzene**
(*p*-Xylene)

FIGURE B-6 The three xylene molecules IUPAC names appear outside of parentheses, whereas trivial names appear inside parentheses.

CONNECTIONS Xylene (Fig. B-6) makes up almost 1% of crude oil. One of its main uses is as a feedstock in the production of polymers like polyethylene terephthalate (PET, used to make plastic bottles) and polyester (used for clothing). Xylene is also used as a cleaning agent and a solvent. In dentistry, for example, xylene is the solvent for gutta percha, which is a material used to fill the canals of a tooth after a root canal.

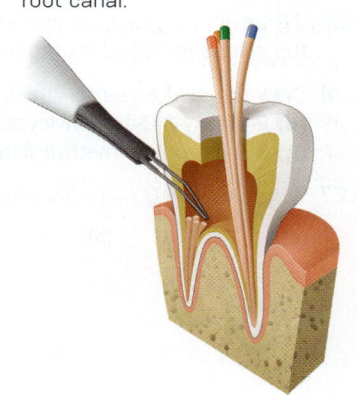

Problems

B.1 Alkenes, Alkynes, Cycloalkenes, and Cycloalkynes: Molecules with One C=C or C≡C

B.19 Write the IUPAC name for each of the following molecules.

(a)

$H_2C=CH_2$

(b) **(c)**

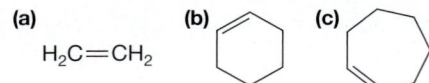

B.20 What is the IUPAC name of each of the following compounds?

(a) **(b)** **(c)** **(d)** **(e)**

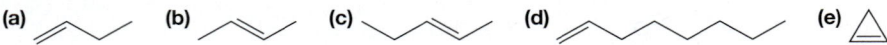

B.21 Draw the structures for **(a)** hex-2-ene; **(b)** hex-3-ene; **(c)** hept-1-ene; **(d)** oct-2-yne; **(e)** cycloheptene.

B.22 Name the following structures.

(a) **(b)** **(c)** Cl Cl

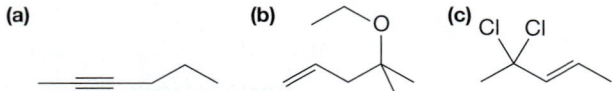

B.23 Provide correct names for the following compounds.

(a) **(b)** Cl **(c)** **(d)** OCH₃ **(e)** **(f)**

OCH₃

Br
Br

NO₂

B.24 Provide the IUPAC names for the following compounds.

(a) **(b)** **(c)**

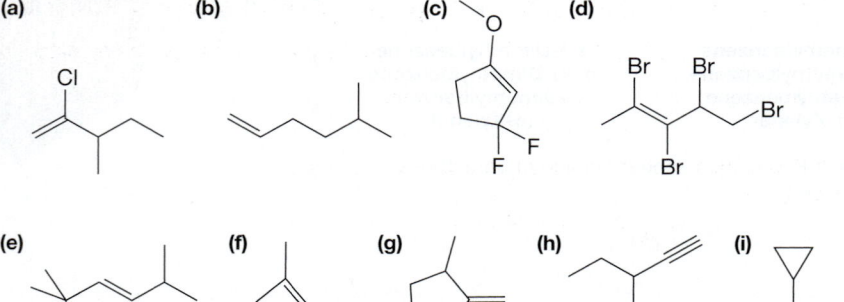

B.25 2-Propylbut-1-ene is an incorrect name under the IUPAC system. What is the correct name for this compound? *Hint:* Draw the structure based on the incorrect name and rename it.

B.26 Draw the structures for these molecules. **(a)** 2-chloropropene; **(b)** 3-methylbut-1-ene; **(c)** 2,3-dimethyl-2-butene; **(d)** 2-ethoxy-3,3-dimethylcyclohexene; **(e)** 3,4,5-trimethoxycycloheptene; **(f)** 3-bromo-2-methyl-4-nitrocyclopentene; **(g)** 3,3-dibromo-4-methylcyclopentene; **(h)** 4-methyl-2-pentyne

B.27 Write the IUPAC name for each of the following molecules.

(a) **(b)** **(c)** **(d)**

Cl

Br Br

Br

Br

(e) **(f)** **(g)** **(h)** **(i)**

O

F

B.2 Molecules with Multiple C=C or C≡C Bonds

B.28 Draw the structure of 1,4-cyclohexadiene. The "1,4" can appear in another place in the name. Write that name.

B.29 Draw the structures for penta-1,4-diene and cyclopenta-1,3-diene.

B.30 What is the name for 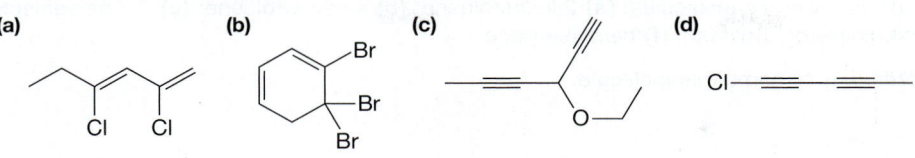 ?

B.31 Draw the structures for **(a)** 2-methyl-1,3,5-hexatriene and **(b)** 1,6-dimethoxyhexa-1,5-diene.

B.32 Determine the IUPAC name for each of the following molecules.

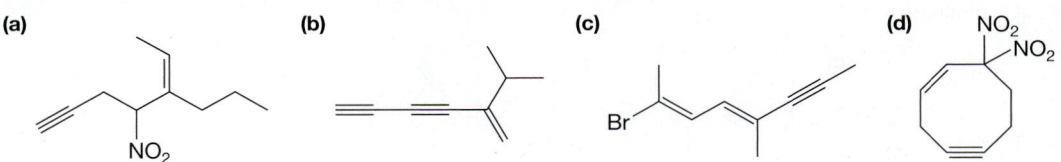

(a)   **(b)**   **(c)**   **(d)**

(e)   **(f)**   **(g)**

B.33 Determine the IUPAC name for each of the following molecules.

(a)   **(b)**   **(c)**   **(d)**

B.3 Benzene and Benzene Derivatives

B.34 Draw the structures for **(a)** hexylbenzene and **(b)** bromobenzene.

B.35 What is the name of this compound?

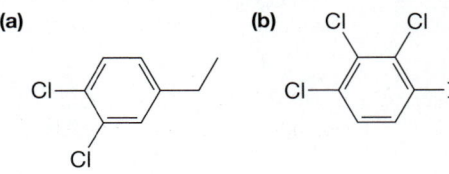

B.36 What are the IUPAC names for the following molecules?

(a)   **(b)**

B.37 Draw the structures for each of the following molecules. **(a)** fluorobenzene; **(b)** 1-chloro-2-fluorobenzene; **(c)** 1-iodo-4-nitrobenzene; **(d)** 1,3-dibromobenzene; **(e)** 2,3-dimethyl-1-cyclopentylbenzene; **(f)** 4-ethoxy-1,2-dinitrobenzene

B.38 Give the IUPAC name for each of the following molecules. Which of these compounds could be named using *ortho*, *meta*, or *para* as a prefix?

(a)   **(b)**   **(c)**   **(d)**

(e)   **(f)**   **(g)**

B.4 Trivial Names Involving Alkenes, Alkynes, and Benzene Derivatives

B.39 Draw the structure for each of the following molecules. **(a)** 1,3-divinylcyclohexane; **(b)** 3-allyl-4-vinylcyclopentene; **(c)** dimethylacetylene; **(d)** divinyl ether; **(e)** 4-vinylocta-1,3,7-triene; **(f)** 2-allylcyclohexa-1,3-diene

B.40 Draw the structures for each of the following molecules. **(a)** 2-fluorotoluene; **(b)** 4-ethoxytoluene; **(c)** 2-ethoxyanisole; **(d)** 1,3-diphenylheptane; **(e)** 4,4-diphenyl-1-octene; **(f)** benzylbenzene

B.41 Use trivial names, where appropriate, to name this molecule.

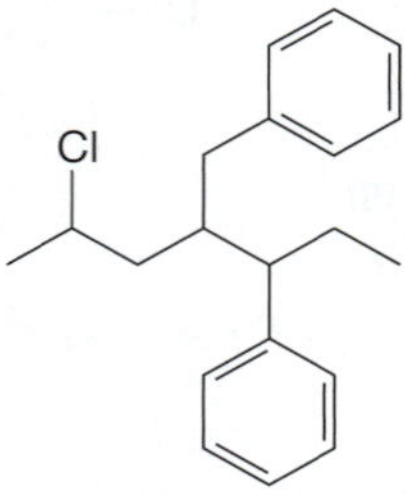

B.42 Name the following molecules.

(a)   (b) 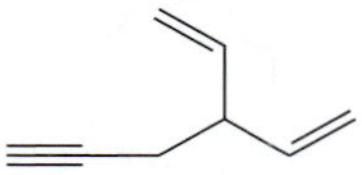   (c) 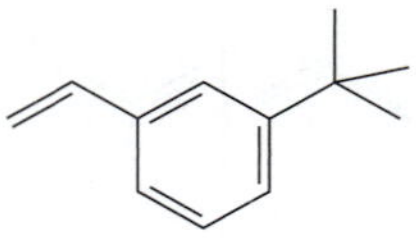   (d)

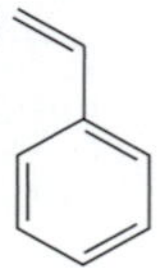

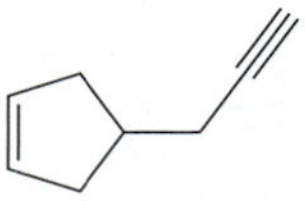

Integrated Problems

B.43 Provide the IUPAC name for each of the following molecules.

(a)   (b) 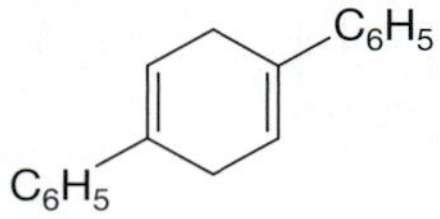   (c) 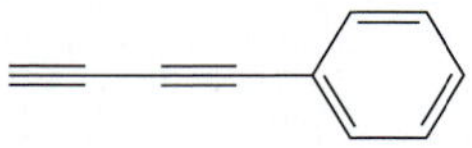   (d)

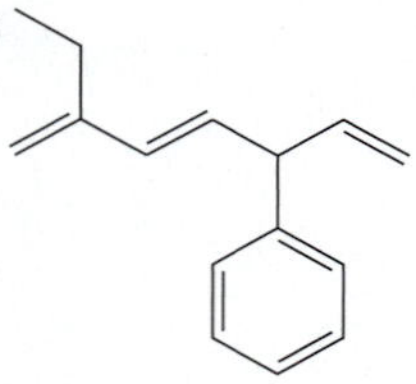

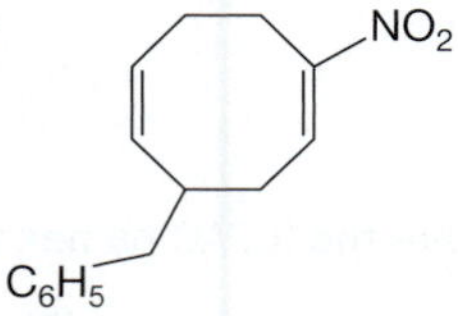

B.44 Provide the IUPAC name for each of the following molecules.

(a)   (b) 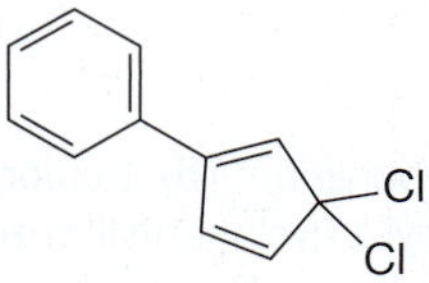   (c)

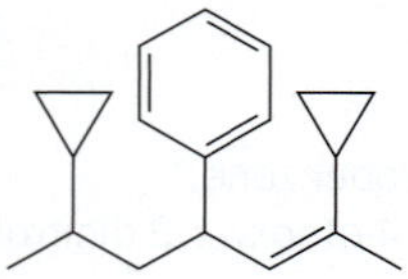

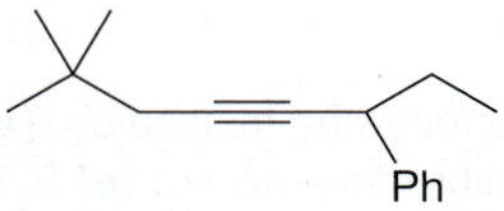

B.45 Draw the structure for these molecules. **(a)** 3-phenylmethylhex-4-en-1-yne; **(b)** 7-phenylcyclohepta-1,3,5-triene; **(c)** 3,5,6-trinitro-4-phenylmethylhepta-1,3,5-triene; **(d)** 5,5-dichloro-6-ethenyl-7-phenylcyclooct-3-en-1-yne

When a roller coaster car travels up and down along its rails, it gains and loses potential energy, much like a molecule's energy changes as it rotates about its single bonds. These energy changes within a molecule give rise to conformers, a topic of this chapter.

4

Isomerism 1
Conformers and Constitutional Isomers

There are tens of millions of known organic compounds, and more are being isolated, synthesized, and identified every year! Fortunately, the chemical behavior of every compound is *not* entirely unique. Instead, *molecules that are similar in composition and structure tend to have similar reactivity*. We first encountered this principle in Chapter 1, where we saw that identical functional groups in different molecules tend to react similarly due to the specific arrangement of their atoms.

Here in Chapter 4 and later in Chapter 5, we focus more closely on the relationships between similar molecules, in particular those with the same molecular formulas: *isomers*. In Chapter 4 we concentrate on the distinctive characteristics of two classes of these: *conformers* and *constitutional isomers*. In Chapter 5 we explore two other types of isomers: *enantiomers* and *diastereomers*.

4.1 Isomerism: A Relationship

Two molecules are said to be **isomers** of each other if they have the same molecular formula but are different in some way.

There are a variety of ways in which two molecules with the same molecular formula can differ from each other, and these differences give rise to the different types

Chapter Objectives

On completing Chapter 4 you should be able to:

1. Define isomers and distinguish between conformers and constitutional isomers; identify pairs of molecules as one or the other type of isomer.
2. Draw and interpret Newman projections for various conformations about a single bond.
3. Determine the relative energies of conformations about single bonds based on their torsional strain and steric strain and explain what enables conformations to interconvert.
4. Identify gauche and anti conformations about a single bond.
5. Explain what gives rise to ring strain, and predict relative amounts of ring strain, given only Lewis

structures; describe how heats of combustion are used experimentally to determine ring strain.

6. Draw the most stable conformations of cyclohexane and cyclopentane rings, and explain why they attain their respective geometries.
7. Describe how chair conformations of cyclohexane interconvert and how envelope conformations of cyclopentane interconvert.
8. Predict the more stable chair conformation for substituted cyclohexanes.
9. Compute the index of hydrogen deficiency (IHD) for a given molecular formula.
10. Efficiently draw various constitutional isomers of a given molecular formula.

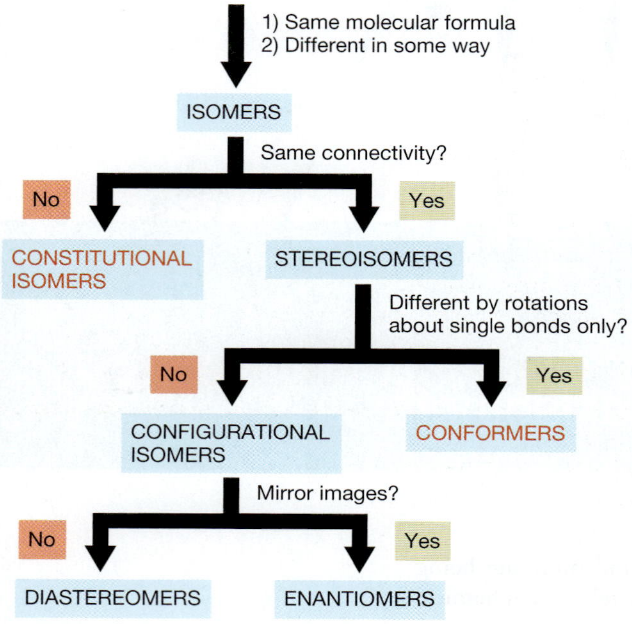

FIGURE 4-1 Flowchart for categorizing types of isomers Constitutional isomers and conformers (red) are discussed in depth here in Chapter 4. Diastereomers and enantiomers are discussed in Chapter 5.

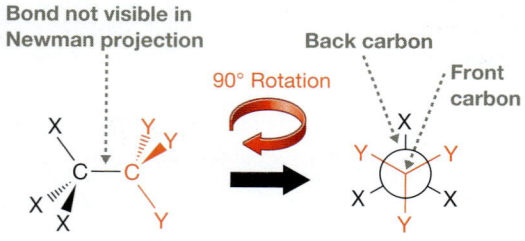

FIGURE 4-2 Interpretation of a Newman projection The generic molecule on the left can be depicted by the Newman projection on the right. Each carbon has three substituents pointing outward.

of *isomerism* shown in the flowchart in Figure 4-1. Constitutional isomers and conformers (i.e., the categories highlighted in red in Fig. 4-1) are examined in depth here in Chapter 4; configurational isomers, which include diastereomers and enantiomers, are discussed in Chapter 5.

> Each type of **isomerism** is a *specific relationship* between molecules.

Therefore, the term "isomer" cannot be applied to a single molecule. It is a mistake, for example, to refer to a single molecule as a constitutional isomer, a diastereomer, or an enantiomer without reference to a second molecule. This is no different from saying, "John is a cousin." A cousin of whom? We *can* say, however, that two molecules are constitutional isomers *of each other*, or that one molecule is an enantiomer *of a second*. Similarly, we can say, "John is Jane's cousin," or, "John and Jane are cousins."

4.2 Conformers: Rotational Conformations, Newman Projections, and Dihedral Angles

According to the flowchart in Figure 4-1, a pair of **conformers** differ only by rotations about single bonds. (Conformers are sometimes referred to as conformational isomers, but we will avoid that term in this book because conformers generally cannot be isolated from each other.) To more easily study conformers, it is helpful to have a systematic way to depict molecules having different angles of rotation about single bonds—that is, different **rotational conformations**. One convenient way to illustrate rotational conformations is with a **Newman projection** (Fig. 4-2), which is a two-dimensional representation of a molecule *viewed down* *the bond of interest*. The following conventions are used when drawing Newman projections:

- In a Newman projection, the two atoms directly connected by the bond of interest are shown explicitly.
  - The nearer atom is depicted as a point.
  - The more distant atom is depicted as a circle.
- Bonds to the front atom converge at the point, whereas bonds to the back atom connect to the circle.

YOUR TURN **4.1**

Based on the Newman projection given, label the front carbon and the back carbon in the structure on the left.

90° Rotation

Answers to Your Turns are in the back of the book.

The Newman projection in Figure 4-2 shows that *the bond of interest is not visible; instead, it must be imagined as connecting the front and back carbons*. A Newman projection does, however, highlight the other bonds to the front and back atoms.

YOUR TURN **4.2**

Construct a molecule of 1,1,1-tribromoethane (H_3C—CBr_3) using a molecular modeling kit. Rotate the CH_3 and CBr_3 groups relative to each other until the molecule looks like the structure shown in dash–wedge notation on the left side of Figure 4-2. Look down the C—C bonding axis and fill in the missing atoms in the incomplete Newman projection provided here.

Figure 4-3 (next page) shows the Newman projections for a series of rotations about the C—C bond of a generic molecule XCH_2—CH_2Y. In the figure, we have arbitrarily chosen to leave the C atom in front (the CH_2Y group) frozen in place and to rotate the back C (the CH_2X group).

Each angle of rotation defines a particular **dihedral angle**, θ, corresponding to the angle between the C—X and C—Y bonds as they appear in the Newman projection. By convention, θ can assume values between $-180°$ and $+180°$, where the conformations at $-180°$ and $+180°$ are exactly the same because they are 360° apart.

YOUR TURN **4.3**

Add the substituents to the incomplete Newman projections to represent the first molecule after +60° and +120° rotations of the back carbon.

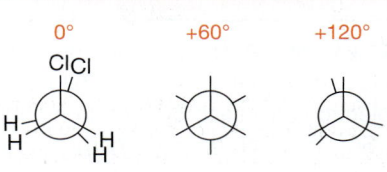

Dihedral Angle, θ	−180°	−120°	−60°	0°	+60°	+120°	+180°

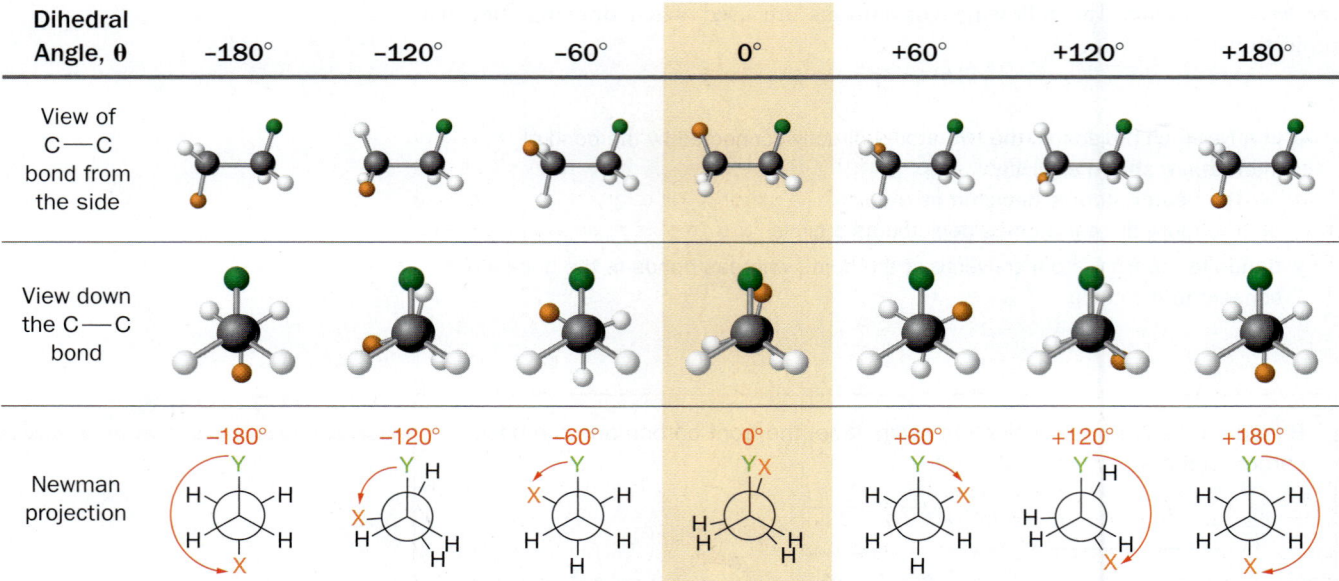

FIGURE 4-3 Newman projections and bond rotations A generic molecule of XCH$_2$—CH$_2$Y is shown at various angles of rotation (dihedral angles, θ) about the C—C bond. (*Top*) View of the molecule from the side. (*Middle*) View of the molecule down the C—C bonding axis (from the right end). (*Bottom*) Newman projection of the molecule that corresponds to the view down the C—C bond.

SOLVED PROBLEM 4.1

Draw the Lewis structure of the molecule whose Newman projection is given.

Think Which atoms are connected by the bond not observable in the Newman projection? What atoms or groups are attached to each of those atoms?

Solve The bond not shown is a C—C single bond. The C atom in the front (represented by the point) is bonded to two H atoms and a CH$_3$ group, whereas the C in the back (represented by a circle) is bonded to three H atoms. Thus, the structure consists of three C atoms bonded together. The compound is propane.

H H ··· Front carbon of Newman projection (point)

H$_3$C

H

··· Back carbon of Newman projection (circle)

H H

Propane

PROBLEM 4.2 Draw the Lewis structure that corresponds to each of the following Newman projections.

(a)   (b)   (c)

PROBLEM 4.3 The Newman projection at the right is of the same generic molecule used in Figure 4-3. Taking this dihedral angle to be −180°, draw Newman projections for each 60° rotation about the C—C single bond, from −180° to +180°, in which *the back carbon remains frozen in place*.

4.3 Conformers: Energy Changes and Conformational Analysis

To better understand the nature of a rotation about a given bond, we can perform a **conformational analysis**, which is a plot of a molecule's energy as a function of that bond's dihedral angle. The energy values for various angles of rotation are relative to that of the lowest-energy conformation. In this section, we examine the conformational analyses of ethane and 1,2-dibromoethane.

4.3a Conformational Analysis of Ethane: Torsional Strain, Eclipsed Conformations, and Staggered Conformations

Figure 4-4 depicts the conformational analysis of ethane ($H_3C—CH_3$) about the C—C bond. There are three rotational conformations in Figure 4-4 in which ethane's energy is at a *maximum* (i.e., $\theta = 0°$, $+120°$, and $-120°$). These are called **eclipsed conformations** because the C—H bonds on the front carbon atom cover, or *eclipse*, those on the rear carbon atom in the Newman projections. Because the hydrogen substituents are all identical, these three conformations are indistinguishable.

There are also three rotational conformations of $H_3C—CH_3$ in which the energy is at a minimum (i.e., $\theta = \pm180°$, $-60°$, and $+60°$). In the Newman projections of these conformations, each C—H bond on the front carbon atom bisects a pair of C—H bonds on the rear carbon, so the bonds to the front and rear carbon atoms alternate around the circle. These are called **staggered conformations**.

YOUR TURN **4.4**

In Figure 4-4, label each of the unlabeled conformations as either eclipsed or staggered.

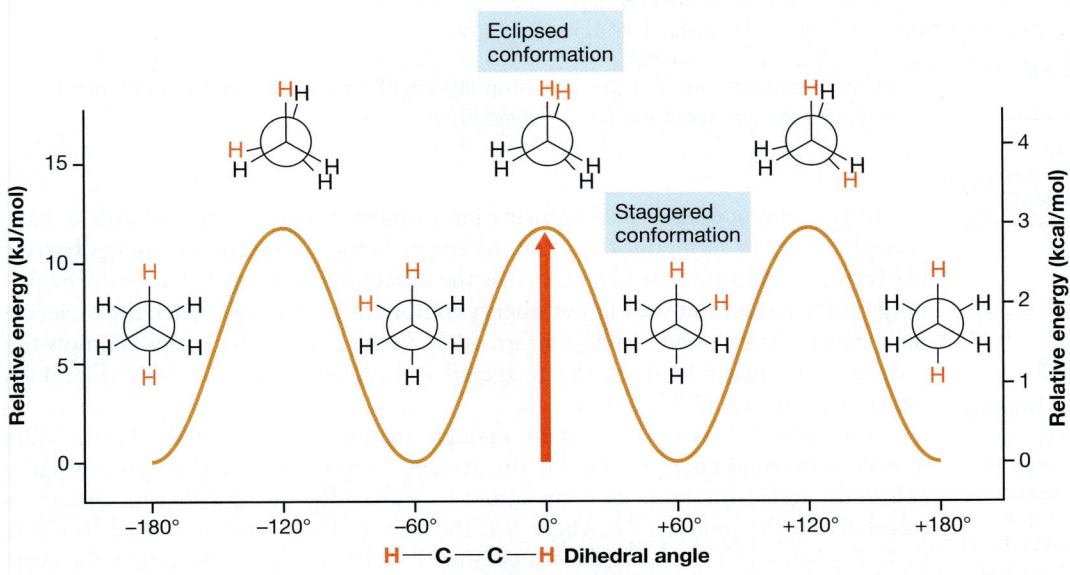

FIGURE 4-4 Conformational analysis of ethane, $H_3C—CH_3$ The plot shows the relative energy of ethane as a function of the H—C—C—H dihedral angle. Energies are relative to the lowest-energy conformation. The thick, red arrow represents the energy barrier for rotation about that bond.

More stable because electron repulsion is at a minimum

(a) Staggered

Less stable because electron repulsion is at a maximum

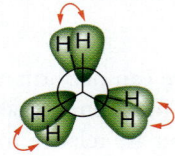

(b) Eclipsed

FIGURE 4-5 Origin of torsional strain (a) In the staggered conformation, repulsion among the electrons in the C—H bonds is at a minimum, making this the more stable conformation. (b) In the eclipsed conformation, that repulsion is at a maximum, making it the less stable conformation.

What is the source of the increase in energy in eclipsed conformations relative to staggered conformations? As we can see in Figure 4-5, the electrons making up the C—H bonds on the adjacent C atoms are closer in an eclipsed conformation, and the resulting electron repulsion causes energy to rise. Because this rise in energy, or *strain*, occurs on rotation of the single bond connecting the C atoms, we call it **torsional strain**. Also, because the C—H electron groups are closest when the conformation is eclipsed and are the farthest when it is staggered, we can say the following:

> Eclipsed conformations possess torsional strain and staggered conformations do not.

YOUR TURN 4.5

Construct a molecule of ethane (H_3C—CH_3) using a molecular modeling kit. Rotate the C—C bond so that the molecule is in an eclipsed conformation. Now rotate the C—C bond so that it is in a staggered conformation. Are the C—H bonds closer in the eclipsed conformation or in the staggered conformation?

YOUR TURN 4.6

Use Figure 4-4 to estimate the torsional strain in a molecule of ethane in an eclipsed conformation. _____ kJ/mol _____ kcal/mol

Although they appear in the conformational analysis of ethane, eclipsed conformations do not exist for significant amounts of time because they represent energy maxima (see Fig. 4-4). Instead, essentially all molecules of ethane exist in staggered conformations, which appear at energy minima. Nevertheless, ethane molecules are not locked in the *same* staggered conformation indefinitely.

> At room temperature, staggered conformations of ethane constantly interconvert through rotation about the C—C single bond.

To get from one staggered conformation to another requires the molecule to have enough energy to surmount the *rotational energy barrier*. A **rotational energy barrier** between two conformations, in general, is the difference in energy between the beginning conformation and the highest-energy conformation through which the molecule must pass to arrive at the ending conformation. In the case of ethane, this is simply the difference in energy between the staggered and eclipsed conformations (Fig. 4-4), which is 12 kJ/mol (2.9 kcal/mol).

From where does ethane acquire enough energy to surmount the barrier? The answer is **thermal energy**, which is the average energy available through molecular collisions, and it increases as temperature increases. The average thermal energy is calculated as the product RT, where R is the universal gas constant (8.314 J/mol·K or 1.987 cal/mol·K) and T is the temperature in kelvins. If T is 298 K, then the average thermal energy is ~2.5 kJ/mol (~0.6 kcal/mol). Although this is less than the 12 kJ/mol (2.9 kcal/mol) rotational energy barrier in ethane, we can see from Figure 4-6 that, at any given time, 2% of molecules have at least that much energy. In a mole of ethane, that is approximately 1.2×10^{22} molecules! We can extend this idea:

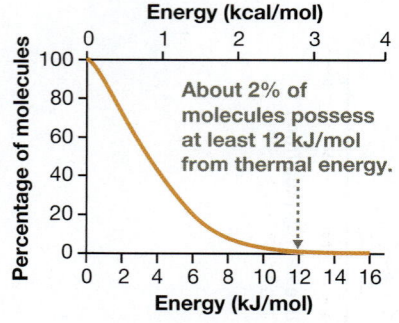

FIGURE 4-6 Distribution of thermal energy in molecules at 298 K For each energy plotted on the *x* axis, the percentage of molecules possessing at least that much energy at 298 K is plotted on the *y* axis. At any given time, about 2% of molecules possess enough energy to surmount an energy barrier of 12 kJ/mol (2.9 kcal/mol), the rotational energy barrier for ethane.

In general, two conformations will rapidly interconvert if the rotational energy barrier is of the same order of magnitude as thermal energy.

4.3b Conformational Analysis of 1,2-Dibromoethane: Steric Strain, Gauche Conformations, and Anti Conformations

If one H atom of each CH_3 group in ethane (H_3C—CH_3) is replaced by another substituent, the energies of the three staggered conformations are no longer the same. Neither are the energies of the three eclipsed conformations. This can be seen in the conformational analysis of 1,2-dibromoethane (Br—CH_2—CH_2—Br) shown in Figure 4-7.

YOUR TURN 4.7

Label each unlabeled structure in Figure 4-7 as either eclipsed or staggered.

YOUR TURN 4.8

Use Figure 4-7 to estimate how much higher in energy one eclipsed conformation of 1,2-dibromoethane (BrCH$_2$—CH$_2$Br) is than the other two.

_____ kJ/mol   _____ kcal/mol

Additionally, estimate how much lower in energy one staggered conformation is than the other two.

_____ kJ/mol   _____ kcal/mol

FIGURE 4-7 Conformational analysis of 1,2-dibromoethane, Br—CH$_2$—CH$_2$—Br The energy of 1,2-dibromoethane is plotted as a function of the Br—C—C—Br dihedral angle. Energies are relative to that of the most stable conformation. The gauche and anti conformations are labeled.

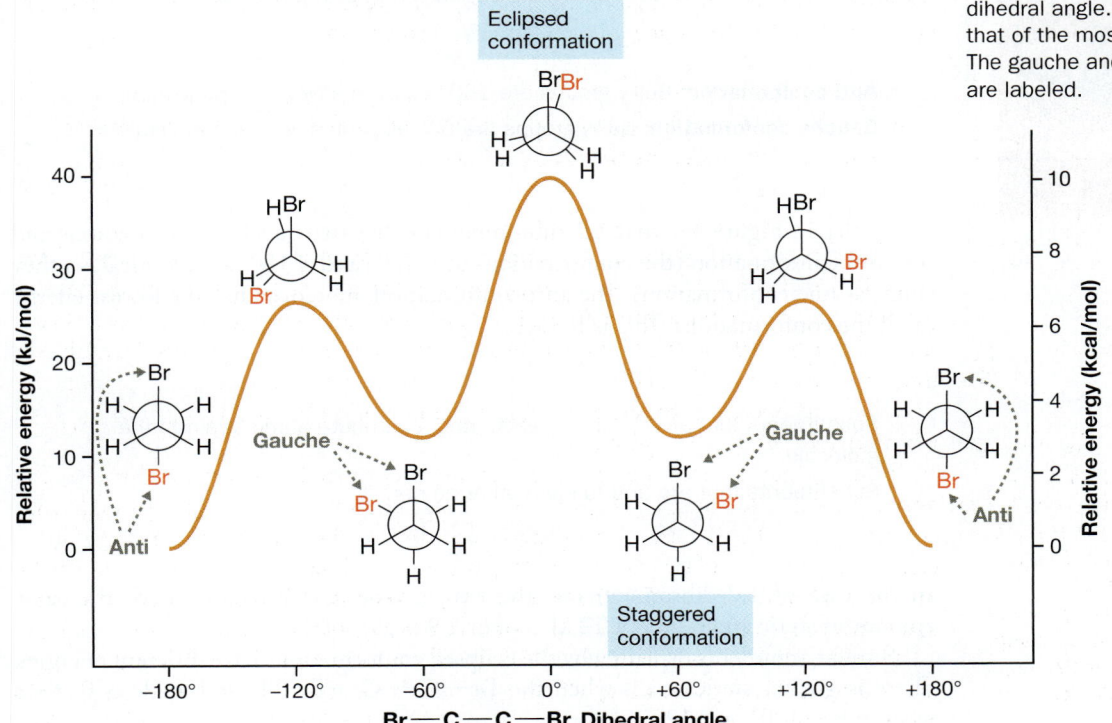

Gauche

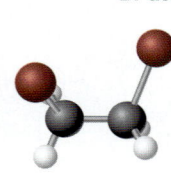 

Br atoms crashing into each other

Anti

 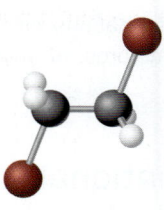 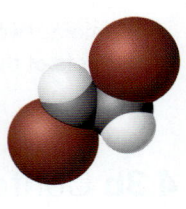

(a)

(b)

FIGURE 4-8 **Steric strain in 1,2-dibromoethane, BrCH$_2$CH$_2$Br** (a) 1,2-Dibromoethane in its gauche conformation, shown as a Newman projection (left), a ball-and-stick model (middle), and a space-filling model (right). Because the two Br atoms are quite large, their electrons begin to occupy the same space, causing steric strain. (b) 1,2-Dibromoethane in its anti conformation. Because the Br atoms are 180° apart, this conformation does not possess steric strain.

CONNECTIONS

1,2-Dibromoethane (Fig. 4-8), also called ethylene dibromide, has use in the synthesis of other compounds in the organic laboratory. It is also used as a fumigant for the control of bark beetles and termites in trees that have fallen, and also for the control of wax moths in beehives.

The differences in energy that you estimated in Your Turn 4.8 are mainly due to the much larger size of a bromine atom compared to a hydrogen atom. As the Br atoms are brought closer together through rotation about the C—C bond, they begin to crash into each other, and their electrons, forced to occupy the same space, repel one another. This is a form of strain called *steric strain* and is depicted in Figure 4-8. More generally:

Steric strain is an increase in energy that results from electron repulsion between atoms or groups of atoms that are not directly bonded together but occupy the same space.

Because of the energy changes caused by bulky groups like bromine atoms, the staggered conformations are further distinguished as anti and gauche:

- **Anti conformation:** Bulky groups are 180° apart in a Newman projection.
- **Gauche conformation:** Bulky groups are 60° apart in a Newman projection.

Notice in Figure 4-7 that 1,2-dibromoethane has two gauche conformations and one anti conformation (the conformations at −180° and +180° are identical, so they count as one conformation). The anti conformation, moreover, has the lowest energy of all the conformations. This is to say:

- Substituents that are gauche to each other contribute steric strain to the molecule.
- Substituents that are anti to each other do not.

In the case of 1,2-dibromoethane, the two gauche conformations have the same amount of steric strain (about 12 kJ/mol or 2.9 kcal/mol).

Similar arguments explain why the eclipsed conformations have different energies. There is greater steric strain when the Br—C—C—Br dihedral angle is 0° than when it is +120° or −120°.

Identify which of these conformations is anti and which is gauche.

The three staggered (i.e., the two gauche and one anti) conformations of $Br—CH_2—CH_2—Br$ occur at energy minima, so they are stable relative to the eclipsed conformations. Furthermore, these three conformations have the same molecular formula yet are not exactly the same. Therefore, according to Figure 4-1:

> Gauche and anti conformations are *conformers* of each other.

Like those in ethane, all three staggered conformations of 1,2-dibromoethane interconvert rapidly at room temperature, because the rotational energy barrier between the anti and gauche conformations (see Your Turn 4.10) is similar in magnitude to the thermal energy available at room temperature (2.5 kJ/mol; 0.60 kcal/mol). Consequently, just as in ethane, a significant (but smaller) percentage of 1,2-dibromoethane molecules possess sufficient energy to surmount that barrier.

In Figure 4-7, draw an arrow, similar to the one in Figure 4-4, to indicate the rotational energy barrier on going from the anti conformation to one of the gauche conformations. Estimate that energy barrier here:

_____ kJ/mol   _____ kcal/mol

Even though the gauche and anti conformations of 1,2-dibromoethane rapidly interconvert, they do not exist in equal abundance.

> An anti conformation is lower in energy (more stable) than the corresponding gauche conformation, so an anti conformation is preferred.

In fact, $Br—CH_2—CH_2—Br$ molecules at room temperature are 98% anti and 2% gauche.

PROBLEM 4.4 Perform a conformational analysis of 1,2-dichloroethane ($ClCH_2—CH_2Cl$) by sketching its energy as a function of the dihedral angle about the C—C bond. Make sure the *relative* energies are correct, but do not concern yourself with the exact values. Draw Newman projections for each staggered and eclipsed conformation, and identify the anti and gauche conformations.

PROBLEM 4.5 Which molecule do you think has a larger rotational energy barrier about the C—C bond: 1,2-dibromoethane or 1,2-difluoroethane? Why?

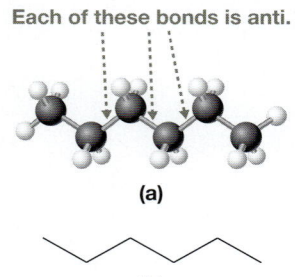

Each of these bonds is anti.

(a)

(b)

FIGURE 4-9 **The all-anti conformation of hexane**
(a) Ball-and-stick model of hexane. The all-anti conformation is most stable, giving rise to a zigzag structure. (b) Line structure of hexane.

4.3c Longer Molecules and the Zigzag Conformation

For longer molecules, such as hexane ($CH_3CH_2CH_2CH_2CH_2CH_3$), there are several single bonds about which rotation can occur, giving rise to many possible conformations. Each of the three interior C—C bonds in hexane has the form R—CH_2—CH_2—R, in which R is an alkyl group. As a result, each of these bonds can exist in either a gauche or an anti conformation. Just as we saw with 1,2-dibromoethane, anti conformations are favored over gauche, so the most stable conformation of hexane is the **all-anti conformation**, often called the **zigzag conformation** (Fig. 4-9a). This is the basis for the zigzag convention we use to depict alkyl chains in line structures (Fig. 4-9b).

4.4 Conformers: Cyclic Alkanes and Ring Strain

Thus far, we have limited our conformational analyses to single bonds in acyclic compounds. Single bonds, however, may also be part of ring structures (Fig. 4-10), so ring structures, too, can have conformations associated with them. We examine these conformations in Sections 4.5 and 4.6. Here, we introduce ring structures in general and discuss their relative stabilities.

FIGURE 4-10 **Cycloalkanes**
Because their rings consist of single bonds, some cycloalkanes can attain different conformations.

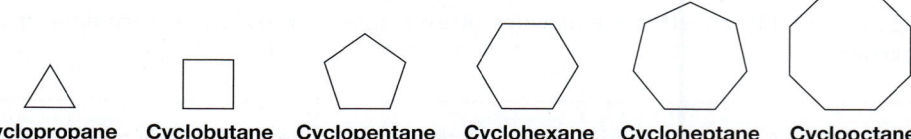

Cyclopropane   Cyclobutane   Cyclopentane   Cyclohexane   Cycloheptane   Cyclooctane

Ring structures consisting only of single bonds are abundant in nature (Fig. 4-11). Menthol, for example, is a natural oil that contains a six-membered ring. Androsterone, a steroid, consists of three six-membered rings and one five-membered ring. Diamond, one of the hardest substances known, is an extended network of six-membered rings. And a principal component of RNA and DNA are five-membered rings called ribose and deoxyribose, respectively.

Although compounds can exist with rings of any size, five- and six-membered rings are most prevalent in nature, suggesting that rings of these sizes are particularly stable. We attribute the instability of other ring sizes to **ring strain**, the increase in energy due to geometric constraints on the ring.

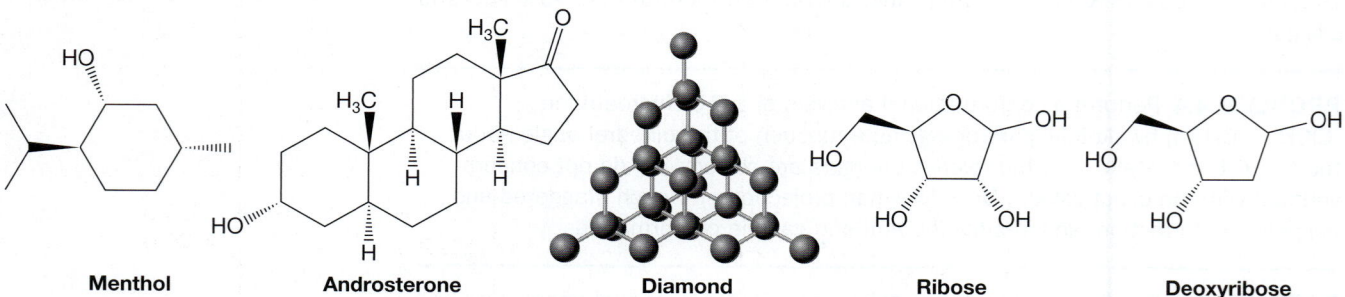

Menthol   Androsterone   Diamond   Ribose   Deoxyribose

FIGURE 4-11 **Rings in nature** Menthol, androsterone, and diamond have rings made of only carbon. Ribose and deoxyribose have rings that consist of carbon and oxygen.

Ring strain can be quantified using **heats of combustion**, the energy given off in the form of heat ($\Delta H°$) during a combustion reaction. The heats of combustion for some cyclic alkanes are listed in Table 4-1.

We can extract information about strain from heats of combustion because the *balanced* chemical equation for the combustion of each of these compounds has the same form; all that varies is n, the number of carbon atoms:

$$(CH_2)_n + (3n/2)\, O_2 \rightarrow n\, CO_2 + n\, H_2O + \text{heat} \qquad \text{(4-1)}$$

The number of moles of each combustion product is also n, so the amount of product depends on the size of the ring. To account for this, we can think of dividing the above terms by n to yield the balanced chemical equation in Equation 4-2, which contains each ring's heat of combustion *per CH_2 group*: heat/n.

$$\text{``}(CH_2)\text{''} + (3/2)\, O_2 \rightarrow CO_2 + H_2O + \text{heat}/n \qquad \text{(4-2)}$$

Values of heat/n are listed for the C_3 through C_7 rings in Table 4-1.

TABLE 4-1 Heats of Combustion and Ring Strain for Cyclic Alkanes

Cycloalkane	Number of CH_2 Units	HEAT OF COMBUSTION, $\Delta H°$		HEAT OF COMBUSTION, $\Delta H°$, PER CH_2 UNIT		RING STRAIN PER CH_2 UNIT, RELATIVE TO CYCLOHEXANE		TOTAL RING STRAIN RELATIVE TO CYCLOHEXANE	
		kJ/mol	kcal/mol	kJ/mol	kcal/mol	kJ/mol	kcal/mol	kJ/mol	kcal/mol
Cyclopropane	3	1961.0	468.7	653.7	156.2	653.7 − 615.1 = 38.6	156.2 − 147.0 = 9.2	3(38.6) = 115.8	3(9.2) = 27.6
Cyclobutane	4	2570.2	614.3	642.6	153.6	642.6 − 615.1 = 27.5	153.6 − 147.0 = 6.6	4(27.5) = 110.0	4(6.6) = 26.4
Cyclopentane	5	3102.5	741.5	620.5	148.3	620.5 − 615.1 = 5.4	148.3 − 147.0 = 1.3	5(5.4) = 27.0	5(1.3) = 6.5
Cyclohexane	6	3690.7	882.1	615.1	147.0	615.1 − 615.1 = 0	147.0 − 147.0 = 0	6(0) = 0	6(0) = 0
Cycloheptane	7	4332.1	1035.4	618.9	147.9	618.9 − 615.1 = 3.8	147.9 − 147.0 = 0.9	7(3.8) = 26.6	7(0.9) = 6.3

Because the products of combustion are otherwise identical, we may assume the following for a cyclic alkane:

- Any difference in the heat of combustion per CH_2 group for two cyclic alkanes reflects a difference in *ring strain* per CH_2 group.
- The total ring strain is obtained by multiplying the strain per CH_2 group by the number of CH_2 groups, n.

According to Table 4-1, ring strain is smallest for cyclohexane, the six-membered ring. In fact, as we will see in Section 4.5, cyclohexane is considered to have no ring strain at all. Cyclopentane (the five-membered ring) and cycloheptane (the seven-membered ring), on the other hand, both have small amounts of ring strain. Cyclopropane (the three-membered ring) and cyclobutane (the four-membered ring) are highly strained.

YOUR TURN 4.11

Cyclooctane

The heat of combustion of cyclooctane is 4962.2 kJ/mol (1186.0 kcal/mol).

How many CH_2 groups does cyclooctane have? _____

Compute the heat of combustion per CH_2 group. _____

Compute the ring strain per CH_2 group by subtracting cyclohexane's heat of combustion per CH_2 group from that determined for cyclooctane. _____

Compute the total ring strain for cyclooctane by taking into account the total number of CH_2 groups in the molecule. _____

How do these values compare to those for the other molecules in Table 4-1?

_____

SOLVED PROBLEM 4.6

Rank the following cycloalkanes in order from lowest heat of combustion to highest.

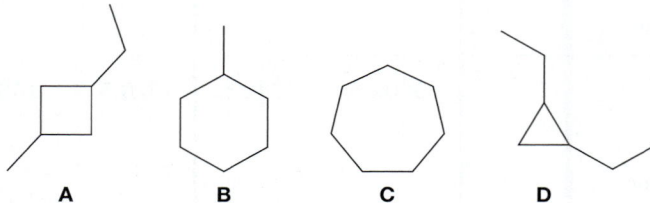

A   **B**   **C**   **D**

Think What is the same among these molecules? What is different? How do those differences translate into heats of combustion?

Solve All four of these compounds have the molecular formula C_7H_{14} and contain only C—C and C—H single bonds. Differences in heats of combustion, therefore, will largely reflect differences in ring strain. The three-membered ring (**D**) is the most strained, so it will give off the most heat during combustion—it will have the highest heat of combustion. Next comes the four-membered ring (**A**), followed by the seven-membered ring (**C**). The six-membered ring (**B**) is the least strained, so it will have the lowest heat of combustion. In order, then, the heats of combustion are as follows: **B** < **C** < **A** < **D**.

E F G H

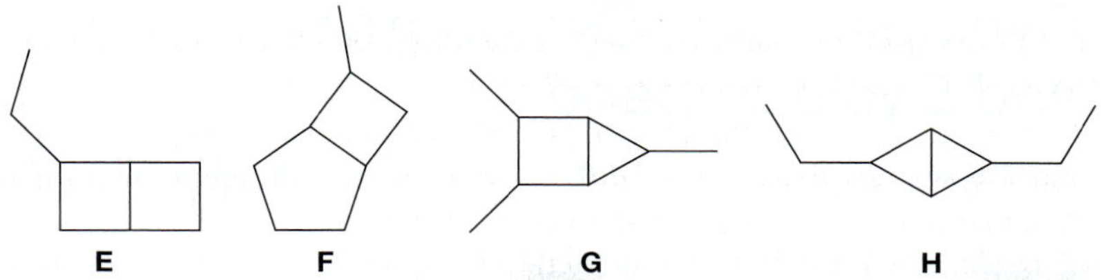

An All-Gauche Alkane

How can a linear alkane be made to adopt an all-gauche conformation? The answer is to trap it inside a molecular capsule whose cavity is shorter than the stretched alkane in its zigzag conformation. Recall that a linear alkane, free of any external influences, prefers the all-anti conformation. But in 2004, Julius Rebek, Jr., at The Scripps Research Institute, trapped linear tetradecane ($C_{14}H_{30}$) inside a capsule that self-assembles via hydrogen bonds between its two halves.

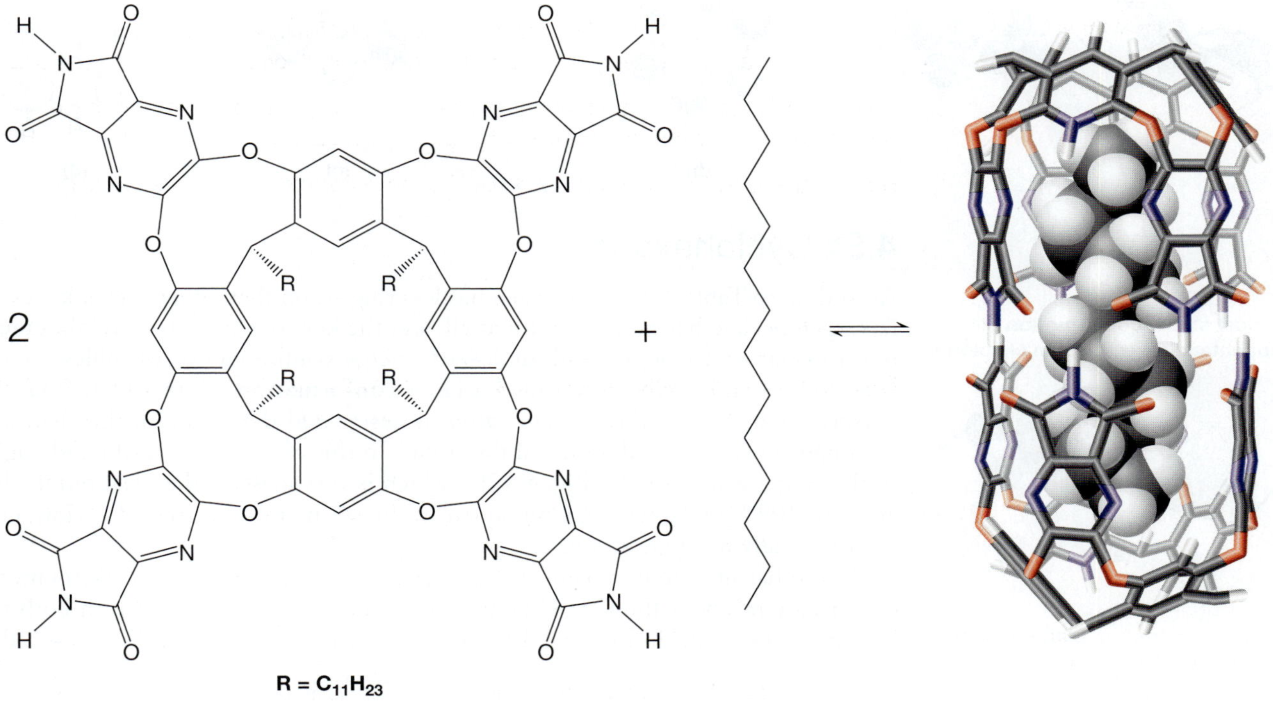

$R = C_{11}H_{23}$

To fit inside the cavity, the molecule must twist into a helical structure to become shorter. In that helical structure, all C—C bonds adopt a gauche conformation. Each gauche conformation raises the energy of, and thus destabilizes, the alkane, but that destabilization is more than compensated by the hydrogen bonding between the two halves of the capsule and by the favorable interactions between the alkane and the capsule walls.

Studies such as this are more than just a novelty. They can, in fact, help us better understand the way that some medicines function. Medicines are typically small molecules that operate inside a relatively small and confining cavity of a protein or nucleic acid. And it is known that the severe confinement of a molecule can have a dramatic effect on its chemical behavior.

4.5 Conformers: The Most Stable Conformations of Cyclohexane, Cyclopentane, Cyclobutane, and Cyclopropane

In Section 4.4, we saw that rings of different size possess different amounts of *ring strain*. What structural factors give rise to ring strain?

Angle strain, which we first encountered in Chapter 2, makes a significant contribution to ring strain. In the cyclic alkanes discussed so far, the atoms of the ring are sp^3 hybridized, with an *ideal* bond angle of 109.5°. Geometric constraints imposed by the ring, however, can force significant deviations in these bond angles, thereby causing an increase in energy. Additionally, *torsional strain* and *steric strain* can contribute to ring strain. This should come as no surprise, because torsional strain and steric strain can exist for any single bond, whether or not the bond is part of a ring structure.

Here in Section 4.5, we study these contributions toward ring strain in rings of various sizes.

FIGURE 4-12 Various representations of cyclohexane (a) Ball-and-stick model of cyclohexane viewed from the side, illustrating its chair conformation. (b) Ball-and-stick model of cyclohexane viewed from the top. All C—C—C bond angles are about 111°. (c) Ball-and-stick model of cyclohexane viewed down two C—C bonds, illustrating that those bonds are in staggered conformations. (d) Newman projection of cyclohexane.

The most stable conformation of cyclohexane resembles a chair.

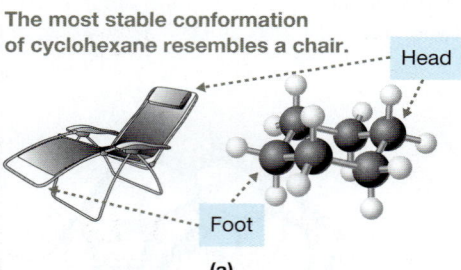
Head
Foot
(a)

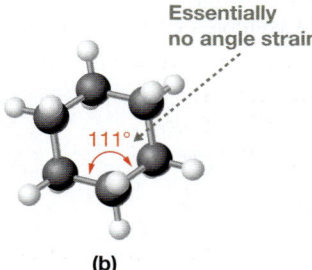
Essentially no angle strain
111°
(b)

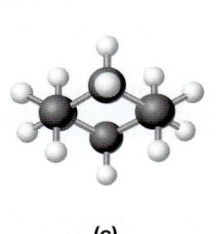
(c)

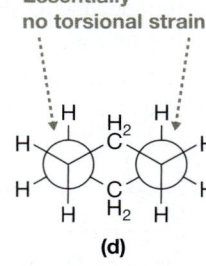
Essentially no torsional strain
(d)

4.5a Cyclohexane

According to Table 4-1, cyclohexane has less ring strain than other cycloalkanes. In fact, cyclohexane has no ring strain at all, and the key to why is that cyclohexane is not a planar molecule; instead, its lowest-energy conformation resembles a chair (Fig. 4-12a) and is therefore called a **chair conformation**. Notice that, from the perspective in Figure 4-12a, one C atom is designated the "head" of the chair and the opposite C atom is designated the "foot." In this conformation, all bond angles of the ring are about 111° (Fig. 4-12b), which is very close to the ideal tetrahedral angle of 109.5° (Chapter 2). Consequently, the six-membered ring of cyclohexane has essentially no angle strain.

It also has little to no torsional strain (Section 4.3a), because all of the rotational conformations about the C—C bonds are staggered. This can be seen in both the ball-and-stick model in Figure 4-12c and the Newman projection in Figure 4-12d.

The most stable conformation of cyclopentane resembles an envelope.

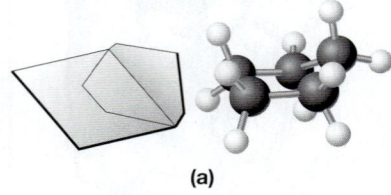
(a)

Bond in front of plane of paper and parallel to paper.

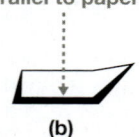
(b)

FIGURE 4-13 The envelope conformation of cyclopentane (a) Ball-and-stick representation of the envelope conformation of cyclopentane. (b) Line structure of the envelope conformation with dash–wedge notation. The thicker bond on the bottom indicates that it is in front of, and parallel to, the plane of the paper.

4.5b Cyclopentane

Like cyclohexane, the lowest-energy conformation of cyclopentane is not entirely planar, as shown in Figure 4-13. Four of its five carbon atoms lie essentially in one plane, but the fifth carbon is outside of that plane. If you imagine cyclopentane's carbon atoms located at the five corners of the envelope in Figure 4.13a, you can see why this geometry is referred to as an **envelope conformation**.

Unlike cyclohexane, we saw from Table 4-1 that cyclopentane has a small amount of ring strain. It possesses some angle strain, because its bond angles range from 102° to 106°, which are somewhat farther from the ideal tetrahedral bond angle of 109.5° than the 111° angles of cyclohexane. Furthermore, as shown in Figure 4-14, cyclopentane's C—C—C—C dihedral angles are slightly eclipsed.

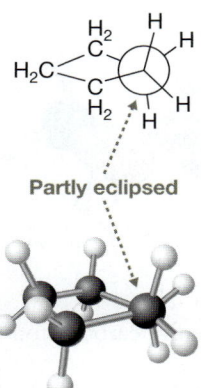

Partly eclipsed

FIGURE 4-14 Torsional strain in cyclopentane Ball-and-stick model (*bottom*) and Newman projection (*top*) of cyclopentane. The dihedral angle shown is partly eclipsed, giving rise to a small amount of torsional strain.

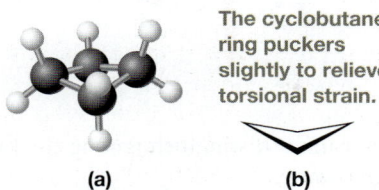

The cyclobutane ring puckers slightly to relieve torsional strain.

(a)   (b)

FIGURE 4-15 Puckered conformation of cyclobutane (a) Ball-and-stick representation of cyclobutane. (b) Line structure. The most stable conformation of cyclobutane has a slightly puckered ring.

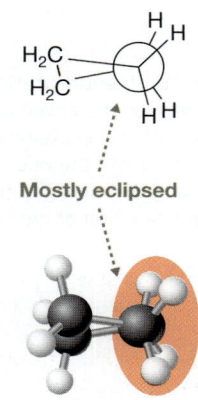

Mostly eclipsed

FIGURE 4-16 Torsional strain in cyclobutane Ball-and-stick model (*bottom*) and Newman projection (*top*) of puckered cyclobutane, illustrating that the C—C bond is mostly eclipsed.

YOUR TURN **4.12**

Using a molecular modeling kit, build a molecule of cyclohexane in its chair conformation, then rotate it in space until it appears as in Figure 4-12c. Observe the staggered conformation about each C—C bond.

YOUR TURN **4.13**

Identify the C—C bond in Figure 4-13 that has the most torsional strain. It may help if you build a model of cyclopentane in its envelope conformation using a modeling kit.

4.5c Cyclobutane and Cyclopropane

The most stable conformation for cyclobutane is close to square, but not exactly; instead, the ring is slightly puckered, with interior angles of about 88° (Fig. 4-15). This large deviation from the ideal bond angle of 109.5° leaves cyclobutane with a substantial amount of angle strain. Furthermore, as we can see in Figure 4-16, the conformation of each C—C bond remains mostly eclipsed, giving cyclobutane a substantial amount of torsional strain, too.

For cyclopropane, there is no alternative to having all three carbon atoms in the same plane, because three points define a plane. All three angles of the ring are exactly 60° and all three C—C bonds are the same length, so the ring forms an equilateral triangle. As a result, cyclopropane has more angle strain than the other cyclic molecules we have examined. Furthermore, the conformation at all of the C—C bonds is fully eclipsed, as shown in Figure 4-17, resulting in significant torsional strain.

4.6 Conformers: Cyclopentane, Cyclohexane, Pseudorotation, and Chair Flips

In the envelope conformation of cyclopentane (Fig. 4-13), four of the carbon atoms lie in the same plane, with the remaining carbon outside that plane. It appears, therefore, that one carbon atom is distinct from the other four. Experiments, to the

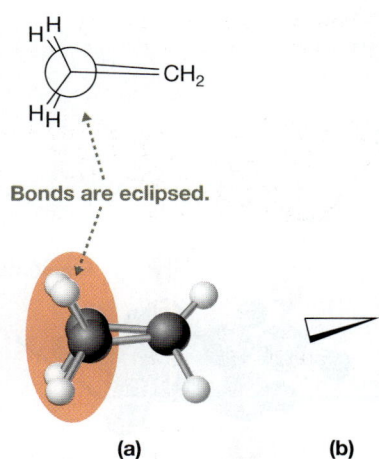

Bonds are eclipsed.

(a)   (b)

FIGURE 4-17 Cyclopropane (a) Ball-and-stick model (bottom) and Newman projection (top) of cyclopropane, showing that the C—C bonds are eclipsed. (b) Line structure of cyclopropane.

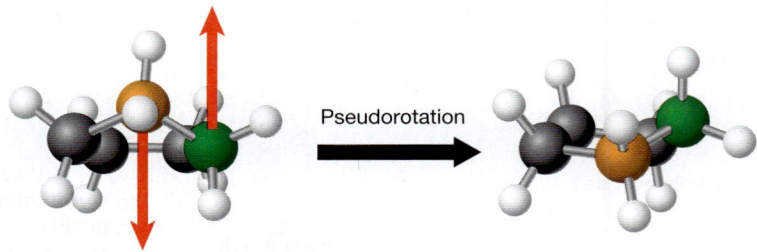

FIGURE 4-18 Pseudorotation in cyclopentane The envelope conformation of cyclopentane on the left is converted into the one on the right by moving the highlighted carbon atoms in the direction of the red arrows.

Pseudorotation

contrary, cannot distinguish among the five carbons, so the five carbon atoms are said to be *equivalent*.

How can we explain this discrepancy? The answer is **pseudorotation**, which enables all five possible envelope conformations—each with a different carbon atom out of the plane—to interconvert. The energy barrier for pseudorotation is very small, which means that these interconversions take place constantly and very rapidly. It is impossible, therefore, to isolate any one of the envelope conformations from the others.

The mechanism by which pseudorotation occurs in cyclopentane is shown in Figure 4-18. The out-of-plane carbon atom (orange) moves into the plane, whereas one of the carbon atoms in the plane (green) moves out of the plane. Although it may be difficult to see from the figure:

> Pseudorotation occurs by *partial rotations* about the C—C bonds.

This explains, at least in part, why the energy barrier for the interconversion is so small—namely, covalent bonds are never broken. You may find a molecular modeling kit useful in illustrating these bond rotations (see Your Turn 4.14).

YOUR TURN 4.14

> Use a molecular modeling kit to build a molecule of cyclopentane in its envelope conformation. View it so that it appears as shown on the left in Figure 4-18 and move the atoms as indicated in the figure. Identify which C—C bonds rotate the most and circle those bonds on the drawing on the left in Figure 4-18.

In cyclohexane, all six carbon atoms are completely indistinguishable, but there are two different types of hydrogen atoms, as shown in Figure 4-19. Six hydrogen atoms occupy *equatorial* positions and six occupy *axial* positions. Each carbon atom in cyclohexane is bonded to one of each.

> - **Equatorial** bonds lie almost in the plane that is roughly defined by the ring (i.e., the *equator* of the molecule; see Fig. 4-19) and point outward from the center of the ring.
> - **Axial** bonds are perpendicular to the plane.

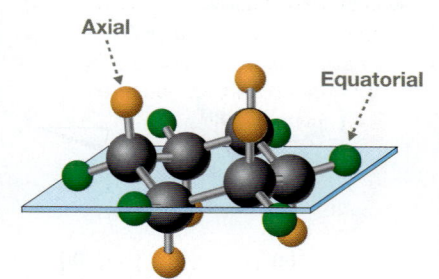

Axial

Equatorial

FIGURE 4-19 The axial and equatorial hydrogens in cyclohexane This chair conformation of cyclohexane shows that there are six axial H atoms (orange) and six equatorial H atoms (green). Bonds to axial H atoms are perpendicular to the plane indicated, whereas bonds to equatorial H atoms are nearly in the plane.

In most experiments, however, the axial and equatorial hydrogens in cyclohexane are indistinguishable. This is because there are two chair conformations that interconvert very rapidly—on the order of millions of times per second at room temperature—through a process called a **chair flip** or **ring flip**. As shown in Figure 4-20, the head of the chair becomes the foot after a chair flip, and vice versa. Moreover:

> A chair flip converts axial hydrogens into equatorial hydrogens, and vice versa.

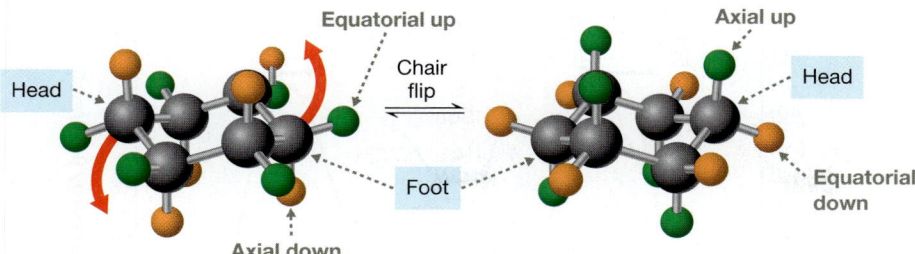

Equatorial up Axial up

Head Head

Chair flip

Foot

Axial down Equatorial down

FIGURE 4-20 Chair flip of a cyclohexane ring The chair conformation on the left can be converted into the chair conformation on the right by moving the two specified C atoms in the directions indicated by the red arrows. Notice that all the equatorial positions in the structure on the left (green) become axial after the chair flip, and all the axial positions (orange) become equatorial.

The rate at which the two chair conformations interconvert is so high because the energy barrier between the two chair conformations is quite low. The energy barrier is low because a chair flip does not involve breaking or forming bonds. Instead, like pseudorotation in cyclopentane, a chair flip in cyclohexane involves only rotations about bonds that make up the ring.

YOUR TURN 4.15

Use a molecular modeling kit to construct a model of cyclohexane and view it so that it looks just like the structure on the left in Figure 4-20. Perform a chair flip by moving the two carbon atoms indicated by the red arrows to obtain the conformation on the right in the figure. Reverse and repeat this procedure several more times. As you flip the chair back and forth, identify the bonds of the ring that rotate and the directions in which they rotate. Circle and label those bonds in the drawing on the left in Figure 4-20.

Even though a chair flip interconverts axial and equatorial positions on a cyclohexane ring, it does *not* allow substituents to switch sides of the ring's plane. During a chair flip, a hydrogen atom in an axial position on one side of the ring's plane becomes an equatorial hydrogen on the same side. Likewise, an equatorial hydrogen on one side of the plane becomes an axial hydrogen on that same side. Viewing the cyclohexane ring from the side, as in Figure 4-20:

- A chair flip converts axial-up positions to equatorial-up (and vice versa).
- A chair flip converts axial-down positions to equatorial-down (and vice versa).

Rather than occurring in a single step, a cyclohexane chair flip involves the multiple independent steps shown in Figure 4-21 (next page). Throughout such a chair flip, cyclohexane assumes key conformations known as the *half-chair*, the *twist–boat*, and the *boat*.

As can be seen from Figure 4-21:

The half-chair, twist-boat, and boat conformations in a chair flip are higher in energy than the chair conformation itself, due to added ring strain.

In the boat conformation (Fig. 4-22), for example, two of the C—C bonds are in an eclipsed conformation (Fig. 4-22b). In addition, there is a **flagpole interaction** between hydrogen atoms on an opposing pair of carbon atoms (Fig. 4-22a). This is a form of steric strain that is absent in the chair conformation.

CONNECTIONS When cyclohexane is cooled to well below room temperature, its chair flips are slowed substantially, allowing the two types of hydrogens to be detected more easily. Nuclear magnetic resonance (NMR) spectroscopy (Chapter 16) is one technique that has been successful in doing so.

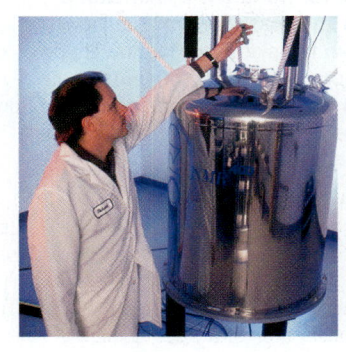

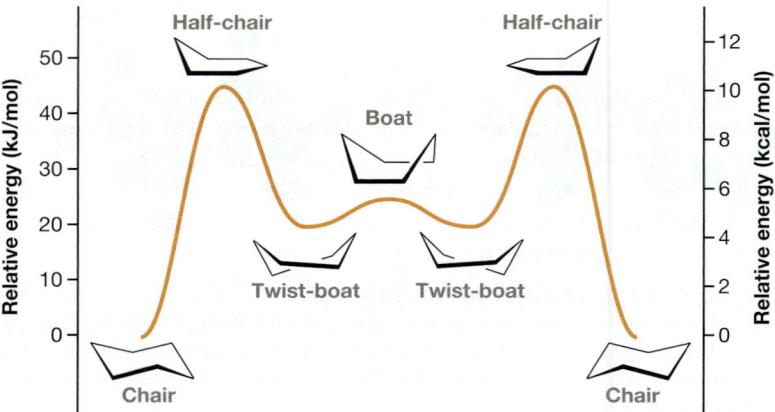

FIGURE 4-21 Energy diagram of a chair flip A chair flip that converts the conformation on the left into the one on the right goes through several key conformations, including the half-chair, twist-boat, and boat. Energies are relative to the chair conformation.

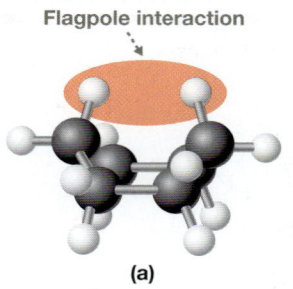

Flagpole interaction

(a)

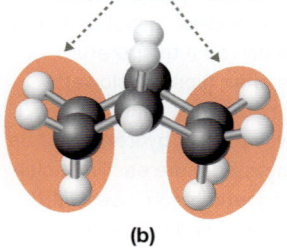

Eclipsed conformations

(b)

FIGURE 4-22 Strain in the boat conformation of cyclohexane
(a) The boat conformation viewed from the side, showing the flagpole interaction, a form of steric strain.
(b) The boat conformation viewed from one end, showing the two eclipsed conformations.

Use a molecular modeling kit to build this half-chair conformation. Examine the model from different points of view, and note the different types of strain. In the figure provided here, indicate where the strain exists, as well as the type of strain (i.e., angle, steric, or torsional). With these observations, can you justify why the half-chair conformation is so high in energy?

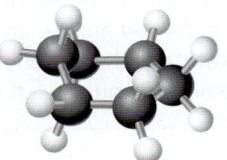

4.7 Strategies for Success: Drawing Chair Conformations of Cyclohexane

Given the abundance of cyclohexane rings, it would soon become cumbersome if we always had to represent chair conformations three-dimensionally as ball-and-stick models (Fig. 4-23a) or in dash–wedge notation (Fig. 4-23b). Chemists, therefore, have devised the shorthand notation for drawing chair conformations shown in Figure 4-23c. Working with these structures is not trivial, so we devote the rest of this section to

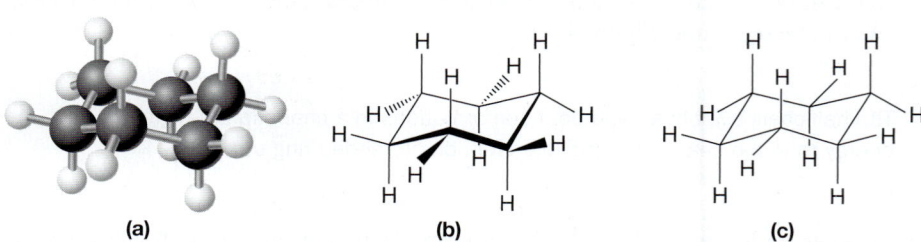

(a) (b) (c)

FIGURE 4-23 Various representations of the chair conformation of cyclohexane
(a) Ball-and-stick model. (b) Dash–wedge notation. (c) Shorthand notation.

Cubane: A Useful "Impossible" Compound?

Cubane, C_8H_8, is an exotic molecule in which the eight carbon atoms are located at the corners of a cube. It was once thought to be an impossible compound, unable to exist due to excessive strain. In 1964, however, Philip Eaton and Thomas Cole, Jr., at the University of Chicago, successfully carried out its synthesis. At the time, such a synthesis was more a novelty than anything else. But since then, cubane and its derivatives have been finding widespread potential applications. One of the longest-studied applications is its potential use as a high-energy fuel. With the large amount of strain it has, combined with its relatively high density (nearly twice that of gasoline), cubane can store energy more efficiently than conventional fuels.

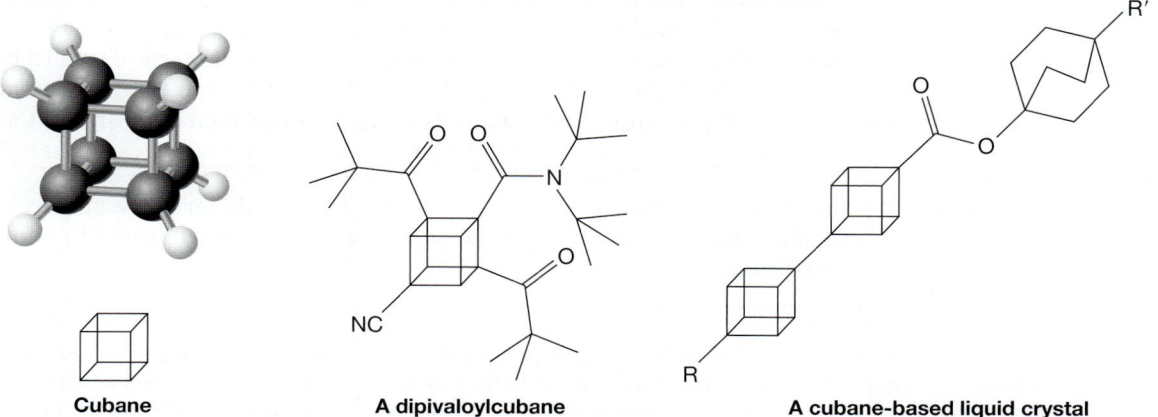

Cubane   **A dipivaloylcubane**   **A cubane-based liquid crystal**

Derivatives of cubane also have potential use in medicine, owing to the rigid, lipophilic framework to which up to eight, independent, functional groups can be attached at eight specific locations. Consider, for example, the dipivaloylcubane shown above, which has been shown to exhibit moderate activity against the human immunodeficiency virus (HIV).

Cubane derivatives also have applications in materials science. For example, liquid crystals have been synthesized with cubane as the central structural component, and individual cubane units have been linked together to form polymers with interesting properties. With each passing year, more applications of cubane and its derivatives are sure to be found.

---

both drawing and interpreting chair structures using this shorthand method. Before we begin, note the following features of the shorthand notation:

- By convention, the C—C bonds toward the bottom of the ring are interpreted as being in front of the plane of the paper (Fig. 4-24a).
- When the ring is oriented as it is in Figure 4-23, all of the axial bonds (red) are perfectly vertical, *alternating up and down* around the ring (Fig. 4-24b).
- All equatorial bonds (blue) are either slightly up or slightly down, *alternating around the ring*; on a carbon where the axial bond is down, there is an equatorial bond slightly up, and vice versa (Fig. 4-23c).

Axial bonds alternate up and down around the ring.

Equatorial bonds alternate slightly up and down around the ring.

These bonds are in front.

(a)   **(b)**   **(c)**

FIGURE 4-24 Interpreting the shorthand notation for the chair conformation of cyclohexane (a) By convention, the bonds on the bottom are in front of the plane of the paper. (b) Axial bonds (red) are vertical and alternate up and down around the ring. (c) Equatorial bonds (blue) alternate slightly up and slightly down around the ring.

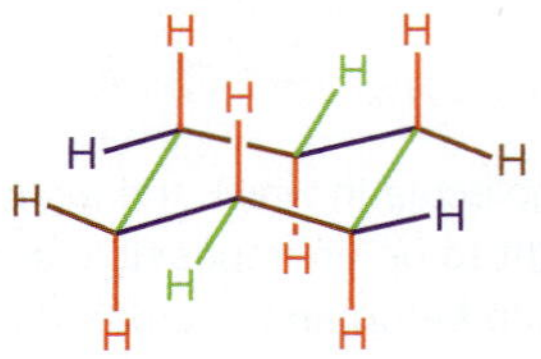

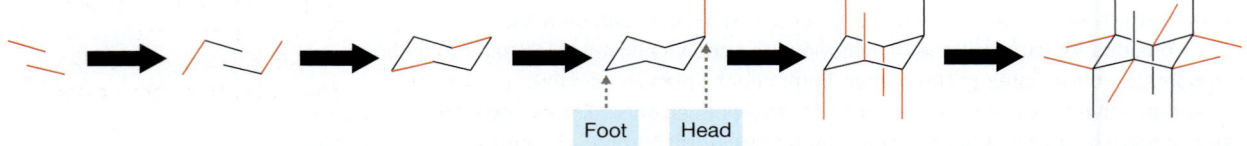

FIGURE 4-25 Parallel bonds in the shorthand notation for cyclohexane

The chair conformation of cyclohexane is color-coded to show its four sets of parallel bonds. Each C—C bond is parallel to the one opposite it in the ring and to a pair of equatorial C—H bonds. The six axial C—H bonds are also parallel to one another.

Figure 4-25 uses color coding to illustrate which sets of bonds are parallel in the shorthand notation for cyclohexane. All six axial C—H bonds (red) are parallel to one another, since they are all drawn vertically. Each equatorial C—H bond (brown, purple, or green) is parallel to one equatorial C—H bond on the opposite side of the ring and to two C—C bonds that are part of the ring. With these sets of parallel lines in mind, examine Figure 4-26 to learn one way to draw a complete chair structure from scratch. (If you learn another way that suits you better, then use that.)

YOUR TURN 4.17

Draw the chair conformation of cyclohexane in Figure 4-26 using the steps described there. Practice doing so until you can draw it without having to refer to the figure.

Draw the front and back bonds so they are parallel.	Draw two more sets of parallel lines to complete the ring. Bonds should all zigzag around the ring.	Add the axial bond to the C at the head of the chair so it points up. Then add the rest of the axial bonds alternating up and down.	Draw equatorial bonds alternating up and down. The three on the left should point left and the three on the right should point right.

Foot   Head

FIGURE 4-26 **The progression in drawing a chair conformation of cyclohexane** The steps proceed from left to right. The lines that are added in each step are indicated in red.

YOUR TURN 4.18

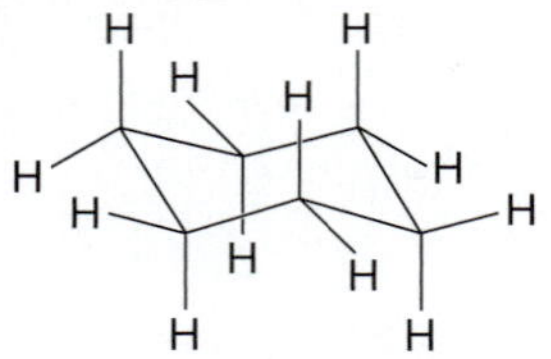

This chair conformation of cyclohexane is obtained after the one in Figure 4-26 has undergone a chair flip. Use the steps in Figure 4-26 to practice drawing this chair conformation until you can draw it without having to refer to the figure.

4.8 Conformers: Monosubstituted Cyclohexanes

If one of the hydrogen atoms in cyclohexane is replaced by a substituent such as CH_3, the result is a **monosubstituted cyclohexane**.

The two chair conformations of a monosubstituted cyclohexane are *not* equivalent.

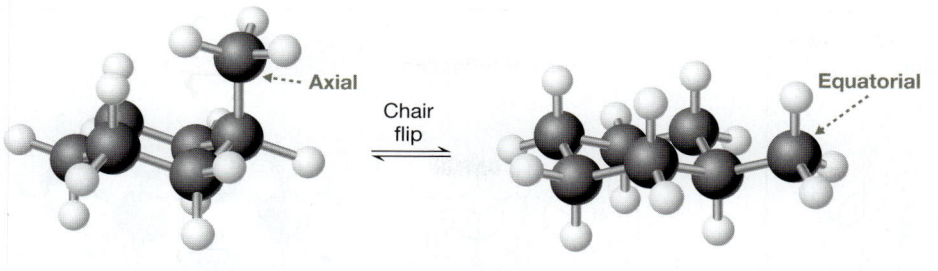

Axial

Chair flip

Equatorial

For example, the CH_3 group of methylcyclohexane is axial in one chair form (Fig. 4-27, left), whereas it is equatorial in the other (Fig. 4-27, right).

YOUR TURN 4.19

In both chair conformations in Figure 4-27, label each of the 11 hydrogen atoms bonded to the ring as either axial or equatorial.

Because they have the same molecular formulas but differ due to rotations about single bonds:

Two nonequivalent chair forms are *conformers* of each other.

The two chair conformations of methylcyclohexane rapidly interconvert, but they are not equally favored. At any given time, about 95% of the molecules exist in the form with an equatorial CH_3 group, and the remaining 5% exist in the form with an axial CH_3 group. Other monosubstituted cyclohexanes exhibit the same trend:

A monosubstituted cyclohexane is lower in energy (more stable) when the substituent occupies an equatorial position.

The equatorial conformer is more stable than the axial conformer because there is *more room for the substituent in the equatorial position.* To see why this is so, compare methylcyclohexane with an axial CH_3 group (Fig. 4-28a, next page) to methylcyclohexane with an equatorial CH_3 group (Fig. 4-28b).

In the axial position, the CH_3 group experiences significant steric strain from gauche interactions, one of which is illustrated in the Newman projection in Figure 4-28a. Notice that the CH_2 group containing C3 of the ring is gauche to the CH_3 substituent (red) bonded to C1. Repulsion between the electrons from the respective CH_3 and CH_2 groups gives rise to the strain.

A major contribution to the strain in these gauche interactions comes specifically from repulsion between the electrons on the axial CH_3 group bonded to C1 and those on the axial H atoms two positions away on the ring (i.e., at positions 3 and 5). There are two of these **1,3-diaxial interactions** (the "1,3" identifies the *relative* positions on the ring for the interacting substituents), as indicated in Figure 4-28a.

No such steric strain exists when the CH_3 group is in the equatorial position (Fig. 4-28b). In the equatorial position, the CH_2 group at position 3 on the ring is anti to the CH_3 group, so no steric strain arises from their interaction. More specifically, the 1,3-diaxial interactions are eliminated because the CH_3 group is relatively far away from the axial H atoms at positions 3 and 5.

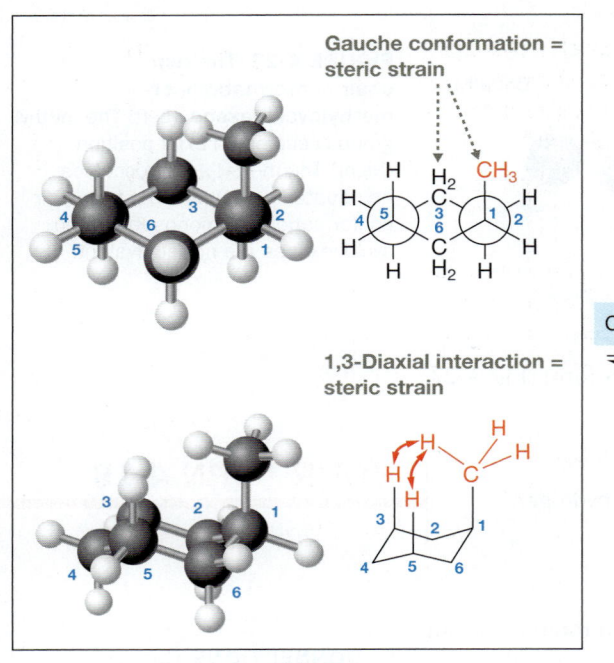

Gauche conformation = steric strain

1,3-Diaxial interaction = steric strain

Chair flip

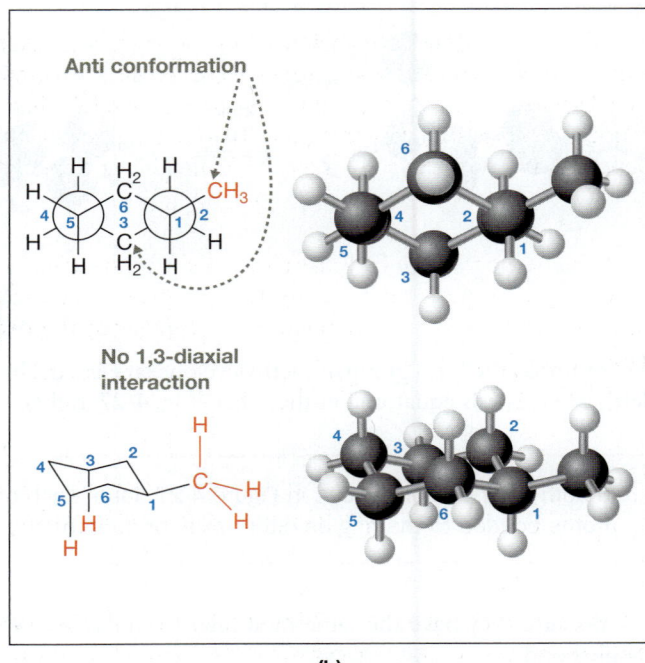

Anti conformation

No 1,3-diaxial interaction

(a)

(b)

FIGURE 4-28 Strain from axial versus equatorial substituents on cyclohexane
(a) The methyl group is in an axial position. A Newman projection looking down the C1—C2 and C5—C4 bonds (top) shows strain from a gauche interaction between the CH_3 substituent on the ring and a CH_2 group of the ring. A chair representation (bottom) shows strain from 1,3-diaxial interactions. Ball-and-stick models are also provided. (b) The methyl group is in an equatorial position. A Newman projection (top) shows that the CH_3 substituent on the ring is anti to the CH_2 group of the ring. A chair representation (bottom) shows that no 1,3-diaxial interactions are present.

YOUR TURN 4.20

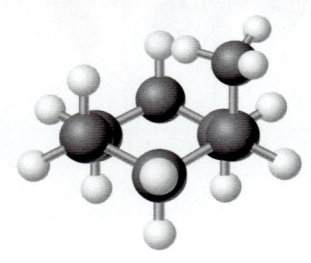

This model of methylcyclohexane corresponds to the top structure in Figure 4-28a. It shows two hydrogen atoms on the ring involved in 1,3-diaxial interactions with the CH_3 group. Identify these two hydrogens. Also, *two different* CH_2 groups are gauche to the CH_3 group. One is indicated in Figure 4-28a. Identify the second one in the structure shown. *Hint:* Build a molecular model of methylcyclohexane.

SOLVED PROBLEM 4.8

Which species, **A** or **B**, is more stable?

Think How are the two molecules related? How are they different? How do those differences translate into stabilities?

Solve **A** and **B** are related by a simple chair flip. The Br atom occupies an equatorial position in **A**, whereas it occupies an axial position in **B**. Bromine is much bulkier than any of the 11 H atoms also bonded to the cyclohexane ring, so it requires more room. Because the equatorial position provides more room than the axial position, conformation **A** is more stable.

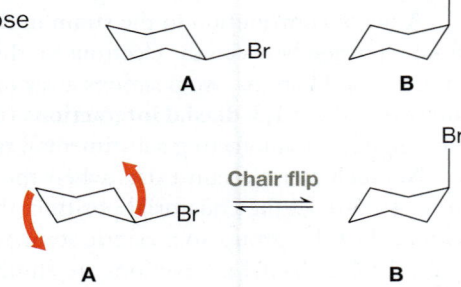

Chair flip

A   B

PROBLEM 4.9 Draw both chair conformations of cyclohexane-d_1, in which a hydrogen atom on cyclohexane has been replaced by a deuterium atom. Which conformation would you expect to be in greater abundance, if any? Explain. *Hint:* Deuterium is an isotope of hydrogen, possessing one more neutron than hydrogen. Recall from Chapter 1 that the size of an atom is dictated by the size of its electron cloud.

Cyclohexane-d_1

Even though all nonhydrogen groups prefer the equatorial position over the axial position, some groups have a stronger preference than others. This can be seen in Table 4-2, which shows the relative percentages of the two chair conformations of various alkylcyclohexanes.

As can be seen from Table 4-2:

> The bulkier the substituent on a cyclohexane ring, the more favored is the chair conformation with the substituent in the equatorial position.

Notice, especially, that the *tert*-butyl group [$C(CH_3)_3$] is so bulky that only a very small percentage of molecules exist with the group in the axial position.

PROBLEM 4.10 For which compound—triiodomethylcyclohexane or trifluoromethylcyclohexane—would you expect to find a greater percentage of molecules that have the substituent in the axial position? Explain.

Triiodomethylcyclohexane   **Trifluoromethylcyclohexane**

TABLE 4-2 Relative Percentages of Chair Conformations of Alkylcyclohexanes

Substituted Cyclohexane	Percent Axial	Percent Equatorial
R = H	50.0	50.0
R = CH_3	5.3	94.7
R = CH_2CH_3	4.5	95.5
R = $CH(CH_3)_2$	2.8	97.2
R = $C(CH_3)_3$	0.02	99.98

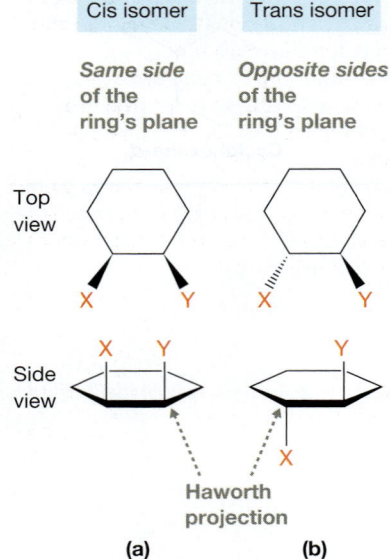

FIGURE 4-29 Haworth projections of disubstituted cyclohexanes (a) A cis isomer of a generic 1,2-disubstituted cyclohexane shown in its dash–wedge notation (top) and as a Haworth projection (bottom). (b) A trans isomer of a generic 1,2-disubstituted cyclohexane.

4.9 Conformers: Disubstituted Cyclohexanes, Cis and Trans Isomers, and Haworth Projections

With **disubstituted** cyclohexanes (i.e., those with two substituents), we have to take into account the relationship of each substituent to the plane of the ring. Are they on the same side of the ring (i.e., *cis* to each other) or are they on opposite sides (i.e., *trans*)? A chair flip does *not* switch a substituent from one side of the plane to the other, so *the cis–trans relationship between any pair of substituents on a cyclohexane ring is independent of the particular chair conformation the species is in.*

> Substituents that are cis to each other on a cyclohexane ring remain cis after a chair flip; substituents that are trans remain trans.

Because the cis–trans relationship of any pair of substituents is unaffected by a chair flip, chemists often find it more convenient to represent substituted cyclohexanes (and rings of other size, for that matter) using *Haworth projections*. In a **Haworth projection**, the ring is depicted as being planar, and bonds to substituents are drawn perpendicular to that plane (Fig. 4-29). Despite their convenience, Haworth projections are inaccurate representations of the true structure and should be used with caution. In particular, they are incapable of portraying axial versus equatorial positions, so they do not accurately represent steric interactions.

YOUR TURN 4.21

> Draw each line structure as a Haworth projection and each Haworth projection as a line structure including dash–wedge notation.
>
>   OH   OH CH_3 CH_3 CH_3
>
> HO   HO CH_3
>
> (a)   (b)   (c)   (d)

Because each substituent on a cyclohexane ring is more stable in an equatorial position than in an axial position, we can reasonably predict the more stable chair conformation of a number of disubstituted cyclohexanes. In *cis*-1,3-dimethylcyclohexane (Fig. 4-30), for example, one chair conformation has two equatorial CH_3 groups, whereas the other has both groups axial. The diequatorial conformer is favored over the diaxial conformer.

In *trans*-1,3-dimethylcyclohexane (Fig. 4-31), both chair conformations are equally favored because both have one axial and one equatorial CH_3 group. With a chair flip, the axial CH_3 group becomes equatorial and the equatorial group becomes axial, thus yielding a conformation identical to the first.

FIGURE 4-30 Relative stabilities of *cis*-1,3-dimethylcyclohexane chair conformations (a) Haworth projection of *cis*-1,3-dimethylcyclohexane. (b) The two chair conformations of *cis*-1,3-dimethylcyclohexane. The one on the left has both methyl groups in equatorial positions and the one on the right has them both axial. The diequatorial conformation is favored over the diaxial.

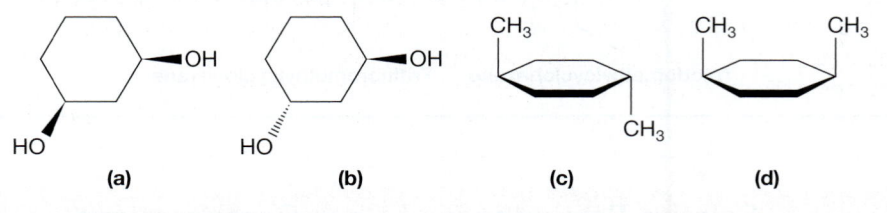

cis-1,3-Dimethylcyclohexane

(a)

Favored chair conformation

(b)

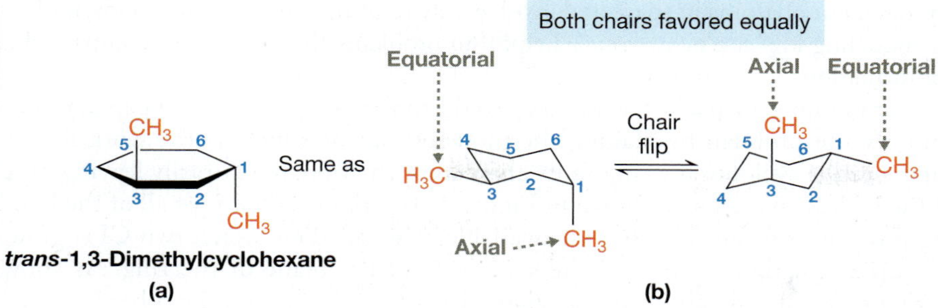

Both chairs favored equally

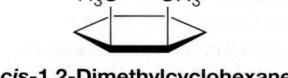

Equatorial

CH₃ 6
5
4   1
3   2
CH₃

Same as

Equatorial

H₃C
4   5   6
3   2   1

Chair flip ⇌

Axial   Equatorial

CH₃
5   6   1
4   3   2
CH₃

Axial ·····▸ CH₃

trans-1,3-Dimethylcyclohexane
(a)

(b)

FIGURE 4-31 **Relative stabilities of *trans*-1,3-dimethylcyclohexane chair conformations** (a) Haworth projection of *trans*-1,3-dimethylcyclohexane. (b) The two chair conformations of *trans*-1,3-dimethylcyclohexane. Both chair conformations have one methyl group in an equatorial position and one in an axial position, so both conformations are favored equally.

YOUR TURN 4.22

Use a molecular modeling kit to build a model of *trans*-1,3-dimethylcyclohexane so that it looks exactly like the chair conformation shown on the left in Figure 4-31b. Without flipping the chair into a different conformation, simply rotate the molecule again until it looks like the chair conformation on the right.

SOLVED PROBLEM 4.11

Draw both chair conformations of *cis*-1,2-dimethylcyclohexane and determine which one, if any, is more stable.

Think Which substituents require the most room? Which position, axial or equatorial, offers more room? Can both substituents achieve that position?

Solve The two methyl groups are the largest substituents on the ring and each is more stable in an equatorial position. To determine whether they can both occupy equatorial positions, begin by drawing a chair conformation with one equatorial methyl group. This methyl group is pointed up, and for the two methyl groups to be cis, the methyl group on the adjacent carbon must also point up. As shown in the first conformation at the right, that bond is axial. If we perform a chair flip, the equatorial methyl group becomes axial and vice versa, as shown in the second conformation. The two chairs have precisely the same stability, so both conformations are favored equally.

H₃C   CH₃

cis-1,2-Dimethylcyclohexane

H₃C

CH₃
H₃C

Chair flip ⇌

CH₃
CH₃

PROBLEM 4.12 Draw the most stable conformation of *trans*-1,2-dimethylcyclohexane.

PROBLEM 4.13 Draw the most stable conformation of the molecule shown here.

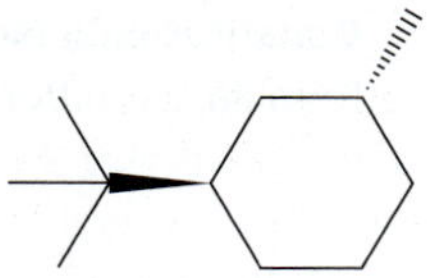

PROBLEM 4.14 Draw the most stable conformation of *cis*-1-methyl-4-trichloromethylcyclohexane.

4.10 Strategies for Success: Molecular Modeling Kits and Chair Flips

Molecular modeling kits can be really useful because they help us "see" molecules in three dimensions from different vantage points and they accurately portray the rotational characteristics of single bonds. Because chair flips affect the three-dimensional arrangement of atoms in space and involve only rotations about single bonds, molecular modeling kits can be *extremely* helpful in problems that ask you to compare chair conformations.

For example, instead of working Solved Problem 4.11 entirely on paper, you can simplify the problem by making the modeling kit do much of the work for you. First, build a cyclohexane ring in its chair conformation, temporarily leaving off all of the hydrogen atoms, as shown in Figure 4-32a. Being able to see all of the bonding positions on the ring that are available, you can then attach two CH_3 groups on adjacent carbon atoms, on the same side of the plane of the ring. As shown in Figure 4-32b, one position is axial, pointing straight up, and the other is equatorial, pointing slightly up. After adding the remaining hydrogen atoms, one of the two chair conformations of *cis*-1,2-dimethylcyclohexane is complete. Flipping the chair as indicated in Figure 4-32b, you arrive at the second chair conformation in Figure 4-32c, which also has one axial and one equatorial CH_3 group. Hence, the two conformations are equivalent.

FIGURE 4-32 Model kits and chair flips (a) A cyclohexane ring without hydrogen atoms. Bonds on the same side of the ring's plane and on adjacent C atoms are identified. (b) CH_3 groups have been added. One is axial and the other equatorial. Twisting the C atoms on the left and right sides according to the red arrows flips the chair. (c) The model after the chair flip. Both CH_3 groups remain up, with one axial and the other equatorial.

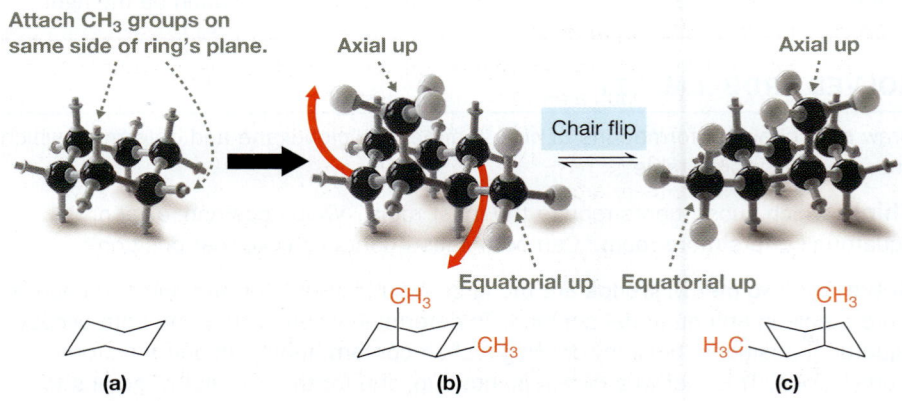

Attach CH_3 groups on same side of ring's plane.   Axial up   Chair flip   Axial up

Equatorial up   Equatorial up

(a)   (b)   (c)

4.11 Constitutional Isomerism: Identifying Constitutional Isomers

Constitutional isomers are the second type of isomers we discuss in this chapter. As you can see from the portion of the Figure 4-1 flowchart reproduced in Figure 4-33:

1) Same molecular formula
2) Different in some way

ISOMERS

Same connectivity?

No

CONSTITUTIONAL ISOMERS

FIGURE 4-33 Constitutional isomers Constitutional isomers have the same formula but different connectivity.

> **Constitutional isomers**, also called **structural isomers**, have the same molecular formula but *differ in their connectivity*.

Unlike conformers, constitutional isomers generally do not interconvert and can be separated from each other. Recall from Chapter 1 that the *connectivity* of a molecule describes its *bonding scheme*, including which atoms are bonded together and by what type of bond (e.g., single, double, or triple). Cyclobutane and but-1-ene are constitutional isomers, for example, because both have the molecular formula C_4H_8, but their connectivities are different.

These are constitutional isomers because they have different connectivities.

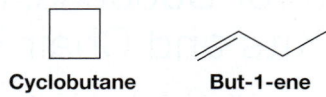

Cyclobutane   **But-1-ene**

Cyclobutane has only single bonds, whereas but-1-ene contains a C=C double bond.

Furthermore, the C—C single bonds in cyclobutane form a ring, whereas but-1-ene is acyclic.

Both of the following structures have the molecular formula C_8H_{18}, but are they constitutional isomers?

These two molecules are not constitutional isomers of each other!

Although they are drawn differently, they actually have the same connectivity, so they are *not* constitutional isomers of each other. How can we tell?

Perhaps the most straightforward method to determine whether two molecules with the same formula are constitutional isomers is one that takes advantage of some aspects of nomenclature that were discussed in Interchapters A and B.

Identifying Constitutional Isomers

1. For each molecule, identify the longest continuous chain or ring of carbons that contains any C═C double and C≡C triple bonds.
2. Number the carbons in the chain or ring sequentially so that:
   - The carbon atoms involved in the double or triple bonds receive the lowest numbers, or, if there are no such multiple bonds,
   - The first substituent is attached to the lowest-numbered carbon.
3. If there is a difference in any of the following, the molecules must have different connectivities, and must therefore be constitutional isomers:
   - The size of the longest continuous chain or ring.
   - The numbers assigned to the carbons involved in the multiple bonds.
   - The numbers assigned to the carbons to which any substituent is attached.
   - The identities of the substituents attached to the same-numbered carbon.
4. Otherwise, the molecules have the same connectivity and are not constitutional isomers.

Let's now apply this method to the two molecules above. For Step 1, the longest continuous chain of carbon atoms in each molecule has six carbons. For Step 2, the carbons are numbered 1 through 6 so that the first methyl group is encountered on C2, as shown below. Because the methyl groups are attached to C2 and C4 in both cases, the molecules have the same connectivity and therefore are not constitutional isomers.

Longest continuous chain has six carbon atoms.

Longest continuous chain has six carbon atoms.

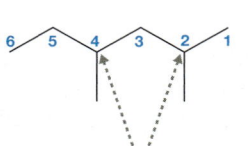

Methyl groups on C2 and C4

Methyl groups on C2 and C4

YOUR TURN 4.23

Using the above method, show that this molecule is *not* a constitutional isomer of the previous two molecules.

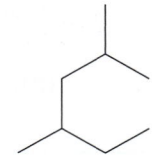

SOLVED PROBLEM 4.15

Are these two molecules constitutional isomers of each other?

Think Do both compounds have the same molecular formula? Is the largest continuous chain or ring in each molecule the same? Are the C atoms that are involved in the double bonds assigned the same numbers in each molecule? In each molecule, is the C atom to which the methyl group is attached assigned the same number?

Solve Both compounds have a molecular formula of C_6H_{10}. In each molecule, the largest continuous ring has five C atoms. As shown at the right, the numbers assigned to the doubly bonded C atoms are the same in each molecule. However, the molecules differ in the number assigned to the C atom to which the methyl group is attached. In the first molecule, the methyl group is attached to C3, whereas it is attached to C4 in the second. Thus, the molecules have different connectivities, making them constitutional isomers.

Methyl groups attached to different carbons

PROBLEM 4.16 For each pair of molecules, determine whether they are constitutional isomers.

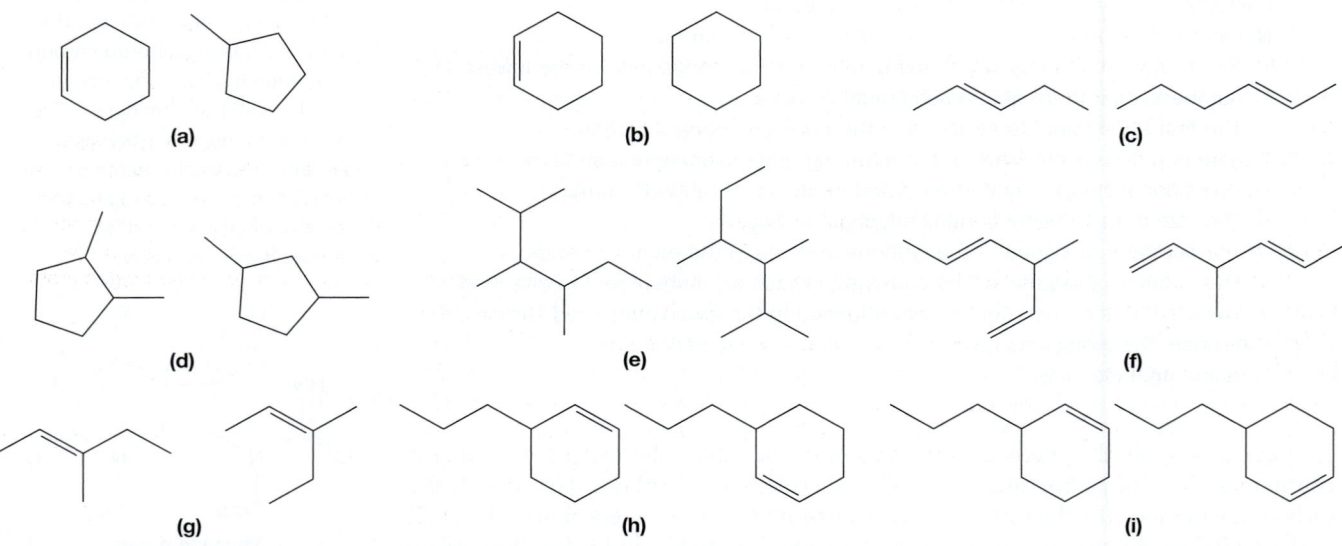

(a)   (b)   (c)

(d)   (e)   (f)

(g)   (h)   (i)

PROBLEM 4.17 For each pair of molecules, determine whether they are constitutional isomers.

(a)   (b)   (c)   (d)

(e)   (f)   (g)

4.12 Constitutional Isomers: Index of Hydrogen Deficiency (Degree of Unsaturation)

Looking ahead, Section 4.13 is devoted to drawing constitutional isomers of a given formula. There are a variety of contexts where being able to do so is useful, including in *spectroscopy*, which we discuss in Chapters 15 and 16.

To help you address these kinds of issues, we introduce here in Section 4.12 a concept called the *index of hydrogen deficiency (IHD)*, also known as *degree of unsaturation*. Before we examine IHD, however, you must understand what it means for a molecule to be *saturated*.

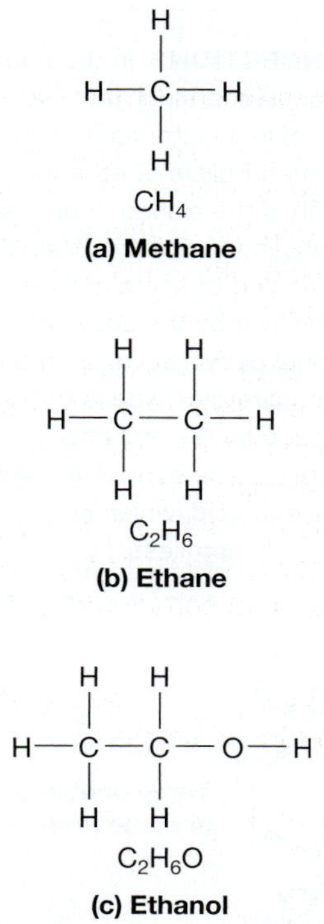

A molecule is said to be **saturated** if it *has the maximum number of hydrogen atoms possible*, consistent with:

- The number and type of each nonhydrogen atom in the molecule.
- The octet rule and the duet rule.

Each of the molecules in Figure 4-34, for example, is saturated. Methane (CH_4) is saturated because a single carbon atom can bond at most to four other atoms; in this case, they are all hydrogen atoms. Ethane (CH_3CH_3) is also saturated because both carbon atoms are bonded to four other atoms (i.e., one carbon and three hydrogens). Ethanol (CH_3CH_2OH) is saturated because each of its carbon atoms is bonded to four other atoms, and the oxygen atom, being uncharged, is bonded to its maximum of two other atoms.

Molecules that contain double bonds, triple bonds, or rings are said to be **unsaturated** because they have fewer than the maximum number of hydrogen atoms possible (Fig. 4-35). Ethene (ethylene, $H_2C{=}CH_2$), for example, has the same number of carbon atoms as ethane, but has two fewer hydrogen atoms. Ethyne (acetylene, $HC{\equiv}CH$) has the same number of carbon atoms, too, but has four fewer hydrogen atoms than ethane. Ethanal (acetaldehyde, $CH_3CH{=}O$) and oxirane each have the same number of carbon and oxygen atoms as ethanol, but have two fewer hydrogen atoms.

FIGURE 4-34 Some saturated molecules These molecules contain the most number of hydrogen atoms possible, given (a) one carbon atom, (b) two carbon atoms, and (c) two carbon atoms and one oxygen atom.

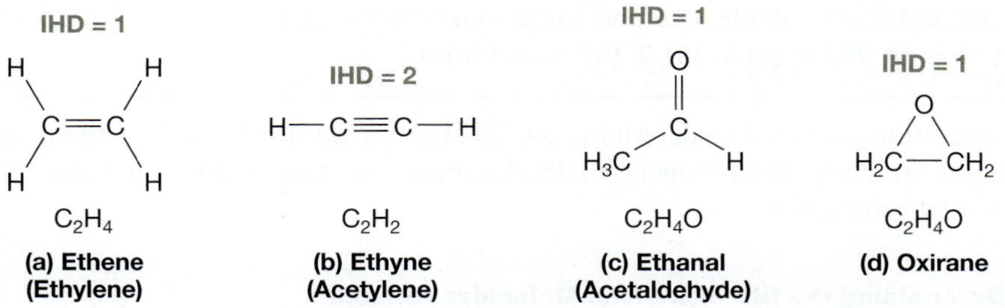

FIGURE 4-35 Some unsaturated molecules Ethene (a) and ethyne (b) are unsaturated because they have fewer hydrogen atoms than ethane. Ethanal (c) and oxirane (d) are unsaturated because they have fewer hydrogen atoms than ethanol.

YOUR TURN 4.24

Draw the structure of a *saturated* molecule that corresponds to each of the unsaturated molecules in Figure 4-35.

The *extent* to which a molecule is unsaturated is indicated by its index of hydrogen deficiency:

A molecule's **index of hydrogen deficiency (IHD)**, or **degree of unsaturation**, is defined as *half* the number of hydrogen atoms missing from that molecule compared to an analogous, completely saturated molecule.

Alternatively, it is the number of H_2 *molecules* that are missing from the molecule compared to an analogous saturated structure. Based on this definition, ethene, ethanal, and oxirane all have an IHD of 1, whereas ethyne has an IHD of 2. In other words:

- Each double bond contributes 1 to a molecule's IHD.
- Each triple bond contributes 2 to a molecule's IHD.
- Each ring contributes 1 to a molecule's IHD.

Benzene (C_6H_6), therefore, has an IHD of 4, because its Lewis structure has three double bonds and one ring, each of which contributes 1 to its IHD (Fig. 4-36).

IHD = 4

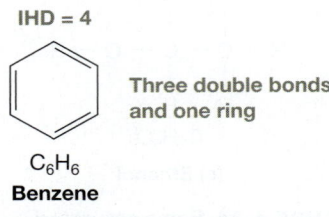

Three double bonds and one ring

C_6H_6
Benzene

FIGURE 4-36 The IHD of benzene Each of the three double bonds contributes 1 to the IHD, and the ring also contributes 1 to the IHD, for a total of 4.

SOLVED PROBLEM 4.18

What is the IHD of naphthalene?

Think How many double bonds are there? How many triple bonds are there? How many rings are there? How much does each contribute to the overall IHD?

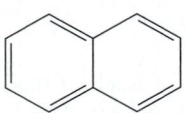

Naphthalene

Solve There are five double bonds, each contributing 1 to the IHD, and there are two rings, each contributing 1 to the IHD. There are no triple bonds. The total IHD, therefore, is 5(1) + 2(1) = 7.

PROBLEM 4.19 Determine the IHD for each of the following molecules.

(a) (b) (c) (d) (e)

PROBLEM 4.20 Identify all of the compound classes listed in Table 1-6 (p. 35) that have an IHD of **(a)** 1, **(b)** 2, **(c)** 3, and **(d)** 4.

Sometimes we need to determine the IHD of a molecular formula without being given its structure. The following method outlines one way to do so, but you might encounter others, too.

Determining the IHD of a Given Molecular Formula

1. Draw any saturated molecule that has the same number of each nonhydrogen atom as in the formula you are given.
   - The saturated molecule must contain no double bonds, no triple bonds, and no rings.
   - Each atom should have its "normal" number of bonds and lone pairs to avoid formal charges.
2. Determine how many more hydrogen atoms are in this saturated molecule compared to the formula you were given.
3. Compute the IHD by dividing that number of additional hydrogen atoms by 2.

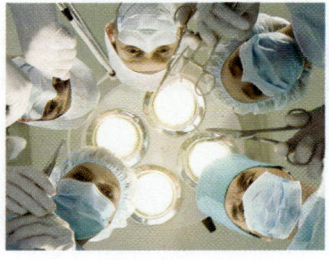

To see how these steps are applied, study Solved Problem 4.21 and then work through the problem that follows it.

SOLVED PROBLEM 4.21

Determine the IHD for a compound whose molecular formula is $C_7H_7NO_2$.

Think What is the formula for an analogous saturated compound? How many hydrogen atoms are missing from the formula we are given?

Solve We can construct a saturated compound containing seven carbon atoms, one nitrogen atom, and two oxygen atoms simply by connecting all of these atoms in a row using single bonds only. It takes 17 hydrogen atoms in total to saturate each carbon, nitrogen, and oxygen in this compound. Thus, the formula we were given is missing 10 H atoms, for an IHD of 5.

A saturated compound

$$H_3C-\underset{H_2}{C}-\underset{H_2}{C}-\underset{H_2}{C}-\underset{H_2}{C}-\underset{H_2}{C}-\underset{H_2}{C}-\underset{H}{N}-O-OH$$

$$C_7H_{17}NO_2$$

PROBLEM 4.22 Compute the IHD for a compound whose molecular formula is $C_4N_2OH_7F$.

Computing a formula's IHD can be sped up considerably by realizing that a saturated hydrocarbon with n carbon atoms has $2n + 2$ hydrogen atoms—that is, its formula is C_nH_{2n+2}. See Table 4-3, which lists a variety of saturated hydrocarbons.

Problems 4.23 and 4.24 show that adjustments can be made to the generic formula of C_nH_{2n+2} for saturated compounds containing atoms such as nitrogen, oxygen, and the halogens.

PROBLEM 4.23 If a saturated compound contains four carbon atoms as the only nonhydrogen atoms, how many hydrogen atoms does it contain? How many hydrogen atoms are in a saturated compound containing four carbon atoms and one oxygen atom? How many hydrogen atoms are in a saturated compound containing four carbon atoms and two oxygen atoms? What can you conclude about the effect that each oxygen atom has on the formula for a completely saturated molecule?

PROBLEM 4.24 **(a)** Repeat Problem 4.23, adding nitrogen atoms instead of oxygen atoms. **(b)** Repeat it once again, adding fluorine atoms instead of oxygen atoms.

TABLE 4-3 Number of Hydrogen Atoms in a Saturated Hydrocarbon with n Carbon Atoms

Hydrocarbon	Number of H Atoms	$2n + 2$
CH_4 ($n = 1$)	4	4
CH_3CH_3 ($n = 2$)	6	6
$CH_3CH_2CH_3$ ($n = 3$)	8	8
$CH_3(CH_2)_2CH_3$ ($n = 4$)	10	10
$CH_3(CH_2)_3CH_3$ ($n = 5$)	12	12
$CH_3(CH_2)_4CH_3$ ($n = 6$)	14	14

4.13 Strategies for Success: Drawing All Constitutional Isomers of a Given Formula

Being able to draw all constitutional isomers of a given molecular formula can be useful, especially when trying to determine a compound's structure using results from spectroscopy (Chapters 15 and 16). More immediately, however, the time you spend drawing constitutional isomers will deepen your understanding of connectivity.

It helps to have a systematic method to tackle these kinds of problems. Here we present one method, and you may even develop your own.

1. Determine the formula's IHD. This will tell you the possible combinations of double bonds, triple bonds, and rings required in each isomer you draw.
2. Draw all possible isomers that omit double bonds, triple bonds, and halogen atoms. (It is most convenient to work with line structures so that the H atoms are accounted for appropriately when features are added in Steps 3 and 4.)
   - Double bonds, triple bonds, and halogen atoms will be added later.
   - Include rings. The number of rings must not exceed the IHD computed from Step 1.
3. For each structure generated in Step 2, add double or triple bonds to satisfy the total IHD calculated in Step 1. Try to add the multiple bonds at various locations to generate as many unique connectivities as possible.
4. For each structure generated in Step 3, add halogen atoms at various locations to generate as many unique connectivities as possible.

How we apply these steps depends specifically on the nature of the formula we are given. We present two examples.

4.13a Drawing All Constitutional Isomers of $C_4H_8F_2$

Let's draw all possible constitutional isomers having the formula $C_4H_8F_2$. According to Step 1, we first determine the IHD for this formula. To do so, we construct a saturated molecule having four C atoms and two F atoms:

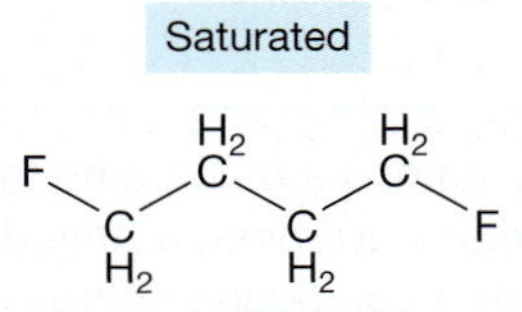

Saturated

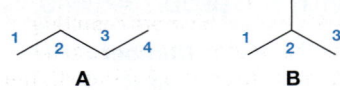

$C_4H_8F_2$

Because this compound has eight H atoms, *the formula we are given has an IHD of 0.* That means that every constitutional isomer we draw must have *no double bonds,* *no triple bonds,* and *no rings.*

For Step 2, we draw all isomers that have four C atoms only and just single bonds, disregarding the F atoms for now. Because the IHD is 0, these structures also should contain no rings. There are only two such structures, **A** and **B**, shown in Figure 4-37.

Normally, for Step 3, we would add double or triple bonds to satisfy the IHD. We skip Step 3 in this case, however, because the IHD = 0.

For Step 4, we have two F atoms to add. As shown in Figure 4-38a, we can add the first F atom to C1 or C2 of structure **A** to produce structures **C** and **D** having different connectivities. We do not add the F atom to C3 or C4 because the resulting structures would be redundant to **D** and **C**, respectively. Similarly, in structure **B**, we can add the first F atom to C1 or C2 to produce structures **E** and **F** having two unique connectivities (Fig. 4-38b).

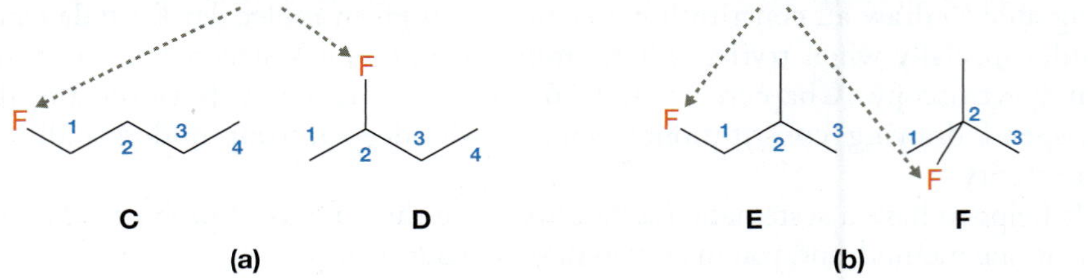

- **Four C atoms**
- **No double bonds, triple bonds, or halogens**
- **No rings**

FIGURE 4-37 Four-carbon isomers (Step 2) These isomers have four C atoms, the same as in the formula we were given: $C_4H_8F_2$. Step 2 requires them to have no double bonds, triple bonds, or halogen atoms. There are no rings because the IHD = 0.

FIGURE 4-38 Isomers resulting from the addition of the first F atom (Step 4) (a) Structures **C** and **D** were produced by adding F to C1 and C2, respectively, of structure **A**. (b) Structures **E** and **F** were produced by adding F to C1 and C2, respectively, of structure **B**.

F atom was added to structure A.

C   D

(a)

F atom was added to structure B.

E   F

(b)

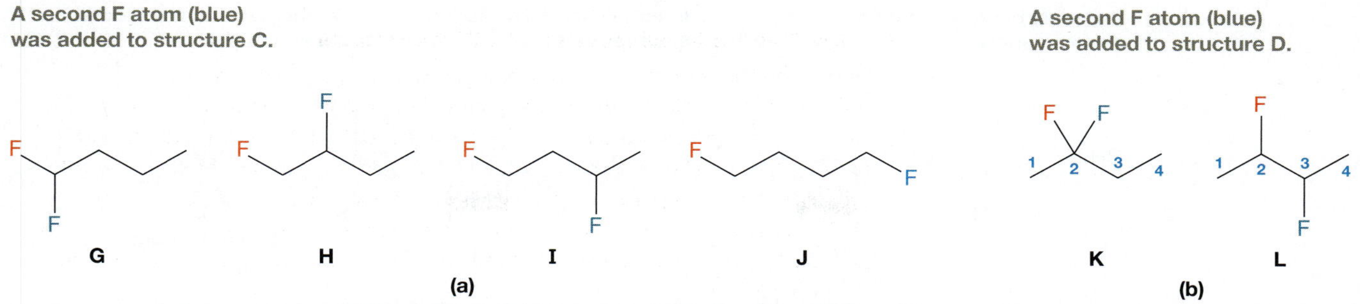

A second F atom (blue) was added to structure C.

A second F atom (blue) was added to structure D.

G   H   I   J   K   L

(a)   (b)

FIGURE 4-39 Isomers resulting from the addition of the second F atom to structures C and D (Step 4)
(a) The second F atom (blue) has been added to each of the four C atoms in structure **C**, producing structures **G–J**. (b) The second F atom (blue) has been added to C2 and C3 in structure **D**, producing structures **K** and **L**, respectively. It is not added to C1 or C4 because the resulting molecules would be redundant.

We can now add the second F atom separately to structures **C–F**. First let's do so for structures **C** and **D**, as shown in Figure 4-39. In Figure 4-39a, the second F atom (blue) has been added to each of the four different C atoms in structure **C** to yield structures **G–J**, which have the desired formula of $C_4H_8F_2$. In Figure 4-39b, we add the second F atom only to C2 and C3 in structure **D** because adding it to C1 or C4 would repeat structures **H** and **I**, respectively.

Now let's add the second F atom to structure **E**, as shown in Figure 4-40, to yield molecules with the desired formula $C_4H_8F_2$. We add the second F atom only to C1, C2, and C3 to produce molecules **M, N,** and **O**. We do not add the second F atom to the unnumbered carbon in structure **E** because it would repeat molecule **O**.

Finally, we should consider adding the second F atom to structure **F**. However, adding the second F atom to any one of its available carbon atoms would repeat structure **N**. Therefore, there are a total of nine isomers having the formula $C_4H_8F_2$, structures **G–O**.

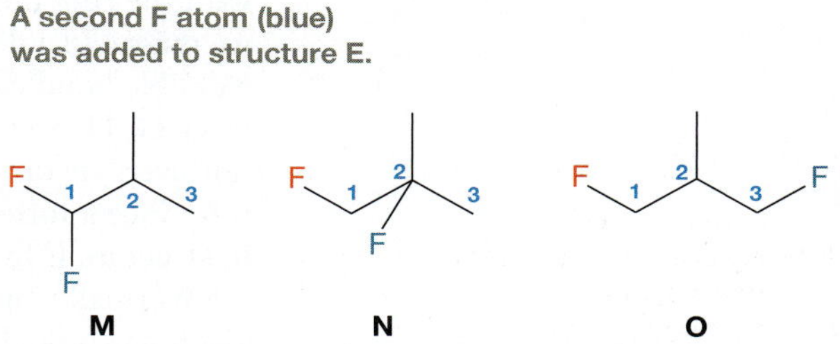

A second F atom (blue) was added to structure E.

M   N   O

FIGURE 4-40 Isomers resulting from the addition of the second F atom to structure E (Step 4) The second F atom (blue) has been added to C1, C2, and C3 in structure **E**, producing structures **M, N,** and **O**.

4.13b Drawing All Constitutional Isomers of C_3H_6O

Let us now consider the formula C_3H_6O. Using the same systematic approach, we first determine the IHD. We begin by comparing the formula we are given with a completely saturated molecule containing three C atoms and one O atom. One possibility is $CH_3CH_2CH_2OH$, which has a formula C_3H_8O. Consequently, our formula has an IHD of 1, which can be due to either a double bond or a ring.

In Step 2, we draw all unique line structures containing three C atoms and one O atom, leaving out double bonds and triple bonds. As shown in Figure 4-41, there are six such structures (**A–F**), which can contain up to one ring without exceeding an IHD of 1.

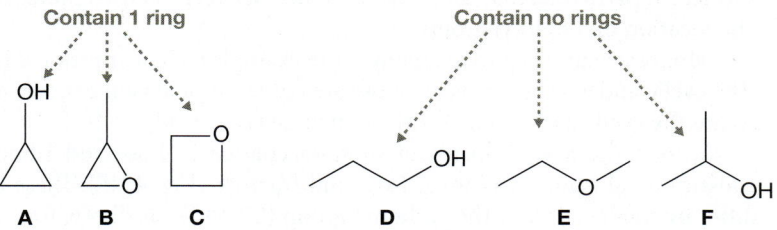

Contain 1 ring   Contain no rings

A   B   C   D   E   F

FIGURE 4-41 Isomers containing three C atoms and one O atom (Step 2) These isomers have three C atoms and one O atom, the same as in the formula we were given: C_3H_6O. Step 2 requires them to have no double bonds or triple bonds. These isomers can contain up to one ring (structures **A–C**) because the formula we were given corresponds to an IHD of 1.

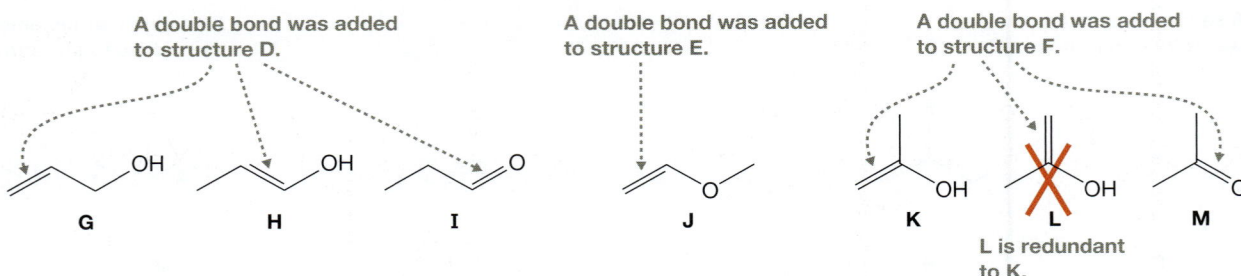

A double bond was added to structure D.

A double bond was added to structure E.

A double bond was added to structure F.

G H I J K L M

L is redundant to K.

FIGURE 4-42 **Isomers resulting from the addition of a double bond to structures D–F (Step 3)** Structures **G–I** were produced by adding a double bond to various locations in structure **D**. Structure **J** was produced by adding a double bond to structure **E**. Structures **K–M** were produced by adding a double bond to various locations in structure **F**. Structure **L** is redundant to **K**.

For Step 3, we add double or triple bonds to obtain a total IHD of 1 (calculated in Step 1). We cannot add double or triple bonds to structures **A–C**, however, because with the ring present, their IHD is already 1. For structures **D–F**, on the other hand, we must add a double bond to various locations, as shown in Figure 4-42. Notice that a double bond can be added to structure **D** at three different locations to yield structures **G**, **H**, and **I**. A double bond can be added to only one location in structure **E**, however, yielding structure **J**; adding the double bond to any other location would introduce a formal charge on O. A double bond also can be added to three locations in structure **F** to yield structures **K**, **L**, and **M**, but, as indicated, **L** is redundant to **K**.

Normally, Step 4 would have us add halogens, but because the formula we were given has no halogen atoms, the structures we have constructed thus far all have the formula C_3H_6O. Therefore, there are nine constitutional isomers of C_3H_6O: **A**, **B**, **C**, **G**, **H**, **I**, **J**, **K**, and **M**.

YOUR TURN 4.25

Identify the functional groups present in *each* molecule in Figure 4-42.

Even though we can draw these nine constitutional isomers, molecules **H** and **K** (Fig. 4-42) are unstable. As a class of compounds, they are called *enols*: They have a carbon that is bonded to an OH group and is also part of a C=C double bond. We explain in Section 7.9 that these two molecules will undergo a rearrangement reaction in solution to form the more stable isomers **I** and **M**, respectively.

THE ORGANIC CHEMISTRY OF BIOMOLECULES

4.14 Constitutional Isomers and Biomolecules: Amino Acids and Monosaccharides

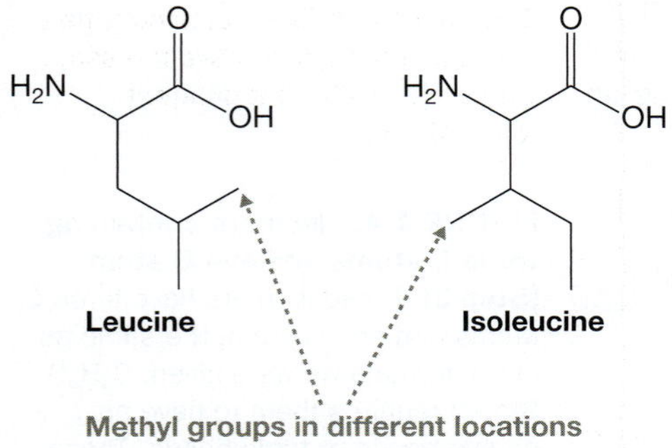

Leucine and isoleucine (Fig. 4-43), two naturally occurring amino acids, are constitutional isomers because they have the same molecular formula but differ in their connectivities. (The prefix *iso*, in fact, stands for "isomer.") Notice, in particular, that the only difference between the two molecules is in the location of a methyl group.

Monosaccharides provide many more examples of constitutional isomers. The cyclic and acyclic forms of ribose are constitutional isomers, for instance, as are the cyclic and acyclic forms of glucose (Fig. 4-44).

Among the acyclic forms of monosaccharides, ribose and ribulose are constitutional isomers, as are glucose and fructose (Fig. 4-45). These isomers differ by the location of the carbonyl group (C=O). In ribose, for example, the carbonyl group involves a terminal carbon, characteristic of an aldehyde,

Leucine

Isoleucine

Methyl groups in different locations

FIGURE 4-43 **Isomeric amino acids** Leucine and isoleucine are naturally occurring amino acids that have the same formula but different connectivities, so they are constitutional isomers.

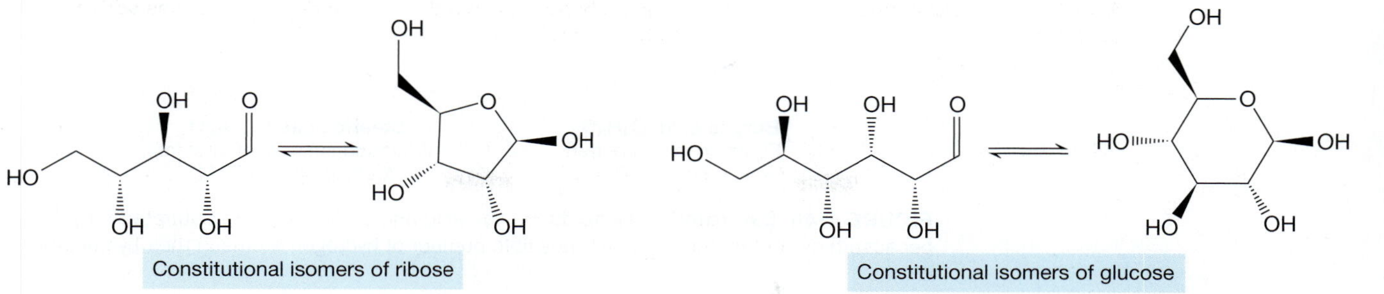

Constitutional isomers of ribose

Constitutional isomers of glucose

FIGURE 4-44 Acyclic and cyclic forms of sugars as constitutional isomers Acyclic and cyclic ribose have the same formula ($C_5H_{10}O_5$) but different connectivities. Acyclic and cyclic glucose have the same formula ($C_6H_{12}O_6$) but different connectivities.

whereas in ribulose, it involves an internal carbon, characteristic of a ketone. Thus, ribose is classified as an **aldose**, whereas ribulose is a **ketose**. Glucose is similarly classified as an aldose, whereas fructose is a ketose.

To further distinguish sugars on the basis of their carbon atoms, ribose and ribulose are **pentoses**, because they both contain five carbons, whereas glucose and fructose are **hexoses**, containing six carbons. Combining these terminologies, we say that ribose is an **aldopentose**, ribulose is a **ketopentose**, glucose is an **aldohexose**, and fructose is a **ketohexose**.

PROBLEM 4.25 Draw the Lewis structure of each of the following: **(a)** an aldotetrose; **(b)** a ketotetrose; **(c)** an aldotriose; **(d)** a ketotriose; **(e)** a ketohexose different from fructose.

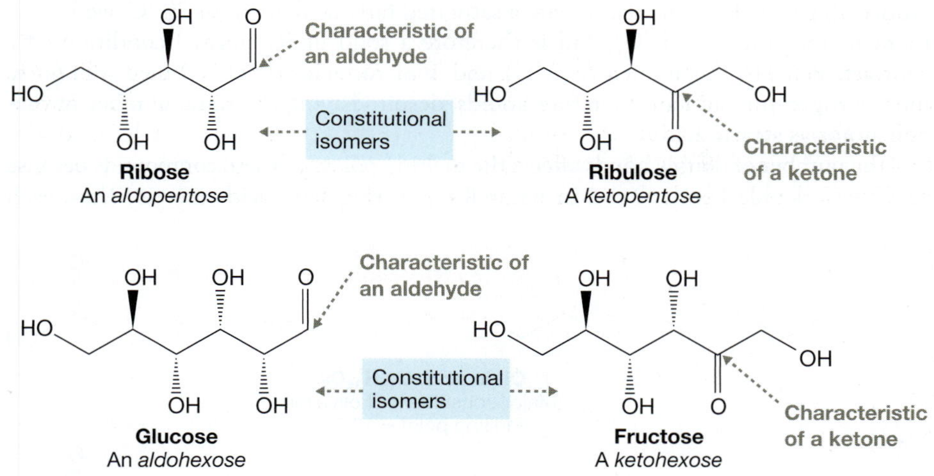

FIGURE 4-45 Acyclic sugars as constitutional isomers Ribose and ribulose have the same formula ($C_5H_{10}O_5$) but different connectivities. Glucose and fructose have the same formula ($C_6H_{12}O_6$) but different connectivities.

4.15 Saturation and Unsaturation in Fats and Oils

Diets high in saturated fats, typically of animal origin, are unhealthy because they increase the risk of coronary heart disease. Conversely, unsaturated fats, typically derived from plants, are a healthy part of our diets.

The concepts of saturation and unsaturation apply to the hydrocarbon chains of fats and oils in the same way they do to other organic compounds (see Section 4.12). That is, each C═C double bond that is present adds 1 unit of unsaturation (or IHD)

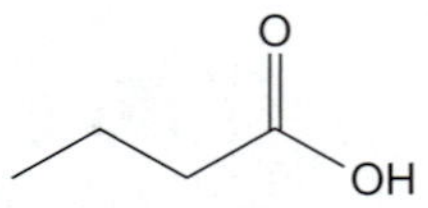

Butyric acid, $C_4H_8O_2$
(Found in rancid butter)
Melting point = −7.9 °C

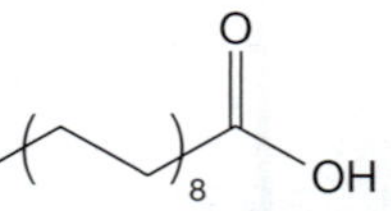

Stearic acid, $C_{18}H_{36}O_2$
(Major constituent of beef fat)
Melting point = 70 °C

FIGURE 4-46 Saturated fatty acids Butyric acid and stearic acid are saturated fatty acids because they contain the maximum possible number of hydrogen atoms in their hydrocarbon chains.

to the molecule. Thus, fats that contain fatty acids such as butyric acid or stearic acid (Fig. 4-46) are *saturated*, because they contain no C=C double bonds, whereas fats that contain oleic, linoleic, or linolenic acids (Fig. 4-47) are *unsaturated*, because these fatty acids have one or more C=C double bonds.

We can also distinguish the extent of unsaturation on the basis of the number of C=C double bonds present. For example, oleic acid is a **monounsaturated** fatty acid, because it has just one C=C double bond, whereas linolenic acid is a **polyunsaturated** fatty acid, because it has three.

Linoleic and linolenic acids are further classified as **essential fatty acids** because these are the only naturally occurring fatty acids that cannot be synthesized in the human body by any known chemical pathways. Instead, they must be consumed as part of our diet.

The different health effects of saturated and unsaturated fats seem to correlate with their different physical properties. Namely:

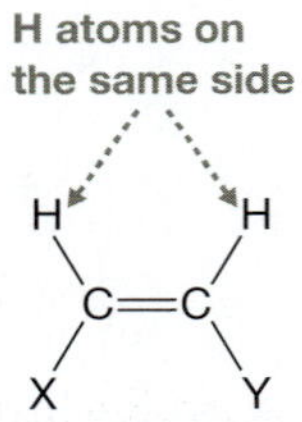

Each cis C=C double bond that is present in a fatty acid lowers the melting point.

Notice, in particular, that stearic acid, a saturated fatty acid, melts at 70 °C, well *above* room temperature (~25 °C), and is therefore a solid under normal conditions. By contrast, oleic (16 °C), linoleic (−7 °C), and linolenic acids (−11 °C) melt well *below* room temperature and are therefore liquids, despite having the same number of carbon atoms as stearic acid.

The number of double bonds affects the melting points of these compounds because all C=C double bonds found in naturally occurring fatty acids are cis. Thus, each

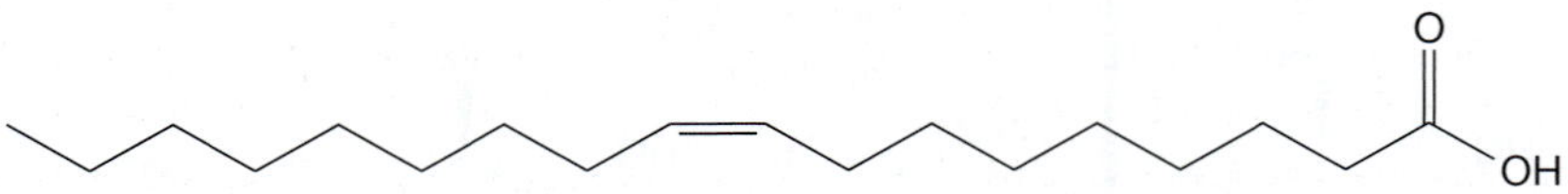

H atoms on the same side

A cis double bond

Oleic acid, $C_{18}H_{34}O_2$
(Major constituent of olive oil)
Melting point = 16 °C

Essential fatty acids

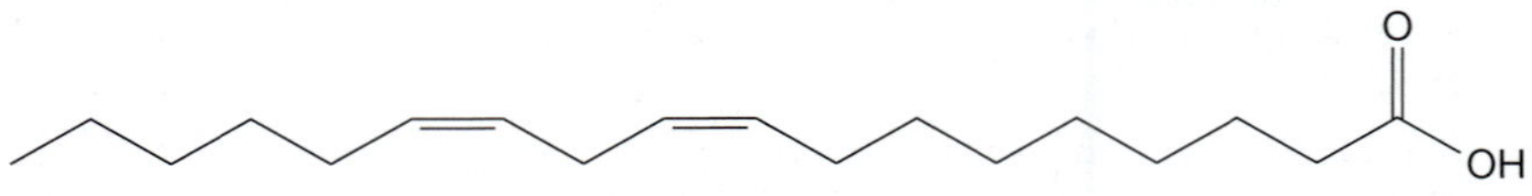

Linoleic acid, $C_{18}H_{32}O_2$
(Major constituent of safflower oil)
Melting point = −7 °C

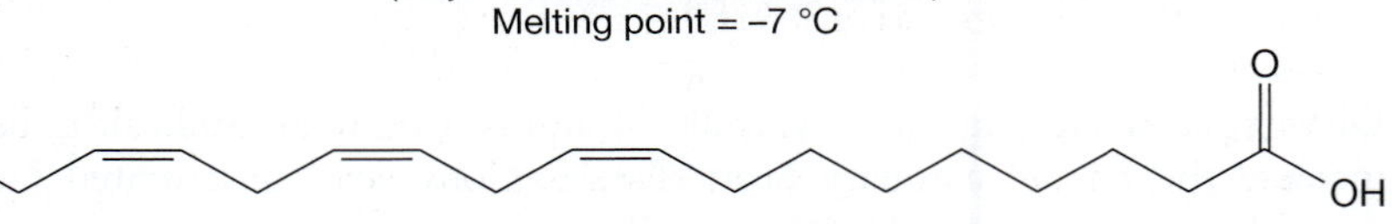

Linolenic acid, $C_{18}H_{30}O_2$
(Major constituent of flaxseed oil)
Melting point = −11 °C

FIGURE 4-47 Unsaturated fatty acids Oleic acid, linoleic acid, and linolenic acid are unsaturated fatty acids because they contain units of unsaturation (i.e., C=C bonds) in their carbon chains.

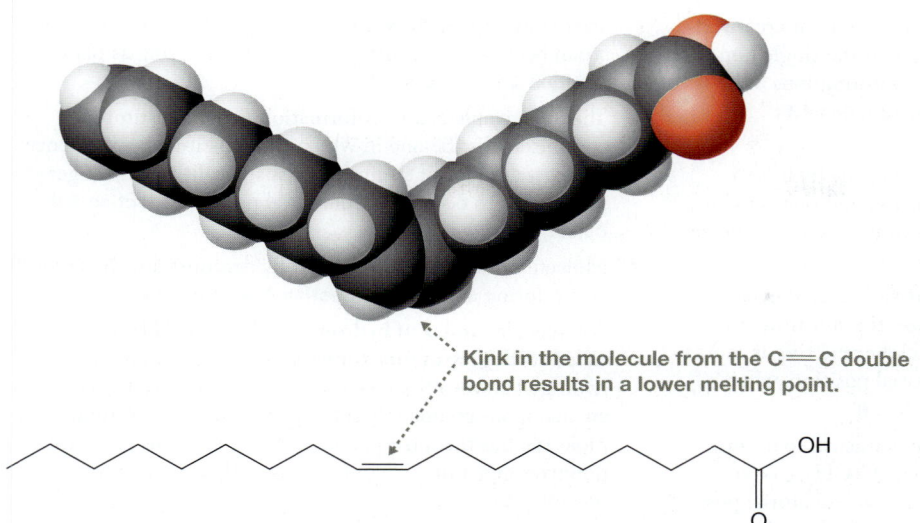

Kink in the molecule from the C=C double bond results in a lower melting point.

FIGURE 4-48 Space-filling model of oleic acid A cis double bond introduces a kink in the carbon chain of a naturally occurring fatty acid such as oleic acid. This decreases the contact surface area among separate fatty acid molecules, which reduces the strength of their intermolecular interactions and thus decreases the melting point.

double bond introduces a kink in the carbon chain, as shown for oleic acid in Figure 4-48. Such a kink makes it more difficult for separate fatty acid molecules to align, so the contact surface area among the molecules is decreased. In turn, this weakens the induced dipole–induced dipole interactions (Section 2.6), resulting in a lower melting point.

Chapter Summary and Key Terms

- **Isomerism** is a relationship between two or more molecular species. Molecules are **isomers** of each other if they have the same molecular formula but are different in some way. (Section 4.1; Objective 1)

- **Conformers** differ by rotations about single bonds, whereas **constitutional isomers** have different connectivities. (Sections 4.2 and 4.11; Objective 1)

- **Newman projections** are used to show conformations about bonds. They depict the view down a bonding axis. The atom in front is represented as a point and the atom in back is represented as a circle. (Section 4.2; Objective 2)

- In a **conformational analysis** of ethane, energy is plotted as a function of the H—C—C—H **dihedral angle (θ)**. In a 360° rotation, we observe three equivalent **staggered conformations**, each at an energy minimum, and three equivalent **eclipsed conformations**, each at an energy maximum. (Section 4.3a; Objective 3)

- **Torsional strain** is the energy increase that appears in an eclipsed conformation. (Section 4.3a; Objective 3)

- Staggered conformations in ethane rapidly interconvert because the **rotational energy barrier** is comparable to the **thermal energy** available. (Section 4.3a; Objective 3)

- Based on the conformational analysis of $BrCH_2$—CH_2Br, one staggered conformation, called the **anti conformation**, is lower in energy than the other two, called **gauche conformations**. This is because there is less **steric strain** between

the two Br atoms in the anti conformation. Similarly, one eclipsed conformation is higher in energy than the other two. (Section 4.3b; Objective 4)

- Although the anti conformation is more stable than the gauche, the two conformers rapidly interconvert because the energy barrier between them is comparable to the available thermal energy. (Section 4.3b; Objectives 3 and 4)

- The lowest-energy conformation of an alkyl chain is the **zigzag** or **all-anti conformation**, which has the anti conformation at each C—C bond. (Section 4.3c; Objectives 3 and 4)

- **Heats of combustion** correlate with the **ring strain** for rings of various sizes. Cyclohexane (a six-membered ring) has essentially no ring strain. Cyclopentane and cycloheptane (five- and seven-membered rings, respectively) have mild ring strain. Cyclopropane and cyclobutane (three- and four-membered rings, respectively) are highly strained. (Section 4.4; Objective 5)

- Cyclohexane has no ring strain because it adopts a **chair conformation**, in which all angles are about 111° (close to the ideal tetrahedral angle of 109.5°) and all C—C bonds are staggered. Cyclopentane adopts an **envelope conformation**, which has slightly more angle strain and torsional strain. Cyclobutane and cyclopropane rings are highly strained due to substantial torsional and angle strain. (Section 4.5; Objective 6)

- Envelope conformations of cyclopentane rapidly interconvert via **pseudorotation**—slight rotations about the single bonds that compose the ring. Similarly, chair conformations of cyclohexane interconvert via **chair flips**. (Section 4.6; Objective 7)

- In each chair conformation of cyclohexane, one hydrogen atom on each carbon is in an **axial position** and one is in an **equatorial position**. The two positions interconvert during a chair flip. (Section 4.6; Objective 7)

- The two chair conformations of a **monosubstituted cyclohexane** are no longer equivalent. Because the substituent is larger than the hydrogen atoms, the cyclohexane ring is more stable with the substituent in an equatorial position, where there is more room. (Section 4.8; Objective 8)

- The bulkier a substituent, the greater its tendency to occupy an equatorial position. A *tert*-butyl group, C(CH₃)₃, is so bulky that it is almost exclusively found in an equatorial position. (Section 4.8; Objective 8)

- **Disubstituted** cyclohexanes introduce **cis** and **trans** relationships relative to the plane of the ring. **Haworth projections** illustrate cis and trans relationships well, but they do not

- accurately convey three-dimensional relationships or steric strain because they portray the cyclohexane ring as planar. (Section 4.9; Objective 8)

- The most stable chair conformation of a disubstituted cyclohexane is the one in which the substituents experience the least amount of strain. This usually calls for the larger substituent to occupy an equatorial position. (Section 4.9; Objective 8)

- Molecular modeling kits effectively demonstrate changes that occur during a chair flip. (Section 4.10; Objective 7)

- A molecule's **index of hydrogen deficiency (IHD)**, also called its **degree of unsaturation**, is half the number of hydrogen atoms missing from that molecule compared to an analogous completely **saturated** molecule. A saturated molecule has the most possible hydrogen atoms given the nonhydrogen atoms it contains; its IHD is 0. (Section 4.12; Objective 9)

- Each double bond a molecule possesses contributes 1 to its IHD. Each triple bond contributes 2, and each ring contributes 1. (Section 4.12; Objective 9)

Problems

4.2 Newman Projections

4.26 Draw the Newman projection for each of the following species, looking down the bond that is indicated in red.

(a)   (b)   (c)   (d)

(e)   (f)   (g)   (h)

4.27 Draw the corresponding dash–wedge structure for each of the following Newman projections.

(a)   (b)   (c)   (d)

4.28 Identify which C—H bonds are axial and which are equatorial in the following Newman projection for cyclohexane.

4.29 **(a)** Draw the Newman projection for each molecule shown here, looking down the C—C bond indicated by the arrow. **(b)** Which configuration do you think is more stable? Explain.

cis-1,2-Dimethylcyclopropane   *trans*-1,2-Dimethylcyclopropane

4.3 Conformational Analysis

4.30 Rank the following conformations in order from least stable to most stable.

A B C D E F

4.31 Perform a conformational analysis of propane, $CH_3CH_2CH_3$. Pay attention to the relative energies of the various conformations, but do not concern yourself with the actual energy values.

4.32 Perform a conformational analysis of butane, $CH_3CH_2CH_2CH_3$, looking down the C2—C3 bond. Pay attention to the relative energies of the various conformations, but do not concern yourself with the actual energy values.

4.33 Perform a conformational analysis of 1-bromo-2-chloroethane, $BrCH_2CH_2Cl$. Pay attention to the relative energies of the various conformations, but do not concern yourself with the actual energy values.

4.34 Perform a conformational analysis of 2-methylbutane, $(CH_3)_2CHCH_2CH_3$, looking down the C2—C3 bond. Pay attention to the relative energies of the various conformations, but do not concern yourself with the actual energy values.

4.35 Perform a conformational analysis of 1,2-dibromo-1-fluoroethane, $BrFCH—CH_2Br$. Pay attention to the relative energies of the various conformations, but do not concern yourself with the actual energy values.

4.4–4.8 Ring Strain, Stable Conformations of Rings, and Chair Conformations of Monosubstituted Cyclohexanes

4.36 The heat of combustion of cyclononane is 5586 kJ/mol (1335 kcal/mol). Calculate the ring strain per CH_2 group and the total ring strain of cyclononane. Which compound has more ring strain, cyclononane or cycloheptane?

4.37 Rank the following compounds in order from smallest heat of combustion to largest heat of combustion.

A B C D E

4.38 There are three distinct chair conformations for cyclohexylcyclohexane. **(a)** Draw all three and **(b)** determine which one is the most stable.

Cyclohexylcyclohexane

4.39 These two compounds are each in their more stable chair conformation. **(a)** Which occupies more space, a lone pair of electrons or a N—H bond? **(b)** Which occupies more space, a lone pair of electrons or a CH_3 group?

4.40 For which isomer would you expect a greater equilibrium percentage of molecules with the alkyl group in the axial position, isopropylcyclohexane or propylcyclohexane? Explain.

Isopropylcyclohexane　　　　**Propylcyclohexane**

4.9 Chair Conformations and Chair Flips with Two or More Substituents

4.41 Draw the more stable chair conformation of each of the following molecules.

(a) 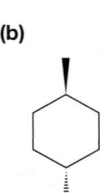 (b) 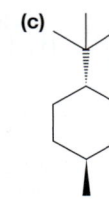 (c) 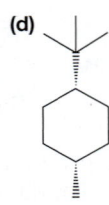 (d) 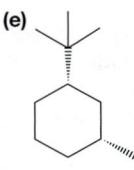 (e) 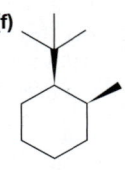 (f) 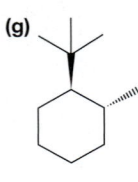 (g)

4.42 Draw a chair conformation of this molecule with **(a)** all CH_3 groups in axial positions and **(b)** all CH_3 groups in equatorial positions.

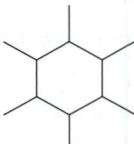

4.43 Identify each of the following disubstituted cyclohexanes as either a cis or a trans isomer.

(a)    (b)    (c)   (d)   (e)   (f)

4.44 For each of the following disubstituted cyclohexanes, determine whether the cis or trans isomer is more stable.

(a) 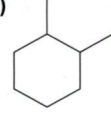   (b) 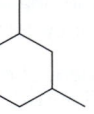   (c)

4.45 Both cis and trans isomers exist for this molecule. Which one is more stable?

4.46 Draw the more stable chair conformation for each of the following disubstituted cyclohexanes.

(a) OH 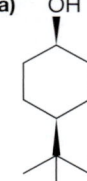   (b) OH 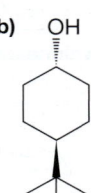   (c) 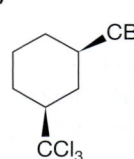   (d) 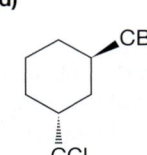   (e) 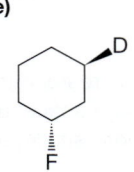   (f)

4.12 Index of Hydrogen Deficiency

4.47 Determine the IHD for each of the following compounds.

(a) 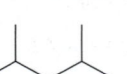   (b) 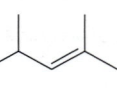   (c) N 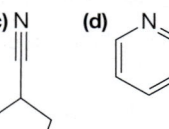   (d) N   (e) 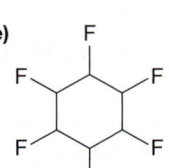   (f) O   (g) O 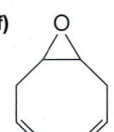   (h)

4.48 Determine the number of hydrogen atoms in each compound, given the number and type of nonhydrogen atoms it contains and its IHD.

(a) Four carbon atoms; IHD = 0
(b) Four carbon atoms; IHD = 2
(c) Three carbon atoms; two oxygen atoms; IHD = 1
(d) Five carbon atoms; two chlorine atoms; one nitrogen atom; IHD = 3
(e) One carbon atom; one nitrogen atom; IHD = 2
(f) Six carbon atoms; two nitrogen atoms; one oxygen atom; three fluorine atoms; IHD = 4

4.49 Calculate the IHD for each of the following molecular formulas. (a) C_6H_6; (b) $C_6H_5NO_2$; (c) $C_8H_{13}F_2NO$; (d) $C_4H_{12}Si$; (e) $C_6H_5O^-$; (f) $C_4H_6O_3S$

4.11 and 4.13 Identifying and Drawing Constitutional Isomers

4.50 Which pairs of structures are constitutional isomers?

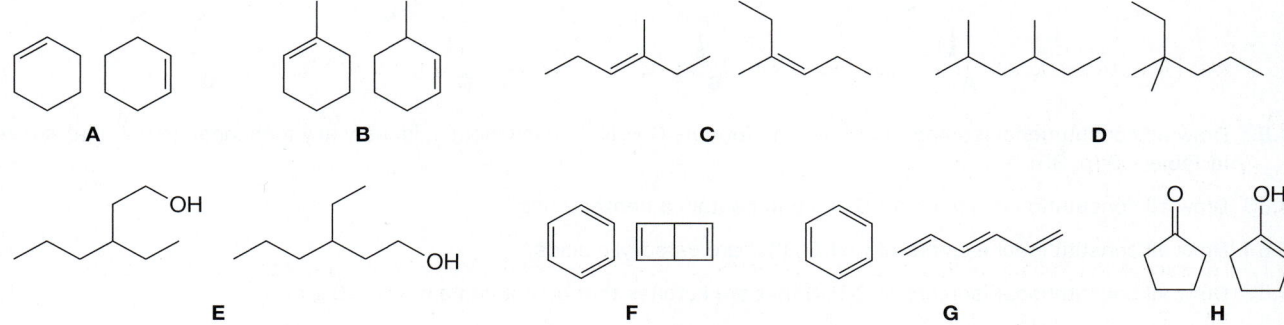

4.51 Draw all constitutional isomers that have the formula $C_5H_{11}Br$.

4.52 Draw all constitutional isomers that have the molecular formula $C_3H_6F_2O$, in which the oxygen is bonded to only one carbon atom. (There are 14 isomers.)

4.53 Draw all constitutional isomers that have the molecular formula $C_3H_6F_2O$, in which the oxygen is bonded to two carbon atoms.

4.54 Draw all constitutional isomers that have the molecular formula C_4H_6. (There are nine isomers.)

4.14 and 4.15 The Organic Chemistry of Biomolecules

4.55 Behenic acid and erucic acid are two fatty acids isolated from rapeseed oil. Which fatty acid has a higher melting point?

Behenic acid

Erucic acid

4.56 As we saw in Section 4.15, oleic acid has one C=C double bond that is in the cis configuration. Elaidic acid is identical to oleic acid, but the C=C is in the trans configuration. Which fatty acid has the higher melting point? Explain.

4.57 How many constitutional isomers of acyclic monosaccharides are there that can be classified as aldohexoses? As ketohexoses?

4.58 Mannoheptulose is a monosaccharide found in avocados and, by blocking the enzyme hexokinase, it inhibits glucose phosphorylation. Classify mannoheptulose according to the distinctions made in Section 4.14 (e.g., aldohexose).

Mannoheptulose

4.59 Olive oil melts around −6 °C and palm oil melts around +35 °C. What does this say about the relative amount of unsaturation in the fatty acids that make up these oils?

Integrated Problems

4.60 For each of the following substituted cyclohexane rings, **(a)** draw the corresponding dash–wedge structure, **(b)** draw the corresponding Haworth projection, **(c)** determine whether the given conformation is the most stable one, and **(d)** if it is *not* the most stable conformation, draw the Newman projection of the one that is.

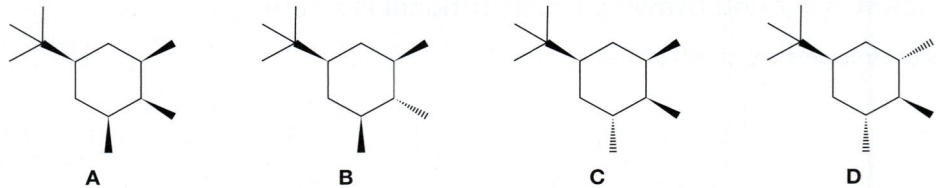

4.61 Rank the following compounds in order from smallest heat of combustion to largest heat of combustion.

A **B** **C** **D**

4.62 Draw all constitutional isomers that have the formula C_3H_5N. In each isomer, identify any functional groups that are listed in Table 1-6 (p. 35).

4.63 Draw all constitutional isomers of C_9H_{12} that contain a benzene ring.

4.64 Draw all constitutional isomers of $C_5H_8O_2$ that are carboxylic acids.

4.65 Draw all constitutional isomers of C_5H_8O that are ketones that do *not* contain a C=C group.

4.66 Draw all constitutional isomers of all-cis ethylmethylisopropylcyclohexane—that is, in which a methyl group (CH_3), an ethyl group (CH_2CH_3), and an isopropyl group [$CH(CH_3)_2$] are all bonded to a cyclohexane ring on the same side of the ring's plane. Which of those isomers do you think is the most stable? Explain.

4.67 Glucose, a monosaccharide, has both acyclic and cyclic forms. One of the cyclic forms is called β-D-glucopyranose. Draw the more stable chair conformation of β-D-glucopyranose.

β-D-**Glucopyranose**

4.68 In addition to β-D-glucopyranose (see Problem 4.67), glucose can exist in another cyclic form, called β-D-glucofuranose. Which form is more stable, β-D-glucopyranose or β-D-glucofuranose? Explain.

β-D-**Glucofuranose**

4.69 Draw the more stable chair conformation of each of the following compounds in which an sp^3-hybridized heteroatom is part of the ring.

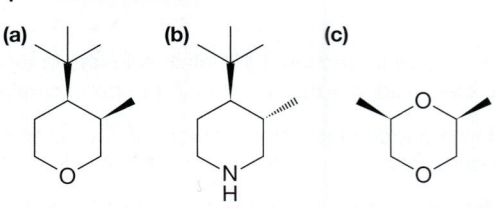

(a) (b) (c)

4.70 Draw the more stable chair conformation of each of the following compounds in which an sp^2-hybridized carbon atom is part of the ring.

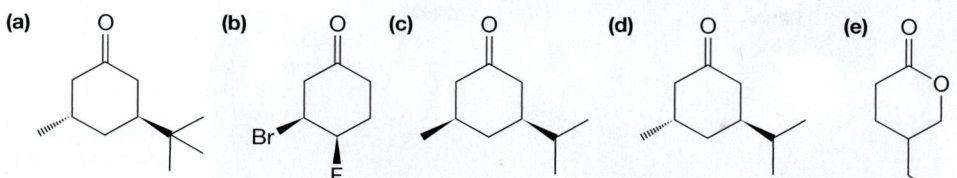

(a)   (b)   (c)   (d)   (e)

4.71 Which monosaccharide has a greater heat of combustion, β-D-glucopyranose or β-D-allopyranose? Explain.

β-D-**Glucopyranose**   β-D-**Allopyranose**

4.72 Even though an iodine atom is larger in size than a bromine atom, both bromocyclohexane and iodocyclohexane exist with 31% of molecules having the halogen atom in the axial position. Explain why.

4.73 5-Hydroxy-1,3-dioxane is more stable with the OH group in the axial position than in the equatorial position. Explain why.

OH

5-Hydroxy-1,3-dioxane

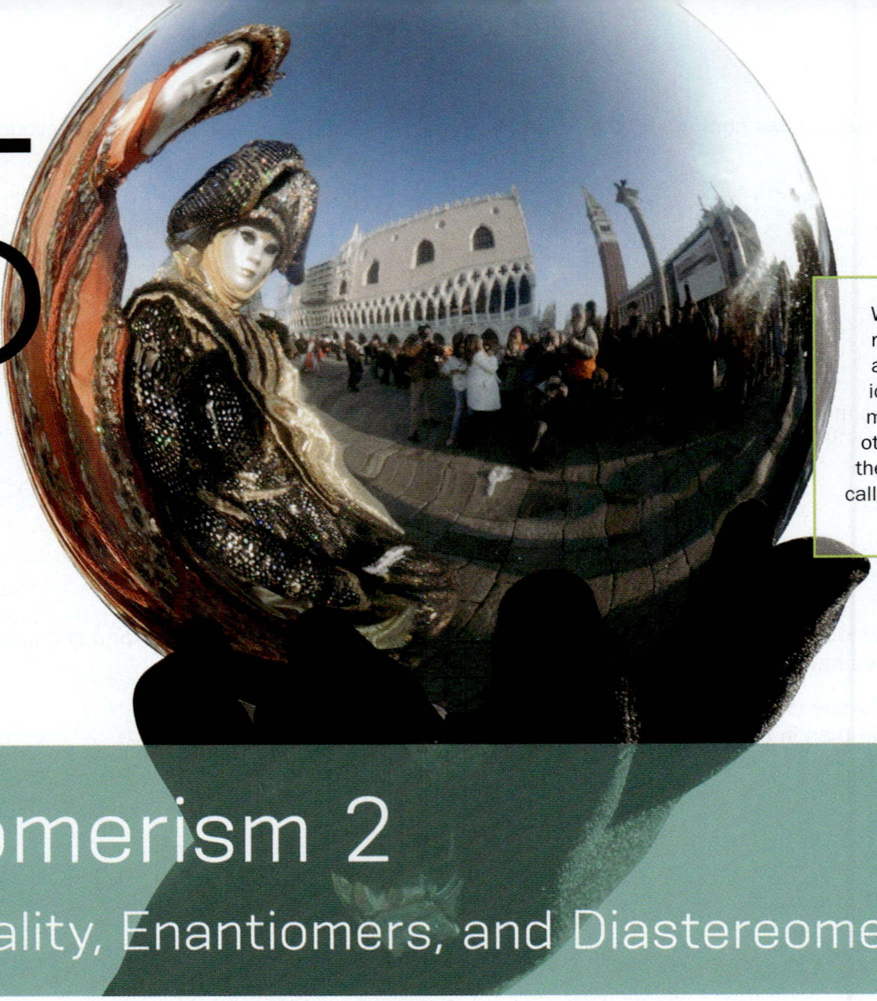

5

When objects are reflected through a mirror, some are identical to their mirror image and others are not. This is the basis for what is called *chirality*, a major topic of this chapter.

Isomerism 2
Chirality, Enantiomers, and Diastereomers

In Chapter 4, we examined *conformers* and *constitutional isomers* in detail. Here in Chapter 5, we examine *configurational isomers*, of which there are two types: *enantiomers* and *diastereomers*. We study in depth the structural relationships among these types of isomers, and we begin to examine how those relationships affect their respective physical and chemical behaviors. Whether molecules are enantiomers or diastereomers can have dramatic consequences in chemical reactions, as we will study in greater depth in Chapter 8 and beyond.

5.1 Defining Configurational Isomers, Enantiomers, and Diastereomers

To formally define *configurational isomers*, *enantiomers*, and *diastereomers*, review the flowchart shown in both Figure 4-1 and Figure 5-1. The types of isomers we are interested in here are highlighted in red in Figure 5-1. According to the flowchart:

Configurational isomers have the same connectivity, making them a type of **stereoisomers**, but differ in a way *other* than by rotations about single bonds. They include two types:

- **Enantiomers:** Configurational isomers that are mirror images of each other.
- **Diastereomers:** Configurational isomers that are *not* mirror images of each other.

Chapter Objectives

On completing Chapter 5 you should be able to:

1. Define the structural relationships that characterize and distinguish pairs of configurational isomers, enantiomers, and diastereomers.

2. Determine whether a pair of molecules are enantiomers or diastereomers.

3. Identify a molecule as either chiral or achiral and draw any chiral molecule's enantiomer.

4. Identify chiral centers and explain their relevance to chirality.

5. Determine whether a compound is meso.

6. Recognize whether a given nitrogen atom is a chiral center.

7. Understand the importance of stereochemical configuration as it pertains to the structural relationships of enantiomers and diastereomers.

8. Draw and interpret Fischer projections.

9. Predict whether two molecules will have the same or different physical and chemical properties based on their structural relationship.

10. Predict the relative stability of alkenes based on the alkyl substitution of the carbon atoms involved in the double bond.

11. Explain the general principles behind methods for separating enantiomers and diastereomers.

12. Describe the relationship between chirality and optical activity.

13. Compute a chiral compound's specific angle of rotation, measured angle of rotation, concentration, and path length, given three of these values.

14. For a mixture of enantiomers, compute enantiomeric excess, the specific angle of rotation of the mixture, and the specific angle of rotation of its components, given two of these values.

Whereas conformers are related solely by *rotations* about single bonds, converting from one configurational isomer to another usually requires the breaking of a covalent bond followed by the formation of a new one—something that generally does not readily take place at room temperature. Therefore, *configurational isomers can usually be isolated from one another.*

Even though the structural differences between a pair of configurational isomers may seem subtle, the differences in their behavior may not be. For example, one of two enantiomers of the drug thalidomide acts as a sedative and antinausea medication for pregnant women suffering from morning sickness, whereas the other enantiomer causes terrible birth defects in newborns. Widely marketed in Europe during the late 1950s and early 1960s, thalidomide is now used to treat multiple myeloma (a kind of cancer) and erythema nodosum leprosum (an inflammation of fat cells under the skin). We will have more to say about this phenomenon later in the chapter.

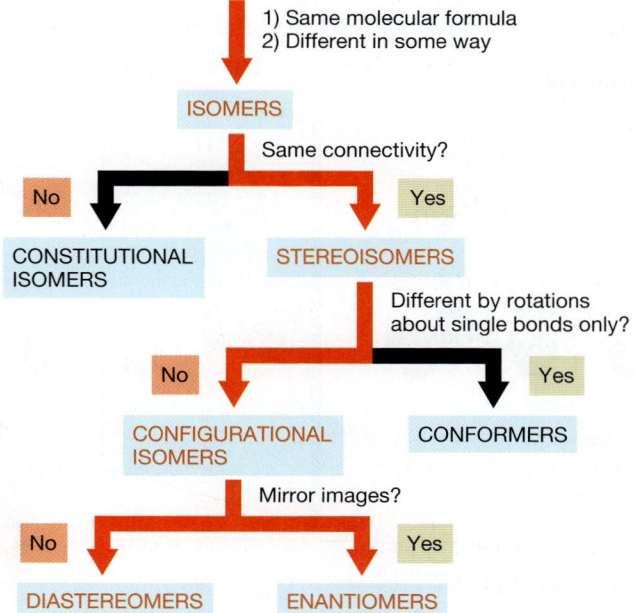

FIGURE 5-1 Flowchart illustrating the subcategories of isomers Those in red are examined in detail in this chapter, whereas those in black were discussed in Chapter 4.

5.2 Enantiomers, Mirror Images, and Superimposability

Enantiomers are mirror images of each other. To be *isomers* of each other, however, they must be *different* in some way. Enantiomers, therefore, are *nonsuperimposable mirror images*.

> Molecules are **nonsuperimposable** if there is no orientation in which *all* atoms of both molecules can be lined up perfectly (i.e., superimposed).

The two molecules of CHBrClF in Figure 5-2a are enantiomers. They are mirror images of each other—that is, if we looked at the reflection in the mirror of the molecule on the right (Fig. 5-2b), we would see an image that is identical to the molecule on the left. There is no orientation of the two molecules, however, that allows *all* of the atoms to be superimposed (see Your Turn 5.1).

YOUR TURN 5.1

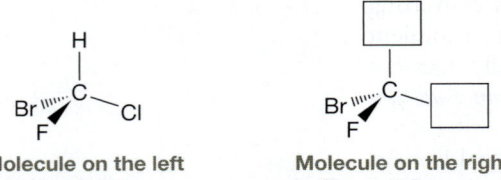

Molecule on the left in Figure 5-2a

Molecule on the right in Figure 5-2a

The molecule on the left in Figure 5-2a and part of the one on the right in Figure 5-2a are redrawn here. Using a molecular modeling kit, construct the molecule on the right in Figure 5-2a and orient it so that the C, Br, and F atoms occupy the positions shown in the second structure here. Complete the drawing of the molecule by adding the H and Cl atoms in the positions where they appear in your model. Comparing these two structures, are the molecules superimposable?

Answers to Your Turns are in the back of the book.

Unlike CHBrClF, a molecule of CH_2ClF does *not* have an enantiomer. CH_2ClF has a mirror image (Fig. 5-3a), but its mirror image is exactly the same as itself

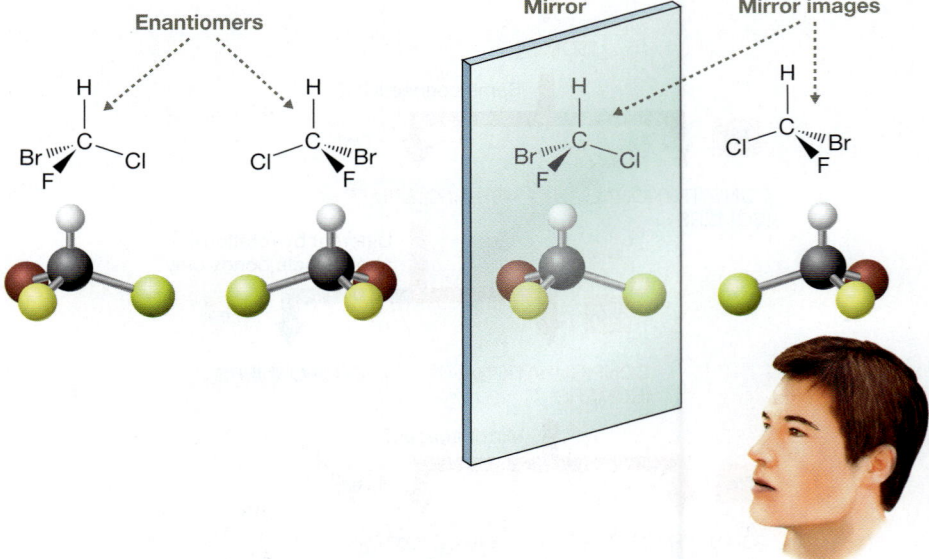

Enantiomers   **Mirror**   **Mirror images**

FIGURE 5-2 Enantiomers The two molecules of CHBrClF (a) are enantiomers because, as shown in (b), they are nonsuperimposable mirror images of each other. Ball-and-stick models are shown below each dash–wedge structure.

(a)   (b)

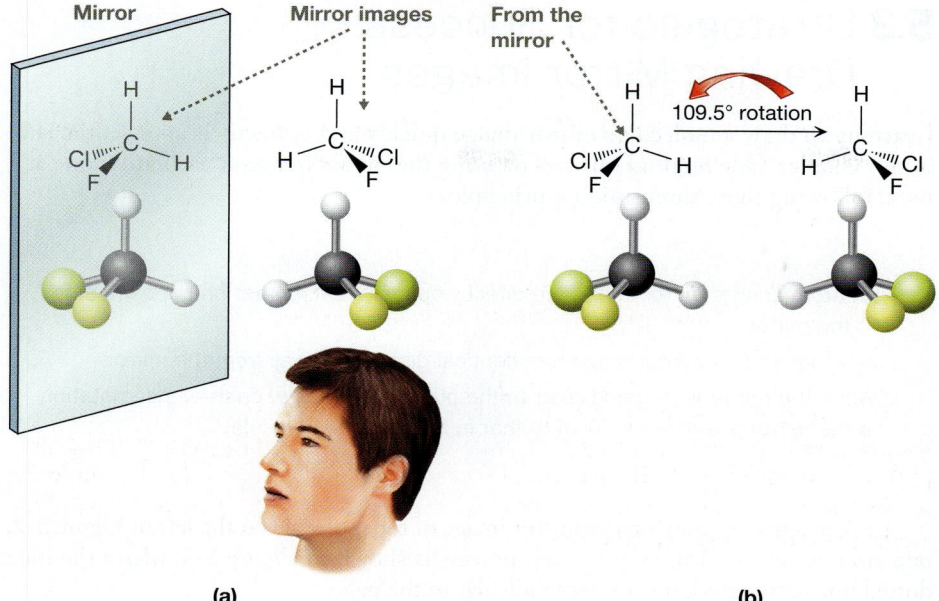

FIGURE 5-3 Superimposable mirror images CH_2ClF does not have an enantiomer because the molecule and its mirror image (a) are superimposable. If the image from the mirror is rotated by 109.5° (b), then the molecule and its mirror image are exactly the same. Ball-and-stick models are shown below each dash–wedge structure.

(Fig. 5-3b). If the image in the mirror is rotated, it can be superimposed on the original molecule (see Your Turn 5.2). Remember:

> Every molecule has a mirror image, but not every molecule has a *nonsuperimposable* mirror image.

YOUR TURN 5.2

Use a molecular modeling kit to construct both the original molecule and the mirror image shown in Figure 5-3a. Rotate the mirror image structure as indicated in Figure 5-3b to verify that the two molecules are superimposable.

PROBLEM 5.1 For each pair, determine whether the molecules are *superimposable* or *nonsuperimposable*. *Hint:* It may help to build models.

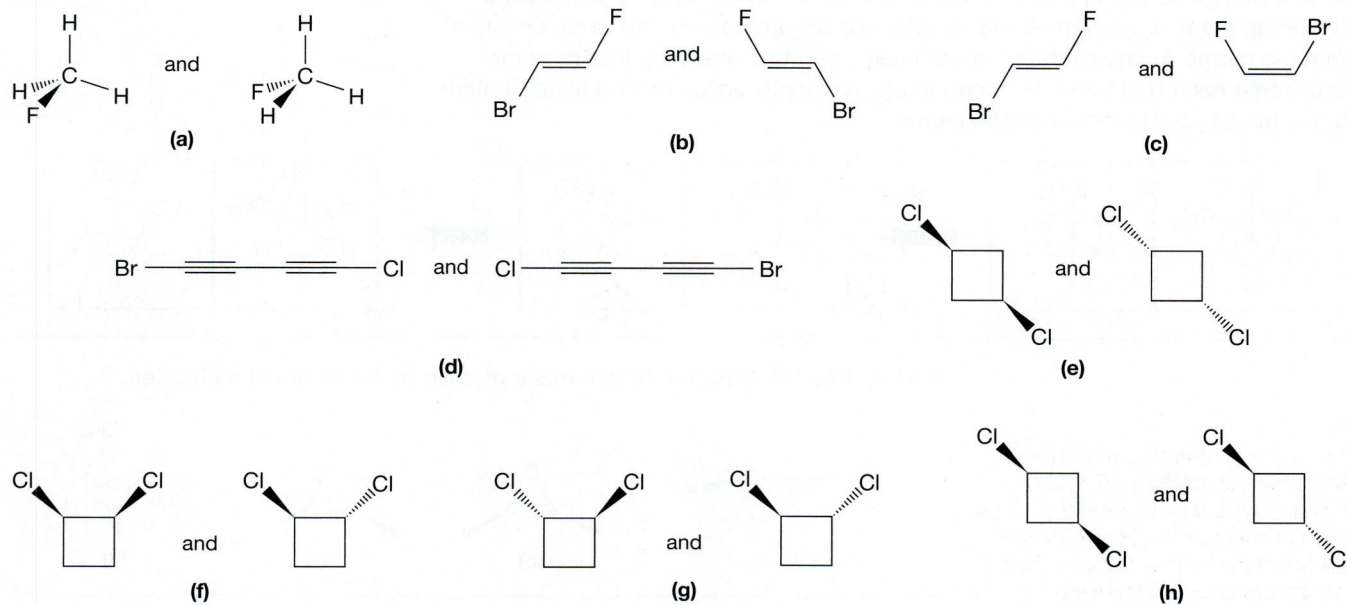

5.3 Strategies for Success: Drawing Mirror Images

Learning to draw a molecule's mirror image quickly and correctly is an essential skill in this chapter. One method involves drawing the mirror image of each atom one at a time, following these three guiding principles:

- An atom and its mirror image are directly opposite each other on opposite sides of the mirror.
- An atom and its mirror image are identical distances *away from* the mirror.
- When the mirror is perpendicular to the plane of the page, dash–wedge notation in the mirror image is identical to that in the original molecule.

Let's practice by drawing the mirror image of the molecule on the left in Figure 5-2, one atom at a time. This step-by-step process is shown in Figure 5-4, where the blue dotted line represents a mirror perpendicular to the page.

SOLVED PROBLEM 5.2

Given the Newman projection of butane in a gauche conformation, draw the Newman projection of its mirror image. The blue dotted line next to the molecule represents a mirror perpendicular to the page.

Think For each atom or group in the original molecule, where should its mirror image be with respect to the length of the mirror? With respect to the distance from the mirror?

Solve Six substituents must be added to the Newman projection on the right to complete the mirror image—two CH_3 groups and four H atoms. Begin by drawing the mirror image of the topmost CH_3 group, which must be directly opposite the original CH_3 group and must also be the same distance from the mirror. This is shown in the first frame below. Next, as shown in the second frame, add the second CH_3 group to the mirror image so that it is directly opposite the second CH_3 group in the original molecule and the two CH_3 groups are the same distance from the mirror. To complete the mirror image, add the remaining four H atoms. Notice that each H atom in the mirror image is directly opposite its original H atom and is the same distance from the mirror.

PROBLEM 5.3 Draw the mirror image of each of the following molecules.

(a)   (b)   (c)   (d)

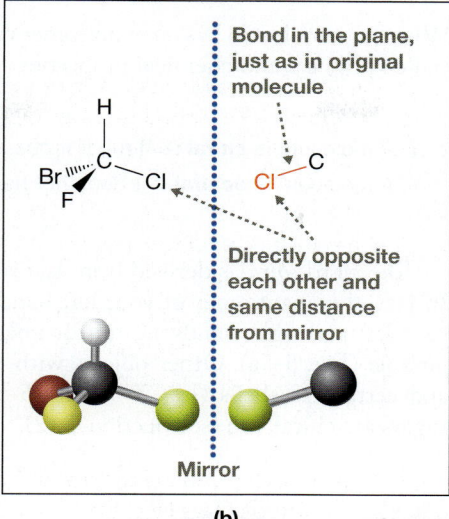

Draw the mirror image of C.

Same distances from the mirror

Directly opposite original C

Mirror

(a)

Add Cl and its bond.

Bond in the plane, just as in original molecule

Directly opposite each other and same distance from mirror

Mirror

(b)

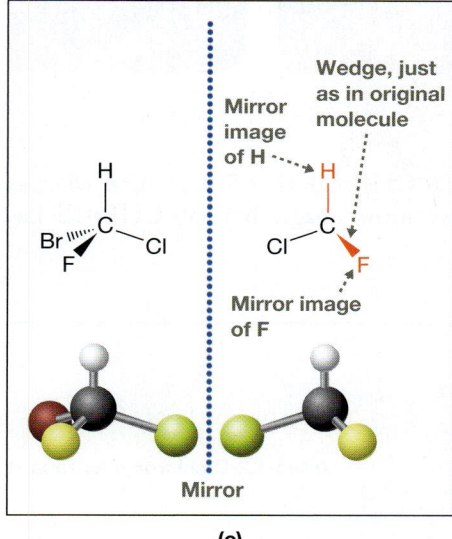

Add H, F, and their bonds.

Mirror image of H

Wedge, just as in original molecule

Mirror image of F

Mirror

(c)

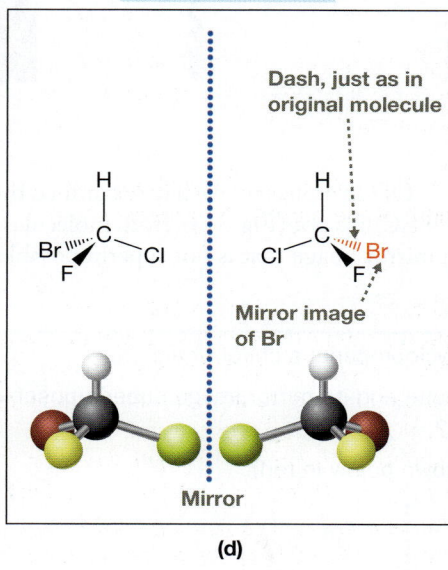

Add Br and its bond.

Dash, just as in original molecule

Mirror image of Br

Mirror

(d)

FIGURE 5-4 Progression of drawing the mirror image of CHFClBr In each frame, the mirror is represented by the blue dotted line, the original molecule is on the left, and the mirror image is on the right. The atoms and bonds added to the mirror image of the dash–wedge structure are highlighted in red. A ball-and-stick model is shown below each dash–wedge structure.

PROBLEM 5.4 For each molecule in Problem 5.3, determine whether the mirror image is superimposable on the original molecule.

PROBLEM 5.5 Determine which pairs of molecules are mirror images.

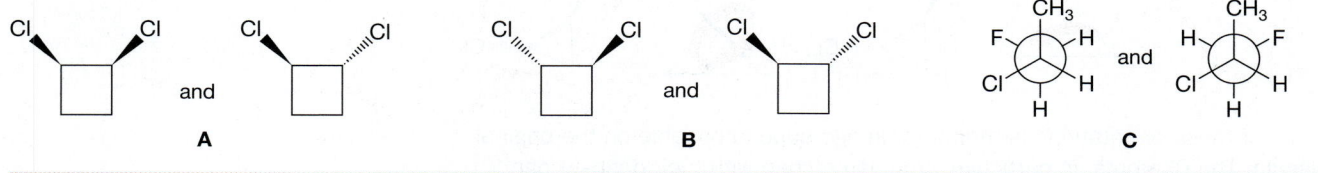

and   **A**

and   **B**

and   **C**

5.4 Chirality

Whether a molecule has an enantiomer can have important implications for the molecule's physical and chemical properties.

- A molecule is **chiral** (KAI-rul) if it *has* an enantiomer.
- A molecule is **achiral** if it does *not* have an enantiomer.

The word *chiral* is derived from Latin for "hand," given that your hands are chiral. In fact, the enantiomer of your left hand is your right hand, and vice versa, because your left and right hands are mirror images of each other but they are not superimposable (Fig. 5-5a). Other objects with handedness, or chirality, include corkscrews and certain seashells (Fig. 5-5b and 5-5c). Biomolecules such as amino acids and sugars are chiral, too (see Section 5.12).

 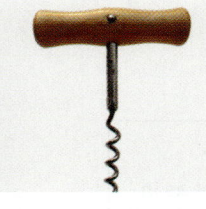   

| (a) | (b) | (c) |

FIGURE 5-5 Familiar objects that are chiral (a) Hands. (b) A corkscrew. (c) A seashell.

Of the molecules we have examined thus far, CHBrClF (Fig. 5-2) is chiral, whereas CH$_2$ClF is not (Fig. 5-3). Both molecules have mirror images, but only CHBrClF has a mirror image that is not superimposable.

SOLVED PROBLEM 5.6

Determine whether *trans*-1,2-dichlorocyclopropane is chiral or achiral.

Think Are *trans*-1,2-dichlorocyclopropane and its mirror image superimposable or nonsuperimposable? How can you tell?

Solve First draw the mirror image (shown below in red).

trans-**1,2-Dichlorocyclopropane**

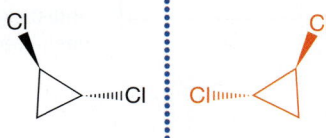

It then becomes a matter of trying to line up the two molecules with every orientation possible to see if they are superimposable:

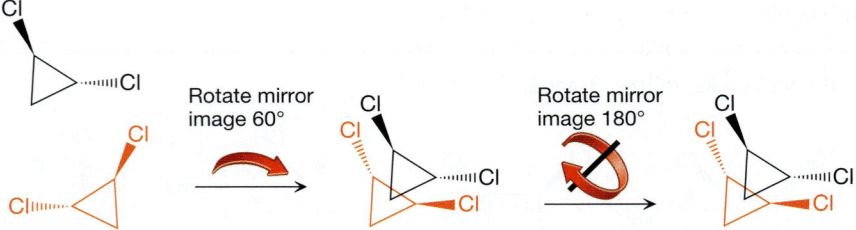

In none of these orientations is the mirror image superimposable on the original molecule. The Cl atoms, in particular, are mismatched with their dash–wedge notation. There are other orientations you can try, but you will find that none of them allows the mirror image to be superimposable on the original molecule. Therefore, *trans*-1,2-dichlorocyclopropane is chiral. If you are unconvinced, construct a model of each molecule and try to superimpose them (see Your Turn 5.3).

YOUR TURN **5.3**

Build a molecular model of both the original molecule and the mirror image in Solved Problem 5.6. Verify that the two molecules are nonsuperimposable by performing the rotations indicated in Solved Problem 5.6. Are there other rotations you can imagine to determine whether the two molecules line up perfectly?

YOUR TURN **5.4**

Is the mirror image of the molecule in Solved Problem 5.6 (i.e., the molecule in red) chiral or achiral?

PROBLEM 5.7 Is *trans*-1,2-dichloroethene chiral or achiral?

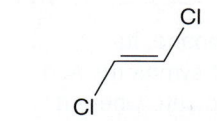

***trans*-1,2-Dichloroethene**

5.4a The Plane of Symmetry Test for Chirality

A convenient way to determine whether a molecule is achiral is to search for a *plane of symmetry*. A molecule has a **plane of symmetry** if it can be bisected in such a way that *one half of the molecule is the mirror image of the other half*. If a molecule has at least one plane of symmetry, then its mirror image is the same as itself. Thus:

A molecule that possesses a plane of symmetry must be achiral.

Most molecules that do not possess a plane of symmetry are chiral, though there are some exceptions. (See Problem 5.79 at the end of the chapter.)

 CH_2ClF is achiral (Fig. 5-3, p. 211), and Figure 5-6a (next page) shows that it has a plane of symmetry that bisects the H—C—H bond angle. CHBrClF is chiral (Fig. 5-2, p. 210), on the other hand, so it must *not* have a plane of symmetry. Notice that what was a plane of symmetry in CH_2ClF (Fig. 5-6a) is no longer one in CHBrClF (Fig. 5-6b).

YOUR TURN **5.5**

In Solved Problem 5.6, we showed that *trans*-1,2-dichlorocyclopropane is chiral. Can it have a plane of symmetry? (You should build a model of it and look for one to verify your answer.)

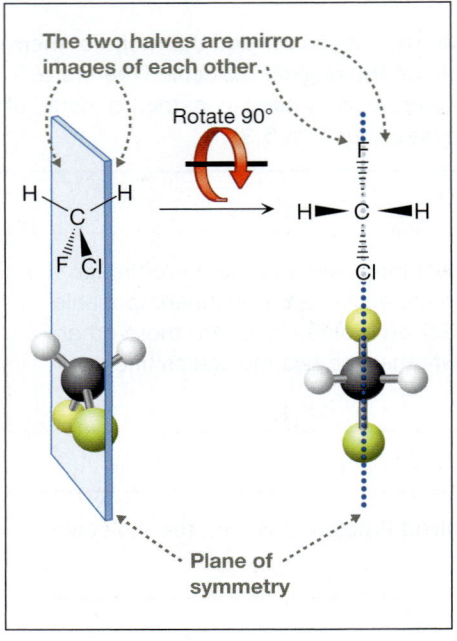

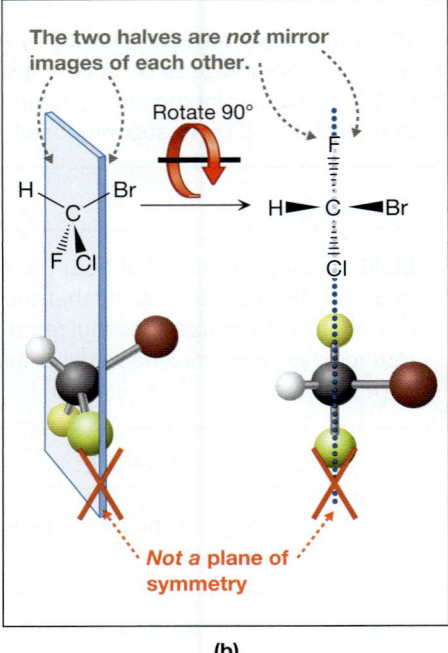

FIGURE 5-6 **Plane of symmetry test for chirality** (a) Two orientations of CH_2ClF. To obtain the orientation on the right, the molecule on the left is rotated 90° about the axis indicated. In both orientations, the molecule possesses a plane of symmetry (indicated in blue) that bisects the H—C—H angle. As a result, CH_2ClF is achiral. (b) Two orientations of CHBrClF that differ by a rotation of 90° about the axis. Replacing one H atom from CH_2ClF by a Br atom destroys the plane of symmetry that exists in CH_2ClF. This is consistent with CHBrClF being chiral.

YOUR TURN 5.6

Unlike *trans*-1,2-dichlorocyclopropane, its cis isomer does have a plane of symmetry. Add a dashed line to the figure to indicate where it is. Is *cis*-1,2-dichlorocyclopropane chiral or achiral?

cis-**1,2-Dichlorocyclopropane**

PROBLEM 5.8 Determine whether each of the following molecules possesses a plane of symmetry. If it does, indicate the plane of symmetry using a dashed line.

(a)   (b)   (c)   (d)   (e)   (f)

5.4b Stereocenters, Asymmetric Atoms, and Stereochemical Configurations

Thus far, we have shown how a molecule's chirality depends on whether it contains a plane of symmetry. However, another structural feature, a *tetrahedral stereocenter*, is also related to chirality.

- A **stereocenter** is an atom with the property that interchanging any two of its attached groups produces a different stereoisomer.
- A tetrahedral stereocenter, called a **chiral center**, is bonded to four *different groups*.

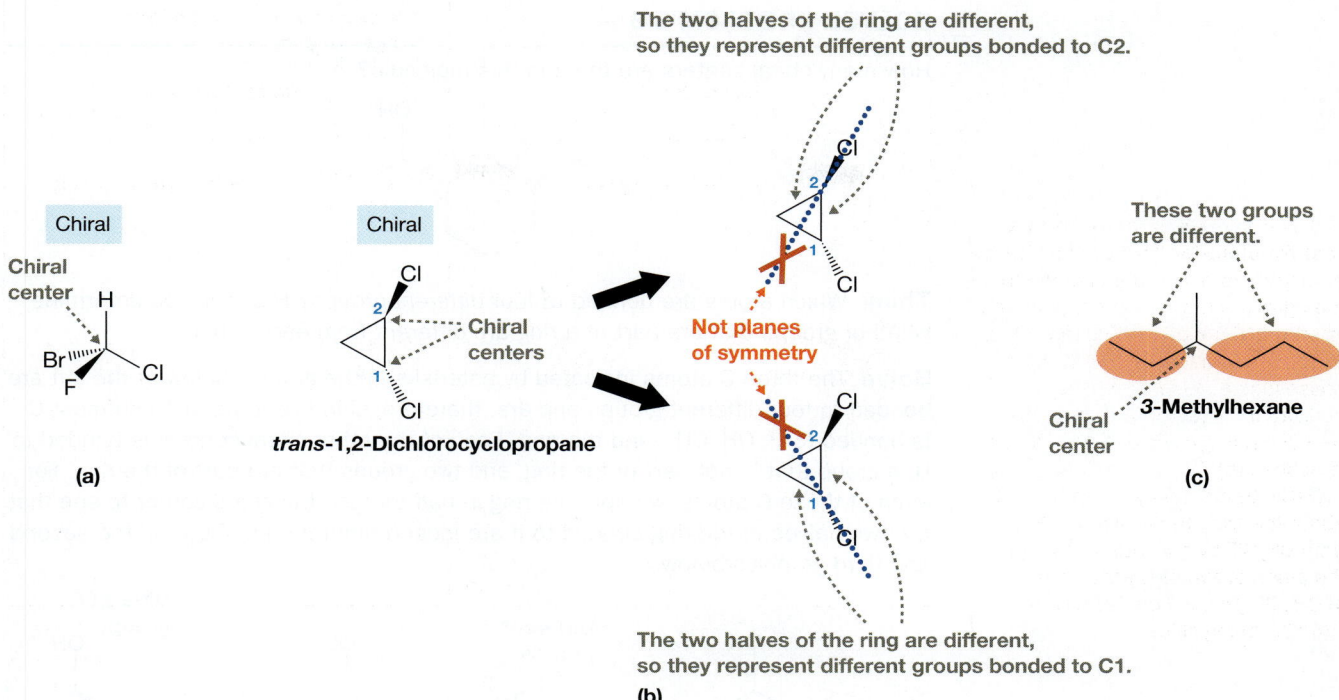

The two halves of the ring are different, so they represent different groups bonded to C2.

Chiral

Chiral center

Chiral

Chiral centers

trans-1,2-Dichlorocyclopropane

Not planes of symmetry

These two groups are different.

3-Methylhexane

Chiral center

(a)

(b)

(c)

The two halves of the ring are different, so they represent different groups bonded to C1.

FIGURE 5-7 Chiral centers (a) The C atom of CHBrClF is a chiral center because it is bonded to four different atoms. (b) Both C1 and C2 of *trans*-1,2-dichlorocyclopropane are chiral centers (left) because each is bonded to four different groups. For each chiral center, the groups that make up the ring are shown to be different (right). (c) In 3-methylhexane, C3 is a chiral center because it is bonded to four different groups.

When the chiral center is a carbon atom, in particular, it is referred to as an **asymmetric carbon**.

Each of the molecules we have identified thus far as being chiral possesses at least one chiral center. The C atom in CHBrClF, for example, is a chiral center, because it is bonded to four different atoms (Fig. 5-7a). *trans*-1,2-Dichlorocyclopropane contains two chiral centers—namely, each C that is bonded to a Cl (Fig. 5-7b). While it should be straightforward that both of those C atoms in Figure 5-7b are tetrahedral, it may be less straightforward that both are also bonded to four *different* groups. On each C atom, one bond is to H and another is to Cl. The other two bonds are part of the ring. To see that those last two bonds are to different groups, we bisect the vertex made by those bonds, as indicated by each dashed line on the right in Figure 5-7b. In each case, the two halves of the ring are different from each other.

When you look for stereocenters, make sure you take into account *entire* substituents, not just the immediate atoms at the points of attachment. In 3-methylhexane (Fig. 5-7c), for example, C3 is a chiral center because it is attached to four different groups: H, CH_3, CH_2CH_3, and $CH_2CH_2CH_3$. The last two groups are distinct, even though they both have CH_2 at their points of attachment.

Chiral centers are not limited to just carbon atoms. Nitrogen atoms can also form four bonds. Therefore:

A nitrogen atom that is bonded to four different groups is a chiral center.

An example is shown in Figure 5-8. Notice that, to be a chiral center, a nitrogen atom, unlike carbon, must have a +1 formal charge.

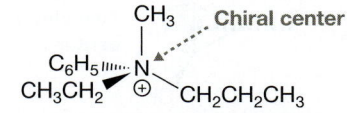

CH_3 Chiral center

C_6H_5—N
CH_3CH_2 ⊕ $CH_2CH_2CH_3$

A quaternary ammonium ion

FIGURE 5-8 A nitrogen chiral center The nitrogen atom is tetrahedral and is bonded to four different groups, so it is a chiral center.

SOLVED PROBLEM 5.9

How many chiral centers are there in this molecule?

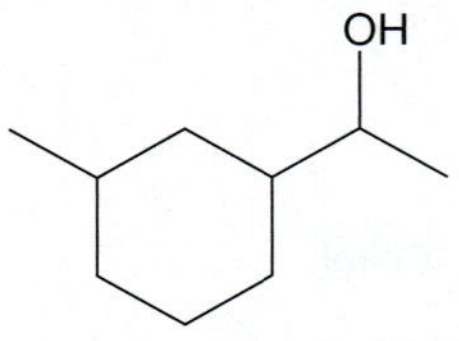

Think Which atoms are bonded to four *different* groups? How can you determine whether groups that are part of a ring are different from each other?

Solve The three C atoms indicated by asterisks in the graphic below at the left are bonded to four different groups and are, therefore, chiral centers. The rightmost C is bonded to H, OH, CH_3, and the ring. Each of the other chiral centers is bonded to H, a group that is not part of the ring, and two groups that are part of the ring. For each of those C atoms, we split the ring in half through the chiral center to see that the two halves of the ring bonded to it are indeed different, as shown in the second and third graphics below.

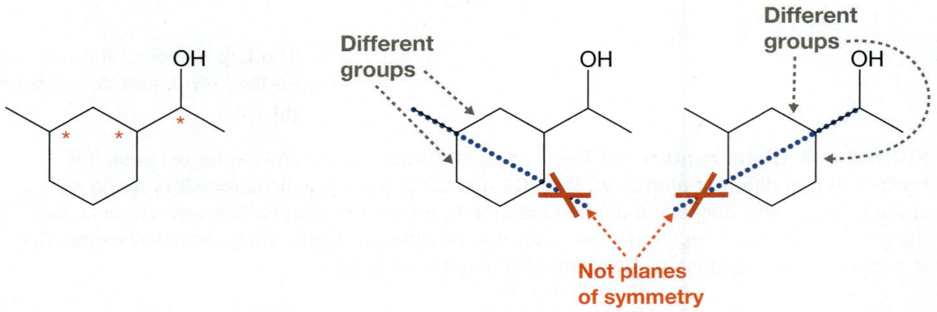

PROBLEM 5.10 Determine how many chiral centers exist in each of the following molecules.

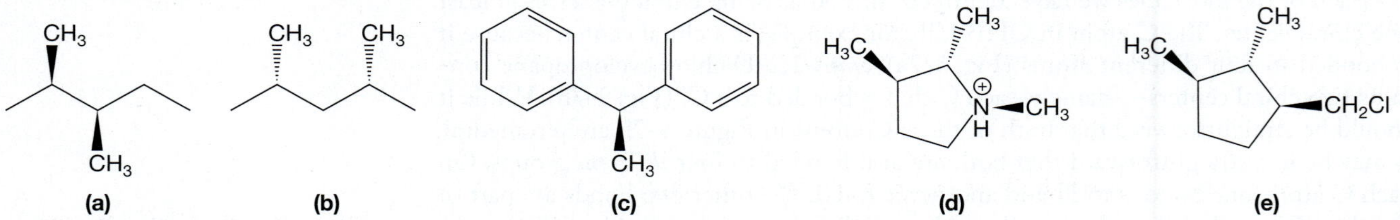

(a)   (b)   (c)   (d)   (e)

Although chiral centers are related to the concept of chirality, be careful:

> The presence of chiral centers does not guarantee that a molecule is chiral.

For example, *cis*-1,2-dichlorocyclopropane contains two chiral centers, but it is achiral because the molecule has a plane of symmetry (Fig. 5-9).

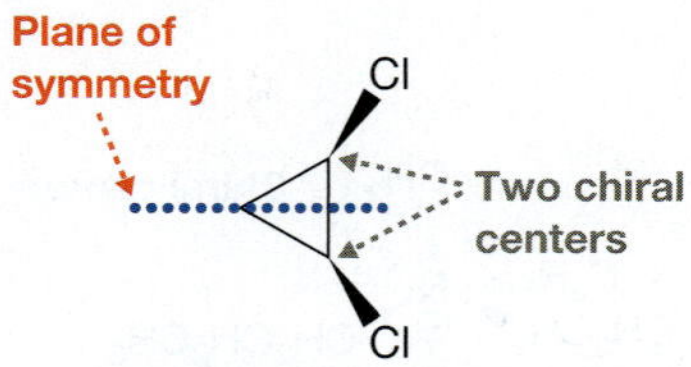

> A molecule is **meso** if it contains at least two chiral centers but has a plane of symmetry that makes it achiral *overall*.

The term comes from the Greek word for "middle," and refers to the fact that the molecule reflects about its middle—its plane of symmetry.

Plane of symmetry

Two chiral centers

cis-1,2-Dichlorocyclopropane
(*meso*-1,2-Dichlorocyclopropane)

FIGURE 5-9 A meso compound This molecule is meso because it contains chiral centers but, due to the plane of symmetry, is achiral.

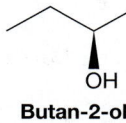

FIGURE 5-10 A chiral molecule with no chiral centers HFC=C=CHF contains no chiral centers, yet it is chiral. It does not have a plane of symmetry.

PROBLEM 5.11 Which of the following molecules are *meso*? *Hint:* Consider rotations about single bonds.

A **B** **C** **D** **E**

In most cases, a molecule (e.g., CH_2ClF) is achiral if it does not have a chiral center. However:

> The absence of chiral centers does not guarantee that a molecule is achiral.

1,3-Difluoropropa-1,2-diene (HFC=C=CHF; Fig. 5-10), for example, has no chiral centers (it has no tetrahedral atoms), yet it is chiral.

There is, however, a strict rule that relates the number of chiral centers and chirality:

> Any molecule that contains exactly one chiral center must be chiral.

We have previously seen (Fig. 5-2) that CHBrClF has a single chiral center and is chiral. The same would be true if H, F, Cl, and Br were replaced by any four distinct groups. Examples include butan-2-ol and 2-bromo-1,1-dimethylcyclobutane (Fig. 5-11).

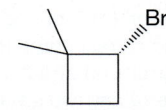

Butan-2-ol

2-Bromo-1,1-dimethylcyclobutane

FIGURE 5-11 Chiral centers and chirality Each molecule has exactly one chiral center and is therefore chiral.

YOUR TURN **5.7**

Place an asterisk at the chiral center in butan-2-ol and in 2-bromo-1,1-dimethylcyclobutane in Figure 5-11.

YOUR TURN **5.8**

Prove to yourself that butan-2-ol is chiral by (1) drawing the mirror image, (2) building models of both the original molecule and the mirror image, and (3) lining the models up with all orientations you can imagine to determine whether they are superimposable or nonsuperimposable.

There are two possible **stereochemical configurations** for any given chiral center. That is, there are two possible ways in which to arrange the four atoms or groups about the stereocenter in a tetrahedral fashion. One configuration is called "*R*" and the other is "*S*."

> The convention for naming a stereochemical configuration as *R* or *S* is discussed in Section C.1 of Interchapter C, which immediately follows this chapter.

Learning this nomenclature convention could provide insight into some aspects of stereochemistry discussed here in Chapter 5. However, the remainder of this chapter does *not* rely on having already covered that nomenclature section. What is critical to this chapter is understanding that, for any chiral center, one configuration is the *opposite* or *inverse* of the other, and that the two configurations are related as follows:

- One configuration of a chiral center is the mirror image of the other.
- Exchanging any two of the four groups on a chiral center gives the opposite stereochemical configuration.

These relationships are illustrated in Figure 5-12. The black structures on the left in Figure 5-12a and 5-12b are identical; hence, the stereochemical configuration at the C atom (the chiral center) is identical in both cases. In Figure 5-12a, the structure is reflected through a mirror plane, giving the structure shown in red. In Figure 5-12b, groups W and X in the black structure are interchanged, giving the structure shown in blue. Accordingly, the stereochemical configuration about the C atom in the red and blue structures is opposite that in the black structure, and the red and blue structures have the same configuration.

FIGURE 5-12 Two ways to invert stereochemical configurations (a) Reflection of the chiral center through a mirror. (b) Exchanging two groups on the chiral center. Reflection through a mirror (red structure) and the exchange of any two groups (blue structure) give exactly the same stereochemical configuration, which is opposite that in the original (black) structure.

PROBLEM 5.12 Figure 5-12b shows the result of exchanging groups W and X on the black structure, giving the blue structure. Repeat this exercise, but instead, exchange groups W and Y on the black structure and determine whether the resulting structure is superimposable on the blue structure in Figure 5-12.

5.4c Mirror Images That Rapidly Interconvert: Single-Bond Rotation and Nitrogen Inversion

In our discussion about chirality so far, the nonsuperimposable mirror images we have examined do not interconvert; doing so would require breaking covalent bonds. But in some cases, nonsuperimposable mirror images do rapidly interconvert, making it impossible to separate the mirror images.

> If a molecule and its mirror image rapidly interconvert, then the molecule is effectively *achiral*.

This is the case, for example, with conformers. Consider the gauche conformation of 1,2-dibromoethane shown in black in Figure 5-13a. Its mirror image is the gauche conformation shown in red. The two gauche conformations are nonsuperimposable, but they rapidly interconvert through rotation about the C—C bond (Fig. 5-13b), so 1,2-dibromoethane is achiral (Fig. 5-13c).

YOUR TURN **5.9**

Build molecular models of the gauche 1,2-dibromoethane shown on the left in Figure 5-13a and its mirror image. By overlaying the two molecules, verify that they are nonsuperimposable. Next, rotate the bond as in Figure 5-13b and show that the resulting molecules are superimposable.

PROBLEM 5.13 Determine whether each of the following molecules is chiral or achiral.

(a)   (b)   (c)

The same principle applies to conformers of cyclohexane chairs. Figure 5-14a (next page), for example, shows that one chair conformation of *cis*-1,2-difluorocyclohexane is nonsuperimposable on its mirror image. Figure 5-14b shows, however, that the

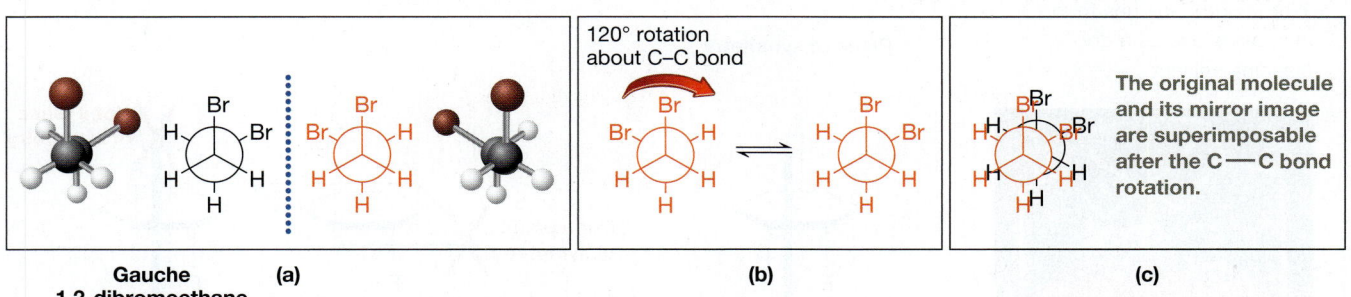

Gauche   (a)
1,2-dibromoethane

120° rotation about C–C bond

(b)

The original molecule and its mirror image are superimposable after the C—C bond rotation.

(c)

FIGURE 5-13 Bond rotations and chirality (a) Newman projection of 1,2-dibromoethane in a gauche conformation (black), along with its mirror image (red). The mirror is indicated by the blue dotted line. Ball-and-stick models are provided next to the corresponding Newman projections. (b) The mirror image after a 120° rotation of the rear CH$_2$Br group about the C—C single bond. (c) Overlaying the conformer from (b) with the original molecule from (a) shows that the two are superimposable.

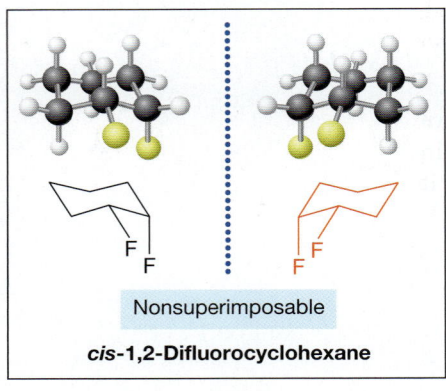

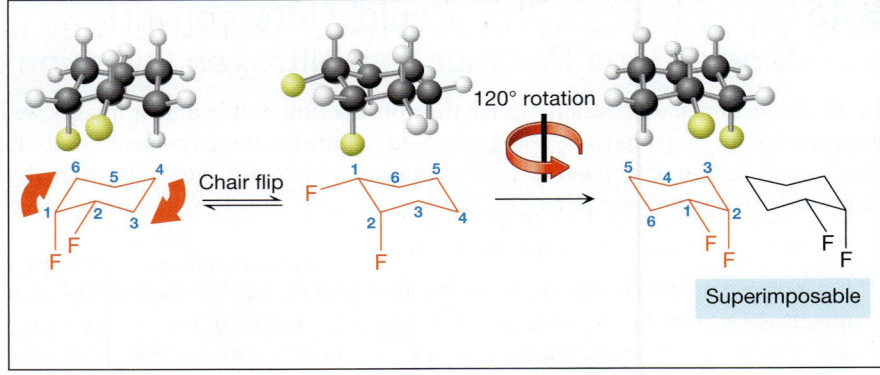

(a)   (b)

FIGURE 5-14 Chair conformations and chirality (a) The chair conformation of *cis*-1,2-difluorocyclohexane (black) is depicted with its mirror image (red), along with their respective ball-and-stick models. The mirror is represented by the blue dotted line. As indicated, the two structures are nonsuperimposable. (b) The mirror image (red) undergoes a chair flip, followed by a rotation of 120°. The resulting orientation is superimposable on the original molecule (black). Therefore, *cis*-1,2-difluorocyclohexane is achiral.

mirror image is simply the *other* chair conformation. And, because the two chair conformations rapidly interconvert, *cis*-1,2-difluorocyclohexane is *achiral*.

YOUR TURN 5.10

Build molecular models of *cis*-1,2-difluorocyclohexane and its mirror image (as indicated in Fig. 5-14a). Align the two molecules with every orientation you can imagine to convince yourself that they are *not* superimposable. Then perform a chair flip on the mirror image and again align the molecules to convince yourself that the resulting structure *is* superimposable with the original molecule.

The preceding example suggests that, when determining the chirality of substituted cyclohexanes, we do not need to consider the details surrounding the conformation of the chair. Rather, what matters is the *relative* positioning of the substituents with respect to the plane of the ring (i.e., cis or trans). Therefore, we can simplify the problem by working with structures that assume a flat ring, such as Haworth projections or regular line structures that incorporate dash–wedge notation. For example, the Haworth projection of *cis*-1,2-difluorocyclohexane (Fig. 5-15a) has a plane of

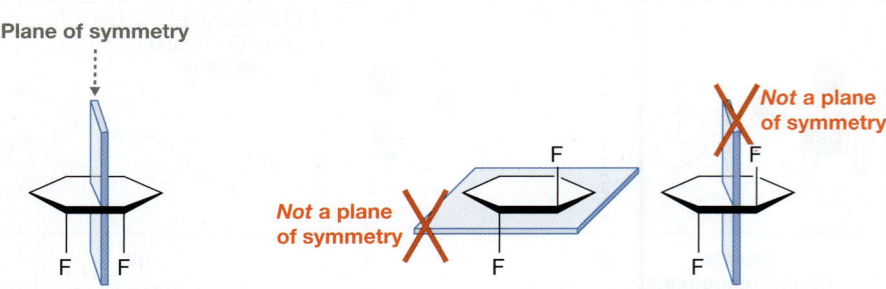

(a) *cis*-1,2-Difluorocyclohexane   (b) *trans*-1,2-Difluorocyclohexane

FIGURE 5-15 Haworth projections and chirality Haworth projections accurately depict the chirality of a substituted cyclohexane ring. (a) The Haworth projection of *cis*-1,2-difluorocyclohexane has a plane of symmetry, so it is achiral. (b) *trans*-1,2-Difluorocyclohexane is chiral, so it does not have a plane of symmetry.

symmetry, so it is achiral. *trans*-1,2-Difluorocyclohexane, on the other hand, is chiral, and this is in agreement with the fact that its Haworth projection (Fig. 5-15b) has no plane of symmetry.

Use molecular models to prove to yourself that *trans*-1,2-difluorocyclohexane is chiral, as predicted using Haworth projections in Figure 5-15. The molecule and its mirror image are shown.

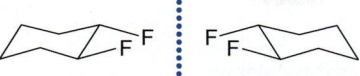

trans-1,2-Difluorocyclohexane

(a) Build a model of the original molecule and a model of its mirror image.
(b) Rotate the two molecules in space to show that they are *not* superimposable.
(c) Carry out a chair flip on the model of the mirror image and once again rotate the molecules in space to show that they are still *not* superimposable.

PROBLEM 5.14 Determine whether each of the following molecules is chiral or achiral.

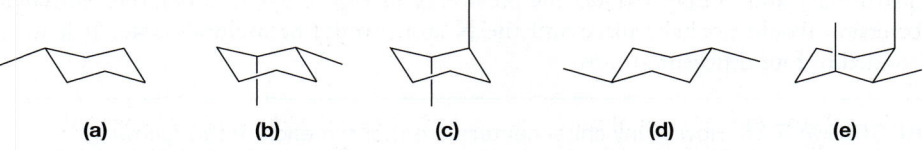

(a)   (b)   (c)   (d)   (e)

Nonsuperimposable mirror images that rapidly interconvert are not limited to cases where the mirror images are conformers. Another example involves uncharged nitrogen atoms in molecules such as *N*-methylethanamine (ethylmethylamine), shown in Figure 5-16. Notice that the molecule and its mirror image are nonsuperimposable and therefore appear to be enantiomers. The N atom, furthermore, appears to be a chiral center because it is tetrahedral and has four different groups: H, CH_3, CH_2CH_3, and a lone pair of electrons. However, the mirror images rapidly interconvert through a process called **nitrogen inversion**, the dynamics of which are depicted in Figure 5-17 (next page). Because of this rapid interconversion, the mirror images cannot be separated and *N*-methylethanamine is achiral. Thus, the N atom cannot be counted as a chiral center. In general:

Uncharged nitrogen atoms that undergo *nitrogen inversion* are not chiral centers.

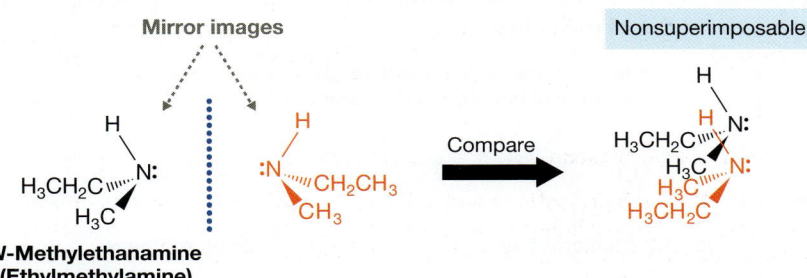

**N-Methylethanamine
(Ethylmethylamine)**

Mirror images   Nonsuperimposable   Compare

FIGURE 5-16 Chiral centers and uncharged nitrogen atoms Counting the lone pair of electrons as a group, *N*-methylethanamine has a tetrahedral nitrogen atom with four different groups, suggesting that *N* should be a chiral center. Additionally, the mirror image is nonsuperimposable on the original molecule, suggesting that ethylmethylamine should be chiral.

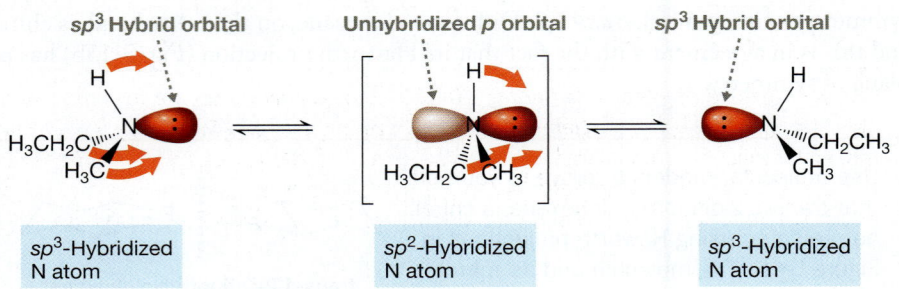

FIGURE 5-17 Nitrogen inversion
N-Methylethanamine is achiral because the two pyramidal mirror images rapidly interconvert via a planar intermediate with an sp^2-hybridized N atom. Consequently, the uncharged nitrogen atom is not a chiral center.

Nitrogen inversion is analogous to an umbrella that turns inside out in a gust of wind. As shown on the left in Figure 5-17, the N atom is initially pyramidal with its three bonds pointing toward the left. Those bonds and the attached groups swing to the right to proceed through a planar species and continue on to ultimately produce the pyramidal structure shown at the right of the figure.

This process occurs with relative ease in large part because one of the groups on the N atom is a lone pair of electrons. In the two pyramidal structures, the N atom is sp^3 hybridized and the lone pair occupies an sp^3 hybrid AO. In the planar species, the N atom is sp^2 hybridized and the lone pair occupies a p AO. This temporary rehybridization facilitates the migration of the lone pair from one side of the molecule to the other. If that lone pair were instead tied up in a covalent bond, as is the case with the quarternary ammonium ion we saw previously in Figure 5-8 (p. 217), then nitrogen inversion would not take place and the N atom would be a chiral center if it were bonded to four different groups.

PROBLEM 5.15 How many chiral centers are there in each of the following species?

(a)   (b)   (c)   (d)

5.5 Diastereomers

Recall that *diastereomers* are *stereoisomers that are not mirror images of each other* (review the flowchart in Fig. 5-1, p. 209). To be stereoisomers, they must have the same molecular formula, be different molecules, and have the same connectivity.

Cis and trans alkenes, such as *cis*-1,2-dichloroethene and *trans*-1,2-dichloroethene, are diastereomers. They have the same molecular formula and the same connectivity, but they are different molecules, given that one has chlorine atoms on the same side of the double bond, and the other has chlorine atoms on opposite sides. Moreover, they are *not* mirror images of each other.

Cis and trans isomers with respect to a double bond are diastereomers of each other.

cis-1,2-Dichloroethene   *trans*-1,2-Dichloroethene

Nanocars

Can you imagine an electric car about 10,000 times smaller than the thickness of a human hair? Dr. Ben Feringa, who shared the Nobel Prize in chemistry for the design and synthesis of molecular machines, could imagine such a car, and in 2011, he and his research team at the University of Groningen (Netherlands) constructed one! It is a single molecule, roughly 1 nm long, as shown below on the left. Each of the four "wheels" comprises a planar three-ring system.

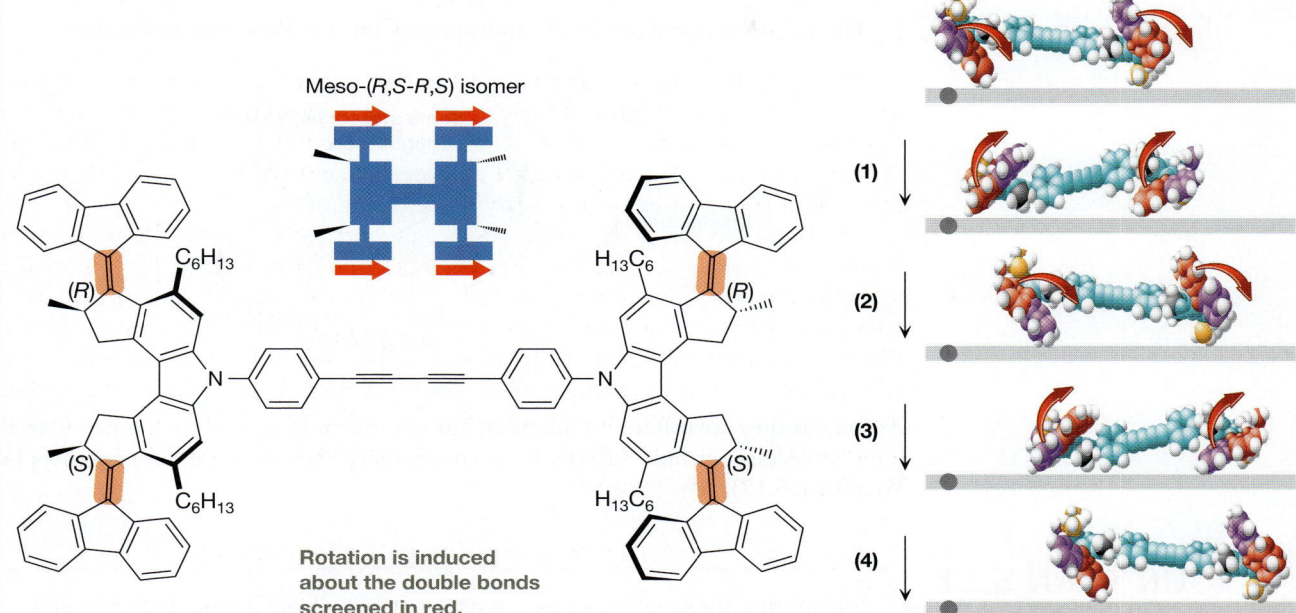

Meso-(R,S-R,S) isomer

Rotation is induced about the double bonds screened in red.

(1)
(2)
(3)
(4)

After the car is sublimed onto a copper surface, it is powered by electrons from a scanning tunneling microscope (STM). Each of the wheels functions independently of the others, making the nanocar analogous to a four-wheel-drive vehicle. As shown above on the right, a full rotation of each wheel consists of four separate steps. First, excitation by a pulse from the STM provides energy to temporarily break the π bond of the double bond that connects the wheel. This isomerizes the double bond, resulting in a partial turn. The wheel is left in a sterically strained conformation as a result, and in Step 2 it relieves that strain by relaxing into a more stable conformation. Steps 3 and 4 are the same as Steps 1 and 2.

The direction in which each wheel rotates is governed by the specific configuration of its four chiral centers. With the configurations shown, the four wheels work in concert to propel the nanocar forward. Other configurations can lead to the wheels working against each other, resulting either in no net motion or in the car turning.

Powering a nanocar in a single direction like this provides a proof of concept for more advanced nanomachines that could carry out specific tasks within our bodies. One task that scientists envision is targeting and killing cancer cells.

The compounds *cis*- and *trans*-cyclopentane-1,3-diol are diastereomers, too. Unlike the previous example, however, *cis*- and *trans*-cyclopentane-1,3-diol do not differ by a 180° rotation about a double bond. In fact, neither compound has a double bond. Nevertheless, the two have the same molecular formula and the same connectivity, but they are different molecules and are not mirror images of each other.

Cis and trans isomers with respect to a ring are diastereomers of each other.

cis-Cyclopentane-1,3-diol   *trans*-Cyclopentane-1,3-diol

Diastereomers need not be cis–trans pairs. Consider these two molecules:

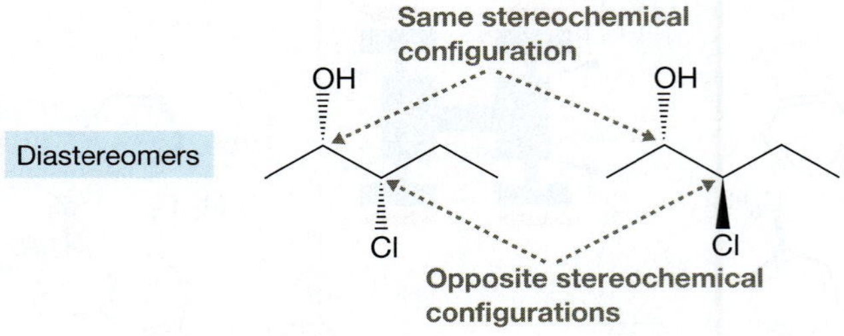

Same stereochemical configuration

Diastereomers

Opposite stereochemical configurations

As long as they are different molecules, but not mirror images of each other, with the same molecular formula and the same connectivity, then they are diastereomers (see Your Turn 5.12).

YOUR TURN 5.12

Confirm that these compounds are different by building a model of each and rotating one to determine if it is superimposable on the other.

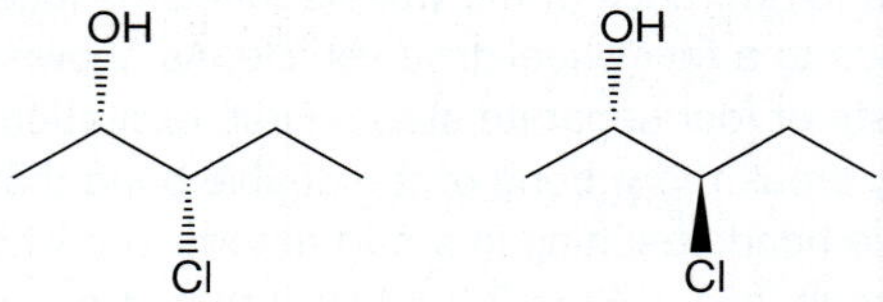

It can be quite helpful to use the relative configurations at the chiral centers to determine whether two molecules are diastereomers or enantiomers:

- If two isomers are related by the inversion of some, but not all, chiral centers, then the two molecules are *diastereomers* of each other.
- If two isomers are related by the inversion of *all* chiral centers, then the two molecules are *enantiomers*.

These criteria apply regardless of the number of chiral centers a molecule contains. For example, if two molecules have opposite configurations at two out of three chiral centers, then they are *diastereomers* (Fig. 5-18a); if they are nonsuperimposable and have opposite configurations at all three chiral centers, then they are *enantiomers* (Fig. 5-18b).

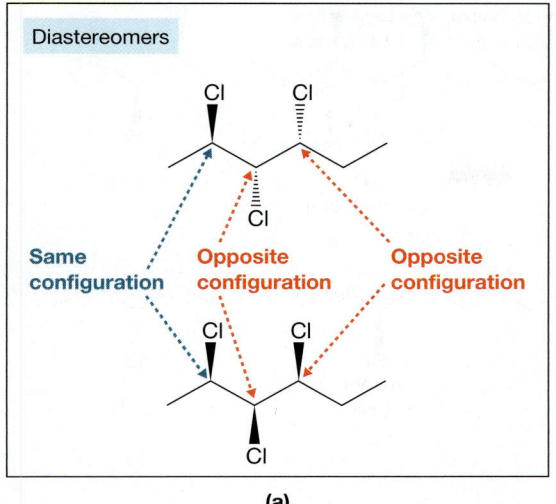

Diasteromers

Same configuration / Opposite configuration / Opposite configuration

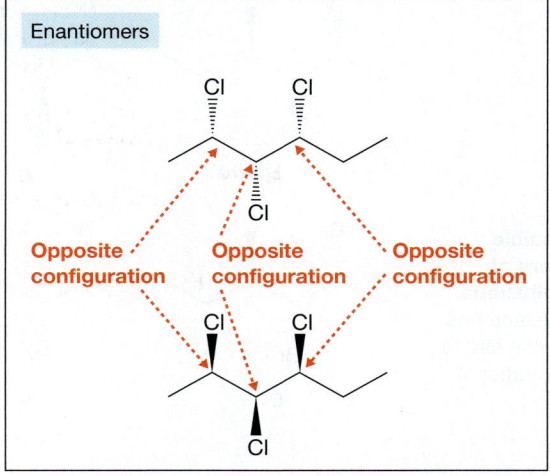

Enantiomers

Opposite configuration / Opposite configuration / Opposite configuration

(a)   (b)

FIGURE 5-18 Stereoisomers and relative configurations about chiral centers (a) Two of the three chiral centers in the top molecule have configurations that are opposite those in the bottom molecule, so the molecules are diastereomers. (b) All three chiral centers in the top molecule have configurations that are opposite those in the bottom molecule, so the molecules are enantiomers.

SOLVED PROBLEM 5.16

Given this configurational isomer of 1-chloro-2,3-dimethylcyclopentane, draw its enantiomer and one of its diastereomers.

Think How many chiral centers are there? To obtain its enantiomer, how many of those configurations must be reversed? To obtain one of its diastereomers, how many of those configurations must be reversed?

Solve There are three chiral centers, indicated by asterisks below. To obtain its enantiomer, we must reverse *all* three of those configurations. To obtain a diastereomer, we must reverse at least one configuration, but not all of them. The enantiomer and one diastereomer are shown.

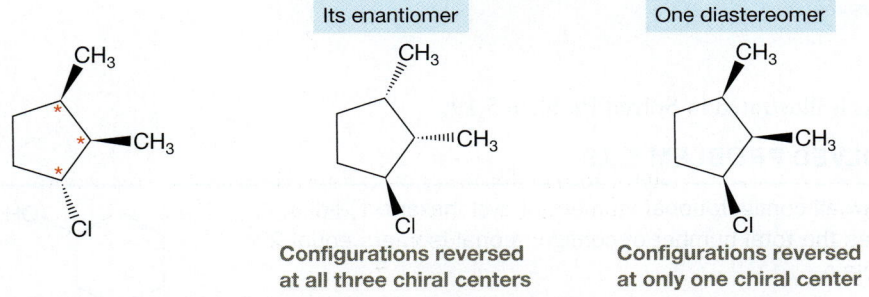

Its enantiomer

One diastereomer

Configurations reversed at all three chiral centers

Configurations reversed at only one chiral center

PROBLEM 5.17 Draw two more diastereomers of the molecule given in Solved Problem 5.16.

Because a new configurational isomer can be obtained for *each* inversion of a chiral center's configuration, the number of configurational isomers that exist can double with the addition of each chiral center. Therefore:

The maximum number of configurational isomers that can exist for a molecule with n chiral centers is 2^n.

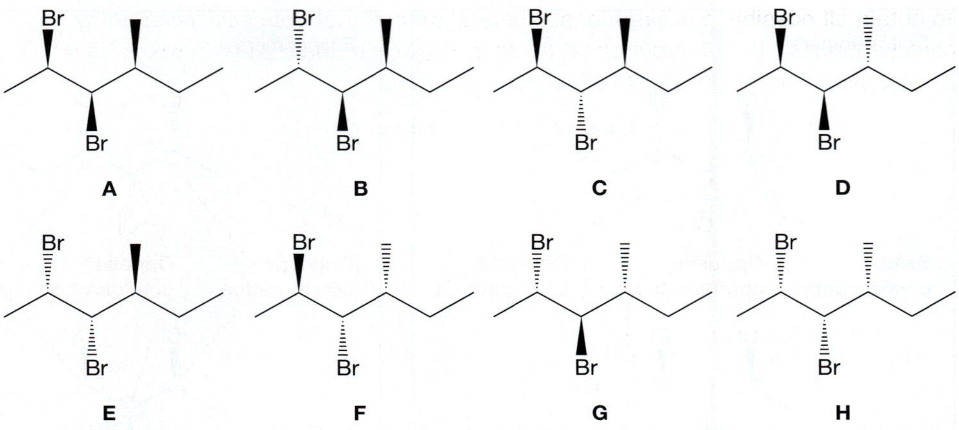

FIGURE 5-19 **All possible configurational isomers of 2,3-dibromo-4-methylhexane**
2,3-Dibromo-4-methylhexane has three chiral centers, giving rise to a total of $2^3 = 8$ configurational isomers.

2,3-Dibromo-4-methylhexane, for example, has three chiral centers, so it has the $2^3 = 8$ configurational isomers shown in Figure 5-19.

YOUR TURN 5.13

In molecule **A** in Figure 5-19, indicate each chiral center with an asterisk.

PROBLEM 5.18 Which molecules in Figure 5-19 are enantiomers of molecule **A**? Which molecules are diastereomers? *Hint:* How many chiral centers are there? In each structure, how many chiral centers are inverted relative to those in molecule **A**?

It is important to realize that 2^n represents a *maximum* number.

Fewer than 2^n configurational isomers can exist when at least one of the isomers is *meso*.

This is illustrated in Solved Problem 5.19.

SOLVED PROBLEM 5.19

Draw all configurational isomers of cyclohexane-1,3-diol. Does the total number of configurational isomers equal 2^n? Explain.

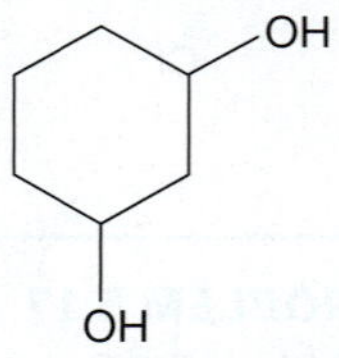

Think How many chiral centers are there? How can we change each chiral center to convert from one configurational isomer to another?

Cyclohexane-1,3-diol

Solve There are two chiral centers, indicated by the asterisks.

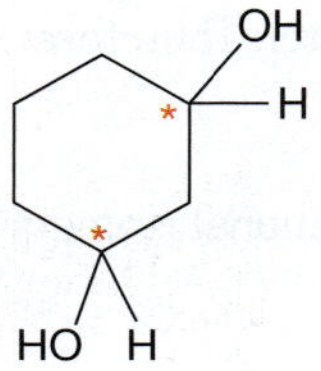

To obtain all possible configurational isomers, we can systematically reverse the configurations at the different chiral centers. Begin with the isomer in which both OH groups point toward us (structure **A**):

Structures A and D are the same molecule.

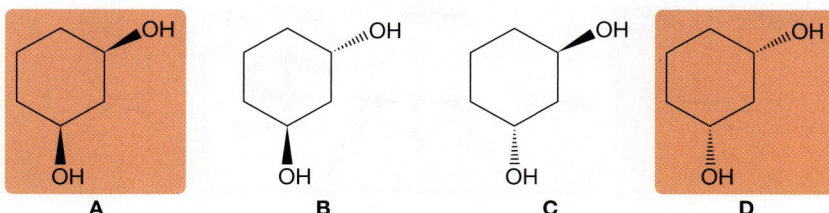

It may appear that there are four total isomers, which does equal 2^n with n (the number of chiral centers) being two. Structures **A** and **D** are exactly the same, however, so there are only three total isomers. There are fewer than 2^n isomers because one of the isomers is meso. Although it has two chiral centers, it has a plane of symmetry, too, and is therefore achiral—its mirror image is no different from itself.

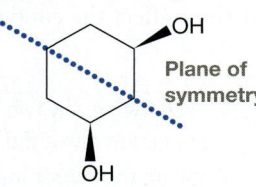

PROBLEM 5.20 Draw all possible configurational isomers of each molecule. How many chiral centers are there? Does the number of configurational isomers equal 2^n? Explain.

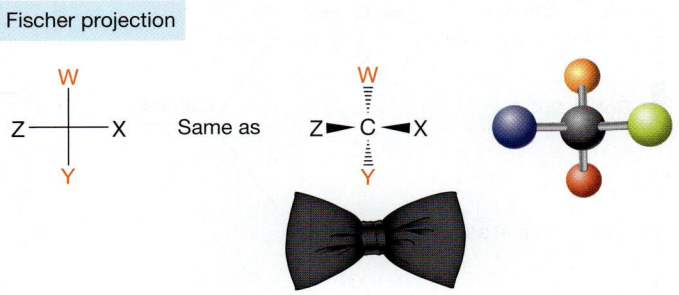

5.6 Fischer Projections and Stereochemistry

In his study of simple sugars (saccharides) in the late 19th century, Emil Fischer found himself working with several asymmetric carbons in a given molecule and several molecules at a time. This led Fischer to develop a quicker and more convenient way to depict configurations about these chiral centers, now known as **Fischer projections**. There are two conventions for drawing Fischer projections (Fig. 5-20):

- The intersection of a horizontal line and a vertical line indicates a carbon atom—typically an asymmetric carbon.
- The substituents on the horizontal bonds are understood to point toward you, reminiscent of a bow tie, whereas the substituents on the vertical bonds are understood to point away from you.

FIGURE 5-20 Fischer projections (*Left*) A generic Fischer projection of an asymmetric carbon, where W, X, Y, and Z are different substituents. (*Middle*) The dash–wedge structure that corresponds to the Fischer projection. The substituents on the horizontal point toward you, reminiscent of a bow tie, whereas the substituents on the vertical point away from you. (*Right*) A ball-and-stick model of the molecule.

Build a model of the molecule represented by the Fischer projection in Figure 5-20 (shown again on the left below). Use different colored balls to represent the four different substituents. View the molecule from the vantage point indicated on the right and fill in the atoms in the boxes provided.

Fischer projection

Rotate 90° about axis

Z——X   Same as   Z►C◄X

W

Y

To work comfortably with Fischer projections, we must know how certain manipulations affect the configurations about their chiral centers.

- Exchanging any two substituents on an asymmetric carbon in a Fischer projection gives the opposite stereochemical configuration.
- Taking the mirror image of a Fischer projection gives the opposite stereochemical configuration.
- Rotating a Fischer projection 90° in the plane of the paper gives the opposite stereochemical configuration.

The first two manipulations apply to dash–wedge representations, too, whereas the third manipulation is peculiar to Fischer projections. When a Fischer projection is rotated 90°, the bonds that were pointing toward us instead point away from us, and vice versa. As illustrated in Figure 5-21, this results in a configuration that is nonsuperimposable on the original molecule.

Using the same logic, *a 180° rotation of a Fischer projection results in no change in the configuration*, because this is the same as two 90° rotations. The first 90° rotation inverts the configurations of all asymmetric carbons and the second 90° rotation inverts them again, thus restoring them to their original configurations.

The convenience of Fischer projections is fully realized when multiple asymmetric carbons exist in the same molecule, as in the case of simple sugars. D-Allose, for example, is a molecule with six adjacent C atoms, four of which are asymmetric carbons (Fig. 5-22). Notice that the two C atoms that are not asymmetric are *not* represented by the intersection of perpendicular lines.

D-Glucose has the same molecular formula and the same connectivity as D-allose. Based on their Fischer projections, however, they are diastereomers of each other (not enantiomers), because their stereochemical configurations are opposite at only one of

FIGURE 5-21 Rotation of a Fischer projection by 90° (*Top left*) A generic Fischer projection and its dash–wedge representation. (*Bottom left*) Rotation of the Fischer projection by 90° and the resulting dash–wedge structure. (*Right*) The two structures are nonsuperimposable, so rotation of a Fischer projection by 90° gives the opposite configuration at an asymmetric carbon.

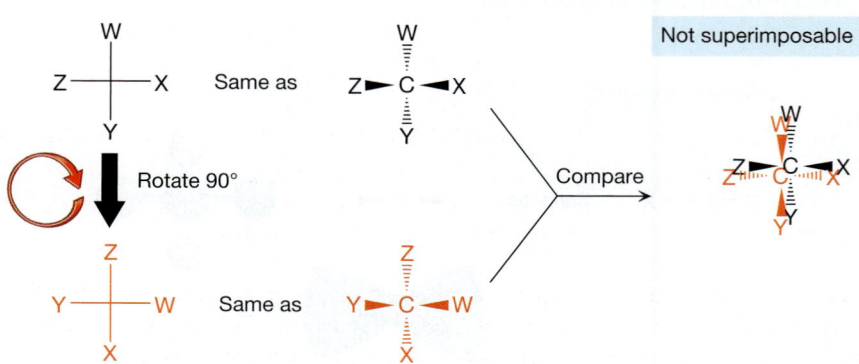

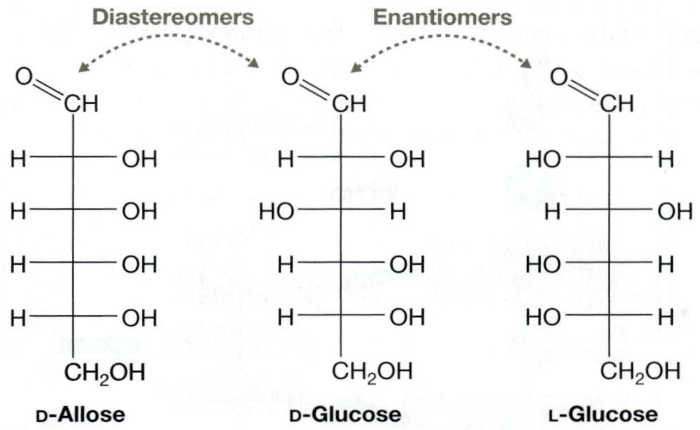

FIGURE 5-22 Fischer projections with multiple stereocenters Each Fischer projection represents four asymmetric carbons. D-Allose and D-glucose differ in the configuration at one of these carbons, so they are diastereomers. D-Glucose and L-glucose are nonsuperimposable mirror images, so they are enantiomers.

the four asymmetric carbons. Similarly, we can see from their Fischer projections that L-glucose and D-glucose are enantiomers of each other: They are mirror images that are nonsuperimposable.

YOUR TURN **5.15**

> Identify the C atoms at which the stereochemical configurations are different in D-allose and D-glucose.

PROBLEM 5.21 What is the stereochemical relationship between D-allose and L-glucose? Explain.

PROBLEM 5.22 Draw the enantiomer of D-allose as a Fischer projection.

5.7 Strategies for Success: Converting between Fischer Projections and Zigzag Conformations

A molecule containing a chain of carbon atoms is frequently represented in its zigzag conformation. If it contains multiple asymmetric carbons, however, it may be more convenient to work with its Fischer projection. How does one convert from a zigzag conformation to a Fischer projection?

Consider, for example, the following molecule in its zigzag conformation on the left:

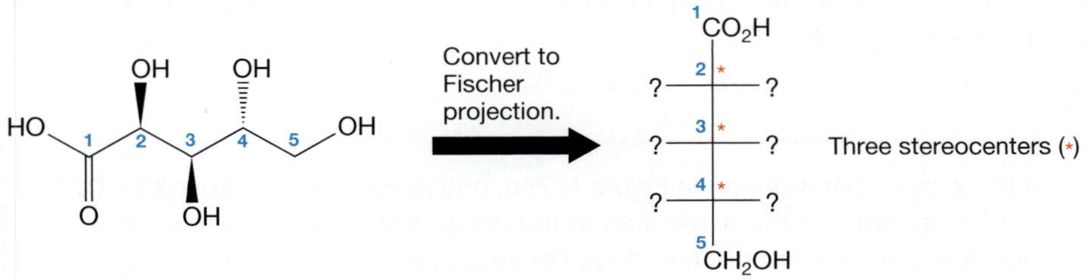

It consists of a chain of five C atoms, each of which is numbered; C2, C3, and C4 are asymmetric carbons. We can therefore begin to draw the Fischer projection with the framework shown above on the right, in which the three adjacent asymmetric carbons

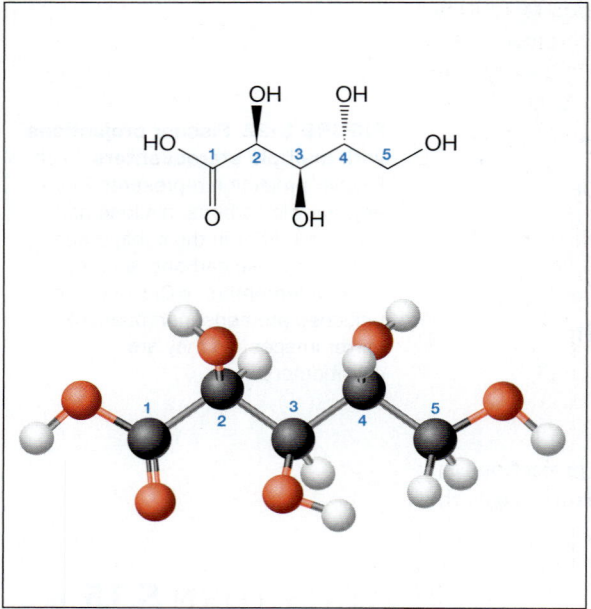

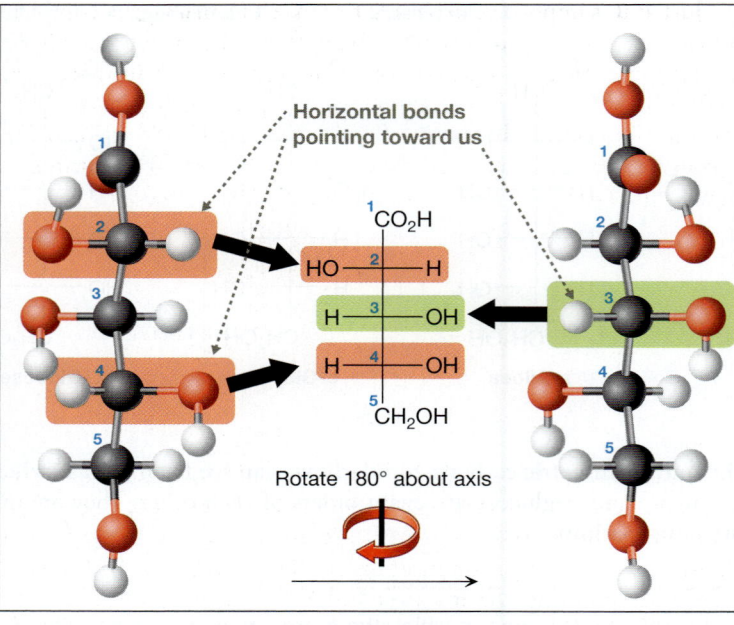

(a)

(b)

FIGURE 5-23 Converting a zigzag conformation to a Fischer projection (a) Ball-and-stick model of the molecule in question in its zigzag conformation. (b) The molecular models to the left and right of the Fischer projection are related by a 180° rotation. On the left, the model is oriented such that the asymmetric carbons C2 and C4 have the horizontal bonds pointing toward us, highlighted in red. This allows us to complete the Fischer projection at C2 and C4. On the right, the model is oriented such that the horizontal substituents at C3 are pointing toward us, highlighted in green, allowing us to complete the Fischer projection at C3.

are denoted by asterisks. Two of the bonds on each asymmetric carbon have been left with a question mark because two substituents must still be added to each—in this case, H and OH—and we must add them so that the stereochemical configuration at each of those carbons in the Fischer projection matches what we were given in the dash–wedge notation.

One way to complete the Fischer projection is to build a molecular model of the molecule in its zigzag conformation, paying particular attention to the dash–wedge notation on each asymmetric carbon (Fig. 5-23a). Then, orient the model vertically (Fig. 5-23b) so that the carbon chain matches the Fischer projection (i.e., C1 on top and C5 on the bottom).

With the molecular model as it appears on the left of Figure 5-23b, notice that the horizontal bonds on C2 and C4 (highlighted in red) are pointing toward us, just as the convention of Fischer projections demands. For those carbons, we simply add the H and OH to the Fischer projection on the same sides as they appear in the model. Thus, in the Fischer projection, we add OH to the left side on C2 and to the right side on C4.

To add the remaining substituents to C3 in the Fischer projection, we cannot do so as they appear in the molecular model on the left of Figure 5-23b because the horizontal bonds are pointed *away* from us, which is opposite of what is demanded by the convention. Therefore, we flip the model over, as shown on the right of Figure 5-23b, so the horizontal bonds on C3 point toward us. Then we add the H and OH substituents as they appear in the model: H on the left and OH on the right. This completes the Fischer projection.

YOUR TURN 5.16

In the zigzag conformation in Figure 5-23a, notice that the OH groups on C2 and C3 are both on the same side of the plane of the page—in this case, they both point toward us. Are those OH groups on the same side of the Fischer projection in Figure 5-23b? Notice that the OH groups on C2 and C4 are on opposite sides of the plane of the page. Are they on opposite sides of the Fischer projection, too? Do you see a pattern?

Just as it is important to be able to convert from a zigzag conformation to a Fischer projection, it is also important to be able to convert from a Fischer projection to a zigzag conformation. Suppose, for example, that we want to convert the Fischer projection below on the left into the zigzag conformation on the right. There are four carbon atoms in the chain, two of which are asymmetric: C2 and C3. To complete the zigzag structure, we must add the remaining H and OH substituents so that the configurations agree with those given in the Fischer projection.

Two stereocenters (*)

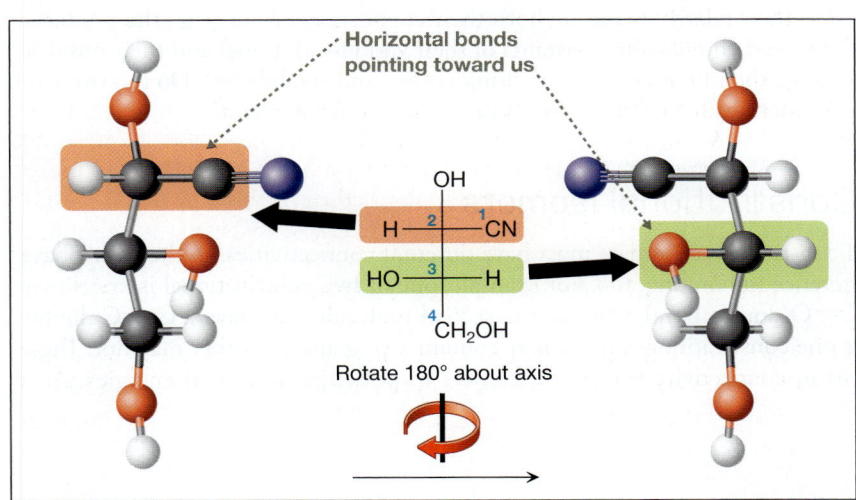

Convert to zigzag.

Again, model kits can help. Begin by building a molecular model of the zigzag structure, temporarily leaving off the H and OH substituents on C2 and C3. Orient that model vertically so it appears like the Fischer projection, as shown in Figure 5-24a. With the orientation on the left in Figure 5-24a, the horizontal bonds on C2 are pointing toward us (highlighted in red), so we add the remaining substituents on C2 just as they appear in the Fischer projection: H on the left and CN on the right. Before adding the remaining substituents to C3, we turn the molecular model around so the horizontal bonds on C3 point toward us (highlighted in green). Then the substituents are added just as they appear in the Fischer projection: OH on the left and H on the right. The molecular model is now complete and accurate. Looking at it from the side, as shown in Figure 5-24b (you will need to rotate around the C2—C3 bond), we then add the H and OH substituents to the dash and wedge bonds exactly as they appear in the model: OH toward us on C2 and away from us on C3.

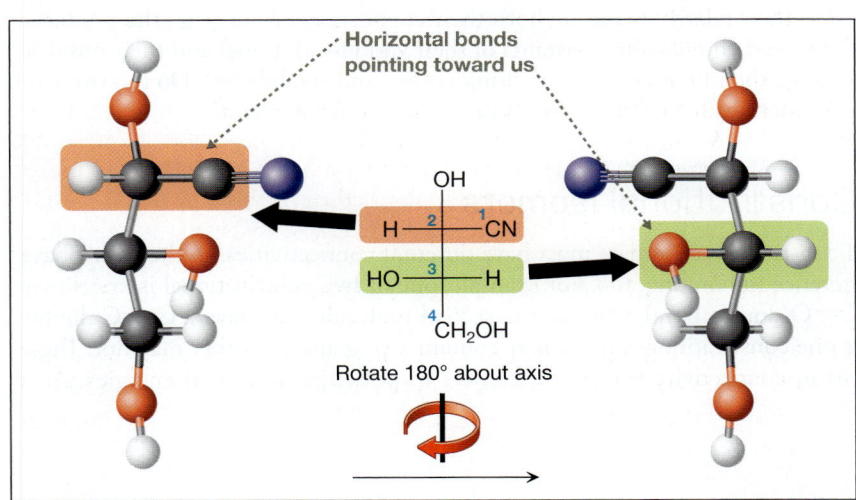

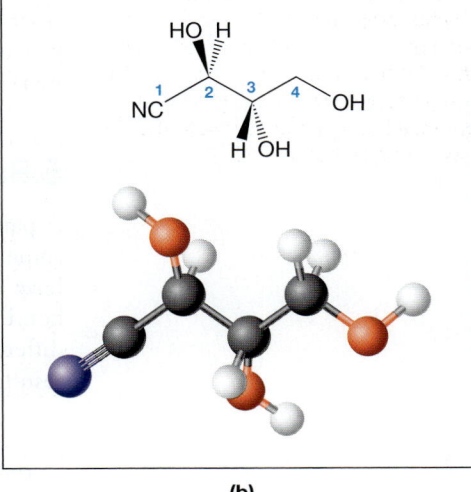

(a) (b)

FIGURE 5-24 Converting a Fischer projection to a zigzag conformation (a) The molecular models to the left and right of the Fischer projection differ by a 180° rotation. In the molecular model on the left, the horizontal bonds on C2 point toward us (highlighted in red), so the substituents on C2 must appear in the same locations as in the Fischer projection. In the model on the right, the horizontal bonds on C3 (highlighted in green) point toward us, so the substituents on C3 are added just as they appear in the Fischer projection. (b) The completed molecular model is viewed from the side and the H and OH substituents on C2 and C3 are added to the zigzag structure just as they appear in the model.

PROBLEM 5.23 Draw each of the following molecules as Fischer projections so that the carbon indicated is at the top.

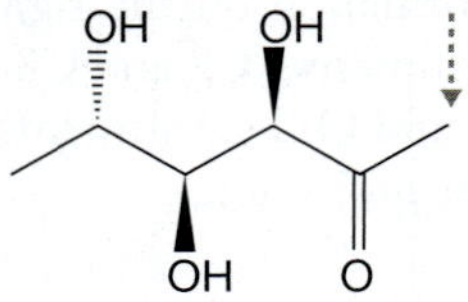

(a)

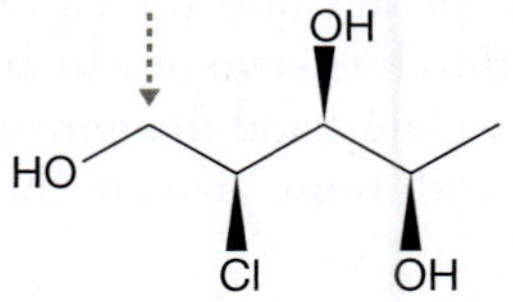

(b)

PROBLEM 5.24 Draw the following molecules in their zigzag conformations.

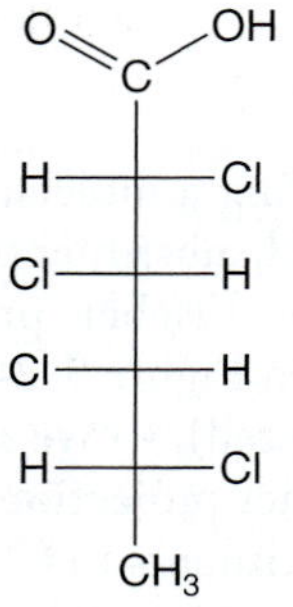

(a)

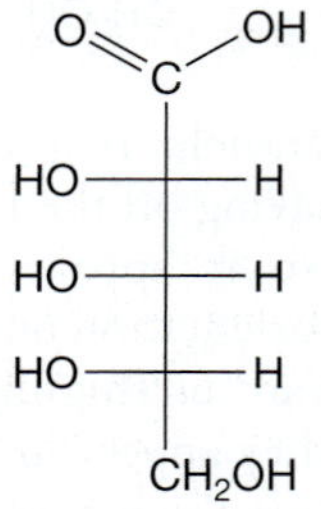

(b)

5.8 Physical and Chemical Properties of Isomers

One benefit of knowing the specific relationship between two isomers is the insight it provides into their relative behavior, both their chemical behavior (e.g., the products, reaction rates, and equilibrium constants of their various reactions) and their physical properties (e.g., their boiling points, melting points, and solubilities). Do the two isomers behave identically or differently? If differently, to what extent?

5.8a Constitutional Isomers

A pair of constitutional isomers must have different connectivities, so they must have some difference in bonding, too. For example, one of two constitutional isomers may have a C=O double bond, whereas the second molecule may have a C=C double bond. Or one constitutional isomer may contain a ring and the other may not. These differences in connectivity lead to differences in polarities and bond energies. As a result:

> Constitutional isomers must have different physical and chemical properties.

How differently constitutional isomers behave depends largely on how different their connectivities are. For example, pent-1-ene and *trans*-pent-2-ene (shown at the left on the next page) are two constitutional isomers of C_5H_{10}. They both contain one C=C double bond in addition to C—C and C—H single bonds. Although they have different connectivities, the differences are not great—the double bonds are simply found at different locations within the molecules. As a result, these two molecules behave similarly, both physically and chemically.

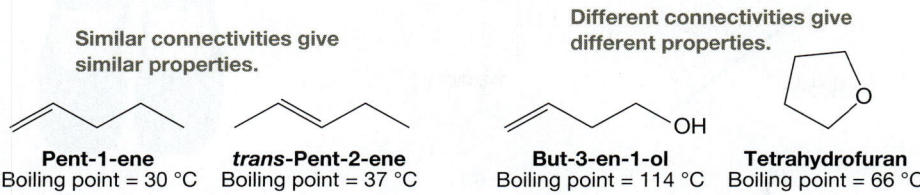

Similar connectivities give similar properties.

Pent-1-ene
Boiling point = 30 °C

***trans*-Pent-2-ene**
Boiling point = 37 °C

Different connectivities give different properties.

But-3-en-1-ol
Boiling point = 114 °C

Tetrahydrofuran
Boiling point = 66 °C

On the other hand, but-3-en-1-ol and tetrahydrofuran (shown above at the right) are two constitutional isomers of C_4H_8O that have quite different physical and chemical behavior because of their different functional groups.

5.8b Enantiomers

Enantiomers are mirror images of each other, so they have exactly the same connectivity and precisely the same polarity. For these reasons, it might seem that enantiomers should behave identically. Indeed, both enantiomers of butan-2-ol boil at 99 °C.

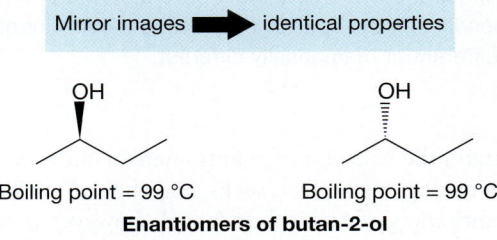

Mirror images ➡ identical properties

OH

OH

Boiling point = 99 °C

Boiling point = 99 °C

Enantiomers of butan-2-ol

In general, however, whether enantiomers have identical properties depends on if they are in a *chiral environment* or an *achiral environment*.

- A **chiral environment** is one that is nonsuperimposable on its mirror image.
  - Chiral species must be present, other than the enantiomers of interest.
  - At least one of those chiral species must be in unequal proportions of its enantiomers.
- An **achiral environment** is one that is superimposable on its mirror image.
  - This can occur if no chiral species are present other than the enantiomers of interest.
  - This can also occur if chiral species (other than the enantiomers of interest) are present in equal proportions of their enantiomers.

Most environments we encounter in the laboratory are *achiral*. This is the case, for example, when a pure enantiomer boils. Furthermore, most solvents, solutes, and reactants an enantiomer might encounter are achiral.

Chiral environments can occur in a variety of ways. In chromatography, for example, a chiral stationary phase can be used. In chemical reactions, we can use chiral

CONNECTIONS

Tetrahydrofuran is principally used as a solvent, both industrially and in the organic chemistry laboratory. When it is treated with a strong acid, it forms the polymer poly(tetramethylene ether) glycol, which is used to make elastic fibers such as Spandex.

FIGURE 5-25 Chiral and achiral environments (a) A sock has a plane of symmetry and is therefore achiral. (b) Because a sock is an achiral environment for feet, socks fit both the left and right feet equally well. (c) A shoe is chiral and is therefore a chiral environment for feet. This is why a left shoe fits a left foot far better than a right shoe and why a right shoe fits a right foot far better than a left shoe.

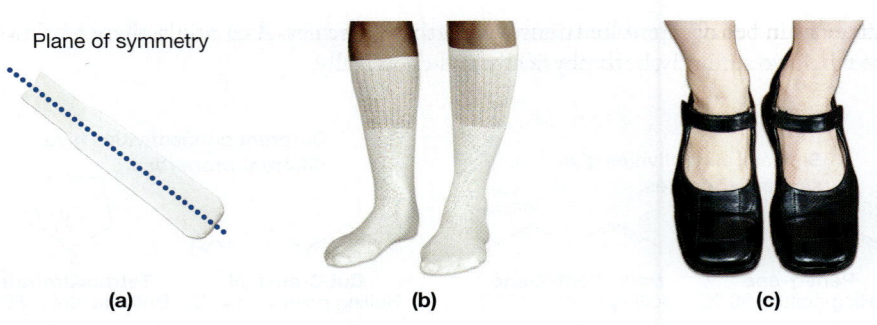

Plane of symmetry

(a) (b) (c)

catalysts. Perhaps more importantly, biological systems are chiral environments because biomolecules such as proteins, sugars, and DNA are chiral and, in the body, are each present in exclusively one of their enantiomers.

With an understanding of chiral and achiral environments, we can now be more precise in our statement regarding the relative behavior of enantiomers.

- In an *achiral environment*, enantiomers have exactly the same physical and chemical properties.
- In a *chiral environment*, enantiomers must have different physical and chemical properties. Depending on the specific situation, the behavior of the enantiomers can be slightly different or dramatically different.

To better understand the behavior of enantiomers in chiral versus achiral environments, we present an analogy using feet, socks, and shoes. Your left and right feet are chiral—they are enantiomers of each other. A sock, however, is achiral because it has a plane of symmetry along its length (Fig. 5-25a). Thus, both a left foot and right foot should "behave" identically when they interact with the achiral environment of a sock. From experience we know this to be the case; a sock will fit either foot equally well (Fig. 5-25b).

Shoes, however, are chiral objects—left and right shoes are enantiomers of each other. Left and right feet, therefore, should "behave" differently in the presence of the chiral environment of a given shoe. Indeed, a right shoe fits the right foot better than the left (Fig. 5-25c) and vice versa.

5.8c Diastereomers

Like enantiomers, a pair of diastereomers have the same connectivity. They are *not* mirror images of each other, however, so they must behave differently.

Diastereomers must have different physical and chemical properties.

This can be illustrated most straightforwardly with a pair of cis and trans isomers, which, as we learned in Section 5.5, constitute one category of diastereomers. *trans*-1,2-Dichloroethene, for example, is a nonpolar molecule, whereas *cis*-1,2-dichloroethene, its diastereomer, has a substantial dipole moment. This difference in polarity results in physical properties that can be quite different, as shown on the next page at the left. The difference in behavior between diastereomers isn't always so dramatic, as shown below on the right with the diastereomers of butane-2,3-diol.

Diastereomers with significantly different properties

Diastereomers with similar properties

Polar

Nonpolar

cis-1,2-Dichloroethene
Boiling point = 60 °C

trans-1,2-Dichloroethene
Boiling point = 47.5 °C

(2*S*,3*S*)-Butane-2,3-diol
Boiling point = 179–182 °C

meso-Butane-2,3-diol
Boiling point = 183–184 °C

SOLVED PROBLEM 5.25

In Chapter 8, you will learn that 2-bromo-4-methylhexane (molecule **Z**) can undergo the substitution reaction at the right when treated with NaCl.

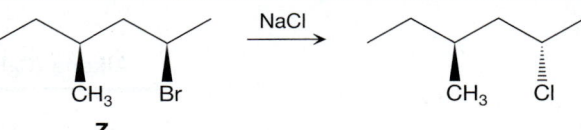

CH_3 Br

Z

NaCl →

CH_3 Cl

The following molecules, **A–E**, undergo a similar substitution reaction with NaCl.

CH_3 I
A

CH_3 Br
B

CH_3 Br
C

CH_3 Br
D

CH_3
Br
E

For which of these molecules will the *rate* of the reaction be precisely the same as that of the reaction involving **Z**? Explain.

Think How is each molecule **A** through **E** related to **Z**? How do those relationships translate into relative behavior?

Solve **A** and **B** are unrelated to **Z** because they have different molecular formulas. **C**, **D**, and **E** are all isomers of **Z**. **C** is its enantiomer, **D** is a diastereomer of it, and **E** is one of its constitutional isomers. Only enantiomers have precisely the same behavior, so the correct answer is **C**.

PROBLEM 5.26 Molecule **Y** is a carboxylic acid (RCO_2H), so it is moderately acidic (see Chapter 6 for details). Which of the molecules **A–E** have an acidity that is *different* from that of **Y**? Explain.

CH_3
H_3C CO_2H

Y

CH_3
CH_3 CO_2H
A

CH_3
H_3C CO_2H
B

CO_2H
H_3C CH_3
C

CH_3
H_3C CO_2H
D

CH_3
HO_2C CH_3
E

5.9 Stability of Double Bonds and Chemical Properties of Isomers

Recall from Section 5.8 that constitutional isomers and diastereomers *must* have different chemical properties. This can be seen in the heats of combustion of the six C_6H_{12} isomeric alkenes listed in Table 5-1.

The C=C double bonds of the isomers in Table 5-1 have different *alkyl substitutions*. **Alkyl substitution** is the number of alkyl groups bonded to the alkene carbon

TABLE 5-1	Heats of Combustion of Six Representative Isomeric Alkenes with the Formula C_6H_{12}					

Alkene (C_6H_{12})	Type of Alkene	HEAT OF COMBUSTION		DIFFERENCE	
		kJ/mol	kcal/mol	kJ/mol	kcal/mol
2,3-Dimethylbut-2-ene	Tetrasubstituted	–3741.5	–894.2	—	—
3-Methylpent-2-ene	Trisubstituted	–3749.9	–896.2	8.4	2.0
2-Ethylbut-1-ene	Disubstituted	–3755.9	–897.7	14.4	3.5
trans-Hex-3-ene	Disubstituted	–3762.6	–899.3	21.1	5.1
cis-Hex-3-ene	Disubstituted	–3766.3	–900.2	24.8	6.0
Hex-1-ene	Monosubstituted	–3769.7	–901.0	28.2	6.8

atoms. 2,3-Dimethylbut-2-ene, for example, has a *tetrasubstituted* double bond because the carbons of the C=C double bond are connected to four alkyl groups (in this case, all CH_3 groups). On the other hand, the double bond in hex-1-ene is *monosubstituted* because the carbons of the C=C double bond are connected to one alkyl group (in this case, a butyl group, $CH_2CH_2CH_2CH_3$).

Because all of the molecules in Table 5-1 are isomers of one another, their combustion products are identical. That is, each mole of C_6H_{12} that undergoes combustion produces 6 moles of CO_2 and 6 moles of H_2O, as shown in Equation 5-1.

$$C_6H_{12} + 9\,O_2 \rightarrow 6\,CO_2 + 6\,H_2O \qquad (5\text{-}1)$$

As a result, any difference in the heat of combustion must be attributed to differences in stabilities among the reactant alkenes. Because combustion releases heat, *the compound with the smallest heat of combustion is the most stable.* The stability of a C=C double bond decreases as follows in going from tetrasubstituted to monosubstituted, where R represents a generic alkyl group:

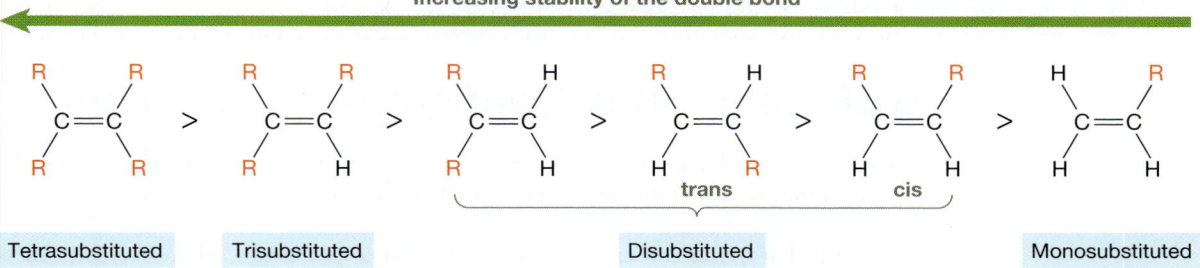

Two features of this sequence are particularly worth noting:

- Double bond stability increases as the amount of *alkyl substitution* increases.
- Trans alkenes are more stable than cis alkenes.

Trans isomers are more stable than cis isomers due to *steric strain* (Chapter 4). The *bulkiness* of the two alkyl substituents causes electrons from those groups to occupy the same space in the cis isomer (Fig. 5-26), whereas they are on opposite sides of the double bond in the trans isomer. Repulsion of those electrons decreases the stability of the cis isomer.

Double bonds become more stable as the number of attached R groups increases due to **hyperconjugation**. A more detailed discussion of hyperconjugation is presented in Interchapter D. Briefly, the electrons in the single bonds of an R group can interact with the empty π^* MO of the double bond. As a result, the electrons from the R group are *delocalized* to some extent, similar to what we observe in resonance. *Hyperconjugation leads, therefore, to an energy lowering (i.e., to stabilization), and each additional alkyl group present leads to additional hyperconjugation.*

SOLVED PROBLEM 5.27

Which of these molecules do you think will have the smaller heat of combustion? Explain.

Think Which of the C=C double bonds is more highly alkyl substituted? More stable? What is the relationship between relative stability and relative heats of combustion?

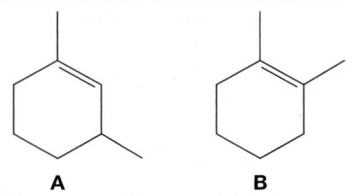

A   B

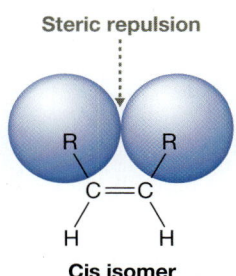

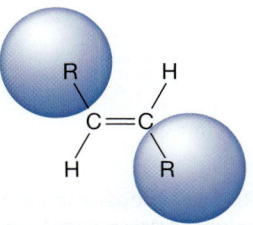

FIGURE 5-26 Steric strain in cis alkenes Steric repulsion between the alkyl groups causes the cis isomer of an alkene to be less stable than the trans isomer.

Solve The two molecules are isomers of each other, each with the formula C_8H_{14}. As a result, they have precisely the same combustion products, in which case any difference in the heat of combustion must come from differences in the stability of the reactants. The C=C double bond in molecule **A** is trisubstituted (one CH_3 group and two alkyl groups that are part of the ring), whereas the C=C double bond in molecule **B** is tetrasubstituted (two CH_3 groups and two alkyl groups that are part of the ring). Thus, **B** is more stable than **A** and will have a *smaller* heat of combustion (i.e., less heat is released).

PROBLEM 5.28 Which of these molecules do you think will have the greatest heat of combustion? The smallest? Explain.

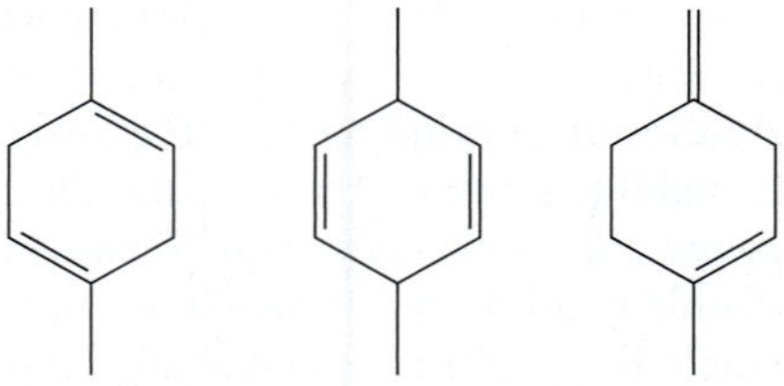

C   D   E

5.10 Separating Configurational Isomers

As we discuss in Chapter 8, if a chemical reaction forms a chiral product, it usually forms a mixture of stereoisomers. How, then, do we separate stereoisomers from one another?

Recall that diastereomers have different physical properties, whereas enantiomers have identical properties in achiral environments. Consequently:

- Diastereomers often can be separated by common laboratory techniques such as fractional distillation, crystallization, and simple chromatography.
- Enantiomers generally cannot be separated by these methods.

Louis Pasteur (1822–1895) was the first to isolate a pair of enantiomers from each other. The enantiomers he separated were those of sodium ammonium tartrate, an ionic compound that forms crystals (Fig. 5-27). As Pasteur noted, the crystals appeared

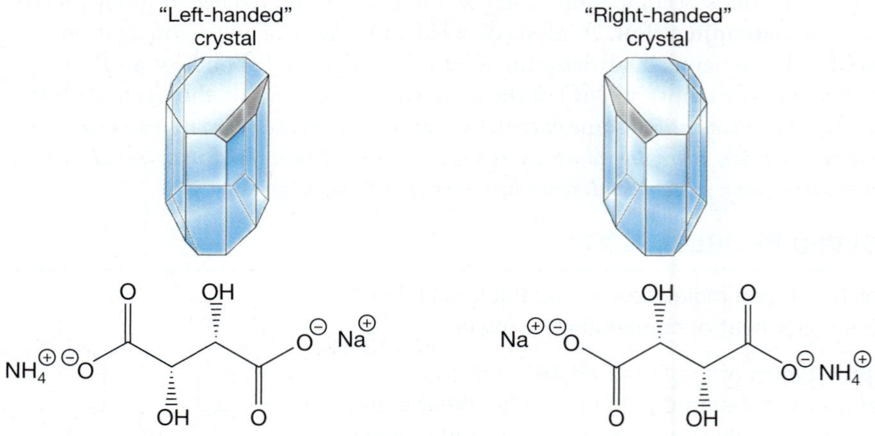

"Left-handed" crystal   "Right-handed" crystal

FIGURE 5-27 Separation of sodium ammonium tartrate enantiomers Right- and left-handed crystals of sodium ammonium tartrate that Louis Pasteur separated by hand are depicted at the top. The two crystals are mirror images of each other because their molecular structures are enantiomers.

Sodium ammonium tartrate

to grow in one of two varieties—left-handed crystals and right-handed crystals—that are mirror images of each other. Using nothing more than a microscope and a pair of tweezers, he physically separated the two types of crystals.

Most enantiomers cannot be separated using tweezers, so today, other techniques are used. Chromatography, for example, can exploit the fact that *enantiomers have different physical properties in a chiral environment*. A sample containing a mixture of enantiomers is passed through a chiral stationary medium, for which the enantiomers have different affinities. Traveling through the chiral medium at different rates allows them to be collected separately.

A second method of separating enantiomers takes advantage of the fact that diastereomers are readily separable, as mentioned previously. The key to this method involves three steps:

1. Temporarily converting the enantiomers into a pair of diastereomers.
2. Separating those diastereomers from each other by exploiting their different physical and chemical properties.
3. Regenerating the enantiomers from the separated diastereomers.

See Problem 5.57 for a specific example of how this method is applied.

5.11 Optical Activity

Although enantiomers have identical physical and chemical properties in an achiral environment, they behave differently in a chiral environment (Section 5.8b). They also interact differently with **plane-polarized light**.

Light can be regarded as both a particle and a wave. When treated as a particle, we think of light as consisting of **photons**, each of which carries a specific quantity of energy that can be associated with its frequency and wavelength (this is discussed further in Chapters 15 and 16). When treated as a wave, we think of light as consisting of oscillating electric and magnetic fields. The frequency of oscillation of those fields is what defines the frequency of light.

As a ray of light travels through space, its electric and magnetic fields oscillate in planes perpendicular to each other and also perpendicular to the direction it travels (Fig. 5-28). If all photons from a light source have their electric fields oscillating in the same plane, then the light is *plane polarized*. In these cases, we can represent that plane of polarization using a double-headed arrow, as shown in red on the right in Figure 5-28. Note that for simplicity, the plane in which the magnetic field oscillates is not shown in that representation; it is understood, though, to be perpendicular to the electric field's plane of oscillation.

Most light sources emit light that is **unpolarized**. That is, if we could view all of the photons traveling in the same direction, we would see each one's electric field oscillating in a different plane. A **polarizer** (Fig. 5-29) generates plane-polarized light by allowing through only those photons whose electric field is oscillating in a

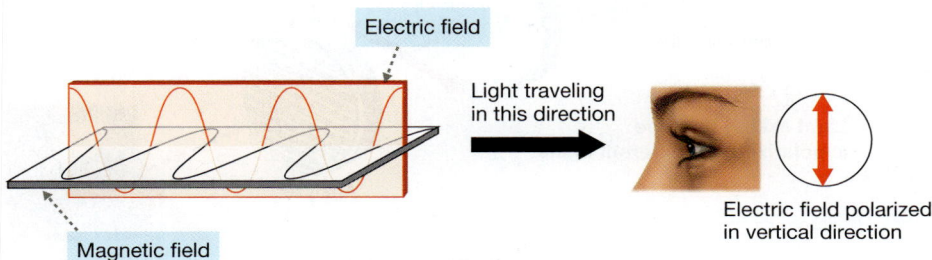

Electric field

Light traveling in this direction

Electric field polarized in vertical direction

Magnetic field

FIGURE 5-28 Plane-polarized light (*Left*) The electric field (red) oscillates in a vertical direction and the magnetic field (black) oscillates in the plane perpendicular to the page as the light wave travels to the right. (*Right*) A red, double-headed arrow represents the electric field's plane of oscillation.

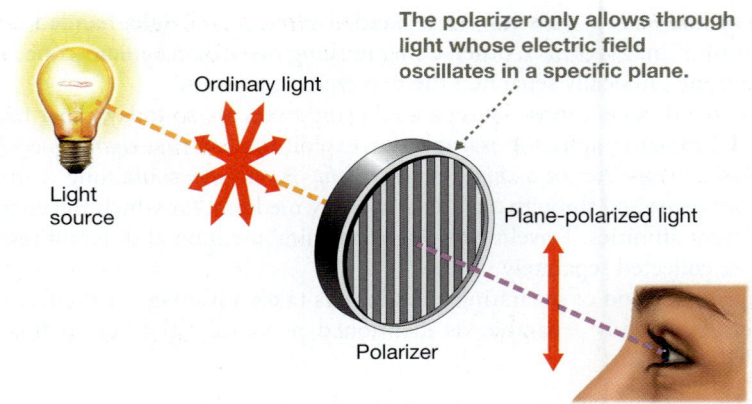

FIGURE 5-29 Function of a polarizer (*Left*) Light emitted from most sources is unpolarized. The electric-field vectors of its photons oscillate in all planes perpendicular to the direction of travel. (*Middle*) A polarizer effectively filters out photons whose electric-field vector does not oscillate in the specified plane (in this case, the vertical direction). (*Right*) The light that passes through the polarizer is plane polarized.

Ordinary light

Light source

The polarizer only allows through light whose electric field oscillates in a specific plane.

Plane-polarized light

Polarizer

specified plane, effectively filtering out light whose electric field oscillates in any other plane.

If plane-polarized light passes through a sample of a compound (Fig. 5-30), the plane in which the light is polarized can change, depending on whether the compound is chiral or achiral.

- Enantiomerically pure chiral compounds are said to be **optically active** because they rotate the plane of polarization.
- Achiral compounds are said to be **optically inactive** because they leave the plane of polarization unchanged.

The angle by which a chiral compound rotates plane-polarized light can be measured using an analyzer. Light enters the analyzer after it exits the sample. Some chiral compounds rotate light clockwise (in the + direction) and are called **dextrorotatory** (from Latin, meaning "rotating to the right"). Others rotate light counterclockwise (in the − direction) and are called **levorotatory** (meaning "rotating to the left"). *The*

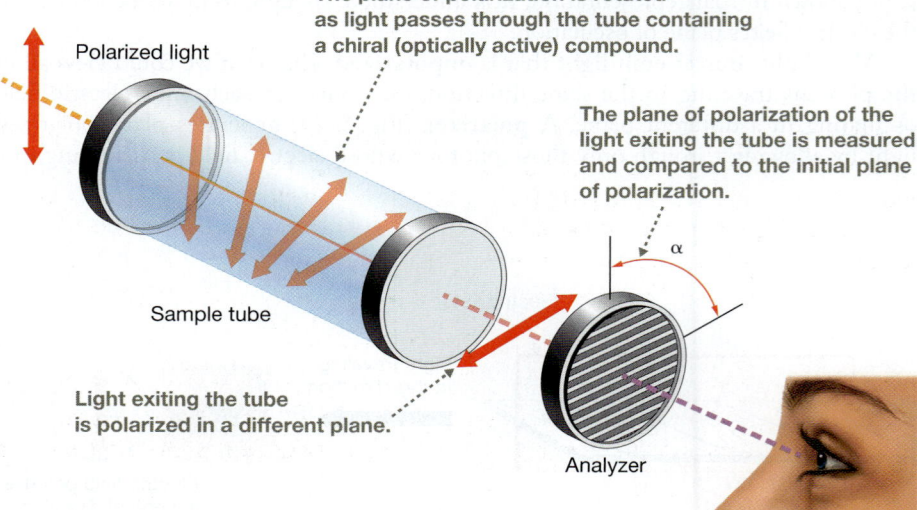

FIGURE 5-30 Optical activity (*Left*) Plane-polarized light enters a tube containing a solution of a compound being studied. If the compound is optically active, the plane of polarization is rotated as the light passes through the tube. (*Right*) The angle of rotation, α, is measured using an analyzer.

Polarized light

The plane of polarization is rotated as light passes through the tube containing a chiral (optically active) compound.

The plane of polarization of the light exiting the tube is measured and compared to the initial plane of polarization.

Sample tube

Light exiting the tube is polarized in a different plane.

α

Analyzer

direction of rotation generally cannot be known without performing the experiment. In fact, two chiral compounds that are structurally very similar may rotate light in opposite directions!

The amount by which the plane-polarized light is rotated on passing through a sample of a chiral compound depends on both the *concentration* of the chiral compound and the *length of the sample* through which the light travels:

- As the concentration of a chiral compound increases, so does the angle of rotation.
- As the length of the sample tube increases, so does the angle of rotation.

This should make sense, because the number of molecules the light encounters increases as the concentration of the sample or the length of sample tube increases.

The **measured angle of rotation, α,** can thus be expressed by Equation 5-2, where c is the concentration of the sample in units of grams per milliliter (g/mL) and l is the length of the sample tube in units of decimeters (dm) (1 dm $= \frac{1}{10}$ meter).

$$\alpha = ([\alpha]_{\lambda}^{T})(l)(c) \tag{5-2}$$

The $[\alpha]_{\lambda}^{T}$ term is the **specific rotation**, and it is a constant that is *unique for a given chiral compound.* It is the angle of rotation of light of a given wavelength (λ, in nanometers), if it passes through a sample whose concentration is 1 g/mL, whose length is 1 dm, and whose temperature is $T\,°C$. Most often, the light used for measurement is the sodium D line (589.6 nm), abbreviated simply D, and the temperature is 20 °C. Therefore, the specific rotation is usually reported as $[\alpha]_{D}^{20}$.

SOLVED PROBLEM 5.29

Suppose that 20.00 g of a chiral compound is dissolved in 0.1000 L of solution and is placed in a tube that is 20.00 cm long. What is its specific rotation of light if the observed rotation is determined experimentally to be +45.00°?

Think Which variable in Equation 5-2 are we solving for? The equation calls for concentration—how do we calculate it? Are the units correct for the length of the tube, l?

Solve We are asked for $[\alpha]_{D}^{20}$ and are given the value for α, which is +45.00°. Equation 5-2 must therefore be rearranged as follows:

$$[\alpha]_{D}^{20} = \frac{\alpha}{(l)(c)}$$

To solve this equation, we must have values for c and l in their correct units. We can calculate concentration in units of g/mL by dividing the 20.00 g of sample by the 100.0 mL of solution, to yield 0.2000 g/mL. The length of the tube, l, is given to us as 20.00 cm, but must be converted to dm. Because 1 dm = 10 cm, the length of our tube is 2.000 dm. Therefore, the specific rotation is

$$[\alpha]_{D}^{20} = \frac{(+45.00°)}{(2.000\ \text{dm})(0.2000\ \text{g/mL})} = 112.5°$$

PROBLEM 5.30 Penicillin V has a specific rotation of +223°. What would the measured angle of rotation be of a 0.00300 g/mL solution, if it were measured in a tube 10.0 cm long?

Recall that enantiomers interact differently with plane-polarized light. In fact:

> Enantiomers have equal but opposite specific rotations.

Just as enantiomers are mirror images of each other, the mirror image of a rotation in the clockwise direction is an identical rotation in the counterclockwise direction. For this reason, one enantiomer can always be designated as the (+) enantiomer and the other as the (−) enantiomer.

A **racemic mixture** (pronounced ruh-SEE-mik) contains equal amounts of the (+) and (−) enantiomers of a chiral molecule. That is, light traveling through a racemic mixture encounters an equal number of molecules of each enantiomer. Therefore, the tendency of one enantiomer to rotate the light in one direction is exactly balanced by the tendency of the other enantiomer to rotate the light in the opposite direction. The net result is zero rotation of the light. Consequently:

> A racemic mixture of enantiomers is optically inactive, despite being made up of chiral molecules.

If a mixture of enantiomers is not racemic, then it will be optically active, but it will not rotate light as much as one of the pure enantiomers. Such a mixture can be viewed as being a certain percentage racemic, with the remaining percentage, called the **enantiomeric excess (ee)**, viewed as being composed of one of the pure enantiomers. The percentage that is racemic will not contribute toward the rotation of plane-polarized light, but the enantiomeric excess will. This idea is summarized in Equation 5-3.

$$\text{(Specific rotation of mixture)} = (\% \text{ ee})(\text{specific rotation of pure enantiomer})/(100) \quad \textbf{(5-3)}$$

If you know the relative amounts of two enantiomers in solution, you can solve for the enantiomeric excess by first determining the percentage of the solution that is racemic. Suppose, for example, that a solution consists of 70% enantiomer A and 30% enantiomer B. The percent of the solution that is racemic is determined by combining all of the enantiomer in the smaller amount (B) with an equal amount of the other enantiomer (A). In this case, that would be 30% B + 30% A = 60% racemic. The remaining 40% of the solution is entirely A and is, therefore, the enantiomeric excess of A.

YOUR TURN 5.17

Determine the enantiomeric excess of a solution that consists of 30% A and 70% B.

SOLVED PROBLEM 5.31

Suppose a solution of a pure chiral molecule has a specific rotation of −32°. What is the specific rotation of a solution that is 90% of the (+) enantiomer and 10% of the (−) enantiomer?

Think Which enantiomer is in excess and what is its ee? How does that govern the mixture's ability to rotate plane-polarized light?

Solve Because the solution is excess in the (+) enantiomer, the specific rotation of the mixture should be in the (+) direction. The 10% that is the (−) enantiomer can be combined with 10% of the solution that is the (+) enantiomer, such that 20% of the solution is effectively racemic and the other 80% is excess in the (+) enantiomer. Thus, 20% of the solution does not contribute toward the rotation of plane-polarized light and 80% contributes toward a specific rotation of +32° [the value for the pure (+) enantiomer]. We then substitute the appropriate numbers into Equation 5-3.

(Specific rotation of mixture) = (% ee)(specific rotation of pure enantiomer)/(100)

= (80)(+32°)/(100)

= 26°

PROBLEM 5.32 Suppose that a pure compound has a specific rotation of +49°. In the laboratory, a solution in which the compound is mixed with its enantiomer is found to have a specific rotation of +12°. What is the ee of the mixture? What percentage of the mixture is the (+) enantiomer and what percentage is the (−) enantiomer?

Mapping the Earth with Polarimetry

The polarimetry technique we learned here in Chapter 5 takes advantage of the fact that, as plane-polarized light travels through a chiral compound, the polarization angle of the light is rotated. Another polarimetry technique, called radar polarimetry, takes advantage of the polarization angle changing when polarized light *scatters* off of a surface. Airplanes and satellites equipped with a specialized polarimeter transmit a radar wave with one type of polarization toward Earth's surface, and the polarization of the wave that is bounced back to the detector is analyzed. Transmitted waves can be polarized vertically (V) or horizontally (H), and each type of transmitted wave can produce a scattered wave that has V or H polarization, giving rise to four possible outcomes: VV, VH, HV, or HH. With a detection channel for each of these outcomes, the signature of the scattered light can be matched to specific types of materials.

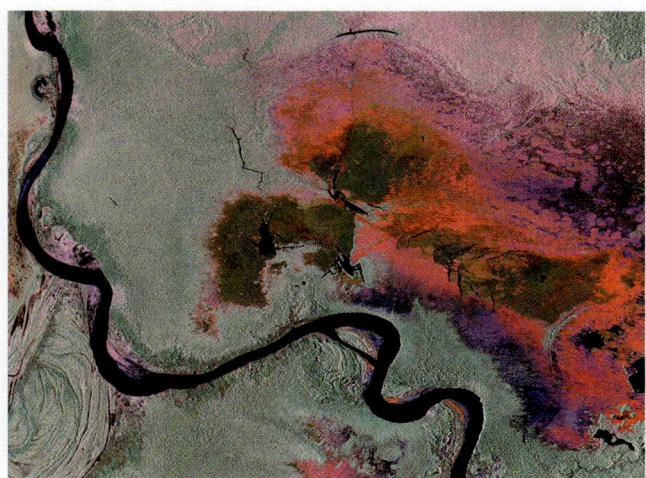

One of the major applications of radar polarimetry is in detecting different kinds of vegetation, both in agriculture and in natural ecosystems. The image shown here, for example, is of the Pacaya Samiria National Reserve in Peru. Black is open water, grayish-green is tropical forest, dark green is low vegetation, and red and pink are wetlands. Radar polarimetry has been used in other important applications, too, including monitoring glaciers and sea ice, as well as tracking crude oil spills.

THE ORGANIC CHEMISTRY OF BIOMOLECULES

5.12 The Chirality of Biomolecules

The tragedy of thalidomide was mentioned briefly in Section 5.1. In the 1950s and 1960s, thalidomide was prescribed as an antinausea medication for pregnant women with morning sickness. Unfortunately, thalidomide is teratogenic: It causes birth defects. As a direct result of their mothers' taking the drug, it is estimated that more than 10,000 children worldwide were born with deformed or missing limbs.

Like many drugs, thalidomide is chiral and was sold as a *racemic mixture* of its two enantiomers:

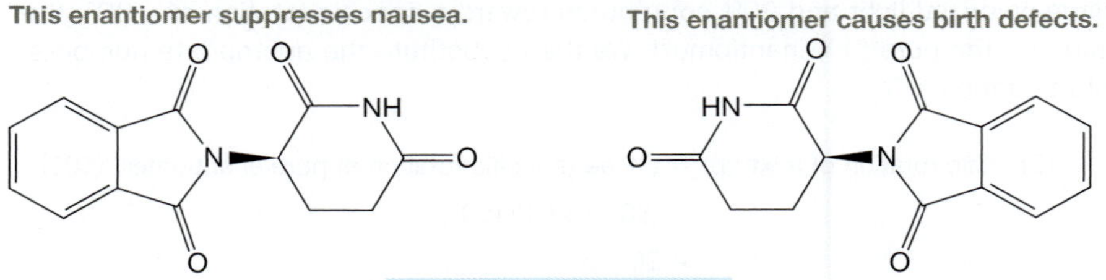

This enantiomer suppresses nausea.   This enantiomer causes birth defects.

Enantiomers of thalidomide

Later testing on mice showed that the enantiomer on the left is primarily responsible for suppressing nausea, whereas the one on the right is primarily responsible for the teratogenic properties. (It turns out, however, that administering only the enantiomer on the left would not have solved the problem because the two enantiomers interconvert in the body.)

How can enantiomers—molecules that are mirror images of each other—behave so differently? Recall from Section 5.8 that *enantiomers have different physical and chemical properties in a chiral environment*. In other words:

The body acts as a chiral environment.

Most biomolecules, including those encountered in previous chapters, are chiral. For example, as shown in Figure 5-31, a typical amino acid has a single chiral center (marked by an asterisk), and is thus chiral. There is one exception (see Problem 5.63). Glucose, in its acyclic form, has four chiral centers—each C atom that is bonded to an H and an OH group. A nucleotide in DNA has three chiral centers, and testosterone, a steroid, has six chiral centers.

Despite the presence of these chiral compounds, the body would remain an achiral environment if each pair of enantiomers were present in equal amounts. (This is analogous to a racemic mixture being optically inactive, as discussed in Section 5.11.) Instead:

Natural amino acids and monosaccharides appear in the body exclusively in one enantiomeric form.

FIGURE 5-31 Chiral centers in biomolecules In each of these biomolecules, chiral centers are marked by asterisks.

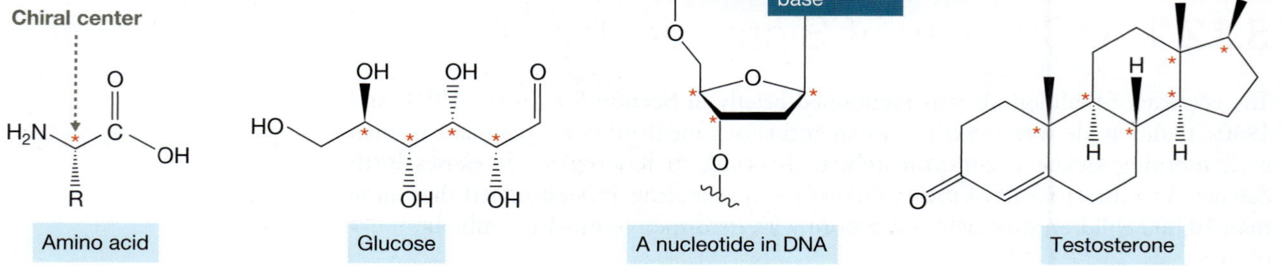

Amino acid   Glucose   A nucleotide in DNA   Testosterone

For amino acids, that form is the L enantiomer, and for monosaccharides, it is the D enantiomer. (These designations are discussed in greater detail in Section 5.13.) Reasons why these compounds appear exclusively in these forms are not known and are the subject of debate.

5.13 The D/L System for Classifying Monosaccharides and Amino Acids

Each chiral amino acid and monosaccharide has two enantiomers, specified using the D/L system. The system was established around 1910, before the advent of the IUPAC system of nomenclature and before the technology existed to determine the specific location of atoms in three-dimensional space.

The basis of the D/L system is the optical rotation of glyceraldehyde. Glyceraldehyde is an *aldotriose*, a three-carbon monosaccharide possessing a $CH{=}O$ group characteristic of an aldehyde (see Section 4.14). It has a single asymmetric carbon, and thus has enantiomers that rotate plane-polarized light in equal but opposite directions (Section 5.11). The enantiomer that rotates plane-polarized light in the clockwise direction was designated as D-glyceraldehyde, because it is dextrorotatory. (Recall that dextrorotatory derives from Latin, and means "rotating to the right.") The other enantiomer, which rotates plane-polarized light in the counterclockwise direction, was designated as L-glyceraldehyde, because it is levorotatory ("rotating to the left").

At the time the D/L system was established, other sugars could be synthesized from glyceraldehyde by lengthening the molecule at the $CH{=}O$ end of the molecule, while leaving unchanged the configuration of glyceraldehyde's chiral center. As a result, sugars synthesized in this way from D-glyceraldehyde were designated as D sugars, and sugars synthesized from L-glyceraldehyde were designated as L sugars.

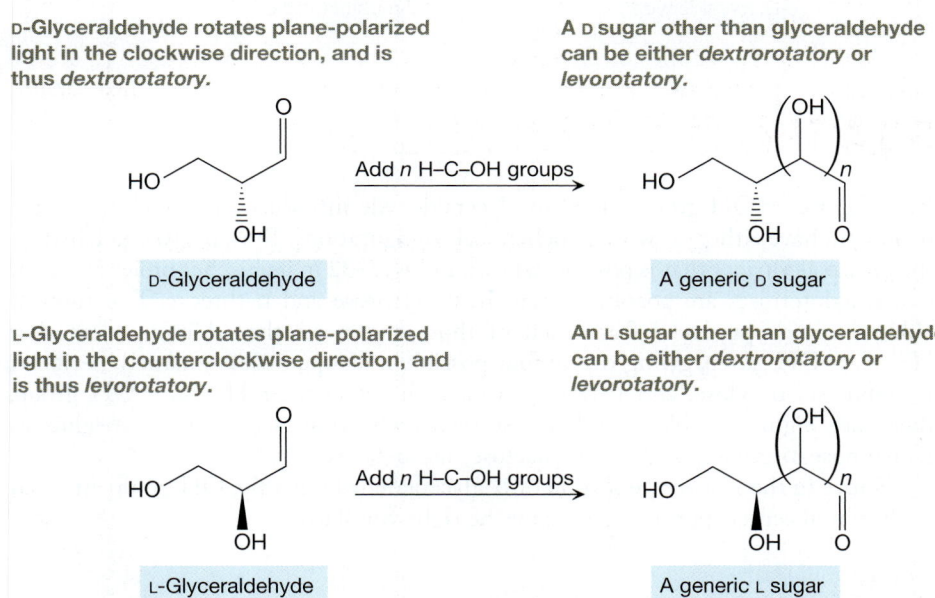

D-Glyceraldehyde rotates plane-polarized light in the clockwise direction, and is thus *dextrorotatory*.

Add *n* H–C–OH groups

A D sugar other than glyceraldehyde can be either *dextrorotatory* or *levorotatory*.

D-Glyceraldehyde

A generic D sugar

L-Glyceraldehyde rotates plane-polarized light in the counterclockwise direction, and is thus *levorotatory*.

Add *n* H–C–OH groups

An L sugar other than glyceraldehyde can be either *dextrorotatory* or *levorotatory*.

L-Glyceraldehyde

A generic L sugar

Based on its origin from either D- or L-glyceraldehyde, a sugar *other* than glyceraldehyde is assigned a D/L designation only as part of the *name*. In those cases, the D/L designation does not have any connection to the direction in which the sugar rotates plane-polarized light.

- Some D sugars rotate plane-polarized light in the clockwise direction, and others rotate it in the counterclockwise direction.
- Some L sugars rotate plane-polarized light in the clockwise direction, and others rotate it in the counterclockwise direction.

The D and L designations for amino acids are assigned by analogy. The second conformation of each glyceraldehyde enantiomer below resembles the configuration of the amino acid next to it—specifically, glyceraldehyde's HC═O, OH, and CH_2OH groups are analogous to the amino acid's CO_2H, NH_2, and R groups. Thus, the top amino acid is the D enantiomer, because its R group points toward you, just as D-glyceraldehyde's CH_2OH group does. For similar reasons, the bottom amino acid is the L enantiomer.

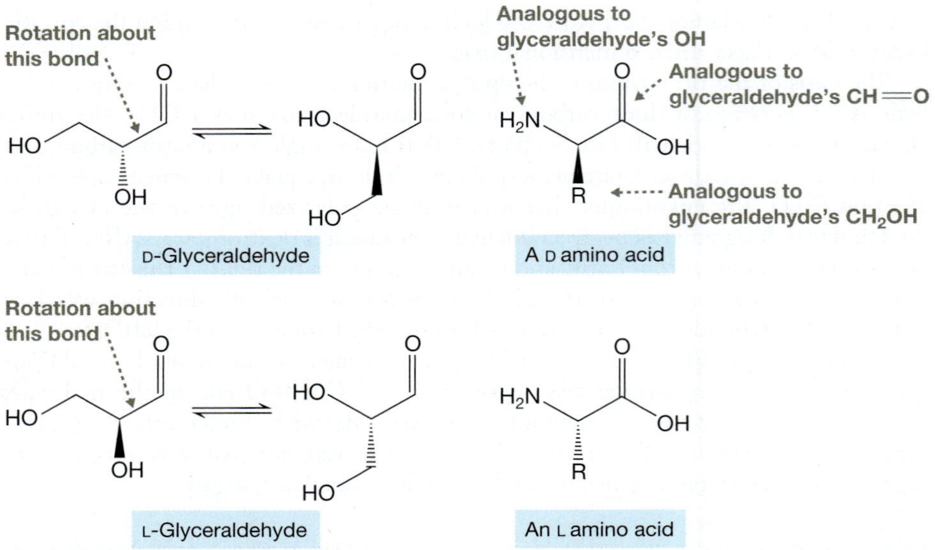

5.14 The D Family of Aldoses

Each H—C—OH group added to glyceraldehyde introduces a new chiral center, which can have either of two stereochemical configurations. Thus, D-glyceraldehyde is the only D aldotriose that is possible (shown in Fig. 5-32 in its Fischer projection), but two D aldotetroses are possible—namely, D-erythrose and D-threose. Two more D aldoses can be produced from each of those sugars on the addition of another H—C—OH group, giving rise to four possible D aldopentoses—namely, D-ribose, D-arabinose, D-xylose, and D-lyxose. And, with yet another H—C—OH group, there are eight possible D aldohexoses—namely, D-allose, D-altrose, D-glucose, D-mannose, D-gulose, D-idose, D-galactose, and D-talose.

Notice in the Fischer projection of D-glyceraldehyde that the OH group attached to the chiral center (purple) appears on the right. Similarly:

Any D sugar is distinguished by having the OH group of the highest-numbered asymmetric carbon appear on the right in its Fischer projection.

Notice, too, that D-glyceraldehyde rotates plane-polarized light in the clockwise direction, denoted by (+), but this is not true of all D sugars. For example, D-erythrose

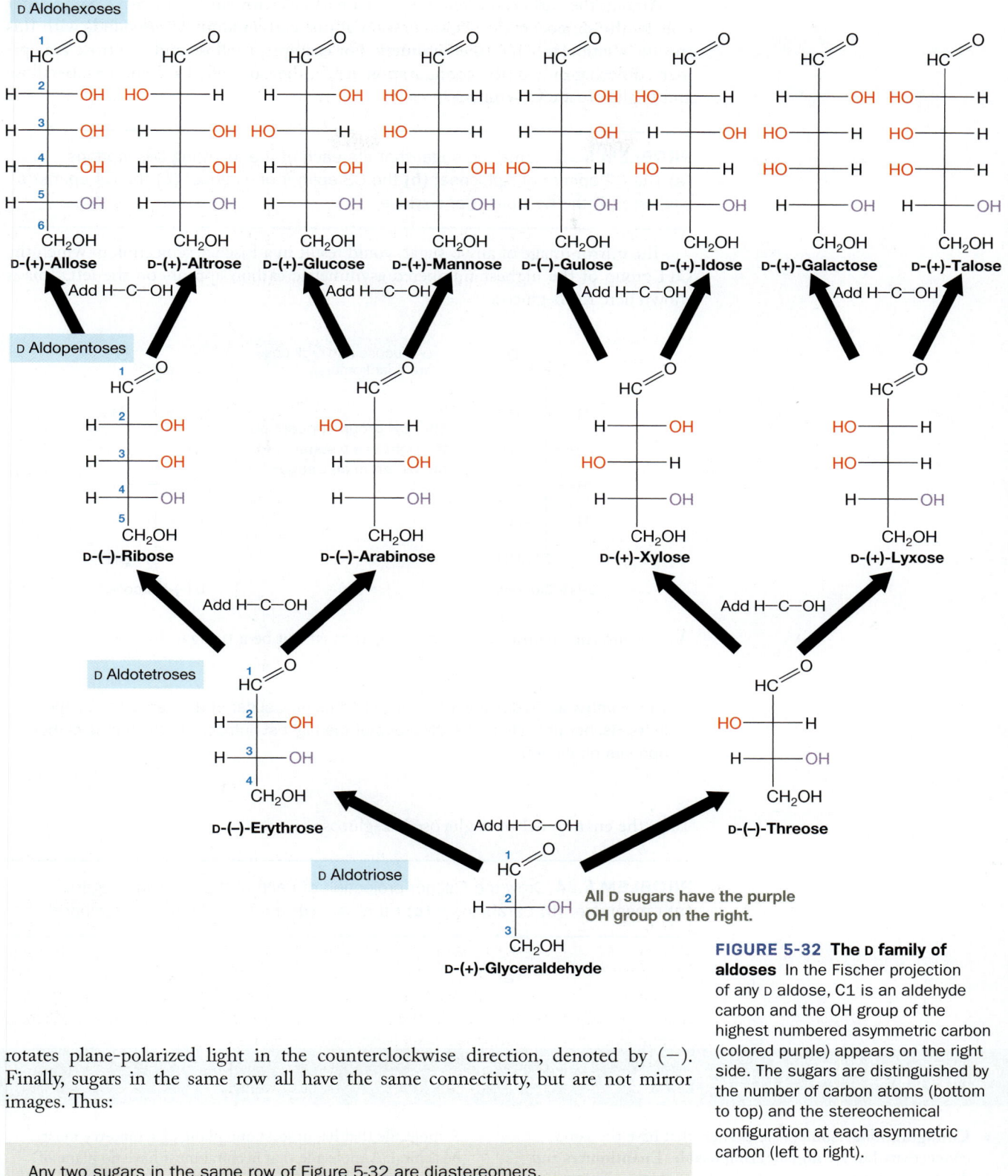

D Aldohexoses

D-(+)-Allose D-(+)-Altrose D-(+)-Glucose D-(+)-Mannose D-(−)-Gulose D-(+)-Idose D-(+)-Galactose D-(+)-Talose

Add H—C—OH   Add H—C—OH   Add H—C—OH   Add H—C—OH

D Aldopentoses

D-(−)-Ribose   D-(−)-Arabinose   D-(+)-Xylose   D-(+)-Lyxose

Add H—C—OH   Add H—C—OH

D Aldotetroses

D-(−)-Erythrose   D-(−)-Threose

Add H—C—OH

D Aldotriose

D-(+)-Glyceraldehyde

All D sugars have the purple OH group on the right.

FIGURE 5-32 The D family of aldoses In the Fischer projection of any D aldose, C1 is an aldehyde carbon and the OH group of the highest numbered asymmetric carbon (colored purple) appears on the right side. The sugars are distinguished by the number of carbon atoms (bottom to top) and the stereochemical configuration at each asymmetric carbon (left to right).

rotates plane-polarized light in the counterclockwise direction, denoted by (−). Finally, sugars in the same row all have the same connectivity, but are not mirror images. Thus:

Any two sugars in the same row of Figure 5-32 are diastereomers.

Recall from Section 5.8 that diastereomers have different physical and chemical properties, which is why no two sugars in the same row have the same name.

Among the various diastereomers in a particular row in Figure 5-32, some differ only by the stereochemical configuration at one carbon atom. Compounds with this specific relationship are called **epimers**. For example, D-allose and D-altrose are epimers, differing only in the configuration at C2; they are called C2 epimers. D-Allose and D-glucose are C3 epimers.

PROBLEM 5.33 Name the sugar that fits each of the following descriptions.
(a) The C2 epimer of D-glucose; **(b)** the C3 epimer of D-talose; **(c)** the C4 epimer of D-talose; **(d)** the C3 epimer of D-xylose.

The mirror image of any D sugar would result in a Fischer projection in which the OH group of the highest-numbered asymmetric carbon appears on the left. This is shown here for D-glucose:

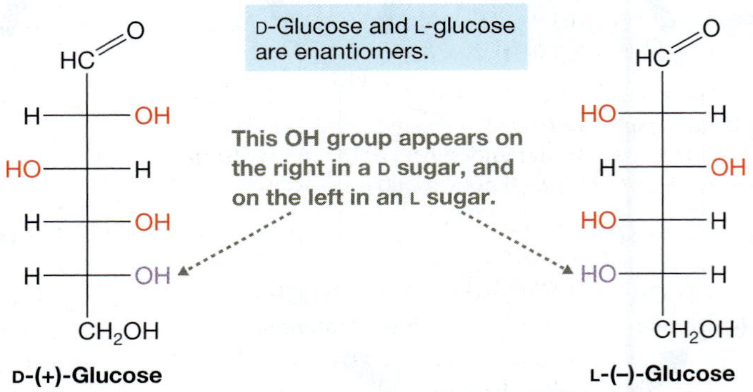

D-Glucose and L-glucose are enantiomers.

This OH group appears on the right in a D sugar, and on the left in an L sugar.

D-(+)-Glucose   L-(−)-Glucose

As a result, the enantiomer of any D sugar must not be a D sugar. Instead:

The enantiomer of a D sugar is designated as an L sugar of the same name, and in its Fischer projection, the OH group of the highest-numbered asymmetric carbon appears on the left.

Thus, the enantiomer of D-glucose is L-glucose.

PROBLEM 5.34 Draw the Fischer projection of each of the following sugars:
(a) L-mannose; **(b)** L-arabinose; **(c)** L-threose; **(d)** the C2 epimer of L-arabinose.

Chapter Summary and Key Terms

- **Configurational isomers** are isomers that have the same connectivity but are **nonsuperimposable**. **Enantiomers** and **diastereomers** are types of configurational isomers; enantiomers are mirror images of each other, whereas diastereomers are not. **(Sections 5.1 and 5.2; Objectives 1 and 2)**

- A molecule is **chiral** if it has an enantiomer. Otherwise it is **achiral**. **(Section 5.4; Objective 3)**

- A molecule that has at least one **plane of symmetry** must be achiral. A molecule that is chiral must have no plane of symmetry. **(Section 5.4a; Objective 3)**

- A tetrahedral **stereocenter**, or a **chiral center**, is bonded to four *different* substituents. When the chiral center is carbon, it is called an **asymmetric carbon**. **(Section 5.4b; Objective 4)**

- Every chiral center has two different configurations possible, related to each other by either (1) reflection through a mirror or (2) the interchange of any two groups. (Section 5.4b; Objective 4)

- A molecule that contains exactly one chiral center must be chiral. (Section 5.4b; Objective 4)

- A molecule with no chiral centers may or may not be chiral. Similarly, a molecule with two or more chiral centers may or may not be chiral. (Section 5.4b; Objective 4)

- A molecule with at least two chiral centers is **meso** if it contains a plane of symmetry. (Section 5.4b; Objective 5)

- If a molecule and its mirror image rapidly interconvert, the molecule is achiral. This can occur through single-bond rotations or nitrogen inversion. (Section 5.4c; Objective 3)

- Because chair conformations rapidly interconvert, Haworth projections can be used to evaluate the chirality of a cyclohexane ring. (Section 5.4c; Objective 3)

- A nitrogen atom may be a chiral center if it is bonded to four different groups (and thus bears a +1 formal charge). An uncharged nitrogen atom that undergoes **nitrogen inversion** is not a chiral center. (Section 5.4c; Objective 6)

- Cis–trans pairs are diastereomers. If a molecule contains two or more chiral centers, then inversion of some, but not all, of their configurations gives a different diastereomer. (Section 5.5; Objective 7)

- Inversion of configuration at all chiral centers in a chiral molecule gives the molecule's enantiomer. (Section 5.5; Objective 7)

- For a molecule that contains n chiral centers, there are, at most, 2^n configurational isomers. (Section 5.5; Objective 7)

- **Fischer projections** are shorthand notations used to represent the configurations about asymmetric carbons in a molecule. Perpendicular intersecting lines represent a carbon atom—typically a chiral center. Horizontal bonds point toward the viewer, whereas vertical bonds point away from the viewer. (Section 5.6; Objective 8)

- Rotation of a Fischer projection 90° in the plane of the page gives the opposite configuration at all asymmetric carbons. Rotation of 180° leaves all stereochemical configurations unchanged. Taking a mirror image of a chiral Fischer projection gives the molecule's enantiomer. (Section 5.6; Objective 8)

- Constitutional isomers have different chemical and physical properties. Similarly, diastereomers have different chemical and physical properties. In an **achiral environment**, enantiomers have *identical* physical and chemical properties. In a **chiral environment**, enantiomers have *different* properties. (Section 5.8; Objective 9)

- Heats of combustion can be used to determine the relative stabilities of isomeric alkenes. The stability of C=C double bonds increases as the amount of **alkyl substitution** increases; trans alkenes are more stable than cis alkenes. (Section 5.9; Objective 10)

- Diastereomers can be separated based on their physical properties. Enantiomers can be separated in a chiral environment, such as a chromatography column containing a chiral stationary phase. (Section 5.10; Objective 11)

- Chiral compounds rotate **plane-polarized light** and are therefore **optically active**. Achiral compounds are **optically inactive**. (Section 5.11; Objective 12)

- A compound's **specific rotation** (i.e., $[\alpha]_D^{20}$) characterizes its ability to rotate plane-polarized light and is a constant that is unique to every chiral compound. (Section 5.11; Objectives 12 and 13)

- Enantiomers have equal but opposite specific rotations. A **racemic mixture** of enantiomers is optically inactive. (Section 5.11; Objectives 12 and 13)

- The **enantiomeric excess (ee)** is the fraction of a mixture that is not racemic. It is the fraction of a mixture that contributes to the rotation of plane-polarized light. (Section 5.11; Objective 14)

Problems

5.2–5.4 Enantiomers, Chirality, and Stereocenters

5.35 Determine whether each of the following objects is chiral or achiral. (Assume that there are no graphics on any of these objects.)

(a) A coffee mug	(b) Your ears	(c) A bowling ball	(d) An automobile	(e) A pair of scissors
(f) A t-shirt	(g) Eyeglasses	(h) A piano	(i) Golf clubs	(j) A tennis racquet

5.36 Draw the Newman projection of 1,2-dichloroethane in its anti conformation and in each of its gauche conformations. Determine which of these conformers possess a plane of symmetry. For those that do, indicate the plane of symmetry using a dashed line.

5.37 For each molecule below, determine whether it is the same enantiomer as the one shown at the right.

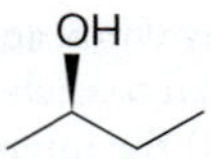

(a)

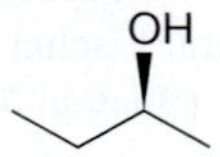

(b)

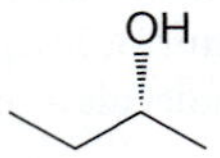

(c)

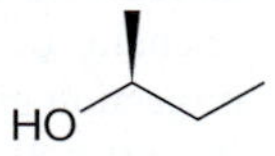

(d)

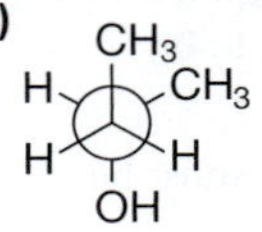

(e)

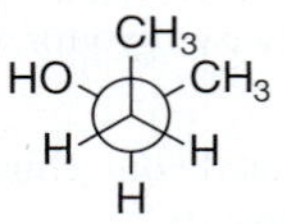

(f)

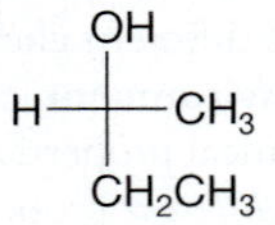

(g)

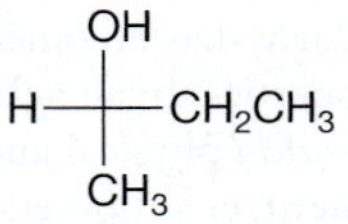

(h)

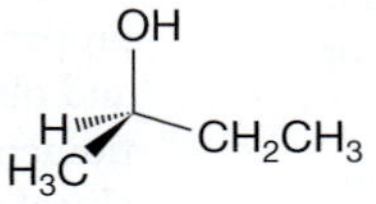

5.38 Identify all of the following species that are chiral.

(a)

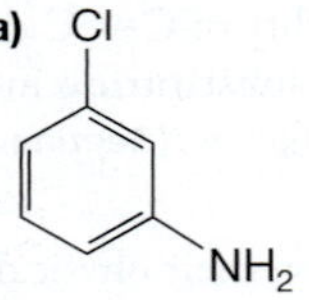

(b)

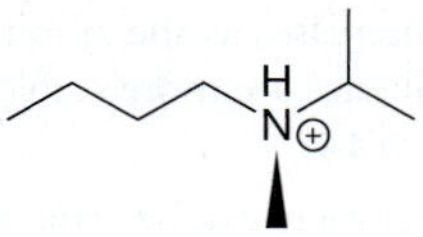

(c)

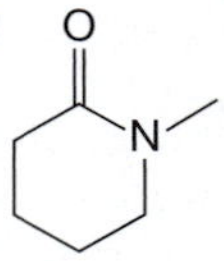

(d)

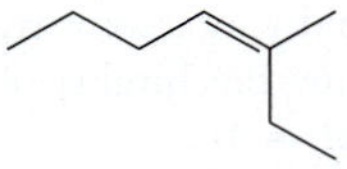

(e)

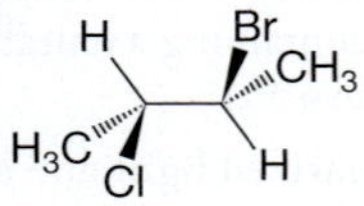

(f)

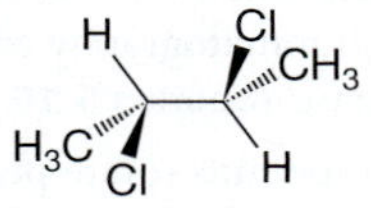

(g)

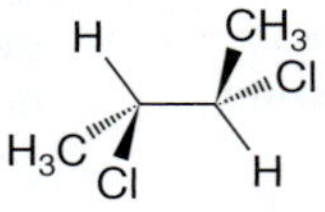

(h)

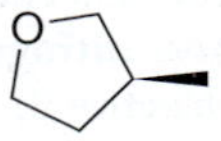

5.39 For each molecule in Problem 5.38, identify all of the chiral centers that exist. Which of those molecules, if any, are meso?

5.40 Which of the following molecules are chiral? *Hint:* Do these Lewis structures accurately depict the three-dimensional geometry?

(a)

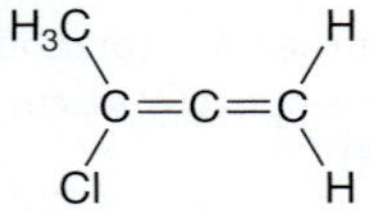

(b)

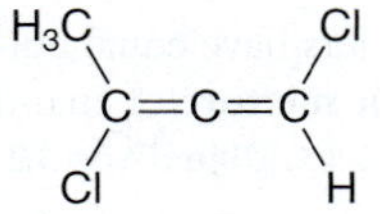

(c)

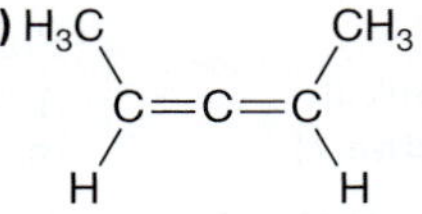

(d)

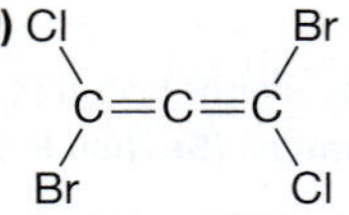

(e)

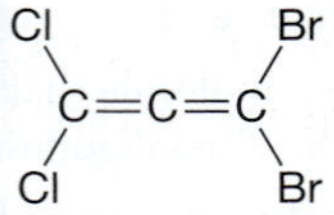

5.41 Is it possible for a meso compound to contain three chiral centers? Why or why not?

5.42 How many chiral centers are in each of the following molecules? Mark each one with an asterisk.

(a)

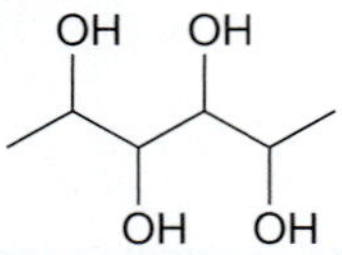

(b)

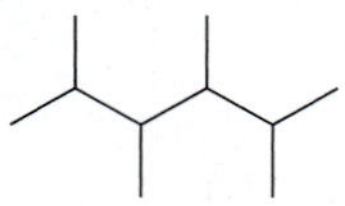

(c)

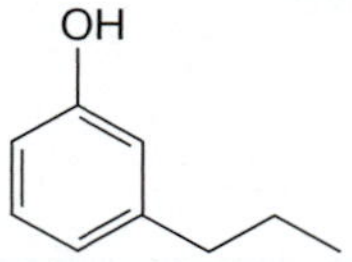

(d)

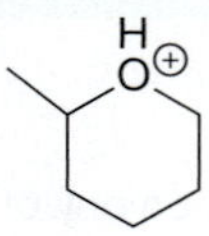

(e)

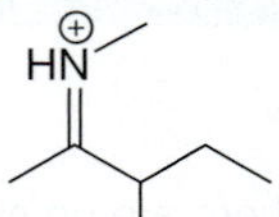

(f)

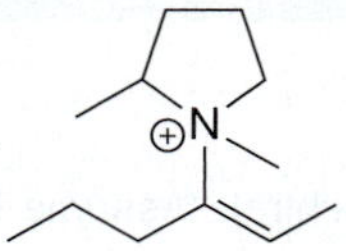

(g)

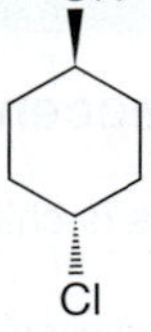

5.43 Aldosterone, a steroid involved in regulating blood pressure, is shown without dash–wedge notation. How many chiral centers does aldosterone have?

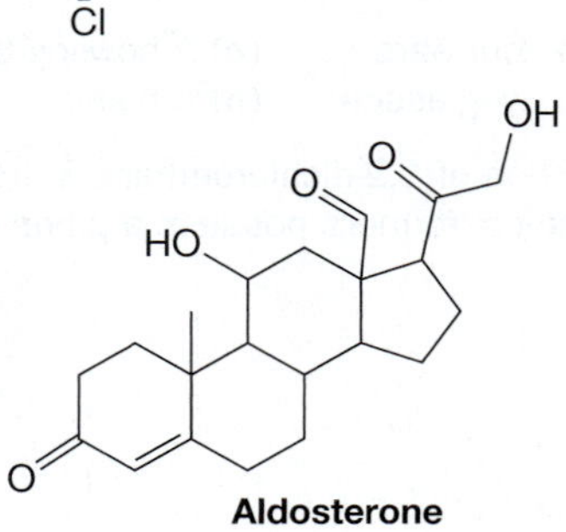

Aldosterone

5.44 Taxol, an anticancer drug, is shown here without dash–wedge notation. How many chiral centers does Taxol have?

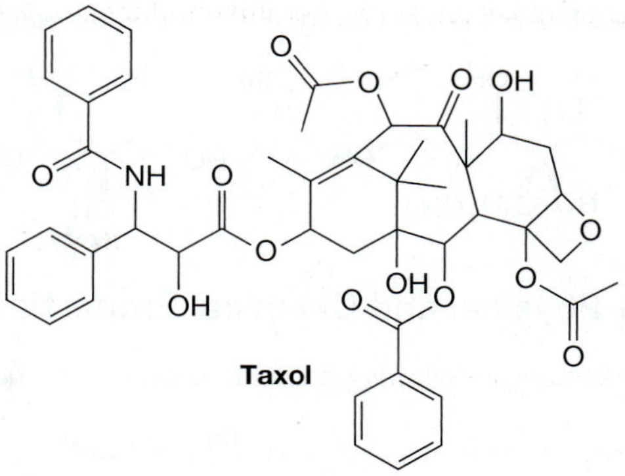

Taxol

5.45 Although we learned in Section 5.4 that uncharged nitrogen atoms generally cannot be chiral centers (due to nitrogen inversion), an exception is Tröger's base. Tröger's base has two enantiomers that can be separated from each other. They are different in their configurations at the N atoms. One enantiomer is shown on the right, viewed from two different perspectives. **(a)** Draw the second enantiomer of Tröger's base. **(b)** Explain why the two enantiomers do not interconvert.

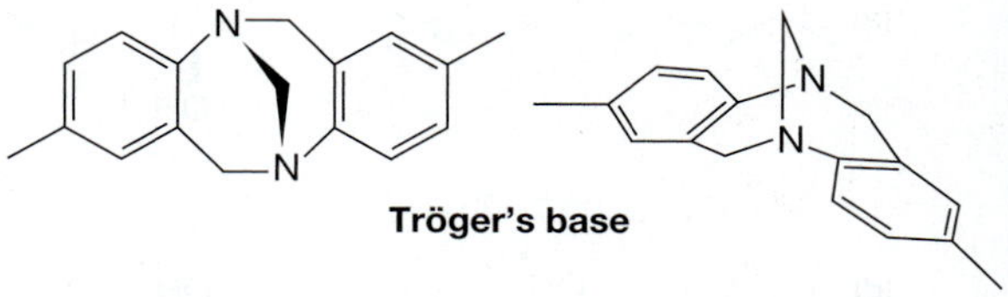

Tröger's base

5.5–5.7 Diastereomers and Fischer Projections

5.46 Consider the molecule shown here. **(a)** How many chiral centers does it have? **(b)** How many total configurational isomers are possible? *Hint:* Determine whether it is possible for any of the configurational isomers to be meso.

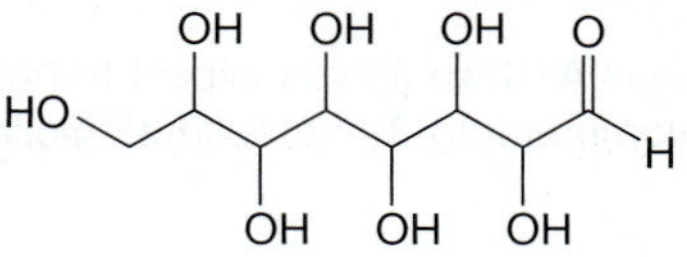

5.47 Draw all possible configurational isomers of the molecule shown here. Which ones are meso?

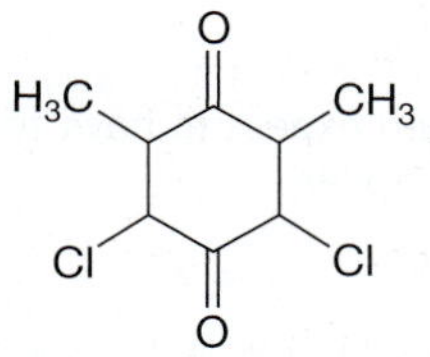

5.48 Draw all possible configurational isomers of the molecule shown here.

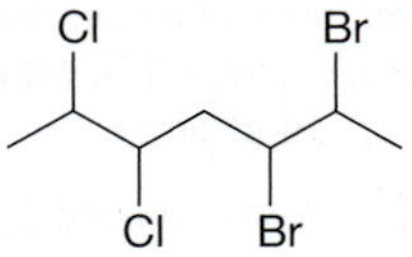

5.49 Which of the following species are chiral?

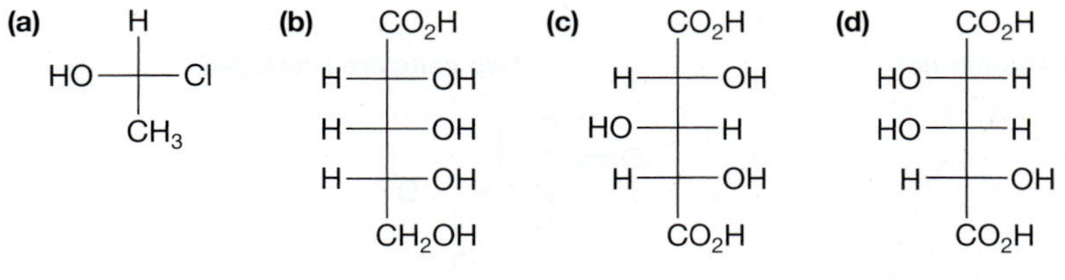

5.50 For each molecule in Problem 5.49, identify all chiral centers that exist. Which of those molecules, if any, are meso?

5.51 Draw each of the following molecules in a zigzag conformation.

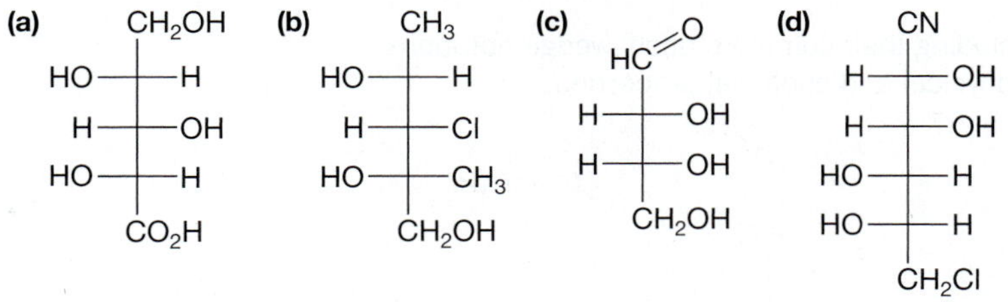

5.52 Draw a Fischer projection for each of the following molecules shown in its zigzag conformation.

(a)

(b)

(c)

5.8–5.10 Physical and Chemical Properties of Isomers

5.53 For each of the following pairs, determine if the compounds have the same boiling point or different boiling points.

(a)

(b)

(c)

CH₂OH	CH₂OH
H——OH	H——OH
H——OH	HO——H
H——OH	H——OH
CH₂OH	CH₂OH

The (c) structures are:

$$\begin{array}{cc} CH_2OH & CH_2OH \\ H\!-\!\!-\!OH & H\!-\!\!-\!OH \\ H\!-\!\!-\!OH & HO\!-\!\!-\!H \\ H\!-\!\!-\!OH & H\!-\!\!-\!OH \\ CH_2OH & CH_2OH \end{array}$$

(d)

(e)

5.54 Which of compounds **A–C** would you expect to have the greatest heat of combustion? The smallest? Explain.

A   **B**   **C**

5.55 Which isomer, **A** or **B**, would you expect to have the greater heat of combustion? Explain.

A   **B**

5.56 The heat of combustion of either but-1-yne or but-2-yne is −2577 kJ/mol (−615.9 kcal/mol), whereas the other's is −2597 kJ/mol (−620.7 kcal/mol). Match each compound to its heat of combustion.

But-1-yne   **But-2-yne**

5.57 Molecule **A** is an acid that has a single chiral center and is therefore chiral. As with any chiral compound, the enantiomers of **A** have identical physical and chemical properties and therefore cannot be separated in an achiral environment. However, when a racemic mixture of **A** is reacted with an enantiomerically pure chiral base such as **B**, then two salts of the form **C** are produced.

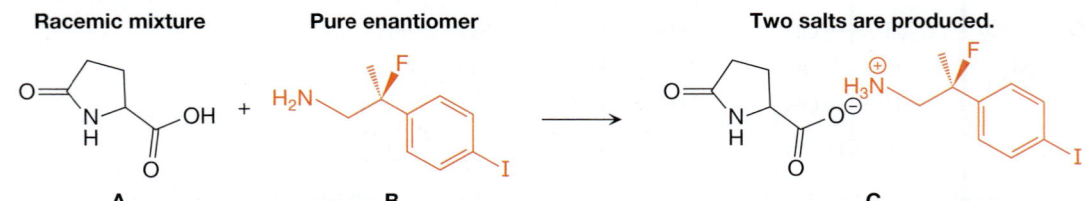

Racemic mixture   Pure enantiomer   Two salts are produced.

A   **B**   **C**

The two product salts have *different* physical and chemical properties, which allow them to be separated in an achiral environment.
(a) Draw the two salts that are produced, including their complete dash–wedge notations.
(b) Explain why the two salts have different physical and chemical properties.

5.11 Optical Activity

5.58 Which of the molecules you drew in Problem 5.48 are optically active?

5.59 If Taxol (see Problem 5.44) has a specific rotation of −49°, then what is the specific rotation of its enantiomer?

5.60 Consider a mixture that is 60% (+) enantiomer and 40% (−) enantiomer. In which direction will the mixture rotate plane-polarized light? What is the enantiomeric excess of the mixture?

5.61 (S)-2-Bromobutane has a specific rotation of +23.1°. Suppose that a solution is made by dissolving 20.00 g of (S)-2-bromobutane in 1.00 L total solution. What would be the measured rotation of plane-polarized light sent through a 10.00-cm tube containing that solution?

5.62 Ibuprofen is a drug used to manage mild pain, fever, and inflammation. It is a chiral drug, and only the S enantiomer, whose specific rotation is +25.0°, is effective. The R enantiomer exhibits no biological activity. If a particular process is capable of producing ibuprofen in 84% enantiomeric excess of the S enantiomer, then what is the specific rotation of that mixture? What is the percentage S enantiomer in that mixture?

5.12–5.14 Organic Chemistry of Biomolecules

5.63 In this chapter, we saw that amino acids typically have a single chiral center and are therefore chiral. One amino acid, however, is achiral. Which one? Explain. *Hint:* Review Table 1-7, page 39.

5.64 How many D-aldoheptoses are possible? Draw the Fischer projection of one of them and its enantiomer.

5.65 Identify which D-aldopentoses are C2 epimers. Identify which ones are C3 epimers.

5.66 Draw the Fischer projection of **(a)** L-lyxose and **(b)** L-talose.

5.67 In the late 1800s, Emil Fischer was able to determine the stereochemical relationships among the four asymmetric carbons in D-glucose by strategically carrying out reactions on various known sugars and studying their optical activities. His proof, known today as the Fischer proof, is a very elegant application of logic in chemistry. As one piece of the proof, Fischer carried out a Wohl degradation reaction on D-glucose to produce D-arabinose. He then oxidized D-arabinose with warm nitric acid, which converted both the CH=O at C1 and the C—OH at C5 to CO_2H functional groups. He found that the resulting compound, a type of *aldaric acid*, was optically active. Based on this result, which aldopentose structures could be ruled out for D-arabinose? Which aldohexose structures could be ruled out for D-glucose?

5.68 As another piece to the Fischer proof (see Problem 5.67), D-glucose was reacted in such a way as to convert the CH=O at C1 to a CH_2OH group and to convert the CH_2OH group at C6 to a CH=O group. The same was done to D-mannose. When this transformation was carried out on D-glucose, a sugar other than D-glucose was produced, but when it was carried out on D-mannose, D-mannose was returned as the product. How do these results agree with the currently known structures of the two sugars? What sugar was produced when D-glucose was reacted?

5.69 Carvone is present in many essential oils. It is chiral, and its two enantiomers are shown here. Even though they are enantiomers, they have different odors. The (−) enantiomer smells like spearmint, whereas the (+) enantiomer smells like caraway seeds. What does this say about the olfactory receptors that detect carvone?

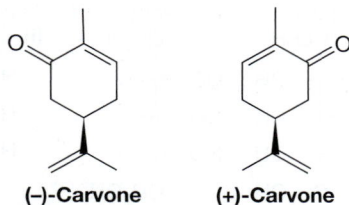

(−)-Carvone   (+)-Carvone

Integrated Problems

5.70 If molecules **A** and **B** are isomers of each other, then what kinds of isomers could they be (i.e., *enantiomers*, *diastereomers*, or *constitutional isomers*) under each of the following conditions?
(a) Both molecules have the same IHD.
(b) Molecule **A** has a ring but molecule **B** does not.
(c) Molecules **A** and **B** contain different functional groups.
(d) Molecules **A** and **B** share exactly the same functional groups.
(e) Molecule **A** has a plane of symmetry but molecule **B** does not.

5.71 Identify the specific type of relationship between each of the following pairs of molecules (i.e., either *same, constitutional isomers, enantiomers, diastereomers, conformers,* or *unrelated*).

(a)

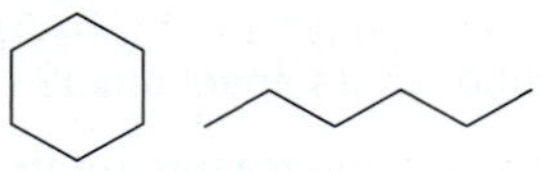

(b)

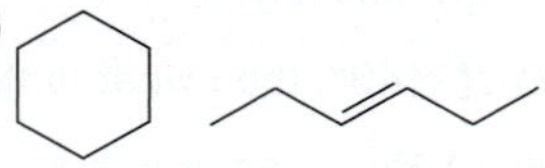

(c)

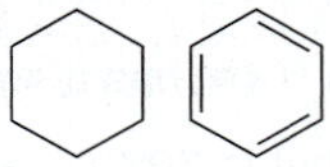

(d)

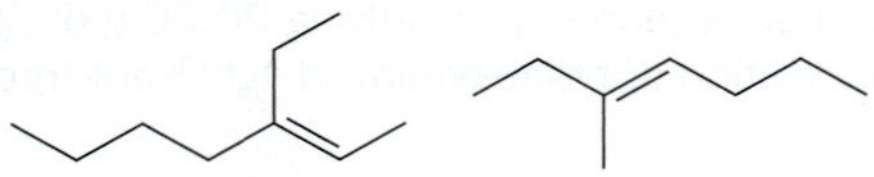

(e)

(f)

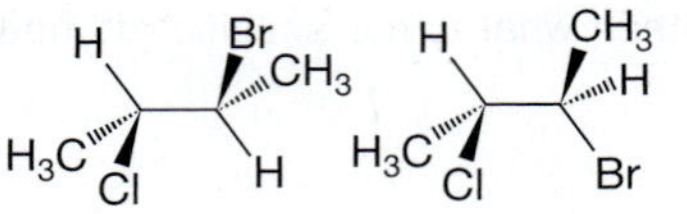

(g)

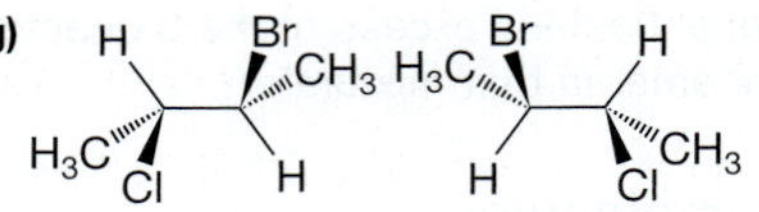

(h)

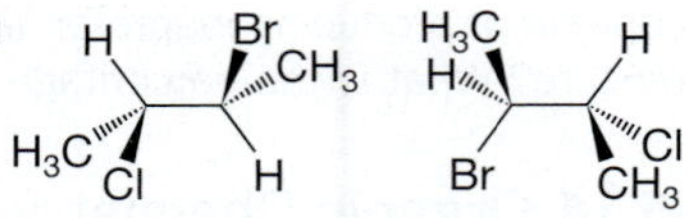

5.72 Identify the specific relationship between each of the following pairs of molecules (i.e., either *same, constitutional isomers, enantiomers, diastereomers, conformers,* or *unrelated*).

(a)

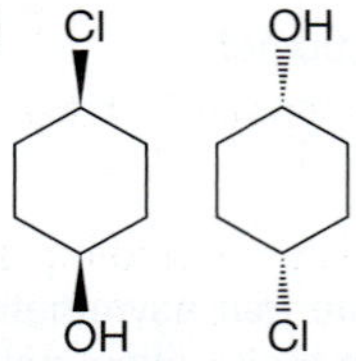

(b)

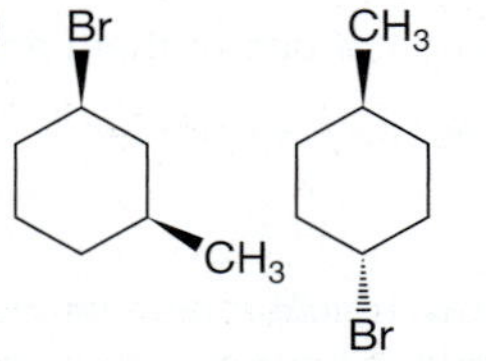

(c)

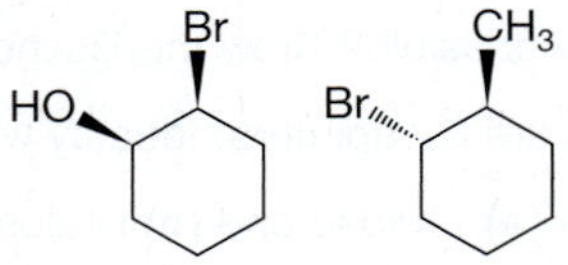

(d)

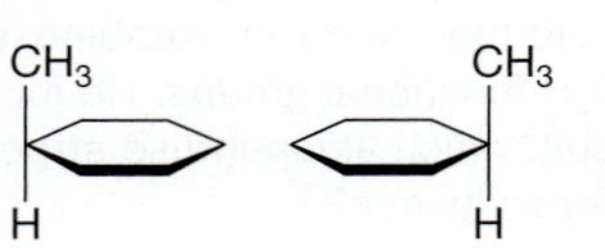

(e)

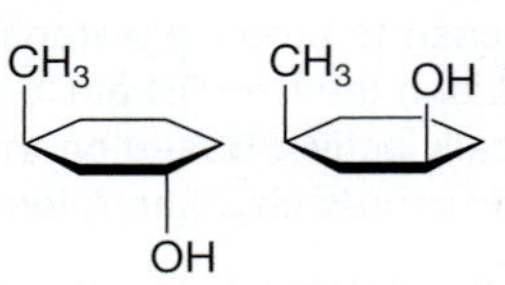

(f)

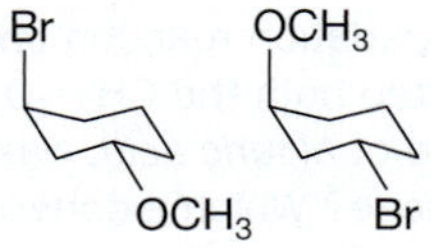

5.73 Identify the specific relationship between each of the following pairs of molecules (i.e., either *same, constitutional isomers, enantiomers, diastereomers, conformers,* or *unrelated*).

(a)

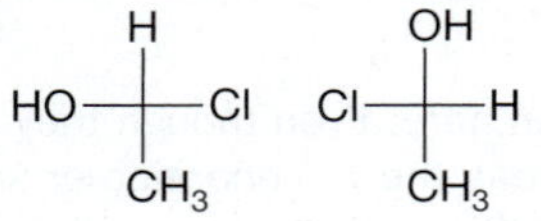

(b)

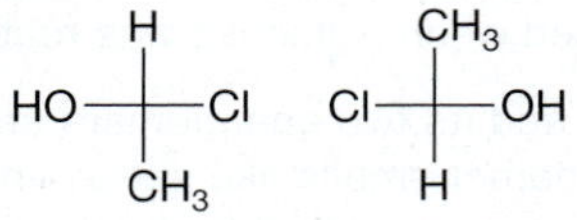

(c)

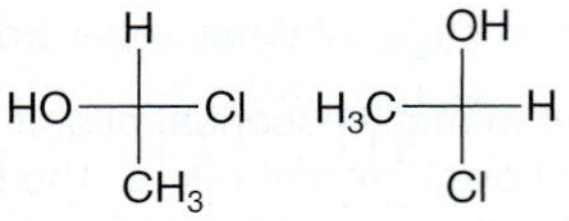

(d)

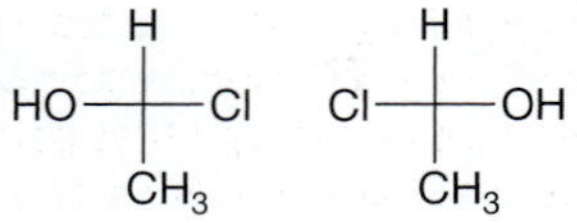

(e)

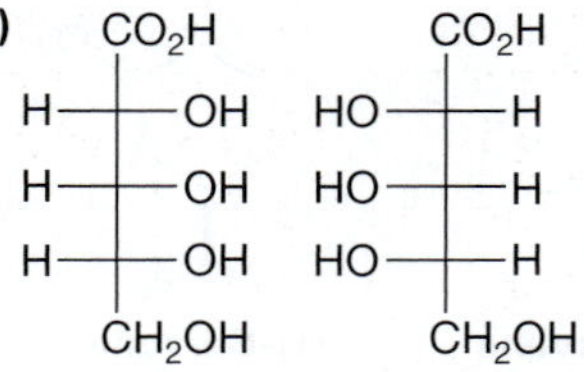

(f)

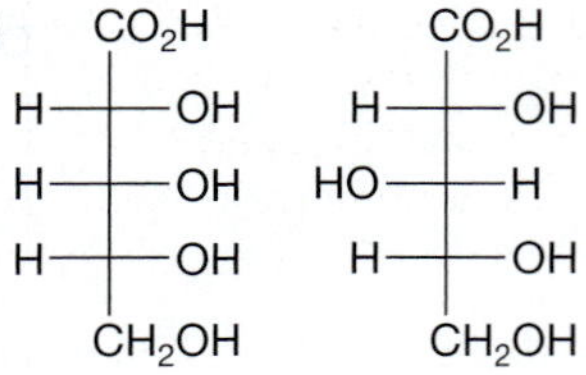

(g)

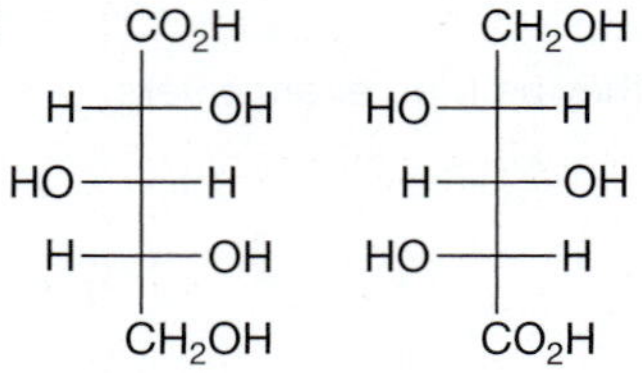

(h)

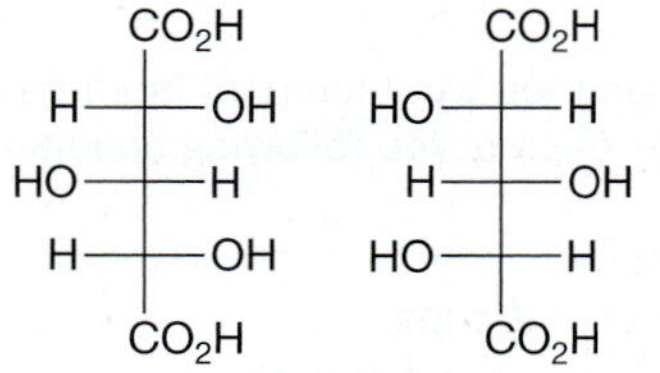

5.74 Draw all configurational isomers of $C_4H_{11}N$ that are optically active.

5.75 Butanal (C_4H_8O) has eight H atoms. Suppose that any of these H atoms can be replaced by a Cl atom to yield a molecule with the formula C_4H_7ClO. **(a)** Identify two H atoms where this substitution would yield *constitutional isomers* of C_4H_7ClO; **(b)** *enantiomers* of C_4H_7ClO; **(c)** *conformers* of C_4H_7ClO.

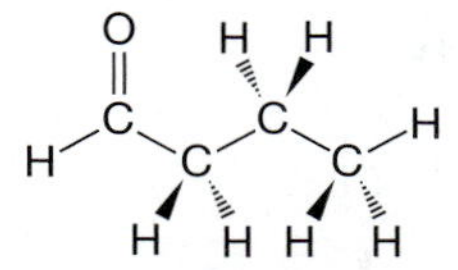

Butanal

5.76 2,3-Dibromoprop-1-ene ($C_3H_4Br_2$) has four H atoms. Suppose that any of these H atoms can be replaced by a Cl atom to yield a molecule with the formula $C_3H_3Br_2Cl$. **(a)** Identify two H atoms where this substitution would yield *constitutional isomers* of $C_3H_3Br_2Cl$; **(b)** *enantiomers* of $C_3H_3Br_2Cl$; **(c)** *diastereomers* of $C_3H_3Br_2Cl$.

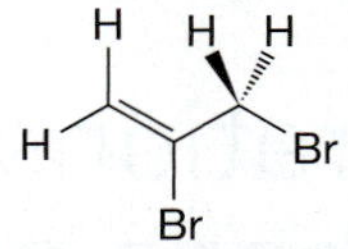

2,3-Dibromoprop-1-ene

5.77 An unknown compound is optically active and has the molecular formula C_6H_{12}. Draw all possible isomers of the compound.

5.78 1,1′-Bi-2-naphthol is a chiral compound, and its enantiomers can be separated from each other. The enantiomer shown here has a specific rotation of $-32.70°$ in tetrahydrofuran. (Notice that the two fused-ring systems are not in the same plane. In the top ring system, the left side is in front of the plane of the page, whereas in the bottom ring system, the right side is in front.)

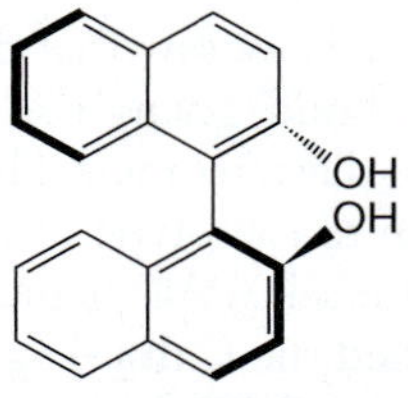

1,1′-Bi-2-naphthol

(a) How many chiral centers does 1,1′-bi-2-naphthol have?
(b) Draw the enantiomer of the molecule shown. What is its specific rotation?
(c) Aside from being mirror images of each other, how else are the two enantiomers related? Why do you think they do not readily interconvert? *Hint:* It may help to construct a molecular model.

5.79 In this chapter, you learned that a molecule that possesses a plane of symmetry must be achiral. It is possible, however, for a molecule to be achiral without possessing a plane of symmetry, such as in cases where the molecule possesses a *point of symmetry*. If a point of symmetry exists for a molecule, then each atom can be reflected through that point (rather than a plane) to arrive at an identical atom. This molecule, for example, is achiral, yet it possesses no plane of symmetry.

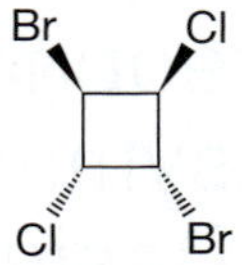

(a) Convince yourself that the molecule has no plane of symmetry.
(b) Draw the molecule's mirror image and show that the molecule and its mirror image are superimposable.
(c) Where is the point of symmetry located?

Stereochemistry in Nomenclature

R and *S* Configurations about Asymmetric Carbons and *Z* and *E* Configurations about Double Bonds

In Chapter 5, we saw that different configurations can exist about a chiral center or a double bond, giving rise to different configurational isomers—namely, enantiomers and diastereomers. Here in Interchapter C, we explain how to assign a specific designation to each type of configuration and to incorporate these designations into the molecule's IUPAC name. The collection of rules used to assign these configurations is called the **Cahn–Ingold–Prelog convention**.[1,2]

The basis of this convention involves first assigning priorities to the various substituents attached to the chiral center or double bond and then examining the spatial arrangement of a subset of those substituents. In Section C.1, we show how the Cahn–Ingold–Prelog rules are used to determine the configuration about an asymmetric carbon, and in Section C.2, we do the same for the configuration about a double bond.

C.1 Priority of Substituents and Stereochemical Configurations at Asymmetric Carbons: *R/S* Designations

The IUPAC rules for assigning the configuration at a particular asymmetric carbon involve three basic steps:

[1]Cahn, R. S.; Ingold, C. K.; Prelog, V. *Angew. Chem.* **1966**, 78, 413–447.
[2]Prelog, V.; Helmchen, G. *Angew. Chem. Int. Ed.* **1982**, 21, 567–583.

Basic Steps for Assigning an *R* or *S* Configuration to an Asymmetric Carbon

1. Assign a priority, 1 through 4 (where 1 is the highest and 4 is the lowest), to each of the four substituents bonded to the asymmetric carbon.
2. Orient the molecule so that *the lowest-priority substituent is pointed away from you*.
3. If the substituents having priorities 1 through 3 are arranged *clockwise*, then the stereocenter is assigned the *R* configuration (Fig. C-1a). If they are arranged *counterclockwise*, then the stereocenter is assigned the *S* configuration (Fig. C-1b). (The *R* and *S* designations derive from Latin—*rectus* means right and *sinister* means left.)

Turning a steering wheel in the *clockwise* direction turns a car to the *right*.

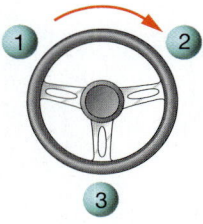

(a) *R* configuration

Turning a steering wheel in the *counterclockwise* direction turns a car to the *left*.

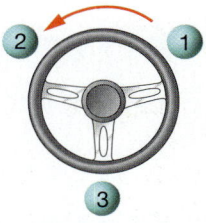

(b) *S* configuration

As you can see, determining the relative priorities of substituents is a key part of assigning an *R* or *S* configuration to an asymmetric carbon. Because of the complexity introduced when substituents contain double or triple bonds, we first present the basic system involving substituents that contain only single bonds (Section C.1a). We then look at the modifications required to deal with double and triple bonds in a substituent (Section C.1b).

FIGURE C-1 *R* and *S* configurations With the lowest-priority substituent pointed away (as if down the column of a steering wheel), a clockwise arrangement of the top-three-priority substituents defines an *R* configuration (a) and a counterclockwise arrangement defines an *S* configuration (b).

C.1a Substituents Involving Only Single Bonds

To assign priorities to substituents, we use a system of tie-breaking rules that have us examine one atom at a time from each substituent rather than substituents as a whole. When we compare atoms from different substituents, we establish priorities of the atoms as follows:

Relative Priorities of Atoms

- The atom with the greater atomic number has the higher priority.
- If two isotopes are being compared, the one with the greater atomic mass has the higher priority.

For example, Br has a higher priority than Cl because Br has the greater atomic number. The ^{2}H (i.e., deuterium) and ^{1}H isotopes have the same atomic number, but ^{2}H has the higher priority because it has the greater atomic mass.

PROBLEM C.1 Which atom in each pair has the higher priority?
(a) F or O; **(b)** P or F; **(c)** ^{13}C or ^{12}C

With this in mind, the first tiebreaker is as follows:

First Tiebreaker for Establishing Substituent Priorities

Examine each substituent's atom at the point of attachment. If those atoms are different, assign the higher priority to the substituent that has the higher-priority atom.

Let's apply this to the following stereoisomer of 1-bromo-1-chloroethane.

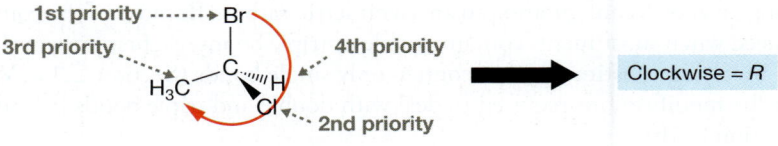

1-Bromo-1-chloroethane

There is one chiral center, which is at C1, and the substituents attached to it are CH_3, Br, H, and Cl. The atoms at the points of attachment are C, Br, H, and Cl. Because these atoms are all different, their atomic numbers establish the priorities of the substituents as follows: Br > Cl > CH_3 > H. As shown below, the fourth-priority substituent, H, is pointing away (it is attached by a dash bond) and the first-, second-, and third-priority substituents are arranged clockwise. Therefore, this configuration is *R*.

1st priority ······→ Br
3rd priority ····
4th priority
H_3C
C
H
Cl ···· 2nd priority

Clockwise = *R*

First-, second-, and third-priority substituents
are arranged clockwise, and the fourth-priority
substituent points away from you.

PROBLEM C.2 The atoms in each of the following sets are arranged alphabetically. Reorder the substituents in each set from *highest* to *lowest* priority. **(a)** Br, CH_3, Cl, F; **(b)** Br, CH_3, H, I; **(c)** CH_3, F, O, N

When you are presented with a molecular structure, the fourth-priority substituent will not always point away from you. If it is not, you can build a model and reorient the molecule so the fourth-priority substituent points away. Alternatively, you can use one of the following tricks to help.

Assigning *R* and *S* Configurations When the Fourth-Priority Substituent Does Not Point Away

- If the fourth-priority substituent points toward you, determine whether the first-, second-, and third-priority substituents are arranged clockwise or counterclockwise and reverse that arrangement before assigning *R* or *S*.
- If the fourth-priority substituent is in the plane of the page, switch the fourth-priority substituent with the substituent that points away, then determine whether the first-, second-, and third-priority substituents are arranged clockwise or counterclockwise and reverse that arrangement before assigning *R* or *S*.

Suppose the exact molecule above were instead given to us as shown below.

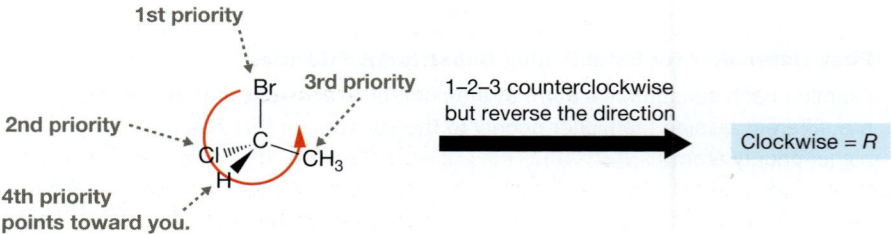

1st priority
2nd priority ····
Br
3rd priority
Cl
C
CH₃
H
4th priority
points toward you.

1–2–3 counterclockwise
but reverse the direction

Clockwise = *R*

In this case, the fourth-priority substituent points toward us. The first-, second-, and third-priority substituents are arranged counterclockwise, but we must reverse that arrangement to clockwise before assigning the configuration. This gives us the R configuration, the same as we determined before.

Once we know the stereochemical configuration of a particular chiral center, we can incorporate it into the IUPAC name according to the following rules:

> **Incorporating *R* and *S* Configurations into an IUPAC Name**
> - Add an *R* or *S* designation in parentheses for each asymmetric carbon in the molecule and use hyphens to separate those designations from the rest of the IUPAC name.
> - The *R* and *S* designations can be placed in the IUPAC name in one of two ways:
>   - Each can be placed immediately before the first number used to locate a substituent on the corresponding asymmetric carbon.
>   - Alternatively, all *R* and *S* designations can be placed together at the front of the name. If there is more than one asymmetric carbon, the locator number for each one must appear before its corresponding *R* or *S* designation, and the designations must be separated from each other by commas.

For the stereoisomer of 1-bromo-1-chloroethane we have been working with, which has one asymmetric carbon whose configuration is R, the complete IUPAC name is (*R*)-1-bromo-1-chloroethane.

PROBLEM C.3 Designate the configuration of each stereocenter in these molecules as *R* or *S*. What is the complete IUPAC name for each molecule?

(a)   (b)   (c)

PROBLEM C.4 Draw the specific stereoisomer that corresponds to each name. **(a)** (S)-1-methoxy-1-nitrobutane; **(b)** (R)-1-methoxy-1-nitrobutane; **(c)** (R)-1-fluoro-1-methoxy-1-nitropropane; **(d)** (S)-3,3-dichloro-1-ethoxy-1-fluorohexane

---

PROBLEM C.5 Designate the configuration of each stereocenter in these molecules as *R* or *S*.

(a)   (b)   (c)

The molecule below, whose name is given without stereochemical designations, requires us to consider two chiral centers: one at C1 and the other at C4.

1-Bromo-1-chloro-4-fluoro-4-nitrobutane

The *R/S* designation for *each* chiral center is determined just as before. The substituents on C1 are all attached by different atoms, whose priorities decrease in the order: Br > Cl > C > H. As shown on the next page at the left, the fourth-priority substituent is pointing away and the top-three-priority substituents are arranged clockwise, so the configuration at C1 is *R*.

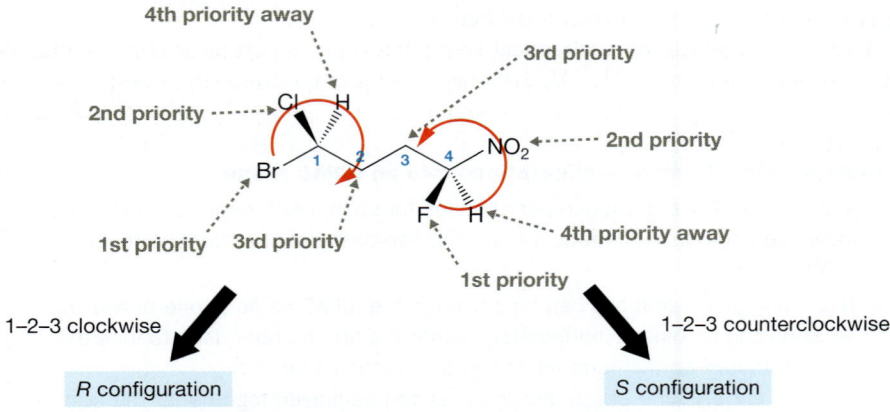

(R)-1-Bromo-1-chloro-(S)-4-fluoro-4-nitrobutane
(1R,4S)-1-Bromo-1-chloro-4-fluoro-4-nitrobutane

The substituents on C4 are attached by different atoms, too, and, as shown above at the right, their priorities decrease in the order: F > N > C > H. Again, the fourth-priority substituent points away, but this time the top-three-priority substituents are arranged counterclockwise, so the configuration at C4 is *S*.

To add the *R* and *S* designations to the IUPAC name, we can place them each immediately before their respective locator numbers already present: (*R*)-1-bromo-1-chloro-(*S*)-4-fluoro-4-nitrobutane. Alternatively, we can group them at the beginning of the name, but we must add an additional locator number immediately before each designation: (1*R*,4*S*)-1-bromo-1-chloro-4-fluoro-4-nitrobutane.

So far, when assigning priorities to substituents, the substituents have all differed in the atoms at their points of attachment. Often, however, we must distinguish substituents that are attached by the same atom. In these cases, we must apply the next tiebreaker.

Second Tiebreaker for Establishing Substituent Priorities

For each substituent, identify the set of three atoms one bond away from its point of attachment. In each set, arrange the three atoms from highest priority to lowest.

- Compare each set's highest-priority atom. If they are different, then the atom that has the higher priority corresponds to the higher-priority substituent.
- If the highest-priority atoms from each set are the same, then compare each set's second-highest-priority atom to break the tie.
- If the second-highest-priority atoms from each set are the same, then compare each set's lowest-priority atom to break the tie.

The second tiebreaker is necessary to establish the configuration of the chiral center in the following stereoisomer of 2-chlorobutane.

1st priority ······▶ Cl

4th priority ◀······ H

The first tiebreaker results in a tie because both substituents are attached by C.

By considering the atoms at the points of attachment, we use the first tiebreaker to assign Cl as the first-priority substituent and H as the fourth-priority substituent. The priorities of the CH_3 and CH_2CH_3 substituents, however, remain tied because both are attached by C.

To break the tie, we proceed to the second tiebreaker, which has us examine the set of atoms one bond away from the point of attachment, beginning with the highest-priority atom. For the CH_2CH_3 substituent, that set of atoms is {C, H, H} and for the CH_3 group it is {H, H, H}:

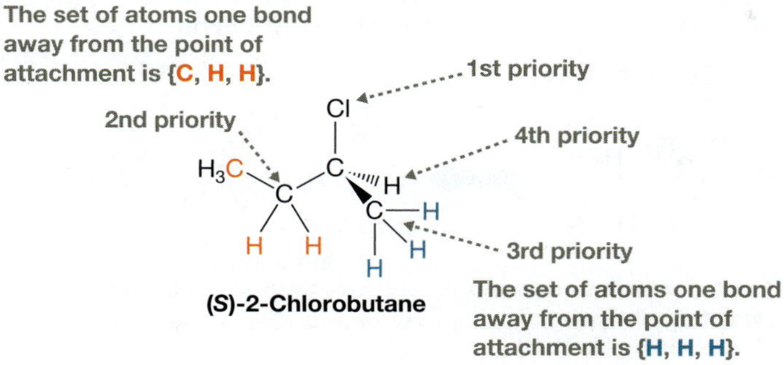

(S)-2-Chlorobutane

Comparing the highest-priority atoms in each set (C vs. H), we assign the higher priority to the CH_2CH_3 substituent, so CH_2CH_3 is second priority and CH_3 is third. Notice that the fourth-priority substituent (H) is pointing away and the first-, second-, and third-priority substituents are arranged counterclockwise, so the configuration is *S*. The complete name of the molecule is therefore (*S*)-2-chlorobutane.

PROBLEM C.6 Which substituent in each pair has the higher priority?
(a) $CH_2CH_2CH_3$ or $CH(CH_3)_2$; **(b)** OH or OCH_3

The second tiebreaker is not always sufficient to break the tie between substituents. If it fails to break the tie, we must apply the third tiebreaker.

Third Tiebreaker for Establishing Substituent Priorities

- If sets of atoms one bond away from the point of attachment are identical, then apply the second tiebreaker to the sets of atoms *one additional bond away* from the point of attachment.

- If a tie persists, continue to move farther away from the point of attachment until there is a difference.
  - If the substituent's backbone is branched, then follow the chain on which the higher-priority atoms are encountered first.
  - If two atoms being compared are the same and one of them is a substituent's terminal atom, then the substituent to which it belongs has the lower priority.
  - If a point of difference is never found, then the substituents must be identical.

The following example shows how to apply the third tiebreaker.

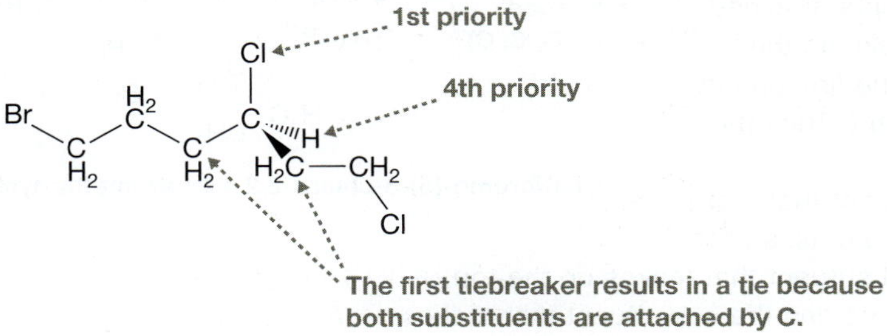

By examining the atoms at the points of attachment, we can apply the first tiebreaker to assign Cl as the first-priority substituent and H as the fourth-priority substituent. The remaining two substituents are attached by C, so we proceed to the second tiebreaker. For each of those two substituents, we identify the sets of atoms one bond away from the points of attachment as {C, H, H}, as shown below on the left, so the tie remains unbroken.

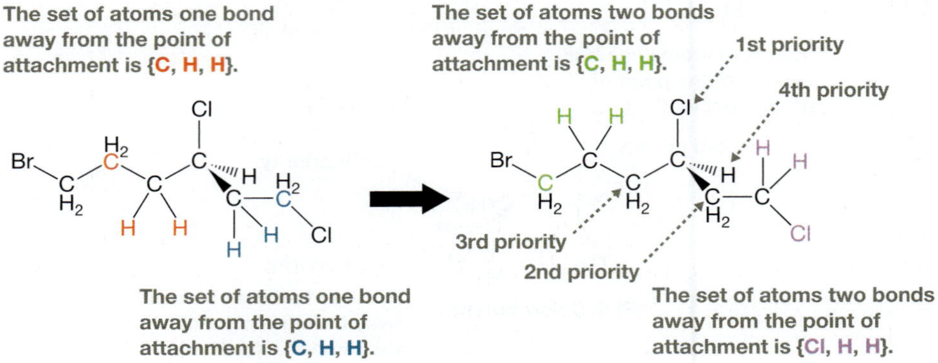

The set of atoms one bond away from the point of attachment is {C, H, H}.

The set of atoms two bonds away from the point of attachment is {C, H, H}.

1st priority
4th priority
3rd priority
2nd priority

The set of atoms one bond away from the point of attachment is {C, H, H}.

The set of atoms two bonds away from the point of attachment is {Cl, H, H}.

To apply the third tiebreaker, we examine the sets of atoms one bond farther away from the points of attachment. As shown on the right above, that set of atoms for the Br-containing substituent is {C, H, H} and that set of atoms for the Cl-containing substituent is {Cl, H, H}. Because the atomic number of Cl is higher than that of C, the Cl-containing substituent wins. Therefore, the Cl-containing substituent is assigned second priority and the Br-containing substituent is assigned third priority. Notice in this case that the Br atom, whose atomic number is higher than any other atom in the molecule, never figured in the evaluation of assignments.

Now that the priorities of the four substituents are established, we can assign the configuration of the stereocenter and write the complete name of the molecule. With the fourth-priority substituent pointing away, the first-, second-, and third-priority substituents are arranged clockwise, so the configuration is R. Thus, the complete IUPAC name of this molecule is (R)-6-bromo-1,3-dichlorohexane.

SOLVED PROBLEM C.7

What is the complete IUPAC name of the molecule at the right?

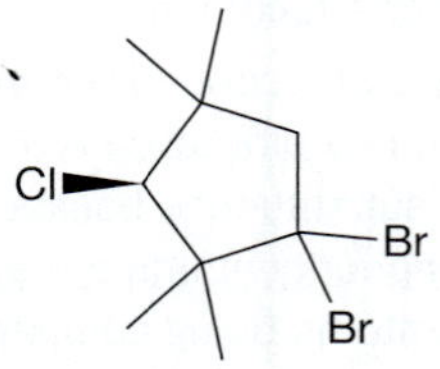

Think What is the name of the molecule without considering stereochemical configurations? How many chiral centers are there? What are the attached substituents? In determining the priorities of those substituents, what distinctions can be made using the first tiebreaker? The second? The third? Once you determine the stereochemical configuration, how do you incorporate that designation into the IUPAC name?

Solve Without taking stereochemistry into consideration, the molecule's IUPAC name is 1,1-dibromo-3-chloro-2,2,4,4-tetramethylcyclopentane. There is a single chiral center at C3, highlighted in red at the right. The four substituents attached to that carbon are: Cl, H, and two others that make up the ring. Using the first tiebreaker, we assign Cl as the first-priority substituent and H as the fourth-priority substituent. The other two are both attached by carbon.

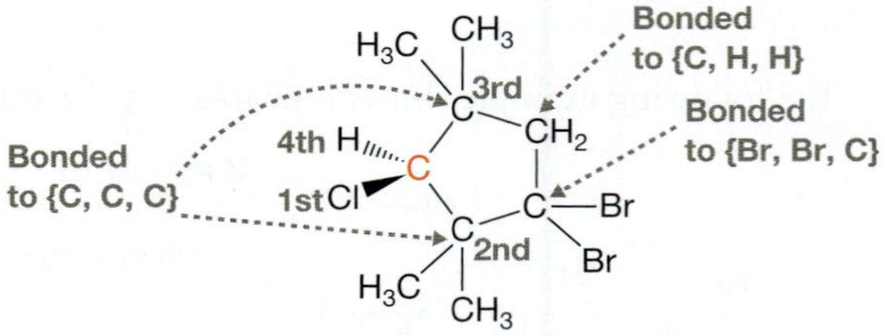

Applying the second tiebreaker, each substituent that makes up the ring gives us {C, C, C} as the set of three atoms, so the tie remains. Applying the third tiebreaker, the substituent that makes up the top of the ring gives {C, H, H} as the set of three atoms and the substituent that makes

Bonded to {C, H, H}
Bonded to {Br, Br, C}
Bonded to {C, C, C}
4th
3rd
1st
2nd

1,1-Dibromo-(S)-3-chloro-2,2,4,4-tetramethylcyclopentane

up the bottom of the ring gives us {Br, Br, C}. The bottom substituent therefore wins, making it the second-priority substituent. Finally, we see that, with the fourth-priority substituent pointing away, the top-three-priority substituents are arranged counterclockwise, so the configuration is S. That S designation can be placed immediately before the stereocenter's locator number, as shown in the complete IUPAC name on the preceding page.

PROBLEM C.8 What is the complete IUPAC name for each molecule?

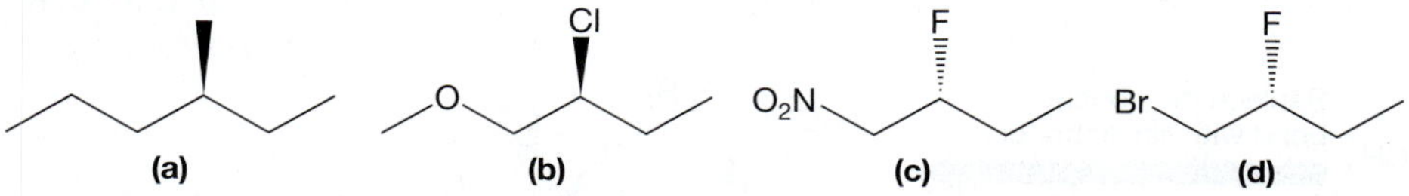

(a)   (b)   (c)   (d)

PROBLEM C.9 Draw the specific stereoisomer that corresponds to each name.
(a) (*R*)-2-bromohexane; **(b)** (*S*)-4-ethoxy-1,1,1,2-tetrafluorobutane; **(c)** (*R*)-1,4-dinitropentane; **(d)** (*S*)-2-ethoxy-2-methoxypentane

---

C.1b Substituents with Double or Triple Bonds

Substituents with double or triple bonds are treated somewhat differently from substituents that contain only single bonds.

How to Deal with Substituents That Have a Double or Triple Bond

- If atoms X and Y in a substituent are connected by a *double bond*, then we treat X as being singly bonded to two Y atoms. Likewise, we treat Y as being singly bonded to two X atoms.

- If atoms X and Y in a substituent are connected by a *triple bond*, then we treat X as being singly bonded to three Y atoms, and we treat Y as being singly bonded to three X atoms.

To apply these rules, it is helpful to consider adding *imaginary* atoms when replacing double and triple bonds with single bonds. This is shown in Figure C-2 for a

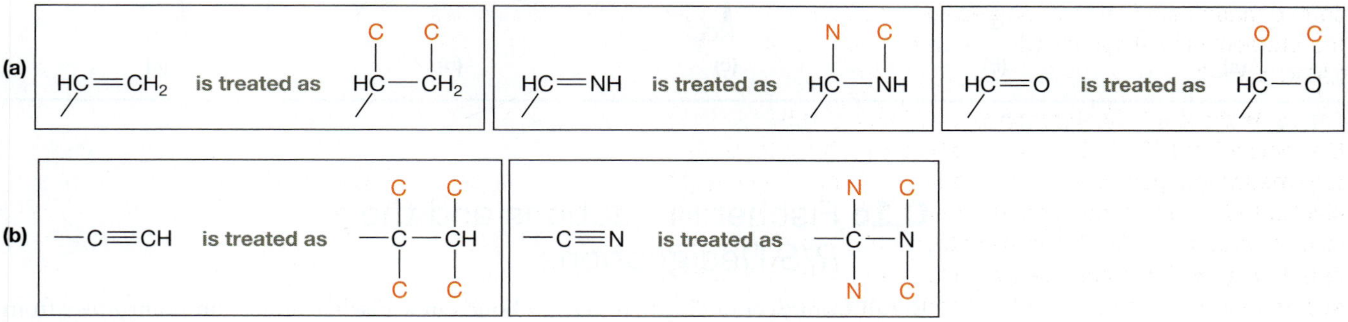

FIGURE C-2 Priorities of substituents with double and triple bonds (a) An atom that is doubly bonded to another atom is treated as having two single bonds to the atom—one real atom (black) and one imaginary (red). (b) An atom that is triply bonded to another atom is treated as having three single bonds to the atom—one real atom (black) and two imaginary (red).

variety of cases, where the atoms in red are the imaginary atoms that have been added.

With these rules in mind, let's determine the configuration of the asymmetric carbon in the following isomer of 3-methylhex-1-ene:

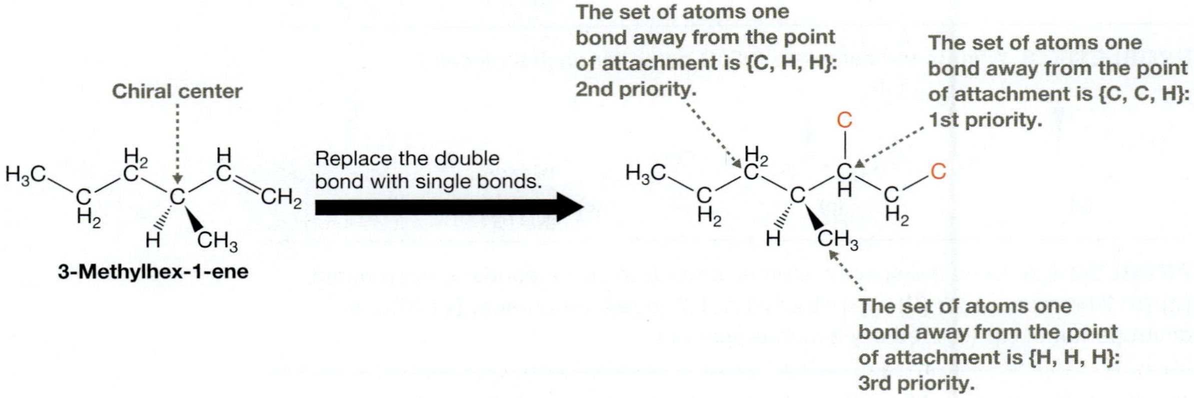

In the structure on the right above, the C=C double bond has been replaced by a single bond, and single bonds to two imaginary C atoms (red) have been added. The H atom is assigned fourth priority because the other three substituents are attached by C; C has a higher atomic number than H. The priorities of the remaining substituents are determined by the sets of atoms one bond away from the respective points of attachment: {H, H, H} for the CH₃ substituent, {C, H, H} for the CH₂CH₂CH₃ substituent, and {C, C, H} for the CH=CH₂ substituent. Therefore, CH=CH₂ is first priority, CH₂CH₂CH₃ is second priority, and CH₃ is third priority. Given that these substituents are arranged counterclockwise with the fourth-priority substituent pointing away, the configuration is assigned *S* and the complete name of the molecule is (*S*)-3-methylhex-1-ene.

PROBLEM C.10 What is the configuration, *R* or *S*, of the chiral center in each of the following molecules?

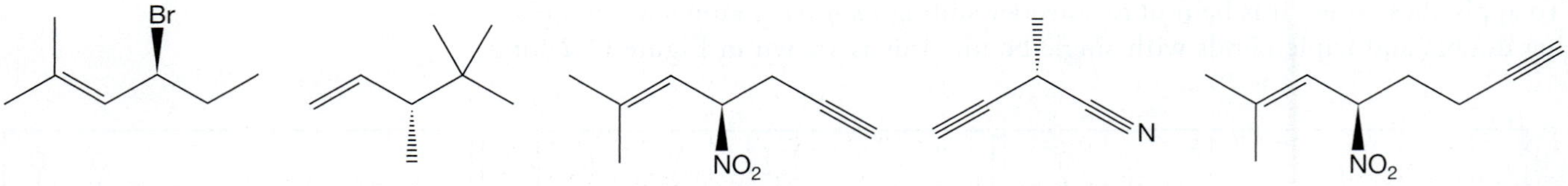

C.1c Fischer Projections and the *R/S* Designations

Recall from Section 5.6 that *vertical* bonds in a Fischer projection point *away* from you and horizontal bonds point *toward* you. Therefore, if the lowest-priority group is on a vertical bond, then the projection represents a properly oriented model. If the lowest-priority substituent is on a horizontal bond, instead, then it points *toward* you and, prior to assigning the configuration, we can simply reverse the order in which the first-, second-, and third-priority substituents are arranged. As an example, consider the Fischer projection shown below on the left.

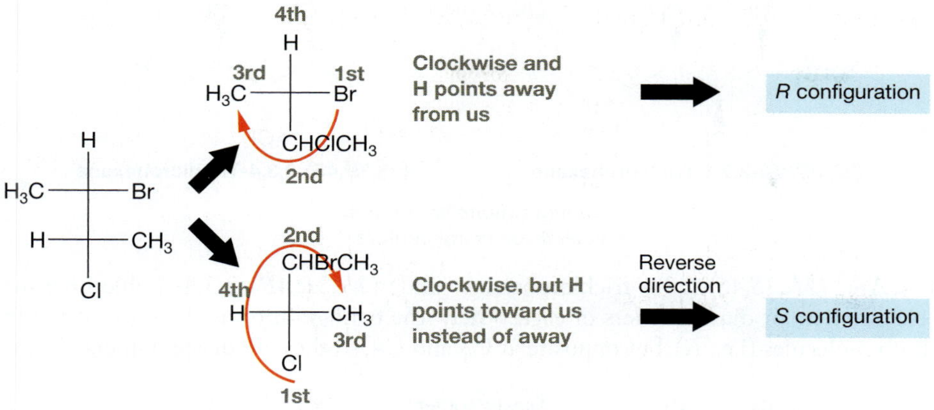

Each asymmetric carbon has its first-, second-, and third-priority groups arranged clockwise, as indicated above by the red curved arrows. In the top asymmetric carbon, the fourth-priority substituent is on a vertical bond, so it points away and we can simply assign the configuration as R. In the bottom asymmetric carbon, on the other hand, the fourth-priority substituent is on a horizontal bond, so it points toward us. We must therefore reverse the arrangement to counterclockwise, giving us the S configuration.

PROBLEM C.11 Determine the IUPAC name of each of the following compounds.

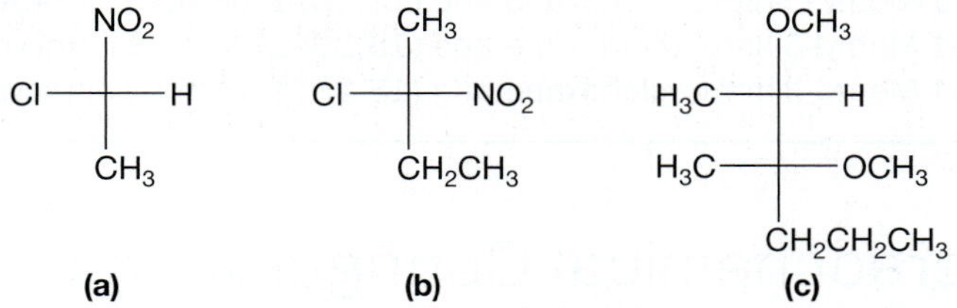

C.1d Using the *R/S* Designations to Identify Enantiomers and Diastereomers

The *R/S* designation in the names of molecules can be used to determine the specific relationship between a pair of configurational isomers—that is, whether they are enantiomers or diastereomers of each other. This is done by applying the following rules we learned in Section 5.5.

- If two stereoisomers are related by the inversion of some, but not all chiral centers, then they are *diastereomers* of each other.
- If two stereoisomers are related by the inversion of *all* chiral centers, then they are *enantiomers* of each other.

For example, (2*R*,3*S*,4*S*)-2,3,4-trichlorohexane and (2*S*,3*R*,4*R*)-2,3,4-trichlorohexane are enantiomers of each other because each chiral center's configuration in one is the reverse of the corresponding configuration in the other. In other words, wherever there is an R configuration in one, there is an S configuration in the other.

(2R,3S,4S)-2,3,4-Trichlorohexane   **(2S,3R,4R)-2,3,4-Trichlorohexane**

Configurations are opposite
at all three chiral centers.

Are (2R,3S,4S)-2,3,4-trichlorohexane and (2R,3R,4R)-2,3,4-trichlorohexane enantiomers or diastereomers of each other? The configuration is the same at C2 in both molecules (i.e., R), but opposite at C3 and C4, so they are diastereomers.

(2R,3S,4S)-2,3,4-Trichlorohexane   **(2R,3R,4R)-2,3,4-Trichlorohexane**

Configurations are opposite at
two of the three chiral centers.

PROBLEM C.12 Considering only the IUPAC names, determine whether each of the following pairs of molecules are enantiomers, diastereomers, or neither.
(a) (1R,3S)-1-Methyl-3-nitrocyclohexane and (1S,3R)-1-methyl-3-nitrocyclohexane
(b) (1R,3S)-1-Methyl-3-nitrocyclohexane and (1S,3S)-1-methyl-3-nitrocyclohexane
(c) (1R,3S)-1-Methyl-3-nitrocyclohexane and (1S,2R)-1-methyl-2-nitrocyclohexane

---

C.2 Stereochemical Configurations of Alkenes: *Z/E* Designations

In Chapter 5, we learned that molecules differing by the exchange of two substituents attached to an atom of a double bond are an example of diastereomers. We first encountered this idea of different configurations about double bonds in Chapter 3. There we learned we can designate these configurations as cis or trans if each atom connected by the double bond has one hydrogen atom attached to it. However, that requirement limits us to a relatively narrow set of cases. In this section, we present the Z/E system for configurations about double bonds, which allows us to assign a configuration to any double bond.

C.2a Determining the Configuration of a C═C Double Bond

To determine the configuration about a double bond, we apply the following rules.

Rules for Determining a Z or E Configuration
- Examine the pair of substituents attached to one atom of the double bond and, using the tiebreakers from Section C.1a, determine which substituent has the higher priority.

- Examine the pair of substituents on the other atom of the double bond and determine which substituent has the higher priority.
- If the two higher-priority groups are on the same side of the double bond, as shown at the right (top), then the alkene is assigned the *Z* configuration, and the molecule's name is given the prefix (*Z*).
- If the two higher-priority groups are on opposite sides of the double bond, as shown at the right (bottom), then the alkene is assigned the *E* configuration, and the molecule's name is given the prefix (*E*).

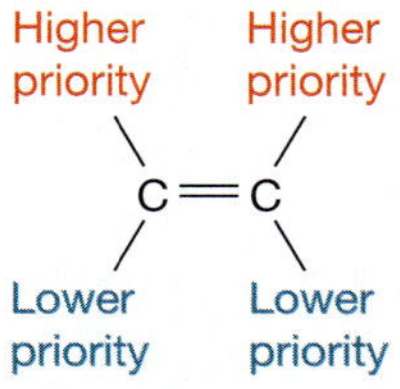

Higher priority   Higher priority

Lower priority   Lower priority

Z configuration

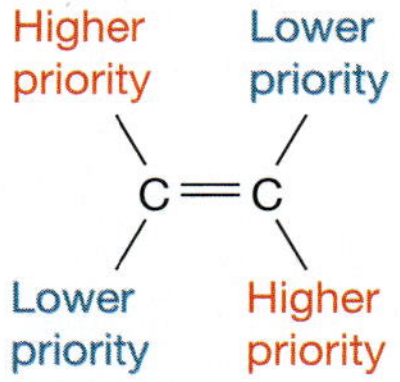

Higher priority   Lower priority

Lower priority   Higher priority

E configuration

The *Z*/*E* notation derives from German, in which *zusammen* means "together" and *entgegen* means "opposed." Alternatively, you can remember this mnemonic, noting whether the two higher-priority groups are on the "Zame" side of the double bond or on "Epposite" sides.

Let's see how this system is applied to the two isomers of pent-2-ene:

Higher-priority substituents are on the same side.

C wins over H.   C wins over H.

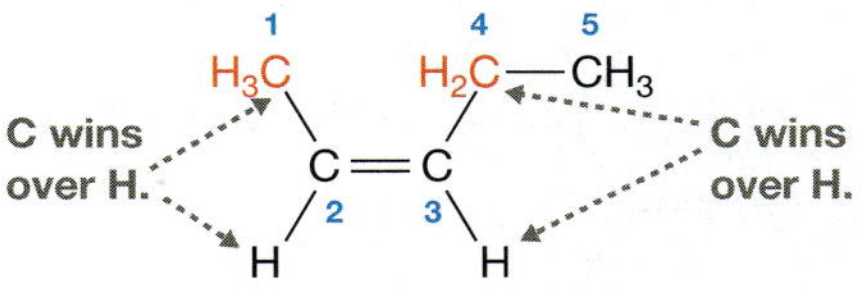

(*Z*)-Pent-2-ene

Higher-priority substituents are on opposite sides.

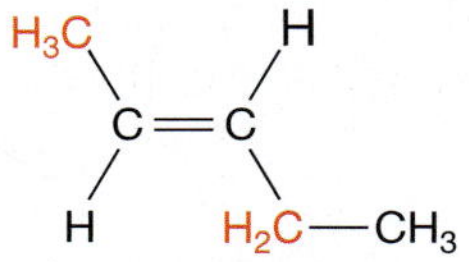

(*E*)-Pent-2-ene

Notice in the molecule above on the left that C2 and C3 are the atoms connected by the double bond. The substituents on C2 are H and CH_3, and CH_3 beats H because its atom at the point of attachment, C, has the higher atomic number. Similarly, on C3, the CH_2CH_3 substituent beats H. Notice in the isomer above on the left that the two higher-priority substituents are on the same side of the double bond, so the configuration is *Z*. Thus, the name of the molecule is (*Z*)-pent-2-ene. In the isomer above on the right, the two higher-priority substituents are on opposite sides of the double bond, so the configuration is *E* and the name of the molecule is (*E*)-pent-2-ene.

SOLVED PROBLEM C.13

What is the configuration, *Z* or *E*, of the double bond in the molecule at the right?

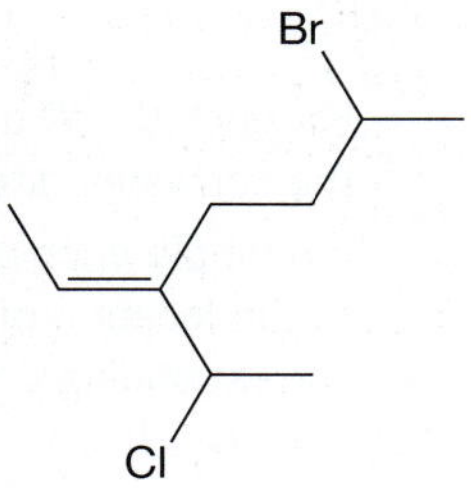

Think What are the two substituents attached to one atom of the double bond? Which of those substituents has the higher priority? What are the two substituents attached to the other atom of the double bond? Which of those substituents has the higher priority? Are the two higher-priority substituents on the same side or opposite sides of the double bond?

Solve The atoms connected by the double bond are C2 and C3 (see next page). The two substituents attached to C2 are H and CH_3. Of those, the CH_3 substituent has the higher priority because its atom at the point of attachment (C) has the higher atomic number. On C3, the two substituents are both attached by C, so to break the tie, we examine the set of atoms one bond farther away. For the Cl-containing substituent, that set of atoms is {Cl, C, H}, as shown on the next page, and for the Br-containing substituent it is {C, H, H}. Because Cl has a higher priority than C, the Cl-containing substituent is assigned the higher priority on C3.

Thus, the higher-priority substituents on C2 and C3 are on opposite sides of the double bond, making the configuration *E*.

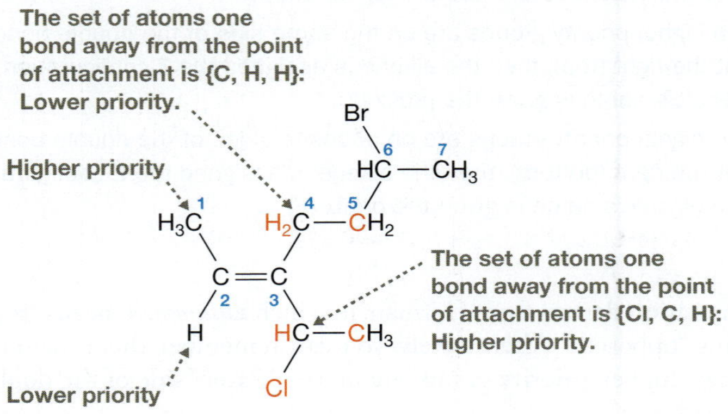

The set of atoms one bond away from the point of attachment is {C, H, H}: Lower priority.

Higher priority

The set of atoms one bond away from the point of attachment is {Cl, C, H}: Higher priority.

Lower priority

PROBLEM C.14 What is the configuration, *Z* or *E*, of each of the following double bonds?

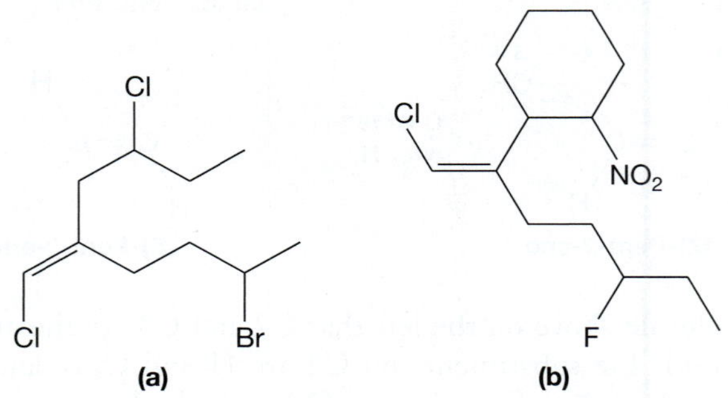

(a)

(b)

C.2b Assigning Configurations to Molecules with More than One C=C Group

When a molecule has two or more C=C groups that can be assigned either an *E* or a *Z* configuration, the name of the molecule must do so unambiguously.

- For each C=C double bond that can be assigned either an *E* or *Z* configuration, the corresponding designation must appear in the name.
- If multiple such designations appear in the name, the number that corresponds to the location of each double bond must appear immediately before the corresponding *E* or *Z* designation.

In the penta-1,3-diene isomer on the next page at the left, there are two double bonds but only one *E/Z* configuration is assigned. The double bond at C1 cannot have an *E/Z* configuration because two H atoms are bonded to C1. In the 1-chloropenta-1,3-diene isomer on the next page at the right, however, a *Z/E* designation is required for both double bonds. In this case, the double bond involving C1 has the *E* configuration and the one involving C3 has the *Z* configuration. In the IUPAC name, the number corresponding to each double bond appears immediately before its *Z/E* designation, even though those numbers are duplicated later in the name.

C3 is *E*; C1 is neither.   C1 is *E*; C3 is *Z*.

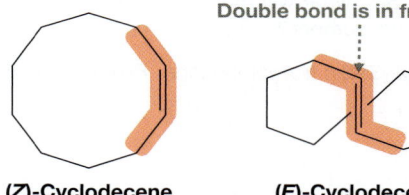

(*E*)-Penta-1,3-diene   **(1*E*,3*Z*)-1-Chloropenta-1,3-diene**

PROBLEM C.15 What is the complete IUPAC name for each of the following compounds?

(a)   (b)   (c)

C.2c Configurations of C=C Bonds as Part of Rings

A double bond that is part of a ring, such as that in cyclodecene, can have both *Z* and *E* configurations.

Double bond is in front of the single bond.

(*Z*)-Cyclodecene   **(*E*)-Cyclodecene**

For rings consisting of seven or fewer carbons, however, the *E* configuration about the double bond is unstable due to excessive ring strain (Section 4.4). Consequently, the *Z/E* convention for small rings is usually omitted; it is understood that the double bond is in the configuration in which the ring carbons are cis to each other.

PROBLEM C.16 Given each of the following names, draw the corresponding structure. **(a)** 3,3-dichlorocyclohexene; **(b)** 1-chlorocyclohexene; **(c)** (*E*)-4,4-dinitrocyclodecene; **(d)** (*R*)-3-fluorocycloheptene; **(e)** cyclohepta-1,3-diene

---

PROBLEM C.17 What is the complete IUPAC name for each of the following molecules?

(a)   (b)   (c)

Problems

C.1 *R* and *S* Configurations

C.18 Designate the configuration of each chiral center in the following molecules as *R* or *S*.

(a)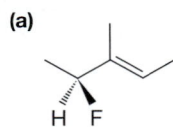

(b)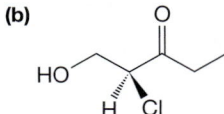

(c)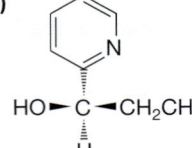

(d)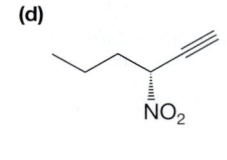

C.19 Assign the configuration to each asymmetric carbon as *R* or *S*.

(a)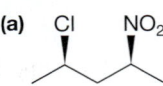

(b)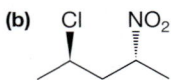

(c)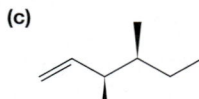

(d)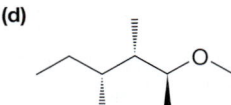

C.20 Assign the configuration to each asymmetric carbon as *R* or *S*.

(a)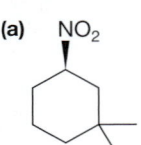

(b)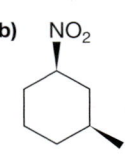

(c)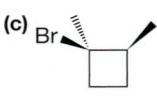

(d)

C.21 If the asymmetric carbons in a molecule have been designated (2*R*,3*S*,5*R*), what will the designations be in the molecule's enantiomer? What will the designations be in *one* of its diastereomers?

C.22 The IUPAC name of a particular stereoisomer is (*R*)-1-chloro-(*R*)-4-methyl-(*S*)-2-nitrocyclohexane. Rewrite this IUPAC name using another way to incorporate the *R* and *S* designations.

C.2 *E* and *Z* Configurations

C.23 What is the configuration, *Z* or *E*, of each of the following double bonds?

(a)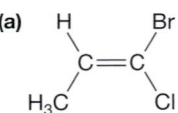

(b)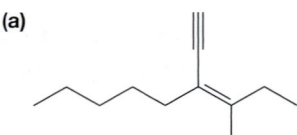

(c)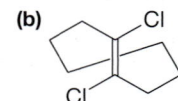

(d)

C.24 What is the configuration, *Z* or *E*, of each of the following double bonds?

C.25 What is the configuration, *Z* or *E*, of each of the following double bonds?

(a)

(b)

Note: Make a model; the diagonal bond is *behind* the double bond.

C.26 Assign the configuration *Z* or *E* to each double bond, where appropriate, in the following molecules.

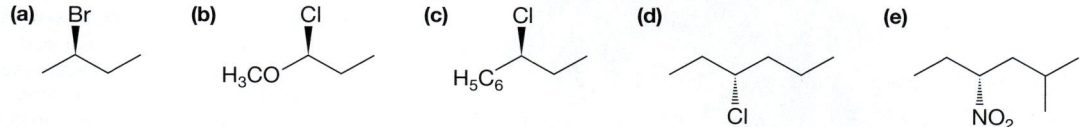

Integrated Problems

C.27 Determine the IUPAC name of each of the following molecules.

(a) ![structure] (b) ![structure] (c) ![structure] (d) ![structure] (e) ![structure]

C.28 Draw the structure of each of the following molecules. **(a)** (S)-2-chloro-(S)-3-ethoxypentane; **(b)** (4R,5S)-2,4-dimethyl-5-nitrohex-2-ene; **(c)** (4R,5R,6S)-4,5,6-trichloro-2-methyl-3-phenylhept-2-ene

C.29 Draw the structure of each molecule. **(a)** (S)-1-chloro-2,2-dimethyl-1-phenylcyclopentane; **(b)** (1R,2S)-1-methyl-1,2-dinitrocyclopropane; **(c)** (R)-4-ethoxycyclohexene; **(d)** (3S,4S)-3-chloro-4-fluoro-2-methylhepta-1,6-diene

C.30 Determine the IUPAC name of each of the following molecules.

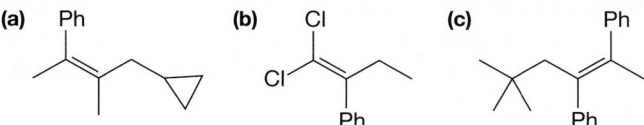

C.31 Draw the structure of each of the following molecules. **(a)** (R)-1-chloro-1-fluorobutane; **(b)** (S)-2-chloropentane; **(c)** (R)-2-chloro-2-methoxypentane; **(d)** (R)-2,2,3-trichlorobutane; **(e)** (S)-3-methylhexane; **(f)** (S)-2-bromo-1-nitropentane

C.32 Draw the structure of each of the following molecules. **(a)** (R)-3-chloropent-1-ene; **(b)** (2S,3S)-2-bromo-3-chloropentane; **(c)** (R)-1-bromo-(R)-2-iodocyclopentane; **(d)** (S)-3-chlorocyclohexene; **(e)** (1R,2S)-1,2-dibromocyclopentane

C.33 What is the complete IUPAC name for each of the following molecules?

(a) ![structure] (b) ![structure] (c) ![structure] (d) ![structure]

C.34 What is the complete IUPAC name for each of the following molecules?

(a) ![structure] (b) ![structure] (c) ![structure]

C.35 Given each of the following names, draw the corresponding structure. **(a)** (Z)-2-methoxypent-2-ene; **(b)** (E)-3-methylpent-2-ene; **(c)** (Z)-1-chloro-2-methylpent-1-ene; **(d)** (E)-2-chloro-3-methoxybut-2-ene; **(e)** (Z)-1-bromo-1-chloropent-1-ene; **(f)** (Z)-3-methylpent-2-ene

C.36 Given each of the following names, draw the corresponding structure. **(a)** (Z)-3-phenylhex-2-ene; **(b)** 1,2-dichlorocyclopentene; **(c)** (2E,4E)-2-ethoxyhexa-2,4-diene; **(d)** (1E,3E,5E)-1,3,4,6-tetrachlorohexa-1,3,5-triene

C.37 Determine whether each of the following names describes a single stereoisomer unambiguously. For each one that does, rewrite the name using the *R* and *S* designations for the chiral centers if appropriate. **(a)** *cis*-1,2-difluorocyclohexane; **(b)** *trans*-1,2-difluorocyclohexane; **(c)** *trans*-1,4-difluorocyclohexane; **(d)** *cis*-1-chloro-2-fluorocyclohexane; **(e)** *trans*-1,4-dimethylcycloheptane

6

This statue of George Washington (Washington Square Park, New York City) exhibits damage from acid rain. Sulfuric acid in rainwater (acid rain) reacts with calcium carbonate in the stone via a proton transfer reaction, the same type of reaction that is the primary focus of this chapter.

The Proton Transfer Reaction

An Introduction to Mechanisms, Thermodynamics, and Charge Stability

In Chapters 1 through 5, we focused primarily on structural aspects of atoms, molecules, and ions. We asked questions about the nature of the chemical bond, about how electrons are distributed in a given molecule or ion, and about the similarities and differences among the various types of isomers. Here in Chapter 6, we shift our focus to *chemical reactions*. A **chemical reaction** is the transformation of one substance (a reactant) into another substance (a product), typically through changes in chemical bonds. Although this definition is quite broad, in this book the term *reaction* will usually be applied only when the starting material is converted into another substance that can be isolated.

When we watch a chemical reaction take place in the laboratory, it may seem to proceed smoothly, with the reactants disappearing and the products appearing in one continuous process. If we could observe the reaction on the molecular level, however, we would see that it actually takes place in distinct events called **elementary steps**, each of which involves the breaking and/or formation of specific bonds. The precise sequence of elementary steps that results in the conversion of the original reactants to the final products is called the **mechanism** of the reaction. Depending on the nature of the particular reaction, the mechanism may consist of a single step (i.e., Reactants → Products) or multiple steps (e.g., Reactants → A → B → C → Products).

As a result, a reaction mechanism can be used to illustrate *how* a reaction takes place. Much more importantly, however, mechanisms can help us understand *why* a

Chapter Objectives

On completing Chapter 6 you should be able to:

1. Describe a proton transfer reaction and draw its curved arrow notation, given the reactants and products.

2. Define the equilibrium constant, K_{eq}, and write the expression for the acidity constant, K_a, for any acid, and determine the relative strengths of acids from their K_a or pK_a values.

3. Determine which side of a proton transfer reaction is favored given only the pK_a values of the acids involved.

4. Establish whether a solvent is suitable for a reactant, with respect to the leveling effect.

5. Calculate the percent dissociation of an acid at various solution pH values given the acid's pK_a.

6. Estimate an acid's pK_a value by comparing its structure to that of an acid with known pK_a.

7. Calculate ΔG°_{rxn} for a reaction from its K_a value and explain how ΔG°_{rxn} relates to spontaneity and to the relative stability of products versus reactants.

8. Interpret a free energy diagram to identify the locations of reactants, products, and transition states and infer relative values for ΔG°_{rxn} and $\Delta G^{\circ\ddagger}$.

9. Determine relative stabilities of ions from the pK_a values of acids.

10. Explain how the stability of an ion is governed by the identity and hybridization of the atom bearing the charge, the extent of charge delocalization via resonance, and inductive effects from nearby electron-donating and electron-withdrawing substituents.

11. Predict relative acid and base strengths based on charge stability.

12. Predict the relative contribution of a resonance structure toward the resonance hybrid on the basis of charge stability.

reaction takes place, and ultimately can help us make *predictions* about reactions we may not have seen before. Therefore, each time a new reaction is introduced, we will focus on its mechanism.

Another important aspect of reaction mechanisms is the way in which they simplify organic chemistry. As we will see throughout the rest of the book, there are many seemingly different organic reactions. However, the mechanisms for most of those reactions are constructed from just a few types of elementary steps. Here in Chapter 6, we examine one such elementary step—the *proton transfer*—and in Chapter 7 we examine others.

We begin with the proton transfer reaction because it is one of the simplest reactions that you will encounter in organic chemistry and you are likely already familiar with it from general chemistry. Many of the concepts we learn here will be applied to the other elementary steps discussed in Chapter 7.

6.1 An Introduction to Reaction Mechanisms: The Proton Transfer Reaction and Curved Arrow Notation

A **proton transfer reaction** (or a **Brønsted–Lowry acid–base reaction**) is one in which a *Brønsted–Lowry base* reacts with a *Brønsted–Lowry acid*:

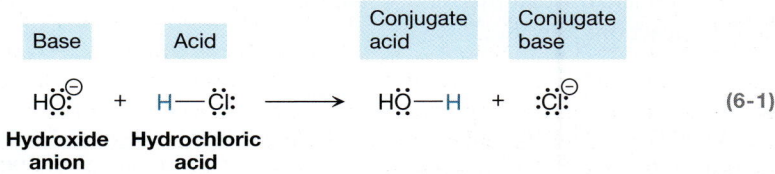

In Equation 6-1, HO^- is the **Brønsted–Lowry base** because it *accepts a proton* (H^+), and HCl is the **Brønsted–Lowry acid** because it *donates a proton*. Overall, a proton

(H^+), highlighted in blue, is transferred from the acid to the base; the formal charge of the base, therefore, increases by 1, whereas the formal charge of the acid decreases by 1. The **conjugate acid** is the species that the base becomes after picking up a proton, and the **conjugate base** is the species that the acid becomes after losing a proton.

Proton transfer reactions take place as a single event, which is to say:

> A proton transfer reaction consists of a single *elementary step*.

As a result, all of the changes that occur in the reaction at the molecular level do so simultaneously; they are said to be **concerted**.

Organic chemists follow these changes by keeping track of the arrangement of the electrons. As shown in Equation 6-2, a single bond between H and Cl in the reactants is broken, while a single bond between H and O in the products is formed:

$$\text{HO}^{\ominus} \ + \ \text{H}\!-\!\text{Cl:} \ \longrightarrow \ \text{HO}\!-\!\text{H} \ + \ \text{:Cl:}^{\ominus} \tag{6-2}$$

Bond is broken.   Bond is formed.

Moreover, organic chemists use **curved arrow notation** (also called **arrow pushing**) to keep track of the valence electrons as they move throughout a mechanism. There are essentially four rules to using curved arrow notation:

> 1. A curved arrow represents the movement of *valence electrons*, not *atoms*.
> 2. Each *double-barbed curved arrow* (⌒) represents the movement of two valence electrons.
> 3. To show bond breaking, the tail of the arrow originates from the center of a bond.
> 4. To show bond formation, the head of the arrow points to:
>   • an atom if the new bond is a σ bond, or
>   • the region where the bond is formed if the new bond is a π bond.

The curved arrow notation for the reaction in Equation 6-1 is illustrated in Equation 6-3.

$$\text{HO}^{\ominus} \ + \ \text{H}\!-\!\text{Cl:} \ \longrightarrow \ \text{HO}\!-\!\text{H} \ + \ \text{:Cl:}^{\ominus} \tag{6-3}$$

Two curved arrows are required because the reaction involves a total of four electrons. The curved arrow on the left uses a lone pair on O to form an O—H bond. The curved arrow on the right represents the breaking of the H—Cl bond. The pair of electrons from that bond ends up as an additional lone pair on Cl. You will learn in Solved Problem 6.1 that this set of curved arrows can be used to describe other proton transfer reactions as well.

YOUR TURN 6.1

> In Equation 6-3, circle all of the electrons that are involved in the chemical reaction. Label each curved arrow as either "bond breaking" or "bond formation."
>
> *Answers to Your Turns are in the back of the book.*

SOLVED PROBLEM 6.1

Draw the curved arrow notation for the proton transfer between ammonia (NH_3) and water, where water acts as the acid and ammonia acts as the base.

Think What does it mean to be an acid? A base? What are the important electrons to keep track of during the course of the reaction? What bonds are broken? What bonds are formed?

Solve Because water acts as an acid (a proton donor) and ammonia acts as a base (a proton acceptor), we may write the reaction as follows:

Acid = H⁺ donor   Conjugate base
Base = H⁺ acceptor   Conjugate acid

$$H_2\ddot{O} \; + \; \ddot{N}H_3 \longrightarrow H\ddot{O}^{\ominus} \; + \; {}^{\oplus}NH_4$$

Bond breaking and bond formation involve only valence electrons, so draw all valence electrons in the two reactants. That way, we can clearly see which electrons are involved in the reaction, both from the reactants and from the products.

Becomes a lone pair on O   Becomes a bond to H

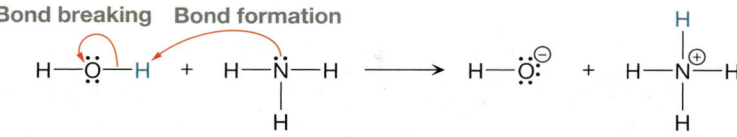

It then becomes a matter of using curved arrows to show the appropriate movement of those electrons. One curved arrow is drawn from the lone pair on N to the H on water to illustrate the formation of the H—N bond. A second curved arrow originates from the center of the O—H bond to illustrate the breaking of that bond, and points to the O atom, showing that those electrons end up as a lone pair on O.

Bond breaking   Bond formation

$$H-\ddot{O}-H \; + \; H-\overset{\underset{|}{H}}{N}-H \longrightarrow H-\ddot{O}^{\ominus} \; + \; H-\overset{\overset{H}{|}}{\underset{\underset{H}{|}}{N}}{}^{\oplus}-H$$

PROBLEM 6.2 Draw the curved arrow notation for the reverse of the reaction in Solved Problem 6.1.

PROBLEM 6.3 Draw the curved arrow notation for the proton transfer reaction between NH_3 and H_2O, in which NH_3 acts as the acid and H_2O acts as the base.

6.2 Chemical Equilibrium and the Equilibrium Constant, K_{eq}

Even though a proton transfer reaction can be *written* for any combination of acid and base, not every reaction will actually occur to a significant extent. A proton transfer reaction between HCl and HO^- will take place readily, for example, producing Cl^- and H_2O, but essentially no proton transfer will occur between HO^- and NH_3.

A reaction's tendency to form products is described by its **equilibrium constant**, K_{eq}. An equilibrium constant for a given reaction can be obtained experimentally by allowing that reaction to come to equilibrium (the point at which the net concentrations of reactants and products no longer change) and then measuring the concentrations of all products and all reactants. Those equilibrium concentrations are then substituted into the reaction's equilibrium constant expression, which is shown for a generic reaction in Equation 6-4.

$$a\,A + b\,B + c\,C + \cdots \rightleftharpoons w\,W + x\,X + y\,Y + \cdots$$

$$K_{eq} = \frac{[W]_{eq}^{w}[X]_{eq}^{x}[Y]_{eq}^{y}\cdots}{[A]_{eq}^{a}[B]_{eq}^{b}[C]_{eq}^{c}\cdots} \tag{6-4}$$

In this reaction, the uppercase letters are the reactants and products, and the lowercase letters are the stoichiometric coefficients used to balance the equation. The numerator of the K_{eq} expression contains the equilibrium concentrations of the products, whereas the denominator contains the equilibrium concentrations of the reactants. Each concentration is raised to a power specified by the exponent corresponding to its stoichiometric coefficient (i.e., $a, b, c, \ldots$ and $w, x, y, \ldots$).

For a proton transfer reaction between HA (a generic acid) and B^{-} (a generic base), the equilibrium constant expression is written as in Equation 6-5.

$$HA + B^{-} \rightleftharpoons A^{-} + HB \qquad K_{eq} = \frac{[A^{-}]_{eq}[HB]_{eq}}{[HA]_{eq}[B^{-}]_{eq}} \tag{6-5}$$

If the equilibrium concentrations of the products are high (and hence the equilibrium concentrations of the reactants are low), then the value of the numerator will be large and the value of the denominator will be small. The converse is true, as well. Thus:

- A very large K_{eq} value (e.g., 10^{10}) heavily favors products.
- A very small K_{eq} value (e.g., 10^{-10}) heavily favors reactants.
- A K_{eq} value close to 1 favors significant concentrations of both reactants and products.

YOUR TURN 6.2

Add the curved arrow notation to each of the following reactions. Based on the K_{eq} values, which reaction tends to form more products at equilibrium?

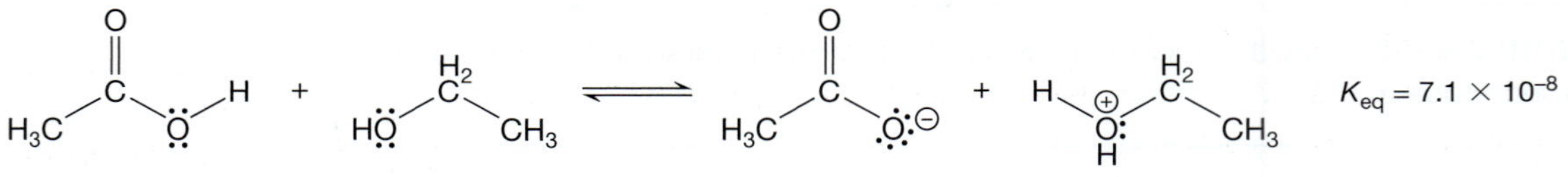

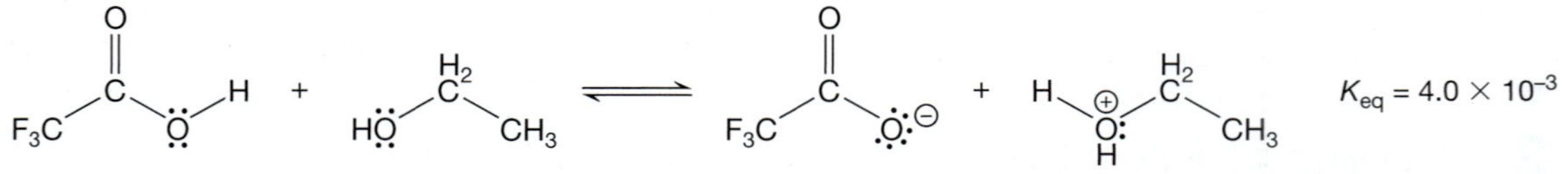

6.2a Acid Strengths: K_a and pK_a

The strength of an acid is defined as its *tendency to donate a proton*. Acid strengths can be obtained experimentally via the equilibrium that is established between a particular acid (HA) and water, producing the acid's conjugate base (A^{-}) and the *hydronium ion* (H$_3$O^{+}), as shown in Equation 6-6. This equilibrium is convenient because water acts as the base *and* the solvent.

$$HA(aq) + H_2O(\ell) \rightleftharpoons A^-(aq) + H\!\!-\!\!OH_2^+(aq) \qquad K_{eq} = \frac{[A^-]_{eq}[H_3O^+]_{eq}}{[HA]_{eq}[H_2O]_{eq}} \qquad \text{(6-6)}$$

Water is in such a huge excess in this equilibrium that its concentration is effectively constant, at 55.5 mol/L. Therefore, $[H_2O]_{eq}$ can be eliminated from the equilibrium expression without any loss of information, yielding what is called the **acidity constant, K_a.**

$$K_a = \frac{[A^-]_{eq}[H_3O^+]_{eq}}{[HA]_{eq}}$$

Chemists generally prefer to work with values of K_a instead of K_{eq}.

Because all K_a values are obtained with water acting as the base, any difference in K_a for two compounds reflects a difference in the strength of the acid.

When comparing two acids, the one with the larger K_a value is the stronger acid.

Values of K_a for several common acids encountered in organic chemistry are listed in Table 6-1. (A more comprehensive table can be found in Appendix A.) Notice the incredibly wide range of values—from about 10^{-50} for ethane (CH_3CH_3) to about 10^{13} for trifluoromethanesulfonic acid (CF_3SO_3H). In other words, CF_3SO_3H is 10^{63} times stronger than CH_3CH_3—a number that dwarfs even Avogadro's number (6.02×10^{23})!

Due to the immensely large range of values for K_a, chemists frequently work with values of **pK_a,** which is related to K_a through Equation 6-7.

$$pK_a = -\log K_a \qquad \text{(6-7)}$$

The log function can make very large and very small numbers easier to work with. For example, the pK_a values for the acids in Table 6-1 range from 50 to -13, a difference of only 63.

The negative sign in front of the log function in Equation 6-7 has the effect of reversing the relative values of K_a and pK_a. That is, a higher (i.e., more positive) value of K_a corresponds to a lower (i.e., more negative) value of pK_a. Therefore, when comparing pK_a values of two different acids:

- The acid with the lower pK_a value is the stronger acid.
- Each difference of 1 in pK_a values represents a factor of 10 difference in acid strength.

SOLVED PROBLEM 6.4

Which of these is a stronger acid? How much stronger (i.e., by what factor?)

Think Is the stronger acid the one with the higher or lower pK_a value? What is the difference in their pK_a values and how can that difference be used to calculate the difference in their acid strengths?

4-Methylphenol
$pK_a = 10.26$

4-Chlorophenol
$pK_a = 9.43$

Solve The compound with the lower pK_a value, 4-chlorophenol, is the stronger acid. The difference in pK_a values is $10.26 - 9.43 = 0.83$, which corresponds to a difference in acid strength of $10^{0.83} = 6.8$. Thus, 4-chlorophenol is 6.8 times stronger an acid than 4-methylphenol.

CONNECTIONS
4-Methylphenol, also called *para*-cresol, is one of the compounds responsible for the odor of pigs and is also found in human sweat. One of the main uses of 4-methylphenol is in the production of antioxidants.

TABLE 6-1 Values of K_a and pK_a for Various Acids[a]

Acid	Conjugate Base	K_a	pK_a	Acid	Conjugate Base	K_a	pK_a
F₃C—S—OH (Trifluoromethanesulfonic acid)	F₃C—S—O⁻	1×10^{13}	−13	CH₃OH Methanol	CH₃O⁻	3.2×10^{-16}	15.5
HO—S—OH (Sulfuric acid)	HO—S—O⁻	1×10^{9}	−9	H₂O Water	HO⁻	2×10^{-16}	15.7
HCl Hydrochloric acid	Cl⁻	1×10^{7}	−7	Ethanol (OH)	(O⁻)	1×10^{-16}	16
H₃O⁺ Hydronium ion	H₂O	55	−1.7	Propan-2-ol (Isopropyl alcohol)		3.2×10^{-17}	16.5
Cl₃C—OH Trichloroethanoic acid (Trichloroacetic acid)	Cl₃C—O⁻	0.17	0.77	Methylpropan-2-ol (tert-Butyl alcohol)		1×10^{-19}	19
HF Hydrofluoric acid	F⁻	6.3×10^{-4}	3.2	Propanone (Acetone)		1×10^{-20}	20
Benzoic acid		6.3×10^{-5}	4.2	HC≡CH Ethyne (Acetylene)	HC≡C⁻	1×10^{-25}	25
Ethanoic acid (Acetic acid)		1.8×10^{-5}	4.75	Aniline (Phenylamine) NH₂	NH⁻	1×10^{-27}	27
H₂S Hydrogen sulfide	HS⁻	6.3×10^{-8}	7.2	H₂ Hydrogen gas	H⁻	1×10^{-35}	35
H₄N⁺ Ammonium ion	NH₃	4×10^{-10}	9.4	N-Methylmethanamine (Dimethylamine) H₃C—N(H)—CH₃	H₃C—N⁻—CH₃	1×10^{-38}	38
Phenol (OH)	(O⁻)	1×10^{-10}	10.0	H₂C=CH₂ Ethene (Ethylene)	H₂C=CH⁻	1×10^{-44}	44
H₃C—NH₃⁺ Methylammonium ion	H₃C—NH₂	2.3×10^{-11}	10.63	Ethoxyethane (Diethyl ether)		$\sim 1 \times 10^{-45}$	~45
2,2,2-Trifluoroethanol (CF₃CH₂OH)	(CF₃CH₂O⁻)	4×10^{-13}	12.4	CH₄ Methane	H₃C⁻	1×10^{-48}	48
2-Chloroethanol (ClCH₂CH₂OH)	(ClCH₂CH₂O⁻)	5.0×10^{-15}	14.3	CH₃CH₃ Ethane	CH₃CH₂⁻	1×10^{-50}	50

[a] pK_a = −log K_a. The less positive (or more negative) the pK_a value, the stronger the acid relative to another acid.

PROBLEM 6.5 Which is a stronger acid, phenol or 4-methylphenol? By what numerical factor? *Hint:* Consult Table 6-1 and Solved Problem 6.4.

Phenol

4-Methylphenol

To determine relative *base* strengths, compare the pK_a values of the bases' respective *conjugate acids* and apply the following rule:

The strength of a base *decreases* as the strength of its conjugate acid *increases*.

For example, what are the relative base strengths of the hydroxide anion (HO^-) and ammonia (H_3N)? The conjugate acids of these bases are H_2O and H_4N^+, whose pK_a values are 15.7 and 9.4, respectively (Table 6-1). Because H_4N^+ is the stronger of these two acids (it has the lower pK_a value), H_3N must be a weaker base than HO^-. Furthermore, because the difference in those pK_a values is $15.7 - 9.4 = 6.3$, H_3N is weaker by a factor of $10^{6.3} = 2.0 \times 10^6$.

PROBLEM 6.6 Which is a stronger base, Cl^- or the phenoxide anion ($C_6H_5O^-$)? By what numerical factor? Explain. *Hint:* Consult Table 6-1.

6.2b Predicting the Outcome of a Proton Transfer Reaction Using pK_a Values

Although pK_a values can be found for a wide variety of acids, many proton transfer reactions do not involve water acting as a base. In these cases, pK_a values can still be used to predict the outcome of a proton transfer reaction—that is, to determine which side of the reaction is favored at equilibrium and by how much.

To make use of pK_a values in this way, keep in mind that there are two acids involved in an equilibrium for a proton transfer reaction: the acid on the left side of the equation (a reactant) and the conjugate acid of the base on the right side (a product) (Equation 6-8).

HA drives the reaction to the right.   HB drives the reaction to the left.

$$B:^{\ominus} \quad + \quad H-A \quad \rightleftharpoons \quad B-H \quad + \quad :A^{\ominus} \qquad (6\text{-}8)$$

Base   **Acid**   **Conjugate acid**   **Conjugate base**

The two acids can be thought of as being in competition with each other. The acid on the left drives the reaction to the right, while the acid on the right drives the reaction to the left. The outcome of this competition is dictated by the stronger acid:

- Proton transfer reactions favor the side *opposite* the stronger acid (i.e., opposite the lower pK_a value).
- Each difference of 1 in pK_a values between the two acids corresponds to a power of 10 by which that side of the reaction is favored.

For example, consider the reaction in which CH_3CO_2H donates a proton to CH_3NH_2 (Equation 6-9).

$$pK_a = 4.75$$

Equilibrium favors the side
opposite the stronger acid.

Product side favored
by 7.6×10^5

$$pK_a = 10.63$$

CONNECTIONS Methanamine $(CH_3NH_2$, Equation 6-9) is a common reagent used in the production of a wide variety of compounds, including some pesticides, pharmaceuticals, and surfactants. Biologically, methanamine is produced in putrefaction, one of the stages in the decomposition of dead animals.

The conjugate acid on the product side is $CH_3NH_3^+$ and the conjugate base is $CH_3CO_2^-$. The pK_a of CH_3CO_2H is 4.75 and the pK_a of $CH_3NH_3^+$ is 10.63 (Table 6-1). Because CH_3CO_2H has the lower pK_a value, it is a stronger acid than $CH_3NH_3^+$. Thus, the product side of this proton transfer reaction is favored over the reactant side. This is signified in Equation 6-9 with an equilibrium arrow that is longer in the forward direction than in the reverse direction ($\rightleftharpoons$).

To determine the extent to which the product side is favored, notice that the pK_a values of the acids, 10.63 and 4.75, differ by 5.88. Therefore, the product side is favored by 5.88 powers of 10, or $10^{5.88} = 7.6 \times 10^5$.

SOLVED PROBLEM 6.7

Predict which side of the following reaction is favored. By what numerical factor is that side favored?

Think What acid is present on each side of the reaction? Which one is stronger? What is the difference in their pK_a values?

Solve The acid on the left is propanone (acetone), $(CH_3)_2C{=}O$, whose pK_a value is 20. The conjugate acid on the right is 2-methylpropan-2-ol (*tert*-butyl alcohol), $(CH_3)_3COH$, whose pK_a is 19. Therefore, the acid on the right is stronger (lower pK_a), making the left side of the equilibrium favored. The difference between the two pK_a values is 1, so the left side of the reaction is favored over the right by a factor of 10^1, or simply 10.

PROBLEM 6.8 Repeat Solved Problem 6.7 using $(CH_3)_2N^-$ as the base.

6.2c The Leveling Effect

When there is a *desired reaction* to carry out, R → P, we usually do not want the solvent to react with the species we add, R. If the species R is too strong as an acid or base, however, it can react with the solvent in an *undesired proton transfer reaction*, in which case R would not remain intact to undergo the desired reaction to produce P. Such an undesired proton transfer reaction is the result of what is called the **leveling effect**, which can be stated as follows:

- At equilibrium, the strongest acid that can exist in solution to any appreciable concentration is the *protonated solvent*.
- The strongest base that can exist in solution is the *deprotonated solvent*.

In water, for example, H_3O^+ is the strongest acid that can exist and HO^- is the strongest base that can exist.

To see why this is so, let's examine what would happen if an acid stronger than H_3O^+, such as HCl, were placed in water. Being an acid, HCl could donate its proton to water according to the proton transfer reaction in Equation 6-10:

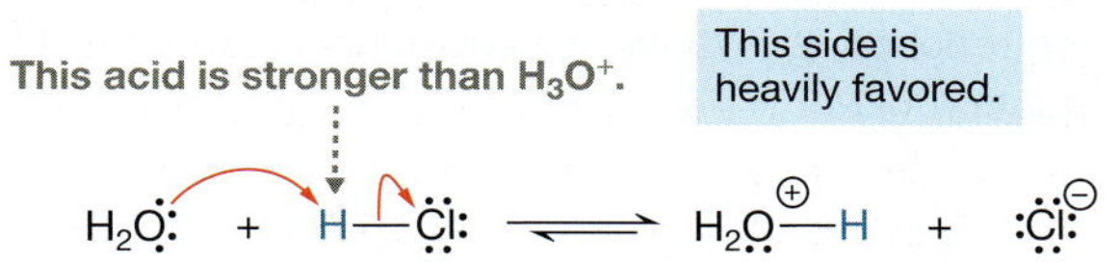

$$H_2\ddot{O}: \; + \; H{-}\ddot{\underset{..}{C}}l: \; \rightleftharpoons \; H_2\overset{\oplus}{O}{-}H \; + \; :\ddot{\underset{..}{C}}l:^{\ominus} \tag{6-10}$$

This acid is stronger than H_3O^+. · This side is heavily favored.

Because HCl is a significantly stronger acid than H_3O^+, HCl wins the competition and the product side is heavily favored. This means that, when HCl is added to water, HCl will be consumed by the above reaction, producing H_3O^+ instead.

YOUR TURN 6.3

What are the pK_a values of HCl and H_3O^+ in Equation 6-10, and do they verify that HCl is a stronger acid than H_3O^+?

A similar story happens when a base stronger than HO^-, such as $(CH_3)_2N^-$, is added to water. Being a base, $(CH_3)_2N^-$ could remove a proton from water, according to the proton transfer reaction in Equation 6-11.

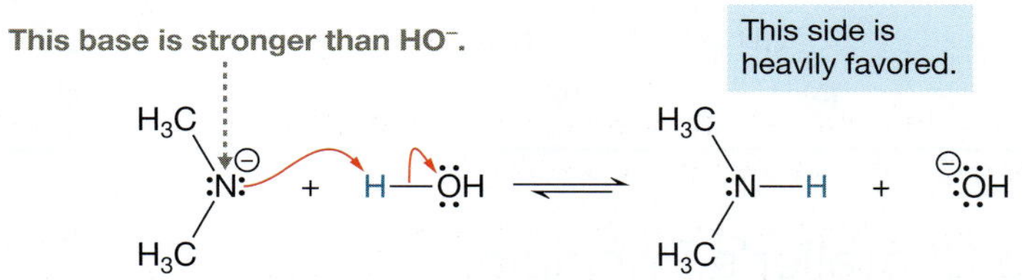

$$\underset{H_3C}{\overset{H_3C}{{>}}}\ddot{N}:^{\ominus} \; + \; H{-}\ddot{O}H \; \rightleftharpoons \; \underset{H_3C}{\overset{H_3C}{{>}}}N{-}H \; + \; ^{\ominus}:\ddot{O}H \tag{6-11}$$

This base is stronger than HO^-. · This side is heavily favored.

$(CH_3)_2N^-$ is a much stronger base than HO^-, consistent with the fact that H_2O is a much stronger acid than $(CH_3)_2NH$, so the product side of Equation 6-11 is heavily favored. Therefore, when $(CH_3)_2N^-$ is added to water, it is consumed by the above reaction, producing HO^- instead.

YOUR TURN 6.4

What are the pK_a values of H_2O and $(CH_3)_2NH$ in Equation 6-11, and do they verify that H_2O is a stronger acid than $(CH_3)_2NH$?

The important lesson to take away from these examples is that, if we want to carry out a particular reaction involving either HCl or $(CH_3)_2N^-$ as a reactant, then water is a poor choice of solvent because each of these species will undergo an *undesired proton transfer* with water. We would say that, with respect to the leveling effect, water would be an *unsuitable* solvent. In general:

With respect to the leveling effect, a solvent is unsuitable for a particular reactant R if:

- R is a stronger acid than the solvent's conjugate acid (i.e., R has the lower pK_a).
- R is a stronger base than the solvent's conjugate base (i.e., the conjugate acid of R has a higher pK_a than the solvent).

If one solvent is unsuitable with respect to the leveling effect, we would need to choose another solvent. For example, diethyl ether ($CH_3CH_2OCH_2CH_3$) would be a suitable solvent for $(CH_3)_2N^-$. (See Your Turn 6.5.)

YOUR TURN 6.5

Verify the preceding statement that diethyl ether would be a suitable solvent for $(CH_3)_2N^-$. To do so, use Table 6-1 to fill in the boxes below with the appropriate pK_a values and label which one is the stronger acid. Indicate which side of the reaction is favored at equilibrium. Is this the same side that is favored in Equation 6-11?

PROBLEM 6.9 With respect to the leveling effect, determine whether each of the following solvents would be suitable for a reaction involving $HC\equiv C:^{\ominus}$ as a reactant. *Hint:* You might need pK_a values from Table 6-1 or Appendix A.

H_2O OH   NH₂ (over C=O)   S (over O)   O

(a)   (b)   (c)   (d)   (e)

6.2d Le Châtelier's Principle, pH, and Percent Dissociation

A chemical equilibrium is a *stationary state* of a reaction, in which the net concentrations of reactants and products do not change with time. Even after equilibrium has been established, however, the reaction can be driven in the forward or reverse direction by altering the concentrations of the species involved in the reaction. This response by a chemical equilibrium is one consequence of **Le Châtelier's principle**, which states that *if a reaction at equilibrium experiences a change in reaction conditions (e.g., concentrations, temperature, pressure, or volume), then the equilibrium will shift to counteract that change.* Specifically:

- A reaction at equilibrium can be shifted in the forward direction (i.e., to form more products) by the addition of reactants or the removal of products.
- A reaction at equilibrium can be shifted in the reverse direction (i.e., to form more reactants) by the addition of products or the removal of reactants.

In an acid's equilibrium with water (Equation 6-6, p. 279), for example, the concentration of $H_3O^+(aq)$—a product of the reaction—directly affects the relative amounts of HA and A^- at equilibrium. Increasing the concentration of H_3O^+ (e.g., by adding a strong acid) will cause the equilibrium to shift toward reactants. At the new equilibrium, there will be more HA and less A^-, so the **percent dissociation** of HA (Equation 6-12) is said to decrease.

$$\% \text{ Dissociation} = \frac{[A^-]_{eq}}{[HA]_{initial}} \times 100 = \frac{[A^-]_{eq}}{[HA]_{eq} + [A^-]_{eq}} \times 100 \qquad (6\text{-}12)$$

Likewise, decreasing the concentration of H_3O^+ (e.g., by adding a strong base) will cause the equilibrium to shift toward products, leading to an increase in the percent dissociation of HA.

The concentration of H_3O^+ in solution is commonly reported as pH:

$$pH = -\log[H_3O^+(aq)] \qquad (6\text{-}13)$$

A solution is considered *acidic* if its pH is <7, *basic* if its pH is >7, and *neutral* if its pH equals 7. Furthermore, pH increases as the concentration of H_3O^+ decreases. Thus:

> In the equilibrium between an acid (HA) and water, increasing the solution's pH (i.e., decreasing $[H_3O^+]$) causes an increase in the percent dissociation of the acid into its conjugate base (A^-).

This can be seen graphically in Figure 6-1.

Of particular significance in Figure 6-1 is the pH at which the acid dissociates 50%. This significance can best be seen with the **Henderson–Hasselbalch equation** (Equation 6-14), which is commonly applied in general chemistry toward problems dealing with buffers:

Henderson–Hasselbalch equation

$$pH = pK_a + \log\left(\frac{[A^-(aq)]_{eq}}{[HA(aq)]_{eq}}\right) \qquad (6\text{-}14)$$

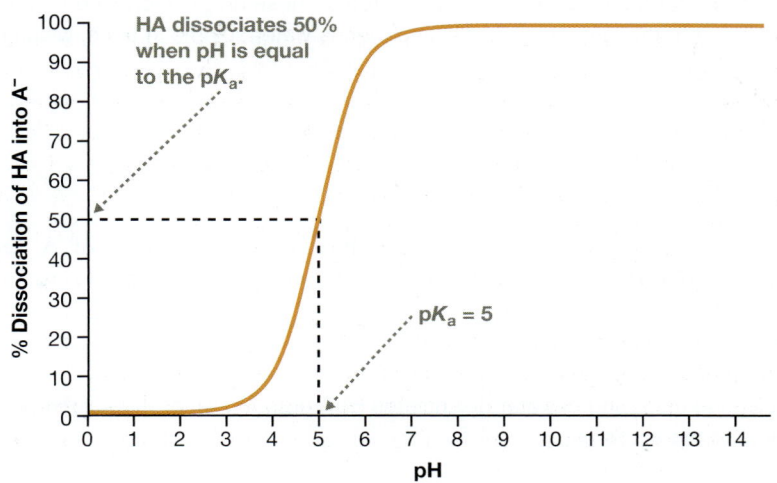

FIGURE 6-1 Percent dissociation of an acid with $pK_a = 5$ as a function of pH As the pH of the solution increases (i.e., as the solution becomes more basic), the percent dissociation of the acid (HA) increases. When the pH of the solution is equal to the pK_a of the acid, the acid has dissociated 50%.

pK_a and the Absorption and Secretion of Drugs

Ibuprofen and morphine are both drugs that are taken to relieve pain, but when they are taken orally, they have different routes for absorption into the blood. Ibuprofen is absorbed when it is in the stomach, whereas morphine is absorbed in the upper bowel. Their excretion routes are different, too. Whereas ibuprofen is excreted in the urine, morphine is excreted via the stomach. What gives rise to these different absorption and excretion routes?

Ibuprofen — pK_a = 4.9

Morphine — pK_a = 8.2

The answer has to do with the different pK_a values of the two drugs. Ibuprofen has a CO_2H group and its pK_a is 4.9. Morphine, on the other hand, has an R_3NH^+ group, whose pK_a is 8.2. In the stomach, where the pH of gastric juice is roughly 2, both drugs retain their proton because their respective pK_a values are above the pH. Ibuprofen, therefore, is uncharged in the stomach, which allows it to diffuse across a cell's lipophilic membrane. Morphine, on the other hand, is positively charged in the stomach, which makes it hydrophilic. Morphine therefore remains soluble in the aqueous environment of the stomach until it reaches the large intestine, where the pH can be as high as about 7. With the higher pH, a greater percentage of morphine exists in the deprotonated, uncharged state, in which case it can diffuse across the cell membrane.

The same ideas account for the different excretion routes of these drugs. Morphine is water soluble in the stomach due to the low pH, whereas ibuprofen is water soluble in urine, which has a pH of 6.5–8.

At 50% dissociation of the acid, the equilibrium concentrations of A$^-$(aq) and HA(aq) are equal. This makes the last term in Equation 6-14 equal to zero, in which case pH = pK_a.

An acid (HA) dissolved in water dissociates 50% (into A$^-$) when the solution's pH is equal to the acid's pK_a.

Thus, the graph in Figure 6-1 represents an acid whose pK_a is 5. We will find in Section 6.10 that this general relationship is useful in *gel electrophoresis*, a method used to analyze mixtures of amino acids and proteins.

YOUR TURN 6.6

In Figure 6-1, how many pH units above the acid's pK_a must the solution be to cause the acid to dissociate nearly 100%? _____

How many pH units below the acid's pK_a must the solution be for the acid to be nearly 100% associated? _____

YOUR TURN 6.7

In Figure 6-1, sketch a plot of the percent dissociation of an acid with $pK_a = 9$ as a function of pH.

6.3 Thermodynamics and Gibbs Free Energy

In Section 6.2, we saw that the value of a reaction's equilibrium constant (K_{eq}) is larger for reactions in which more products are formed at equilibrium. As you may recall from general chemistry, the equilibrium constant is also related to the **standard Gibbs free energy difference** between reactants and products, ΔG°_{rxn}, as shown in Equation 6-15.

$$G^\circ_{products} - G^\circ_{reactants} = \Delta G^\circ_{rxn} = -RT \ln K_{eq} \qquad \text{(6-15)}$$

Here R is the universal gas constant (8.314 J/mol·K or 1.987 cal/mol·K), T is the temperature in kelvin, and K_{eq} is the equilibrium constant calculated from the expression in Equation 6-4. The naught (°) signifies *standard conditions*, in which all pure substances are in their most stable states at 298 K, the partial pressures of all gases are 1 atm, and the concentrations of all solutions are 1 mol/L.

With the relationship in Equation 6-15, we can associate ΔG°_{rxn} with the reaction's tendency to favor products at equilibrium (i.e., to be *spontaneous*):

- A reaction tends to be spontaneous if $\Delta G^\circ_{rxn} < 0$. Such a reaction is said to be **exergonic** (energy out), because it releases free energy.
- A reaction tends to be nonspontaneous (i.e., the reverse reaction is spontaneous) if $\Delta G^\circ_{rxn} > 0$. Such a reaction is said to be **endergonic** (energy in), because it absorbs free energy.

Because R and T are both positive, a negative value for ΔG°_{rxn} results in $K_{eq} > 1$, in which case products are favored over reactants. On the other hand, a positive value for ΔG°_{rxn} means that $K_{eq} < 1$, in which case reactants are favored over products.

Knowing the value of ΔG°_{rxn} for a given reaction can also be helpful because it conveys information about the *stability* of the products relative to the reactants. If $\Delta G^\circ_{rxn} < 0$, then the standard Gibbs free energy of the products ($G^\circ_{products}$) must be lower than that of the reactants ($G^\circ_{reactants}$). Conversely, if $\Delta G^\circ_{rxn} > 0$, then $G^\circ_{products}$ is higher than $G^\circ_{reactants}$. Thus:

- If $\Delta G^\circ_{rxn} < 0$, then products are more stable than reactants thermodynamically.
- If $\Delta G^\circ_{rxn} > 0$, then products are less stable than reactants thermodynamically.

6.3a Enthalpy and Entropy

The ΔG°_{rxn} can be expressed in terms of the reaction's *enthalpy change* and *entropy change*, as follows:

$$\Delta G^{\circ}_{rxn} = \Delta H^{\circ}_{rxn} - T\Delta S^{\circ}_{rxn} \tag{6-16}$$

The ΔH°_{rxn} term in Equation 6-16 is the **standard enthalpy difference** between the reactants and products. At constant pressure, ΔH°_{rxn} equals the heat absorbed or released by the reaction. If ΔH°_{rxn} is positive, then the reaction absorbs heat and is said to be **endothermic**; if ΔH°_{rxn} is negative, then the reaction releases heat and is **exothermic**.

In the second term on the right side of Equation 6-16, T is the temperature in kelvin and ΔS°_{rxn} is the **standard entropy difference** between the reactants and products. As mentioned in Section 2.7, entropy is related to the number of different states available to a system and is often thought of as a "measure of disorder."

For a proton transfer reaction, the entropy term in Equation 6-16 is usually quite small. This is because the most significant factor that governs ΔS°_{rxn} is the number of independent species on each side of the reaction, and, as we can see from the general proton transfer reaction below, both the reactant side and the product side contain two separate species. Therefore, the enthalpy term in Equation 6-16 generally dominates, and $\Delta G^{\circ}_{rxn} \approx \Delta H^{\circ}_{rxn}$.

ΔS°_{rxn} is small because both reactants and products have the same number of species.

$\Delta G^{\circ}_{rxn} \approx \Delta H^{\circ}_{rxn}$

For most other reactions we will encounter, ΔH°_{rxn} also governs whether ΔG°_{rxn} is positive or negative. In situations where ΔS°_{rxn} is significant, however, the sign of ΔG°_{rxn} depends on the signs of both ΔH°_{rxn} and ΔS°_{rxn}, as well as the temperature. We will discuss this idea further in the context of the reactions to which it is pertinent.

6.3b The Reaction Free Energy Diagram

So far, we have considered only the end points of the proton transfer reaction—namely, the reactants and the products. However, the changes that the reactants undergo to become products are continuous, not instantaneous. To help us discuss these changes, chemists often use free energy diagrams.

In a **reaction free energy diagram**, Gibbs free energy is plotted as a function of the *reaction coordinate*. A **reaction coordinate** is a variable that corresponds to the changes in geometry, on a molecular level, as reactants are converted into products. The way in which a reaction coordinate is quantified is unique to each individual reaction and can be a complicated function of bond angles and bond lengths. However, we can think about it in simpler terms:

> As the reaction coordinate increases, the geometries of the species involved in the reaction increasingly resemble those of the products.

This becomes clearer if we examine the specific examples in Figure 6-2. Figure 6-2a is the free energy diagram for the proton transfer between HO^- and HCl, whereas Figure 6-2b is the diagram for the reverse reaction, between Cl^- and H_2O.

For the reaction in Figure 6-2a, an H—Cl bond is broken during the course of the reaction and an O—H bond is formed. Therefore, we can say that as the reaction coordinate increases, the distance between the H and Cl atoms increases and the distance between the H and O atoms decreases.

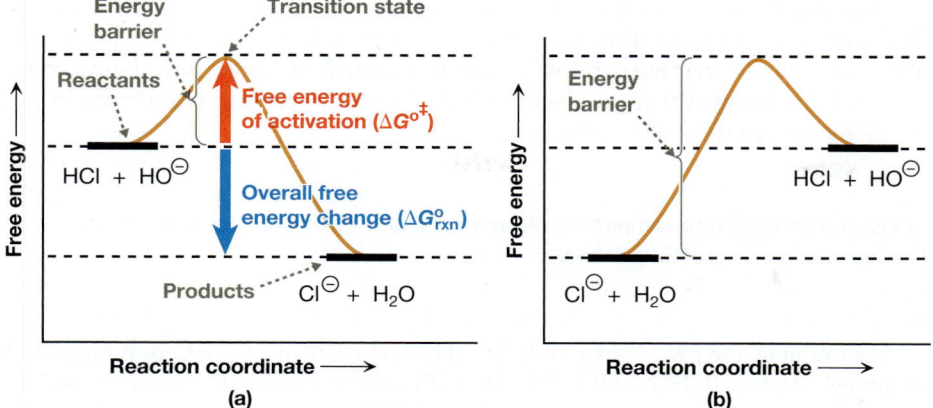

FIGURE 6-2 **Free energy diagrams**
(a) Free energy diagram for the proton transfer between HCl and HO^-. The reaction is exergonic ($\Delta G^\circ_{rxn} < 0$), but it still has an energy barrier. (b) Free energy diagram for the proton transfer between Cl^- and H_2O, an endergonic reaction ($\Delta G^\circ_{rxn} > 0$).

YOUR TURN 6.8

In Figure 6-2b, as the reaction coordinate increases, does the distance between Cl and H increase or does it decrease? _____ Does the distance between the HO and H increase or decrease? _____

One of the main benefits of free energy diagrams is that they allow us to see quickly whether the products are more stable or less stable than the reactants. In Figure 6-2a, for example, the products are lower in energy than the reactants (the reaction is exergonic), signifying that the products are more stable; this difference in energy, ΔG°_{rxn}, is indicated by the thick blue arrow. For the reverse reaction (Fig. 6-2b), the reactants are more stable than the products.

Notice in Figure 6-2a and 6-2b that there is an *energy maximum along the reaction coordinate*. The structure that corresponds to this energy maximum is the elementary step's **transition state**. Because there is a maximum in energy between the reactants and products, an *energy barrier* must be surmounted to form products. *This is true regardless of whether the elementary step is exergonic (Fig. 6-2a) or endergonic (Fig. 6-2b).* This energy barrier, called the **free energy of activation** ($\Delta G^{\circ\ddagger}$, "delta-*G*-naught-double-dagger"), is the difference in standard free energy between the reactants and the transition state. $\Delta G^{\circ\ddagger}$ is an important factor that governs reaction rates:

As the free energy of activation decreases, the rate of the reaction increases and the reactants are said to be more *reactive*.

We will see how this applies to reactions in Chapter 9.

YOUR TURN 6.9

Indicate ΔG°_{rxn} and $\Delta G^{\circ\ddagger}$ in Figure 6-2b the way it is done in Figure 6-2a. Which reaction has a greater $\Delta G^{\circ\ddagger}$, the one in Figure 6-2a or the one in Figure 6-2b? _____

6.4 Strategies for Success: Functional Groups and Acidity

Although the pK_a values for several different compounds are listed in Table 6-1, they represent only a small fraction of the millions of compounds known. There is a good chance, then, that the compound for which you need to know the pK_a value is not included in the table. How can we obtain the pK_a values for those other compounds?

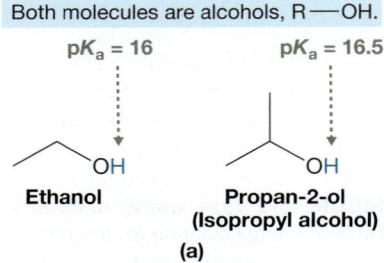

Both molecules are alcohols, R—OH.

$pK_a = 16$ $pK_a = 16.5$

Ethanol   **Propan-2-ol (Isopropyl alcohol)**

(a)

Both molecules are carboxylic acids, R—CO₂H.

$pK_a = 4.75$ $pK_a = 4.2$

Ethanoic acid (Acetic acid)   **Benzoic acid**

(b)

FIGURE 6-3 Functional groups and pK_a (a) The acidic proton (in blue) in both molecules is part of an OH functional group, and the pK_a values are similar. (b) The acidic proton (in blue) in both molecules is part of a CO₂H functional group, and the pK_a values are similar. The pK_a value of a carboxylic acid, however, is significantly different from that of an alcohol.

One way is to *estimate* the value based on structural similarities to a compound that is listed in the table. This may seem to be a formidable task at first, due to the many structural variations of the compounds listed in Table 6-1. Recall from Section 1.13, however, that the chemical behavior of a compound is governed by the functional groups it possesses. Therefore:

> The acidity (pK_a) of a compound is governed largely by the functional group on which the acidic proton is found.

For example, the pK_a of ethanol (CH₃CH₂OH) is 16 and the pK_a of propan-2-ol [isopropyl alcohol, (CH₃)₂CHOH] is 16.5 (Fig. 6-3a). Their pK_a values are similar because their acidic protons are both part of an O—H functional group characteristic of alcohols. Likewise, ethanoic acid (acetic acid, CH₃CO₂H) and benzoic acid (C₆H₅CO₂H) have similar pK_a values, at 4.75 and 4.2, respectively, because their acidic protons are both part of a CO₂H functional group characteristic of carboxylic acids (Fig. 6-3b).

Carboxylic acids are substantially more acidic than alcohols, even though both have their acidic protons on an oxygen atom. Ethanoic acid (acetic acid, CH₃CO₂H) has a pK_a of 4.75, whereas ethanol (CH₃CH₂OH) has a pK_a of 16—a difference in acid strength of $>10^{11}$. Thus, the presence of a carbonyl group (C=O) significantly increases the acidity of an adjacent OH group.

Although acidity is governed primarily by which functional group the acidic proton is on, some nearby structural features, such as a highly electronegative substituent, can alter the acidity significantly. Trichloroacetic acid (Cl₃CCO₂H; $pK_a = 0.77$) is a stronger acid than acetic acid (H₃CCO₂H; $pK_a = 4.75$) by nearly four pK_a units—a factor of 10^4 in acid strength (Fig. 6-4a). The presence of a double bond attached to the atom containing the acidic proton can significantly increase the acidity, too. Phenol (C₆H₅OH; $pK_a = 10.0$), for example, is more acidic than ethanol (CH₃CH₂OH; $pK_a = 16$) by a factor of about 10^6 (Fig. 6-4b).

The reasons for these structural effects on pK_a are discussed in greater detail in Sections 6.5 and 6.6. Based on these few examples, however, we can quickly recognize when *not* to expect acidic protons on the same atom to have similar acidities.

FIGURE 6-4 Effects on pK_a from nearby structural features (a) The nearby electronegative Cl atoms significantly lower the pK_a of the proton in blue. (b) The adjacent double bond significantly lowers the pK_a of the proton in blue.

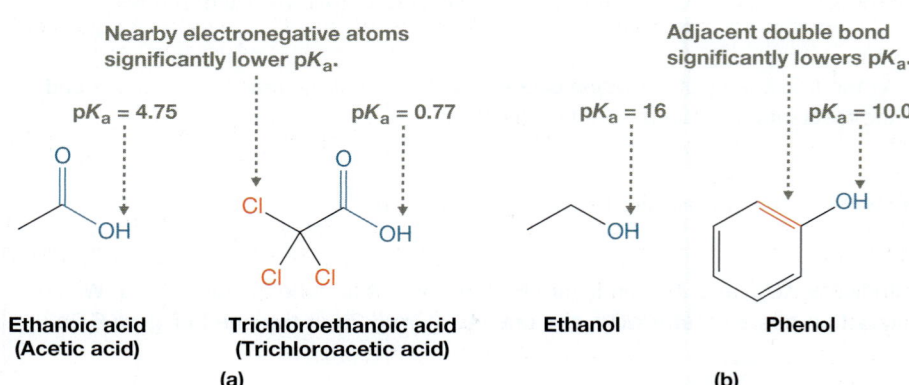

Nearby electronegative atoms significantly lower pK_a.

$pK_a = 4.75$ $pK_a = 0.77$

Adjacent double bond significantly lowers pK_a.

$pK_a = 16$ $pK_a = 10.0$

Ethanoic acid (Acetic acid)   **Trichloroethanoic acid (Trichloroacetic acid)**   **Ethanol**   **Phenol**

(a)   **(b)**

CONNECTIONS Isopropyl alcohol (Fig. 6-3a), as a concentrated aqueous solution, is sold commercially as rubbing alcohol, a topical antiseptic. It is also useful as a solvent in the laboratory because it can dissolve a variety of nonpolar compounds. Isopropyl alcohol has multiple uses in the automotive industry, including as a gasoline additive to combat the problems that arise when water enters the fuel lines.

SOLVED PROBLEM 6.10

Using Table 6-1, estimate the pK_a for the NH₂ protons in cyclohexylamine.

Think On what functional group do the H atoms appear? What molecule(s) in Table 6-1 have the same functional group? Are there any nearby electronegative atoms or adjacent double bonds?

Cyclohexylamine

Solve The H atom in question is part of a C—N functional group, characteristic of amines. According to Table 6-1, the pK_a of $(CH_3)_2NH$, another amine, is 38. Because cyclohexylamine does not have any nearby electronegative atoms or adjacent double bonds, we estimate its pK_a to be 38, too.

PROBLEM 6.11 Using Table 6-1 and/or Appendix A, estimate the pK_a for the proton explicitly shown on each of the following compounds.

(a)   (b)   (c)   (d)   (e)

CONNECTIONS In biochemistry, trichloroacetic acid (Fig. 6-4a) is used to precipitate large biomolecules, such as DNA and proteins. It is also used in cosmetic treatments, such as in wart removal and chemical peels.

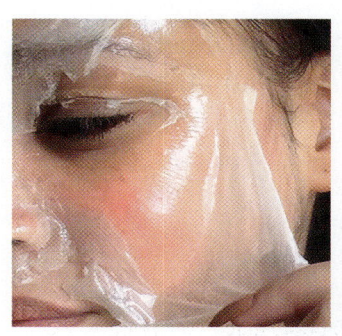

6.5 Relative Strengths of Charged and Uncharged Acids: The Reactivity of Charged Species

Perhaps the most conspicuous trend from Table 6-1 involves the charge on the atom to which a proton is attached.

> A proton is significantly more acidic when the atom to which it is attached is positively charged than when that atom is uncharged.

For example, the pK_a of H_3O^+ is -1.7, whereas that of H_2O is 15.7. Additionally, the pK_a of H_4N^+ is 9.4, whereas that of H_3N is 36. Why should the presence of a charge have this effect?

To begin to answer this question, consider Equation 6-17, the *autoionization* equilibrium for water.

$$H_2\ddot{O}{:}(\ell) + H{-}\overset{..}{\underset{..}{O}}H(aq) \rightleftharpoons H_2\overset{\oplus}{\underset{..}{O}}{-}H(aq) + {:}\overset{..}{\underset{..}{O}}H^{\ominus}(aq) \qquad \begin{array}{l} K_{eq} \approx 10^{-14} \\ \Delta G^{\circ}_{rxn} \approx +80 \text{ kJ/mol} \end{array} \qquad (6\text{-}17)$$

The equilibrium constant for the autoionization of water is very small ($K_{eq} \approx 10^{-14}$), which corresponds (via Equation 6-15) to a substantially positive change in Gibbs free energy ($\Delta G^{\circ}_{rxn} \approx +80$ kJ/mol). See the free energy diagram in Figure 6-5. The H_3O^+ and HO^- products, which are charged, are less stable than the uncharged reactants by 80 kJ/mol.

> Generally speaking, a charged species is significantly higher in energy (and thus less stable) than its uncharged counterpart.

With this in mind, we can understand the trend among charged and uncharged acids, mentioned previously. Consider

FIGURE 6-5 Free energy diagram for the autoionization of H_2O (Equation 6-17) The charged products are much higher in energy than the uncharged reactants, meaning that the charged species are significantly less stable.

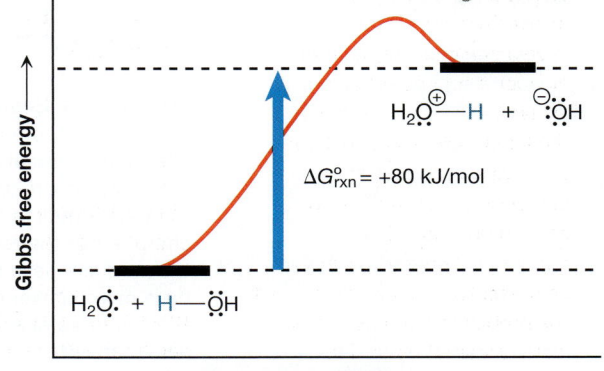

the reactions shown in Equation 6-18a and 6-18b, in which water deprotonates NH_3 and H_4N^+, respectively.

pKa = 36 (6-18a)

pKa = 9.4 (6-18b)

Two additional charges

No additional charges

When NH_3 is deprotonated (Equation 6-18a), two new charges appear. Therefore, in the corresponding free energy diagram in Figure 6-6 (red curve), the products are considerably higher in energy than the reactants. When H_4N^+ is deprotonated (Equation 6-18b), on the other hand, no new charges appear, so in the corresponding free energy diagram in Figure 6-6 (blue curve), the reactants and products are more similar in energy. Therefore, it is energetically easier to deprotonate H_4N^+, making H_4N^+ the stronger acid.

Similar ideas explain the relative energies of the two sets of reactants in Figure 6-6, as well as the relative energies of the two sets of products. The reactants in the deprotonation of H_4N^+ (blue curve) exhibit one charge, so they are higher in energy than

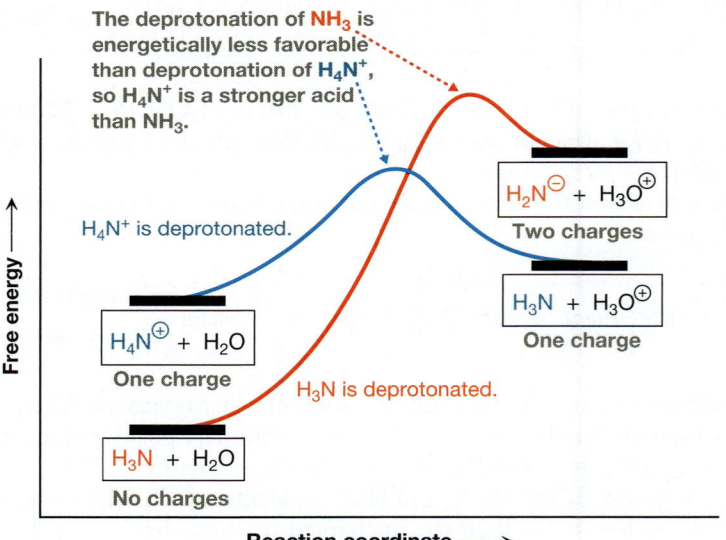

FIGURE 6-6 Energy diagrams for the deprotonation of NH_3 (red) and H_4N^+ (blue) by water The reactants in Equation 6-18a ($NH_3 + H_2O$; red curve) are lower in energy than the reactants in Equation 6-18b ($H_4N^+ + H_2O$; blue curve) because they have fewer charges. Similarly, the products in Equation 6-18b ($NH_3 + H_3O^+$; blue curve) are lower in energy than the products in Equation 6-18a ($H_2N^- + H_3O^+$; red curve) because they have fewer charges. For the red curve, the products are much less stable than the reactants because the products have two additional charges. For the blue curve, the reactants and products have more similar energies because they each have one total charge. The deprotonation of H_4N^+ is more energetically favorable than the deprotonation of NH_3, so H_4N^+ is a stronger acid than NH_3.

the reactants in the deprotonation of NH_3 (red curve), which have no charges. Likewise, the products in the deprotonation of H_4N^+ (blue curve) exhibit one charge, so they are lower in energy than the products in the deprotonation of NH_3 (red curve), which have two charges.

PROBLEM 6.12 Draw an energy diagram similar to Figure 6-6, using H_3O^+ and H_2O as the acids. Based on the energy diagram, which acid is predicted to be stronger? Is this consistent with their relative pK_a values?

6.6 Relative Acidities of Protons on Atoms with Like Charges

The analysis in Section 6.5 explains the greater acidity of a positively charged species relative to a comparable uncharged molecule, but we have yet to explain the relative acidities of species bearing the *same* charge. As we explain here in Section 6.6, there are a variety of trends to examine. In Sections 6.6a and 6.6b, for example, we discuss how the identity of an atom affects the acidity of protons attached to it; in Section 6.6c, we describe how hybridization affects acidity; and in Sections 6.6d and 6.6e, we explain the effects of nearby π bonds and nearby atoms.

6.6a Protons on Different Atoms in the Same Row of the Periodic Table

The acidic protons in CH_4, H_3N, H_2O, and HF are bonded to different atoms in the second row of the periodic table — namely, C, N, O, and F, respectively. The pK_a values of these acids are 48, 36, 15.7, and 3.2, respectively.

> The farther to the right an atom appears in the periodic table, the more acidic the protons that are attached to it.

To better understand the basis for such a trend, consider the equilibria in which H_3N and CH_4 are deprotonated by water.

$$H_2N-H \;+\; H_2O \;\rightleftharpoons\; H_2N^{\ominus} \;+\; H_3O^{\oplus} \qquad \text{(6-19a)}$$

$$H_3C-H \;+\; H_2O \;\rightleftharpoons\; H_3C^{\ominus} \;+\; H_3O^{\oplus} \qquad \text{(6-19b)}$$

The free energy diagrams for these reactions are shown in Figure 6-7 (next page). Unlike what we saw in Figure 6-6, the reactants for both reactions in Figure 6-7 appear at the same location, which is an outcome of the following assumption:

> When constructing free energy diagrams for reactions that involve both charged and uncharged species, assume that the various uncharged species have similar stabilities.

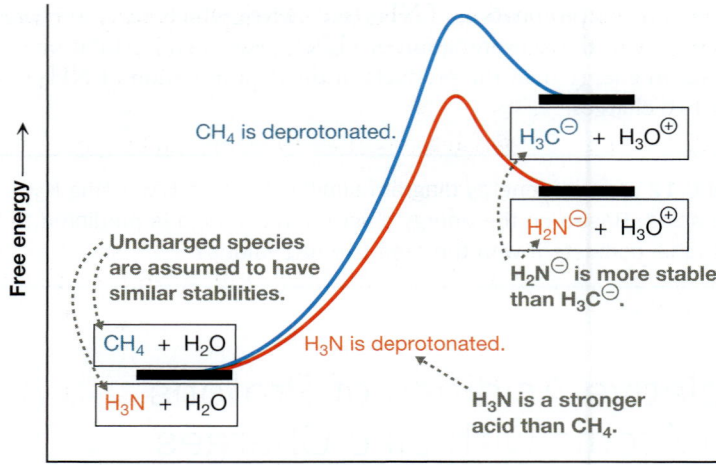

FIGURE 6-7 Relative stabilities of H_3C^- and H_2N^- Because H_3N is a stronger acid than CH_4, the reaction in Equation 6-19a (red curve) is more favorable than the reaction in Equation 6-19b (blue curve). Thus, H_2N^- is more stable than H_3C^-.

In reality, NH_3 and CH_4 have different free energies, but the above assumption works because the free energy difference between uncharged species is typically much smaller than the free energy difference between their charged counterparts.

Next, notice that, for both reactions represented in Figure 6-7, the products have two more charges than the reactants. Therefore, the free energy diagrams for these reactions have the products significantly higher in energy than the reactants. NH_3 is a stronger acid than CH_4, however, so the deprotonation of NH_3 (red curve) must be more energetically favorable than the deprotonation of CH_4 (blue curve). This places the products of NH_3 deprotonation lower in energy than the products of CH_4 deprotonation, which means that H_2N^- is more stable than H_3C^-.

The way we determined the relative stabilities of H_2N^- and H_3C^- leads to the following general rule:

> When comparing two uncharged acids, the stronger acid is the one whose negatively charged conjugate base is more stable.

Why should H_2N^- be more stable than H_3C^-? The answer lies in the difference between the atoms on which the charge resides. As we know from Section 1.7, N is more electronegative than C, meaning that N has a stronger attraction for electrons (i.e., for negative charge). Thus, although both species are destabilized by the presence of a negative charge, N is less destabilized than C.

A similar analysis shows that H_2N^- is less stable than HO^-, which, in turn, is less stable than F^-. Thus:

Negative charges on atoms in the same row of the periodic table

Increasing electronegativity of atom bearing the negative charge →

$H_3C^{\ominus}$ $H_2N^{\ominus}$ $HO^{\ominus}$ $F^{\ominus}$

Increasing stability of the anion →

> An anion becomes more stable as the electronegativity of the atom bearing the negative charge increases.

The same trend is observed among acids with acidic protons bonded to atoms in other rows of the periodic table. For example, HCl is a stronger acid than H_2S because Cl is more electronegative than S, allowing Cl to better accommodate the negative charge that appears when the proton is lost.

YOUR TURN 6.10

Verify that HCl is a stronger acid than H_2S by looking up their pK_a values in Table 6-1. HCl _____ H_2S _____

The preceding analysis involving uncharged acids applies equally well to positively charged acids. H_3O^+, for example, is a stronger acid than H_4N^+.

YOUR TURN **6.11**

Verify that H_3O^+ is a stronger acid than H_4N^+ by looking up their pK_a values in Table 6-1. H_3O^+ _____ H_4N^+ _____

The proton transfer equilibria for H_3O^+ and H_4N^+ with water are shown in Equation 6-20a and 6-20b.

Stronger acid H_2O—H + H_2O ⇌ H_2O + H_3O^+ (6-20a)

Weaker acid H_3N—H + H_2O ⇌ NH_3 + H_3O^+ (6-20b)

On the product side, the difference in these reactions is H_2O versus NH_3. Relative to the charged species on the reactant sides, we assume that these uncharged molecules have similar stabilities. Therefore, in the free energy diagrams for these reactions shown in Figure 6-8, the products are placed at the same height.

Next, because H_3O^+ is more acidic than H_4N^+, we know that the reaction in Equation 6-20a (red curve in Fig. 6-8) is more energetically favorable than the reaction in Equation 6-20b (blue curve in Fig. 6-8). Therefore, in the free energy diagrams, the reactants for Equation 6-20a appear above the reactants for Equation 6-20b. This allows us to conclude that H_3O^+ is less stable than H_4N^+.

In general:

When comparing two positively charged acids, the stronger acid is the one that is less stable.

Why is H_3O^+ less stable than H_4N^+? Once again, it can be attributed to a difference in electronegativity. The O atom bearing the positive charge in H_3O^+ is more electronegative than the N atom bearing the positive charge in H_4N^+. Thus, for the

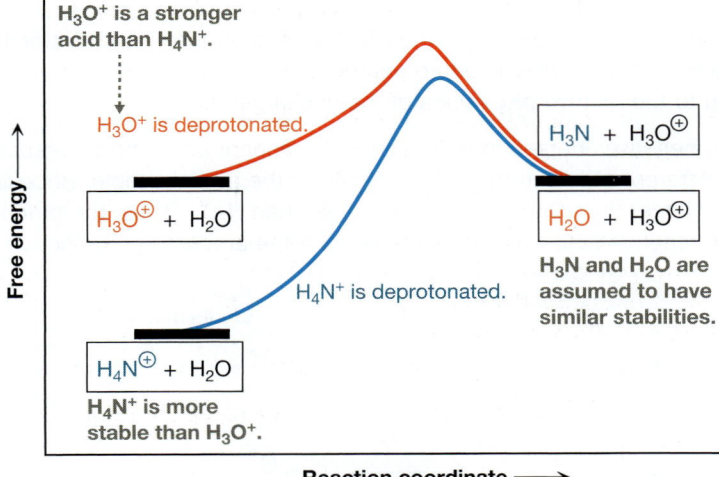

FIGURE 6-8 **Relative stabilities of H_3O^+ and H_4N^+** H_3O^+ is a stronger acid than H_4N^+, so the reaction in Equation 6-20a (red curve) is more favorable than the reaction in Equation 6-20b (blue curve). Thus, H_4N^+ is more stable than H_3O^+.

same reason that O can accommodate a *negative* charge better than N, O accommodates a *positive* charge *worse* than N.

The relative stabilities of H_3O^+ and H_4N^+ are consistent with the following periodic table trend for the stabilities of positively charged ions:

> The stability of a cation decreases as the electronegativity of the atom bearing the positive charge increases.

6.6b Protons on Different Atoms in the Same Column of the Periodic Table

HCl ($pK_a = -7$) is a *stronger* acid than HF ($pK_a = 3.2$), which means that Cl^- is *more* stable than F^-. F is *more* electronegative than Cl, however, in which case you might expect that F could accommodate a negative charge better than Cl.

To untangle this apparent discrepancy, recall that *Cl is a substantially larger atom than F*, because Cl is one row below F in the periodic table. A negative charge on Cl is therefore less concentrated than a negative charge on F because the same -1 charge is spread out over a larger volume in Cl. This lower concentration of charge contributes to the greater stability of Cl^- than F^-.

The electrostatic potential maps of F^- and Cl^-, shown in Figure 6-9, are consistent with this analysis. The electron cloud of F^- is a deep red, whereas the electron cloud of Cl^- is orange, which suggests that the concentration of charge is higher in F^- than in Cl^-.

This trend continues down group 17 of the periodic table, because HBr ($pK_a = -9$) is a stronger acid than HCl ($pK_a = -7$). Br^- is more stable than Cl^- because Br is larger than Cl, so the negative charge on Br^- is more spread out than it is on Cl^-. The following general trend applies to other columns of the periodic table as well, as shown in Solved Problem 6.13:

> The stability of an anion tends to increase when the negative charge is on an atom farther down a column of the periodic table.

SOLVED PROBLEM 6.13

Predict which has a lower pK_a: CH_4 or SiH_4.

Think Which is more stable, H_3C^- or H_3Si^-? Based on their relative stabilities, which anion's conjugate acid is deprotonated more favorably? How does that correspond to the relative pK_a values of the uncharged acids?

Solve The negative charges in H_3C^- and H_3Si^- appear on C and Si, respectively, which are different atoms in the same column of the periodic table. Because Si is significantly larger than C, H_3Si^- is more stable than H_3C^-. Thus, the products of the second reaction below are more stable than the products of the first.

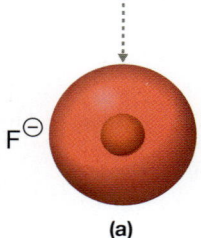
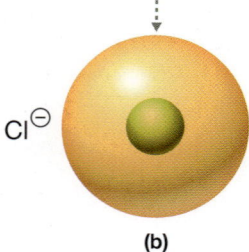

FIGURE 6-9 Atomic size and concentration of negative charge The electrostatic potential map of F^- is shown in (a) and that of Cl^- is shown in (b). Because Cl is larger than F, the negative charge on Cl^- is less concentrated, which is why it appears less red than F^-. This lower concentration of charge helps make Cl^- more stable than F^-.

- Smaller volume
- More concentrated negative charge
- Less stable

$F^{\ominus}$

(a)

- Larger volume
- Less concentrated negative charge
- More stable

$Cl^{\ominus}$

(b)

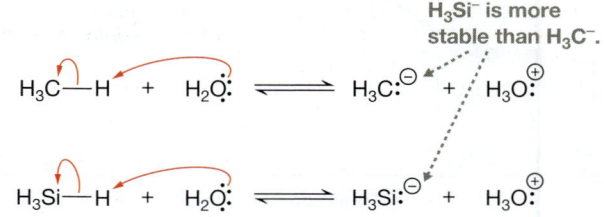

H_3Si^- is more stable than H_3C^-.

Consequently, SiH_4 is deprotonated more favorably than CH_4, making SiH_4 the stronger acid. SiH_4 has a lower pK_a than CH_4 (34 vs. 48).

PROBLEM 6.14 Which is a stronger acid, HBr or HI? Explain.

PROBLEM 6.15 Which is a stronger acid, CH_4 or PH_3? Explain.

6.6c Hybridization of the Atom to Which the Proton Is Attached

According to Table 6-1, $H_3C—CH_3$ is an extremely weak acid (among the weakest known), $H_2C=CH_2$ is somewhat stronger, and $HC\equiv CH$ is nearly as acidic as some alcohols (R—OH).

YOUR TURN **6.12**

Verify the relative acidities of ethane, ethene, and ethyne by looking up their pK_a values in Table 6-1.

pK_a: $H_3C—CH_3$ _____ $H_2C=CH_2$ _____ $HC\equiv CH$ _____

Because ethane, ethene, and ethyne are uncharged acids, their differences in acidity must be caused by differences in the stability of their negatively charged conjugate bases. In other words, $HC\equiv C^-$ must be more stable than $H_2C=CH^-$, which, in turn, must be more stable than $H_3C—CH_2^-$.

These differences in stability cannot come from differences in the size of the charged atom, because the negative charge is on carbon in each conjugate base. Instead, they arise from differences in the carbon atom's *effective electronegativity* in each of these species. As we saw in Section 3.11, the effective electronegativity of an atom depends on its hybridization, increasing in the order: $sp^3 < sp^2 < sp$. The effective electronegativity has the same effect on charge stability that true electronegativity does:

> The stability of a charged species increases as the *effective electronegativity* of an atom bearing a *negative* charge increases.

The C atom bearing the negative charge in $HC\equiv C^-$ is *sp* hybridized (highest effective electronegativity = most stable); that in $H_2C=CH^-$ is sp^2 hybridized (lower effective electronegativity = less stable); and that in $H_3C—CH_2^-$ is sp^3 hybridized (lowest effective electronegativity = least stable).

As you might expect, the effective electronegativity has the opposite effect on the stability of positively charged species.

> The stability of a charged species decreases as the *effective electronegativity* of an atom bearing a *positive* charge increases.

All else being equal, a positive charge on an sp^2-hybridized atom is *less* stable than a positive charge on an sp^3-hybridized atom (see Solved Problem 6.16). The greater effective electronegativity of the sp^2-hybridized atom makes it less able to accommodate a positive charge.

Predict which acid is stronger: $(CH_3)_2C$=OH^+ or $(CH_3)_2CH$—OH_2^+.

Think Which atom bears the positive charge in each acid? What is the hybridization of each of those atoms, and which atom has a greater effective electronegativity? How do their relative effective electronegativities correspond to the relative stabilities of the positively charged species? Based on those stabilities, which acid undergoes deprotonation more favorably?

Solve Each acid is positively charged, so we know that the one that is less stable is the stronger acid. In both cases, the positive charge appears on O. The O atom in $(CH_3)_2C$=OH^+ is sp^2 hybridized, whereas the O atom in $(CH_3)_2CH$—OH_2^+ is sp^3 hybridized, giving the O atom in $(CH_3)_2C$=OH^+ a higher effective electronegativity. Consequently, $(CH_3)_2C$=OH^+ cannot accommodate a positive charge on O as well as $(CH_3)_2CH$—OH_2^+ can, making $(CH_3)_2C$=OH^+ less stable, and therefore a stronger acid, than $(CH_3)_2CH$—OH_2^+.

PROBLEM 6.17 Which is a stronger acid: CH_3—NH_3^+ or HC≡NH^+?

6.6d Effects from Adjacent Double and Triple Bonds: Resonance Effects

Recall from Section 6.4 that the pK_a of an acid can be altered significantly by the presence of a double or triple bond *adjacent* to the atom bonded to the acidic proton. For example, ethanoic acid (acetic acid, CH_3CO_2H), which has a C=O bond adjacent to the acidic OH group, is more acidic than ethanol (CH_3CH_2OH) by about 11 pK_a units.

Ethanoic acid
(Acetic acid)

Ethanol

YOUR TURN 6.13

Use Table 6-1 to look up the pK_a values of ethanoic acid and ethanol.

pK_a: Ethanoic acid _____ Ethanol _____

One of the main reasons ethanoic acid is so much more acidic than ethanol has to do with **resonance effects**. Both of these acids are uncharged (of the type in Equation 6-19), so the difference in their acidities is largely due to differences in the stability of their negatively charged conjugate bases, $CH_3CO_2^-$ and $CH_3CH_2O^-$ (Equation 6-21a and 6-21b). Because ethanoic acid is more acidic, we can say that $CH_3CO_2^-$ is more stable than $CH_3CH_2O^-$ (review Fig. 6-7, p. 294).

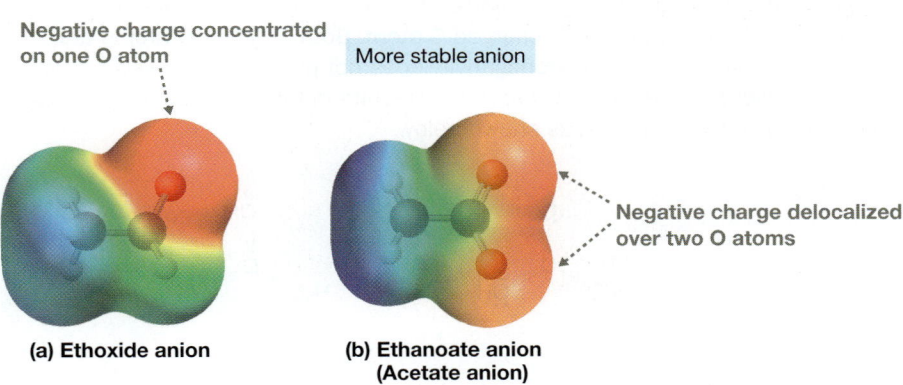

More stable anion

Negative charge delocalized over two O atoms

Ethanoic acid (Acetic acid) → Loss of H⊕ → **Ethanoate anion (Acetate anion)**   (6-21a)

Negative charge localized on one O atom

Ethanol → Loss of H⊕ → **Ethoxide anion**   (6-21b)

As we can see in Equation 6-21a, ethanoic acid's conjugate base, the ethanoate anion, has two resonance structures, each of which has the negative charge on a different O atom. We say that the negative charge is *delocalized* over these two O atoms.

On the other hand, the negative charge is *localized* on a single O atom in ethanol's conjugate base (the ethoxide anion), shown in Equation 6-21b. In other words, the negative charge is *less concentrated* in the ethanoate anion than it is in the ethoxide anion. This is confirmed in the electrostatic potential maps of the two anions in Figure 6-10. With lower concentration of charge comes greater charge stability.

Delocalizing a charge via resonance lowers the *concentration* of the charge and increases the stability of the charged species.

Negative charge concentrated on one O atom

More stable anion

Negative charge delocalized over two O atoms

(a) Ethoxide anion

(b) Ethanoate anion (Acetate anion)

FIGURE 6-10 Resonance delocalization of a negative charge (a) An electrostatic potential map of the ethoxide anion ($CH_3CH_2O^-$) shows that the negative charge is localized on just the one O atom. (b) An electrostatic potential map of the ethanoate anion ($CH_3CO_2^-$) shows that the negative charge is delocalized onto both O atoms, consistent with its resonance hybrid. The concentration of negative charge is lower (less red) in the ethanoate anion, so the ethanoate anion is more stable than the ethoxide anion.

YOUR TURN 6.14

In the space provided here, draw the curved arrow notation for the conversion of one of the ethanoate anion's resonance structures into the other and draw the resonance hybrid. (You may want to review Section 1.10.)

Resonance hybrid

SOLVED PROBLEM 6.18

Which species, **A** or **B**, is the stronger acid?

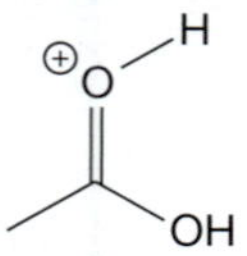

A

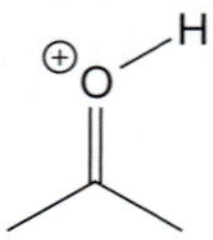

B

Think Write reactions that depict each acid being deprotonated by a base. Do you expect a significant difference in energy between the reactants of one and the reactants of the other? Between the products of one and the products of the other? Are the charge-bearing atoms different in these two acids? Do they have different effective electronegativities? Do the ions differ in resonance delocalization of the charge?

Solve The acids can be deprotonated by water according to the following reactions:

More stable cation

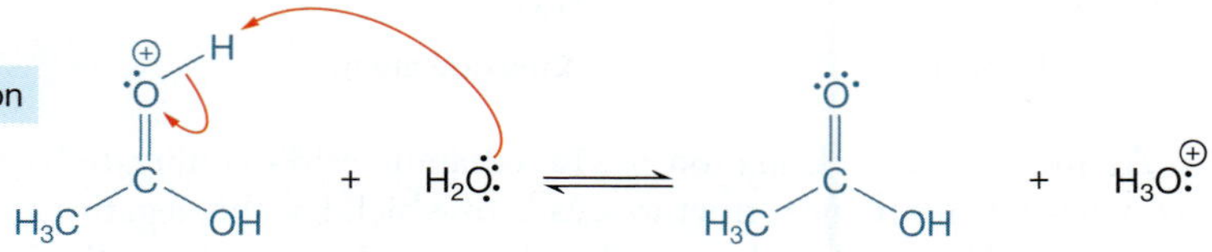

A

Less stable cation

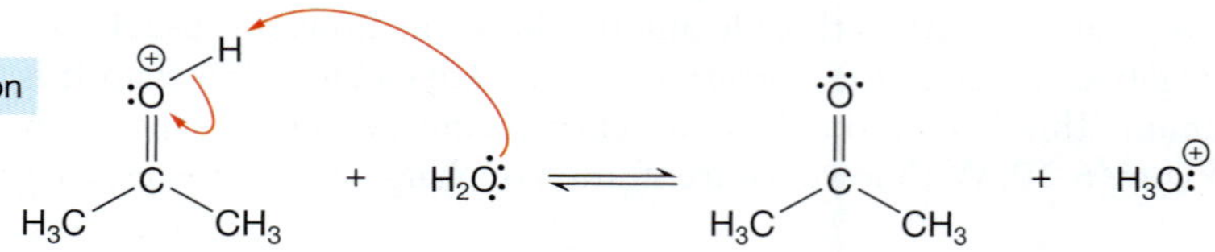

B

Being positively charged, these acids have the same form as H_3O^+ and H_4N^+ in Equation 6-20a and 6-20b. So, as we learned from Figure 6-8, the energetically more favorable deprotonation will involve the less stable reactant ion. Thus, the stronger acid is the one that is *less stable*. In both of these acids, the acidic proton is attached to a positively charged O atom. Both of those O atoms are sp^2 hybridized, so effective electronegativity does not play a role. However, *acid* ***A*** *has two major resonance structures*, which results in the positive charge being delocalized over two O atoms, as shown below.

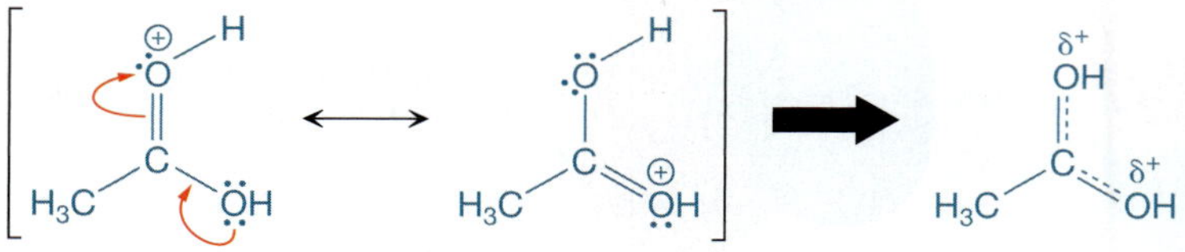

Positive charge delocalized over two O atoms

A

Acid **B**, by contrast, does not have another strongly contributing resonance structure like **A**, so the positive charge is more *localized* on one oxygen atom. As a result, acid **B** is the stronger acid because it is less stable than the first.

PROBLEM 6.19 Which compound, **C** or **D**, is the stronger acid? Explain.

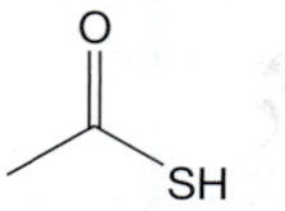

C

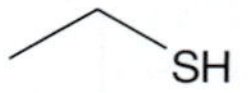

D

The number of atoms over which a charge is delocalized by resonance affects charge stability.

Stabilization via resonance generally increases as the number of atoms over which a charge is delocalized increases.

For example, sulfuric acid (H_2SO_4; $pK_a = -9$) is a stronger acid than ethanoic acid (CH_3CO_2H; $pK_a = 4.75$). Both acids are uncharged, so the conjugate base of the stronger acid must be more stable than that of the weaker acid (recall Equation 6-19 and Fig. 6-7). In HSO_4^-, the negative charge is delocalized over three O atoms, whereas in $CH_3CO_2^-$, the negative charge is delocalized over only two. Delocalizing over more atoms in HSO_4^- means the negative charge is less concentrated, leading to greater stability. We can see that the negative charge is more delocalized in HSO_4^- than in $CH_3CO_2^-$, by comparing their electrostatic potential maps in Figure 6-11.

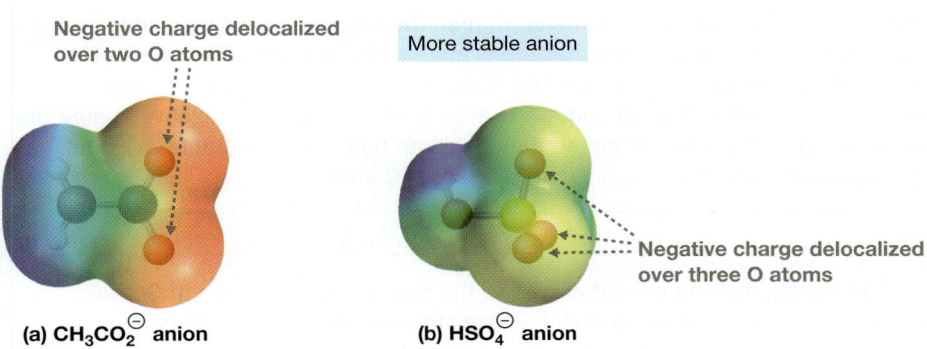

Negative charge delocalized over two O atoms

More stable anion

Negative charge delocalized over three O atoms

(a) $CH_3CO_2^\ominus$ anion

(b) $HSO_4^\ominus$ anion

FIGURE 6-11 Resonance delocalization of a negative charge onto different numbers of atoms (a) An electrostatic potential map of the ethanoate anion ($CH_3CO_2^-$) shows that the negative charge is delocalized by resonance over two O atoms. (b) An electrostatic potential map of the hydrogen sulfate anion (HSO_4^-) shows that the negative charge is delocalized over three O atoms. The concentration of negative charge is lower in HSO_4^- (is less red around the O atoms), so HSO_4^- is more stable than $CH_3CO_2^-$.

YOUR TURN 6.15

In the space provided, draw the two remaining resonance structures of HSO_4^- that illustrate the sharing of its negative charge. Be sure to include the appropriate curved arrows. Then, draw the corresponding resonance hybrid.

Resonance hybrid

SOLVED PROBLEM 6.20

Deprotonation in pentane-2,4-dione may occur at a terminal C atom or at the central C atom. Which site is more acidic? Explain.

Pentane-2,4-dione

Think Does charge stability play a role in the acid or in the conjugate bases? What, specifically, leads to differences in charge stability? The type of atom? Effective electronegativity? Resonance delocalization of the charge? How are the answers to these questions related to the relative strengths of the acids?

Solve The acid is uncharged, so we must look for differences in charge stability in the possible conjugate bases. Since both the CH$_2$ and CH$_3$ protons are attached to sp^3-hybridized C atoms, neither the type of atom nor effective electronegativity should play a role. We do find, however, that there is a difference in resonance stabilization of the resulting charge.

If deprotonation occurs at a terminal C, then the resulting negative charge in the conjugate base is delocalized over the C atom and one O atom via resonance. If deprotonation occurs at the central C, however, then the negative charge is delocalized over the C atom and *two* O atoms. As a result, the negative charge that develops is less concentrated, so the conjugate base is more stable. This makes the central C atom more acidic than a terminal C atom.

YOUR TURN 6.16

Add the appropriate curved arrows to the resonance structures in Solved Problem 6.20 to show how each is transformed into the one on its right.

PROBLEM 6.21 Based on differences in resonance delocalization, predict whether HNO$_3$ or CH$_3$CO$_2$H is the stronger acid. Explain. *Hint:* In this case, you can ignore the fact that the O atoms are bonded to different atoms (i.e., N vs. C).

6.6e Effects from Nearby Atoms: Inductive Effects

Recall from Section 6.4 that another of the structural differences responsible for differences in acid strength is *the presence of electronegative atoms near the acidic proton*. For example, 2-chloroethanol (ClCH$_2$CH$_2$OH; pK_a = 14.3) is more acidic than ethanol (CH$_3$CH$_2$OH; pK_a = 16) by almost two pK_a units. Both acids are uncharged, so the conjugate base of 2-chloroethanol (ClCH$_2$CH$_2$O$^-$) must be more stable than that of ethanol (CH$_3$CH$_2$O$^-$).

In both conjugate bases, the negative charge is on the same type of atom (oxygen) with the same hybridization (sp^3). Furthermore, neither of the conjugate bases has any resonance structures that place the negative charge on other atoms. Therefore, the difference in charge stability must come from another factor.

The electrostatic potential maps of the two anions are shown in Figure 6-12. Notice that some of the negative charge (red) surrounding the O atom has been removed when the Cl atom is present. In other words, the O atom bears less of a negative charge in ClCH$_2$CH$_2$O$^-$ than it does in CH$_3$CH$_2$O$^-$. Just as with charge delocalization via resonance, this decrease in the concentration of charge increases the stability of the anion.

Why does the presence of the Cl atom remove electron density from the nearby O atom? Cl is more electronegative than H, so Cl is **electron withdrawing** relative to H. Therefore, if Cl replaces an H atom on CH$_3$ in CH$_3$CH$_2$O$^-$, then electron density on

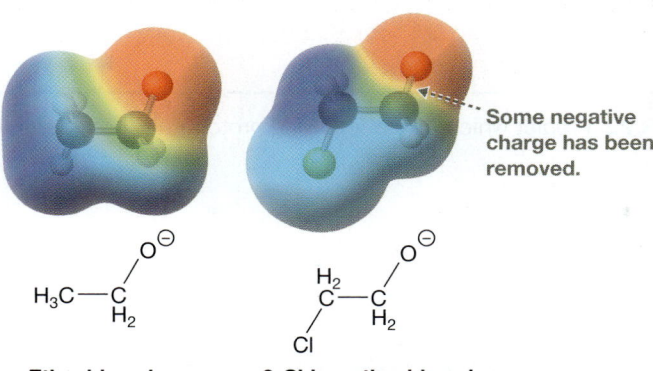

More stable anion

Some negative charge has been removed.

Ethoxide anion   2-Chloroethoxide anion

FIGURE 6-12 Delocalization of a negative charge via induction Electrostatic potential maps of the ethoxide anion ($CH_3CH_2O^-$, *left*) and the 2-chloroethoxide anion ($ClCH_2CH_2O^-$, *right*). The presence of the Cl atom in place of a H atom removes some electron density (red) from the negatively charged O atom. This decrease in charge concentration stabilizes the anion.

the leftmost C atom would be shifted toward the Cl atom, along the C—Cl bond. As a result, the leftmost C atom would develop a deficiency of electrons, and to compensate, it would draw electron density away from other atoms to which it is bonded. This effect would continue down the chain until, ultimately, electron density was removed from the O atom bearing the negative charge:

More stable anion

Electron density is removed from the O^-, thus *stabilizing the negative charge.*

Cl is more electronegative than H and is thus *electron withdrawing.*

> In general, anions are *stabilized* by electron-withdrawing groups near the negative charge.

This phenomenon is called **induction**, and it can be defined as the distortion of electron density along covalent bonds, brought about by the replacement of a H atom with another substituent. The effect that induction has on stability is called an **inductive effect**. In $ClCH_2CH_2O^-$, for example, the Cl atom is *inductively stabilizing*.

The presence of a nearby electron-withdrawing substituent does not always lead to stabilization. If electron density (i.e., negative charge) is drawn away from an atom that has a *positive* charge, then the concentration of positive charge on that atom *increases*, which is destabilizing. This is what we observe in $ClCH_2CH_2OH_2^+$, the conjugate acid of 2-chloroethanol. In this case, the Cl atom is *inductively destabilizing*.

Less stable cation

Electron density is removed from the O^+, thus *destabilizing the positive charge.*

Cl is more electronegative than H and is thus *electron withdrawing.*

In general, cations are *destabilized* by electron-withdrawing groups near the positive charge.

PROBLEM 6.22 Predict which of the following protonated alcohols is the stronger acid.

A   B

Most substituents, like chlorine, are inductively electron-withdrawing groups because most atoms common in organic compounds are more electronegative than hydrogen. However, there are a handful of substituents that are inductively **electron donating**. A silicon atom, for example, is electron donating relative to hydrogen because silicon is less electronegative than hydrogen (1.90 vs. 2.20).

The most common electron-donating groups in organic chemistry are *alkyl groups*. $(CH_3)_3COH$ (pK_a = 19), for example, is about three pK_a units *less acidic* than CH_3CH_2OH (pK_a = 16), suggesting that $(CH_3)_3CO^-$ is less stable than $CH_3CH_2O^-$. $(CH_3)_3CO^-$ is less stable because the electron-donating ability of the CH_3 groups compared to H atoms increases the concentration of negative charge on O.

In general, anions are *destabilized* by electron-donating groups near the negative charge.

The opposite is true for cations. $CH_3NH_3^+$ (pK_a = 10.63), for example, is a weaker acid than NH_4^+ (pK_a = 9.4), suggesting that $CH_3NH_3^+$ is the more stable cation. $CH_3NH_3^+$ is more stable because the electron-donating ability of the CH_3 group compared to H decreases the concentration of positive charge on N.

In general, cations are *stabilized* by electron-donating groups near the positive charge.

Draw the arrow in $CH_3NH_3^+$ in the previous graphic to represent the inductive effect of the CH_3 group.

PROBLEM 6.23 Predict which compound, **A** or **B**, is the stronger acid. Explain.

SH H_2S

A   B

This effect is particularly important for **carbocations**—species that contain a positively charged carbon atom (C^+)—which are key reactive intermediates in a variety of chemical reactions.

> Carbocation stability generally increases with each additional alkyl group attached to the positively charged carbon atom.

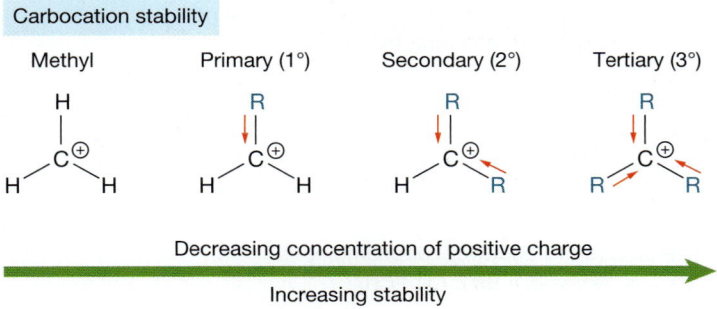

Carbocation stability

Methyl   Primary (1°)   Secondary (2°)   Tertiary (3°)

Decreasing concentration of positive charge

Increasing stability

The inductively electron-donating ability of each alkyl group helps to diminish the concentration of positive charge on C^+. (Alkyl groups attached to C^+ also stabilize carbocations through hyperconjugation, a topic discussed in Interchapter D.)

Because of the impact on carbocation stability, carbocations are distinguished based on the number of alkyl groups attached to C^+. A **methyl cation** (H_3C^+) has no attached alkyl groups; a **primary (1°) carbocation**, RCH_2^+, has one alkyl group attached to C^+; a **secondary (2°) carbocation**, R_2CH^+, has two alkyl groups; and a **tertiary (3°) carbocation**, R_3C^+, has three alkyl groups. Carbocation stability increases in the following order: methyl < 1° < 2° < 3°.

The difference in electronegativity between carbon and hydrogen is quite small—2.55 for carbon compared to 2.20 for hydrogen—but it is enough to make alkyl groups inductively electron donating relative to hydrogen, even though they are made up of only C—H and C—C single bonds. The reason becomes clearer if we closely examine CH_3, the simplest alkyl group (Fig. 6-13). In each of the C—H bonds, a small bond dipole points toward the C atom, giving the carbon atom a partial negative charge (δ^-). This buildup of negative charge enables that carbon atom to donate some electron density to a group to which it is bonded.

Alkyl groups of different sizes have different electron-donating capabilities, but the difference is typically insubstantial. For example, the alkyl group attached to the acidic OH in ethanol (CH_3CH_2OH) is twice the size of that in methanol (CH_3OH), but both have pK_a values of about 16.

Thus far, we have discussed inductive effects on ion stability only in a completely qualitative sense—that is, electron-withdrawing groups stabilize nearby negative charges and destabilize nearby positive charges, whereas electron-donating groups destabilize nearby negative charges and stabilize nearby positive charges. However,

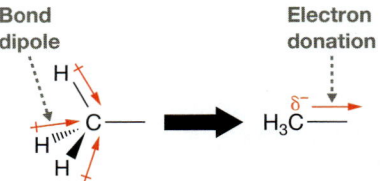

Bond dipole   Electron donation

FIGURE 6-13 The electron-donating ability of an alkyl group (*Left*) C is slightly more electronegative than H, so each C—H bond dipole points toward C. (*Right*) The electron density (δ^-) built up on C can be donated to adjacent atoms.

we can be somewhat quantitative as well, by considering a few straightforward principles.

---

1. The inductive electron-withdrawing ability of an uncharged atom increases as its electronegativity increases, whereas an atom's electron-donating ability increases as its electronegativity decreases. For example,
   - Fluorine is more electronegative than chlorine, so fluorine is more inductively electron withdrawing than chlorine.
   - Silicon is less electronegative than carbon, so the H_3Si group is more electron donating than the H_3C group.

These relationships occur because inductive effects arise from a distortion of electron density along covalent bonds. Thus, there is a greater shift of electron density toward the more electronegative group and away from the less electronegative group.

---

2. Charged substituents have more pronounced inductive effects than uncharged substituents. For example,
   - $(CH_3)_3N^+$ is more electron withdrawing than fluorine.
   - CO_2^- is more electron donating than CH_3.

A full positive charge signifies a substituent that is very highly electron deficient. A full negative charge, on the other hand, signifies a large excess of electron density.

---

3. Inductive effects are additive. For example,
   - $CF_3CH_2^+$ is *less stable* than $CH_2FCH_2^+$.
   - $(CH_3CH_2)_2NH_2^+$ is *more stable* than $CH_3CH_2NH_3^+$.

In $CF_3CH_2^+$, three F atoms withdraw electron density from the positively charged C atom, whereas only one F atom does in $CH_2FCH_2^+$. As a result, there is a greater concentration of positive charge in $CF_3CH_2^+$ than in $CH_2FCH_2^+$. In $(CH_3CH_2)_2NH_2^+$, two alkyl groups donate electron density to the positively charged N atom, whereas only one alkyl group does in $CH_3CH_2NH_3^+$. Thus, there is a smaller concentration of charge in $(CH_3CH_2)_2NH_2^+$.

---

4. Inductive effects fall off very quickly with distance. For example,
   - $CH_3CH_2CHFCH_2NH^-$ is *more stable* than $CH_3CHFCH_2CH_2NH^-$.

The F atom is separated from the negatively charged N by one fewer bond in $CH_3CH_2CHFCH_2NH^-$ than in $CH_3CHFCH_2CH_2NH^-$. In $CH_3CH_2CHFCH_2NH^-$, therefore, the electron-withdrawing effect of F is transmitted more efficiently to the negatively charged N, resulting in a lower concentration of negative charge on N.

SOLVED PROBLEM 6.24

Predict which carboxylic acid, **A** or **B**, is more acidic.

Think For each acid, does the stability of the acid or the stability of the conjugate base dictate pK_a? Do electron-donating or electron-withdrawing effects stabilize those species? Are the Br and I substituents electron donating or electron withdrawing? Which substituent invokes stronger inductive effects?

Solve **A** and **B** are uncharged acids, so their pK_a values are dictated by the stability of their negatively charged conjugate bases, in which the OH group has become O^-. Both Br and I are electron-withdrawing substituents, so they stabilize negatively charged species. Br is more electronegative than I, so Br better stabilizes the conjugate base, in which case **A** is a stronger acid than **B**. (Notice that the larger size of I compared to Br does *not* come into play because in neither case does the negative charge appear on those atoms.)

PROBLEM 6.25 Which acid is stronger, $O_2NCH_2CH_2OH$ or $H_2NCH_2CH_2OH$? Explain. *Hint:* Draw out the complete Lewis structure for each.

Superacids: How Strong Can an Acid Be?

Sulfuric acid (H_2SO_4) is arguably the most important industrial compound—nearly 200 million tons of it are produced in huge facilities (see photo) each year, about 30% of that in the United States alone. It is so important, in part, because it is among the strongest acids that exist ($pK_a = -9$), and strong acids catalyze numerous chemical reactions.

According to Table 6-1, however, trifluoromethanesulfonic acid (CF_3SO_3H; $pK_a = -13$) is about four pK_a units (a factor of 10,000) stronger than sulfuric acid. As a result, CF_3SO_3H is one of a variety of acids classified as a *superacid*. A few others are shown below, along with their pK_a values. The acid at the far right—the strongest acid known—is some 16 pK_a units lower than sulfuric acid, making it 10 quadrillion times stronger!

| $pK_a = -10.7$ | $pK_a = -13.3$ | $pK_a = -18$ | $pK_a = -25$ |

What all of these superacids have in common is very heavy stabilization of the negative charge that appears in the conjugate base. That stabilization results from extensive resonance delocalization of the charge, multiple inductively electron-withdrawing groups, or both.

What is the purpose of superacids, beyond simply pushing the limits of chemistry? One application is to make acid catalysis "greener" by carrying out certain reactions using smaller quantities of strong acid at lower temperatures. A second use is as a medium in which to generate and sustain highly reactive species, such as carbocations, so they can be studied. And a third application, used in the petroleum industry, is to convert long-chain hydrocarbons into smaller ones for the production of high-octane fuels.

PROBLEM 6.26 Which carboxylic acid is more acidic? Explain.

6.7 Strategies for Success: Ranking Acid and Base Strengths — The Relative Importance of Effects on Charge

Thus far in this chapter, we have analyzed relative acid strengths by considering only one charge-stability factor at a time. Often in organic chemistry, however, we must consider multiple factors simultaneously. To do so, we need to understand the relative importance of each factor in evaluating the stability of a particular species.

With relatively few exceptions, the order of priority for these factors follows the "CARDIN"-al rule:

Charge > Atom > Resonance Delocalization > Inductive effects

That is to say: Whether a species is *charged or uncharged* is typically the most important factor; the next-most-important factor is the type of *atom* on which the charge appears; this is followed by the extent of charge delocalization via *resonance*; and *inductive effects* are usually the least important factor. When evaluating the relative stabilities of two species, therefore, ask the following questions in order. The first question to which the answer is "yes" will likely correspond to the factor that most influences the species' relative stabilities.

1. Do the species have different charges?

A charged species is generally more reactive than one that bears no formal charge.

2. Do the charges appear on different atoms?

- A charge, positive or negative, is better accommodated on a larger atom (i.e., one farther down the periodic table).
- A negative charge is better accommodated on a more electronegative atom; a positive charge is better accommodated on a less electronegative atom.

3. Are the charges delocalized differently via resonance?

All else being equal, stability increases as the number of atoms over which a charge is shared increases.

4. Are there differences in inductive effects?

- Electron-withdrawing groups stabilize nearby negative charges but destabilize nearby positive charges.
- Electron-donating groups stabilize nearby positive charges but destabilize nearby negative charges.

The relative importance of these factors is reflected in the magnitude of their effect on pK_a. The appearance of a charge has quite a significant effect. A protonated amine ($R-NH_3^+$), for example, is nearly 30 pK_a units more acidic than a comparable uncharged amine ($R-NH_2$). A protonated alcohol ($R-OH_2^+$) is about 18 pK_a units more acidic than an uncharged alcohol ($R-OH$).

YOUR TURN **6.18**

Verify the differences in pK_a between the protonated and uncharged amines and alcohols just mentioned by estimating the pK_a values of the generic species shown below based on actual compounds in Table 6-1. Write them in the spaces provided here.

pK_a: $R-NH_3^+$ _____ $R-NH_2$ _____ $R-OH_2^+$ _____ $R-OH$ _____

The type of atom on which an acidic proton is found also has a dramatic effect on pK_a. A typical alcohol ($R-OH$), for example, is about 20 pK_a units more acidic than a typical amine ($R-NH_2$). The main difference is that the negative charge resides on an oxygen atom in the $R-O^-$ conjugate base and on the nitrogen atom in the $R-NH^-$ conjugate base. When we consider differences in type of atom, we should also consider effective electronegativity because an acidic hydrogen on an sp-hybridized carbon atom is over 20 pK_a units more acidic than a comparable hydrogen on an sp^3-hybridized carbon atom.

YOUR TURN **6.19**

Verify the preceding statements by estimating the pK_a values of the generic alcohol, amine, alkane, and terminal alkyne species shown below based on actual compounds in Table 6-1. Write them in the spaces provided here.

pK_a: $R-OH$ _____ $R-NH_2$ _____ $R-CH_3$ _____ $RC\equiv CH$ _____

Resonance effects have less of an impact on pK_a values than the type of atom on which the acidic hydrogen is found. Nevertheless, they are quite important because a carboxylic acid ($R-CO_2H$) is about 11 pK_a units more acidic than an alcohol ($R-OH$). Similarly, the negative charge on N in the conjugate base of aniline ($C_6H_5NH^-$) can be delocalized by resonance, as shown below, making aniline about 11 pK_a units more acidic than a typical amine ($R-NH_2$).

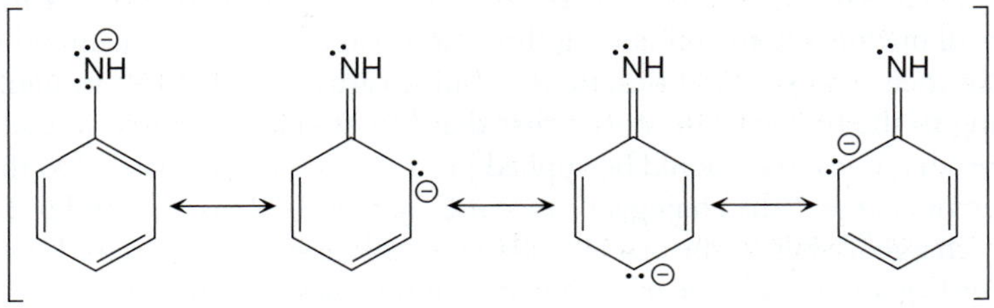

In the preceding graphic, draw the curved arrows needed to show how each resonance structure of $C_6H_5NH^-$ is converted to the one on its right. Then, draw the resonance hybrid to the right of the resonance structures.

Inductive effects generally have the least effect on pK_a values. For example, the pK_a of 2,2,2-trifluoroethanol (F_3CCH_2OH) is 12.4, whereas the pK_a of ethanol (CH_3CH_2OH) is 16. With its three F atoms near the acidic O—H group, 2,2,2-trifluoroethanol exhibits relatively strong inductive effects, yet it is only about four pK_a units more acidic than ethanol.

Let's now apply this knowledge toward ranking the strengths of the following four acids:

A   **B**   **C**   **D**

Because pK_a is defined in terms of each acid's reaction with water, we begin by writing out their reactions, as shown in Equation 6-22.

$$\text{(6-22a)}$$

$$\text{(6-22b)}$$

$$\text{(6-22c)}$$

$$\text{(6-22d)}$$

CONNECTIONS Aniline ($C_6H_5NH_2$, previous page) is a precursor used to make acetaminophen, the pain reliever sold as the brand name Tylenol. Aniline is also an important compound in the dye industry, where it is a precursor to indigo, the dye commonly associated with blue jeans.

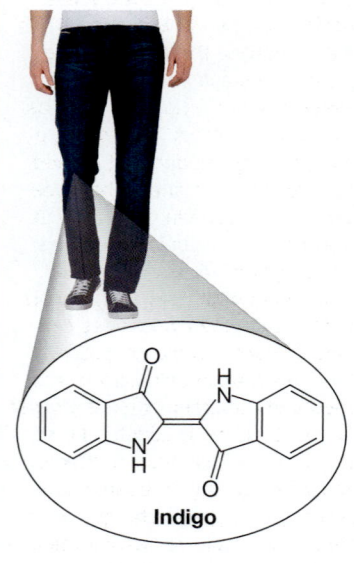

Indigo

To help determine how favorable each reaction is (the more favorable the reaction, the stronger the acid), construct a free energy diagram of all four reactions on a single plot (see Fig. 6-14). We do so by determining the relative stabilities of the various reactant and product species. H_2O (a reactant) can be ignored because it appears in all cases. The remaining reactants can be evaluated using the tie-breaking questions. Do the remaining reactants have different charges? The acid in Equation 6-22c (red) is charged, whereas those in Equation 6-22a, 6-22b, and 6-22d are uncharged. In Figure 6-14, therefore, the reactants in Equation 6-22a, 6-22b, and 6-22d are lower in energy than the reactants in Equation 6-22c. Because the acids in Equation 6-22a, 6-22b, and 6-22d are all uncharged, the remaining three tie-breaking questions cannot be used to distinguish their relative stabilities; their stabilities are assumed to be similar.

Moving to the product side, notice first that H_3O^+ can be ignored in each case, so the tie-breaking questions should be applied just to the conjugate bases. Do the species have different charges? The conjugate bases in Equation 6-22a, 6-22b, and 6-22d are all negatively charged, whereas the one in Equation 6-22c is uncharged (red). Therefore, the products of Equation 6-22c are more stable than the ones from the other reactions.

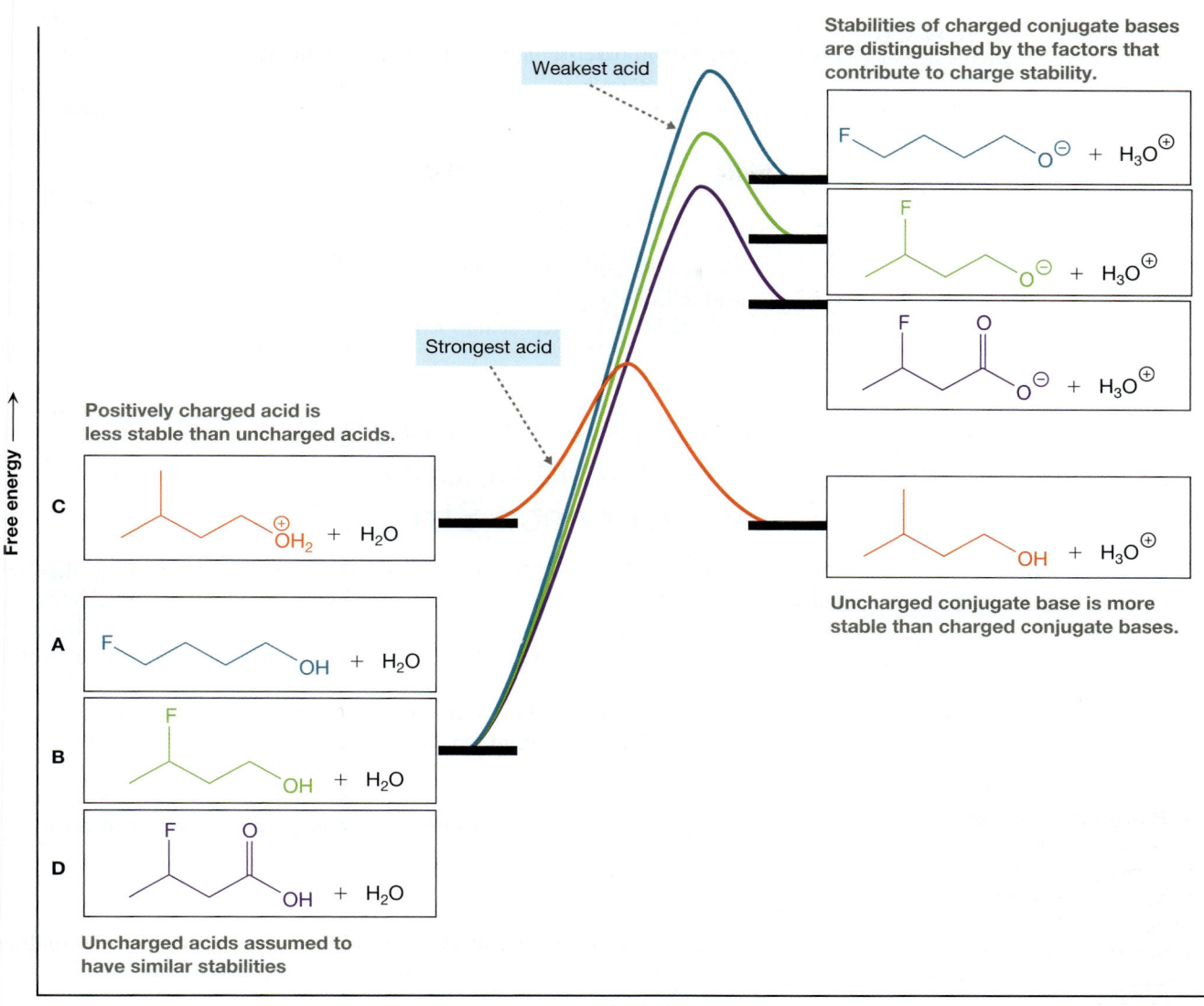

Weakest acid

Stabilities of charged conjugate bases are distinguished by the factors that contribute to charge stability.

Strongest acid

Positively charged acid is less stable than uncharged acids.

Uncharged conjugate base is more stable than charged conjugate bases.

Uncharged acids assumed to have similar stabilities

Reaction coordinate →

Free energy →

FIGURE 6-14 Free energy diagrams for the proton transfer reactions shown in Equation 6-22 The reactants of the red curve are highest in energy because the acid in that reaction is charged, whereas the ones in the other reactions are uncharged. The products of the red curve are lowest in energy because the conjugate base in that reaction is uncharged; the conjugate bases in the other reactions are charged. The products of the purple curve are next lowest in energy because the negative charge on the conjugate base is resonance delocalized. The products of the green curve are lower in energy than the products of the blue curve due to inductive effects. In the green conjugate base, F is closer to the negative charge, which more substantially decreases the concentration of negative charge on O⁻.

For the charged conjugate bases in Equation 6-22a, 6-22b, and 6-22d, do the charges appear on different atoms? The −1 charge appears on O in all three cases, so the answer is no.

Are the charges delocalized differently via resonance? The conjugate base in Equation 6-22d (purple) has two resonance structures that delocalize the negative charge over the two O atoms, making it more stable than the negatively charged conjugate bases in Equation 6-22a (blue) and 6-22b (green).

Are there differences in inductive effects for the conjugate bases in Equation 6-22a and 6-22b? Both have a highly electron-withdrawing F atom that inductively stabilizes the negative charge, but the one in Equation 6-22b (green) is closer to the negative charge, so the conjugate base in Equation 6-22b is more stable than the one in Equation 6-22a.

By establishing the relative stabilities of all four sets of products, we can see in Figure 6-14 that deprotonation of the acids becomes more favorable (less positive ΔG°_{rxn}) in the order $A < B < D < C$. This, then, is the order of increasing acid strength.

If you know how to rank species according to acid strength, then you can rank species according to base strength by applying the concept from Section 6.2a: *The stronger a base's conjugate acid, the weaker the base.* Thus, the conjugate bases in Equation 6-22 increase in base strength in the following order:

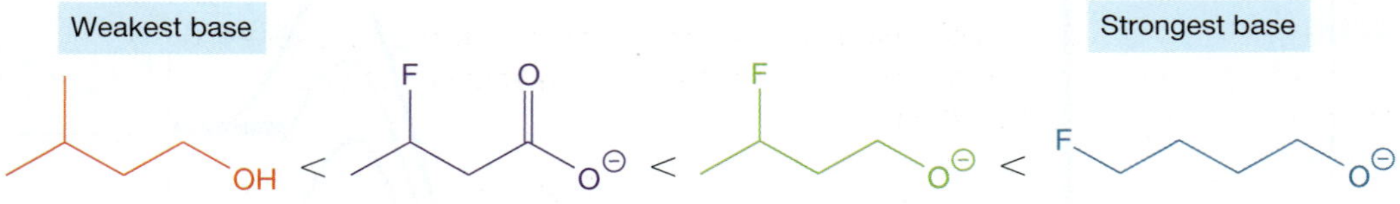

The molecule on the left is the weakest base because it is the conjugate base of the strongest acid in Equation 6-22. The species on the right is the strongest base because its conjugate acid is the weakest acid in Equation 6-22.

6.8 Strategies for Success: Determining Relative Contributions by Resonance Structures

Our knowledge of charge stability is applicable beyond simply predicting the outcomes of proton transfer reactions. In this section, we discuss how charge stability can be used to determine the relative contributions by resonance structures. In Chapter 7, we explain how charge stability can be applied to elementary steps other than proton transfers.

Although the resonance hybrid is an *average* of all the resonance contributors, not all resonance contributors are weighted equally. As discussed in Section 1.10:

> The resonance hybrid looks most like the lowest-energy (most stable) resonance structure.

We also learned in Chapter 1 that a variety of factors govern the relative stabilities of resonance structures. In particular, a resonance structure is more stable with:

- a greater number of atoms with octets
- more covalent bonds
- fewer atoms with nonzero formal charges

Based on what you have learned here in Chapter 6, a resonance structure is more stable with fewer atoms possessing nonzero formal charges because charged species are inherently less stable than the corresponding uncharged molecules. Consider, for example, the following resonance contributors to $H_2C{=}CH{-}OCH_3$:

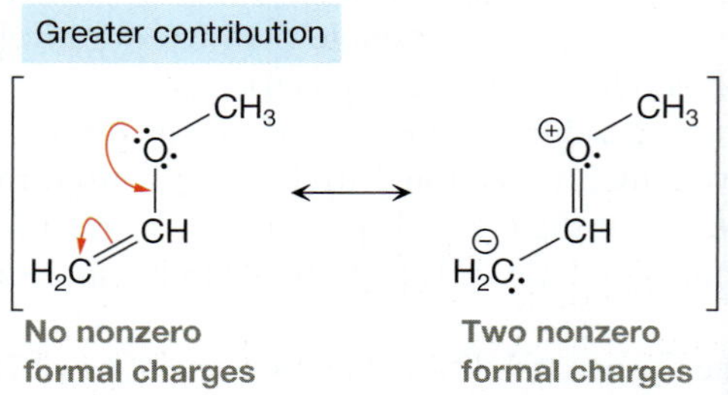

No atoms in the structure on the left possess a nonzero formal charge, but the O atom has a +1 formal charge and one C atom has a −1 formal charge in the structure on the right. The structure on the right is less stable, therefore, so it makes a smaller contribution to the resonance hybrid. Thus, a molecule of $H_2C=CH-OCH_3$ looks more like the resonance contributor on the left.

PROBLEM 6.27 (a) Draw the curved arrows necessary to convert the resonance structure on the left into the one on the right. (b) Which resonance structure contributes more to the hybrid? Explain.

$$\left[H_3C-C\equiv C-\ddot{\underset{..}{C}}l: \longleftrightarrow H_3C-\overset{\ominus}{\underset{..}{C}}=C=\overset{\oplus}{\underset{..}{C}}l: \right]$$

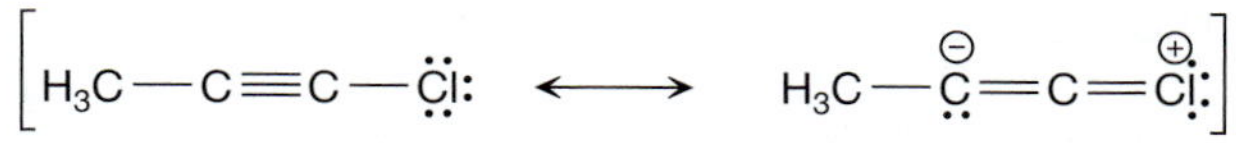

Frequently, the relative stabilities of resonance structures cannot be determined on the basis of the octet rule, the total number of bonds, or the number of atoms bearing a formal charge, but they can be determined on the basis of charge stability. In these cases, first examine the atoms on which the formal charges appear (the more important factor) and then examine any inductive effects. Because we are considering a single resonance contributor only, charge delocalization via resonance does not apply.

In both resonance contributors to an enolate anion, for example, each atom has its octet or duet, both resonance structures have the same total number of bonds, and each has a single atom with a -1 formal charge.

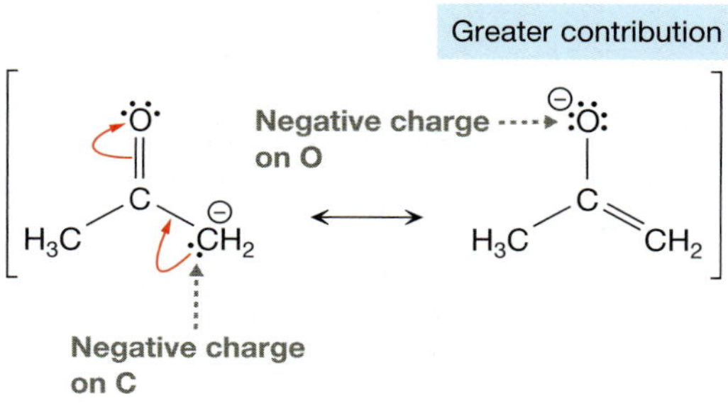

Greater contribution

Negative charge on O

Negative charge on C

The main difference between these structures is the atom on which the formal charge appears — namely, C in the structure on the left and O in the structure on the right. O is more electronegative than C, so O can better accommodate the negative charge than C. The structure on the right, therefore, is more stable and makes the greater contribution to the resonance hybrid.

PROBLEM 6.28 Which resonance structure contributes more to the hybrid? Explain.

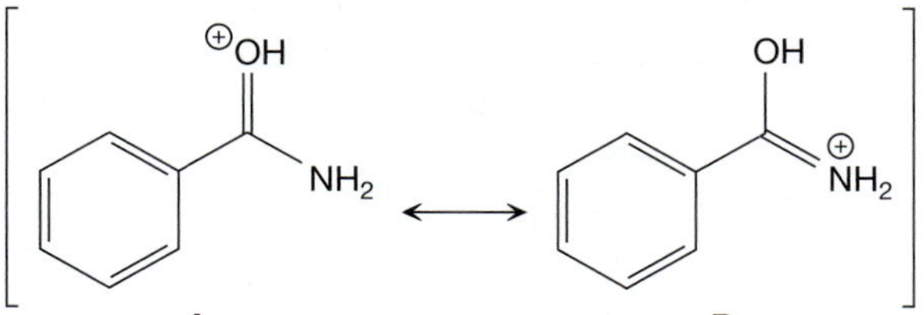

A   B

PROBLEM 6.29 Of these two resonance structures, the greater contribution is from the second one. (a) Explain why this is counterintuitive based on charge stability. (b) Why does the structure on the right contribute more?

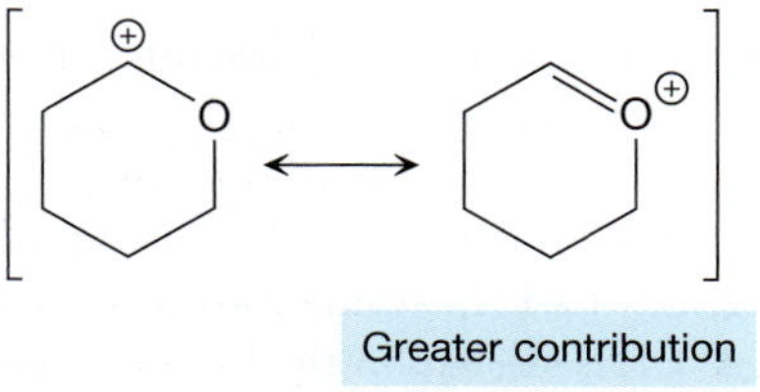

Greater contribution

In the following two resonance contributors, there is a +1 formal charge on a different C atom in each.

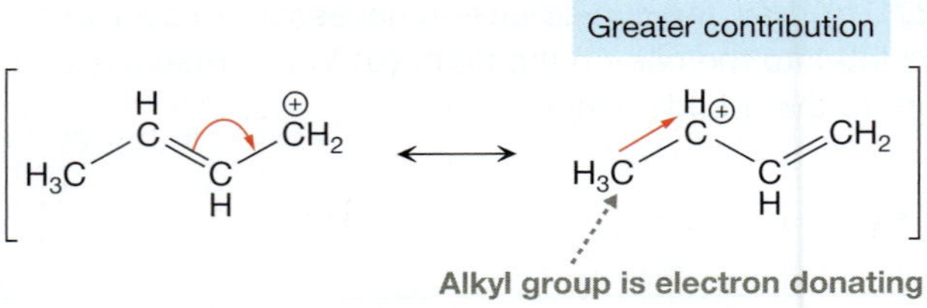

Greater contribution

Alkyl group is electron donating

In the structure on the left, the C atom bearing the positive charge is bonded to two H atoms and a C atom that is part of the double bond. In the structure on the right, the positively charged C atom is bonded to one H atom, a CH_3 group, and a C atom that is part of the double bond. Because the CH_3 group is electron donating compared to H, the positive charge on the C atom is reduced in the structure on the right, making this structure more stable and a more significant resonance contributor.

PROBLEM 6.30 Which is the more important resonance contributor, **A** or **B**? Explain.

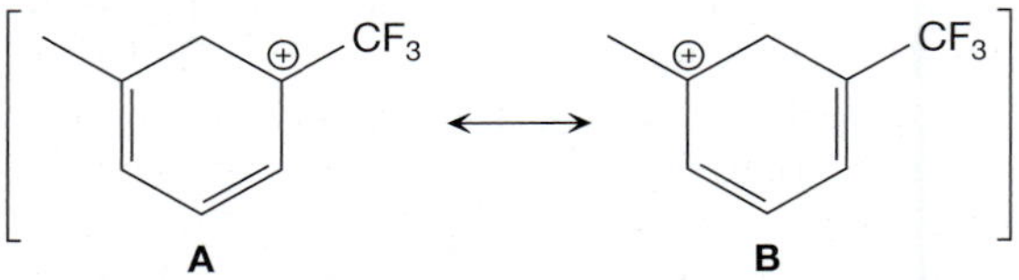

A   B

6.9 The Structure of Amino Acids in Solution as a Function of pH

Recall from Section 1.14 that the general form of an α-amino acid contains both an amino group (NH_2) and a carboxyl group (CO_2H) bonded to the same C atom:

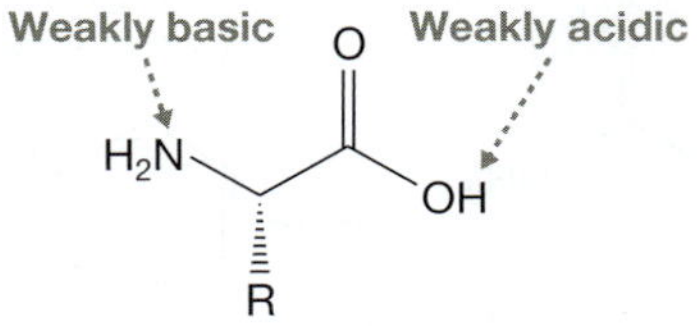

Weakly basic   Weakly acidic

An α-amino acid

It turns out, however, that this is never the dominant form in aqueous solution because the carboxyl group is weakly acidic and the amino group is weakly basic.

What form, then, does an α-amino acid take in aqueous solution? The answer depends on the pH of the solution. Under highly acidic (pH < 2) conditions, the weakly basic N atom is protonated. The resulting species, which bears an overall charge of +1, is shown on the left in Equation 6-23.

This is a zwitterion, because it has a separated positive and negative charge, but its net charge is zero.

$pK_a \approx 2$ $pK_a \approx 9$ to 10

$$(6\text{-}23)$$

pH << 2   pH between ~2 and ~9   pH >> 9

As the solution becomes more basic (i.e., as the pH of the solution increases), a proton can be removed. It does *not* come from the NH_3^+, however, because the most acidic proton in the cationic species is the one that belongs to the CO_2H; its pK_a is around 2, whereas that of the NH_3^+ group is around 9 to 10. (The exact pK_a values depend on the specific amino acid, characterized by the side group R.) Therefore, when the pH of the solution has risen significantly above 2, the CO_2H proton is lost and the dominant form is the middle species in Equation 6-23. This species, called a **zwitterion** (pronounced ZVITTER-eye-on), has both a positive and a negative formal charge, but a net charge of zero.

As the solution becomes more basic still, the proton of the NH_3^+ group is lost, yielding the species on the right in Equation 6-23. This second deprotonation takes place when the pH of the solution is significantly above 9 or 10, the pK_a of the NH_3^+.

Table 6-2 (next page) lists the pK_a values associated with the 20 naturally occurring amino acids. For the majority of these amino acids, only two pK_a values are listed—one for the CO_2H group, and one for the NH_3^+. For seven of them, however, a third pK_a value is given. Those amino acids have side chains that are **ionizable**, meaning that an atom in the side chain can gain or lose a proton to become charged.

PROBLEM 6.31 Draw the dominant form of alanine in solutions whose pH values are 1, 4, 8, and 11.

Aspartic acid is one of the seven amino acids with an ionizable side chain. Its side chain (shown in blue in Equation 6-24) contains a second CO_2H.

Aspartic acid

$pK_a = 2.10$ $pK_a = 9.82$

$$(6\text{-}24)$$

The zwitterion

pH < 2.10   2.10 < pH < 3.86   3.86 < pH < 9.82   pH > 9.82

At low pH (i.e., strongly acidic conditions), aspartic acid is in its fully protonated form, in which both CO_2H groups are uncharged and the NH_3^+ is positively charged. As the pH is increased, the first proton that is lost comes from the CO_2H group that is part of the amino acid backbone ($pK_a = 2.10$), thus yielding the zwitterion. Increasing the pH further causes the CO_2H on the side chain ($pK_a = 3.86$) to lose

TABLE 6-2 pK_a Values of the 20 Naturally Occurring Amino Acids

Amino Acid	Side Chain	pK_a of $-CO_2H$	pK_a of $-NH_3^+$	pK_a of Side Chain	Amino Acid	Side Chain	pK_a of $-CO_2H$	pK_a of $-NH_3^+$	pK_a of Side Chain
Alanine	—CH$_3$	2.35	9.87	N/A	Leucine	—C$_{H_2}$—CH(CH$_3$)CH$_3$	2.33	9.74	N/A
Arginine	—C$_{H_2}$(C$_{H_2}$)$_2$—N$_H$—C(=$^\oplus$NH$_2$)—NH$_2$	2.01	9.04	12.48	Lysine	—C$_{H_2}$(C$_{H_2}$)$_3$—$^\oplus$NH$_3$	2.18	8.95	10.53
Asparagine	—C$_{H_2}$—C(=O)—NH$_2$	2.02	8.80	N/A	Methionine	—C$_{H_2}$—C$_{H_2}$—S—CH$_3$	2.28	9.21	N/A
Aspartic acid	—C$_{H_2}$—C(=O)—OH	2.10	9.82	3.86	Phenylalanine	—C$_{H_2}$—C$_6$H$_5$	2.58	9.24	N/A
Cysteine	—C$_{H_2}$—SH	2.05	10.25	8.00	Proline[a]	(ring structure)	2.00	10.60	N/A
Glutamic acid	—C$_{H_2}$—C$_{H_2}$—C(=O)—OH	2.10	9.47	4.07	Serine	—C$_{H_2}$—OH	2.21	9.15	N/A
Glutamine	—C$_{H_2}$—C$_{H_2}$—C(=O)—NH$_2$	2.17	9.13	N/A	Threonine	—CH(CH$_3$)OH	2.09	9.10	N/A
Glycine	—H	2.35	9.78	N/A	Tryptophan	—C$_{H_2}$—(indole)	2.38	9.39	N/A
Histidine	—C$_{H_2}$—(imidazole)	1.77	9.18	6.10	Tyrosine	—C$_{H_2}$—C$_6$H$_4$—OH	2.20	9.11	10.07
Isoleucine	—CH(CH$_3$)CH$_2$CH$_3$	2.32	9.76	N/A	Valine	—CH(CH$_3$)CH$_3$	2.29	9.72	N/A

[a]For proline, the side chain is shown in black.

its proton. Under more strongly basic conditions, the NH_3^+ group ($pK_a = 9.82$) loses its proton.

PROBLEM 6.32 Draw the structure of the most abundant form of glutamic acid in solutions whose pH values are 1, 3, 5, and 11.

Lysine is another amino acid that has an ionizable side chain, though it has a N atom that can be protonated, rather than a CO_2H that can be deprotonated. As indicated in Equation 6-25, the CO_2H group is uncharged and both N atoms are positively charged (i.e., are protonated) at very low pH. As the pH of the solution is increased, the first group to lose its proton is the CO_2H ($pK_a = 2.18$). Next is the NH_3^+ group of the amino acid backbone ($pK_a = 8.95$), and last is the NH_3^+ group of the side chain ($pK_a = 10.53$). With lysine, the zwitterion is the product of the second deprotonation, not the first.

PROBLEM 6.33 Draw the structure of the most abundant form of arginine in solutions whose pH values are 1, 4, 8, 10, and 14.

6.10 Electrophoresis and Isoelectric Focusing

Knowing the pK_a values of amino acids is quite helpful in **electrophoresis**, one of the most common ways of separating a mixture of amino acids or proteins. In electrophoresis, the mixture is spotted on a gel or strip of paper that has been buffered to a specific pH (Fig. 6-15). A high voltage (50–1000 V) is applied by positively and negatively charged poles—the **anode** and **cathode**, respectively—situated at opposite ends of the apparatus. Any ion that has a net positive charge at that pH will migrate toward the negatively charged terminal, whereas any ion having a net negative charge will migrate toward the positively charged terminal. If the net charge is zero (as in a zwitterion), then the species will not move. To help you remember this, use the following mnemonic:

- **Cat**ions migrate toward the **cat**hode.
- **An**ions migrate toward the **an**ode.

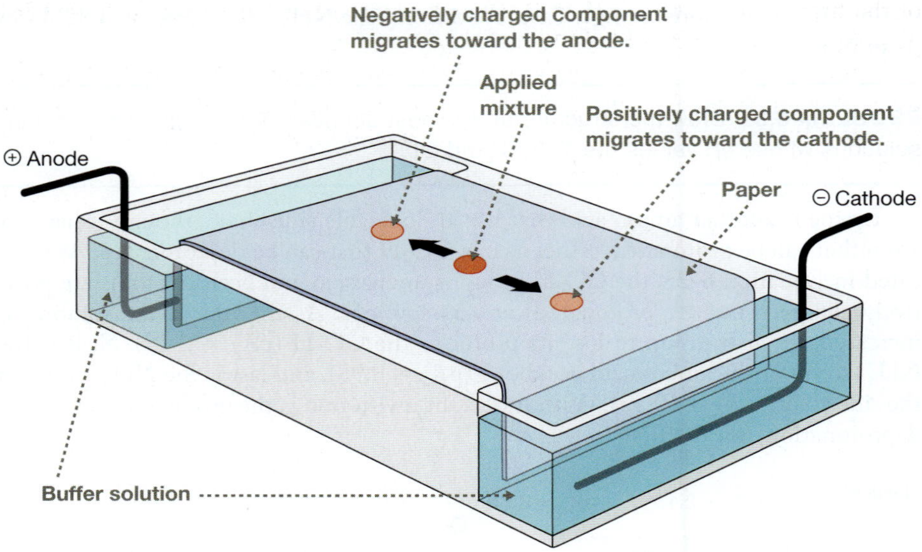

FIGURE 6-15 Paper electrophoresis Any substance carrying a positive charge will migrate toward the negatively charged terminal (the cathode) and any substance carrying a negative charge will migrate toward the positively charged terminal (the anode). An uncharged substance will not migrate.

An amino acid's **isoelectric pH**, or **isoelectric point (pI)**, is *the pH at which the substance has a charge of zero*, and it is unique for each of the 20 amino acids. If the pH of the solution is higher (i.e., more basic) than the pI, then the substance will, on average, be deprotonated to a greater extent and will carry a net negative charge. Conversely, if the pH is lower (i.e., more acidic) than the pI, then the substance will, on average, be protonated to a greater extent and will carry a net positive charge.

An amino acid's pI can be computed straightforwardly from its pK_a values:

An amino acid's pI is the average of its two pK_a values associated with the zwitterion—namely, the pK_a value corresponding to the proton transfer reaction in which the zwitterion is a product and the pK_a value corresponding to the proton transfer reaction in which the zwitterion is a reactant.

For an amino acid with only two ionizable groups (i.e., no ionizable side chains), there are only two pK_a values in all. As shown previously in Equation 6-23, the first pK_a corresponds to the proton transfer reaction in which the zwitterion is produced, and the second corresponds to the proton transfer reaction in which the zwitterion reacts. Therefore, the pI is simply the average of those two pK_a values. For example, for glycine, pI = (2.35 + 9.78)/2 = 6.07.

PROBLEM 6.34 In an electrophoresis experiment where the pH of the gel is 7, will glycine migrate toward the anode or the cathode? Explain.

PROBLEM 6.35 Calculate the pI of alanine. In an electrophoresis experiment where the pH of the gel is 4, will alanine migrate toward the anode or the cathode? Explain.

If an amino acid has an ionizable side chain, then there are three ionizable groups in all and three pK_a values to consider. The way in which pI is computed does not change, however, because it is still the average of the two pK_a values associated with the reactions that involve the zwitterion. Which two pK_a values those are, however, depends on whether the side chain is acidic or basic.

Aspartic acid, for example, has an ionizable side chain that contains a CO_2H and is therefore acidic. According to Equation 6-24, aspartic acid's zwitterion is a product of the first deprotonation (pK_a = 2.10) and a reactant in the second (pK_a = 3.86). Therefore, aspartic acid's pI is (2.10 + 3.86)/2 = 2.98.

Lysine, on the other hand, has a side chain that contains an NH_2 and is therefore basic. According to Equation 6-25, lysine's zwitterion is a product in the second deprotonation (pK_a = 8.95) and a reactant in the third (pK_a = 10.53). As a result, the pI for lysine is (8.95 + 10.53)/2 = 9.74.

The three examples we have discussed show how an amino acid's pI depends on the nature of the side chain:

- Amino acids with *acidic* side chains have pI values that lie between about 2 and 3.
- Amino acids with *neutral* side chains have pI values that lie between about 5 and 6.
- Amino acids with *basic* side chains have pI values >7.

PROBLEM 6.36 Compute the pI value of glutamic acid. If the gel in an electrophoresis experiment is at pH 7, in which direction will glutamic acid migrate, toward the cathode or toward the anode? Explain.

Could you use simple electrophoresis to separate a solution containing glycine (pI = 6.07), aspartic acid (pI = 2.98), lysine (pI = 9.74), and asparagine (pI = 5.41)? If the pH of the gel is buffered to 5.41, then asparagine will not migrate (because pH = pI), aspartic acid will migrate toward the positively charged terminal (because the gel is more basic than its pI), but both glycine and lysine will migrate toward the negatively charged terminal (because the gel is more acidic than their pIs). It is difficult, therefore, to separate glycine from lysine in this manner.

In cases like this, chemists often turn to **isoelectric focusing**, a particularly elegant way of separating amino acids or proteins. Isoelectric focusing is similar to the electrophoresis experiments described previously, except that a *pH gradient* is used. That is, the pH increases toward the cathode along the paper or gel (Fig. 6-16), rather than being constant throughout. If an amino acid is at a location where the pH is higher (more basic) than its pI value, then it will have a net negative charge and will move

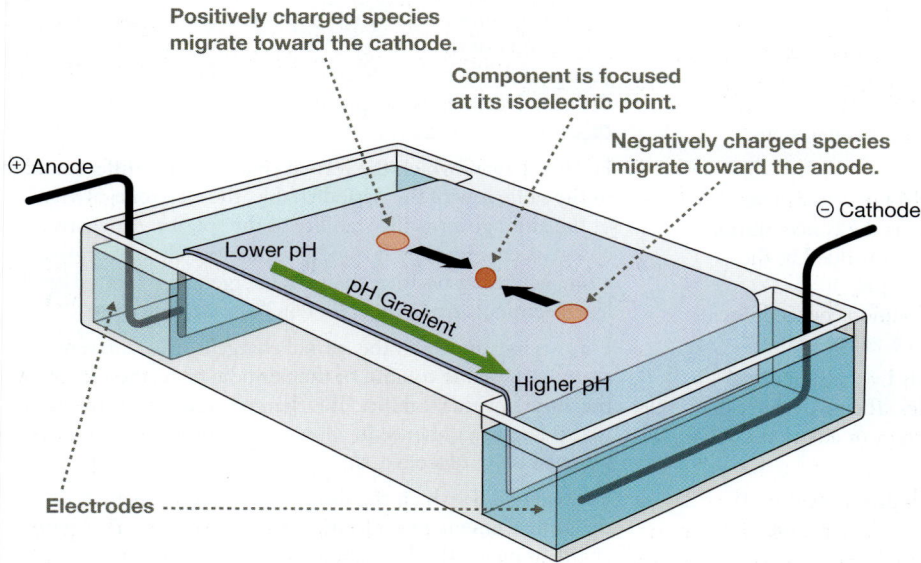

Positively charged species migrate toward the cathode.

Component is focused at its isoelectric point.

Negatively charged species migrate toward the anode.

⊕ Anode

⊖ Cathode

Lower pH

pH Gradient

Higher pH

Electrodes

FIGURE 6-16 Isoelectric focusing of peptides and proteins The paper or gel has a pH gradient in which pH increases toward the cathode and decreases toward the anode. Each peptide "seeks" the point on the paper or gel that corresponds to its pI.

away from the cathode, which is in the direction of lower pH. It will continue to migrate until pH = pI, at which point the amino acid will no longer migrate.

Another species at a location where the pH is lower than its pI will have a net positive charge, and will move toward the cathode, which is in the direction of increasing pH. It, too, will continue to migrate until pH = pI. In this way it is possible to separate amino acids or proteins whose isoelectric points differ by only a few hundredths of a pH unit.

PROBLEM 6.37 The pI for lysine is 9.74. In an isoelectric focusing experiment, toward which terminal will lysine migrate if it is placed on the gel at a pH of 12? If it is placed on the gel at a pH of 1? If it is placed on the gel at a pH of 7? At what pH will lysine remain stationary?

Chapter Summary and Key Terms

- In a **proton transfer reaction**, a proton is transferred from a **Brønsted–Lowry acid** to a **Brønsted–Lowry base** in a single **elementary step**; that is, one bond is broken and another is formed simultaneously. (Section 6.1; Objective 1)

- **Curved arrow notation** describes the movement of electrons in an elementary step of a mechanism, showing explicitly the breaking and/or forming of bonds. A *double-barbed* curved arrow (⟶) represents the movement of a pair of valence electrons. (Section 6.1; Objective 1)

- A reaction's **equilibrium constant**, K_{eq}, reflects the tendency of that reaction to form products. A reaction with $K_{eq} > 1$ favors products and one with $K_{eq} < 1$ favors reactants. (Section 6.2; Objective 2)

- The equilibrium between an acid (HA) and water is described by the **acidity constant**, K_a, where $K_a = [A^-]_{eq}[H_3O^+]_{eq}/[HA]_{eq}$. A compound's K_a reflects its strength as an acid; for convenience, acid strength is usually reported as pK_a, where $pK_a = -\log K_a$. (Section 6.2a; Objective 2)

- In a proton transfer reaction, the equilibrium favors the side *opposite* the stronger acid. (Section 6.2b; Objective 3)

- A solvent's **leveling effect** dictates the maximum strength of an acid or a base that can exist in solution. The strongest acid that can exist is the protonated solvent; the strongest base that can exist is the deprotonated solvent. (Section 6.2c; Objective 4)

- The **percent dissociation** of an acid (HA) in water increases as the pH of the solution increases (i.e., as the concentration of H_3O^+ decreases). This observation is embedded in the Henderson–Hasselbalch equation: $pH = pK_a + \log([A^-]/[HA])$. When the pH of the solution equals the acid's pK_a, the acid is 50% dissociated. (Section 6.2d; Objective 5)

- A compound's pK_a is governed primarily by the functional group on which the acidic proton resides. The pK_a value can be affected by nearby electronegative atoms or adjacent double bonds. (Section 6.4; Objective 6)

- A reaction's **standard Gibbs free energy difference**, ΔG°_{rxn}, is related to the reaction's equilibrium constant: $\Delta G^\circ_{rxn} = -RT \ln K_{eq}$. A reaction tends to be spontaneous if ΔG°_{rxn} is negative (**exergonic**). If ΔG°_{rxn} is positive (**endergonic**), then the reverse reaction tends to be spontaneous. (Section 6.3; Objective 7)

- ΔG°_{rxn} consists of a **standard enthalpy difference** term and a **standard entropy difference** term: $\Delta G^\circ_{rxn} = \Delta H^\circ_{rxn} - T\Delta S^\circ_{rxn}$. For many reactions, including the proton transfer reaction, $\Delta G^\circ_{rxn} \approx \Delta H^\circ_{rxn}$ because the change in entropy is so small. (Section 6.3a; Objective 7)

- A **reaction free energy diagram** plots free energy as a function of the **reaction coordinate**: a measure of geometric changes of the species involved in a reaction as reactants are transformed into products. The maximum in free energy between reactants and products, which corresponds to the **transition state**, creates an energy barrier for the reaction called the **free energy of activation**, $\Delta G^{\circ\ddagger}$. (Section 6.3b; Objective 8)

- Positively charged acids are stronger acids than their uncharged counterparts, indicating that charged species are generally high in energy and tend to be unstable and reactive. (Section 6.5; Objective 9)

- Relative pK_a values for acids reflect the relative *charge stability* of the reactants and products. For an uncharged acid (HA), the stability of the conjugate base (A^-) increases as the pK_a decreases. For a positively charged acid (HA^+), the stability of the acid decreases as the pK_a decreases. (Section 6.6; Objectives 9 and 11)

- For two ions in which the formal charge is on a different atom in the same row of the periodic table, the electronegativity of the atom governs the stability of the species. A negative charge is energetically favored on the more electronegative atom, whereas a positive charge is energetically favored on the less electronegative atom. (Section 6.6a; Objective 10)

- For two ions in which the formal charge is on a different atom in the same column of the periodic table, the size of the atom governs stability. The charge is favored on the atom that is larger, and hence in a lower row of the periodic table. (Section 6.6b; Objective 10)

- For two ions in which the formal charge is on an atom of the same element, hybridization governs stability. A negative charge is energetically favored on the atom with the greater effective electronegativity (i.e., $sp^3 < sp^2 < sp$). A positive charge is energetically favored on the atom with the lower effective electronegativity. (Section 6.6c; Objective 10)

- **Resonance effects** can stabilize a charged species. A species in which a charge is *delocalized* by resonance is more stable than one in which a charge is *localized*. All else being equal, the stability of the charged species increases as the number of atoms over which the charge is delocalized increases. (Section 6.6d; Objective 10)

- **Inductive effects** can affect the stability of a charged species by shifting electron density through covalent bonds. An atom that is more electronegative than hydrogen is considered to be **electron withdrawing,** so it stabilizes a nearby negative charge but destabilizes a nearby positive charge. An **electron-donating** group, such as an alkyl group or an atom that is less electronegative than hydrogen, stabilizes a nearby positive charge but destabilizes a nearby negative charge. (Section 6.6e; Objective 10)

- Inductive effects are additive. The greater the number of groups that contribute to an inductive effect, the greater the effect. (Section 6.6e; Objective 10)

- Inductive effects fall off quickly with distance. (Section 6.6e; Objective 10)

- In general, the order of importance for factors affecting charge stability is (Section 6.7; Objective 10):
  1. the presence of formal charges,
  2. the type of atom on which the charge resides,
  3. resonance effects, and
  4. inductive effects.

- Charge stability is an excellent predictor of a resonance structure's relative contribution to the overall resonance hybrid. All else being equal, the contribution of a resonance structure increases as the stability of the charge increases. (Section 6.8; Objective 12)

Problems

6.1–6.3 The Proton Transfer Reaction, Equilibrium, and Thermodynamics

6.38 Given the curved arrow notation for each of the following proton transfer reactions, draw the appropriate products.

(a)

(b)

(c)

(d)

6.39 Given the reactants and products of each of the following proton transfer reactions, supply the missing curved arrows. Add relevant electrons if they are not shown.

(a)

(b)

(c)

(d)

6.40 If 0.100 mol of phenol, C_6H_5OH, were dissolved in pure water to make 1.000 L of total solution, what would the concentration of $C_6H_5O^-$ be at equilibrium? What would the acid's percent dissociation be?

6.41 For each of the following proton transfer reactions, **(a)** draw the products, **(b)** determine which side of the equilibrium is favored, and **(c)** compute the numerical factor by which that side is favored. *Hint:* For acids not in Table 6-1 or in Appendix A, you will need to estimate pK_a values based on charge stability and the functional group in which the acidic proton appears.

(i)

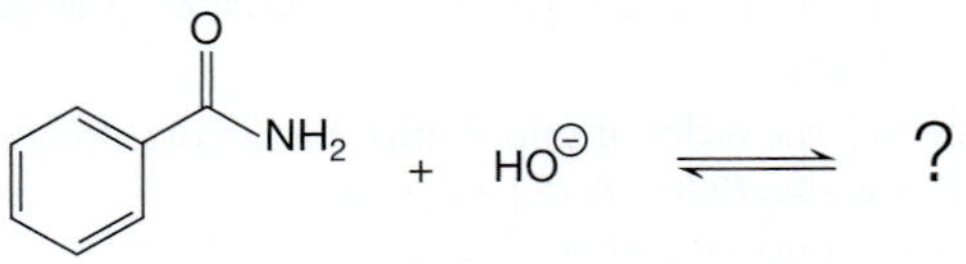

(ii)

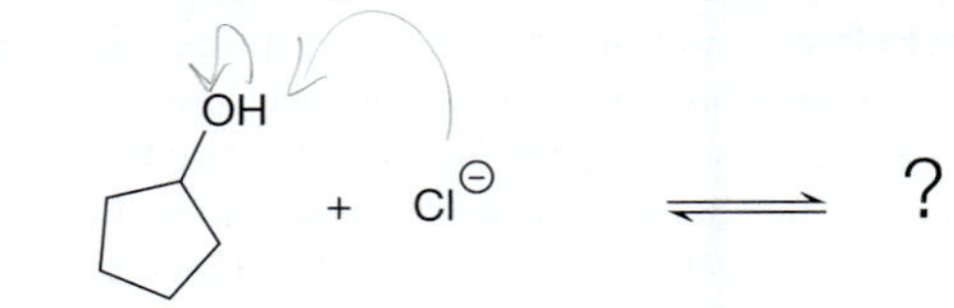

(iii)

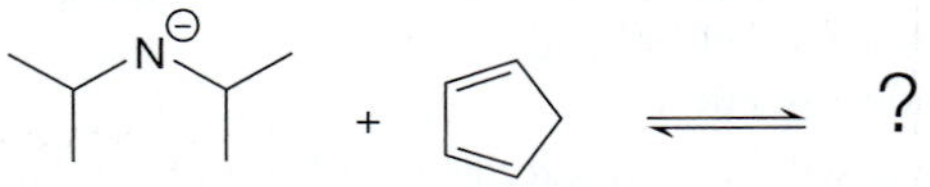

(iv)

(v)

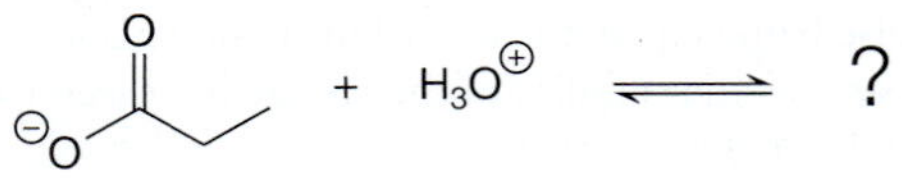

(vi)

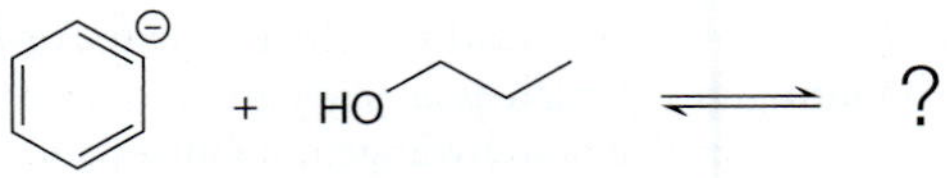

(vii)

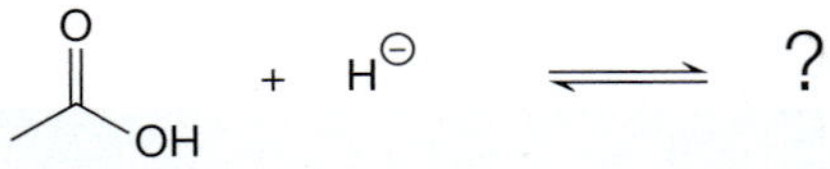

6.42 Draw the curved arrow notation for the proton transfer reaction between the hydride anion (H⁻) and ethanol (CH$_3$CH$_2$OH). Using the pK_a values listed in Table 6-1, predict which side of this reaction is favored. By what numerical factor is that side favored?

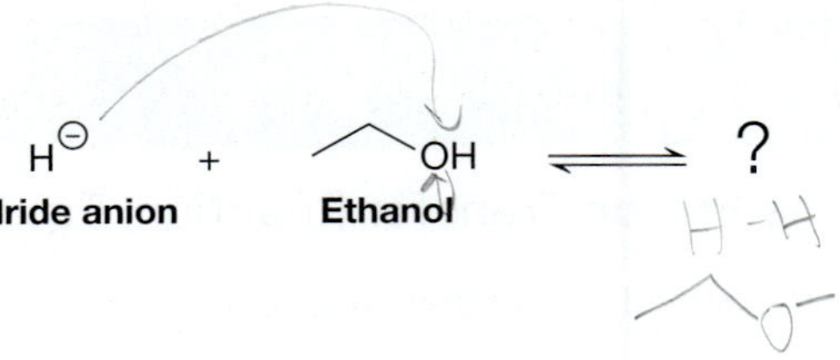

Hydride anion   **Ethanol**

6.43 At what pH will Cl$_3$CCO$_2$H dissociate 50% in water? At what pH will it dissociate 90%? At what pH will it dissociate 10%?

6.44 The protonated form of aniline has a pK_a of about 4.6. At what pH would you expect the species to exist predominantly in the protonated (cationic) form? At what pH would you expect it to exist predominantly in the uncharged form? At what pH would you expect an equal mixture of the two forms?

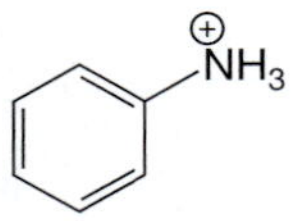

pK_a = 4.6

6.45 Keeping in mind the leveling effect, can the following species be used as a reactant in ethanol?

(a)   (b) 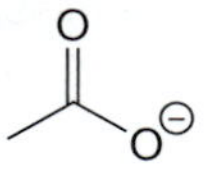   (c)   (d) 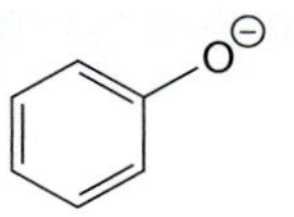   (e)   (f)

6.46 Keeping in mind the leveling effect, can the following species be used as a reactant in ethanamine (CH$_3$CH$_2$NH$_2$)?

(a) Cl⁻   (b) HCl   (c) ⁻CH$_3$ (d) ⁻NH$_2$ (e) ⁺NH$_4$ (f) ⁻OH

6.47 **(a)** Draw the products of the proton transfer reaction shown here. **(b)** Draw a free energy diagram for this reaction, indicating whether it is endothermic or exothermic.

6.48 **(a)** Draw the products of the proton transfer reaction shown here. **(b)** Draw a free energy diagram for this reaction, indicating whether it is endothermic or exothermic.

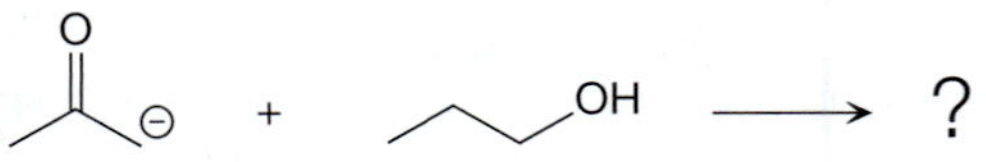

6.49 A reactant **W** can undergo two separate reactions to yield either **X** or **Y**. If the free energy diagram on the left were to describe these competing reactions, which product would be in greater abundance at equilibrium? If the free energy diagram on the right were to describe these competing reactions, which product would be in greater abundance at equilibrium?

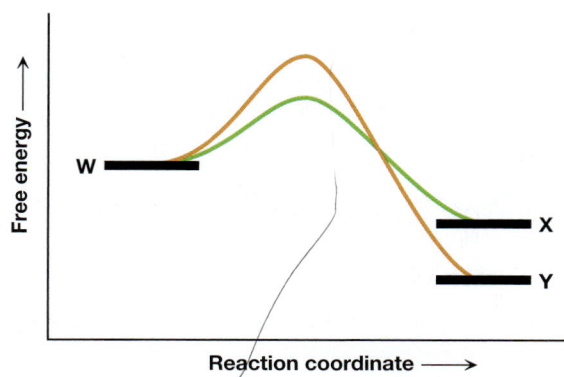
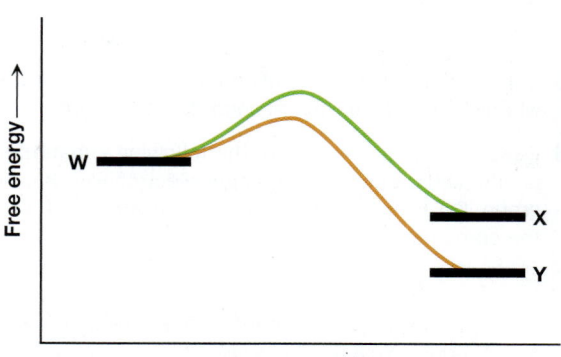

6.4 Strategies for Success: Functional Groups and Acidity

6.50 In each of the following species, identify the most acidic proton and estimate its pK_a.

(a) CH₃ ... OH, HO

(b) O ... NH₂ 38

(c) O ... HO ... SH

(d) N—H ... N

(e) O=S=O ... OH, HO

(f) O ... OH, H₂O

(g) H₃N⁺ ... N—H

(h) NH₂

6.51 For each of the following species, identify the most basic site.

(a) O⁻ ... NH₂

(b) OH ... NH₂

(c) O ... S⁻

(d) O ... O⁻ ... O⁻ ... O

(e) O⁻ ... NH⁻

6.52 For which of the following molecules, **A–D**, do you expect the pK_a value of the OH proton to be the most similar to that of cyclohexanol? Which do you expect to be the most different? Explain.

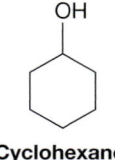

OH

Cyclohexanol

A — OH

B — O ... OH

C — Cl ... OH

D — OH ... O

6.5–6.6 Relative Acidities and Factors of Charge Stability

6.53 Students are often taught in general chemistry that HCl, HBr, and HI are all "strong acids" and no distinction is made among them. Based on charge stability, rank these acids from least acidic to most acidic.

6.54 For each pair of molecules, predict which is the stronger acid and explain your reasoning.

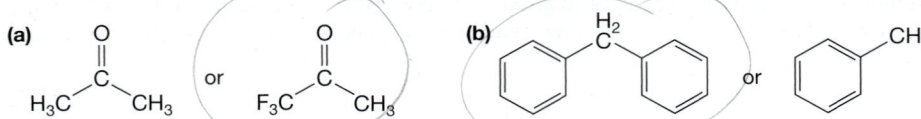

(a)

(b)

6.55 Sulfuric acid (H_2SO_4) is called a *diprotic acid* because it has two acidic protons. The pK_a for the first deprotonation is −9, whereas the pK_a for the second deprotonation is 2. Explain these relative acid strengths.

6.56 Based on the pK_a values of the following substituted acetic acids, which is a stronger electron-withdrawing group, CO_2H or NO_2? Can you explain why? *Hint:* Write out the complete Lewis structures.

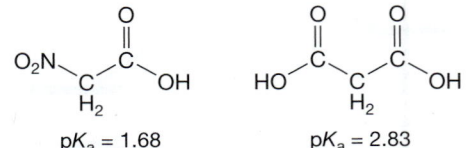

pK_a = 1.68   pK_a = 2.83

6.57 Use the electrostatic potential maps provided to predict whether C≡N or CF_3 is a stronger electron-withdrawing substituent. Explain.

6.58 Which do you expect to be the stronger base: HCN or HNC? Explain. *Hint:* Draw out the complete Lewis structure for each molecule.

6.59 The pK_a values of acids **A–C** are 9.0, 9.1, and 9.2. Match each of these values with its structure.

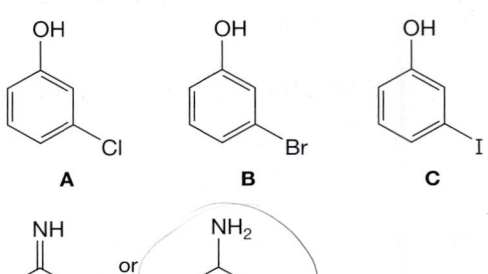

A   **B**   **C**

6.60 Which of the compounds shown here do you expect to be the stronger base? Explain.

6.7 Strategies for Success: Ranking Acid and Base Strengths — The Relative Importance of Effects on Charge

6.61 Based on the following pK_a values, which do you think is more important in determining inductive effects: electronegativity or distance from the reaction center? Explain.

pK_a = 2.68   pK_a = 3.85   pK_a = 2.95   pK_a = 3.97

6.62 Based on charge stability, rank species **A–G** from weakest base to strongest base.

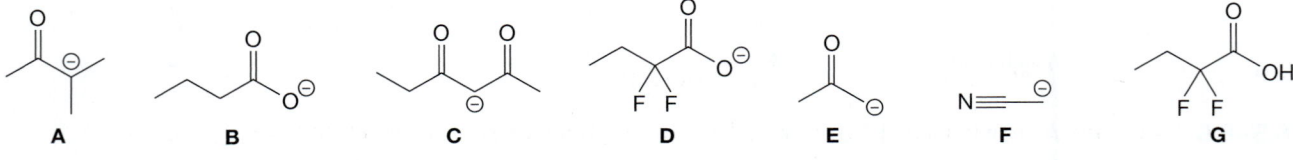

A   **B**   **C**   **D**   **E**   **F**   **G**

6.63 Explain why phenol (C$_6$H$_5$OH) is substantially more acidic than methanol (CH$_3$OH), but benzoic acid (C$_6$H$_5$CO$_2$H) is not much more acidic than acetic acid (CH$_3$CO$_2$H).

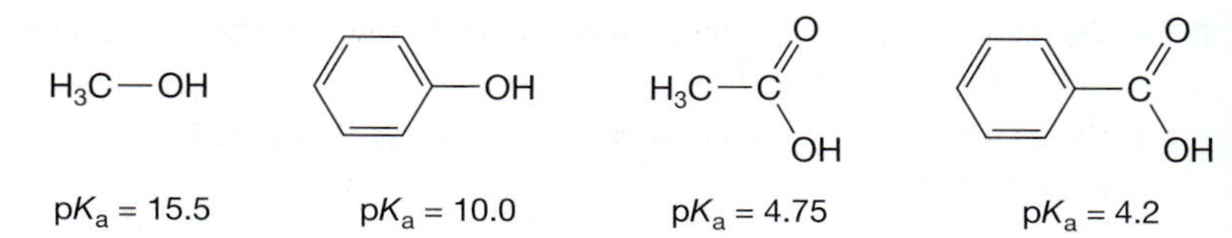

H$_3$C—OH

pK$_a$ = 15.5

(phenol) —OH

pK$_a$ = 10.0

H$_3$C—C(=O)OH

pK$_a$ = 4.75

(benzoic acid) C(=O)OH

pK$_a$ = 4.2

6.64 For each pair of species, predict which is the stronger base. Explain.

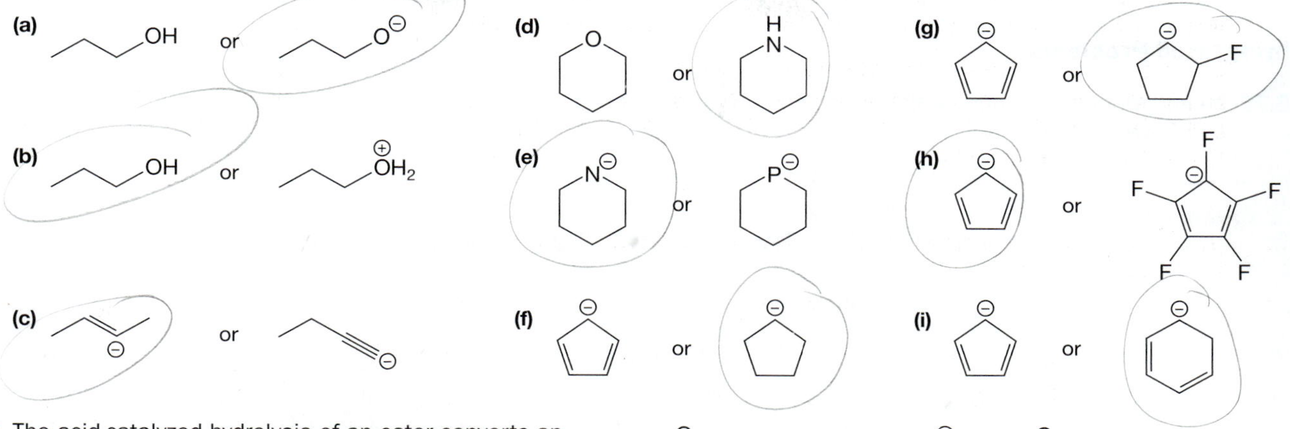

(a) ⌒⌒OH   or   ⌒⌒O$^{\ominus}$

(b) ⌒⌒OH   or   ⌒⌒$\overset{\oplus}{O}$H$_2$

(c) ⌒=⌒$^{\ominus}$ or   ⌒≡⌒$^{\ominus}$

(d) (pyran O)   or   (piperidine, H-N)

(e) (N$^{\ominus}$ ring)   or   (P$^{\ominus}$ ring)

(f) (cyclopentadienyl $^{\ominus}$)   or   (cyclopentyl $^{\ominus}$)

(g) (cyclopentadienyl $^{\ominus}$)   or   (cyclopentyl-F $^{\ominus}$)

(h) (cyclopentadienyl $^{\ominus}$)   or   (perfluorocyclopentadienyl $^{\ominus}$, F F F F)

(i) (cyclopentadienyl $^{\ominus}$)   or   (cyclohexadienyl $^{\ominus}$)

6.65 The acid-catalyzed hydrolysis of an ester converts an ester into a carboxylic acid. Although there are two O atoms that can be protonated, the first step in the mechanism is believed to be protonation of the oxygen in the C=O group. Based on charge stability, why is it favorable to protonate that oxygen? *Hint:* Draw out the products of each protonation.

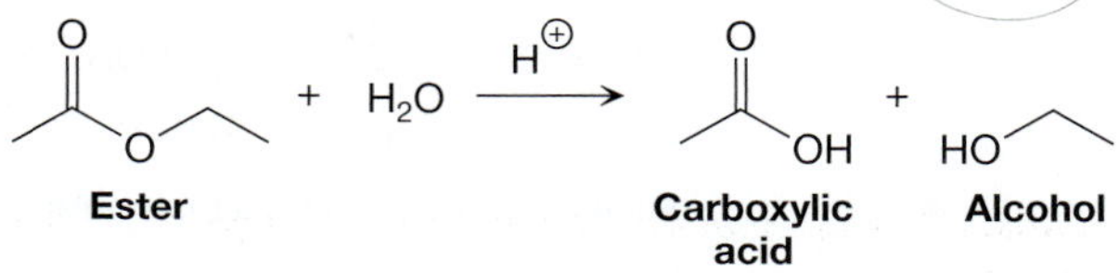

O‖C—O—CH$_2$CH$_3$ +   H$_2$O $\xrightarrow{\overset{\oplus}{H}}$ O‖C—OH   +   HO—⌒

Ester   **Carboxylic acid**   **Alcohol**

6.8 Strategies for Success: Determining Relative Contributions by Resonance Structures

6.66 Which of the following resonance structures has the greatest contribution to the hybrid? Explain.

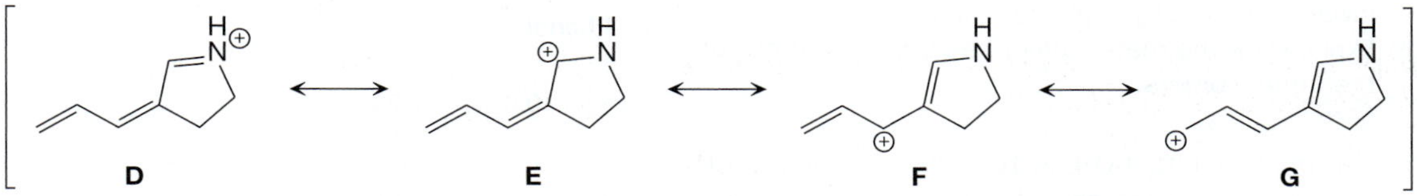

D ↔ E ↔ F ↔ G

6.67 Draw all resonance structures of the ion shown here and determine which is the strongest contributor to its resonance hybrid.

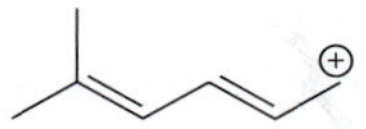

6.68 Draw all resonance structures of the ion shown here and rank them in order from strongest to weakest contributor.

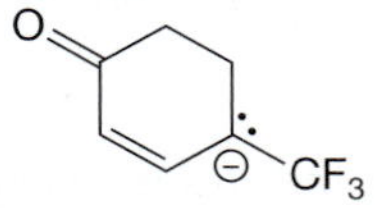

6.69 Each of molecules **A–C** has a resonance structure that exhibits separated positive and negative charges. Which molecule will resemble that resonance structure the most? The least? Explain.

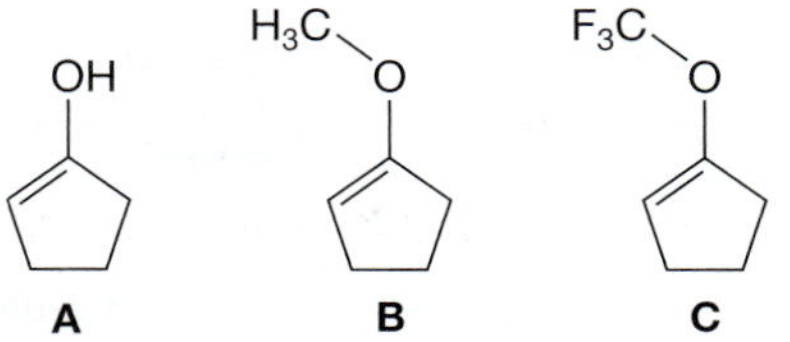

A   B   C

6.9–6.10 The Organic Chemistry of Biomolecules

6.70 Draw the structure of the most abundant form of cysteine in solutions whose pH values are 1, 4, 6, 9, and 11.

6.71 Draw the structure of the most abundant form of histidine in solutions whose pH values are 1, 3, 5, 7, and 11.

6.72 Calculate the pI of proline. In an electrophoresis experiment, will proline migrate at a pH of 6? If so, toward which terminal will it migrate?

6.73 Calculate the pI of serine. In an electrophoresis experiment, will serine migrate at a pH of 6? If so, toward which terminal will it migrate?

6.74 Compute the pI value of tyrosine. If the gel in an electrophoresis experiment is at pH 7, in which direction will tyrosine migrate—toward the cathode or toward the anode? Explain.

Integrated Problems

6.75 An important step in one synthesis of carboxylic acids is the deprotonation of diethyl malonate and its alkyl-substituted derivative:

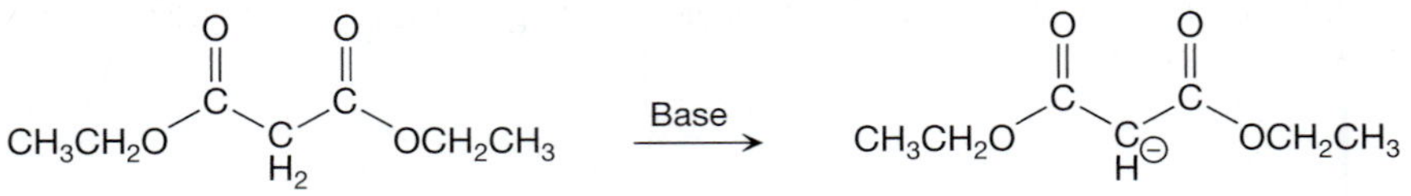

Diethyl malonate

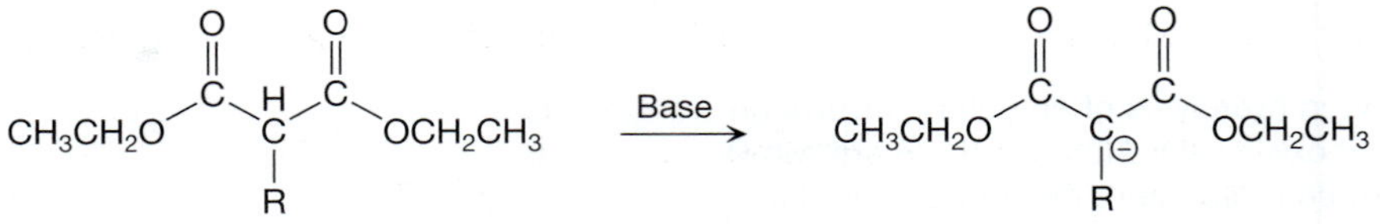

Alkyl substituted diethyl malonate

NaOH can deprotonate diethyl malonate effectively, but $NaOC(CH_3)_3$ is typically used to deprotonate the alkyl-substituted derivative. Explain why.

6.76 The pK_a of phenol (C_6H_5OH) is 10.0. When a nitro group (NO_2) is attached to the ring, the pK_a decreases, as shown for the ortho, meta, and para isomers.

(a) Explain why the pK_a values of all three isomers are lower than the pK_a of phenol itself.

(b) Explain why the meta isomer has the highest pK_a of the three isomers.

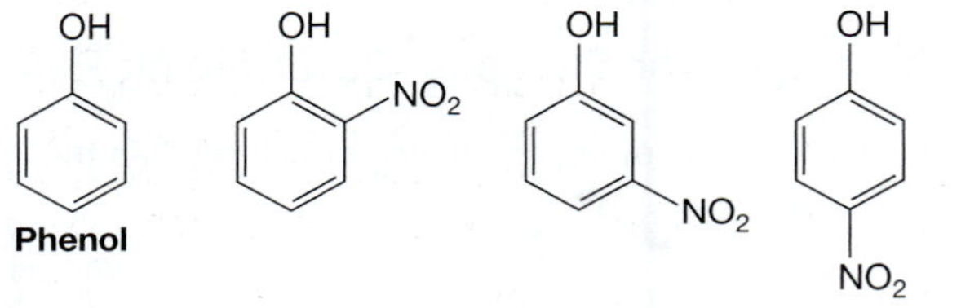

Phenol

$pK_a = 10.0$ 7.23   8.35   7.14

6.77 The pK_a of phenol (C_6H_5OH) is 10.0. When a methyl (CH_3) group is attached to the ring, the pK_a increases, as shown here for the ortho, meta, and para isomers.

(a) Explain why the pK_a values of all three isomers are higher than the pK_a of phenol itself.

(b) Explain why the meta isomer has the lowest pK_a of the three isomers.

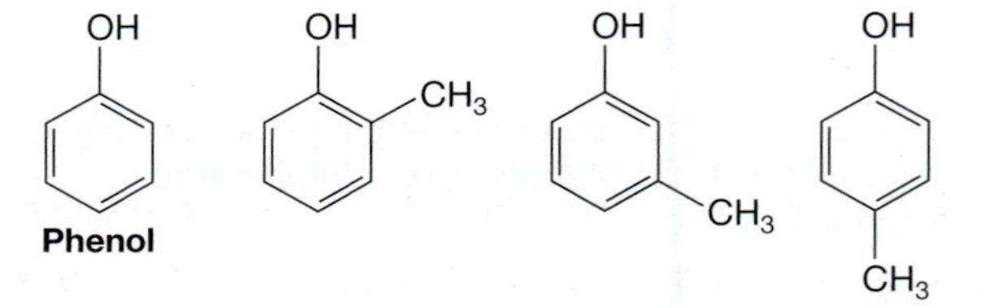

Phenol

$pK_a = 10.0$ 10.3   10.1   10.2

6.78 *cis*-Cyclohexane-1,2-diol is more acidic than *trans*-cyclohexane-1,2-diol. Based on charge stability, explain why this is the case.

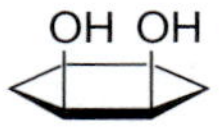

cis-**Cyclohexane-1,2-diol** 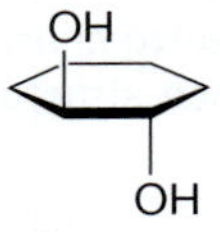   *trans*-**Cyclohexane-1,2-diol**

6.79 The pK_a of a typical ketone is 20, whereas the pK_a of a typical ester is 25. Explain why.

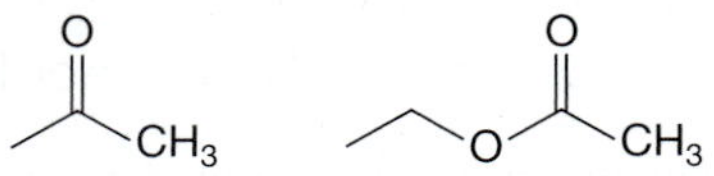

$pK_a = 20$ $pK_a = 25$

6.80 Which do you predict will be a stronger acid: *m*-hydroxybenzaldehyde or *p*-hydroxybenzaldehyde? Explain.

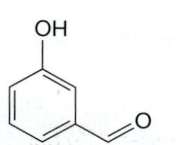

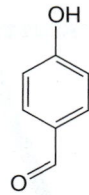

m-Hydroxybenzaldehyde   **p-Hydroxybenzaldehyde**

6.81 Which do you expect to be the stronger acid: CH_3CN or CH_3NC? Explain. *Hint:* Draw out the complete Lewis structure for each molecule.

6.82 The pK_a of formaldehyde ($H_2C=O$) is not listed in Table 6-1.
   (a) Based on your understanding of charge stability, which of the following compounds (**A–F**) would you expect to have a pK_a most similar to that of formaldehyde?

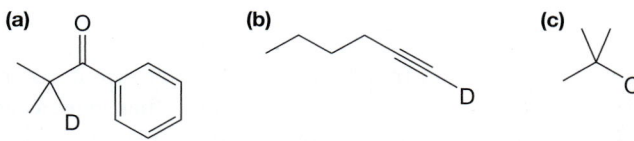

(b) Using Table 6-1, as well as your knowledge of the factors that affect charge stability, estimate the pK_a of formaldehyde.

6.83 Two possible proton transfer reactions can take place between the reactants shown here.

$\triangle\!\!\!\!\diagdown OH + \overset{H}{\underset{}{N}}\diagdown \longrightarrow ?$

   (a) Write the products of each possible proton transfer reaction.
   (b) Determine which reaction is more energetically favorable.

6.84 Deuterium (D) is an isotope of H. Both D and H have one proton and one electron; H has no neutrons and D has one. Consequently, D and H have nearly identical behavior, but they can be distinguished from each other experimentally due to their different masses. Therefore, replacing an H with a D in a molecule—*deuterium isotope labeling*—can provide valuable information about a mechanism. With this in mind, how would you synthesize each of the following deuterium-labeled compounds from the analogous unlabeled compound, using D_2O as your only source of deuterium? *Hint:* You will have to use two separate proton transfer reactions to synthesize each one.

(a)   **(b)**   **(c)**

6.85 When water is the solvent, the pK_a of acetic acid (CH_3CO_2H) is 4.75, but when DMSO is the solvent, the pK_a is 12.6. Explain why. *Hint:* Review Section 2.9 and consider the ability of each solvent to solvate cations and anions.

6.86 When water is the solvent, the pK_a of NH_4^+ is 9.4, but when DMSO is the solvent, the pK_a is 10.5. Explain why the acid strength of NH_4^+ is similar in the two solvents, but, as shown in Problem 6.85, the acid strength of acetic acid is much different in the two solvents. *Hint:* Review Section 2.9 and consider the ability of each solvent to solvate cations and anions.

6.87 In which of solvents **A–E** will CH_3NH_2 be the *weakest* base? Explain.

A   B   C   D   CCl₄ E

6.88 The pK_a value for a protonated amine (R_3NH^+) depends on the number of alkyl groups attached to N, as shown here. This order disagrees with what we would predict using charge stability.

$\overset{\oplus}{N}H_4$ $pK_a = 9.25$

$H_3C\overset{\oplus}{-}NH_3$ 10.66

$H_3C\overset{\oplus H_2}{\underset{}{N}}CH_3$ 10.73

$H_3C\overset{CH_3}{\underset{\oplus N}{\underset{H}{}}}CH_3$ 9.81

   (a) From least acidic to most acidic, what is the order that would be predicted using charge stability?
   (b) Can you explain the reason for the discrepancy? *Hint:* These pK_a values are all measured in water.

7

An Overview of the Most Common Elementary Steps

Pollen sticks to this bee due to electrostatic attraction. When it flies, a bee loses electrons and builds up a positive electrical charge. The flower's pollen, on the other hand, is negatively charged, so when the bee lands, the pollen experiences a driving force from the flower to the bee. Similarly, the elementary steps we examine here in Chapter 7 are driven by the flow of electrons from a site with excess negative charge toward a site with excess positive charge.

In Chapter 6, we learned that a proton transfer is an *elementary step*: it occurs as a single event. If a proton transfer takes place in isolation from other steps, as we saw in the examples in Chapter 6, then it constitutes an *overall reaction*. Frequently, however, a proton transfer makes up an individual step of a *multistep mechanism*—something we will discuss more extensively in Chapter 8.

There are a handful of other quite common elementary steps as well, which can be combined in various ways to produce mechanisms for numerous reactions. Chapter 7 provides an overview of nine of these elementary steps:

1. Bimolecular nucleophilic substitution (S_N2, Section 7.2)
2. Coordination (Section 7.3)
3. Heterolysis (Section 7.3)
4. Nucleophilic addition (Section 7.4)
5. Nucleophile elimination to form a polar π bond (Section 7.4)
6. Bimolecular elimination (E2, Section 7.5)
7. Electrophilic addition (Section 7.6)
8. Electrophile elimination to form a nonpolar π bond (Section 7.6)
9. Carbocation rearrangements (Section 7.7)

Coordination and heterolysis (Section 7.3), nucleophilic addition and nucleophile elimination (Section 7.4), and electrophilic addition and electrophile elimination (Section 7.6) are discussed in pairs in their respective sections because, in each case, one is simply the reverse of the other. This means both are governed by the same factors.

On completing Chapter 7 you should be able to:

1. Describe how the curved arrow notation for a proton transfer step reflects the flow of electrons from an electron-rich site to an electron-poor site.
2. Simplify metal-containing species in terms of their electron-rich behavior.
3. Identify and draw the curved arrows for the following common elementary steps:
   - Proton transfer
   - Bimolecular nucleophilic substitution (S_N2)
   - Coordination
   - Heterolysis
   - Nucleophilic addition
   - Nucleophile elimination
   - Bimolecular elimination (E2)
   - Electrophilic addition
   - Electrophile elimination
   - Carbocation rearrangement
4. Describe the electron flow in each of the above steps in terms of electron-rich and electron-poor sites.
5. Rationalize the driving force for each of the above elementary steps in terms of charge stability and bond energies.
6. Draw the keto and enol forms of a keto–enol tautomerization, and explain the driving force for such a reaction.

Each of these nine elementary steps can be depicted using curved arrow notation, much as proton transfer steps were depicted in Chapter 6. However, beyond simply describing how each step takes place, we need to know *why* it would (or would not) take place. Therefore, we devote Section 7.8 to the *driving force* for elementary steps. Understanding the driving force for each step will ultimately enable us to use mechanisms to make predictions about the outcomes of reactions.

The nine new elementary steps we learn here in Chapter 7, along with proton transfer steps (10 in all), make up nearly all of the reaction mechanisms you will encounter through Chapter 23. This knowledge, that the same 10 elementary steps constitute the mechanisms for a large number of reactions, can be very powerful. It allows you to be confident that *mechanisms simplify organic chemistry*. The time and effort you spend mastering these elementary steps will be rewarded as you continue to learn reactions throughout the rest of this book.

To help you see the similarities and differences among the various elementary steps:

General forms of all 10 elementary steps are provided in the chapter summary on pages 355–356.

You should visit those pages frequently. As you do, challenge yourself to become familiar with three aspects of each elementary step: (1) the types of species that appear as reactants and products, (2) the curved arrows that describe the electron movement, and (3) the driving force. You will especially benefit in later chapters when we study reactions that have mechanisms constructed from these steps.

7.1 Mechanisms as Predictive Tools: The Proton Transfer Step Revisited

A reaction mechanism can be very helpful as a *predictive* tool. In this section we begin to see how mechanisms help, by revisiting *curved arrow notation* for a proton transfer step and examining specifically how it can be used to make predictions about bond formation and bond breaking.

Even though our discussion here is in the context of proton transfer steps, we can apply the same logic to other elementary steps as well. Effectively, then, this section establishes a process by which to *analyze* an elementary step. The utility of the process becomes apparent when we discuss new elementary steps later in this chapter.

7.1a Curved Arrow Notation: Electron Rich to Electron Poor

Curved arrow notation was introduced in Section 6.1 as a means for keeping track of valence electrons in an elementary step. It can be far more powerful than that, though, if we recall two concepts:

1. *Opposite charges attract; like charges repel.*
2. *Atoms in the first and second rows of the periodic table must obey the duet and octet rules, respectively.*

With these ideas in mind, examine Equation 7-1, which shows the curved arrows for the proton transfer between HCl and HO^-.

Curved arrow drawn from
"electron rich to electron poor"

Electron-rich site   Electron-poor site

$$HO^{\ominus} \quad + \quad H\!-\!\overset{\delta^+}{C}l\!: \quad \longrightarrow \quad HO\!-\!H \quad + \quad :\overset{..}{C}l\!:^{\ominus} \qquad (7\text{-}1)$$

Notice that HO^- bears a full negative charge. This excess electron density on O means the electrons on O are somewhat destabilized due to their mutual repulsion. The proton on HCl bears a partial positive charge, δ^+, because Cl is more electronegative than H. As a result, the electrons on O are attracted to the proton on HCl. This simultaneous charge repulsion among the electrons on O and their attraction to H facilitates the flow of electrons from O to H and results in the formation of the new O—H bond.

This example illustrates one of the most important guidelines for drawing curved arrows:

> In an elementary step, electrons tend to flow from an *electron-rich* site to an *electron-poor* site.

In Equation 7-1, the HO^- anion is relatively electron rich and the H atom on HCl is relatively electron poor, denoted by the light red screen behind the negatively charged O atom and the light blue screen behind the H bearing a partial positive charge. (This is the same color scheme used in electrostatic potential maps to represent areas of more and less electron density, respectively.) To help you keep track of those atoms, we have kept the red and blue screens the same in the products.

The curved arrow in Equation 7-1 that goes from the H—Cl bond to the Cl atom is needed to maintain H's duet of electrons. Without that curved arrow, and hence without breaking that bond, H would end up with two bonds, which is one too many.

YOUR TURN 7.1

In this proton transfer step, identify the electron-rich and electron-poor sites, and label the curved arrow that connects the two as "electron rich to electron poor."

Answers to Your Turns are in the back of the book.

The curved arrow notation for the following proton transfer is faulty. What is unacceptable about it? In the space provided here, make the necessary corrections to the curved arrow notation.

$$HO^{\ominus} + H-Cl: \longrightarrow HO-H + :Cl:^{\ominus}$$

SOLVED PROBLEM 7.1

Identify the electron-poor H atom in methanol. Draw the mechanism by which methanol acts as an acid in a proton transfer reaction with H_2N^-.

H—C—O—H structure with H atoms
Methanol

Think What kinds of charges characterize an electron-poor atom? Are there any formal charges present? Any strong partial charges? When H_2N^- and CH_3OH are combined, what curved arrow can we draw to depict the flow of electrons from an electron-rich site to an electron-poor site?

Solve CH_3OH does not bear any full charges, but the highly electronegative O places a strong partial positive charge on the H to which it is bonded. That H is therefore electron poor, and a blue screen is placed behind it as a reminder. H_2N^- bears a full negative charge and is therefore electron rich. As a reminder, a red screen is placed behind N. To represent the flow of electrons from the electron-rich site (H_2N^-) to the electron-poor site (CH_3OH), draw a curved arrow from a lone pair on N to the H atom on O. A second curved arrow is needed to make sure H has only one bond, not two.

Electron rich to electron poor

$$H-C-O-H + :NH_2^{\ominus} \longrightarrow H-C-O:^{\ominus} + H-NH_2$$
δ^+

CONNECTIONS
Trimethylamine [$(CH_3)_3N$, Problem 7.2] is a gas that is often associated with the odor of rotting fish. Sensors have been developed to test for trimethylamine to assess the freshness of fish.

PROBLEM 7.2 Identify the electron-rich and electron-poor sites in each of the reactant molecules shown here. Draw the curved arrows and the products for the proton transfer between these two molecules and label the curved arrow that represents the flow of electrons from an electron-rich site to an electron-poor site.

$$H_3C-N(CH_3)-CH_3 + H_2O \longrightarrow ?$$

7.1b Simplifying Assumptions Regarding Electron-Rich and Electron-Poor Species

Contrary to what is suggested in Equation 7-1, we cannot simply add hydroxide anion (HO^-) to HCl to initiate a proton transfer reaction, because of the following restriction:

> Anions do not exist in the solid or liquid phase without the presence of cations, and vice versa.

We can, however, add a *source* of HO^-, such as NaOH; NaOH is an ionic compound, so in solution it dissolves as Na^+ and HO^-.

This initially may seem to complicate the picture, because Na^+ is electron poor. We can thus envision Na^+ reacting with an electron-rich site of another species. However, Na^+ is relatively inert in solution and tends not to react. It behaves instead as a **spectator ion**.

Consequently, when we envision the flow of electrons from an electron-rich site to an electron-poor site, we can disregard such metal cations.

SOLVED PROBLEM 7.3

Draw the necessary curved arrows for the proton transfer between KOCH$_3$ and HCN in solution.

Think What are the electron-rich and electron-poor sites in each compound? Are there any simplifying assumptions we can make?

Solve KOCH$_3$ is an ionic compound that dissolves in solution as K$^+$ and CH$_3$O$^-$. We can therefore treat K$^+$ as a spectator ion and consider just CH$_3$O$^-$ as the reactive species it contributes. HCN has an electron-poor H atom due to the high effective electronegativity of the sp-hybridized C atom. Thus, a curved arrow is drawn from the electron-rich O to the electron-poor H to initiate the proton transfer. A second curved arrow is drawn to break the H—C bond to avoid two bonds to H.

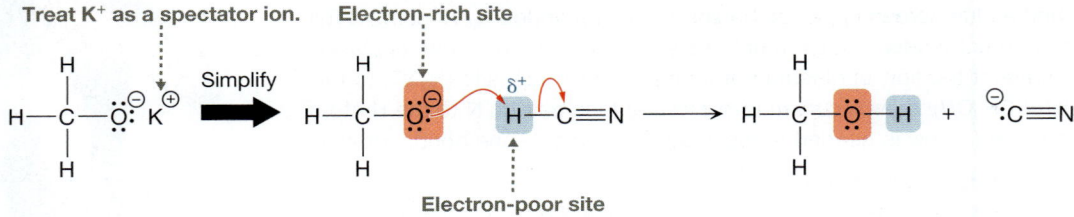

PROBLEM 7.4 Use curved arrow notation to indicate the proton transfer between NaSH and CH$_3$CO$_2$H.

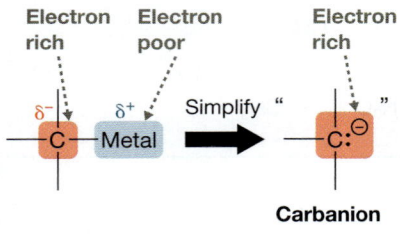

FIGURE 7-1 Simplifying assumptions in organometallic compounds (*Left*) Because of the high polarity in a carbon–metal bond, organometallic compounds contain an electron-rich site on C and an electron-poor site on the metal. (*Right*) We can usually ignore the reactivity of the metal-containing portion and treat the organometallic compound simply as a carbanion, which is electron rich. The quotation marks remind us that the carbanion does not actually exist in solution.

Similar ideas allow us to make simplifying assumptions for **organometallic** compounds, which contain *a metal atom bonded directly to a carbon atom*. Examples of organometallic compounds include alkyllithium (R—Li); alkylmagnesium halide (R—MgX, where X = Cl, Br, or I), also called a **Grignard reagent**; and lithium dialkyl cuprate [Li$^+$(R—Cu—R)$^-$]. These kinds of organometallic compounds are useful reagents for forming new carbon–carbon bonds (see Chapter 10).

To simplify how we treat organometallic compounds in mechanisms, we must first recognize that the carbon–metal bond is a polar covalent bond. The carbon atom's electronegativity (2.55) is significantly greater than that of the metal (Li = 0.98, Mg = 1.31, and Cu = 1.90), so there is a large partial negative charge on carbon (making it electron rich) and a large partial positive charge on the metal atom (making it electron poor), as shown in Figure 7-1.

In most reactions involving organometallic compounds, the product that we are interested in isolating contains the organic portion of the organometallic compound and not the metal-containing portion. Therefore, even though the carbon–metal bond is covalent, it often helps to simplify organometallic compounds in a way that allows us to disregard the metal-containing portion, in much the same way as we did with the group 1A metal cations:

FIGURE 7-2 Simplifying some specific organometallic compounds (a) $LiCH_2CH_3$ is simplified to $^-CH_2CH_3$. (b) C_6H_5MgBr is simplified to $C_6H_5^-$. (c) $(CH_3)_2CuLi$ is simplified to H_3C^-.

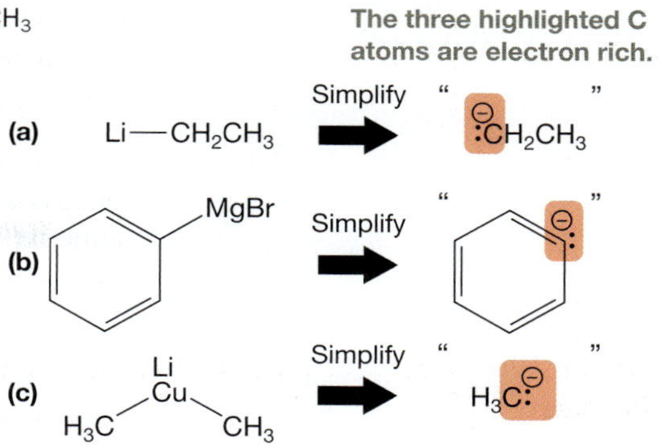

The three highlighted C atoms are electron rich.

(a) $Li—CH_2CH_3$ Simplify ➡ " $:\overset{\ominus}{C}H_2CH_3$ "

Therefore, as shown on the right in Figure 7-1, we can often treat organometallic compounds simply as electron-rich **carbanions**—compounds in which a negative formal charge appears on C. Applying this simplifying assumption allows us, for example, to treat CH_3CH_2Li as a source of $CH_3CH_2^-$, C_6H_5MgBr as a source of $C_6H_5^-$, and $(CH_3)_2CuLi$ as a source of CH_3^-, as shown in Figure 7-2.

(b) [benzene ring with MgBr]   Simplify ➡ " [benzene ring with $\overset{\ominus}{:}$] "

(c) [structure with Li, Cu, H_3C, CH_3]   Simplify ➡ " $H_3C\overset{\ominus}{:}$ "

SOLVED PROBLEM 7.5

What are the products of the proton transfer step between C_6H_5MgBr and H_2O?

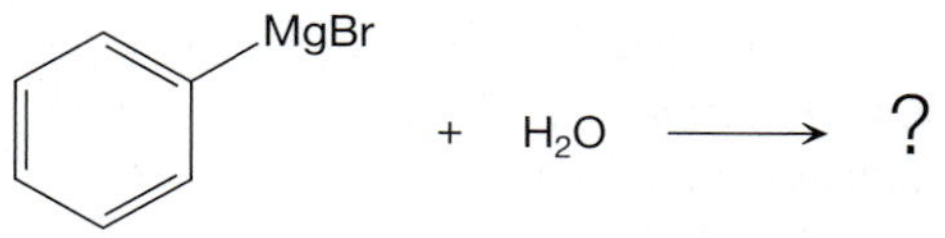

[benzene ring with MgBr] $+ \; H_2O \longrightarrow$?

Think What are the electron-rich sites? What are the electron-poor sites? What can be disregarded?

Solve In H_2O, O is electron rich, and both H atoms are electron poor. In C_6H_5MgBr, we can disregard the MgBr portion because it contains the metal, and treat the reactive species as $C_6H_5^-$. The proton transfer is initiated by drawing a curved arrow from the electron-rich C on $C_6H_5^-$ to an electron-poor H on H_2O.

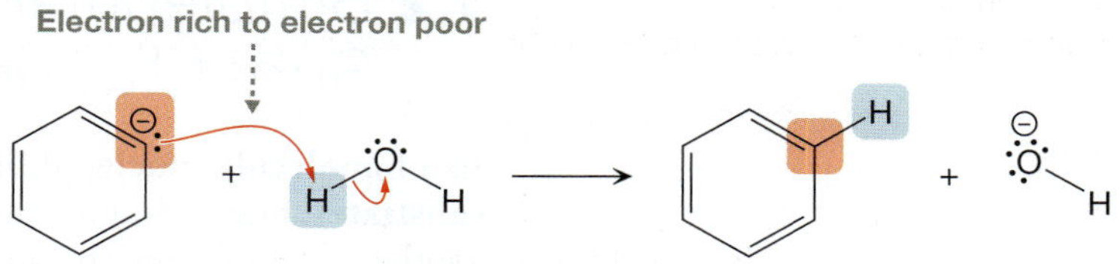

Electron rich to electron poor

[reaction scheme: benzene ring with $\overset{\ominus}{:}$ + H—O—H $\longrightarrow$ benzene ring with H + $\overset{\ominus}{:}O$—H]

PROBLEM 7.6 Use curved arrow notation to show the proton transfer step that occurs between CH_3Li and CH_3OH. Predict the products of this reaction.

The strategy we applied to simplifying organometallic compounds can also be used to simplify **hydride reagents**, such as lithium aluminum hydride ($LiAlH_4$) and sodium borohydride ($NaBH_4$), which are commonly used as *reducing agents* (see Chapter 17). $LiAlH_4$ consists of Li^+ ions and AlH_4^- ions, as shown on the left in Figure 7-3a. We can treat Li^+ as a spectator ion, leaving AlH_4^- as the reactive species (Fig. 7-3a, middle). According to the Lewis structure of AlH_4^-, the H atoms [electronegativity (EN) = 2.20] are covalently bonded to an Al metal atom (EN = 1.61). The excess negative charge built up on each H atom means that we can treat $LiAlH_4$

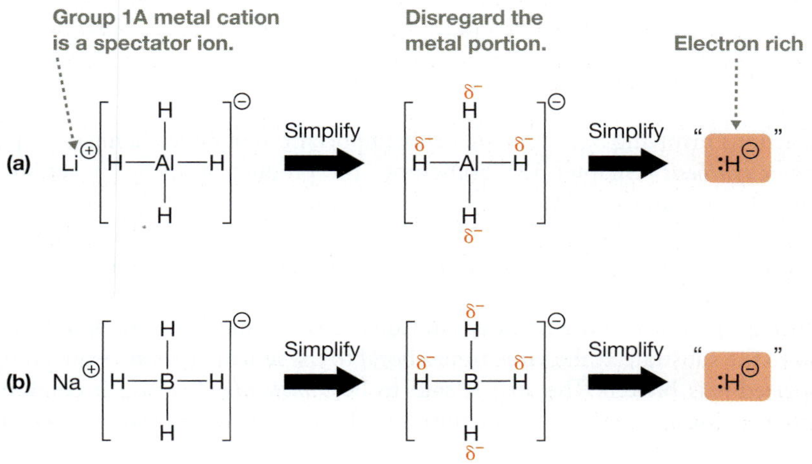

Group 1A metal cation is a spectator ion.

Disregard the metal portion.

Electron rich

(a) $Li^{\oplus}$ [H—Al—H structure with H top and bottom, $\ominus$]   Simplify ➡ [δ^- H—Al—H δ^- structure, $\ominus$]   Simplify ➡ " $:H^{\ominus}$ "

(b) $Na^{\oplus}$ [H—B—H structure with H top and bottom, $\ominus$]   Simplify ➡ [δ^- H—B—H δ^- structure, $\ominus$]   Simplify ➡ " $:H^{\ominus}$ "

FIGURE 7-3 Simplifying assumptions in hydride reagents (a) Lithium aluminum hydride, $LiAlH_4$, consists of Li^+ and AlH_4^- ions. The reactive species is AlH_4^- (middle), which can be treated simply as $:H^-$. The quotation marks indicate that $:H^-$ does not actually exist in solution. (b) Sodium borohydride, $NaBH_4$, consists of Na^+ and BH_4^- ions. The reactive species is BH_4^- (middle), which can be treated as $:H^-$, too.

as a source of hydride, :H⁻, as indicated on the right in Figure 7-3a. For similar reasons, we can treat NaBH₄ as a source of :H⁻, too.

PROBLEM 7.7 Use curved arrow notation to show the proton transfer step that takes place between LiAlH₄ and water. Predict the products of the reaction. Do the same for the proton transfer step between NaBH₄ and phenol (C_6H_5OH).

Later, we will need to be careful when making these simplifying assumptions, particularly when treating organometallic compounds as carbanions, R⁻, and hydride reagents as hydride anions, H⁻. Experimentally, significant differences in reactivity are observed among the different organometallic compounds and among the different hydride reagents. These differences are discussed more extensively in the context of nucleophilic addition reactions in Chapters 17 and 18. For now, you should take the time to commit to memory the simplifications in Figures 7-2 and 7-3.

7.2 Bimolecular Nucleophilic Substitution (S$_N$2) Steps

In a **bimolecular nucleophilic substitution (S$_N$2) step**, a molecular species, called a **substrate**, undergoes substitution in which one atom or group of atoms is replaced by another. Examples are shown in Equations 7-2 and 7-3.

HO^- substitutes for Cl^- in Equation 7-2, and H_3N substitutes for $CH_3SO_3^-$ in Equation 7-3.

Reactions involving S$_N$2 steps are really important in organic chemistry, especially in *organic synthesis* (Chapter 13). Sometimes the product of an S$_N$2 reaction is the compound you want to synthesize. In other cases, an S$_N$2 reaction, by substituting one atom or group for another, can be used to alter the reactivity of a molecule in ways that make further reactions possible. Because of this central role in synthesis, we revisit S$_N$2 reactions in Chapters 8–10.

During the course of the S$_N$2 steps in Equations 7-2 and 7-3, a **nucleophile** forms a bond to the substrate at the same time a bond to the **leaving group** — the group that is displaced — is broken. The step is said to be *bimolecular* because it contains two separate reacting species in an elementary step. In other words, the step's **molecularity** is 2. It is called *nucleophilic* because a nucleophile is the species that reacts with the substrate.

Equations 7-2 and 7-3 also show that a leaving group often comes off in the form of a negatively charged species. *Common leaving groups are relatively stable with a negative charge.* Applying what we learned in Sections 6.6 and 6.7, we can say that:

Leaving groups are typically *conjugate bases of strong acids.*

Leaving groups thus include Cl^-, Br^-, and I^- (conjugate bases of the strong acids HCl, HBr, and HI, respectively) because, in each case, the negative charge is on a relatively large and/or electronegative atom. They also include anions in which there is substantial resonance and inductive stabilization, such as an alkylsulfonate anion, RSO_3^- (the conjugate base of RSO_3H). Water and alcohols (ROH) are also leaving groups because they, too, are the conjugate bases of strong acids (H_3O^+ and ROH_2^+, respectively).

A nucleophile tends to be attracted by and form a bond to a nonhydrogen atom that bears a partial or full positive charge.

The term *nucleophile* literally means "nucleus loving." It is given this name because the nucleus of an atom bears a positive charge, the type of charge to which a nucleophile is attracted.

Species that act as nucleophiles generally have the following two attributes:

1. A nucleophile has an atom that carries a full negative charge or a partial negative charge. The charge is necessary for it to be attracted to an atom bearing a positive charge.
2. The atom with the negative charge on the nucleophile has a pair of electrons that can be used to form a bond to an atom in the substrate. As we saw in Equations 7-2 and 7-3, those electrons are usually lone pairs.

In the nucleophile in Equation 7-2 (HO^-), the O atom possesses a full negative charge and three lone pairs of electrons, as indicated below. In Equation 7-3, on the other hand, the nucleophile (H_3N) has a N atom bearing a *partial* negative charge and one lone pair of electrons.

Other common, negatively charged nucleophiles include CH_3O^-, Cl^-, Br^-, I^-, H_2N^-, CH_3NH^-, HS^-, CH_3S^-, $N\equiv C^-$, and N_3^-. Other common uncharged nucleophiles include H_2O, CH_3OH, CH_3NH_2, H_2S, and CH_3SH.

YOUR TURN 7.3

Draw the complete Lewis structures for three of the *negatively charged* nucleophiles listed in the previous paragraph (other than HO^-) and for three of the uncharged nucleophiles (other than H_3N). Include all lone pairs. For each of the uncharged nucleophiles, write "δ^-" next to the atom bearing a partial negative charge.

SOLVED PROBLEM 7.8

Should CH_4 act as a nucleophile? Why or why not?

Think Does CH_4 have an atom that carries a partial or full negative charge? Does that atom have a pair of electrons that can be used to form a bond to another atom?

Solve The complete Lewis structure for CH_4 is shown here. The C atom has a small partial negative charge (because C is slightly more electronegative than H), but it does not possess a lone pair of electrons that can be used to form a bond with another atom. Thus, CH_4 should *not* act as a nucleophile.

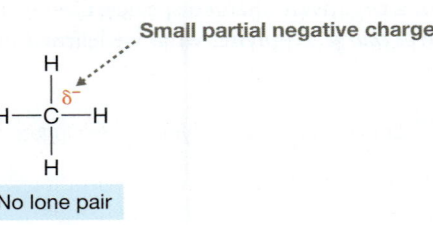

Small partial negative charge

No lone pair

PROBLEM 7.9 Which of the following species can behave as a nucleophile? Explain. **(a)** SiH_4; **(b)** NaSCN; **(c)** NH_4^+; **(d)** CH_3Li

Recall from Section 7.1a that the electrons in an elementary step tend to flow from an *electron-rich* site to an *electron-poor* site. In the proton transfer in Equation 7-1 (shown again in Equation 7-4a), for example, where HO^- acts as a base, a curved arrow is drawn from a lone pair on the O in HO^- to the H in HCl. A second curved arrow is drawn to show the initial bond to H being broken, with its electrons becoming a lone pair on Cl^-.

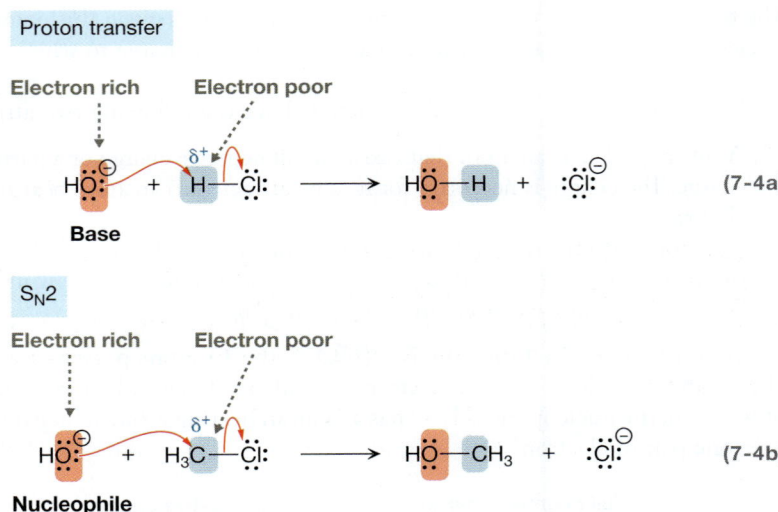

A similar thing happens in the S_N2 step shown in Equation 7-4b. In this case, HO^- is still relatively electron rich, but the C atom of CH_3Cl is relatively electron poor because it carries a partial positive charge. Thus, HO^- acts as a nucleophile, and a curved arrow is drawn from a pair of electrons on O to the C atom to signify bond formation. To avoid five bonds to C (which would exceed the C atom's octet), a second curved arrow shows that the pair of electrons initially composing the C—Cl bond becomes a lone pair on Cl^-.

YOUR TURN 7.4

The following S_N2 step is similar to the one in Equation 7-4b:

$$:\overset{..}{\underset{..}{Cl}}:^{\ominus} \;+\; H_3C—\overset{..}{\underset{..}{Br}}: \;\longrightarrow\; :\overset{..}{\underset{..}{Cl}}—CH_3 \;+\; :\overset{..}{\underset{..}{Br}}:^{\ominus}$$

Label the appropriate reacting species as "electron rich" or "electron poor" and draw in the correct curved arrows.

Draw the S_N2 step that would occur between $C_6H_5CH_2I$ and CH_3SNa.

Think Which species is the nucleophile? Which is the substrate? What do we do with the metal atom to simplify our treatment of CH_3SNa? Which species is electron rich? Electron poor?

Solve $C_6H_5CH_2I$ will behave as the substrate because it possesses an I, a good leaving group that departs as I^-. The conjugate acid of I^-, HI, is a very strong acid. CH_3SNa has a metal atom that can be treated as a spectator ion and can thus be ignored. The nucleophile is therefore CH_3S^-. In an S_N2 step, a curved arrow is drawn from the lone pair of electrons on the electron-rich S atom to the electron-poor C atom bonded to I. A second curved arrow must be drawn to indicate that the C—I bond is broken (otherwise that C would end up with five bonds).

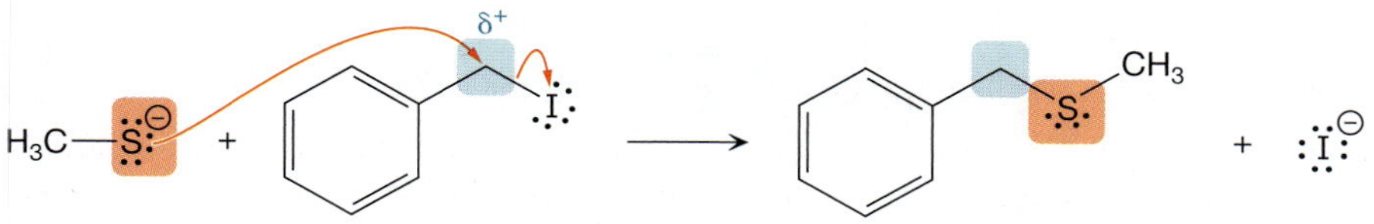

PROBLEM 7.11 Draw the S_N2 step that would occur between the two compounds at the right.

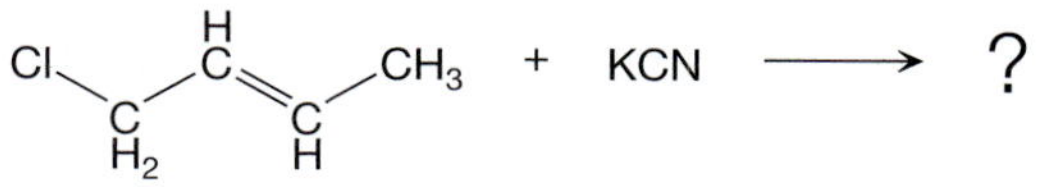

7.3 Bond-Forming (Coordination) and Bond-Breaking (Heterolysis) Steps

In both the proton transfer and the S_N2 steps we have examined so far, a bond is formed and a separate bond is broken simultaneously. It is possible, however, for bond formation and bond breaking to occur as independent steps. In Equations 7-5 and 7-6, for example, only a single covalent bond is formed. These are called **coordination steps**.

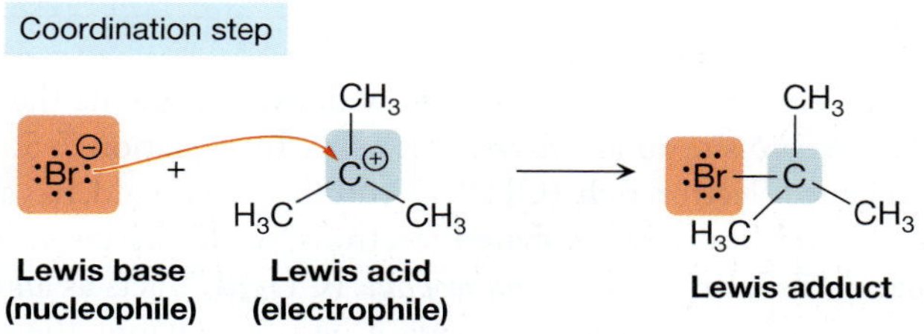

(7-5)

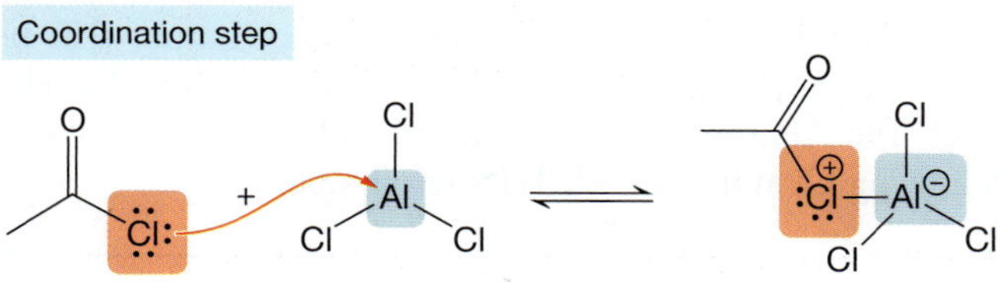

(7-6)

You may have learned in general chemistry that coordination steps are also called **Lewis acid–base reactions**. A **Lewis acid** is an electron-pair acceptor, having an atom that lacks an octet. A **Lewis base**, on the other hand, is an electron-pair donor. Thus, in Equation 7-5, $(CH_3)_3C^+$ is the Lewis acid and Br^- is the Lewis base. The product, $Br-C(CH_3)_3$, is called the **Lewis adduct**.

> Label each species in Equation 7-6 as a Lewis acid, Lewis base, or Lewis adduct.

An elementary step can also occur in which only a single bond is broken and *both electrons from that bond end up on one of the atoms initially involved in the bond*, as shown in Equations 7-7 and 7-8. These are called **heterolytic bond dissociation steps**, or **heterolysis steps** (*hetero* = different; *lysis* = break), to emphasize that, once the bond is broken, the two electrons are not distributed equally to the atoms initially involved in the bond. One of those atoms, in particular, will be left without an octet. Viewed in this way, *heterolysis steps are the reverse of coordination steps.*

Heterolysis step

(7-7)

Heterolysis step

(7-8)

Unlike proton transfer and S_N2 steps, coordination and heterolysis generally do not take place in isolation. This is because one of the reactant or product species lacks an octet and is therefore highly unstable and reactive. Instead, coordination and heterolysis steps usually compose one step of a mechanism involving two or more elementary steps—a so-called *multistep mechanism* (Chapter 8). We will see, for example, that coordination and heterolysis occur in mechanisms for S_N1 and E1 reactions (Chapter 8), electrophilic addition reactions (Chapter 11), and Friedel–Crafts reactions (Chapter 22).

> In Equations 7-5 through 7-8, identify all atoms lacking an octet.

In a coordination step, the single curved arrow drawn represents the flow of electrons from an *electron-rich* site to an *electron-poor* site. In Equation 7-5, Br^- is negatively charged and hence electron rich. $(CH_3)_3C^+$, on the other hand, has a C atom that is positively charged and has only six shared electrons, so it is electron poor. Thus, as indicated in Equation 7-5, Br^- acts as a *nucleophile*. $(CH_3)_3C^+$ acts as an **electrophile**, because it forms a bond with a species bearing a negative charge, the same kind of charge an electron has.

In Equation 7-6, CH_3COCl has an electron-rich Cl atom that bears a partial negative charge, whereas the Al atom in $AlCl_3$ is electron poor, largely because it is short of an octet. The Al atom is made even more electron poor by the surrounding electron-withdrawing Cl atoms on $AlCl_3$ (Section 6.6e).

PROBLEM 7.12 **(a)** Draw the appropriate curved arrows for the coordination step between $FeCl_3$ and Cl^-. Draw the reaction products. Identify which reactant species is electron rich and which is electron poor. **(b)** Use curved arrow notation to show the product from part (a) undergoing heterolysis to regenerate $FeCl_3$ and Cl^-.

7.4 Nucleophilic Addition and Nucleophile Elimination Steps

In Section 7.2, we introduced the S_N2 step, in which a nucleophile bonds to an atom containing a suitable leaving group. A nucleophile can also bond to an atom that is involved in a *polar π bond*—that is, a π bond that is part of a double or triple bond connecting atoms with significantly different electronegativities, such as those in carbonyl groups (C=O), imine groups (C=N), and cyano groups (C≡N). Specifically, as shown in Equations 7-9 through 7-11, the nucleophile forms a bond to the less electronegative atom and the π bond breaks, becoming a lone pair on the more electronegative atom. A nucleophile adds to the polar π bond in these steps, so they are called **nucleophilic addition steps**.

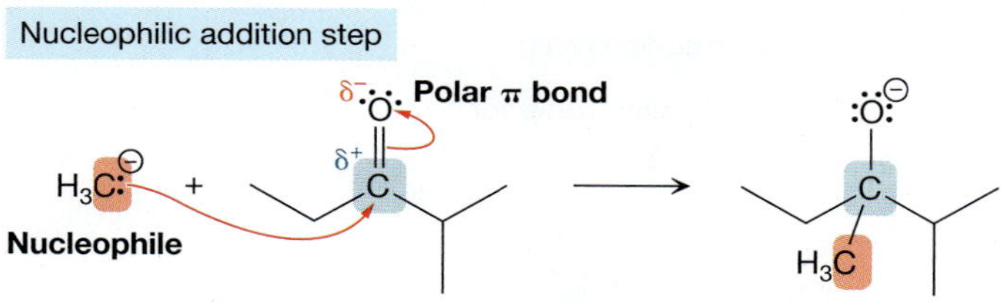

Nucleophilic addition step

(7-9)

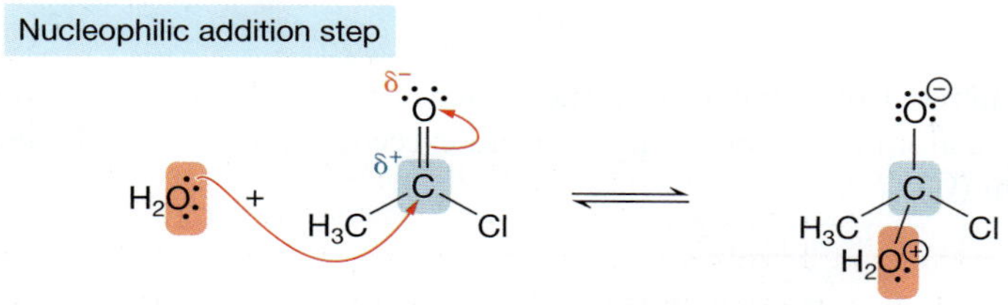

Nucleophilic addition step

(7-10)

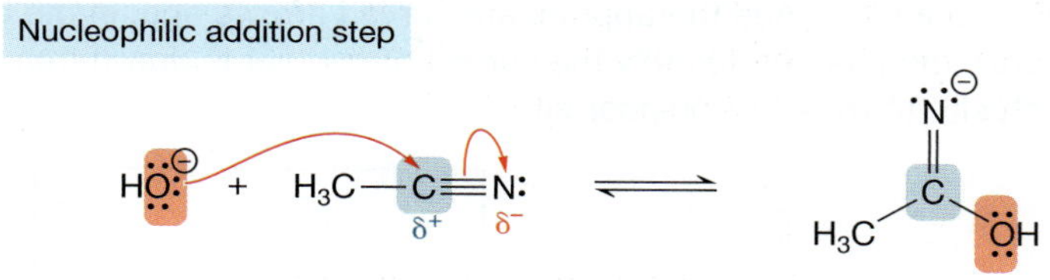

Nucleophilic addition step

(7-11)

The reverse of nucleophilic addition (Equations 7-12 and 7-13) is also commonly encountered in organic chemistry.

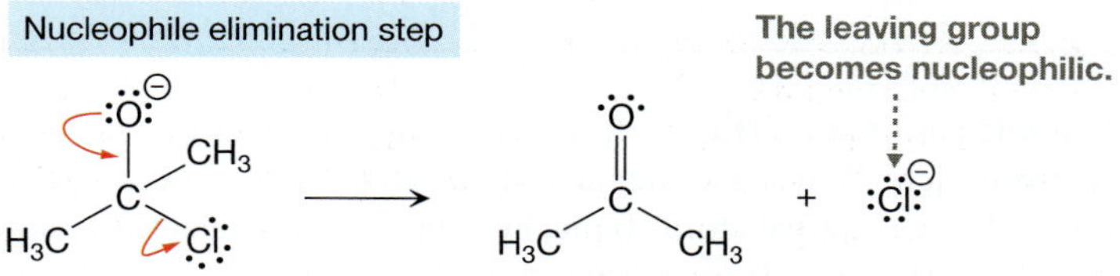

Nucleophile elimination step

The leaving group becomes nucleophilic.

(7-12)

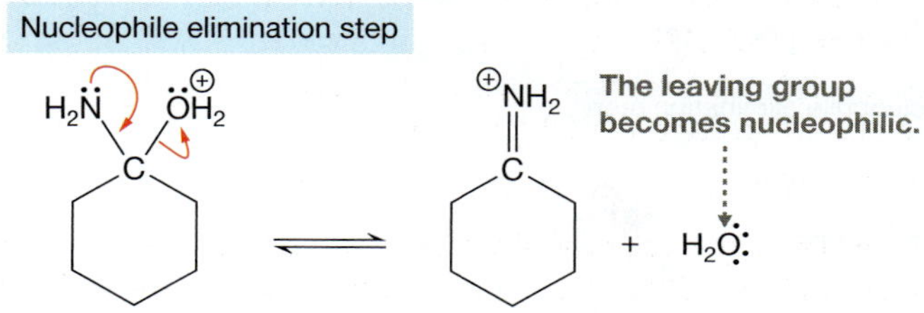

Nucleophile elimination step

The leaving group becomes nucleophilic.

(7-13)

In both examples, a lone pair of electrons from a more electronegative atom forms a π bond to a less electronegative atom. A *leaving group* is simultaneously expelled to avoid exceeding an octet on the less electronegative atom. In the products, the leaving group ends up with a lone pair of electrons and an excess of negative charge, both of which are characteristics of a nucleophile. As a result, we call these **nucleophile elimination steps**.

Nucleophilic addition and nucleophile elimination steps are typically found in multistep mechanisms. In Chapters 17 and 18, for example, we will learn reactions in which proton transfer steps are coupled with nucleophilic addition steps. In Chapters 20 and 21, we will see how nucleophilic addition and nucleophile elimination steps in the same mechanism can result in a substitution.

The *electron-rich to electron-poor* nature of a nucleophilic addition step is fairly straightforward. The nucleophile in a nucleophilic addition step, which has an excess of negative charge, is relatively electron rich, and the less electronegative atom of the polar π bond is relatively electron poor. Thus, the curved arrow drawn from the nucleophile to the polar π bond represents the flow of electrons from an electron-rich site to an electron-poor site:

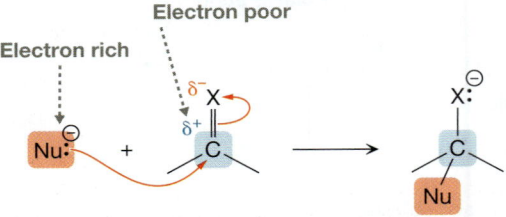

The second curved arrow, drawn from the center of the double (or triple) bond to the electronegative atom (X), is necessary to avoid exceeding an octet on the less electronegative atom (C).

YOUR TURN 7.7

For the following nucleophilic addition step, label the pertinent electron-rich and electron-poor sites. Add the appropriate curved arrows and, in the box provided, draw the product. Identify the curved arrow that is drawn from the electron-rich site to the electron-poor site.

In nucleophile elimination (the reverse step), the more electronegative atom (X) is relatively electron rich: It bears either a full negative charge or a partial negative charge. The less electronegative atom (typically C) is relatively electron poor. Thus, the curved arrow that originates from a lone pair of electrons on X and points to the bonding region between C and X represents the flow of electrons from an electron-rich site to an electron-poor site:

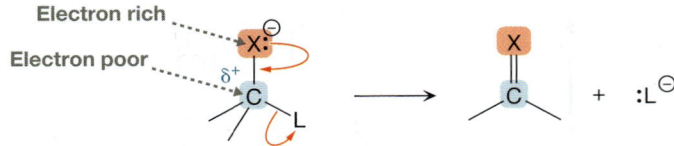

Once again, a second curved arrow, to represent the breaking of the C—L bond to the leaving group (L), is necessary to avoid exceeding an octet on the less electronegative atom (C).

For the following nucleophile elimination step, label the pertinent electron-rich and electron-poor sites. Add the appropriate curved arrows and identify the curved arrow that is drawn from the electron-rich site to the electron-poor site.

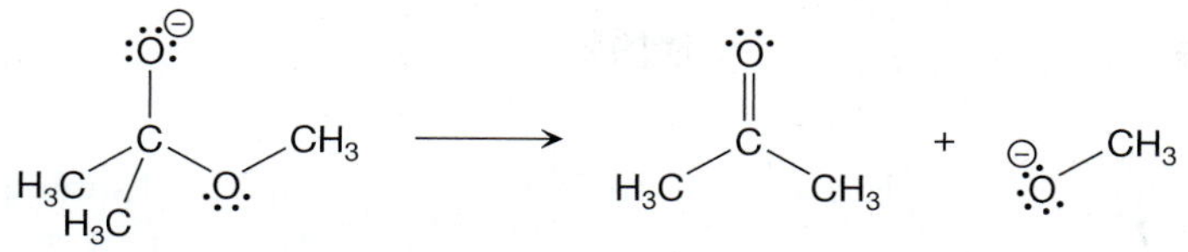

PROBLEM 7.13 Draw the appropriate curved arrows and the products for each of the following nucleophilic addition steps.

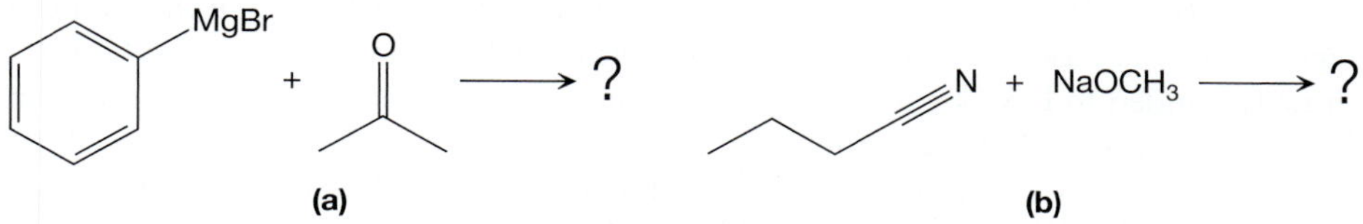

(a) (b)

PROBLEM 7.14 Draw the appropriate curved arrows necessary for each of the following nucleophile elimination steps to produce a ketone, and draw the resulting ketone.

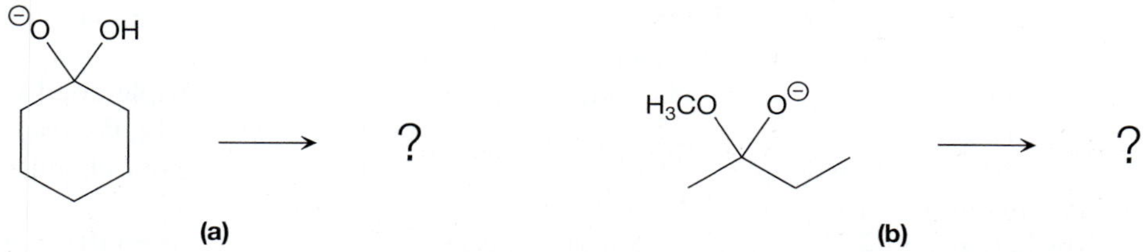

(a) (b)

7.5 Bimolecular Elimination (E2) Steps

Each of the elementary steps we have examined so far involve only one or two curved arrows. A **bimolecular elimination (E2) step**, however, is an example of an elementary step that requires three curved arrows, as shown in Equations 7-14 through 7-16.

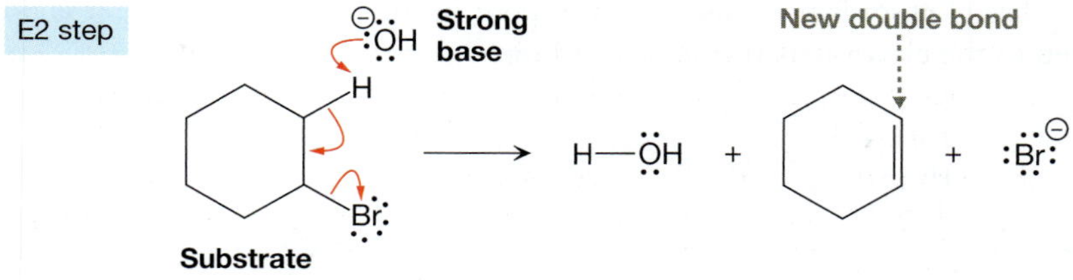

(7-14)

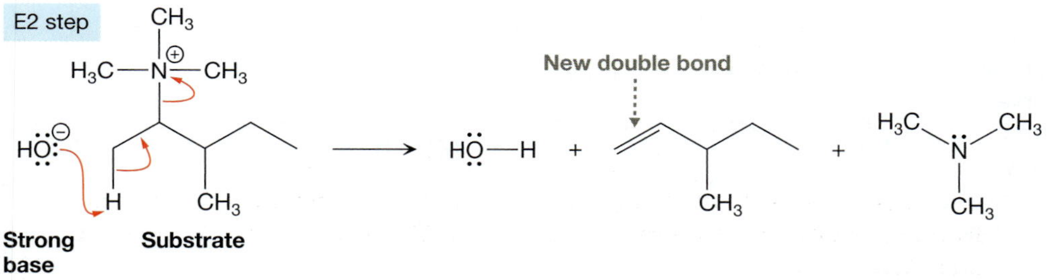

(7-15)

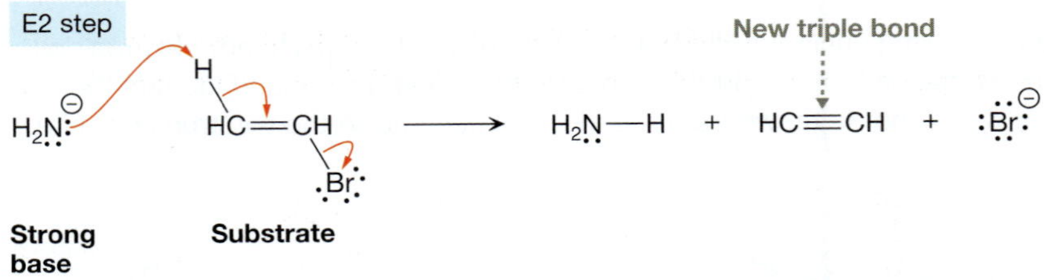

New triple bond

$$H_2N: \quad + \quad HC=CH \quad \xrightarrow{} \quad H_2N-H \quad + \quad HC\equiv CH \quad + \quad :\overset{\ominus}{\underset{..}{Br}}: \quad \text{(7-16)}$$

Strong base   **Substrate**

An E2 step can take place when a strong base reacts with a substrate in which *a leaving group (L) and a hydrogen atom are on adjacent carbon atoms.* That is:

> In an E2 step the substrate generally has the form $H-\overset{\beta}{C}-\overset{\alpha}{C}-L$ or $H-\overset{\beta}{C}=\overset{\alpha}{C}-L$.

It is called an elimination step because both the H atom and L are *eliminated* from the substrate. It is a *bimolecular* step because there are two reactant species in the elementary step—the base and the substrate. Furthermore, by assigning α to the C atom initially bonded to L and β to the adjacent C atom (shown above), we can see why an E2 step is a type of **β elimination**.

Equations 7-14, 7-15, and 7-16 illustrate the diversity of reactants and products that can be involved in E2 steps. The base is HO^- in both Equations 7-14 and 7-15, whereas it is H_2N^- in Equation 7-16. The leaving group can come off as a negatively charged species, as in Equations 7-14 and 7-16, or it can be uncharged, as in Equation 7-15. And the C atoms containing the H atom and the leaving group in the substrate may be joined by either a single bond (Equations 7-14 and 7-15) or a double bond (Equation 7-16).

The primary importance of E2 steps is the incorporation of a carbon–carbon multiple bond into a molecule at a particular site. In Equations 7-14 and 7-15, for example, a $C=C$ double bond is generated. And, in Equation 7-16, a $C\equiv C$ triple bond is formed. As we discuss later in this chapter, as well as in Chapters 11 and 12, the reactivity of carbon–carbon double and triple bonds is quite important in organic chemistry. We therefore revisit E2 steps in greater detail in Chapters 8, 9, and 10.

The base in an E2 step is the electron-rich species, but the hydrogen atom that the base attacks is not particularly electron poor; instead, the electron-poor atom is the carbon atom bonded to the leaving group. Thus, the movement of electrons from the *electron-rich* site to the *electron-poor* site is depicted with two curved arrows (Equation 7-17). One curved arrow is drawn from the negatively charged atom in the base to the hydrogen in the substrate. The second curved arrow is then drawn from the C—H bond to the bonding region between the two C atoms, and the third curved arrow is necessary to depict the departure of L to avoid exceeding the octet on carbon. Overall, the electron-poor C atom gains a share of electrons from the C—H bond. Furthermore, that C atom is no longer electron poor in the products because it is no longer bonded to the electronegative leaving group.

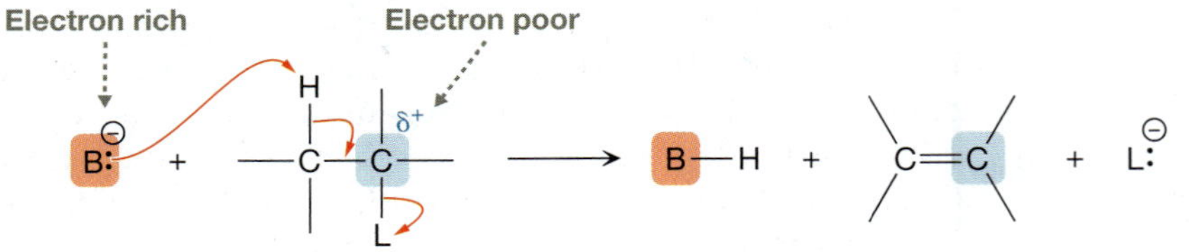

Electron rich   **Electron poor**

$$B:^{\ominus} \quad + \quad \overset{H}{\underset{L}{-C-C-}}^{\delta+} \quad \xrightarrow{} \quad B-H \quad + \quad \underset{\diagdown}{\overset{\diagup}{C}}=\underset{\diagup}{\overset{\diagdown}{C}} \quad + \quad L:^{\ominus} \quad \text{(7-17)}$$

YOUR TURN 7.9

The reactants and products for this E2 step are shown, but the curved arrow notation has been omitted. Supply the missing curved arrow notation and identify the pertinent electron-rich and electron-poor sites.

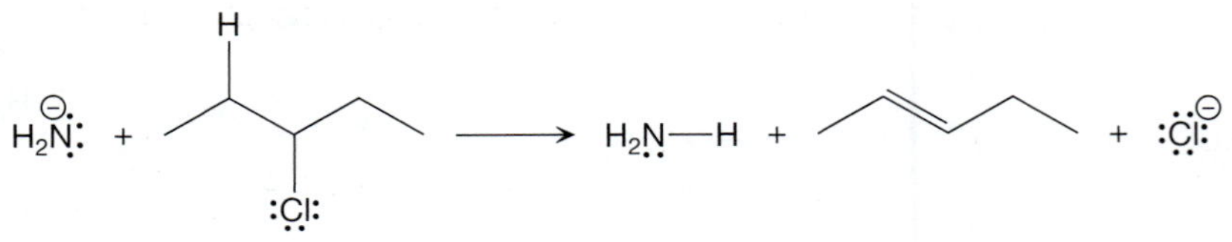

$$H_2\overset{\ominus}{N}: \quad + \quad \text{[substrate structure with H and :Cl:]} \quad \xrightarrow{} \quad H_2\underset{..}{N}-H \quad + \quad \text{[alkene structure]} \quad + \quad :\overset{\ominus}{\underset{..}{Cl}}:$$

PROBLEM 7.15 Supply the appropriate curved arrows and the products for each of the following E2 steps. *Hint:* Consider simplifying the electron-rich species.

NaOH +

(a)

CH$_3$Li + Cl

(b)

7.6 Electrophilic Addition and Electrophile Elimination Steps

An **electrophilic addition step** occurs when a species containing a nonpolar π bond (as part of a double or triple bond) approaches a strongly electron-deficient species—an *electrophile*—and a bond forms between an atom of the π bond and the electrophile (Equations 7-18 and 7-19).

Electrophilic addition step

$$H_3C-C\equiv C-CH_3 \quad H-\overset{\delta+}{\underset{}{C}}\overset{\delta-}{\underset{}{l}} \longrightarrow H_3C-\overset{\oplus}{C}=C\overset{H}{\underset{CH_3}{}} + \; :\overset{..}{\underset{..}{Cl}}:^{\ominus} \quad (7\text{-}18)$$

Nonpolar π bond   **Electrophile**

Electrophilic addition step

$$(7\text{-}19)$$

Nonpolar π bond   **Electrophile**

The nonpolar π bonds involved in electrophilic addition steps are typically ones that join a pair of carbon atoms. The electrophile, on the other hand, can have a variety of different forms. For example, the electrophile can be H$^+$ from a Brønsted acid, such as HCl in Equation 7-18. Alternatively, the electrophile can exist on its own, as shown for the NO$_2^+$ species in Equation 7-19.

The product of each of these electrophilic addition steps is a *carbocation*, which is highly unstable and will react further because it has a positive charge and lacks an octet. Electrophilic additions, therefore, are generally part of multistep mechanisms. Equation 7-18, for example, is the first step in the electrophilic addition of an acid across a multiple bond (Chapters 11 and 12). Equation 7-19, on the other hand, is the first step in electrophilic aromatic substitution (Chapters 22 and 23).

YOUR TURN 7.10

Add the appropriate curved arrows for the following electrophilic addition step.

$$H_3C\overset{H}{\underset{H_2}{C}}-\overset{H_2}{\underset{H}{C}}=\overset{C^2}{\underset{}{C}}-CH_3 + H-\overset{..}{\underset{..}{Br}}: \longrightarrow H_3C\overset{H}{\underset{H_2}{C}}-\overset{H}{\underset{H}{C}}\overset{\oplus}{\underset{H}{C}}-\overset{H_2}{\underset{}{C}}-CH_3 + :\overset{..}{\underset{..}{Br}}:^{\ominus}$$

CONNECTIONS

Nitrobenzene ($C_6H_5NO_2$, Equation 7-21) is primarily used in the production of aniline, which is a precursor to a variety of compounds, such as explosives, dyes, and pharmaceutical drugs. Nitrobenzene has an odor that resembles almonds, making it useful in the fragrance industry.

Carbocations are typically quite unstable, so the reverse of electrophilic addition is also a common elementary step in organic reactions. In the reverse step, called **electrophile elimination**, an electrophile is *eliminated* from the carbocation, generating a stable, uncharged, organic species. Equations 7-20 and 7-21 show examples in which H^+ is the electrophile that is eliminated.

Electrophile elimination

(7-20)

Electrophile elimination

(7-21)

Equation 7-20 is the second step of an E1 reaction (Chapter 8) and Equation 7-21 is the second step of an electrophilic aromatic substitution reaction (Chapters 22 and 23).

YOUR TURN 7.11

Add the appropriate curved arrow(s) to the following electrophile elimination step, which is essentially the reverse of the addition step in Your Turn 7.10.

Even though Equations 7-20 and 7-21 show that H^+ is the electrophile that is eliminated, *a proton does not exist on its own in solution.* Rather, it must be associated with a base. Any base that is present in solution will therefore assist in the removal of a proton in an electrophile elimination step. If water is present, for example, then Equation 7-20 would more appropriately be written as follows in Equation 7-22:

Electrophile elimination

(7-22)

As shown in Equation 7-23, the electrophile (E^+) in an electrophilic addition step is relatively electron poor because it either carries a full positive charge (Equation 7-19) or has an atom with a significant partial positive charge (Equation 7-18). The double or triple bond, on the other hand, is relatively electron rich. In a $C=C$ double bond, for example, four electrons are localized in the region between two atoms, and in a $C\equiv C$ triple bond, six electrons are localized between two atoms. Therefore, the movement of electrons from *electron rich to electron poor* is indicated by a curved arrow that originates from the center of the multiple bond and terminates at the electrophile.

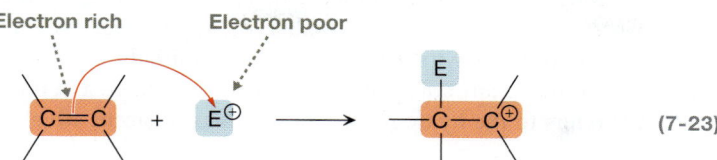

In electrophile elimination (Equation 7-24), the positively charged C atom is relatively electron poor, whereas the C—E single bond is relatively electron rich. Therefore, the curved arrow originates from the C—E single bond and points to the bond between C and C⁺.

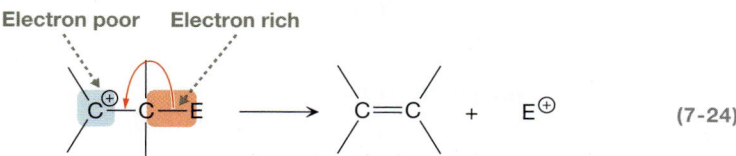

YOUR TURN 7.12

In both Your Turn 7.10 and 7.11, label the pertinent *electron-rich* and *electron-poor* sites.

PROBLEM 7.16 Supply the appropriate curved arrows and draw the product of each of the following electrophilic addition steps.

(a)   (b)

PROBLEM 7.17 Supply the appropriate curved arrows and draw the product of each of these electrophile elimination steps. (You may assume that a weak base is present.)

(a)   (b)

7.7 Carbocation Rearrangements: 1,2-Hydride Shifts and 1,2-Alkyl Shifts

Carbocations typically appear as intermediates in multistep mechanisms. They are usually too unstable to exist for a prolonged time because they are extremely electron deficient due to (1) the carbon atom's +1 formal charge and (2) its lack of an octet. As we saw in Section 7.3, carbocations commonly behave as Lewis acids to form a bond with an electron-rich Lewis base. Carbocations can also eliminate H⁺ (Section 7.6) to yield an alkene or alkyne. Given a chance, however, carbocations can also undergo a **rearrangement** before taking part in one of these steps with another species.

Equations 7-25 and 7-26 show two **carbocation rearrangements**. Equation 7-25 is a **1,2-hydride shift**. A *hydride anion* (H^-) is said to shift because a hydrogen atom migrates along with the pair of electrons initially making up the C—H bond. The numbering system denotes the number of atoms over which the hydride anion migrates; the atom to which the hydrogen is initially bonded is designated as the number 1 atom and any adjacent atom can be designated as a number 2 atom. Thus, a "1,2" shift refers to the hydrogen migrating to an adjacent atom.

A 1,2-hydride shift

H migrates with
two electrons.

$$\tag{7-25}$$

A 1,2-methyl shift

CH_3 migrates with
two electrons.

$$\tag{7-26}$$

Equation 7-26 shows a **1,2-alkyl shift**; more specifically, a methyl group is transferred, so this rearrangement is called a **1,2-methyl shift**. The numbering system is no different from that of the hydride shift because the migrating group—the methyl group—is transferred to an adjacent atom.

Both Equations 7-25 and 7-26 share identical curved arrow notation. In both cases, a single curved arrow indicates that the initial bond between the C atom and the migrating group is broken, and those electrons are used to form a bond to the adjacent C. Meanwhile, the +1 formal charge is also shifted over one atom to the atom that was initially bonded to the migrating group.

Carbocation rearrangements are important to consider whenever carbocations are formed in a particular reaction. These reactions include S_N1 and E1 reactions (Chapters 8 and 9) and electrophilic addition reactions (Chapter 11).

In a carbocation rearrangement, the positively charged carbon atom of the carbocation is very electron poor because it carries a full positive charge and it has less than an octet of electrons. On the other hand, a single bond to hydrogen or carbon on an adjacent atom is relatively electron rich because two electrons are localized in the bonding region. Therefore, the single curved arrow that is used to depict a carbocation rearrangement in Equation 7-27 represents the flow of electrons from an electron-rich site to an electron-poor site.

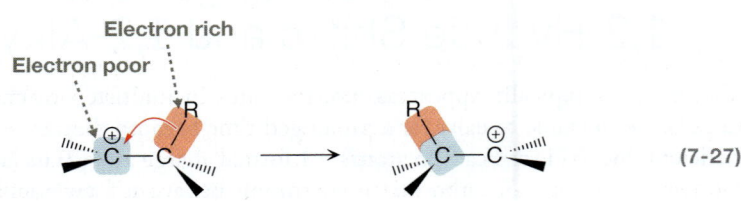

$$\tag{7-27}$$

"Watching" a Bond Break

An elementary step can be extremely fast—on the order of picoseconds (1 ps = 10^{-12} s) or femtoseconds (1 fs = 10^{-15} s). Is it possible, then, to actually observe one taking place? A few decades ago, the answer would have been no, but now we have the ability to do so, at least with some types of reactions. How is it done? In essence, snapshots of a reaction are taken using a really fast camera. But not a camera in the traditional sense. In this case, the "camera" is constructed from lasers capable of producing light pulses lasting <1000 fs. Such a femtosecond laser is shown here.

The decomposition of gaseous ICN into I + CN, carried out in 1988 by Stewart O. Williams and Dan G. Imre of the University of Washington, has been studied using this technology. Two lasers were used—one to provide a ~125-fs pulse of 306-nm light to break the I—C bond and a second to provide a ~125-fs pulse of ~389-nm to 433-nm light at various delay times after the first pulse. The second pulse excited the newly forming CN species, causing it to emit fluorescence that could be detected and correlated with the I—C distance over time. They found that the I—C bond was nearly completely broken at ~60 fs.

Even though studies such as this one focus on specific chemical reactions in the gas phase, the knowledge we gain can be used to develop a deeper understanding of reaction dynamics in general.

YOUR TURN 7.13

Supply the curved arrow notation for the carbocation rearrangement shown here.

PROBLEM 7.18 Supply the appropriate curved arrows and draw the product for this carbocation undergoing **(a)** a 1,2-hydride shift and **(b)** a 1,2-methyl shift.

7.8 The Driving Force for Chemical Reactions

So far, our focus on elementary steps has been on *how* they occur, using curved arrows to account for bonds breaking and bonds forming. But why should these elementary steps occur in the first place? That is, what is their *driving force*?

The **driving force** for a reaction reflects the extent to which the reaction favors products over reactants, and that tendency increases with increasing stability of the products relative to the reactants. Thus, to understand a reaction's driving force, it is a matter of identifying and evaluating the factors that help stabilize the products, destabilize the reactants, or both.

Although there are various factors that dictate the stability of a species, we can often make reasonable predictions about a reaction's driving force by examining just two factors. One is *charge stability*, which, as we saw in Chapter 6, can vary dramatically from one species to another. The other is *total bond energy*. As we saw in Chapter 1, each covalent bond provides stability to a molecule, and the amount of stabilization depends heavily on the type of bond that is formed—that is, which atoms are bound together, and whether it is by a single, double, or triple bond. (Review the tables of bond energies in Tables 1-2 and 1-3 on p. 10.)

A reaction's driving force generally increases with:

- Greater charge stabilization in the products relative to the reactants.
- Greater total bond energy in the products relative to the reactants.

With this in mind, we can see that a coordination step such as the one in Equation 7-28 is unambiguously driven toward products.

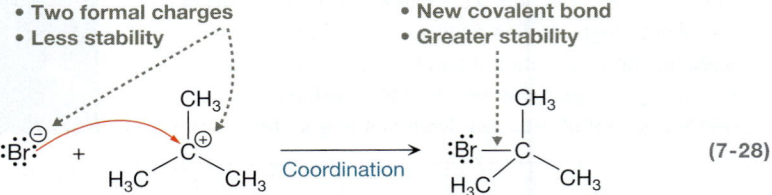

(7-28)

Charge stability heavily favors products because there are two formal charges in the reactants but no formal charges in the products. Furthermore, total bond energy favors products because, during the course of the reaction, one covalent bond is formed, giving carbon an octet, and no bonds are broken.

Charge stability and bond energy do not always work in the same direction. Consider the proton transfer step in Equation 7-29.

The H—Cl bond broken is stronger than the H—N bond formed.

The negative charge is better accommodated on Cl than N.

$$H_2\overset{\ominus}{N}: \ + \ H-\overset{..}{\underset{..}{Cl}}: \ \xrightarrow{\text{Proton transfer}} \ H_2N-H \ + \ :\overset{..}{\underset{..}{Cl}}:^{\ominus} \quad K_{eq} = 1 \times 10^{43} \quad (7\text{-}29)$$

In this case, a negative charge appears in both the reactants and the products. In the reactants, the negative charge is on N, whereas it is on Cl in the products. Because Cl is significantly larger (i.e., lower in the periodic table) than N, the negative charge is better accommodated on Cl than on N, so we say that charge stability favors products. Bond energy, however, favors the reactants because the H—Cl bond that appears on the reactant side (431 kJ/mol; 103 kcal/mol) is stronger than the H—N bond that appears on the product side (389 kJ/mol; 93 kcal/mol). Despite the disagreement, notice that the products are heavily favored; HCl is a much stronger acid than NH_3, giving rise to a very large K_{eq} of 1×10^{43}.

YOUR TURN 7.14

Consult Appendix A to verify that HCl is a stronger acid than NH_3. Write the pK_a values of the two acids below. Do these pK_a values agree with the fact that the reaction in Equation 7-29 heavily favors products?

pK_a of HCl _____ pK_a of NH_3 _____

The previous example leads to the following general rule:

When charge stability and bond energy favor opposite sides of a chemical reaction, charge stability usually wins.

This idea is very useful when considering the E2 step in Equation 7-30.

(7-30)

Charge stability favors the products because the negative charge is better accommodated on Cl than on O. Bond energy, however, favors the reactants because, on the product side, a σ and a π bond are formed, while two σ bonds are broken. Effectively, there is a net gain of one σ bond and a net loss of one π bond and, as we learned in Chapter 3, a σ bond is generally stronger. Experimentally, we know that the product side of the reaction is heavily favored, in agreement with the prediction obtained using charge stability.

SOLVED PROBLEM 7.19

Determine which side of this carbocation rearrangement is favored.

Think On the two sides of the reaction, is there a difference in charge stability? Is there a difference in total bond energy?

Solve The positive charge is on a tertiary carbon in the reactant and is on a secondary carbon in the product. Because a tertiary carbocation is more stable (Section 6.6e), charge stability favors the reactant side. Total bond energy is not a significant factor in this reaction because a C—C σ bond is broken in the reactant and another one is formed in the product. Overall, then, this reaction will favor the reactant side.

PROBLEM 7.20 Determine which side of each of the following carbocation rearrangements is favored.

(a)

(b)

Caution must be taken when using charge stability and total bond energy to predict the driving force for an elementary step. One reason is that a reaction that is product favored is not guaranteed to occur at a rate that is practical. For example, both charge stability and total bond energy heavily favor the products of the S_N2 reaction in Equation 7-31, but essentially no reaction occurs.

(7-31)

Understanding the reason for this requires reaction kinetics, a topic discussed in Chapter 9.

A second reason to be cautious is that other factors can come into play. Consider, for example, the heterolysis step in Equation 7-32, which is the reverse of the coordination step in Equation 7-28.

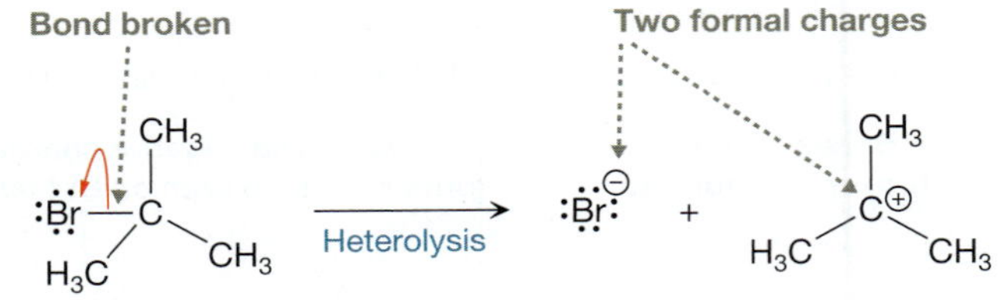

It might seem that this step shouldn't take place at all because both charge stability and total bond energy heavily disfavor the products. However, heterolysis steps such as this one often appear in multistep mechanisms. As it turns out, factors that influence the driving force include effects from the solvent, as well as entropy—topics that we discuss in Chapter 9.

By and large, charge stability and total bond energy are the dominant contributors to the driving force of elementary steps in most situations. These are the ones you should focus on the most. We will discuss other factors in the context of the reactions in which they are relevant.

7.9 Keto–Enol Tautomerization: An Example of Bond Energies as the Major Driving Force

In aqueous basic or acidic conditions, ketones and aldehydes exist in rapid equilibrium with a rearranged form, called an *enol*:

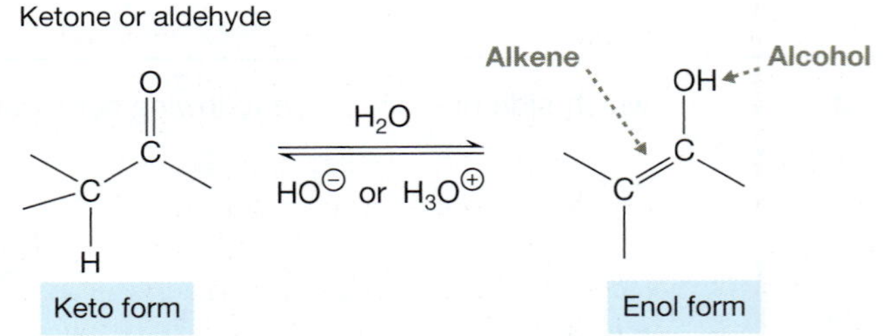

As a ketone or aldehyde, the species is called the **keto form**. In the **enol form**, the species has a carbon atom that is simultaneously part of a C=C bond characteristic of an alk**ene** and is bonded to OH, characteristic of an alcoh**ol**. Because the keto and enol forms are constitutional isomers in equilibrium, they are called **tautomers** (Greek: *tauto* = same; *mer* = part), and the equilibrium is called **keto–enol tautomerization**. In later chapters, we will see that this equilibrium has important consequences in a variety of chemical reactions.

In the equilibrium in Equation 7-33, one of the main differences between the keto and enol forms is the location of a hydrogen atom. In the keto form, there is a hydrogen atom on the carbon atom that is adjacent to the C=O group, the so-called **α (alpha) carbon**, whereas in the enol form, the hydrogen appears on the oxygen atom instead. Realizing that a proton transfer step serves to add or remove a proton from a particular species, we can account for this transformation with a mechanism consisting of back-to-back proton transfer steps. Such a mechanism depends, however, on whether a strong acid or strong base is present.

Equation 7-34 is the mechanism for the reaction in Equation 7-33 under basic conditions. The strong base removes a proton from the α carbon in Step 1, producing an **enolate anion**, which has a resonance-delocalized negative charge. In the resonance structure on the right, the O atom is electron-rich and, in Step 2 of the mechanism, picks up a proton from water, which acts as the acid.

Mechanism for Equation 7-33 under basic conditions

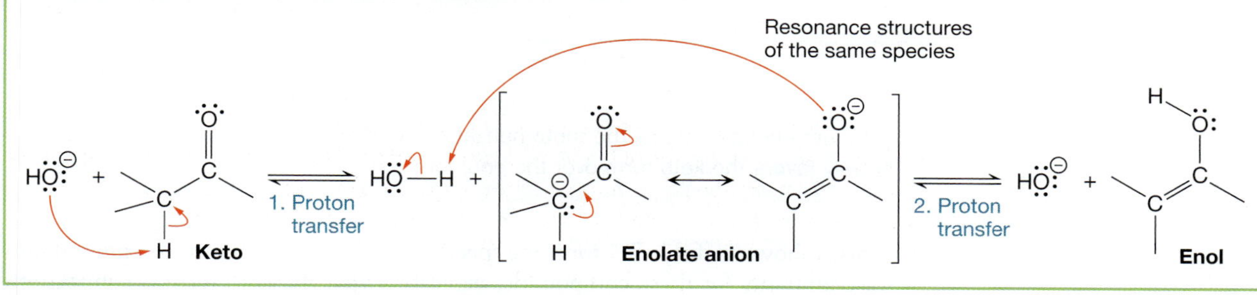

Mechanism for Equation 7-33 under acidic conditions

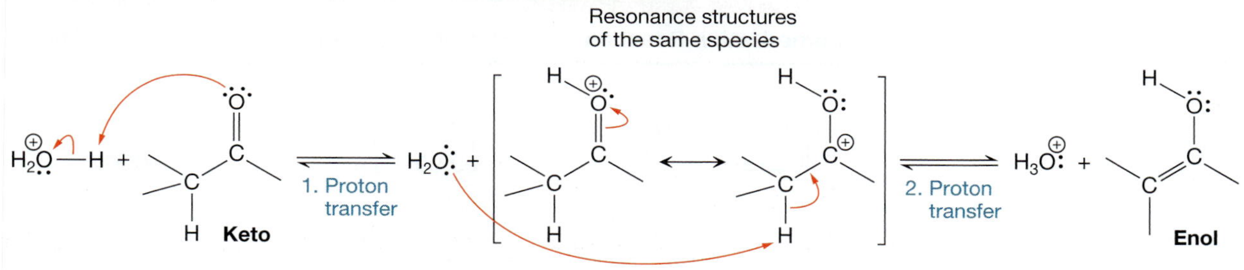

The mechanism for the reaction under acidic conditions is shown in Equation 7-35. The strong acid that is present donates a proton to the O atom of the C=O group in Step 1. In Step 2, water (a weak base) removes the proton from the α carbon to produce the uncharged enol product.

Because it is an equilibrium, the tautomerization reaction in Equation 7-33 takes place in the reverse direction, too. Equations 7-36 and 7-37 are the mechanisms showing how the enol form produces the keto form under basic and acidic conditions, respectively. Notice how the steps in Equations 7-36 and 7-37 are the same as in Equations 7-34 and 7-35, respectively, but in reverse order.

Mechanism for the reverse of Equation 7-33 under basic conditions

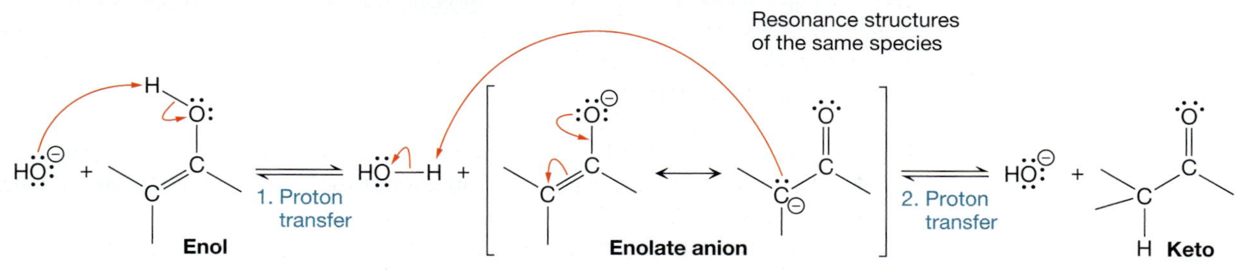

Resonance structures of the same species

(7-37)

For most enolate anions, the tautomerization equilibrium (Equation 7-33, p. 350) heavily favors the keto form over the enol form.

This is shown in Table 7-1 for some specific examples. The enol form is present only in trace amounts for these compounds, suggesting that the keto form is significantly more stable. In other words, the driving force for these reactions favors the keto form.

TABLE 7-1 Relative Percentages of Keto and Enol Forms

Tautomerization Reaction	Percent Keto Form	Percent Enol Form
	99.99994%	0.00006%
	99.986%	0.014%
	99.9999995%	0.0000005%
	99.9999988%	0.0000012%
	99.99996%	0.00004%

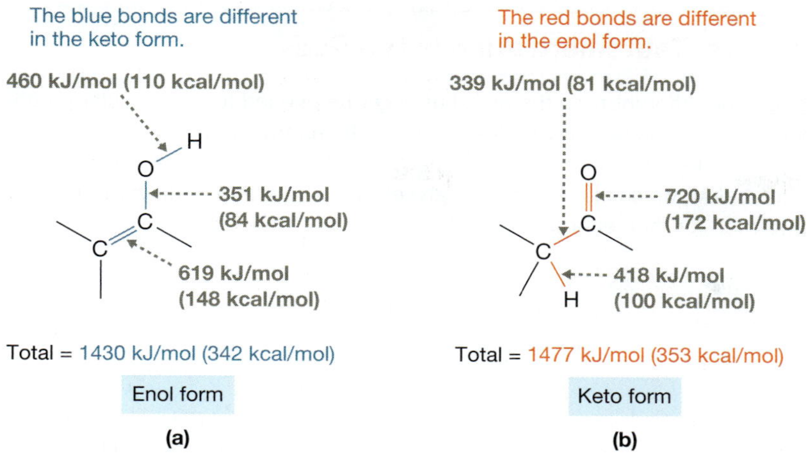

The blue bonds are different in the keto form.

460 kJ/mol (110 kcal/mol)

351 kJ/mol (84 kcal/mol)

619 kJ/mol (148 kcal/mol)

Total = 1430 kJ/mol (342 kcal/mol)

Enol form

(a)

The red bonds are different in the enol form.

339 kJ/mol (81 kcal/mol)

720 kJ/mol (172 kcal/mol)

418 kJ/mol (100 kcal/mol)

Total = 1477 kJ/mol (353 kcal/mol)

Keto form

(b)

FIGURE 7-4 Relative stabilities of enol and keto forms (a) Energies of the bonds that appear in the enol form but not in the keto form. The sum of the energies is 1430 kJ/mol (342 kcal/mol). (b) Energies of the bonds that appear in the keto form but not in the enol form. The sum of the energies is 1477 kJ/mol (353 kcal/mol). Because of its greater total bond energy, the keto form is more stable than the enol form.

Unlike the reactions we saw in the previous section, the reactant and product of the tautomerization equilibrium are both uncharged. Therefore, charge stability cannot contribute to the driving force. Instead, the keto form is favored because it has a greater total bond energy than the enol form. This is shown explicitly in Figure 7-4, which tallies the energies of the bonds (see Tables 1-2 and 1-3, p. 10) that appear in one form but not the other. As we can see, the total bond energy of the keto form is roughly 47 kJ/mol (11 kcal/mol) greater than that of the enol form, making the keto form more stable.

YOUR TURN 7.15

In Figure 7-4, there is a similar bond in red in the keto form for each bond in blue in the enol form. Therefore, we can pair these bonds as follows:

Bond in Keto Form	Bond in Enol Form	Difference in Bond Energy
C=O	C=C	
C—C	C—O	
C—H	O—H	

For each pair, compute the *difference* in bond energy and enter it into the table. Based on these data, which bond is most responsible for the additional stability of the keto form? _____

PROBLEM 7.21 Decarboxylation (i.e., elimination of CO_2) occurs when a β-ketoacid is heated under acidic conditions.

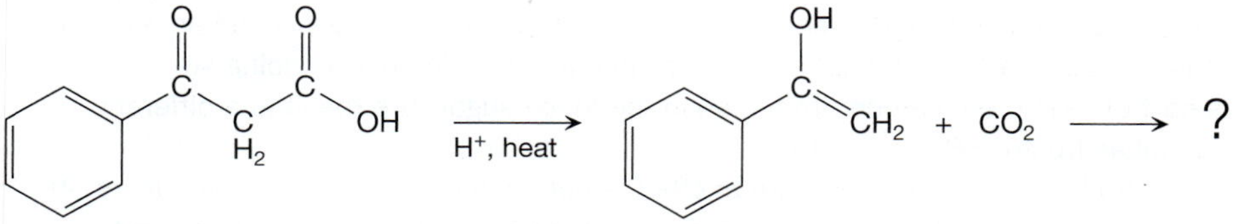

The immediate product of decarboxylation is an enol, which quickly rearranges. Draw the overall product of the rearrangement.

Sugar Transformers: Tautomerization in the Body

A sugar inside a cell can be different from the one that might be needed for a particular purpose, but the body has developed an elegant way to deal with this: It can transform one sugar into another! It does so using keto–enol tautomerization reactions (Section 7.9). This is exemplified in the reaction scheme below, which is a key part of **glycolysis**, a metabolic pathway that breaks down simple carbohydrates for their energy.

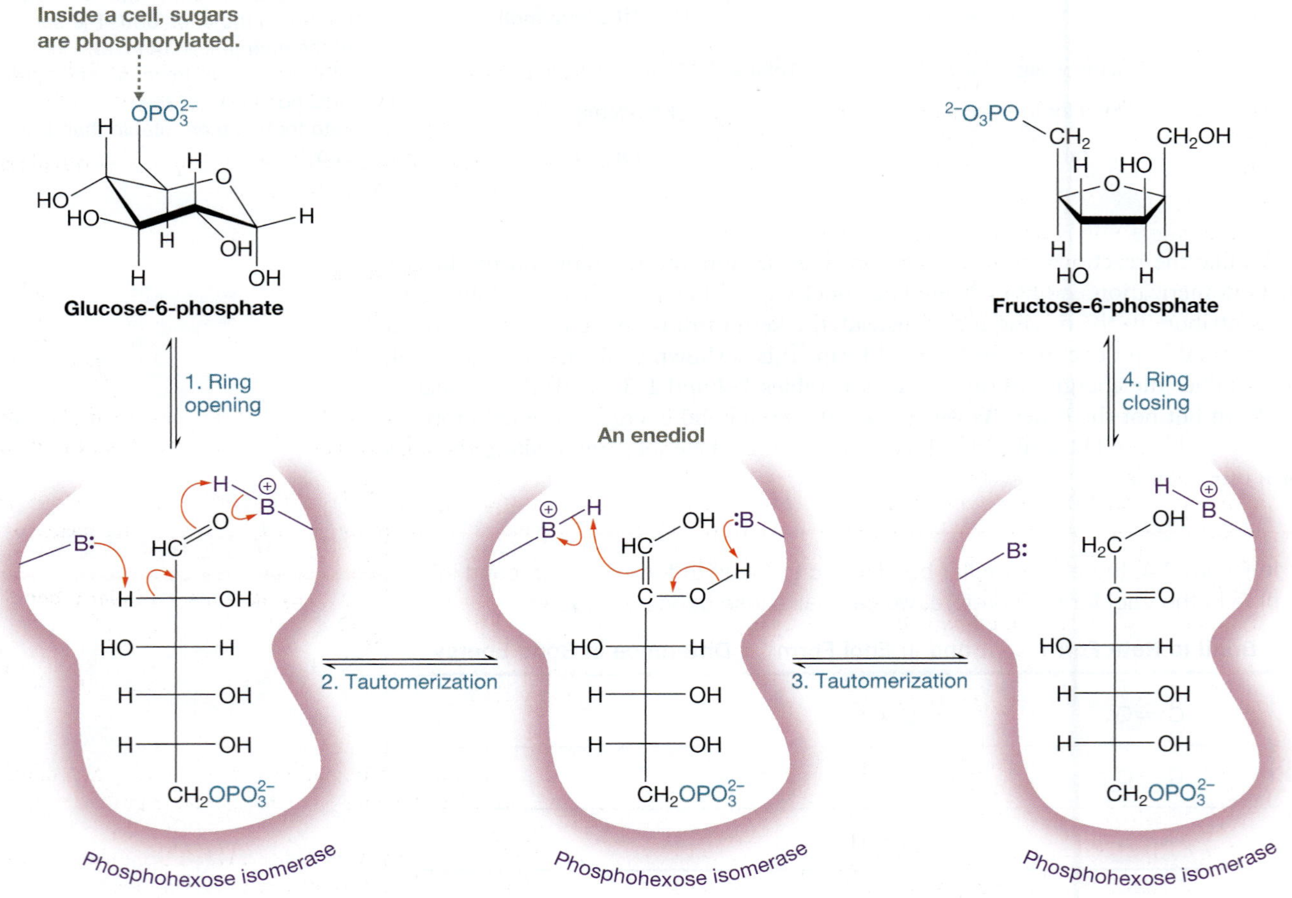

Inside a cell, sugars are phosphorylated.

Glucose-6-phosphate

Fructose-6-phosphate

1. Ring opening

4. Ring closing

An enediol

2. Tautomerization

3. Tautomerization

Phosphohexose isomerase

Phosphohexose isomerase

Phosphohexose isomerase

When D-glucose enters a cell, it is *phosphorylated* to become glucose-6-phosphate (shown above at left), in which the hydroxyl group on C6 (the bottommost carbon in the Fischer projection) has been replaced by a phosphate ($-OPO_3^{2-}$) group. Before it can be broken down, however, glucose-6-phosphate must be converted into fructose-6-phosphate (shown at right). This conversion involves back-to-back tautomerization reactions catalyzed by the enzyme phosphohexose isomerase (the active site is shown in purple). The enzyme supplies the basic (B:) and acidic sites (B^+—H) necessary for the proton transfer steps. The first tautomerization produces an enol that contains two different OH groups, so it is more precisely called an enediol. The second tautomerization converts the enediol back into a keto form, but on doing so, the C=O bond is part of a different carbon atom than in the initial sugar. The result is a different phosphorylated sugar, fructose-6-phosphate.

This process of transforming one sugar into another is not limited to just six-carbon sugars. At a later stage in glycolysis, an enzyme called *phosphotriose isomerase* converts one three-carbon sugar into another. These kinds of processes truly are a testament to how efficient biological organisms are.

- The curved arrow notation for an elementary step reflects the flow of electrons from an *electron-rich* site to an *electron-poor* site. **(Section 7.1a; Objectives 1, 3, and 4)**

- Metal cations from group 1A of the periodic table behave as spectator ions, so they can be disregarded when identifying electron-rich sites in an elementary step. **(Section 7.1b; Objective 2)**

- **Organometallic** reagents, such as alkyllithium reagents (RLi), Grignard reagents (RMgX), and lithium dialkylcuprates (R_2CuLi), can be treated as sources of R^-. Hydride reagents, such as lithium aluminum hydride ($LiAlH_4$) and sodium borohydride ($NaBH_4$), can be treated as sources of H^-. **(Section 7.1b; Objective 2)**

- Charge stability and total bond energy are two major factors that contribute to a reaction's **driving force**. For a reaction or elementary step involving both ions and uncharged molecules, the side that is favored generally exhibits greater charge stability. **(Section 7.8; Objective 5)**

- For a reaction or elementary step involving only uncharged species, the side that is favored generally has the greater bond energies. **(Section 7.9; Objective 5)**

- In a **keto–enol tautomerization**, the **keto form** is in equilibrium with its **enol form** via proton transfer steps. In general, the keto form is much more stable than, and therefore favored over, the enol form because it has a greater total bond energy. **(Section 7.9; Objectives 5 and 6)**

- **Bimolecular nucleophilic substitution (S_N2) steps.** **(Section 7.2; Objectives 3 and 4)**

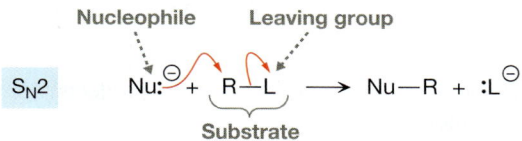

- A **substrate** (R—L) contains a **leaving group** (L). Good leaving groups have strong conjugate acids. A **nucleophile** contains an atom that has a full or partial negative charge and possesses a lone pair of electrons.
- The nucleophile is relatively *electron rich* and the atom attached to the leaving group is relatively *electron poor*.

- **Coordination steps** and **heterolytic bond dissociation (heterolysis) steps.** **(Section 7.3; Objectives 3 and 4)**

Lewis base (nucleophile)   Lewis acid (electrophile)

Coordination   B:⁻ + ↑A⁺ → B—A—

- In a coordination step, the Lewis acid is usually deficient of an octet, and the Lewis base has an atom with a partial or full negative charge and a lone pair of electrons.

- The Lewis base is relatively *electron rich*, whereas the Lewis acid is relatively *electron poor*—it is an **electrophile**.

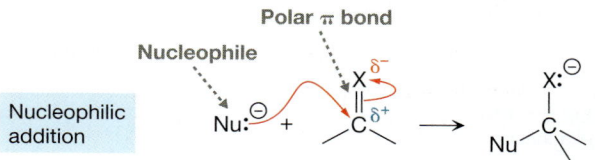

- In a heterolysis step, the bond to the substrate is broken and the bonding electrons become a lone pair on the leaving group.

- **Nucleophilic addition steps** and **nucleophile elimination steps.** **(Section 7.4; Objectives 3 and 4)**

Polar π bond

Nucleophile

Nucleophilic addition

- A nucleophile forms a bond to the positive end of a polar C—X π bond, forcing a pair of electrons from the π bond onto X.
- The nucleophile is relatively *electron rich*, and the atom at the positive end of the polar C—X π bond is relatively *electron poor*.

Leaving group   Polar π bond

Nucleophile elimination

- In a nucleophile elimination step, a new C—X π bond is formed at the same time that a leaving group is expelled.
- The X atom of the C—X bond is relatively *electron rich*, whereas the C atom is relatively *electron poor*.

- **Bimolecular elimination (E2) steps.** **(Section 7.5; Objectives 3 and 4)**

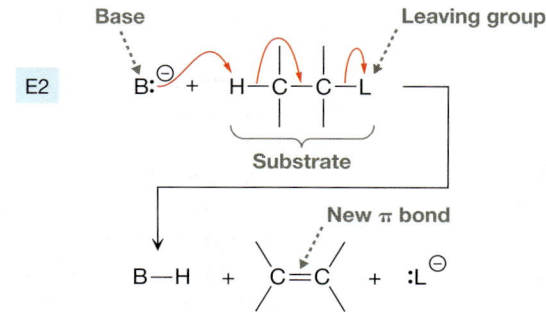

- In an E2 step, a base deprotonates an atom on the substrate at the same time that a leaving group is expelled, forming an additional π bond between the atoms to which the hydrogen and the leaving group were initially bonded.
- The base is relatively *electron rich*; the carbon atom bonded to the leaving group is relatively *electron poor*.

- **Electrophilic addition steps** and **electrophile elimination steps**. (Section 7.6; Objectives 3 and 4)

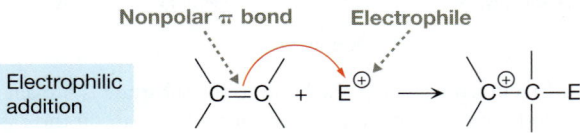

- In electrophilic addition, a pair of electrons from a nonpolar π bond forms a bond to an *electrophile*.
- The nonpolar π bond is relatively *electron rich*, whereas the electrophile is relatively *electron poor*.

- In an electrophile elimination step, an electrophile is eliminated from a carbocation species and a nonpolar π bond is formed simultaneously.
- The C—E single bond is relatively *electron rich*; the positively charged C atom is relatively *electron poor*.

- **Carbocation rearrangements.** (Section 7.7; Objectives 3 and 4)

- In a **1,2-hydride shift** or **1,2-alkyl shift**, a C—H or C—C bond adjacent to a carbocation is broken, and the bond is reformed to the C atom initially with the positive charge. The positive charge moves to the C atom whose bond is broken.
- The positively charged C atom is relatively *electron poor*; the C—H or C—C single bond that is broken is relatively *electron rich*.

Problems

7.1–7.7 Curved Arrow Notation and Elementary Steps

7.22 Predict the product of the reaction between phenol (C_6H_5OH) and each of the compounds shown here. *Hint:* First determine which elementary step is likely to occur.

(a) CH_3OK (b) CH_3Li (c) [phenyl]—MgBr (d) $LiAlH_4$

7.23 Determine whether each of the following elementary steps is acceptable. For those that are, draw the products. For those that are not, explain why. *Hint:* Explaining why may involve drawing the products.

(a)

(b)

(c)

(d)

(e)

7.24 Draw the curved arrows and the product for each of the following S_N2 steps.

(a) + NaSH ⟶ ?

(b) + KCN ⟶ ?

7.25 **(a)** Draw the appropriate curved arrows and products for each set of reactants undergoing a coordination step. Identify each reactant species as either a Lewis acid or a Lewis base. **(b)** Use curved arrow notation to show each product undergoing heterolysis to regenerate reactants.

(i)

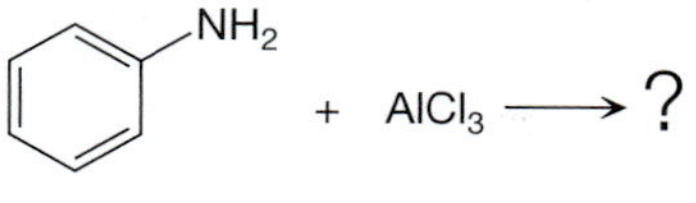

$+ \ AlCl_3 \longrightarrow$?

(ii)

$H_2O \ + \ BF_3 \longrightarrow$?

(iii)

$+ \ $ $\longrightarrow$?

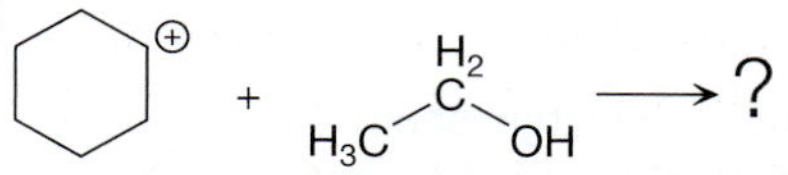

7.26 Draw the curved arrows and the product for each of the following nucleophilic addition steps.

(a)

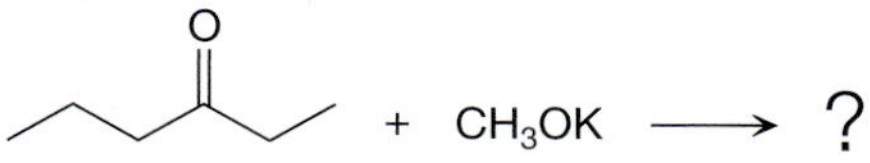

$+ \ CH_3OK \longrightarrow$?

(b)

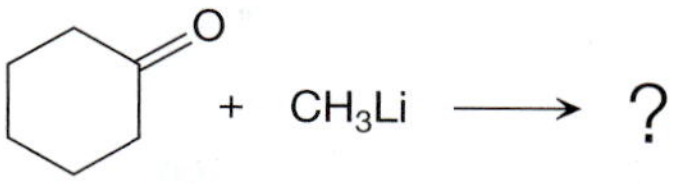

$+ \ CH_3Li \longrightarrow$?

(c)

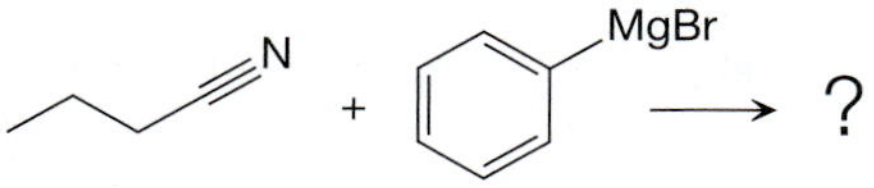

$+$ $\longrightarrow$?

(d)

$+ \ NaBH_4 \longrightarrow$?

(e)

$+ \ NaOH \longrightarrow$?

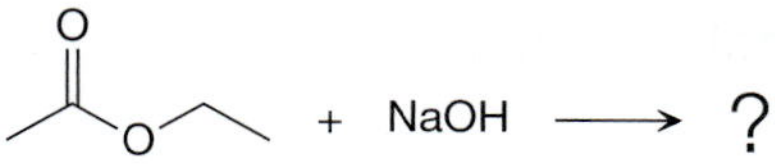

(f)

$\longrightarrow$?

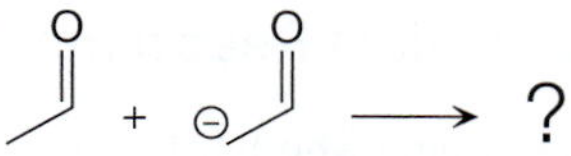

7.27 For each of the steps in Problem 7.26, determine whether the product can eliminate a leaving group to produce a compound that is different from the reactants. For those that can, draw the appropriate curved arrows and the new product that forms.

7.28 If the anionic species shown were to eliminate a leaving group, the three possibilities would be H_3C^-, Cl^-, or CH_3O^-. Draw the curved arrow notation and the products for each of these elimination steps. Which is the major product? Why?

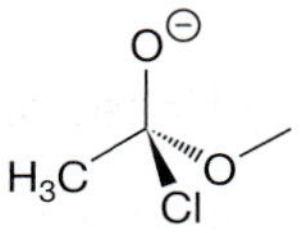

7.29 Which of the following substrates can undergo an E2 step with H_2N^- as the base? For those that can, draw the curved arrow notation and the products.

(a)

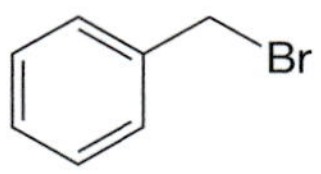

(b)

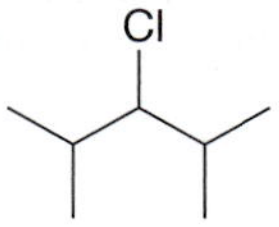

(c)

(d)

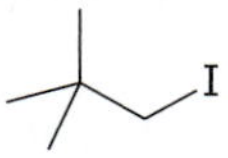

(e)

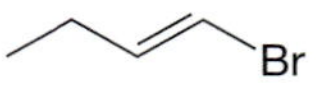

7.30 There are two possible products in the electrophilic addition step shown here. Draw both possible products and predict which one is more stable.

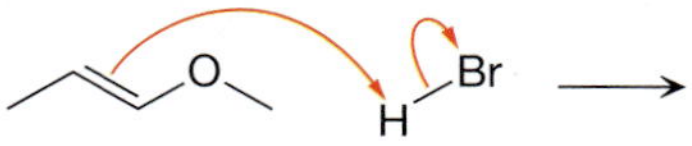

7.31 If the H colored red is eliminated as H^+, then two possible diastereomers can form. Draw the curved arrows for this electrophile elimination step, and draw each of the diastereomeric products.

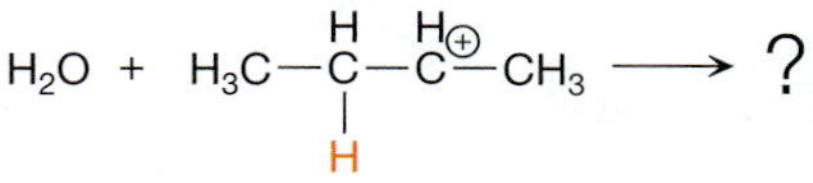

7.32 If H^+ is eliminated from the carbocation shown here in an electrophile elimination step, then three possible constitutional isomers can form. Draw the mechanism for the formation of all three of those products.

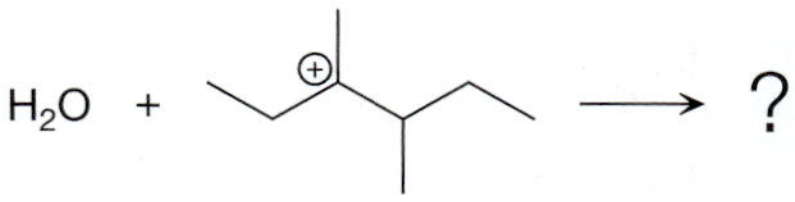

7.33 In the electrophilic addition step shown here, where NO_2^+ adds to phenol, the electrophile can add either ortho, meta, or para to the OH substituent on the ring. Draw the curved arrows and products of each electrophilic addition step.

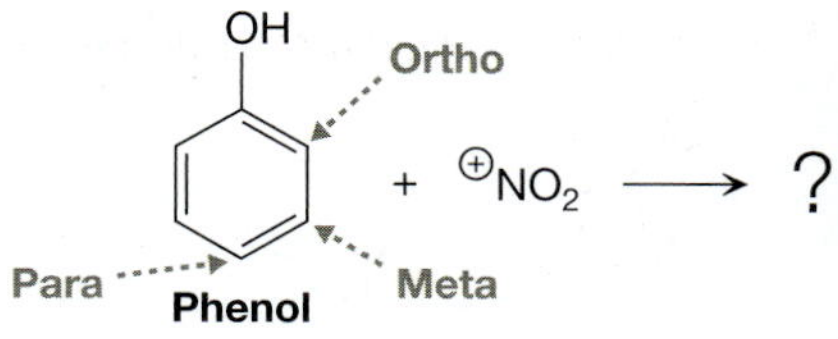

7.34 A proton (H^+) from trifluoromethanesulfonic acid, CF_3SO_2OH, can add to the alkyne shown to yield two different carbocation products. **(a)** Draw the mechanism for each of these steps, along with the corresponding products. **(b)** Which carbocation is more stable?

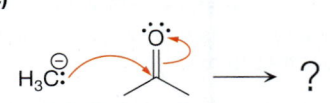

7.35 The carbocation shown here is formed in one step of an electrophilic aromatic substitution reaction (discussed in Chapter 22). **(a)** Draw the curved arrow notation and the product for the elimination of H^+. **(b)** Do the same for the elimination of SO_3H^+.

7.36 For each of these carbocations, draw the curved arrow notation and product for all possible 1,2-hydride shifts and all possible 1,2-methyl shifts.

(a) (b) (c)

(d) (e) (f)

7.8–7.9 The Driving Force for Chemical Reactions and Keto–Enol Tautomerization

7.37 For each elementary step given, determine whether the reactant or product side is favored.

(a) (b) (c)

(d) (e) (f)

7.38 Each heterolysis step on the left does not readily occur, but the corresponding one on the right does. Explain why. *Hint:* Draw the products of each heterolysis and determine the contributions to their driving force.

(a)

(b)

(c)

(d)

7.39 The first of the two heterolysis reactions below takes place readily, but the second one does not. Explain why.

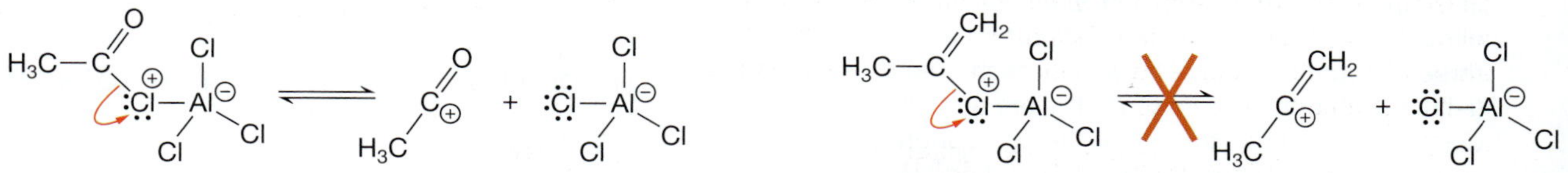

7.40 Determine which carbocation rearrangements you drew for Problem 7.36 produce a significantly more stable carbocation than the reactant.

7.41 According to Table 7-1, the equilibrium percentage of the first molecule in its keto form is lower than that of the second molecule. Explain why. *Hint:* What do you know about the stability of C=C double bonds?

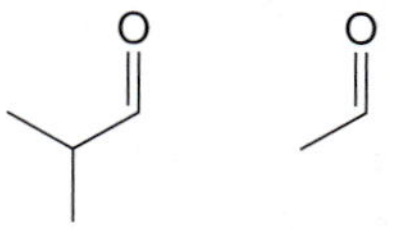

7.42 According to Table 7-1, the equilibrium percentage of the first molecule in its keto form is lower than that of the second molecule. Explain why.

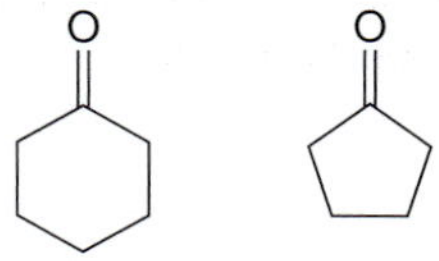

7.43 Recall from Section 7.9 that most ketones and aldehydes exist primarily in their keto form, as shown in Table 7-1. Propanedial, however, exists primarily (>99%) in its enol form. Explain why. *Hint:* Examine its structure in the enol form.

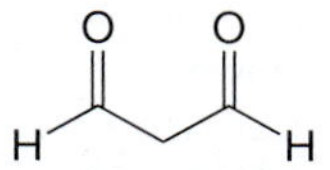

Propanedial

7.44 Draw the mechanism for the conversion of propanedial (see Problem 7.43) in its keto form to its enol form under basic and acidic conditions. Do the same for the conversion from the enol form to the keto form.

7.45 A tautomerization reaction can occur with an imine in a way analogous to that of a ketone or aldehyde. Using the appropriate bond energies from Section 1.4, estimate ΔH°_{rxn} for this reaction and determine which form is more stable, the imine or enamine.

Imine   Enamine

Integrated Problems

7.46 The following reaction, which is discussed in Chapter 8, is an example of a unimolecular nucleophilic substitution (S_N1) reaction. It consists of the four elementary steps shown here. For each step (i–iv), **(a)** identify all electron-rich sites and all electron-poor sites, **(b)** draw in the appropriate curved arrows to show the bond formation and bond breaking that occur, and **(c)** name the elementary step.

(i)

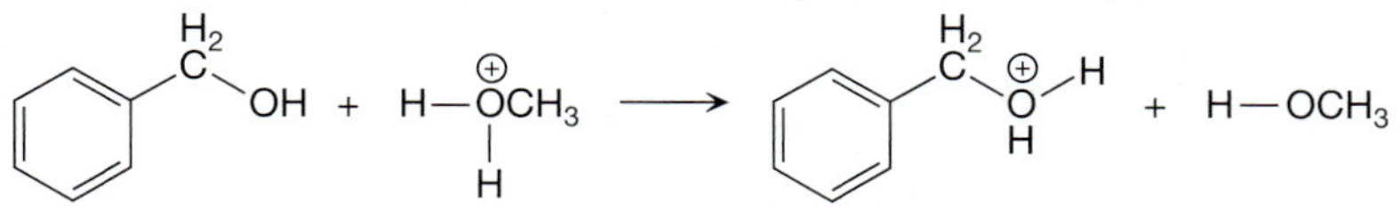

(ii)

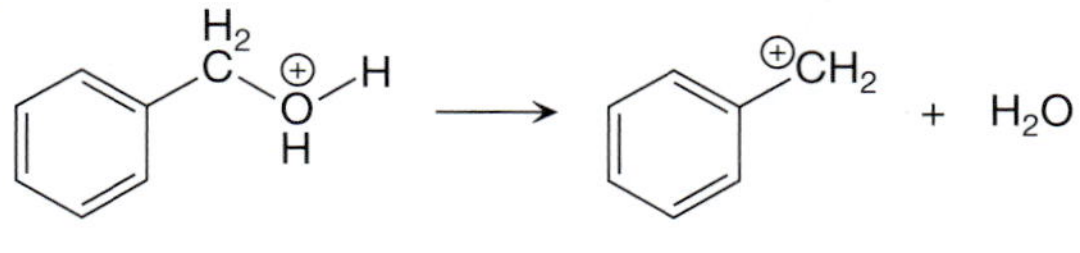

(iii)

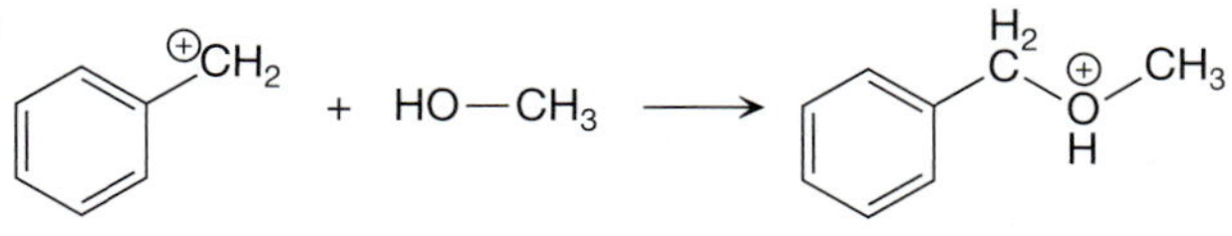

(iv)

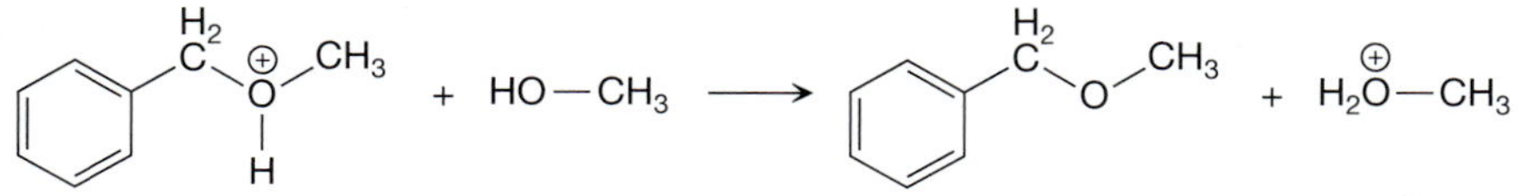

7.47 The reaction shown here, which is discussed in Chapter 8, is an example of a unimolecular elimination (E1) reaction and consists of the three elementary steps shown. For each step (i–iii), **(a)** identify all electron-rich sites and all electron-poor sites, **(b)** draw in the appropriate curved arrows to show the bond formation and bond breaking that occur, and **(c)** name the elementary step.

(i)

(ii)

(iii)

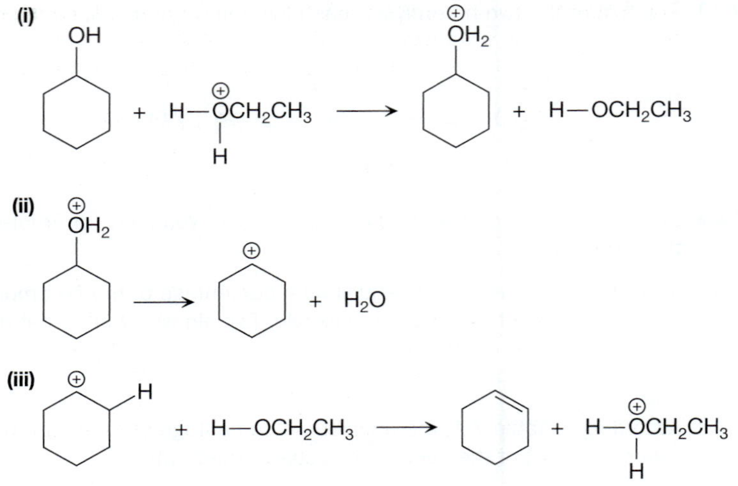

7.48 The reaction shown here, which converts an epoxide into a bromohydrin, is discussed in Chapter 10. It consists of the two elementary steps shown. For each step (i and ii), **(a)** identify all electron-rich sites and all electron-poor sites, **(b)** draw in the appropriate curved arrows to show the bond formation and bond breaking that occur, and **(c)** name the elementary step.

(i)

(ii)

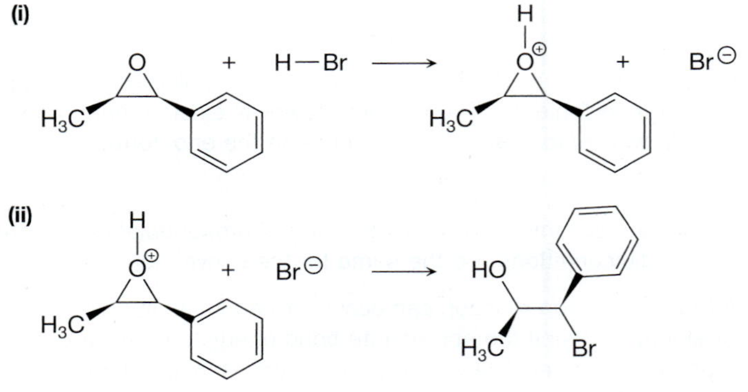

7.49 The reaction shown here, which is discussed in Chapter 10, consists of the two elementary steps shown. For each step (i and ii), **(a)** identify all electron-rich sites and all electron-poor sites, **(b)** draw in the appropriate curved arrows to show the bond formation and bond breaking that occur, and **(c)** name the elementary step.

(i)

(ii)

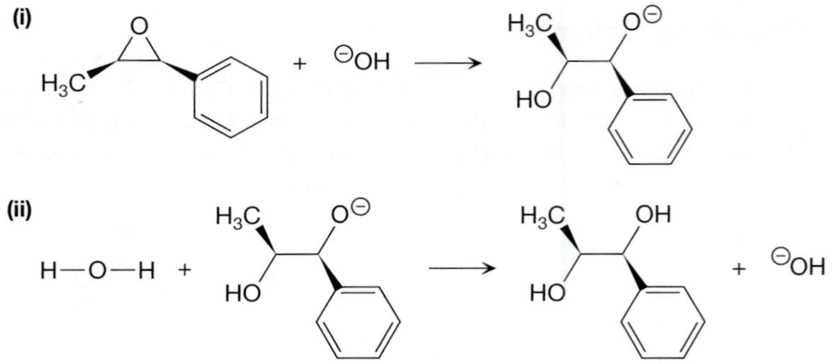

7.50 The following reaction, which converts a cyclic ether into a diol, is discussed in Chapter 8. It consists of the three elementary steps shown. For each step (i–iii), **(a)** identify all electron-rich sites and all electron-poor sites, **(b)** draw in the appropriate curved arrows to show the bond formation and bond breaking that occur, and **(c)** name the elementary step.

(i)

(ii)

(iii)

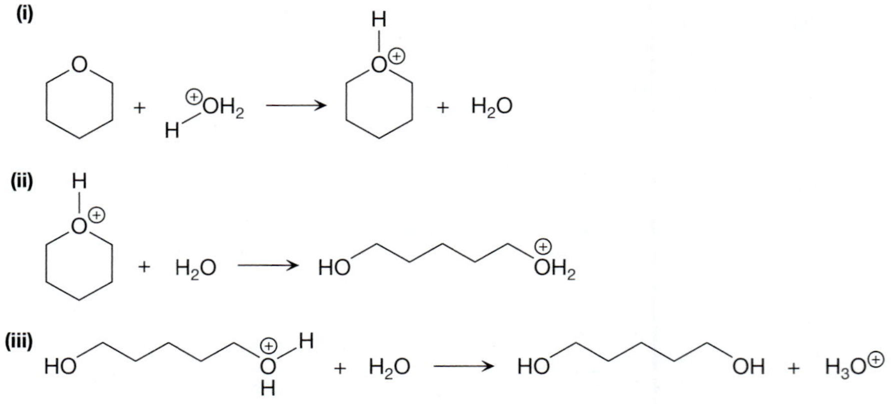

7.51 The following reaction, which is discussed in Chapter 21, consists of the three elementary steps shown. For each step (i–iii), **(a)** identify all electron-rich sites and all electron-poor sites, **(b)** draw in the appropriate curved arrows to show the bond formation and bond breaking that occur, and **(c)** name the elementary step.

(i)

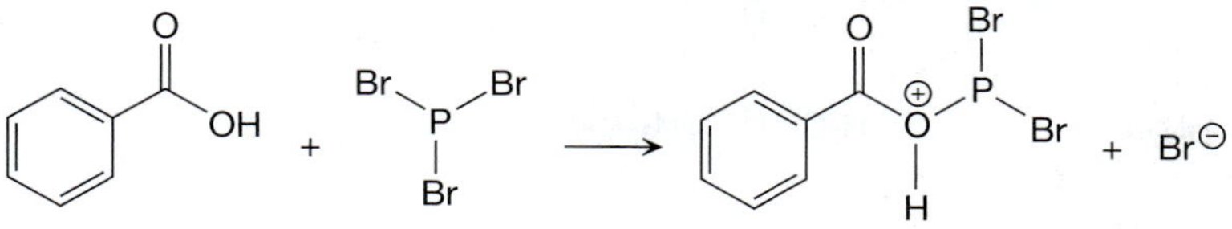

(ii)

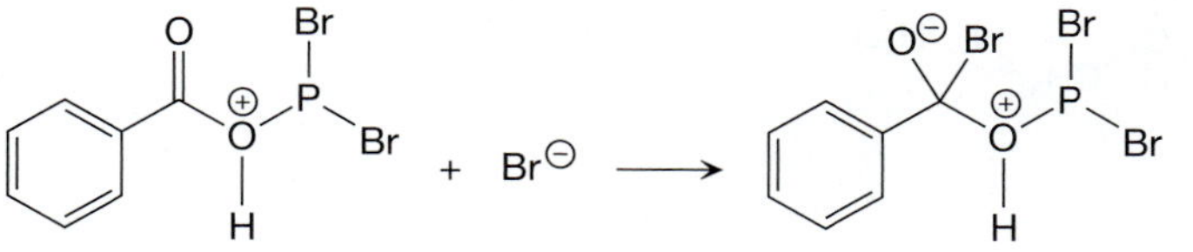

(iii)

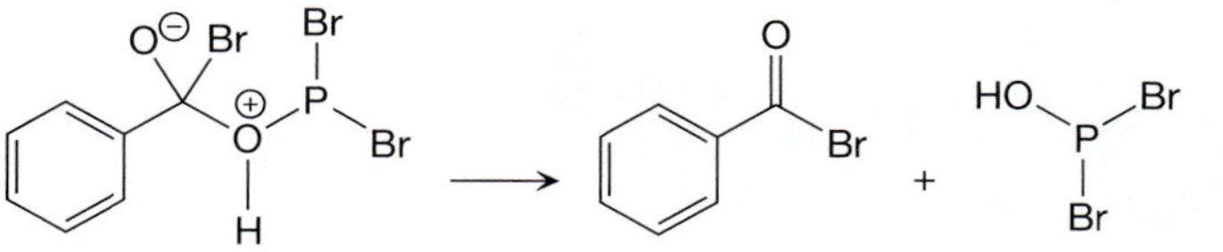

7.52 The following reaction, which is discussed in Chapter 17, consists of the two elementary steps shown. For each step (i and ii), **(a)** identify all electron-rich sites and all electron-poor sites, **(b)** draw in the appropriate curved arrows to show the bond formation and bond breaking that occur, and **(c)** name the elementary step.

(i)

(ii)

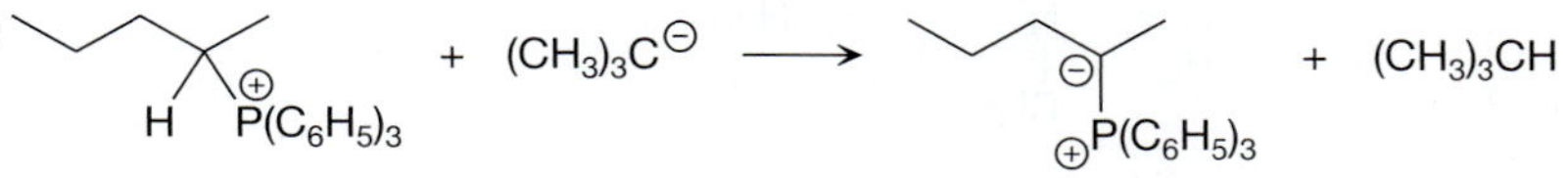

7.53 The following is an example of a *Fischer esterification* reaction, which is discussed in Chapter 21. The mechanism consists of the six elementary steps shown. For each step (i–vi), **(a)** identify all electron-rich sites and all electron-poor sites, **(b)** draw in the appropriate curved arrows to show the bond formation and bond breaking that occur, and **(c)** name the elementary step.

(i)

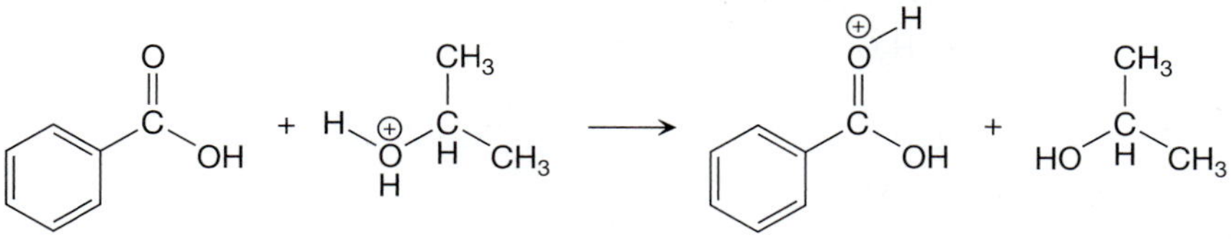

(ii)

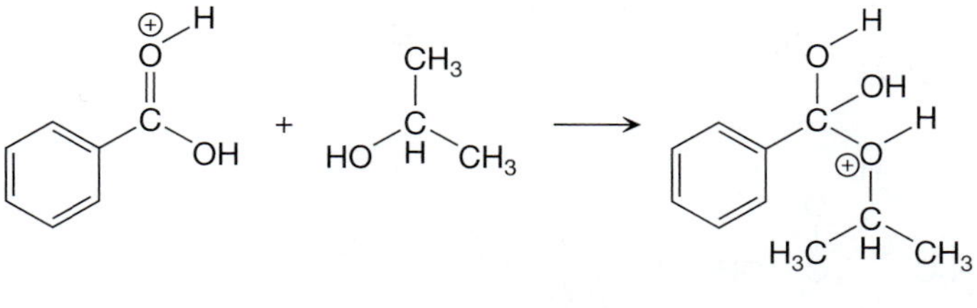

(iii)

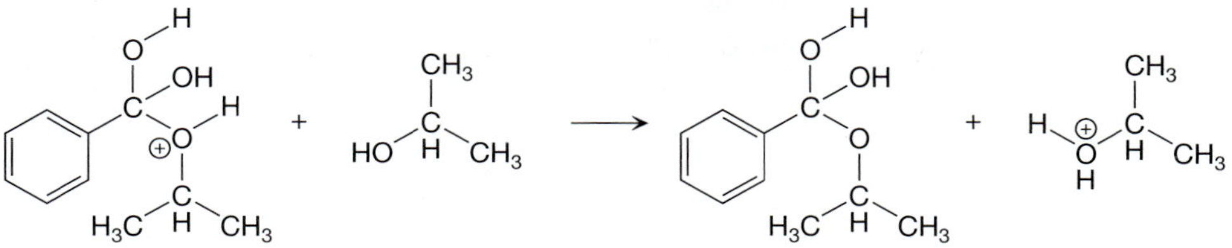

(iv)

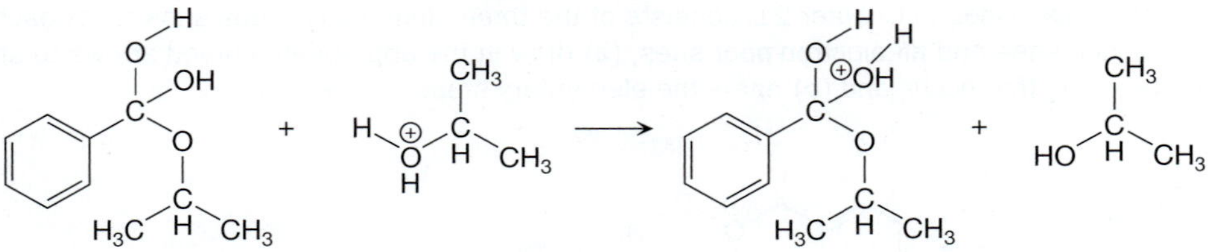

(v)

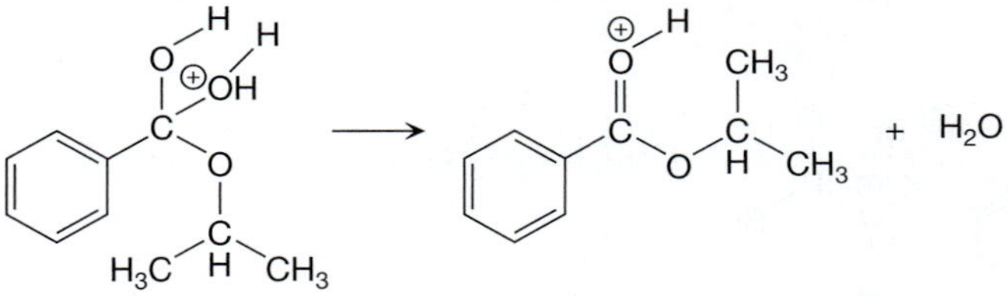

(vi)

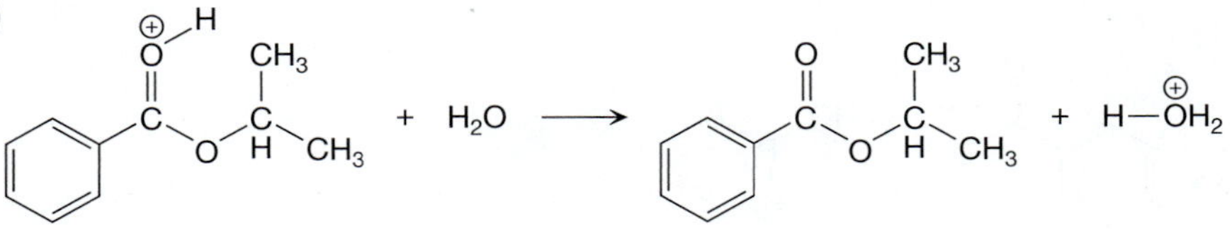

7.54 The following reaction is an example of an *acid-catalyzed hydrolysis* of an amide. This reaction, which is the same one that breaks down proteins into amino acids in your stomach, is discussed in Chapter 21. The mechanism consists of the six elementary steps shown. For each step (i–vi), **(a)** identify all electron-rich sites and all electron-poor sites, **(b)** draw in the appropriate curved arrows to show the bond formation and bond breaking that occur, and **(c)** name the elementary step.

(i)

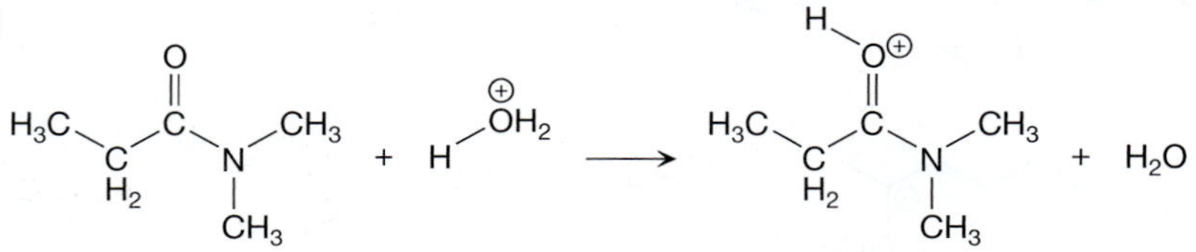

(ii)

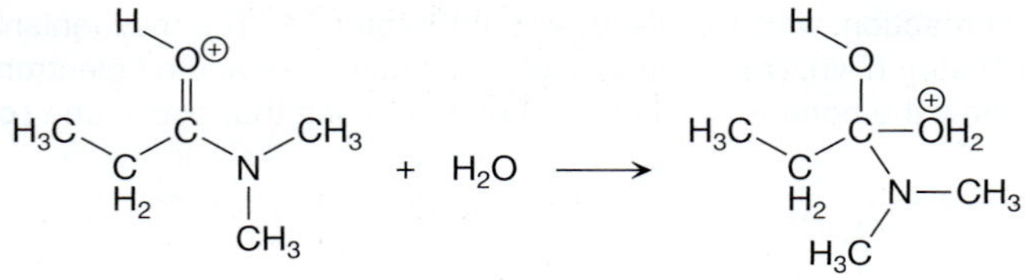

(iii)

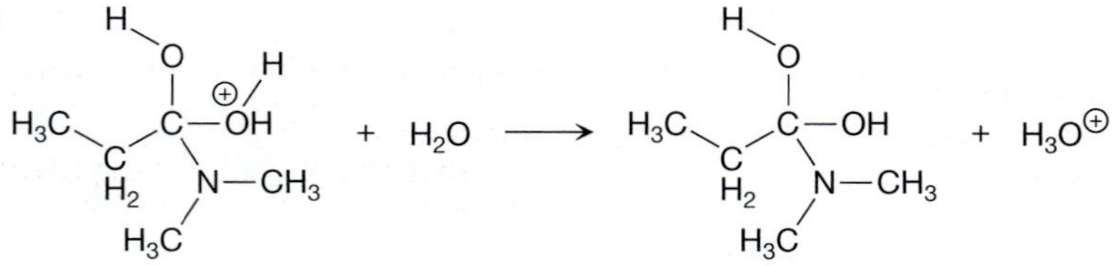

(iv)

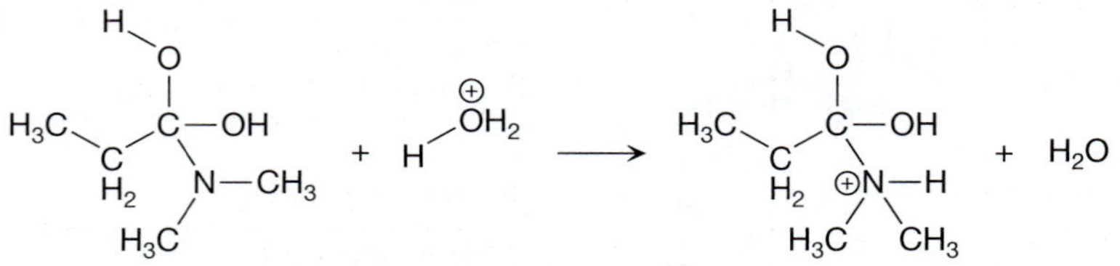

(v)

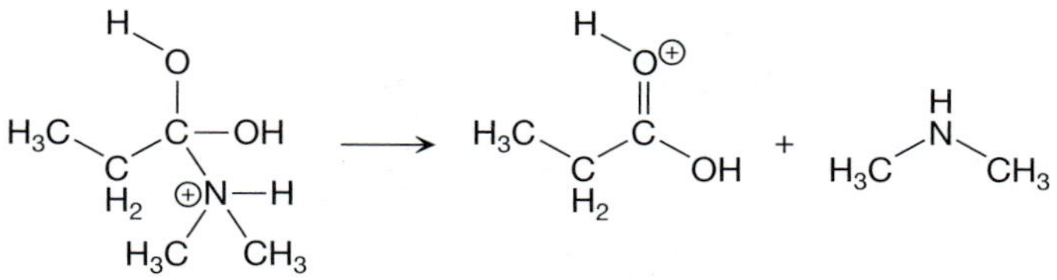

(vi)

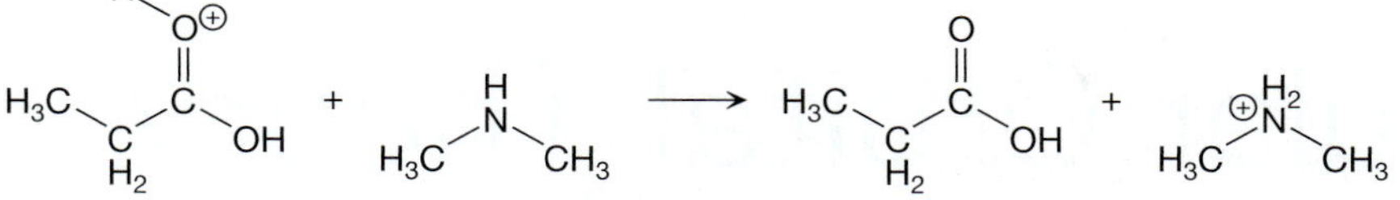

7.55 Shown here is an example of an *electrophilic aromatic substitution* reaction, which we examine in Chapter 22. The mechanism consists of the four elementary steps shown. For each step (i–iv), **(a)** identify all electron-rich sites and all electron-poor sites, **(b)** draw in the appropriate curved arrows to show the bond formation and bond breaking that occur, and **(c)** name the elementary step.

(i)

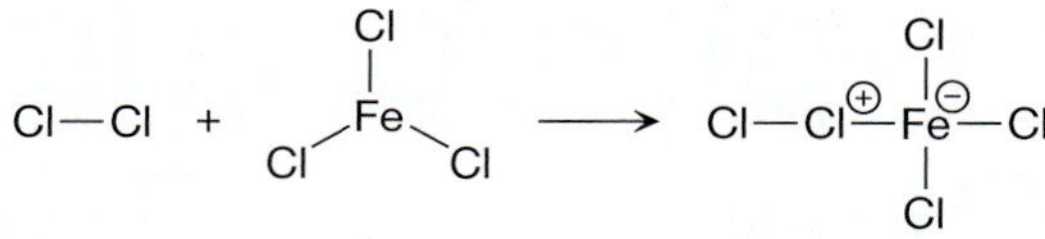

(ii)

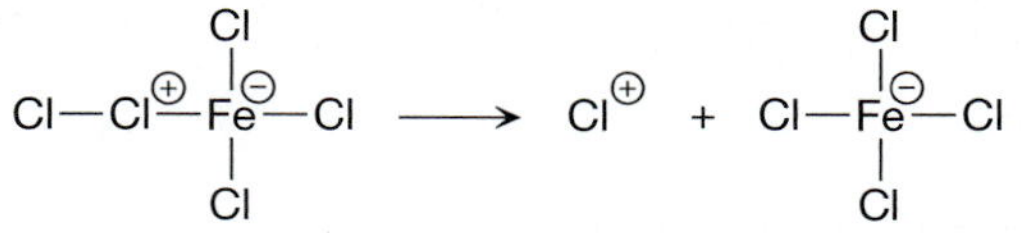

(iii)

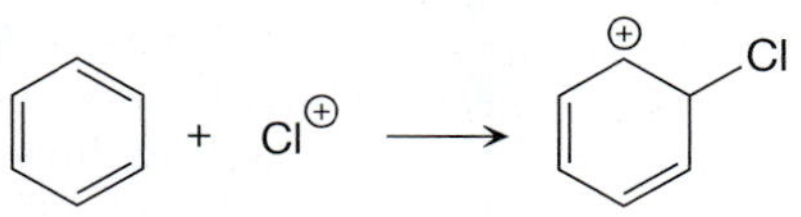

(iv)

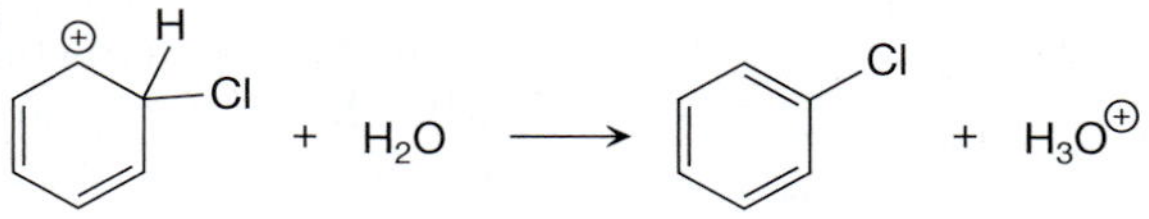

7.56 Draw the curved arrow notation and products for each elementary step described by the sequence shown here. *Note:* The products of the first step should be used as reactants in the second step.

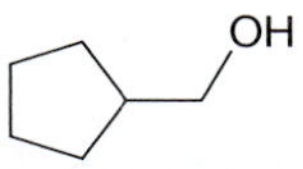

1. Proton transfer involving HCl
2. S_N2 involving $Cl^{\ominus}$

?

7.57 Draw the curved arrow notation and products for each elementary step described by the sequence shown here. *Note:* The products of each step should be used as reactants in the subsequent step.

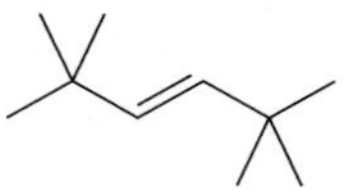

1. Electrophilic addition of $H^{\oplus}$
2. 1,2-Methyl shift
3. Electrophile elimination of $H^{\oplus}$

?

7.58 Draw the curved arrow notation and products for each elementary step described by the sequence shown here. *Note:* The products of each step should be used as reactants in the subsequent step.

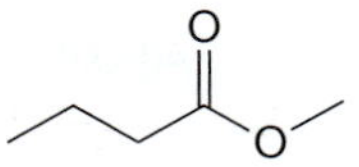

1. Nucleophilic addition involving CH_3MgBr
2. Nucleophile elimination
3. Nucleophilic addition involving CH_3MgBr
4. Proton transfer involving $H_3O^{\oplus}$

?

7.59 Draw the curved arrow notation and products for each elementary step described by the sequence shown here. *Note:* The products of each step should be used as reactants in the subsequent step.

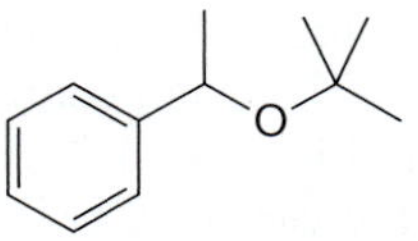

1. Proton transfer involving $H_3O^{\oplus}$
2. Heterolysis
3. Coordination involving H_2O
4. Proton transfer involving H_2O

?

7.60 Draw the curved arrow notation and products for each elementary step described by the sequence shown here. *Note:* The products of each step should be used as reactants in the subsequent step.

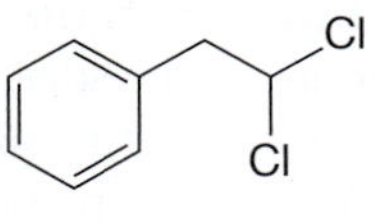

1. E2 involving $NaNH_2$
2. E2 involving $NaNH_2$
3. Proton transfer involving $NaNH_2$
4. Proton transfer involving $H_3O^{\oplus}$

?

Molecular Orbital Theory, Hyperconjugation, and Chemical Reactions

In Chapter 3, we saw how hybridization and molecular orbital (MO) theory help us understand several fundamental characteristics of a molecule's or molecular ion's structure. We saw, for example, that an atom's geometry determines its hybridization, and we saw how pure and hybrid atomic orbitals (AOs) can mix to form new bonding and antibonding molecular orbitals with either σ or π symmetry.

Here in Interchapter D, we see how MO theory can be used to provide insights beyond just the structural characteristics of a molecular species. In Section D.1, we use MO theory to explain how the degree of alkyl substitution affects the stabilities of carbocations and alkenes. In Section D.2, we use MO theory to better understand the elementary steps introduced in Chapter 7.

D.1 Relative Stabilities of Carbocations and Alkenes: Hyperconjugation

In Section 6.6e, we learned that the stability of a carbocation increases as the number of alkyl groups bonded to the positively charged carbon increases:

Carbocation stability: CH_3^+ < RCH_2^+ < R_2CH^+ < R_3C^+
(Methyl)   (1°)   (2°)   (3°)

These relative stabilities were explained by inductive effects: Each alkyl group is electron donating through its attached σ bond and therefore reduces the amount of positive charge that is localized on the C^+. Inductive effects, however, are only part of the story. The major contribution, in fact, comes from *hyperconjugation*, which is illustrated in Figure D-1.

In the case of CH_3^+ (Fig. D-1a), the empty p AO is perpendicular to all three C—H bonds, so it does not interact with any other orbitals, making it effectively isolated. By contrast, the p AO in $CH_3CH_2^+$ (Fig. D-1b) is aligned with one of the C—H bonds in the adjacent CH_3 group, which represents a pair of electrons occupying a C—H σ bonding orbital. The empty p AO and that σ bonding orbital therefore interact. As with any pair of interacting orbitals, two new orbitals are produced: In this case, constructive interference results in a new orbital that is more stable than the σ bonding orbital and destructive interference results in a new orbital that is less stable

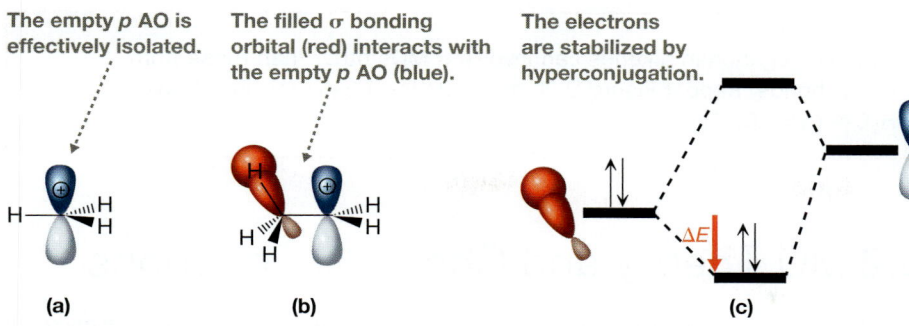

The empty *p* AO is effectively isolated.

The filled σ bonding orbital (red) interacts with the empty *p* AO (blue).

The electrons are stabilized by hyperconjugation.

(a)   (b)   (c)

FIGURE D-1 Hyperconjugation in carbocations (a) In CH_3^+, the *p* AO on carbon is empty and does not interact with any adjacent orbitals. (b) In $CH_3CH_2^+$, the empty *p* AO overlaps with the σ orbital of the adjacent C—H bond, so the two orbitals can interact. (c) The orbital interaction in $CH_3CH_2^+$ (b), called hyperconjugation, stabilizes the electrons from the σ bonding orbital.

than the *p* AO. The two electrons from the σ bonding MO end up in the lower-energy of the two new orbitals and are thus stabilized, as shown in Figure D-1c. Such a stabilizing effect would occur for each additional alkyl group attached to C^+.

Notice in Figure D-1b and D-1c that the mixing of the orbitals effectively transfers some electron density from the alkyl group into the empty *p* AO. In this way, the alkyl groups are electron donating, in agreement with how we view alkyl groups from the perspective of inductive effects.

YOUR TURN D.1

Hyperconjugation can involve σ bonding orbitals other than those from C—H bonds. Repeat Figure D-1b and D-1c for the 2,2-dimethylpropyl cation, $(CH_3)_3CCH_2^+$.

Answers to Your Turns are in the back of the book.

Let's now turn our attention to alkenes. Recall from Section 5.9 that the stability of an alkene generally increases with each additional alkyl group bonded to the alkene carbons:

Alkene stability:

Unsubstituted < Monosubstituted < Disubstituted < Trisubstituted < Tetrasubstituted

Just as with carbocation stability, this trend can be explained by *hyperconjugation* involving orbitals from the attached alkyl groups. Figure D-2a shows the π* antibonding MO from ethene, an unsubstituted alkene. There are no adjacent orbitals with which the π* MO can interact.

In propene (Fig. D-2b), which has a methyl group attached to an alkene carbon, there are C—H σ bonding orbitals from the CH_3 group, just as we saw with hyperconjugation in carbocations. The empty π* MO (red) can interact with the filled σ MO from the CH_3 group, and the result, as shown in Figure D-2c, is the stabilization of those σ electrons. This type of stabilization increases with each additional alkyl group attached to the C=C bond.

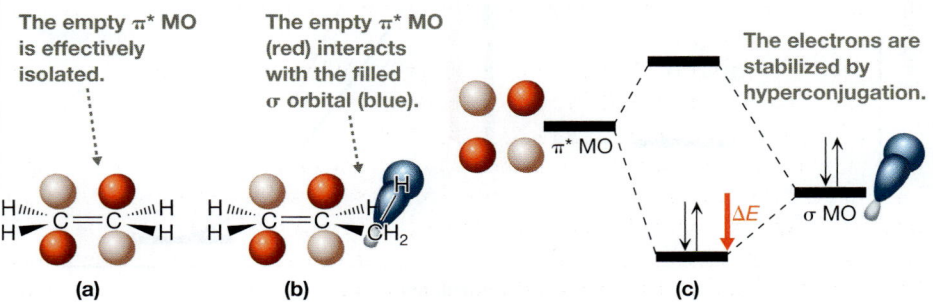

The empty π* MO is effectively isolated.

The empty π* MO (red) interacts with the filled σ orbital (blue).

The electrons are stabilized by hyperconjugation.

(a)   (b)   (c)

FIGURE D-2 Hyperconjugation in alkenes (a) In $H_2C=CH_2$, the π* MO does not interact with any adjacent orbitals. (b) In $H_2C=CH—CH_3$, the empty π* MO interacts with the filled σ MO of the C—H bond in CH_3. (c) The orbital interaction in $H_2C=CH—CH_3$, an example of hyperconjugation, stabilizes the electrons from the σ MO.

Hyperconjugation in alkenes can involve σ MOs other than those from C—H bonds. Repeat Figure D-2b and D-2c for 3,3-dimethylbut-1-ene, $H_2C\!=\!CH\!-\!C(CH_3)_3$.

D.2 MO Theory and Chemical Reactions

Chapters 6 and 7 introduced 10 of the most common elementary steps that make up organic reaction mechanisms. For each step, we examined the curved arrow notation that enables us to depict the changes that take place among valence electrons throughout the course of the reaction. Furthermore, we discussed how such changes in valence electrons tend to represent a flow of electrons from an electron-rich site to an electron-poor site. Thus far, however, these aspects have been discussed only in terms of the Lewis structure model—there has been no mention of orbitals.

Here in Section D.2, each of the elementary steps we have previously learned is presented again from the perspective of MO theory. First, however, we give an overview of what is called *frontier molecular orbital (FMO) theory*. Then, we apply FMO theory to each of the elementary steps discussed in Chapter 7. In so doing, we will be able to see how FMO theory allows for each step to occur and accounts for the flow of electrons from an electron-rich site to an electron-poor one. Additionally, we will see how FMO theory justifies the *stereochemistry* of certain elementary steps, a topic discussed more fully in Chapter 8.

D.2a An Overview of Frontier Molecular Orbital Theory

The application of **frontier molecular orbital (FMO) theory** to chemical reactions begins with the idea that the reactants in an elementary step must typically surmount a significant energy barrier—the free energy of activation—to get to products. The energy barrier arises from the relative instability of the transition state of the reaction relative to the reactants. Without a source of significant stabilization for that transition state, many elementary steps would have very large energy barriers (red curve, Fig. D-3),

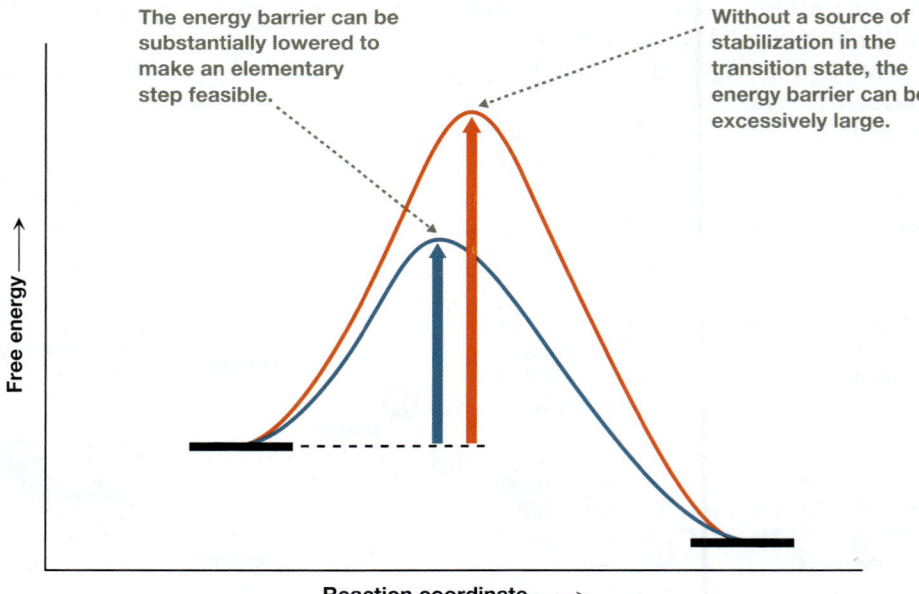

The energy barrier can be substantially lowered to make an elementary step feasible.

Without a source of stabilization in the transition state, the energy barrier can be excessively large.

FIGURE D-3 Energy barriers and transition-state stabilization The red curve represents an elementary step in which the transition state is not significantly stabilized (large energy barrier), making the reaction excessively slow. When the transition state is significantly stabilized (small energy barrier), as indicated in the blue curve, the reaction rate is substantially increased.

making their rates excessively slow. Conversely, stabilization of the transition state can lower the energy barrier to allow the elementary step to take place at a reasonable rate (blue curve, Fig. D-3).

What source of stabilization in the transition state can lower the energy barrier? One of the main contributions comes from the **frontier molecular orbitals** of the reacting species, which are defined as its *highest occupied* and *lowest unoccupied* MOs (i.e., the HOMO and LUMO, respectively; review Section 3.3, if necessary).

- If the HOMO of one reactant and the LUMO of another have substantial net overlap in the transition state, then the transition state tends to be significantly stabilized, and the reaction of interest is said to be *allowed*.
- Otherwise, the transition state tends *not* to be sufficiently stabilized, and the reaction of interest is said to be *forbidden*.

The frontier orbitals of the reactants are the focus for two reasons. One is that, of all imaginable interactions of a filled orbital with an empty orbital, the HOMO–LUMO interaction is generally the one that involves orbitals that are closest in energy (Fig. D-4a). This maximizes any interaction that takes place between the orbitals (Section 3.3).

The second reason we focus on the frontier orbitals is that if the HOMO and LUMO orbitals can interact, then the interaction *must* stabilize the entire species, as shown in Figure D-4b. This interaction must be stabilizing because, by definition, the HOMO contains electrons and the LUMO is empty. When the orbitals mix (via constructive and destructive interference; Section 3.2), therefore, two new orbitals must be produced—one that has been lowered in energy and one that has been raised in energy—and the electrons from the HOMO will end up in the lower of the two orbitals in the transition state.

With what we have seen so far, we can make a very useful association between FMO interactions and the tendency of electrons to flow from an electron-rich site to an electron-poor one. Notice in Figure D-4b that the pertinent electrons originate

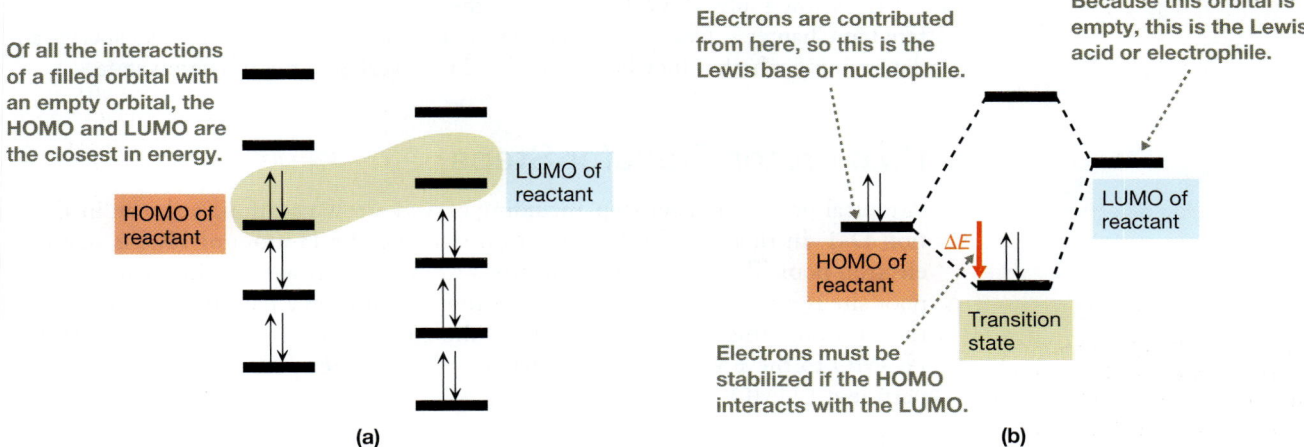

FIGURE D-4 HOMO–LUMO interactions (a) The similar energies of the HOMO and LUMO maximize any interaction between the two orbitals. (b) If an interaction takes place between the HOMO of one reactant and the LUMO of another, then stabilization is guaranteed. When the two orbitals interact, two new orbitals in the transition state (center) are produced—one that is lower in energy than either the HOMO or LUMO, and one that is higher in energy. The electrons contributed by the HOMO end up in the lower of the two orbitals.

from the HOMO of one species and interact with the empty LUMO of the other. Thus, there is an effective flow of electrons from the HOMO to the LUMO:

- The species that contributes the pertinent HOMO in the FMO interaction is relatively electron rich; it is the Lewis base or the nucleophile.
- The species that contributes the pertinent LUMO is relatively electron poor; it is the Lewis acid or the electrophile.

One of the important messages from Chapter 7 is that some elementary steps might simply be the reverse of each other, but receive different names. Examples include coordination and heterolysis (Section 7.3), nucleophilic addition and nucleophile elimination (Section 7.4), and electrophilic addition and electrophile elimination (Section 7.6). In such cases, we can imagine applying FMO theory separately to the elementary steps in each pair, and indeed this can be done. However, it is much more convenient to take advantage of the following idea.

- If an elementary step is allowed in the forward direction, then it will be allowed in the reverse direction as well.
- If an elementary step is forbidden in the forward direction, then it will be forbidden in the reverse direction as well.

This idea is related to the **principle of microscopic reversibility**, which demands that if one reaction is the reverse of another, then the two reactions must proceed along the same path in the energy diagram and, therefore, must proceed through the same transition state. Thus, the FMO interactions that might contribute to the stabilization of the elementary step's transition state in the forward direction will be identical to the ones for the elementary step in the reverse direction.

Keep in mind that FMO theory focuses only on the *energy barrier* of an elementary step—that is, on the *kinetics*. Even if an elementary step is allowed, *thermodynamics* (i.e., the relative energies of the reactants and products) will still dictate whether the elementary step is favorable at equilibrium.

Let's now apply FMO theory toward the various elementary steps that we encountered in Chapter 7. Doing so entails identifying the FMOs in the reacting species and determining whether they have substantial net overlap in the transition state.

D.2b Proton Transfer Steps

A typical proton transfer step, including curved arrow notation, is shown in Equation D-1. In this case, HO⁻ is electron rich and the H atom of HCl is relatively electron poor. Thus, the flow of electrons from an electron-rich site to an electron-poor site is represented by the curved arrow drawn from a pair of electrons on O to the H atom. This means that HO⁻ is the species that contributes the pertinent HOMO to the FMO interaction, whereas HCl is the species that contributes the pertinent LUMO.

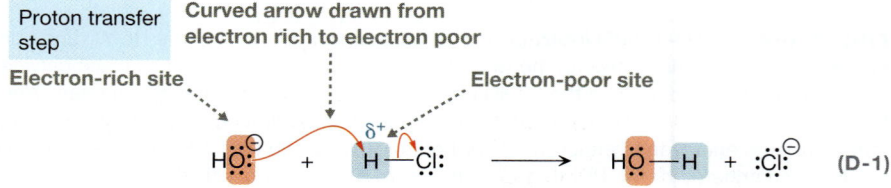

The HOMO and LUMO have
substantial net overlap,
so this reaction is *allowed*.

The HOMO and LUMO do *not* have
substantial net overlap,
so this reaction is *forbidden*.

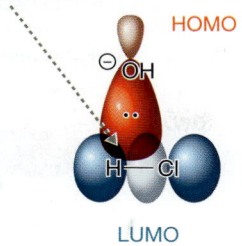

HOMO   LUMO

(a)

HOMO

LUMO

(b)

**FIGURE D-5 Frontier orbital
interactions for the proton
transfer involving HO⁻ and HCl**
(a) The approach of HO⁻ from
the end opposite the Cl atom. As
indicated, the HOMO and LUMO
have substantial net overlap. The
specific phases shown would lead
primarily to constructive interference,
whereas reversing the phase of one
of the orbitals would lead primarily
to destructive interference. (b) The
approach of HO⁻ from the side of the
H—Cl bond. In this transition state,
the HOMO and LUMO do not have a
significant net interaction. Whereas
the left side in this depiction (shaded
dark red and dark blue) indicates
constructive interference, the right
side (shaded dark red and light blue)
contributes to destructive interference.

It is now a matter of identifying the HOMO of HO⁻ and the LUMO of HCl and determining whether they have substantial net overlap in the transition state. In the two-center MO approach we used in Chapter 3, the HOMO of HO⁻ is a nonbonding pair of electrons, as shown in Figure D-5, and the LUMO of HCl is a σ^* antibonding MO. When HO⁻ approaches HCl from the end opposite the Cl atom (Fig. D-5a), the HOMO and LUMO will undergo substantial interaction in the transition state. The specific phases shown in the figure will lead to a net constructive interference, whereas destructive interference would result from reversing the phase of one of the orbitals. As we learned in the previous section (Fig. D-4b), this kind of HOMO–LUMO interaction will stabilize the transition state, making the reaction allowed.

YOUR TURN D.3

Construct the atomic orbital overlap and MO energy diagrams separately for HO⁻ and HCl (review Figs. 3-13 through 3-25). From these diagrams, determine the HOMO of HO⁻ and the LUMO of HCl. Do they agree with the ones shown in Figure D-5?

Is there a transition state in which these frontier orbitals do *not* undergo significant interaction? One example is shown in Figure D-5b, in which HO⁻ approaches HCl from the side of the H—Cl bond. In that transition state, one portion of the orbital overlap leads to constructive interference (in this case, the left side), whereas the other portion (the right side) leads to destructive interference. Thus, there is very little net interaction, so the transition state remains unstabilized and the reaction is forbidden.

YOUR TURN D.4

In Figure D-5a and D-5b, label every region of constructive interference and destructive interference between the HOMO and LUMO.

D.2c Bimolecular Nucleophilic Substitution (S$_N$2) Steps

The S$_N$2 step we first encountered in Chapter 7 is shown once again in Equation D-2.

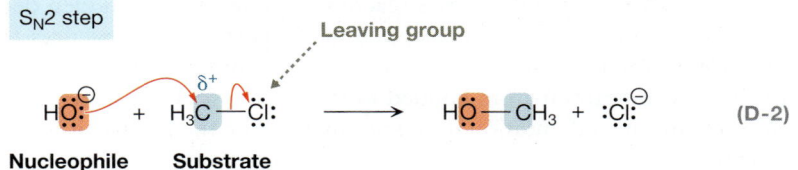

Nucleophile   **Substrate**   (D-2)

The HOMO and LUMO have
substantial net overlap,
so this reaction is *allowed*.

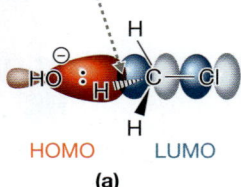

HOMO   LUMO

(a)

The HOMO and LUMO do *not* have
substantial net overlap, so this
reaction is *forbidden*.

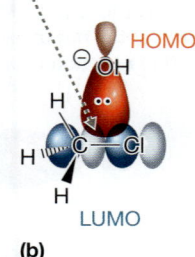

HOMO

LUMO

(b)

FIGURE D-6 **Frontier orbital interactions for the S$_N$2 step involving HO$^-$ and CH$_3$Cl**
(a) The approach of HO$^-$ from the end opposite the Cl leaving group. As indicated, the
HOMO and LUMO have substantial net overlap. (b) The approach of HO$^-$ from the side of
the C—Cl bond. In this transition state, the HOMO and LUMO do not have a significant
net interaction. Whereas the right side in this depiction (shaded dark red and dark blue)
indicates constructive interference, the left side (shaded dark red and light blue) contributes
to destructive interference.

The curved arrow notation for this reaction is essentially the same as for the proton
transfer step in Equation D-1, so the FMO picture should be similar, too. Indeed,
HO$^-$ is once again the electron-rich species, so it will contribute the relevant HOMO
to the FMO interaction. CH$_3$Cl has a relatively electron-poor C atom, so it will contribute the relevant LUMO.

Just as we saw in Figure D-5, the HOMO that is contributed by HO$^-$ in Equation D-2 contains the nonbonded electrons. The LUMO of CH$_3$Cl, once again, is
a σ* MO, this time of the C—Cl bond. These orbitals are shown in Figure D-6. If
the HO$^-$ nucleophile approaches the CH$_3$Cl substrate from the end opposite the
Cl leaving group (Fig. D-6a), then the resulting overlap of the HOMO and LUMO
orbitals will lead to significant interaction, the transition state will be stabilized, and
the reaction will be allowed. Also, similar to what we saw in Figure D-5b, if the
approach of the HO$^-$ nucleophile is from the side of the C—Cl bond, then there
will be no significant interaction between the FMOs, making the reaction
forbidden.

YOUR TURN D.5

Construct the atomic orbital overlap and MO energy diagrams for CH$_3$Cl
(review Figs. 3-13 through 3-25) and use these diagrams to determine its
LUMO. Does your answer agree with the one shown in Figure D-6?

YOUR TURN D.6

In Figure D-6a and D-6b, label every region of constructive interference and
destructive interference between the HOMO and LUMO.

The fact that whether an S$_N$2 reaction is allowed or forbidden depends on the
particular approach of the nucleophile has important consequences for the *stereochemistry* of the reaction—a topic that is more completely discussed in Chapter 8.
In short, if the carbon atom bonded to the leaving group is a chiral center, then its
stereochemical configuration in the product will be the reverse of what it was in the
reactant.

D.2d Bond-Formation (Coordination) and Bond-Breaking (Heterolysis) Steps

The coordination step depicted in Equation D-3 was shown previously in Section 7.3.

Coordination step

(D-3)

Lewis base or nucleophile   Lewis acid or electrophile   Lewis adduct

Br^- is relatively electron rich and is the Lewis base or nucleophile, so it will contribute a nonbonding orbital as the relevant HOMO to the FMO interaction. By contrast, the $(CH_3)_3C^+$ carbocation is electron poor, so it serves as the Lewis acid or electrophile, and it will contribute the relevant LUMO (the empty p orbital in Fig. D-7) to the FMO interaction.

Figure D-7a shows that the HOMO and LUMO have substantial net overlap when the Br^- nucleophile approaches from the left face of the planar C atom. Figure D-7b shows that the same is true for the approach of the nucleophile from the right face of the planar C atom, too. (Notice that we reversed the phases of the orbitals to illustrate this.) Thus, coordination is allowed from either face.

The HOMO and LUMO have substantial net overlap, so this reaction is *allowed*.

The HOMO and LUMO have substantial net overlap, so this reaction is *allowed*.

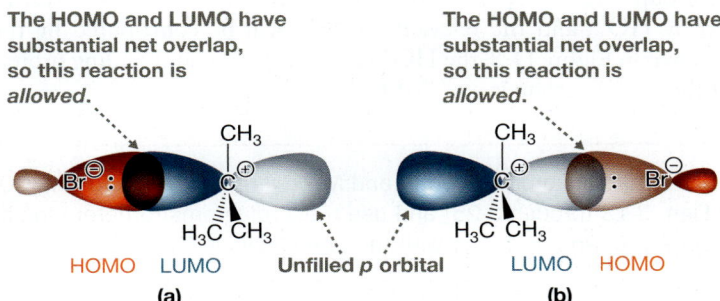

HOMO   LUMO   Unfilled p orbital   LUMO   HOMO

(a)   (b)

FIGURE D-7 Frontier orbital interactions for the coordination of Br^- and $(CH_3)_3C^+$ (a) The approach of Br^- from one face of the planar C atom. (b) The approach of Br^- from the other face of the planar C atom. As indicated, the HOMO and LUMO for both approaches have substantial net overlap, so the elementary step is allowed for either approach.

YOUR TURN D.7

Construct the atomic orbital overlap and MO energy diagrams for $(CH_3)_3C^+$ (review Figs. 3-13 through 3-25) and use these diagrams to determine its LUMO. Does your answer agree with the one shown in Figure D-7?

YOUR TURN D.8

In Figure D-7a and D-7b, label every region of constructive interference and destructive interference between the HOMO and LUMO.

Let's now turn our attention to heterolysis, the reverse of coordination. An example that we previously encountered in Chapter 7 is shown again in Equation D-4.

Heterolysis step

(D-4)

We don't need to begin an FMO analysis of such a reaction from scratch to determine whether it is allowed or forbidden, however, because we can use the principle of microscopic reversibility (Section D.2a) instead. Thus, because coordination steps are allowed, heterolysis steps must be allowed, too.

D.2e Nucleophilic Addition and Nucleophile Elimination Steps

Recall from Section 7.4 that a nucleophilic addition step involves the formation of a bond between a nucleophile and the atom at the positive end of a polar π bond. The π bond breaks during this process, and its electrons end up as a lone pair on the other atom that was initially part of the polar π bond. An example is shown in Equation D-5, in which the HO^- nucleophile adds to the C atom of a C=O group.

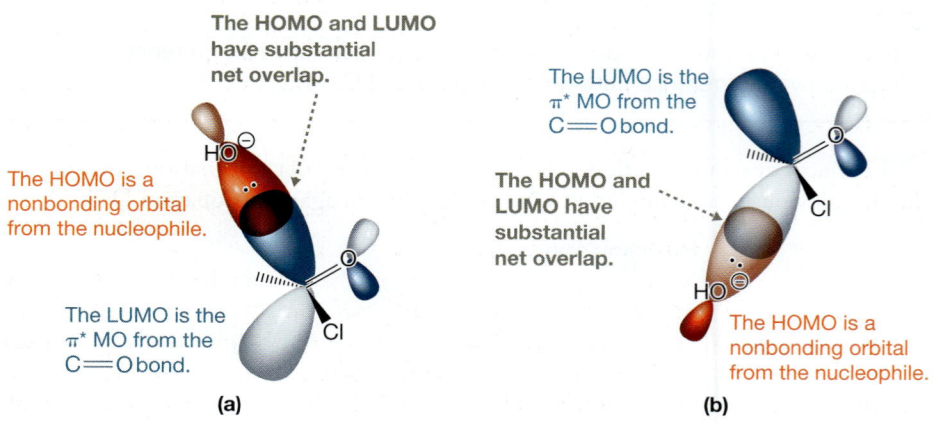

Nucleophilic addition

Electron rich   Electron poor

(D-5)

Once again, the HO^- nucleophile is relatively electron rich, whereas the C atom of the C=O group is relatively electron poor. As a result, the relevant HOMO will be contributed by HO^-, and the relevant LUMO will be contributed by the C=O group. As shown in Figure D-8, the HOMO of HO^- is a nonbonding orbital, and the LUMO of the C=O group is a π^* MO.

YOUR TURN D.9

Construct the atomic orbital overlap and MO energy diagrams for CH_3COCl (review Figs. 3-13 through 3-25) and use these diagrams to determine its LUMO. Does your answer agree with the one shown in Figure D-8a?

YOUR TURN D.10

In Figure D-8a and D-8b, label every region of constructive interference and destructive interference between the HOMO and LUMO.

FIGURE D-8 Frontier orbital interaction for the nucleophilic addition of HO^- to CH_3COCl The nonbonding orbital of the HO^- nucleophile (red; HOMO) has substantial net overlap with the π^* orbital of the C=O bond (blue; LUMO), so this nucleophilic addition is allowed. *Note:* The π^* MO on O is smaller than on C because, being more electronegative, O contributes more of its p AO to the π MO and less to the π^* MO.

The HOMO and LUMO have substantial net overlap.

The HOMO is a nonbonding orbital from the nucleophile.

The LUMO is the π^* MO from the C=O bond.

(a)

The LUMO is the π^* MO from the C=O bond.

The HOMO and LUMO have substantial net overlap.

The HOMO is a nonbonding orbital from the nucleophile.

(b)

Figure D-8 shows the approach of the nucleophile from either of the two faces of the planar carbon. Regardless of the nucleophile's approach, the HOMO and LUMO have substantial net overlap, so either approach leads to an allowed nucleophilic addition.

The reverse of nucleophilic addition is nucleophile elimination. An example is shown in Equation D-6, in which Cl^- is eliminated from a reactant species.

Nucleophile elimination

$$ (D-6) $$

According to the principle of microscopic reversibility, nucleophile elimination steps are allowed because nucleophilic addition steps are allowed.

D.2f Bimolecular Elimination (E2) Steps

In Section 7.5, we learned that a bimolecular elimination (E2) step involves a base and a substrate. As shown again in Equation D-7, the base removes a proton at the same time a leaving group is displaced from an adjacent C atom, thus producing a new π bond.

E2 step

$$ (D-7) $$

Electron rich   Electron poor

Unlike the previous examples, an E2 step involves the formation of two separate bonds (in this case, the HO—H bond and the π bond) and the breaking of another two bonds (in this case, the H—C and C—Br bonds). To simplify the picture using FMO theory, we can treat the entire process as two separate parts that take place at essentially the same time—one that forms the HO—H bond and breaks the H—C bond, and the second that forms the π bond and breaks the C—Br bond.

In the first of these parts of the E2 step—namely, forming the HO—H bond and breaking the H—C bond—the HO⁻ base is electron rich and the H—C bond of the substrate is relatively electron poor. Thus, HO⁻ will contribute its nonbonding MO as the HOMO, and the substrate will contribute its σ* orbital of the H—C bond as the LUMO, as shown in Figure D-9. Similar to the proton transfer step in Figure D-5, the HOMO and LUMO have substantial net overlap.

In the second part of the E2 step—namely, forming the π bond and breaking the C—Br bond—the electrons originate from the H—C bond that is undergoing bond breaking. Thus, the H—C bond is treated as electron rich, while the carbon of the C—Br bond is electron poor. As such, the HOMO that is contributed to the FMO interaction is the σ bonding orbital of the H—C bond, and the LUMO is the σ* orbital of the C—Br bond. This is shown in Figure D-10.

Notice, however, that there are two orientations in Figure D-10 in which the HOMO and LUMO have substantial net overlap: One has the H—C and C—Br bonds anti to each other (Fig. D-10a), whereas the other has the two bonds syn to each other (Fig. 10-10b), in which case they eclipse each other. Thus, an E2 step is allowed in either of these conformations. The HOMO and LUMO overlap to a greater extent,

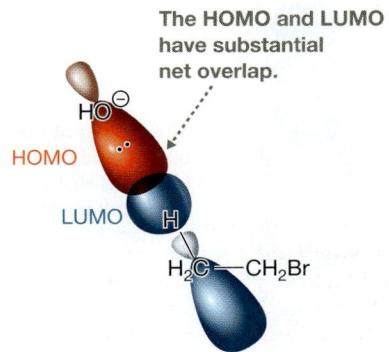

The HOMO and LUMO have substantial net overlap.

HOMO

LUMO

FIGURE D-9 The first part of the frontier orbital interactions of an E2 step The nonbonding orbital of the HO⁻ base (red; HOMO) has substantial net overlap with the σ* orbital of the H—C bond (blue; LUMO), so this part of the E2 step is allowed.

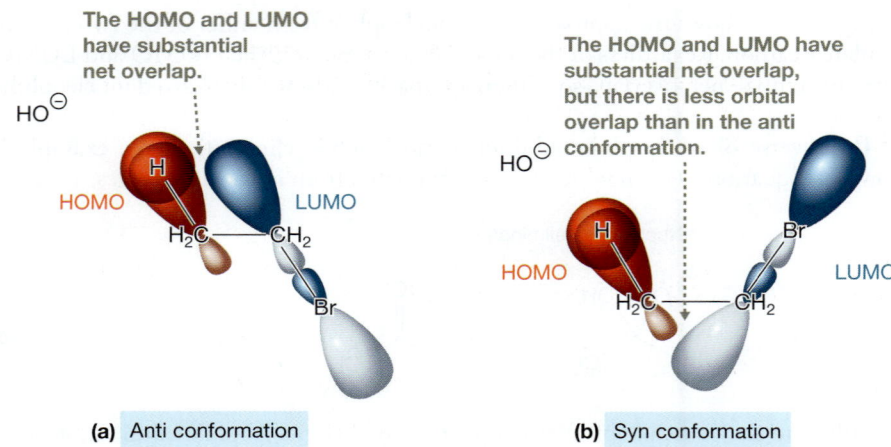

The HOMO and LUMO have substantial net overlap.

HO⊖

HOMO   **LUMO**

H_2C —— CH_2

Br

(a) Anti conformation

The HOMO and LUMO have substantial net overlap, but there is less orbital overlap than in the anti conformation.

HO⊖

HOMO   **LUMO**

H_2C —— CH_2

(b) Syn conformation

FIGURE D-10 The second part of the frontier orbital interactions of an E2 step In both the (a) anti and (b) syn conformations, the filled C—H σ orbital (red; HOMO) has substantial net overlap with the empty C—Br σ* orbital (blue; LUMO), so the E2 step is allowed with either conformation. The anti conformation is favored, however, due to the greater extent of orbital overlap.

however, when the two bonds are anti to each other. As a result, an E2 step tends to proceed faster when the hydrogen and the leaving group on the adjacent carbon atom in the substrate are anti to each other than when they are syn. This kind of preference for the particular conformation of the substrate greatly affects the *stereochemistry* of the reaction, often favoring one diastereomeric product over another (see Chapter 8).

YOUR TURN D.11

Construct the atomic orbital overlap and MO energy diagrams for CH_3CH_2Br (review Figs. 3-13 through 3-25) and use these diagrams to determine its HOMO and LUMO. Do your answers agree with the ones shown in Figure D-10a?

YOUR TURN D.12

In Figure D-10a and D-10b, label every region of constructive interference and destructive interference between the HOMO and LUMO.

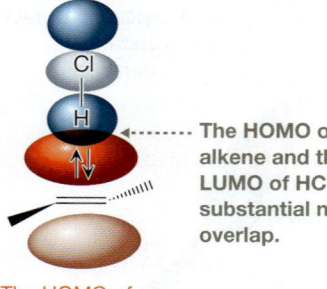

The LUMO of HCl is a σ* MO.

Cl

H

The HOMO of the alkene and the LUMO of HCl have substantial net overlap.

The HOMO of an alkene is a π MO.

FIGURE D-11 Frontier orbital interaction for the electrophilic addition involving an alkene and HCl The π MO of the alkene (red; HOMO) has substantial net overlap with the σ* orbital of the H—Cl bond (blue; LUMO), so this electrophilic addition step is allowed.

D.2g Electrophilic Addition and Electrophile Elimination Steps

Electrophilic addition was first discussed in Section 7.6. In a typical electrophilic addition step, a carbon atom of an alkene or alkyne group forms a new σ bond to an electrophile. An example is shown in Equation D-8, in which a C atom of a C=C bond forms a new bond to the proton of H—Cl.

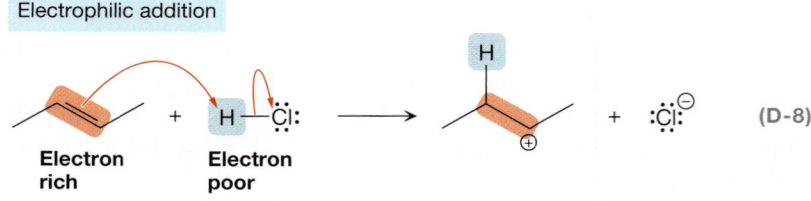

Electrophilic addition

Electron rich   Electron poor

(D-8)

Recall that the C=C double bond of an alkene is relatively electron rich, whereas the proton of HCl is electron poor. Therefore, the alkene donates the relevant HOMO to the FMO interaction, and HCl donates the relevant LUMO. As shown in Figure D-11, the HOMO of the alkene is a π MO, and the LUMO of H—Cl is a σ* orbital. Notice that those two MOs have substantial overlap, so the elementary step is allowed.

YOUR TURN **D.13**

Construct the atomic orbital overlap and MO energy diagrams for $CH_3CH\!=\!CHCH_3$ (review Figs. 3-13 through 3-25) and use these diagrams to determine its HOMO. Does your answer agree with the one shown in Figure D-11?

YOUR TURN **D.14**

In Figure D-11, label every region of constructive interference and destructive interference between the HOMO and LUMO.

Electrophile elimination is the reverse of electrophilic addition, so electrophile elimination is allowed, too, according to the principle of microscopic reversibility. A typical electrophile elimination step is shown in Equation D-9, in which a H^+ electrophile is eliminated from a carbocation species, producing a new π bond. Simultaneously, a base forms a bond to the departing H^+ electrophile.

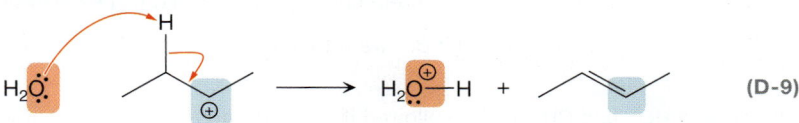

Electrophile elimination

$(D-9)$

D.2h Carbocation Rearrangements

Carbocation rearrangements were discussed in Section 7.7. Recall that a hydrogen or methyl group can migrate from one carbon atom to an adjacent one to increase carbocation stability. An example of a 1,2-hydride shift is shown in Equation D-10.

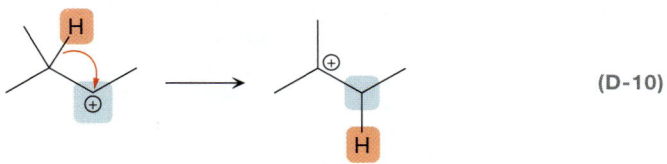

1,2-Hydride shift

$(D-10)$

This reaction involves just one species, so the carbocation must provide both the HOMO and the LUMO to the FMO interaction. As shown in Figure D-12, the HOMO is the σ orbital that connects to the migrating group, and the LUMO is the empty p orbital of the positively charged C atom.

Figure D-12 shows that the HOMO and LUMO do, indeed, have substantial net overlap, so these carbocation rearrangements are allowed.

The HOMO and LUMO have substantial net overlap, so this reaction is *allowed*.

The HOMO is the bonding σ orbital of the C—H bond.

The LUMO is the empty p orbital of the positively charged C atom.

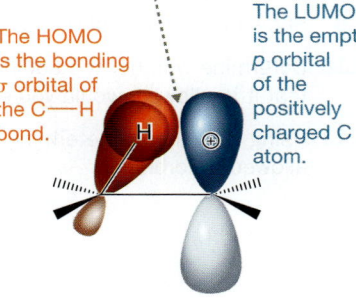

FIGURE D-12 The frontier orbital interaction for a 1,2-hydride shift
The σ orbital of the C—H bond (red; HOMO) has substantial net overlap with the empty p orbital of the positively charged carbon atom (blue; LUMO), so this carbocation rearrangement step is allowed.

YOUR TURN **D.15**

Construct the atomic orbital overlap and MO energy diagrams for $(CH_3)_2CHC^+HCH_3$ (review Figs. 3-13 through 3-25) and use these diagrams to determine its HOMO and LUMO. Do your answers agree with the ones shown in Figure D-12?

YOUR TURN **D.16**

In Figure D-12, label every region of constructive interference and destructive interference between the HOMO and LUMO.

Problems

D.1 Relative Stabilities of Carbocations and Alkenes: Hyperconjugation

D.1 Draw the orbital interaction that represents hyperconjugation in $CF_3CH_2^+$. Would you expect hyperconjugation to lead to greater stabilization in this cation or in $CH_3CH_2^+$? Explain. *Hint:* Because F is substantially more electronegative than H, the C—F σ bonding orbital is lower in energy than that of C—H.

D.2 A type of hyperconjugation occurs in the $F—CH_2—NH^-$ anion even though there are no double bonds, triple bonds, or empty p AOs. Draw the orbital interaction that illustrates the most significant hyperconjugation interaction in this species. *Hint:* What is the HOMO in this species? Which adjacent σ^* orbital is the lowest in energy?

D.3 Draw the orbital interaction that illustrates the hyperconjugation that takes place in propyne.

D.2 MO Theory and Chemical Reactions

D.4 Would the proton transfer between HO^- and HCl be allowed if HO^- were to approach HCl from the same end as Cl (i.e., directly opposite the H)? If your answer is yes, would you expect that elementary step to take place? Why or why not?

D.5 Would the S_N2 step between HO^- and CH_3Cl be allowed if HO^- were to approach CH_3Cl from the same end as Cl (i.e., directly opposite the CH_3)? If your answer is yes, would you expect that elementary step to take place? Why or why not?

D.6 Would the coordination step between Br^- and $(CH_3)_3C^+$ be allowed if Br^- were to approach from within the plane of the central C? Explain.

D.7 Would the nucleophilic addition step between HO^- and CH_3COCl be allowed if HO^- were to approach the carbonyl C atom directly along the C=O bond? Explain.

D.8 Show that the electrophilic addition involving $CH_3CH=CHCH_3$ and $(CH_3)_3C^+$ is allowed.

D.9 Show that a 1,2-methyl shift involving $(CH_3)_3CC^+HCH_3$ is allowed.

D.10 Determine whether the addition of a nucleophile to the O atom of a C=O group is allowed or forbidden. If you determine that it is allowed, would you expect that elementary step to take place? Why or why not?

D.11 Using FMO theory, determine whether a carbanion rearrangement, analogous to the 1,2-hydride shift in Equation D-10, is allowed or forbidden.

Naming Compounds with a Functional Group That Calls for a Suffix 1

Alcohols, Amines, Ketones, and Aldehydes

In previous nomenclature interchapters (A–C), we have seen that the presence of a functional group requires a very specific modification to the name. When the compound is an alkene or alkyne, for example, the corresponding alkane's root (parent compound) is changed to *ene* or *yne*, respectively. Groups such as halogen atoms, nitro groups, and ethers, on the other hand, are indicated by appropriate prefixes. As we see here in Interchapter E, a *suffix* must be added to the appropriate root if functional groups other than the ones just mentioned are present in a molecule. These ideas are captured in the following diagram.

The root (parent compound) indicates
the longest chain or ring of carbons, and
the presence of C═C and C≡C bonds.

Prefixes indicate the presence
and locations of substituents.

The suffix indicates the presence of
a functional group other than halo,
nitro, ether, C═C, or C≡C.

prefixesrootsuffix

In Section E.1, we provide the basic rules for naming compounds that require adding a suffix, beginning with alcohols and amines. In Section E.2, we extend these rules to aldehydes and ketones, and in Section E.3, we present trivial names that are in common use for alcohols, amines, ketones, and aldehydes.

We focus on alcohols, amines, ketones, and aldehydes here in Interchapter E because they appear as reactants in some important reactions you will encounter in Chapters 8–10. Other functional groups that call for a suffix are explained in Interchapter F.

E.1 The Basic System for Naming Compounds Having a Functional Group That Calls for a Suffix: Alcohols and Amines

To name a compound with a functional group that requires a suffix, all of the rules we learned in Interchapters A, B, and C remain in effect and the following additional rules are applied.

Basic Rules for Naming Compounds with a Functional Group That Requires a Suffix

- Determine the highest-priority functional group present that calls for a suffix.
- Establish the main chain or ring as the one that contains that functional group.
- Remove the "e" from the normal *ane, ene,* or *yne* ending and add the suffix that corresponds to the highest-priority functional group.
  - Nitriles are a notable exception to this rule.
- Number the main chain or ring so the carbon atom involving the highest-priority functional group receives the lowest number possible.
- Add the locator number (locant) for the highest-priority functional group immediately before the suffix, unless it is redundant.
- All other functional groups in the molecule are treated as substituents and appear in the name as a prefix.

Based on these rules, you must know the relative priorities of functional groups to establish which one appearing in the molecule has the highest priority. You must also know the proper suffix and prefix that correspond to each of those functional groups. These pieces of information are presented in Table E-1. Notice that Table E-1 includes the functional groups we focus on in this interchapter (characteristic of alcohols, amines, ketones, and aldehydes) as well as the functional groups discussed in Interchapter F (characteristic of carboxylic acids, acid anhydrides, esters, acid chlorides, amides, and nitriles).

Let's apply these rules to name the following alcohols, which have just one OH functional group.

OH group located on C1	OH group located on C2	No locator number added	No locator number added
Propan-1-ol	Propan-2-ol	Ethanol	Cyclohexanol

If the OH groups were absent, the first two molecules would be *propane*, the third would be *ethane*, and the fourth would be *cyclohexane*. According to Table E-1, we therefore drop the final "e" and add the *ol* suffix, so the first two molecules become *propanol*, the third becomes *ethanol*, and the fourth becomes *cyclohexanol*. Locator numbers are added to the names of the first two molecules, immediately before the *ol* suffix, to establish where the OH group is in the molecule. In the third and fourth molecules, locator atoms are not added to the name because the carbon atoms are equivalent without the OH; a locator number would be redundant.

CONNECTIONS Millions of tons of cyclohexanol are produced annually, primarily as a precursor in the production of nylon. Nylon is an important synthetic fiber used in a variety of consumer products, such as this tent.

TABLE E-1 Common Functional Groups, Their Relative Priorities, and Corresponding Suffixes and Prefixes

Functional Group Priority	Compound Class	SUFFIX Drop "e" from Root?	Add	Prefix	Functional Group Priority	Compound Class	SUFFIX Drop "e" from Root?	Add	Prefix
1	Carboxylic acid	Yes	oic acid	carboxy	6 R—C≡N	Nitrile	No	nitrile	cyano
2	Acid anhydride	Yes	oic anhydride	—	7	Aldehyde	Yes	al	oxo
3	Ester	—	—	—	8	Ketone	Yes	one	oxo
4	Acid chloride	Yes	oyl chloride	chlorocarbonyl	9 R—OH	Alcohol	Yes	ol	hydroxy
5	Amide	Yes	amide	carbamoyl	10 R—N	Amine	Yes	amine	amino

Note: R indicates an alkyl group, and the absence of an atom at the end of a bond indicates that either R or H may be attached.

PROBLEM E.1 Write the IUPAC name for each of the following compounds.

(a)   (b)   (c)

PROBLEM E.2 Draw the structure for each of the following molecules.
(a) pentan-3-ol; **(b)** cyclobutanol; **(c)** hexan-1-ol

In each of the following molecules, the highest-priority functional group is NH_2, characteristic of amines. According to Table E-1, then, they each receive the suffix *amine* after removing the final "e" from the corresponding alkane.

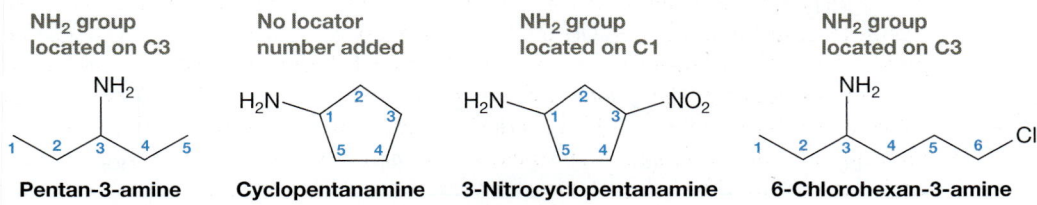

NH₂ group located on C3	No locator number added	NH₂ group located on C1	NH₂ group located on C3
Pentan-3-amine	**Cyclopentanamine**	**3-Nitrocyclopentanamine**	**6-Chlorohexan-3-amine**

For the first molecule above, the *pentane* root becomes *pentanamine* and the locator number 3 is included immediately before the *amine* suffix. The second molecule, cyclopentanamine, does not require a locator number in the name. In the third molecule, C1 is assigned to the carbon attached to NH_2 and numbering increases clockwise around the ring so the nitro group is encountered the earliest. Even though the locator number for the NH_2 group is not included in this molecule's name, the one for the nitro group is necessary. In the fourth molecule, numbering starts from the left end of the chain to give the lowest locator number to the carbon attached to NH_2, which is the highest-priority group in the molecule.

PROBLEM E.3 Write the IUPAC name for each of the following compounds.

(a)   (b)   (c)

PROBLEM E.4 Draw the structure for each of the following molecules.
(a) 4,4-dibromocyclohexanamine; **(b)** 2-nitroethanamine;
(c) 2-methylpropan-1-amine; **(d)** 5-cyclopropylheptan-2-amine

The following molecules contain one or more C=C or C≡C bonds in addition to an OH or NH_2 group.

C=C bond beginning at C2

4-Chlorocyclohex-2-en-1-amine

C=C bonds beginning at C4 and C6

Hepta-4,6-dien-2-ol

C≡C bond beginning at C1

Hept-1-yn-4-ol

To name these molecules, we must add the appropriate *ol* or *amine* suffix after first dropping the final "e" from the normal *ene* or *yne* ending. We must also apply the rules we learned in Interchapter B to establish the presence and location of the C=C and C≡C bonds. In the first molecule above, C1 is the carbon attached to NH_2 and numbering proceeds counterclockwise around the ring to arrive at the C=C bond the earliest. Locator numbers are required for both the chloro substituent and the C=C bond. In the second molecule, numbering begins on the right to give the carbon attached to OH the lowest possible number, and locator numbers are added to account for the C=C bonds at C4 and C6. Remember that *di* is added to specify that there are two C=C bonds. In the third molecule, the carbon attached to OH would receive the same locator number regardless of the end at which numbering begins, so we begin at the right to encounter the C≡C bond the earliest.

PROBLEM E.5 Write the IUPAC name for each of the following compounds. You may disregard stereochemistry.

(a)   (b)   (c)   (d)

PROBLEM E.6 Draw the structure for each of the following molecules.
(a) pent-4-en-1-amine; **(b)** 3-cyclopropylcyclopent-3-en-1-amine;
(c) 6-chlorohexa-1,4-diyn-3-amine; **(d)** 2-methylcycloocta-3,6-dien-1-ol

The following molecules contain both an OH and an NH_2 group:

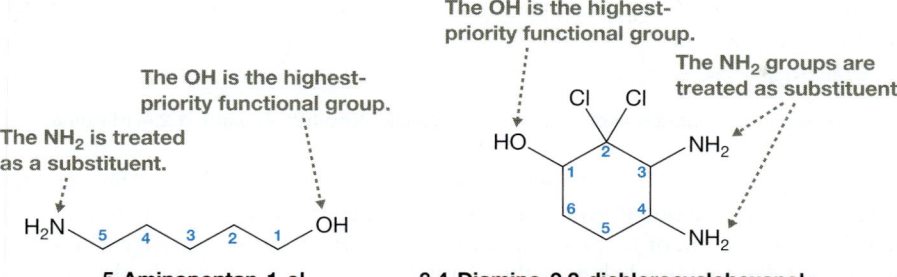

The NH₂ is treated as a substituent.

The OH is the highest-priority functional group.

The OH is the highest-priority functional group.

The NH₂ groups are treated as substituents.

5-Aminopentan-1-ol   **3,4-Diamino-2,2-dichlorocyclohexanol**

CONNECTIONS
5-Aminopentan-1-ol is used in the synthesis of manzamines, a class of compounds that exhibit antitumor activity. Manzamines were originally isolated from marine sponges like the one shown here.

Both the NH_2 and OH functional groups require adding a suffix to the IUPAC name, but only one suffix can be added. According to Table E-1, the higher priority goes to the OH, so each molecule receives the suffix *ol* and numbering is established to give the carbon attached to OH the lowest possible number. In the first molecule, then, the numbering must begin at the right, and in the second molecule, the carbon attached to OH is C1 and numbering increases clockwise around the ring. The NH_2 groups are then treated as substituents and, according to Table E-1, are named *amino* in the prefix.

PROBLEM E.7 Write the IUPAC name for each of the following compounds.

(a)   **(b)**   **(c)**

PROBLEM E.8 Draw the structure for each of the following molecules.
(a) 4,5-diaminoheptan-2-ol; **(b)** 2,3,4-triaminocycloheptanol

E.1a Molecules with Two or More of the Highest-Priority Functional Group

In all of the molecules we have examined so far here in Interchapter E, the highest-priority functional group appears just once. If there are two or more of that functional group, we must apply the following additional rules:

Naming a Molecule with Two or More of the Highest-Priority Functional Groups

- Do not remove the final "e" of *ane*, *ene*, or *yne* prior to adding the suffix.
- Add *di*, *tri*, etc. (p. 57) immediately before the suffix to specify how many of the highest-priority functional groups appear.
- Add one locator number for *each* of the highest-priority functional groups immediately before the *di*, *tri*, etc.

Let's apply these rules to the following examples:

The highest-priority functional group is OH, and there are two of them located at C2 and C4.

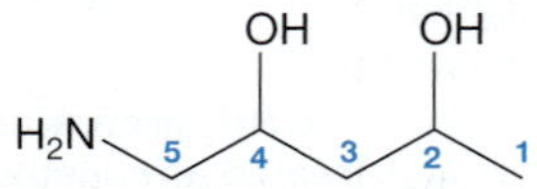

5-Aminopentane-2,4-diol

The highest-priority functional group is NH₂, and there are three of them located at C1, C2, and C4.

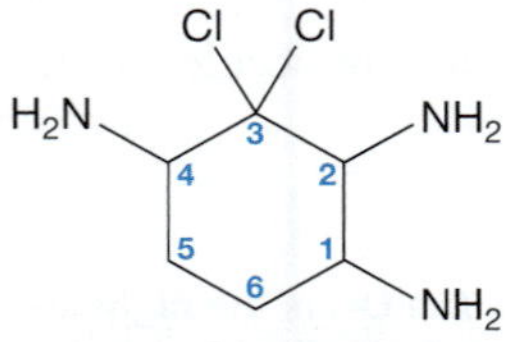

3,3-Dichlorocyclohexane-1,2,4-triamine

In the first molecule, the OH groups establish the suffix because they are the highest-priority functional groups in the molecule. There are two of them, so we keep the "e" at the end of *pentane*, add locator numbers 2 and 4 to specify where they are, and add *di* immediately before the *ol* suffix. In the second molecule, the highest-priority functional group is an NH₂ and there are three of them. Because the molecule is cyclic, one of the carbons attached to NH₂ must be assigned C1. The correct one is at the bottom-right because that allows the second NH₂ to be encountered the earliest.

SOLVED PROBLEM E.9

Write the IUPAC name for the compound shown here.

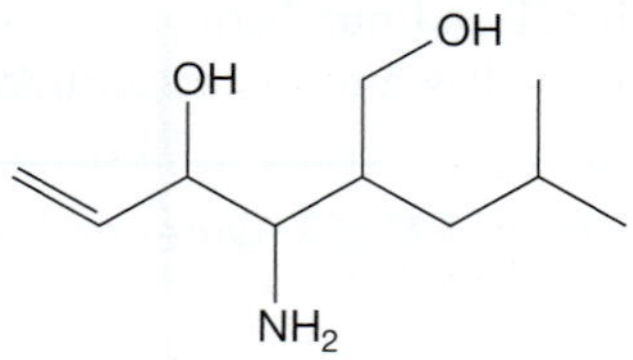

Think Are there any functional groups that require adding a suffix to the name? Which one has the highest priority and what is its suffix? How many are there of that functional group? What is the longest carbon chain that contains those functional groups? Should you drop the final "e" before adding the suffix? How should you number the main chain so the highest-priority functional groups are encountered the earliest? Are there any C═C or C≡C bonds within that main chain?

Solve The molecule contains OH and NH₂ groups, both of which call for a suffix to be added. An OH group has a higher priority than NH₂, so the suffix is *ol*. There are two OH groups, and the longest carbon chain that contains both of them has six carbons, as shown with the numbering system here. The chain is numbered so the first OH encountered receives the lowest possible number—in this case, C1. There is one C═C bond that is part of that chain, located at C5. Because there are two OH groups, we retain the final "e" in *ene* before adding the suffix, we specify the location of the OH groups using the locator numbers 1 and 4, and we add *di* before *ol* to indicate how many OH goups there are. Finally, we use prefixes to account for the amino substituent at C3 and the (2-methylpropyl) substituent at C2.

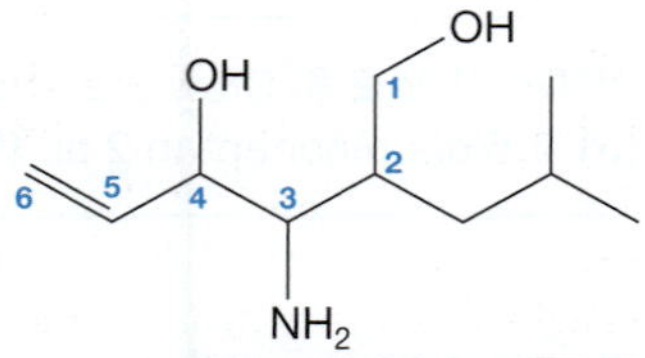

3-Amino-2-(2-methylpropyl)-
hex-5-ene-1,4-diol

PROBLEM E.10 Write the IUPAC name for each of the following compounds.

(a)

(b)

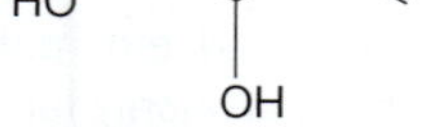

(c)

PROBLEM E.11 Draw the structure for each of the following molecules.
(a) propane-1,2-diol; **(b)** 1-methoxypentane-2,3-diamine;
(c) but-2-ene-1,4-diol

CONNECTIONS Propane-1,2-diol (Problem E.11a), more commonly known as propylene glycol, is the principal ingredient used in antifreeze for automobiles.

E.1b More Than One Alkyl Group Bonded to the Amine Nitrogen

In the amines we have examined so far, the amine nitrogen has been bonded to two H atoms, appearing as R—NH_2. An amine N can have up to three alkyl groups bonded to it, however, so to account for these additional alkyl groups attached to N, we apply the following rules:

Accounting for Alkyl Groups Bonded to an Amine Nitrogen

- If the N atom of an amine is bonded to more than one alkyl group, then the root is established by the alkyl group that makes up the longest carbon chain or largest carbon ring.
- Any other alkyl groups attached to the amine N are treated as substituents and each of them is given an italic *N* as a locator instead of a number.

Consider the following amines as examples.

The N is located at C1 and there is a methyl group attached to it.

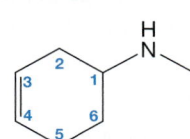

***N*-Methylcyclohex-3-en-1-amine**

The N is located at C2. There is one ethyl group attached to N, and there are two methyl groups: one attached to N and the other to C3.

***N*-Ethyl-5-methoxy-*N*,3-dimethylpentan-2-amine**

The first molecule has two alkyl groups attached to N and the six-carbon ring establishes the root: It is a cyclohexenamine. The methyl group that is attached to N is treated as a substituent and is given *N* as its locator. In the second molecule, the longest carbon chain that is attached to N has five carbons, so it is a pentanamine. There is one ethyl substituent that is attached to N, so it appears as a prefix and is given the locator *N*. There are two methyl substituents in the molecule, one attached to N and one attached to C3, so *dimethyl* appears in the prefix along with the locators "*N*,3."

PROBLEM E.12 Write the IUPAC name for each of the following compounds.

(a)   (b)   (c)

PROBLEM E.13 Draw the structure for each of the following molecules.
(a) *N*-methylhepta-1,6-dien-4-amine; **(b)** *N*,*N*,3-tricyclopropylcyclobutanamine

---

E.1c Stereochemistry and Functional Groups That Require a Suffix

Interchapter C discussed how to incorporate the *R/S* stereochemical configurations for an asymmetric carbon into a molecule's name. The focus in Interchapter C was on the configurations for asymmetric carbons to which substituents are attached, but the *R/S* designation also applies to situations in which the functional group corresponding to the suffix involves an asymmetric carbon. In these situations, the rules we learned in Interchapter C are applied no differently. This can be seen in the following examples. (Revisit Interchapter C if you need to review the rules for determining stereochemical configurations.)

(2S,3E,5R)-5-Aminohex-3-en-2-ol   **(1R,2S)-Cyclopentane-1,2-diol**

In the first molecule, the carbon attached to NH$_2$ at C5 has the *R* configuration and the carbon attached to OH at C2 has the *S* configuration. The C=C bond at C3 has the *E* configuration. These stereochemical designations and their locator numbers appear in parentheses at the very beginning of the name in increasing order of the corresponding locator number. Similarly, in the second molecule, the carbon attached to OH at C1 has the *R* configuration and the one at C2 has the *S*. Again, these stereochemical designations and their locators are enclosed in parentheses at the very beginning of the name.

PROBLEM E.14 Write the IUPAC name for each of the following compounds, including stereochemical designations.

(a)   **(b)**   **(c)**

PROBLEM E.15 Draw the structure for each of the following molecules.
(a) (1R,3S)-cyclooct-6-ene-1,3-diol; **(b)** (3S,4Z)-6-amino-5-chlorohex-4-en-3-ol

---

E.2 Naming Ketones and Aldehydes

Both ketones and aldehydes contain a C=O bond, the *carbonyl group*. In a ketone, the carbonyl carbon is bonded to two alkyl and/or aryl groups, whereas in an aldehyde, the carbonyl carbon is bonded on one side to an H atom and on the other side to an alkyl group, an aryl group, or a second H.

Ketones and aldehydes require adding a suffix to the IUPAC name and therefore follow the rules we learned in Section E.1. Let's first see how those rules are applied to the following ketones.

The C=O group can only be at C2, so no locator number is added.

Propanone
(Acetone)

The OH group is treated as a substituent because a C=O has a higher priority than an OH.

4-Hydroxypentan-2-one

The main chain is chosen to include both C=O groups.

3-Propylpentane-2,4-dione

Cyclohexane-1,2,4-trione

The longest carbon chain in the first molecule has three carbons and no C=C or C≡C bonds, so we begin with *propane*. According to Table E-1 (p. 379), we drop the "e" at the end and add the suffix *one* because the C=O group is part of a ketone. We do not add a locator number because the carbonyl carbon must be at C2 for the compound to be a ketone rather than an aldehyde. The name, therefore, is just propanone.

The second molecule has two functional groups that require a suffix: a C=O group that establishes a ketone and an OH. According to Table E-1, priority goes to the C=O, so we name the compound as a ketone and treat the OH group as a substituent, adding the prefix *hydroxy*. The ketone has five carbons in its longest carbon chain, making it a *pentanone*. Unlike propanone, a locator number is necessary for this molecule because a pentanone can have the C=O group at more than one location. In this case, we begin numbering from the right to give the C=O the lowest number—C2 instead of C4. The name, therefore, is 4-hydroxypentan-2-one.

In the third molecule, the longest carbon chain has six carbon atoms, but the longest carbon chain that contains *both* C=O functional groups has just five carbon atoms. Therefore, it is a pentane-2,4-dione. Because there is more than one C=O, we do not drop the "e" at the end of *pentane*. Furthermore, locator numbers are required for each C=O. Finally, the propyl group at C3 is treated as a substituent.

The fourth molecule is a cyclic ketone. To give the lowest locator number to the first C=O group, we begin numbering at one of the C=O carbons. To encounter the second C=O the earliest, we must begin with the C=O on the right and number the carbons counterclockwise. Again, we retain the "e" at the end of *ane* and include a locator number for each C=O.

To see how the rules are applied for aldehydes, let's name the following molecules.

No locator number is added because the C=O group must be at a terminal carbon.

Butanal
(Butyraldehyde)

The aldehyde C=O has priority over the ketone C=O.

3-Oxopentanal

The main chain must include both aldehyde C=O groups.

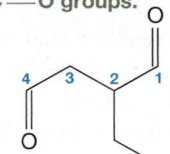

2-Ethylbutanedial

A cyclic aldehyde

Cyclopentanecarbaldehyde

The first molecule has four carbons in the main chain and, according to Table E-1 (p. 379), an aldehyde requires dropping the final "e" and adding the *al* suffix. Therefore, *butane* becomes *butanal*. Notice that no locator number is added because an aldehyde C=O must be at a terminal carbon.

The second molecule has both a ketone C=O and an aldehyde C=O, both of which call for a suffix. According to Table E-1, priority goes to the aldehyde, so the =O of the ketone is treated as an *oxo* substituent. A locator number is required for the ketone C=O but not for the aldehyde C=O.

The third molecule has five carbons in the longest carbon chain, but the main chain is assigned as the four-carbon chain that contains both aldehyde C=O groups. A locator number is required for the ethyl substituent, but not the aldehyde C=O groups, so the name is 2-ethylbutanedial. Notice that *di* is added to account for the two aldehyde C=O groups, and the "e" in *ane* is retained.

The fourth molecule is a cyclic aldehyde. Unlike the functional groups we have dealt with previously, the carbon of an aldehyde C=O cannot be part of the carbon ring. Rather, it is external to the ring but is attached directly to it. In situations like this, we apply the following rule:

Naming Cyclic Aldehydes

The name of a cyclic aldehyde takes the form *cycloalkanecarbaldehyde*.

In the case at hand, the cycloalkane ring is a cyclopentane, so the molecule is cyclopentanecarbaldehyde.

PROBLEM E.16 Write the IUPAC name for each of the following compounds.

(a)   (b)   (c)   (d)

PROBLEM E.17 Draw the structure for each of the following molecules.
(a) cyclopentane-1,2,3-trione; **(b)** 3,3,3-trichloro-2-oxopropanal;
(c) 2,4,4-trinitrocycloheptanecarbaldehyde; **(d)** 4-hydroxyhept-2-ynedial

---

E.3 Trivial Names of Alcohols, Amines, Ketones, and Aldehydes

In Section A.7, we learned that many trivial names of alkanes, alkyl halides, and ethers remain in common use. In Section B.4, we encountered a variety of these trivial names involving alkenes, alkynes, and substituted benzenes. The types of compounds we focus on here in Interchapter E—namely, alcohols, amines, ketones, and aldehydes—have trivial names that remain in common use, too. We examine several of them in this section.

E.3a Trivial Names of Alcohols

Because alcohols consist of an alkyl group attached to a hydroxyl (OH) group, many trivial names of alcohols are derived simply by identifying the specific alkyl group present and adding the separate word *alcohol*.

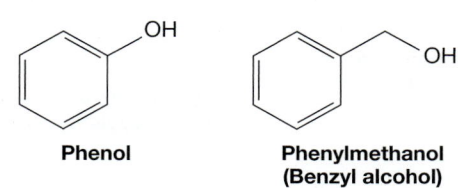

Alcohols

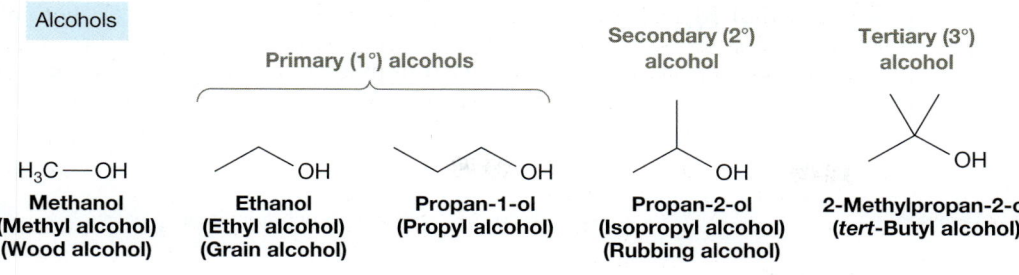

Primary (1°) alcohols

Secondary (2°) alcohol

Tertiary (3°) alcohol

H₃C—OH

Methanol
(Methyl alcohol)
(Wood alcohol)

Ethanol
(Ethyl alcohol)
(Grain alcohol)

Propan-1-ol
(Propyl alcohol)

Propan-2-ol
(Isopropyl alcohol)
(Rubbing alcohol)

2-Methylpropan-2-ol
(*tert*-Butyl alcohol)

Phenol

Phenylmethanol
(Benzyl alcohol)

CONNECTIONS Benzyl alcohol is largely used as a solvent for inks and paints, as well as for lacquers and epoxy resin coatings. It occurs naturally in foods such as cranberries, cocoa, and apricots, as well as in essential oils like jasmine. It has therefore found use as food and perfume additives.

For example, methanol (sometimes referred to as wood alcohol because it was first isolated from the pyrolysis of wood) consists of a methyl group attached to OH, so its trivial name is methyl alcohol. Similarly, the trivial name of ethanol (sometimes called grain alcohol because of its production during the fermentation of grains) is ethyl alcohol, and the trivial name of propan-1-ol is propyl alcohol. Examples that incorporate trivial names for the alkyl group include isopropyl alcohol (also known as rubbing alcohol) and *tert*-butyl alcohol. Phenol and benzyl alcohol represent examples that contain a phenyl ring.

Phenol does *not* appear in parentheses in the above graphic because it has been adopted by the IUPAC, much like toluene ($C_6H_5CH_3$) and anisole ($C_6H_5OCH_3$) (see Interchapter B). The term *phenol*, in fact, generally refers to the compound class describing molecules in which an OH group is directly attached to a benzene ring.

Alcohols are often classified as either primary (1°), secondary (2°), or tertiary (3°), depending on the degree of alkyl substitution of the carbon to which the OH group is attached.

- In a *primary* (1°) alcohol, the OH group is attached to a primary carbon.
- In a *secondary* (2°) alcohol, the OH group is attached to a secondary carbon.
- In a *tertiary* (3°) alcohol, the OH group is attached to a tertiary carbon.

Recall from Section A.7 that a carbon's degree of substitution is defined by the number of alkyl or aryl groups to which it is bonded. A primary carbon is bonded to one alkyl or aryl group, a secondary carbon is bonded to two alkyl or aryl groups, and a tertiary carbon is bonded to three alkyl or aryl groups. Therefore, as indicated in the preceding figure, ethanol (ethyl alcohol) and propan-1-ol (propyl alcohol) are primary alcohols, whereas propan-2-ol (isopropyl alcohol) and methylpropan-2-ol (*tert*-butyl alcohol) are secondary and tertiary alcohols, respectively.

PROBLEM E.18 For each trivial name, draw the complete structure and provide the IUPAC name. **(a)** trichloromethyl alcohol; **(b)** isobutyl alcohol; **(c)** pentyl alcohol; **(d)** *sec*-butyl alcohol

PROBLEM E.19 Determine whether each alcohol in Problem E.18 is a primary, secondary, or tertiary alcohol.

E.3b Trivial Names of Amines

The trivial names for amines are constructed in much the same way as they are for alcohols. The alkyl groups attached to the amine N are named first, followed by the suffix *amine*.

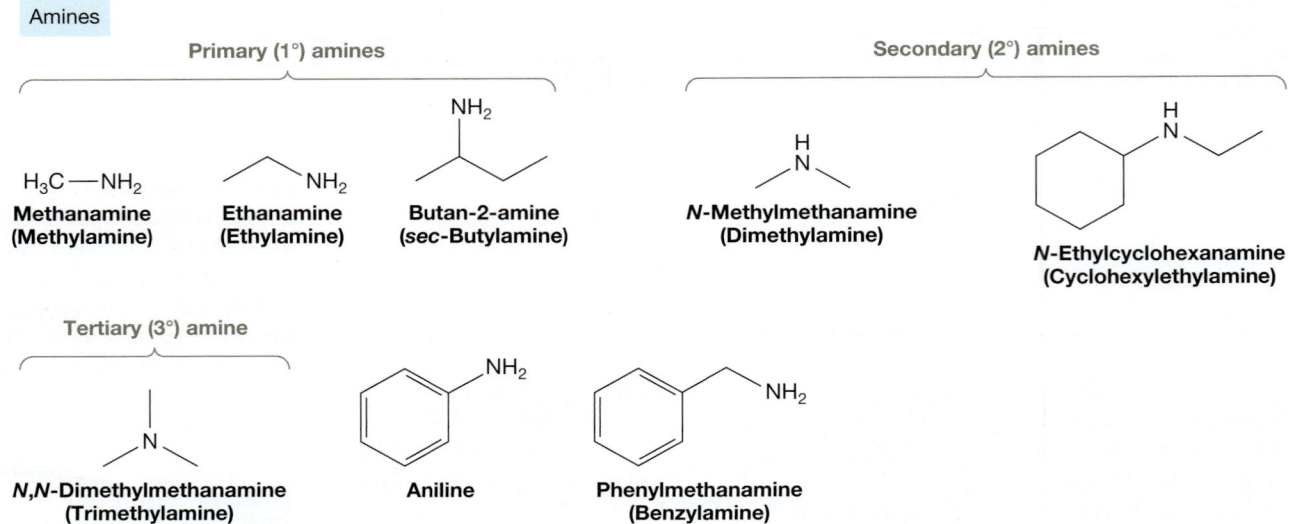

Amines

Primary (1°) amines

Methanamine
(Methylamine)

Ethanamine
(Ethylamine)

Butan-2-amine
(*sec*-Butylamine)

Secondary (2°) amines

***N*-Methylmethanamine**
(Dimethylamine)

***N*-Ethylcyclohexanamine**
(Cyclohexylethylamine)

Tertiary (3°) amine

***N*,*N*-Dimethylmethanamine**
(Trimethylamine)

Aniline

Phenylmethanamine
(Benzylamine)

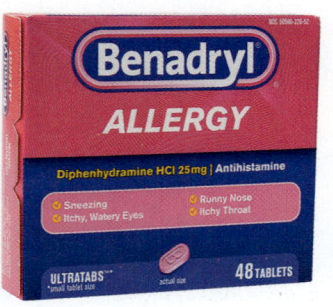

The main difference between naming amines and naming alcohols is in the number of alkyl groups that must be identified. Whereas an alcohol oxygen is bonded to only one alkyl group, the nitrogen atom of an amine can be bonded to one, two, or three alkyl groups.

In methylamine, for example, the N atom is bonded to just a methyl group, so the other two bonds to N are N—H bonds. The names ethylamine and *sec*-butylamine indicate that the N atom is attached to an ethyl group and a *sec*-butyl group, respectively. Dimethylamine and cyclohexylethylamine are examples in which the N atom is bonded to two alkyl groups, and trimethylamine is an example in which the N atom is bonded to three alkyl groups. Aniline and benzylamine involve a phenyl ring. Just like we saw previously for phenol (C_6H_5OH), aniline has been adopted by the IUPAC, which is why the name does not appear in parentheses in the preceding graphic.

Amines are classified as either primary (1°), secondary (2°), or tertiary (3°). Although this terminology is identical to that used for classifying alcohols (Section E.3a), the way in which amines are classified is different from that for alcohols. Whereas alcohols are classified by the number of alkyl groups bonded to the alcohol carbon, an amine is classified by the number of alkyl groups to which the N atom is attached.

- In a *primary* (1°) amine, the N atom is bonded to one alkyl or aryl group.
- In a *secondary* (2°) amine, the N atom is bonded to two alkyl or aryl groups.
- In a *tertiary* (3°) amine, the N atom is bonded to three alkyl or aryl groups.

Therefore, methylamine, ethylamine, and *sec*-butylamine are all primary amines, whereas dimethylamine and cyclohexylethylamine are secondary amines, and trimethylamine is a tertiary amine.

PROBLEM E.20 For each trivial name, draw the complete structure and provide the IUPAC name. **(a)** diisopropylamine; **(b)** *sec*-butylisopropylamine; **(c)** *tert*-butyldimethylamine; **(d)** triethylamine; **(e)** diphenylamine

PROBLEM E.21 Identify each molecule in Problem E.20 as either a primary, secondary, or tertiary amine.

E.3c Trivial Names of Aldehydes

Some of the more common trivial names for aldehydes are as follows:

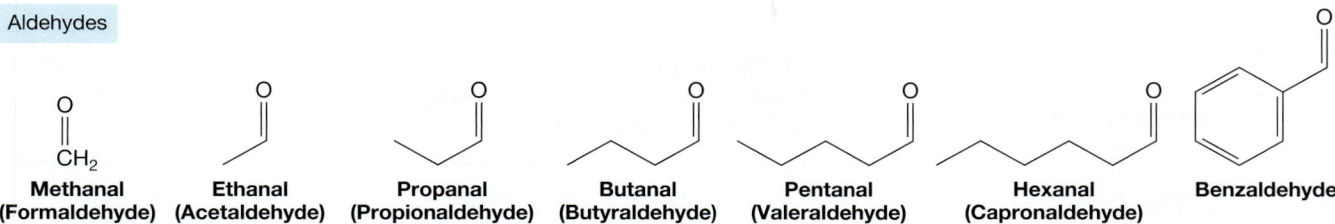

Aldehydes

Methanal
(Formaldehyde)

Ethanal
(Acetaldehyde)

Propanal
(Propionaldehyde)

Butanal
(Butyraldehyde)

Pentanal
(Valeraldehyde)

Hexanal
(Capronaldehyde)

Benzaldehyde

Many of these trivial names derive from trivial names for the analogous carboxylic acids, a topic in Interchapter F. For example, the trivial name for methanal, which has one carbon atom, is formaldehyde. This is analogous to formic acid, the carboxylic acid that contains just one carbon atom. Similarly acetaldehyde, the trivial name for the two-carbon aldehyde, derives from acetic acid, the two-carbon carboxylic acid.

PROBLEM E.22 For each trivial name, draw the complete structure and provide the IUPAC name. **(a)** phenylacetaldehyde; **(b)** 2,3-dichloropropionaldehyde; **(c)** 4-nitrobutyraldehyde

CONNECTIONS
Formaldehyde is used widely in industry to make resins, plastics, and polyurethane paints. It also has medical uses, such as for killing bacteria and fungi, and for treating warts. You are perhaps most familiar with the use of formaldehyde in embalming and in preserving organs and entire organisms.

E.3d Trivial Names of Ketones

Because ketones consist of two alkyl or aryl groups attached to a carbonyl (C=O) group, their trivial names consist of identifying the alkyl or aryl groups and listing them both before the word *ketone*. Therefore, the trivial name for propanone, $(CH_3)_2C$=O, is dimethyl ketone, the trivial name for pentan-3-one is diethyl ketone, and the trivial name for diphenylmethanone, $(C_6H_5)_2C$=O, is diphenyl ketone.

CONNECTIONS
Benzaldehyde is used as the starting material for some synthetic dyes and as an intermediate in the synthesis of some pharmaceuticals. It is also used widely as a flavoring and fragrance agent because it has the taste and smell of almonds.

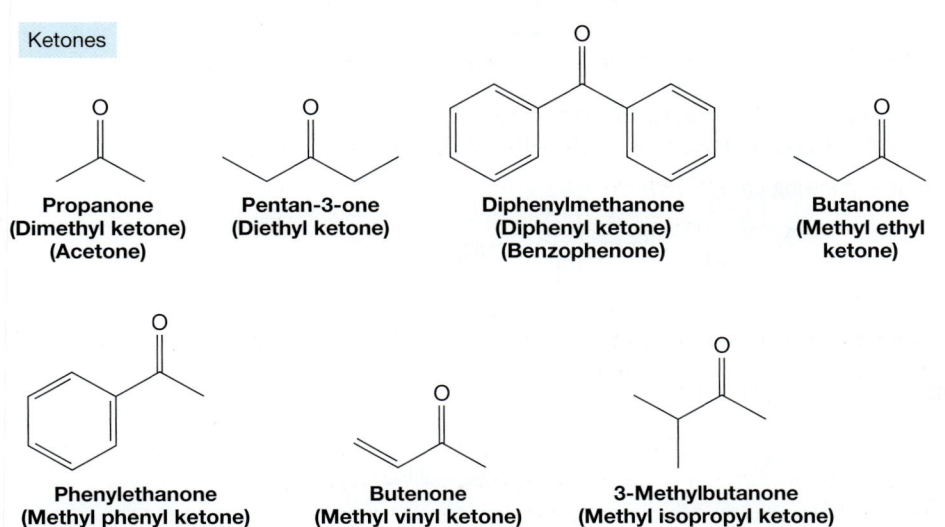

Ketones

Propanone
(Dimethyl ketone)
(Acetone)

Pentan-3-one
(Diethyl ketone)

Diphenylmethanone
(Diphenyl ketone)
(Benzophenone)

Butanone
(Methyl ethyl ketone)

Phenylethanone
(Methyl phenyl ketone)
(Acetophenone)

Butenone
(Methyl vinyl ketone)

3-Methylbutanone
(Methyl isopropyl ketone)

CONNECTIONS Butanone (methyl ethyl ketone) is often used as an industrial solvent in manufacturing plastics and textiles, as well as many household products. It is also commonly used as the solvent in dry-erase markers.

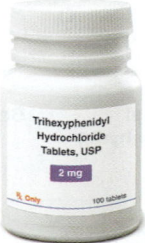

When the two alkyl groups are different, each alkyl group is written and they are separated by a space. Examples include methyl ethyl ketone (commonly abbreviated MEK), methyl phenyl ketone (MPK), methyl vinyl ketone (MVK), and methyl isopropyl ketone. Notice in the last two examples that the trivial names vinyl and isopropyl are used to identify the alkyl groups.

Some ketones have more than one trivial name—for example, propanone is also known as dimethyl ketone and acetone. Diphenylmethanone (trivial name: diphenyl ketone) is also known as benzophenone, and phenylethanone (trivial name: methyl phenyl ketone) is also known as acetophenone.

PROBLEM E.23 For each trivial name, draw the complete structure and provide the IUPAC name. **(a)** divinyl ketone; **(b)** benzyl isopropyl ketone; **(c)** cyclohexyl methyl ketone; **(d)** diisopropyl ketone; **(e)** isobutyl phenyl ketone

---

Problems

E.1 Alcohols and Amines

E.24 Provide the IUPAC name for each of the following alcohols.

(a)   **(b)**   **(c)**   **(d)**

E.25 Provide the IUPAC name for each of the following amines.

(a)   **(b)**   **(c)**   **(d)**

E.26 Draw the molecule that corresponds to each IUPAC name. **(a)** 3,3-dipropoxypentan-1-amine; **(b)** 2,3,4-trichlorocyclohexanol; **(c)** 3-cyclopropylpentan-1-ol; **(d)** 3-(1-methylethyl)cycloheptanamine

E.27 Provide the IUPAC name for each of the following compounds.

(a)   **(b)**   **(c)**

E.28 Provide the IUPAC name for each of the following compounds.

(a)   **(b)**   **(c)**

E.29 Draw the molecule that corresponds to each IUPAC name. **(a)** 5-amino-2,3,4-trimethylpentan-1-ol; **(b)** 3-amino-4,5-diethoxyoctan-1-ol; **(c)** 3,4-diamino-5-bromocyclohexanol; **(d)** 4-amino-3,3-diethylhexan-1-ol

E.30 Provide the IUPAC name for each of the following compounds.

(a)   **(b)**   **(c)**

E.31 Provide the IUPAC name for each of the following compounds.

(a)   **(b)**   **(c)**

E.32 Provide the IUPAC name for each of the following compounds. Pay close attention to stereochemistry.

(a)   **(b)**   **(c)**

E.33 Draw the molecule that corresponds to each IUPAC name. **(a)** (S)-hexan-3-amine; **(b)** (2R,4R)-4-aminopentan-2-ol; **(c)** (1S,3R,4S)-4-nitrocycloheptane-1,3-diamine

E.2 Ketones and Aldehydes

E.34 Provide the IUPAC name for each of the following ketones.

(a)   **(b)**   **(c)**   **(d)**

E.35 Provide the IUPAC name for each of the following aldehydes.

(a)   **(b)**   **(c)**   **(d)**   **(e)**

E.36 Draw the molecule that corresponds to each IUPAC name. **(a)** 2,3-dimethylcyclopentanone; **(b)** 4,4-difluoroheptanal; **(c)** 1,1,1-trichloropentan-3-one; **(d)** (1S,3S)-3-ethoxycyclohexanecarbaldehyde

E.37 Provide the IUPAC name for each of the following compounds.

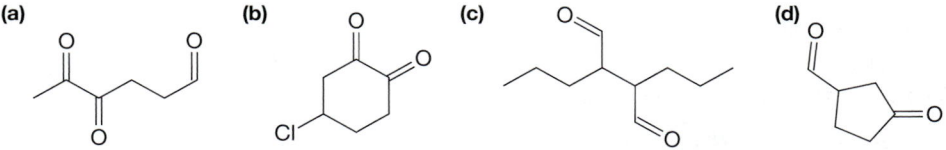

(a)   **(b)**   **(c)**   **(d)**

E.38 Provide the IUPAC name for each of the following compounds.

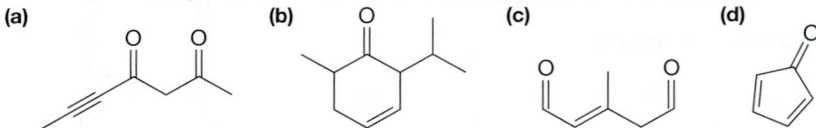

(a)   (b)   (c)   (d)

E.39 Draw the molecule that corresponds to each IUPAC name. **(a)** (Z)-hept-3-enedial; **(b)** 2,2-diethoxy-4-oxopentanal; **(c)** cyclohept-4-ene-1,3-dione; **(d)** pent-4-ynal

E.3 Trivial Names

E.40 Draw the molecule that corresponds to each trivial name and determine whether the alcohol is primary, secondary, or tertiary. **(a)** hexyl alcohol; **(b)** neopentyl alcohol; **(c)** pentafluoroethyl alcohol; **(d)** cyclohexyl alcohol

E.41 Draw the molecule that corresponds to each trivial name and determine whether the amine is primary, secondary, or tertiary. **(a)** diethylamine; **(b)** diethylpropylamine; **(c)** allylmethylamine; **(d)** cyclopentylamine; **(e)** dibenzylamine

E.42 Draw the molecule that corresponds to each trivial name. **(a)** di-*tert*-butyl ketone; **(b)** *tert*-butyl vinyl ketone; **(c)** dicyclopentyl ketone; **(d)** chloroacetone; **(e)** isopropyl phenyl ketone

E.43 Draw the molecule that corresponds to each trivial name. **(a)** trichloroacetaldehyde; **(b)** pentafluorobenzaldehyde; **(c)** isobutyraldehyde

Integrated Problems

E.44 Provide the IUPAC name for each of the following compounds.

(a)   (b)   (c)

E.45 Provide the IUPAC name for each of the following compounds.

(a)   (b)   (c)

E.46 Provide the IUPAC name for each of the following compounds. Pay attention to stereochemistry.

(a)   (b)   (c)

E.47 Provide the IUPAC name for each of the following compounds.

(a)   (b)

E.48 For each IUPAC name, draw the complete structure. **(a)** (2S,3R,4R)-2,3-diamino-4-hydroxycyclopentanone; **(b)** (2R,3R)-2-butyl-3-hydroxypentanedial; **(c)** (4S,5R)-4,5-diaminoheptane-2,3,6-trione; **(d)** (3E,5R,6Z)-5-hydroxy-7-methoxyocta-3,6-dien-2-one

Eventually, all of these dominos will fall, but they must do so in a particular order. Analogously, the overall reactions we will learn in this chapter—S_N1 and E1 reactions—take place via multistep mechanisms constructed from precise sequences of elementary steps.

8

An Introduction to Multistep Mechanisms

S_N1 and E1 Reactions and Their Comparisons to S_N2 and E2 Reactions

Many organic reactions take place by mechanisms that involve several steps. Fortunately, such **multistep mechanisms** are typically built up from just a handful of different *elementary steps*. We have already studied 10 of these elementary steps in Chapters 6 and 7. Just as letters of the alphabet can be combined in various ways to create a large number of different words, these elementary steps can be combined in various ways to create a large number of different reaction mechanisms.

Here in Chapter 8, we introduce two of the simplest multistep mechanisms: the unimolecular nucleophilic substitution (S_N1) reaction and the unimolecular elimination (E1) reaction. In doing so, we will also revisit the single-step S_N2 and E2 reactions to see the similarities and differences. Even though S_N1 and E1 reactions consist of just two steps each, by studying them we can learn much about multistep mechanisms in general, including those with three, four, five, or more steps.

We begin Chapter 8 by examining key aspects of S_N1 and E1 reactions—namely, their curved arrow notation, reaction energy diagrams, chemical kinetics (reaction rates), and stereoselectivity (the tendency to form one stereoisomer over another). We then present general guidelines for reasonable multistep mechanisms, which are intended to help you develop a certain level of "chemical intuition" that can be applied to other reaction mechanisms in subsequent chapters.

Chapter Objectives

On completing Chapter 8 you should be able to:

1. Draw the curved arrow notation and the products of a unimolecular nucleophilic substitution (S_N1) reaction, given a substrate and nucleophile.

2. Draw the curved arrow notation and the products of a unimolecular elimination (E1) reaction, given a substrate and base.

3. Distinguish intermediates from the overall reactants and overall products of a given reaction.

4. Construct a free energy diagram for an S_N1 or E1 reaction and identify key structures, such as overall reactants, overall products, intermediates, and transition states.

5. Describe the general types of evidence that can support a mechanism.

6. Distinguish S_N1 reactions from S_N2 reactions and E1 reactions from E2 reactions on the basis of their rate laws.

7. Explain how the mechanisms of S_N2, S_N1, E2, and E1 reactions are consistent with their respective rate laws.

8. Explain the differences in stereochemistry between S_N1 and S_N2 reactions and between E1 and E2 reactions, and predict which stereoisomers can be produced from each reaction.

9. Draw reasonable mechanisms for S_N2, S_N1, E2, and E1 reactions that include proton transfer steps.

10. Predict when a carbocation rearrangement will take place in a nucleophilic substitution or elimination reaction.

11. Incorporate resonance structures of a resonance-delocalized intermediate into a mechanism.

8.1 The Unimolecular Nucleophilic Substitution (S_N1) Reaction

In Section 7.2, we saw that a nucleophile (Nu^-) can replace a leaving group (L) on a substrate (R—L) in a single step via a bimolecular nucleophilic substitution (S_N2) reaction (Equation 8-1). However, we can also envision a nucleophilic substitution reaction taking place in two steps, as shown in Equation 8-2. This is an example of a **unimolecular nucleophilic substitution (S_N1) reaction** mechanism. (Note that the numbers in the names of the mechanisms do *not* correspond to the number of steps in the mechanism.)

S_N2 mechanism

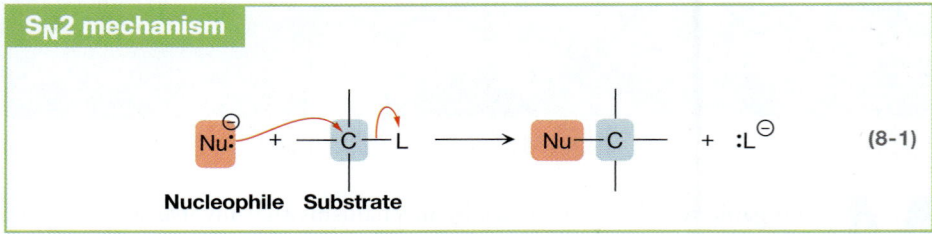

Nucleophile   Substrate

$$(8\text{-}1)$$

S_N1 mechanism

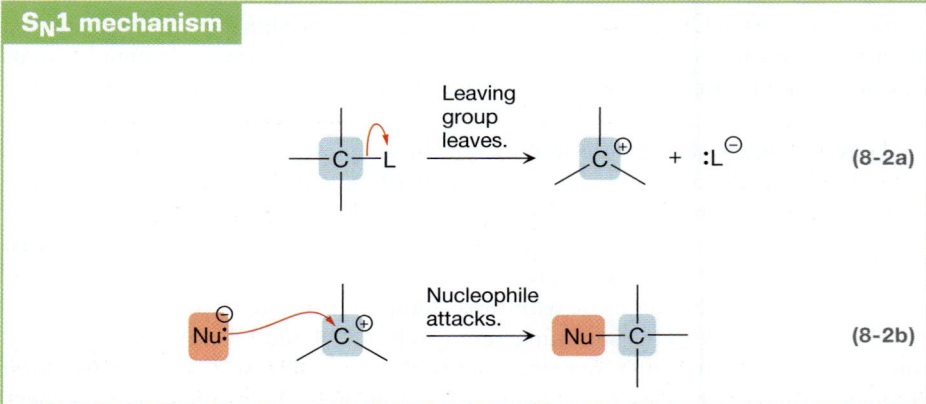

Leaving group leaves.

$$(8\text{-}2a)$$

Nucleophile attacks.

$$(8\text{-}2b)$$

Equation 8-2a shows that the leaving group (L) simply leaves in the first step of an S_N1 mechanism, producing L^- and a carbocation. In the second step (Equation 8-2b), the nucleophile (Nu^-) forms a bond to the carbocation (which is a very reactive electrophile) to complete the reaction. Even though this mechanism is different from what we have seen before, it is composed of elementary steps with which we are familiar (see Your Turn 8.1).

YOUR TURN 8.1

Under each arrow in Equation 8-2, write the name of the elementary step that is occurring. *Hint:* See Chapter 7.

Answers to Your Turns are in the back of the book.

Although a nucleophile replaces a leaving group in both the S_N1 and S_N2 mechanisms, there are important differences in their respective outcomes. We will see, for example, that S_N2 reactions tend to produce a single stereoisomer, whereas S_N1 reactions tend to produce mixtures of stereoisomers. We will also see that the carbon backbone can rearrange in an S_N1 mechanism, but it cannot in an S_N2 mechanism.

PROBLEM 8.1 Using curved arrow notation, draw **(a)** an S_N2 mechanism and **(b)** an S_N1 mechanism for the substitution reaction shown here.

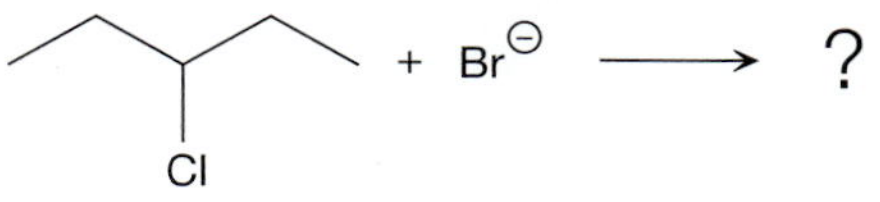

8.1a Overall Reactants, Overall Products, and Intermediates

In any reaction that takes place in multiple steps, we must be able to distinguish the *overall reaction* from the elementary steps of its mechanism. The **overall reaction** (or **net reaction**) describes the *net changes* that occur, and can be obtained by simply adding together all of the elementary steps.

An S_N1 mechanism

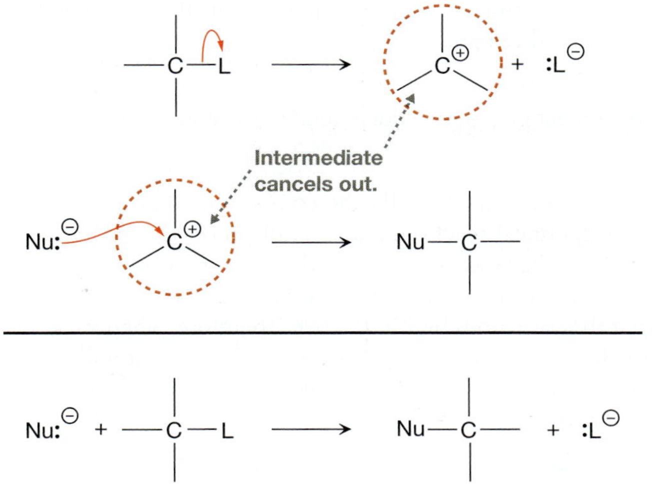

(8-2a)

(8-2b)

Overall reaction (8-3)

When we add together the steps of an S_N1 mechanism in Equation 8-2, for example, we cancel the carbocation species because it is a *product* in Equation 8-2a and then a *reactant* in Equation 8-2b. As a result, there is no net consumption or net production of the carbocation. The species that remain appear in the overall reaction, as shown in Equation 8-3.

Use the following definitions to distinguish between *overall reactants*, *overall products*, and *intermediates*.

- A species is an **overall reactant** or an **overall product** if it appears in the overall (i.e., net) reaction.
- A species is an **intermediate** if it is *produced* in a mechanism but does *not* appear in the overall reaction.

The overall reactants in Equation 8-3 are R—L and Nu⁻, whereas the overall products are Nu—R and L⁻. The carbocation (R⁺) that appears in the reaction mechanism (Equation 8-2) is the only intermediate.

YOUR TURN 8.2

In the S_N1 mechanism shown here, **(a)** identify each species as an *overall reactant*, *overall product*, or *intermediate*, then **(b)** sum the steps to yield the overall reaction.

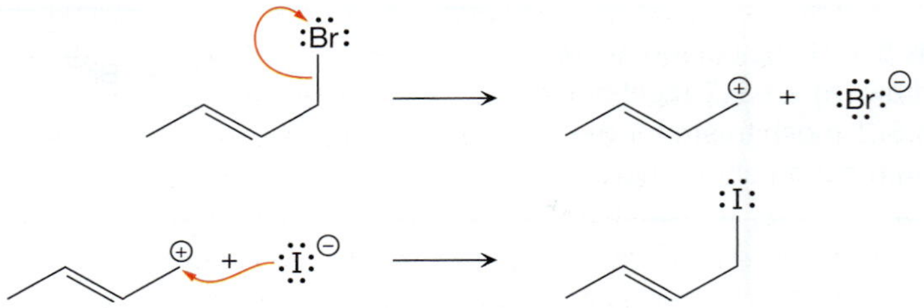

One reason to distinguish intermediates from overall reactants and products is to know which species in the mechanism can be isolated.

Intermediates usually are too reactive to be isolated.

They are intermediates in the first place because they tend to be highly reactive: As soon as one is formed as a product in one step, it is consumed as a reactant in the next step. On the other hand:

Overall reactants and products tend to be stable and can be isolated.

Overall reactants are typically the compounds that are physically added together to initiate a chemical reaction, and overall products are the newly formed compounds remaining in the reaction mixture when the reaction has come to completion.

Another reason it is important to be able to identify intermediates stems from the fact that they are often highly reactive. Therefore, when we know a species is a reactive intermediate, we often do not need to ask *whether* it will react, but instead can focus on *how* it will react. This, for example, is the focus of Section 8.6d on predicting carbocation rearrangements.

PROBLEM 8.2 For the S_N1 reaction in Problem 8.1, identify the overall reactants, overall products, and any intermediates.

8.1b Free Energy Diagram of an S_N1 Reaction

The free energy diagram for the S_N1 mechanism in Equation 8-2 is shown in Figure 8-1. Just as in any reaction free energy diagram, the Gibbs free energy of the

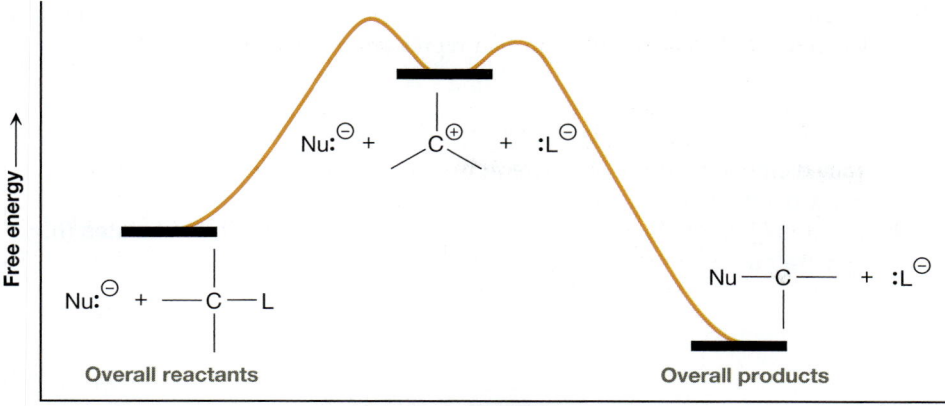

FIGURE 8-1 Free energy diagram for an S$_N$1 reaction The overall reactants are on the far left, the overall products are on the far right, and the intermediate appears at the local energy minimum in between.

species involved in the reaction is plotted as a function of the *reaction coordinate* (Section 6.3b). As the reaction coordinate increases from left to right, the species that are involved in the reaction less closely resemble the overall reactants and more closely resemble the intermediate or overall products. That is, the *overall reactants* are located on the far left and the *overall products* are located on the far right.

Unlike energy diagrams we have seen before, the one in Figure 8-1 has two humps connecting reactants to products, not one. This is because there are two separate elementary steps for the S$_N$1 mechanism. Each step is an individual reaction that has its own reactants and products and proceeds through a *transition state*; each transition state occurs at a local energy maximum along the reaction coordinate.

There are two characteristics of the intermediate worth noting:

1. The intermediate is higher in energy (less stable) than the overall reactants and the overall products.
2. The intermediate occurs at a **local minimum** in energy along the reaction coordinate, so energy increases when going either forward or backward from the intermediate along the reaction coordinate.

The energy of the reaction intermediate in Equation 8-2 is high because the C atom loses its octet and also because two additional charges are created in the process. Although the intermediate is itself highly unstable, even more destabilization is created when *partial bonds* are introduced as the reaction proceeds either forward or backward from the intermediate along the reaction coordinate.

YOUR TURN 8.3

In Figure 8-1, label the transition state for the first step of the mechanism "TS 1" and label the one for the second step "TS 2." Also, label the intermediate.

Because they are situated at local energy minima, intermediates can *theoretically* be isolated. We can think of the intermediate as "trapped" between the two energy barriers on either side. As noted previously, however, isolating an intermediate is very difficult because, compared to the energy difference between the intermediate and either the overall reactants or overall products, those energy barriers are typically quite small.

These observations can be generalized for any mechanism involving any number of steps.

For a mechanism that contains *n* total elementary steps, there must be *n* total transition states.

Because an intermediate is located in between two transition states, there must be $n - 1$ locations in the energy diagram that represent intermediates.

YOUR TURN 8.4

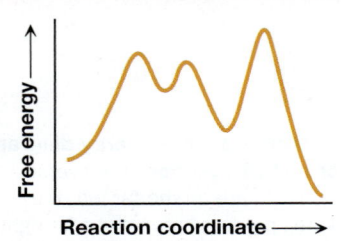

How many elementary steps are there in the mechanism represented by this free energy diagram? How many locations in the energy diagram represent intermediates? Mark the locations of the overall reactants (R), intermediates (I), transition states (TS), and overall products (P) on the diagram.

PROBLEM 8.3 Draw a detailed free energy diagram for the S_N1 reaction in Your Turn 8.2. Include and label the overall reactants, the overall products, the intermediate(s), and the transition state(s).

8.2 The Unimolecular Elimination (E1) Reaction

In Section 7.5, we saw that a leaving group and a proton (H^+) are eliminated from adjacent carbon atoms in a *bimolecular elimination (E2) reaction*, resulting in an additional π bond between those two carbon atoms. Equation 8-4 shows an E2 reaction between a generic base (B^-) and a generic substrate (R—L).

E2 mechanism

$$B:^{\ominus} + H-C-C-L \longrightarrow B-H + \;C=C\; + \;:L^{\ominus} \qquad (8\text{-}4)$$

Base   Substrate

Elimination reactions can also take place in two steps, via a **unimolecular elimination (E1) reaction**, as shown in Equation 8-5. As with S_N1 and S_N2 reactions, the numbers in the names of the elimination reactions do *not* refer to the number of steps in each mechanism.

E1 mechanism

Leaving group leaves.

$$H-C-C-L \longrightarrow H-C-C^{\oplus} + \;:L^{\ominus} \qquad (8\text{-}5a)$$

Base attacks.

$$B:^{\ominus} + H-C-C^{\oplus} \longrightarrow B-H + \;C=C\; \qquad (8\text{-}5b)$$

Notice that the first step in an E1 reaction (Equation 8-5a) is precisely the same as the first step in an S_N1 reaction (Equation 8-2a)—namely, the leaving group leaves. The difference between the S_N1 and E1 mechanisms is in the second step. Whereas the second step of an S_N1 mechanism is a *coordination* step (Equation 8-2b), H^+ is eliminated in the second step of an E1 mechanism (Equation 8-5b), resulting in a π bond.

Just as we did with the S_N1 mechanism, we can obtain the overall reactants and products of an E1 mechanism by adding together the steps in Equation 8-5, yielding Equation 8-6. Notice once again that the carbocation is an intermediate in the E1 mechanism because it does not appear in the overall reaction.

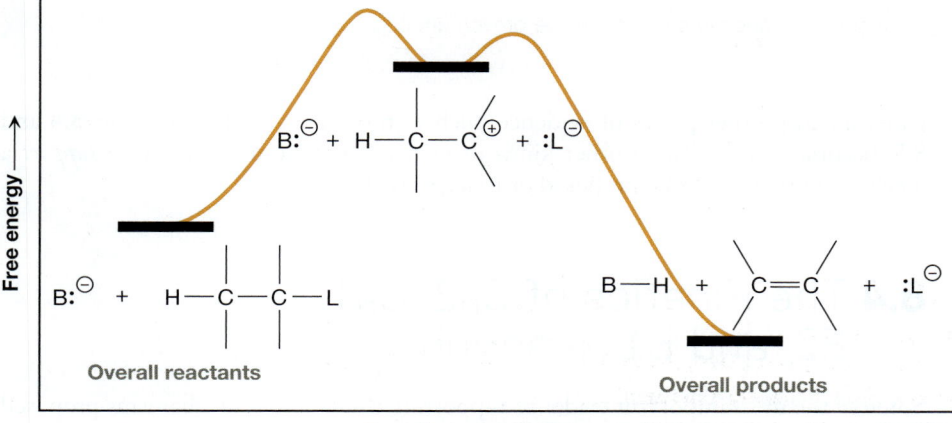

PROBLEM 8.4 Draw the mechanism for the E1 reaction that would take place between the two species shown here. Label each species in the mechanism as an overall reactant, an overall product, or an intermediate.

The free energy diagram for an E1 reaction is shown in Figure 8-2. It is strikingly similar to the one in Figure 8-1 for an S_N1 reaction. Notice that there are two transition states (energy maxima), consistent with the fact that the E1 mechanism consists

FIGURE 8-2 Free energy diagram for an E1 reaction The overall reactants are on the far left, the overall products are on the far right, and the intermediate appears at the local energy minimum in between.

of two separate steps. Furthermore, there is a single intermediate, appearing at a *local energy minimum* between the two transition states. Finally, the energy of the intermediate is substantially higher than either the reactants or products due to the loss of an octet on carbon and the appearance of charges.

YOUR TURN 8.7

In Figure 8-2, label the transition state for the first step of the mechanism "TS 1" and label the one for the second step "TS 2." Also, label the intermediate.

PROBLEM 8.5 Draw the energy diagram for the E1 reaction in Problem 8.4. Include and label the overall reactants, the overall products, the transition state(s), and the intermediate(s).

8.3 Direct Experimental Evidence for Reaction Mechanisms

So far in this chapter we have seen that S_N1 products are similar to S_N2 products and that E1 products are similar to E2 products. What evidence, then, do we have for each mechanism? How do we know that the mechanisms we have presented are "correct"?

Perhaps the most powerful evidence for the correctness of a reaction mechanism is the direct observation of the intermediates that are proposed. For both S_N1 and E1 reactions, for example, we would try to directly observe the carbocation that is presumed to be an intermediate. As mentioned previously, however, intermediates (e.g., carbocations) tend to be very reactive and have very short lifetimes. The direct observation of intermediates, therefore, is a formidable task and quite rare.

In 1962, however, George A. Olah, a Hungarian-born American chemist, directly observed the *tert*-butyl carbocation, $(CH_3)_3C^+$, using *nuclear magnetic resonance spectroscopy* (an important technique that we discuss in Chapter 16). Olah dissolved 2-fluoro-2-methylpropane, $(CH_3)_3C$—F, in a *superacid*, which provided the extreme conditions necessary for the carbocation to be long-lived enough for observation.

Key intermediates in a few other reactions have been directly observed as well. Most intermediates, however, do not lend themselves to direct observation, so:

In general, mechanisms cannot be proven directly.

Consequently, other pieces of evidence, such as those described in Sections 8.4 and 8.5, become critical. These other kinds of evidence can be used either to *support* a mechanism that might be proposed or to *disprove* it.

8.4 The Kinetics of S_N2, S_N1, E2, and E1 Reactions

Some of the most important evidence supporting the different mechanisms proposed for S_N1 and S_N2 nucleophilic substitution reactions, as well as for the E1 and E2 elimination reactions, comes from **reaction kinetics**, the study of *reaction rates*. Here in Section 8.4, we discuss differences in the *rate laws* for the different reactions, which, as we will see, correspond to the differences in their respective free energy diagrams.

8.4a Empirical and Theoretical Rate Laws for the S_N2, S_N1, E2, and E1 Reactions

The rates of S_N1 and S_N2 reactions can be determined experimentally by measuring the amount of product that forms over time. As it turns out, the S_N1 and S_N2 reaction rates differ in the way they depend on reactant concentrations. We can see this difference from the experimentally determined rate laws, or **empirical rate laws**, for the two reactions, shown in Equations 8-7 and 8-8.

$$\text{Rate } (S_N2) = k_{S_N2}[Nu^-][R-L] \tag{8-7}$$

$$\text{Rate } (S_N1) = k_{S_N1}[R-L] \tag{8-8}$$

Notice in Equation 8-7 that the rate of an S_N2 reaction is directly proportional to both the nucleophile ($[Nu^-]$) and substrate ($[R-L]$) concentrations. The rate of an S_N1 reaction, on the other hand, is directly proportional only to the substrate concentration ($[R-L]$). In other words:

> The rate of an S_N2 reaction is *directly proportional to* the concentration of the nucleophile, whereas the rate of an S_N1 reaction is *independent of* the concentration of the nucleophile!

The proportionality constant in Equation 8-7, k_{S_N2}, is the **rate constant** for the S_N2 reaction. Similarly, the rate constant for the S_N1 reaction is denoted k_{S_N1} in Equation 8-8.

YOUR TURN 8.8

In the S_N2 reaction of $BrCH_2CH_2CH_2CH_3$ with $NaSCH_3$, how would the reaction rate be affected if the concentration of $NaSCH_3$ were doubled?

SOLVED PROBLEM 8.6

A chemist wants to determine whether the reaction shown here proceeds via an S_N1 or S_N2 reaction. She therefore decided to carry out kinetics experiments, measuring the dependence of the reaction rate on the initial concentrations of each reactant. Her results are provided in the following table:

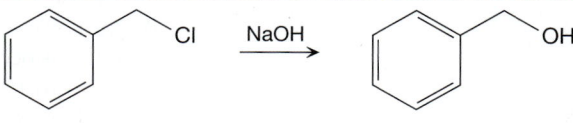

Trial Number	$[C_6H_5CH_2Cl]$	$[HO^-]$	Rate (M/s)
1	0.10 M	0.10 M	2.6×10^{-6}
2	0.10 M	0.20 M	5.1×10^{-6}
3	0.20 M	0.20 M	1.0×10^{-5}

Based on these results, does the reaction proceed by the S_N1 or S_N2 mechanism?

Think What is the dependence of the reaction rate on each of the reactants? What is the corresponding empirical rate law? Is it consistent with the rate law for an S_N1 or an S_N2 reaction?

Solve In going from Trial 1 to Trial 2, the concentration of HO$^-$ (the nucleophile) doubles, but the concentration of $C_6H_5CH_2Cl$ remains the same. This causes the reaction rate to double—from 2.6×10^{-6} M/s to 5.1×10^{-6} M/s—suggesting that the rate is directly proportional to the concentration of the nucleophile. Similarly, in going from Trial 2 to Trial 3, the concentration of $C_6H_5CH_2Cl$ (the substrate) doubles, but the concentration of HO$^-$ remains the same, and the rate also doubles. Thus, the rate is directly proportional to the concentration of the substrate as well. These results indicate an empirical rate law of the form Rate $= k[C_6H_5CH_2Cl][HO^-]$, which is consistent with an S_N2 reaction.

PROBLEM 8.7 Rate data for the following substitution reaction are presented in the accompanying table.

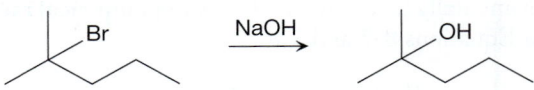

Trial Number	[R—Br]	[HO$^-$]	Rate (M/s)
1	0.10 M	0.10 M	5.5×10^{-7}
2	0.10 M	0.05 M	5.5×10^{-7}
3	0.20 M	0.10 M	1.0×10^{-6}

Are the data consistent with an S_N1 reaction or an S_N2 reaction?

Why do the rates of S_N2 and S_N1 reactions have different dependences on reactant concentrations? To help answer this question, we introduce one of the foundational principles of chemical kinetics:

For an *elementary step*, $aA + bB \rightarrow$ products, where A and B are reactants and a and b are the coefficients required to balance the chemical equation, the **theoretical rate law** is:

$$\text{Rate}_{\text{elementary step}} = k_{\text{elementary step}} [A]^a[B]^b \qquad (8\text{-}9)$$

The quantities [A] and [B] are the concentrations of the reactants, and their exponents are the **orders** of the reaction with respect to the reactants A and B. That is, a is the order of the reaction with respect to A, and b is the order of the reaction with respect to B. Importantly, a and b are the same as the coefficients used to balance the elementary step. This is because the reactants must collide at effectively the same time for an elementary step to take place, and the frequency with which that happens is proportional to $[A]^a[B]^b$. Note that Equation 8-9 is referred to as a *theoretical* rate law (not an *empirical* rate law) because it is derived from a proposed elementary step, not from an experiment.

For an S_N2 reaction, which is an elementary step itself, there are two reactant species—the nucleophile and the substrate. According to Equation 8-9, therefore, the concentrations of both reactants should appear in the theoretical rate law. Moreover, the exponent for each reactant concentration should be 1, the coefficient required to balance the corresponding species in the reaction. Thus, the theoretical rate law should appear as in Equation 8-10, which matches the rate law obtained experimentally (Equation 8-7).

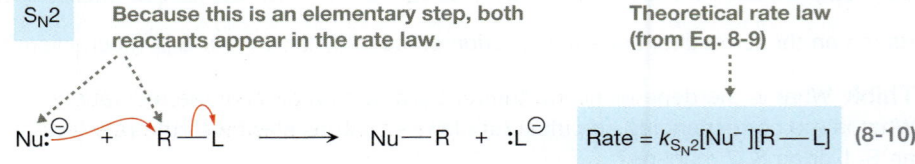

For an S_N1 reaction, Equation 8-9 can be applied to each of the two elementary steps, as shown in Equations 8-11a and 8-11b.

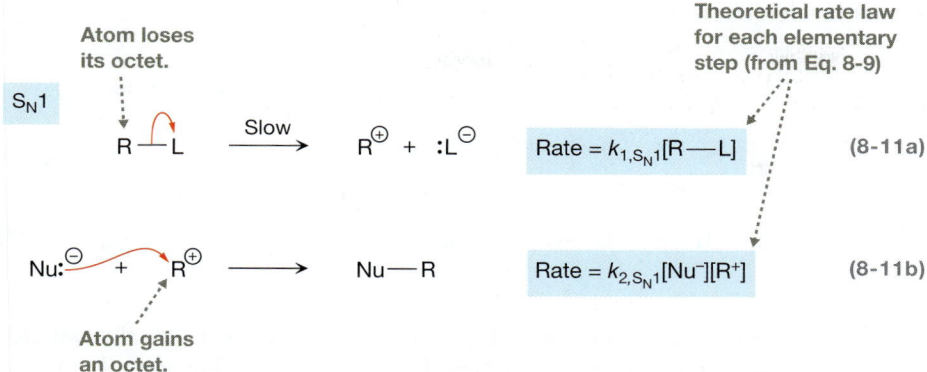

The first step is **unimolecular** (its *molecularity* is 1) because only one reactant species appears—the substrate, R—L. According to Equation 8-9, the rate for that step should therefore depend on the concentration of R—L only, as indicated by its theoretical rate law in Equation 8-11a. The second step, on the other hand, is *bimolecular* (its molecularity is 2), because it involves the nucleophile and the newly formed carbocation as reactants. Consequently, the rate for that step depends on the concentrations of both species, as we can see from its theoretical rate law in Equation 8-11b.

The theoretical rate law for the first step of an S_N1 reaction (Equation 8-11a) is essentially identical to the empirical rate law, shown previously in Equation 8-8. In other words:

> The first step of an S_N1 reaction establishes the rate of the overall reaction, and is thus the **rate-determining step**.

In fact, the overall reaction is called a *unimolecular* nucleophilic substitution reaction because the rate-determining step is unimolecular.

To say that the first step of the S_N1 mechanism is rate determining means that the second step converts the carbocation intermediate into products essentially as fast as the first step produces the intermediate. Often, chemists will describe this situation by saying that the first step is the "slow" step, as indicated in Equation 8-11.

To understand this idea better, consider the analogy of water flowing through a system of pipes. If a narrower pipe is connected to a wider pipe, as shown in Figure 8-3, then it is the width of the narrower pipe that dictates the rate at which water can flow through the entire system. The narrower pipe is analogous to the slow step—that is, to the rate-determining step.

Like S_N2 and S_N1 reactions, E2 and E1 reactions differ in their empirical rate laws. An E2 reaction is first order with respect to both the base and the substrate (Equation 8-12), whereas an E1 reaction is first order with respect to the substrate only (Equation 8-13).

$$\text{Rate (E2)} = k_{E2}[\text{Base}][\text{R—L}] \qquad (8\text{-}12)$$

$$\text{Rate (E1)} = k_{E1}[\text{R—L}] \qquad (8\text{-}13)$$

In other words:

> The rate of an E2 reaction is directly proportional to the concentration of the base, whereas the rate of an E1 reaction is independent of the concentration of the base.

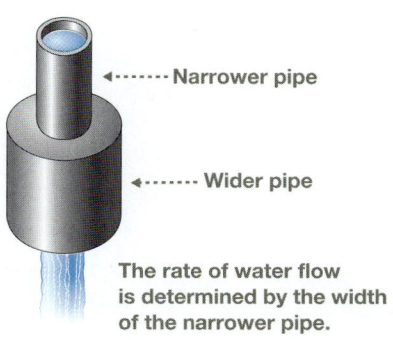

FIGURE 8-3 **Rate-determining steps** The narrower and wider pipes represent slower and faster elementary steps, respectively. Just as the rate that water can flow through the system of pipes is governed by the narrower pipe, the rate of an overall reaction is governed by its slowest step—that is, the rate-determining step.

The rate law for the E2 reaction in Equation 8-12 is consistent with a single-step mechanism involving both the base and the substrate, as indicated in Equation 8-14. According to Equation 8-9, the rate should be directly proportional to the concentration of each reactant.

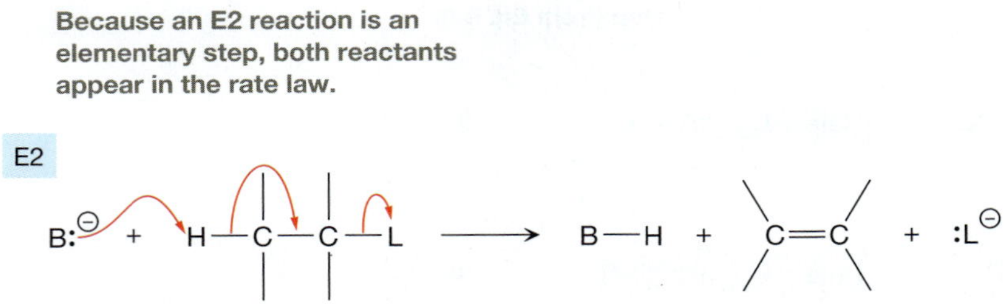

Because an E2 reaction is an elementary step, both reactants appear in the rate law.

Theoretical rate law (from Eq. 8-9)

E2

$$\text{Rate} = k_{E2}[B^-][R\!-\!L] \qquad (8\text{-}14)$$

The rate law for the E1 reaction in Equation 8-13 is consistent with the two-step mechanism shown in Equation 8-15, where the first step is the slow step. That is:

The first step of an E1 reaction is the rate-determining step.

E1

Atom loses its octet.

Theoretical rate law for each elementary step (from Eq. 8-9)

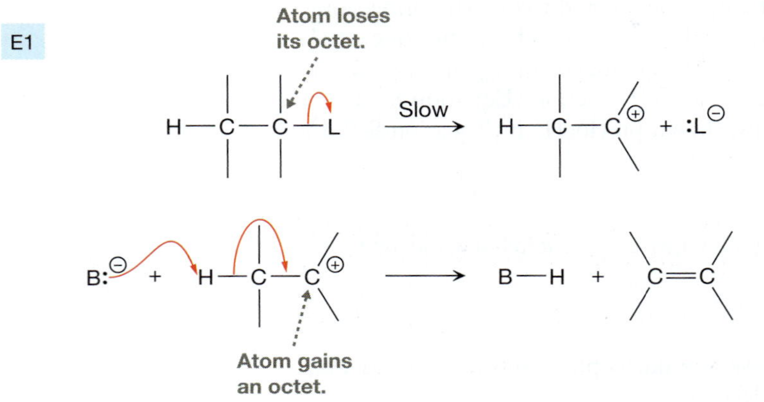

$$\text{Rate} = k_{1,E1}[R\!-\!L] \qquad (8\text{-}15a)$$

$$\text{Rate} = k_{2,E1}[B^-][R^+] \qquad (8\text{-}15b)$$

Atom gains an octet.

As with the S_N1 reaction, the E1 reaction is said to be *unimolecular* because its rate-determining step is unimolecular.

SOLVED PROBLEM 8.8

Suppose that the first step of an S_N1 reaction were much faster than the second step. Would the resulting theoretical rate law for the overall reaction agree or disagree with the observed dependence of the S_N1 reaction on the concentration of the nucleophile?

Think What would the rate-determining step be under this assumption? What is the theoretical rate law for that step?

Solve If the first step of an S_N1 reaction (Equation 8-11a) were much faster than the second step (Equation 8-11b), then the second step would be the rate-determining step for the reaction. The theoretical rate law for the overall reaction would simply be that for the second step—namely:

$$\text{Rate} = k_{2,S_N1}[Nu^-][R^+]$$

Thus, the reaction rate would be directly proportional to $[Nu^-]$. This contradicts the observed S_N1 reaction rate, which is independent of $[Nu^-]$.

PROBLEM 8.9 Suppose that the first step of an E1 reaction were much faster than the second step. Would the resulting theoretical rate law for the overall reaction agree or disagree with the observed dependence of the E1 reaction on the base concentration? Explain.

8.4b Energy Barriers and Rate Constants: Transition State Theory

In Section 8.4a, we learned that the first step of an S_N1 reaction tends to be the slow step. Likewise, the first step of an E1 reaction tends to be the slow step. We can see why this should be so from each reaction's free energy diagram, shown previously in Figures 8-1 and 8-2. Both diagrams resemble Figure 8-4, which depicts two separate steps, the first of which leads to the formation of a high-energy (unstable) intermediate.

To form that intermediate, the first step proceeds through a transition state that is particularly unstable, which gives rise to a large energy barrier, or *standard free energy of activation* ($\Delta G^{\circ\ddagger}$, Section 6.3b), and makes the step rather slow. Moreover, because the transition state for the first step is the highest energy point along the path from reactants to products, $\Delta G^{\circ\ddagger}$ for Step 1 essentially establishes the energy barrier for the overall reaction, making Step 1 the rate-determining step.

We can gain better insight into this idea by examining Equation 8-16, a result from **transition state theory**.

$$k_{\text{elementary step}} = CT\, e^{-(\Delta G^{\circ\ddagger}/RT)} \tag{8-16}$$

Here, $k_{\text{elementary step}}$ is the rate constant for an elementary step, C is a constant, T is the absolute temperature (in kelvin), $\Delta G^{\circ\ddagger}$ is the standard Gibbs free energy of activation, and R is the universal gas constant.

Because $\Delta G^{\circ\ddagger}$ appears in the numerator of a negative exponent in Equation 8-16:

The rate constant ($k_{\text{elementary step}}$) decreases as $\Delta G^{\circ\ddagger}$ increases, so the corresponding reaction rate decreases.

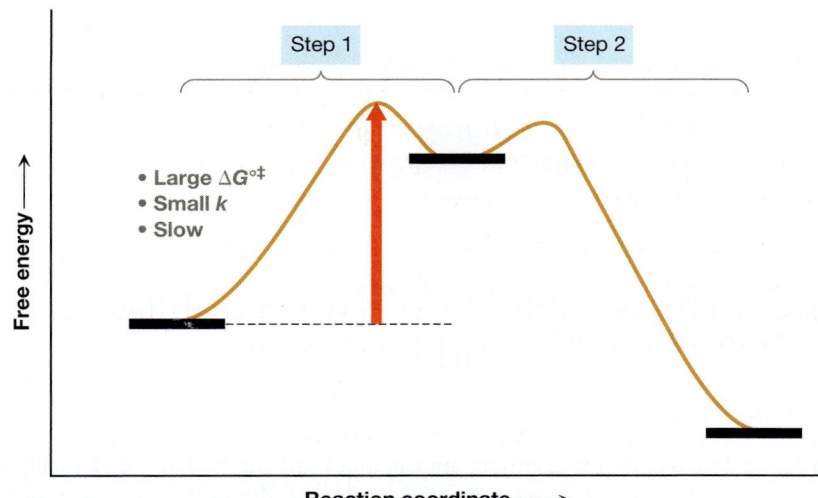

- Large $\Delta G^{\circ\ddagger}$
- Small k
- Slow

Step 1   Step 2

Free energy →

Reaction coordinate →

FIGURE 8-4 $\Delta G^{\circ\ddagger}$ and k The transition state for Step 1 is very high in energy, resulting in a really large energy barrier, $\Delta G^{\circ\ddagger}$. Step 1, therefore, has a small rate constant, k, and is slow.

FIGURE 8-5 **Percentage of molecules able to surmount an energy barrier ($\Delta G^{\circ\ddagger}$) as a function of the size of the energy barrier and the temperature** The percentage of molecules able to surmount an energy barrier is plotted against the size of the energy barrier for two different temperatures: 0 °C (blue curve) and 100 °C (red curve). The region inside the box is expanded in the inset. Notice that as the size of the energy barrier increases, the percentage of molecules having enough energy to react decreases dramatically. Notice also that for a given energy barrier, increasing the temperature significantly increases the percentage of molecules able to react.

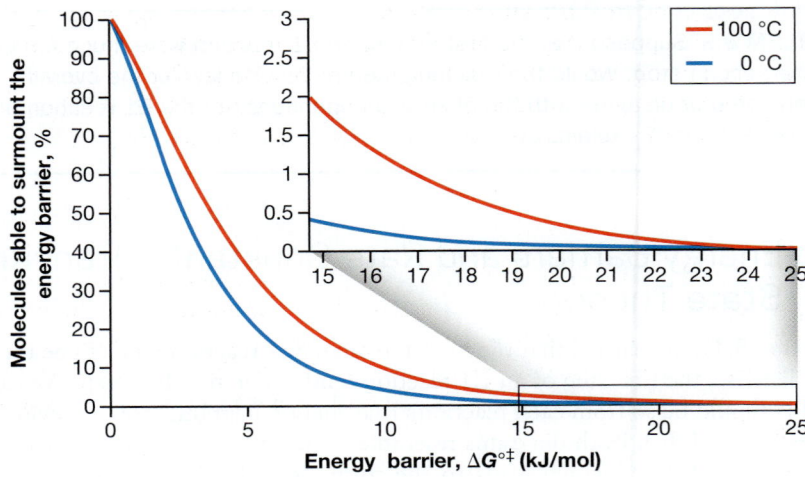

Temperature (T), on the other hand, appears in the denominator of the negative exponent and as part of the *pre-exponential factor*—the collection of terms (i.e., CT) that multiply the exponential factor. Consequently:

The rate constant ($k_{\text{elementary step}}$) increases as T increases, so the reaction rate increases.

The rate constant depends on $\Delta G^{\circ\ddagger}$ and T in these ways because, at a particular temperature, only a certain percentage of molecules possess enough energy to surmount an energy barrier of a given size (Section 4.3). As shown in Figure 8-5, the percentage of molecules with sufficient energy decreases as the size of the energy barrier increases (left to right). That percentage increases as temperature increases (blue curve to red curve), on the other hand, because a higher temperature supplies the reactants with greater kinetic energy.

YOUR TURN 8.9

Using Figure 8-5, estimate the percentage of molecules at 100 °C having enough energy to surmount an energy barrier of **(a)** 15 kJ/mol and **(b)** 25 kJ/mol.

YOUR TURN 8.10

Using Figure 8-5, estimate the percentage of molecules having enough energy to surmount an energy barrier of 20 kJ/mol at **(a)** 0 °C and **(b)** 100 °C.

8.5 Stereochemistry of Nucleophilic Substitution and Elimination Reactions

From Equations 8-1 and 8-2, it may appear that both the S_N2 and S_N1 mechanisms yield the same products. Likewise, from Equations 8-4 and 8-5, it may appear that

E2 and E1 reactions yield the same products. However, depending on the particular nature of the reaction, there can be subtle but important differences. One way in which the reaction products can differ is in their stereochemical configurations. Thus, the **stereochemistry** of an S_N1 reaction differs from that of an S_N2 reaction, and the stereochemistry of an E1 reaction differs from that of an E2 reaction.

8.5a Stereochemistry of an S_N1 Reaction

Stereochemistry is pertinent to an S_N1 reaction when the atom that is attached to the leaving group is a *chiral center*, because the breaking and formation of bonds to the chiral center can impact the stereochemical configuration of that atom.

If an S_N1 reaction is carried out on a stereochemically pure substrate, such as the one shown in Equation 8-17, then a *mixture* of both the *R* and *S* enantiomers is produced.

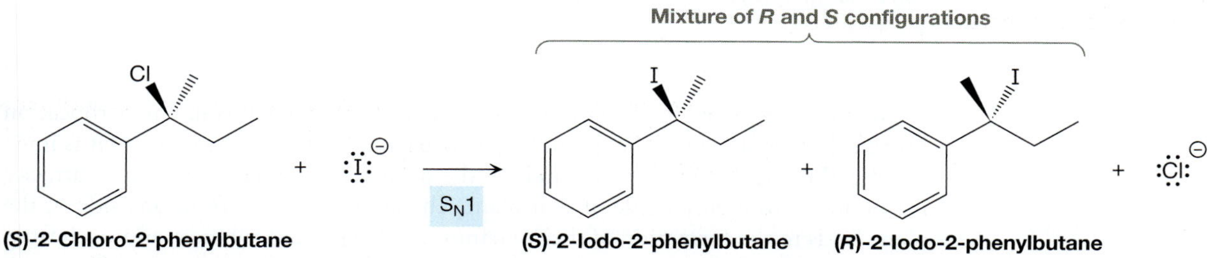

(S)-2-Chloro-2-phenylbutane   **(S)-2-Iodo-2-phenylbutane**   **(R)-2-Iodo-2-phenylbutane**

Why does the reaction produce both configurations of the chiral center?

To answer this question, let's look at the mechanism, shown in Equation 8-18.

Mechanism for Equation 8-17

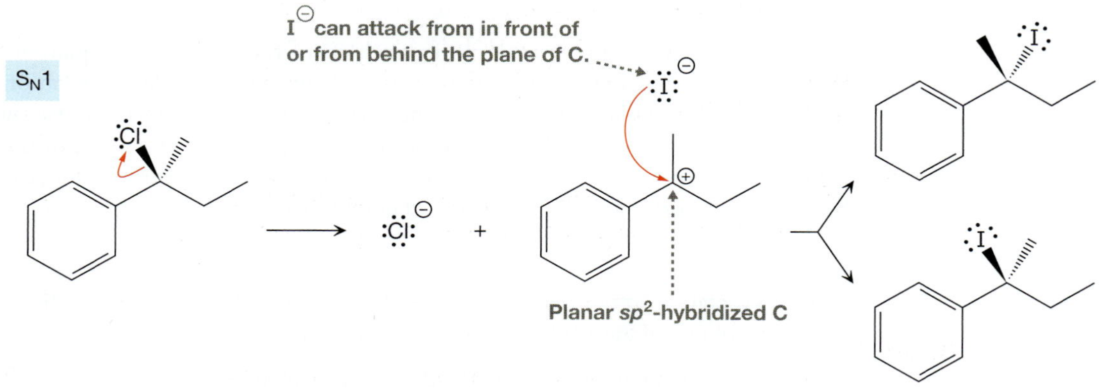

In the first step, Cl⁻ simply departs, leaving behind a planar carbocation. The C atom that was initially bonded to Cl becomes sp^2 hybridized in the carbocation and is no longer a chiral center (i.e., the stereochemistry of that C atom has been lost). In the second step, in which I⁻ forms a bond to the carbocation, that same C atom becomes a chiral center once again.

YOUR TURN 8.11

Label each organic species in Equation 8-18 as either *chiral* or *achiral*.

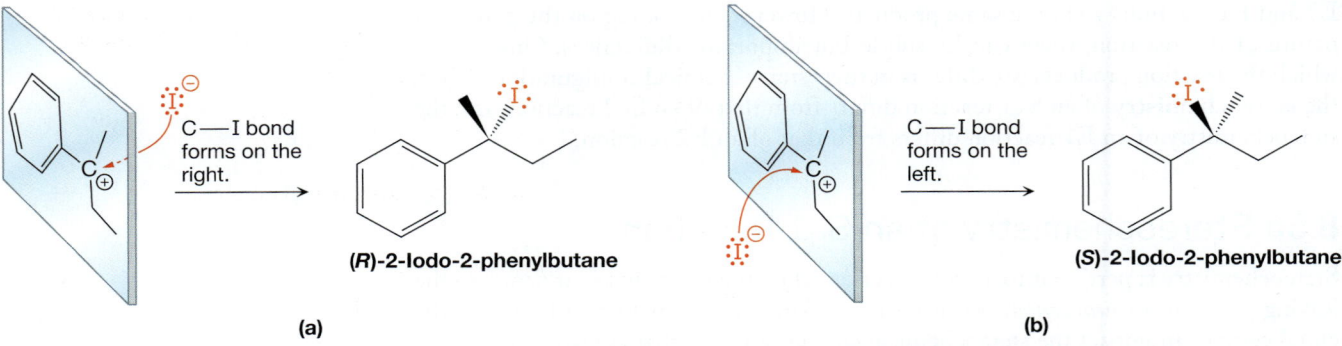

(R)-2-Iodo-2-phenylbutane

(S)-2-Iodo-2-phenylbutane

(a)

(b)

FIGURE 8-6 Generating a mixture of stereoisomers in an S_N1 reaction The specific stereoisomer that is formed in the reaction in Equation 8-17 is determined by the approach of the I^- nucleophile. (a) I^- approaches from the right, generating the new C—I bond shown, thus producing the *R* configuration. (b) I^- approaches from the left, generating the new C—I bond shown, thus producing the S enantiomer.

In Figure 8-6, we focus on the second step of the mechanism. The carbocation intermediate is shown at the left in Figure 8-6a and 8-6b, and the plane that is indicated is the one defined by the bonds to the positively charged C. When I^- attacks, it can do so from either side of that plane. The approach of I^- from one side of the plane leads to the formation of the *R* enantiomer (Fig. 8-6a), whereas approach from the other side of the plane leads to the formation of the *S* enantiomer (Fig. 8-6b). This idea can be summarized as follows:

> If a chiral center is generated in a single elementary step from an atom that is not a chiral center, then both *R* and S configurations will usually be produced.

You might expect the enantiomeric products of the reaction in Equation 8-17 to be produced in equal amounts—that is, as a *racemic mixture*. The carbocation shown in Figure 8-6 has a plane of symmetry and is therefore achiral, in which case the attack of the I^- nucleophile would appear to be equally likely from either side of the carbocation's plane. Indeed, if the nucleophile were to attack the *free* carbocation, this would be the case, according to the following general rules:

> If a new chiral center is produced in an elementary step, then the *R* and S configurations of the chiral center are produced in:
>
> - *Equal* amounts if the reactants and the environment are *achiral*.
> - *Unequal* amounts if the reactants or the environment are *chiral*.

However, the nucleophile does not attack the free carbocation in S_N1 reactions like this one. After the bond to the leaving group is broken in the first step, the leaving group remains associated with the side of the carbon from which it was initially attached, forming what is called an **ion pair**. The ion pair produced in the reaction in Equation 8-17 retains some of the chiral character from the original substrate, so effectively there is no plane of symmetry when the nucleophile attacks. This results in an *unequal* likelihood of nucleophilic attack from the two sides of the C atom, producing a mixture that is close to, but not completely, racemic.

The stereochemistry of the S_N1 reaction in Equation 8-19 is more clear-cut.

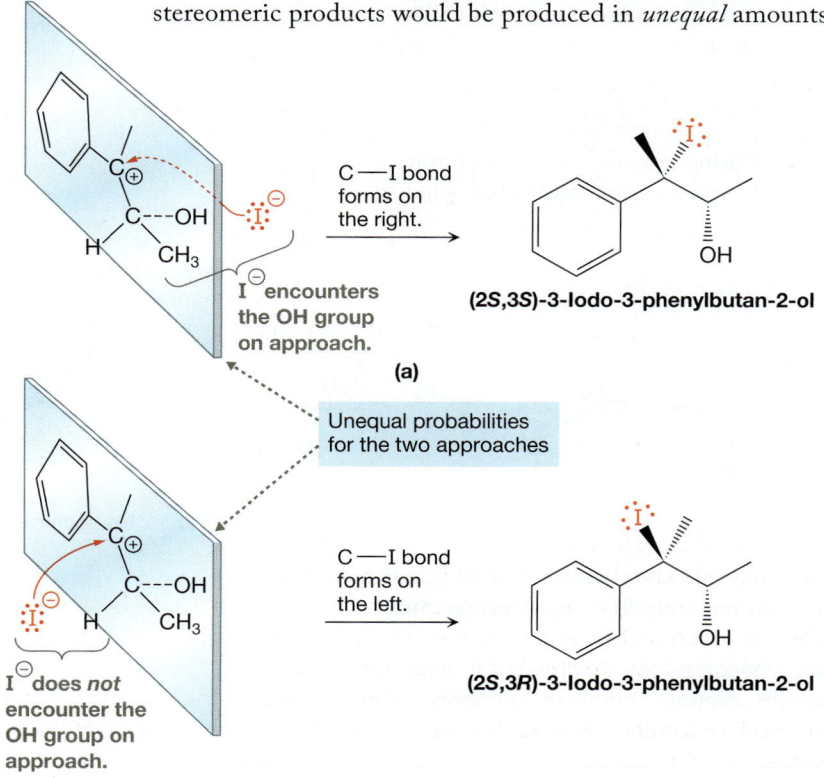

S_N1

I⊖ can attack from in front of or from behind the plane of C...

Planar C

Chiral carbocation

Unequal mixture of diastereomers

(2S,3S)-3-Iodo-3-phenylbutan-2-ol   (8-19a)

(2S,3R)-3-Iodo-3-phenylbutan-2-ol   (8-19b)

Here, the overall reactant has two chiral centers: the C atom bonded to the Cl leaving group and the C atom bonded to OH. The stereochemical configuration of the C bonded to OH remains unchanged because the reaction takes place only at the C atom bonded to Cl. Just as in Equation 8-18, though, the C atom bonded to Cl reacts to form both the *R* and *S* configurations in the product. As indicated in Equation 8-19, the result is a mixture of two *diastereomers*.

Unlike the previous situation, the carbocation intermediate shown in Equation 8-19 has a single chiral center, so it is unambiguously chiral and has no plane of symmetry. Without a plane of symmetry in the carbocation, the nucleophile will be influenced differently depending on which side of the carbocation it approaches. For the conformation shown in Figure 8-7, the nucleophile would encounter the OH group on the right side (Fig. 8.7a) but not the left (Fig. 8.7b), so the likelihood of each approach would be unequal and the diastereomeric products would be produced in *unequal* amounts.

C—I bond forms on the right.

I⊖ encounters the OH group on approach.

(2S,3S)-3-Iodo-3-phenylbutan-2-ol

(a)

Unequal probabilities for the two approaches

C—I bond forms on the left.

I⊖ does *not* encounter the OH group on approach.

(2S,3R)-3-Iodo-3-phenylbutan-2-ol

(b)

FIGURE 8-7 Formation of an unequal mixture of diastereomers in an S_N1 reaction The OH group is present only on one side of the plane, so whether it interacts with an approaching I⁻ depends on the direction from which the ion arrives. (a) If the I⁻ nucleophile comes from the right, it encounters the OH group. (b) If the I⁻ nucleophile comes from the left—the side opposite the OH group—it avoids the OH group. The two possible resulting reactions thus occur with different probabilities, producing an unequal mixture of diastereomers.

Determine which of carbocations **A–C** will undergo coordination with I⁻ to produce a single product, an equal mixture of stereoisomers, or an unequal mixture of stereoisomers. (Assume each carbocation is free from any leaving group.)

A   **B**   **C**

You may be tempted to try to predict which diastereomer is produced in greater abundance. We caution against it, however, because the factors that favor the production of one configuration over another can be subtle, involving the interplay between steric effects and electronic effects (such as hydrogen bonding, inductive effects, and polarizability). As a result, we limit such discussions to reactions in which these factors are clear-cut.

SOLVED PROBLEM 8.10

Draw the complete, detailed mechanism for this reaction, assuming that it takes place via the S_N1 mechanism. Will the reaction produce a single stereoisomer, or will it produce a mixture of stereoisomers? If it produces a mixture, then will the isomers be produced in equal amounts?

Think What can act as the leaving group? What can act as the nucleophile? What are the steps in an S_N1 mechanism? Does the reaction take place at a chiral center?

Solve The Br substituent can act as a leaving group, coming off as Br⁻, which is a relatively stable species. The HS⁻ ion can act as a nucleophile. The S_N1 mechanism takes place in two steps. First, the leaving group leaves via heterolysis, yielding a planar carbocation, then HS⁻ attacks C⁺ via a coordination step, yielding the overall products.

Only this stereoisomer is formed.

Chiral center   Not a chiral center   Chiral center   Not a chiral center

Notice in the alkyl halide reactant that there is a single chiral center, but no bonds to it are affected throughout the course of the reaction. Therefore, its stereochemical configuration in the product is the same as in the reactant. Notice also in the second step of the reaction that the planar C⁺ becomes tetrahedral, but it does *not* become a chiral center—it is bonded to two CH_3 groups. Therefore, stereochemistry is not a concern at that C, and the only stereoisomer formed is the one shown.

PROBLEM 8.11 Draw the complete mechanism and products for an S_N1 reaction between the following reactants. Will the reaction produce a single stereoisomer, or will it produce a mixture of stereoisomers? If it produces a mixture, will those isomers be produced in equal amounts?

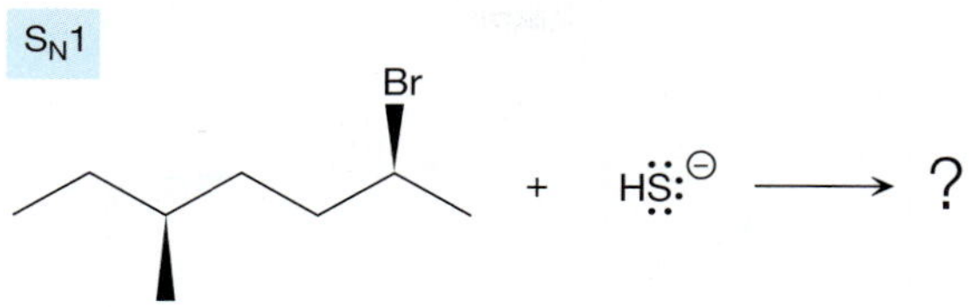

8.5b Stereospecificity of an S_N2 Reaction

Previously in Section 8.5 we said that the S_N1 and S_N2 reactions have different stereo-chemistries. This is evident in Equation 8-20, which shows the outcome of an S_N2 reaction between I$^-$ and (S)-1-chloro-1-phenylethane.

Only this stereoisomer is formed.

(S)-1-Chloro-1-phenylethane   (R)-1-Iodo-1-phenylethane

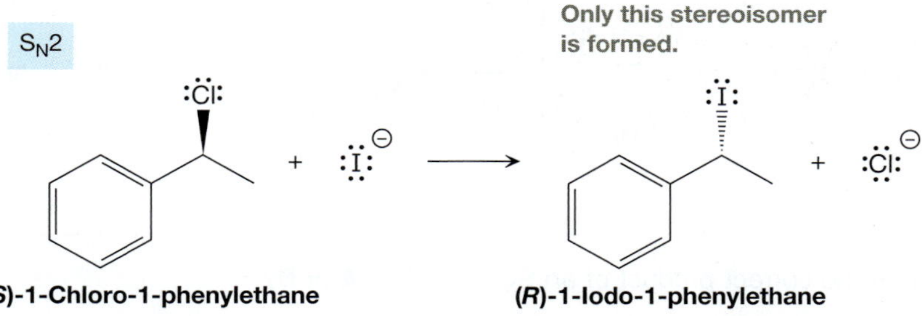

(8-20)

Whereas the S_N1 mechanism involving similar reactants (Equation 8-17) produces a mixture of enantiomers, the S_N2 mechanism yields only one of those enantiomers. In general:

A stereochemically pure substrate that undergoes an S_N2 reaction yields only one stereoisomer, which depends on the specific configuration of the substrate, so S_N2 reactions are **stereospecific**.

More specifically, as shown in Equation 8-20, the C atom's bond to the nucleophile in the products is on the side of the substrate *opposite* the initial bond to the leaving group in the reactants.

The S_N2 reaction takes place in a single step, so this stereospecificity suggests that the nucleophile attacks the substrate exclusively from the side opposite the leaving group (Equation 8-21a). In other words:

S_N2 reactions require **backside attack** of the substrate by the nucleophile.

The stereoisomer that is *not* produced would be generated by attack of the nucleophile on the *same* side as the leaving group (Equation 8-21b), a process called **frontside attack**.

Backside attack

The three R groups must flip over to the other side.

This stereoisomer is formed exclusively.

$$ \text{(8-21a)} $$

Frontside attack

This stereoisomer is *not* formed.

$$ \text{(8-21b)} $$

YOUR TURN 8.13

Which is the correct product of an S_N2 reaction, **A** or **B**?

Br →(NaOH)→ OH (**A**) or OH (**B**)

As shown in Equation 8-21a, the backside attack requires the remaining three groups of the substrate to flip over to the other side. This is known as a **Walden inversion**, and is analogous to an umbrella inverting in a windstorm. In a frontside attack, the three groups would remain on the same side in the products.

One factor that contributes to the stereospecificity of an S_N2 reaction is *steric hindrance*. The leaving group is often large (otherwise it could not accommodate the negative charge with which it leaves), so it would crash into any nucleophile approaching in a frontside attack.

Frontside attack is disfavored, too, due to *charge repulsion* (Fig. 8-8). The atom of the leaving group bonded to the substrate is typically highly electronegative, so it usually bears a significant partial negative charge. The nucleophile, which itself bears either a partial or a full negative charge, is thus repelled from that side of the substrate.

The stereospecificity of an S_N2 reaction can be understood instead from a molecular orbital point of view. For such a discussion, see Section D.2c in the interchapter that follows Chapter 7.

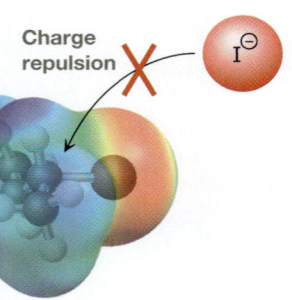

Charge repulsion

(a)

Attraction between opposite charges

(b)

FIGURE 8-8 Stereospecificity of an S_N2 reaction (a) *Frontside attack* of a nucleophile on a substrate is disfavored due to charge repulsion between the incoming nucleophile and the leaving group. (b) *Backside attack* of a nucleophile on a substrate is favored because the negatively charged nucleophile is attracted to the partially positively charged carbon that has the leaving group.

SOLVED PROBLEM 8.12

Draw the complete, detailed mechanism for the following reaction, assuming that it takes place via an S_N2 mechanism. Pay attention to stereochemistry.

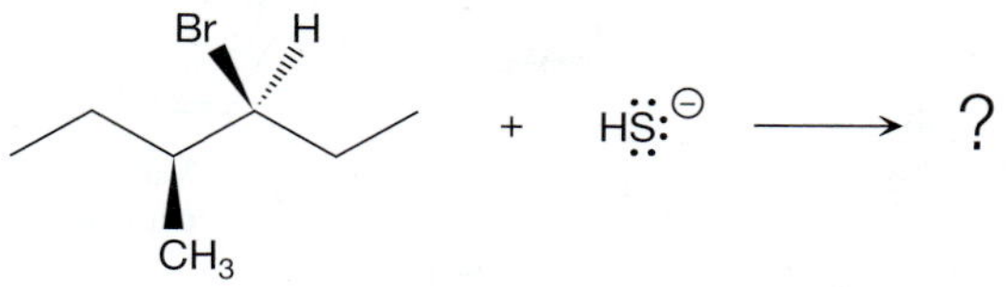

Think What can act as the leaving group? What can act as the nucleophile? How many steps make up an S_N2 mechanism? How does the nucleophile approach the substrate during attack?

Solve The leaving group is Br^- and the nucleophile is HS^-. In an S_N2 reaction, the HS^- nucleophile should attack from the side opposite the C—Br bond—in this case, from behind the plane of the page. Thus, the new C—S bond remains behind the plane of the page.

The nucleophile attacks from behind the plane of the page.

The new C—S bond remains behind the plane of the page.

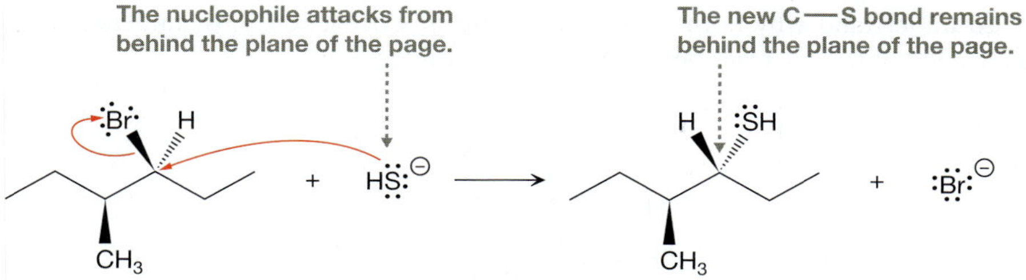

Notice that the stereochemical configuration at the other chiral center—the one bonded to CH_3—remains unchanged because no bonds to it were broken or formed.

PROBLEM 8.13 Draw the complete mechanism and the products for each of the following S_N2 reactions, paying close attention to stereochemistry.

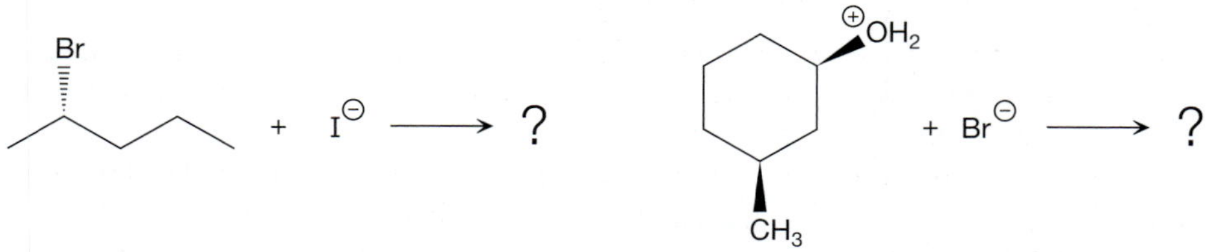

(a)

(b)

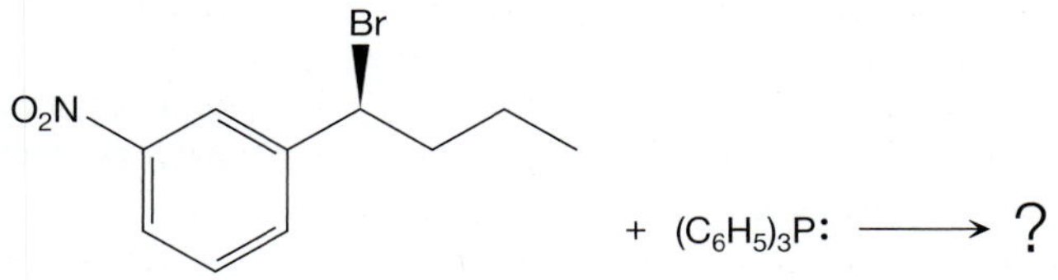

(c)

8.5c Stereochemistry of an E1 Reaction

When an E1 reaction produces a new double bond, stereochemistry is an issue if both E and Z configurations about the double bond exist. An example is shown in Equation 8-22.

Both *E* and *Z* isomers are produced in an E1 mechanism.

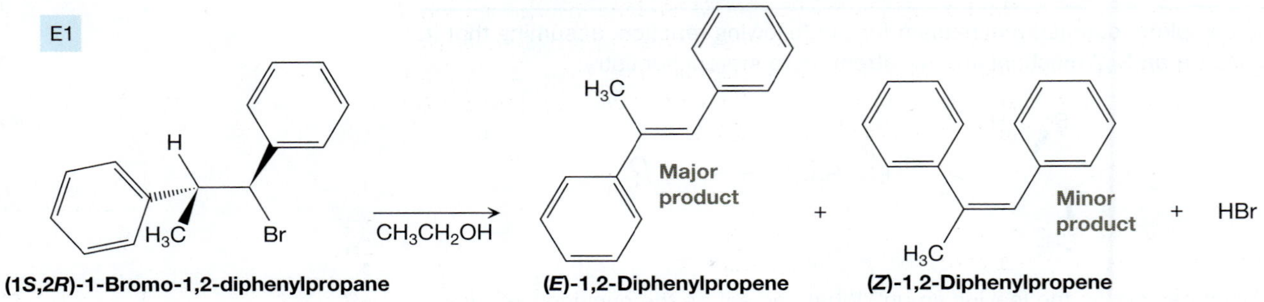

(1*S*,2*R*)-1-Bromo-1,2-diphenylpropane → (CH₃CH₂OH) → **(*E*)-1,2-Diphenylpropene** (Major product) + **(*Z*)-1,2-Diphenylpropene** (Minor product) + HBr (8-22)

As indicated, both of these diastereomers are produced in the reaction. This is a general result for all E1 reactions:

> An E1 reaction produces a mixture of the *E* and *Z* configurations about a double bond formed in the products.

To understand why both stereoisomers are produced, you must understand the mechanism of the E1 reaction, shown in Equation 8-23.

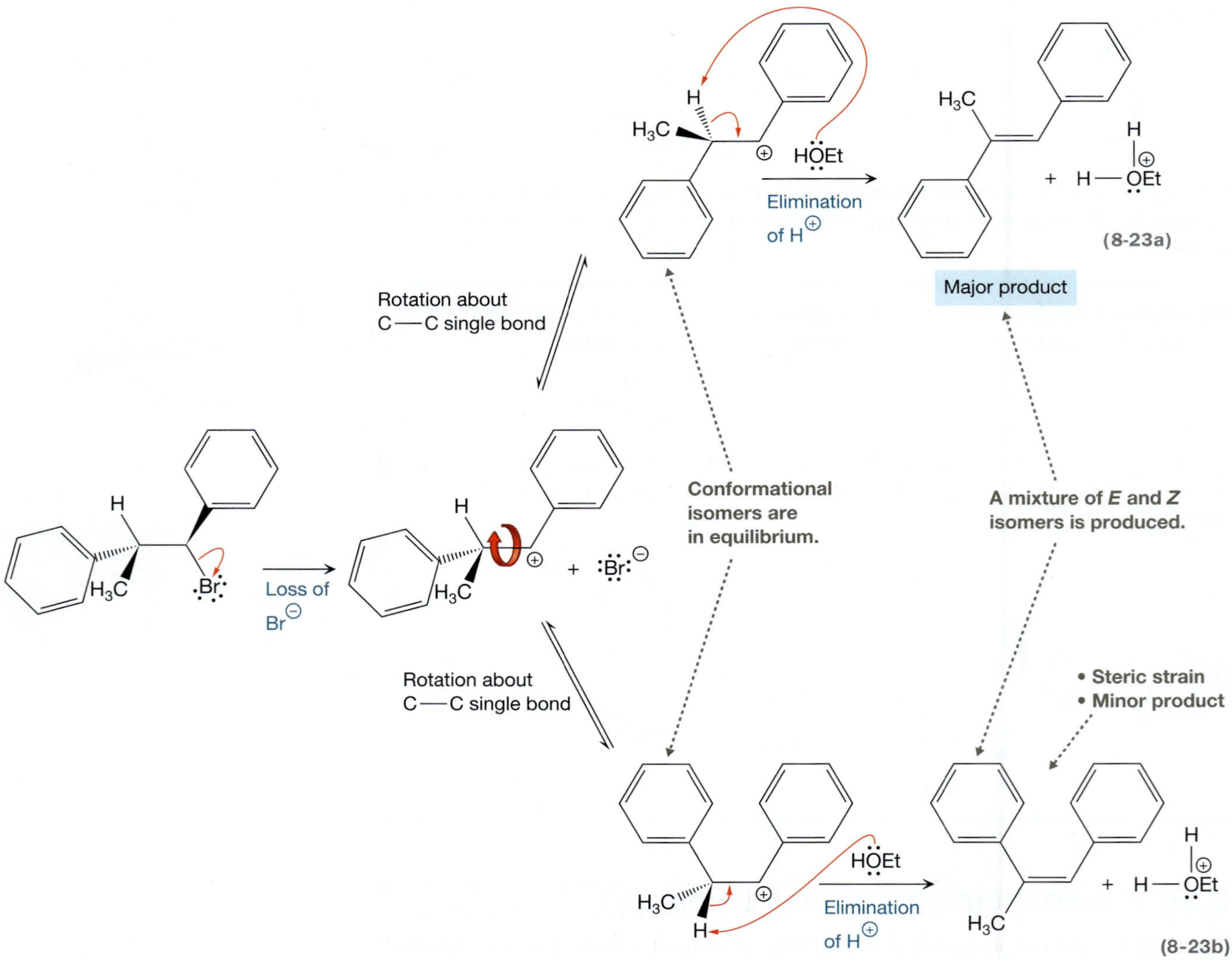

Rotation about C—C single bond

Elimination of H⊕ (8-23a)

Major product

Loss of Br⊖

Rotation about C—C single bond

Conformational isomers are in equilibrium.

A mixture of *E* and *Z* isomers is produced.

Elimination of H⊕ (8-23b)

- Steric strain
- Minor product

In the carbocation intermediate that is formed, a single bond connects the C atoms that were initially bonded to the leaving group and the H$^+$ undergoing elimination. Thus, an equilibrium is established among the various conformers about that single bond. Depending on the specific conformation of the carbocation when the double bond is formed in the second step of the mechanism, the product can be either the *E* isomer (Equation 8-23a) or the *Z* isomer (Equation 8-23b).

SOLVED PROBLEM 8.14

Draw the complete, detailed mechanism for this reaction, assuming it proceeds via an E1 mechanism. Pay attention to stereochemistry.

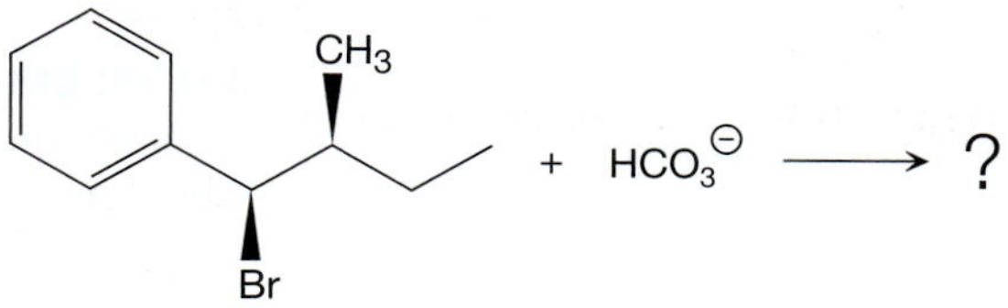

Think What can act as the leaving group? What can act as the base? What proton can be removed in the second step of an E1 mechanism? Do *E* and *Z* configurations exist for the product? If so, can both be formed?

Solve As shown below, Br$^-$ can act as the leaving group in the first step, thus generating a carbocation intermediate. In the second step, HCO$_3^-$ can act as a base to remove a proton from the adjacent carbon and produce the alkene product. Both *E* and *Z* configurations exist about that double bond, and because the bond indicated in the carbocation can undergo rotation, both the *E* and *Z* isomers are formed.

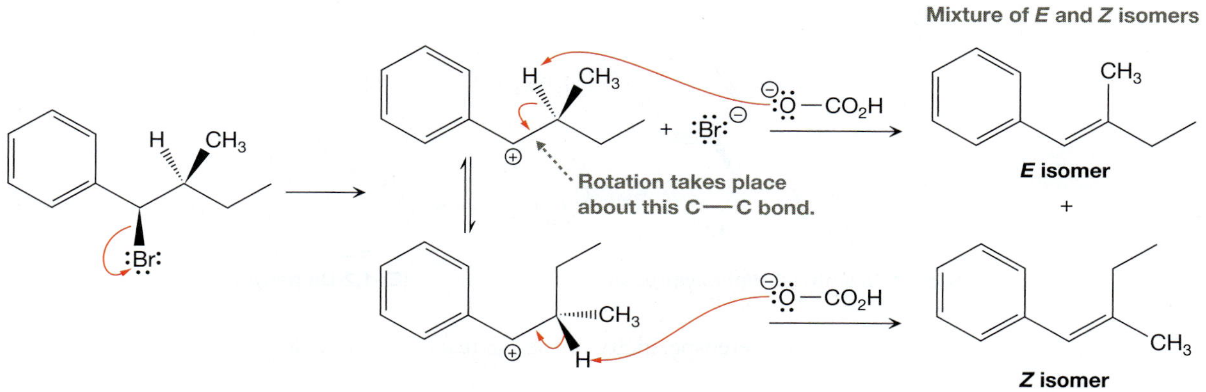

PROBLEM 8.15 Draw the complete, detailed mechanism for each of the following reactions, assuming they proceed via E1 mechanisms. Pay attention to stereochemistry.

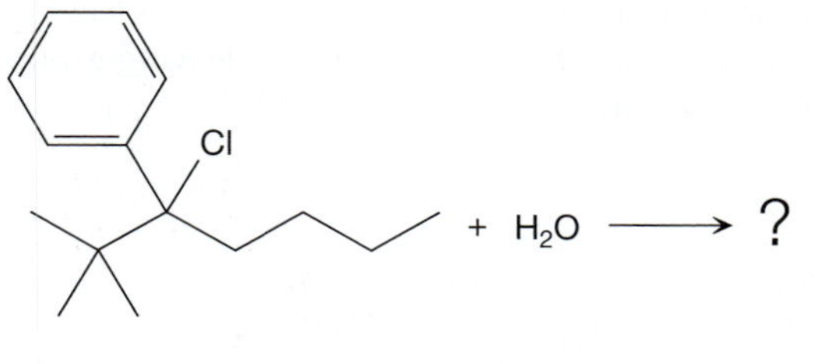

(a)

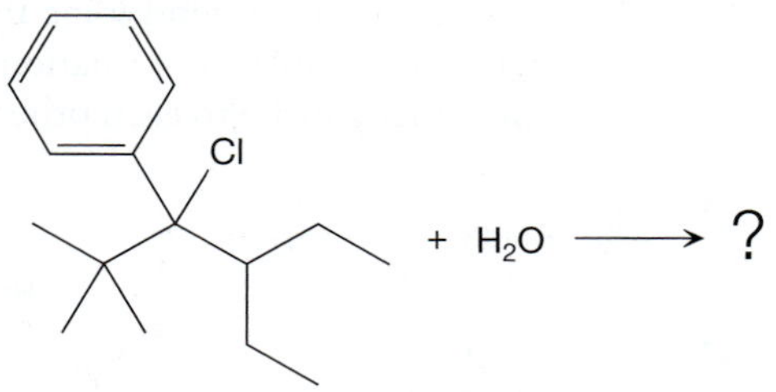

(b)

Notice that the *E* isomer in Equation 8-22 is favored over the *Z* isomer. This is because bulky phenyl rings cause steric strain. Thus, the conformer from which the *E* isomer is produced (Equation 8-23a) will be in greater abundance than the one from which the *Z* isomer is produced (Equation 8-23b). This idea can be generalized for other E1 reactions as well.

> If an E1 reaction produces both *E* and *Z* isomers, the isomer with less steric strain will generally be favored.

PROBLEM 8.16 For each reaction in Problem 8.15 that produces a mixture of diastereomers, predict which diastereomer will be produced in greater abundance.

8.5d Stereospecificity of an E2 Reaction

Unlike the E1 reaction in Equation 8-22, which yields a mixture of both the *E* and *Z* alkene products, an E2 reaction involving the same substrate yields only the diastereomer shown in Equation 8-24, making the E2 reaction *stereospecific*.

(1S,2R)-1-Bromo-1,2-diphenylpropane   **(E)-1,2-Diphenylpropene**   (8-24)

The stereospecificity of the E2 reaction can be described as follows:

> E2 reactions are favored by the substrate conformation in which the leaving group and the hydrogen atom that are eliminated are anti to each other.

Because a single bond joins the C atoms that are bonded to the H and the leaving group, the stable conformations are those in which the H atom and the leaving group are either gauche to each other or anti to each other (see Section 4.3b):

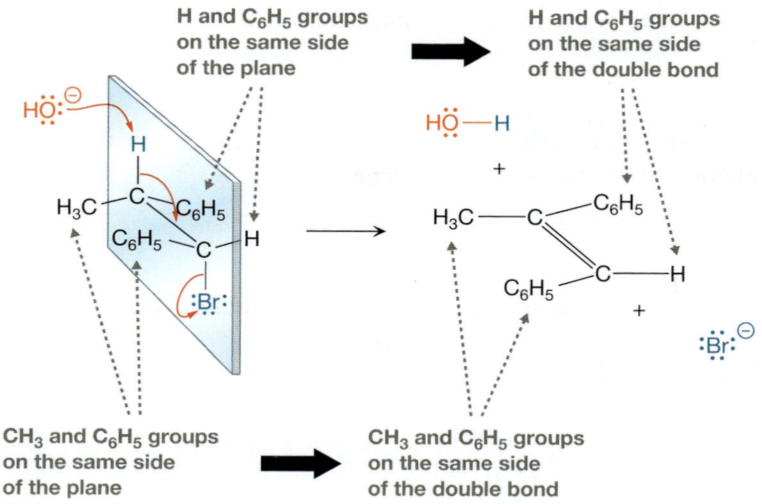

H and C₆H₅ groups on the same side of the plane

H and C₆H₅ groups on the same side of the double bond

CH₃ and C₆H₅ groups on the same side of the plane

CH₃ and C₆H₅ groups on the same side of the double bond

FIGURE 8-9 Stereospecificity in E2 reactions An E2 reaction is favored when the substrate is in the *anticoplanar* conformation (left), in which the H atom and the leaving group (in this case, Br) on adjacent C atoms are anti to each other and in the same plane. Because the CH₃ and the C₆H₅ groups are on the same side of the plane in the substrate, they end up on the same side of the double bond in the product. Similarly, the H and C₆H₅ groups are on the same side of the plane in the substrate and are on the same side of the double bond in the product.

The different conformations interconvert via rotation about that C—C bond.

In the conformation in which the H and the leaving group are anti to each other, the H, the leaving group, and the C atoms to which they are attached all reside in a single plane (Fig. 8-9). This conformation is therefore sometimes referred to as **anticoplanar** or **antiperiplanar**.

With the substrate in the anticoplanar conformation, the stereoisomer that is formed is determined by the location of the other substituents on the carbon atoms of the substrate. Notice in Figure 8-9 that the CH₃ and C₆H₅ substituents are on the same side of the plane in the substrate, so they are on the same side of the double bond in the product, too. Similarly, the H and C₆H₅ substituents are on the same side of the double bond in the product because they began on the same side of the plane in the substrate.

PROBLEM 8.17 Draw the substrate that would undergo an E2 reaction to yield the diastereomer of the product alkene in Equation 8-24. Be sure to include an accurate dash–wedge structure.

YOUR TURN 8.14

Using a model kit, construct the alkyl halide substrate in Equation 8-24, paying particular attention to the dash–wedge notation. Orient the molecule so you are looking down the bond connecting the two tetrahedral carbons and rotate about that single bond until the H and Br atoms that are eliminated are anti to each other. In that conformation, what do you notice about the relative positions of the two phenyl rings? How does that compare to the product shown in Equation 8-24?

If there are two H atoms that can be eliminated from the same C atom, then a *mixture* of diastereomers can form in an E2 reaction (see Equation 8-25).

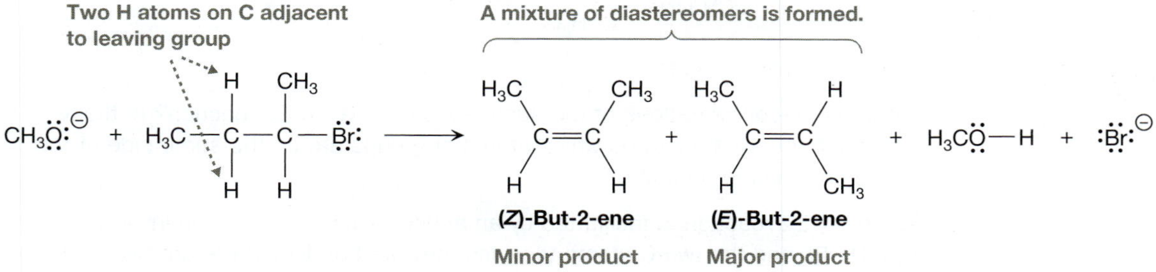

Two H atoms on C adjacent to leaving group

A mixture of diastereomers is formed.

$$CH_3\ddot{O}:^{\ominus} + H_3C-\overset{H}{\underset{H}{C}}-\overset{CH_3}{\underset{H}{C}}-\ddot{Br}: \longrightarrow$$

(Z)-But-2-ene
Minor product

(E)-But-2-ene
Major product

$+ \ H_3C\ddot{O}-H \ + \ :\ddot{Br}:^{\ominus}$ (8-25)

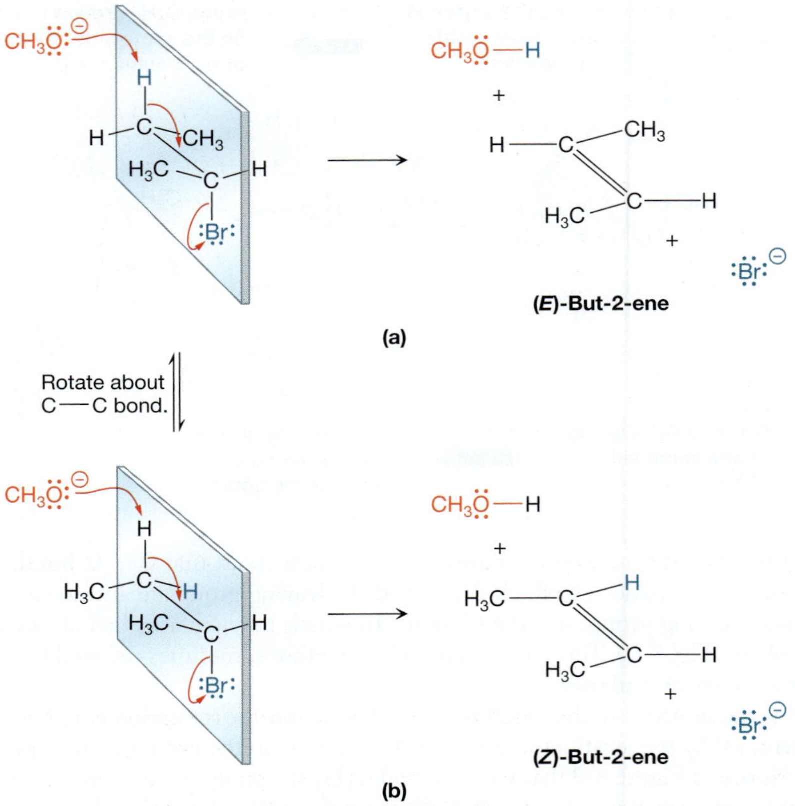

FIGURE 8-10 Formation of diastereomers in an E2 reaction The substrate in Equation 8-25 (left) has two different conformers in which an H and the Br are anti. The *E* diastereomer is formed from the conformer shown in (a), whereas the *Z* diastereomer is formed from the conformer shown in (b).

Each of the products in Equation 8-25 is the result of elimination from the substrate in an anticoplanar conformation. Figure 8-10 shows that there are two different anticoplanar conformers, which differ by rotation about the C—C single bond. One of them leads to the *E* configuration about the C═C double bond (Fig. 8-10a), whereas the other leads to the *Z* configuration (Fig. 8-10b).

SOLVED PROBLEM 8.18

Draw the products of E2 elimination involving **(a)** the D atom and **(b)** the H atom indicated. Pay attention to stereochemistry, and note that D is an isotope of H, so the two atoms exhibit nearly identical chemical behavior.

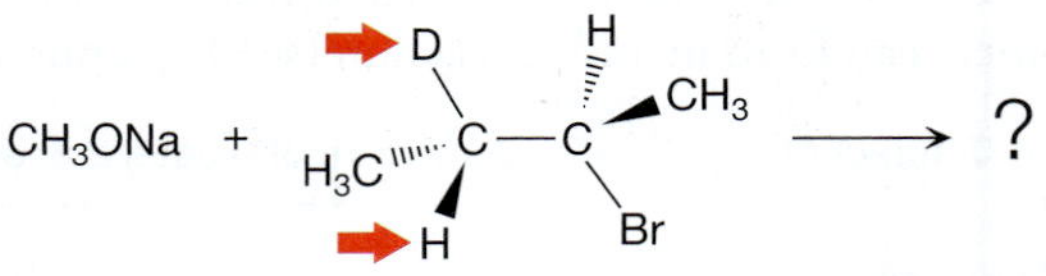

Think What conformations of the substrate facilitate an E2 reaction? In those conformations, what dictates which atoms or groups are on the same side of the double bond in the products?

Solve An E2 reaction is facilitated by an anticoplanar conformation involving H (or D), Br, and the two C atoms to which they are bonded. There are two such conformations, as shown on the left in each of the following reactions: one involves D and the other involves H.

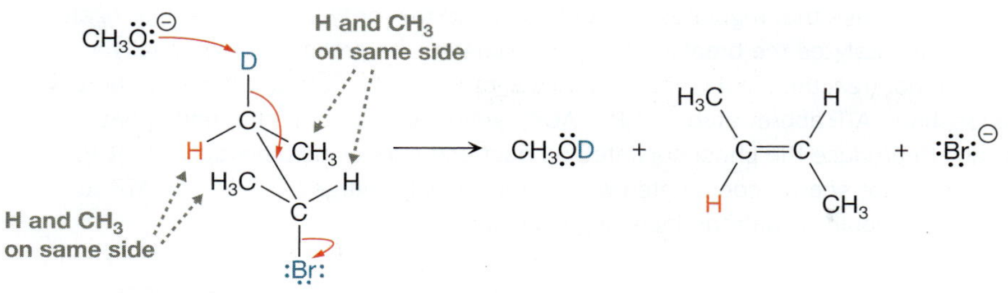

(a)

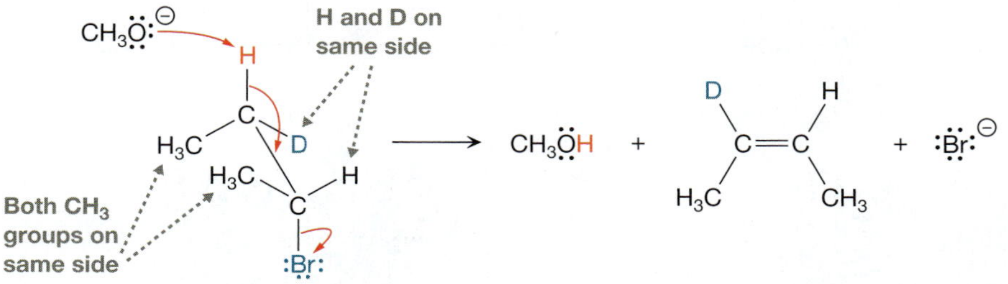

(b)

In the first reaction, D is eliminated along with Br. Because H and a CH_3 group appear on each side of the D—C—C—Br plane in the substrate, H and a CH_3 group appear on each side of the double bond in the products. In the second reaction, H is eliminated along with Br. Because two CH_3 groups appear on one side of the H—C—C—Br plane in the substrate, two CH_3 groups appear on the same side of the double bond in the products. Similarly, because H and D appear on the same side of that plane in the substrate, they also appear on the same side of the double bond in the product.

PROBLEM 8.19 Draw the products of E2 elimination involving **(a)** the H atom and **(b)** the D atom indicated, and compare these products to the ones in Solved Problem 8.18. Pay attention to stereochemistry.

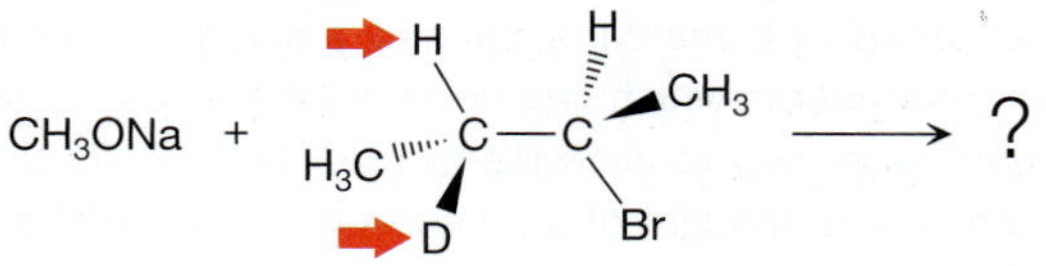

PROBLEM 8.20 Which compound, **A** or **B**, would you expect to undergo E2 elimination more readily? Why? *Hint:* Can the H and leaving group attain an anticoplanar conformation in each structure?

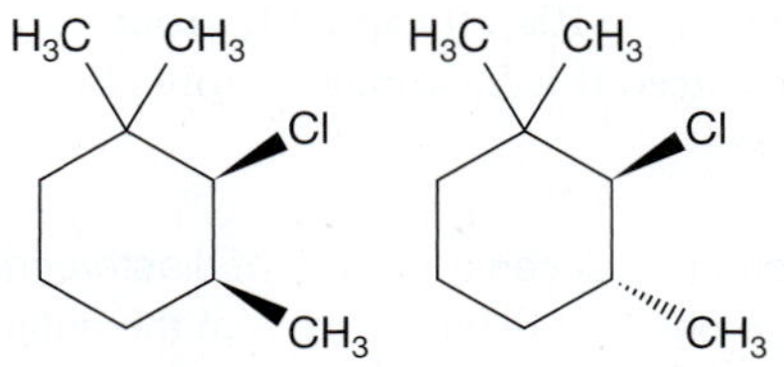

Looking back at Equation 8-25, notice that the *E* isomer is favored over the *Z* isomer. Once again, this is primarily due to steric strain. In the anticoplanar

Phosphorylation: An Enzyme's On/Off Switch

Phosphorylation is a process that regulates the function of certain enzymes, such as glycogen phosphorylase, which catalyzes the breaking down of glycogen. The process is essentially a nucleophilic substitution reaction, and is facilitated by another enzyme called a *kinase*, whereby adenosine triphosphate (ATP, abbreviated $^{2-}O_3P$—ADP), acting as the substrate, undergoes nucleophilic attack to produce the phosphorylated product and adenosine diphosphate (ADP). In this reaction, Mg^{2+} (not shown) coordinates with the negatively charged O atoms of ATP to minimize electrostatic repulsion with the incoming nucleophile.

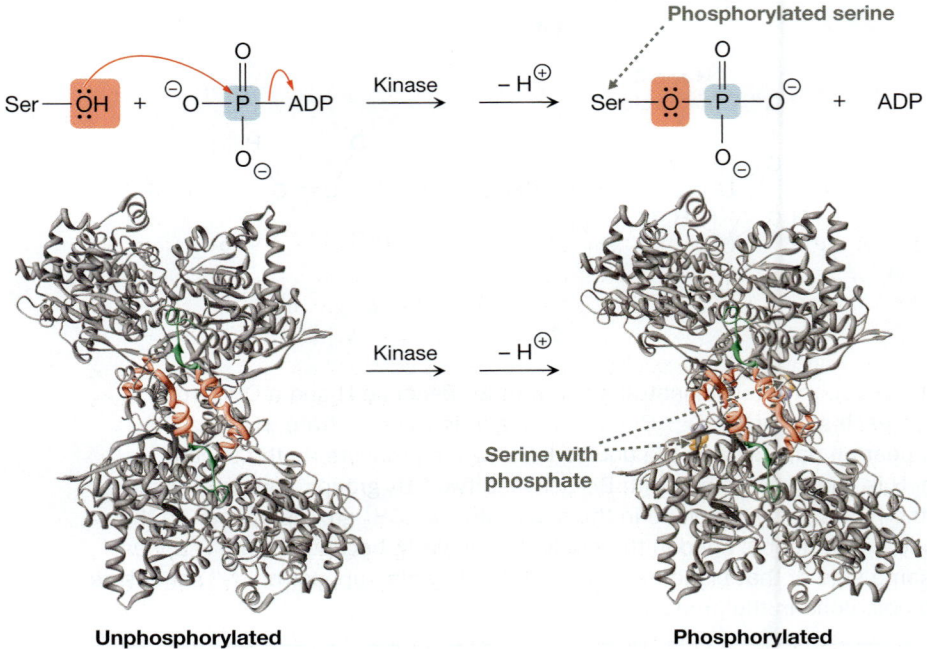

Unphosphorylated   **Phosphorylated**

In glycogen phosphorylase, in particular, a serine amino acid in the enzyme's active site is subject to phosphorylation, owing to the nucleophilic character of the OH group in serine's side group. The introduction of the negatively charged phosphate group onto that residue significantly changes the interactions with other amino acid residues, causing a change in the conformation of the enzyme. Such a change in conformation, which can be seen above (particularly the region highlighted in red), causes the enzyme's activity to increase by roughly 25%. For this and other enzymes, therefore, phosphorylation can be thought of as a convenient "on/off" switch.

conformation from which the *Z* isomer is produced (Fig. 8-10b), the bulky CH_3 groups are gauche to each other, whereas in the anticoplanar conformation from which the *E* isomer is produced (Fig. 8-10a), those CH_3 groups are anti to each other. Thus, the conformation that produced the *E* isomer is more abundant.

> In general, if an E2 reaction produces a mixture of diastereomers, the diastereomer that is favored will be the one that is produced from the anticoplanar conformation that is more stable.

Why is it that the E2 reaction is favored by the anticoplanar orientation of the substrate? One important factor has to do with electrostatic attraction and repulsion.

As in the S_N2 reaction, the bond between the leaving group and the C atom to which it is attached is polar—the atom bonded to C bears a partial negative charge, whereas the C atom bears a partial positive charge. Therefore, if the strong base (which is negatively charged) attacks the H atom anti to the leaving group, then it would be attracted to the partial positive charge on C (Fig. 8-11a). On the other hand, if the base attacks the H atom that is gauche to the leaving group, then the incoming base would be repelled by the partial negative charge on the leaving group (Fig. 8-11b).

Just as with S_N2 reactions, the stereospecificity of an E2 reaction can be understood instead from a molecular orbital point of view. For this discussion, see Section D.2f in the interchapter that follows Chapter 7.

8.6 The Reasonableness of a Mechanism: Proton Transfers and Carbocation Rearrangements

So far, we have considered just the simplest of nucleophilic substitution and elimination reactions. Each of the S_N2 and E2 mechanisms that have been presented consists of precisely one step, and each of the S_N1 and E1 reactions that have been presented consists of precisely two steps. However, you will encounter many instances where other elementary steps are incorporated into these rudimentary mechanisms, making the mechanisms slightly longer and more complex. The most common of these steps are *proton transfer reactions* and *carbocation rearrangements*.

Because of the greater complexity resulting from including these additional steps, it is important for you to gain a sense of how they can be incorporated into a mechanism in a *reasonable* way. To help you acquire this "chemical intuition," we present four general rules. The first three pertain to proton transfer reactions and are discussed in Sections 8.6a–8.6c. The fourth rule, discussed in Section 8.6d, pertains to carbocation rearrangements.

Although these rules are introduced in the context of S_N2, S_N1, E2, and E1 reactions, they apply generally to all reaction mechanisms. Therefore, you should try to apply the lessons you learn here each time you encounter a new reaction mechanism.

8.6a Acidic and Basic Conditions: Proton Transfer Reactions Are Fast

The first general rule addresses the conditions under which a reaction takes place:

Proton transfer reactions should be incorporated in such a way as to avoid the appearance of species that are incompatible with the conditions of the solution. Specifically:

- Strong acids are compatible with acidic and neutral conditions, but are incompatible with basic conditions.
- Strong bases are compatible with basic conditions and neutral conditions, but are incompatible with acidic conditions.

Under basic conditions, the equilibrium concentration of strongly acidic species is extremely small, and under acidic conditions, the equilibrium concentration of strongly basic species is extremely small.

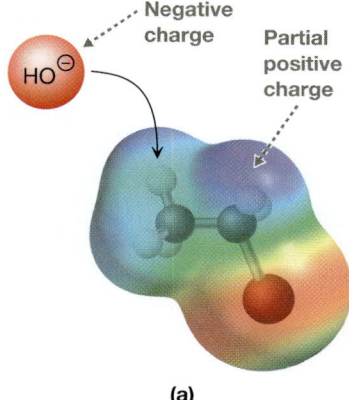

Attraction between opposite charges

Negative charge

Partial positive charge

HO⊖

(a)

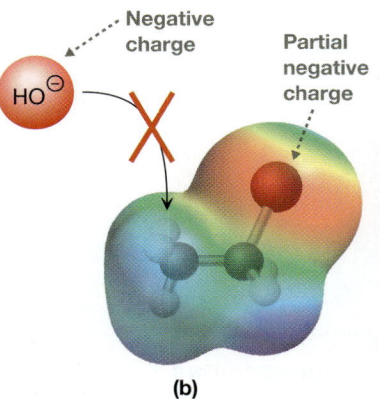

Repulsion between like charges

Negative charge

Partial negative charge

HO⊖

(b)

FIGURE 8-11 Stereospecificity in E2 reactions E2 reactions with the substrate in different conformations. (a) The H and leaving group are anti to each other. In this conformation, the incoming base is attracted by the positively charged end of the substrate. (b) The H and leaving group are gauche to each other. In this conformation, the base is repelled by the partial negative charge on the leaving group.

Proton transfer reactions are fast, so they can be incorporated before or after another elementary step, as necessary, to avoid incompatible species. Recall, for example, the acid–base titrations you have carried out in lab. Changes in the pH of a solution occur almost instantaneously on the addition of acid or base. On the other hand, many organic reactions require hours to reach completion, even at an elevated temperature.

To apply this rule correctly, we must be able to recognize strong acids and strong bases:

- Examples of strong bases: HO^-, H_2N^-, H_3C^-, H^-
- Examples of weak bases: H_2O, NH_3, Cl^-, Br^-, I^-, HSO_4^-, HCO_2^-

- Examples of strong acids: H_3O^+, $CH_3OH_2^+$, $H_3CCH_2^+$
- Examples of weak acids: H_2O, NH_3, H_4N^+, $CH_3NH_3^+$

Strong bases generally have a negative charge localized on an atom from the first or second row of the periodic table. As we learned in Chapter 6, these relatively small atoms do not accommodate the negative charge very well. Weak bases are generally uncharged but can be negatively charged if the charge is heavily stabilized. In the above halides, for example, the negative charge appears on a large atom, and in HSO_4^- and HCO_2^-, the negative charge is stabilized by resonance and inductive effects.

Strong acids generally have a positive charge, but some positively charged acids, such as H_4N^+, are weak. A nitrogen atom can handle the positive charge decently well because it is not very highly electronegative. You might think that a carbocation should be a weak acid, too, because carbon is even less electronegative than nitrogen, but remember that a carbon atom in a carbocation lacks an octet. This makes the carbocation very highly reactive and a strong acid.

With this in mind, consider the substitution reaction in Equation 8-26, in which phenylmethanol is converted to methoxyphenylmethane under *basic* conditions (indicated by the presence of KOH).

CONNECTIONS

Methoxyphenylmethane (benzyl methyl ether) is a component of lilac and is used as a fragrance.

**Phenylmethanol
(Benzyl alcohol)**

Basic conditions

Methoxyphenylmethane
75%

(8-26)

In the proposed mechanism in Equation 8-27, the alcohol acts as a nucleophile in an S_N2 reaction, followed by a proton transfer step.

Proposed mechanism for Equation 8-26

Unreasonable

1. S_N2

2. Proton transfer

A strong acid is incompatible with basic conditions.

(8-27)

This mechanism, however, is unreasonable because the product of the first step has a highly acidic proton, which is incompatible with the basic conditions of the reaction.

A more reasonable mechanism is shown in Equation 8-28, in which HO^- first deprotonates phenylmethanol to produce an alkoxide anion, RO^-. The alkoxide anion then acts as the nucleophile to displace I^-. In contrast to Equation 8-27, no strongly acidic species appear in this mechanism.

Proposed mechanism for Equation 8-26

Reasonable

No strong acids appear.

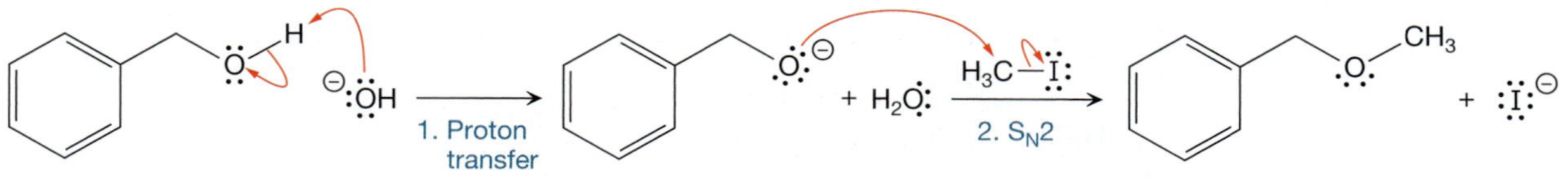

1. Proton transfer

2. S_N2

(8-28)

YOUR TURN 8.15

Consider this reaction, which takes place under *basic* conditions. In the following mechanism, label the incompatible species that makes the mechanism unreasonable.

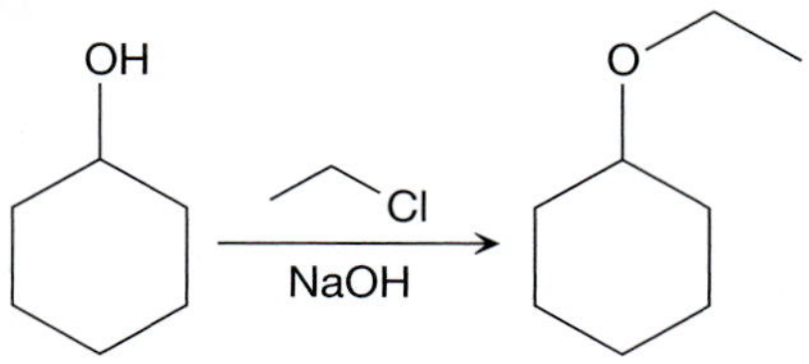

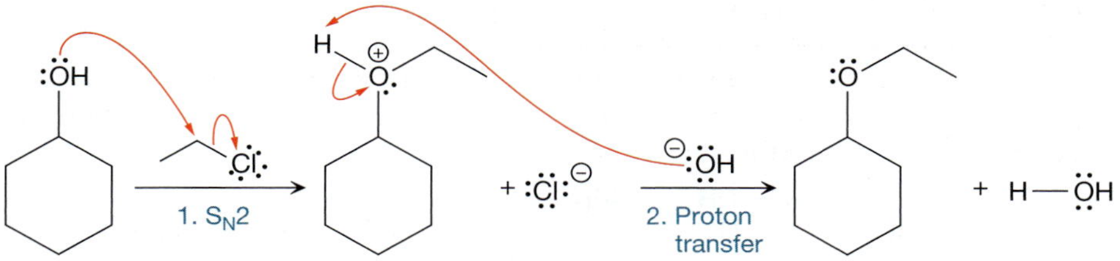

1. S_N2

2. Proton transfer

PROBLEM 8.21 Propose a reasonable mechanism for the reaction in Your Turn 8.15.

The substitution reaction in Equation 8-29 takes place under *acidic* conditions.

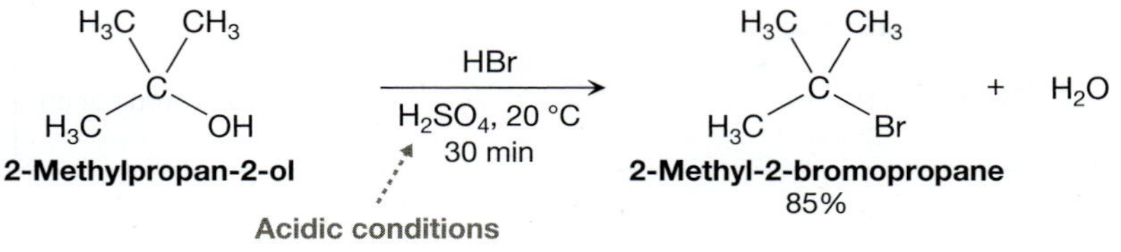

2-Methylpropan-2-ol

$\xrightarrow[\text{H}_2\text{SO}_4, \, 20\,°\text{C}]{\text{HBr}}$ 30 min

Acidic conditions

2-Methyl-2-bromopropane 85%

+ H_2O (8-29)

The proposed S_N1 mechanism shown in Equation 8-30 accounts for the formation of the product, but it is unreasonable. Notice that the product of the first step is HO^-, which is a strong base and is thus incompatible with the acidic conditions of the reaction. (In Chapter 9, we will also learn that this mechanism is unreasonable because strong bases are poor *leaving groups*.)

Proposed mechanism for Equation 8-29

Unreasonable

A strong base is incompatible with the acidic conditions.

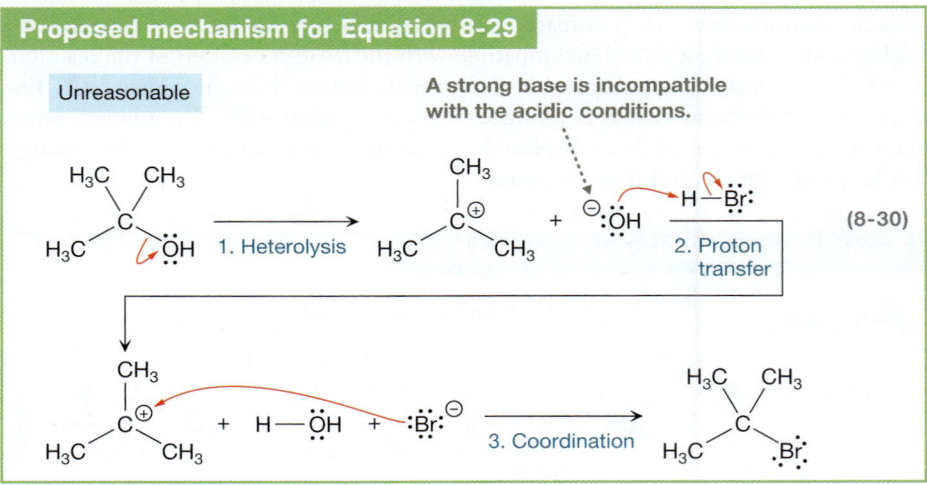

(8-30)

1. Heterolysis
2. Proton transfer
3. Coordination

A more reasonable mechanism is shown in Equation 8-31, in which the OH group is protonated prior to the S_N1 mechanism. Thus, the leaving group departs as H_2O instead of HO^-, and no strongly basic species appear at any stage of the mechanism.

Proposed mechanism for Equation 8-29

Reasonable

1. Proton transfer
2. Heterolysis
(8-31)
3. Coordination

YOUR TURN 8.16

The reaction shown here takes place under *acidic* conditions. In the following proposed mechanism, label the incompatible species that makes the mechanism unreasonable.

1. Heterolysis
2. Proton transfer
3. Coordination
4. Proton transfer

PROBLEM 8.22 Propose a reasonable mechanism for the reaction in Your Turn 8.16.

SOLVED PROBLEM 8.23

Draw a reasonable mechanism for the elimination reaction shown here, assuming it takes place via an E1 mechanism.

Think What are the first and second steps of a typical two-step E1 mechanism? What incompatible species appear in such a mechanism for this reaction? How can you incorporate a proton transfer reaction in a reasonable way to avoid the formation of such a species?

Solve The normal two-step E1 mechanism includes a heterolysis step, followed by elimination of H^+ to form the double bond.

A strong base is incompatible
with acidic conditions.

This requires the formation of CH_3O^-, however, which is a strong base and is thus incompatible with the acidic conditions given. To avoid the formation of this species, the strong acid can protonate oxygen prior to the heterolysis step (remember, proton transfer reactions are *fast*). Therefore, in the step where the C—O bond is broken, a weakly basic CH_3OH molecule is formed instead.

1. Proton transfer 2. Heterolysis 3. Elimination of H^+

PROBLEM 8.24 Draw a reasonable mechanism for the reaction in Solved Problem 8.23, assuming that it proceeds via an E2 mechanism.

8.6b Intramolecular versus Solvent-Mediated Proton Transfer Reactions

Some mechanisms must account for the removal of a proton at one site within a molecular species and the addition of a proton at another site within the *same* species. Consider, for example, the reaction shown in Equation 8-32, in which ammonia attacks oxirane to produce 2-aminoethanol. (We discuss this type of ring-opening reaction in greater detail in Chapter 10.)

Oxirane
(Ethylene oxide)

2-Aminoethanol
70%

(8-32)

CONNECTIONS One of the main uses of cyclohexene (the product in Solved Problem 8.23) is as a precursor of adipic acid, a monomer of the polymer nylon-66. Nylon-66 can be used to make synthetic fibers, such as the ones in this rope.

A reasonable first step is shown in Equation 8-33, in which an S$_N$2 reaction opens the ring to produce a species with a positively charged ammonium ion (R—NH$_3^+$) and a negatively charged alkoxide anion (RO$^-$). From there, the N atom must be deprotonated and the negatively charged O atom must be protonated to arrive at the final, uncharged product. But how does this happen?

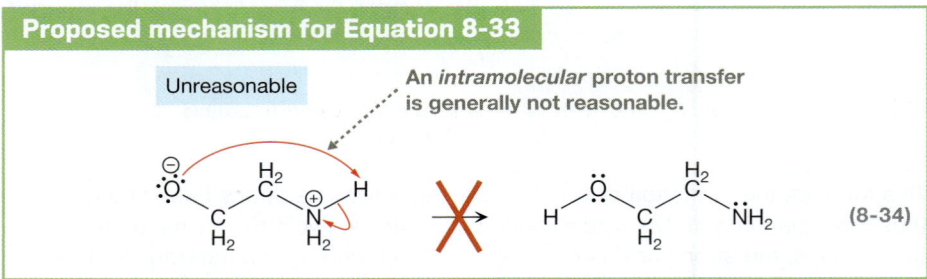

We could imagine that the proton is transferred directly from the N to the O, through an *intra*molecular proton transfer reaction. The curved arrow notation for that process is shown in Equation 8-34.

Proposed mechanism for Equation 8-33

Unreasonable

An *intramolecular* proton transfer is generally not reasonable.

(8-34)

CONNECTIONS

β-Propiolactone (see Your Turn 8.17) has had use in the medical field as a sterilizing agent for blood plasma, tissue grafts, and vaccines like this flu vaccine.

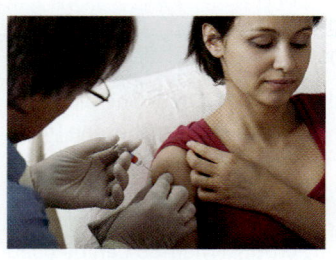

In general, however:

> Intramolecular proton transfer reactions are unreasonable.

This is because there are typically several solvent molecules that reside between the acidic and basic sites at any given time, making it difficult for the *direct* transfer of the proton from one site to the other.

Instead, if these solvent molecules are sufficiently acidic or basic, they invariably participate in the transfer of a proton from one site to another in a particular species, via a **solvent-mediated proton transfer**. In Equation 8-32, the solvent (water) is both weakly acidic and weakly basic. Therefore, the solvent-mediated proton transfer in Equation 8-35 is reasonable.

Proposed mechanism for Equation 8-33

Reasonable

Water assists the transfer of a proton from N to O via a *solvent-mediated proton transfer*.

Proton transfer   Proton transfer   (8-35)

Methanethiol (HSCH$_3$) attacks β-propiolactone to open the ring:

β-**Propiolactone**   +   HS—CH$_3$ $\xrightarrow{\text{H}_2\text{O}}$ 3-**Methylthiopropanoic acid**

The following proposed mechanism for this reaction is unreasonable. Label the step that is unreasonable and explain why.

PROBLEM 8.25 Draw a reasonable mechanism for the reaction in Your Turn 8.17.

Using Proton Transfer Reactions to Discover New Drugs

Proton transfer reactions may be relatively simple, but they can be used to identify new drugs. This is possible because proton transfer reactions tend to be quite fast, and because there are several weakly acidic protons throughout the structure of a protein, both in the amide groups that make up the protein's backbone and in the side groups of certain amino acids. If a protein is dissolved in deuterated water (D$_2$O), these protons can exchange with the D atoms of the solvent via simple proton transfer reactions. The rate of this H/D exchange can be monitored with mass spectrometry (see Chapter 16), because the atomic mass of D is greater than that of H.

How can this help us discover new drugs? Drugs are typically designed to bind to target proteins that are in their *folded* state, as shown below. A potentially viable drug, therefore, will help keep the protein folded, preventing D$_2$O from exchanging with protons on the interior of the protein. In these cases, the rate of H/D exchange will be slowed in the protein's interior and the protein will have a lighter mass.

A drug bound to a protein stabilizes the protein in its folded state.

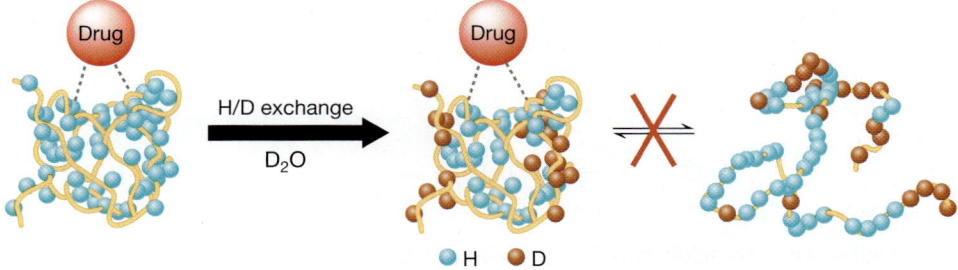

This technique is especially attractive because it requires only picomole amounts of protein, it can be carried out even in the presence of impurities, and it can be automated. As many as 10,000 potential drugs can be tested in a single day!

8.6c Molecularity

A third general rule addresses the number of reactant species in a particular elementary step. Equation 8-36, for example, accomplishes the same proton transfers we saw previously in Equation 8-35 but is unreasonable.

Proposed mechanism for Equation 8-33

Unreasonable   A *termolecular step* is, in general, highly unlikely.

$$HO-H \; + \; {}^{\ominus}O-CH_2-C\overset{H_2}{C}-\overset{+}{N}H_2-H \quad {}^{\ominus}OH \longrightarrow HO:^{\ominus} \; + \; HO-CH_2-\overset{H_2}{C}-NH_2 \; + \; H_2O: \qquad (8\text{-}36)$$

Neither of the rules we have encountered so far is violated in Equation 8-36—namely, there are no incompatible species together and the proton transfers are solvent-mediated, not intramolecular. Why, then, is Equation 8-36 unreasonable?

Equation 8-36 is unreasonable because it is a **termolecular** elementary step—that is, it involves three reactant species simultaneously.

> In general, termolecular steps (and steps of higher molecularity) are unreasonable.

By definition, an elementary step takes place in a *single event*—that is, the breaking and/or forming of bonds occur simultaneously. Thus, a termolecular step would require the collision of all three reactant species at precisely the same moment, the probability of which is vanishingly small. By contrast, bimolecular and unimolecular steps are reasonable because a bimolecular step requires the collision of just two species, and a unimolecular step is a spontaneous transformation that does not require a collision with another reactant species at all.

SOLVED PROBLEM 8.26

Consider this overall reaction. Why is the following mechanism unreasonable? Suggest an alternate mechanism that would be more reasonable.

$$H_2O_3PO-H \; + \quad \text{(cyclopentanol with OH)} -H \; + \; H_2O: \longrightarrow H_2O_3PO:^{\ominus} \; + \quad \text{(cyclopentene)} \; + \; H_2O: \; + \; H_3O:^{\oplus}$$

Think Are all of the species compatible with the acidic conditions under which the reaction takes place? Does an intramolecular proton transfer appear in the mechanism? Does a termolecular step appear?

Solve The reaction takes place under acidic conditions. The species that appear are either strongly acidic or neutral, so they are compatible with the reaction conditions. No intramolecular proton transfer appears, but the step shown is termolecular, because it involves the alcohol, H_2O, and H_3PO_4. To avoid this, the single step can be split into two separate steps.

PROBLEM 8.27 Suggest why the following S_N2 step is unreasonable and provide an alternate mechanism that is reasonable.

8.6d Carbocation Rearrangements

Carbocations, because of their net positive charge and lack of an octet, are inherently quite reactive. We saw in Section 8.1 that they can achieve greater stability by undergoing a coordination reaction, thus forming S_N1 products. As we saw in Section 8.2, they can also increase their stability by eliminating H^+, thus forming E1 products. Furthermore, as we learned in Section 7.7, they can undergo *carbocation rearrangements*, in which both the reactant and the product carbocations are isomers of each other.

S_N1 reactions provide clear evidence of these carbocation rearrangements. Consider the S_N1 reaction in Equation 8-37 between 2-iodo-3-methylbutane and water. It *appears* that substitution occurs at a C atom without the leaving group!

S_N1

Substitution appears to occur at the C *without* the leaving group.

2-Iodo-3-methylbutane   **2-Methylbutan-2-ol**   (8-37)

The mechanism in Equation 8-38 (next page) shows that a carbocation rearrangement takes place. In Step 1, the leaving group leaves as in the usual S_N1 reaction, yielding I^- and a secondary carbocation intermediate. In Step 2, a 1,2-hydride shift converts the secondary carbocation into the tertiary carbocation. In Step 3, the normal second step of an S_N1 reaction—the coordination step—takes place between H_2O and the carbocation. A final proton transfer yields the uncharged alcohol product.

CONNECTIONS
2-Methylbutan-2-ol (also called *tert*-amyl alcohol) is a by-product of grain fermentation and has physiological effects similar to ethanol. It was once used as an anesthetic but has been replaced by safer drugs.

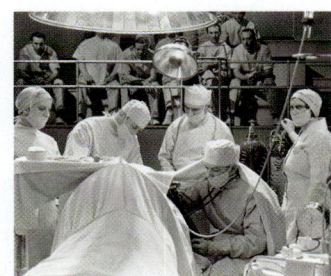

1. Heterolysis

Less stable
secondary carbocation

**2. 1,2-Hydride
shift**

More stable
tertiary carbocation

3. Coordination

(8-38)

**4. Proton
transfer**

$+ \; H_3O:^{\oplus}$

Why should this carbocation rearrangement take place? Recall from Section 7.8 that there are two major factors that can contribute to the driving force for an elementary step: charge stability and bond energies. In this case, bond energies will not contribute significantly because one σ bond is broken and a second σ bond is formed. Charge stability, however, significantly favors the tertiary carbocation over the secondary one, because, as we learned in Section 6.6e, the additional alkyl group stabilizes the positive charge.

Carbocation rearrangements that result in significantly greater stability tend to be quite rapid. George Olah and Joachim Lukas experimentally measured the rate constant for these steps to be on the order of $10^4 \; s^{-1}$ or faster, which, under normal conditions, makes them faster than most other elementary steps we encounter. Therefore:

> If an energetically favorable 1,2-hydride shift or 1,2-methyl shift competes with another possible elementary step, the carbocation rearrangement will usually win.

In the mechanism in Equation 8-38, for example, such a carbocation rearrangement beats out the coordination step that would otherwise take place. (Recall from Section 8.6a that proton transfer reactions are fast, too, but we seldom have to consider competition between a simple proton transfer and a carbocation rearrangement.)

Because carbocation rearrangements tend to be so fast, it is important that you pay attention whenever you see a carbocation produced in a mechanism. In such cases, you should consider every possible 1,2-hydride shift and 1,2-methyl shift. Consider, for example, an S_N1 reaction involving the substrate and nucleophile in Equation 8-39.

S_N1

? (8-39)

Once the carbocation is produced, there are two possible carbocation rearrangements to consider. Equation 8-40 shows a 1,2-hydride shift, and Equation 8-41 shows a 1,2-methyl shift.

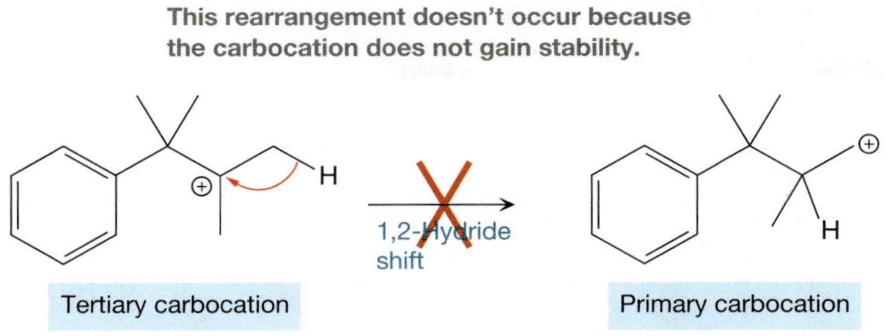

This rearrangement doesn't occur because the carbocation does not gain stability.

Tertiary carbocation   Primary carbocation   (8-40)

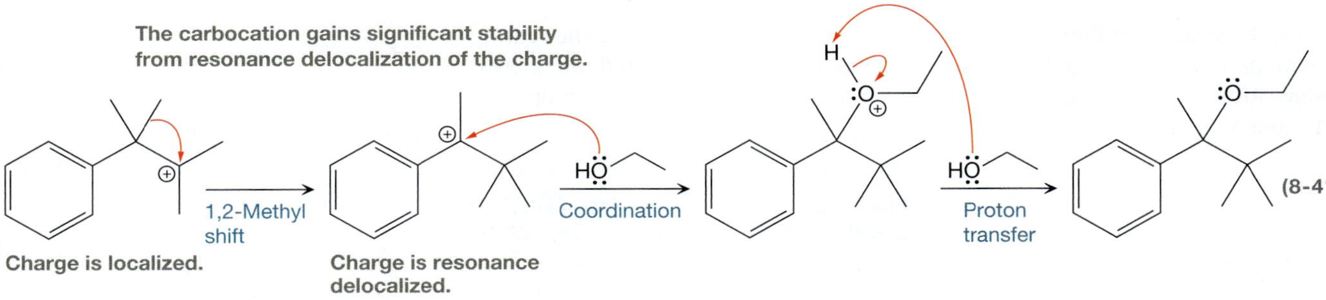

The carbocation gains significant stability from resonance delocalization of the charge.

Charge is localized.   Charge is resonance delocalized.

1,2-Methyl shift   Coordination   Proton transfer   (8-41)

As indicated, the 1,2-hydride shift does not take place because a more stable tertiary carbocation is converted to a less stable primary one. On the other hand, the 1,2-methyl shift is expected to take place rapidly because it produces a carbocation that has gained significant stability via resonance delocalization of the charge. From there, a coordination step and a proton transfer step complete the S_N1 mechanism.

YOUR TURN 8.18

Draw all resonance structures of each carbocation intermediate in Equation 8-41.

YOUR TURN 8.19

Draw free energy diagrams for the carbocation rearrangements in Equations 8-40 and 8-41. Include only the initial and final carbocations.

E1 reactions are also susceptible to carbocation rearrangements, as illustrated in Solved Problem 8.28. Later in this book, we discuss other mechanisms that involve carbocations. Keep in mind the possibility of carbocation rearrangements in those reactions as well.

SOLVED PROBLEM 8.28

Predict the products of the reaction shown here, which takes place via an E1 mechanism.

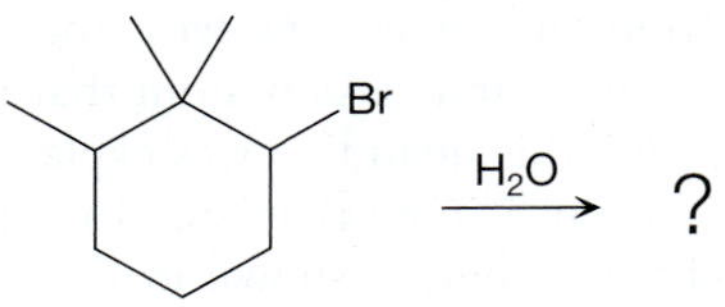

Think What are the normal steps of an E1 mechanism? Is a carbocation generated as an intermediate? If so, can it undergo a 1,2-hydride shift or a 1,2-methyl shift to become significantly more stable?

Solve The normal E1 reaction takes place in two steps. The leaving group leaves in the first step, and a proton is eliminated with the help of a base in the second step.

Notice, however, that there is a secondary carbocation intermediate that can rearrange to become significantly more stable via a 1,2-methyl shift, as shown below. After this rearrangement, elimination of an H$^+$ takes place to complete the E1 mechanism.

PROBLEM 8.29 Draw the complete, detailed mechanism for the S$_N$1 reaction between the reactants in Solved Problem 8.28.

CONNECTIONS
2-Methylbut-3-en-2-ol is a pheromone for the bark beetle, a species that has been responsible for destroying millions of acres of forest in the western United States since 2005.

8.7 Resonance-Delocalized Intermediates in Mechanisms

When 2-methylbut-3-en-2-ol is treated with concentrated hydrochloric acid, 1-chloro-3-methylbut-2-ene is produced, as shown in Equation 8-42.

$$\text{(8-42)}$$

2-Methylbut-3-en-2-ol   **1-Chloro-3-methylbut-2-ene**
                                              96%

This is a substitution reaction and, as we saw with the examples in Section 8.6d, substitution appears to take place at a carbon atom that initially does *not* have the leaving group. In Section 8.6d, this quandary was explained by a carbocation rearrangement incorporated into an S$_N$1 mechanism. The reaction in Equation 8-42 does undergo an S$_N$1 mechanism, but, as shown in Equation 8-43, no carbocation rearrangement takes place.

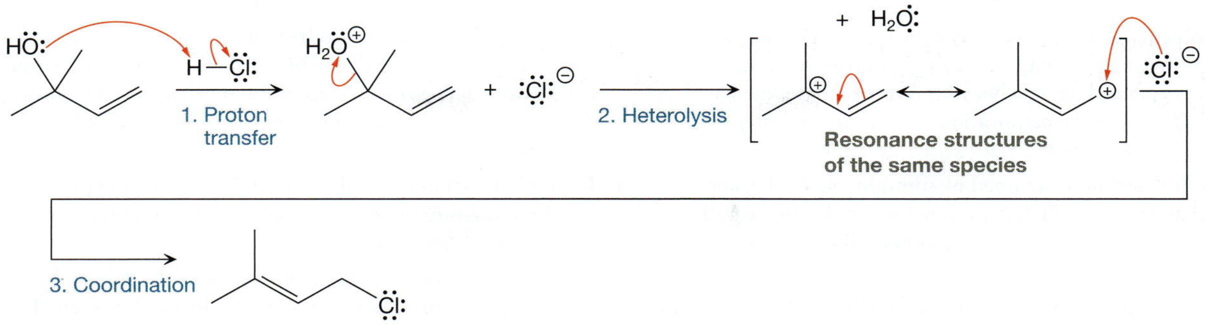

(8-43)

Instead, the carbocation that is produced in Step 2 has two resonance structures, showing that the positive charge is shared over two carbon atoms: the initial carbon to which the leaving group was attached and the terminal carbon. To produce the particular alkyl chloride shown in Equation 8-42, Cl⁻ must attack the terminal carbon, as indicated by the curved arrow notation in Step 3.

When resonance structures are incorporated into a mechanism like this, there are two important things you must keep in mind. First:

> The conversion of one resonance structure to another is *NOT* an elementary step.

An individual resonance structure is hypothetical and the one, true species is most accurately represented by the resonance hybrid. Therefore, resonance structures are just different depictions of the *same* species. This idea leads to the second important thing you must keep in mind:

> When an intermediate has two or more resonance structures, *any* resonance structure can be shown to participate as a reactant in the next step of the mechanism.

In Equation 8-43, the second resonance structure was used, but the first resonance structure could have been used instead, as shown in Equation 8-44.

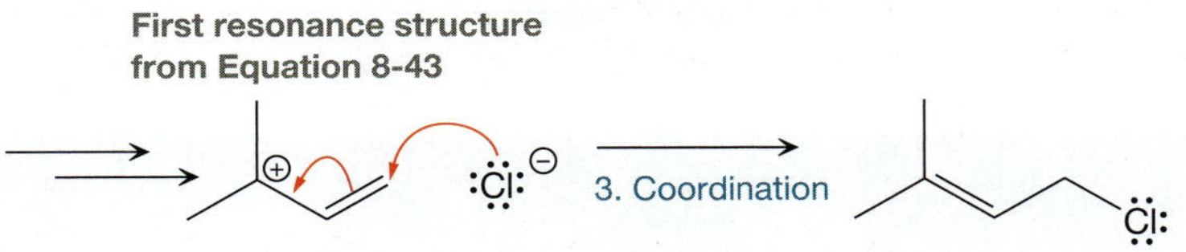

(8-44)

Notice, however, that the choice of resonance structure will impact the curved arrow notation for the subsequent elementary step. In Equation 8-43, a single curved arrow was used to depict the final coordination step. In Equation 8-44, two curved arrows must be used.

PROBLEM 8.30 Draw the complete, detailed mechanism for the following reaction.

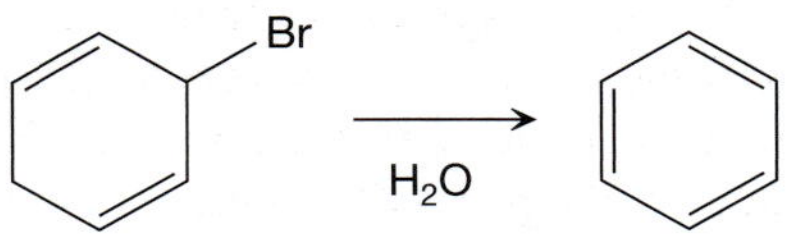

Chapter Summary and Key Terms

- A **unimolecular nucleophilic substitution (S_N1) reaction** consists of two steps. First, the leaving group leaves in a *heterolysis* step, yielding a carbocation **intermediate**, then a nucleophile attacks the carbocation in a *coordination* step. **(Section 8.1; Objective 1)**

- An **overall reaction** is obtained by summing all of the steps in a mechanism. Intermediates do not appear in the overall reaction—just **overall reactants** and **overall products**. **(Section 8.1a; Objective 3)**

- The reaction free energy diagram of an S_N1 reaction (Fig. 8-1, p. 397) contains two transition states—one for each step. Between the transition states is the intermediate, which occurs at a **local energy minimum**. **(Section 8.1b; Objective 4)**

- The **unimolecular elimination (E1) reaction** also consists of two steps. First the leaving group leaves, generating a carbocation intermediate, then H^+ is eliminated with the aid of a base to yield a double bond. **(Section 8.2; Objective 2)**

- The free energy diagram of an E1 reaction (Fig. 8-2, p. 399), like that of an S_N1 reaction, shows two transition states flanking the local energy minimum, which represents the carbocation intermediate. **(Section 8.2; Objective 4)**

- Direct observation of an intermediate is powerful evidence for a mechanism, but it is rare because intermediates tend to be unstable. Thus, a mechanism generally cannot be proven directly. **(Section 8.3; Objective 5)**

- Most of what we know about mechanisms comes from **reaction kinetics**. A *proposed mechanism* yields a **theoretical rate law**. Agreement between the theoretical rate law and the **empirical rate law** lends support to a mechanism. **(Section 8.4; Objective 5)**

- S_N2 and E2 reactions are both *second-order* reactions. In each case, the reaction rate is directly proportional to the concentration of both the nucleophile/base and the substrate. **(Section 8.4; Objectives 6, 7)**

- S_N1 and E1 reactions are both *first order*. In each case, the reaction rate depends only on the concentration of the substrate; it is independent of the concentration of the nucleophile or the base, respectively. **(Section 8.4; Objectives 6, 7)**

- For S_N1 and E1 reactions to be first order, the first step of each one—that is, the departure of the leaving group—must be the **rate-determining step** of the mechanism. As a result, the rate of the entire reaction is essentially the rate of the first step. **(Section 8.4; Objectives 6, 7)**

- If an S_N1 reaction takes place at a chiral center, the products contain a mixture of both stereochemical configurations. **(Section 8.5a; Objective 8)**

- In an S_N2 reaction, the nucleophile attacks the substrate only from the side opposite the leaving group—so-called **backside attack**. The substituents that remain on the atom being attacked undergo **Walden inversion**, and if the atom is a chiral center, a single stereoisomer is produced, making the S_N2 reaction **stereospecific**. **(Section 8.5b; Objective 8)**

- Because an E1 reaction takes place in two steps, both *E* and *Z* alkenes are produced. **(Section 8.5c; Objective 8)**

- An E2 reaction is favored by the **anticoplanar** conformation of the substrate, in which the H atom and the leaving group that are eliminated are anti to each other about the C—C bond involved in the reaction. Thus, the E2 reaction is stereo-specific. **(Section 8.5d; Objective 8)**

- Under acidic conditions, strong bases should not appear in a mechanism and, under basic conditions, strong acids should not appear. **(Section 8.6a; Objective 9)**

- **Intramolecular proton transfer** reactions are generally *unreasonable*. Instead, a proton is usually transferred from one part of a molecule to another by a **solvent-mediated proton transfer**. **(Section 8.6b; Objective 9)**

- **Termolecular** elementary steps are generally *unreasonable* because they require three species to collide at precisely the same time. **(Section 8.6c; Objective 9)**

- *Carbocation rearrangements* are fast. If a 1,2-hydride shift or a 1,2-methyl shift is energetically favorable, it will generally occur before any other step. **(Section 8.6d; Objective 10)**

- When an intermediate has resonance structures, any of the resonance structures can be used in the curved arrow notation for the subsequent step. The conversion of one resonance structure to another does not constitute an elementary step. **(Section 8.7; Objective 11)**

Problems

Problems that are related to synthesis are denoted (SYN).

8.1 and 8.2 S_N1 and E1 Mechanisms; Free Energy Diagrams

8.31 Draw the complete, detailed S_N1 mechanism for each of the following reactions.

8.32 Draw the free energy diagram for each of the reactions in Problem 8.31. For each diagram, include and label the overall reactants, overall products, all intermediates, and all transition states.

8.33 Draw the complete, detailed E1 mechanism for each of the following reactions.

(a)

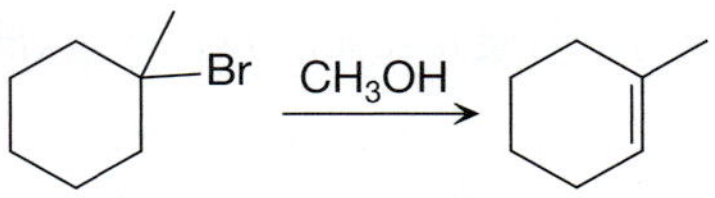

(b)

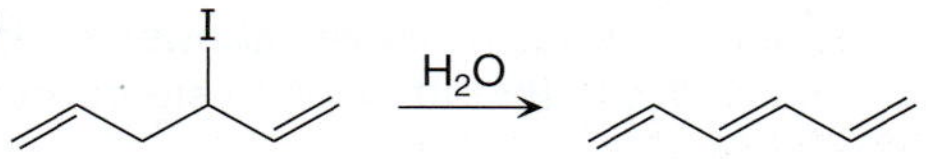

8.34 Draw the free energy diagram for each of the reactions in Problem 8.33. For each diagram, include and label the overall reactants, overall products, all intermediates, and all transition states.

8.35 (SYN) For each of these compounds, draw an alkyl halide that can be used to produce it in an S_N1 reaction. Then, determine whether the alkyl halide would need to be treated with water or methanol and draw the corresponding mechanism.

(a) OH

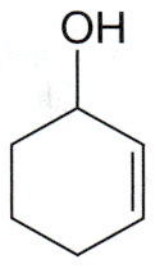

(b)

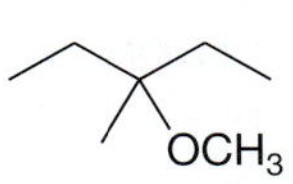

8.36 (SYN) For each of these compounds, draw an alkyl halide that can be used to produce it in an E1 reaction. Then, draw the E1 mechanism that would take place when the alkyl halide is treated with water.

(a)

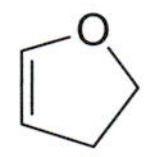

(b)

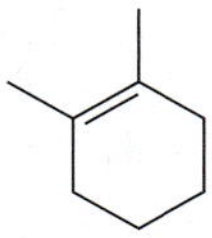

8.37 Draw all possible E1 mechanisms and products involving this alkyl halide and water.

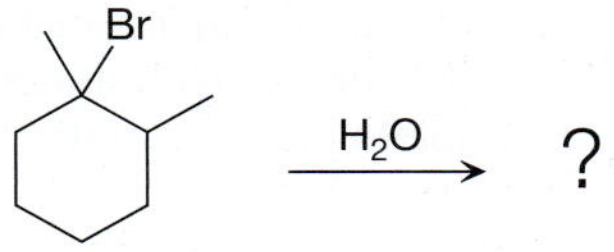

8.4 The Kinetics of S_N2, S_N1, E2, and E1 Reactions

8.38 When benzyl bromide is treated separately with KI and CH_3OH, the substitution products are different but the reaction rates are about the same.

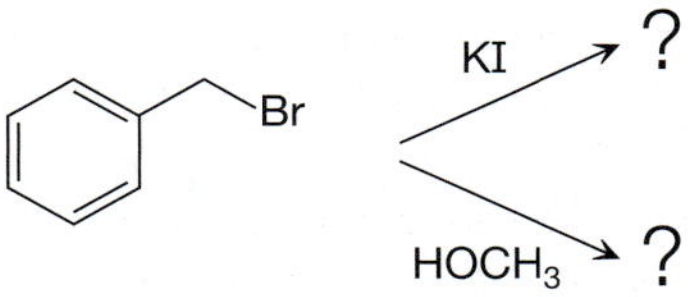

(a) What does this suggest about the mechanism—is it S_N1 or S_N2? Explain.

(b) Draw the complete mechanism (including curved arrows) for the formation of each product.

(c) If the concentration of KI were doubled, what would happen to the rate of the substitution reaction?

8.39 The initial rates for the following elimination reaction were measured under different concentrations of the substrate and base (water); the data are tabulated at the right. Do the data suggest an E1 reaction or an E2 reaction?

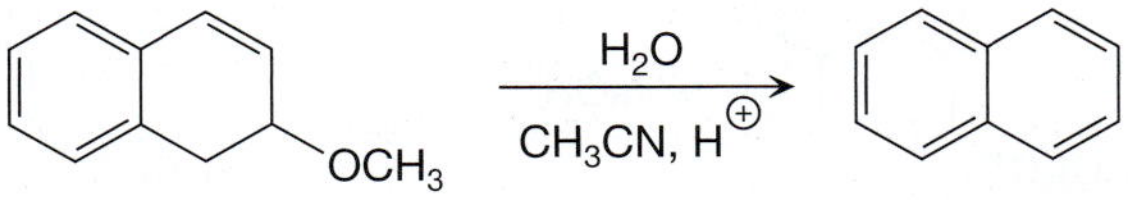

Trial Number	[R—OCH₃]	[H₂O]	Rate (M/s)
1	0.010 M	0.45 M	9.50×10^{-4}
2	0.020 M	0.45 M	1.85×10^{-3}
3	0.020 M	0.22 M	1.85×10^{-3}

8.40 Write the rate law for the reaction in Problem 8.39.

8.41 The initial rates for the following elimination reaction were measured under different concentrations of the substrate and base; the data are tabulated at the right. Do the data suggest an E1 reaction or an E2 reaction?

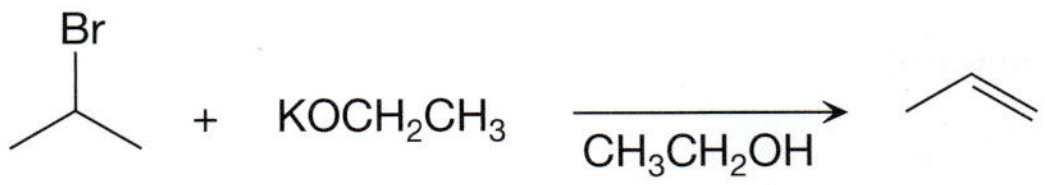

Trial Number	[R—Br]	[KOCH₂CH₃]	Rate (M/s)
1	1.0 M	1.0 M	2.35×10^{-6}
2	0.50 M	0.50 M	5.9×10^{-7}
3	0.50 M	1.0 M	1.20×10^{-6}

8.42 Write the rate law for the reaction in Problem 8.41.

8.5 Stereochemistry of Nucleophilic Substitution and Elimination Reactions

8.43 Draw the complete, detailed mechanism (including curved arrows) for each of the following reactions occurring via **(a)** an S_N2 mechanism and **(b)** an S_N1 mechanism. Pay attention to stereochemistry.

(i)

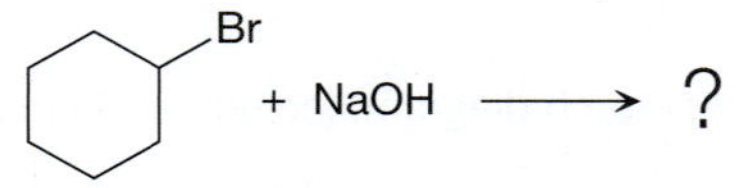

+ NaOH ⟶ ?

(ii)

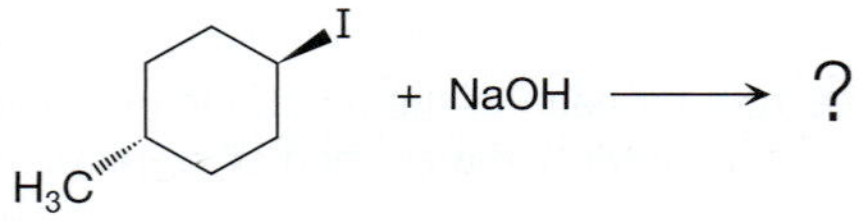

+ NaOH ⟶ ?

(iii)

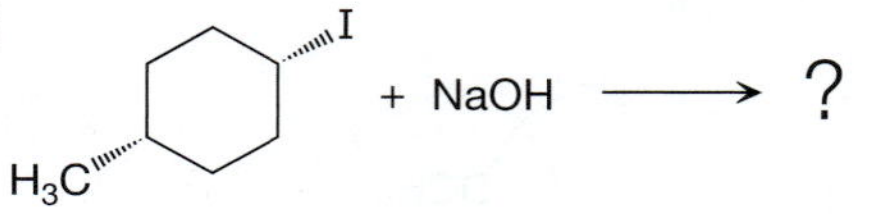

+ NaOH ⟶ ?

(iv)

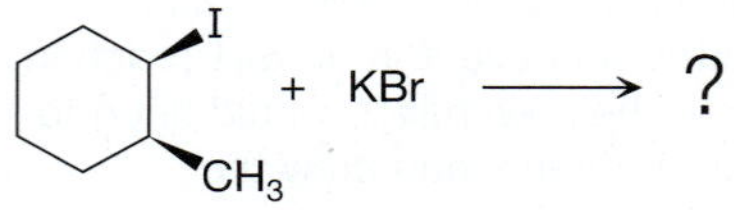

+ KBr ⟶ ?

(v)

+ NaOCH₃ ⟶ ?

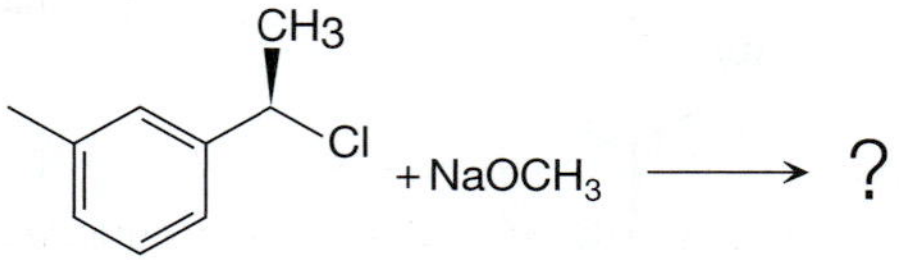

8.44 Draw the complete, detailed mechanism (including curved arrows) for each of the following reactions occurring via **(a)** an E2 mechanism and **(b)** an E1 mechanism. If more than one possible product can be produced from the same type of mechanism, draw the complete mechanism that leads to each one. Pay attention to stereochemistry.

(i)

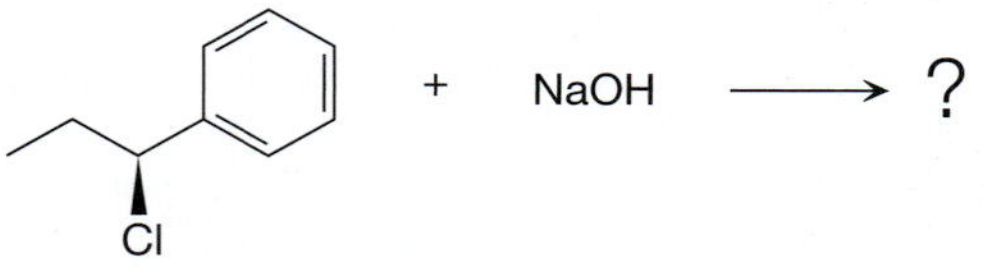

+ NaOH ⟶ ?

(ii)

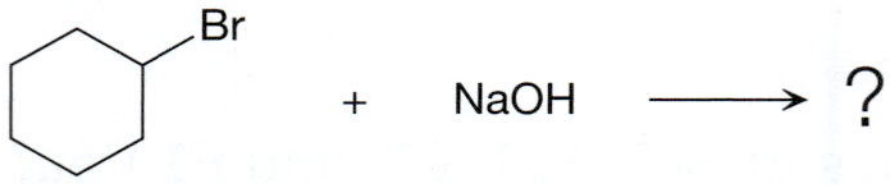

+ NaOH ⟶ ?

(iii)

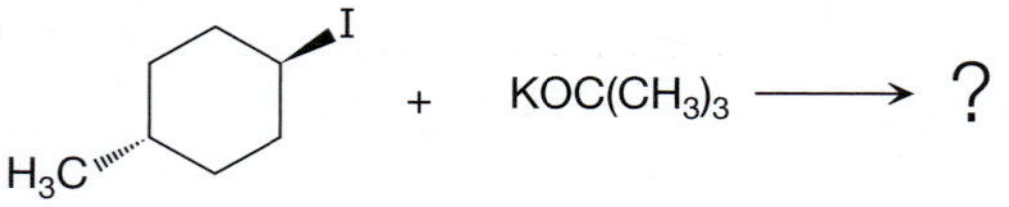

+ KOC(CH₃)₃ ⟶ ?

(iv)

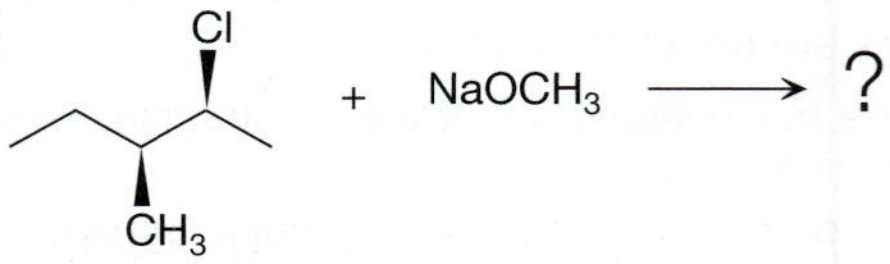

+ NaOCH₃ ⟶ ?

(v)

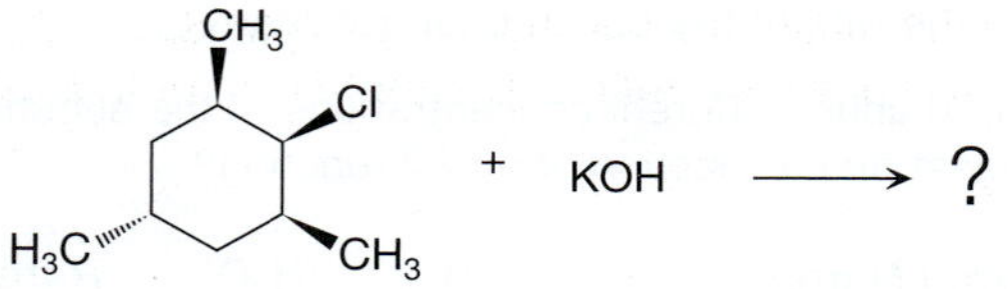

+ KOH ⟶ ?

(vi)

+ KOH ⟶ ?

8.45 The cis isomer of 1-bromo-4–*tert*-butylcyclohexane undergoes E2 elimination about 1,000 times faster than the trans isomer. Explain why the cis isomer reacts faster. *Hint: It is not because of steric hindrance.*

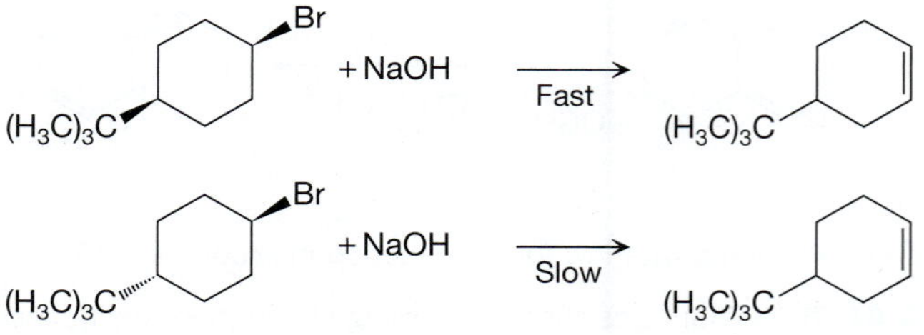

8.46 Racemization occurs when (S)-3,3-dimethylcyclohexanol is dissolved in dilute acid. Draw a mechanism to account for this, including curved arrows.

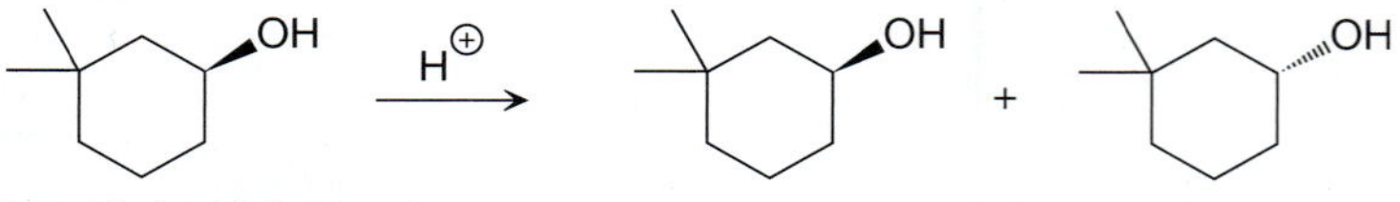

(S)-3,3-Dimethylcyclohexanol

8.47 Consider the *intramolecular* nucleophilic substitution reaction shown here. Does the stereochemistry of the product suggest an S_N1 or S_N2 mechanism? Draw the complete mechanism for this reaction, including curved arrows.

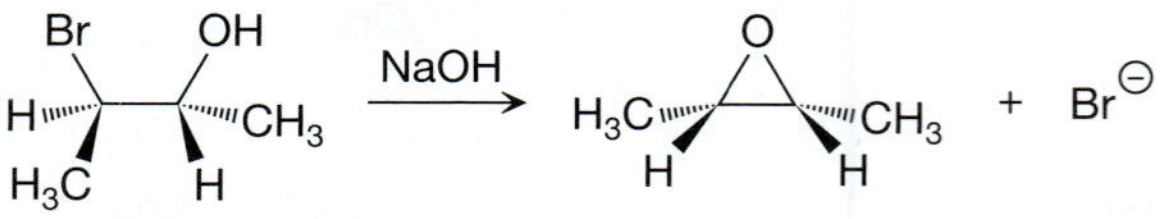

8.48 Consider the elimination reaction shown here, which produces a mixture of diastereomers. Based on the stereochemistry of this reaction alone, is it possible to tell whether the reaction takes place via an E1 or an E2 reaction? Explain.

8.49 Consider the nucleophilic substitution reaction shown here. Based on the stereochemistry, does it proceed by an S_N1 or S_N2 mechanism? Explain.

8.50 (E)-Anethole is the major component of anise oil, which is used as artificial licorice flavoring and has potential antimicrobial and antifungal properties. It can be synthesized from an alkyl halide precursor, as shown here.

(a) Will an E2 reaction produce (E)-anethole exclusively, or will the reaction produce a mixture of stereoisomers?
(b) Will an E1 reaction produce the pure stereoisomer or a mixture?
(c) For each of these reactions that produces a mixture, which stereoisomer will be produced in greater abundance?

8.51 The following nucleophilic substitution reaction is monitored by measuring the optical rotation of the solution as a function of time. Based on the results graphed on the right, suggest whether the reaction takes place by the S_N1 or S_N2 mechanism. *Hint:* Review optical rotation in Chapter 5.

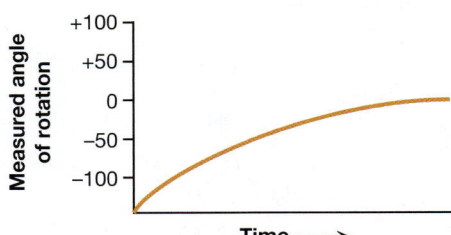

8.6 and 8.7 The Reasonableness of Mechanisms

8.52 For each of the following substrates, predict whether a carbocation rearrangement will take place in an S_N1 or E1 mechanism. Explain. Draw the curved arrow notation illustrating the carbocation rearrangement that is likely to occur.

(a)

(b)

(c)

(d)

(e)

(f)

(g)

(h) H₃CH₂C CH₂CH₃

(i)

8.53 Consider the following *overall* reaction, which will be discussed in Chapter 20.

$$H_3C-C\equiv N \quad \xrightarrow[HO^{\ominus}]{H_2O} \quad H_3C-\overset{\overset{\displaystyle O}{\|}}{C}-O^{\ominus} \quad + \quad NH_3$$

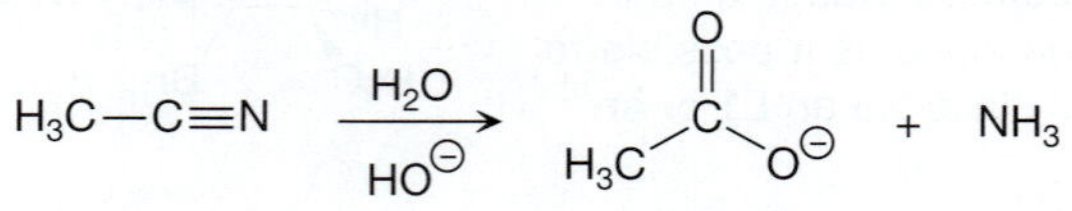

Below is a proposed mechanism for this reaction. Use the rules we learned in this chapter to evaluate whether each step in the proposed mechanism is reasonable. For each step that is *not* reasonable, explain why.

(i)

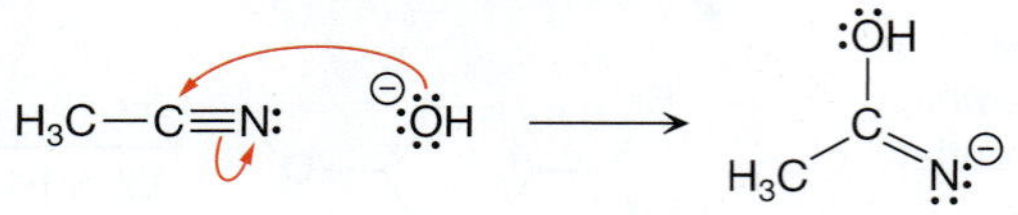

(ii)

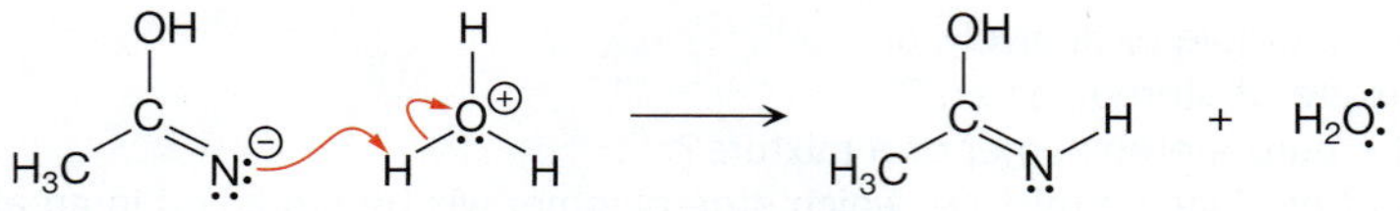

(iii)

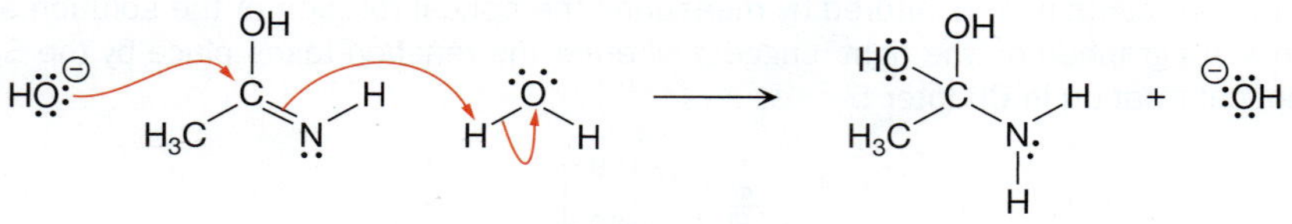

(iv)

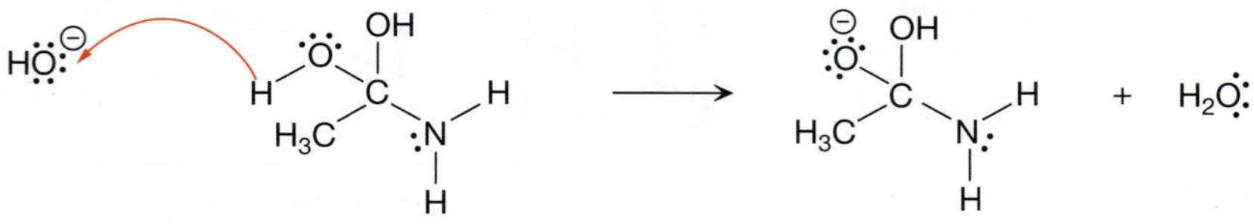

(v)

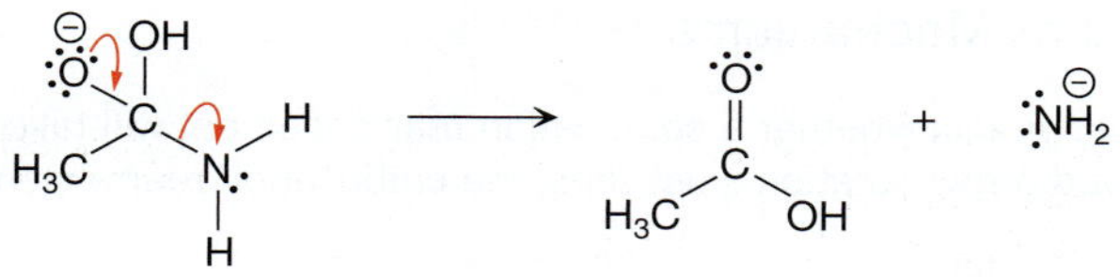

(vi)

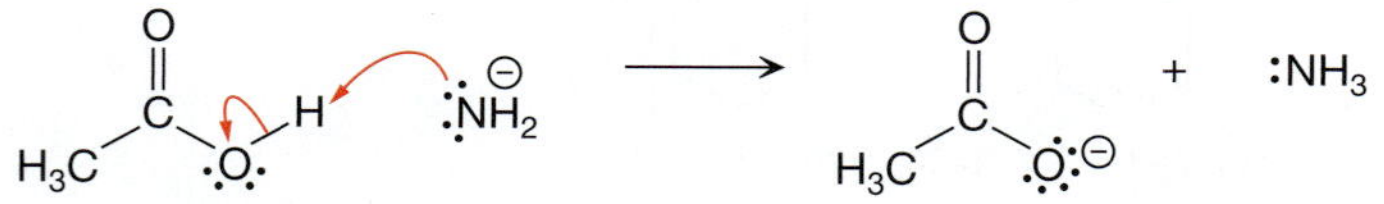

8.54 Consider the following *overall* reaction, which will be discussed in Chapter 20.

$$2 \quad \overset{\overset{\displaystyle O}{\|}}{\underset{H}{C}}\diagdown_{CH_3} \quad \xrightarrow[H_2O]{H^{\oplus}} \quad H-\overset{\overset{\displaystyle O}{\|}}{C}-\overset{H_2}{C}-\overset{\overset{\displaystyle OH}{|}}{CH}-CH_3$$

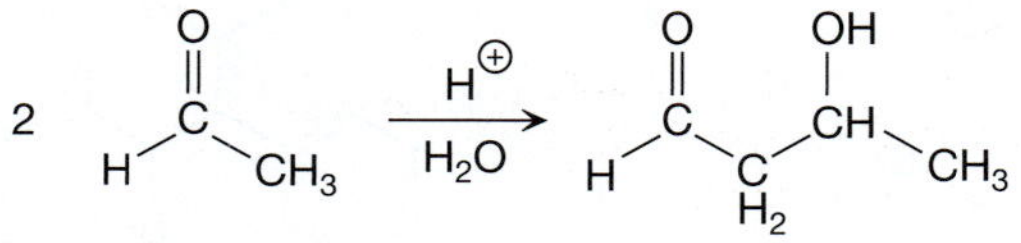

Below is a proposed mechanism. Evaluate whether each step of the mechanism is reasonable. For each step that is *not* reasonable, explain why.

(i)

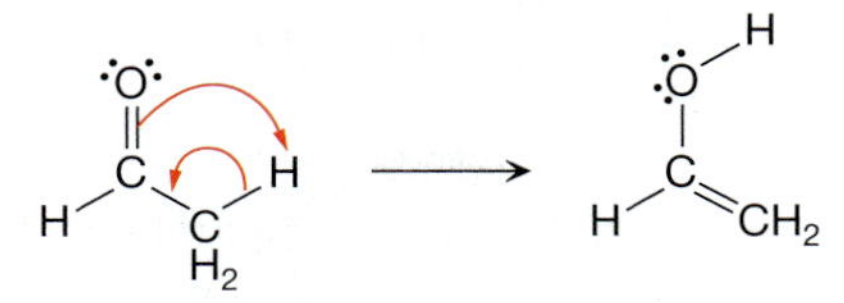

(ii)

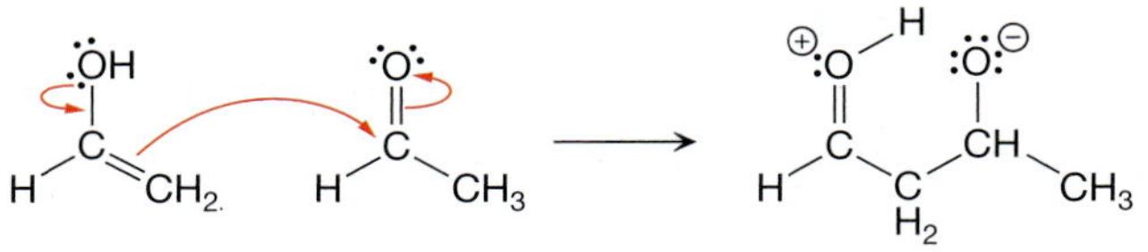

(iii)

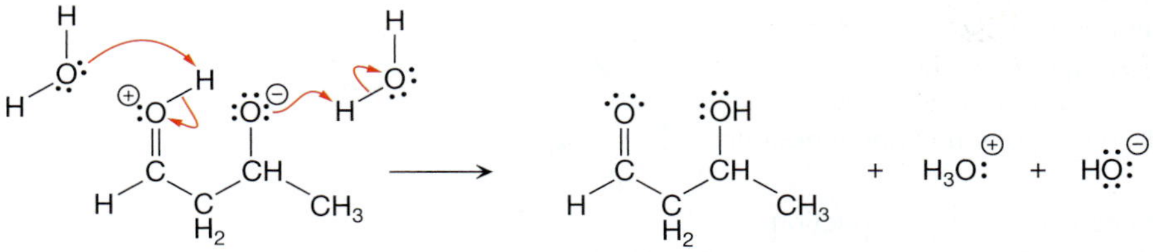

8.55 Consider the nucleophilic substitution reaction shown here, which yields a mixture of constitutional isomers. **(a)** Does this occur via an S_N1 or an S_N2 mechanism? How do you know? **(b)** Propose a mechanism that accounts for the formation of *each* product.

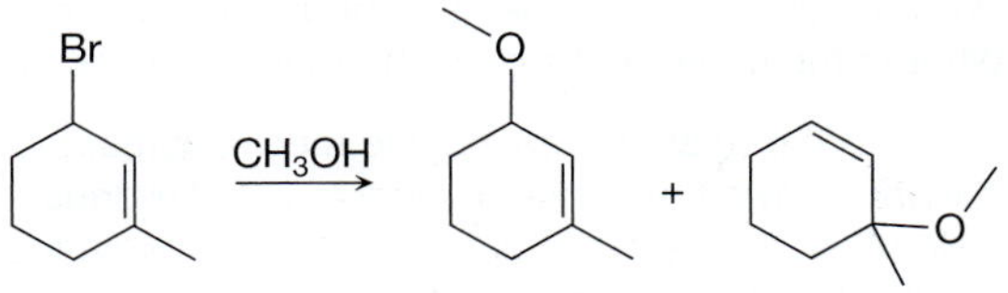

8.56 Consider the nucleophilic substitution reaction shown here. **(a)** Argue whether this reaction takes place via an S_N1 or an S_N2 reaction. **(b)** Draw the complete mechanism (including curved arrows) for this reaction.

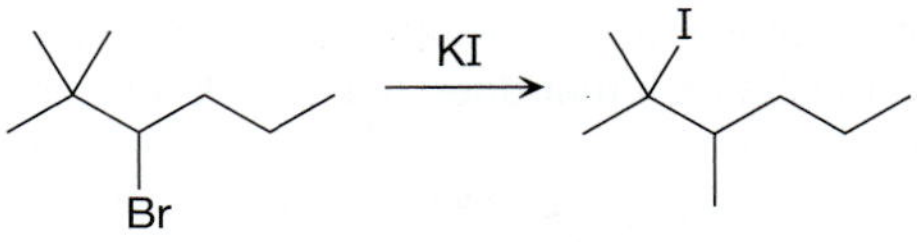

8.57 The reaction shown here yields three different nucleophilic substitution products that are constitutional isomers of one another. **(a)** Does this suggest an S_N1 or an S_N2 mechanism? **(b)** Draw the mechanism for the formation of each of these products.

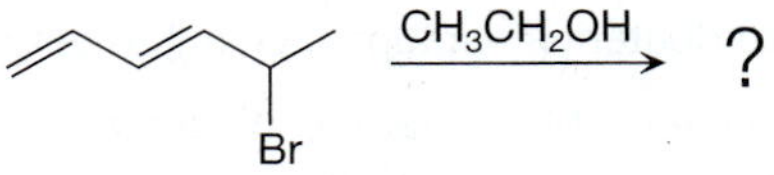

8.58 The reaction shown here proceeds via a carbocation rearrangement. Draw a complete, detailed mechanism to account for the product. Explain why the carbocation rearrangement is favorable.

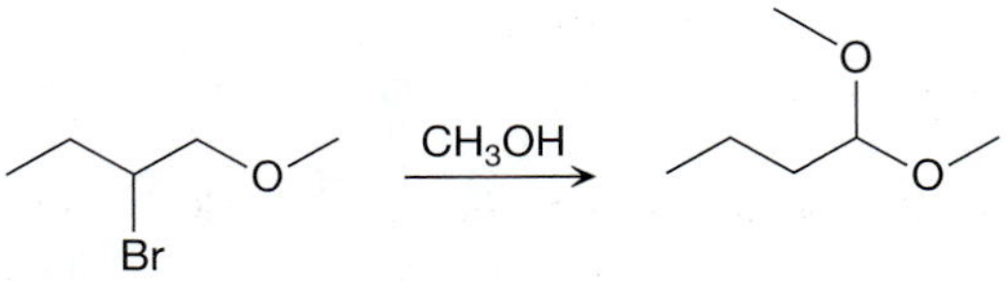

8.59 One way to synthesize diethyl ether is to heat ethanol in the presence of a strong acid, as shown here. Draw a complete, detailed mechanism for this reaction.

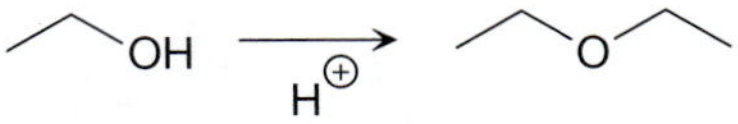

8.60 Suggest a reasonable mechanism for the reaction shown here.

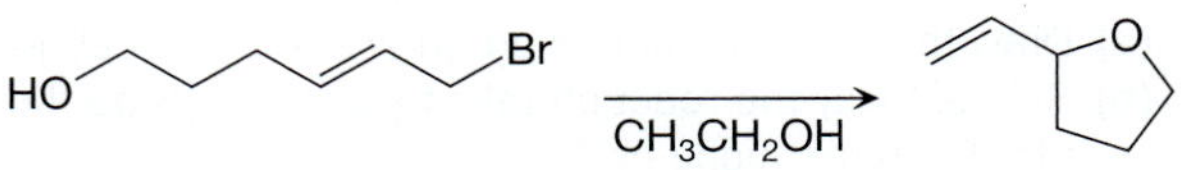

Integrated Problems

8.61 Consider this E1 reaction. **(a)** Draw a complete, detailed mechanism for the reaction. **(b)** Draw a reaction free energy diagram that agrees with that mechanism, labeling overall reactants, overall products, all transition states, and all intermediates.

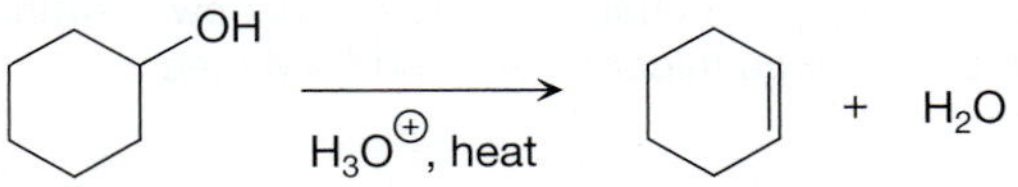

8.62 Draw a reasonable, detailed mechanism that shows this racemization at the α (alpha) carbon. *Note:* The reaction takes place under basic conditions.

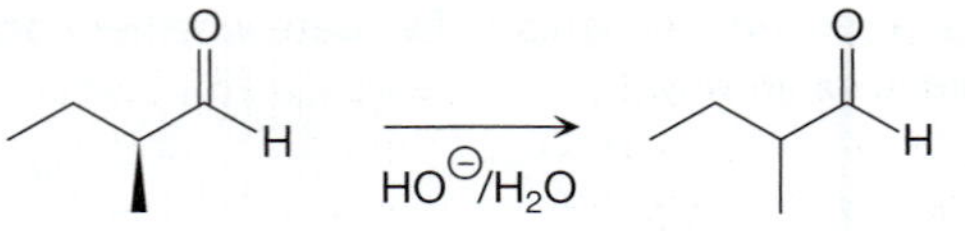

Racemic mixture

8.63 Draw a reasonable, detailed mechanism that shows this racemization at the α (alpha) carbon. *Note:* The reaction takes place under acidic conditions.

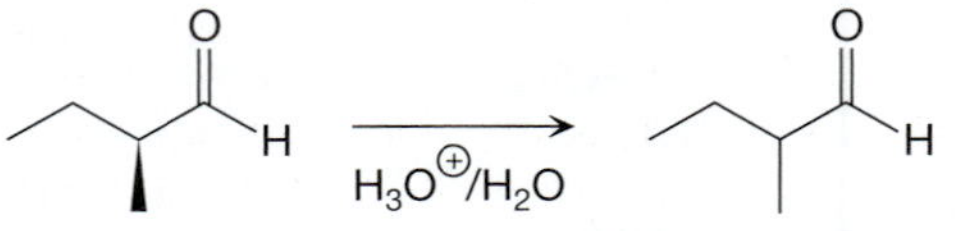

Racemic mixture

8.64 The elimination reaction at the right (top) yields the same alkene product independent of whether it proceeds by the E2 or E1 mechanism. The mechanism by which the reaction proceeds could be determined if the D-labeled substrate shown at the right (bottom) were used instead. *Note:* Deuterium (D) is an isotope of H. They both have one proton and one electron, so they have nearly identical chemical properties, but D, having one additional neutron, is heavier by 1 u.

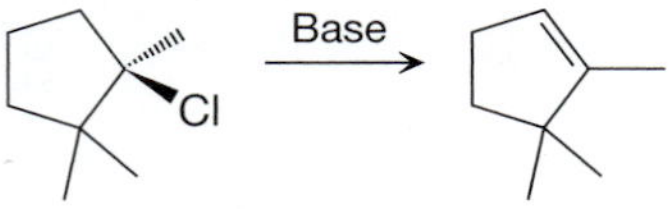

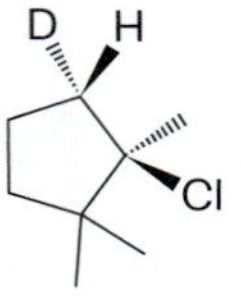

(a) Draw the complete mechanism for an E2 reaction involving the D-labeled substrate and predict the major product(s).
(b) Draw the complete mechanism for an E1 reaction involving the D-labeled substrate and predict the major product(s).
(c) What is the molar mass of each product from (a) and (b)?

8.65 According to the rules for reasonable mechanisms, the E1 reaction shown here should undergo a 1,2-hydride shift. However, the same product is produced regardless of whether the rearrangement occurs.

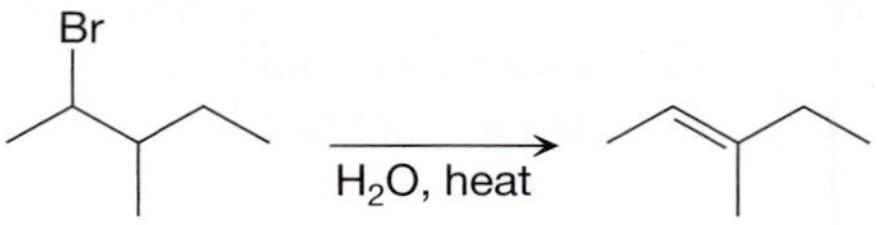

(a) Draw the mechanism that includes that carbocation rearrangement.
(b) Draw the mechanism that does not include the rearrangement.
(c) Experimentally, how can we use ^{13}C isotope labeling to determine whether the rearrangement occurs? In other words, can a ^{12}C atom in the substrate be replaced by a ^{13}C atom so that the E1 products would depend on whether the rearrangement takes place?
(d) How can we use deuterium isotope labeling to determine whether the rearrangement occurs?

8.66 (S)-Adenosylmethionine (SAM) is a cosubstrate that is involved in biological methyl group transfers. SAM is believed to be produced by an S_N2 type of reaction between methionine and ATP, as shown below.

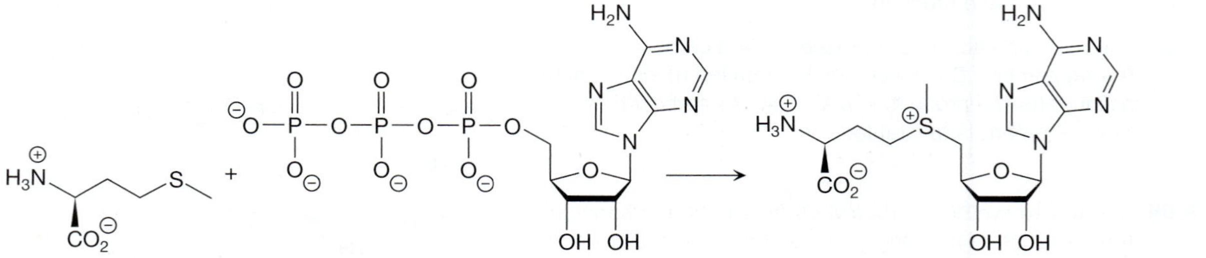

Methionine   **ATP**   **SAM**

(a) Draw the appropriate curved arrows for this reaction.
(b) Suggest why the reaction takes place with S as the nucleophilic atom instead of one of the negatively charged O atoms on methionine.

8.67 Creatine is a naturally occurring compound that helps provide energy to cells in the body, especially muscle cells. Draw an S_N2 mechanism that shows how creatine is produced from guanidoacetate and SAM (see Problem 8.66).

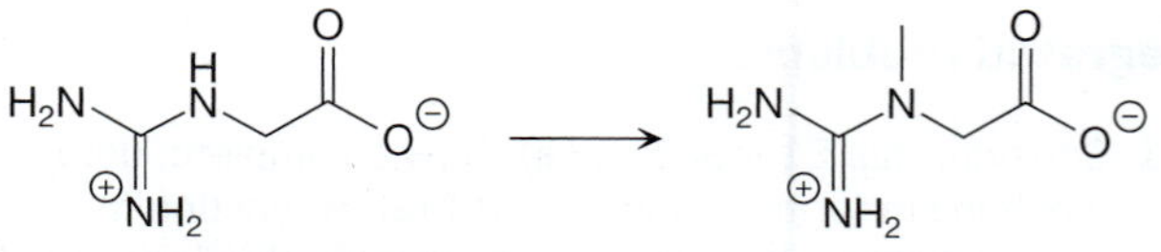

Guanidoacetate   **Creatine**

8.68 One of the ways in which an L α-amino acid can be synthesized is to carry out an S$_N$2 reaction between an α-bromo acid and ammonia. (The wavy line indicates that the bond could be a dash or a wedge.) Draw the stereoisomer of the α-bromo acid that would be necessary to produce L-alanine, in which the R group is CH$_3$.

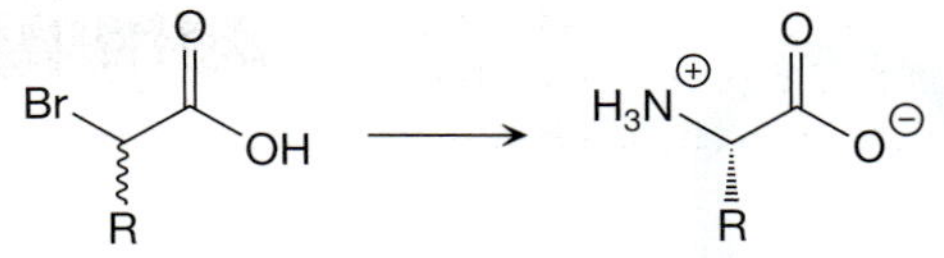

An α-bromo acid   **An L α-amino acid**

8.69 Structures **A** and **B** are intermediates in the biosynthesis of steroids. **(a)** Draw the mechanism (including curved arrows) that shows how **A** can be converted to **B** through two 1,2-hydride shifts followed by two 1,2-methyl shifts. **(b)** Draw the mechanism (including curved arrows) that shows how **B** is converted to lanosterol.

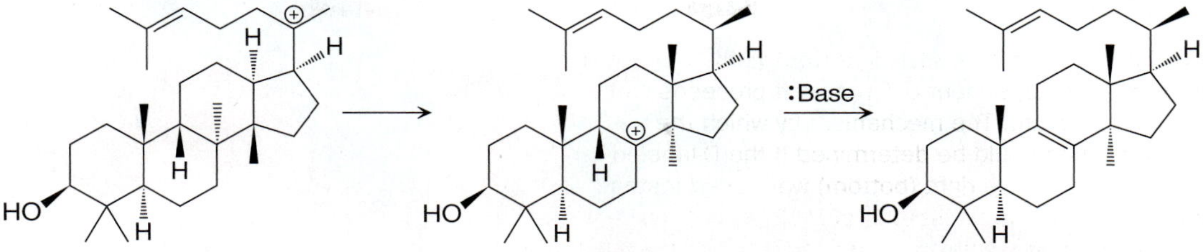

Intermediate **A**   Intermediate **B**   **Lanosterol**

8.70 The reaction shown here is called the pinacol rearrangement. A carbocation rearrangement is believed to be involved. **(a)** Propose a reasonable mechanism for this reaction. **(b)** Suggest why the carbocation rearrangement is favorable.

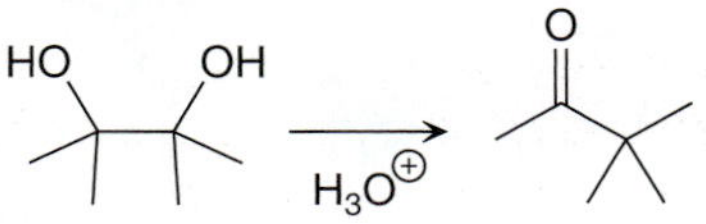

8.71 Propose a mechanism for the reaction shown here, which produces 1,4-dioxane.

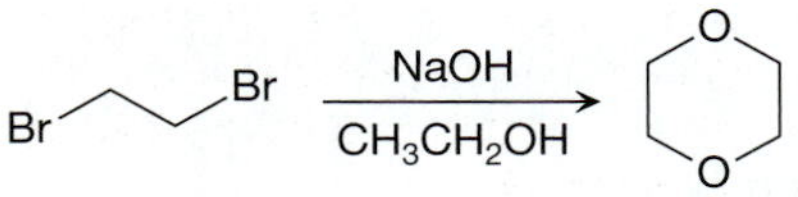

1,4-Dioxane

8.72 The specific angle of rotation of (R)-2-bromobutane is −23.1°. Treatment of (R)-2-bromobutane with potassium bromide produces a racemic mixture of (R)- and (S)-2-bromobutane, which is optically inactive. The rate at which the product's angle of rotation decreases is directly proportional to the concentration of KBr. Does this suggest that the reaction takes place by an S$_N$1 or an S$_N$2 mechanism? Explain.

8.73 Propose a mechanism for the reaction shown here, which takes place under conditions that favor an S$_N$1 reaction.

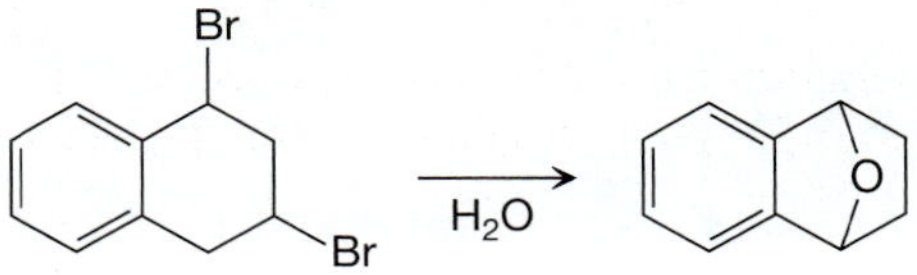

8.74 Propose a mechanism for the reaction shown here, which takes place under conditions that favor an S$_N$1 reaction. Based on the mechanism, do you think that the products will be formed in a mixture of stereoisomers?

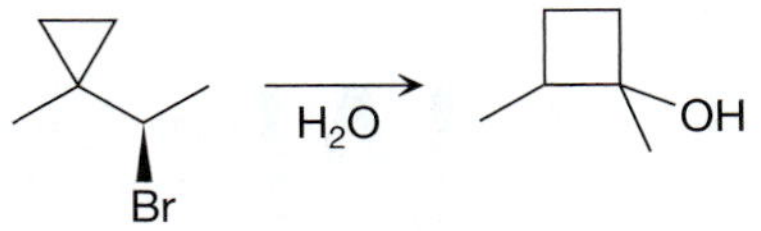

8.75 A chemist proposes that the reaction shown here proceeds by an S$_N$2 mechanism. She carries out the reaction with oxygen-18 (^{18}O)-labeled hydroxide in ^{18}O-labeled water (i.e., ^{18}OH$^-$/H$_2$^{18}O). When analyzing the products, she finds that the ^{18}O isotope appeared only in CH$_3$CO$_2^-$ and not in CH$_3$OH. What does this result suggest about her hypothesis?

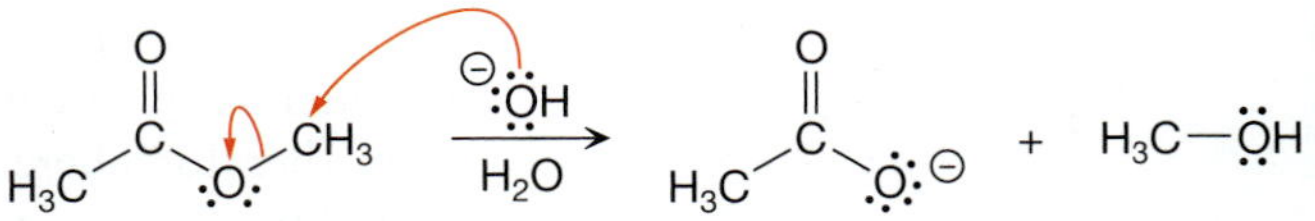

8.76 A chemist proposes that the following reaction occurs via an S$_N$2 mechanism.

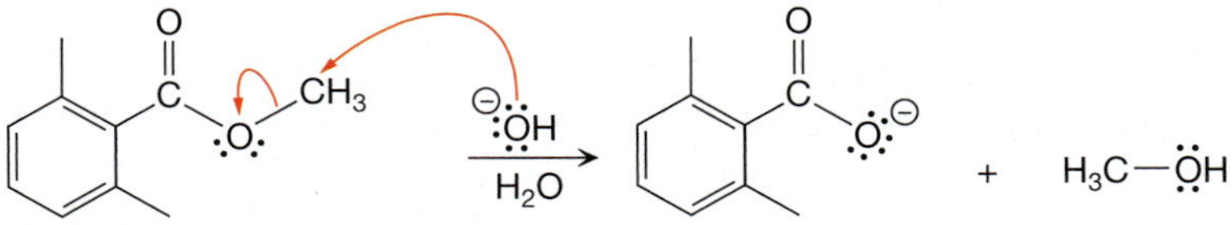

On carrying out the reaction using ^{18}O-labeled hydroxide anion in ^{18}O-labeled water, she finds that ^{18}O-labeled methanol is produced. What does this suggest about her hypothesis?

9

In this children's game, players compete to accumulate the most marbles in their respective bins. Analogously, S_N2, S_N1, E2, and E1 reactions compete to establish the major product.

Nucleophilic Substitution and Elimination Reactions 1

Competition among S_N2, S_N1, E2, and E1 Reactions

W e have studied nucleophilic substitution and elimination reactions extensively in the last two chapters: S_N2 and E2 in Chapter 7, and S_N1 and E1 in Chapter 8. Up to this point, we have considered these four reactions as if they were independent of one another. They are generally in competition, however, so under most circumstances, *if one reaction is feasible, we must consider all four of them*. Therefore, here in Chapter 9 we first examine how a number of factors affect each of the four reactions, then we see how to use that knowledge to predict reaction products.

Sometimes a competition can take place between reactions that proceed by the same mechanism. One example we examine here in Chapter 9 deals with *regioselectivity*—the tendency of a particular reaction to be favored at one site within a molecule over another. Another example we examine involves the competition between an *intermolecular reaction* (involving functional groups on *separate* reactant species) with an *intramolecular reaction* (involving functional groups on the *same* reactant).

Many of the reactions we will encounter in subsequent chapters also involve competitions. Thus, the ideas we learn in Chapter 9 pertaining to nucleophilic substitution and elimination reactions will be applied throughout the book.

Chapter Objectives

On completing Chapter 9 you should be able to:

1. Recognize suitable *substrates* for nucleophilic substitution and elimination reactions and draw the S_N2, S_N1, E2, and E1 mechanisms that compete with each other when a substrate is treated with an attacking species.

2. Explain why S_N2, S_N1, E2, and E1 reactions tend to compete under kinetic control and apply the principles behind the Hammond postulate to rationalize the impacts that certain factors have on the rates of these reactions.

3. Determine the strength of an attacking species as a nucleophile and as a base, and on that basis predict whether an S_N2, S_N1, E2, or E1 reaction is favored.

4. Determine which reactions—S_N2, S_N1, E2, or E1—are favored by high concentration of the attacking species and which are favored by low concentration of the attacking species.

5. Distinguish among good, moderate, and poor leaving groups, and specify which substitution and elimination reactions each type of leaving group favors, if any at all.

6. Establish whether nucleophilic substitution and elimination reactions are feasible based on the hybridization of the carbon atom bonded to the leaving group.

7. Classify the carbon atom bonded to the leaving group as either primary (1°), secondary (2°), or tertiary (3°), and on that basis predict which substitution and elimination reactions are favored.

8. Identify a solvent as either protic or aprotic, and on that basis determine which reactions—S_N2, S_N1, E2, or E1—are favored.

9. Justify why the relative strengths of *some* nucleophiles are reversed in protic solvents versus aprotic solvents.

10. Explain the role of heat in nucleophilic substitution and elimination reactions.

11. Predict the major product(s) of a given nucleophilic substitution or elimination reaction by systematically evaluating the factors affecting the reaction.

12. Recognize when an elimination reaction can produce two or more alkene products, and predict which is the major product.

13. Predict the major product of competing intermolecular and intramolecular substitution reactions.

9.1 The Competition among S_N2, S_N1, E2, and E1 Reactions

A competition usually exists among S_N2, S_N1, E2, and E1 reactions. In the four reactions shown in Equations 9-1 through 9-4, for example, the reactants are identical, but the mechanisms are different, which can lead to different products.

There are essentially two reasons that this competition occurs:

1. S_N2, S_N1, E2, and E1 reactions all involve a substrate containing a *leaving group*.
2. Any species that can act as a nucleophile also has the potential to act as a base, and vice versa. Such a species is called an **attacking species**.

Acting as a nucleophile, an attacking species uses a lone pair of electrons to form a bond to an electron-poor *nonhydrogen atom*. Acting as a base, on the other hand, an attacking species forms a bond to a *proton*. Notice that $CH_3CO_2^-$ acts as a nucleophile in Equations 9-1 and 9-2, but acts as a base in Equations 9-3 and 9-4.

S_N2 mechanism

The attacking species acts as a nucleophile.

(9-1)

SN1 mechanism

The attacking species acts as a nucleophile.

(9-2)

E2 mechanism

The attacking species acts as a base.

(9-3)

E1 mechanism

The attacking species acts as a base.

(9-4)

YOUR TURN 9.1

Which of the following reactions, **A** or **B**, shows NH_3 acting as a base and which shows NH_3 acting as a nucleophile?

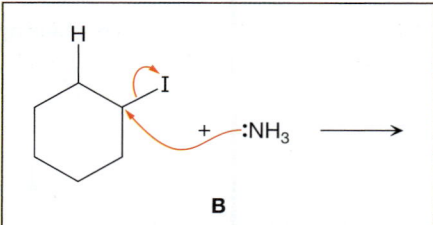

A

B

Answers to Your Turns are in the back of the book.

PROBLEM 9.1 Draw the complete, detailed mechanisms for the S_N2, S_N1, E2, and E1 reactions between iodocyclohexane and ammonia. Include the necessary curved arrows and draw the products.

How can we predict the major products of $S_N2/S_N1/E2/E1$ competitions? We first need to know whether the competition takes place under *kinetic control* or *thermodynamic control*.

- In a competition that takes place under **kinetic control**, the major product is the one that is produced the *fastest*.
- Under **thermodynamic control**, the major product is the one that is the *most stable* (i.e., the lowest energy).

As we discuss more fully in Section 9.12, whether reactions compete under kinetic control or thermodynamic control depends on the nature of the reactants and products and on the reaction conditions—topics of Sections 9.3–9.8 for S_N2, S_N1, E2, and E1 reactions. For now, suffice it to say that:

The $S_N2/S_N1/E2/E1$ competition usually takes place under *kinetic control*.

Thus, we must turn our attention to the various factors that influence the relative *rates* of the S_N2, S_N1, E2, and E1 reactions.

9.2 Rate-Determining Steps Revisited: Simplified Pictures of the S_N2, S_N1, E2, and E1 Reactions

As we learned in Section 9.1, the outcome of an $S_N2/S_N1/E2/E1$ competition is dictated by the relative rates of the reactions. Moreover, recall from Section 8.4 that the rate of an overall reaction is essentially the same as the rate of the rate-determining step.

The S_N2 mechanism consists of just a single step, so that single step must be rate determining:

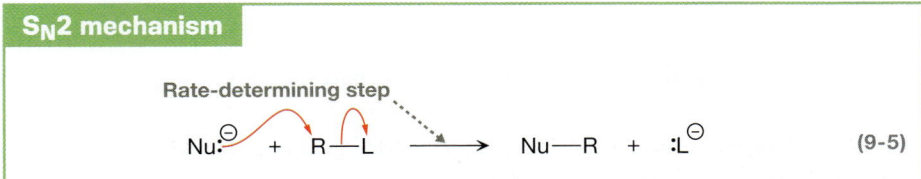

S_N2 mechanism

Rate-determining step

$$Nu:^{\ominus} + R-L \longrightarrow Nu-R + :L^{\ominus} \qquad (9\text{-}5)$$

The S_N1 reaction, on the other hand, takes place in two steps. The first step is rate determining because formation of the carbocation is so slow:

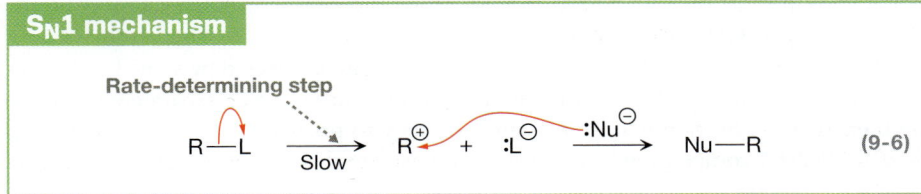

S_N1 mechanism

Rate-determining step

$$R-L \xrightarrow{\text{Slow}} R^{\oplus} + :L^{\ominus} \xrightarrow{:Nu^{\ominus}} Nu-R \qquad (9\text{-}6)$$

In the rate-determining step of an S_N2 reaction, the nucleophile forms a bond to the substrate at the same time the leaving group leaves. In the rate-determining step of an S_N1 reaction, on the other hand, the leaving group leaves on its own. The nucleophile is present in solution throughout the reaction, even though it does not enter the mechanism until the second step. Therefore, the role of the nucleophile differs in S_N2 and S_N1 reactions:

- In an S_N2 reaction, the nucleophile forces the leaving group out.
- In an S_N1 reaction, the nucleophile waits until the leaving group has left.

As a result, the rate of an S_N2 reaction is highly sensitive to factors that affect the nucleophile's ability to attack the substrate and to factors that affect the ability of the leaving group to depart. The rate of an S_N1 reaction, on the other hand, is highly sensitive only to factors that help the leaving group depart.

Like an S_N2 reaction, an E2 reaction (Equation 9-7) consists of a single step that must be rate determining. And, as in an S_N1 reaction, the rate-determining step of an E1 reaction (Equation 9-8) is the first of its two steps.

E2 mechanism

Rate-determining step

(9-7)

E1 mechanism

Rate-determining step

(9-8)

In the E2 reaction, the substrate is deprotonated at the same time the leaving group leaves. In an E1 reaction, the base does not enter the picture until the second step, even though it is in solution throughout the reaction. That is:

- In an E2 reaction, the base pulls off the proton, thus forcing the leaving group to leave.
- In an E1 reaction, the base waits until the leaving group has left.

Consequently, the rate of an E2 reaction is highly sensitive to factors that affect the ability of the base to pull off the proton and to factors that affect the ability of the leaving group to leave. And, like an S_N1 reaction, the rate of an E1 reaction is highly sensitive only to factors that help the leaving group to depart.

We are now ready to consider how each reaction rate is affected by some key factors. In Sections 9.3 through 9.8 we thoroughly examine those factors separately from one another. Then, in Section 9.9, we present a strategy to predict the major product of an $S_N2/S_N1/E2/E1$ competition by bringing together all of these factors in a systematic way.

9.3 Factor 1: Strength of the Attacking Species

Because the S_N2, S_N1, E2, and E1 reactions are sensitive to the attacking species in different ways, the *identity* of the attacking species can play a major role in the outcome of the $S_N2/S_N1/E2/E1$ competition. Here in Section 9.3, we examine how the attacking species affects each of these reactions differently. First we discuss the nature of the attacking species in S_N2 and S_N1 reactions, in which it behaves as a nucleophile. Then we turn our attention to E2 and E1 reactions, in which the attacking species behaves as a base.

9.3a The Nucleophile Strength in S_N2 and S_N1 Reactions: An Introduction to the Hammond Postulate

Table 9-1 (next page) lists the S_N2 reaction rates for various nucleophiles attacking CH_3I in the solvent *N,N*-dimethylformamide (DMF). These rate differences reflect differences in *nucleophile strength*, or **nucleophilicity**.

> The stronger nucleophile promotes a faster S_N2 reaction.

Thus, the relative nucleophilicities of Br^-, Cl^-, and NC^- in the solvent DMF are 1, 2, and 250, respectively. A smaller energy barrier leads to a faster reaction (Section 8.4), so the energy barrier for an S_N2 reaction must be smaller when Cl^- is the nucleophile than when Br^- is. The energy barrier must be smaller still when NC^- is the nucleophile.

To better understand these variations in the size of the energy barrier, we turn to the **Hammond postulate**, proposed in 1955 by the American chemist George S. Hammond: "If two states . . . occur consecutively during a reaction process and have nearly the same energy content, their interconversion will involve only a small reorganization of the molecular structures." One of the main lessons of the Hammond postulate comes from its application to transition states:

- For an elementary step whose $\Delta G°_{rxn}$ is negative (i.e., is *exergonic*), the transition state resembles the reactants more than it does the products, in both structure and energy (Fig. 9-1a).
- For an elementary step whose $\Delta G°_{rxn}$ is positive (i.e., is *endergonic*), the transition state resembles the products more than it does the reactants, in both structure and energy (Fig. 9-1b).

If the energy diagrams in Figure 9-1a and 9-1b have the same energy scale on the vertical axis and represent reactant species undergoing very similar structural changes, the energy barrier appears to be smaller when the value of $\Delta G°_{rxn}$ is negative instead of positive. Thus, we arrive at a very useful guideline that allows us to understand and make predictions about relative reaction rates:

> For two elementary steps that are of the same type (e.g., both S_N2 or both E2), the one with the more negative (less positive) value of $\Delta G°_{rxn}$ tends to have the smaller energy barrier, and thus tends to be faster.

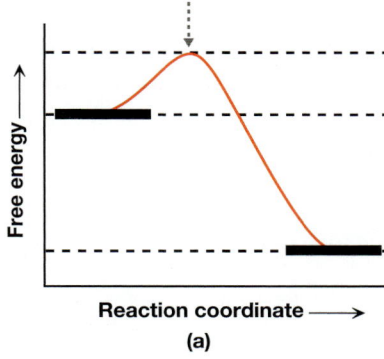

An exergonic reaction

The transition state lies closer to the reactants than to the products, both along the energy axis and along the reaction coordinate.

(a)

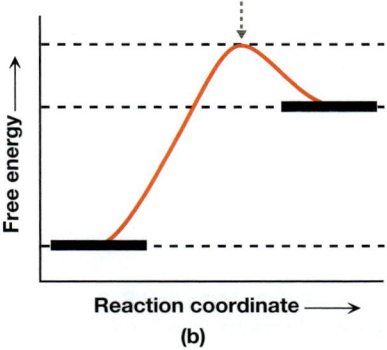

An endergonic reaction

The transition state lies closer to the products than to the reactants, both along the energy axis and along the reaction coordinate.

(b)

FIGURE 9-1 The Hammond postulate and $\Delta G°_{rxn}$ (a) Free energy diagram for an exergonic reaction, showing that the transition state lies closer in energy to the reactants than to the products, and that its structure resembles reactants more than it does products. (b) Free energy diagram for an endergonic reaction, showing that the transition state lies closer in energy to the products than to the reactants, and that its structure resembles products more than it does reactants.

| TABLE 9-1 | S_N2 Reaction Rate in DMF of the Reaction: |

$$Nu^{\ominus} + CH_3I \longrightarrow NuCH_3 + I^{\ominus}$$

Nucleophile	H_2O	pyridine $\overset{..}{N}$:	$NCS^{\ominus}$	$Br^{\ominus}$	$Cl^{\ominus}$	$N_3^{\ominus}$	$C_6H_5S^{\ominus}$	$CH_3CO_2^{\ominus}$	$NC^{\ominus}$
Relative Reaction Rate	~0	~0.0005	0.06	1	2	3	12	15	250

Nucleophile strength increases ⟶

YOUR TURN 9.2

Draw an arrow in Figure 9-1a to indicate the energy barrier of the exergonic reaction. Do the same for the energonic reaction in Figure 9-1b. Assuming that the energy diagrams have the same energy scale on the vertical axis, what do you notice?

To see how this guideline applies to S_N2 reactions, consider the reactions in Equations 9-9 and 9-10, involving Cl^- and Br^- as nucleophiles.

$$:\overset{..}{\underset{..}{Cl}}{}^{\ominus} \quad H_3C \overset{}{\longrightarrow} \overset{..}{\underset{..}{I}}: \quad \longrightarrow \quad :\overset{..}{\underset{..}{Cl}} \text{—} CH_3 \;+\; :\overset{..}{\underset{..}{I}}:{}^{\ominus} \qquad (9\text{-}9)$$

$$:\overset{..}{\underset{..}{Br}}:{}^{\ominus} \quad H_3C \overset{}{\longrightarrow} \overset{..}{\underset{..}{I}}: \quad \longrightarrow \quad :\overset{..}{\underset{..}{Br}} \text{—} CH_3 \;+\; :\overset{..}{\underset{..}{I}}:{}^{\ominus} \qquad (9\text{-}10)$$

The free energy diagrams for these two reactions are shown together in Figure 9-2.

As we learned in Section 7.8, the relative energies of the reactants and products are governed largely by charge stability. In both cases, ΔG_{rxn}° is negative because I^- is more stable than either Br^- or Cl^-. (Recall from Section 6.6b that larger atoms can

FIGURE 9-2 ΔG_{rxn}° **and S_N2 energy barriers** Free energy diagrams for the S_N2 reactions of Cl^- (red) and Br^- (blue) with CH_3I. With Cl^- as the nucleophile, the reaction is energetically more favorable. Consequently, the transition state involving Cl^- as the nucleophile lies closer in energy to the reactants than does the transition state involving Br^- as the nucleophile. This corresponds to a smaller energy barrier, and thus a faster reaction, when Cl^- is the nucleophile.

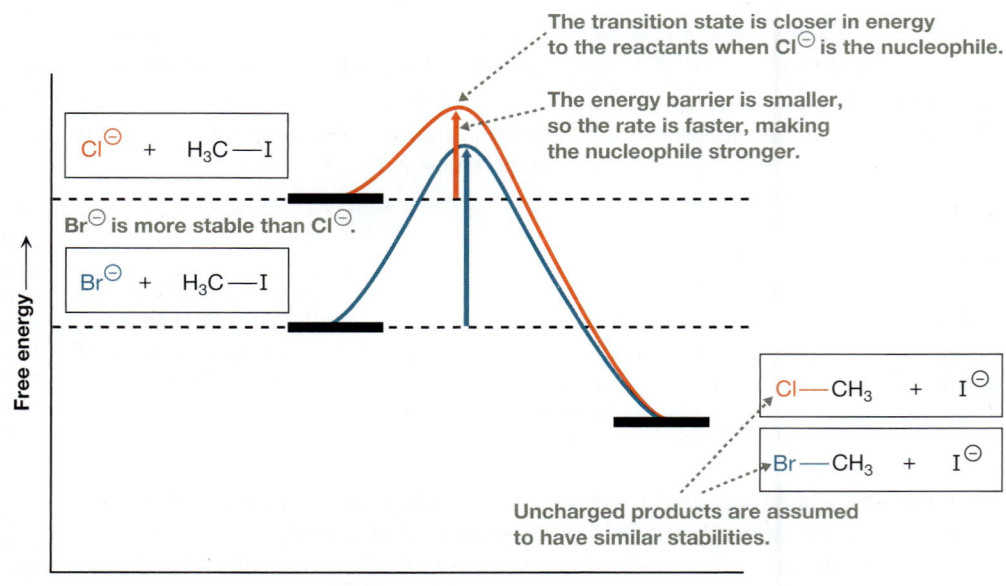

The transition state is closer in energy to the reactants when $Cl^{\ominus}$ is the nucleophile.

The energy barrier is smaller, so the rate is faster, making the nucleophile stronger.

$Cl^{\ominus} + H_3C\text{—}I$

$Br^{\ominus}$ is more stable than $Cl^{\ominus}$.

$Br^{\ominus} + H_3C\text{—}I$

Free energy ⟶

$Cl\text{—}CH_3 + I^{\ominus}$

$Br\text{—}CH_3 + I^{\ominus}$

Uncharged products are assumed to have similar stabilities.

Reaction coordinate ⟶

accommodate charges better.) Moreover, Br$^-$ is more stable than Cl$^-$ (Br is larger), making the set of reactants belonging to the blue curve lower in energy than the set of reactants belonging to the red curve. Thus, the reaction involving Cl$^-$ as the nucleophile (the red curve) has the more negative value for ΔG°_{rxn}. According to the guideline just discussed, this corresponds to a smaller energy barrier—and thus a faster rate—for the Cl$^-$ reaction, which agrees with the relative rates we saw in Table 9-1.

PROBLEM 9.2 Use the values in Table 9-1 to draw a free energy diagram, similar to the one in Figure 9-2, comparing the S$_N$2 reactions of NC$^-$ and Br$^-$ with CH$_3$I.

The principles we have learned in this section are quite helpful in predicting the relative strengths of nucleophiles *not* listed in Table 9-1, including uncharged nucleophiles (see Solved Problem 9.3). Moreover, as we will see on several occasions, both in this chapter and throughout the book, these principles represent powerful tools that can help us understand the relative reactivities of species in other reactions.

SOLVED PROBLEM 9.3

Which nucleophile will react faster with CH$_3$I in an S$_N$2 reaction: H$_2$O or H$_2$S? What does this indicate about their relative nucleophilicities?

Think What are the products of the two reactions? Are the products higher or lower in energy than the reactants? Do the reactants of the two reactions differ significantly in energy? Do the products? Which reaction has a more negative (less positive) value for ΔG°_{rxn}, and how should that guide the way you draw the free energy diagrams?

Solve The two reactions are as follows:

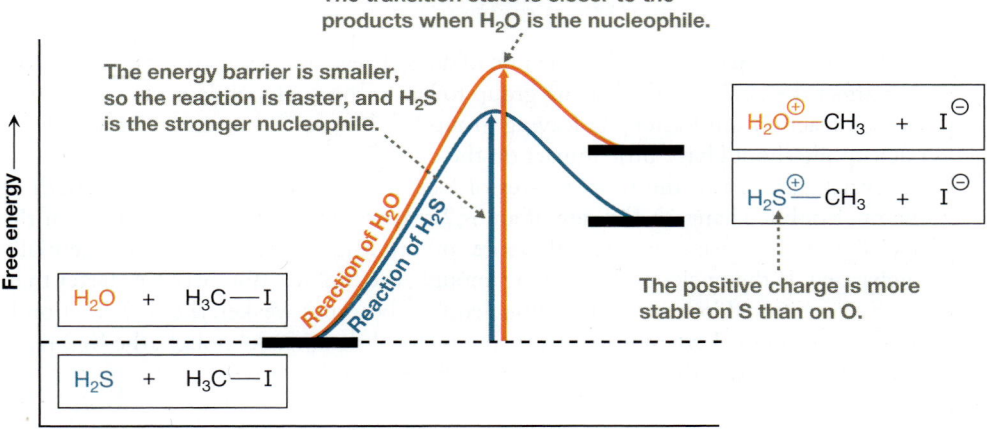

The free energy diagrams for these reactions are as follows:

The transition state is closer to the products when H$_2$O is the nucleophile.

The energy barrier is smaller, so the reaction is faster, and H$_2$S is the stronger nucleophile.

Reaction of H$_2$O

Reaction of H$_2$S

Free energy →

H$_2$O + H$_3$C — I

H$_2$S + H$_3$C — I

H$_2$O$^{⊕}$— CH$_3$ + I$^{⊖}$

H$_2$S$^{⊕}$— CH$_3$ + I$^{⊖}$

The positive charge is more stable on S than on O.

Reaction coordinate →

Both reactions are endergonic because two new charges are produced. Both sets of reactants are uncharged, so they are placed at the same energy. The set of products from the H$_2$S reaction (the blue curve) is lower in energy than the set of products from the H$_2$O reaction (the red curve) because a positive charge is more stable on the larger S atom than it is on the smaller O atom. Consequently, the reaction in which H_2S is the nucleophile has a less positive value for ΔG°_{rxn} and, according to the guideline derived from the Hammond postulate, should have a smaller activation energy. Thus, H_2S is a stronger nucleophile than H_2O.

PROBLEM 9.4 **(a)** For each pair of nucleophiles, predict which will react faster with CH_3I in an S_N2 reaction: **(i)** CH_3O^- or $CH_3CO_2^-$; **(ii)** H_3N or H_3P. **(b)** Which of each pair is the stronger nucleophile?

How, then, does the nucleophile influence the rate of an S_N1 reaction, such as those shown in Equation 9-11a and 9-11b? The substrate in both reactions is chlorodiphenylmethane, but the nucleophiles are the thiocyanate ion (NCS^-) in Equation 9-11a and the azide anion (N_3^-) in Equation 9-11b.

S_N1 reactions

The rates of both reactions are nearly the same.

(9-11a)

(9-11b)

Recall from Table 9-1 that N_3^- is a stronger nucleophile than NCS^- by a factor of roughly 50. The rates of the two reactions in Equation 9-11, however, are about the same. It turns out that:

The rate of an S_N1 reaction is essentially independent of the strength of the nucleophile.

Recall from Section 9.2 that the rate of an S_N1 reaction is sensitive only to factors that affect the ability of the leaving group to leave. The nucleophile is not involved in an S_N1 reaction until after the leaving group has left, so the specific identity of the nucleophile should have little impact on the overall rate.

What happens to the *relative* rates of S_N2 and S_N1 reactions as the strength of the nucleophile changes? The rate of an S_N2 reaction increases as the strength of the nucleophile increases, whereas the rate of an S_N1 reaction remains essentially unchanged. If the nucleophile is strong enough, the S_N2 reaction becomes faster than the S_N1 reaction. Conversely, as the nucleophile becomes weaker, the S_N2 reaction is slowed, but the S_N1 reaction is not. If the nucleophile is weak enough, the S_N2 reaction becomes slower than the S_N1 reaction. Stated another way:

- Strong nucleophiles tend to favor S_N2 reactions.
- Weak nucleophiles tend to favor S_N1 reactions.

The distinction between "strong" and "weak" nucleophiles is purely empirical. In practice, with all else being equal, S_N2 reactions are usually favored by nucleophiles

$$\text{Weak} \qquad\qquad\qquad\qquad\qquad\qquad\qquad\qquad\qquad \text{Strong}$$

$$H_2O, ROH < H_3N, R_2NH < I^\ominus < Br^\ominus < Cl^\ominus < HS^\ominus, RS^\ominus \approx RCO_2^\ominus < HO^\ominus, RO^\ominus < H_2N^\ominus, R_2N^\ominus < H_3C^\ominus, R^\ominus$$

Intrinsic nucleophile strength increases

→

Stability of negative charge decreases

FIGURE 9-3 Intrinsic nucleophile strength Uncharged nucleophiles tend to be intrinsically weak, whereas negatively charged nucleophiles tend to be strong.

bearing a total -1 charge, whereas S_N1 reactions are usually favored by uncharged nucleophiles. Thus:

- Strong nucleophiles tend to have a total -1 charge.
- Weak nucleophiles tend to be uncharged.

Examples of strong nucleophiles include the halide anions Cl^-, Br^-, and I^-, as well as HO^-, alkoxide anions (RO^-), thiolate anions (RS^-), and deprotonated amines (R_2N^-) (Fig. 9-3). Organometallic reagents, which behave as R^-, are strong nucleophiles if the partial negative charge on carbon is sufficiently concentrated. These include alkyllithium (R—Li) and Grignard (R—MgX) reagents. Examples of weak nucleophiles include H_2O, alcohols (ROH), and amines (R_2NH).

PROBLEM 9.5 Is H_2P^- a strong nucleophile or a weak nucleophile? Will it favor the S_N2 or the S_N1 mechanism? Explain.

9.3b Generating Carbon Nucleophiles

As we will see in Chapter 13, nucleophilic carbon atoms are often important in synthesis, especially in the formation of carbon–carbon bonds. In uncharged molecules, however, carbon atoms are typically non-nucleophilic for two reasons: (1) They do not possess a lone pair of electrons, and (2) they are rarely considered electron rich because they are generally bonded to atoms with electronegativities comparable to their own (e.g., hydrogen and other carbon atoms), if not greater (e.g., oxygen, nitrogen, and halogens).

A carbon atom is quite nucleophilic, however, when it bears a -1 formal charge—that is, when it is a *carbanion* (Fig. 9-4). These carbon atoms not only are electron rich, but also possess a lone pair of electrons that can be used to form a bond.

It is often beneficial, then, to generate carbanions from uncharged carbon atoms. The simplest way to do so would be to deprotonate the uncharged carbon, but carbon atoms typically do not possess acidic hydrogens. Alkanes, for example, have pK_a values around 50, so they are such weak acids that deprotonation is unfeasible.

Some carbon atoms, on the other hand, are deprotonated much more readily. A terminal alkyne (RC≡C—H), for example, is weakly acidic ($pK_a \approx 25$) because the alkyne C atom is sp hybridized. Recall from Section 3.11 that an sp-hybridized atom has a greater effective electronegativity than its sp^3-hybridized counterpart, so it can better stabilize a negative charge.

Once deprotonated (Equation 9-12a), the resulting **alkynide anion** (RC≡C⁻) can behave as a strong nucleophile, as shown in Equation 9-12b. Note the new C—C bond that is formed.

Not nucleophilic   A strong nucleophile

FIGURE 9-4 Carbon nucleophiles An uncharged, tetrahedral C atom (*left*) is not nucleophilic. A carbanion (*right*) is a strong nucleophile because of the negative charge and the lone pair of electrons.

$$R-C\equiv CH \ + \ NaH \ \longrightarrow \ R-C\equiv C:^{\ominus} \ + \ H-H \ + \ Na^{\oplus} \qquad (9\text{-}12a)$$

Alkynide anion

$$R-C\equiv C:^{\ominus} \ + \ R'-Br \ \xrightarrow{\ S_N2\ } \ R-C\equiv C-R' \ + \ Br^{\ominus} \qquad (9\text{-}12b)$$

Strong nucleophile   **New C—C bond**

Other types of compounds also have weakly acidic protons on carbon atoms. Ketones (R—CR=O), aldehydes (R—CH=O), and nitriles (R—C≡N) all have acidic α hydrogens (i.e., on carbons that are *adjacent* to the C=O and C≡N group), and compounds containing these functional groups can be converted into strong carbon nucleophiles as well. This is discussed in greater detail in Section 10.3.

YOUR TURN 9.3

Hydrocyanic acid (HCN), like a terminal alkyne, can be converted into a carbon nucleophile by treatment with a sufficiently strong base such as HO⁻, as shown here. Draw the products of this reaction and label the nucleophilic atom that is produced.

$$N\equiv C-H \ + \ :\!\ddot{O}H^{\ominus} \ \longrightarrow$$

Hydrocyanic acid

CONNECTIONS HCN is found naturally in very small concentrations in the pits or seeds of some fruits, such as cherries and apples. The compound has a number of industrial uses, including as a precursor to sodium cyanide, NaCN, which is used in gold and silver mining to separate the metal from the ore.

PROBLEM 9.6 Draw the complete, detailed mechanism for the S$_N$2 reaction that takes place when hex-1-yne is treated with NaH, followed by treatment with bromoethane.

9.3c The Base Strength in E2 and E1 Reactions

Table 9-2 lists the relative rates of E2 reactions involving 1,2-dichloroethane as the substrate. These data show that the relative rates of E2 reactions depend on the *strength* of the base.

TABLE 9-2 E2 Reaction Rates in H$_2$O for the Reaction:

$$Base^{\ominus} \ + \ ClHC\underset{\underset{Cl}{|}}{\overset{\overset{H}{|}}{-}}CH_2 \ \longrightarrow \ Base-H \ + \ ClHC=CH_2 \ + \ Cl^{\ominus}$$

Base	H₃C–C(=O)–O⁻	pyridine	C₆H₅O⁻ (phenoxide)	(CH₃)₂N–CH₃ (trimethylamine)	HO⁻
Relative Reaction Rate	1	3	40	60	353
pK_a of Base—H	4.75	5.2	10.0	9.8	15.7

The rate of an E2 reaction generally increases as the strength of the base increases (i.e., as the pK_a of the base's conjugate acid increases).

The base in an E2 reaction pulls off the proton, allowing the electrons from the C—H bond to form a new π bond to the neighboring carbon, thereby forcing the leaving group to leave. The stronger the base, the faster this process can occur.

Free energy diagrams can provide insight into how the rate of an E2 reaction depends on the strength of the base. Consider the E2 reactions in Equations 9-13 and 9-14, in which 1,2-dichloroethane is the substrate, and the acetate anion ($CH_3CO_2^-$) and hydroxide anion (HO^-), respectively, are the bases.

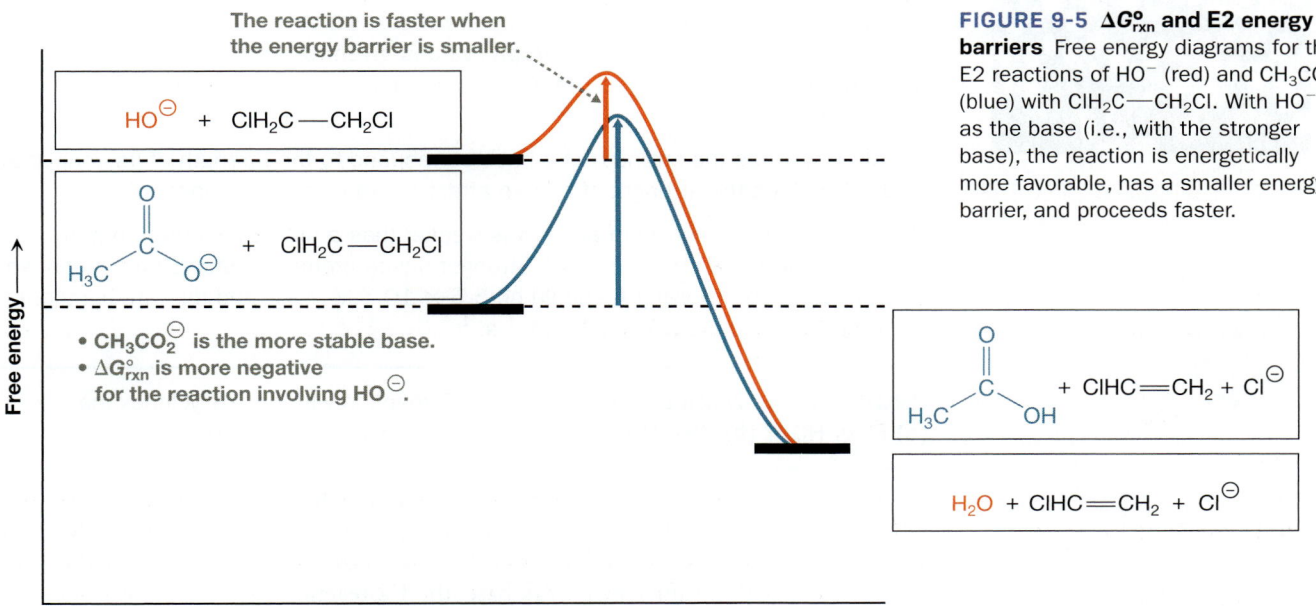

The free energy diagrams for these two reactions are shown in Figure 9-5. For both reactions, ΔG_{rxn}° is negative because a negative charge is better stabilized on Cl than it is on O. Also, ΔG_{rxn}° is more negative when HO^- is the base than when $CH_3CO_2^-$ is, because the negative charge is resonance delocalized in $CH_3CO_2^-$. Thus,

The reaction is faster when the energy barrier is smaller.

FIGURE 9-5 ΔG_{rxn}° and E2 energy barriers Free energy diagrams for the E2 reactions of HO^- (red) and $CH_3CO_2^-$ (blue) with ClH_2C—CH_2Cl. With HO^- as the base (i.e., with the stronger base), the reaction is energetically more favorable, has a smaller energy barrier, and proceeds faster.

Free energy

Reaction coordinate

$HO^\ominus$ + ClH_2C—CH_2Cl

+ ClH_2C—CH_2Cl

• $CH_3CO_2^\ominus$ is the more stable base.
• ΔG_{rxn}° is more negative for the reaction involving $HO^\ominus$.

+ $ClHC$=CH_2 + $Cl^\ominus$

H_2O + $ClHC$=CH_2 + $Cl^\ominus$

using the guideline derived from the Hammond postulate (Section 9.3a), the reaction involving HO⁻ as the base has a smaller energy barrier and proceeds faster, consistent with the data in Table 9-2.

Similar to what we saw with S$_N$2 and S$_N$1 reactions, there is a stark contrast between the E2 and E1 reactions when it comes to the attacking species.

> The rate of an E2 reaction depends on the strength of the base, whereas the rate of an E1 reaction is essentially independent of the identity of the base.

The base does not participate in an E1 reaction until the leaving group has left (i.e., until after the rate-determining step), at which point it removes the proton from the carbocation intermediate. Thus, because the base does *not* help the leaving group to leave, the strength of the base has little effect on the reaction rate.

SOLVED PROBLEM 9.7

CONNECTIONS One of the uses of bromocyclohexane (C$_6$H$_{11}$Br) is as a solvent to match the refractive index of poly(methylmethacrylate) (PMMA). This refractive index matching strategy helps improve the resolution in confocal microscopy, a three-dimensional optical imaging technique used to obtain high-resolution images, such as the one of PMMA microspheres below.

F. G. Bordwell and S. R. Mrozack carried out the following E2 elimination reaction in DMSO, using a variety of bases.

Base: + [structure: Bromocyclohexane with H and Br] → Base—H + [cyclohexene] + :Br:⁻

Bromocyclohexane

Two of the bases they used are shown below.

[structure A: naphthalenol anion O⁻]   [structure B: brominated naphthalenol anion O⁻ with Br]

pK$_a$(Base—H) = 17.1   pK$_a$(Base—H) = 16.2

A   **B**

The rate constants for these reactions are 5.5×10^{-2} M^{-1} s^{-1} and 1.6×10^{-2} M^{-1} s^{-1}. Based on the pK$_a$ values given for each base's conjugate acid, Base—H (measured in DMSO), match each rate constant to the appropriate base.

Think How does the strength of a base correspond to the strength of its conjugate acid? How does the strength of a base affect the rate of an E2 reaction?

Solve The conjugate acid of anion **A** is weaker (has a higher pK$_a$) than that of anion **B**, so **A** is the stronger base. Stronger bases promote faster E2 reactions, so the rate constant involving **A** should be 5.5×10^{-2} M^{-1} s^{-1} and the rate constant involving **B** should be 1.6×10^{-2} M^{-1} s^{-1}.

PROBLEM 9.8 Which promotes a faster E2 reaction with bromocyclohexane: **(a)** F⁻ or HO⁻? **(b)** CH$_3$CH$_2$O⁻ or CF$_3$CH$_2$O⁻? Explain your answers.

The E2 rate is highly sensitive to base strength, whereas the E1 rate is not, but what impact does that have on the competition between the two reactions? With a sufficiently strong base, the E2 reaction becomes faster than the corresponding E1 reaction, and with a sufficiently weak base, the E2 reaction becomes slower than the corresponding E1 reaction:

$$H_2O, ROH < H_3N, R_2NH < I^{\ominus} < Br^{\ominus} < Cl^{\ominus} < F^{\ominus} < RCO_2^{\ominus} < HS^{\ominus}, RS^{\ominus} < HO^{\ominus}, RO^{\ominus} < H_2N^{\ominus}, R_2N^{\ominus} < H_3C^{\ominus}, R^{\ominus}$$

Base strength generally increases

→

Stability of negative charge decreases

FIGURE 9-6 Intrinsic base strength
Strong bases are at least as strong as HO^-. Weak bases are significantly weaker than HO^-.

- Strong bases tend to favor E2 reactions.
- Weak bases tend to favor E1 reactions.

To use these concepts to predict products, we must be able to distinguish strong bases from weak bases. In practice, with all else being equal, E2 reactions are usually favored by bases that are as strong or stronger than HO^-. Otherwise, E1 reactions tend to be favored. Thus:

- Strong bases are at least as strong as HO^-.
- Weak bases are significantly weaker than HO^-.

Alkoxides (RO^-), then, are strong bases, as are deprotonated amines (R_2N^-) and organometallic reagents such as alkyllithium (RLi) and Grignard (RMgX) reagents (Fig. 9-6). Weak bases include halide anions (i.e., F^-, Cl^-, Br^-, and I^-), thiolate anions (RS^-), and carboxylate anions (RCO_2^-). Similarly, uncharged species like H_2O, alcohols (ROH), and amines (R_2NH) are weak bases.

SOLVED PROBLEM 9.9

Is the phenoxide anion, $C_6H_5O^-$, a strong base or a weak base? Will it tend to favor the E1 or the E2 mechanism?

Think What is the pK_a of C_6H_5OH, the conjugate acid of $C_6H_5O^-$? How does this compare to the pK_a of H_2O, the conjugate acid of HO^-? Based on these relative pK_a values, is $C_6H_5O^-$ a stronger or weaker base than HO^-?

Solve The pK_a values of C_6H_5OH and H_2O are 10.0 and 15.7, respectively. C_6H_5OH, therefore, is a significantly stronger acid than H_2O, making $C_6H_5O^-$ a significantly weaker base than HO^-. Thus, $C_6H_5O^-$ is a weak base and weak bases tend to favor the E1 mechanism over the E2.

PROBLEM 9.10 Is NC^- a strong base or a weak base? Will it tend to favor the E1 or the E2 mechanism? Explain. *Hint:* Consult Appendix A.

9.3d Strong, Bulky Bases

Based on the guidelines just described, the *tert*-butoxide anion, $(CH_3)_3CO^-$, should be both a strong nucleophile (because it has a full negative charge) and a strong base (because it is stronger than HO^-). Thus, it should favor both S_N2 and E2 reactions. In the competition between substitution and elimination, however, the *tert*-butoxide

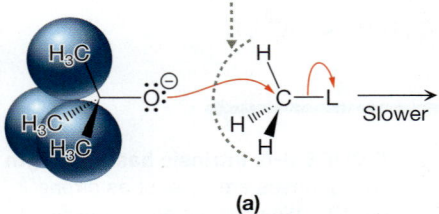

Difficult for the O⁻ to approach the C to form a bond

(a)

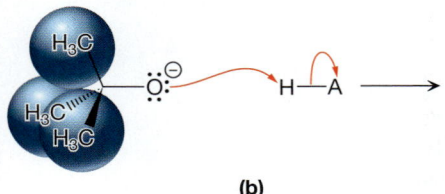

No significant steric hindrance in a proton transfer

(b)

FIGURE 9-7 Steric repulsion from the *tert*-butoxide anion (a) Steric hindrance by the CH₃ groups diminishes the nucleophilicity of (CH₃)₃CO⁻. (b) The basicity of (CH₃)₃CO⁻ is unaffected because steric hindrance by the CH₃ groups does not come into play—protons are very small and usually well exposed.

anion usually favors just E2 products because *(CH₃)₃CO⁻ is a much weaker nucleo-phile than we would expect based on charge stability.*

Figure 9-7a shows that the diminished nucleophilicity of the *tert*-butoxide anion stems from the bulkiness of the methyl groups surrounding the nucleophilic O⁻. They are so bulky that it is difficult for the O⁻ to form a bond to a C atom and to displace a leaving group, which slows the reaction. Thus, the S$_N$2 reaction suffers from *steric hindrance* caused by the methyl groups. The ability of the anion to act as a base, however, is essentially unaffected by steric hindrance because protons are very small and are usually well exposed. Consequently, protons are not encumbered by the methyl groups surrounding the O⁻ (Fig. 9-7b) and can therefore be pulled off by O⁻ easily.

> Strong, bulky bases (such as the *tert*-butoxide anion) tend to favor E2 reactions over S$_N$2 reactions.

In addition to the *tert*-butoxide anion, other common, strong, bulky bases include the neopentoxide anion and the diisopropylamide anion:

Strong, bulky bases

The *tert*-butoxide anion   **The neopentoxide anion**   **Lithium diisopropylamide (LDA)**

YOUR TURN 9.5

In the three bases above, identify the groups that contribute to significant steric hindrance.

PROBLEM 9.11 Rank alkoxide nucleophiles **A–C** in order of increasing S$_N$2 reaction rate.

A   **B**   **C**

9.4 Factor 2: Concentration of the Attacking Species

The dependence of each reaction on the concentration of the substrate and the attacking species is summarized in its respective *empirical rate law* (i.e., its experimentally derived rate law). These were first introduced in Chapter 8 and are shown again in Equations 9-15 through 9-18.

$$\text{Rate (S}_N2) = k_{S_N2}[\text{Att}^-][\text{R—L}] \qquad (9\text{-}15)$$

$$\text{Rate (E2)} = k_{E2}[\text{Att}^-][\text{R—L}] \qquad (9\text{-}16)$$

$$\text{Rate (S}_N1) = k_{S_N1}[\text{R—L}] \qquad (9\text{-}17)$$

$$\text{Rate (E1)} = k_{E1}[\text{R—L}] \qquad (9\text{-}18)$$

[Att$^-$] represents the concentration of the attacking species, whether it is acting as a nucleophile (in the substitution reactions) or as a base (in the elimination reactions). [R—L] is the substrate concentration. The rate constants for the various reactions are different and are therefore denoted as k_{S_N2}, k_{E2}, k_{S_N1}, and k_{E1}.

Both the S$_N$2 and E2 reaction rates depend on the concentration of the attacking species, [Att$^-$], which agrees with our simplified pictures of their rate-determining steps (Section 9.2). In an S$_N$2 or E2 reaction, the attacking species forces off the leaving group, so with a greater number of attacking species (higher concentration), the rate-determining step occurs more frequently and the rate increases. The S$_N$1 and E1 reaction rates, on the other hand, are independent of [Att$^-$], because the attacking species in an S$_N$1 or E1 reaction must *wait* until the leaving group has departed. Therefore, a higher concentration of the attacking species simply means more attacking species waiting for the leaving group to come off, but the S$_N$1 or E1 reaction rate does not change.

Because the S$_N$2, S$_N$1, E2, and E1 reaction rates have different dependencies on the concentration of the attacking species, these concentrations can be used to control the outcome of the competition as follows:

- A high concentration of a strong attacking species will promote fast S$_N$2 and E2 reactions, favoring both S$_N$2 and E2 reactions over S$_N$1 and E1 reactions.
- A low concentration of the attacking species will promote slow S$_N$2 and E2 reactions, which can leave the S$_N$1 and E1 reactions favored over S$_N$2 and E2 reactions.

In practice, *we may assume that a high concentration of the attacking species is present unless it has been deliberately made dilute.* If the concentration is indeed dilute, it should be indicated in the reaction conditions—for example,

$$\xrightarrow{\text{dil Br}^{\ominus}}$$

We must be careful when drawing these conclusions about high and low concentrations of the attacking species. Weak nucleophiles and bases are essentially incapable of forcing off the leaving group, regardless of their number present in solution. Therefore:

If the attacking species is *weak*, its concentration does not significantly affect S$_N$2 or E2 reaction rates.

SOLVED PROBLEM 9.12

State which reaction(s)—S$_N$2, S$_N$1, E2, or E1—are favored by: **(a)** a high concentration of HS$^-$; **(b)** a low concentration of HS$^-$.

Think Is HS$^-$ a strong nucleophile or a weak nucleophile? A strong base or a weak base? What relative concentrations favor each reaction? What are the exceptions?

Solve HS$^-$ is a strong nucleophile because it possesses a -1 charge. At a high concentration of HS$^-$, therefore, S$_N$2 is favored over S$_N$1, whereas at low concentrations, S$_N$1 is favored over S$_N$2. HS$^-$ is a weak base because it is significantly weaker than HO$^-$. Therefore, its concentration does not substantially affect the E2 rate, so E1 is favored over E2 at both low and high concentrations of HS$^-$. Overall, then: **(a)** at a high concentration of HS$^-$, S$_N$2 and E1 are favored; **(b)** at a low concentration of HS$^-$, S$_N$1 and E1 are favored.

PROBLEM 9.13 For each of the following conditions, state whether an S_N1 or S_N2 reaction is favored and whether an E1 or E2 reaction is favored. **(a)** A high concentration of HO^-. **(b)** A low concentration of HO^-. **(c)** A high concentration of Br^-. **(d)** A low concentration of Br^-. **(e)** A high concentration of $(CH_3)_3CO^-$. **(f)** A low concentration of $(CH_3)_3CO^-$.

An alkyl tosylate (R—OTs)

FIGURE 9-8 The tosylate leaving group A tosylate group (red), abbreviated OTs, has an excellent leaving group ability.

9.5 Factor 3: Leaving Group Ability

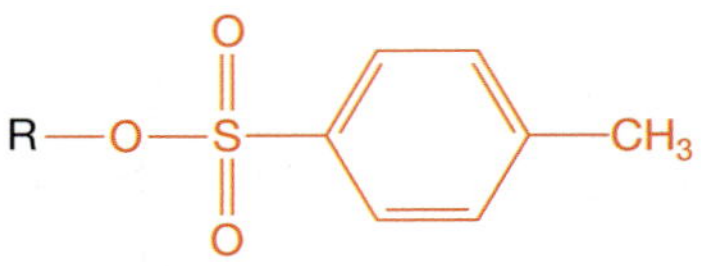

A leaving group must leave in the rate-determining step whether the reaction is S_N2, S_N1, E2, or E1. Not surprisingly, then, the identity of the leaving group has an effect on the rate of each reaction. As shown in Table 9-3, for example, an alkyl tosylate (R—OTs) (Fig. 9-8) undergoes an S_N2 reaction about 300 times faster than a comparable alkyl chloride (R—Cl). We say, therefore, that the **leaving group ability** of the tosylate anion (TsO^-) in an S_N2 reaction is 300 times greater than that of the chloride anion. More dramatically, Table 9-4 shows that the leaving group ability of TsO^- in an S_N1 reaction is about 10,000 times greater than that of Cl^-!

TABLE 9-3 Effects of Leaving Group Ability on the Relative Rate of the S_N2 Reaction:

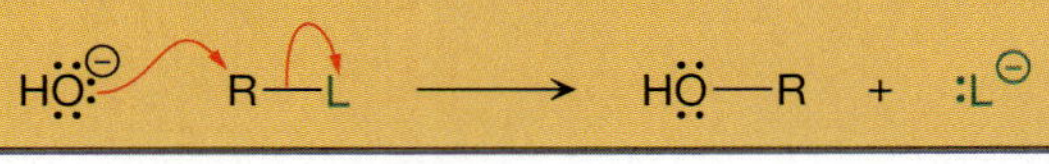

Leaving Group (L^-)	HO^-, H_2N^-, RO^-	F^-	Cl^-	Br^-	I^-	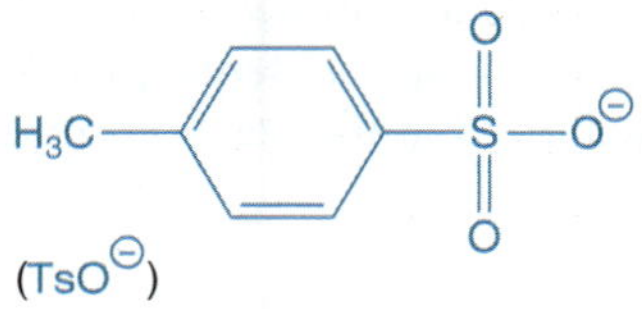 (TsO$^-$)
Relative S_N2 Reaction Rate	~0	0.005	1	50	150	300
pK_a of Conjugate Acid	> ~16	3.2	−7	−9	−10	−2.8

TABLE 9-4 Effects of Leaving Group Ability on the Relative Rate of the S_N1 Reaction:

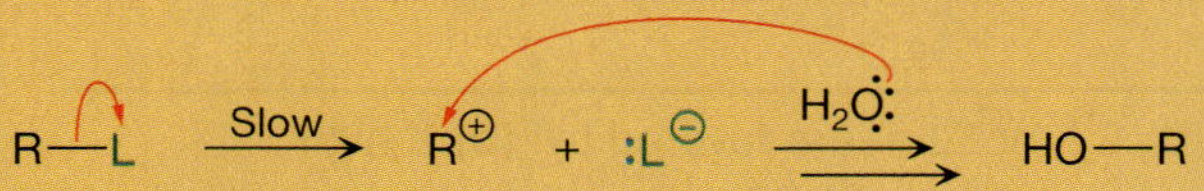

Leaving Group (L^-)	HO^-, RO^-, H_2N^-	Cl^-	Br^-	H_2O	I^-	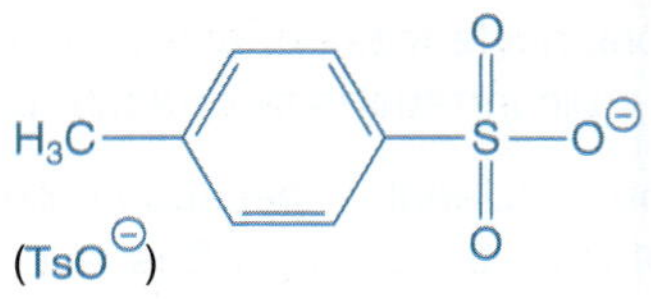 (TsO$^-$)	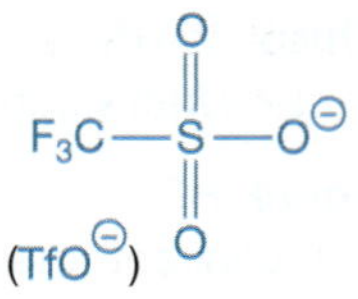 (TfO$^-$)
Relative S_N1 Reaction Rate	~0	1	~10	~10	~100	~10,000	~100,000,000
pK_a of Conjugate Acid	> ~16	−7	−9	−1.7	−10	−2.8	−13

S_N1 reactions are more sensitive to leaving group ability than S_N2 reactions are.

What are the reasons for these relative leaving group abilities? And why should the reactions have different sensitivities to leaving group ability? We answer these questions here in Section 9.5.

9.5a Leaving Group Ability, Charge Stability, and Base Strength

Relative leaving group abilities are governed largely by the stabilities of the leaving groups in the form in which they have come off the substrate.

The more stable the leaving group is (in the form in which it has left), the better its leaving group ability.

For example, H_2O is a much better leaving group than HO^- because water is uncharged. (Recall from Section 6.5 that negatively charged species are intrinsically less stable than analogous uncharged ones.) Similarly, Br^- is a much better leaving group than Cl^- because the larger Br atom can better stabilize the negative charge.

Leaving group ability tends to increase as the stability of the leaving group increases because the leaving group (in the form in which it departs) is a *product* in the rate-determining step. As those products become lower in energy, the value of ΔG°_{rxn} for the rate-determining step becomes more negative (less positive). Thus, as we learned in Section 9.3a, the energy barrier decreases and the rate increases.

Conveniently, relative leaving group abilities correlate rather well with relative base strengths:

Better leaving groups tend to be weaker bases.

Br^-, for example, is a weaker base than Cl^-, and it is a better leaving group, too.

The correlation between leaving group ability and base strength is particularly useful when it comes to understanding why the tosylate anion (TsO^-), the mesylate anion (MsO^-), and the triflate anion (TfO^-) are among the best leaving groups (Fig. 9-9). Structurally, they are very similar to the bisulfate anion (HSO_4^-), which is the conjugate base of sulfuric acid. Sulfuric acid is among the strongest acids known, which means the HSO_4^- anion is among the weakest bases.

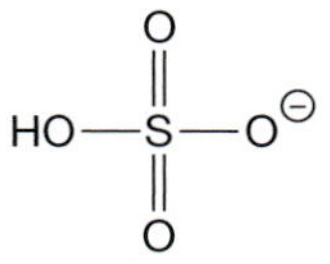

Bisulfate anion

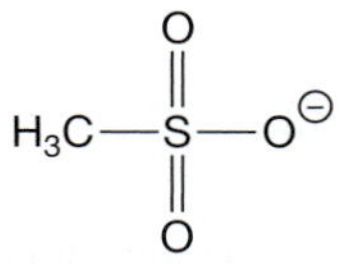

Methyl sulfonate (Mesylate) ($MsO^\ominus$)

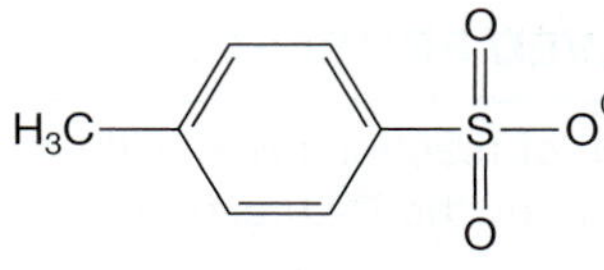

***p*-Toluene sulfonate (Tosylate) ($TsO^\ominus$)**

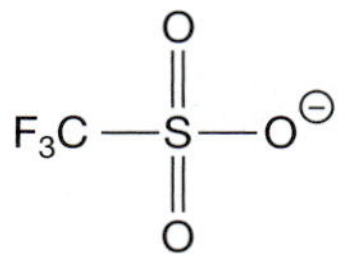

Trifluoromethyl sulfonate (Triflate) ($TfO^\ominus$)

FIGURE 9-9 Sulfonate leaving groups The excellent leaving group abilities of the mesylate, tosylate, and triflate groups derive from the same type of resonance and inductive stabilization exhibited by HSO_4^-.

YOUR TURN 9.6

Based on the pK_a value of its conjugate acid, identify the leaving group in Table 9-4 that is the *weakest base*. How does the S_N1 reaction rate involving that leaving group compare with the others?

YOUR TURN 9.7

Sulfonate anions owe some of their stability to delocalization of the negative charge. Draw all of the resonance structures of the MsO^- anion.

We can apply this reasoning to understand why the rate of nucleophilic substitution and elimination is essentially zero when the leaving group is HO^-. HO^- is a strong base (it has a weak conjugate acid, H_2O), so it is a very poor leaving group. Stronger bases, such as RO^-, H_2N^-, H^-, and R^-, are even poorer leaving groups. H_2O, on the other hand, is an excellent leaving group, because it is a weak base (it has a strong conjugate acid, H_3O^+).

PROBLEM 9.14 Which is a better leaving group, HCO_2^- or $C_6H_5O^-$? Explain.

As mentioned previously, *S_N1 reactions are much more sensitive to leaving group ability than S_N2 reactions.* S_N1 reactions are so sensitive to leaving group ability because the leaving group must depart on its own in the rate-determining step. By contrast, in an S_N2 reaction, the departure of the leaving group is greatly assisted by the attacking nucleophile, so the stability of the leaving group becomes less important.

The effects of leaving group abilities on E2 and E1 reaction rates parallel those of S_N2 and S_N1 reactions. Both E2 and E1 reaction rates are enhanced by better leaving group abilities, but E1 reactions are much more sensitive to leaving group ability than E2 reactions are. (See Solved Problem 9.15.)

SOLVED PROBLEM 9.15

Which of reactions **A–C** will proceed *fastest* by the E1 mechanism? Which will proceed *slowest* by the E2 mechanism? (Assume the same initial concentrations in **A–C**.)

| A | B | C |

Think What are the relative strengths of the bases in **A–C**, and how are the E1 and E2 reactions affected by base strength? What are the relative leaving group abilities of the leaving groups that appear in **A–C**, and how are the E1 and E2 reactions affected by leaving group ability?

Solve The E1 reaction is sensitive to leaving group ability, but is relatively insensitive to base strength. Because TsO^- is a better leaving group than Br^-, **A** proceeds by the E1 mechanism faster than either **B** or **C**. The E2 reaction is sensitive to both leaving group ability and base strength. Because Br^- is a worse leaving group than TsO^- and NH_3 is a weaker base than HO^-, **C** proceeds slower by the E2 mechanism than either **A** or **B**.

PROBLEM 9.16 Which of reactions **D–F** will proceed the *slowest* by the E1 mechanism? Which will proceed the *fastest* by the E2 mechanism? (Assume the same initial concentrations in **D–F**.)

| D | E | F |

Poor .... Moderate .... Good

$$H^{\ominus}, H_3C^{\ominus} < H_2N^{\ominus} < HO^{\ominus}, RO^{\ominus} < F^{\ominus} < RCO_2^{\ominus} < Cl^{\ominus} < Br^{\ominus}, H_2O, ROH < I^{\ominus} < MsO^{\ominus} < TsO^{\ominus} < TfO^{\ominus}$$

Leaving group ability increases →

Charge stability generally increases

FIGURE 9-10 Leaving group ability Good leaving groups tend to accommodate a developing negative charge rather well. Poor leaving groups do not.

Given the much greater sensitivity of S_N1 and E1 reactions to leaving group ability than S_N2 and E2 reactions, we arrive at the following general rule:

> Excellent leaving groups favor S_N1 and E1 reactions over the corresponding S_N2 and E2 reactions.

It thus becomes important to recognize whether a leaving group is good, moderate, or poor. Some examples of each are shown in Figure 9-10.

Poor leaving groups favor S_N2 and E2 reactions over S_N1 and E1 reactions. According to Tables 9-3 and 9-4, however:

> The rates of nucleophilic substitution and elimination reactions under normal conditions are essentially zero unless the leaving group is at least as stable as F^-.

PROBLEM 9.17 With C_6H_5OH as the leaving group, which reaction would be favored: S_N2 or S_N1? E2 or E1? Explain.

9.5b Converting a Poor Leaving Group into a Good Leaving Group

Frequently, a desired nucleophilic substitution or elimination reaction is unfeasible because the leaving group is unsuitable—that is, it is not stable enough in the form in which it leaves. Consider, for example, Equation 9-19, which shows that no reaction takes place when Br^- is added to butan-1-ol under normal conditions.

$$\text{OH} + \text{NaBr} \xrightarrow{H_2O} \text{No reaction} \qquad (9\text{-}19)$$

This result should not be surprising, though, because the leaving group in this case is HO^-, which is a very poor leaving group in substitution and elimination reactions. Under acidic conditions, however, a reaction does take place, as shown in Equation 9-20.

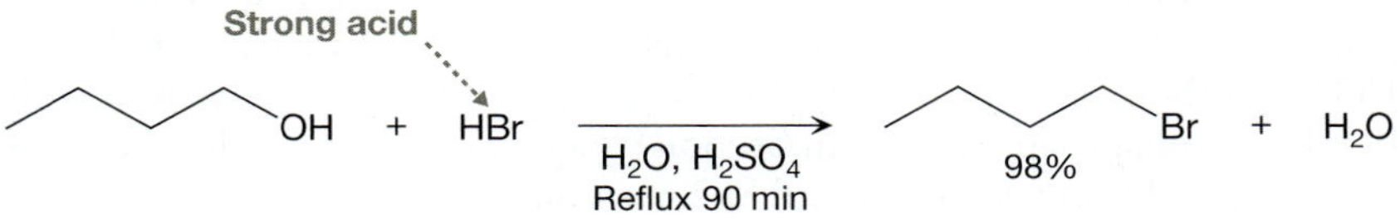

Strong acid

$$\text{OH} + \text{HBr} \xrightarrow[\substack{\text{Reflux 90 min}}]{H_2O, H_2SO_4} \underset{98\%}{\text{Br}} + H_2O \qquad (9\text{-}20)$$

The acidic conditions in Equation 9-20 facilitate the substitution reaction because the O atom of the OH group is weakly basic and therefore becomes protonated (remember, proton transfer reactions are *fast*), as shown in Equation 9-21. The leaving group then leaves as H_2O, not HO^-. Being uncharged, H_2O is much more stable than HO^-, and is a *good* leaving group (review Fig. 9-10). This is consistent with the fact that H_2O is a much weaker base than HO^-: The pK_a of H_3O^+ is -1.7, whereas the pK_a of H_2O is 15.7.

Mechanism for Equation 9-20

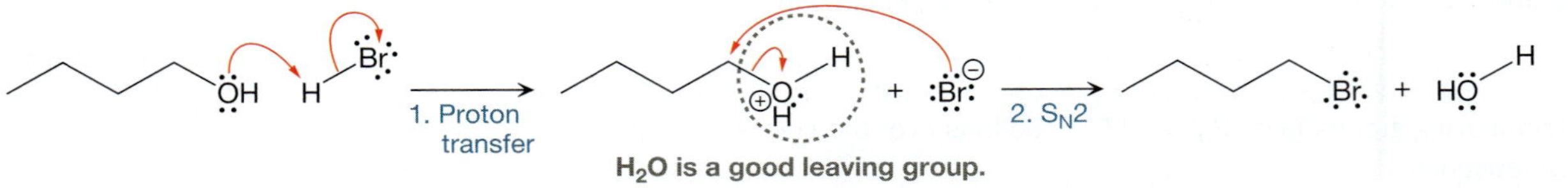

H_2O is a good leaving group. (9-21)

YOUR TURN 9.8

Circle the potential leaving group in both the reactant and product of the following proton transfer reaction:

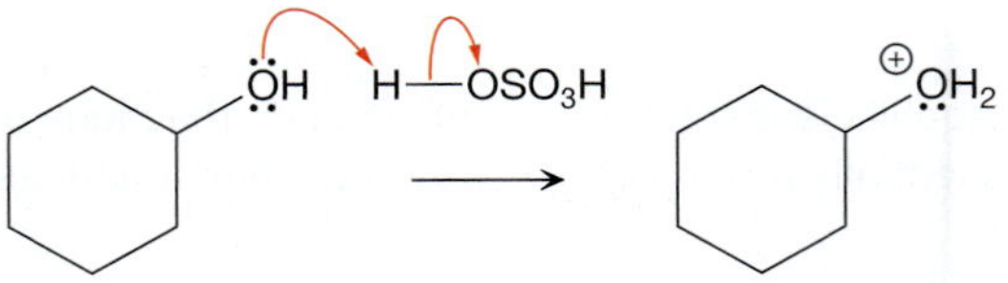

Next to each, write the form of the leaving group in which it would leave, and label it as either a good leaving group or a poor leaving group.

CONNECTIONS
2-Methoxyphenol ($HOC_6H_4OCH_3$), commonly called guaiacol, is one of the pheromones that cause desert locusts to swarm.

The protonation of ethers works in much the same way. Under normal conditions (Equation 9-22), 2-methoxyphenol ($HOC_6H_4OCH_3$) does not undergo nucleophilic substitution or elimination reactions because the leaving group would be an RO^- anion, which is a poor leaving group. A reaction does take place, however, under acidic conditions, as shown in Equation 9-23.

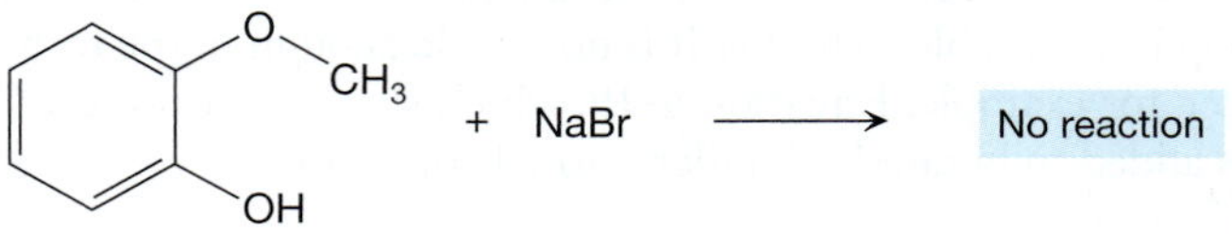

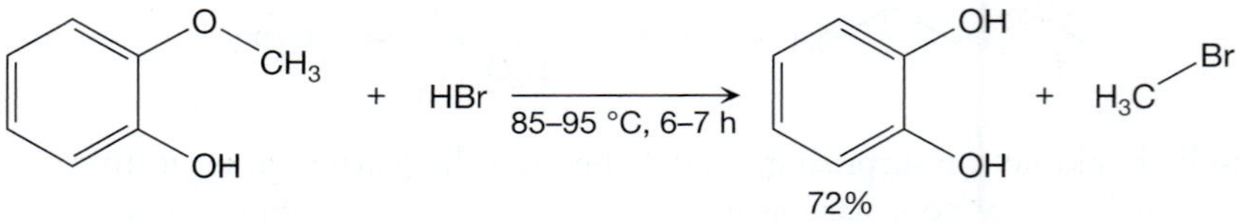

Just as we saw in Equation 9-21, protonation of the O atom (Equation 9-24) generates an excellent leaving group: an uncharged and very weakly basic molecule, HOC_6H_4OH. (Although the protonated intermediate could instead be viewed with CH_3OH as the leaving group, we learn in Section 9.6a that the sp^2-hybridized carbon to which it is attached is unreactive toward S_N2 reactions.)

Mechanism for Equation 9-23

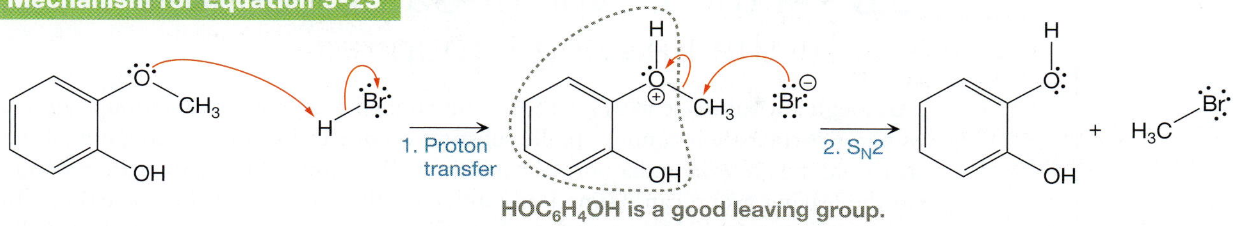

HOC₆H₄OH is a good leaving group.

$$\text{HOC}_6\text{H}_4\text{OH is a good leaving group.}$$ (9-24)

Generating a good leaving group in this fashion can also facilitate elimination reactions, as shown in the **dehydration** reaction in Equation 9-25.

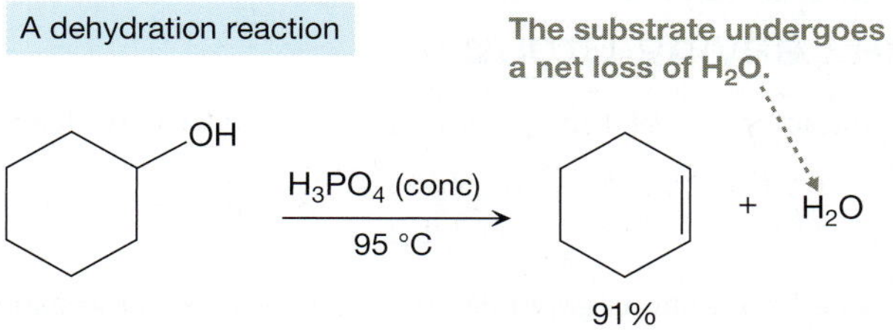

A dehydration reaction

The substrate undergoes a net loss of H_2O.

H_3PO_4 (conc) 95 °C

91%

(9-25)

HO^- would be the leaving group under neutral conditions, but acidic conditions generate a water leaving group, as shown in Equation 9-26.

Mechanism for Equation 9-25

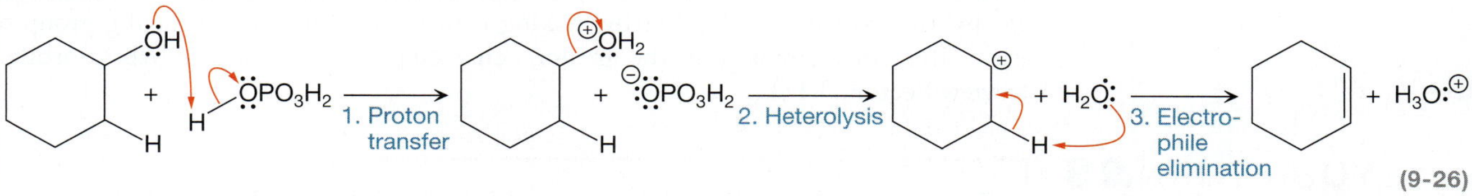

1. Proton transfer   2. Heterolysis   3. Electrophile elimination

(9-26)

Like alcohols and ethers, amines tend *not* to undergo nucleophilic substitution or elimination reactions under normal conditions (Equation 9-27). To occur, these reactions would require the departure of a very poor leaving group—namely, H_2N^-, HRN^-, or R_2N^-. Protonation of the mildly basic nitrogen of the amino group would make the leaving group better, but *even under acidic conditions, amines tend not to act as substrates in nucleophilic substitution or elimination reactions*. Instead, the reaction stops at the formation of the ammonium ion (Equation 9-28).

NH₂ + NaBr ⟶ No reaction (9-27)

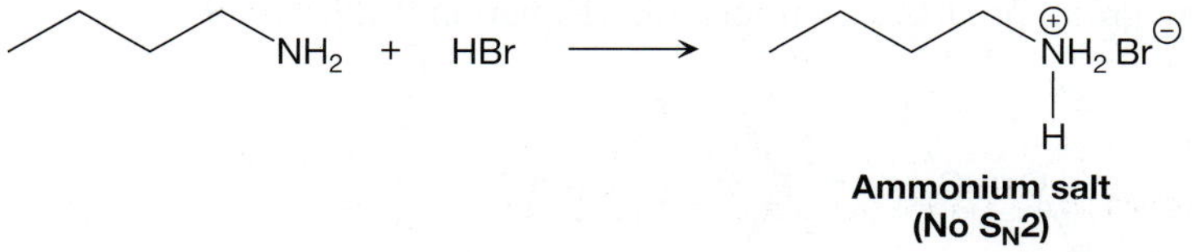

NH₂ + HBr ⟶ (9-28)

Ammonium salt (No S_N2)

Why do protonated alcohols tend to undergo nucleophilic substitution and elimination reactions but protonated amines do not? Evidently, the leaving group in a protonated alcohol (i.e., H_2O) is much better than the leaving group in a protonated primary amine (i.e., NH_3). This can be explained by the much greater basicity of NH_3 relative to that of H_2O (the pK_a of NH_4^+ is about 10, and the pK_a of H_3O^+ is -1.7).

9.6 Factor 4: Type of Carbon Bonded to the Leaving Group

Although any substrate with a carbon atom bonded to a good leaving group can *theoretically* participate in a nucleophilic substitution or elimination reaction, not all such reactions are *practical* under normal conditions. The nature of the carbon atom bonded to the leaving group can dramatically influence the outcome of these reactions. The relevant factors include the particular hybridization of the carbon, the number of alkyl groups to which it is bonded, and the proximity of double or triple bonds.

9.6a Hybridization of the Carbon Atom Bonded to the Leaving Group

An important characteristic of the carbon atom bonded to the leaving group is its hybridization.

> S_N2, S_N1, E2, and E1 reactions generally do *not* occur unless the carbon atom bonded to the leaving group is sp^3 hybridized.

(We will see an exception to this general rule in Chapter 10.)

One of the main reasons the leaving group needs to be bonded to an sp^3-hybridized carbon atom is that sp^2- and sp-hybridized carbons form stronger σ bonds than sp^3-hybridized carbons do, thereby making it more difficult for the leaving group to leave. This effect stems from the greater *s* character in the sp^2 and sp hybrid orbitals (review Section 3.11).

YOUR TURN 9.9

To verify that bond strength increases as the percent s character in the hybridized orbital used to form the bond increases, find the following C—H bond energies in Section 3.11 and write them in the boxes provided here.

$$sp^3 \text{C—H} \quad sp^2 \text{C—H} \quad sp \text{C—H}$$

S_N1 and E1 reactions are further hindered when the leaving group is on an *sp*- or sp^2-hybridized carbon atom due to the carbon atom's effective electronegativity (Section 3.11). In the rate-determining step of such proposed reactions, departure of the leaving group produces a carbocation (Equation 9-29a and 9-29b).

(9-29a)

Excessively unstable carbocations

(9-29b)

Because *sp*- and *sp²*-hybridized carbon atoms possess greater *s* character than an *sp³*-hybridized carbon, these atoms have a higher effective electronegativity, which makes the resulting positive charge excessively unstable.

S_N2 reactions are further hindered with the leaving group on an *sp*- or *sp²*-hybridized carbon due to electrostatic repulsion. As we learned in Section 7.6, C=C and C≡C bonds are relatively electron rich, so they repel an incoming negatively charged nucleophile (Fig. 9-11).

The π electrons in the C=C and C≡C bonds repel negatively charged nucleophiles.

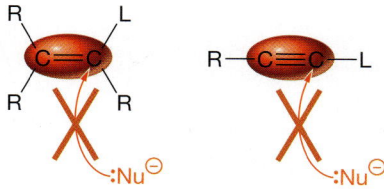

PROBLEM 9.18 Which carbon atom in the molecule shown here would undergo nucleophilic substitution most readily? Explain.

FIGURE 9-11 Electrostatic repulsion between π electrons and a nucleophile When a leaving group is bonded to an alkene or an alkyne carbon, S_N2 reactions are hindered by repulsion between the negatively charged π electron cloud and the incoming nucleophile.

9.6b The Number of Alkyl Groups on the Carbon Bonded to the Leaving Group

Even with a good leaving group on an *sp³*-hybridized carbon atom, nucleophilic substitution and elimination reactions can be strongly influenced by the number of alkyl groups bonded to that carbon. Recall from Interchapter A (Section A.7) that a carbon atom bonded to one alkyl group is called a *primary (1°) carbon*; if it is bonded to two alkyl groups, then it is called a *secondary (2°) carbon*; and if it is bonded to three alkyl groups, then it is called a *tertiary (3°) carbon*. Therefore, substrates of the form RCH_2—L, R_2CH—L, and R_3C—L are called primary, secondary, and tertiary substrates, respectively. Alternatively, if the carbon is bonded only to hydrogen atoms, then it is called a *methyl carbon* and the methyl substrate takes the form CH_3—L.

The data in Tables 9-5 and 9-6 show that both the S_N1 and S_N2 reaction rates are very sensitive to the type of carbon atom bonded to the leaving group. Specifically:

As the number of alkyl groups on the carbon atom to which the leaving group is bonded increases, the S_N1 reaction rate sharply increases, whereas the S_N2 reaction rate sharply decreases.

TABLE 9-5 Relative Reaction Rates in the S_N1 Reaction:

$$H_2O \ + \ \overset{|}{\underset{|}{C}}\!-\!Br \ \longrightarrow \ HO\!-\!\overset{|}{\underset{|}{C}}\!- \ + \ HBr$$

Substrate	H_3C–Br	H_3C–CH_2–Br	H_3C–CH(CH_3)–Br	H_3C–C(CH_3)(CH_3)–Br
	Methyl	1°	2°	3°
Relative S_N1 Reaction Rate	~0	0.08	1	100,000

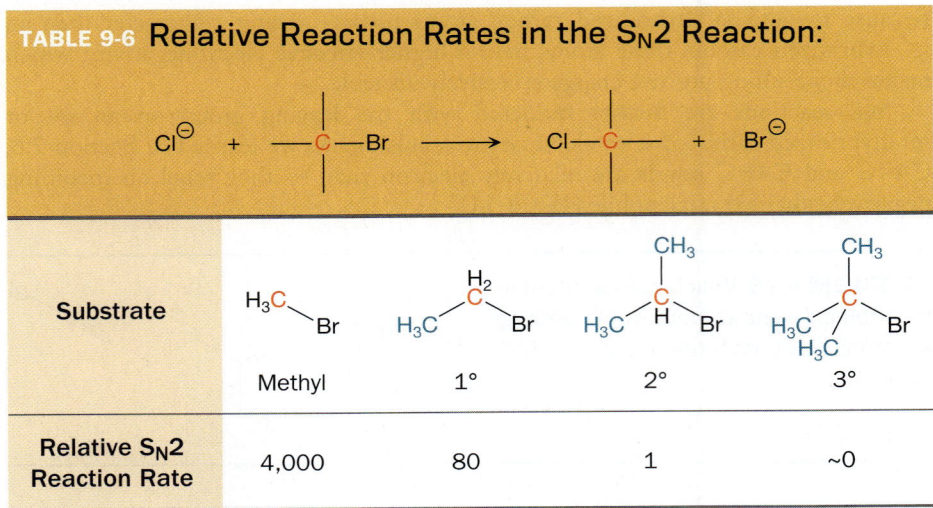

TABLE 9-6 Relative Reaction Rates in the S$_N$2 Reaction:

$$Cl^{\ominus} + \underset{|}{\overset{|}{-}}C-Br \longrightarrow Cl-\underset{|}{\overset{|}{C}}- + Br^{\ominus}$$

Substrate	H_3C-Br Methyl	H_3C-CH_2-Br 1°	$H_3C-CH(CH_3)-Br$ 2°	$(H_3C)_3C-Br$ 3°
Relative S$_N$2 Reaction Rate	4,000	80	1	~0

To understand why the number of alkyl groups affects reaction rates in these ways, we must revisit their mechanisms. For an S$_N$2 reaction, recall that the nucleophile *forces off the leaving group*. Each alkyl group surrounding the carbon atom bonded to the leaving group adds *steric hindrance* (Fig. 9-12), similar to what we saw with the *tert*-butoxide anion, $(CH_3)_3CO^-$, in Section 9.3d. With greater steric hindrance surrounding the C—L carbon, it becomes more difficult for the nucleophile to attack and the S$_N$2 rate decreases. That steric hindrance becomes excessive with three alkyl groups:

S$_N$2 reactions are unfeasible for tertiary substrates, R$_3$C—L.

In an S$_N$1 reaction, the reaction rate is dictated by the ability of the leaving group to leave, so steric hindrance of the nucleophile does not come into play. Instead, *each additional alkyl group, which is electron donating, stabilizes the carbocation intermediate that is produced and thus helps the leaving group leave* (Fig. 9-13). A tertiary substrate reacts the fastest because the carbocation it produces has the maximum number of

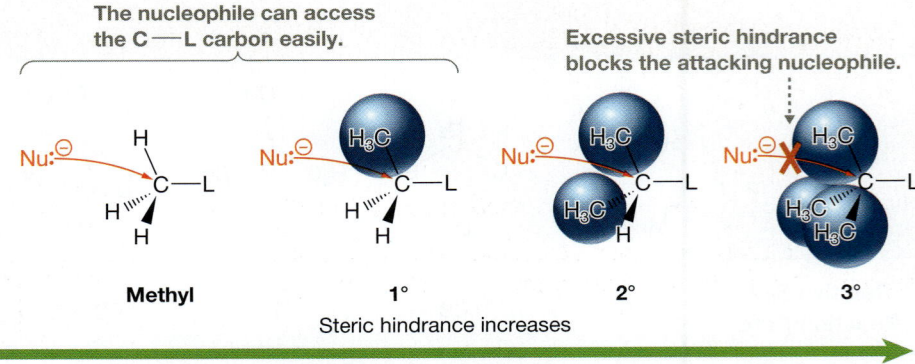

FIGURE 9-12 How steric hindrance affects the rate of an S$_N$2 reaction Steric hindrance of the nucleophile in an S$_N$2 reaction increases as the number of alkyl groups surrounding the C atom bonded to the leaving group increases. As steric hindrance increases, the S$_N$2 reaction rate decreases.

The nucleophile can access the C—L carbon easily.

Excessive steric hindrance blocks the attacking nucleophile.

Methyl   1°   2°   3°

Steric hindrance increases

S$_N$2 reaction rate decreases

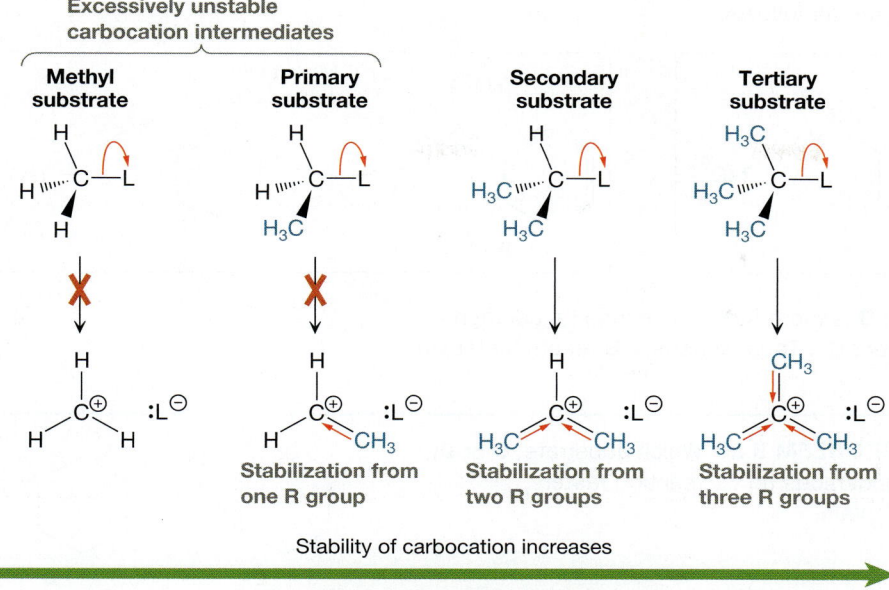

Excessively unstable carbocation intermediates

Methyl substrate	Primary substrate	Secondary substrate	Tertiary substrate

Stabilization from one R group

Stabilization from two R groups

Stabilization from three R groups

Stability of carbocation increases

Rate of S_N1 and E1 reactions increases

FIGURE 9-13 How S_N1 and E1 reaction rates depend on carbocation stability Carbocation stability increases as the number of alkyl groups bonded to the positively charged C of a carbocation increases. In the substrate, therefore, each alkyl group surrounding the C atom bonded to the leaving group increases the rate of S_N1 and E1 reactions.

alkyl groups (three), making it the most stable. Secondary substrates react slower because the carbocation that is produced has only two stabilizing alkyl groups. Methyl and primary substrates produce carbocations that are even less stable.

> S_N1 reactions are unfeasible for methyl substrates, H_3C—L, and most primary substrates, RCH_2—L.

E1 reaction rates depend on the number of alkyl groups in the same way as S_N1 reaction rates:

> The rate of an E1 reaction increases as the number of alkyl groups on the carbon bonded to the leaving group increases.

E1 and S_N1 reaction rates depend on carbocation stability in the same way because they both have *exactly* the same rate-determining step. (See Solved Problem 9.19.) Therefore, like S_N1 reactions:

> E1 reactions are unfeasible for methyl substrates, H_3C—L, and most primary substrates, RCH_2—L.

SOLVED PROBLEM 9.19

Which substrate, **A** or **B**, undergoes an E1 reaction faster?

Think Which rate-determining step produces a more stable carbocation? How does this affect the E1 reaction rate?

A

B

Solve The two rate-determining steps are as follows:

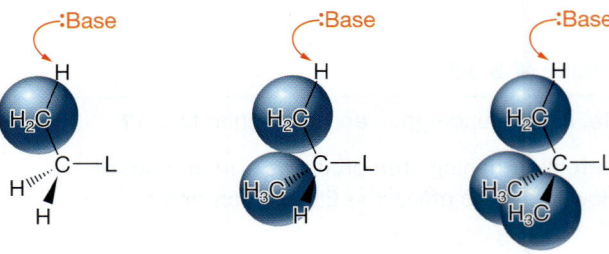

The carbocation formed from substrate **B** is more stable, because its additional CH_3 group stabilizes the electron-deficient C^+. Thus, substrate **B** reacts faster via the E1 mechanism.

PROBLEM 9.20 Which substrate, **C** or **D**, undergoes an E1 reaction faster? Explain.

C

D

Unlike S_N2, S_N1, and E1 reactions, *E2 reactions are relatively insensitive to the number of alkyl groups on the carbon bonded to the leaving group.* That is:

> Substrates with leaving groups on primary, secondary, and tertiary carbons can usually undergo E2 reactions quickly.

This may be surprising because the attacking species in an E2 reaction *must force off the leaving group*, just as in an S_N2 reaction. In an E2 reaction, however, the attacking species does so by pulling off a proton from a carbon *adjacent* to the one bonded to the leaving group. Because a proton is small and generally well exposed, *steric hindrance tends to be a minor factor in E2 reactions*, even with the leaving group on a 3° carbon (Fig. 9-14).

Table 9-7 summarizes much of the preceding discussion regarding the feasibility of various types of substrates undergoing S_N2, S_N1, E2, and E1 reactions. Specifically, each entry that is shaded darkly represents an unfeasible reaction.[1] This information will be very useful when you are asked to predict the products of an $S_N2/S_N1/E2/E1$ competition, so take the time to review why the reactions are unfeasible in those cases.

PROBLEM 9.21 According to Table 9-7, E2 reactions are feasible for substrates in which the leaving group is bonded to a primary, secondary, or tertiary carbon, but not for substrates in which the leaving group is bonded to a methyl carbon. Explain why.

TABLE 9-7 Feasibility of S_N2, S_N1, E2, and E1 Reactions

Type of Substrate (R—L)	S_N2	S_N1	E2	E1
CH_3—L (methyl)	✓			
RCH_2—L (1°)	✓		✓	
R_2CH—L (2°)	✓	✓	✓	✓
R_3C—L (3°)		✓	✓	✓

FIGURE 9-14 Accessibility of the proton in E2 reactions Unlike in S_N2 reactions, an increase in the number of alkyl groups attached to the carbon bonded to the leaving group does not cause much steric hindrance in E2 reactions. Even with tertiary C atoms, the protons on adjacent C atoms remain well exposed to the base.

Protons are typically well exposed on 1°, 2°, and 3° substrates.

:Base   :Base   :Base

---

[1]According to Table 9-7, S_N1 and E1 reactions are feasible for secondary substrates, R_2CH—L, but this is not entirely clear. For a discussion on the matter, see Murphy, T. J.; *J. Chem. Educ.* **2009**, 84(4), 519.

PROBLEM 9.22 Which of substrates **X**, **Y**, and **Z** will undergo an S$_N$2 reaction the fastest? Which will undergo an S$_N$1 reaction the fastest? Which will undergo an E1 reaction the fastest? Explain.

X　　　　　**Y**　　　　　**Z**

An exception to Table 9-7 arises when the reaction involves an **allyl substrate** (CH$_2$=CH—CH$_2$—L) or a **benzyl substrate** (C$_6$H$_5$—CH$_2$—L), as shown in Equations 9-30 and 9-31.

1° Allyl carbon

Resonance–stabilized charge

Allyl cation

$$ + \; :L^{\ominus} \quad (9\text{-}30)$$

Resonance-stabilized charge

1° Benzyl carbon　　**Benzyl cation**

$$ + \; :L^{\ominus} \qquad (9\text{-}31)$$

CONNECTIONS Allyl halides, such as allyl chloride (CH$_2$=CH—CH$_2$Cl) and allyl bromide (CH$_2$=CH—CH$_2$Br), are used as alkylating agents, such as in the manufacture of pharmaceuticals and polymers. Allyl chloride, furthermore, is epoxidized to epichlorohydrin, which is used to produce epoxy resins for protective coatings, like the one on this boat.

> If the leaving group is bonded to a primary allyl or benzyl carbon, then both S$_N$2 and S$_N$1 reactions are feasible.

Normally, S$_N$1 mechanisms are unavailable to substrates in which the leaving group is bonded to a primary carbon because the resulting primary carbocation is too unstable. However, if the leaving group is bonded to a primary allyl or benzyl carbon, then the resulting carbocation is heavily stabilized by resonance with the adjacent π electrons.

YOUR TURN 9.10

Draw all possible resonance structures for the benzyl cation to illustrate how resonance stabilizes the charge.

SOLVED PROBLEM 9.23

Which substrate, **A** or **B**, would undergo an S$_N$1 reaction more rapidly?

Think What is the carbocation intermediate involved in each S$_N$1 mechanism? Which one is more stable? How does this affect the S$_N$1 reaction rate?

A　　　　　**B**

Solve In each case, the carbocation is generated by loss of the leaving group, Cl⁻. The carbocation from **A** is resonance stabilized by a lone pair of electrons on the neighboring O atom, yielding a strong resonance contributor with all atoms having an octet. The carbocation from **B** is *not* resonance stabilized, so it is less stable. Thus, the carbocation from **A** is generated faster than the carbocation from **B**. Since these are the rate-determining steps of the respective S$_N$1 reactions, **A** undergoes the S$_N$1 reaction faster than **B**.

Strong resonance contributor

A

No resonance stabilization

B

PROBLEM 9.24 Which substrate, **C** or **D**, will undergo an E1 reaction faster? Explain.

C

D

9.7 Factor 5: Solvent Effects

S$_N$2, S$_N$1, E2, and E1 reactions can take place in *polar protic solvents* and *polar aprotic solvents*. The solvent must be polar to dissolve the reactants efficiently, given that the reactants are typically polar or ionic. As we learned in Section 2.9, protic solvents

Rotaxanes: Exploiting Steric Hindrance

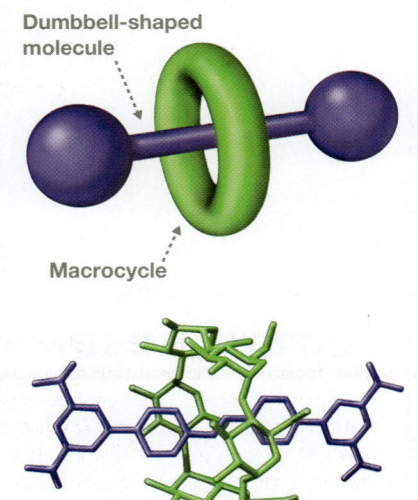

Dumbbell-shaped molecule

Macrocycle

As we have just seen, steric hindrance is a major factor that helps govern the outcome of an S$_N$2/S$_N$1/E2/E1 competition. Steric hindrance generally has a negative connotation, though, because the role of steric hindrance is to *prevent* a process from occurring. However, steric hindrance can, in fact, be exploited to make possible the fabrication of new molecules such as *rotaxanes*. **Rotaxanes** are a class of compounds in which a dumbbell-shaped molecule is threaded through a large cyclic molecule called a *macrocycle*, as shown schematically on the left. The graphic on the bottom is the crystal structure of an actual rotaxane.

Once this particular rotaxane is formed, the dumbbell-shaped molecule can rotate about its axis relative to the macrocycle, much like a wheel and axle. The macrocycle can also slide back and forth along the axis of the dumbbell. The dumbbell-shaped molecule doesn't slip out of the macrocycle entirely, however, because of the steric hindrance that is induced by the bulky end groups.

By applying an external stimulus (e.g., a voltage), the geometries of some rotaxanes can be manipulated in specific ways, giving rise to a variety of interesting potential applications. These include molecular switches, memory storage devices, and molecular muscles.

The rotaxane motif has also been shown to exist in biological systems. One example is microcin J25, which is produced by *Escherichia coli* during periods of nutrient depletion. Microcin J25 belongs to a class of peptides called lasso peptides, in which a C-terminal tail is threaded through an N-terminal cyclized ring.

(e.g., water and alcohols) possess hydrogen-bond donors, whereas aprotic solvents such as dimethyl sulfoxide (DMSO), *N,N*-dimethylformamide (DMF), and acetone do not. Here in Section 9.7, we examine the impact that the choice of solvent has on the competition between S_N2, S_N1, E2, and E1 reactions, as well as on relative nucleophile strengths.

9.7a Protic Solvents, Aprotic Solvents, and the $S_N2/S_N1/E2/E1$ Competition

Experimentally, the choice of solvent can have a significant influence on the outcome of nucleophilic substitution and elimination reactions.

- Polar *aprotic* solvents tend to favor S_N2 and E2 reactions.
- Polar *protic* solvents tend to favor S_N1 and E1 reactions.

These effects are illustrated with the data listed in Tables 9-8 and 9-9.

These solvent effects are explained, in part, by the solvation of the *attacking species*. As we learned in Section 2.9, *anions are solvated very strongly by protic solvents* because the large partial positive charge on H in the solvent molecule is well exposed (Fig. 9-15a). This stabilizes the nucleophile, which, as we learned in Section 9.3, slows an S_N2 reaction, but essentially leaves the S_N1 rate unchanged. Aprotic solvents, however, do not solvate anions nearly as strongly because the positive end of the solvent molecule's dipole is much less accessible (Fig. 9-15b). *Steric hindrance* keeps the anion away from that partial positive charge. As a result, aprotic solvents do not stabilize nucleophiles as much as polar protic solvents do, allowing S_N2 reactions to remain faster than S_N1 reactions. Putting these ideas together:

Protic solvents weaken nucleophiles substantially via solvation; aprotic solvents do not.

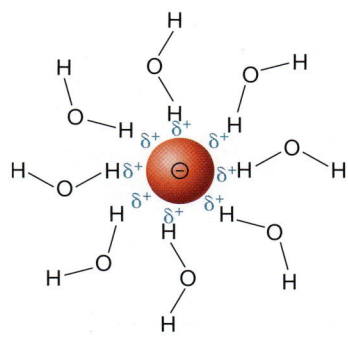

(a) Protic solvent

Anion very strongly solvated (heavily stabilized)

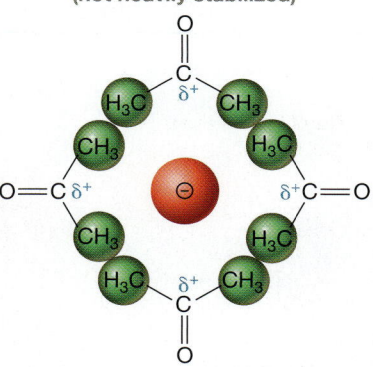

(b) Aprotic solvent

Anion weakly solvated (not heavily stabilized)

FIGURE 9-15 Comparing solvation of anions in protic and aprotic solvents (a) Solvation of an anion by water, a polar protic solvent. The large concentration of positive charge on a well-exposed hydrogen atom enables water to strongly solvate negative charges. (b) Solvation of an anion by acetone, a polar aprotic solvent. The partial positive end of the net dipole is buried inside the solvent molecule, which severely decreases acetone's ability to solvate negative charges.

TABLE 9-8 **Reaction Rates in Various Solvents for the S_N2 Reaction:**

Solvent	Type of Solvent	Relative S_N2 Reaction Rate
CH_3OH	Protic	1
H_2O	Protic	7
$(CH_3)_2S{=}O$, DMSO	Aprotic	1,300
$HCON(CH_3)_2$, DMF	Aprotic	2,800
CH_3CN	Aprotic	5,000

TABLE 9-9 **Reaction Rates in Various Solvents for the S$_N$1 Reaction:**

Solvent	Type of Solvent	Relative S$_N$1 Reaction Rate
CH$_3$OH	Protic	7,400,000
HCONH$_2$	Protic	1,200,000
HCON(CH$_3$)$_2$, DMF	Aprotic	12.5
CH$_3$CON(CH$_3$)$_2$	Aprotic	1

The strong solvation by polar protic solvents further weakens nucleophiles as a result of the tight "cage" that is formed by the solvent molecules around the nucleophile. To form a bond with a substrate, the nucleophile must first shed some of those solvent molecules.

PROBLEM 9.25 In which solvent—dimethyl sulfoxide or ethanol— does this reaction proceed faster via an S$_N$1 mechanism? In which solvent does it proceed faster via an S$_N$2 mechanism? Explain.

The solvent effects shown in Tables 9-8 and 9-9 also derive from the different abilities of the solvent to solvate the *leaving group*. Recall that the S$_N$1 reaction rate is dictated entirely by the departure of the leaving group, which is often negatively charged. In a protic solvent, the anionic leaving group is stabilized heavily by the strong solvation that ensues, but in an aprotic solvent it is not. Therefore:

> A protic solvent will dramatically speed up the rate-determining step of an S$_N$1 reaction by solvating the leaving group; an aprotic solvent will not.

The influence that the solvent has on the leaving group departing does not affect S$_N$2 reaction rates as greatly, due to the assistance that the attacking species provides.

PROBLEM 9.26 In which solvent—ethanol or acetone—does this reaction proceed faster by an S$_N$2 mechanism? Explain.

Similar reasons explain why protic solvents tend to favor E1 reactions over E2 reactions, whereas aprotic solvents tend to favor E2 over E1. Strong solvation of the base by a protic solvent weakens the base, thereby slowing the E2 reaction rate. At the same time, protic solvents enhance the ability of the leaving group to leave in the rate-determining step of an E1 reaction. Conversely, aprotic solvents favor E2 reactions over E1 because the base is not weakened as much. The leaving group is slow to depart on its own, moreover, because it is so poorly solvated by the aprotic solvent.

PROBLEM 9.27 In which solvent, **Y** or **Z**, does the reaction in Problem 9.26 proceed faster by the E1 mechanism? Explain.

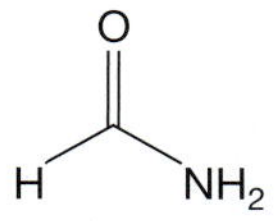

9.7b Relative Nucleophilicities in Protic and Aprotic Solvents

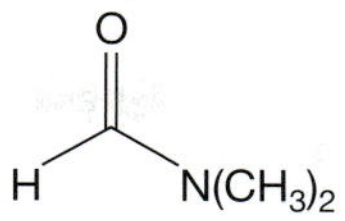

We have just seen that solvation in protic solvents has the effect of weakening nucleophiles substantially. The solvation of anions by protic solvents is so dramatic that it can *reverse* the relative strengths of some nucleophiles. This can be seen in Table 9-10, which lists the S_N2 reaction rates of various nucleophiles in ethanol (a protic solvent).

Note, for example, that the nucleophilicity of Br^- in ethanol is about 30 times greater than that of Cl^- (i.e., 620,000 vs. 23,000). This is contrary to DMF, an aprotic solvent (Table 9-1, p. 448), in which Cl^- is a stronger nucleophile than Br^- (i.e., 2 vs. 1). This reversal occurs in protic solvents because Cl^- is substantially smaller than Br^- (Cl is above Br in group 7A of the periodic table), which means the negative charge is more concentrated when it is on Cl^-. With a higher concentration of negative charge, Cl^- is solvated much more strongly in a protic solvent, as indicated in Figure 9-16 (next page), and is thus more weakened as a nucleophile.

YOUR TURN 9.11

Compare Tables 9-1 and 9-10 to find another pair of nucleophiles (other than Cl^- and Br^-) whose *relative* nucleophilicities in ethanol are the reverse of what they are in DMF. What does this observation suggest about which of the two species is more strongly solvated in ethanol?

In general:

We see a reversal of nucleophilicity in protic versus aprotic solvents when we compare two nucleophiles that have negative charges localized on atoms in different rows of the periodic table.

This is illustrated further in Solved Problem 9.28.

TABLE 9-10 S_N2 Reaction Rate in Ethanol for the Reaction:

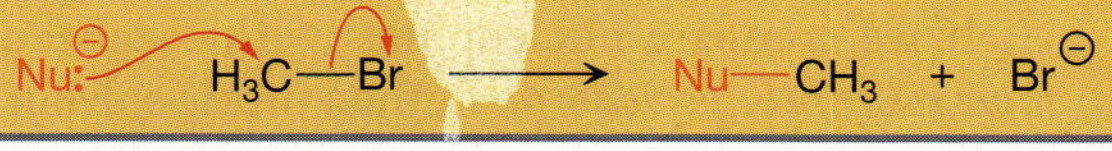

$$Nu{:}^{\ominus} \quad H_3C{-}Br \longrightarrow Nu{-}CH_3 \ + \ Br^{\ominus}$$

Nucleophile	CH_3OH	$F^{\ominus}$	$CH_3CO_2^{\ominus}$	$Cl^{\ominus}$	NH_3	$N_3^{\ominus}$
Relative Reaction Rate	1	500	20,000	23,000	320,000	600,000

Nucleophile	$Br^{\ominus}$	$CH_3O^{\ominus}$	$NC^{\ominus}$	Ph_3P	$HS^{\ominus}$
Relative Reaction Rate	620,000	2,000,000	5,000,000	10,000,000	100,000,000

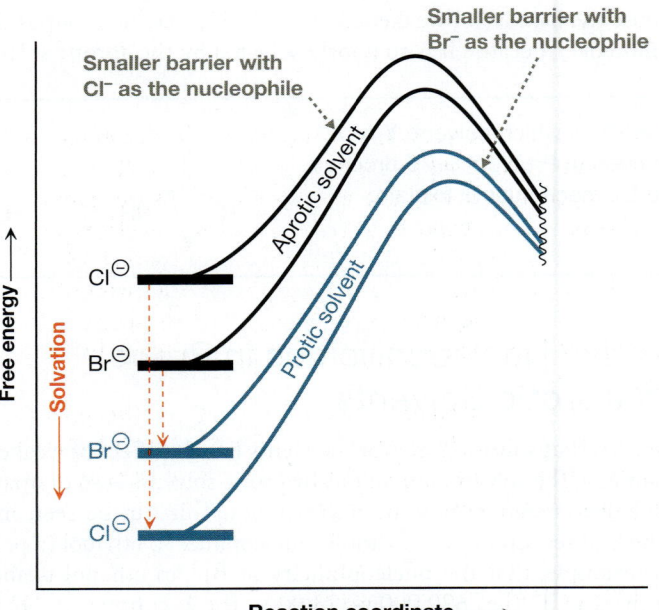

FIGURE 9-16 **Solvation and the reversal of nucleophile strengths** Cl^- is a stronger nucleophile than Br^- in an aprotic solvent (black curves) because Cl^-, being less stable, undergoes an S_N2 reaction having the smaller energy barrier. In a protic solvent (blue curves), however, Cl^- is solvated (red arrows) much more strongly than Br^-, which results in Br^- undergoing an S_N2 reaction having the smaller energy barrier.

In the figure: "Smaller barrier with Cl^- as the nucleophile", "Smaller barrier with Br^- as the nucleophile", "Aprotic solvent", "Protic solvent", "Free energy", "Solvation", "Reaction coordinate", labels $Cl^\ominus$, $Br^\ominus$, $Br^\ominus$, $Cl^\ominus$.

SOLVED PROBLEM 9.28

Is HS^- or HO^- a stronger nucleophile in water?

Think Which is a stronger nucleophile *intrinsically* (i.e., without considerations of solvation)? Is water a protic or an aprotic solvent? Is the solvation of HO^- in water much different from the solvation of HS^-?

Solve Without considering solvation, HO^- is a stronger nucleophile because O does not stabilize the negative charge as well as S. This is what we would expect in an aprotic solvent, too, in which solvation is not dramatic. Water is a *protic solvent*, however, so solvation is very important. The solvation of HO^- is much stronger than that of HS^- because the atoms on which the negative charge is localized (O and S) are in different rows of the periodic table. We therefore expect a reversal of the nucleophile strengths in a protic solvent versus an aprotic solvent. Thus, HS^- is a stronger nucleophile than HO^- in water.

PROBLEM 9.29 Is F^- or I^- a stronger nucleophile in ethanol? Explain.

The tremendous solvation that occurs in protic solvents causes some negatively charged nucleophiles to be weaker than some uncharged nucleophiles. Notice in Table 9-10, for example, that NH_3 is more than 10 times stronger than Cl^- in ethanol (i.e., 320,000 vs. 23,000), making NH_3 a moderately strong nucleophile. More impressively, triphenylphosphine (Ph_3P) is more than 400 times stronger than Cl^- (i.e., 10,000,000 vs. 23,000).

If we compare nucleophiles with localized negative charges on atoms in the *same row of the periodic table*, then protic solvents do *not* reverse nucleophilicities. In both protic and aprotic solvents, for example, nucleophilicities increase in the order $F^- < CH_3O^- < NC^-$. Even though each ion is solvated quite strongly by the protic solvent, they are not solvated very *differently* from each other because the negatively charged atoms are similar in size.

PROBLEM 9.30 Would you expect H_2S to be a stronger or weaker nucleophile than H_3N in ethanol? Explain.

How an Enzyme Can Manipulate the Reactivity of a Nucleophile and Substrate

Pseudomonas putida PP3, a type of gram-negative bacteria, carries out the conversion shown here. This may seem surprising to you, because the Cl substituent is replaced by OH with inversion of configuration, making this an S_N2 reaction. As we learned here in Chapter 9, S_N2 reactions are promoted by strong nucleophiles, so in the laboratory, we might carry out the reaction by treating the substrate with a source of HO⁻, such as NaOH. The involvement of HO⁻, however, would be unfeasible at physiological pH because a strong base like HO⁻ would be kept at *very* low concentration. What is the bacteria's secret?

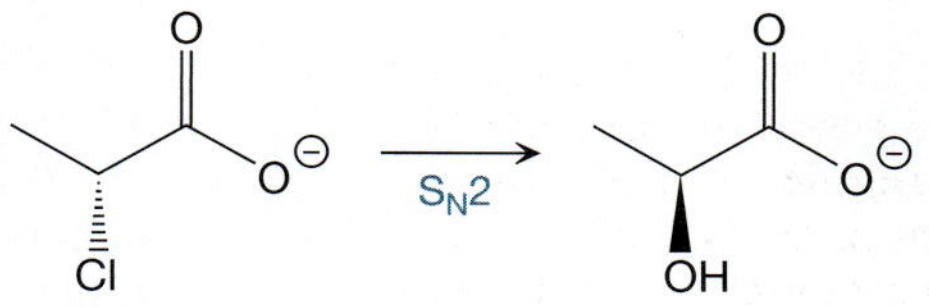

The key is an enzyme called haloacid dehalogenase (shown below at the left), which catalyzes the reaction with *water* as the nucleophile, not HO⁻. But how does the enzyme enable water, a weak nucleophile, to facilitate an S_N2 reaction on a secondary chloride substrate?

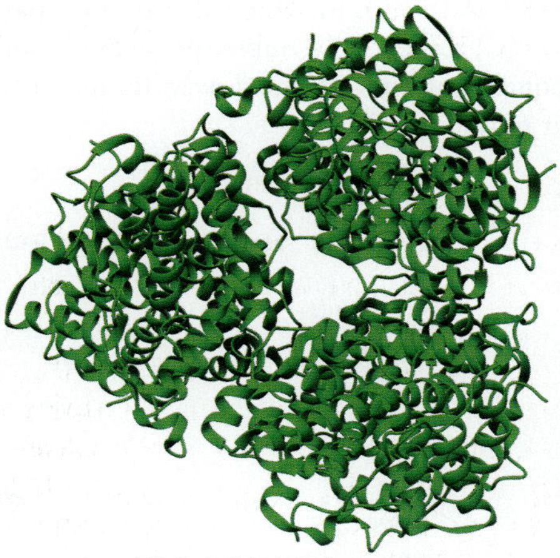

Haloacid dehalogenase

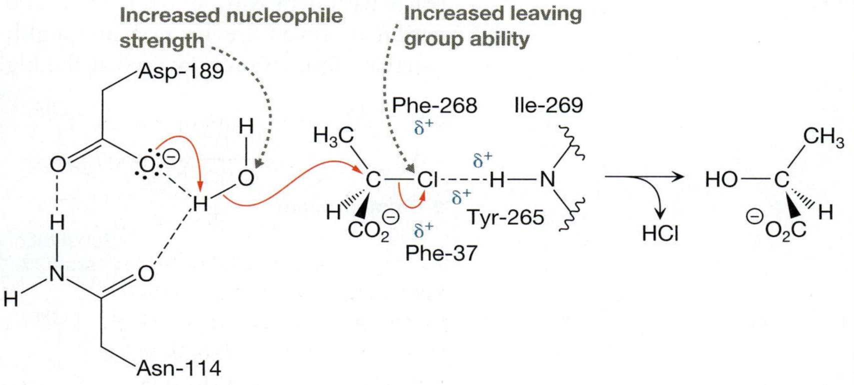

As shown above at the right, water is initially held in place in the enzyme's active site by hydrogen bonds involving the side groups of the aspartate-189 and asparagine-114 amino acid residues. Meanwhile, the substrate is situated so that its negatively charged regions benefit from the electrostatic attractions to the partially positive H of an amide group from the protein's backbone (Ile-269) and also to the partially positive aromatic hydrogens of the phenylalanine-37, tyrosine-265, and phenylalanine-268 residues.

The hydrogen bonding involving water sets up the loss of a proton from water during nucleophilic attack, which adds to the electron density at water's O atom and effectively increases water's nucleophilicity. Simultaneously, the negative charge that develops on Cl during its departure is stabilized by the surrounding partially positive environment, which effectively enhances the leaving group ability. Both of these modes of action together make it easier for the nucleophile to force off the leaving group and facilitate an S_N2 reaction that would otherwise not take place.

The idea of manipulating the reactivities of species involved in a chemical reaction is not unique to haloacid dehalogenase, but rather is a common strategy employed by enzymes to deal with various limitations that physiological conditions might bring. This goes a long way toward enzymes being so efficient at carrying out specific reactions vital to an organism.

9.8 Factor 6: Heat

Products from an S_N2 reaction are often formed along with abundant E2 products. Similarly, S_N1 and E1 reaction products are often formed together. This occurs because conditions that favor S_N2 reactions—such as a high concentration of a strong attacking species, a leaving group bonded to a primary carbon, and a polar aprotic solvent—generally favor E2 reactions as well. S_N1 and E1 reactions are favored by the same conditions, too—namely, a weak attacking species (or a low concentration of a strong attacking species); a good leaving group; a leaving group on a tertiary, allylic, or benzylic carbon; and a polar protic solvent.

When substitution and elimination reactions are both favored under a specific set of conditions, it is often possible to influence the outcome by changing the temperature under which the reactions take place. Specifically:

> Increasing the temperature of the reaction tends to promote elimination more than substitution.

This effect can be seen in Equations 9-32 and 9-33, in which 2-bromopropane reacts with hydroxide anion at 45 °C and 100 °C, respectively. Substitution and elimination products are formed in roughly equal amounts at the lower temperature, whereas elimination is favored at the higher temperature.

$$\text{2-Bromopropane} \xrightarrow[\text{CH}_3\text{OH/H}_2\text{O} \quad 45\,°C]{\text{NaOH}} \text{(47% Substitution product)} + \text{(53% Elimination product)} + \text{NaBr} + \text{H}_2\text{O} \quad (9\text{-}32)$$

$$\text{2-Bromopropane} \xrightarrow[\text{CH}_3\text{OH/H}_2\text{O} \quad 100\,°C]{\text{NaOH}} \text{(29%)} + \text{(71%)} + \text{NaBr} + \text{H}_2\text{O} \quad (9\text{-}33)$$

At higher temperatures, elimination becomes more favored.

This temperature effect can be understood by considering *entropy*, which, as we discussed in Section 2.7, is often thought of as a measure of disorder. If we examine an *overall* nucleophilic substitution reaction alongside an *overall* elimination reaction, as in Equation 9-34a and 9-34b, then we can see that their reactants are identical, but they differ in the number of product species: There are two product species in a substitution reaction and three in an elimination reaction. As a result, *there is significantly more disorder in the products of an elimination reaction than in the products of a competing substitution reaction*. In other words, the standard change in entropy of a reaction, $\Delta S^{\circ}_{\text{rxn}}$, is more positive for an elimination reaction than for a substitution reaction.

Two product species = less entropy

$$\text{2-Bromopropane} \xrightarrow{\ominus\text{OH}} \text{Substitution} \quad \text{OH} + \text{Br}^{\ominus} \quad (9\text{-}34a)$$

$$\xrightarrow{\text{Elimination}} + \text{H—OH} + \text{Br}^{\ominus} \quad (9\text{-}34b)$$

Three product species = greater entropy

For the reaction shown here, identify the products that have greater entropy. Label the products of substitution and the products of elimination.

Styrene

Recall from Section 6.3 (Equation 6-16) that $\Delta G^{\circ}_{rxn} = \Delta H^{\circ}_{rxn} - T\Delta S^{\circ}_{rxn}$. Because ΔS°_{rxn} is more positive for elimination than substitution, an increase in T causes a larger *decrease* in ΔG°_{rxn} for an elimination reaction. The Hammond postulate (Section 9.3a) would then suggest that $\Delta G^{\circ\ddagger}$ is lowered—and reaction rate is increased—more for elimination than for substitution.

There are a variety of ways to indicate that a reaction is run at high temperatures. One is simply to write the temperature at which the reaction is run underneath the reaction arrow (e.g., $\xrightarrow[90\,°C]{}$). Presumably that temperature will be significantly above room temperature. Alternatively, you can indicate that heat has been added to the reaction mixture by writing either "Heat" or its shorthand notation, Δ, underneath the reaction arrow (e.g., $\xrightarrow[Heat]{}$ or $\xrightarrow[\Delta]{}$).

PROBLEM 9.31 Suppose the following reaction produces a substantial amount of both substitution and elimination products at room temperature. If the temperature at which the reaction is run is raised, then which product will be the major product? If the temperature is lowered, then which product will be the major product?

9.9 Predicting the Outcome of an $S_N2/S_N1/E2/E1$ Competition

If all of the factors we have discussed were to favor the same reaction, then predicting the outcome of the competition between the S_N2, S_N1, E2, and E1 reactions would be pretty easy. Generally speaking, however, those factors do *not* all pull in the same direction, so predicting the major products is not always clear-cut. In many cases, definitively knowing the major product requires carrying out the reaction in question and measuring the relative concentrations of each product. However, by systematically evaluating all of the factors we have examined thus far, we can often reasonably *predict* the major products by applying the following four steps.

Step 1: Determine whether the substrate has a suitable leaving group.

- For an S_N1, S_N2, E1, or E2 reaction to take place readily, the leaving group should be at least as stable as F^-. [Leaving groups that are *not* suitable include HO^-, RO^-, H_2N^-, H^-, and R^- (Tables 9-3 and 9-4).]
- If the leaving group is suitable, then proceed to Step 2.

Step 2: Evaluate the type of carbon bonded to the leaving group.

- For an S_N1, S_N2, E1, or E2 reaction to take place readily, the carbon atom needs to be sp^3 hybridized.
- Recall from Table 9-7 that if the leaving group is on:
  - a *primary* carbon, then the S_N1 and E1 reactions are *not* feasible unless the resulting carbocation is resonance stabilized.
  - a *tertiary* carbon, then S_N2 reactions are *not* feasible.
  - a *secondary* carbon, then all four reactions must be considered.

Step 3: Examine the influences of the four factors:

- the strength of the attacking species as a nucleophile and as a base,
- the concentration of the attacking species,
- the leaving group ability, and
- the effects of the solvent.

Temporarily ignoring the influence of heat, we essentially give equal weight to each of those factors and tally up a score to determine the "winner" of the competition.

Step 4: If both substitution and elimination reactions appear to be favored, then factor in the influence of heat.

High temperatures tend to favor elimination, whereas low temperatures tend to favor substitution.

To illustrate this systematic method, let's first try to predict the outcome of the reaction in Equation 9-35.

$$\text{(S)-2-Chloropentane} \quad + \quad \text{NaBr} \quad \xrightarrow{\text{DMSO}} \quad ? \qquad (9\text{-}35)$$

The attacking species is Br^-, the substrate is (*S*)-2-chloropentane, the solvent is DMSO, and the leaving group would be Cl^-.

In Step 1, we need to determine whether the substrate has a suitable leaving group. Cl^- is more stable than F^-, so Cl^- has adequate leaving group ability (Tables 9-3 and 9-4) and we proceed to Step 2.

In Step 2, we confirm that the leaving group is on an sp^3-hybridized carbon and we need to determine the number of alkyl groups to which that carbon atom is attached. In this case, the carbon is 2°, so we must consider all four mechanisms (review Table 9-7).

For Step 3, we can construct a table (Table 9-11) to keep track of the reactions favored by each factor. We begin by considering the strength of the attacking species, Br⁻. Because it has a full negative charge, Br⁻ is a strong nucleophile, thereby favoring S_N2 reactions over S_N1 reactions. This is recorded in Table 9-11 with a check mark. On the other hand, Br⁻ is a weak base because it is substantially weaker than HO⁻. Therefore, the E1 reaction is favored over E2. Again, this is recorded in Table 9-11 with a check mark.

Next, what influence does the attacking species' concentration have on the competing reactions? Because we are not explicitly told that Br⁻ is dilute, we may assume that it is concentrated. Given that Br⁻ is a strong nucleophile, this high concentration will favor the S_N1 reaction over the S_N1, and we record this in Table 9-11 with another check mark. Br⁻ is a weak base, however, so its concentration does not matter significantly. Therefore, neither the E1 nor the E2 column receives a check mark.

The leaving group ability of Cl⁻ is considered next. Cl⁻ is a moderate leaving group (Tables 9-3 and 9-4), so it does not discriminate greatly between S_N1 and S_N2 or between E1 and E2. We therefore record a check in all four columns in Table 9-11 (alternatively, we could leave all four columns blank).

Finally, what are the solvent effects? DMSO is a polar *aprotic* solvent, so it favors S_N2 and E2 reactions over their corresponding S_N1 and E1 reactions. Once again, this is recorded in Table 9-11.

We essentially give equal weight to these four factors, so tallying up the number of factors that favor each of the four reactions gives four "votes" to the S_N2 reaction, and the other three reactions receive two or fewer. As a result, the S_N2 reaction is heavily favored, which means we can skip Step 4. The effect of heat must be considered only if there is a tie between substitution and elimination.

To predict the major product, enter the specific reactants into the S_N2 mechanism, as is shown in Equation 9-36. Notice that *stereochemistry is important* here, given that the C atom bonded to the leaving group is a chiral center. Backside attack of the nucleophile ensures that the only stereoisomer produced is the one shown, which is (R)-2-bromopentane.

TABLE 9-11	Summary Table for the Reaction in Equation 9-35			
Factor	**S_N1**	**S_N2**	**E1**	**E2**
Strength		✓	✓	
Concentration		✓		
Leaving group	✓	✓	✓	✓
Solvent		✓		✓
Total	**1**	**4**	**2**	**2**

Mechanism for Equation 9-35

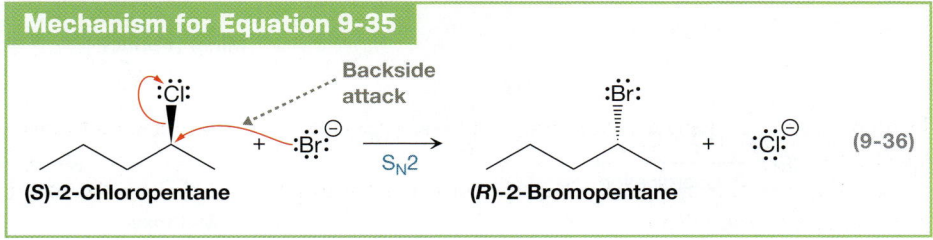

Now, predict the major product(s) of the reaction in Equation 9-37.

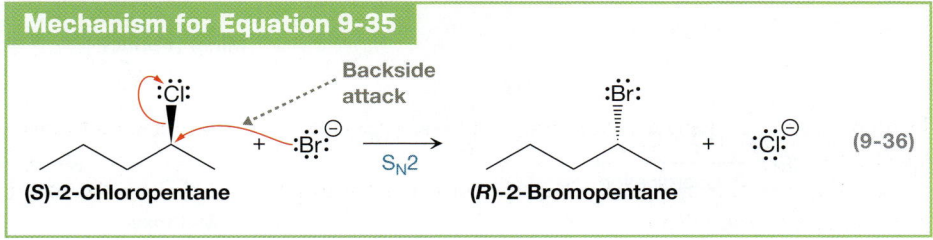

3-Bromo-3-ethylpentane $\xrightarrow{CH_3OH,\ \Delta}$ **?**   (9-37)

In this reaction, 3-bromo-3-ethylpentane is the substrate and Br⁻ is the leaving group. The substrate is dissolved in methanol, so methanol must be the attacking species as well as the solvent. This type of reaction is called **solvolysis** because the solvent acts as a reactant.

For Step 1 of our analysis, Br⁻ is a better leaving group than F⁻, so it is indeed a suitable leaving group for nucleophilic substitution and elimination. For Step 2, the

TABLE 9-12	Summary Table for the Reaction in Equation 9-37			
Factor	S_N1	S_N2	E1	E2
Strength	✓		✓	
Concentration				
Leaving group	✓		✓	
Solvent	✓		✓	
Total	3		3	0

leaving group is on a tertiary carbon, so S_N2 reactions can be ruled out (Table 9-7). The S_N2 column in Table 9-12 is therefore grayed out.

For Step 3, we determine which reactions are favored by the remaining factors (except heat), and enter the results into Table 9-12.

The attacking species, CH_3OH, is a weak nucleophile and a weak base, so S_N1 and E1 are favored over S_N2 and E2. The concentration of the nucleophile appears to be high (it is, after all, the solvent), which normally would favor S_N2 and E2. However, the attacking species is weak, both as a nucleophile and as a base, so its concentration does *not* matter significantly (Section 9.4), and we leave those entries in Table 9-12 without check marks. Next, Br^- is a good leaving group, so this factor favors S_N1 and E1 over S_N2 and E2. Finally, methanol is a polar protic solvent, which favors the S_N1 and E1 reactions over the S_N2 and E2.

The final tally of the results in Table 9-12 shows that both the S_N1 and E1 reactions are heavily favored. This suggests that both the S_N1 and E1 products should be formed in significant amounts. To break the tie, we must consider the effect of heat (Step 4). The Δ symbol beneath the reaction arrow indicates that the reaction is heated, which tips the balance in the favor of elimination over substitution (Section 9.8). We therefore conclude that the E1 reaction (Equation 9-38) leads to the major product, and the S_N1 reaction (Equation 9-39) leads to a minor product.

E1 mechanism for Equation 9-37 (Major product)

(9-38)

3-Ethylpent-2-ene

S_N1 mechanism for Equation 9-37 (Minor product)

(9-39)

3-Methoxy-3-ethylpentane

A proton can be removed from one of three possible carbons of 3-bromo-3-ethylpentane in an E1 reaction. In this case, however, those carbons are chemically equivalent, so removal of any one of them leads to the same alkene product (Equation 9-38). (In Section 9.10 we discuss how to predict the major elimination product when such protons are chemically distinct.)

Equation 9-39 shows that there are three steps for this S_N1 mechanism instead of two. The nucleophile that attacks in the second step of the S_N1 reaction, CH_3OH, is uncharged, so it develops a positive charge on oxygen. That positive charge is removed in the third step, which is a simple proton transfer involving the solvent.

What is the major product of the reaction in Equation 9-40? The substrate contains a tosylate (TsO^-) leaving group, the attacking species is the *tert*-butoxide anion, $(CH_3)_3CO^-$, and the solvent is *N,N*-dimethylformamide (DMF).

$$\text{(9-40)}$$

For Step 1 of our analysis, the TsO$^-$ leaving group is suitable for nucleophilic substitution and elimination; in fact, it is among the best leaving groups possible. For Step 2, the leaving group is bonded to a 2° carbon, so all four reactions must be considered. (Table 9-13 is constructed but is left for you to fill in as an exercise in Your Turn 9.13.)

For Step 3, consider all of the remaining factors except heat. The attacking species, $(CH_3)_3CO^-$, is a strong base, so it favors the E2 reaction over the E1. We would normally expect alkoxides (RO^-) to favor S_N2 reactions as well, but because of the excessive bulkiness of $(CH_3)_3CO^-$, it favors only the E2 reaction. The attacking species' concentration is assumed to be high because we are not told otherwise. This favors the E2 reaction because $(CH_3)_3CO^-$ is a strong base, but it does *not* favor the S_N2 reaction because of the diminished nucleophilicity of $(CH_3)_3CO^-$. TsO$^-$ is an excellent leaving group, so it favors the S_N1 reaction over the S_N2 and the E1 reaction over the E2. Finally, DMF is a polar aprotic solvent, which favors the S_N2 and E2 reactions over the S_N1 and E1 reactions. With these factors tallied at the bottom of Table 9-13, we can conclude that the E2 reaction will win.

TABLE 9-13	Summary Table for the Reaction in Equation 9-40			
Factor	**S_N1**	**S_N2**	**E1**	**E2**
Strength				
Concentration				
Leaving group				
Solvent				
Total	**1**	**1**	**1**	**3**

YOUR TURN 9.13

Complete Table 9-13 by placing check marks in the appropriate boxes, using the information provided above.

Notice in the substrate that there is a H atom on each C adjacent to the C atom bonded to the leaving group. Thus, it may appear that there are two possible E2 products, depending on which of those protons is removed. The E2 reaction is *stereospecific*, however, favoring the anticoplanar conformation of the substrate (review Section 8.5d). In this substrate, only the proton indicated in Equation 9-41 can be anti to the leaving group, giving rise to only one E2 product.

Mechanism for Equation 9-40

This H is anti to the leaving group.

This H is *not* anti to the leaving group.

E2 Reaction

Major product

$$\text{(9-41)}$$

SOLVED PROBLEM 9.32

Draw the complete mechanism and predict the products for the reaction shown here. In this reaction, (R)-(bromomethyl-d)-benzene is dissolved in methanol.

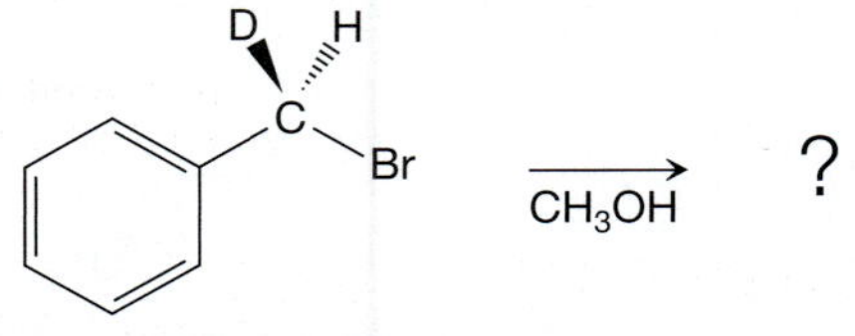

Think Is there a suitable leaving group? Does the type of carbon bonded to the leaving group rule out any of the four reactions? What is the attacking species? What is the solvent? Which reaction does each of the factors favor?

(R)-(Bromomethyl-d)-benzene

Solve The substrate is (R)-(bromomethyl-d)-benzene. This is a solvolysis reaction because methanol is both the attacking species and the solvent. The leaving group, Br^-, is suitable for nucleophilic substitution and elimination (it has much better leaving group ability than F^-). The leaving group is on a 1° carbon, suggesting that we might not have to consider the S_N1 and E1 options. The leaving group is bonded to a *benzylic carbon*, however, which makes the S_N1 and E1 reactions possible. Thus, all four reactions must be considered. Looking ahead, though, we can omit both the E1 and E2 options because we do not have a hydrogen and a suitable leaving group on *adjacent* carbons. That leaves only S_N1 and S_N2 to consider.

The attacking species (CH_3OH) is a weak nucleophile, favoring S_N1 reactions. Although the concentration of the attacking species is high, we do not add check marks to that row because the attacking species is a weak nucleophile and its concentration does not matter significantly. Br^- is a very good leaving group, favoring S_N1. Finally, the solvent (CH_3OH) is polar protic, which also favors S_N1. Tallying up the scores using the table in the margin, the S_N1 mechanism should be the predominant mechanism.

The mechanism is as follows:

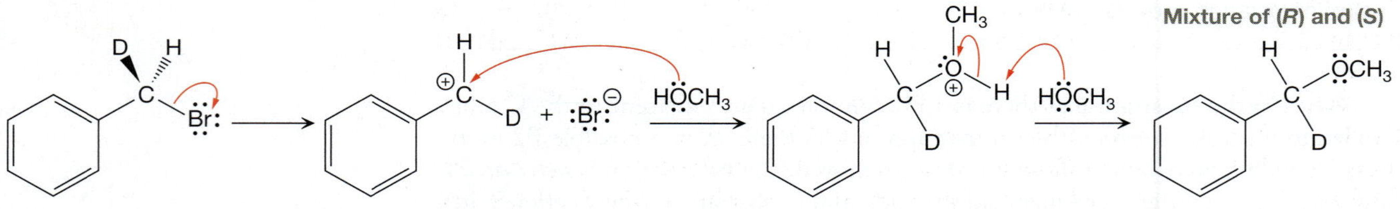

Factor	S_N1	S_N2	E1	E2
Strength	✓			
Concentration				
Leaving group	✓			
Solvent	✓			
Total	**3**	0		

The loss of the leaving group produces an achiral carbocation, so the attack of the nucleophile produces a mixture of enantiomers. The S_N1 product is then deprotonated to yield an uncharged ether.

PROBLEM 9.33 Predict the major product of the reaction shown here. In this reaction, the alkyl bromide is treated with sodium methoxide in acetone.

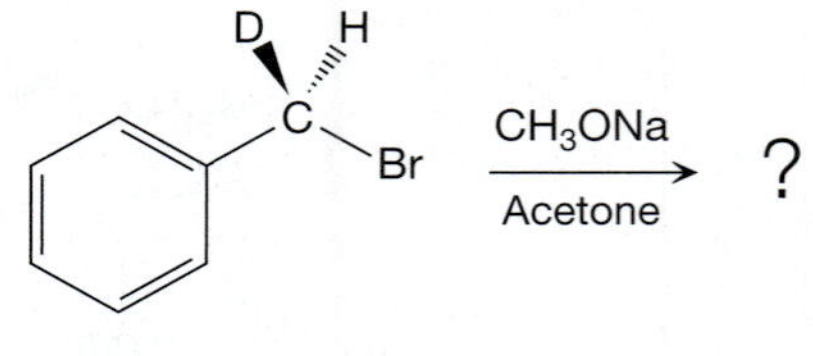

9.10 Regioselectivity in Elimination Reactions: Zaitsev's Rule

A substrate can possess two or more distinct hydrogen atoms that can be removed as protons in an elimination reaction, leading to two or more possible alkene products. This is the case for 2-iodohexane, shown in Equation 9-42. Removing a proton from

C3 produces hex-2-ene, whereas removing a proton from C1 produces hex-1-ene. With CH_3O^- as the base, the major product is hex-2-ene.

The major product is the more highly substituted alkene.

$$\overset{6\quad5\quad4\quad3\quad2\quad1}{CH_3CH_2CH_2CH_2CHCH_3} \quad \overset{CH_3ONa}{\underset{\underset{100\ °C}{CH_3OH}}{\longrightarrow}} \quad CH_3CH_2CH_2CH=CHCH_3 \quad + \quad CH_3CH_2CH_2CH_2CH=CH_2 \quad \textbf{(9-42)}$$

2-Iodohexane $\quad\quad\quad\quad\quad\quad\quad\quad\quad\quad$ 68% $\quad\quad\quad\quad\quad\quad\quad$ 16%

$\overset{\quad\quad\quad\quad\quad\quad\quad\quad\quad}{\underset{I}{\cdot\cdot}}$ $\quad\quad\quad\quad\quad\quad\quad\quad\quad\quad\quad\quad\quad\quad$ **Hex-2-ene** $\quad\quad\quad\quad\quad$ **Hex-1-ene**

This elimination reaction exhibits **regioselectivity**, which is the preference of a reaction to take place at one site within a molecule over another.

Why is hex-2-ene the major product, not hex-1-ene? The conditions in Equation 9-42 favor E2 reactions, so each product is formed from a competing E2 reaction. The free energy diagram for each of these reactions is shown in Figure 9-17.

Hex-2-ene is more stable than hex-1-ene because it is more highly alkyl substituted (Section 5.9), so the reaction that leads to the formation of hex-2-ene has a more negative ΔG°_{rxn} than that leading to the formation of hex-1-ene. Thus, as we learned in Section 9.3a, the reaction that produces hex-2-ene should have a lower energy barrier and proceed faster than the reaction that produces hex-1-ene.

The lesson we learn from this example can be generalized:

> Elimination usually takes place so as to produce the most stable (e.g., the most highly alkyl-substituted) alkene.

It applies to E1 reactions, too, such as those in Equation 9-43. A proton on 2-methylbutan-2-ol can be removed from C1 to produce 2-methylbut-1-ene, or from C3 to produce 2-methylbut-2-ene. The major product is 2-methylbut-2-ene, because, once again, it is the more highly substituted and more stable alkene product.

Before the elimination mechanisms were fully understood, Alexander M. Zaitsev (1841–1910), a Russian chemist, summarized the regioselectivity of elimination reactions as follows: *The major elimination product is the one produced by deprotonating the carbon atom initially attached to the fewest hydrogen atoms.* Eliminating a proton from the C atom that has the fewest H atoms leads to the most highly alkyl-substituted alkene product, which, we now know, is often the most stable. Zaitsev's summary from his empirical observations came to be known as **Zaitsev's rule**, and the most highly substituted alkene product is called the **Zaitsev product**.

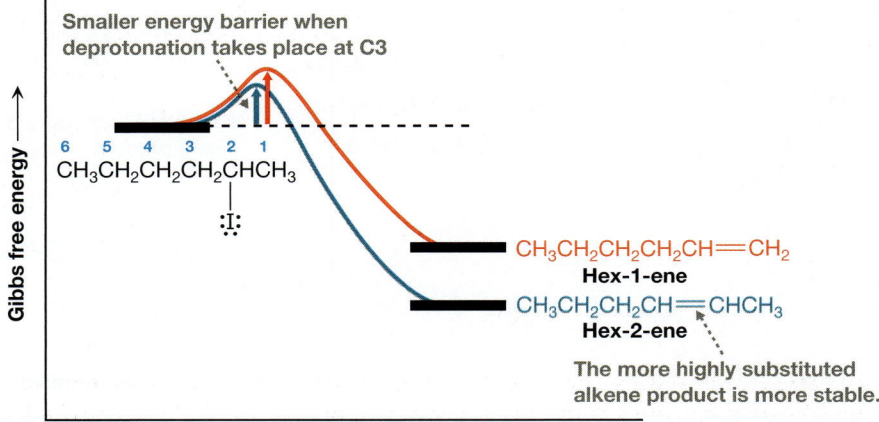

FIGURE 9-17 Zaitsev's rule The free energy diagram is shown for the reaction leading to each product in Equation 9-42. The more stable alkene product is hex-2-ene because it is more highly substituted. According to the Hammond postulate, then, the transition state leading to hex-2-ene is lower in energy than that leading to hex-1-ene, so the energy barrier leading to hex-2-ene is smaller, too.

Smaller energy barrier when deprotonation takes place at C3

$\overset{6\quad5\quad4\quad3\quad2\quad1}{CH_3CH_2CH_2CH_2CHCH_3}$

CH₃CH₂CH₂CH₂CH=CH₂
Hex-1-ene

CH₃CH₂CH₂CH=CHCH₃
Hex-2-ene

The more highly substituted alkene product is more stable.

Gibbs free energy

Reaction coordinate

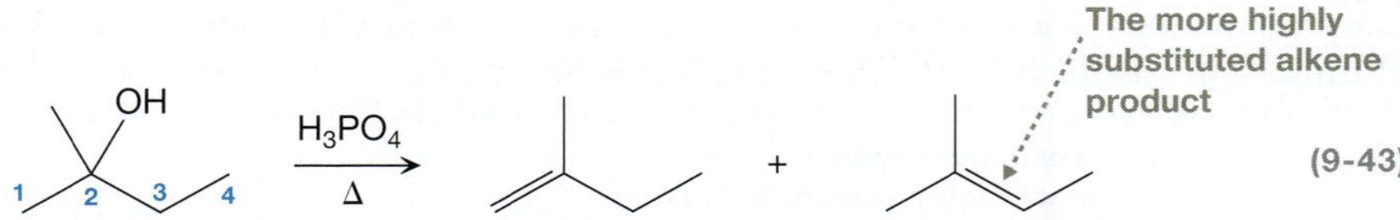

$$\text{2-Methylbutan-2-ol} \xrightarrow[\Delta]{H_3PO_4} \text{2-Methylbut-1-ene (21\%)} + \text{2-Methylbut-2-ene (78\%)} \quad (9\text{-}43)$$

The more highly substituted alkene product

YOUR TURN 9.14

For each of the two products in Equation 9-42, circle all of the alkyl groups that are attached to the alkene group and specify the degree of alkyl substitution (i.e., mono, di, tri, tetra, or unsubstituted). Do the same for each product in Equation 9-43.

One noteworthy exception to Zaitsev's rule involves a strong, bulky base, such as the *tert*-butoxide ion, $(CH_3)_3CO^-$. When 2-iodohexane is treated with potassium *tert*-butoxide, the major product is hex-1-ene (Equation 9-44), which is the *less* highly substituted alkene product—that is, the **anti-Zaitsev product**.

There is less steric hindrance surrounding these H atoms.

Strong bulky base

The major product is the less highly substituted alkene.

$$\underset{\text{2-Iodohexane}}{CH_3CH_2CH_2CH_2CHCH_3} \xrightarrow[\substack{(CH_3)_3COH \\ 99\,°C}]{(CH_3)_3COK} \underset{\substack{\text{Hex-2-ene} \\ 27\%}}{CH_3CH_2CH_2CH=CHCH_3} + \underset{\substack{\text{Hex-1-ene} \\ 60\%}}{CH_3CH_2CH_2CH_2CH=CH_2} \quad (9\text{-}44)$$

Although the H atoms at both C1 and C3 are relatively well exposed to the base (Section 9.6b), there is more steric hindrance surrounding those at C3. Thus, the base, which itself is quite bulky, favors abstraction of the proton at C1.

SOLVED PROBLEM 9.34

Predict the major product of the E2 reaction at the right.

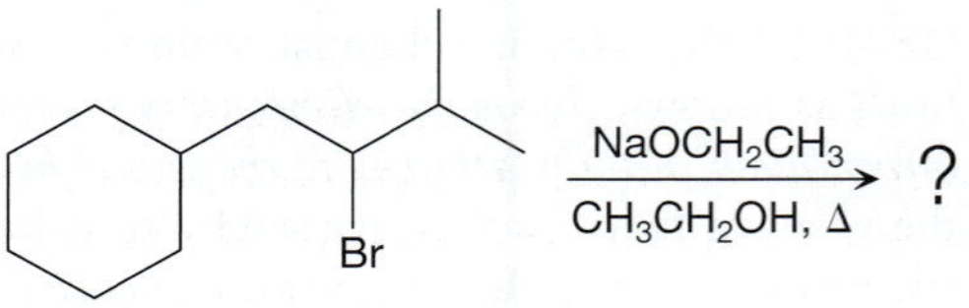

Think What are the possible alkene products? Draw them out. Which is the most stable? Is the base very bulky?

Solve The possible E2 products are obtained by eliminating the leaving group (Br⁻) and a proton on a carbon *adjacent* to the one bonded to the leaving group. These protons are highlighted on the reactant side below and the corresponding products are shown on the right.

Alkene **B** is more highly substituted than alkene **A**, so **B** is the more stable alkene and is thus favored as long as the base is not very bulky. The base in this reaction, $CH_3CH_2O^-$, is not bulky, so the major product is **B**.

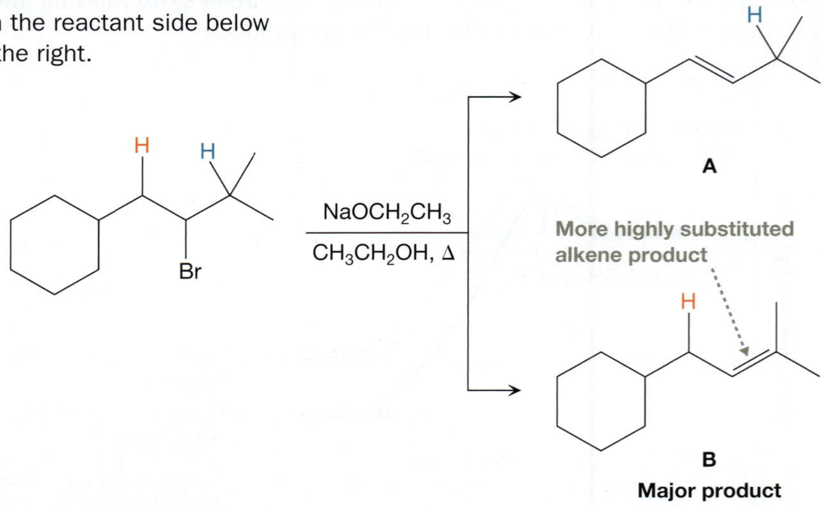

$$\xrightarrow[CH_3CH_2OH, \Delta]{NaOCH_2CH_3} \quad ?$$

A

More highly substituted alkene product

B

Major product

PROBLEM 9.35 For each of the following substrates, draw *all* possible E2 products using CH_3ONa as the base and determine which of them is the major product. (Disregard stereochemistry.)

(a)　　　　　　　　(b)　　　　　　　　(c)

9.11 Intermolecular Reactions versus Intramolecular Cyclizations

We generally think of a chemical reaction being between two *separate species*—a so-called **intermolecular** (between molecule) reaction. A reaction typically requires two separate *functional groups*, but those functional groups need not be on separate molecules. Instead, they can be part of the *same molecule*, attached at different places on the molecule's backbone. In that case, the reaction is said to be **intramolecular** (within the same molecule).

Equation 9-45 shows an intramolecular S_N2 reaction. It is intramolecular because the nucleophile (the negatively charged O atom) and the leaving group (departing as Cl^-) are on the same molecule, resulting in the formation of a ring.

An *intramolecular* reaction

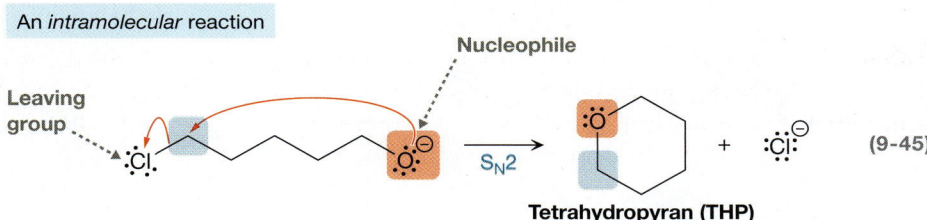

$$(9\text{-}45)$$

Tetrahydropyran (THP)

Whenever an intramolecular reaction can take place, an intermolecular reaction is also possible—that is, the two reactions compete with each other. This can be seen in Equation 9-46, in which the reactants are the same as those in Equation 9-45.

An *intermolecular* reaction

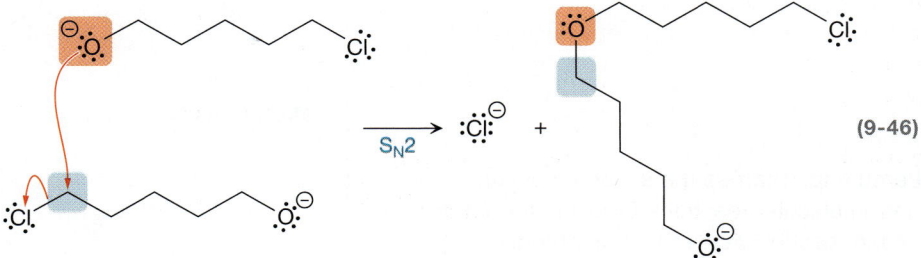

$$(9\text{-}46)$$

Which is the predominant reaction? Experimentally:

An *intramolecular* reaction typically "wins out" over its competing intermolecular reaction when the formation of a five- or six-membered ring is possible.

This is an outcome of the interplay between the changes in entropy and strain during the formation of a ring. As the chain becomes longer, it becomes less likely for the ends to meet to form a ring; there is a greater loss of entropy in the process. On the other hand, recall from Section 4.4 that ring strain decreases sharply to zero on going from a three-membered ring to a six-membered ring, and larger rings have a small amount of ring strain. Therefore, as the size of the ring being formed increases from three to six members, the process of forming the ring becomes easier energetically. These opposing factors—entropy and strain energy—reach an optimum in the formation of a five- or six-membered ring.

SOLVED PROBLEM 9.36

Two different intramolecular nucleophilic substitution reactions are possible with the substrate shown here. Draw the complete, detailed mechanism that leads to each product. Predict which one is the major product and explain why.

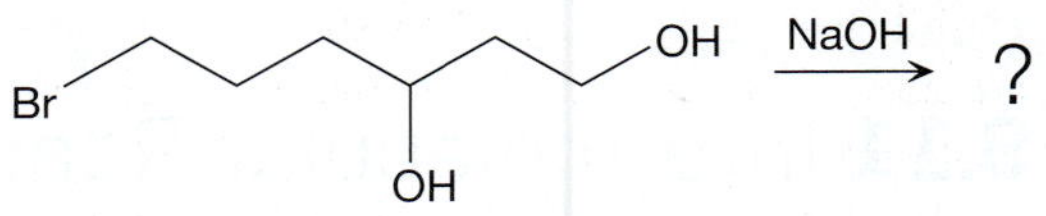

Think Under what conditions does the reaction take place? For an intramolecular nucleophilic substitution reaction under these conditions, what can act as the nucleophile? What can act as the leaving group? For each intramolecular nucleophilic substitution reaction, what is the size of the ring that is formed?

Solve The reaction takes place under basic conditions, so deprotonation of an OH group on the organic species would generate a strongly nucleophilic O⁻ site. The leaving group departs as Br⁻. The two possible intramolecular nucleophilic substitution reactions that can take place are as follows:

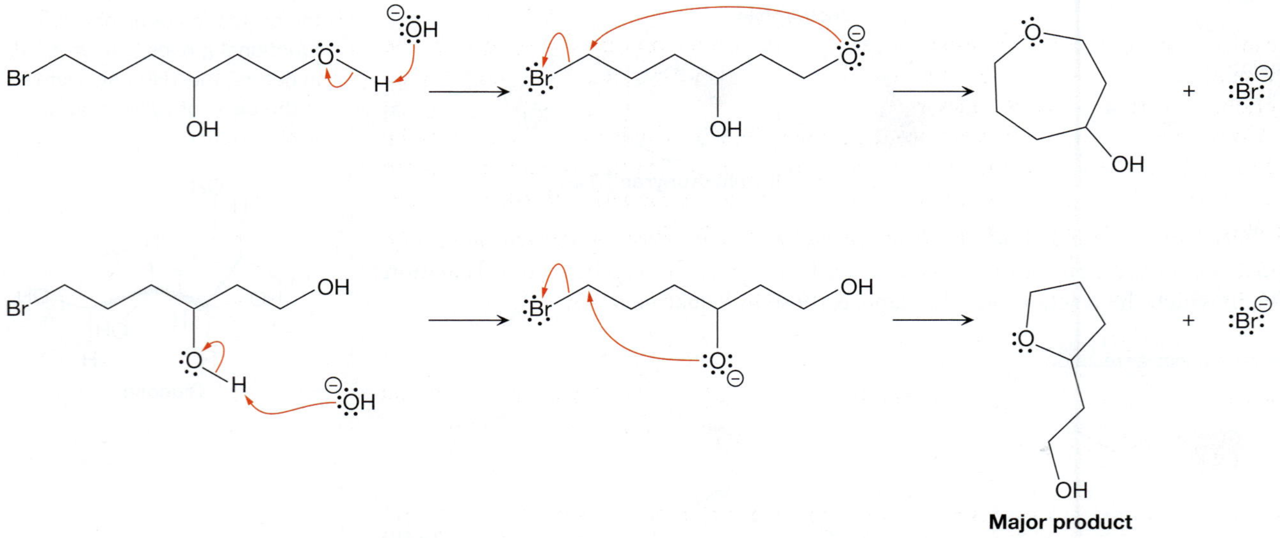

The first reaction yields a seven-membered ring, whereas the second reaction yields a five-membered ring. Because intramolecular reactions favor the formation of five- and six-membered rings, the second reaction yields the major product.

PROBLEM 9.37 What is the major product of the nucleophilic substitution reaction shown here?

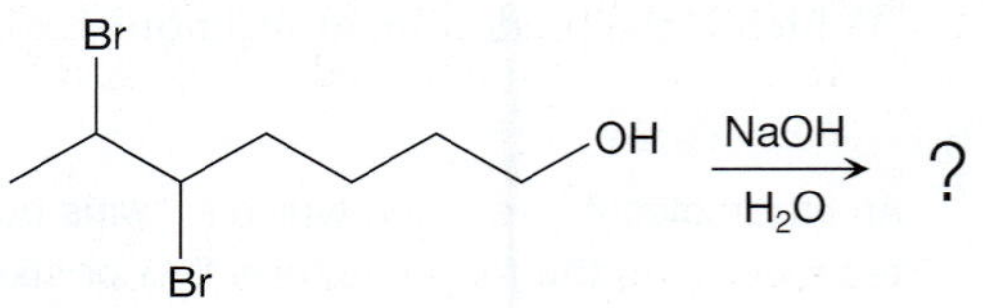

9.12 Kinetic Control, Thermodynamic Control, and Reversibility

Recall from Section 9.1 that competing reactions can take place under *kinetic control*, in which case the major product is the one that forms the fastest, or under *thermodynamic control*, in which case the major product is the one that is most stable. Recall, too, that nucleophilic substitution and elimination reactions generally compete under kinetic control. Why?

Kinetic control and thermodynamic control differ depending on whether the products formed from each of the competing reactions are in equilibrium with each other. If an equilibrium involving the various products is established, then, as with any equilibrium, the species with the lowest free energy *must* be present in the greatest amount (Section 6.3). This describes thermodynamic control for the competing reactions.

For products from competing reactions to be in equilibrium, there must be a way that those products interconvert. One typical scheme, shown in Equation 9-47, is **reversible** competing reactions. That is, as suggested by the **reversible reaction arrows** ($\rightleftharpoons$), each reaction can take place readily in both the forward and reverse directions.

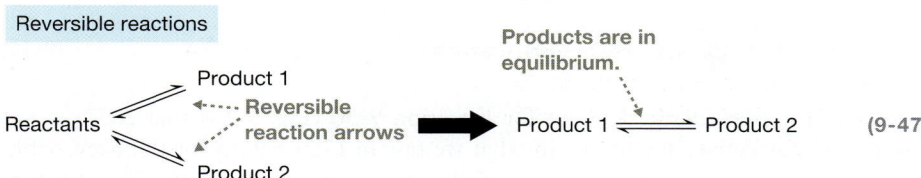

Because of this reversibility, a molecule of Product 1, once produced, can react in the reverse direction to regenerate reactants. Then, the regenerated reactants can undergo reaction again to produce Product 2.

Conversely, if competing reactions are **irreversible**, in which case they do *not* take place readily in the reverse direction, then equilibrium is *not* established between the products from the respective reactions. This is illustrated in Equation 9-48, using **irreversible reaction arrows** ($\rightarrow$) to connect reactants to products.

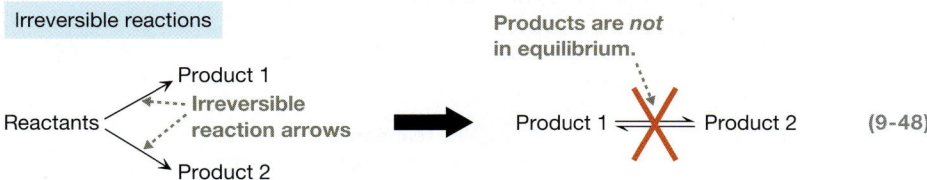

If the respective products are not in equilibrium, then the major product is *not* necessarily the most stable product. Rather, the major product is the one that is produced the fastest, in which case the competing reactions take place under kinetic control.

These ideas can be summarized as follows:

- Reversible reactions tend to take place under thermodynamic control.
- Irreversible reactions tend to take place under kinetic control.

Thus, competing S_N2, S_N1, E2, and E1 reactions usually take place under kinetic control because *these reactions tend to be irreversible*.

Whether a reaction is reversible or irreversible can often be determined by carefully examining its free energy diagram. For example, consider a typical S_N2 reaction, such as the one in Equation 9-49.

An *irreversible* S_N2 reaction

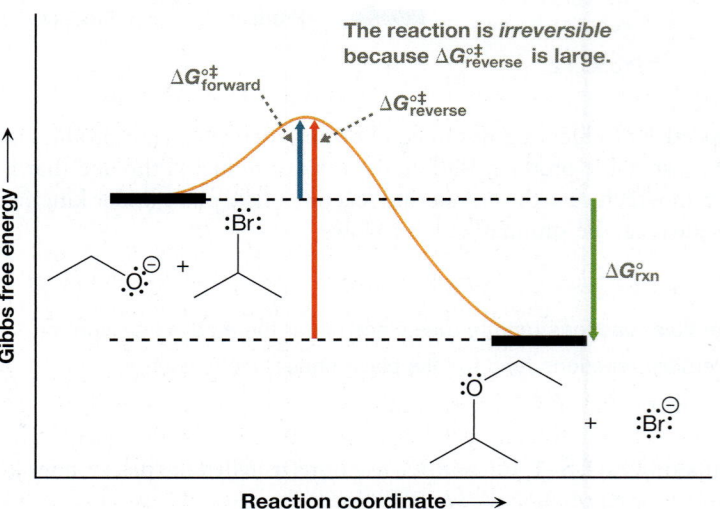

2-Bromopropane   **2-Ethoxypropane**   (9-49)

In its free energy diagram in Figure 9-18, the products are much lower in energy than the reactants, making ΔG°_{rxn} substantially negative. This is mainly due to the greater charge stability on the product side; the Br atom, because it is substantially larger, can better accommodate the negative charge than the O atom can (Section 6.6b). Consequently, the sizes of the energy barriers in the forward and reverse directions—$\Delta G^{\circ\ddagger}_{forward}$ and $\Delta G^{\circ\ddagger}_{reverse}$, respectively—differ significantly. In fact, $\Delta G^{\circ\ddagger}_{reverse}$ is so large that the reaction in the reverse direction is very slow, thus making the reaction virtually irreversible.

Some S_N2 reactions, including that in Equation 9-50, are reversible.

A *reversible* S_N2 reaction

$$\ddot{\ddot{I}}{:}^{\ominus} \ + \ H_3C-\ddot{B}\ddot{r}{:} \ \rightleftharpoons \ \ddot{\ddot{I}}{:}-CH_3 \ + \ {:}\ddot{B}\ddot{r}{:}^{\ominus} \qquad (9\text{-}50)$$

Notice in the free energy diagram for Equation 9-50 (Fig. 9-19) that ΔG°_{rxn} is *not* substantially negative, in contrast to what we saw in Figure 9-18 for the irreversible reaction in Equation 9-49. In fact, the products are somewhat higher in energy than the reactants, making ΔG°_{rxn} somewhat positive. This is primarily because Br^- is less stable than I^-. Consequently, $\Delta G^{\circ\ddagger}_{reverse}$, is not very large, allowing the reaction to proceed in the reverse direction at a significant rate.

The connection between ΔG°_{rxn} and reversibility extends to other reactions as well.

- A reaction tends to be *irreversible* if its ΔG°_{rxn} is substantially negative (i.e., if the products are much more stable than the reactants).
- Otherwise, the reaction tends to be *reversible*.

FIGURE 9-18 Standard Gibbs free energy change, ΔG°_{rxn}, and irreversible reactions The free energy diagram represents the substitution reaction in Equation 9-49. The products are much more stable than the reactants, so the reaction has a large negative value for ΔG°_{rxn}. As a result, the energy barrier in the reverse direction, $\Delta G^{\circ\ddagger}_{reverse}$, is large. This makes the reverse reaction excessively slow, in which case the reaction is *irreversible*.

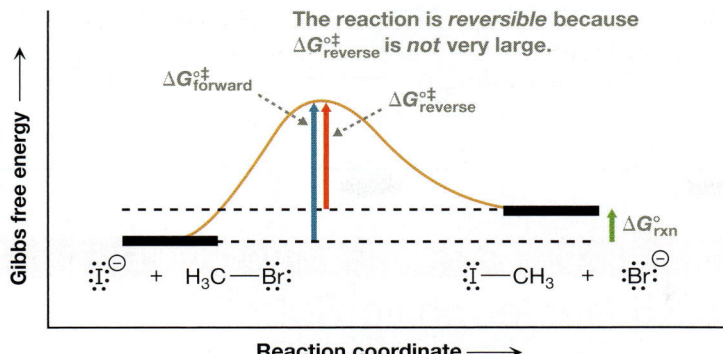

The reaction is *reversible* because $\Delta G^{\circ\ddagger}_{reverse}$ is *not* very large.

$\Delta G^{\circ\ddagger}_{forward}$ $\Delta G^{\circ\ddagger}_{reverse}$

Gibbs free energy

ΔG°_{rxn}

$:\!\overset{\ominus}{\ddot{I}}\!:$ + $H_3C\!-\!\overset{..}{\underset{..}{Br}}\!:$ $\overset{..}{\ddot{I}}\!-\!CH_3$ + $:\!\overset{..}{\underset{..}{Br}}\!:^{\ominus}$

Reaction coordinate $\longrightarrow$

FIGURE 9-19 Standard Gibbs free energy change, ΔG°_{rxn}, and reversible reactions The free energy diagram for the substitution reaction in Equation 9-50. The products are *not* dramatically more stable than the reactants, so the energy barrier in the reverse direction, $\Delta G^{\circ\ddagger}_{reverse}$, is comparable to that in the forward direction, $\Delta G^{\circ\ddagger}_{forward}$. Thus, the rate in the reverse direction is significant, and the reaction is *reversible*.

How, then, can S_N1 and E1 reactions, such as those in Equations 9-51 and 9-52, participate in the $S_N2/S_N1/E2/E1$ competition under kinetic control? Kinetic control of these reactions might seem counterintuitive because both of them *directly* produce charged products, which are higher in energy than the uncharged reactants.

This product is irreversibly removed from the equilibrium, thus driving the equilibrium more toward products.

The large amount of H₂O makes this step effectively irreversible.

S_N1 Reversible

H_2O + [cyclohexyl–Br] ⇌ [cyclohexyl–$\overset{\oplus}{O}H_2$] + $Br^{\ominus}$ $\xrightarrow{H_2O}$ [cyclohexyl–OH] + $H\!-\!\overset{\oplus}{O}H_2$ (9-51)

E1   Reversible   Irreversible

H_2O + [cyclohexyl–Br] $\underset{Heat}{\rightleftharpoons}$ [cyclohexene] + $H\!-\!\overset{\oplus}{O}H_2$ + $Br^{\ominus}$ → $H\!-\!\overset{\oplus}{O}H_2$ + $Br^{\ominus}$ (9-52)

This product can be distilled away from the reaction mixture as the reaction progresses, thus driving the equilibrium more toward products.

Indeed, as indicated, the reactions leading to the immediate substitution and elimination products are intrinsically reversible. However, typical conditions for nucleophilic substitution and elimination reactions quite often prevent such reactions from reaching equilibrium. For example, the substitution reaction in Equation 9-51 is a solvolysis reaction, with water as the solvent. Because water is present in such a large amount, Le Châtelier's principle dictates that the deprotonation of the $C_6H_{11}OH_2^+$ product is shifted heavily to the right. Once $C_6H_{11}OH_2^+$ is formed, therefore, it is rapidly deprotonated and removed from the equilibrium. Without the necessary products present for the reaction to take place in the reverse direction, the S_N1 reaction is effectively *irreversible*.

A similar phenomenon occurs with the elimination reaction in Equation 9-52. As we saw in Section 9.8, elimination reactions such as this are heated substantially. The relatively low-boiling product, cyclohexene, can thus be distilled away from the product mixture as the reaction progresses. Once again, without this product present, the reverse reaction cannot take place, effectively making the E1 reaction *irreversible*.

PROBLEM 9.38 The E2 reaction shown here competes with the S$_N$2 reaction from Equation 9-49. Is this reaction *reversible* or *irreversible*? Explain.

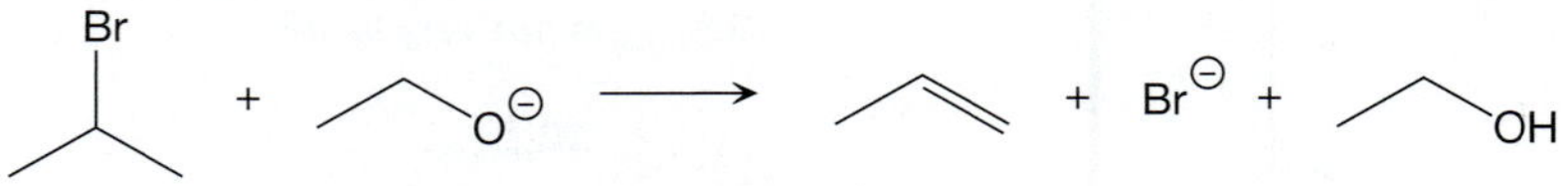

9.13 Nucleophilic Substitution Reactions and Monosaccharides: The Formation and Hydrolysis of Glycosides

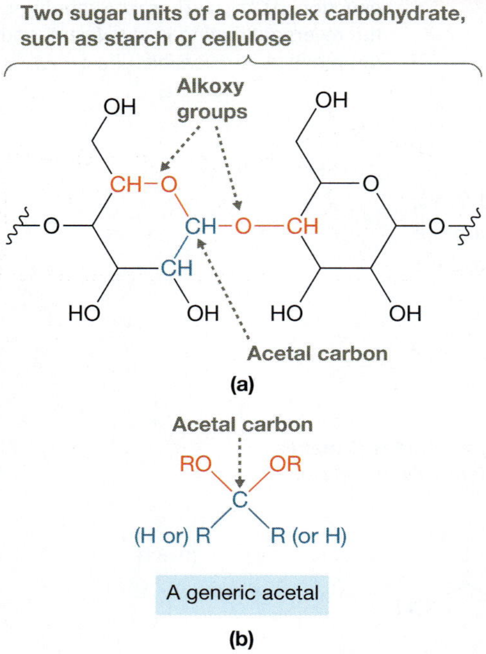

As we saw in previous topics on biomolecules (Sections 1.14b, 4.14, and 5.14), monosaccharides are relatively small carbohydrates and are often shown in their acyclic form. In nature, however, carbohydrates largely exist as more complex structures, such as starch and cellulose, in which numerous simple sugars (frequently many thousands) in their ring forms are connected together. A portion of one such carbohydrate is shown in Figure 9-20a. (We describe these macroscopic structures in greater detail in Chapter 26.)

As indicated in red, an *acetal* functional group connects the sugar units together. First introduced in Table 1-6, an acetal group (shown in its generic form in Fig. 9-20b) is characterized by a C atom bonded to two alkyl (or H) groups and two alkoxy (RO−) groups. In the case of sugar units connected together, the alkoxy groups belong to different sugar units.

An acetal group involving a sugar can be produced in the laboratory simply by treating a monosaccharide with an alcohol under acidic conditions. An example is shown in Equation 9-53, in which β-D-glucopyranose is treated with methanol and hydrochloric acid. In this case, only one alkoxy group in the product belongs to a sugar unit. The second alkoxy group is simply a methoxy (CH$_3$O−) group from the alcohol used.

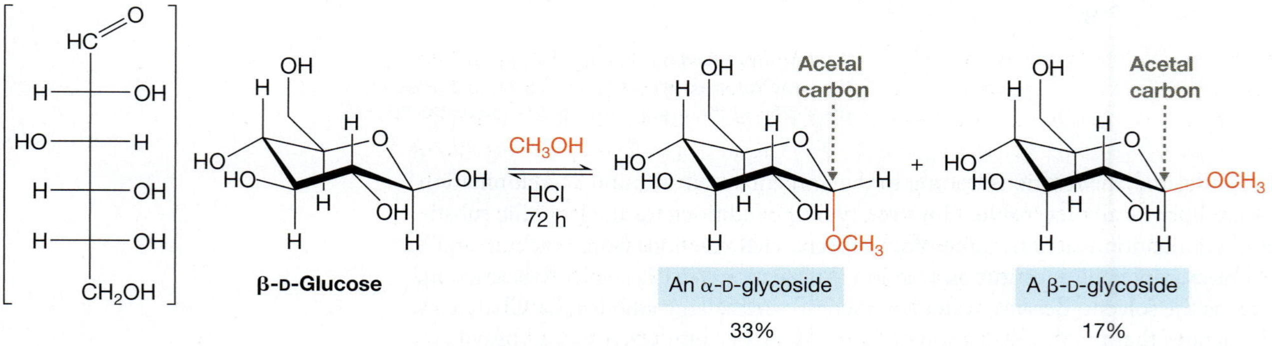

Two sugar units of a complex carbohydrate, such as starch or cellulose

FIGURE 9-20 Acetal groups in carbohydrates (a) In a complex carbohydrate, one cyclic sugar unit is connected to another via an acetal functional group, highlighted in red. (b) An acetal carbon is bonded to two RO− groups.

(9-53)

β-D-Glucose   An α-D-glycoside (33%)   A β-D-glycoside (17%)

The products in Equation 9-53, in which OH from a sugar molecule has been replaced by OR, are called **glycosides**, and they are produced as a mixture of diastereomers. The first diastereomer, in which the alkoxy group is in the axial position, is designated as α, and the second one, in which the alkoxy group is in the equatorial position, is designated as β. (The α and β designations are explained in greater detail in Section 18.13.)

To understand why a mixture of diastereomers is produced, study the mechanism of the reaction, which is shown in Equation 9-54.

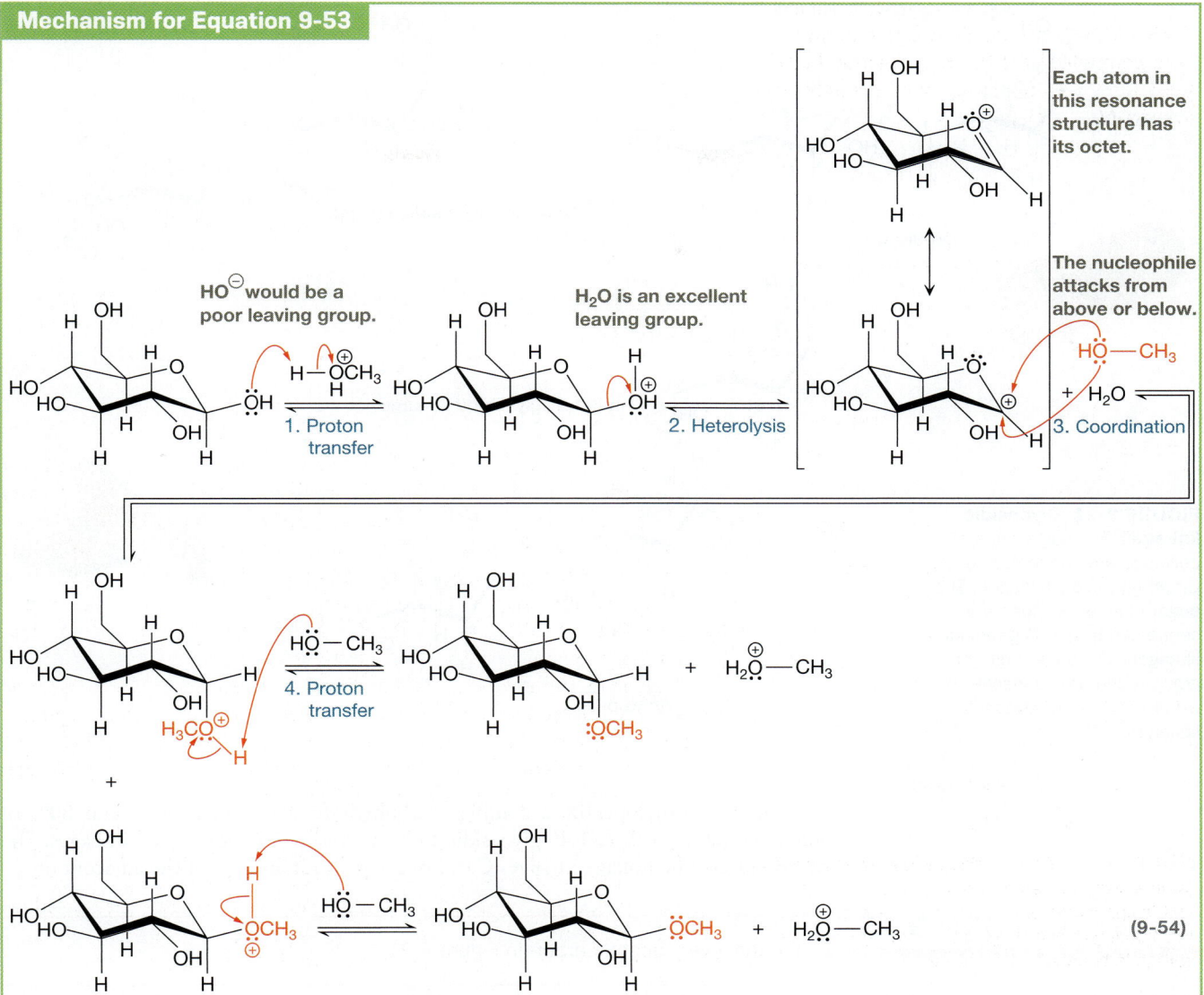

(9-54)

This is really an S_N1 mechanism, similar to the one in Equation 9-39. In Step 1, protonation converts the poor HO^- leaving group into a good leaving group, which departs as H_2O in Step 2, resulting in a carbocation that is stabilized by resonance with the neighboring O atom. In Step 3, the newly formed carbocation is attacked by the methanol nucleophile. As indicated, this can take place on either side of the plane of the carbocation C. Finally, in Step 4, deprotonation removes the positive charge from O.

The reaction in Equation 9-53 favors an S_N1 mechanism because the nucleophile, methanol, is weak; the leaving group, H_2O, is excellent; the carbocation that is produced is resonance stabilized (as shown in the mechanism); and the solvent, methanol, is protic.

In nature, complex carbohydrates consisting only of D-glucose subunits can differ by the way in which the acetal groups connect the sugars together (Fig. 9-21, next page). In cellulose (which makes up about half of wood), for example, the acetal group involves C1 on one sugar unit and C4 on the adjacent one, and the C—O—C bond that connects the sugars together, called a **glycosidic linkage**, is in the equatorial position of the first sugar unit. Therefore, cellulose is said to have β-1,4'-glycosidic linkages, where the prime (') indicates that C4 and C1 are on different sugar units.

By contrast, amylose, a complex carbohydrate that constitutes about 20% of starch (as in corn), consists of D-glucose units connected by α-1,4'-glycosidic linkages. In this case, the C—O—C bond resides in the axial position of the sugar unit that

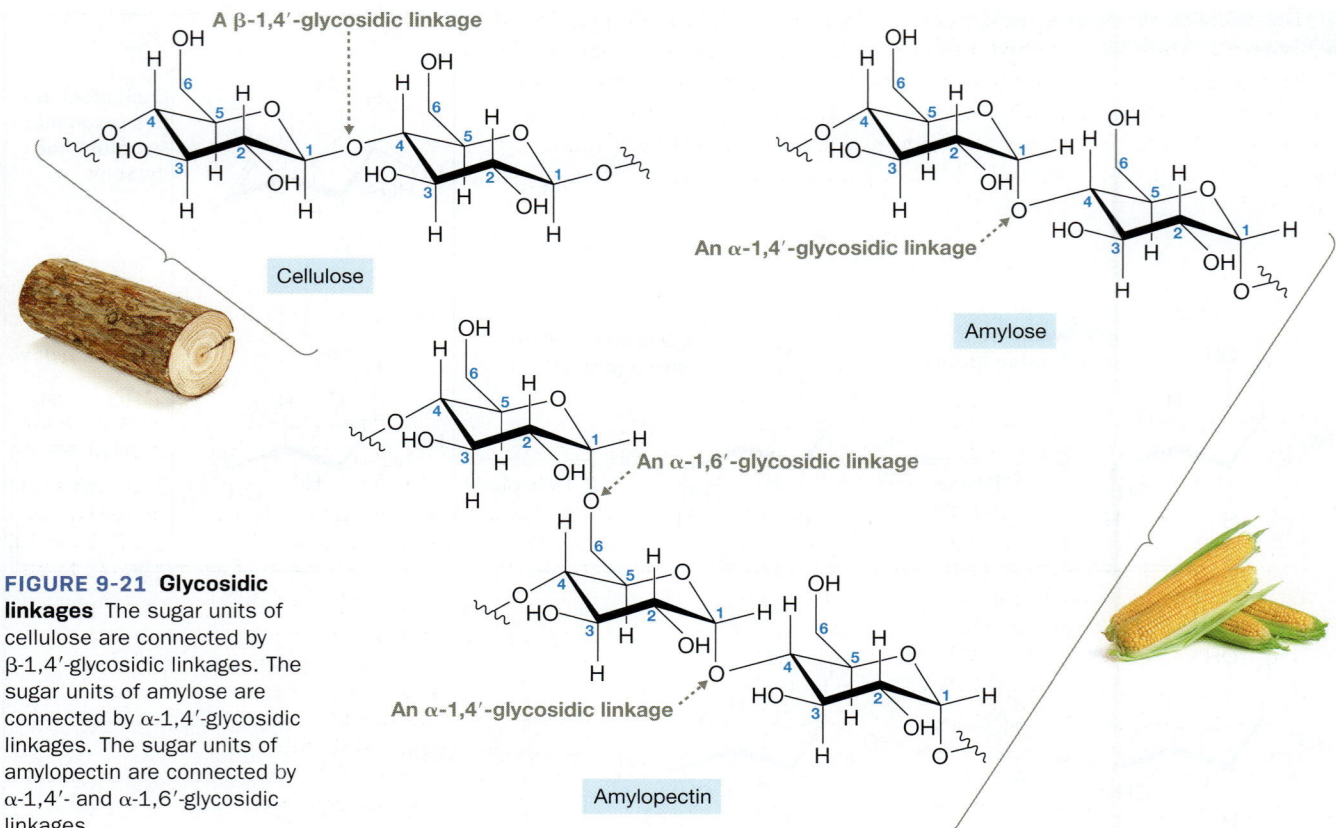

A β-1,4'-glycosidic linkage

Cellulose

An α-1,4'-glycosidic linkage

Amylose

An α-1,6'-glycosidic linkage

An α-1,4'-glycosidic linkage

Amylopectin

FIGURE 9-21 Glycosidic linkages The sugar units of cellulose are connected by β-1,4'-glycosidic linkages. The sugar units of amylose are connected by α-1,4'-glycosidic linkages. The sugar units of amylopectin are connected by α-1,4'- and α-1,6'-glycosidic linkages.

contains C1. Amylopectin, a complex carbohydrate that constitutes about 80% of starch, contains both α-1,4'-glycosidic linkages and α-1,6'-glycosidic linkages. The α-1,6'-glycosidic linkage involves C1 of one sugar unit and C6 of the adjacent one.

YOUR TURN 9.15

Identify every acetal carbon in Figure 9-21.

These relatively small structural differences among complex carbohydrates can have a large impact on macroscopic behavior, and therefore dictate the specific function of the carbohydrate. These consequences are discussed in greater detail in Chapter 26.

Notice in Equation 9-53 that glycoside formation is reversible, indicated by the equilibrium arrow. Thus, as shown in Equation 9-55, a glycoside can undergo acid-catalyzed hydrolysis to produce the monosaccharide. The mechanism for this reaction is identical to the one in Equation 9-54, except the roles of water and methanol are reversed—namely, water is the nucleophile and the alcohol is the leaving group in the hydrolysis of a glycoside.

Glycoside hydrolysis

$$\text{Methyl β-D-glucopyranoside} \xrightleftharpoons[\text{HCl}]{\text{H}_2\text{O}} \text{α-D-Glucopyranose} + \text{β-D-Glucopyranose}$$ (9-55)

The stomach provides an acidic environment, so it might seem plausible that the breakdown of complex carbohydrates into their monosaccharide subunits takes place in the stomach. Acid hydrolysis is too slow, however, for the body to make use of complex carbohydrates as quick sources of energy. Instead, hydrolysis depends on various enzymes located in the saliva as well as in the small intestine. These enzymes are highly specialized, and target only the α-1,4′-glycosidic linkages. As a result, humans can metabolize starches but not cellulose.

Chapter Summary and Key Terms

- S_N2, S_N1, E2, and E1 reactions generally compete with one another under **kinetic control**. The fastest reaction yields the major product. (Section 9.1; Objectives 1, 2)

- According to the **Hammond postulate**, the transition state for a reaction resembles reactants more than products if ΔG°_{rxn} is negative, and it resembles products more than reactants if ΔG°_{rxn} is positive. If two elementary steps are of the same type, then the one with the more negative (less positive) value for ΔG°_{rxn} tends to be faster. (Section 9.3a; Objective 2)

- S_N2 reaction rates are highly sensitive to **nucleophilicity**, whereas S_N1 reactions are not. Thus, strong nucleophiles favor S_N2 reactions and weak nucleophiles favor S_N1 reactions. (Section 9.3a; Objectives 3, 11)

- E2 reaction rates are highly sensitive to the strength of the attacking base, whereas E1 reactions are not. Thus, strong bases favor E2 reactions and weak bases favor E1 reactions. (Section 9.3c; Objectives 3, 11)

- Nucleophilicity can be weakened significantly by bulky alkyl groups surrounding the nucleophilic site. Base strength, however, is not significantly affected. Thus, strong, bulky bases favor E2 over S_N2 reactions. (Section 9.3d; Objectives 3, 11)

- If a nucleophile is strong, then high concentration favors S_N2 and low concentration favors S_N1. If a base is strong, then high concentration favors E2 and low concentration favors E1. (Section 9.4; Objectives 4, 11)

- All four reaction rates are affected by **leaving group ability**. However, S_N1 and E1 reactions are more sensitive to this factor than S_N2 and E2 reactions are. Excellent leaving groups favor S_N1 and E1, whereas poor leaving groups favor S_N2 and E2. (Section 9.5; Objectives 5, 11)

- Leaving group ability is determined largely by charge stability. The stronger the leaving group's conjugate acid, the better the leaving group. (Section 9.5; Objective 5)

- Substrates with very poor leaving groups, such as HO^-, RO^-, H_2N^-, H^-, and R^-, generally cannot undergo any of the four reactions. (Section 9.5; Objectives 5, 11)

- Nucleophilic substitution and elimination reactions generally do not occur when leaving groups are on sp^2- or sp-hybridized carbon atoms. (Section 9.6a; Objectives 6, 11)

- S_N2, S_N1, and E1 reactions are highly sensitive to the type of carbon to which the leaving group is bonded (i.e., 1°, 2°, or 3°), whereas E2 reactions are not. (Section 9.6b; Objectives 7, 11)

  - S_N2 reactions are unfeasible when the leaving group is on a tertiary carbon because steric hindrance is excessive.
  - S_N1 and E1 reactions are unfeasible when the leaving group is on a primary carbon, because the carbocation intermediate is too unstable. Exceptions arise for primary **benzylic substrates** and **allylic substrates**.
  - E2 reactions are feasible for all types of carbons because hydrogen atoms are well exposed.

- *Aprotic solvents* favor S_N2 and E2; *protic solvents* favor S_N1 and E1. (Section 9.7; Objectives 8, 11)

- Nucleophilicity in a protic solvent is reversed from that in an aprotic solvent when the nucleophilic atoms are negatively charged and are in different rows of the periodic table. These reversals are due to dramatic differences in solvation. (Section 9.7b; Objective 9)

- Heat favors elimination over substitution due to the greater *entropy* in the elimination products. (Section 9.8; Objectives 10, 11)

- When multiple elimination products are possible, the major product is usually the one that is the most stable, in accordance with **Zaitsev's rule**. The **anti-Zaitsev product** can be obtained using a strong, bulky base like $(CH_3)_3CO^-$. (Section 9.10; Objective 12)

- **Intramolecular** cyclization reactions are favored over their corresponding **intermolecular** reactions when a five- or six-membered ring can be formed. (Section 9.11; Objective 13)

- A reaction with a substantially negative ΔG°_{rxn} tends to be **irreversible** and takes place under *kinetic control*. Otherwise, the reaction tends to be **reversible** and takes place under *thermodynamic control*. (Section 9.12; Objective 2)

Reaction Tables

This is the first of several end-of-chapter sections in which the reactions from the chapter are summarized in tabular form. For each reaction, the starting compound class, the compound class formed, and the typical reagents and reaction conditions are provided. The sections in which the reaction is discussed are also listed. Similar tables will be provided at the ends of future chapters in which new reactions are encountered.

There are two other important features of these tables to note. First, each entry identifies the key electron-rich and electron-poor species that appear in the mechanism of the reaction (recall from Section 7.1a that these species tend to react with each other). Knowing key electron-rich and electron-poor species for various reactions can help you focus on the mechanisms as you learn new reactions. Second, we have collected the reactions into two tables (Tables 9-14 and 9-15), depending on whether the reaction leads to the formation and/or breaking of a carbon–carbon σ bond. Reactions that do *not* are designated as *functional group transformations*, whereas reactions that do are said to *alter the carbon skeleton*. We discuss the importance of these designations in Chapter 13 in the context of *organic synthesis*.

TABLE 9-14 Functional Group Transformations[a]

	Starting Compound Class	Typical Reagents and Reaction Conditions	Compound Class Formed	Key Electron-Rich Species	Key Electron-Poor Species	Comments	Discussed in Section(s)
(1)	$R-CH_2-X$ Primary alkyl halide	NaOH	$R-CH_2-OH$ 1° Alcohol	$HO^{\ominus}$	$R-CH_2-X$ δ^+	S_N2 reaction	7.2, 8.4, 8.5, 9.9
(2)	$R-CRR-X$ Tertiary alkyl halide	H_2O	$R-CRR-OH$ 3° Alcohol	H_2O	$R-C^{\oplus}RR$	S_N1 reaction	7.3, 8.1, 8.4, 8.5, 9.5a, 9.9
(3)	R—O—R Ether	H_2O $H^{\oplus}$	R—OH Alcohol	H_2O	$R^{\oplus}$ or $ROH_2^{\oplus}$	S_N1 or S_N2 reaction	7.2, 7.3, 8.1, 8.4, 8.5, 9.5a, 9.9
(4)	$R-CH_2-X$ Primary alkyl halide	R'ONa	$R-CH_2-O-R'$ Ether	$R'O^{\ominus}$	$R-CH_2-X$ δ^+	S_N2 reaction	7.2, 8.4, 8.5, 9.9
(5)	R—OH Alcohol	H_3PO_4 or H_2SO_4 Δ	R—O—R Ether	ROH	$R^{\oplus}$ or $ROH_2^{\oplus}$	S_N1 or S_N2 reaction	7.2, 7.3, 8.1, 8.4, 8.5, 9.5a, 9.9
(6)	R—L L = Cl, Br, I, OTs, OMs, or OTf	NaX	R—X Alkyl halide	$X^{\ominus}$	$R-CH_2-L$ δ^+ or $R^{\oplus}$	S_N1 or S_N2 reaction	7.2, 7.3, 8.1, 8.4, 8.5, 9.9

[a] X = Cl, Br, I.

TABLE 9-14 Functional Group Transformations[a] (continued)

Starting Compound Class	Typical Reagents and Reaction Conditions	Compound Class Formed	Key Electron-Rich Species	Key Electron-Poor Species	Comments	Discussed in Section(s)
(7) R—OH Alcohol	HX	R—X Alkyl halide	X$^{\ominus}$	R$^{\oplus}$ or R$\overset{\oplus}{O}H_2$	S_N1 or S_N2 reaction	7.2, 7.3, 8.1, 8.4, 8.5, 9.5a, 9.9
(8) R—CH H—CH X—R′ Alkyl halide	$(CH_3)_3CONa$, Δ	R—C=C—R′ Alkene	$(CH_3)_3CO^{\ominus}$	R—CH H—$\overset{\delta+}{C}$—R′ (H X)	E2 reaction	7.5, 8.4, 8.5, 9.3d, 9.9
(9) R—CH H—CH OH—R′ Alcohol	H_3PO_4 or H_2SO_4, Δ	R—C=C—R′ Alkene	H_2O	R—CH H—$\overset{\oplus}{C}$—R′ (H)	E1 reaction	7.3, 7.6, 8.2, 8.4, 8.5, 9.5a, 9.9

[a]X = Cl, Br, I.

TABLE 9-15 Reactions That Alter the Carbon Skeleton[a]

Reactant	Typical Reagents and Reaction Conditions	Product Formed	Key Electron-Rich Species	Key Electron-Poor Species	Comments	Discussed in Section(s)
(1) R—C≡C—H Alkyne	1. NaH 2. R′—X	R—C≡C—R′ Alkyne	R—C≡C:$^{\ominus}$ Alkynide anion	$\overset{\delta+}{R'}$—X Alkyl halide	S_N2 reaction	7.2, 8.4, 8.5, 9.3b, 9.9
(2) R—X Alkyl halide	NaCN	R—C≡N Nitrile	N≡C:$^{\ominus}$ Cyanide anion	$\overset{\delta+}{R}$—X Alkyl halide	S_N2 reaction	7.2, 8.4, 8.5, 9.9

[a]X = Cl, Br, I.

Problems

Problems that are related to synthesis are denoted (SYN).

9.1 The Competition among S_N2, S_N1, E2, and E1 Reactions

9.39 When bromocyclohexane is treated with sodium cyanide, the S_N2, S_N1, E2, and E1 reactions compete. Draw the mechanism for each of these reactions.

9.40 (SYN) Suggest an alkyl bromide that can be treated with CH_3SNa to form each of the following products exclusively. By what mechanism should the respective reactions proceed?

(a)

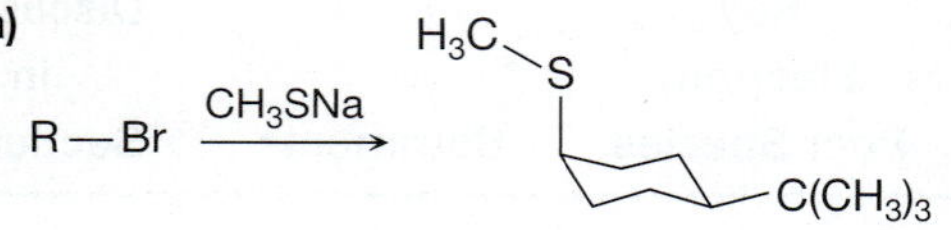

R—Br $\xrightarrow{CH_3SNa}$

(b)

R—Br $\xrightarrow{CH_3SNa}$

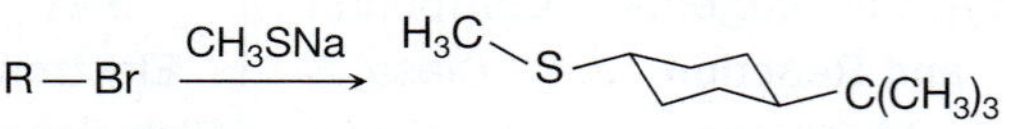

9.41 Did the following overall reaction occur by an S_N2, S_N1, E2, or E1 mechanism? How do you know? Draw the complete, detailed mechanism to account for the formation of both products.

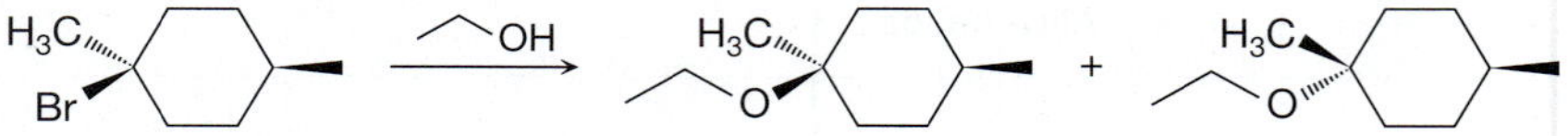

9.42 Did the overall reaction shown here occur by an S_N2, S_N1, E2, or E1 mechanism? How do you know? Draw the complete, detailed mechanism to account for the formation of both products.

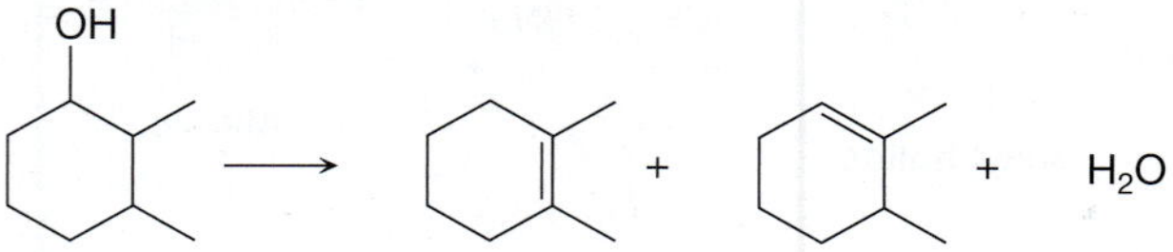

9.43 The formula of the precursor is given for each of the following reactions. Draw its structure, paying attention to stereochemistry, if appropriate.

(a)

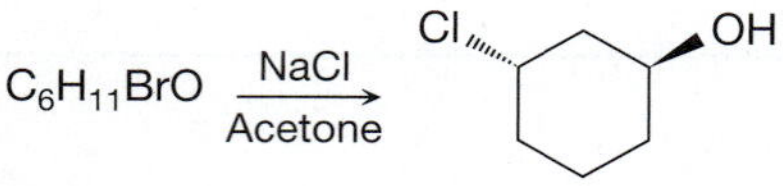

$C_6H_{11}BrO$ $\xrightarrow[\text{Acetone}]{NaCl}$

(b)

$C_{12}H_{17}Cl$ $\xrightarrow[\text{DMSO}]{(CH_3)_3CONa}$

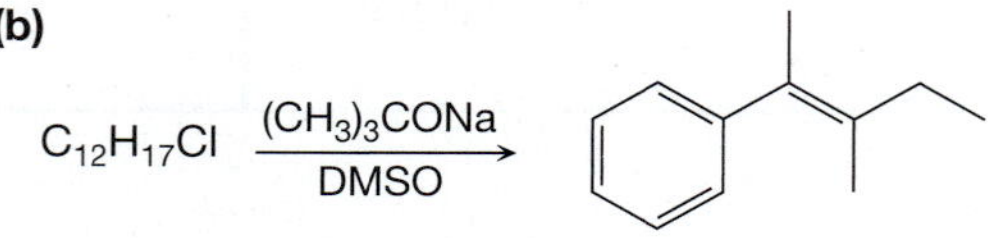

9.3–9.5 Strength and Concentration of the Attacking Species; Leaving Group Ability

9.44 Rank the following bases in order from slowest E2 reaction rate to fastest.

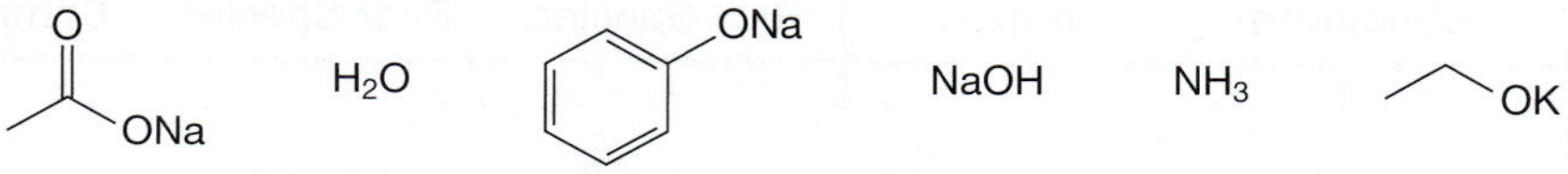

A   B   C   D   E   F

H_2O NaOH NH_3

9.45 Rank the following nucleophiles in order from slowest S_N2 reaction rate to fastest if DMSO is the solvent.

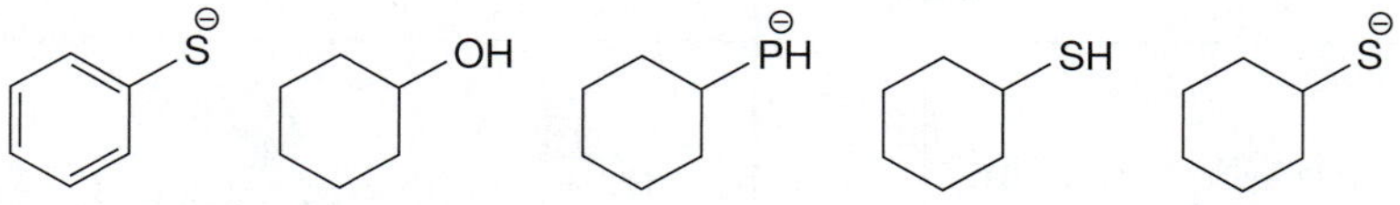

A   B   C   D   E

9.46 (SYN) Show how pent-2-yne can be made from two different alkyl bromides.

9.47 When acetic acid is treated with a strong base, followed by benzyl bromide, a compound is formed whose formula is $C_9H_{10}O_2$. Draw the structure of this product, and draw the mechanism leading to its formation.

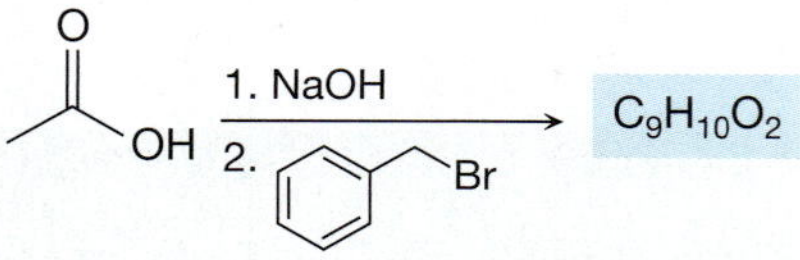

$\xrightarrow[\text{2.}]{\text{1. NaOH}}$ $C_9H_{10}O_2$

9.48 When the following deuterium-labeled compound is treated with potassium *tert*-butoxide in *N,N*-dimethylformamide, a single product is observed.

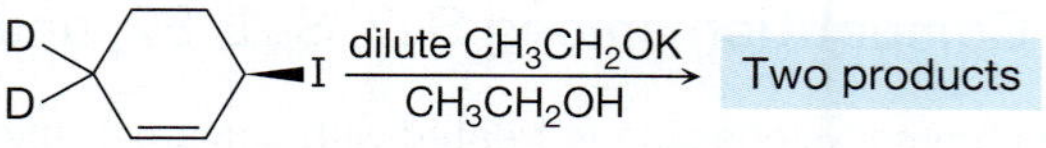

$\xrightarrow[\text{DMF}]{(CH_3)_3COK}$ One product

$\xrightarrow[\text{CH}_3\text{CH}_2\text{OH}]{\text{dilute CH}_3\text{CH}_2\text{OK}}$ Two products

When the same substrate is heated in the presence of dilute potassium ethoxide in ethanol, a mixture of two products is formed. Provide the complete, detailed mechanism for each reaction and explain these results.

9.49 Rank the following substrates in order of increasing rate of the S_N2 reaction.

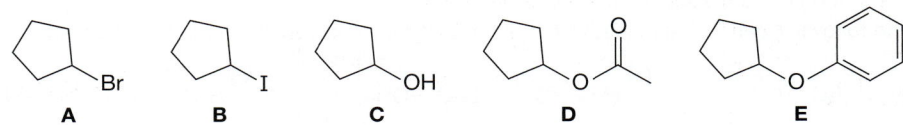

A   B   C   D   E

9.50 Rank the following substrates in order of increasing rate of the E1 reaction.

A   B   C   D

9.51 Rank the following E2 reactions in order of increasing rate.

A   B   C

D   E

9.6 Type of Carbon

9.52 Rank the following substrates in order from slowest E1 reaction rate to fastest.

A   B   C   D

9.53 Rank the following substrates in order from slowest S_N2 reaction rate to fastest.

A   B   C   D

9.54 Rank the following substrates in order from slowest S_N1 reaction rate to fastest.

A   B   C   D

9.55 Both of the following reactions will give the same S_N2 product. Draw the mechanism for each of these reactions and show the product. Which reaction is more efficient?

(a) H_3C-Br $\xrightarrow{(CH_3)_2CHONa}$?

(b) H_3C-ONa $\xrightarrow{(CH_3)_2CHBr}$?

9.7, 9.8 Effects of Solvent and Heat

9.56 For each of the following pairs of species, which is the stronger nucleophile in acetone? Explain.

(a) H₃C—OH or H₃C—O⊖

(b) H₃C—O⊖ or H₃C—O⊕H₂

(c) H₃C—OH or H₃C—NH₂

(d)

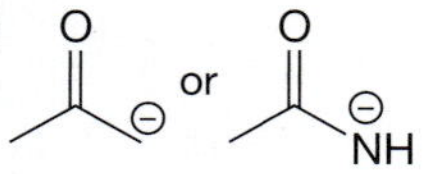

(e)

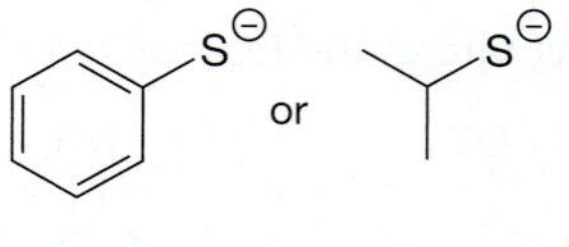

(f)

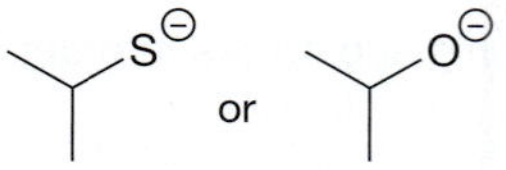

(g) 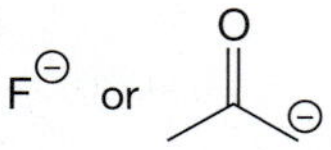 F⊖ or

(h)

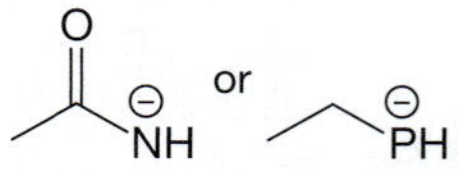

(i) CH₃S⊖ or CH₃Se⊖

(j) CH₃Se⊖ or Br⊖

9.57 For each of the following pairs of species, which is the stronger nucleophile in ethanol? Explain.

(a) H₃C—OH or H₃C—O⊖

(b) H₃C—OH or H₃C—NH₂

(c)

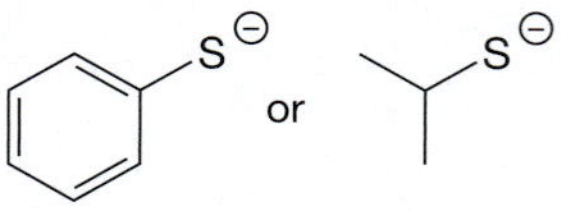

(d)

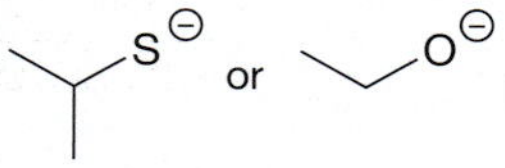

(e) CH₃S⊖ or CH₃Se⊖

(f) CH₃Se⊖ or Br⊖

9.58 (SYN) Which solvent—ethanol or dimethyl sulfoxide—would be better to use to carry out the reaction shown here? Why?

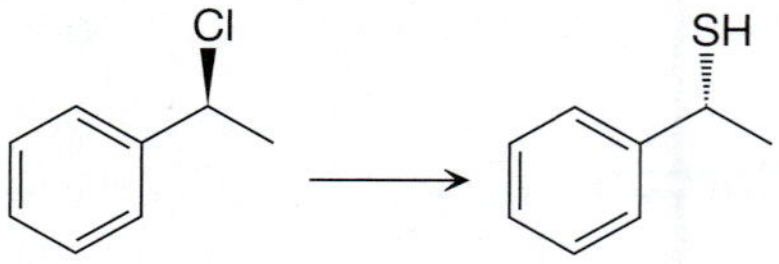

9.59 (SYN) Which solvent—acetone or *tert*-butyl alcohol—would be better to use to carry out the reaction shown here? Why?

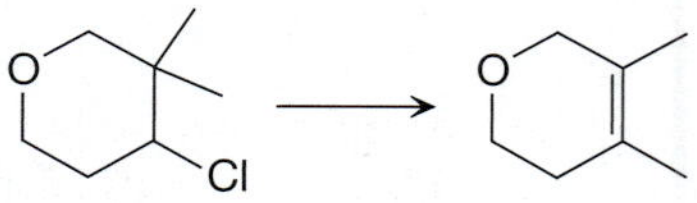

9.60 Would heating each of the following reactions increase the yield of the product shown?

(a)

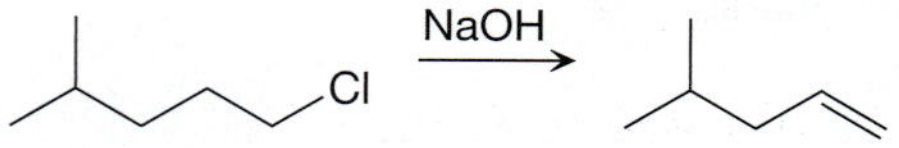

(b)

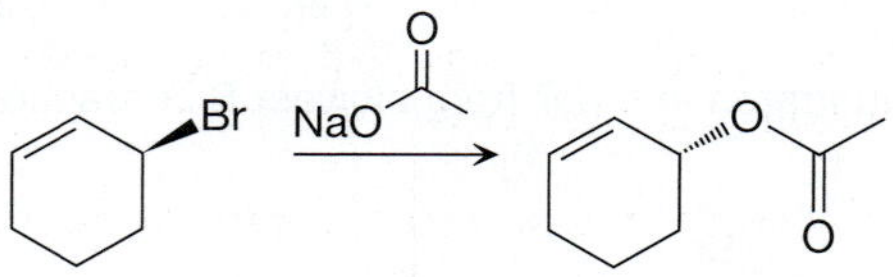

9.9–9.11 Predicting the Outcome of Nucleophilic Substitution and Elimination Reactions; Zaitzev's Rule; Intramolecular versus Intermolecular Reactions

9.61 (SYN) Suggest how each of the reactions shown here could be carried out, focusing in particular on the identity of the nucleophile and the choice of solvent.

(a)

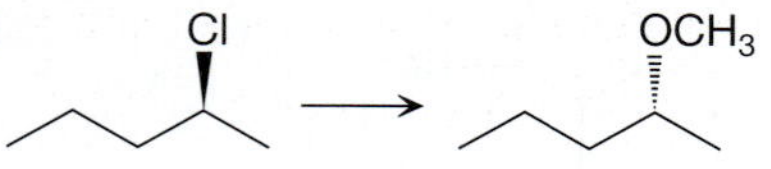

(b) 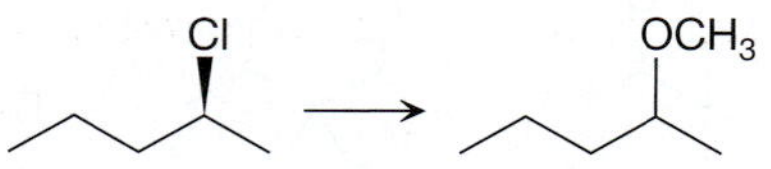 Racemic

9.62 Provide a complete, detailed mechanism for the reaction shown here.

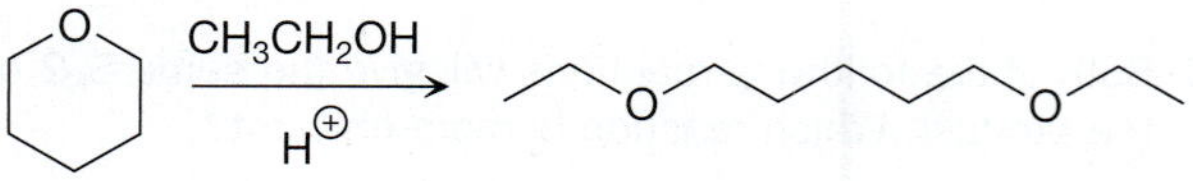

9.63 For each of the following reactions, provide a complete, detailed mechanism and predict the products, including stereochemistry where appropriate. Determine whether the reaction will yield exclusively one product or a mixture of products. For each reaction that yields a mixture, determine which is the major product.

(a)
(S)-3-Chlorooctane
CH3CH2OK
DMSO
?
(b)
(S)-3-Methyl-3-chlorooctane
CH3CH2OK
CH3CH2OH
?
(c)
OH
conc H3PO4
Δ
?
(d)
Cl
(CH3)3CONa
DMF
?
(e)
TfO
O
NaN3
Ethanol
?
(f)
OMs
CH3CH2OH
?

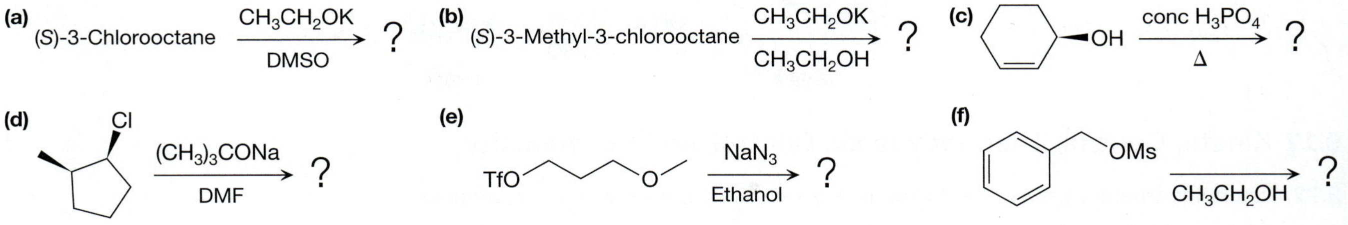

9.64 For each of the following reactions, provide a complete, detailed mechanism and predict the products, including stereochemistry where appropriate. Determine whether the reaction will yield exclusively one product or a mixture of products. For each reaction that yields a mixture, determine which is the major product.

(a)
H3C
OTs
CH3CH2OH
?
(b)
H3C
OTs
NaCN
DMF
?
(c)
H3C
OTs
(CH3)3CONa
DMSO
?
(d)
H3C
OTs
Na2CO3
CH3CH2OH, Δ
?
(e)
Cl CH3
CH3CH2OH, Δ
?
(f)
Cl
CH3
NaOCH2CH3
DMSO, 95 °C
?

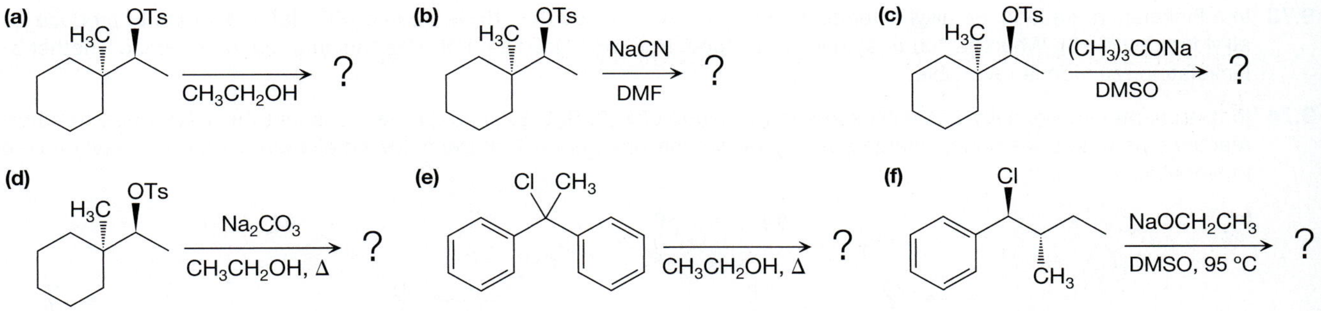

9.65 Predict the major products of each of the following reactions.

(a)
Cl
Cl
NaBr
DMSO
?
(b)
Br
Cl
F
(CH3)3COK
DMF
?

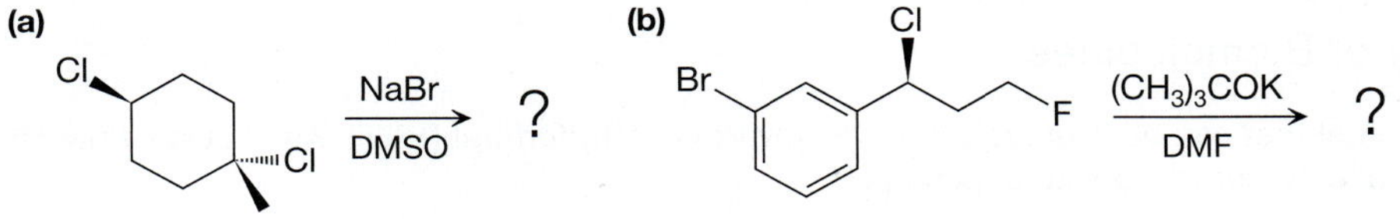

9.66 The reaction shown here produces two isomers with the formula $C_7H_{11}Br$. **(a)** Draw each product. **(b)** Draw the mechanism that accounts for the formation of each product.

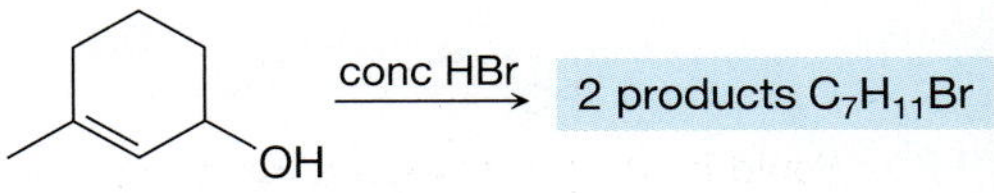

OH
conc HBr
2 products C7H11Br

9.67 When the reaction mixture in Problem 9.66 is heated, the compounds shown here are produced. Draw the complete, detailed mechanism that accounts for the formation of each of these products.

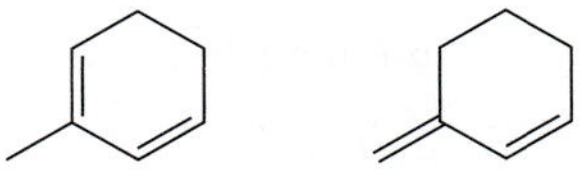

9.68 For each of the following substrates, draw the major E2 product when NaOH is used as the base.

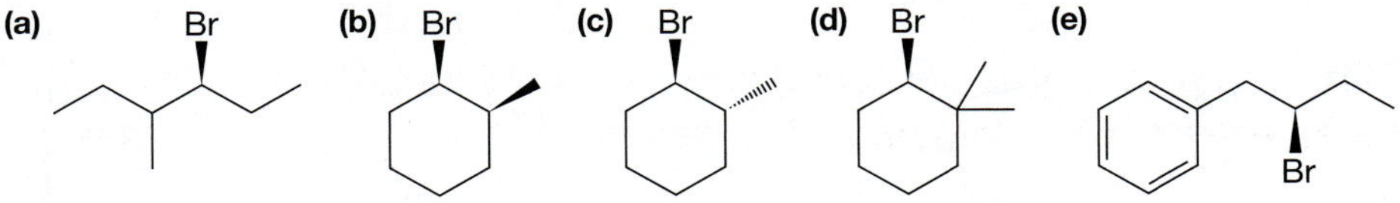

(a)
Br
(b)
Br
(c)
Br
(d)
Br
(e)
Br

9.69 For each of the substrates in Problem 9.68, draw the major E1 product.

9.70 The following isomers react separately with sodium hydroxide to give different products with the formulas shown.

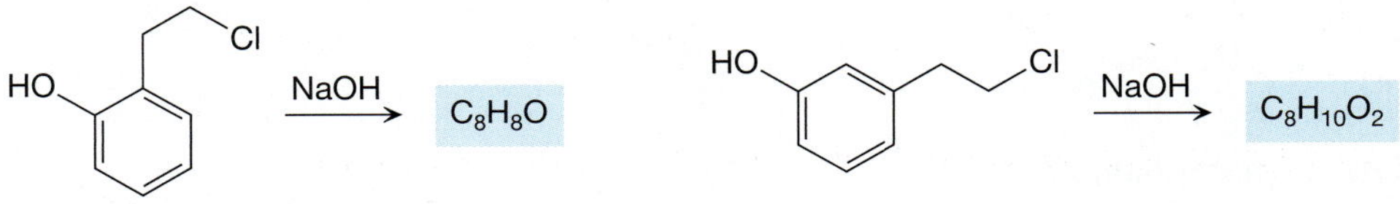

Cl
HO
NaOH
C8H8O
HO
Cl
NaOH
C8H10O2

(a) Draw the structure of each product.
(b) Draw the mechanism that accounts for the formation of each of those products.
(c) Explain why the isomeric reactants lead to different products.

9.71 Draw the mechanism for the following reaction.

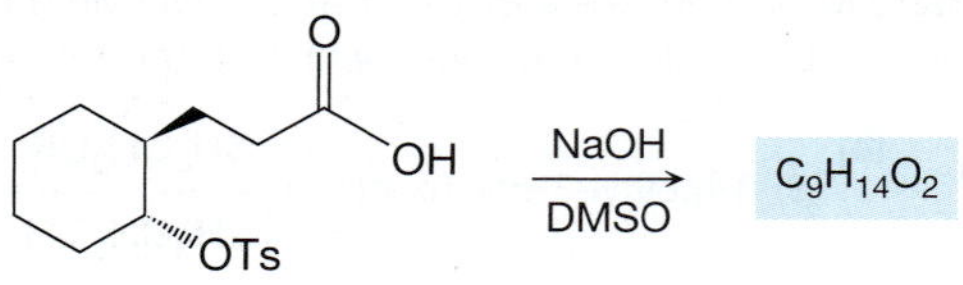

$$\xrightarrow[\text{DMSO}]{\text{NaOH}} \quad C_9H_{14}O_2$$

9.12 Kinetic Control, Thermodynamic Control, and Reversibility

9.72 Determine whether each of the following S_N2 reactions is reversible or irreversible.

(a)

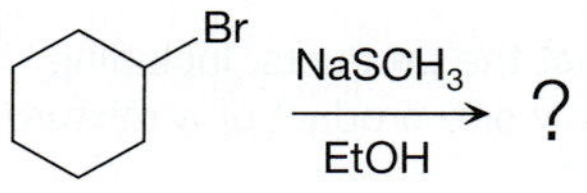

$$\xrightarrow[\text{EtOH}]{\text{NaSCH}_3} \quad ?$$

(b)

$$\xrightarrow[\text{EtOH}]{\text{KBr}} \quad ?$$

(c)

$$\xrightarrow[\text{EtOH}]{\text{NaCN}} \quad ?$$

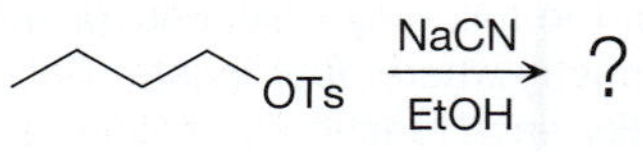

9.73 In a Finkelstein reaction, an alkyl chloride (R—Cl) or alkyl bromide (R—Br) is treated with NaI in acetone to produce an alkyl iodide (R—I). Whereas NaI is soluble in acetone, NaCl and NaBr are not. Use this information to argue whether a Finkelstein reaction is reversible.

9.74 In this chapter, reversibility was discussed only in terms of S_N2, S_N1, E2, and E1 reactions, but the ideas apply to other reactions as well. Considering charge stability, determine whether each of the following elementary steps is reversible or irreversible.

(a)

$\longrightarrow \quad ?$

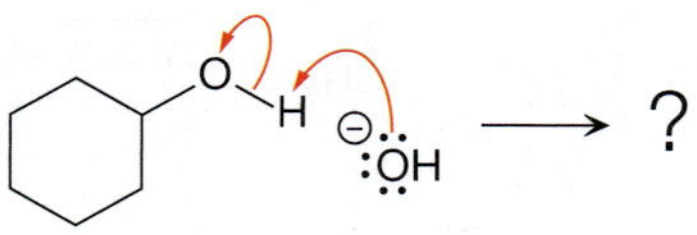

(b)

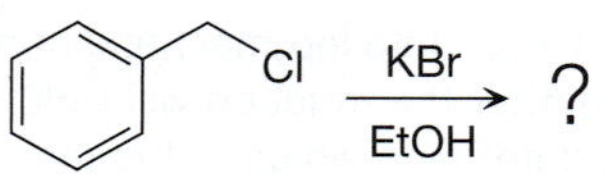

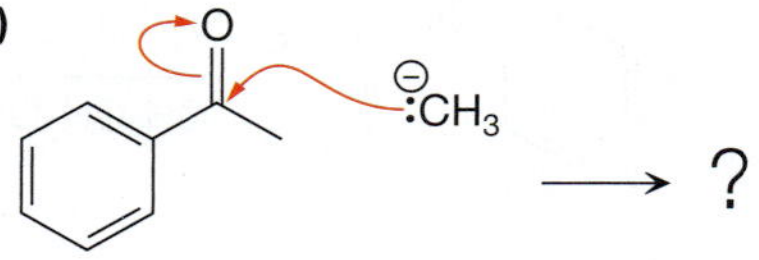

$\longrightarrow \quad ?$

(c)

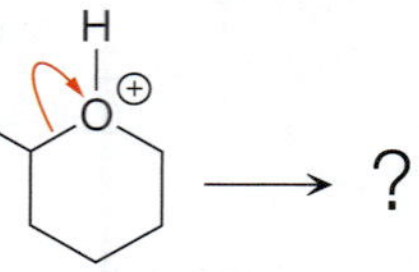

$\longrightarrow \quad ?$

9.13 The Organic Chemistry of Biomolecules

9.75 2,5-Dimethylfuran is a liquid biofuel that can be synthesized from 5-hydroxymethylfurfural (HMF). As shown below, HMF can be synthesized from D-fructose by treatment with sulfuric acid.

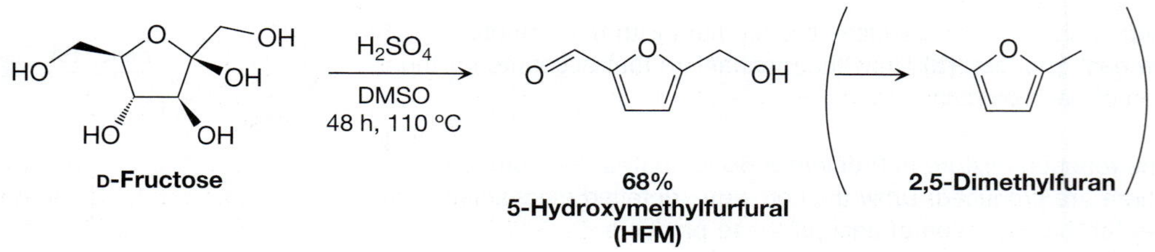

$$\xrightarrow[\substack{\text{DMSO} \\ \text{48 h, 110 °C}}]{\text{H}_2\text{SO}_4}$$

D-Fructose

68%
5-Hydroxymethylfurfural
(HFM)

2,5-Dimethylfuran

The mechanism for the formation of HMF from D-fructose is believed to involve the following series of dehydration reactions:

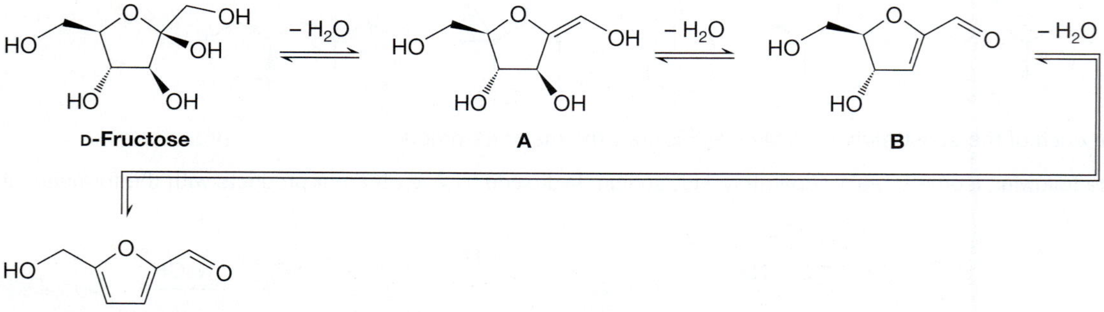

D-Fructose $- H_2O$ **A** $- H_2O$ **B** $- H_2O$

5-Hydroxymethylfurfural

Draw the complete, detailed mechanism for each of these dehydration reactions.

9.76 Draw the complete, detailed mechanism and the products for the glycoside formation shown here. Pay attention to stereochemistry.

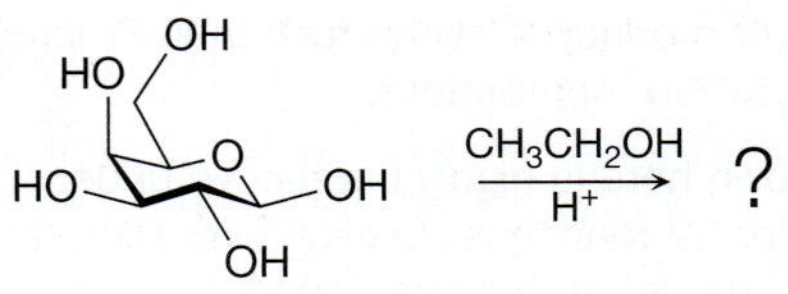

Galactose

$$\xrightarrow[\text{H}^+]{\text{CH}_3\text{CH}_2\text{OH}} \quad ?$$

9.77 Lactose is a disaccharide in which a glycosidic linkage connects the monosaccharides galactose and glucose.

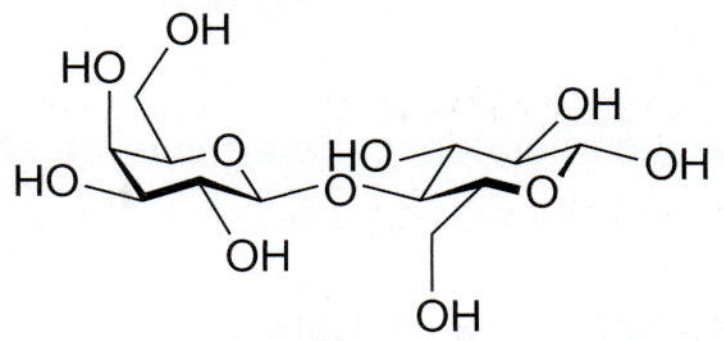

(a) Identify the glycosidic linkage and the acetal carbon in lactose.

(b) What type of glycosidic linkage does lactose have (i.e., is it 1,1'-, 1,2'-, etc., and is it α or β)?

(c) People who are lactose intolerant are deficient in the enzyme lactase, and therefore cannot efficiently break down the disaccharide into its monosaccharides. When lactose is treated with aqueous acid, however, this hydrolysis can take place, though relatively slowly. Draw the complete, detailed mechanism and the products of the acid-catalyzed hydrolysis of lactose.

Lactose

9.78 DNA is damaged when a base from the DNA chain is removed after an alkylation has occurred. In a *depurination reaction*, the purine nitrogenous base is displaced from its sugar as shown in this reaction. Draw the mechanism for this reaction and suggest a reason why it occurs so easily.

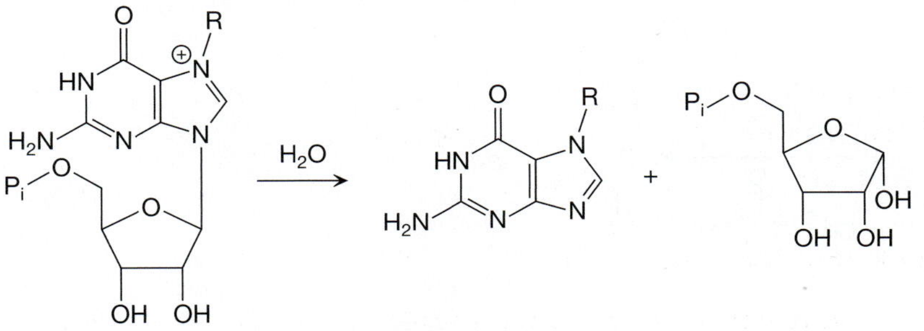

Integrated Problems

9.79 Draw the complete, detailed mechanism for the reaction shown here and predict the major products, including stereochemistry.

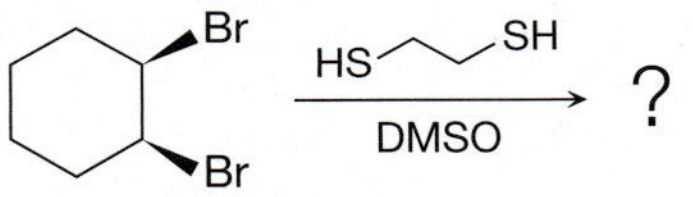

9.80 In which of the following reactions would you expect a carbocation rearrangement? Explain. *Hint:* You will need to figure out whether the predominant mechanism is S_N1, S_N2, E1, or E2.

(a)

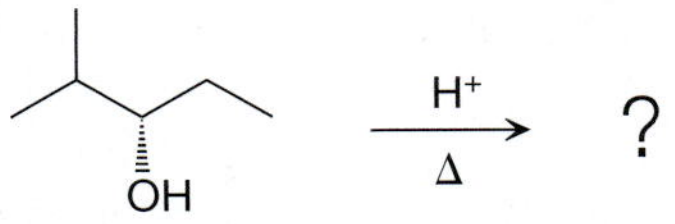

$$\xrightarrow[\Delta]{\text{H}^+} \quad ?$$

(b)

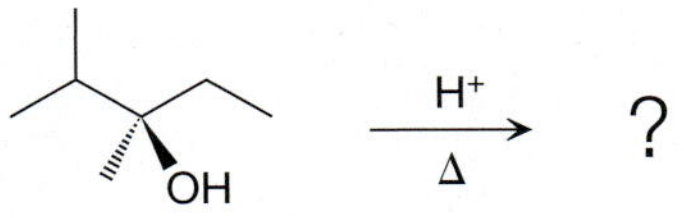

$$\xrightarrow[\Delta]{\text{H}^+} \quad ?$$

(c)

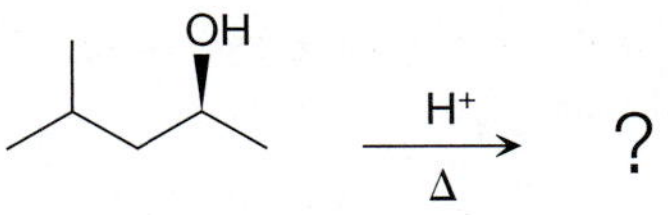

$$\xrightarrow[\Delta]{\text{H}^+} \quad ?$$

(d)

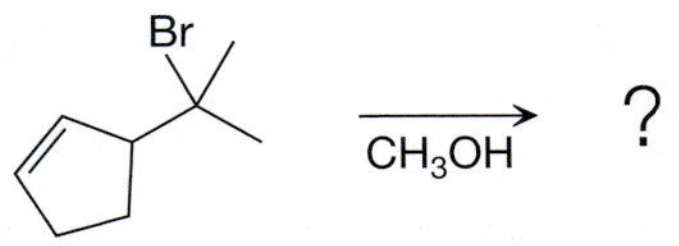

$$\xrightarrow{\text{CH}_3\text{OH}} \quad ?$$

(e)

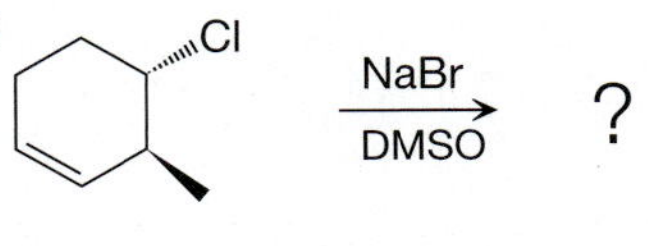

$$\xrightarrow[\text{DMSO}]{\text{NaBr}} \quad ?$$

(f)

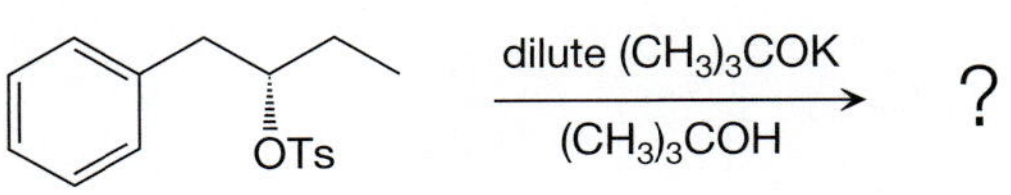

$$\xrightarrow[\text{(CH}_3)_3\text{COH}]{\text{dilute (CH}_3)_3\text{COK}} \quad ?$$

(g)

$$\xrightarrow{\text{Ethanol}} \quad ?$$

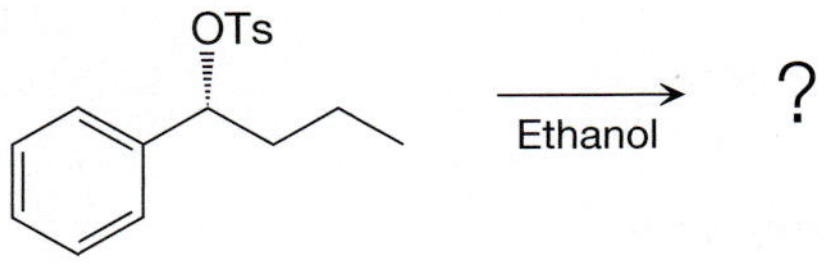

9.81 Determine the major product of each reaction in Problem 9.80 and draw the complete, detailed mechanism. Pay attention to stereochemistry where appropriate.

9.82 The compound shown here is highly unreactive under conditions that favor E2 reactions. Explain why. *Hint:* It may help to build a model of this compound.

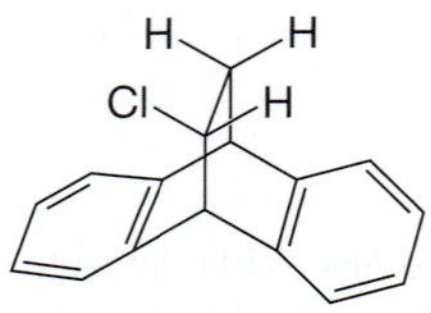

9.83 Given the following reaction sequence, determine the structures of **A** and **B**, including proper stereochemistry.

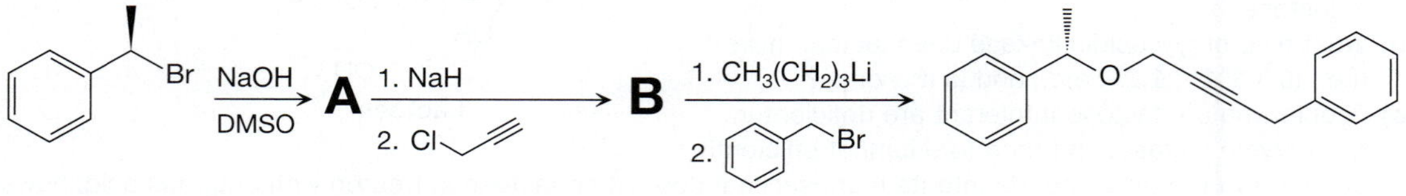

9.84 Suggest why each of the following reactions will not occur as indicated.

(a)

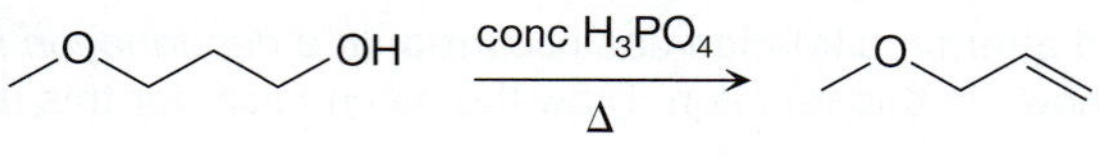

(b)

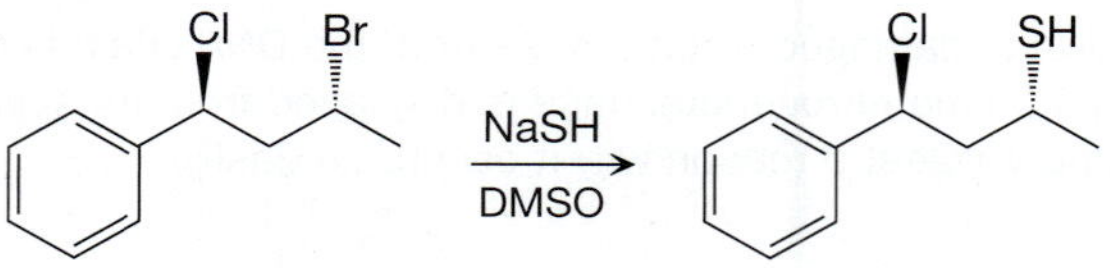

(c)

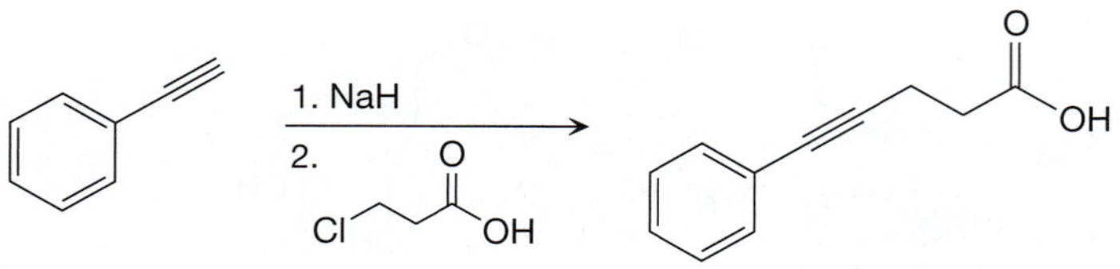

(d)

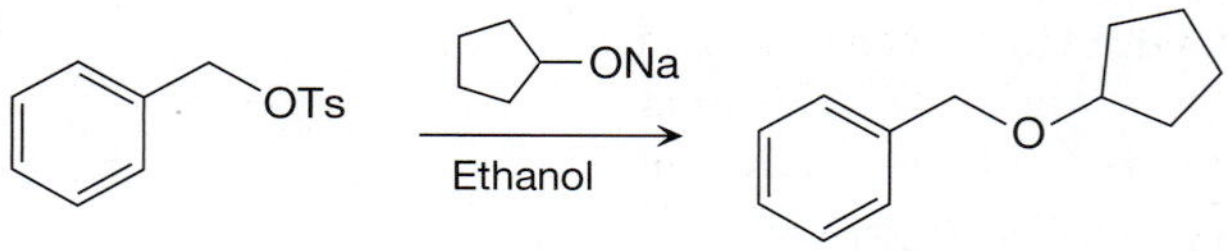

Naming Compounds with a Functional Group That Calls for a Suffix 2
Carboxylic Acids and Their Derivatives

Interchapter E explained how to name alcohols, amines, ketones, and aldehydes, all of which have a functional group that requires adding a suffix to the IUPAC name to indicate the highest-priority functional group present. Those are not the only functional groups that require adding a suffix, however, as shown in Table F-1 (which previously appeared as Table E-1 in Interchapter E). Other functional groups include the ones characteristic of carboxylic acids (RCO_2H) and **carboxylic acid derivatives**: acid anhydrides, esters, acid chlorides, amides, and nitriles. Of these, carboxylic acids, acid chlorides, amides, and nitriles are named using essentially the same rules we learned in Interchapter E, so we deal with those functional groups first here in Interchapter F. Then we learn how to name acid anhydrides and esters, which require different rules.

F.1 Naming Carboxylic Acids, Acid Chlorides, Amides, and Nitriles

We name carboxylic acids, acid chlorides, amides, and nitriles according to the following basic rules:

Basic Rules for Naming Carboxylic Acids, Acid Chlorides, Amides, and Nitriles

- Establish the root (parent compound) according to the longest carbon chain that contains the highest-priority functional group.
- Choose the numbering system that gives the C atom of the highest-priority functional group the lowest locator number (locant).
- Remove the final "e" of *ane*, *ene*, or *yne* prior to adding the suffix unless the suffix is *nitrile* (see Table F-1) or unless there are two or more of the highest-priority functional group present.
- Account for the presence and location of all other functional groups, as well as stereochemistry, using the rules presented in the previous nomenclature units (Interchapters A–C, E).

TABLE F-1 Common Functional Groups, Their Relative Priorities, and Corresponding Suffixes and Prefixes

Priority	Functional Group	Compound Class	Drop "e" from Root?	Add	Prefix	Priority	Functional Group	Compound Class	Drop "e" from Root?	Add	Prefix
1		Carboxylic acid	Yes	oic acid	carboxy	6	R—C≡N	Nitrile	No	nitrile	cyano
2		Acid anhydride	Yes	oic anhydride	—	7		Aldehyde	Yes	al	oxo
3		Ester	See Section F.2a		—	8		Ketone	Yes	one	oxo
4		Acid chloride	Yes	oyl chloride	chloro-carbonyl	9	R—OH	Alcohol	Yes	ol	hydroxy
5		Amide	Yes	amide	carbamoyl	10	R—N	Amine	Yes	amine	amino

Note: R indicates an alkyl group, and the absence of an atom at the end of a bond indicates that either R or H may be attached.

Let's apply these rules to name the following molecules.

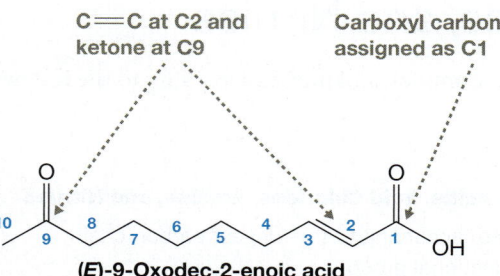

C=C at C2 and ketone at C9

Carboxyl carbon assigned as C1

The longest carbon chain containing both carboxyl groups has four carbons.

(E)-9-Oxodec-2-enoic acid

(R)-2-Pentylbutanedioic acid

In the first molecule, the highest-priority functional group is a carboxyl group, so numbering begins at the right to give its carbon the lowest number. The longest carbon chain containing the carboxyl group has 10 carbons and a C=C bond, making the root decene. After removing the final "e" and adding the suffix *oic acid*, the root becomes decenoic acid. The location of the carboxyl group is not included in the name because it must be at C1, but the locator numbers for the C=C bond at C2 and the carbonyl group of the ketone at C9 are included. Those groups are named according to the rules we have encountered previously. Notice, specifically, that the ketone's C=O group is given the prefix *oxo*. Finally, the *E* configuration about the double bond is indicated in parentheses at the beginning of the name.

In the second molecule on the previous page, there are two carboxyl groups and the longest carbon chain that contains both of them has four carbons. In this case, the final "e" is retained, so the molecule is a *butanedioic acid*. The pentyl group is indicated at C2 and the *R* configuration of the asymmetric carbon appears at the beginning of the name. Notice that no numbers are used to locate the two carboxyl groups because they must be at the ends of the chain.

Both of the following molecules contain a CN group. In the first molecule, that group is the highest-priority functional group present, so the molecule is given the suffix *nitrile* and its carbon is assigned C1. In the second molecule, the CN group is treated as the substituent *cyano* because the highest-priority functional group present is O=C—Cl. Notice, in this case, that the carbon belonging to the CN group is not counted as part of the longest carbon chain that establishes the root because, when treated as a substituent, the cyano group already accounts for that carbon.

CN is the highest-priority functional group so the suffix is *nitrile* and its carbon is assigned C1.

O=C—Cl is the highest-priority functional group so the suffix is *oyl chloride*. The CN group is a substituent, so its C atom is not numbered.

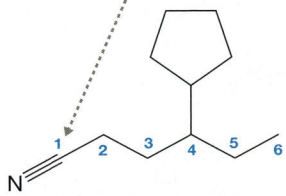

4-Cyclopentylhexanenitrile

3-Cyanopropanoyl chloride

The next two molecules both contain an O=C—N functional group, characteristic of an amide. In the first molecule below, O=C—N is the highest-priority functional group, so we add the suffix *amide*. Notice that the ethyl and methyl groups attached to the amide N are indicated using the same rules we learned in Interchapter E to specify alkyl groups attached to an amine N. The number of alkyl groups attached to the amide N, moreover, characterizes the amide as primary, secondary, or tertiary, as shown in Figure F-1.

The highest-priority functional group is O=C—N. An ethyl and a methyl group are attached to the amide N.

The O=C—N group is treated as a substituent because the highest-priority functional group is CO₂H.

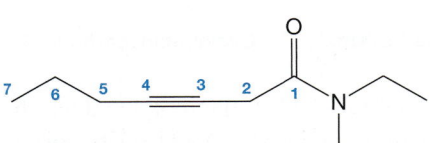

N-Ethyl-N-methylhept-3-ynamide

2-Carbamoylbutanedioic acid

In the second molecule above, the highest-priority functional group is CO_2H, so the O=C—N group is treated as a substituent, indicated by the prefix *carbamoyl*. Just as we saw with CN groups, *the carbon atom of an O=C—N group is not numbered when the O=C—N is treated as a substituent.*

1° Amide

2° Amide

3° Amide

FIGURE F-1 Amide classifications Just as we saw for amines in Interchapter E, amides are classified as primary (1°), secondary (2°), or tertiary (3°) according to the number of bonds to carbon the amide N has.

SOLVED PROBLEM F.1

Write the IUPAC name for L-asparagine, one of the naturally occurring amino acids, which is shown here in its unionized form.

Think What is the highest-priority functional group present and the corresponding suffix that must be added? How many carbon atoms are in the longest carbon chain? Where should the numbering of that chain begin? What other functional groups are present and how are those named as substituents? Are there stereochemical configurations that must be specified?

Solve The highest-priority functional group in the molecule is CO_2H, so the suffix is *oic acid* and numbering begins at the CO_2H carbon. The other functional groups present are NH_2 and $O=C-N$, which must be treated as substituents and receive the prefixes *amino* and *carbamoyl*, respectively. The prefix carbamoyl already accounts for the $O=C-N$ carbon, so there are three carbon atoms in the longest carbon chain containing the CO_2H group, making the molecule a *propanoic acid*. The amino group is attached to C2 and the carbamoyl group is attached to C3. C2 is an asymmetric carbon, which has an S configuration.

The carbamoyl carbon is not numbered.

(S)-2-Amino-3-carbamoyl-propanoic acid

PROBLEM F.2 Write the IUPAC name for each of the following compounds.

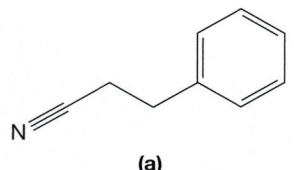

(a)

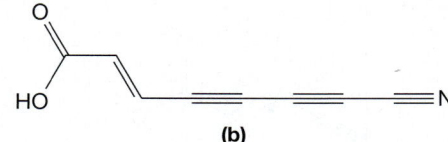

(b)

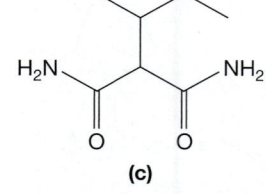

(c)

CONNECTIONS The compound in Problem F.2(a) is isolated from watercress. The compound in Problem F.2(b) is a fungal metabolite isolated from *Lepista diemii*, shown here.

PROBLEM F.3 Draw the structure for each of the following molecules.
(a) 4-chlorocarbonylbutanoic acid; **(b)** (*R*)-4-hydroxy-*N*-methyl-5-phenylpentanamide; **(c)** (*Z*)-2,3-dimethylhex-2-enedinitrile; **(d)** 2-carboxy-1,4-butanedioic acid

The functional groups we have been studying in this section cannot be incorporated into a ring structure. They can be *attached* directly to a ring, however, as shown with the following carboxylic acid, amide, and nitrile:

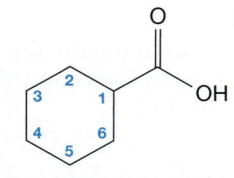

Cyclohexanecarboxylic acid

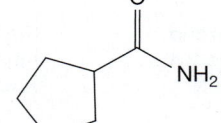

Cyclopentanecarboxamide

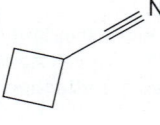

Cyclobutanecarbonitrile

In such cases, both the ring and the functional group establish the root, in which the ring is named first, followed by *carboxylic acid*, *carboxamide*, or *carbonitrile* to specify the CO_2H, $CONH_2$, or CN functional group, respectively. Furthermore, the carbon atoms of the ring are numbered, not the carbon atom of the functional group.

SOLVED PROBLEM F.4

Write the IUPAC name for shikimic acid.

Think What is the highest-priority functional group present? Can it be numbered as part of the root? What other functional groups are present and how should they

Shikimic acid

be incorporated into the molecule's name? Are there stereochemical configurations that must be specified?

Solve The highest-priority functional group present is CO_2H and, because it is directly attached to a cyclohexene ring, the molecule is a *cyclohexenecarboxylic acid*. The carbon atoms of the ring are numbered, with C1 being the carbon that is attached to the carboxyl group. The C=C bond begins at C1 and three hydroxy substituents are attached at C3, C4, and C5, which are asymmetric carbons that have *R*, *S*, and *R* configurations, respectively.

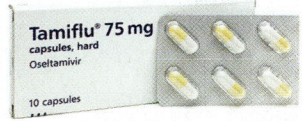

(3R,4S,5R)-3,4,5-Trihydroxycyclohex-1-enecarboxylic acid

PROBLEM F.5 Write the IUPAC name for each of the following compounds.

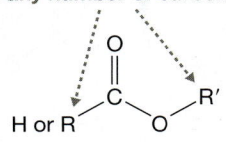

(a)   (b)   (c)

PROBLEM F.6 Draw the structure for each of the following molecules.
(a) 2,2-dimethylcyclopentanecarboxylic acid; **(b)** cyclohepta-3,5-diene-1-carboxamide;
(c) (1*R*,4*S*)-4-cyanocyclooctane-1-carboxylic acid;
(d) (1*S*,2*S*)-2-hydroxy-*N*-(1-methylethyl)cyclobutane-1-carboxamide;
(e) (*R*)-cyclohexa-2,4-diene-1-carbonitrile

F.2 Naming Esters and Acid Anhydrides

The rules for naming esters and acid anhydrides differ from those we just learned in Section F.1 for naming carboxylic acids and their derivatives.

F.2a Nomenclature Rules for Esters

An ester consists of an O=C—O group with H or an alkyl group (R) attached to the carbonyl carbon and another alkyl group (R′) attached to the singly bonded O.

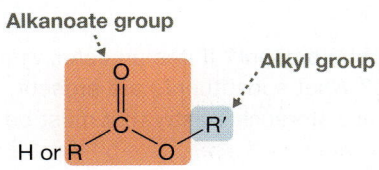

An ester

Because the two groups attached to O=C—O can consist of any number of carbon atoms, the name of the ester must accurately account for both groups. We begin by dividing the molecule into two parts, as shown on the previous page at the right. One part is simply the R′ alkyl group bonded to the O atom, highlighted in blue. The other is the RCO_2 group, called the **alkanoate group**, which is highlighted in red and contains the carbonyl (C=O) group and the singly bonded O atom. Esters are then named as follows:

Rules for Naming Esters

- The name of an ester has the general form *alkyl alkanoate*.
- The alkyl group is named precisely as described in Section A.4 (i.e., methyl, ethyl, propyl, etc.)
- The alkanoate portion derives from the analogous alkane having the same number of carbon atoms.

In ethyl butanoate (shown below), for example, an ethyl group is attached to the singly bonded O atom. The alkanoate group is named butanoate because it is made up of a four-carbon chain. In propyl ethanoate (propyl acetate), the alkyl group is a propyl group and the alkanoate group has two C atoms, so it is the ethanoate group (acetate group). (The name acetate derives from the trivial name for the analogous carboxylic acid, acetic acid; Section F.3.)

CONNECTIONS Ethyl butanoate tastes like oranges and is commonly used as a flavoring agent, even in many brands of orange juice.

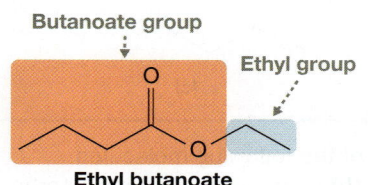

Ethyl butanoate

Butanoate group · Ethyl group

Propyl ethanoate (Propyl acetate)

Ethanoate group (Acetate group) · Propyl group

Both the alkyl and alkanoate groups of an ester can have attached substituents, so each carbon chain can have a numbering system, as shown in the following examples:

1-Methylpropyl 3-oxobutanoate

Phenyl (3S,4S)-3,4-dihydroxypentanoate

In the first molecule above, a methyl group is attached to C1 of the propyl group, and C3 of the butanoate portion is part of a carbonyl group. In the second molecule, hydroxy substituents are attached to C3 and C4 of the pentanoate portion.

SOLVED PROBLEM F.7

Write the IUPAC name for the molecule shown here.

Think What is the highest-priority functional group present? If it is an ester, what are the names of the alkyl and alkanoate parts? What substituents are present, and how do you number the carbons? Is there any stereochemistry that must be specified?

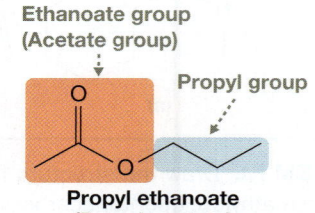

Solve The highest-priority functional group present is CO_2R, characteristic of an ester. We first name the alkyl group attached to O as a cyclopentyl group. The alkanoate part is propanoate, as shown with the numbering system here, and attached to C3 are a cyano and a hydroxy group. C3 is an asymmetric carbon with an S configuration.

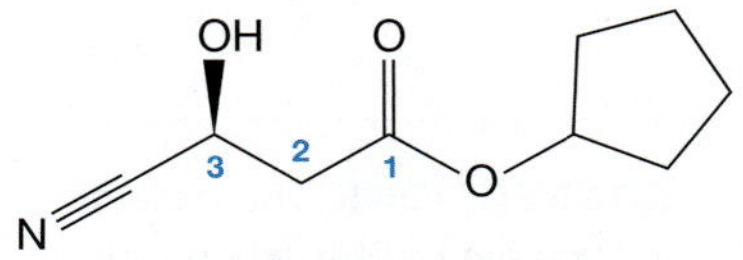

Cyclopentyl (*S*)-3-cyano-
3-hydroxypropanoate

PROBLEM F.8 What is the IUPAC name for each of the following molecules?

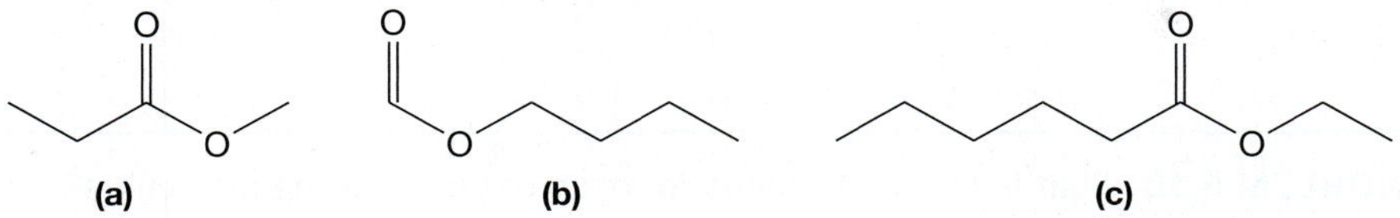

(a)   (b)   (c)

PROBLEM F.9 Draw the molecules that correspond to the following IUPAC names. **(a)** Pentyl pentanoate; **(b)** propyl butanoate; **(c)** ethyl methanoate; **(d)** ethyl (*Z*)-3-methylpent-2-enoate; **(e)** (*R*)-3-chlorobutyl (*S*)-2-hydroxypropanoate

F.2b Nomenclature Rules for Acid Anhydrides

The rules for naming acid anhydrides are derived from the fact that an acid anhydride can be produced from two carboxylic acids in what is called a dehydration reaction, as shown below. (This reaction is described in detail in Chapter 21.)

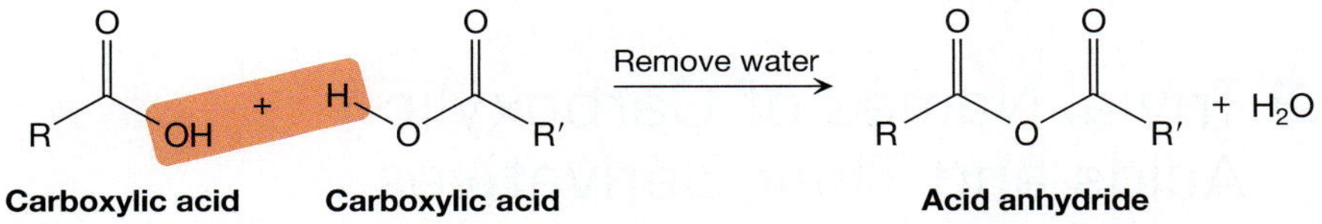

Remove water

Carboxylic acid   **Carboxylic acid**   **Acid anhydride** $+ H_2O$

The R and R′ groups can be the same, in which case the acid anhydride would be *symmetric*, or they can be different, in which case the acid anhydride would be *unsymmetric*.

Rules for Naming Acid Anhydrides

- For a symmetric acid anhydride, name the molecule according to the general form *alkanoic anhydride*. Here, *alkanoic* corresponds to the specific carboxylic acid that could undergo dehydration to produce the anhydride.
- If the acid anhydride is unsymmetric, then name the molecule according to the general form *alkanoic alkanoic anhydride*.
  - The two instances of *alkanoic* will be different, each one corresponding to the different carboxylic acids that would be required to form the anhydride via dehydration.
  - The two instances of *alkanoic* should appear in alphabetical order.

The first molecule on the next page, for example, is a symmetric anhydride that would be produced on dehydration of propanoic acid. Therefore, its name is propanoic anhydride. The second molecule is an unsymmetric anhydride, which would be produced on the dehydration of ethanoic acid (acetic acid) and benzoic acid. It is named benzoic ethanoic anhydride.

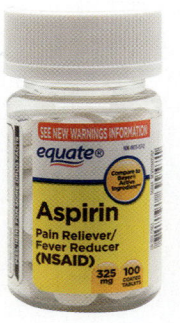

Each of these carbon-containing groups is the same as in propanoic acid.

This is the same carbon-containing group as in ethanoic acid.

This is the same carbon-containing group as in benzoic acid.

Propanoic anhydride

Benzoic ethanoic anhydride

PROBLEM F.10 What is the IUPAC name for each of the following molecules?

(a)   (b)   (c)

PROBLEM F.11 Draw the molecules that correspond to the following IUPAC names. **(a)** butanoic anhydride; **(b)** butanoic propanoic anhydride; **(c)** 2-methylbutanoic anhydride; **(d)** benzoic 2-methylbutanoic anhydride

F.3 Trivial Names of Carboxylic Acids and Their Derivatives

Just as we have seen with other compound classes, trivial names of carboxylic acids and their derivatives are frequently used.

F.3a Trivial Names of Carboxylic Acids

Trivial names of some common carboxylic acids are shown below. They include mono-carboxylic acids as well as dicarboxylic acids.

Monocarboxylic acids

Methanoic acid (Formic acid)   **Ethanoic acid (Acetic acid)**   **Propanoic acid (Propionic acid)**   **Butanoic acid (Butyric acid)**   **Pentanoic acid (Valeric acid)**

Hexanoic acid (Caproic acid)   **Propenoic acid (Acrylic acid)**   **Benzenecarboxylic acid Benzoic acid**

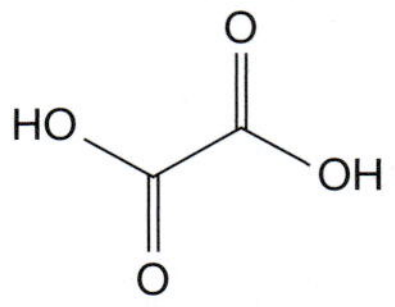

Ethanedioic acid (Oxalic acid)	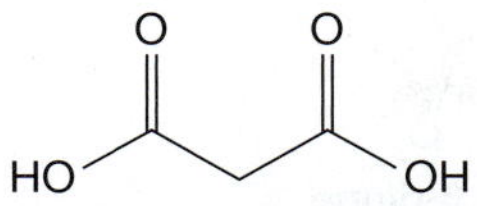 **Propanedioic acid** (Malonic acid)	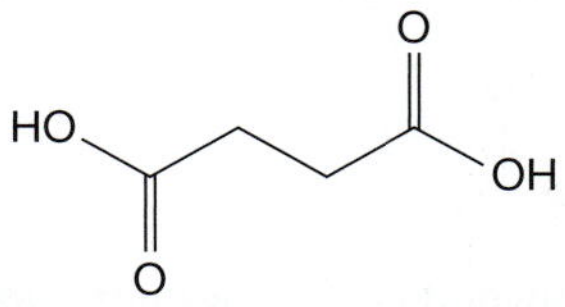 **Butanedioic acid** (Succinic acid)	**(Z)-Butenedioic acid** (Maleic acid)

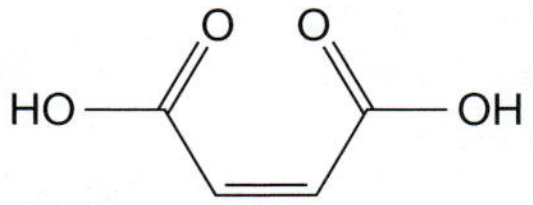

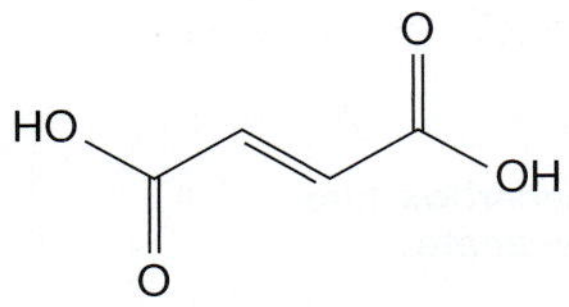

(E)-Butenedioic acid (Fumaric acid)	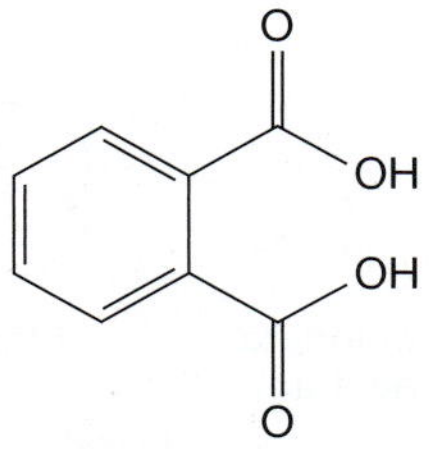 **Benzene-1,2-dicarboxylic acid** (Phthalic acid)

Several of the preceding trivial names are used almost to the exclusion of their IUPAC names. Acetic acid, the principal component of vinegar, is one example. Another is formic acid, a main constituent of ant venom (*formica* is the Latin word for ant). The name *benzoic acid*, in fact, has been accepted by IUPAC, which is why it appears without parentheses.

PROBLEM F.12 For each trivial name, draw the complete structure and provide the IUPAC name. **(a)** trichloroacetic acid; **(b)** 2,2-dimethylbutyric acid; **(c)** 2-aminopropionic acid; **(d)** dimethylmaleic acid; **(e)** diethylmalonic acid

---

With trivial names, a system of Greek letters is used to locate substituents relative to the carboxyl group. The carbon atom adjacent to the carboxyl group is designated α (alpha), and the atoms that are two, three, four, and five carbons away from the carboxyl carbon atom are designated β (beta), γ (gamma), δ (delta), and ε (epsilon), respectively.

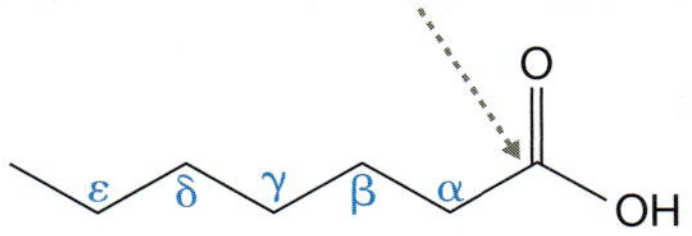

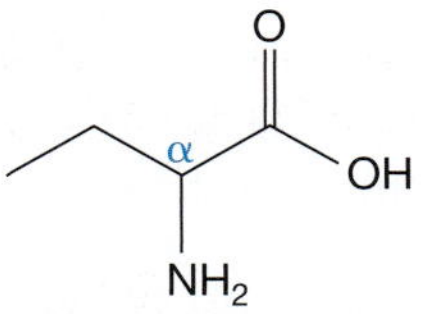

	An α-amino acid	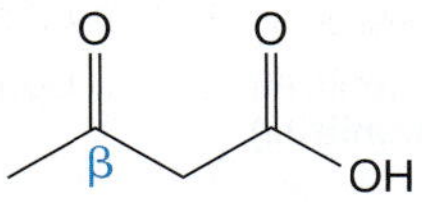 **A β-keto acid**

Thus, a compound that has an amino group attached to a carbon atom adjacent to the carboxyl group is called an α-amino acid, and a compound in which a ketone group is two carbons removed from the carboxyl carbon is called a β-keto acid.

F.3b Trivial Names of Carboxylic Acid Derivatives: Acid Anhydrides, Esters, Acid Chlorides, Amides, and Nitriles

The trivial names of many carboxylic acid derivatives derive from the trivial names of the analogous carboxylic acids.

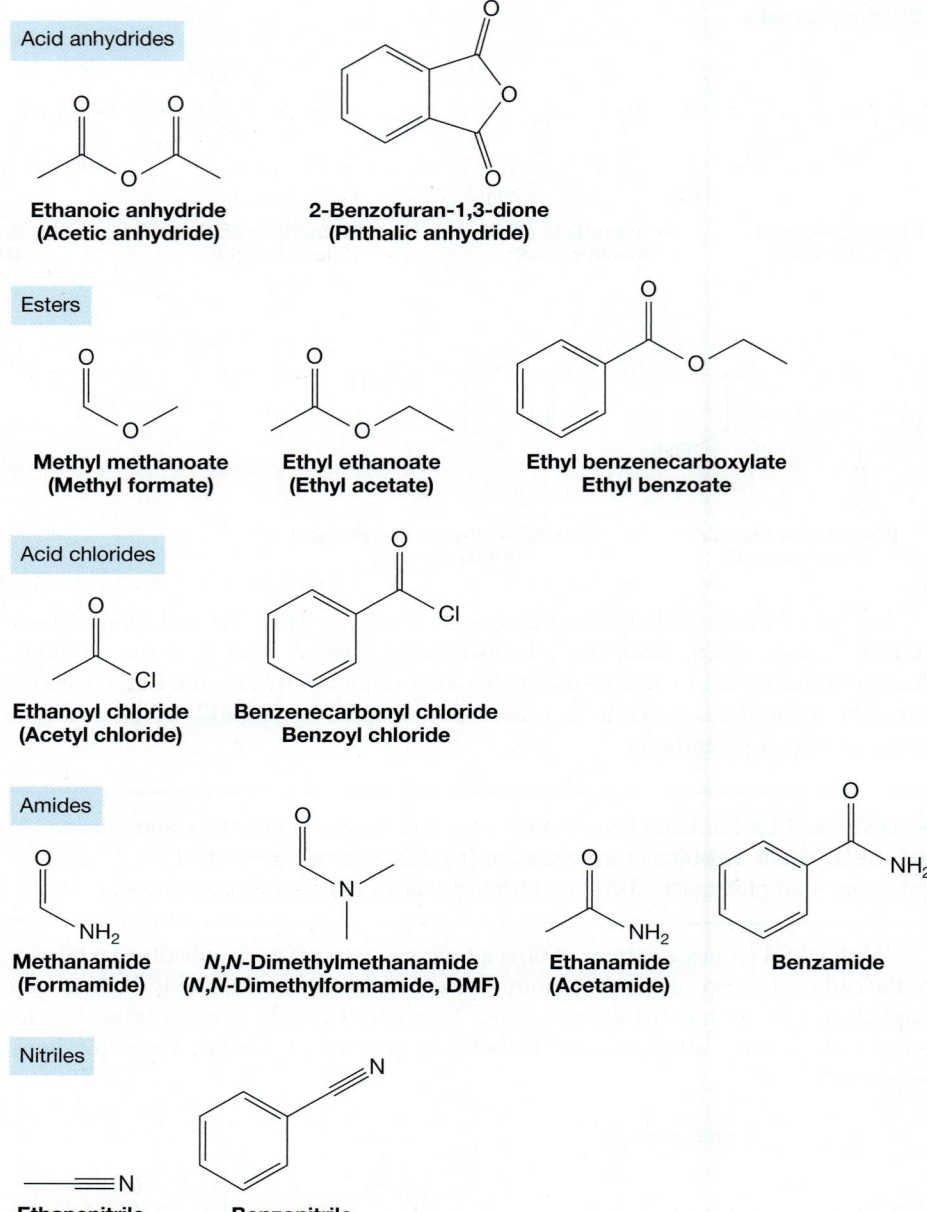

Acid anhydrides

Ethanoic anhydride
(Acetic anhydride)

2-Benzofuran-1,3-dione
(Phthalic anhydride)

Esters

Methyl methanoate
(Methyl formate)

Ethyl ethanoate
(Ethyl acetate)

Ethyl benzenecarboxylate
Ethyl benzoate

Acid chlorides

Ethanoyl chloride
(Acetyl chloride)

Benzenecarbonyl chloride
Benzoyl chloride

Amides

Methanamide
(Formamide)

N,N-Dimethylmethanamide
(*N,N*-Dimethylformamide, DMF)

Ethanamide
(Acetamide)

Benzamide

Nitriles

Ethanenitrile
(Acetonitrile)

Benzonitrile

Notice, in particular, the prevalence of the roots *form*, *acet*, and *benz*, which derive from formic acid, acetic acid, and benzoic acid, respectively.

PROBLEM F.13 For each trivial name, draw the complete structure and provide the IUPAC name. **(a)** *N,N*-dimethylacetamide; **(b)** methoxyacetonitrile; **(c)** ethyl trichloroacetate; **(d)** isopropyl formate; **(e)** *N,N*-diphenylbenzamide

---

F.1 Naming Carboxylic Acids, Acid Chlorides, Amides, and Nitriles

F.14 Provide the IUPAC name for each of the following carboxylic acids.

(a)

(b)

(c)

F.15 Draw the structure of each of the following molecules. **(a)** 2,2-dimethylcyclopentane-1-carboxylic acid; **(b)** (R)-3-chloropentanoic acid; **(c)** (2R,3S)-2,3-dinitrobutanedioic acid

F.16 Provide the IUPAC name for each of the following acid chlorides.

(a)

(b)

(c)

F.17 Draw the structure of each of the following molecules. **(a)** pentanoyl chloride; **(b)** 4-(2-methylpropyl)heptanedioyl chloride; **(c)** (S)-5-phenyloctanoyl chloride

F.18 Provide the IUPAC name for each of the following amides.

(a)

(b)

(c)

F.19 Draw the structure of each of the following molecules. **(a)** 5-phenylpentanamide; **(b)** (2S,3S)-2,3-dimethoxyhexanediamide; **(c)** N-phenylcyclobutanecarboxamide

F.20 Provide the IUPAC name for each of the following nitriles.

(a)

(b)

(c)

F.21 Draw the structure of each of the following molecules. **(a)** hexanedinitrile; **(b)** (S)-4-nitroheptanenitrile; **(c)** 4,4-diethylcyclohexanecarbonitrile

F.2 Naming Esters and Acid Anhydrides

F.22 Provide the IUPAC name for each of the following esters.

(a)

(b)

(c)

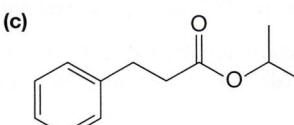

F.23 Draw the structure of each of the following molecules. **(a)** cyclohexyl butanoate; **(b)** 1,1-dimethylethyl hexanoate; **(c)** phenyl 4,4-dinitroheptanoate

F.24 Provide the IUPAC name for each of the following acid anhydrides.

(a)

(b)

(c)

F.25 Draw the structure of each of the following molecules. **(a)** pentanoic anhydride; **(b)** hexanoic propanoic anhydride; **(c)** ethanoic 3-methylpentanoic anhydride

F.3 Trivial Names of Carboxylic Acids and Their Derivatives

F.26 For each of the following trivial names, draw the structure and write the correct IUPAC name. **(a)** pentachloropropionic acid; **(b)** trimethylacrylic acid; **(c)** phenylmalonic acid; **(d)** α-hydroxycaproic acid

F.27 For each of the following trivial names, draw the structure and write the correct IUPAC name. **(a)** trichloroacetic anhydride; **(b)** valeric anhydride; **(c)** acetic butyric anhydride

F.28 For each of the following trivial names, draw the structure and write the correct IUPAC name. **(a)** isobutyl benzoate; **(b)** *tert*-butyl acetate; **(c)** isopropyl formate

F.29 For each of the following trivial names, draw the structure and write the correct IUPAC name. **(a)** N,N-diphenylformamide; **(b)** acrylamide; **(c)** N-isopropylbenzamide; **(d)** oxalamide

F.30 For each of the following trivial names, draw the structure and write the correct IUPAC name. **(a)** acrylonitrile; **(b)** fumaronitrile; **(c)** malononitrile; **(d)** phthalonitrile; **(e)** valeronitrile

Integrated Problems

F.31 Provide the IUPAC name for each of the following molecules.

F.32 Provide the IUPAC name for each of the following molecules.

F.33 Provide the IUPAC name for each of the following molecules.

F.34 Draw the structure of each molecule. **(a)** (E)-4-carbamoylbut-3-enoic acid; **(b)** 3-carbamoylpentanediamide; **(c)** 4,6-dioxohexanenitrile; **(d)** (1S,2S)-2-methoxycyclopent-3-ene-1-carbonitrile; **(e)** cyclohexa-3,6-diene-1,3-dicarboxylic acid

F.35 Draw the structure of each molecule. **(a)** 2,2-dimethylpropyl hex-3-ynoate; **(b)** cyanoethanoic anhydride; **(c)** cyanomethyl 5,5-dibromopentanoate; **(d)** hex-3-ynoic anhydride; **(e)** butyl (R)-4-carbamoylhexanoate

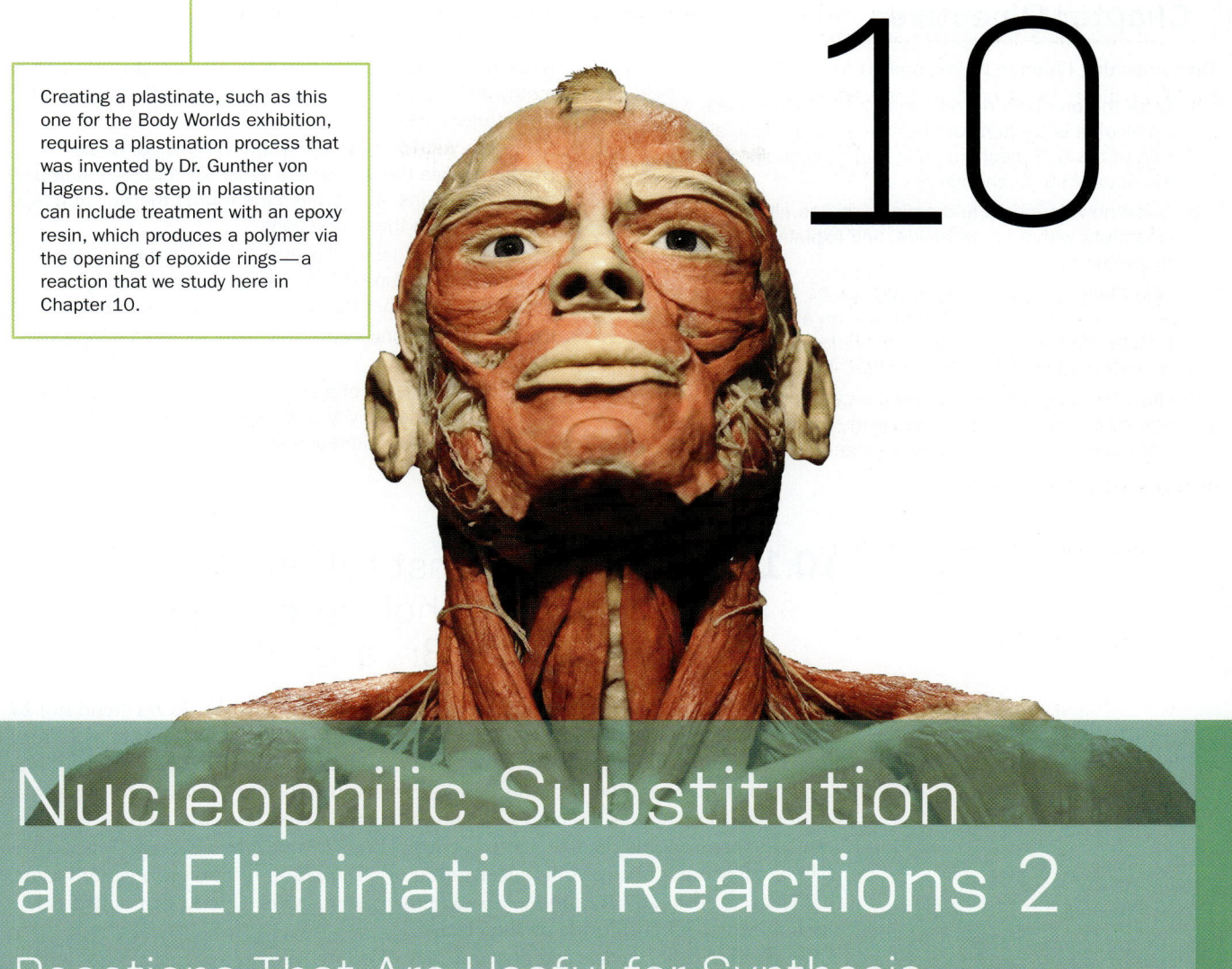

Creating a plastinate, such as this one for the Body Worlds exhibition, requires a plastination process that was invented by Dr. Gunther von Hagens. One step in plastination can include treatment with an epoxy resin, which produces a polymer via the opening of epoxide rings—a reaction that we study here in Chapter 10.

10

Nucleophilic Substitution and Elimination Reactions 2

Reactions That Are Useful for Synthesis

In the last few chapters we have focused a great deal on S_N2, S_N1, E2, and E1 reactions. In Chapters 7 and 8, the mechanisms of these four reactions were introduced, including their stereochemistries. In Chapter 9, we learned that all four reactions are generally in competition with one another, and we analyzed the main factors that can be used to predict the outcome of that competition.

Thus far, however, our focus on such reactions has primarily been on their fundamental mechanisms, which has limited us to a relatively narrow range of reactions. Here in Chapter 10, we learn how to use nucleophilic substitution and elimination reactions to carry out specific transformations that are useful in synthesis (the topic of Chapter 13). In doing so, we hope to illustrate the importance of these reactions throughout organic chemistry and to reinforce important aspects of nucleophilic substitution and elimination in general.

Because both nucleophilic substitution and elimination reactions are presented in this chapter, we have grouped the reactions according to their mechanisms. Sections 10.1 through 10.7 discuss reactions involving nucleophilic substitution, then Sections 10.8 and 10.9 discuss reactions involving elimination.

Chapter Objectives

On completing Chapter 10 you should be able to:

1. Draw the products of the reaction that takes place when an alcohol is treated with PBr_3 or PCl_3, and explain the role of the S_N2 mechanism in these reactions, including aspects of stereochemistry.

2. Show how ammonia and amines can be alkylated by treatment with an alkyl halide, and explain the potential problem of polyalkylation.

3. Show how a ketone or aldehyde can be alkylated at an α (alpha) carbon by treatment with an alkyl halide under basic conditions, and explain how the regioselectivity can be governed by the choice of base.

4. Draw the products of halogenation at an α carbon of a ketone or aldehyde, and explain the effect that acidic or basic conditions have on polyhalogenation.

5. Show how diazomethane converts a carboxylic acid into a methyl ester via an S_N2 reaction, and explain the role of the nitrogen leaving group.

6. Show how an ether can be synthesized under basic conditions (via the Williamson ether synthesis) and under acidic conditions, and explain the role of the S_N2 and S_N1 mechanisms in these reactions.

7. Recognize epoxides and oxetanes as suitable substrates in S_N2 reactions and explain how the regioselectivity depends on whether the reaction conditions are acidic.

8. Explain how to convert a vinylic halide into an alkyne via an E2 reaction.

9. Predict the products of a Hofmann elimination reaction, including regioselectivity, and explain the role of the E2 mechanism in this reaction.

10.1 Nucleophilic Substitution: Converting Alcohols into Alkyl Halides Using PBr_3 and PCl_3

In Section 9.5, we saw that under normal conditions, alcohols (R—OH) tend not to undergo nucleophilic substitution and elimination reactions, but alkyl chlorides (R—Cl) and alkyl bromides (R—Br) do so quite readily. This is because HO^- is a rather poor leaving group for nucleophilic substitution and elimination reactions, whereas halides such as Br^- and Cl^- are quite good. (Recall that Cl and Br are substantially larger atoms than O and can better stabilize a negative charge.) Thus, if we wish to carry out a nucleophilic substitution or elimination reaction at a carbon atom that is attached to an OH group, it is extremely useful to be able to first convert that alcohol into an alkyl chloride or alkyl bromide.

In Section 9.5b, we learned that one way of doing so is to treat an alcohol with a strong acid like HCl, HBr, or HI (Equation 10-1).

Alcohol → **Alkyl halide**

conc HBr
Δ, 4 h

76%

(10-1)

Under these acidic conditions, a poor leaving group (HO^-) is converted into a very good leaving group (H_2O), as shown in the mechanism in Equation 10-2. The Br^- that is generated in the process can then act as a nucleophile to displace H_2O. If the substrate is a primary alcohol, as in Equation 10-1, then this displacement takes place via an S_N2 reaction.

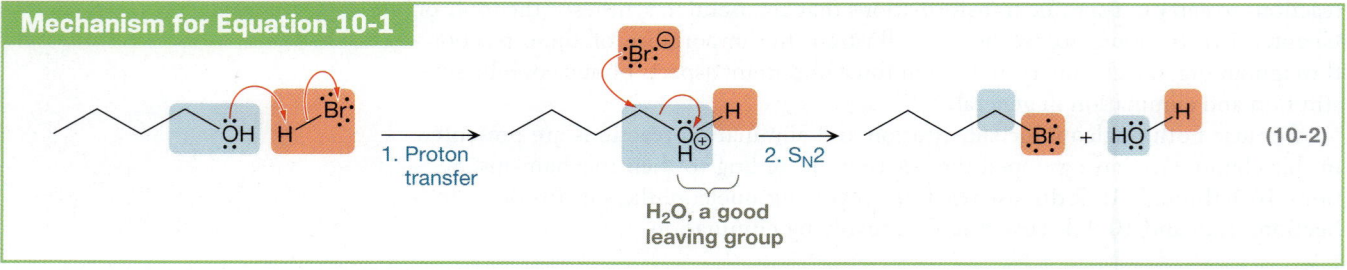

Mechanism for Equation 10-1

1. Proton transfer

H_2O, a good leaving group

2. S_N2

(10-2)

516   **CHAPTER 10** Nucleophilic Substitution and Elimination Reactions 2

However, a variety of problems are often associated with this kind of reaction. For example, there may be other functional groups in the substrate (not shown in Equation 10-1) that are sensitive to strongly acidic conditions, leading to unwanted side reactions. If the substrate is a secondary or tertiary alcohol, moreover, then these reaction conditions would favor both S_N1 and E1 mechanisms (see Section 9.6b). Because some elimination products would form, the yield of the intended substitution product would be compromised. Also, S_N1 reactions proceed through a planar *carbocation intermediate*, so any stereochemistry that might exist at the carbon atom bonded to the leaving group would be lost, as shown in Equations 10-3 and 10-4 (see Problem 10.1).

Mixture of stereoisomers

$$\text{(10-3)}$$

conc HBr / S_N1

$$\text{(10-4)}$$

conc HCl / S_N1

38%   +   38%

PROBLEM 10.1 Draw the complete, detailed mechanism for the reaction in **(a)** Equation 10-3 and **(b)** Equation 10-4.

Finally, generating a carbocation means a carbocation rearrangement may be possible (review Section 8.6d), as shown in Equation 10-5 (see also Problem 10.2).

Carbocation rearrangement

conc HBr / S_N1

$$\text{(10-5)}$$

PROBLEM 10.2 Draw the complete detailed mechanism for the reaction in Equation 10-5.

A much better way to convert an alcohol into an alkyl halide is to use phosphorus tribromide, PBr_3 (Equation 10-6), or phosphorus trichloride, PCl_3 (Equation 10-7). (Another reagent, $SOCl_2$, is even better than PCl_3, as we discuss in Chapter 20.)

Reaction is *stereospecific* (inversion of configuration).

PBr_3 / Diethyl ether

$$\text{(10-6)}$$

No carbocation rearrangement

PCl_3 / Diethyl ether

$$\text{(10-7)}$$

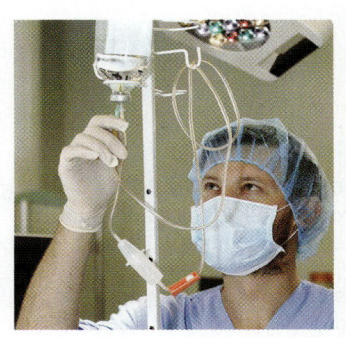

These reagents have none of the disadvantages that arise from using HBr or HCl:

- PBr$_3$ and PCl$_3$ convert an alcohol to an alkyl halide under relatively mild conditions.
- No elimination products are generated.
- The conversion is *stereospecific*. It takes place with *inversion of configuration* at the alcohol carbon atom (see Equation 10-6).
- No carbocation rearrangements occur (compare Equations 10-5 and 10-7).

These observations are explained by the two-step mechanism shown in Equation 10-8, which consists of back-to-back S$_N$2 steps.

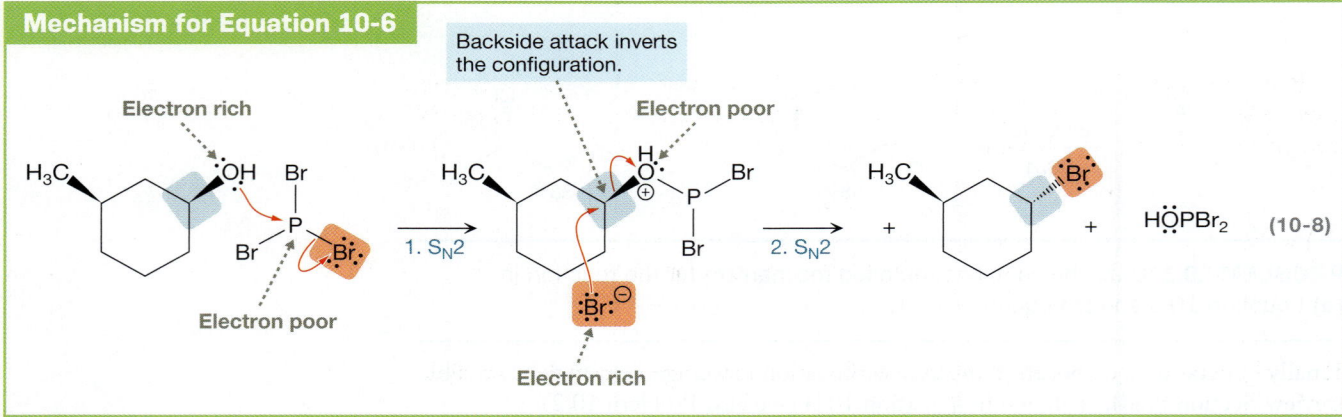

Mechanism for Equation 10-6

In Step 1, the O atom of the alcohol is electron rich and acts as the nucleophile, whereas the P atom of PBr$_3$ is electron poor and acts as the substrate. A Br$^-$ anion is the leaving group in this step and is therefore liberated. In Step 2, the Br$^-$ anion generated in Step 1 acts as the nucleophile and the phosphorus-containing species acts as the substrate. The leaving group is HOPBr$_2$.

The phosphorus-containing species in the second step of the mechanism resembles a protonated alcohol, ROH$_2^+$. The leaving group (HOPBr$_2$) resembles a stable, uncharged H$_2$O molecule, which is an excellent leaving group (Fig. 10-1).

YOUR TURN 10.1

Repeat the mechanism shown in Equation 10-8, using PCl$_3$ instead of PBr$_3$.

Answers to Your Turns are in the back of the book.

How does the mechanism in Equation 10-8 account for the stereospecificity of the reaction? In Step 2, the asymmetric carbon undergoes S$_N$2 attack, resulting in the

FIGURE 10-1 **The HOPBr$_2$ leaving group** The HOPBr$_2$ leaving group in Step 2 of Equation 10-8 is highlighted in red (*left*). This leaving group resembles the excellent H$_2$O leaving group in a protonated alcohol (*right*).

HOPBr$_2$ leaving group

H$_2$O leaving group

Protonated alcohol

inversion of the stereochemical configuration. In Step 1, the stereochemical configuration of that carbon is unaffected because no bonds to that carbon are broken or formed. There is no carbocation rearrangement, moreover, because no carbocations are ever generated!

PROBLEM 10.3 Draw the complete, detailed mechanism for each of the following reactions and predict the product.

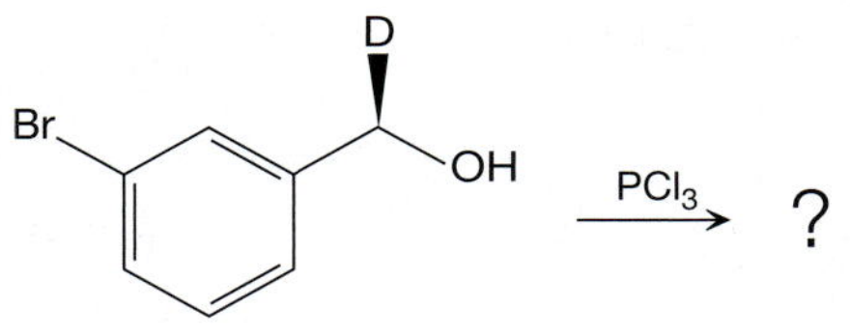

(a) 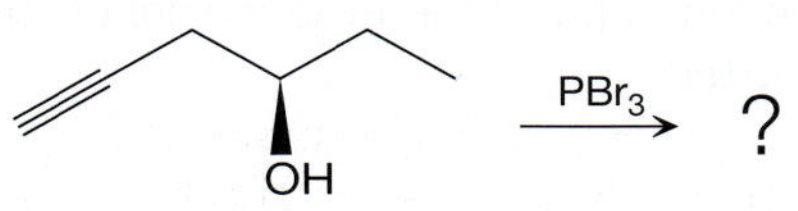   (b)

Recall from Section 9.6a that nucleophilic substitution reactions generally do *not* occur when a leaving group is on an sp^2- or sp-hybridized carbon atom. The same is true of these reactions involving phosphorus trihalides:

> The conversion of an alcohol into an alkyl halide using PBr_3 or PCl_3 occurs only when the OH group is bonded to an sp^3-hybridized carbon atom.

The OH group in Equation 10-9 is on an sp^2-hybridized C, not sp^3, so the halide does *not* form.

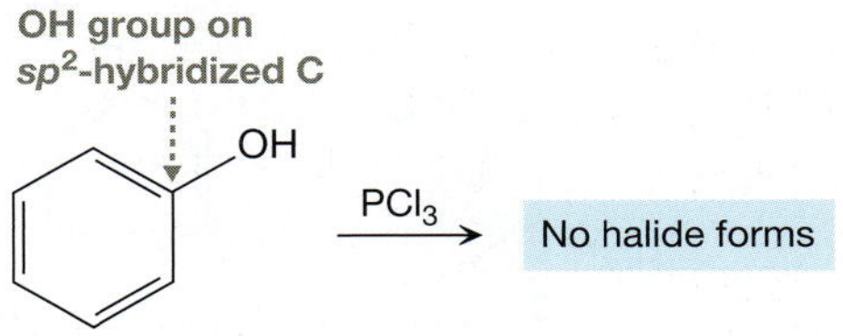

Recall from Section 9.6b, too, that S_N2 reactions effectively do not occur when the leaving group is bonded to a tertiary carbon (i.e., a carbon bonded to three R groups), due to excessive steric hindrance. Similarly, the intermediate that is formed when a tertiary alcohol (R_3COH) is treated with PBr_3 or PCl_3 does not readily undergo an S_N2 reaction, though it more feasibly can undergo S_N1. Thus:

> The reaction of a tertiary alcohol (R_3COH) with PBr_3 or PCl_3 often gives poor yield of the alkyl halide.

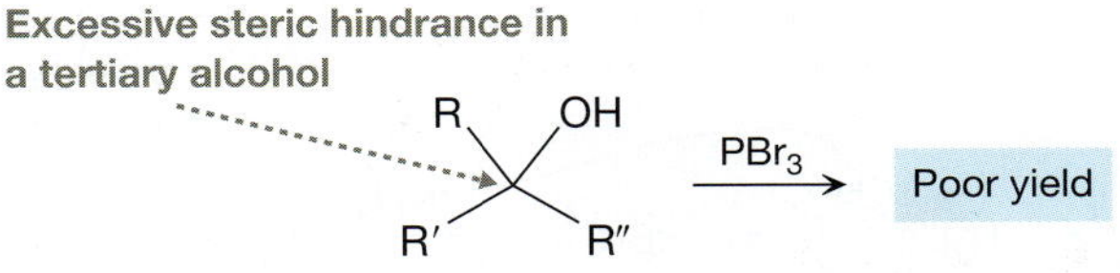

As we can see in Solved Problem 10.4, these restrictions on the type of carbon atom bonded to OH make it possible to carry out a selective conversion into the alkyl halide.

SOLVED PROBLEM 10.4

Predict the products of this reaction and draw its complete, detailed mechanism. Pay attention to stereochemistry.

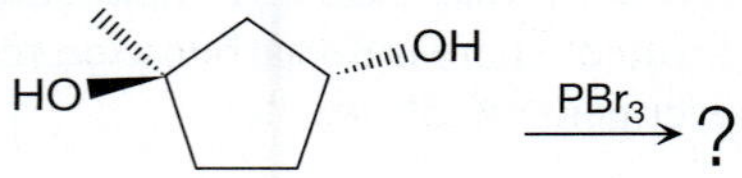

Think Which OH group is more susceptible to reaction with PBr$_3$? What type of functional group is produced? How is the stereochemistry of the molecule affected?

Solve PBr$_3$ converts an OH group into a Br substituent via back-to-back S$_N$2 steps. Of the two OH groups in the molecule given, the one on the right is part of a secondary alcohol, and the one on the left is part of a tertiary alcohol. Therefore, the OH group on the right will be converted.

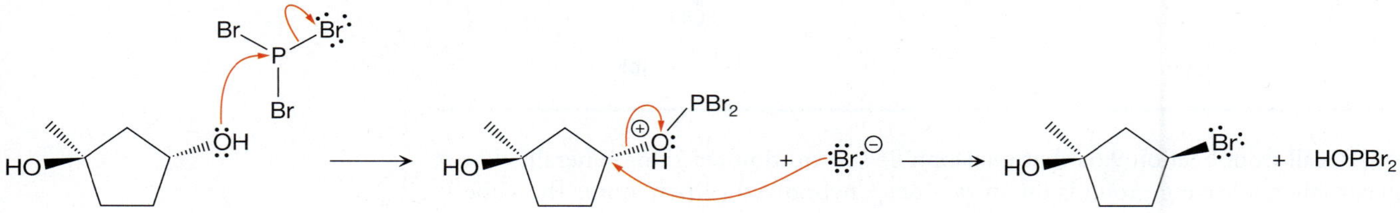

According to the mechanism shown, inversion of configuration at the C atom bonded to OH results in a C—Br bond that points toward you because the initial bond to OH pointed away.

PROBLEM 10.5 Which of the following compounds will readily react with PCl$_3$? For those that will, **(a)** draw the products, including stereochemistry, and **(b)** draw the complete, detailed mechanism.

A   B   C   D

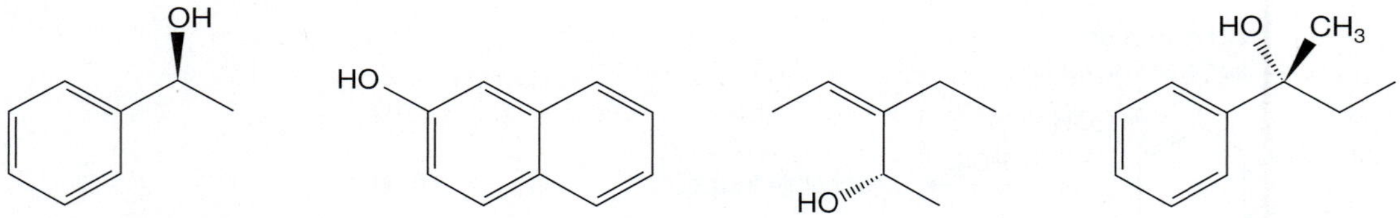

10.2 Nucleophilic Substitution: Alkylation of Ammonia and Amines

Although ammonia (NH$_3$) is uncharged, it is a moderately strong nucleophile (Section 9.7b). Therefore, when an alkyl halide such as bromoethane is treated with ammonia, an S$_N$2 reaction can occur, producing a protonated amine (Equation 10-11). Ammonia is also weakly basic, so it can deprotonate the S$_N$2 product to yield an uncharged amine. Overall, ammonia has undergone a single **alkylation**, a reaction in which a hydrogen atom is replaced by an alkyl group.

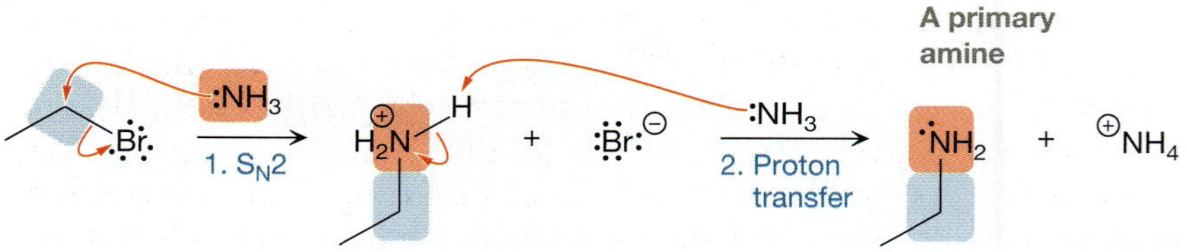

A primary amine

(10-11)

Verify that ammonia is a moderately good nucleophile by looking up its relative nucleophilicity and that of Cl⁻ in Section 9.7. Write their relative nucleophilicities here:

NH_3 _____   Cl⁻ _____

This reaction is *not* an effective method of synthesizing primary amines ($R—NH_2$), however, due to the mixture of other products formed (Equation 10-12). (We present a more efficient synthesis of primary amines in Chapter 20.)

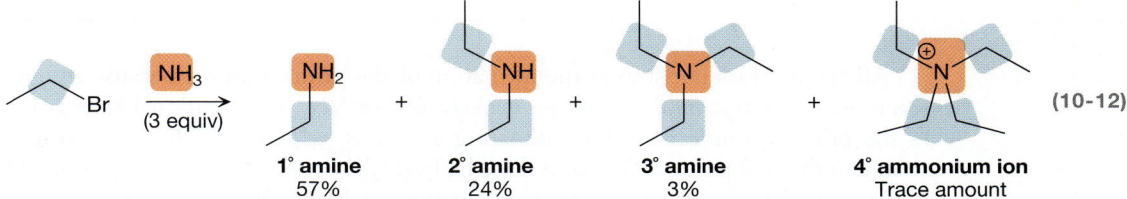

1° amine	2° amine	3° amine	4° ammonium ion
57%	24%	3%	Trace amount

(10-12)

In other words:

> If an alkyl halide is treated with an excess of ammonia, then the 1° amine (RNH_2) is formed in a mixture that also contains a 2° amine (R_2NH), a 3° amine (R_3N), and a 4° ammonium ion (R_4N^+).

The primary amine that is formed from the first alkylation is also nucleophilic at the N atom. It can therefore react with any unreacted alkyl halide species remaining in solution (Equation 10-13), leading to further alkylations.

Partial Mechanism for Equation 10-12

1° Amine product from first alkylation
(Eq. 10-11, steps 1 and 2)

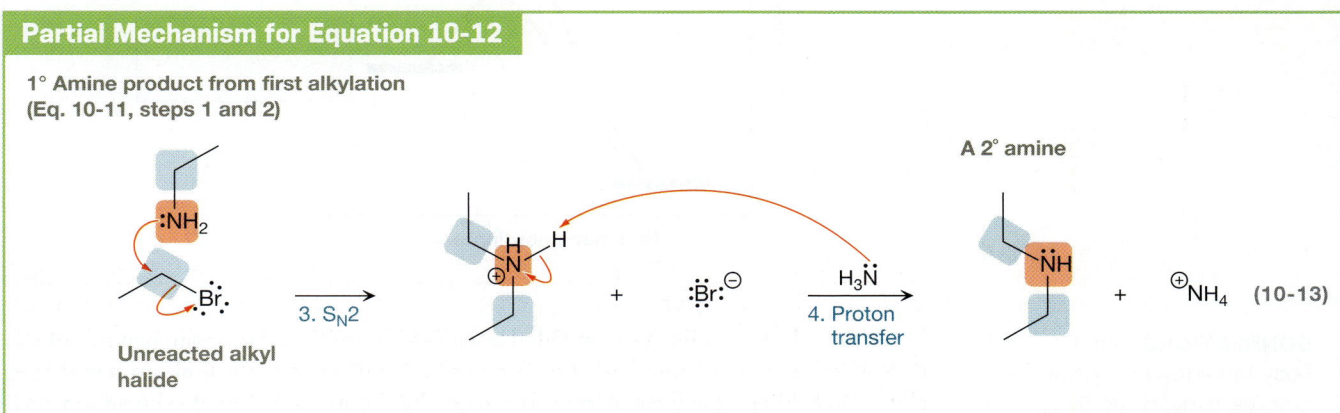

A 2° amine

(10-13)

The same process—an S_N2 reaction followed by deprotonation—can occur a third time to yield the tertiary amine, $(CH_3CH_2)_3N$ (see Your Turn 10.3). The tertiary amine is nucleophilic, too, so it can react in a fourth S_N2 reaction to yield the quaternary ammonium ion, $(CH_3CH_2)_4N^+$. Unlike the first three S_N2 reactions, the quaternary ammonium ion cannot be deprotonated because there is no acidic hydrogen left on the N atom.

The following reaction scheme shows the mechanism for the formation of the quaternary ammonium ion in Equation 10-12 from the secondary amine shown in Equation 10-13. Draw in the necessary curved arrows, and below each reaction arrow label each step as either "proton transfer" or "S$_N$2."

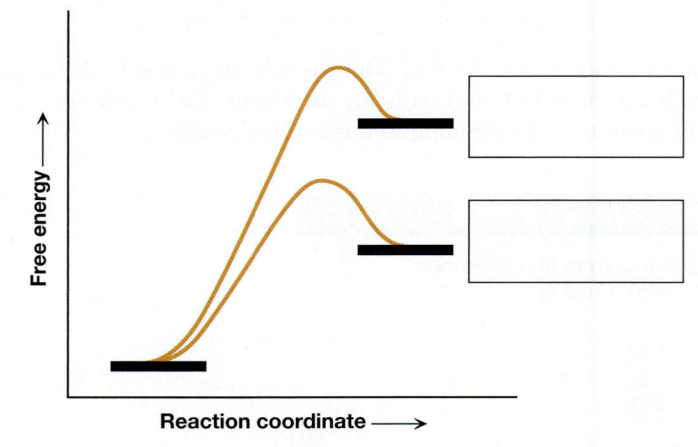

Alkylation does not stop at the formation of the primary amine because *the primary amine that is formed is a stronger nucleophile than NH$_3$*. The additional alkyl group in the primary amine is electron donating and thus helps to stabilize the positive charge in the S$_N$2 product [i.e., in $(CH_3CH_2)_2NH_2^+$]. Consequently, the S$_N$2 reaction in Equation 10-13 is *faster* than that in Equation 10-11 (see Your Turn 10.4). Similar arguments explain the subsequent alkylations as well.

This figure compares the free energy diagrams for the S$_N$2 steps in Equations 10-11 and 10-13. The uncharged reactants in each reaction are assumed to have the same stability, which is why they appear at the same energy. Draw the products of the S$_N$2 steps using the boxes provided, and draw arrows along each C—N bond to indicate the electron donation by the alkyl groups.

Free energy →

Reaction coordinate →

CONNECTIONS In the body, tetraethylammonium bromide, $(CH_3CH_2)_4N^+Br^-$, is a ganglionic blocker, which inhibits the transmission of nerve signals in the autonomic nervous system. Clinically, this gave tetraethylammonium bromide use in the treatment of hypertension, but it has been replaced by other, safer drugs.

Although the stabilization provided by each additional alkyl group makes the alkylation of ammonia a poor method of synthesizing primary amines, those electron-donating effects do facilitate the formation of the quaternary ammonium ion. As shown in Equation 10-14, this can be done simply by adding a large excess of the alkyl halide.

Large excess of alkyl halide

The 4° ammonium ion is the major product.

$$ \text{Br} \xrightarrow[\text{>8 h, 100 °C}]{NH_3} \quad + \quad Br^{\ominus} \qquad (10\text{-}14) $$

87%

If desired, the quaternary ammonium ion can be isolated with an anion as a **quaternary ammonium salt**.

This alkylation process is not limited to just ammonia. As shown in Solved Problem 10.6, other amines can be alkylated as well, which is particularly useful in Hofmann elimination reactions (discussed in Section 10.9).

SOLVED PROBLEM 10.6

Draw the complete, detailed mechanism for the reaction shown here and predict the major product.

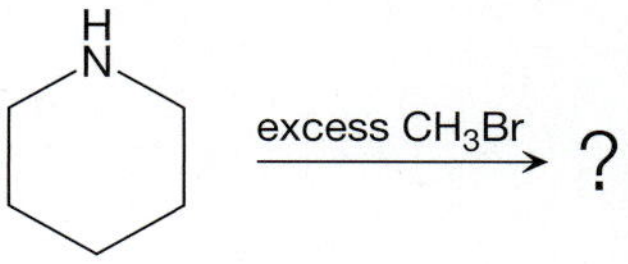

Think Which species is electron rich? Which is electron poor? What kind of reaction will take place? Can that reaction take place a second time?

Solve The amine is electron rich at the N atom, and CH_3Br is electron poor at the C atom. The resulting reaction is an S_N2, yielding a protonated tertiary amine. The product can be deprotonated by another molecule of the amine to yield an uncharged tertiary amine. Because there is excess CH_3Br, a further alkylation generates a quaternary ammonium ion as the major product. No further reaction can take place because there are no acidic H atoms on the quaternary ammonium ion.

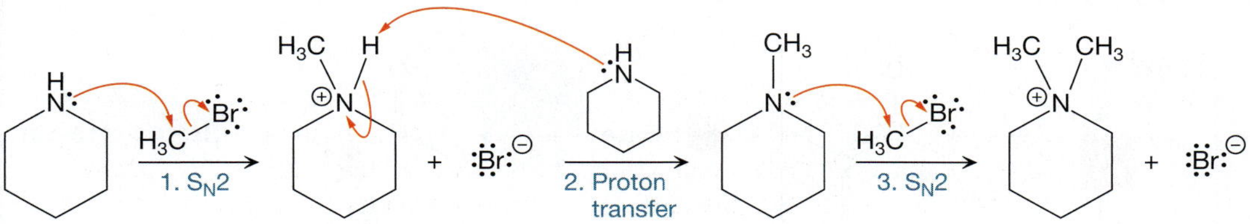

PROBLEM 10.7 Draw the complete, detailed mechanism for the reaction shown here and predict the major product.

10.3 Nucleophilic Substitution: Alkylation of α Carbons

No reaction occurs when an aldehyde (RCH=O) or ketone (R_2C=O) is treated with an alkyl halide alone:

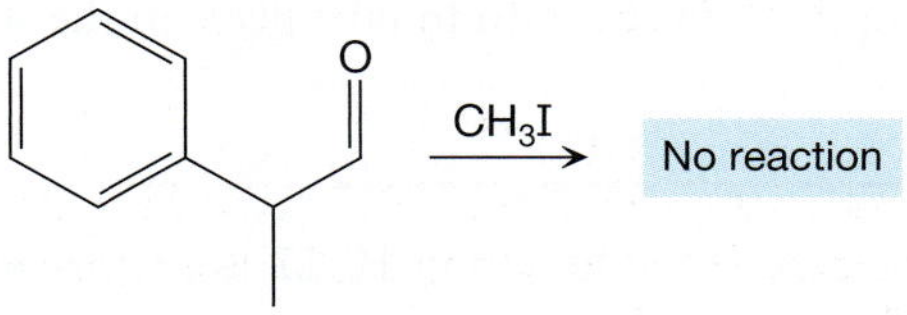

$$(10\text{-}15)$$

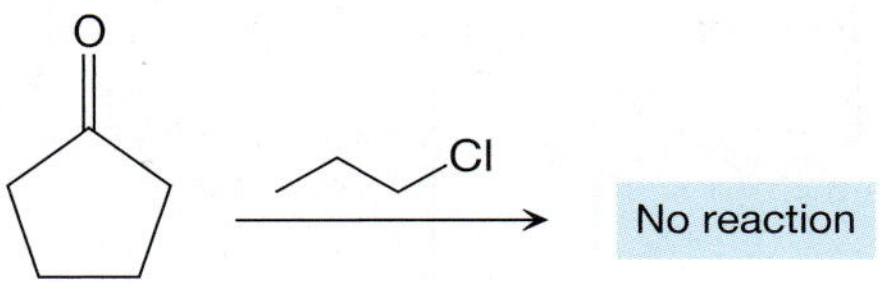

$$(10\text{-}16)$$

When a strong base like sodium hydride (NaH) or sodium amide ($NaNH_2$) is added first, however, *alkylation* can take place at the *α carbon* (i.e., at the C atom adjacent to the C=O group). Examples are shown in Equations 10-17 and 10-18.

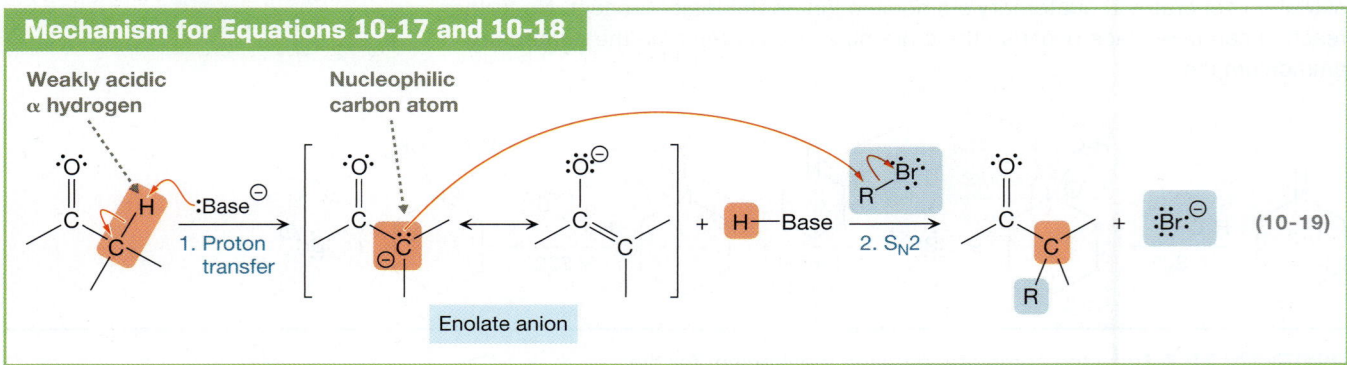

(10-17)

The α carbon
is alkylated.

77%

(10-18)

1. NaNH₂

2. Br

72%

The mechanism for these reactions, shown in Equation 10-19, is essentially a proton transfer step followed by an S_N2 step.

Mechanism for Equations 10-17 and 10-18

Weakly acidic
α hydrogen

Nucleophilic
carbon atom

:Base

1. Proton
transfer

Enolate anion

+ H—Base

2. S_N2

+ :Br:⁻

(10-19)

A base is required because an α carbon is non-nucleophilic in the uncharged form of a ketone or aldehyde. It is non-nucleophilic because the α carbon does not have a lone pair of electrons, nor does it bear a significant partial or full negative charge (review Section 9.3b). As we learned in Chapter 6, however, a hydrogen atom on such a carbon is weakly acidic ($pK_a \approx 20$), so it can be deprotonated by a sufficiently strong base to generate an *enolate anion*. In the first resonance structure in Equation 10-19, the α carbon possesses a lone pair and a −1 formal charge. Thus:

> An enolate anion is nucleophilic at the α carbon.

This enables the nucleophilic substitution to take place in the second step.

YOUR TURN 10.5

The mechanism for the reaction in Equation 10-17 is as follows:

:H⁻

+ H₂

CH₃—I:

+ I⁻

Provide the appropriate curved arrows, label the enolate anion, and identify the type of elementary step by writing "proton transfer" or "S_N2" below each reaction arrow.

PROBLEM 10.8 Draw the complete, detailed mechanism, including curved arrows, for the reaction in Equation 10-18.

10.3a Regioselectivity in α Alkylations

Alkylation at an α carbon is rather straightforward if there is only one distinct type of α carbon to consider. This is the case for the aldehyde in Equation 10-17, which contains only one α carbon altogether. It is also the case for the ketone in Equation 10-18, because the two α carbons are indistinguishable (Fig. 10-2).

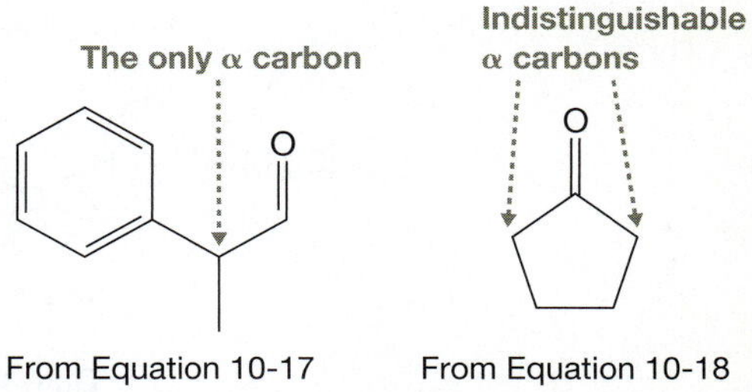

For many ketones, however, such as 2-methylcyclohexanone, the α carbons are distinct, so alkylation of the different C atoms leads to different products (Fig. 10-3). *Regioselectivity*, therefore, is a concern for the alkylation of ketones that have distinct α carbons, but regioselectivity can be controlled by the choice of base and reaction conditions, as shown in Equation 10-20.

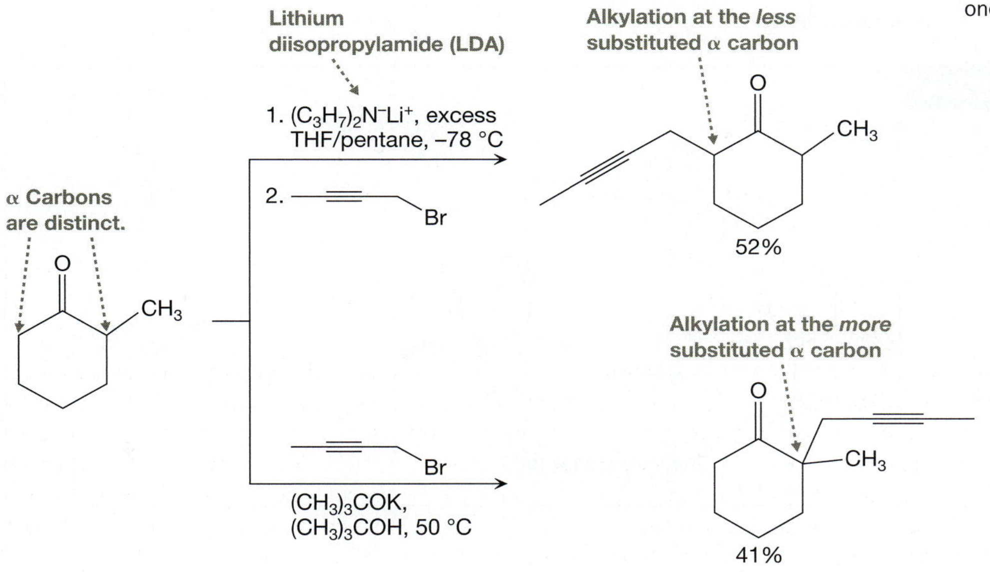

The only α carbon

Indistinguishable α carbons

From Equation 10-17   From Equation 10-18

FIGURE 10-2 Species with one distinct α carbon Regiochemistry is not an issue in the α alkylation of these species because each has just one distinct α carbon.

Lithium diisopropylamide (LDA)

α Carbons are distinct.

1. (C₃H₇)₂N⁻Li⁺, excess THF/pentane, −78 °C

2. ═══ Br

Alkylation at the *less* substituted α carbon

52%   (10-20a)

═══ Br
(CH₃)₃COK, (CH₃)₃COH, 50 °C

Alkylation at the *more* substituted α carbon

41%   (10-20b)

- An excess of lithium diisopropylamide (LDA) at low temperatures causes alkylation to occur at the *less* substituted α carbon of a ketone.
- Alkoxide bases (RO⁻) cause alkylation to occur at the *more* highly substituted α carbon of a ketone.

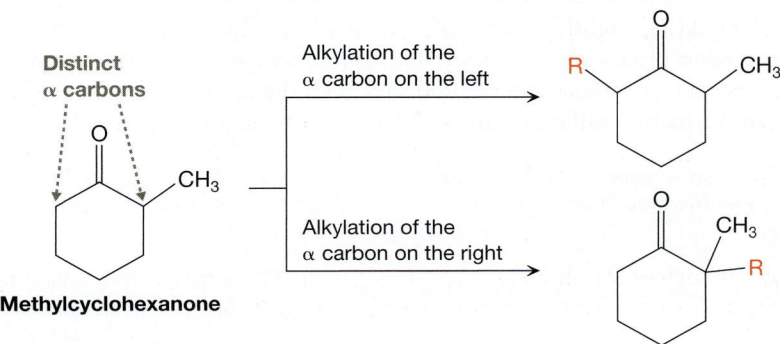

Distinct α carbons

Alkylation of the α carbon on the left

Alkylation of the α carbon on the right

2-Methylcyclohexanone

FIGURE 10-3 Regiochemistry in α alkylation 2-Methylcyclohexanone has two distinct α carbons, so α alkylation can produce different constitutional isomers.

The regioselectivity demonstrated by LDA versus RO⁻ is an outcome of whether deprotonation at the α carbon takes place *reversibly* or *irreversibly*. When LDA is used as the base (Equation 10-21), products are strongly favored ($K_{eq} \gg 1$), making $\Delta G°_{rxn}$ substantially negative.

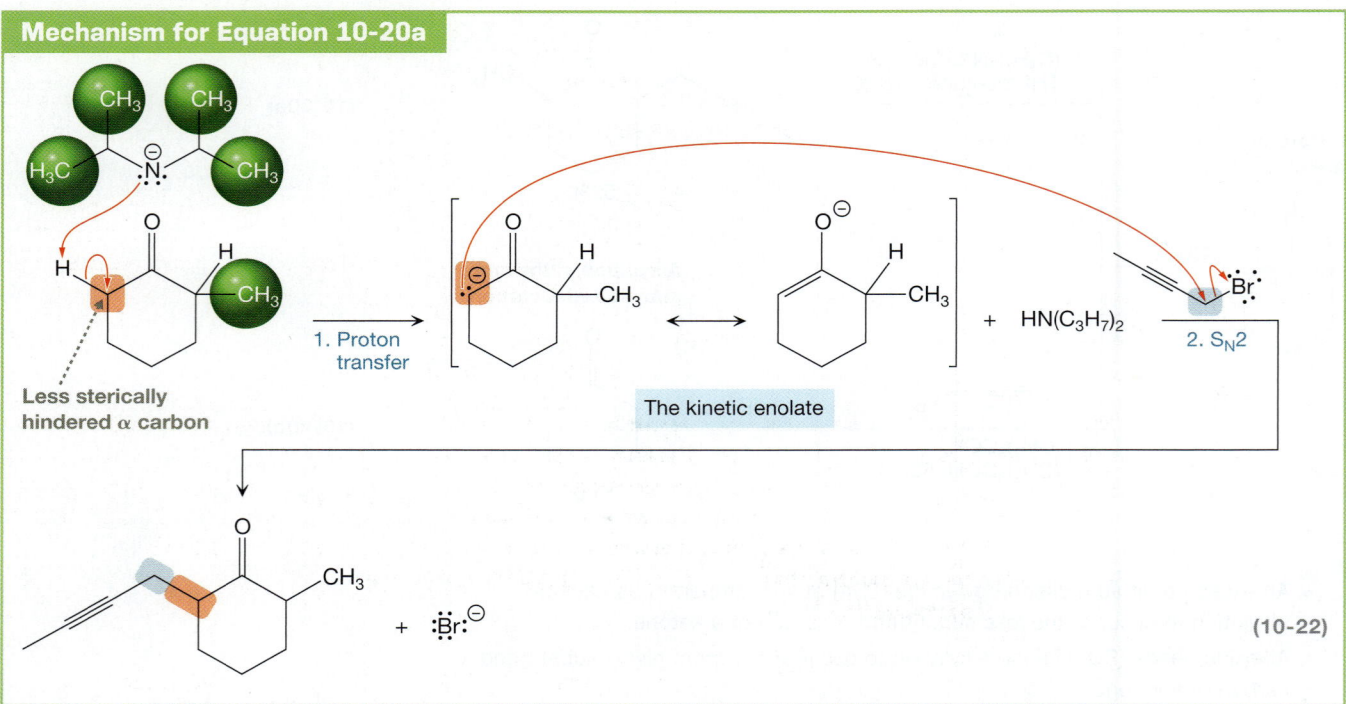

From LDA

Deprotonation is *irreversible*, so it takes place under *kinetic control*.

$\Delta G°_{rxn}$ is very negative.

$$K_{eq} \approx 10^{18}$$
$$\Delta G°_{rxn} \approx -103 \text{ kJ/mol}$$
$$\approx -25 \text{ kcal/mol}$$

(10-21)

$pK_a \approx 20$ $pK_a = 38$

Therefore, applying what we learned from Section 9.12, we can say that deprotonation by LDA is *irreversible* and hence takes place under *kinetic control*. In other words, LDA will predominantly deprotonate the α hydrogen that can be removed the fastest, yielding the **kinetic enolate anion**. As shown in Equation 10-22, that is the proton on the less-substituted α carbon, because LDA, a very bulky base, will encounter less steric hindrance there.

Mechanism for Equation 10-20a

Less sterically hindered α carbon

1. Proton transfer

The kinetic enolate

+ HN(C₃H₇)₂

2. S_N2

+ :Br:⁻

(10-22)

When $(CH_3)_3CO^-$ is used as the base, K_{eq} for the proton transfer step is somewhat <1, making $\Delta G°_{rxn}$ slightly positive (Equation 10-23). The reaction is *reversible*, therefore, so it takes place under *thermodynamic control* and an *equilibrium* is established. If two distinct enolate anions can be produced, then the more stable of the two—called the **thermodynamic enolate anion**—will have greater abundance prior to the S_N2 step.

Deprotonation is *reversible*, so it takes place under *thermodynamic control*.

$\Delta G°_{rxn}$ is slightly positive.

$$K_{eq} \approx 10^{-1}$$
$$\Delta G°_{rxn} \approx +6 \text{ kJ/mol}$$
$$\approx +1.4 \text{ kcal/mol}$$

(10-23)

$pK_a \approx 20$ $pK_a = 19$

To identify the thermodynamic enolate anion, let's examine both enolate anions that can be formed.

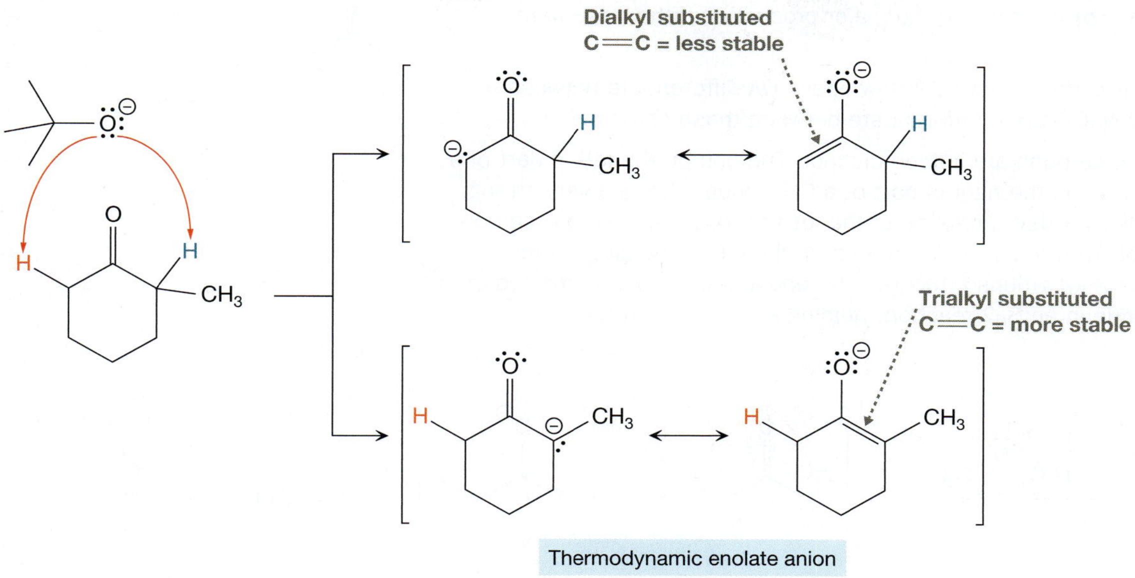

Dialkyl substituted C=C = less stable

(10-24a)

Trialkyl substituted C=C = more stable

(10-24b)

Thermodynamic enolate anion

Both enolate anions stabilize the negative charge roughly equally because the negative charge in each case is resonance-delocalized over a C atom and an O atom. The difference is in the stability contributed by the C=C double bond in the respective resonance forms in Equation 10-24. The C=C double bond in Equation 10-24a is dialkyl substituted, whereas that in Equation 10-24b is trialkyl substituted, so the enolate anion in Equation 10-24b is more stable (Section 5.9), making it the *thermodynamic enolate anion.*

Equation 10-25 shows the mechanism for the reaction in Equation 10-20b. In Step 1, $(CH_3)_3CO^-$ deprotonates the ketone reversibly to produce the thermodynamic enolate anion. In Step 2, the thermodynamic enolate anion displaces Br^- from the alkyl halide in an S_N2 reaction.

Mechanism for Equation 10-20b

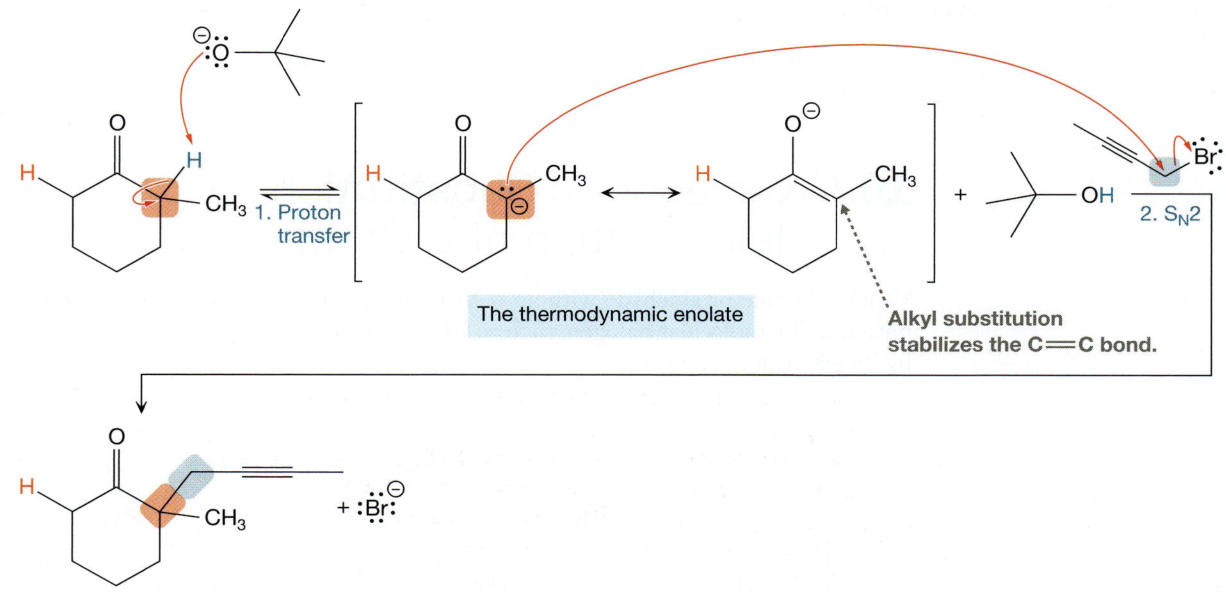

1. Proton transfer

The thermodynamic enolate

Alkyl substitution stabilizes the C=C bond.

2. S_N2

(10-25)

(a) Predict the major product for the reaction shown here and draw the complete, detailed mechanism. **(b)** What would the major product be if the base were $NaOC(CH_3)_3$ instead?

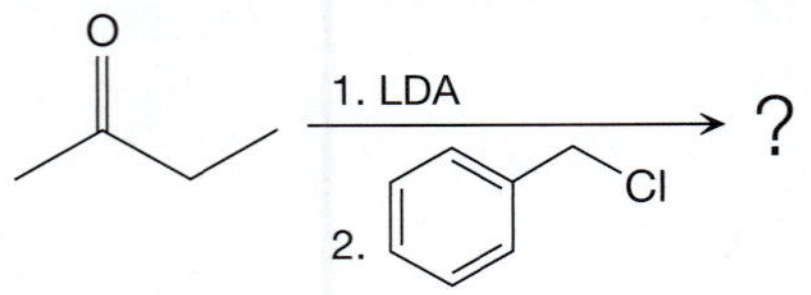

Think Are the two α carbons distinct? How does LDA differentiate between α carbons? How does $NaOC(CH_3)_3$ differentiate between those carbons?

Solve (a) The two α carbons are indeed distinct. The one on the left is part of a CH_3 group, and the one on the right is part of a CH_2 group. LDA is a very strong, bulky base, so it will favor deprotonation of the less sterically hindered α carbon under kinetic control. That α carbon is the one on the left in the given ketone because it has fewer alkyl groups attached. The enolate anion that is formed then attacks benzyl chloride in an S_N2 reaction, yielding an alkylated ketone.

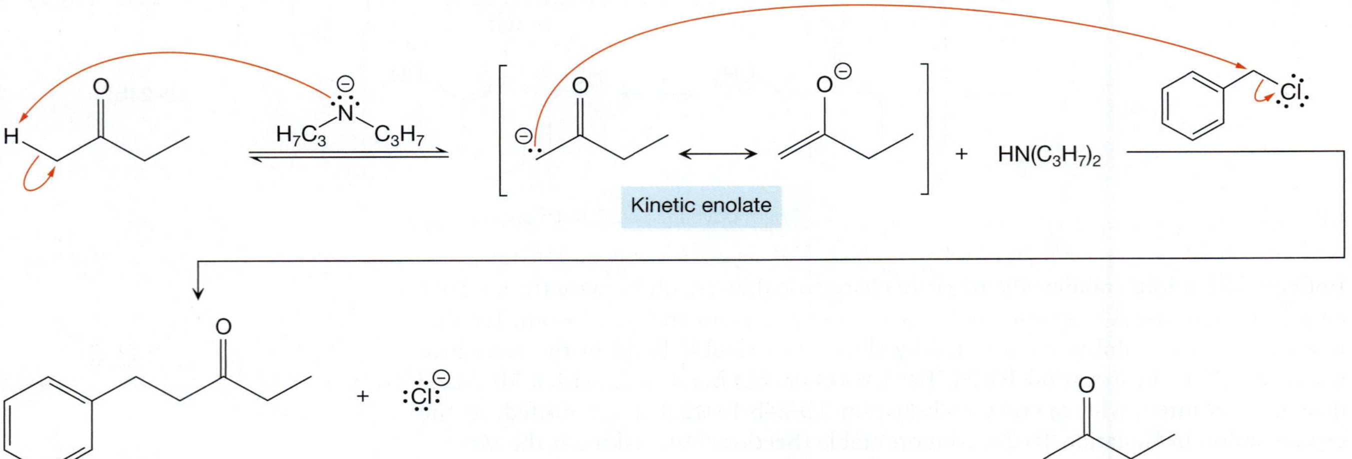

Kinetic enolate

(b) If $NaOC(CH_3)_3$ were used as the base instead, then the α carbon to the right of the C=O group would be deprotonated reversibly to yield the more stable (thermodynamic) enolate anion. Therefore, alkylation would take place at that α carbon yielding the product at the right.

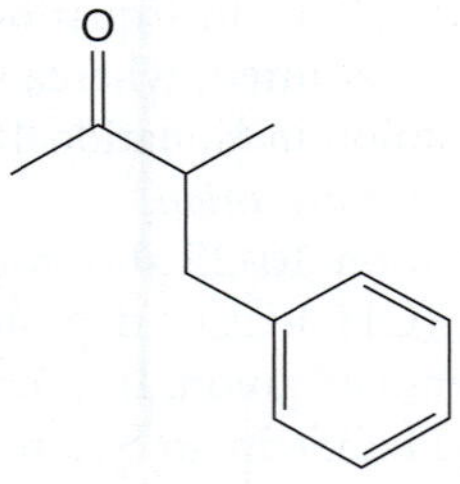

PROBLEM 10.10 (a) Show the complete mechanism for this reaction and predict the major product. **(b)** Do the same for the reaction in which $KOC(CH_3)_3$ is used as the base instead of LDA.

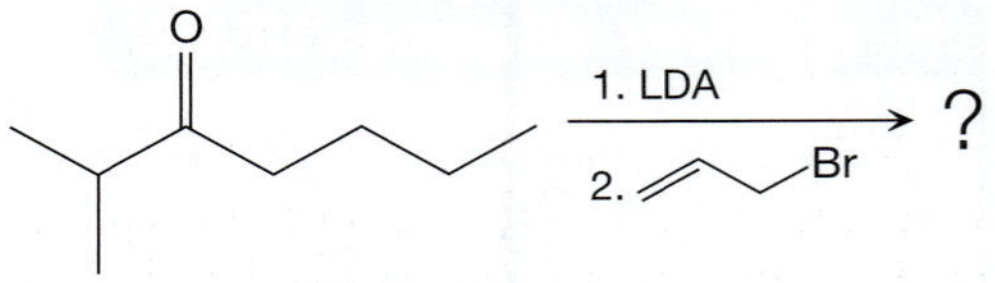

10.4 Nucleophilic Substitution: Halogenation of α Carbons

When a ketone or aldehyde with an α hydrogen is treated with a strong base in the presence of a molecular halogen, such as Cl_2, Br_2, or I_2, a halogen atom replaces the α hydrogen (Equation 10-26).

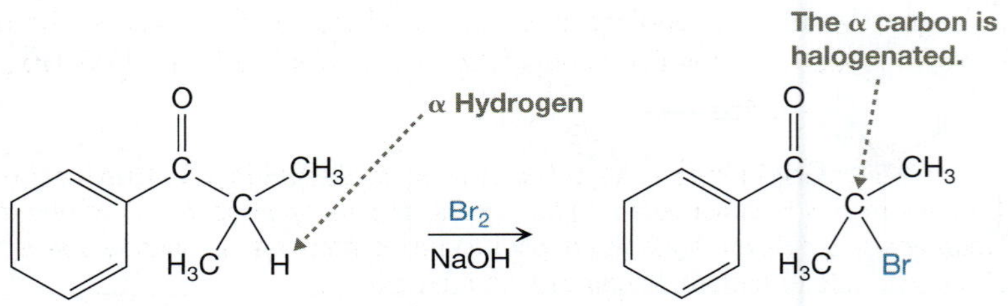

(10-26)

The mechanism for this **halogenation** reaction is shown in Equation 10-27.

Mechanism for Equation 10-26

(10-27)

It is essentially the same as the mechanism for *alkylation* at an α carbon (Section 10.3). In Step 1, the base deprotonates the α carbon, producing a nucleophilic enolate anion. In Step 2, the enolate anion attacks the molecular halogen in an S$_N$2 reaction.

YOUR TURN 10.6

The reaction shown here is nearly identical to that in Equation 10-27. The only difference is that I$_2$ is the molecular halogen instead of Br$_2$. Draw its complete, detailed mechanism, including any curved arrows.

$$\xrightarrow[\text{NaOH}]{\text{I}_2}$$

Why can a molecular halogen species act as a substrate in a nucleophilic substitution reaction, given that it does not appear to possess an electron-poor site that a nucleophilic enolate anion would seek? The halogen molecule is nonpolar, and the three lone pairs of electrons surrounding each halogen atom may even suggest that the substrate is electron rich. However:

Halogens like Cl$_2$, Br$_2$, and I$_2$ can behave as substrates in S$_N$2 reactions in part because of their *polarizability*.

Cl$_2$, Br$_2$, and I$_2$ are relatively highly *polarizable* due to their abundance of electrons, especially nonbonded electrons (review Section 2.6d). Consequently, when an electron-rich species like the enolate anion approaches, the electrons on the halogen molecule are repelled quite readily, generating a substantial *induced dipole* (Fig. 10-4).

FIGURE 10-4 Molecular halogens as substrates in nucleophilic substitution (a) An isolated bromine molecule is nonpolar. (b) As the nucleophile approaches, electron repulsion forces electron density around Br$_2$ to the opposite side, generating an induced dipole moment. The near side, therefore, becomes electron poor.

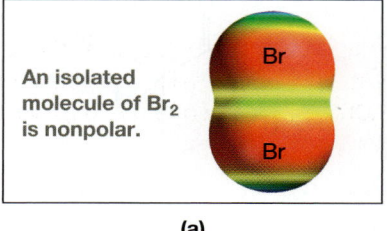

An isolated molecule of Br$_2$ is nonpolar.

(a)

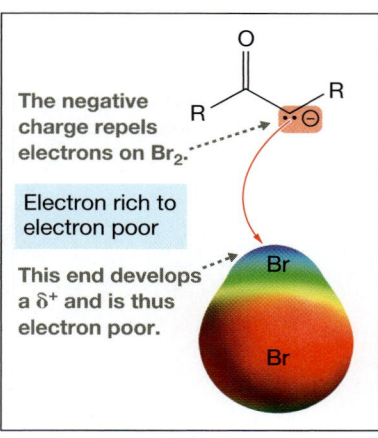

The negative charge repels electrons on Br$_2$.

Electron rich to electron poor

This end develops a δ$^+$ and is thus electron poor.

(b)

The halogen atom that is nearer the nucleophile, therefore, becomes electron poor. As a result, a curved arrow can be drawn from the electron-rich nucleophile to the electron-poor halogen atom.

Notice in the halogenation shown in Equation 10-27 that the ketone has only a single α hydrogen. The ketone in Equation 10-28 contains two α hydrogens, and *both* α hydrogens are replaced.

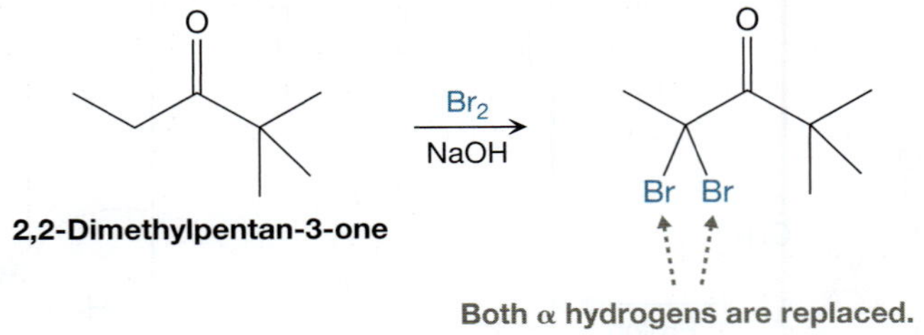

Br2
NaOH
2,2-Dimethylpentan-3-one
Br Br
Both α hydrogens are replaced.

(10-28)

Under *basic* conditions, **polyhalogenation** generally occurs, in which every α hydrogen is replaced by a halogen atom.

In the mechanism for Equation 10-28, which is shown in Equation 10-29, the mechanism for a single halogenation (Equation 10-27) essentially occurs twice.

Mechanism for Equation 10-28

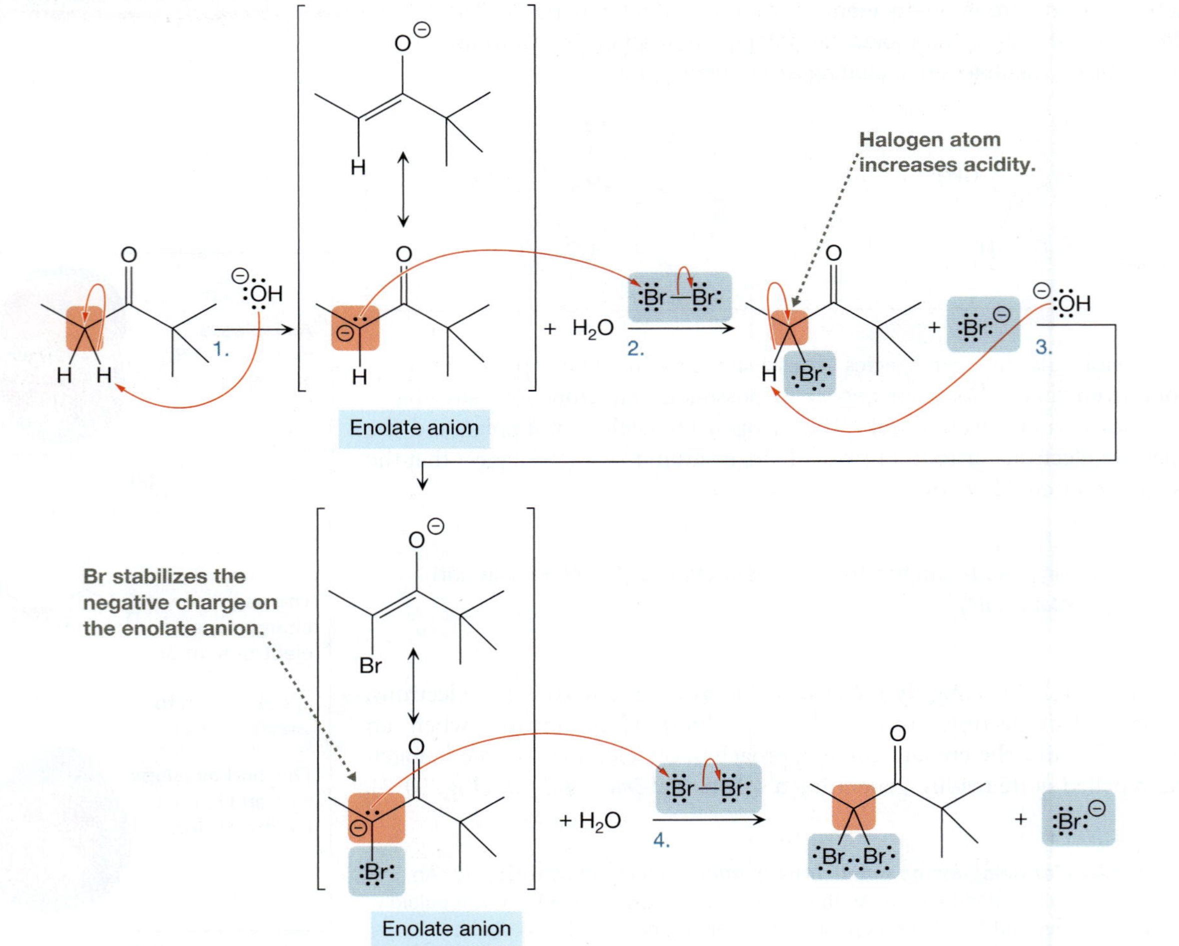

Enolate anion
Halogen atom increases acidity.
+ H2O
1.
2.
3.
4.
Br stabilizes the negative charge on the enolate anion.
Enolate anion
+ H2O

(10-29)

Under each *reaction arrow* in the mechanism in Equation 10-29, label the step as either *proton transfer* or *S$_N$2*.

Halogenation does not stop after a single substitution because the ketone product after the first halogenation is more acidic at the α carbon than the original ketone was. As indicated in Equation 10-29, this greater acidity is due to the increased stability of the halogenated ketone's enolate anion, which, in turn, is a result of the halogen atom's ability to withdraw electron density from the negative charge. Therefore, the second enolate is formed more easily than the first, so *the second halogenation is faster than the first halogenation.*

Which ketone, **A** or **B**, will undergo chlorination faster under basic conditions?

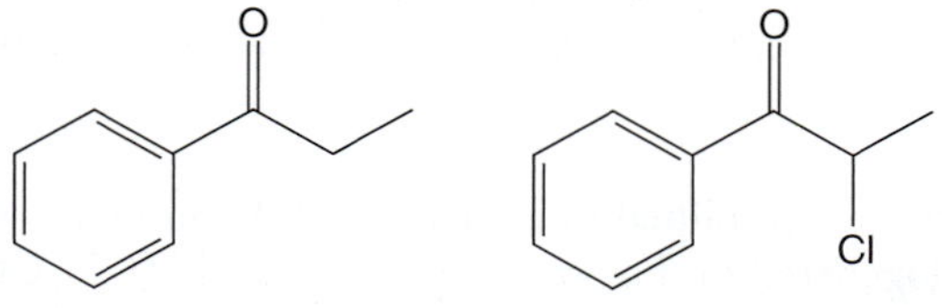

A   **B**

SOLVED PROBLEM 10.11

Draw the complete, detailed mechanism for the chlorination of butanal.

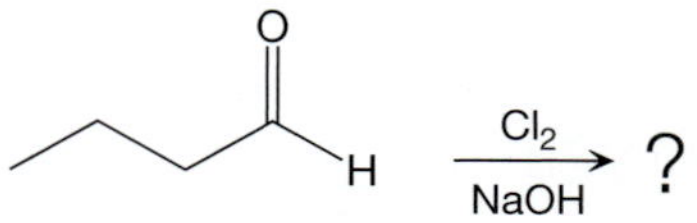

Think In the presence of a strong base, what reaction takes place at an α carbon? How does this affect the chemical properties at that carbon? What species will be attacked as a result? How many times will such a reaction take place at that α carbon?

Butanal

Solve In the presence of a strong base, an α carbon can be deprotonated to generate a nucleophilic C atom. Subsequently, Cl$_2$ will be attacked in an S$_N$2 reaction. Because this reaction takes place under basic conditions, each subsequent halogenation will be faster, so all α hydrogens will be replaced. In this case, there are two.

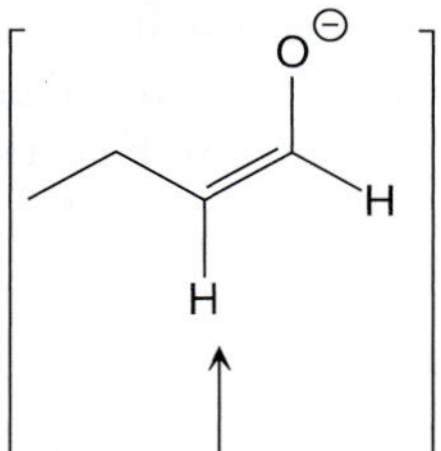

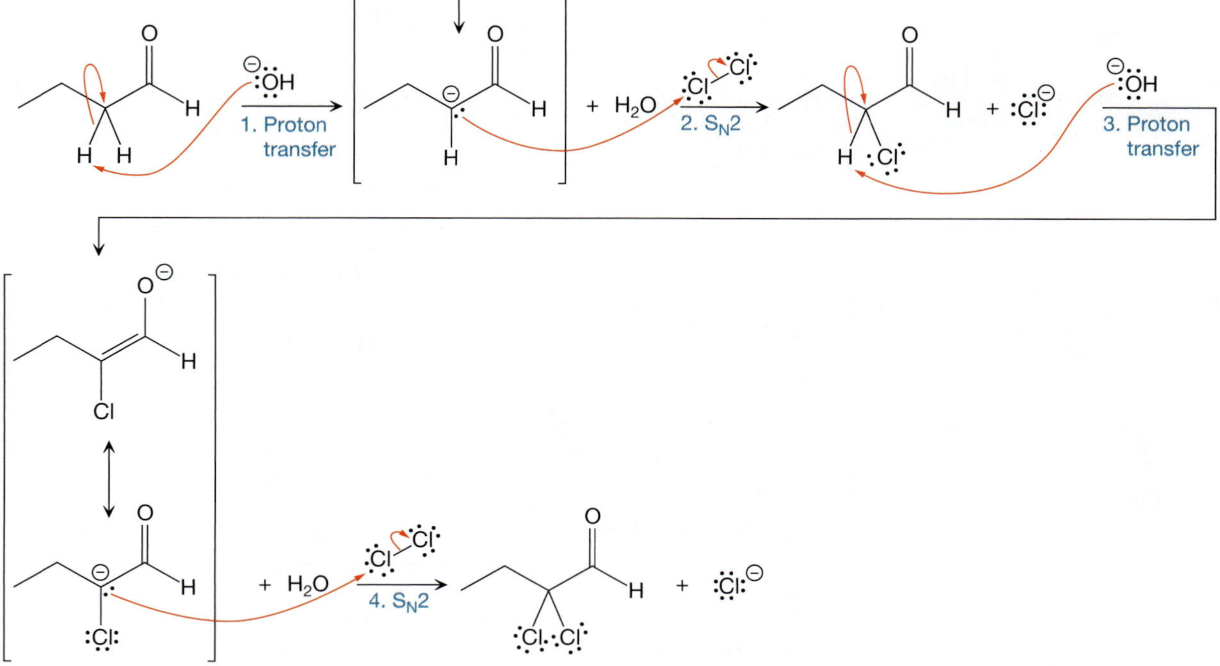

PROBLEM 10.12 Draw the complete mechanism for the iodination of 2,6-dimethylcyclohexanone, which takes place under basic conditions, and predict the major product.

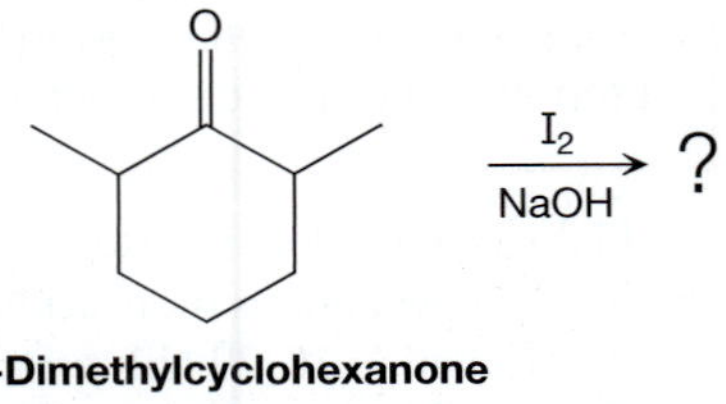

O
I2
NaOH
?
2,6-Dimethylcyclohexanone

If polyhalogenation is *not* desired, then we can change the conditions in which the reaction is carried out.

Under *acidic conditions*, only a single halogenation takes place.

This is exemplified in Equation 10-30, in which the ketone from Equation 10-28 is halogenated in the presence of acetic acid (CH_3CO_2H), abbreviated as HOAc.

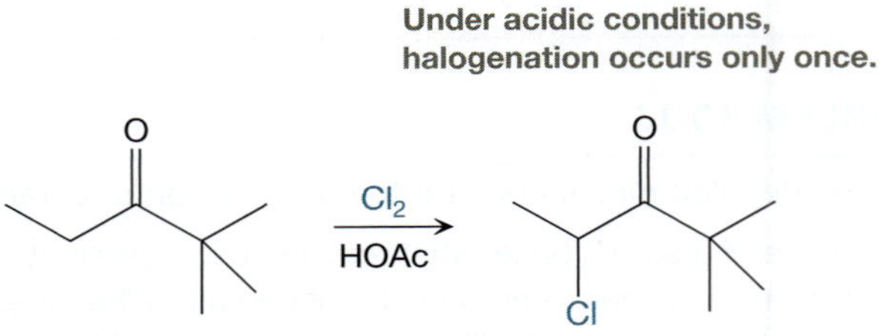

Under acidic conditions, halogenation occurs only once.
O
Cl2
HOAc
O
Cl

(10-30)

Under these conditions, a negatively charged enolate anion cannot exist at any substantial concentration because it is strongly basic (the pK_a of a ketone or aldehyde is ~20). However, the α carbon can still become electron rich via acid-catalyzed keto–enol tautomerization, as shown in Equation 10-31.

Mechanism for Equation 10-30

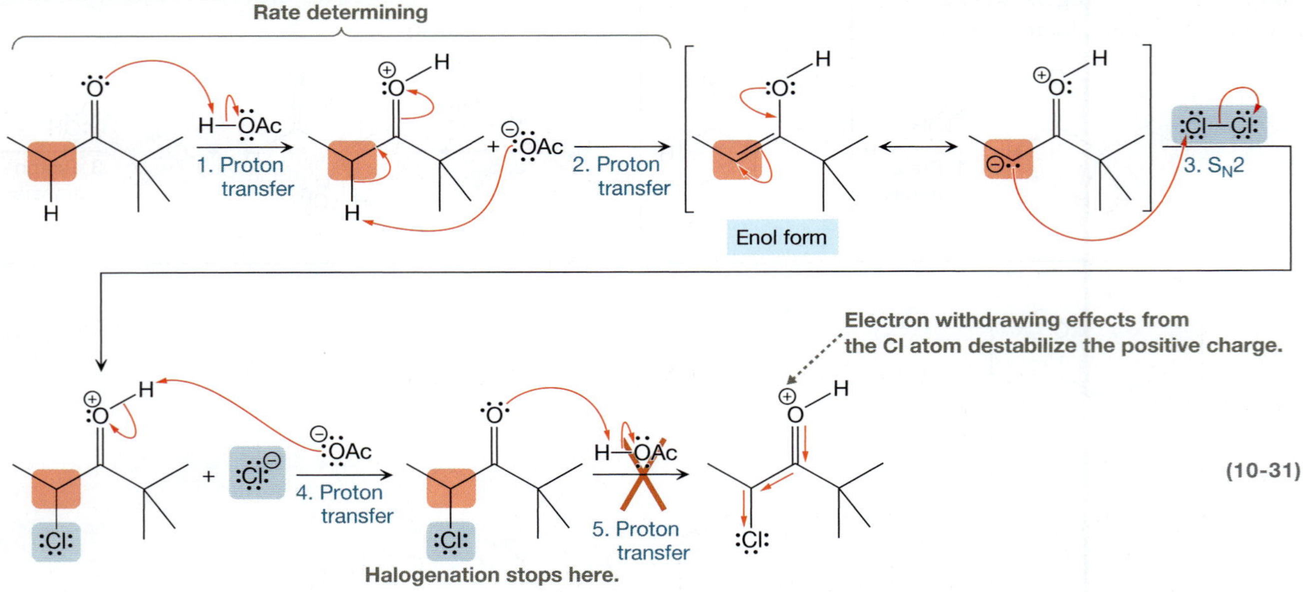

Rate determining
H—ÖAc
1. Proton transfer
H
+ :ÖAc
2. Proton transfer
Enol form
:Cl—Cl:
3. SN2
H
+ :Cl:
:ÖAc
4. Proton transfer
:Cl:
H—ÖAc
5. Proton transfer
Halogenation stops here.
Electron withdrawing effects from the Cl atom destabilize the positive charge.
:Cl:

(10-31)

In the enol form, the α carbon is electron rich, due to the small contribution by the resonance structure in which a negative charge and a lone pair of electrons are located on the α carbon. As before, one end of the halogen molecule becomes electron poor when the enol approaches, thus setting up the *electron-rich to electron-poor* driving force for the subsequent nucleophilic substitution.

The reaction shown here is identical to the one in Equation 10-30, except Br$_2$ takes the place of Cl$_2$. Draw the complete, detailed mechanism for this reaction.

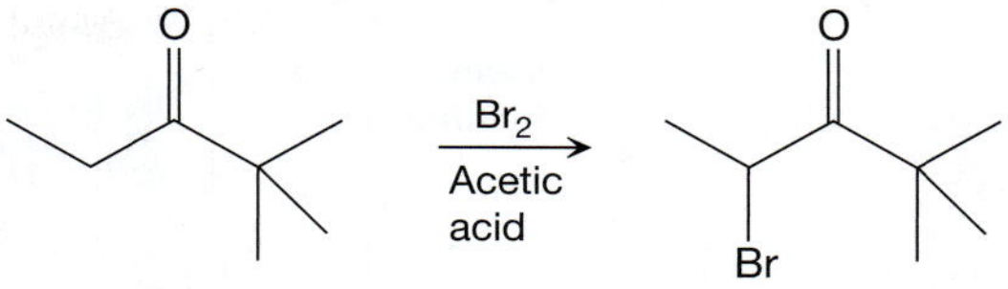

Why does halogenation take place only once under acidic conditions? As indicated in Equation 10-31, formation of the enol—the key nucleophilic species in the mechanism—is rate determining, and it requires protonation of the carbonyl group. The resulting positive charge on O is inductively *destabilized* by the presence of the halogen atom, so the formation of the second enol is more difficult than the formation of the first. Therefore, in contrast to what occurs under basic conditions, halogenation becomes *slower* with each additional halogen.

Which ketone, **A** or **B**, will undergo bromination faster under acidic conditions?

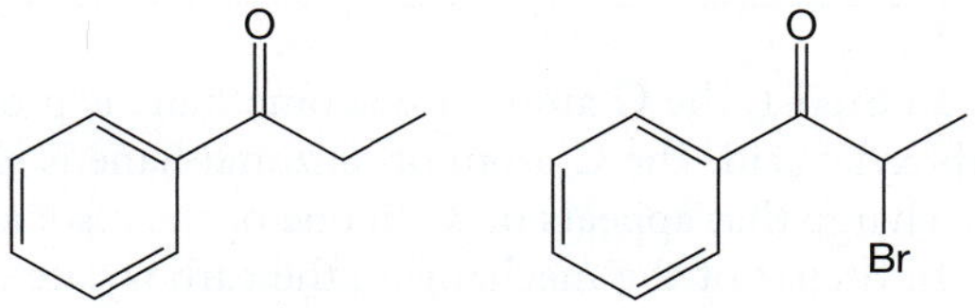

A

B

PROBLEM 10.13 Draw the complete, detailed mechanism for the following iodination under acidic conditions and predict the major product.

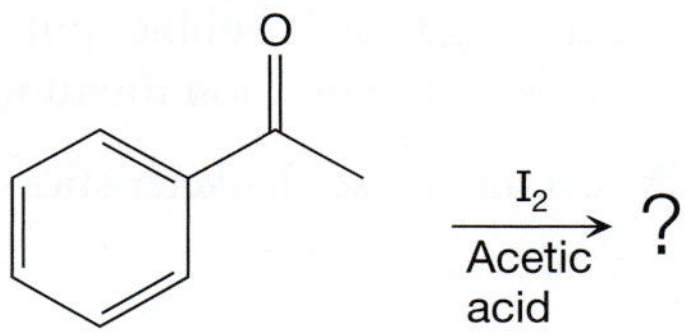

$$\xrightarrow[\text{Acetic acid}]{I_2} \ ?$$

10.5 Nucleophilic Substitution: Diazomethane Formation of Methyl Esters

When a carboxylic acid is treated with **diazomethane** (CH$_2$N$_2$), the acidic H atom is replaced by a CH$_3$ group to produce a methyl ester:

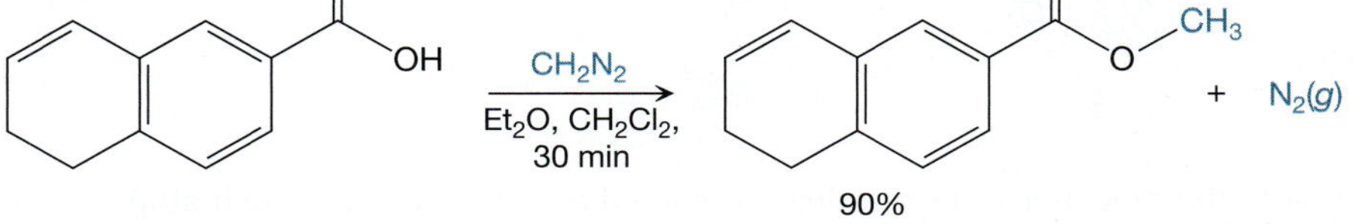

$$\xrightarrow[\substack{\text{Et}_2\text{O, CH}_2\text{Cl}_2, \\ 30 \text{ min}}]{\text{CH}_2\text{N}_2} \quad + \ N_2(g) \quad \textbf{(10-32)}$$

90%

The mechanism for this reaction consists of a proton transfer followed by a nucleophilic substitution, as shown in Equation 10-33.

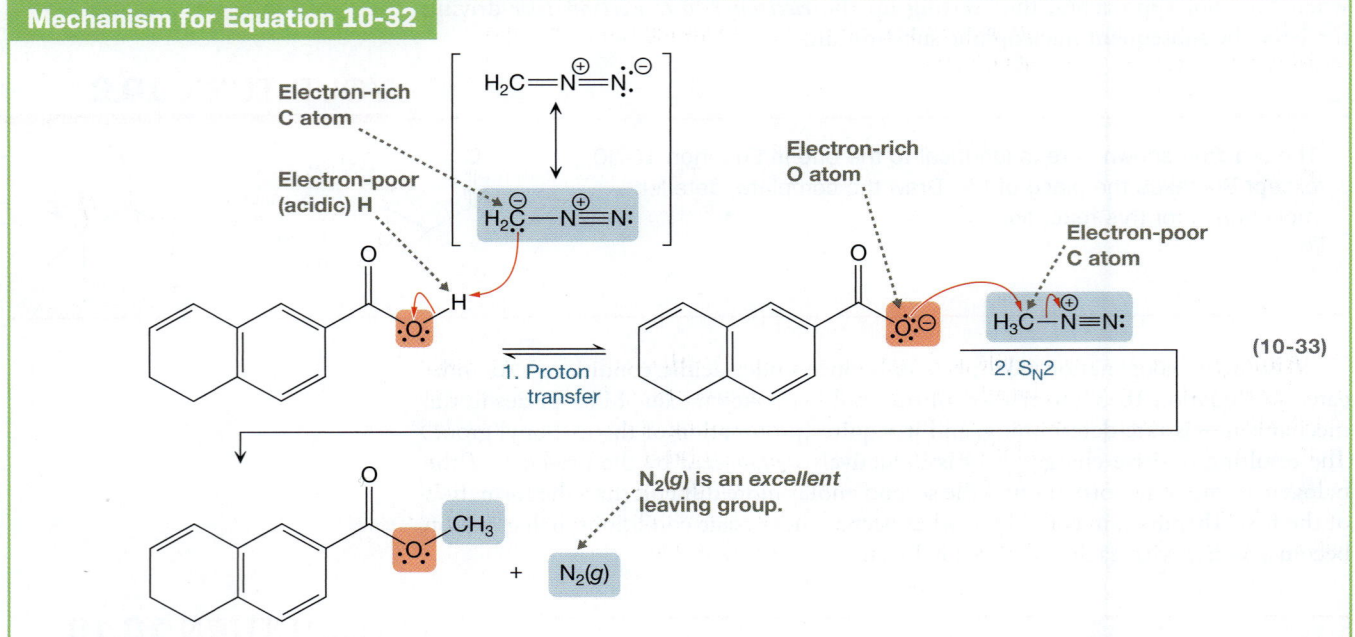

Electron-rich C atom

Electron-poor (acidic) H

Electron-rich O atom

Electron-poor C atom

1. Proton transfer

2. S_N2

(10-33)

$N_2(g)$ is an *excellent* leaving group.

$+$ $N_2(g)$

In Step 1, the C atom of diazomethane is protonated by the acidic proton of the carboxylic acid. The C atom of diazomethane is electron rich, indicated by the negative charge that appears on C in one of the resonance structures.

In Step 2 of the mechanism, the carboxylate anion (RCO_2^-) is electron rich, and acts as the nucleophile in the subsequent S_N2 reaction. The protonated form of diazomethane is electron poor and acts as the substrate. The leaving group is nitrogen gas, which is an *excellent* leaving group for two reasons:

1. $N_2(g)$ is extremely stable—one of the most inert compounds known.
2. It is a gas, so it bubbles out of solution as it is formed, which permanently removes it from the reaction mixture and drives the reaction to completion.

As a result, these characteristics of the $N_2(g)$ leaving group make the reaction *irreversible*.

YOUR TURN 10.11

Diazomethane can be used to convert acetic acid into methyl acetate:

$$CH_2N_2$$

Ethanoic acid (Acetic acid)

Methyl ethanoate (Methyl acetate)

The mechanism for this reaction is as follows, with the curved arrows omitted:

$+$ $H_3C-N\equiv N:$ $\longrightarrow$ $+$ $N_2(g)$

Complete the mechanism by drawing the curved arrows, and label each step under its corresponding reaction arrow as either *proton transfer* or S_N2.

PROBLEM 10.14 Predict the products of each of the following reactions and provide complete, detailed mechanisms:

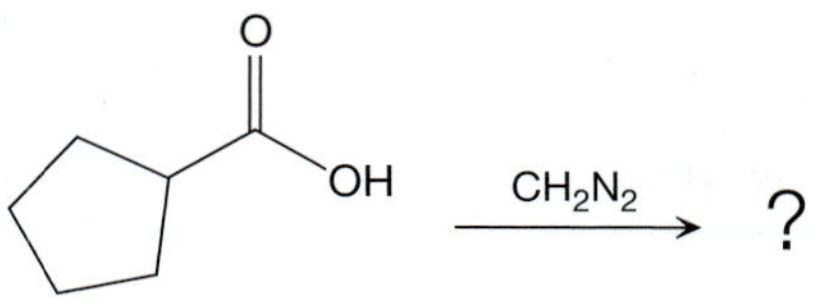

(a)

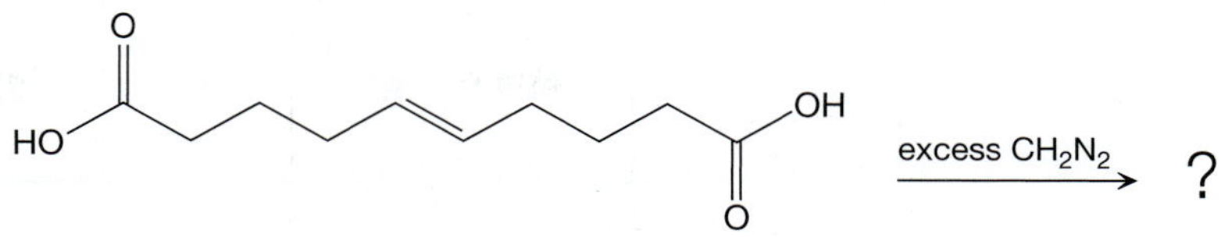

(b)

Although turning a carboxylic acid into a methyl ester with CH_2N_2 is a very clean reaction on paper, care must be taken whenever using diazomethane in the lab because it is toxic and explosive! As a result, other methods for forming methyl esters are often much more attractive. Trimethylsilyldiazomethane [$(CH_3)_3SiCHN_2$], for example, is a less explosive alternative to CH_2N_2. And the *Fischer esterification reaction*, presented in detail in Chapter 21, is a much milder, safer, and altogether different reaction that allows us to produce a wide variety of esters from carboxylic acids.

10.6 Nucleophilic Substitution: Formation of Ethers and Epoxides

One of the most convenient ways of forming an ether (R—O—R′) is via the **Williamson ether synthesis**, in which an alkyl halide is treated with a salt of an alkoxide anion. Examples are shown in Equations 10-34 and 10-35.

Williamson ether synthesis

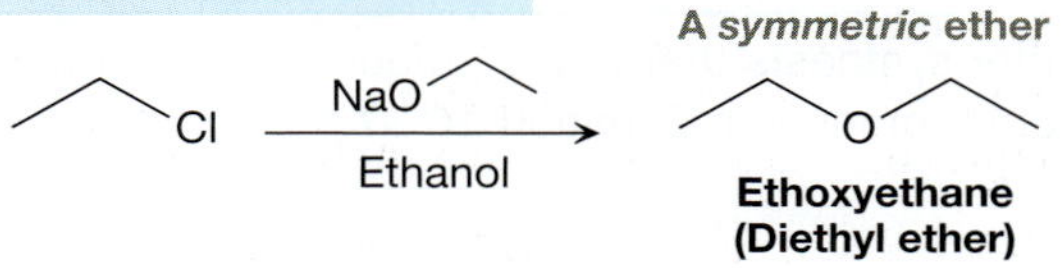

(10-34)

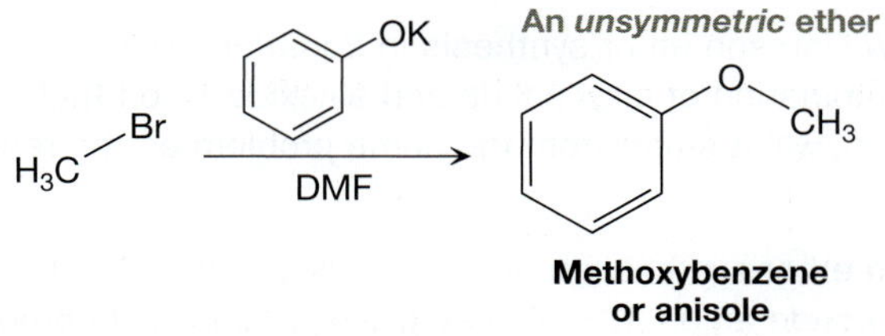

(10-35)

The Williamson ether synthesis makes it possible to synthesize a wide variety of ethers. In particular:

> The *Williamson ether synthesis* can be used to synthesize both *symmetric* ethers (in which the alkyl groups bonded to O are the same) and *unsymmetric* ethers (in which the alkyl groups bonded to O are different).

In solution, the alkoxide salts dissolve to form alkoxide anions (RO^-), which are strong nucleophiles. In the presence of an alkyl halide, then, an S_N2 reaction takes place, as shown in Equation 10-36.

X = Cl, Br, I

$$R \overset{\ominus}{\underset{..}{\text{O}}} \quad R' \!-\! X \quad \xrightarrow{\;S_N2\;} \quad R \overset{\text{O}}{\underset{}{}} R' \;+\; X^{\ominus} \qquad (10\text{-}36)$$

An ether

As in any S_N2 reaction, the reactivity of the substrate is very important in the success of a Williamson ether synthesis. For example, reaction of sodium ethoxide with 2-bromo-2-methylpropane (Equation 10-37) yields essentially no ether product. The C atom bonded to the leaving group is tertiary, and, as we learned in Section 9.6, tertiary substrates have excessive steric hindrance and are not susceptible to nucleophilic attack in an S_N2 reaction. Instead, the major product is methylpropene, which is the product of E2 elimination. (Recall from Section 9.6 that steric hindrance plays a much smaller role in E2 reactions.)

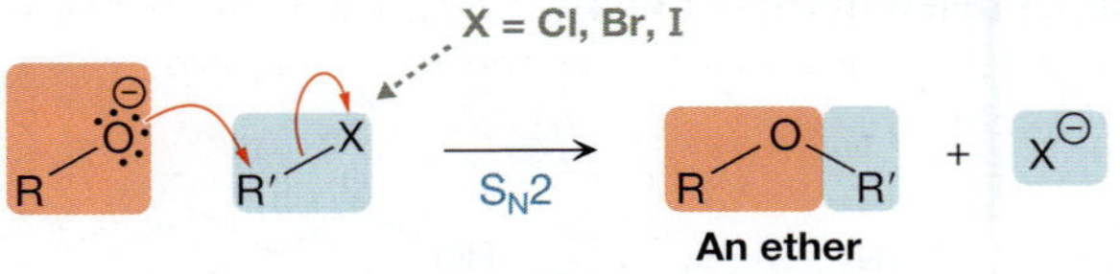

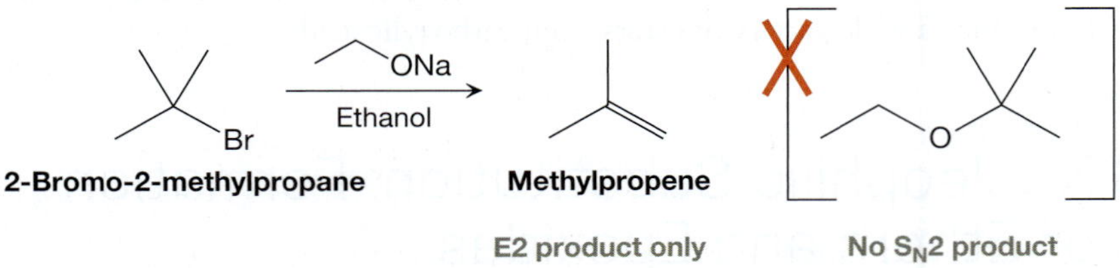

2-Bromo-2-methylpropane   **Methylpropene**   (10-37)

E2 product only   No S_N2 product

Fortunately, there are two possible choices for the Williamson synthesis of an unsymmetric ether, which differ by the alkyl groups that make up the substrate and alkoxide. If one choice is unfeasible, the other might be feasible (see Solved Problem 10.15).

SOLVED PROBLEM 10.15

Devise a Williamson ether synthesis that would produce the ether shown here, which is the intended ether product in Equation 10-37.

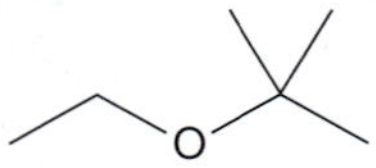

Think Why is the Williamson ether synthesis in Equation 10-37 unfeasible? Is there a different combination of alkyl halide and alkoxide anion that would produce the same ether? If so, will it suffer from the same problem as the reaction in Equation 10-37?

Solve A Williamson ether synthesis requires an alkyl halide (R—X) as the substrate and an alkoxide anion (R'—O⁻) as the nucleophile. To produce an unsymmetric ether like the one desired here, there are two such combinations, which differ by the choice of R and R'. In Equation 10-37, R is the *tert*-butyl group and R' is the ethyl group, but the reaction is unfeasible because the substrate is tertiary. The other combination has R as the ethyl group and R' as the *tert*-butyl group. Now the substrate is primary, so the S_N2 reaction is feasible.

This is a primary carbon, so the S_N2 reaction is feasible.

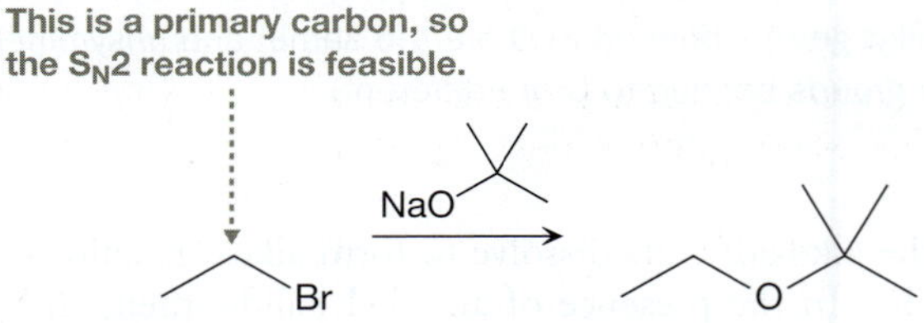

PROBLEM 10.16 Determine which of the following ethers can be produced from a Williamson ether synthesis. For those that can, show the synthesis that would be successful. If there are two feasible syntheses for a particular ether, determine which one is preferable.

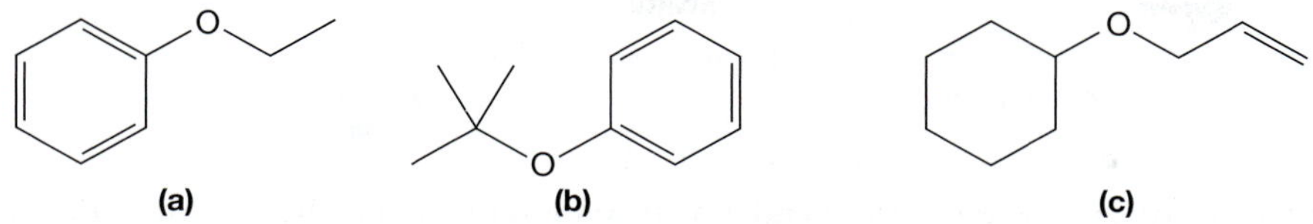

(a) (b) (c)

As we learned in Section 9.11, an S_N2 reaction between an alkoxide anion and an alkyl halide can also take place *intramolecularly*, yielding cyclic ethers. For example, treating 4-bromobutan-1-ol with a strong base produces tetrahydrofuran (THF) (Equation 10-38):

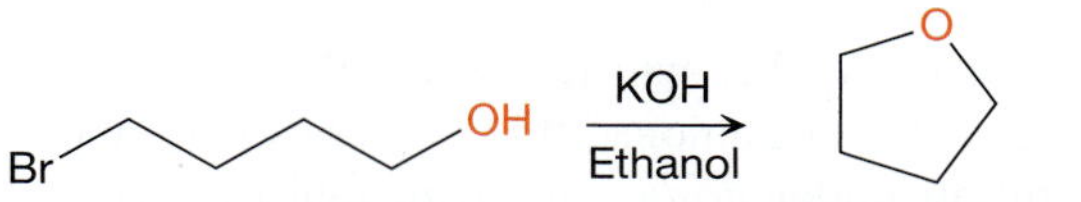

4-Bromobutan-1-ol   **Tetrahydrofuran (THF)**   (10-38)

Equation 10-38 proceeds by a two-step mechanism, as shown in Equation 10-39. In Step 1, the hydroxyl group is deprotonated to produce an alkoxide nucleophile. Step 2 is an intramolecular S_N2 reaction. In this case, the reaction is favored in large part by the formation of a five-membered ring, which, as we learned in Section 9.11, tends to favor intramolecular reactions over intermolecular ones.

Mechanism for Equation 10-38

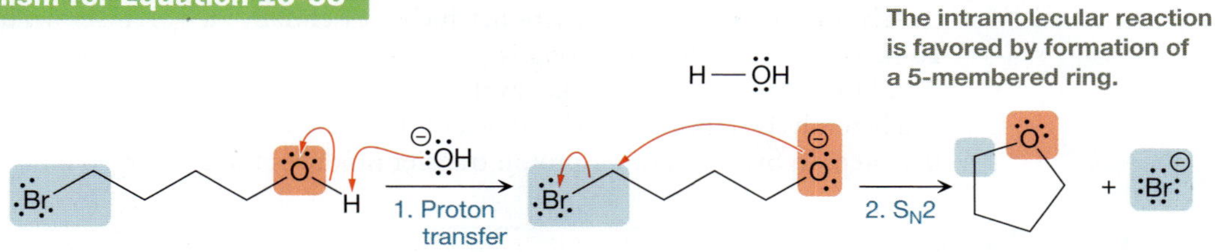

The intramolecular reaction is favored by formation of a 5-membered ring.

(10-39)

YOUR TURN 10.12

Tetrahydropyran (THP) can be made with the reaction shown here, which is analogous to Equation 10-38. If the mechanism mimics the one in Equation 10-39, draw the complete, detailed mechanism for this reaction.

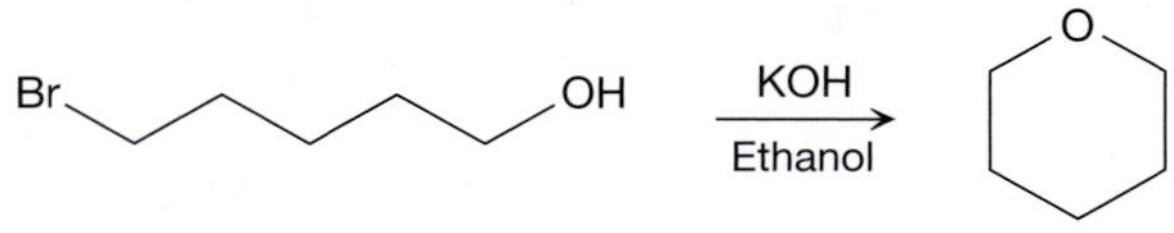

Tetrahydropyran (THP)

PROBLEM 10.17 Draw the complete, detailed mechanism for the following reaction and predict the major product.

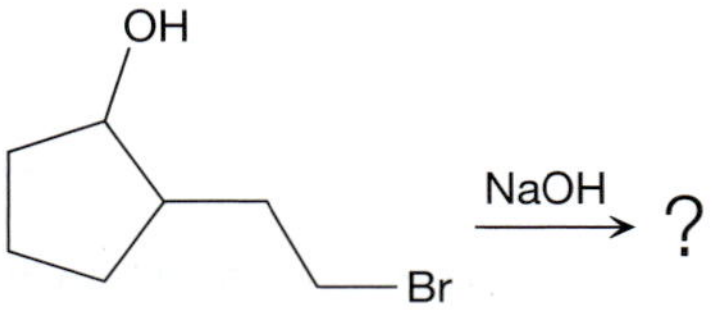

Epoxides—three-membered ring ethers—can be made in a similar fashion using **halohydrins**, as shown in Equation 10-40. (Reactions of epoxides are discussed in Section 10.7.)

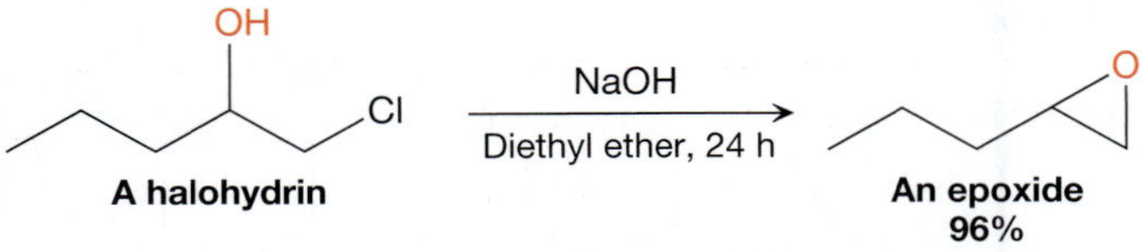

A halohydrin → NaOH, Diethyl ether, 24 h → **An epoxide** 96% (10-40)

In a halohydrin, a halogen atom and a hydroxyl group are on adjacent carbon atoms. (Halohydrins are typically made via *electrophilic addition* to an alkene, described in Chapter 12.)

YOUR TURN 10.13

Draw the complete, detailed mechanism for the reaction in Equation 10-40.

Whereas the Williamson ether synthesis takes place under *basic* conditions (indicated by the presence of alkoxide anions, RO^-), ether formation can also take place via nucleophilic substitution under *acidic conditions*. Diisopropyl ether, for example, is formed by heating propan-2-ol in the presence of H_2SO_4 (Equation 10-41).

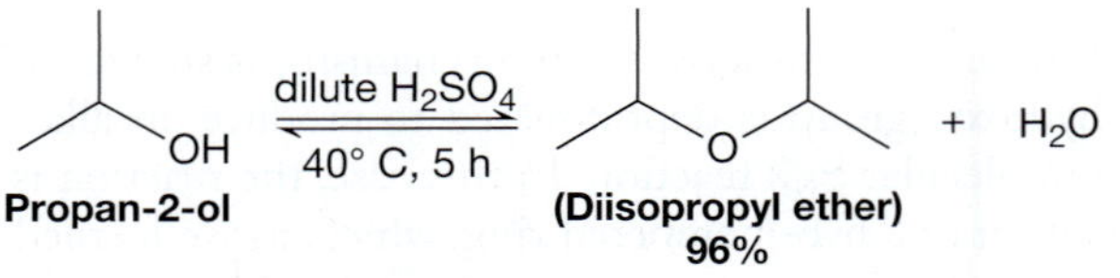

Propan-2-ol → dilute H_2SO_4, 40° C, 5 h → **(Diisopropyl ether)** 96% + H_2O (10-41)

Equation 10-41 is an example of a *dehydration* reaction (review Section 9.5b), so called because water is lost in the process.

The mechanism for this reaction, which is essentially an S_N1, is shown in Equation 10-42. In Step 1, the OH group is protonated to become a very good leaving group (i.e., H_2O). Steps 2 and 3 make up the two-step S_N1 reaction—namely, water leaves in a heterolysis step, followed by attack of the nucleophile, propan-2-ol, in a coordination step. In Step 4, another proton transfer produces the uncharged ether.

Mechanism for Equation 10-41

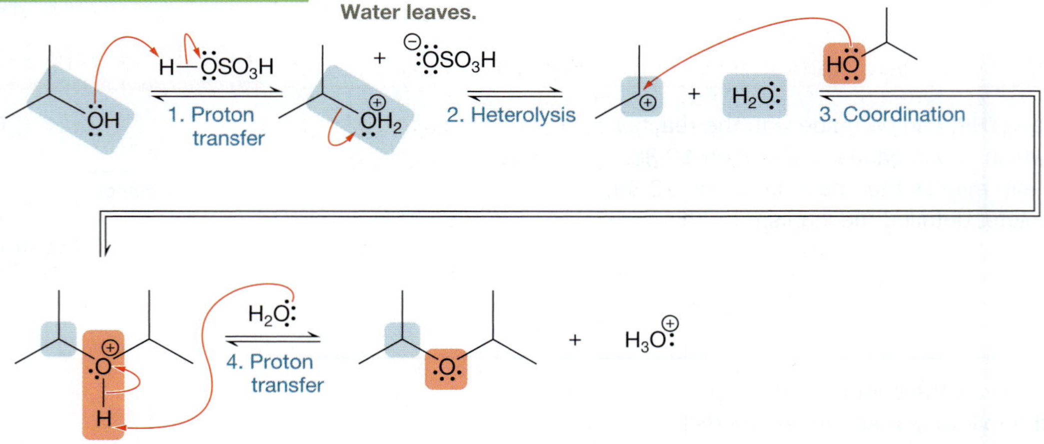

Water leaves.

1. Proton transfer
2. Heterolysis
3. Coordination
4. Proton transfer

(10-42)

Propan-1-ol also forms an ether under acidic conditions (Equation 10-43), but the reaction proceeds through an S_N2 reaction instead of an S_N1 (see Your Turn 10.14).

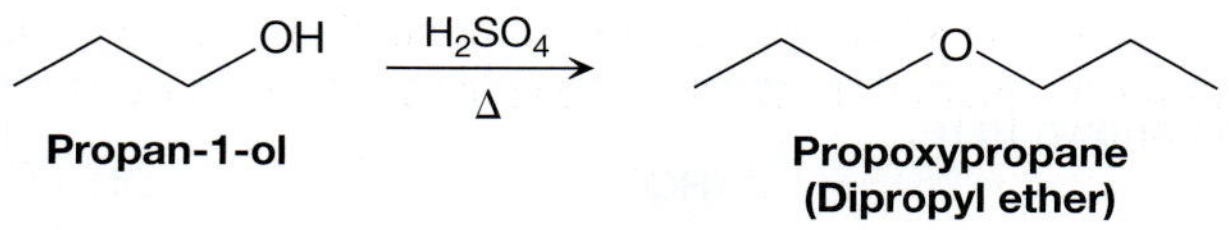

Propan-1-ol → Propoxypropane (Dipropyl ether) (10-43)

Despite the very good leaving group (H_2O), primary carbocations are too unstable for the leaving group to leave on its own.

The mechanism for the reaction in Equation 10-43 is as follows:

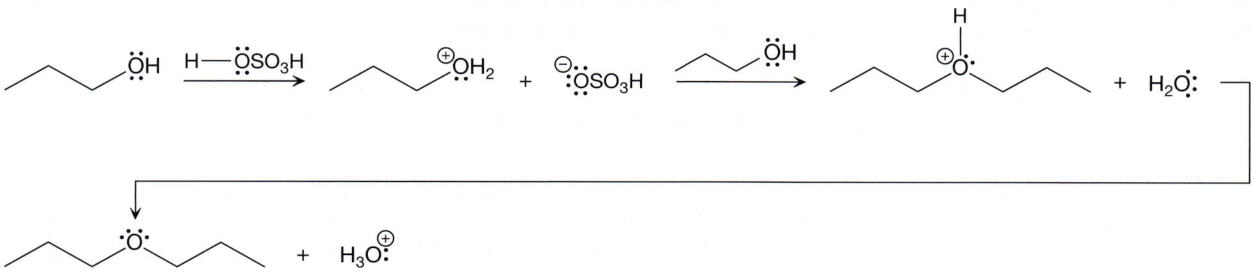

Supply the appropriate curved arrows, and below each reaction arrow label the step as *proton transfer* or S_N2.

PROBLEM 10.18 Draw the complete, detailed mechanism for the reaction shown here and predict the major product.

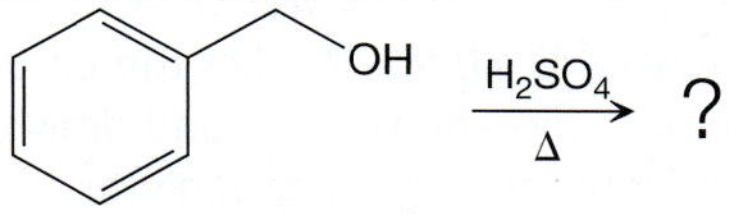

Whereas the Williamson ether synthesis can be used to make either symmetric or unsymmetric ethers, the formation of an ether via dehydration is more restrictive.

> The synthesis of an ether via the dehydration of an alcohol is useful only when the target is a *symmetric* ether.

If you try to synthesize an unsymmetric ether via dehydration, a mixture of ethers is produced, since the reactants provide two different potential nucleophiles and two different potential substrates. For example, if propan-1-ol is treated with propan-2-ol under acidic conditions, the three ethers shown in Equation 10-44 are produced.

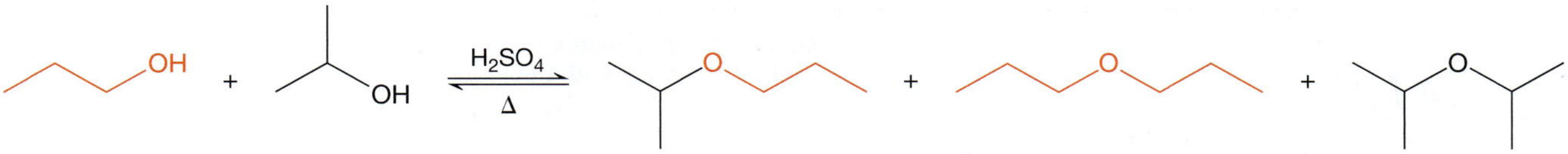

(10-44)

PROBLEM 10.19 Mechanisms for the formation of the two symmetric ethers in Equation 10-44 (i.e., the second and third products) have been shown previously. Draw a complete, detailed mechanism for the formation of the unsymmetric ether.

Predict the ether that would be formed from the reaction shown here, and then draw the complete, detailed mechanism.

HO⁓⁓⁓OH $\xrightarrow[\Delta]{H_2SO_4}$ **?**

Think In a substitution reaction, what could act as the nucleophile? What could act as the substrate? Is an S_N1 or S_N2 mechanism favored? Can an intramolecular reaction take place? If so, is it more favorable or less favorable than the corresponding intermolecular reaction?

Solve The acidic conditions cause an OH group to be protonated, generating a good leaving group (H_2O). The second OH group in the molecule can act as a nucleophile in an intramolecular S_N2 reaction. The intramolecular reaction is favored over the intermolecular one, due to the formation of the six-membered ring.

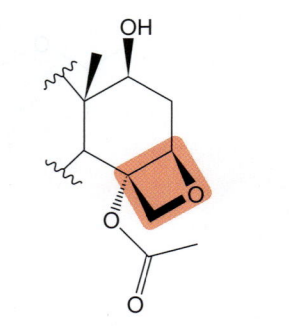

PROBLEM 10.21 Predict the product of the reaction shown here and draw the complete, detailed mechanism.

$\xrightarrow[\Delta]{H_2SO_4}$ **?**

10.7 Nucleophilic Substitution: Epoxides and Oxetanes as Substrates

In Section 9.5, we learned that ethers (R—O—R′) are resistant to nucleophilic substitution and elimination reactions under normal conditions because the leaving group would have to be an alkoxide anion (RO⁻), a very poor leaving group. Thus, if diethyl ether is treated with a strong nucleophile such as the methoxide anion (CH₃O⁻), then no reaction takes place, as shown in Equation 10-45.

Under normal conditions, ethers tend not to react with nucleophiles.

$\xrightarrow{NaOCH_3}$ No reaction   (10-45)

An exception arises with ethers that are part of small rings, such as oxirane and other **epoxides** (three-membered ring ethers), as well as **oxetanes** (four-membered ring ethers) (Fig. 10-5). Because epoxides have the smaller ring size, they are less stable and therefore react with a greater variety of nucleophiles than oxetanes do. Thus, we spend the bulk of this section discussing the reactivity of epoxides. We first discuss their reactivity under neutral and basic conditions (Section 10.7a), and then under acidic conditions (Section 10.7b). In each of these sections, we examine not only the mechanisms by which these types of compounds react, but also aspects of *stereochemistry* and *regiochemistry*. Finally, we conclude this section by introducing some of the reactions that oxetanes undergo.

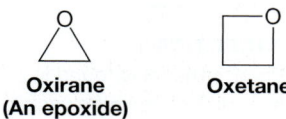

Oxirane
(An epoxide)

Oxetane

10.7a Reactions of Epoxides under Neutral and Basic Conditions

Equations 10-46 and 10-47 show that oxirane reacts readily under neutral and basic conditions, respectively.

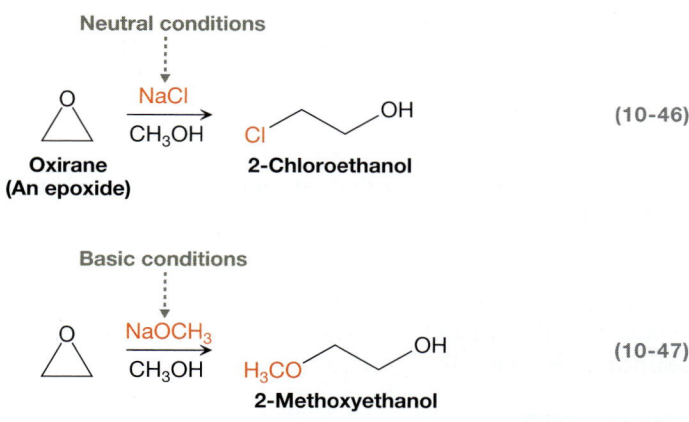

Neutral conditions

$$\text{Oxirane} \xrightarrow[\text{CH}_3\text{OH}]{\text{NaCl}} \text{2-Chloroethanol} \qquad (10\text{-}46)$$

Oxirane (An epoxide)

2-Chloroethanol

Basic conditions

$$\xrightarrow[\text{CH}_3\text{OH}]{\text{NaOCH}_3} \qquad (10\text{-}47)$$

2-Methoxyethanol

The mechanism for the reaction in Equation 10-47 consists of an S$_N$2 step followed by a proton transfer, as shown in Equation 10-48.

CONNECTIONS
2-Methoxyethanol is used as an additive in airplane deicing solutions.

Mechanism for Equation 10-47

Ring opening relieves ring strain.

Poorly stabilized leaving group

1° C

$$\text{CH}_3\ddot{\text{O}}^{\ominus} + \quad \xrightarrow{\text{1. S}_N2} \quad \text{H}_3\text{CO} \quad \ddot{\text{O}}^{\ominus} \quad \text{H}-\ddot{\text{O}}\text{CH}_3 \quad \xrightarrow[\text{transfer}]{\text{2. Proton}} \quad \text{H}_3\text{CO} \quad \ddot{\text{O}}\text{H} \quad + \quad {}^{\ominus}\ddot{\text{O}}\text{CH}_3 \quad (10\text{-}48)$$

YOUR TURN 10.15

Draw the complete, detailed mechanism for the reaction in Equation 10-46, which is nearly identical to that of Equation 10-47.

As we can see in the mechanism, the S$_N$2 reaction causes the highly strained epoxide ring to open, which compensates for the poor leaving group ability of alkoxides (RO$^-$).

Epoxides can undergo S$_N$2 reactions in neutral or basic solution due to the relief of ring strain.

The ring opening of epoxides can be carried out with a variety of different nucleophiles. Equations 10-49 and 10-50 show examples of treating an epoxide with an alkyllithium reagent (R—Li) and a Grignard reagent (R—MgX), respectively.

New C—C bond

$$
\text{(epoxide)} \xrightarrow[\text{2. HCl, H}_2\text{O}]{\text{1. CH}_3\text{CH}_2\text{Li, ether}} \text{CH}_3\text{CH}_2\text{—CH}_2\text{CH}_2\text{OH} \quad \textbf{(10-49)}
$$

Butan-1-ol

Acid workup

$$
\text{(epoxide)} \xrightarrow[\text{2. HCl, H}_2\text{O}]{\text{1. PhMgCl, ether}} \quad \textbf{(10-50)}
$$

New C—C bond

2-Phenylethanol
75%

Acid workup

> Treatment of an epoxide with R—Li or R—MgX results in the formation of a new C—C bond.

The general mechanism for these types of reactions is shown in Equation 10-51. Notice that R—Li and R—MgX are treated as strong R⁻ nucleophiles, just as we learned in Section 7.1b.

Mechanism for Equations 10-49 and 10-50

Add aqueous acid in an acid workup.

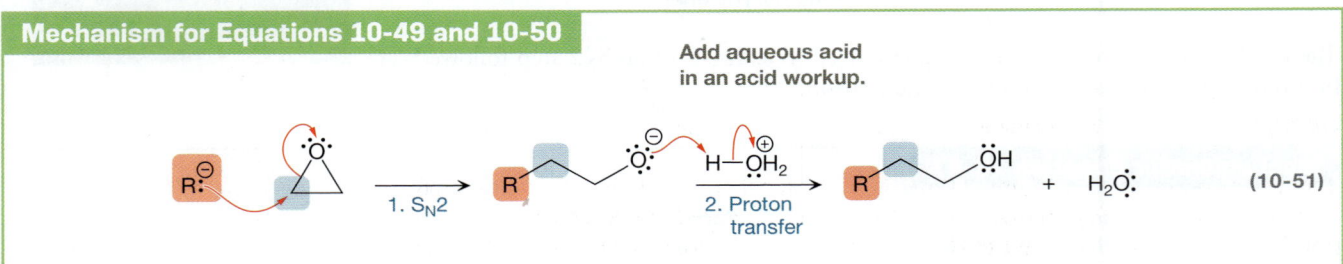

$$\textbf{(10-51)}$$

1. S$_N$2   2. Proton transfer

Alkyllithium and Grignard reagents are extremely strong bases, so Step 1 must take place in the absence of any acidic species, including protic solvents such as water. As shown in Equations 10-49 and 10-50, the acid must be added in Step 2, only after the S$_N$2 reaction is complete, in what is called an **acid workup**.

Other carbon nucleophiles can also react with epoxides to form a new C—C bond. Examples include the cyanide anion (NC⁻; Equation 10-52) and alkynide anions (RC≡C⁻; Equation 10-53). (Recall from Section 9.3b that a very strong base, such as H⁻, will deprotonate a terminal alkyne to produce an alkynide anion.)

New C—C bond

$$
\text{(epoxide)} \xrightarrow[\text{H}_2\text{O}]{\text{NaCN}} \text{NC—CH}_2\text{CH}_2\text{OH} \quad \textbf{(10-52)}
$$

3-Hydroxypropanenitrile

New C—C bond

$$
\text{RC≡CH} \xrightarrow[\substack{\text{2. (epoxide)} \\ \text{3. H}_2\text{SO}_4,\ \text{H}_2\text{O}}]{\text{1. NaH}} \text{RC≡C—CH}_2\text{CH}_2\text{OH} \quad \textbf{(10-53)}
$$

Acid workup

In a similar fashion, *hydride nucleophiles* can open an epoxide ring via an S$_N$2 reaction. Equation 10-54, for example, shows that LiAlH$_4$ or NaBH$_4$ can behave as H$^-$ (review Section 7.1b) to open oxirane rings. Acid workup of the resulting alkoxide produces ethanol.

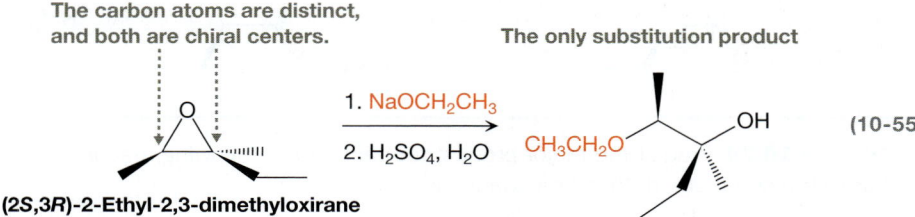

$$(10\text{-}54)$$

PROBLEM 10.22 Draw the complete, detailed mechanisms for the reactions in Equations 10-52 through 10-54.

With oxirane as the substrate in these ring-opening reactions, only H atoms are bonded to the C atoms. Therefore, we do *not* need to consider *stereochemistry* (because the C atom that is attacked is not a chiral center) or *regiochemistry* (because the two C atoms that can be attacked are indistinguishable). Therefore, the product is the same, regardless of which C atom is attacked.

This is not the case for (2*S*,3*R*)-2-ethyl-2,3-dimethyloxirane (Equation 10-55).

The carbon atoms are distinct,
and both are chiral centers.   The only substitution product

$$(10\text{-}55)$$

(2*S*,3*R*)-2-Ethyl-2,3-dimethyloxirane

The C atoms attached to O are distinct from each other, so different constitutional isomers are possible, depending on which C atom is attacked. Furthermore, the reaction involves attack at a chiral center, so different stereoisomers are imaginable. As shown in Equation 10-55, however, only one substitution product is formed.

These results are consistent with the following rule:

> Under neutral or basic conditions, a nucleophile attacks an epoxide at the *less highly alkyl-substituted C atom* of the ring, from the side *opposite the O atom*.

The mechanism in Equation 10-56 shows how this rule applies to the reaction in Equation 10-55. Specifically, the C atom on the left is attacked because it has one fewer alkyl group than the one on the right.

Mechanism for Equation 10-55

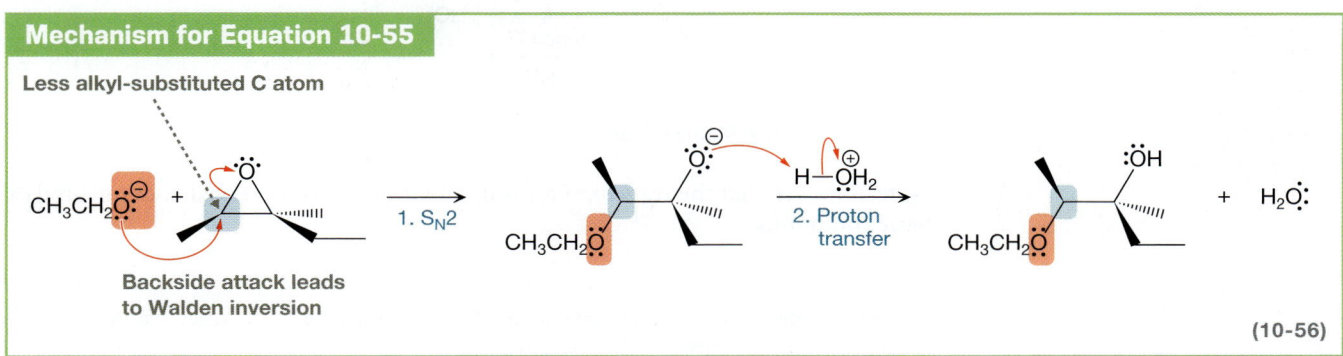

$$(10\text{-}56)$$

These observations are consistent with aspects of the S_N2 reaction we have encountered previously. In Section 9.6b, we saw that with less alkyl substitution on the carbon bonded to the leaving group, there is less *steric hindrance*, which enables an S_N2 reaction to proceed faster. Also, as we learned in Section 8.5b, backside attack is required of all S_N2 reactions, and the substituents that remain bonded to the carbon atom that was attacked undergo *Walden inversion*.

SOLVED PROBLEM 10.23

Predict the major product of the following reaction.

Think What is the nucleophile? Which C atom of the epoxide ring will it attack? What is the stereochemistry of such a reaction?

(2S,3S)-2-Ethyl-2,3-dimethyloxirane

Solve The nucleophile is the ethoxide anion, $CH_3CH_2O^-$. It will undergo an S_N2 reaction with the epoxide ring, attacking the less sterically hindered C, which is the one on the left. The nucleophile attacks from the bottom of the epoxide (i.e., opposite the O atom of the ring), forcing the H and CH_3 substituents upward. Notice that because the epoxide here differs from that in Equation 10-55 only by the configuration at the C atom that is attacked, the major product also differs only by the configuration at that C atom.

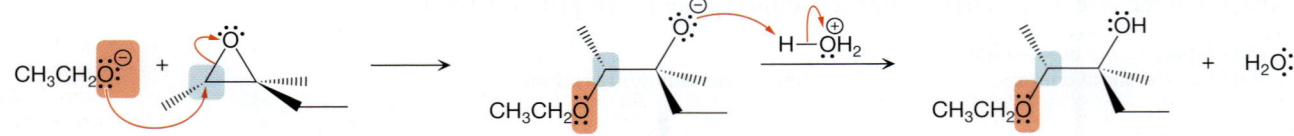

PROBLEM 10.24 Predict the major product for each of the following reactions, and draw the complete, detailed mechanisms.

(a)

NaCN / H_2O → ?

(b)

NaCN / H_2O → ?

10.7b Reactions of Epoxides under Acidic Conditions

The ring opening of epoxides occurs not only under neutral or basic conditions, but under acidic conditions as well. For example, (R)-2-methyloxirane reacts with HBr to form the bromoalcohol shown in Equation 10-57.

Acidic conditions

Nucleophilic attack at the more *highly* alkyl-substituted C

conc HBr / H_2O

(R)-2-Methyloxirane

(10-57)

Note, however, that the *regiochemistry* is different from that observed under neutral or basic conditions.

> Under *acidic conditions*, a nucleophile attacks an epoxide at the *more highly alkyl-substituted carbon*.

We can understand this regiochemistry by studying the mechanism illustrated in Equation 10-58, which is different from that in Equation 10-56 for neutral or basic conditions.

Mechanism for Equation 10-57

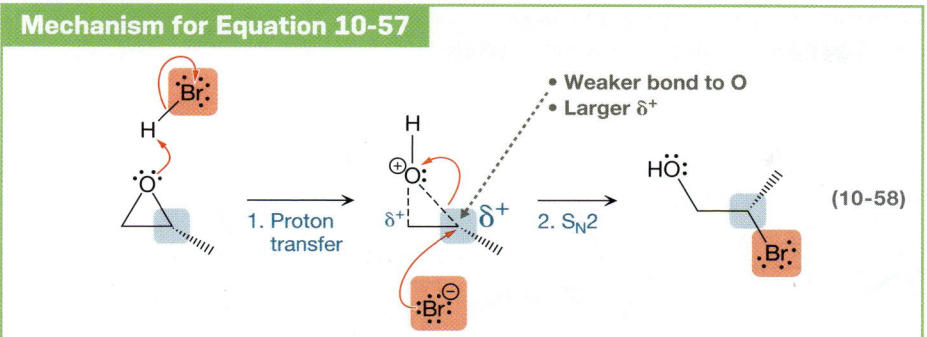

(10-58)

Prior to the attack of the nucleophile, a fast proton transfer occurs, generating an intermediate with a positive charge on the O atom in the ring.

To help delocalize the positive charge on the protonated O, both C—O bonds *partially* break in the intermediate, indicated by the dashed bonds in Equation 10-58. As a result, both C atoms of the epoxide gain a partial positive charge. The additional alkyl group attached to the C atom on the right is electron donating and helps stabilize the developing positive charge. Thus, the C—O bond involving the more highly alkyl-substituted C is broken to a greater extent, indicated in Figure 10-6 by the fact that it is slightly longer than the other C—O bond. With the larger partial positive charge generated on the side of the ring with the greater alkyl substitution, the nucleophile is attracted more strongly to that side of the ring in the subsequent S_N2 step.

FIGURE 10-6 Effect of protonation on epoxide C—O bond lengths The C—O bonds in an epoxide (*left*) partially break on protonation (*right*), as shown by the lengthening of the C—O bonds. As a result, both C atoms develop a partial positive charge, but a larger partial positive charge develops on the side of the ring with greater alkyl substitution, drawing the incoming nucleophile to that side.

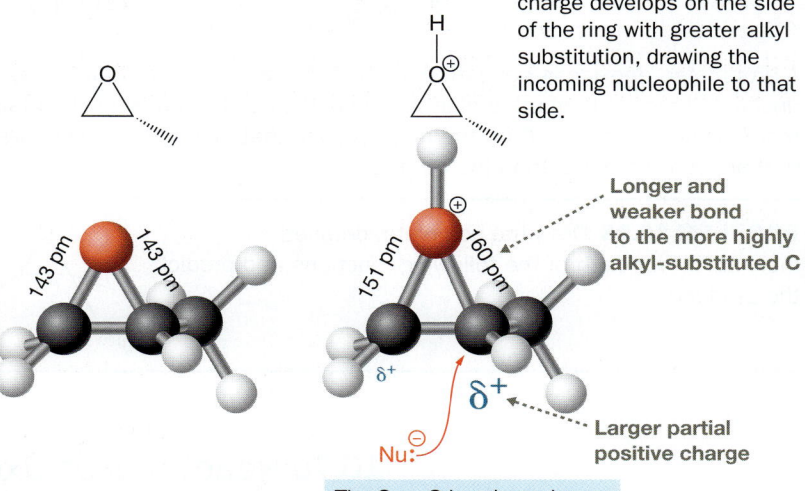

Longer and weaker bond to the more highly alkyl-substituted C

Larger partial positive charge

The C—O bonds are longer in the protonated form.

SOLVED PROBLEM 10.25

(S)-2-Methyloxirane undergoes a chemical reaction when it is dissolved in ethanol. However, what products are formed depends on whether the reaction takes place under acidic or basic conditions. Draw the complete, detailed mechanism for each reaction and predict the products.

(S)-2-Methyloxirane

dilute HCl
CH₃CH₂OH → ?

CH₃CH₂ONa
CH₃CH₂OH → ?

Think What types of species are allowed under acidic conditions and what types are not allowed? Under basic conditions? Under each set of conditions, what is the substrate being attacked? What is the nucleophile?

Solve Under basic conditions, the nucleophile is $CH_3CH_2O^-$ and the substrate is the uncharged epoxide:

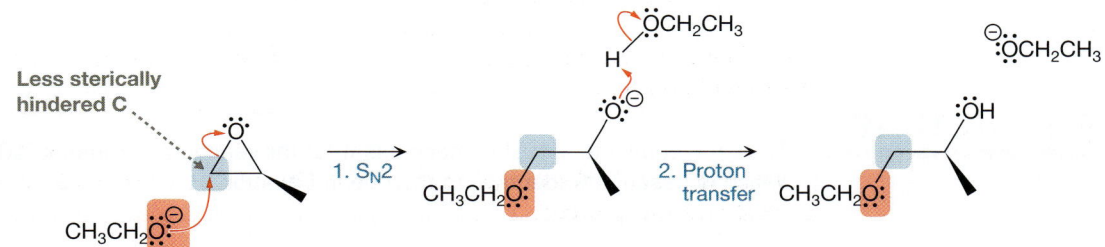

Less sterically hindered C

The mechanism follows the one in Equation 10-56. Because the epoxide being attacked is uncharged, steric hindrance guides the nucleophile to the less-substituted carbon of the ring.

Under acidic conditions, however, strong bases should not appear in a mechanism, so we may *not* include species with a localized negative charge on O. Instead, the mechanism in Equation 10-58 takes place, as shown:

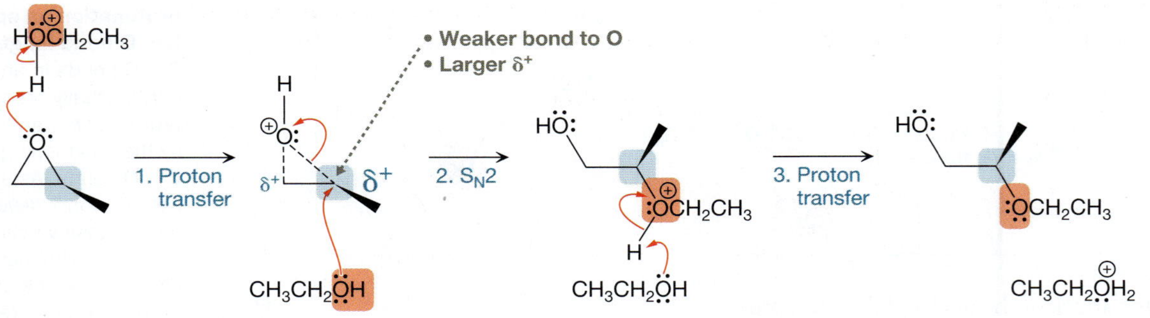

A protonated epoxide is generated in the first step, which is attacked by an uncharged molecule of ethanol in the second step. The nucleophile attacks the more highly substituted C on the ring because that side of the ring bears a larger partial positive charge than the other side.

PROBLEM 10.26 Draw the complete, detailed mechanism for each of the following reactions and predict the products.

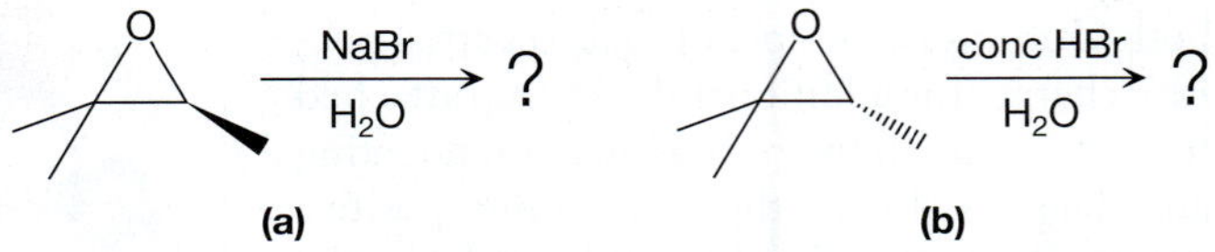

(a)   (b)

10.7c Reactions of Oxetanes and Cyclic Ethers That Are Part of Larger Rings

Oxetanes, which are four-membered ring ethers, can also open via nucleophilic attack under neutral or basic conditions, aided by the relief of ring strain. Oxetanes are larger rings than epoxides, however, so they have less ring strain, giving these reactions less of a driving force.

Consequently, reactions that open oxetanes are generally limited to ones involving *very* strong nucleophiles, such as alkyllithium reagents (R—Li) and Grignard reagents (R—MgX). As shown in Equation 10-59, these reactions result in the formation of a new C—C bond, just as we saw with the analogous reactions involving epoxides (Equations 10-49 and 10-50, p. 542).

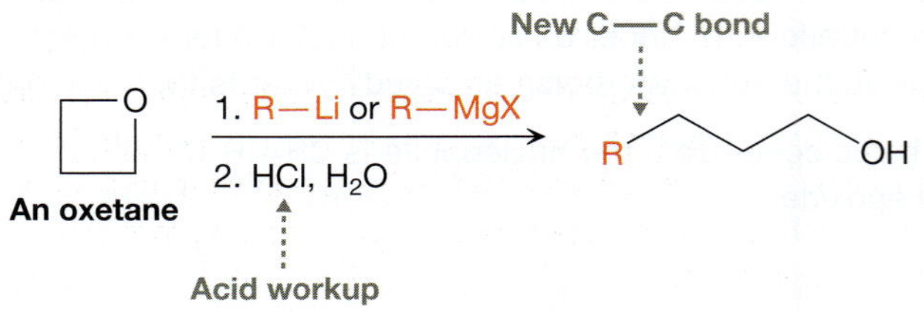

(10-59)

YOUR TURN 10.16

Draw the complete, detailed mechanism for the reaction in Equation 10-59, which is essentially identical to the one in Equation 10-51 (p. 542).

DNA Alkylation: Cancer Causing and Cancer Curing

DNA bases (adenine, guanine, cytosine, and thymine) contain nucleophilic nitrogen atoms, which is why many halogenated compounds are carcinogenic. The moderate-to-good leaving group abilities of halogen atoms facilitate a nucleophilic substitution reaction, leaving the DNA base *alkylated*, much like the way in which ammonia and amines become alkylated on treatment with an alkyl halide (Section 10.2). Alkylated DNA can still function in its process of replication, though it will do so abnormally, resulting in mutations in the DNA and, ultimately, cancerous cells.

Guanine

Damaging DNA via alkylation can also be used to *treat* cancer. The key is that cancer cells grow and divide more rapidly than normal cells, and thus are more susceptible to mechanisms that damage DNA and impair its function. A number of successful chemotherapeutic drugs, such as mechlorethamine, have a bis(2-chloroethyl)amino motif to carry out these alkylations.

With a leaving group at two locations, a single molecule of the drug will alkylate two DNA bases and tether them together, disrupting DNA function even more severely. Each alkylation by mechlorethamine involves the formation of an aziridinium ring followed by ring opening, analogous to the formation and opening of an epoxide ring that we learned about in Sections 10.6 and 10.7, respectively.

Unfortunately, alkylating agents like mechlorethamine cannot differentiate the DNA in cancer cells from the DNA in normal cells. Other cells that grow and divide rapidly, such as cells that grow hair and cells in the epithelial lining of the gastrointestinal tract, will also be targeted. This accounts for some of the side effects associated with chemotherapy, such as hair loss, nausea and vomiting, and diarrhea.

Ethers that are part of larger rings do not readily undergo ring opening under neutral or basic conditions, because five-membered rings and larger have little or no ring strain, and thus behave much like acyclic ethers. As shown in Equations 10-60 and 10-61, for example, treating tetrahydrofuran (THF) or tetrahydropyran (THP) with a very strong R^- nucleophile results in no reaction. (In fact, because of this lack of reactivity, THF and THP are excellent solvents for reactions that involve such strong R^- nucleophiles.)

Tetrahydrofuran (THF) $\xrightarrow{\text{R—Li or R—MgX}}$ No reaction (10-60)

Tetrahydropyran (THP) $\xrightarrow{\text{R—Li or R—MgX}}$ No reaction (10-61)

10.8 Elimination: Generating Alkynes via Elimination Reactions

Although most of the instances in which we discuss elimination reactions lead to the formation of alkenes, it is also possible to generate an alkyne from an elimination reaction (Equation 10-62). As with the other E2 reactions we have examined, a proton and a leaving group are required on adjacent atoms.

The leaving group is in a *vinylic* position (i.e., attached to a C=C double bond), so these substrates are quite resistant to nucleophilic substitution and elimination reactions. As we learned in Section 9.6a, the C—L bond strength is significantly greater with an sp^2-hybridized carbon atom than with an sp^3-hybridized carbon atom. Furthermore, because the four electrons of the C=C bond establish a region that is relatively electron rich, bases, which are themselves electron rich, tend to be repelled. Under normal conditions, therefore, even a strong base such as HO^- does not facilitate such a reaction. As a result, these reactions are carried out under more extreme conditions, often at temperatures >200 °C. An example is shown in Equation 10-63 with a vinylic chloride.

If a much stronger base is used, such as NaH or $NaNH_2$, then the elimination can take place without extreme temperatures. An example is shown in Equation 10-64, which takes place at room temperature.

When a very strong base like NaH or $NaNH_2$ is used and the elimination product is a terminal alkyne, as in Equation 10-65, an *acid workup* is necessary.

The mechanism for Equation 10-65 is shown in Equation 10-66. Step 1 is the E2 elimination of H^+ and Br^-, which *initially* forms the uncharged terminal alkyne. In Step 2, this terminal alkyne is *irreversibly* deprotonated because the pK_a of H_2 (the conjugate acid of H^-) is substantially higher than that of a terminal alkyne (see Your Turn 10.17). The reaction stops there, so an *acid workup* is carried out to replace that proton on the terminal C atom and recover the final uncharged product.

Mechanism for Equation 10-65

YOUR TURN 10.17

Look up (Table 6-1) or estimate the pK_a values of H_2 and a terminal alkyne to verify that Step 2 in Equation 10-66 is irreversible. To what extent (i.e., by what numerical factor) is that step favored?

H_2 ________________ R—C≡C—H ________________

Internal alkynes, such as the ones produced in Equations 10-63 and 10-64, do not contain an acidic proton. Therefore, the mechanisms for those reactions would not include the proton transfer steps that appear in Equation 10-66 (see Your Turn 10.18).

YOUR TURN 10.18

Draw the complete mechanisms for the reactions in Equations 10-63 and 10-64.

These elimination reactions are particularly useful when we begin with a dihalide such as 1,1-dichloropentane (Equation 10-67) or 1,2-dibromo-1-phenylethane (Equation 10-68).

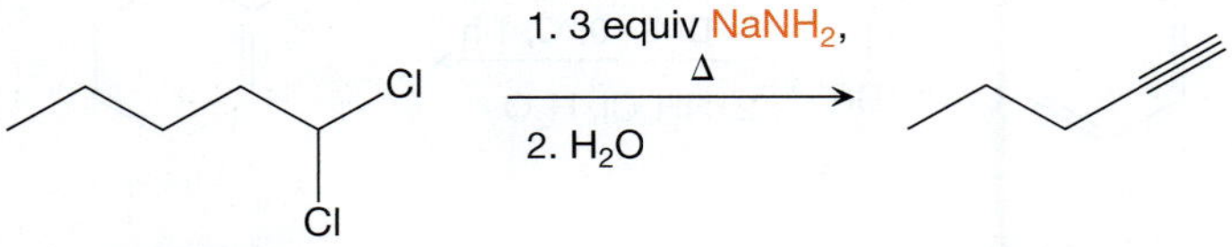
Cl
1. 3 equiv $NaNH_2$, Δ
2. H_2O
Cl

1,1-Dichloropentane **Pent-1-yne** **(10-67)**

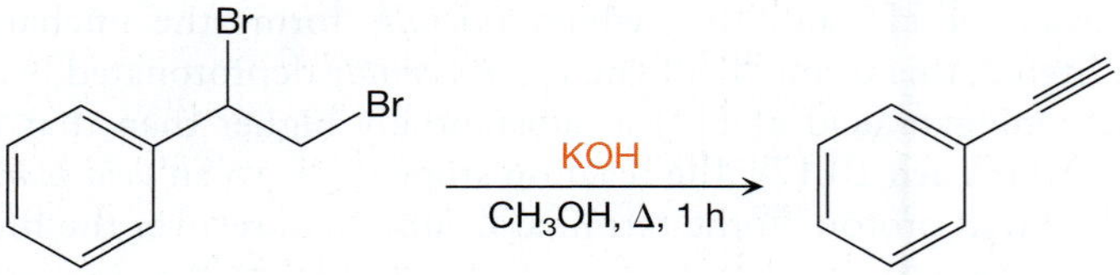
Br
Br
KOH
CH_3OH, Δ, 1 h

1,2-Dibromo-1-phenylethane **Phenylethyne (Phenylacetylene)** 40–50% **(10-68)**

Three equivalents of H_2N^- are used in Equation 10-67 to convert the dihalide into an alkyne. The first two equivalents bring about two separate E2 reactions, removing H and Cl each time. The third equivalent of base deprotonates the acidic proton on the newly formed terminal alkyne. *Acid workup* replaces that proton. The reaction in Equation 10-68, on the other hand, does not require acid workup because HO^- is not a strong enough base to deprotonate the terminal alkyne that forms.

PROBLEM 10.27 Draw the complete, detailed mechanisms for the reactions in Equations 10-67 and 10-68.

PROBLEM 10.28 Predict the major product of each of the following reactions, and draw their complete, detailed mechanisms.

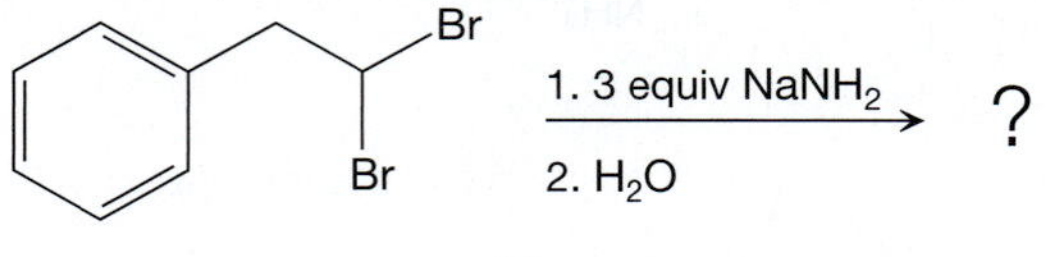
Br
Br
1. 3 equiv $NaNH_2$
2. H_2O
?

(a)

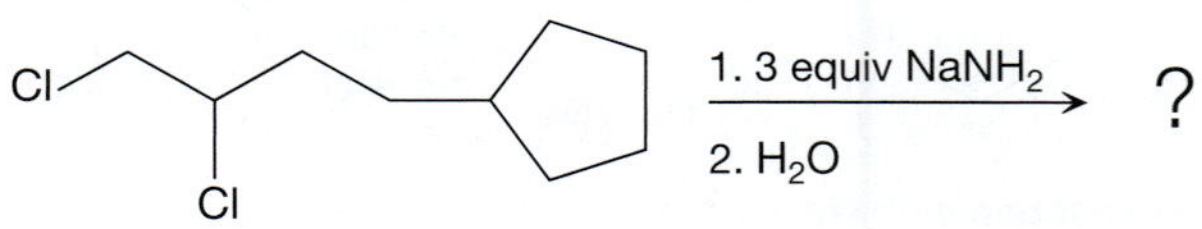
Cl
Cl
1. 3 equiv $NaNH_2$
2. H_2O
?

(b)

10.9 Elimination: Hofmann Elimination

But-1-ene is produced when butan-2-amine is treated first with excess iodomethane, then with silver oxide (Ag_2O), and is finally heated:

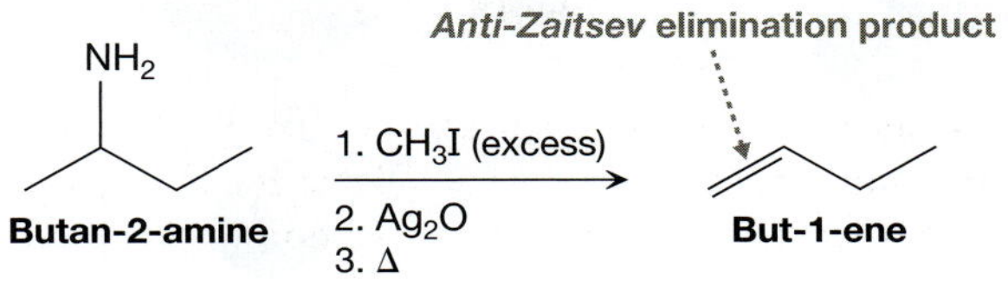

Anti-Zaitsev elimination product

1. CH_3I (excess)
2. Ag_2O
3. Δ

Butan-2-amine   **But-1-ene**   (10-69)

Equation 10-69 is an example of a **Hofmann elimination reaction**. The formation of a C=C double bond strongly suggests that an elimination reaction has occurred in which the leaving group contains the N atom. Two problems, however, must be reconciled: (1) amino groups are terrible leaving groups and therefore do not participate as substrates in elimination reactions, and (2) the major product is a terminal alkene, which is the less alkyl substituted, and therefore the less stable, of two possible elimination products (review Section 9.10). In other words:

> The major product of Hofmann elimination is the *anti-Zaitsev product* (or **Hofmann product**).

The more stable elimination product (i.e., the *Zaitsev product*) would be but-2-ene, in which the alkene group is more highly alkyl substituted.

The presence of CH_3I ensures that the leaving group is not simply H_2N^-. As we saw in Section 10.2, an amine reacts with multiple equivalents of CH_3I to yield the quaternary ammonium salt—in this case, the iodide salt (Equation 10-70). The leaving group is therefore a stable, uncharged amine, $N(CH_3)_3$. Silver oxide replaces I^- with HO^-, and then heating causes the E2 reaction to occur.

Mechanism for Equation 10-69

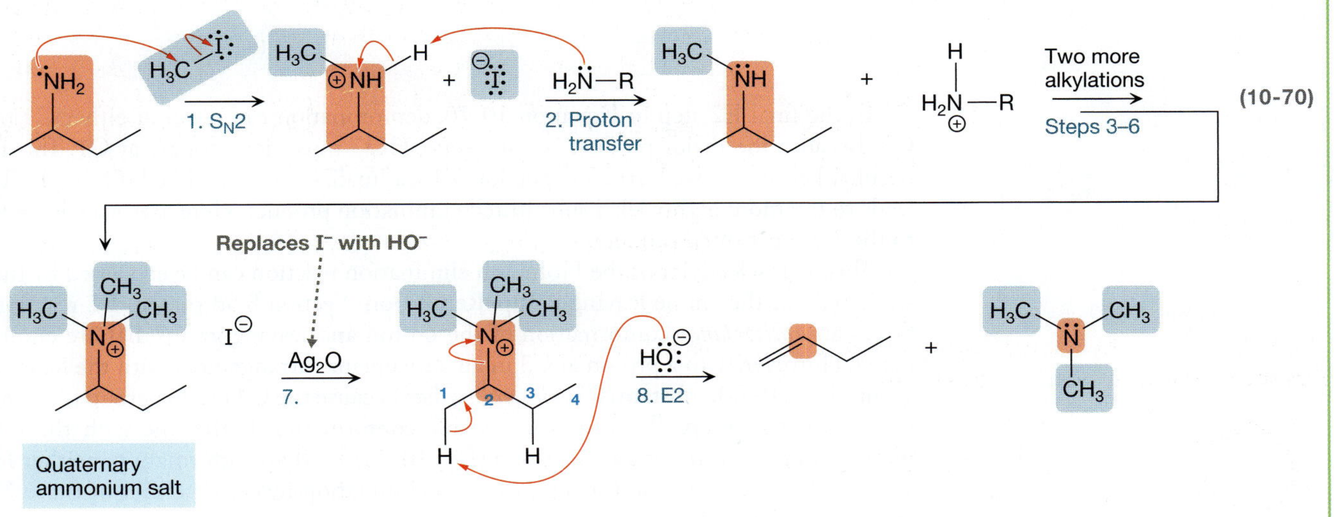

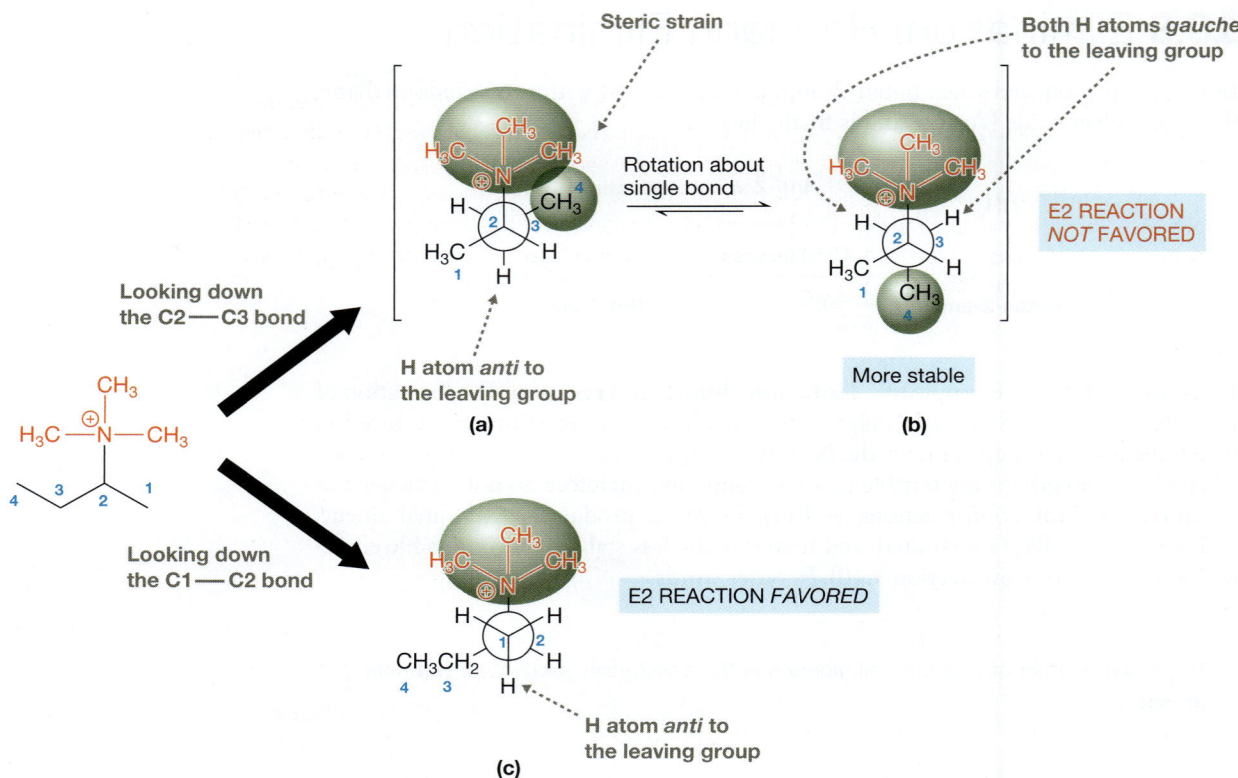

FIGURE 10-7 Regioselectivity in Hofmann elimination reactions Newman projections looking down the C2—C3 bond (*top*) and the C1—C2 bond (*bottom*) of the tetraalkyl ammonium ion shown at the left. (a) Conformation necessary for formation of the Zaitsev product, where the H on C3 is *anti* to the amine leaving group. In this conformation, there is substantial steric strain between the C4 methyl group and the leaving group. (b) The more stable conformation is the one in which the C4 methyl group is *anti* to the leaving group. This conformation does not favor the E2 reaction because the H and the amine leaving group are not in the anticoplanar conformation. (c) A H atom is *anti* to the leaving group, and no substantial steric strain exists, so an E2 reaction favors elimination of a proton on C1.

In the final E2 step in Equation 10-70, deprotonation can occur at either C1 or C3. Because the major product is but-1-ene, HO^- must deprotonate at C1. This is peculiar because, as we learned in Section 9.10, a small, strong base like HO^- typically leads to the more highly alkyl substituted elimination product. Here, however, it leads to the *less* substituted product.

This regioselectivity of the Hofmann elimination reaction can be explained by the steric bulk of the amine leaving group. Recall from Section 8.5d that an E2 reaction favors an *anticoplanar* conformation of the proton and leaving group that are eliminated. However, with a proton at C3 in an anticoplanar arrangement with the leaving group (Fig. 10-7a), substantial *steric strain* exists because the C4 methyl group is *gauche* to the leaving group. Thus, the more stable conformation is the one with the C4 methyl group *anti* to the leaving group (Fig. 10-7b). In that conformation, neither H atom on C3 is *anti* to the leaving group, so elimination involving either of those H atoms is *not* favored.

As shown in Figure 10-7c, on the other hand, a H atom on C1 can be *anti* to the leaving group without causing substantial steric strain. In this case, elimination involving the H atom is favored.

Mechanically Generated Acid and Self-Healing Polymers

When you bend or pull on a piece of plastic, it can crack or tear. And for typical plastics, that damage is essentially permanent, without some outside intervention. But imagine if that plastic could, instead, repair the damage on its own. This might seem like fiction, but it is, in fact, possible. Plastics consist of very long-chain molecules called polymers, and a crack or tear in a plastic represents bonds in the polymer having been severed. For the polymer to "heal" itself—and thus for the plastic to return to its original form—a chemical reaction might be initiated automatically, which would reconnect the broken bonds.

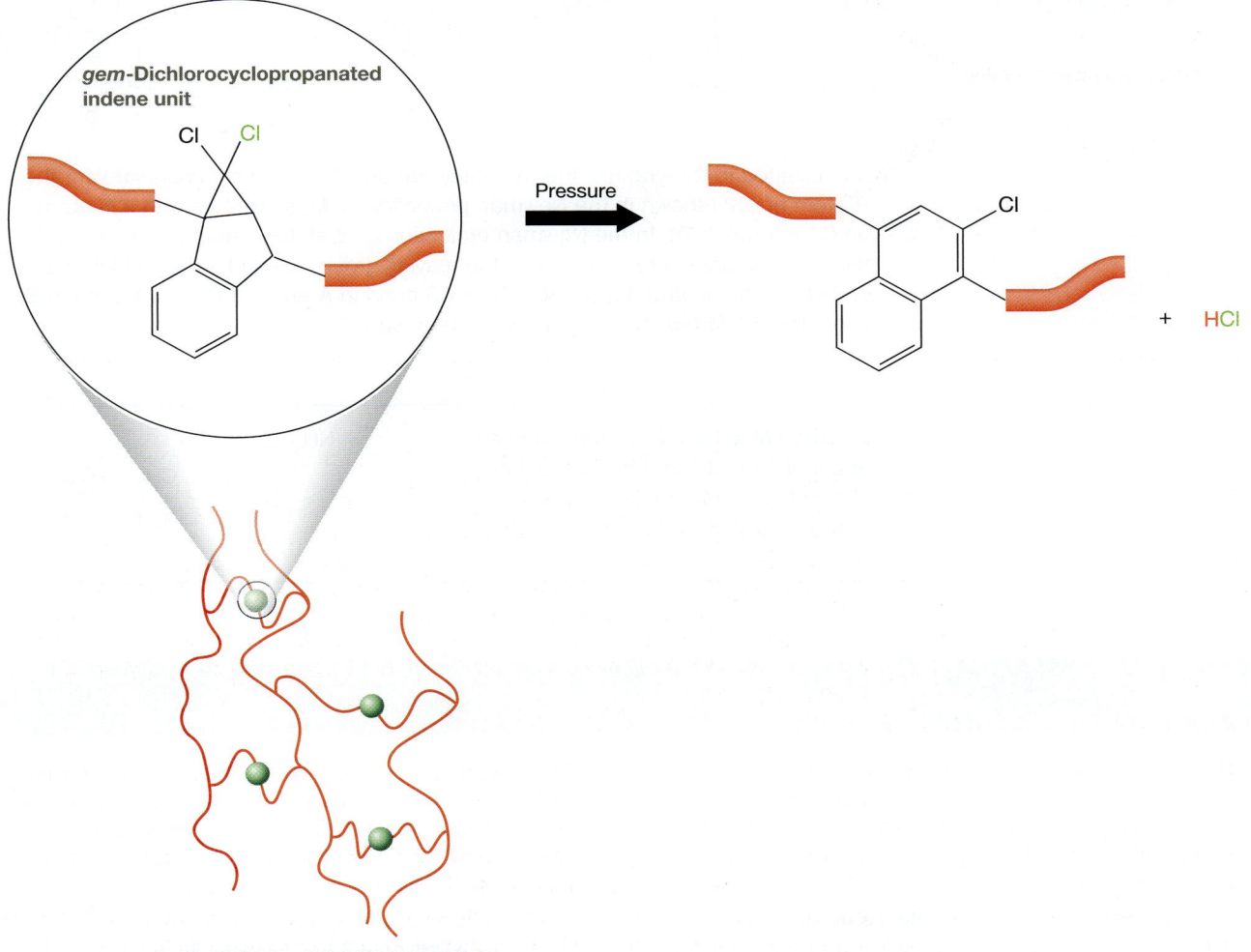

Cross-linked poly(methyl acrylate)

How can such a reaction be initiated? Jeffrey Moore at the University of Illinois at Urbana-Champaign envisions that the physical stress placed on the polymer, which would have caused the polymer's damage in the first place, would produce a significant amount of acid at the location where the damage occurred. The acid could behave, in turn, as a catalyst to start a reaction that repairs the polymer molecule.

Moore developed a polymer system that produces this acid when pressure is applied—so-called mechanically generated acid. The key was to covalently bond *gem*-dichlorocyclopropanated indene units into cross-linked polymer chains of poly(methyl acrylate).

When pressure is applied, the induced strain causes HCl to be eliminated, analogous to the elimination reactions we have seen here in Chapter 10 and previously in Chapter 9. This reaction is driven by the additional stabilization gained by the opening of a three-membered ring and the resonance stabilization involving the newly formed double bonds.

2-Methylpentan-3-amine undergoes complete alkylation to give the quaternary ammonium ion shown below in the middle.

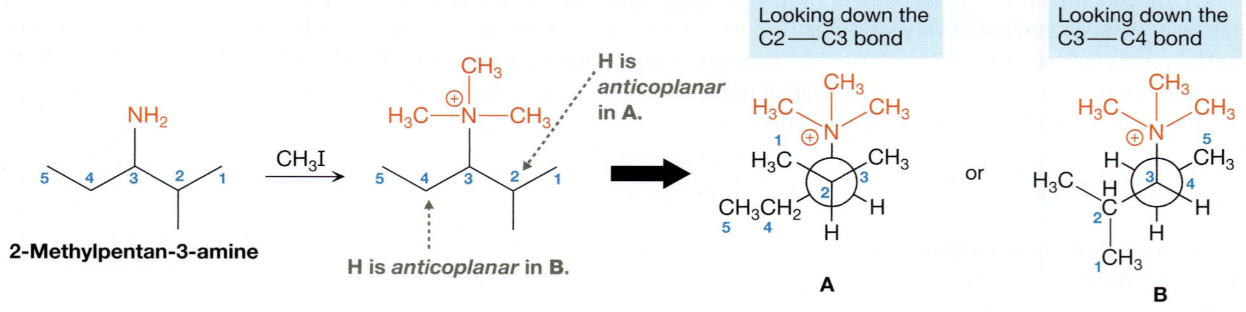

As indicated in the graphic, the leaving group on C3 can be *anticoplanar* with the H atom on C2 (shown in the Newman projection in **A**), as well as with a H atom on C4 (shown in **B**). In the Newman projections, label the H atoms that are in the *anticoplanar* conformation with the leaving group. Also, identify and label all sources of steric strain along the C2—C3 bond in **A** and the C3—C4 bond in **B**. Label the conformation with the least steric strain.

PROBLEM 10.29 Draw the complete, detailed mechanism for the reaction shown here and predict the major product. *Hint:* See Your Turn 10.19.

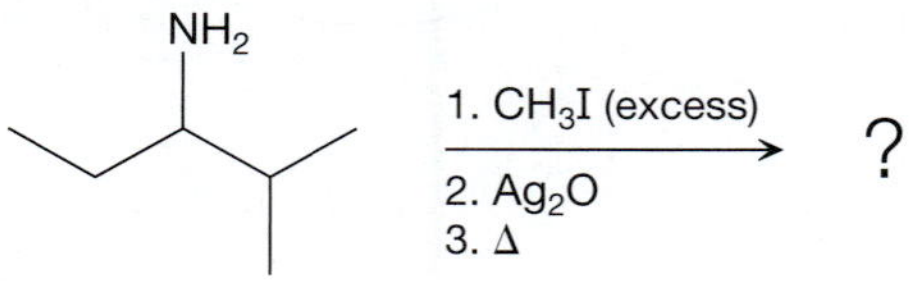

1. CH_3I (excess)
2. Ag_2O
3. Δ

?

Chapter Summary and Key Terms

- Phosphorus tribromide (PBr_3) and phosphorus trichloride (PCl_3) convert primary and secondary alcohols into alkyl halides *stereospecifically* via back-to-back S_N2 reactions. The configuration of the alkyl halide is opposite that of the initial alcohol. (**Section 10.1; Objective 1**)

- Ammonia and amines undergo **alkylation** when treated with an alkyl halide that has a good leaving group. Multiple alkylations occur because the product amine is often more nucleophilic than the reactant. Alkylation of ammonia is a poor method to synthesize a primary amine, but it is an effective way of generating a **quaternary ammonium salt**. (**Section 10.2; Objective 2**)

- Alkylation can take place at an α carbon of a ketone or aldehyde. Under basic conditions, the α carbon is deprotonated and becomes nucleophilic. The resulting enolate anion may then attack an alkyl halide in an S_N2 reaction. The choice of base can dictate the regiochemistry in the alkylation of a ketone. Reaction with LDA leads to alkylation at the *less* highly alkyl substituted α carbon, whereas reaction with an RO^- base leads to alkylation at the *more* highly alkyl substituted carbon. (**Section 10.3; Objective 3**)

- **Halogenation** may occur at an α carbon of aldehydes and ketones under either acidic or basic conditions. The reaction stops after just a single halogenation under acidic conditions, whereas multiple halogenations occur under basic conditions. (**Section 10.4; Objective 4**)

- **Diazomethane** (CH_2N_2) converts a carboxylic acid (RCO_2H) into a methyl ester (RCO_2CH_3). After an initial proton transfer, an S_N2 reaction occurs, displacing $N_2(g)$ as the leaving group. (**Section 10.5; Objective 5**)

- The **Williamson ether synthesis** can be used to synthesize either symmetric or unsymmetric ethers via an S_N2 reaction between an alkoxide anion ($R'O^-$) and an alkyl halide (RX). The Williamson synthesis takes place under basic conditions. Under acidic conditions, a dehydration reaction can be used to synthesize a symmetric ether. (**Section 10.6; Objective 6**)

- **Epoxides** and **oxetanes** are three- and four-membered ring ethers, respectively, that can be opened in an S_N2 reaction that relieves their substantial ring strain. Under neutral or basic conditions, a nucleophile attacks the less alkyl-substituted carbon of the epoxide ring. Under acidic conditions, the nucleophile attacks the more highly alkyl-substituted carbon of the ring. (Section 10.7; Objective 7)

- A *vinylic* halide can undergo an E2 reaction to produce an alkyne. Because bonds to an sp^2-hybridized carbon atom are quite strong, and because alkenes are electron rich, these reactions require strong bases and often elevated temperatures. (Section 10.8; Objective 8)

- The **Hofmann elimination reaction** leads to an *anti-Zaitsev* elimination product, or **Hofmann product**. An amine with a poor leaving group is converted into a tetraalkyl ammonium ion, which has a moderately good leaving group, using excess of an alkyl halide like CH_3I. Subsequent treatment with Ag_2O followed by heat induces elimination. The bulkiness of the leaving group in the tetraalkyl ammonium ion causes the anti-Zaitsev regiochemistry. (Section 10.9; Objective 9)

Reaction Tables

The functional group transformations from this chapter have been collected in Table 10-1; reactions that alter the carbon skeleton are listed in Table 10-2.

TABLE 10-1 Functional Group Transformations[a]

	Starting Compound Class	Typical Reagents and Reaction Conditions	Compound Class Formed	Key Electron-Rich Species	Key Electron-Poor Species	Comments	Discussed in Section
(1)	1° or 2° alcohol	PBr_3	1° or 2° alkyl halide	$Br^{\ominus}$		Back-to-back S_N2 reactions	10.1
(2)	R—X 1° alkyl halide	1 equiv NH_3	R—NH_2 + other amines 1° amine	NH_3	R—X	S_N2 reaction (Not useful in synthesis)	10.2
(3)	R—NH_2 Amine	excess R'—X	Quaternary ammonium salt	R—NH_2	R'—X	Multiple S_N2 reactions	10.2
(4)	Ketone or aldehyde	X_2 / NaOH	α-Halogenated ketone or aldehyde	Enolate anion	X—X	Multiple S_N2 reactions	10.4
(5)	Ketone or aldehyde	X_2 / Acid	α-Halogenated ketone or aldehyde	Enol	X—X	Single S_N2 reaction	10.4

[a]X = Cl, Br, I.

(continued)

TABLE 10-1 Functional Group Transformations[a] (continued)

	Starting Compound Class	Typical Reagents and Reaction Conditions	Compound Class Formed	Key Electron-Rich Species	Key Electron-Poor Species	Comments	Discussed in Section
(6)	Carboxylic acid	CH_2N_2	Methyl ester	Carboxylate anion	$H_3C-N_2^{\oplus}$	S_N2	10.5
(7)	Alkyl halide $R-X$	$NaOR'$	Ether (symmetric or unsymmetric)	$R'O^{\ominus}$	$\overset{\delta+}{R}-X$	Williamson ether synthesis, S_N2	10.6
(8)	Halohydrin	$NaOH$	Epoxide			Intramolecular S_N2	10.6
(9)	Alcohol $R-OH$	H_2SO_4, Δ	Ether (symmetric)	$R-OH$	$R-\overset{\oplus}{O}H_2$	S_N1 or S_N2 (dehydration)	10.6
(10)	Epoxide	$:Nu^{\ominus}$ Neutral or basic	Alcohol (2-substituted)	$:Nu^{\ominus}$		S_N2	10.7a
(11)	Epoxide	$H-Nu$ Acidic	Alcohol (2-substituted)	$:Nu^{\ominus}$		S_N2	10.7b
(12)	Vinylic halide	$NaNH_2$ or NaH or $NaOH$, Δ	Internal alkyne $R-C\equiv C-R'$	$^{\ominus}NH_2$ or $^{\ominus}H$ or $^{\ominus}OH$		E2	10.8
(13)	Vinylic halide	1. $NaNH_2$ or NaH 2. H_2O	Terminal alkyne $R-C\equiv CH$	$^{\ominus}NH_2$ or $^{\ominus}H$		E2	10.8
(14)	Amine	1. CH_3I (excess) 2. Ag_2O 3. Δ	Alkene	$^{\ominus}OH$		E2	10.9

[a] X = Cl, Br, I.

TABLE 10-2 **Reactions That Alter the Carbon Skeleton[a]**

	Starting Compound Class	Typical Reagents and Reaction Conditions	Compound Class Formed	Key Electron-Rich Species	Key Electron-Poor Species	Comments	Discussed in Section
(1)	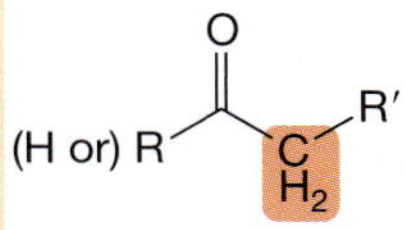 Ketone or aldehyde	1. Base⁻ 2. R″—X	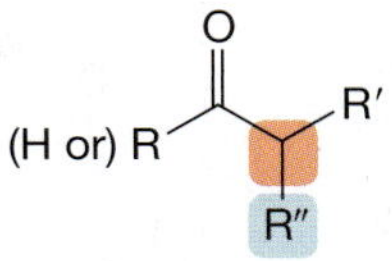 α-Alkylated ketone or aldehyde	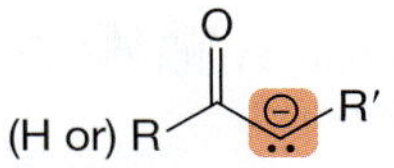 Enolate anion	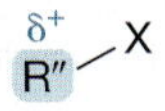 Alkyl halide	S$_N$2	10.3
(2)	Epoxide	1. R″—Li or R″—MgX 2. H$_3$O⁺	Alcohol	R″⁻		S$_N$2	10.7a
(3)	Epoxide	NaCN H$_2$O	Nitrile (β-hydroxy)	NC⁻		S$_N$2	10.7a
(4)	R″C≡CH Alkyne (terminal)	1. NaH 2. 3. H$_3$O⁺	Alcohol (3-Alkyn-1-ol)	R″C≡C⁻		S$_N$2	10.7a
(5)	Oxetane	1. R″—Li or R″—MgX 2. H$_3$O⁺	Alcohol	R″⁻		S$_N$2	10.7c

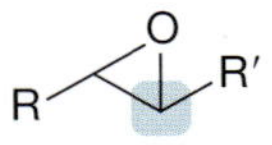

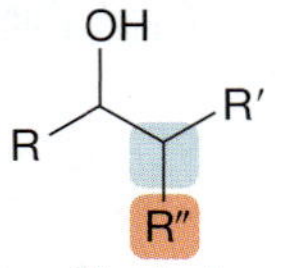

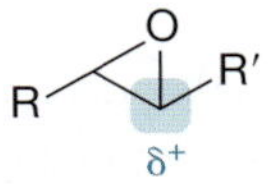

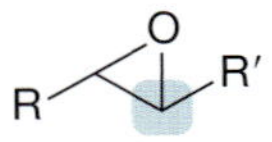

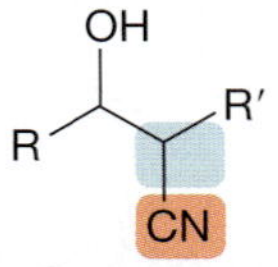

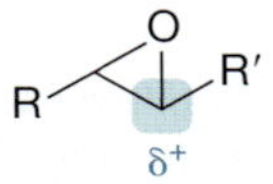

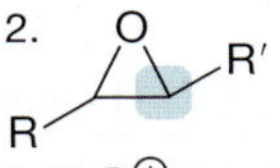

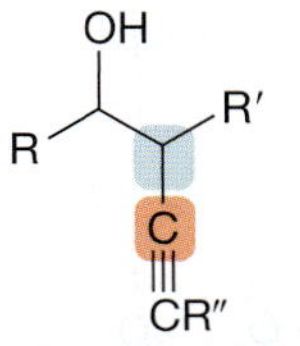

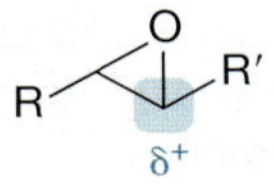

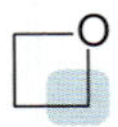

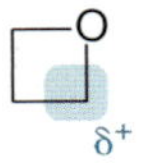

[a]X = Cl, Br, I.

Problems

Problems that are related to synthesis are denoted (SYN).

10.1 Converting Alcohols to Alkyl Halides Using PBr$_3$ and PCl$_3$

10.30 For each reaction, draw the complete, detailed mechanism and the major organic product.

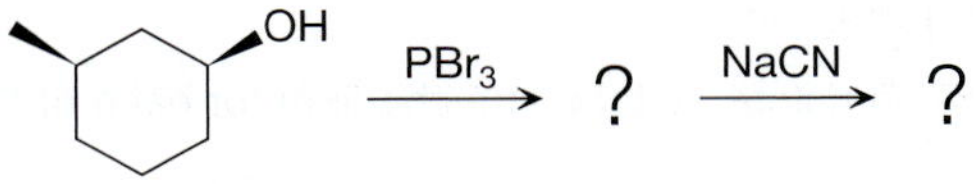

10.31 (SYN) (a) Draw the alcohol that would be required to form the alkyl chloride shown here using PCl$_3$. **(b)** Draw the complete, detailed mechanism by which this transformation would occur.

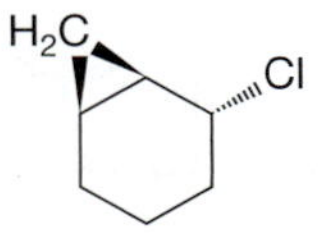

10.32 (SYN) Show how to carry out the transformation at the right, and draw the complete, detailed mechanism for that reaction.

HO
HO
OH
?
Cl
HO
OH

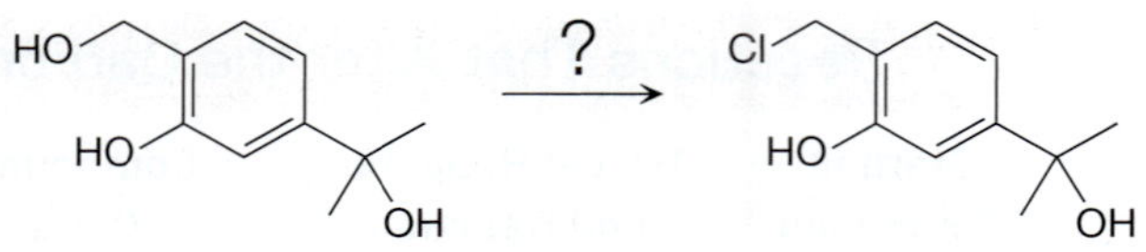

10.2 Alkylation of Ammonia and Amines

10.33 Draw the complete, detailed mechanism for the reaction shown here and give the major product.

O
NH2
CH3I (excess)
?

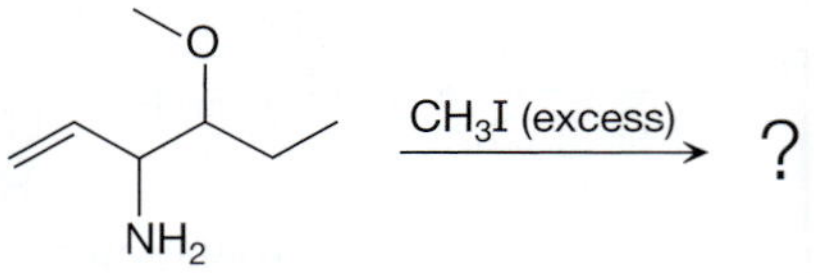

10.34 Draw the reactant of this reaction, whose formula is $C_{10}H_{16}ClN$.

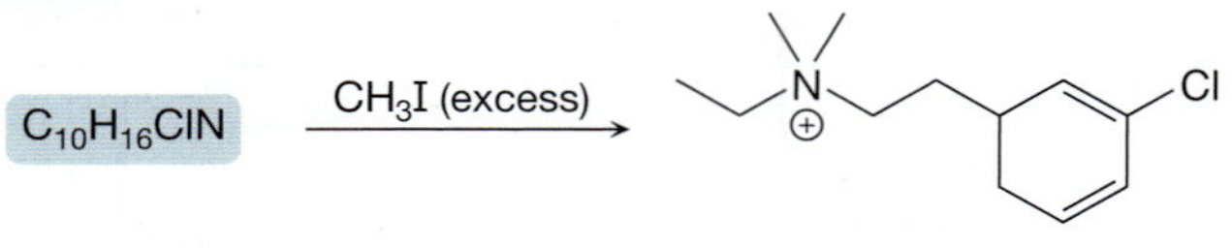

C10H16ClN
CH3I (excess)
N
Cl

10.35 (SYN) Suggest a reagent that can be used to carry out the transformation shown here, and draw the complete, detailed mechanism for the reaction.

NH2
?
N

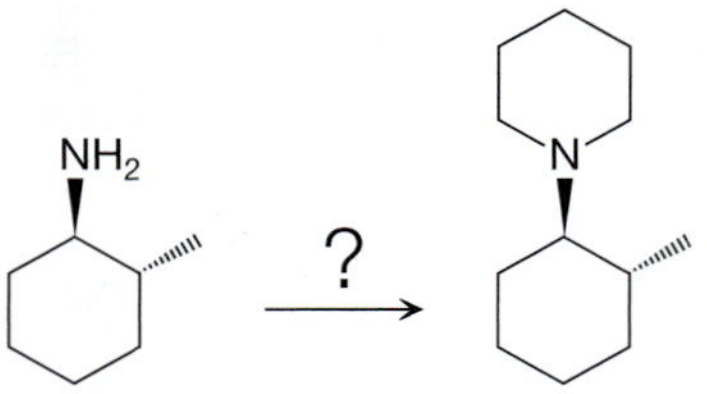

10.36 Draw the complete, detailed mechanism for the reaction shown here and draw the major organic product.

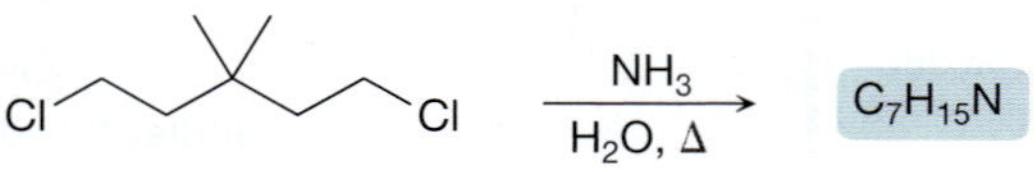

Cl
Cl
NH3
H2O, Δ
C7H15N

10.3 and 10.4 Alkylation and Halogenation of α Carbons

10.37 When pentane-2,4-dione is treated with one molar equivalent of sodium carbonate and bromoethane, 3-ethylpentane-2,4-dione is the major product. If $NaNH_2$ is the base and two molar equivalents are used, however, then heptane-2,4-dione is the major product. Explain these observations.

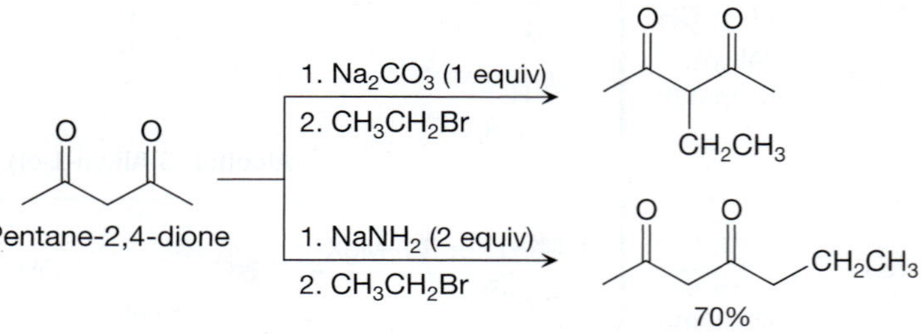

O
O
1. Na2CO3 (1 equiv)
2. CH3CH2Br
O
O
CH2CH3
1. NaNH2 (2 equiv)
2. CH3CH2Br
O
O
CH2CH3
70%

Pentane-2,4-dione

10.38 An α carbon of a ketone or aldehyde can be alkylated or halogenated under basic conditions. Recall that multiple halogenations take place under these conditions, whereas polyalkylation is generally not a concern. Suggest why.

10.39 Draw the complete, detailed mechanism for the reaction shown here.

O
Cl
Cl
NaOH
H2O, CH3CH2OH
O

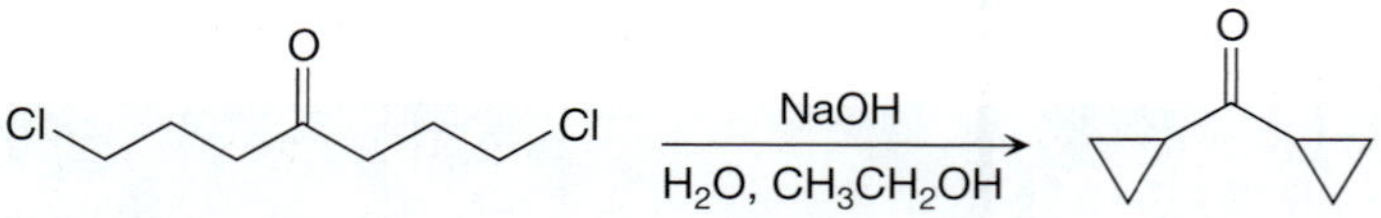

10.40 Amides are moderately acidic at the N atom, so they can be alkylated in a fashion that is quite similar to the alkylation of ketones and aldehydes. Predict the product of the reaction shown here, and draw its complete, detailed mechanism.

O
N
H
1. NaH
2.
Cl
?

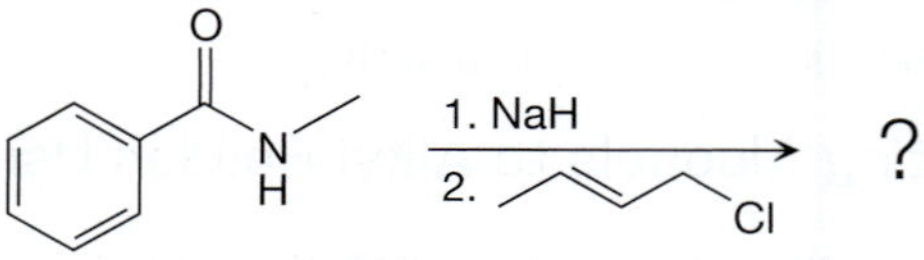

10.41 Draw the complete, detailed mechanism for each of the following reactions and predict the major product(s).

(a)

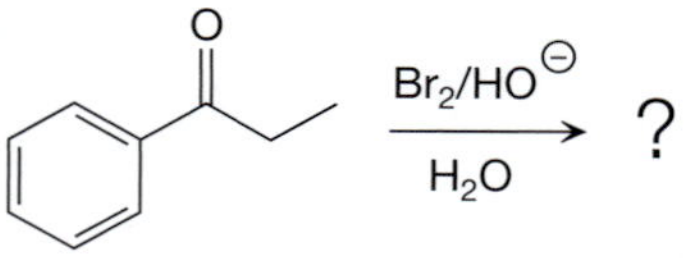

O
Br2/HO⊖
H2O
?

(b)

O
Br2/H⊕
H2O
?

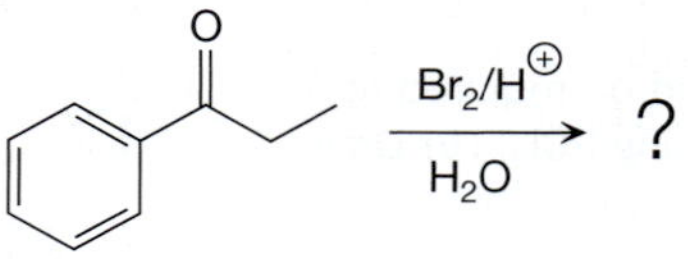

10.42 Halogenation readily takes place at an α carbon of a ketone or aldehyde under basic conditions if the halogen is Cl_2, Br_2, or I_2. Explain why it does *not* readily take place with F_2.

10.5 Diazomethane Formation of Methyl Esters

10.43 Predict the products for each of the following reactions and draw the complete, detailed mechanisms.

(a)

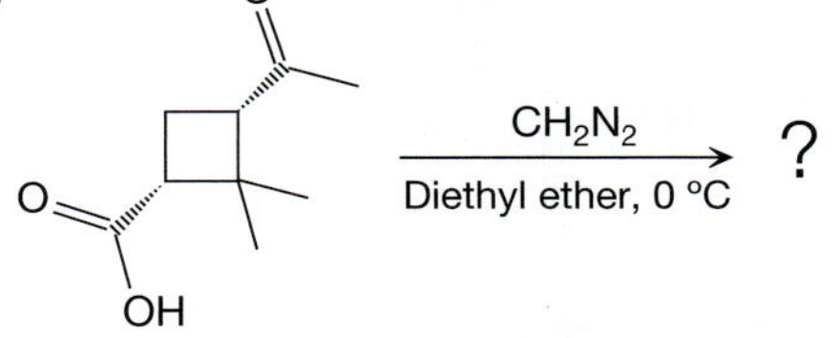

$$\xrightarrow[\text{Diethyl ether, 0 °C}]{\text{CH}_2\text{N}_2}$$?

(b)

$$\xrightarrow[\text{CH}_3\text{OH, benzene, 0 °C}]{\text{CH}_2\text{N}_2}$$?

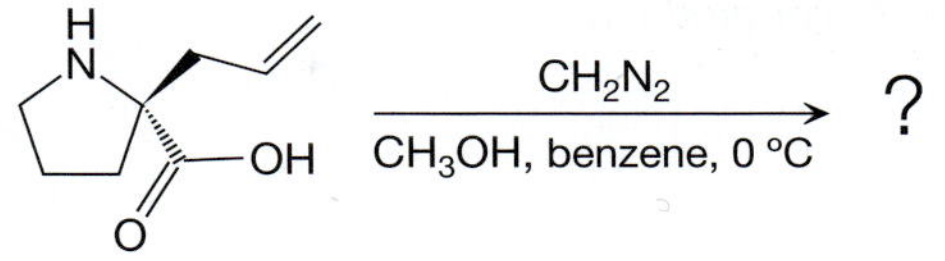

10.44 (SYN) Draw the structure of the carboxylic acid that can be reacted with diazomethane to form each of the following compounds.

(a)

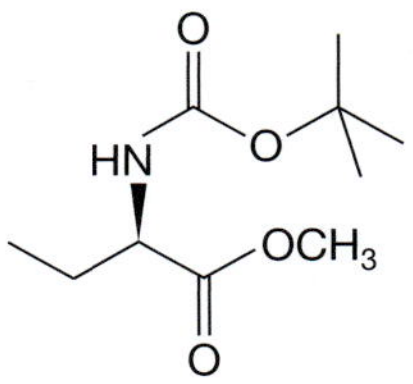

(b)

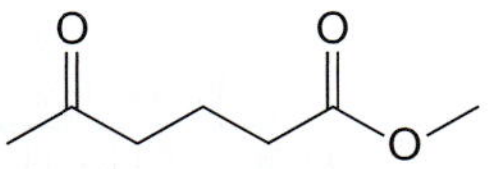

(c)

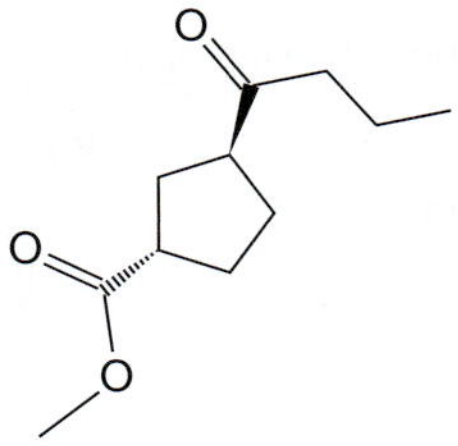

10.45 A compound, $C_2H_2O_4$, reacts with excess diazomethane to produce $C_4H_6O_4$. Draw the structure of each compound.

$$C_2H_2O_4 \xrightarrow{\text{CH}_2\text{N}_2 \text{ (excess)}} C_4H_6O_4$$

10.46 The pK_a of a compound, $C_6H_{12}O_2$, is measured to be 25. Do you expect this compound to react with diazomethane? Explain.

10.6 Formation of Ethers and Epoxides

10.47 Draw a complete, detailed mechanism for the reaction shown here.

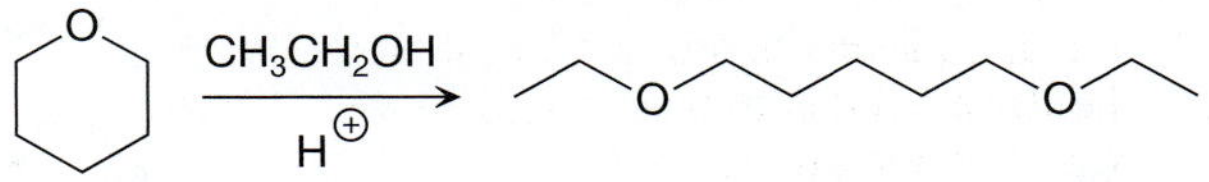

$$\xrightarrow[\text{H}^{\oplus}]{\text{CH}_3\text{CH}_2\text{OH}}$$

10.48 A student wanted to prepare (1-Ethylpropoxy)cyclopentane by heating cyclopentanol and pentan-3-ol under acidic conditions. On carrying out the reaction, however, she found that there were three ethers produced, each containing 10 carbon atoms. One of the ethers was (1-Ethylpropoxy)cyclopentane. Draw the structures of the other two ethers, and draw the complete, detailed mechanisms showing how all three ethers are produced.

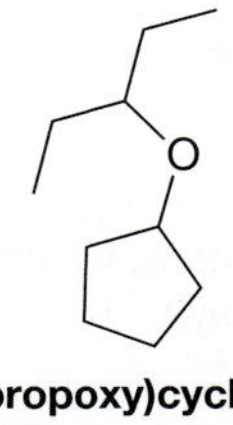

(1-Ethylpropoxy)cyclopentane

10.49 One stereoisomer of 2,6-dibromocyclohexanol is labeled with a ^{13}C isotope (indicated by an asterisk) at one of the C atoms bonded to Br. When this compound is treated with a strong base and heated, only the product shown is formed. Draw the stereoisomer of 2,6-dibromocyclohexanol that is consistent with these results. Explain.

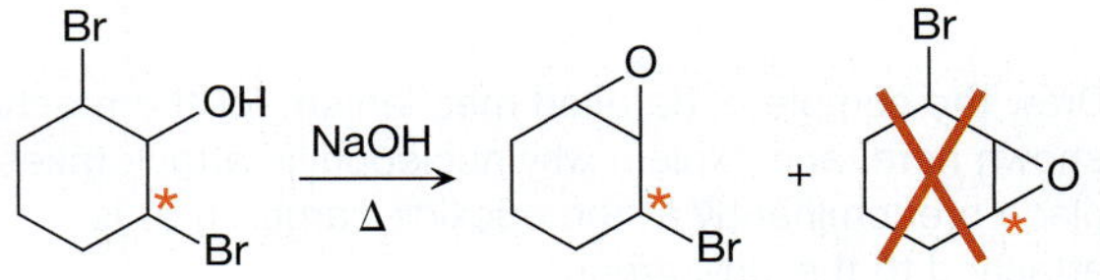

10.50 Draw and name the stereoisomer of 3-bromobutan-2-ol that, on heating in the presence of sodium hydroxide, will produce an epoxide that has no optical activity.

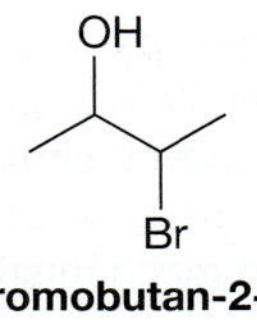

3-Bromobutan-2-ol

10.51 A student attempted to synthesize an epoxide according to the reaction scheme shown here, but no epoxide was formed. Explain why. *Hint:* It may be helpful to build a model of the reactant molecule.

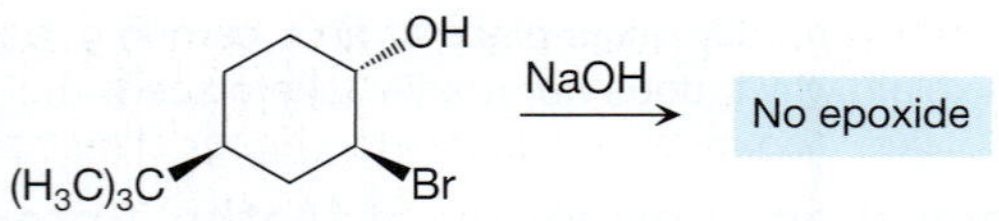

$$(H_3C)_3C \quad \overset{OH}{\cdots} \quad Br \quad \xrightarrow{NaOH} \quad \boxed{\text{No epoxide}}$$

10.7 Epoxides and Oxetanes as Substrates

10.52 For each reaction, draw the complete, detailed mechanism and the major product.

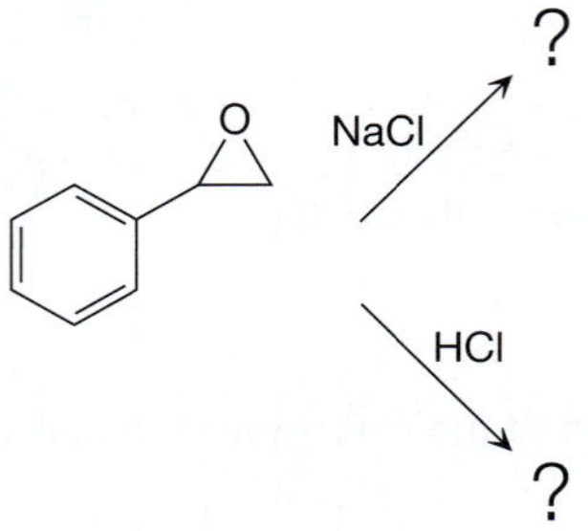

$$\xrightarrow{NaCl} \quad ?$$
$$\xrightarrow{HCl} \quad ?$$

10.53 Draw the product of each of the following reactions.

(a)
$$\xrightarrow[H_2O]{NaN_3} \quad ?$$

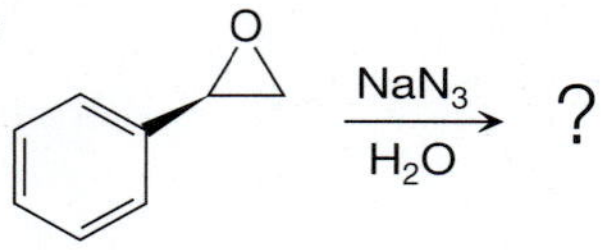

(b)
$$\xrightarrow[HOCH_2CH_2OH]{KCN} \quad ?$$

(c)
$$\xrightarrow[HOCH_2CH_2OH]{HCN} \quad ?$$

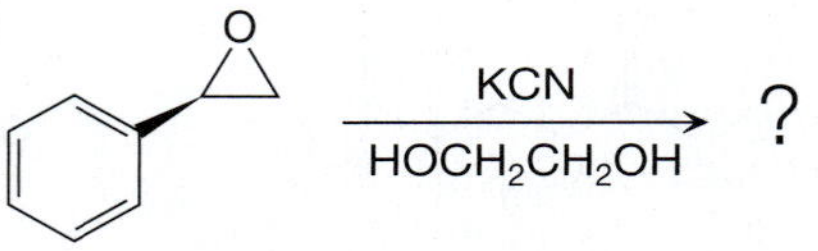

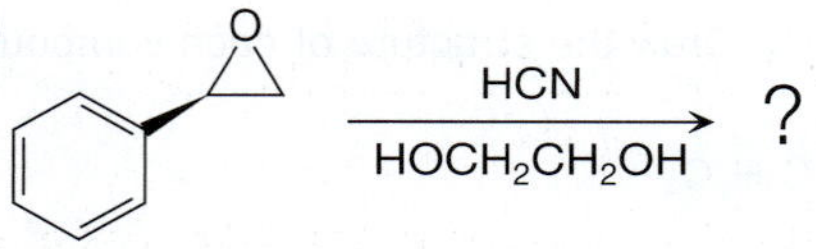

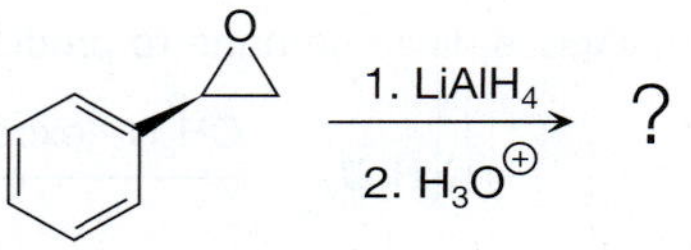

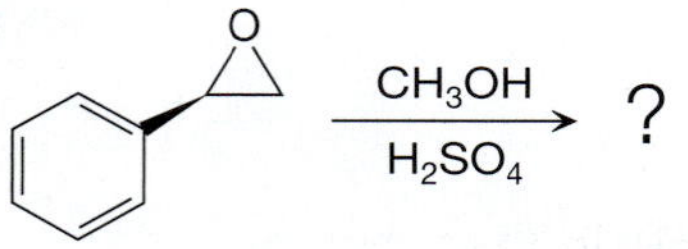

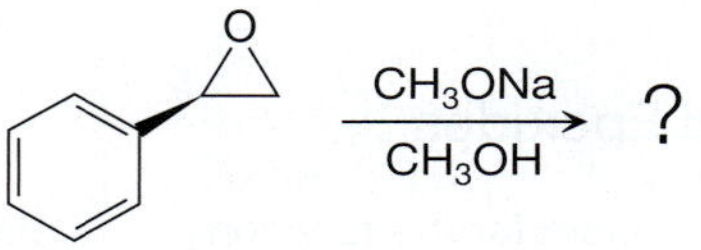

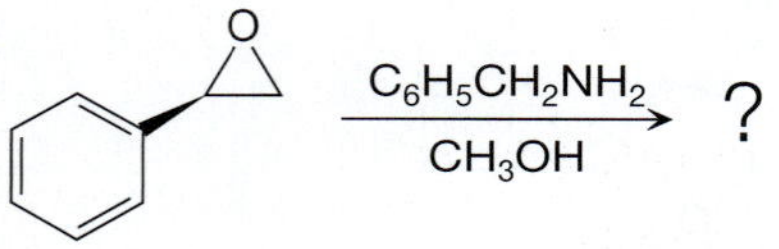

(d)
$$\xrightarrow[2.\ H_3O^{\oplus}]{1.\ LiAlH_4} \quad ?$$

(e)
$$\xrightarrow[H_2SO_4]{CH_3OH} \quad ?$$

(f)
$$\xrightarrow[CH_3OH]{CH_3ONa} \quad ?$$

(g)
$$\xrightarrow[CH_3OH]{C_6H_5CH_2NH_2} \quad ?$$

(h)
$$\xrightarrow[2.\ H_3O^{\oplus}]{1.\ CH_3Li} \quad ?$$

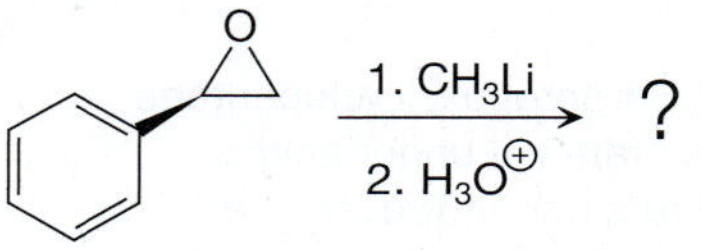

10.54 In the protonated epoxide shown here, which C—O bond would you expect to be longer? Why?

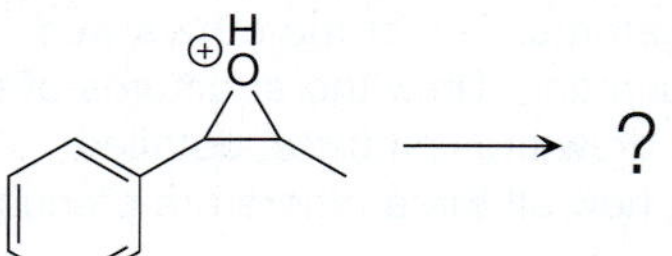

$$\xrightarrow{\quad} \quad ?$$

10.55 Draw the complete, detailed mechanism for the reaction shown here, and explain why nucleophilic attack takes place predominantly at the epoxide carbon that is attached to the vinyl group.

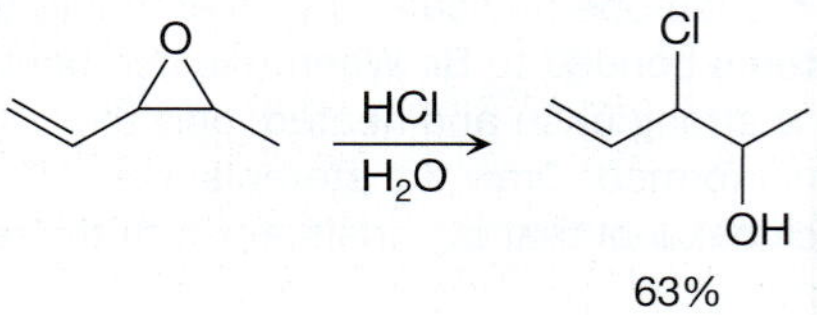

$$\xrightarrow[H_2O]{HCl} \quad \overset{Cl}{\underset{OH}{\diagup}}$$
63%

10.56 An *aziridine* is a compound that contains a three-membered ring consisting of two carbon atoms and a nitrogen atom. Because of the strain in the ring, it behaves similarly to an epoxide. Predict the product of the reaction shown here, and draw the complete, detailed mechanism.

$$\xrightarrow{H_2S} \quad ?$$

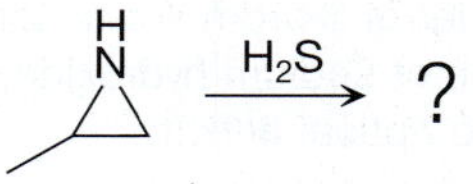

10.8 Generating Alkynes via Elimination Reactions

10.57 For each reaction, draw the complete, detailed mechanism and the major organic product.

(a)

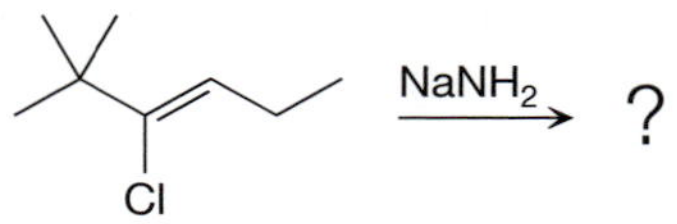

NaNH₂ → ?

(b)

1. NaNH₂ (excess)
2. H₂O
→ ?

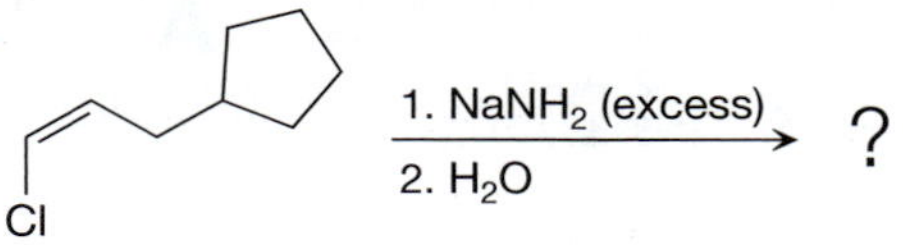

10.58 When hydroxide is used as the base to carry out an E2 reaction on a vinylic halide, the reaction usually needs to be heated significantly. As shown below, such a reaction involving the *E* isomer typically requires much higher temperatures. Why is this so?

Z isomer

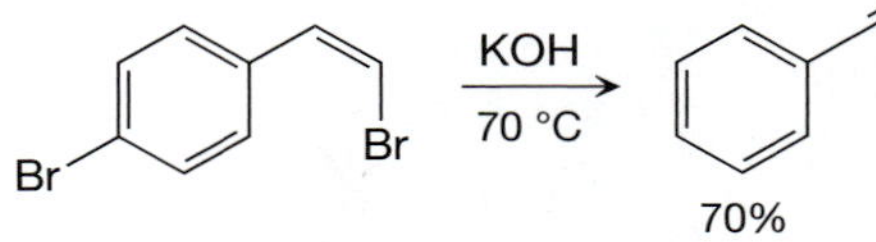

KOH
70 °C
→

70%

E isomer

KOH
200–230 °C
→

67%

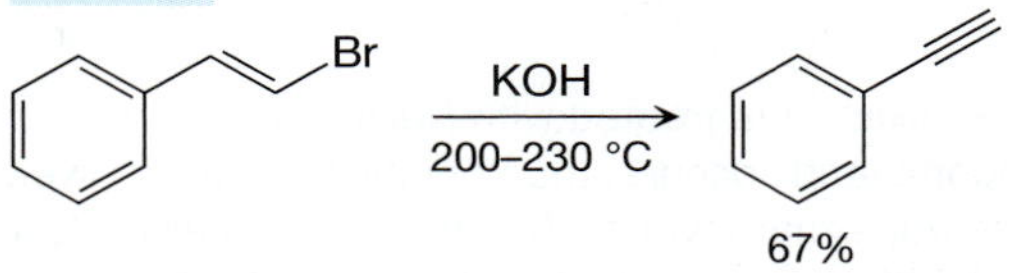

10.59 In the elimination of a vicinal dihalide to yield an internal alkyne, two equivalents of NaNH₂ are required—one for each equivalent of HX that is eliminated. When the product is a terminal alkyne, however, three equivalents of NaNH₂ are required. Explain why.

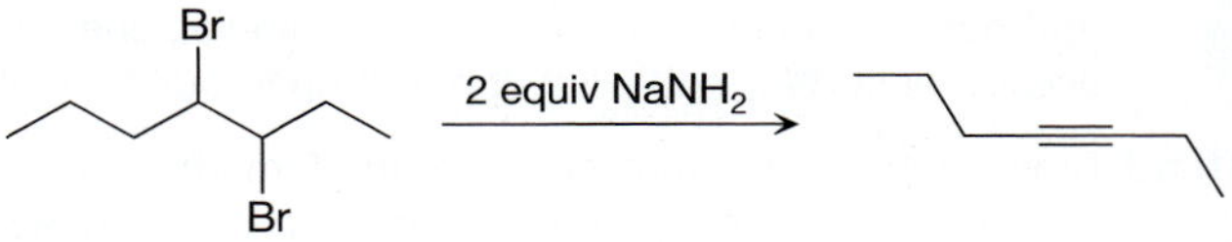

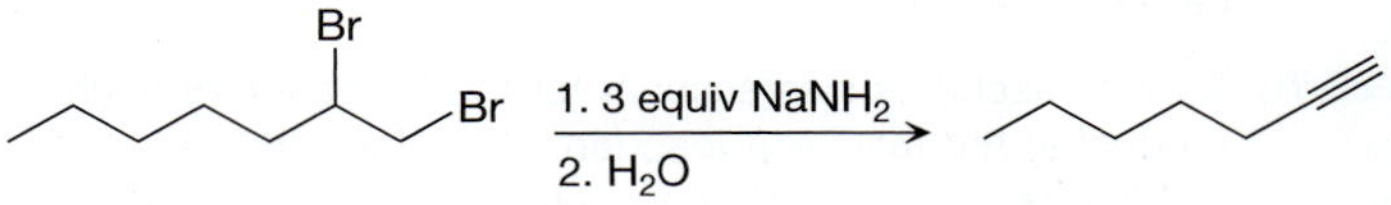

2 equiv NaNH₂ →

1. 3 equiv NaNH₂
2. H₂O
→

10.60 Elimination occurs when (*Z*)-3-bromohex-3-ene is treated with NaNH₂. Under the same conditions, 1-bromocyclohexene undergoes elimination much more sluggishly. Explain why.

10.9 Hofmann Elimination

10.61 For the reaction shown here, draw the complete, detailed mechanism and major product.

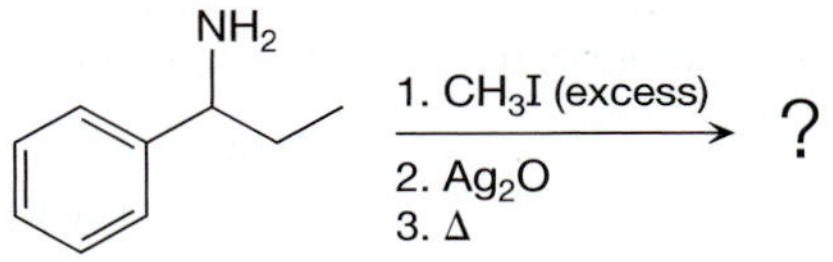

1. CH₃I (excess)
2. Ag₂O
3. Δ
→ ?

10.62 The Hofmann elimination reaction shown here is significantly slower than one involving pentan-3-amine. Explain why. *Hint:* You may want to build a molecular model of this molecule and the intermediates formed in the mechanism.

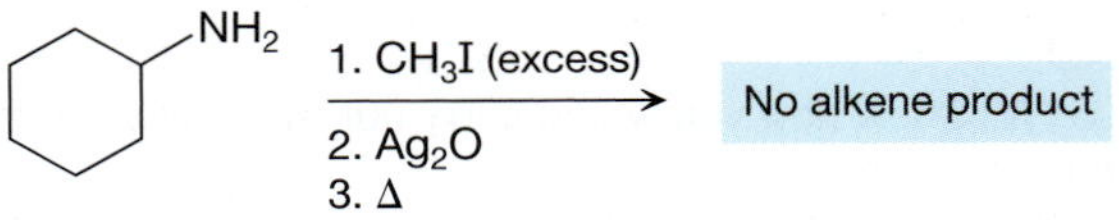

1. CH₃I (excess)
2. Ag₂O
3. Δ
→ No alkene product

10.63 Draw the complete, detailed mechanism for the reaction shown here.

1. CH₃I (excess)
2. Ag₂O
3. Δ
→

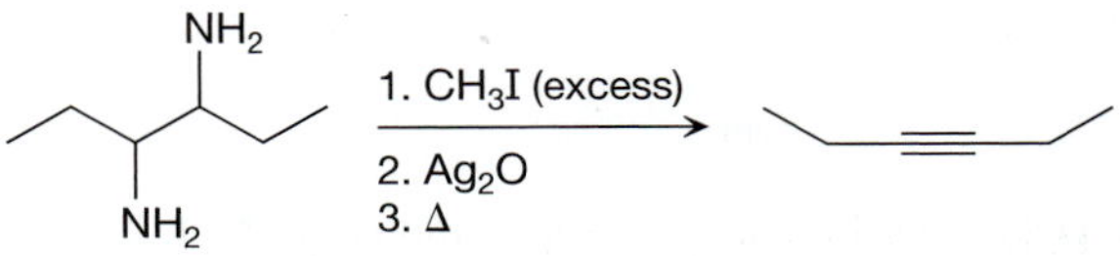

10.64 Under conditions that favor Hofmann elimination, *N*-ethylhexan-3-amine can lead to three different alkene products. Draw the complete mechanism that leads to each alkene product, and predict the major product.

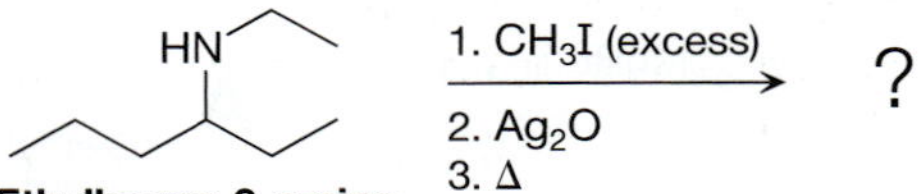

***N*-Ethylhexan-3-amine**

1. CH₃I (excess)
2. Ag₂O
3. Δ
→ ?

10.65 An amine, whose formula is C₉H₁₃N, was treated first with excess iodomethane and then silver oxide, and was finally heated. When the product mixture was analyzed, two isomeric alkenes were found, along with trimethylamine, as indicated here. Draw the structure of the initial amine.

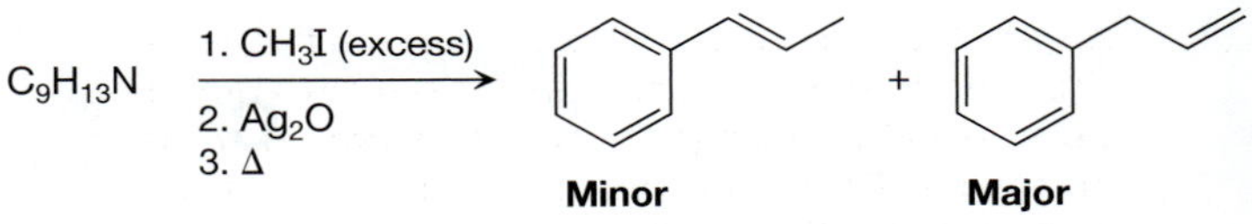

C₉H₁₃N

1. CH₃I (excess)
2. Ag₂O
3. Δ
→

Minor

+

Major

10.66 What are the structures of **A** and **B** in the following reaction sequence?

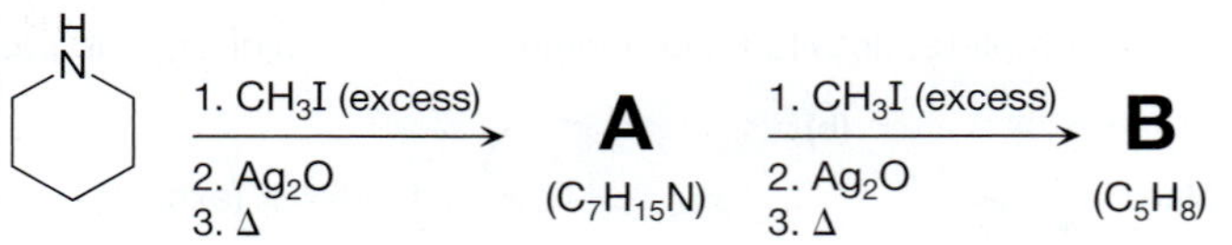

Integrated Problems

10.67 (SYN) Propose how you would carry out the transformation shown here. *Hint:* It may take more than a single reaction.

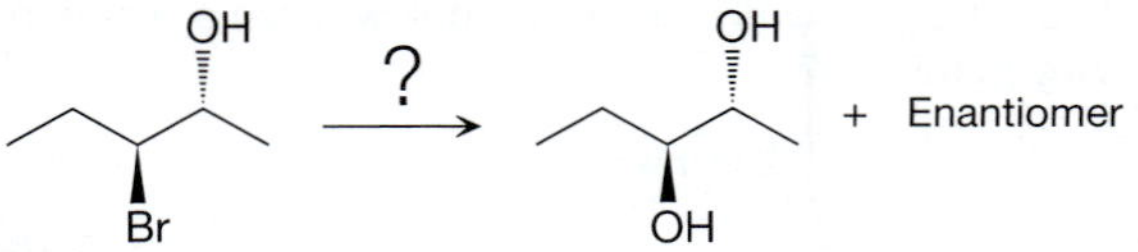

10.68 When oxirane is treated with NaOH, an S_N2 reaction predominantly occurs, thus opening the ring. Given that conditions that favor an S_N2 reaction generally favor an E2 reaction, too, we can also write an E2 mechanism that opens the ring. The E2 reaction, however, does *not* occur readily. Why not? *Hint:* It may help to build a molecular model of the epoxide.

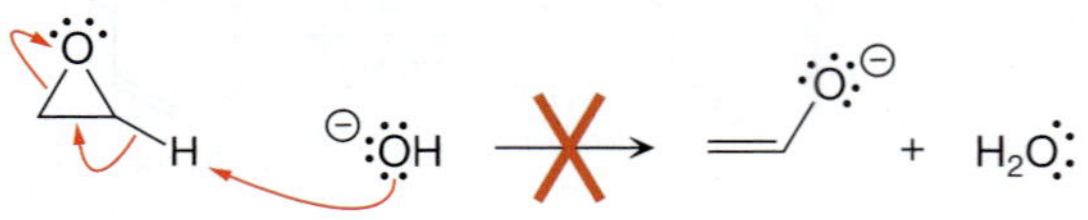

10.69 Draw a mechanism to account for the formation of the product in the reaction shown here. *Hint:* Under these conditions, deprotonation of a propargylic (C≡C—CH) carbon is reversible.

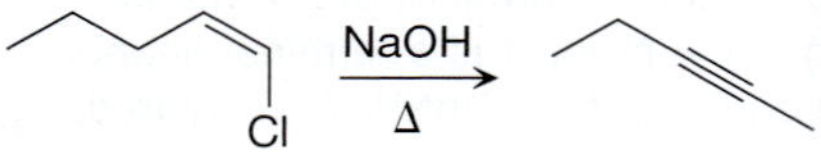

10.70 Draw a mechanism to account for the formation of each product in the following reaction. *Hint:* Under these conditions, deprotonation of a propargylic (C≡C—CH) carbon is reversible.

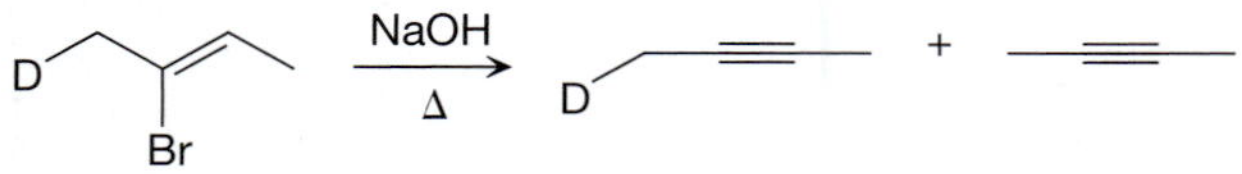

10.71 (SYN) The reaction in Problem 10.48 produces a mixture of ethers rather than the desired ether exclusively. Starting with the same two alcohols given in Problem 10.48, show the sequence of reactions that would need to be carried out to produce (1-Ethylpropoxy)cyclopentane as the exclusive ether.

10.72 When the following 2,3-epoxyalcohol is dissolved in aqueous base, an equal mixture of the two compounds shown is produced. Draw a complete, detailed mechanism for this reaction.

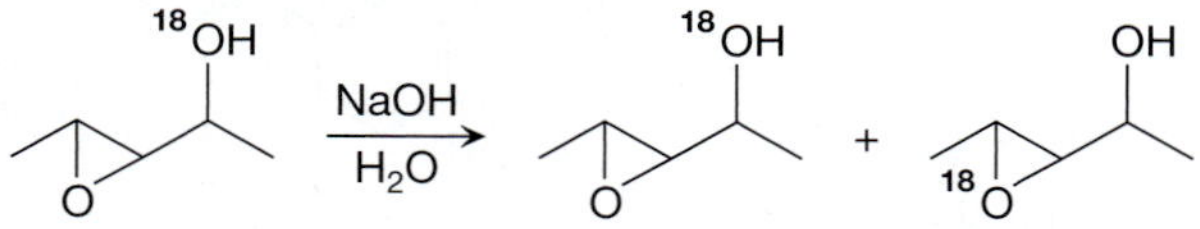

10.73 (SYN) Show how you would carry out the following transformations. *Hint:* Each transformation may require more than one reaction.

(a)

(b)

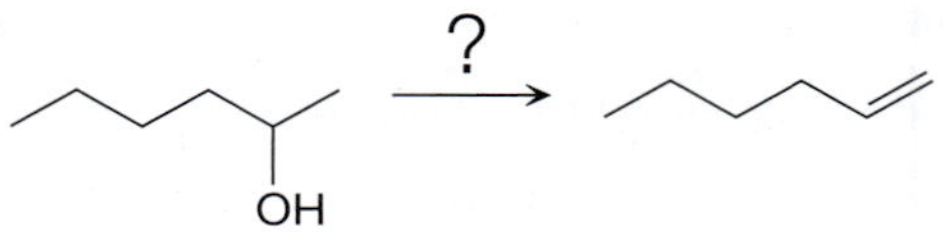

10.74 Determine the structures of compounds **A**, **B**, **C**, and **D** in the following reaction sequences.

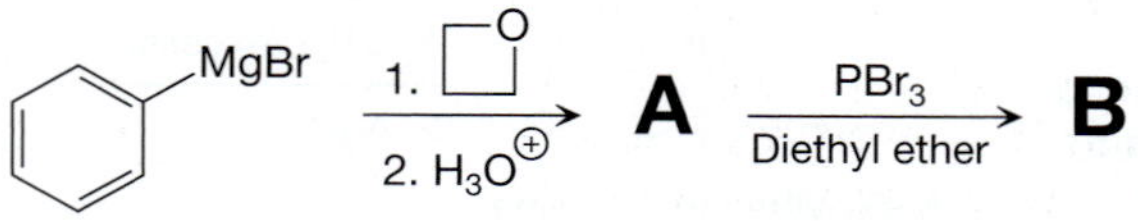

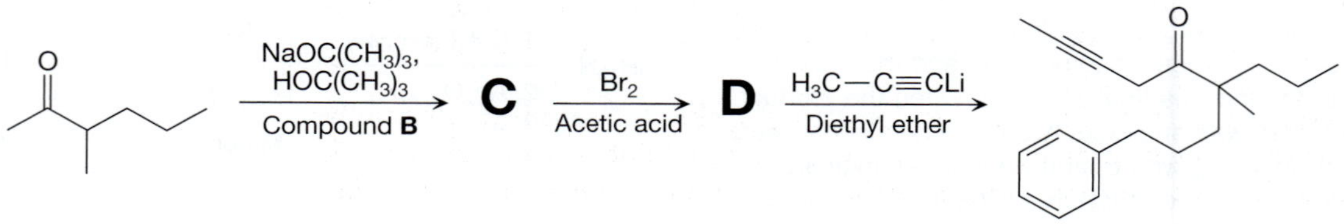

Mycobacterium tuberculosis is the bacterial species responsible for most cases of tuberculosis, which attacks the lungs. These bacteria use an electrophilic addition reaction to produce branched-chain fatty acids, which become incorporated into their cell walls.

Electrophilic Addition to Nonpolar π Bonds 1

Addition of a Brønsted Acid

In the last several chapters, we have seen that bond formation is frequently driven by the flow of electrons from an electron-rich species to an electron-poor species. In the reactions we have examined in depth so far, the electron-rich species, acting as a nucleophile or base, typically possesses an atom that has not only a partial or full negative charge, but also a lone pair of electrons. In such cases, those electrons are the ones used to form the new bond.

Here in Chapter 11 and later in Chapter 12, we discuss **electrophilic addition reactions**, whose mechanisms contain an *electrophilic addition step* (or a variation of it). Recall from Section 7.6 that the electrons used to form a new bond in an electrophilic addition step originate from a nonpolar π bond, such as the C=C double bond of an alkene or the C≡C triple bond of an alkyne. These bonds are relatively electron rich because there are multiple electrons confined to the region between two atoms (Fig. 11-1a). Thus, alkenes and alkynes tend to react with electrophiles (i.e., with electron-poor species).

Two other factors contribute to the reactivity of the π electrons of a nonpolar double or triple bond compared to an analogous single bond. One is the fact that π electrons are, on average, located farther away from a molecule's nuclei than

Chapter Objectives

On completing Chapter 11 you should be able to:

1. Draw the general mechanism for the addition of a strong Brønsted acid to an alkene or alkyne and explain which step is rate determining.

2. Specify why benzene rings tend *not* to participate in electrophilic addition reactions.

3. Predict the major products of the addition of a strong Brønsted acid to an alkene or alkyne, including both regiochemistry and stereochemistry.

4. Explain the role of an acid catalyst in the addition of a weak Brønsted acid to an alkene or alkyne.

5. Predict the major product of hydration of an alkyne and draw the mechanism for its formation.

6. Identify the products of 1,2- and 1,4-addition to a conjugated diene and draw the mechanisms for their formation.

7. Predict the major product of electrophilic addition to a conjugated diene under kinetic control and under thermodynamic control, and explain why they can differ.

σ electrons are, as shown in Figure 11-1b. Therefore, π electrons are more accessible spatially. The second factor is the relative energy of the π electrons. As indicated by Figure 11-1c, π electrons are higher in energy, and thus less stable than σ electrons. Consequently:

> It is easier to break the π bond of a double bond than the σ bond of an analogous single bond.

We examine a variety of electrophilic addition reactions here in Chapter 11 and in Chapter 12. The complexity of these reactions can vary greatly. Some are stepwise additions, whose mechanisms consist of two or more steps, while others are concerted, taking place in a single step. Some even result in the formation of rings. Realizing the complexity that electrophilic addition reactions can have, we begin with the addition of a Brønsted acid (HA) to an alkene, a prototype of electrophilic addition reactions. We then explore the intricacies of the other electrophilic addition mechanisms.

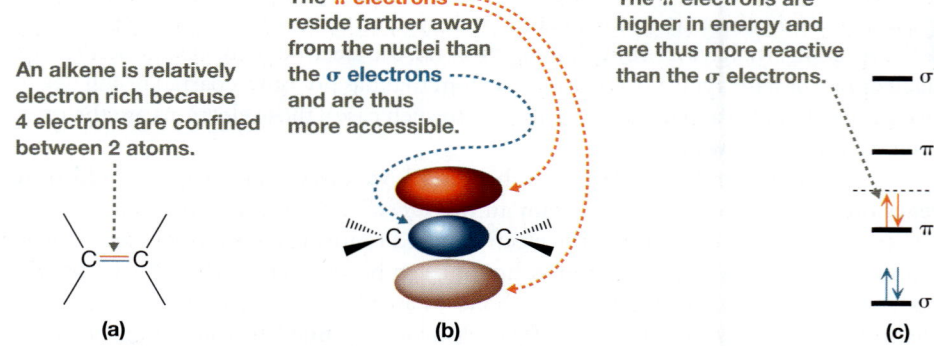

An alkene is relatively electron rich because 4 electrons are confined between 2 atoms.

The π electrons reside farther away from the nuclei than the σ electrons and are thus more accessible.

The π electrons are higher in energy and are thus more reactive than the σ electrons.

(a)   (b)   (c)

FIGURE 11-1 Relative reactivity of π electrons (a) The electrons that make up a nonpolar π bond are relatively electron rich because there are multiple electrons confined to the region between two atoms. In general, π electrons are more reactive than σ electrons because π electrons are (b) more easily accessed spatially and (c) higher in energy.

11.1 The General Electrophilic Addition Mechanism: Addition of a Strong Brønsted Acid to an Alkene

Equation 11-1 shows a prototypical electrophilic addition reaction, in which cyclohexene reacts with HCl, a strong Brønsted acid, to produce chlorocyclohexane, the **adduct** (i.e., the addition product). As we can see, one alkene carbon forms a new bond to H, and the other one forms a new bond to Cl. The HCl molecule is said to *add across* the C=C double bond. Similarly, in the reaction in Equation 11-2, H_2SO_4 adds across the C=C double bond in 2,3-dimethylbut-2-ene.

Cyclohexene → HCl → **Chlorocyclohexane** 95%

HCl adds across the double bond. (11-1)

2,3-Dimethylbut-2-ene → conc H_2SO_4 → OSO_3H / H

H_2SO_4 adds across the double bond. (11-2)

YOUR TURN **11.1**

Describe what takes place in the following reaction:

(E)-But-2-ene + HBr → Br / H **2-Bromobutane**

Answers to Your Turns are in the back of the book.

These reactions take place via the two-step mechanism shown in Equation 11-3 for the addition of HCl. Step 1 is an electrophilic addition step, in which a pair of electrons from the electron-rich π bond forms a bond to the acid's electron-poor H atom (review Section 7.6). This leaves one of the initial alkene C atoms with only three bonds, giving it a +1 formal charge. Cl^- is produced as a result of the concurrent breaking of the H—Cl bond. Step 2 is a coordination step, whereby Cl^- forms a bond to the carbocation. It is identical to the second step of the S_N1 mechanism discussed in Chapter 8.

Mechanism for Equation 11-1

The electron-rich alkene attacks the electron-poor H.

The electron-rich chloride anion attacks the electron-poor carbocation.

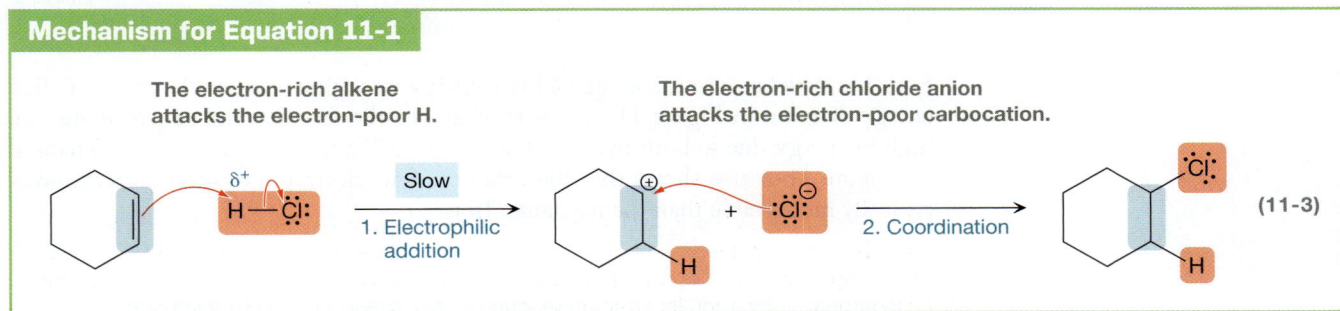

Slow
1. Electrophilic addition

2. Coordination

(11-3)

The mechanism for the addition of HBr across the double bond in cyclohexene is as follows, but the curved arrows have been omitted:

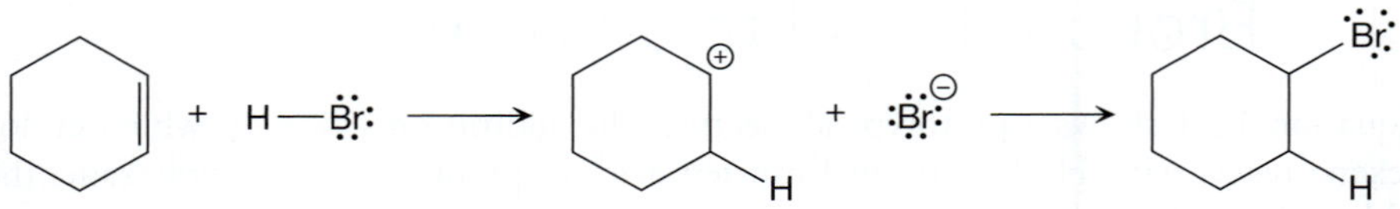

The mechanism is essentially identical to the one in Equation 11-3. Supply the missing curved arrows and write the name of each elementary step below the appropriate reaction arrow. Also, label the appropriate electron-rich and electron-poor sites in each step.

PROBLEM 11.1 Draw the complete, detailed mechanism and predict the major product for each of the following reactions. Neglect stereochemistry.

(a)

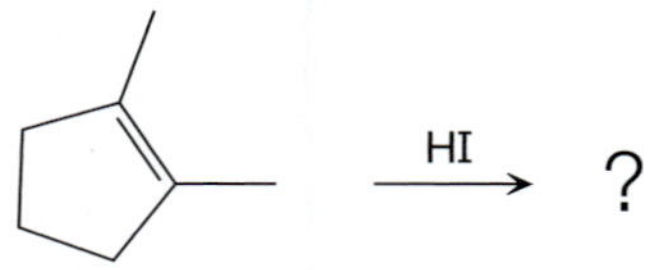

(b)

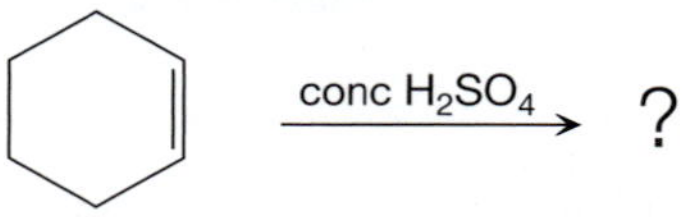

(c)

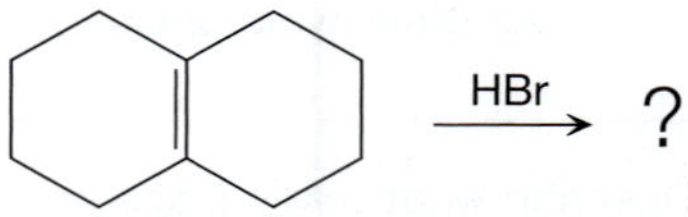

(d)

PROBLEM 11.2 Show how each of the following compounds can be produced from an alkene.

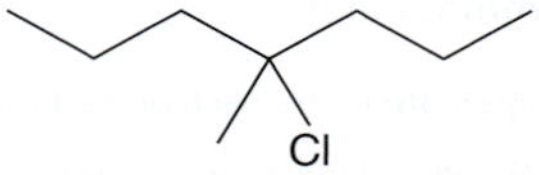

(a)

(b)

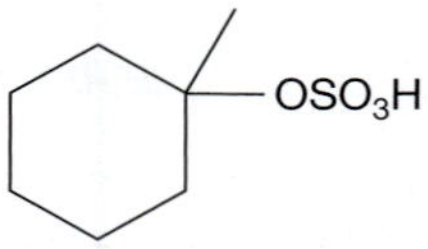

(c)

Step 1 in Equation 11-3 is the slow step, so:

> In an electrophilic addition of a Brønsted acid to an alkene, the rate-determining step is Step 1: addition of the H^+ electrophile.

Step 1 is slow because it has such a high-energy transition state, as shown in the free energy diagram in Figure 11-2. This is related to the fact that the intermediates are high in energy, due to both the lack of an octet on C and the creation of two charges.

Figure 11-2 also shows that the product of an electrophilic addition reaction is typically more stable than the reactants. Thus:

> In general, electrophilic addition reactions are energetically quite favorable.

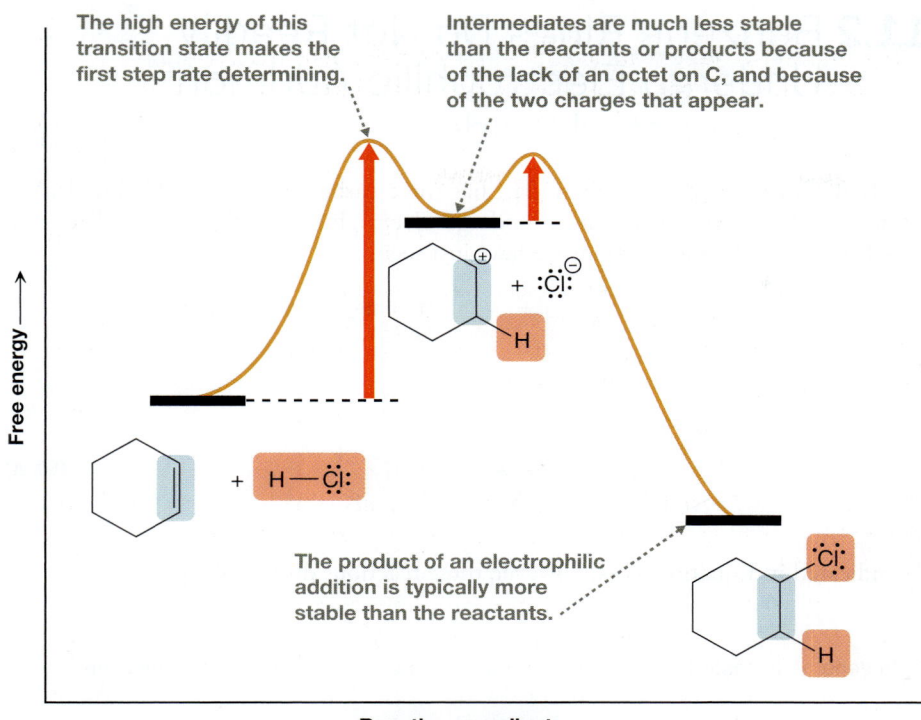

The high energy of this transition state makes the first step rate determining.

Intermediates are much less stable than the reactants or products because of the lack of an octet on C, and because of the two charges that appear.

The product of an electrophilic addition is typically more stable than the reactants.

Free energy →

Reaction coordinate →

FIGURE 11-2 Reaction energy diagram for the electrophilic addition of a strong acid to an alkene The intermediates are much higher in energy than the reactants or products, due to the lack of an octet on C and the presence of two charges. The transition state of the first step is higher in energy than that of the second step, which makes the first step rate determining.

These reactions tend to be favorable because the sum of the bond energies in the product is greater than that in the reactants. Figure 11-3 shows that a C=C double bond and a single bond to H are replaced by three single bonds. Thus, the overall change in bond energy is largely accounted for by the gain of a σ bond at the expense of the loss of a π bond. Because the σ bond is stronger than the π bond (Section 3.6), the total bond strength increases throughout the course of the reaction.

YOUR TURN 11.3

Verify that the addition of HCl across a C=C double bond is energetically favorable by using the table of bond energies from Chapter 1 (p. 10) to estimate ΔH°_{rxn}. Is the value significantly positive or significantly negative?

$\Delta H^\circ_{rxn} = ($ _____ kJ/mol + _____ kJ/mol) − (_____ kJ/mol + _____ kJ/mol + _____ kJ/mol)

⎵ C=C   H—Cl ⎵   ⎵ C—C   H—C   C—Cl ⎵

Sum of energies of bonds lost

Sum of energies of bonds gained

= _____ kJ/mol

σ Bond + π bond   σ Bond   σ Bond   σ Bond   σ Bond

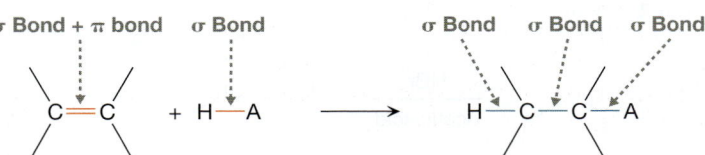

A σ bond is stronger than a π bond, so the total bond strength is greater in the product than in the reactants.

FIGURE 11-3 Driving force for electrophilic addition Two σ bonds and a π bond are lost (red) and three σ bonds are gained (blue). Because of the greater strength of a σ bond, the products are lower in energy.

11.2 Benzene Rings Do Not Readily Undergo Electrophilic Addition of Brønsted Acids

The Lewis structure of benzene, C_6H_6, has three carbon–carbon double bonds. We might therefore expect benzene to undergo electrophilic addition with a Brønsted acid, HA, similar to the reactions we have just seen.

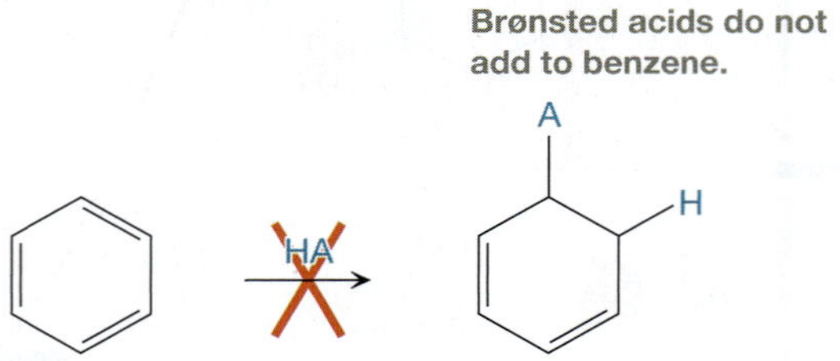

Brønsted acids do not add to benzene.

$$ \text{(11-4)} $$

As indicated in Equation 11-4, however, these reactions do *not* take place.

> In general, Brønsted acids do *not* add across a C=C double bond of benzene.

The π electrons of benzene are too heavily stabilized.

The source of this stability is a phenomenon called *aromaticity*, a topic of Chapter 14. Although we will not fully discuss aromaticity here, it is helpful to know that benzene's aromaticity results in the six π electrons being completely delocalized around the ring, just as we would expect in the hybrid of benzene's two resonance structures:

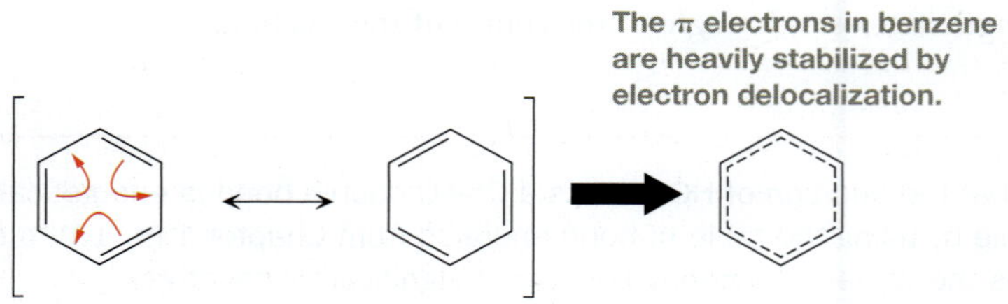

The π electrons in benzene are heavily stabilized by electron delocalization.

If HA were to add across one of the C=C double bonds, as shown previously in Equation 11-4, then this cyclic delocalization of the π electrons would cease to exist, and the resulting stabilization would be destroyed.

If a species contains both a benzene ring and a separate C=C, then treatment with a Brønsted acid will lead to a reaction at the separate C=C double bond, as shown in Equation 11-5.

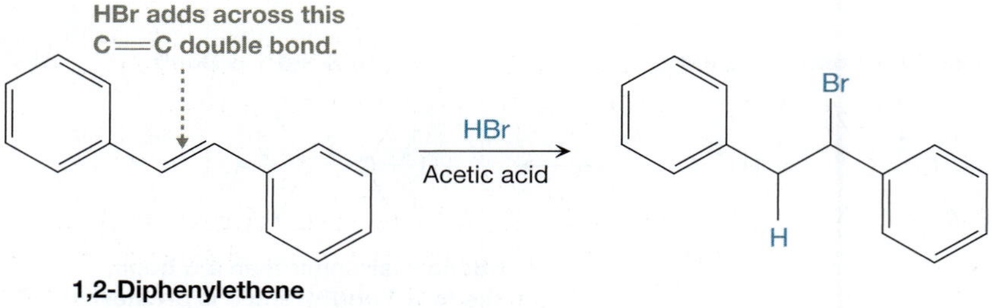

HBr adds across this C=C double bond.

$$ \xrightarrow[\text{Acetic acid}]{\text{HBr}} \qquad \text{(11-5)} $$

1,2-Diphenylethene

11.3 Regiochemistry: Production of the More Stable Carbocation and Markovnikov's Rule

In the addition of a Brønsted acid across a double bond, the proton can bond to one of two possible carbon atoms. In each of the reactions we have examined thus far, the alkene has been *symmetric* (i.e., the groups attached to one of the alkene carbons are identical to the groups attached to the other), so the same adduct is produced regardless of which alkene carbon gains the proton. With an *unsymmetric* alkene, on the other hand, two constitutional isomers can be produced, as shown in Equation 11-6 for the reaction of propene with HCl. In this case, 2-chloropropane is the major product, not 1-chloropropane.

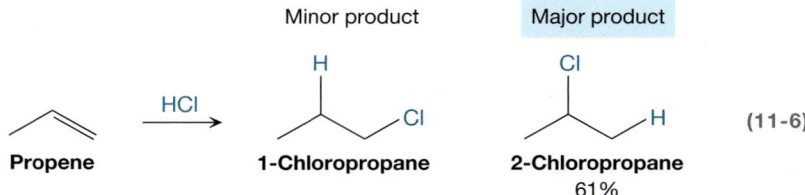

Addition to an unsymmetric
alkene can produce two products.

Minor product   Major product

Propene   1-Chloropropane   2-Chloropropane
61%

(11-6)

Why is 2-chloropropane the major product in Equation 11-6? To begin, recall from Section 11.1 that the overall product of an electrophilic addition reaction is significantly more stable than the overall reactants. Thus, applying what we learned from Section 9.12:

> Electrophilic addition reactions tend to be irreversible and generally take place under *kinetic control*.

As a result, the major product is the one that is produced the *fastest*.

Also recall from Section 11.1 that the first step of the mechanism—formation of the carbocation intermediate—is the rate-determining step of the overall reaction. Thus, the major product derives from the *carbocation intermediate* that is produced most rapidly.

With this in mind, let's examine Figure 11-4, the free energy diagrams for the formation of 1-chloropropane (minor product; red curve) and 2-chloropropane (major product; blue curve) from propene. The carbocation intermediate that is produced by the addition of H^+ to the terminal C (blue curve) is *secondary*, so it is lower in energy than the *primary* carbocation intermediate that is produced by the addition of H^+ to the central C (red curve). Consequently, according to the Hammond postulate (Section 9.3a), production of the secondary carbocation involves a smaller energy barrier. Thus, the secondary carbocation is formed faster and must lead to the major product, consistent with what we saw in Equation 11-6.

The lesson we learn here from the addition of HCl to propene can be generalized:

> The major product of an electrophilic addition of a Brønsted acid to an alkene is the one that proceeds through the more stable carbocation intermediate.

Applying this idea to the addition of H_2SO_4 to dodec-1-ene (Equation 11-7), we can see why the major product is the one in which the proton has added to the terminal carbon (see Your Turn 11.4). Moreover, this concept allows us to *predict* the major

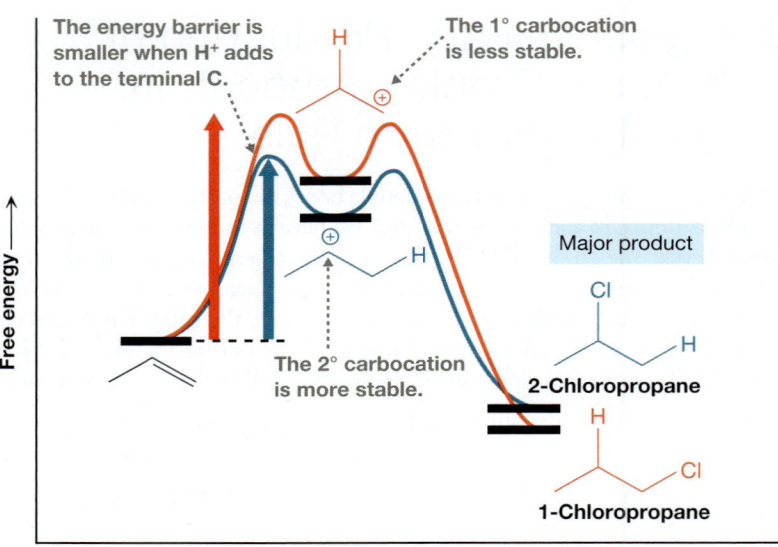

FIGURE 11-4 Regiochemistry and relative energy barriers in electrophilic addition Electrophilic addition of HCl to propene to form 2-chloropropane is represented by the blue curve, whereas formation of 1-chloropropane via the same reaction is represented by the red curve. The faster rate of the blue reaction is consistent with a smaller energy barrier, which, in turn, is consistent with a more stable carbocation intermediate.

The energy barrier is smaller when H⁺ adds to the terminal C.

The 1° carbocation is less stable.

The 2° carbocation is more stable.

Major product

2-Chloropropane

1-Chloropropane

Free energy →

Reaction coordinate →

products of other electrophilic addition reactions of Brønsted acids to alkenes, exemplified by Solved Problem 11.3.

The more stable carbocation intermediate is produced when H⁺ adds to the terminal C atom.

Dodec-1-ene → H_2SO_4 → **78%** (11-7)

YOUR TURN 11.4

The free energy diagrams for the reactions that produce the isomeric adducts in Equation 11-7 are shown below. Complete the diagram by drawing the carbocation intermediates in the appropriate boxes provided.

Free energy →

+ H_2SO_4

OSO_3H

OSO_3H

Reaction coordinate →

SOLVED PROBLEM 11.3

Predict the major product when indene is treated with HCl.

Indene

HCl → ?

CONNECTIONS Indene is used in industry to make coumarone-indene resin, which is applied as an edible, protective coating to citrus fruits and apples.

Think Which C=C double bond will undergo electrophilic addition? What are the *possible* products and the corresponding carbocation intermediates from which they are produced? Which carbocation intermediate is more stable?

Solve The rightmost C=C double bond is the one that will undergo electrophilic addition. The others make up a benzene ring and are much too stable to react under these conditions. The two possible products of HCl addition differ by which C atom gains the H⁺ and which gains the Cl⁻:

A

The benzylic carbocation is resonance stabilized.

B

Major product

We can predict the major product by identifying the more stable carbocation intermediate. As indicated in the reaction scheme, carbocation **B** is benzylic and therefore is resonance stabilized, giving it significantly greater stability than carbocation **A**. Thus, the product that derives from **B** is the major product.

YOUR TURN 11.5

In Solved Problem 11.3, carbocation **B** is said to be resonance stabilized. Draw all resonance structures of that carbocation.

PROBLEM 11.4 Draw the detailed mechanism for the reaction of each of the following with HCl and predict the major product.

(a) (b) (c) (d) NO₂

Electrophilic Addition and Laser Printers

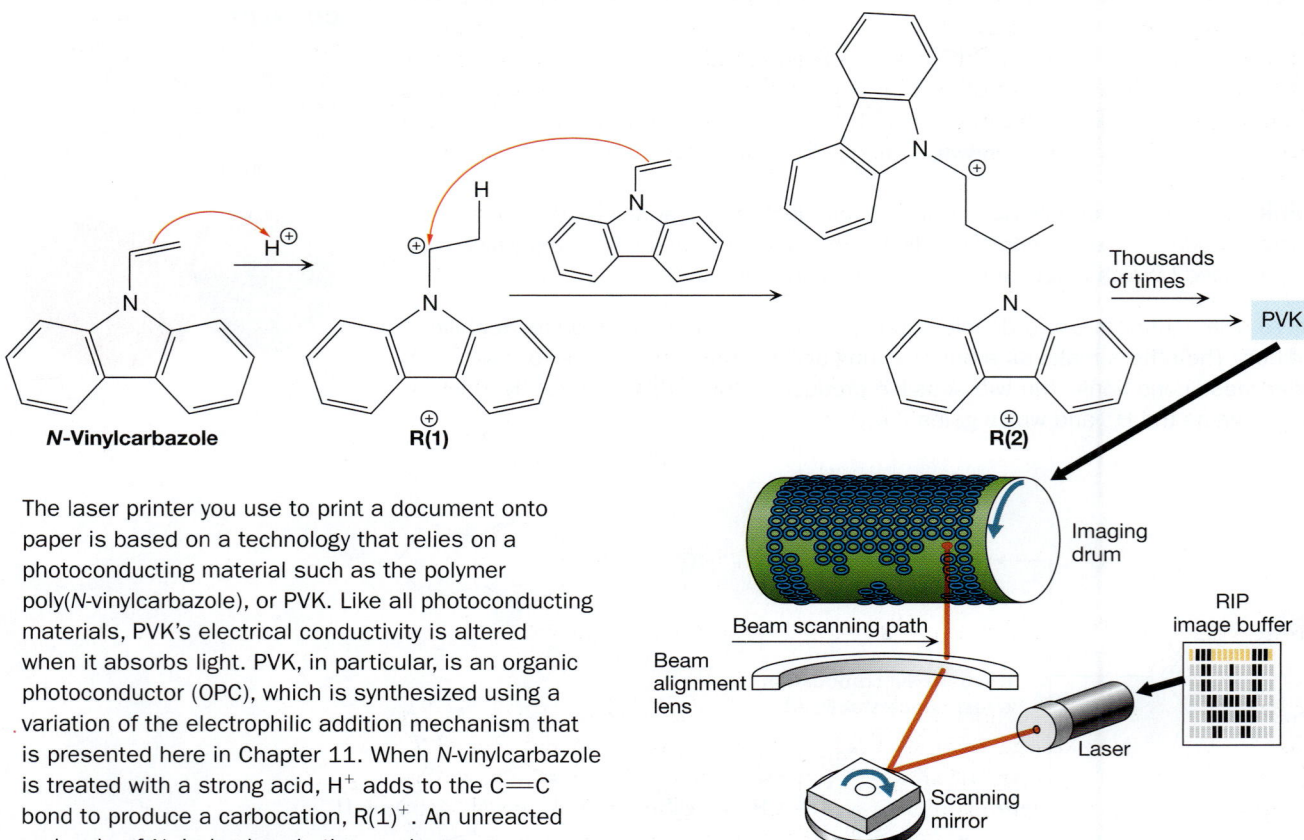

N-Vinylcarbazole

R(1)

R(2)

Thousands of times

PVK

Imaging drum

Beam scanning path

Beam alignment lens

Scanning mirror

Laser

RIP image buffer

The laser printer you use to print a document onto paper is based on a technology that relies on a photoconducting material such as the polymer poly(N-vinylcarbazole), or PVK. Like all photoconducting materials, PVK's electrical conductivity is altered when it absorbs light. PVK, in particular, is an organic photoconductor (OPC), which is synthesized using a variation of the electrophilic addition mechanism that is presented here in Chapter 11. When N-vinylcarbazole is treated with a strong acid, H^+ adds to the C=C bond to produce a carbocation, $R(1)^+$. An unreacted molecule of N-vinylcarbazole then undergoes electrophilic addition of $R(1)^+$ to produce a larger carbocation, $R(2)^+$. This process is repeated thousands of times, resulting in PVK.

In a laser printer, an imaging drum coated with an OPC picks up charged particles when it is exposed to a high voltage. As the drum spins, an image to be printed is scanned onto the drum using a light source such as a laser or an array of light-emitting diodes. Because of the OPC's properties, areas of the drum that are exposed to that image of light eject their charged particles. The surface of the drum is then treated with toner (dry ink), which adheres to the remaining charged particles. As paper is fed through the printer and comes in contact with the drum's surface, the toner is deposited and burned onto the paper. Once the drum's surface is cleaned, the cycle repeats.

PROBLEM 11.5 Show how each of the given compounds can be synthesized from two different alkenes.

(a)

(b)

(c)

Before the mechanism for electrophilic addition reactions was known, the Russian chemist Vladimir Markovnikov (1838–1904) made the following generalization to describe the reaction's regioselectivity: *The addition of a hydrogen halide to an alkene favors the product in which the proton adds to the alkene carbon that is initially bonded to the greater number of hydrogen atoms.* Thus, the H^+ forms a bond to the terminal alkene C when

HCl adds to propene (Equation 11-6, p. 569); the terminal alkene C atom is initially bonded to two H atoms, whereas the central C atom is initially bonded to only one.

We now know that Markovnikov's generalization is just the outcome of the reaction favoring the more stable carbocation intermediate. Nonetheless, his generalization has come to be known as **Markovnikov's rule**, and the type of regioselectivity it describes is called **Markovnikov addition**. In Chapter 12, we present examples of electrophilic addition reactions with the opposite regioselectivity—so-called **anti-Markovnikov addition**—whose mechanisms do not involve carbocation intermediates at all.

11.4 Carbocation Rearrangements

When 3-methylbut-1-ene is treated with hydrochloric acid (Equation 11-8), significant amounts of both 2-chloro-3-methylbutane and 2-chloro-2-methylbutane are produced.

The production of this adduct involves a carbocation rearrangement.

| 3-Methylbut-1-ene | → HCl → | 2-Chloro-3-methylbutane 45% | + | 2-Chloro-2-methylbutane 45% | (11-8) |

2-Chloro-3-methylbutane is the product of a normal Markovnikov addition of HCl across the C=C double bond. Specifically, H⁺ adds to the terminal alkene C and Cl⁻ adds to the adjacent, secondary alkene C. By contrast, 2-chloro-2-methylbutane does not appear to be the product of addition of HCl across the double bond because the C atom to which the Cl⁻ attaches was not initially part of the double bond.

As with any mechanism that proceeds through a carbocation intermediate:

> Electrophilic addition of a Brønsted acid across a C=C double bond is susceptible to carbocation rearrangements.

Recall from Chapter 8 that carbocation rearrangements are fast! In this case, a 1,2-hydride shift can account for the formation of 2-chloro-2-methylbutane, as shown in the mechanism in Equation 11-9. This 1,2-hydride shift transforms a less stable secondary carbocation into a more stable tertiary carbocation prior to the attack of the Cl⁻ nucleophile.

1,2-Methyl shifts also take place quickly, so they can appear in the mechanism of an electrophilic addition reaction, too. An example is illustrated in Solved Problem 11.6.

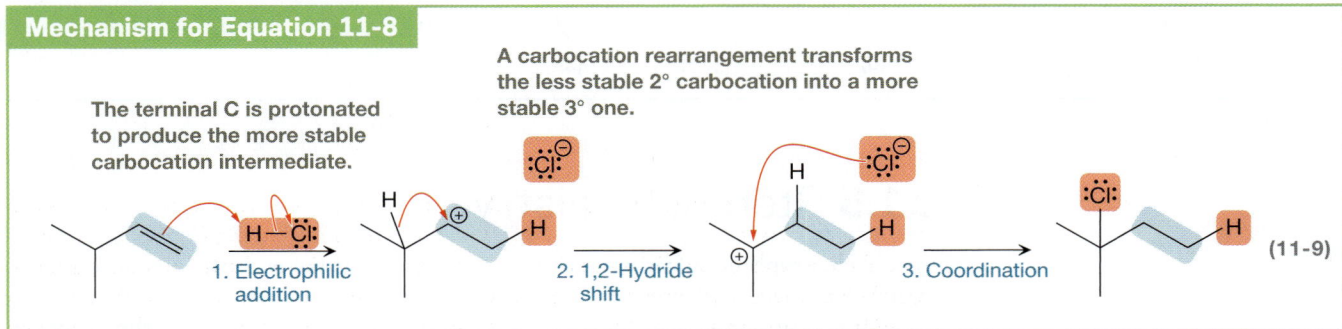

Mechanism for Equation 11-8

The terminal C is protonated to produce the more stable carbocation intermediate.

A carbocation rearrangement transforms the less stable 2° carbocation into a more stable 3° one.

1. Electrophilic addition   2. 1,2-Hydride shift   3. Coordination   (11-9)

SOLVED PROBLEM 11.6

Draw the complete, detailed mechanism and predict the major product of the reaction shown here.

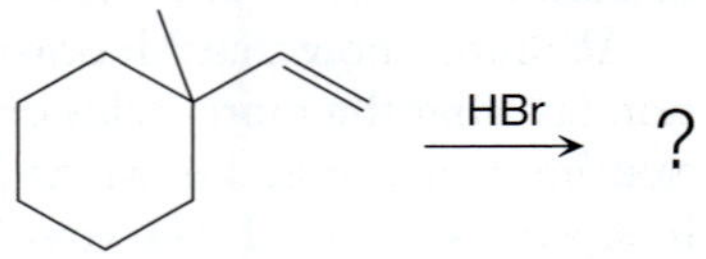

Think What carbocation intermediate is produced on addition of H^+? Can that carbocation intermediate undergo a 1,2-hydride shift or a 1,2-methyl shift to attain greater stability?

Solve In Step 1 of this electrophilic addition reaction, H^+ adds to the terminal alkene C to produce a secondary carbocation intermediate, as shown below. (Addition of H^+ to the internal alkene C would instead produce a less stable primary carbocation intermediate.)

The more stable carbocation intermediate is produced when H^+ adds to the terminal C.

The less stable 2° carbocation intermediate transforms into a more stable 3° one via a 1,2-methyl shift.

1. Electrophilic addition

2. 1,2-Methyl shift

3. Coordination

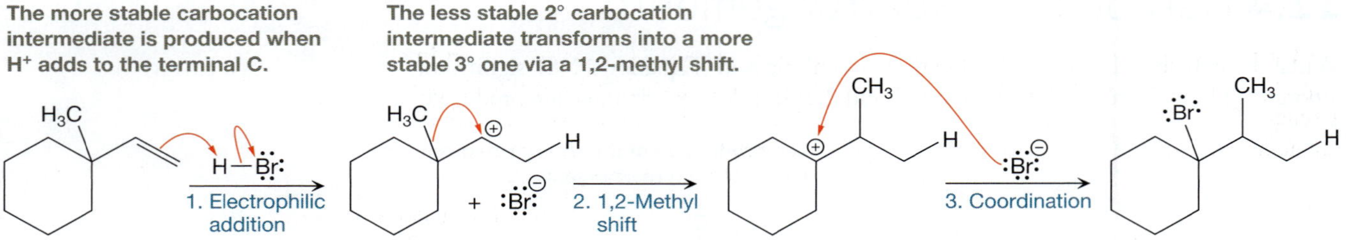

The secondary carbocation intermediate rapidly converts to a more stable tertiary one via a 1,2-methyl shift in Step 2. After coordination of Br^- in Step 3, the reaction is complete.

PROBLEM 11.7 Draw the complete, detailed mechanism for each of the following reactions and predict the major product.

(a)   (b)   (c)

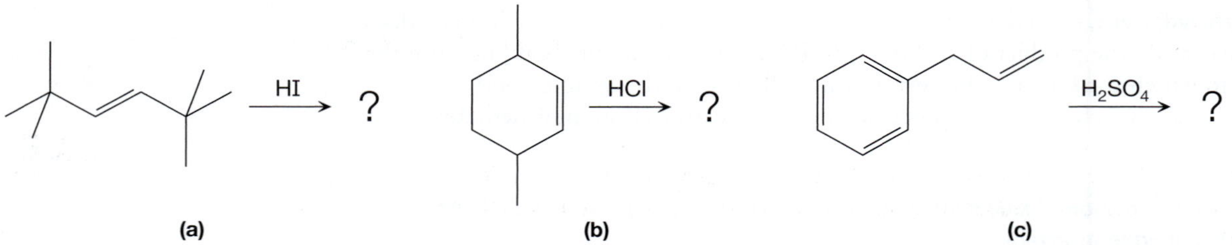

PROBLEM 11.8 Each of the following deuterated bromoalkanes can be produced by treating an alkene with deuterium bromide (DBr). Draw the corresponding alkenes that could have been used.

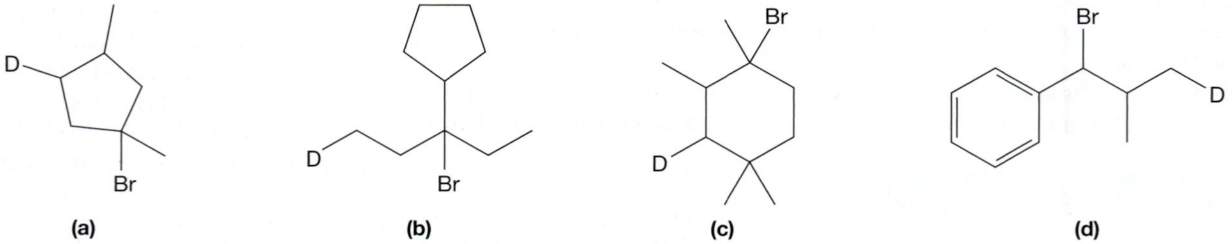

(a)   (b)   (c)   (d)

11.5 Stereochemistry

In an electrophilic addition reaction involving an alkene, both alkene carbons, which are initially planar, become tetrahedral in the product. It is possible, therefore, for new chiral centers to be produced during the course of the reaction.

Depending on the symmetry of the product, stereochemistry may or may not be an issue.

In the reaction shown previously in Equation 11-1 (shown again in Equation 11-10), stereochemistry is *not* a concern because neither alkene C becomes a chiral center, and the product is achiral; in the product, none of the C atoms is bonded to four *different* groups.

$$\text{Cyclohexene} \xrightarrow{\text{HCl}} \text{Chlorocyclohexane}$$

Neither C becomes a chiral center. (11-10)

By contrast, in the reaction shown previously in Equation 11-5 (shown again in Equation 11-11), a single chiral center is formed, so each product molecule is chiral. Because the starting material and the conditions under which the reaction takes place are achiral, a racemic mixture of enantiomers is produced.

A new chiral center is produced.

$$\text{1,2-Diphenylethene} \xrightarrow[\text{Acetic acid}]{\text{HBr}} \quad + \quad \text{(11-11)}$$

Racemic mixture of products

YOUR TURN **11.6**

In the electrophilic addition reaction shown previously in Equation 11-7 (p. 570), the major product has gained a chiral center. Mark that chiral center with an asterisk.

Equation 11-12 shows an example of an electrophilic addition reaction in which two new chiral centers are produced.

Two new chiral centers (*) are produced.

$$\xrightarrow{\text{HBr}} \quad \text{(11-12)}$$

There are $2^n = 2^2 = 4$ stereoisomers of the product that exist: two enantiomers in which the H and Br are cis to each other and two enantiomers in which they are trans. It is tempting to say that the reaction should produce all four of these stereoisomers in significant amounts. Indeed, the mechanism in Equation 11-3 (p. 565) can be applied to arrive at each one of them. Frequently, however, the solvent plays a major role in the stereoselectivity of these kinds of reactions, heavily favoring the trans products over the cis.

CONNECTIONS
1,2-Diphenylethene (Equation 11-11), commonly called stilbene, provides the framework for several useful derivatives. One derivative, 4,4'-diamino-2,2'-stilbenedisulfonic acid, is used as an optical brightener in some laundry detergents.

PROBLEM 11.9 Draw the complete mechanism for the formation of all products formed in this reaction, paying attention to stereochemistry.

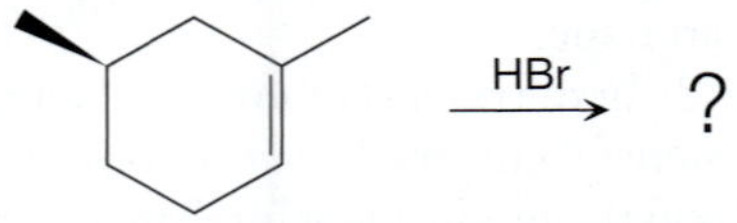

$\xrightarrow{\text{HBr}}$?

11.6 Addition of a Weak Acid: Acid Catalysis

Thus far, we have examined the addition of only strong acids to alkenes. What happens when weak acids are added to alkenes?

As indicated in Equation 11-13, effectively no reaction occurs if an alkene is treated with water—a weak acid—under neutral conditions.

$\xrightarrow{\text{H}_2\text{O}}$ | No reaction | (11-13)

With water as the acid, the first step of the general electrophilic addition mechanism (Equation 11-3, p. 565) is too unfavorable to take place at a reasonable rate. Not only would the first step produce a carbocation, it would also produce HO^-, in which the negative charge is relatively poorly stabilized. (See Your Turn 11.7.)

YOUR TURN 11.7

Draw the hypothetical two-step mechanism just mentioned, in which water adds across the double bond in Equation 11-13 according to the mechanism in Equation 11-3 (p. 565). Identify the HO^- intermediate. Why is this mechanism unfeasible, whereas the same mechanism with HCl as the acid is feasible?

Equation 11-14 shows, on the other hand, that:

The addition of water to an alkene takes place readily in the presence of a strong acid, such as sulfuric acid, producing an alcohol.

Equation 11-14 is an example of an **acid-catalyzed hydration reaction**.

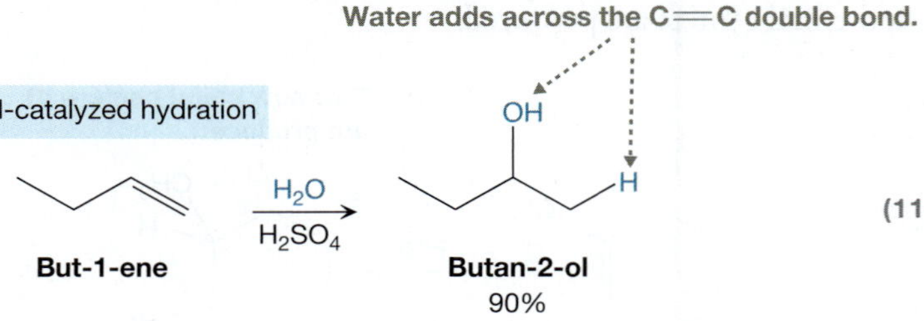

Water adds across the C≡C double bond.

Acid-catalyzed hydration

But-1-ene $\xrightarrow[\text{H}_2\text{SO}_4]{\text{H}_2\text{O}}$ **Butan-2-ol** 90% (11-14)

The mechanism for the hydration reaction is shown in Equation 11-15. Due to the leveling effect (Section 6.2c), H_3O^+ is the strongest acid that can exist in significant amounts in water. Being a strong Brønsted acid, H_3O^+ can protonate the alkene to produce a carbocation intermediate, as illustrated in Step 1. Notice in Step 1 that the proton adds to the terminal C to yield the more stable carbocation intermediate. In Step 2, that carbocation intermediate is attacked by a weak H_2O nucleophile. A final deprotonation in Step 3 removes the positive charge from the adduct, yielding an uncharged product. Step 3 also regenerates H_3O^+, a reactant in Step 1. Therefore, H_3O^+ speeds up the reaction but is not consumed overall—characteristics required of a *catalyst*.

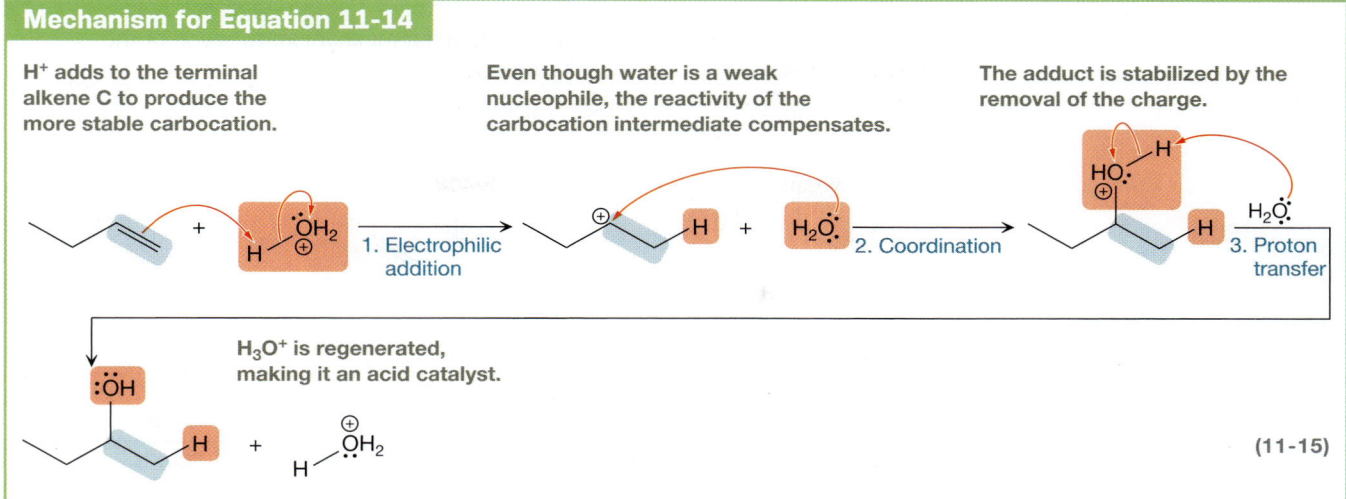

Mechanism for Equation 11-14

H⁺ adds to the terminal alkene C to produce the more stable carbocation.

Even though water is a weak nucleophile, the reactivity of the carbocation intermediate compensates.

The adduct is stabilized by the removal of the charge.

1. Electrophilic addition

2. Coordination

3. Proton transfer

H_3O^+ is regenerated, making it an acid catalyst.

(11-15)

Notice that H_2SO_4 appears in both Equations 11-14 and 11-2 (p. 565), but the two reactions have different outcomes. H_2O adds across the double bond in Equation 11-14, whereas H_2SO_4 adds across the double bond in Equation 11-2. In Equation 11-2, very little water is present, so H_2SO_4 acts as the strong acid and HSO_4^- is the nucleophile. In Equation 11-14, on the other hand, water is the solvent, so H_3O^+ acts as the strong acid and H_2O is the nucleophile.

Water is not the only weak nucleophile that can add across a $C=C$ double bond via acid catalysis. When an alcohol adds, as shown in Solved Problem 11.10, it is called an **acid-catalyzed alkoxylation reaction**.

SOLVED PROBLEM 11.10

Draw the complete, detailed mechanism for this reaction and predict the products.

Think What is the electrophile? What is the nucleophile? To which alkene carbon does the electrophile add?

$$\xrightarrow[H_2SO_4]{CH_3OH} \quad ?$$

Solve The strongest acid that can exist in methanol is $CH_3OH_2^+$, which can react with the alkene in an electrophilic addition step. Formally, H^+ from this acid adds to the terminal C in Step 1 to produce the more stable carbocation intermediate, which is secondary and resonance stabilized.

1. Electrophilic addition

2. Coordination

3. Proton transfer

In Step 2, CH_3OH behaves as a nucleophile, forming a bond to the positively charged C. In Step 3, CH_3OH acts as a base to remove a proton from the adduct, resulting in an uncharged product. Overall, this is the same mechanism as for the acid-catalyzed hydration in Equation 11-15. Unlike Equation 11-15, however, the alkoxylation product here is an *ether*, not an alcohol.

PROBLEM 11.11 Draw the complete, detailed mechanism for each of the following reactions and predict the products.

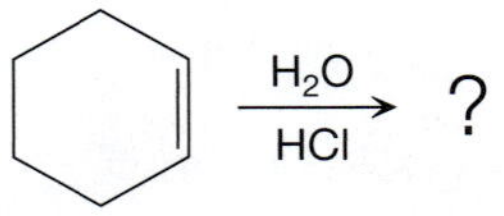

(a)

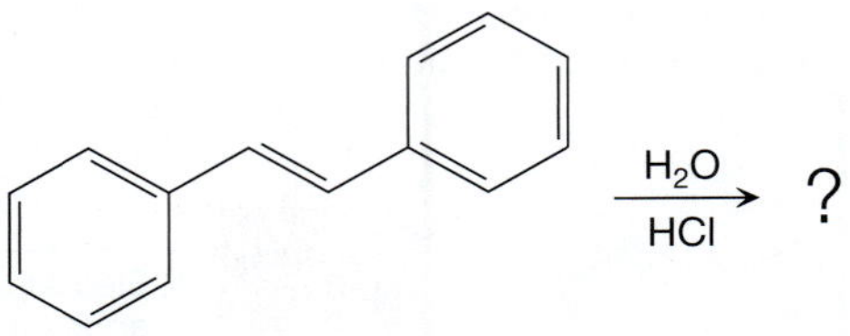

(b)

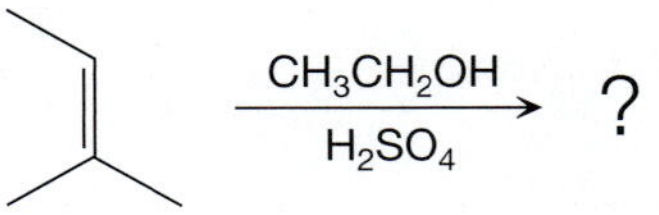

(c)

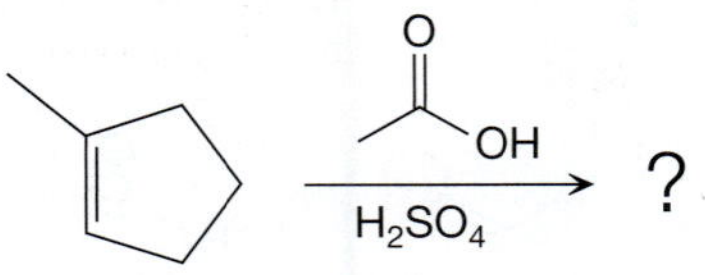

(d)

PROBLEM 11.12 Show how each of the following compounds can be produced from an alkene.

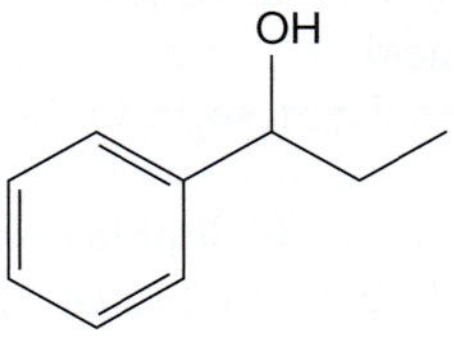

(a)

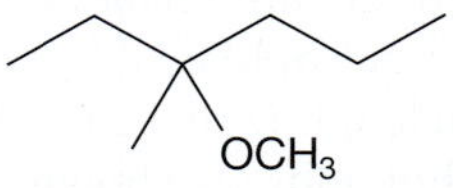

(b)

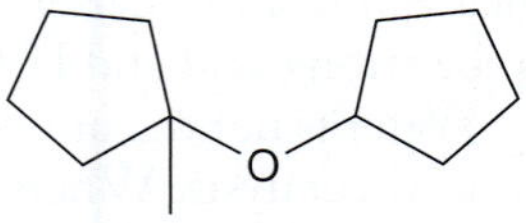

(c)

11.7 Electrophilic Addition of a Strong Brønsted Acid to an Alkyne

Like the C=C double bond of an alkene, the C≡C triple bond of an alkyne is relatively electron rich—that is, the six electrons of the C≡C triple bond reside largely in the region between the two carbon atoms, and the π electrons are relatively easily accessible. As a result, alkynes undergo electrophilic addition with strong Brønsted acids in much the same way alkenes do.

When ethynylbenzene is treated with excess HCl, addition does occur, as shown in Equation 11-16, but the major product is the result of two additions of HCl, not one.

Two additions of HCl produce a geminal dichloride.

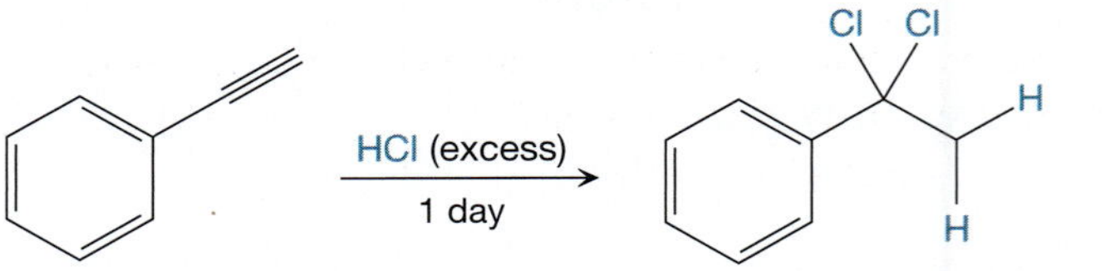

(11-16)

Ethynylbenzene

1,1-Dichloro-1-phenylethane
82%

In general:

> The reaction of an alkyne with a hydrogen halide produces a **geminal dihalide** as the major product, in which both halogen atoms appear on the same carbon.

The mechanism shown in Equation 11-17 accounts for the formation of this product.

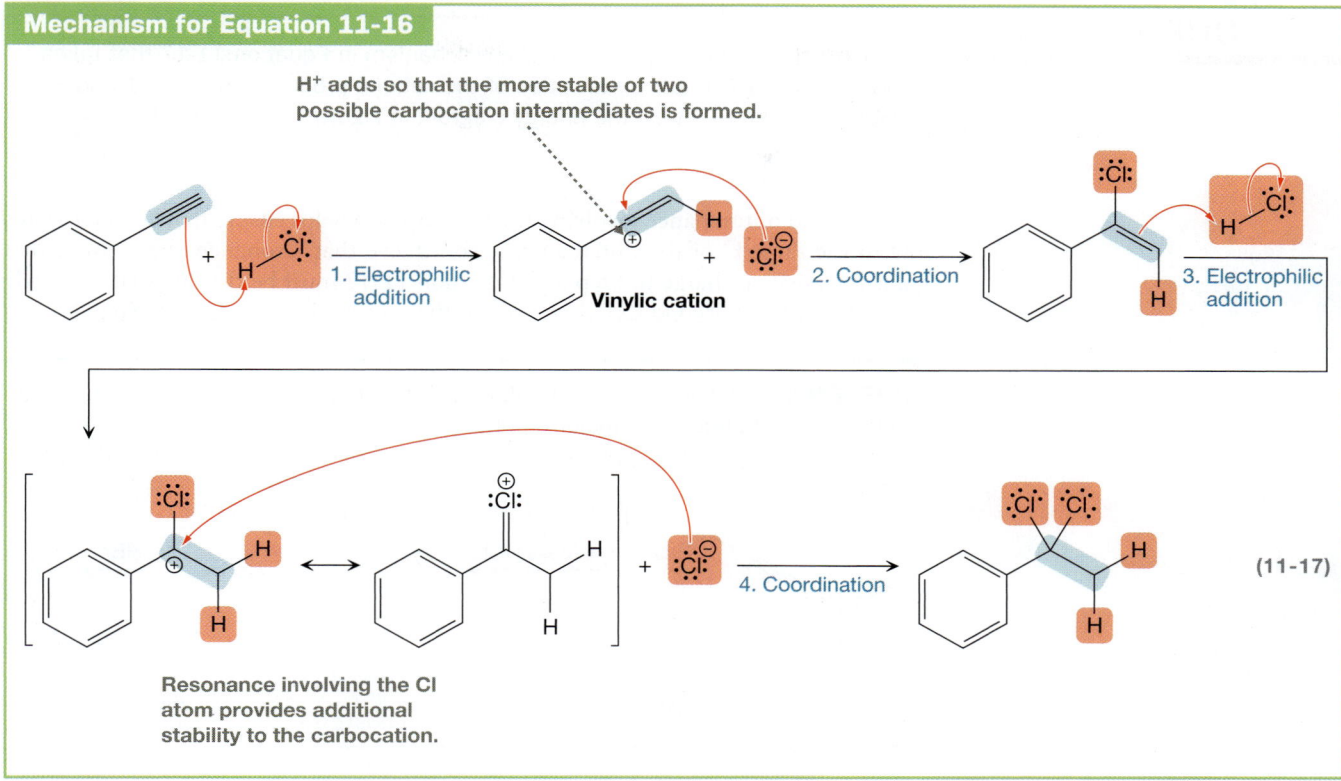

Mechanism for Equation 11-16

H⁺ adds so that the more stable of two possible carbocation intermediates is formed.

1. Electrophilic addition

Vinylic cation

2. Coordination

3. Electrophilic addition

4. Coordination

(11-17)

Resonance involving the Cl atom provides additional stability to the carbocation.

Step 1 is electrophilic addition of H⁺ to the C≡C triple bond, producing a vinylic cation. Just as we saw with alkenes, the more stable carbocation is produced. In this case, the proton adds to the terminal C to produce a secondary carbocation that is stabilized by resonance involving the phenyl ring. In Step 2, Cl⁻ attacks the positively charged C in a coordination step to produce a vinylic chloride in which Cl is attached directly to an alkene C. That vinylic chloride, however, reacts further because it contains a C=C double bond, to which another H⁺ adds in Step 3. Finally, in Step 4, a second Cl⁻ undergoes coordination with the carbocation to produce the geminal dichloride, which is the major product in Equation 11-16.

YOUR TURN 11.8

Draw the resonance structures of the vinylic cation in Equation 11-17 to explain the regiochemistry of the first addition.

The major product is the result of two additions, not one, because the second addition is easier than the first. It is easier because the carbocation that is produced in the rate-determining step of the second addition (Step 3) is more stable than the one produced in the rate-determining step of the first addition (Step 1).

The greater stability of the carbocation produced in Step 3 comes from two main factors. One is effective electronegativity (Section 3.11). An alkyne C is *sp* hybridized, so it has a higher effective electronegativity than an alkene C, which is *sp²* hybridized. The alkene C, therefore, handles the positive charge better than an alkyne C.

The second factor is resonance involving the lone pair from the Cl atom that added in Step 2, as shown in Equation 11-17. Without the Cl atom present, this resonance stabilization is unavailable to the carbocation produced in Step 1.

YOUR TURN 11.9

Construct an energy diagram for the mechanism in Equation 11-17 that takes into account the relative stabilities of the cations produced in Steps 1 and 3. (You may wish to review the energy diagram in Figure 11-2, p. 567.)

This kind of resonance involving the Cl atom also helps ensure that both Cl atoms add to the same C of the initial alkyne. To achieve that resonance, the C atom that gains the positive charge in Step 3 of Equation 11-17 must be the one that is already bonded to Cl. The second Cl⁻ then adds to that positively charged C in Step 4.

PROBLEM 11.13 For each of the following reactions, draw the complete, detailed mechanism and predict the major product.

excess HCl → ?

HBr → ?

(a)   (b)

Step 1 of the mechanism in Equation 11-17 becomes unreasonable when the positive charge that develops on the vinylic cation is no longer delocalized by resonance. In these cases, the vinylic cation is too unstable to be produced. Even without that resonance delocalization available, alkynes can still undergo addition of a hydrogen halide to produce a geminal dihalide, as shown in Equation 11-18.

Addition of two equivalents of HBr to an alkyne produces a *geminal* dibromide, in which both Br atoms are on the same C atom.

HBr (excess)
4 days

Propyne   2,2-Dibromopropane
                              100%

(11-18)

How does the first addition of HBr take place without proceeding through an unstable vinylic cation intermediate? One way is for HBr to initially form a π complex with the alkyne instead of fully transferring its proton, as shown in Equation 11-19. This π complex is stabilized, in part, by the attraction between the electron-rich C≡C triple bond and the electron-poor proton of the acid. Then, the proton is transferred at the same time Br⁻ attacks, producing the vinylic halide that continues with the addition of a second equivalent of HBr, just as we saw in Steps 3 and 4 of Equation 11-17.

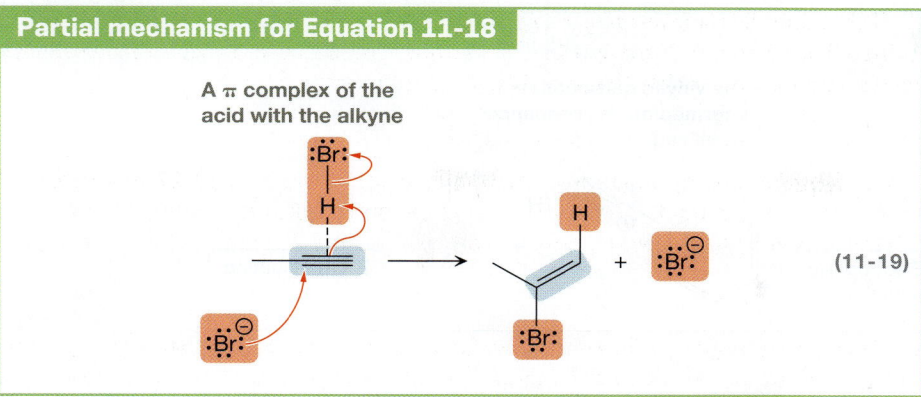

Partial mechanism for Equation 11-18

A π complex of the
acid with the alkyne

(11-19)

PROBLEM 11.14 Show how to synthesize each of the following compounds from an alkyne.

(a)   (b)

11.8 Acid-Catalyzed Hydration of an Alkyne: Synthesis of a Ketone

Recall from Section 11.6 that, under acidic conditions, water adds across the double bond of an alkene to produce an alcohol. It makes sense, therefore, that an alkyne would react in a similar way under the same conditions. Indeed, Equation 11-20 shows that a reaction does take place, but:

Acid-catalyzed hydration of an alkyne produces a ketone, not an alcohol.

Acid-catalyzed hydration of an
alkyne produces a ketone.

$$\underset{\textbf{Ethynylbenzene}}{} \xrightarrow[\text{H}_2\text{SO}_4,\ 90\ \text{h}]{\text{H}_2\text{O}} \underset{\substack{\textbf{Phenylethanone}\\\text{(trace amount)}}}{} \quad (11\text{-}20)$$

The mechanism for Equation 11-20 is shown in Equation 11-21.

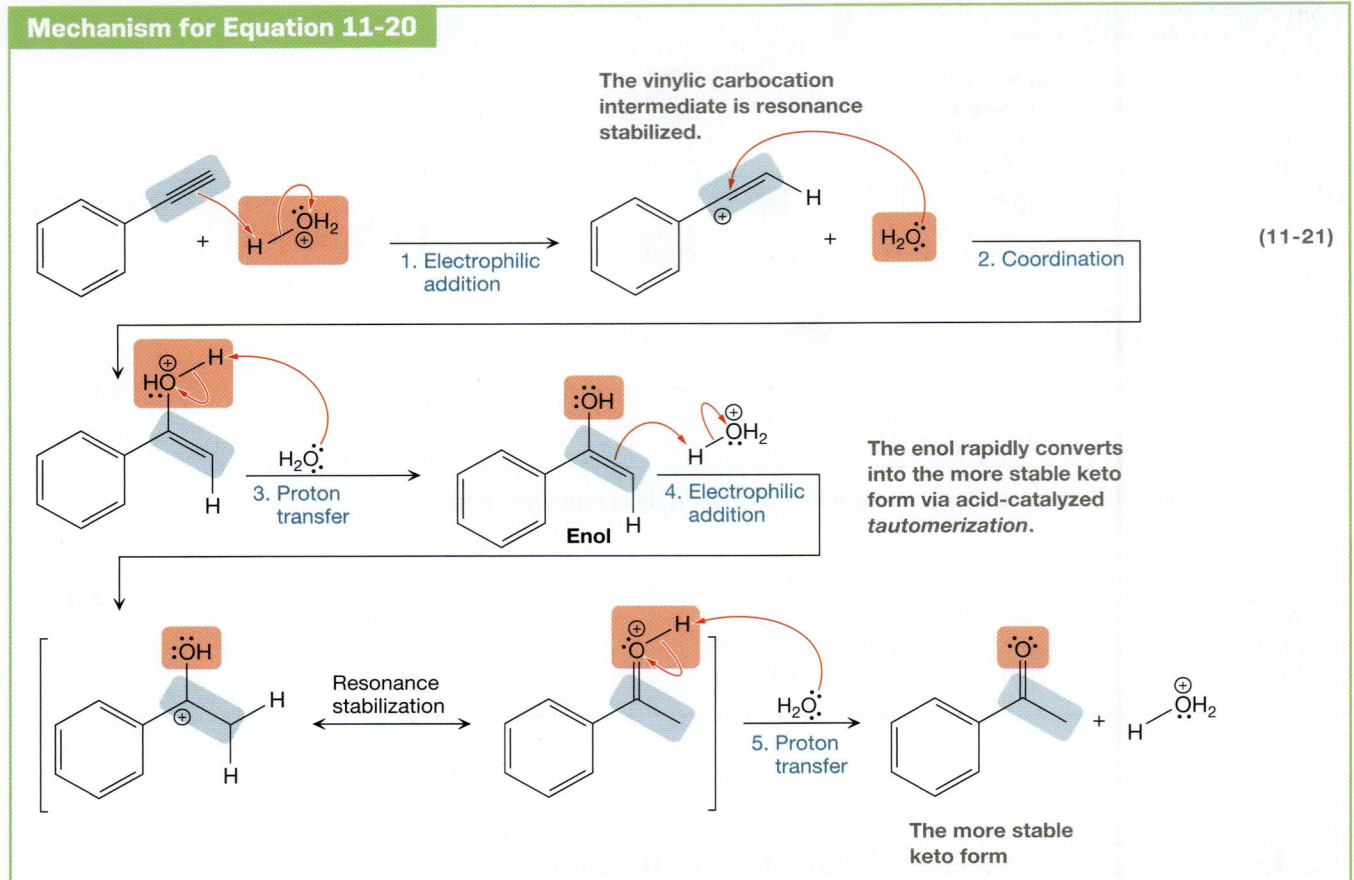

(11-21)

Steps 1–3 are the same as the ones that make up the mechanism for the acid-catalyzed hydration of an alkene, shown previously in Equation 11-15 (p. 577). Step 1 is the electrophilic addition of H^+, which produces a resonance-stabilized vinylic carbo-cation intermediate. In Step 2, water behaves as a nucleophile, attacking the positively charged C of that intermediate. Deprotonation in Step 3 produces the uncharged OH group, which is attached to an alkene C. The product of Step 3 is an *enol*.

Under these acidic conditions, the enol rapidly tautomerizes (review Section 7.9) to the more stable *keto* form in two steps. First, in Step 4, a proton adds to the terminal C atom. Then, in Step 5, the O atom is deprotonated, yielding the final uncharged ketone.

There are two important aspects of this reaction that you need to understand. First, the product is a ketone, not an aldehyde. This is an outcome of H^+ adding preferentially to the terminal C in Step 1 to give the more stable carbocation intermediate. Water must then add to the internal C atom in the second step. An aldehyde could form if the proton added to the internal C atom in the first step, but that would produce the less stable carbocation intermediate (see Your Turn 11.10).

YOUR TURN 11.10

Draw the carbocation intermediate that must be generated to produce an aldehyde from the acid-catalyzed hydration of ethynylbenzene. Explain why it is less stable than the carbocation intermediate shown in Equation 11-21.

The second aspect of this reaction worth noticing is that the product yield of the reaction in Equation 11-20 is very low; in fact, only a trace amount of ketone is produced, so this reaction is not very useful synthetically. The yield is so low because the addition of H^+ in the first step of the mechanism produces a vinylic carbocation intermediate, which, as mentioned in Section 11.7, is generally less stable than a carbocation in which the positive charge is on an sp^2-hybridized C. The first step of the mechanism, therefore, is excessively unfavorable with H_3O^+ as the acid.

Using conditions that can support an acid stronger than H_3O^+, though, makes the first step of the mechanism more favorable. An example using trifluoromethanesulfonic acid (CF_3SO_3H, also known as triflic acid, TfOH) and 2,2,2-trifluoroethanol (CF_3CH_2OH) is shown in Equation 11-22.

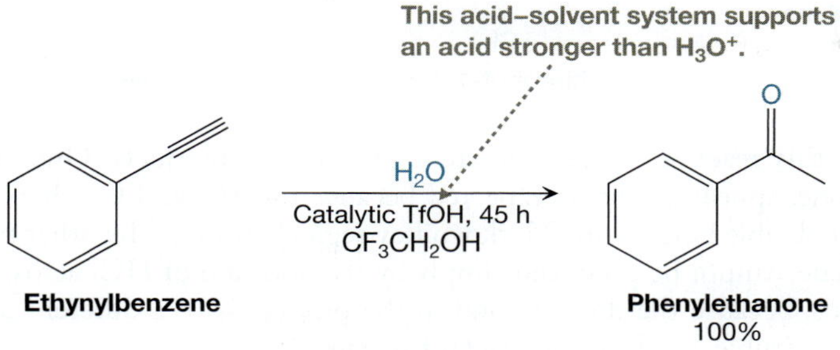

This acid–solvent system supports an acid stronger than H_3O^+.

$$\text{(11-22)}$$

Ethynylbenzene

Phenylethanone
100%

YOUR TURN **11.11**

What is the strongest acid that can exist in solution when CF_3CH_2OH is the solvent, as in Equation 11-22? Is that acid stronger or weaker than H_3O^+? Why?

Another way to circumvent the problem associated with the acid-catalyzed hydration of an alkyne is to use a mercury(II) catalyst, Hg^{2+}, as shown in Equation 11-23. The presence of a mercury(II) catalyst changes the mechanism, which is discussed in Chapter 12.

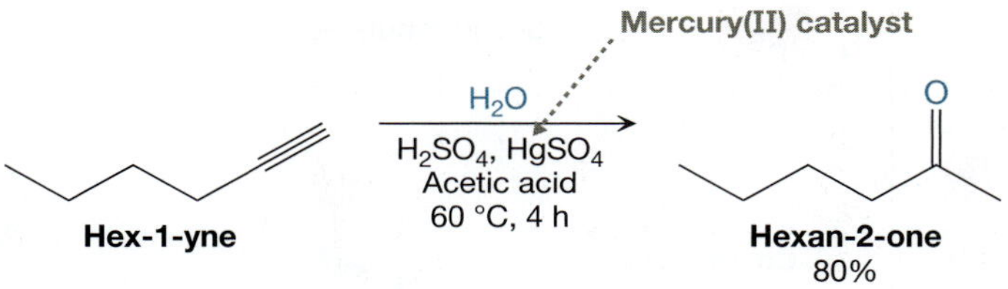

Mercury(II) catalyst

$$\text{(11-23)}$$

Hex-1-yne

Hexan-2-one
80%

PROBLEM 11.15 Draw the complete, detailed mechanism for the reaction shown here, and predict the major product.

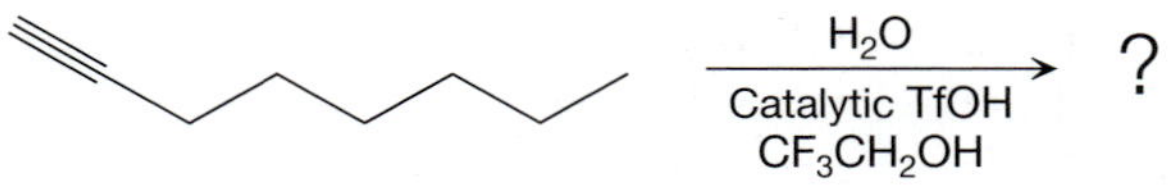

11.9 Electrophilic Addition of a Brønsted Acid to a Conjugated Diene: 1,2-Addition and 1,4-Addition

A molecule such as buta-1,3-diene is said to have **conjugated** double bonds because the two double bonds are separated by another bond (in this case, a single bond). A conjugated diene such as buta-1,3-diene is electron rich, much like the previous

alkenes we have studied in this chapter, so it undergoes electrophilic addition with Brønsted acids (an example is shown in Equation 11-24 with HCl as the acid).

$$\text{(11-24)}$$

Notice that this reaction yields a mixture of isomeric products. One product, 3-chlorobut-1-ene, appears to be as expected because the H^+ and Cl^- have added across one of the double bonds with Markovnikov regiochemistry. The other product, 1-chlorobut-2-ene, cannot be produced simply by the addition of HCl across one of the double bonds, because the double bond in the product is in a different location than either of the double bonds in the reactant species.

Both reaction products are produced from the same general mechanism, as shown in Equation 11-25. In Step 1, H^+ adds to a π bond, and in Step 2, the Cl^- anion attacks the newly formed carbocation intermediate in a coordination step.

Mechanism for Equation 11-24

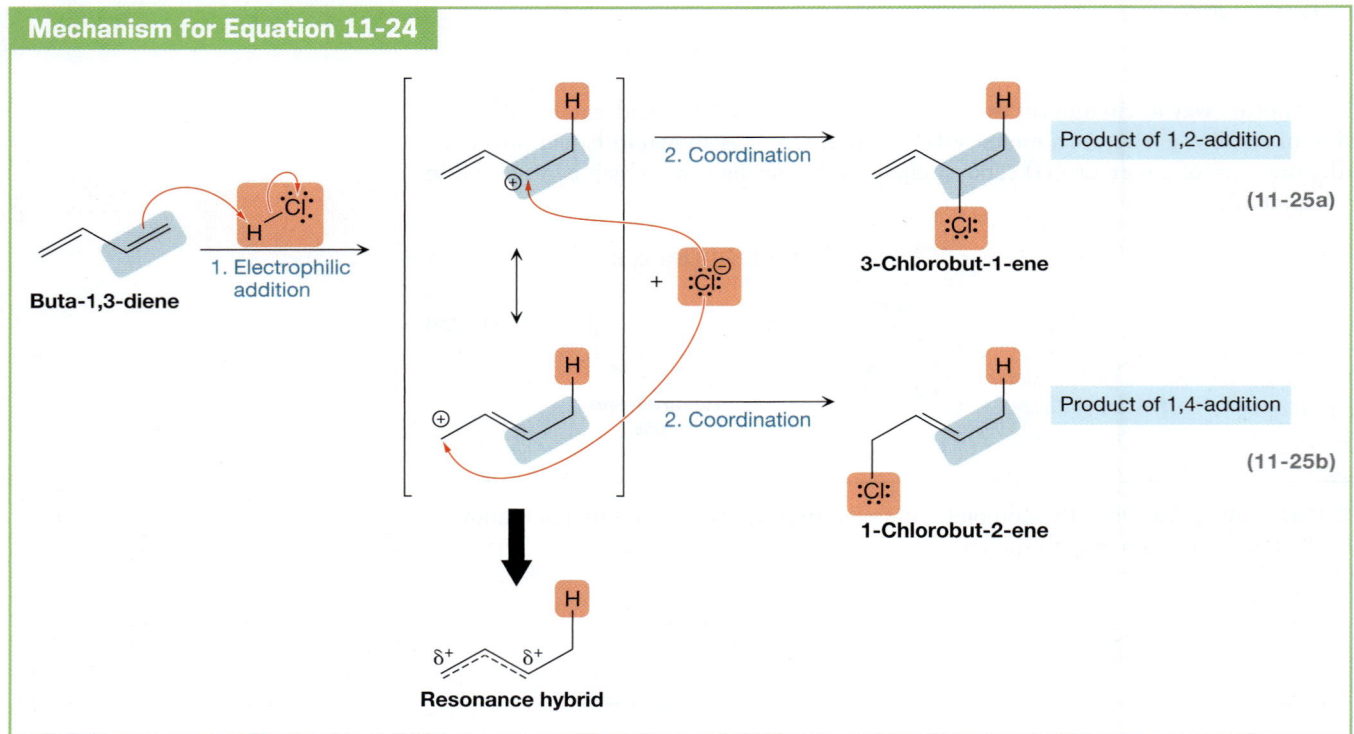

Both products are produced from the *same carbocation intermediate*. As indicated in Equation 11-25, this intermediate has two resonance structures—one with the positive charge on C1 and one with the positive charge on C3. In the resonance hybrid, therefore, each of those two C atoms bears a partial positive charge, so each can be attacked by Cl^-. Attack on one of those C atoms (Equation 11-25a) yields 3-chlorobut-1-ene, whereas attack on the other carbon atom (Equation 11-25b) yields 1-chlorobut-2-ene.

The H^+ and Cl^- that added to the diene to produce 3-chlorobut-1-ene are separated by two C atoms, making it the product of **1,2-addition**. On the other hand, 1-chlorobut-2-ene is the product of **1,4-addition** because the H^+ and Cl^- that added to the diene are separated by four C atoms.

Which of the products shown here is produced from 1,2-addition, and which is produced from 1,4-addition?

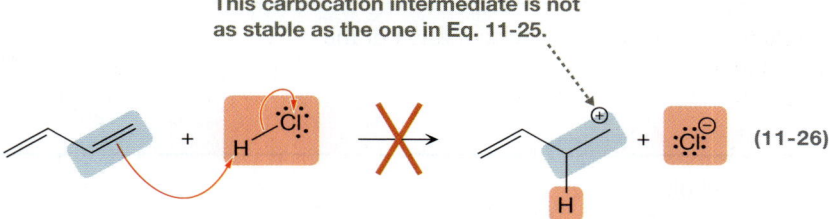

Addition of the proton in the first step of the mechanism occurs to give *the more stable carbocation intermediate*, just as we saw previously in the electrophilic addition to propene (Equation 11-6, p. 569). The carbocation intermediate shown in Equation 11-25, which is produced by the addition of H^+ to a terminal C, is *resonance stabilized*. If one of the internal C atoms were to gain the H^+ instead, then the positive charge in the resulting carbocation would be *localized* on a primary C, as shown in Equation 11-26.

This carbocation intermediate is not as stable as the one in Eq. 11-25.

(11-26)

SOLVED PROBLEM 11.16

Draw the major 1,2-addition and 1,4-addition products of this reaction.

Think How many distinct carbocation intermediates are possible from protonation of a double bond? Which one is the most stable? What are the species that can be produced on nucleophilic attack of that carbocation intermediate?

Solve Each of the alkene groups can gain a proton at either of its C atoms, giving rise to four possible carbocation intermediates **A–D**.

Intermediates **A** and **D** are more stable than **B** and **C** due to resonance delocalization of the positive charge. Intermediate **A** is more stable than **D** (and is the most stable of all four) because its positive charge is shared on a tertiary C atom.

A   B   C   D

The positive charge is resonance delocalized over 2° and 3° C atoms.

Most stable carbocation intermediate

The positive charge is resonance delocalized over 2° and 1° C atoms.

With the most stable intermediate identified, the 1,2- and 1,4-addition products are obtained by attack of the nucleophile, H_2O, on the carbon atoms sharing the positive charge.

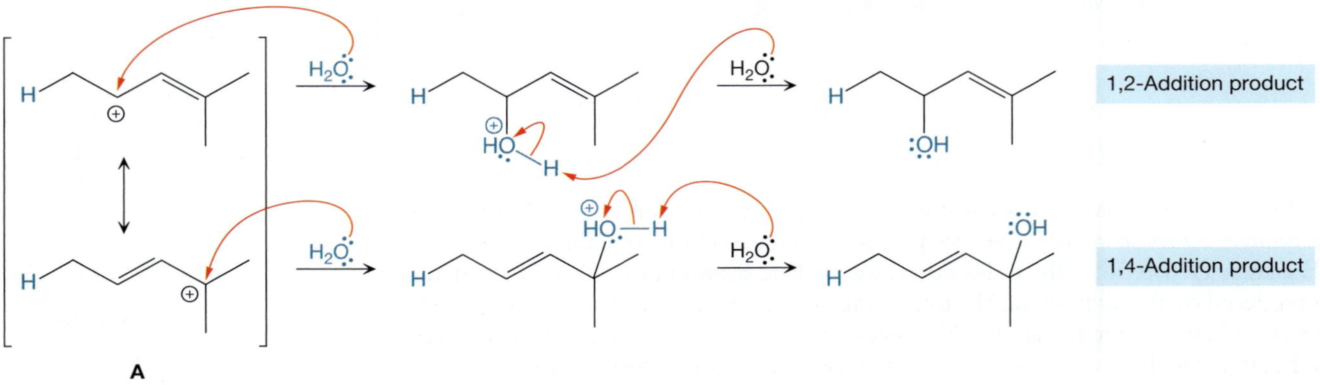

PROBLEM 11.17 Draw the complete, detailed mechanism for the formation of the major 1,2- and 1,4-addition products of this reaction.

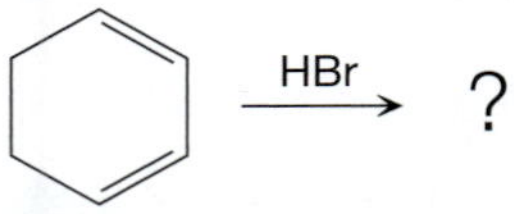

PROBLEM 11.18 When an unknown conjugated diene is treated with HCl, a mixture of the two chloroalkenes at the right is produced. Draw the conjugated diene that was used as the reactant and draw the complete, detailed mechanism that leads to the formation of each product.

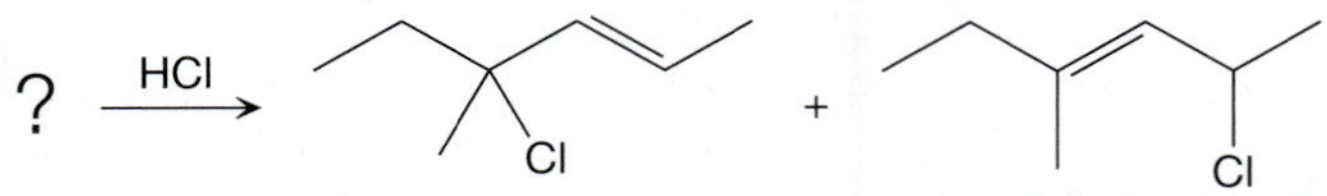

11.10 Kinetic versus Thermodynamic Control in Electrophilic Addition to a Conjugated Diene

As we discussed in Section 11.9, electrophilic addition to buta-1,3-diene produces a mixture of addition products—a 1,2-adduct and a 1,4-adduct—from the same carbocation intermediate. Which adduct is the major product? The answer to that question can depend on the *temperature* at which the reaction is carried out. If the electrophilic addition of HCl to buta-1,3-diene is carried out at room temperature, for example, then the 1,4-adduct is the major product (Equation 11-27).

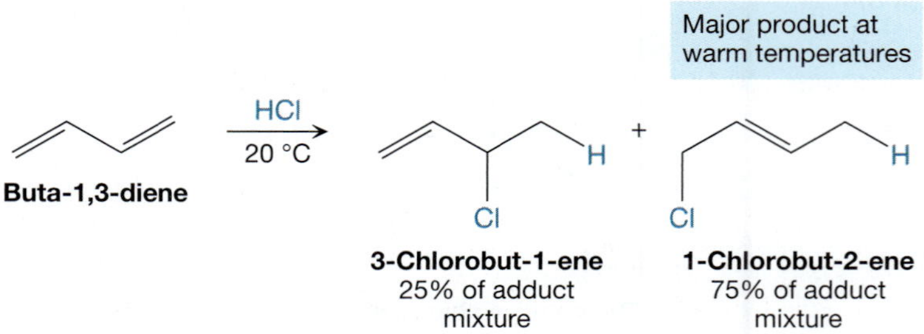

Major product at warm temperatures

(11-27)

Buta-1,3-diene

3-Chlorobut-1-ene
25% of adduct mixture

1-Chlorobut-2-ene
75% of adduct mixture

If the reaction is carried out at cold temperatures, however, then the 1,2-adduct is the major product (Equation 11-28).

$$\text{Buta-1,3-diene} \xrightarrow[-80\,°\text{C}]{\text{HCl}} \text{3-Chlorobut-1-ene} + \text{1-Chlorobut-2-ene} \quad (11\text{-}28)$$

Major product at cold temperatures

3-Chlorobut-1-ene
80% of adduct mixture

1-Chlorobut-2-ene
20% of adduct mixture

We observe these temperature effects because the temperature at which the reaction is run governs whether the reaction is *reversible* or *irreversible*. At low temperatures, the product molecules do not possess enough energy to climb over the energy barrier in the reverse direction to reform reactants at a significant rate (review Fig. 11-2, p. 567), effectively making electrophilic addition to the diene irreversible. At high temperatures, on the other hand, the product molecules do possess enough energy to make the reaction reversible. Combining this information with what we learned in Section 9.12:

- At low temperatures, electrophilic addition to a conjugated diene takes place under *kinetic control*, so the major product is the one that is produced most rapidly.
- At high temperatures, electrophilic addition to a conjugated diene takes place under *thermodynamic control*, so the major product is the one that is lowest in energy.

Based on the temperatures reported in Equations 11-27 and 11-28, therefore, the 1,4-adduct must be the thermodynamic product and the 1,2-adduct must be the kinetic product.

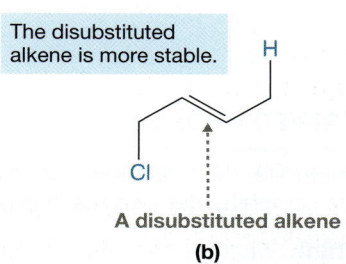

A monosubstituted alkene
(a)

The disubstituted alkene is more stable.

A disubstituted alkene
(b)

FIGURE 11-5 Stability of 1,2- and 1,4-adducts (a) The 1,2-adduct and (b) the 1,4-adduct of Equations 11-27 and 11-28. The 1,4-adduct is more stable because its double bond is more highly substituted.

YOUR TURN 11.13

In Equations 11-27 and 11-28, label each product as either the "kinetic product" or the "thermodynamic product."

We can rationalize the 1,4-adduct being the thermodynamic product by the stability of the C=C group (Fig. 11-5). Notice that the C=C group in the 1,4-adduct is *disubstituted* (i.e., bonded to two alkyl groups), whereas the C=C group in the 1,2-adduct is only *monosubstituted*. Then recall from Section 5.9 that the more highly alkyl substituted a C=C is, the more stable the alkene is.

YOUR TURN 11.14

Identify the thermodynamic product in the following electrophilic addition reaction.

The 1,2-adduct is the kinetic product due to the *location* of Cl^- at the instant it is formed from the first step of the mechanism. On addition of the H^+ to the diene, Cl^- appears closer to C2 than to C4 (Equation 11-29).

As a result, Cl^- can attack C2 more quickly than it can attack C4.

In the case of electrophilic addition to buta-1,3-diene, the kinetic and thermodynamic products are different—the 1,2-adduct is the kinetic product and the 1,4-adduct is the thermodynamic product. When the reaction involves conjugated dienes other than buta-1,3-diene, the kinetic product is generally the 1,2-adduct, too. The thermodynamic product, however, is not always the 1,4-adduct, as shown in Solved Problem 11.19.

SOLVED PROBLEM 11.19

Determine the major thermodynamic and kinetic products in this reaction and draw the complete, detailed mechanism for the formation of each.

Think What are the possible carbocation intermediates? What is the most stable carbocation intermediate? What are the adducts that can be produced on nucleophilic attack of that carbocation intermediate? Which of those adducts is produced the fastest? Which is the more stable adduct?

Solve Two carbocations can be produced if the π bond at the top is used to form a bond to H^+. (Because the diene is symmetric, addition of H^+ to the bottom π bond would produce the same carbocation intermediates.)

The more stable carbocation intermediate is **A**, because it is resonance stabilized:

The 1,2-adduct is formed faster, and is also the more stable alkene product.

The 1,4-adduct

Adding I⁻ yields the 1,2- and 1,4-adducts. As usual, the 1,2-adduct is the kinetic product. It is also the more stable alkene product because it is more highly alkyl substituted; the C=C in the 1,2-adduct is trisubstituted, whereas it is only disubstituted in the 1,4-adduct. Therefore, the 1,2-adduct is the thermodynamic product, too.

PROBLEM 11.20 Determine the major thermodynamic and kinetic products in the reaction shown here and draw the complete, detailed mechanism for the formation of each.

$$\text{(structure)} \xrightarrow{\text{HCl}} ?$$

Kinetic Control, Thermodynamic Control, and Mad Cow Disease

Mad cow disease and its human form, Creutzfeldt–Jakob disease (CJD), are neurological disorders that are incurable, degenerative, and fatal. They are classified as prion (short for "protein infection") diseases because of the mode by which they are transmitted and proliferate: A protein in its native conformation (i.e., the conformation in which the protein functions normally) becomes misfolded, and that diseased protein promotes the refolding of other native proteins into their diseased states. In the case of CJD, the protein that undergoes this refolding (i.e., the prion protein) is water soluble and is believed to be responsible for transmembrane transport or cell signaling when in its native state. In its diseased state, that prion protein forms aggregates and becomes water *insoluble*.

 The mechanism by which prion proteins refold into their diseased states is not exactly clear, but a key piece of the puzzle has to do with kinetic control versus thermodynamic control in the protein folding process—the same type of phenomenon that governs electrophilic addition to conjugated dienes that we studied here in Chapter 11. In their native states, many proteins are believed to adopt their most energetically stable conformations—that is, they fold under thermodynamic control. But F. E. Cohen and coworkers at the University of California, San Francisco, showed that the native state of the mouse recombinant prion protein is *not* in the most stable conformation; it folds under kinetic control. Understanding the specific factors that lead to such a kinetically controlled process could someday lead to treatments or cures for prion diseases.

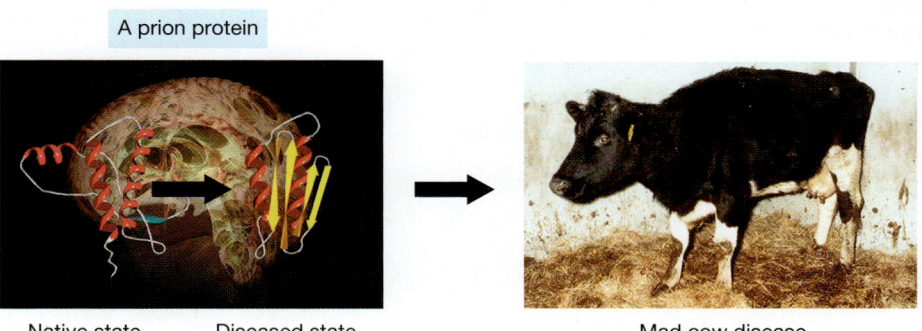

A prion protein

Native state   Diseased state

Mad cow disease

THE ORGANIC CHEMISTRY OF BIOMOLECULES

11.11 Terpene Biosynthesis: Carbocation Chemistry in Nature

Recall from Section 2.11c that many natural products are *terpenes* or *terpenoids*, including essential oils from plants and steroids. More importantly, recall that terpenes and terpenoids are distinguished from other natural products by the structure of their carbon backbone—namely, the carbon backbone of a terpene or terpenoid consists of multiple *isoprene units* linked by their terminal carbons (Fig. 11-6). One of the keys to

FIGURE 11-6 An isoprene unit
Four carbons make up a chain and a fifth carbon is connected to C2 of that chain.

Isoprene unit

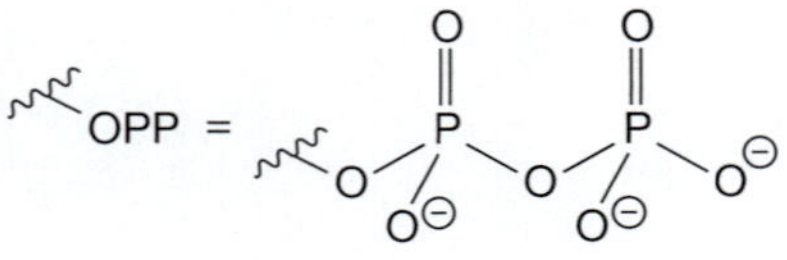

terpene synthesis in nature is isopentenyl pyrophosphate, which itself is produced in several steps from acetic acid (Equation 11-30). As you can see, isopentenyl pyrophosphate has the same carbon structure as the fundamental isoprene unit.

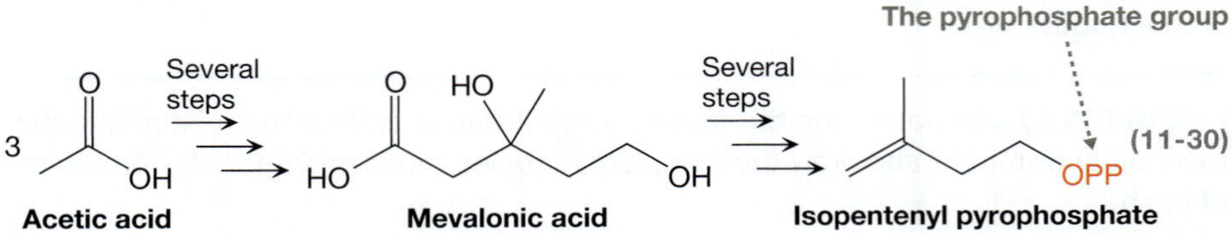

The pyrophosphate group

(11-30)

Acetic acid   **Mevalonic acid**   **Isopentenyl pyrophosphate**

Some fraction of isopentenyl pyrophosphate reacts with an enzyme (a protein catalyst), isomerizing to dimethylallyl pyrophosphate. As shown in Equation 11-31, the first step is electrophilic addition of a proton, and the second step is the elimination of a different proton.

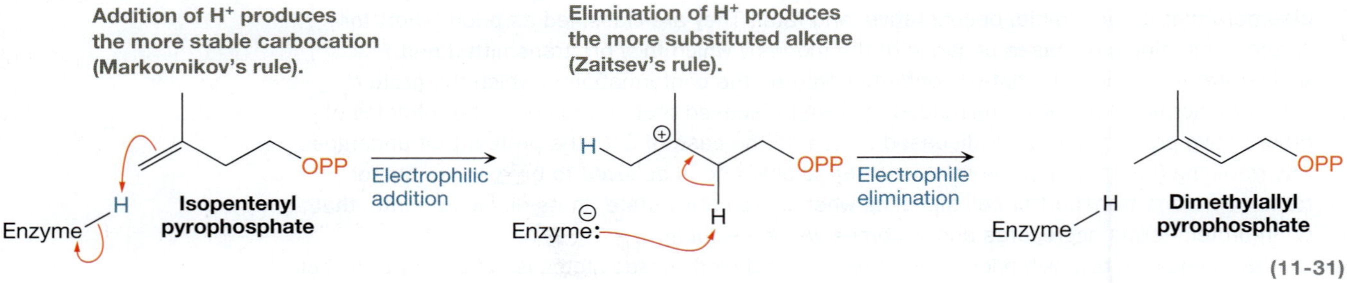

Addition of H⁺ produces the more stable carbocation (Markovnikov's rule).

Elimination of H⁺ produces the more substituted alkene (Zaitsev's rule).

(11-31)

Notice that the electrophilic addition step produces the more stable carbocation intermediate, thus adhering to Markovnikov's rule (Section 11.3), and that the elimination step produces the more highly substituted alkene product, consistent with Zaitsev's rule (Section 9.10).

Dimethylallyl pyrophosphate then loses its pyrophosphate group (a rather good leaving group) in a heterolysis step, as shown in Equation 11-32, resulting in an allylic carbocation.

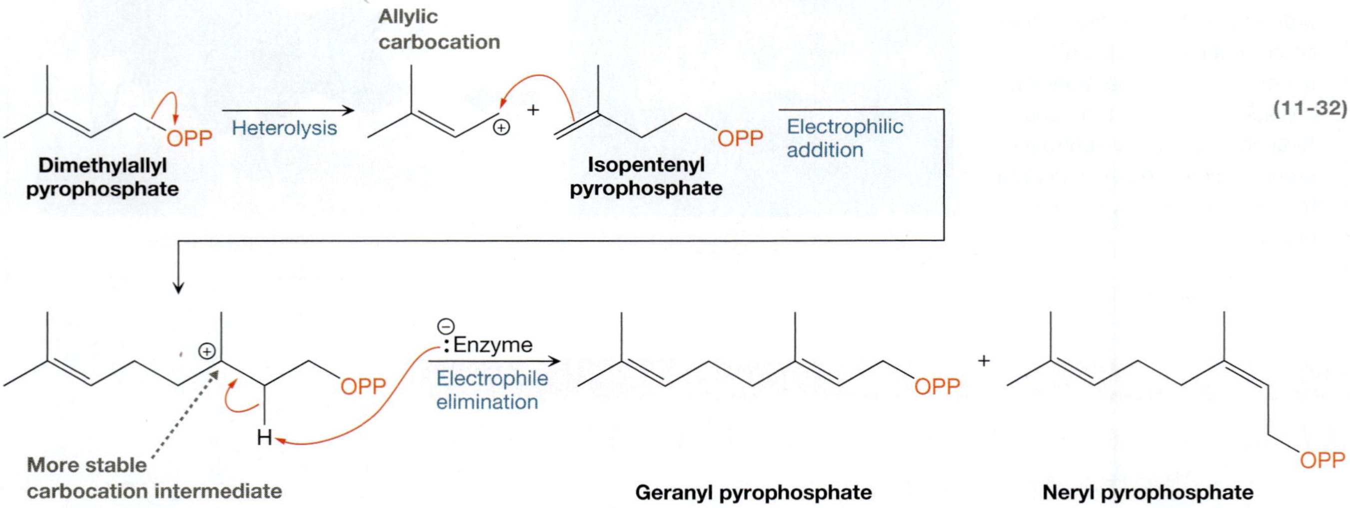

Allylic carbocation

(11-32)

Dimethylallyl pyrophosphate   **Isopentenyl pyrophosphate**

More stable carbocation intermediate   **Geranyl pyrophosphate**   **Neryl pyrophosphate**

The allylic carbocation then reacts with isopentenyl pyrophosphate in an electrophilic addition step, producing another carbocation intermediate, which now contains 10 carbon atoms. Subsequent elimination of H^+ produces geranyl pyrophosphate and neryl pyrophosphate, which are E/Z isomers.

Neryl pyrophosphate can then react in a number of ways to produce various terpenes and terpenoids containing 10 carbons. For example, Equation 11-33 shows it can undergo an intramolecular electrophilic addition reaction to produce a carbocation intermediate with a six-membered ring.

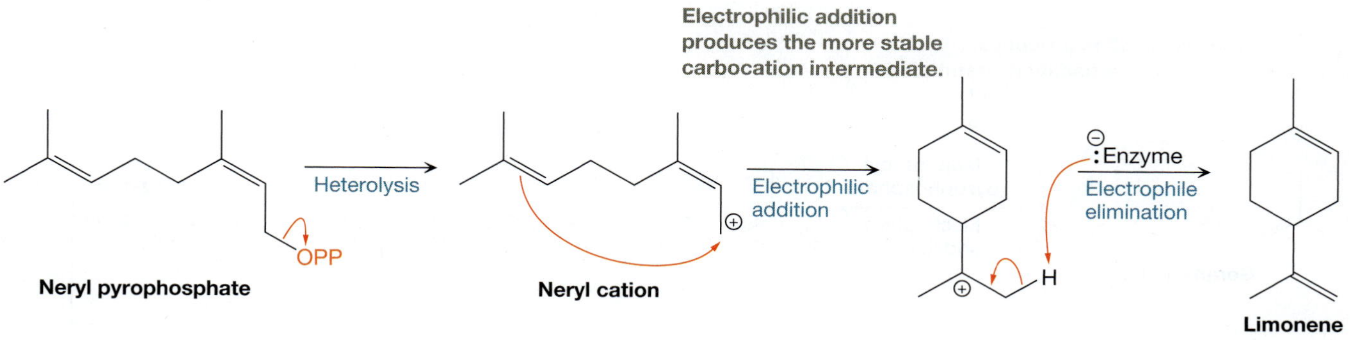

Electrophilic addition produces the more stable carbocation intermediate.

Neryl pyrophosphate → (Heterolysis) → Neryl cation → (Electrophilic addition) → :Enzyme (Electrophile elimination) → Limonene

(11-33)

Subsequent elimination of H^+ yields limonene, a monoterpene that is a constituent of citrus oils. Alternatively, geranyl pyrophosphate can undergo nucleophilic substitution with water to produce geraniol (Equation 11-34), found in oils of the rose and geranium plants.

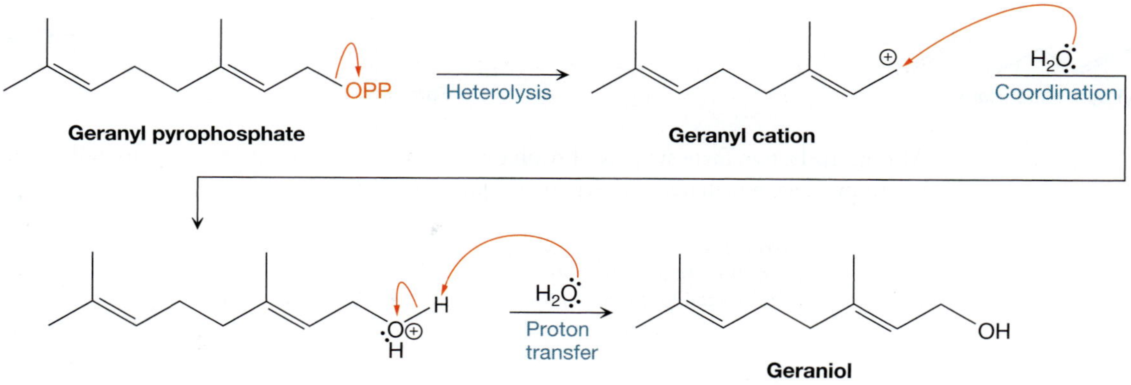

Geranyl pyrophosphate → (Heterolysis) → Geranyl cation → H_2O (Coordination) (11-34)

→ H_2O (Proton transfer) → Geraniol

PROBLEM 11.21 Neryl pyrophosphate can react with water to produce α-terpineol and terpin hydrate, terpenoids found in natural oils. Draw the mechanism for each of these reactions.

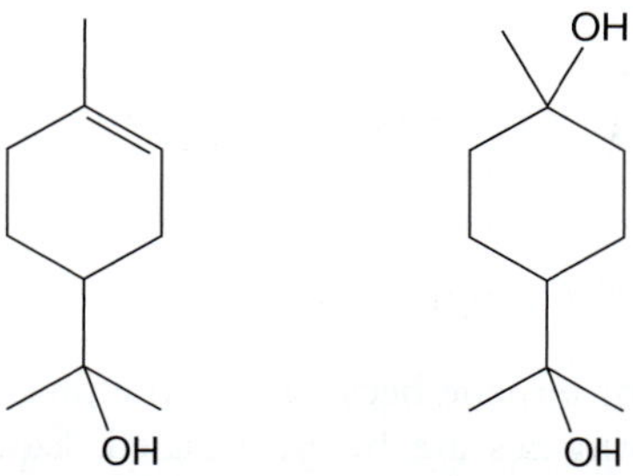

α-Terpineol   Terpin hydrate

Geranyl pyrophosphate can be used to further grow the carbon chain in the synthesis of terpenes and terpenoids. For example, after the pyrophosphate group has left, the geranyl cation can react with isopentenyl pyrophosphate to produce a carbocation intermediate containing 15 carbon atoms, and subsequent elimination of a proton produces farnesyl pyrophosphate, as shown in Equation 11-35.

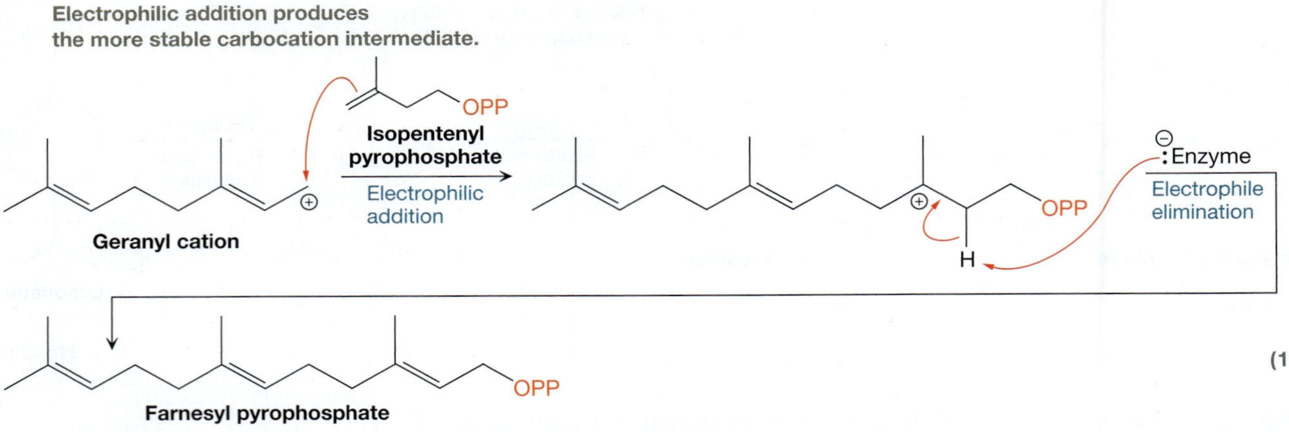

Electrophilic addition produces the more stable carbocation intermediate.

Isopentenyl pyrophosphate

Electrophilic addition

Geranyl cation

:Enzyme

Electrophile elimination

Farnesyl pyrophosphate

(11-35)

Farnesyl pyrophosphate can be hydrolyzed with water to produce farnesol (Equation 11-36)—a pheromone for some insects, and an oil used to enhance the aroma of perfumes for some humans.

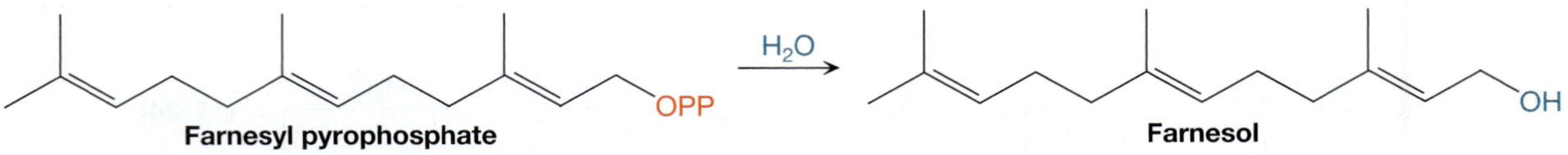

Farnesyl pyrophosphate $\xrightarrow{H_2O}$ **Farnesol**

(11-36)

Alternatively, two farnesyl pyrophosphate molecules can couple in a "tail-to-tail" fashion to produce squalene, a triterpene (Equation 11-37).

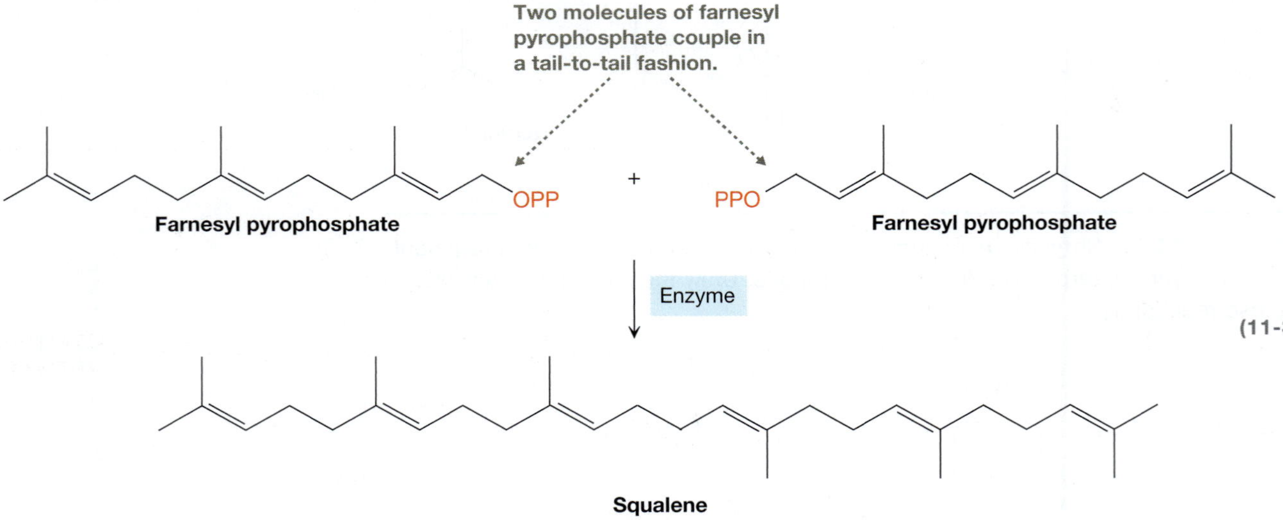

Two molecules of farnesyl pyrophosphate couple in a tail-to-tail fashion.

Farnesyl pyrophosphate   +   **Farnesyl pyrophosphate**

Enzyme

(11-37)

Squalene

Squalene is an important terpene because it is the precursor to cholesterol, from which all other steroid hormones are biosynthesized. Equation 11-38 shows that squalene is oxidized by the enzyme squalene epoxidase in the presence of oxygen, producing squalene oxide.

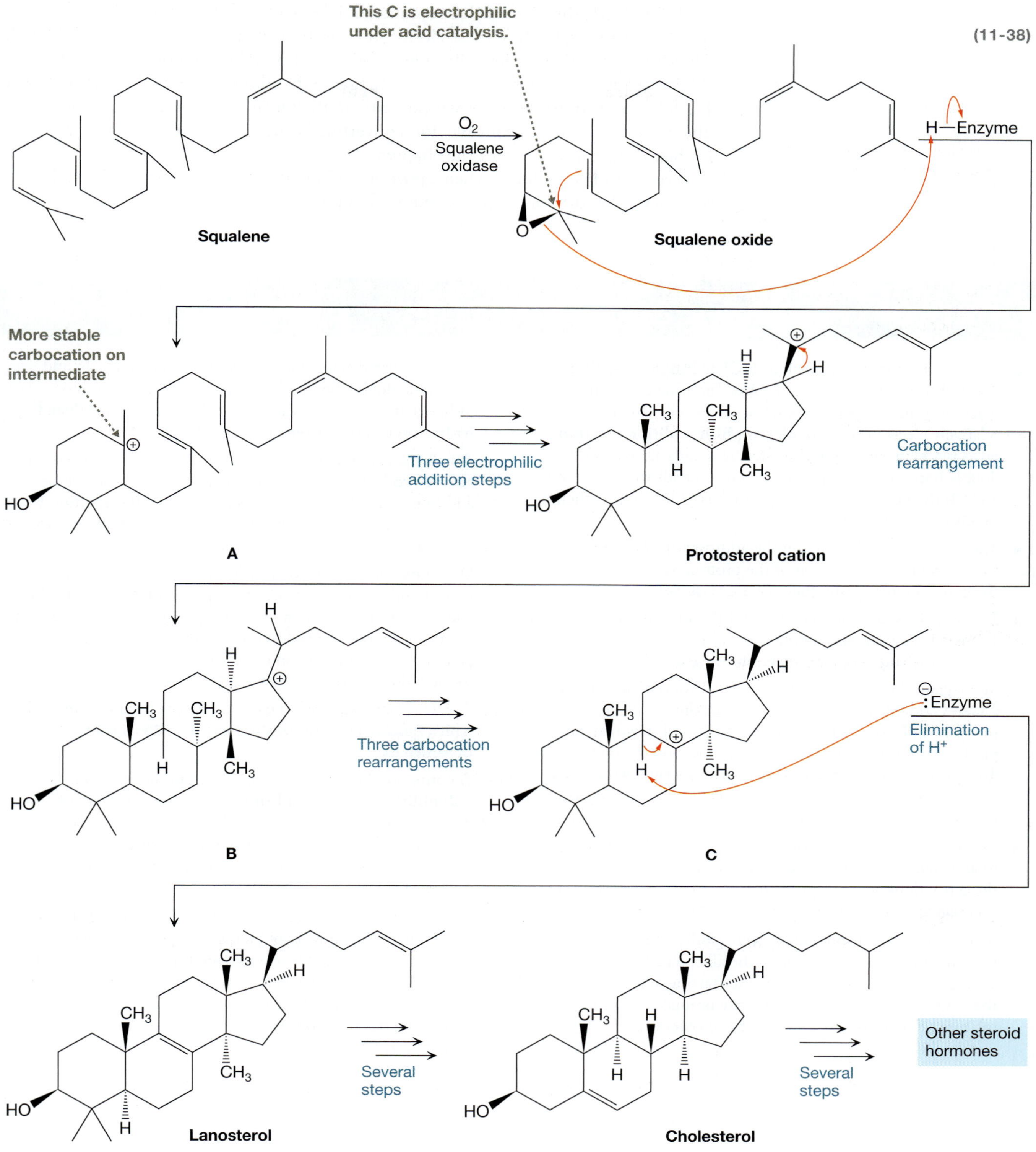

This C is electrophilic under acid catalysis.

Squalene

O₂
Squalene oxidase

Squalene oxide

H—Enzyme

More stable carbocation on intermediate

HO

A

Three electrophilic addition steps

CH₃ CH₃

H

CH₃

HO

Protosterol cation

Carbocation rearrangement

H

CH₃ CH₃

H CH₃

HO

B

Three carbocation rearrangements

CH₃

H

CH₃

H CH₃

HO

C

⊖:Enzyme

Elimination of H⁺

CH₃

H

CH₃

CH₃

HO H

Lanosterol

Several steps

CH₃

H

CH₃ H

H H

HO

Cholesterol

Several steps

Other steroid hormones

Under acid catalysis, the epoxide ring in squalene oxide is electrophilic at the tertiary C that is indicated, which facilitates the electrophilic addition step to produce structure **A**. One new C—C bond forms, resulting in the formation of a new six-membered ring. In doing so, notice that the more stable carbocation is produced, which is what we would expect in a Markovnikov addition.

From there, three more electrophilic addition reactions take place to produce the protosterol cation. Each of these electrophilic addition steps forms a new C—C bond, which produces another new ring. (See Problem 11.50 at the end of the chapter.)

Subsequently, four carbocation rearrangements take place. The first of those is a 1,2-hydride shift, which is shown explicitly in the mechanism, producing structure **B**. The next three are responsible for converting structure **B** into structure **C**. (See Problem 11.51 at the end of the chapter.)

Lanosterol is produced when a proton is eliminated from structure **C**. From lanosterol, several additional steps are required to produce cholesterol.

Chapter Summary and Key Terms

- A strong Brønsted acid, such as HCl or HBr, has an electron-poor proton, which is an *electrophile*. These acids can therefore react with the nonpolar π bond of an alkene, which is relatively electron rich. An **electrophilic addition reaction** takes place, in which the Brønsted acid adds across a C=C double bond in two steps: (1) the proton adds to one carbon, and (2) the newly formed conjugate base adds to the other. **(Section 11.1; Objective 1)**

- The rate-determining step in the addition of a Brønsted acid to an alkene is the addition of the proton, which is Step 1 of the general mechanism. **(Section 11.1; Objective 1)**

- Benzene does not undergo electrophilic addition with Brønsted acids, because doing so would destroy the aromaticity of the ring. **(Section 11.2; Objective 2)**

- The *regioselectivity* of the addition of a Brønsted acid across a C=C double bond is an outcome of Step 1 of the mechanism (protonation of the C=C bond) favoring the production of the more stable carbocation intermediate. This principle accounts for **Markovnikov's rule**, which was determined empirically before the mechanism was known. **(Section 11.3; Objective 3)**

- Because the mechanism for the addition of a Brønsted acid to an alkene proceeds through a carbocation intermediate, these reactions are susceptible to carbocation rearrangements. **(Section 11.4; Objective 3)**

- Stereochemistry is an issue in electrophilic addition reactions involving alkenes if one of the alkene carbons becomes a chiral center in the product. If the starting alkene is achiral, then any chiral products are typically produced as a racemic mixture of enantiomers. **(Section 11.5; Objective 3)**

- Weak Brønsted acids, such as water and alcohols, can add across the double bond of an alkene under acid catalysis. When water adds, the reaction is called an **acid-catalyzed hydration reaction**. **(Section 11.6; Objective 4)**

- An alkyne can undergo electrophilic addition of a Brønsted acid to produce a substituted alkene. As with electrophilic addition to an alkene, the regiochemistry of the addition to an alkyne is an outcome of protonation favoring the more stable carbocation intermediate. **(Section 11.7; Objective 3)**

- When two equivalents of a hydrogen halide add to a C≡C triple bond, the major product is a **geminal dihalide**. In the addition of the second equivalent of the acid, the carbocation intermediate that is produced is resonance stabilized if the positive charge appears on the carbon atom already attached to the halogen atom. **(Section 11.7; Objective 3)**

- The acid-catalyzed hydration of an alkyne produces an enol, which quickly tautomerizes to a ketone. **(Section 11.8; Objective 5)**

- A **conjugated** diene undergoes electrophilic addition via **1,2-addition** and **1,4-addition**, producing a mixture of products. Both mechanisms involve precisely the same carbocation intermediate, in which the positive charge is resonance delocalized over two carbon atoms. **(Section 11.9; Objective 6)**

- Electrophilic addition to a conjugated diene takes place under kinetic control at cold temperatures, and under thermodynamic control at warm temperatures. The 1,2-adduct is typically the kinetic product. The thermodynamic product could be either the 1,2- or the 1,4-adduct, depending on which one has the more stable C=C. **(Section 11.10; Objective 7)**

Reaction Table

TABLE 11-1 Functional Group Transformations[a]

	Starting Compound Class	Typical Reagents Required	Compound Class Formed	Key Electron-Rich Species	Key Electron-Poor Species	Comments	Discussed in Section(s)
(1)	Alkene ($C=C$)	HX	Alkyl halide	$:\!\ddot{X}\!:^{\ominus}$	carbocation (H–C–C⊕)	Markovnikov addition	11.1
(2)	Alkene ($C=C$)	H_2O, $H^{\oplus}$	Alcohol	$H_2\ddot{O}:$	carbocation	Markovnikov addition, acid catalysis	11.6
(3)	Alkene ($C=C$)	ROH, $H^{\oplus}$	Ether	$R\ddot{O}H$	carbocation	Markovnikov addition, acid catalysis	11.6
(4)	Alkyne ($C\equiv C$)	HX (1 equiv)	Vinylic halide	$:\!\ddot{X}\!:^{\ominus}$	vinyl cation	Markovnikov addition, predominantly trans; not useful for synthesis	11.7
(5)	Alkyne ($C\equiv C$)	HX (2 equiv)	Geminal dihalide	$:\!\ddot{X}\!:^{\ominus}$	carbocation	Markovnikov addition twice	11.7
(6)	Alkyne ($C\equiv C$)	H_2O, TfOH, CF_3CH_2OH	Ketone	$H_2\ddot{O}:$	vinyl cation	Markovnikov addition of H_2O, keto–enol tautomerization	11.8
(7)	Conjugated diene	HX (1 equiv), cold	1,2-Adduct	$:\!\ddot{X}\!:^{\ominus}$	allylic cation	Kinetic control	11.9; 11.10
(8)	Conjugated diene	HX (1 equiv), warm	1,4-Adduct	$:\!\ddot{X}\!:^{\ominus}$	allylic cation	Thermodynamic control	11.9; 11.10

[a] X = Cl, Br, I.

Problems

Problems that are related to synthesis are denoted (SYN).

11.1 The General Electrophilic Addition Mechanism: Strong Brønsted Acids

11.22 Cyclohexene can react with hydrogen halides, HX, to yield the various halocyclohexanes, $C_6H_{11}X$. Rank HF, HCl, HBr, and HI in order from slowest reaction rate to fastest. Explain. *Hint:* What is the rate-determining step?

11.23 Rank alkenes **A–D** in order from slowest rate of electrophilic addition of HCl to fastest. Explain. *Hint:* What is the rate-determining step?

$H_2C = CH_2$

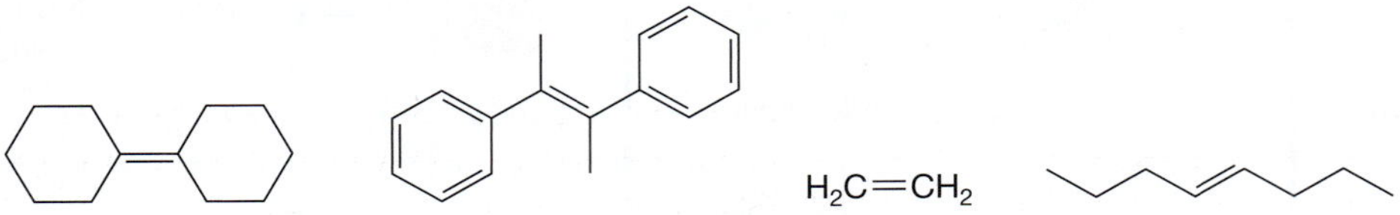

A **B** **C** **D**

11.24 Draw the complete, detailed mechanism for each of the following reactions, including the major organic product.

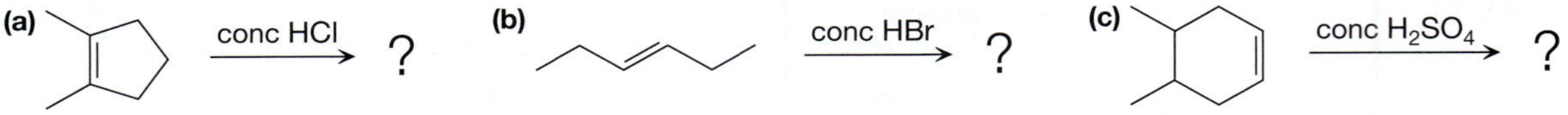

(a) conc HCl ? **(b)** conc HBr ? **(c)** conc H_2SO_4 ?

11.25 A compound, C_5H_8, is treated with excess HCl to produce a compound, C_5H_9Cl. Draw all possible structures for the initial compound.

11.2–11.5 Benzene Rings, Regiochemistry, Carbocation Rearrangements, and Stereochemistry

11.26 Which of alkenes **E–I** will produce 2-chloro-3-methyl-2-phenylbutane as the major product when treated with HCl? Explain.

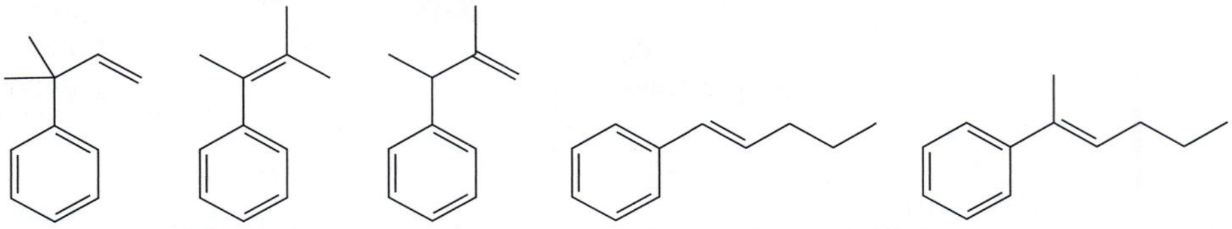

E **F** **G** **H** **I**

11.27 (SYN) Each of the following compounds can be produced from an alkene, using a single electrophilic addition reaction. Write that reaction and draw its complete, detailed mechanism. **(a)** 4-chloro-1,2-dimethylcyclohexane; **(b)** 1-chloro-1,2-dimethylcyclohexane; **(c)** 1-bromo-1,1-diphenylbutane; **(d)** 2,2-dichloropentane

11.28 Treatment of (R)-4-chlorocyclohexene with HCl produces a mixture of four products. Draw the mechanism that accounts for the formation of each product, and identify which products are optically active.

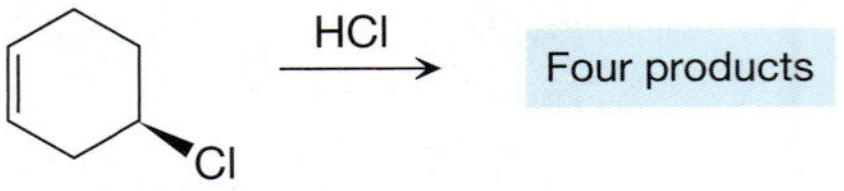

HCl → Four products

11.29 Draw the complete, detailed mechanism for the addition of HCl to dihydropyran and predict the major product.

+ HCl ⟶ ?

11.30 When penta-1,4-diene is heated in the presence of HCl, the major product is 1-chloropent-2-ene. Draw the complete, detailed mechanism that accounts for this reaction.

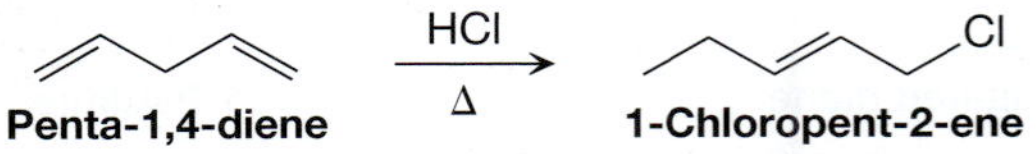

Penta-1,4-diene $\xrightarrow{\text{HCl}}{\Delta}$ **1-Chloropent-2-ene**

11.31 The regiochemistry in the electrophilic addition reaction shown here does not adhere to the original generalization put forth by Markovnikov. Draw the complete mechanism for this reaction and explain its regiochemistry.

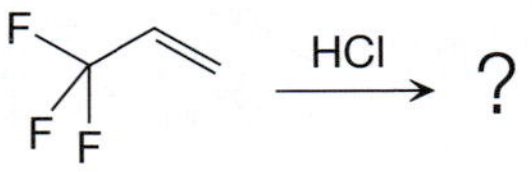

$\xrightarrow{\text{HCl}}$?

11.32 In the biosynthesis of aromatic amino acids, erythrose-4-phosphate undergoes electrophilic addition to phosphoenolpyruvate (PEP). Draw the products of this step, paying particular attention to regiochemistry.

Phosphoenolpyruvate (PEP)   **Erythrose 4-phosphate**

11.6 Addition of a Weak Acid: Acid Catalysis

11.33 (SYN) Each of the following compounds can be produced from an alkene, using a single electrophilic addition reaction. Write that reaction and draw its complete, detailed mechanism. **(a)** pentan-2-ol; **(b)** 3-methylpentan-3-ol; **(c)** 1-methoxy-1,4-dimethylcyclohexane; **(d)** (cyclopentyldimethoxymethyl)benzene

11.34 (SYN) Show how 1-methylcyclohexanol can be produced from two *different* alkenes.

11.35 Draw the complete, detailed mechanism for the addition of hexan-3-ol to dihydropyran and predict the major product.

11.36 Draw the mechanism for this reaction.

11.7 and 11.8 Electrophilic Addition to an Alkyne

11.37 For each reaction, draw the complete mechanism and the major organic product.

(a)

excess HCl   ?

(b)

excess HBr   ?

11.38 For each reaction, draw the complete mechanism and the major organic product. (Recall that each of these reactions is expected to have a low yield.)

(a) $HC\equiv CH$ $\xrightarrow[H_2O]{H_2SO_4}$?   **(b)** $HC\equiv CH$ $\xrightarrow[D_2O]{D_3PO_4}$?   **(c)**

$\xrightarrow[H_2O]{H_2SO_4}$?

11.39 (SYN) Each of the following compounds can be produced from an alkyne, via electrophilic addition. Write the reactions and draw complete, detailed mechanisms.

(a) (1,1-Dichloro-2-cyclopentylethyl)benzene   **(b)** 3,3-Dibromohexane   **(c)**

(d)

11.40 (SYN) Each of the following compounds can be produced (in low yield) from an alkyne, using a single electrophilic addition reaction. Write the reactions and draw complete, detailed mechanisms.

(a)

(b)

11.41 The acid-catalyzed hydration of the internal alkyne shown here leads to only a single adduct (in low yield). Draw the product and explain why a mixture of adducts is not produced.

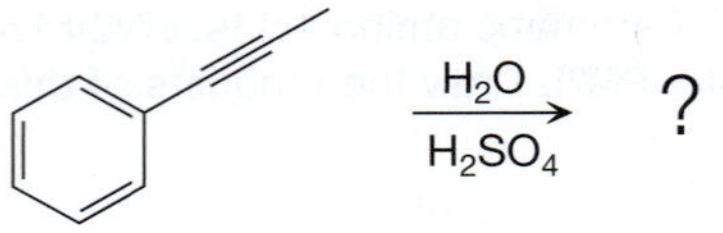

11.42 Draw the complete mechanism for this reaction.

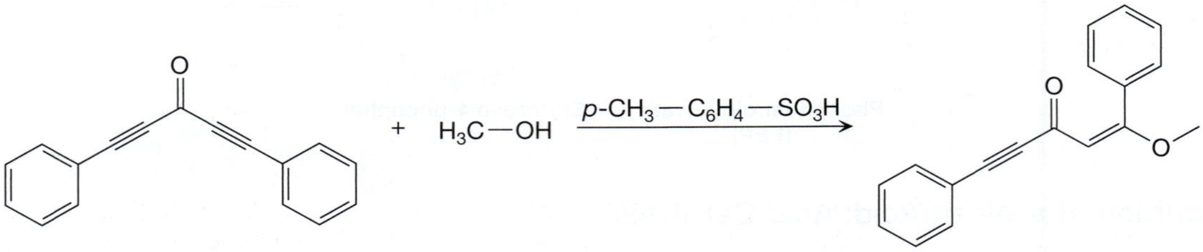

11.9 and 11.10 Electrophilic Addition to a Conjugated Diene; Kinetic versus Thermodynamic Control

11.43 Which product is the result of 1,2-addition and which one is the result of 1,4-addition?

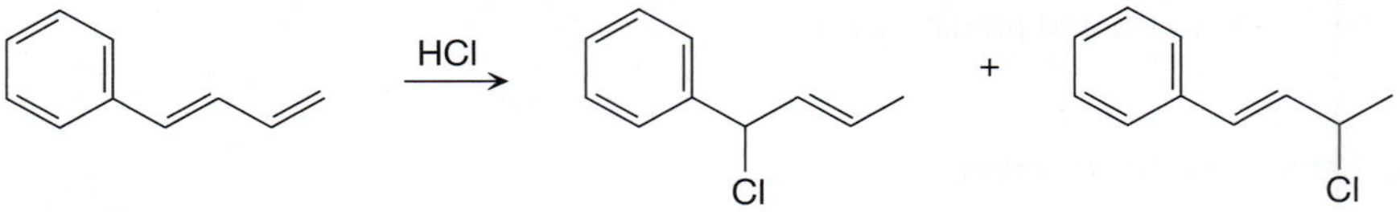

11.44 Consider the addition of HBr shown here.

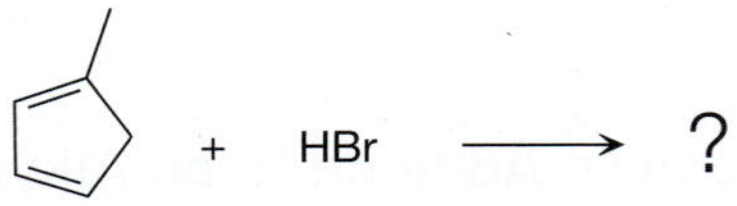

(a) Draw all four carbocation intermediates possible from protonation of the diene and identify the most stable one.
(b) Draw both halogenated products formed by attack of Br^- on that carbocation.
(c) Which of those products would you expect to be formed in the greatest amount at low temperatures?
(d) Which would you expect to be formed in the greatest amount at high temperatures?

11.45 Consider the addition of HBr shown here.

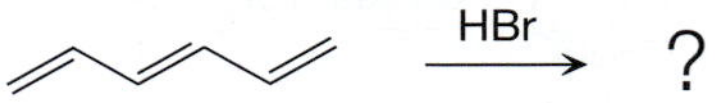

(a) There are three carbocation intermediates possible from the protonation of this triene. Draw all three of them and identify the most stable one.
(b) Draw all halogenated products formed by attack of Br^- on the most stable carbocation.
(c) Which of those products would you expect to be formed in the greatest amount at low temperatures?
(d) Which would you expect to be formed in the greatest amount at high temperatures?

11.46 The addition of HBr to buta-1,3-diene results in 1,2-addition at cold temperatures and 1,4-addition at warm temperatures. If the 1,2-adduct is first formed at cold temperatures and then warmed up, the 1,4-adduct is formed, as shown here. Draw a mechanism for this isomerization.

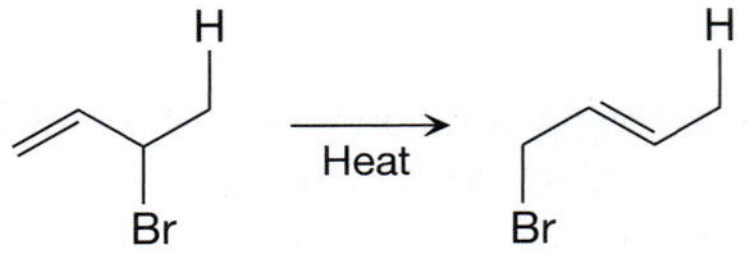

1,2-Adduct **1,4-Adduct**

11.11 The Organic Chemistry of Biomolecules

11.47 α-Terpineol, a naturally occurring monoterpene alcohol, isomerizes to 1,8-cineole and 1,4-cineole when treated with acid. Draw a complete, detailed mechanism to account for the formation of each product.

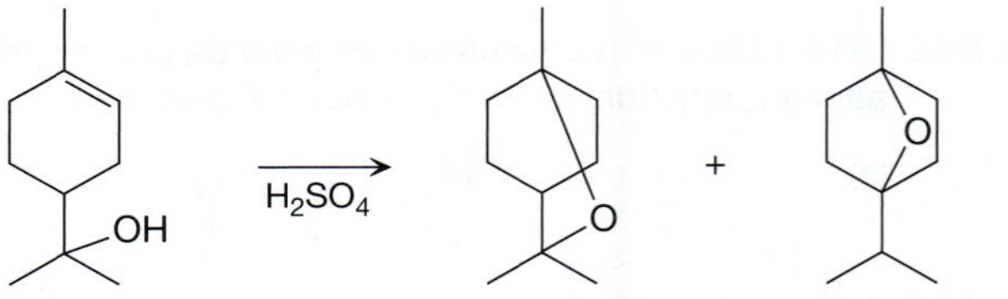

α-Terpineol **1,8-Cineole** **1,4-Cineole**

11.48 Draw the mechanism for how α-pinene is biosynthesized from geranyl pyrophosphate.

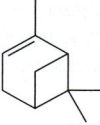

α-Pinene

11.49 Draw the mechanism for how γ-curcumene is biosynthesized from farnesyl pyrophosphate.

γ-Curcumene

11.50 In the partial mechanism for the formation of cholesterol from squalene (Equation 11-38, p. 593), the steps for the following transformation were not shown explicitly. As indicated, the transformation involves three electrophilic addition steps. Draw each step and its curved arrow notation explicitly.

A   Three electrophilic addition steps   **Protosterol cation**

11.51 In the partial mechanism for the formation of cholesterol from squalene (Equation 11-38, p. 593), the steps for the following transformation were not shown explicitly. As indicated, the transformation involves three carbocation rearrangement steps. Draw each step and its curved arrow notation explicitly.

B   Three carbocation rearrangements   **C**

Integrated Problems

11.52 Predict the major product(s) for each of the following reactions.

(a) 4-Chlorobut-1-ene + HBr ⟶ ?

(b) 1-Chlorobut-1-ene + HBr ⟶ ?

(c) 4,4-Dimethylcyclopentene + H₂O, H⁺ ⟶ ?

(d) Propyne + 2 HCl ⟶ ?

(e) Cyclopentylethene $\xrightarrow{\text{H}_3\text{O}^+}$?

11.53 (SYN) Show how you would carry out each of the following transformations. *Hint:* Two or more separate reactions may be required.

(a) ⟶ ?

(b) ⟶ ?

11.54 As discussed in greater detail in Chapter 22, AlCl$_3$ is a powerful Lewis acid that effectively catalyzes the dissociation of an alkyl chloride, RCl, into its respective ions, R$^+$ and Cl$^-$, as shown at the right.

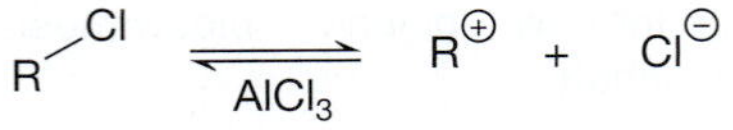

(a) Propose a mechanism for the addition of RCl across a double bond, as shown at the right.

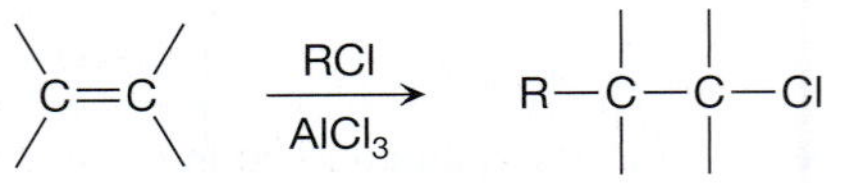

(b) Using that mechanism, what are the two possible isomers that can be formed when 2-methylpropene is treated with 2-chloropropane in the presence of AlCl$_3$? Which one is the major product? Explain.

11.55 As discussed in Chapters 25 and 26, a polymer is a very large molecule that contains many repeating units called monomers. The reaction here shows, for example, how styrene reacts to form polystyrene. The reaction is *initiated* by the electrophilic addition of H$^+$ from an acid like sulfuric acid, which generates an initial carbocation. Afterward, that carbocation behaves as an electrophile in the presence of another molecule of styrene, resulting in yet another carbocation. This reaction can repeat many thousands of times to build up the polymer. With this in mind, draw the detailed mechanism that illustrates the initiation of the polymerization reaction and the addition of the first two monomers, as shown at the right.

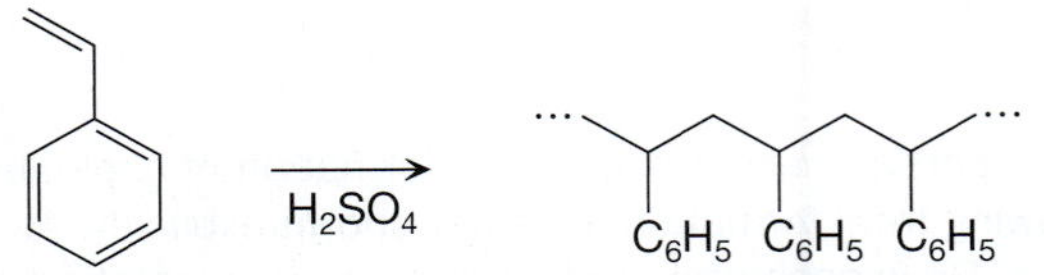

Styrene   **Polystyrene**

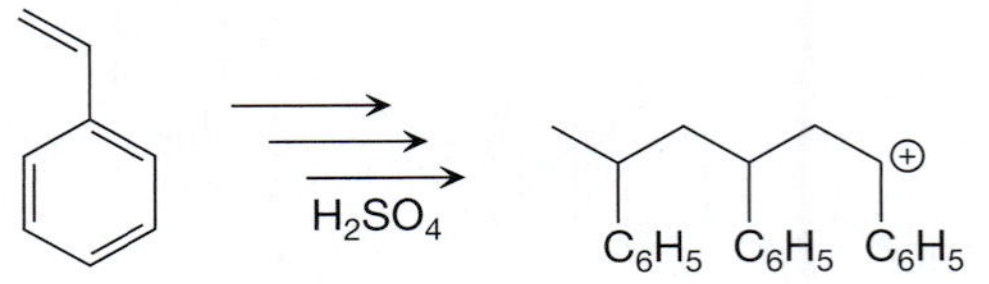

11.56 The treatment of but-1-en-3-yne with HBr produces 4-bromobuta-1,2-diene, which is an allene. Draw the complete, detailed mechanism for this reaction.

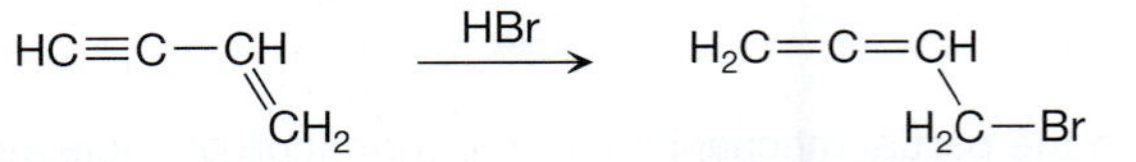

But-1-en-3-yne   **4-Bromobuta-1,2-diene**

11.57 Propose a mechanism for the reaction shown here, in which HCl adds to hepta-1,6-diene.

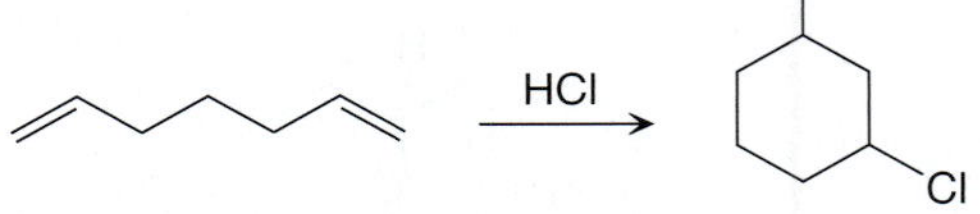

11.58 Propose a reasonable mechanism that would account for the reaction shown here.

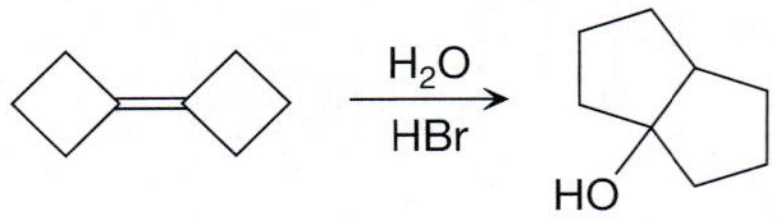

11.59 Determine the structures of compounds **A** through **D** in the following reaction scheme:

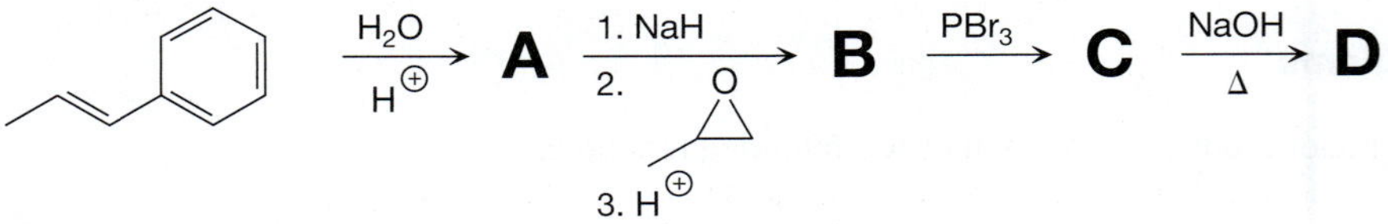

11.60 Determine the structures of compounds **E** through **I** in the following reaction scheme:

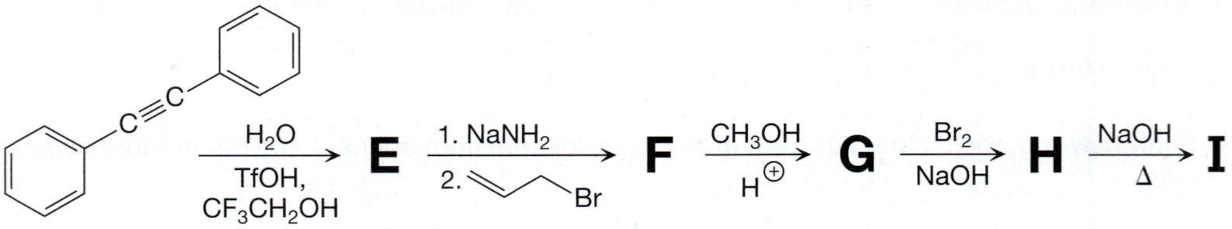

11.61 Determine the structures of compounds **J** through **N** in the following reaction scheme:

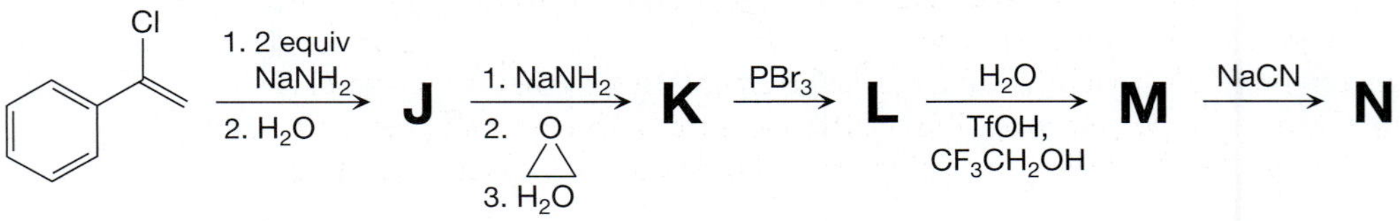

A pyrethrin, which has a characteristic cyclopropane ring, is a natural, potent insecticide that can be isolated from chrysanthemum flowers. Pyrethroids are synthetic variations of pyrethrin and constitute the bulk of household insecticides. Many of these pyrethroids are synthesized using a cyclopropanation reaction analogous to the one we present here in Chapter 12.

A pyrethrin

12

Electrophilic Addition to Nonpolar π Bonds 2

Reactions Involving Cyclic Transition States

Chapter 11 discussed reactions in which a Brønsted acid adds across a C=C double bond of an alkene or a C≡C triple bond of an alkyne. The π bond of the alkene or alkyne is relatively electron rich in those reactions, whereas the proton (H$^+$) from the Brønsted acid is electron poor. Consequently, the proton acts as an *electrophile* and, as it is picked up by the alkene or alkyne, a carbocation intermediate is produced. Then, that carbocation intermediate is attacked by a nucleophile.

Here in Chapter 12, we will see that species other than protons can act as electrophiles, and thus can also add to an alkene or alkyne. Examples include molecular halogens like Cl$_2$ and Br$_2$, peroxyacids (RCO$_3$H), carbenes (R$_2$C:), and borane (BH$_3$). Unlike the electrophilic addition of a Brønsted acid, all of these reaction mechanisms involve a step whose transition state is cyclic. This avoids the formation of a carbocation intermediate and therefore has significant consequences for the outcomes of the reactions, both regiochemically and stereochemically.

We begin Chapter 12 with a brief look at a key elementary step that is involved in several of the above reactions: electrophilic addition to form a three-membered ring. We then examine the details of several reactions that proceed by such a step, including

On completing Chapter 12 you should be able to:

1. Recognize when electrophilic addition to an alkene or alkyne favors the formation of a three-membered ring instead of a carbocation intermediate, and draw the mechanism for such a step.

2. Show how carbene and dichlorocarbene are produced from their respective precursors, and draw the mechanism of their reaction with an alkene or alkyne.

3. Describe the role of the halonium ion intermediate in the addition of a molecular halogen to an alkene or alkyne, and predict the products of such reactions, including stereochemistry and regiochemistry.

4. Explain the importance of the oxymercuration–reduction reaction, and draw its complete mechanism.

5. Show how an epoxide is produced in the reaction of a peroxyacid with an alkene.

6. Draw the detailed mechanism of hydroboration–oxidation of an alkene or alkyne, and predict the products of such a reaction, including stereochemistry and regiochemistry.

7. Explain the role of disiamylborane and other bulky dialkylboranes in the hydroboration–oxidation of an alkyne.

carbene addition, halogenation, oxymercuration, and epoxide formation. Following that, we examine hydroboration, which proceeds through a four-membered-ring transition state.

12.1 Electrophilic Addition via a Three-Membered Ring: The General Mechanism

The mechanisms of all of the electrophilic addition reactions we saw in Chapter 11 include a step in which a carbocation is produced. Recall that this step is generally highly unfavorable, due in large part to the loss of an octet on a C atom, as indicated in Equation 12-1.

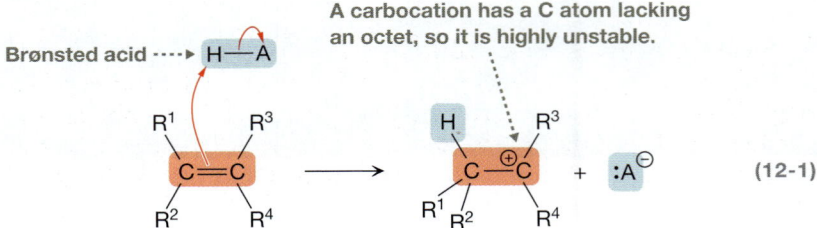

(12-1)

However, if the electrophilic atom on the electrophile has a lone pair of electrons, then addition can take place in a way that avoids losing an octet. This is illustrated in Equation 12-2, in which E: represents a generic electrophile that possesses a lone pair of electrons.

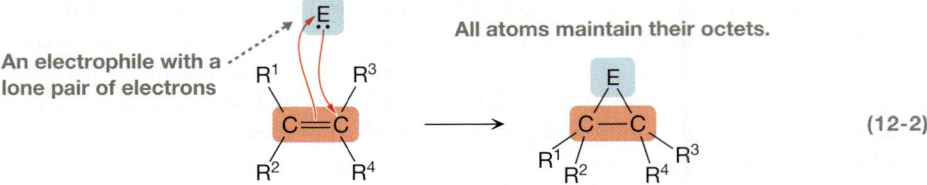

(12-2)

In this step, two new E—C σ bonds are formed simultaneously to produce a three-membered ring. One of those bonds is formed by the electrons from the initial carbon–carbon π bond, and the other is formed by the lone pair of electrons from the electrophile.

In the cyclic structure in Equation 12-2, how many bonds does each of the highlighted C atoms have? How does that compare to the number of bonds each of those C atoms has in the initial alkene?

Answers to Your Turns are in the back of the book.

Notice in Equation 12-2 that as the alkene reacts, two C atoms rehybridize from sp^2 to sp^3, so *stereochemistry* can be an issue. When it is, stereochemistry is governed by the following rules.

- The cis/trans relationship is conserved for the groups attached to the alkene carbons in Equation 12-2. Groups that are on the same side of the double bond in the reactant must end up on the same side of the plane of the three-membered ring in the product.
- If the cyclic product in Equation 12-2 is chiral, then a mixture of stereoisomers is produced: a racemic mixture of enantiomers if the original alkene and other reagents are achiral, or an unequal mixture of stereoisomers if otherwise.

These rules are exemplified by Equation 12-3.

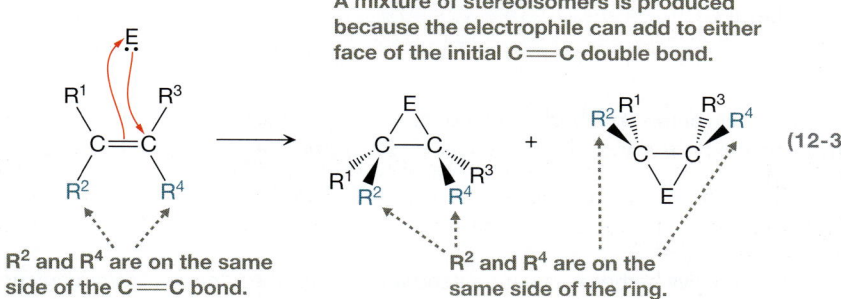

A mixture of stereoisomers is produced because the electrophile can add to either face of the initial C=C double bond.

R^2 and R^4 are on the same side of the C=C bond.

R^2 and R^4 are on the same side of the ring.

(12-3)

Stereochemistry is conserved in such reactions because both C—E bonds are formed *simultaneously*. Normally, if a C=C double bond is converted into a C—C single bond, rotation about that single bond can lead to scrambling of the cis/trans relationship among the groups attached to those carbon atoms. In this case, however, the C—C single bond that remains is part of a ring, so free rotation cannot take place.

The reaction produces a mixture of stereoisomers because an alkene is planar, and the electrophile can add to either face of the alkene. One stereoisomer is produced by the addition of the electrophile to one face, and the second is produced by addition to the other face. If the initial alkene is achiral (e.g., by possessing a plane of symmetry), then addition to each face of the alkene is equally likely. If the initial alkene is chiral, however, then it possesses no plane of symmetry, making electrophilic addition to one face more likely than addition to the other.

Match each curved arrow formalism on the left with the appropriate product on the right. Write "racemic" next to each product that will be produced as a racemic mixture.

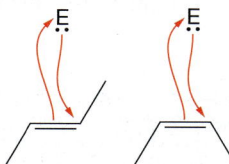

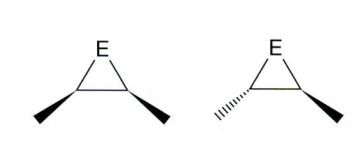

Having examined this elementary step in a generic fashion, let's now turn our attention toward four specific reaction mechanisms in which it appears. Section 12.2 discusses the electrophilic addition of a carbene to an alkene; Section 12.3 discusses electrophilic addition involving molecular halogens; Section 12.4 discusses oxymercuration–reduction; and Section 12.5 discusses epoxide formation.

12.2 Electrophilic Addition of Carbenes: Formation of Cyclopropane Rings

Mechanistically, the simplest of the reactions we discuss here in Chapter 12 occurs between an alkene and a **carbene**. A carbene is a species containing a carbon atom that has two bonds and a lone pair of electrons,[1] as shown in Figure 12-1. Having three total electron groups, that C atom lacks an octet and is sp^2 hybridized; the lone pair of electrons occupies a hybrid orbital, leaving the p orbital empty. Thus, these carbenes resemble carbocations and are generally *highly* electron poor. Unlike a carbocation, however, the electron-poor C atom of a carbene has a formal charge of 0.

YOUR TURN 12.3

Using the method we learned in Chapter 1, calculate the formal charge on the carbene C.

Most carbenes are highly reactive, so they typically have very short lifetimes and cannot be isolated. In these cases, *we cannot simply add a carbene as a reagent.* Instead:

> Highly reactive carbenes must be generated in situ (i.e., "on site") from precursors that can be added as reagents.

Diazomethane, CH_2N_2, is one precursor from which a carbene can be made. An example of how it is used is shown in Equation 12-4. Notice that it produces a new cyclopropane ring.

No octet = highly electron deficient

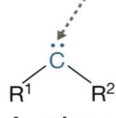

A carbene

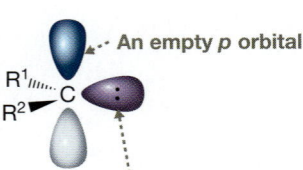

An empty p orbital

The lone pair occupies an sp^2 hybrid orbital.

FIGURE 12-1 A generic carbene (*Top*) A carbene is characterized by a C atom with two bonds and a lone pair of electrons. Even though the C atom is uncharged, carbenes are typically highly electrophilic due to the lack of an octet. (*Bottom*) Carbenes are sp^2 hybridized. The lone pair occupies an sp^2 hybrid AO, whereas the p AO is empty.

New cyclopropane ring ----> CH_2

$+ CH_2N_2 \xrightarrow{h\nu}$ $+ N_2(g)$ (12-4)

80%

The symbol $h\nu$ indicates energy from light.[2]

The complete mechanism for this reaction is shown in Equation 12-5.

---

[1] These are called *singlet carbenes* when the nonbonding electrons occupy the same orbital. In a *triplet carbene*, one nonbonding electron is in each of two separate orbitals. In this book, we discuss only singlet carbenes.

[2] In the symbol $h\nu$, h is Planck's constant and ν is the photon frequency, and the product of these values is the photon's energy.

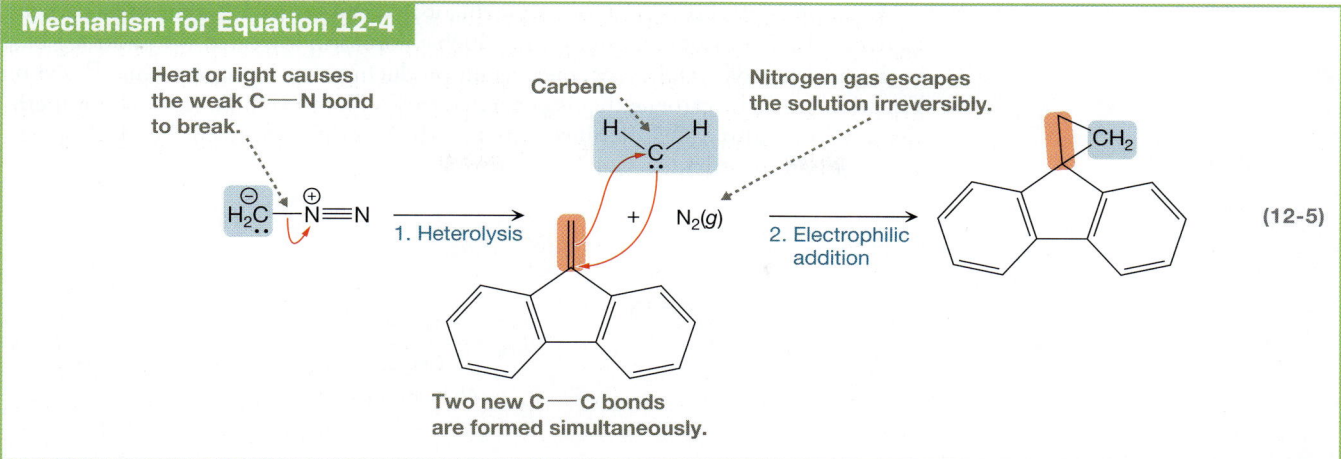

Heat or light causes the weak C—N bond to break.

Carbene

Nitrogen gas escapes the solution irreversibly.

1. Heterolysis

+ $N_2(g)$

2. Electrophilic addition

(12-5)

Two new C—C bonds are formed simultaneously.

In Step 1, the C—N bond in diazomethane is broken when energy is absorbed from light (but the energy could instead come from added heat), and $H_2C\colon$ is produced. Even though $H_2C\colon$ is highly unstable, this step is helped by the production of $N_2(g)$, an excellent leaving group. Recall from Section 10.5 that $N_2(g)$ is very stable and, being a gas, escapes the reaction mixture *irreversibly*. In Step 2, $H_2C\colon$ reacts with the alkene to generate the three-membered ring via the same mechanism as in Equation 12-2.

PROBLEM 12.1 Draw the complete, detailed mechanism for the reaction shown here and predict the major products.

$$\text{(diagram)} \xrightarrow[\Delta]{CH_2N_2} \quad ?$$

PROBLEM 12.2 Show how each of the following compounds can be produced from an alkene.

(a)   (b)   (c)   (d)

Although the formation of carbene from diazomethane can lead to a good yield of the cyclopropane-containing product, the reaction is of rather limited use in synthesis because diazomethane is explosive and requires extreme caution in the laboratory. A safer way to produce a cyclopropane ring from an alkene is with the Simmons–Smith reaction, an example of which is shown in Equation 12-6. (The mechanism for this reaction is explored in Problem 12.57.)

Simmons–Smith reaction

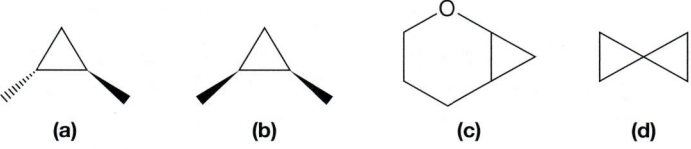

$$\xrightarrow[(CH_3CH_2)_2Zn]{CH_2I_2}$$

CH_2 + Enantiomer   (12-6)

86%

Chloroform (CHCl₃) is another precursor that can be used to generate a carbene. Equation 12-7 provides an example in which chloroform is treated with a strong base in the presence of cyclohexene, once again producing a cyclopropane ring. Based on Equation 12-7, the carbene that is generated from chloroform must be dichloromethylene (Cl₂C:), also called **dichlorocarbene**, which is different from the H₂C: generated from diazomethane.

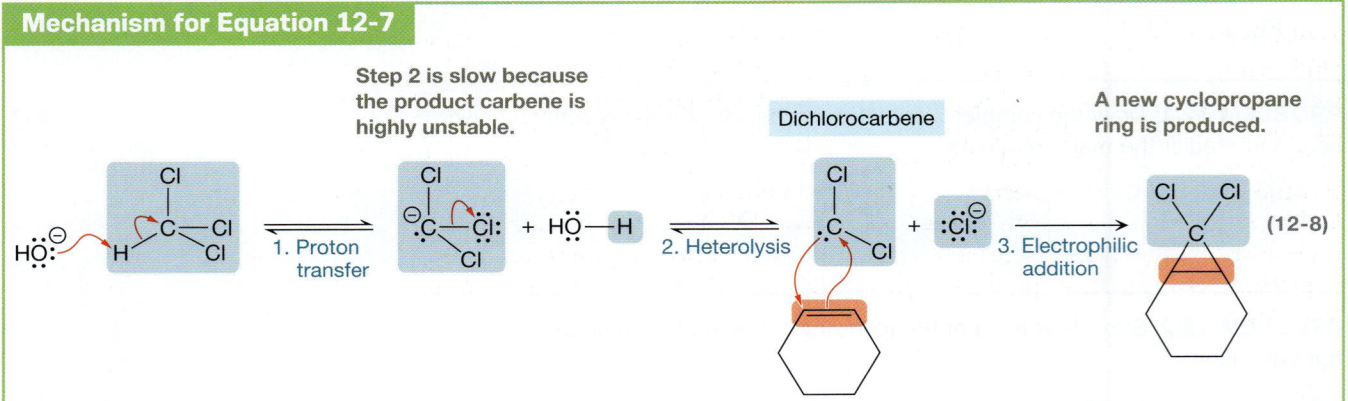

$$+ \quad CHCl_3 \quad \xrightarrow[\substack{CH_2Cl_2 \\ 45\ h,\ 40\text{--}45\ °C}]{NaOH} \quad \quad 52\% \quad \quad (12\text{-}7)$$

As shown in the mechanism in Equation 12-8, dichlorocarbene is produced in two steps via **α-elimination**. The base first deprotonates CHCl₃, which is weakly acidic (pK_a = 24). In Step 2, Cl⁻ departs to generate the Cl₂C: carbene. Although Cl⁻ is a moderately good leaving group, this step occurs quite slowly because the carbene that is produced is very unstable. In Step 3, Cl₂C: adds to the alkene, producing the cyclopropane ring.

Mechanism for Equation 12-7

Step 2 is slow because the product carbene is highly unstable.

Dichlorocarbene

A new cyclopropane ring is produced.

1. Proton transfer   2. Heterolysis   3. Electrophilic addition   (12-8)

YOUR TURN 12.4

The reaction shown here is similar to the one in Equation 12-7. The mechanism for this reaction is given below, but the curved arrows have been omitted. Complete the mechanism by adding the curved arrows, and write the name of each elementary step below the appropriate reaction arrow.

PROBLEM 12.3 Draw the complete, detailed mechanism for the reaction shown here. Using the mechanism, predict the products, paying close attention to stereochemistry.

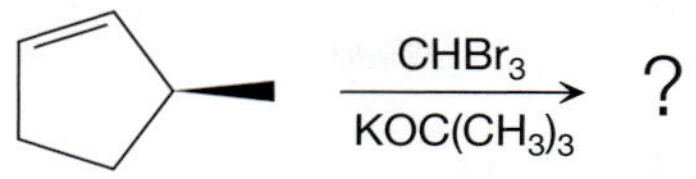

PROBLEM 12.4 Provide two different ways of synthesizing the compound shown here via carbene addition, each using a different starting alkene.

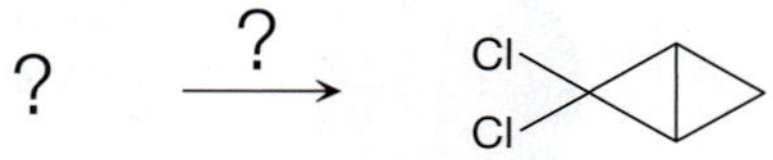

12.3 Electrophilic Addition Involving Molecular Halogens: Synthesis of 1,2-Dihalides and Halohydrins

Section 12.2 highlighted the reactions of carbenes with alkenes. The highly electrophilic carbon atom of the carbene possesses a lone pair of electrons, which is what facilitates the formation of a three-membered ring. Carbenes are not the only species in which an electrophilic atom possesses a lone pair of electrons. As we see here in Section 12.3, molecular halogens, such as Cl_2 and Br_2, share that feature with carbenes, and can therefore react with an alkene to produce a three-membered ring. Unlike what we observe with carbenes, however, the three-membered ring that is produced from a molecular halogen is an unstable intermediate and reacts further to produce relatively stable products such as a *1,2-dihalide* (i.e., a *vicinal dihalide*) or a *halohydrin*.

12.3a Synthesis of 1,2-Dihalides

When cyclohexene is treated with molecular bromine (Br_2) in tetrachloromethane (CCl_4), also called carbon tetrachloride, a racemic mixture of *trans*-1,2-dibromocyclohexane is produced, as shown in Equation 12-9.

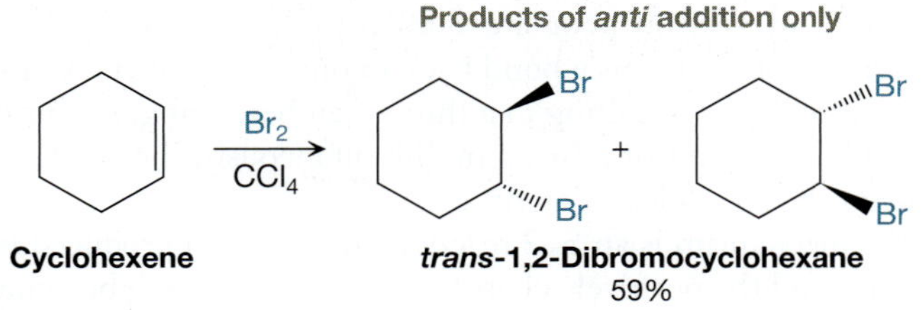

Products of *anti* addition only

(12-9)

Cyclohexene   ***trans*-1,2-Dibromocyclohexane**
                              59%

In other words:

Molecular bromine undergoes anti addition across a C=C double bond.

Although Br_2 does not appear to possess an electron-poor atom, recall that we have seen Br_2 behave as an electrophile once before—in the S_N2 bromination of enols

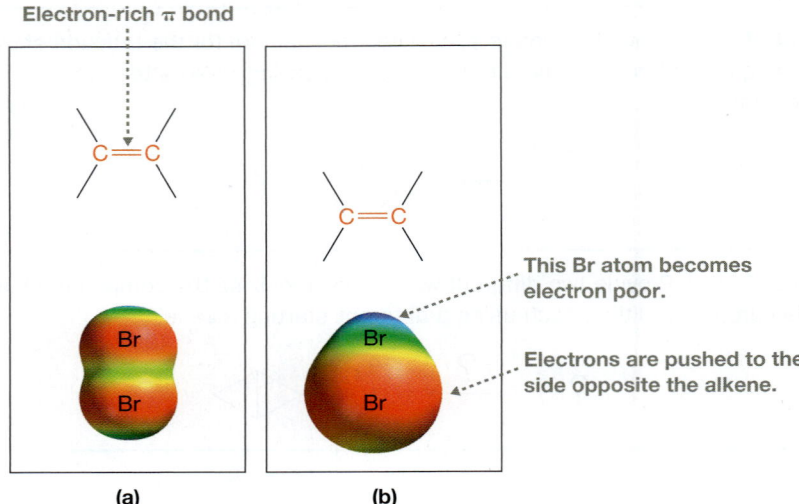

Electron-rich π bond

This Br atom becomes electron poor.

Electrons are pushed to the side opposite the alkene.

(a)   (b)

FIGURE 12-2 The electrophilic nature of a molecular halogen in the presence of an alkene (a) When it is isolated from other species, Br_2 is not electron poor. (b) As the electron-rich alkene approaches the Br_2 molecule, however, electrons on Br_2 are repelled to the side opposite the alkene, temporarily generating an electron-poor site on the Br atom closer to the alkene. Subsequent flow of electrons takes place between the electron-rich alkene and the electron-poor Br atom.

and enolate anions (Section 10.4). As the electron-rich π bond approaches Br_2, the Br_2 molecule becomes *polarized*, temporarily making one of the Br atoms electron poor (Fig. 12-2).

To account for the stereochemistry of the reaction in Equation 12-9, the mechanism must *not* proceed through a carbocation intermediate; otherwise, both syn and anti addition would take place (review Section 11.5). Instead, as shown in Equation 12-10, the mechanism proceeds through a **bromonium ion intermediate** (possessing a positively charged bromine atom), which is produced in Step 1.

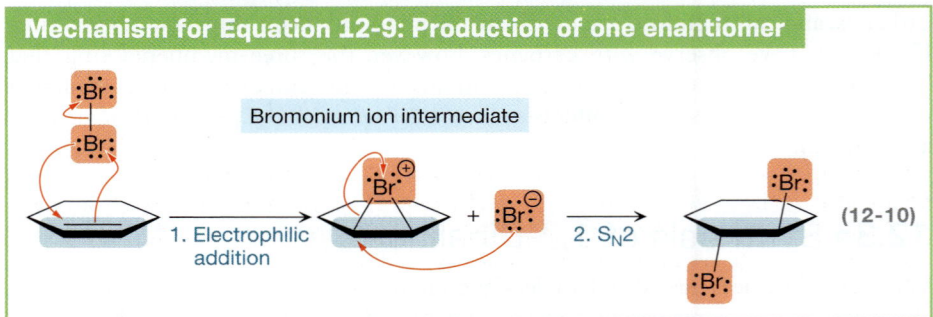

Mechanism for Equation 12-9: Production of one enantiomer

Bromonium ion intermediate

1. Electrophilic addition   2. S_N2 (12-10)

The curved arrow from the π bond to Br represents the flow of electrons from an electron-rich site to an electron-poor site. Similar to the reaction in Equation 12-2, a lone pair of electrons on Br forms a bond back to one of the alkene C atoms to avoid breaking the C atom's octet, resulting in a three-membered ring. In this case, the ring consists of two C atoms and one Br atom. Simultaneously, the weak Br—Br bond breaks, and one of the atoms leaves as Br^-.

Step 2 of the mechanism is an S_N2 reaction: The Br^- ion produced in Step 1 acts as the nucleophile, and the positively charged Br atom in the ring becomes the leaving group. This is exactly the same step we saw in the opening of a protonated epoxide ring in Section 10.7b, where a positively charged oxygen atom behaves as the leaving group.

Notice in Equation 12-10 that the S_N2 reaction in Step 2 is precisely what requires the two Br atoms to be anti to each other in the product. Recall from Section 8.5b that the nucleophile in an S_N2 reaction must attack from the side *opposite the leaving group*. The only way to achieve syn addition would be for the Br^- nucleophile to attack from the same side on which the leaving group leaves — something that is forbidden in an S_N2 reaction.

Notice also in Step 2 of Equation 12-10 that nucleophilic attack is shown to occur at just the left-hand C atom of the three-membered ring. Equation 12-11 shows that nucleophilic attack can also occur at the other C atom of the three-membered ring, resulting in the enantiomer of the product shown in Equation 12-10. Because this particular bromonium ion is achiral, the mixture must be *racemic*.

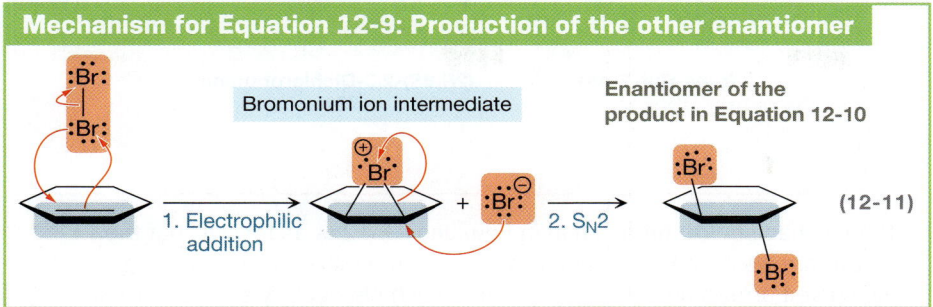

Mechanism for Equation 12-9: Production of the other enantiomer

Bromonium ion intermediate

Enantiomer of the product in Equation 12-10

1. Electrophilic addition

2. S_N2

(12-11)

SOLVED PROBLEM 12.5

Predict the products of the reaction shown here. Do you expect the product mixture to be optically active? Why or why not?

$$\xrightarrow[CCl_4]{Br_2} \quad ?$$

Think What reaction takes place between Br_2 and an alkene? Are the product molecules chiral? How does chirality relate to optical activity?

Solve Br_2 undergoes anti addition across the double bond, yielding the following mixture of products.

Each product molecule is chiral. Whereas the products in Equation 12-9 are enantiomers, here the two products are diastereomers. They are formed in unequal amounts, therefore, and do not have equal and opposite values of $[\alpha]_D^{20}$. As a result, the product solution will be optically active.

PROBLEM 12.6 Draw the product(s) that would be produced from reacting each of the following compounds with molecular bromine in carbon tetrachloride. Will the product mixture be optically active?

HO_2C CO_2H

(a)   (b)   (c)   (d)

Molecular chlorine (Cl_2) also undergoes addition to alkenes to produce vicinal dichlorides (Equation 12-12). In the particular example shown in Equation 12-12, Cl_2 adds in an anti fashion to the double bond, just as we saw when Br_2 adds to an alkene. The mechanism for this chlorination reaction, therefore, should be similar to bromination, proceeding through a **chloronium ion intermediate** to avoid breaking the C atom's octet (see Your Turn 12.5). Not all chlorination reactions exhibit anti addition exclusively, however, suggesting that the mechanism for chlorination depends on the structure of the alkene. (See Problem 12.56 at the end of the chapter.)

Anti addition of Cl₂

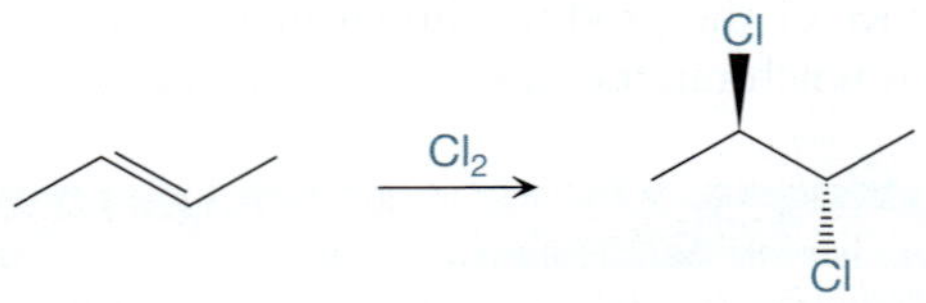

(12-12)

trans-But-2-ene → **(2R,3S)-2,3-Dichlorobutane**
73%

YOUR TURN 12.5

The mechanism for the reaction in Equation 12-12 is as follows, but the curved arrows have been omitted. Complete the mechanism by adding the curved arrows and write the name of each elementary step under the appropriate reaction arrow. Label the chloronium ion intermediate.

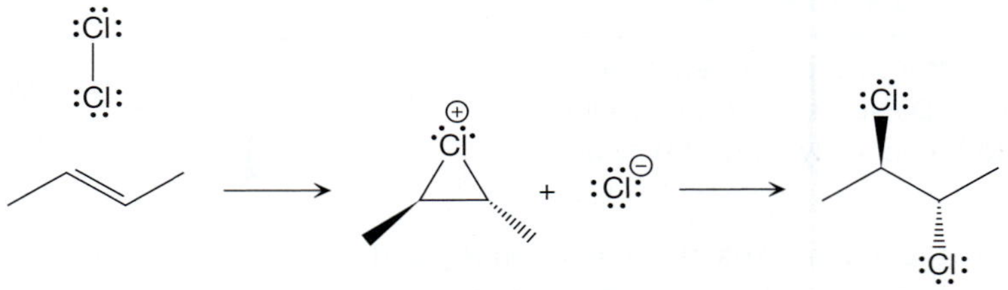

Molecular fluorine (F₂) and iodine (I₂) react with alkenes as well. These reactions are not as useful in organic synthesis, however, because F₂ reacts explosively with alkenes, whereas the reaction with I₂ is energetically unfavorable. Therefore, we will not discuss these reactions any further.

PROBLEM 12.7 Treatment of an alkene with molecular chlorine in carbon tetrachloride yields a racemic mixture of (R,R)-3,4-dichloro-2-methylhexane and its enantiomer. Draw the structure of the initial alkene.

Br₂ and Cl₂ add to alkynes in much the same way they add to alkenes. For example, oct-1-yne reacts with *one equivalent* of Br₂ to yield 1,2-dibromooct-1-ene (Equation 12-13) and with *excess* Br₂ to yield the tetrabromide (Equation 12-14).

Br₂ adds both syn and anti to the C≡C bond.

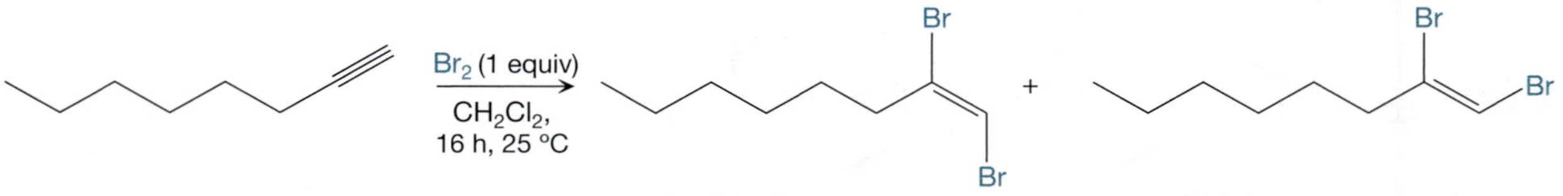

(12-13)

Oct-1-yne

(E)-1,2-Dibromooct-1-ene
65%

(Z)-1,2-Dibromooct-1-ene
19%

Two equivalents of Br₂ add to an alkyne.

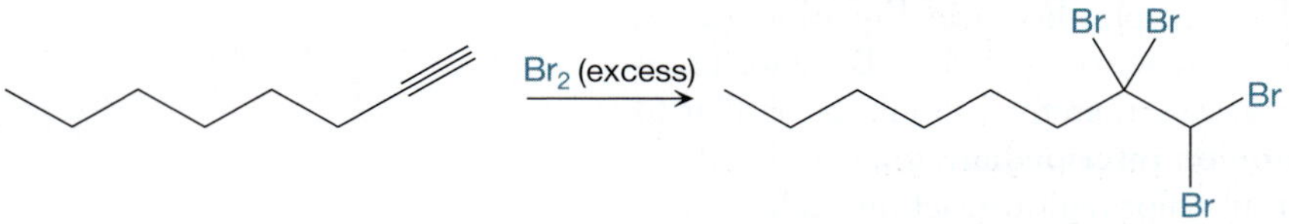

(12-14)

Oct-1-yne

1,1,2,2-Tetrabromooctane

Notice in Equation 12-13 that a mixture of both *E* and *Z* isomers is formed. In other words:

Br$_2$ adds to an alkyne via both syn and anti addition.

This means that the reaction must not proceed cleanly through a bromonium ion intermediate—otherwise, only anti addition would take place, producing the *E* isomer exclusively. Why is this so?

Essentially, it is because of ring strain (Equation 12-15). When a bromonium ion intermediate is produced from an alkyne, the resulting three-membered ring consists of two sp^2-hybridized C atoms, whose ideal angles are 120°. By contrast, when a bromonium ion intermediate is produced from an alkene, those two C atoms are sp^3 hybridized, and their ideal angles are 109.5°. As a result, the sp^3 C atoms from the alkene reaction can better handle the small angles required for the three-membered ring, which are near 60°.

A bromonium ion is not produced when Br$_2$ reacts with an alkyne, because the sp^2-hybridized C atoms would be too highly strained in the ring.

(12-15)

Because of this ring strain, chemists believe that the reaction proceeds through a carbocation-like intermediate instead, much as in the addition of a Brønsted acid. As we saw in Section 11.5, nucleophilic attack of a carbocation intermediate is not stereospecific.

12.3b Synthesis of Halohydrins

In Section 12.3a (Equation 12-9), we saw that a vicinal dibromide is produced if an alkene reacts with Br$_2$ in carbon tetrachloride. If the same reaction is carried out in water, however, as shown in Equation 12-16, a **bromohydrin** is produced instead. We can view the formation of a bromohydrin as the *net* addition of HO—Br across the double bond. Similarly, a **chlorohydrin** is produced when an alkene is treated with Cl$_2$ in water (Equation 12-17).

Water is the solvent. Anti addition of HO and Br

Cyclohexene → **trans-2-Bromocyclohexanol**

A bromohydrin

(12-16)

Water is the solvent. Anti addition of HO and Cl

(E)-But-2-ene → **(2S,3R)- and (2R,3S)-3-Chlorobutan-2-ol**

A chlorohydrin

(12-17)

These two reactions are *stereospecific*. Namely:

> In the formation of a halohydrin from an alkene, the OH group and the halogen atom add anti to each other.

This stereochemistry is the same as in the addition of Br_2 or Cl_2 across a double bond (recall Equations 12-9 and 12-12), strongly suggesting that halohydrin formation, too, proceeds through a *halonium ion intermediate*. This is illustrated in the mechanism in Equation 12-18.

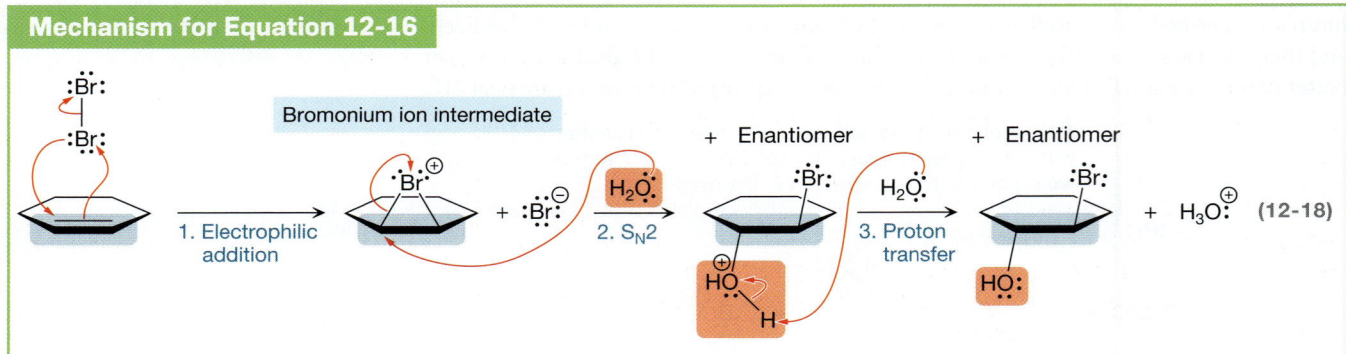

Mechanism for Equation 12-16

Bromonium ion intermediate

1. Electrophilic addition   2. S_N2 + Enantiomer   3. Proton transfer   + Enantiomer   (12-18)

The first two steps are essentially the same as the ones in Equation 12-10, in which Br_2 adds to an alkene. Step 1 is electrophilic addition that produces the bromonium ion intermediate, and Step 2 is an S_N2 reaction that produces a racemic mixture of enantiomers—the enantiomers are produced from attack of the two different C atoms bonded to the Br leaving group. In contrast to the mechanism in Equation 12-10, however, the ring opening in Step 2 involves H_2O as the nucleophile instead of Br^-. Finally, in Step 3, a proton transfer produces the uncharged bromohydrin.

It might seem counterintuitive that H_2O acts as the nucleophile in Step 2, even though the much stronger Br^- nucleophile is present in solution. Water is the solvent, however, and is therefore quite abundant, whereas Br^- is produced in Step 1 of the mechanism and is therefore maintained at much lower concentrations. Consequently, the bromonium ion intermediate is much more likely to encounter a molecule of H_2O than Br^-.

With CCl_4 as the solvent instead, we saw that Br^- is the nucleophile that attacks the bromonium ion intermediate (Equation 12-10). Even though CCl_4 is much more abundant than Br^-, CCl_4 is a non-nucleophilic solvent.

YOUR TURN 12.6

The mechanism for the reaction in Equation 12-17 is as follows, but the curved arrows have been omitted. Complete the mechanism by drawing in the curved arrows. Also, write the name of each elementary step below the appropriate reaction arrow.

+ Enantiomer   + Enantiomer

PROBLEM 12.8 Draw the complete, detailed mechanism leading to the formation of the second enantiomer shown in Equation 12-16, which is not shown explicitly in Equation 12-18.

SOLVED PROBLEM 12.9

When the reaction in Equation 12-16 is carried out in an aqueous NaCl solution instead of pure water, racemic *trans*-1-bromo-2-chlorocyclohexane is produced along with the bromohydrin. Draw a complete mechanism that accounts for the formation of the dihalo compound.

Think When an alkene reacts with Br_2, what is electron rich and what is electron poor? Will Cl^- behave as an electrophile or a nucleophile? How can you account for the stereochemistry of the reaction—that is, the formation of only the trans isomer?

Solve To account for the production of the trans isomers exclusively, the reaction must proceed through a bromonium ion intermediate. In the formation of the bromonium ion, the alkene is relatively electron rich and one of the bromine atoms from Br_2 is relatively electron poor. To open the ring and form a C—Cl bond, Cl^- in solution must act as the nucleophile.

Cyclohexene

trans-1-Bromo-2-chlorocyclohexane

(+ Enantiomer)

PROBLEM 12.10 Draw the complete, detailed mechanism of the reaction in Equation 12-16, assuming it was carried out in ethanol instead of water.

Regiochemistry becomes an issue with halohydrin formation if the alkene reactant is *unsymmetrical*. With distinct C=C atoms, two possible constitutional isomers can be produced, depending on which carbon atom ends up bonded to the halogen atom and which carbon atom ends up bonded to the OH group. This is the case, for example, with the reaction in Equation 12-19.

These C atoms are distinct.

Major product

Phenylethene

Br_2
H_2O

2-Bromo-1-phenylethan-1-ol
78%

2-Bromo-2-phenylethan-1-ol

(12-19)

Why is the first of these products the major product? The answer can be understood by studying the mechanism, which is presented in Equation 12-20.

Mechanism for Equation 12-19

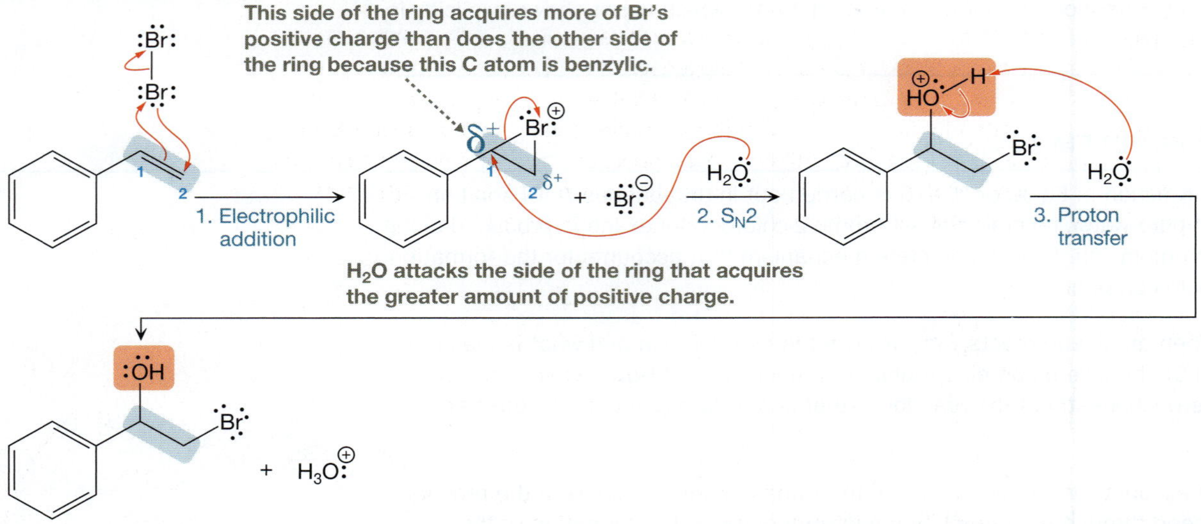

This side of the ring acquires more of Br's positive charge than does the other side of the ring because this C atom is benzylic.

1. Electrophilic addition

H₂O attacks the side of the ring that acquires the greater amount of positive charge.

2. S_N2

3. Proton transfer

(12-20)

The specific isomer that is formed is dictated by Step 2 of the mechanism—namely, nucleophilic attack by water. Although attack can occur at either C atom of the three-membered ring, attack is favored at C1 instead of C2. Recall that we saw a very similar mechanistic step in Equation 10-58 of Section 10.7b (p. 545), in which a protonated epoxide acts as a substrate in an S_N2 reaction. Even though the positive charge is formally represented on the heteroatom—in this case, Br—it is actually shared among the three atoms of the ring. Given that C1 is benzylic, it can better handle a positive charge than C2, an isolated primary C atom. Therefore, the side of the ring that contains C1 acquires the greater amount of positive charge and draws the H₂O nucleophile to that side, even though C1 is more sterically hindered.

PROBLEM 12.11 Draw a complete, detailed mechanism for the reaction shown here and predict the major product, paying careful attention to both regiochemistry and stereochemistry.

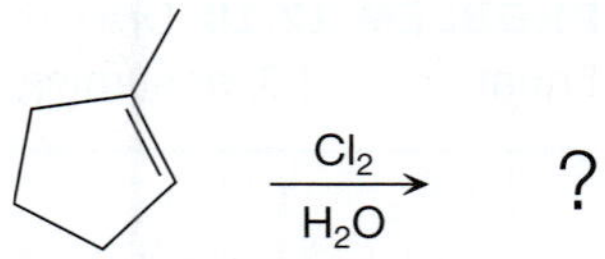

$$\xrightarrow[\text{H}_2\text{O}]{\text{Cl}_2} \quad ?$$

12.4 Oxymercuration–Reduction: Addition of Water

Recall from Chapter 11 that water can undergo an acid-catalyzed addition to an alkene or alkyne, producing an alcohol or ketone, respectively. In these reactions, the electrophilic addition of H⁺ produces the more stable carbocation intermediate, then H₂O attacks the carbocation in a coordination step. Thus, we observe *Markovnikov addition* of water. One drawback of these reactions, however, is that carbocation rearrangements are also possible. An example is shown in Equation 12-21.

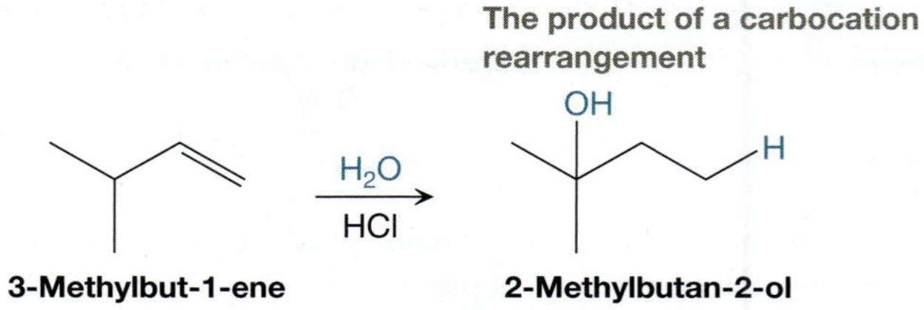

The product of a carbocation rearrangement

$$\xrightarrow[\text{HCl}]{\text{H}_2\text{O}}$$

(12-21)

3-Methylbut-1-ene   **2-Methylbutan-2-ol**

Halogenated Metabolites: True Sea Treasures

Natural products isolated from marine organisms have a wide range of potential applications in medicine and beyond. Of particular interest are halogenated compounds that serve as metabolites for many of these organisms. Compounds **A–D** below, for example, are some of the halogenated compounds that have been isolated from marine red algae. Compound **A** has antitumor and cytotoxic properties; **B** has anthelminthic properties (i.e., it expels parasitic worms from the body); **C** has antifouling properties (i.e., it prevents the accumulation of organisms on wet surfaces); and **D** has antimicrobial properties.

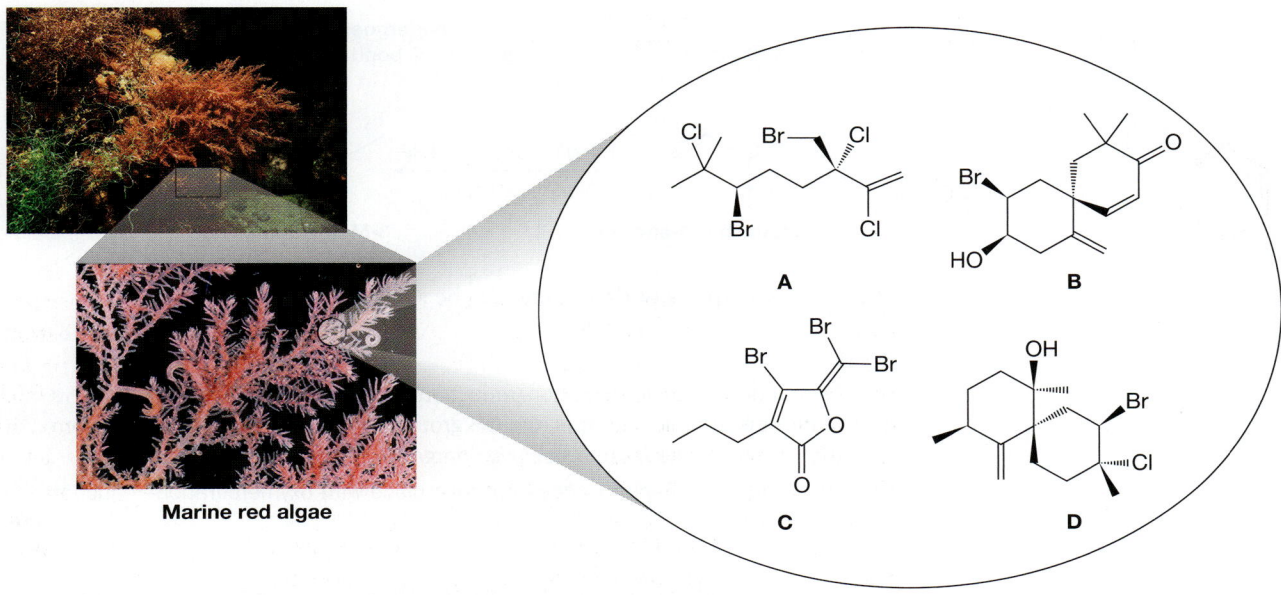

Marine red algae

A

B

C

D

(E)-(+)-Nerolidol

A bromonium ion intermediate

J. N. Carter-Franklin and A. Butler, from the University of California, Santa Barbara, have shown that the biosynthesis of these halogenated metabolites probably involves a bromonium ion intermediate, much like we see in the bromination of alkenes here in Chapter 12. Production of the bromonium ion intermediate in these biosynthetic pathways, however, does not involve molecular Br_2. Rather, an alkene is believed to react with Br^+, which is produced from Br^- (an abundant ion in seawater) in the presence of hydrogen peroxide (H_2O_2), catalyzed by the enzyme vanadium bromoperoxidase (V—BrPO). Br^- undergoes a two-electron oxidation in the active site of the enzyme to make Br^+ (or its equivalent), which then adds to the C=C double bond. In many cases, the opening of the three-membered ring results in a new carbocycle, as shown with (E)-(+)-nerolidol in the example above.

Draw the mechanism that accounts for the formation of the product in Equation 12-21.

An alternate method to add water across a double bond is **oxymercuration–reduction** (also called **oxymercuration–demercuration**), an example of which is shown in Equation 12-22. The alkene is first treated with mercury(II) acetate, $Hg(OAc)_2$, in a water–tetrahydrofuran (THF) solution, and that is followed by reduction with sodium borohydride.

Oxymercuration–reduction

Water undergoes Markovnikov addition across the C=C bond with no rearrangement.

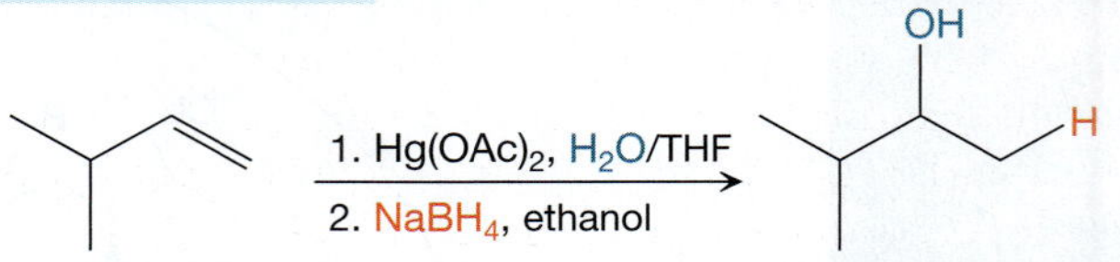

1. $Hg(OAc)_2$, H_2O/THF
2. $NaBH_4$, ethanol

(12-22)

3-Methylbut-1-ene   **3-Methylbutan-2-ol**

There are two aspects of this reaction to be aware of:

- The product of oxymercuration–reduction is the one expected from Markovnikov's rule—that is, the OH group forms a bond to the carbon atom that can better stabilize a positive charge.
- Rearrangement generally does *not* take place with oxymercuration–reduction.

To better understand the reasons for these outcomes, study the mechanism shown in Equation 12-23.

Mechanism for Equation 12-22

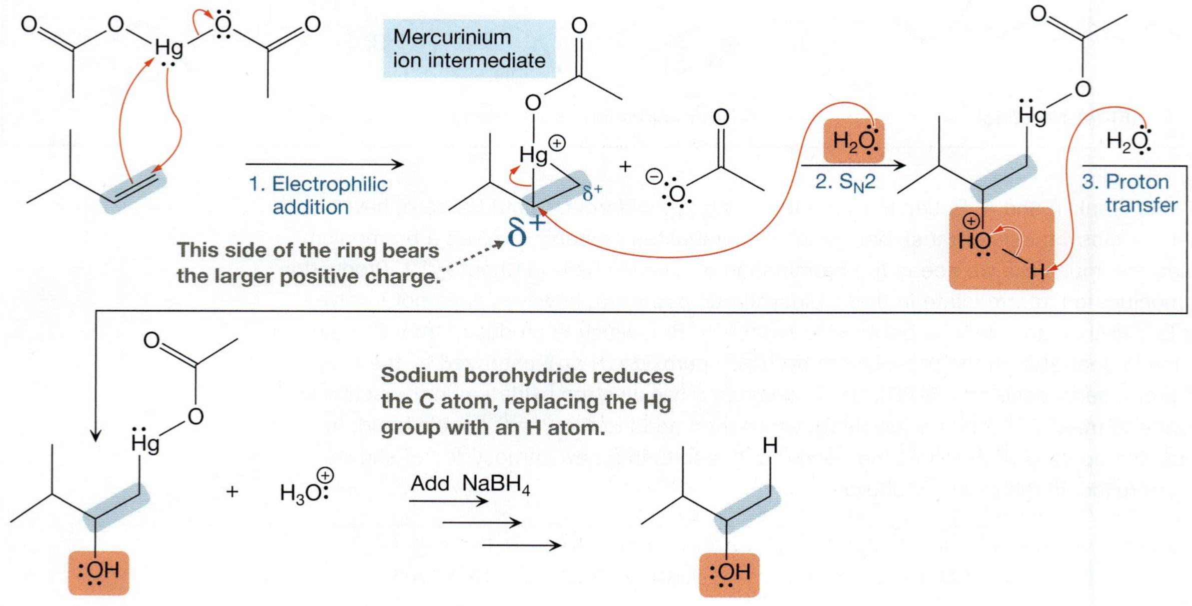

Mercurinium ion intermediate

1. Electrophilic addition

This side of the ring bears the larger positive charge.

2. S_N2

3. Proton transfer

(12-23)

Sodium borohydride reduces the C atom, replacing the Hg group with an H atom.

Add $NaBH_4$

The first three steps of the mechanism are identical to those in the formation of a halohydrin, shown previously in Equation 12-20. In Step 1, the Hg atom is electron poor, given that it is bonded to two highly electronegative O atoms. It is therefore attacked by the electron-rich double bond, and simultaneously a lone pair of electrons on Hg forms a bond to C to produce the three-membered ring. The result is a **mercurinium ion intermediate**, which is analogous to the bromonium (or chloronium) ion intermediate we encountered previously. In Step 2 of the mechanism, H_2O acts as a nucleophile to open the three-membered ring, and in Step 3, the positively charged O atom is deprotonated.

Reduction then occurs when sodium borohydride ($NaBH_4$) is added, in which the Hg-containing substituent is replaced by H. Although this may appear to be a simple nucleophilic substitution reaction with H^- as the nucleophile, it actually proceeds through a more complex mechanism that is believed to involve free radicals (species with unpaired electrons; see Chapter 25). Consequently, even though oxymercuration takes place with anti addition, any stereochemistry of the C atom bonded to Hg is scrambled during the reduction step, giving a mixture of both syn and anti addition of water. This is exemplified in the oxymercuration–reduction of methyl (E)-2-methylbut-2-enoate shown in Equation 12-24.

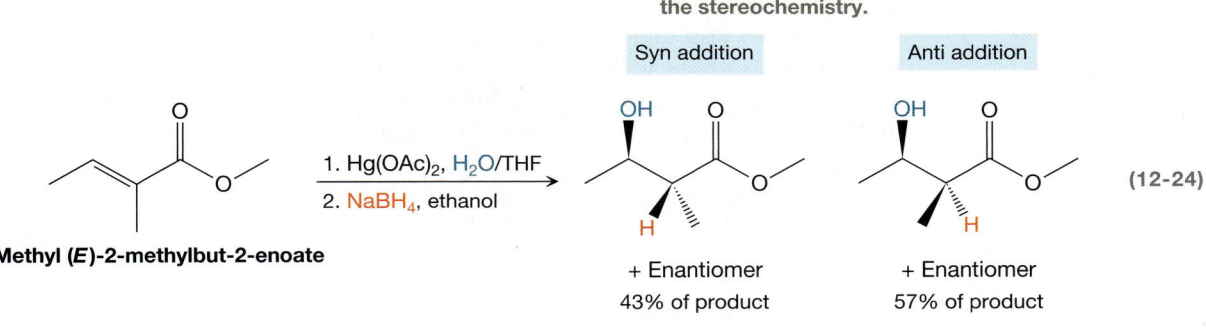

(12-24)

PROBLEM 12.12 Draw a detailed mechanism for each of the following reactions and identify the major products.

Why does the addition of H_2O in Step 2 of Equation 12-23 follow Markovnikov's rule? Like we saw with nucleophilic attack on a protonated epoxide or a bromonium ion, the positive charge on Hg in the mercurinium ion is shared over the two C atoms of the ring. The side of the ring that can handle the positive charge better will acquire more of the charge and will more strongly draw in the nucleophile. In this case, the side of the ring that has the secondary C atom acquires more of the positive charge, not the side that has the primary C. Notice that the secondary C atom is the one that would gain the full positive charge if a proton were to add to the alkene, and thus is the same carbon atom that would be attacked by water in an acid-catalyzed hydration.

The mechanism also shows why rearrangements tend *not* to take place in oxymercuration–reduction, in contrast to acid-catalyzed hydration. In the acid-catalyzed hydration of an alkene, a *full* positive charge develops on C in the carbocation intermediate, whereas only a *partial* positive charge develops on C in the mercurinium ion intermediate of the oxymercuration reaction. With a smaller concentration of positive charge on C in a mercurinium ion, a rearrangement would not lead to as much of an increase in stability compared to a carbocation.

PROBLEM 12.13 Draw two possible alkenes that can undergo oxymercuration–reduction to yield 2-methylpentan-2-ol as the major product.

PROBLEM 12.14 Propose how to carry out the following transformation.

Alkynes can also undergo Markovnikov addition of water via oxymercuration. However, just as we saw with the acid-catalyzed hydration of an alkyne (Section 11.8), an unstable enol is produced initially. Subsequent tautomerization converts the enol into the more stable keto form, as shown in Equation 12-25.

Hydration of an alkyne leads to an initial enol, which tautomerizes into the more stable keto form.

The oxymercuration reaction in Equation 12-25 does not require reduction with NaBH$_4$ to remove the mercury(II) substituent. Instead, as shown in Equation 12-26, the mercurinium ion intermediate opens to produce a mercuric enol, which, after tautomerizing to a mercuric ketone, is hydrolyzed by water to produce the enol in Equation 12-25.

Mercury(II) acetate is not the only source of Hg^{2+} used for these reactions. Equations 12-27 and 12-28 show examples in which HgCl$_2$ and HgSO$_4$ are used instead.

Phenylethyne → Mercury(II) catalyst, H_2O, $HgCl_2$ → **Phenylethanone** 82% (12-27)

Hex-1-yne → H_2O, $HgSO_4$, H_2SO_4, Acetic acid, Mercury(II) catalyst → **Hexan-2-one** 80% (12-28)

YOUR TURN **12.8**

Equations 12-27 and 12-28 show the overall products of hydration of the alkynes. For each reaction, draw the enol that is produced prior to tautomerization to the more stable keto form.

Internal alkynes can also be converted to ketones. A mixture of isomeric ketones will be produced, however, unless the alkyne is symmetric. This is shown in Equation 12-29.

Hydration of an internal alkyne leads to a mixture of isomeric ketones.

Hept-2-yne → H_2O, HgO, H_2SO_4, Methanol → **Heptan-2-one** ~67% + **Heptan-3-one** ~33% (12-29)

SOLVED PROBLEM 12.15

In an **alkoxymercuration–reduction** reaction, an alcohol is used as the solvent in the first step instead of water. Draw the complete, detailed mechanism for the alkoxymercuration part of this reaction and predict the major product.

1. $Hg(OAc)_2$, EtOH/THF
2. $NaBH_4$
?

Think How does the substitution of EtOH for water change the mechanism in Equation 12-23? What considerations should be made about *regiochemistry*? What considerations should be made about rearrangements?

Solve The mechanism still proceeds through a cyclic mercurinium ion intermediate. Normally, H_2O acts as the nucleophile to open the ring. Here, however, this is the role of ethanol. Because the alkene is unsymmetric, the alcohol can attack either C atom of the ring, but predominantly attacks the more highly substituted one because it can acquire more of the positive charge from Hg. Finally,

CONNECTIONS
Heptan-2-one (Equation 12-29) is believed to be used as an anesthetic by honeybees when they bite small insects and larvae that enter the hive. It is also responsible for the odor of gorgonzola cheese.

just as we saw with oxymercuration–reduction, carbocation rearrangements are not an issue here.

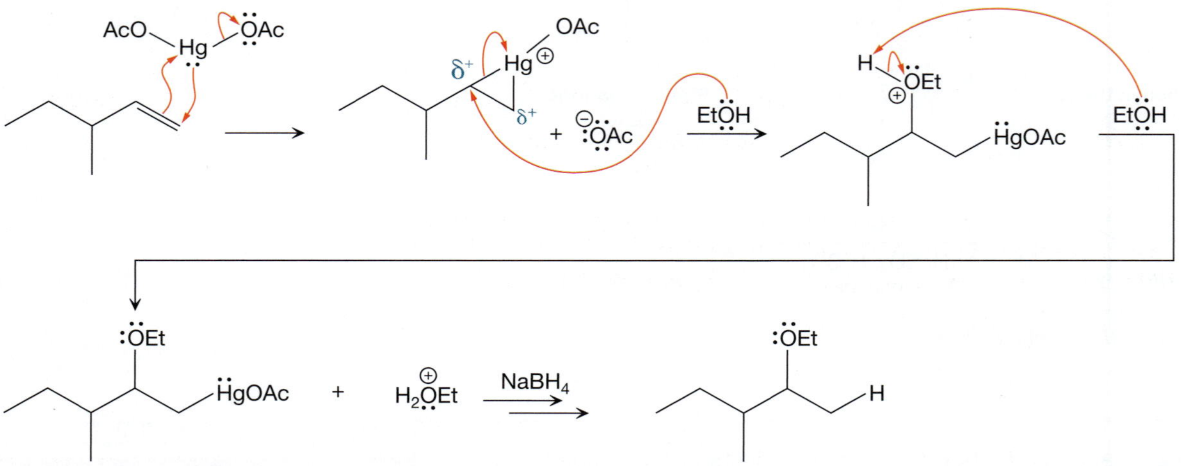

AcO
Hg
OAc
δ+
Hg
OAc
δ+
OAc
EtOH
H
OEt
HgOAc
EtOH
OEt
HgOAc
H2OEt
NaBH4
OEt
H

PROBLEM 12.16 Predict the product of this reaction.

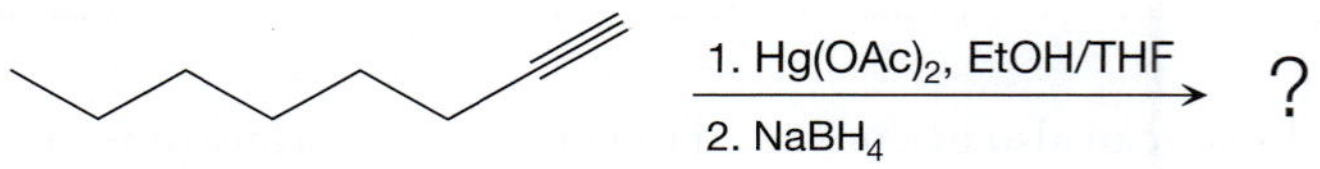

1. Hg(OAc)2, EtOH/THF
2. NaBH4
?

12.5 Epoxide Formation Using Peroxyacids

Recall from Section 10.6 that an epoxide can be produced by treating a halohydrin with a strong base, as shown in Equation 12-30.

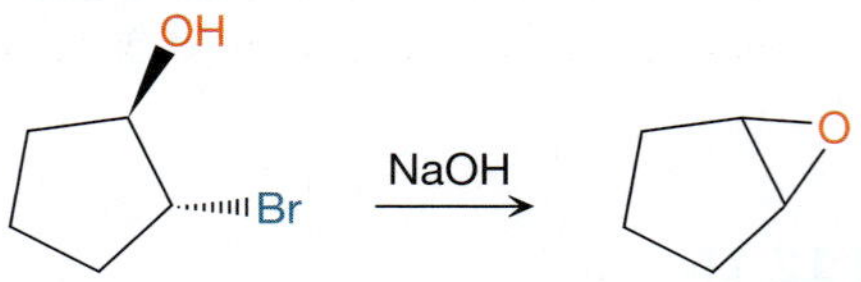

OH
Br
NaOH
O

(12-30)

Unfortunately, synthesizing an epoxide in this way is often impractical, in part because HO^-, a strong base and strong nucleophile, could lead to unwanted side reactions. Moreover, halohydrins are not very common, so they may not be readily available to be used as precursors.

Fortunately, an epoxide can be produced from an alkene using a **peroxyacid** (RCO_3H), also called a *peracid*. Examples of such **epoxidation reactions** are shown in Equations 12-31 and 12-32, using ***meta*-chloroperbenzoic acid (MCPBA)** as the peroxyacid.

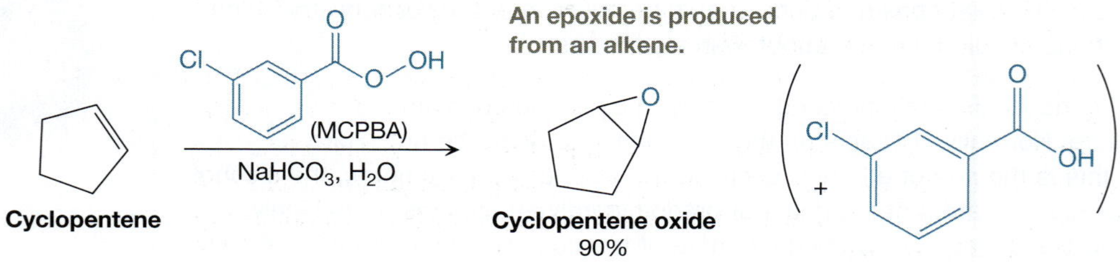

Cl
O
OH
(MCPBA)
NaHCO3, H2O
An epoxide is produced from an alkene.
O
Cl
O
OH
+
Cyclopentene
Cyclopentene oxide
90%

(12-31)

The CH$_3$ groups are on opposite sides of the C=C double bond.

The CH$_3$ groups are on opposite sides of the plane of the epoxide ring.

trans-But-2-ene   (MCPBA) / CH$_2$Cl$_2$ →   trans-But-2-ene oxide 60%   (12-32)

Synthesizing epoxides in this way is advantageous because the reaction conditions are relatively mild and many alkenes are common or easily synthesized.

The mechanism for these epoxidation reactions is shown in Equation 12-33. Notice that it takes place in a single step—that is, it is *concerted*.

Mechanism for Equation 12-32

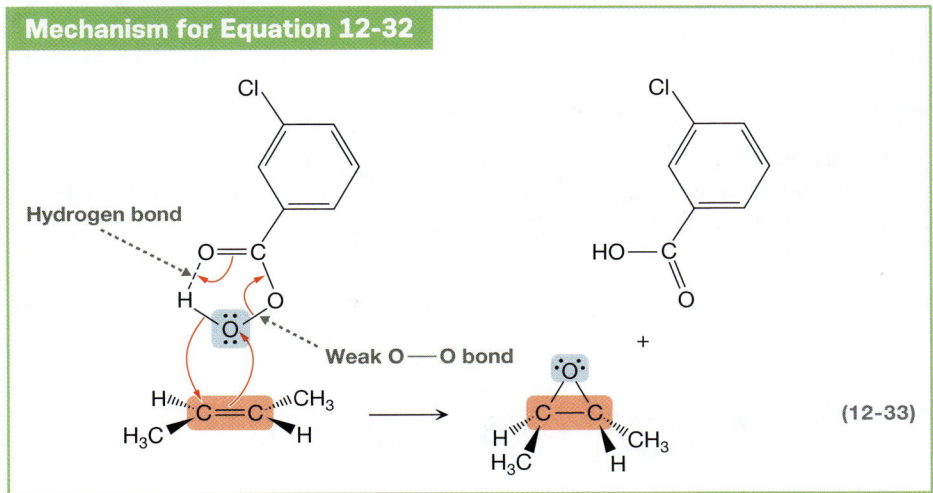

Hydrogen bond

Weak O—O bond

(12-33)

The O atom of the OH group is the one that is transferred to the alkene because both of its original covalent bonds are relatively weak. The O—O single bond is inherently very weak. On the other hand, an O—H bond is normally quite strong, but in this case the O's bond to H is weakened by the internal hydrogen bond involving the OH and the carbonyl O atom.

YOUR TURN **12.9**

Verify that the O—O single bond is weak by looking up its average value (Table 1-2, p. 10). For comparison, do the same for an average C—C single bond.

O—O _____ kJ/mol   C—C _____ kJ/mol

With its weakened bonds, the O atom of the OH group is highly reactive and is attacked by the π electrons from the alkene. To avoid generating a carbocation on one of the C atoms from the initial alkene, the pair of electrons from the OH bond is used to form an O—C bond, thereby completing the three-membered epoxide ring. The carbonyl's O atom acquires the H, and the pair of electrons from the initial O—O bond goes to make a C=O double bond.

Because epoxidation takes place in a single step, the stereochemical requirements presented in Section 12.1 apply. As shown previously in Equation 12-32, for example, stereochemistry is conserved, so the trans relationship of the CH$_3$ groups about the C=C double bond in the reactants establishes a trans relationship about the plane of the epoxide ring in the products. Furthermore, because the products are chiral, a mixture of stereoisomers is produced.

PROBLEM 12.17 Draw the complete, detailed mechanism for the reaction shown here and predict the major product(s), paying particular attention to the stereochemistry.

MCPBA → ?

CONNECTIONS
Cyclohexane-1,2-diol, the final product in Solved Problem 12.19, is found in castoreum, an excretion by North American beavers that is used to mark their territories by scent.

PROBLEM 12.18 What alkene can be epoxidized using MCPBA to yield the following compound?

? → MCPBA →

Because an epoxide readily undergoes ring opening (Section 10.7), the product of an epoxidation reaction is often used as a reactant in a subsequent reaction. This is exemplified in Solved Problem 12.19.

SOLVED PROBLEM 12.19

Identify the missing compounds **A–C**.

OH → conc H₃PO₄, Δ → **A** → MCPBA, CH₂Cl₂ → **B** → NaOH, H₂O → **C**

Think In the first reaction, what is the nature of the leaving group? The attacking species? The solvent? What type of reaction does heat promote? For the second reaction, what type of reaction takes place between an alkene and a peroxyacid? In the third reaction, what acts as the nucleophile? The substrate?

Solve The missing products are as follows:

OH → conc H₃PO₄, Δ → **A** → MCPBA, CH₂Cl₂ → **B** → NaOH, H₂O → **C** + Enantiomer

The first reaction is an acid-catalyzed dehydration to form an alkene, which is an E1 reaction. Under the acidic conditions, the leaving group is H_2O, the attacking species is weak, and the solvent is protic. Heat favors elimination over substitution. The second reaction involves an alkene and MCPBA, which is a peroxyacid, leading to the formation of an epoxide. In the third reaction, HO^- acts as the nucleophile to open the epoxide in an S_N2 reaction. The trans stereochemistry is governed by the attack of the nucleophile from the side of the ring opposite the epoxide O.

PROBLEM 12.20 Identify the missing compounds **D–F**.

Br → NaOCH₃, Δ → **D** → MCPBA → **E** → 1. CH₃MgBr 2. H₃O⁺ → **F**

Benzo[a]pyrene: Smoking, Epoxidation, and Cancer

Benzo[a]pyrene is found in smoke that is produced from the combustion of organic compounds (such as the tobacco in cigarettes), and several different studies have linked benzo[a]pyrene to various forms of cancer. Benzo[a]pyrene is classified as a procarcinogen because it is a *precursor* to the specific compound that is directly responsible for causing cancer. In this case, in the detoxification process in the liver, benzo[a]pyrene reacts with molecular oxygen in the presence of the enzymes cytochrome P450 1A1 (CYP1A1) and cytochrome P450 1B1 (CYP1B1). This is an epoxidation of the C=C bond between C7 and C8, analogous to the MCPBA epoxidation of alkenes we have studied here in Chapter 12, resulting in (+)-benzo[a]pyrene-7,8-epoxide (**A**).

Adduct of
benzo[a]pyrene

The epoxide in **A** then undergoes ring opening with H_2O in the presence of the enzyme epoxide hydrolase, which produces a trans diol, much like we saw in Section 10.7. A second epoxidation targets the C=C bond between C9 and C10, resulting in (+)-benzo[a]pyrene-7,8-dihydrodiol-9,10-epoxide (**B**). **B** can intercalate into the DNA double helix, where it covalently bonds to a nucleophilic guanine nitrogenous base. As you can see above on the right, this adduct distorts the double helix, which interferes with the normal process by which DNA is copied, thus inducing mutations.

12.6 Hydroboration–Oxidation: Anti-Markovnikov Syn Addition of Water to an Alkene

So far, we have seen two reactions that serve to add water across a carbon–carbon double or triple bond. One is *acid-catalyzed hydration* (Chapter 11), which proceeds through a carbocation intermediate. The second is *oxymercuration–reduction* (Section 12.4), which proceeds through a cyclic mercurinium ion intermediate, thereby avoiding carbocation rearrangements. Both of these reactions add water in a *Markovnikov* fashion, in which the carbon atom that gains the OH group is the one that is better able to stabilize a positive charge. Also, neither of these reactions is stereospecific—a mixture of stereoisomers from both syn addition and anti addition is produced.

Hydroboration–oxidation provides a third way to add water across a nonpolar π bond, as shown in Equations 12-34 and 12-35. In each case, the alkene first undergoes **hydroboration**, in which borane, BH_3 (from either $BH_3 \cdot THF$ or B_2H_6), adds across the double bond. The product is then *oxidized* with a basic solution of hydrogen peroxide, H_2O_2.

Anti-Markovnikov addition of H and OH

$$\text{Hept-1-ene} \xrightarrow[\text{2. } H_2O_2,\ NaOH,\ H_2O]{\text{1. } BH_3 \cdot THF} \text{Heptan-1-ol}$$

Heptan-1-ol
92%

(12-34)

H and OH add with syn stereochemistry and anti-Markovnikov regiochemistry.

$$\text{1-Methylcyclopentene} \xrightarrow[\substack{\text{2. } H_2O_2,\ NaOH,\ H_2O \\ 0\,°C,\ 2\ h}]{\substack{\text{1. } B_2H_6,\ THF \\ 0\,°C,\ 2\ h,\ then\ 20\,°C,\ 1\ h}} \text{}$$

trans-2-Methylcyclopentanol
85%

(12-35)

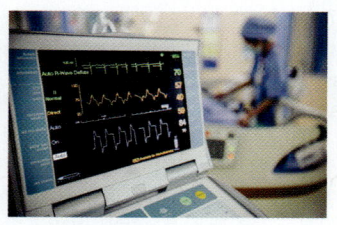

Notice the regiochemistry that is exhibited by these reactions. Specifically, in both Equations 12-34 and 12-35, addition of OH is favored at the alkene C atom that can *least* stabilize a positive charge—namely, a primary C atom in Equation 12-34 and a secondary C atom in Equation 12-35. This is opposite to what we see with Markovnikov regiochemistry exhibited by acid-catalyzed hydration and oxymercuration–reduction. Thus:

Hydroboration–oxidation adds H and OH to an alkene with **anti-Markovnikov regiochemistry**.

Notice also that hydroboration–oxidation is stereospecific. In Equation 12-35, for example, both the H and OH groups add to the same face of the planar alkene functional group. In other words:

In hydroboration–oxidation, an alkene undergoes *syn addition* of H and OH.

To understand these aspects of hydroboration–oxidation reactions, we must understand their mechanisms. First, we study the mechanism of hydroboration in Section 12.6a, then we examine the mechanism of the subsequent oxidation in Section 12.6b.

12.6a Hydroboration: Addition of BH_3 across a C=C Double Bond

Hydroboration is a very useful reaction because it is the starting point for a variety of synthetic sequences. It is the first of two sequential reactions in the anti-Markovnikov hydration of alkenes, for example, and it can be used as a starting point for hydrogenating alkenes and alkynes, thereby reducing them to alkanes and alkenes, respectively. For these reasons, Herbert C. Brown (1912–2004), who pioneered hydroboration in the field of organic synthesis, shared the 1979 Nobel Prize in Chemistry.

Borane (BH_3) is the reactive species in the hydroboration of an alkene. Borane is highly unstable, however, because the central B atom does not have an octet. Therefore, BH_3 cannot be isolated. Instead, in its pure form it exists as a gaseous dimer, **diborane (B_2H_6)**, in which two H atoms constitute a bridge between the B atoms (shown in Fig. 12-3a). Those H atoms are involved in what are called **three-center, two-electron bonds**. Although these bonds provide some stability, B_2H_6 is highly reactive, and thus is both *toxic* and *explosive*. A more stable variation of BH_3 is sold commercially as a one-to-one complex with tetrahydrofuran (THF; shown in Fig. 12-3b), denoted $BH_3 \cdot THF$, where THF acts as a Lewis base to give the B atom its octet. Similarly, a one-to-one complex between BH_3 and dimethyl sulfide (DMS; CH_3—S—CH_3), denoted $BH_3 \cdot DMS$, can be used as a source of borane.

The mechanism for the hydroboration reaction is shown in Equation 12-36.

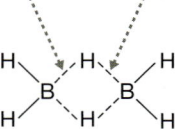

A 3-center, 2-electron bond provides some stability to the B atom.

(a) Diborane, B_2H_6

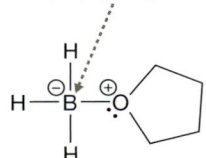

Coordination of THF to BH_3 provides some stability by giving B an octet.

(b) Borane·THF complex ($BH_3 \cdot THF$)

FIGURE 12-3 Sources of BH_3
(a) Diborane, B_2H_6, is a dimer of BH_3, formed by two separate, three-center, two-electron bonds indicated by the dashed lines. (b) In $BH_3 \cdot THF$, all nonhydrogen atoms have an octet.

Partial mechanism for Equation 12-35

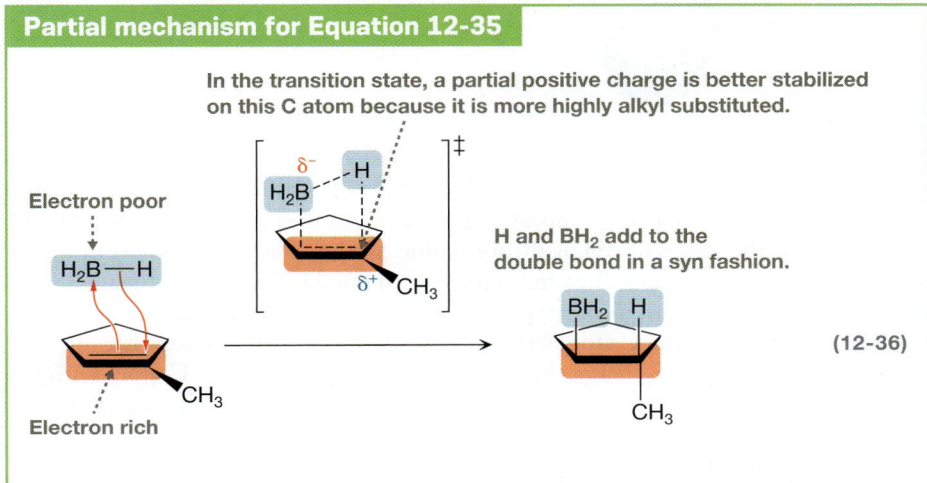

In the transition state, a partial positive charge is better stabilized on this C atom because it is more highly alkyl substituted.

Electron poor

Electron rich

H and BH_2 add to the double bond in a syn fashion.

(12-36)

It is driven primarily by the flow of electrons from the electron-rich π bond to the electron-poor boron atom of BH_3. Similar to the mechanisms we have seen previously in this chapter, this reaction avoids the formation of a highly unstable carbocation by simultaneously forming a bond back to the second C atom of the double bond. Whereas a lone pair was used in previous reactions to form that second bond to C (Section 12.1), in hydroboration that bond comes from electrons originally part of a B—H bond in BH_3. Overall, then, two bonds are broken and two bonds are formed in a *concerted* fashion—that is, without the formation of intermediates.

According to Equation 12-36:

Hydroboration is *stereospecific*, with the H and BH_2 groups adding to the alkene in a syn fashion.

Hydroboration is also *regioselective*. The H atom primarily adds to the C with the greater number of alkyl groups, whereas the BH_2 group adds to the C with the fewer number of alkyl groups. This, too, can be explained by the fact that the reaction is concerted. With no intermediates formed along the reaction coordinate, no formal charge is generated. Charge stability, therefore, does not provide as much of a driving force as it does in other reactions we have seen in this chapter and Chapter 11, allowing *steric hindrance* to play a more significant role.

CONNECTIONS Borane will also form a complex with ammonia, resulting in $BH_3 \cdot NH_3$. This borane–ammonia complex is being studied as an air-stable form of hydrogen storage, which can have a wide variety of energy applications, including a source of H_2 gas for fuel cells in automobiles.

When BH_3 adds to an alkene, *steric repulsion* directs the larger group, BH_2, to the carbon atom with the fewer number of alkyl groups.

Equation 12-37 shows the steric repulsion that would occur if the BH_2 group were to add to the more highly substituted alkene carbon.

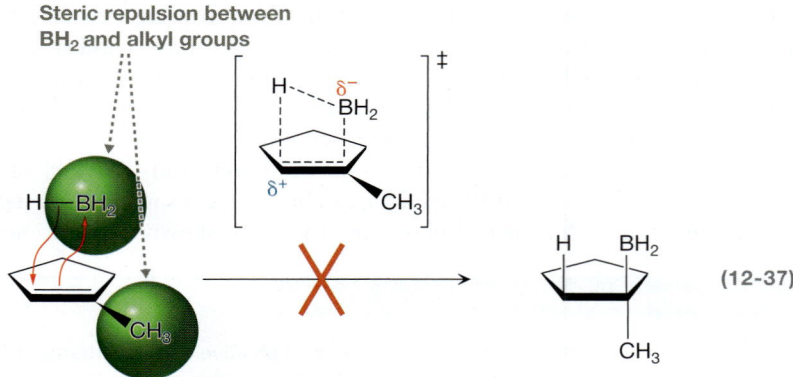

(12-37)

Even though no formal charges appear, charge stability is also a factor in the regioselectivity of hydroboration. In the transition states shown in Equations 12-36 and 12-37, a *partial* positive charge appears on the C atom that has a partial bond to H. This is because the step is driven by the formation of a bond between the electron-rich alkene and the electron-poor B atom, so the C—B bond is formed to a greater extent in the transition state than the C—H bond. The transition state is more stable when the BH_2 group adds to the C atom with the fewer number of alkyl groups because that allows the partial positive charge that is generated to appear on the C atom with the greater number of alkyl groups. The additional alkyl groups are electron donating and stabilize that partial positive charge.

By considering this charge buildup during hydroboration, we can say the following about regiochemistry:

Both the Markovnikov addition of a Brønsted acid to an alkene and the anti-Markovnikov addition of BH_3 to an alkene are an outcome of the same factor: which alkene C can better handle positive charge.

In both the addition of a Brønsted acid and hydroboration, one alkene C forms a bond to the electrophilic (i.e., electron-poor) atom, and the other alkene C is the one that gains the partial or full positive charge. The difference, as we explore in Your Turn 12.10, is the atom that is electrophilic: a proton in the case of a Brønsted acid and the boron atom in the case of BH_3.

YOUR TURN 12.10

Consider the hydroboration and protonation steps shown here. In each case, label (a) the electrophilic atom and (b) the alkene C that acquires positive or partial positive charge. Do the electrophilic atoms add to the same alkene C in each case?

Borane can add to propene to produce two different products, as shown below. Complete each of the hydroboration steps by adding the curved arrows. Also, indicate the pertinent steric repulsion that is present in each reaction, as well as the partial charges that develop in the transition state, similar to Equations 12-36 and 12-37, and determine which reaction is favored.

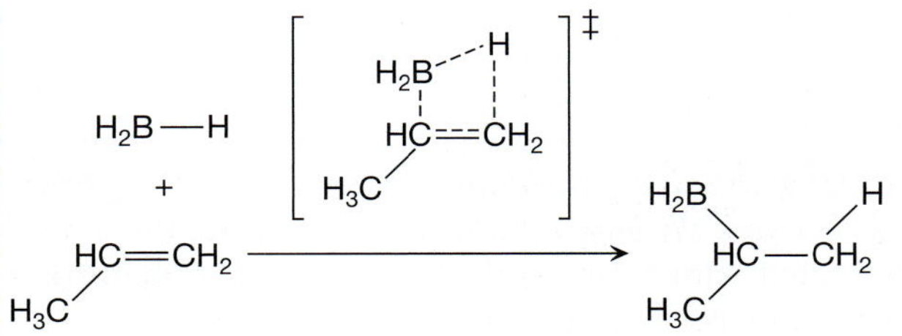

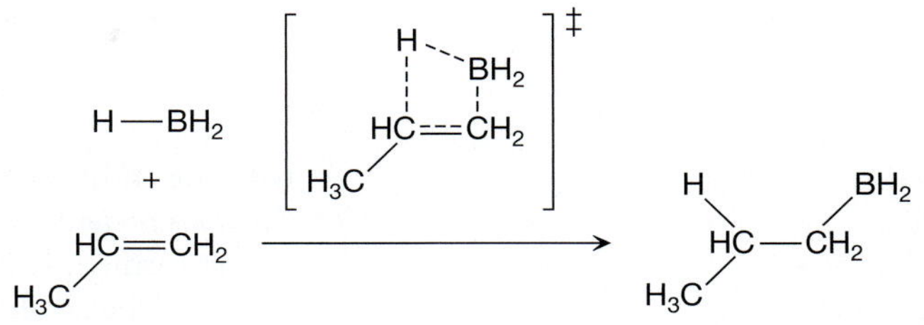

The product of hydroboration is an **alkylborane**, R—BH$_2$ (Equation 12-38a), which has two B—H bonds remaining.

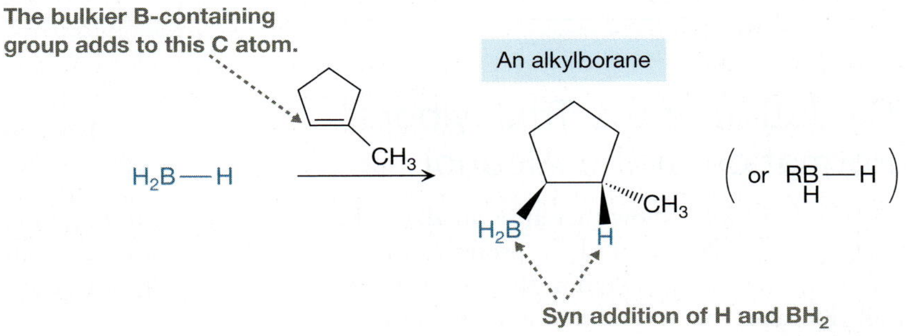

The bulkier B-containing group adds to this C atom.

An alkylborane

$$\left(\text{or} \quad RB\underset{H}{-}H \right) \quad (12\text{-}38a)$$

Syn addition of H and BH$_2$

That alkylborane can therefore react with an additional unreacted alkene, as shown in Equation 12-38b.

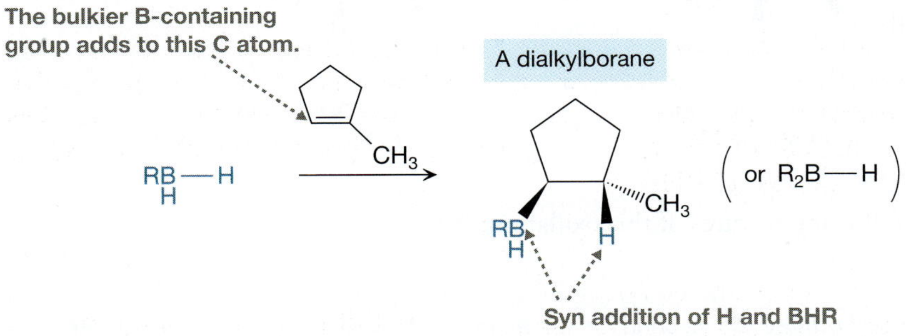

The bulkier B-containing group adds to this C atom.

A dialkylborane

$$\left(\text{or} \quad R_2B-H \right) \quad (12\text{-}38b)$$

Syn addition of H and BHR

That is, RHB—H adds across the double bond of the alkene to produce a **dialkylborane**, R$_2$B—H. In turn, the dialkylborane product still has a B—H bond and adds across yet another equivalent of the unreacted alkene to produce a **trialkylborane**, R$_3$B, as shown in Equation 12-38c.

The bulkier B-containing
group adds to this C atom.

A trialkylborane

$$R_2B-H \longrightarrow$$

$$\left(\text{or } R_3B \right) \quad (12\text{-}38c)$$

Syn addition of H and BR$_2$

As with the addition of BH$_3$, *each of these is a syn addition, and the bulkier B-containing portion adds preferentially to the less sterically hindered alkene C atom.* Once the trialkylborane is formed, it is then treated with a basic solution of hydrogen peroxide to convert it to the alcohol, which is discussed in the next section.

PROBLEM 12.21 Draw the detailed mechanism for the formation of the monoalkylborane in the reaction shown here. Also draw the trialkylborane that is ultimately produced.

$$\xrightarrow{\text{BH}_3 \cdot \text{THF}} \text{?}$$

12.6b Oxidation of the Trialkylborane: Formation of the Alcohol

Equations 12-34 and 12-35 show that after an alkene has undergone hydroboration, treatment with a basic solution of H$_2$O$_2$ produces the alcohol. We can now see from Equation 12-38 in the previous section that the actual species that undergoes oxidation is a trialkylborane. The net reaction is shown in Equation 12-39.

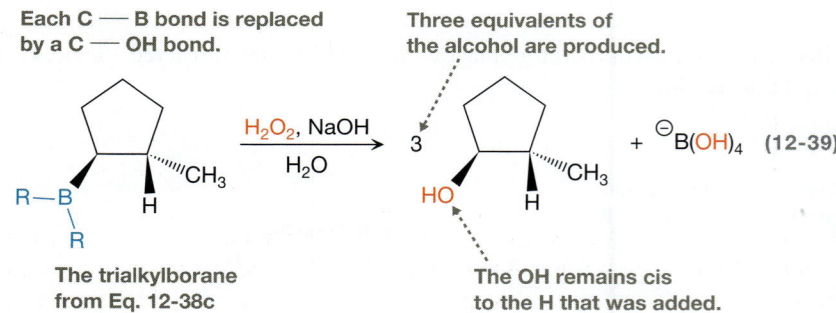

Each C — B bond is replaced
by a C — OH bond.

Three equivalents of
the alcohol are produced.

$$\xrightarrow[\text{H}_2\text{O}]{\text{H}_2\text{O}_2,\ \text{NaOH}}$$

$$+ \ ^{\ominus}\text{B(OH)}_4 \quad (12\text{-}39)$$

The trialkylborane
from Eq. 12-38c

The OH remains cis
to the H that was added.

Note the following features of this oxidation:

- Each of the three C—B bonds is replaced by a C—OH bond, producing three equivalents of the alcohol.
- The OH group in the product is syn to the hydrogen atom added from the previous hydroboration reaction. Thus, this oxidation takes place with *retention of configuration* at each C atom bonded to B.

These aspects of the oxidation reaction can be better understood by studying the mechanism, which is shown in Equation 12-40.

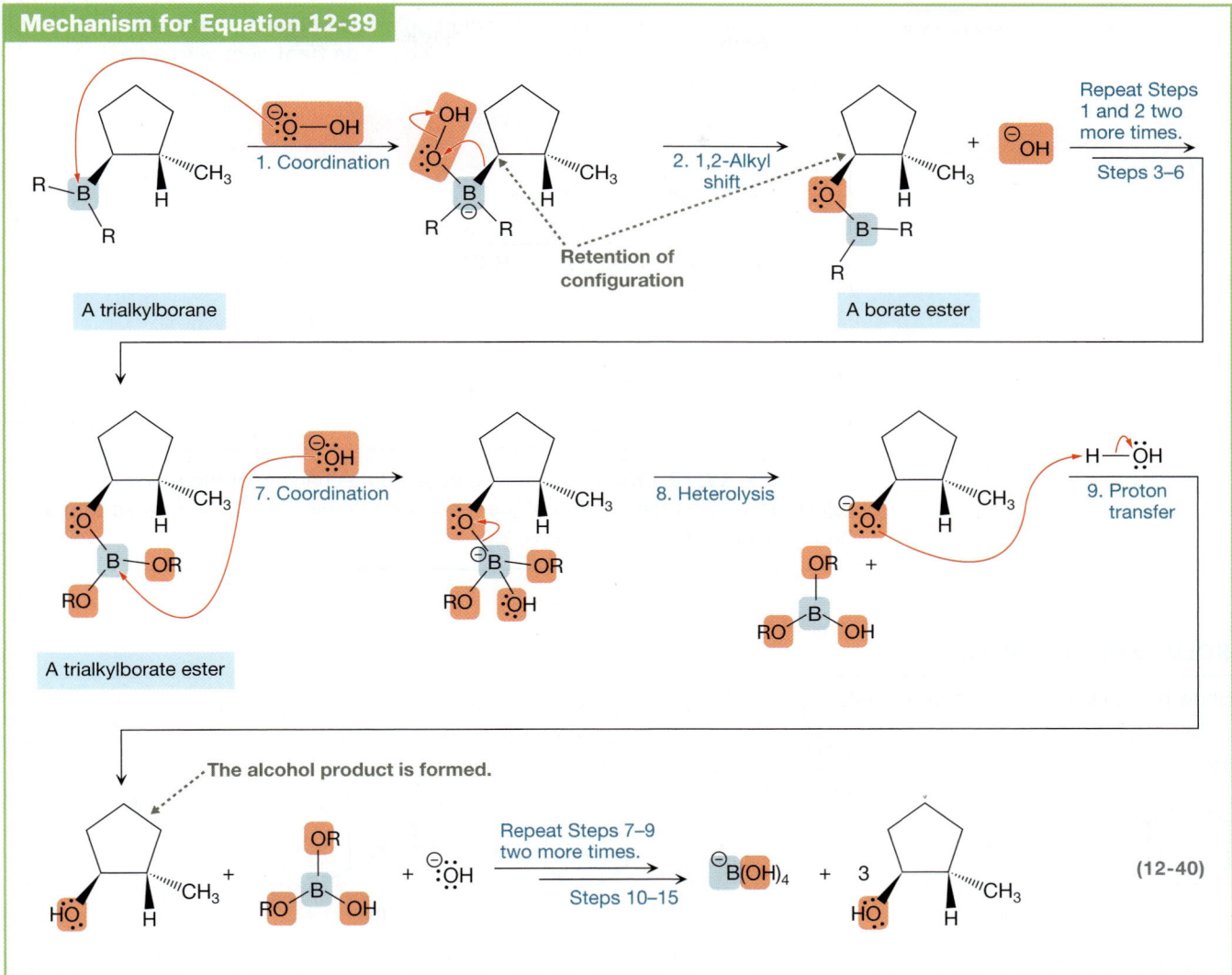

Mechanism for Equation 12-39

A trialkylborane

Retention of configuration

A borate ester

Repeat Steps 1 and 2 two more times.
Steps 3–6

1. Coordination
2. 1,2-Alkyl shift

A trialkylborate ester

7. Coordination
8. Heterolysis
9. Proton transfer

The alcohol product is formed.

Repeat Steps 7–9 two more times.
Steps 10–15

(12-40)

Under basic conditions, there is an equilibrium amount of the deprotonated peroxide, HOO⁻, called the **hydroperoxide ion**. The mechanism begins with coordination of HOO⁻ to the electron-deficient B atom of the trialkylborane, thereby producing an unstable tetrahedral intermediate. In Step 2, the breaking of a weak peroxide (O—O) bond drives a 1,2-alkyl shift that yields a **borate ester**. This pair of steps occurs twice more (Steps 3–6), resulting in a **trialkylborate ester**. In Step 7, hydroxide anion (HO⁻) coordinates to the B atom of the trialkylborate ester, and in Step 8, heterolysis occurs to liberate an alkoxide anion (RO⁻) leaving group. In Step 9, the strongly basic alkoxide anion gains a proton from H_2O to produce the first equivalent of the final alcohol. This trio of steps is then repeated twice (Steps 10–15).

With this mechanism, we can now see how the conversion of the trialkylborane to the alcohol takes place with retention of configuration. The critical step is Step 2, where a C—B bond breaks at the same time a C—O bond forms. Because of geometric constraints during this *concerted* process, the O-containing group simply assumes the position that was originally occupied by the B-containing group.

YOUR TURN 12.12

This trialkylborane is an intermediate in a hydroboration–oxidation reaction. Draw the alcohol that is produced on treatment with a basic solution of H_2O_2.

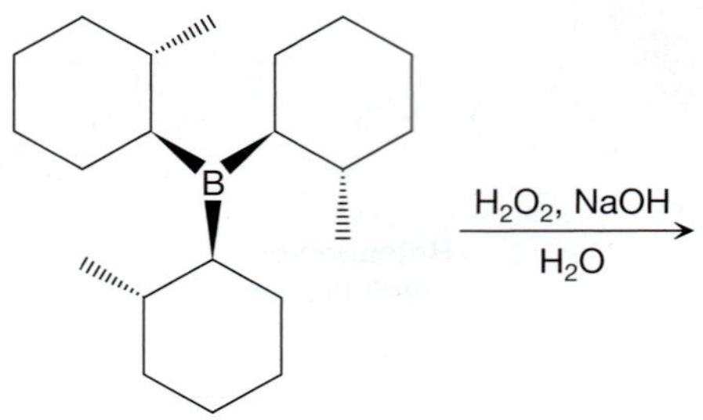

$$\xrightarrow[\text{H}_2\text{O}]{\text{H}_2\text{O}_2,\ \text{NaOH}}$$

PROBLEM 12.22 Draw the complete detailed mechanism of the reaction that takes place when the product of the reaction in Problem 12.21 is treated with a basic solution of hydrogen peroxide.

SOLVED PROBLEM 12.23

Show how to carry out each of the following transformations.

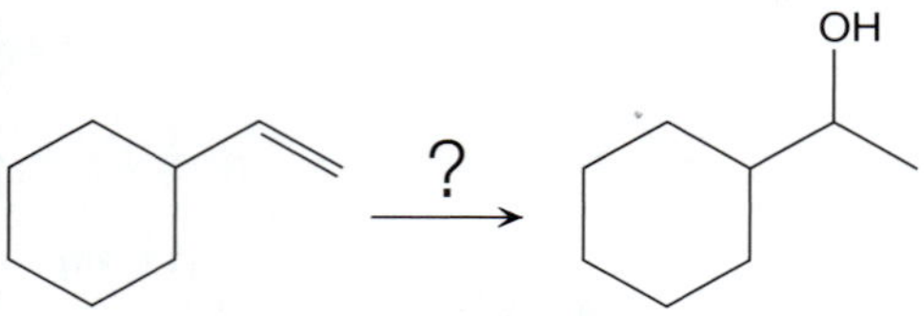

(a)

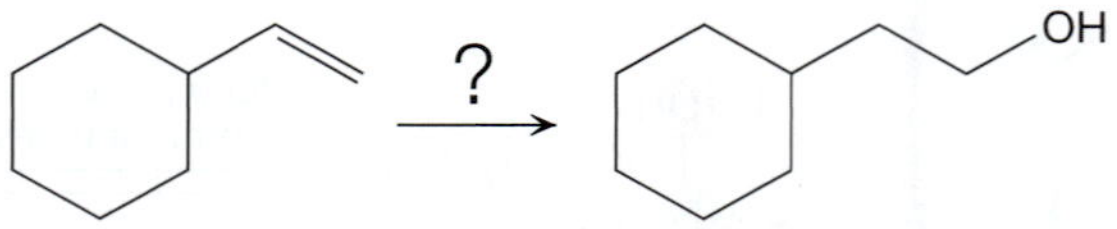

(b)

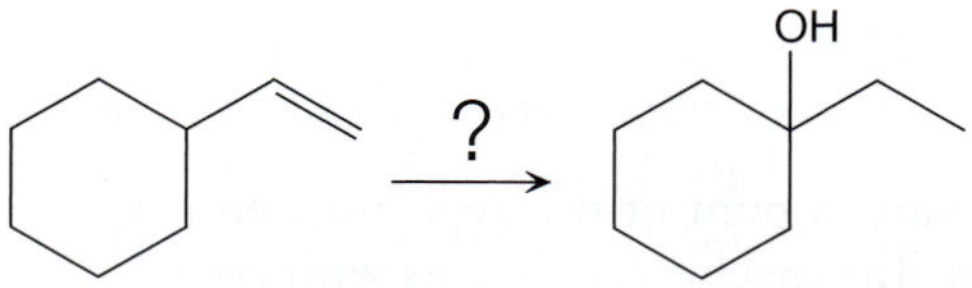

(c)

Think In each case, there is a net addition of what molecule across the double bond? What is the regiochemistry required in each reaction? Are carbocation rearrangements a concern?

Solve In each case, H_2O is being added across the double bond. In **(a)**, the addition takes place with Markovnikov regiochemistry, so there is a choice between acid-catalyzed hydration and oxymercuration–reduction. In this case, the carbocation that would be produced under an acid-catalyzed hydration would undergo a carbocation rearrangement, which is undesirable. Therefore, oxymercuration–reduction would be the better choice. In **(b)**, the regiochemistry is anti-Markovnikov, which requires hydroboration–oxidation. In **(c)**, a rearrangement takes place, so the reaction must proceed through a carbocation intermediate. The reaction therefore requires aqueous acid.

(a)
1. $Hg(OAc)_2$, H_2O
2. $NaBH_4$

(b)
1. $BH_3 \cdot THF$
2. H_2O_2, NaOH, H_2O

(c)
H_2O
HCl

PROBLEM 12.24 Show how to carry out each of the following transformations.

(a)

?

(b)

?

(c)

?

12.7 Hydroboration–Oxidation of Alkynes

The addition of BH_3 to an alkyne takes place in much the same way as it does to an alkene, with BH_2 adding to the less sterically hindered carbon and H adding to the more sterically hindered one (Equation 12-41). An alkene is produced, however, and if that alkene is terminal, a second hydroboration takes place readily. As a result, such a reaction would lead to a mixture of products, making it a fairly useless reaction.

Syn addition of H and BH₂

$$R-C\equiv C-H \xrightarrow{BH_3} \underset{\underset{R}{\mid}\;\;\underset{H}{\mid}}{C=C} \xrightarrow[\text{further with } BH_3.]{\text{The alkene reacts}} \boxed{\text{Mixture of products}} \qquad (12\text{-}41)$$

CONNECTIONS Hexanal (Equation 12-42) is an oxidation product of linoleic acid and is thought to be partly responsible for the distinct flavor of cooked meats.

Chemists avoid this problem by using a bulky dialkylborane, such as **disiamylborane** [$(C_5H_{11})_2BH$], instead of BH_3, as shown in Equation 12-42. Disiamylborane reacts with an alkyne in the usual way, with the R_2B group adding to the less sterically hindered C atom, just as we saw previously in Equation 12-38c. With the bulky R_2B

group attached to the alkene, *a second addition of disiamylborane does not occur*, so hydroboration stops at the alkene stage.

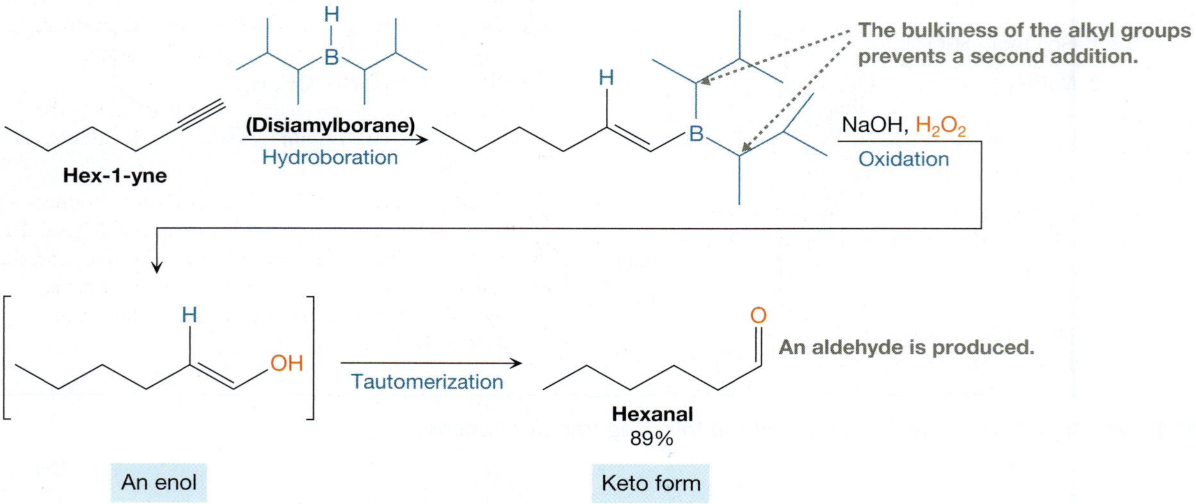

(12-42)

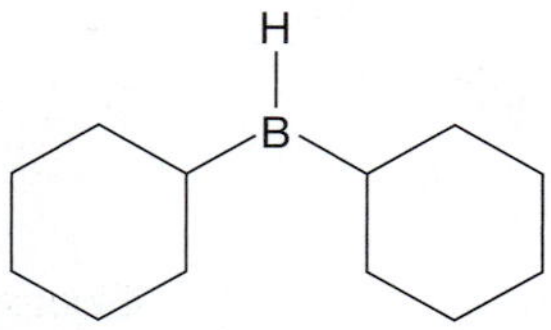

The bulkiness of the alkyl groups prevents a second addition.

An aldehyde is produced.

Hexanal
89%

An enol — Keto form

(a) Dicyclohexylborane

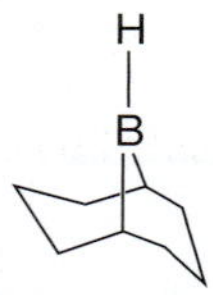

(b) 9-Borabicyclo[3.3.1]nonane (9-BBN)

FIGURE 12-4 Bulky dialkylboranes
(a) Dicyclohexylborane and
(b) 9-borabicyclo[3.3.1]nonane (9-BBN)
add just once to a terminal alkyne.

Subsequent oxidation with a basic solution of H_2O_2 converts the R_2B substituent on the alkene into an OH group, similar to Equation 12-39. In this case, however, the product is an *enol*, which tautomerizes into the more stable keto form.

Disiamylborane is not the only bulky dialkylborane used to carry out these kinds of conversions. Other examples include **dicyclohexylborane** and **9-borabicyclo[3.3.1] nonane** (abbreviated **9-BBN**), as shown in Figure 12-4.

As we can see from Equation 12-42, hydroboration–oxidation is a useful way to convert a *terminal alkyne* into an aldehyde. This reaction can also be used to convert an internal alkyne into a ketone, but a mixture of products results unless the alkyne is symmetric.

PROBLEM 12.25 Alkyne **A** is treated with disiamylborane followed by a basic solution of H_2O_2. The overall product is an aldehyde. Draw the structures of the initial alkyne **A** and intermediate **B**.

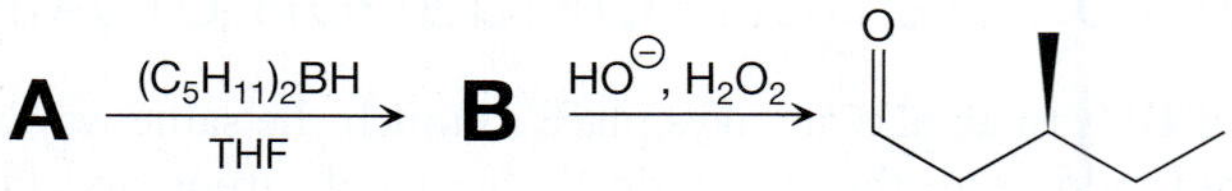

Chapter Summary and Key Terms

- Addition of an electrophile to an alkene or alkyne tends to proceed through a three-membered ring transition state if the electron-poor atom of the electrophile possesses a lone pair of electrons. Thus, no carbocation is produced. (**Section 12.1; Objective 1**)

- A **carbene** is characterized by an uncharged carbon atom that possesses a lone pair of electrons, and is typically produced in situ from an appropriate precursor. A carbene adds to an alkene, via a three-membered-ring transition state, to produce a cyclopropane ring. This reaction preserves the cis/trans relationships of the groups attached to the alkene carbons. (**Section 12.2; Objective 2**)

- In a non-nucleophilic solvent (such as carbon tetrachloride), Br_2 or Cl_2 adds to an alkene to produce a 1,2-dihalide. This reaction takes place with anti addition of the halogens (i. e., produces a vicinal dihalide), signifying the initial production of a **bromonium** or **chloronium ion intermediate**. (Section 12.3a; Objective 3)

- In water, treatment of an alkene with a molecular halogen produces a *halohydrin*, in which a halogen atom and an OH group add in an anti fashion to the alkene carbon atoms. The OH group ends up on the carbon that can better handle a positive charge. (Section 12.3b; Objective 3)

- **Oxymercuration–reduction** proceeds through a **mercurinium ion intermediate** and results in the addition of water to an alkene or alkyne with Markovnikov regiochemistry. (Section 12.4; Objective 4)

- Treatment of an alkene with a **peroxyacid** (or peracid) produces an epoxide. This reaction preserves the cis/trans relationships of the groups attached to the alkene. (Section 12.5; Objective 5)

- In a **hydroboration–oxidation reaction**, H—OH adds to an alkene in a syn fashion, with **anti-Markovnikov regiochemistry**, to produce an alcohol. (Section 12.6; Objective 6)
  - The hydroboration step involves the addition of BH_3 to the alkene to produce a **trialkylborane** and proceeds through a four-membered-ring transition state. (Section 12.6a; Objective 6)
  - The oxidation step converts the trialkylborane to the alcohol, with retention of configuration. (Section 12.6b; Objective 6)

- A single hydroboration–oxidation of an alkyne produces an enol that tautomerizes into a ketone or an aldehyde. To ensure that an alkyne undergoes only a single hydroboration, a bulky dialkylborane is used, such as **disiamylborane, dicyclohexylborane,** or **9-borabicyclo[3.3.1]nonane (9-BBN)**. (Section 12.7; Objective 7)

Reaction Tables

TABLE 12-1 Functional Group Transformations[a]

	Starting Compound Class	Typical Reagent Required	Compound Class Formed	Key Electron-Rich Species	Key Electron-Poor Species	Comments	Discussed in Section
(1)	Alkene	X_2 / CCl_4	Vicinal dihalide			Anti addition	12.3a
(2)	Alkyne	X_2 2 equiv / CCl_4	1,1,2,2-Tetrahalide			2 equivalents of halogen	12.3a
(3)	Alkene	X_2 / H_2O	Halohydrin			OH ends up on more highly substituted carbon; anti addition	12.3b
(4)	Alkene	1. $Hg(OAc)_2$, H_2O / 2. $NaBH_4$	Alcohol			Markovnikov addition of water; no carbocation rearrangements	12.4

[a]X = Cl, Br.

(continued)

TABLE 12-1 Functional Group Transformations[a] (continued)

Starting Compound Class	Typical Reagent Required	Compound Class Formed	Key Electron-Rich Species	Key Electron-Poor Species	Comments	Discussed in Section
(5) C=C Alkene	1. B_2H_6 or BH_3·THF 2. H_2O_2, NaOH, H_2O	H OH C–C Alcohol	C=C	BH_3	Anti-Markovnikov syn addition of water	12.6
(6) C=C Alkene	O‖C R–O–OH	O C–C Epoxide	C=C	O‖ R–C–O–O–H	Conservation of cis/trans configuration	12.5
(7) –C≡C– Alkyne	Hg(OAc)₂, H_2O	O‖ C–C H H Ketone	$H_2\ddot{O}$:	OAc ⊕Hg C=C	Markovnikov addition of water	12.4
(8) –C≡C– Alkyne	1. $(C_5H_{11})_2BH$ 2. H_2O_2, NaOH, H_2O	H O‖ C–C H Ketone or aldehyde	–C≡C–	$(C_5H_{11})_2BH$	Anti-Markovnikov addition of water	12.7

[a]X = Cl, Br.

TABLE 12-2 Reactions That Alter the Carbon Skeleton

Starting Compound Class	Typical Reagent Required	Compound Class Formed	Key Electron-Rich Species	Key Electron-Poor Species	Comments	Discussed in Section
(1) C=C Alkene	CH_2N_2 Δ or hν	H₂ C C–C Cyclopropane ring	C=C	:CH_2	Syn addition; retention of cis/trans configuration; not very useful in synthesis	12.2
(2) C=C Alkene	$CHCl_3$ NaOH	Cl Cl C C–C Dichlorocyclopropane ring	C=C	:CCl_2	Syn addition; retention of cis/trans configuration	12.2

Problems

Problems that are related to synthesis are denoted (SYN).

12.2 Electrophilic Addition of Carbenes: Formation of Cyclopropane Rings

12.26 Draw the mechanism for the reaction that would take place when each of the following compounds is treated with CH_2N_2 and irradiated with ultraviolet light.

(a)

(R)-1,6-Dimethylcyclohexene

(b)

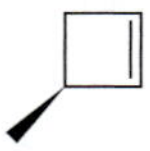

(c)

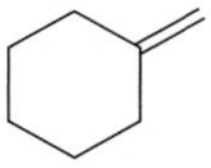

(d)

(e)

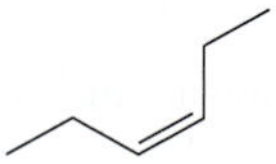

12.27 For each reaction, draw the complete mechanism and the major organic product(s).

(a)

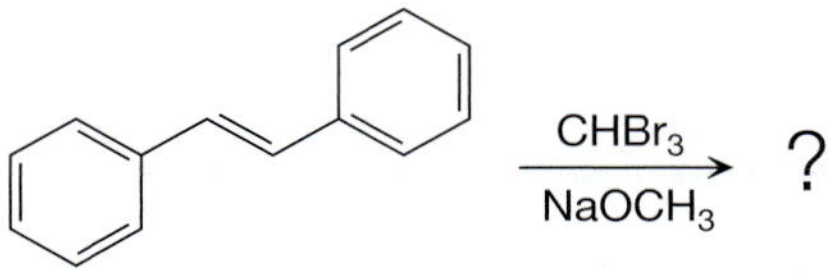

$\xrightarrow[\text{NaOCH}_3]{\text{CHBr}_3}$?

(b)

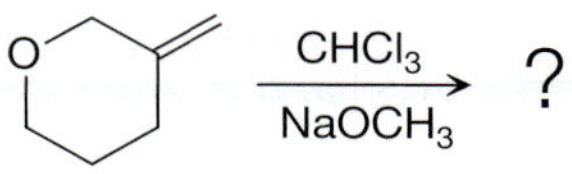

$\xrightarrow[\text{NaOCH}_3]{\text{CHCl}_3}$?

(c)

$\xrightarrow[\text{NaOCH}_3]{\text{CHCl}_3}$?

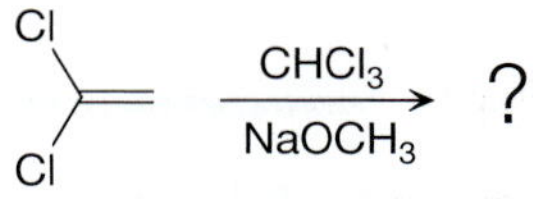

12.28 **(SYN)** The high reactivity of carbenes can facilitate the synthesis of some unusual compounds. Show how each of the following can be synthesized from acyclic compounds.

(a)

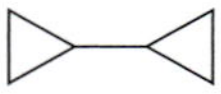

(b)

12.29 **(SYN)** Show how each of these compounds can be produced from an alkene. In each case, include the alkene, the reagents, and any special reaction conditions. Pay attention to stereochemistry.

(a)

(b)

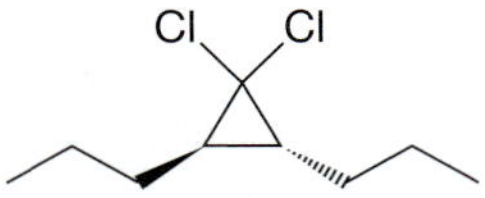

(c)

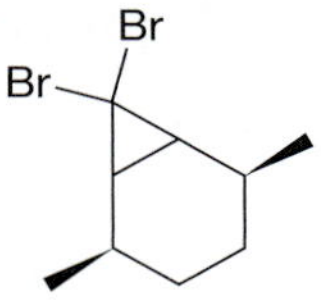

12.30 **(SYN)** Draw the alkyne that, when treated with diazomethane and irradiated with ultraviolet light, will produce the compound shown here. Draw the complete mechanism for that reaction, too.

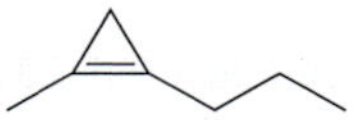

12.31 Dichloromethane (CH_2Cl_2) can be treated with butyllithium, $CH_3CH_2CH_2CH_2$—Li (Bu—Li), to make a carbene in situ, analogous to the way a carbene is generated from trichloromethane ($CHCl_3$) using HO^-.
(a) Show the mechanism for the generation of a carbene from CH_2Cl_2.
(b) Why is butyllithium used instead of HO^-?
(c) The reaction shown here leads to a mixture of four products. Draw all four products.

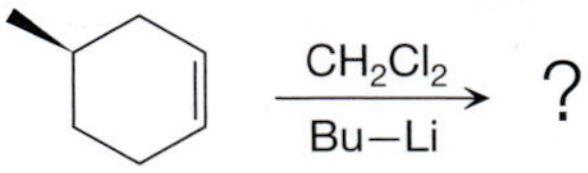

$\xrightarrow[\text{Bu—Li}]{\text{CH}_2\text{Cl}_2}$?

12.32 Draw the mechanism for this reaction.

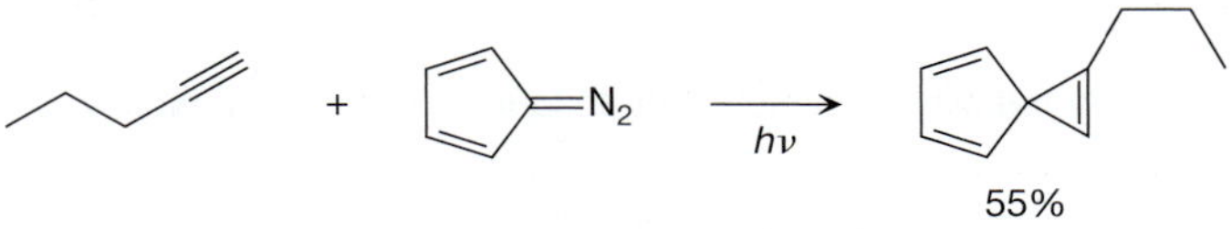

$+$ $=N_2$ $\xrightarrow{h\nu}$

55%

12.33 Draw all stereoisomers that can be produced in this reaction.

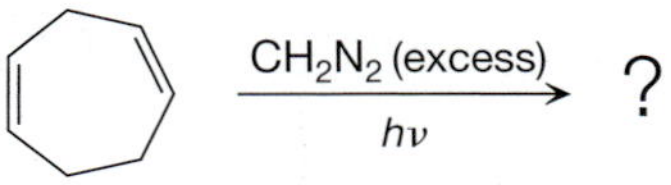

$\xrightarrow[h\nu]{\text{CH}_2\text{N}_2 \text{ (excess)}}$?

12.3 Electrophilic Addition Involving Molecular Halogens: Synthesis of 1,2-Dihalides and Halohydrins

12.34 Br_2 undergoes electrophilic addition to maleic anhydride as shown here. Explain why this reaction is much slower than the analogous one with cyclopentene.

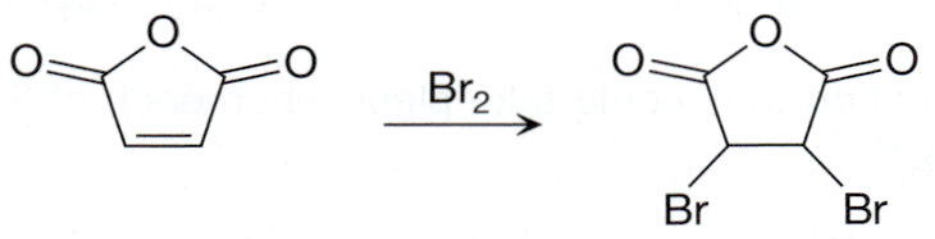

12.35 The electrophilic addition of Br_2 to several alkenes was examined. Explain why the relative reaction rates are as follows:

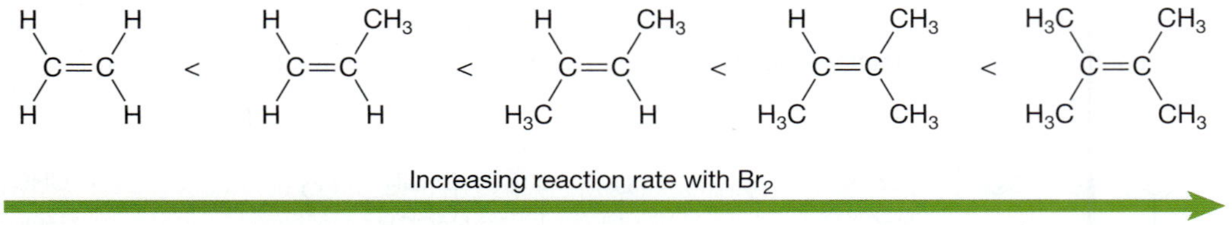

Increasing reaction rate with Br_2

12.36 For each of the following reactions, draw a complete, detailed mechanism and predict the major products, paying close attention to regiochemistry and stereochemistry.

(a)

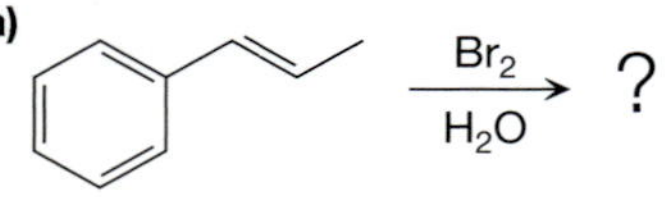

$\xrightarrow[H_2O]{Br_2}$?

(b)

$\xrightarrow[H_2O]{Cl_2}$?

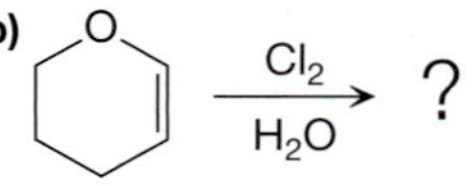

12.37 For each of the following reactions, draw a complete, detailed mechanism and predict the major organic products.

(a) 2-Methylbut-2-ene $\xrightarrow[CCl_4]{Br_2}$?

(b) 2-Methylbut-2-ene $\xrightarrow[H_2O]{Br_2}$?

(c) Hexa-1,5-diene $\xrightarrow[CCl_4]{Cl_2 \text{ (excess)}}$?

(d) 3-Ethylpent-1-yne $\xrightarrow[CCl_4]{Br_2 \text{ (excess)}}$?

12.38 Bromination can occur in a 1,4 fashion across conjugated double bonds, as shown here for cyclohexa-1,3-diene. One mechanism that has been proposed involves a five-membered ring, bromonium ion intermediate, as shown below.
(a) According to this mechanism, what should the stereochemistry be for the products—namely, all cis, all trans, or a mixture of both?
(b) Observations from experiment show that both cis and trans products are formed. Does this support or discredit the proposed mechanism shown here?

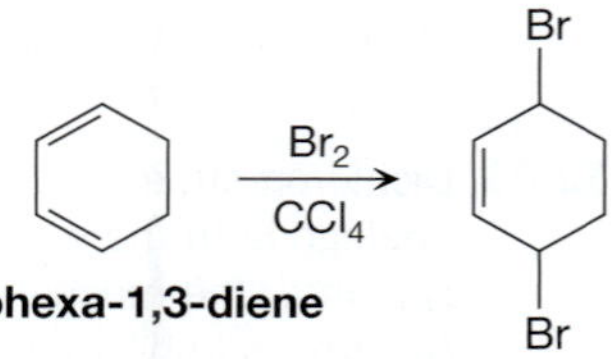

Cyclohexa-1,3-diene

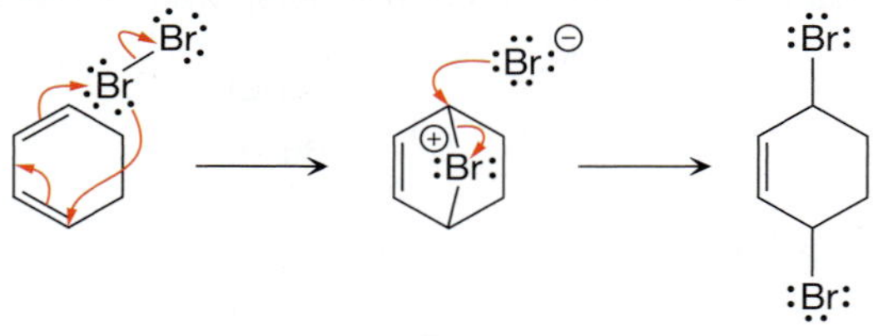

12.39 Propose a mechanism for the following reaction that accounts for the observed stereochemistry.

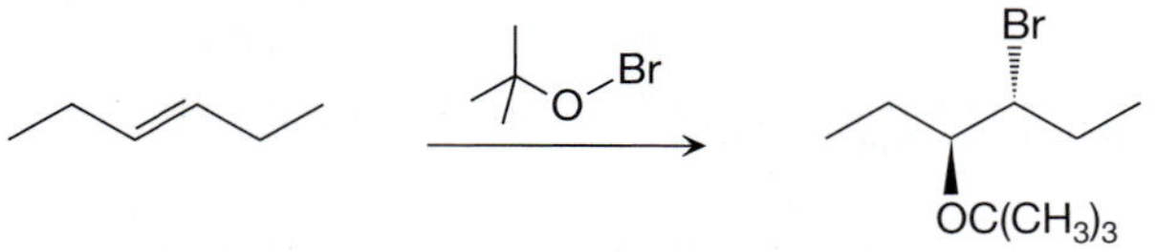

+ Enantiomer

12.40 A student attempted to brominate the double bond in pent-4-en-1-ol, but ended up with the following cyclic ether instead. Propose a mechanism for the formation of this product.

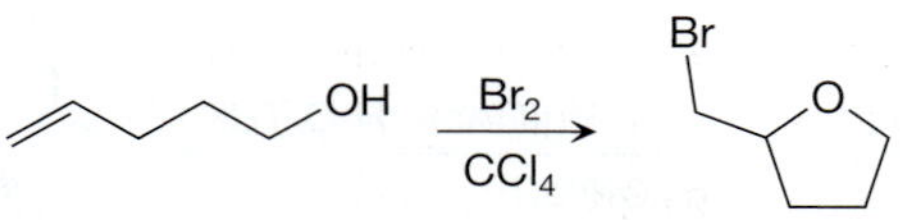

12.41 Iodine monochloride (ICl) is a mixed halogen that adds to an alkene via the same mechanism by which bromination takes place. With that in mind, propose a mechanism for the following reaction, and use that mechanism to predict the products, paying attention to both *regiochemistry* and *stereochemistry*. *Hint:* In ICl, one atom is more electrophilic than the other.

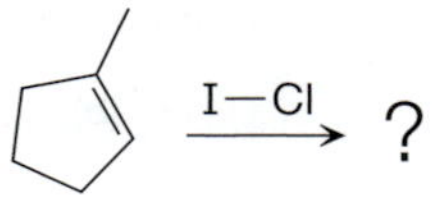

12.4 Oxymercuration–Reduction: Addition of Water

12.42 Draw the mechanism and the major organic product(s) for each of the following reactions.

(a)

Cyclopentylethene $\xrightarrow{\text{H}_3\text{O}^{\oplus}}$?

(b)

Cyclopentylethene $\xrightarrow[\text{2. NaBH}_4,\ \text{ethanol}]{\text{1. Hg(OAc)}_2,\ \text{H}_2\text{O}}$?

(c)

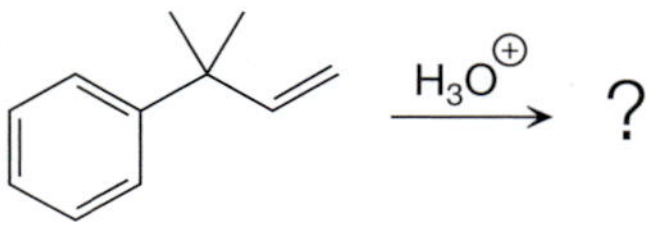

$\xrightarrow{\text{H}_3\text{O}^{\oplus}}$?

(d)

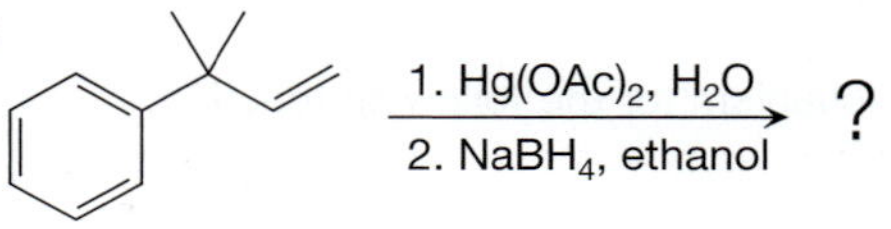

$\xrightarrow[\text{2. NaBH}_4,\ \text{ethanol}]{\text{1. Hg(OAc)}_2,\ \text{H}_2\text{O}}$?

12.43 (SYN) Show how to make each compound from an alkene.

(a)

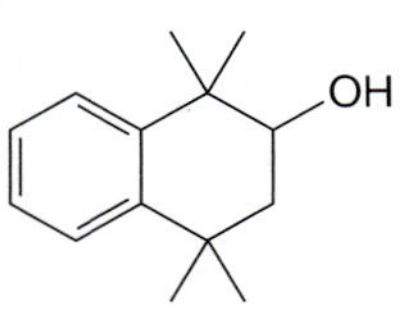

(b)

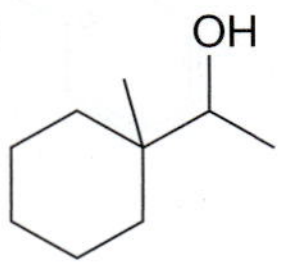

12.44 Which of these transformations would be the result of acid-catalyzed hydration, and which would be the result of oxymercuration–reduction?

(a)

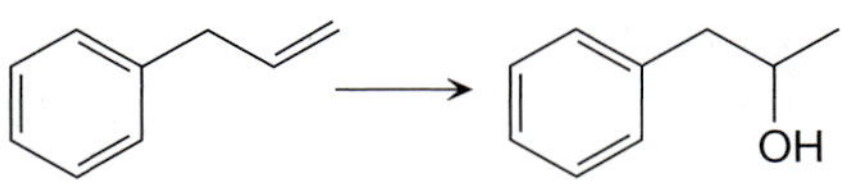

(b)

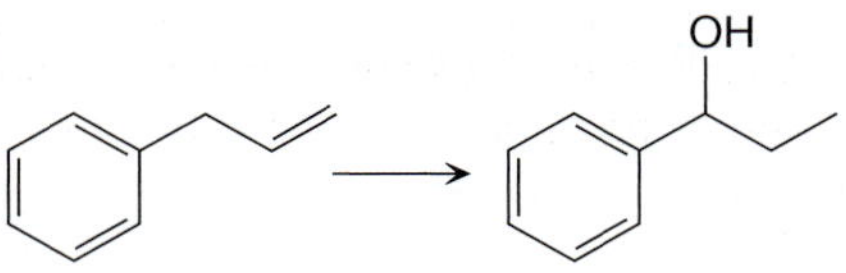

(c)

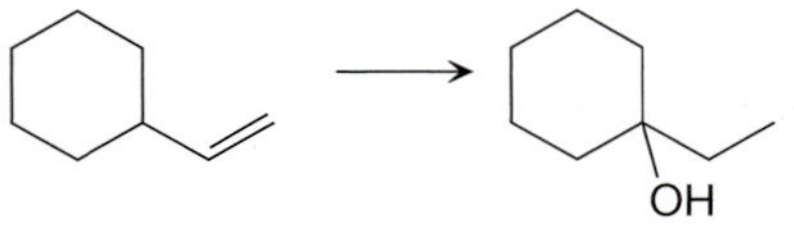

(d)

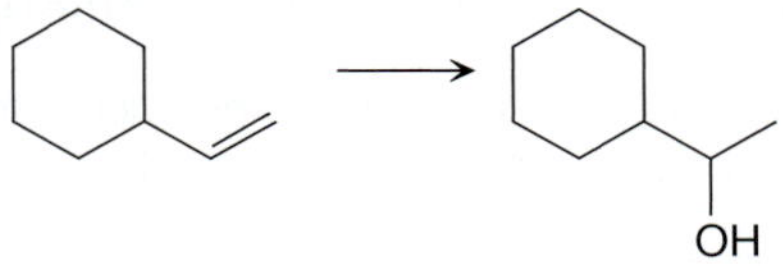

12.45 Draw the mechanism and the major organic product(s) for each of the following reactions.

(a)

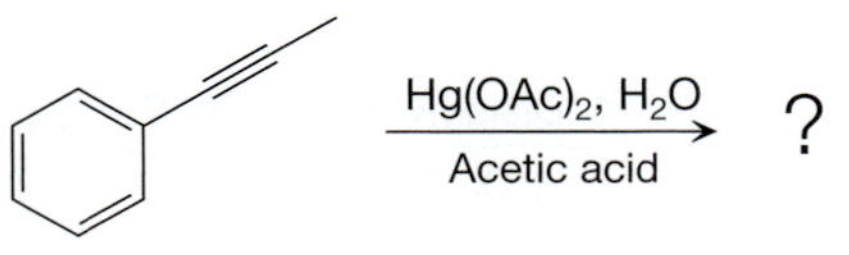

$\xrightarrow[\text{Acetic acid}]{\text{Hg(OAc)}_2,\ \text{H}_2\text{O}}$?

(b)

$\xrightarrow[\text{Acetic acid}]{\text{Hg(OAc)}_2,\ \text{H}_2\text{O}}$?

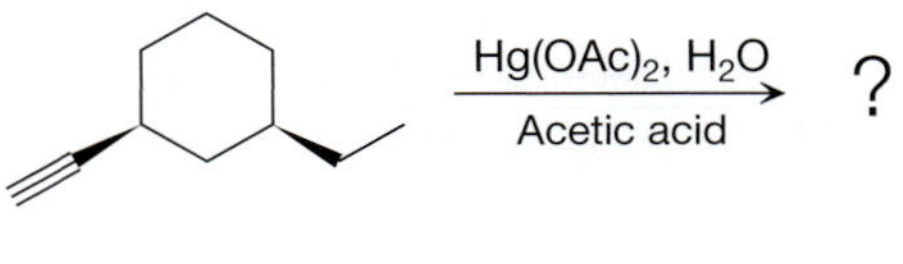

12.46 In the following oxymercuration–reduction reaction, H adds to the more highly substituted C atom, which might appear to violate Markovnikov's rule. Explain why this reaction exhibits this regiochemistry.

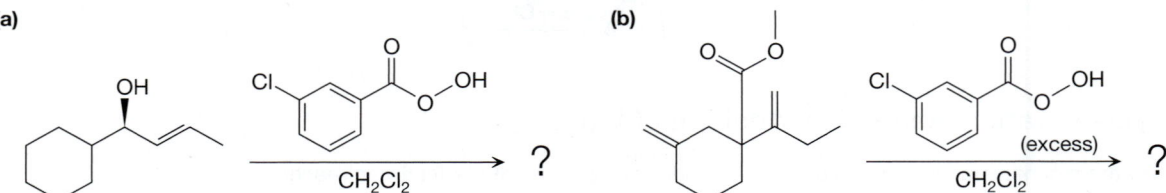

Methyl (E)-2-methylbut-2-enoate

1. Hg(OAc)₂, H₂O/THF
2. NaBH₄, ethanol

+ Enantiomer   + Enantiomer

12.5 Epoxide Formation Using Peroxyacids

12.47 For each reaction, draw the complete mechanism and the major organic product(s), paying attention to stereochemistry.

(a)

OH

CH_2Cl_2 ?

(b)

(excess)
CH_2Cl_2 ?

12.48 (SYN) Show how each epoxide can be produced from an alkene.

(a)   (b)   (c)

12.49 Draw the complete mechanism for the following reaction.

$NaHCO_3$
CH_2Cl_2, H_2O

12.50 When cyclohexene is treated with *m*-chloroperbenzoic acid and H₂O, *trans*-cyclohexane-1,2-diol is produced. Propose a mechanism for this reaction, accounting for the observed stereochemistry. *Hint:* Recall what a peroxyacid does to an alkene.

MCPBA
H_2O

+   Enantiomer

12.51 Why do you think the epoxidation reaction below favors the C=C bond on the left?

MCPBA
CH_2Cl_2, 0 °C

96%

12.6 and 12.7 Hydroboration–Oxidation of Alkenes and Alkynes

12.52 Draw the mechanism and the major organic product(s) for each of the following reactions.

(a)
Cyclopentylethene

1. BH₃·THF
2. H₂O₂, NaOH, H₂O   ?

(b)

1. B₂H₆
2. H₂O₂, NaOH, H₂O   ?

(c)

1. Disiamylborane, THF
2. H₂O₂, NaOH, H₂O   ?

(d)

1. Disiamylborane, THF
2. H₂O₂, NaOH, H₂O   ?

12.53 (SYN) Show how each of these compounds can be produced from an alkene or alkyne. Draw the appropriate alkene or alkyne and include any necessary reagents and special reaction conditions.

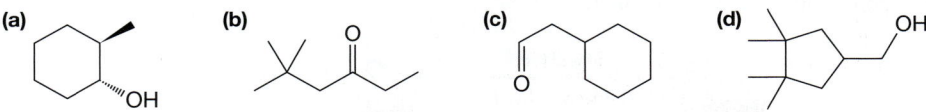

(a)   (b)   (c)   (d)

12.54 (SYN) Hydroboration–oxidation can be carried out with deuterated forms of the reagents and solvent. For example, BD₃·THF can be used instead of BH₃·THF, and D₂O could be used instead of H₂O. With this in mind, show how each of the following compounds can be produced from an alkene.

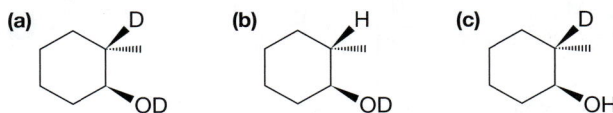

(a)   (b)   (c)

12.55 Predict the product of the reaction shown here.

$$\text{Buta-1,3-diene} \xrightarrow[\text{2. NaOH, H}_2\text{O}_2, \text{H}_2\text{O}]{\text{1. BH}_3\cdot\text{THF (excess)}} \ ?$$

Integrated Problems

12.56 In Section 12.3, we learned that Cl_2 undergoes anti addition to an alkene such as but-2-ene. Under similar conditions, Cl_2 undergoes both syn and anti addition to (E)-1-phenylprop-1-ene to produce the isomers shown below, plus their enantiomers. **(a)** Draw the mechanism to account for this mixture of products. **(b)** Explain why the mechanism for 1-phenylprop-1-ene is different from the one for but-2-ene.

12.57 In Section 12.2, we learned that the Simmons–Smith reaction produces a cyclopropane ring from an alkene. Diiodomethane (CH_2I_2) is treated with a source of zinc to produce the Simmons–Smith reagent (ICH_2ZnI), which reacts with the alkene in a single elementary step. Complete the mechanism below by adding the necessary curved arrows for this elementary step. *Hint:* The curved arrow notation is very similar to that for epoxidation involving a peroxyacid.)

Simmons–Smith reagent

12.58 When benzene is treated with diazomethane and irradiated with light, cyclohepta-1,3,5-triene is produced. Propose a mechanism for this reaction.

$$\xrightarrow[h\nu]{\text{CH}_2\text{N}_2}$$

12.59 Draw the mechanism for the first reaction in the following sequence.

$$\xrightarrow[\text{2. NaBH}_4, \text{ ethanol}]{\text{1. Hg(OAc)}_2, \text{ H}_2\text{O/THF}}$$

60%

12.60 When norbornene undergoes hydroboration–oxidation, a mixture of two stereoisomers is produced in a roughly 6:1 ratio. **(a)** Draw both of these isomeric products. **(b)** Which product is favored? *Hint:* You should build a model of norbornene and consider the transition state leading to each product.

$$\text{Norbornene} \xrightarrow[\text{2. H}_2\text{O}_2,\text{ NaOH, H}_2\text{O}]{\text{1. BH}_3\cdot\text{THF}} \text{?}$$

Norbornene

12.61 Supply the missing compounds in the following sequence of reactions.

$$\bigcirc \xrightarrow{\text{MCPBA}} \mathbf{A} \xrightarrow[\text{2. H}_3\text{O}^{\oplus}]{\text{1. CH}_3\text{CH}_2\text{Li}} \mathbf{B} \xrightarrow[\Delta]{\text{H}_3\text{O}^{\oplus}} \mathbf{C} \xrightarrow{\text{MCPBA}} \mathbf{D} \xrightarrow[\text{CH}_3\text{CH}_2\text{OH}]{\text{NaBr}} \mathbf{E} \xrightarrow[\Delta]{\text{NaOH}} \mathbf{F} \xrightarrow[h\nu]{\text{CH}_2\text{N}_2} \mathbf{G}$$

12.62 Supply the missing compounds in the following synthesis scheme.

$$\xrightarrow[\text{CCl}_4]{\text{Br}_2} \mathbf{H} \xrightarrow[\text{2. H}_2\text{O}]{\text{1. NaNH}_2 \text{ (3 equiv)}} \mathbf{I} \xrightarrow[\text{2. NaBH}_4]{\text{1. Hg(OAc)}_2, \text{H}_2\text{O}} \mathbf{J} \xrightarrow[\text{Acetic acid}]{\text{Br}_2} \mathbf{K} \xrightarrow[\Delta]{\text{NaOH}} \mathbf{L} \xrightarrow[\text{H}_2\text{O}]{\text{Br}_2} \mathbf{M}$$

12.63 (SYN) Propose how to convert hex-1-yne into **(a)** 2,2-dibromohexane and **(b)** 1,2-dibromohexane. *Hint:* Each conversion might require carrying out more than one reaction.

12.64 (SYN) Show two different ways to convert 2-methylbut-2-ene to 3-bromo-2-methylbutan-2-ol. *Hint:* Each conversion might require carrying out more than one reaction.

Taxol, used in cancer chemotherapy, was first isolated from the bark of the Pacific yew tree. For many years, taxol was produced industrially from readily available starting compounds via organic synthesis. Currently it is produced from fresh leaf tissue via plant cell fermentation and extraction, a much greener process. We discuss some beginning concepts of multistep organic synthesis and green chemistry here in Chapter 13.

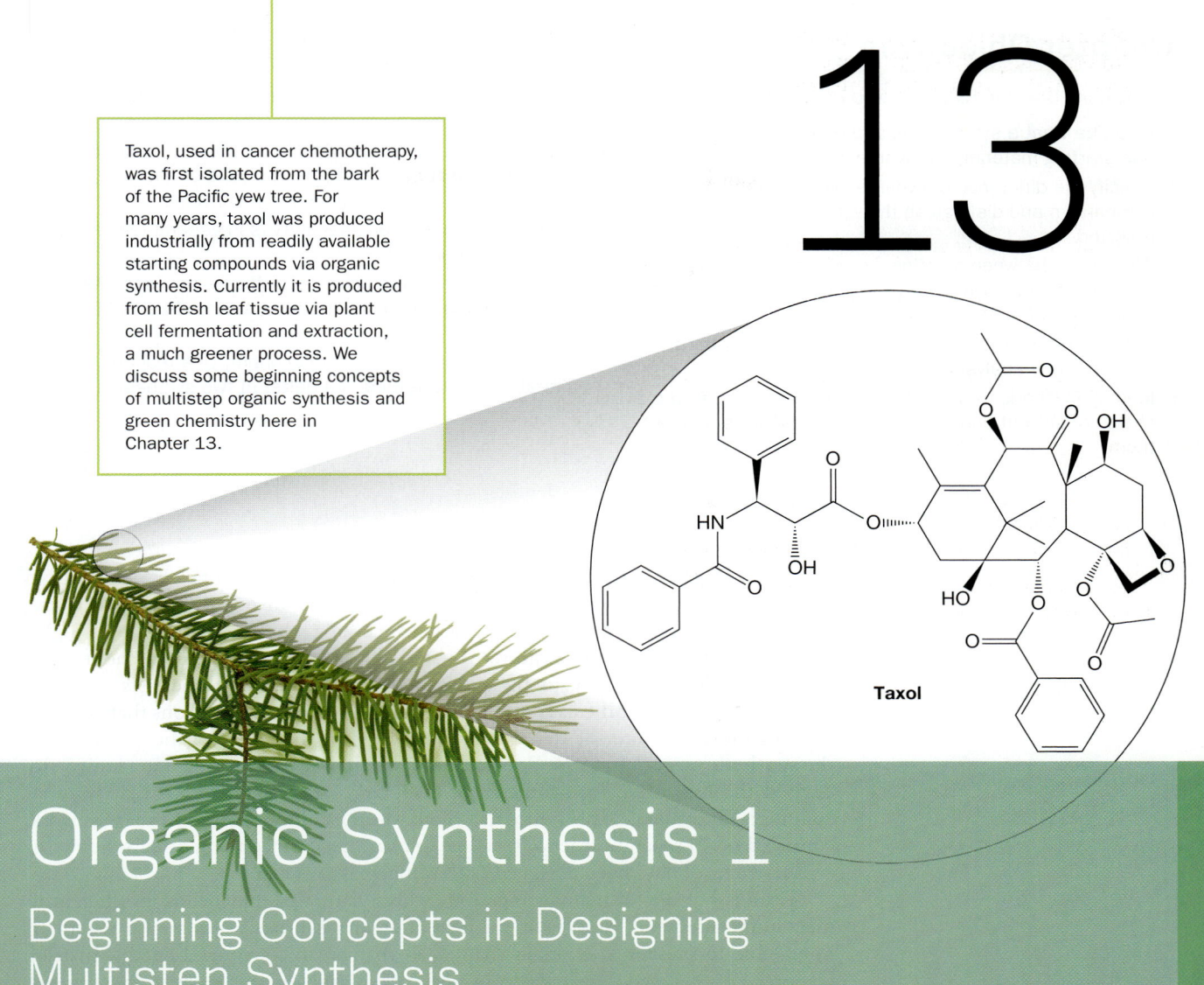

Taxol

Organic Synthesis 1
Beginning Concepts in Designing Multistep Synthesis

Before chemistry developed into a scientific discipline, alchemists sought various *elixirs* to cure diseases and lengthen life. These elixirs were mixtures of compounds, often produced by carrying out extractions or distillations on samples of plant or animal material, and they contained what was called the "essence" of that material. Examples include peppermint oil and clove oil, produced by the steam distillation of peppermint and cloves, respectively. As the discipline of chemistry became more sophisticated throughout the 19th and 20th centuries, scientists realized that these elixirs were made up of individual compounds—now called **natural products**—and that only certain natural products are responsible for the elixir's bioactivity. Menthol and eugenol, for example, which are natural products that have local anesthetic properties, are found in peppermint oil and clove oil, respectively.

Because of the potential benefits they offer, many natural products are in high demand today. Whenever we find a use for a natural product, however, demand almost invariably exceeds nature's supply. One notable example of this is taxol.

On completing Chapter 13 you should be able to:

1. Describe what a synthesis is and define both the target and starting material of a synthesis.
2. Specify the difference between a synthesis and a mechanism and distinguish the ways in which each is reported.
3. Distinguish between functional group transformations and reactions that alter the carbon skeleton, and explain the importance of each.
4. Define a transform and use transforms to carry out a retrosynthetic analysis.
5. Identify synthetic traps and potential issues involving the choice of solvent and explain some of the ways to avoid them.
6. Design a synthesis that incorporates stereospecific reactions.
7. Explain how the overall yield in a synthesis is affected by the total number of steps, the yield of each step, and whether the synthesis is linear or convergent.
8. Describe what green chemistry is in general terms and why it is important.
9. Explain why, in green chemistry, the use of reagents and solvents that are toxic or otherwise unsafe should be avoided.
10. Compute the percent atom economy of a reaction and, on that basis, determine the better of two syntheses.

Originally isolated from the bark of the Pacific yew tree (*Taxus brevifolia*) in 1966, taxol is an effective anticancer drug. Unfortunately, the amount of taxol available from the bark of one tree can provide only one 300-mg dose for one person. Even worse, isolating taxol requires harvesting the bark, which kills the tree.

In order to make taxol more readily available to cancer patients, without over-harvesting the Pacific yew to the point of extinction, organic chemists sought a *synthesis* of taxol. A **synthesis** is a specific sequence of chemical reactions that converts **starting materials** into the desired compound, called the **target** of the synthesis (or the **synthetic target**). In 1994, Robert A. Holton (b. 1944) of Florida State University devised the first route for synthesizing taxol from starting materials that were all commercially available. This synthesis, which was the culmination of 12 years of work, consisted of 46 separate reactions, called *synthetic steps*. Since then, other schemes to synthesize taxol have been designed with somewhat fewer steps.

Taxol is but one of many compounds whose natural abundance is insufficient to meet human demand. As a result, the field of organic synthesis is immense, making it impossible to present in a single chapter all that organic synthesis entails. Instead, Chapter 13 is devoted to *introductory* aspects of designing multistep organic syntheses, focusing on targets that are much simpler than taxol. First, we present proper ways to write the reactions of a synthesis. Then we explore the basic thinking that goes into planning an efficient synthesis of an organic molecule, including *retrosynthetic analysis*, a powerful tool that helps organize the search for synthetic pathways. Finally, we discuss *green chemistry*, which describes a responsible approach to designing and carrying out a synthesis, especially as it pertains to taking care of our planet.

In Chapter 19, we tackle some of the more complex aspects of organic synthesis, such as the proper placement of functional groups within a molecule, and how to use selective reagents and protecting groups.

13.1 Writing the Reactions of an Organic Synthesis

Although formalizing an organic synthesis on paper is typically the last step in designing a synthesis scheme, we begin by examining some of the conventions used to do so. These conventions are not new to you; they have been used consistently in the last several chapters when reactions have been introduced and discussed, and you have been using them to answer problems asking you to show how to produce a particular compound. We take the time to review the conventions here, however, so that they are at the forefront of your mind when you write out a synthesis. They help you *communicate* effectively, ensuring that a synthesis scheme you have devised is interpreted by others in the way that you intend.

There are essentially three main conventions routinely used in writing a synthesis scheme. The first stems from the fact that a synthesis is an abbreviated *recipe*. As such:

For each **synthetic step**, provide just:

- the overall reactants,
- the reagents added,
- the reaction conditions, and
- the overall products.

In particular, *do not include the details of any reaction mechanism.* The mechanisms that you have seen throughout the last several chapters enable you to understand *how* and *why* each reaction takes place as it does, but when conveying information about a synthesis to others, the focus changes to just the sequence of actions that must be done in the laboratory to produce the target from the starting material.

Analogously, when writing a recipe for chocolate chip cookies, it is not important to explain the details of the chemical processes that take place when an egg is added; it is important, however, to know *that* an egg is needed, as well as *when* and *how* to add the egg.

For example, a synthetic step showing how to convert 2-phenyl-2-tosyloxypropane into 2-bromo-2-phenylpropane might be written like this:

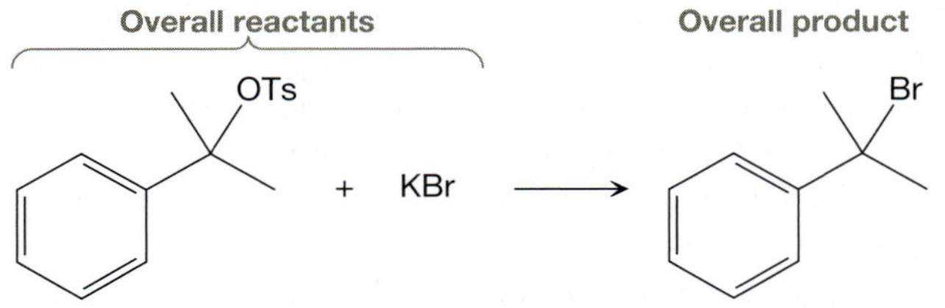

Overall reactants   **Overall product**

OTs   Br

+ KBr →   (13-1)

2-Phenyl-2-tosyloxypropane   **2-Bromo-2-phenylpropane**

This synthetic step implies that when 2-phenyl-2-tosyloxypropane and potassium bromide are combined, 2-bromo-2-phenylpropane is generated as an overall product. By contrast, the mechanism for this reaction, which is an S_N1, would be written as follows:

Mechanism for Equation 13-1

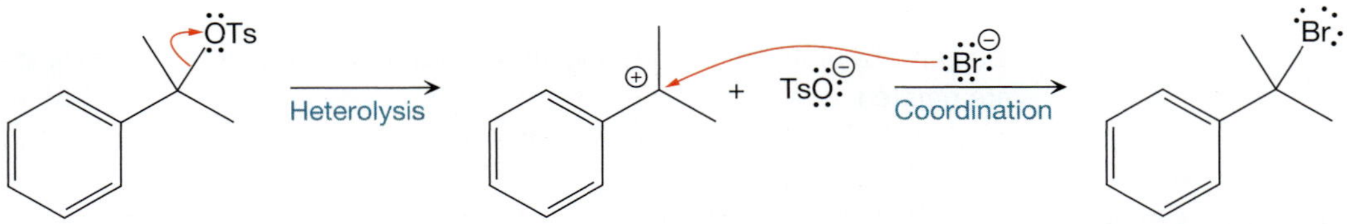

:ÖTs   :Br:

⊕   :Br:⊖

+ TsÖ:⊖   (13-2)

Heterolysis   Coordination

Even though KBr dissolves to form K^+ and Br^-, and even though Br^- is the active nucleophile in the mechanism in Equation 13-2, it is inappropriate to write Br^- as a reagent in a *synthetic step*. Br^- cannot exist in a pure form, due to the charge that it carries, so the synthetic step proposed in Equation 13-3 is technically incorrect.

Not an appropriate synthetic step
because Br$^\ominus$ does not exist in pure form

$$\text{(structure: 2-phenyl-2-tosyloxypropane with OTs)} \quad + \quad \text{Br}^\ominus \quad \longrightarrow \quad \text{(structure: 2-bromo-2-phenylpropane with Br)} \qquad (13\text{-}3)$$

In general:

> Reagents must be written in the form in which they can be added, not as they appear in the mechanism.

For this reason, Equation 13-4 is also unacceptable as a synthetic step.

Not an appropriate synthetic step
because :CH$_2$ does not exist in pure form

$$\text{(cyclohexene)} \quad + \quad :CH_2 \quad \longrightarrow \quad \text{(bicyclic with CH}_2) \qquad (13\text{-}4)$$

Recall from Section 12.2 that :CH$_2$, a carbene, is a very highly reactive species and does not exist in pure form. Rather, :CH$_2$ is produced in situ from a precursor that can exist in pure form, such as diazomethane (CH_2N_2). Thus, Equation 13-5 would be an acceptable synthetic step to accomplish the electrophilic addition of carbene to cyclohexene.

CH_2N_2 *does* exist in pure form and
is a precursor of :CH$_2$, so this is an
acceptable synthetic step.

$$\text{(cyclohexene)} \quad + \quad CH_2N_2 \quad \xrightarrow{\Delta} \quad \text{(bicyclic with CH}_2) \qquad (13\text{-}5)$$

Notice in Equations 13-1 and 13-5 that not all of the products are included in the synthesis scheme—the product containing the tosylate anion (TsO$^-$), in particular, has been omitted from Equation 13-1 and the N$_2$(g) product has been omitted from Equation 13-5. This simply suggests that TsO$^-$ and N$_2$ are by-products that we are not interested in collecting. Whereas organic products are usually of interest:

> Inorganic by-products and leaving groups are often irrelevant to the synthesis and are omitted.

YOUR TURN 13.1

(a) Write the synthetic step that shows 2-phenyl-2-tosyloxypropane reacting with NaCl to produce 2-chloro-2-phenylpropane.

(b) Draw the mechanism for this reaction.

Answers to Your Turns are in the back of the book.

The second convention for writing synthesis schemes relates to how the information in a chemical reaction is presented:

> Reagents added in a particular synthetic step can be written *above* the reaction arrow (→) that connects reactants to products, whereas reaction conditions (including solvent, temperature, pH, time of reaction, etc.) are usually written *below* the arrow.

For example, consider the synthetic step in Equation 13-6, which shows the conversion of 7-chlorohept-1-ene to 7-iodohept-1-ene.

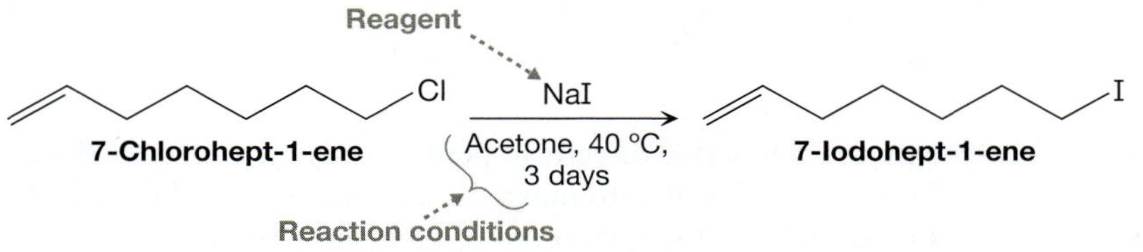

7-Chlorohept-1-ene   **7-Iodohept-1-ene**   (13-6)

Sodium iodide is the reagent that is added to 7-chlorohept-1-ene in this synthetic step, so it is written above the arrow. The reaction takes place in acetone at an elevated temperature (40 °C) for 3 days, so these conditions are written below the arrow.

YOUR TURN 13.2

Complete the following synthetic step by indicating that water is used as the solvent and that the reaction is carried out at 70 °C.

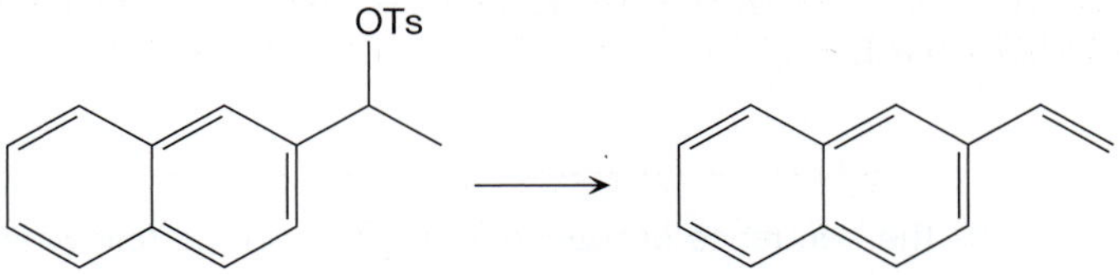

The convention for writing reagents above a reaction arrow and reaction conditions below the arrow is not rigorous. In a solvolysis reaction, for example, such as the one shown in Equation 13-7, CH_3OH acts as both the solvent and a reagent. Therefore, CH_3OH is written just once, below the arrow.

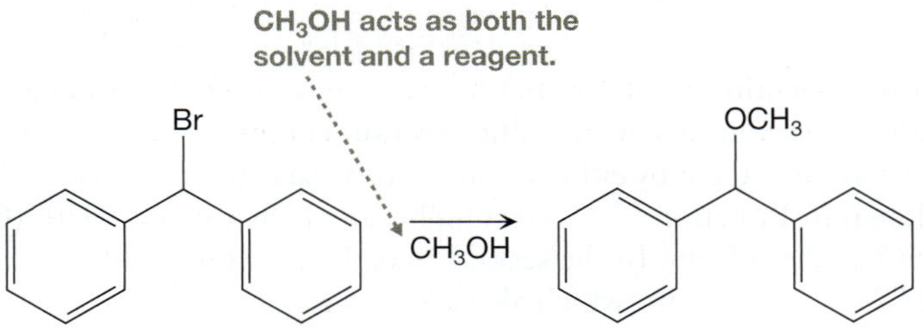

CH_3OH acts as both the solvent and a reagent.

(13-7)

The third convention provides some flexibility as to how we can write a synthesis that has multiple steps.

> Synthetic steps that are customarily carried out sequentially may be combined, using numbers above and below the reaction arrow to distinguish the steps.

Numbering the steps above and below the reaction arrow is often done when one of the reactions is a proton transfer. It can also be done when the product of a reaction is difficult to isolate and purify. In that case, the next set of reagents can simply be added to the crude product mixture, as long as the components of that mixture do not cause any unwanted side reactions.

An example is presented in Equation 13-8, which shows how hex-2-yne can be synthesized from pent-1-yne in a two-step process.

The numbers indicate that the reaction with NaH is carried out to completion before CH_3I is added.

$$\text{Pent-1-yne} \quad \xrightarrow[\text{2. } CH_3I]{\text{1. NaH}} \quad \text{Hex-2-yne} \qquad (13\text{-}8)$$

According to this scheme, NaH is added directly to pent-1-yne in the first reaction, in which the terminal alkyne is deprotonated to make the alkynide nucleophile. CH_3I is then added, and the subsequent reaction goes on to produce hex-2-yne.

A critical part of this numbering convention is the following:

> In a synthetic step, each reaction that is denoted by a number is understood to go to completion before the next reagent is added.

Therefore, in the case of the reaction sequence in Equation 13-8, it is possible to carry out the pair of reactions so that the two reagents—NaH and CH_3I—never come into contact with each other.

YOUR TURN 13.3

Describe in words the sequence of reactions depicted in the following transformation.

$$\text{Butanoic acid} \quad \xrightarrow[\text{2. } CH_3CH_2Br]{\text{1. NaOH}} \quad \text{Ethyl butanoate}$$

CONNECTIONS Butanoic acid is produced when butter turns rancid, giving the butter an unpleasant odor. In fact, the trivial name of butanoic acid is butyric acid, which derives from the Greek word that means butter.

Using this convention for sequential steps, reaction conditions can be written after the reagent for each numbered step. The reaction conditions are typically separated from the reactant or reagent by either a comma or a slash. For example, in the two-step synthesis shown in Equation 13-9, 1-bromohexane is first treated with $(CH_3)_3COK$ using DMSO as the solvent. In the second step, the resulting product is treated with Br_2, using H_2O as the solvent (which also acts as a reagent).

Reagent   Conditions

$$\text{1-Bromohexane} \quad \xrightarrow[\text{2. } Br_2/H_2O]{\text{1. } (CH_3)_3COK/DMSO} \qquad (13\text{-}9)$$

Reagent   Conditions

Using the number convention for common sequential steps, show how the following synthetic step would be represented: (Z)-Hex-3-ene is treated with mercury(II) acetate in water, and on completion of that reaction, the product is treated with sodium borohydride to yield hexan-3-ol as the overall product.

13.2 Cataloging Reactions: Functional Group Transformations and Carbon–Carbon Bond-Forming/Breaking Reactions

How challenging it is to design a practical synthesis partly depends on the nature of the target molecule and on the types of compounds available as starting materials. It also depends on how familiar you are with the wide variety of existing reactions, and how specifically you know what each reaction can accomplish. Therefore, it becomes important to *catalog* reactions you have encountered according to their utility in a synthesis. Specifically:

For each reaction you encounter, you should ask yourself the following questions:

- What functional groups in the reactants are involved?
- What functional groups are produced in the products?
- What are the major structural changes that occur?

To help you answer these questions quickly, reaction summary tables are included at the end of every chapter in which new reactions are introduced, beginning with Chapter 9. As shown in the following example, each entry in these tables provides the reactants, typical reagents and reaction conditions, products, and the section of the book in which the reaction is discussed. Also, to better enable you to apply your knowledge of mechanisms in synthesis, each entry provides key electron-rich and electron-poor species, along with the reaction type.

	Starting Compound Class	Typical Reagent and Reaction Conditions	Compound Class Formed	Key Electron-Rich Species	Key Electron-Poor Species	Comments	Discussed in Section
(1)	(H or) R O R' Ketone or aldehyde	1. Base⊖ 2. R''—X (X = Cl, Br, I)	(H or) R O R' R'' α-Alkylated ketone or aldehyde	(H or) R O ⊖ R' Enolate anion	δ+ R''—X Alkyl halide	S_N2	10.3

In general, you will find two of these tables at the end of a chapter. One contains **functional group transformations** or **functional group conversions**, which, as their name suggests, simply transform one functional group into another. They do so by leaving alone the **carbon skeleton** (i.e., the bonding arrangement of the carbon atoms).

The second of these tables contains reactions that do, indeed, bring about changes in the carbon skeleton. These reactions *require that carbon–carbon σ bonds be broken and/or formed*. (The breaking or formation of a carbon–carbon π bond alone is not considered to be among these reactions, because the σ bond between those atoms remains intact, thus preserving the carbon skeleton.)

YOUR TURN 13.5

To design a synthesis successfully, you must become very familiar with a variety of reactions. Therefore, take the time now to assess how familiar you are with the reactions we have learned so far, which appear in the reaction tables at the ends of Chapters 9–12 (Table 9-14, p. 494; Table 9-15, p. 495; Table 10-1, p. 555; Table 10-2, p. 557; Table 11-1, p. 595; Table 12-1, p. 633; and Table 12-2, p. 634).

(a) Go through the tables first to see if you can determine the typical reagents and reaction conditions for each reaction by examining just the reactants and products.

(b) Go through the reaction tables again to see if you can determine the reactants for each reaction by examining just the products and the typical reagents and reaction conditions.

If you encounter difficulty with a particular reaction, make sure to review the sections where the reaction is discussed and study the mechanism.

Equation 13-10 shows the various ways in which the two types of reactions are used in synthesis.

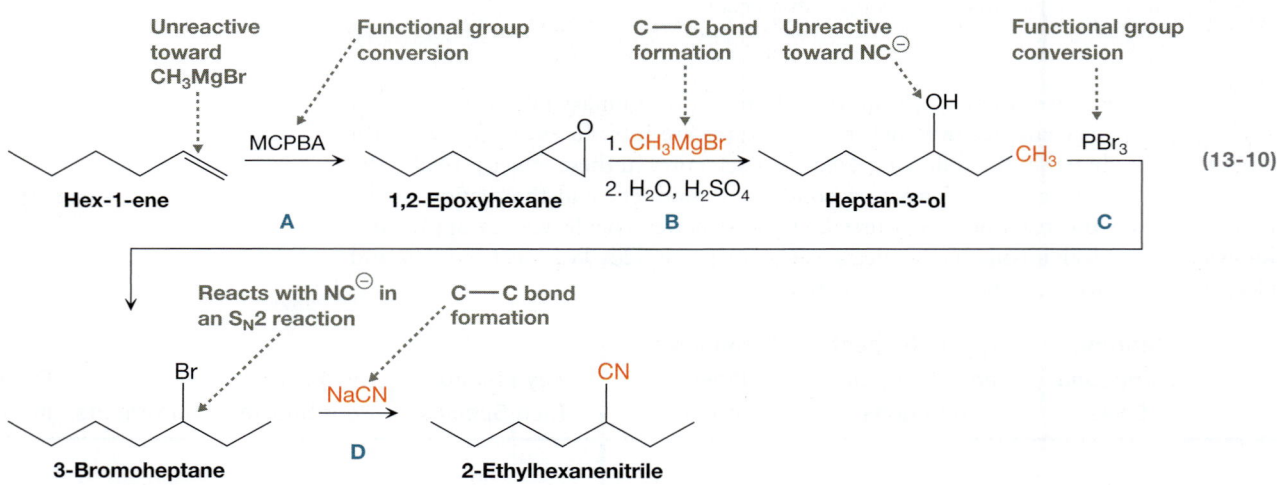

(13-10)

Synthetic steps **A** and **C** are functional group conversions, whereas **B** and **D** are carbon–carbon bond-forming reactions that alter the carbon skeleton.

Notice in Equation 13-10 that functional group conversions do not necessarily produce a functional group that appears in the target of the synthesis. In synthetic step **A**, for example, the epoxide that is produced sets up the Grignard reaction in the subsequent step. And the reaction in synthetic step **C** converts a poor HO^- leaving group into a good Br^- one, and therefore sets up the S_N2 reaction in **D**.

Although Equation 13-10 illustrates how reactions can be used to alter a carbon skeleton and to convert one functional group into another, our ability to use these schemes is quite limited right now because we have not yet learned many reactions. As you encounter more reactions in subsequent chapters, you will acquire a better feel for how these types of reactions can and should be used in synthesis.

Label each step in the following synthesis as either a functional group conversion or a reaction that alters the carbon skeleton.

PROBLEM 13.1 Supply the missing reagents for each step in Your Turn 13.6.

As the collection of reactions you have learned continues to grow, you should consider using Appendixes C and D to search for a reaction that you may need in a synthesis but does not come to mind immediately. These appendixes contain all the reactions presented in this book. Appendix C contains reactions that alter the carbon skeleton and Appendix D contains functional group transformations. Within Appendix D, you will find one table devoted to each functional group that you might be trying to produce. This organization should help you find that reaction more quickly and efficiently.

Use Appendixes C and D to find **(a)** synthetic step **B** and **(b)** synthetic step **C** shown in the synthesis scheme in Equation 13-10. For each reaction, identify the table number and the entry number.

13.3 Retrosynthetic Analysis: Thinking Backward to Go Forward

When presented with a synthesis problem, the natural tendency is to think in the *forward* direction, first deciding which compound(s) to use as the starting material, and then determining the specific sequence of reactions to carry out. Solving the problem in this way can be straightforward for syntheses that require only one or two synthetic steps, but it becomes impractical as the number of steps in an organic synthesis increases. In many cases, it is not obvious what the starting material should be, especially since the starting material does not need to resemble the target. Moreover, even if you do establish a viable starting compound, you will find that there are many dozens or even hundreds of different reactions that can be carried out for each synthetic step. Therefore, even for a three-step synthesis, there might be more than a million different imaginable schemes to consider!

The problem is greatly streamlined by a method called **retrosynthetic analysis**. The concept of retrosynthetic analysis was developed in the first half of the 20th century, but some of the most significant advances came in the 1960s when Elias J. Corey (b. 1928) of Harvard University applied it to construct complex synthesis schemes. This changed the face of organic synthesis and helped earn Professor Corey the 1990 Nobel Prize in Chemistry.

The basis of retrosynthetic analysis is the **transform**, which is the *proposed undoing* of a single reaction or set of reactions. To begin the design of a synthesis scheme, the chemist applies a transform to the target molecule to dissect it into smaller and/or simpler **precursors**, often without consideration of the starting material. An open arrow ($\Longrightarrow$), called a **retrosynthetic arrow**, is the convention used to indicate a transform, and is drawn from the target to the precursor. Therefore:

When considering a reaction or set of reactions that would produce a target **A** from a precursor **B**, a transform can be written as follows:	Target ⟍ Precursor ⟍ **A** $\Longrightarrow$ **B**

Think of the retrosynthetic arrow as meaning "can be made from." Thus, the above notation would be read as "**A** can be made from **B**." Also, notice that there are no reagents or conditions written above or below the retrosynthetic arrow. This allows us to focus on just the structural differences between the target and precursor.

Each precursor is then dissected in the same way. The chemist asks what reaction can be used to produce that precursor and then applies a transform to undo one of those reactions to arrive at another precursor. The cycle is repeated until the chemist arrives at precursors that are either readily available compounds or are easy to produce.

To see how retrosynthetic analysis can be used, let's design a synthesis of 1-methoxypent-2-yne, limiting ourselves to starting compounds that contain three or fewer carbons.

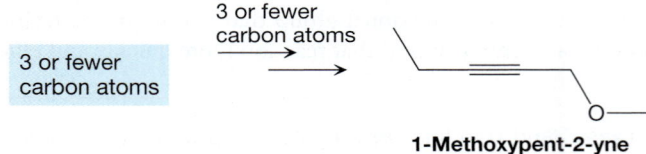

Because the target contains a continuous chain of five C atoms, and the compounds we may use as our starting material can contain at most three C atoms, the synthesis of 1-methoxypent-2-yne must contain at least one reaction that forms a C—C bond.

There are essentially nine reactions we have encountered so far that produce a new C—C bond, but only two leave a C≡C bond in the product, as shown in Table 13-1. (Entry 1 is taken from Table 9-15 and entry 2 is taken from Table 10-2.) We should therefore consider using one of these reactions.

Of these two reactions, the first (entry 1 of Table 13-1) is the better choice because the product of the second reaction contains an arrangement of functional groups that does *not* appear in our target molecule: an OH group separated by two carbons from the C≡C bond. We can therefore apply a transform to our target molecule that undoes the first reaction in the table, so that it *disconnects* the single bond between C3 and C4. This is indicated in Equation 13-11 by the wavy line.

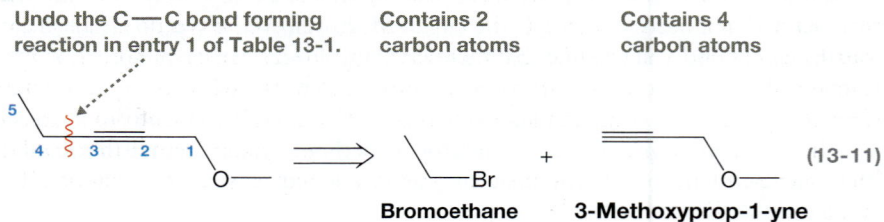

TABLE 13-1 Reactions That Alter the Carbon Skeleton and Leave a C≡C Bond

	Starting Compound Class	Typical Reagent and Reaction Conditions	Compound Class Formed	Key Electron-Rich Species	Key Electron-Poor Species	Comments	Discussed in Section
(1)	RC≡CH **Alkyne**	1. NaH 2. R'—X (X = Cl, Br, I)	RC≡C—R' **Alkyne**	RC≡C:$^{\ominus}$	δ^+ R'—X	S_N2	9.3b
(2)	R"C≡CH **Alkyne (terminal)**	1. NaH 2. 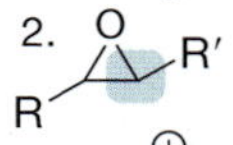 $\overset{O}{R\triangle R'}$ 3. $H_3O^{\oplus}$	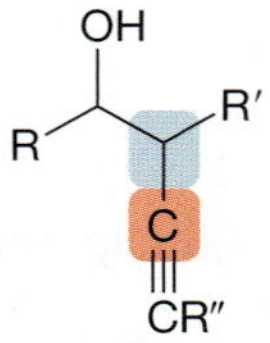 OH / R—\overset{}{\underset{\overset{|}{C}}{}}—R' / C≡CR" **Alcohol (3-Alkyn-1-ol)**	R"C≡C:$^{\ominus}$	$\underset{\delta^+}{R\overset{O}{\triangle}R'}$	S_N2	10.7a

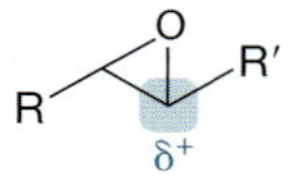

According to Table 13-1, the two necessary precursors for such a reaction are an alkyl halide and a terminal alkyne. In this case, the alkyl halide precursor contains two C atoms, and the other precursor, 3-methoxyprop-1-yne, contains the terminal alkyne.

YOUR TURN 13.8

Draw the appropriate precursors indicated by the transform shown here, similar to what was done in Equation 13-11.

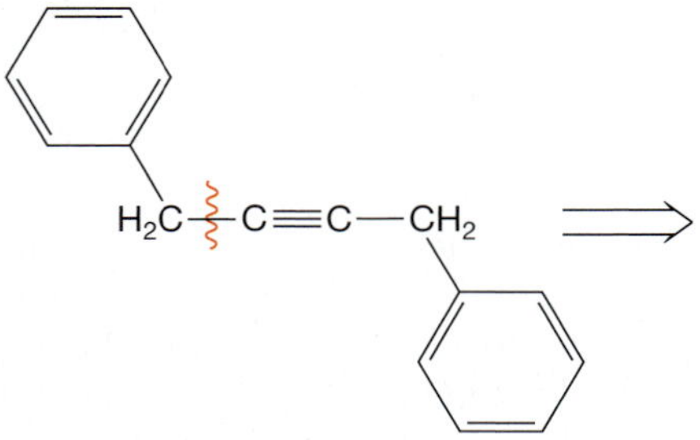

$$H_2C\overset{\xi}{-}C≡C—CH_2 \Longrightarrow$$

PROBLEM 13.2 Draw the appropriate precursors to the target shown here by applying a transform that undoes the reaction in entry 2 of Table 13-1. *Hint:* Which C≡C bond should be disconnected?

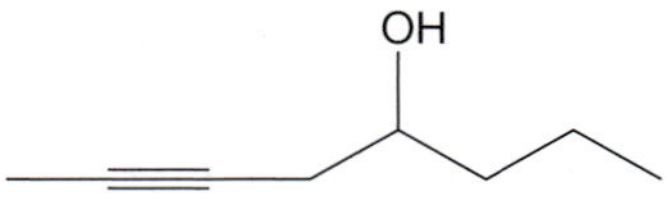

Of those two precursors, only bromoethane is acceptable for our starting material, because it contains three or fewer C atoms. 3-Methoxyprop-1-yne contains four C atoms, so it cannot be used as starting material. Instead, we must apply a transform to dissect it into smaller precursors. 3-Methoxyprop-1-yne contains a C—O—C group characteristic of ethers, so we can apply a transform that undoes an ether-forming reaction. According to Table 13-2 (taken from Table 10-1), one such reaction—the Williamson ether synthesis—requires an alkyl halide (R—X) and a sodium alkoxide (NaOR) as precursors.

TABLE 13-2 Functional Group Transformation

Starting Compound Class	Typical Reagent and Reaction Conditions	Compound Class Formed	Key Electron-Rich Species	Key Electron-Poor Species	Comments	Discussed in Section
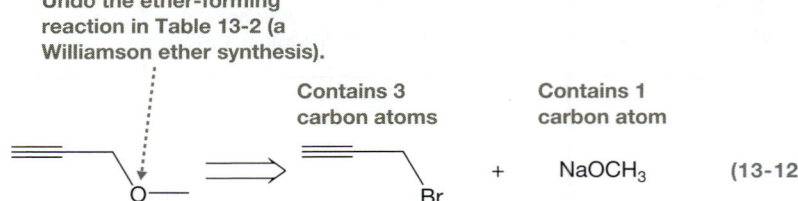 R—X (X = Cl, Br, I) **Alkyl halide**	NaOR′ →	R—O—R′ **Ether (symmetric or unsymmetric)**	R′O⁻	R—X δ^+	Williamson ether synthesis, S_N2	10.6

The resulting transform might appear as follows:

Undo the ether-forming reaction in Table 13-2 (a Williamson ether synthesis).

Contains 3 carbon atoms   **Contains 1 carbon atom**

NaOCH₃   (13-12)

Both of these precursors now contain three or fewer carbons and can be used as starting materials.

What remains to complete the synthesis is to reverse the transforms and to include the necessary reagents and conditions that will accomplish each reaction, as shown in Equation 13-13. In other words, write the synthesis in the *forward* direction.

NaOCH₃ →   1. NaH   2. CH₃CH₂Br →   (13-13)

Notice in each of the previous transforms that the reactions could have been undone in a different way from the one that was shown. For example, instead of disconnecting the C3—C4 bond to the left of the triple bond in the target (Equation 13-11), we could begin by disconnecting the C1—C2 bond to the right of the triple bond, as shown in Equation 13-14.

Undo the C—C bond-forming reaction in entry 1 of Table 13-1.   **Undo the C—C bond-forming reaction in entry 1 of Table 13-1.**

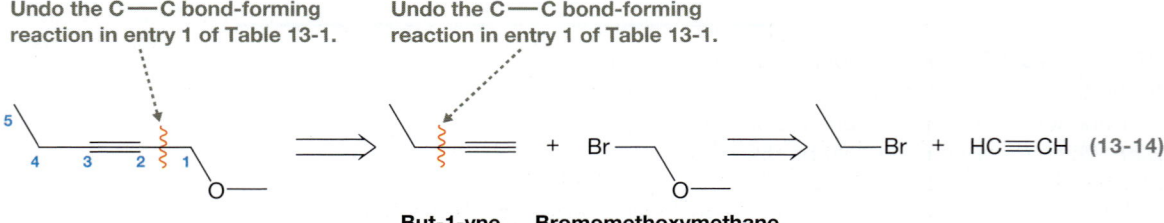

+ Br—   Br + HC≡CH   (13-14)

But-1-yne   **Bromomethoxymethane**

One of the precursors, bromomethoxymethane, contains two C atoms, so it is acceptable as starting material. The other precursor, but-1-yne, however, contains four C atoms, which exceeds the maximum of three, so another transform is applied. The C—C bond to the left of the triple bond is disconnected by undoing the reaction in entry 1 of Table 13-1, resulting in two precursors that each have two C atoms and are therefore acceptable as starting material.

The synthesis can then be written as shown in Equation 13-15.

$$HC\equiv CH \xrightarrow[\text{2. } CH_3CH_2Br]{\text{1. NaH}} \diagdown\!\!\!\!\!\!\equiv \xrightarrow[\text{2. } BrCH_2OCH_3]{\text{1. NaH}} \diagdown\!\!\!\!\!\!\equiv\!\!\!\diagup\!\!\diagdown_{O\diagdown} \quad \text{(13-15)}$$

The previous example illustrates an important point:

A synthesis can usually be designed in multiple different ways.

There are a variety of factors that can make one synthesis better than another, and we will examine some of these factors throughout the rest of our discussion of synthesis here in Chapter 13 and in Chapter 19.

YOUR TURN **13.9**

Show how to carry out the transform in Equation 13-12 by disconnecting the other O—C bond, and modify the synthesis in Equation 13-13 accordingly.

SOLVED PROBLEM 13.3

Propose a way to carry out the synthesis shown here.

Think Is there a reaction we have encountered that can transform an alcohol into a nitrile directly? If not, then what other precursor can be transformed into a nitrile? Can that precursor be made directly from an alcohol?

Solve There are no reactions we have learned that convert an alcohol into a nitrile directly. Notice, however, that a C—C single bond must be formed to carry out this synthesis. Only the S_N2 reaction shown here (taken from entry 2 in Table 9-15) accomplishes this while leaving a CN group in the product. We can thus undo this reaction, as shown in the first transform.

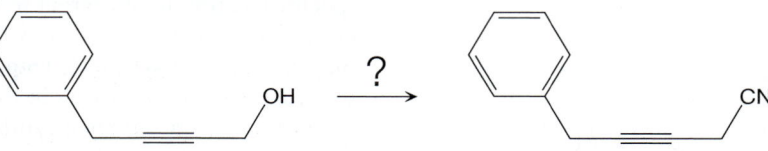

| Undo the S_N2 reaction in entry 2 of Table 9-15. | Undo the PBr₃ bromination in entry 1 of Table 10-1. |

This yields a primary alkyl halide as a precursor, which can be made directly from an alcohol using PBr₃ (entry 1 of Table 10-1), as shown in the second transform. To complete the synthesis, we reverse the transforms and apply the appropriate reagents.

PROBLEM 13.4 Propose a synthesis of pent-2-yne from compounds containing two or fewer carbons.

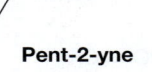

Pent-2-yne

PROBLEM 13.5 Propose a synthesis of 2-benzyl-3-phenylpropanal from (bromomethyl)benzene (benzyl bromide), using any reagents containing two or fewer carbon atoms.

**(Bromomethyl)benzene
(Benzyl bromide)**   ?   **2-Benzyl-3-phenylpropanal**

CONNECTIONS

(Bromomethyl)benzene (Problem 13.5) is a strong irritant and lachrymator (induces tears) and has been used in chemical warfare.

13.4 Synthetic Traps

When we undo a particular reaction by applying a transform, we tend to focus on the specific location in our target molecule where we want those changes to occur. However, if we restrict ourselves in these ways while considering various transforms, we may encounter a **synthetic trap**, in which some factor we have overlooked prevents the reaction from occurring in the *forward* direction as planned.

When we encounter a synthetic trap, we must alter our synthesis in some way. Sometimes a synthetic trap can be circumvented with only minor changes to the synthesis scheme, perhaps simply by reordering the steps. At other times, the synthesis might require a specialized reaction, many of which we will learn in the chapters to come. Occasionally, however, there will be no apparent way around a synthetic trap, forcing us to search for an altogether different synthetic route. As you gain more experience with organic synthesis, and as you learn more reactions in subsequent chapters, you will be better equipped to make these kinds of decisions.

13.4a Reactants with More than One Reactive Functional Group

Among the most conspicuous synthetic traps is one in which there are two or more functional groups on a precursor that can react in a particular synthetic step. Consider, for example, the transform proposed in Equation 13-16, which undoes a C—C bond-forming reaction from entry 1 of Table 13-1.

Undo the C—C bond-forming
reaction in entry 1 of Table 13-1.

Hex-4-yn-1-ol   **Propyne** + **3-Bromopropan-1-ol**   (13-16)

In the forward direction, this synthetic step would appear as follows:

1. NaH
2. Br⌒⌒OH   (13-17)

When NaH is added, the terminal alkyne is deprotonated to generate the alkynide anion, $CH_3C\equiv C^-$. A problem arises in the second of these sequential reactions, however, because the alkynide anion is *intended* to behave as a nucleophile and displace the Br leaving group, as shown in Equation 13-18a. Unfortunately, it can also behave as a strong base, deprotonating the OH group, as shown in Equation 13-18b. Because proton transfer reactions tend to be much faster than nucleophilic substitution reactions (Section 8.6a), the reaction in Equation 13-18b dominates, and the synthesis does not proceed as intended.

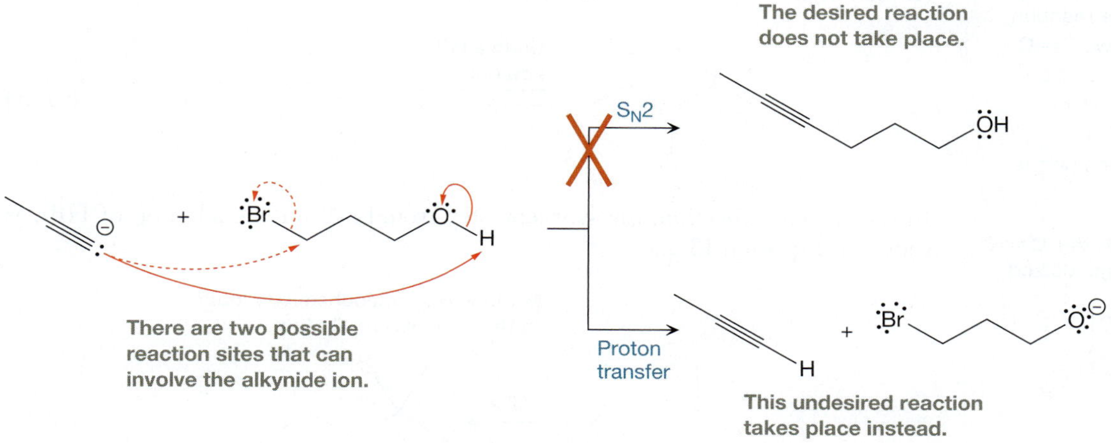

The desired reaction does not take place.

(13-18a)

There are two possible reaction sites that can involve the alkynide ion.

Proton transfer

This undesired reaction takes place instead.

(13-18b)

One way to circumvent this kind of problem involves what are called *protecting groups*. In this case, the OH group can be protected to prevent the proton transfer reaction from occurring, leaving the S_N2 reaction as the only possibility. We discuss the use of protecting groups in synthesis in greater detail in Chapter 19. For now, we simply focus on identifying synthetic traps so that we can avoid them.

PROBLEM 13.6 For each of the following proposed transforms, determine whether it leads to a synthetic trap (i.e., it will not proceed as planned in the forward direction). Explain your reasoning.

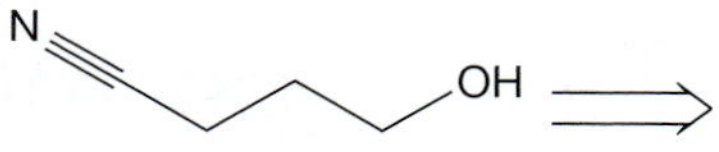

(a)

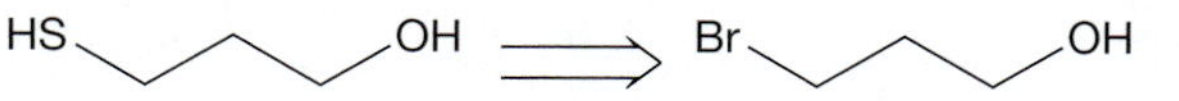

(b)

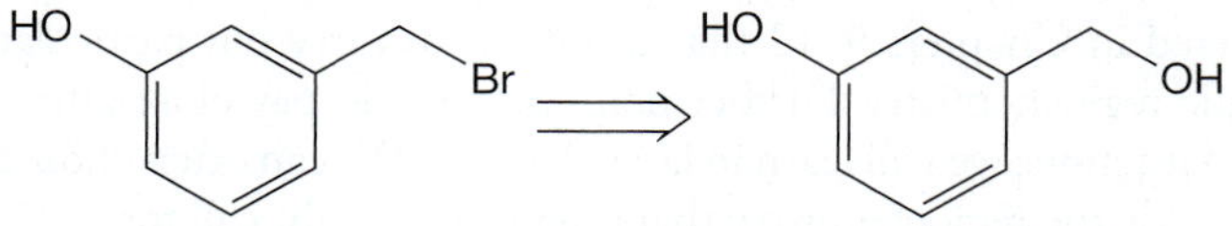

(c)

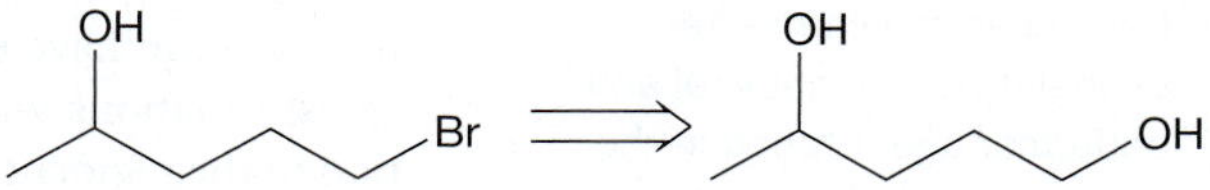

(d)

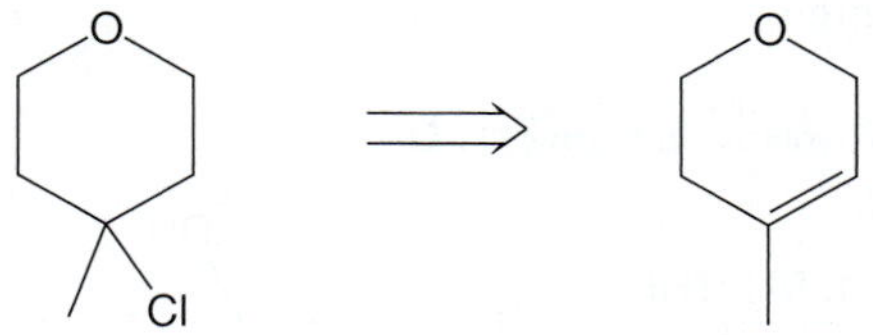

(e)

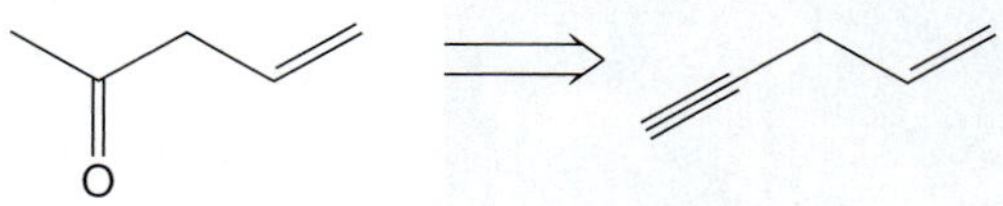

(f)

13.4b Synthetic Traps and Regiochemistry

CONNECTIONS To deal with the problem of multiple reactive functional groups in biochemical reactions, specialized enzymes are often recruited to carry out the necessary reaction with high selectivity. In the reaction below, for example, two C=C groups (highlighted) could be epoxidized, but the enzyme chloroperoxidase (CPO) epoxidizes just the one on the right.

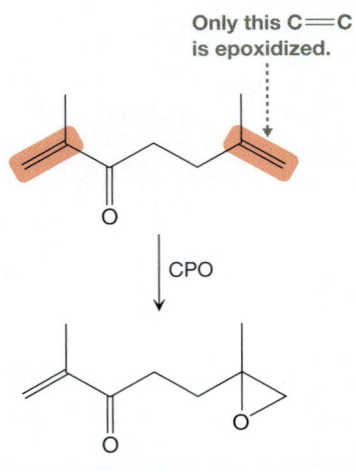

Only this C=C is epoxidized.

CPO

In Section 13.4a, we learned that a synthetic trap can arise when a reactant has two or more reactive functional groups. Even when there is only one reactive functional group, however, a synthetic trap can still arise if a reaction can take place at two or more sites *within* that functional group—that is, if *regiochemistry* is an issue. Thus, running such a reaction in the forward direction might result in a different product than we envisioned when we undid the reaction in the transform.

Consider, for example, a transform that undoes the hydrobromination of an alkene (Equation 13-19), which we learned in Section 11.1:

Undo a HBr addition.

(13-19)

Br

In the forward direction, the synthetic step would call for the addition of HBr, as shown in Equation 13-20.

Markovnikov regiochemistry leads to Br on the wrong C atom.

HBr

Br

(13-20)

However, the major product of this reaction is not the desired one because, as we saw in Section 11.3, the reaction would proceed through the more stable carbocation (tertiary in this case), resulting in Markovnikov regiochemistry.

One way to solve this problem is to consider a different reaction to obtain the desired regiochemistry. We have not yet encountered a reaction that directly adds HBr to an alkene with anti-Markovnikov regiochemistry, but we learned in Section 12.6 that hydroboration–oxidation results in the anti-Markovnikov addition of H_2O. As shown in Equation 13-21, the alcohol product from this reaction can then be converted to the bromide using PBr_3, resulting in the desired product from Equation 13-19.

Anti-Markovnikov addition of H_2O   **OH is converted to Br.**

1. BH_3·THF
2. H_2O_2, NaOH, H_2O

OH

PBr_3

Br

(13-21)

CONNECTIONS
Methylenecyclopentane has been used to synthesize cephalotoxin, an antiviral and antitumor agent found in the salivary gland of the common octopus.

Because of the importance of regiochemistry in synthesis, you should review the reactions we have encountered in Chapters 9–12 that enable us to carry out particular transformations with specific regiochemistry. Furthermore, you should pay close attention to the regiochemistry of reactions we will learn in later chapters. We can often choose one of these reactions to provide the regiochemistry that a synthesis might call for.

For example, Equations 13-22 through 13-24 are three different reactions that add H_2O across a C=C double bond.

Anti-Markovnikov addition of H_2O

1. BH_3·THF
2. H_2O_2, NaOH, H_2O

OH

(13-22)

Methylenecyclopentane

Markovnikov addition of H₂O

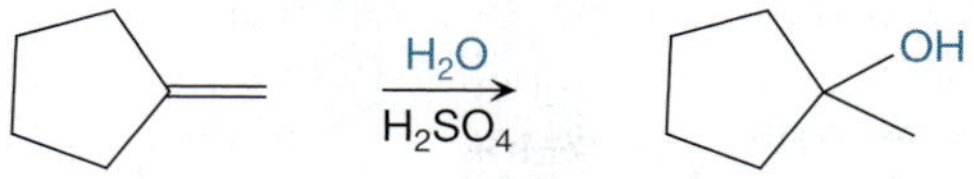

$$\text{(13-23)}$$

Markovnikov addition of H₂O

1. Hg(OAc)₂,
   H₂O
2. NaBH₄

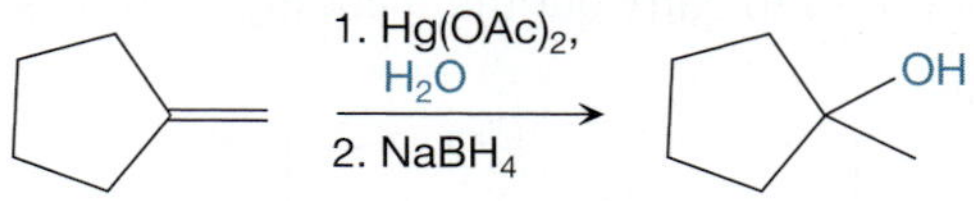

$$\text{(13-24)}$$

Equation 13-22 is the hydroboration–oxidation reaction we just reviewed, which takes place with anti-Markovnikov regiochemistry. Equations 13-23 and 13-24, on the other hand, take place with Markovnikov regiochemistry.

SOLVED PROBLEM 13.7

Show how to carry out the synthesis at the right.

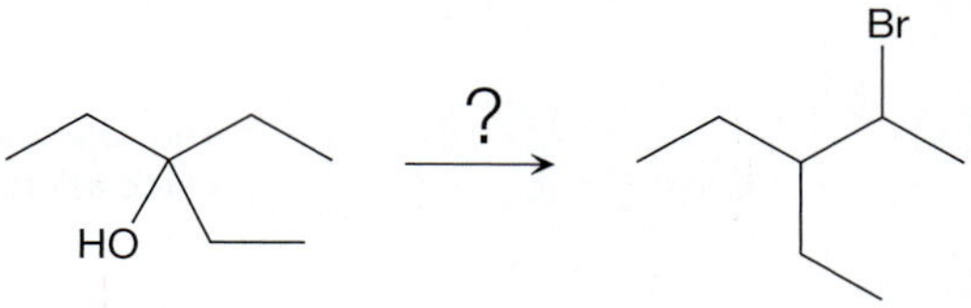

Think What precursor can be used to produce the alkyl bromide target? Is regiochemistry a concern? If so, can the alkyl bromide be produced directly with the desired regiochemistry? If not, what other reaction can be carried out to give the desired regiochemistry?

Solve As shown in the transform below, the alkyl bromide target can be produced from a hydrobromination of an alkene.

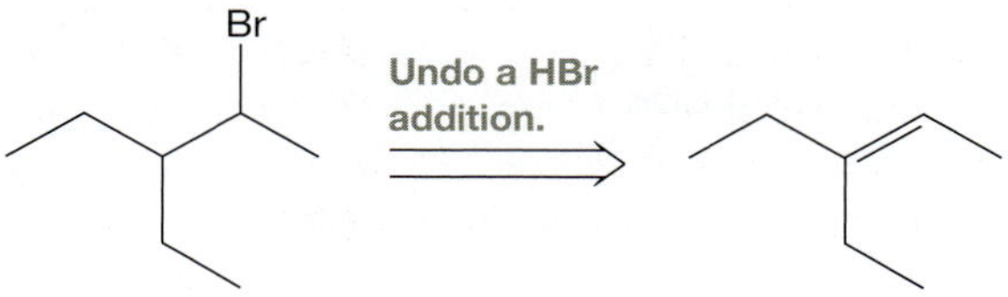

Undo a HBr
addition.

However, the HBr addition in the forward direction has Markovnikov regiochemistry, which would produce an undesired product.

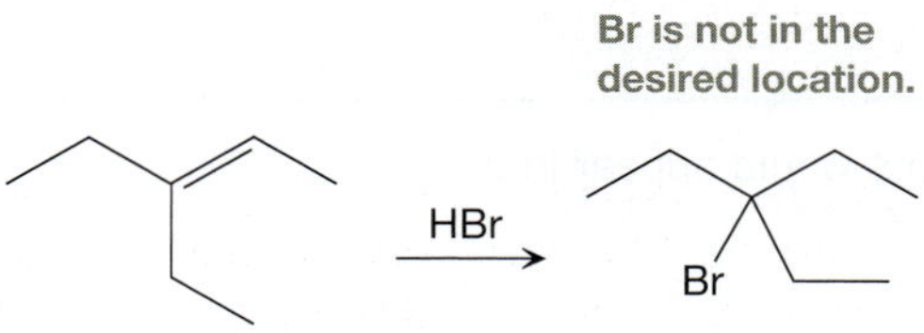

Br is not in the
desired location.

HBr

Instead, the alkyl bromide target can be produced from the corresponding alcohol, which can be produced from an alkene, in turn, by an anti-Markovnikov addition of H₂O. The alkene can be produced from the starting compound by dehydration.

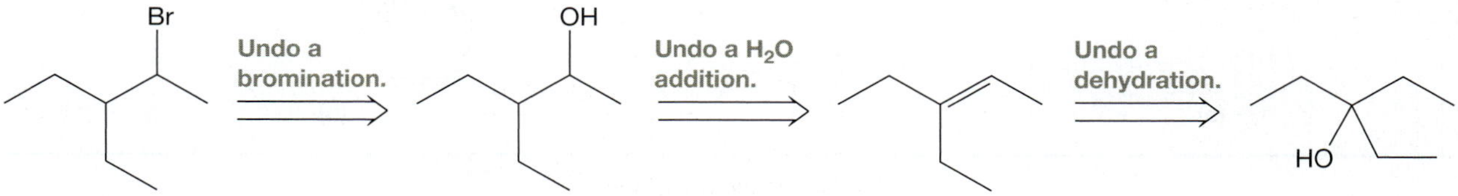

Undo a
bromination.

Undo a H₂O
addition.

Undo a
dehydration.

The synthesis in the forward direction would then be written as follows:

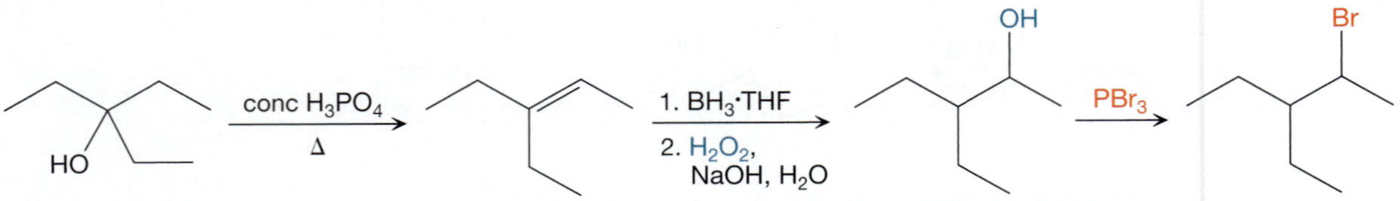

PROBLEM 13.8 Show how to carry out the following synthesis.

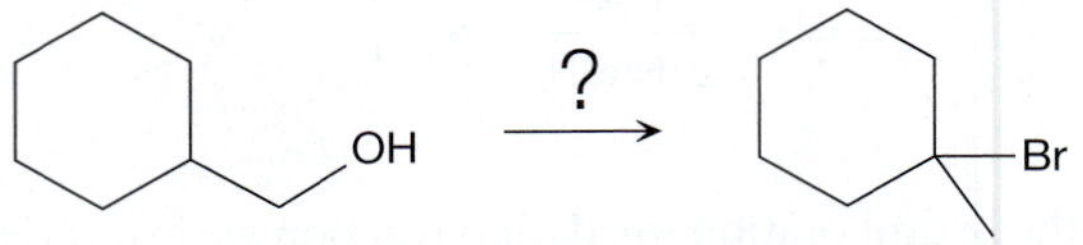

The alkylation of an α carbon (Section 10.3) is another example of a reaction whose regiochemistry we can control, as shown in Equation 13-25.

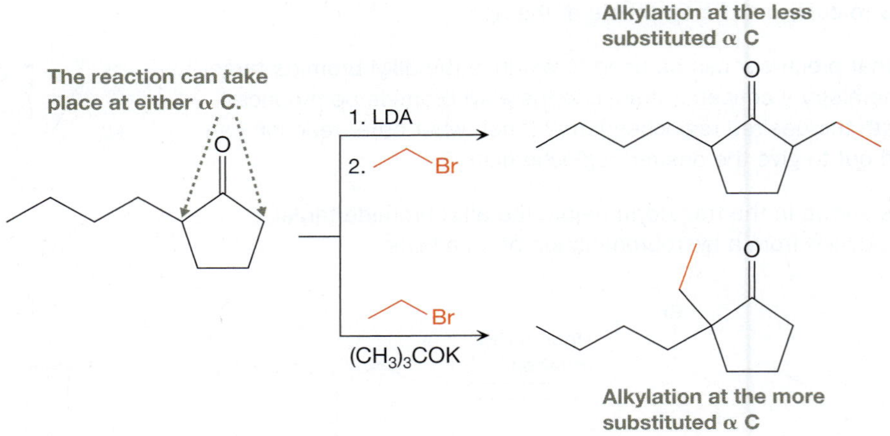

In this case, the choice of base can dictate which α carbon undergoes reaction. A very strong base such as LDA leads to alkylation at the less substituted α C (Equation 13-25a), whereas a moderately strong base such as the *tert*-butoxide anion leads to alkylation at the more highly substituted α C (Equation 13-25b).

PROBLEM 13.9 Show how each of the following transforms would appear in a synthesis.

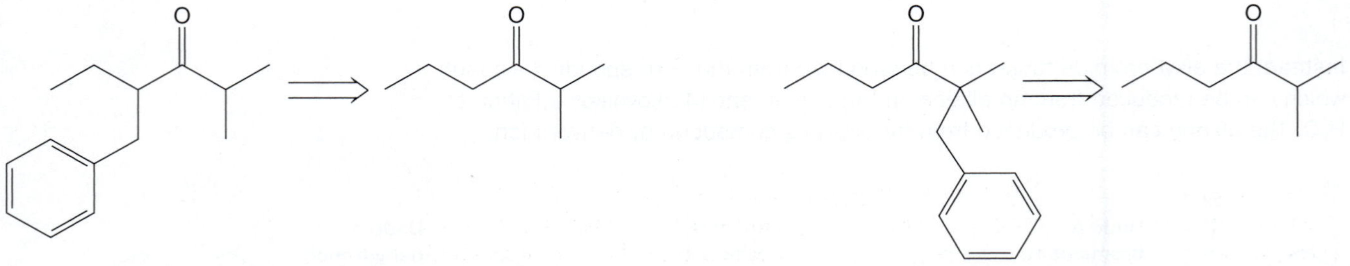

(a)

(b)

Epoxide ring opening (Equation 13-26) is another reaction whose regiochemistry can be controlled.

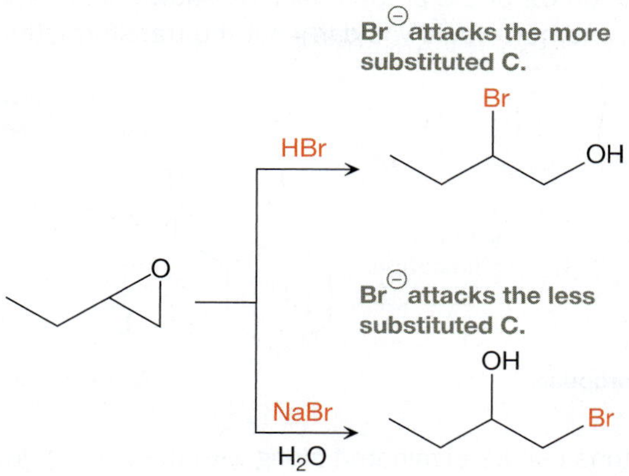

As we learned in Section 10.7, the nucleophile tends to attack the more highly substituted C atom under acidic conditions (Equation 13-26a), and tends to attack the less highly substituted C under neutral or basic conditions (Equation 13-26b).

PROBLEM 13.10 Show how each of the following transforms would appear in a synthesis.

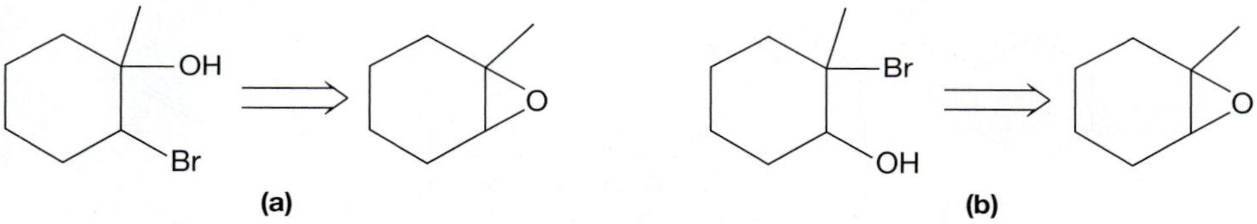

Regiochemistry can also be controlled in elimination reactions, as shown in Equations 13-27 and 13-28.

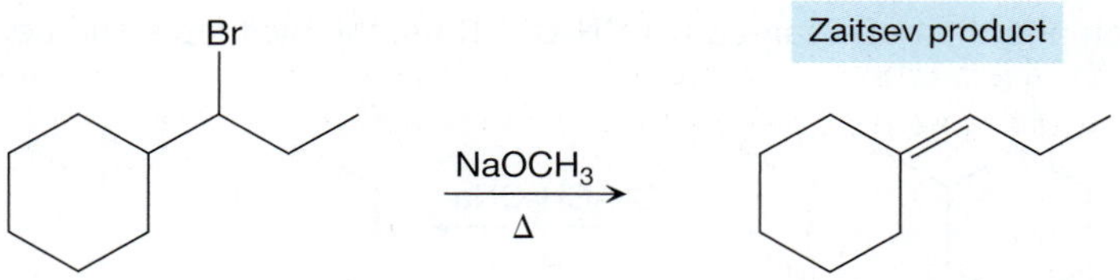

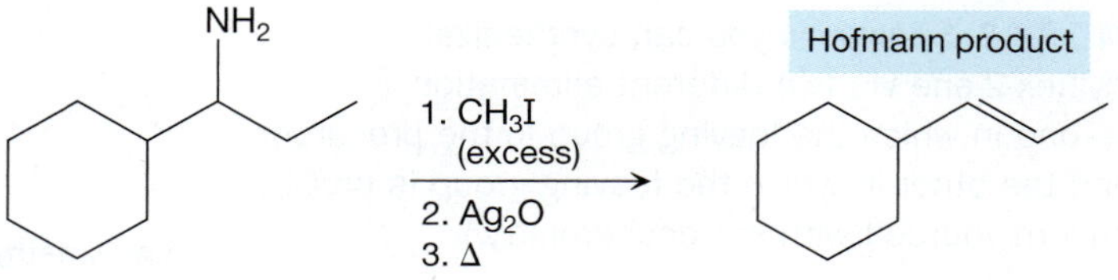

As we learned in Section 9.10, the elimination reaction in Equation 13-27 tends to produce the most highly substituted alkene product, or the Zaitsev product, because it is the most stable. On the other hand, as we saw in Section 10.9, the Hofmann elimination reaction in Equation 13-28 tends to produce the less substituted alkene product.

SOLVED PROBLEM 13.11

The alkene in Equation 13-28 can also be synthesized from a precursor in which the leaving group is on C2 of the propyl chain, as shown here. What leaving group would you choose, and how would you carry out the transformation?

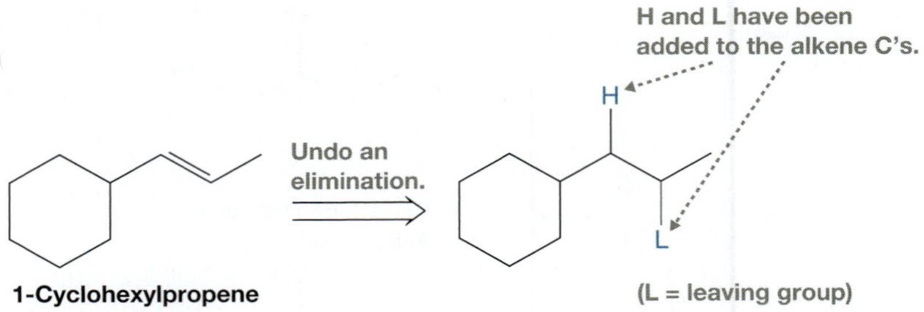

H and L have been added to the alkene C's.

Undo an elimination.

1-Cyclohexylpropene

(L = leaving group)

Think Which protons can be eliminated along with the leaving group? What would be the elimination product in each case? Which is the Zaitsev product, which is the Hofmann product, and which corresponds to the target?

Solve The protons that can be eliminated are on C1 and C3 of the propyl chain.

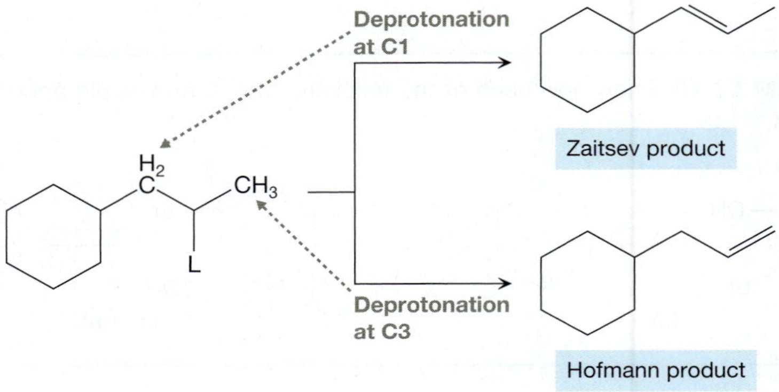

Deprotonation at C1

Zaitsev product

Deprotonation at C3

Hofmann product

Deprotonation at C1 gives the desired product, which, in this case, is the Zaitsev product. This deprotonation can be achieved by choosing a good leaving group and a strong base, neither of which should be bulky. The leaving group can be a halide anion and the base can be $CH_3CH_2O^-$. Thus, the synthesis can be written as follows:

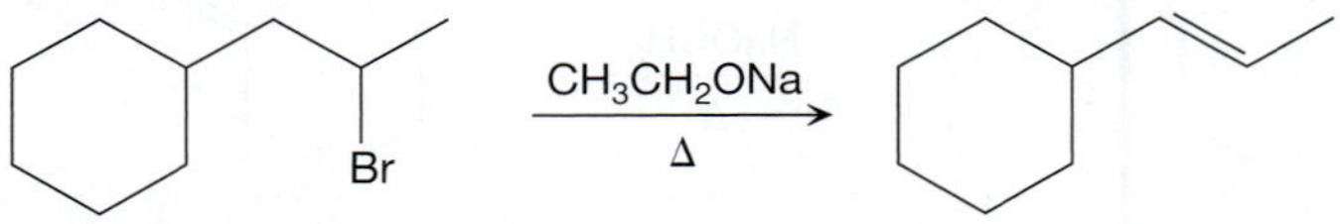

CH_3CH_2ONa

Δ

PROBLEM 13.12 Show how you can synthesize 3,4-dimethylhex-2-ene via two different elimination reactions—one in which the leaving group in the precursor is on C2 and the other in which the leaving group is on C3. (Do not concern yourself with stereochemistry.)

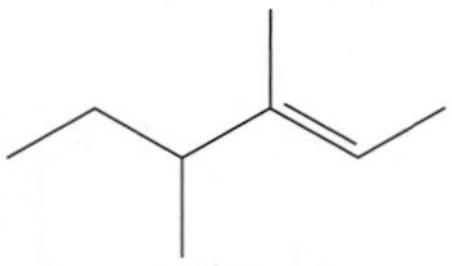

3,4-Dimethylhex-2-ene

Finally, in the electrophilic addition to a conjugated diene (Section 11.9), regiochemistry can be governed by whether the conditions lead to kinetic or thermodynamic control of the reaction. For example, HCl adds to buta-1,3-diene in a 1,2 fashion at cold temperatures (Equation 13-29), but in a 1,4 fashion at warm temperatures (Equation 13-30).

1,2-Addition

$$\text{Buta-1,3-diene} \xrightarrow[\text{Cold } T]{\text{HCl}}$$

(13-29)

1,4-Addition

$$\xrightarrow[\text{Warm } T]{\text{HCl}}$$

(13-30)

Manipulating Atoms One at a Time: Single-Molecule Engineering

As we have seen here in Chapter 13, the basic strategy of organic synthesis is to use known reactions in precise sequences to construct a particular target. Despite its tremendous success, traditional organic synthesis is limited by the specific structural changes that reactions can bring about. Indeed, these limits were the basis for our discussion on *synthetic traps* (Section 13.4). Are there ways around these limitations? Perhaps.

In 1989, Don Eigler and Erhard Schweizer, working for IBM, used scanning tunneling microscopy (STM) to move 35 xenon atoms on a nickel surface to spell out the company's acronym. STM works by locating a very small metal tip narrowly above a sample surface (typically within about 1 nm) and passing a voltage to the tip. That voltage can be used to control the interaction between the STM tip and an individual atom on the sample surface, thereby giving the STM the ability to move atoms.

Though making the IBM acronym out of atoms was a novelty, scientists see the technology as having tremendous potential for creating nanostructures that have valuable functions. In recent years, the technology has even been applied toward organic molecules. In early 2016, for example, Leo Gross and coworkers, working at IBM Research–Switzerland, reported using single-atom manipulation to carry out the following transformation. After depositing the initial dibromo compound on a copper surface, they applied 1.6 V to the STM tip to remove the first Br atom, and 3.3 V to remove the second one. Then, applying 1.7 V, they carried out a retro-Bergman cyclization to produce the final product.

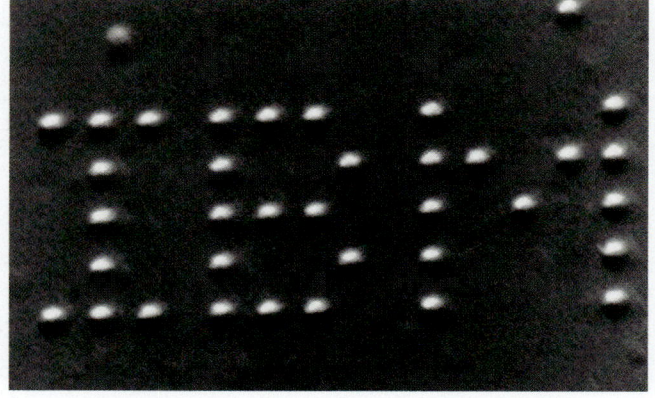

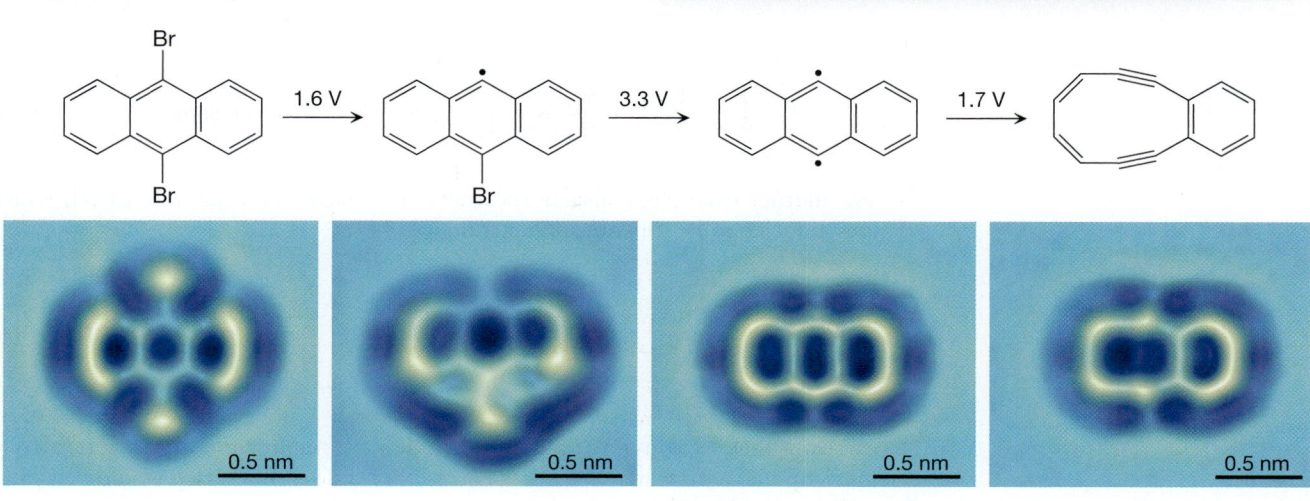

PROBLEM 13.13 Show how to produce each of the following compounds beginning with buta-1,3-diene. (Do not concern yourself with stereochemistry.)

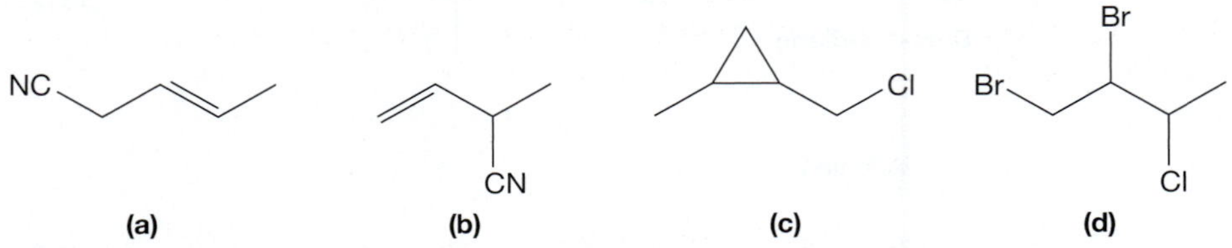

(a) (b) (c) (d)

13.5 Choice of the Solvent

With a variety of factors to consider in devising a synthesis, it is easy to overlook the importance of the solvent. However, the nature of the solvent can have a dramatic effect on the outcome of a reaction. For example, consider the transform shown in Equation 13-31, which undoes the bromination of an alkene.

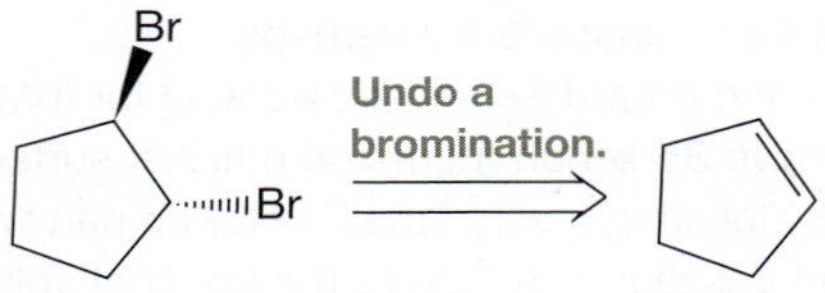

Undo a bromination. (13-31)

To carry out the reaction in the forward direction, Br_2 must add across the double bond as discussed in Section 12.3a. If we choose water as the solvent, however, then the unwanted bromohydrin shown in Equation 13-32 would be produced, as we learned in Section 12.3b.

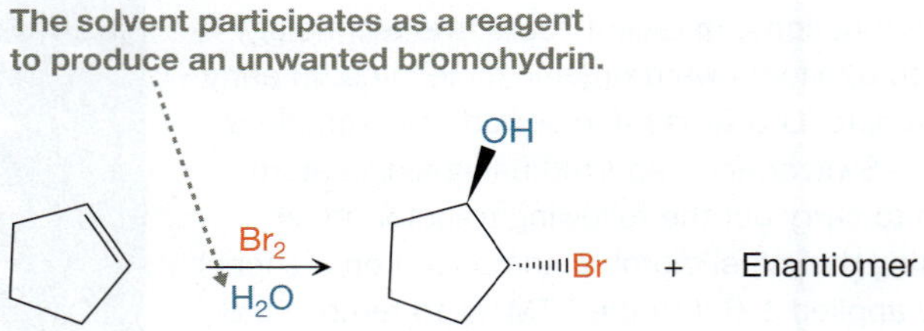

The solvent participates as a reagent to produce an unwanted bromohydrin.

Br_2 / H_2O + Enantiomer (13-32)

To produce the desired 1,2-dibromide instead, the solvent must be non-nucleophilic, such as CCl_4. This is shown in Equation 13-33.

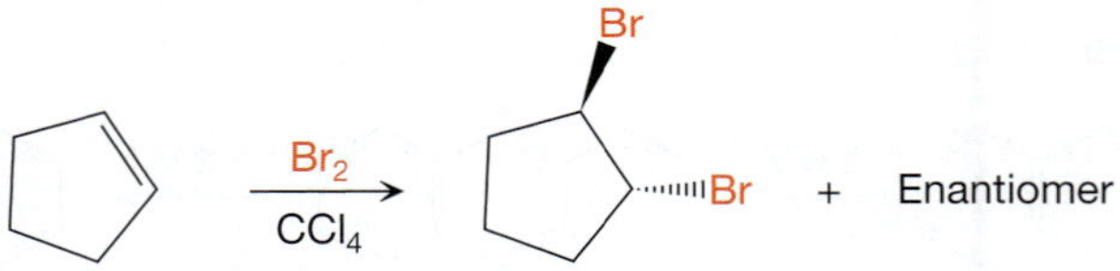

Br_2 / CCl_4 + Enantiomer (13-33)

As another example, consider the transform shown in Equation 13-34, which undoes a nucleophilic substitution reaction.

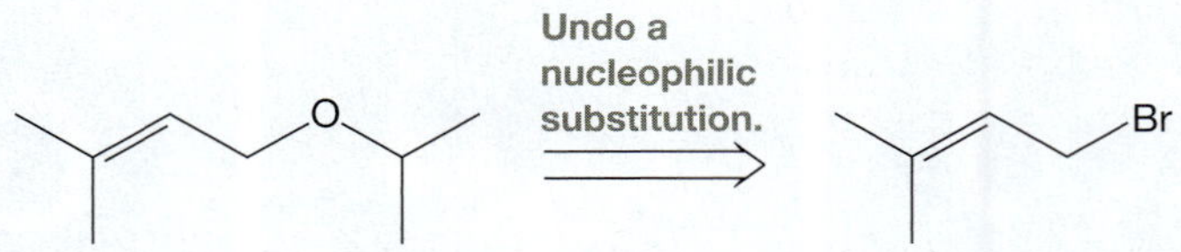

Undo a nucleophilic substitution. (13-34)

To carry out the reaction in the forward direction, we might choose $^-OCH(CH_3)_2$ as the nucleophile, with water as the solvent, so that the synthetic step would be as follows (Equation 13-35):

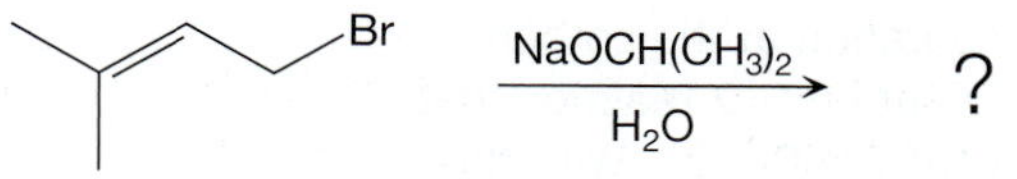

$$\text{(13-35)}$$

Keep in mind, however, that water has a weakly acidic hydrogen that can be deprotonated by $^-\text{OCH(CH}_3)_2$, according to Equation 13-36.

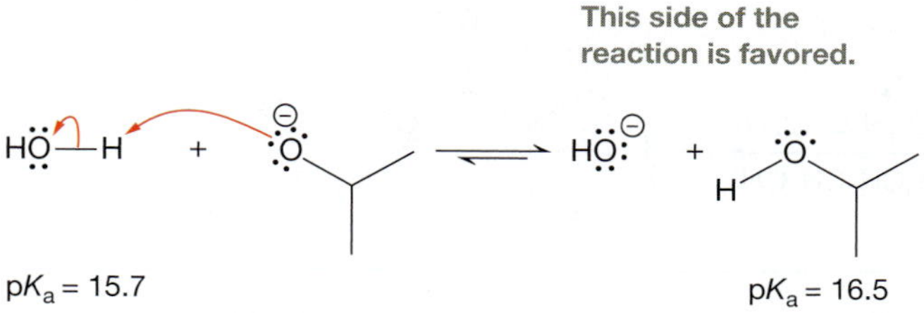

This side of the reaction is favored.

$$\text{(13-36)}$$

$pK_a = 15.7$ $pK_a = 16.5$

This proton transfer reaction effectively removes $^-\text{OCH(CH}_3)_2$ and replaces it with HO^-, so the *actual* nucleophile is primarily HO^-. Thus, even though the *intended* product is an ether, the major product of the reaction would be an alcohol, as shown in Equation 13-37.

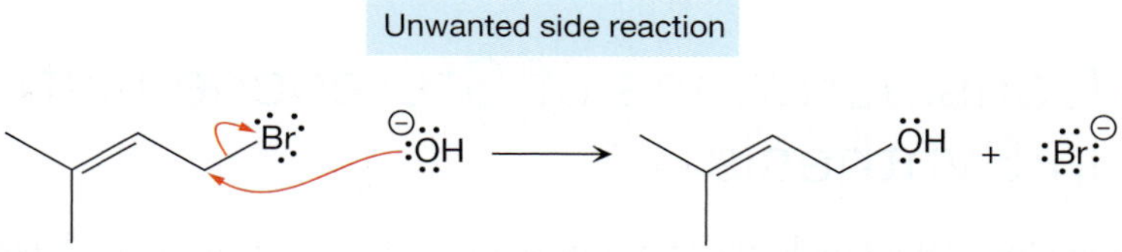

Unwanted side reaction

$$\text{(13-37)}$$

One way to solve this problem is to choose an aprotic solvent whose protons are not acidic enough to be deprotonated, such as dimethyl sulfoxide (DMSO) or *N,N*-dimethylformamide (DMF). A second way is to use a solvent whose conjugate base is the same as the intended nucleophile. In the example at hand, an appropriate solvent would be propan-2-ol, $(\text{CH}_3)_2\text{CHOH}$. The solvent would still be deprotonated by the nucleophile (Equation 13-38), but this would not result in any new species.

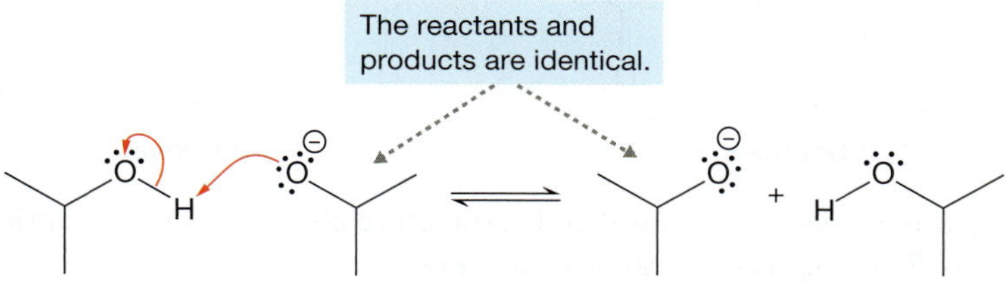

The reactants and products are identical.

$$\text{(13-38)}$$

YOUR TURN 13.10

If $\text{CH}_3\text{CH}_2\text{O}^-$ is to be used as a nucleophile, then what alcohol would be an appropriate solvent?

SOLVED PROBLEM 13.14

The transform shown here undoes a nucleophilic substitution reaction. How would you carry out the reaction in the forward direction using a *protic* solvent?

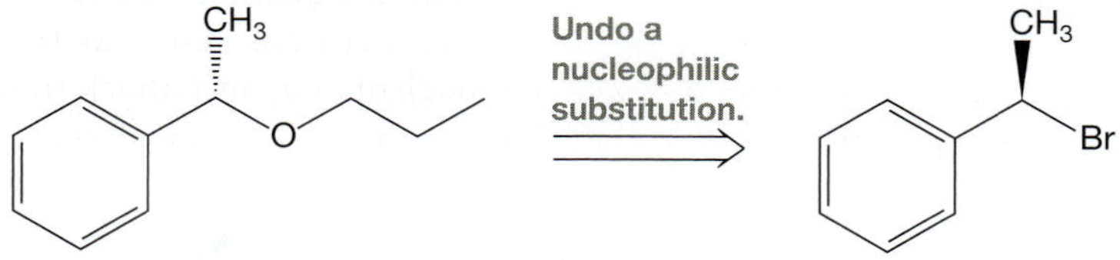

Undo a nucleophilic substitution.

Think Should the reaction be S_N1 or S_N2? To ensure that this reaction takes place, should you use a strong nucleophile or a weak nucleophile? What protic solvent could be used that would not react with that nucleophile to form a new nucleophile?

Solve Because there is inversion of stereochemical configuration at the carbon atom where substitution must take place, we want to carry out an S_N2 reaction, not an S_N1. To do so, we should choose a strong nucleophile, $CH_3CH_2CH_2O^-$. With this nucleophile, the proper choice of protic solvent would be $CH_3CH_2CH_2OH$, because deprotonating it would produce $CH_3CH_2CH_2O^-$, our desired nucleophile. Therefore, the synthetic step would be written as follows:

$$\text{(structure with } CH_3, Br) \xrightarrow[CH_3CH_2CH_2OH]{NaOCH_2CH_2CH_3} \text{(structure with } CH_3, O\text{-propyl)}$$

PROBLEM 13.15 The transform shown here undoes a nucleophilic substitution reaction. What nucleophile would you use to carry out this reaction in the forward direction? What *protic* solvent would you use?

Undo a nucleophilic substitution.

13.6 Considerations of Stereochemistry in Synthesis

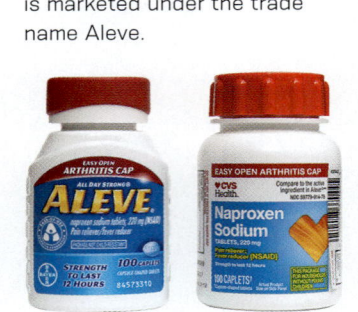

It is common for a target molecule to be one of multiple stereoisomers, each with different behavior. This is particularly true with biologically active compounds such as *naproxen*, a drug that is used for pain relief and the reduction of fever and inflammation.

Synthetic drug that relieves pain, fever, and inflammation

Causes liver damage

(S)-Naproxen

(R)-Naproxen

Naproxen has one asymmetric atom and is therefore chiral—its two enantiomers are designated as *R* and *S*. Whereas the *S* enantiomer is relatively safe to use, the *R* enantiomer causes liver damage! It would therefore be extremely useful to devise a synthesis in which the *desired S* stereoisomer was predominantly produced. Syntheses that accomplish this are called **asymmetric syntheses**, and they constitute the subject of a major area of ongoing research in organic synthesis.

A key to asymmetric synthesis is using a reaction that favors one type of product stereochemistry at the stage at which it is possible to generate more than one stereoisomer. Recall that *stereospecific* reactions accomplish this by producing one type of configuration exclusively, which depends on the reactant configuration. The S_N2 reaction in Equation 13-39, for example, is stereospecific. As we learned in Section 8.5b, S_N2 reactions result in the inversion of stereochemical configuration because the nucleophile must attack from the side opposite the leaving group.

An S_N2 reaction reverses the configuration.

$$\text{(cyclopentane with Br)} \xrightarrow{NaN_3} \text{(cyclopentane with } N_3)$$

(13-39)

A particularly useful stereospecific reaction is the conversion of an alcohol to an alkyl halide using PBr_3 or PCl_3, an example of which is shown in Equation 13-40.

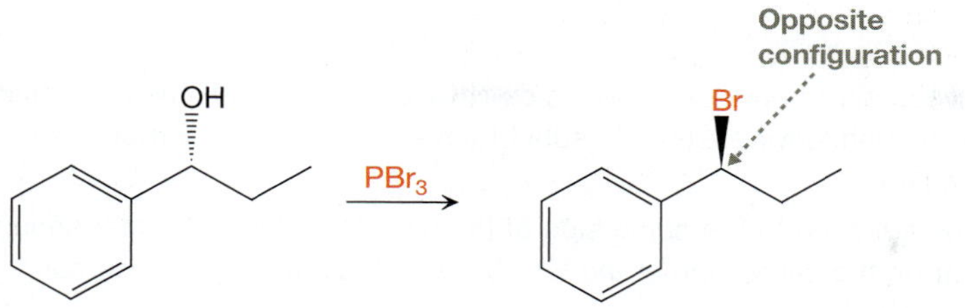

Opposite configuration

(13-40)

As we learned in Section 10.1, the stereospecificity of this reaction is accounted for by the mechanism, which consists of back-to-back S_N2 steps. Only the second of those S_N2 steps involves the C atom initially bonded to the OH group, so the reaction leads to inversion of configuration at that C.

To successfully implement an S_N2 reaction in an organic synthesis, we must take stereochemistry into account. It is especially helpful to do so in the planning stages, as part of a retrosynthetic analysis. Suppose, for example, that we want to synthesize (*S*)-2-ethylpentanenitrile via an S_N2 reaction. Because S_N2 reactions cause inversion of configuration, the precursor should have the opposite configuration at the atom bonded to the leaving group. In this case, the stereospecific transform might be written as shown in Equation 13-41.

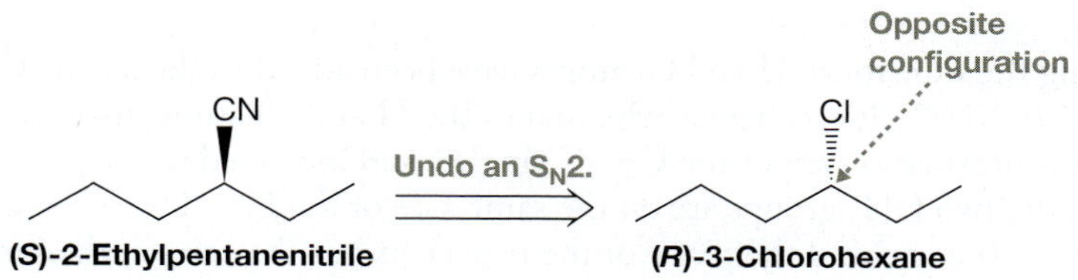

Opposite configuration

Undo an S_N2.

(13-41)

(S)-2-Ethylpentanenitrile

(R)-3-Chlorohexane

PROBLEM 13.16 The transform shown here, which is intended to undo an S_N2 reaction, does not take stereochemistry into consideration. Draw the precursor with the appropriate stereochemical configuration.

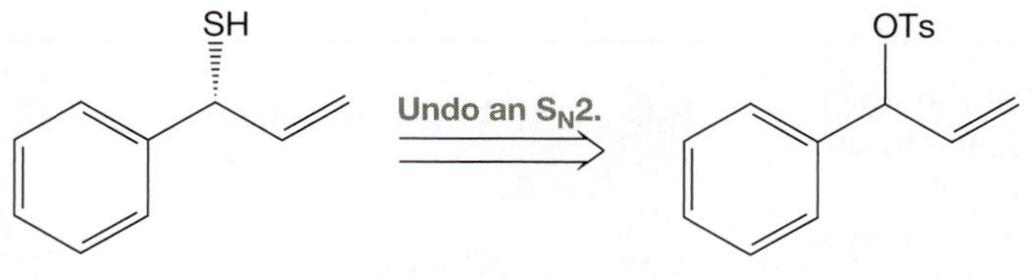

Undo an S_N2.

An E2 reaction is also stereospecific. As we learned in Section 8.5d, an E2 reaction is favored by the anticoplanar conformation of the substrate, in which a hydrogen atom and a leaving group on adjacent carbon atoms are anti to each other (Equation 13-42):

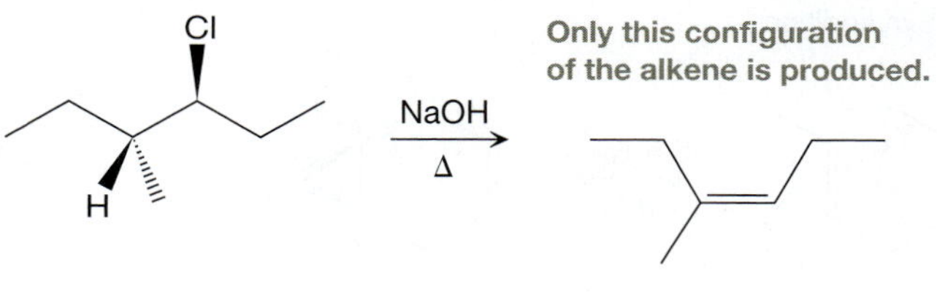

Only this configuration of the alkene is produced.

NaOH
Δ

(13-42)

To take stereochemistry into account in a transform that undoes an E2 reaction, the hydrogen atom and the leaving group must be anti to each other before they are eliminated from the substrate. Thus:

- The precursor of an E2 reaction is constructed from an alkene by adding the hydrogen and leaving group (L) substituents to the alkene carbon atoms in an anti fashion.
- Groups attached to the same side of the C=C bond in the target should appear on the same side of the H—C—C—L plane in the precursor.

An example of such a transform is shown in Equation 13-43, in which (E)-3-methyl-pent-2-ene is the target.

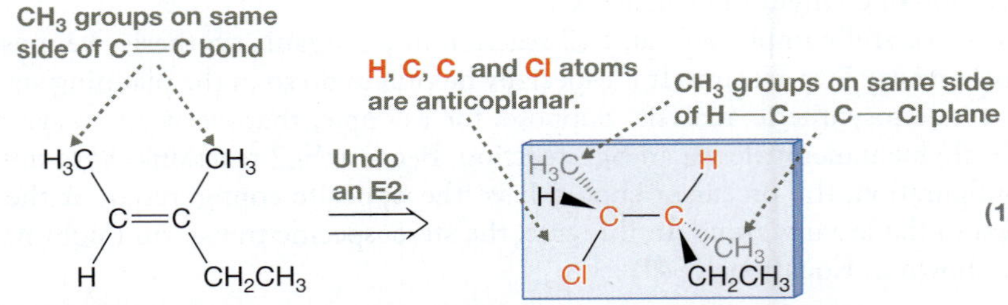

(13-43)

(E)-3-Methylpent-2-ene   **(2R,3R)-2-Chloro-3-methylpentane**

In applying the transform, H and Cl atoms have been added to the alkene C atoms to arrive at (2R,3R)-2-chloro-3-methylpentane. The H and Cl atoms that are added are in an anti conformation about the C—C single bond indicated in the precursor. Furthermore, the two CH_3 groups are on the same side of the H—C—C—Cl plane in the precursor (i.e., behind the plane of the paper), just as they are on the same side of the C=C double bond in the target.

YOUR TURN 13.11

Draw a Newman projection, looking down the C2—C3 bond, for both the target and the precursor in Equation 13-43.

PROBLEM 13.17 This transform shows another way in which H and Cl atoms can be added to (E)-3-methylpent-2-ene to undo an E2 reaction. The H and Cl atoms are in an anticoplanar conformation about the indicated C—C bond. Supply the three substituents that are missing.

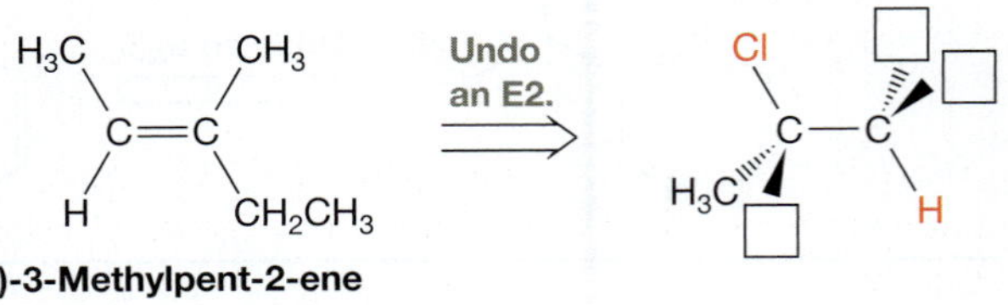

(E)-3-Methylpent-2-ene

Electrophilic addition reactions can be stereospecific, too, taking place in a syn or anti fashion. As we saw in Chapter 12, carbene addition, epoxidation, and hydroboration–oxidation reactions (Equations 13-44 through 13-46) result in syn addition.

Syn addition

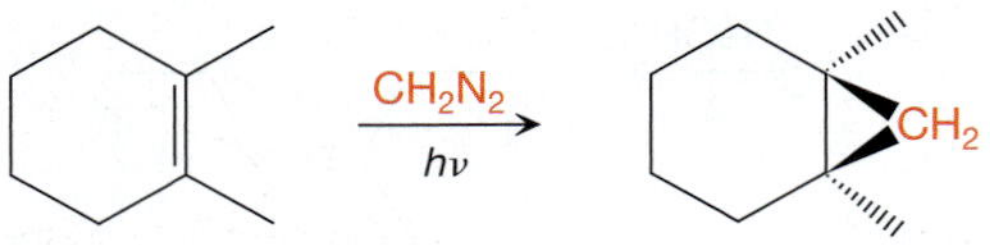

(13-44)

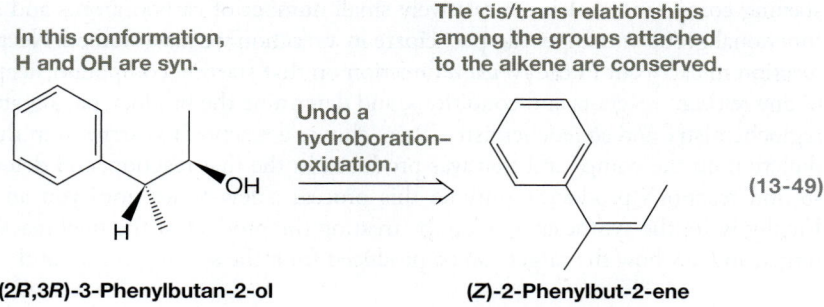

$$\text{(13-45)}$$

$$\text{(13-46)}$$

By contrast, the formation of 1,2-dihalides and halohydrins (Equations 13-47 and 13-48) results in anti addition.

Anti addition

$$\text{(13-47)}$$

$$\text{(13-48)}$$

Recall that when chirality is introduced in the products (e.g., Equations 13-46 through 13-48), these reactions result in a *mixture* of stereoisomers.

To take stereochemistry into account when incorporating these reactions in a synthesis, remember:

- In a syn addition to an alkene, the cis/trans relationships are conserved among the groups initially attached to the alkene carbons.
- In an anti addition, these relationships are inverted.

With this in mind, suppose that we want to apply a transform to (2*R*,3*R*)-3-phenylbutan-2-ol that undoes a hydroboration–oxidation reaction, as shown in Equation 13-49.

In this conformation, H and OH are syn.

The cis/trans relationships among the groups attached to the alkene are conserved.

Undo a hydroboration–oxidation.

$$\text{(13-49)}$$

(2R,3R)-3-Phenylbutan-2-ol   **(Z)-2-Phenylbut-2-ene**

Ensuring that the H and OH groups in the target are syn to each other, we can determine the precursor by removing those substituents and adding a π bond between the two C atoms to which they are initially attached. The cis/trans relationships of the groups attached to the alkene C atoms in the precursor remain the same as those in the target.

PROBLEM 13.18 Apply a transform to (2S,3R)-3-phenylbutan-2-ol that undoes a hydroboration–oxidation reaction.

(2S,3R)-3-Phenylbutan-2-ol

Undo a hydroboration–oxidation.

?

Even though a transform like the one in Equation 13-49 yields a single precursor, keep in mind that the reaction in the forward direction produces a racemic mixture of enantiomers because both the reactants and the environment are achiral. It is often possible, however, to carry out these kinds of reactions in a way that would favor one enantiomer over the other—a so-called **enantioselective synthesis**. Commonly, this is done by carrying out the reaction in the presence of a chiral reagent or catalyst.

The reactions that we have revisited here in Section 13.6 are not the only reactions with characteristic stereochemistry that we will encounter in this book. When you encounter other such reactions in subsequent chapters, be sure to pay particular attention to the aspects of their mechanisms that dictate their stereochemistry. Having a keen sense of the mechanisms of stereospecific reactions will better equip you to design more efficient syntheses.

CONNECTIONS One way to carry out an enantioselective hydroboration–oxidation is to use a chiral borane, such as diisopinocampheylborane (Ipc₂BH), which can be prepared from borane and α-pinene (p. 110), a natural product found in pine trees.

Diisopinocampheylborane (Ipc₂BH)

13.7 Strategies for Success: Improving Your Proficiency with Solving Multistep Syntheses

Your success in solving multistep syntheses will depend on your mastery of various reactions, and on your ability carry out a retrosynthetic analysis. Developing these skills requires practice, and there are several end-of-chapter problems in this and later chapters that will help you become more proficient. But what should you do if you have exhausted all of those end-of-chapter problems or if you are just looking for more challenge?

One of the best ways to improve is to *construct your own synthesis problems*. Choose a starting compound that has a relatively small number of carbon atoms and also has a functional group you know can participate in a reaction we have studied. Next, choose a reaction to carry out in the *forward* direction on that starting compound, keeping track of any relevant reagents and conditions, and determine the product, paying attention to regiochemistry and stereochemistry. Then, choose a reaction to carry out in the forward direction on the compound that was produced in the first reaction, and determine the second reaction's product. Continue this process a few times until you are satisfied. Finally, write the synthesis problem by treating the product of the final reaction as the target, and ask how the target can be produced from the starting compound.

For example, let's choose a terminal alkene as our starting compound and carry out a bromination (Section 12.3a):

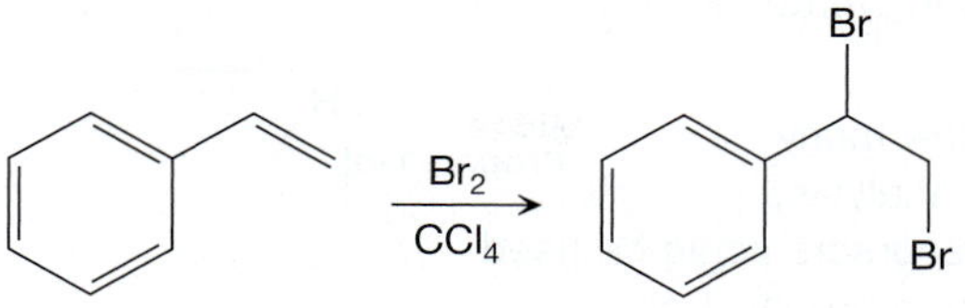

The vicinal dibromide that is produced can then undergo back-to-back E2 eliminations, followed by acid workup, to produce a terminal alkyne (Section 10.8):

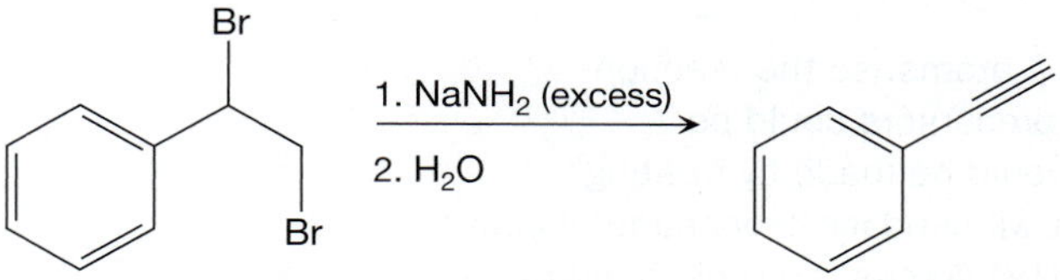

The terminal alkyne can be deprotonated to convert it to an alkynide anion, which is a strong nucleophile. Then, the alkynide anion can react with an alkyl halide substrate, producing a new C—C bond (Section 9.3b):

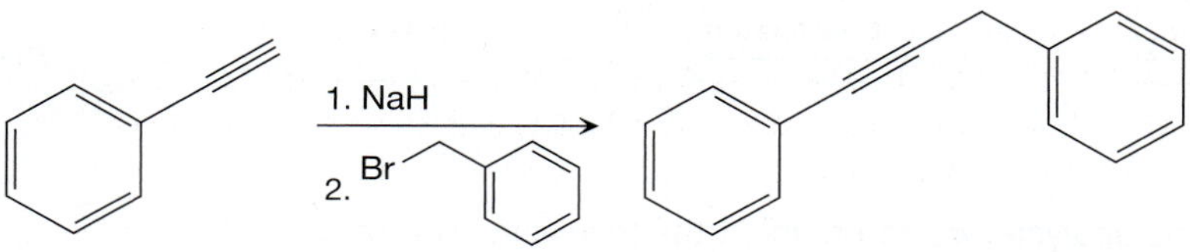

Although we could continue with more reactions, let's stop here and set up the synthesis problem. We treat the final product as the target and simply ask how to synthesize it from the starting material, as follows:

SYNTHESIS PROBLEM Show how to carry out the following synthesis:

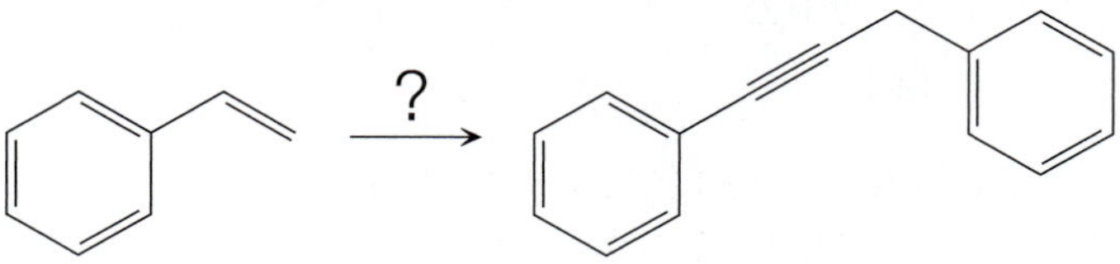

Constructing a synthesis problem like this will help you become more familiar with the reactions you have previously encountered. It won't benefit you much to try to solve it, though, because you already know which reactions to use and in what order.

Instead, work with a study partner. You should each construct separate synthesis problems and then swap. To solve the problem your study partner gave you, make sure to practice the retrosynthetic analysis strategies you learned here in Chapter 13. When you finish, check if your solution agrees with the reaction scheme your partner used to construct the problem. If they disagree, perhaps one solution is flawed. Or perhaps both are good!

As you learn new reactions throughout this book, revisit this exercise of constructing your own synthesis problems and swapping with a partner. Each time you do, you are sure to improve.

SOLVED PROBLEM 13.19

Show how to carry out the synthesis at the right, which your study partner gave to you to solve. Propan-1-ol is your only starting material that contributes carbon atoms to the target.

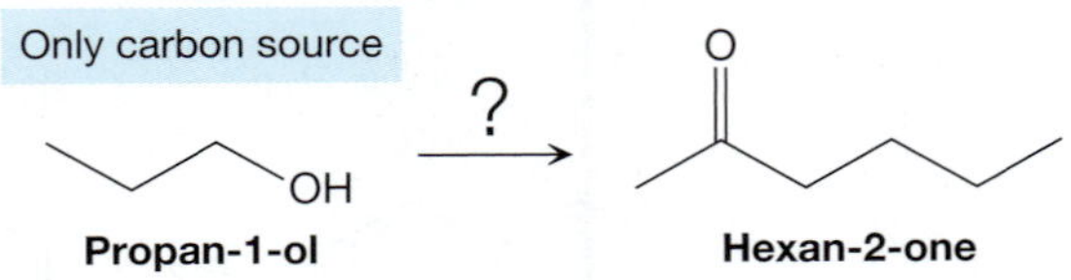

Think How many carbon atoms does the target have? How many does the starting compound have? In your retrosynthetic analysis, which C—C bond should you consider undoing? What precursors could be used to form that bond? How can those precursors be made from propan-1-ol?

Solve The target has six C atoms but the starting compound has only three, so let's begin by applying a transform that undoes the formation of the C—C bond shown here.

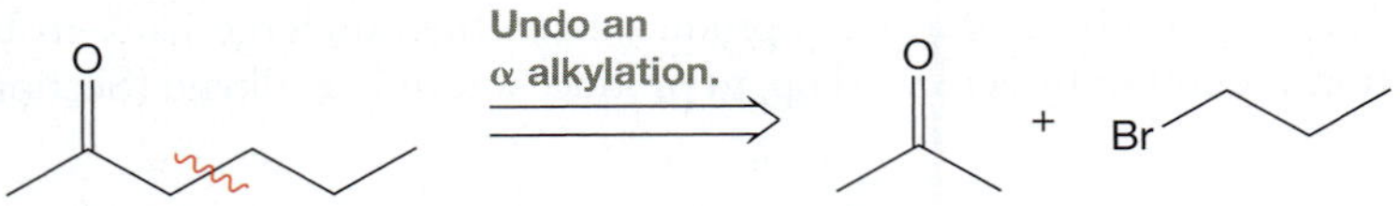

That C—C bond involves one of the ketone's α C atoms, so the reaction we should consider undoing is an α alkylation. The precursors could be 1-bromopropane and propanone. 1-Bromopropane could be made by treating propan-1-ol with PBr$_3$. We don't have a reaction that will produce propanone directly from propan-1-ol, but it can be produced by adding H$_2$O across the triple bond of propyne, as shown in the retrosynthesis below. Propyne can be produced from propene by carrying out a bromination followed by elimination. Finally, propene can be produced by dehydrating propan-1-ol.

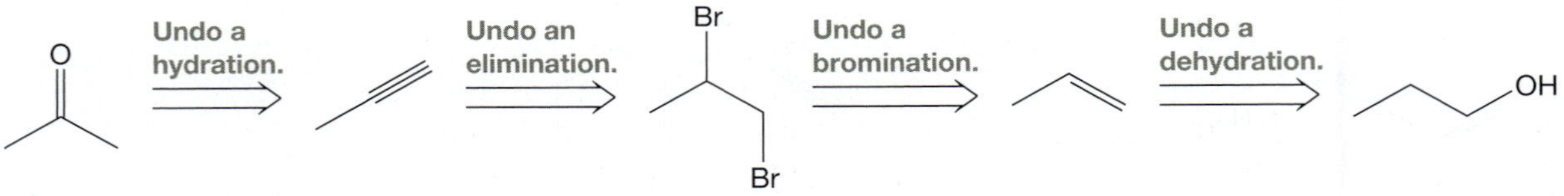

Now that we have completed our retrosynthetic analysis, we can present the following synthesis:

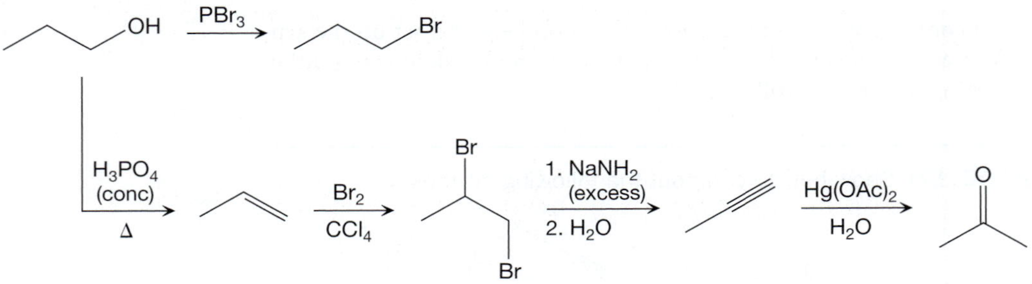

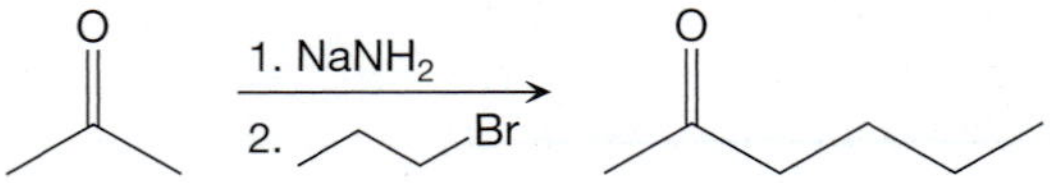

Notice that Markovnikov addition is necessary for the hydration of propyne, which is why oxymercuration–reduction is chosen. Also, notice how 1-bromopropane and propanone are separately made from propan-1-ol and are reacted together in the final step to produce hexan-2-one.

PROBLEM 13.20 Show how to carry out the following synthesis, which your study partner gave to you to solve. Propan-1-ol is your only starting material that contributes carbon atoms to the target.

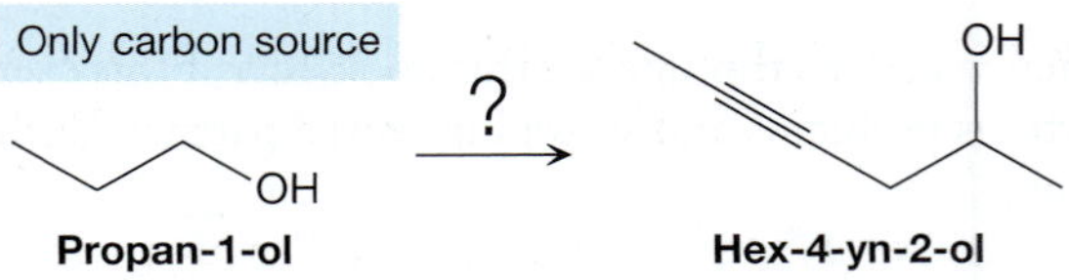

13.8 Choosing the Best Synthesis Scheme

Often, you will find that multiple synthesis schemes can be devised for the same target. What makes one better or worse than another? Generally speaking, a synthesis scheme is better if it is:

- less costly or
- more environmentally friendly (i.e., more "green").

Considering each of these ideas thoroughly for a particular synthesis can be an involved process and, indeed, would be necessary if you were to carry out a synthesis in the laboratory. Here, however, we discuss just some of the most important aspects. Section 13.8a deals with percent yield, one of the main factors contributing to cost, whereas Section 13.8b discusses some of the guiding principles behind what is called *green chemistry*.

Although the rules we examine are very important in the laboratory, their importance in the classroom can vary significantly from one instructor to the next. If it has not already been made clear to you by your instructor, you should ask whether the application of these rules will impact the credit you are awarded for a particular synthesis problem.

13.8a Considerations of Percent Yield

There are a number of factors that contribute to the cost of a synthesis, such as the purchase price of the starting materials, the time the synthesis takes, and the costs associated with the disposal of waste generated in each synthetic step. For multi-step syntheses, these costs typically decrease significantly as the percent yield of the target increases because less starting material is used, less time is spent in the laboratory, and less waste is generated. Therefore, *we should seek to maximize the percent yield of the synthesis* and, to do so, there are two basic rules that should always be applied.

1. The number of steps in a synthesis scheme should be minimized.
2. Each step in the synthesis should proceed with the highest possible percent yield.

These rules are essentially an outcome of how percent yield is computed for a **linear synthesis** (i.e., a synthesis composed of sequential steps): *For a linear synthesis, the overall percent yield is equal to the product of the yields of the individual steps*. For example, compare a three-step linear synthesis to one that is six steps long, and suppose that each individual step in both syntheses proceeds with an 80% yield of product (which is often considered quite good). The three-step synthesis will have an *overall* yield of $(0.80) \times (0.80) \times (0.80) = (0.80)^3 = 0.51$, or 51%, whereas the six-step synthesis will have an overall yield of 26%. Thus, the synthesis with the fewer number of steps has the greater yield, consistent with the first rule.

YOUR TURN 13.12

Show how 26% was obtained as the overall yield for the six-step synthesis.

Now consider another pair of syntheses, each of which consists of three steps. Suppose that each step in the first synthesis proceeds with a 90% yield, to give an

TABLE 13-3	How Overall Yield in a Synthesis Is Affected by the Number and Yield of Individual Steps	
Number of Steps in a Synthesis	Percent Yield per Step	Overall Percent Yield
5	90%	59%
5	80%	33%
5	70%	17%
10	90%	35%
10	80%	11%
10	70%	2.8%
20	90%	12%
20	80%	1.2%
20	70%	0.08%

overall yield of $(0.90) \times (0.90) \times (0.90) = 0.73$, or 73%. Suppose, in the second synthesis, that each step proceeds with a 70% yield, so the overall yield is $(0.70) \times (0.70) \times (0.70) = 0.34$, or 34%. Thus, as suggested in rule 2, the first synthesis has the higher overall yield because each of its steps has a higher yield than the individual steps in the second synthesis.

Table 13-3 illustrates how dramatically the overall yield of a synthesis can change with the number of steps it has and the yield of each step. The published syntheses of natural products commonly consist of 15 to 30 steps, with an overall yield of <5%! In the synthesis of taxol highlighted in the chapter opener (p. 641), considered a triumph at the time, the overall yield was <2%.

PROBLEM 13.21 Calculate the percent yield of a linear synthesis consisting of eight steps, each of which proceeds with 75% yield.

PROBLEM 13.22 Calculate the percent yield of the following seven-step synthesis that converts the starting material **S** into the target **T**.

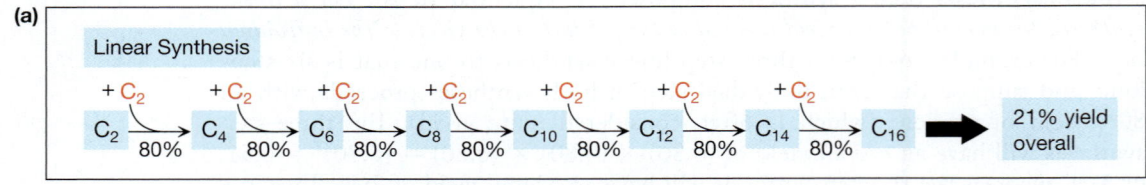

So far, we have considered percent yield in a *linear synthesis* only. In a **convergent synthesis**, portions of a target molecule are synthesized separately and are assembled together at a later stage.

> In general, the yield of a multistep synthesis is higher if a target can be produced from a *convergent synthesis* instead of a linear one.

To demonstrate this point, consider Figure 13-1. Both Figure 13-1a and 13-1b depict an overly simplified synthesis of a target that contains 16 carbons, assuming that our only carbon source contains two carbons. In Figure 13-1a, we see a *linear synthesis* where two

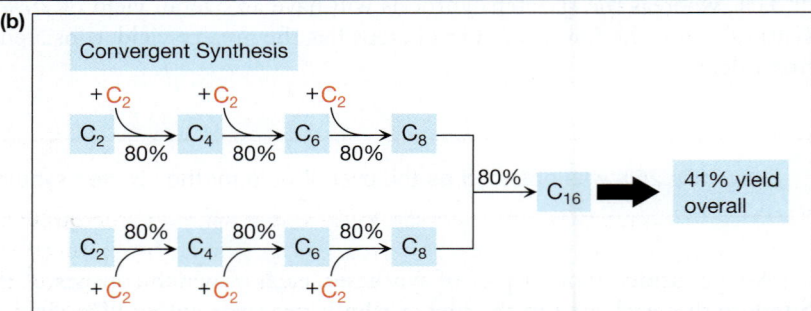

FIGURE 13-1 Linear versus convergent synthesis schemes Hypothetical synthesis of a 16-carbon target from two-carbon pieces, using (a) a linear synthesis scheme and (b) a convergent synthesis scheme. Assuming each step has an 80% yield, the overall yield of the linear synthesis is 21%, whereas that of the convergent synthesis is 41%.

carbons are added at a time. Seven steps are required and, if we assume an 80% yield for each step, then the overall yield would be $(0.80)^7 = 0.21 = 21\%$. Figure 13-1b depicts a convergent synthesis, where two eight-carbon pieces are connected together in the final step, with each of those pieces having been constructed separately from our two-carbon source. Reaching the final target still requires seven separate steps, but the longest branch of the synthesis is only four steps, so the overall yield would be $(0.80)^4 = 0.41 = 41\%$. This is roughly twice that from the linear synthesis!

PROBLEM 13.23 Consider synthesis schemes **I** and **II** below, both of which produce the same eight-carbon target. For each synthesis scheme, **(a)** determine whether it is linear or convergent, and **(b)** compute the overall percent yield assuming that each synthetic step proceeds with an 80% yield.

Scheme I

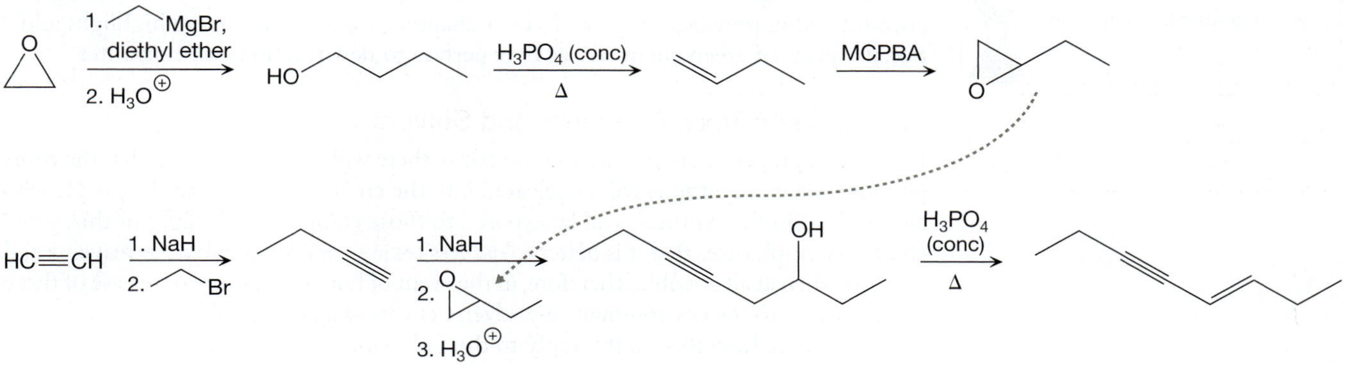

Scheme II

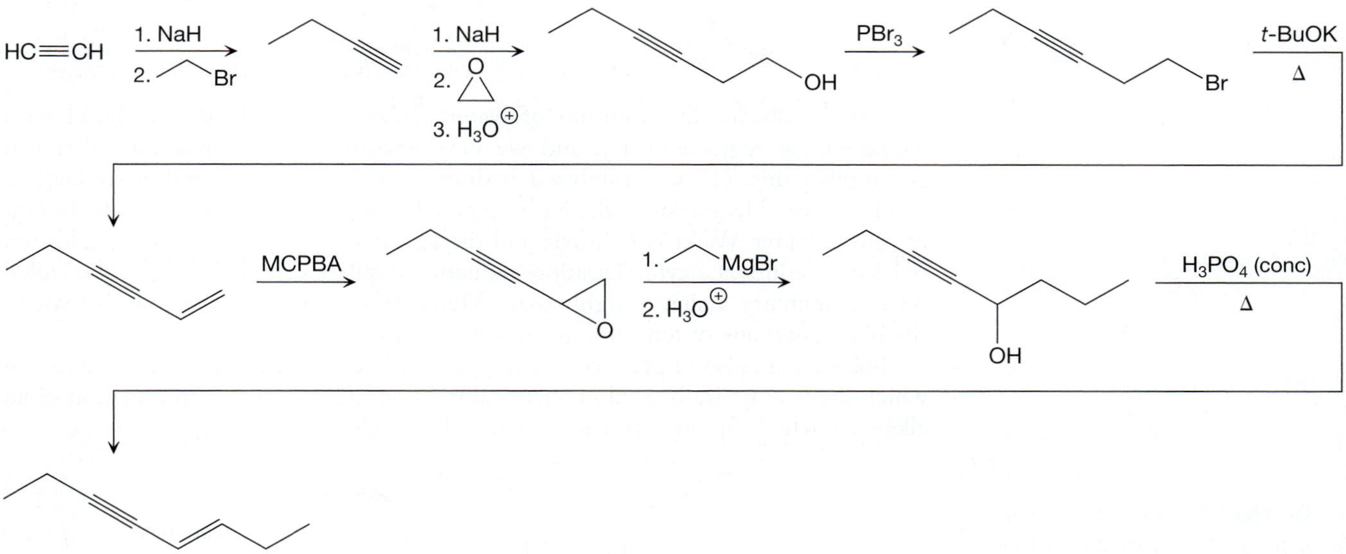

13.8b Green Chemistry

Over the last century or more, organic synthesis has offered numerous benefits to our society, including safer and more effective medicines, better crop protection, enhanced energy production and storage, and new materials for consumer products. Unfortunately, considerations for the environment were often an afterthought and, by the middle of the 20th century, the adverse effects of the chemicals industry on the environment were apparent. Acid rain was polluting our waters and destroying our forests, "holes" developed in our ozone layer, and compounds linked to cancer were in common use. In response to these issues, *green chemistry* evolved in the 1990s.

In short, **green chemistry** is a set of guiding principles established to help prevent various forms of pollution and other hazards. Rather than focus on the proper handling of materials after a synthesis has been carried out, the goal of green chemistry is to *avoid* the production and accumulation of hazardous materials in the first place, and it depends on the innovation of new chemical reactions, new techniques, and new technologies, which can be called **green alternatives**.

There are 12 distinct principles of green chemistry, which you can find on the American Chemical Society's website at: acs.org/content/acs/en/greenchemistry/what-is-green-chemistry.html. A thorough discussion of all of the principles would be inappropriate for this book. Instead, in the discussion that follows, we focus on some of the overarching ideas, showing how they might apply to a handful of reactions we have encountered in previous chapters. In later chapters, we will continue to highlight additional aspects of green chemistry as they pertain to new reactions we encounter.

GREEN CHEMISTRY When we learn new reactions in later chapters, look for this symbol to learn about ways in which green chemistry is applied.

13.8b.1 Less Toxic Reagents and Solvents

Even when proper safety precautions are taken, there will still be some risk that the compounds used in a synthesis will be released into the environment or that the people who are involved in the synthesis will be exposed to those compounds. In light of this, green chemistry emphasizes that it is better to use less toxic—or even nontoxic—reagents and solvents when at all possible. Therefore, in the event of human exposure or release of those compounds into the environment, any adverse effects will be minimized.

We can see how this might apply to the following transformation.

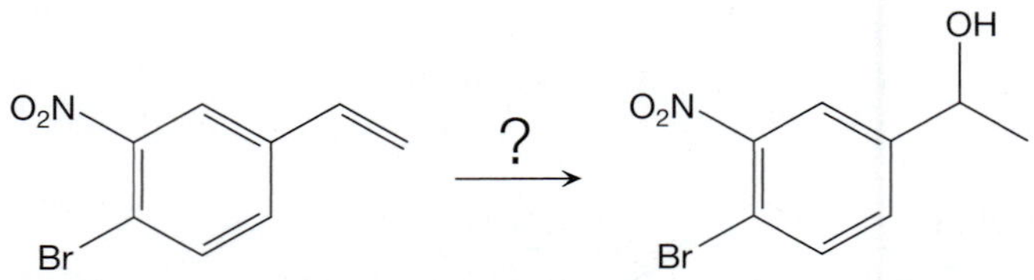

(13-50)

This step calls for the addition of water across the C=C double bond with Markovnikov regiochemistry, and we have encountered two reactions that can accomplish this: (1) acid-catalyzed hydration and (2) oxymercuration–reduction. Acid-catalyzed hydration is the far better choice because it requires a small amount of acid in water. Water is nontoxic and the acid can be neutralized by the addition of base. By contrast, oxymercuration–reduction requires a source of Hg^{2+}, also called divalent mercury, which is highly toxic. Mercury(II) acetate, for example, is toxic to the blood, nervous system, and mucous membranes.

Solvents are also of great concern, especially because of the large abundances in which they are typically used in chemical reactions. Consider the bromination of an alkene, in which Br_2 adds anti across a C=C double bond.

CONNECTIONS Cyclooctene is the smallest cycloalkene that can exist in a trans isomer (see p. 271), with the cis isomer, shown here, being more stable.

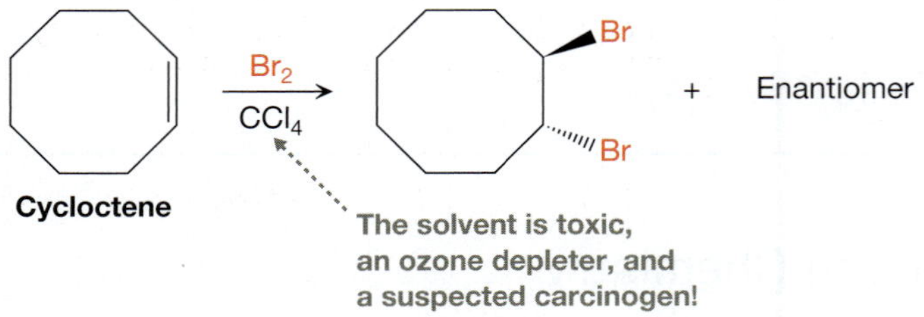

(13-51)

As we learned in Section 12.3, this reaction calls for a non-nucleophilic solvent such as CCl_4. A nucleophilic solvent such as water would lead to the production of a halohydrin instead. However, CCl_4 is toxic, causes ozone depletion in the stratosphere, and is a suspected carcinogen. Fortunately, this is not the only method of producing these kinds of 1,2-dibromides. Another method uses a 5:1 ratio of $NaBr:NaBrO_3$ as the source of bromine, with acetic acid as the solvent. Acetic acid is significantly less toxic than CCl_4, so the reaction that uses $NaBr:NaBrO_3$ can be viewed as a *green alternative* to the traditional reaction that uses CCl_4.

PROBLEM 13.24 Recall from Section 9.7 that S_N2 reactions are promoted by aprotic solvents, such as DMSO or DMF. All else being equal, which solvent should you use? *Hint:* You can search the Internet to learn about each solvent's toxicity.

13.8b.2 Safer Synthesis Routes

Sometimes a synthesis can be green in terms of the by-products and other waste that it produces but should be avoided if it poses a high risk for accidents, including fires and explosions. Consider the reactions in which we used diazomethane, CH_2N_2. In Chapter 10, we learned that it can be used to convert a carboxylic acid into a methyl ester (Equation 13-52), and in Chapter 12 we learned that it can be used as a source of carbene to convert the C=C double bond of an alkene into a cyclopropane ring (Equation 13-53).

In both cases, the only by-product is $N_2(g)$, which is nontoxic. However, diazomethane is *explosive* and should be avoided if there are safer alternatives. As we will learn in Chapter 21, the esterification reaction in Equation 13-52 can be accomplished with methanol and a catalytic amount of acid. And, as we saw in Section 12.2, the cyclopropanation in Equation 13-53 can be carried out with a Simmons–Smith reaction, which uses CH_2I_2 in the presence of a zinc source. Both of these alternatives are safer than using diazomethane.

13.8b.3 Minimize By-products and Other Waste

For any reagents used in a synthesis, there is risk of release into the environment as well as human exposure. The same is true for any by-products and other accumulated waste. Therefore, green chemistry places value on incorporating as much of the starting material into the desired products as possible. One way to do so is to maximize the percent yield, which we discussed in Section 13.8a. Even if a reaction proceeds with 100% yield, however, it can still be quite wasteful. Consider, for example, the following epoxidation reaction:

The carboxylic acid by-product is unusable in the synthesis and would probably be discarded. It represents a significant amount of material that would go to waste.

By contrast, consider the following hydrochlorination reaction:

All of the material that reacts ends up in the desired product.

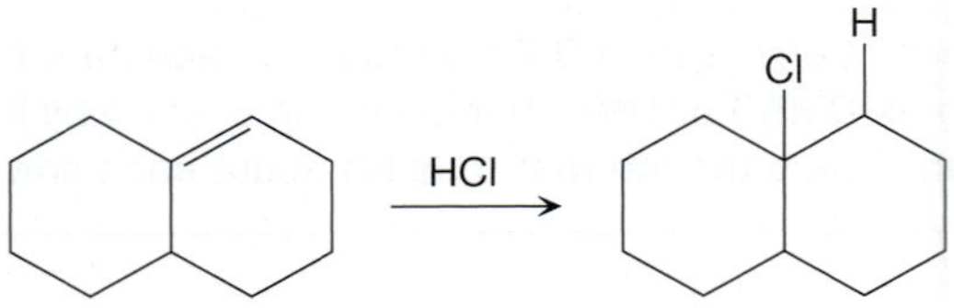

(13-55)

In this case, *all* of the material that reacts ends up in the desired product. If the reaction were to proceed with 100% yield, then *none* of the starting materials would go to waste. Inherently, then, this reaction is more *efficient* than the one in Equation 13-54.

This type of inherent efficiency of a reaction is called **percent atom economy**, and it is calculated according to Equation 13-56.

$$\% \text{ Atom economy} = \frac{\text{Mass of atoms in the desired product}}{\text{Mass of atoms in all reagents}} \times 100\% \quad \text{(13-56)}$$

When the numerator and denominator are the same, as is the case for the hydrochlorination reaction in Equation 13-55, the atom economy is 100%. On the other hand, when there are by-products that are not usable in the synthesis, the atom economy falls below 100%. This is the case for the epoxidation reaction in Equation 13-54, as we can see more explicitly in Solved Problem 13.25.

SOLVED PROBLEM 13.25

Compute the percent atom economy of the epoxidation reaction in Equation 13-54.

Think What is the desired product? What is the mass of its atoms? What is the mass of all the reactant atoms?

Solve In Equation 13-54, the desired product is the epoxide ($C_{10}H_{16}O$), whose molecular weight is 152.2 u. There are two reactant molecules: the alkene ($C_{10}H_{16}$) and MCPBA ($C_7H_5ClO_3$). Their molecular weights are 136.2 u and 172.6 u, respectively, for a total mass of 308.8 u. The percent atom economy is therefore: [(152.2 u)/(308.8 u)] × 100% = 49.3%.

PROBLEM 13.26 Compute the percent atom economy for the following substitution reaction.

Chapter Summary and Key Terms

- A **synthesis** scheme is an abbreviated recipe that tells how **starting material** is converted into a particular **target**. (Introduction; Objective 1)

- In each **synthetic step**, only the reactants, reagents added, reaction conditions, and products are included. Reagents are usually written above the reaction arrow, whereas reaction conditions are usually written below. (Section 13.1; Objective 2)

- Reactions that form and/or break C—C σ bonds alter the **carbon skeleton**, whereas **functional group conversions** do not. (Section 13.2; Objective 3)

- **Retrosynthetic analysis** streamlines the process of designing an organic synthesis. **Transforms** are applied to the target molecule to arrive at smaller or simpler **precursors**, and the cycle is repeated on the precursors until a suitable starting material is reached. (Section 13.3; Objective 4)

- Some transforms cannot be carried out in the forward direction. These **synthetic traps** can occur if a precursor has multiple reactive functional groups or if one reactive functional group has two reactive sites. (Section 13.4; Objective 5)
- Often the choice of solvent can have a dramatic impact on the outcome of a reaction. (Section 13.5; Objective 5)
- When incorporating a stereospecific reaction into a synthesis, the stereochemistry of the reaction should be considered when applying a transform. (Section 13.6; Objective 6)
- For the most efficient synthesis, the percent yield should be maximized. This often calls for minimizing the number of steps and using a **convergent synthesis** scheme instead of a **linear synthesis**. (Section 13.8a; Objective 7)

- **Green chemistry** is a set of guiding principles for chemists to apply when designing a synthesis, largely for the purpose of preventing environmental pollution and reducing health and safety risks. (Section 13.8b; Objective 8)
- Green chemistry promotes minimizing the use of toxic and otherwise hazardous substances because of the risk of release into the environment as well as human exposure. (Section 13.8b; Objective 9)
- Green chemistry also promotes maximizing the **percent atom economy**, which is a measure of how much of the starting material is incorporated into the desired product. (Section 13.8b; Objective 10)

Problems

Problems that are related to synthesis are denoted (SYN).

13.1–13.3 Writing the Reactions of an Organic Synthesis, Cataloging Reactions, and Retrosynthetic Analysis

13.27 Each of the following is a set of directions for carrying out a reaction or sequence of reactions. Rewrite each set of directions in the form of a synthesis.
 (a) To 2-ethylcyclohexanone, add lithium diisopropylamide, and when that reaction is complete, add bromoethane to yield 2,6-diethylcyclohexanone.
 (b) Add molecular bromine to 2,2-dimethylcyclohexanone in the presence of acetic acid to yield 6-bromo-2,2-dimethyl-cyclohexanone. To the resulting mixture, add sodium cyanide to yield 6-cyano-2,2-dimethyl-cyclohexanone.
 (c) Treat pent-4-ynoic acid with diazomethane to produce methyl pent-4-ynoate. Next, add sodium hydride, followed by (bromomethyl)benzene, to yield methyl 6-phenylhex-4-ynoate.

13.28 Convert each synthesis scheme into words that can be used as instructions in the laboratory, similar to what you see in Problem 13.27.

(a)

(b)

(c)

13.29 Determine whether each of the following syntheses requires a reaction that alters the carbon skeleton.

(a)

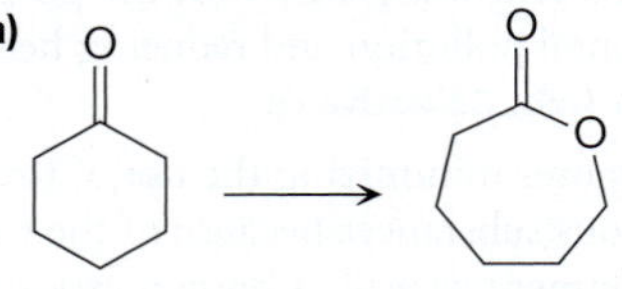

(b)

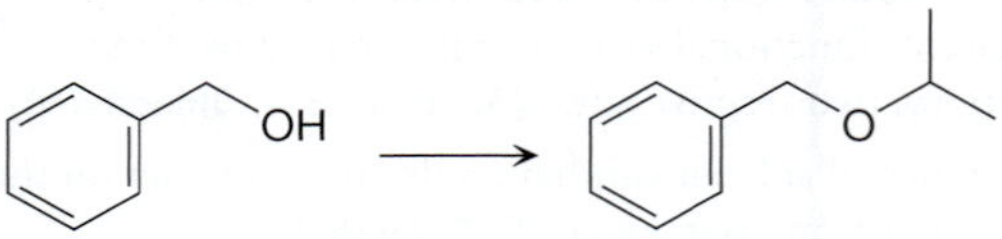

(c)

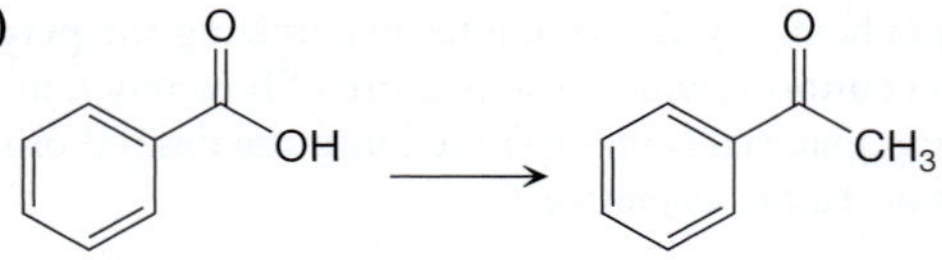

(d)

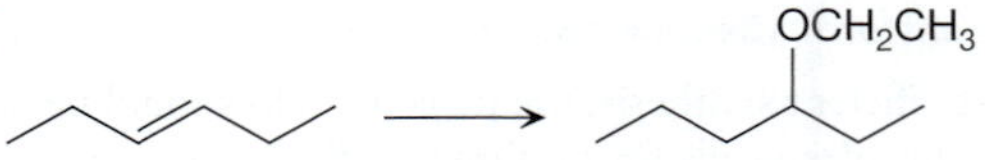

(e)

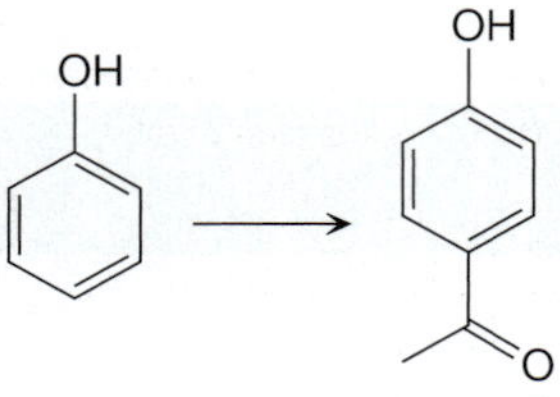

(f)

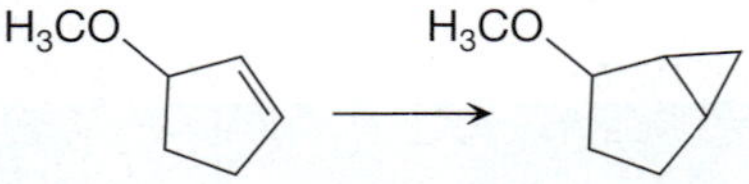

(g)

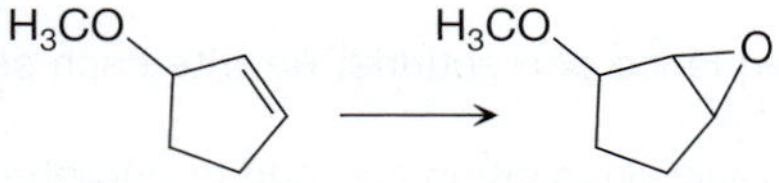

(h)

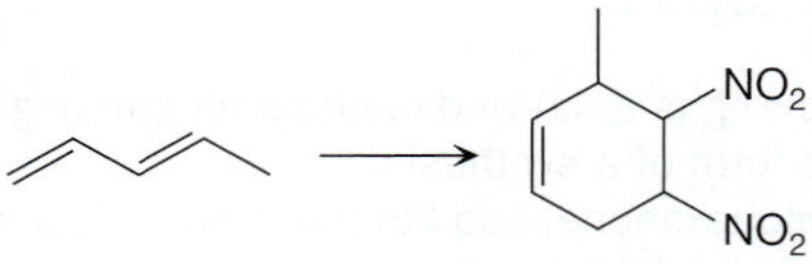

13.30 Show how a retrosynthetic analysis might be constructed for each synthesis in Problems 13.27 and 13.28.

13.31 Rewrite each of these transforms as a synthetic step in the forward direction, including reagents and any special reaction conditions.

(a)

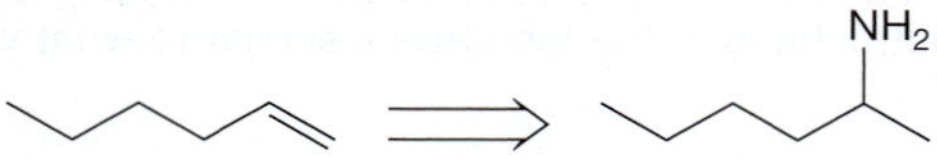

(b)

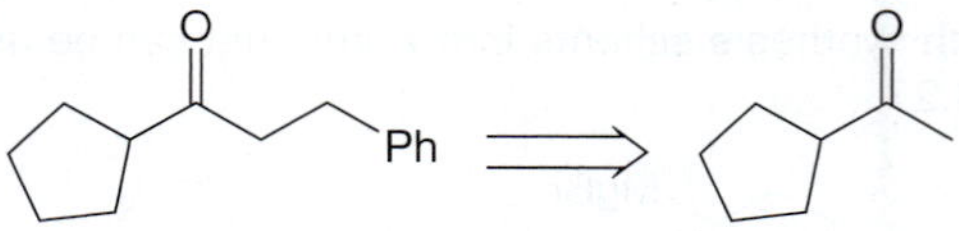

(c)

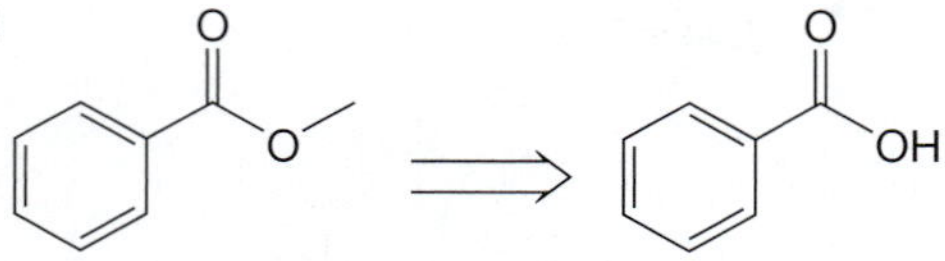

(d)

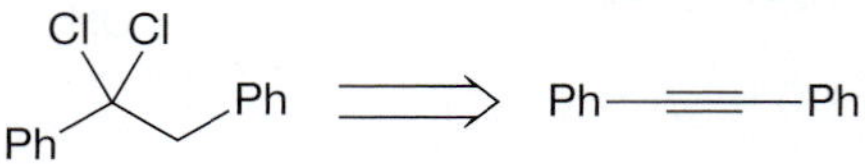

13.4–13.6 Synthetic Traps, the Choice of Solvent, and Considerations of Stereochemistry

13.32 Determine whether each of the following proposed transforms represents a synthetic trap (i.e., it will not proceed as planned in the forward direction). Explain your reasoning.

(a)

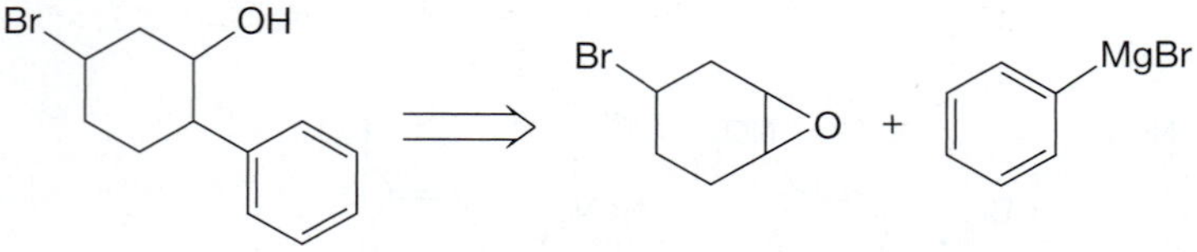

(b)

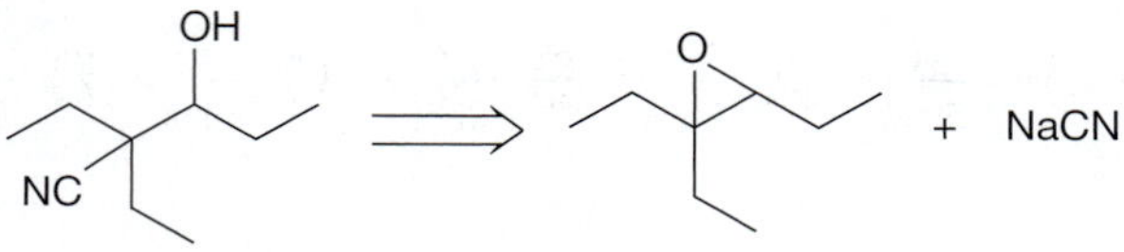

+ NaCN

(c)

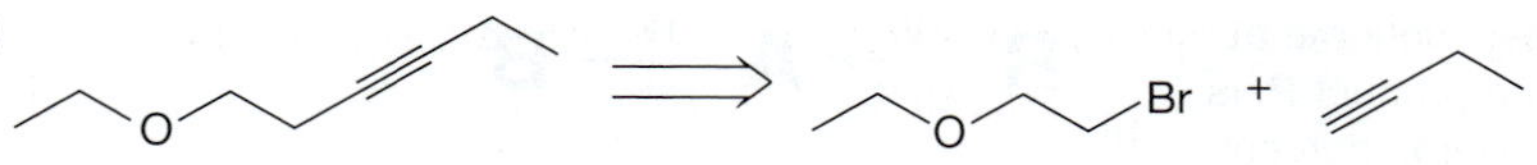

O
O
Br
+

(d)

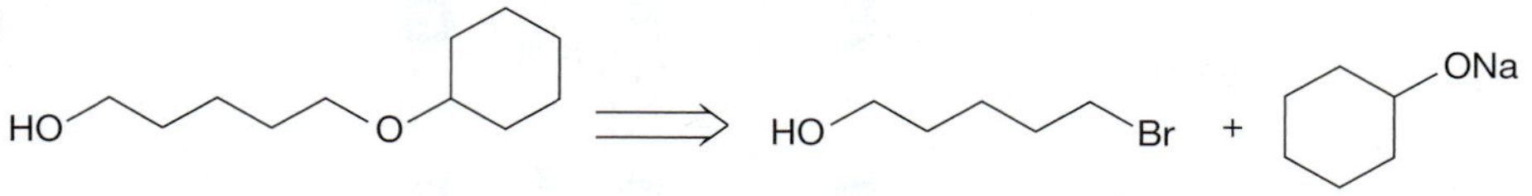

HO
O
HO
Br
+
ONa

(e)

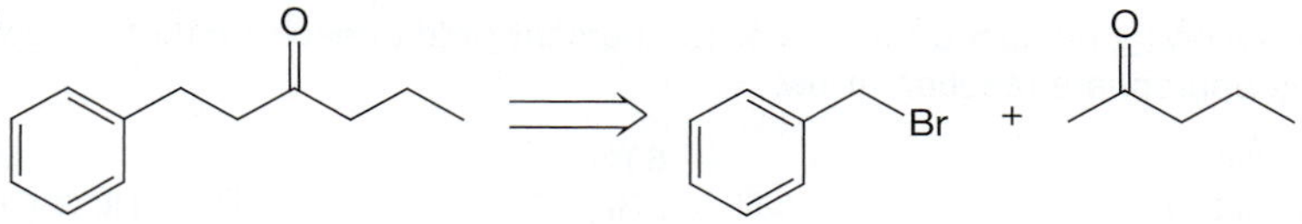

O
Br
+
O

13.33 For each of the following proposed synthetic steps, determine whether the solvent would interfere with producing the intended target. For those that would, explain why and suggest another solvent that could be used instead.

(a)

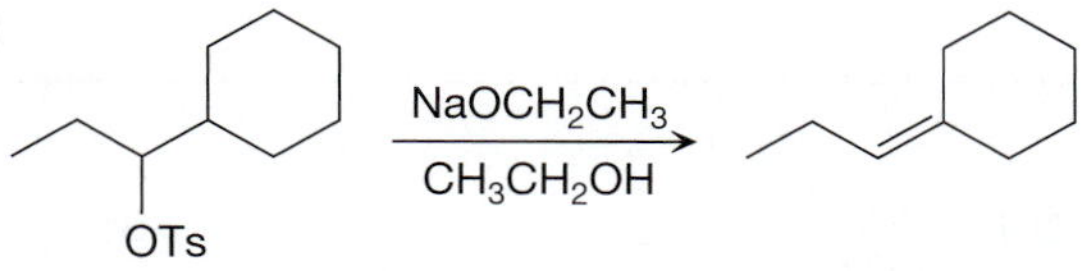

NaOCH2CH3
CH3CH2OH
OTs

(b)

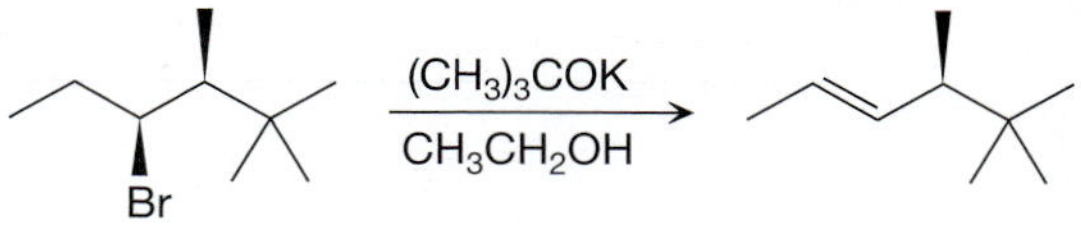

(CH3)3COK
CH3CH2OH
Br

(c)

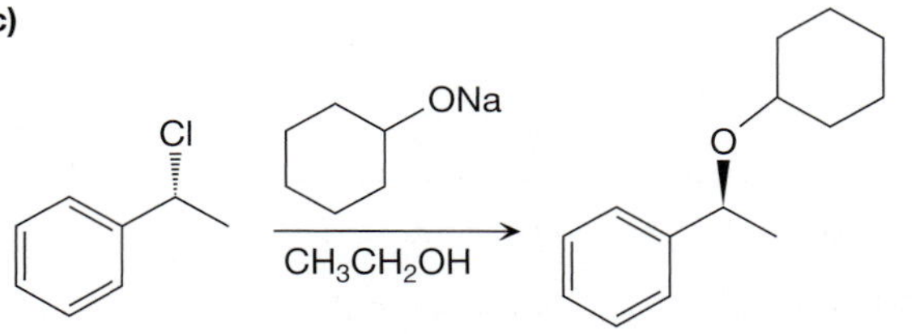

Cl
ONa
CH3CH2OH
O

(d)

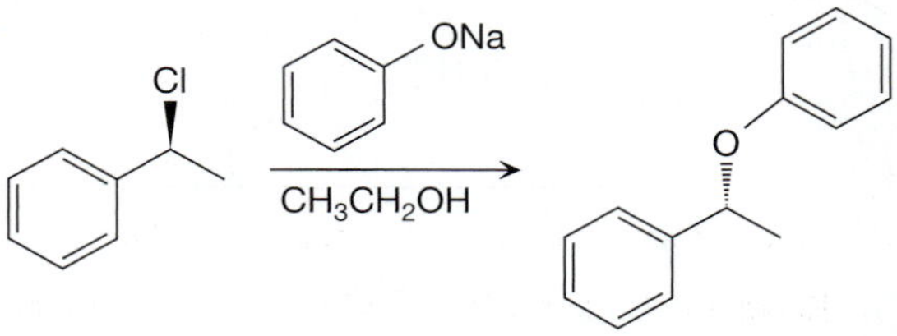

Cl
ONa
CH3CH2OH
O

13.34 Rewrite each of the following stereospecific transforms as a synthetic step in the forward direction, including reagents and any special reaction conditions.

(a)

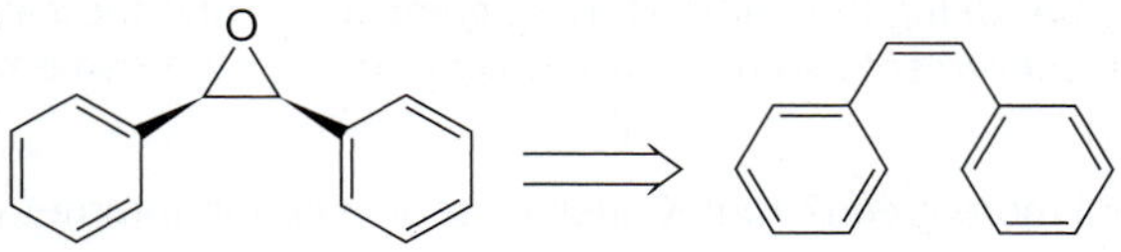

O

(b)

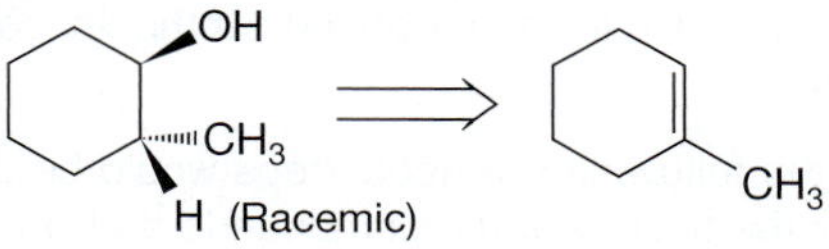

OH
CH3
H (Racemic)
CH3

(c)

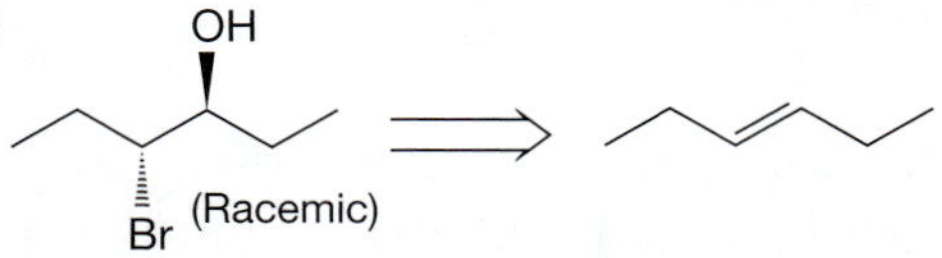

OH
Br
(Racemic)

(d)

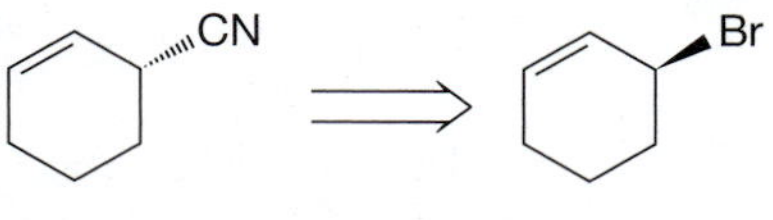

CN
Br

13.35 (SYN) For each of the following alkenes, provide an alkyl halide that can be used to synthesize it exclusively via an E2 reaction. Pay attention to stereochemistry.

(a)

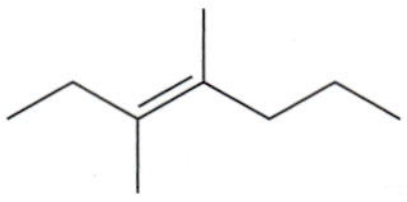

(b)

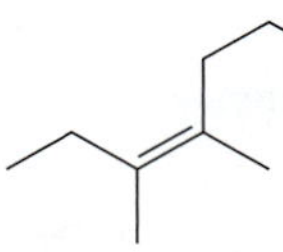

(c)

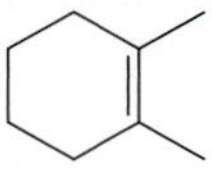

(d)

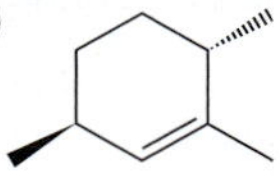

(e)

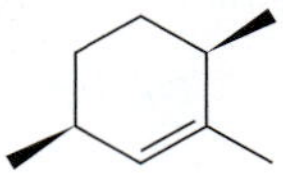

13.8 Percent Yield and Green Chemistry

13.36 Determine whether this generic synthesis scheme, which converts reactant **R** into product **P**, is linear or convergent and compute the overall percent yield.

$$\mathbf{R} \xrightarrow[59\%]{V} \mathbf{A} \xrightarrow[78\%]{W} \mathbf{B} \xrightarrow[92\%]{X} \mathbf{C} \xrightarrow[66\%]{Y} \mathbf{D} \xrightarrow[85\%]{Z} \mathbf{P}$$

13.37 Determine whether the generic synthesis scheme at the right, which converts reactant **R** into product **P**, is linear or convergent and compute the overall percent yield.

$$R \xrightarrow[59\%]{V} A \xrightarrow[78\%]{W} B$$

$$E \xrightarrow[66\%]{X} F \xrightarrow[85\%]{Y} G$$

$$B + G \xrightarrow[81\%]{Z} P$$

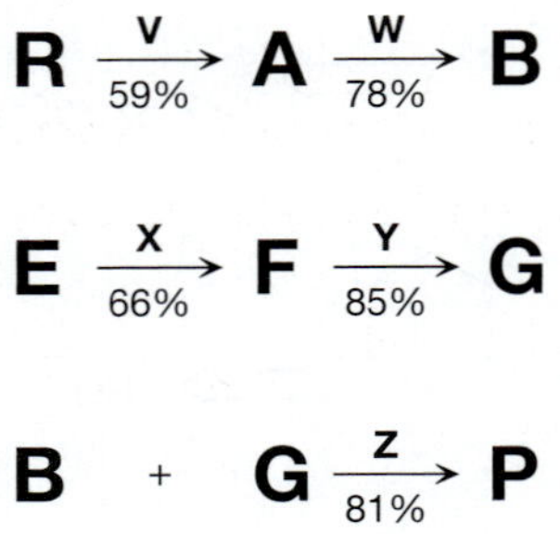

13.38 Compute the *overall* percent yield of the following proposed synthesis. The percent yield of each synthetic step or sequence of steps is provided above the appropriate reaction arrow.

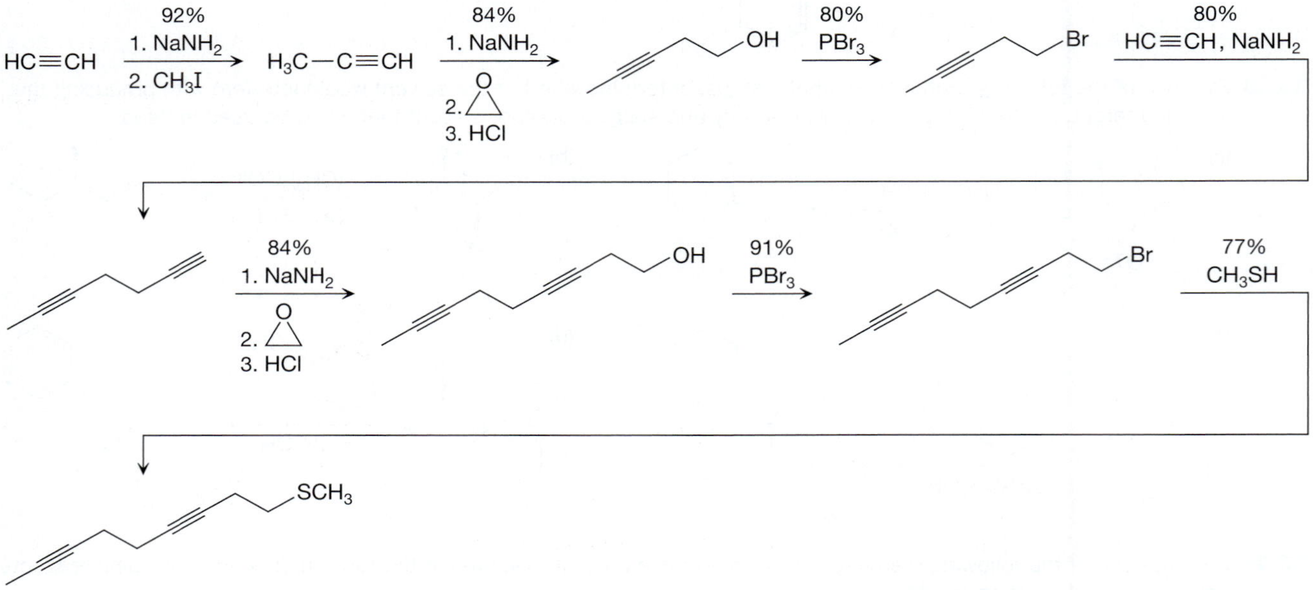

13.39 The proposed synthesis in Problem 13.38 is a *linear* scheme. Using the same starting compounds and the same types of reactions shown in the proposed synthesis, construct a *convergent* synthesis that might produce the same target in a higher yield.

13.40 Which of the following synthetic steps would be considered more green? *Hint:* Consider searching the Internet to determine the toxicity of the compounds that are involved.

Synthetic step I

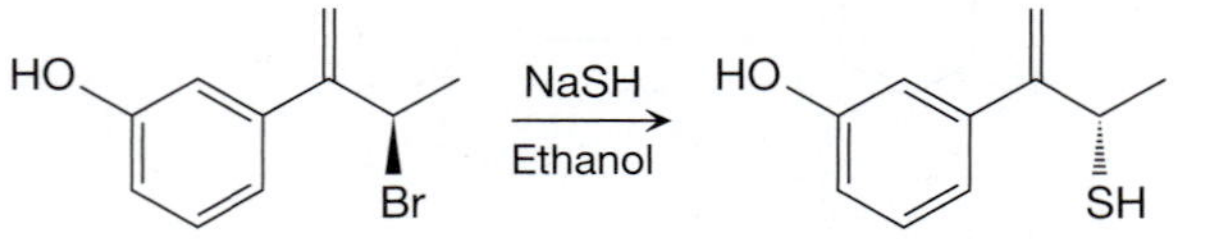

Synthetic step II

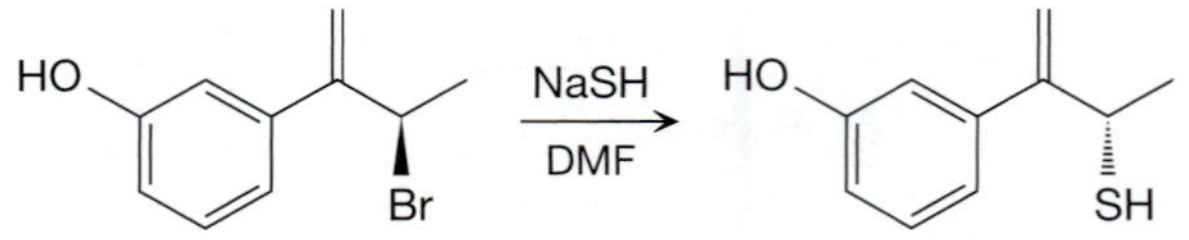

13.41 Which of the following synthetic steps would be considered more green? *Hint:* Consider searching the Internet to determine the toxicity of the compounds that are involved.

Synthetic step III

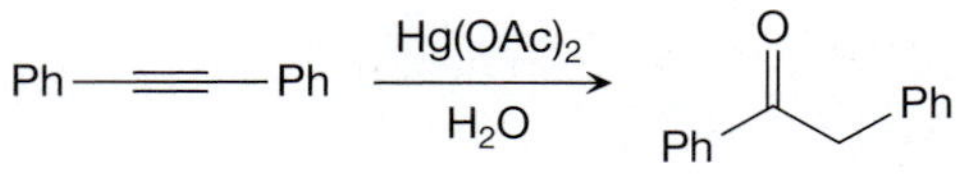

Synthetic step IV

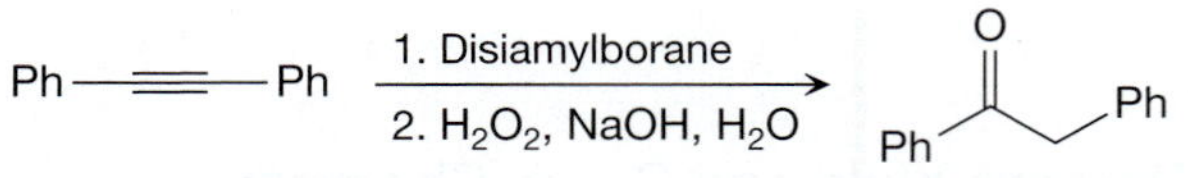

13.42 Compute the percent atom economy for the following reaction.

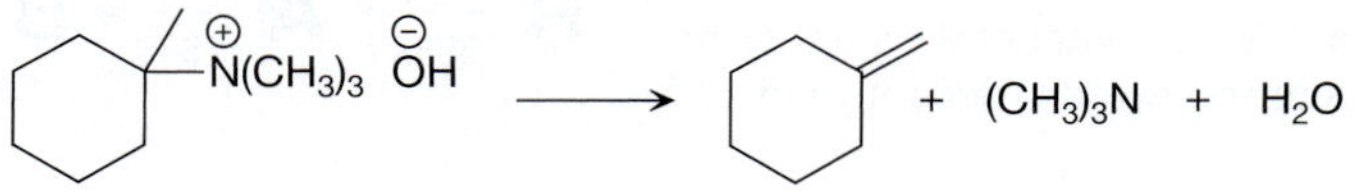

Integrated Problems

13.43 (SYN) Show how pent-2-yne can be made from two different alkyl bromides.

13.44 (SYN) Show how propoxybenzene can be synthesized using phenol and alcohols as your only source of carbon atoms that appear in the target.

13.45 (SYN) Show how you would synthesize each of the following compounds from the materials specified.

(a)

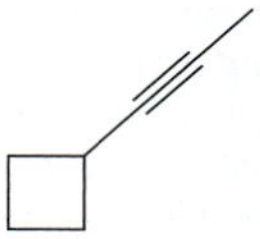

From reagents containing
4 or fewer carbons

(b)

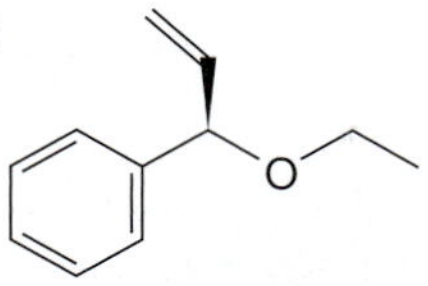

From any alcohols

(c)

From reagents containing
5 or fewer carbons

13.46 (SYN) Show how you would carry out the following synthesis from the starting material given, using any other compound containing, at most, one carbon atom.

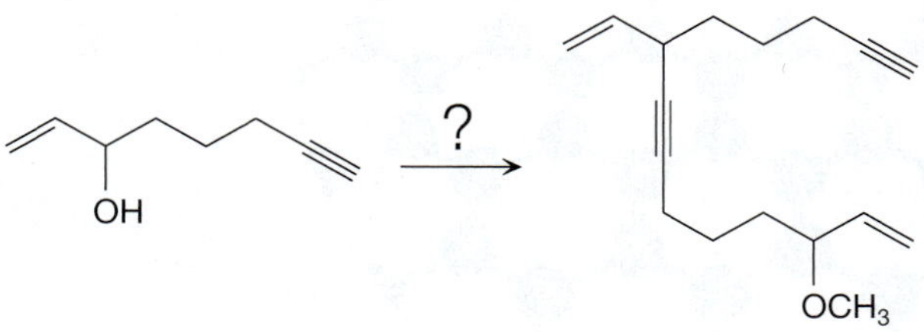

13.47 (SYN) Show how you would carry out the synthesis at the right using the starting material given, plus propyne, and any other inorganic reagents (i.e., reagents containing no carbon).

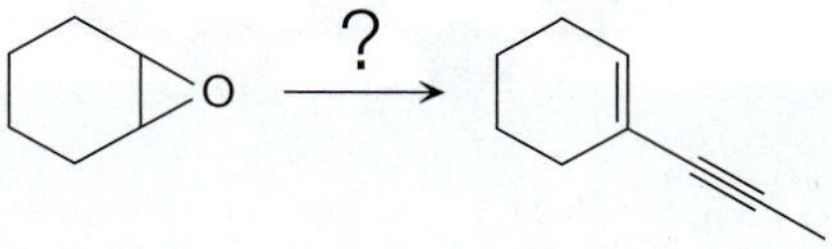

13.48 (SYN) Suggest how you would carry out each of the following syntheses.

(a)

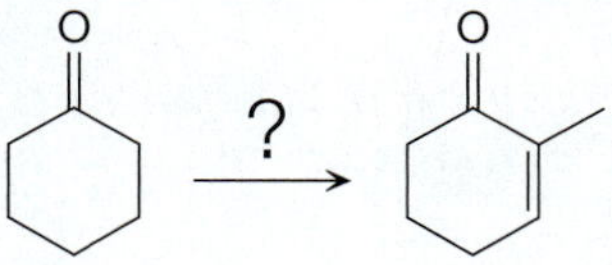

(b)

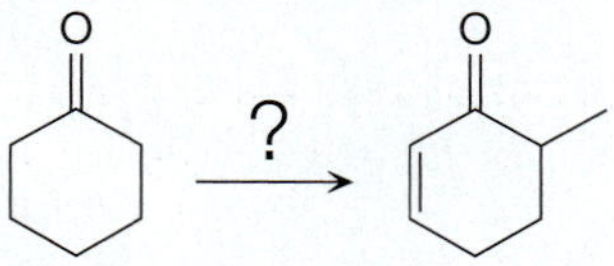

13.49 (SYN) Propose how to carry out the synthesis shown here using any reagents necessary.

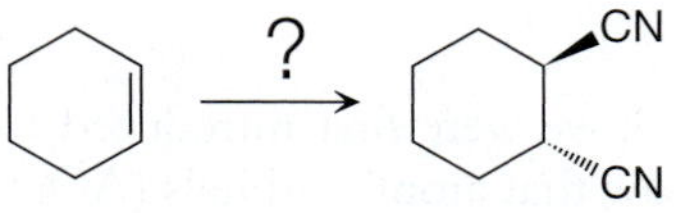

+ Enantiomer

13.50 (SYN) Propose how to carry out the synthesis shown here using any reagents necessary.

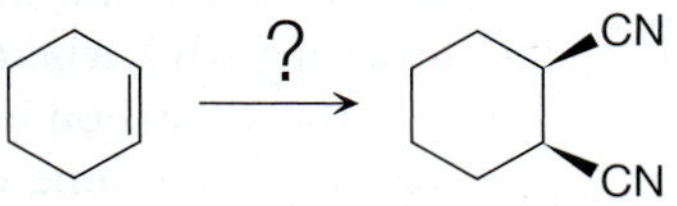

13.51 (SYN) Propose how to carry out the following synthesis using any reagents necessary. Pay attention to the stereochemistry in the target to determine the required stereochemistry of the starting material.

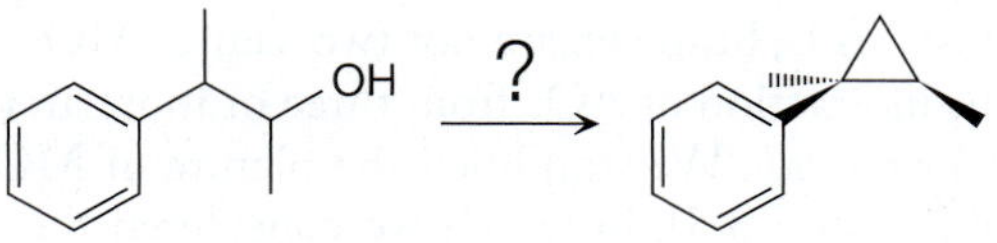

+ Enantiomer

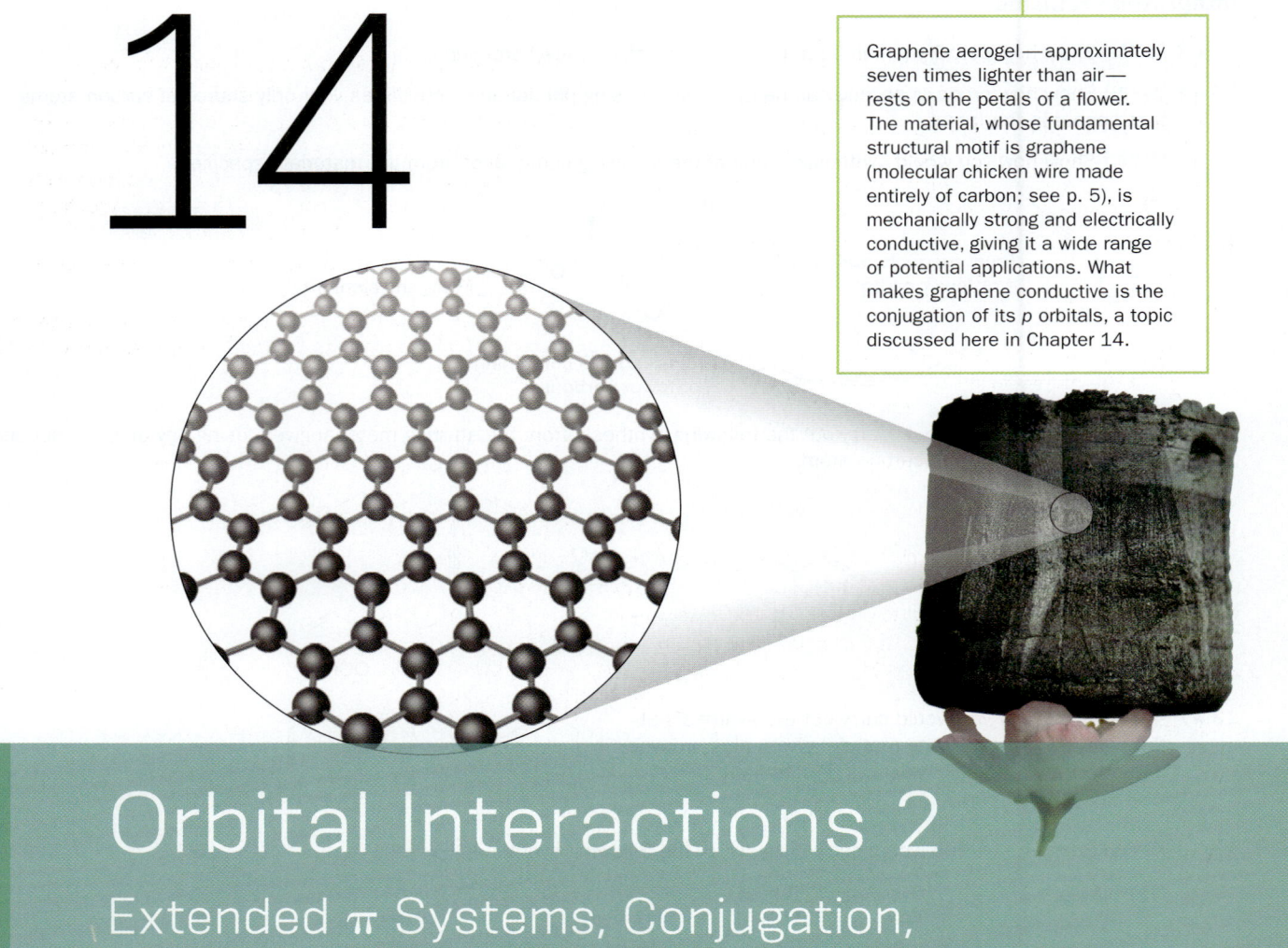

14

Orbital Interactions 2
Extended π Systems, Conjugation, and Aromaticity

I n Chapter 3, we were first introduced to the concept of *molecular orbitals (MOs)*. There, we saw that atomic orbitals (AOs) from separate atoms can overlap in space to generate new orbitals. Regions of *constructive interference* lower the energy of the resulting MO, whereas regions of *destructive interference* raise the energy.

The discussion in Chapter 3 dealt only with the MOs generated by the interaction of two AOs—one orbital from each of two atoms. These kinds of MOs are called *two-center molecular orbitals*, and include the σ bonding and σ* antibonding MOs produced by the interaction of two *s* orbitals, as well as the π bonding and π* antibonding MOs produced by the interaction of two *p* orbitals.

Most MOs, however, are not two-center MOs, but instead result from the simultaneous interaction of AOs from three or more atoms. These are called *multiple-center molecular orbitals*. We simplified the picture of MOs by applying valence bond (VB) theory (Section 3.5), in which we considered the interactions of just two AOs at a time—that is, one from each of two adjacent atoms. VB theory can account for a number of important aspects of molecular structure, but it does have significant shortcomings. Therefore, here in Chapter 14, we extend our discussion of MO theory by examining multiple-center MOs more carefully.

Chapter Objectives

On completing Chapter 14 you should be able to:

1. Explain why the two-center orbital picture of VB theory does *not* account for aspects of electron delocalization.
2. Draw qualitatively correct representations of multiple-center π MOs that result from the interaction of several parallel *p* orbitals on adjacent atoms.
3. Construct MO energy diagrams for, and determine the ground-state electron configurations of, species that have such multiple-center MOs, given only their Lewis structures.
4. Explain how multiple-center MOs account for electron delocalization over multiple atoms.
5. Identify a species as aromatic, antiaromatic, or nonaromatic, given only its Lewis structure.
6. Explain the consequences of a molecule's aromatic or antiaromatic character in terms of energy stabilization.
7. Construct a π MO energy diagram for any species with a cyclic π system and derive its ground-state electron configuration, given only its Lewis structure.
8. Determine whether a species is aromatic based on its ground-state electron configuration.
9. Determine the number of separate π systems in a given molecular species and the number of electrons occupying each system.

Although multiple-center MOs can have σ or π symmetry (or even other symmetries), we restrict the scope of our discussion solely to π MOs. We do so because of their importance in *conjugation* and *aromaticity*, concepts we learn about here in Chapter 14 that are key to our understanding of chemical structure and stability. Moreover, we often find the HOMO (highest occupied MO) and LUMO (lowest unoccupied MO) of a species to be MOs of π symmetry, giving them important roles in chemical reactions.

What we learn about multiple-center MOs here is quite useful in the chapters that follow. Important concepts of spectroscopy (the study of how molecules interact with light, discussed in Chapters 15 and 16) can be better understood with a solid foundation in multiple-center MOs. Moreover, the energetic consequences of conjugation and aromaticity can significantly affect the reactivity of particular species and the outcome of chemical reactions in which they take part. We introduce examples of this beginning in Chapter 17.

14.1 The Shortcomings of VB Theory

Some chemical and structural features of molecules are unaccounted for by VB theory—most notably *electron delocalization*, or *resonance* (Section 1.10), over three or more nuclei. Here in Section 14.1, we examine these issues, using as examples the allyl cation ($H_2C{=}CH{-}CH_2^+$) and buta-1,3-diene ($H_2C{=}CH{-}CH{=}CH_2$).

14.1a The Allyl Cation

The allyl cation ($H_2C{=}CH{-}CH_2^+$) has the following two resonance structures:

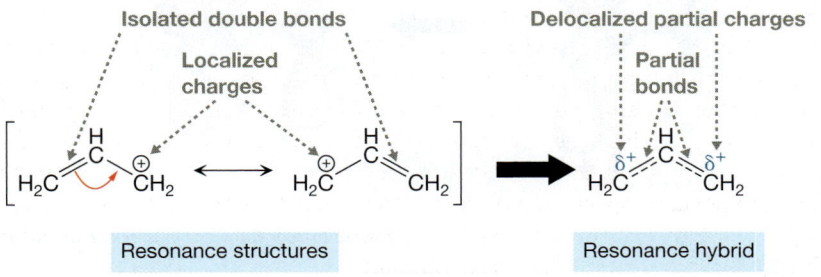

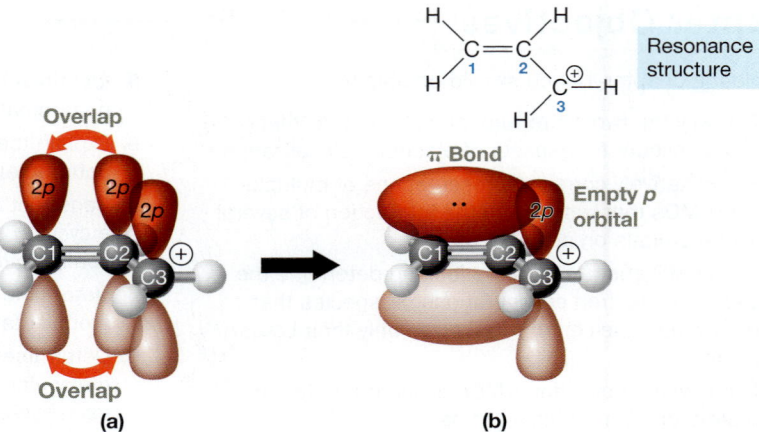

Resonance structure

FIGURE 14-1 VB theory and the allyl cation (a) Atomic *p* orbitals on the C atoms in the allyl cation. (b) A π bond has been generated by the interaction between the *p* AOs on C1 and C2. This VB theory picture represents the resonance structure shown above it, which is an inaccurate description of the species.

The one true structure is better represented by the *resonance hybrid*. The hybrid is entirely planar, with all three C atoms sp^2 hybridized. It has identical C—C bonds that are intermediate in character between single and double bonds, and it has identical partial positive charges that appear on the terminal C atoms.

VB theory does not generate this picture of the allyl cation, however, as shown in Figure 14-1. Each C atom is sp^2 hybridized, so each has a single valence *p* AO. If we suppose that the *p* orbital on C1 interacts with the one on C2, as indicated in Figure 14-1a, then the result is a double bond between C1 and C2 and a positive charge on C3. As shown in Figure 14-1b, this VB theory picture is analogous to the first of the two resonance structures shown previously, not the resonance hybrid. A similar picture emerges if we suppose that the *p* orbitals on C2 and C3 interact instead (see Your Turn 14.1). In general:

> The VB theory picture of a species that has resonance is inaccurate because it represents a single resonance structure, not the hybrid.

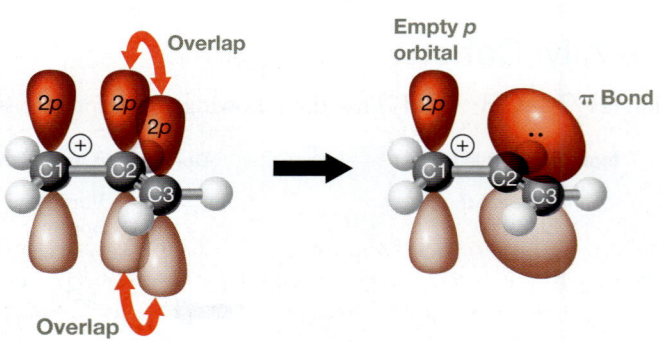

133.8 pm is slightly longer than 132 pm.

135.4 pm is shorter than 154 pm.

Ethene

Ethane

Buta-1,3-diene

FIGURE 14-2 **Bond lengths in buta-1,3-diene** The C—C single bond in buta-1,3-diene is significantly shorter than the one in ethane. Each of the C=C double bonds in buta-1,3-diene is slightly longer than the one in ethene.

PROBLEM 14.1 Draw a VB picture for the π orbitals of the allyl anion, $CH_2=CH—CH_2^-$ (all C atoms are sp^2 hybridized). Draw the resonance hybrid of the allyl anion. Do the two pictures agree or disagree? Explain.

14.1b Buta-1,3-diene

This shortcoming of VB theory is not limited to just charged species like the allyl cation. Buta-1,3-diene ($H_2C=CH—CH=CH_2$) also has structural features that VB theory does not account for. As shown in Figure 14-2, the double bonds indicated at either end of buta-1,3-diene are found experimentally to be slightly longer than the C=C double bond in ethene ($H_2C=CH_2$), whereas the C—C single bond in the middle of the molecule is found to be substantially shorter than that in ethane ($H_3C—CH_3$). Moreover, the favored geometry of buta-1,3-diene is entirely planar.

Both of these anomalies can be explained by the following weak resonance contributors:

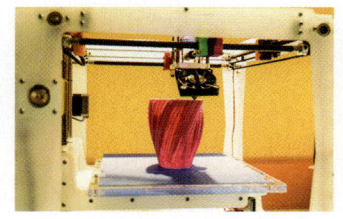

Weak resonance contributors

Buta-1,3-diene

Partial single-bond character

Partial double-bond character

These weak resonance contributors give some single-bond character to the two terminal C=C bonds, resulting in the lengthening of those bonds. The resonance contributors also give some double-bond character to the central C—C bond, resulting in the shortening of that bond, as well as the molecule's preference for an all-planar conformation (Section 3.9).

Figure 14-3 (next page) shows, however, that a VB picture fails to account for these observations. Each of the C atoms in Figure 14-3a, which are sp^2 hybridized, has one unhybridized p orbital. Interaction between the p orbitals on C1 and C2 results in a π bond between those two carbons. A second π bond results from the interaction between the p orbitals on C3 and C4. Thus, the bond between C2 and C3 remains a single bond, and the VB picture resembles the resonance structure shown (Fig. 14-3b), not the resonance hybrid.

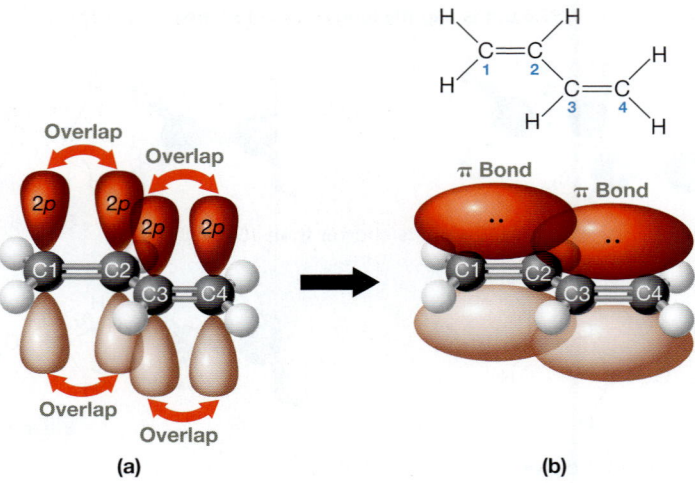

FIGURE 14-3 VB theory and buta-1,3-diene (a) The atomic *p* orbitals that contribute to the MOs. (b) The interaction between the *p* orbitals on C1 and C2 results in a π bond between C1 and C2. The interaction between the *p* orbitals on C3 and C4 results in a π bond between C3 and C4. The VB picture shown corresponds to the Lewis structure above it.

14.2 Multiple-Center MOs

To accurately describe a species like the allyl cation or buta-1,3-diene, we must use **multiple-center molecular orbitals**, which are MOs that span *multiple atoms* and involve the simultaneous interaction of *multiple AOs*.

To begin to understand the nature of these orbitals, we can apply some general results from our treatment of two-center MOs previously discussed in Section 3.3:

1. MOs arise from interacting AOs on different atoms.
2. The number of MOs formed must equal the number of AOs that contribute to their creation (this is known as the *conservation of number of orbitals*).
3. Differences in the shapes of the resulting MOs are due to the different phase relationships of the contributing AOs.
   a. Overlapping AOs with the same phase leads to *constructive interference* and a buildup of the MO.
   b. Overlapping AOs with opposite phase leads to *destructive interference* and a *node*, and the resulting MO must have opposite phases on either side of that node.
4. MO energies differ as a result of constructive and destructive interference among contributing AOs.
   a. Each pair of overlapping AOs with the same phase lowers the MO energy and is a **bonding contribution** to the MO.
   b. Each pair of overlapping AOs with opposite phases (i.e., each node) raises the MO energy and is an **antibonding contribution** to the MO.
   c. A *nonbonding MO* is similar in energy to its contributing AOs; a *bonding MO* is significantly lower in energy than its AOs; an *antibonding MO* is significantly higher in energy than its AOs.

With these rules in mind, let us see how the MOs of π symmetry are constructed for the allyl cation and buta-1,3-diene.

14.2a The Allyl Cation (H_2C=CH—CH_2^+)

There are three unhybridized *p* orbitals in the allyl cation that are **conjugated**, meaning that they are *adjacent* and *overlap in a side-by-side fashion* (Fig. 14-4). From the simultaneous interaction of these three orbitals, *three MOs must be formed*. What do those MOs look like, and how do the *p* AOs mix to form them?

Three conjugated atomic *p* orbitals

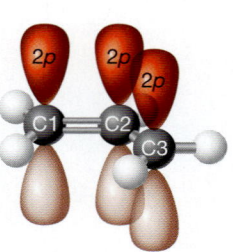

FIGURE 14-4 Conjugation in the allyl system These three *p* orbitals are conjugated because they are adjacent and parallel.

The following rule from quantum mechanics helps us answer these questions:

> For a linear system of conjugated *p* orbitals, all nodal planes in a resulting π MO must be positioned symmetrically about the center of that set of *p* orbitals.

The lowest-energy π MO, called π_1, has no nodal planes perpendicular to the bonding axes, so all three *p* AOs must mix together with the same phase, as shown on the left in the following graphic.

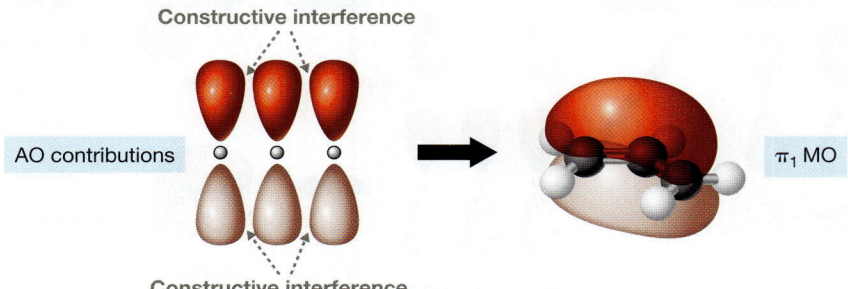

Notice that only constructive interference takes place above and below the bonding axes. Mixing the AOs therefore results in one large lobe above the bonding axes and one large lobe below, as indicated by the picture of the MO above on the right.

The second π MO, called π_2, has one nodal plane perpendicular to the bonding axes, so π_2 will appear at a higher energy than π_1. The nodal plane must appear on the central C atom because it must be positioned symmetrically about the center of the set of *p* orbitals:

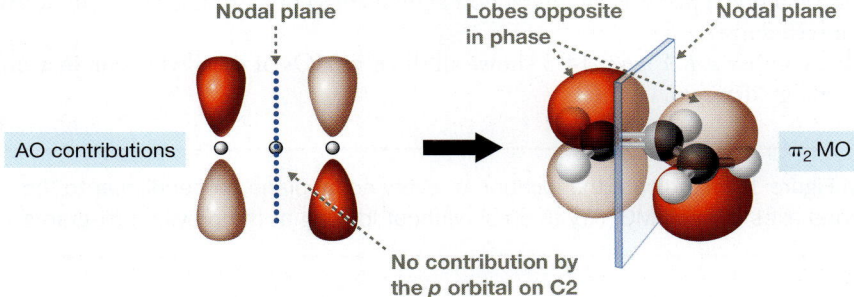

The presence of the nodal plane precisely at the central C atom means that its atomic *p* orbital must make no contribution to the MO. This ensures that an electron in that MO has zero probability of being found in that plane. The remaining two *p* orbitals contribute with opposite phase because they appear on opposite sides of the nodal plane.

The third π MO, called π_3, has two nodal planes perpendicular to the bonding axes, so π_3 has a higher energy than either π_1 or π_2. To be symmetric, the nodal planes must appear on either side of the central C atom, as shown below on the left:

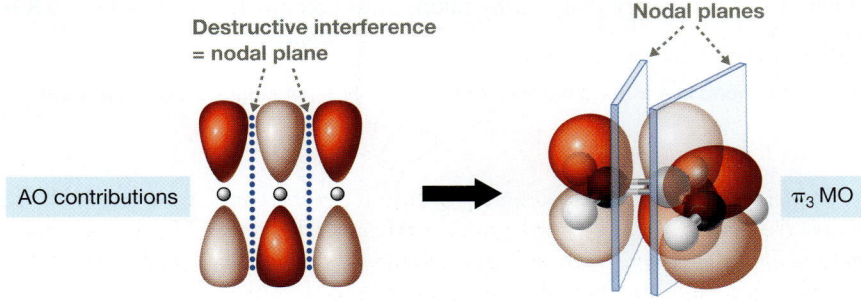

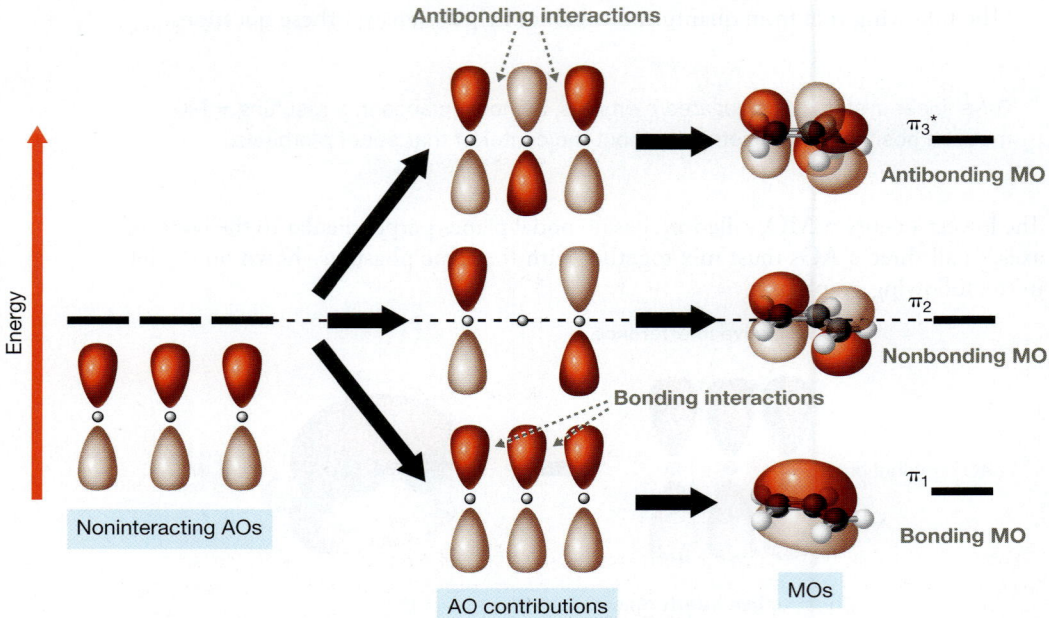

Antibonding interactions

Energy

Noninteracting AOs

AO contributions

MOs

π_3^*
Antibonding MO

π_2
Nonbonding MO

Bonding interactions

π_1
Bonding MO

FIGURE 14-5 π MOs of the allyl cation The energy of the three noninteracting p AOs is indicated on the left. The phase relationships of those AOs that give rise to the three MOs are shown in the center, and the MOs themselves (designated π_1, π_2, and π_3^*), with their respective energies, are shown on the right.

All three p AOs, therefore, contribute to the MO. Because the phases change going across each nodal plane, each interaction among adjacent p orbitals results in destructive interference.

For comparison, Figure 14-5 shows all three π MOs of the allyl cation in a single MO energy diagram.

YOUR TURN 14.2

In Figure 14-5, indicate the location of every nodal plane perpendicular to the bond axes in each MO. (Try to do so without looking at the previous diagrams.)

The π_1 MO is lower in energy than the isolated p AOs (represented by the horizontal dashed line in the middle of Fig. 14-5) because the p AOs exhibit two bonding contributions and no antibonding contributions, so it is a *bonding MO* (or, more simply, a π MO). By contrast, π_3^* is an *antibonding MO* (or, more simply, a π^* MO), because the p AOs exhibit two antibonding contributions and no bonding contributions. Finally, the energy of π_2 is roughly the same as that of the noninteracting p AOs, because the p AOs shown are not adjacent, so they essentially do not interact. This makes π_2 a *nonbonding MO*.

Having established the relative energies of the π MOs of the allyl cation, we can complete the MO energy diagram by taking into account the allyl cation's MOs of σ symmetry.

A double bond is a σ bond plus a π bond.   A single bond is a σ bond.

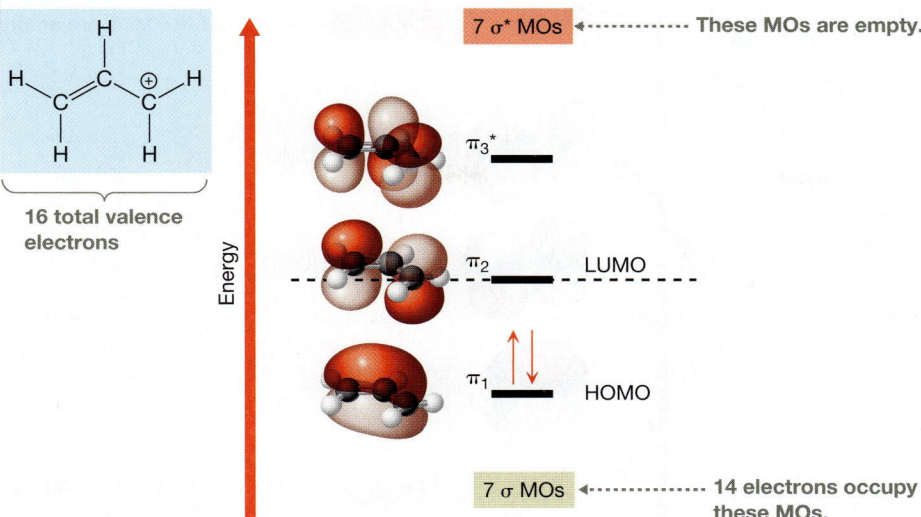

16 total valence electrons

7 σ* MOs ·········· **These MOs are empty.**

π_3^*

π_2 LUMO

π_1 HOMO

7 σ MOs ·········· **14 electrons occupy these MOs.**

Energy

FIGURE 14-6 Complete orbital energy diagram of the allyl cation The σ MOs are lower in energy than the MOs of π symmetry, whereas the σ* MOs are higher in energy. The 16 total valence electrons fill the eight MOs that are lowest in energy, making π_1 the HOMO and π_2 the LUMO.

Recall from Section 3.5 that each single bond depicted in the Lewis structure represents a pair of electrons in a σ MO. The double bond, moreover, which consists of one σ bond and one π bond, represents another pair of electrons occupying a σ MO. Thus, in the allyl cation, there are seven σ MOs in all. As we learned in Section 3.6, these σ MOs should be lower in energy than the π MOs, as shown in Figure 14-6.

As we learned in Section 3.5, too, the number of σ* MOs should equal the number of σ MOs. The σ* MOs should appear higher in energy than the MO of π symmetry.

The 16 valence electrons (depicted as eight total bonds in the Lewis structure) occupy these MOs according to the aufbau principle. The first 14 electrons completely fill all seven of the σ MOs. The remaining two electrons are placed into π_1, leaving the π_2 and π_3^* MOs empty. As a result, π_1 is the HOMO and π_2 is the LUMO.

SOLVED PROBLEM 14.2

Draw the complete MO picture and energy diagram for the allyl anion ($H_2C{=}CH{-}CH_2^-$), which is similar to that shown in Figure 14-6. Identify the HOMO and the LUMO. (All carbon atoms are sp^2 hybridized, which accounts for the double-bond character involving the carbon atoms in the resonance hybrid; Problem 14.1.)

Think How many p orbitals are conjugated? How many π MOs should result? How do those π MOs differ from each other? How many total σ and σ* MOs should there be? How many total valence electrons are there, and how should they be arranged in the orbitals?

Solve Each sp^2-hybridized carbon contributes one p orbital, so there are three p orbitals that are conjugated. Their simultaneous overlap forms three π MOs, the same as those in the allyl cation. Additionally, as we can see from their Lewis structures, the allyl anion has the same types of σ MOs as the allyl cation. The difference between the two species is that the allyl anion has two more electrons than the allyl cation. In this case, the last two electrons are placed in π_2, making π_2 the HOMO and π_3^* the LUMO.

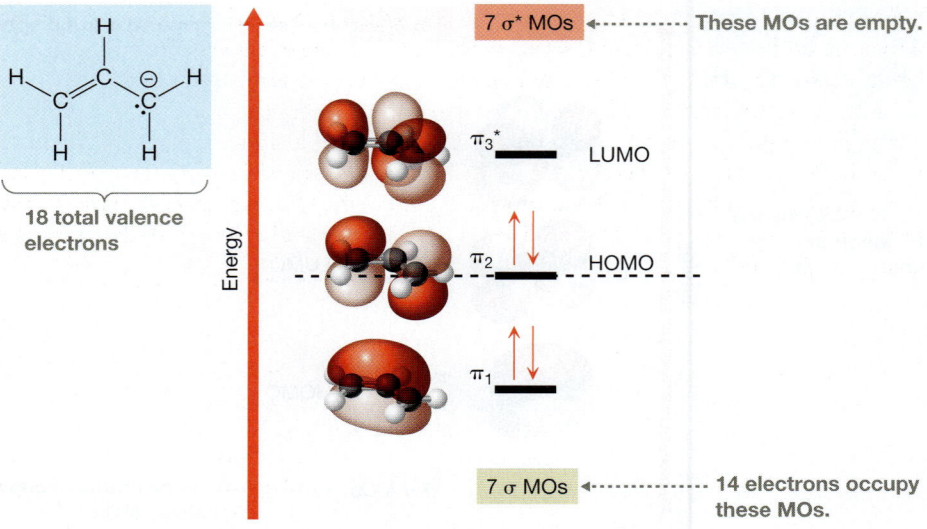

7 σ* MOs ←-------- These MOs are empty.

π_3^* — LUMO

π_2 — HOMO

π_1

7 σ MOs ←------- 14 electrons occupy these MOs.

H₂C=CH—C:̈H

18 total valence electrons

Energy

PROBLEM 14.3 Draw the complete MO picture and energy diagram for the allyl *radical* (H₂C=CH—CH₂˙), which has one unpaired electron. Identify the HOMO and the LUMO. (All carbon atoms are sp^2 hybridized.)

How does this MO treatment account for the characteristics of the allyl cation that VB theory could not? That is, how does MO theory account for the allyl cation's planarity, its equivalent carbon–carbon bonds, and the identical partial positive charges on the terminal carbons?

The planarity of the allyl cation is largely due to π_1, which contains both π electrons. Recall that π_1 is the result of two bonding contributions (shown again in Fig. 14-7, left). When a terminal CH₂ group is rotated 90° (Fig. 14-7, right), its p AO is no longer conjugated to the other two p AOs, leaving only one bonding contribution. This raises the energy of the π MO, and of the π electrons, too.

The π_1 MO also accounts for the equivalent carbon–carbon bonds. As shown in Figure 14-8, the electrons in π_1 are distributed symmetrically over both the C1—C2 and the C2—C3 internuclear regions. These two electrons correspond

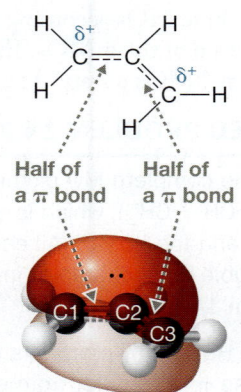

Half of a π bond   Half of a π bond

The π_1 MO extends over all three C atoms.

FIGURE 14-8 Symmetric carbon–carbon bonds in the allyl cation The π electrons that occupy the π_1 MO are symmetrically distributed over the two carbon–carbon bonding regions. Because these two electrons account for one bond total, each pair of C atoms effectively receives half of a π bond.

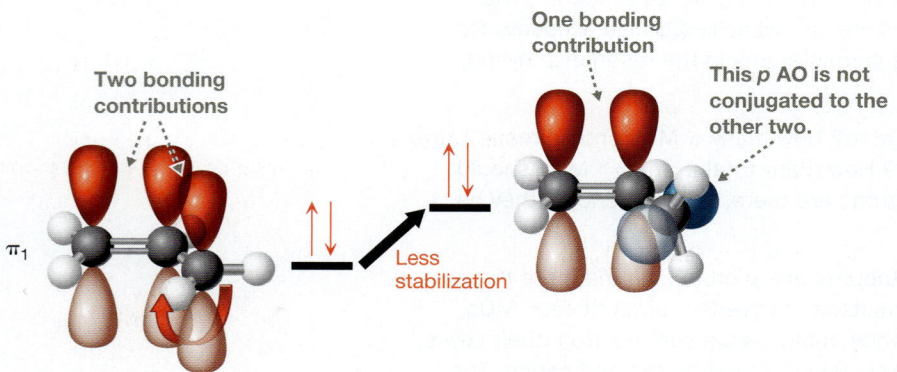

Two bonding contributions

One bonding contribution

This p AO is not conjugated to the other two.

π_1

Less stabilization

FIGURE 14-7 The planarity of the allyl cation When the allyl cation is entirely planar (*left*), two bonding contributions stabilize the electrons in π_1. When the terminal CH₂ group is rotated 90° (*right*), only one bonding contribution remains, which raises the energy of the orbital and its electrons.

to one π bond in total, so C1—C2 and C2—C3 each receive half of a π bond. Taking into account the C1—C2 and C2—C3 σ bonds, there are effectively one and one-half bonds connecting each pair of C atoms, consistent with the resonance hybrid.

Finally, to understand why each terminal carbon bears the same partial positive charge, examine the orbitals in Figure 14-9. Figure 14-9a shows the empty p orbital from the VB theory picture (Fig. 14-1), which we associate with a full positive charge. The analogous orbital in MO theory is the empty π_2 MO, which is equally shared between C1 and C3 (Fig. 14-9b). This results in an equal sharing of that positive charge on C1 and C3, or a partial charge of $+\frac{1}{2}$ on each.

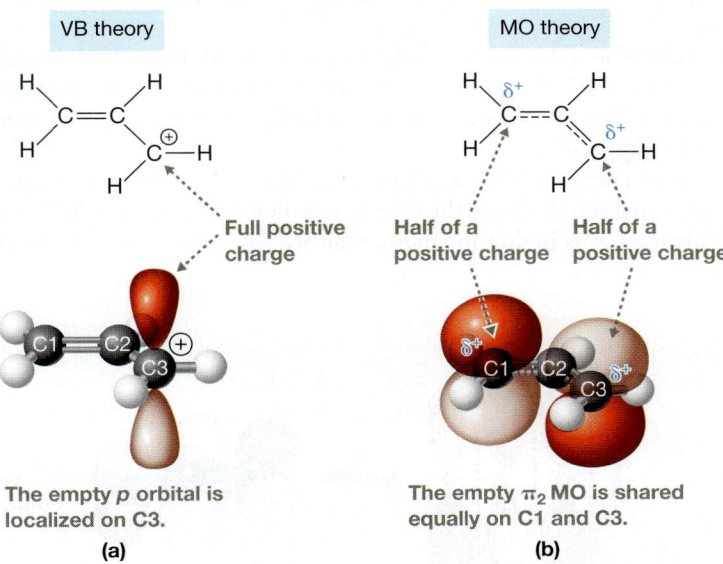

FIGURE 14-9 Equivalent partial positive charges in the allyl cation (a) In the VB picture the localized positive charge is associated with an empty p AO. (b) The corresponding MO is symmetrically distributed over C1 and C3, so the positive charge is equally shared by those two atoms.

SOLVED PROBLEM 14.4

Using the MO picture of the allyl anion ($H_2C\!=\!CH\!-\!CH_2^-$), explain why each terminal carbon atom bears the same partial negative charge. (If necessary, review Solved Problem 14.2.)

Think There are two more electrons in the allyl anion than in the allyl cation. Which orbital do those electrons occupy in the VB theory picture? What kind of charge do we associate with an occupied orbital like that? Which orbital do those two electrons occupy in the MO picture? How is the distribution of charge affected when that multiple-center MO is occupied?

Solve In the VB theory picture, the two additional electrons constitute a lone pair of electrons occupying a p orbital, which we associate with a full negative charge for carbon. In the MO theory picture, those electrons occupy the π_2 MO, which delocalizes the electrons onto C1 and C3. Thus, the negative charge is shared equally on C1 and C3, for a partial charge of $-\frac{1}{2}$ on each.

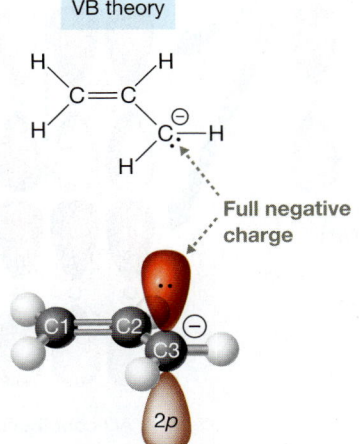

The p orbital localizes the nonbonding electrons on C3.

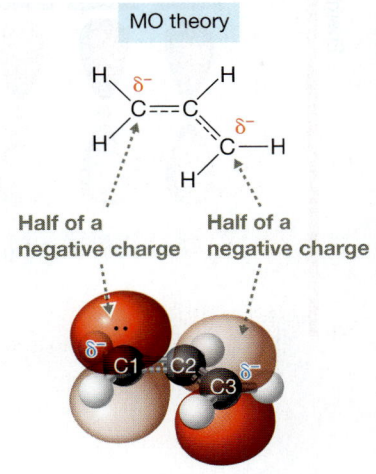

The π_2 MO delocalizes the nonbonding electrons onto C1 and C3.

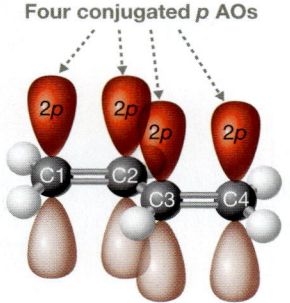

Four conjugated *p* AOs

FIGURE 14-10 Conjugation in buta-1,3-diene These four *p* orbitals are conjugated because they are adjacent and parallel.

14.2b Buta-1,3-diene (H₂C=CH—CH=CH₂)

To arrive at the MO picture and electron configuration of buta-1,3-diene, we can apply the same principles we just used for the allyl cation. With an all-planar conformation of the molecule, the four *p* AOs are parallel and, therefore, *conjugated* to one another (Fig. 14-10). As a result, they interact simultaneously to form *four MOs of π symmetry.*

The π MOs of buta-1,3-diene are shown in Figure 14-11. As with the allyl cation, the π MOs of buta-1,3-diene differ in the number of nodal planes perpendicular to the bonding axes. These differences are due, in turn, to the various phase combinations of the contributing *p* AOs.

In the lowest-energy π MO, π_1, all of the *p* AOs have the same phase, so there are no nodal planes perpendicular to the C—C bonding axes. The next MO, π_2, contains one nodal plane, so it is higher in energy than π_1. Just as we saw with π_2 of the allyl

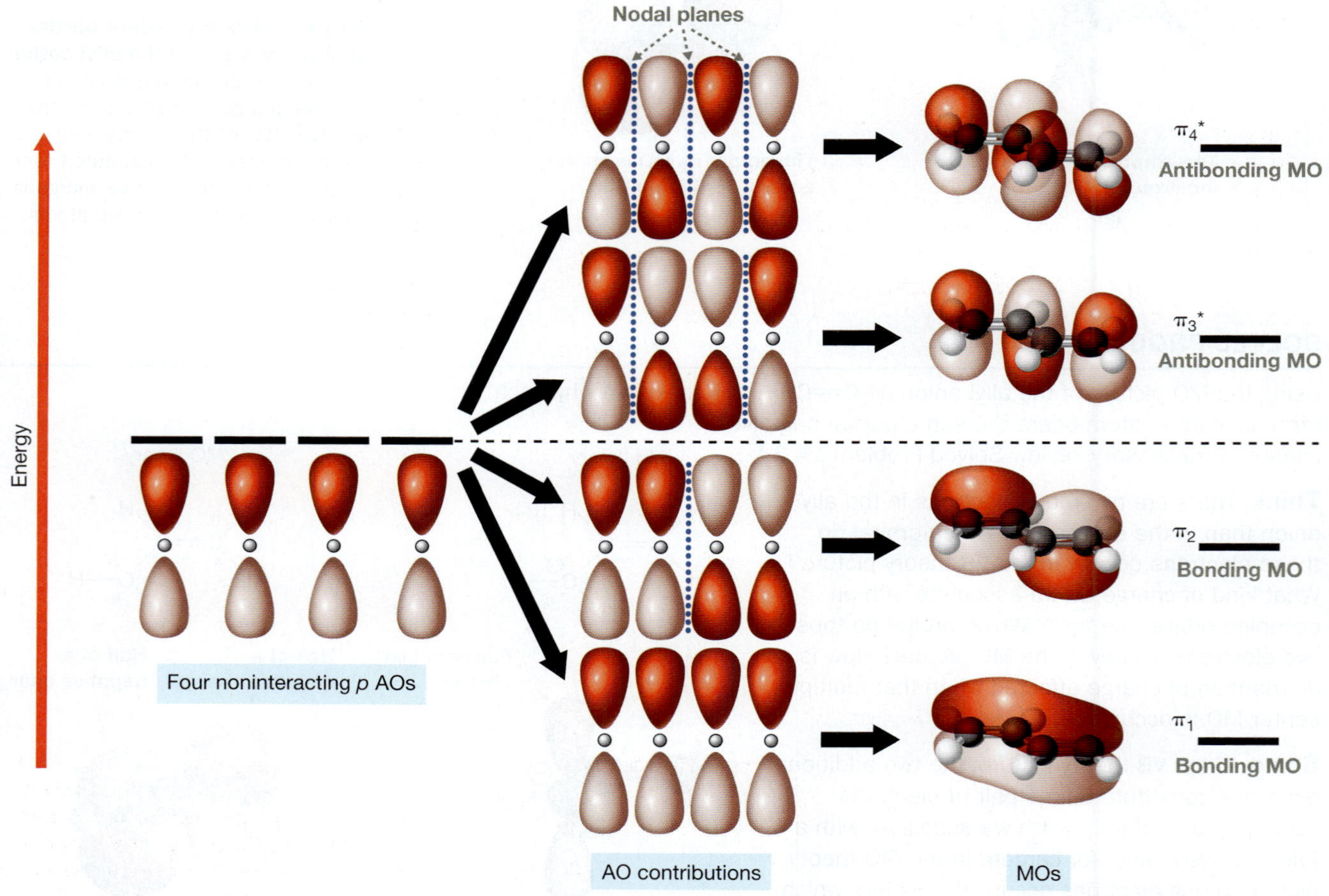

Nodal planes

Energy

Four noninteracting *p* AOs

AO contributions

MOs

π_4* Antibonding MO

π_3* Antibonding MO

π_2 Bonding MO

π_1 Bonding MO

FIGURE 14-11 π MOs of buta-1,3-diene The energy of the four noninteracting *p* orbitals is indicated on the left. The phase relationships of those AOs that give rise to the four MOs are shown in the center, and the MOs themselves (designated π_1, π_2, π_3*, and π_4*), with their respective energies, are shown on the right. The blue dotted lines represent nodal planes.

cation, that single nodal plane must appear at the center of the system of p orbitals. The π_3^* MO has two nodal planes and is thus higher in energy than π_2. Notice that those two nodal planes are positioned symmetrically about the center of the molecule—one between C1 and C2, and the other between C3 and C4. Finally, π_4^*, the highest-energy MO, contains three nodal planes positioned symmetrically about the center of the molecule.

Unlike the allyl cation, buta-1,3-diene has no nonbonding π MOs. For both π_1 and π_2, the contributing p AOs exhibit more bonding contributions than antibonding ones. This results in a significant lowering of the energy, making both of them *bonding MOs*. The opposite is true for both π_3^* and π_4^*.

YOUR TURN 14.3

For the AO contributions of each MO in Figure 14-11, label all regions that give rise to bonding contributions and all regions that give rise to antibonding contributions.

PROBLEM 14.6 The drawing shown here indicates the p AO contribution to a single MO of π symmetry. Locate the regions that give rise to bonding contributions and to antibonding contributions. How many are there of each? Will the contributions of the p AOs in this fashion produce a bonding, nonbonding, or antibonding MO? Explain.

The complete orbital energy diagram of buta-1,3-diene is shown in Figure 14-12. There are nine total σ MOs, one for each of the seven single bonds shown in the Lewis structure and one for each of the two double bonds. There are also nine total σ^* MOs, one for each bonding MO. Just as with the allyl cation, the σ MOs are lower in energy than the π MOs, and the σ^* MOs are higher in energy than the π^* MOs.

Looking again at the Lewis structure of buta-1,3-diene, notice that there are 22 total valence electrons. The first 18 of those electrons occupy the nine σ MOs. The remaining four electrons occupy π_1 and π_2, the two lowest-energy π MOs. The remaining two MOs of π symmetry, π_3^* and π_4^*, are empty. Therefore, π_2 is the HOMO, and π_3 is the LUMO.

FIGURE 14-12 Complete orbital energy diagram of buta-1,3-diene The σ MOs are lower in energy than the MOs of π symmetry, whereas the σ^* MOs are higher in energy. The 22 total valence electrons occupy the 11 MOs that are lowest in energy, making π_2 the HOMO and π_3^* the LUMO.

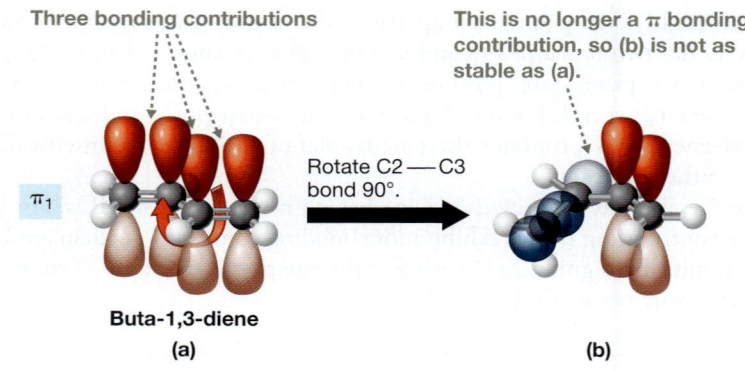

Three bonding contributions

This is no longer a π bonding contribution, so (b) is not as stable as (a).

π_1

Rotate C2—C3 bond 90°.

Buta-1,3-diene

(a)

(b)

FIGURE 14-13 Planarity of buta-1,3-diene (a) With buta-1,3-diene in an all-planar conformation, the four *p* atomic orbitals contribute to the π_1 MO to give rise to three bonding contributions. (b) Rotating 90° about the C2—C3 bond destroys one of the bonding contributions, resulting in a less stable molecule.

PROBLEM 14.7 Draw the complete MO energy diagram for the butadienyl dication, $[H_2C—CH=CH—CH_2]^{2+}$, similar to the one in Figure 14-12. Draw in all of the valence electrons and identify the HOMO and the LUMO. *Note:* All carbons are sp^2 hybridized.

Now that we have developed the MO picture of buta-1,3-diene, we can see why the molecule favors the all-planar conformation about the C2—C3 bond. Planarity is favored by the electrons occupying π_1, which exhibits a bonding contribution between C2 and C3, as shown in Figure 14-13a.[1] When the C2—C3 bond is rotated 90°, however, as shown in Figure 14-13b, that bonding interaction is no longer intact, resulting in an increase in the energy of two π electrons.

The increase in energy that comes about when conjugation is disrupted between two pairs of *p* AOs is evident in some chemical properties, too. For example, the heat of combustion of hexa-1,4-diene is greater (more exothermic) than that of hexa-1,3-diene by about 20 kJ/mol (5 kcal/mol).

These double bonds are *conjugated* and more thermodynamically stable.

These double bonds are *isolated* and less thermodynamically stable.

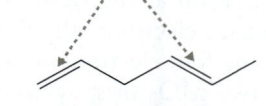

Hexa-1,3-diene
Heat of combustion = −3624 kJ/mol
(−866 kcal/mol)

Hexa-1,4-diene
Heat of combustion = −3644 kJ/mol
(−871 kcal/mol)

In hexa-1,3-diene, all four *p* AOs are conjugated, so the molecule is said to have *conjugated double bonds*. In hexa-1,4-diene, the double bonds are separated from each other by an sp^3-hybridized carbon, making these *isolated* double bonds. The greater heat of combustion of hexa-1,4-diene means that the molecule is higher in energy and less thermodynamically stable. In general:

> Double bonds are more thermodynamically stable when they are conjugated than when they are isolated.

The unusually short C2—C3 bond length and the unusually long C1—C2 and C3—C4 bond lengths in buta-1,3-diene are also accounted for by its π_1 MO. As

---

[1] In π_2, there is an *antibonding* contribution between C2 and C3, which partly cancels the bonding contribution in π_1. The bonding interaction in π_1 is not fully canceled, however, because the *p* AOs on C2 and C3 contribute more to π_1 than to π_2. Why this is so is beyond the scope of this book.

shown in Figure 14-14, the two electrons in π_1 are delocalized over the entire molecule. Thus, the electrons that occupy that MO give the C2—C3 bond some π bond character, at the expense of some π bond character of the C1—C2 and the C3—C4 bonds. This makes the central bond slightly stronger and shorter than a normal C—C single bond and makes each terminal bond weaker and longer than a normal C=C double bo.. [1]

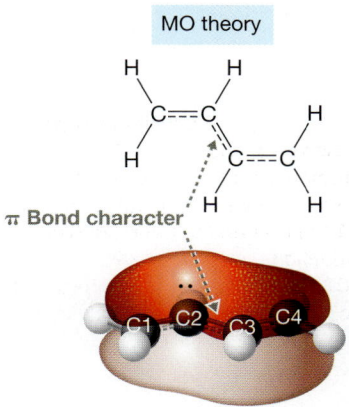

MO theory

π Bond character

The π_1 MO delocalizes electrons over the C1—C2, C2—C3, and C3—C4 bonding regions.

14.3 Aromaticity and Hückel's Rules

Sections 14.1 and 14.2 discussed aspects of conjugation pertaining to the allyl cation and buta-1,3-diene. Conjugation is also apparent in benzene (C_6H_6), a cyclic, six-carbon molecule whose Lewis structure depicts alternating single and double bonds. Based on the Lewis structure, benzene should have alternating shorter and longer carbon–carbon bonds. Experimentally, however, all six of them are identical in length, intermediate between that of a normal single bond and a normal double bond, as shown in Figure 14-15a. Furthermore, the entire molecule is perfectly planar and all six C—C—C bond angles are 120°. This high degree of symmetry is consistent with benzene's resonance hybrid, shown in Figure 14-15b.

FIGURE 14-14 The unusual carbon–carbon bond lengths of buta-1,3-diene The π_1 MO is delocalized over the entire molecule, which gives the C2—C3 bond some π bond character at the expense of some π bond character from the C1—C2 and C3—C4 bonds. This shortens the C2—C3 bond and lengthens the other two.

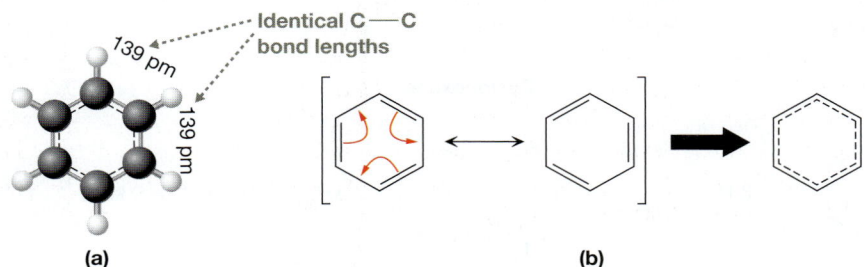

Identical C—C bond lengths

139 pm

139 pm

(a)   (b)

FIGURE 14-15 The structure of benzene (a) The carbon–carbon bond lengths in benzene are all identical. (b) Benzene has equivalent resonance structures (left), the hybrid of which (right) suggests a symmetric molecule.

Perhaps more importantly, from a chemical standpoint:

Benzene's system of π electrons is *unusually stable*.

For example, whereas Br_2 adds to a normal alkene under relatively mild conditions (Equation 14-1; Section 12.3), no reaction occurs with benzene under these same conditions (Equation 14-2).

Br $\overset{Br_2}{\longrightarrow}$ Br ⟍⟍ Br ⟍⟍⟍   +   Enantiomer   (14-1)

$\overset{Br_2}{\longrightarrow}$ No reaction   (14-2)

If a strong Lewis acid catalyst such as $FeBr_3$ is present (Equation 14-3), then a reaction does take place, but rather than addition, benzene undergoes substitution to preserve the π system. (We discuss these kinds of reactions in greater detail in Chapters 22 and 23.)

Benzene undergoes substitution rather than addition.

$$(14\text{-}3)$$

We can gain a sense of benzene's stability from **heat of hydrogenation** (ΔH°_{hyd}) data. As Equation 14-4 shows, 1 mol of benzene reacts with 3 mol of H_2 in the presence of a catalyst, under high temperature and pressure, to produce 1 mol of cyclohexane (this reaction is discussed in greater detail in Chapter 19).

$\Delta H^{\circ}_{hyd} = -208$ kJ (−49.7 kcal) $(14\text{-}4)$

$\Delta H^{\circ}_{hyd} = -360$ kJ (−86.0 kcal) $(14\text{-}5)$

Similarly, 3 mol of cyclohexene will undergo hydrogenation in the presence of a catalyst (Equation 14-5) to produce 3 mol of cyclohexane, but under much milder conditions. Both of these reactions can be simplified to Equation 14-6, in which 3 mol of C=C double bonds react with 3 mol of H_2 to produce 3 mol of C—C single bonds and 6 mol of C—H bonds.

$$3\ C{=}C\ +\ 3\ H_2\ \longrightarrow\ 3\ C{-}C\ +\ 6\ C{-}H \qquad (14\text{-}6)$$

However, ΔH°_{hyd} for the two reactions differs significantly. Whereas the hydrogenation of 1 mol of benzene (Equation 14-4) releases 208 kJ (49.7 kcal), the hydrogenation of 3 mol of cyclohexene (Equation 14-5) releases 360 kJ (86.0 kcal). Thus, as shown in the energy diagram in Figure 14-16, benzene's π system is more thermodynamically stable by 360 kJ − 208 kJ = 152 kJ (36.3 kcal). This "extra" stability is much more than we would expect from the normal conjugation of double bonds, such as in hexa-1,3-diene (~20 kJ/mol or 5 kcal/mol; p. 694) and is attributed to **aromaticity**. It is sometimes referred to as benzene's *resonance energy*.

YOUR TURN 14.4

What fraction of an average C—C single bond energy is benzene's resonance energy? *Hint:* You may need to review Section 1.4.

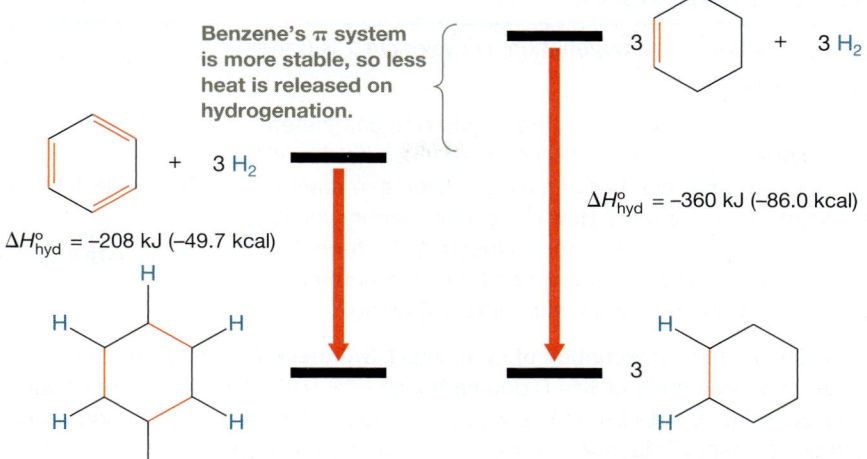

Benzene's π system is more stable, so less heat is released on hydrogenation.

$\Delta H^{\circ}_{hyd} = -208$ kJ (–49.7 kcal)

$\Delta H^{\circ}_{hyd} = -360$ kJ (–86.0 kcal)

FIGURE 14-16 The stability of benzene The hydrogenation of one mole of benzene (*left*) and three moles of cyclohexene (*right*) are both exothermic. The π system of benzene is stabilized by aromaticity, however, so it releases significantly less heat on hydrogenation.

Conjugated Linoleic Acids

Unsaturated and polyunsaturated fatty acids (PUFAs) have received a lot of attention over the years for their health benefits, including lowering the risk of heart attacks and cardiovascular disease. Some of these fatty acids—oleic acid, linoleic acid, and linolenic acid—were examined in Section 4.15. In the biosynthesis of these fatty acids, the cis configurations of the double bonds are produced almost exclusively, and the double bonds themselves are separated by a single CH_2 group. These double bonds, therefore, are isolated from each other.

In the stomach of ruminant animals (such as cattle), certain bacteria carry out biohydrogenation to convert dietary linoleic acid to stearic acid, its saturated form. Conjugated linoleic acids (CLAs), which are isomers of linoleic acid, are intermediates in this process, and the predominant CLA is the 9-cis, 11-trans isomer.

Isolated double bonds

Linoleic acid

Conjugated double bonds

9-Cis, 11-trans CLA

These CLAs have significantly different bioactivity than PUFAs, and some studies suggest that CLAs exhibit remarkable health benefits of their own. CLAs have had success complementing some cancer treatments, they are used to treat inflammatory bowel disease and Crohn's disease, and they have even been shown to be effective weight-loss supplements.

It is not entirely clear what gives rise to the unique bioactivity of these CLAs, but at least in some cases, the additional rigidity brought about by the conjugation of the double bonds is believed to play a role. Just as we saw with buta-1,3-diene here in Chapter 14, the conjugation in 9-cis, 11-trans CLA favors a coplanar arrangement of the double bonds, and this rigidity can affect the ability of the fatty acid to dock with specific enzyme active sites or receptors.

SOLVED PROBLEM 14.8

Based on its heat of hydrogenation, is cycloocta-1,3,6-triene aromatic? Explain.

Cycloocta-1,3,6-triene

$\Delta H^{\circ}_{hyd} = -334$ kJ/mol
(−79.8 kcal/mol)

Think How many moles of H_2 are required to completely hydrogenate cycloocta-1,3,6-triene? How much heat is released when that number of moles of hydrogen reacts with cyclohexene instead? How does that quantity compare to the heat of hydrogenation for cycloocta-1,3,6-triene? Can the extra stability be accounted for by the normal conjugation of double bonds as in hexa-1,3-diene?

Solve Complete hydrogenation of cycloocta-1,3,6-triene requires 3 mol of H_2—one for each mole of C=C double bonds. If 3 mol of H_2 were to react with cyclohexene instead, then (3)(120 kJ/mol) = 360 kJ/mol (86.0 kcal/mol) of heat would be released. This is 26 kJ/mol (6.2 kcal/mol) more heat than is released when hydrogenating cycloocta-1,3,6-triene, suggesting that the three double bonds in cycloocta-1,3,6-triene are about 26 kJ/mol (6.2 kcal/mol) more stable than in cyclohexene. This is much smaller than the 152 kJ/mol (36.3 kcal/mol) of stabilization provided by the three double bonds in benzene, and is similar to the stability from conjugation we saw for hexa-1,3-diene, so cycloocta-1,3,6-triene is *not* aromatic.

PROBLEM 14.9 Based on their heats of hydrogenation (ΔH°_{hyd}), predict whether each of these compounds is aromatic.

(a) Cyclopenta-1,3-diene

$\Delta H^{\circ}_{hyd} = -211$ kJ/mol
(−50 kcal/mol)

(b) Naphthalene

$\Delta H^{\circ}_{hyd} = -318$ kJ/mol
(−76 kcal/mol)

The unusual stability of benzene's π system can be explained, in part, by electron delocalization through *resonance* (Chapter 1). The two resonance structures of benzene, called **Kekulé structures** after the German chemist, August Kekulé (1829–1896), who proposed them, are shown in Figure 14-15b (p. 695). Because the resonance structures are *equivalent*, they should contribute equally to the hybrid structure, allowing the six π electrons to be delocalized completely around the ring.

If resonance fully explained this phenomenon, however, then we should expect a similar stability in the π system of cyclobutadiene. Just like benzene, cyclobutadiene can be represented as a fully conjugated ring and two equivalent resonance structures can be drawn (Fig. 14-17). Unlike benzene, however:

Cyclobutadiene's system of π electrons is *unusually unstable*.

FIGURE 14-17 The structure of cyclobutadiene (a) Cyclobutadiene has two different bond lengths, one that is characteristic of a C=C double bond and one that is characteristic of a C—C single bond. (b) Cyclobutadiene does not enjoy the same resonance stabilization that benzene does.

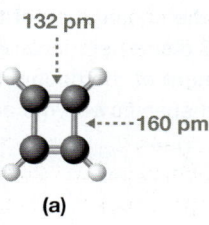

132 pm

·····160 pm

(a)

Not resonance structures because atoms move in response to differing bond lengths

(b)

Cyclobutadiene's π system is so unstable, in fact, that the compound has only been isolated and studied in an argon matrix at extremely low temperatures. Moreover, those studies reveal that cyclobutadiene consists of *two different carbon–carbon bonds*: one that is consistent with a C=C double bond and one that is consistent with a C—C single bond; the molecule is rectangular! In other words, cyclobutadiene does not exist as a resonance hybrid, but rather equilibrates between two different structures.

YOUR TURN 14.5

Compare the carbon–carbon bond lengths in ethane and ethene to those in cyclobutadiene. (See Fig. 14-2, p. 685.)

Benzene and cyclobutadiene are members of two separate *classes* of compounds that exhibit different characteristic behavior. Benzene is classified as *aromatic*, whereas cyclobutadiene is *antiaromatic*. In general:

- **Aromatic** compounds have cyclic π systems that are unusually *stable*.
- **Antiaromatic** compounds have cyclic π systems that are unusually *unstable*.
- All other compounds are classified as **nonaromatic** (i.e., as neither unusually stable nor unusually unstable).

From his observations on the structural similarities of compounds in each of these classes, Erich Hückel (1896–1980), a German physicist and physical chemist, proposed what are now known as **Hückel's rules** for aromaticity:

Hückel's Rules for Aromaticity

- If a species possesses a π system of MOs constructed from *p* atomic orbitals that are fully conjugated around a ring, then the species is:
  - *Aromatic* if the number of electrons in that cyclic π system is either 2, 6, 10, 14, 18, and so on. (These are called **Hückel numbers**.)
  - *Antiaromatic* if the number of electrons in that cyclic π system is either 4, 8, 12, 16, 20, and so on. (These are called **anti-Hückel numbers**.)
- All other species are *nonaromatic*.

For *p* AOs to be fully conjugated around a ring, *the atoms that contribute those orbitals must lie entirely in the same plane.*

To help you remember the above numbers, notice that each Hückel number is an *odd number of pairs* (e.g., 6 is the same as three pairs), whereas each anti-Hückel number is an *even number of pairs* (e.g., 4 is the same as two pairs). Alternatively, the Hückel numbers correspond to $4n + 2$, where n is any integer ≥ 0, and the anti-Hückel numbers correspond to $4n$, where n is any integer ≥ 1.

Benzene satisfies the criteria for an aromatic species, as shown in Figure 14-18a. All six C atoms lie in the same plane and each C atom has an unhybridized *p* atomic orbital (because each is sp^2 hybridized). The six *p* orbitals are all adjacent and parallel, making them conjugated in a complete ring. Additionally, there are a total of six π electrons, which is a Hückel number.

Cyclobutadiene (Fig. 14-18b) also has a fully conjugated ring of *p* atomic orbitals because all of its C atoms, too, lie in the same plane and are sp^2 hybridized. Cyclobutadiene's π system contains four electrons, however, which is an anti-Hückel number.

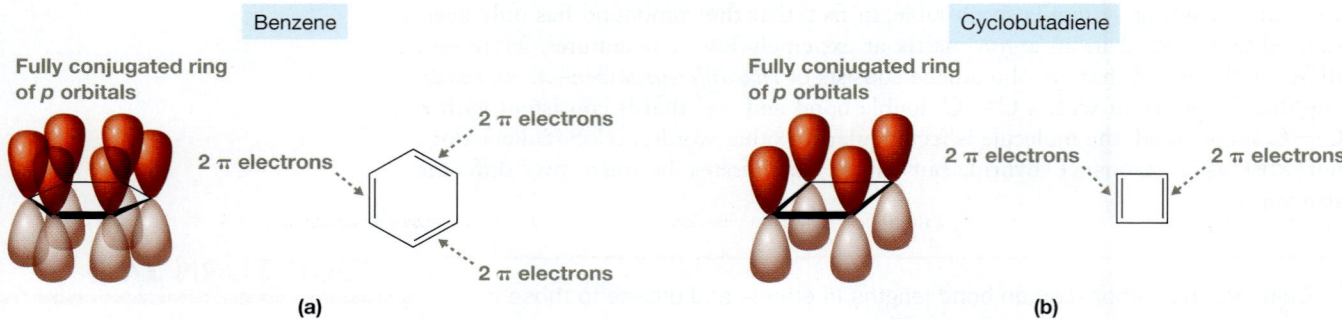

Benzene

Fully conjugated ring of p orbitals

2 π electrons

2 π electrons

2 π electrons

(a)

Cyclobutadiene

Fully conjugated ring of p orbitals

2 π electrons

2 π electrons

(b)

FIGURE 14-18 Benzene, cyclobutadiene, and Hückel's rules (a) Benzene has a cyclic π system that is fully conjugated (left), which is occupied by six electrons (right). (b) Cyclobutadiene has a cyclic π system that is fully conjugated (left), which is occupied by four electrons (right).

SOLVED PROBLEM 14.10

Assuming the molecule is entirely planar, would cycloocta-1,3,5,7-tetraene be aromatic, antiaromatic, or nonaromatic? Explain your reasoning.

Think Does planar cycloocta-1,3,5,7-tetraene contain a set of *p* atomic orbitals that are fully conjugated around a ring? If so, does it contain a Hückel or an anti-Hückel number of π electrons?

Solve All eight C atoms are sp^2 hybridized, so each contributes a *p* atomic orbital. Consequently, if the molecule is planar, then the eight *p* orbitals are fully conjugated around the ring. In that π system, there are eight electrons (two from each of the four double bonds), which is an anti-Hückel number. Thus, this compound would be antiaromatic. (In actuality, cycloocta-1,3,5,7-tetraene is *not* planar, as discussed later in Section 14.6.)

PROBLEM 14.11 Predict whether each of the following compounds is aromatic, antiaromatic, or nonaromatic.

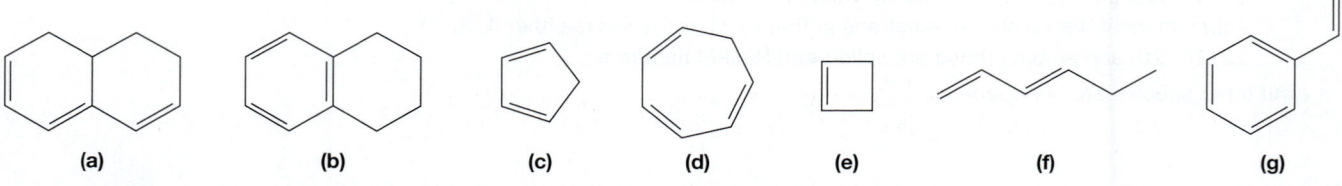

(a)   (b)   (c)   (d)   (e)   (f)   (g)

14.4 The MO Picture of Benzene: Why It's Aromatic

Given the difficulties of resonance theory in explaining the unusual stability of benzene, as well as the unusual instability of cyclobutadiene, we must turn to MO theory. In benzene, there are six unhybridized *p* orbitals fully conjugated around the ring (Fig. 14-19a). The simultaneous interaction among benzene's *p* atomic orbitals must yield six MOs of π symmetry.

We can derive the relative energies of benzene's π MOs using the **Frost method**, a shortcut developed by the American chemist Arthur A. Frost (1909–2002) that can be applied to any species with a fully conjugated, cyclic π system:

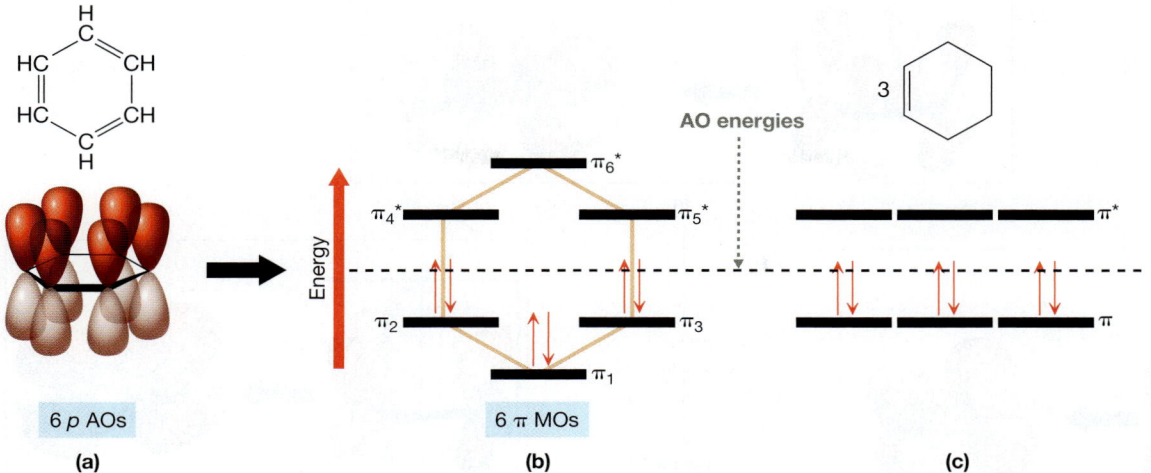

FIGURE 14-19 Energy diagram for the π MOs of benzene (a) The six p atomic orbitals of benzene. (b) The energies of the six MOs of benzene. (c) The π MOs of three molecules of cyclohexene. The dashed line indicates the energy of the noninteracting p atomic orbitals.

The Frost Method for Energies of Cyclic π MOs

1. Draw the polygon that represents the cyclic compound's line structure, with one of the vertices pointed directly downward.
2. Place the energy of the noninteracting p atomic orbitals at the center of the polygon.
3. Place the energies of the π MOs at the vertices of the polygon.

Thus, Figure 14-19b shows that benzene has a single π MO (π_1) that is lowest in energy. At a somewhat higher energy, there are the π_2 and π_3 MOs—they have identical energies and are thus called **degenerate orbitals**. A second pair of degenerate orbitals, π_4^* and π_5^*, is found at a higher energy still, and π_6^* has the highest energy. Benzene's six π electrons fill the lowest-energy π MOs first—namely π_1, π_2, and π_3.

YOUR TURN 14.6

In Figure 14-19b, label each of the six MOs as either *bonding*, *nonbonding*, or *antibonding*. Also, identify each *pair* of degenerate orbitals.

PROBLEM 14.12 Draw the complete MO energy diagram for benzene, including the MOs of σ symmetry, similar to Figures 14-6 and 14-12.

For comparison, the energy of the six π electrons in three molecules of cyclohexene, whose double bond is isolated, is shown in Figure 14-19c. Whereas benzene's π_2 and π_3 MOs are similar in energy to the π MO in cyclohexene, benzene's π_1 MO is significantly lower in energy. This provides insight into why benzene's π system of electrons is so stable.

Figure 14-20 (next page) shows how each of benzene's π MOs is produced from the contributing p AOs. As we have seen before, one way in which these MOs differ is in the phases of the contributing p orbitals. Some also differ in the number of nodal planes perpendicular to the bonding axes.

- Each pair of degenerate MOs has the same number of nodal planes perpendicular to the bonding axes.
- The energy of each MO rises with each additional nodal plane.

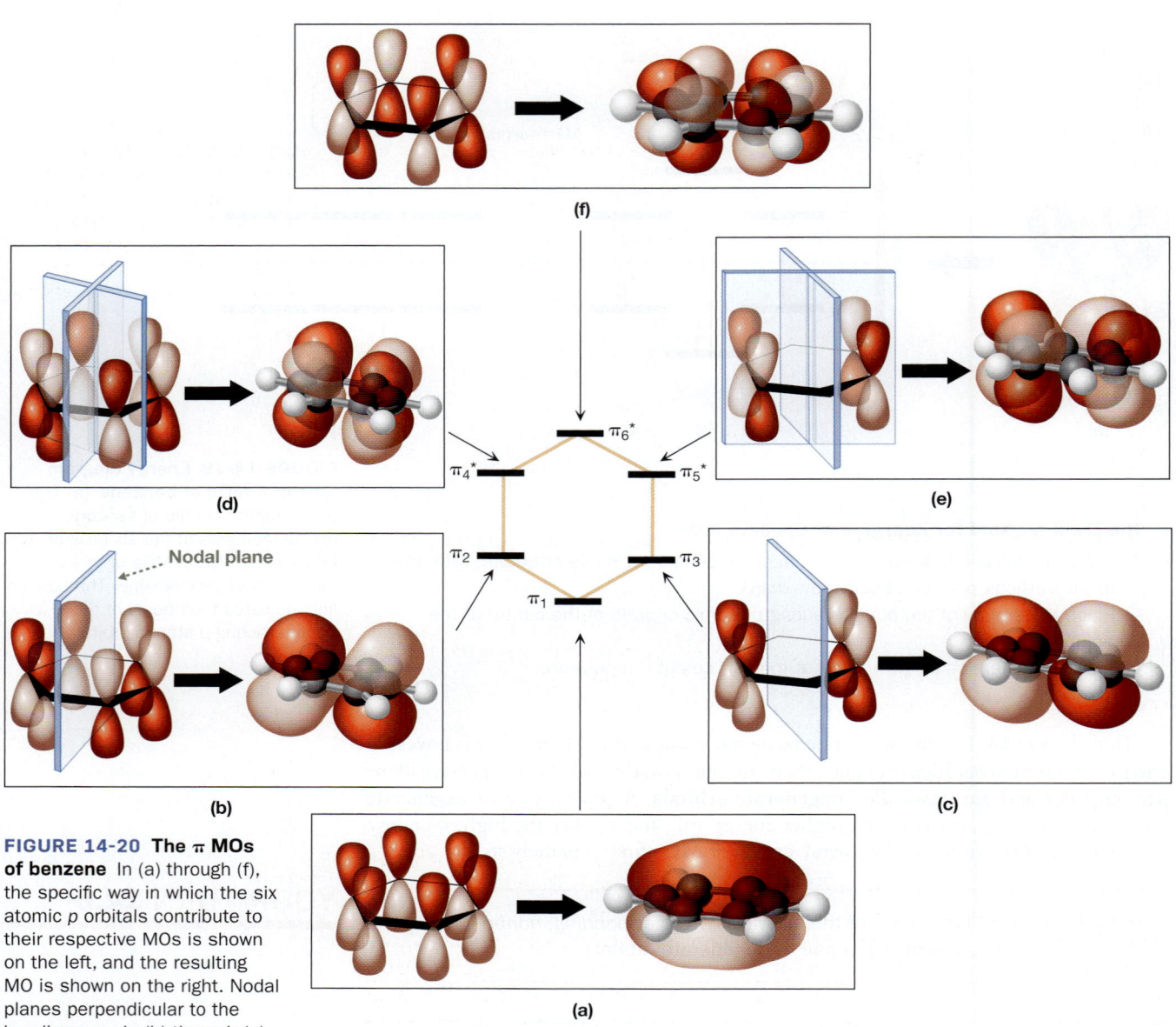

FIGURE 14-20 The π MOs of benzene In (a) through (f), the specific way in which the six atomic p orbitals contribute to their respective MOs is shown on the left, and the resulting MO is shown on the right. Nodal planes perpendicular to the bonding axes in (b) through (e) are shown in blue.

In particular, there are zero nodes for π_1, one for π_2 and π_3, and two for π_4^* and π_5^*.

YOUR TURN 14.7

In Figure 14-20, how many nodal planes perpendicular to the bonding axes should π_6^* have? Can you locate them?

14.5 The MO Picture of Cyclobutadiene: Why It's Antiaromatic

Cyclobutadiene presents a somewhat different situation. In cyclobutadiene, there must be four MOs of π symmetry generated from the four unhybridized p AOs that are fully conjugated around the ring. To obtain their relative energies, we can use the Frost method (Section 14.4) again. We orient a square so that one vertex is pointed

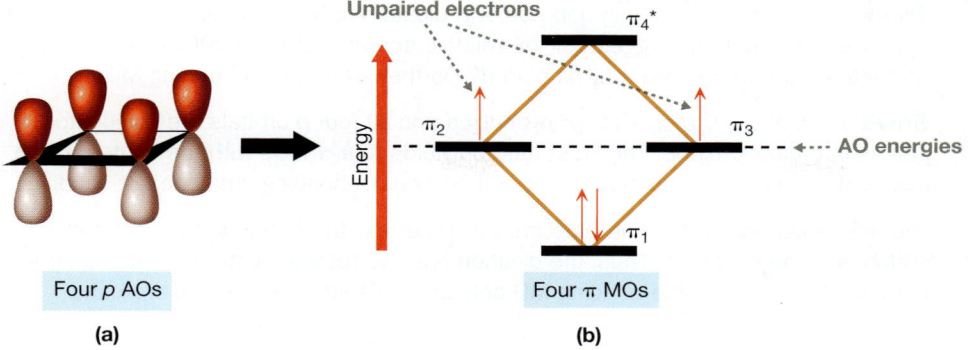

Unpaired electrons

Energy

π_4^*

π_2

π_3

AO energies

π_1

Four p AOs

Four π MOs

(a)

(b)

FIGURE 14-21 Energy diagram for the π MOs of square cyclobutadiene (a) Four unhybridized p atomic orbitals on the C atoms of cyclobutadiene. (b) Energy diagram of the four π MOs of cyclobutadiene. The dashed line represents the energy of the unhybridized p AOs.

downward (Fig. 14-21). Each of the four vertices then represents the energy of a π MO (π_1, π_2, π_3, and π_4^*), and the center of the square represents the energy of each unhybridized p AO.

YOUR TURN **14.8**

In Figure 14-21, label each MO as either bonding, antibonding, or nonbonding.

Figure 14-21 shows that the four π electrons in cyclobutadiene occupy π_1, π_2, and π_3. The π_1 MO is completely filled, whereas π_2 and π_3 contain only one electron each. This is in accordance with *Hund's rule* (Section 1.3c)—namely, orbitals of the same energy are not doubly occupied unless it is absolutely necessary.

PROBLEM 14.13 Draw the complete MO energy diagram for cyclobutadiene, including the MOs of σ symmetry, similar to Figures 14-6 and 14-12.

We can see why cyclobutadiene's π system is so much less stable than benzene's by comparing the MO energy diagram of square cyclobutadiene (Fig. 14-21) to that of benzene (Fig. 14-19). In benzene, all valence electrons occupy bonding MOs, so they are highly stabilized relative to their energies in unhybridized AOs. In cyclobutadiene, on the other hand, only two of the four π electrons are found in a bonding MO; the other two are in higher-energy nonbonding MOs. Furthermore, whereas all of benzene's π electrons are paired, cyclobutadiene has two *unpaired* electrons. As we discuss in greater detail in Chapter 25, species with unpaired electrons, known as *free radicals,* tend to be quite unstable.

In fact, the two unpaired electrons in nonbonding MOs makes square cyclobutadiene significantly less stable than if the two double bonds were isolated and the species were nonaromatic. If the two double bonds were isolated, then each double bond would give rise to one π MO and one π* MO, and all four π electrons would occupy π bonding MOs.

Figure 14-22 (next page) shows how each of cyclobutadiene's π MOs is produced from the contributing p AOs. As with the other π systems we have seen, the lowest-energy MO (π_1) has no nodal planes perpendicular to the bonding axes, and each additional nodal plane raises the energy of the other MOs.

SOLVED PROBLEM 14.14

Based on the characteristics of its MO diagram, would you expect the cyclobutadienyl dication to be aromatic, antiaromatic, or nonaromatic? Explain. *Hint:* How would Figure 14-21 change on going from the uncharged molecule to the species with a +2 charge?

$$\left[\begin{array}{c} HC-CH \\ \| \quad | \\ HC-CH \end{array} \right]^{2+}$$

The cyclobutadienyl dication

Think Are the *p* AOs in the cyclobutadienyl dication fully conjugated in a ring? What does the Frost method suggest for the relative energies of the π MOs? How many π electrons are there? Are they all paired? Do they all reside in bonding MOs?

Solve Each of the C atoms is sp^2 hybridized and all four *p* orbitals are conjugated, just as in cyclobutadiene. The Frost method yields four π MOs with the same relative energies as in cyclobutadiene—one bonding, two nonbonding, and one antibonding.

The +2 charge indicates that there are two fewer electrons than in the uncharged cyclobutadiene molecule. Thus, the dication has two total π electrons. Both electrons are paired and occupy the bonding MO only, so this species should be aromatic.

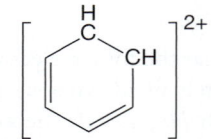

All π electrons are paired and reside in a bonding MO.

PROBLEM 14.15 Based on the characteristics of its MO diagram, do you think the benzene dication should be aromatic, antiaromatic, or nonaromatic? What would you expect for the benzene dianion? Explain. (Assume that each ion is planar.)

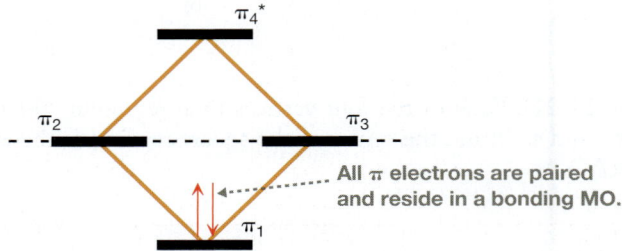

The benzene dication   The benzene dianion

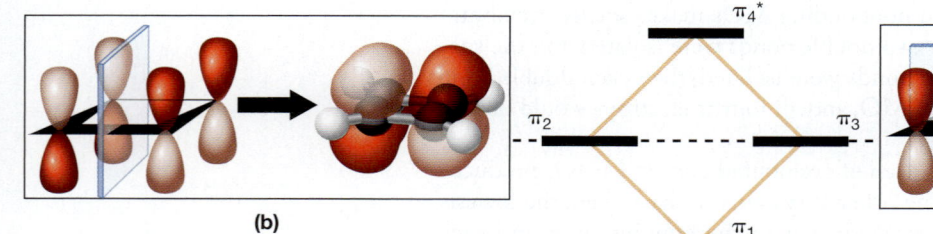

(b)

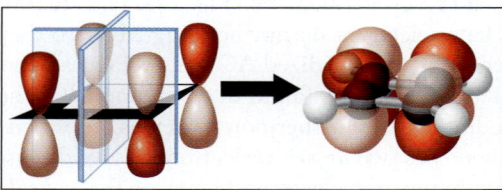

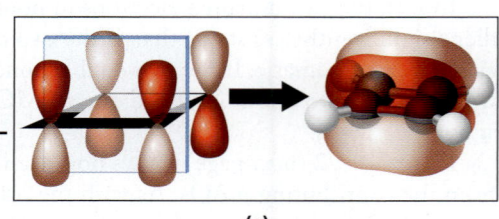

(c)

FIGURE 14-22 The π MOs of cyclobutadiene In (a) through (d), the specific way in which the four *p* atomic orbitals contribute to their respective MOs is shown on the left, and the resulting MO is shown on the right. Nodal planes perpendicular to the bonding axes are shown in blue.

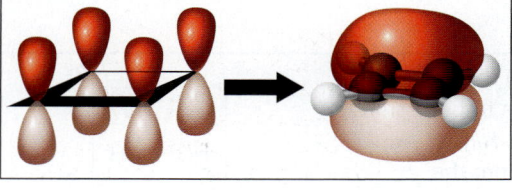

(a)

14.6 Aromaticity in Larger Rings: [*n*]Annulenes

The two species we have examined thus far with cyclic π systems—benzene and cyclobutadiene—are *monocyclic* (one ring) hydrocarbons with C=C and C—C bonds alternating around the ring. These molecules belong to a class of compounds collectively called *annulenes*. They are commonly named **[*n*]annulenes**, where *n* is the number of carbon atoms in the ring. Therefore, cyclobutadiene is [4]annulene and benzene is [6]annulene.

Cyclobutadiene and benzene are not the only [*n*]annulenes. For example, cycloocta-1,3,5,7-tetraene (Fig. 14-23a) is [8]annulene. According to Hückel's rules, [8]annulene should be antiaromatic if it were planar (review Solved Problem 14.10). This would give it a fully conjugated, cyclic π system, and it has eight π electrons, which is an anti-Hückel number. Because of the instability associated with antiaromaticity, however, [8]annulene resists planarity. Its most stable conformation is *tub shaped* (Fig. 14-23b), in which the *p* AOs are not all parallel to one another (Fig. 14-23c), and are therefore not fully conjugated. Thus, [8]annulene is better characterized as nonaromatic.

Like benzene, each carbon atom in [10]annulene, [14]annulene, and [18]annulene (Fig. 14-24) contributes a *p* AO, and each molecule has a Hückel number of π electrons—10, 14, and 18, respectively. As expected, [14]annulene and [18]annulene are aromatic. [10]Annulene is not aromatic, however, due to the excessive ring strain it would have if it were completely planar. With all of the double bonds in a cis configuration (Fig. 14-24a), there would be excessive *angle strain*; the average interior angle would have to be 144°, compared to the ideal angle of 120° for an sp^2-hybridized carbon atom. The angle strain would be eliminated by introducing two trans double bonds into the ring (Fig. 14-24b), but that molecule would have excessive *steric strain*, due to the H atoms crashing into each other in the center of the ring.

Tub-shaped conformation

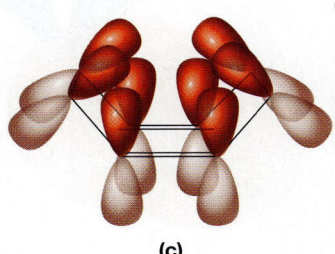

Cycloocta-1,3,5,7-tetraene ([8]Annulene)
(a)

(b)

The *p* AOs are *not* fully conjugated around the ring.

(c)

FIGURE 14-23 Cycloocta-1,3,5,7-tetraene ([8]annulene) (a) Line structure and (b) ball-and-stick model of [8]annulene. (c) In its tub shape, the *p* AOs do not make a fully conjugated ring, so the compound is nonaromatic.

YOUR TURN 14.9

In the molecule of [18]annulene shown in Figure 14-24d, draw bonds to the H atoms on the inside of the ring. Is there significant steric strain?

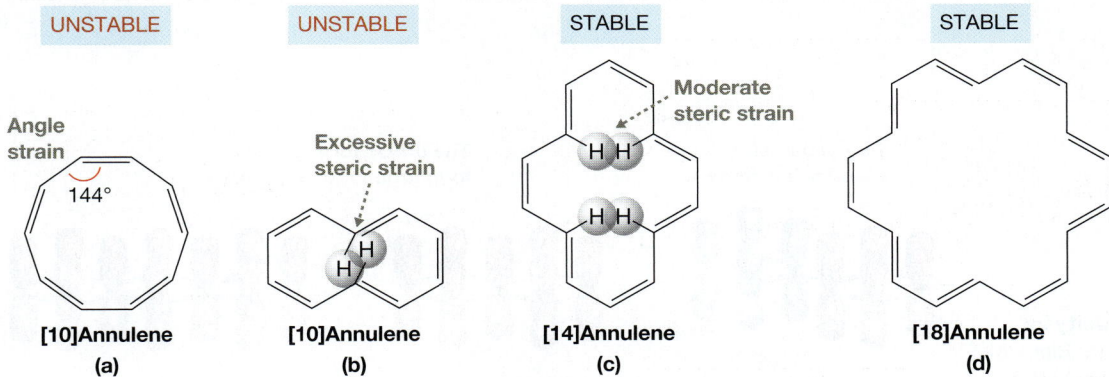

UNSTABLE UNSTABLE STABLE STABLE

Angle strain
144°

Excessive steric strain

Moderate steric strain

H H
H H

[10]Annulene
(a)

[10]Annulene
(b)

[14]Annulene
(c)

[18]Annulene
(d)

FIGURE 14-24 Aromaticity in [10]annulene, [14]annulene, and [18]annulene
[10]Annulene, (a) and (b), should be aromatic according to Hückel's rules, but it is unstable. The all-cis configuration (a) has too much angle strain, whereas the configuration with two trans double bonds (b) has too much steric strain. (c) [14]Annulene is aromatic, but is less stable than benzene due to the steric strain of the H atoms on the inside of the ring. (d) [18]Annulene is aromatic and is stable because the ring is large enough to comfortably accommodate trans double bonds without substantial steric strain.

PROBLEM 14.16 Draw the most stable configuration of [16]annulene and suggest whether it should be aromatic, antiaromatic, or nonaromatic. Explain.

14.7 Aromaticity and Multiple Rings

Compounds with two or more rings can also be aromatic. Biphenyl (Fig. 14-25a) is a somewhat trivial example, consisting of two separate benzene rings connected by a single bond. Although the two rings are conjugated to each other, each π system forms its own ring with six π electrons. Therefore, each essentially behaves as an independent aromatic system.

Naphthalene (Fig. 14-25b) and anthracene (Fig. 14-25c) contain two and three **fused rings**, respectively—so called because *the rings have bonds in common*. These molecules are aromatic, and as a class are called **polycyclic aromatic hydrocarbons (PAHs)**. Like benzene, each of these molecules is planar and possesses a system of conjugated *p* AOs that form a single loop and contain a Hückel number of π electrons. Respectively, they contain 10 and 14 π electrons.

The Lewis structures of PAHs can be misleading. In the Lewis structures of naphthalene and anthracene, for example, the bond lines might make it appear that there are separate cyclic π systems. The atomic *p* orbitals do indeed make a single loop, however, so they should be viewed as a single aromatic π system.

YOUR TURN 14.10

In Figure 14-25b and 14-25c, connect all of the darkly shaded lobes on the *p* orbitals to illustrate the single loop that exists in each.

YOUR TURN 14.11

In the resonance structures of naphthalene and anthracene in Figure 14-25, the double bonds appear to make one ring around the periphery of the molecules. Draw a resonance structure of each compound that shows the double bonds forming *two* separate rings. Do these resonance structures have an effect on the locations of the *p* AOs in Figure 14-25?

FIGURE 14-25 Aromaticity in polycyclic compounds (a) Biphenyl can be viewed as two independent aromatic systems, given that the *p* atomic orbitals form two different closed loops. (b) Naphthalene and (c) anthracene, on the other hand, have *p* orbitals that form one complete ring around the periphery.

Two separate aromatic systems

The *p* AOs form a single loop around the periphery.

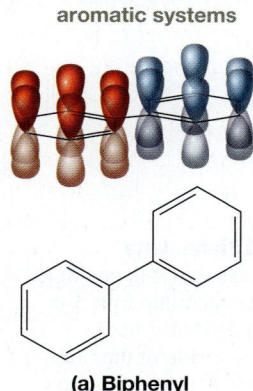
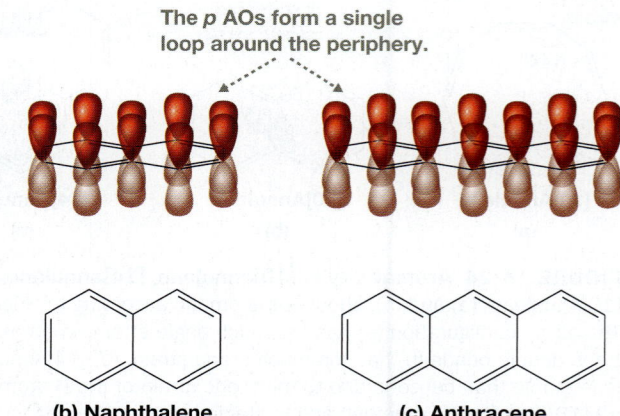

(a) Biphenyl   **(b) Naphthalene**   **(c) Anthracene**

SOLVED PROBLEM 14.17

Do you think that the compound shown here is aromatic, antiaromatic, or nonaromatic? Explain.

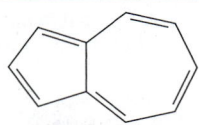

Think Does the molecule contain a π system formed from a ring of fully conjugated *p* orbitals? How many electrons are in that π system?

Solve Because the C atoms are all sp^2 hybridized and lie in the same plane, there is a fully conjugated system of *p* AOs around the periphery of the molecule, as shown by the blue dashed line. In that π system, there are 10 π electrons (indicated by the five double bonds), which is a Hückel number (five pairs), so the molecule should be aromatic.

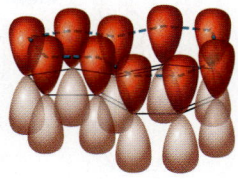

PROBLEM 14.18 Are the compounds shown here aromatic, antiaromatic, or nonaromatic? Explain.

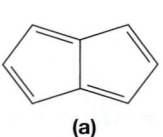 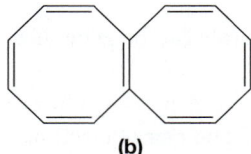 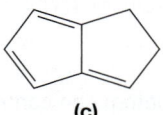

(a)   (b)   (c)

CONNECTIONS Anthracene (Fig. 14-25c) has insecticidal and fungicidal properties and has been used as a wood preservative. It is also used to make the mixture (along with potassium perchlorate and sulfur) that is burned to produce black smoke out of the chimney of the Sistine Chapel, indicating that a new pope has not yet been chosen.

14.8 Heterocyclic Aromatic Compounds

So far, we have only studied examples of aromatic hydrocarbons. There are many other compounds that have been experimentally determined to be aromatic, however, in which atoms other than carbon (i.e., heteroatoms) are part of the aromatic ring. Examples of these **heterocyclic aromatic compounds** include pyridine, pyrrole, and furan (Fig. 14-26).

Pyridine has a six-membered ring containing one N atom and resembles benzene. The main difference is that the N atom possesses a lone pair of electrons instead of a bond to H.

According to Hückel's rules, pyridine will be aromatic if there is a complete ring of conjugated *p* AOs possessing a Hückel number of electrons. Just as in benzene, all six atoms in pyridine (the N atom included) lie in the same plane and are sp^2 hybridized, so each possesses an unhybridized *p* AO. Furthermore, pyridine contains a total of six π electrons, as shown by the three formal double bonds in the Lewis structure.

This must mean that the lone pair of electrons on the N atom in pyridine is not part of the π system. If it were, then that system would contain eight total π electrons, making it *antiaromatic*. Instead, that lone pair resides in a MO that has contribution from the sp^2 hybrid AO on N, as shown in Figure 14-27a (next page). *Because that hybrid orbital and the π system are perpendicular to each other, the two cannot interact.*

Pyrrole is slightly more complex. According to VSEPR theory, the N atom should be sp^3 hybridized (four electron groups), but the molecule is entirely planar, so N is actually sp^2 hybridized. Although this imposes some angle strain on the N atom, the sp^2 hybridization allows the molecule to gain much more stability by becoming aromatic. Notice in Figure 14-27b that the N atom contributes its unhybridized *p* orbital to the π system of the ring and the lone pair of electrons on N is part of that π system. Consequently, there is a fully conjugated ring of *p* orbitals containing six electrons—a Hückel number—two from the lone pair and four from the two π bonds.

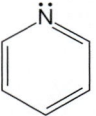 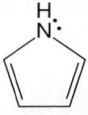 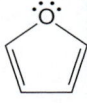

Pyridine   **Pyrrole**   **Furan**

FIGURE 14-26 Heterocyclic aromatic compounds In each compound, a noncarbon atom contributes a *p* AO to the aromatic π system.

CONNECTIONS Pyridine is an important compound in numerous chemical applications, which is why roughly 100,000 tons of it are produced annually. It is used as a solvent, as a precursor in the synthesis of certain herbicides and pesticides, and as a starting material for pyridinium chlorochromate (PCC), which is a specialized oxidizing agent (Section 19.6a). Pyridine is not found much in nature, but one natural source is the marshmallow plant.

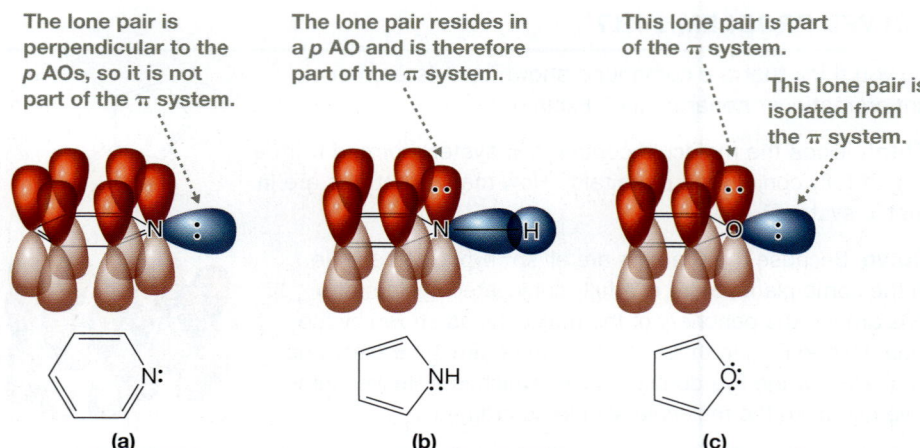

FIGURE 14-27 **Aromaticity in heterocyclic compounds** (a) The orbital picture of pyridine, showing that the lone pair of electrons on N is isolated from the aromatic π system. (b) The orbital picture of pyrrole, showing that the lone pair is part of the π system. (c) The orbital picture of furan, showing that one lone pair is part of the aromatic π system and the other is not.

The lone pair is perpendicular to the *p* AOs, so it is not part of the π system.

The lone pair resides in a *p* AO and is therefore part of the π system.

This lone pair is part of the π system.

This lone pair is isolated from the π system.

(a)

(b)

(c)

CONNECTIONS Furan is found in foods that have been heat processed because it is produced when some natural components undergo thermal degradation. It is found, for example, in foods that are canned and jarred, as well as in espresso coffee.

The lesson we learn from pyrrole can be generalized:

A heteroatom can contribute a lone pair of electrons to the π system to attain aromaticity.

This is evident in furan, whose O atom has two lone pairs of electrons. Only one of those lone pairs, however, is part of the π system, giving furan a total of six π electrons: the two electrons from the lone pair and the four electrons from the two π bonds. If both lone pairs were part of the π system, then there would be eight total π electrons, in which case furan would be antiaromatic. Instead, as shown in Figure 14-27c, the O atom is sp^2 hybridized with one lone pair residing in a *p* atomic orbital and the other lone pair *isolated* from the aromatic π system.

SOLVED PROBLEM 14.19

Do you think this compound is aromatic, antiaromatic, or nonaromatic? Explain.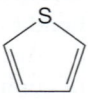

Think How many lone pairs from S are part of the π system? How many total π electrons are there?

Solve Just like the O atom in furan, the S atom here can contribute one of its two lone pairs (a total of two electrons) to the π system involving the double bonds. Including the four π electrons from the two double bonds, this makes a total of six electrons (a Hückel number) in a completely conjugated, cyclic π system. The compound is aromatic.

PROBLEM 14.20 Identify each of the following compounds as aromatic, antiaromatic, or nonaromatic.

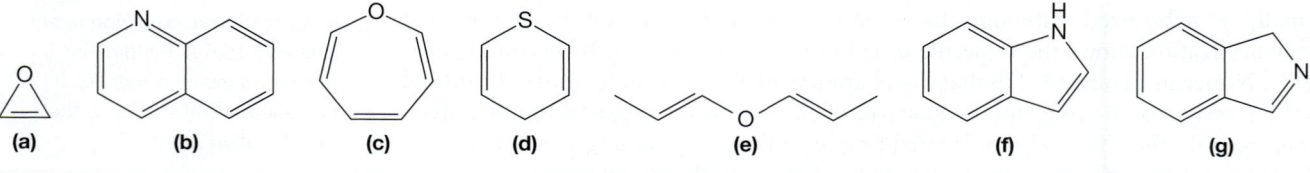

(a)   (b)   (c)   (d)   (e)   (f)   (g)

Aromaticity Helping Us Breathe: A Look at Hemoglobin

Hemoglobin is the protein responsible for carrying oxygen from our lungs to the tissues that need it. In humans and other mammals, the most common type of hemoglobin has four subunits that are associated noncovalently, and each subunit has an embedded heme group. It is this heme group that is directly responsible for binding and releasing O_2 molecules.

A heme group is made up of a *porphyrin*, which has a planar, cyclic arrangement of four pyrrole-type rings (see Fig. 14-26), covalently bonded to each other by single-carbon bridges. At the center of the porphyrin is an iron atom in the $+2$ oxidation state. The iron atom has six sites available for coordination in an essentially octahedral geometry, and four of those sites are occupied by the N atoms of the porphyrin ring, all in the same plane. On one side of that plane, a fifth coordination site is used to covalently bond the iron to a N atom in the side chain of a histidine residue of the protein. On the other side of that plane, the sixth coordination site is used to temporarily bind a molecule of O_2 for transport.

The geometry about the iron atom is vital to the proper functioning of the heme group, and the rigidity of the porphyrin ensures the proper geometry. Where does this rigidity come from? Notice that the porphyrin consists of several sp^2-hybiridized atoms, each of which contributes a valence p AO to make a fully conjugated ring. Furthermore, there are 18 electrons occupying that π system, which are highlighted. This is a Hückel number, so the porphyrin is, in fact, aromatic! As with any aromatic species, planarity is heavily favored to maximize the overlap among the contributing p AOs.

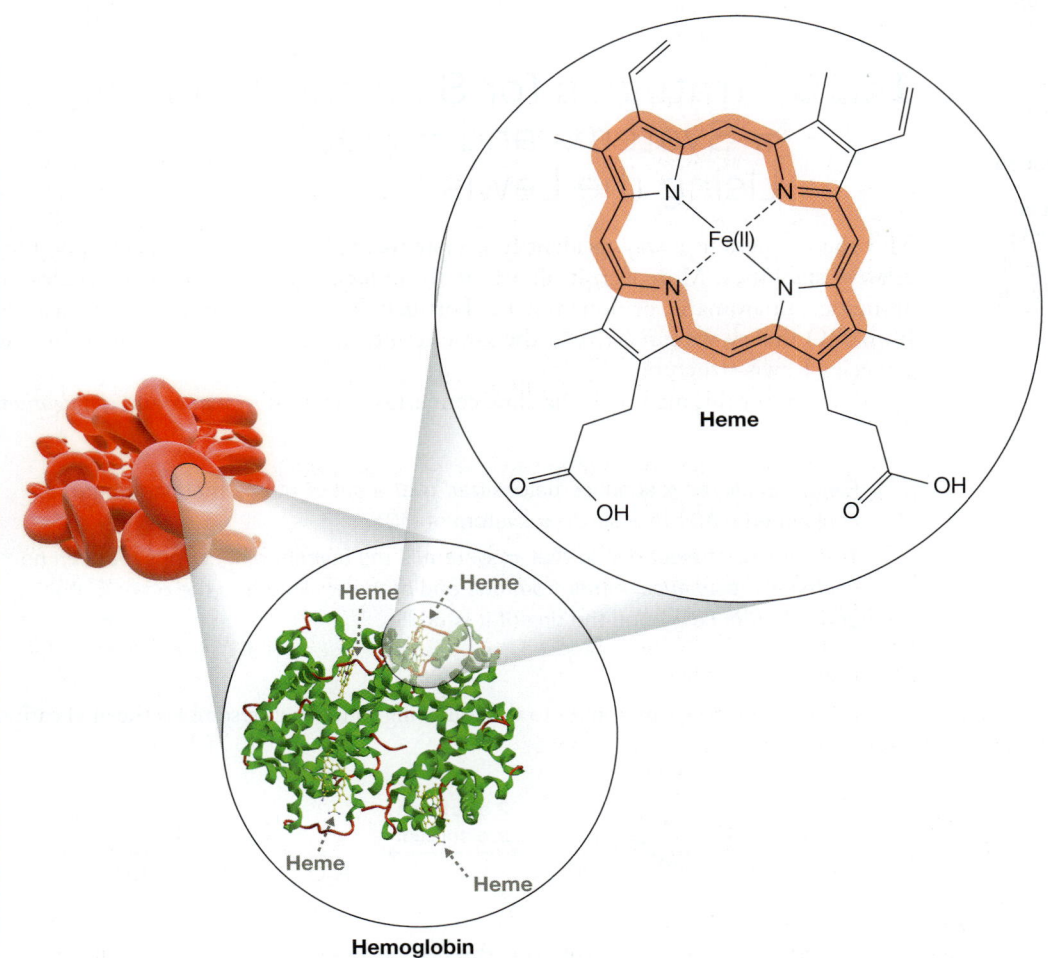

Heme

Hemoglobin

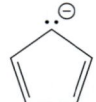

The cyclopentadienyl anion

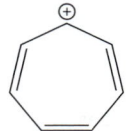

**The cycloheptatrienyl cation
(The tropylium ion)**

FIGURE 14-28 Aromatic ions In each ion, the π system is made of a fully conjugated ring of *p* AOs and contains a Hückel number of electrons.

The cyclopropenyl anion

The cyclopentadienyl cation

FIGURE 14-29 Antiaromatic ions In each ion, the π system is made of a fully conjugated ring of *p* AOs and contains an anti-Hückel number of electrons.

14.9 Aromatic Ions

Aromaticity is not limited to uncharged molecules, because several molecular *ions* that conform to Hückel's rules are also aromatic. The cyclopentadienyl anion and the cyclo-heptatrienyl cation (also called the tropylium ion), are two common examples (Fig. 14-28).

The cyclopentadienyl anion is similar to pyrrole because it is a five-membered ring with two conjugated double bonds and a lone pair of electrons. Therefore, it has six π electrons. The cycloheptatrienyl cation, on the other hand, is a seven-membered ring that does not possess any lone pairs of electrons. Each C atom is sp^2 hybridized, and the three conjugated double bonds provide the six π electrons.

Antiaromatic ions can also exist (Fig. 14-29). One example is the cyclopropenyl anion, which contains a fully conjugated ring of *p* atomic orbitals with a total of four π electrons—an anti-Hückel number. Additionally, the cyclopentadienyl cation contains a total of four π electrons, making it antiaromatic as well. Unlike [8]annulene (Fig. 14-23), these three- and five-membered rings are not large enough to bend out of plane to allow the π system to become nonaromatic.

PROBLEM 14.21 Based on Hückel's rules, determine whether each of the species shown here is aromatic, antiaromatic, or nonaromatic. Explain.

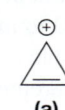

(a)

(b)

14.10 Strategies for Success: Counting π Systems and π Electrons Using the Lewis Structure

MO theory gives us a way to identify π systems of electrons and to determine their relative stabilities. Moreover, it allows us to understand *why* certain molecules are aromatic, antiaromatic, or nonaromatic. Fortunately, applying the lessons we learned from MO theory, we can arrive at the same results more quickly and easily using just a species' Lewis structure.

The basis for this method is the close connection between *resonance* and *conjugation*:

- Electrons can be resonance delocalized over a set of atoms that contribute conjugated *p* AOs to a single π system of MOs.
- The number of electrons in that π system is *the number of electrons that can be shifted via resonance*, either from one end of the π system to the other (if it is acyclic), or once around the ring (if it is cyclic).

We can show how this applies to the following ion, which resembles the allyl cation.

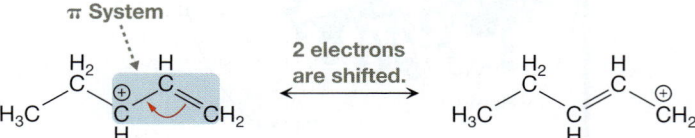

A single π system encompasses the three C atoms indicated by the blue screen, because those are the atoms over which electrons can be resonance delocalized. Furthermore, two electrons can be shifted from the right side of that π system to the left, so the π system contains two electrons, the same as we saw in the allyl cation in Section 14.2a.

In the following allylic anion, there is also a single π system that encompasses three C atoms:

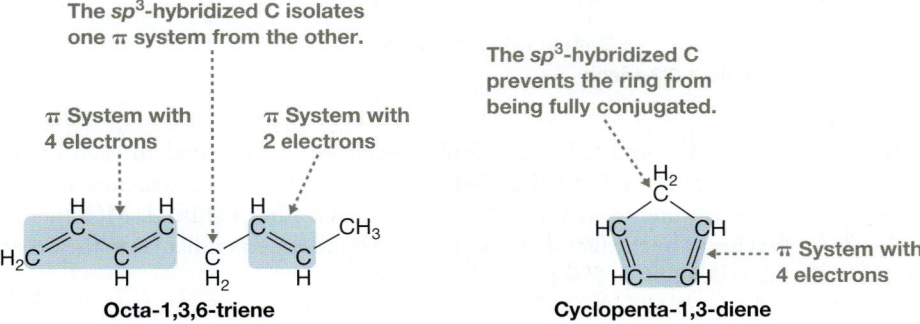

In this case, however, four electrons can be shifted to the right via resonance, so the π system contains four total electrons.

We can also apply this method to hexa-2,4-diene, which resembles buta-1,3-diene.

Hexa-2,4-diene

There is a single π system that encompasses the four C atoms indicated. And, because four electrons can be shifted toward the right via resonance, that system contains a total of four electrons, the same as we obtained for buta-1,3-diene (Section 14.2b).

This method can also be applied to both hexa-1,3,5-triene and benzene.

Hexa-1,3,5-triene

Benzene

Each compound has a single π system that encompasses six C atoms and contains six electrons. In the case of hexa-1,3,5-triene, six electrons can be shifted to the right. In the case of benzene, six electrons can be shifted cyclically around the ring.

From all of these examples, we can make some very useful generalizations:

- A single π system cannot encompass *sp³*-hybridized carbon atoms.
- Two double bonds are part of the same π system if they are separated by exactly one bond, making the double bonds *conjugated*.

With these generalizations in mind, notice that octa-1,3,6-triene has six total π electrons in two separate π systems:

The *sp³*-hybridized C isolates one π system from the other.

π System with 4 electrons

π System with 2 electrons

Octa-1,3,6-triene

The *sp³*-hybridized C prevents the ring from being fully conjugated.

π System with 4 electrons

Cyclopenta-1,3-diene

One π system consists of two conjugated double bonds, whereas the other is a lone double bond. The two systems are isolated from each other by the sp^3-hybridized C that separates them. Notice, too, that the π system of cyclopenta-1,3-diene is *not* fully conjugated around the ring. Although the two double bonds are conjugated to each other, the sp^3-hybridized C atom isolates one end of that π system from the other. Moreover, the top two C—C bonds are both single bonds, so the ring does not consist *entirely* of alternating double and single bonds.

PROBLEM 14.22 How many π systems does β-carotene contain? How many electrons are in each?

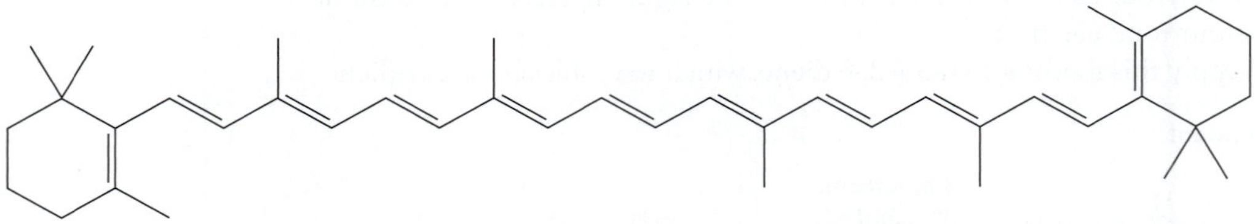

β-Carotene

Using this method of counting π systems and π electrons is particularly helpful when triple bonds are involved, as shown with octa-3,5-dien-1-yne:

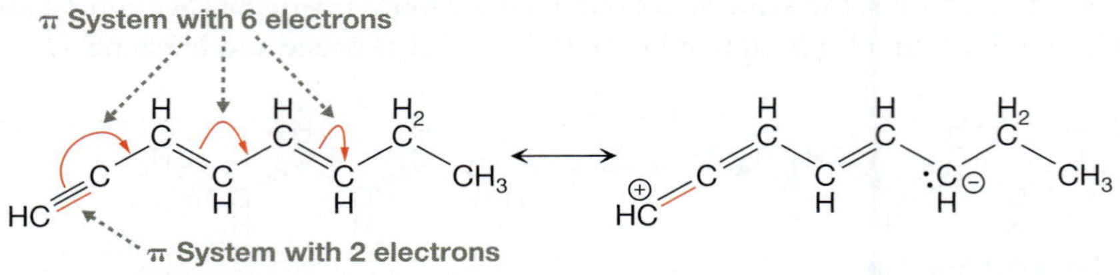

Octa-3,5-dien-1-yne

According to the drawing above on the left, there are eight total π electrons in the molecule. They reside in two different π systems, however—namely, one that contains six electrons and one that contains two—because, according to MO theory, the two π bonds of a triple bond are perpendicular to each other (Section 3.8). Therefore, only one of those π bonds can be conjugated with a neighboring π system.

A similar situation arises in penta-1,2,3-triene:

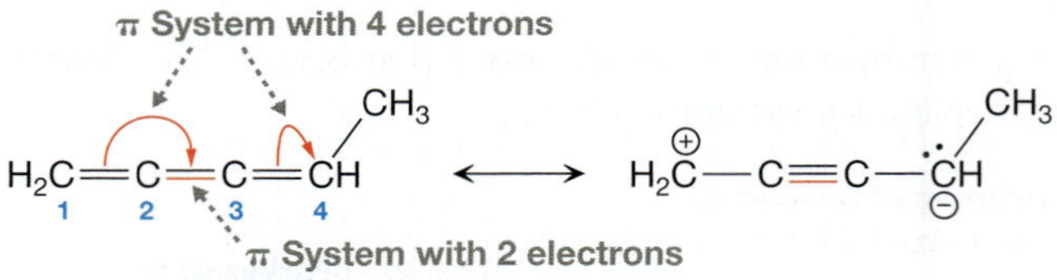

Penta-1,2,3-triene

Although the C1—C2 and C3—C4 double bonds are conjugated to each other, neither is conjugated to the C2—C3 double bond. No resonance structure can be drawn in which electrons from the C2—C3 bond have been shifted. MO theory accounts for this result by the fact that the central π bond is perpendicular to the other two. (See Problem 3.42 on p. 150.)

The previous two examples can be generalized as follows:

Two π bonds to the same atom must belong to different π systems.

SOLVED PROBLEM 14.23

If it is planar, should the compound shown at the right be aromatic, antiaromatic, or nonaromatic? Explain.

Think Can we draw a resonance structure by shifting electrons fully around the ring? If so, how many electrons must be shifted?

Solve We *can* shift electrons fully around the ring via resonance:

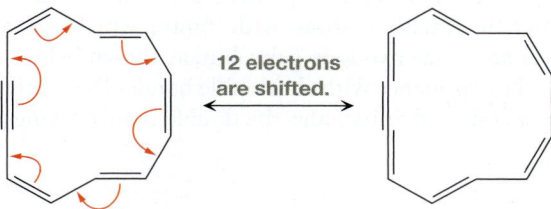

12 electrons are shifted.

Doing so involves two electrons from each of the double bonds, as well as two electrons from the triple bond, for a total of 12 electrons. Thus, there is a fully conjugated cyclic π system that contains 12 electrons (an anti-Hückel number), so the compound should be antiaromatic if it is planar. Notice that one pair of electrons of the triple bond is not shifted, indicating that those two electrons are in a separate π system.

PROBLEM 14.24 If it is planar, should the compound shown here be aromatic, antiaromatic, or nonaromatic? Explain.

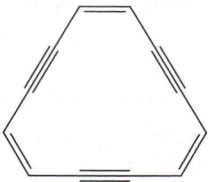

Finally, using resonance structures can be quite useful when counting electrons in a cyclic π system that involves atoms with lone pairs. Consider pyrrole, which we examined in Section 14.8. As shown below at the left, a resonance structure can be drawn by shifting six electrons around the ring. Thus, the π system contains six electrons, including the lone pair on N.

The lone pair is part of the π system.

Pyrrole

6 π electrons are shifted.

In pyridine, however, the lone pair on N is *not* part of the cyclic π system (Section 14.8). This is consistent with the fact that the lone pair is not involved in resonance with the double bonds of the ring, as shown on the left below. The π system thus contains a total of six electrons—namely, those that are part of the double bonds only.

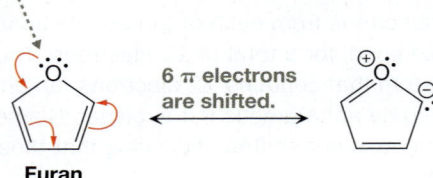

Notice that, as shown in parentheses above on the right, an attempt can be made to include N's lone pair in resonance. In the resulting resonance structure, however, the N atom retains its lone pair, so that lone pair is effectively unchanged.

A slightly different situation arises with furan, which we also examined in Section 14.8. The O atom has two lone pairs, but, as shown below, only one pair at a time can be involved in resonance with the double bonds. Thus, only one of those lone pairs is part of the π system that includes the double bonds, giving the system a total of six electrons.

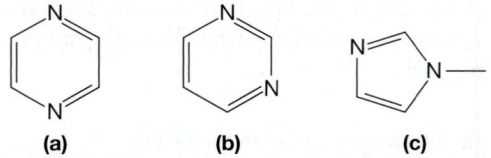

What we observe with the lone pairs in furan can be generalized as follows:

A single π system can include at most one lone pair of electrons on a particular atom.

PROBLEM 14.25 Identify each of these compounds as aromatic, antiaromatic, or nonaromatic.

(a)   (b)   (c)

14.11 Aromaticity and DNA

Although the discussion of aromaticity here in Chapter 14 predominantly focuses on prototypical molecules, the phenomenon is observed in biological molecules as well. Aromaticity, for example, affects the structure and properties of DNA. Recall from Section 1.14c that a nucleic acid is a long chain of nucleotides and that each nucleotide has three components: a sugar group, a phosphate group, and a nitrogenous base.

Furthermore, a nucleic acid's backbone consists of alternating sugar and phosphate groups, and the nitrogenous bases are bonded specifically to the sugar units of the backbone.

One nucleic acid is distinguished from another by the specific sequence of nitrogenous bases attached to the sugar–phosphate backbone. The four nitrogenous bases found in DNA are guanine (G), adenine (A), cytosine (C), and thymine (T).

10 π electrons	10 π electrons	6 π electrons	6 π electrons
Guanine (G)	**Adenine (A)**	**Cytosine (C)**	**Thymine (T)**

Each of these ring systems consists of atoms that have planar geometries and are sp^2 hybridized. Thus, each nitrogenous base has a fully conjugated, cyclic system of p AOs. Additionally, each of these π systems contains a Hückel number of electrons—10 π electrons for guanine and adenine, and 6 for cytosine and thymine—which are highlighted in red in the preceding structures. According to Hückel's rules (Section 14.3), then:

All nitrogenous bases in nucleic acids are aromatic.

YOUR TURN **14.12**

Draw resonance structures of guanine and adenine, to which the methods from Section 14.10 can be applied to show 10 electrons in a single π system. Do the same for cytosine and thymine to show 6 electrons in a single π system.

To understand the implications of the aromatic nature of these nitrogenous bases, we must consider the general structure of DNA shown in Figure 14-30. DNA consists of two nucleic acid strands, wrapped around each other to form a *double helix*. The nitrogenous bases from each nucleic acid point toward the center of the cylinder that is created by the two sugar–phosphate backbones. Moreover, the specific sequence of nitrogenous bases in one nucleic acid strand dictates the sequence in the other, according to the following rules for base pairing:

- Guanine in one strand of DNA is matched with cytosine in the other.
- Adenine in one strand of DNA is matched with thymine in the other.

Thus, the strands are said to be **complementary**.

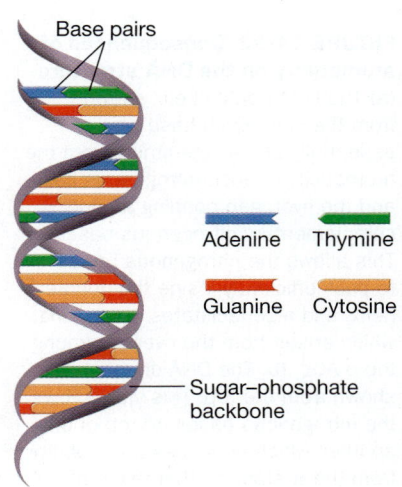

FIGURE 14-30 The general structure of DNA DNA consists of two nucleic acid strands that are wrapped around each other in a double helix structure. The sugar–phosphate backbones of the strands define a cylinder, inside of which the nitrogenous bases are located. The strands have *complementary* sequences of nitrogenous bases—that is, adenine in one strand is matched with thymine in the other, and guanine in one strand is matched with cytosine in the other.

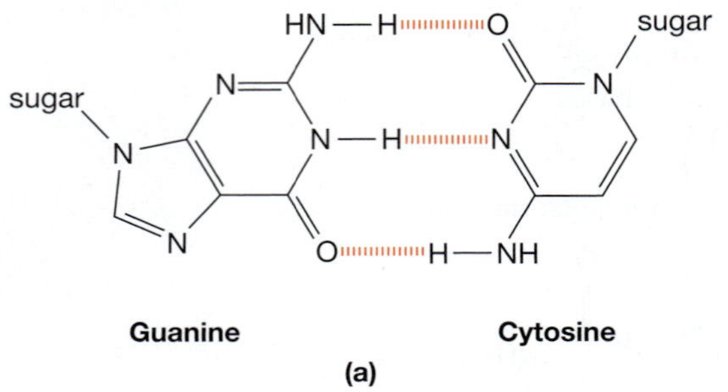

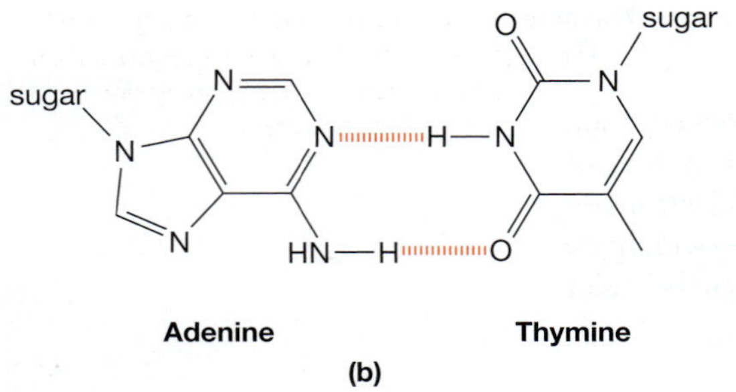

The G and C bases are paired because they form three hydrogen bonds.

Guanine   **Cytosine**

(a)

The A and T bases are paired because they form two hydrogen bonds.

Adenine   **Thymine**

(b)

FIGURE 14-31 Complementarity among nitrogenous bases (a) G and C bases are complementary and (b) A and T bases are complementary because each hydrogen-bond donor in one nitrogenous base is aligned with a hydrogen-bond acceptor in the other.

The G–C/A–T base pairing is heavily favored because the two strands are held together relatively strongly by hydrogen bonding among the nitrogenous bases, as shown in Figure 14-31, and this hydrogen bonding is maximized when the strands are complementary.

Figure 14-31a shows that guanine has two hydrogen-bond donors and one hydrogen-bond acceptor that are ideally located to establish hydrogen bonds with two acceptors and one donor on cytosine. Thus, guanine and cytosine form three hydrogen bonds. Similarly, adenine and thymine form two hydrogen bonds, as shown in Figure 14-31b.

Considering again the double helix structure of DNA, we can see how the aromaticity of the nitrogenous bases is advantageous. First, it allows each base pair to be planar, as shown in Figure 14-32a. As we learned earlier in Chapter 14, aromatic rings favor a planar structure to maximize the overlap of the p orbitals that make up the π system of MOs. Furthermore, the hydrogen bonding essentially locks complementary bases in the same plane. Thus, as shown in Figure 14-32, the DNA base pairs are "stacked" on the inside of DNA, much like sheets of paper, which keeps steric crowding to a minimum. As a result, the nitrogenous bases are packed in a highly efficient way, allowing DNA to store a large amount of genetic information in relatively little space.

The second way in which the aromaticity of the nitrogenous bases is advantageous is in the stability it provides to the DNA double helix. Because nitrogenous bases are stacked on top of one another, the p AOs from the aromatic system of one nitrogenous base overlap with the p AOs from another nitrogenous base attached to an adjacent sugar. This is a stabilizing interaction called π **stacking**, and is illustrated in Figure 14-32a. Individually, these interactions are relatively weak, but, viewing DNA from the top (Fig. 14-32b), notice that there is a high density of these interactions inside the double helix. The sum of these interactions along the entire length of DNA can provide significant stability.

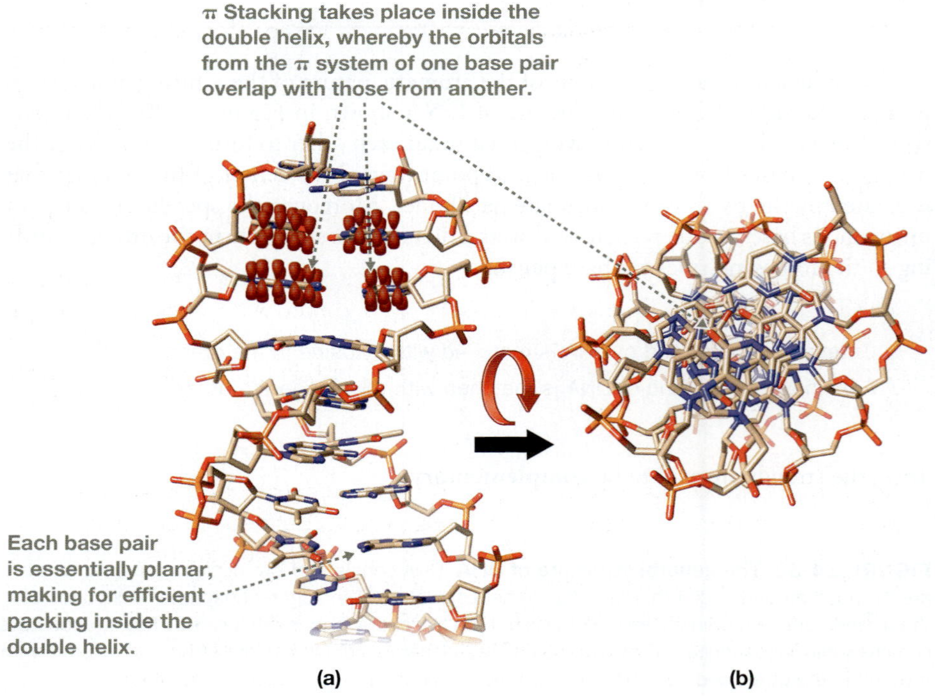

FIGURE 14-32 Consequences of aromaticity on the DNA structure (a) The DNA double helix shown from the side. Each base pair is essentially planar, stemming from the aromaticity of each nitrogenous base and the hydrogen bonding among complementary nitrogenous bases. This allows the nitrogenous bases to pack efficiently inside the double helix, and also facilitates π stacking, which arises from the overlap among the p AOs. (b) The DNA double helix shown from the top. This view shows the nitrogenous bases on top of one another, which enhances the stability from the π stacking that takes place.

π Stacking takes place inside the double helix, whereby the orbitals from the π system of one base pair overlap with those from another.

Each base pair is essentially planar, making for efficient packing inside the double helix.

(a)   **(b)**

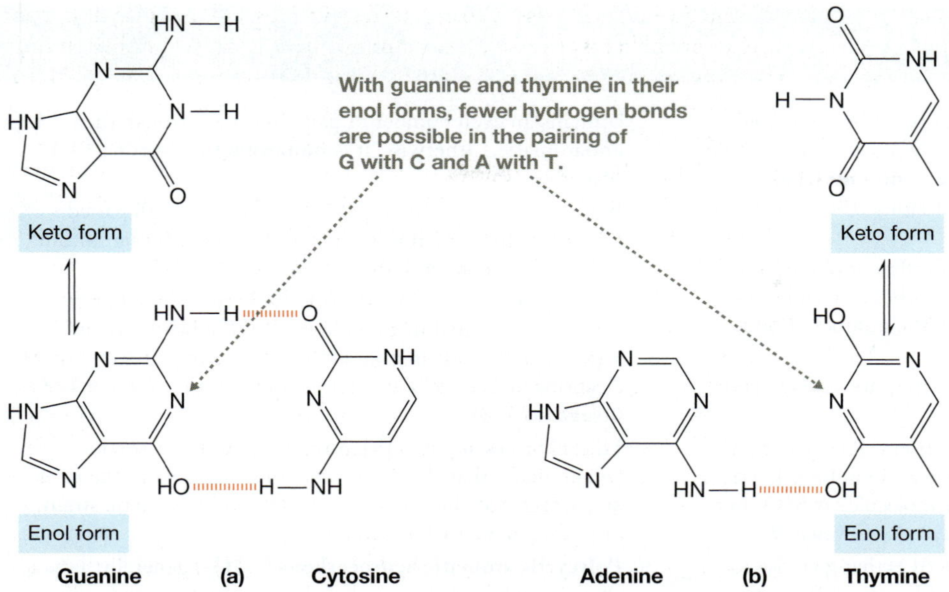

With guanine and thymine in their enol forms, fewer hydrogen bonds are possible in the pairing of G with C and A with T.

Keto form

Enol form

Guanine **(a)** **Cytosine**

Keto form

Enol form

Adenine **(b)** **Thymine**

FIGURE 14-33 **The enol forms of nitrogenous bases** In its enol form (a) G forms only two hydrogen bonds with C, and (b) T forms only one hydrogen bond with A.

The fact that the nitrogenous bases in DNA are aromatic delayed the elucidation of its structure, which was eventually published in 1953 by James Watson (b. 1928), an American biologist, and Francis Crick (1916–2004), a British physicist. Watson and Crick had two important pieces of information with which to work. One was the parity relationship among the nitrogenous bases, discovered by the Austrian chemist Erwin Chargaff (1905–2002), which stated that the amounts of guanine and cytosine in DNA are equal, as are the amounts of adenine and thymine. The second piece of information came from X-ray diffraction images of DNA taken by Rosalind Franklin (1920–1958), a British biophysicist, which suggested that the structure was helical.

Watson and Crick used physical models to determine how the base pairing could occur. They were working with the incorrect structures for guanine and thymine, however, because these bases were assumed to be more stable in their enol forms (Section 7.9) instead of their keto forms.

This assumption is understandable, because the guanine and thymine rings in their enol forms (Fig. 14-33) have alternating single and double bonds, which resemble a benzene ring. In their enol forms, however, guanine and cytosine cannot participate in as many hydrogen bonds with their complementary bases. The enol form of guanine can participate in only two hydrogen bonds with cytosine, and the enol form of thymine can participate in only one hydrogen bond with adenine. Thus, Crick and Watson could not devise a model that was consistent with the complementarity among the nitrogenous bases.

Fortunately, Jerry Donohue (1920–1985), an American chemist, was studying at Cambridge on a six-month grant and was sharing an office with Crick and Watson. Donohue had expertise with small organic molecules and suggested that guanine and thymine were more stable in their keto forms. This suggestion turned out to be the final piece of the puzzle that Crick and Watson needed to elucidate the structure of DNA.

YOUR TURN 14.13

Figure 14-31 shows the possible hydrogen bonding between G–C base pairs when both bases are in their keto forms. Figure 14-33 shows the possible hydrogen bonding when one base is in its enol form and the other is in its keto form. Without changing the positions of the bases, draw the possible hydrogen bonding between G–C base pairs when both bases are in their enol forms. Which of the three scenarios maximizes hydrogen bonding?

Chapter Summary and Key Terms

- Electron delocalization via resonance cannot be accounted for by the two-center orbital picture of VB theory. The orbital picture of resonance requires **multiple-center molecular orbitals**, which span multiple adjacent atoms. **(Sections 14.1 and 14.2; Objective 1)**

- MOs are created by the interaction of different phases of the contributing AOs, the total number of orbitals must be conserved, and MO energy increases with the number of nodes. **(Section 14.2; Objective 2)**

- *p* AOs are **conjugated** to each other if they are *adjacent* and *parallel*. **(Section 14.2a; Objective 2)**

- Three conjugated *p* AOs in the allyl cation yield three π MOs: one bonding, one nonbonding, and one antibonding. These π MOs are higher in energy than the molecule's σ MOs but lower than the σ* MOs. **(Section 14.2a; Objectives 2, 3)**

- Buta-1,3-diene has four conjugated *p* AOs that give rise to four π MOs: two bonding and two antibonding. These π MOs are higher in energy than the molecule's σ MOs but lower than the σ* MOs. **(Section 14.2b; Objectives 2, 3)**

- The allyl cation and buta-1,3-diene favor an all-planar conformation to allow all of their *p* AOs to be conjugated. **(Sections 14.2a and 14.2b; Objectives 2, 3)**

- MO theory accounts for the characteristics of resonance hybrids, including those for the allyl cation and buta-1,3-diene. **(Sections 14.2a and 14.2b; Objective 4)**

- Benzene (C_6H_6) is symmetric, with six identical carbon–carbon bonds, and it has an unusually stable π system of electrons. These characteristics are attributable to **aromaticity**. **Heat of hydrogenation** data can be used to determine the stabilities of π systems. From these experiments, we know that aromaticity provides benzene with ~150 kJ/mol (~36 kcal/mol) of *extra* stability. **(Section 14.3; Objectives 5, 6)**

- Cyclobutadiene (C_4H_4) is highly unstable and is classified as **antiaromatic**. **(Section 14.3; Objectives 5, 6)**

- According to **Hückel's rules**, a species is **aromatic** if it has a cyclic system of conjugated *p* orbitals that contains an odd number of pairs of electrons ($4n + 2$ π electrons). If it contains an even number of pairs ($4n$ π electrons), then it is antiaromatic. Otherwise, it is **nonaromatic**. **(Section 14.3; Objective 5)**

- Benzene has six π MOs. Its six π electrons are all paired, filling the three π MOs, which gives benzene its significant stability. **(Section 14.4; Objectives 7, 8)**

- Cyclobutadiene has four π MOs. Only two of its four π electrons are paired in a bonding MO. The other two are *unpaired* and reside in nonbonding MOs. This distribution of electrons makes cyclobutadiene very unstable. **(Section 14.5; Objectives 7, 8)**

- Hückel's rules apply to **[*n*]annulenes**. However, some [*n*]annulenes that should be aromatic according to the rules are, in fact, not, due to excessive angle strain or steric strain, or both. **(Section 14.6; Objective 5)**

- **Polycyclic aromatic hydrocarbons (PAHs)** generally have a single π system, although their Lewis structures show them as separate rings. **(Section 14.7; Objective 5)**

- Heteroatoms can be part of an aromatic ring, giving rise to **heterocyclic aromatic compounds**. In these rings, the heteroatoms can contribute a lone pair of electrons to achieve a Hückel number of π electrons. **(Section 14.8; Objective 5)**

- Ions follow the same rules that Hückel outlined and can therefore be aromatic, antiaromatic, or nonaromatic. **(Section 14.9; Objective 5)**

- Electron delocalization via resonance is closely related to conjugation in π systems. **(Section 14.10; Objective 9)**
  - The atoms over which electrons can be shifted via resonance encompass a single π system of MOs. The number of electrons that can be shifted is the same number of electrons that occupy that π system.
  - Tetrahedral carbon atoms disrupt a π system.
  - At most one lone pair of electrons from a given atom can contribute to the same π system.
  - At most one π bond from a given triple bond can contribute to the same π system.

Problems

14.1 and 14.2 The Shortcomings of VB Theory; Multiple-Center MOs

14.26 The following are three π MOs for the heptatrienyl cation. For each MO, **(a)** determine the number of nodal planes perpendicular to the bonding axes and rank them in order of increasing energy; **(b)** draw the *p* AO contributions on each C atom that would give rise to the MO; **(c)** identify each internuclear region as having either a *bonding* or an *antibonding* type of contribution; and **(d)** based on your answer to part (c), determine whether each MO is overall *bonding*, *nonbonding*, or *antibonding*.

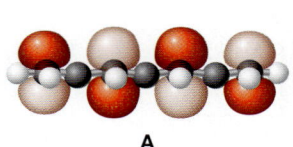
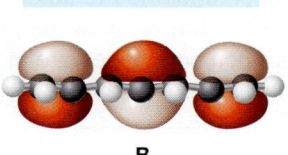
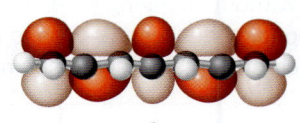

A   B   C

14.27 The following are three π MOs of octa-1,3,5,7-tetraene. Repeat Problem 14.26 for these orbitals.

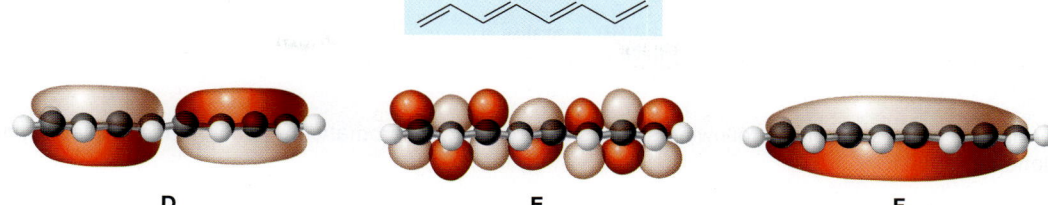

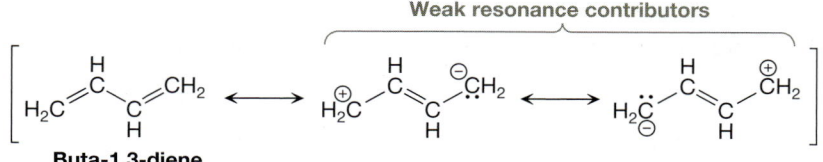

| D | E | F |

14.28 Buta-1,3-diene has one strong and two weak resonance contributors. The two-center orbital picture that describes the strong resonance contributor was previously shown in Figure 14-3. Draw the two-center orbital picture that describes each of the two weak contributors.

Weak resonance contributors

Buta-1,3-diene

14.29 An *excited state* of buta-1,3-diene is achieved when one electron from π_2 has been promoted to π_3^*. How do you think this electronic transition affects the bond length between C1 and C2? Does it increase, decrease, or stay about the same? Explain. What about the C2—C3 bond length?

14.30 For the penta-1,3-dienyl anion ($H_2C=CH-CH=CH-CH_2^-$), draw the π MOs and the MO energy diagram similar to what is shown in Figures 14-5 and 14-6 for the allyl cation. Identify each MO as either bonding, nonbonding, or antibonding. Label the HOMO and the LUMO.

14.31 For the penta-1,3-dienyl cation ($H_2C=CH-CH=CH-CH_2^+$), draw the π MOs and the MO energy diagram similar to what is shown in Figures 14-5 and 14-6 for the allyl cation. Identify each MO as either bonding, nonbonding, or antibonding. Label the HOMO and the LUMO.

14.32 For hexa-1,3,5-triene ($H_2C=CH-CH=CH-CH=CH_2$), draw the π MOs and the MO energy diagram similar to what is shown in Figures 14-5 and 14-6 for the allyl cation. Identify each MO as either bonding, nonbonding, or antibonding. Label the HOMO and the LUMO.

14.33 The bond length of the indicated C—C single bond in hexa-1,3,5-triene is about 146 pm, which is considerably shorter than that in ethane, H_3C-CH_3.
(a) How much shorter is it? *Hint:* See Figure 14-2, page 685.
(b) Which occupied π MOs (obtained in Problem 14.32) help decrease this bond length?
(c) If an electron were promoted from the HOMO to the LUMO, would the length of the C—C single bond increase, decrease, or stay roughly the same? Explain.

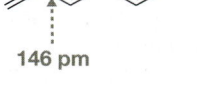

146 pm

14.34 How many total π electrons are in the molecule shown here? In how many separate π systems do they reside? How many electrons are in each π system?

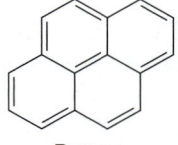

14.3–14.9 Aromaticity and Molecular Orbitals in Cyclic π Systems

14.35 Pyrene has been determined experimentally to be aromatic. At first glance, however, its structure appears to break Hückel's rule. How so? Can you explain why pyrene exhibits aromaticity? *Hint:* What are the characteristics of the π system on the periphery of the molecule?

Pyrene

14.36 Draw the π MO energy diagram for [10]annulene, similar to those in Figures 14-19 and 14-21. Fill up the orbitals with the appropriate number of π electrons. Based on this diagram, should [10]annulene be aromatic or antiaromatic? Explain.

14.37 Repeat Problem 14.36 for [8]annulene.

14.38 Repeat Problem 14.36 for the following species:

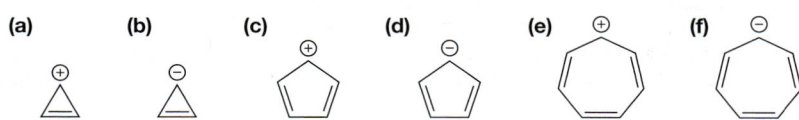

14.39 The molecule shown at the right, B_3N_3, has been synthesized. Should it be aromatic, antiaromatic, or nonaromatic? Explain.

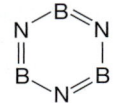

14.40 Each N atom in the molecule in Problem 14.39 contains a lone pair of electrons. In what kind of orbital does each one reside?

14.41 Determine whether each of the following species is aromatic, antiaromatic, or nonaromatic. Explain. *Hint:* Don't forget the lone pairs.

(a) 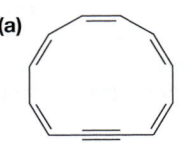 (b) 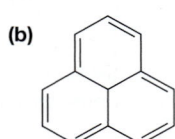 (c) 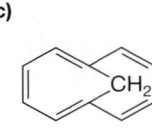 (d) 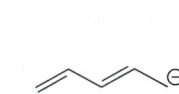 (e)   (f)   (g)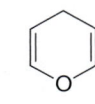

14.42 Rank the following in order of increasingly exothermic heat of hydrogenation. Explain.

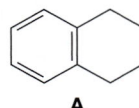 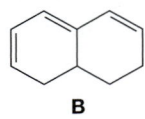 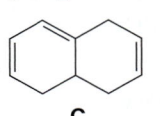 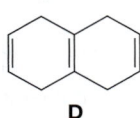

A   B   C   D

14.43 Which ketone, **E** or **F**, do you think has the more stable π system? Why? *Hint:* Consider various resonance structures of each species.

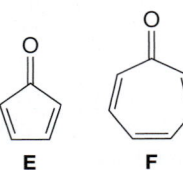

E   F

14.44 Molecules **G** and **H** are isomers of each other, and they have the same number of C=C double bonds. Which one has the more exothermic heat of hydrogenation? Why?

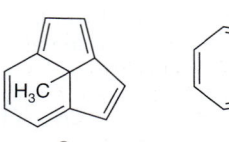

G   H

14.45 Coronene is a polycyclic aromatic hydrocarbon. **(a)** How many total π electrons does it have? **(b)** Is that number consistent with the fact that the compound is aromatic? Explain.

Coronene

14.46 As we saw here in Chapter 14, the tropylium ion is aromatic. Ion **I**, however, is not. Explain.

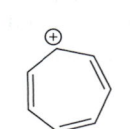 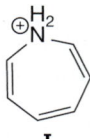

The tropylium ion   **I**

14.47 Most dianions of hydrocarbons are very unstable. However, if cyclooctatetraene is treated with potassium, the cyclooctatetraenyl dianion is readily formed via a redox reaction. **(a)** Explain why $C_8H_8^{2-}$ is easy to make. **(b)** As we saw in the chapter, cyclooctatetraene is tub shaped (Fig. 14-23). What do you think the geometry of $C_8H_8^{2-}$ is? Explain.

$$\text{(cyclooctatetraene)} + 2\text{ K(s)} \longrightarrow C_8H_8^{2-} + 2\text{ K}^+$$

14.11 The Organic Chemistry of Biomolecules: Aromaticity and DNA

14.48 A strand of nucleic acid is defined by its sequence of nucleotides: A, C, T, and G. How many different sequences are possible for a nucleic acid that is 200 nucleotides long? How does that number compare to the estimated number of atoms in the universe, which is approximately 10^{80}?

14.49 Figure 14-31a (p. 716) shows how hydrogen bonding accounts for the fact that guanine is complementary to cytosine, giving rise to three simultaneous hydrogen bonds in the interaction. Draw the optimal alignment for the interaction between guanine and adenine. How many simultaneous hydrogen bonds exist in that interaction?

14.50 One nucleic acid strand in a particular segment of DNA has the following nucleotide sequence: CGGATACATTTGC. In the same segment of DNA, what is the sequence of nucleotides in the other strand?

14.51 Doxorubicin, shown here, is an important chemotherapy drug used to treat a variety of cancers, including bladder cancer, breast cancer, and certain forms of leukemia. Doxorubicin works by binding to DNA in such a way that a portion of it penetrates the DNA double helix—a process called *intercalation*. During transcription—the process that forms RNA—portions of the DNA strands are temporarily separated for the base sequence to be read and then are reconnected. With bound doxorubicin, however, the double helix does not reform properly after the strands are separated, which disrupts replication— the process that forms an identical copy of DNA. Which portion of doxorubicin do you think intercalates into the DNA double helix, and why do you think it has little difficulty doing so?

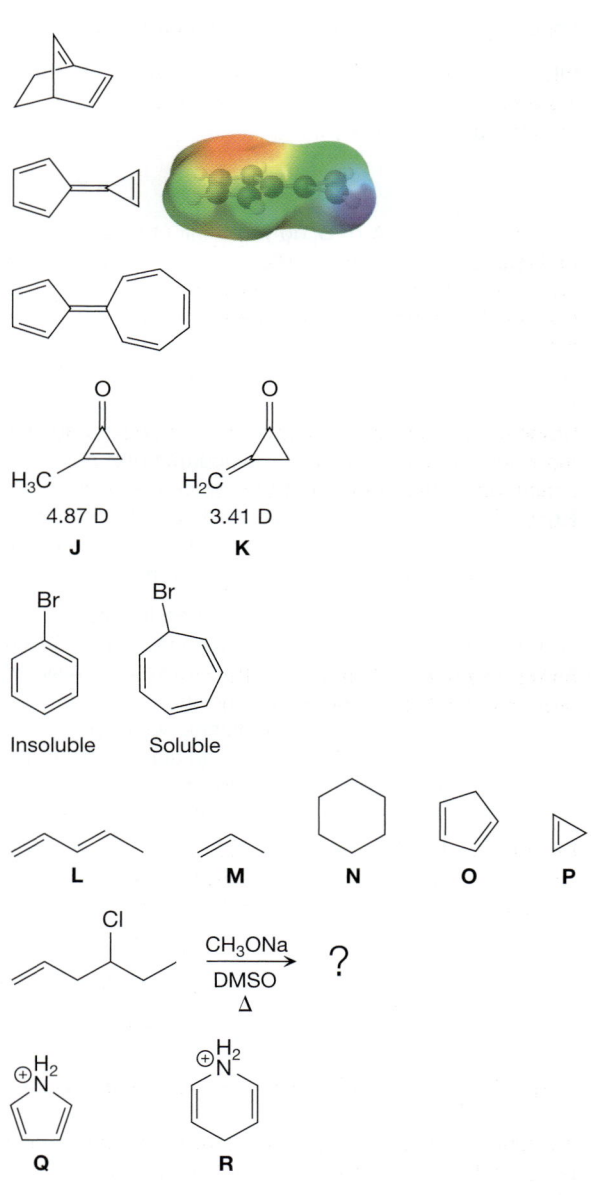

Doxorubicin

Integrated Problems

14.52 Although the hypothetical molecule shown here has alternating single and double bonds, those double bonds are not considered to be conjugated. Why not?

14.53 The molecule shown here has quite a large dipole, as indicated in its electrostatic potential map. Explain why. *Hint:* Consider various resonance structures.

14.54 Based on your answer to Problem 14.53, do you think the compound shown here should have a significant dipole moment? If so, in which direction does it point?

14.55 Using your knowledge of aromaticity, explain the difference in the calculated dipole moments of compounds **J** and **K**. *Hint:* Consider various resonance structures.

H_3C

4.87 D

J

H_2C

3.41 D

K

14.56 Bromobenzene is insoluble in water, but 7-bromocyclohepta-1,3,5-triene is water soluble. Explain why. *Hint:* In order for compounds to have dramatic differences in solubility, do you think the type of intermolecular forces between each compound and water can be the same?

Br

Br

Insoluble   Soluble

14.57 Rank compounds **L–P** in order of decreasing acid strength, from lowest pK_a to highest pK_a. Explain.

L   **M**   **N**   **O**   **P**

14.58 Draw the complete, detailed mechanism for the reaction shown here and predict the major product. Explain.

Cl

$\xrightarrow[\substack{\text{DMSO} \\ \Delta}]{\text{CH}_3\text{ONa}}$?

14.59 Which cation, **Q** or **R**, do you think is the stronger acid? Why?

$\overset{\oplus}{N}H_2$

Q

$\overset{\oplus}{N}H_2$

R

14.60 Which O atom of 4-pyrone do you think is more *basic*? Explain.

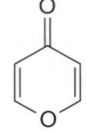

4-Pyrone

14.61 For each of the following pairs of molecules, which do you think should have the more acidic α hydrogen? Explain why.

(a)
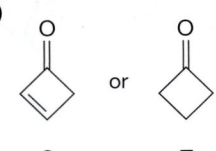

S **T**

(b)
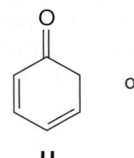

U **V**

14.62 Which do you think is a stronger nucleophile in an S_N2 reaction—pyridine or pyrrole? Explain.

Pyridine **Pyrrole**

14.63 For each of the following pairs of substrates, which do you think will undergo an S_N1 reaction faster? Explain.

(a)
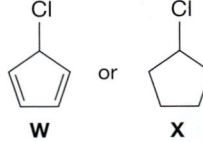

W **X**

(b)

Y **Z**

14.64 Which compound, **A** or **B**, do you think undergoes dehydration more quickly? Why?

A

B

14.65 Treatment of propadiene (an allene) with hydrogen bromide produces 2-bromopropene as the major product. This suggests that the more stable carbocation intermediate is produced by the addition of a proton to a terminal carbon rather than to the central carbon.

$$H_2C=C=CH_2 \xrightarrow{HBr} H_3C-C(Br)=CH_2$$

Propadiene **2-Bromopropene**

(a) Draw both carbocation intermediates that can be produced by the addition of a proton to the allene.
(b) Explain the relative stabilities of those intermediates. *Hint:* Draw the orbital picture of the intermediates and consider whether the CH₂ groups in propadiene are in the same plane.

14.66 Alkynes behave quite like alkenes when it comes to carbene addition. For example, a carbene can add to an alkyne to yield a cyclopropene, as shown here. Alkynes behave differently from alkenes, however, when treated with a peroxyacid. Whereas an alkene would be converted into an epoxide by the addition of an oxygen atom, this addition product is not observed for alkynes under normal conditions. Suggest why. *Hint:* Pay special attention to the lone pairs.

$$-C\equiv C- \xrightarrow[hv]{CH_2N_2} \quad C=C \text{ (cyclopropene)}$$

$$-C\equiv C- \xrightarrow{RCO_3H} \quad$$

Chloroplasts

A plant cell

Plants harness the sun's energy to convert carbon dioxide and water into sugar for storable energy—a process called photosynthesis that takes place inside chloroplasts. Chlorophyll, which gives plants their green color, is the key molecule responsible for absorbing visible light from the sun. As we see here in Chapter 15, this type of interaction with light is the basis for spectroscopy—a laboratory technique that provides information about a molecule's structure.

15

Structure Determination 1
Ultraviolet–Visible and Infrared Spectroscopies

By now you have seen the structures of many hundreds of different molecules throughout this book. What assurance do you have that those structures are accurate? When organic chemistry was a relatively immature field, chemists would derive molecular structure by measuring a compound's physical properties, and by carrying out different reactions designed to provide information about the presence and relative positions of specific functional groups. As you might imagine, these processes were quite painstaking and often unreliable.

Today, much of our knowledge about chemical structure comes from *spectroscopy*, the study of how electromagnetic radiation (such as visible light) interacts with matter. In a spectroscopic experiment, we *shine light (or other radiation) on molecules and carefully observe the outcome* using specialized instruments. There are many different types of spectroscopy, but the ones we focus on in this book (Chapters 15 and 16) are *ultraviolet–visible (UV–vis), infrared (IR),* and *nuclear magnetic resonance (NMR).* Because molecules interact with different kinds of electromagnetic radiation in characteristic ways, each type of spectroscopy reveals different aspects of how a molecule is put together.

Another important tool for elucidating molecular structure (which we discuss in Chapter 16) is *mass spectrometry.* Rather than using electromagnetic radiation, it uses various means to ionize molecules and even shatter them into fragments. A detector then measures the masses of the resulting charged species.

Chapter Objectives

On completing Chapter 15 you should be able to:

1. Define spectroscopy and specify the general features of a spectrum.

2. Describe how a UV–vis spectrum is acquired.

3. Explain what a photon is and how its energy is related to its frequency and wavelength.

4. Identify the longest-wavelength λ_{max} in a UV–vis spectrum and relate it to the energy difference between a molecule's HOMO and LUMO.

5. Determine when the longest-wavelength λ_{max} corresponds to a π-to-π* transition and when it corresponds to an n-to-π* transition, and explain why the former generally occurs at shorter wavelength than the latter.

6. Describe the relationship between the longest-wavelength λ_{max} and the extent of conjugation in a molecule.

7. Relate the frequency of light that is absorbed in IR spectroscopy to the frequency of a molecule's particular type of vibrational motion.

8. Identify the general regions in an IR spectrum where stretching modes and bending modes of vibration appear.

9. Specify the approximate frequencies of IR absorptions for stretching modes of a bond between hydrogen and a heavy atom, and explain why these stretching modes correspond to the highest-frequency IR absorptions.

10. Outline the factors that cause variations in the stretching frequency of a bond between hydrogen and a heavy atom.

11. Explain why stretching frequencies generally decrease in the following order: triple bond > double bond > single bond.

12. Specify the factors that dictate the intensity of a stretching-mode absorption band.

13. Know the characteristic frequencies and features of common IR absorption bands.

14. Obtain structural information about molecules from their UV–vis and IR spectra.

Nowadays, spectroscopy and mass spectrometry are used quite heavily in the area of synthesis. They often help us to identify newly discovered compounds in nature (so-called natural products) so that we can synthesize them from simpler compounds in the laboratory (e.g., as described in Chapters 13 and 19). Additionally, we routinely rely on these techniques to ensure that the products we make in the laboratory are in fact the ones we *intended* to make.

Spectroscopy and mass spectrometry even have many everyday applications. In criminal investigations, for example, an unknown substance encountered at a crime scene can be brought to a lab and analyzed via spectroscopy or mass spectrometry. Often the identity of the compound (or at least certain important characteristics) can be revealed in seconds or minutes.

Here in Chapter 15, we discuss UV–vis and IR spectroscopy. For each type of spectroscopy, we examine the characteristic changes that take place within a molecule when it interacts with that particular electromagnetic radiation, then we explain how to interpret the results of each type of spectroscopic experiment to derive aspects of a molecule's structure. In Chapter 16, we discuss NMR spectroscopy and provide a brief introduction to mass spectrometry.

15.1 An Overview of Ultraviolet–Visible Spectroscopy

Spectroscopy is the study of the interaction of *electromagnetic radiation* with matter. As shown in Figure 15-1, electromagnetic radiation is categorized according to its wavelength, from very-short-wavelength gamma rays to relatively long-wavelength radio waves.

Ultraviolet–visible spectroscopy (UV–vis), in particular, uses light from the UV and visible regions of the spectrum, which are probably more familiar to you than the other regions. The visible region includes wavelengths we associate with colors we can see with our eyes, ranging from red to violet. UV light has shorter wavelengths

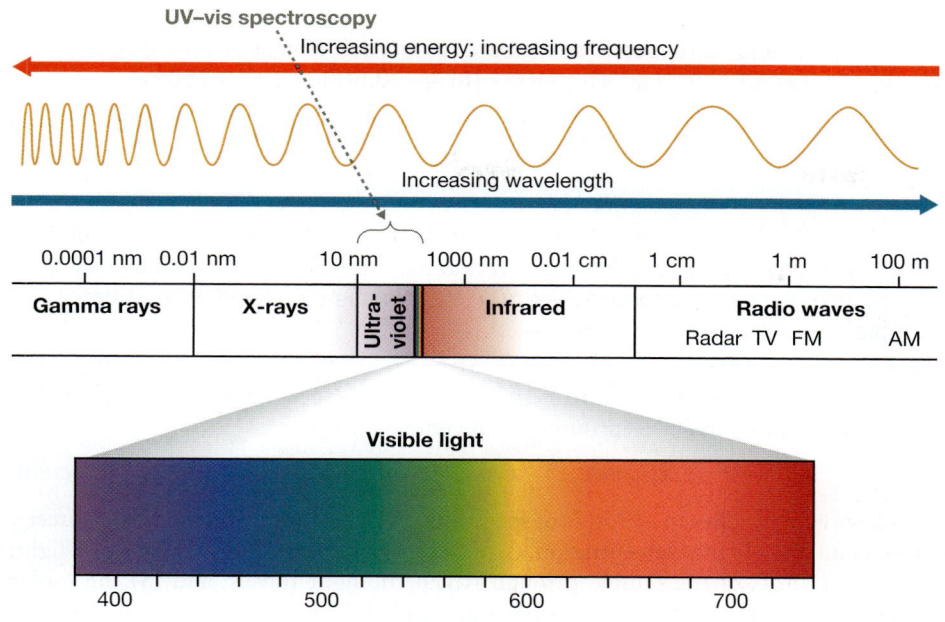

FIGURE 15-1 The electromagnetic spectrum Different types of spectroscopy use different wavelengths of electromagnetic radiation.

than visible light and contains the harmful radiation associated with sunburns and skin cancer.

In a typical UV–vis spectroscopy experiment (Fig. 15-2a), a range of wavelengths (usually 200–800 nm) is sent through an **analyte**—a sample we wish to analyze. At each wavelength, the intensity of light that reaches the detector ($I_{detected}$) is measured, and so is the intensity of light from the source (I_{source}). At certain wavelengths, the analyte might interact with the light from the source as it passes through, in which case $I_{detected}$ would be smaller than I_{source}. The light's efficiency in reaching the detector is the **transmittance** (%T) and is computed according to Equation 15-1.

$$\%T = (I_{detected}/I_{source}) \times 100 \tag{15-1}$$

Although intensity is a quantity that has units such as watts (W), these units cancel when $I_{detected}$ is divided by I_{source}.

Equation 15-2 is used to convert transmittance to **absorbance (A)**, which ranges from 0 (no absorption) to infinity (complete absorption).

$$A = -\log(I_{detected}/I_{source}) = 2 - \log(\%T) \tag{15-2}$$

Finally, a **UV–vis spectrum** is produced by plotting absorbance against **wavelength, λ** (the Greek letter lambda), as shown for a generic case in Figure 15-2b. Features of the spectrum where absorbance is high are called **absorption bands** or **peaks**.

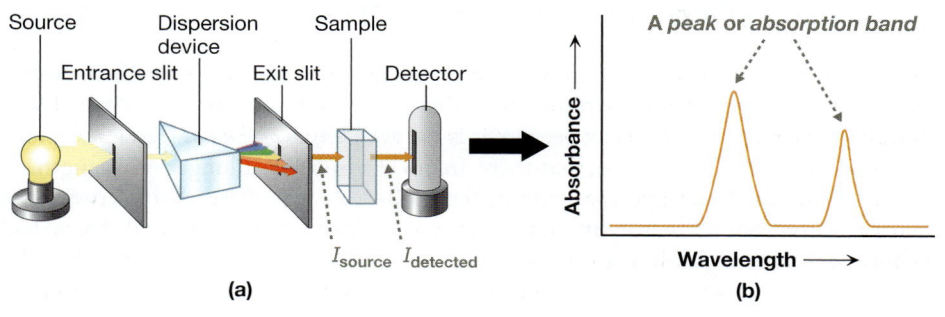

FIGURE 15-2 UV–vis spectroscopy (a) A general schematic of a UV–vis spectrophotometer. Specific wavelengths of light are selected to enter a sample. The intensity of light entering the sample is I_{source}, whereas the intensity of light that exits the sample and is detected by the detector is $I_{detected}$. (b) A generic UV–vis spectrum in which wavelength is plotted on the x axis and absorbance is plotted on the y axis.

Which scenario corresponds to a greater proportion of light absorbed by a sample: **(a)** %T = 20% or %T = 40%? **(b)** A = 0.500 or A = 0.750?

Answers to Your Turns are in the back of the book.

PROBLEM 15.1 What is the percent transmittance if the intensity from a radiation source is measured to be 1.0×10^{-4} W and the intensity of the radiation exiting the sample is measured to be 2.5×10^{-5} W? What is the absorbance of the sample?

The magnitude of absorbance is governed by the **Beer–Lambert law** (Equation 15-3), which you may have encountered in general chemistry:

$$A = \varepsilon \cdot l \cdot C \qquad (15\text{-}3)$$

This law, which applies to *all* forms of spectroscopy, states that absorbance, A, is directly proportional to (1) the concentration, C, of the species responsible for absorbing light; (2) the length, l, of the sample through which the light travels; and (3) the **molar absorptivity**, ε (the Greek letter epsilon), also called the *extinction coefficient*, which is an experimentally derived quantity that is characteristic of a given species at a given wavelength of radiation.

The Beer–Lambert law takes the form it does because the amount of light that a sample absorbs is proportional to the number of light-absorbing molecules the light encounters as it travels through the sample; the number of light-absorbing molecules, in turn, increases with both C and l. The molar absorptivity reflects *the probability that light of a given wavelength will be absorbed when it encounters light-absorbing molecules.* Later here in Chapter 15, and again in Chapter 16, we discuss some of the factors that govern the relative magnitude of ε for various types of spectroscopy. This makes it possible to better understand certain characteristics of the different spectra we encounter.

If increasing the wavelength of light causes the molar absorptivity of a sample to increase by a factor of 3, what happens to the measured absorbance?

PROBLEM 15.2 Calculate the molar absorptivity if a sample's absorbance is 0.78, the concentration of the sample is 6.00×10^{-6} M, and the length of the sample the light travels through is 1.00 cm. What are the units?

15.2 The UV–Vis Spectrum: Photon Absorption and Electron Transitions

The UV–vis absorption spectrum of buta-1,3-diene is shown in Figure 15-3. Notice that there is a peak in the spectrum, centered at 217 nm. This wavelength is called $\boldsymbol{\lambda_{max}}$ (i.e., lambda-max) for that peak because it is *the wavelength of light at which absorbance is a local maximum.* Notice, too, that there are no absorptions at longer wavelength.

To understand why the spectrum of buta-1,3-diene has these characteristics, we must first understand that light (and electromagnetic radiation in general) has a dual nature, behaving as both a wave and a particle. As a wave, light can be assigned a *wavelength* and a *frequency.* The wavelength (λ) is the distance between two successive maxima or minima in the wave's oscillating amplitude. **Frequency, $\boldsymbol{\nu}$** (the Greek letter

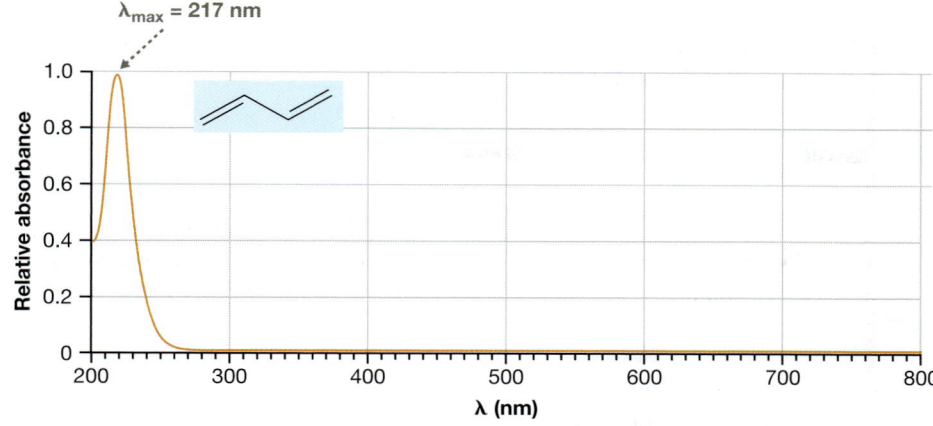

$\lambda_{max} = 217$ nm

FIGURE 15-3 UV–vis spectrum of buta-1,3-diene Relative absorbance is plotted against wavelength. The wavelength of maximum absorption, λ_{max}, is 217 nm.

nu), on the other hand, is the number of complete oscillations that occur in a given time. Its units can be hertz (Hz), cycles/second, or s^{-1}, all of which are equivalent. Wavelength and frequency are related by Equation 15-4:

$$c = \lambda\nu \quad \text{or} \quad \nu = c/\lambda \qquad (15\text{-}4)$$

Here c stands for the speed of light, a universal constant: $c = 2.9979 \times 10^8$ m/s. According to this equation, *wavelength and frequency are inversely related*, so as one increases, the other must decrease.

Behaving as a particle, electromagnetic radiation exists as *photons*. Each photon possesses a characteristic energy that depends only on its frequency (or wavelength), according to Equation 15-5:

$$E_{photon} = h\nu_{photon} = hc/\lambda_{photon} \qquad (15\text{-}5)$$

In this equation, E_{photon} is the energy of a *single* photon, and h is **Planck's constant** $(6.626 \times 10^{-34} \text{ J} \cdot \text{s})$. Thus:

- *The energy of a given photon is directly proportional to its frequency*: As one increases, so does the other.
- *A photon's energy and wavelength are inversely proportional*: As one increases, the other decreases.

YOUR TURN **15.3**

Which wavelength of light corresponds to photons with a greater energy: 375 nm or 530 nm?

According to the **law of conservation of energy**, energy cannot be created or destroyed, so when a molecule absorbs a UV–vis photon, it acquires the photon's energy, too. More specifically, the photon's energy is transferred to an electron:

The absorption of a UV–vis photon causes an **electron transition**, in which an electron in the molecule is promoted to a higher-energy molecular orbital (MO).

The blue arrows in Figure 15-4 depict examples of these kinds of transitions. Typically, many are possible, even for simple molecules.

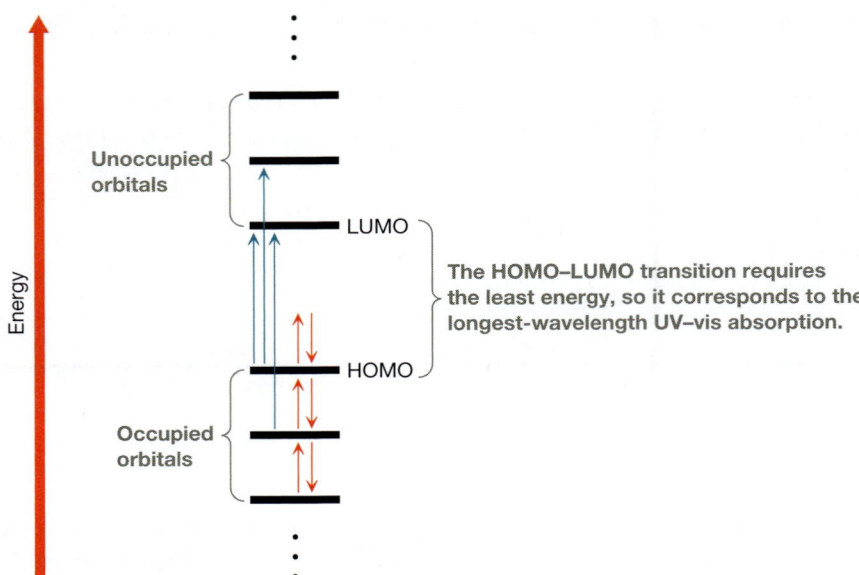

FIGURE 15-4 **Electron transitions in UV–vis spectroscopy** In UV–vis spectroscopy, the absorption of a photon typically corresponds to the promotion of an electron from an occupied MO to an unoccupied one. These electron transitions are represented by the blue arrows.

The HOMO–LUMO transition requires the least energy, so it corresponds to the longest-wavelength UV–vis absorption.

YOUR TURN 15.4

In Figure 15-4, draw arrows to represent five more electron transitions involving the six MOs that are given.

Which MOs, in particular, are involved in the transition that corresponds to the absorption band at 217 nm for buta-1,3-diene? Because no absorption bands appear at longer wavelength in the UV–vis spectrum, the electron transition must be the one that requires the *least* amount of energy, which, as indicated in Figure 15-4, is the one that promotes an electron from the HOMO to the LUMO. Thus, in general:

The longest-wavelength UV–vis absorption band corresponds to the HOMO–LUMO transition of the analyte.

YOUR TURN 15.5

How do the energies required for the transitions you drew in Your Turn 15.4 compare to the energy required for the HOMO–LUMO transition?

SOLVED PROBLEM 15.3

Given the UV–vis spectrum shown here, calculate the approximate energy difference between the HOMO and LUMO.

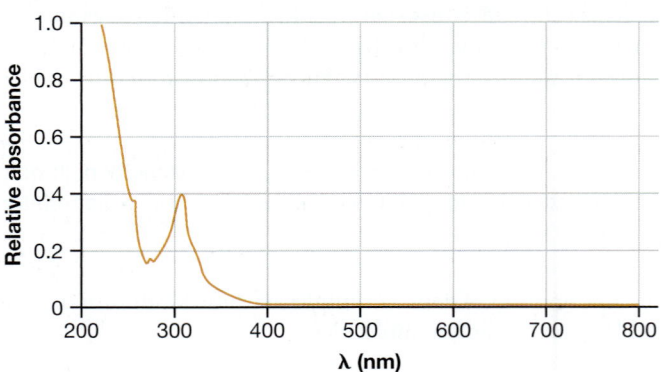

Think Which is the longest-wavelength absorption band? What is its λ_{max}? What energy does a photon with that wavelength possess?

Solve Although there may be a peak that is partially shown around 230 nm, the longest-wavelength absorption band has a λ_{max} of 310 nm. This corresponds to an energy of:

$$E = h\nu = h\left(\frac{c}{\lambda}\right) = (6.626 \times 10^{-34} \text{ J}\cdot\text{s})\left(\frac{2.9979 \times 10^8 \text{ m/s}}{310 \times 10^{-9} \text{ m}}\right)$$

$$= 6.41 \times 10^{-19} \text{ J}$$

PROBLEM 15.4 Given the UV–vis spectrum shown here, calculate the approximate energy difference between the HOMO and LUMO of the unspecified organic compound.

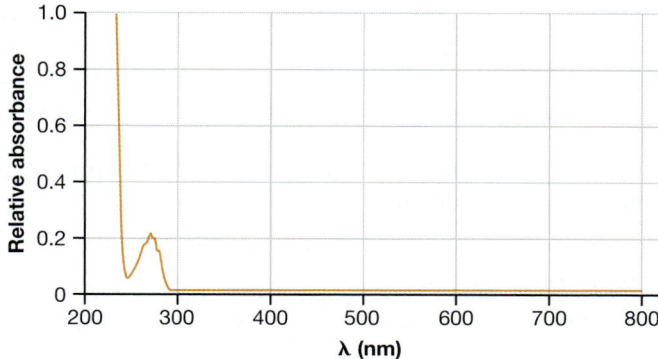

To gain a better feel for the HOMO–LUMO transition in buta-1,3-diene, consider Figure 15-5a, its MO energy diagram (shown previously in Fig. 14-12). The HOMO is π_2, which is a bonding MO, and the LUMO is $\pi_3{}^*$, an antibonding MO. When an incoming photon with the appropriate energy (represented by the wavy red arrow) is absorbed, the molecule undergoes what is called a **$\pi \rightarrow \pi^*$** ("pi-to-pi-star") **transition**, producing an *excited state* of the molecule.

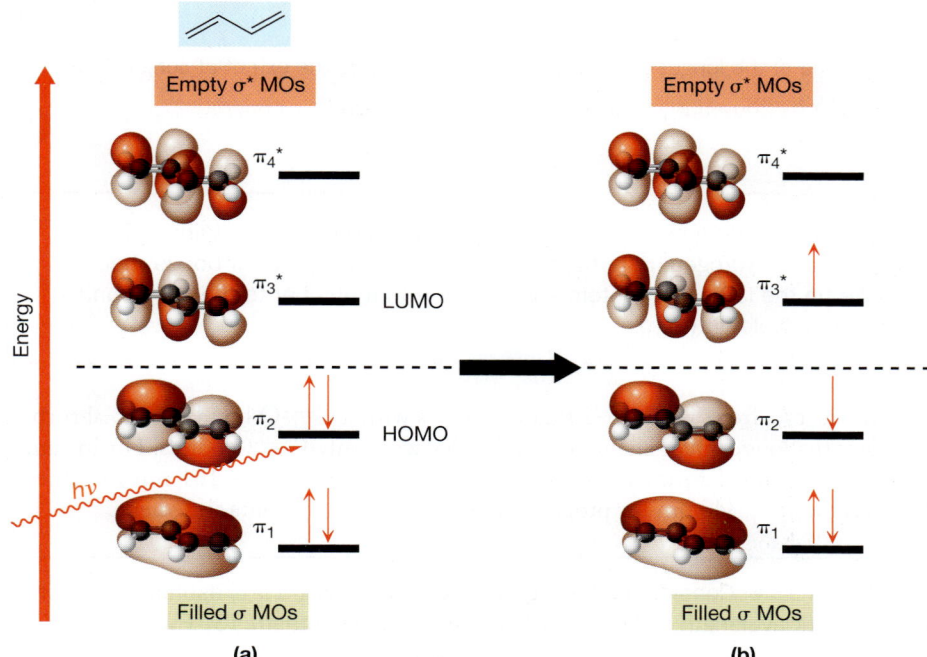

FIGURE 15-5 Absorption of an UV–vis photon in buta-1,3-diene (a) Ground-state electron configuration of buta-1,3-diene. The red wavy arrow represents a photon with energy $h\nu$. (b) Excited state electron configuration of the same molecule after absorption of the photon. Notice that an electron has been promoted from the HOMO (π_2) to the LUMO ($\pi_3{}^*$).

In Figure 15-5a, circle the electron that gains the energy of the incoming photon and draw an arrow from it to the orbital to which it is promoted. Label each of the π MOs as either bonding, nonbonding, or antibonding.

PROBLEM 15.5 Construct a diagram, similar to Figure 15-5, that depicts the longest-wavelength UV–vis absorption for ethene (H_2C=CH_2).

15.3 Effects of Structure on λ_{max}

As we saw in Section 15.2, λ_{max} for the longest-wavelength UV–vis absorption band of buta-1,3-diene is 217 nm (Fig. 15-3). The specific structure of a molecule, however, can have significant effects on the λ_{max} for these kinds of absorption bands, as shown in Table 15-1. The range is wide, from ~161 nm (in the UV region) to >450 nm (in the visible region).

Table 15-1 shows that λ_{max} for the longest-wavelength absorption depends on the nature of bonding within the molecule:

- Molecules that contain only σ bonds (e.g., alkanes and cycloalkanes) do not absorb in the UV or visible regions—their λ_{max} is much too short.
- Molecules that contain at least one π bond (e.g., ethene or hex-1-ene) have a λ_{max} greater than about ~160 nm.

Of the molecules containing π bonds, λ_{max} *depends heavily on the extent of conjugation.* For example, the λ_{max} of H_2C=CH_2 and the λ_{max} of cyclohexene, each of which contains an isolated double bond, are similar at 161 nm and 182 nm, respectively. The values for λ_{max} of *cis*-penta-1,3-diene and cyclopentadiene, each of which contains two conjugated double bonds, are similar, too, but they are at 223 nm and 239 nm, respectively. In general:

The value of the longest-wavelength λ_{max} increases as the extent of conjugation increases.

Examine the compounds in Table 15-1. For each one that contains only carbon and hydrogen, determine the number of conjugated π bonds that make up the largest π system and compare that number to the compound's longest-wavelength λ_{max}.

Notice, too, from Table 15-1 that molecules with a C=O bond tend to absorb at wavelengths longer than analogous molecules with only C=C bonds. For instance, λ_{max} = 280 nm for formaldehyde (H_2C=O), whereas λ_{max} = 161 nm for ethene. Similarly, $\lambda_{max} \approx 340$ nm for propenal, but only 217 nm for buta-1,3-diene.

PROBLEM 15.6 Considering the trend observed in Table 15-1, estimate λ_{max} for the longest-wavelength absorption of deca-1,3,5,7,9-pentaene ($C_{10}H_{12}$).

TABLE 15-1 λ_{max} for the Longest-Wavelength UV–Vis Absorptions of a Variety of Organic Compounds[a]

Compound	λ_{max} (nm)	Compound	λ_{max} (nm)	Compound	λ_{max} (nm)
Alkanes and cycloalkanes	<150				
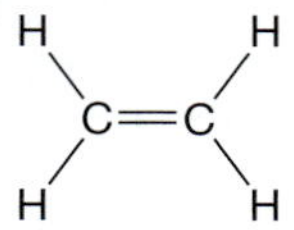 Ethene	161	 Buta-1,3-diene	217	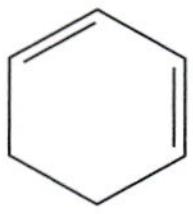 Cyclohexa-1,3-diene	256
 Hex-1-ene	177	 cis-Penta-1,3-diene	223	 Hexa-1,3,5-triene	274
 Penta-1,4-diene	178	 trans-Penta-1,3-diene	223.5	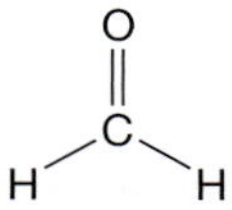 Methanal (Formaldehyde)	280
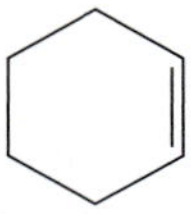 Cyclohexene	182	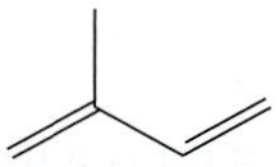 2-Methylbuta-1,3-diene (Isoprene)	224	 Octa-1,3,5,7-tetraene	290
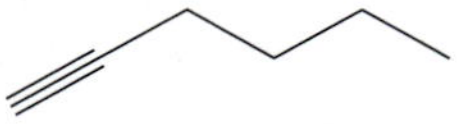 Hex-1-yne	185	 Cyclopentadiene	239	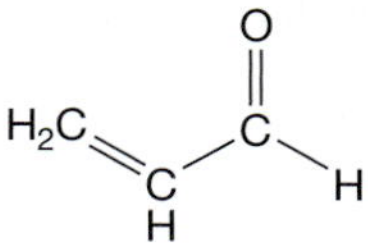 Propenal (Acrolein)	340

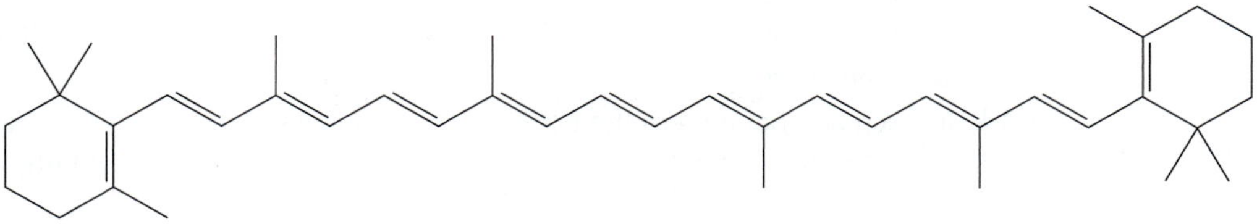

β-Carotene — 455

[a]Values <200nm are measured in the vapor phase to avoid absorption by a cuvette, solvent, or air.

PROBLEM 15.7 Considering the trend observed in Table 15-1, estimate λ_{max} for the longest-wavelength absorption of penta-2,4-dienal ($H_2C=CH-CH=CH-CH=O$).

15.3a Effect of Conjugation on λ_{max}

As conjugation increases, λ_{max} for the longest-wavelength absorption increases, too. We can see why in Figure 15-6, which shows the specific electron transitions that occur when a photon is absorbed by ethene, buta-1,3-diene, and hexa-1,3,5-triene—molecules

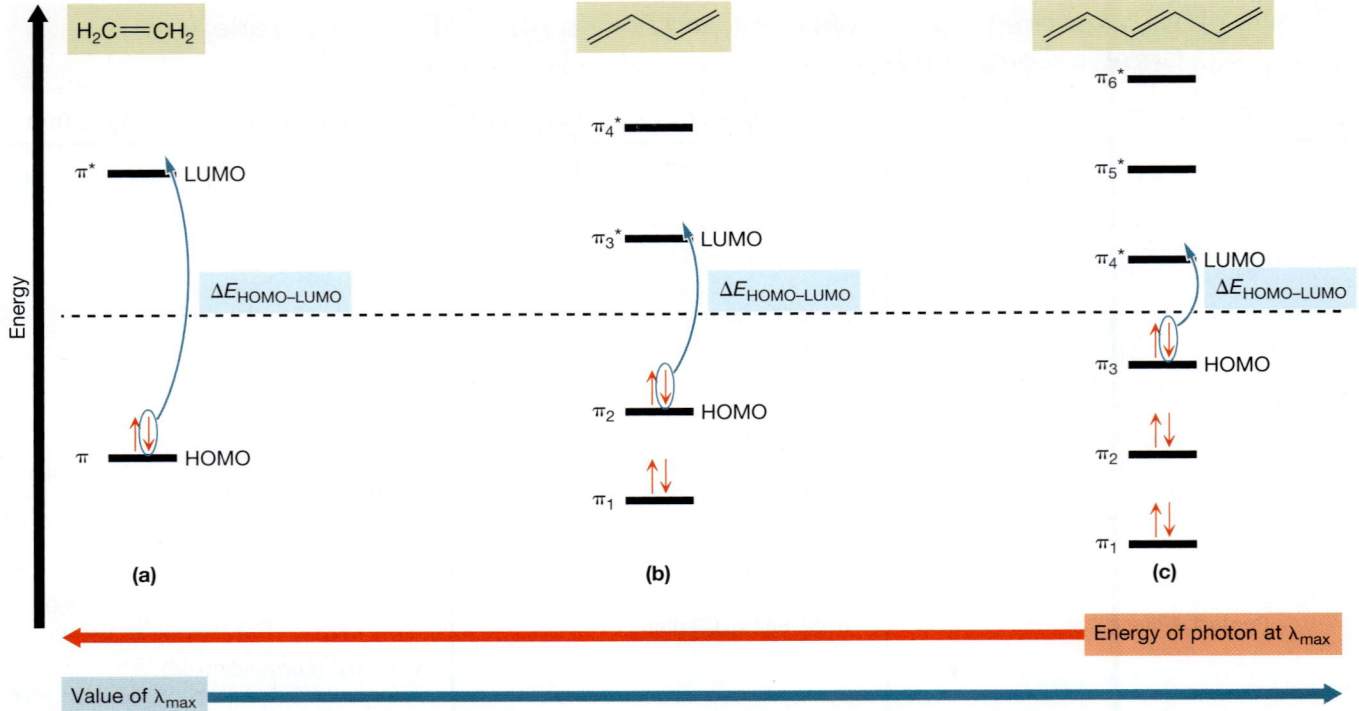

FIGURE 15-6 Conjugation and HOMO–LUMO transitions MO energy diagram for the π MOs of (a) ethene, (b) buta-1,3-diene, and (c) hexa-1,3,5-triene. As conjugation increases, the HOMO–LUMO energy difference ($\Delta E_{HOMO-LUMO}$) decreases, so the value of λ_{max} must increase.

that differ in the extent of the conjugation of their π systems. In all three cases, the lower-energy σ MOs are all filled, so the HOMO–LUMO transition is from a π MO to a π* MO (a π → π* transition).

According to Figure 15-6, the HOMO–LUMO energy difference ($\Delta E_{HOMO-LUMO}$) *decreases* as the extent of conjugation *increases*. Because this energy difference corresponds to the energy of the photon that is absorbed, the photon's energy must also decrease. And, according to Equation 15-5, that means the photon's wavelength must *increase*.

SOLVED PROBLEM 15.8

Both *trans*-penta-1,3-diene and penta-1,4-diene have a total of four MOs of π symmetry, yet according to Table 15-1, penta-1,3-diene absorbs at a much longer wavelength. Explain why.

Think In each compound, are the double bonds conjugated or isolated? What do the resulting MO energy diagrams look like? What are the relative energy differences between the HOMO and LUMO in each molecule?

Solve The structures of the compounds are shown here. The double bonds in *trans*-penta-1,3-diene are conjugated, so all four π electrons are in the same π system, resembling that shown in Figure 15-6b. In penta-1,4-diene, however, the double bonds are isolated from each other, so each π system looks like that in Figure 15-6a. The HOMO–LUMO energy difference is smaller in Figure 15-6b, so the longest λ_{max} in *trans*-penta-1,3-diene is longer than that in penta-1,4-diene.

trans-**Penta-1,3-diene**   **Penta-1,4-diene**

PROBLEM 15.9 Which of the compounds shown here do you expect to have the longer λ_{max}? Explain.

CONNECTIONS *trans*-Penta-1,3-diene is produced in soft drinks and energy drinks that contain the preservative sorbic acid when a spoilage yeast, such as *Saccharomyces cerevisiae*, degrades sorbic acid.

15.3b Effect of Lone Pairs on λ_{max}

A molecule such as formaldehyde, $H_2C{=}O$, does not at first seem to be consistent with the dependence of λ_{max} on the extent of conjugation. Its λ_{max} is 280 nm, which is significantly longer than that of its carbon analog, $H_2C{=}CH_2$ ($\lambda_{max} = 161$ nm). In fact, the λ_{max} of formaldehyde is about the same as that of hexa-1,3,5-triene ($\lambda_{max} = 274$ nm), which has three conjugated double bonds!

In formaldehyde, the HOMO is a nonbonding MO (i.e., *not* a π MO), which holds a lone pair of electrons on O. As a result, the transition that occurs in formaldehyde on absorption of a UV–vis photon is *not* a $\pi \to \pi^*$ transition. Instead, it is from the nonbonding orbital to the π^* orbital, which is called an **n → π*** ("n-to-pi-star") **transition**. A simple valence bond picture suggests that formaldehyde's nonbonding orbitals are intermediate in energy between the π and π^* MOs (Fig. 15-7b). Therefore, the n → π* transition requires substantially *less* energy than the π → π* transition (Fig. 15-7a), in which case λ_{max} is *longer*. In general:

> The longest λ_{max} for a species that has an available n → π* transition (due to the presence of a lone pair) will be significantly longer than a comparable species that does not have such a transition available.

SOLVED PROBLEM 15.10

Which of the following has a longer λ_{max}: $H_2C{=}CH_2$ or $H_2C{=}NH$? Explain.

Think For each compound, does the HOMO–LUMO transition correspond to a π → π* transition or an n → π* transition? In general, what are the relative energies of these types of transitions?

Solve The HOMO–LUMO transition for $H_2C{=}CH_2$ is π → π*, as shown in Figure 15-7a. The transition for $H_2C{=}NH$, however, is n → π*, similar to Figure 15-7b, given the presence of both a lone pair on N and the π bond. Accordingly, the HOMO–LUMO transition for $H_2C{=}NH$ should require less energy, and will thus appear at a longer wavelength.

PROBLEM 15.11 Which of the following do you think has a longer λ_{max}: $H_2C{=}NH_2^+$ or $H_2C{=}OH^+$? Explain.

CONNECTIONS Methanimine, $H_2C{=}NH$, has been detected in distant galaxies. This is significant because methanimine is viewed as a possible precursor to amino acids, so its existence in distant galaxies supports the idea that life can exist elsewhere in the universe.

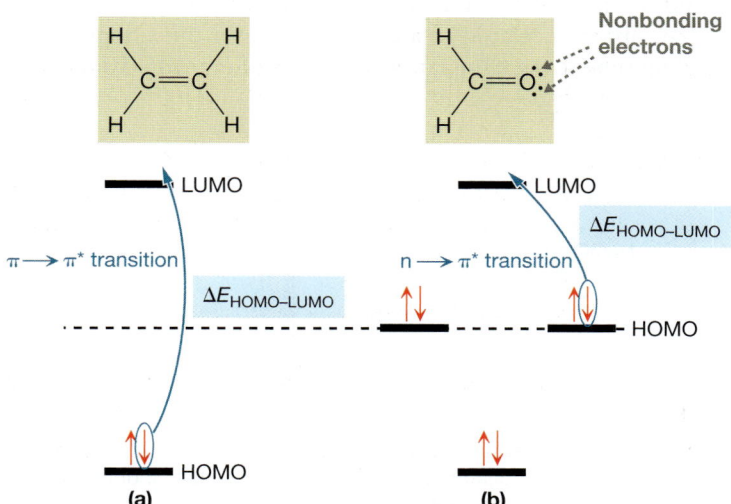

FIGURE 15-7 Nonbonding electrons and HOMO–LUMO transitions MO energy diagram for (a) ethene and (b) formaldehyde. The presence of the nonbonding electrons on O significantly decreases the energy difference between the HOMO and LUMO ($\Delta E_{HOMO–LUMO}$).

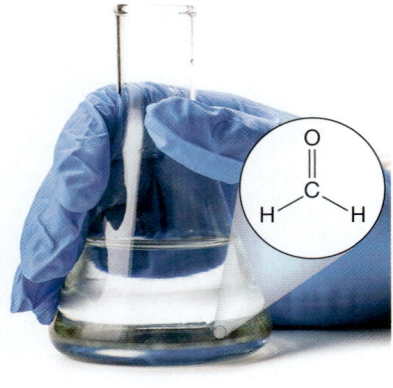

(a)

(b)

FIGURE 15-8 UV–vis absorption and color (a) Formaldehyde is colorless because it absorbs in the UV portion of the spectrum. (b) Absorption by β-carotene in the visible portion of the electromagnetic spectrum gives carrots their characteristic color.

β-Carotene absorbs in this blue region.

Our eye registers the complementary color, which is orange.

FIGURE 15-9 Complementary colors of visible light Complementary colors of visible light are found on opposite sides of this color wheel. When a particular color of light is removed from white light, our eye registers the complementary color.

15.3c UV–Vis Absorption and Color

Liquid formaldehyde is colorless (Fig. 15-8a), whereas β-carotene, the compound responsible for the color of carrots, is orange (Fig. 15-8b). What makes these compounds so different in appearance?

The answer stems from the fact that white light is a mixture of all colors from the visible spectrum (about 400 nm to 700 nm). As we can see from Table 15-1 (p. 731), formaldehyde's longest-wavelength absorption is at 280 nm, which is in the UV region of the spectrum. Therefore, all the colors of white light pass through formaldehyde and reach our eye, so the white light appears unaffected.

β-Carotene, on the other hand, absorbs at 455 nm, which is in the visible part of the spectrum. Therefore, when white light impinges on the compound, not all colors reach our eye, giving the appearance of a color that is nonwhite.

Why, specifically, should β-carotene appear orange? When a particular color of light is removed from the spectrum, our eye registers the *complementary* color. In the case of β-carotene, 455 nm light is blue and its complementary color is orange. This is illustrated in Figure 15-9, a color wheel in which complementary colors are located on opposite sides.

PROBLEM 15.12 Crystal violet is a dye that is used as a pH indicator. At a pH of –1, a solution of the dye absorbs at 420 nm. What color does that solution appear? *Hint:* Use Figure 15-1 (p. 725) to determine the color to which 420 nm roughly corresponds.

15.3d UV–Vis Spectroscopy in Quantitative Measurements

So far, we have focused on how UV–vis spectra relate to the structures of various molecular species. In practice, however, chemists typically turn to other methods to determine a species' structure, such as infrared spectroscopy (Section 15.4), nuclear magnetic resonance spectroscopy (Chapter 16), and mass spectrometry (Chapter 16)—all of which provide more structural information than UV–vis spectroscopy.

Chemists most often use UV–vis spectroscopy to *quantify the amount* of one or more compounds present in solution. The idea is based on the Beer–Lambert law (review Equation 15-3, p. 726). The path length (l) will be known—it's typically a standard length such as 1.0 cm. A particular λ will be chosen to monitor, so the molar absorptivity (ε) will be known, too. Therefore, the measured absorbance (A) is directly proportional to the concentration (C) of the analyte.

Applying UV–vis spectroscopy toward quantitative measurements like this is particularly useful in kinetics experiments, where relative reactant and/or product concentrations (i.e., absorbance at λmax) are monitored over time. Suppose, for example, we want to determine if the elimination reaction in Figure 15-10 takes place by the

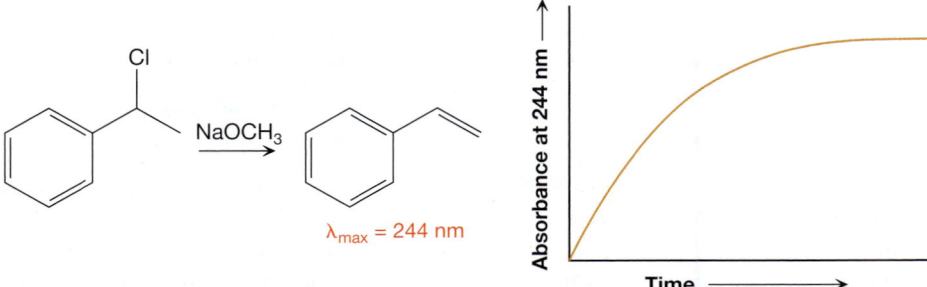

λmax = 244 nm

FIGURE 15-10 UV–vis spectroscopy and kinetics The kinetics of the elimination reaction at the left can be investigated by monitoring the absorbance of the product's λmax as a function of time. As shown at the right, as the concentration of the product increases, so does the absorbance.

E2 or E1 mechanism. The product has a λ_{max} of 244 nm, so as the reaction progresses, the absorbance at that wavelength increases. Because absorbance is directly proportional to concentration, the initial reaction rate is simply the change in absorbance divided by time. Therefore, if the concentration of the base, $NaOCH_3$, is doubled and the absorbance at 244 nm that is measured over an initial period also doubles, we can conclude that the reaction proceeds by the E2 mechanism, not the E1.

YOUR TURN **15.8**

For the reaction in Figure 15-10, suppose that the concentration of the base is doubled and the absorbance at 244 nm that is measured over an initial period remains unchanged. What would we conclude about the mechanism? Explain.

UV–Vis Spectroscopy and DNA Melting Points

Under normal conditions, two complementary strands of DNA associate into a double helix, held together by hydrogen bonding among complementary bases (A with T and C with G; review Section 14.11, p. 714). At high enough temperatures, however, that hydrogen bonding can be overcome and the double-stranded DNA will *denature*, unwinding into its single strands. The temperature at which half the DNA exists as single strands is called the melting temperature, T_m. T_m is an important piece of information about a sample of DNA that can be determined using UV–vis spectroscopy.

UV–vis spectroscopy can be used to obtain T_m because DNA (due to the π systems of its nitrogenous bases) has an absorption peak at ~260 nm, both as a double helix and as single strands, but single-stranded DNA has a significantly stronger absorbance—a phenomenon called *hyperchromicity*. Therefore, as a sample of DNA is heated and becomes denatured, the absorbance measured at 260 nm will increase, as shown here in the graph. The steepest point on the plot of absorbance versus temperature indicates T_m.

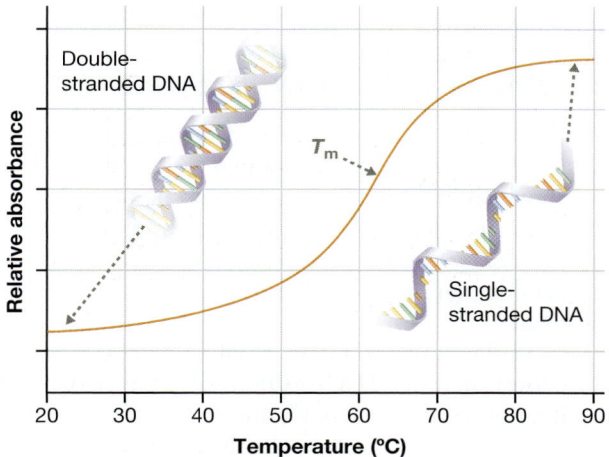

The value of T_m is somewhat sensitive to the DNA sequence, so this method of obtaining T_m is often employed to determine the purity of a sample of DNA that has been synthesized in the laboratory. This technique is also used to determine the GC content of DNA (i.e., the percentage of nucleotides that are either G or C), which can vary dramatically among organisms. Because there is more extensive hydrogen bonding between G and C than there is between A and T (three hydrogen bonds versus two), a higher value of T_m indicates higher GC content. Finally, T_m can be used to detect DNA mutations because mutated DNA tends to have a lower value of T_m compared to the wild type.

λ_{max} = 244 nm

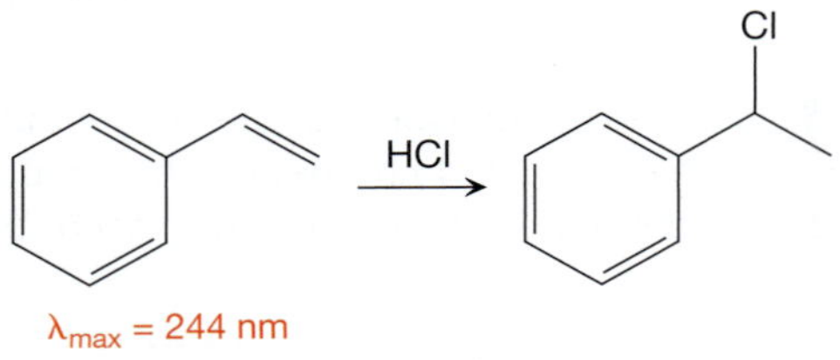

15.4 IR Spectroscopy

In IR spectroscopy, radiation from the infrared region of the electromagnetic spectrum is sent through the analyte, similar to what we saw with UV–vis spectroscopy (Fig. 15-2). IR radiation, which includes wavelengths from ~800 nm to ~10^6 nm (1 mm), is invisible to our eyes, but we can feel it—it feels warm. It is the radiation that emanates from lamps used at fast-food restaurants to keep your food warm, and it is responsible for the heat you can feel when you stand several meters away from a campfire, or when sunlight shines on your face. IR radiation feels warm to us because of the molecular property that is affected:

IR absorption causes excitations in the vibrational motions of molecules.

That includes molecules in our skin.

15.4a General Theory of IR Spectroscopy

Recall from Chapter 1 that chemical bonds behave like springs. Thus, the atoms of a molecule, like masses connected by springs, can undergo different types of regular oscillations in which interatomic distances and bond angles vary rhythmically. There are two basic, independent types of vibrational motion: **stretching** and **bending**. In stretching, the *distance* between two atoms in a chemical bond grows longer and shorter; in bending, an *angle*—either a bond angle or a dihedral angle—becomes larger and smaller.

Figure 15-11 shows several modes of vibration that take place simultaneously in a molecule of formaldehyde (H_2C=O). The carbonyl (C=O) group undergoes a stretching vibration, as indicated in Figure 15-11a. The C—H bonds can also undergo stretching vibrations, but not independently of each other. They undergo a **symmetric stretch** (Fig. 15-11b) and an **asymmetric stretch** (Fig. 15-11c).

Two types of bending modes are also shown for formaldehyde—namely, an **in-plane bending vibration** (Fig. 15-11d) and an **out-of-plane bending vibration** (Fig. 15-11e).

PROBLEM 15.14 Shown here is another type of vibration that formaldehyde undergoes, one that is not shown in Figure 15-11. Classify it as a symmetric stretch, an asymmetric stretch, an in-plane bend, or an out-of-plane bend.

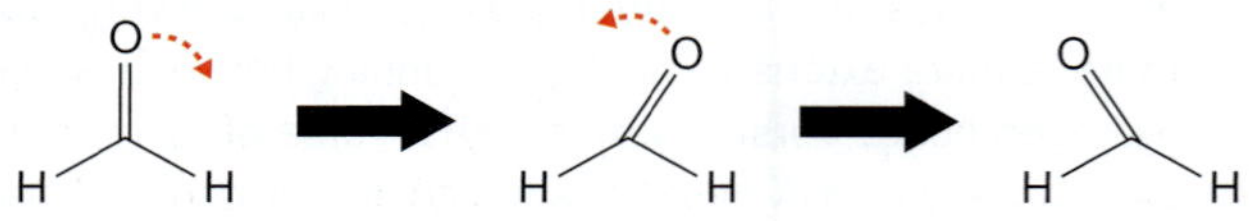

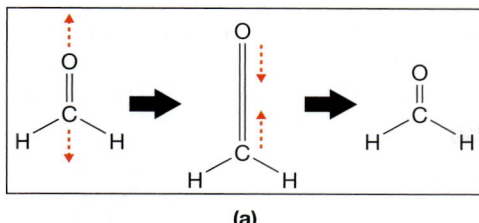

C=O stretch

(a)

FIGURE 15-11 Types of vibration Various types of vibration are shown for a molecule of formaldehyde. (a) The C=O bond stretches and compresses. (b) Both C—H bonds stretch and compress in unison. (c) While one C—H bond stretches, the other compresses. (d) The H—C—H bond angle closes and opens. (e) The two H atoms and the O atom shift downward and upward. *Note:* The magnitudes of displacement are exaggerated for illustration.

Symmetric C—H stretch

(b)

Asymmetric C—H stretch

(c)

In-plane bend

(d)

Out-of-plane bend

(e)

Quantum mechanics dictates that *each type of vibrational motion is quantized.* For each type of vibration, therefore, only certain energy levels are allowed, so only photons of certain energies can be absorbed. Molecules at room temperature, in particular, tend to absorb an IR photon when the photon's frequency equals the frequency of vibration for a particular type of motion. Therefore:

> The frequency at which a peak appears in an IR spectrum is often the same as the frequency of the vibration that is responsible for the absorption of the photon.

A typical IR spectrum is shown in Figure 15-12 (next page), in which percent transmittance is plotted against the frequency of light in units of **cm^{-1}**, called **reciprocal centimeters**, **inverse centimeters**, or **wavenumbers**, $\bar{\nu}$. This unit, the reciprocal of the wavelength, is calculated by dividing 1 by the wavelength in centimeters (Equation 15-6) and, like any unit of frequency, is directly proportional to photon energy. Physically, it corresponds to the number of waves that fit in 1 cm. The two common units of frequency—cm^{-1} and Hz—are related by the speed of light, as shown in Equation 15-7. (The factor of 100 accounts for different units of length used for wavenumbers and the speed of light: cm for wave numbers and m for the speed of light.)

$$\bar{\nu} \text{ (in cm}^{-1}) = 1/\text{wavelength (in cm)} \qquad (15\text{-}6)$$

$$\bar{\nu} \text{ (in cm}^{-1}) = \nu \text{ (in Hz)}/(100\,c) \qquad (15\text{-}7)$$

For historical reasons, frequency increases from right to left along the *x* axis. Commonly, the frequency range in an IR spectrum is from ~400 cm^{-1} to ~4000 cm^{-1}.

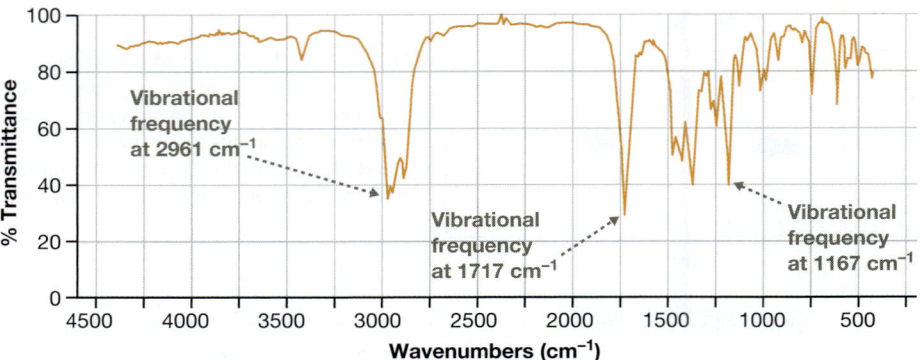

FIGURE 15-12 A typical infrared spectrum *Percent transmittance* is plotted on the *y* axis and frequency of light in wavenumbers (cm⁻¹) is plotted on the *x* axis. Each peak drops down from the top and represents the frequency at which various vibrations occur within the molecule.

Because percent transmittance is plotted on the *y* axis, the baseline (no absorption) is at 100% and *peaks in the spectrum appear as dips in percent transmittance*. Therefore, a **strong absorption** is one whose percent transmittance is near zero (near the bottom of the spectrum), and a **weak absorption** is one whose percent transmittance is near 100% (near the top of the spectrum). In Figure 15-12, for example, we can identify peaks at 2961, 1717, and 1167 cm⁻¹. Applying what we learned earlier, we can say that the molecule that produces this IR spectrum has types of vibrational motion with these same frequencies.

YOUR TURN 15.9

> Choose a fourth peak in Figure 15-12 and annotate it with the approximate frequency of the type of vibration it represents.

PROBLEM 15.15 Calculate the wavelengths of the photons required to excite the three vibrations indicated in Figure 15-12.

15.4b Location of Peaks in an Infrared Spectrum

One of the greatest advantages of IR spectroscopy is that the frequency of a particular type of vibration is typically found within a characteristic range of frequencies, regardless of the functional group with which the vibration is associated. Thus:

> A given type of bond has an IR absorption frequency that does not change dramatically from one molecule to another.

For example, compare the IR spectra of hexan-2-one and 2-ethylhexanal, shown in Figure 15-13. Both compounds contain a C=O bond—hexan-2-one as part of a ketone and 2-ethylhexanal as part of an aldehyde. Even though the molecules belong to different compound classes, the absorptions corresponding to the C=O stretching frequencies are quite similar, appearing at 1717 cm⁻¹ and 1723 cm⁻¹, respectively.

The fact that absorptions by certain vibrations appear with characteristic frequency ranges is what enables us to use IR spectroscopy to obtain structural information about a molecule. For example, based on the spectra in Figure 15-13, we can safely say that, in the IR spectrum of an unknown compound, the appearance of a strong, sharp

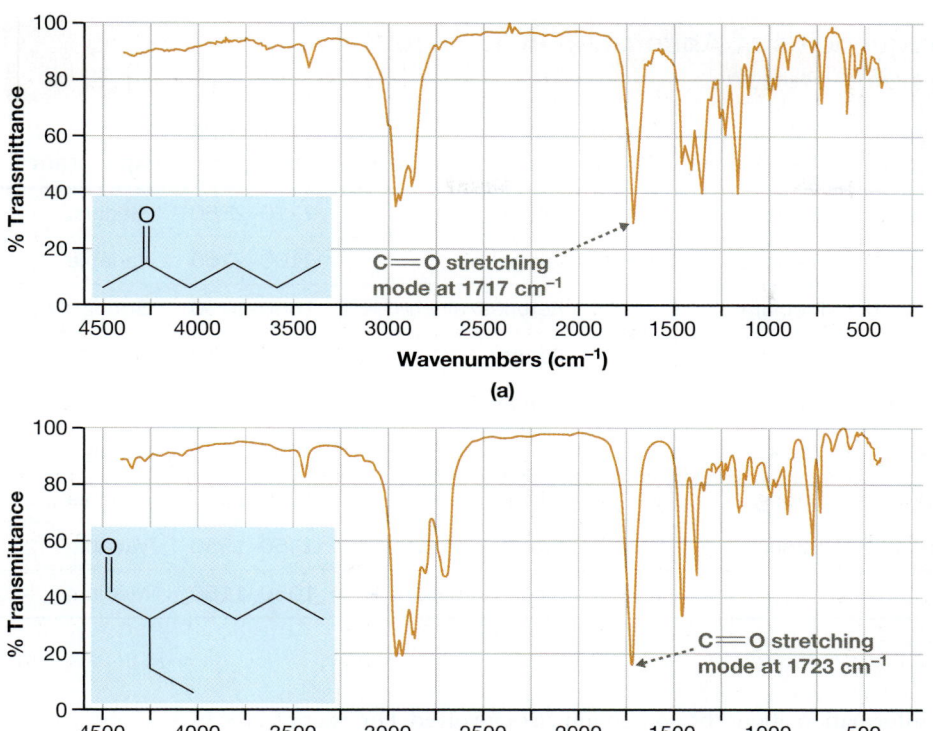

(a)

(b)

FIGURE 15-13 Characteristic absorption frequency of the C=O stretching mode of vibration The IR spectra of (a) hexan-2-one and (b) 2-ethylhexanal are shown. Although the molecules belong to different compound classes—hexan-2-one is a ketone and 2-ethylhexanal is an aldehyde—the absorptions corresponding to the C=O stretching modes are quite similar.

peak around 1720 cm⁻¹ strongly suggests that the compound contains a carbonyl (C=O) group.

YOUR TURN **15.10**

Benzaldehyde, whose IR spectrum is shown below, contains a C=O bond. Identify the absorption that corresponds to the C=O bond.

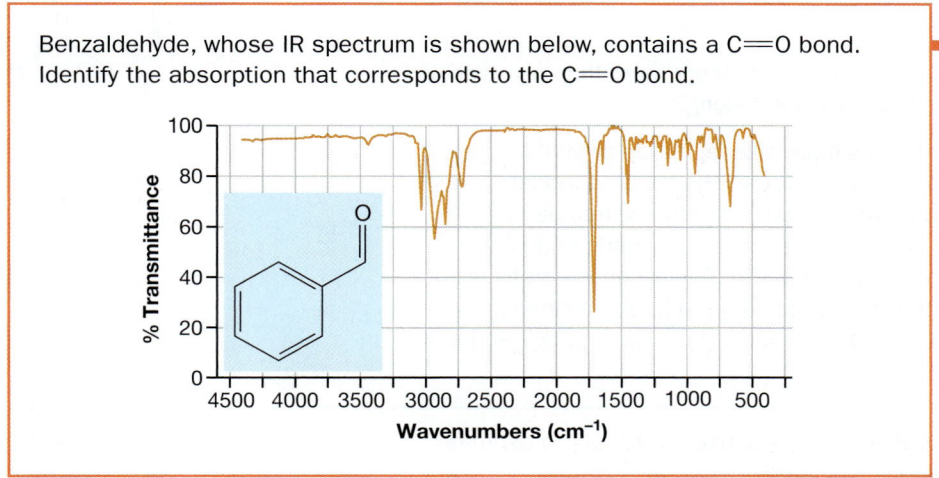

Absorptions for other types of vibrational modes appear within characteristic frequency ranges, too, as shown in Table 15-2. These are all for stretching modes of vibration (we examine some for bending modes of vibration in Section 15.5g). Notice that these stretching modes have frequencies from ~1000 cm⁻¹ to a few thousand cm⁻¹ (bending modes generally occur from ~500 cm⁻¹ to ~1000 cm⁻¹).

Although the absorptions for a particular mode of vibration appear within a relatively narrow range of frequencies, small differences in frequency can provide valuable

TABLE 15-2 Characteristic Frequencies of Absorption in IR Spectroscopy: Stretching Modes of Vibration

Type of Bond	Compound Class	Frequency Range (cm^{-1})	Appearance	Type of Bond	Compound Class	Frequency Range (cm^{-1})	Appearance
O—H	Alcohol and phenol	3200–3600	Broad, strong	C≡N	Nitrile	2210–2260	Medium
	Carboxylic acid	2500–3000	Broad, strong	C≡C	Alkyne	2100–2260	Variable
N—H	Amine	3300–3500	Medium	C=O	Ketones/aldehydes	1680–1750	Strong
	Amide	3350–3500	Medium		Esters	1730–1750	Strong
C—H	Alkane	2800–3000	Variable		Carboxylic acid	1710–1780	Strong
	Alkene	3000–3100	Weak		Amide	1630–1690	Strong
	Alkyne	~3300	Strong	C=C	Alkene	1620–1680	Variable
	Aldehyde	2720 and 2820	Strong		Aromatic	1450–1550	Variable
				C—O	Alcohol, ester, ether	1050–1150	Medium

information about the compound class involved. For example, the C=O absorptions for ketones, aldehydes, esters, carboxylic acids, and amides all appear between ~1630 and ~1780 cm^{-1}. However, as we see in Section 15.5c, an amide's C=O absorption tends to appear at the low end of that range, whereas that for a carboxylic acid tends to appear at the high end.

SOLVED PROBLEM 15.16

Use Table 15-2 to estimate the frequencies of the stretching vibrations indicated in the molecule shown here.

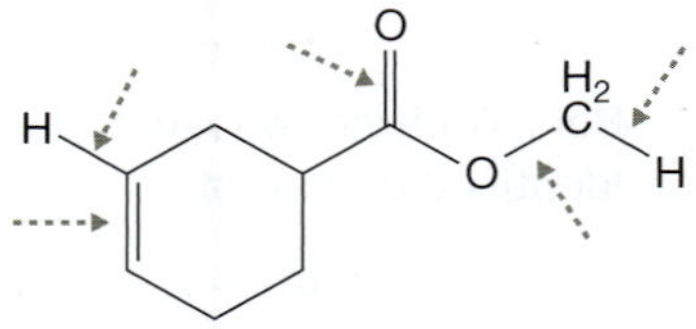

Think What type of bond does each arrow indicate? Is there more than one type of compound class to which each of those bonds could belong?

Solve The C=C bond resembles that of a simple alkene, not an aromatic ring, so its stretching frequency is ~1620 to 1680 cm^{-1}. The C—H bond on the left, which is attached to that C=C bond, should therefore have a stretching frequency ~3050 cm^{-1}. The C=O bond is part of a CO$_2$R group, characteristic of an ester, so its stretching frequency is ~1730 to 1750 cm^{-1}. The C—O bond's stretching frequency will be ~1100 cm^{-1}. And the C—H bond on the right most closely resembles that from an alkane (rather than an alkene, alkyne, or aldehyde), so its frequency is ~2800 to 3000 cm^{-1}.

PROBLEM 15.17 For each of the following molecules, use Table 15-2 to estimate the frequencies of the stretching vibrations indicated.

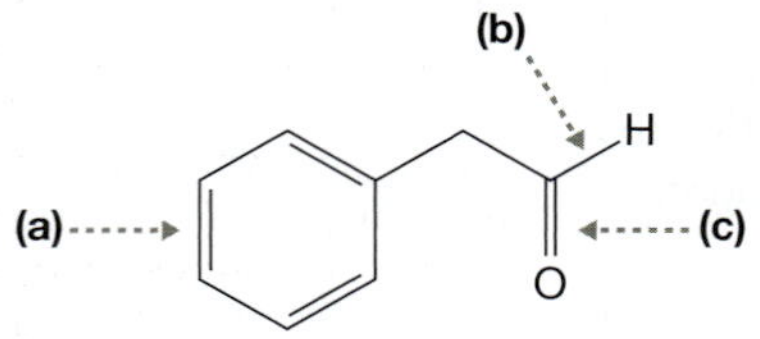

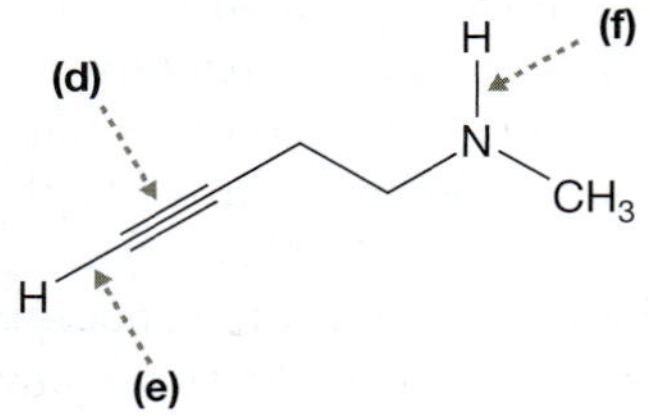

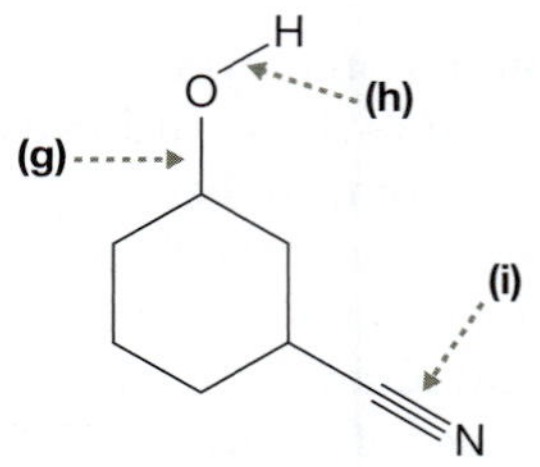

In Table 15-2, there are a number of trends worth noting. The first defines five major regions in which absorptions appear. These five regions, shown in Figure 15-14, are as follows:

The Five Major Regions in an IR Spectrum

- Stretching vibrations of Q—H bonds (where Q is a "heavy" atom such as C, N, or O) occur in the region between ~2500 and 4000 cm^{-1}.
- Absorptions by triple bonds (C≡C or C≡N) appear between 2000 and 2500 cm^{-1}.
- Double bonds appear between ~1500 and 2000 cm^{-1}.
- Single bonds between two heavy atoms appear between ~1000 and 1500 cm^{-1}.
- The frequencies of bending vibrational modes appear below 1000 cm^{-1}.

The region below ~1400 cm^{-1} is called the **fingerprint region**. Numerous peaks typically appear in this region, even for relatively simple molecules, and many of those peaks overlap. Therefore, we focus primarily on identifying absorptions at greater than ~1400 cm^{-1}. Suffice it to say, however, that the collection of peaks in a fingerprint region is unique for each molecular species, just as every human has unique fingerprints.

YOUR TURN 15.11

Identify the two regions in Figure 15-14 that contain the fingerprint region.

Two other trends that are useful to know involve bonds between hydrogen and a heavy atom—that is, absorptions that appear in the region above 2500 cm^{-1}. First, a periodic-table trend is in effect for these stretching frequencies:

- C—H stretches appear between ~2700 and 3300 cm^{-1}.
- N—H stretches appear around 3300 to 3500 cm^{-1}.
- Alcohol O—H stretches creep up a bit higher to ~3600 cm^{-1}.

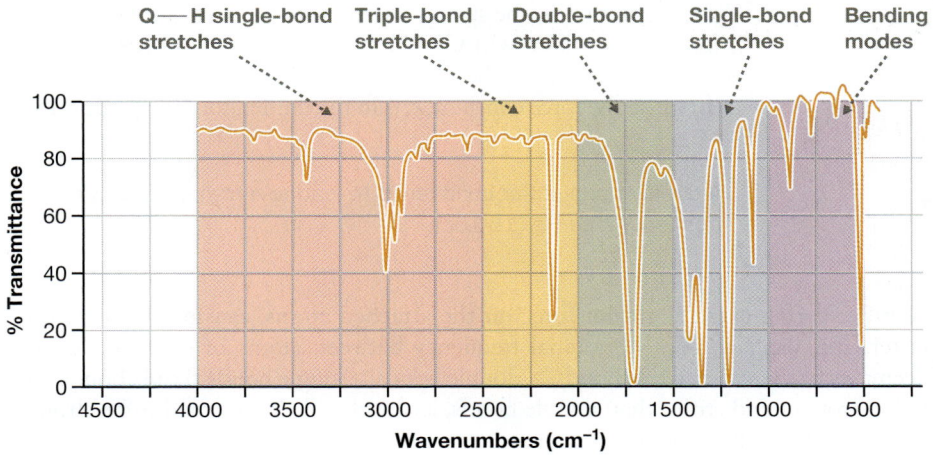

FIGURE 15-14 The five major regions of absorption in IR spectroscopy Absorptions corresponding to the stretching frequencies of O—H, N—H, and C—H bonds appear between 2500 and 4000 cm^{-1}. Those for triple bonds generally appear between 2000 and 2500 cm^{-1}. Those for double bonds generally appear between 1500 and 2000 cm^{-1}. Those for single bonds not involving hydrogen appear between 1000 and 1500 cm^{-1}. Absorptions corresponding to bending modes appear below 1000 cm^{-1}.

Second, there is a trend involving hybridization for C—H stretches:

- Alkane (sp^3) C—H stretches appear below 3000 cm^{-1}.
- Alkene (sp^2) C—H stretches appear between ~3000 and 3100 cm^{-1}.
- Alkyne (sp) C—H stretches appear at ~3300 cm^{-1}.

Reasons for these trends are discussed next in Section 15.4c.

15.4c The Ball-and-Spring Model for Explaining Peak Location

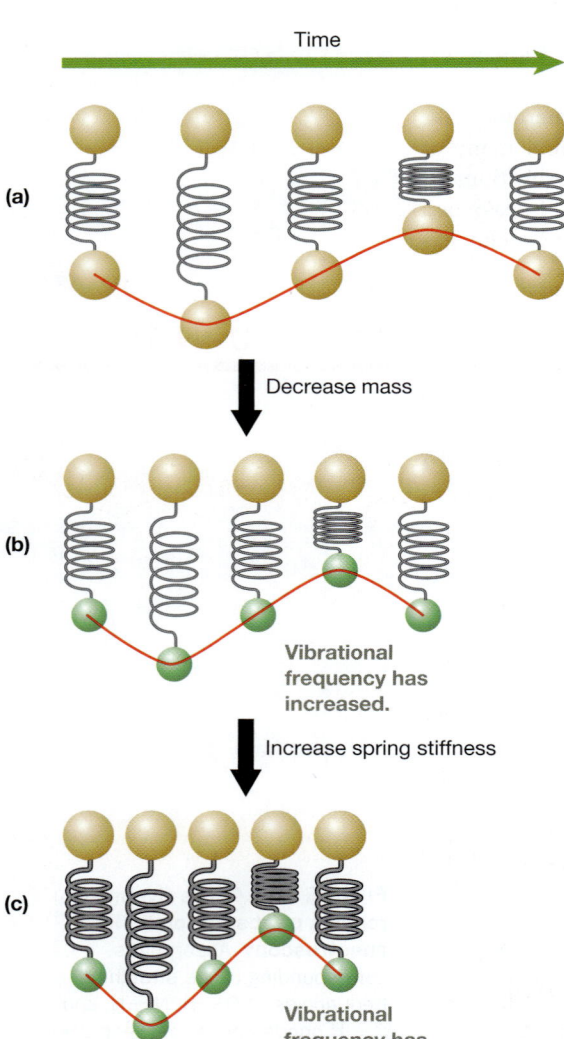

Time

(a)

Decrease mass

(b)

Vibrational frequency has increased.

Increase spring stiffness

(c)

Vibrational frequency has increased.

FIGURE 15-15 Vibrational frequency, mass, and spring stiffness (a) Masses connected by a spring undergo vibration. (b) With a lighter mass, the vibrational frequency increases. (c) With a stiffer spring, the vibrational frequency increases, too.

To interpret an IR spectrum, you must be able to determine the types of bonds that individual IR peaks represent. Table 15-2 shows these correlations for several IR peaks, but why should those peaks appear at the frequencies they do? To begin to answer this question, recall that an IR peak's frequency is generally equal to the frequency of the molecular vibration that is responsible for producing the peak. Therefore, if we understand the factors that control molecular vibrational frequency, we can make sense out of the information in Table 15-2.

We simplify the picture of molecular vibrations by considering the *ball-and-spring model*, which treats bonds as simple springs that connect atoms together (Fig. 15-15a). You might remember from physics that the behavior of such a system is described by **Hooke's law**. According to Hooke's law, the spring vibrates at a particular frequency (v_{spring}) that depends on the mass (m) of the object connected to the spring (in this case, an atom) and the spring's force constant (k), which is often thought of as the *stiffness* of the spring. The stiffer the spring, the more difficult it is to compress and extend it. Mathematically, these values are related according to Equation 15-8.

$$v_{spring} = \sqrt{\frac{k}{m}} \tag{15-8}$$

Because mass is in the denominator:

A decrease in the mass of an atom leads to an increase in vibrational frequency (see Fig. 15-15b).

With a lighter mass, the spring can move the atom more easily—that is, *faster*. This explains why C—H, N—H, and O—H stretches have such high frequencies: The H atom is *very light*.

Because the force constant, k, is in the numerator of Equation 15-8:

A stronger and stiffer bond tends to lead to a higher vibrational frequency (see Fig. 15-15c).

A stiffer spring exerts a greater force on the attached atoms, causing them to move faster. This explains why vibrational frequency between atoms of comparable mass decreases in the order: triple bonds > double bonds > single bonds. Triple bonds tend to be stronger and stiffer than double bonds, just as double bonds tend to be stronger and stiffer than single bonds.

The generalization about spring stiffness also accounts for the relationship between the C—H stretching frequency and the hybridization of C. We learned in Chapter 3 that the C—H bond energy increases as the s character of the C atom's hybridization increases. Thus, a bond between hydrogen and an sp-hybridized carbon atom is stronger and stiffer than one with an sp^2-hybridized carbon, which is stronger and stiffer, in turn, than one involving an sp^3-hybridized carbon. For this reason, the C—H stretching frequency decreases in the same order: alkyne C—H > alkene C—H > alkane C—H.

SOLVED PROBLEM 15.18

Deuterium (D), an isotope of hydrogen, has roughly twice the mass of hydrogen, but the two have nearly identical chemical properties. A C—H bond, therefore, has essentially the same strength as a C—D bond. Which vibrational mode, a C—H stretch or a C—D stretch, absorbs at a higher frequency in the IR region? Explain.

Think Considering the model of masses connected by a spring, are the masses the same? Is the stiffness of each spring the same? How do these factors govern vibrational frequency? How does vibrational frequency affect IR absorption frequencies?

Solve We are told that C—H and C—D bonds are essentially identical in strength, so each one's spring stiffness must be about the same, too. Spring stiffness, therefore, does not come into play, but the greater mass of D contributes to a lower frequency of vibration. Because the frequency of vibration is the same as the photon frequency that can be absorbed, a C—D stretch must absorb lower-frequency photons than a C—H stretch.

PROBLEM 15.19 For each of the following pairs of compounds, indicate which CN stretching mode absorbs the higher-frequency IR photons. Explain.

(a)   (b)   (c)

15.4d Amplitudes of Peaks in an IR Spectrum

As shown in Table 15-2, some absorptions have characteristically strong intensities, whereas others are typically weak. Still others are variable, depending on the specific molecules in which they are found. Why do these absorption peaks have the intensities they do? The answer has to do with a result from quantum mechanics, which states that a photon is more likely to be absorbed if a particular vibrational mode causes oscillations in the molecule's dipole moment. In turn, an oscillating dipole moment tends to be more pronounced for a stretching vibration when the bond is polar. Therefore:

- A stretching mode involving a highly polar bond (such as C=O and O—H) tends to have a strong IR absorption.
- A stretching mode involving a nonpolar bond (such as C=C) tends to have a weak or nonexistent IR absorption unless the two portions of the molecule connected by that bond are highly dissimilar.

To illustrate the second point, examine the IR spectra of hept-1-ene and hept-3-ene in Figure 15-16 (next page). In the IR spectrum of hept-1-ene (Fig. 15-16a), the C=C stretching band at $1643 \ cm^{-1}$ is moderately intense because the two portions of the

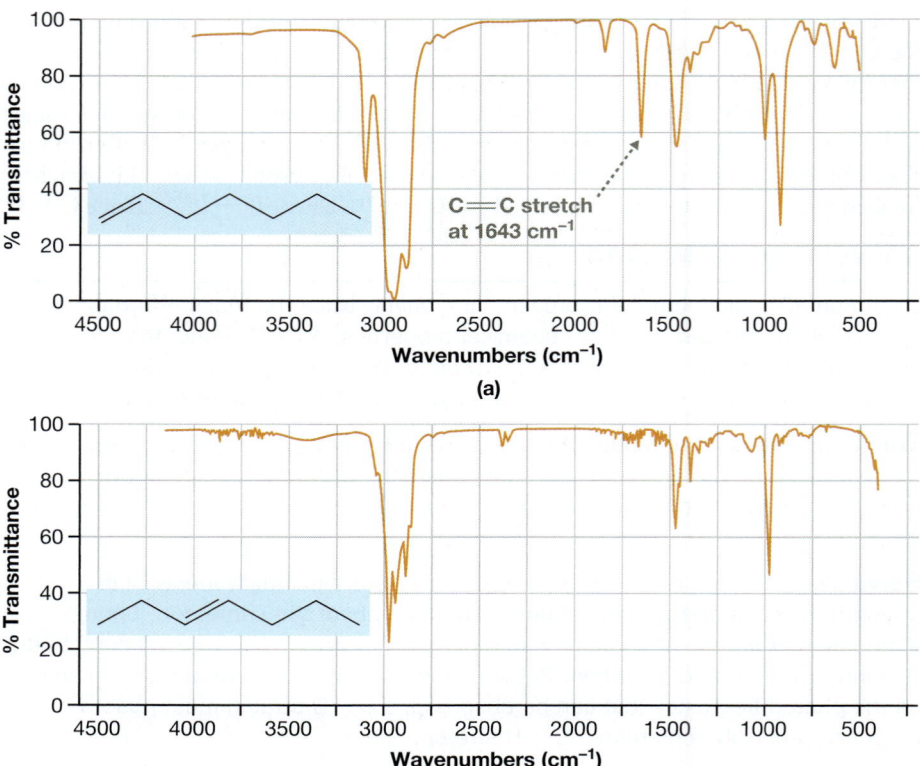

FIGURE 15-16 Molecular symmetry and intensity of stretching absorptions (a) IR spectrum of hept-1-ene, illustrating the presence of the C=C stretching mode. Because of the considerable lack of symmetry about the C=C double bond, that absorption is moderately intense. (b) IR spectrum of hept-3-ene. The C=C stretch band is essentially absent because of the greater symmetry about the C=C double bond.

molecule connected by the C=C bond (i.e., the CH₂ and C₆H₁₂ portions) are rather dissimilar. The C=C stretch is essentially absent in the spectrum of hept-3-ene (Fig. 15-16b), however, because of the much greater symmetry about the C=C bond.

YOUR TURN 15.12

If an absorption corresponding to a C=C stretch were to have appeared in the spectrum of hept-3-ene in Figure 15-16b, indicate roughly where it would have been located.

PROBLEM 15.20 Based on the intensities of the C=C stretching bands, match each of the compounds shown here with its IR spectrum.

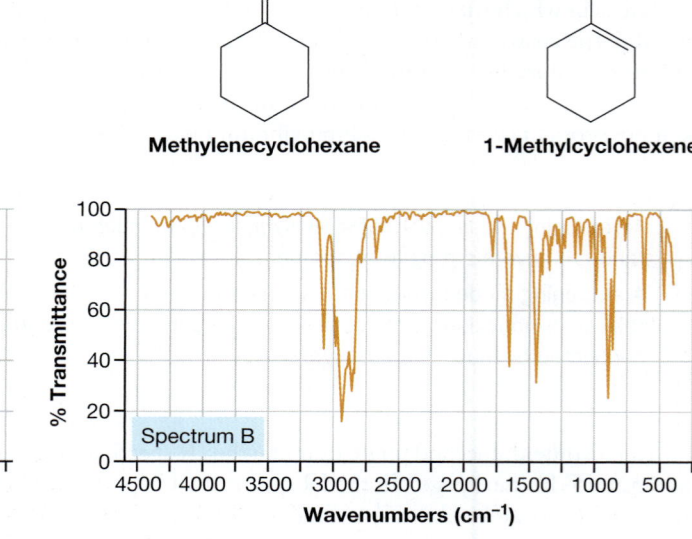

Methylenecyclohexane

1-Methylcyclohexene

Spectrum A

Spectrum B

15.5 A Closer Look at Some Important IR Absorption Bands

When you encounter an IR spectrum of an unknown compound, it may seem daunting at first. Even a simple organic molecule can have dozens of peaks in its IR spectrum. As you gain experience interpreting these spectra, though, these problems will become less and less intimidating. But where do we begin?

The best way to begin is by examining in detail the regions of the spectrum above ~1400 cm^{-1}, where peaks tend to be well separated, are relatively easy to spot, and typically provide unambiguous information. Then look for the presence or absence of specific peaks in the fingerprint region (less than ~1400 cm^{-1}) if the information provided by the spectrum above 1400 cm^{-1} needs to be clarified.

The sections that follow will help you identify and interpret several of these absorptions, which, in turn, will help you determine aspects of a molecule's structure.

15.5a The O—H Stretch

Among the easiest IR absorption bands to spot are those associated with the OH stretch.

> OH stretch bands are intense and are usually rather broad, centered above 3000 cm^{-1}.

Such absorption bands are evident in the IR spectra of cyclohexanol (Fig. 15-17) and 3,3-dimethylbutanoic acid (Fig. 15-18).

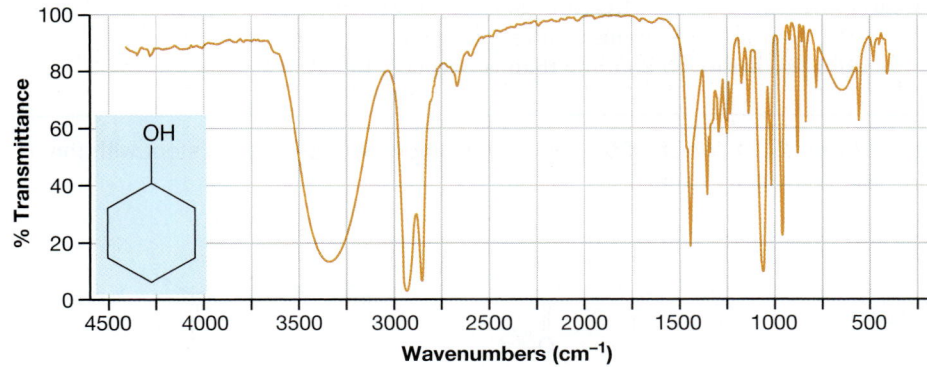

FIGURE 15-17 **IR spectrum of cyclohexanol** Notice the broad OH stretch band centered at ~3350 cm^{-1}.

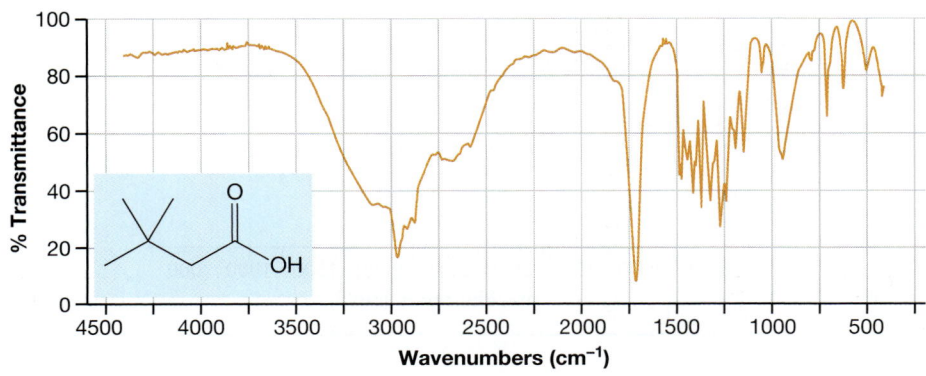

FIGURE 15-18 **IR spectrum of 3,3-dimethylbutanoic acid** Notice the broad OH stretch band centered at ~3000 cm^{-1}.

Identify the peaks in Figures 15-17 and 15-18 that correspond to the O—H stretches.

The broadening of an OH stretch band is due to hydrogen bonding involving the OH groups of the sample. A network of hydrogen bonding in a sample facilitates a rapid exchange of hydrogen atoms from one OH group to another, since each OH hydrogen atom is already partially bonded to two different oxygen atoms.

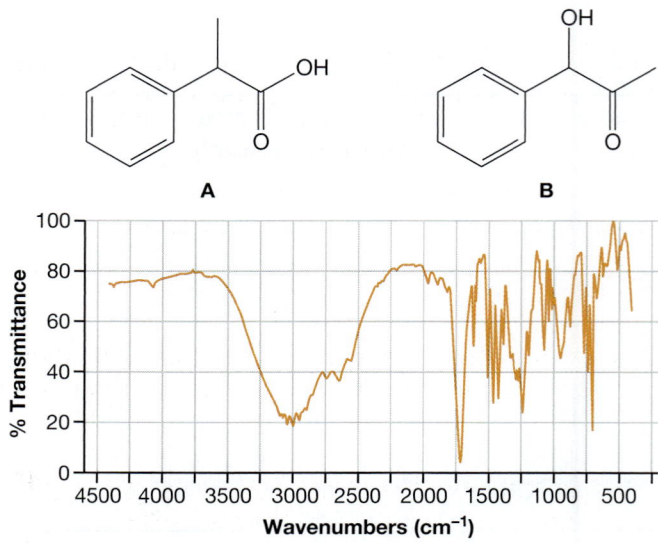

This process gives rise to a lot of variation in the O—H bond strengths. Thus, a wide range of O—H stretching frequencies overlap in the spectrum, resulting in what appears to us as a single broad peak.

Notice that the OH stretch bands in Figures 15-17 and 15-18 are markedly different.

- An alcohol O—H stretch is typically centered near 3300 cm^{-1} and is usually well resolved from the alkane C—H stretch bands between 2800 and 3000 cm^{-1}.
- A carboxylic acid O—H stretch is centered at around 3000 cm^{-1} and can extend down to ~2500 cm^{-1}, thus overlapping the alkane C—H stretches.

A carboxylic acid O—H stretch is generally lower in frequency and much broader than that of an alcohol because carboxylic acids can form dimers in which there is more extensive hydrogen bonding than is normally found in alcohols.

PROBLEM 15.21 Which of the following isomers, **A** or **B**, is consistent with the IR spectrum provided? Explain.

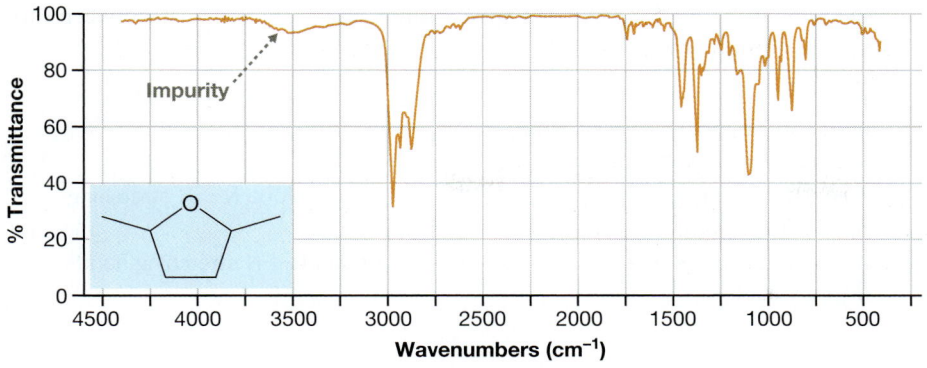

FIGURE 15-19 IR spectrum of 2,5-dimethyltetrahydrofuran Because the compound itself does not contain an OH group, the weak, broad absorption near 3500 cm^{-1} must indicate the presence of an impurity, such as water or an alcohol.

An O—H stretching band in an IR spectrum can also indicate the presence of an impurity such as water or an alcohol—solvents that are often difficult to eliminate entirely from a sample. This is the case with the IR spectrum of 2,5-dimethyltetrahydrofuran shown in Figure 15-19. No OH group appears in the molecule, but a weak, broad absorption appears around 3500 cm^{-1}. If the compound itself were to contain an OH group, that peak would be *much* more intense, as we saw in Figures 15-17 and 15-18.

If a sample is contaminated with a *substantial* amount of water or an alcohol solvent, then the OH absorption can be intense, making it difficult to determine whether the OH group is an impurity or is part of the analyte molecule. For reasons like this, it is usually important to remove solvents (and other contaminants) from a sample before acquiring the spectrum.

YOUR TURN 15.14

The IR spectrum of hept-3-ene shown previously in Figure 15-16b (p. 744) exhibits a water or alcohol impurity. Identify that peak.

15.5b The N—H Stretch

Bands representing N—H stretches from an amine or amide share many similarities with those of O—H stretches, including being generally easy to identify.

- N—H stretches appear between 3300 and 3500 cm^{-1}.
- Hydrogen bonding involving N—H bonds usually causes these bands to be moderately broad.

There are often noticeable differences between N—H and O—H absorption bands, because N—H bonds are less polar and tend to undergo weaker hydrogen bonding.

- The intensity of an N—H stretch is usually less than that of an O—H stretch.
- An N—H absorption band is usually not as broad as an O—H one.

The number of N—H peaks that appear can be used to distinguish between various types of amines or amides.

- A primary amine (RNH$_2$) or amide (RCONH$_2$) exhibits two separate (but closely spaced) N—H stretching bands (Fig. 15-20a).
- A secondary amine (R$_2$NH) or amide (RCONHR) exhibits one N—H stretching band (Fig. 15-20b).
- A tertiary amine (R$_3$N) or amide (RCONR$_2$) exhibits no N—H stretching bands (Fig. 15-20c).

YOUR TURN 15.15

Identify *all* of the peaks in Figure 15-20 that correspond to N—H stretching modes of vibration.

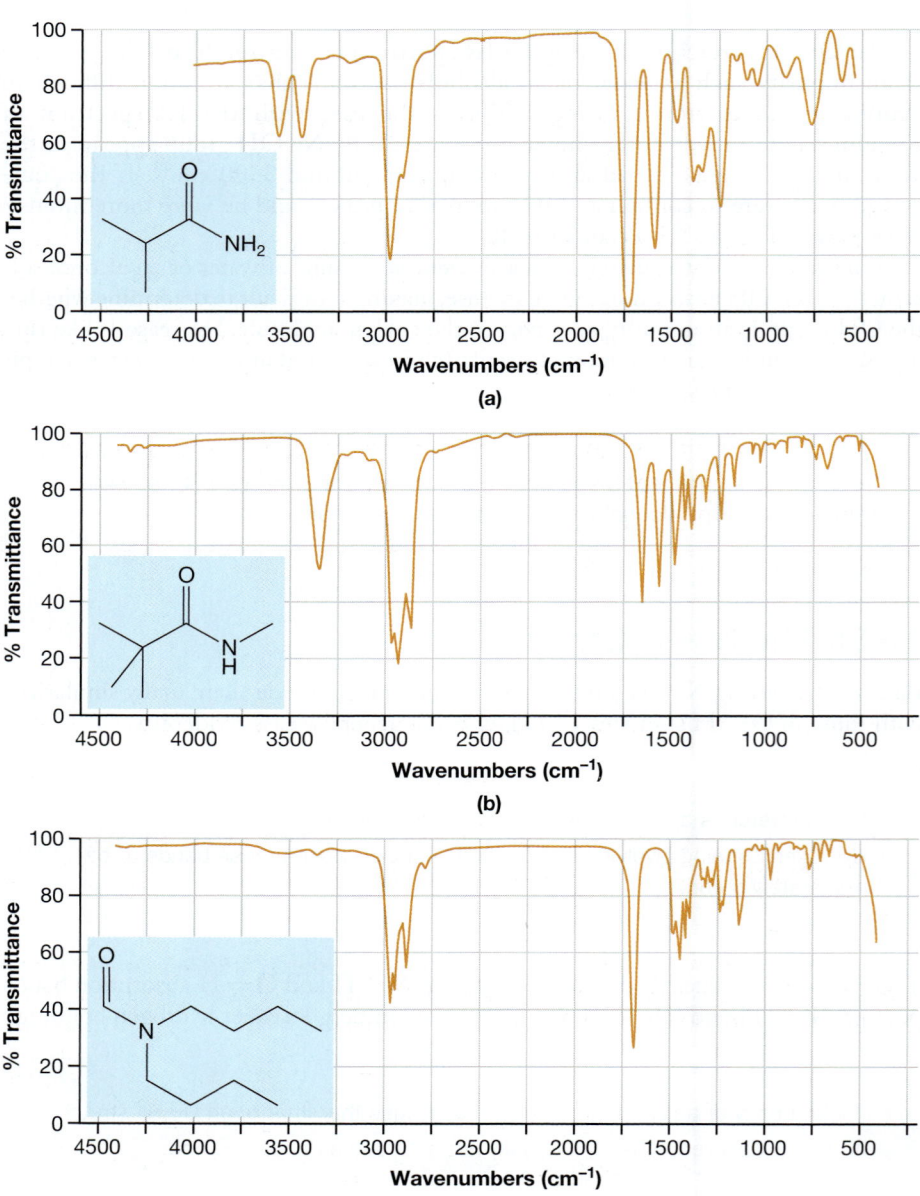

(a)

(b)

(c)

FIGURE 15-20 N–H stretching bands in IR spectra IR spectra of (a) 2-methylpropanamide, a primary amide; (b) *N*-methyl-2,2-dimethylpropanamide, a secondary amide; and (c) *N,N*-dibutylformamide, a tertiary amide, are shown.

Two N—H stretching bands appear for a primary amine or amide because an NH_2 group has two different stretching modes of vibration: a symmetric stretch (both N—H bonds stretch simultaneously) and an asymmetric stretch (one N—H bond lengthens while the other shortens). A secondary amine or amide has just one N—H bond, and a tertiary amine or amide has none.

YOUR TURN 15.16

In the generic primary amine on the left, add arrows to illustrate the motion of the *symmetric* N—H stretch. Illustrate the motion of the *asymmetric* N—H stretch in the molecule on the right. *Hint:* Review Figure 15-11.

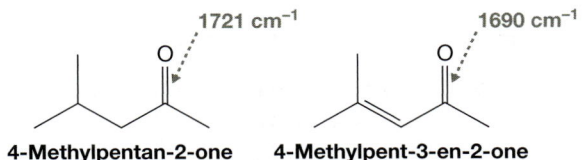

Symmetric stretch   Asymmetric stretch

15.5c The Carbonyl (C=O) Stretch

The carbonyl group (C=O) is found in quite a few classes of compounds, including simple ketones and aldehydes, carboxylic acids, esters, and amides. A carbonyl stretch absorption is quite strong and the band is relatively narrow, usually making it very easy to spot.

Given the variety of compound classes that contain a carbonyl group, we find a moderately large range of frequencies at which the carbonyl stretch absorption can appear—generally between 1600 and 1800 cm^{-1}. Within that range, however, the C=O stretch absorption tends to appear at frequencies that are somewhat characteristic of the compound class to which the C=O group belongs:

For each of the following compound classes, the C=O stretch typically appears in the corresponding range:

- Ester = 1730–1750 cm^{-1}
- Aldehyde = 1720–1740 cm^{-1}
- Ketone = 1710–1730 cm^{-1}
- Amide = 1630–1690 cm^{-1}

Moreover, conjugation can have a significant impact on the location of a C=O stretch.

When a C=O group is conjugated to a C=C or C≡C bond, the frequency of the C=O absorption is typically lowered by 20–40 cm^{-1}.

In 4-methylpentan-2-one, the C=O bond is *not* conjugated and its absorption appears at 1721 cm^{-1}. For 4-methylpent-3-en-2-one, however, the conjugated C=C double bond lowers the C=O frequency to 1690 cm^{-1}.

1721 cm^{-1} 1690 cm^{-1}

4-Methylpentan-2-one   **4-Methylpent-3-en-2-one**

CONNECTIONS

4-Methylpentan-2-one, more commonly called methyl isobutyl ketone (MIBK), is used to extract gold and silver from aqueous cyanide solutions as part of an analytical technique used during certain mining processes.

These differences in absorption frequencies reflect differences in *stiffness* of the C=O bond (Section 15.4c). Specifically, the stiffness of the C=O bond in general increases in the order: amide < ketone < aldehyde < ester. Furthermore, conjugation decreases the stiffness of the C=O bond. The reasons for this are explored in Problems 15.43–15.46 at the end of the chapter.

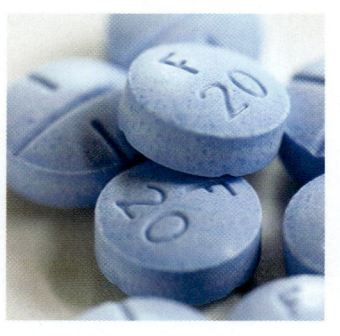
PROBLEM 15.22 The IR spectrum of an unknown compound, whose molecular formula is $C_9H_{10}O$, has a strong absorption band at 1686 cm^{-1}. Which of these compounds is consistent with that spectrum? Explain.

A   **B**

15.5d The C=C Stretch

Alkenes and aromatic rings both contain C=C double bonds in their Lewis structures, but their C=C stretching modes of vibration appear differently in an IR spectrum.

- An alkene C=C stretch generally appears as a single peak in the range 1620–1680 cm^{-1}. (Review Fig. 15-16, p. 744.)
- Aromatic C=C stretches generally appear as two or three peaks in the range 1450–1600 cm^{-1}. (See Fig. 15-21.)

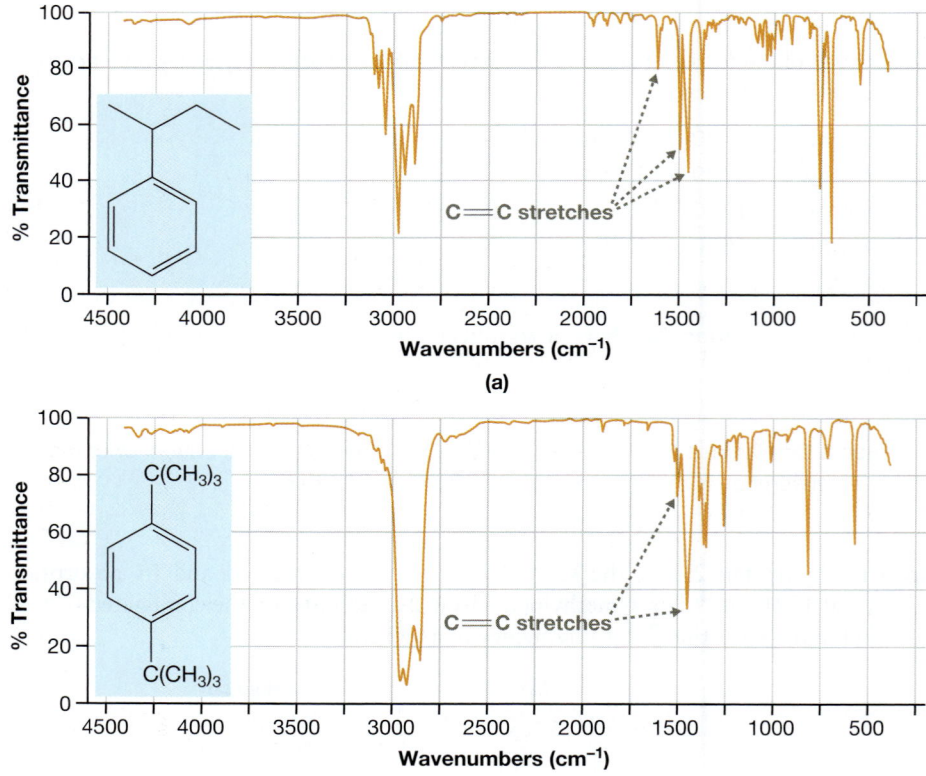

FIGURE 15-21 Aromatic C=C stretches Bands that appear around 1600, 1500, and just below 1500 cm^{-1} are characteristic of aromatic C=C stretches. All three are present in the spectrum of *sec*-butylbenzene (a), but only two of the three are present in the spectrum of 1,4-di-*tert*-butylbenzene (b).

An alkene C=C stretch can sometimes be difficult to identify in an IR spectrum for two reasons. First, an alkene C=C stretch can appear in the same region as some C=O stretches, in which case the absorption by the C=C stretch can be obscured. Second, in some highly symmetric alkenes, such as *trans*-hept-3-ene, the C=C stretch has very little absorption (review Fig. 15-16b, p. 744).

Absorptions from aromatic C=C stretches, on the other hand, tend to be distinct and are often prominent. In Figure 15-21a, for example, three peaks are present between roughly 1450 and 1600 cm^{-1} and two of them are relatively intense; in Figure 15-21b, two peaks appear in that range, one of which is relatively intense.

PROBLEM 15.23 Match isomers **C** and **D** with the correct IR spectrum. Explain.

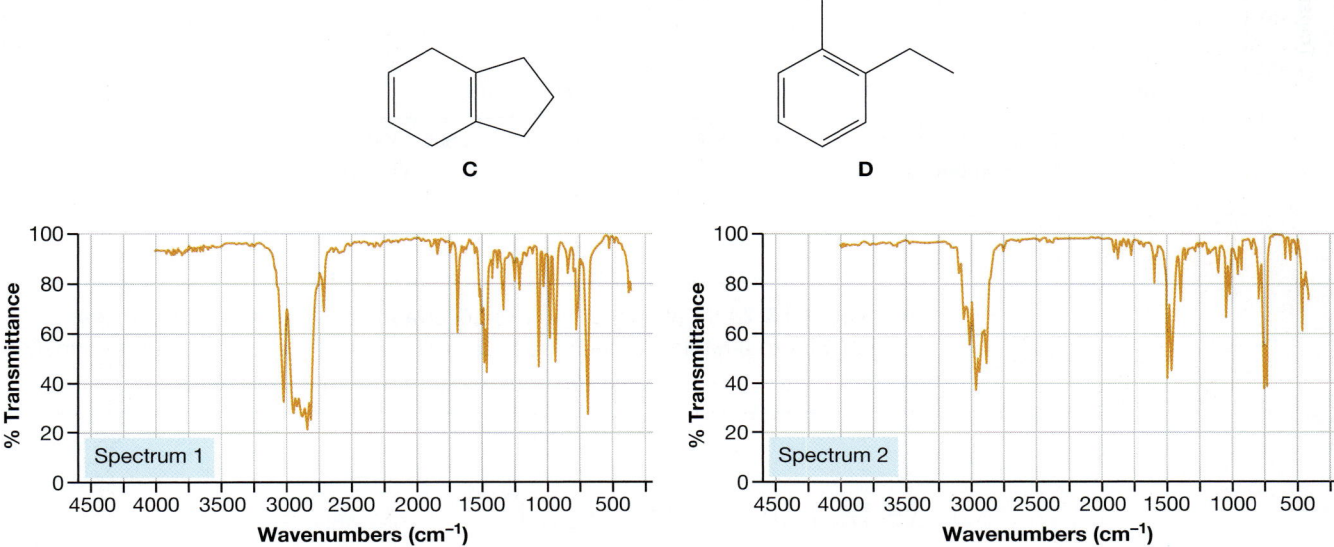

15.5e Alkyne (C≡C) and Nitrile (C≡N) Stretches

Triple bonds are generally among the easiest to identify in an IR spectrum, largely because no other absorption bands appear where they do. Unfortunately, however, C≡C and C≡N bonds both appear as sharp bands at roughly the same position, with C≡C between 2100 and 2260 cm^{-1} and C≡N between 2210 and 2260 cm^{-1}. Without additional information, therefore, such as whether the compound contains nitrogen, it can be difficult to decide between these two functional groups based solely on the frequency of the triple-bond absorption band.

Probably the best way to distinguish between the two functional groups is to search for an alkyne C—H peak at ~3300 cm^{-1}. The presence of one, which appears as a *sharp* peak (in contrast to a broad peak representative of an N—H or O—H stretch), is consistent with a terminal alkyne (C≡C—H). The absence of one, however, may still leave us the choice between a C≡N or an internal alkyne (R—C≡C—R).

In these cases, the intensity of the absorption band for the triple-bond stretch can help you distinguish between functional groups:

Whereas RC≡CR stretches typically have weak absorption intensities, RC≡N absorptions have moderate intensities.

This is because internal alkynes tend to have significant symmetry about the triple bond, whereas the C≡N group has a significant bond dipole (Section 15.4d).

YOUR TURN 15.17

In each of the following spectra, identify the band corresponding to a triple-bond stretching mode.

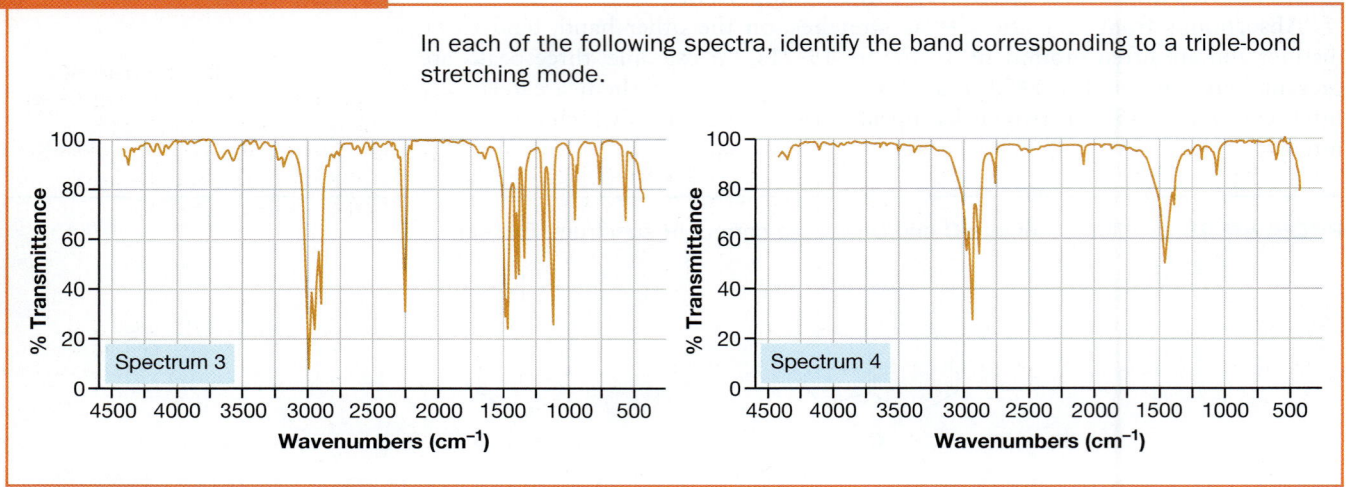

PROBLEM 15.24 (a) Which of the two spectra in Your Turn 15.17 most likely corresponds to a nitrile? Explain. (b) Is the alkyne most likely internal or terminal? Explain.

15.5f The C—H Stretch

All C—H stretches appear between 2700 and ~3300 cm^{-1} (Table 15-2). Within that range, however, the location of the C—H absorption band is often characteristic of a particular compound class:

- Alkyne C—H, ~3300 cm^{-1} (moderate intensity)
- Alkene C—H, between 3000 and 3100 cm^{-1} (variable intensity)
- Aromatic C—H, 3000–3100 cm^{-1} (variable intensity)
- Alkane C—H, 2800–3000 cm^{-1} (variable intensity)
- Aldehyde C—H, two peaks: ~2720 and ~2820 cm^{-1} (moderate intensity)

CONNECTIONS Heptanal (Fig. 15-22) can be isolated from castor oil and has a fruity odor, making it useful as a flavoring agent or as an ingredient in some perfumes and cosmetics.

Often, the presence or absence of one of these absorptions can help you narrow your choices of compounds when another region of the spectrum is difficult to interpret unambiguously. For example, a C=O stretch appearing in an IR spectrum at 1720 cm^{-1} could be consistent with either a ketone or an aldehyde. If you also see peaks at 2720 and 2820 cm^{-1}, however, then the compound is almost certainly an aldehyde, as you can see in the IR spectrum for heptanal (Fig. 15-22). As another example, if you are uncertain as to whether an IR spectrum exhibits a C=C stretch, you should examine the 3000–3100 cm^{-1} region to see if there is evidence of an alkene C—H stretch. This idea is explored in Problem 15.25.

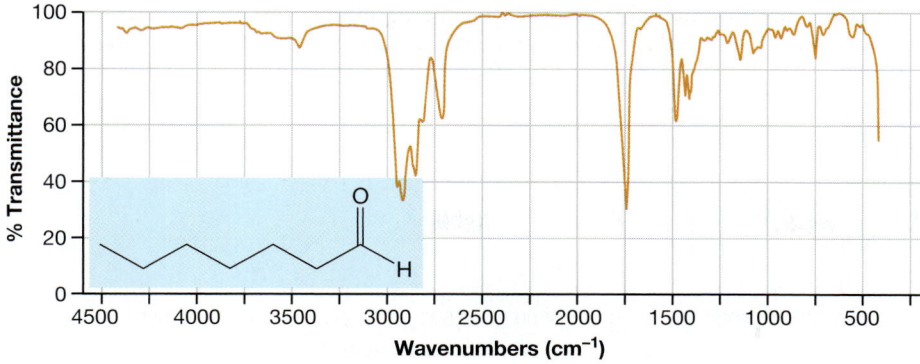

FIGURE 15-22 **IR spectrum of heptanal** Notice the aldehyde C—H stretch peaks at ~2820 cm^{-1} and ~2720 cm^{-1}. The one at ~2820 cm^{-1} is partly obscured by the alkane C—H band.

Identify the bands in Figure 15-22 that correspond to the aldehyde C—H stretch.

PROBLEM 15.25 Which spectrum, 5 or 6, is more likely of a compound that contains a C=C double bond? Explain.

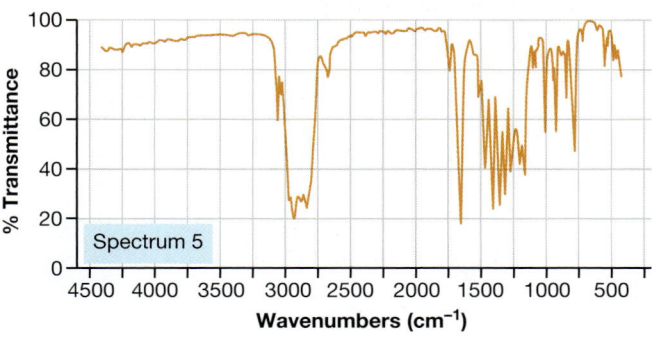

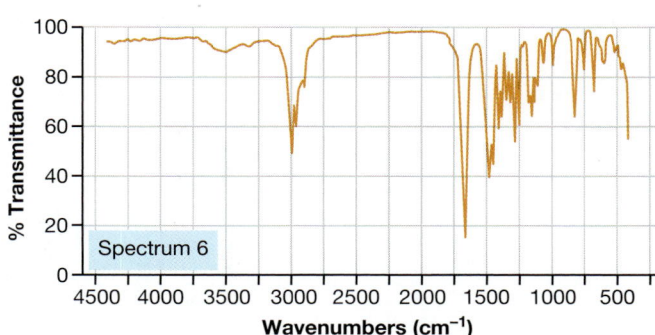

Alkane C—H stretches, which appear between 2800 and 3000 cm^{-1}, are so commonplace in organic molecules that they are generally *not* useful in structure elucidation. Trying to use these bands to elucidate a structure is like trying to identify a car based on its tires.

The alkane C—H stretching band is perhaps most useful when it is low in intensity or absent altogether. This is because:

> The relative intensity of the alkane C—H band decreases as the relative number of alkane C—H bonds in the molecule decreases.

In the structure of 2-ethylhexan-1-ol, for example, there are 17 C—H bonds, making the C—H stretching band even more intense than the O—H stretch (Fig. 15-23a). In ethanol's IR spectrum (Fig. 15-23b), on the other hand, the alkane C—H stretching band is less intense than the O—H stretch because there are only five alkane

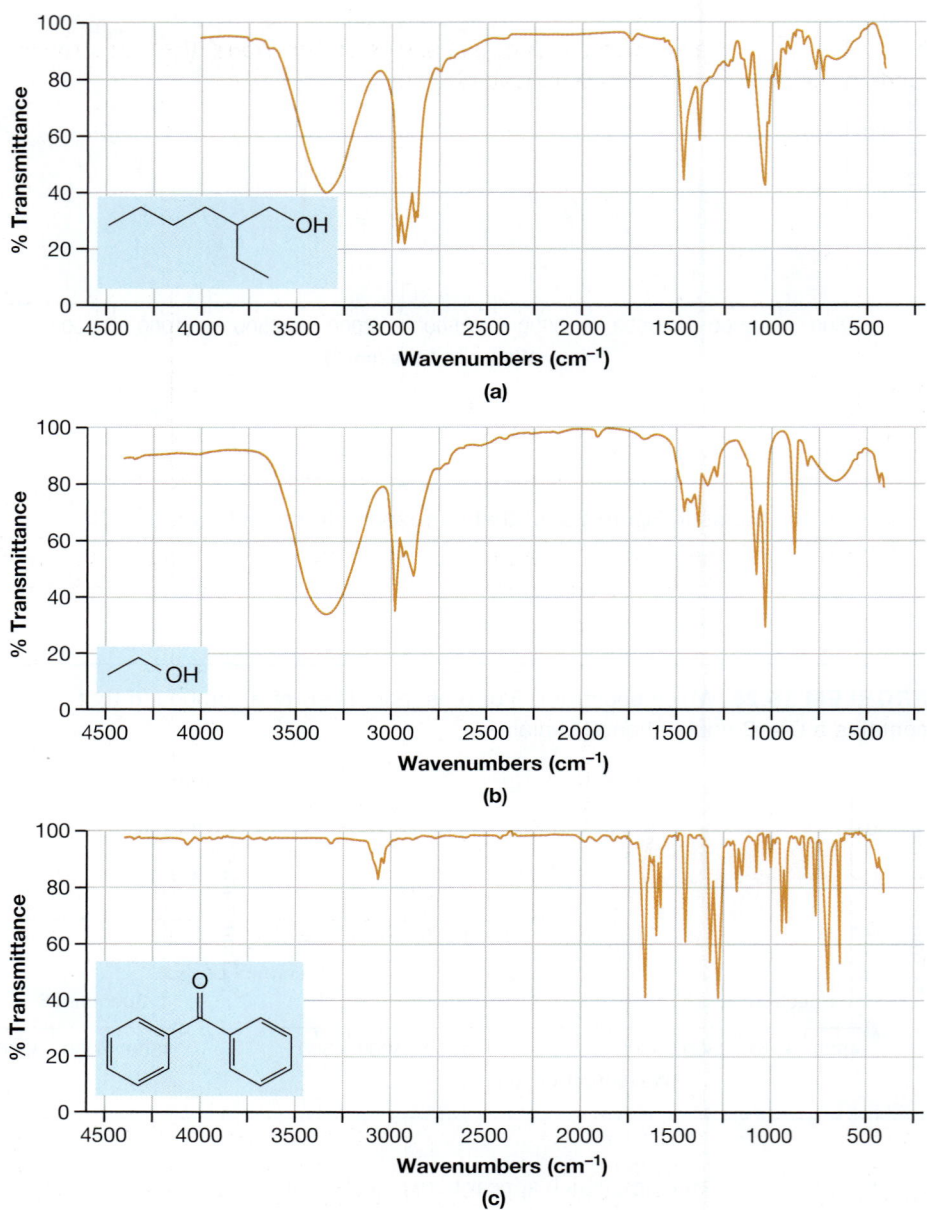

FIGURE 15-23 Relative intensities of the alkane C—H bands IR spectra of (a) 2-ethylhexan-1-ol, (b) ethanol, and (c) diphenylmethanone are shown. As the number of alkane C—H bonds decreases, so does the intensity of the alkane C—H absorption band.

CONNECTIONS
Benzophenone (Fig. 15-23c) has been used as an additive in plastic packaging to help prevent photodegradation.

C—H bonds. Diphenylmethanone (benzophenone) has no alkane C—H bonds at all. Its IR spectrum (Fig. 15-23c), therefore, has no absorption band between 2800 and 3000 cm^{-1}.

YOUR TURN 15.19

Identify the alkane C—H stretching absorption bands in Figure 15-23a and 15-23b. In Figure 15-23c, indicate where alkane C—H stretching bands would normally appear.

PROBLEM 15.26 Based on the relative intensities of the alkane C—H absorption bands, match compounds **E–G** with spectra 7–9.

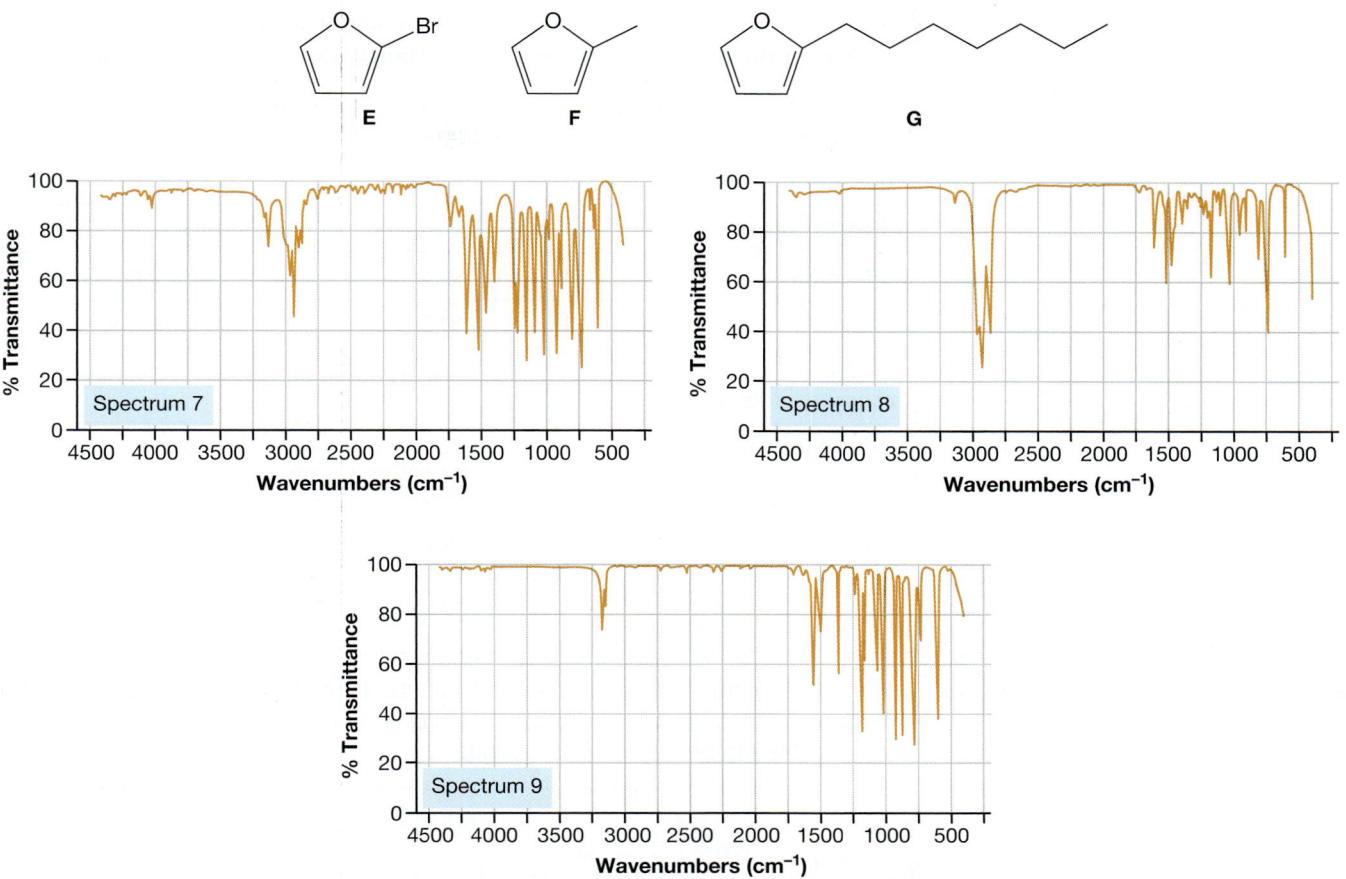

15.5g Bending Vibrations

Recall from Figure 15-14 (p. 741) that IR absorptions corresponding to bending modes of vibration appear below 1000 cm^{-1}. Although these peaks can provide useful information about a molecule's structure, the region below 1000 cm^{-1} falls within the fingerprint region, which tends to have a large number of peaks, even for relatively small molecules. Therefore, we generally don't turn our attention to the region below 1000 cm^{-1} unless we expect a specific type of absorption to be there.

For example, if we find an absorption at ~1620 cm^{-1}, corresponding to an alkene C=C stretch, we can search for bending modes to help us determine what kind of alkene the molecule is. As shown in Table 15-3, these bending modes can help us determine if the alkene is mono-, di-, or trisubstituted, or if the alkene is cis or trans. Similarly, if we find multiple peaks between about 1450 and 1600 cm^{-1}, indicating an aromatic C=C, we could look for bending modes to help us determine if the compound has a monosubstituted benzene ring, or if it has an ortho-, meta-, or para-disubstituted benzene ring.

To see how this might work, consider the spectrum of *sec*-butylbenzene (Fig. 15-21a, p. 750). Notice the relatively strong absorptions at 730 and 759 cm^{-1}, which represent the bending modes of a monosubstituted benzene, consistent with Table 15-3. In the spectrum of di-*tert*-butylbenzene (Fig. 15-21b), the bending mode at 833 cm^{-1} indicates a para-disubstituted benzene, again consistent with Table 15-3.

TABLE 15-3 Characteristic Frequencies of Absorption in IR Spectroscopy: C—H Bending Modes of Vibration

Type of Bond	Number of Bends	Frequency Range (cm^{-1})	Appearance
R—CH=CH$_2$	2	910 and 990 (two peaks)	Strong
R$_2$C=CH$_2$	1	890	Strong
RCH=CHR (cis)	1	660–730	Strong
RCH=CHR (trans)	1	970	Strong
R$_2$C=CHR	1	815	Moderate
Aromatic (monosubstituted)	2	700 and 750	Strong
Aromatic (ortho)	1	750	Strong
Aromatic (meta)	2	780 and 880	Strong
Aromatic (para)	1	830	Strong

YOUR TURN 15.20

In the IR spectra in Figure 15.21a and 15.21b, label the bending-mode absorptions described on the preceding page and identify the specific entry in Table 15-3 to which they correspond.

15.6 Structure Elucidation Using IR Spectroscopy

Interpreting an IR spectrum requires practice applying the concepts presented in Sections 15.4 and 15.5. For this reason, the IR spectra of three unknown compounds are presented here in Section 15.6 and we take the time to analyze them to obtain structural information.

There are many ways to approach the analysis of an IR spectrum, and you will probably develop a strategy that works particularly well for you. To begin, however, it is a good idea to ask the questions below, which we will reference frequently as we analyze the IR spectra that follow.

1. What stretching modes (i.e., above 1400 cm^{-1}) are evident in the spectrum?
   - Look especially for the presence or absence of peaks highlighted in Section 15.5: O—H, N—H, C=O, C=C, C≡C, C≡N, and C—H.
2. For the peaks you just identified in question 1, is it possible that the corresponding bonds are consistent with a particular compound class? For example:
   - (C=O stretch) + (O—H stretch) = carboxylic acid?
   - (C=O stretch) + (N—H stretch) = amide?
   - (C=O stretch) + (two C—H stretches at 2720 and 2820 cm^{-1}) = aldehyde?
   - (C=C stretch) + (C—H stretch between 3000 and 3100 cm^{-1}) = alkene or aromatic ring?
   - (C≡C stretch) + (C—H stretch ~3300 cm^{-1}) = terminal alkyne?

3. Are there special features or specific frequencies of the stretching absorptions you have identified that give you additional insight into the structure of the molecule? For example:
   - If you identify an O—H stretch, is it consistent with an alcohol or a carboxylic acid? (Section 15.5a)
   - If you identify an N—H stretch, is it consistent with an NH or an NH_2 feature? (Section 15.5b)
   - If you identify a C=O stretch, does its specific frequency suggest a carboxylic acid or ester? A ketone or aldehyde? An amide? (Section 15.5c)
   - If you identify a C=C stretch, is it consistent with an alkene or an aromatic ring? (Section 15.5d)
   - If you identify a triple bond, is it consistent with C≡C or C≡N? (Section 15.5e)
   - Is the intensity of the alkane C—H stretch band noticeably weak? What might that suggest? (Section 15.5f)
4. If you determine that a C=O bond is present as well as a C=C or C≡C bond, is there evidence that they are conjugated?
   - Conjugation typically lowers the stretching frequency of a C=O bond by 20–40 cm^{-1}. (Section 15.5c)
5. Does the presence of an alkene or aromatic C=C stretch warrant searching the fingerprint region for particular bending modes of vibration? (Section 15.5g)
   - Are there bending modes that suggest a mono-, di-, or trisubstituted alkene? Cis or trans?
   - Are there bending modes that suggest a monosubstituted benzene ring? An ortho-, meta-, or para-disubstituted benzene ring?

Even after extracting all of this information from an IR spectrum, you will be unable to determine the compound's structure unambiguously without some significant additional information, such as the molecular formula or results from other forms of spectroscopy. That is because IR spectra allow you to determine the types of bonds and functional groups present, but they provide little structural information beyond that.

Therefore, if you are given the formula of the compound that generated the IR spectrum, take full advantage of it. For example, if there is a single O atom in the formula and the IR spectrum indicates a C=O stretch, the compound *cannot* be a carboxylic acid or ester; those types of compounds each require two O atoms.

Furthermore, use the formula to calculate the molecule's index of hydrogen deficiency (IHD; Section 4.12) and make sure that your answers to the above questions are consistent with it. For example, if the IHD = 0, the compound cannot contain C=O, C=C, C≡C, or C≡N bonds. As another example, if the IHD ≤ 3, the compound cannot contain a benzene ring, which would require an IHD of 4 (three double bonds and a ring).

15.6a Unknown 1

The IR spectrum of Unknown 1 is shown in Figure 15-24. Its molecular formula is C_7H_8O, so its IHD is 4, because a completely saturated molecule with 7 C atoms and 1 O atom would have 16 H atoms, not just 8. Unknown 1, therefore, could contain up to four rings, four double bonds, or two triple bonds.

To answer question 1 from Section 15.6, there is an intense, broad O—H stretch centered at ~3300 cm^{-1}, an alkene or aromatic C—H stretch at ~3050 cm^{-1}, alkane C—H stretches between 2800 and 3000 cm^{-1}, and C=C stretches between ~1450 and 1600 cm^{-1}.

YOUR TURN **15.21**

Label the O—H stretching absorption that appears in Figure 15-24 and indicate where a C=O stretch would normally appear.

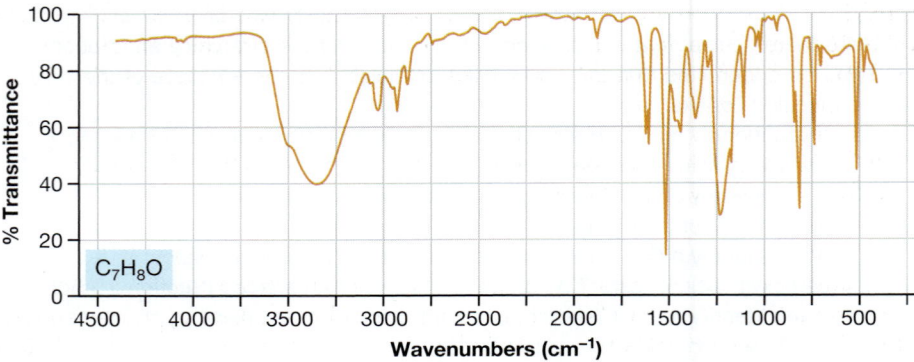

FIGURE 15-24 IR spectrum of Unknown 1, C_7H_8O

To answer question 2 from Section 15.6, the C—H stretch around 3050 cm^{-1} and the C=C stretches between ~1450 and 1600 cm^{-1} are both consistent with an alkene or aromatic ring.

To answer question 3, we should establish whether the C=C stretch is consistent with an alkene or aromatic ring. An alkene typically exhibits one peak above

IR Spectroscopy and the Search for Extraterrestrial Life

Our culture is obsessed with extraterrestrial life. Even though we have yet to find any hard evidence that extraterrestrial life exists, we continue to invest significant time and resources looking for it. One way this is done by the SETI (Search for Extraterrestrial Intelligence) Institute is to use radio telescopes to detect radio waves emanating from deep space. The data collected are then analyzed for patterns that might suggest the radio waves were produced by intelligent life.

But radio telescopes are not the only tool used to search for life on other planets; IR spectroscopy is a pivotal tool in this quest as well. Researchers are able to record an IR spectrum produced by the atmosphere of a distant planet, and with such a spectrum in hand, they look for absorbances that are characteristic of gases generally believed to be conducive to life. Some of these gases include H_2O, CO_2, CH_4, and O_3 (ozone), as can be seen in the IR spectrum of Earth's atmosphere.

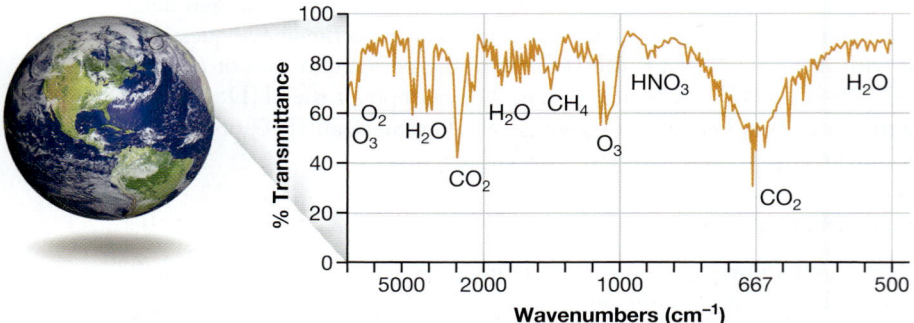

Of these gases, researchers are particularly interested in the existence of O_3. Not only is O_3 produced from atmospheric O_2 (which sustains life), but it also provides an important layer of protection from harmful UV radiation.

Whereas an IR spectrum can indicate whether an atmosphere might be able to support life, researchers believe that the detection of IR radiation in the range of 750–1000 nm (13,000–10,000 cm^{-1}) would provide more direct evidence of life. That's because, on Earth, photosynthetic plants tend to reflect IR radiation in that range with relatively high efficiency.

~ 1620 cm^{-1}, whereas an aromatic ring typically exhibits two or three peaks in the 1450–1600 cm^{-1} range. In this case, we appear to have an aromatic ring. With the IHD of 4 that we previously calculated, this could indicate a benzene ring.

YOUR TURN **15.22**

In Figure 15-24, label the aromatic C—H stretch bands and all three bands corresponding to aromatic C=C stretches.

Continuing with question 3, notice the absence of an intense alkane C—H band between 2800 and 3000 cm^{-1}. There are alkane C—H peaks present, but their intensity is low (less than the O—H band). This suggests that there are relatively few alkane C—H bonds and would be consistent with six out of the seven C atoms being part of an aromatic ring.

YOUR TURN **15.23**

Identify the alkane C—H stretching band in Figure 15-24.

Because there is no evidence of a C=O bond, we can skip question 4 and address question 5. Because we believe that Unknown 1 likely contains a substituted benzene ring, we can look in the region below 1000 cm^{-1} for characteristic bending modes. There is a relatively intense peak around 830 cm^{-1}, suggesting that the benzene ring is para disubstituted (Table 15-3).

YOUR TURN **15.24**

Identify the C—H bending mode in Figure 15-24 that indicates a para-disubstituted benzene ring.

At this stage, we know that the compound has the general form G(1)—C$_6$H$_4$—G(2), where the C$_6$H$_4$ portion is the para-disubstituted benzene ring and G(1) and G(2) represent the attached groups. Given that the molecular formula is C$_7$H$_8$O, the G(1) and G(2) portions must contain a total of one carbon, four hydrogens, and one oxygen. We also know that there is an OH group, leaving one carbon and three hydrogens yet to be accounted for. That could simply be a CH$_3$ group, in which case the unknown would be CH$_3$—C$_6$H$_4$—OH, as shown at the right.

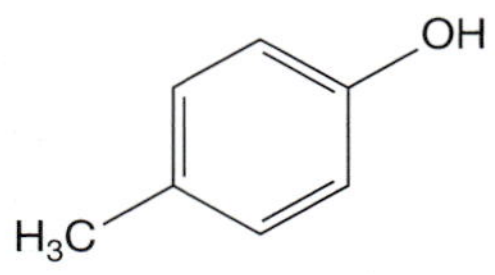

Unknown 1

15.6b Unknown 2

The IR spectrum of Unknown 2 (molecular formula C$_8$H$_{14}$O) is shown in Figure 15-25. We begin again with determining the compound's IHD. A completely saturated molecule with 8 C and 1 O would have 18 H, not 14, so Unknown 2 has an IHD of 2. Thus, Unknown 2 could have one triple bond, two double bonds, two rings, or a double bond and a ring.

To answer question 1 from Section 15.6, notice the following stretching absorption bands: alkene or aromatic C—H at ~ 3050 cm^{-1}, alkane C—H stretches between 2800 and 3000 cm^{-1}, a C=O stretch at ~ 1690 cm^{-1}, and a C=C stretch at ~ 1650 cm^{-1}.

YOUR TURN **15.25**

Label the above stretching absorptions in Figure 15-25.

For question 2, let's consider the compound class in which the C=O bond might be involved. We can rule out a carboxylic acid or ester because these require

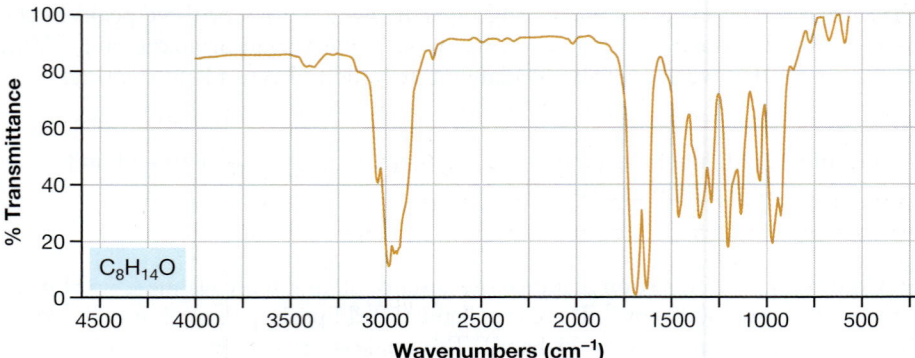

FIGURE 15-25 IR spectrum of Unknown 2, $C_8H_{14}O$

two O atoms each, but the formula we were given has just one. We can also rule out an amide because the formula has no N atoms. Furthermore, there are no O—H or N—H stretches in the IR spectrum, which might have supported a carboxylic acid or amide. We can also rule out an aldehyde because the spectrum doesn't exhibit aldehyde C—H stretches at ~2720 and ~2820 cm⁻¹. Therefore, the C=O bond is probably part of a ketone.

YOUR TURN 15.26

In Figure 15-25, mark the regions where a carboxylic acid O—H stretch, an N—H stretch, and aldehyde C—H stretches would appear.

Continuing with question 2, notice that the C—H stretch at ~3050 cm⁻¹ and the C=C stretch at ~1650 cm⁻¹ might be consistent with an alkene or aromatic ring. However, we can rule out an aromatic ring because the formula's IHD is 2, which is insufficient. The molecule therefore is an alkene.

For question 3, we would normally try to assign the C=C stretch to an alkene or aromatic ring, but we have already established that it indicates an alkene. To lend further support to that conclusion, notice that we do *not* see two or three peaks in the 1450–1600 cm⁻¹ region, which is what we would expect of an aromatic ring.

YOUR TURN 15.27

In Figure 15-25, circle the region where multiple C=C stretch absorptions would appear if the molecule contained an aromatic ring.

For question 4, notice that a C=O stretch indicating a ketone normally appears at ~1720 cm⁻¹, but in this case it appears at ~1690 cm⁻¹. This suggests that the C=O and C=C bonds are conjugated.

Finally, for question 5, we can turn to the fingerprint region to search for C—H bending modes that would help us determine what type of C=C bond is present (Table 15-3). There appear to be two bands, around 920 and 980 cm⁻¹, indicative of a terminal alkene of the form R—CH=CH₂.

YOUR TURN 15.28

Label the C—H bending bands in Figure 15-25 that indicate the presence of a terminal alkene.

Knowing that the molecule is a conjugated ketone, and knowing that the C=C is terminal, the general structure for Unknown 2 is C—CO—CH=CH₂. But we still have four carbon atoms and 11 hydrogen atoms as yet unaccounted for. The remaining carbon atoms are sp^3 hybridized, but lacking any other information we cannot determine definitively which of multiple possible isomers our unknown compound is. Two possibilities are shown here.

The IR spectrum of Unknown 2 is an excellent example of the limitations of IR spectroscopy. Specific absorption bands in an IR spectrum provide structural information about only *pieces* of a molecule. Frequently, there is insufficient explicit information to allow you to converge on a single structure.

Two possibilities for Unknown 2

SOLVED PROBLEM 15.27

Draw a third possible structure that is consistent with the IR spectrum in Figure 15-25.

Think What is the basic structural unit that we derived from the IR spectrum? How can the remaining atoms be attached without disturbing that basic structure? What kinds of bonds must connect those atoms?

Solve The first structure is the basic structural unit we derived from the spectrum, which accounts for four C, three H, and one O atom. That leaves four C and 11 H atoms yet to add, all of which can be attached only by single bonds—otherwise, we would exceed the compound's IHD of 2. To preserve the basic structure, we must add C_4H_{11} to the left-hand C atom. One way to do so, which does not repeat either of the previous two structures, is as shown in the second structure.

PROBLEM 15.28 Draw two additional structures consistent with the IR spectrum in Figure 15-25, each of which is different from the previous three.

15.6c Unknown 3

The IR spectrum for Unknown 3 (molecular formula C_6H_7N) is shown in Figure 15-26. Suppose, too, that the longest-wavelength absorption in its UV–vis spectrum has λ_{max} around 190 nm.

Again we begin by determining the IHD, which is 4. This corresponds to a variety of combinations of double bonds, triple bonds, and rings (including a single aromatic ring).

YOUR TURN 15.29

How many hydrogen atoms are there in a completely saturated compound with six carbon atoms and one nitrogen atom? Is that consistent with Unknown 3 having an IHD of 4?

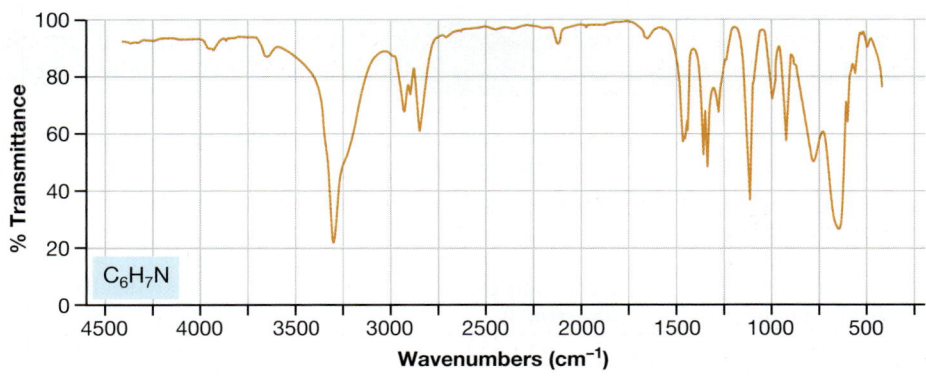

FIGURE 15-26 IR spectrum of Unknown 3, C_6H_7N

To answer question 1 from Section 15.6, there is an absorption at ~3300 cm^{-1}, which is roughly where O—H, N—H, and alkyne C—H stretches appear; a small, sharp peak at ~2100 cm^{-1}, which is where C≡C or C≡N triple bonds appear; and an alkane C—H band between 2800 and 3000 cm^{-1}. For the band at ~3300 cm^{-1}, we can rule out O—H because the molecular formula does not contain O, leaving N—H and alkyne C—H stretches as possibilities.

YOUR TURN 15.30

Label the stretching bands in Figure 15-26.

For question 2, notice that an alkyne C—H stretch at ~3300 cm^{-1} would agree with the C≡C stretch at ~2100 cm^{-1}.

For question 3, we can rule out a C≡N triple bond because a polar C≡N bond would have produced a much more intense peak at ~2100 cm^{-1}. Therefore, the N atom must be part of an amine.

Is there evidence of an N—H stretch at ~3300 cm^{-1} to lend support to an amine? We already identified an alkyne C—H stretch there, but if you examine that absorption carefully, you will see that two peaks seem to be overlapping. As shown at the left, the broader portion of that feature would be consistent with an N—H stretch and the skinnier portion would be consistent with the alkyne C—H stretch. Having identified an N—H stretch, we can conclude that the N atom is likely part of a secondary amine (R$_2$NH) and not a primary amine (RNH$_2$). A primary amine would have produced two separate peaks in that region, but the spectrum exhibits just one.

Finally, for question 3, notice that the alkane C—H band is relatively weak. This suggests that several C atoms are not sp^3 hybridized, so the majority of the C atoms could be alkyne carbons.

We can skip question 4 because the spectrum does not exhibit a C=O peak, and we can skip question 5 because we don't have any evidence of an alkene or aromatic ring.

So far, we have evidence of a terminal alkyne, which would account for an IHD of 2, and a secondary amine. But how do we achieve the remaining IHD of 2 to arrive at a total IHD of 4?

Unknown 3 could have a second C≡C triple bond. If so, this would account for the relatively small alkane C—H absorption band. The two C≡C bonds, however, must not be conjugated to each other because, according to Table 15-1, a λ_{max} of ~190 nm in a UV–vis spectrum suggests an isolated π bond. One possibility that fits all of these criteria is shown in the structure here in the margin.

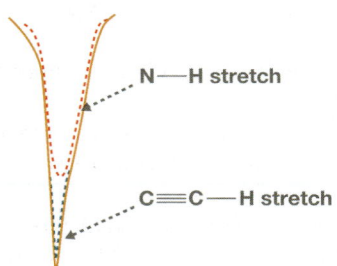

Possible structure for Unknown 3

PROBLEM 15.29 Draw another molecule that is consistent with the IR spectrum in Figure 15-26.

Chapter Summary and Key Terms

- **Spectroscopy**, in general, is the study of how electromagnetic radiation interacts with matter. **(Introduction and Section 15.1; Objective 1)**

- In **UV–vis spectroscopy**, a range of wavelengths of UV and visible light from a source is sent through an **analyte**. At each wavelength, the intensity of light that is transmitted through the analyte is compared to the intensity of light from the source. **(Section 15.1; Objective 2)**

- A **UV–vis spectrum** is obtained by plotting **absorbance** against the wavelength of light. **(Section 15.1; Objective 2)**

- Electromagnetic radiation has a dual nature, behaving as both a wave and a particle. As a wave, we can assign it a wavelength and frequency. As a particle, or *photon*, radiation carries only certain, discrete amounts of energy. **(Section 15.2; Objective 3)**

- In UV–vis spectroscopy, absorption of a photon promotes an electron from a lower-energy MO to a higher-energy MO. The λ_{max} for the longest-wavelength (lowest-energy) absorption in the spectrum generally corresponds to the HOMO–LUMO transition. (Section 15.2; Objective 4)

- In a $\pi \rightarrow \pi^*$ ("pi-to-pi-star") **transition**, an electron from a π MO is promoted to a π^* MO. (Section 15.2; Objective 5)

- In an $n \rightarrow \pi^*$ ("n-to-pi-star") **transition**, an electron from a nonbonding MO is promoted to a π^* MO. (Section 15.3b; Objective 5)

- Conjugation decreases the energy difference between the HOMO and LUMO, and thus increases the wavelength of the photons absorbed. (Section 15.3a; Objective 6)

- Compounds with lone pairs tend to have longer-wavelength absorptions than analogous compounds without lone pairs. (Section 15.3b; Objective 5)

- Absorption of an IR photon excites a particular type of vibration within a molecule. Absorption occurs when the frequency of the photon equals the frequency of that vibration. (Section 15.4a; Objective 7)

- In an IR spectrum, transmittance is plotted against IR frequency, in units of **wavenumbers (cm^{-1})**. Each peak in the spectrum corresponds to a vibrational frequency of the molecule. (Section 15.4a; Objectives 7, 8)

- The highest-frequency vibrations in an organic molecule are those of Q—H bonds (where Q is a heavy atom like C, N, or O), because H is very light. (Sections 15.4b and 15.4c; Objectives 8, 9)

- For a bond between two heavy atoms, the vibrational frequency depends on the strength and stiffness of the bond. Thus, vibrational frequency decreases in the order: triple bond > double bond > single bond. Vibrational frequency

- also decreases in the order: alkyne (sp) C—H > alkene (sp^2) C—H > alkane (sp^3) C—H. (Sections 15.4b and 15.4c; Objectives 10, 11)

- Bending modes of vibration are low in frequency, appearing below 1000 cm^{-1}. (Sections 15.4b and 15.5g; Objective 8)

- The intensity of an IR stretching absorption increases, in general, as the magnitude of the dipole undergoing vibration increases. (Section 15.4d; Objective 12)

- The O—H stretch typically appears as a broad absorption due to extensive hydrogen bonding. (Section 15.5a; Objective 13)

- The N—H stretch appears as two peaks for a primary amide or amine and one peak for a secondary amide or amine. No N—H peaks appear for a tertiary amide or amine. (Section 15.5b; Objective 13)

- The C=O stretch of an amide is lower in frequency than that of a ketone or aldehyde, whereas an ester's C=O stretch is at a higher frequency than that of a ketone or aldehyde. Conjugation of a C=O bond tends to lower the absorption frequency by ~20–40 cm^{-1}. (Section 15.5c; Objective 13)

- The C=C stretch of an alkene generally appears above 1620 cm^{-1}, whereas that of an aromatic ring usually appears as two or three peaks in the 1450–1600 cm^{-1} range. (Section 15.5d; Objective 13)

- A C≡N stretch is usually more intense than that of a C≡C, due to the greater bond dipole of C≡N. (Section 15.5e; Objective 13)

- The intensity of an alkane C—H stretching band (2800–3000 cm^{-1}) is indicative of the number of C—H bonds that contribute to it. (Section 15.5f; Objective 13)

- Absorptions that are characteristic of bending modes of vibration can help distinguish the substitution pattern of an alkene or aromatic ring. (Section 15.5g; Objective 13)

Problems

15.1–15.3 UV–Vis Spectroscopy

15.30 Naturally occurring carotene primarily exists in two different forms, called α-carotene and β-carotene. Which has the longer-wavelength UV–vis absorption? Explain.

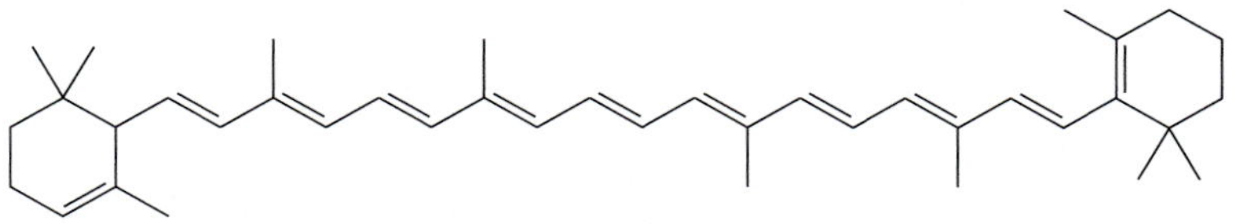

α-Carotene

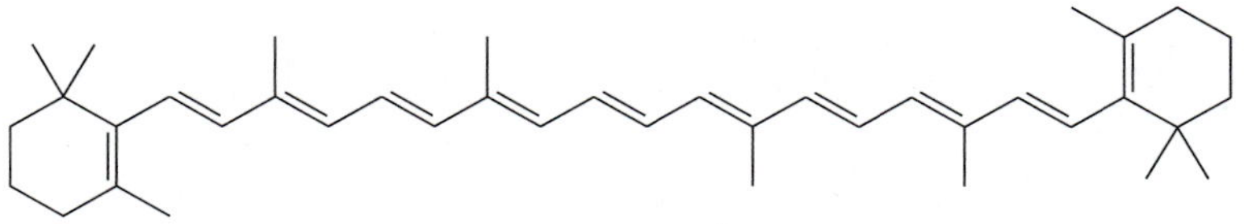

β-Carotene

15.31 In the UV–vis spectrum of buta-1,2-diene (CH$_3$CH=C=CH$_2$), the longest-wavelength absorption appears at 178 nm. Compare this to the longest-wavelength absorption in buta-1,3-diene (see Table 15-1) and explain the significant difference.

15.32 An unknown compound has the formula C_5H_6. Its longest-wavelength UV–vis absorption is centered at 215 nm. Draw four isomers that are consistent with these results.

15.33 A compound, whose formula is C_7H_{12}, is known to have a six-membered ring. In its UV–vis spectrum, the longest-wavelength λ_{max} appears at 191 nm. Draw four isomers that are consistent with these results.

15.34 Phenolphthalein is often used as an indicator in acid–base titration experiments because its color depends on the pH of the solution. When the solution is acidic or near neutral (pH < 8), it is colorless. Under mildly basic conditions (pH 9–13), the solution is red. Under strongly basic conditions (pH > 14), the solution is colorless again. Given the following structures of phenolphthalein under the various pH conditions indicated, explain the color dependence on pH.

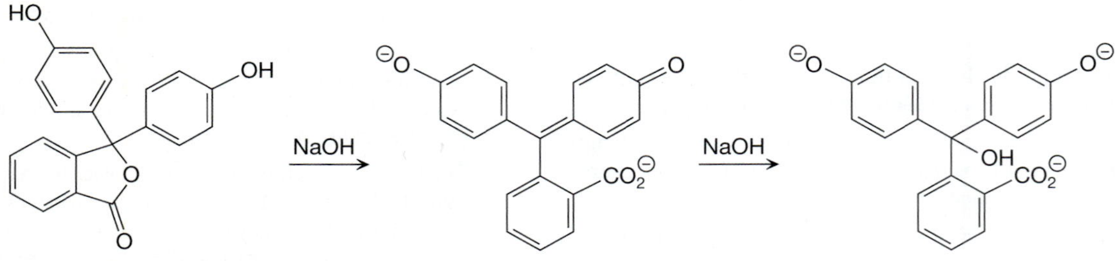

Phenolphthalein
pH < 8

pH 9–13

pH > 14

15.35 Which compound has the longer-wavelength λ_{max}: propyne or acetonitrile ($CH_3C\equiv N$)? Explain.

15.36 Suggest how UV–vis spectroscopy could be used to determine whether each of the following reactions actually took place. Explain your reasoning.

(a) Base

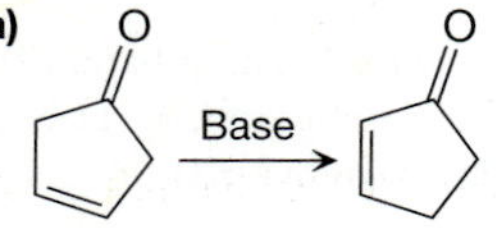

(b) 1. LiAlH₄ 2. H₃O⁺

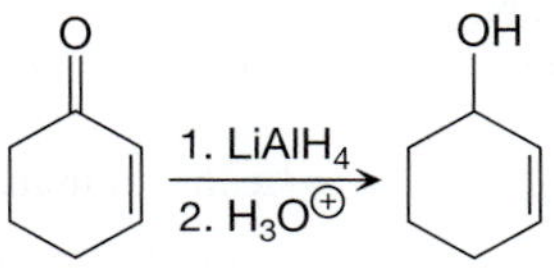

(c) Li / NH₃/EtOH

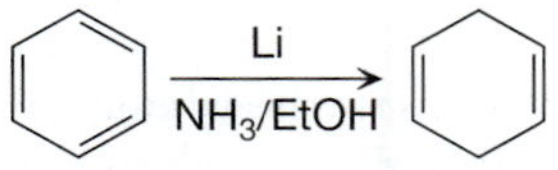

(d) Fe / HCl

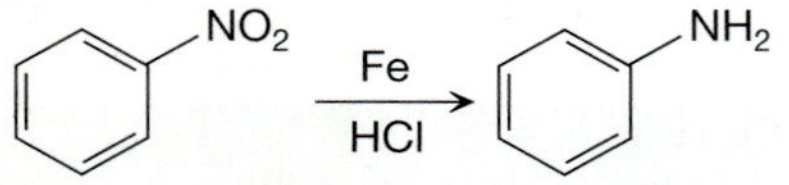

(e) AlCl₃

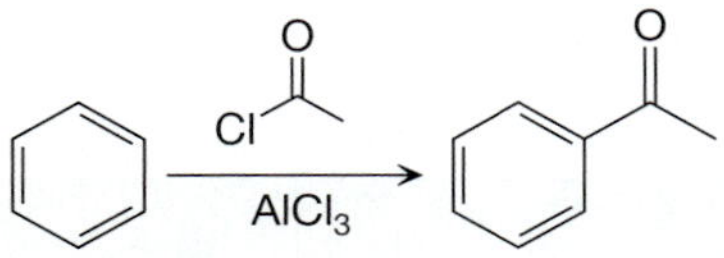

15.37 According to Table 15-1, propenal (acrolein) has a λ_{max} at 340 nm. In the UV–vis spectrum shown here, the region that contains that absorption peak is magnified and shown in the inset. A more intense absorption appears at 202 nm. What kind of electron transition does the absorption at 202 nm correspond to? How do you know?

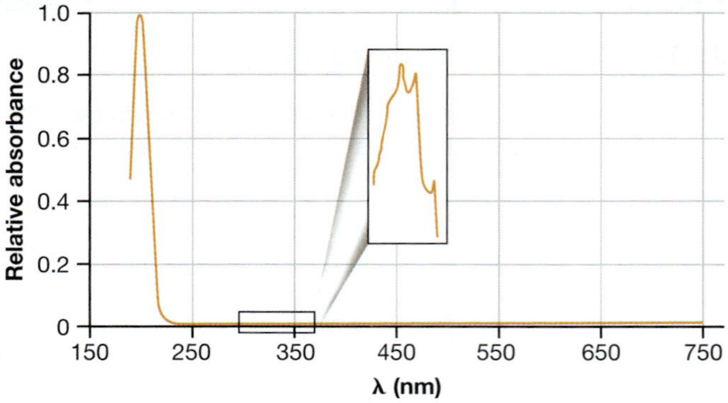

15.38 In the UV–vis spectrum of benzene, the absorption that corresponds to the HOMO–LUMO transition occurs at 184 nm. How does this compare to the corresponding electron transition in hexa-1,3,5-triene (see Table 15-1)? Explain why there is a significant difference.

15.39 In Section 6.2d, we saw that the pK_a of an acid is equal to the pH of the solution at which half the acid has dissociated into its conjugate base. UV–vis spectroscopy can be used to measure the relative concentrations involved if the acid or conjugate base absorbs UV–vis light. With this in mind, suppose that a particular acid has a λ_{max} of 312 nm. The figure here shows the absorbance at that wavelength as a function of pH. What is the pK_a of the acid?

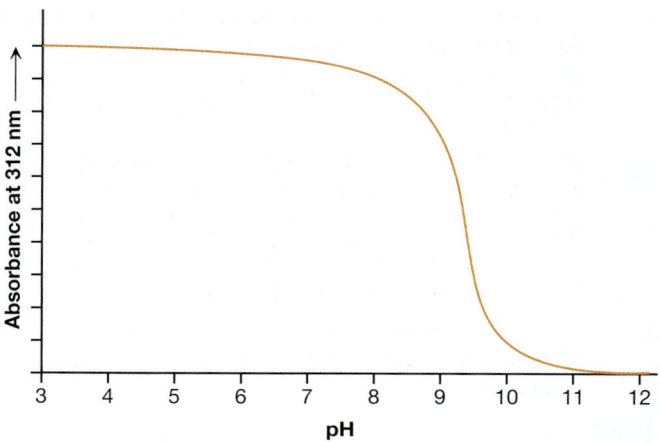

15.4 and 15.5 IR Spectroscopy

15.40 For each of the following compounds, estimate the IR stretching frequencies for the bonds indicated by the arrows.

15.41 What differences in the IR spectra of the reactant and product would enable you to tell that each of the following reactions took place?

(a) cyclohexanol $\xrightarrow[\Delta]{85\% \text{ H}_3\text{PO}_4}$ cyclohexene

(b) acetic acid $\xrightarrow{\text{CH}_2\text{N}_2}$ methyl acetate

(c) butylamine + butyl bromide → dibutylamine

(d) $\xrightarrow{\text{CH}_3\text{I}}$ trimethylamine

(e) propylene oxide $\xrightarrow[\text{H}_2\text{O}]{\text{NaOH}}$ diol

(f) cyclohexenone $\xrightarrow[\text{Pd}]{\text{H}_2}$ cyclohexanone

(g) benzene $\xrightarrow{\text{Li} / \text{NH}_3/\text{EtOH}}$ 1,4-cyclohexadiene

(h) $\xrightarrow[\Delta]{\text{NaOH}}$

(i) cyclohexenone $\xrightarrow[\text{2. H}_3\text{O}^\oplus]{\text{1. LiAlH}_4}$ cyclohexenol

15.42 How could you use IR spectroscopy to distinguish between compounds **A** and **B**?

A

B

15.43 In Section 15.5c, we learned that the frequency of the C=O stretch for an amide is lower than it is for a ketone, suggesting that the C=O bond is weaker in an amide. This weakening of the C=O bond can be explained by the significant contribution from one of its resonance structures toward the resonance hybrid. Draw that resonance structure for *N,N*-dimethylacetamide [CH₃CON(CH₃)₂] and explain how it accounts for the weakening of the C=O bond.

15.44 Problem 15.43 calls attention to a resonance structure of an amide that accounts for the lowering of the C=O stretch frequency. An analogous resonance structure can be drawn for an ester. **(a)** Draw that resonance structure for methyl acetate (CH₃CO₂CH₃). **(b)** Based on the fact that an ester's C=O stretch frequency is higher than an amide's, does that resonance structure have a larger contribution toward the resonance hybrid of an ester or an amide? Using arguments of charge stability, explain why that should be so.

15.45 The carbonyl stretch of acetyl chloride (CH₃COCl) is found at 1806 cm⁻¹. Is the C=O bond stronger or weaker than the one in an amide? What does this suggest about the resonance contribution of a lone pair on Cl in an acid chloride compared to that of the lone pair on N in an amide? Explain. *Hint:* See Problems 15.43 and 15.44.

15.46 In Section 15.5c, we learned that the frequency of a C=O stretch decreases when the C=O bond is conjugated to a C=C bond. Draw the pertinent resonance contributor of a conjugated carbonyl (C=C—C=O) and, based on the resulting resonance hybrid, explain why the frequency decreases.

15.47 A student acquires the IR spectra of cyclohepta-1,3-diene and 3-methylcyclohexa-1,4-diene, which are isomers. One has an absorption band at 1648 cm⁻¹ and the other at 1618 cm⁻¹. Which spectrum belongs to which compound?

15.48 The C=C stretch frequency of an isolated C=C bond is normally ~1620 cm⁻¹. The C=C stretch frequencies for an aromatic ring are typically found between 1450 and 1600 cm⁻¹. Explain why the frequencies are lower in an aromatic ring.

15.49 In an IR spectrum, the C—O stretch for an alcohol (ROH) or an ether (ROR) appears near 1050 cm⁻¹, but for a carboxylic acid (RCO₂H) or an ester (RCO₂R), the C—O stretch appears near 1250 cm⁻¹. Explain why.

15.50 The C=O stretching frequency for butanal is centered at 1746 cm⁻¹, whereas the one for butan-2-one is centered at 1735 cm⁻¹. Explain the difference.

15.51 In which compound, **C** or **D**, would you expect the C—O stretch band to be higher in frequency? Explain.

15.52 In which compound, **E** or **F**, would you expect the C=C stretch absorption to be more intense? Explain.

15.53 The IR spectra for cyclohexanone and cyclobutanone are shown. In which compound is the C=O bond stronger? How do you know?

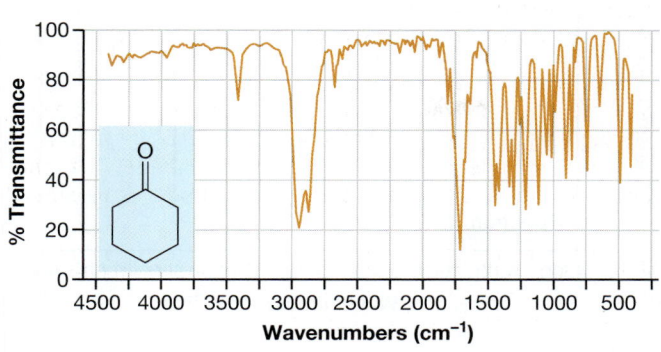

15.54 Which compound, **G** or **H**, do you expect to exhibit the lower C=O stretching frequency? Explain.

G   H

15.55 Sodium acetate has a strong, sharp IR peak appearing at 1569 cm^{-1}, as shown in the spectrum here. To what kind of stretching mode does this band correspond? Why is its frequency so different from that of an ester?

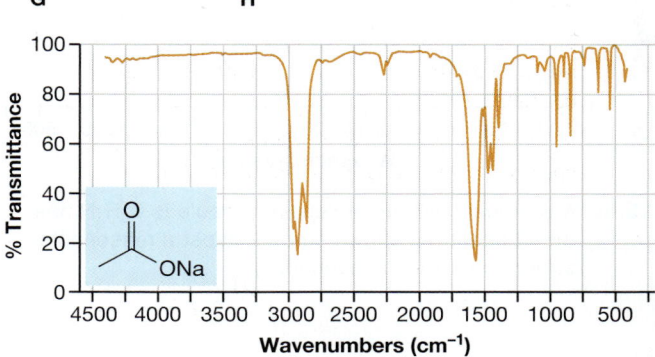

15.56 For each compound below, sketch the IR spectrum it would produce. Have the x axis range from 400 to 4000 cm^{-1} and pay attention to each absorption's frequency, breadth, and intensity.

(a)   (b)   (c)   (d)

15.6 Structure Elucidation Using IR Spectroscopy; **Integrated Problems**

15.57 IR spectroscopy does not distinguish very well between isomers **I** and **J**. Explain why. How could UV–vis spectroscopy be used to distinguish between them?

I   **J**

15.58 A student ran the reaction shown here separately using two different bases: once with NaOH and again with LDA. When NaOH was used, the organic product's UV–vis spectrum had a λ_{max} of ~220 nm. When LDA was used, the product's spectrum had a λ_{max} of ~180 nm. In both cases, the formula of the organic product was C_7H_{10}. Explain.

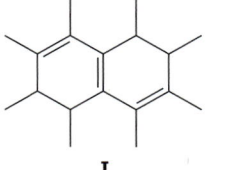

15.59 Match each IR spectrum provided with either compound **K**, **L**, or **M**.

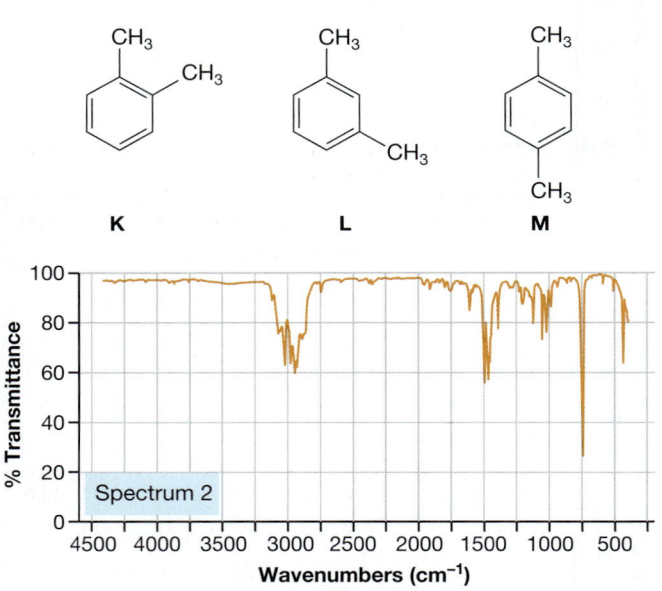

K   L   M

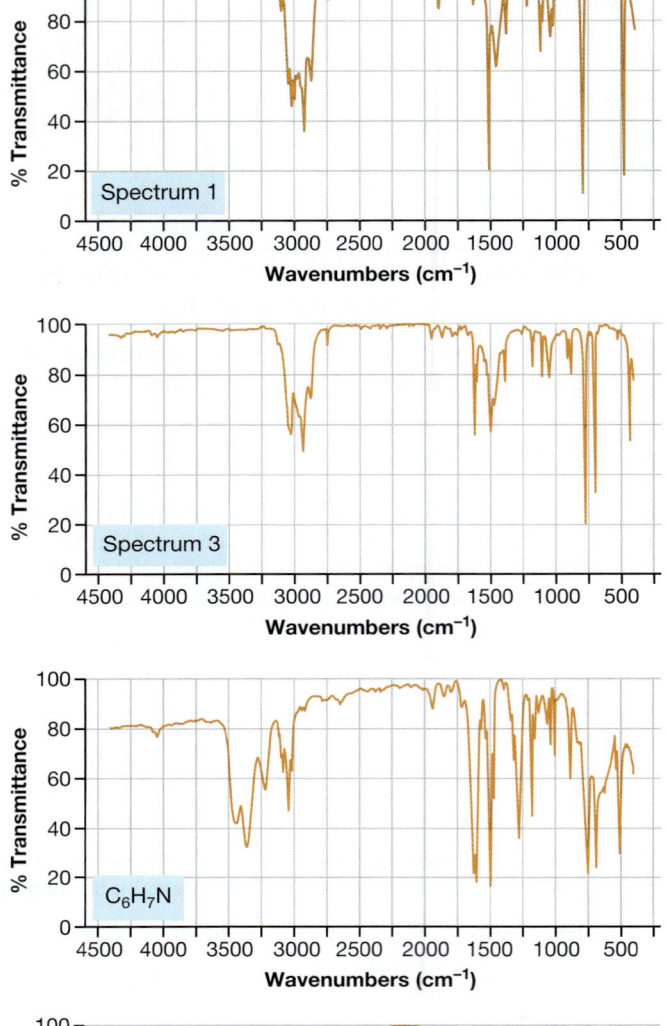

Spectrum 1

Spectrum 2

Spectrum 3

15.60 A compound whose molecular formula is C_6H_7N has the IR spectrum shown here. Suggest a reasonable structure for this compound.

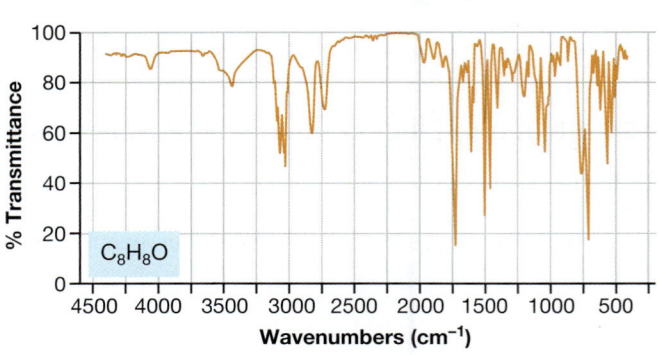

C_6H_7N

15.61 A compound whose molecular formula is C_8H_8O has the IR spectrum shown. Suggest a reasonable structure for this compound.

C_8H_8O

15.62 A compound whose molecular formula is $C_3H_2O_2$ has the IR spectrum shown. Propose a structure for this compound and estimate its pK_a.

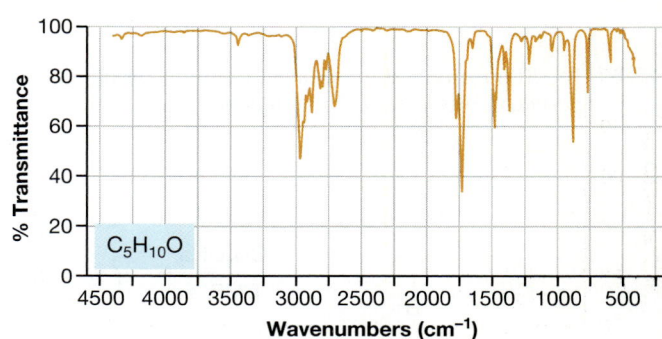

15.63 A compound whose molecular formula is $C_5H_{10}O$ has the IR spectrum shown. The compound's pK_a is >40. Propose a structure for this compound that is consistent with these data.

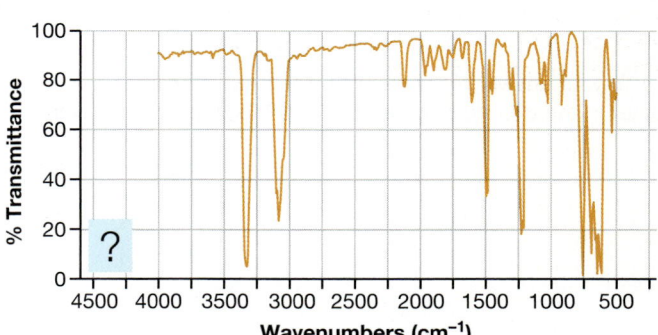

15.64 When 1,2-dibromo-1-phenylethane is heated with sodium hydroxide, a compound with the IR spectrum provided is produced. Propose a structure for this product.

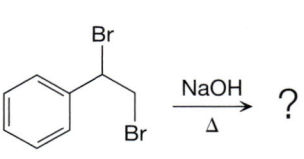

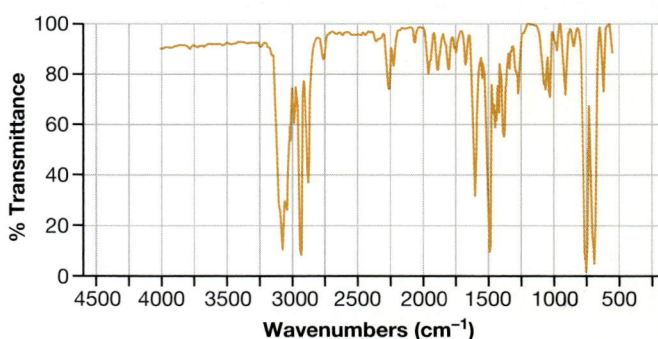

15.65 The product obtained from the reaction in Problem 15.64 is then treated with sodium hydride, followed by treatment with iodomethane. The IR spectrum of the resulting compound is provided. Propose a structure for this product.

15.66 The IR spectrum of the product of the following reaction is shown below. Propose a structure for this compound.

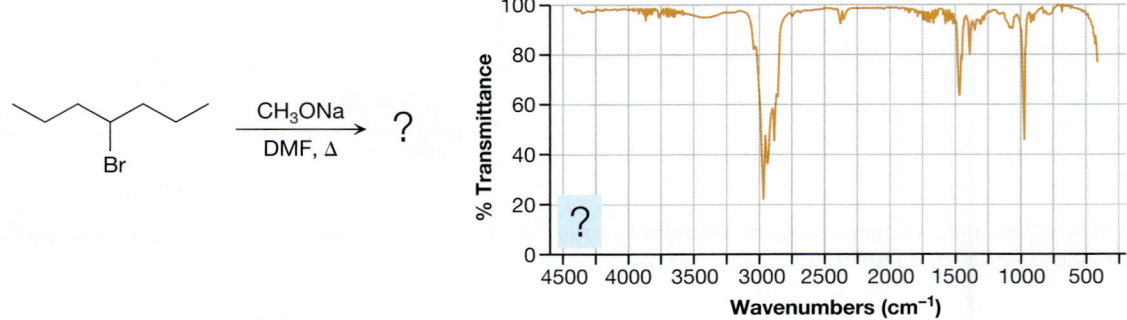

15.67 The IR spectrum of the product of the following reaction is shown below. Propose a structure for this compound, paying particular attention to its stereochemistry.

15.68 Based on your answer to Problem 15.53, in which compound—cyclohexanone or cyclobutanone—is there greater s character in the σ bond between C and O? Explain your answer.

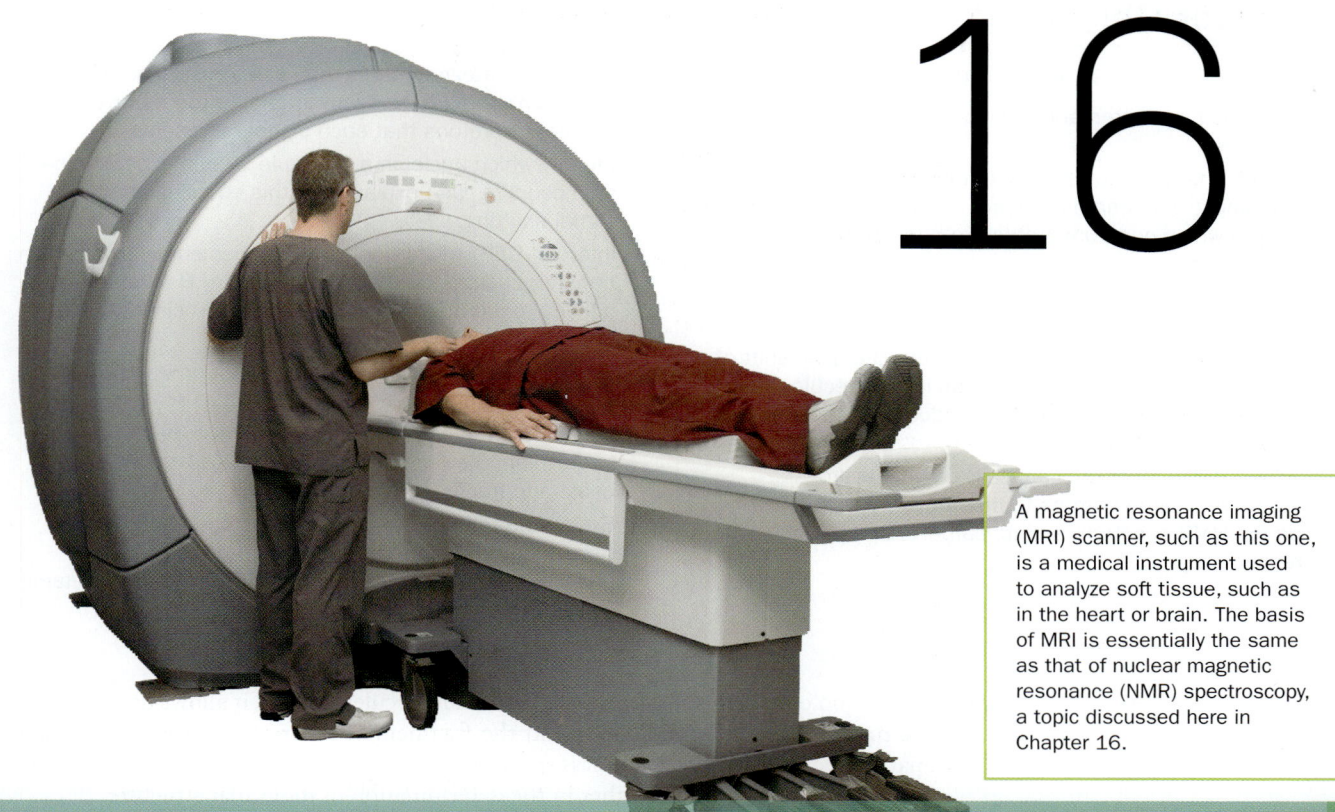

A magnetic resonance imaging (MRI) scanner, such as this one, is a medical instrument used to analyze soft tissue, such as in the heart or brain. The basis of MRI is essentially the same as that of nuclear magnetic resonance (NMR) spectroscopy, a topic discussed here in Chapter 16.

Structure Determination 2
Nuclear Magnetic Resonance Spectroscopy and Mass Spectrometry

Chapter 15 discussed UV–vis and IR spectroscopy. Here in Chapter 16, we continue the discussion of spectroscopy, focusing on **nuclear magnetic resonance (NMR) spectroscopy**. Toward the end of this chapter, we also introduce *mass spectrometry*.

Because of the kind of information it provides, NMR spectroscopy is generally considered the single most powerful tool in organic chemistry for the elucidation of molecular structure. NMR spectroscopy can tell us about the number of distinct types of hydrogen and carbon atoms in a given molecule, as well as each atom's specific environment. This, in turn, allows us to determine the *connectivity* of atoms within the molecule. By comparison, IR spectroscopy provides relatively little information about connectivity, and UV–vis spectroscopy provides even less.

Understanding NMR spectroscopy, however, requires a grasp of principles that are somewhat more complex than those underlying UV–vis or IR spectroscopy. Moreover, an NMR spectrum often has more features to analyze than a typical UV–vis or IR spectrum, so interpreting an NMR spectrum is slightly more involved. For these reasons, we devote nearly an entire chapter to NMR spectroscopy.

Mass spectrometry is another very powerful tool for structural elucidation. Unlike UV–vis, IR, and NMR spectroscopy, however, mass spectrometry does not involve the

On completing Chapter 16 you should be able to:

1. Describe how proton and carbon-13 NMR spectra are generated.

2. Identify the region of the electromagnetic spectrum containing wavelengths that are absorbed by nuclei, and specify which quantum states are affected by such an absorption.

3. Explain why the signal frequency of a nucleus increases when the nucleus is exposed to a stronger magnetic field.

4. Define chemical shift and estimate the chemical shift of each hydrogen and carbon nucleus in a molecule, taking into account the roles played by inductive effects and magnetic anisotropy.

5. Relate the number of chemically distinct hydrogen atoms and carbon atoms to the number of signals in its proton NMR spectrum and its carbon-13 NMR spectrum, respectively.

6. Define the integration of a signal, and use a proton spectrum's integral trace to determine the relative number of protons that each signal represents.

7. Identify which nuclei in a molecule should exhibit spin–spin coupling, and explain how this coupling affects signal splitting.

8. Interpret DEPT (distortionless enhancement by polarization transfer) spectra to determine how many H atoms are bonded to the C atom giving rise to a particular carbon signal.

9. Describe the features of a mass spectrum, and explain how one is generated.

10. Identify the molecular ion (M^+) peak in a mass spectrum, and explain the origin of the M + 1 peak, the M + 2 peak, and fragment peaks.

11. Exploit the relative intensities of isotope peaks to determine the number of carbon atoms present in a molecule, and to determine whether a molecule contains bromine, chlorine, or sulfur.

interaction of electromagnetic radiation with molecules. Rather, it allows us to measure the masses of the molecule itself and the fragments that are created when it is split apart. Nevertheless, we group mass spectrometry with the three spectroscopies because all four complement one another in the determination of molecular structure.

16.1 NMR Spectroscopy: An Overview

Recall from Chapter 15 that UV–vis and IR spectroscopies are characterized by the kind of radiation that they use. UV–vis spectroscopy uses radiation from the ultraviolet and visible regions of the electromagnetic spectrum, whereas IR spectroscopy uses infrared radiation. NMR spectroscopy, on the other hand, uses electromagnetic radiation from the **radio frequency (RF)** region of the spectrum, also called *radio waves* (review Fig. 15-1, p. 725).

As its name suggests, RF radiation is largely used for communication purposes, including AM and FM radio signals and wireless network connections for computers. Photons in this region are less energetic than those in the UV–vis or IR regions, encompassing frequencies between roughly 3 Hz and 3×10^{11} Hz (i.e., wavelengths between 100,000 km and 1 mm). Most modern NMR spectrometers, however, work in the relatively narrow region between about 10^8 and 10^9 Hz (i.e., wavelengths between 3 m and 30 cm).

A typical setup for an NMR experiment is shown in Figure 16-1a. A solution containing a small amount (on the order of 5 mg) of the sample is placed in a glass tube and the tube is lowered into the hollow bore of a superconducting magnet, where it is irradiated with a short pulse (several microseconds) of RF radiation emitted from an NMR probe. The probe delivers a range of frequencies and, as in any spectroscopic experiment, we want to obtain a *spectrum* that can tell us which of those frequencies are absorbed and how strong each absorption is.

The general setup for NMR spectroscopy, however, differs from the typical setup for UV–vis or IR spectroscopy in two major ways. The first is the use of a superconducting magnet, which subjects the sample to a very strong **external magnetic field**, abbreviated as B_{ext}. The second is in the way that each absorbed frequency is detected. In UV–vis and IR spectroscopy, absorbance at a particular frequency is obtained by

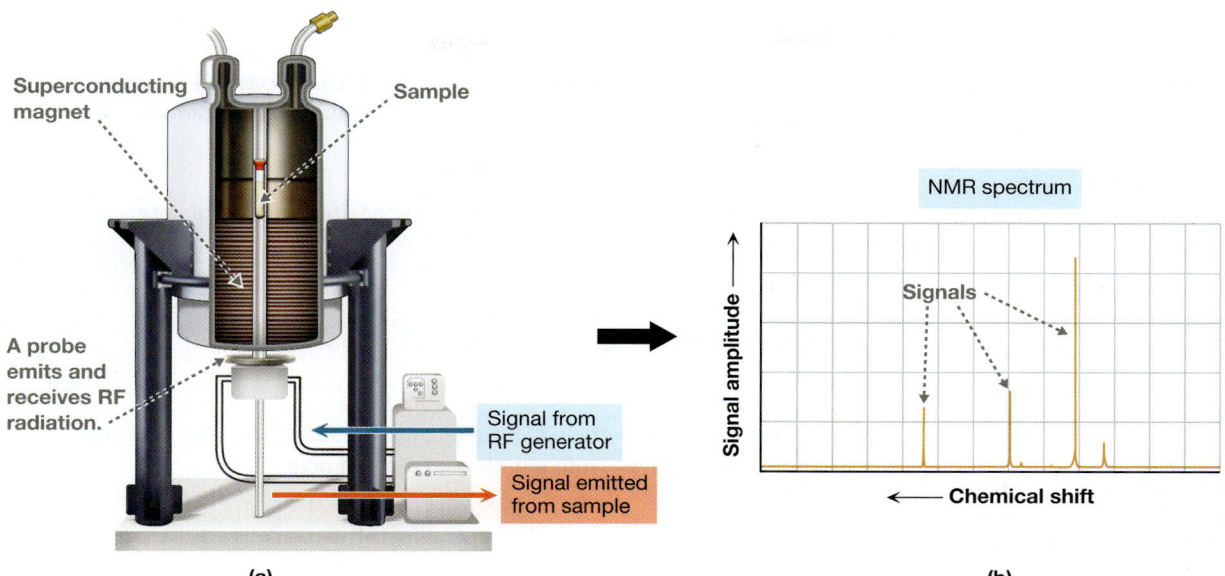

(a)

(b)

FIGURE 16-1 Typical nuclear magnetic resonance experiment
(a) A modern superconducting NMR spectrometer. A sample is placed in the hollow bore of a superconducting magnet. An NMR probe irradiates the sample with RF radiation, and detects RF signals emitted by the sample.
(b) The RF radiation emitted by the sample is recorded and converted into an NMR spectrum, in which amplitude is plotted against chemical shift.

comparing the intensity of radiation that has passed through the sample, $I_{detected}$, to the intensity of radiation from a source, I_{source} (review Equations 15-1 and 15-2, p. 725).

By contrast, most NMR spectrometers take advantage of the fact that a portion of the RF radiation absorbed by the sample at each frequency is *re-emitted* at the same frequency. The magnitude of the oscillating RF radiation re-emitted from the sample is recorded over a period of a few seconds and is digitized, producing what is called a **free induction decay (FID)**. Many FIDs are typically acquired and are averaged to improve the quality, after which a mathematical algorithm, called a **Fourier transform**, decomposes the FID into its individual frequencies, called **signals**. The *amplitude* of each signal is proportional to the amount of radiation that was originally absorbed. Thus, the recorded signals are converted automatically into a spectrum analogous to other spectra we have seen, in which relative absorbance is plotted against a characteristic of radiation (Fig. 16-1b). In this case, the *x* axis of the NMR spectrum is *chemical shift*, which we can think of for now as representing *relative frequency*. (We discuss chemical shift in greater detail in Section 16.6.)

With this overview of NMR spectroscopy in mind, we ask the following questions:

- Why must a large magnetic field be applied to the sample?
- Why are only certain frequencies of RF radiation absorbed by a given sample?
- How can an NMR spectrum be used to interpret molecular structure?

We answer these questions in the sections that follow.

16.2 Nuclear Spin and the NMR Signal

The sample is placed in a strong magnetic field in an NMR experiment because:

> Without being subjected to a strong magnetic field, a sample does not absorb RF radiation and therefore cannot produce an NMR spectrum.

Why is this?

The answer stems from the fact that some atomic nuclei possess a property called **nuclear spin**. The most common example is the nucleus of a ^1H atom (i.e., a proton), whose spin is analogous to that of an electron. Recall from Section 1.3c that electrons can assume one of two spin states. A proton's spin is quantized in the same way.

A proton has a spin of $+\frac{1}{2}$ or $-\frac{1}{2}$ au, which corresponds to the **α spin state** or the **β spin state**, respectively.

Nuclear spin generates a small magnetic field. Consequently:

Protons (and other nuclei with spin) have quantized magnetic dipoles and thus behave as tiny bar magnets, with a north and a south pole.

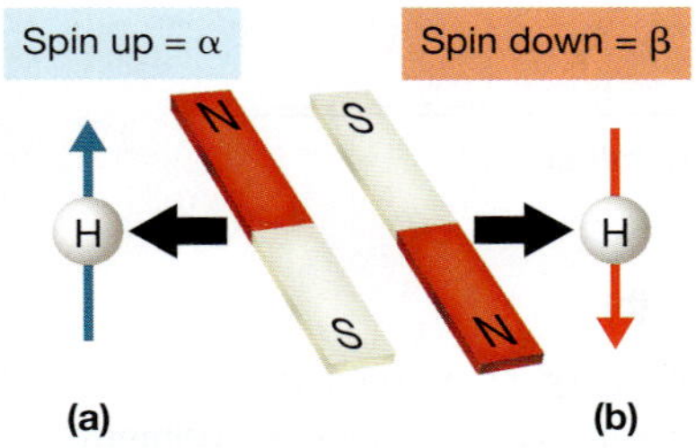

These nuclei are thus often represented as arrows, indicating the direction in which their magnetic dipoles point (Fig. 16-2).

Outside of an external magnetic field, there is no energy difference between the α and β states, so there is an equal probability of a proton being in either state (Fig. 16-3, left). Moreover, the magnetic dipoles of the protons point in random directions. The situation is similar to what we would observe if we were to grab several bar magnets and toss them onto the floor. Unless two magnets lie close enough to affect each other, they will end up with random orientations.

In an external magnetic field, B_{ext}, however, the situation is different. Each nucleus with spin has its average magnetic dipole aligned parallel to B_{ext}. As shown on the right in Figure 16-3:

- When exposed to B_{ext}, the magnetic dipoles of nuclei in the α spin state become aligned *with* B_{ext}, whereas those of nuclei in the β spin state become aligned *against* it.
- In these orientations, the energy of the α state is slightly lower than the energy of the β state.

Spin up = α **Spin down = β**

(a) **(b)**

FIGURE 16-2 Quantized magnetic dipoles of the hydrogen nucleus (a) The α spin or spin-up state. (b) The β spin or spin-down state.

FIGURE 16-3 Effect of an external magnetic field on nuclear spin states In the absence of an external magnetic field, the α and β states are at exactly the same energy, and the nuclear magnetic dipoles are oriented in random directions (*left*). In an external magnetic field, represented by the green arrow, the α spins align with the external magnetic field and the β spins align against it (*right*). This makes the energy of the α state slightly lower than that of the β state, by a difference ΔE_{spin}.

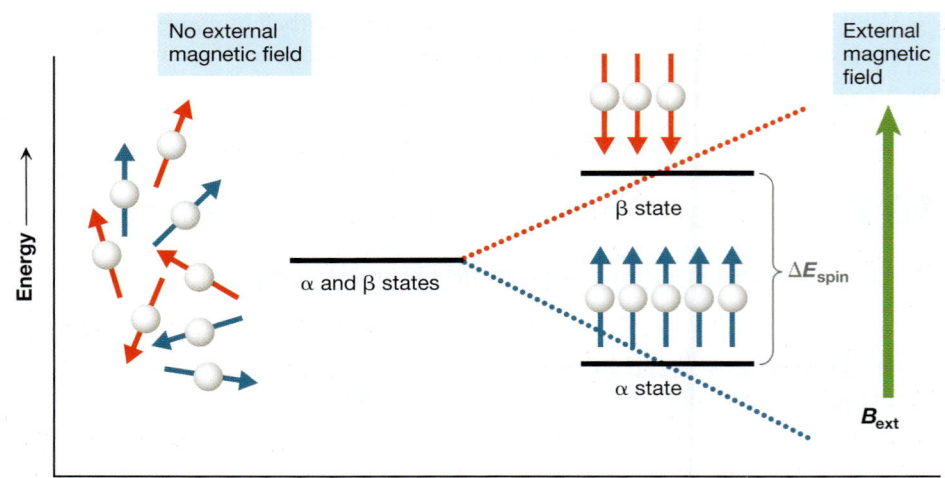

No external magnetic field

External magnetic field

α and β states

β state

α state

ΔE_{spin}

B_{ext}

Energy

Magnitude of external magnetic field, B_{ext}

This energy difference is like being in a canoe on a river — paddling with the current requires less energy than paddling against it.

Equation 16-1 shows that the difference in energy between the two spin states of a proton, ΔE_{spin}, is directly proportional to B_{ext}.

$$\Delta E_{spin} = \frac{\gamma h B_{ext}}{2\pi} = (constant) \times B_{ext} \qquad (16\text{-}1)$$

Here, B_{ext} is in units of **tesla (T)**, h is Planck's constant (6.626×10^{-34} J·s), and γ (the Greek letter gamma) is the **gyromagnetic ratio**. Every nucleus that has spin has a characteristic value for the gyromagnetic ratio; for a proton, its value is 2.67512×10^8 T^{-1} s^{-1}.

YOUR TURN 16.1

In Figure 16-3, choose a position on the x axis representing a B_{ext} that is stronger than the one indicated by the separated spin states shown. At that magnetic field, draw horizontal lines representing the energies of the α and β states and draw a vertical curly brace (similar to the one already drawn) representing the energy difference between the two states. Next to that curly brace, write "larger ΔE_{spin}."

Answers to Your Turns are in the back of the book.

In a typical NMR experiment, values of ΔE_{spin} are on the order of 10^{-25} J (see Solved Problem 16.2), which is about 0.01% of the thermal energy available at room temperature. Therefore, nuclei in the lower-energy α spin state are in small excess, but that excess is significant.

PROBLEM 16.1 At which magnetic field experienced by a hydrogen nucleus is the energy difference between the α and β states greater: 0.5 T or 5.0 T? Explain.

SOLVED PROBLEM 16.2

A common magnetic field strength of modern NMR instruments is 7.046 T. In this magnetic field, calculate the energy difference between the α and β spin states of a proton.

Think In Equation 16-1, what value should be substituted for γ? For h? For B_{ext}?

Solve As mentioned in the text, γ for a proton is 2.67512×10^8 T^{-1} s^{-1} and h is 6.626×10^{-34} J·s. In the problem, the value for B_{ext} is given as 7.046 T. Therefore,

$$\Delta E_{spin} = \frac{\gamma h B_{ext}}{2\pi}$$

$$= \frac{(2.67512 \times 10^8 \text{ T}^{-1} \text{s}^{-1})(6.626 \times 10^{-34} \text{ J·s})(7.046 \text{ T})}{2\pi}$$

$$= 1.988 \times 10^{-25} \text{ J}$$

The units s and T both cancel, leaving energy in units of J.

PROBLEM 16.3 Calculate the energy difference between the α and β spin states of a proton in a magnetic field of 2.2 T.

Typical values of ΔE_{spin}, moreover, correspond to photon energies from the RF region of the electromagnetic spectrum. Therefore, a nucleus in the α state can absorb an RF photon and be promoted to the β state — a so-called **spin flip** (Fig. 16-4).

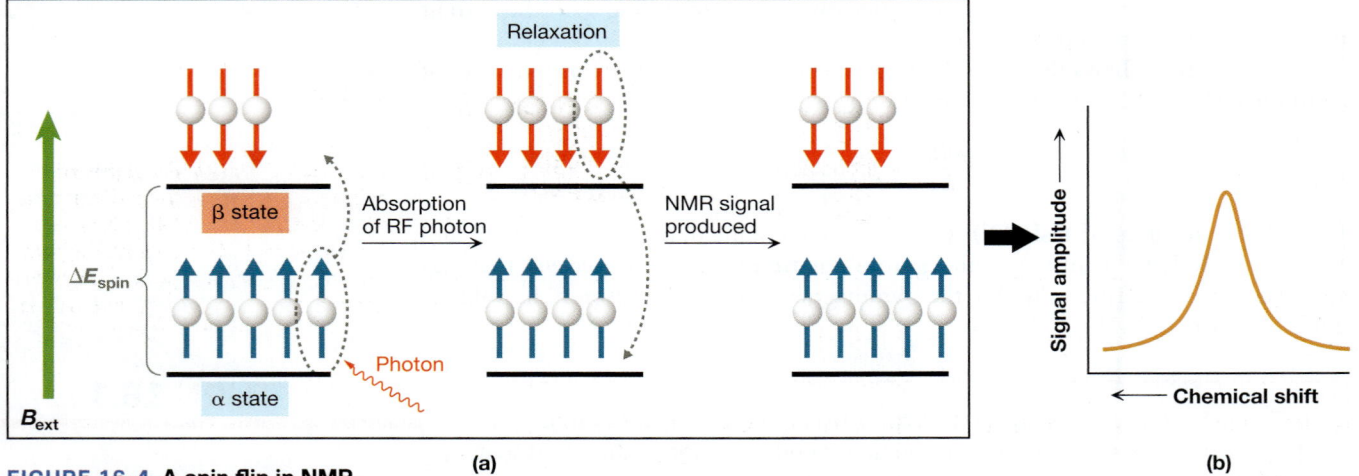

(a)

(b)

When a nucleus in the β state relaxes back to the α state, it emits RF radiation that we detect as an NMR signal.

We can now understand why no NMR spectrum can be acquired in the absence of a B_{ext}. As we saw in Figure 16-3, the α and β states of a proton have exactly the same energy in the absence of a B_{ext}. Under these circumstances, there is no higher-energy spin state to which a nucleus can be promoted, so no RF photon can be absorbed.

Most of this chapter focuses on **proton NMR** (or **^{1}H NMR**) **spectroscopy**, in which the RF radiation used causes spin flips in the nuclei of hydrogen atoms. NMR spectroscopy, however, can also be used to probe other nuclei that possess spin. In general:

> A nucleus possesses spin and thus can be studied with NMR spectroscopy, making it *NMR active*, if it has an odd number of protons, an odd number of neutrons, or both.

This is consistent with ^{1}H being NMR active, because its nucleus has just one proton (an odd number). ^{2}H (D, deuterium) is NMR active, too, because its nucleus has one proton and one neutron (both odd numbers). In addition, the carbon-13 (^{13}C) nucleus, which has six protons (an even number) and seven neutrons (an odd number), is NMR active, but the carbon-12 (^{12}C) nucleus, which has six protons (even) and six neutrons (even), is not.

Many other nuclei are NMR active, including ^{14}N, ^{15}N, ^{17}O, ^{19}F, and ^{31}P. Organic molecules are principally composed of hydrogen and carbon, however, so we will discuss only ^{1}H NMR spectroscopy (through Section 16.12) and ^{13}C NMR spectroscopy (Sections 16.13 and 16.14).

16.3 Chemical Distinction and the Number of NMR Signals

All protons in a molecule typically do *not* absorb at the same frequency; if they did, then we would be unable to use NMR spectroscopy to determine aspects of molecular structure. Instead, protons surrounded by different electron distributions—that

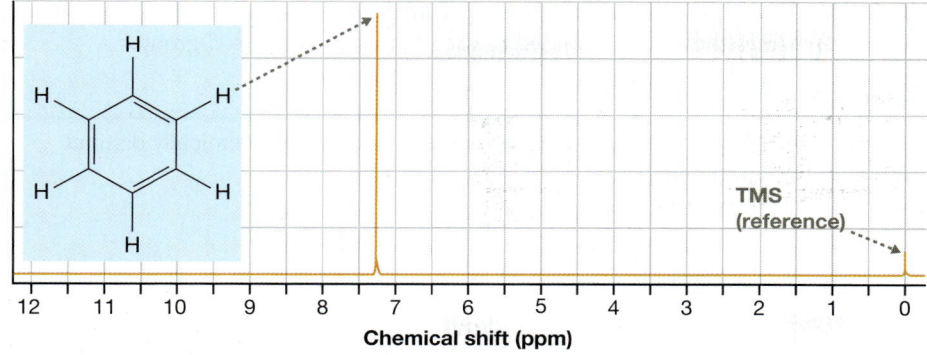

FIGURE 16-5 ¹H NMR signal of benzene The six H atoms of benzene are chemically equivalent, giving rise to a single ¹H NMR signal at 7.3 ppm. The small signal at 0 ppm arises from tetramethylsilane (TMS) that is added to provide a reference signal.

is, ones that reside in different **chemical environments**—absorb at different frequencies. These protons are called **heterotopic protons**, or **chemically distinct protons**. By the same token, protons that reside in identical chemical environments, called **homotopic protons**, or **chemically equivalent protons**, absorb at the same RF frequency.

Because a proton's absorption frequency depends on its chemical environment, we arrive at one of the most important rules for ¹H NMR spectroscopy:

> Each chemically distinct type of hydrogen atom in a molecule gives rise to an independent signal in a proton NMR spectrum.

In benzene, for example, the electron distribution about each proton is identical, so all six H atoms are chemically equivalent. This gives rise to a single ¹H NMR signal, which appears at 7.3 ppm in the NMR spectrum (Fig. 16-5). (The unit ppm stands for "parts per million," which is discussed in Section 16.6.)

The second signal in the spectrum, which appears at 0 ppm, is due to a small amount of **tetramethylsilane (TMS)**, $(CH_3)_4Si$, which can be added to samples to provide a reference signal.

PROBLEM 16.4 Based on the positions of the two signals in Figure 16-5, do you think that the protons in benzene are chemically equivalent to those in TMS? Explain.

Propynoic acid ($HC\equiv C-CO_2H$) gives rise to two ¹H NMR signals, as shown in the spectrum in Figure 16-6. (Once again, the signal at 0 ppm corresponds to TMS that is added to the sample.) Thus, there are two chemically distinct H atoms in

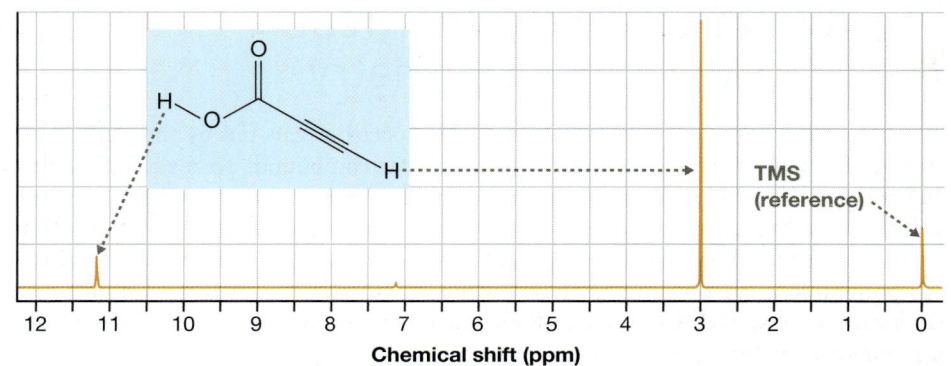

FIGURE 16-6 ¹H NMR spectrum of propynoic acid The two signals on the left correspond to the protons in propynoic acid, indicating that propynoic acid has two chemically distinct protons. The signal at 0 ppm corresponds to TMS, the reference compound.

$HC\equiv C-CO_2H$. This should not be surprising, because one proton is bonded to an alkyne C, whereas the other is bonded to the O atom of a carboxyl group.

SOLVED PROBLEM 16.5

From the ^1H NMR spectrum given, determine the number of chemically distinct protons in dichloroacetic acid.

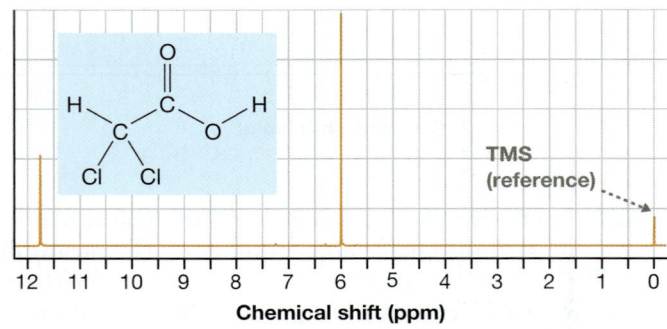

Think How many signals appear in the proton NMR spectrum? Of those signals, which are generated by dichloroacetic acid?

Solve Although there are three peaks in the ^1H NMR spectrum, only the ones at 6.0 ppm and 11.8 ppm correspond to protons in dichloroacetic acid. The peak at 0 ppm corresponds to added TMS. Thus, there are two chemically distinct types of protons in dichloroacetic acid.

PROBLEM 16.6 From the ^1H NMR spectrum given, determine the number of chemically distinct protons in *trans*-1,2-dichloroethene.

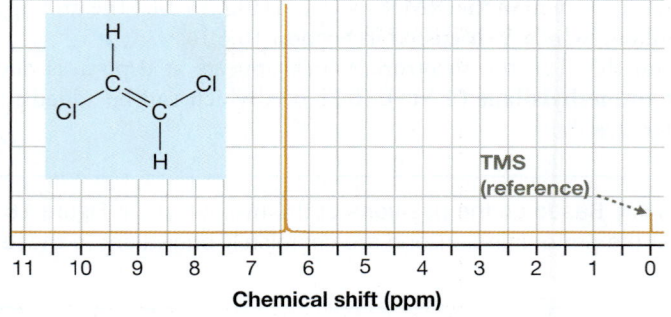

16.4 Strategies for Success: The Chemical Distinction Test and Molecular Symmetry

As discussed in Section 16.3, a proton NMR spectrum can tell us the number of chemically distinct hydrogen atoms in a particular compound. To make use of this information, however, it is important to be able to determine the number of chemically distinct hydrogen atoms solely from a molecule's structure. The ability to do so helps us to establish whether a compound is consistent with the NMR spectrum.

Sometimes, it is not immediately obvious whether certain hydrogen atoms in a molecule are chemically distinct. In these situations, we can use a protocol called the **chemical distinction test**.

The Chemical Distinction Test for Hydrogen Atoms

1. For each hydrogen atom in question, draw the complete structure of the molecule in which *just that hydrogen atom* is replaced by an imaginary atom, "X." There should be one X-substituted molecule for each hydrogen atom being tested.
2. Determine which of those X-substituted molecules are either *identical* or *enantiomers*. The corresponding hydrogen atoms in the original molecule are *chemically equivalent*.
3. Determine which of those X-substituted molecules are either *constitutional isomers* or *diastereomers*. The corresponding hydrogen atoms in the original molecule are *chemically distinct*.

If the chemical distinction test yields *enantiomers*, then the corresponding hydrogen atoms are said to be **enantiotopic**. If the test yields *diastereomers*, then the corresponding hydrogen atoms are said to be **diastereotopic**. As we can see from the above steps, enantiotopic hydrogen atoms are chemically equivalent whereas diastereotopic hydrogen atoms are chemically distinct.

Let's apply the chemical distinction test to the six H atoms in benzene. If we want to test all six hydrogen atoms, then Step 1 results in the following six molecules shown on the right:

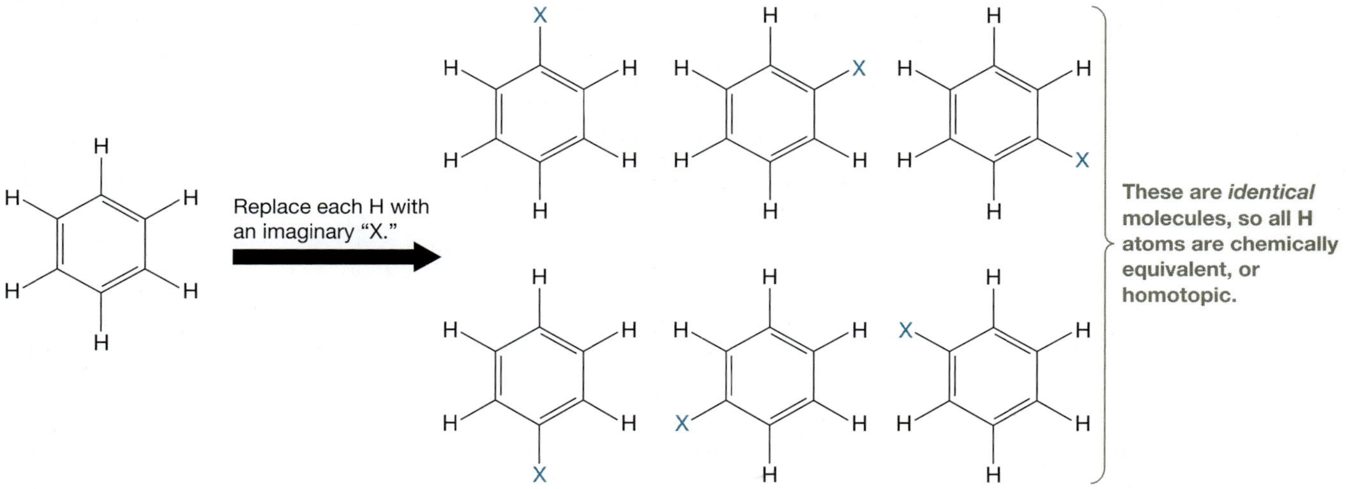

These are *identical* molecules, so all **H** atoms are chemically equivalent, or homotopic.

All six of the X-substituted molecules are identical to each other, so according to Step 2, all six H atoms must be chemically equivalent. This conclusion is consistent with the fact that benzene gives rise to just one signal in its ^{1}H NMR spectrum (Fig. 16-5, p. 777).

Now consider propynoic acid, $HC{\equiv}C{-}CO_2H$. When each H atom is separately replaced by "X," we get the following two structures shown on the right:

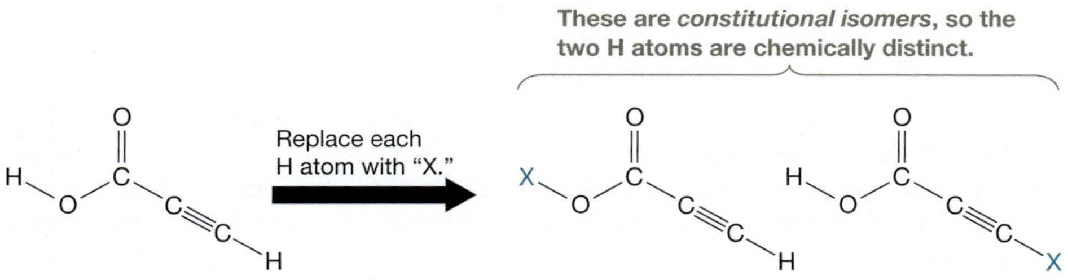

These are *constitutional isomers*, so the two H atoms are chemically distinct.

The resulting molecules are *constitutional isomers* because they have the same molecular formula but different connectivities. Thus, the two H atoms are chemically distinct, consistent with the fact that the compound gives rise to two signals in its ¹H NMR spectrum (Fig. 16-6, p. 777).

SOLVED PROBLEM 16.7

How many chemically distinct H atoms are there in chloroethene, $H_2C=CHCl$?

Think What molecules are generated by substituting each H atom separately by "X"? Which of those molecules are identical? Enantiomers? Diastereomers? Constitutional isomers?

Solve The three X-substituted molecules are as follows:

Molecules **A** and **B** are diastereomers—they are different molecules that have the same connectivity but are not mirror images. Molecules **B** and **C** are constitutional isomers. Therefore, according to Step 3 in the chemical distinction test, all three H atoms are chemically distinct.

PROBLEM 16.8 How many chemically distinct H atoms are in each of the following molecules?

(a)   (b)   (c)   (d)   (e)   (f)

PROBLEM 16.9 A compound gives rise to two signals in its ¹H NMR spectrum. Which of compounds **D–G** can it be? Explain.

D   E   F   G

After you have practiced the chemical distinction test with a variety of different molecules, you should expect to be able to do some of the steps in your head and this will help speed up the process. You will also be able to speed up the process by noticing that molecules with greater symmetry tend to have fewer chemically distinct hydrogens. That's because a molecule's symmetry can place demands on the chemical equivalence of certain atoms.

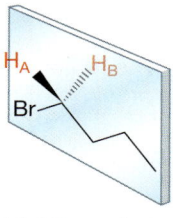

H_A and H_B are mirror images of each other, so they are equivalent.

(a) 1-Bromobutane

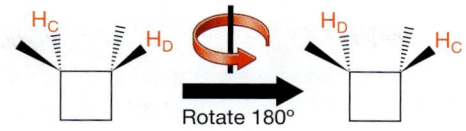

This 180° rotation of the molecule exchanges H_C and H_D, but the molecule otherwise appears unchanged, so they are equivalent.

Rotate 180°

(b) (1R,2R)-1,2-Dimethylcyclobutane

FIGURE 16-7 Molecular symmetry and chemical equivalence (a) H_A and H_B are mirror images of each other through the molecule's plane of symmetry, so the two H atoms are equivalent. (b) The 180° rotation of the molecule causes H_C and H_D to exchange locations but the molecule otherwise appears unchanged, so the two H atoms are equivalent.

Two atoms must be chemically equivalent if:

- The atoms are mirror images with respect to a plane of symmetry of the molecule; or,
- A rotation of the entire molecule causes the atoms to exchange locations, but the molecule otherwise appears unchanged.

In 1-bromobutane (Fig. 16-7a), for example, H_A and H_B are mirror images with respect to the plane of symmetry, so they are equivalent. In (1R,2R)-1,2-dimethylcyclobutane (Fig. 16-7b), the 180° rotation of the molecule results in the exchange of H_C and H_D, but the molecule otherwise appears unchanged, so the two H atoms are equivalent.

YOUR TURN 16.2

In benzene (Fig. 16-5, p. 777), all six H atoms are chemically equivalent.

(a) Can you identify a plane of symmetry through which one H atom is the mirror image of another?

(b) Can benzene be rotated in such a way as to exchange the locations of H atoms, but otherwise leave the molecule apparently unchanged?

16.5 The Time Scale of NMR Spectroscopy

In its chair conformation, cyclohexane has two types of hydrogens: axial and equatorial. The two types of hydrogen atoms are in different chemical environments, so you might expect cyclohexane to give rise to two ^1H NMR signals. However, at room temperature, cyclohexane's ^1H NMR spectrum (Fig. 16-8) exhibits only one signal. How can we explain this?

Cyclohexane gives rise to just one proton signal because the axial and equatorial hydrogens rapidly interchange environments via chair flips. More specifically, *the time it takes for a chair flip ($\sim 10^{-5}$ s) is much shorter than the time it takes to acquire an NMR spectrum (~ 1 s).* Thus, the different environments of the interchanging protons in cyclohexane are blurred, in much the same way that a camera with a

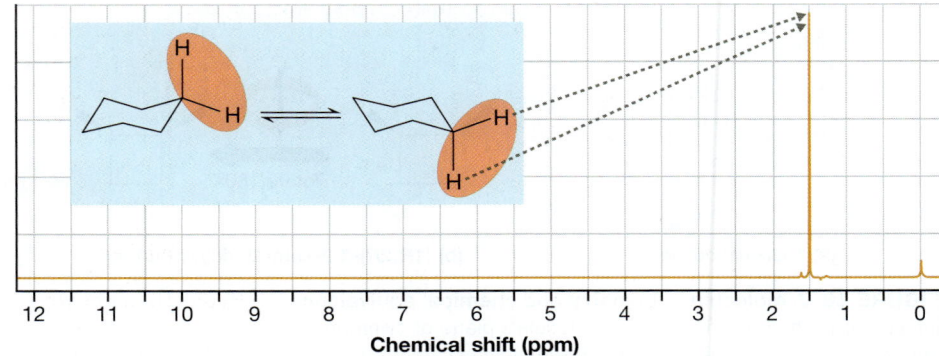

Axial and equatorial H atoms rapidly interconvert.

Chemical shift (ppm)

FIGURE 16-8 **¹H NMR spectrum of cyclohexane** The axial and equatorial H atoms give rise to the same signal because they rapidly interconvert via a chair flip.

slow shutter speed blurs objects in motion. The NMR signal, as a result, reflects the *average* proton environment of each proton during its chair flip. (At low temperatures, however, two signals can be observed; see Problem 16.41 at the end of the chapter.)

This phenomenon is not limited to just cyclohexane. In general:

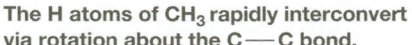

If protons in different chemical environments interchange rapidly, then they give rise to the same NMR signal that reflects the *average* of those environments.

Another example involves acetic acid, CH_3CO_2H, whose proton NMR spectrum (Fig. 16-9) exhibits two signals. One signal is from the carboxyl H and the other is from the CH_3 protons. The CH_3 protons all give rise to the same signal because they exchange positions rapidly via rotation of the CH_3 group. The time it takes to do so ($\sim 10^{-10}$ s) is even shorter (much shorter) than the time it takes for a cyclohexane chair flip.

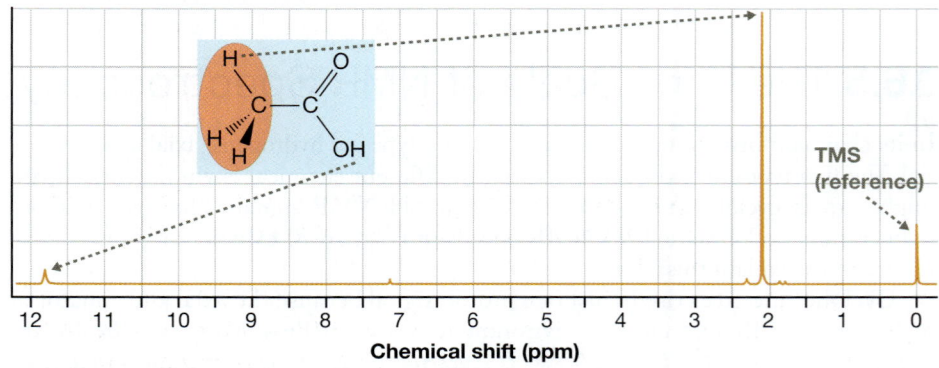

The H atoms of CH₃ rapidly interconvert via rotation about the C—C bond.

TMS (reference)

Chemical shift (ppm)

FIGURE 16-9 **¹H NMR spectrum of acetic acid** The three H atoms of CH_3 give rise to the same signal because they rapidly interconvert via rotation of the CH_3 group.

16.6 Chemical Shift

Chemical shift, abbreviated as the Greek letter δ (delta), is a measure of the extent to which a signal's frequency differs from that of a reference compound. The reference compound used for ^{1}H NMR spectroscopy is TMS. Formally, *chemical shift*, in units of *parts per million (ppm)*, is defined by the mathematical expression in Equation 16-2:

$$\text{Chemical shift (ppm)} = \frac{\nu_{\text{sample}} - \nu_{\text{TMS}}}{\nu_{\text{op}}} \times 10^6 \qquad \text{(16-2)}$$

Here, ν_{sample} is the frequency of the signal of interest, ν_{TMS} is the frequency of the signal from the protons in TMS, and ν_{op} is the *operating frequency* of the NMR spectrometer, all of which are in units of hertz (Hz).

The spectrometer's **operating frequency** is the frequency at which a bare proton absorbs radiation when subjected to the spectrometer's magnetic field (Equation 16-3).

$$\nu_{\text{op}} = \frac{\gamma B_{\text{ext}}}{2\pi} \qquad \text{(16-3)}$$

Because the gyromagnetic ratio (γ) is a constant, the operating frequency is directly proportional to B_{ext}. Thus, the operating frequency is constant for a given magnetic field strength. If the strength of the spectrometer's magnet is 7.046 T, for example, then the operating frequency is calculated as follows:

$$\nu_{\text{op}} = \frac{(2.67512 \times 10^8 \ \text{T}^{-1} \ \text{s}^{-1})(7.046 \ \text{T})}{2(3.14159)} = 3.000 \times 10^8 \ \text{s}^{-1}$$

Alternatively, because the unit s^{-1} is equivalent to the unit Hz, and because 1 MHz (i.e., megahertz) equals 10^6 Hz, a spectrometer with a 7.046-T magnet has an operating frequency of 300 MHz. We say that the spectrometer is a 300-MHz instrument.

PROBLEM 16.10 What is the operating frequency of an NMR spectrometer using a magnet whose strength is 11.74 T? Give your answer in units of MHz.

Chemical shift is plotted on the x axis instead of frequency, because signal frequency depends on B_{ext} (review Equation 16-1, p. 775). Therefore, signal frequency can vary from one NMR instrument to the next, depending on the strength of its magnet. By contrast, a proton's *chemical shift is independent of B_{ext}*, because the operating frequency that appears in the denominator of Equation 16-2 has the same dependence on magnetic field strength as the frequency terms in the numerator. The magnetic field dependence thus cancels out, so:

> A proton's chemical shift is the same regardless of the NMR instrument used to measure it.

PROBLEM 16.11 Suppose that the signal of a specific hydrogen nucleus appears 2200 Hz higher than that of the protons in TMS. If the NMR spectrometer uses a 7.046-T magnet, what is the chemical shift of that proton?

PROBLEM 16.12 Suppose that a proton's chemical shift is 2.4 ppm in a 300-MHz NMR spectrometer. What is the difference between the signal frequency of that proton and the signal frequency of a proton in TMS?

By convention, chemical shift increases from right to left in an NMR spectrum. The chemical shift of the protons in TMS always appears at 0 ppm. (We can solve for this value by substituting ν_{TMS} for ν_{sample} in Equation 16-2, which makes the numerator zero.) The farther to the left an absorption peak appears in a spectrum (i.e., the more positive the chemical shift), the more the signal is said to be shifted **downfield**, whereas the farther to the right it appears, the more the signal is said to be shifted **upfield**.[1] These situations correspond to protons that are more *deshielded* or more *shielded*, respectively, as we explain in Section 16.7.

16.6a Deuterated Solvents for NMR Samples

Recall from Section 16.1 that an NMR spectrum is typically acquired using a solution that has a small amount (on the order of 5 mg) of dissolved sample. Most of the solution, therefore, is the solvent, so if the solvent contains any protons, then the resulting ^{1}H NMR spectrum would be dominated by solvent signals. To avoid this problem, we use solvents that have no protons, such as deuterochloroform ($CDCl_3$), carbon tetrachloride (CCl_4), deuterium oxide (D_2O), or acetone-d_6 (CD_3—CO—CD_3).

Many such NMR solvents contain deuterium (D, or ^{2}H), which is NMR active (Section 16.2). This is not a concern, however, because the gyromagnetic ratio of D is 0.41065×10^8 T^{-1} s^{-1}, which is very different from that of a proton, 2.67512×10^8 T^{-1} s^{-1}. This means that the deuterium signals will have very different chemical shifts from protons and will not appear in a ^{1}H NMR spectrum.

Deuterated solvents typically contain a small percentage of the ^{1}H analog, which will produce a solvent peak in the spectrum. $CDCl_3$, for example, contains a small percentage of $CHCl_3$, whose proton's chemical shift is about 7.2 ppm. Consequently, a ^{1}H NMR spectrum acquired using $CDCl_3$ as the solvent will often exhibit a signal there. (Can you find evidence that $CDCl_3$ was used to acquire the spectrum in Figure 16-9 on page 782?)

16.7 Characteristic Chemical Shifts, Inductive Effects, and Magnetic Anisotropy

Recall from Section 16.3 that protons in different chemical environments give rise to different signals in an NMR spectrum. Table 16-1, which lists the chemical shifts observed in ^{1}H NMR spectroscopy, shows that a proton's chemical shift is governed largely by the identity of the surrounding atoms.

YOUR TURN 16.3

Identify two different types of protons in Table 16-1 whose absorption peaks appear *downfield* and two that appear *upfield* from those produced by the protons in $ClCH_3$.

To begin to understand why each type of proton in Table 16-1 has the characteristic chemical shift it does, keep in mind that H atoms that are part of molecules are not bare protons—they are surrounded by electrons. When subjected to B_{ext}, those electrons move

---

[1]The terms *downfield* and *upfield* originate from older types of NMR spectrometers, in which a sample was irradiated with a fixed frequency of RF radiation, and the magnetic field was adjusted to make ΔE_{spin} equal E_{photon}. The more positive a proton's chemical shift, the stronger the magnetic field that was required, and vice versa.

TABLE 16-1 **Chemical Shifts in ¹H NMR Spectroscopy**

Type of Proton	Chemical Shift (ppm)	Type of Proton	Chemical Shift (ppm)	Type of Proton[a]	Chemical Shift (ppm)
1. H₃C—Si—CH₂ with CH₃, H, CH₂, CH₃ (TMS)	0	8. R—C≡C—H	2.4	15. benzene ring with H	7.3
2. R—CH₂ with H	0.9	9. Br—CH₂ with H₂, H	2.7	16. R—C(=O)—H	9–10
3. R₂HC—H	1.3	10. Cl—CH₂ with H₂, H	3.1	17. R—NH with H	1.5–4
4. R₃C—H	1.4	11. R—O—CH₂ with H₂, H	3.3	18. R—O with H	2–5
5. R₂C=CH—H₂C—H	1.7	12. F—CH₂ with H₂, H	4.1	19. Ar—O with H	4–7
6. R—C(=O)—CH₂—H	2.1	13. R₂C=CH—H	4.7	20. R—C(=O)—O—H	10–12
7. benzene ring—CH₂—H	2.3	14. R₂C=CR—H	5.3		

[a]Protons in boxes shaded in green undergo hydrogen bonding.

in a concerted fashion and create a **local magnetic field (B_{loc})** that opposes B_{ext}. The magnetic field that is "felt" by the nucleus, called the **effective magnetic field (B_{eff})**, is somewhat less than B_{ext}, so protons in a molecule are said to be **shielded**. This idea is captured in Equation 16-4, where B_{ext} and B_{loc}, which oppose each other, have opposite signs.

$$B_{eff} = B_{ext} + B_{loc}$$

(16-4)

Chemically distinct protons are surrounded by different electron distributions and therefore have different values of B_{loc}. Consequently, chemically distinct protons are shielded from B_{ext} to a different extent and will absorb photons at different frequencies—they will have different chemical shifts.

YOUR TURN 16.4

The following diagram (not to scale) represents the spin states of two different protons (proton 1 and proton 2) exposed to the same B_{ext}.

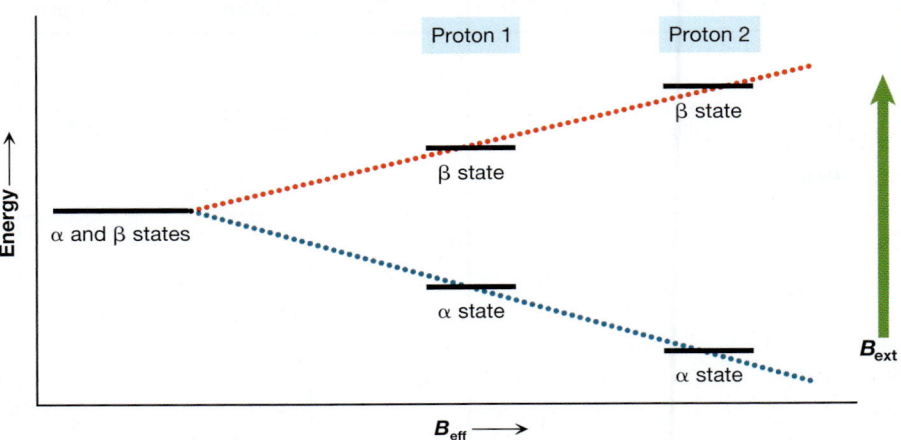

The B_{eff} for one proton differs from the B_{eff} of the other. Which proton is shielded to a greater extent? Which proton requires higher frequency radio waves to cause a spin flip?

Two major factors influence the extent to which a hydrogen nucleus is shielded: *inductive effects* and *magnetic anisotropy*. The more straightforward of the two is inductive effects (review Section 6.6e). Magnetic anisotropy, on the other hand, is a phenomenon we have not previously encountered.

16.7a Inductive Effects

The role of inductive effects on chemical shift can be seen by comparing entries 2 and 12 in Table 16-1 (RCH_2—H and FCH_2—H, respectively). In entry 2, no significantly electronegative atoms are present and the proton has a very low chemical shift of 0.9 ppm. In entry 12, the C atom to which the H atom is attached is itself bonded to a highly electronegative F atom. The presence of that F atom causes a substantial downfield shift of the protons' absorption peak, to 4.1 ppm. A similar phenomenon occurs with other highly electronegative atoms. In general, then:

A proton's absorption peak is shifted downfield (i.e., to higher chemical shift) by nearby electronegative atoms.

This downfield shift occurs because electronegative atoms are inductively *electron withdrawing*, so they remove electron density from the nearby proton. Consequently, the shielding that those electrons provide is diminished, making B_{loc} less effective at canceling B_{ext} (review Equation 16-4). The proton is said to be **deshielded**, resulting in a higher signal frequency, and thus a larger chemical shift.

The extent of this inductive deshielding depends, in part, on the electronegativity of the nearby atom. Notice in Table 16-1, for example, that the chemical shifts of the protons in FCH_3 appear downfield from those in $ClCH_3$. The F atom is more electronegative than Cl, so it more effectively deshields the CH_3 protons.

Look up the chemical shifts for the protons in FCH_3, $ClCH_3$, and $BrCH_3$ in Table 16-1 and compare them to the electronegativities of fluorine, chlorine, and bromine in Figure 1-16 (p. 16). Are the trends the same or different?

PROBLEM 16.13 In which compound will the proton chemical shift be greater, CH_3Br or CH_3I? Explain.

16.7b Magnetic Anisotropy

The second factor that affects B_{eff} is called **magnetic anisotropy**. The word *anisotropy* derives from Greek (*anisos* = unequal; *tropos* = way) and refers to a characteristic that is directionally dependent. Magnetic anisotropy, in particular, is responsible for the large chemical shift in an aromatic compound like benzene, whose protons have a chemical shift of 7.3 ppm. Inductive effects cannot explain such a large downfield shift because benzene (C_6H_6) contains no highly electronegative atoms.

Benzene's six π electrons occupy a π system that looks like two donuts—one above the plane of the ring and one below (Fig. 16-10). The B_{ext} forces those π electrons to move in a circular path parallel to the plane of the ring—a so-called **ring current** (Fig. 16-10a). The movement of those charged particles creates an additional magnetic field (Fig. 16-10b). Notice that the H atoms are located where those additional magnetic field lines (represented by the blue arrows) are in the same direction as B_{ext}, so B_{eff} is increased and the hydrogen nuclei are *deshielded*.

In Figure 16-10, identify a *local* magnetic field line oriented in the same direction as the B_{ext}.

This phenomenon applies to other aromatic rings, too, not just the benzene ring. That is:

Protons on the outside of an aromatic ring are substantially deshielded as a result of the local magnetic field induced by the ring current.

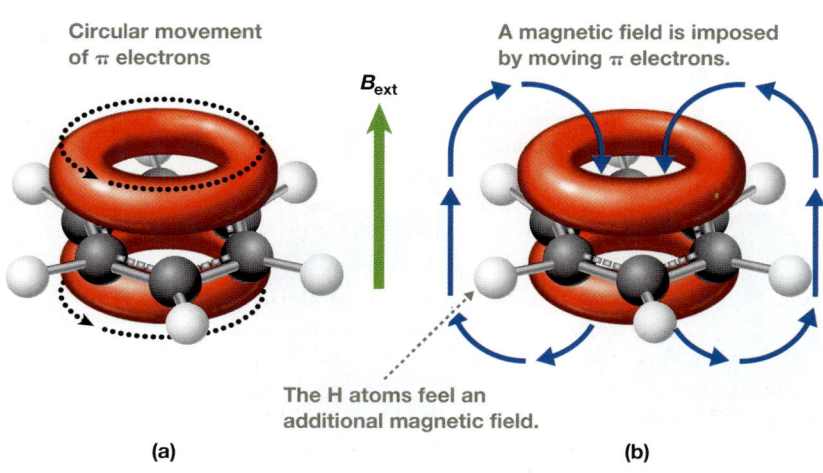

Circular movement of π electrons

B_{ext}

A magnetic field is imposed by moving π electrons.

The H atoms feel an additional magnetic field.

(a)　　　(b)

FIGURE 16-10 Deshielding of hydrogens in benzene (a) The B_{ext} forces the π electrons of benzene to move in a circular trajectory above and below the plane of the ring, creating a ring current (black dotted arrows). (b) The ring current gives rise to a local magnetic field (blue arrows) that is in the same direction as the B_{ext} at the location of the hydrogen atoms. The hydrogen nuclei are therefore deshielded so much that their signal appears well downfield of TMS.

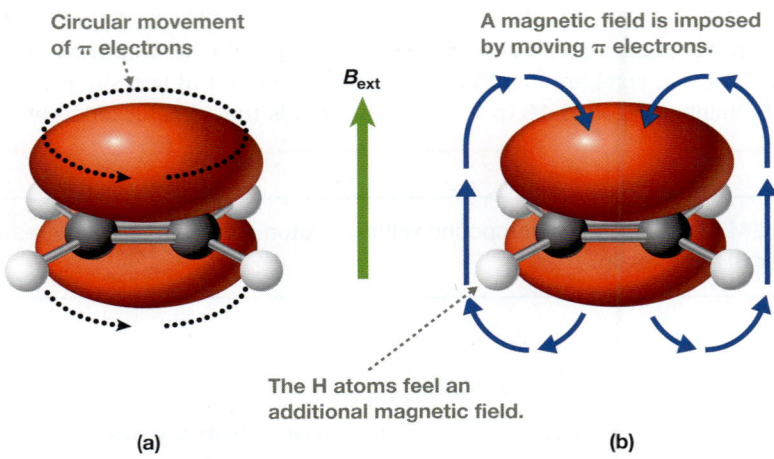

FIGURE 16-11 Deshielding of vinylic hydrogens (a) The B_{ext} forces the π electrons of a double bond to move in a circular trajectory above and below the plane of the double bond, creating a ring current (black dotted arrows). (b) The ring current gives rise to a local magnetic field (blue arrows) that is in the same direction as the B_{ext} at the location of the hydrogen atoms. The hydrogen nuclei are therefore deshielded.

Circular movement of π electrons

A magnetic field is imposed by moving π electrons.

B_{ext}

The H atoms feel an additional magnetic field.

(a)

(b)

Just like the H atoms in benzene, vinylic H atoms (C═C—H) have relatively high chemical shifts at around 5 ppm, despite not having any highly electronegative atoms. Although the π electron density is not donut shaped, the π electrons still undergo a coherent circular motion on either side of the molecular plane, causing an increase in B_{eff} where the vinylic H atom is located (Fig. 16-11). Consequently:

Magnetic anisotropy deshields hydrogen nuclei associated with simple alkenes and aldehydes.

We might also expect alkyne hydrogens (RC≡CH) to have relatively high chemical shifts, but they do not. Their chemical shifts tend to be relatively low, around 2.4 ppm, indicating that they are much less deshielded than alkene hydrogens. They are less deshielded because the electron density of a triple bond has *cylindrical symmetry* about the bonding axis (Fig. 16-12). As a result, the B_{ext} causes electrons in a triple bond to move in a circular path around the bonding axis. As shown in Figure 16-12b, an alkyne proton lies where B_{loc} opposes B_{ext}, which results in a shielding effect, not a deshielding effect.

FIGURE 16-12 Cylindrical symmetry of a triple bond The two π bonding MOs of ethyne (HC≡CH) are shown. The electron density of a triple bond has cylindrical symmetry about the bonding axis. When placed in a magnetic field, the motion of those electrons (black dotted arrow in a) generates magnetic field lines (blue arrows in b) that shield the hydrogen nuclei.

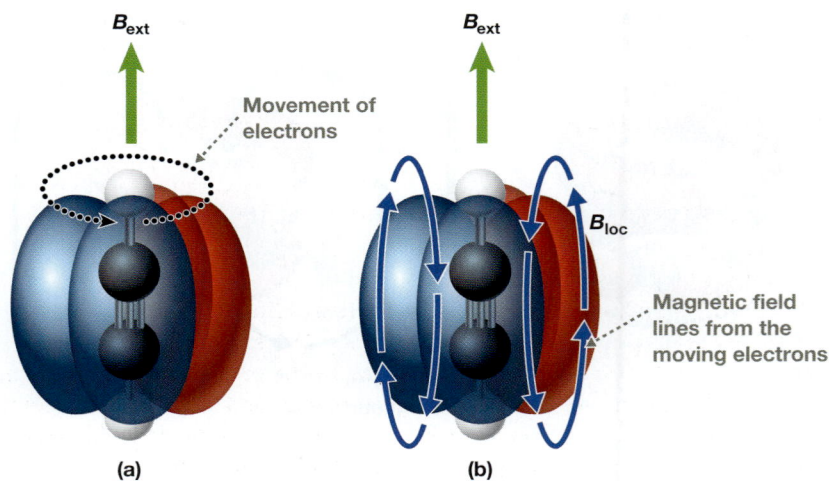

B_{ext}

B_{ext}

Movement of electrons

B_{loc}

Magnetic field lines from the moving electrons

(a)

(b)

In Figure 16-12b, draw an arrow on top of each hydrogen atom, indicating the direction of the local magnetic field there. *Hint:* Each arrow should be either with or against B_{ext}.

16.8 Trends in Chemical Shift

Table 16-1 provides chemical shifts of representative protons; it is not an exhaustive table of chemical shifts. How do we deal with protons in molecules that don't closely match the ones in the table?

We can take advantage of two useful trends. The first can be stated as follows:

The effects that cause deshielding are essentially additive.

We can see the additivity of inductive effects by examining the chemical shifts of the protons in the following molecules:

Increasing number of electronegative atoms

Increasing chemical shift

$\delta =$ 0.9 ppm   3.1 ppm   5.3 ppm

Notice that the chemical shift increases by roughly the same amount as each Cl atom is added to the C bearing the H.

Will the chemical shift of the proton be greater in CH_2Cl_2 or $CHCl_3$?

PROBLEM 16.14 Rank protons **A–C** in order of increasing chemical shift.

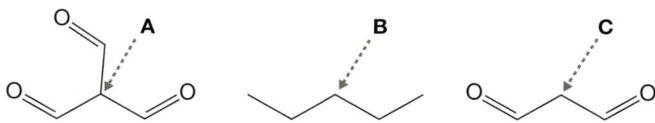

The second useful trend is as follows:

Deshielding effects fall off very rapidly with distance.

When the atom or group of atoms that causes deshielding is more than two carbon atoms away, the effect on chemical shift is generally very small.

In ethylbenzene, for example, the protons that are directly attached to the aromatic ring are deshielded the most—their chemical shift is around 7.2 ppm. The proton on C1 of the ethyl group is farther away from the aromatic ring, so it is deshielded less—its chemical shift is 2.6 ppm. And the proton on C2 is the farthest away from the ring, so it is deshielded the least. In fact, notice that the chemical shift of the proton on C2 is not much greater than 0.9 ppm, the chemical shift of a CH_3 group that is part of an alkane.

CONNECTIONS Ethylbenzene is used primarily in the production of styrene, the precursor to the polymer polystyrene. It is also a component of fluids that can be injected underground as part of a process to recover natural gas.

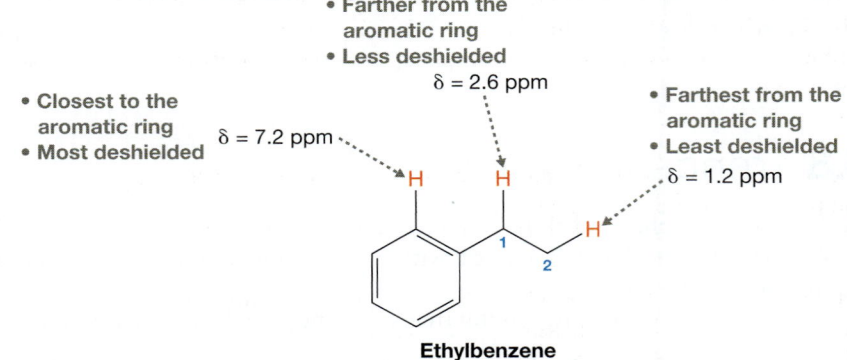

• Farther from the aromatic ring
• Less deshielded
δ = 2.6 ppm

• Closest to the aromatic ring
• Most deshielded
δ = 7.2 ppm

• Farthest from the aromatic ring
• Least deshielded
δ = 1.2 ppm

Ethylbenzene

SOLVED PROBLEM 16.15

Rank protons **D–H** in order from largest chemical shift to smallest.

Think What groups present are responsible for deshielding nearby protons? Do those groups deshield protons to the same extent? How severely do those deshielding effects fall off with distance?

Solve There are two groups present that cause significant deshielding: the C=O group (through magnetic anisotropy and inductive effects) and the Cl atom (through inductive effects only). Therefore, protons **D** and **G**, which are closest to those groups, will have the highest chemical shifts. The remaining protons will have smaller chemical shifts because deshielding falls off rapidly with distance. Of protons **D** and **G**, proton **D** will have the higher chemical shift because, as we can see from Table 16-1, the chemical shift of an aldehyde proton (H—C=O) is around 9–10 ppm, whereas that of a proton on a C that is attached to a Cl atom is around 3 ppm. Proton **E** is one additional carbon atom away from the C=O group, and protons **F** and **H** are one additional bond away from the Cl atom. Because of the greater deshielding effects by the C=O group, proton **E** will have a higher chemical shift than protons **F** or **H**. Finally, protons **F** and **H** will be deshielded similarly by the Cl atom, but because protons **F** are closer to the C=O group, they will be deshielded slightly more. The chemical shifts of these protons, then, decrease in the order: **D** > **G** > **E** > **F** > **H**.

PROBLEM 16.16 Rank protons **I–M** in order from largest chemical shift to smallest.

16.9 Integration of Signals

One of the most important features of NMR spectrometers is their ability to compute the *area under an absorption peak*, called the signal's **integration**. Signal integration is important because:

> The area under an absorption peak is proportional to the number of protons that generate that peak.

In modern instruments, the integration is done digitally, and the results can be displayed in a variety of ways.

One way to display the integration is as an **integral trace** superimposed on the spectrum, as shown in blue in Figure 16-13. The integral trace for a single peak resembles a stair step and the height of that stair step is proportional to the total area under the peak. Therefore:

> The higher an integral trace's stair step rises for a particular signal, the greater the number of protons that contribute to that signal.

With this in mind, consider the ^{1}H NMR spectrum of 1,4-dimethylbenzene (*p*-xylene) shown in Figure 16-14. The integral trace is drawn in blue.

The signal at 2.3 ppm represents the six CH_3 protons, and the signal at 7.1 ppm represents the four aromatic protons. The ratio of these numbers of protons is 6 : 4, or 1.5 : 1, which is essentially the same as the ratio of the stair-step heights.

> Verify that the integrations of the two signals in Figure 16-14 have a ratio of 1.5 : 1. To do so, use a ruler to measure the height of each stair step (e.g., in millimeters). Is the one on the right roughly 1.5 times the height of the one on the left?

PROBLEM 16.17 Which signal in this proton NMR spectrum represents the greatest number of protons? Which signal represents the fewest? Roughly how many times more protons contribute to the former signal than to the latter?

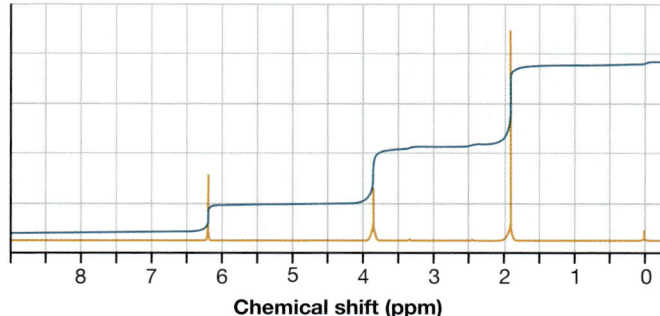

Chemical shift (ppm)

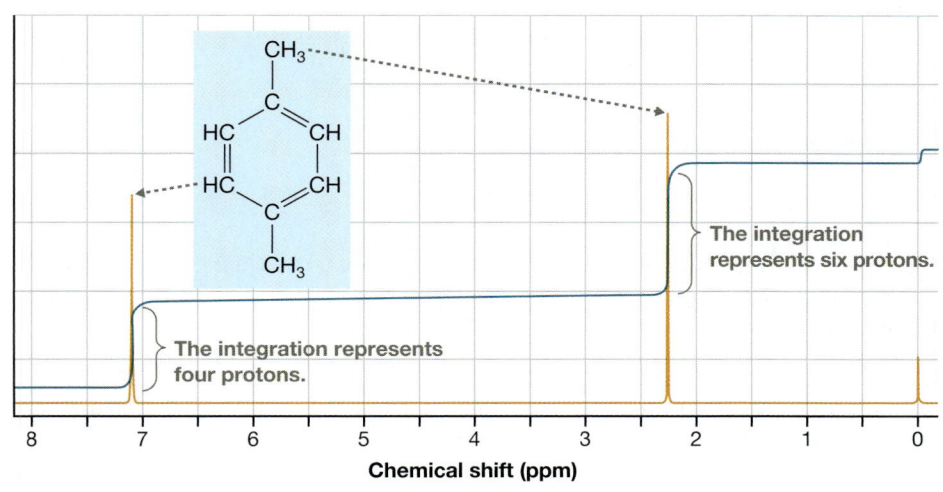

The integration represents four protons.

The integration represents six protons.

CH_3

HC   CH

HC   CH

CH_3

Chemical shift (ppm)

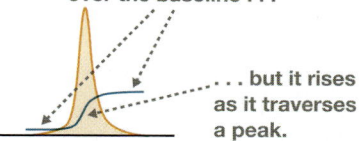

The integral trace is horizontal over the baseline . . .

. . . but it rises as it traverses a peak.

FIGURE 16-13 An integral trace
A proton signal is represented by the orange peak. The blue stair-step line is the integral trace, which represents the cumulative area under the peak (represented by the shading) going from left to right.

YOUR TURN 16.9

CONNECTIONS
1,4-Dimethylbenzene, also called *para*-xylene, is used as a precursor to terephthalic acid, which is a starting material for poly(ethylene terephthalate), or PET, a polyester from which drinking bottles are made.

FIGURE 16-14 ^{1}**H NMR spectrum of 1,4-dimethylbenzene** The integral trace, shown in blue, indicates that the ratio of CH_3 hydrogens (2.3 ppm) to aromatic hydrogens (7.1 ppm) is 1.5 : 1.

Often, it is convenient to convert the output from integration to the smallest set of whole-number ratios because the number of hydrogen atoms each signal represents *must* be a whole number. The actual number of hydrogen atoms each signal represents is either the same as that in the ratio or a multiple of it. For example, suppose that an NMR spectrum contains three signals and the stair-step heights measure 10 mm, 10 mm, and 15 mm, respectively. We can divide each height by the smallest number, yielding $1:1:1.5$. The smallest set of whole number ratios is double that, or $2:2:3$. The actual number of hydrogen atoms contributing to each signal could be 2, 2, and 3; or 4, 4, and 6; or 6, 6, and 9; and so on.

In spectral problems involving NMR, the smallest set of whole number ratios is sometimes given to you. If the integration is from a proton NMR spectrum, it is common to place an "H" after the numbers. A specific signal might be said to have an integration of 1 H, 2 H, 3 H, and so on.

PROBLEM 16.18 If the molecule giving the spectrum in Problem 16.17 contains a total of six protons, then how many protons does each signal represent? What if the molecule contains a total of 12 protons instead?

16.10 Splitting of the Signal by Spin–Spin Coupling: The *N* + 1 Rule

1,1,2-Trichloroethane (Cl_2CHCH_2Cl) has only two chemically distinct H atoms, so its ^1H NMR spectrum (Fig. 16-15) exhibits two signals. Notice in the magnifications, however, that each signal is *split* into more than one peak. Why does this happen, and how can we predict these splitting patterns? We answer these questions here in Section 16.10.

16.10a Spin–Spin Coupling and the *N* + 1 Rule

This **signal splitting** occurs because every proton behaves like a tiny bar magnet that can be in either the α state or the β state. The magnetic field that any particular proton experiences (i.e., B_{eff}) can therefore be altered by the magnetic fields of nearby protons, and these alterations can affect the proton's apparent chemical shift. In these

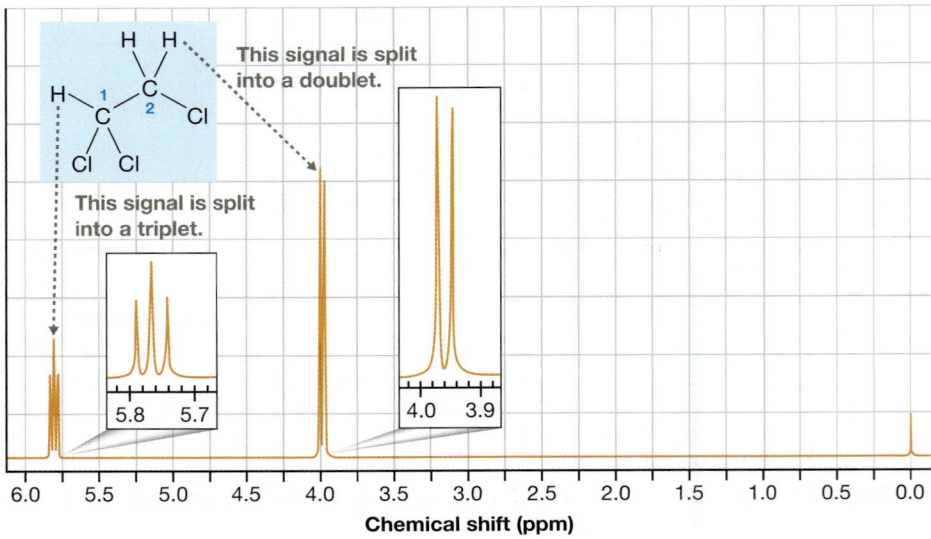

FIGURE 16-15 Proton NMR spectrum of 1,1,2-trichloroethane The signal at δ ≈ 4.0 ppm is split into a doublet and the one at δ ≈ 5.8 ppm is split into a triplet. Magnifications (insets) show these splitting patterns more clearly.

situations, protons are said to interact with each other through **spin–spin coupling**, and the protons involved are said to be **coupled** to each other. In general:

> Protons that are coupled are chemically distinct from each other and are generally separated by three or fewer single bonds.

(In some cases, protons that are separated by more than three bonds can exhibit coupling. This situation is discussed in Section 16.11.)

Relatively simple splitting patterns like the ones in Figure 16-15 can be predicted by the $N + 1$ **rule**:

> If proton **A** is coupled to proton **B** and there are N equivalent protons **B**, then the signal from proton **A** is split into $N + 1$ peaks.

It is important to keep in mind that N is not the number of protons responsible for generating a particular signal, but rather it is the number of *coupled* protons.

Various types of splitting patterns that arise from the $N + 1$ rule are summarized in Table 16-2. When there are zero, one, two, three, or four coupled protons, the signal appears as a **singlet (s)**, **doublet (d)**, **triplet (t)**, **quartet (q)**, or **quintet (qn)**, respectively. With a significantly greater number of coupled protons, the splitting pattern is described as a **multiplet (m)**.

Notice how the splitting patterns in Figure 16-15 are consistent with the $N + 1$ rule. The H that is bonded to C1 is coupled to the two H atoms bonded to C2, so $N = 2$ and $N + 1 = 3$. Therefore, the signal that arises from the proton on C1 ($\delta \approx 5.8$ ppm) is a triplet. The H atoms that are bonded to C2 are coupled to the lone H bonded to C1, so $N = 1$ and $N + 1 = 2$ for the corresponding signal ($\delta \approx 4.0$), giving rise to a doublet.

TABLE 16-2 Common Splitting Patterns and Relative Peak Heights

Number of Coupled Protons, N	Number of Peaks in Splitting Pattern, $N + 1$	Description of Splitting Pattern	Relative Peak Heights
0	1	Singlet (s)	1
1	2	Doublet (d)	1:1
2	3	Triplet (t)	1:2:1
3	4	Quartet (q)	1:3:3:1
4	5	Quintet (qn)	1:4:6:4:1
Several	Several	Multiplet (m)	—

YOUR TURN 16.10

In Figure 16-15, circle the proton(s) responsible for splitting the signal at ~4.0 ppm and box the proton(s) responsible for splitting the signal at ~5.8 ppm.

YOUR TURN 16.11

How many protons are coupled to the one that gives rise to each of the signals shown here? Explain.

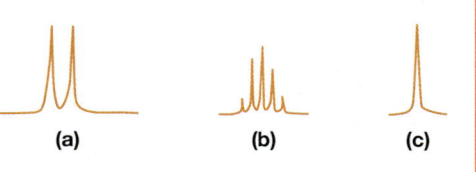

(a)   (b)   (c)

According to the $N + 1$ rule, the protons that are responsible for splitting a signal must be distinct from the protons that give rise to the signal. In other words:

> Chemically equivalent protons do not show the effects of being coupled together.

This is why the six H atoms of benzene give rise to a singlet (Fig. 16-5, p. 777). If the adjacent protons split each other's signal, then benzene's ^{1}H NMR spectrum would be quite a bit more complex.

SOLVED PROBLEM 16.19

An unknown compound produces a ^{1}H NMR spectrum that has just two signals, each of which is a singlet. Which of compounds **1–4** could the unknown be?

Think How many chemically distinct protons are indicated by the spectrum? Which of those protons are coupled to other protons that are chemically distinct from themselves?

Solve The chemically distinct protons are indicated as **A–G** in the structures shown here. The unknown has two chemically distinct protons, because there are two signals in the spectrum. This rules out molecule **4**, because all six of its protons **G** are chemically equivalent. For the remaining molecules, we must consider the splitting patterns of their proton signals.

In **1**, proton **A** has two coupled protons **B**, so N = 2 and N + 1 =3, making signal **A** a triplet. Likewise, proton **B** has two coupled protons **A** (N = 2), making signal **B** a triplet. In **2**, proton **C** has three coupled protons **D** (N = 3), making signal **C** a quartet. Proton **D** has two coupled protons **C** (N = 2), so signal **D** is a triplet. Finally, in **3**, protons **E** and **F** each have no coupled protons (N = 0), so each signal is a singlet. Therefore, the spectrum is consistent with **3**, but not with any of the other molecules.

PROBLEM 16.20 Determine the number of signals that would be generated in the ^{1}H NMR spectrum of each of the following compounds, and predict the splitting pattern of each signal.

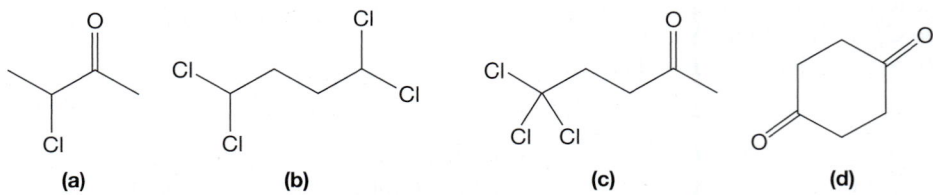

The magnetic field from an α proton causes B_{eff} to increase slightly, which increases δ.

The magnetic field from a β proton causes B_{eff} to decrease slightly, which decreases δ.

One coupled proton

B_{ext}

← Increasing chemical shift

The origin of the N + 1 rule can be understood by considering the various magnetic fields (B_{loc}) generated by a set of protons responsible for splitting a particular signal. Figure 16-16, for example, shows how a doublet arises when a signal is split by one coupled proton. The peak at the greater chemical shift (on the left) is produced when the coupled proton is in the α state, and the peak at the smaller chemical shift is produced when the coupled proton is in the β state. When in the α state, the coupled proton exerts a magnetic field that increases B_{eff} for the proton giving rise to the signal. When in the β state, the magnetic field from the coupled β proton decreases B_{eff}.

Figure 16-17 shows how a triplet arises when a signal is split by two coupled protons. The peak on the left is produced when both coupled protons are in the α state,

FIGURE 16-16 The origin of a doublet A doublet arises when a signal is split by one coupled proton. The peak at the greater chemical shift (on the left) is due to the coupled proton in the α state, which causes B_{eff} to increase. The peak at lower chemical shift (on the right) is due to the coupled proton in the β state, which decreases B_{eff}.

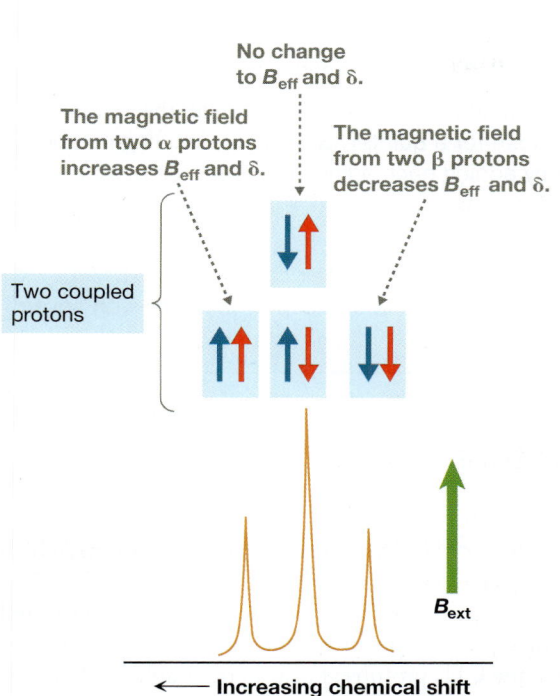

The magnetic field from two α protons increases B_{eff} and δ.

No change to B_{eff} and δ.

The magnetic field from two β protons decreases B_{eff} and δ.

Two coupled protons

B_{ext}

← **Increasing chemical shift**

The magnetic field from three α protons increases B_{eff} and δ.

The magnetic field from one excess α proton increases B_{eff} and δ.

The magnetic field from one excess β proton decreases B_{eff} and δ.

The magnetic field from three β protons decreases B_{eff} and δ.

Three coupled protons

B_{ext}

← **Increasing chemical shift**

FIGURE 16-17 The origin of a triplet A triplet arises when a signal is split by two coupled protons. The peak on the far left is due to both coupled protons in the α state, the one on the right is due to both coupled protons in the β state, and the one in the middle is due to one coupled proton in each state.

FIGURE 16-18 The origin of a quartet A quartet arises when a signal is split by three coupled protons. From left to right, the four peaks are due to the three coupled protons in the following states: all three α, two α and one β, one α and two β, and all three β.

which leads to an increase in B_{eff}. The peak on the right is produced when both coupled protons are in the β state, which leads to a decrease in B_{eff}. And the tall peak in the middle is produced when one coupled proton is in the α state and one is in the β state, resulting in no change to B_{eff}.

Finally, Figure 16-18 shows how a quartet arises when a signal is split by three coupled protons, which can produce four different magnetic fields: one in which all three coupled protons are α, a second in which there is one excess α proton, a third in which there is one excess β proton, and a fourth in which all three protons are β. These scenarios correspond to decreasing magnitudes of B_{eff} and, therefore, decreasing chemical shift.

Notice in Table 16-2 (p. 793) that the peaks in each splitting pattern appear in characteristic *height ratios*. The peak heights of a doublet are in a 1:1 ratio. Those of a triplet are in a 1:2:1 ratio, and those of a quartet are in a 1:3:3:1 ratio. The reason for these patterns has to do with the fact that each spin combination of a set of coupled protons is nearly equally likely. Therefore, the more spin combinations that contribute to an individual peak in a splitting pattern, the larger the peak. One spin combination contributes to each peak of a doublet (Fig. 16-16), which is why the peaks are roughly equal in height. For a triplet (Fig. 16-17), one spin combination contributes to each of the outer peaks, and two contribute to the middle peak, which is why the middle peak has roughly twice the height. And for a quartet (Fig. 16-18), one spin combination contributes to each outer peak, and three contribute to the inner peaks, giving rise to the 1:3:3:1 height ratio.

These ratios describing the relative peak heights in splitting patterns form a pattern known as **Pascal's triangle** (Fig. 16-19). The first and last numbers in each row of numbers is 1. Each of the remaining numbers in a particular row is the sum of the two closest numbers in the row above it. For example, consider the relative peak heights for a quintet, 1:4:6:4:1. As shown in Figure 16-19, 4 is the sum of 1 and 3, the two closest numbers in the previous line, and 6 is the sum of the two 3s that appear in the line above it.

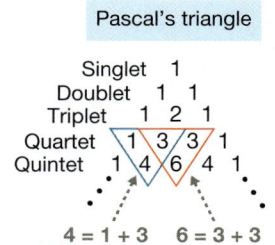

Pascal's triangle

Singlet	1	
Doublet	1 1	
Triplet	1 2 1	
Quartet	1 3 3 1	
Quintet	1 4 6 4 1	

4 = 1 + 3   6 = 3 + 3

FIGURE 16-19 Pascal's triangle The numbers in each row correspond to the height ratios in various splitting patterns.

SOLVED PROBLEM 16.21

In what ratios would the peaks of a sextet (a signal with six peaks) appear?

Think What should the first and last numbers in the set of ratios be? How is each remaining number derived from the line above it in Pascal's triangle?

Solve The first and last numbers should be 1. In Pascal's triangle, the ratios for a sextet would appear just below those for a quintet. So, each remaining number in the sextet's ratios is computed by adding each adjacent pair of numbers in the quintet's ratios. That is, $1 + 4 = 5$, $4 + 6 = 10$, $6 + 4 = 10$, and $4 + 1 = 5$. Therefore, the ratios are $1:5:10:10:5:1$.

PROBLEM 16.22 In what ratios would the peaks of a septet (a signal with seven peaks) appear?

16.10b OH and NH Protons Often Don't Exhibit Coupling

The $N + 1$ rule seems to be violated in the ^1H NMR spectrum of ethanamine, which is shown in Figure 16-20. The amino (NH_2) protons ($\delta \approx 3.7$ ppm) and the CH_2 protons ($\delta \approx 2.7$ ppm) don't appear to be coupled together, even though they are separated by only three bonds. The NH_2 protons give rise to a broad singlet, so they appear to be uncoupled to any protons. Moreover, the CH_2 protons appear to be coupled only to the three CH_3 protons, making the signal a quartet. This phenomenon is not limited to ethanamine, but rather occurs commonly with protons bonded to oxygen or nitrogen.

> O—H and N—H protons generally give rise to broad singlets, and thus do not appear to couple to other protons.

The reason for this is rapid *proton exchange*, which we first encountered in Section 15.5a. Protons on nitrogen and oxygen undergo extensive hydrogen bonding, and thus can hop rapidly from one nitrogen or oxygen atom to another. This proton exchange is usually much faster than the time it takes to acquire an NMR spectrum,

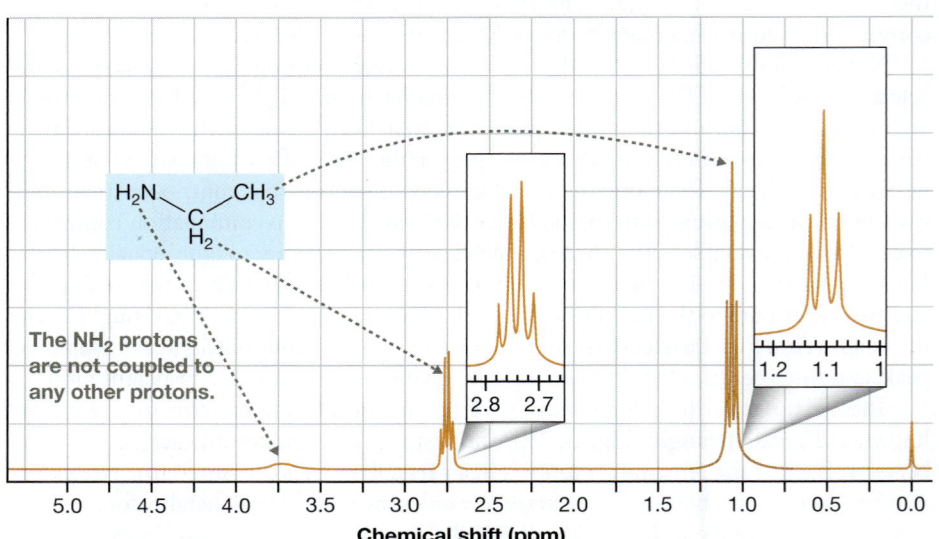

FIGURE 16-20 Proton NMR spectrum of ethanamine The NH_2 protons and the CH_2 protons do not appear to be coupled together.

so the signal that we observe reflects the *average* magnetic environment from all the molecules on which it spends time, resulting in an unsplit singlet. Moreover, protons bonded to oxygen or nitrogen have less well-defined chemical environments than protons bonded to carbon, so they tend to absorb a larger range of frequencies, which also broadens their signals.

We can take advantage of this proton exchange to identify O—H or N—H signals. If a sample is treated with D_2O, the sample's O—H and N—H protons are rapidly replaced by D (i.e., by ^2H), whose signals do not appear in ^1H NMR spectra (Section 16.6a). Thus, the intensities of the O—H and N—H signals decrease dramatically, or disappear entirely.

PROBLEM 16.23 What similarities would you expect between proton NMR spectra of compounds **A** and **B**? What differences would you expect? How would each spectrum be affected by the addition of D_2O?

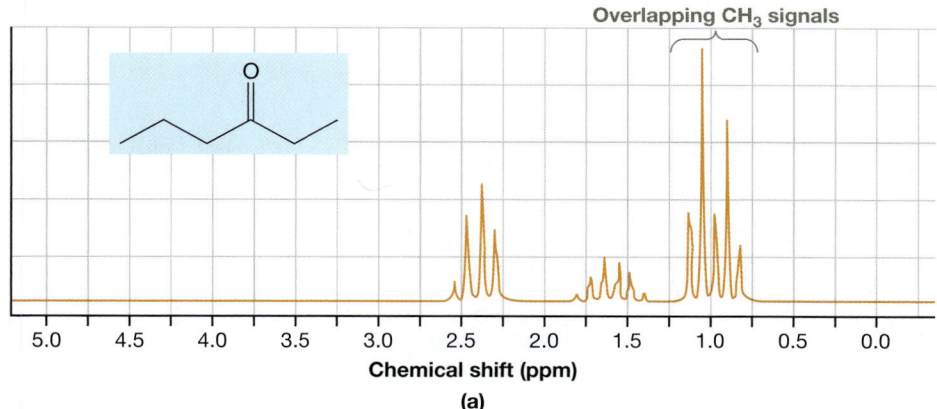

16.11 Coupling Constants and Signal Resolution

Figure 16-21 shows the ^1H NMR spectrum of hexan-3-one taken with a 90-MHz (Fig. 16-21a) and a 300-MHz (Fig. 16-21b) spectrometer. Based on the structure of

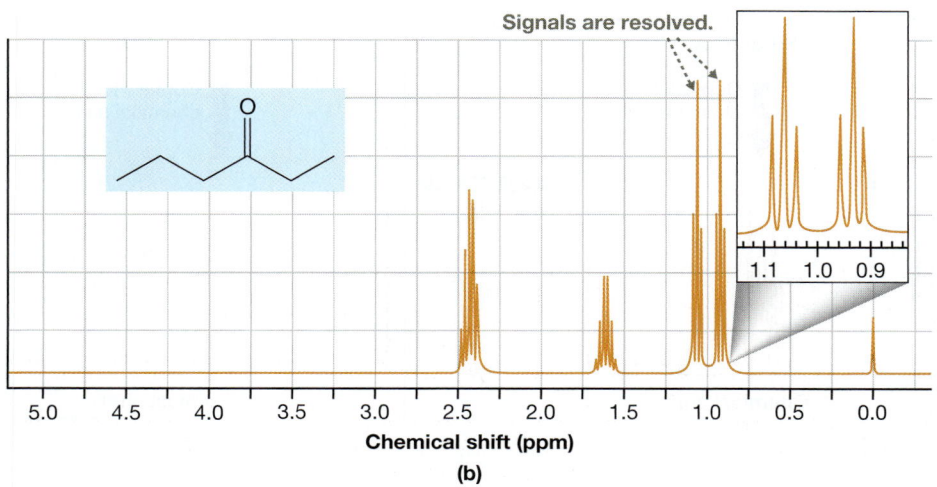

FIGURE 16-21 Proton NMR spectra of hexan-3-one (a) Spectrum taken with a 90-MHz spectrometer. (b) Spectrum taken with a 300-MHz spectrometer. The stronger magnet resolves the signals much better.

hexan-3-one, the two sets of CH_3 hydrogens are chemically distinct and should give rise to two different signals, each a triplet. However, the spectrum taken with the 90-MHz spectrometer does not clearly show this. Instead, the two signals overlap, and are therefore not **resolved** from each other. In cases such as this, interpreting the spectrum can be a more formidable task. Taken with the 300-MHz spectrometer, on the other hand, the spectrum clearly shows two distinct triplets.

The 90-MHz and 300-MHz ^1H NMR spectra of hexan-3-one demonstrate the following important point:

> The signals in an NMR spectrum tend to be better resolved as the magnet of the spectrometer becomes more powerful.

This is true primarily because the magnetic fields that cause signal splitting—that is, those produced by the various spin combinations of nearby nuclei—are independent of the B_{ext}. Therefore, regardless of the strength of B_{ext}, the *frequency difference* between peaks of a split signal, called the **coupling constant (J)**, remains unchanged. By contrast, recall from Figure 16-3 (p. 774) that for an unsplit signal, increasing B_{ext} causes an increase in ΔE_{spin}, and hence an increase in the signal frequency.

With these two concepts in mind, we can now see how the magnet strength affects the quality of the spectrum. At the top of Figure 16-22a, two unresolved signals are plotted, with frequency on the x axis. As the strength of the magnet increases (Fig. 16-22, top), the centers of two signals move apart, but the frequency separation between peaks of each triplet remains essentially the same. Thus, the signals are resolved from each other.

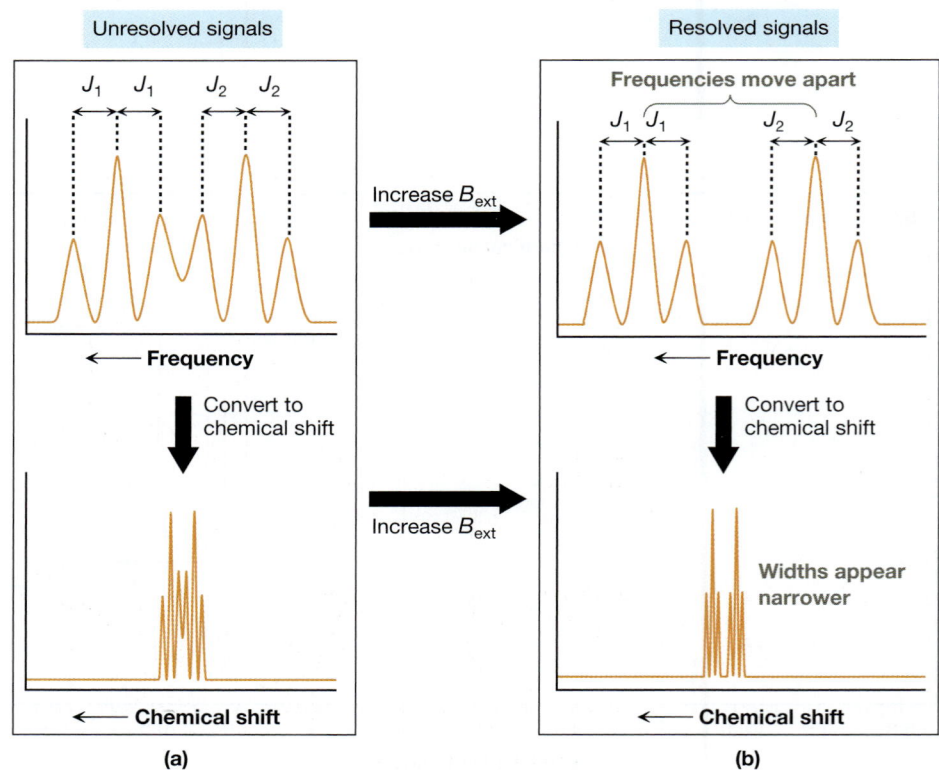

FIGURE 16-22 Dependence of signal resolution on the spectrometer's magnet strength (a) A representation of two triplets overlapping in an NMR spectrum. The figure at the top has frequency on the x axis, whereas the figure at the bottom has chemical shift on the x axis. J_1 and J_2 are the coupling constants for the respective signals. (b) NMR spectrum of the same compound using a spectrometer with a higher magnetic field strength. The frequency separation between the two signals is increased, but the separation between peaks within the same signal (i.e., the coupling constants J_1 and J_2) remains the same. When we convert to chemical shift, therefore, the signals appear narrower.

Recall that to convert signal frequencies to chemical shifts (which are independent of the instrument's magnet strength), we must divide those frequencies by the instrument's *operating frequency* (Equation 16-3, p. 783). This is shown at the bottom of Figure 16-22. When we do this, the signal width (in units of ppm) becomes narrower with the stronger magnet.

Coupling constants can be used to interpret a ^{1}H NMR spectrum if you remember the following:

> Signals of protons that are coupled together exhibit the same coupling constant.

This is because the energetic effects of one magnetic dipole on a second are the same as the energetic effects of the second magnetic dipole on the first.

For example, in the spectrum of 1-bromo-4-ethylbenzene (Fig. 16-23), there are four distinguishable signals. Signals **A** and **B** each have two major peaks, the centers of which are separated by 8.5 Hz, so each signal has an apparent coupling constant of 8.5 Hz. The quartet (signal **C**) and the triplet (signal **D**), on the other hand, have a coupling constant of 7.6 Hz. This suggests that the protons that give rise to signal **A** are coupled to the protons that give rise to signal **B**. Likewise, the protons that give rise to signal **C** are coupled to those that give rise to signal **D**. Indeed, looking at the structure, we see that protons **A** and **B** are coupled together, and protons **C** and **D** are coupled together. (In actuality, the coupling constants that describe signals **A** and **B** are more complicated because of long-range coupling, described later in this section.)

YOUR TURN 16.12

In Figure 16-23, do the coupling constants suggest that protons **B** and **C** are coupled? Does your answer agree with the molecule's structure?

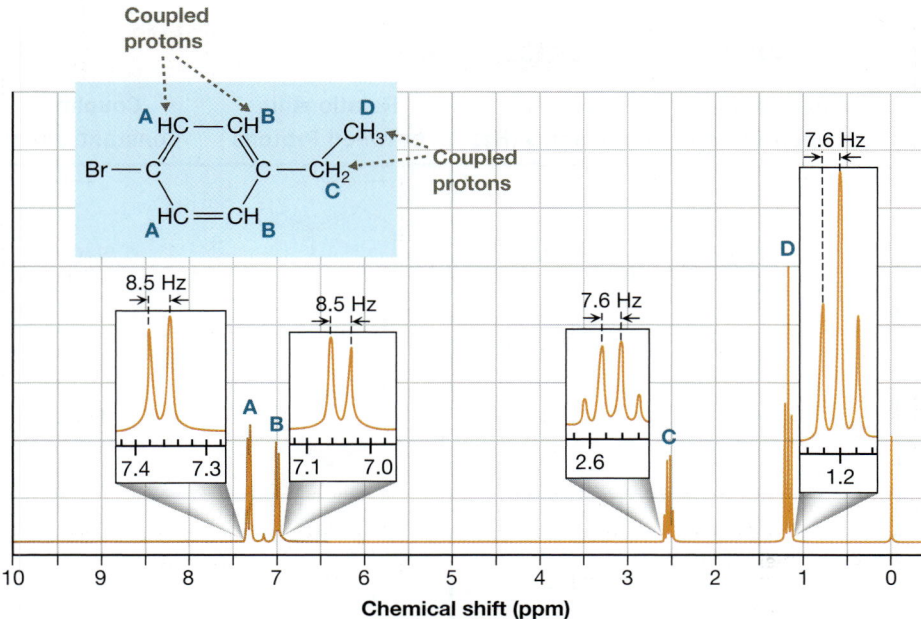

FIGURE 16-23 Proton NMR spectrum of 1-bromo-4-ethylbenzene Protons **A** and **B** are coupled together and their signals have the same apparent coupling constant (i.e., $J = 8.5$ Hz). Protons **C** and **D** are coupled together, too, and their signals have the same coupling constant (i.e., $J = 7.6$ Hz).

YOUR TURN 16.13

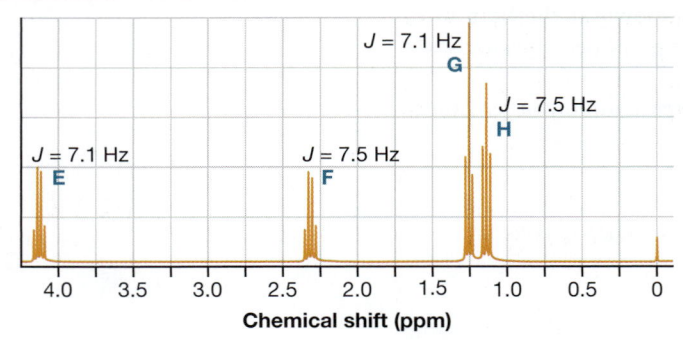

In the ^{1}H NMR spectrum shown here, identify each pair of signals that represent coupled protons.

The *magnitude* of the coupling constant can provide structural information about the protons coupled together. As shown in Table 16-3, coupling constants generally range up to about 18 Hz.

There are two examples listed in Table 16-3 in which protons separated by more than three bonds exhibit weak coupling. These situations are said to be **long-range coupling**, and they tend to occur when the protons are connected by a rigid carbon framework, established by π bonds.

PROBLEM 16.24 Using the ^{1}H NMR spectrum provided, along with the relative frequencies of absorption noted in Hz, determine which signal or signals represent protons coupled to the protons with δ = 4.1 ppm.

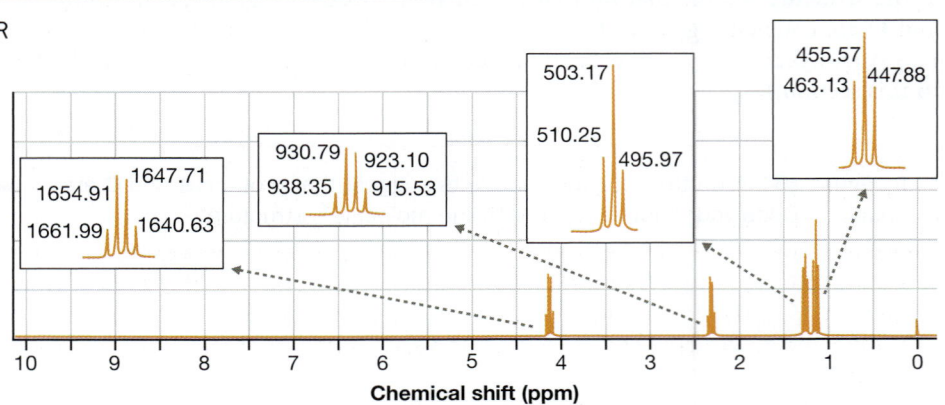

TABLE 16-3 Commonly Encountered Coupling Constants

Relationship between Protons	Coupling Constant, J (Hz)	Relationship between Protons	Coupling Constant, J (Hz)	Relationship between Protons	Coupling Constant, J (Hz)
H–C–C–H	6–9	C=C (H cis type, both H same side)	13–18	benzene ortho H	6–9
H–C–C–C–H	~0	C=C (geminal)	1–3	benzene meta H	1–3
C=C (H H same carbon side)	7–12	aldehyde C(=O)–C–H	1–3	benzene para H	0–1

16.12 Complex Signal Splitting

The signal splitting we have discussed so far has been due to the coupling of a given proton to only one chemically distinct type of proton. It is common, however, for proton signals to be coupled to two or more chemically distinct types of protons. One of the simplest examples is bromoethene, whose ^1H NMR spectrum is shown in Figure 16-24. The $N + 1$ rule does not *seem* to hold for molecules of this type.

YOUR TURN **16.14**

> According to the $N + 1$ rule, what should the splitting pattern be for the signal of the H_C proton in Figure 16-24 if the two protons to which H_C is coupled (H_A and H_B) were equivalent? Which signals, if any, are consistent with that splitting pattern?

We observe **complex splitting** patterns in these kinds of situations because the strength of spin–spin coupling is different for each type of chemically distinct proton. This results in different coupling constants. When a given proton is coupled to two types of protons that are distinct from each other, therefore, two independent splitting patterns appear simultaneously.

In the case of bromoethene, there are three chemically distinct types of protons, and each is coupled to the other two. It helps to view the coupling to each chemically distinct proton *one at a time* and to construct what is called a **splitting diagram**, an example of which is shown in Figure 16-25 (next page) for the H_B proton ($\delta = 5.97$ ppm).

Figure 16-25a considers first the splitting of the H_B signal by proton H_C, temporarily ignoring the splitting by proton H_A. H_B is coupled to H_C by a coupling constant of 7.1 Hz, which would cause the unsplit signal (Fig. 16-25a, top) to be split into a doublet of peaks separated by 7.1 Hz (Fig. 16-25a, middle). Now, if we take into account the coupling between H_B and H_A, for which the coupling constant is 1.8 Hz, then each of the peaks in the doublet brought about by the first coupling is split again into a doublet, this time with the peaks separated by 1.8 Hz. The result is a **doublet of doublets** (Fig. 16-25a, bottom), consistent with our observations in the spectrum.

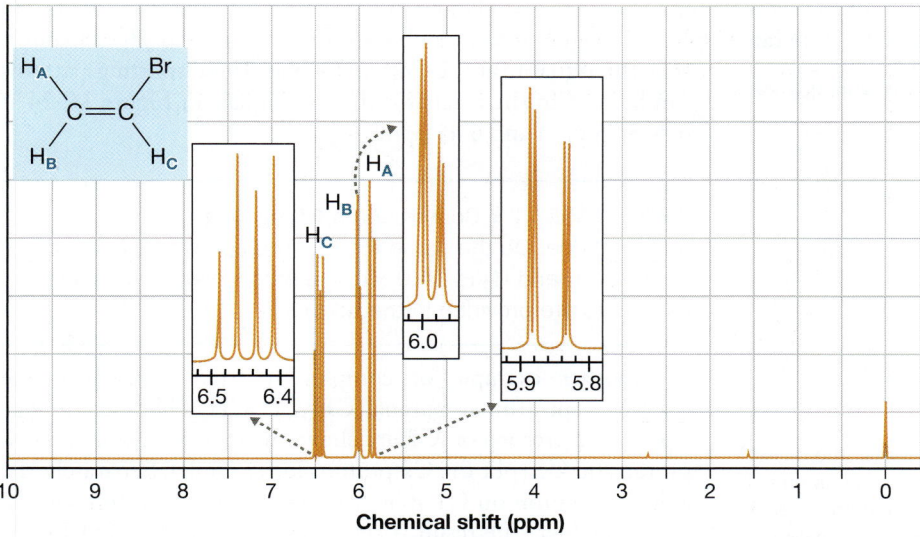

FIGURE 16-24 Proton NMR spectrum of bromoethene The spectrum would be less complicated if H_A and H_B were equivalent.

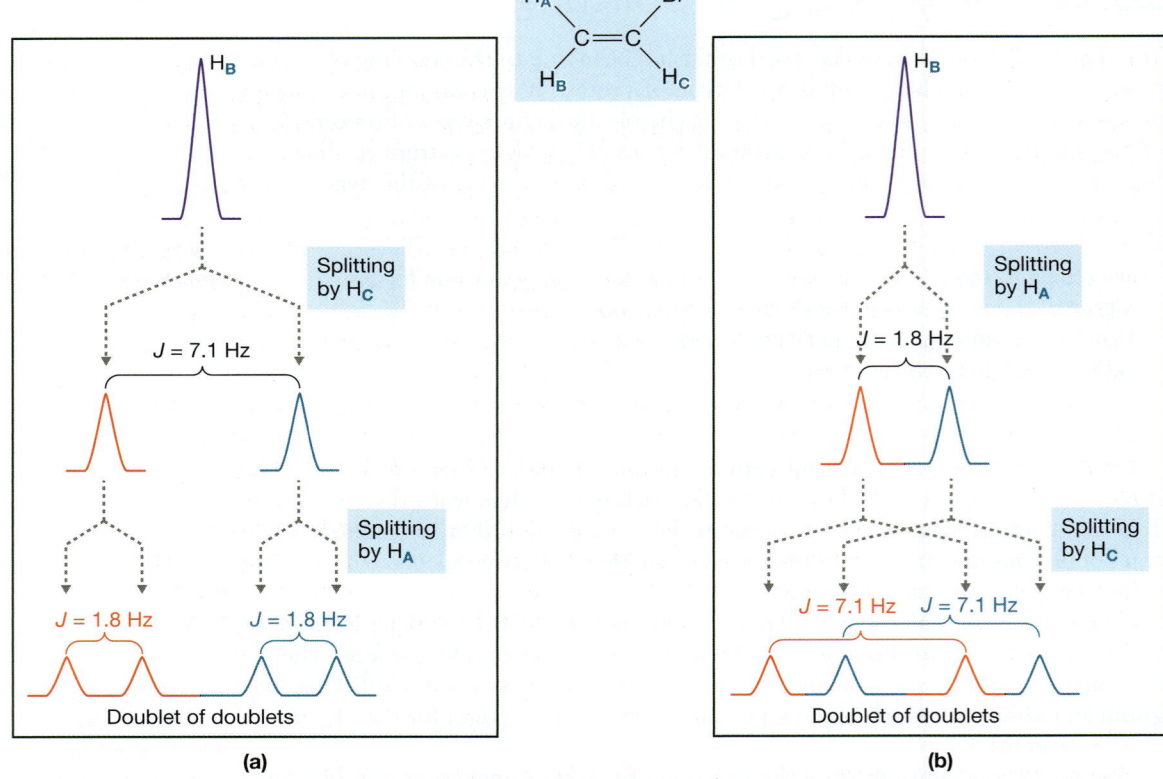

FIGURE 16-25 Splitting diagram for proton B in bromoethene (a) The signal from proton **B** is split first by proton **C** and then by proton **A**. (b) The signal from proton **B** is split first by proton **A** and then by proton **C**. The peak locations in the resulting doublet of doublets are independent of the order in which we think about the splitting taking place.

The same splitting pattern, a doublet of doublets, is obtained if we reverse the order in which we perform the splitting. As shown in Figure 16-25b, H_B's signal is first split by H_A, and the resulting peaks are then split by H_C. It is generally easier, however, to construct these splitting diagrams by working in order from largest coupling constant to smallest.

For the same reasons, the signals of H_A and H_C are each a doublet of doublets. H_B and H_C separately split H_A's signal into a doublet. Likewise, H_A and H_B separately split the signal from H_C into a doublet. These splitting patterns can be seen in the insets for the two signals in Figure 16-24 at $\delta = 5.85$ ppm and 6.45 ppm.

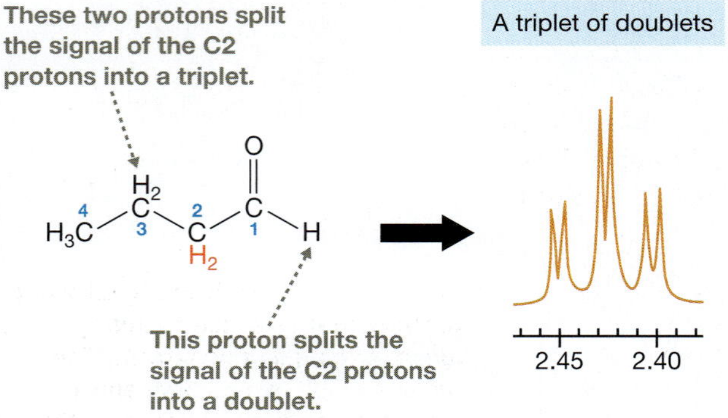

PROBLEM 16.25 Construct a splitting diagram for H_A in bromoethene. Do the same for H_C. (The coupling constant between H_A and H_C is 14.9 Hz. The other relevant coupling constants are provided in the text.)

Another example of complex splitting appears in the ^{1}H NMR spectrum of butanal, $CH_3CH_2CH_2CH{=}O$. The signal for the protons on C2 are shown in Figure 16-26. The two protons on C3 split the C2 protons into a triplet. Separately, the aldehyde proton on C1 splits the signal for the protons on C2 into a doublet. The result is six peaks in a 1:1:2:2:1:1 height ratio, or a **triplet of doublets**.

These two protons split the signal of the C2 protons into a triplet.

A triplet of doublets

This proton splits the signal of the C2 protons into a doublet.

FIGURE 16-26 Origin of a triplet of doublets The signal from the protons on C2 (in red) is split by the protons on C3 into a triplet, and each of the three peaks is separately split by the proton on C1 into a doublet. The result is a triplet of doublets, as shown on the right.

Magnetic Resonance Imaging

The figure shown here is a magnetic resonance imaging (MRI) scan of a human brain. As you may know, MRI is a valuable, noninvasive tool used in medicine to analyze tissues, and it is particularly useful for diagnosing cancer. You might not know, however, that MRI is based on the same fundamental concepts that we have applied toward NMR spectroscopy.

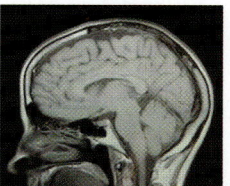

In the chamber of an MRI scanner (see photo on p. 771), a patient is exposed to a very strong magnetic field and subjected to pulses of RF radiation. As we learned here in Chapter 16, a strong magnetic field separates the nuclei of hydrogen atoms into different energy levels, which allows them to absorb RF radiation. In the body, the hydrogen atoms from water are principally responsible for absorption of that RF radiation because water is so abundant in our bodies. This RF absorption causes spin flips, producing a nonequilibrium distribution of nuclear spin. When the RF pulse is turned off, those nuclei produce an RF signal as they relax back into their equilibrium distribution, and that signal is detected. The MRI image that is generated uses gray scale to depict the signal intensity at various locations in space—the higher the water content, the stronger the signal and the brighter the spot. Therefore, an MRI scan can be thought of as a map of water density in the body.

Additional information obtained from the signal is the time it takes for hydrogen nuclei to relax back into an equilibrium distribution of spin states—that is, the relaxation time. This relaxation time is affected by the specific environment in which hydrogen nuclei are found, which depends, in turn, on the type of tissue in which the water is located. As a result, different types of tissues can be mapped differently in an MRI scan.

YOUR TURN **16.15**

How would you describe the splitting pattern for the CH₂ proton signal in propanal? Sketch the signal, which should be similar to the one in Figure 16-26.

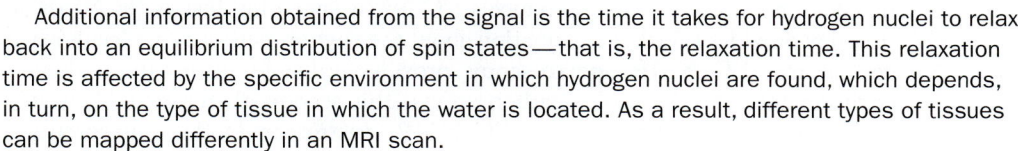

CONNECTIONS Propanal (Your Turn 16.15) is used to synthesize trimethylolethane, $CH_3C(CH_2OH)_3$, which is important in the manufacture of alkyd resins.

Sometimes, when two chemically distinct types of protons split the signal of the same proton, the coupling constants for the two different interactions are very nearly equal. When this is the case, the resulting splitting pattern is essentially the same as we would obtain by treating the two chemically distinct types of hydrogens as equivalent and using the $N + 1$ rule. For example, the protons highlighted on C2 in Figure 16-27 are flanked by five protons of two chemically distinct types. Theoretically, then, the signal of the protons on C2 are split twice—once into a triplet by the CH_2 protons on C1 and once into a quartet by the CH_3 protons on C3. In theory, then, this should give a **quartet of triplets** (or a **triplet of quartets**)—12 peaks in all—for the protons on C2. In actuality, however, the protons on C2 in propylbenzene give rise to a signal that appears to be split into a sextet (Fig. 16-27, right),

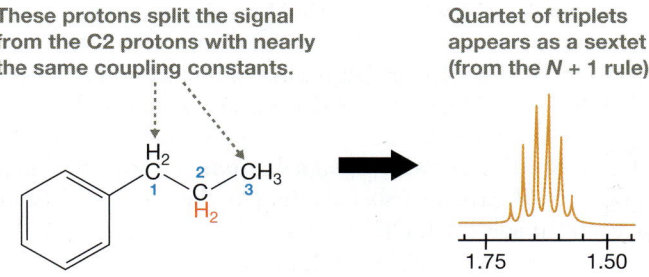

FIGURE 16-27 Splitting by distinct protons that give rise to nearly identical coupling constants The protons on C2 (in red) have about the same coupling constant with the protons on C1 as they do with the protons on C3. As a result, what is theoretically a quartet of triplets (or a triplet of quartets) actually appears as a sextet.

consistent with the $N + 1$ rule when $N = 5$. Because of the nearly identical coupling constants, some of the peaks that would normally result from such complex splitting overlap instead, thus merging into a single peak (see Problem 16.26).

PROBLEM 16.26 Construct a splitting diagram similar to the one in Figure 16-25 for a quartet of triplets in which both the quartet and the triplet have the same coupling constant.

16.13 ^{13}C NMR Spectroscopy

We have focused so far on ^{1}H NMR spectroscopy, but the nucleus of the ^{13}C isotope (six protons, seven neutrons) can have spin states of $+\frac{1}{2}$ and $-\frac{1}{2}$ au, too, just like ^{1}H. As a result, ^{13}C nuclei can absorb radiation in the RF region when placed in a strong magnetic field, and can thus produce a ^{13}C NMR spectrum. Just as ^{1}H NMR spectroscopy provides valuable information about the environments of a molecule's hydrogen atoms, **^{13}C NMR spectroscopy** provides valuable information about the environments of a molecule's carbon atoms. This section, therefore, is dedicated to explaining some of the details of ^{13}C NMR spectroscopy and to interpreting ^{13}C NMR spectra.

16.13a The ^{13}C NMR Signal

Because the ^{1}H and ^{13}C nuclei each have the same available spin states, the two types of spectroscopy have a variety of characteristics in common. In ^{13}C NMR spectroscopy, however, the frequency of the RF radiation that is used to irradiate a sample is significantly lower than in ^{1}H NMR spectroscopy, because the *gyromagnetic ratio* (γ) of a ^{13}C nucleus is about one-fourth that of a proton. This ensures that proton signals do not appear on carbon spectra, and vice versa.

One of the challenges of ^{13}C NMR spectroscopy is the low natural abundance of the ^{13}C isotope, which is only 1.1%; the remaining 98.9% of carbon is ^{12}C, which is NMR inactive. This results in poor signal strength, yielding noisy spectra, as shown in the ^{13}C NMR spectrum of 1-chloropropane (Fig. 16-28a). Therefore, in ^{13}C NMR spectroscopy, obtaining a good quality spectrum often requires averaging a large number of acquisitions—sometimes thousands. Figure 16-28b shows the improvement that this kind of averaging can produce.

Just as a ^{1}H NMR spectrum can tell us the number of chemically distinct protons in a molecule, a ^{13}C NMR spectrum can tell us the number of chemically distinct carbons.

Every chemically distinct carbon atom produces one ^{13}C NMR signal.

In 1-chloropropane, there are three chemically distinct carbon atoms, which give rise to three ^{13}C NMR signals (Fig. 16-28b). The peak at 0 ppm is due to TMS, which is added as a reference.

Similar to ^{1}H NMR spectroscopy, signals from the solvent typically appear in ^{13}C NMR spectra. In Figure 16-28b, for example, the peak at 77 ppm is due to the carbon signal from the solvent, $CDCl_3$.

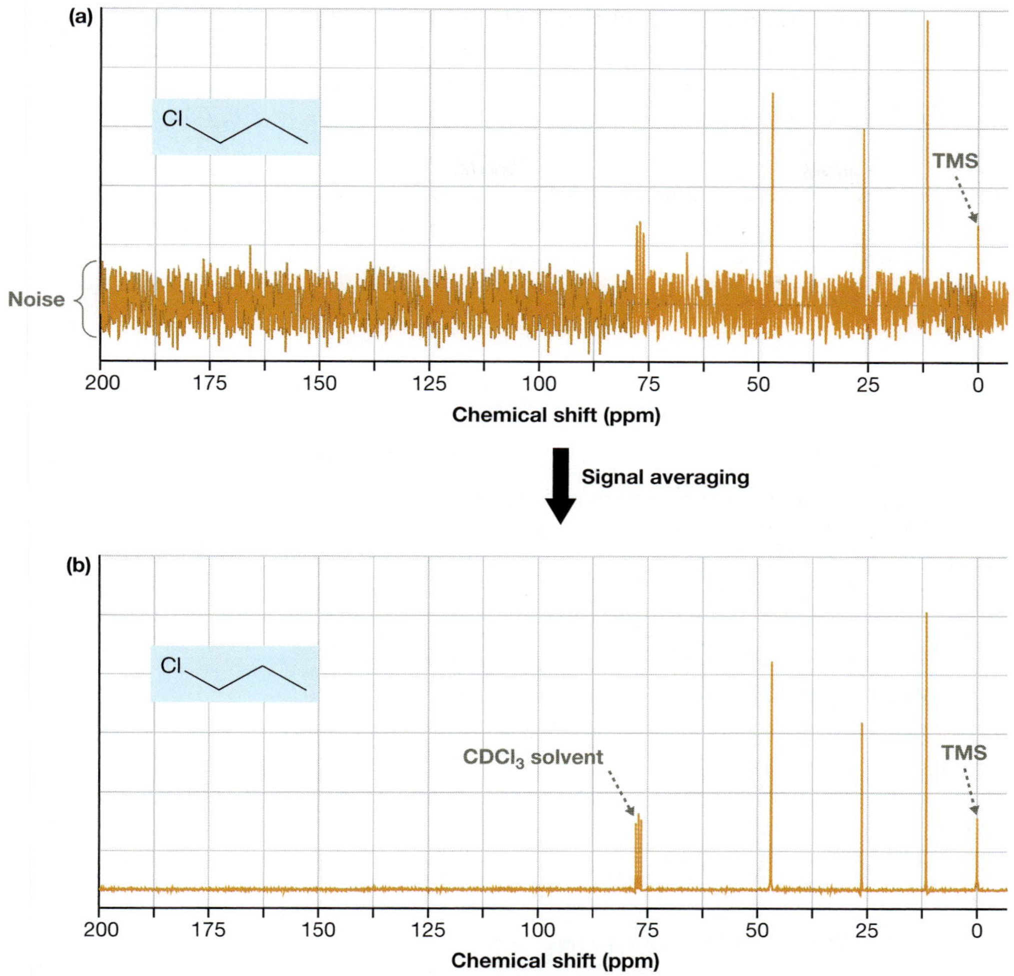

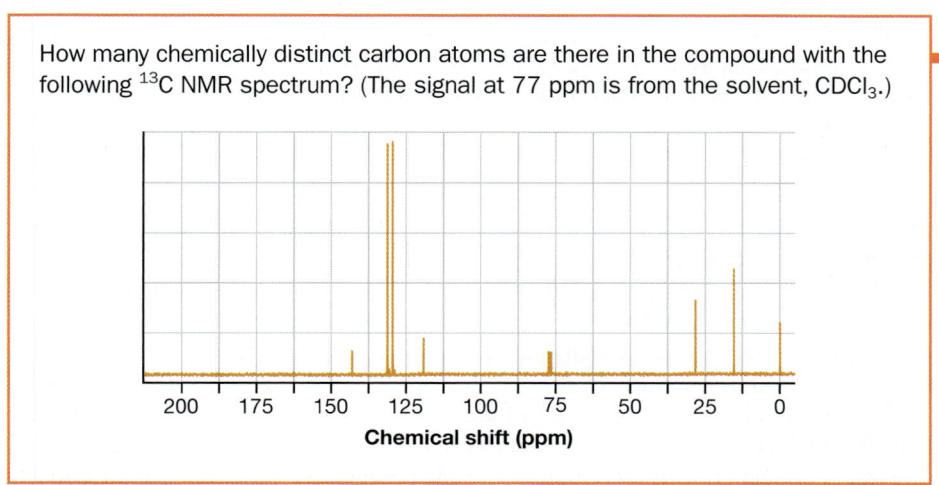

(a)

Cl

Noise

200 175 150 125 100 75 50 25 0
Chemical shift (ppm)

TMS

Signal averaging

(b)

Cl

CDCl₃ solvent

TMS

200 175 150 125 100 75 50 25 0
Chemical shift (ppm)

FIGURE 16-28 **¹³C NMR spectrum of 1-chloropropane** (a) Without signal averaging. (b) With signal averaging.

YOUR TURN 16.16

How many chemically distinct carbon atoms are there in the compound with the following ¹³C NMR spectrum? (The signal at 77 ppm is from the solvent, CDCl₃.)

200 175 150 125 100 75 50 25 0
Chemical shift (ppm)

To determine the number of chemically distinct carbon atoms in a molecule, we can apply the chemical distinction test we learned for the hydrogen atom, but substitute an imaginary atom "X" for each carbon atom instead of hydrogen. This is illustrated in Solved Problem 16.27.

How many ^{13}C NMR signals would you expect for this molecule?

Think What molecules are obtained by separately replacing each C atom with "X"? What are the relationships among the resulting molecules?

Solve Replacing each of the carbon atoms with X, we get compounds **A–F** as follows:

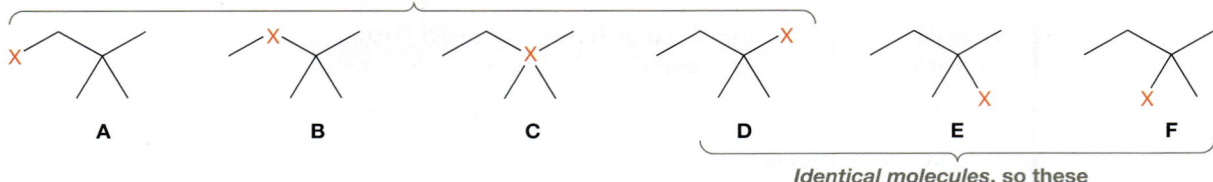

Compounds **A–D** are constitutional isomers, so the corresponding four C atoms are chemically distinct. Compounds **E** and **F** are identical to **D**, so the C atoms substituted in **E** and **F** are chemically equivalent to the one substituted in **D**. In all, there are four chemically distinct C atoms in the molecule, so we expect four ^{13}C NMR signals.

PROBLEM 16.28 How many ^{13}C NMR signals would you expect for the molecule shown here? Explain.

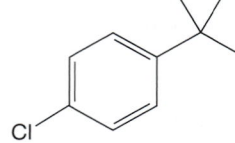

16.13b Signal Splitting

Because ^{13}C nuclei have spin, they undergo coupling with other nearby nuclei that have spin, such as protons or other ^{13}C nuclei. However, under the conditions that the spectra are acquired, we do not observe signal splitting. In other words:

> In most of the ^{13}C NMR spectra you will encounter, all of the signals will be singlets.

In the ^{13}C NMR spectrum of 1-chloropropane in Figure 16-28 (p. 805), for example, all three signals are singlets.

We do not observe coupling between two ^{13}C nuclei because the natural abundance of ^{13}C is so low. With ^{13}C comprising approximately 1 out of every 100 carbon atoms, there is only about a 1% chance that a given ^{13}C nucleus being detected will be adjacent to another ^{13}C nucleus.

We do not observe the splitting of a carbon signal by attached protons because **broadband decoupling** is usually used to essentially "turn off" the coupling between ^{13}C and ^{1}H nuclei. In this technique, a specific range of RF radiation is continuously sent through the sample, forcing all protons to flip back and forth rapidly between the α and β spin states. Therefore, the average magnetic field each proton produces is zero.

A major advantage of broadband decoupling is that it causes the signal intensity to increase, which helps with the interpretation of the spectrum. A disadvantage, however, is that we lose valuable information about the number of protons attached to each carbon. This type of information is obtained in other ways, such as with DEPT spectroscopy (see Section 16.14).

16.13c Chemical Shifts

As in ^1H NMR spectroscopy, ^{13}C absorptions are reported as chemical shifts, with TMS as the reference compound (its carbon atom is assigned a chemical shift of 0 ppm).

Chemical shifts for a variety of carbon atoms are listed in Table 16-4. Notice that their range is different from that for protons (see Table 16-1, p. 785). That is:

Whereas the chemical shifts in ^1H NMR generally range from 0 to 12 ppm, those for carbon atoms generally range from 0 to 220 ppm.

TABLE 16-4 ^{13}C NMR Chemical Shifts

Type of Carbon	Chemical Shift (ppm)	Type of Carbon	Chemical Shift (ppm)	Type of Carbon	Chemical Shift (ppm)
1. 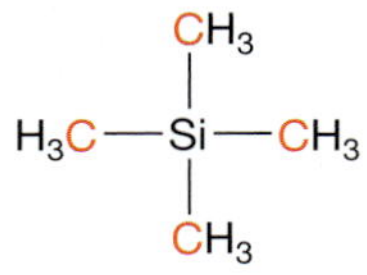 $H_3C-Si(CH_3)(CH_3)-CH_3$	0	6. 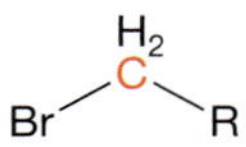 $Br-CH_2-R$	25–35	11. 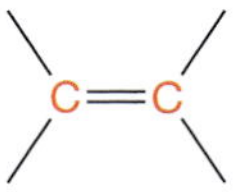 $C=C$	105–150
2. $R-CH_3$	10–25	7. 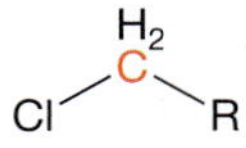 $Cl-CH_2-R$	35–55	12. 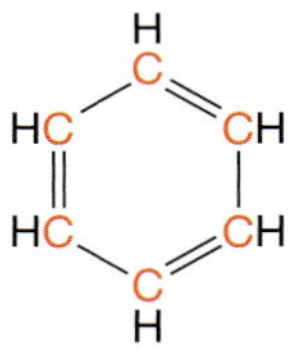 benzene ring	128
3. 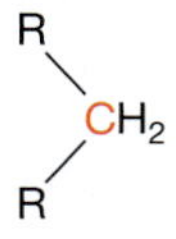 R_2CH_2	20–45	8. 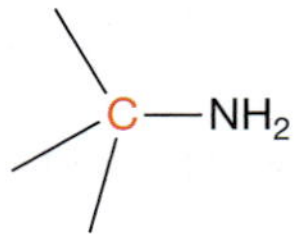 $C-NH_2$	35–40	13. 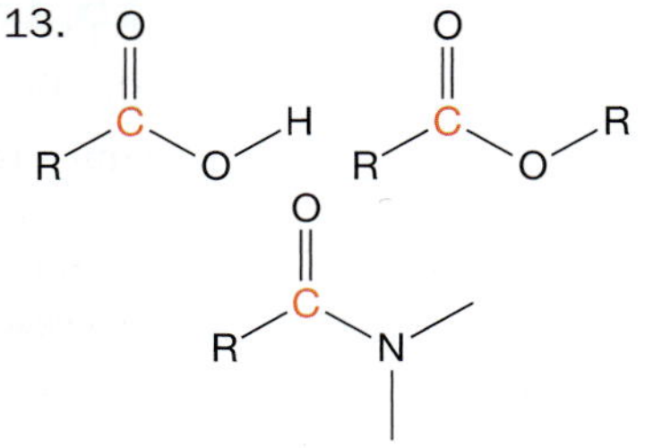 $RCOOH$, $RCOOR$, $RCON<$	160–185
4. 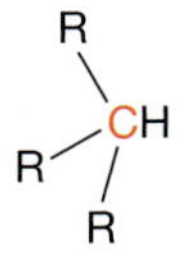 R_3CH	25–45	9. 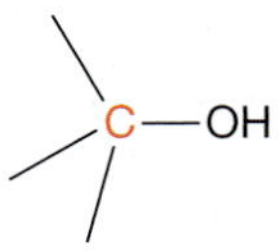 $C-OH$	55–70	14. 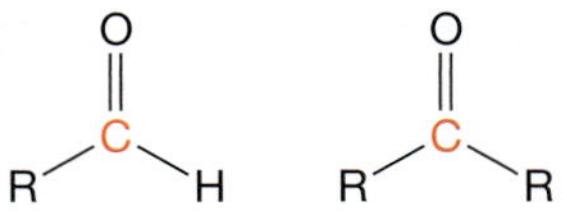 $RCHO$, $RCOR$	190–220
5. 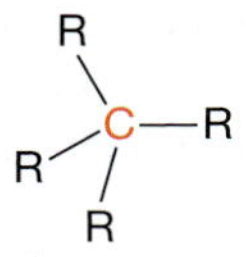 R_4C	30–35	10. $R-C\equiv C-R$	65–85		

Despite the different values for ^1H and ^{13}C chemical shifts:

> The order of the functional groups for ^{13}C chemical shifts in Table 16-4 is roughly the same as the order for ^1H NMR chemical shifts in Table 16-1.

Saturated alkane carbons have among the lowest chemical shifts (0–50 ppm); carbons attached to hydroxy, halo, and amino groups have chemical shifts that are somewhat farther downfield (10–90 ppm), followed by the carbon atoms of alkenes and aromatics (100–170 ppm) and carbonyl carbons (150–220 ppm).

The trends in chemical shift are the same for both carbon and proton signals because shielding and deshielding phenomena affect both types of nuclei in roughly the same way. Sigma (σ) bonding electrons shield the carbon nucleus, and the greater the electron density contributed by those electrons, the more the carbon nucleus is shielded. Nearby electron-withdrawing substituents (such as halogen, oxygen, or nitrogen atoms) remove electron density from the carbon atom, thereby deshielding it. The result is a downfield shift. In addition, ring current from aromatic rings or double bonds adds to the magnetic field experienced by the carbon atom, resulting, once again, in a downfield shift.

YOUR TURN 16.17

> Identify two pairs of carbon atoms in Table 16-4 whose chemical shifts are in the same order as the chemical shifts of the analogous protons in Table 16-1.

CONNECTIONS Benzyl chloride (Problem 16.29) is used in a wide variety of organic syntheses because of how easily its Cl substituent can be replaced in S_N2 or S_N1 reactions. In particular, it can be used in the synthesis of a number of pharmaceuticals, including amphetamines. Benzyl chloride sales, therefore, are monitored by the U.S. Drug Enforcement Agency.

Amphetamine

PROBLEM 16.29 Predict which C atom in benzyl chloride would have the highest chemical shift. Explain.

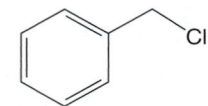

16.13d Integration of ^{13}C NMR Signals

The ^{13}C NMR spectrum of 2-methylbutanal (Fig. 16-29) shows five signals, each of which is different in height. If the number of carbon atoms contributing to each signal were proportional to the area, then all five peaks would have essentially the same height, because each signal represents exactly one carbon atom in the molecule. However:

> In typical ^{13}C NMR spectra, the area under a peak is *not* proportional to the number of carbon atoms contributing to that peak.

There is poor correlation because of the broadband decoupling used to produce only singlets. One of the outcomes of this technique is a magnification of each carbon signal via what is called the nuclear Overhauser effect. That magnification differs for different types of carbon atoms, so the integration is of relatively little value in ^{13}C NMR spectroscopy.

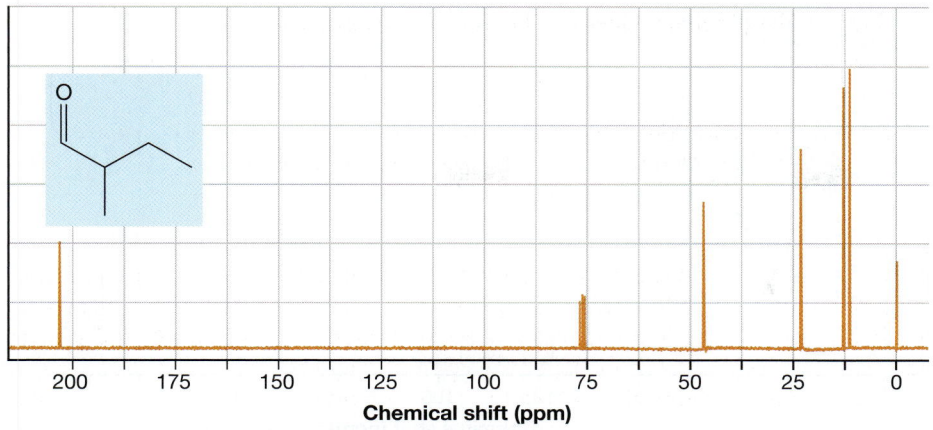

Chemical shift (ppm)

FIGURE 16-29 **^{13}C NMR spectrum of 2-methylbutanal** Although one carbon atom contributes to each signal, the heights of the signals differ markedly.

16.14 DEPT ^{13}C NMR Spectroscopy

In Section 16.13b, we saw that proton decoupling is used in ^{13}C NMR spectroscopy to make all signals appear as singlets. This makes it easier for us to interpret spectra, but we lose valuable information about the number of protons attached to each carbon. Fortunately, a separate experiment can be used to obtain this information: *distortionless enhancement by polarization transfer (DEPT) spectroscopy*. In **DEPT ^{13}C NMR spectroscopy**, three separate spectra are acquired, as shown for ethylbenzene in Figure 16-30 (next page). One spectrum is the normal broadband-decoupled spectrum (Fig. 16-30a), in which all carbon signals appear. The other two spectra are acquired after the sample has been subjected to specific pulse sequences of RF radiation, so that the signals given off by the ^{13}C nuclei depend on the number of attached protons. In one of those spectra, called a DEPT-90 (Fig. 16-30b), the only signals are those of carbon atoms bonded to a single proton (CH). Signals of carbons with zero, two, or three protons do not appear in a DEPT-90. In the other spectrum, called a DEPT-135 (Fig. 16-30c), carbon atoms with one and three protons (CH and CH$_3$) produce normal signals, whereas carbons with two protons (CH$_2$) produce negative signals—that is, *below* the baseline. As in DEPT-90, signals from carbon atoms with no protons are absent in a DEPT-135.

With DEPT spectra, all four types of carbon atoms are readily identified:

- CH carbon signals appear normally in the DEPT-90 spectrum.
- CH$_2$ carbon signals appear as negative signals in the DEPT-135 spectrum.
- CH$_3$ carbon signals appear as positive signals in the DEPT-135 but do not appear in the DEPT-90 spectrum.
- Signals of carbons without any protons appear in the broadband-decoupled spectrum but not in any of the DEPT spectra.

Frequently, information from DEPT spectroscopy is simply summarized in the normal broadband-decoupled ^{13}C NMR spectrum, as shown in Figure 16-31 for the C and CH carbons of ethylbenzene. Providing all three spectra each time can become quite cumbersome.

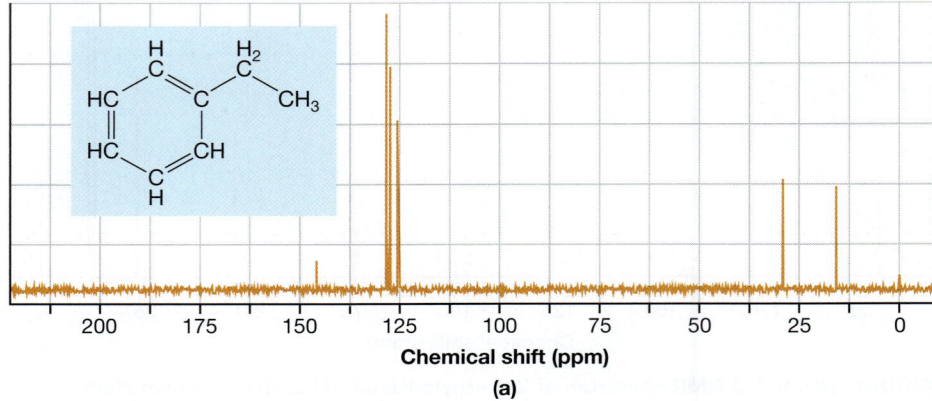

Normal broadband-decoupled
¹³C NMR spectrum

(a)

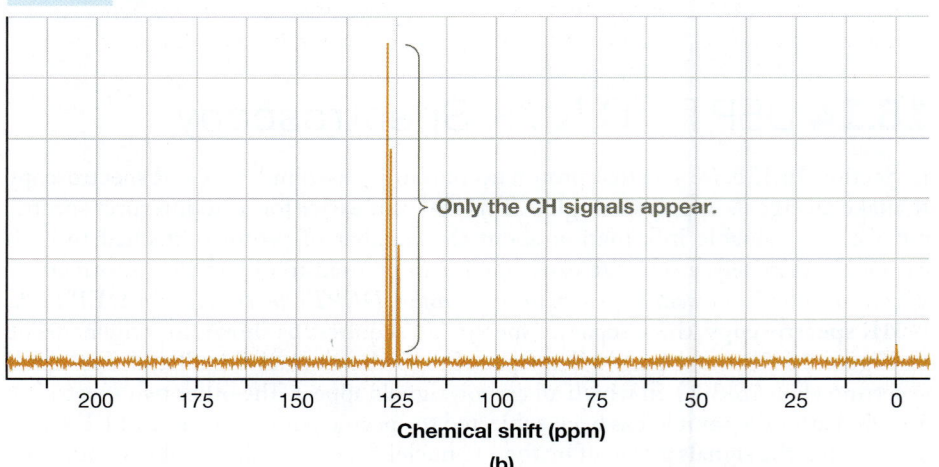

DEPT-90

Only the CH signals appear.

(b)

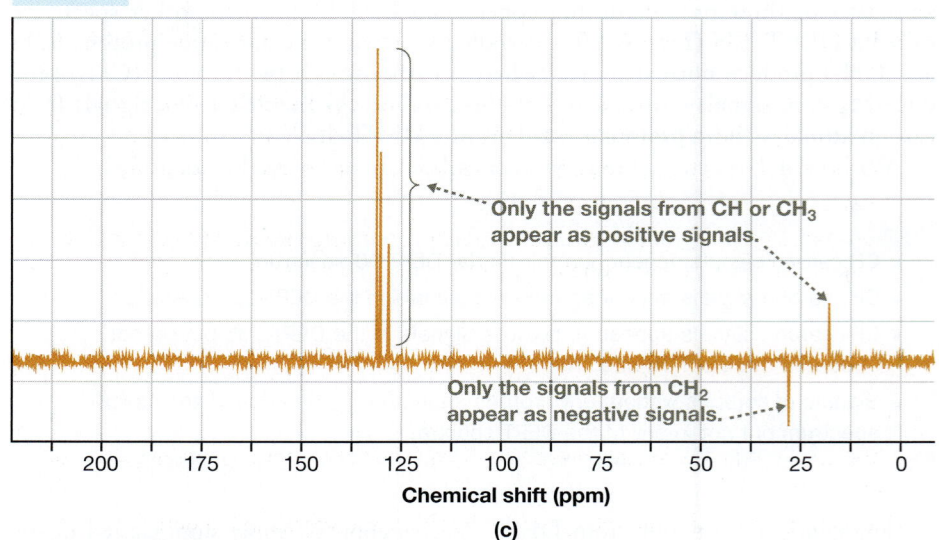

DEPT-135

Only the signals from CH or CH₃
appear as positive signals.

Only the signals from CH₂
appear as negative signals.

(c)

FIGURE 16-30 DEPT ¹³C NMR spectroscopy (a) The normal broadband-decoupled ¹³C NMR spectrum of ethylbenzene. (b) The DEPT-90 spectrum of ethylbenzene. Only CH carbon signals appear. (c) The DEPT-135 spectrum of ethylbenzene. CH and CH₃ carbons appear as positive signals, whereas CH₂ carbons appear as negative signals.

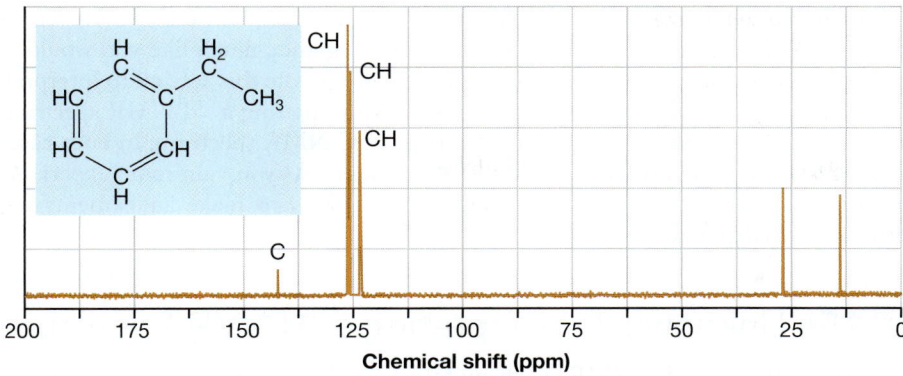

FIGURE 16-31 **The normal broadband ^{13}C NMR spectrum of ethylbenzene** Partial results from a DEPT experiment have been provided (see Your Turn 16.18).

YOUR TURN 16.18

In Figure 16-31, identify the CH_2 signal as the one that appears as a negative signal in the DEPT-135 spectrum. Identify the CH_3 signal as the one that appears as a positive signal in the DEPT-135 spectrum but does not appear in the DEPT-90 spectrum.

16.15 Structure Elucidation Using NMR Spectroscopy

To this point in Chapter 16, we have dealt with various aspects of NMR spectra somewhat independently of one another. Each aspect of an NMR spectrum tells us a certain amount of information about the structure of the compound that produced the spectrum.

In ^1H NMR spectroscopy:

- The total number of signals gives you the number of chemically distinct types of protons.
- A signal's chemical shift tells you whether aromatic rings, double bonds, or electronegative atoms are nearby the protons responsible for that signal.
- The integration of each signal tells you the number of protons responsible for that signal.
- A signal's splitting pattern tells you the number of neighboring protons that are distinct from the protons responsible for that signal.

In ^{13}C NMR spectroscopy:

- The total number of signals gives you the number of chemically distinct types of carbons.
- A signal's chemical shift tells you whether aromatic rings, double bonds, or electronegative atoms are nearby the carbon atoms responsible for that signal.
- The results from the DEPT spectra tell you the number of protons attached to each type of carbon.

To determine a more complete picture of the molecule's structure, however, we must bring these individual pieces of information together, much like you would fit pieces of a jigsaw puzzle together. For this reason, we devote this section to interpreting NMR spectra. In Section 16.15a we focus on interpreting a ^{1}H NMR spectrum, and in Section 16.15b we focus on interpreting a ^{13}C NMR spectrum. In both cases, we suggest ways to approach these kinds of problems. As you gain more experience interpreting NMR spectra, however, you will learn how to make adjustments and settle on a strategy that works best for you.

16.15a Unknown 1: Interpreting a ^{1}H NMR Spectrum

It helps to approach a ^{1}H NMR spectrum by asking the following questions:

1. How many signals are apparent in the spectrum?
2. For each signal:
   a. What is the value of the chemical shift? What kinds of aromatic rings, double bonds, or electronegative atoms, if any, does that chemical shift suggest might be nearby? Can you use the molecular formula or information from other spectra to provide insight into the presence of electronegative atoms or double bonds?
   b. What is the relative integration? If you know the total number of protons in the molecule (such as from the molecular formula), can you determine how many protons each signal represents?
   c. What is the splitting pattern? What does it say about the number of neighboring protons distinct from the ones responsible for the signal?
3. Can you combine the information from question 2 to build molecular fragments with multiple carbon atoms?
4. Can you assemble those molecular fragments in a way that is consistent with every aspect of the ^{1}H NMR spectrum?

With this in mind, suppose you are given the ^{1}H NMR spectrum in Figure 16-32, which is generated from a compound (Unknown 1) having the formula $C_{10}H_{12}O_2$.

To answer question 1, the spectrum appears to have at least four signals: one at $\delta \approx 1.1$ ppm, one at $\delta \approx 1.8$ ppm, one at $\delta \approx 4.2$ ppm, and a set of signals between $\delta \approx 7.4$ ppm and $\delta \approx 8.1$ ppm. Therefore, there are at least four distinct types of protons.

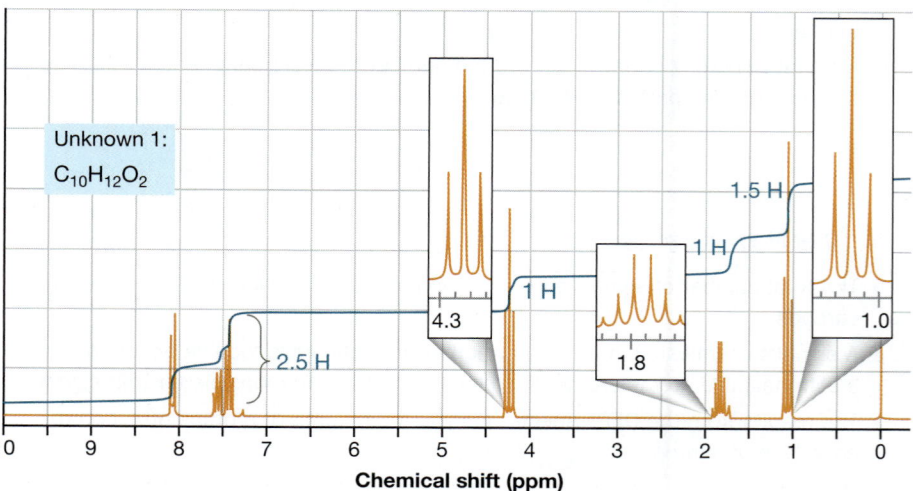

FIGURE 16-32 Proton NMR spectrum of Unknown 1, $C_{10}H_{12}O_2$

To organize our thoughts as we work through questions 2 and 3, let's construct a table that has a row for each signal, as shown in Table 16-5. For the signal at $\delta \approx 1.1$ ppm (Entry 1), the answer to question 2a is that the protons are not immediately attached to a double bond, nor are those protons on a C atom that is attached to an electronegative atom. These scenarios would have required chemical shifts that are significantly higher.

YOUR TURN 16.19

> According to Table 16-1 (p. 785), what is the lowest chemical shift you should expect to observe for a proton that is attached to a double bond or is on a carbon attached to an electronegative atom?

To answer question 2b for the signal at $\delta \approx 1.1$ ppm, we are given an integration of 1.5 H, which is relative to the smallest integration of 1 H. Summing the integrations given in the spectrum, we arrive at a total integration of 6 H (i.e., 1.5 H + 1 H + 1 H + 2.5 H), but the molecular formula shows that there are actually 12 H atoms in the molecule. Therefore, each of the relative integration values given to us must represent two H atoms, so the signal at $\delta \approx 1.1$ ppm represents three equivalent H atoms.

To answer question 2c for the signal at $\delta \approx 1.1$ ppm, we see that the signal is a triplet. This suggests that there are two neighboring protons distinct from the ones responsible for the signal.

Putting all of this information together for the signal at $\delta \approx 1.1$ ppm, we can propose that the signal is due to a CH_3 group that is attached to a CH_2 group, as shown in the rightmost column in Table 16-5. In that column, the protons in red are the ones that are responsible for generating the signal of interest and the protons on the neighboring group in black and in the parentheses are responsible for generating *another* signal.

Let's now move to the signal at $\delta \approx 1.8$ ppm, the information for which is summarized in Entry 2 of Table 16-5. As before, the low chemical shift tells us that the

TABLE 16-5 Summary of Information from Figure 16-32

	QUESTION 2a		QUESTION 2b		QUESTION 2c		QUESTION 3
Entry	Chemical Shift (ppm)	Nearby Double Bonds or EN Atoms?	Relative Integration	Number of Protons	Splitting Pattern	Number of Neighboring Distinct H	Molecular Fragment
1	~1.1	None	1.5	3	Triplet	2	$(H_2C)\!-\!CH_3$
2	~1.8	None	1	2	Sextet	5	$(H_2C)\!-\!\underset{H_2}{C}\!-\!(CH_3)$
3	~4.2	H—C—O	1	2	Triplet	2	$(O)\!-\!\underset{}{\overset{H_2}{C}}\!-\!(CH_2)$
4	~7.4–8.1	H—Aromatic	2.5	5	Complex	?	aromatic ring fragment

protons responsible for this signal are not directly attached to an atom that is part of a double bond, nor are the protons on a carbon atom that has an attached electronegative atom. The relative integration of 1 H tells us that the signal represents two protons, and the apparent sextet splitting pattern tells us that there are five neighboring protons. Putting this information together, we can propose that the signal is due to a CH_2 group that is bonded to a CH_3 group on one side and a CH_2 group on the other. (Recall from Section 16.12 that this would actually produce a quartet of triplets that would appear as a sextet because each splitting occurs with nearly the same coupling constant.)

The signal at $\delta \approx 4.2$ ppm (Entry 3 in Table 16-5) is consistent with a proton on a carbon that has an attached electronegative atom. According to the molecular formula, the electronegative atom would need to be oxygen. The relative integration of 1 H tells us that the signal represents two protons, and the fact that the signal is split into a triplet tells us that there are two neighboring protons. As shown at the right in Table 16-5, this signal can be from a CH_2 group that is bonded to O on one side and a CH_2 group on the other.

YOUR TURN 16.20

In Table 16-1 (p. 785), what is the normal chemical shift of a proton that is part of a fragment H—C—O in an ether or alcohol? How does that compare to the chemical shift in Entry 3 of Table 16-5? What could be the source of that difference in chemical shift?

The final set of signals between $\delta \approx 7.4$ ppm and $\delta \approx 8.1$ ppm are consistent with aromatic protons, as shown in Entry 4 of Table 16-5. The relative integration of 2.5 H tells us that there are five such protons, which would be consistent with a monosubstituted benzene, as shown in the rightmost column. Because these signals appear at different chemical shifts and complex splitting is apparent, we will analyze the splitting pattern later.

YOUR TURN 16.21

In Table 16-1 (p. 785), what is the normal chemical shift of a proton in benzene? How does that compare to the chemical shifts in Entry 4 of Table 16-5? What could be the source of those differences in chemical shift?

Question 4 asks us to consider how to assemble the proposed molecular fragments to complete the entire molecule. To do so, let's try to find portions of one fragment that overlap with portions of another fragment. For example, the red CH_3 group in Entry 1 could be the same as the black CH_3 group in parentheses in Entry 2:

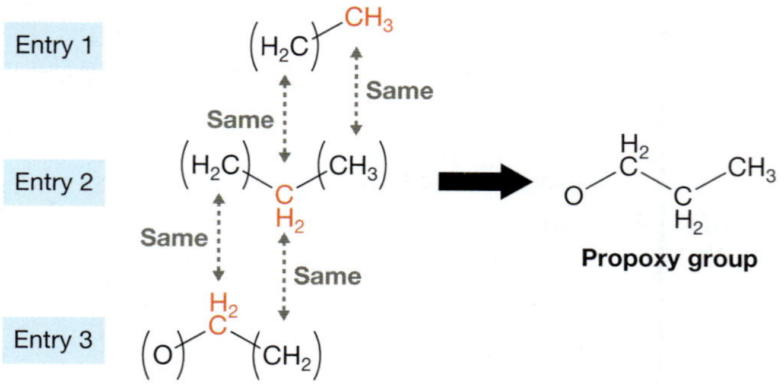

Similarly, the red CH_2 group in Entry 3 could be the same as the black CH_2 group in parentheses in Entry 2. Seeing these areas of overlap allows us to arrive at the larger structure on the right, which is a propoxy group.

Can the propoxy group we just derived be attached directly to the benzene ring? The answer, in this case, is no because the resulting molecule would be propoxybenzene,

$C_6H_5OCH_2CH_2CH_3$, which has the formula $C_9H_{12}O$. This would be one carbon atom and one oxygen atom less than the formula we were given, $C_{10}H_{12}O_2$. These two atoms could be added as a carbonyl group (C=O) between the propoxy group and the benzene ring, giving propyl benzoate, as shown here. All aspects of propyl benzoate are consistent with the ^1H NMR spectrum.

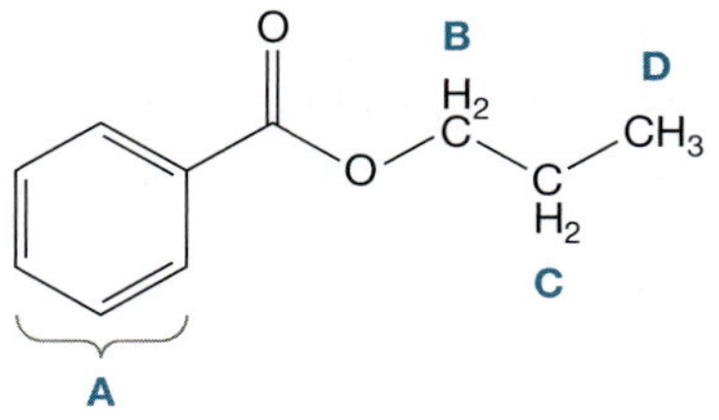

Unknown 1: $C_{10}H_{12}O_2$

YOUR TURN 16.22

Write **A**, **B**, **C**, or **D** above the signals in Figure 16-32 to match up with each chemically distinct H atom in propyl benzoate.

PROBLEM 16.30 Based on the NMR spectrum in Figure 16-32, can Unknown 1 be $CH_3CH_2CH_2CO_2C_6H_5$ or $CH_3CH_2OCH_2(C=O)C_6H_5$? Why or why not?

Let's now return to the features of the aromatic signals, which we temporarily ignored, to verify that they agree with our completed structure. In the completed structure, there are three distinct aromatic protons: two ortho protons, two meta protons, and one para. Being the closest to the carbonyl group, the ortho protons would be expected to be deshielded the most. This is consistent with the peaks at $\delta = 8.1$ ppm, which have the highest chemical shift and an integration of 1 H, representing two protons. The meta and para protons must therefore give rise to the overlapping peaks centered at ~7.5 ppm, which have an integration of ~1.5 H, representing three protons.

16.15b Unknown 2: Interpreting a ^{13}C NMR Spectrum

Similar to a ^1H NMR spectrum, it helps to approach a ^{13}C NMR spectrum by asking the following questions:

1. How many signals are apparent in the spectrum? Can you compare the number of signals to the number of carbons in the molecular formula to determine whether there are equivalent carbons?
2. For each signal:
   a. What is the value of the chemical shift? What kinds of aromatic rings, double bonds, or electronegative atoms, if any, does that chemical shift suggest might be nearby? Can you use the molecular formula or information from other spectra to provide insight into the presence of aromatic rings, double bonds, or electronegative atoms?
   b. Can you apply results from DEPT spectroscopy to determine which type of carbon generated the signal (i.e., C, CH, CH_2, or CH_3)?
3. Can you combine the information from question 2 to build molecular fragments with multiple carbon atoms?
4. Can you assemble those molecular fragments in a way that is consistent with every aspect of the ^{13}C NMR spectrum?

With these questions in mind, let's consider the ^{13}C NMR (with DEPT) and IR spectra for Unknown 2 ($C_5H_{12}O$), shown in Figure 16-33.

To answer question 1, the ^{13}C spectrum has three signals: one at $\delta \approx 26$ ppm, one at $\delta \approx 33$ ppm, and one at $\delta \approx 73$ ppm (remember that the signal at 77 ppm is from the $CDCl_3$ solvent and the one at 0 ppm is from the TMS reference). Thus, only three of the five total C atoms in Unknown 2 are distinct. This means that two of the C atoms are equivalent to others in the molecule.

As we did for the ^1H NMR spectrum of Unknown 1 in Section 16.15a, let's construct a table (Table 16-6) to answer questions 2 and 3 in an organized fashion, with a separate row for each signal. We begin with the signal at ~26 ppm (Entry 1). To answer question 2a, the chemical shift is quite low and suggests that the carbon responsible for generating that signal is not part of a double bond or attached to an electronegative atom.

YOUR TURN 16.23

According to Table 16-4 (p. 807), what is the lowest chemical shift you should expect to observe for a carbon that is part of a double bond or is attached to an electronegative atom?

To answer question 2b, the DEPT results show that the carbon is part of a CH_3 group. To answer question 3, we can combine this information to arrive at a fragment in which the CH_3 group is attached to a C atom, as shown in the rightmost column in Table 16-6. The carbon in red is responsible for generating the signal of interest and the carbon in black and in the parentheses is responsible for generating *another* signal.

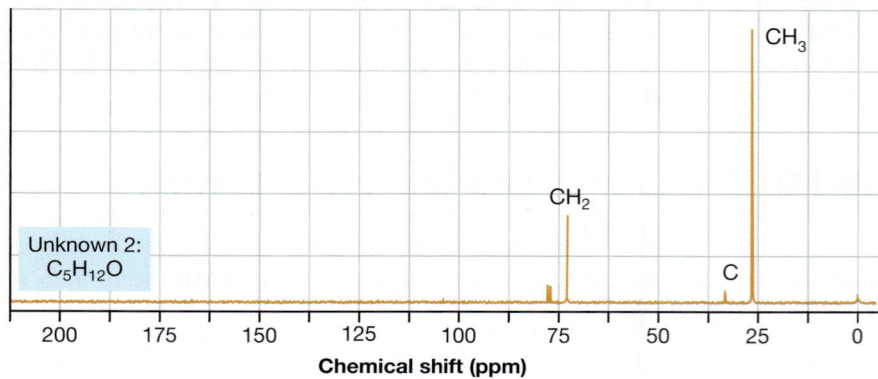

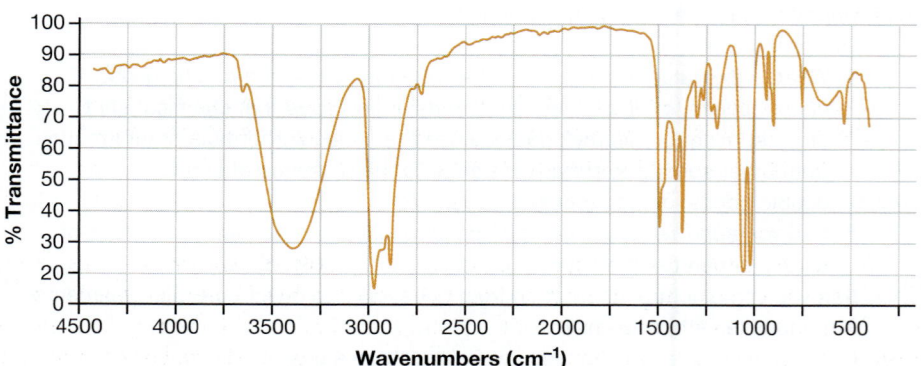

FIGURE 16-33 ^{13}C NMR and IR spectra of Unknown 2, $C_5H_{12}O$

TABLE 16-6 **Summary of Information from Figure 16-33**

	QUESTION 2a		QUESTION 2b	QUESTION 3
Entry	Chemical Shift (ppm)	Nearby Double Bonds or EN Atoms?	DEPT Spectrum	Molecular Fragment
1	~26	None	CH_3	$(C)\!-\!CH_3$
2	~33	None	C	$(C)\,(C)\!-\!C\!-\!(C)\,(C)$
3	~73	C—O	CH_2	$(HO)\!-\!\overset{H_2}{C}\!-\!(C)$

Next, to answer question 2 about the signal at ~33 ppm (Entry 2), the low chemical shift again suggests that the C atom is neither part of a double bond nor bonded to an electronegative atom. The DEPT results show that it is not bonded to any H atoms, either, suggesting that it is singly bonded to other C atoms only. To answer question 3, we arrive at the fragment shown in the rightmost column.

To answer question 2a for the signal at ~73 ppm (Entry 3), notice that the chemical shift is moderately high. The chemical shift is not high enough for the C atom to be part of a double bond, but it could be part of an alkyne or it could be singly bonded to an O atom.

YOUR TURN 16.24

Use Table 16-4 (p. 807) to verify that the signal at ~73 ppm could be consistent with an alkyne carbon or a carbon that is singly bonded to O, but is inconsistent with a carbon that is part of a double bond.

The IHD for a molecular formula of $C_5H_{12}O$ is 0, however, which rules out an alkyne C. Furthermore, the IR spectrum indicates the presence of an OH group but not an alkyne, so that C atom is likely part of an alcohol. To answer question 2b, the DEPT results show that the C atom is attached to two H atoms, suggesting the fragment at the far right of Entry 3.

YOUR TURN 16.25

Identify the IR band in Figure 16-33 that indicates an OH group. In addition, identify the locations where you would expect to see absorptions corresponding to a C≡C and an alkyne C—H stretch.

To tackle question 4, let's try to assemble the fragments shown in Table 16-6. The red C atom in Entry 2 must be bonded to four other C atoms. One of them could be the red CH_3 group from Entry 1 and another could be the red CH_2 group from Entry 3, as shown on the next page. We thus end up with the larger fragment shown to the right of the arrow, where the two C atoms in parentheses are not yet fully characterized.

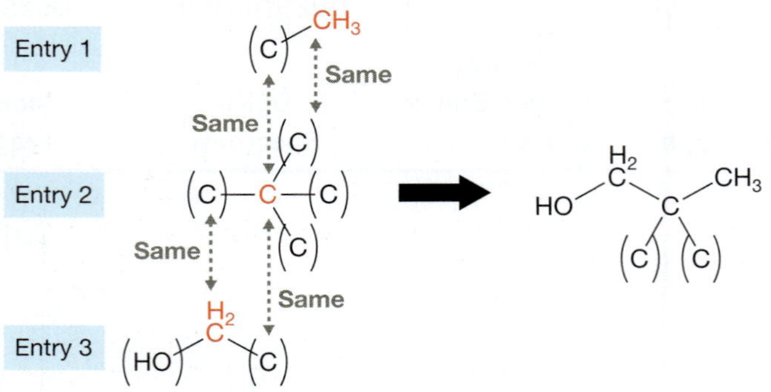

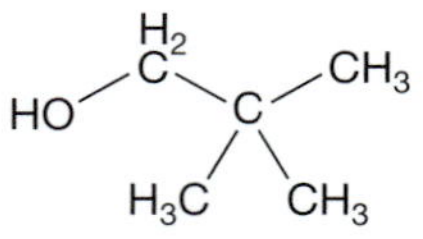

Entry 1

Entry 2

Entry 3

HO $-\overset{H_2}{C}-\overset{CH_3}{C}-CH_3$
$H_3C \quad CH_3$

Unknown 2: $C_5H_{12}O$

Each of the C atoms in parentheses in the partial structure above on the right must be part of one of the fragments already in Table 16-6. Otherwise, we would have more than three distinct carbons. Neither C can be part of another fragment from Entry 2, however, because that would require bonds to additional carbon atoms and we would exceed the five C atoms specified by the molecular formula. Moreover, neither C can be part of another fragment from Entry 3 because that would add another O atom, and the formula specifies just one. Both C atoms must therefore be part of additional CH_3 groups, as shown here in the margin.

PROBLEM 16.31 An unknown compound has the formula $C_9H_{10}O_2$. Four signals appear in its ^1H NMR spectrum: (1) singlet, $\delta = 2.3$ ppm, 6 H; (2) doublet, $\delta = 7.0$ ppm, 2 H; (3) triplet, $\delta = 7.2$ ppm, 1 H; and (4) very broad singlet, $\delta = 12.9$ ppm, 1 H. Six signals appear in its ^{13}C NMR spectrum at $\delta = 19.4, 127.2, 128.5, 133.6, 135.3$, and 170.9 ppm. Propose a structure for this compound.

PROBLEM 16.32 An unknown compound has the formula $C_8H_{10}O$. There are four distinct signals in its ^1H NMR spectrum: (1) doublet, $\delta = 1.4$ ppm, 3 H; (2) singlet, $\delta = 2.4$ ppm, 1 H; (3) quartet, $\delta = 4.8$ ppm, 1 H; and (4) overlapping signals, $\delta = 7.2–7.4$ ppm, 5 H. When the sample is treated with D_2O, the signal at 2.4 ppm disappears. Propose a structure for this compound.

16.16 Mass Spectrometry: An Overview

Mass spectrometry provides insight into the mass of a molecule and the fragments that compose it. As with UV–vis, IR, and NMR spectroscopy, this information about a molecule is obtained by interpreting a spectrum, in this case a *mass spectrum*. Unlike spectroscopy, however, mass spectrometry does *not* involve electromagnetic radiation.

Before we examine a mass spectrum, you need to understand how one is generated. One way to do so is shown in Figure 16-34. A small sample (typically on the order of $\leq 1 \ \mu L$ of a dilute solution) of a compound M is injected into the spectrometer, where it is immediately vaporized. This *vaporization* can be represented by Equation 16-5.

$$M(\ell) \rightarrow M(g) \tag{16-5}$$

Once vaporized, these gaseous molecules, $M(g)$, drift through a beam of fast-moving electrons. When an electron from that beam impacts a molecule of $M(g)$, an electron from $M(g)$ is knocked off, producing a gaseous species that has one fewer electron, and therefore has a positive charge. The process of producing this **molecular** ion, $M^+(g)$, is called **electron impact ionization** and can be represented by Equation 16-6.

$$M(g) \xrightarrow{\text{Electron beam}} M^+(g) + e^- \qquad \text{(16-6)}$$

A charged species like $M^+(g)$ can be guided into a detector through a curved tube using a magnetic field to bend the ion's path. The extent that a given ion's path is bent depends on the strength of the magnetic field that is applied and the **mass-to-charge ratio** of the particle, represented as *m/z*. If the magnetic field strength is held fixed, then only ions of a specific value of *m/z* can reach the detector. Ions with *m/z* different from that value will collide with the wall of the tube and will be destroyed before reaching the detector—those that are too light will be deflected too much, whereas those that are too massive will not be deflected enough. If multiple ions with different *m/z* values are present, then the magnetic field strength can be adjusted to allow the various ions to be detected.

YOUR TURN 16.26

In Figure 16-34, draw a line to represent the path of an ion that is too light to reach the detector and label it "too light." Do the same for an ion that is too massive to reach the detector and label it "too massive."

The detector itself is designed to keep track of the number of charged species that collide with it; this number is translated into what is called the **relative abundance** of the ion. The greater the number of charged species detected, the greater the ion's relative abundance. If multiple ions are produced with different values of *m/z*, then the magnetic field strength can be scanned to determine the relative abundance at *each m/z*.

The process represented in Figure 16-34 is not the only one that can be used to generate a mass spectrum. A variety of other types of mass spectrometers exist as well, each using its own particular method for generating a mass spectrum. All types of mass spectrometers, however, must have three basic components, providing the means to (1) produce gaseous ions from an uncharged sample, (2) separate ions by their *m/z* values, and (3) detect the relative abundance of ions having a particular value of *m/z*. It is worth noting, in particular, that modern instruments use a *quadrupole* (four charged poles) to separate ions, which causes the ions to move in a spiral path, rather than a bent path.

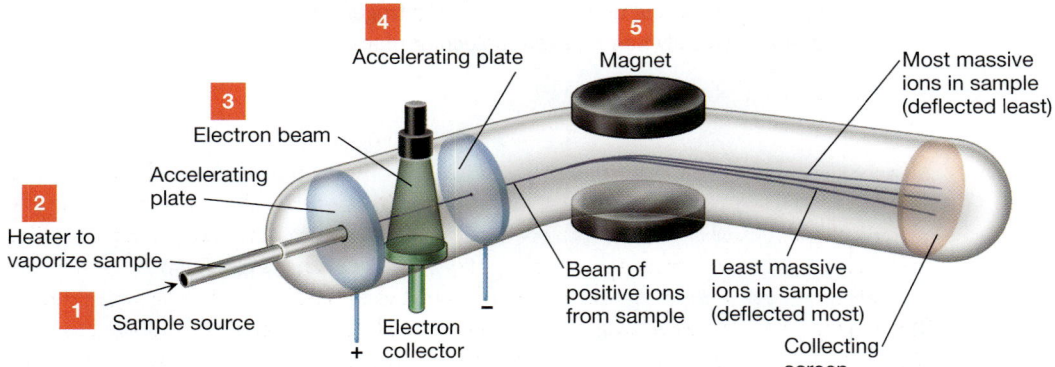

FIGURE 16-34 Schematic of a mass spectrometer An injected sample (1) is converted to vapor (2) and passed through a beam of highly energetic electrons (3). Collisions with the electrons generate positively charged ions of the entire molecule and of its fragments. These ions are accelerated (4) through a magnet that separates them (5) on the basis of their mass and charge. As the ions impact the detector, a signal is registered. Varying the magnetic field allows the spectrometer to measure the relative abundance of charged particles with different mass-to-charge ratios.

16.17 Features of a Mass Spectrum, the Nitrogen Rule, and Fragmentation

The **mass spectrum** of a compound is often represented as a bar graph with m/z on the x axis (where m is in atomic mass units, u) and the relative abundance of ions at each m/z on the y axis. When a mass spectrometer uses electron impact to ionize the compound (Equation 16-6), the charge (z) on each ion that is produced is typically $+1$. In that case, then, m/z is simply m. That is:

> The x axis of a mass spectrum can usually be interpreted as the mass of each ion detected.

In turn, the mass of each ion is just the mass of the atoms that make it up because the mass of the electron that was lost from the uncharged molecule is negligible.

The mass spectrum of hexane is shown in Figure 16-35. The mass of hexane (C_6H_{14}) is 86 u (or 86 g/mol), so the peak representing the molecular ion, M^+, is found at $m/z = 86$. In general:

> The M^+ peak of a mass spectrum is taken to be the compound's molecular mass.

The molecular mass of a compound can help us determine the compound's molecular formula, because the molecular mass equals the sum of the masses of its individual atoms.

The value of the M^+ peak can also provide insight into whether the compound contains nitrogen, based on the **nitrogen rule**:

> **The Nitrogen Rule**
>
> - A compound containing an odd number of nitrogen atoms typically has an odd molecular mass.
> - A compound containing an even number of nitrogen atoms, or no nitrogen atoms at all, typically has an even molecular mass.

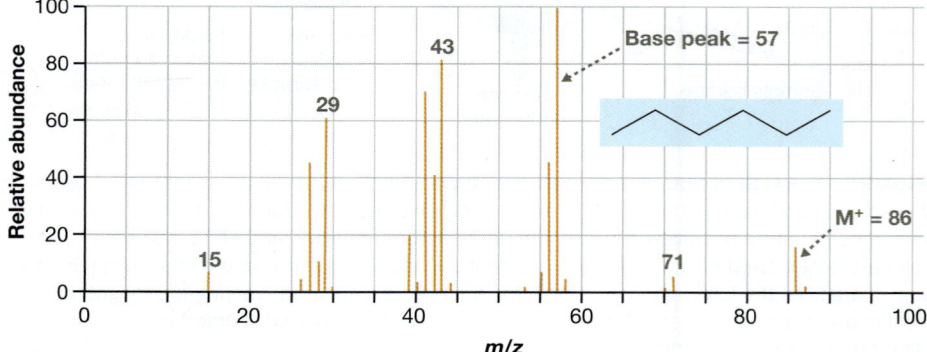

FIGURE 16-35 Mass spectrum of hexane The M^+ peak and the base peak are identified. The values of m/z for selected peaks are provided above the peaks.

For example, NH_3 contains a single nitrogen atom and its mass is 17 u, whereas ethane (H_3CCH_3) and hydrazine (H_2NNH_2) contain zero and two nitrogen atoms, respectively, and have masses of 30 and 32 u, respectively.

YOUR TURN 16.27

Compute the molecular mass of *N,N*-dimethylacetamide, CH_3—CO—$N(CH_3)_2$. Is it consistent with the nitrogen rule?

Notice in Figure 16-35 that there are several peaks present other than the M^+ peak. The relative abundance of each peak is assigned a value relative to that of the *base peak*.

The **base peak** in a mass spectrum is the one that is most intense and is assigned a relative abundance of 100%.

In Figure 16-35, the base peak is at $m/z = 57$, which is *not* the same as the M^+ peak, but don't be surprised if they are the same in mass spectra of some other compounds.

There are multiple mass peaks in Figure 16-35, despite the fact that the spectrum was produced from a pure sample of hexane, because **fragmentation** occurs after M^+ is formed. Typically, an electron from the electron beam is moving sufficiently fast that it not only knocks an electron off of the uncharged molecule when it collides, but also *breaks bonds*! In a molecule such as hexane, there are several different bonds that can be broken; hence, there are several different fragmentation pathways. Three of the fragmentation pathways that involve the breaking of carbon–carbon bonds are shown in Figure 16-36.

When a molecular ion undergoes fragmentation, there must be conservation of charge. As indicated in Figure 16-36, a molecular ion whose charge is +1 invariably fragments into one species bearing a +1 charge and another that is uncharged. *Only the charged fragments can be detected in mass spectrometry!*

FIGURE 16-36 Fragmentation pathways of hexane's molecular ion involving the breaking of carbon–carbon bonds Each fragmentation produces a cation and an uncharged species, ensuring that the total charge after fragmentation is the same as it was before fragmentation—that is, +1.

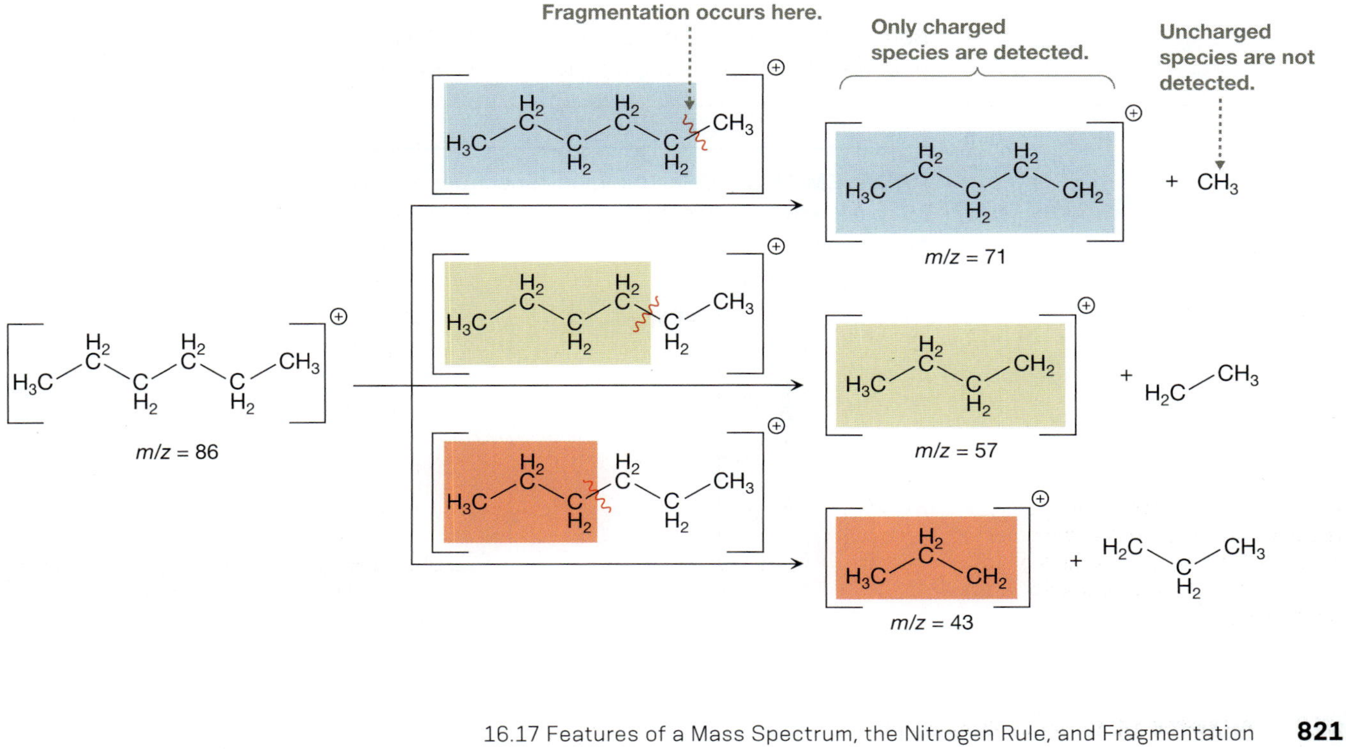

Each of the fragment ions in Figure 16-36 appears in hexane's mass spectrum and each is indicated in Figure 16-35. Thus, if we understand fragmentation pathways, then the fragment peaks that appear can help us to determine molecular structure.

Fragmentation pathways can be quite complex. We can begin to appreciate this complexity by noting that many more ion peaks appear in the spectrum of hexane than the ones we examined in Figure 16-36. The complexity of fragmentation processes is due, in part, to the fact that the pathways involve very high-energy species in the gas phase. Additionally, the molecular ion contains an odd number of electrons, making it a *radical cation*. As discussed in Chapter 25, the chemistry of radicals is somewhat different from that of *closed-shell species* in which all electrons are paired. Therefore, we leave a more in-depth discussion of fragmentation in mass spectrometry until Interchapter G, which comes after we have discussed radicals and radical reactions more fully in Chapter 25.

YOUR TURN 16.28

Hexane's molecular ion can also undergo the following two fragmentation pathways (not shown in Fig. 16-36):

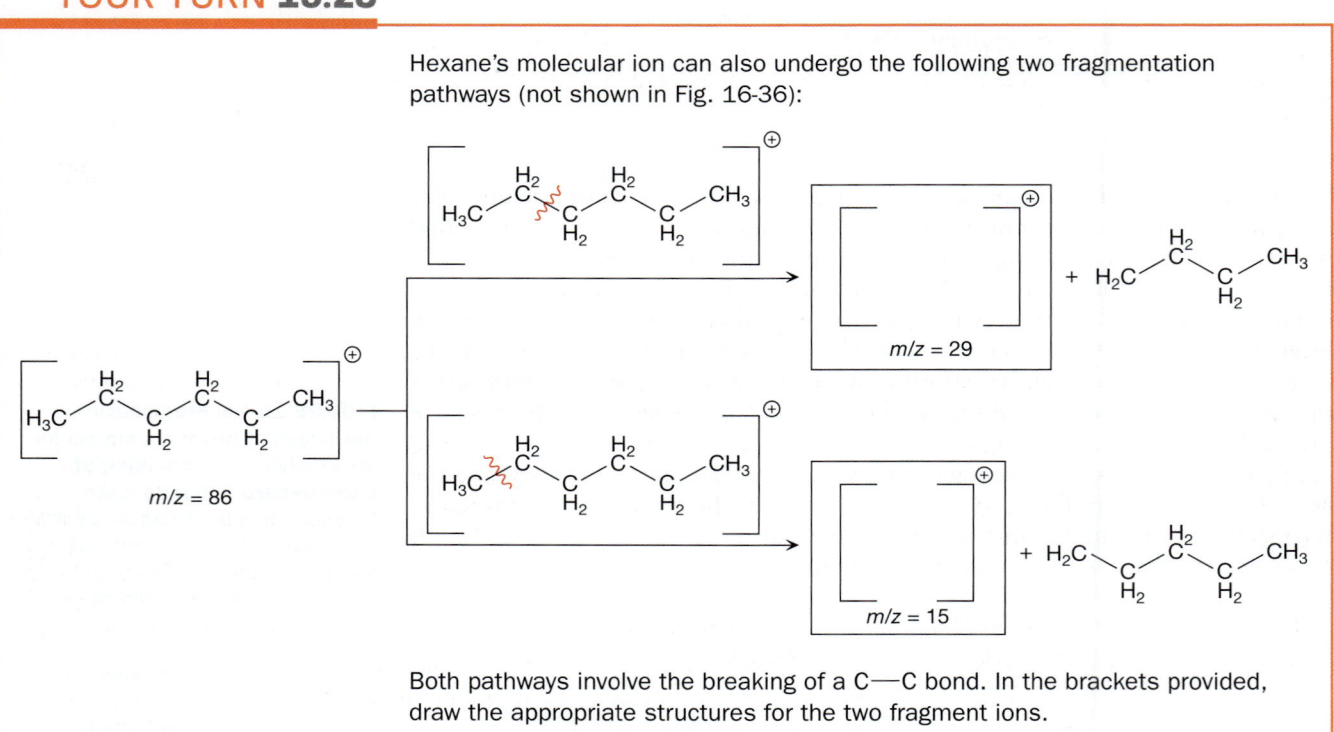

Both pathways involve the breaking of a C—C bond. In the brackets provided, draw the appropriate structures for the two fragment ions.

PROBLEM 16.33 Using the mass spectrum of ethylbenzene provided, identify the M$^+$ peak and the base peak. What is the formula of the ion responsible for the base peak?

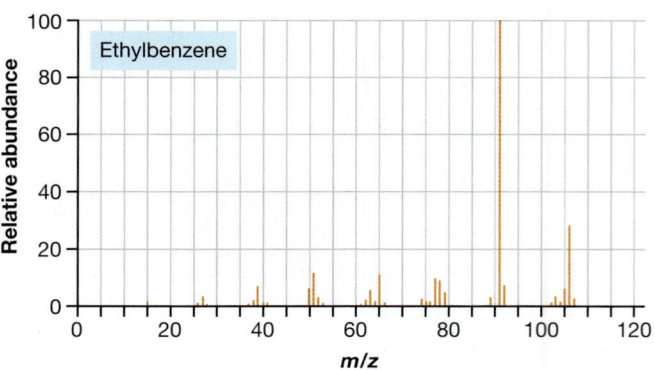

16.18 Isotope Effects: M + 1 and M + 2 Peaks

Although the M^+ peak of hexane appears at $m/z = 86$, there is a small peak at $m/z = 87$ in Figure 16-35, too. This peak is called the **M + 1 peak**.

YOUR TURN 16.29

Label the M + 1 peak in Figure 16-35. Also, identify and label the M + 1 peak in the mass spectrum of ethylbenzene in Problem 16.33.

The M + 1 peak represents a molecular ion containing a heavy isotope. Hexane contains only carbon and hydrogen atoms, so in this case it must be due to the appearance of either an atom of ^2H (deuterium) or ^{13}C. According to Table 16-7, however, the natural abundance of ^2H is negligible, whereas that of ^{13}C is about 1.1%, so hexane's M + 1 peak must be due primarily to an ion that contains five ^{12}C atoms, one ^{13}C atom, and 14 ^1H atoms.

YOUR TURN 16.30

Compute the molecular mass of $^{12}C_5{}^{13}C^1H_{14}$. How does this compare to m/z for the M + 1 peak in the mass spectrum of hexane?

The fact that an M + 1 peak appears in a spectrum is not particularly useful by itself, but the *intensity* of that peak relative to the intensity of the M^+ peak can be very useful. In the mass spectrum of hexane, for example, the relative intensity of the M + 1 peak is 1.0% and that of the M^+ peak is 15.5%. The ratio of these two peaks is $(1.0)/(15.5) = 0.065$, or 6.5%. Consider, now, the mass spectrum of dodecane, $C_{12}H_{26}$, which is shown in Figure 16-37. The relative intensity of the M + 1 peak is 0.8%, and that of the M^+ peak is 5.9%. The ratio of these intensities is $(0.8)/(5.9) = 0.136$, or 13.6%. In other words:

CONNECTIONS Dodecane is considered to be a substitute for traditional jet fuels.

The ratio of the M + 1 peak intensity to the M^+ peak intensity increases as the number of carbon atoms in the molecule increases.

TABLE 16-7 Relative Isotopic Abundance of Naturally Occurring Elements Common in Organic Molecules

Element	Most Abundant Isotope	Abundance	Heavy Isotopes	Abundance[a]	Element	Most Abundant Isotope	Abundance	Heavy Isotopes	Abundance[a]
Carbon	^{12}C	98.90%	^{13}C	1.10%	Chlorine	^{35}Cl	75.77%	^{37}Cl	24.23%
Hydrogen	^1H	99.985%	^2H (D)	0.015%	Bromine	^{79}Br	50.69%	^{81}Br	49.31%
Nitrogen	^{14}N	99.634%	^{15}N	0.366%	Sulfur	^{32}S	95.02%	^{33}S	0.75%
Oxygen	^{16}O	99.76%	^{17}O	0.038%				^{34}S	4.21%
			^{18}O	0.20%	Silicon	^{28}Si	92.23%	^{29}Si	4.67%
								^{30}Si	3.10%

[a]Abundances shown in red are the heavy isotopes most likely observable in a mass spectrum.

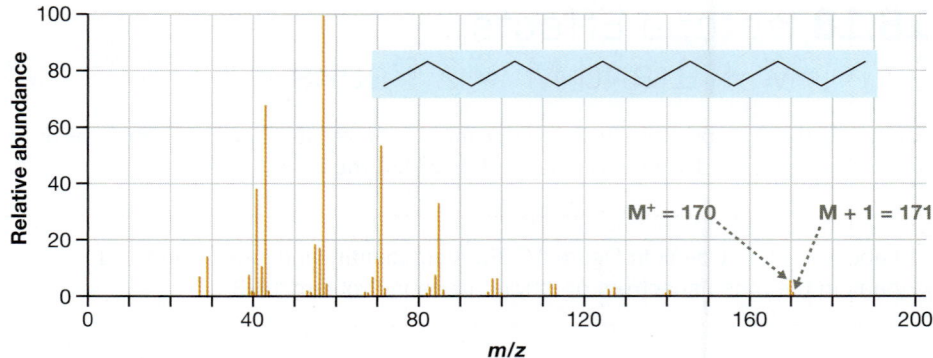

FIGURE 16-37 **Mass spectrum of dodecane, $C_{12}H_{26}$** The M^+ and the $M + 1$ peaks are labeled.

This is because the likelihood that a molecular ion contains one ^{13}C nucleus increases when the molecule contains more carbons.

This idea can be used to *estimate* the number of carbon atoms, according to Equation 16-7.

$$\text{Number of C atoms} \approx \frac{\text{Intensity of M} + 1}{\text{Intensity of M}^+} \times \frac{100\%}{1.1\%} \tag{16-7}$$

The 1.1% appears in the equation because that is the probability that a given carbon atom is ^{13}C (Table 16-7). Part of the reason that Equation 16-7 can only be used as an estimate is that atoms other than carbon, such as hydrogen and nitrogen, have isotopes that contribute a small amount to the $M + 1$ peak.

YOUR TURN 16.31

Use Equation 16-7 to verify that hexane has six carbons and dodecane has 12.

PROBLEM 16.34 Estimate the number of carbon atoms present in the compound that produced the following mass spectrum.

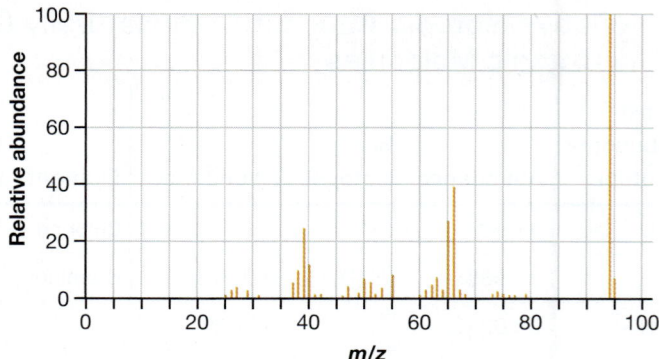

Based on the relative abundances in Table 16-7, carbon contributes significantly to an $M + 1$ peak. Other elements, however, have relatively abundant isotopes that are 2 u heavier than their most common isotopes. These include sulfur, chlorine, and

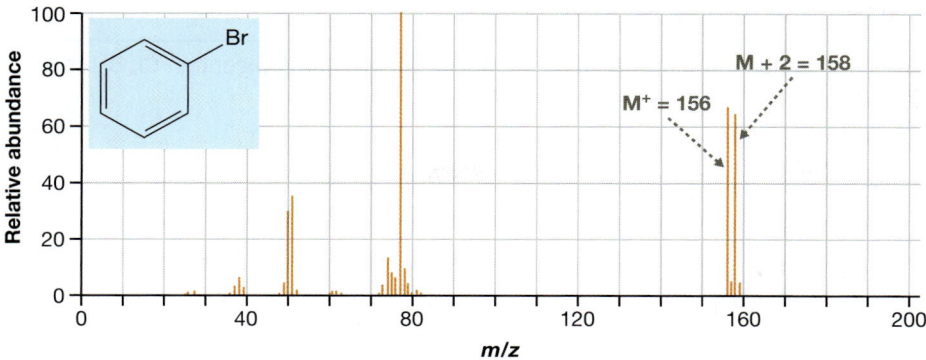

FIGURE 16-38 **Mass spectrum of bromobenzene** The M^+ and M + 2 peaks correspond to molecular ions with the ^{79}Br and ^{81}Br isotopes, respectively.

bromine. The appearance of an **M + 2 peak**, therefore, may indicate the presence of one of these elements, but which one?

With good certainty, we can identify which of these elements is present by the intensity of the M + 2 peak relative to that of the M^+ peak. For example, notice in the mass spectrum of bromobenzene (C_6H_5Br; Fig. 16-38) that the M^+ peak and the M + 2 peak have roughly equal intensities. The M^+ peak represents the molecular ion $[C_6H_5Br]^+$ that contains the ^{79}Br isotope, whereas the M + 2 peak represents the molecular ion $[C_6H_5Br]^+$ that contains the ^{81}Br isotope. The similar intensities of these peaks are consistent with the fact that the ^{79}Br and ^{81}Br isotopes have nearly equal abundances in nature (i.e., 50.69% and 49.31%, respectively).

YOUR TURN **16.32**

Verify that the values of m/z for the M^+ and M + 2 peaks in Figure 16-38 are consistent with the presence of the ^{79}Br and ^{81}Br isotopes by calculating the molecular masses of both species. Also, what molecular formula (including isotope designations) would account for the small peak at $m/z = 157$? What formula would account for the peak at $m/z = 159$?

Fragment ions that contain a bromine atom also appear as pairs of mass peaks of roughly equal magnitude, which differ by 2 in their values of m/z. This is quite evident in the mass spectrum of 1-bromo-4-(1-methylethyl)benzene, shown in Figure 16-39.

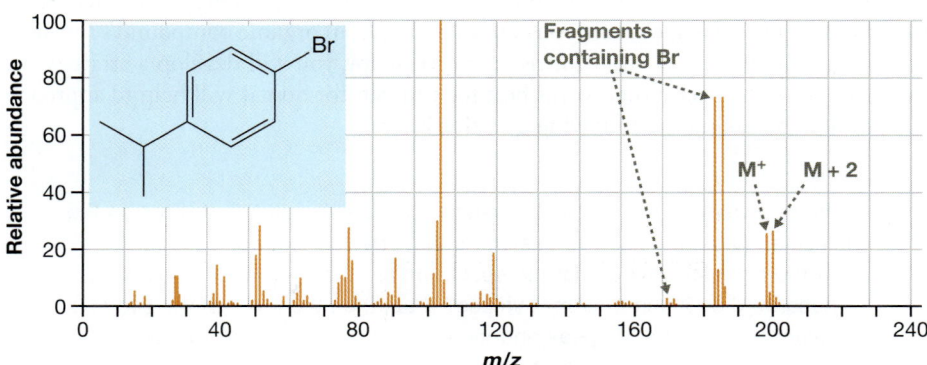

FIGURE 16-39 **Mass spectrum of 1-bromo-4-(1-methylethyl)benzene** Two fragments containing Br are indicated, as evidenced by pairs of peaks of roughly equal intensity, differing by two units of m/z.

SOLVED PROBLEM 16.35

Does the compound that produced the spectrum shown here contain Cl, Br, or S? Explain.

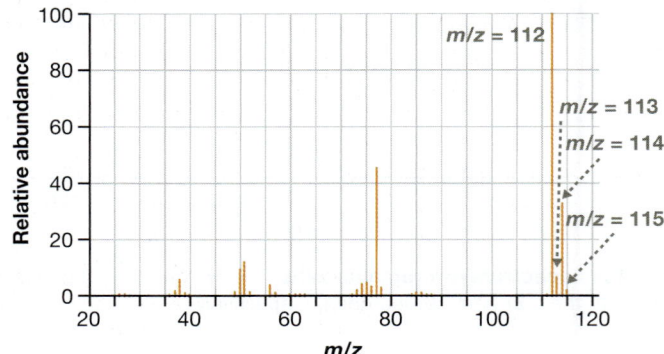

Think Can you identify an M^+ and an $M + 2$ peak in the spectrum? If so, what is the ratio of their intensities? Does that match the relative abundance of either Cl, Br, or S from Table 16-7 (p. 823)?

Solve There are two significant mass peaks separated by 2 u at the high-mass part of the spectrum: one at $m/z = 112$ and one at $m/z = 114$. The peak at $m/z = 112$, therefore, is M^+ and the one at $m/z = 114$ is $M + 2$. The intensities of the two peaks are in roughly a 3:1 ratio, which matches the relative abundances of the ^{35}Cl and ^{37}Cl isotopes. The compound that generated this mass spectrum, therefore, more likely contains Cl than Br or S.

PROBLEM 16.36 At what value of m/z do you expect the M^+ peak of C_6H_5SH to appear? At what value of m/z do you expect the $M + 2$ peak to appear? If the M^+ peak is also the base peak, then what intensity would you expect the $M + 2$ peak to have?

16.19 Determining a Molecular Formula of an Organic Compound from the Mass Spectrum

One of the fundamental pieces of information you can obtain from a mass spectrum is a compound's molecular mass. Knowing the molecular mass, along with other information from the mass spectrum, you can often derive an organic compound's formula. Just as we mentioned for other forms of spectroscopy, you will develop a strategy for interpreting mass spectra that works best for you, but for now it will help to approach the problem by asking the following questions:

1. Is the m/z value of the M^+ peak given to you? If so, assign that value as the molecular mass. If not, try to assign the M^+ peak as follows:
   a. At the highest-mass end of the spectrum, are there two major peaks separated by 2 u in their m/z values? If so, then the lower of the two m/z values is likely the M^+ peak and the higher one is the $M + 2$ peak.
   b. If no $M + 2$ peak is apparent, then find the two highest-mass peaks that are separated by 1 u in their values of m/z. If the lower-mass peak is significantly more intense, it is likely the M^+ peak and the higher-mass peak is the $M + 1$ peak.

2. Can you calculate the ratio of the M$^+$ and M + 1 peak intensities and estimate the number of C atoms present using Equation 16-7 (p. 824)?
3. Is there evidence that the molecule contains any heteroatoms?
    a. Does the nitrogen rule suggest the presence of any nitrogen atoms?
    b. If an M + 2 peak is present, do the relative intensities of the M$^+$ and M + 2 peaks suggest the presence of Br, Cl, or S?
    c. Is there other spectral information (such as from an IR spectrum) that suggests the presence of a heteroatom?
4. Subtract the masses of the ^{12}C atoms you determined in question 2 and the heteroatoms you determined in question 3 from the molecular mass. Can the remaining mass reasonably be made up by only hydrogens?
    a. If so, then you have a complete molecular formula that could be feasible.
    b. If that number of H atoms is unreasonably high, then consider adding another non-hydrogen atom to the formula and repeat question 4.

Let's apply this strategy toward determining the molecular formula of Unknown 3 from its mass spectrum, which is presented in Figure 16-40. To answer question 1, we are not given the M$^+$ peak, but notice that the two highest-mass peaks are at $m/z = 129$ and $m/z = 130$ and the peak at $m/z = 129$ is significantly more intense. We therefore assign $m/z = 129$ to the M$^+$ peak and $m/z = 130$ to the M + 1 peak, in which case the molecular mass of Unknown 3 is 129 u.

For question 2, we can estimate that the compound has seven carbons, and for question 3a, the nitrogen rule suggests that the molecule has an odd number of N atoms. We can skip questions 3b and 3c because the spectrum does not contain an M + 2 peak and we are not given other spectral information.

YOUR TURN 16.33

Plug the values of the M$^+$ and M + 1 peaks into Equation 16-7 to verify the estimate of seven carbons in Unknown 3.

For question 4, we subtract the mass of seven ^{12}C atoms (weighing 84 u) from the molecular mass of 129 u, leaving us with 45 u that must be accounted for by non-carbon atoms. We also subtract the mass of an odd number of N atoms because of what we concluded from the nitrogen rule. If we assume that the compound contains three N atoms (weighing 42 u), then the remaining mass of 3 u could be accounted for by three H atoms. It is not very reasonable, however, for a compound containing seven C atoms and three N atoms to contain only three H atoms. Therefore, Unknown 3 probably contains just one N atom, leaving a mass of 31 u yet to be accounted for.

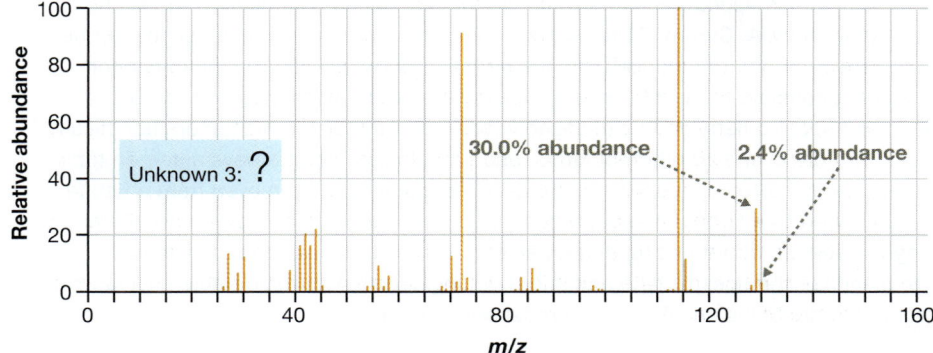

FIGURE 16-40 Mass spectrum of Unknown 3

Although that remaining mass of 31 u could be accounted for by 31 H atoms, that would be unreasonable, too, exceeding the number of H atoms for a completely saturated molecule containing seven C atoms and one N atom. Therefore, the molecule must contain another heteroatom, such as O or F. Formulas that are consistent with this mass spectrum could therefore be $C_7H_{15}NO$ or $C_7H_{12}NF$.

YOUR TURN 16.34

To confirm that it would be unreasonable for Unknown 3 to contain 31 H atoms, calculate the number of H atoms a saturated molecule with seven C atoms and one N atom would have.

PROBLEM 16.37 Suppose that the IR spectrum of Unknown 3 contains a strong absorption at 1670 cm^{-1} and no absorptions above 3000 cm^{-1}. Draw three structures consistent with these data.

PROBLEM 16.38 An unknown compound has the mass spectrum shown here. Its ^{13}C NMR spectrum has four signals and its IR spectrum has an intense peak near 1720 cm^{-1}. Propose a structure for this compound.

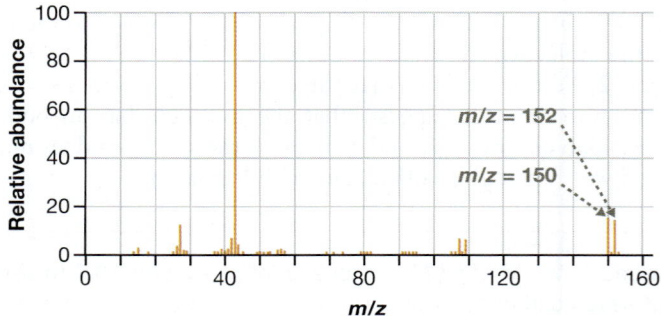

Mass Spectrometry, *CSI*, and *Grey's Anatomy*

Mass spectrometry is not just an analytical tool in chemistry; it can help investigators solve crimes and can help doctors identify and treat cancer!

Investigating a crime involving explosives is often challenging when the explosive has been nearly entirely consumed. But a mass spectrum of the residue that is produced can help investigators trace the explosive back to its origin. When investigators suspect arson, they can use mass spectrometry to analyze the partially charred wood for the presence of trace amounts of an accelerant such as gasoline, kerosene, or mineral spirits. And scientists are looking into the possibility of using mass spectrometry as evidence that can place a criminal at the scene of a crime. In their 2010 study, A. Curran, P. Prada, and K. Furton showed that human scent can be analyzed by mass spectrometry to produce a unique bar code of an individual's "primary odor" compounds, making it possible to identify an individual by the odor that they leave behind.

In medicine, mass spectrometry can help diagnose brain tumors. Detecting tumors is difficult, in part, because there are over 125 different kinds, and pathologists don't always agree on their diagnoses. To help identify a tumor, a spectrum produced by compounds removed from a tissue sample can be compared to a library of spectra generated from various types of tumors. Perhaps more interestingly, researchers are finding that a patient's breath can be analyzed by mass spectrometry to diagnose the specific type and stage of lung cancer with remarkable accuracy. And other breath tests are being developed for breast and colon cancers.

Chapter Summary and Key Terms

- In **nuclear magnetic resonance (NMR) spectroscopy**, a sample is placed in a strong **external magnetic field (B_{ext})** and irradiated with radiation from the **radio frequency (RF)** portion of the electromagnetic spectrum. The re-emitted frequencies are recorded as a **free induction decay (FID)**, and multiple FIDs are averaged before a **Fourier transform** is applied to obtain the separate NMR signals. An NMR spectrum plots the intensity of each signal emitted by the sample against *chemical shift*, a quantity related to relative frequency. **(Section 16.1; Objectives 1, 2)**

- Absorption of an RF photon causes a nucleus to undergo a **spin flip**. In **^{1}H NMR spectroscopy**, the nuclei that undergo spin flips are those of hydrogen atoms (protons, ^{1}H); in **^{13}C NMR spectroscopy**, carbon-13 nuclei undergo spin flips. **(Sections 16.2 and 16.3; Objective 2)**

- Signal frequency depends on the identity of the nucleus and increases with increasing B_{ext} due to an increasing energy separation between nuclear spin states. **(Section 16.2; Objective 3)**

- Atoms that are **chemically distinct** have nuclei that reside in different **chemical environments**. Chemically distinct nuclei differ in the extent to which they are **shielded** from B_{ext} and thus generate signals at different frequencies. **(Sections 16.3 and 16.13a; Objectives 4, 5)**

- If nuclei in different chemical environments interchange rapidly, they tend to give rise to the same averaged signal. **(Section 16.5; Objective 5)**

- **Chemical shift** is a measure of a signal's frequency relative to the signal frequency generated by **tetramethylsilane (TMS)**. Whereas a signal's frequency is proportional to B_{ext}, chemical shift is independent of B_{ext}. **(Section 16.6; Objective 4)**

- Nearby electron-withdrawing groups **deshield** nuclei inductively, thereby increasing chemical shift. Nearby π electrons from double bonds deshield nuclei via **magnetic anisotropy**. **(Section 16.7; Objective 4)**

- Deshielding from inductive effects and from magnetic anisotropy is additive, and falls off rapidly with distance. A group that is separated from a nucleus by more than two bonds has little effect on the chemical shift of that nucleus. **(Section 16.8; Objective 4)**

- The **integration** of an NMR signal is the area under the peaks in the spectrum and is graphically represented as an **integral trace**. In a ^{1}H NMR spectrum, integration is proportional to the number of protons giving rise to that signal. **(Section 16.9; Objective 6)**

- According to the **$N + 1$ rule**, N protons will split the signal of an adjacent, chemically distinct proton into $N + 1$ peaks. The relative intensities of those peaks are described by **Pascal's triangle**. **(Section 16.10a; Objective 7)**

- Protons on oxygen or nitrogen tend to exhibit none of the effects of coupling, appearing instead as broad singlets. **(Section 16.10b; Objective 7)**

- The **coupling constant (J)** of an NMR signal is the difference in frequency between adjacent peaks belonging to the same split signal. Coupling constants are independent of B_{ext}, so signals become better resolved with increasing B_{ext}. **(Section 16.11; Objective 7)**

- Signals of protons that are coupled together have the same coupling constant. **(Section 16.11; Objective 7)**

- A proton signal can exhibit **complex splitting** if the proton is coupled to two or more protons that are not equivalent to each other. These splitting patterns can be derived from a **splitting diagram**. **(Section 16.12; Objective 7)**

- Carbon signals are generated by ^{13}C nuclei. Signal averaging compensates for the low natural abundance of the ^{13}C nucleus. **(Section 16.13a; Objective 1)**

- All ^{13}C signals appear as singlets due to **broadband decoupling**, and their integrations do *not* correlate precisely with the number of nuclei. **(Sections 16.13b and 16.13d; Objective 7)**

- **DEPT ^{13}C NMR spectroscopy** gives the number of hydrogen atoms on each carbon. **(Section 16.14; Objective 8)**

- A **mass spectrum** plots the relative abundance of gaseous ions against the **mass-to-charge ratio, m/z**. **(Section 16.17; Objective 9)**

- A molecule's mass can be ascertained from the m/z of the **molecular ion, M^+**. **Fragmentation** gives rise to peaks with smaller m/z values. **(Sections 16.16 and 16.17; Objective 10)**

- The presence of ^{13}C gives rise to an **M + 1 peak**. The total number of carbon atoms in a molecule can be computed from the intensity of the M + 1 peak relative to that of the M^+ peak. **(Section 16.18; Objectives 10, 11)**

- The presence of bromine, chlorine, or sulfur can be identified by the relative intensity of an **M + 2 peak**. **(Section 16.18; Objective 11)**

Problems

16.3–16.5 Chemical Distinction, the Number of NMR Signals, and the Time Scale of NMR Spectroscopy

16.39 For each of the following molecules, determine how many signals should appear in its ^1H NMR spectrum.

(a)   (b)   (c)   (d)   (e)

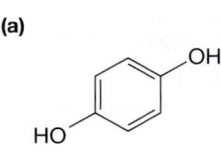 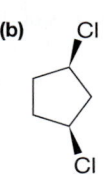 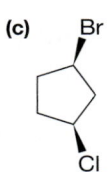

(f)   (g)   (h)   (i)

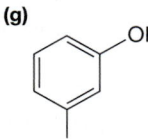 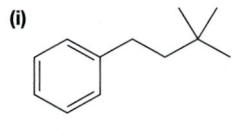

16.40 At room temperature, N,N-dimethylformamide, $HCON(CH_3)_2$, has three ^1H NMR signals, appearing at 2.9, 3.0, and 8.0 ppm. As the temperature is increased, the two signals at 2.9 and 3.0 ppm merge into one signal. Explain. *Hint:* Consider the resonance structures of the compound.

16.41 Cyclohexane-d_{11} (C_6HD_{11}) exhibits one signal in its ^1H NMR spectrum at room temperature. Two signals appear in the spectrum, however, when the temperature is lowered significantly, as shown here. Explain why this happens.

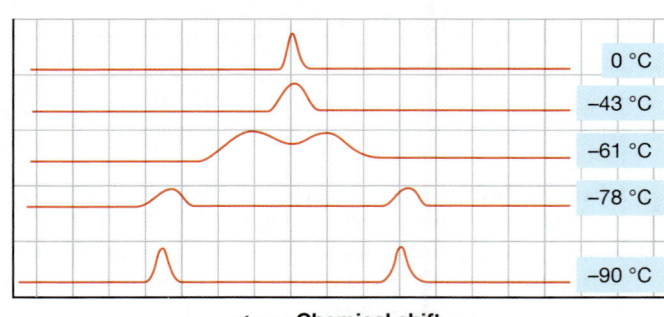

⟵ Chemical shift

16.42 At room temperature, the ^1H NMR spectrum of all-*cis*-1,2,3,4,5,6-cyclohexanehexacarboxylic acid exhibits two signals for the H atoms directly bonded to the ring. Explain why. *Hint:* Draw its chair conformation explicitly.

16.6–16.8 Operating Frequency, Chemical Shift, Shielding, and Deshielding

16.43 When a 300-MHz NMR spectrometer is used, one proton signal produced from a compound is 150 Hz higher than another signal. In a 90-MHz instrument, what would be the frequency difference between the two signals?

16.44 For a particular NMR instrument, the operating frequency is 300 MHz for ^1H NMR spectroscopy and 75 MHz for ^{13}C NMR spectroscopy. Calculate the gyromagnetic ratio for a ^{13}C nucleus.

16.45 In 2015, scientists developed a 1020-MHz NMR instrument. Calculate the strength of the magnet.

16.46 Equation 16-3, shown again below, relates the operating frequency of an NMR spectrophotometer (ν_{op}) to the gyromagnetic ratio of the nucleus of interest (γ) and the magnitude of the external magnetic field (B_{ext}).

$$\nu_{op} = \frac{\gamma B_{ext}}{2\pi}$$

Show how this equation is derived from Equation 16-1 and the condition necessary for photon absorption: $E_{photon} = \Delta E_{spin}$.

16.47 Using the NMR instrument described in Problem 16.44, suppose that a ^{13}C nucleus from a sample generates a signal whose frequency is 11,250 Hz higher than that from the carbons in TMS. What is the chemical shift of that carbon atom from the sample?

16.48 A compound generates two ^1H NMR signals. The frequency of the first is 450 Hz higher than that of the signal from TMS, whereas the frequency of the second is 755 Hz higher than that of TMS. Which signal corresponds to the protons that are shielded to a greater extent? Explain.

16.49 Explain why TMS has a lower chemical shift than its carbon analog, dimethylpropane, $(CH_3)_4C$.

16.50 Estimate the ^1H NMR chemical shift for the protons on C2 in the following compound.

16.51 How would the ^1H NMR spectra of compounds **A** and **B** differ?

A   **B**

16.52 The chemical shifts of two carbon atoms are given for the molecules shown here. It appears that the addition of the OH group *increases* the chemical shift of the C atom to which it is attached, but that it *decreases* the chemical shift of the C atom on the opposite side. How do you account for these observations? *Hint:* Draw all of the pertinent resonance structures.

125 ppm   138 ppm   153 ppm   130 ppm

16.53 Two signals appear in the ^1H NMR spectrum of the compound shown here. One has twice the integration of the other. The signal with greater area corresponds to a chemical shift of 9.3 ppm. The signal with less area corresponds to a chemical shift of −2.9 ppm. Explain. *Hint:* Consider the magnetic field lines from a ring current.

16.9–16.12 Signal Integration and Signal Splitting

16.54 A compound, whose formula is $C_{11}H_{14}$, produces a ^1H NMR spectrum with four signals. The steps made by the integral trace measure 37, 9, 26, and 52 mm. How many protons give rise to each signal?

16.55 Both 1,4-dimethylbenzene and 1,3,5-trimethylbenzene produce a ^1H NMR spectrum that has two signals. In which spectrum do the signal integrations have a 1:3 ratio?

16.56 Suppose that the ^1H NMR spectrum of a compound, $C_5H_{12}O$, exhibits five signals with the relative integration 1:5:5:20:30. What can you say about the signal with the relative integration of 1?

16.57 A ^1H NMR spectrum exhibits the set of absorption peaks shown here. Can those peaks be generated by one chemically distinct type of proton? Why or why not?

6.8   6.7

16.58 Determine the splitting pattern for each type of H highlighted in the following molecules. (Ignore long-range coupling.)

(a)   (b)   (c)

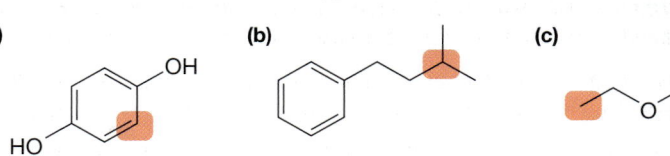

(d)   (e)   (f)

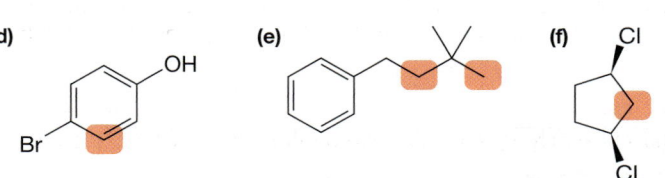

16.59 Suppose that a sextet appears in a ^{1}H NMR spectrum. Is it possible for the protons that produce the signal to be coupled to protons that are all equivalent to each other? Explain.

16.60 Draw the splitting pattern that would be observed for the proton **B**, highlighted here. The coupling constant between protons **A** and **B** is about 7 Hz, whereas the coupling constant between protons **B** and **C** is about 16 Hz.

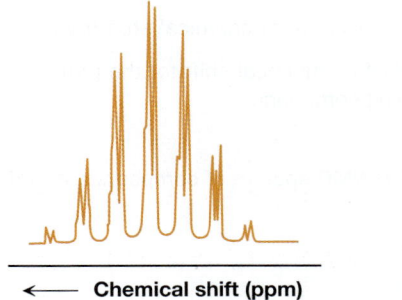

16.61 Based on the splitting pattern, how many total protons are coupled to the proton that gave rise to the signal shown here?

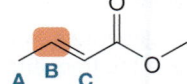

Chemical shift (ppm)

16.13 and 16.14 ^{13}C NMR Spectroscopy and DEPT ^{13}C NMR Spectroscopy

16.62 For each of the molecules in Problem 16.39, determine how many signals should appear in its ^{13}C NMR spectrum.

16.63 ^{13}C NMR spectra of three isomers of C_8H_{18} are provided. Match each of those spectra to one of the following compounds: octane; 2,5-dimethylhexane; and 4-methylheptane. (Remember that the signal at ~77 ppm is due to the solvent.)

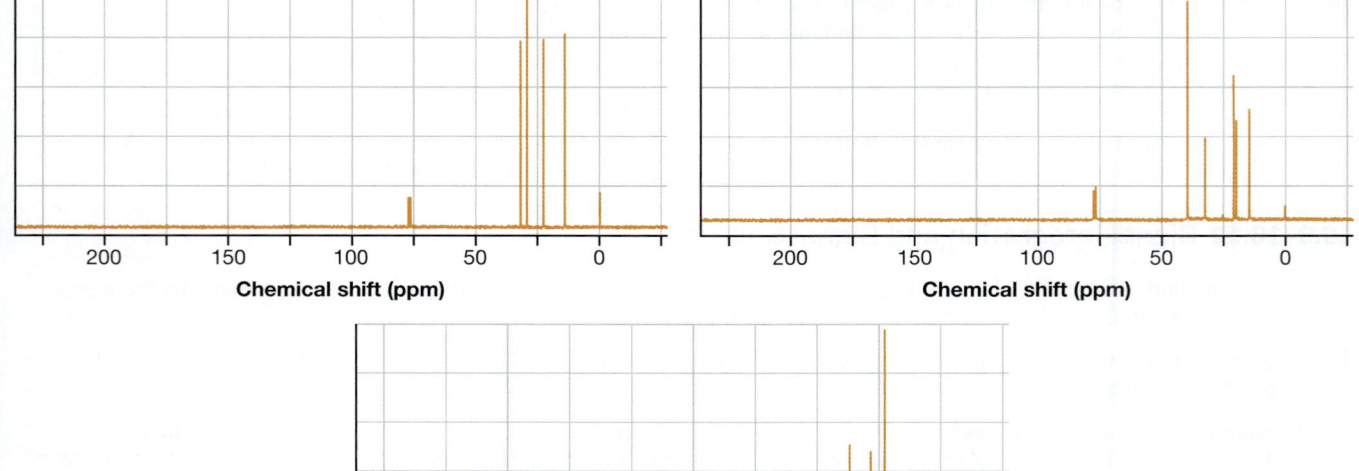

16.64 Which of the two ^{13}C NMR spectra below corresponds to chlorocyclohexane and which corresponds to iodocyclohexane? Explain.

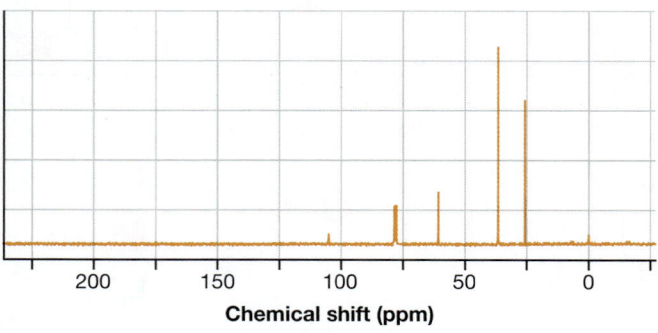

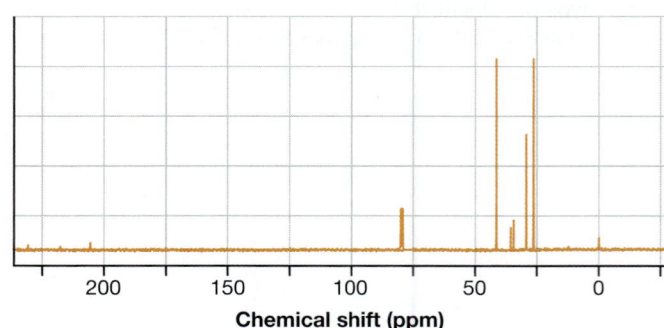

16.65 1-Chloropropane produced the ^{13}C NMR spectrum shown at right. Match each carbon atom in the molecule to the signal to which it corresponds.

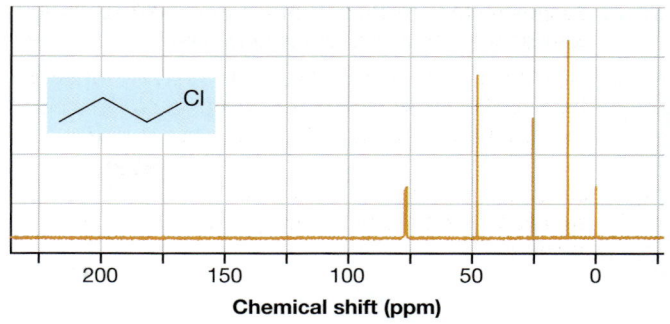

16.66 Which of the signals in the ^{13}C NMR spectrum in Problem 16.65 would appear in the DEPT-90 spectrum? In the DEPT-135 spectrum, which signals would appear as positive signals and which would appear as negative signals?

16.15 Structure Elucidation Using NMR Spectroscopy

16.67 A compound whose formula is $C_9H_{18}O$ produces a ^1H NMR spectrum that contains only one signal—namely, a singlet that appears at 1.25 ppm. What is the structure of the compound?

16.68 Determine the structure of the compound $C_6H_{10}O$ whose ^1H NMR spectrum is shown here.

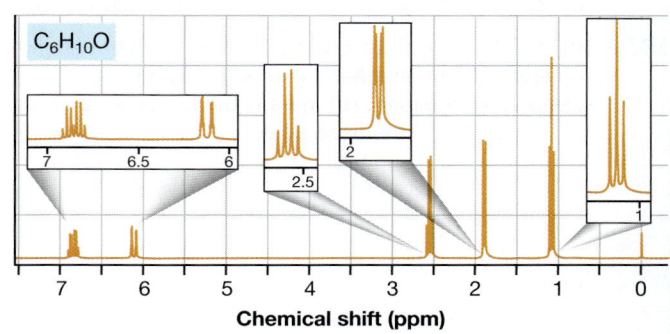

16.69 Determine the structure of the compound $C_7H_{14}O_2$ whose ^1H NMR spectrum is shown here.

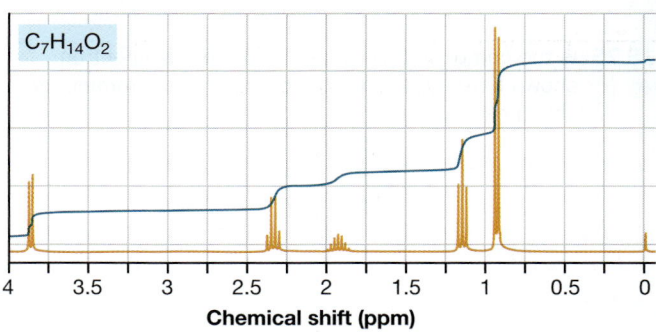

16.70 Determine the structure of the compound $C_4H_9NO_2$ whose ^1H NMR spectrum is shown here.

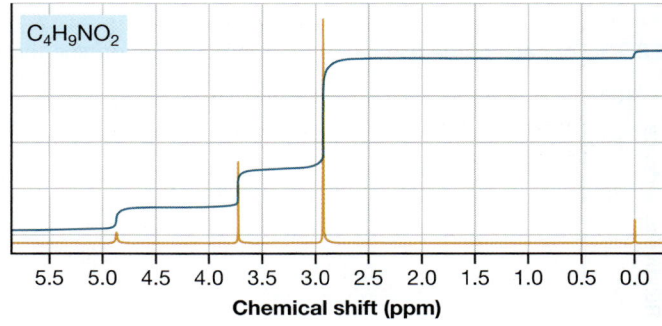

16.71 Determine the structure of the compound $C_{10}H_{14}$ whose ^{13}C NMR spectrum is shown here.

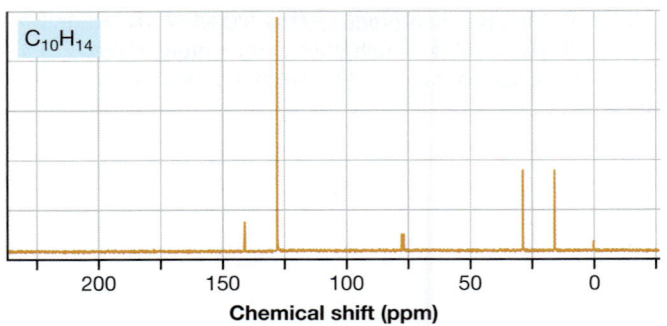

$C_{10}H_{14}$

Chemical shift (ppm)

16.16–16.19 Mass Spectrometry and Determining a Molecular Formula from a Mass Spectrum

16.72 Compounds **X** and **Y** both produce mass spectra in which the M$^+$ peak appears at $m/z = 122$. In the spectrum of compound **X**, the relative intensity of the M$^+$ peak is 83.2% and that of the M + 1 peak is 6.7%. In the mass spectrum of compound **Y**, the analogous intensity values are 16.9% and 1.6%. Which compound contains more carbon atoms?

16.73 A compound containing only carbon, nitrogen, oxygen, and hydrogen contains four carbon atoms. If the M$^+$ peak in its mass spectrum appears at $m/z = 87$, then how many nitrogen atoms does it contain?

16.74 **(a)** Draw the structures of the species that correspond to each of the peaks **A–E** in the following mass spectrum of heptane. **(b)** Identify the M$^+$ peak, the M + 1 peak, and the base peak.

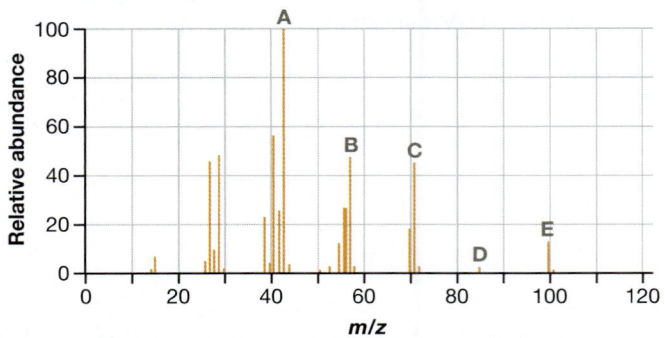

16.75 Is the compound giving rise to the mass spectrum shown here more likely to contain sulfur, bromine, or chlorine? Explain.

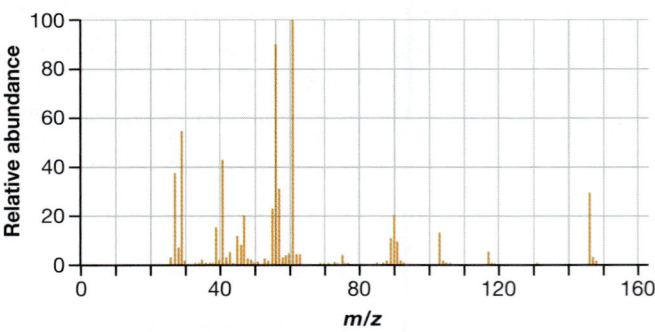

16.76 For each of the two mass spectra below, determine the formula of a compound that can produce it.

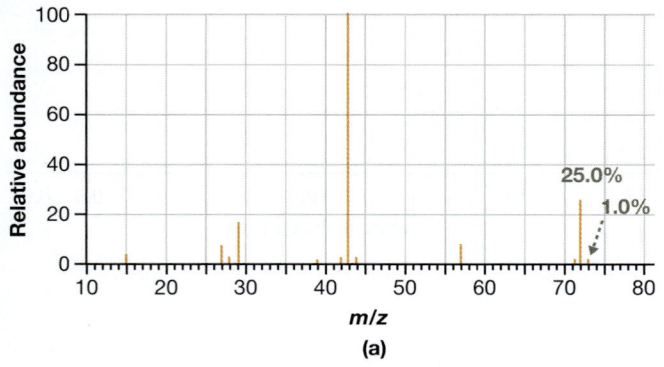

(a)

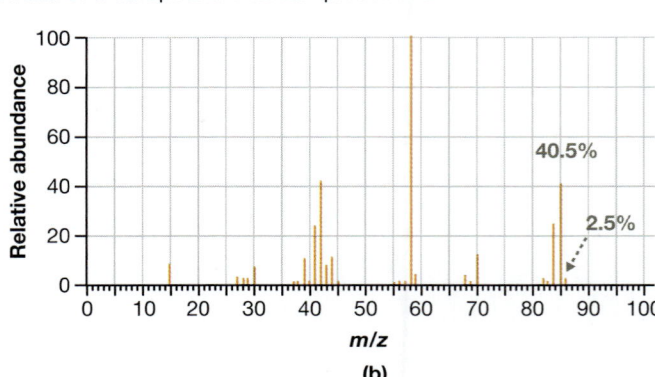

(b)

16.77 Determine the formula of a compound that can give rise to the mass spectrum at right.

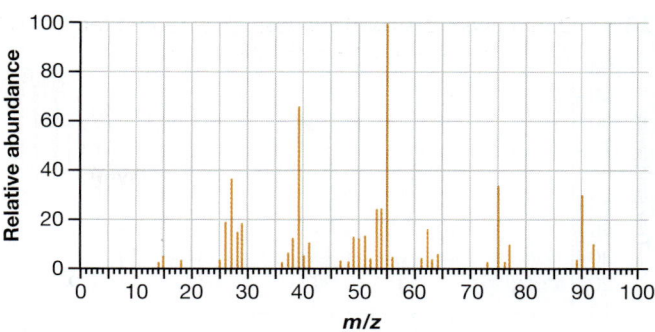

Integrated Problems

16.78 Do you think ^{1}H NMR or ^{13}C NMR spectroscopy would be more suitable for distinguishing among compounds **A**, **B**, and **C**? Explain.

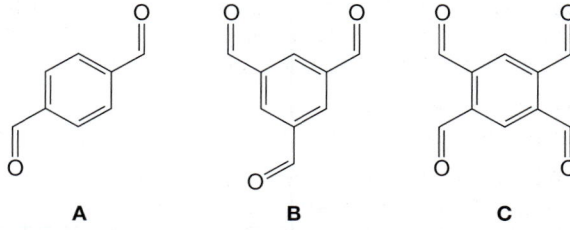

A   **B**   **C**

16.79 For each of the following compounds, sketch both a ^{1}H NMR spectrum and a ^{13}C NMR spectrum. In the ^{1}H NMR spectrum, pay attention to the splitting patterns and include the integral trace. (Ignore long-range coupling.)

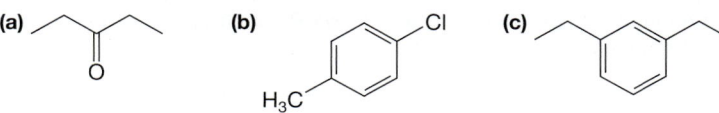

16.80 A compound has a formula of $C_{19}H_{16}$. Five signals appear in its ^{13}C NMR spectrum—one at 57 ppm and the remaining four between 126 and 144 ppm. In its ^{1}H NMR spectrum, an unresolved multiplet (15 H) appears between 6.9 and 7.44 ppm and a singlet (1 H) appears at 5.5 ppm. What is the structure of this compound?

16.81 The ^{1}H NMR spectrum of an unknown compound is shown here. In the compound's mass spectrum, the M^+ peak appears at $m/z = 92$. An M + 2 peak, whose intensity is roughly one-third that of the M^+ peak, also appears. Draw the structure of this compound.

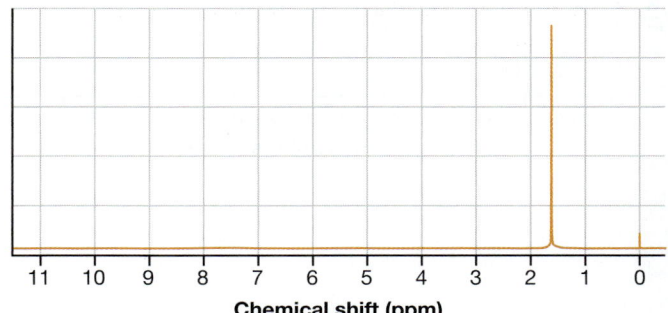

16.82 The formula of a compound is $C_9H_{10}O_3$. Use the IR, ^1H NMR, and ^{13}C NMR spectra provided to determine its structure.

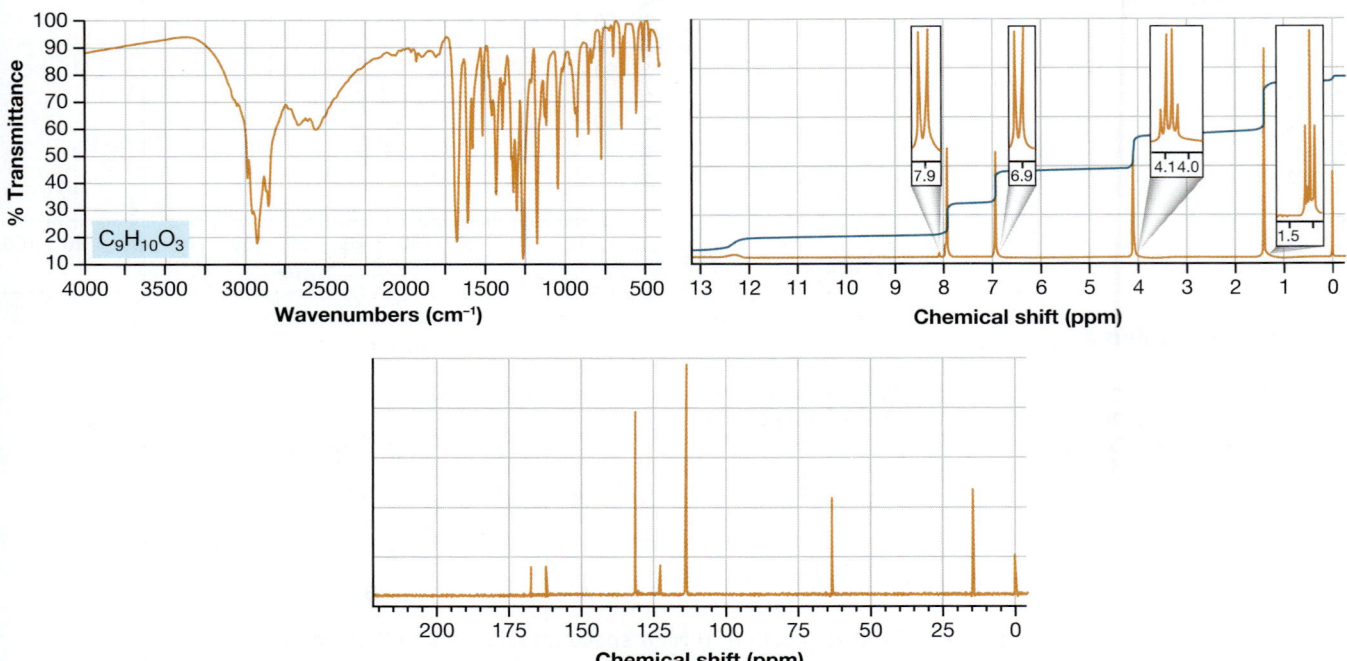

16.83 The formula of a compound is C_9H_8O and its IR and ^1H NMR spectra are provided. If its ^{13}C NMR spectrum has seven signals, then what is the compound's structure?

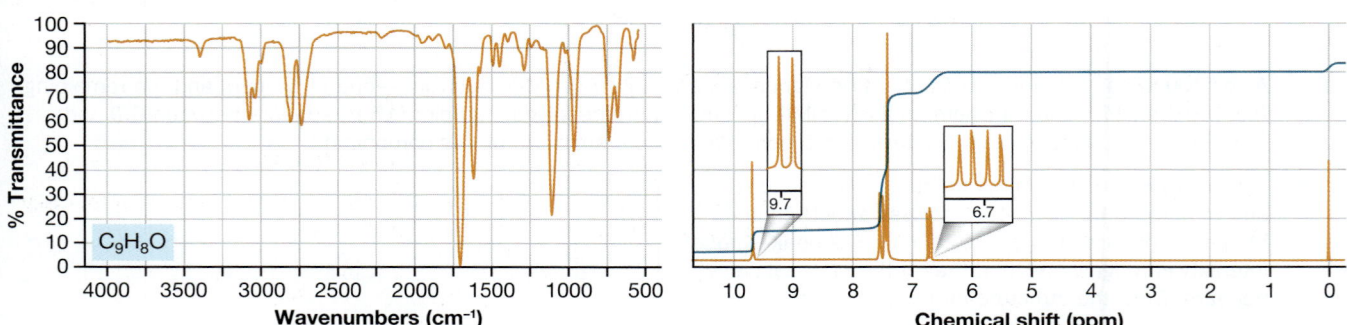

16.84 A compound with the formula $C_4H_{10}O_2$ has the following IR, ^{1}H NMR, and ^{13}C NMR spectra. Determine the structure of the compound.

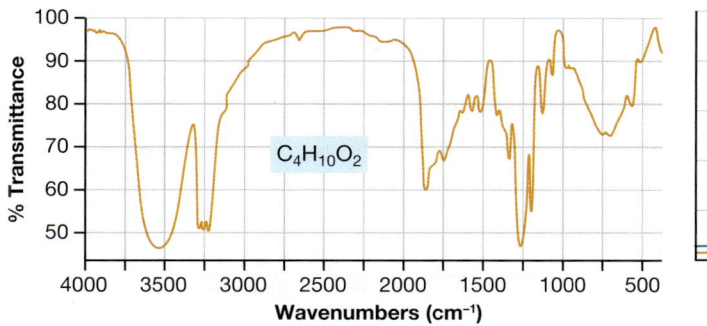

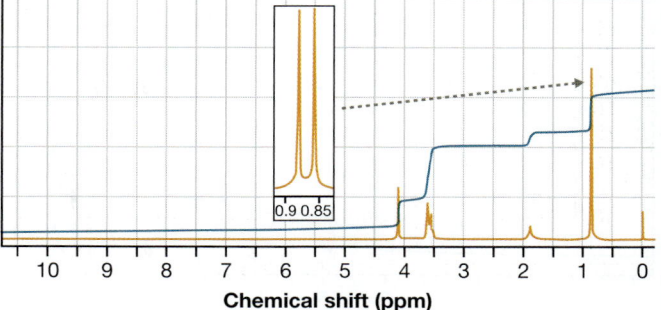

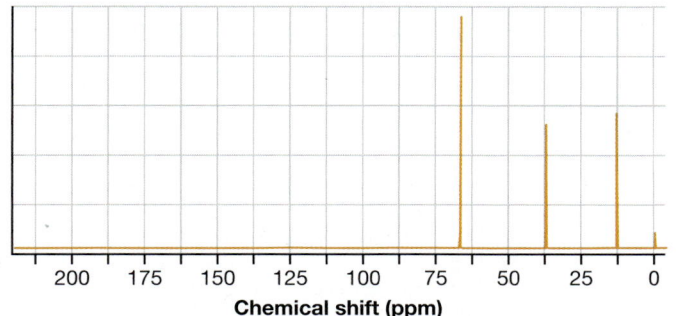

16.85 The chemical shifts of the α protons on cyclohexanone and cyclobutanone are as shown here: **(a)** Which α carbon has a greater effective electronegativity? **(b)** Can you explain why, using arguments of s character and p character?

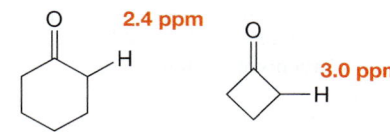

Cyclohexanone   Cyclobutanone

16.86 A student wants to determine if the S_N2 reaction shown here occurs. **(a)** Why would this experiment require isotopically labeled compounds? **(b)** Suggest how she can answer this question using mass spectrometry.

16.87 A student runs the reaction shown here and obtains the ^{1}H NMR spectrum shown. Determine the product of the reaction and draw the complete, detailed mechanism of the reaction.

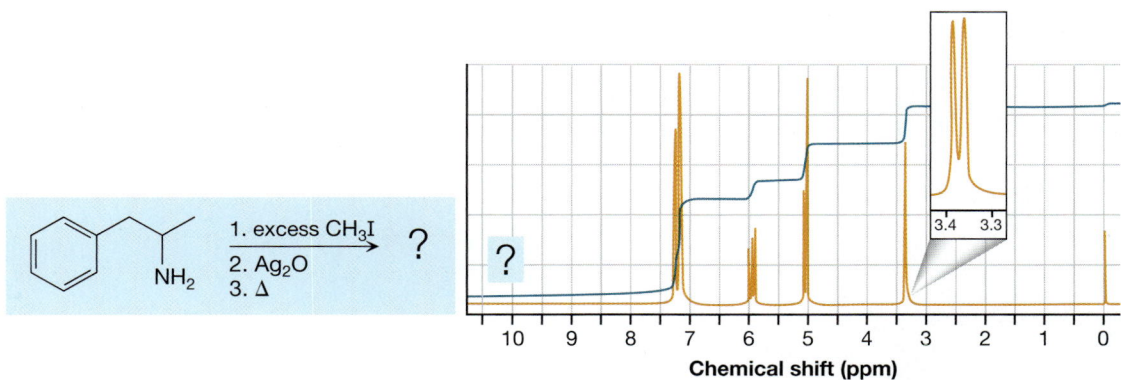

16.88 An alcohol was treated with HBr, yielding a mixture of 1-bromopent-2-ene and 3-bromopent-1-ene. The ^{1}H and ^{13}C NMR spectra of the starting alcohol are given here. Determine the structure of the starting alcohol and draw a complete, detailed mechanism to account for each of the products.

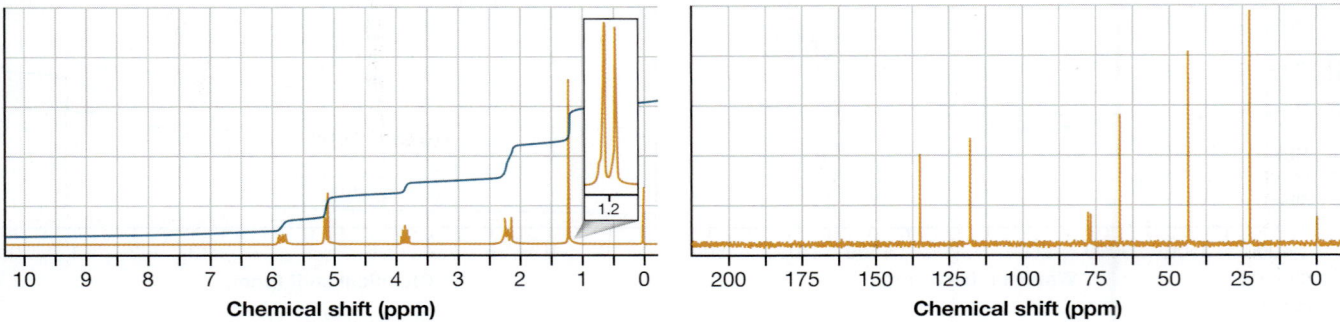

16.89 An alcohol was treated with concentrated sulfuric acid and heated. The product that was obtained had the following IR, ^{1}H NMR, and ^{13}C NMR spectra. **(a)** Determine the structure of the alcohol. **(b)** Draw a complete, detailed mechanism for the reaction that took place.

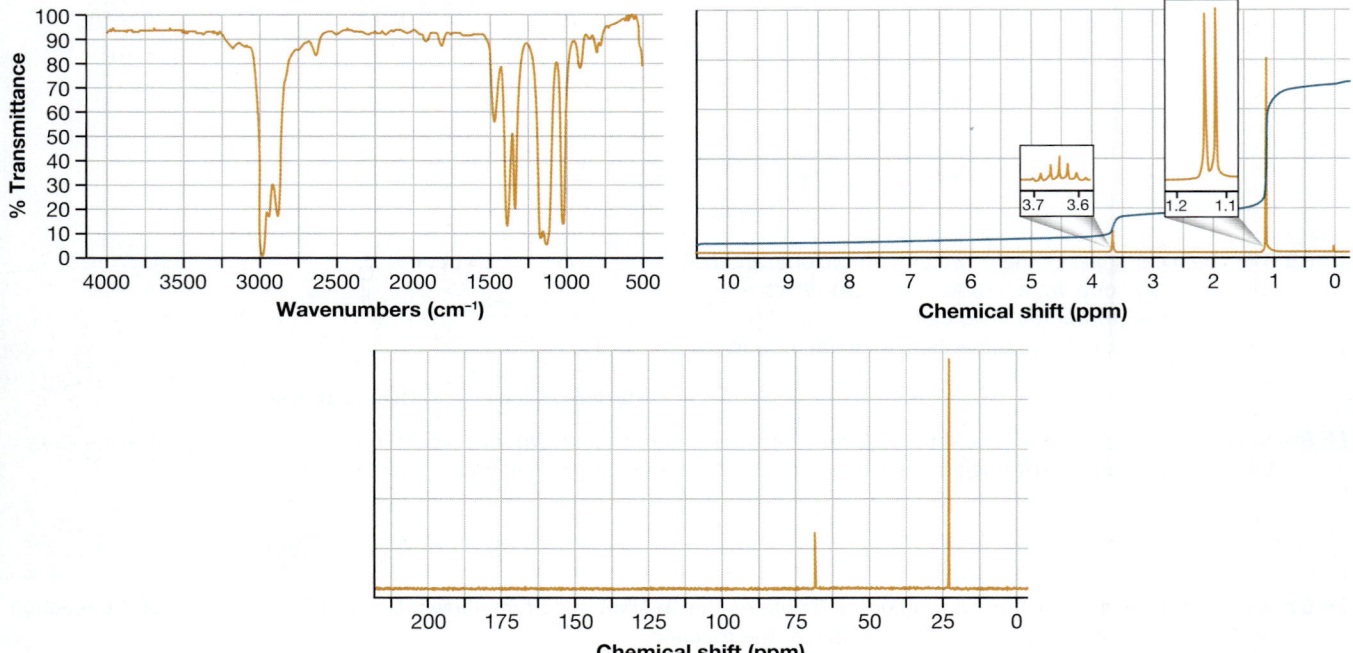

This species of millipede (*Apheloria corrugata*) defends itself by secreting a mixture of hydrogen cyanide and benzaldehyde, produced from a stored cyanohydrin. Cyanohydrins can be made from nucleophilic addition reactions, the topic of Chapter 17.

Nucleophilic Addition to Polar π Bonds 1

Addition of Strong Nucleophiles

In Chapters 11 and 12, we examined reactions involving nonpolar π bonds, such as the C≡C bond of an alkene (Fig. 17-1a). Being electron rich, a *nonpolar* π bond tends to undergo the addition of an electrophile, which is electron poor. Here in Chapter 17, we examine addition reactions involving *polar* π bonds—that is, π bonds joining atoms of significantly different electronegativity—represented as C≡Y in Figure 17-1b. Unlike in an alkene, a polar π bond has an atom that is relatively electron poor (i.e., electrophilic), so it tends to undergo the addition of a nucleophile, which is electron rich. These **nucleophilic addition reactions** are some of the most important reactions in organic chemistry, and they are integral in a variety of biological reactions, too.

A polar π bond appears in several compound classes, as shown in Figure 17-2. Far and away the most common polar π bond that participates in nucleophilic addition reactions is the one in the

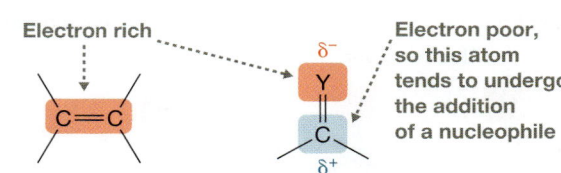

(a) A nonpolar π bond   (b) A polar π bond

FIGURE 17-1 Nonpolar and polar π bonds (a) A nonpolar π bond is relatively electron rich, so it tends to undergo the addition of an electrophile. (b) A polar π bond, where Y is more electronegative than C, has an electron-poor C atom that tends to undergo the addition of a nucleophile.

Chapter Objectives

On completing Chapter 17 you should be able to:

1. Draw the general mechanism for the addition of a strong nucleophile to a polar π bond that does not have an attached leaving group.

2. Account for the difference between aldehydes and ketones in their relative susceptibility to nucleophilic addition.

3. Recognize $NaBH_4$ and $LiAlH_4$ as reducing agents and predict the major product when one of those reducing agents reacts with a ketone, aldehyde, imine, or nitrile.

4. Distinguish the reactivity of NaH from that of $NaBH_4$ or $LiAlH_4$.

5. Predict the major product when an organometallic reagent such as RMgX or RLi reacts with a ketone, aldehyde, imine, or nitrile.

6. Specify which solvents are compatible with hydride reagents, such as $NaBH_4$ and $LiAlH_4$, and which are compatible with organometallic reagents, such as RMgX and RLi.

7. Determine whether the existence of a proposed Grignard or alkyllithium reagent is feasible, based on functional group compatibility.

8. Predict the major organic product of a Wittig reaction and draw its complete, detailed mechanism.

9. Show how to synthesize a Wittig reagent from an alkyl halide precursor.

10. Draw the general mechanism for both direct addition (1,2-addition) and conjugate addition (1,4-addition) to an α,β-unsaturated polar π bond.

11. Determine whether nucleophilic addition to an α,β-unsaturated carbonyl favors the direct addition product or the conjugate addition product, and explain the role of reversibility in such an outcome.

carbonyl (C=O) group. Carbonyl groups are present in ketones (R_2C=O), aldehydes (RCH=O), carboxylic acids (RCO_2H), esters (RCO_2R), amides ($RCONR_2$), acid halides (RCOX), and acid anhydrides (RCO_2COR). Of these, we focus primarily on ketones and aldehydes here in Chapter 17 and in Chapter 18. The carbonyl group in the remaining compound classes behaves somewhat differently because of a *leaving group* attached to the carbonyl C (red type in Fig. 17-2); we study such compounds in Chapters 20 and 21. Although less common, other polar π bonds encountered in organic molecules, such as the C=N bond in imines and the C≡N bond in nitriles, react similarly to the carbonyl group in ketones and aldehydes and are discussed here in Chapter 17, too.

Recall from Section 9.3a that species containing a negatively charged atom with lone pairs tend to be strong nucleophiles while uncharged species tend to be weak. Both types of nucleophiles can add to polar π bonds, and the mechanisms of these reactions are quite similar. As we will see, however, if the nucleophilic reagent that is added to the reaction mixture is weak, the reaction often requires either acid catalysis or base catalysis. Here in Chapter 17, therefore, we focus primarily on reactions in which the nucleophilic reagent that is added is strong, and in Chapter 18, we will study reactions in which the nucleophilic reagent that is added is weak.

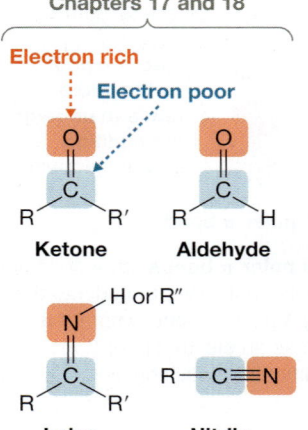

Chapters 17 and 18

Chapters 20 and 21

FIGURE 17-2 Some compound classes containing a polar π bond The compound classes in the first row contain the polar C=O bond. The compound classes in the second row contain polar C=N and C≡N bonds. Ketones, aldehydes, imines, and nitriles do not contain a leaving group and are discussed here in Chapter 17 and in Chapter 18; each of the remaining compound classes do contain a leaving group (red type) and are discussed in Chapters 20 and 21.

17.1 An Overview of the General Mechanism: Addition of Strong Nucleophiles

As we saw in Chapter 7, a nucleophile tends to form a bond with the atom at the positive end of a *polar π bond*. Whereas the nucleophile is relatively electron rich, the partially positive (δ^+) atom of the π bond is relatively electron poor. An example with a generic strong nucleophile (Nu^-) and a ketone is shown in Equation 17-1.

General mechanism for nucleophilic addition to a polar π bond

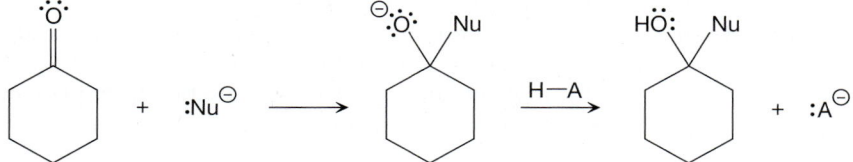

(17-1)

To avoid exceeding the octet on the C atom attacked by the nucleophile, the π bond is broken and the pair of electrons from the π bond becomes a lone pair on the more electronegative O atom.

As shown in Equation 17-1, the immediate product of nucleophilic addition (Step 1) is often a strong base (in this case, an alkoxide anion, RO^-) because it possesses a relatively unstable negative charge. Sometimes, species already present in the reaction mixture, such as the solvent, are acidic enough to protonate that product (Step 2). Otherwise, we can carry out this protonation by adding an acid in a subsequent *acid workup* (Section 10.7a).

The following reaction is similar to the one in Equation 17-1.

Label each reacting species as either "electron rich" or "electron poor," draw in the appropriate curved arrows, and under each reaction arrow name the type of elementary step involved.

Answers to Your Turns are in the back of the book.

YOUR TURN 17.1

CONNECTIONS
3-Methylbutanal, commonly called isovaleraldehyde, is a flavor component described as having a cheesy or malt flavor, and is found in a variety of foods and beverages, including cheese, beer, chicken, and fish. 3-Methylbutanal is also used as feedstock in the manufacture of some pesticides.

SOLVED PROBLEM 17.1

Draw the complete, detailed mechanism and predict the product for the reaction of 3-methylbutanal shown here.

3-Methylbutanal

1. KCN, H$_2$O
2. H$_2$SO$_4$

?

Think What nucleophile is generated when KCN dissolves in water? Which atom will it attack? Which bond is most easily broken in the process? What is the role of the acid?

Solve KCN is ionic, so it dissolves in water as K^+ and ^-CN ions. ^-CN has a localized negative charge on C and is a strong nucleophile that will subsequently attack the electron-poor carbonyl carbon, breaking the π bond of the double bond. The resulting O^- is then protonated by the acid, producing a *cyanohydrin*, which is characterized by CN and OH groups bonded to the same C atom.

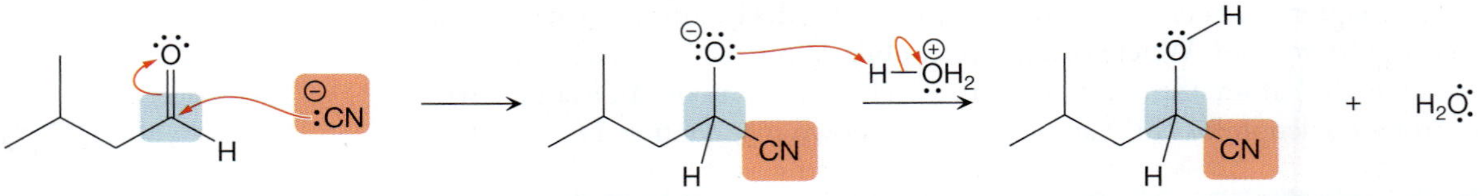

PROBLEM 17.2 Draw the complete, detailed mechanism for the reaction shown here and draw the product.

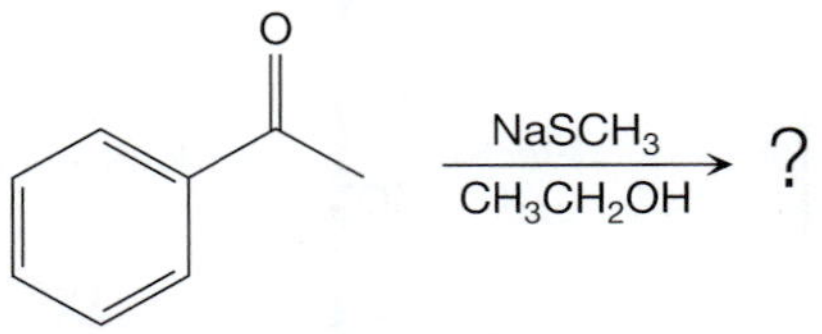

Stereochemistry is something that must be accounted for in nucleophilic addition when the carbon atom that is attacked becomes an asymmetric carbon in the product. An example is shown in Equation 17-2.

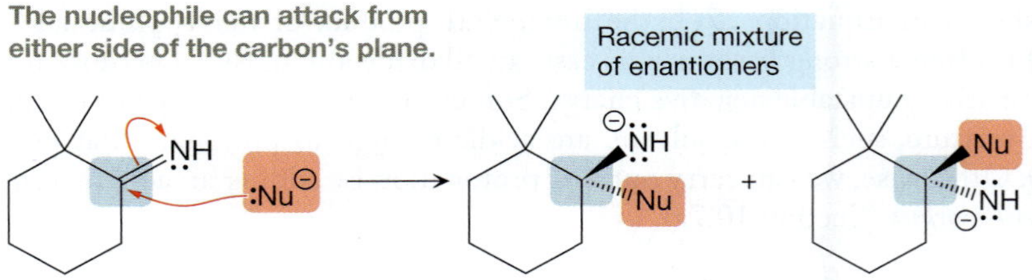

The nucleophile can attack from either side of the carbon's plane.

Racemic mixture of enantiomers

An imine

(17-2)

Because the C atom of the polar π bond has a planar electron geometry, a nucleophile can attack from either side. Thus, as we learned in Chapter 9, the result is a mixture of *R* and *S* stereochemical configurations at the bond formation site. A *racemic mixture* of enantiomers is formed if the reactants are achiral and the reaction takes place in an achiral environment. Otherwise, an unequal mixture of the *R* and *S* configurations will be formed.

YOUR TURN 17.2

The reaction in Solved Problem 17.1 results in a new asymmetric carbon. Identify that atom and draw both stereoisomers that would be produced.

17.2 Substituent Effects: Relative Reactivity of Ketones and Aldehydes in Nucleophilic Addition

Ketones and aldehydes exhibit very similar chemical behavior because they are rather similar structurally, but the nucleophilic addition reactions they undergo differ noticeably both thermodynamically and kinetically:

Nucleophilic attack at a carbonyl carbon tends to be more energetically favorable and faster for aldehydes than for ketones.

These points are exemplified in Table 17-1, which presents both equilibrium and rate data for the conversion of three different carbonyl compounds to their *hydrates* (two OH groups bonded to the same C) under basic conditions.

YOUR TURN 17.3

Draw the mechanism for each of the hydration reactions shown in Table 17-1.

TABLE 17-1 Extent and Rate of Carbonyl Hydration in Ketones and Aldehydes

Reaction	Relative Rate Constant[a]	Percent Hydrate at Equilibrium[b]
(formaldehyde + $^{\ominus}OH$ / H_2O → hydrate)	3×10^6	>99.9%
(acetaldehyde + $^{\ominus}OH$ / H_2O → hydrate)	5×10^4	57%
(acetone + $^{\ominus}OH$ / H_2O → hydrate)	1	<1%

[a]Reactions take place in water.
[b]Reactions take place in methanol.

Notice in Table 17-1 that both the rate of hydration and the extent of hydration at equilibrium are greater for the aldehydes than for the ketone. Both of these results can be explained in part by steric effects associated with the bulky alkyl groups of the ketone, as shown in Figure 17-3. In a ketone, the *steric repulsion* from the two alkyl groups on the carbonyl carbon (Fig. 17-3a, left) makes it more difficult for the nucleophile to attack. Furthermore, the bulky alkyl groups in the hydrate product (Fig. 17-3a, right) overlap with the bulky OH groups, generating *steric strain* and decreasing stability. An aldehyde (Fig. 17-3b), on the other hand, has at most one alkyl group bonded to the carbonyl carbon, so both of these steric effects are diminished.

FIGURE 17-3 Steric effects in nucleophilic addition (a) In a ketone, steric repulsion by each bulky alkyl group (left) makes it more difficult for a nucleophile to attack. Steric strain in the hydrate product (right) decreases stability. (b) In an aldehyde, there is at most one bulky alkyl group, so steric effects are decreased.

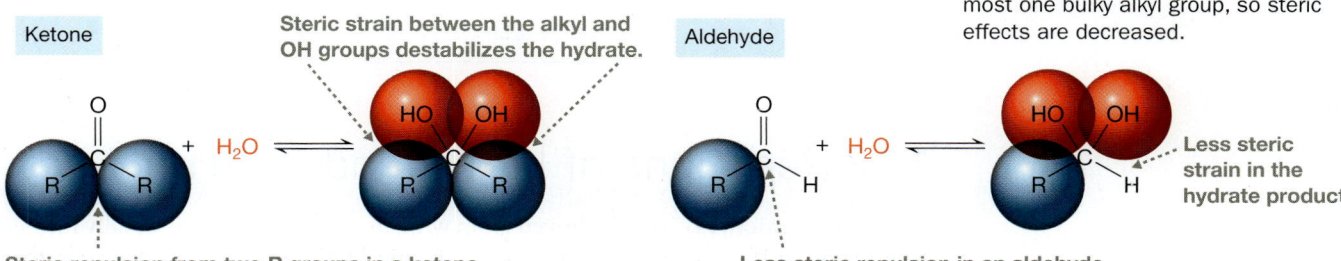

Ketone
Steric strain between the alkyl and OH groups destabilizes the hydrate.
Aldehyde

Less steric strain in the hydrate product

Steric repulsion from two R groups in a ketone makes it more difficult for the nucleophile to attack.
(a)

Less steric repulsion in an aldehyde, so the nucleophile can attack more easily
(b)

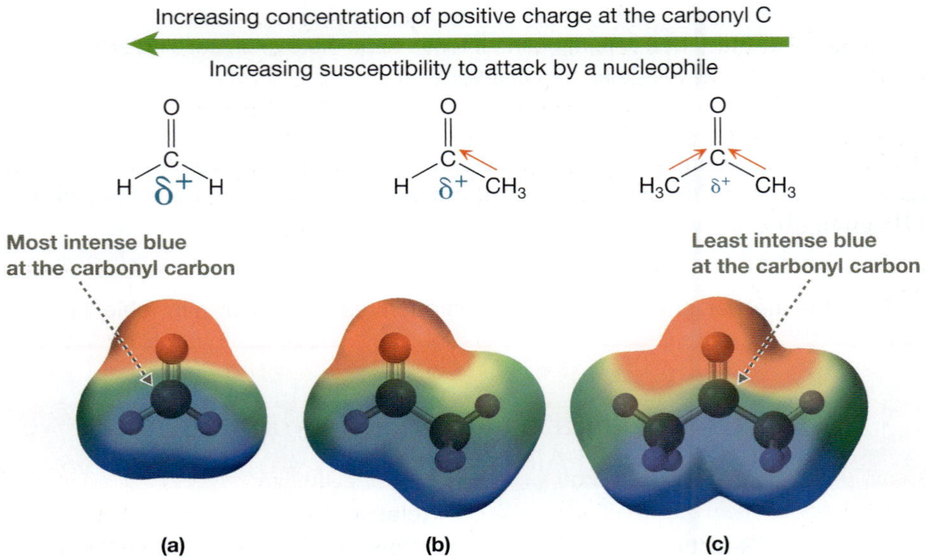

Increasing concentration of positive charge at the carbonyl C

Increasing susceptibility to attack by a nucleophile

Most intense blue at the carbonyl carbon

Least intense blue at the carbonyl carbon

(a)　　(b)　　(c)

FIGURE 17-4 Inductive effects in nucleophilic addition Electrostatic potential maps of (a) formaldehyde, (b) acetaldehyde, and (c) acetone. Additional alkyl groups bonded to the carbonyl carbon decrease its concentration of positive charge by donating electron density, making it less reactive.

Aldehydes are more reactive than ketones, too, because ketones are stabilized more by *inductive effects*, as illustrated in Figure 17-4. Recall from Section 6.6e that alkyl groups are electron donating, so they decrease the concentration of positive charge at the carbonyl carbon. With less concentration of positive charge at that carbon, the carbonyl group is less susceptible to attack by a nucleophile bearing excess negative charge.

YOUR TURN 17.4

Which of imines **A**–**C** has the greatest concentration of positive charge at the C=N carbon, and which has the least concentration of positive charge? Which is the most reactive imine? The least reactive?

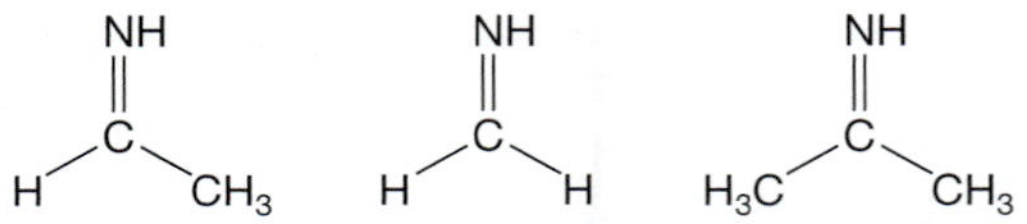

A　　B　　C

PROBLEM 17.3 Chloral, $Cl_3CCH{=}O$, forms a very stable hydrate called chloral hydrate. When dissolved in water, essentially 100% of chloral is hydrated. In contrast, the extent of hydration for ethanal (acetaldehyde) is much smaller (see Table 17-1). Explain why.

PROBLEM 17.4 Which molecule, **D** or **E**, would you expect to undergo hydration to a greater extent? Explain.

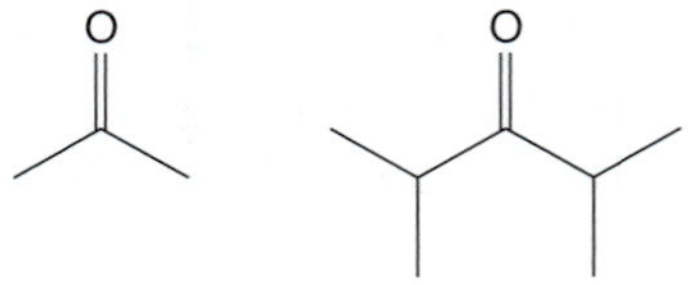

D　　E

17.3 Reactions of LiAlH₄ and NaBH₄

The **hydride anion, H:⁻**, is a hydrogen atom with an extra electron (it is usually further abbreviated to H⁻). Unlike such species as Cl⁻ and HO⁻, however, H⁻ does not exist on its own at any appreciable concentration in solution because it is extremely reactive, both as a nucleophile and as a base, even reacting with many common solvents. H⁻ is so reactive because it is small in size and its nucleus is only moderately electronegative, so it does not accommodate the negative charge well.

As we learned in Section 7.1b, two common hydride sources in organic chemistry are **sodium borohydride ($NaBH_4$)** and **lithium aluminum hydride ($LiAlH_4$, or LAH)**. In both cases, recall that H is covalently bonded to a less electronegative atom, so H has a partial negative charge (Fig. 17-5). In many instances, therefore, we can think of $NaBH_4$ and $LiAlH_4$ as simply H^-. We see how this is done, in greater detail, with the reactions involving polar π bonds we study here in this section.

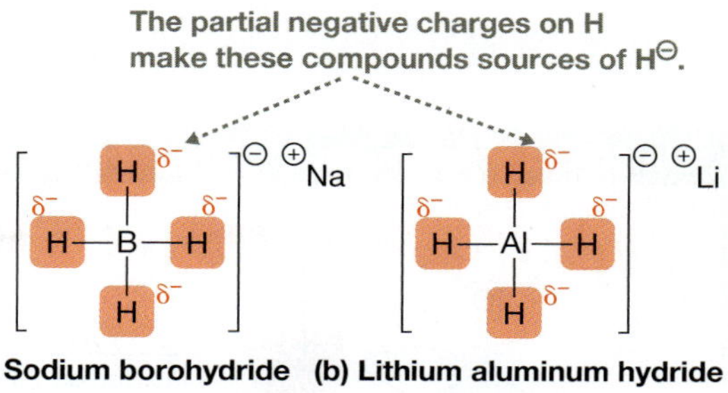

The partial negative charges on H make these compounds sources of $H^\ominus$.

(a) Sodium borohydride   (b) Lithium aluminum hydride

FIGURE 17-5 Common hydride sources (a) Sodium borohydride consists of Na^+ and BH_4^- ions, and (b) lithium aluminum hydride consists of Li^+ and AlH_4^- ions. In both anions, H bears a partial negative charge, so these are sources of hydride ions, H^-.

17.3a Converting Ketones and Aldehydes to Alcohols

Equation 17-3 shows that racemic butan-2-ol is formed when butan-2-one is treated with an aqueous solution of $NaBH_4$. Likewise, Equation 17-4 shows that the same result is achieved when butan-2-one is refluxed with an ether solution of $LiAlH_4$ followed by aqueous acid workup.

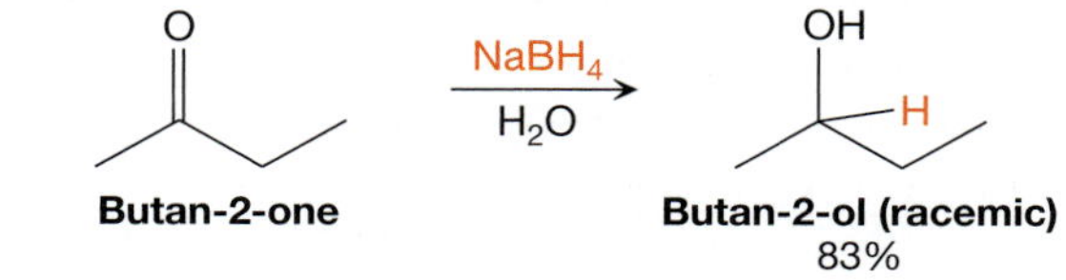

Butan-2-one $\xrightarrow[\text{H}_2\text{O}]{\text{NaBH}_4}$ **Butan-2-ol (racemic)**
83%   (17-3)

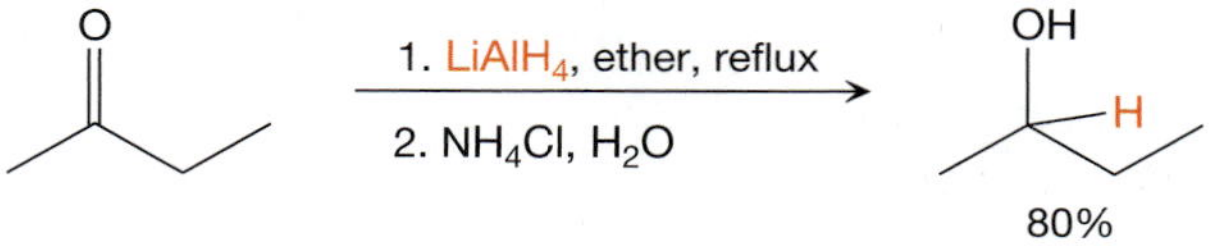

$\xrightarrow[\text{2. NH}_4\text{Cl, H}_2\text{O}]{\text{1. LiAlH}_4\text{, ether, reflux}}$ 80%   (17-4)

Similar reactions, shown in Equations 17-5 and 17-6, convert benzaldehyde to phenylmethanol.

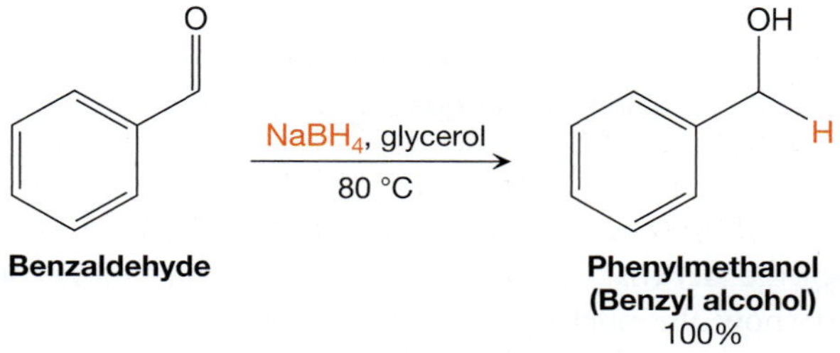

Benzaldehyde $\xrightarrow[\text{80 °C}]{\text{NaBH}_4\text{, glycerol}}$ **Phenylmethanol (Benzyl alcohol)**
100%   (17-5)

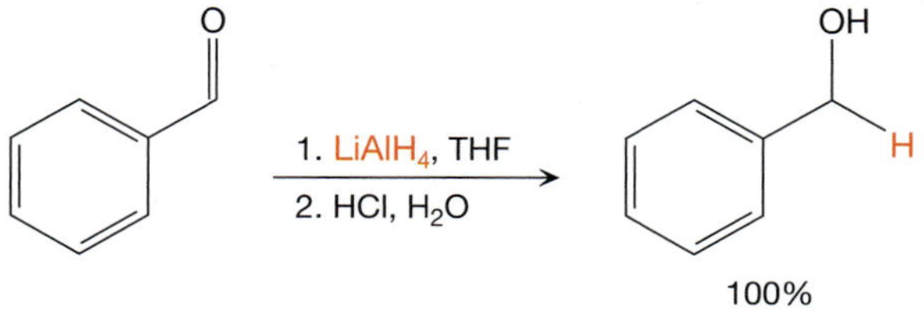

$\xrightarrow[\text{2. HCl, H}_2\text{O}]{\text{1. LiAlH}_4\text{, THF}}$ 100%   (17-6)

In general, $NaBH_4$ or $LiAlH_4$ can be used to convert:

- a ketone to a 2° alcohol.
- an aldehyde to a 1° alcohol.

The simplified mechanism for Equation 17-3, in which NaBH₄ serves as the hydride source, is shown in Equation 17-7.

Simplified mechanism for Equation 17-3

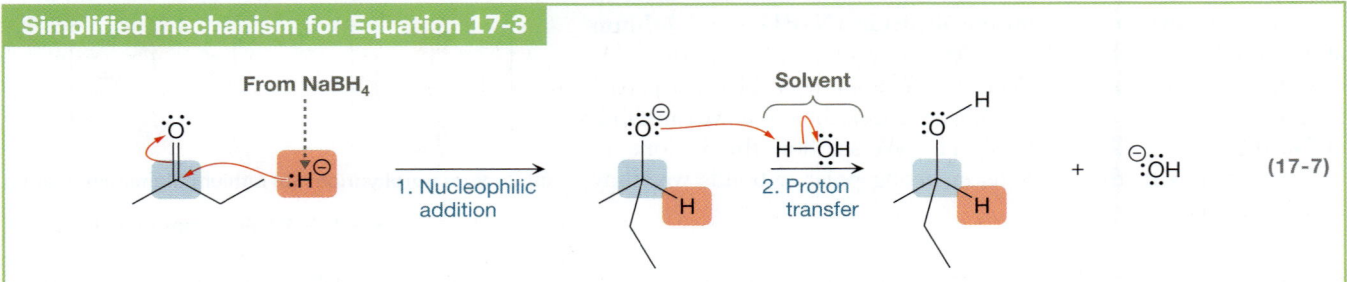

In Step 1, H⁻ undergoes nucleophilic addition to the carbonyl group to produce the strongly basic alkoxide anion. In Step 2, water, the solvent that is already present, protonates the alkoxide anion to produce the final alcohol product.

The simplified mechanism for Equation 17-6, in which LiAlH₄ serves as the hydride source, is shown in Equation 17-8. It is essentially identical to Equation 17-7, in which NaBH₄ is used. The difference is that the reaction with LiAlH₄ occurs in ether, which is not acidic, so the proton transfer in Step 2 does not occur until the acid is added separately.

Simplified mechanism for Equation 17-6

YOUR TURN 17.5

Draw the simplified mechanisms for the reactions involving the ketone in Equation 17-4 and the aldehyde in Equation 17-5. *Hint:* Use Equations 17-7 and 17-8 as your guide.

To account for the fact that H⁻ cannot exist on its own, a more complete, detailed mechanism must show the nucleophilic attack as a *hydride transfer*. That is, the H atom's bond to either B or Al must be broken at the same time the C—H bond is formed. This is illustrated in Equation 17-9 for the NaBH₄ reaction with butan-2-one (Equation 17-3).

Complete mechanism for Equation 17-3

The more complete, detailed mechanism for the LiAlH$_4$ reaction with butan-2-one is very similar (see Your Turn 17.6). In the interest of simplicity, however, we will generally work with the simplified mechanisms depicted in Equations 17-7 and 17-8.

YOUR TURN **17.6**

Draw the complete, detailed mechanism for the reaction of LiAlH$_4$ with butan-2-one (Equation 17-4). *Hint:* It is similar to the one in Equation 17-9.

Nucleophilic addition of NaBH$_4$ to a ketone or aldehyde (Equations 17-3 and 17-5) takes place in solvents such as water or an alcohol. This is advantageous because these solvents are weak acids, so both the addition of H$^-$ to the carbonyl group and the subsequent protonation can occur in the same synthetic step.

Nucleophilic addition involving LiAlH$_4$, on the other hand, cannot take place in water or alcohol because LiAlH$_4$ is too reactive.

LiAlH$_4$ is much more reactive than NaBH$_4$.

The hydride from LiAlH$_4$ deprotonates the weakly acidic proton from water and alcohols very quickly (Equation 17-10), producing hydrogen gas. The reaction is so exothermic that the hydrogen can ignite, causing an explosion!

> **GREEN CHEMISTRY**
> Although a ketone or aldehyde can be converted to an alcohol using either NaBH$_4$ or LiAlH$_4$, the greener choice is NaBH$_4$. NaBH$_4$ is greener because there is essentially no risk of explosion and its solvent—water or alcohol—is less toxic than an ether solvent typically required for LiAlH$_4$.

The H$_2$ gas that is produced is flammable!

$$\text{H}-\underset{\underset{\text{H}}{|}}{\overset{\overset{\text{H}}{|}}{\text{Al}}}{}^{\ominus}\text{H} \quad \text{H}-\overset{\text{H}}{\overset{|}{\ddot{\text{O}}}}-\text{R} \xrightarrow{\text{Fast!}} \underset{\underset{\text{H}}{}}{\overset{\overset{\text{H}}{|}}{\text{Al}}}\text{H} + \text{H}-\text{H}(g) + {}^{\ominus}\ddot{\text{O}}-\text{R} \quad \text{(17-10)}$$

Although NaBH$_4$ can deprotonate water and alcohols, too, this proton transfer is rather slow, especially when the solution is maintained at a slightly basic pH (Equation 17-11).

$$\text{H}-\underset{\underset{\text{H}}{|}}{\overset{\overset{\text{H}}{|}}{\text{B}}}{}^{\ominus}\text{H} \quad \text{H}-\overset{\text{H}}{\overset{|}{\ddot{\text{O}}}}-\text{R} \xrightarrow{\text{Very slow!}} \underset{\underset{\text{H}}{}}{\overset{\overset{\text{H}}{|}}{\text{B}}}\text{H} + \text{H}-\text{H}(g) + {}^{\ominus}\ddot{\text{O}}-\text{R} \quad \text{(17-11)}$$

LiAlH$_4$ is more reactive than NaBH$_4$ because, as shown in Figure 17-6, the electronegativity (EN) of aluminum (1.61) is substantially lower than that of boron (2.04), and both are lower than hydrogen (2.20). As a result, H has a partial negative charge in both the Al—H and the B—H bonds, but the electronegativity *difference* (ΔEN) is greater in the case of Al, giving H a higher concentration of negative charge in LiAlH$_4$ than in NaBH$_4$. As we learned in Chapter 7, a higher concentration of charge like this in LiAlH$_4$ makes it less stable and thus more reactive.

The greater concentration of charge on H makes LiAlH$_4$ more reactive.

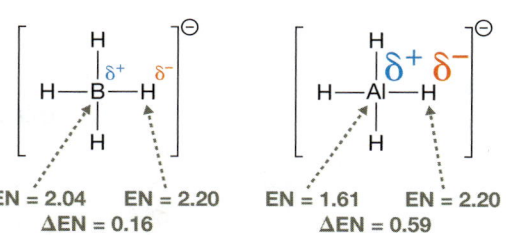

EN = 2.04   EN = 2.20   EN = 1.61   EN = 2.20
ΔEN = 0.16 ΔEN = 0.59

FIGURE 17-6 The relative reactivities of NaBH$_4$ and LiAlH$_4$
The Al—H bond is more polar than the B—H bond, so the H atoms in LiAlH$_4$ bear a larger concentration of negative charge, making LiAlH$_4$ a more reactive source of H$^-$.

SOLVED PROBLEM 17.5

Predict the major organic product of the reaction shown here.

Think Can LiAlH$_4$ be treated as a simpler nucleophile? How does NH$_4$Cl behave in solution, and why must the reaction with LiAlH$_4$ be complete before it is added?

1. LiAlH$_4$, Et$_2$O
2. NH$_4$Cl, H$_2$O
?

Solve LiAlH₄ can be treated simply as H⁻, which attacks the carbonyl carbon, as shown in Step 1 of the following sequence:

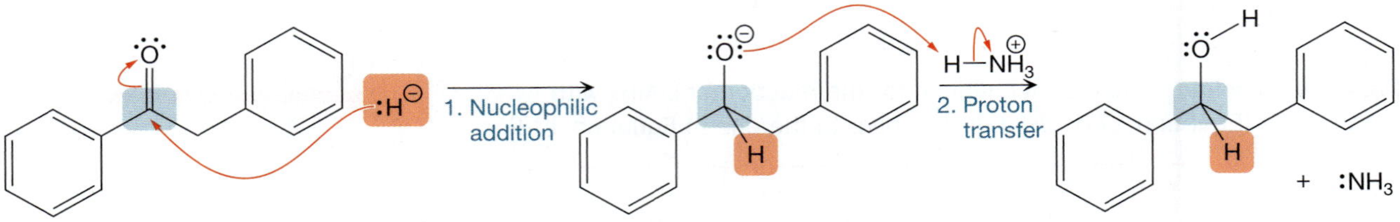

NH₄Cl is ionic, so in aqueous solution it dissociates into NH_4^+ and Cl^- ions. NH_4^+ is weakly acidic, so to avoid destroying LiAlH₄, it is not added until the nucleophilic addition is complete. Instead, in Step 2, NH_4^+ protonates the strongly basic alkoxide anion produced from Step 1.

PROBLEM 17.6 Predict the major organic product in each of these reactions.

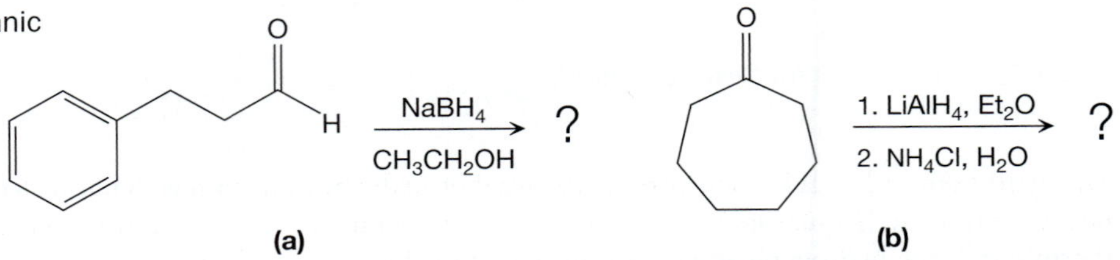

(a)   (b)

17.3b NaBH₄ and LiAlH₄ as Reducing Agents

Recall from general chemistry that a reduction–oxidation (redox) reaction is a reaction in which one species gains electrons and another species loses electrons.

> A species that:
>
> - gains one or more electrons undergoes **reduction** and becomes **reduced**.
> - loses one or more electrons undergoes **oxidation** and becomes **oxidized**.

We can determine whether electrons have been gained or lost by examining changes to an atom's **oxidation state**. To calculate an atom's oxidation state, follow these two steps:

> **Calculating an Atom's Oxidation State**
>
> 1. Assign the proper number of valence electrons to the atom:
>   - Both electrons of a lone pair are assigned to the atom on which they appear.
>   - In a covalent bond, all the electrons are assigned to the *more electronegative* atom. If the atoms are the same, divide the electrons equally.
> 2. Compare the atom's assigned valence electrons to the atom's group number.
>   - Each excess electron contributes −1 to the atom's oxidation state.
>   - Each electron the atom is lacking contributes +1 to the oxidation state.

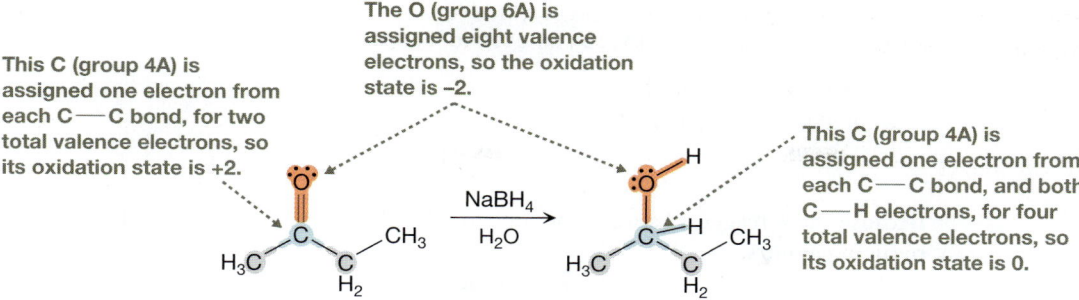

This C (group 4A) is assigned one electron from each C—C bond, for two total valence electrons, so its oxidation state is +2.

The O (group 6A) is assigned eight valence electrons, so the oxidation state is –2.

This C (group 4A) is assigned one electron from each C—C bond, and both C—H electrons, for four total valence electrons, so its oxidation state is 0.

FIGURE 17-7 Calculating oxidation states The valence electrons assigned to the O atom are screened in red, and those assigned to the attached C are screened in blue. In both the reactant and product, the O is assigned eight valence electrons, which is two more than its group number, so the oxidation state of O is −2. The C is assigned two valence electrons in the reactant (two less than C's group number) and four valence electrons in the product (the same as C's group number), so the oxidation state of C is +2 in the reactant and 0 in the product.

This method of calculating oxidation state is very similar to the one we used to calculate an atom's formal charge in Section 1.9. The difference is in how we assign covalently bonded electrons. They are given entirely to the more electronegative atom when calculating oxidation state, whereas they are divided equally when calculating formal charge.

Let's apply this method to the oxidation states of the atoms undergoing bonding changes in Equation 17-3 (p. 845), shown again in Figure 17-7. In the reactant, the O atom is assigned eight total valence electrons (four electrons from the two lone pairs and four electrons from the C=O double bond), which is two more than its group number of 6A, so the oxidation state of O is −2. The attached C atom is assigned just two total valence electrons, one from each of its C—C bonds. In each C—C bond, the two atoms have equal electronegativity, so their covalently bonded electrons are divided equally. The two total valence electrons assigned to C is two less than the group number of 4A, so the oxidation state of C is +2.

In the product, the O atom is still assigned eight valence electrons (the four electrons from the two lone pairs and the four electrons from the O—C and O—H bonds), so its oxidation state remains −2. The attached C atom is again assigned one electron from each of its C—C bonds and, because it is more electronegative than H, it is assigned both electrons from the C—H bond. The C atom therefore has four total valence electrons, which matches its group number, so its oxidation state is 0.

During the course of the reaction, the carbonyl carbon atom gained electrons and its oxidation state became less positive (became more negative). Therefore, the ketone was *reduced* and its conversion to the alcohol is an example of a *reduction*. The same is true when $LiAlH_4$ is used to convert butan-2-one to butan-2-ol (Equation 17-4, p. 845) and in the two reactions in which benzaldehyde is converted to phenylmethanol (Equations 17-5 and 17-6, p. 845).

YOUR TURN 17.7

How many electrons does butan-2-one gain when it is reduced to butan-2-ol?

With this in mind, we can restate our earlier generalization describing the reactions that take place when a ketone or aldehyde is treated with $NaBH_4$ or $LiAlH_4$:

In general, $NaBH_4$ or $LiAlH_4$ can be used to *reduce*:

- a ketone to a 2° alcohol.
- an aldehyde to a 1° alcohol.

Because these reagents cause reductions to take place, $NaBH_4$ and $LiAlH_4$ are called **reducing agents**.

Looking back at Figure 17-7, what accounts for the increase in the number of electrons assigned to the carbonyl C? On going from the reactant to the product species, that C atom loses a bond to O (because the C=O bond was converted to C—O) and gains a bond to H. Losing a bond to O does not change the number of electrons assigned to C because C, being the less electronegative atom, was never assigned any of those bonding electrons. Gaining a bond to H, however, increases the number of electrons assigned to C by two because C, being the more electronegative atom, is assigned both electrons from that new bond.

Seeing these kinds of changes in Figure 17-7 allows us to make the following generalization about reduction reactions:

If a reactant species has a C atom that gains a bond to H or loses a bond to O during the course of a reaction, that species is typically undergoing reduction.

PROBLEM 17.7 Calculate the oxidation state of the carbonyl C and O atoms in both the reactant and product species of Equation 17-5 (p. 845). Which atom's oxidation state is changing? Does this confirm that the reactant is undergoing reduction?

A reduction must have an accompanying oxidation (one can't occur without the other); the reducing agents are the species that are oxidized. Therefore, in the reduction reactions we are examining here, $NaBH_4$ or $LiAlH_4$ must be oxidized. We can verify this by calculating the oxidation states of H before and after the reaction. Prior to reaction, H is part of a H—B or H—Al bond, in which H is the more electronegative atom and is assigned both electrons from the bond. With two valence electrons, H (group 1A) has one excess electron and an oxidation state of -1. In the product, on the other hand, H becomes part of a H—C bond, in which H, being the less electronegative atom, is assigned no valence electrons and has an oxidation state of $+1$. Having lost two electrons, H must have been oxidized.

The conversion of a ketone or aldehyde to an alcohol is not the only example of a reduction. We examine others here in Chapter 17 and many more in later chapters. In later chapters, moreover, we will study reactions in which the organic species is *oxidized*. Because of the importance of reduction and oxidation in organic chemistry, we will highlight these aspects of the reactions when we encounter them.

17.3c Reductions of Imines and Nitriles

Compound classes with polar π bonds other than the carbonyl group react with hydride reagents in much the same way. An imine, for example, is reduced with $NaBH_4$ or $LiAlH_4$ to form an amine, as shown in Equations 17-12 and 17-13, respectively. As before, methanol, a weakly acidic solvent, is suitable for the $NaBH_4$ reduction, but the $LiAlH_4$ reduction must take place in a solvent that has no acidic protons, such as ether.

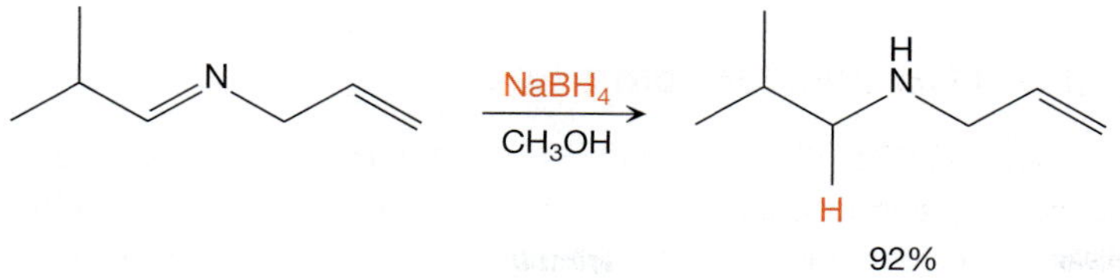

$$(17\text{-}12)$$

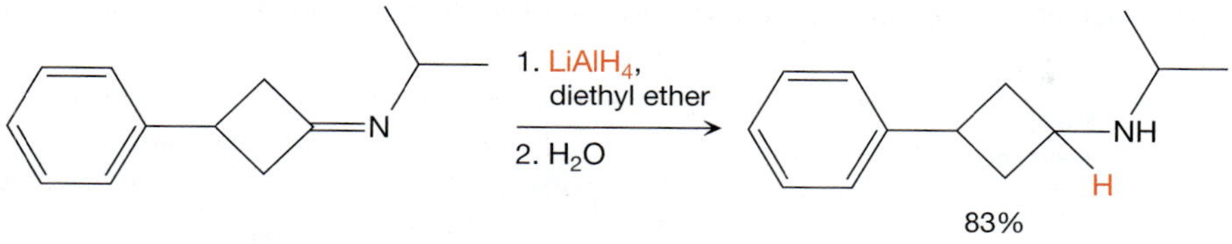

92%

$$(17\text{-}13)$$

83%

$LiAlH_4$ also reduces nitriles to primary amines, as shown in Equation 17-14.

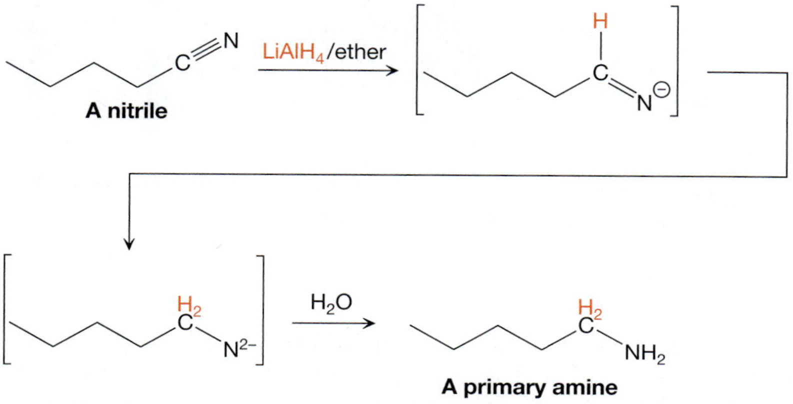

A nitrile

$$(17\text{-}14)$$

A primary amine

The first addition of H^- yields a negatively charged intermediate with a $C=N$ bond, which then undergoes a second addition of H^- to produce a species that resembles a dianion on the N atom. Subsequent acid workup protonates the N atom twice. Weaker reducing agents like $NaBH_4$ normally do not reduce nitriles without the presence of another specialized reagent or catalyst.

The -2 charge shown in the second intermediate of Equation 17-14 is an outcome of our simplification of $LiAlH_4$ to just H^-. Actually, that N atom bonds to Al (similar to what was shown in Equation 17-9), which provides significant stabilization.

PROBLEM 17.8 Draw a complete, detailed mechanism for the reactions in Equations 17-12 and 17-13.

PROBLEM 17.9 Draw a complete, detailed mechanism for the reaction in Equation 17-14.

PROBLEM 17.10 Predict the major organic product in each of the following reactions.

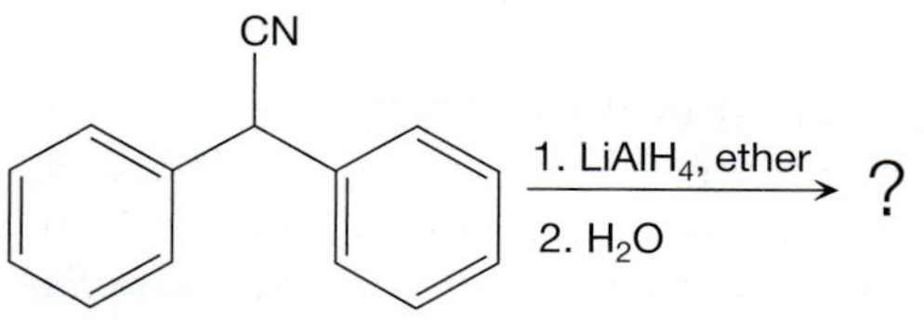

(a)

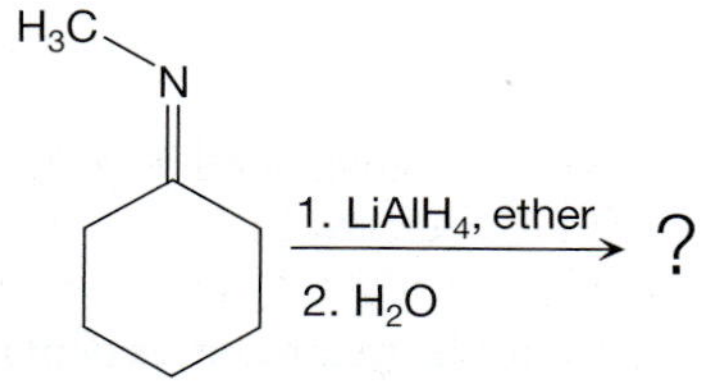

(b)

NADH as a Biological Hydride Reducing Agent

Hydride reductions occur in a variety of biochemical processes, but these processes don't involve the $NaBH_4$ or $LiAlH_4$ hydride reducing agents we have examined here in Chapter 17. Rather, the body uses the reduced form of nicotinamide adenine dinucleotide (NADH). The nicotinamide portion of NADH, shown below on the left, is directly involved in reductions. (R″ represents the remainder of NADH, consisting of two sugar units, two phosphate groups, and adenine.)

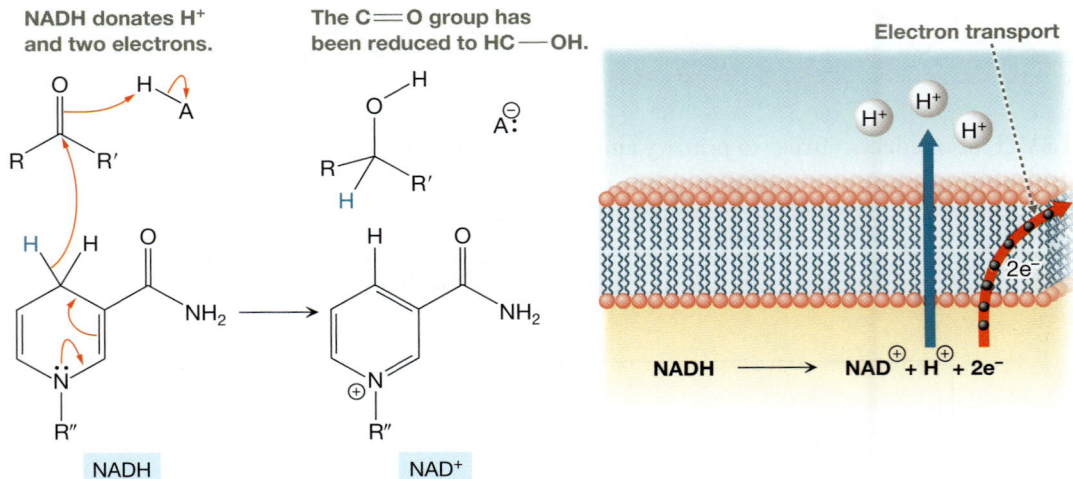

In the simplified mechanism shown above on the left, NADH donates H^+ along with two electrons—the equivalent of an H^- anion—leaving behind NAD^+. Elsewhere, NAD^+ is subsequently used as an oxidizing agent (such as in glycolysis, the citric acid cycle, fatty acid oxidation, and the catabolic elimination of surplus amino acids) and is returned to the NADH form. Thus, NADH and NAD^+ are continuously recycled.

NADH also plays a key role in what is called the electron transport chain in mitochondria, as shown above on the right. When NADH is converted to NAD^+, the H^+ and two electrons that make up the equivalent of H^- are separated from each other. The electrons are passed down the transport chain and ultimately are deposited onto an oxygen atom that becomes reduced to H_2O. In the process, protons are passed to the opposite side of the inner membrane of the mitochondrion, thereby establishing a proton gradient—that is, a difference in proton concentration—that sets up the potential to produce chemical energy in the form of adenosine triphosphate (ATP).

17.4 Sodium Hydride: A Strong Base but a Poor Nucleophile

Sodium hydride (NaH) is another common hydride agent used in organic chemistry. Its behavior, however, is somewhat different from that of $NaBH_4$ and $LiAlH_4$.

Sodium hydride, NaH, is a very strong base, but a poor nucleophile.

For example, treatment of pentan-3-one with NaH yields the enolate anion *quantitatively* (i.e., in 100% yield), as shown in Equation 17-15a. Reduction of the carbonyl (Equation 17-15b) does not occur.

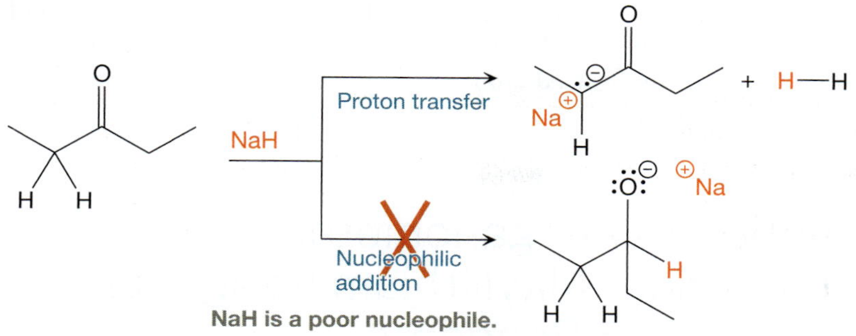

(17-15a)

(17-15b)

NaH is a poor nucleophile.

Because of its properties as a base, NaH is often used to generate carbon nucleophiles, as demonstrated in Solved Problem 17.11.

SOLVED PROBLEM 17.11

Predict the major organic product of the sequence of reactions shown here.

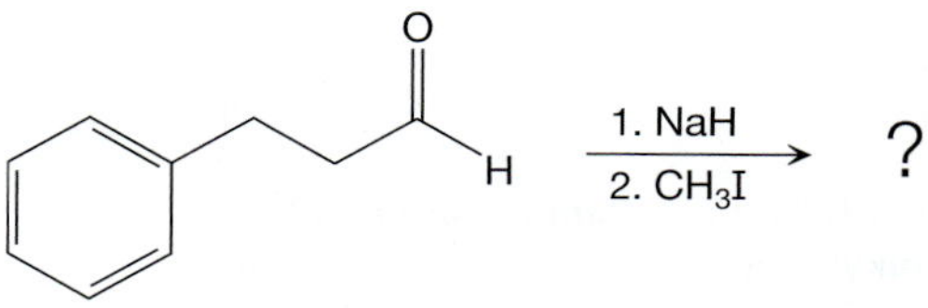

Think Will the hydride anion from NaH act as a base or a nucleophile? Which atom will it attack? How will the resulting species behave in the presence of CH_3I?

Solve NaH is a strong base but a poor nucleophile, so it will deprotonate the α carbon. As we learned in Section 10.3, the resulting enolate anion is a strong nucleophile and will displace I^- from CH_3I in an S_N2 reaction, yielding an α-alkylated aldehyde.

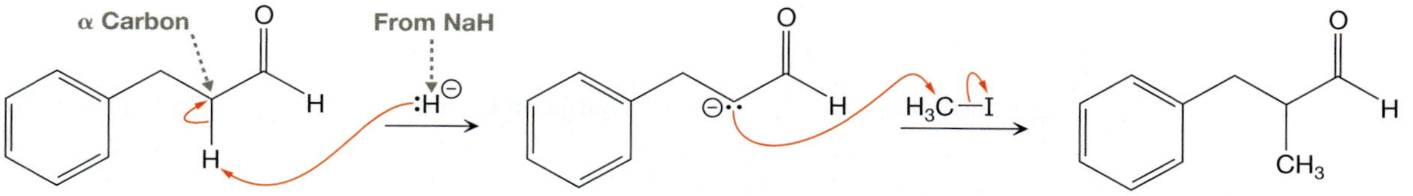

(Racemic)

PROBLEM 17.12 Predict the major organic product in each of the following reactions.

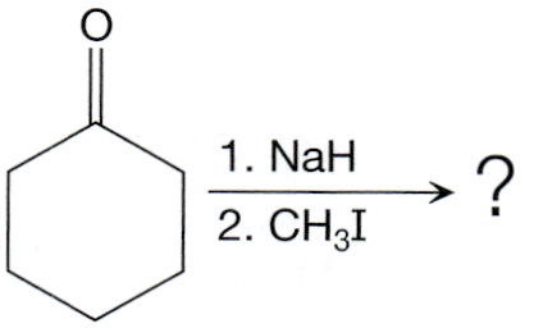

(a)

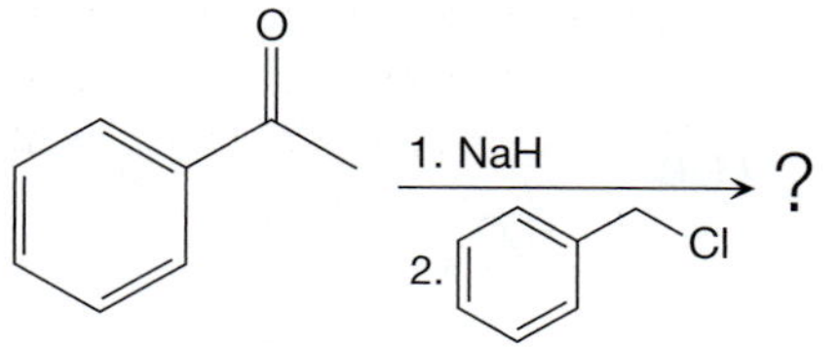

(b)

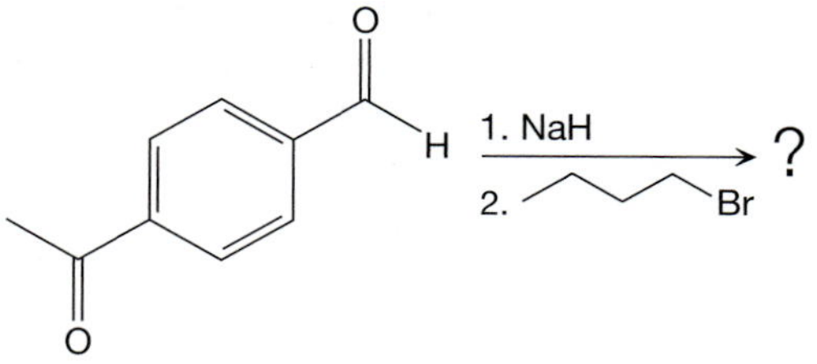

(c)

NaH behaves differently from $NaBH_4$ or $LiAlH_4$ because NaH is an *ionic* hydride, consisting of Na^+ and H^- ions, not a *covalent* hydride. NaH is ionic because there is a large difference in electronegativities between Na and H ($\Delta EN = 1.27$). The B—H and Al—H bonds in $NaBH_4$ and $LiAlH_4$ have more covalent character, on the other hand, because the electronegativity differences in those bonds are significantly smaller ($\Delta EN = 0.16$ and 0.59, respectively).

Because NaH is ionic, it is essentially insoluble in organic solvents; it remains a solid. It is therefore believed that reactions involving NaH take place at the NaH surface. Under these conditions, H^- *directly* participates in reactions, so it could conceivably act as a base or a nucleophile. It acts as a base only, however, because, as we learned in Chapter 8, proton transfers are very fast. In $NaBH_4$ and $LiAlH_4$, on the other hand, H is covalently bonded to B or Al. The B or Al atom forms a bond to the carbonyl O at the same time H^- is transferred to the carbonyl C (review Equation 17-9, p. 846), so the B or Al atom effectively *guides* H^- toward the carbonyl C to act as a nucleophile.

17.5 Reactions of Organometallic Compounds: Alkyllithium Reagents and Grignard Reagents

Recall from Section 7.1b that we can usually treat organometallic compounds such as *Grignard reagents* (RMgX) and **alkyllithium reagents (RLi)** as alkyl anions, $:R^-$ (Fig. 17-8).

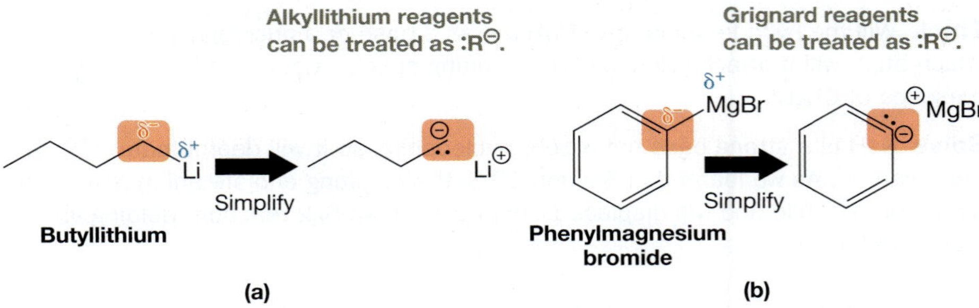

FIGURE 17-8 Simplifications of alkyllithium and Grignard reagents (a) An alkyllithium reagent has a polar covalent C—Li bond, which can be simplified to $C:^-$ and Li^+. (b) A Grignard reagent has a polar C—MgBr bond, which can be treated as $C:^-$ and $(MgBr)^+$.

Because of the poorly stabilized charge on C, these types of reagents behave both as *strong bases* and as *strong nucleophiles*. For example, both alkyllithium reagents and Grignard reagents react rapidly with water in a substantially exothermic proton transfer reaction (Equations 17-16 and 17-17) to produce an alkane and HO^-. The products are much more stable because the negative charge is transferred from a C atom to a more electronegative O atom.

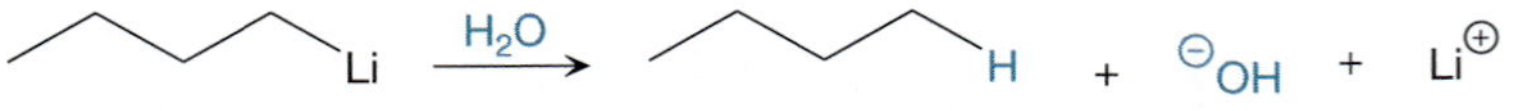

(17-16)

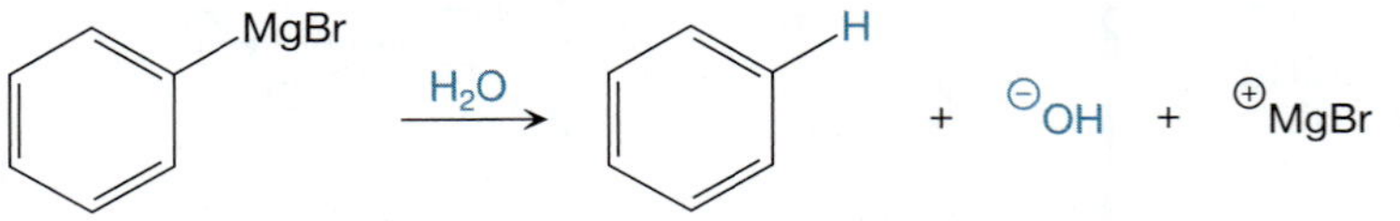

(17-17)

CONNECTIONS Butyllithium is a very powerful base and nucleophile. One of its uses industrially is to initiate polymerization in the production of some types of rubber, such as polybutadiene. Butyllithium and other alkyllithium reagents are pyrophoric, igniting on contact with air, so they are typically stored in a hydrocarbon solvent.

Although these organometallic reagents react as if they are R^-, they are not truly ionic. When dissolved in solution, free R^- does not exist. Instead, think of them as *R^- donors*, in much the same way that we view $LiAlH_4$ and $NaBH_4$ as H^- donors. The reaction that takes place is an R^- transfer from the metal atom to the proton of the acid, as shown for butyllithium in Equation 17-18.

Mechanism for Equation 17-16

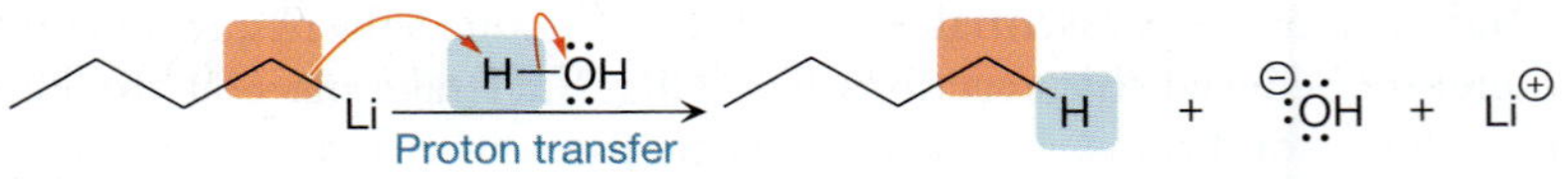

(17-18)

Add the curved arrows to show an R⁻ transfer mechanism for the reaction in Equation 17-17, similar to what is shown in Equation 17-18 for an alkyllithium reagent.

$$\text{PhMgBr} \xrightarrow{\text{H—OH}} \, ?$$

GREEN CHEMISTRY

Grignard reactions traditionally require substantial amounts of ether solvents, which are not environmentally friendly. Green alternatives include reactions that take place in water, such as this one, which facilitates the addition of acetylenic nucleophiles to benzaldehyde.

Here, the catalyst is a three-component system of $RuCl_3$, $In(OAc)_3$, and morpholine.

Even though alkyllithium and Grignard reagents are strong bases, they do not deprotonate at the α carbon of a polar π bond. Instead:

When R—Li or R—MgX reacts with a compound containing a polar π bond, R⁻ acts as a nucleophile.

R⁻ acts as a nucleophile in these reactions because the electron-rich atom of the polar π bond (O or N) coordinates with the electron-poor metal atom during the R⁻ transfer, similar to what we saw with $NaBH_4$ and $LiAlH_4$. Thus, R⁻ is effectively guided to the electron-poor C of the polar π bond to act as a nucleophile. This nucleophilic behavior of R⁻ is exemplified with a ketone in Equation 17-19 and with a nitrile in Equation 17-20. When a Grignard reagent adds in as a nucleophile, it is called a **Grignard reaction**.

(17-19)

(17-20)

Benzonitrile

CONNECTIONS Benzonitrile is a precursor to benzoguanamine, which is a component of some alkyd, acrylic, and formaldehyde resins, and is also an intermediate in the manufacture of some pharmaceuticals.

A new C—C bond is formed in both of these reactions! As we learned in Section 13.2, reactions that alter the carbon framework are distinct from functional group transformations and are, therefore, special.

In both Equations 17-19 and 17-20, the nucleophilic addition takes place in ether, followed by workup with a proton source. In each case, it is vitally important to keep the first reaction free of any acidic compounds, such as water or methanol. Otherwise, the alkyllithium or Grignard reagent will be destroyed via proton transfer reactions like those shown in Equation 17-16 or 17-17.

Notice also that the workup in Equation 17-20 is carried out with methanol. If it is carried out with H_3O^+, instead, then the imine that is produced from nucleophilic addition is hydrolyzed to the analogous carbonyl (C=O) compound—a ketone in this case—as shown in the parentheses. This reaction is discussed in greater detail in Chapter 18.

The mechanisms for these reactions are shown in Equations 17-21 and 17-22. Specifically, R⁻ adds to the electron-poor atom of the polar π bond. Once again, because free R⁻ does not exist in solution, each mechanism would be more accurately shown as a transfer of R⁻.

Mechanism for Equation 17-19

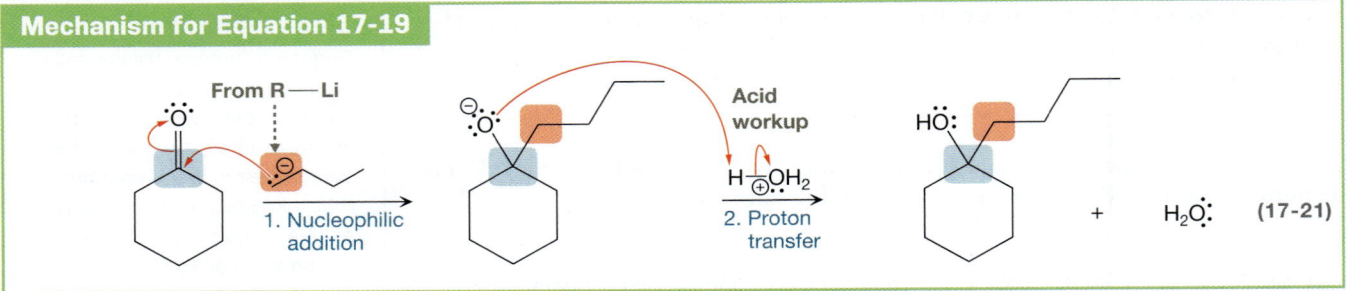

$$+ \quad H_2\ddot{O} \quad \text{(17-21)}$$

Mechanism for Equation 17-20

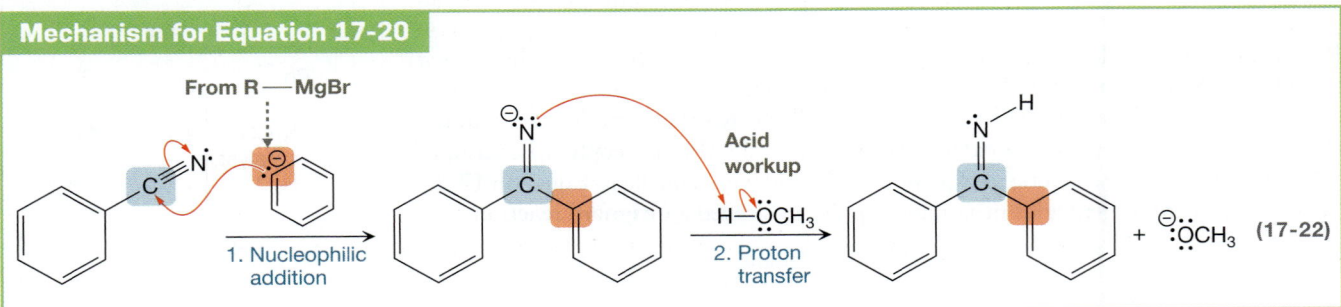

$$+ \quad \overset{\ominus}{\ddot{\text{:}}}\ddot{O}CH_3 \quad \text{(17-22)}$$

The R⁻ nucleophiles from RLi and RMgBr add only once to a nitrile carbon. This is in contrast to what we saw with LiAlH₄ (Equation 17-14), in which H⁻ adds twice. In other words, these R⁻ nucleophiles are not strong enough to generate the effective −2 formal charge on nitrogen that is a necessary outcome of a second nucleophilic addition.

SOLVED PROBLEM 17.13

Predict the major product of the reaction shown here.

Think To what R⁻ nucleophile can the Grignard reagent be simplified? Which atom will it attack? What is the role of the acid?

Solve The C_6H_5MgBr Grignard reagent can be treated simply as $C_6H_5^-$, which will undergo nucleophilic addition at the carbonyl C. Once this is complete, aqueous acid (H_3O^+) is added in an acid workup to protonate the strongly basic O^- generated in the first step.

PROBLEM 17.14 Predict the major organic product in each of the following reactions.

(a)

(b)

(c)

Although CO_2 is a nonpolar compound overall, each C=O bond is highly polar and the central carbon atom is quite electron poor and susceptible to nucleophilic attack. Thus, Grignard reagents can add to CO_2 in what is called a **carboxylation** reaction (Equation 17-23). As with any Grignard reaction, a new C—C bond is formed.

New C — C bond

$$\text{(17-23)}$$

The immediate product of carboxylation—a carboxylate anion—is subsequently protonated through an acid workup to yield a carboxylic acid. Because CO_2 is a gas at room temperature, carrying out this reaction requires either bubbling CO_2 through an ether solution of the Grignard reagent, or pouring the ether solution of the Grignard reagent over dry ice (which is solid CO_2).

PROBLEM 17.15 Draw the complete, detailed mechanism for the reaction in Equation 17-23.

PROBLEM 17.16 Predict the major organic product in each of the following reactions.

(a)

1. $CO_2(s)$
2. HCl, H_2O

?

(b)

1. $CO_2(s)$
2. H_2SO_4, H_2O

?

17.6 Limitations of Alkyllithium and Grignard Reagents

Both alkyllithium and Grignard reagents can be synthesized with a wide variety of R groups. Figure 17-9 illustrates this with Grignard reagents specifically, but the same is typically true of alkyllithium reagents. The R group can be a saturated alkyl group (including the CH_3, CH_2CH_3, and cyclohexyl groups), but it can also be unsaturated. Common unsaturated Grignard reagents include phenylmagnesium bromide (C_6H_5MgBr) and vinylmagnesium bromide ($H_2C=CH—MgBr$).

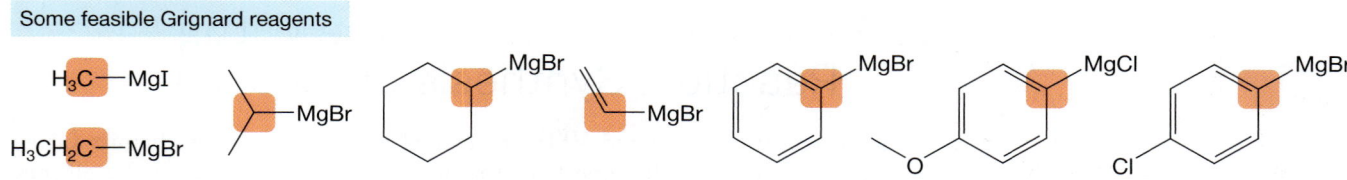

FIGURE 17-9 Feasible Grignard reagents These Grignard reagents are feasible because the strongly nucleophilic and strongly basic carbon, highlighted in red, will not react with other functional groups present in the molecule.

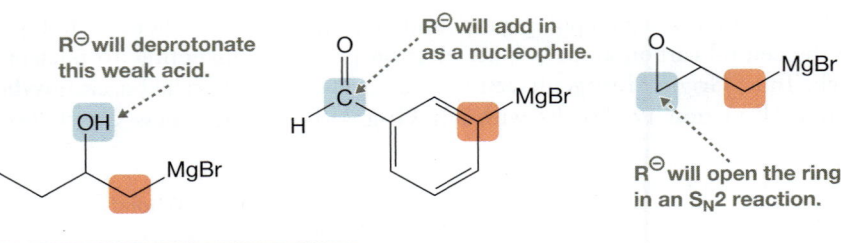

FIGURE 17-10 Unfeasible Grignard reagents These Grignard reagents are unfeasible because the sites highlighted in blue would undergo reaction with the strongly nucleophilic and strongly basic carbon highlighted in red.

$R^{\ominus}$ will deprotonate this weak acid.

$R^{\ominus}$ will add in as a nucleophile.

$R^{\ominus}$ will open the ring in an S_N2 reaction.

Some unfeasible Grignard reagents

These strong organometallic reagents, however, have significant limitations:

> Alkyllithium and Grignard reagents must not contain functional groups that are susceptible to deprotonation or to nucleophilic attack (Fig. 17-10).

For example, the hypothetical Grignard reagent shown on the left in Equation 17-24 is unfeasible.

The Grignard reagent was destroyed.

A *hypothetical* Grignard reagent

(17-24)

The OH group is weakly acidic and can be deprotonated by a strongly basic carbon atom bonded to Mg. As shown in Equation 17-24, this reaction effectively destroys the Grignard reagent; thus, the alkylmagnesium bromide is *incompatible* with the OH group.

Other groups that are incompatible with Grignard reagents or alkyllithium reagents are polar π bonds such as C=O, C=N, and C≡N, because the R⁻ anion will add to the electrophilic carbon. Furthermore, epoxides are incompatible because they are prone to nucleophilic attack, which opens the ring.

PROBLEM 17.17 For each of the Grignard reagents in Figure 17-10, which are *unfeasible*, draw the complete mechanism of the reaction that would take place. The first is shown in Equation 17-24.

17.7 Wittig Reagents and the Wittig Reaction: Synthesis of Alkenes

Wittig reagents (pronounced VIT-tig), named after the German chemist Georg Wittig (1897–1987), are an important class of compounds used to synthesize alkenes. A Wittig reagent (Fig. 17-11) is characterized by a C—P bond in which the C atom bears a −1 formal charge and the P atom bears a +1 formal charge.

Major contributor

Highly nucleophilic carbon

Poor overlap between valence orbitals

Wittig reagent (An ylide)

(H or) R   R' (or H)   (H or) R   R' (or H)

A Wittig reagent is an *ylide* (pronounced IH-lid), which is characterized by adjacent charged atoms that each have a complete octet of electrons. More specifically, because the +1 charge is on P, a Wittig reagent is also called a **phosphonium ylide**. As indicated, the −1 charge on C makes a Wittig reagent highly nucleophilic at the C atom.

Normally, a structure with adjacent positive and negative charges is the weaker contributor of two resonance structures, the stronger contributor being that in which the atoms are connected by a double bond and each atom has a zero formal charge. In the case of a Wittig reagent, however, the C=P resonance structure is the *weaker* contributor, because the π bond of a C=P bond would require the interaction of orbitals from different valence shells—the second shell on carbon and the third shell on phosphorus. This results in a weak interaction and, consequently, little stabilization relative to the atomic orbital energies, leaving the P^+—C^- structure as the more accurate representation.

The primary importance of Wittig reagents is in their reaction with ketones and aldehydes—so-called **Wittig reactions**. An example is shown in Equation 17-25.

Wittig reaction

New C—C bond

40 °C, THF overnight

97%

Triphenylphosphine oxide

(17-25)

As we can see from this example, a Wittig reaction results in the joining of two carbon-containing groups by a C=C bond—one group from the Wittig reagent and the second from the ketone or aldehyde. Notice, in particular, that the newly formed C=C bond is between the original carbonyl carbon and the original P—C carbon. In other words:

In a Wittig reaction, the C=O bond of a ketone or aldehyde is converted into a C=C bond.

The mechanism for the Wittig reaction in Equation 17-25 is shown in Equation 17-26.

(17-26)

A betaine

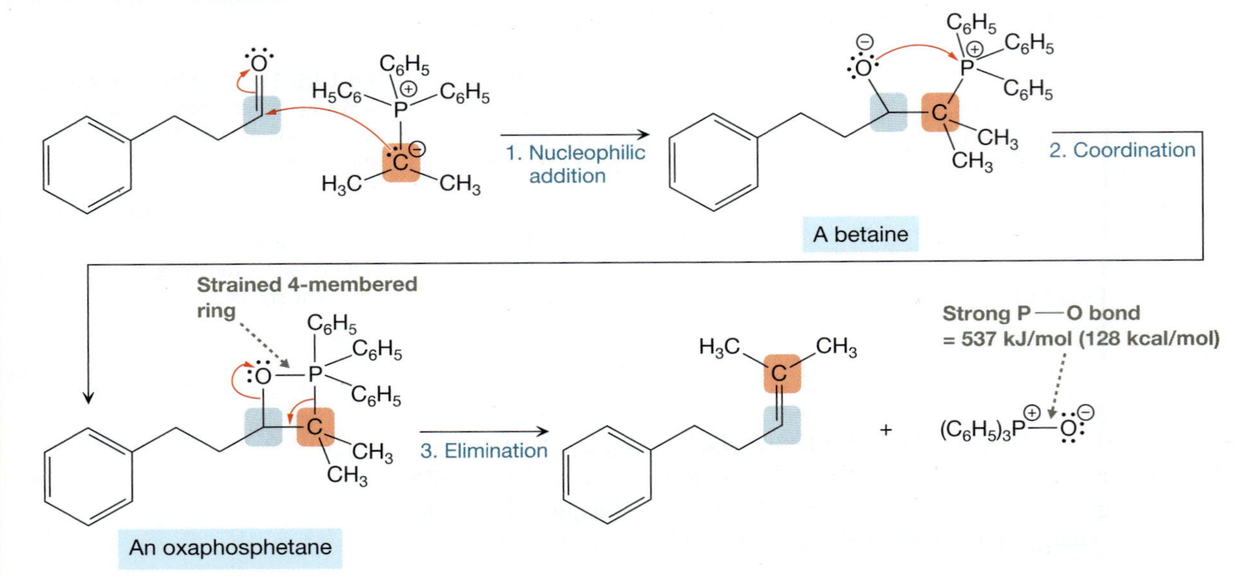

1. Nucleophilic addition

2. Coordination

Strained 4-membered ring

An oxaphosphetane

3. Elimination

Strong P—O bond = 537 kJ/mol (128 kcal/mol)

$(C_6H_5)_3P \overset{\oplus}{—} \overset{\ominus}{O}$

In Step 1, the highly nucleophilic C atom of the Wittig reagent attacks the electrophilic carbonyl C atom, yielding a **betaine** (pronounced BEE-ta-een)—that is, a species in which a positive and a negative charge are separated by two uncharged atoms. Step 2 is a coordination step, in which a bond is formed between the negatively charged O atom and the positively charged P atom. This results in an **oxaphosphetane** that contains a four-membered ring, which, due to the strain, falls apart into the alkene and triphenylphosphine oxide in Step 3.

The first two steps of the Wittig reaction are driven by the principle of *electron rich to electron poor*. The last step, however, is driven primarily by the formation of the very strong P^+—O^- bond, whose bond energy is 537 kJ/mol (128 kcal/mol)—significantly stronger than typical single bonds.

YOUR TURN 17.9

In each of the first two steps of the mechanism in Equation 17-26, label the pertinent electron-rich and electron-poor sites.

For many Wittig reactions, the oxaphosphetane forms in one step rather than two. Which mechanism occurs depends on a number of variables, including the solvent and the stucture of the Wittig reagent. (If the oxaphosphetane forms in a single step, the step would be called cycloaddition, the topic of Chapter 24.)

If E/Z isomerism exists about the C=C bond in the Wittig product, then a mixture of the two diastereomers is produced, as shown in Equation 17-27.

A Wittig reaction produces a mixture of *Z* and *E* isomers.

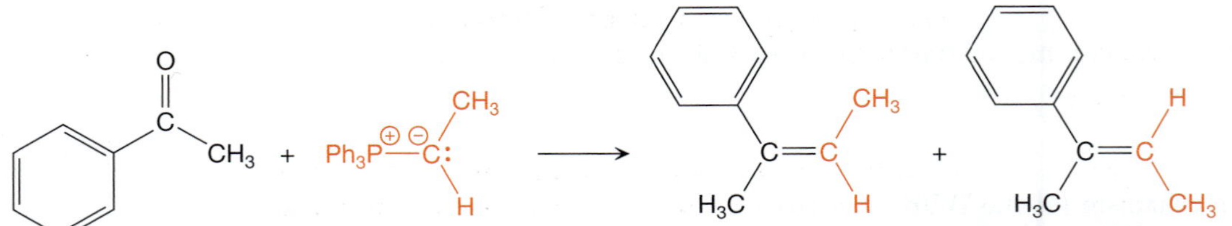

(17-27)

(Z)-2-Phenylbut-2-ene   **(E)-2-Phenylbut-2-ene**

The reason for this stereochemistry can be seen from the first step of the mechanism, in which either face of the carbonyl-containing compound can be attacked. Often, one diastereomer is heavily favored over the other, but predicting which one is beyond the scope of our discussion.

PROBLEM 17.18 Draw the complete, detailed mechanism for the reaction shown here and predict the major organic product(s). Pay attention to stereochemistry.

17.8 Generating Wittig Reagents

One of the reasons that Wittig reactions are so useful is that Wittig reagents can be generated from common precursors—namely, alkyl halides. As shown in Equation 17-28, the Wittig reagent in Equation 17-25 can be synthesized from 2-bromopropane.

$$(17\text{-}28)$$

2-Bromopropane   56%

The alkyl halide is first treated with triphenylphosphine, $P(C_6H_5)_3$, and the product of that reaction is treated with a very strong base such as butyllithium. The mechanism for this reaction is shown in Equation 17-29.

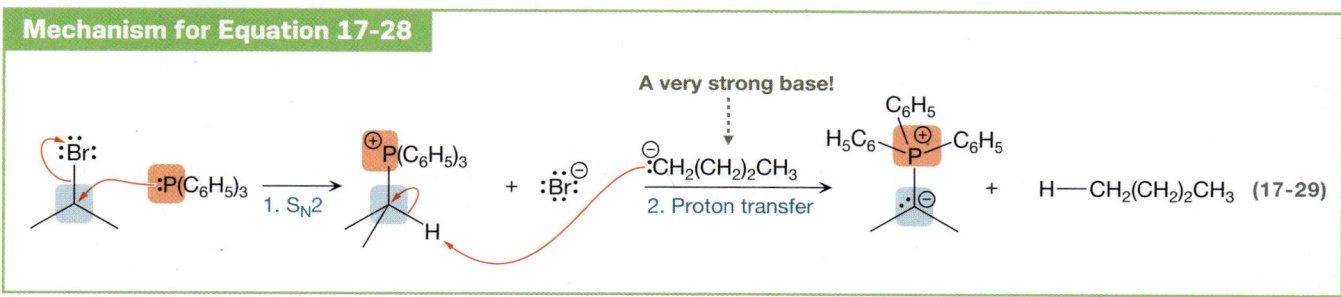

Mechanism for Equation 17-28

A very strong base!

$$(17\text{-}29)$$

Step 1 is an S_N2 reaction in which the nucleophile is Ph_3P and the leaving group is the halide anion. The organic product possesses a positive charge on the P atom. Although Ph_3P is uncharged, it is a good nucleophile because the P atom in the product can accommodate the positive charge rather well, due to both its large size (being in the third row of the periodic table) and its modest electronegativity (EN = 2.19).

YOUR TURN 17.10

Verify that triphenylphosphine is a very good nucleophile by looking up its relative nucleophilicity and that of Br^- in Table 9-10 (p. 473).

Step 2 of Equation 17-29 is deprotonation by the alkyllithium species. A very strong base such as an alkyllithium is necessary because deprotonation occurs at a carbon atom. Carbon acids are in general extremely weak because carbon, due to its small size and moderate electronegativity, does not accommodate a negative charge well. Here, however, deprotonation is facilitated because the C⁻ that is generated is stabilized by the adjacent positive charge on P, and opposite charges attract.

PROBLEM 17.19 Draw the complete, detailed mechanism for the reaction sequence shown here and provide structures for both the product (**B**) and the intermediate (**A**).

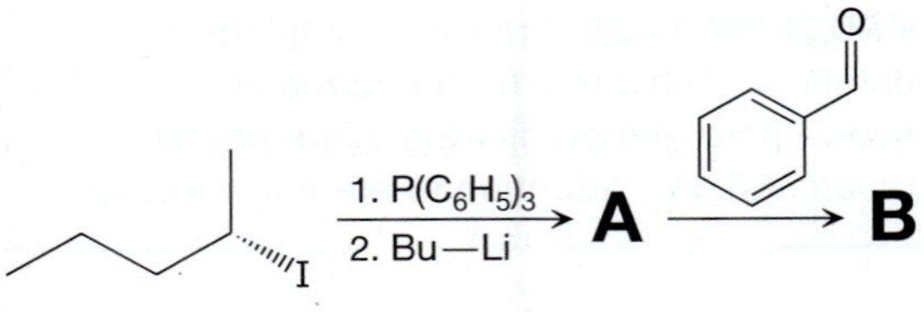

PROBLEM 17.20 Draw an alkyl halide that could be used to synthesize this Wittig reagent.

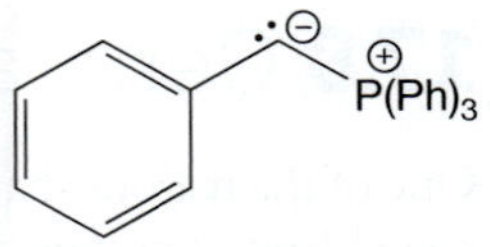

The C atom that is deprotonated in the proton transfer step in Equation 17-29 is the one that was originally bonded to the leaving group in the alkyl halide precursor. Therefore:

> For an alkyl halide to be a suitable precursor for a Wittig reagent, the C atom bonded to the leaving group must possess at least one H.

Moreover, having such a H atom ensures that the halogen leaving group is not on a tertiary carbon—something that, as we learned in Section 9.6b, would preclude the S_N2 reaction from happening in the first place (Equation 17-30).

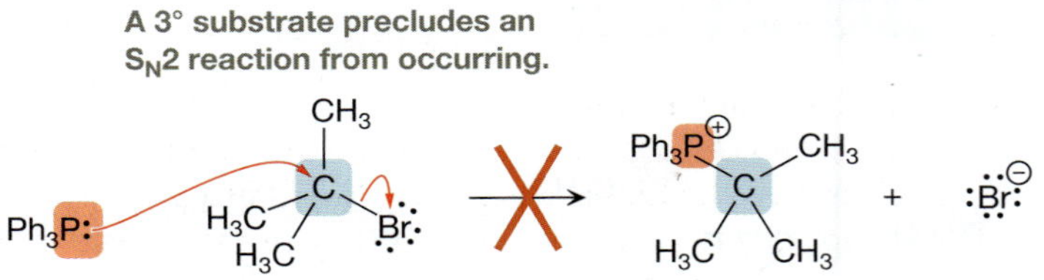

A 3° substrate precludes an
S_N2 reaction from occurring.

(17-30)

SOLVED PROBLEM 17.21

Show how you could synthesize oct-4-ene using butanal as your only source of carbon.

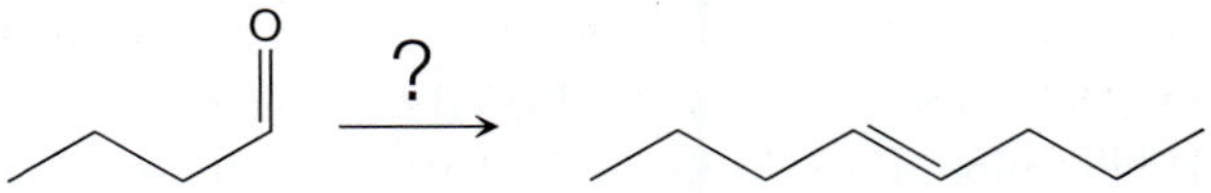

Think What Wittig reagent would you need to carry out this reaction? What alkyl halide could serve as a precursor to that Wittig reagent? How can that alkyl halide be synthesized from butanal?

Solve We can carry out a retrosynthetic analysis on the product, beginning by undoing a Wittig reaction.

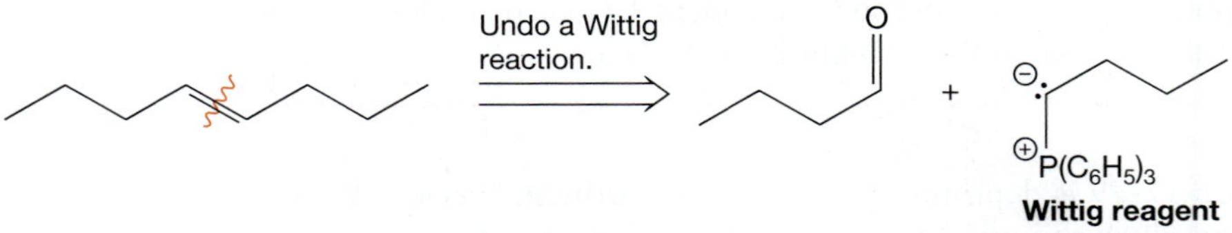

Wittig reagent

The Wittig reagent can be synthesized from a four-carbon alkyl halide. The alkyl halide cannot be produced directly from butanal, but can be produced from butan-1-ol, and butan-1-ol can be synthesized by reducing butanal using a hydride agent.

Undo a Wittig reagent formation.

Undo a substitution.

Undo a reduction.

In the forward direction, the synthesis might appear as follows.

$$\xrightarrow[\text{EtOH}]{\text{NaBH}_4}$$

$$\xrightarrow{\text{PBr}_3}$$

1. P(C₆H₅)₃
2. Bu—Li

(+ Z isomer)

PROBLEM 17.22 Show how you could carry out the synthesis shown here, beginning with benzaldehyde and using any other reagents necessary.

?

17.9 Direct Addition versus Conjugate Addition

When a C=C double bond is conjugated to a polar π bond, as in propenal (Fig. 17-12, next page), two electron-poor sites are present. One of those sites is the carbonyl C atom, given that the highly electronegative O atom is bonded to it, and the other is the C atom *beta* to (i.e., two carbons away from) the carbonyl. We can see why this is so by examining one of its resonance contributors, which places a formal negative charge on the O atom and a formal positive charge on the β C atom (Fig. 17-12a). Although this resonance contributor is somewhat weak, due to the charges present, its contribution to the resonance hybrid does indeed place a small but significant partial positive charge on the β C atom, as indicated in the electrostatic potential map in Figure 17-12b.

YOUR TURN 17.11

Identify the two electron-deficient sites in cyclohex-2-en-1-one, shown here, and label the α and β carbons.

Cyclohex-2-en-1-one

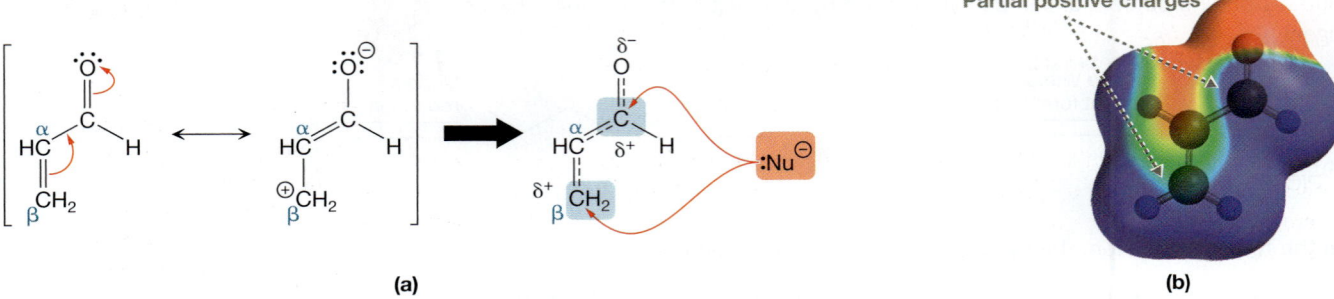

Partial positive charges

(a)   **(b)**

FIGURE 17-12 Reactive sites in an α,β-unsaturated carbonyl (a) Resonance structures of propenal are shown on the left. The weak resonance contributor with separated charges generates a small, partial, positive charge on the β carbon in the resonance hybrid on the right. Thus, a nucleophile can attack at either the carbonyl carbon or the β carbon. (b) An electrostatic potential map of propenal, showing the partial positive charge (blue) on both the carbonyl carbon and the β carbon.

Given these *two electrophilic sites* in an **α,β-unsaturated carbonyl**, nucleophiles in general can attack at two positions: either at the carbonyl carbon itself (Equation 17-31) or at the **β carbon** (Equation 17-32).

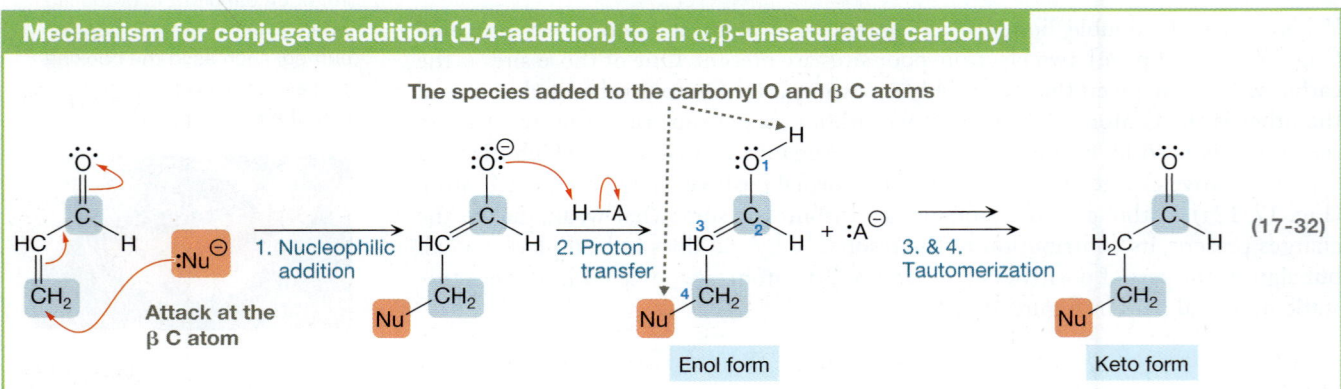

Attack of the nucleophile at the carbonyl carbon is called *1,2-addition*, or *direct addition*, because after protonation in Step 2, species have added to adjacent atoms, which is a 1,2-positioning. Attack at the β C atom, on the other hand, is called *1,4-addition*, or *conjugate addition*. After protonation in Step 2, species have added to atoms separated by three bonds, which is a 1,4-positioning. These 1,2- and 1,4-additions are analogous to electrophiles adding 1,2 or 1,4 to conjugated dienes (Section 11.9).

Notice in Equation 17-32 that the 1,4-addition product is an enol. As we saw in Section 7.9, this enol form will rapidly *tautomerize* to generate the more stable keto form as the major product.

Draw the mechanism that converts the enol in Equation 17-32 into its keto form. *Hint:* Review Section 7.9. You may assume basic conditions.

If a carbon nucleophile adds via conjugate addition, thereby forming a carbon–carbon bond, the process is called a **Michael reaction** or **Michael addition**, after the American chemist Arthur Michael (1853–1942). Over the years, however, the term *Michael reaction* has evolved into a generic description of conjugate additions involving any nucleophile.

PROBLEM 17.23 Draw the complete 1,2- and 1,4-addition mechanisms for each of the reactions shown here.

$$
\text{(a)} \qquad \xrightarrow[\text{H}_2\text{O}]{\text{NaOH}} \quad ? \qquad\qquad \text{(b)} \qquad \xrightarrow[\text{2. H}_2\text{SO}_4]{\text{1. KCN, H}_2\text{O}} \quad ?
$$

(a)   **(b)**

Which of the two possible nucleophilic addition mechanisms dominates (i.e., direct addition or conjugate addition) depends in large part on whether the nucleophile adds to the carbonyl carbon reversibly or irreversibly (Equation 17-33).

$$
\text{C=O} + \quad :\text{Nu}^{\ominus} \quad \underset{\text{or}}{\overset{}{\rightleftharpoons}} \quad \text{C}-\text{O}^{\ominus},\ \text{Nu} \qquad\qquad \textbf{(17-33)}
$$

In general, when an α,β-unsaturated carbonyl undergoes nucleophilic addition:

- Nucleophiles that add reversibly to the carbonyl carbon yield the conjugate addition product as the major product.
- Nucleophiles that add irreversibly to the carbonyl carbon yield the direct addition product as the major product.

Whether the nucleophilic addition in Equation 17-33 is reversible or irreversible depends on the charge stability that each side of the reaction exhibits: *Nucleophilic addition tends to be irreversible if the negative charge that develops in the adduct is substantially better stabilized than it is in the nucleophile.* In practice, these situations are limited to ones in which the nucleophile has the negative charge located on a carbon or hydrogen atom, and the negative charge is not stabilized by resonance or inductive effects. Therefore:

- Nucleophilic addition to a carbonyl carbon tends to be irreversible when it involves a very strong R⁻ or H⁻ nucleophile.
- Otherwise, the nucleophilic addition tends to be reversible.

These ideas are summarized in Table 17-2.

For example, $C_6H_{11}MgBr$ (in which $C_6H_{11}^-$ is the nucleophile) adds irreversibly to a carbonyl C atom, so the major product is from direct addition (Equation 17-34).

CONNECTIONS Cyclohex-2-en-1-one (see Your Turn 17.11) is a useful starting material for the synthesis of pharmaceuticals that contain six-membered rings in the molecular structure. It has been used, for example, in the total synthesis of morphine.

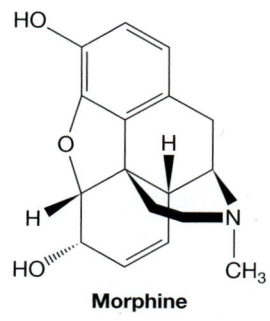

Morphine

TABLE 17-2 **Reversibility in Nucleophilic Addition**	
Nucleophiles That Add *Reversibly*	**Nucleophiles That Add** *Irreversibly*
$HO^{\ominus}$, $RO^{\ominus}$	$R-MgBr$ (:$R^{\ominus}$)
$H_2N^{\ominus}$, $R_2N^{\ominus}$	$R-Li$ (:$R^{\ominus}$)
$Cl^{\ominus}$, $Br^{\ominus}$, $I^{\ominus}$	$H_5C_6-\overset{\oplus}{\underset{C_6H_5}{\overset{C_6H_5}{P}}}-\overset{\ominus}{C}R_2$
$CH_3-\overset{O}{\overset{\|}{C}}-\overset{\ominus}{O}$	$LiAlH_4$ (:$H^{\ominus}$)
$N\equiv C:^{\ominus}$	$NaBH_4$ (:$H^{\ominus}$)
$CH_3-\overset{O}{\overset{\|}{C}}-\overset{\ominus}{\underset{\|}{C}}$	

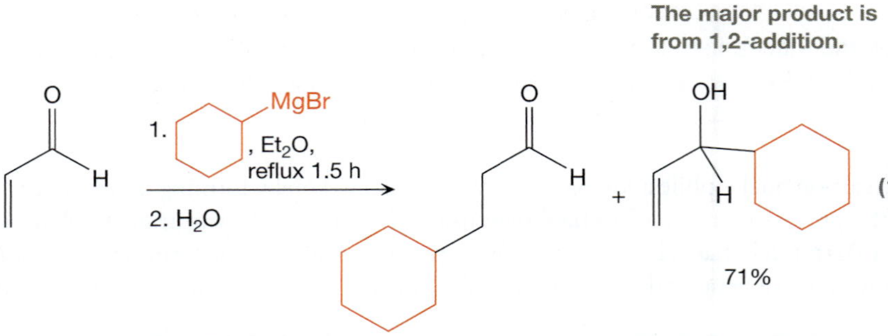

The major product is from 1,2-addition.

(17-34)

71%

On the other hand, CH_3O^- adds reversibly, so the major nucleophilic addition product is from conjugate addition (Equation 17-35).

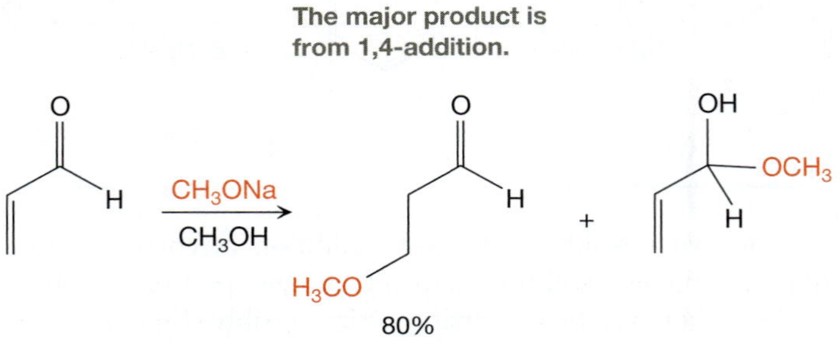

The major product is from 1,4-addition.

(17-35)

80%

PROBLEM 17.24 Draw the major product for the reaction between cyclohex-2-en-1-one and each of the nucleophiles in **(a)–(e)**, and draw the mechanism that leads to each of those products. You may assume an acid workup, if necessary.

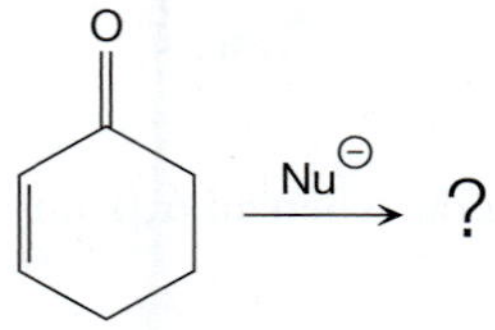

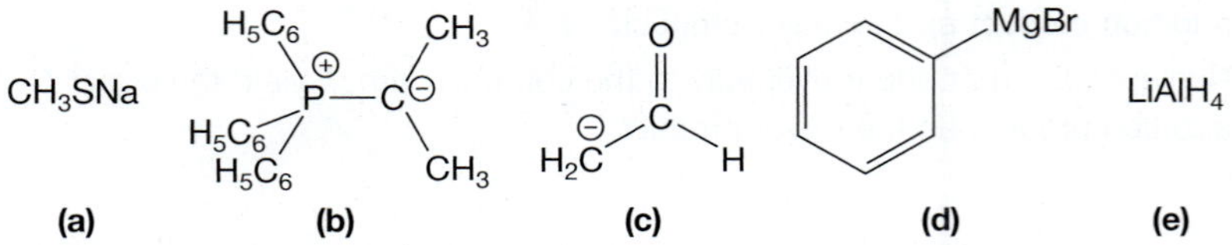

(a) CH_3SNa (b)   (c)   (d)   (e) $LiAlH_4$

Why does this regioselectivity involving 1,2- versus 1,4-addition depend on whether the nucleophile adds reversibly to the carbonyl carbon atom? Just as we saw with electrophilic addition to conjugated dienes (Section 11.10), it is because the reversibility governs whether nucleophilic addition takes place under thermodynamic or kinetic control (Section 9.12).

- If the nucleophile adds reversibly, then the reaction takes place under thermodynamic control and the major product is the thermodynamic product (i.e., the one that is more stable).
- If the nucleophile adds irreversibly, then the reaction takes place under kinetic control and the major product is the kinetic product (i.e., the one that is produced more rapidly).

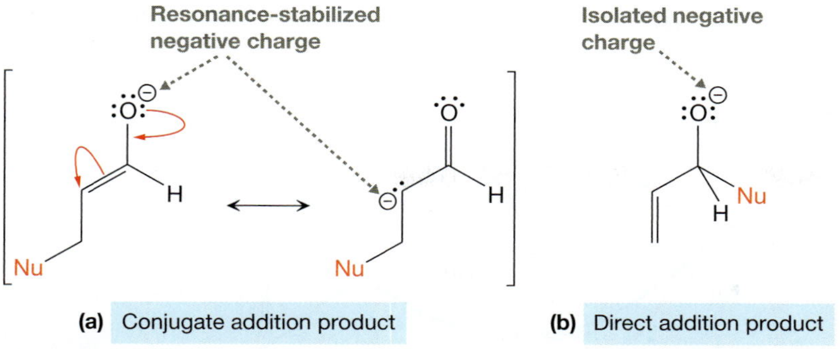

Resonance-stabilized negative charge

Isolated negative charge

(a) Conjugate addition product

(b) Direct addition product

FIGURE 17-13 Charge stability and direct versus conjugate addition The negative charge produced on nucleophilic addition is resonance delocalized in (a) the conjugate addition product but not in (b) the direct addition product, so the conjugate addition product is more stable.

To apply these ideas, we need to know which product is the thermodynamic product and which is the kinetic product. Generally speaking:

- The conjugate addition product is the one that is more stable, making it the thermodynamic product.
- The direct addition product is the one that is produced more rapidly, making it the kinetic product.

The conjugate addition product is more stable in part because of charge stability. Notice from Equation 17-32 that the immediate product of conjugate addition is an enolate anion, which, as shown in Figure 17-13a, has a resonance-stabilized negative charge. The direct addition product (Equation 17-31) shown in Figure 17-13b, on the other hand, possesses an isolated negative charge, and is therefore less stable.

Conjugate addition also leads to the thermodynamic product because the *overall* uncharged product of conjugate addition has greater total bond energy than the one produced from direct addition. The greater total bond energy in the conjugate addition product comes primarily from the type of double bond present. As shown in Figure 17-14, the conjugate addition product (Fig. 17-14a) has a C=O double bond, which, on average, is 101 kJ/mol (24 kcal/mol) stronger than the C=C double bond in the direct addition product (Fig. 17-14b).

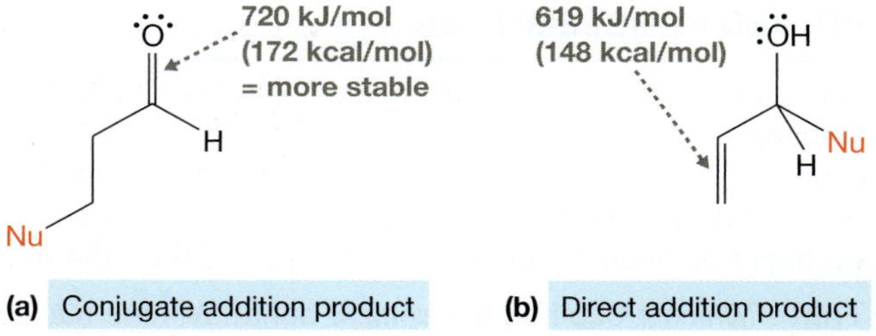

If the direct addition product is the kinetic product, then it must be formed faster, which means the path to form it has a smaller activation energy (Fig. 17-15, next page). Why this is so is due to the greater concentration of positive charge on the carbonyl C atom than on the β C atom. Therefore, in the initial stages of nucleophilic attack, the nucleophile is better stabilized by its attraction to the carbonyl C atom than by its attraction to the β C atom.

720 kJ/mol (172 kcal/mol) = more stable

619 kJ/mol (148 kcal/mol)

(a) Conjugate addition product

(b) Direct addition product

FIGURE 17-14 Bond energies and direct versus conjugate addition The overall product of conjugate addition (a) is more stable than the overall product of direct addition (b) primarily because the average C=O double bond is much stronger than the average C=C double bond.

YOUR TURN 17.13

Which product of the following nucleophilic addition reaction is the *thermodynamic product* and which is the *kinetic product*?

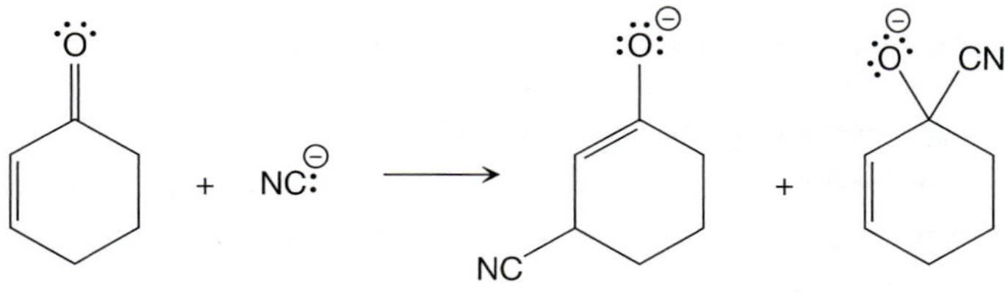

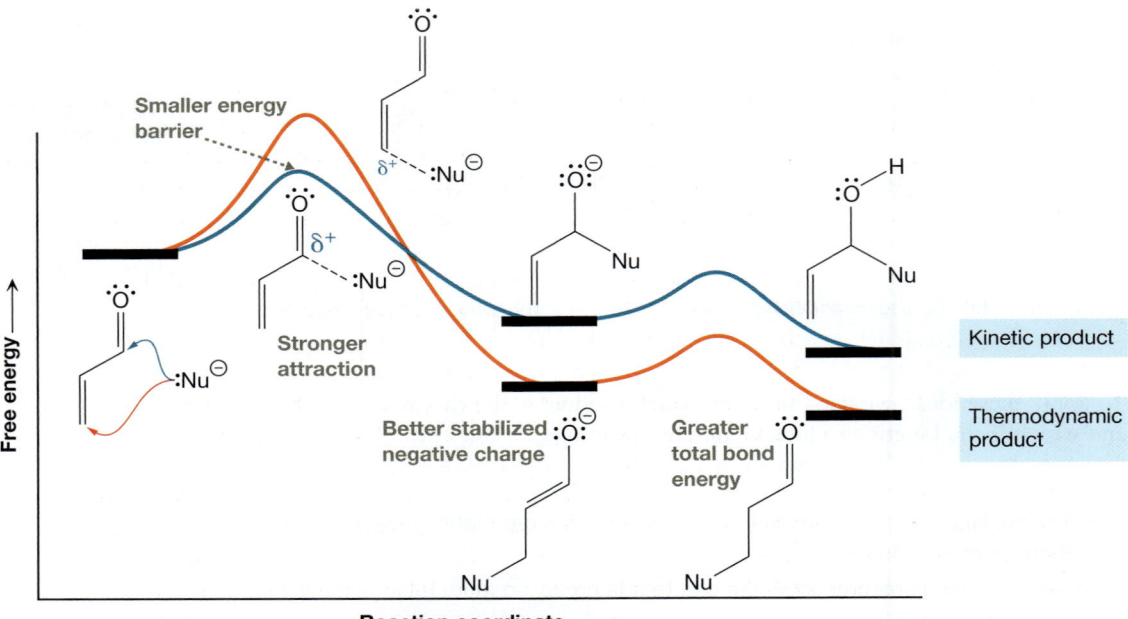

FIGURE 17-15 Direct versus conjugate addition of a generic nucleophile to propenal
The blue curve represents direct addition (Equation 17-31), whereas the red curve represents conjugate addition (Equation 17-32). The energy barrier is lower for direct addition because, in the transition state, the nucleophile is attracted more strongly to the carbonyl carbon than to the β carbon. Thus, direct addition gives the kinetic product. The more stable product is formed from conjugate addition, so conjugate addition gives the thermodynamic product.

SOLVED PROBLEM 17.25

Although thus far we have examined conjugate addition only as it pertains to carbonyl compounds, competition between direct addition and conjugate addition can also occur with other α,β-unsaturated polar π bonds. An example involving an α,β-unsaturated nitrile is shown here. Predict whether the reaction will take place via direct addition or conjugate addition, and draw the major product.

Think Which group has the polar π bond? Will the nucleophile add reversibly or irreversibly? How does that affect whether direct addition or conjugate addition takes place?

Solve The C≡N group has the polar π bond. The nucleophile is CH_3O^-, which, as we know from Table 17-2, adds reversibly. As a result, the thermodynamic product, which is formed via conjugate addition, is favored. The mechanism is essentially the same as that in Equation 17-32.

PROBLEM 17.26 Predict the major product of the reaction shown here.

Several cancer drugs work by taking advantage of the fact that cancer cells grow and divide significantly faster than normal cells. In the box on page 547, for example, we saw how one cancer drug—mechlorethamine—uses S_N2 reactions to tether together two strands of DNA, thereby disrupting DNA function. Eriolangin, isolated from *Eriophyllum lanatum*, is part of another class of cancer drugs that uses Michael additions to combat cancer—reactions that we have studied here in Chapter 17.

Rather than target the DNA itself, eriolangin reacts with DNA polymerase, an enzyme that facilitates the replication of DNA. This is made possible by the two α,β-unsaturated carbonyl groups (highlighted in red below) that can react with the nucleophilic thiol (SH) group of a cysteine amino acid in DNA polymerase. As shown on the right, this reaction is conjugate addition (or Michael addition) instead of direct addition because a thiol group will add to the carbonyl carbon reversibly. The DNA polymerase in the product is incapable of copying additional DNA.

As with other cancer drugs, the way in which eriolangin acts to fight cancer is a double-edged sword. Other α,β-unsaturated carbonyl compounds that enter the body can pose a threat to the normal function of DNA polymerase. One notable example is propenal ($H_2C{=}CH{-}CH{=}O$), a toxin that is found in many foods fried in vegetable oil, such as french fries. In small quantities, however, it does not pose a significant threat, in part because of the presence of glutathione in the body. Glutathione is a type of tripeptide that has a cysteine amino acid, and therefore has a nucleophilic SH group, so it can bind to these kinds of toxins via conjugate addition and safely remove them from the body.

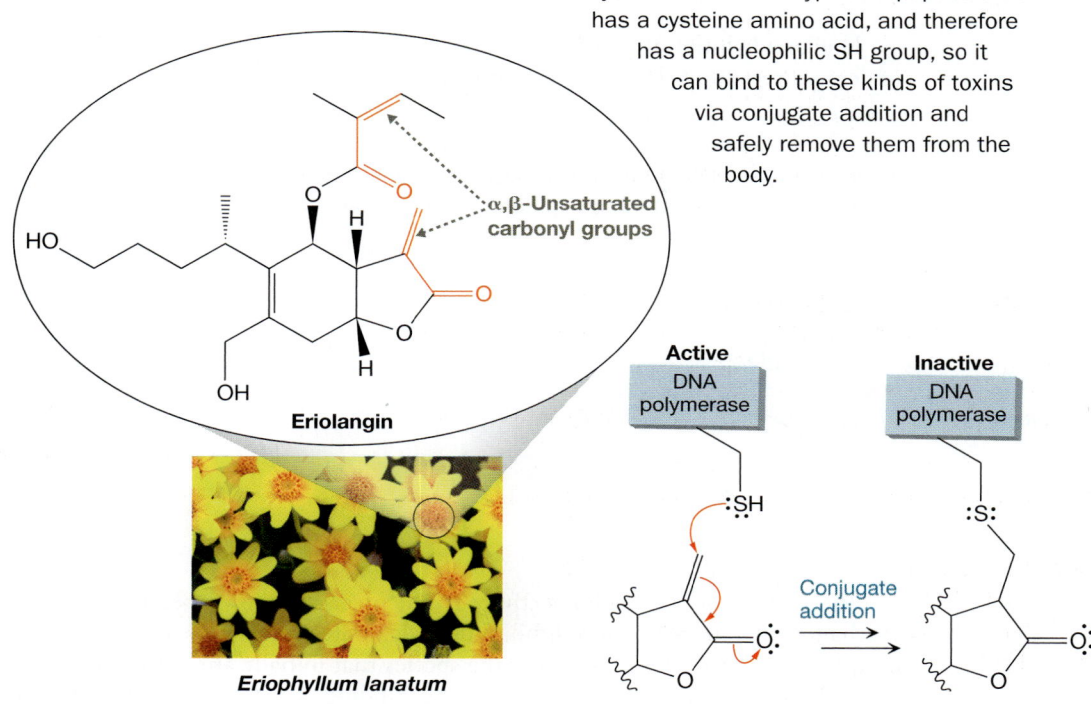

Eriolangin

α,β-Unsaturated carbonyl groups

Eriophyllum lanatum

Active DNA polymerase

Inactive DNA polymerase

Conjugate addition

17.10 Lithium Dialkylcuprates and the Selectivity of Organometallic Reagents

Organocopper reagents, also called organocuprates, constitute an important class of organometallic compounds. Among the most common organocuprates are **lithium dialkylcuprates**, R_2CuLi, sometimes referred to as **Gilman reagents** (after the American chemist Henry Gilman, 1893–1986). As with other organometallic reagents, the carbon atoms of lithium dialkylcuprates are electron rich, because carbon (EN = 2.55) has a higher electronegativity than copper (EN = 1.90). As a result, these reagents, too, can be treated as simply R⁻.

Lithium dialkylcuprates are much less reactive than either alkyllithium reagents (RLi) or Grignard reagents (RMgX). Largely this is because the C—Cu bond is significantly less polar than either the R—Mg bond or the R—Li bond, as shown in Figure 17-16. Because the electronegativity of Cu (1.90) is greater than that of Mg (1.31) or Li (0.98), R₂CuLi has the smallest concentration of negative charge on C.

FIGURE 17-16 Relative reactivities of organometallic reagents As the electronegativity of the metal atom in the metal–carbon bond decreases, the electronegativity difference between the metal and C atoms increases, resulting in a larger partial negative charge on C and a greater reactivity.

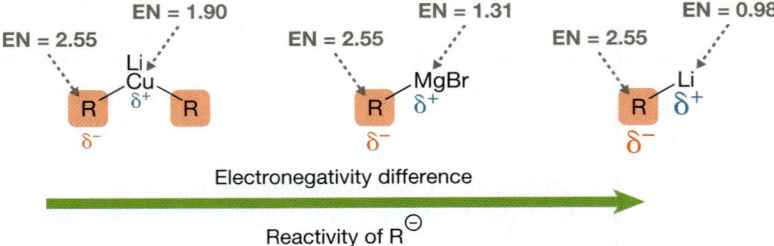

Unlike alkyllithium and Grignard reagents, lithium dialkylcuprates tend *not* to undergo direct addition to a polar π bond.

> When an α,β-unsaturated carbonyl compound is treated with R₂CuLi, R⁻ adds almost exclusively at the β carbon via conjugate addition.

An example is shown in Equation 17-36.

$$
\text{(cyclohexenone)} \xrightarrow[\text{2. NH}_4\text{Cl, H}_2\text{O}]{\substack{\text{1. (CH}_3\text{CH}_2)_2\text{CuLi,} \\ \text{THF, } -78\,°\text{C}}} \text{(3-ethylcyclohexanone, 98\%)} + \text{CH}_3\text{CH}_2\text{Cu} \quad (17\text{-}36)
$$

The reason for this regioselectivity is not precisely known because the specific mechanism for the reaction of lithium dialkylcuprates is not fully understood. However, chemists generally believe that the species that initially attacks as a nucleophile is *not* R⁻. Instead, a key step is the reversible binding of R₂Cu⁻ to the C=C double bond, followed by the transfer of the alkyl group. As we have seen previously with nucleophiles that add reversibly to the carbonyl carbon, the reversible nature of the initial binding may explain the tendency of organocopper reagents to react via conjugate addition instead of direct addition.

SOLVED PROBLEM 17.27

Predict the major organic product of the reaction shown here.

Think To what R⁻ nucleophile can this lithium dialkylcuprate be simplified? Will it add predominantly via direct addition or conjugate addition? What role does NH₄⁺ play after the addition of R⁻ is complete?

$$
\xrightarrow[\text{2. NH}_4\text{Cl, H}_2\text{O}]{\text{1. (CH}_3)_2\text{CuLi}} \quad ?
$$

Solve This lithium dialkylcuprate is a source of H_3C^-, which will add to the α,β-unsaturated carbonyl via conjugate addition. The resulting enolate anion is protonated by NH_4^+.

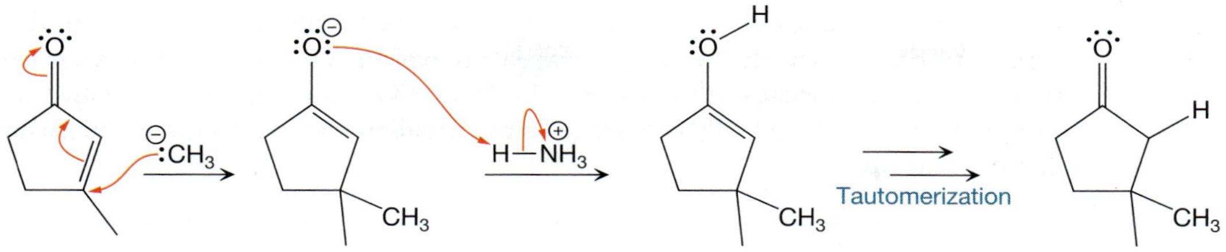

PROBLEM 17.28 Predict the major product of each of the following reactions.

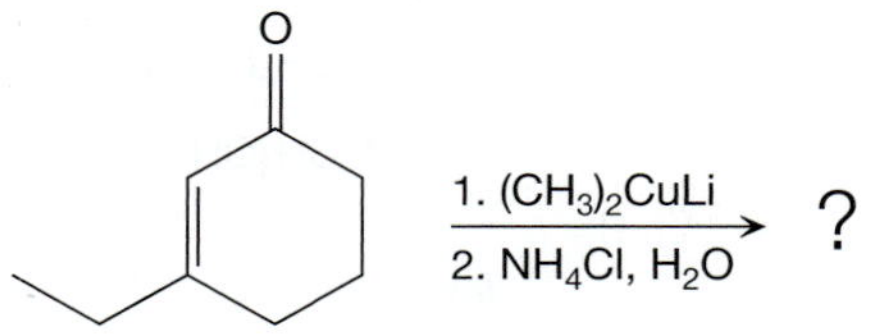

1. $(CH_3)_2CuLi$
2. NH_4Cl, H_2O

?

(a)

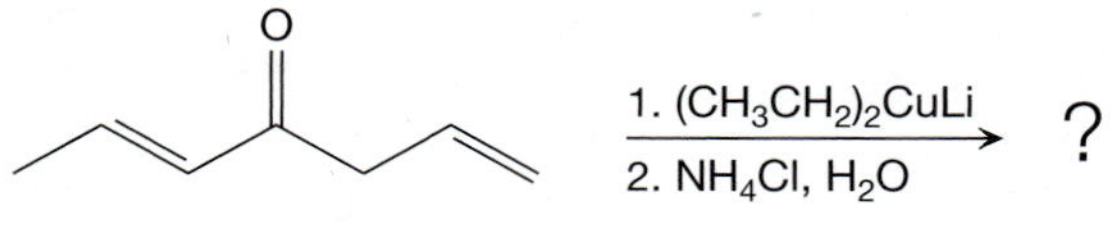

1. $(CH_3CH_2)_2CuLi$
2. NH_4Cl, H_2O

?

(b)

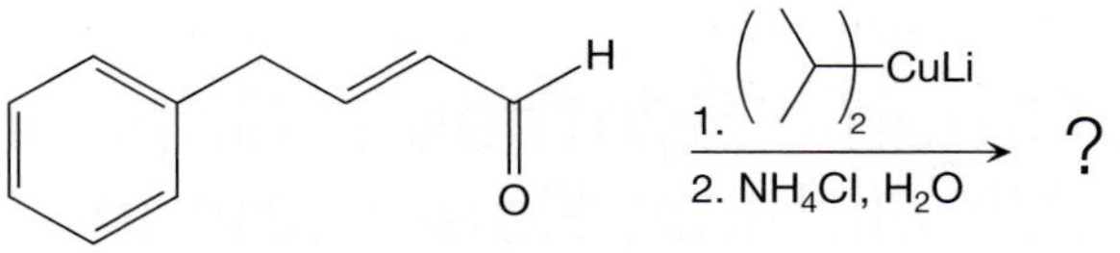

1. $\left(\right)_2$—CuLi
2. NH_4Cl, H_2O

?

(c)

Even though alkyllithium and Grignard reagents predominantly undergo direct addition to α,β-unsaturated carbonyls (Section 17.9), they also produce a minor amount of conjugate addition product, as shown in Equations 17-37 and 17-38.

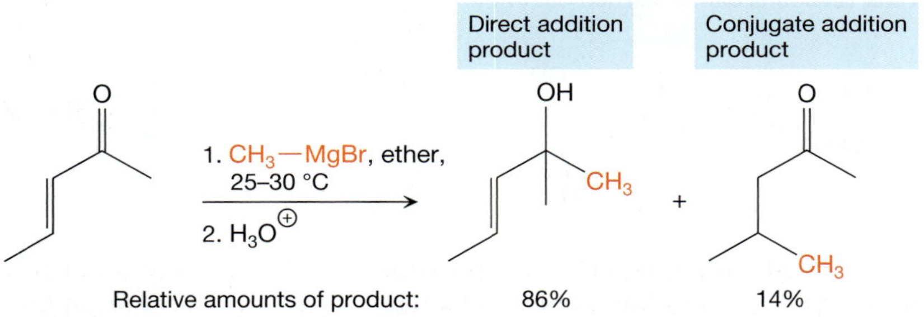

	Direct addition product	Conjugate addition product	
1. CH_3—MgBr, ether, 25–30 °C 2. $H_3O^{\oplus}$	OH ... CH_3	... CH_3	**(17-37)**
Relative amounts of product:	86%	14%	

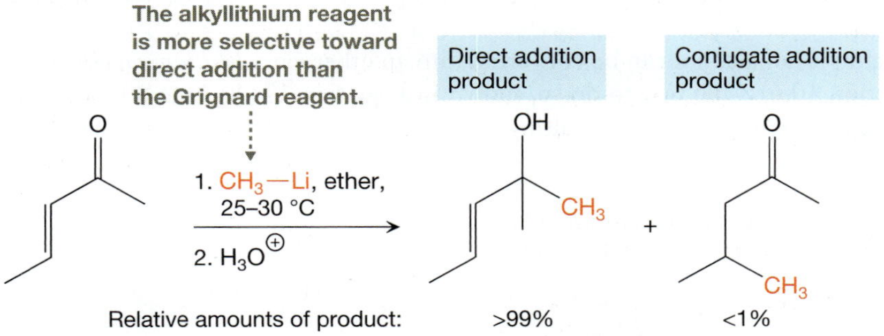

The alkyllithium reagent is more selective toward direct addition than the Grignard reagent.

	Direct addition product	Conjugate addition product	
1. CH_3—Li, ether, 25–30 °C 2. $H_3O^{\oplus}$	OH ... CH_3	... CH_3	**(17-38)**
Relative amounts of product:	>99%	<1%	

Notice, however, that the alkyllithium reagent produces less of the conjugate addition product.

Alkyllithium reagents generally have a greater tendency toward 1,2-addition to α,β-unsaturated carbonyls than Grignard reagents do.

This greater selectivity of alkyllithium reagents is explained by the greater concentration of negative charge on C (review Fig. 17-16, p. 870). Thus, in the kinetically controlled addition of R⁻, alkyllithium reagents react faster with the carbonyl carbon than do Grignard reagents.

PROBLEM 17.29 Predict the major product of each of the following reactions.

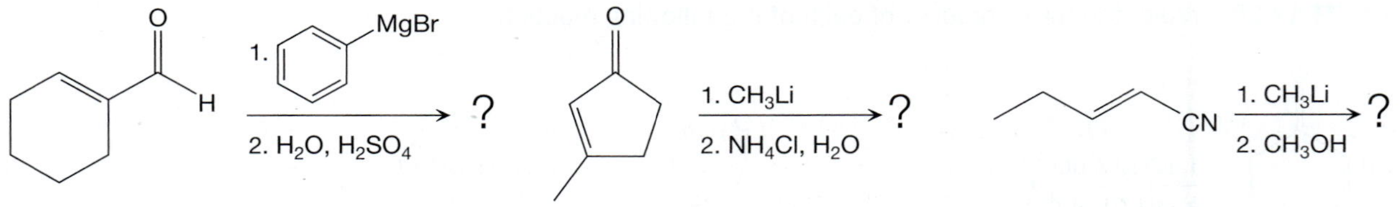

(a) (b) (c)

17.11 Organic Synthesis: Grignard and Alkyllithium Reactions in Synthesis

One of the most important reactions in organic synthesis is the one between a carbonyl compound—such as a ketone or aldehyde—and either a Grignard or an alkyllithium reagent. As shown in Equation 17-39, there are two main reasons why these reactions are so important.

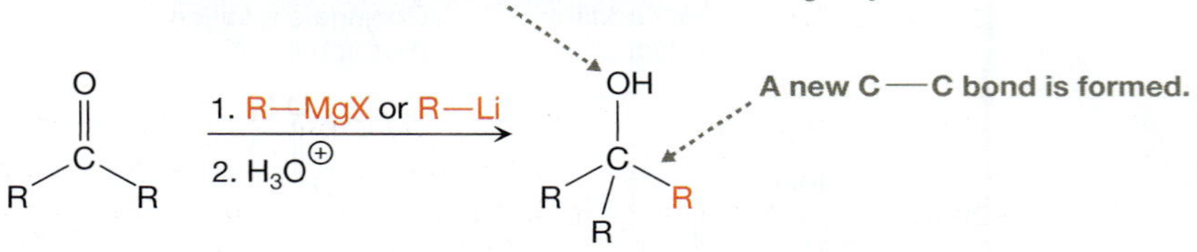

Further reactions can be carried out at the OH group.

A new C—C bond is formed.

(17-39)

First, they are carbon–carbon bond-forming reactions, so they can play a key role in constructing the desired carbon skeleton of a target compound. Second, an OH group is generated at one of the carbon atoms joined by the new bond, so we can use the characteristic reactivity of the OH group to carry out further changes at that carbon.

For example, an alcohol can be converted into an ether via a *Williamson ether synthesis* (Section 10.6). One way to do so, shown in Equation 17-40, is to add a strong base, such as NaH, and treat the resulting alkoxide anion (RO⁻) with an alkyl halide (R—Br).

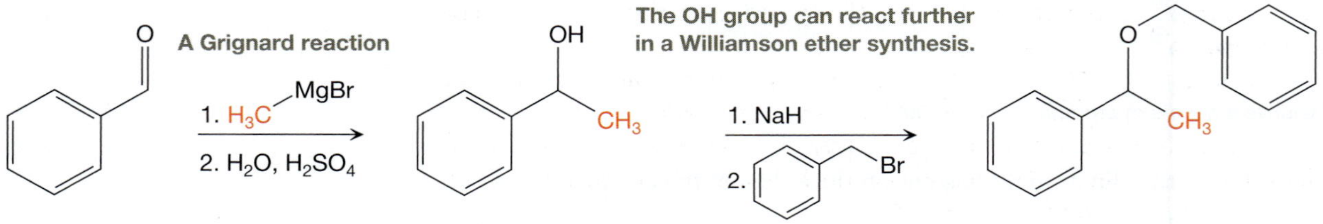

A Grignard reaction

The OH group can react further in a Williamson ether synthesis.

(17-40)

Alternatively, as shown in Equation 17-41, the product alcohol can be converted to an alkyl halide using PBr$_3$ (Section 10.1), and can then be treated with an alkoxide salt (although a substantial amount of E2 product will also form).

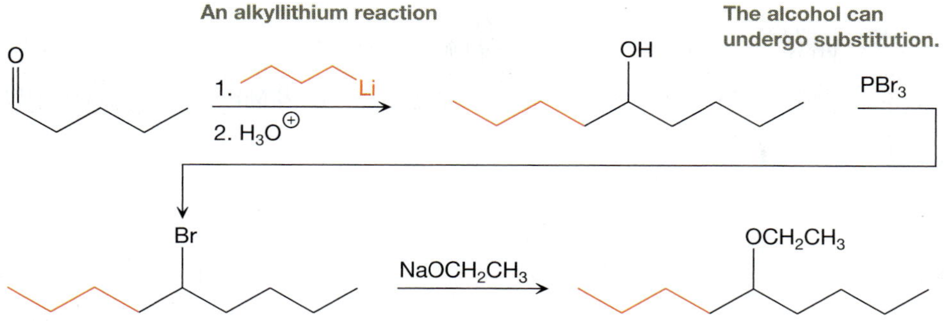

An alkyllithium reaction

1. [pentyl]—Li
2. H$_3$O$^{\oplus}$

The alcohol can undergo substitution.

OH

PBr$_3$

(17-41)

Br

NaOCH$_2$CH$_3$

OCH$_2$CH$_3$

Recall from Sections 17.7 and 17.8, too, that alkyl halides are precursors to Wittig reagents. Therefore, as shown in Equation 17-42, the alkyl halide generated in Equation 17-41 could instead be used to facilitate a Wittig reaction.

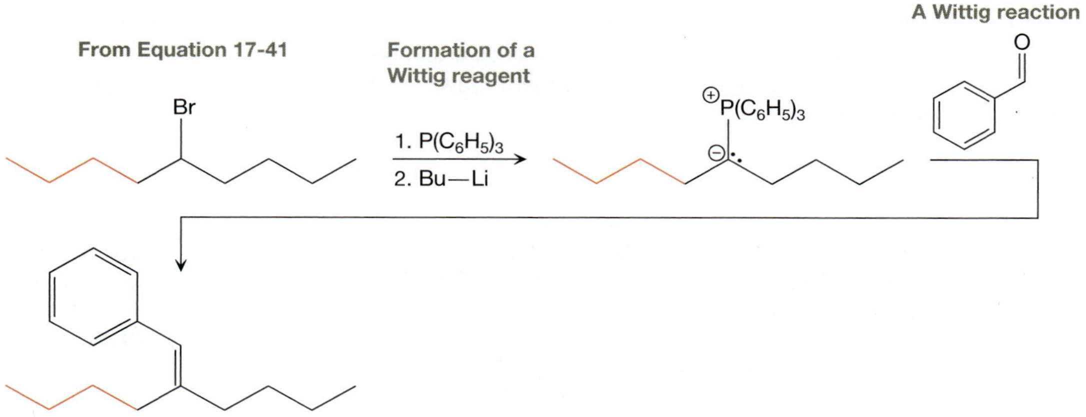

From Equation 17-41

Br

Formation of a Wittig reagent

1. P(C$_6$H$_5$)$_3$
2. Bu—Li

$^{\oplus}$P(C$_6$H$_5$)$_3$

A Wittig reaction

O

(17-42)

Because of the utility of Grignard and alkyllithium reactions in synthesis, it is important to be able to undo these reactions comfortably in your mind to execute transforms efficiently in a retrosynthetic analysis. These transforms can be carried out on essentially any alcohol by disconnecting the bond between the alcohol carbon and an adjacent carbon, as shown in Equation 17-43.

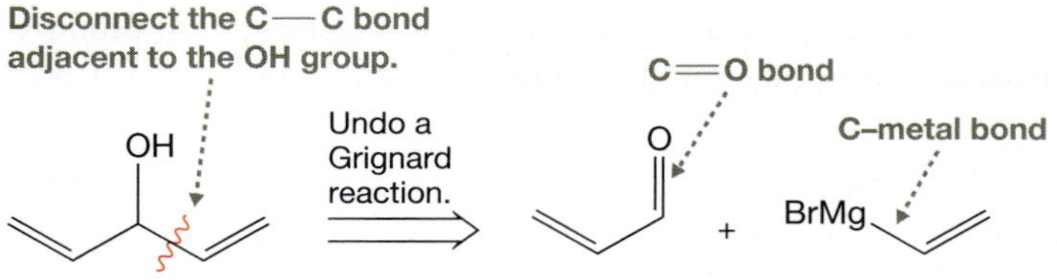

Disconnect the C—C bond adjacent to the OH group.

OH

Undo a Grignard reaction.

C=O bond

O

C–metal bond

BrMg

+

(17-43)

One precursor is a ketone or an aldehyde, whose carbonyl group contains the alcohol C atom from the target. In the other precursor, the second C atom from the disconnected bond is bonded to the metal atom.

Because the target alcohol in Equation 17-43 is symmetric, disconnecting the other C—C bond involving the alcohol carbon would give us the same precursor. With other alcohols, however, disconnecting different C—C bonds can give us different precursors. An example is shown in Equation 17-44, in which the alcohol carbon is involved in three different C—C bonds.

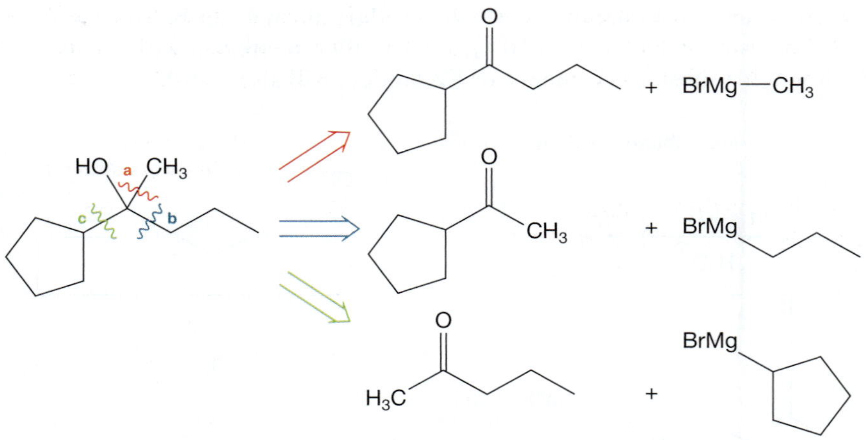

(17-44a)

(17-44b)

(17-44c)

Disconnecting the C—C bond indicated by the wavy line labeled a, b, or c yields the precursors in Equation 17-44a, 17-44b, or 17-44c, respectively.

Having the choice of which C—C bond to disconnect may at first seem daunting. With practice, however, carrying out these kinds of transforms will become quite straightforward, and you will appreciate the options that the Grignard reaction affords.

PROBLEM 17.30 Show how this molecule can be synthesized using each of the Grignard reagents shown in Equation 17-44.

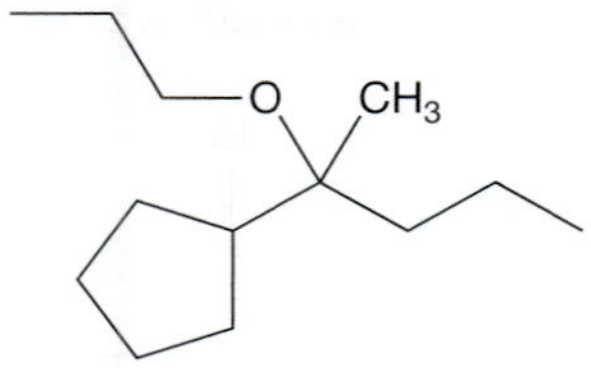

PROBLEM 17.31 Show how this compound can be synthesized from two different alkyllithium reagents.

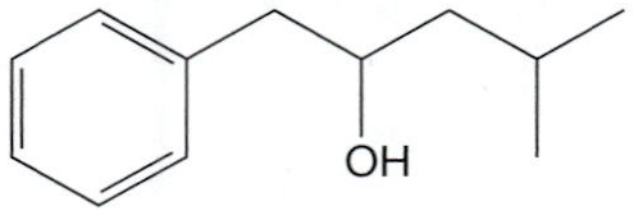

17.12 Organic Synthesis: Considerations of Direct Addition versus Conjugate Addition

As we saw in Section 17.9, a nucleophile can attack an α,β-unsaturated carbonyl group at either the carbonyl carbon, yielding the direct addition product (Equation 17-45a), or the β carbon, yielding the conjugate addition product (Equation 17-45b).

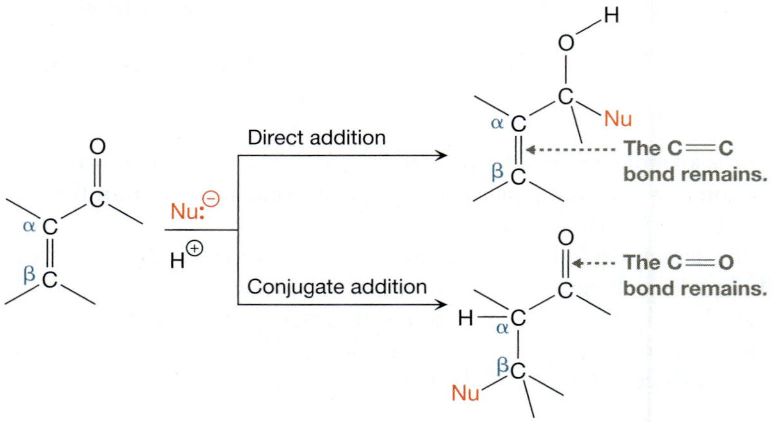

(17-45a)

(17-45b)

Direct addition converts the carbonyl group to an OH group, leaving the C=C double bond unaltered. In conjugate addition, on the other hand, the double bond between the α and β carbons is converted into a single bond, leaving the C=O bond unaltered.

YOUR TURN 17.14

Determine whether each compound shown here is the immediate product of direct or conjugate addition to an α,β-unsaturated carbonyl.

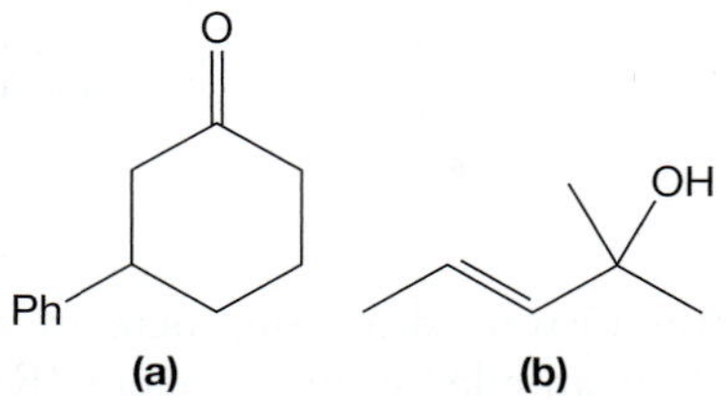

(a)   (b)

When undoing one of these reactions in a retrosynthetic analysis, be sure to look for these clues. In the molecule on the left in Equation 17-46, for example, notice that the cyano group is beta to the C=O bond. Thus, the target can be generated by conjugate addition, in which NC^- attacks the β carbon.

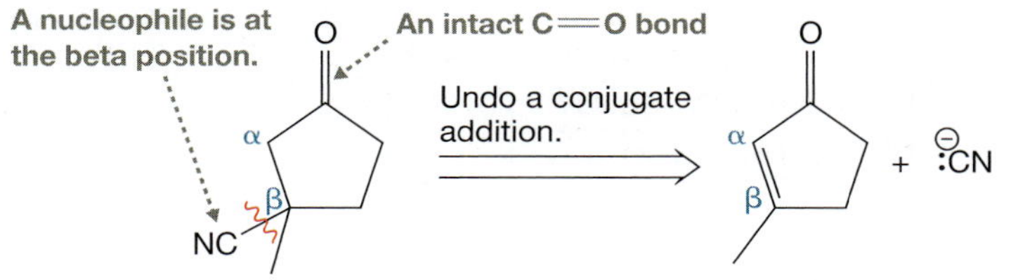

A nucleophile is at the beta position.   An intact C=O bond

Undo a conjugate addition.

$+ \ :CN^{\ominus}$ (17-46)

In the forward direction, the synthetic step might appear as in Equation 17-47.

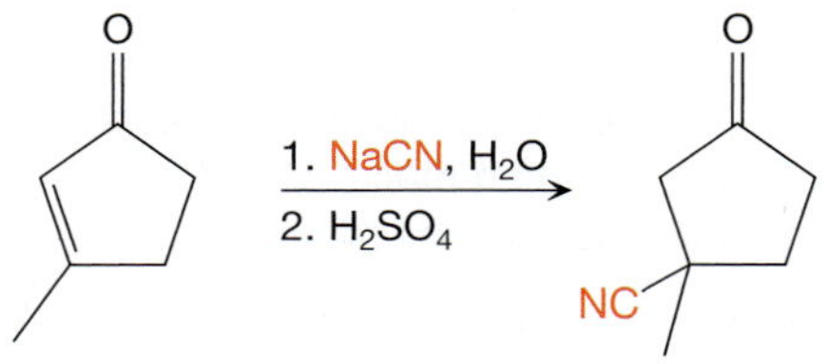

1. NaCN, H_2O
2. H_2SO_4

(17-47)

In the molecule on the left in Equation 17-48, a C=C double bond is adjacent to an OH group, so we can envision it as the result of direct addition, with H^- as the nucleophile.

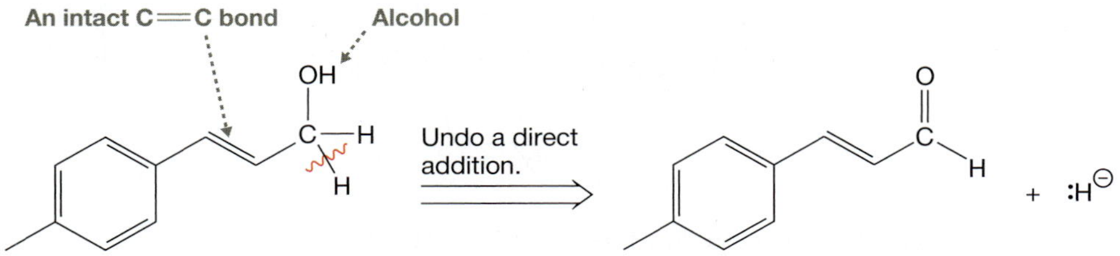

An intact C=C bond   Alcohol

Undo a direct addition.

$+ \ :H^{\ominus}$ (17-48)

Thus, the forward reaction might appear as in Equation 17-49.

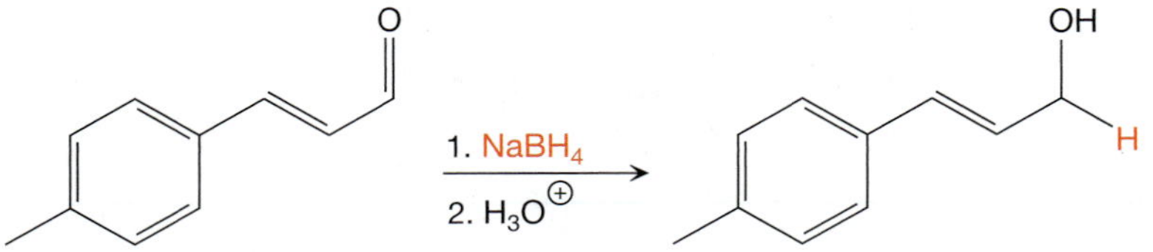

1. $NaBH_4$
2. $H_3O^{\oplus}$

(17-49)

If a particular retrosynthetic analysis suggests the use of an R⁻ nucleophile in the forward direction, then we must also consider *regioselectivity*. Suppose, for example, that we carry out the transform in Equation 17-50, for which the precursors on the right are an α,β-unsaturated ketone and an R⁻ nucleophile.

$$\text{Undo a conjugate addition.} \quad + \quad {}^{\ominus}\!:CH_2CH_3 \quad (17\text{-}50)$$

In the forward direction, there are a variety of forms the R⁻ nucleophile can take, such as an alkyllithium reagent (RLi), a Grignard reagent (RMgX), or a lithium dialkylcuprate (R₂CuLi). We want the nucleophile to attack at the β carbon, however, so we should choose a lithium dialkylcuprate, which leads almost exclusively to the conjugate addition product. Thus, the forward reaction would appear as in Equation 17-51.

R⁻ of R₂CuLi adds via conjugate addition.

New C—C bond

1. (CH₃CH₂)₂CuLi
2. NH₄Cl, H₂O

(17-51)

SOLVED PROBLEM 17.32

Show how to synthesize the compound shown here, using compounds containing five or fewer carbons.

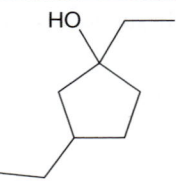

Think Are there C—C bonds that need to be formed? Are there any structural features in the target indicating that a 1,2- or 1,4-addition needs to have taken place? What are those reactions, and how do you undo them in transforms?

Solve The alcohol can be the product of a Grignard reaction, a 1,2-addition. To undo such a reaction in a transform, we can disconnect the ethyl group, as shown below. The 3-ethylcyclopentanone precursor can be the product of conjugate addition, so we can disconnect the ethyl group from C3 to arrive at cyclopent-2-enone as our starting material.

Undo 1,2-addition.   Undo 1,4-addition.

The synthesis in the forward direction might then appear as follows:

1. (⬦)₂CuLi
2. NH₄Cl

1. ⬦MgBr
2. HCl, H₂O

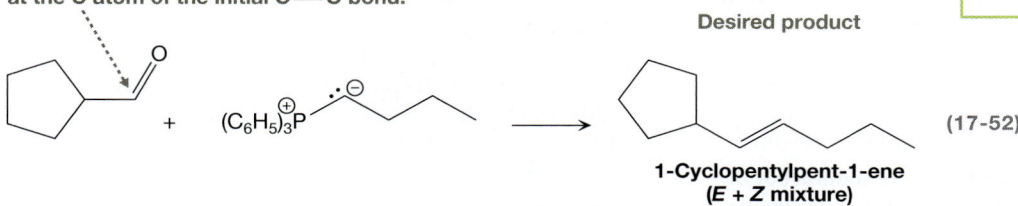

(a)   (b)

17.13 Organic Synthesis: Considerations of Regiochemistry in the Formation of Alkenes

In Section 17.7, we learned that alkenes can be synthesized using the Wittig reaction. In Chapters 8–10, we also saw examples of alkene formation via E1 and E2 reactions. The Wittig reaction is generally regarded as the more synthetically useful reaction, however, due to its *regiospecificity*. In a Wittig reaction, a C=C double bond forms precisely at the location of the initial C=O bond in the ketone or aldehyde. E1 and E2 reactions, on the other hand, typically form a mixture of alkene isomers, and occasionally the desired isomer is the minor product.

Suppose, for example, that we want to synthesize 1-cyclopentylpent-1-ene. As shown in Equation 17-52, this can be accomplished straightforwardly using the Wittig reaction.

In a Wittig reaction, the C=C bond forms at the C atom of the initial C=O bond.

Desired product

(17-52)

1-Cyclopentylpent-1-ene
(*E* + *Z* mixture)

However, attempting to synthesize this compound using an elimination reaction is problematic. One possible precursor is 1-bromo-1-cyclopentylpentane, as shown in Equation 17-53, but the major product is an undesired isomer—a trisubstituted alkene.

An elimination reaction can produce a mixture of isomeric alkenes.

Trisubstituted alkene   **Disubstituted alkene**

CH_3CH_2ONa
Δ

(17-53)

Major product   **Desired product**

Br

1-Bromo-1-cyclopentylpentane

According to Zaitsev's rule (Section 9.10), elimination favors the more highly substituted alkene. In this case, the desired product is the less highly substituted alkene—a disubstituted alkene.

A similar problem arises if 2-bromo-1-cyclopentylpentane is used as the precursor. As shown in Equation 17-54, both possible elimination products are disubstituted alkenes, so a significant amount of each will be produced.

2-Bromo-1-cyclopentylpentane $\xrightarrow[\Delta]{CH_3CH_2ONa}$ **Desired product** (Disubstituted alkene) + (Disubstituted alkene) (17-54)

The main lesson from these examples is that, when given the option of using either a Wittig reaction or an elimination reaction to synthesize an alkene, a Wittig reaction is usually the better choice to maximize percent yield of the desired product.

Chapter Summary and Key Terms

- The general mechanism for the addition of a strong, negatively charged nucleophile (Nu^-) to a *polar π bond* consists of two steps:

Nucleophilic attack occurs in the first step, followed by protonation. (Section 17.1; Objective 1)

- Some carbonyl-containing compound classes, including carboxylic acids (RCO_2H), esters (RCO_2R), amides ($RCONR_2$), acid chlorides ($RCOCl$), and acid anhydrides (RCO_2COR), have a *leaving group* attached to the carbonyl carbon. These compound classes have reactivities that differ from those of ketones and aldehydes; they are discussed in Chapters 20 and 21. (Section 17.1; Objective 1)

- Nucleophilic addition to the carbonyl group of an aldehyde tends to be more energetically favorable and faster than addition to the carbonyl group of a ketone. (Section 17.2; Objective 2)

- **Lithium aluminum hydride ($LiAlH_4$)** and **sodium borohydride ($NaBH_4$)** are sources of H^- that behave as **reducing agents** in the presence of a polar π bond. Both reagents reduce ketones ($R_2C=O$) and aldehydes ($RCH=O$) to alcohols (ROH), and they reduce imines ($R_2C=NR$) to amines ($R_2CH—NHR$). A nitrile (RCN) can be reduced to a primary amine (RCH_2NH_2) by $LiAlH_4$, but not by $NaBH_4$. (Section 17.3; Objective 3)

- $LiAlH_4$ is a very strong base, so it is incompatible with protic solvents like water and alcohols. (Section 17.3; Objective 6)

- **Sodium hydride (NaH)** is an ionic hydride, and thus a strong base, but it is not nucleophilic. (Section 17.4; Objective 4)

- Grignard reagents ($RMgX$) and **alkyllithium reagents (RLi)** are strong nucleophiles and strong bases. They add to the electron-poor C of $C=O$, $C=N$, and $C\equiv N$ bonds, and cannot exist in solvents with functional groups that are acidic or are prone to nucleophilic attack. (Section 17.5; Objectives 5, 6)

- Grignard and alkyllithium reagents cannot possess a functional group that is acidic or susceptible to nucleophilic attack. (Section 17.6; Objective 7)

- A **Wittig reagent**, also called a **phosphonium ylide**, is characterized by a $^+P—C:^-$ bond and is strongly nucleophilic at the negatively charged carbon atom. In a **Wittig reaction**, the phosphonium ylide undergoes nucleophilic addition to the $C=O$ bond of a ketone or aldehyde and, after subsequent elimination, results in the formation of an alkene. (Section 17.7; Objective 8)

- A Wittig reagent is synthesized by treating an alkyl halide with Ph_3P, followed by a very strong base such as an alkyllithium reagent. (Section 17.8; Objective 9)

- A polar π bond that is **α,β-unsaturated** is susceptible to nucleophilic attack at both the electron-poor atom of the polar π bond and the **β carbon**. Attack at the carbonyl carbon yields the *direct addition* (or *1,2-addition*) product, whereas attack at the β carbon yields the *conjugate addition* (or *1,4-addition*) product. (Section 17.9; Objectives 1, 10)

- Direct addition to an α,β-unsaturated polar π bond dominates if the nucleophile adds *irreversibly* to the polar π bond. Otherwise, conjugate addition is favored. (Section 17.9; Objective 11)

- Most nucleophiles add reversibly to a polar π bond. Nucleophiles that add irreversibly include H^- from either $NaBH_4$ or $LiAlH_4$ and R^- from $R—MgX$, $R—Li$, or Wittig reagents. (Section 17.9; Objective 11)

- **Lithium dialkylcuprates (R_2CuLi)** are weak R^- nucleophiles that favor conjugate addition over direct addition. (Section 17.10; Objective 11)

Reaction Tables

TABLE 17-3 Functional Group Conversions

	Starting Compound Class	Typical Reagents and Reaction Conditions	Compound Class Formed	Key Electron-Rich Species	Key Electron-Poor Species	Comments	Discussed in Section
(1)	Ketone or aldehyde	1. $NaBH_4$ or $LiAlH_4$ 2. H_2O, H_2SO_4	Alcohol	Hydride anion		Nucleophilic addition	17.3
(2)	Imine	1. $LiAlH_4$ 2. H_2O	Amine	Hydride anion		Nucleophilic addition	17.3
(3)	$R-C\equiv N$ Nitrile	1. $LiAlH_4$ 2. H_2O	1° Amine	Hydride anion	$-C\equiv N$	Sequential nucleophilic additions	17.3
(4)	α,β–Unsaturated ketone or aldehyde	1. $NaBH_4$ or $LiAlH_4$ 2. NH_4Cl, H_2O	Alcohol	Hydride anion		Nucleophilic addition	17.9
(5)	Ketone or aldehyde	NaH	Enolate anion	Hydride anion		Proton transfer	17.4
(6)	Alkyl halide	1. $P(C_6H_5)_3$ 2. R–Li	Wittig reagent	Triphenyl phosphine		S_N2 followed by proton transfer	17.7

TABLE 17-4 Reactions That Alter the Carbon Skeleton

Starting Compound Class	Typical Reagent and Reaction Conditions	Compound Class Formed	Key Electron-Rich Species	Key Electron-Poor Species	Comments	Discussed in Section
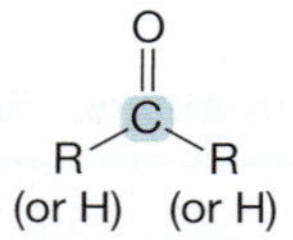 **(1)** Ketone or aldehyde	1. R'—Li or R'—MgX 2. H_2O, H_2SO_4	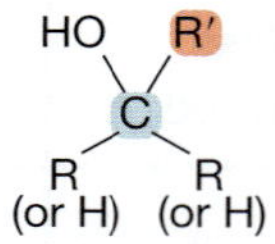 Alcohol	:R'$^{\ominus}$	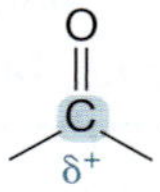	Nucleophilic addition	17.5
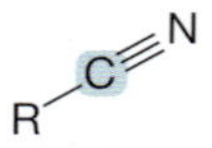 **(2)** Nitrile	1. R'—MgX 2. CH_3OH	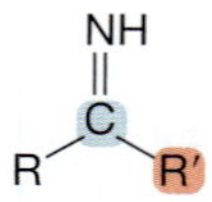 Imine	:R'$^{\ominus}$		Nucleophilic addition	17.5
 **(3)** Grignard reagent	1. CO_2(s) 2. H_2O, H_2SO_4	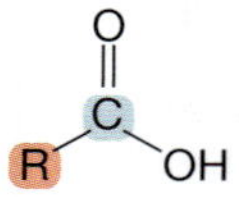 Carboxylic acid	:R$^{\ominus}$	O=C=O	Nucleophilic addition	17.5
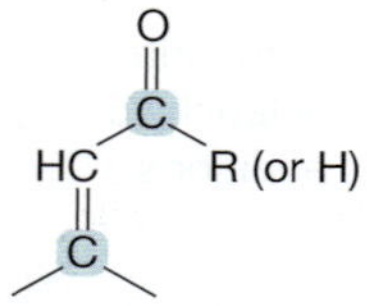 **(4)** α,β–Unsaturated ketone or aldehyde	1. R'$_2$CuLi, THF 2. NH_4Cl, H_2O	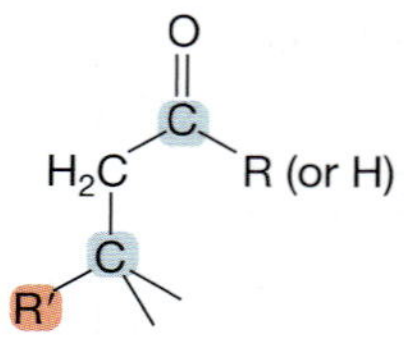 Ketone or aldehyde	:R'$^{\ominus}$	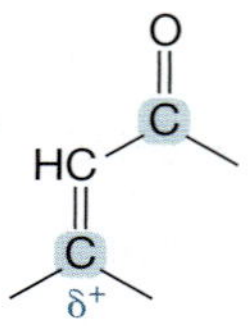	Conjugate nucleophilic addition	17.10
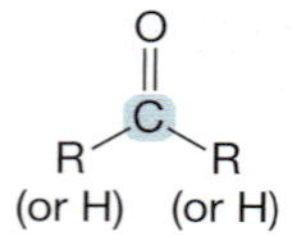 **(5)** Ketone or aldehyde	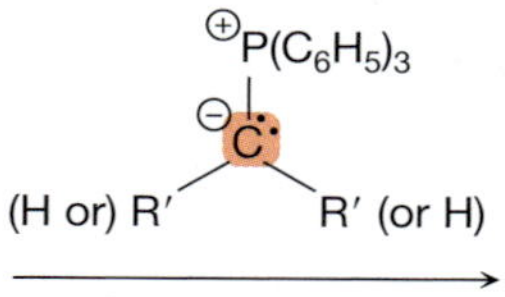 Wittig reagent	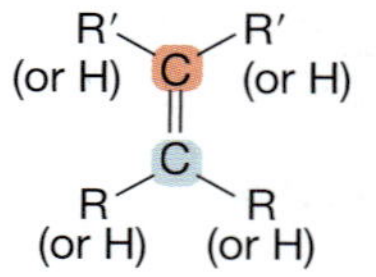 Alkene	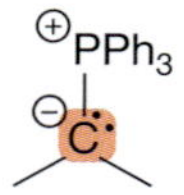	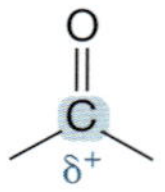	Wittig reaction	17.7

Problems

Problems that are related to synthesis are denoted (SYN).

17.1 and 17.2 Addition of Strong Nucleophiles and Substituent Effects

17.34 For each pair of compounds, which has the polar π bond that will undergo nucleophilic addition more rapidly? Why?

(a)

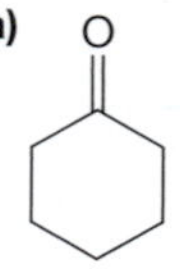

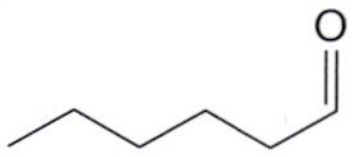

(b)

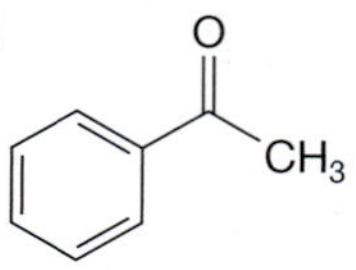

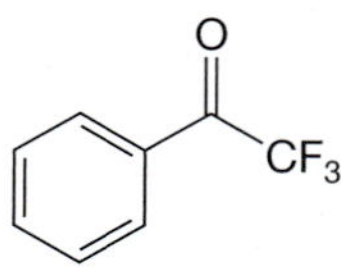

(c)

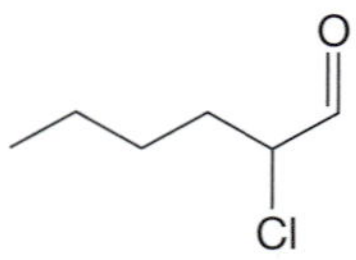

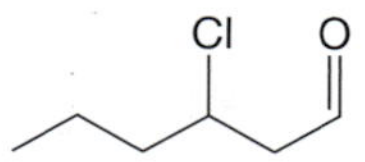

(d) $Cl_3C—C\equiv N$ or $H_3C—C\equiv N$

17.35 Unlike most hydrates, the hydrate of cyclopropanone is stable and can be isolated. Explain why this hydrate is stable.

17.36 Draw the mechanism for each of the following reactions.

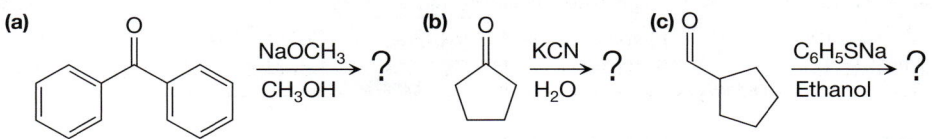

(a) [structure] $\xrightarrow[\text{CH}_3\text{OH}]{\text{NaOCH}_3}$?

(b) [structure] $\xrightarrow[\text{H}_2\text{O}]{\text{KCN}}$?

(c) [structure] $\xrightarrow[\text{Ethanol}]{\text{C}_6\text{H}_5\text{SNa}}$?

17.37 The treatment of a nitrile with cyanide yields an α-aminomalononitrile, as shown here in the reaction. Provide a detailed mechanism for this reaction.

[structure] $\equiv$ N $\xrightarrow[\text{HCN}]{\text{NaCN}}$ [structure with CN, NH₂, CN groups]

An α-aminomalononitrile

17.3 and 17.4 Reactions of Hydride Agents; Oxidation States

17.38 In the chapter, we stated that the imine in this reaction is reduced. Verify that this is the case by calculating the oxidation state of the C and N atoms of the C=N group before and after the reaction has taken place. Does the imine gain electrons? If so, how many?

[structure] $\xrightarrow[\text{CH}_3\text{OH}]{\text{NaBH}_4}$ [structure]

17.39 In the chapter, we stated that the nitrile in this reaction is reduced. Verify that this is the case by calculating the oxidation state of the C and N atoms of the C≡N group before and after the reaction has taken place. Does the nitrile gain electrons? If so, how many?

[structure] N $\xrightarrow[\text{2. H}_3\text{O}^\oplus]{\text{1. LiAlH}_4, \text{ether}}$ [structure] NH₂

17.40 In the chapter, we showed that LiAlH₄ deprotonates weakly acidic solvents such as alcohols, as shown here. Determine which species, if any, are oxidized or reduced.

$R{-}OH + {}^\ominus AlH_4 \longrightarrow R{-}O^\ominus + AlH_3 + H_2$

17.41 Draw the mechanism and predict the major product for each of the following reactions.

(a) [structure] $\xrightarrow[\text{CH}_3\text{OH}]{\text{NaBH}_4}$?

(b) [structure] OCH₃ $\xrightarrow[\text{CH}_3\text{OH}]{\text{NaBH}_4}$?

(c) [structure] $\xrightarrow[\text{2. H}_2\text{SO}_4, \text{H}_2\text{O}]{\text{1. LiAlH}_4, \text{Et}_2\text{O}}$?

17.42 Draw the mechanism and predict the major product for each of the following reactions.

(a) H₃C— [structure] C≡N, CH₃ $\xrightarrow[\text{2. H}_2\text{O}]{\text{1. LiAlH}_4, \text{ether}}$?

(b) [structure] N—CH₃ $\xrightarrow[\text{2. H}_2\text{O}]{\text{1. LiAlH}_4, \text{ether}}$?

(c) [structure] NH $\xrightarrow[\text{H}_2\text{O}]{\text{NaBH}_4}$?

17.43 The chemical behavior of deuterium (D or ²H) is essentially identical to that of hydrogen (¹H). Therefore, D₂O behaves the same as H₂O, and LiAlD₄ behaves the same as LiAlH₄. With this in mind, draw the detailed mechanism of each of the following reactions. Using the mechanism, predict the reaction products.

(a) [structure] $\xrightarrow[\text{2. D}_2\text{O}]{\text{1. LiAlH}_4}$?

(b) [structure] $\xrightarrow[\text{2. H}_2\text{O}]{\text{1. LiAlD}_4}$?

(c) [structure] $\xrightarrow[\text{2. D}_2\text{O}]{\text{1. LiAlD}_4}$?

17.44 Draw the mechanism and predict the major product for the reaction shown here.

Ph [structure] O, Ph $\xrightarrow[\text{2. D}_2\text{O}]{\text{1. NaH}}$?

17.45 Draw the mechanism and predict the major product for each of the following reactions.

(a)

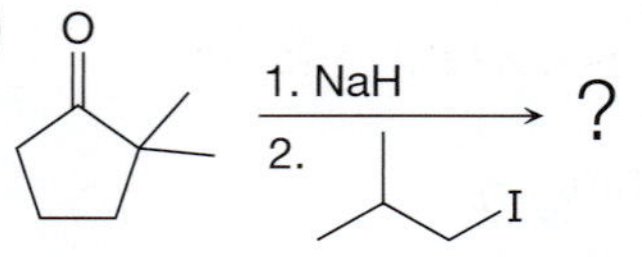

1. NaH
2.

→ ?

(b)

1. NaH
2. Br

→ ?

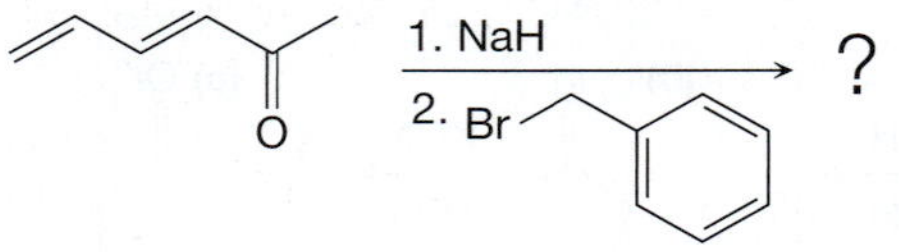

17.46 (SYN) Show how to carry out each of the following transformations.

(a)

? →

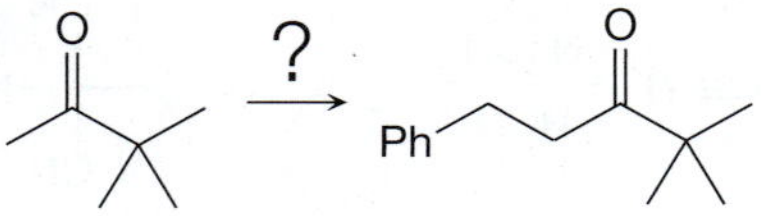

(b)

? →

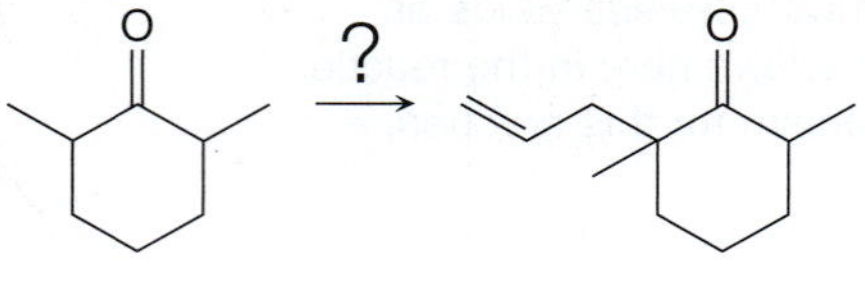

17.5 and 17.6 Reactions of Alkyllithium Reagents and Grignard Reagents

17.47 Draw the mechanism and predict the major product for each of the following reactions.

(a)

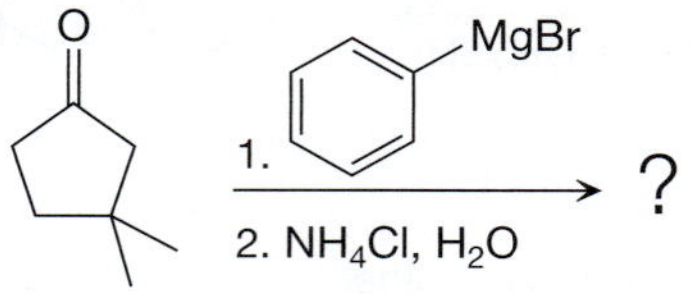

1.   MgBr
2. NH₄Cl, H₂O

→ ?

(b)

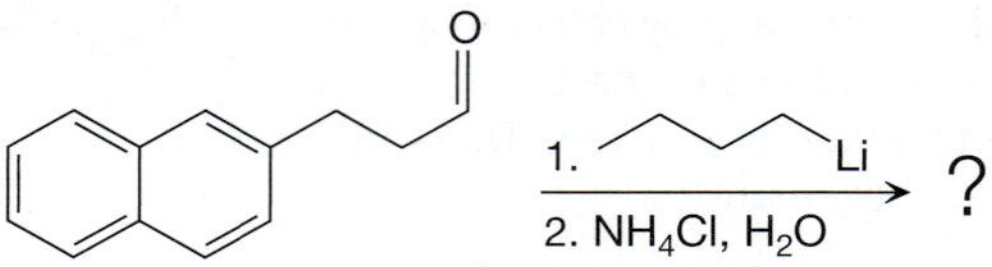

1.   Li
2. NH₄Cl, H₂O

→ ?

(c)

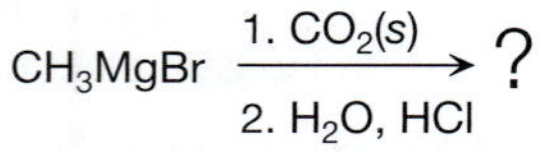

CH₃MgBr

1. CO₂(s)
2. H₂O, HCl

→ ?

17.48 Draw the mechanism and predict the major product for each of the following reactions.

(a)

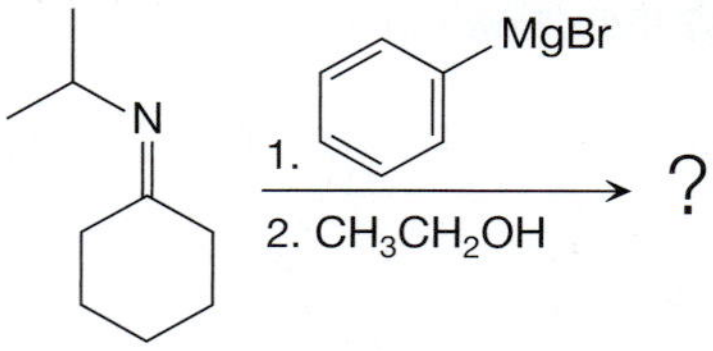

1.   MgBr
2. CH₃CH₂OH

→ ?

(b)

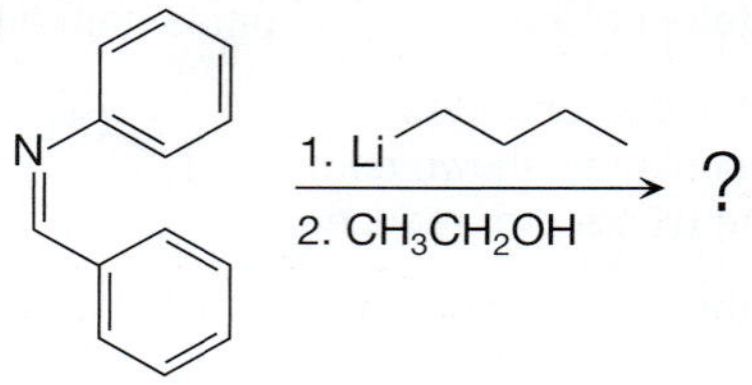

1. Li
2. CH₃CH₂OH

→ ?

(c)

1. CH₃Li
2. CH₃OH

→ ?

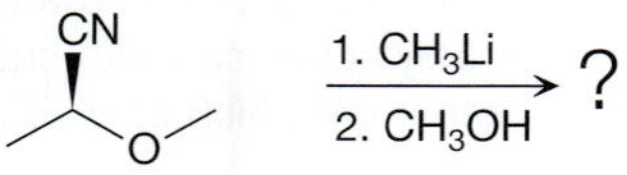

17.49 The chemical behavior of deuterium (D or ^{2}H) is essentially identical to that of hydrogen (^{1}H). Therefore, D₂O behaves the same as H₂O, and CD₃Li behaves the same as CH₃Li. With this in mind, draw the detailed mechanism of each of the following reactions. Using the mechanism, predict the reaction products.

(a)

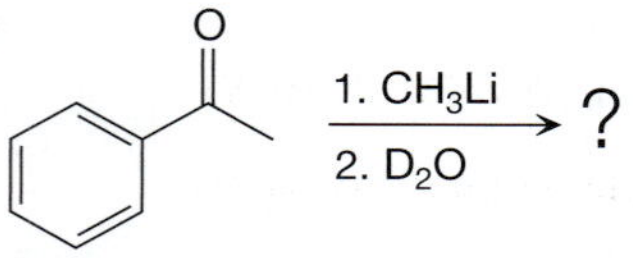

1. CH₃Li
2. D₂O

→ ?

(b)

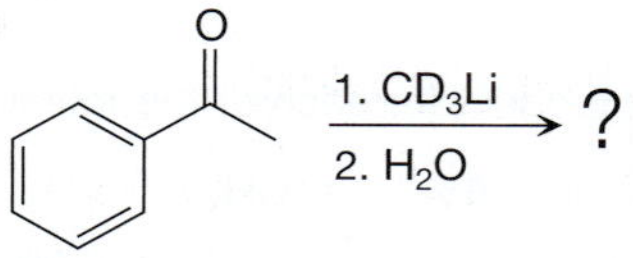

1. CD₃Li
2. H₂O

→ ?

(c)

1. CD₃Li
2. D₂O

→ ?

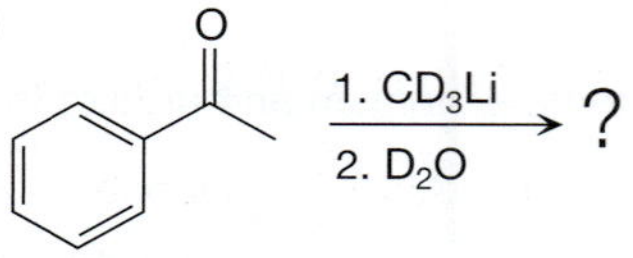

17.50 (SYN) Propose three different syntheses of the alcohol shown here, each using a different Grignard reagent.

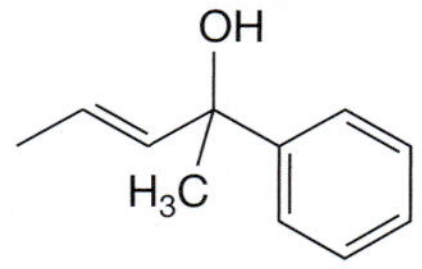

17.51 (SYN) Show how to carry out each of the following transformations.

(a)

? →

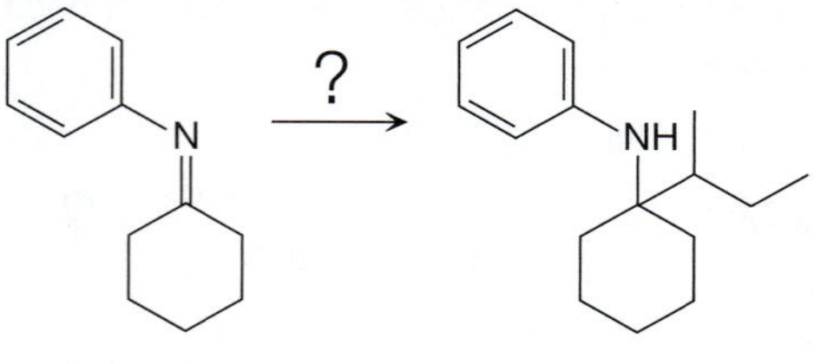

(b)

? →

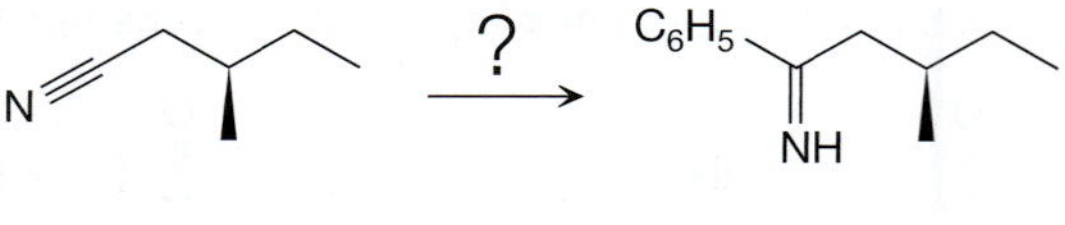

17.52 Alkyllithium and Grignard reagents are highly reactive with protic solvents (e.g., water and alcohols), so they require aprotic solvents (e.g., ethers). Acetone is an aprotic solvent that can be used in nucleophilic substitution and elimination reactions, but it cannot be used as a solvent for reactions involving alkyllithium and Grignard reagents. Explain why.

17.7 and 17.8 Wittig Reagents and the Wittig Reaction

17.53 Draw the complete mechanism that takes place when each of the following species is treated first with triphenylphosphine, followed by butyllithium.

(a) Br with O.O methoxy group on benzene ring   (b) cyclopentyl—Br   (c) pent-2-yne with Cl

17.54 **(SYN)** Show how to synthesize each of these species from an alkyl halide.

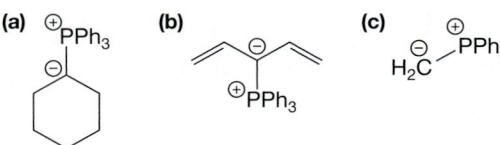

17.55 Draw the complete mechanism for the reaction between benzaldehyde and each of the species from Problem 17.54.

17.56 **(SYN)** Show how to synthesize each of the following compounds from an alkyl halide and a ketone or aldehyde.

(a)   (b) NO₂   (c)

17.57 In the chapter, we learned how a phosphonium ylide is produced from an alkyl halide, and we learned how such a phosphonium ylide reacts with a ketone or aldehyde to produce an alkene. *Sulfonium ylides* can also be produced from alkyl halides, as shown here, but their reaction with a ketone or aldehyde produces an epoxide, not an alkene. Draw the complete mechanism for each reaction in this sequence.

H₃C—S—CH₃ →(1. CH₃I, 2. NaH)→ H₃C—S⁺—CH₂⁻ → [epoxide] + H₃C—S—CH₃

An epoxide

17.58 The reaction of a phosphonium ylide with a ketone or aldehyde produces an alkene, but, as shown in Problem 17.57, the reaction of a sulfonium ylide produces an epoxide. Explain why. *Hint:* What is the major driving force for the Wittig reaction to produce an alkene? The S—O bond energy in dimethyl sulfoxide is 362 kJ/mol.

17.9 and 17.10 Direct Addition versus Conjugate Addition; Lithium Dialkylcuprates

17.59 Which of the following nucleophiles will add *reversibly* to a polar π bond? Which will add *irreversibly*?

(a) Li compound   (b) cyclohexyl—MgBr   (c) benzyl—ONa   (d) cyclopentanone with ONa ester

(e) (chain)₂CuLi   (f) isopropyl—N(Li)—isopropyl   (g) H₂O   (h) cyclohexyl—N(H)—

(i) tetrahydrofuran (O)   (j) pentyl—SNa   (k) benzyl—SH

17.60 Draw the mechanism and predict the major product for each of the following reactions.

(a)

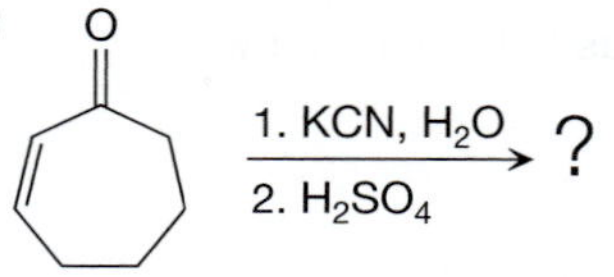

1. KCN, H_2O
2. H_2SO_4

?

(b)

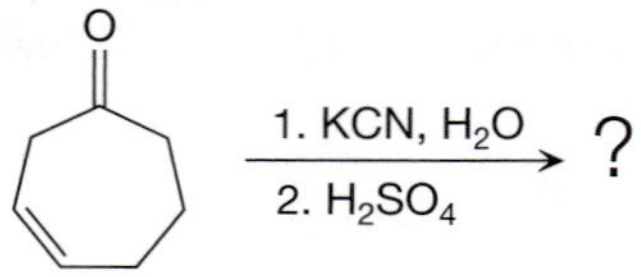

1. KCN, H_2O
2. H_2SO_4

?

(c)

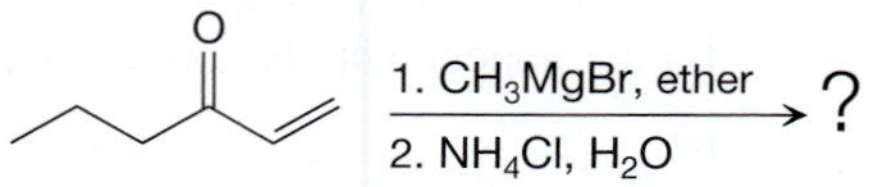

1. CH_3MgBr, ether
2. NH_4Cl, H_2O

?

(d)

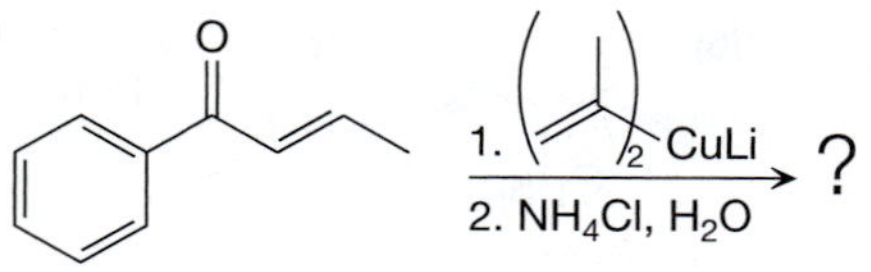

1. (image) $_2$CuLi
2. NH_4Cl, H_2O

?

(e)

1. CH_3CH_2SNa, ethanol
2. NH_4Cl, H_2O

?

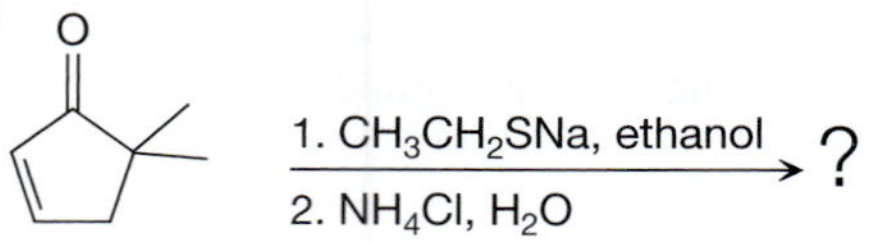

17.61 (SYN) Show two different syntheses for the compound shown here, one using $(CH_3)_2CuLi$ as a reagent and the other using $(CH_3CH_2)_2CuLi$.

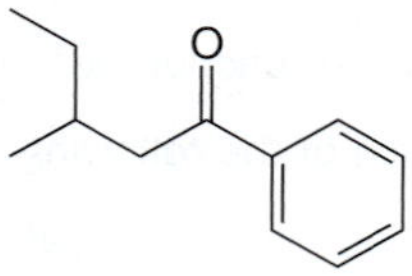

17.62 Both $NaBH_4$ and $LiAlH_4$ favor direct addition to an α,β-unsaturated carbonyl over conjugate addition, but their selectivities differ. Which of these hydride reducing agents would be more selective toward direct addition? Why? *Hint:* Review Section 17.10.

17.63 The conjugate addition of $NaBH_4$ to an α,β-unsaturated carbonyl compound occurs to a small extent, but when it does, it results in two sequential nucleophilic additions. The conjugate addition of a lithium dialkylcuprate (R_2CuLi), on the other hand, results in only a single nucleophilic addition. Explain.

17.64 Draw the complete, detailed mechanism for the reaction shown here. (See Problem 17.63.)

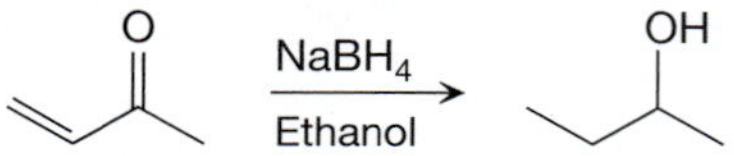

$NaBH_4$ / Ethanol

Integrated Problems

17.65 (SYN) Provide the reagents necessary to perform the following transformation.

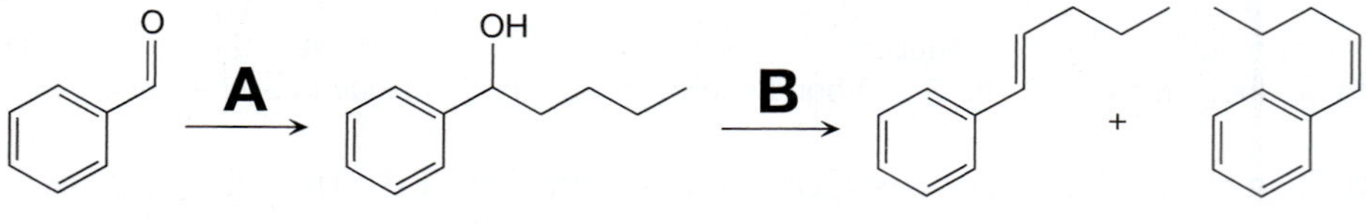

17.66 (SYN) Suggest how you would synthesize each of the following from phenylethanone (acetophenone), using any reagents necessary. *Hint:* Each synthesis may require more than one synthetic step.

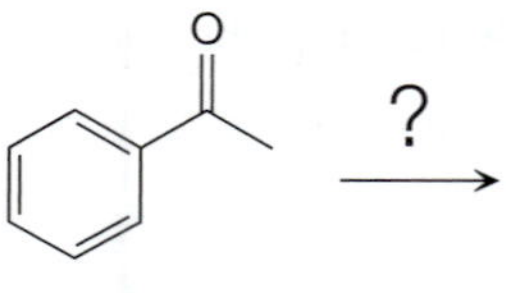

Phenylethanone (Acetophenone)

?

(a)

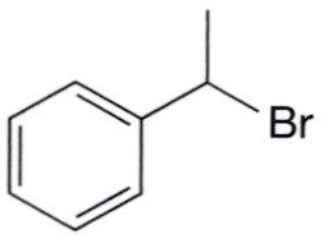

(b)

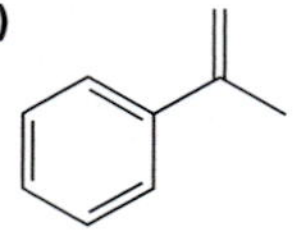

(c)

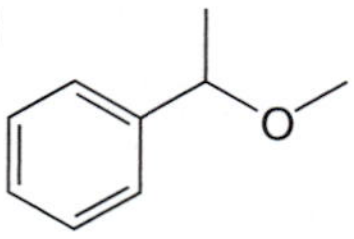

17.67 Predict the major organic product and draw the complete, detailed mechanism for each of the following reactions.

(a)

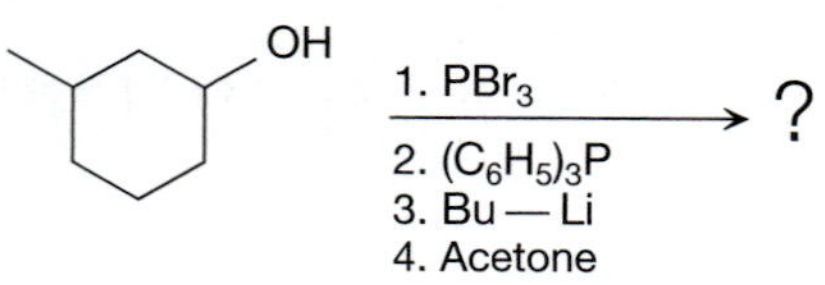

1. PBr_3
2. $(C_6H_5)_3P$
3. Bu—Li
4. Acetone

?

(b)

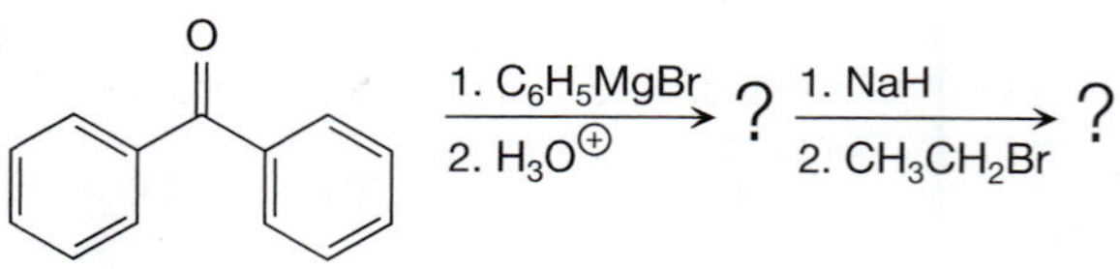

1. C_6H_5MgBr
2. $H_3O^{\oplus}$

?

1. NaH
2. CH_3CH_2Br

?

17.68 When carbon disulfide (CS_2) is treated with an alcohol in the presence of base, the product is a xanthate salt. If an alkyl halide is also present, a xanthate ester is formed. Propose a mechanism for each of these reactions.

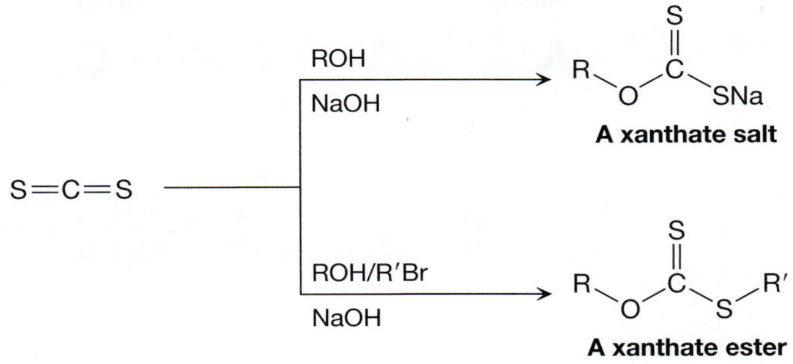

A xanthate salt

A xanthate ester

17.69 Phenylmagnesium bromide reacts with sulfur dioxide to produce a reactive intermediate, which, on further reaction with CH_3Br, produces methylphenylsulfone. Propose a mechanism for this reaction, and propose a structure for the reactive intermediate.

17.70 An α,β-unsaturated ketone reacts with a conjugated Wittig reagent to produce a relatively highly strained bicyclic compound. Draw the complete, detailed mechanism for this reaction. (A key intermediate has been provided.)

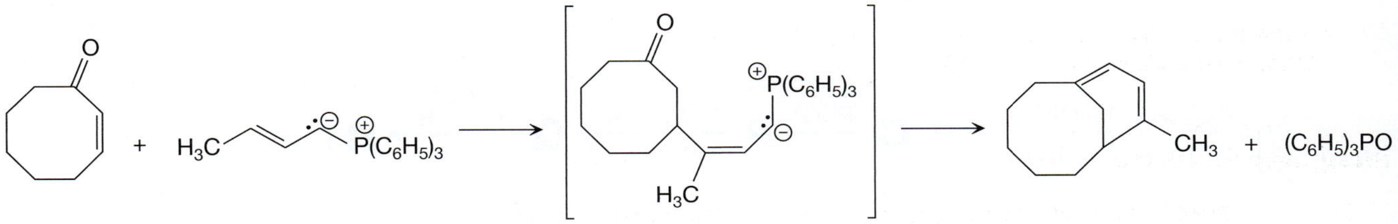

17.71 The reaction shown here is an example of the Corey–Chaykovsky aziridination reaction. Draw its complete, detailed mechanism. *Hint:* See Problem 17.57.

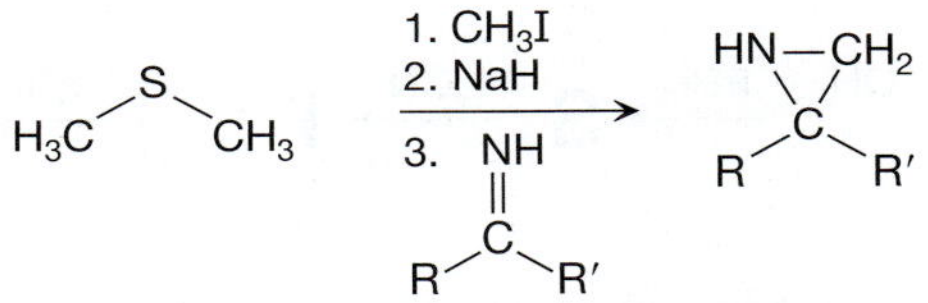

17.72 The following is an example of the Corey–Chaykovsky cyclopropanation. Draw its complete, detailed mechanism. *Hint:* See Problem 17.57.

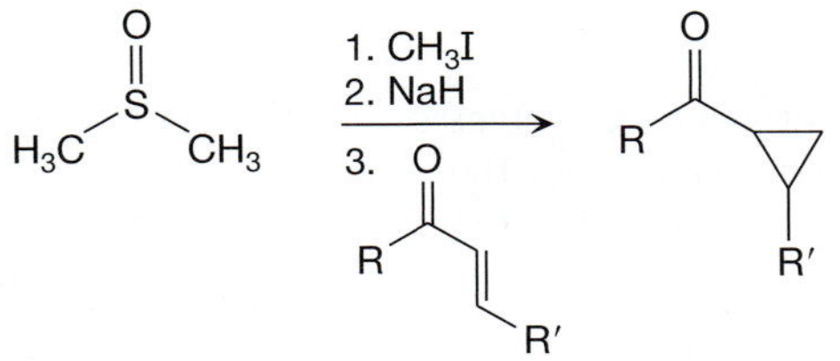

17.73 Although a Wittig reagent can be prepared from 5-bromo-1,3-cyclopentadiene in the usual way, that Wittig reagent is unreactive toward ketones or aldehydes.

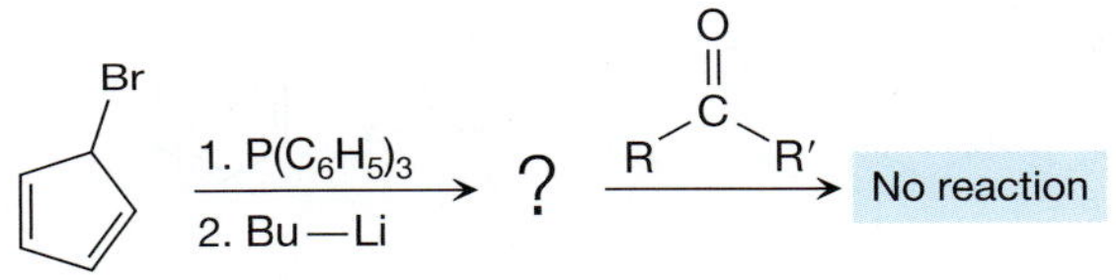

(a) Draw the complete mechanism showing the formation of the Wittig reagent.
(b) Explain why that Wittig reagent does not undergo nucleophilic addition with ketones or aldehydes.

17.74 Determine the structures of compounds **A–J** in the following reaction sequences.

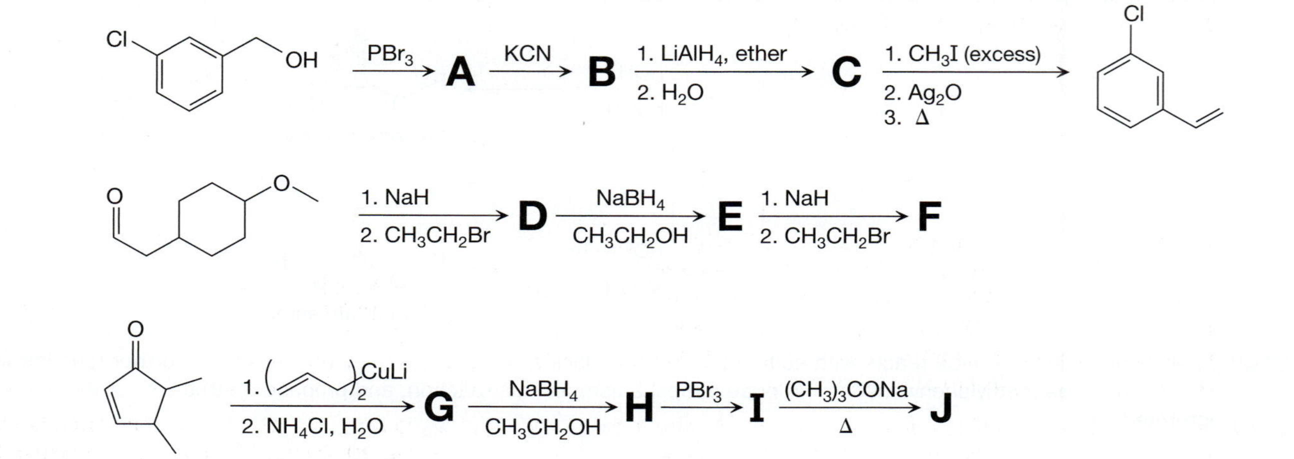

17.75 Determine the structures of compounds **K–V** in the following reaction sequences.

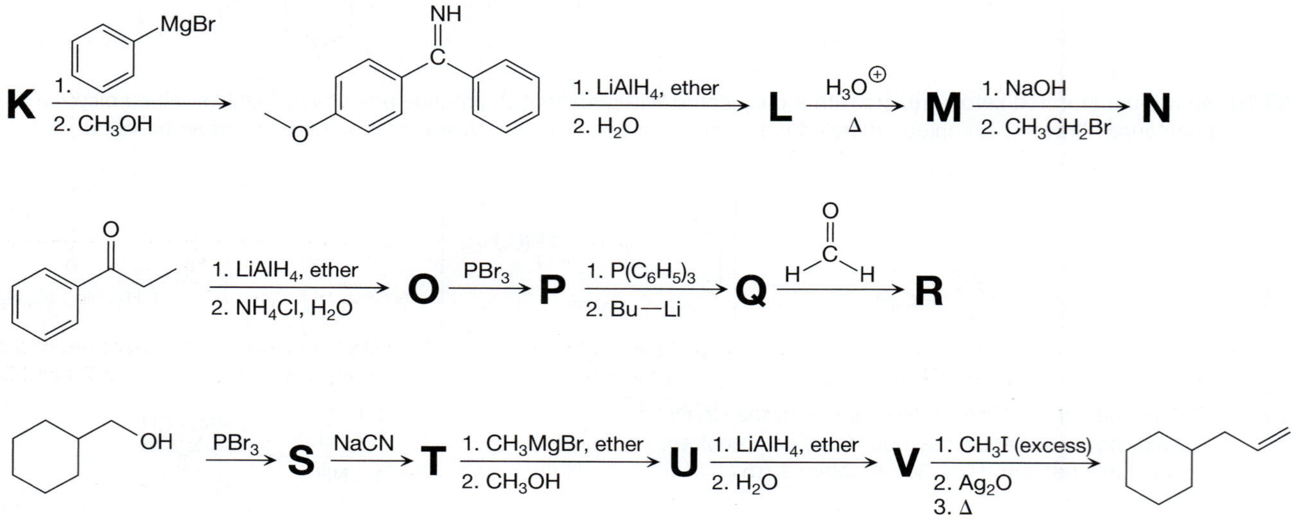

17.76 (SYN) Suggest how you would carry out the synthesis shown here using any reagents necessary. *Hint:* The synthesis may require more than one synthetic step.

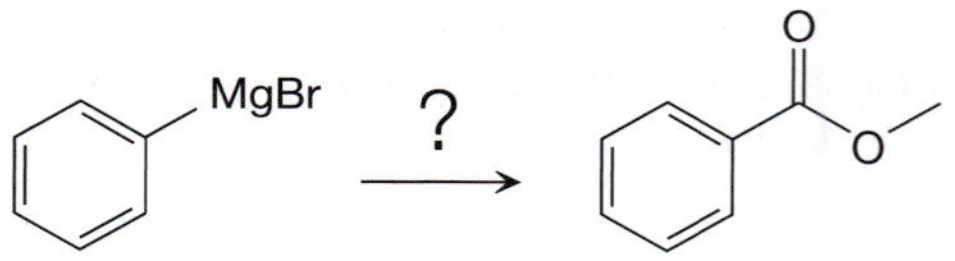

17.77 (SYN) Suggest how you would carry out the synthesis shown here using any reagents necessary. *Hint:* The synthesis may require more than one synthetic step.

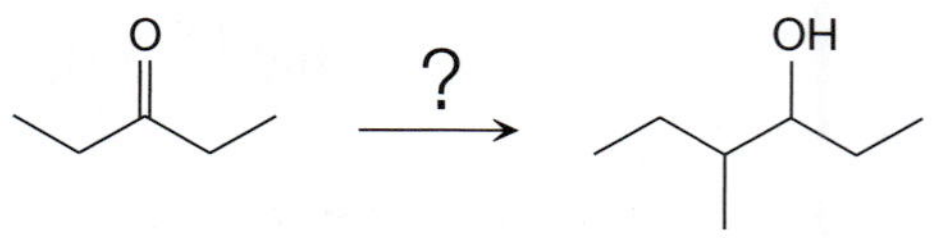

17.78 (SYN) Show how you would carry out the following synthesis using any reagents necessary. *Hint:* The synthesis may require more than one synthetic step.

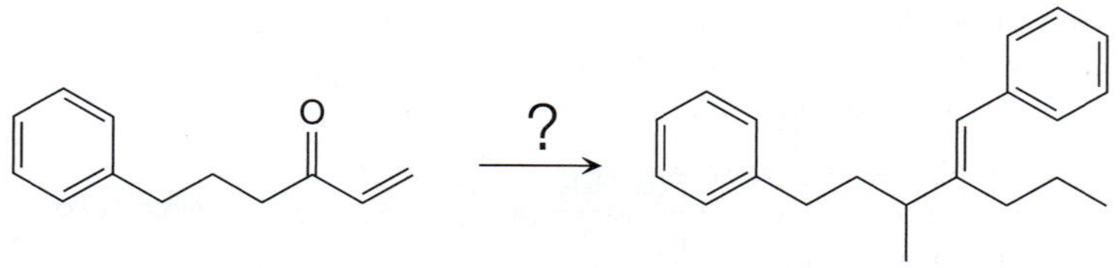

17.79 **(SYN)** Suggest how you would carry out the synthesis shown here using any reagents necessary. *Hint:* The synthesis may require more than one synthetic step.

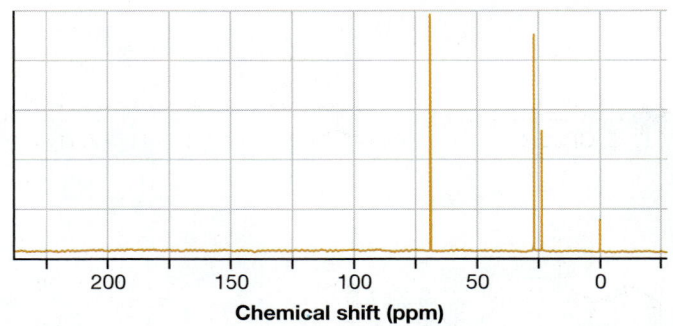

17.80 When propenal is treated with sodium acetylide, a product is formed whose IR spectrum exhibits a broad absorption between 3200 and 3600 cm^{-1}, but shows no absorption near 1700 cm^{-1}. **(a)** Draw the structure of the product. **(b)** Argue whether the nucleophile adds *reversibly* or *irreversibly* to the carbonyl group.

$$\underset{\substack{\|\\ CH_2}}{\overset{\substack{O\\ \|}}{HC}}\!-\!CH \quad \xrightarrow[\text{2. NH}_4\text{Cl}]{\text{1. HC}\equiv\text{CNa}} \quad ?$$

17.81 When 5-bromopentanal is treated with sodium borohydride, a compound is produced whose ^{13}C NMR spectrum is shown here. In its IR spectrum, no absorption bands appear near 1700 cm^{-1} or above 3000 cm^{-1}. Propose a mechanism to account for the formation of this product.

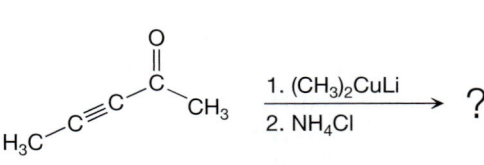

$$\xrightarrow[\text{EtOH}]{\text{NaBH}_4} \quad ?$$

17.82 Lithium dimethylcuprate reacts with the beta alkynyl carbonyl shown in the following reaction. The IR spectrum of the product is shown and the ^{1}H NMR spectrum has the following four signals: 1.9 ppm, 3 H; 2.1 ppm, 3 H; 2.2 ppm, 3 H; and 6.1 ppm, 1 H. What is the structure of the product?

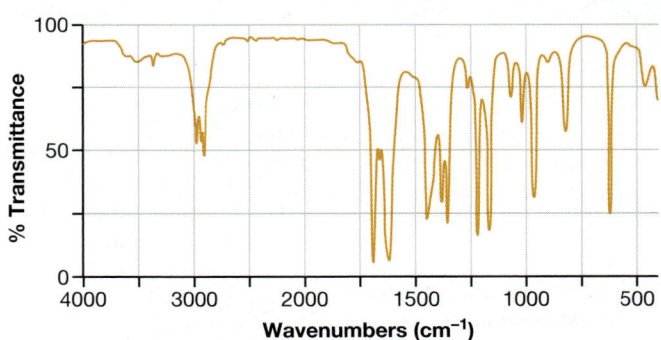

$$\xrightarrow[\text{2. NH}_4\text{Cl}]{\text{1. (CH}_3)_2\text{CuLi}} \quad ?$$

17.83 When phenyl-5-bromopentanone is treated with triphenylphosphine, followed by base, a compound is produced whose formula is C$_{11}$H$_{12}$. Its ^{13}C NMR spectrum has nine signals, six of which appear between 120 and 140 ppm and three of which appear below 50 ppm. For this reaction, draw the complete, detailed mechanism, as well as the major organic product.

$$\xrightarrow[\text{2. NaOCH}_2\text{CH}_3]{\text{1. P(C}_6\text{H}_5)_3} \quad \text{C}_{11}\text{H}_{12}$$

17.84 The reaction shown here produces a compound, C$_{10}$H$_{21}$N, whose IR spectrum exhibits no peaks between 1500 and 2000 cm^{-1}. Draw the complete mechanism for this reaction.

$$\xrightarrow[\substack{\text{Diethyl}\\ \text{ether}}]{\text{LiAlH}_4} \quad \text{C}_{10}\text{H}_{21}\text{N}$$

Our skin can resume its original shape after being stretched in part because of elastin, a protein that exhibits cross-linked bonding between its long-chain molecules. This cross-linking is the result of an aldol condensation reaction, a type of nucleophilic addition reaction we examine here in Chapter 18.

Nucleophilic Addition to Polar π Bonds 2

Weak Nucleophiles and Acid and Base Catalysis

Chapter 17 discussed nucleophilic addition reactions involving *strong* nucleophiles as reagents—for example, NaBH$_4$ and LiAlH$_4$ as sources of H$^-$, organometallic reagents as sources of R$^-$, and Wittig reagents. Here in Chapter 18, we maintain our focus on nucleophilic addition, but the emphasis is now on reactions that involve *weakly* nucleophilic reagents. Such reactions generally require catalysis by a strong acid or a strong base, which slightly alters the nucleophilic addition mechanism we saw in Chapter 17 to provide a lower-energy route to forming products. Of these reactions, we spend the greatest amount of time on the *aldol reaction*, because it is one of the most important and versatile reactions in organic synthesis.

18.1 Weak Nucleophiles as Reagents: Acid and Base Catalysis

Thus far, we have dealt mainly with nucleophilic addition reactions in which the starting materials include a strong nucleophile. In general, those nucleophiles bear a full negative charge or (in the case of hydride reagents and organometallic reagents) can be treated *as if* they bear a full negative charge. As we saw in Chapters 8–10, however,

Chapter Objectives

On completing Chapter 18 you should be able to:

1. Draw the general mechanisms for the addition of a weak nucleophile to a polar π bond under neutral, basic, and acidic conditions.

2. Explain why the addition of weak nucleophiles tends to be catalyzed by strong acids or strong bases.

3. Predict the major product when a weak nucleophile adds to an α,β-unsaturated, polar π bond.

4. Show how to convert a ketone or aldehyde into an acetal, imine, or enamine, and draw the corresponding mechanisms for these reactions.

5. Predict the hydrolysis products of an acetal, imine, enamine, or nitrile, and draw the corresponding mechanisms.

6. Explain the utility of a Wolff–Kishner reduction and draw its mechanism.

7. Draw the mechanisms for the aldol addition and aldol condensation reactions and predict the major products.

8. Incorporate synthetically useful self-aldol and crossed aldol reactions in a synthesis scheme.

9. Recognize when an intramolecular aldol reaction is favorable and predict the major products.

10. Specify the types of reactants that can be used for a Robinson annulation and predict the major products.

a variety of *uncharged* species can act as nucleophiles, including water, alcohols (ROH), amines (RNH$_2$), thiols (RSH), and phosphines (RPH$_2$). These nucleophiles can also add to polar π bonds.

Equation 18-1 shows that ethanol, a weak nucleophile, can add to the C=O group of butan-2-one to produce a **hemiacetal**, a compound in which a carbon atom is bonded to both an OH group and an OR group. This reaction tends to be quite slow, however.

(18-1)

Weak nucleophile

The OH and OR groups are bonded to the same C.

Butan-2-one   **Ethanol**   **A hemiacetal**

YOUR TURN 18.1

Draw the hemiacetal of butan-2-one in which the OR group is a butoxy (OCH$_2$CH$_2$CH$_2$CH$_3$) group.

Answers to Your Turns are in the back of the book.

This reaction can be sped up dramatically when a small amount of a strong base (such as NaOH) or a strong acid (such as H$_2$SO$_4$) is added, as shown in Equation 18-2.

Catalytic amount

Fast

Strong base or acid

(18-2)

In other words:

> The formation of a hemiacetal from a ketone or aldehyde can be base or acid catalyzed; the reaction is said to undergo **base catalysis** or **acid catalysis**.

The reasons these reaction rates increase can be understood by considering the respective mechanisms. The uncatalyzed reaction for Equation 18-1 is shown in Equation 18-3.

Mechanism for Equation 18-1 (uncatalyzed)

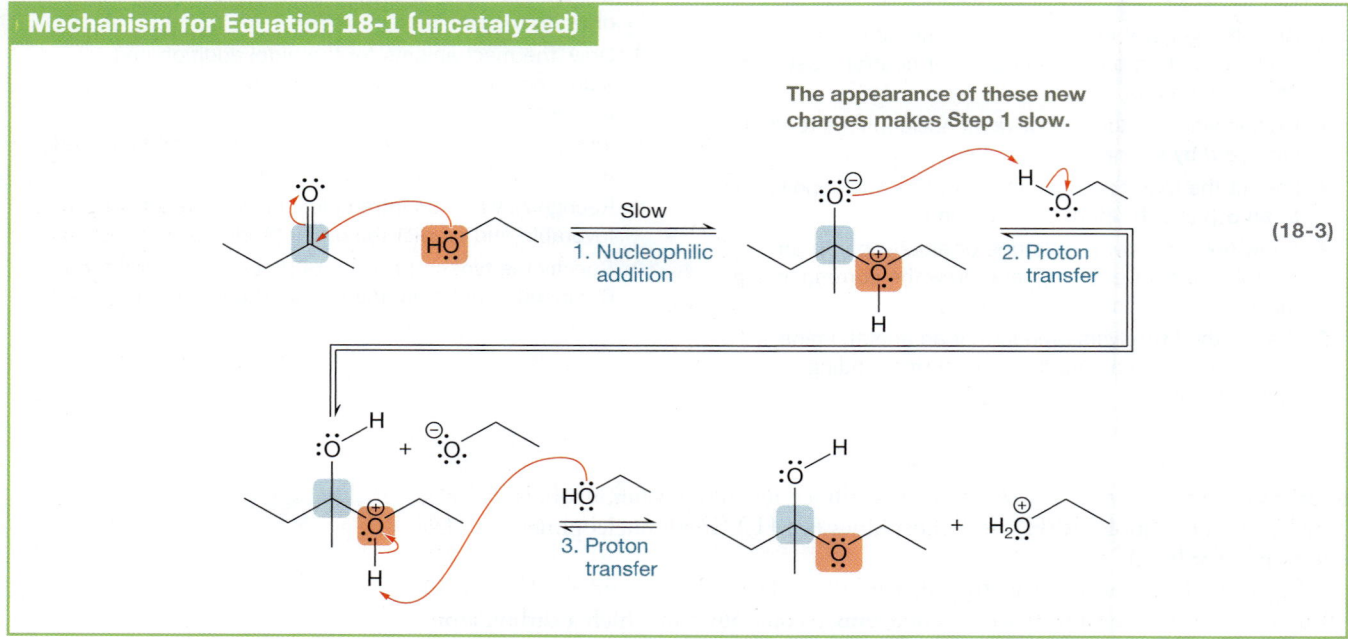

(18-3)

The nucleophile adds to the C=O carbon in Step 1 to produce an intermediate with two formal charges, and proton transfers in Steps 2 and 3 produce the uncharged product. The alcohol is shown as the acid and base in Steps 2 and 3 because it is in much greater abundance than any other acids or bases present.

The mechanism for the base-catalyzed reaction is shown in Equation 18-4.

Mechanism for Equation 18-2 under basic conditions

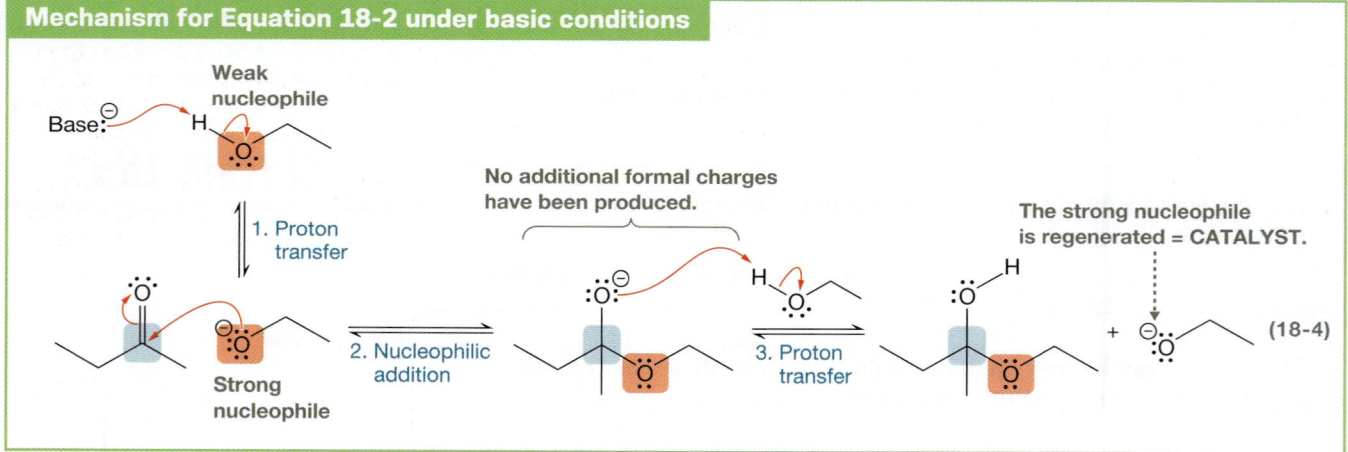

(18-4)

The alcohol (ROH) is still very abundant, so, according to the leveling effect (Section 6.2c), the strongest base that can exist in solution is the deprotonated alcohol, RO^-, produced in Step 1. RO^-, a strong nucleophile, adds to the C=O carbon in Step 2, and the resulting species is protonated by ROH in Step 3. Notice that the RO^- nucleophile is regenerated in Step 3, so it can be recycled in Step 2. Because it is responsible for increasing the reaction rate but is not consumed overall, RO^- is a *catalyst*.

In the acid-catalyzed mechanism in Equation 18-5, the protonated alcohol (ROH_2^+) is the strongest acid that can exist in solution. ROH_2^+ protonates the carbonyl O atom in Step 1, and the nucleophile attacks in Step 2. The deprotonation in Step 3 produces the uncharged product. Similar to the base-catalyzed mechanism, the strong acid (ROH_2^+) is regenerated in Step 3, making it a catalyst that can be recycled in Step 1.

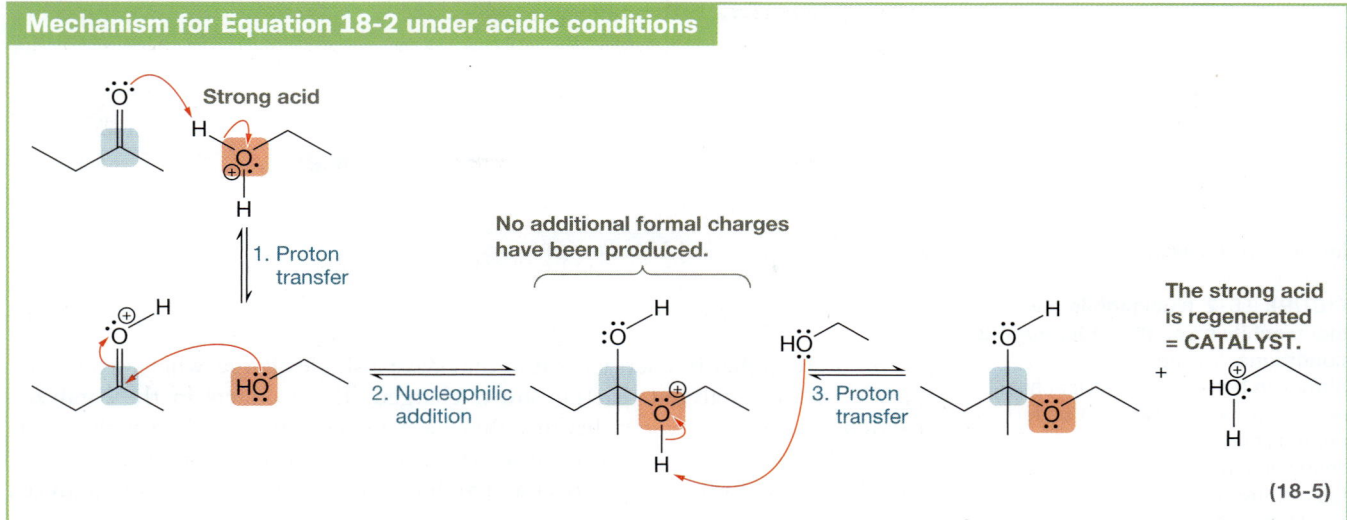

Mechanism for Equation 18-2 under acidic conditions

Strong acid

No additional formal charges have been produced.

1. Proton transfer

2. Nucleophilic addition

3. Proton transfer

The strong acid is regenerated = CATALYST.

(18-5)

The dramatic increase in reaction rate under basic and acidic conditions can be understood by comparing the nucleophilic addition step in each mechanism. In the uncatalyzed reaction (Equation 18-3), the nucleophilic addition step produces *two* new formal charges, whereas in the base- and acid-catalyzed reactions (Equations 18-4 and 18-5), the nucleophilic addition step produces *no* additional formal charges.

> At each stage of the mechanism in Equations 18-3, 18-4, and 18-5, count the total *number* of charges that exist, not the total charge (e.g., one positive charge and one negative charge should be counted as two charges).

FIGURE 18-1 Nucleophilic addition and acid/base catalysis Free energy diagrams for the nucleophilic addition steps of (a) the uncatalyzed mechanism in Equation 18-3, (b) the base-catalyzed mechanism in Equation 18-4, and (c) the acid-catalyzed mechanism in Equation 18-5. For the uncatalyzed mechanism, the two additional charges produced result in a larger energy barrier and a slower reaction.

These differences in the nucleophilic addition steps can be seen in their respective energy diagrams in Figure 18-1. Compared to the uncatalyzed reaction (Fig. 18-1a), the acid- and base-catalyzed reactions (Fig. 18-1b and 18-1c) have higher-energy reactants and lower-energy products. The reactants in the catalyzed

Uncatalyzed

Free energy →

Reaction coordinate →
(a)

Base catalyzed

• Smaller barrier
• Faster reaction

Free energy →

Reaction coordinate →
(b)

Acid catalyzed

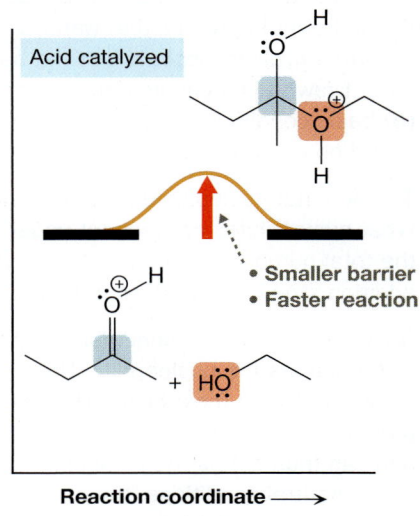

• Smaller barrier
• Faster reaction

Free energy →

Reaction coordinate →
(c)

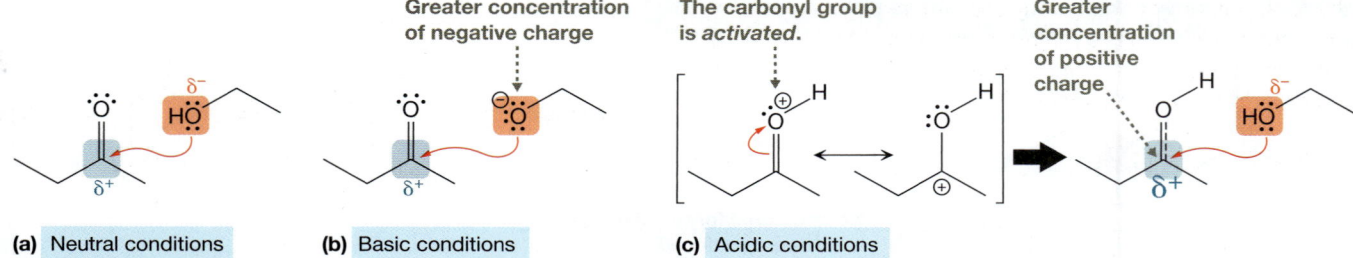

(a) Neutral conditions   **(b)** Basic conditions   **(c)** Acidic conditions

FIGURE 18-2 Nucleophile and electrophile strength under various conditions (a) The electrostatic attraction between the nucleophile and electrophile is weakest under neutral conditions. (b) Under basic conditions, the nucleophile acquires a full negative charge. (c) Under acidic conditions, the partial positive charge on the carbonyl C atom increases due to the second resonance contributor.

reactions are higher in energy because they have a single charge, whereas the reactants in the uncatalyzed reaction are uncharged. The products in the catalyzed reactions are lower in energy because they have a single charge, whereas the products of the uncatalyzed reaction have two charges. Both of these characteristics contribute to a lower energy barrier and an increased rate for each of the catalyzed reactions.

Another way to understand these increases in reaction rate is to see that the electrostatic attraction between the nucleophile and electrophile is increased under basic and acidic conditions, as shown in Figure 18-2. That is:

- Under basic conditions, the nucleophile becomes stronger.
- Under acidic conditions, the electrophile becomes stronger.

Under basic conditions, the electrostatic attraction is increased when the uncharged nucleophile is converted into a negatively charged nucleophile. Under acidic conditions, the protonated carbonyl group acquires a larger partial positive charge on its C atom, stemming from its resonance contributor, and the carbonyl group is said to be **activated** toward nucleophilic attack.

YOUR TURN 18.3

Which of these nitrile groups is activated toward nucleophilic attack?

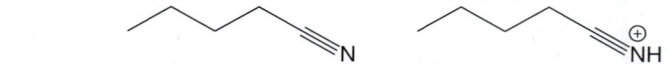

SOLVED PROBLEM 18.1

Recall from Chapter 17 that water can add to a ketone or aldehyde to form a *hydrate*, according to the balanced equation shown here. Draw the mechanism for this reaction under **(a)** neutral and **(b)** basic conditions. Under which conditions do you think this hydration reaction will proceed faster? Why?

O
‖
[cyclohexanone structure] + H₂O ⇌ [cyclohexanone hydrate structure, HO OH]

Cyclohexanone   **Cyclohexanone hydrate**

Think What nucleophiles are present under neutral conditions? What nucleophiles are present under basic conditions? What are the total numbers of charges before and after the nucleophilic addition step in each mechanism?

Solve Under neutral conditions (right), water acts as the nucleophile. Under basic conditions (next page), HO⁻ is present in a significant concentration and can thus act as the nucleophile. Under neutral conditions, the nucleophilic addition step increases the total number of charges by two, whereas the total number of charges remains the same under basic conditions. Thus, nucleophilic addition is faster under basic conditions.

PROBLEM 18.2 Draw the mechanism for the hydration of cyclohexanone under acidic conditions, and argue whether this reaction should be faster or slower than the corresponding hydration under neutral conditions.

PROBLEM 18.3 Draw the complete, detailed mechanism for the reaction shown here under **(a)** basic and **(b)** acidic conditions.

18.1a Addition of HCN: The Formation of Cyanohydrins

When a ketone or aldehyde is treated with aqueous hydrocyanic acid (HCN), the product is a **cyanohydrin**, in which an OH group and a cyano (CN) group are bonded to the same carbon atom. An example is shown in Equation 18-6.

Cyanohydrin formation is important for two reasons. First, it is another valuable *carbon–carbon bond-forming reaction*, giving us an additional means of constructing carbon backbones. Second, the C≡N functional group is readily transformed into CO_2H via a *hydrolysis* reaction (see Chapter 21). The overall product is an α-hydroxy acid. Not only are these compounds useful in synthesis, but some α-hydroxy acids, such as glycolic acid (Fig. 18-3), are used to treat a variety of skin ailments.

In the overall reaction of HCN with the carbonyl group, the cyano group adds to the carbonyl carbon and a hydrogen adds to the carbonyl oxygen. This reaction occurs according to the mechanism shown in Equation 18-7.

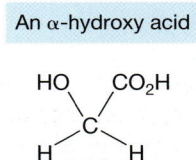

FIGURE 18-3 Glycolic acid Glycolic acid is an α-hydroxy acid used to treat some skin ailments.

Mechanism for Equation 18-6

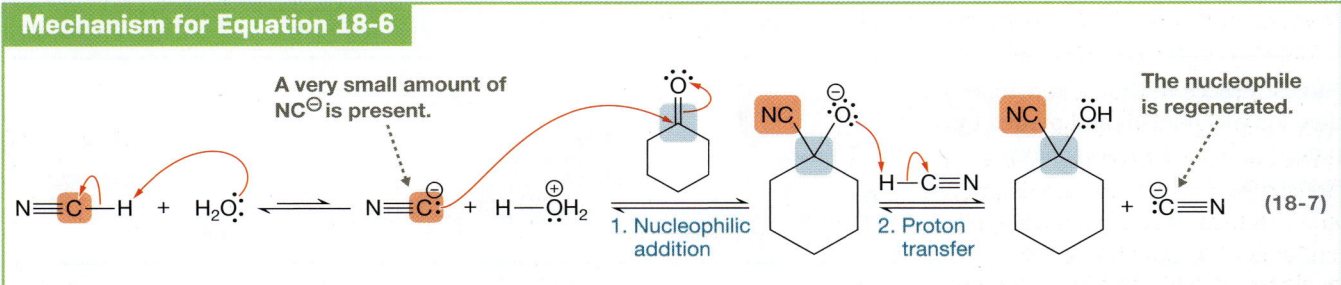

In aqueous HCN, there is a small amount of the cyanide anion (NC^-) present because HCN (pK_a = 9.2) is a weak acid. In Step 1 of the ensuing reaction, NC^- attacks the carbonyl group, generating a negative charge on the O atom, and in Step 2, that O atom is protonated.

The equilibrium proton transfer between HCN and water (Equation 18-7) heavily favors reactants, so only a small amount of NC^- is generated. In fact, only about 0.001% of HCN molecules dissociate into NC^- and H_3O^+. As a result, HCN reacts quite slowly with the carbonyl carbon under normal conditions, even though NC^- is a very good nucleophile.

YOUR TURN 18.4

To verify that the equilibrium between HCN and H_2O in Equation 18-7 heavily favors the reactants, use the appropriate pK_a values (Appendix A) to determine the numerical factor by which the reactant side is favored.

The reaction can be sped up dramatically by the addition of a small amount of either KCN or a strong base such as NaOH (Equation 18-8).

KCN or NaOH will catalyze this reaction.

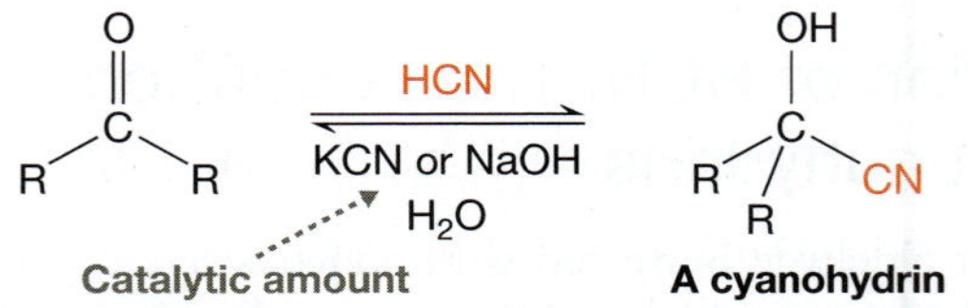

(18-8)

The reaction rate increases because the addition of either KCN or NaOH increases the amount of NC^- present in solution. KCN is an ionic compound that dissociates into K^+ and NC^- in water, while the HO^- from NaOH deprotonates HCN *quantitatively* (i.e., to 100% completion) to produce NC^-. Notice in Equation 18-7 that the NC^- generated in either of these ways is not consumed during the course of the reaction, because the NC^- that is used in Step 1 is regenerated in Step 2. Hence:

Both KCN and NaOH are *catalysts* for the nucleophilic addition of HCN.

Although forming cyanohydrins is very synthetically useful, it is important to know that HCN and KCN are extremely toxic. Proper precautions must be taken to handle these compounds safely.

YOUR TURN 18.5

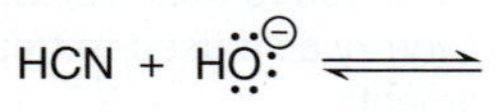

HCN + HO^- ⇌

Verify that HO^- quantitatively deprotonates HCN by **(a)** drawing the products of the reaction shown here, and **(b)** using the appropriate pK_a values to determine the numerical factor by which the product side is favored.

PROBLEM 18.4 Draw the complete, detailed mechanism and predict the products of each of the following reactions.

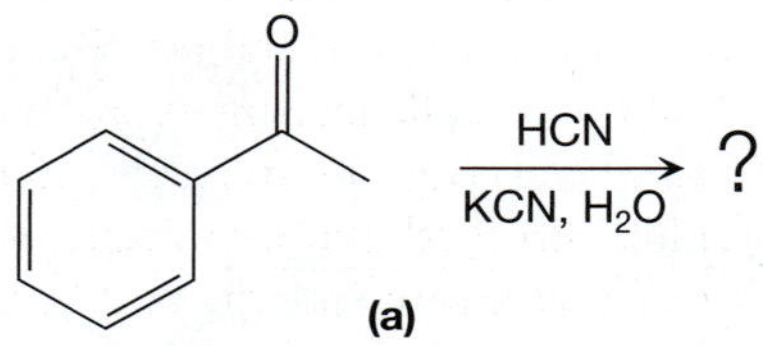

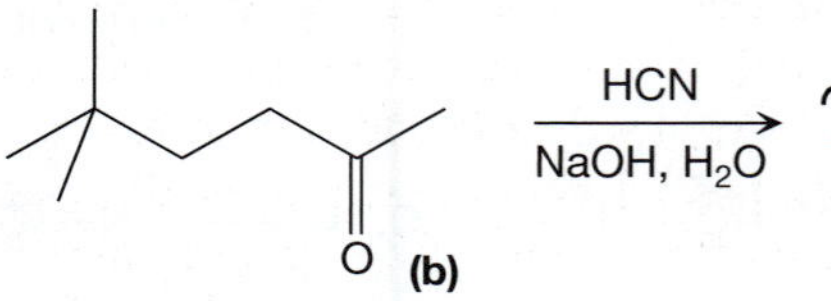

18.1b Direct versus Conjugate Addition of Weak Nucleophiles and HCN

Like negatively charged nucleophiles, uncharged nucleophiles can attack α,β-unsaturated carbonyls at two locations: the carbonyl carbon to give the direct addition product (i.e., the 1,2-addition product) and the β carbon to give the conjugate addition product (i.e., the 1,4-addition product). In general:

> Uncharged nucleophiles heavily favor conjugate addition over direct addition.

An example with ethanethiol (CH_3CH_2SH) as the nucleophile is shown in Equation 18-9.

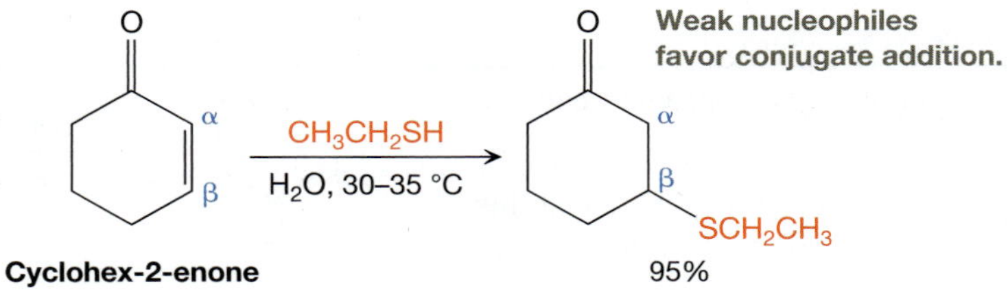

Cyclohex-2-enone

(18-9)

The mechanism for this reaction is shown in Equation 18-10.

Mechanism for Equation 18-9

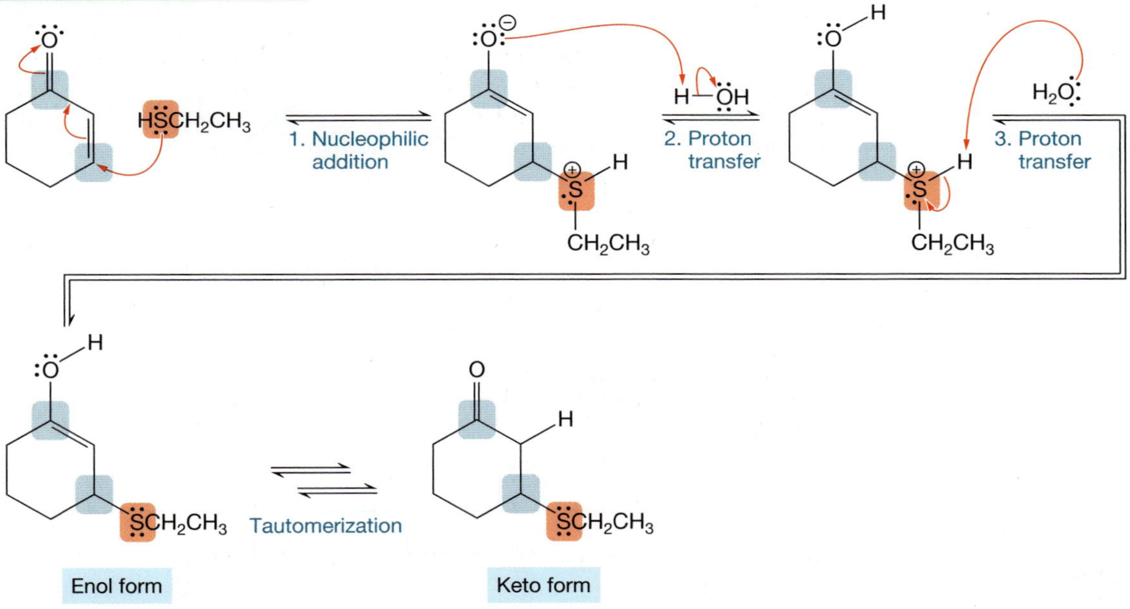

(18-10)

Uncharged nucleophiles favor conjugate addition because Step 1 of the mechanism, the nucleophilic addition step, is reversible, so it takes place under thermodynamic control (Section 17.9). Conjugate addition not only leads to a resonance-stabilized enolate anion, but also leads to greater total bond strength in the overall, uncharged product.

Frequently, the addition of uncharged nucleophiles to α,β-unsaturated carbonyl compounds does not require acidic conditions. Unlike direct addition, conjugate addition of uncharged nucleophiles can proceed at a reasonable rate, in large part because of resonance stabilization in the enolate anion produced in the first step.

HCN similarly adds to α,β-unsaturated carbonyl compounds via conjugate addition because the addition of NC⁻ to the carbonyl group is reversible (Table 17-2, p. 865). An example of one such reaction is shown in Equation 18-11.

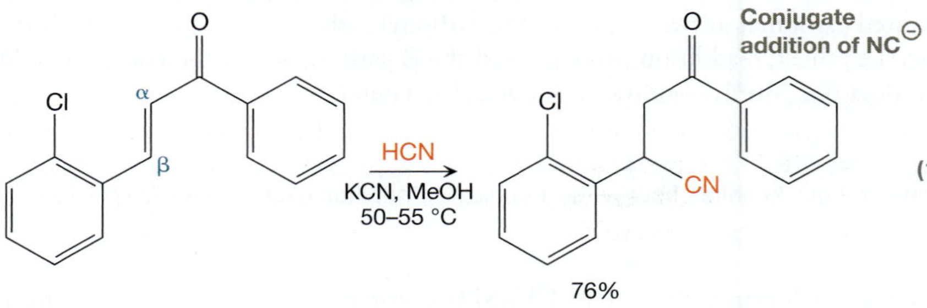

(18-11)

76%

SOLVED PROBLEM 18.5

Predict the product of the reaction shown here and draw its complete, detailed mechanism.

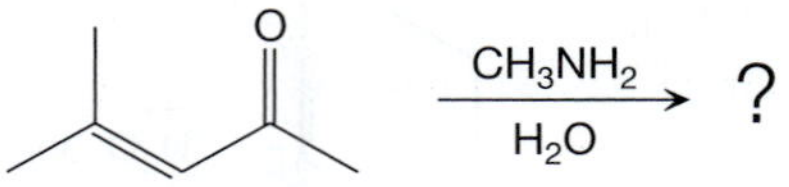

Think Does CH_3NH_2 add to the carbonyl group reversibly or irreversibly? Will this result in direct addition or conjugate addition? What roles do proton transfer steps have in the mechanism?

Solve CH_3NH_2, an uncharged nucleophile, adds reversibly to a carbonyl group, so it will favor conjugate addition on attacking an α,β-unsaturated carbonyl. After CH_3NH_2 adds, two proton transfers take place to produce an uncharged enol, which undergoes tautomerization to the more stable keto form.

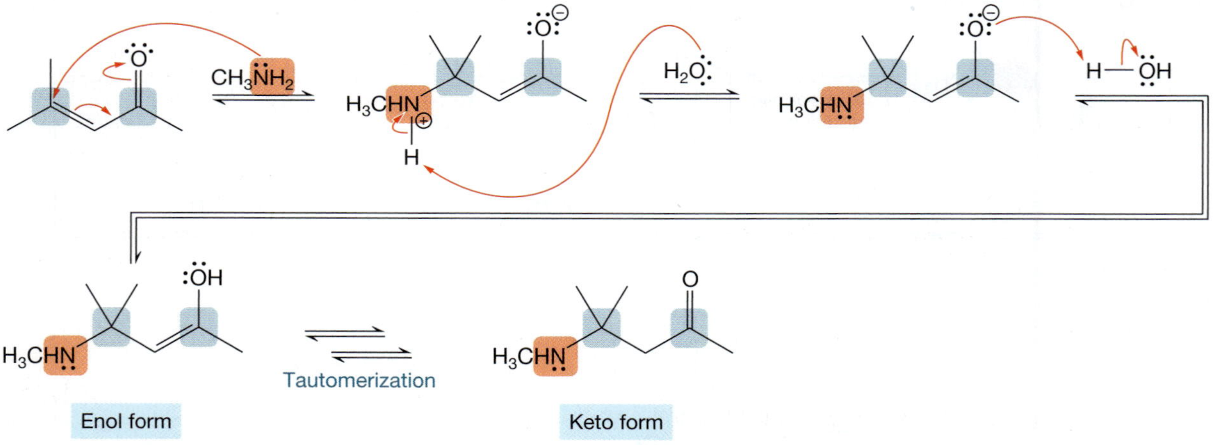

Enol form   Tautomerization   Keto form

PROBLEM 18.6 Predict the major product and draw the complete, detailed mechanism for each of the reactions shown here.

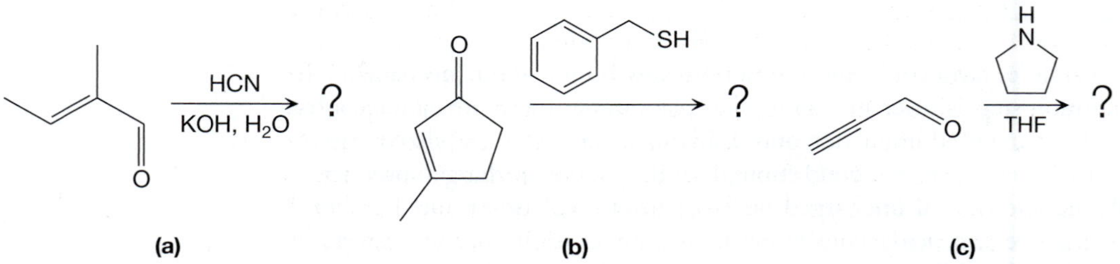

(a)   (b)   (c)

18.2 Formation and Hydrolysis Reactions Involving Acetals, Imines, Enamines, and Nitriles

In Section 18.1, we saw that weak nucleophiles can add to polar π bonds. Frequently, the immediate product of these reactions can react further under the conditions required for nucleophilic addition. We examine some of those subsequent reactions here in Section 18.2—namely, the formation of acetals, imines, and enamines from ketones or aldehydes. We also examine the hydrolysis of acetals, imines, enamines, and nitriles. As we will see, the hydrolyses of acetals, imines, and enamines are simply the reverse of their formation reactions.

18.2a The Formation and Hydrolysis of Acetals

In Equation 18-1, we saw that a *hemiacetal* forms when a ketone or aldehyde is treated with an alcohol, and in Equation 18-2, we saw that the reaction is catalyzed under either basic or acidic conditions.

> If an aldehyde or ketone is treated with a *large excess* of an alcohol under *acidic* conditions, then the hemiacetal produced from nucleophilic addition reacts further to form an *acetal*, in which two alkoxy (RO) groups are bonded to the same carbon.

The overall reaction, shown in Equation 18-12 using pentanal as an example, is reversible.

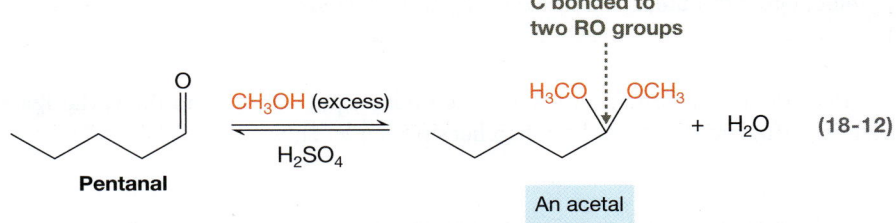

C bonded to two RO groups

Pentanal — CH₃OH (excess) / H₂SO₄ → An acetal + H₂O   (18-12)

YOUR TURN 18.6

Draw the acetal of pentanal in which the two RO groups are propoxy ($CH_3CH_2CH_2O$) groups.

The complete mechanism for this acetal formation reaction is shown in Equation 18-13. The first three steps are identical to the acid-catalyzed nucleophilic addition in Equation 18-5. That is, the C=O group is protonated in Step 1, which *activates* it toward nucleophilic addition by the weak ROH nucleophile in Step 2. In Step 3, deprotonation produces the uncharged hemiacetal. The remaining steps essentially make up an S_N1 reaction. In Step 4, protonation of the OH group generates a good H_2O leaving group, which departs in Step 5 to produce a resonance-stabilized carbocation. That carbocation is subsequently attacked by another ROH nucleophile in Step 6, and the proton transfer in Step 7 results in the uncharged acetal.

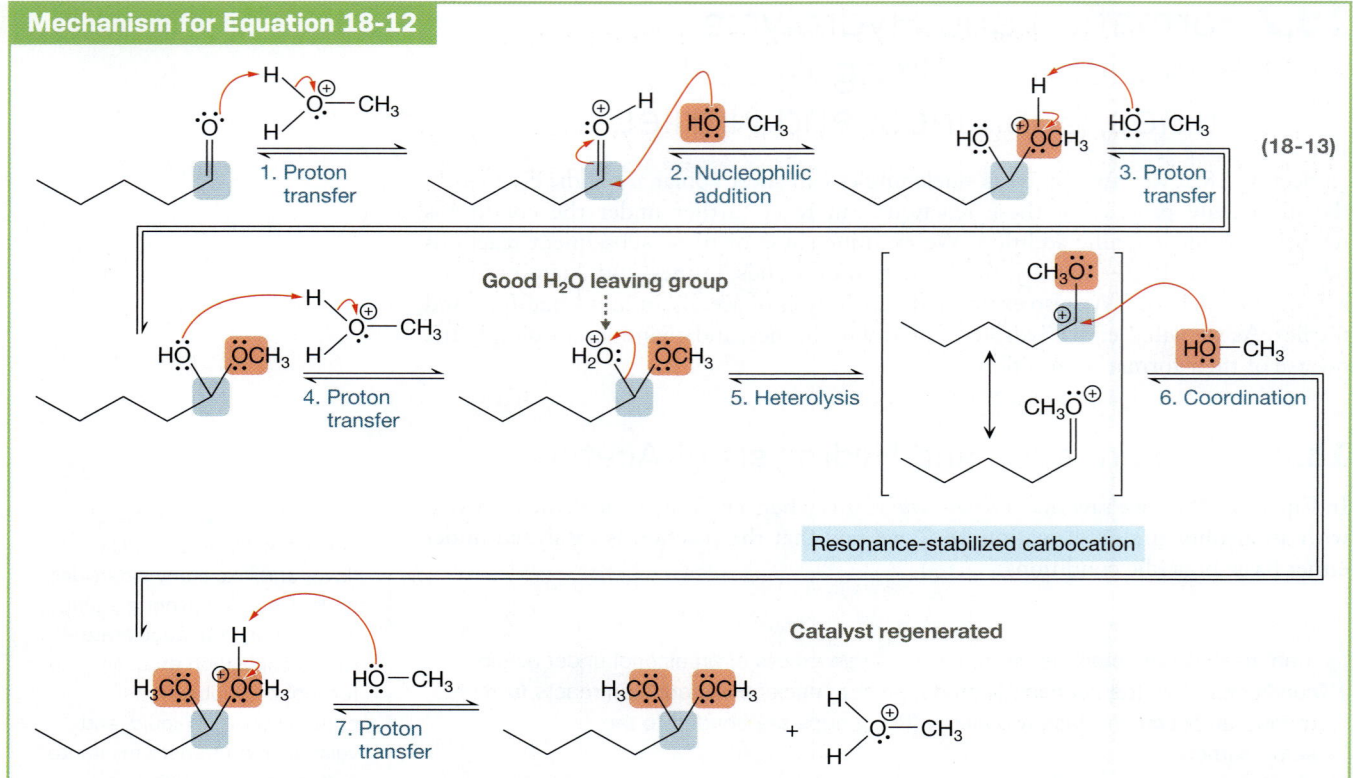

1. Proton transfer
2. Nucleophilic addition
3. Proton transfer

(18-13)

Good H₂O leaving group

4. Proton transfer
5. Heterolysis
6. Coordination

Resonance-stabilized carbocation

Catalyst regenerated

7. Proton transfer

YOUR TURN 18.7

Identify the hemiacetal and the acetal in the mechanism in Equation 18-13. Also, circle the steps the make up an S_N1 reaction.

Because acetal formation takes place under equilibrium conditions, the reaction can be shifted according to Le Châtelier's principle. Thus:

> While formation of the acetal product is favored by using excess alcohol (a reactant), the reverse reaction is favored by using excess water (a product; see Equation 18-12 again) under acidic conditions.

This is exemplified in Equation 18-14, whose mechanism is precisely the reverse of that in Equation 18-13.

The mechanism for this reaction is the reverse of the one in Eq. 18-13.

H_3CO OCH_3 → H_2O (excess) / H_2SO_4 → O + 2 CH_3OH (18-14)

An acetal

Equation 18-14 is a *hydrolysis* reaction, because the addition of water results in the breaking of the C—OCH₃ bonds.

PROBLEM 18.7 Draw the complete, detailed mechanism for the reaction in Equation 18-14.

The reversibility of acetal formation enables us to use acetals as *protecting groups* for ketones and aldehydes (see Chapter 19). Whereas ketones and aldehydes are susceptible to attack by nucleophiles (at the carbonyl carbon) and bases (at the α carbon), acetals are inert to both strong nucleophiles and strong bases.

One popular alcohol used to protect ketones and aldehydes is 1,2-ethanediol (ethylene glycol), as shown in Equation 18-15. 1,2-Ethanediol has two OH functional groups per molecule, so only one equivalent of the diol is needed. It also forms a five-membered ring, which helps favor acetal formation.

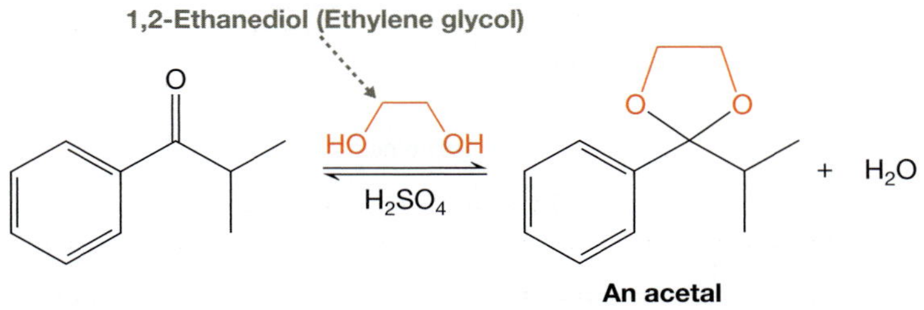

1,2-Ethanediol (Ethylene glycol)

$$+ \ H_2O \quad (18\text{-}15)$$

An acetal

PROBLEM 18.8 Draw a complete, detailed mechanism for the reaction in Equation 18-15.

Although a hemiacetal can form under either basic or acidic conditions (review Equation 18-2, p. 889), the story is somewhat different for acetal formation:

> A ketone or aldehyde readily forms an acetal under *acidic* conditions but not under *basic* conditions.

Acetals do not form under basic conditions because the nucleophilic substitution that would convert the hemiacetal to the acetal would require the leaving group to be HO^- (Equation 18-16). As we learned in Chapter 9, HO^- is an unsuitable leaving group for an S_N1 or S_N2 reaction.

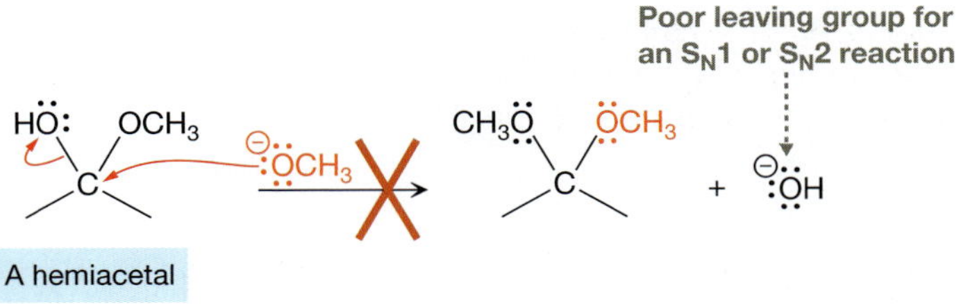

Poor leaving group for an S_N1 or S_N2 reaction

A hemiacetal

$$+ \ \ominus{:}\overset{..}{O}H \quad (18\text{-}16)$$

PROBLEM 18.9 The sulfur analog of an acetal, called a thioacetal, can be produced when a ketone or aldehyde is treated with a thiol (RSH) under acidic conditions. When propane-1,3-dithiol is used, a 1,3-dithiane is produced, as shown here. Dithianes have important applications in organic synthesis. Draw the complete, detailed mechanism for this 1,3-dithiane formation reaction.

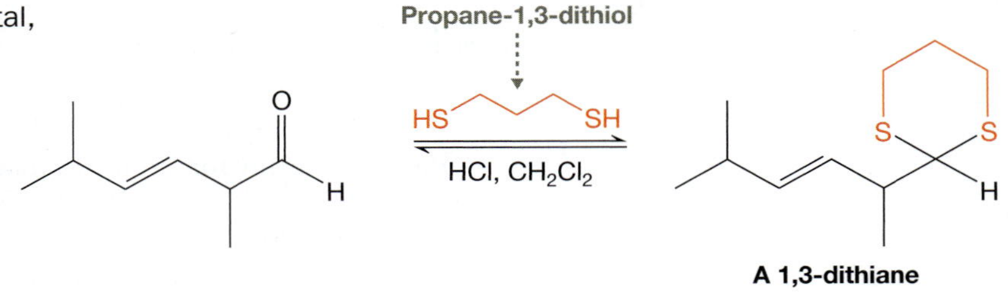

Propane-1,3-dithiol

HS   SH

HCl, CH_2Cl_2

A 1,3-dithiane

18.2b The Formation and Hydrolysis of Imines and Enamines

Ammonia (NH_3) and amines (RNH_2) are uncharged nucleophiles that can add reversibly to carbonyl groups when a small amount of a strong acid is added (i.e., under mildly acidic conditions), as shown in Equations 18-17 and 18-18 for ketones and aldehydes, respectively. (If the conditions are too acidic, no such reaction would occur; see Problem 18.60 at the end of the chapter.) The product is an **imine**, sometimes referred to as a **Schiff base**, which has a characteristic C=N double bond.

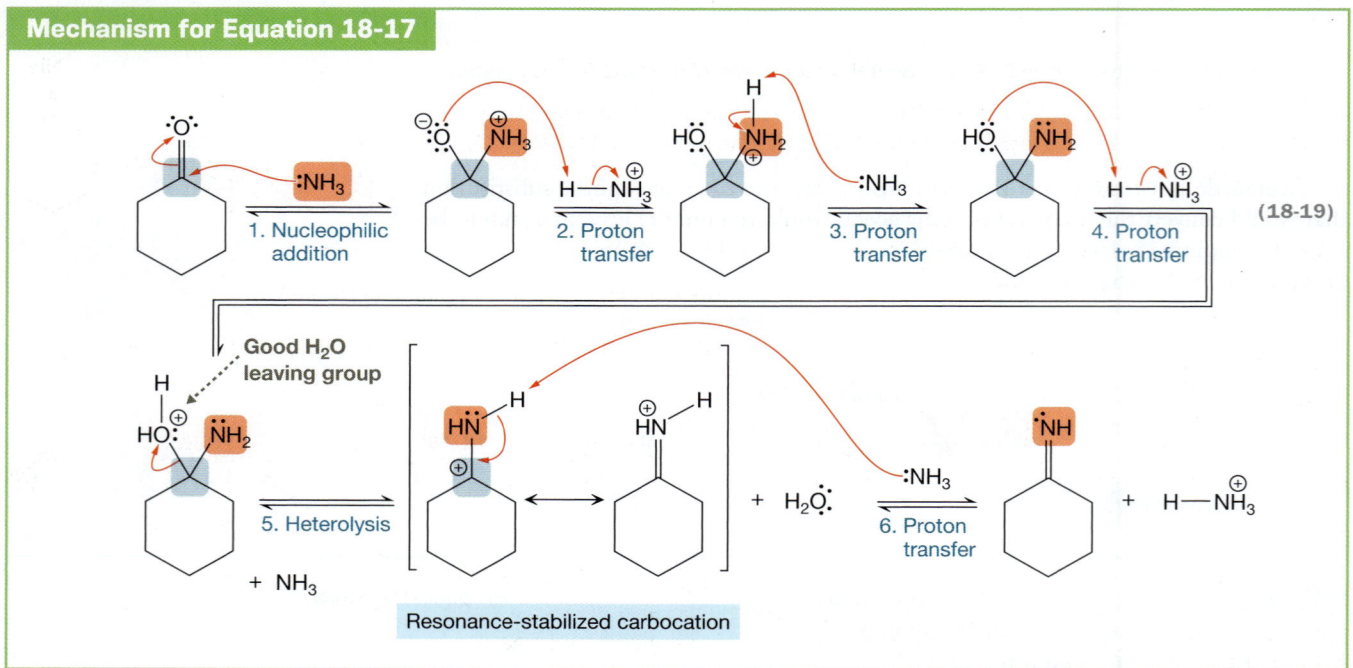

The complete mechanism for the formation of an imine from a ketone is shown in Equation 18-19.

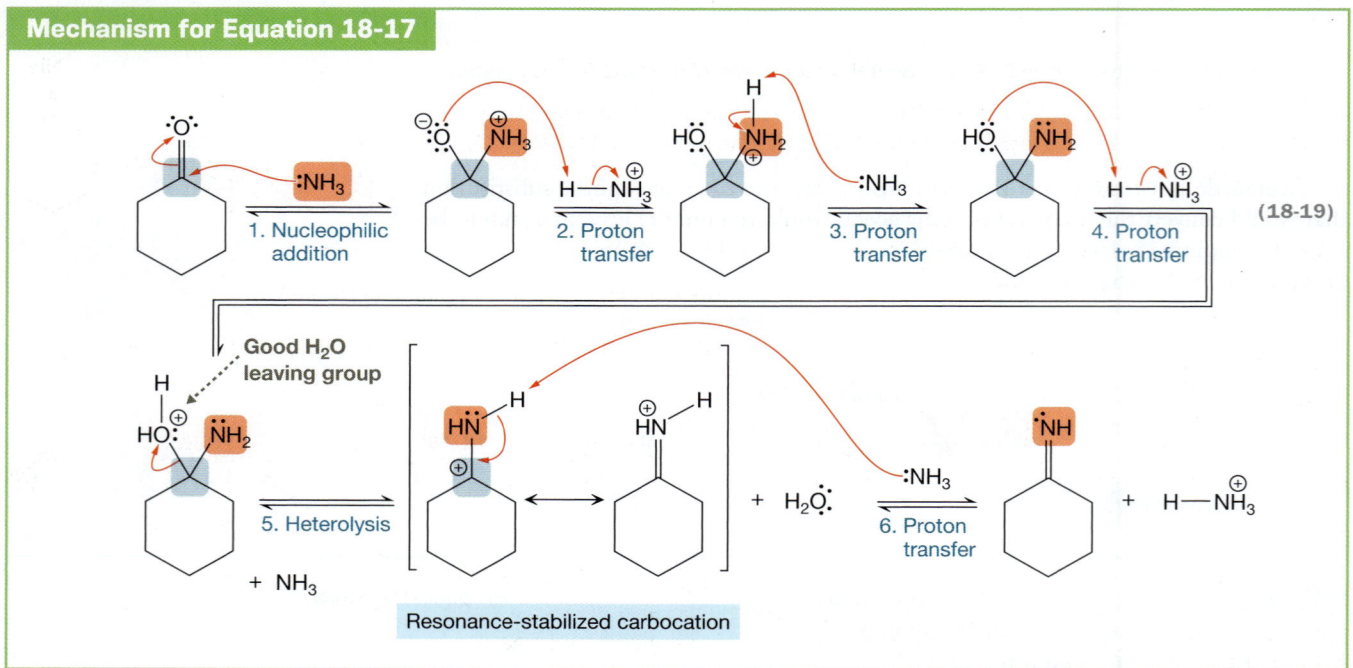

The mechanism for imine formation (Equation 18-19) is similar to that for acetal formation (Equation 18-13). One difference between the mechanisms is the step in which the nucleophile adds. In acetal formation, the weak ROH nucleophile adds in Step 2, after the carbonyl group has been activated. In imine formation, on the other hand, the nucleophile adds in Step 1 because NH_3 is a stronger nucleophile than ROH, and the weakly acidic conditions don't substantially activate the carbonyl group.

The mechanisms for acetal formation and imine formation also differ in Step 6. In imine formation, the N is deprotonated, thus completing the second step of an E1 mechanism (Chapter 8). In acetal formation, an analogous deprotonation is unavailable, because the O that is attached to the C^+ is not bonded to any hydrogen atoms.

PROBLEM 18.10 Draw the complete, detailed mechanism for the formation of the imine from the aldehyde in Equation 18-18.

Just as with acetal formation, the reversibility of imine formation makes it possible to use Le Châtelier's principle to drive the reaction in either direction. Thus:

An imine can be *hydrolyzed* by treating it with excess water under acidic conditions (Equation 18-20).

Under acidic conditions, an imine is hydrolyzed.

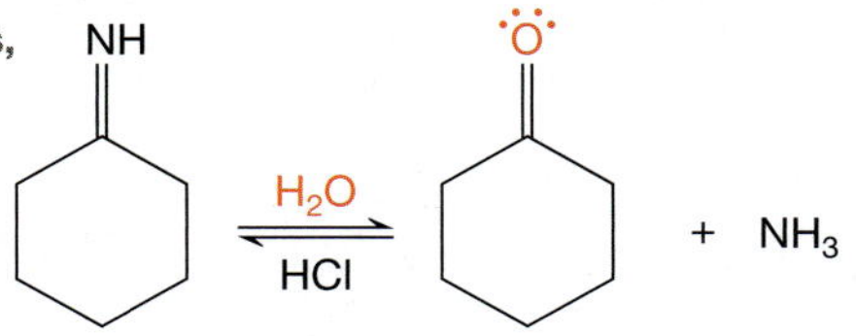

$$\text{imine} \underset{\text{HCl}}{\overset{H_2O}{\rightleftharpoons}} \text{ketone} + NH_3 \qquad \text{(18-20)}$$

PROBLEM 18.11 Draw the complete, detailed mechanism for the hydrolysis reaction in Equation 18-20.

SOLVED PROBLEM 18.12

Can the imine formation shown here take place under *basic* conditions?

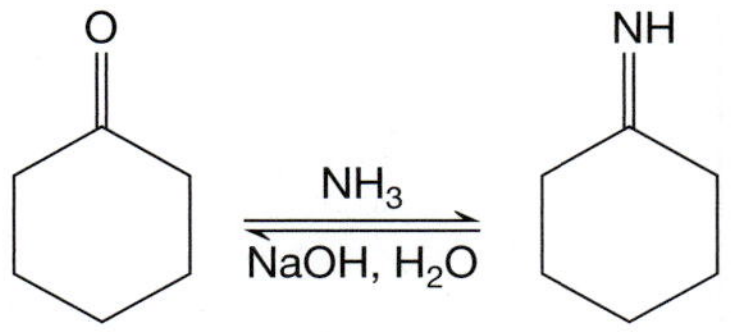

Think Under basic conditions, what types of species should *not* appear in the mechanism? Are all of the steps necessary for imine formation under these conditions reasonable?

$$\text{ketone} \underset{NaOH, H_2O}{\overset{NH_3}{\longrightarrow}} \text{imine}$$

Solve No strong acids should appear in the mechanism under basic conditions. Thus, the mechanism would be as follows:

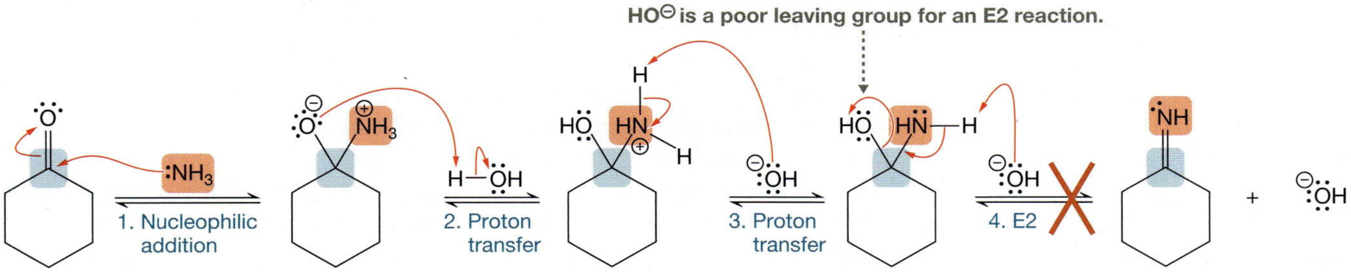

$HO^{\ominus}$ is a poor leaving group for an E2 reaction.

Steps 1–3 are plausible. The appearance of a positive charge on N is reasonable under basic conditions because species like H_4N^+ are only weakly acidic. Step 4, however, is implausible. Recall from Chapter 9 that HO^- is an insufficient leaving group for an E2 reaction. Thus, *imines do not form under basic conditions*.

PROBLEM 18.13 Can the imine hydrolysis shown here take place under *basic* conditions? Explain.

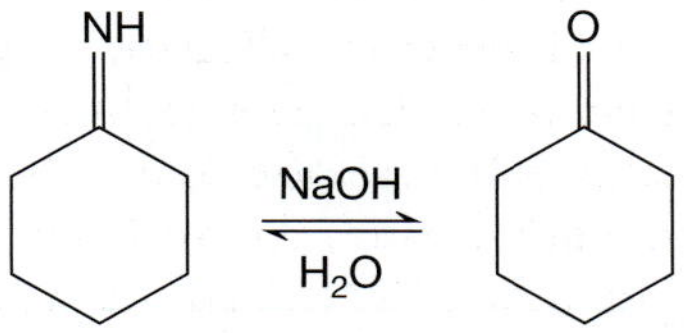

$$\text{imine} \underset{H_2O}{\overset{NaOH}{\longrightarrow}} \text{ketone}$$

Based on the overall reactions in Equations 18-17 and 18-18, *imine formation requires at least two hydrogens on the nucleophilic N atom.* One of those hydrogens is removed in Step 3 of the mechanism (Equation 18-19), and the other is removed in Step 6. Thus:

> Only NH_3 and primary amines (RNH_2) can form imines on reaction with ketones or aldehydes.

Although a secondary amine (R_2NH) has only one hydrogen on nitrogen and thus cannot form an imine, it can react with a ketone or aldehyde to produce an **enamine**, in which an amino group is attached to an alkene carbon (Equation 18-21).

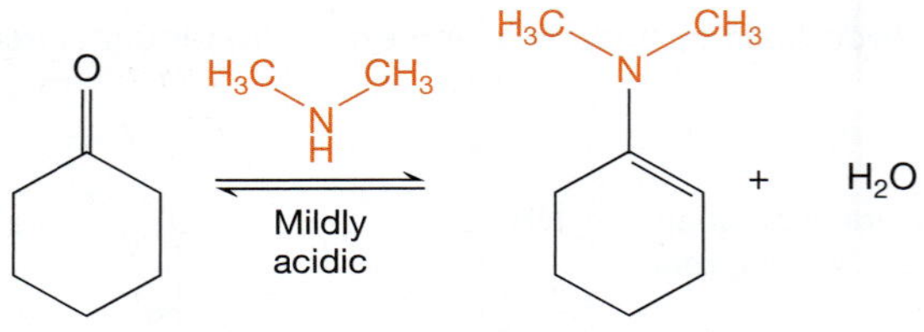

(18-21)

An enamine

The mechanism for enamine formation (Equation 18-22) is identical to the mechanism for imine formation (Equation 18-19) through the first five steps.

Mechanism for Equation 18-21

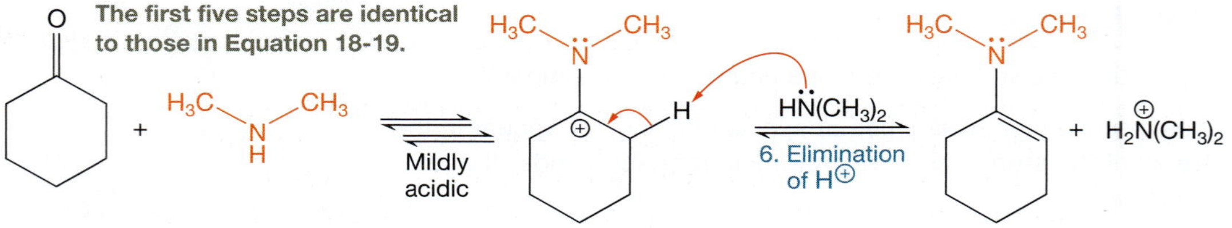

The first five steps are identical to those in Equation 18-19.

6. Elimination of $H^{\oplus}$

(18-22)

The only difference is in the last step. In imine formation (Equation 18-19), the second of two H atoms on N is removed. In enamine formation, no such H exists, so a proton is eliminated from an adjacent C instead (i.e., an α carbon).

Enamines are useful in organic synthesis because the α carbon is nucleophilic and can be used to form new C—C bonds (see Problems 18.81 and 18.82 at the end of the chapter). Furthermore, after an enamine undergoes reaction at the α carbon, it can be hydrolyzed back to the ketone or aldehyde when treated with large amounts of H_2O under acidic conditions (i.e., the reverse of Equation 18-21).

PROBLEM 18.14 Draw the complete mechanism and the major products that are formed in each of the following reactions.

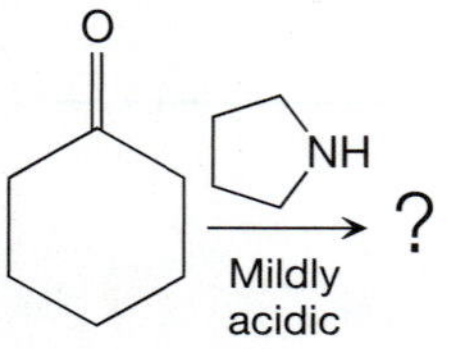

(a) Mildly acidic

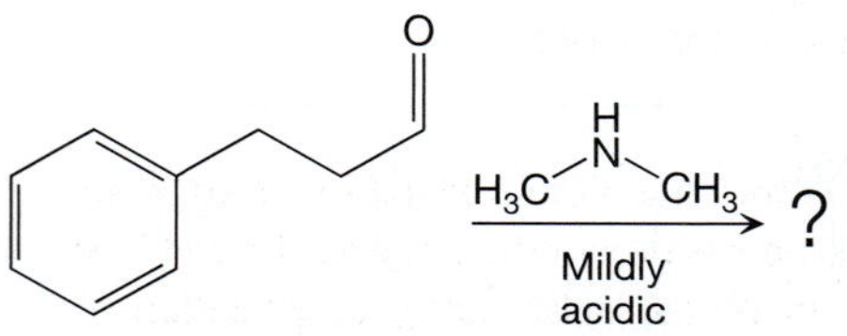

(b) Mildly acidic

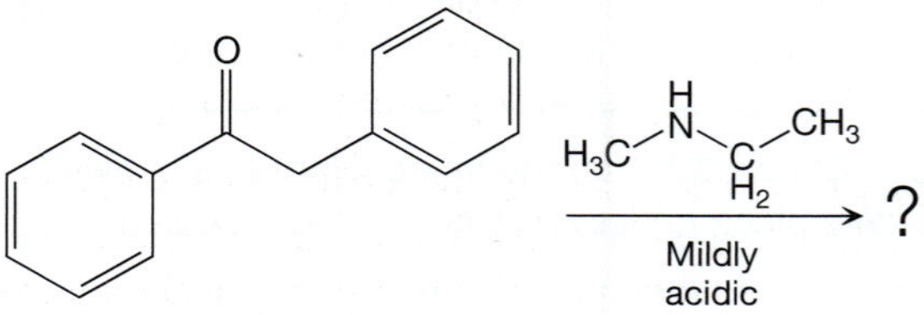

(c) Mildly acidic

Imine Formation and Hydrolysis in Biochemical Reactions

Because of the relative ease with which imines can be formed and hydrolyzed (Section 18.2b), some biochemical processes incorporate imines (or their protonated forms, iminium ions) as a means by which to bind an aldehyde or ketone to a protein for a subsequent reaction. This is integral in the chemistry of vision, for example:

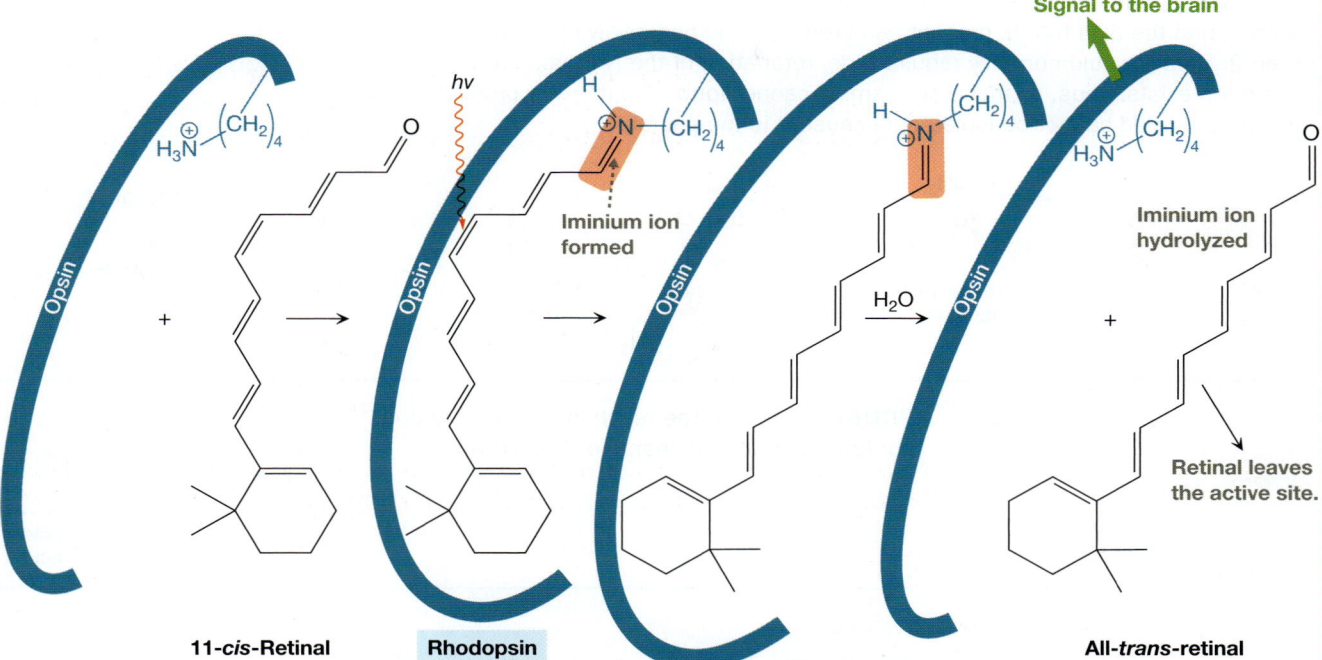

11-*cis*-Retinal   **Rhodopsin**   **All-*trans*-retinal**

The retina in the back of the eye is lined with millions of photoreceptor cells called rods and cones. In cones, the protein opsin binds 11-*cis*-retinal in its active site by forming an iminium ion between the two, producing what is called rhodopsin. Formation of the iminium ion involves the HC=O group from retinal and the NH_3^+ from the side chain of a lysine residue in opsin. Then, when light impinges on rhodopsin, the cis double bond is very quickly converted to trans, which triggers the hydrolysis of the iminium ion. All-*trans*-retinal is ejected and an electrical signal is sent to the brain. Through a series of enzyme-catalyzed steps, the trans form is then recycled back to the cis form.

Another example involves pyridoxal phosphate (PLP), a derivative of vitamin B_6 (OP represents the phosphate group):

Pyridoxal phosphate (PLP)

Many enzymatic reactions are PLP dependent, relying on the formation of an iminium ion between the HC=O group of free PLP and the NH_3^+ group from a lysine residue in the enzyme. Bound PLP can then facilitate a variety of different reactions involving amino acids, including racemization, decarboxylation, transamination, and nucleophilic substitution reactions.

The reaction shown here does *not* form an imine or an enamine. Explain why.

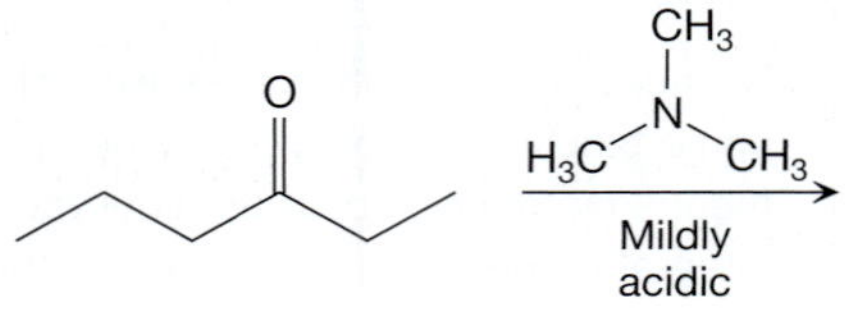

Think Which of the steps in Equations 18-19 or 18-22 can the reaction mechanism include? Are there any steps that are not possible?

Solve Only Steps 1 and 2 of Equation 18-19 (i.e., nucleophilic addition and proton transfer) can take place. Thus, the amine N can attack the carbonyl C in Step 1, and the acid that is present can protonate the negatively charged O in Step 2. Step 3 would normally require a deprotonation of the N atom, but no such proton exists. Thus, an imine or enamine cannot form, and the only product that can be formed cannot be isolated, because it is too unstable.

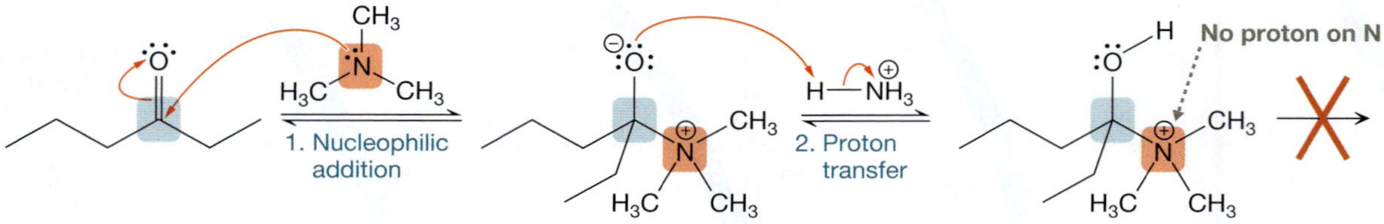

PROBLEM 18.16 The reaction shown here does *not* form an imine or enamine. Explain why.

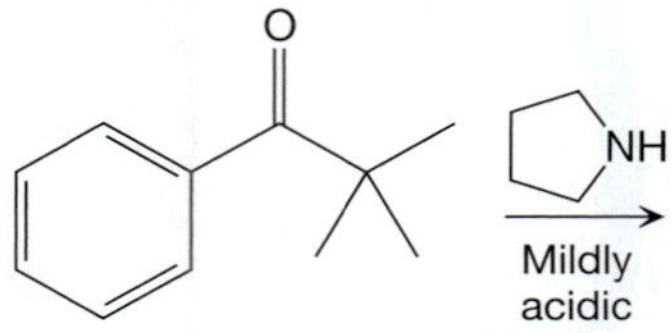

18.2c The Hydrolysis of Nitriles

Closely related to the hydrolysis of acetals and imines is the hydrolysis of nitriles $(R—C\equiv N)$. As shown in Equations 18-23 and 18-24:

> The treatment of a nitrile with water under either acidic or basic conditions produces a primary amide, $R—CO—NH_2$.

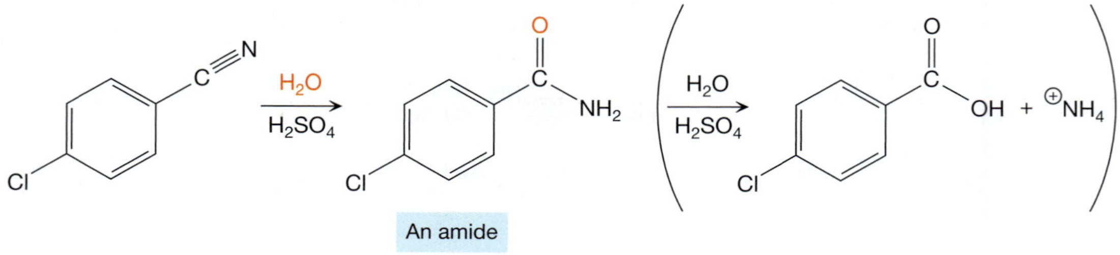

(18-23)

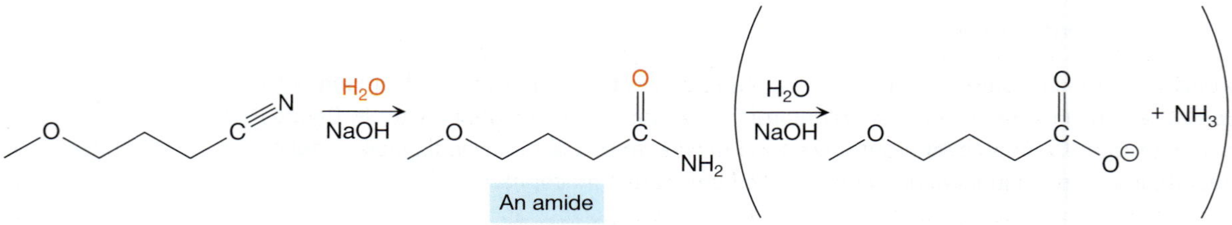

(18-24)

To isolate the amide, the reaction must be monitored carefully because, as shown in parentheses in the two equations, the amide can easily undergo further hydrolysis to produce a carboxylic acid. (Those amide hydrolysis reactions are described in detail in Chapters 20 and 21.)

The mechanism for the hydrolysis of a nitrile under acidic conditions is shown in Equation 18-25.

Mechanism for Equation 18-23

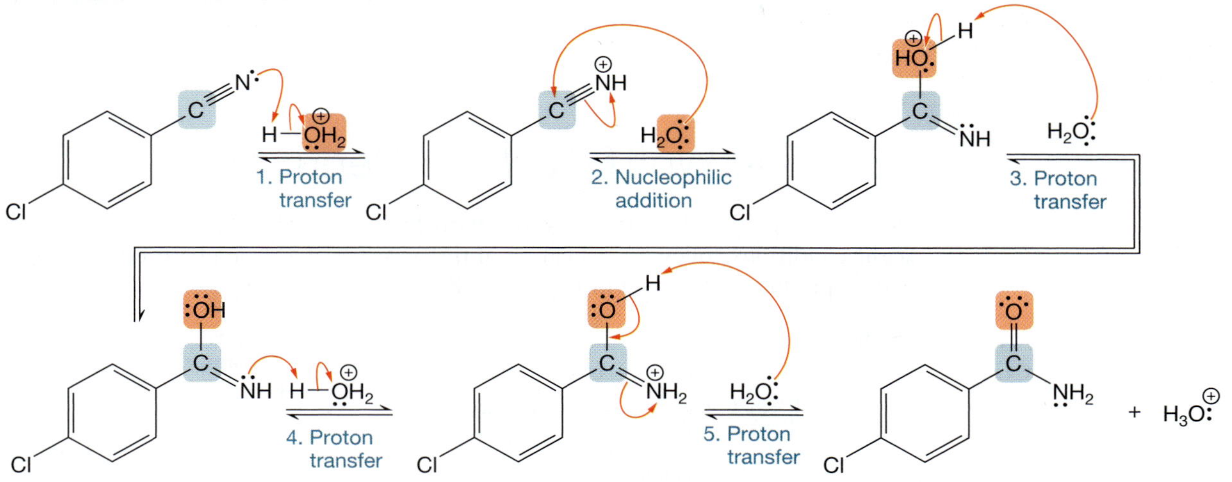

(18-25)

Although this mechanism consists of five elementary steps, the only step that is *not* a proton transfer is the nucleophilic addition step — Step 2. The four proton transfer steps add protons to N and remove them from O, so *no strongly basic species* appear in the mechanism — a necessary condition for a reaction taking place under acidic conditions. Furthermore, notice that Steps 4 and 5 are essentially a *tautomerization* (Section 7.9).

The mechanism for the hydrolysis of a nitrile under basic conditions is shown in Equation 18-26.

Mechanism for Equation 18-24

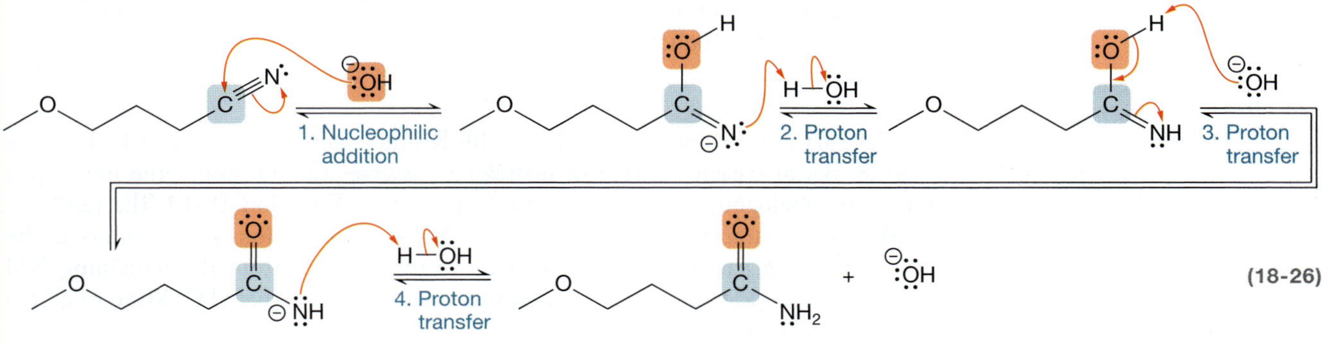

(18-26)

As in hydrolysis under acidic conditions, the only step that is not a proton transfer is the nucleophilic addition step — Step 1 in this case. Similar to the acid-catalyzed mechanism, the proton transfer steps in Equation 18-26 serve to add protons to N and remove one from O. Moreover, Steps 3 and 4 make up a *tautomerization*. Because the reaction takes place under basic conditions, these proton transfer steps allow the reaction to take place without generating any strongly acidic species.

PROBLEM 18.17 Draw the complete, detailed mechanism and predict the major organic product of each of these reactions.

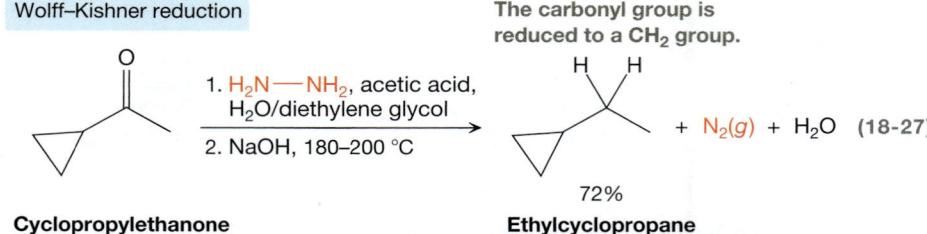

(a)   (b)

18.3 The Wolff–Kishner Reduction

When cyclopropylethanone is treated with hydrazine (H_2N-NH_2) under mildly acidic conditions, followed by heating in the presence of a strong base, ethylcyclopropane is produced, as shown in Equation 18-27.

Wolff–Kishner reduction

The carbonyl group is reduced to a CH_2 group.

1. H_2N-NH_2, acetic acid, H_2O/diethylene glycol
2. NaOH, 180–200 °C

$+ N_2(g) + H_2O$ (18-27)

72%

Cyclopropylethanone   **Ethylcyclopropane**

This is an example of a Wolff–Kishner reduction.

> In a **Wolff–Kishner reduction**, the carbonyl (C=O) group of a ketone or aldehyde is converted to a methylene (CH_2) group.

YOUR TURN 18.8

Verify that the reaction in Equation 18-27 is a reduction by computing the oxidation state of the carbonyl C atom in the reactant and the corresponding C atom in the product.

The partial mechanism for the Wolff–Kishner reduction is shown in Equation 18-28. Because each N atom of hydrazine possesses two H atoms, the first several steps are analogous to those in imine formation in Equation 18-19. The result is a **hydrazone**, characterized by the C=N-NH_2 group. The four steps that occur after the hydrazone has formed serve to remove two protons from the remaining NH_2 group and to add two protons to what was originally the carbonyl C. Moreover, Steps 1 and 2 make up a *tautomerization*.

Notice in Step 1 that HO^- deprotonates a N atom. HO^- is not normally a strong enough base to do this, but the resulting conjugate base is stabilized by resonance (involving the adjacent double bond) and by inductive effects (involving the adjacent electron-withdrawing N atom).

There are two peculiarities about the E2 step (i.e., Step 3) worth noting. First, the proton and the leaving group are removed from adjacent N atoms, whereas we usually see them on adjacent C atoms. Second, the leaving group is quite poor—a negative charge becomes localized on a C atom. Normally this would prevent an E2 reaction from occurring, but the formation of $N_2(g)$ is a substantial driving force. Not only is

$N_2(g)$ very stable, but being a gas, it leaves the reaction system once it is formed. This makes the overall reaction *irreversible*, which, according to Le Châtelier's principle, helps drive the formation of products.

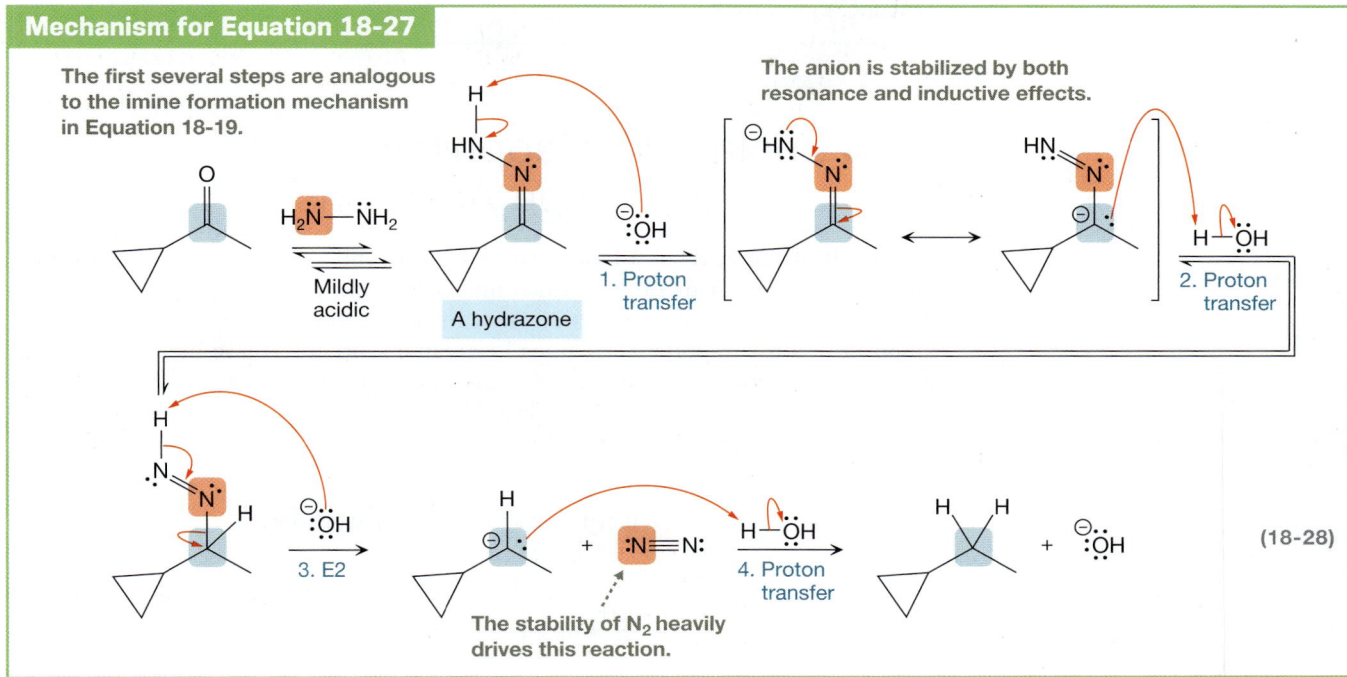

Mechanism for Equation 18-27

The first several steps are analogous to the imine formation mechanism in Equation 18-19.

A hydrazone

1. Proton transfer

The anion is stabilized by both resonance and inductive effects.

2. Proton transfer

3. E2

The stability of N_2 heavily drives this reaction.

4. Proton transfer

(18-28)

The Wolff–Kishner reduction in Equation 18-27 is carried out in two sequential steps: first, the hydrazone forms under mildly acidic conditions, then the hydrazone reacts to form the methylene group under basic conditions. Conveniently, the hydrazone can form under basic conditions, too. Therefore, Wolff–Kishner reductions are frequently carried out in a single step, simply by treating the ketone or aldehyde with hydrazine under strongly basic conditions and high temperature, as shown in Equation 18-29.

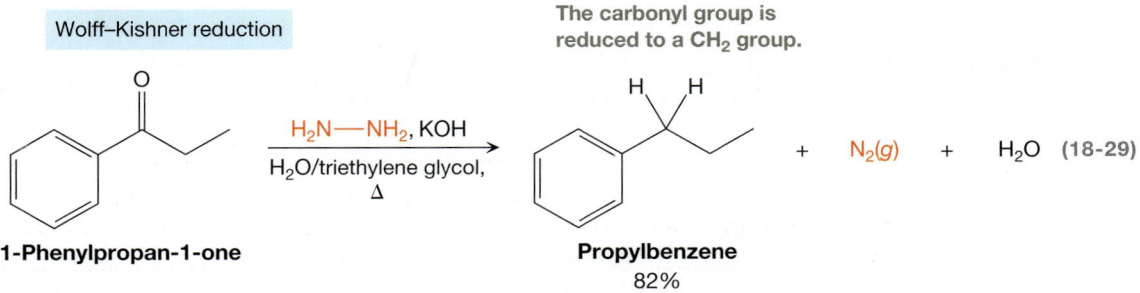

Wolff–Kishner reduction

The carbonyl group is reduced to a CH_2 group.

H_2N-NH_2, KOH

H_2O/triethylene glycol, Δ

$+ \quad N_2(g) \quad + \quad H_2O$ (18-29)

1-Phenylpropan-1-one

Propylbenzene
82%

PROBLEM 18.18 Draw the complete, detailed mechanism and predict the product for each of the following reactions.

1. H_2NNH_2, mildly acidic

2. NaOH/diethylene glycol, Δ

?

1. H_2NNH_2, mildly acidic

2. NaOH/diethylene glycol, Δ

?

(a)

(b)

PROBLEM 18.19 What carbonyl-containing compound could be used as a reactant in the reaction shown here?

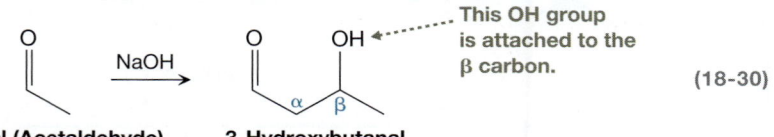

1. H₂NNH₂, mildly acidic
2. NaOH/diethylene glycol, Δ

18.4 Enolate Nucleophiles: Aldol and Aldol-Type Additions

Nearly all of the acid- and base-catalyzed nucleophilic addition reactions we have encountered thus far are ones in which the nucleophilic atom is a heteroatom such as nitrogen, oxygen, or sulfur. Reactions that involve carbon nucleophiles, however, are among the most important reactions in organic synthesis. These include aldol and aldol-type reactions, which we examine throughout the rest of the chapter.

When ethanal (acetaldehyde, $CH_3CH{=}O$) is treated with sodium hydroxide (Equation 18-30), the product is 3-hydroxybutanal, a compound containing a four-carbon chain.

CONNECTIONS 3-Hydroxybutanal is a hypnotic and was once used in medicine as a sedative.

A β-hydroxycarbonyl compound (An aldol)

This OH group is attached to the β carbon.

Ethanal (Acetaldehyde) → NaOH → **3-Hydroxybutanal**   (18-30)

Because the product contains both a $CH{=}O$ group (characteristic of an *ald*ehyde) and an $O{-}H$ group (characteristic of an alcoh*ol*), it is called an **aldol**, and the reaction that forms it is called an **aldol addition**.

Aldol reactions are particularly important because they form a new $C{-}C$ bond. Notice, too, that the carbonyl and the hydroxyl groups in the product are separated by two carbon atoms; the molecule is therefore classified as a **β-hydroxycarbonyl compound**, which is the general form of any aldol addition product.

GREEN CHEMISTRY Aldol addition reactions are highly atom efficient (review Section 13.8b.3) because every atom in the two equivalents of the aldehyde reactant appears in the aldol product.

Every aldol addition produces a β-hydroxycarbonyl compound.

YOUR TURN 18.9

Identify the carbonyl group and the hydroxyl group in the aldol product in Equation 18-30.

PROBLEM 18.20 Which of the following compounds could be the product of an aldol reaction? Explain.

(a)

(b)

(c)

The mechanism for the aldol addition reaction is shown in Equation 18-31.

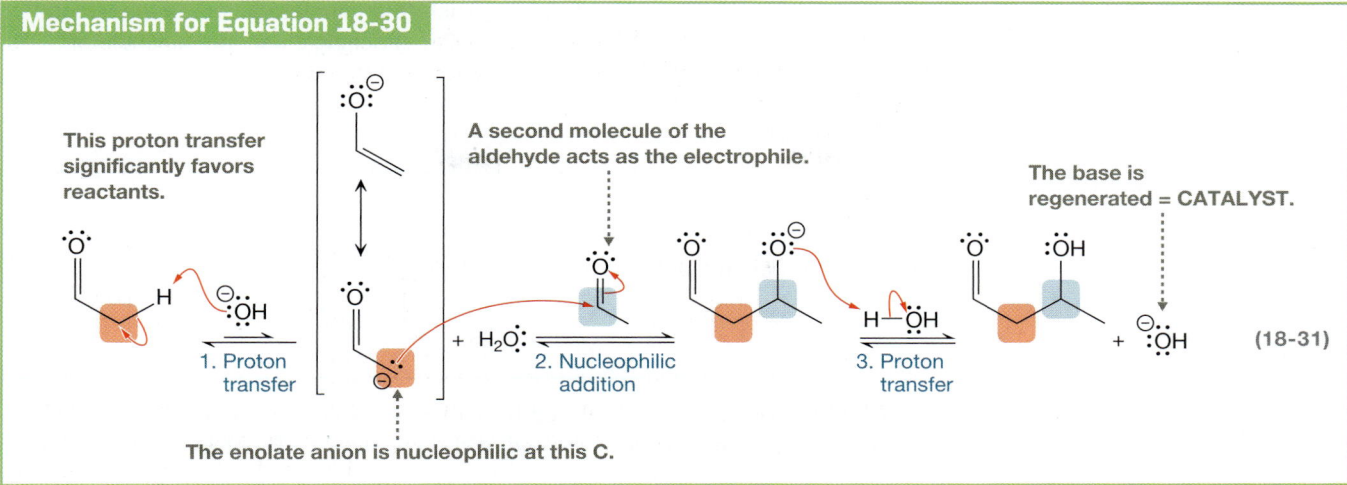

This proton transfer significantly favors reactants.

A second molecule of the aldehyde acts as the electrophile.

The base is regenerated = CATALYST.

1. Proton transfer

2. Nucleophilic addition

3. Proton transfer

(18-31)

The enolate anion is nucleophilic at this C.

In Step 1, HO⁻ *reversibly* deprotonates the α carbon to yield an enolate anion. In fact, that proton transfer favors the reactant side significantly; at equilibrium, only about 0.01% of the aldehyde is deprotonated. In Step 2, the newly formed enolate anion, which is nucleophilic at the α carbon (Sections 10.3 and 10.4), attacks a second molecule of the aldehyde that still has its proton (and hence is uncharged). The immediate product is an alkoxide anion (RO⁻), which is subsequently protonated in Step 3.

YOUR TURN 18.10

To verify that the proton transfer in Step 1 of the mechanism in Equation 18-31 significantly favors the reactant side, use the appropriate pK_a values to calculate the factor by which the reactant side is favored.

The nucleophilic addition step in Equation 18-31 is *reversible*, too (recall Table 17-2, p. 865). It is reversible because the enolate anion is stabilized by resonance and is therefore substantially more stable than the R⁻ nucleophiles such as alkyllithium and Grignard reagents (Section 17.9).

SOLVED PROBLEM 18.21

Draw the complete, detailed mechanism for the reaction shown here and use the mechanism to predict the products.

Think How is this reaction similar to the one in Equation 18-30 (and Equation 18-31)? How is it different? What is the role of NaOH?

Solve Just as in Equation 18-30, an aldehyde is treated with sodium hydroxide. The only difference is the carbon backbone; in this case, we have a four-carbon aldehyde, whereas it was a two-carbon aldehyde in Equation 18-30. The mechanisms are therefore essentially identical.

A β-hydroxy aldehyde

In the first step, HO⁻ acts as a base to generate an enolate anion. In the second step, the enolate anion acts as a nucleophile, attacking a second molecule of butanal that still has its proton. This step forms a C—C bond between the two species. Finally, the O⁻ is protonated by water to yield the β-hydroxy aldehyde and regenerate HO⁻.

PROBLEM 18.22 Draw the complete, detailed mechanism for the reaction shown here and predict the product.

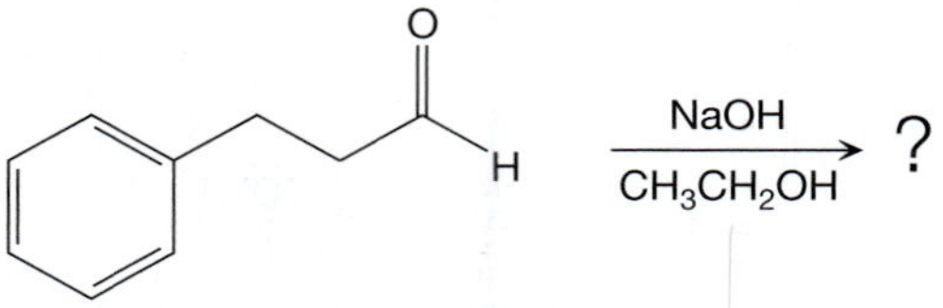

$\xrightarrow[\text{CH}_3\text{CH}_2\text{OH}]{\text{NaOH}}$?

Aldol reactions can be catalyzed by acid, too. Exploring the mechanism for this reaction is left as an exercise at the end of this chapter (see Problem 18.66).

SOLVED PROBLEM 18.23

Show how to synthesize the molecule shown here using an aldol reaction.

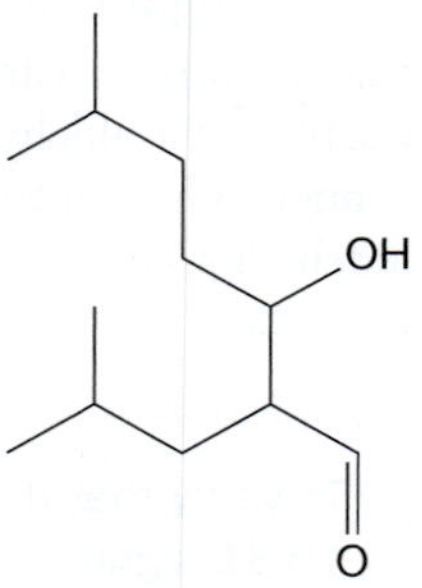

Think Can you identify the portion of the molecule that characterizes it as a β-hydroxy aldehyde? Which carbons of the β-hydroxy aldehyde have been joined in the aldol addition? On disconnecting that C—C bond in a retrosynthesis, what carbon backbones are required in the precursors?

Solve The α and β carbons of the C=O group are labeled in the following structure:

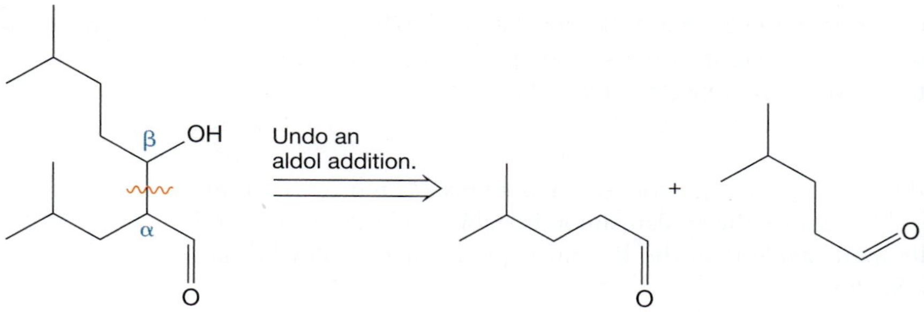

An OH group is attached to the β carbon, so the molecule is indeed a β-hydroxy aldehyde. To undo the aldol reaction, disconnect the bond between the α and β carbons, as indicated. In the precursor, the C—OH bond is a C=O group. The two precursors are the same aldehyde, so in the forward direction, we simply treat that aldehyde with NaOH in a protic solvent.

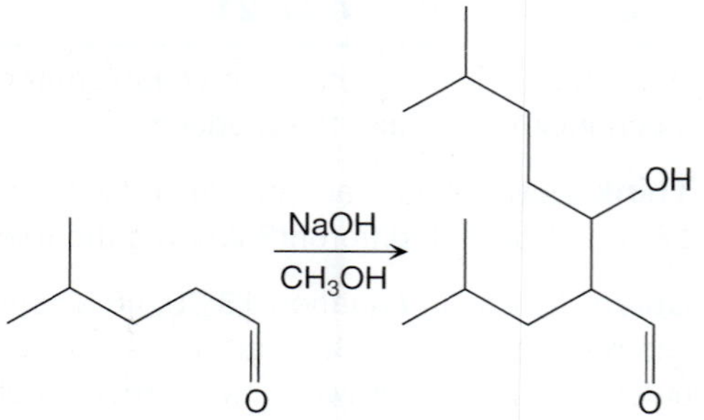

PROBLEM 18.24 Show how to synthesize the molecule shown here using an aldol reaction.

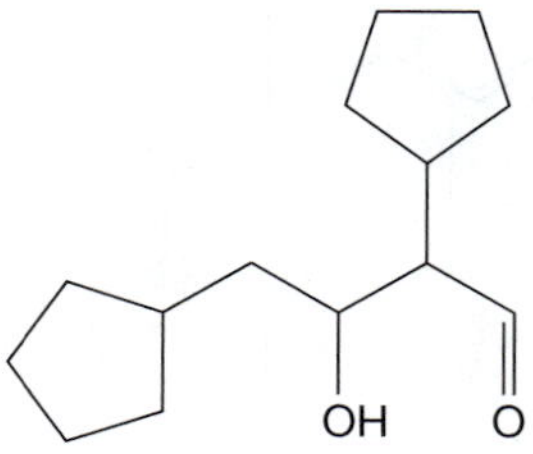

18.5 Aldol Condensations

As we saw in Section 18.4, a β-hydroxy aldehyde is the immediate product of an aldol addition. If the reaction is heated, however, then *dehydration* occurs: Water is eliminated from the β-hydroxy aldehyde, giving the overall product a C=C double bond. An example is shown in Equation 18-32, which takes place under basic conditions.

This overall reaction, in which an aldol addition is followed by dehydration, is called an **aldol condensation**. In general, a *condensation* reaction is one in which two larger molecules bond together with the elimination of a smaller molecule; in this case, that smaller molecule is H_2O.

YOUR TURN **18.11**

To see that Equation 18-32 is truly a condensation reaction, count the total number of C, H, and O atoms in two molecules of the reactant and do the same for the organic product molecule shown. How do they compare?

The newly formed C=C double bond appears between the α and β carbons of the aldehyde in Equation 18-32, so the product is called an *α,β-unsaturated aldehyde*. In general:

The product of an aldol condensation is an α,β-unsaturated carbonyl compound.

Under basic conditions, dehydration of a β-hydroxy aldehyde takes place via an **E1cb mechanism**, which stands for *elimination, unimolecular, conjugate base* (Equation 18-33).

Partial mechanism for Equation 18-32: An E1cb mechanism

Because of the enhanced acidity at the α carbon, Step 1 is a proton transfer, generating a resonance-stabilized enolate anion. In Step 2, the HO⁻ leaving group departs to yield the overall product.

The E1cb mechanism might at first seem to contradict the general rule we encountered for elimination reactions in Section 9.9, which states that an HO⁻ leaving group is generally unsuitable for E1 and E2 reactions. This is not an E1 or E2 mechanism, however. Moreover, the poor leaving group ability of HO⁻ is partly compensated for by the stability that arises from the conjugation between the C=C and C=O double bonds in the overall product.

PROBLEM 18.25 Dehydration occurs if 3-hydroxybutanal is heated under basic conditions, but no dehydration occurs when pent-4-en-2-ol is subjected to the same conditions. Explain.

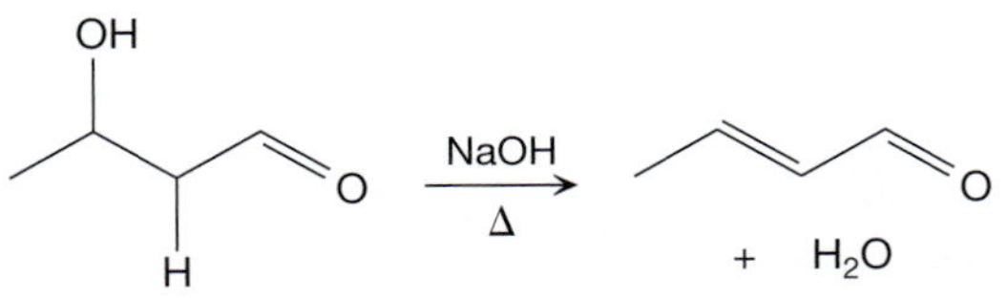

3-Hydroxybutanal 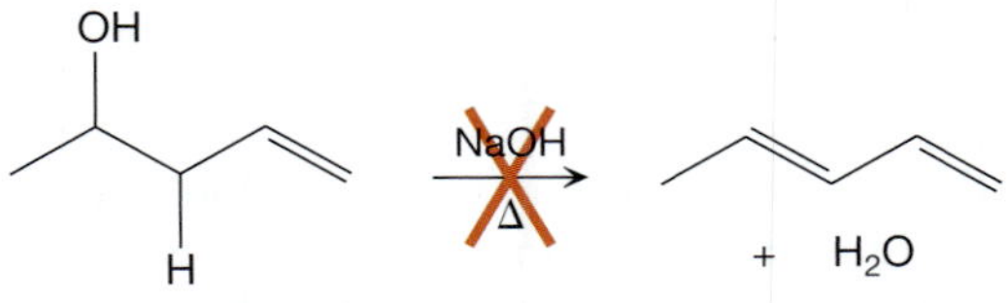   **Pent-4-en-2-ol**

Dehydration of an aldol product can also take place under acidic conditions, again producing an α,β-unsaturated aldehyde (Equation 18-34).

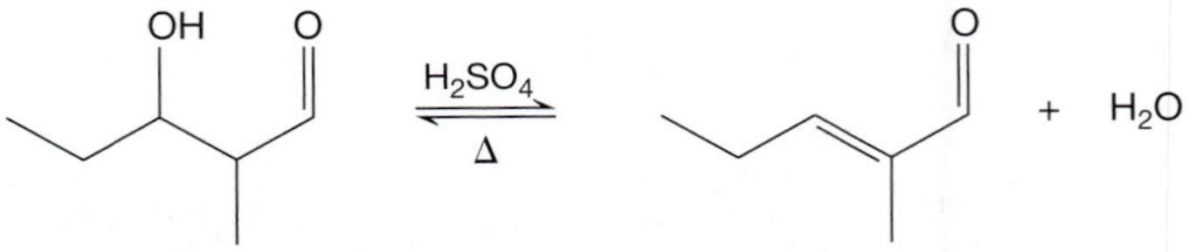

A β-hydroxy aldehyde   **An α,β-unsaturated aldehyde**   (18-34)

Unlike the E1cb mechanism, which takes place under basic conditions, dehydration under acidic conditions proceeds by an E1 mechanism, as shown in Equation 18-35.

Mechanism for the dehydration of an aldol addition product under acidic conditions

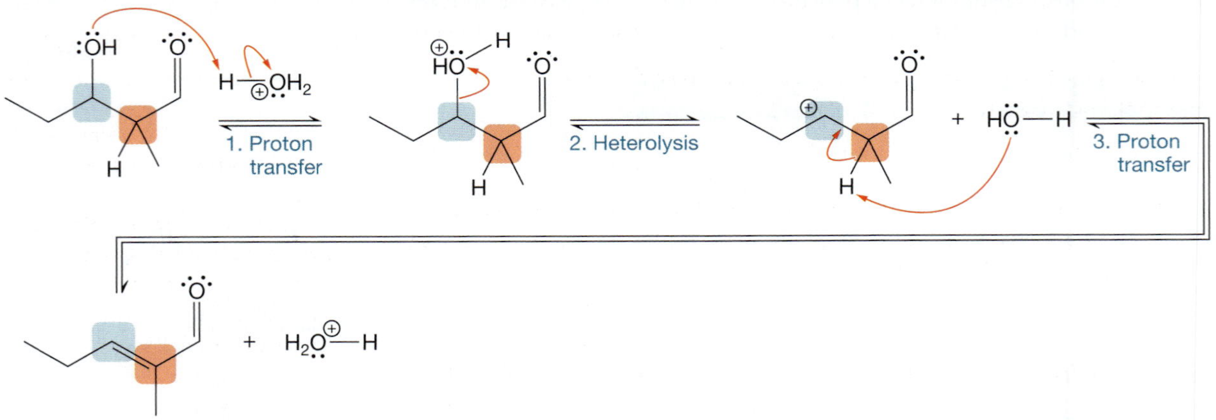

(18-35)

In Step 1, protonation of OH generates a very good water leaving group. The leaving group departs in Step 2, and a final deprotonation yields the overall product.

Notice in Step 3 of Equation 18-35 that the α proton is not the only one that can be detached in a dehydration reaction. As shown in Equation 18-36b, a γ proton can also be removed, yielding a β,γ-unsaturated aldehyde. Essentially none of this product is formed, however, because the α,β-unsaturated aldehyde product (Equation 18-36a) possesses *conjugated double bonds*, and is thus more stable than the β,γ-unsaturated aldehyde.

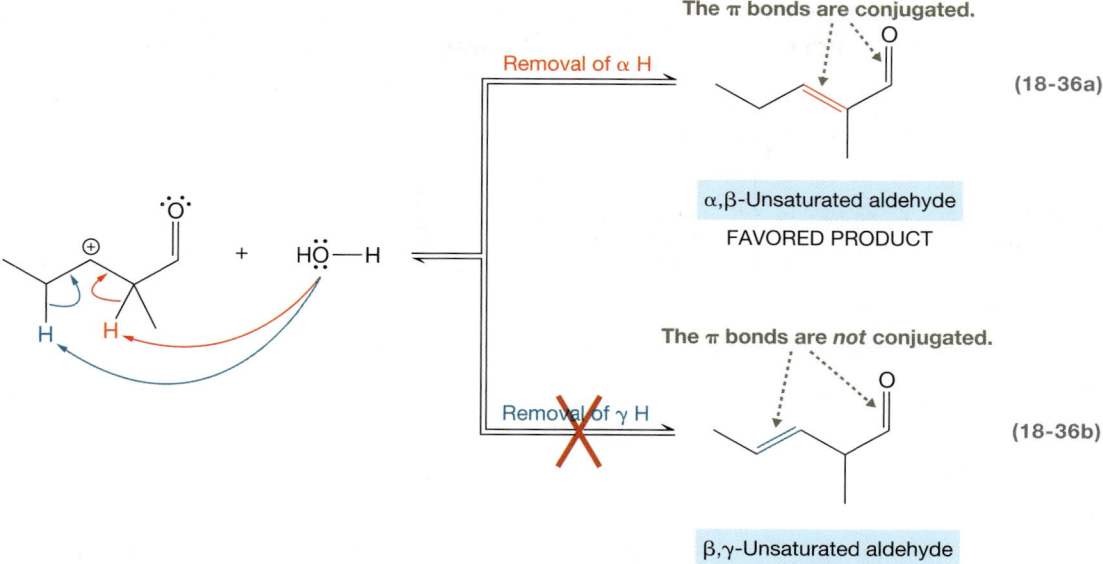

The π bonds are conjugated.

Removal of α H

(18-36a)

α,β-Unsaturated aldehyde
FAVORED PRODUCT

The π bonds are *not* conjugated.

Removal of γ H

(18-36b)

β,γ-Unsaturated aldehyde

18.6 Aldol Reactions Involving Ketones

Ketones with α hydrogens can participate in aldol additions in the same way that aldehydes do, as shown in Equation 18-37 for propanone (acetone).

NaOH

Steric strain surrounding this carbon

(18-37)

Propanone (Acetone)
99%

4-Hydroxy-4-methylpentan-2-one
1%

YOUR TURN 18.12

Label the α and β carbons in the product of Equation 18-37. Is it a β-hydroxycarbonyl compound?

The mechanism for this reaction, shown in Equation 18-38 (next page), is identical to the one for the reaction involving only aldehydes, shown previously in Equation 18-31. According to Equation 18-37, however:

An aldol reaction between ketones tends to form very little product at equilibrium.

Little product is formed because nucleophilic addition to a ketone yields a product with considerable steric strain, making the reaction less favorable than addition to an aldehyde. Additionally, the electron-donating effects from a ketone's alkyl groups diminish the concentration of positive charge on the carbonyl carbon, making it less susceptible to nucleophilic attack. Recall from Section 17.2 that a similar phenomenon occurs with the hydration of ketones and aldehydes.

CONNECTIONS 4-Hydroxy-4-methylpentan-2-one, also called diacetone alcohol, is a component of gravure printing inks. Gravure printing is a type of rotary printing that is often used to manufacture labels for consumer products.

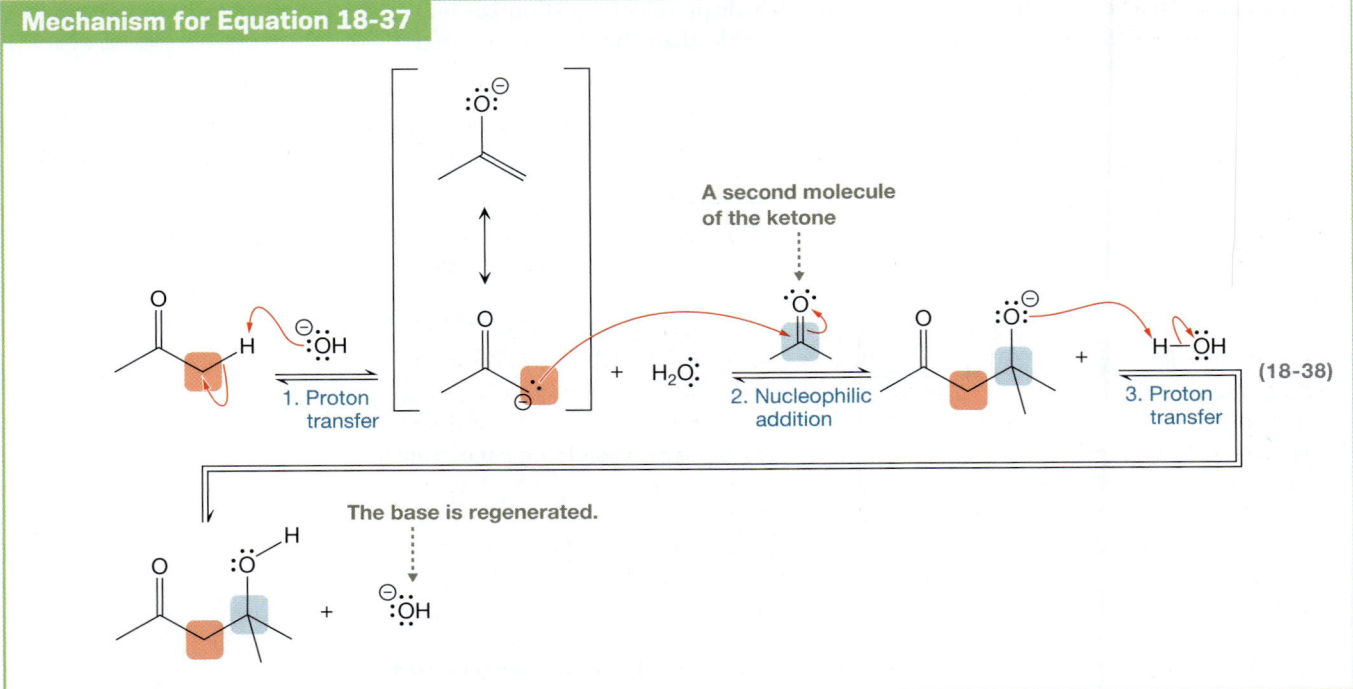

(18-38)

Because aldol additions involving ketones generally favor reactants, chemists must manipulate the equilibrium using Le Châtelier's principle to achieve a reasonable yield. Although we do not discuss the details, one way to do so is to remove the aldol product as it is formed. Alternatively, we can carry out the condensation reaction, which generally makes the α,β-unsaturated ketone product in good yield.

GREEN CHEMISTRY Colin L. Raston and Janet L. Scott (*Green Chem.*, 2000, **2**, 49–52) showed that efficient crossed aldol reactions can be carried out free of solvent, simply by grinding solid reagents together with NaOH. By avoiding the use of solvents, the amount of waste can be dramatically reduced.

18.7 Crossed Aldol Reactions

The aldol additions we have examined thus far have involved only a single aldehyde or a single ketone, in which case the nucleophilic enolate anion is derived from the same aldehyde or ketone it attacks. These kinds of reactions are called **self-aldol additions**. What happens if the nucleophilic species is generated from a *different* aldehyde or ketone than the one that is attacked? Equation 18-39 shows the enolate anion of propanal attacking an uncharged molecule of ethanal (acetaldehyde), yielding 3-hydroxy-2-methylbutanal. This kind of reaction is called a **crossed aldol reaction**.

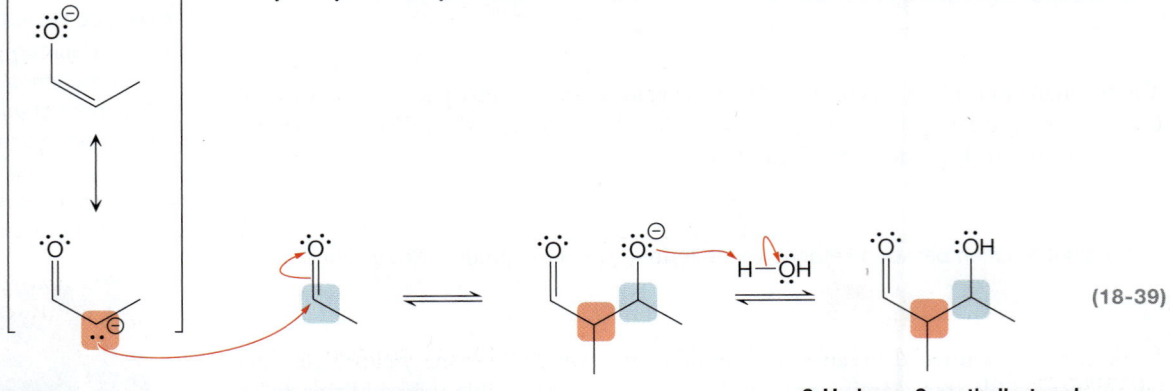

(18-39)

Enolate anion from propanal   **Ethanal (Acetaldehyde)**   **3-Hydroxy-2-methylbutanal**

To carry out a crossed aldol addition, we might imagine treating propanal with NaOH and ethanal (acetaldehyde), as shown in Equation 18-40.

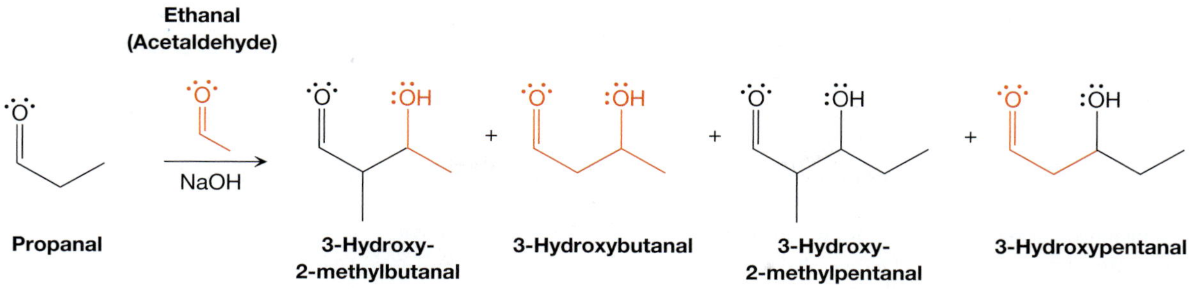

Ethanal
(Acetaldehyde)

(18-40)

Propanal

**3-Hydroxy-
2-methylbutanal**

3-Hydroxybutanal

**3-Hydroxy-
2-methylpentanal**

3-Hydroxypentanal

The hope is that the base will deprotonate propanal, and the resulting enolate anion will attack ethanal to produce 3-hydroxy-2-methylbutanal. Indeed, the desired product is formed, as shown in the equation, but so are three other aldol products. Not only does this compromise the reaction yield, but the four products are difficult to separate due to their similar physical properties. As a result, this kind of reaction is *not* synthetically useful.

The problem stems from the fact that deprotonation of an α hydrogen by NaOH is somewhat unfavorable. Recall from Your Turn 18.10 that an aldehyde (p$K_a \approx 19$) is a weaker acid than water (p$K_a = 15.7$), the conjugate acid of HO⁻. Thus, at any given time throughout the reaction, a significant concentration of HO⁻ exists. When the second aldehyde is added, it can be deprotonated to a small extent by the remaining HO⁻. As a result, the enolate anions from *both* aldehydes are present at the same time, and each enolate anion can attack either of two different uncharged aldehydes. As shown in Equation 18-41, there are four different pairings of an enolate anion with an uncharged aldehyde, and each gives rise to a different aldol product. Two of the products are from *self-aldol reactions* (Equation 18-41a and 18-41d), whereas two are from *crossed aldol reactions* (Equation 18-41b and 18-41c).

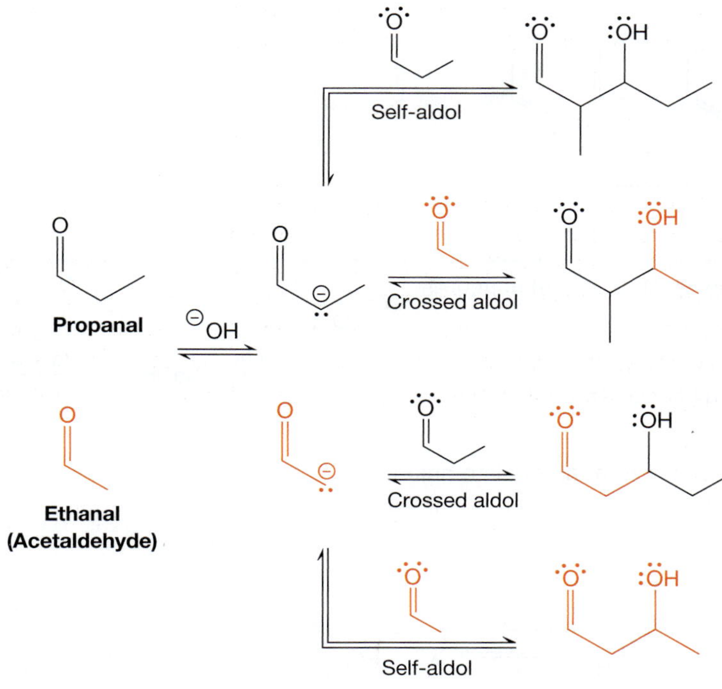

(18-41a)

Self-aldol

Propanal

(18-41b)

Crossed aldol

**Ethanal
(Acetaldehyde)**

(18-41c)

Crossed aldol

(18-41d)

Self-aldol

PROBLEM 18.26 Draw the complete, detailed mechanism for the reaction in Equation 18-41c.

One way around this complication is to use an aldehyde that has no α hydrogens, such as methanal (formaldehyde), benzaldehyde, or dimethylpropanal (Fig. 18-4).

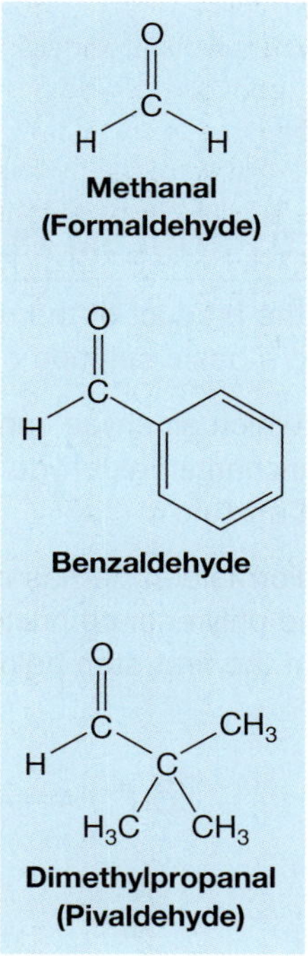

**Methanal
(Formaldehyde)**

Benzaldehyde

**Dimethylpropanal
(Pivaldehyde)**

FIGURE 18-4 Aldehydes with no α hydrogens These aldehydes do not form enolate anions, so they can be used effectively in crossed aldol reactions.

If an aldehyde *with* α hydrogens is added slowly to a basic solution of an aldehyde *without* α hydrogens, primarily one aldol product is formed.

An example is shown in Equation 18-42, in which cinnamaldehyde is produced.

CONNECTIONS As its name suggests, cinnamaldehyde gives cinnamon its characteristic flavor and odor. It is the principal component of the essential oil from cinnamon bark.

No α hydrogens   **Add slowly**

Benzaldehyde + NaOH $\xrightarrow[\text{2. H}_2\text{O, HCl}]{\text{1.}}$ **Cinnamaldehyde** 67%   (18-42)

Benzaldehyde has no α hydrogens, so it cannot be deprotonated by NaOH and does *not* form an enolate anion. In contrast to Equation 18-40, therefore, only one enolate anion is present at any given time. Moreover, if acetaldehyde (which has α hydrogens) is added slowly, we avoid a buildup of the uncharged aldehyde, and thus minimize the amount of product from the self-aldol reaction between acetaldehyde and its own enolate anion.

YOUR TURN 18.13

Which of the following ketones or aldehyde would be the best choice to use for a crossed aldol reaction with butanal?

A   B   C

SOLVED PROBLEM 18.27

Predict the product of the reaction shown here, in which propanal is added slowly to a basic solution of formaldehyde.

Think Which aldehyde can form an enolate anion and which cannot? Which uncharged aldehyde is present in only small concentrations in the reaction mixture?

$$\underset{H}{\overset{O}{\underset{\|}{C}}}\underset{H}{}\ +\ \text{NaOH}\ \xrightarrow[\Delta]{\text{(Add slowly)}}\ ?$$

Solve Formaldehyde has no α hydrogens, so it cannot form an enolate anion. Thus, the only enolate nucleophile that is present is derived from propanal, as shown in the first step below.

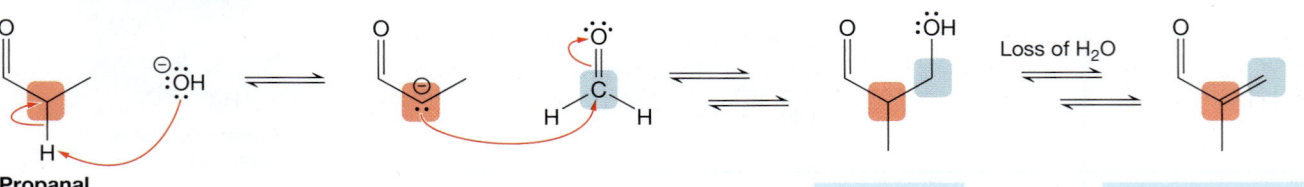

Propanal   Aldol product   Loss of H₂O   Condensation product

Because propanal is added slowly, there is never a substantial concentration of it in its uncharged form, so the major reaction occurs between the propanal enolate anion and formaldehyde. Because the reaction is heated (Δ), dehydration leads to the condensation product.

PROBLEM 18.28 Predict the product of the reaction shown here, in which phenylethanal is added slowly to a basic solution of dimethylpropanal.

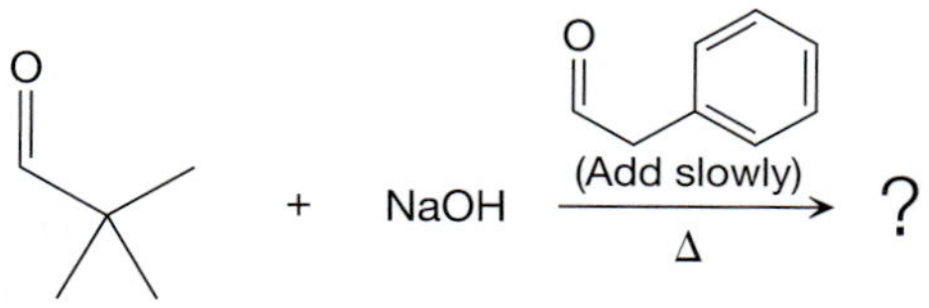

+ NaOH → (Add slowly) Δ → ?

SOLVED PROBLEM 18.29

Show how to synthesize the molecule shown here using an aldol condensation.

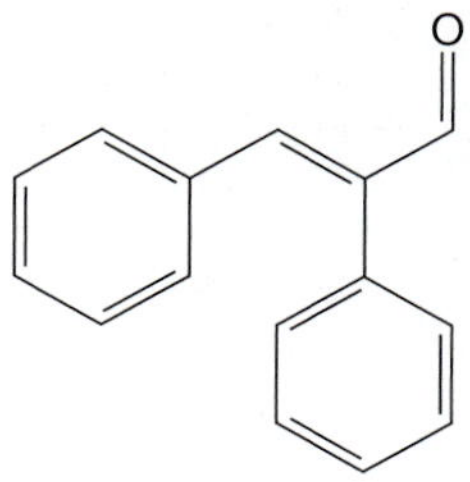

Think Can you identify the portion of the molecule that characterizes it as an α,β-unsaturated carbonyl compound? From what β-hydroxycarbonyl compound could it have been generated? On applying a transform to that β-hydroxycarbonyl compound to undo an aldol addition, which C—C bond should be disconnected?

Solve The carbons that are α and β to the C=O group are labeled in the following structure:

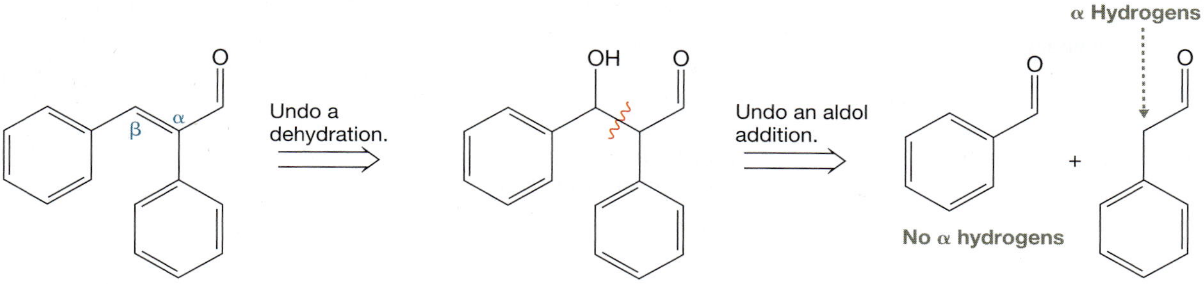

Undo a dehydration.

Undo an aldol addition.

α Hydrogens

No α hydrogens

The target is an α,β-unsaturated aldehyde, so it can be the product of dehydrating the β-hydroxy aldehyde shown. To apply a transform that undoes an aldol addition, we must disconnect the bond between the α and β carbons, as indicated. Because the resulting aldehyde precursors are different from each other, more than one aldol product is possible in the forward direction. To produce only the desired product, we should slowly add the aldehyde *with* α hydrogens to a basic solution of the aldehyde *without* α hydrogens.

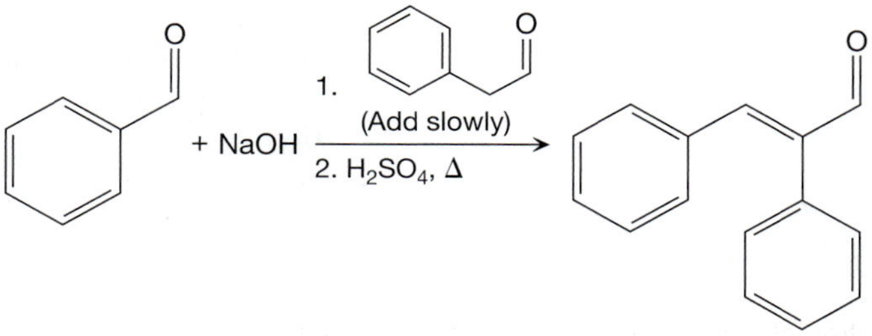

+ NaOH

1. (Add slowly)

2. H₂SO₄, Δ

PROBLEM 18.30 Show how to synthesize the molecule shown here using an aldol reaction.

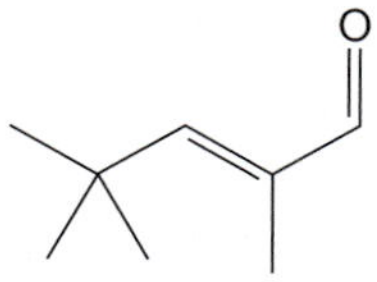

A second way to carry out a crossed aldol reaction selectively is to use a very strong base such as lithium diisopropylamide (LDA) to generate the enolate anion, as shown in Equation 18-43.

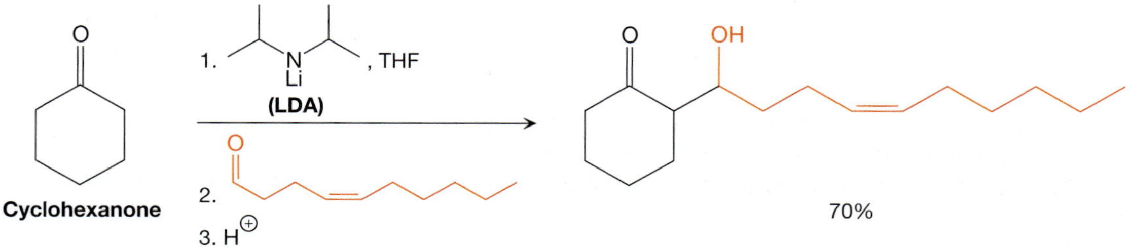

Cyclohexanone

1. [diisopropylamide] , THF
   (LDA)

2. [aldehyde]

3. H⁺

70%

(18-43)

Because LDA is a very strong base, it deprotonates the α carbon rapidly, quantitatively, and irreversibly. Therefore, essentially 100% of the cyclohexanone is converted to its enolate anion prior to the addition of the second carbonyl-containing compound. In this way, only one enolate anion is available to react in the second step of the synthesis.

In an aldol reaction involving ketones, *regiochemistry* becomes a concern when the ketone (e.g., methylbutanone) has two chemically distinct α carbons. As shown in Equation 18-44a and 18-44b, the α C that is deprotonated will lead to one of two enolate anions, and will result in one of two aldol products.

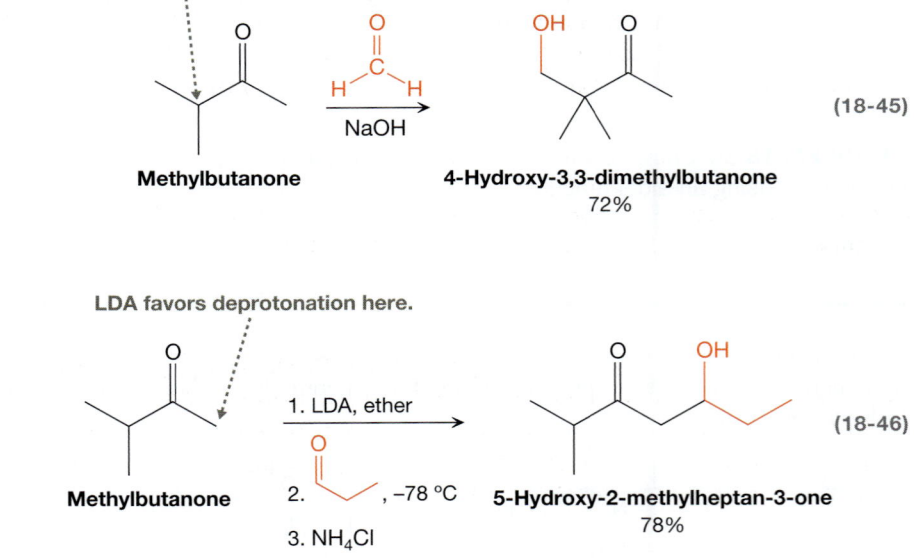

Fortunately, by choosing the base appropriately, we can often control which enolate anion is predominantly formed, and so can control the regiochemistry of these kinds of aldol reactions. For example, Equation 18-45 shows that if NaOH is the base, then the major aldol product has the form that appears in Equation 18-44a. If LDA is the base (Equation 18-46), however, then the major aldol product has the form in Equation 18-44b.

These regiochemistry issues for the aldol reaction are the same ones we previously encountered in Section 10.3a for α-alkylations. When NaOH is used as the base, the favored product is the *thermodynamic* enolate anion, which is formed by deprotonating the *more* highly substituted α carbon. When LDA is used as the base at low temperature, the favored product is the *kinetic* enolate anion, which is formed by deprotonating the *less* highly substituted α carbon.

In each of the following boxes, write the base that would accomplish the respective aldol reactions.

18.8 Intramolecular Aldol Reactions

If a molecule includes two carbonyl groups, as in hexanedial (Equation 18-47), then an *intramolecular* aldol reaction (i.e., one between different parts of the same molecule) is possible. The result is the formation of a *ring*.

Hexanedial   71%   **(18-47)**

Identify the structural features in the product of Equation 18-47 that characterize it as the product of an aldol condensation.

The mechanism for this reaction, shown in Equation 18-48 (next page), is identical to aldol condensations that occur between two separate aldehyde molecules.

In Section 9.11, we learned that cyclization reactions are generally favored over their corresponding intermolecular reactions when the product is a five- or six-membered ring. Aldol reactions are no different.

Intramolecular aldol reactions are favored when five- or six-membered rings are formed.

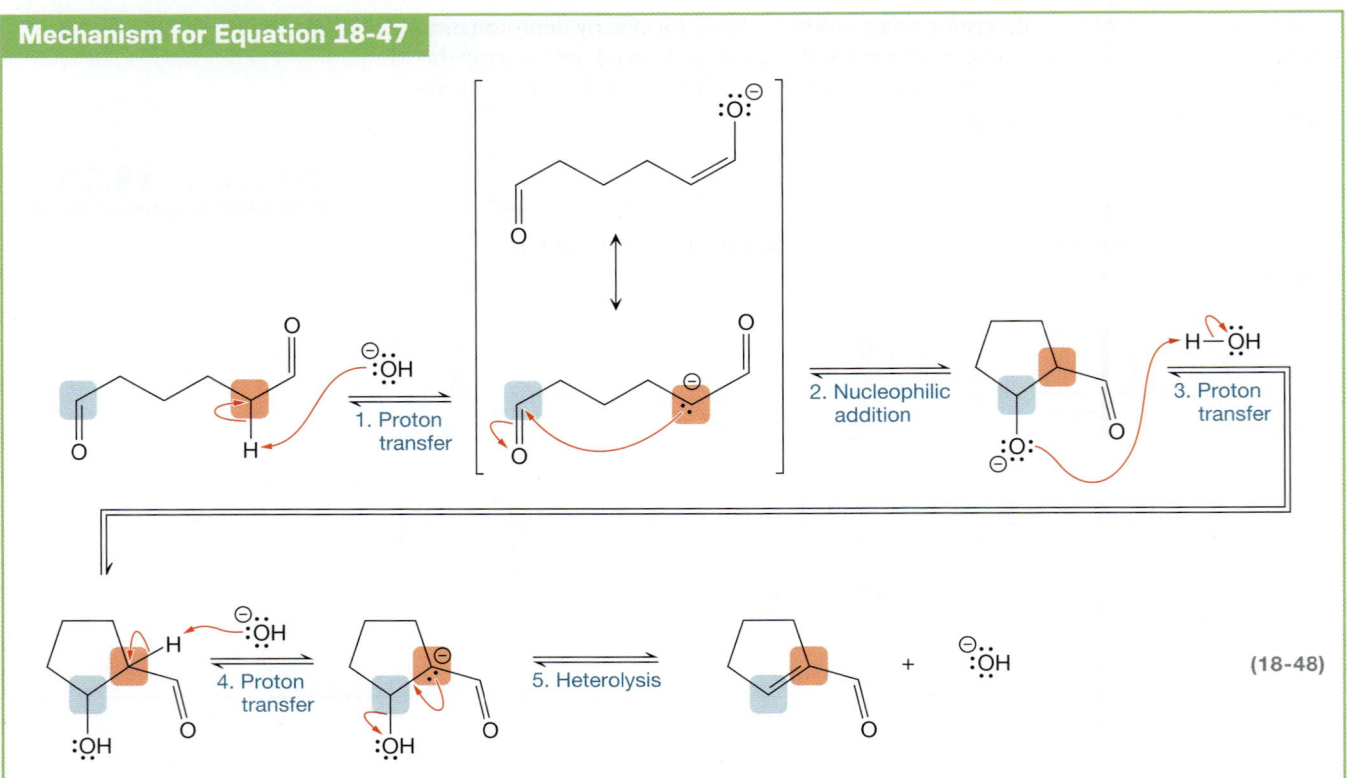

(18-48)

By contrast, pentanedial (Equation 18-49) and octanedial (Equation 18-50) do *not* favor intramolecular aldol reactions because their aldol products would possess four- and seven-membered rings, respectively.

Formation of the 4-membered ring is *not* favorable.

Pentanedial

(18-49)

Formation of the 7-membered ring is *not* favorable.

Octanedial

(18-50)

PROBLEM 18.31 The intramolecular aldol reaction involving heptanedial is favorable. Draw the mechanism for this reaction, along with the major product that is formed. Explain why the reaction is favorable.

Heptanedial

NaOH

?

Equation 18-51 shows a dicarbonyl compound that can form three different enolate anions, in which case we can envision three different cyclic aldol products.

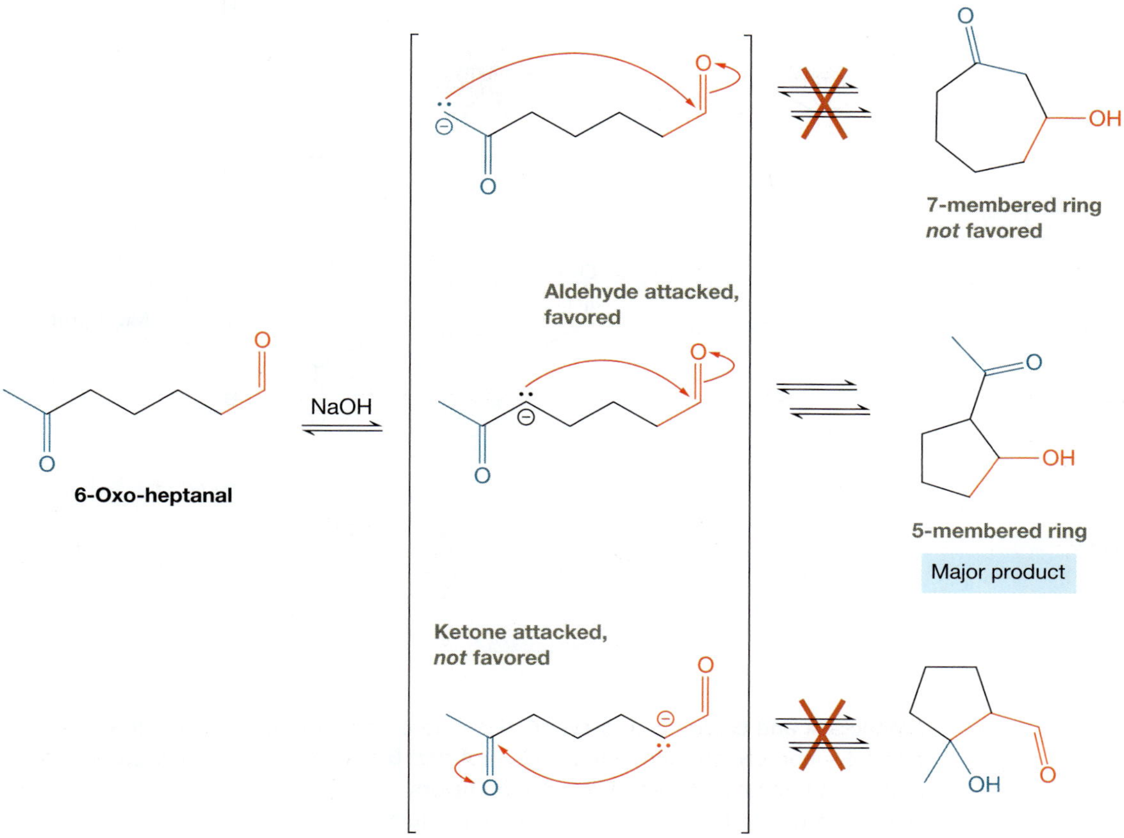

Aldehyde attacked, favored

(18-51a)

7-membered ring
not favored

6-Oxo-heptanal

NaOH

(18-51b)

5-membered ring

Major product

Ketone attacked, *not* favored

(18-51c)

The product in Equation 18-51b is heavily favored over the others. The aldol product in Equation 18-51a is not favored because it forms a seven-membered ring, whereas the reactions in Equation 18-51b and 18-51c both form five-membered rings. The major difference between the two reactions that form five-membered rings is in the carbonyl group that is attacked: In Equation 18-51b, the carbonyl C is attached to one alkyl group, characteristic of an aldehyde, whereas in Equation 18-51c, it is attached to two alkyl groups, characteristic of a ketone. Recall from Section 18.6 that addition to a ketone is generally less favorable than addition to an aldehyde.

SOLVED PROBLEM 18.32

Draw the complete, detailed mechanism for the reaction shown here and predict the major product.

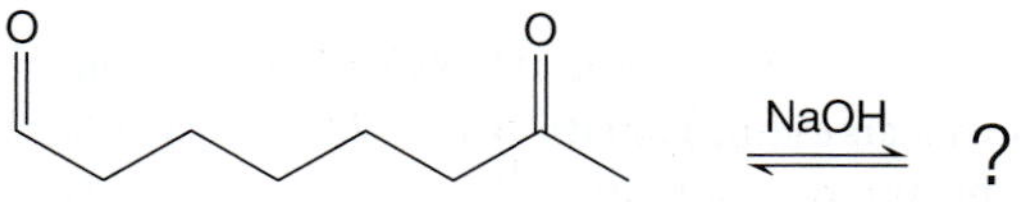

NaOH

?

Think What enolate anions can be formed? Which of those can attack a carbonyl group to form a five- or six-membered ring? In those nucleophilic additions, what kind of carbonyl group is attacked—one characteristic of a ketone or aldehyde?

Solve The three possible enolate anions resulting from deprotonation of an α proton and their corresponding aldol products are as follows:

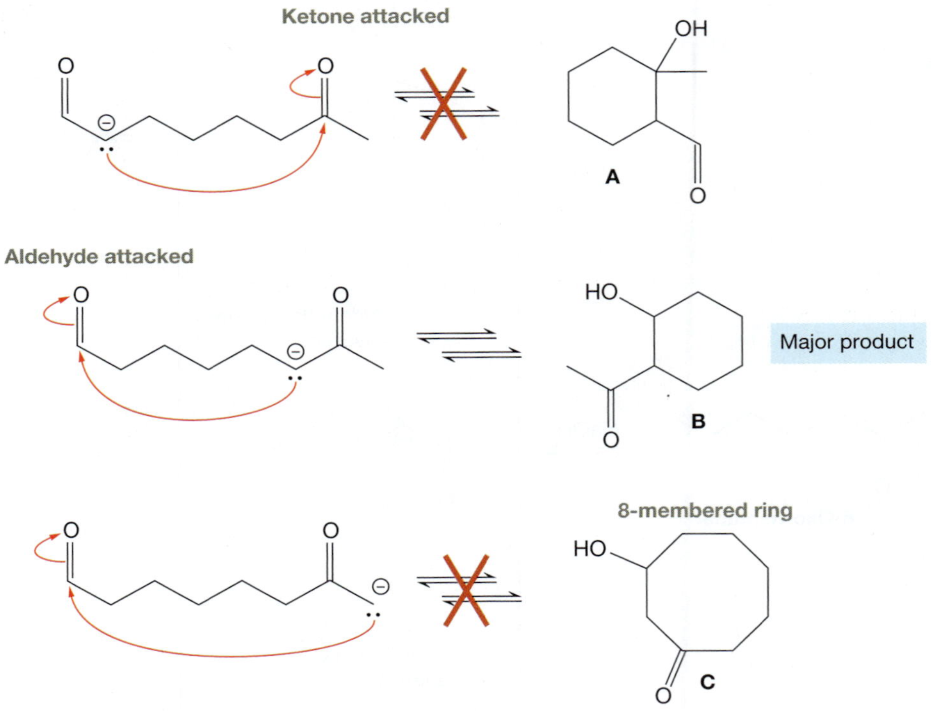

Ketone attacked

A

Aldehyde attacked

HO

Major product

B

8-membered ring

HO

C

Products **A** and **B** are favored over **C** because of the sizes of the rings that are formed—namely, six-membered rings in **A** and **B** and an eight-membered ring in **C**. Moreover, because nucleophilic addition favors attack at an aldehyde over a ketone, **B** is favored over **A**. Thus, **B** is the major product.

PROBLEM 18.33 Predict the major product of the reaction shown here.

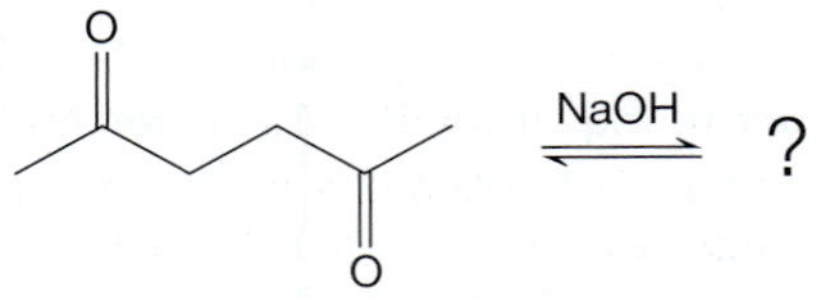

NaOH

?

18.9 Aldol Additions Involving Nitriles and Nitroalkanes

Aldol additions are not limited to just ketones and aldehydes. Other compounds containing a polar π bond have acidic α hydrogens as well. When deprotonated, their resulting enolate anions can act as nucleophiles, just like the enolate anions of ketones and aldehydes.

In nitriles (RC≡N) and nitroalkanes (RNO_2), for example, the α hydrogens are somewhat acidic. As we can see in Equation 18-52a and 18-52b, the pK_a of a nitrile's α hydrogen is ~25, and the pK_a of a nitroalkane's α hydrogen is ~10.

The acidity of a proton alpha to a C≡N or NO_2 group, like the α hydrogen of a ketone or aldehyde, is due to both resonance and inductive stabilization in the conjugate base. Equation 18-52a and 18-52b shows how resonance serves to delocalize the negative charge that develops over two separate atoms: C and N in Equation 18-52a and C and O in Equation 18-52b. Moreover, the CN and NO_2 groups are inductively electron withdrawing, which helps reduce the concentration of negative charge on C.

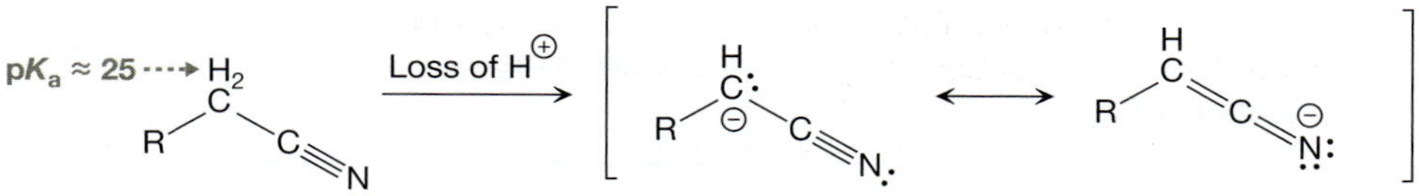

(18-52a)

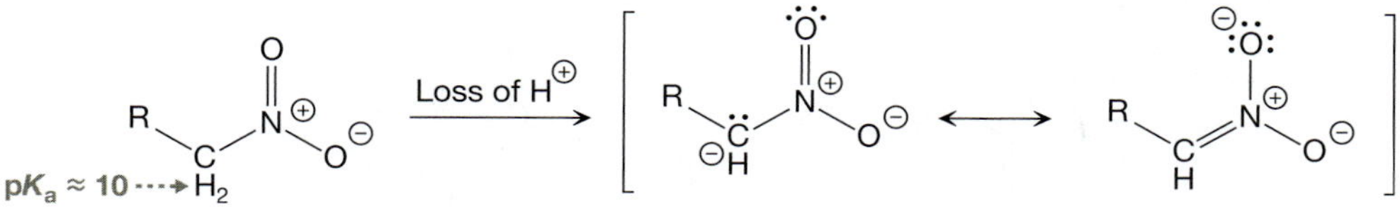

(18-52b)

Equations 18-53 and 18-54 show examples of aldol reactions involving a nitrile and a nitroalkane, respectively.

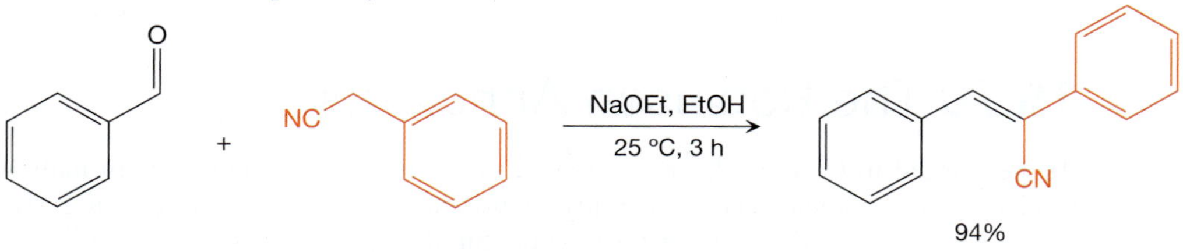

(18-53)

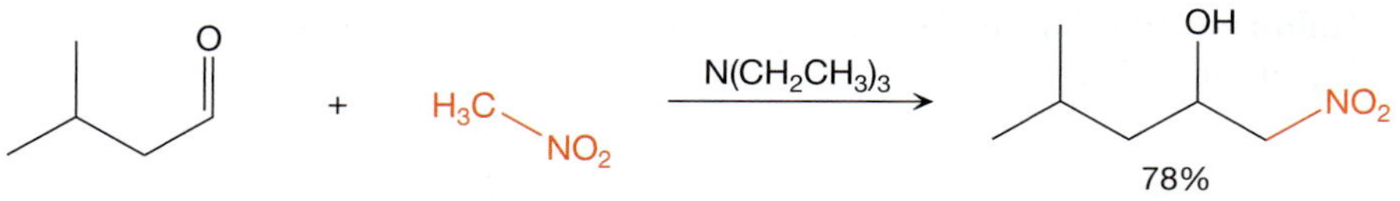

94%

(18-54)

78%

The mechanisms of these reactions are identical to those of aldol reactions involving only ketones or aldehydes (see Solved Problem 18.34 and Problem 18.35). The aldol reaction involving a nitroalkane can be carried out using a weak base such as an amine, however, because the nitroalkane's α proton ($pK_a \approx 10$) is so much more acidic than that of an aldehyde ($pK_a \approx 19$).

SOLVED PROBLEM 18.34

Draw the complete, detailed mechanism for the reaction in Equation 18-53.

Think Which protons are the most acidic? After deprotonation, which polar π bond undergoes nucleophilic addition? How is the C=C double bond formed in an aldol condensation?

Solve The proton alpha to the CN group is the most acidic, so it is removed by the base in the first step. The resulting nitrile anion attacks the carbonyl group of the aldehyde, and the O⁻ generated in that step is subsequently protonated. The C=C bond is then formed by dehydration, which takes place by the E1cb mechanism.

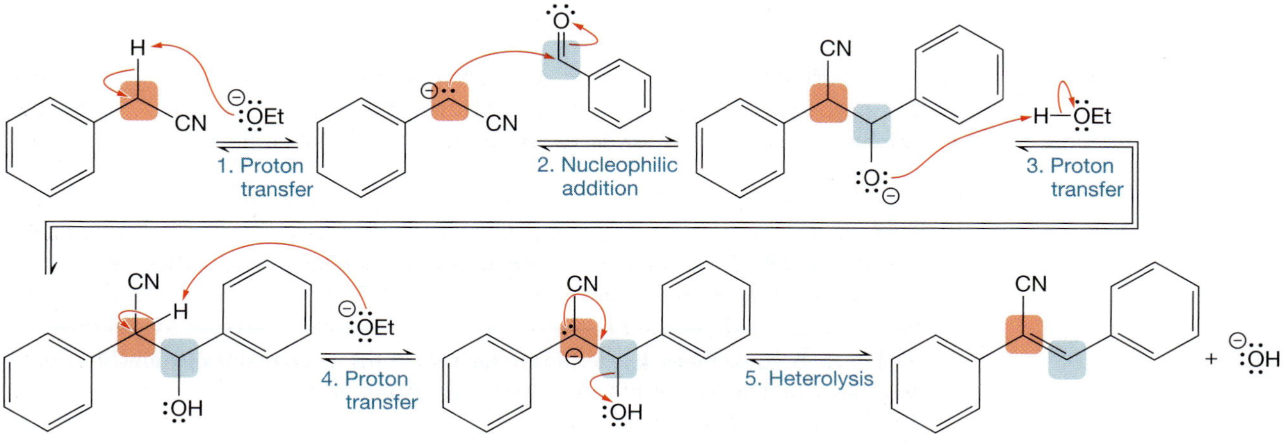

PROBLEM 18.35 Draw the complete, detailed mechanism for the reaction in Equation 18-54.

PROBLEM 18.36 Draw the complete, detailed mechanism and predict the major organic product for the reaction shown here.

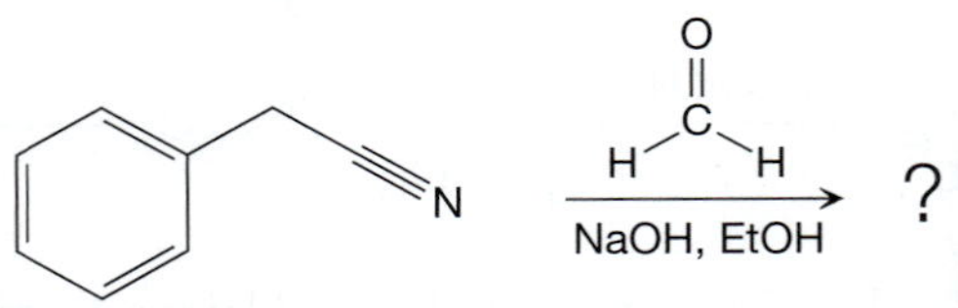

18.10 The Robinson Annulation

As we learned in Chapter 4, six-membered rings are the most abundant in natural products due to their relative stability. Synthetic methods for constructing six-membered rings are therefore of great value. Sir Robert Robinson (1886–1975), an English chemist, developed one such reaction, which has since become known as the **Robinson annulation** (*annulation* means "ring formation"). An example is shown in Equation 18-55.

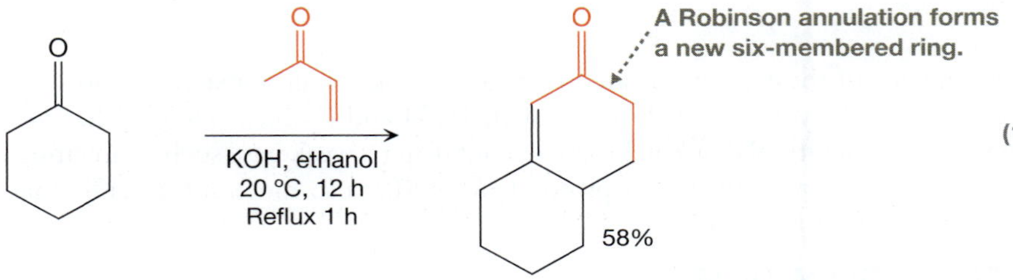

(18-55)

A Robinson annulation reaction is essentially the conjugate addition of an enolate anion (also called a *Michael reaction*; Section 17.9) followed by an intramolecular aldol condensation (i.e., aldol addition plus dehydration), as shown in Equation 18-56.

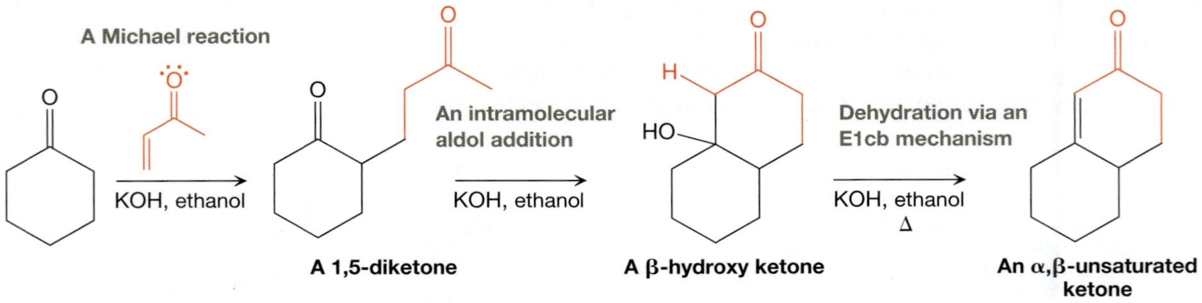

(18-56)

PROBLEM 18.37 Draw the complete, detailed mechanism for the Robinson annulation in Equation 18-55.

For these back-to-back reactions to take place, the reactants of a Robinson annulation must have the general form shown in Equation 18-57.

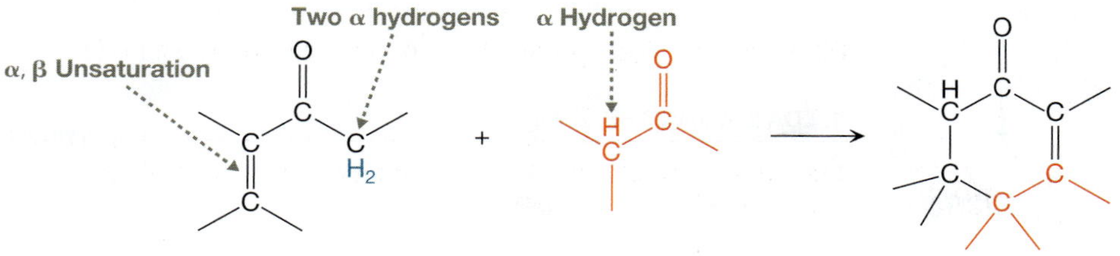

(18-57)

In particular:

> **1.** One of the reactants of a Robinson annulation must be an α,β-unsaturated carbonyl compound and the second must simply be a ketone or aldehyde with an acidic α hydrogen.
> **2.** The α,β-unsaturated carbonyl compound must have two acidic α hydrogens (shown above in blue).

The first of these requirements ensures that an enolate nucleophile can be formed and conjugate addition can take place. The second requirement ensures that the subsequent intramolecular aldol condensation can take place: Deprotonation of the first α hydrogen generates the enolate nucleophile that leads to the formation of a six-membered ring, while the second α hydrogen is removed in the dehydration steps.

PROBLEM 18.38 Based on the above requirements, which of the following pairs of reactants are suitable for a Robinson annulation? For each pair that is suitable, draw the complete, detailed mechanism of the Robinson annulation that could occur.

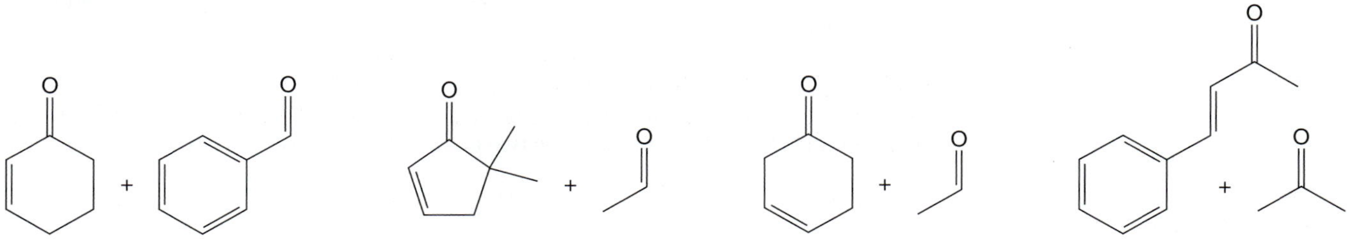

18.11 Organic Synthesis: Aldol Reactions in Synthesis

Aldol reactions are invaluable tools in organic synthesis because they form carbon–carbon bonds and because the compounds they involve—ketones and aldehydes—are quite common. Thus, aldol reactions can be used to produce compounds with a wide variety of structures. In Section 18.8, for example, we saw that aldol reactions can form rings. Perhaps more impressively, aldol reactions can be used to link two compounds with elaborate carbon frameworks, as shown in Equation 18-58, which makes up one of the steps in a synthesis of epothilone B, an anticancer agent.

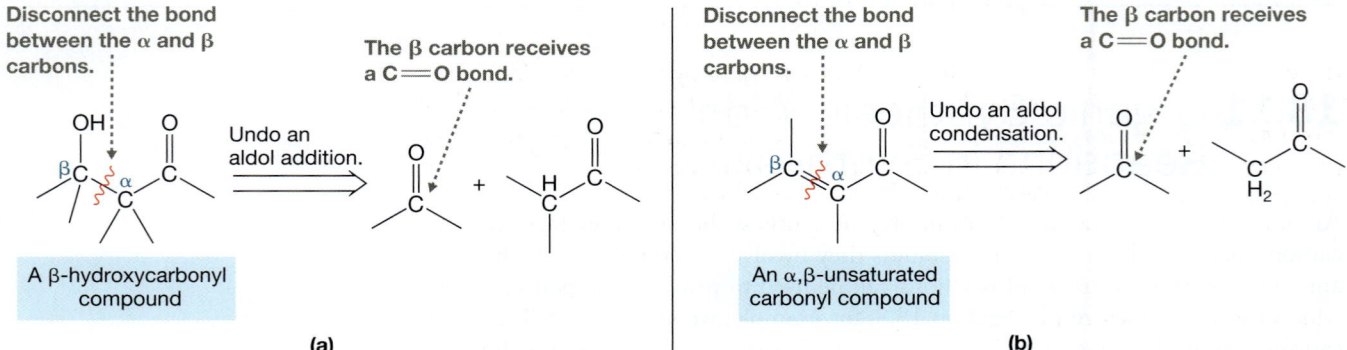

(18-58)

Because of the utility of aldol reactions in synthesis, it is important to be able to efficiently carry out transforms that undo aldol reactions, as part of a retrosynthetic analysis. When doing so, it is helpful to consider the following guidelines:

- The product of an aldol addition is a β-hydroxycarbonyl compound, and the product of an aldol condensation is an α,β-unsaturated carbonyl compound.
- The new C—C bond that is formed in an aldol reaction appears between the α and β carbons in the product, so a transform that undoes an aldol reaction involves disconnecting that bond.
- In constructing the precursors to an aldol reaction, the β carbon from the aldol product receives a C=O bond.

These guidelines are illustrated in the two generic transforms illustrated in Figure 18-5.

FIGURE 18-5 Retrosynthesis involving aldol reactions To carry out a transform that undoes (a) an aldol addition or (b) an aldol condensation, disconnect the α and β carbons and add a C=O bond to the β carbon.

Suppose, for example, that our target compound is 5,5-dimethylcyclohex-2-enone, which is an α,β-unsaturated ketone (Equation 18-59).

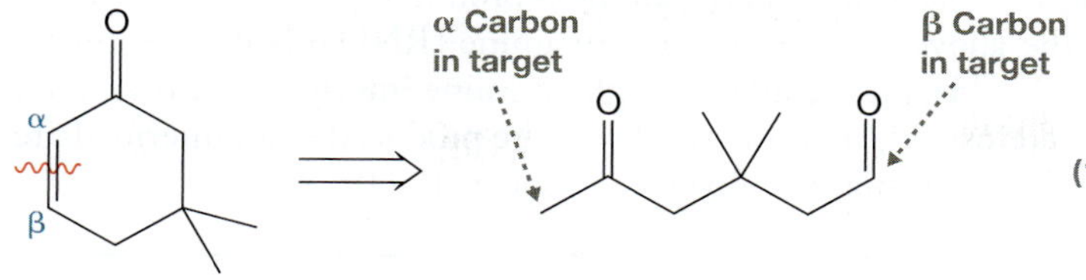

5,5-Dimethylcyclohex-2-enone   **3,3-Dimethyl-5-oxohexanal**

(18-59)

α Carbon
in target

β Carbon
in target

We can undo an aldol condensation by disconnecting the bond indicated by the wavy line and giving the β carbon a C=O bond. Thus, 3,3-dimethyl-5-oxohexanal is a suitable precursor.

In the forward direction, we can simply treat the precursor with NaOH and apply heat (to facilitate dehydration), as shown in Equation 18-60:

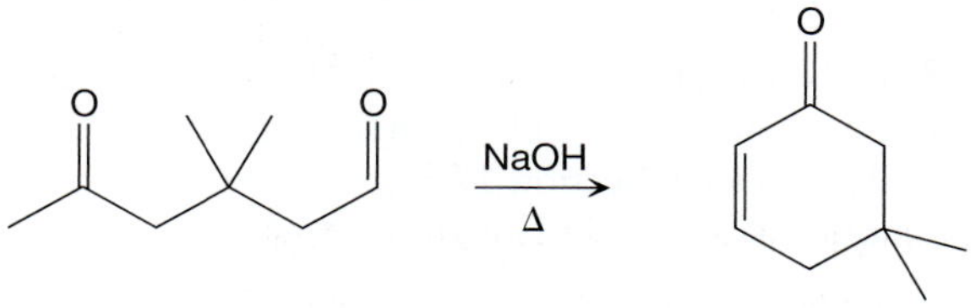

(18-60)

PROBLEM 18.39 Show how to synthesize the following compounds using either an aldol addition or aldol condensation.

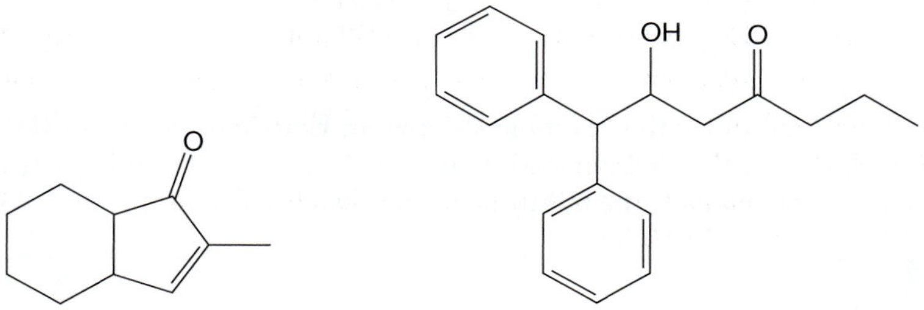

(a)   (b)

18.12 Organic Synthesis: Synthesizing Amines via Reductive Amination

In Section 10.2, we learned that amines can be synthesized by treating an alkyl halide with ammonia or another amine. However, because those reactions tend to produce mixtures of different amines, synthesizing amines that way is generally not very useful. One way around this problem is to carry out a **reductive amination** of a ketone or aldehyde, an example of which is shown in Equation 18-61.

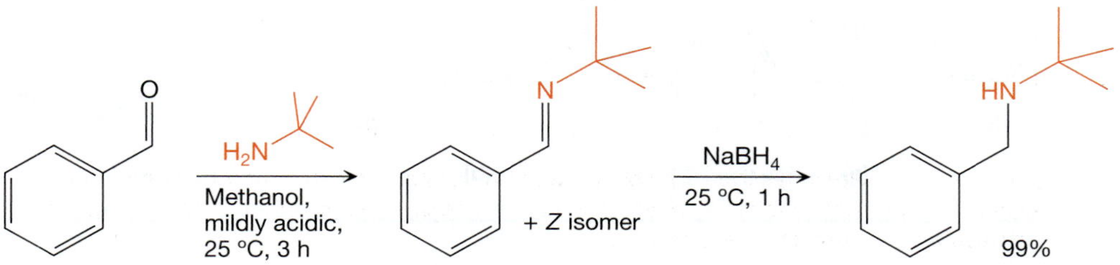

(18-61)

Methanol,
mildly acidic,
25 °C, 3 h

+ Z isomer

NaBH₄
25 °C, 1 h

99%

In this case, the aldehyde is first treated with an amine to produce an imine (review Section 18.2b), after which NaBH$_4$ is added to reduce the imine to the amine (review Section 17.3c). Notice that the amine is produced precisely where the carbonyl O was.

In the above example, a primary amine (RNH$_2$) is used to produce a secondary amine (R$_2$NH), proceeding through an imine intermediate. If ammonia (NH$_3$) were to be used instead, an imine would still be produced as an intermediate, but the product would be a primary amine (see Problem 18.40).

PROBLEM 18.40 For the reaction shown here, draw the intermediate imine and the product.

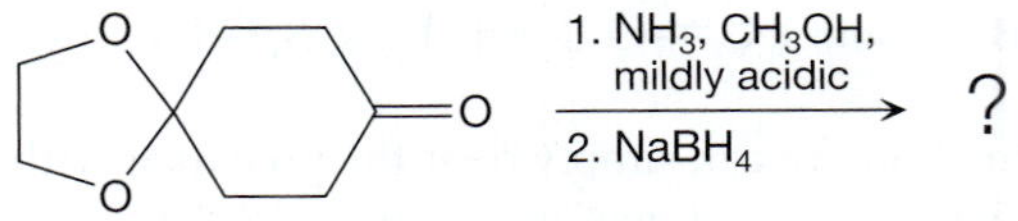

1. NH$_3$, CH$_3$OH, mildly acidic
2. NaBH$_4$

?

In Equation 18-61, NaBH$_4$ is not added until after the imine formation has come to completion. Imine formation takes place under slightly acidic conditions, which would neutralize NaBH$_4$ (review Section 17.3a). A clever solution to this problem is to use NaBH$_3$CN instead, as shown in Equation 18-62.

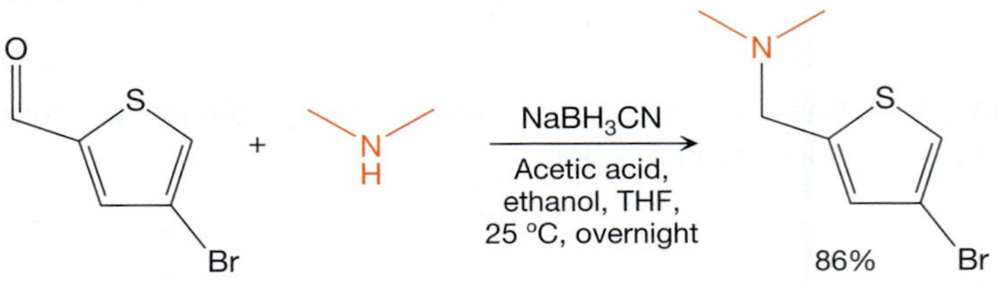

NaBH$_3$CN

Acetic acid, ethanol, THF, 25 °C, overnight

86%

(18-62)

The partial mechanism for this reaction is shown in Equation 18-63. NaBH$_3$CN is a source of hydride, but the electron-withdrawing CN group makes it less basic, allowing it to remain intact under the mildly acidic reaction conditions.

Partial mechanism for Equation 18-62

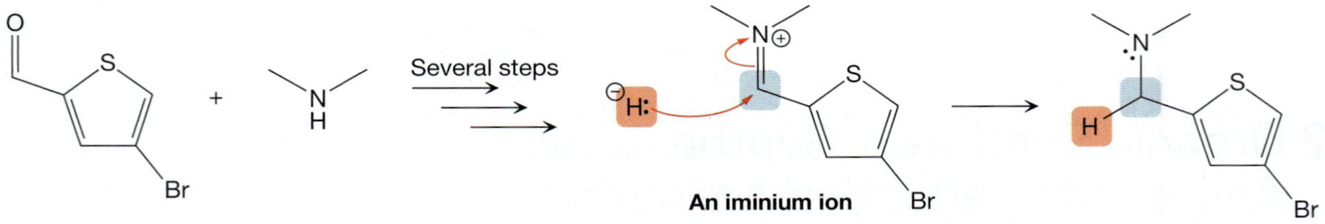

Several steps

An iminium ion

(18-63)

PROBLEM 18.41 Show how to synthesize each of these amines from a ketone or aldehyde.

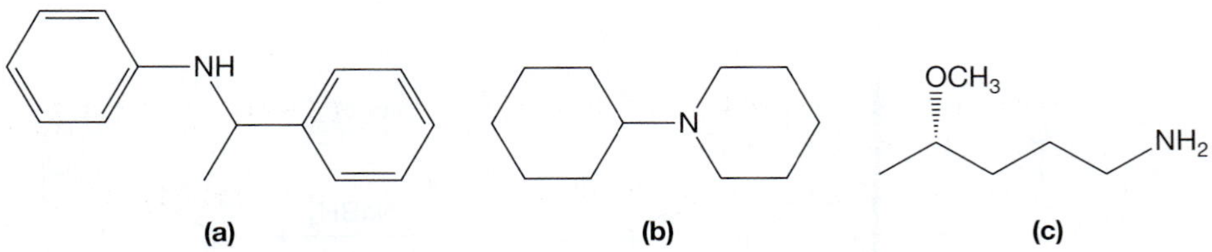

(a)

(b)

(c)

18.13 Ring Opening and Closing of Monosaccharides; Mutarotation

In Section 1.14b, we showed that a monosaccharide can have both open-chain and cyclic forms. In water, the various forms equilibrate, as shown in Equations 18-64 and 18-65 for D-glucose and D-ribose, respectively.

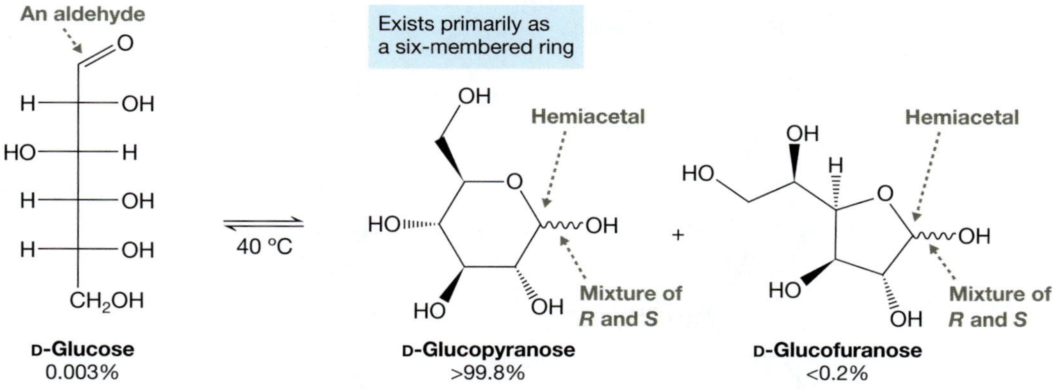

An aldehyde

Exists primarily as a six-membered ring

Hemiacetal   **Hemiacetal**

Mixture of R and S   **Mixture of R and S**

(18-64)

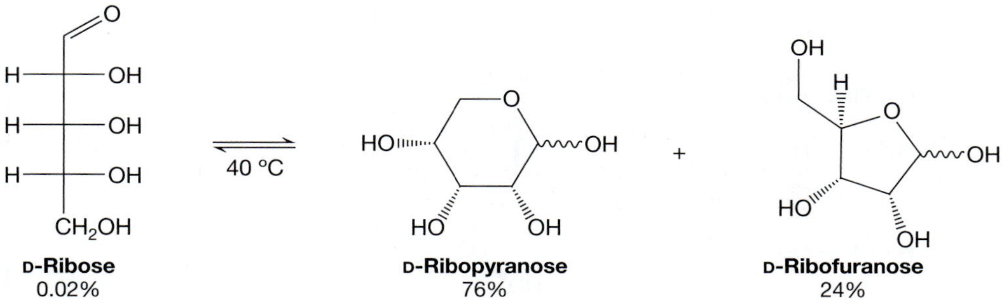

D-Glucose
0.003%

D-Glucopyranose
>99.8%

D-Glucofuranose
<0.2%

(18-65)

D-Ribose
0.02%

D-Ribopyranose
76%

D-Ribofuranose
24%

The mechanisms showing how D-glucose cyclizes to its six- and five-membered rings are shown in Equation 18-66a and 18-66b, respectively. Step 1 is the nucleophilic addition of a hydroxyl group to the carbonyl group, whereas Steps 2 and 3 are both proton transfers. These mechanisms are essentially the same as the one shown previously in Equation 18-3. In each case, the HC=O group reacts with an OH group to produce a hemiacetal. The main difference, however, is that the reaction in Equation 18-3 is *intermolecular*, whereas it is *intramolecular* in Equation 18-66.

Recall from Section 9.11 that cyclization reactions are favored when a five- or six-membered ring is formed, and that the formation of a six-membered ring is usually favored over a five-membered ring. As we can see in Equations 18-64 and 18-65, this is no different for the cyclization of monosaccharides. Specifically:

- At equilibrium, monosaccharides typically favor their five- and six-membered ring forms over their open-chain forms.
- The six-membered ring form is generally favored over the five-membered ring form.

Mechanism showing the cyclization of D-glucose to its six- and five-membered ring forms

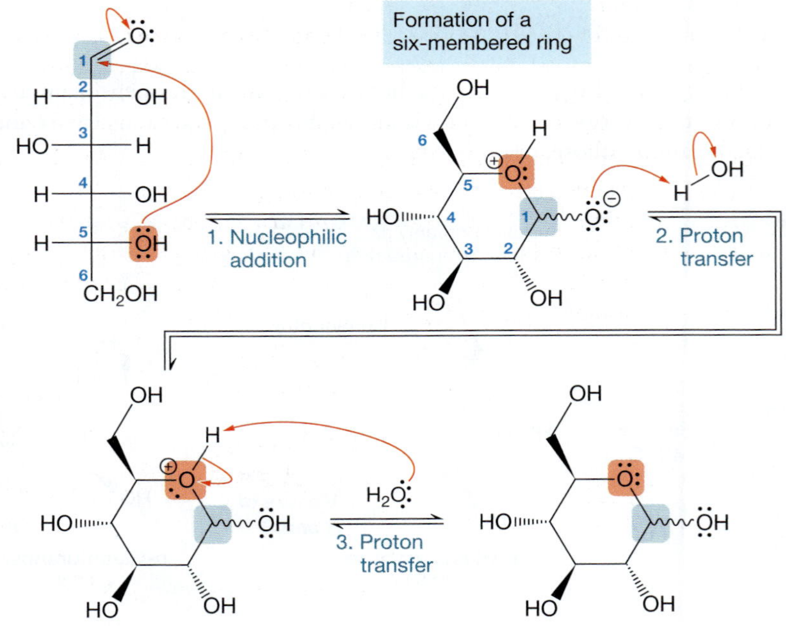

Formation of a six-membered ring

1. Nucleophilic addition

2. Proton transfer

(18-66a)

3. Proton transfer

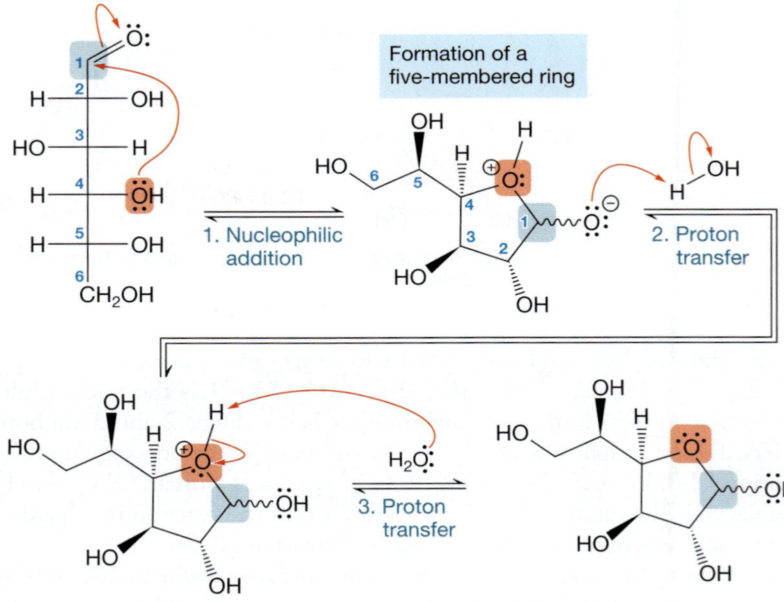

Formation of a five-membered ring

1. Nucleophilic addition

2. Proton transfer

(18-66b)

3. Proton transfer

Because of the importance of the five- and six-membered ring forms of monosaccharides, specific nomenclature has been developed to distinguish the two forms from each other, as well as from the open-chain form. Under this system, the monosaccharide's name (review Sections 4.14 and 5.14) is modified by inserting an "o" followed by either "pyran" or "furan" prior to the *ose* suffix, depending on ring size.

- A monosaccharide that has cyclized to a six-membered ring is designated as a **pyranose**.
- One that has cyclized to a five-membered ring is designated as a **furanose**.

Thus, as indicated previously in Equation 18-64, the cyclic form of D-glucose that has a six-membered ring is called D-glucopyranose, and the one that has a five-membered ring is called D-glucofuranose. The corresponding cyclic forms of D-ribose are D-ribopyranose and D-ribofuranose (Equation 18-65).

Looking back at the reactions in Equations 18-64 and 18-65, notice that the carbonyl carbon of the monosaccharide becomes a *new asymmetric carbon* as a result of the cyclization. As with any reaction in which a new asymmetric atom is generated, a mixture of stereoisomers is produced, one having the *R* configuration at the new asymmetric atom and the other having the *S* configuration. This is why that C—O bond is denoted with a wavy line (∿∿), not a dash or a wedge.

As we can see in Equation 18-66a, the new asymmetric carbon is produced in Step 1, the nucleophilic addition step. In this step, the OH group can attack the carbonyl C atom from either side of that atom's plane, shown in Equation 18-67a and 18-67b for the formation of D-glucopyranose.

OH attack from the top

The new OH ends up on the bottom.

Same as

D-Glucose   α-D-Glucopyranose   (18-67a)

Anomeric carbon

The new OH ends up on the top.

Same as

OH attack from the bottom

β-D-Glucopyranose   (18-67b)

The two stereoisomers differ in configuration at just one of the asymmetric carbons, so they are *diastereomers* of each other. In carbohydrate chemistry, they are more specifically called **anomers** of each other, and the carbon atom that differs in stereochemical configuration—the one that is part of the carbonyl group in the open-chain form—is called the **anomeric carbon**. One anomer is designated the α anomer and the other the β anomer (Fig. 18-6).

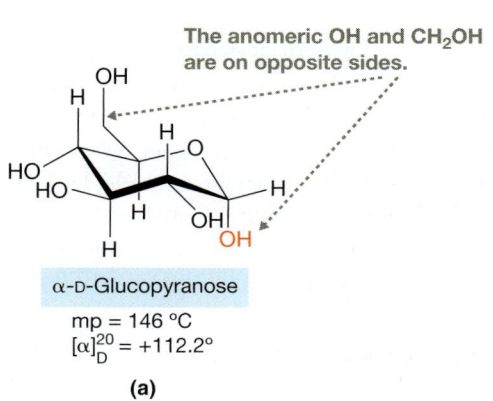

The anomeric OH and CH₂OH are on opposite sides.

The anomeric OH and CH₂OH are on the same side.

α-D-Glucopyranose
mp = 146 °C
$[\alpha]_D^{20}$ = +112.2°

(a)

β-D-Glucopyranose
mp = 150 °C
$[\alpha]_D^{20}$ = +18.7°

(b)

FIGURE 18-6 Anomers of monosaccharides (a) In an α anomer, the anomeric OH is on the side of the ring opposite the CH₂OH group. (b) In a β anomer, the groups are on the same side.

- In the **α anomer** of a cyclic monosaccharide, the anomeric OH and the CH$_2$OH substituents are trans to each other (located on opposite sides of the ring).

- In the **β anomer**, the two substituents are cis to each other (located on the same side of the ring).

PROBLEM 18.42 **(a)** Draw the Haworth projections of α-D-ribofuranose and β-D-ribofuranose. **(b)** Draw the mechanisms that show how each of these anomers is produced from the acyclic form of the monosaccharide.

PROBLEM 18.43 Name each of these monosaccharides. (Consult Fig. 5-32 on p. 249 for the name of each sugar in its acyclic form.)

(a)   (b)

Because they are diastereomers, α and β anomers have different physical properties. The melting point of α-D-glucopyranose is 146 °C, for example, whereas that of β-D-glucopyranose is 150 °C. Additionally, the two have different optical properties. Whereas the specific rotation of α-D-glucopyranose in water is +112.2°, that of β-D-glucopyranose is +18.7°.

If pure crystalline α-D-glucopyranose is dissolved in water, the specific rotation changes over time, from +112.2° to +52.5°. Likewise, the specific rotation of β-D-glucopyranose changes over time, from +18.7° to +52.5°. This phenomenon is called **mutarotation**; in water, each anomer equilibrates with the open-chain form because the formation of the ring is reversible (Equation 18-66a), so the two anomers equilibrate with each other (Equation 18-68).

(18-68)

α-D-Glucopyranose   β-D-Glucopyranose

36%   64%

Using the specific rotation of an equilibrium mixture of two anomers, we can calculate the relative amounts of the two anomers. For D-glucose, for example, let x be the equilibrium fraction of the α anomer, and let the remaining amount in the mixture, $1 - x$, be the equilibrium fraction of the β anomer. The specific rotation of the mixture is the weighted average of their individual specific rotations, according to Equation 18-69.

$$[\alpha]_{D,\text{mixture}}^{20} = x(112.2°) + (1 - x)(18.7°) = 52.5°$$ (18-69)

Solving for x, we find that the equilibrium amount of the α anomer is 36% and that of the β anomer is 64%.

PROBLEM 18.44 The specific rotation of α-D-mannopyranose is +29.3°, and that of β-D-mannopyranose is −16.3°. When either anomer is dissolved in water, the specific rotation slowly changes to +14.5°. Calculate the relative amounts of the two anomers at equilibrium.

Chapter Summary and Key Terms

- Nucleophilic addition of a weak nucleophile to a polar π bond is typically slow. If the nucleophilic atom has a weakly acidic proton, **base catalysis** can increase the rate of the reaction by converting the weak nucleophile into a strong one. Alternatively, **acid catalysis** can increase the reaction rate by making the polar π bond more electrophilic—that is, by **activating** the polar π bond. (Section 18.1; Objectives 1, 2)

- A **cyanohydrin** can be formed via the addition of HCN to the carbonyl group of a ketone or aldehyde, and the reaction can be catalyzed by the addition of a strong base or NC^-. (Section 18.1a; Objectives 1, 2)

- When a weak nucleophile adds to an α,β-unsaturated polar π bond, conjugate addition is generally favored over direct addition. (Section 18.1b; Objective 3)

- *Acetals*, **imines**, and **enamines** can be formed reversibly from ketones or aldehydes under acidic conditions. Formation of each of these products is favored (via Le Châtelier's principle) if there is an excess of the respective nucleophiles. (Sections 18.2a and 18.2b; Objective 4)

- Acetals, imines, and enamines can be hydrolyzed to ketones or aldehydes under acidic conditions using an excess of water. Nitriles can be hydrolyzed to amides under acidic or basic conditions. (Sections 18.2a–c; Objective 5)

- The C=O group of a ketone or aldehyde can be reduced to a methylene (CH_2) group via the **Wolff–Kishner reduction**, in which the ketone or aldehyde is first treated with hydrazine (H_2NNH_2), followed by heating under basic conditions. (Section 18.3; Objective 6)

- When treated with a strong base, a ketone or aldehyde can react via an **aldol addition**, resulting in a **β-hydroxycarbonyl compound**. Under these basic conditions, an enolate anion acts as a nucleophile. (Sections 18.4 and 18.6; Objective 7)

- Heating an aldol reaction facilitates an **aldol condensation**, in which the β-hydroxycarbonyl product undergoes dehydration to yield an α,β-*unsaturated carbonyl compound*. (Section 18.5; Objective 7)

- Aldol reactions involving ketones tend to be unfavorable, and thus require exploiting Le Châtelier's principle to drive the reaction toward products. (Section 18.6; Objectives 7, 8)

- A **crossed aldol reaction** forms a carbon–carbon bond between two different carbonyl compounds. These reactions are not synthetically useful if they form a mixture of aldol products. (Section 18.7; Objective 8)

- A crossed aldol reaction can be synthetically useful if one of the carbonyl-containing reactants possesses no α hydrogens. Alternatively, a crossed aldol reaction can be synthetically useful if it involves a ketone that is first deprotonated quantitatively by a very strong base like LDA. (Section 18.7; Objective 8)

- An intramolecular aldol reaction is favored if it forms a five- or six-membered ring. Intramolecular reactions involving the attack of an aldehyde C=O are favored over ones involving the attack of a ketone C=O. (Section 18.8; Objective 9)

- A nitrile or nitroalkane that possesses an acidic α hydrogen can participate in an aldol-type reaction. (Section 18.9; Objectives 7, 8)

- The **Robinson annulation** produces a six-membered ring from two separate carbonyl compounds, one of which is an α,β-unsaturated carbonyl compound. This reaction consists of a conjugate addition to the α,β-unsaturated carbonyl compound, followed by an intramolecular aldol condensation. (Section 18.10; Objective 10)

Reaction Tables

TABLE 18-1 Functional Group Conversions

Starting Compound Class	Typical Reagents and Reaction Conditions	Compound Class Formed	Key Electron-Rich Species	Key Electron-Poor Species	Comments	Discussed in Section
(1) Ketone or aldehyde (R, R or H, or H)	H—Nu, Strong base; Nu = OR, SR, NR₂	HO, Nu on C (R, R or H, or H)	Nu:⊖	C=O, δ+	Base-catalyzed nucleophilic addition	18.1
(2) Ketone or aldehyde (R, R or H, or H)	H—Nu, Strong acid; Nu = OR, SR, NR₂	HO, Nu on C (R, R or H, or H)	H—Nu:, δ−	⊕OH, C	Acid-catalyzed nucleophilic addition	18.1
(3) α,β-Unsaturated ketone or aldehyde	H—Nu; Nu = OR, SR, NR₂	(conjugate addition product with Nu)	H—Nu:, δ−	HC, C=O, δ+	Conjugate nucleophilic addition	18.1b
(4) Ketone or aldehyde (R, R or H, or H)	R′OH (excess), H₂SO₄	R′O, OR′ on C (R, R or H, or H) **Acetal**	R′—ÖH, δ−	⊕OH, C	Nucleophilic addition, then S_N1	18.2a
(5) Ketone or aldehyde (R, R or H, or H)	Ammonia or 1° amine, NH₃ or R′NH₂ (excess), Mildly acidic	N—R′ (or H), C (R, R or H, or H) **Imine**	R′—ṄH₂, δ−	⊕OH, C	Nucleophilic addition, then E1	18.2b
(6) Ketone or aldehyde (R, or H)(C, R₂, H)	2° Amine, R′₂NH (excess), Mildly acidic	NR′₂ on C (R, or H)(CR₂) **Enamine**	R′₂ṄH, δ−	⊕OH, C	Nucleophilic addition, then E1	18.2b

TABLE 18-1 Functional Group Conversions (continued)

Starting Compound Class	Typical Reagents and Reaction Conditions	Compound Class Formed	Key Electron-Rich Species	Key Electron-Poor Species	Comments	Discussed in Section
(7) $\overset{RO \quad OR}{\underset{R \quad R}{C}}$ (or H) (or H) **Acetal**	$\xrightarrow{\underset{H_2SO_4}{H_2O \text{ (excess)}}}$	$\overset{O}{\underset{R \quad R}{C}}$ (or H) (or H) **Ketone or aldehyde**	$H_2\overset{\delta-}{\ddot{O}}$	$\overset{\oplus OR}{\underset{}{C}}$	S_N1, then E1	18.2a
(8) $\overset{N-R}{\underset{R \quad R}{C}}$ (or H) (or H) **Imine**	$\xrightarrow{\underset{HCl}{H_2O \text{ (excess)}}}$	$\overset{O}{\underset{R \quad R}{C}}$ (or H) (or H) **Ketone or aldehyde**	$H_2\overset{\delta-}{\ddot{O}}$	$\overset{\oplus}{HN}-R$ $\underset{}{C}$	Nucleophilic addition, then E1	18.2b
(9) $R-C{\equiv}N$ **Nitrile**	$\xrightarrow{\underset{H_2SO_4 \text{ or NaOH}}{H_2O \text{ (1 equiv)}}}$	$\overset{O}{\underset{R \quad NH_2}{C}}$ (or H) **Amide**	$H_2\overset{\delta-}{\ddot{O}}$ or $\overset{\ominus}{\ddot{O}H}$	$-C{\equiv}\overset{\oplus}{NH}$ or $-\overset{\delta+}{C}{\equiv}N$	Nucleophilic addition	18.2c
(10) $\overset{O}{\underset{R \quad R}{C}}$ (or H) (or H) **Ketone or aldehyde**	1. H_2N-NH_2, mildly acidic 2. KOH/diethylene glycol, Δ $\xrightarrow{\hspace{2cm}}$ or H_2N-NH_2 $\xrightarrow{KOH/H_2O, \text{ diethylene glycol}, \Delta}$	$\overset{H \quad H}{\underset{R \quad R}{C}}$ **Alkane**	$H_2\overset{\delta-}{N}NH_2$	$\overset{\oplus OH}{\underset{}{C}}$	Wolff–Kishner reduction: nucleophilic addition, then E1, then E2	18.3
(11) $\overset{O}{\underset{R \quad R}{C}}$ (or H) (or H) **Ketone or aldehyde**	(H or) (H or) $\overset{R \quad R}{\underset{H}{N}}$ $\xrightarrow{NaBH_4}$	(H or) (H or) $\overset{R \quad R}{\underset{\underset{(or H) \quad (or H)}{R \quad H \quad R}}{N}}$ **Amine**	(H or) (H or) $\overset{R \quad R}{\underset{H}{N}}$	$\overset{\delta+ O}{\underset{\underset{(or H) \quad (or H)}{R \quad R}}{\overset{}{C}}}{}^{\delta-}$	Reductive amination: imine formation, then reduction	18.12
(12) $\overset{O \quad OH}{\underset{\underset{R}{H \quad C \quad H}}{H \quad C \quad C \quad R}}$ **β-Hydroxycarbonyl compound**	$\xrightarrow{\underset{\Delta}{H_2SO_4 \text{ or NaOH}}}$	$\overset{O \quad H}{\underset{\underset{R}{H \quad C \quad C \quad R}}{}}$ **α,β-Unsaturated carbonyl compound**	—	—	E1 or E1cb	18.5

TABLE 18-2 Reactions That Alter the Carbon Skeleton

Starting Compound Class	Typical Reagents and Reaction Conditions	Compound Class Formed	Key Electron-Rich Species	Key Electron-Poor Species	Comments	Discussed in Section
(1) Ketone or aldehyde	H—CN, NaOH or KCN	Cyanohydrin	NC:⁻		Nucleophilic addition	18.1a
(2) Aldehyde	NaOH	β-Hydroxy aldehyde	Enolate anion		Aldol addition; nucleophilic addition	18.4
(3) Ketone	NaOH	β-Hydroxy ketone	Enolate anion		Aldol addition; nucleophilic addition	18.6
(4) Aldehyde	1. NaOH 2.	β-Hydroxy aldehyde	Enolate anion		Crossed aldol addition; nucleophilic addition	18.7
(5) Ketone	1. LDA 2.	β-Hydroxy ketone	Enolate anion		Crossed aldol addition; nucleophilic addition	18.7
(6) Nitrile or nitroalkane	NaOH		(or NO₂)		Aldol-type addition; nucleophilic addition	18.9
(7) Ketone	KOH		Enolate anion		Robinson annulation	18.10

Problems that are related to synthesis are denoted (SYN).

18.1 Weak Nucleophiles as Reagents: Acid and Base Catalysis

18.45 Draw the mechanism and product for each of the reactions shown.

(a)

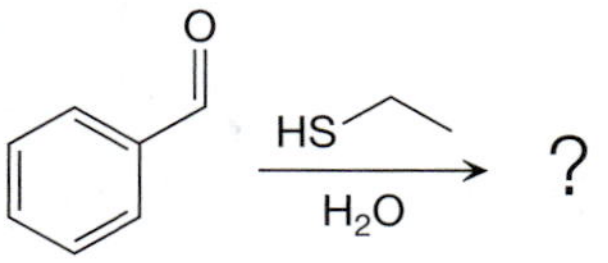

(b)

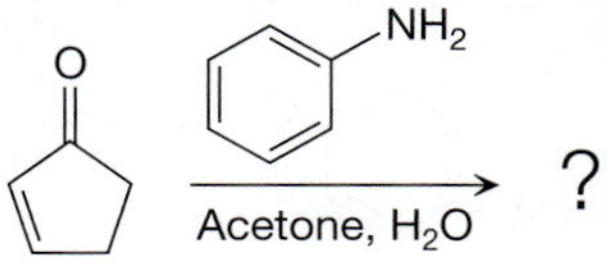

18.46 Draw the mechanism and product for each of the reactions in Problem 18.45 when they are catalyzed by **(a)** base and **(b)** acid.

18.47 Draw the mechanism and product for each of the reactions shown.

(a)

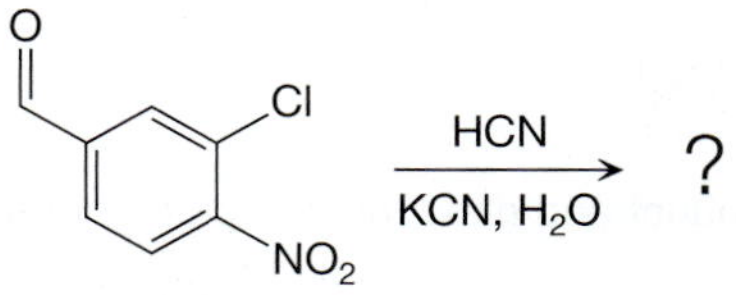

(b)

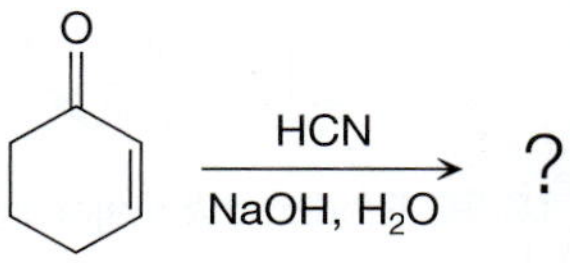

18.48 When acetone is dissolved in either a slightly basic or a slightly acidic solution of oxygen-18 labeled water, $H_2{}^{18}O$, oxygen-18 labeled acetone, $(CH_3)_2C={}^{18}O$, is produced. This is a form of an isotopic exchange reaction between acetone and water. Provide a mechanism to account for this reaction in **(a)** basic solution and **(b)** acidic solution. *Hint:* Is the addition of the nucleophile reversible or irreversible?

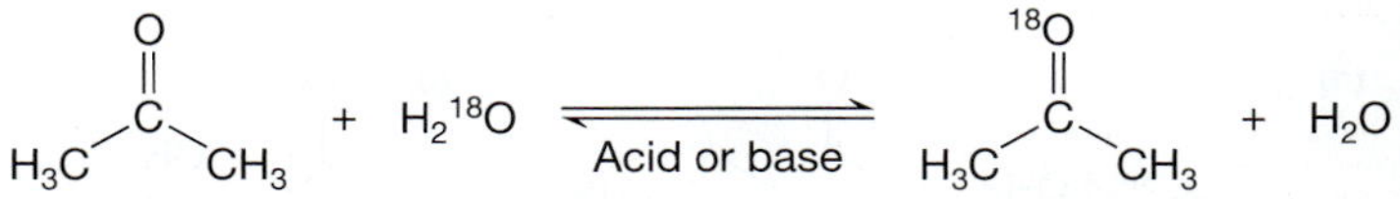

18.49 Explain why the reaction in Problem 18.48 proceeds dramatically more slowly under neutral conditions than under either acidic or basic conditions.

18.50 A carbamate can be prepared by treating an isocyanate with an alcohol, as shown here. This type of reaction is used to synthesize polyurethanes—polymers that have a wide variety of industrial applications, such as surface sealants, high-performance adhesives, and synthetic fibers. Propose a mechanism for this transformation.

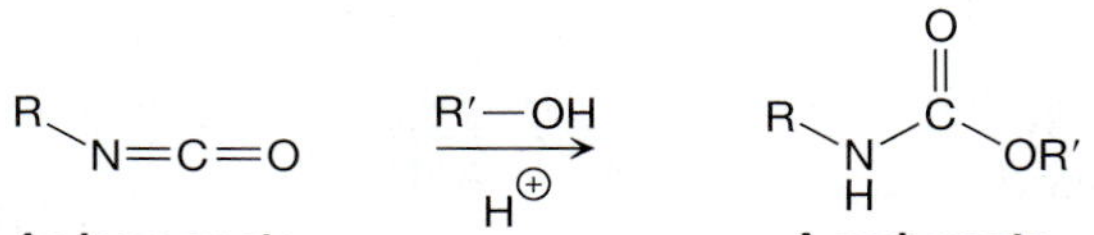

An isocyanate   **A carbamate**
(Substituted urethane)

18.51 An imino ester is formed when a nitrile is treated with an alcohol in the presence of *dry* HCl (i.e., without H_2O), followed by treatment with a weak base such as sodium bicarbonate. Propose a mechanism for this reaction.

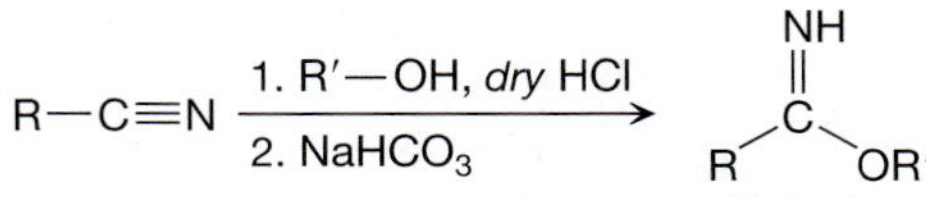

An imino ester

18.52 Draw the complete, detailed mechanism for the reaction shown here, which produces an amidine—a nitrogen analog to an ester.

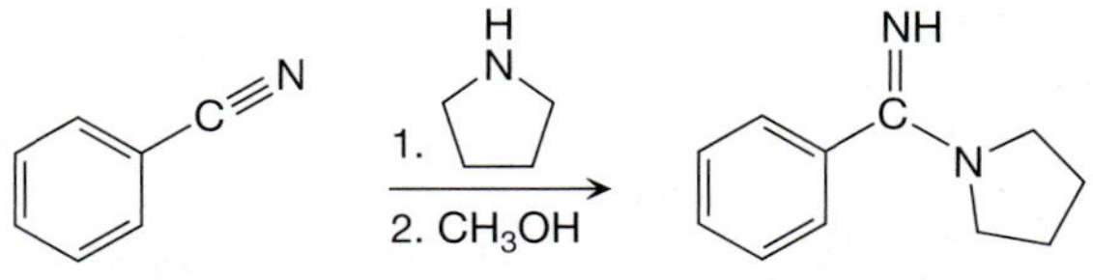

Benzonitrile   **An amidine**

18.53 (SYN) After consulting Problem 18.52, suggest how the amidine shown here can be synthesized from benzonitrile, $C_6H_5C{\equiv}N$. *Hint:* More than one synthetic step may be necessary.

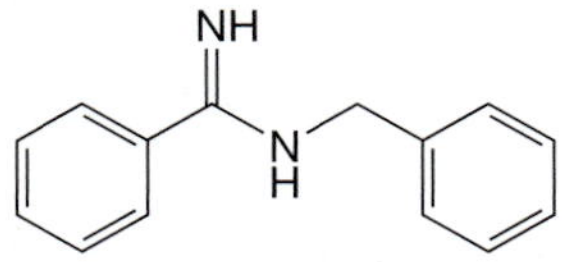

An amidine

18.54 Pentanedinitrile undergoes a cyclization reaction when treated with ammonia under weakly acidic conditions, as shown here. The product is an imidine. Propose a mechanism for this reaction.

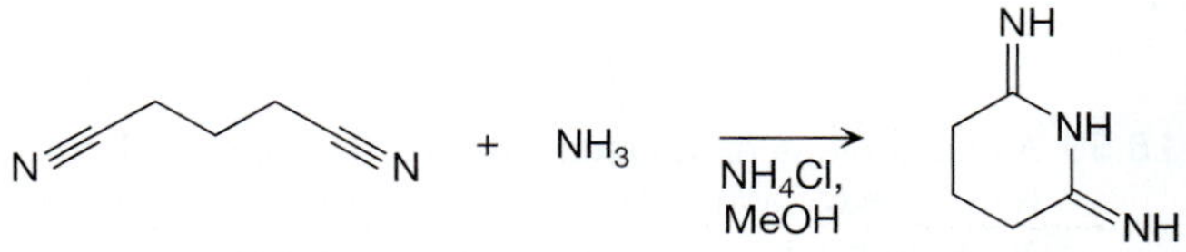

An imidine

18.2 and 18.3 Formation and Hydrolysis Reactions Involving Acetals, Imines, Enamines, and Nitriles; the Wolff–Kishner Reduction

18.55 Predict the major organic product and draw the complete, detailed mechanism for each of the following reactions.

(a)

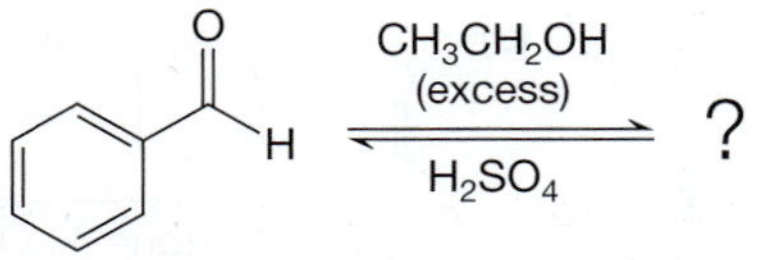

$$\xrightarrow[\text{H}_2\text{SO}_4]{\substack{\text{CH}_3\text{CH}_2\text{OH} \\ \text{(excess)}}} \quad ?$$

(b)

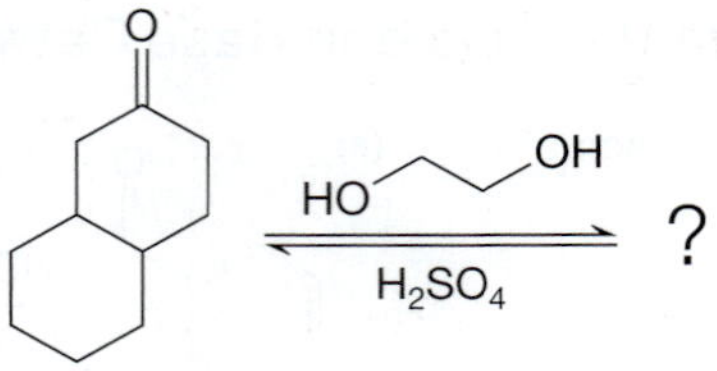

$$\xrightarrow[\text{H}_2\text{SO}_4]{\text{HO}\diagup\text{OH}} \quad ?$$

(c)

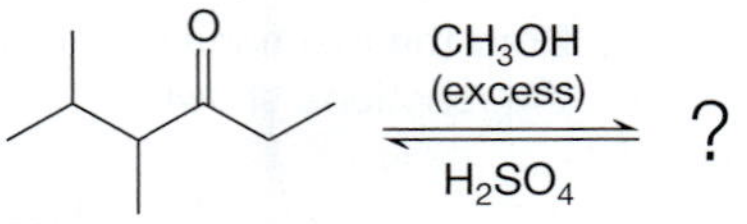

$$\xrightarrow[\text{H}_2\text{SO}_4]{\substack{\text{CH}_3\text{OH} \\ \text{(excess)}}} \quad ?$$

(d)

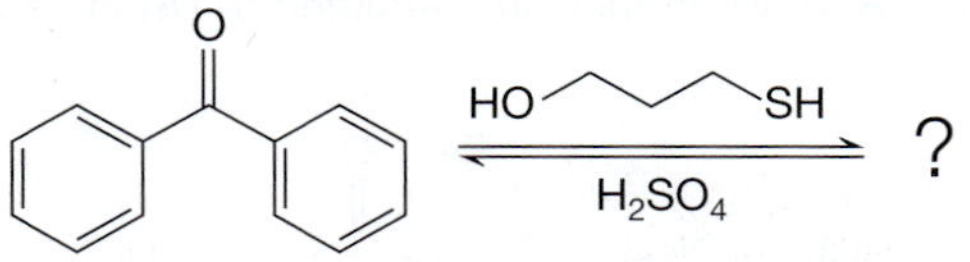

$$\xrightarrow[\text{H}_2\text{SO}_4]{\text{HO}\diagup\diagup\text{SH}} \quad ?$$

(e)

$$\xrightarrow[\text{H}_2\text{SO}_4]{\text{HO}\diagup\diagup\text{OH}} \quad ?$$

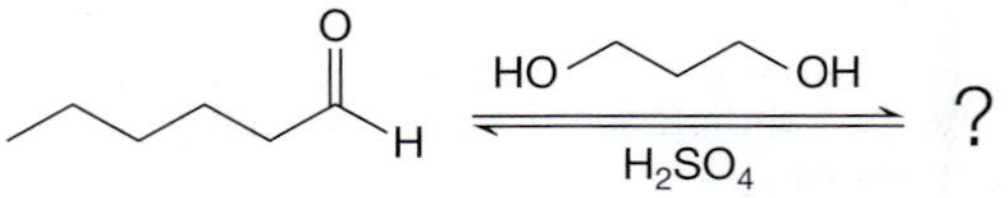

18.56 Predict the major organic product and draw the complete, detailed mechanism for each of the following reactions.

(a)

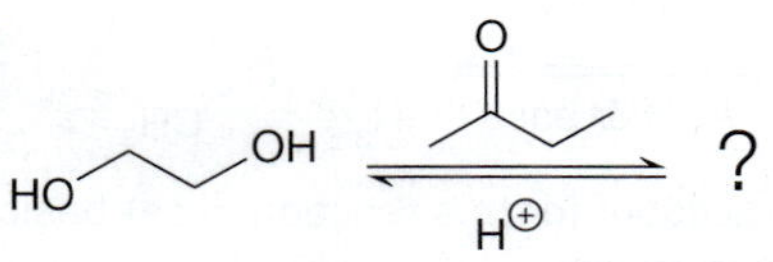

$$\text{HO}\diagup\text{OH} + \text{(ketone)} \xrightleftharpoons[\text{H}^\oplus]{} \quad ?$$

(b)

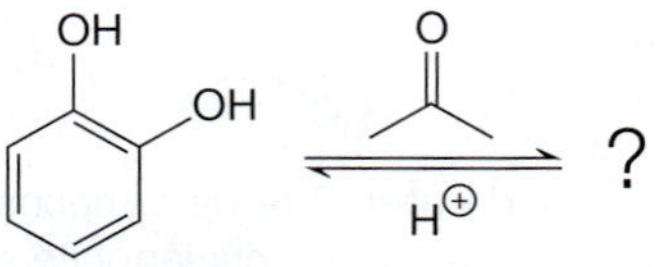

$$\text{(catechol, OH, OH)} + \text{(ketone)} \xrightleftharpoons[\text{H}^\oplus]{} \quad ?$$

(c)

$$\text{HO}\diagup\diagup\text{OH} + \text{(benzaldehyde)} \xrightleftharpoons[\text{H}^\oplus]{} \quad ?$$

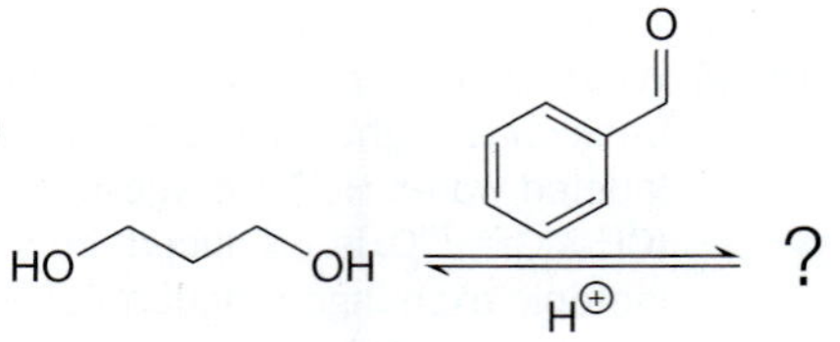

18.57 Predict the major organic product and draw the complete, detailed mechanism for each of the following reactions.

(a)

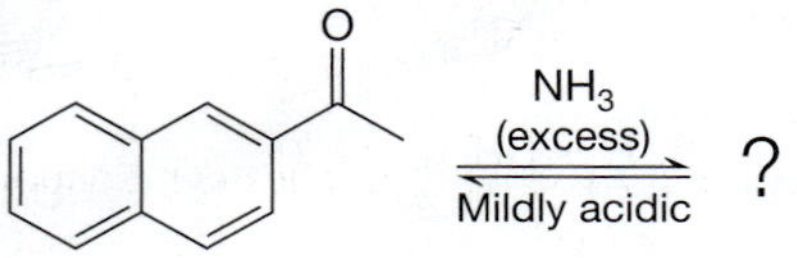

$$\xrightarrow[\text{Mildly acidic}]{\substack{\text{NH}_3 \\ \text{(excess)}}} \quad ?$$

(b)

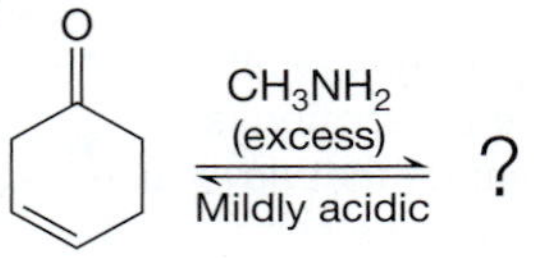

$$\xrightarrow[\text{Mildly acidic}]{\substack{\text{CH}_3\text{NH}_2 \\ \text{(excess)}}} \quad ?$$

(c)

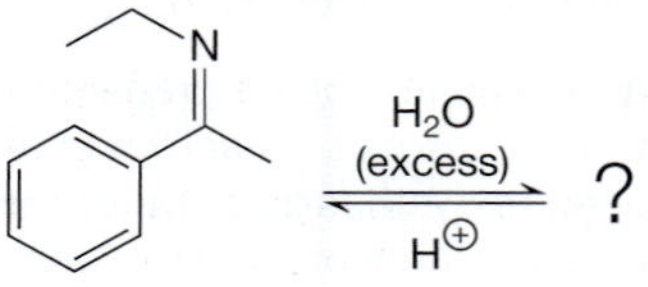

$$\xrightarrow[\text{H}^\oplus]{\substack{\text{H}_2\text{O} \\ \text{(excess)}}} \quad ?$$

(d)

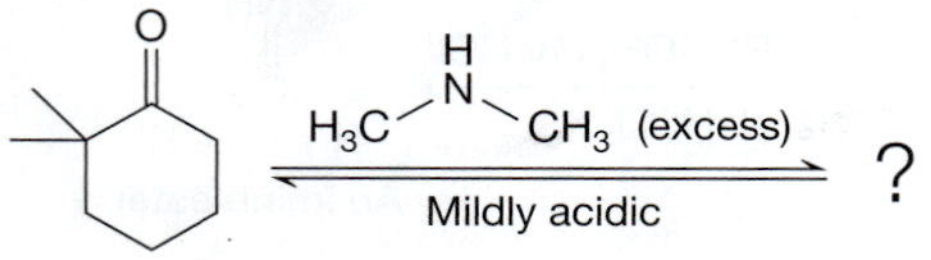

$$\xrightarrow[\text{Mildly acidic}]{\text{H}_3\text{C}-\overset{\text{H}}{\text{N}}-\text{CH}_3 \text{ (excess)}} \quad ?$$

(e)

$$\xrightarrow[\text{H}^\oplus]{\text{H}_2\text{O (excess)}} \quad ?$$

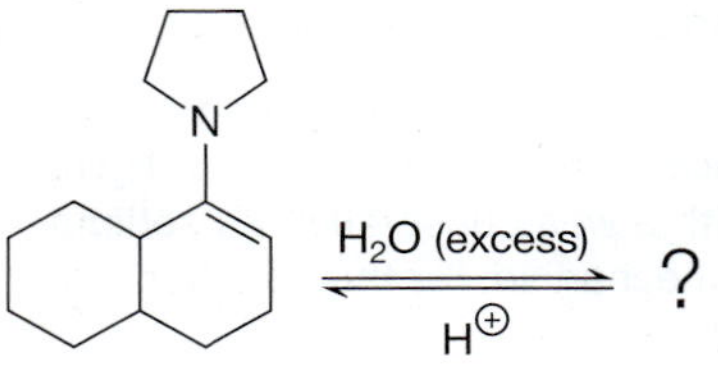

18.58 Predict the major organic product and draw the complete, detailed mechanism for each of the following reactions.

(a)

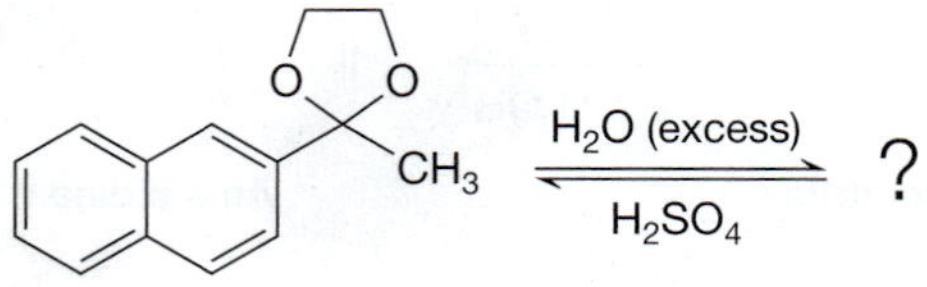

$$\xrightarrow[\text{H}_2\text{SO}_4]{\text{H}_2\text{O (excess)}} \quad ?$$

(b)

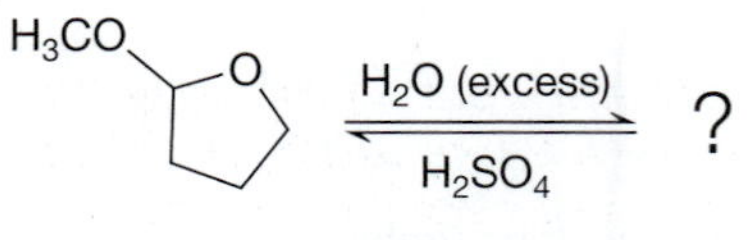

$$\xrightarrow[\text{H}_2\text{SO}_4]{\text{H}_2\text{O (excess)}} \quad ?$$

(c)

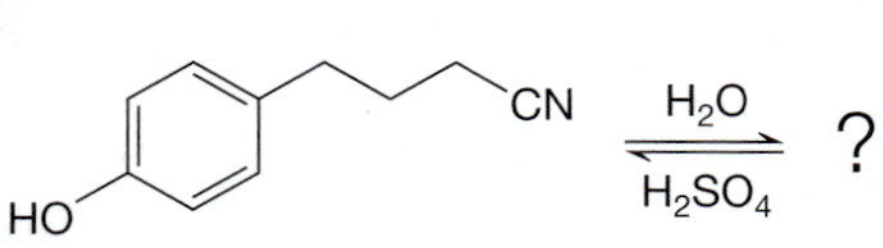

$$\xrightarrow[\text{H}_2\text{SO}_4]{\text{H}_2\text{O}} \quad ?$$

(d) CH₃CH₂O   OCH₂CH₃

$$\xrightarrow[\text{H}_2\text{SO}_4]{\text{H}_2\text{O (excess)}} \quad ?$$

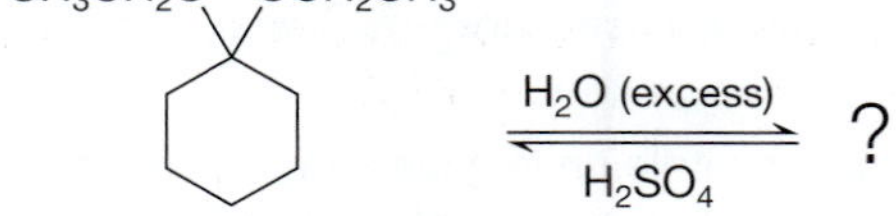

18.59 An isonitrile is an unusual species that has the form $R-^+N\equiv C^-$. When an isonitrile is treated with water, an *N*-alkylformamide is formed, as shown here. Propose a mechanism for this reaction.

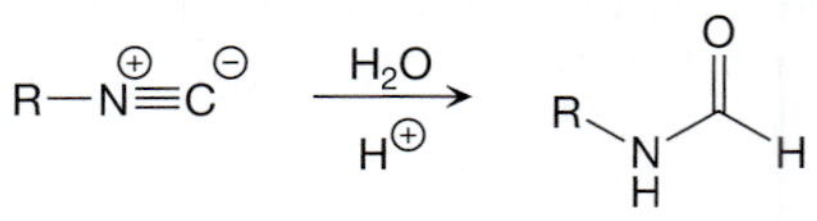

$$R-\overset{\oplus}{N}\equiv\overset{\ominus}{C} \xrightarrow[\text{H}^\oplus]{\text{H}_2\text{O}} R-\underset{\text{H}}{N}-\overset{\text{O}}{\overset{\|}{\text{C}}}-\text{H}$$

18.60 The formation of an acetal from a ketone or aldehyde has a very similar mechanism to the formation of an imine (review Equations 18-13 and 18-19). Both require acidic conditions to generate a water leaving group. At very acidic pH, however, the rate of imine formation slows down (see the accompanying graph), whereas the rate of acetal formation does not. Explain this pH dependence of imine formation.

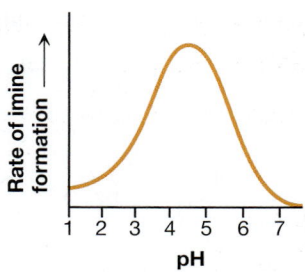

18.61 Hydroxylamine, H_2NOH, has both an OH functional group and an NH_2 functional group, so it can feasibly undergo reaction with a ketone or an aldehyde to produce either an acetal or an imine-like compound called an oxime. **(a)** Draw each of these mechanisms for the reaction of hydroxylamine with acetone. **(b)** Which is the major product? *Hint:* Which step decides the outcome?

18.62 Draw the complete, detailed mechanism for the reaction shown here and, using the mechanism, predict the major product. TsOH is a strong acid.

$$\text{HO} \diagup\diagdown \text{NH}_2$$
$$\xrightarrow{\text{TsOH, C}_6\text{H}_6, \Delta} \quad ?$$

18.63 Predict the major organic product and draw the complete, detailed mechanism for each of the following reactions.

(a)

$$\xrightarrow[\text{2. NaOH/H}_2\text{O, }\Delta]{\text{1. H}_2\text{NNH}_2\text{, mildly acidic}} \quad ?$$

(b)

$$\xrightarrow[\text{2. NaOH/H}_2\text{O, }\Delta]{\text{1. H}_2\text{NNH}_2\text{, mildly acidic}} \quad ?$$

18.64 (SYN) Draw three different ketone or aldehyde precursors that could be used to produce hexane via a Wolff–Kishner reduction.

18.4–18.10 Aldol and Aldol-Type Reactions

18.65 Draw the complete, detailed mechanism and predict the major organic product for each of the following reactions.

(a)

$$\xrightarrow[\Delta]{\text{NaOH}} \quad ?$$

(b)

$$\xrightarrow[\Delta]{\text{NaOH}} \quad ?$$

(c)

$$\xrightarrow{\text{NaOH}} \quad ?$$

(d)

$$\xrightarrow[\Delta]{\text{NaOH}} \quad ?$$

18.66 Draw the mechanism for the aldol addition reaction shown here, which is catalyzed by acid. *Hint:* Under acidic conditions, can an enolate anion act as a nucleophile?

$$\underset{\textbf{Propanal}}{} \quad \xrightleftharpoons{\text{HCl, MeOH}} \quad \underset{\textbf{3-Hydroxy-2-methylpentanal}}{}$$

18.67 Will compound **A** or compound **B** produce more aldol product at equilibrium? Why?

$$\underset{\textbf{A}}{\text{H}_3\text{C}\diagdown\text{CH}_3} \qquad \underset{\textbf{B}}{\text{FH}_2\text{C}\diagdown\text{CH}_2\text{F}}$$

18.68 Draw a complete, detailed mechanism and predict the major product for each of the following reactions.

(a)

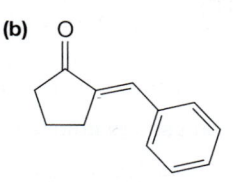

1. LDA

2. (benzaldehyde) → ?

(b)

1. NaOH

2. → ?

(c)

NaOH
Δ
→ ?

(d)

1. LDA

2. → ?

18.69 (SYN) Suggest how you would synthesize each of the following, using cyclopentanone as one of the reagents.

(a)   **(b)**   **(c)**   **(d)**   **(e)**

18.70 When benzaldehyde is treated with a catalytic amount of KCN, the benzoin condensation occurs. Draw a complete, detailed mechanism for this reaction. *Hint:* The reaction does not take place without the presence of NC⁻.

An α-hydroxy ketone

KCN
⇌
HO⁻/H₂O, EtOH

Benzaldehyde

2-Hydroxy-1,2-diphenylethanone (Benzoin)
88%

18.71 No reaction occurs when benzaldehyde and propenenitrile (acrylonitrile) are combined. In the presence of a catalytic amount of NaCN, however, the reaction shown here takes place. Draw a complete, detailed mechanism to account for these results. *Hint:* See Problem 18.70.

+ CN
NaCN
→
HO⁻/H₂O, EtOH

18.72 The crossed aldol reaction shown here can be carried out using a weak base such as pyridine. **(a)** Draw the complete, detailed mechanism for this reaction. **(b)** Explain why a relatively weak base can be used.

EtO   OEt
Diethyl malonate

H—C—H

→ EtO   OEt   OH

18.13 The Organic Chemistry of Biomolecules: Ring Opening and Closing of Monosaccharides; Mutarotation

18.73 Draw Haworth projections for each of the following molecules: (a) α-D-allopyranose, (b) β-D-allopyranose, (c) α-D-allofuranose, (d) β-D-allofuranose. (Consult Fig. 5-32 on p. 249 for the structure of each sugar in its acyclic form.)

18.74 Name each of the following cyclic sugars. (Consult Fig. 5-32 on p. 249 for the name of each sugar in its acyclic form.)

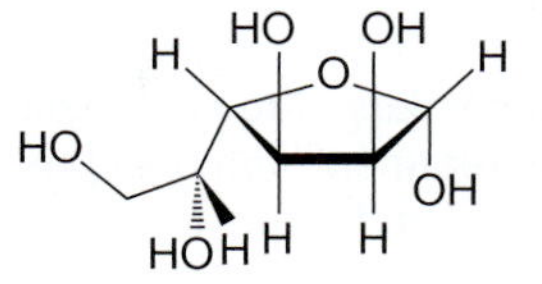

18.75 Identify the anomeric carbon in each molecule in Problems 18.73 and 18.74.

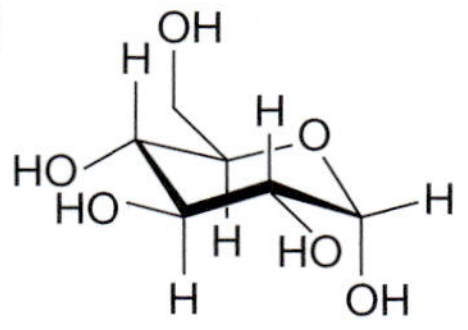

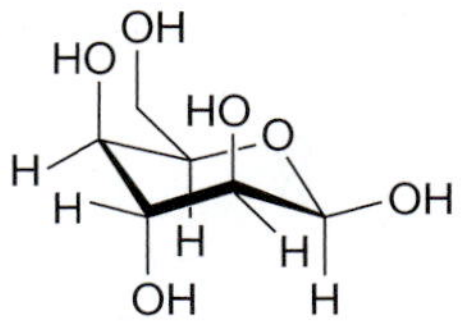

18.76 The acyclic form of D-talose is shown here as its Fischer projection. Identify the anomeric carbon.

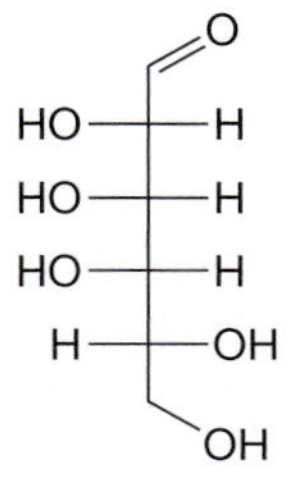

D-Talose

18.77 Draw the mechanism that shows how α-D-allopyranose converts to β-D-allopyranose under acidic conditions.

18.78 Draw the mechanism that shows how α-D-allopyranose converts to α-D-allofuranose under acidic conditions.

18.79 At equilibrium, D-galactose exists almost exclusively in its α and β pyranose forms. Aqueous solutions are freshly prepared for the α and β forms, both at the same concentration and temperature. The solution of the α form rotates plane-polarized light +150.7°, whereas the solution of the β form rotates the light +52.8°. Over time, both solutions have the same measured rotation of +80.2°. How much of the equilibrated solution does each form account for?

Integrated Problems

18.80 Predict the major product(s) of each of the following reactions.

(a)

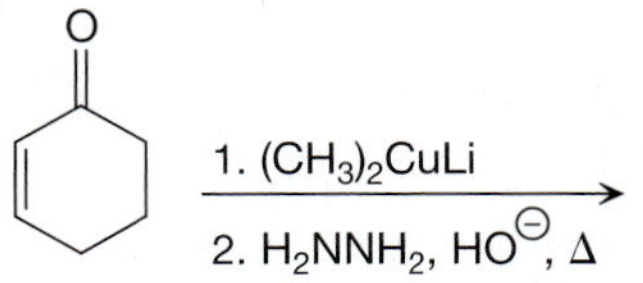

1. (CH$_3$)$_2$CuLi

2. H$_2$NNH$_2$, HO$^\ominus$, Δ

?

(b)

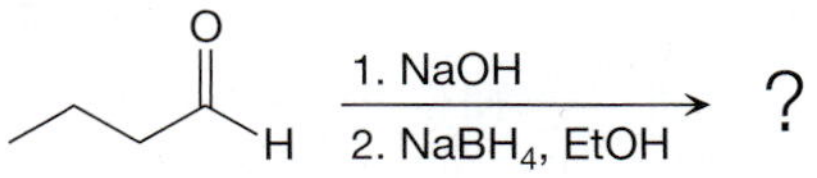

1. NaOH

2. NaBH$_4$, EtOH

?

18.81 An enamine, R$_2$C=C—NR$_2$, behaves as a nucleophile in much the same way that an enolate anion does. This is because an enamine has resonance structures similar to those observed for an enolate anion, as shown here. With this in mind, draw the complete mechanism for the following reaction, and provide the structure of the missing intermediate.

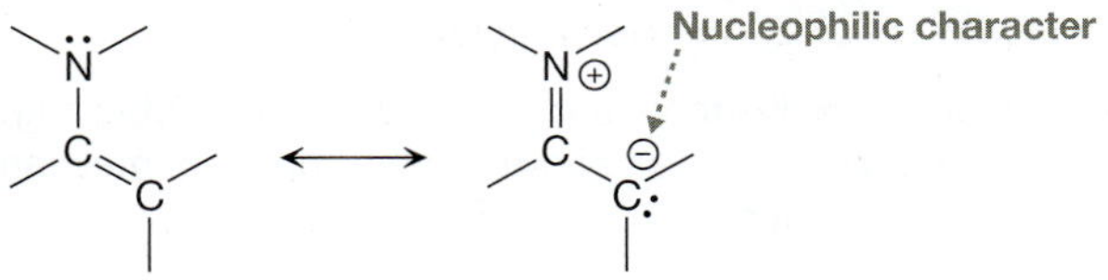

Nucleophilic character

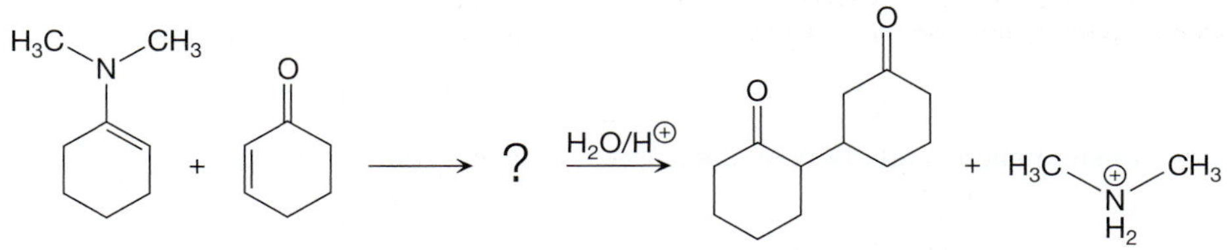

18.82 For each sequence of reactions, draw the complete, detailed mechanism and predict the major organic product. *Hint: See Problem 18.81.*

(a)

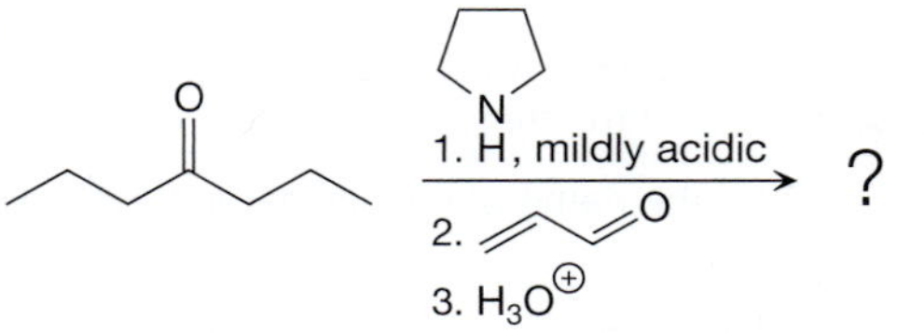

1. H, mildly acidic
2. [allyl vinyl ketone]
3. $H_3O^{\oplus}$

?

(b)

1. H, mildly acidic
2. [propyl bromide]
3. $H_3O^{\oplus}$

?

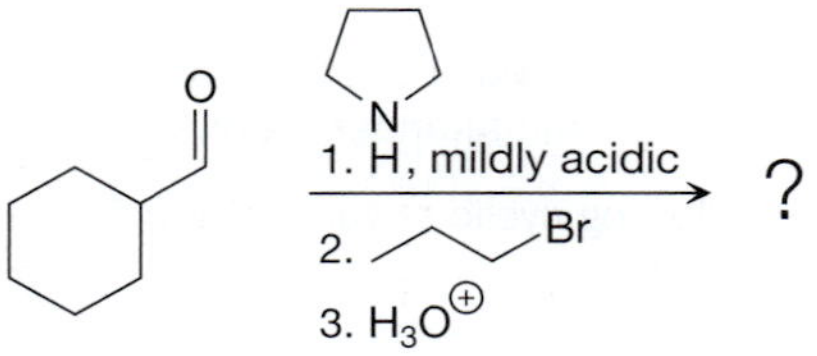

18.83 This reaction shows that a secondary nitro compound can be hydrolyzed to a ketone on treatment with aqueous sulfuric acid. Propose a mechanism for this reaction.

$\xrightarrow[\text{H}_2\text{SO}_4]{\text{H}_2\text{O}}$

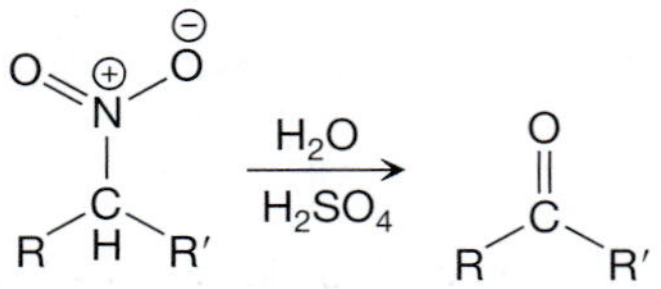

18.84 Draw a complete, detailed mechanism for the following reaction.

$$R-C\equiv N \xrightarrow[\text{2. H}_2\text{O, H}^{\oplus}]{\text{1. R'}-\text{OH, H}^{\oplus}} $$

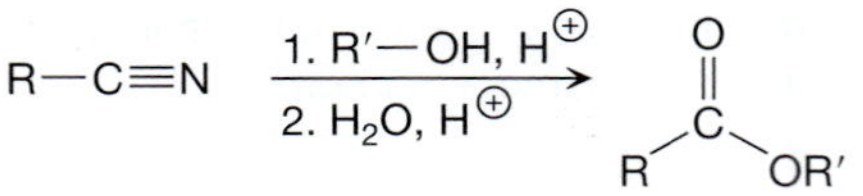

18.85 This reaction shows that an aldol-type reaction can be performed with a deprotonated imine as the nucleophile. On hydrolysis, the product is the normal β-hydroxycarbonyl compound. Notice that LDA is used as the base instead of HO^-.
(a) Explain why HO^- cannot be used as a base for this reaction.
(b) Provide a detailed mechanism for this reaction, including the structure of the intermediate species not shown.

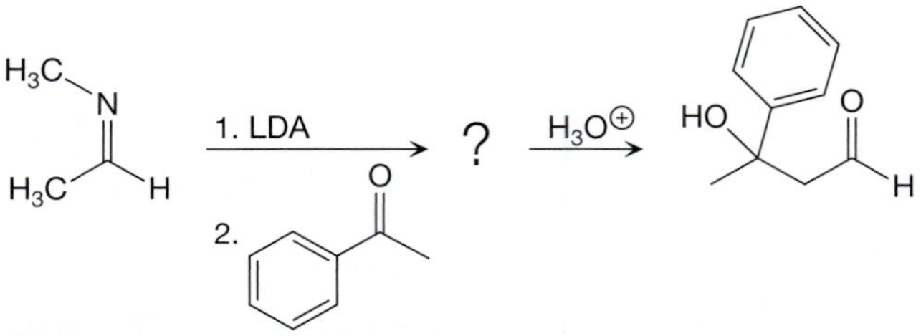

1. LDA
2. [acetophenone]

? $\xrightarrow{H_3O^{\oplus}}$

18.86 Treatment of a nitrile with base, followed by acid hydrolysis, yields a β-keto nitrile, as shown here. Provide a detailed mechanism for this reaction.

$\xrightarrow[\text{2. H}_3\text{O}^{\oplus}]{\text{1. NaOEt}}$

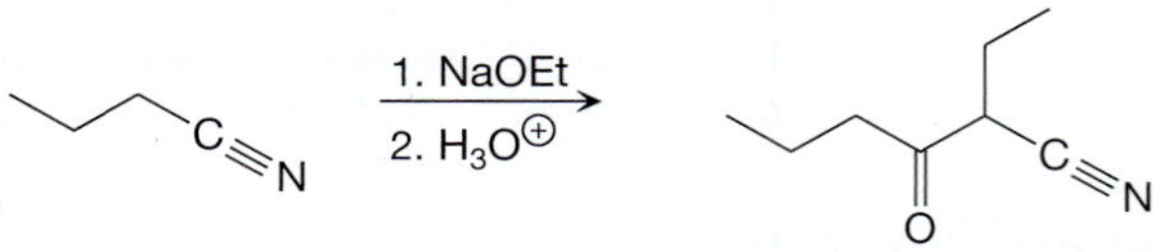

18.87 The following reaction is believed to proceed through the intermediate shown. Draw the complete, detailed mechanism that leads to the formation of that intermediate.

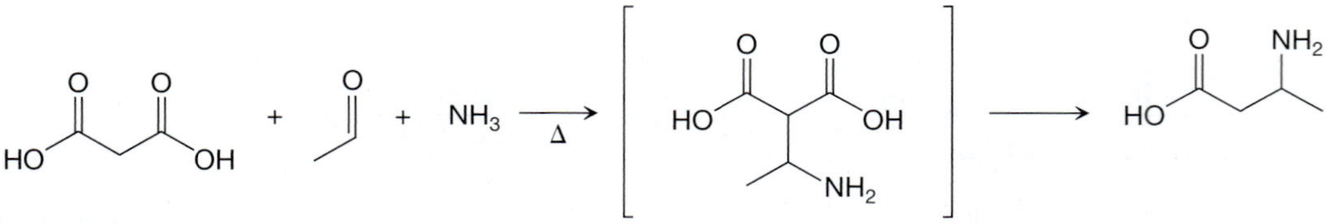

18.88 When a 1,5-diketone is treated with acid, a 2,6-dialkyl pyran is produced, as shown here. Propose a mechanism for this reaction.

$\xrightarrow{H^{\oplus}}$

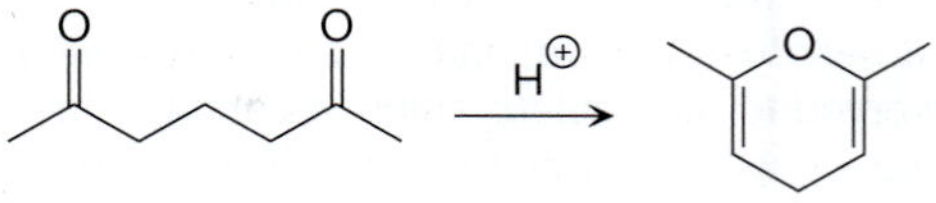

18.89 When an α,β-unsaturated carbonyl is treated with H_2O_2 under basic conditions, the C=C double bond is epoxidized, as shown here. Propose a mechanism for this reaction.

$\xrightarrow[\text{HO}^{\ominus}]{H_2O_2}$

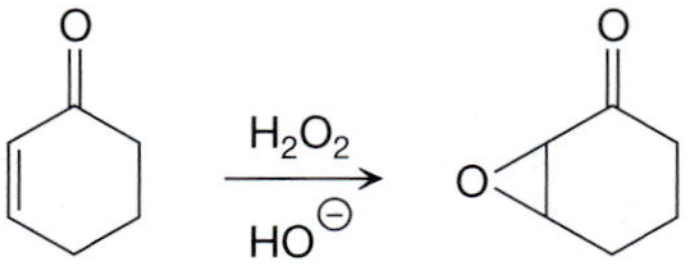

18.90 The following transformation is an example of a Darzens reaction. Draw its complete, detailed mechanism.

$\xrightarrow[\text{HOC(CH}_3)_3]{\text{KOC(CH}_3)_3}$

18.91 The following is an example of a Wittig–Horner reaction. Draw its complete, detailed mechanism. *Hint:* A key intermediate is provided.

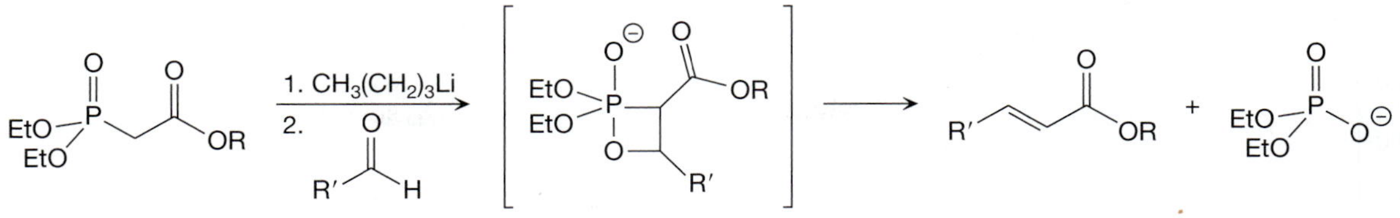

18.92 (SYN) Provide the reagent(s) missing from each of the following reactions.

(a)

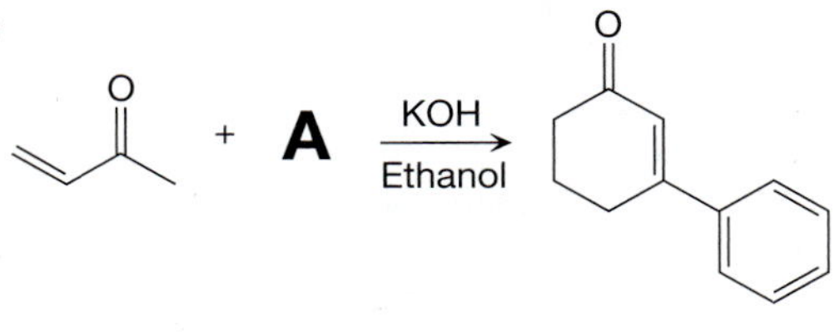

(b)

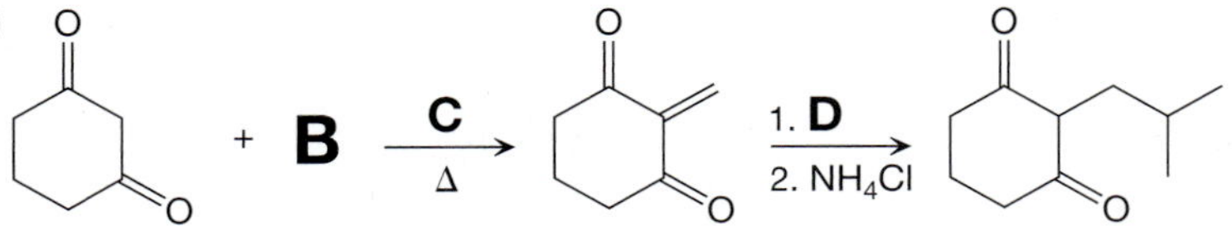

(c)

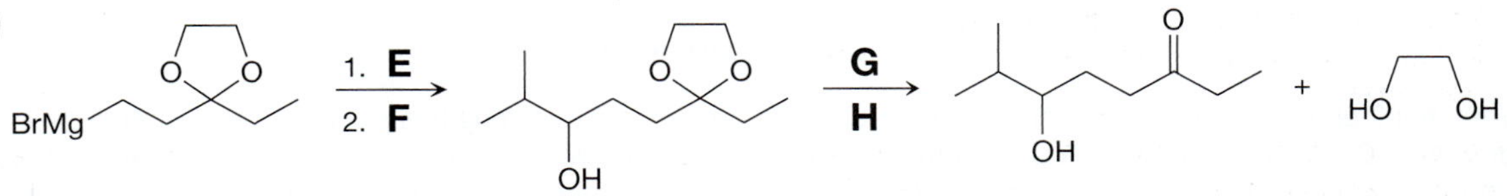

(d)

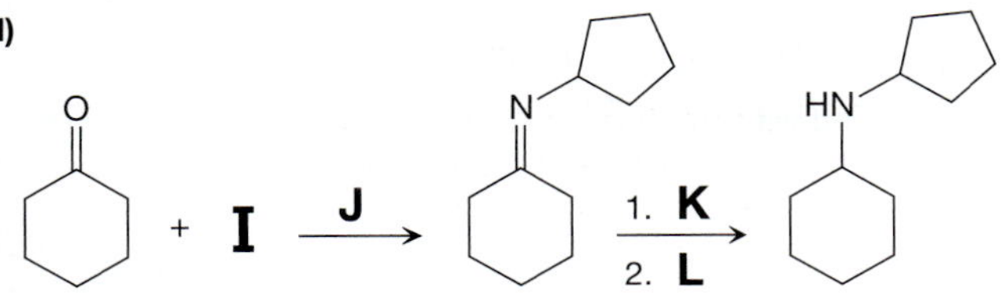

18.93 (SYN) How would you synthesize the compound shown here if, as your starting material, you may use any organic reagents that contain exactly two carbon atoms and any inorganic reagents?

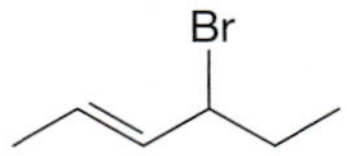

18.94 (SYN) How would you synthesize the compound shown here using phenylethanal as your only source of carbon atoms? You may ignore stereochemistry.

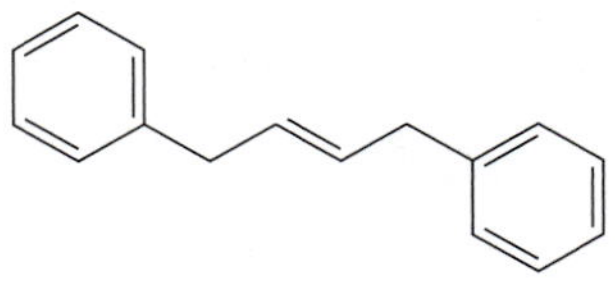

18.95 (SYN) How would you synthesize 2-methylhexane from hex-4-en-3-one and any other reagents necessary?

18.96 (SYN) Using butanal as your only source of carbons, show how to synthesize *N*-butylbutan-1-amine.

18.97 (SYN) Using only compounds containing three or fewer carbons, show how to synthesize 2,3-dimethylpentan-1-amine.

18.98 A compound with formula $C_{10}H_{14}O$ has the IR, ^1H NMR, and ^{13}C NMR spectra shown here. The DEPT spectra reveal that the seven carbon signals farthest upfield are produced from CH_2 carbons. $C_{10}H_{14}O$ can be synthesized using an aldol condensation reaction, simply by heating a compound C_5H_8O in the presence of NaOH. **(a)** Provide the structures of C_5H_8O and $C_{10}H_{14}O$. **(b)** Provide the mechanism for the reaction that is described.

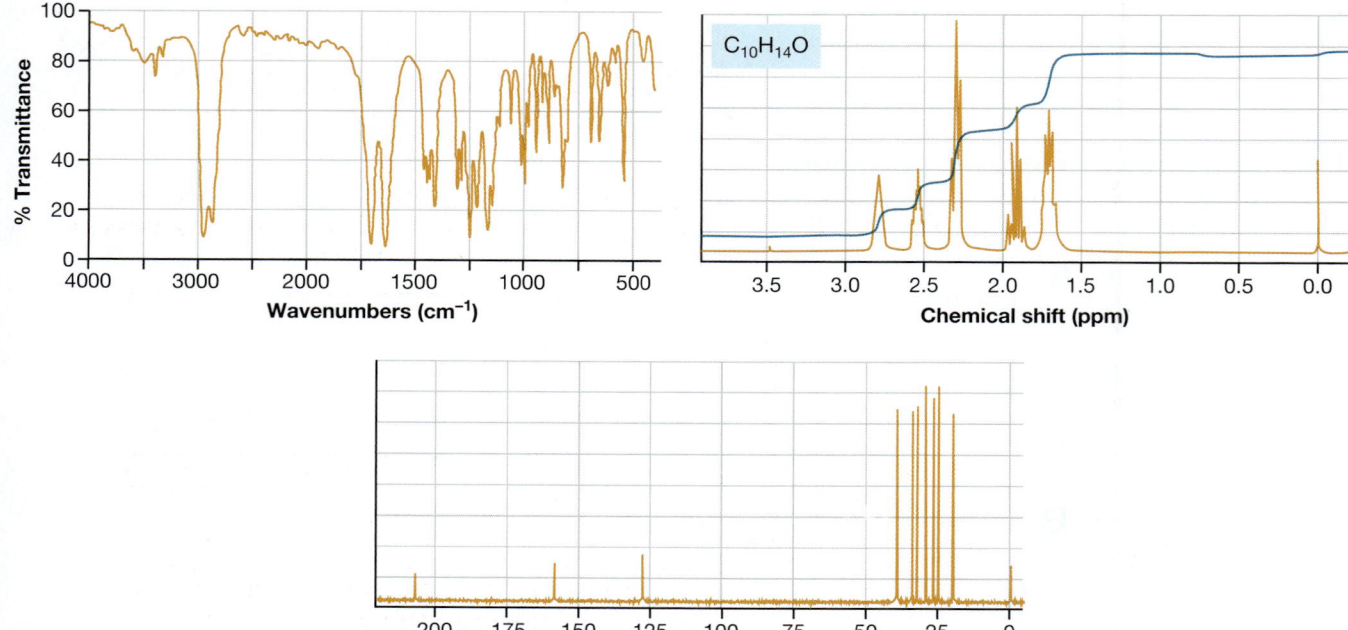

18.99 When but-2-enal is treated with 3-phenylpropenal in the presence of a strong base, a compound is formed whose formula is $C_{13}H_{12}O$. In its ^1H NMR spectrum, one signal has a chemical shift around 10 ppm, several overlapping signals have chemical shifts between 7 and 9 ppm, and several other overlapping signals have chemical shifts between 5 and 6 ppm. Integration of those sets of signals gives a $1:5:6$ ratio.
(a) What is the structure of the product?
(b) Draw a complete, detailed mechanism that accounts for the formation of the product.

18.100 When treated with acid, hexane-2,5-dione forms a compound with the formula C_6H_8O. The ^1H NMR spectrum of the product is shown below. A key intermediate is shown. Identify the structure of the product and propose a mechanism for this reaction.

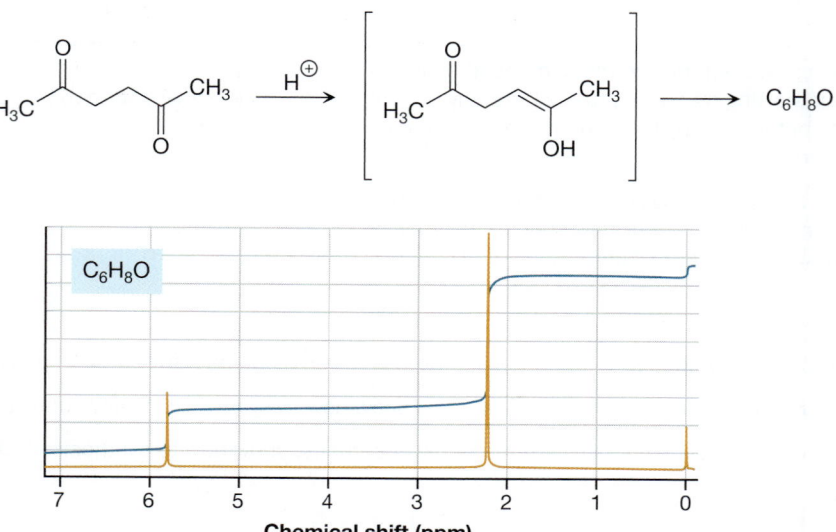

18.101 When acetonitrile is treated with concentrated sulfuric acid and *tert*-butanol, followed by water, a product is formed whose ^{1}H NMR spectrum exhibits the following three signals: singlet, 1.3 ppm, 9 H; singlet, 2.0 ppm, 3 H; and broad singlet, 8.2 ppm, 1 H. Its IR spectrum exhibits one broad absorption of medium intensity between 3300 and 3500 cm^{-1}, and a narrow, intense absorption near 1650 cm^{-1}. A key intermediate is shown. Draw the structure of the product, and draw the complete, detailed mechanism for the reaction.

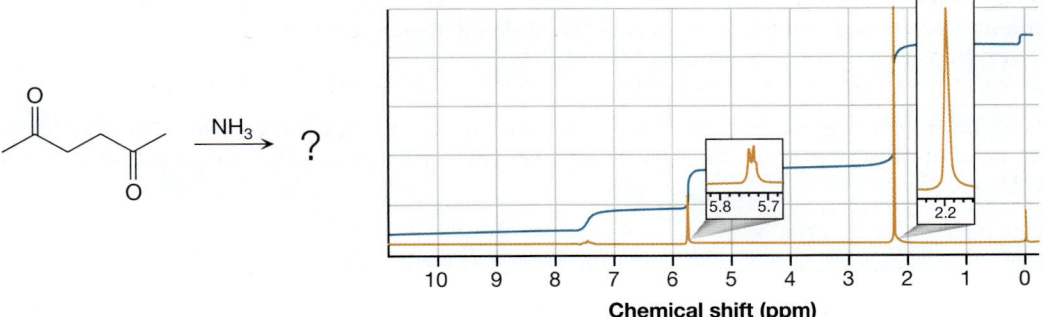

18.102 In mildly acidic conditions, 3-(2-aminophenyl)propenal reacts to form a compound whose molecular weight is 129 g/mol. The ^{1}H NMR spectrum for the product is shown below. Its ^{13}C NMR spectrum contains nine peaks, all between 120 and 150 ppm. Propose a mechanism for this reaction and draw the product.

18.103 Reaction of 2,5-hexanedione with ammonia produces a compound whose ^{1}H NMR spectrum is shown below. Its ^{13}C NMR spectrum exhibits three signals. Draw the complete, detailed mechanism leading to the formation of that product.

Many foods contain synthetic fats derived from naturally occurring oils using a process called catalytic hydrogenation, an important reduction reaction we examine here in Chapter 19.

19

Organic Synthesis 2

Intermediate Topics in Synthesis Design, and Useful Redox and Carbon–Carbon Bond-Forming Reactions

In Chapter 13, we were introduced to the basics of organic synthesis, focusing mainly on how to construct a molecule with the appropriate carbon skeleton and how to convert one functional group into another. Now that we have gained experience with even more reactions, we continue with organic synthesis, focusing on some higher-level topics that enable us to synthesize more-elaborate target molecules.

Much of our focus here is geared toward the formation of new carbon–carbon bonds. Carbon–carbon bonds typically form between carbon atoms with opposite charges, but we are often faced with the challenge of forming these bonds between carbon atoms that are initially of like charge. Therefore, we begin with a strategy to accomplish this. Next, we focus on clues in a target molecule that help us determine the types of carbon–carbon bond-forming reactions that should be considered in a synthesis. We also describe one way to tackle a synthesis in which these kinds of clues do not exist.

Another focus of Chapter 19 is on ways to circumvent *synthetic traps*. (Recall from Section 13.4 that a synthetic trap arises when the conditions necessary for a desired reaction at one site within a molecule will cause an undesired reaction at another site.) Specifically, we discuss the use of selective reactions that target one functional group

On completing Chapter 19 you should be able to:

1. Identify reactions that reverse the polarity at a carbon atom, and incorporate these kinds of reactions in organic syntheses.

2. Analyze the relative positioning of heteroatoms in a target molecule to determine which carbon–carbon bond-forming reactions should be considered in a synthesis.

3. Devise a synthesis that utilizes reactions in which specific functional groups, such as C=O and C=C groups, are removed entirely from a molecule.

4. Identify situations in which a selective reaction is called for, and incorporate these kinds of reactions in a synthesis.

5. Explain the basic principles behind the use of a protecting group, and demonstrate the protection and deprotection of ketones, aldehydes, and alcohols in syntheses.

6. Describe the basic mechanism of catalytic hydrogenation, and predict the products of the catalytic hydrogenation of alkenes, alkynes, aldehydes, and ketones.

7. Predict the products of oxidation reactions involving Cr^{+6} and Mn^{+7} oxidizing agents, and incorporate these reactions effectively in syntheses.

8. Recognize the conditions for coupling and alkene metathesis reactions, and incorporate these reactions effectively in syntheses.

over another. We also discuss strategies for keeping functional groups intact by using protecting groups.

Finally, we introduce some new redox and carbon–carbon bond-forming reactions. These reactions are not only useful in carrying out specific structural changes to a molecule, but they also allow us to apply some of the synthesis strategies just mentioned.

19.1 Umpolung in Organic Synthesis: Forming Bonds between Carbon Atoms Initially Bearing Like Charge; Making Organometallic Reagents

In Section 7.1, we learned that a bond can form between two atoms when one atom bears a partial or full negative charge (and thus is electron rich, or nucleophilic) and the other bears a partial or full positive charge (making it electron poor, or electrophilic). This general idea is applicable to a wide variety of bond-forming reactions, including those that form carbon–carbon bonds. Consider, for example, the reaction in Equation 19-1.

Electron-poor C (electrophilic) Electron-rich C (nucleophilic) A C—C bond tends to form between an electron-rich and an electron-poor C.

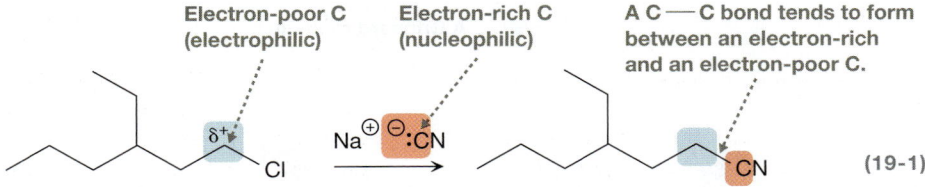

$$(19\text{-}1)$$

The bond that is formed involves the carbon atom from NC^- and the carbon atom attached to Cl in the alkyl halide. The NC^- carbon bears a full negative charge, whereas the alkyl halide carbon bears a partial positive charge.

Because carbon–carbon bond-forming reactions usually involve oppositely charged carbon atoms, a problem arises when we want to form a bond between two carbon atoms bearing like charges. For example, Equation 19-2 shows that no bond readily forms between the carbonyl C atom of a ketone and the C atom attached to Br in an alkyl bromide. This is because both C atoms are attached to atoms with relatively high electronegativities (O and Br, respectively), so both bear partial positive charges.

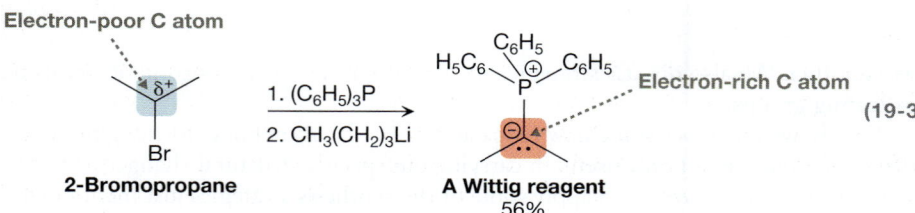

Both C atoms are relatively electron poor.

+   Br   ✗   No reaction   (19-2)

One way to form a bond between two carbons of like charge is to carry out a separate reaction that first reverses the charge (or *polarity*) at one of the carbons. Thus, one carbon atom would become electron rich while the other would remain electron poor. This general idea of reversing a charge at a particular atom, called **umpolung** (a German term meaning "polarity reversal"), is common practice in organic synthesis. In fact, we have already encountered reactions that use umpolung in the formation of a carbon–carbon bond. The synthesis of a Wittig reagent (Section 17.8), shown in Equation 19-3, is one example.

Electron-poor C atom

δ^+

Br

2-Bromopropane

1. $(C_6H_5)_3P$
2. $CH_3(CH_2)_3Li$

H_5C_6 C_6H_5 C_6H_5 P

Electron-rich C atom   (19-3)

A Wittig reagent
56%

In the alkyl halide reactant, the C atom bonded to the halogen atom bears a partial positive charge and is relatively electron poor. By contrast, that C atom has become electron rich in the Wittig reagent that is produced.

Another class of reactions that accomplishes this charge reversal is the formation of organometallic reagents from their corresponding alkyl halides. Equation 19-4 shows, for example, that an alkyl bromide such as bromobenzene can be converted into a *Grignard reagent* (RMgBr) simply by treating it with solid magnesium in an ether solvent such as tetrahydrofuran (THF). Similarly, Equation 19-5 shows that an *alkyllithium reagent* (RLi) can be synthesized from an alkyl bromide by treating it with solid lithium in ether.

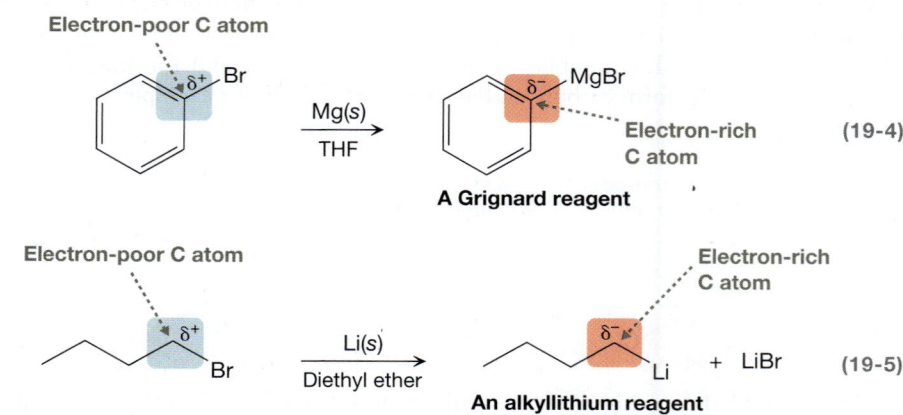

Electron-poor C atom

δ^+ Br

$\xrightarrow[\text{THF}]{\text{Mg(s)}}$

δ^- MgBr

Electron-rich C atom   (19-4)

A Grignard reagent

Electron-poor C atom

δ^+

Br

$\xrightarrow[\text{Diethyl ether}]{\text{Li(s)}}$

δ^-

Li   +   LiBr   (19-5)

Electron-rich C atom

An alkyllithium reagent

In both of these cases, the partial positive charge of the alkyl halide carbon becomes a partial negative charge in the organometallic reagent. Thus, umpolung has taken place, which facilitates the formation of a bond with an electron-poor carbon, such as in an epoxide ring-opening reaction (Section 10.7) or a Grignard reaction (Section 17.5).

Alkyl bromides are not the only alkyl halides that can be used to synthesize a Grignard reagent or alkyllithium reagent. Those organometallic reagents can be synthesized from the corresponding alkyl chloride or alkyl iodide, too, using similar procedures.

For each reaction, write either δ^+ or δ^- next to the C atom bonded to the halogen in the reactant and the C atom bonded to metal atom in the product.

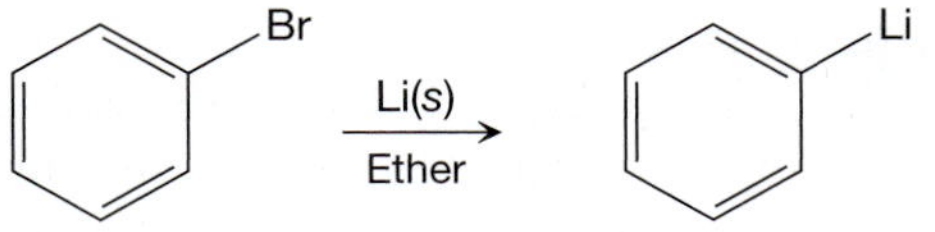

(a)

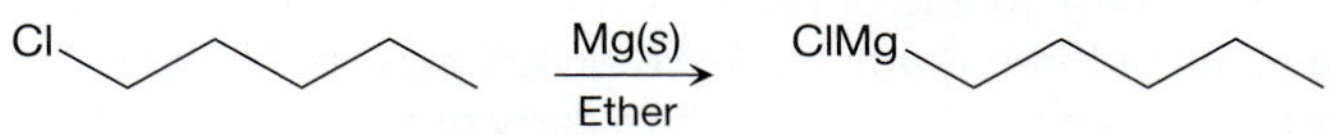

(b)

Answers to Your Turns are in the back of the book.

A related reaction is the synthesis of *lithium dialkylcuprates* (R_2CuLi, also called Gilman reagents), reagents that were first introduced in Section 7.1b. As shown in Equation 19-6, a lithium dialkylcuprate is synthesized from the corresponding alkyl-lithium reagent by treating it with copper(I) iodide, CuI. Because the metal-bonded C atom bears a partial negative charge in both the alkyllithium reagent and the lithium dialkylcuprate, umpolung technically does *not* occur in this reaction, but synthesizing a lithium dialkylcuprate from an alkyl halide does represent polarity reversal at the C atom. Once again, this facilitates the formation of a carbon–carbon bond to an electron-poor carbon, as we have seen previously in conjugate addition to an α,β-unsaturated carbonyl compound (Section 17.10).

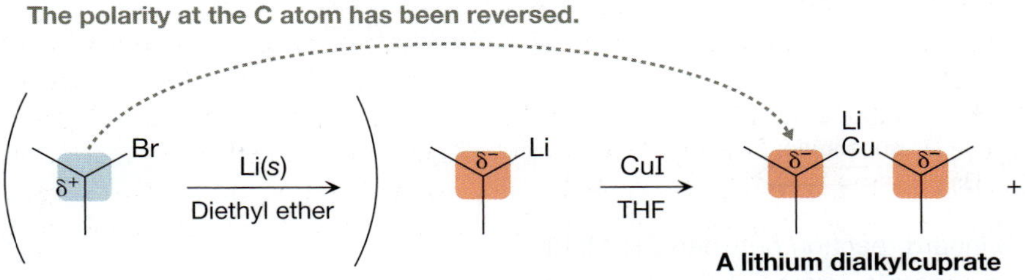

The polarity at the C atom has been reversed.

A lithium dialkylcuprate

$+$ LiI (19-6)

PROBLEM 19.1 Predict the product of each of the following reactions.

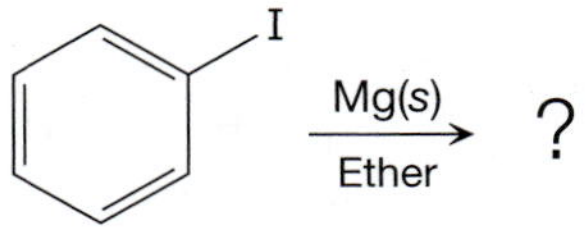

(a)

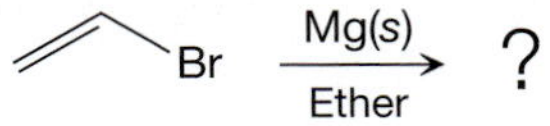

(b)

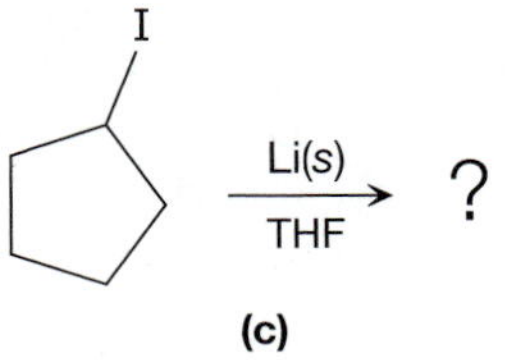

(c)

(d)

PROBLEM 19.2 Show how to synthesize each of the following organometallic compounds from an alkyl halide.

(a)

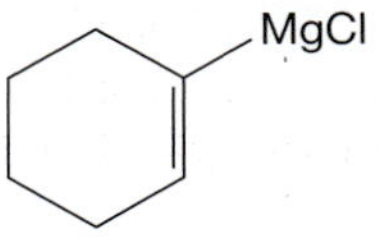

(b)

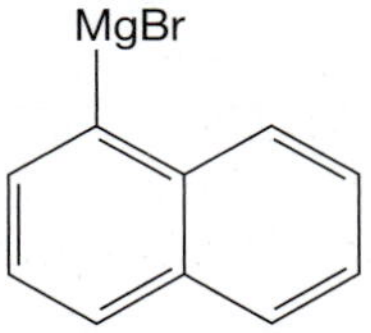

(c)

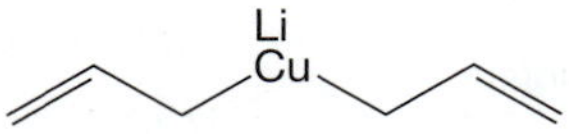

(d)

SOLVED PROBLEM 19.3

Using formaldehyde ($H_2C=O$) as your only source of carbon, show how you would synthesize propan-1-ol.

Think Do C—C bonds have to be formed in the synthesis? If so, what should the electron-rich species be? What should the electron-poor species be? Can the synthesis take place in a single reaction? If not, what precursor should we choose?

Methanal (Formaldehyde) → ? → **Propan-1-ol**

Solve Since our only source of carbon contains a single C atom and the product contains a chain of three C atoms, C—C bonds must be formed. A C—C bond-forming reaction would be difficult to carry out directly between two molecules of formaldehyde, however, because such a reaction would entail the formation of a bond between two carbon atoms bearing a partial positive charge. We must consider, therefore, a precursor that has a carbon atom bearing a significant negative charge. A Grignard reagent would suffice, especially since the target is an alcohol, the product of a Grignard reaction.

An alcohol is the product of a Grignard reaction.

Undo a Grignard reaction.

New target

In our retrosynthetic analysis, undo a Grignard reaction as the first transform, disconnecting the C—C bond involving the C that bears the OH group. This shows that the final product can be made from an ethyl Grignard reagent such as ethylmagnesium bromide (CH_3CH_2MgBr) and formaldehyde. Formaldehyde is an available starting material, but the ethylmagnesium bromide is not. We must, therefore, devise a synthesis for it.

Ethylmagnesium bromide can be generated from CH_3CH_2Br. Once again, CH_3CH_2Br is not an available starting material, but it can be made from ethanol using PBr_3.

New target

MgBr → Undo a Grignard formation. → Br → Undo a substitution. → OH

Ethanol is now the possible product of a Grignard reaction between CH_3MgBr and our starting material, $H_2C=O$. CH_3MgBr is generated from CH_3Br, which can be made from CH_3OH using PBr_3. Finally, CH_3OH can be made by reducing our starting material using a hydride reagent such as $NaBH_4$.

OH → Undo a Grignard reaction. → H_3C—MgBr + formaldehyde → Undo a Grignard formation. → H_3C—Br → Undo a substitution. → H_3C—OH → Undo a reduction. → formaldehyde

The synthesis is reported as follows:

formaldehyde → $NaBH_4$, EtOH → H_3C—OH → PBr_3 → H_3C—Br → Mg(s), Ether → H_3C—MgBr → 1. H—CHO 2. NH_4Cl, H_2O → OH → PBr_3

→ Br → Mg(s), Ether → MgBr → 1. H—CHO 2. NH_4Cl, H_2O → OH

PROBLEM 19.4 Using acetaldehyde ($CH_3CH=O$) as your only source of carbon, show how to synthesize 3-methylpentan-3-ol.

**Ethanal
(Acetaldehyde)**   **3-Methylpentan-3-ol**

PROBLEM 19.5 Show how you would synthesize 4-phenylbutan-1-ol from bromobenzene, using oxirane, , as your only other source of carbon.

Bromobenzene   **4-Phenylbutan-1-ol**

CONNECTIONS
3-Methylpentan-3-ol is a precursor to emylcamate, a drug that was once used to treat anxiety and tension.

Emylcamate

19.2 Relative Positioning of Heteroatoms in Carbon–Carbon Bond-Forming Reactions

In Chapter 13, we learned that carbon–carbon bond-forming reactions are important in organic synthesis because they allow us to alter the carbon framework of a particular molecule. In many of those reactions, two functional groups in the product are left with very specific relative locations along the carbon skeleton. That relative positioning can be critical to a synthesis because, as we have seen many times before, those functional groups govern the sites at which subsequent reactions can take place. Moreover, the relative positioning of heteroatoms within those functional groups can give us a clue as to which C—C bond-forming reactions we might use in the synthesis.

Consider, for example, the formation of a *cyanohydrin*, shown in Equation 19-7.

These heteroatoms are attached to adjacent carbons, so they have a 1,2-positioning.

(19-7)

A cyanohydrin

A C—C bond is formed between the cyanide C and the carbonyl C. Notice, in particular, that the two heteroatoms in the product—N and O atoms—are attached to C atoms that are *adjacent* to each other. In other words, the heteroatoms have a **1,2-positioning** relative to each other along the carbon backbone.

Knowing that this reaction produces a compound with 1,2-positioning of the resulting heteroatoms is particularly useful when our target molecule is *not* a cyanohydrin but still has heteroatoms with 1,2-positioning. Suppose, for example, that we want to carry out the following synthesis starting from the ketone given.

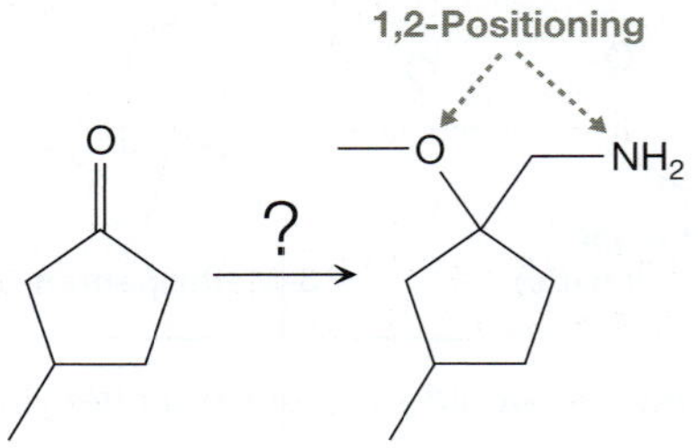

1,2-Positioning

The O and N atoms in the target have a 1,2-positioning along the carbon backbone, so we could consider a cyanohydrin as a synthetic intermediate, as shown in the following retrosynthetic analysis.

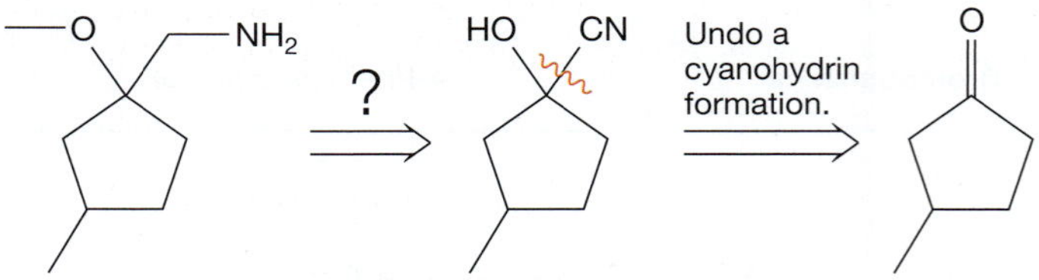

A cyanohydrin

What has yet to be solved is the conversion of the cyanohydrin to the target molecule. This entails converting the O—H to the ether, and converting the C≡N to a primary amine. As shown in the following reaction scheme, the O—H can be converted to the ether using a Williamson synthesis (Section 10.6), and the C≡N can be converted to the primary amine by reduction with LiAlH₄ (Section 17.3).

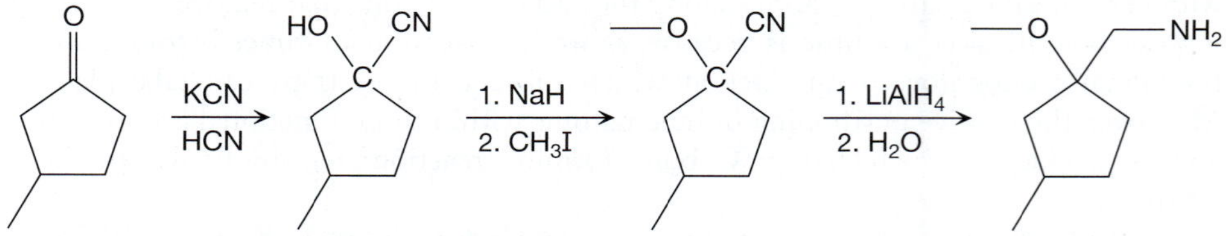

A cyanohydrin

SOLVED PROBLEM 19.6

Show how to synthesize the target shown here, beginning with propanal, $CH_3CH_2CH=O$.

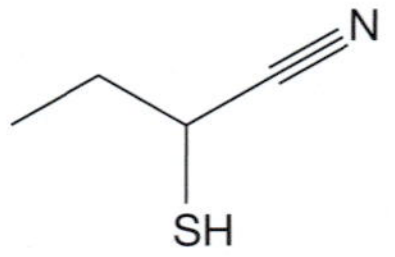

Think Does the target have a 1,2-positioning of the heteroatoms, which could result from the formation of a carbon–carbon bond? If so, from what cyanohydrin could this target be synthesized? How could that cyanohydrin be synthesized from propanal?

Solve The target does indeed have a 1,2-positioning of the heteroatoms involving the CN and SH functional groups, so we can conceive of obtaining it from a cyanohydrin, as shown in the following retrosynthetic analysis.

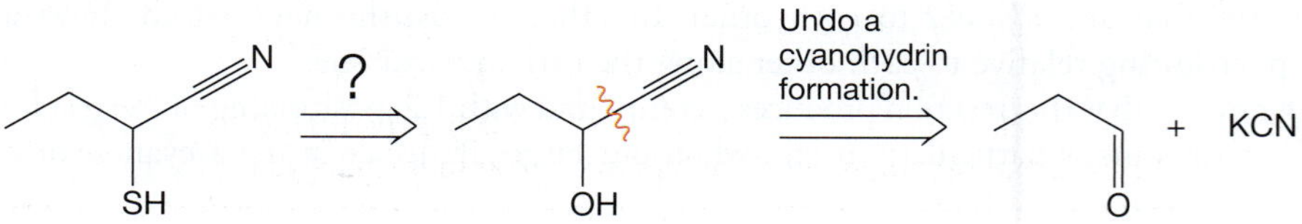

To convert the OH group to an SH group, the cyanohydrin could first be treated with PBr_3, followed by NaSH, as shown in the following synthesis.

PROBLEM 19.7 Show how to synthesize this target, using compounds with seven or fewer carbon atoms.

Cyanohydrin formation is not the only carbon–carbon bond-forming reaction that results in heteroatoms having a specific relative positioning along the carbon backbone. Other such reactions, which we have encountered previously, are listed in Table 19-1.

TABLE 19-1 Relative Positioning of Heteroatoms in Reactions That Form Carbon–Carbon Bonds

Relative Positioning	Product Formed	Electron-Rich Reactant	Electron-Poor Reactant	Discussed in Section
1,2-	Cyanohydrin	Cyanide anion	Ketone/ aldehyde	18.1a
1,3-	β-Hydroxy ketone/ aldehyde	Enolate ion	Ketone/ aldehyde	18.4
1,4-	β-Cyano ketone/ aldehyde	Cyanide anion	α,β-Unsaturated ketone/aldehyde	18.1b
1,5-	1,5-Dicarbonyl compound	Enolate ion	α,β-Unsaturated ketone/aldehyde	18.10

For each entry in Table 19-1, number the carbon atoms in the chains of the respective products so that the numbers assigned to the carbon atoms attached to the heteroatoms agree with the relative positioning listed. *Hint:* The "1" and "2" carbons are not necessarily the atoms that are highlighted.

The products in Table 19-1 have heteroatoms with 1,2-, 1,3-, 1,4-, and 1,5-positioning. Therefore:

- If the heteroatoms in a target have 1,2-, 1,3-, 1,4-, or 1,5- relative positioning and the synthesis calls for a carbon–carbon bond-forming reaction, you should consider using a corresponding reaction from Table 19-1.
- If the functional groups in the target don't match the ones produced from the carbon–carbon bond-forming reaction, then consider implementing functional group transformations after the carbon–carbon bond has been formed.

An example is shown in Solved Problem 19.8.

SOLVED PROBLEM 19.8

Show how you can synthesize 2-methylpentane-1,3-diol from compounds containing five or fewer carbons.

Think Will a carbon–carbon bond-forming reaction be necessary? What is the relative positioning of the heteroatoms in the target? Does the appropriate carbon–carbon bond-forming reaction in Table 19-1 leave us with the correct functional groups, or will we need to carry out an additional functional group conversion?

Solve We will have to use a carbon–carbon bond-forming reaction, because the target's carbon skeleton contains six carbons bonded together and we are allowed to start with compounds containing only five. The 1,3-positioning of the two hydroxyl groups in the product suggests that we should consider using an aldol reaction, which yields a β-hydroxycarbonyl compound. In a retrosynthetic analysis, therefore, our task becomes to apply a transform that takes our target molecule back to a β-hydroxycarbonyl compound. We can do this by undoing a hydride reduction on either of the two OH groups. Below, we show this transform applied to the OH group on C3. From there, we disconnect the appropriate C—C bond to take us back to the aldol reactants, pentan-3-one and formaldehyde.

In the forward direction, the synthesis would appear as follows:

PROBLEM 19.9 Show how you can synthesize 2-methylpentane-1,3-diol from compounds containing three or fewer carbons.

19.3 Reactions That Remove a Functional Group Entirely from a Molecule: Reductions of C=O to CH₂

When we design a synthesis, we typically begin by looking for structural features in the target that might indicate the need to use certain reactions. In Section 19.2, for example, we saw that if a target calls for a carbon–carbon bond-forming reaction and exhibits a specific relative positioning of heteroatoms, we should consider using a reaction from Table 19-1. Frequently, however, a synthesis will call for a carbon–carbon bond-forming reaction, but the functional groups in the target don't provide a clear indication of what that reaction should be.

For cases like this, we can sometimes use reactions that remove a functional group entirely. Before it is removed, however, that functional group could be exploited to form a new C—C bond in a synthesis. Thus, in a retrosynthetic analysis, we might presume that such a functional group *was* present in a precursor.

In this context, we discuss the use of three reactions that convert the C=O group of a ketone or aldehyde into a CH₂ group. In doing so, the functional group is removed entirely. Section 19.3a discusses the Wolff–Kishner reduction, Section 19.3b discusses the Clemmensen reduction, and Section 19.3c discusses the Raney-nickel reduction.

19.3a The Wolff–Kishner Reduction

The *Wolff–Kishner reduction*, first discussed in Section 18.3, converts the carbonyl (C=O) group of a ketone or aldehyde into a CH₂ group. An example is shown in Equation 19-8, which converts 1-phenylpropan-1-one into propylbenzene.

Wolff–Kishner reduction

The carbonyl group has been reduced to a CH₂ group.

$$\text{1-Phenylpropan-1-one} \xrightarrow[\text{H}_2\text{O/triethylene glycol} \atop \Delta]{\text{H}_2\text{N—NH}_2,\ \text{KOH}} \text{Propylbenzene} + \text{N}_2(g) + \text{H}_2\text{O} \quad (19\text{-}8)$$

Propylbenzene
82%

Thus, a Wolff–Kishner reaction removes the C=O functional group of a ketone or aldehyde.

Prior to its removal, however, the carbonyl group can facilitate the formation of a new carbon–carbon bond. As shown in Equation 19-9, for example, alkylation can take place at the α carbon via an S_N2 reaction (Section 10.3).

C—C bond formation at the α carbon

C=O functional group removed

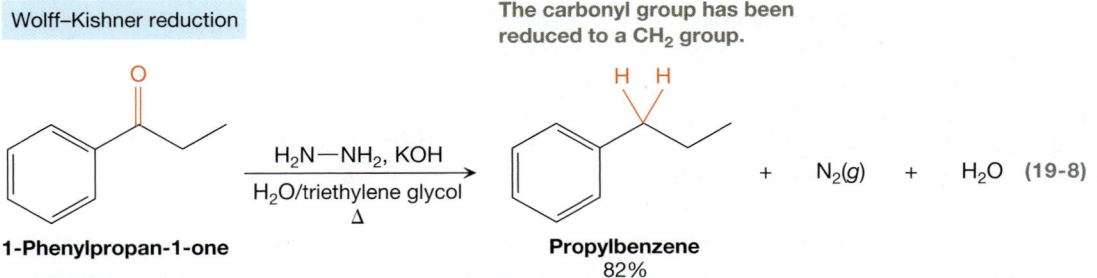

$$\xrightarrow[\text{(CH}_3)_3\text{COK}]{\text{R—Br}} \qquad \xrightarrow[\Delta]{\text{H}_2\text{N—NH}_2,\ \text{KOH}} \qquad (19\text{-}9)$$

Alternatively, Equation 19-10 shows that conjugate addition to an α,β-unsaturated ketone or aldehyde results in the formation of a new C—C bond at the β carbon (Section 17.10).

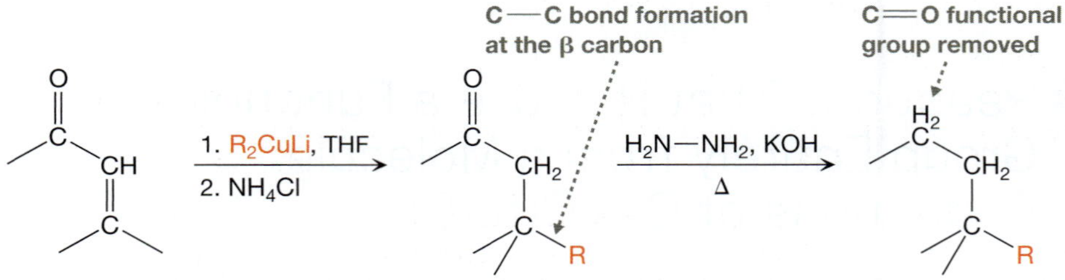

With this in mind, suppose that we want to synthesize propylbenzene from starting materials containing eight or fewer carbon atoms:

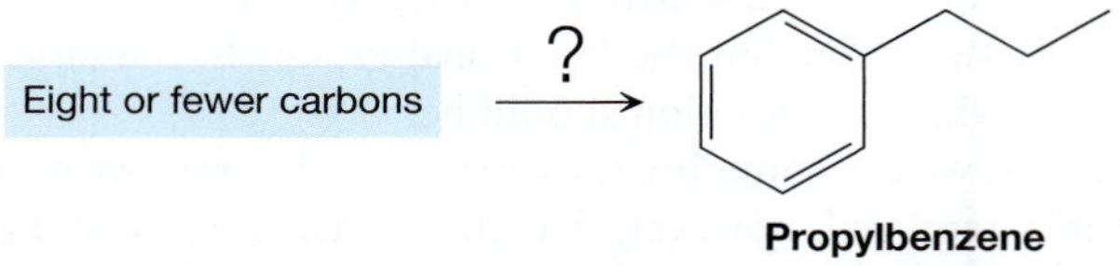

Given that the target contains nine C atoms, we must use a carbon–carbon bond-forming reaction. That reaction could be an α alkylation if we imagine one of the CH_2 groups being produced from a carbonyl group in a precursor, as shown in the following retrosynthetic analysis.

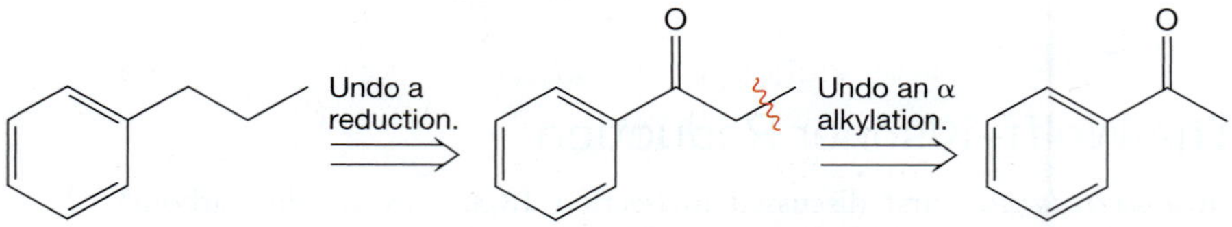

In the forward direction, the synthesis would appear as follows:

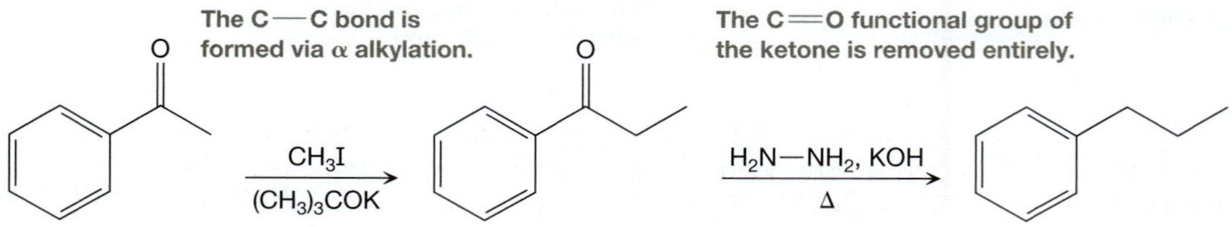

PROBLEM 19.10 Show how to synthesize propylbenzene from starting materials containing seven or fewer carbon atoms.

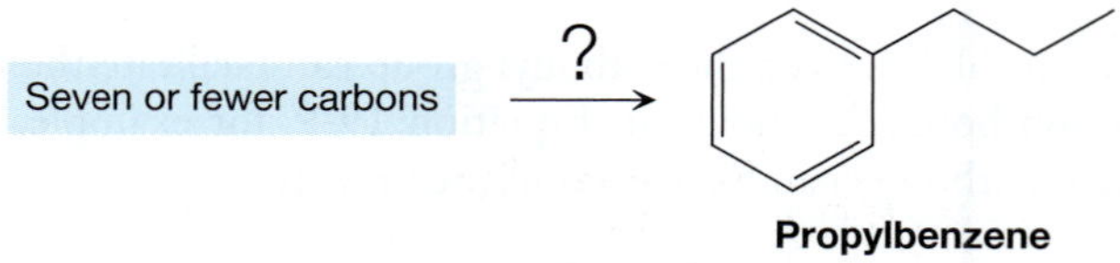

As shown in Solved Problem 19.11, a similar strategy can be applied when the carbon–carbon bond-forming reaction is conjugate addition involving a lithium dialkylcuprate.

SOLVED PROBLEM 19.11

Show how to synthesize this compound, beginning with an α,β-unsaturated ketone or aldehyde.

α,β-Unsaturated ketone or aldehyde $\xrightarrow{?}$

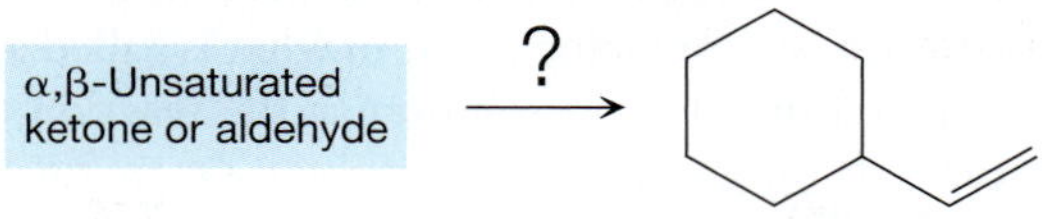

Think What carbon–carbon bond could you disconnect in a transform? If that bond is the result of conjugate addition to an α,β-unsaturated ketone or aldehyde, which CH_2 group could we imagine coming from a C=O group?

Solve The C—C single bond that is not part of the ring could have come from conjugate addition to an α,β-unsaturated ketone, as shown below. This kind of bond formation takes place at the β carbon of an α,β-unsaturated carbonyl compound, so the topmost CH_2 group on the ring could come from a C=O group.

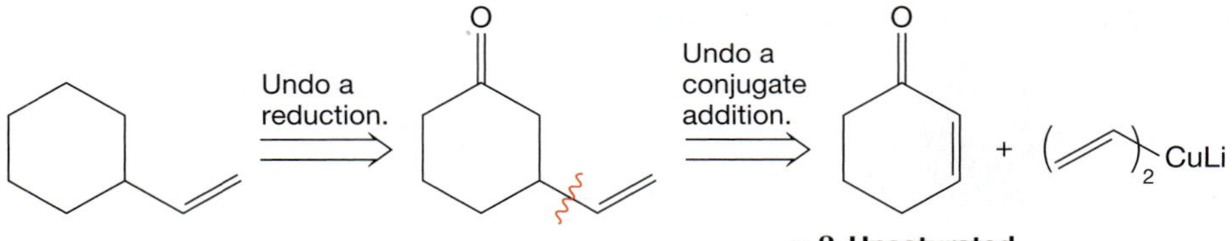

Therefore, the synthesis in the forward direction might appear as follows:

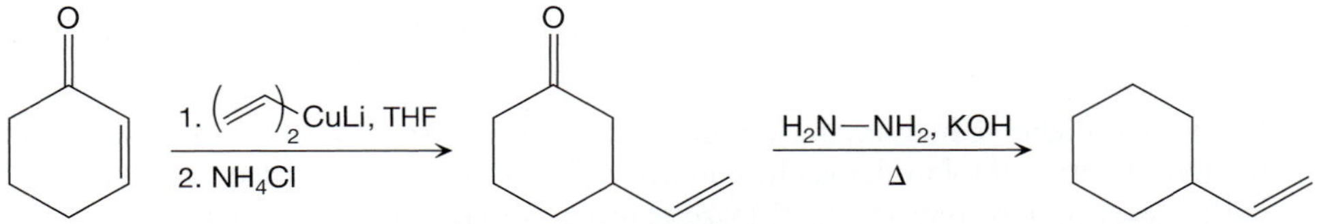

PROBLEM 19.12 Show how this compound can be synthesized separately from four different α,β-unsaturated ketones.

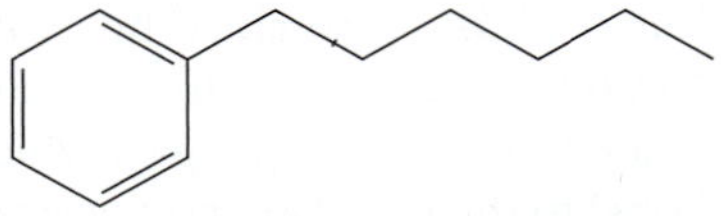

19.3b The Clemmensen Reduction

Section 19.3a focused on the use of the Wolff–Kishner reaction to reduce the C=O group of a ketone or aldehyde to a CH_2 group, but the **Clemmensen reduction**, named after Danish chemist Erik Christian Clemmensen (1876–1941), accomplishes this transformation, too. (So does the Raney-nickel reduction, discussed in Section 19.3c.) An example of a Clemmensen reduction is shown in Equation 19-11.

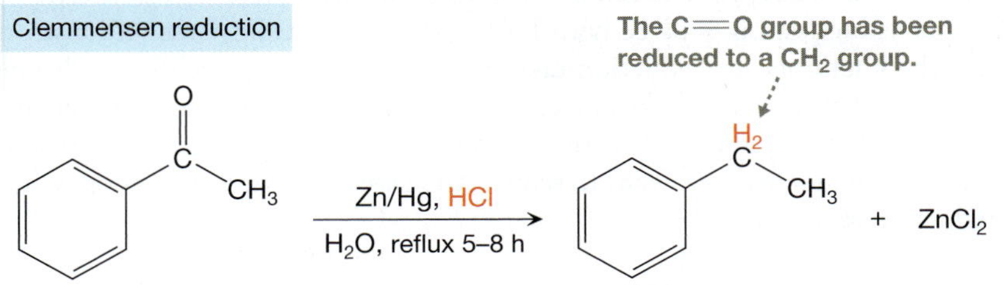

Clemmensen reduction

The C=O group has been reduced to a CH_2 group.

Phenylethanone (Acetophenone) $\xrightarrow[\text{H}_2\text{O, reflux 5-8 h}]{\text{Zn/Hg, HCl}}$ **Ethylbenzene** 90% + $ZnCl_2$ (19-11)

The Clemmensen reduction uses what is called a **zinc amalgam,** which is an alloy (blend) of zinc with mercury. The ketone or aldehyde is refluxed (i.e., continuously evaporated and recondensed under heat) with the amalgam in a concentrated HCl solution. HCl is the source of the protons that form bonds to the carbonyl C. The zinc metal acts as a reducing agent in what is called a **dissolving metal reduction**, whereby the metal dissolves in solution as the reaction ensues. Zinc loses two electrons (it is oxidized) on going from elemental Zn to Zn^{+2} (represented by $ZnCl_2$ in Equation 19-11), whereas the carbonyl C is reduced because it loses the double bond to O and gains two bonds to H.

19.3c The Raney-Nickel Reduction

A third reaction that reduces the carbonyl group of a ketone or aldehyde to a methylene group is the **Raney-nickel reduction**, developed by Murray Raney (1885–1966), an American chemist. An example is shown in Equation 19-12.

In a Raney-nickel reduction, the ketone or aldehyde is first converted to a *thioacetal* by treatment with a thiol under acidic conditions (Section 18.2a). Subsequent treatment with **Raney nickel** converts the thioacetal group into a methylene group.

Raney nickel has two roles in this reaction. First, it is the source of hydrogen. Raney nickel is obtained by treating a 50:50 alloy of aluminum and nickel with hot NaOH, which serves to etch the metal. The spongy network that remains is then treated with hydrogen gas, which adsorbs (accumulates via intermolecular interactions) to the metal in part because of the metal's very large surface area. Second, Raney nickel serves as a *catalyst* for the H_2 reduction, a process that is described in greater detail in Section 19.5.

19.3d Limitations of the Wolff–Kishner, Clemmensen, and Raney-Nickel Reductions

The Wolff–Kishner, Clemmensen, and Raney-nickel reactions reduce the carbonyl group of a ketone or aldehyde to a methylene group, but notice in Equations 19-8, 19-11, and 19-12 that the three reactions take place under different conditions. The Wolff–Kishner reduction takes place with HO^-, which is a strong base and a strong nucleophile; the Clemmensen reduction takes place under strongly acidic conditions; and the Raney-nickel reduction takes place under mild conditions in the presence of a metal catalyst. Thus, if there are functional groups present other than the carbonyl group of a ketone or aldehyde we want to reduce, then we must carefully choose which of these reduction reactions to use.

GREEN CHEMISTRY Raney nickel is very reactive and will ignite spontaneously on contact with atmospheric oxygen, so it must be kept under an inert liquid or an inert gas. For this reason, Wolff–Kishner and Clemmensen reductions are often preferred over Raney-nickel reductions when all else is equal.

- The Wolff–Kishner reduction should be avoided if there is a functional group present that is susceptible to reaction under basic conditions or with strong nucleophiles.

- The Clemmensen reduction should be avoided if there is a functional group present that is susceptible to reaction under acidic conditions.
- The Raney-nickel reduction should be avoided if there is a functional group present that is susceptible to reaction with nucleophilic thiols (RSH) or with H_2 in the presence of a metal catalyst.

Although we have not yet discussed reactions involving H_2 and a metal catalyst, we show in Section 19.5 that these conditions can cause alkenes ($R_2C{=}CR_2$), alkynes ($RC{\equiv}CR$), and aldehydes ($RCH{=}O$) to react readily.

YOUR TURN 19.3

Recall from Section 18.2c that a nitrile ($R{-}C{\equiv}N$) can be hydrolyzed to an amide under either acidic or basic conditions. Do you think a Wolff–Kishner reduction should be carried out on a ketone or aldehyde that contains a $C{\equiv}N$ group? How about a Clemmensen reduction?

SOLVED PROBLEM 19.13

Which reduction reaction(s)—the Wolff–Kishner, Clemmensen, or Raney-nickel—can be used to carry out the following transformation?

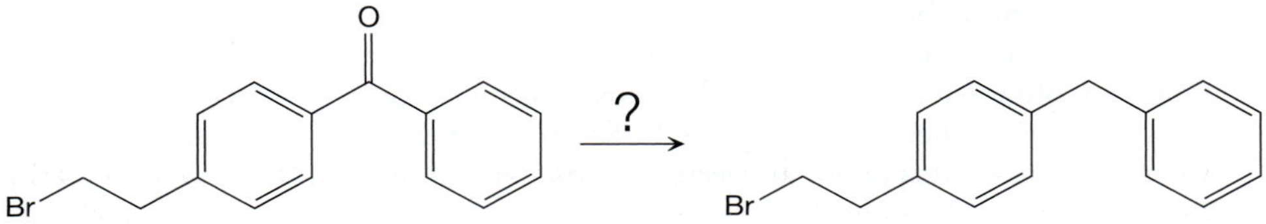

Think Are there any functional groups, aside from the carbonyl group, that are susceptible to reaction in the presence of HO^-? HCl? Thiols? H_2 gas and a metal catalyst?

Solve The primary alkyl halide that is present is susceptible to S_N2 and E2 reactions with HO^- and to S_N2 reactions with hydrazine and thiols, so the Wolff–Kishner and Raney-nickel reductions should be avoided. The aromatic and halo groups are not susceptible to reaction with HCl, so the Clemmensen reduction could be used.

PROBLEM 19.14 Which reduction reaction(s)—the Wolff–Kishner, Clemmensen, or Raney-nickel—can be used for each of the following transformations?

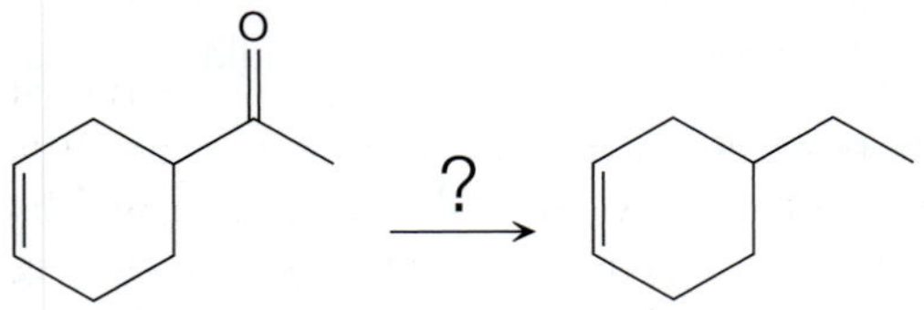

(a)

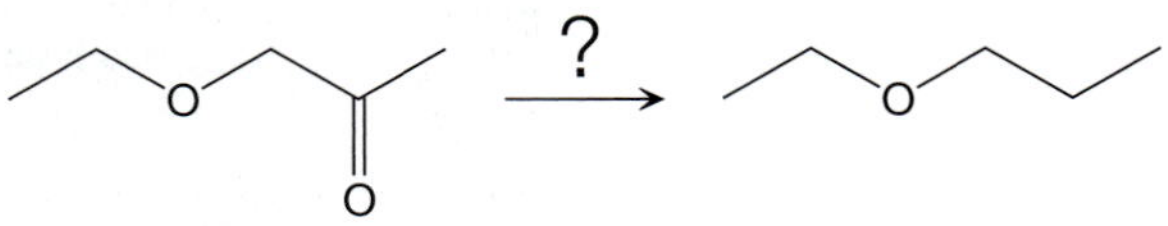

(b)

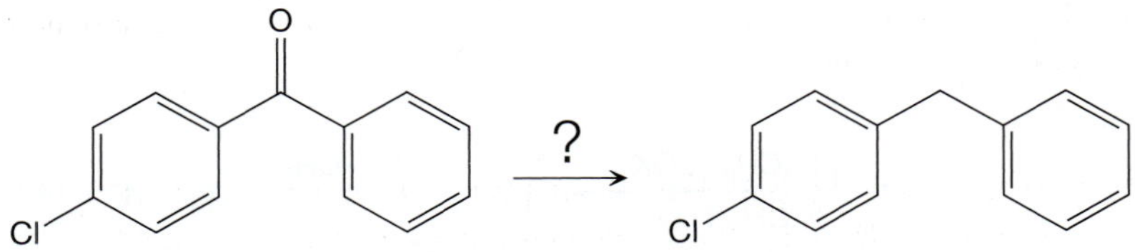

(c)

19.4 Avoiding Synthetic Traps: Selective Reagents and Protecting Groups

As we saw in Section 13.4, a *synthetic trap* arises when a specific transform in a retrosynthesis would lead to an undesired reaction in the *forward* direction. Generally, this is because a functional group we want to leave alone would, in fact, be **labile** under the reaction conditions—that is, the functional group is reactive and readily undergoes a chemical transformation.

Many of the difficulties presented by synthetic traps can be surmounted in one of two ways: (1) by using a *selective reagent* (Section 19.4a) or (2) by using a *protecting group* (Section 19.4b). As discussed in the following sections, both of these strategies exploit differences in reactivity among functional groups.

19.4a Selective Reactions

When two or more outcomes are possible for a given reaction, the actual result often depends on the specific reagents that are used and the specific reaction conditions. In these situations, a reaction that can be chosen to facilitate one outcome over another is called a **selective reaction** and a reagent that is responsible for such selectivity is called a **selective reagent**.

We first encountered an example of a selective reaction in Chapter 9, in the context of the competition between nucleophilic substitution and elimination reactions. Even though S_N2 and E2 reactions tend to compete with each other, given that strong nucleophiles are often strong bases, we can choose specific reagents to selectively favor one reaction over another. For example, Cl^- favors the S_N2 reaction (Equation 19-13a), whereas the *tert*-butoxide anion selectively favors the E2 reaction (Equation 19-13b).

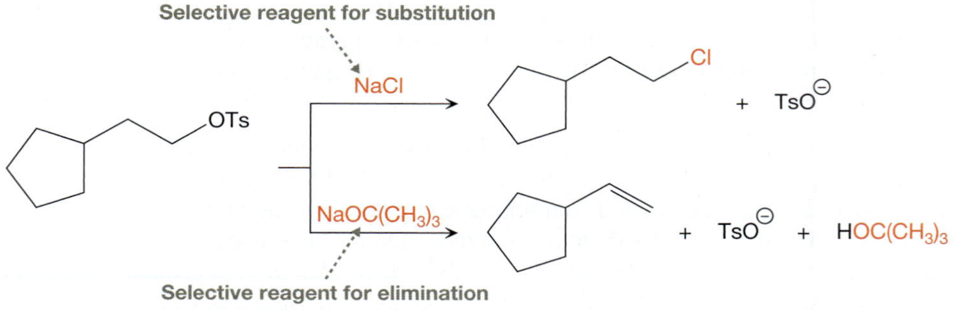

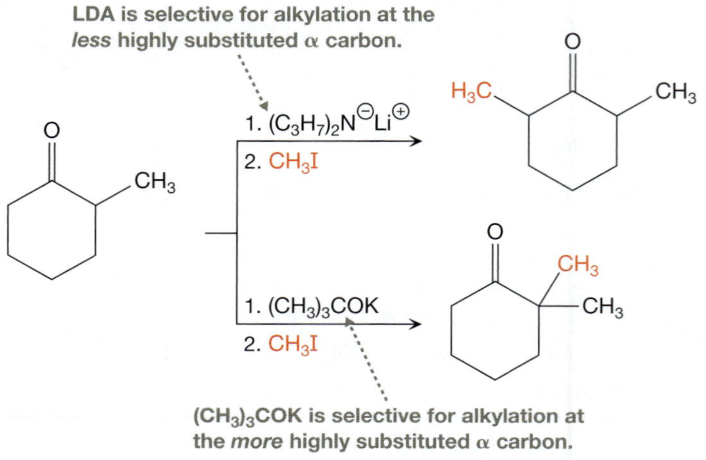

We saw another example of selective reagents in Section 10.3a, when we discussed alkylations at the α carbons of ketones and aldehydes. As mentioned in Section 10.3a and as shown in Equation 19-14, lithium diisopropylamide (LDA) is a bulky base that irreversibly deprotonates ketones and aldehydes, so it is selective for alkylation at the *less* highly substituted α carbon. By contrast, $NaOC(CH_3)_3$ deprotonates ketones and aldehydes reversibly, so it is selective for alkylation at the *more* highly substituted α carbon.

Nucleophilic addition to α,β-unsaturated carbonyl compounds presents us with another case of selectivity in reactions. As we saw in Sections 17.9 and 17.10, very strong nucleophiles like Grignard reagents (RMgX) selectively add to the carbonyl carbon in a *direct addition*, or 1,2-addition (Equation 19-15a). Weaker nucleophiles like lithium dialkylcuprates (R_2CuLi), on the other hand, selectively add to the β carbon in a *conjugate addition*, or 1,4-addition (Equation 19-15b).

Selective for direct addition

1. CH_3MgBr, Et_2O
2. NH_4Cl, H_2O

(19-15a)

1. $(CH_3)_2CuLi$
2. NH_4Cl, H_2O

(19-15b)

Selective for conjugate addition

The selective reactions presented here in Section 19.4a are far from exhaustive; rather, they are intended simply to introduce the concept and to demonstrate how selective reactions can be used to plan a synthesis. As we continue to learn more reactions, we will find several more opportunities for designing syntheses that take advantage of selectivity.

SOLVED PROBLEM 19.15

Show how you would carry out this synthesis, using any reagents necessary.

Think What reagents can be used to form a C—C bond to the carbonyl carbon of a ketone? Which ones do so selectively in an α,β-unsaturated ketone? Will such a reaction produce the necessary functional groups that are in the target, or is a functional group transformation necessary?

Solve The ether target can be produced from an alcohol using a Williamson ether synthesis. The alcohol can be the product of 1,2-addition of an organometallic reagent to the α,β-unsaturated ketone, and Grignard and alkyllithium reagents are selective for 1,2-addition rather than 1,4-addition.

Undo a Williamson synthesis.

Undo a Grignard reaction.

+ BrMg

In the forward direction, the synthesis can be written as follows:

1. BrMg , ether
2. NH_4Cl, H_2O

1. NaH
2. CH_3I

PROBLEM 19.16 Show how you would carry out this synthesis using any reagents necessary.

19.4b Protecting Groups

Selective reactions (Section 19.4a) make it possible to exploit differences in the reactivity of functional groups when we want one group in a molecule to react but leave another alone. There are times, however, when this is unfeasible—when a step in a synthesis may require the use of a reagent that would react with two or more functional groups. In situations like these, we may be able to use a **protecting group** to temporarily make one or more functional groups unreactive under the specific conditions our desired reaction calls for. After making the desired change in the molecule, we must be able to remove the protecting group and restore the original functional group.

Here we focus primarily on the protection of the carbonyl group in ketones and aldehydes, as well as the protection of the hydroxyl group in alcohols. The use of protecting groups has become a mature subdiscipline of organic synthesis, however, and several authors have published books in this area. Suffice it to say that many different functional groups can be protected in a variety of ways, and often there are several factors to consider when choosing a particular protecting group.

The general idea of using a protecting group is outlined in Figure 19-1. Suppose we have a molecule with a functional group A that we want to convert to another functional group B, but there is a labile (i.e., reactive) group that would also be changed. Direct conversion of A to B (i.e., the top reaction in Fig. 19-1) is therefore impossible without affecting the labile group. We can work around this problem by selectively changing the labile group to another functional group that will not react under the conditions that convert A to B. This is called a **protection step**, and the labile group is described as being *protected*. Then, the desired reaction can be carried out and the original labile group can be restored in a subsequent **deprotection step**.

Reversibility is important in reactions that add a protecting group.

> A good protecting group is unreactive under one set of conditions, but is *removable* under another set of mild conditions to restore the original functional group at the same location.

FIGURE 19-1 The general strategy for using a protecting group The starting material (*top left*) cannot be converted to the target (*top right*) directly because the labile group would also react under those conditions. A separate reaction, called a *protection step*, is carried out to convert the labile group into a different group (*bottom left*) that is unreactive under the conditions that convert A to B. Next, the desired reaction is carried out to convert A to B (*bottom right*), and a final *deprotection step* converts the protected group back into the original labile group (*top right*).

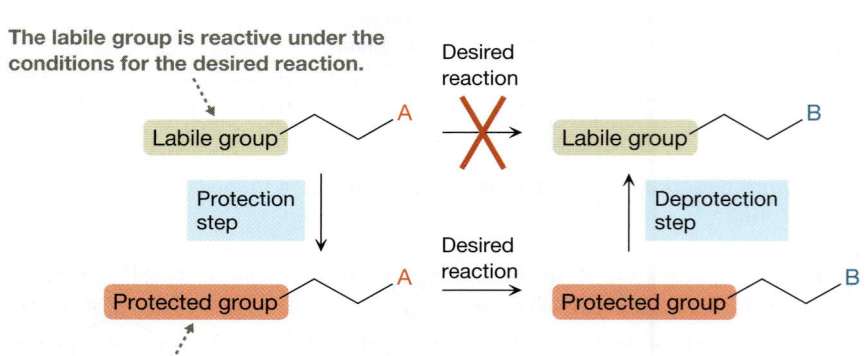

962   **CHAPTER 19** Organic Synthesis 2

It would be a poor choice, for example, to protect the carbonyl group of a ketone by converting it into a methylene (CH$_2$) group using the Clemmensen reduction (Equation 19-16), because we do not know any reactions that would allow us to selectively convert that CH$_2$ group back into a carbonyl group.

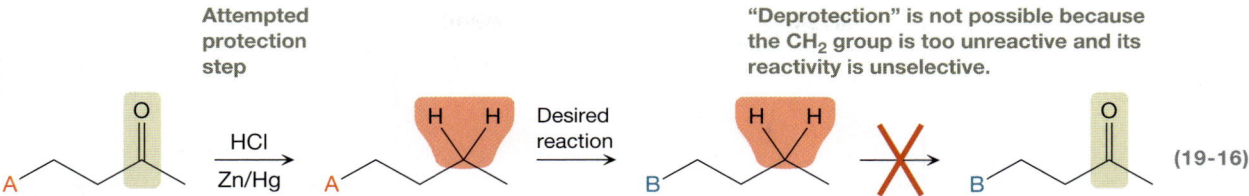

(19-16)

19.4b.1 Protection of the C=O Group in Ketones and Aldehydes

Ketones and aldehydes are most often protected by converting them to *acetals*, as shown in Equation 19-17.

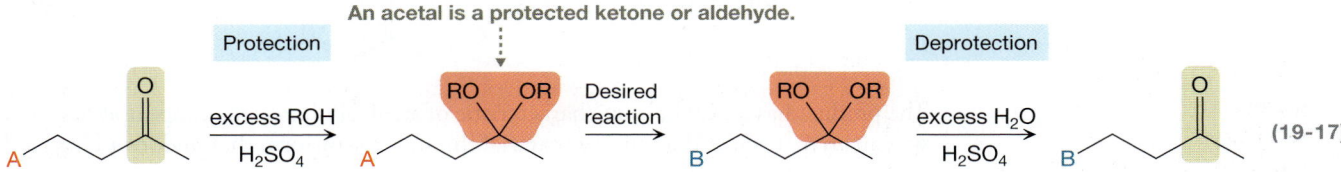

(19-17)

An acetal has the form RO—C—OR, where one C is involved in two C—O—C groups, each resembling an ether.

> Acetals, like ethers, are resistant to strong bases and strong nucleophiles (Section 9.5b). They are also resistant to oxidizing and reducing agents.

An acetal can be formed by reacting a ketone or aldehyde with excess alcohol under acidic conditions (Section 18.2a). Fortunately, the C=O group of a ketone or aldehyde can be recovered by hydrolysis under mildly acidic conditions (review Section 18.2a).

A commonly used alcohol in protecting ketones and aldehydes is ethane-1,2-diol (**ethylene glycol**), which is HOCH$_2$CH$_2$OH (Equation 19-18).

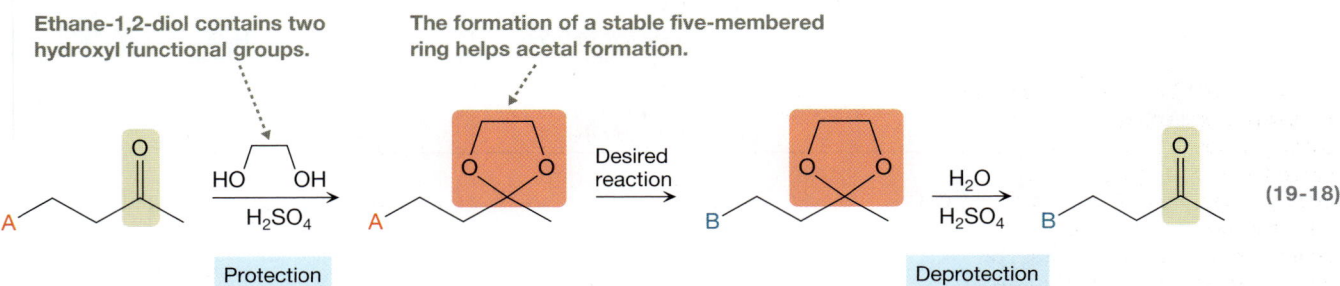

(19-18)

Its popularity is due, in part, to the fact that it contains two hydroxyl groups on the same molecule, so only one molar equivalent is needed to protect a carbonyl (rather than the two molar equivalents that would be required if a monoalcohol were used). Additionally, the acetal that it produces consists of a five-membered ring, which is relatively stable and helps favor the formation of the acetal.

In the reaction shown here, circle the protected carbonyl group and draw the compound that would be produced when it is deprotected.

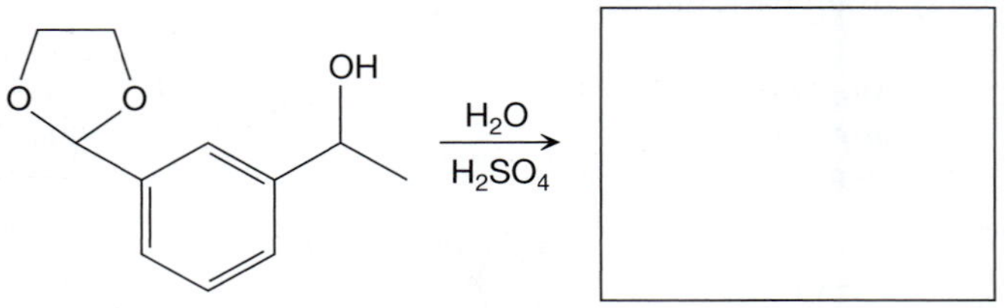

To see how such a protecting group might be used, let's propose a way to carry out the synthesis in Equation 19-19.

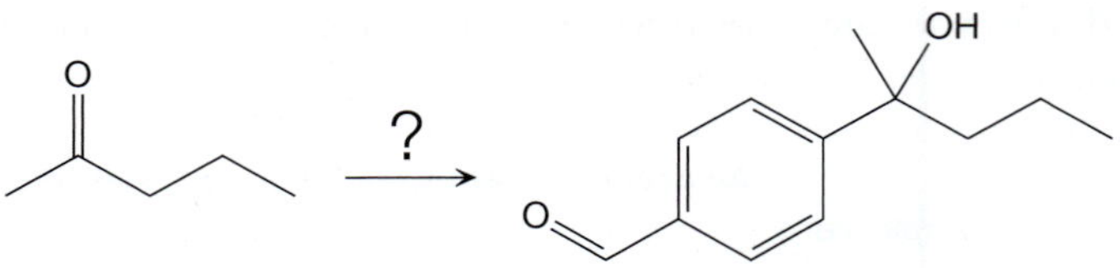

(19-19)

The product has an OH group characteristic of an alcohol, so it may appear at first that we can arrive back at the starting material through a transform that undoes a Grignard reaction, as shown in Equation 19-20.

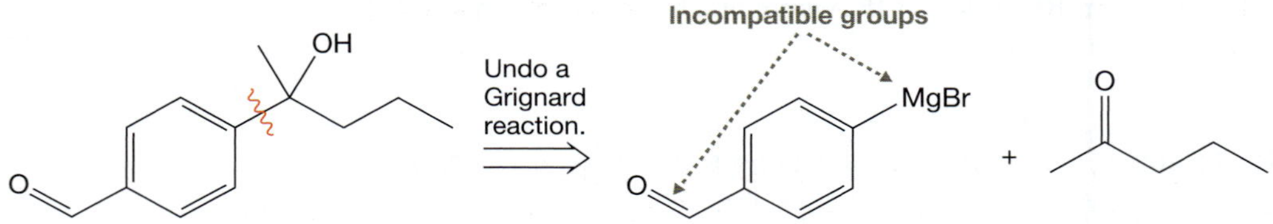

(19-20)

Notice, however, that the necessary Grignard reagent has a carbonyl group on it. As we learned in Section 17.6, carbonyl groups are incompatible with Grignard reagents. To work around this incompatibility, we can instead envision using a Grignard reagent in which that carbonyl group is protected, as shown in Equation 19-21.

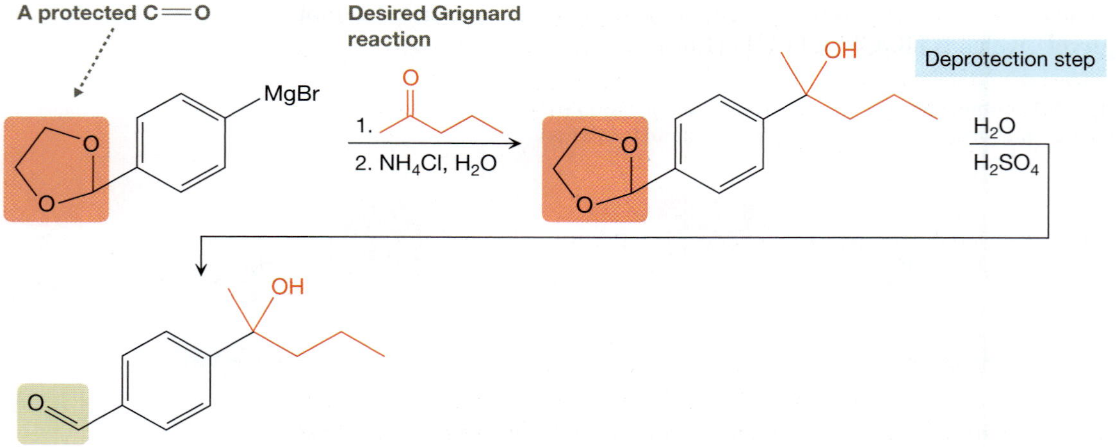

(19-21)

How, then, do we synthesize the protected Grignard reagent? We know from Section 19.1 that a Grignard reagent is produced from an analogous alkyl halide. To avoid the presence of a C=O group on the Grignard reagent, the C=O must be protected prior to the addition of Mg, so the retrosynthetic analysis would appear as shown in Equation 19-22.

The C=O group must be protected *before* the formation of the Grignard reagent.

(19-22)

The synthesis is then reported by reversing the arrows and adding the appropriate reagents (Equation 19-23).

(19-23)

The C=O group of the aldehyde is protected by treating it with ethylene glycol (HOCH$_2$CH$_2$OH) under acidic conditions, and the Grignard reagent is produced by treating the resulting compound with solid Mg in ether.

PROBLEM 19.17 Propose a synthesis of hexane-2,5-dione using acetone as your only source of carbon. *Hint:* Consider how you might use a protecting group.

19.4b.2 Protection of the OH Group in Alcohols

Alcohols are weakly acidic, can react as nucleophiles, are prone to elimination under acidic conditions, and as discussed later here in Chapter 19, are prone to oxidation, too. Depending on the specific step in a synthesis, we may need to suppress the reactivity of an alcohol by temporarily converting its OH group to a less reactive functional group.

Just as we were able to protect the C=O group of a ketone or aldehyde, we can protect the OH group of an alcohol in the form of an *ether* or *acetal*. Table 19-2 (next page) shows three ways to carry out these protection steps, as well as their corresponding deprotection steps.

For Entries 1 and 2, the protection step produces an acetal, but via different reactions. In Entry 1, the protection step is a Williamson ether synthesis, whereas in Entry 2 it is an acid-catalyzed addition of the alcohol to the alkene. For Entry 3, the protection step involves an S$_N$2 reaction to produce a silyl ether, characterized by a C—O—Si group, which, like a regular ether, is unreactive under basic conditions.

YOUR TURN 19.5

Identify the acetal functional groups in Entries 1 and 2 in Table 19-2.

The deprotection steps in Entries 1 and 2 are acid-catalyzed hydrolysis reactions. In Entry 3, the deprotection step is a nucleophilic substitution reaction in which F$^-$ attacks the Si atom. The leaving group is RO$^-$, which is usually a terrible leaving group for substitution reactions, but the Si—F bond that is formed is very strong and thus compensates—its bond energy is 553 kJ/mol (132 kcal/mol), making it about 30% stronger than a C—C single bond!

Even though there are multiple options available to protect an alcohol, they are not always interchangeable. You must consider whether the conditions required for the

TABLE 19-2 Protecting Groups for Alcohols

	Protection Step	Comments	Deprotection Step	Comments
(1)	R—OH $\xrightarrow{\text{1. NaH}}$ R—O⌣O⌣ "MOM" (Cl⌣O⌣)	Williamson ether synthesis (basic conditions, Section 10.6). The product is a methoxymethyl ether (an acetal).	R—O⌣O⌣ $\xrightleftharpoons[\text{H}_2\text{SO}_4]{\text{H}_2\text{O}}$ R—OH + H—C(=O)—H + CH_3OH	S_N1 hydrolysis (acidic conditions, Section 9.5b)
(2)	**Dihydropyran (DHP)** R—OH $\xrightleftharpoons[\text{H}_2\text{SO}_4]{}$ R—O⌣(THP)⌣O	Electrophilic addition of ROH to the alkene (acidic conditions, Section 11.6). The product is an acetal.	R—O⌣⌣O $\xrightleftharpoons[\text{H}_2\text{SO}_4]{\text{H}_2\text{O}}$ R—OH + HO⌣O	S_N1 hydrolysis (acidic conditions, Section 9.5b)
(3)	**tert-Butyldimethylsilyl chloride (TBDMS—Cl)** $\text{Cl}-\overset{\overset{\text{CH}_3}{\|}}{\underset{\underset{\text{CH}_3}{\|}}{\text{Si}}}-\text{C(CH}_3)_3$ R—OH → R—O—$\overset{\overset{\text{CH}_3}{\|}}{\underset{\underset{\text{CH}_3}{\|}}{\text{Si}}}$—C(CH$_3$)$_3$	The product is a silyl ether, characterized by a C—O—Si group. The product is abbreviated R—O—TBDMS.	R—O—$\overset{\overset{\text{CH}_3}{\|}}{\underset{\underset{\text{CH}_3}{\|}}{\text{Si}}}$—C(CH$_3$)$_3$ $\xrightarrow[\text{THF}]{\text{Bu}_4\text{N}^{\oplus}\text{F}^{\ominus}}$ R—OH	Substitution reaction in which $F^{\ominus}$ is the nucleophile and $RO^{\ominus}$ leaves

protection and deprotection of the OH group will interfere with functional groups elsewhere in the molecule. For example, Entry 1 in Table 19-2 calls for basic conditions in the protection step and acidic aqueous conditions in the deprotection step, so that would be a poor choice if other groups in the molecule react under basic or acidic conditions. Entry 2 would be a poor choice if you need to avoid acidic conditions, and Entry 3 should be avoided if the molecule has other sites that are nucleophilic or will react with F^-.

PROBLEM 19.18 Draw the complete, detailed mechanism for both the protection step and the deprotection step of Entries 1 and 2 in Table 19-2.

SOLVED PROBLEM 19.19

A student attempted to carry out the following reaction on 3-hydroxypropanal, but it did not work.

HO⌣⌣O
3-Hydroxypropanal (Reuterin)

$\xrightarrow[\text{2. NH}_4\text{Cl, H}_2\text{O}]{\text{1.}\ \text{(C}_6\text{H}_5)\text{MgBr, THF}}$

HO⌣⌣(C$_6$H$_5$)OH

Suggest why, and propose a synthesis route that would circumvent this problem.

Think What type of reaction does the student intend to carry out? What undesired reaction can take place, and what type of functional group does that reaction involve? Can that functional group be protected?

Solve For the reaction to take place as planned, the Grignard reagent must act as a nucleophile, attacking the carbonyl group. A Grignard reagent is also a very strong base, however, and the OH group on the hydroxyaldehyde is weakly acidic. Instead of the intended nucleophilic addition, the following proton transfer reaction occurs, which destroys the Grignard reagent.

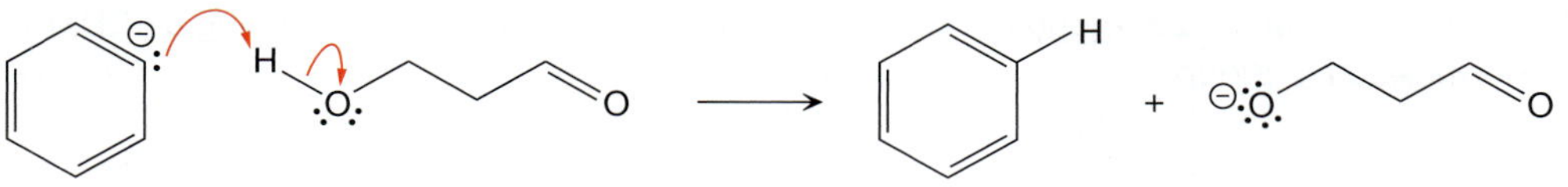

To circumvent this problem, we can protect the OH group before performing the Grignard reaction, as shown below using dihydropyran (DHP). Once the Grignard reaction is complete, the product can be deprotected. Because the workup step of the Grignard reaction requires acidic conditions, the acid workup and deprotection can occur together.

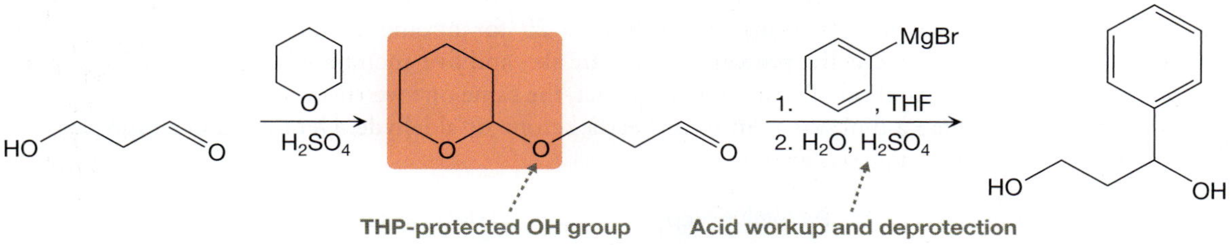

THP-protected OH group   Acid workup and deprotection

PROBLEM 19.20 Show how to carry out the following synthesis.

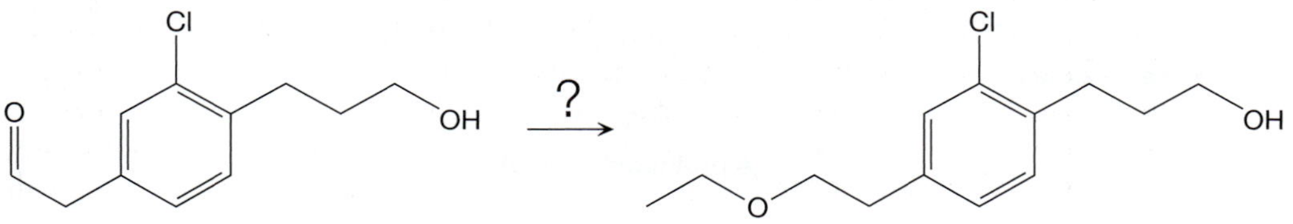

PROBLEM 19.21 A student carried out the following sequence of reactions using *tert*-butyldimethylsilyl chloride (TBDMSCl) to protect the hydroxyl group.

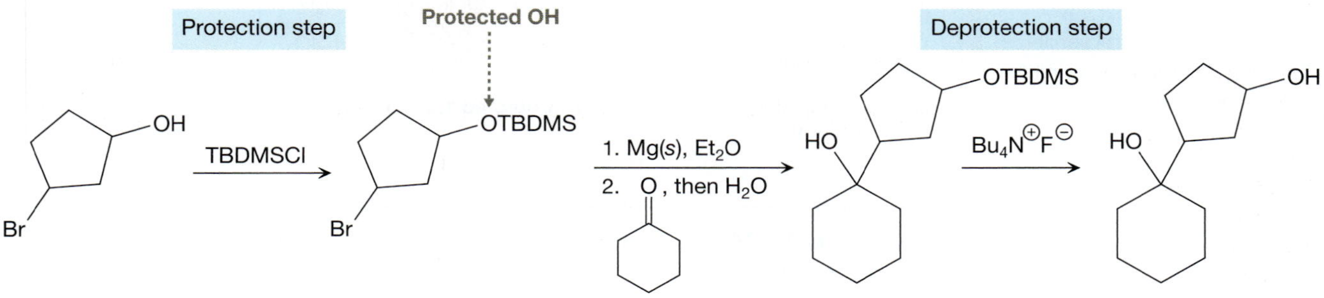

Protection step   Protected OH   Deprotection step

(a) Could this synthesis be carried out without protecting the hydroxyl group? Why or why not?

(b) Which other protecting groups in Table 19-2 could have been used instead? Explain.

PROBLEM 19.22 Show how you would carry out this synthesis.

PROBLEM 19.23 Propose how you would carry out this synthesis using a TBDMS protecting group.

Both 1,2-diols and 1,3-diols can be protected using acetone [(CH₃)₂C═O], as shown in Equations 19-24 and 19-25 for hexane-2,4-diol and benzene-1,2-diol, respectively. The protected forms are six- and five-membered ring acetals, respectively. This method of protection is, in fact, the same one we encountered in Section 19.4b.1, where a diol was used to protect a ketone or aldehyde. Deprotection simply requires dilute aqueous acid.

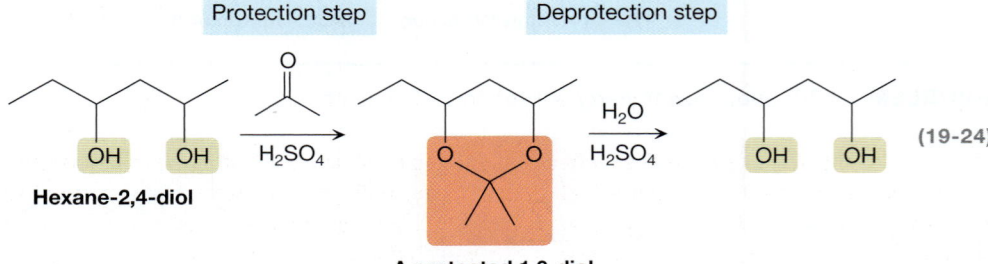

CONNECTIONS Benzene-1,2-diol, also called catechol, is used to manufacture vanillin, which is the principal flavor component of natural vanilla flavoring. Synthetic vanillin is sold commercially as artificial vanilla flavoring.

Vanillin

PROBLEM 19.24 Show how you would synthesize each of the following from 1,2,5-pentanetriol.

(a)

(b)

Protecting Groups in DNA Synthesis

Oligonucleotides are relatively short strands of DNA, on the order of 10–100 nucleotides (i.e., A, C, T, G) long. They have a number of important applications in molecular biology, including DNA sequencing and synthesizing artificial genes. Consequently, it is important to synthesize oligonucleotides in an efficient and accurate way. As it turns out, protecting groups play a major role in these syntheses, as shown in the following simplified scheme.

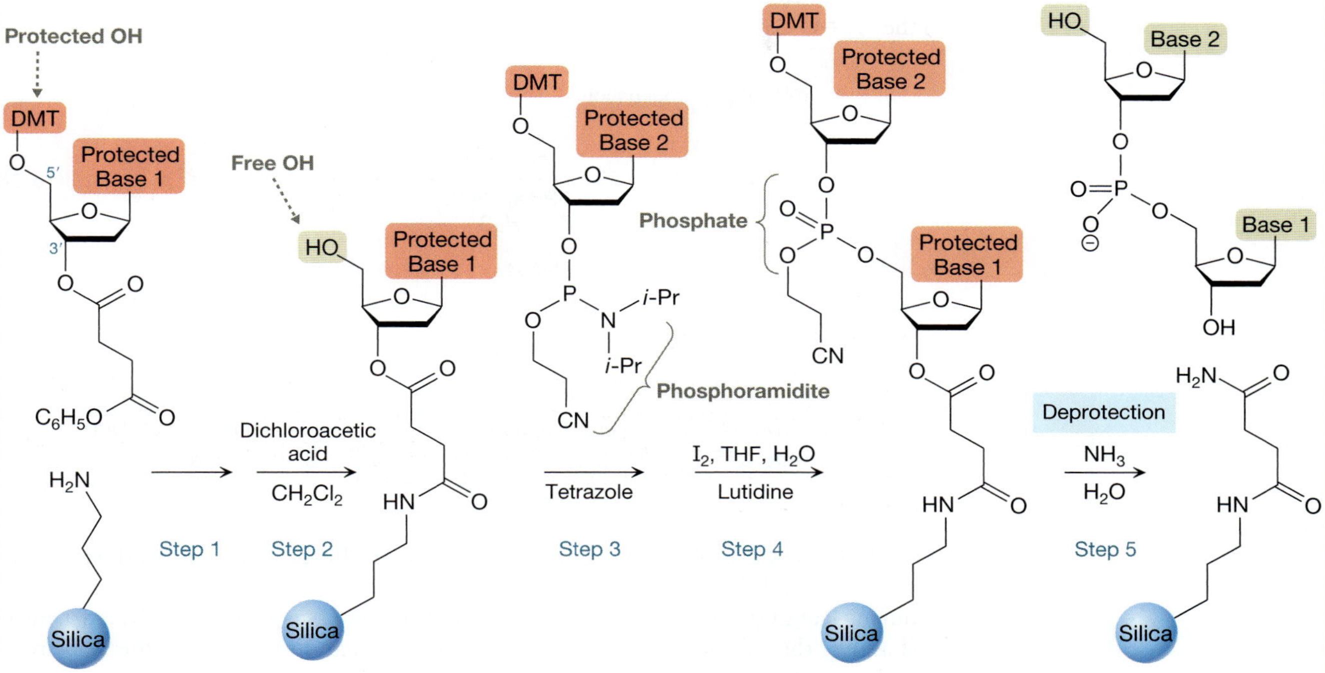

Protecting groups are necessary because each nucleotide has multiple possible reactive sites, including the 5′-OH and the 3′-OH groups of each sugar unit, and the NH_2 group of adenine, guanine, and cytosine bases. By using protecting groups appropriately, we can ensure that the correct groups are free to react at any given time, so that the DNA chain grows in a specific way.

The synthesis consists of the five distinct steps shown above. Step 1 connects the 3′ end of the first nucleoside (i.e., base linked to a sugar) to a support, such as polystyrene, via an amide linkage. Notice that the 5′-OH and the NH_2 group of the base (in the case of A, G, and C) are initially protected. The 5′-OH is protected as a 4,4′-dimethoxytrityl (DMT) ether, and the NH_2 group of the base is protected as an amide. In Step 2, the 5′-OH is deprotected, making it available to react. In Step 3, that 5′-OH connects to the P atom of a phosphoramidite group [characterized by $(RO)_2PNR_2$] at the 3′ end of a second protected nucleoside, and the oxidation in Step 4 produces a phosphate group. Steps 2–4 are repeated to add more nucleosides, and once the desired chain has been constructed, Step 5 deprotects all the groups and releases the DNA chain from the silica support.

19.5 Catalytic Hydrogenation

In Section 19.3, we learned how a functional group can be removed entirely from a molecule via the reduction of the C=O group of a ketone or aldehyde to a CH_2 group. In Section 19.4, we saw how reactions we have encountered previously can be used to selectively target one functional group over another. In this section, we discuss **catalytic hydrogenation**, which can accomplish similar tasks. Catalytic hydrogenation can be used to remove a C=C or C≡C functional group entirely from a molecule. Additionally, it is a selective reaction, involving alkenes, alkynes, and carbonyl-containing groups such as those in ketones and aldehydes.

19.5a Catalytic Hydrogenation of Alkenes

In the presence of a finely divided, solid metal catalyst such as palladium, platinum, or nickel (all elements in the same column of the periodic table), H_2 gas readily adds to the C=C double bond of an alkene. An example is shown in Equation 19-26.

Catalytic hydrogenation

The double bond is converted to a single bond.

$$H_2 \text{ (1 atm)}$$
Pd(s), 25 °C
H_2O

Solid catalyst

82%

(19-26)

Notice, in particular, that:

> Catalytic hydrogenation using a platinum, palladium, or nickel metal catalyst reduces the C=C double bond of an alkene to a C—C single bond.

In these kinds of reactions, the metal is called a **heterogeneous catalyst** because it exists in a different phase from the rest of the reaction mixture. The catalyst used during catalytic hydrogenation is *an insoluble solid*, suspended in (by stirring or shaking) the solution that contains both hydrogen and the organic substance to be hydrogenated.

The detailed mechanism for catalytic hydrogenation is relatively complex. A simplified picture, however, is shown in Figure 19-2.

Initially, the hydrogen and the alkene adsorb onto the surface of the catalyst, as shown in Figure 19-2a through 19-2c. As that occurs, the H—H single bonds are effectively broken, producing individual H atoms that reside on the metal surface. Similarly, adsorption of the alkene partially breaks the π bond of the C=C double bond. Eventually, an adsorbed H atom encounters an adsorbed alkene, and the first of two C—H bonds is formed (Fig. 19-2d). Very shortly thereafter, the second C—H bond is formed through a similar process (Fig. 19-2e). As each C—H bond is formed, the corresponding C atom is released from the metal surface. Because the reaction takes place at the metal's surface, using a finely divided catalyst dramatically increases the surface area of the metal and, hence, the rate of the reaction.

We can take advantage of these alkene reductions in a synthesis that calls for the formation of a new C—C single bond by first incorporating a reaction that forms a new C=C double bond. Suppose, for example, that we need to carry out the following synthesis, beginning with starting materials having six or fewer carbon atoms.

Six or fewer carbon atoms

?

No functional groups appear in the target, and there is no relative positioning of heteroatoms to provide a clear indication of which carbon–carbon bond-forming

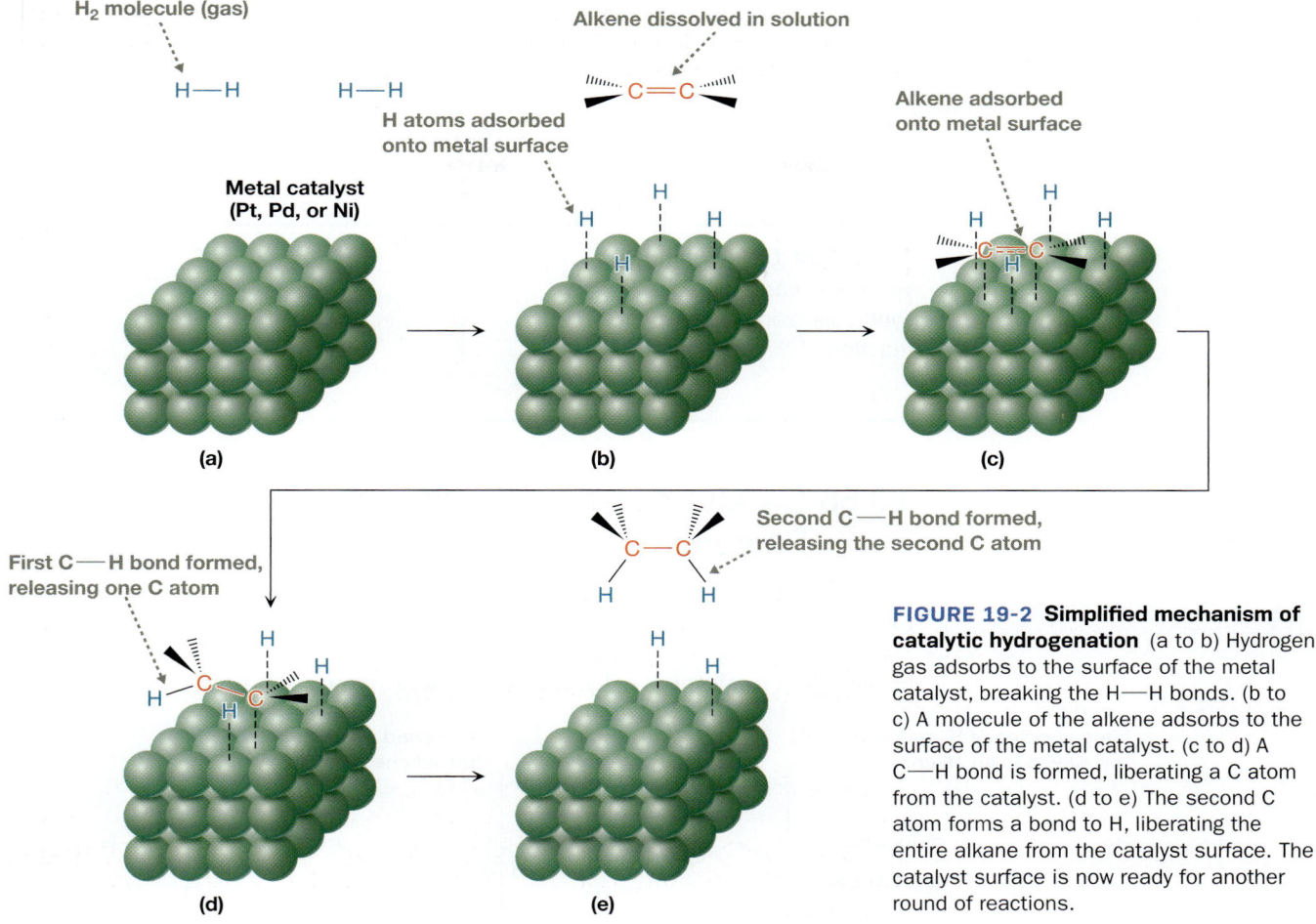

H₂ molecule (gas)

H—H   H—H

Alkene dissolved in solution

C=C

H atoms adsorbed onto metal surface

Alkene adsorbed onto metal surface

Metal catalyst (Pt, Pd, or Ni)

(a)   (b)   (c)

First C—H bond formed, releasing one C atom

Second C—H bond formed, releasing the second C atom

C—C

(d)   (e)

FIGURE 19-2 Simplified mechanism of catalytic hydrogenation (a to b) Hydrogen gas adsorbs to the surface of the metal catalyst, breaking the H—H bonds. (b to c) A molecule of the alkene adsorbs to the surface of the metal catalyst. (c to d) A C—H bond is formed, liberating a C atom from the catalyst. (d to e) The second C atom forms a bond to H, liberating the entire alkane from the catalyst surface. The catalyst surface is now ready for another round of reactions.

reaction should be used. As shown in the following retrosynthetic analysis, however, the target can be produced from an alkene intermediate in which the ring is joined to the alkyl chain by a C=C double bond, because catalytic hydrogenation could reduce the C=C double bond to a C—C single bond.

Undo a hydrogenation. ⟹ Undo a Wittig reaction. ⟹ $\overset{\oplus}{P}(C_6H_5)_3$ Undo a Wittig reagent formation. ⟹ Br

+

O

That C=C double bond, in turn, could be produced using a Wittig reaction (Section 17.7), which requires the ketone and Wittig reagent shown. The Wittig reagent could be produced from 1-bromopropane (Section 17.8), which has three carbon atoms. The synthesis in the forward direction would then appear as follows:

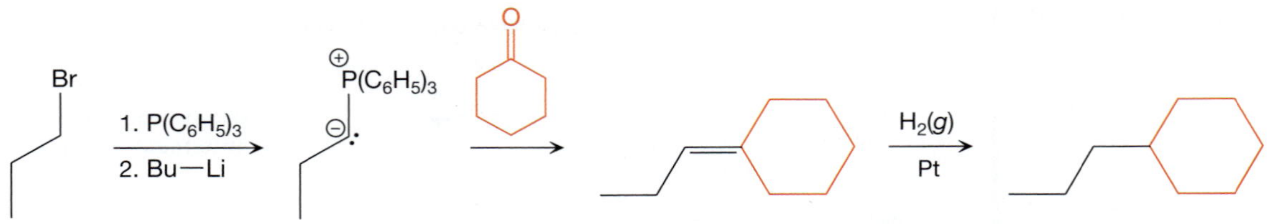

PROBLEM 19.25 Show how you would carry out this synthesis using a Wittig reaction.

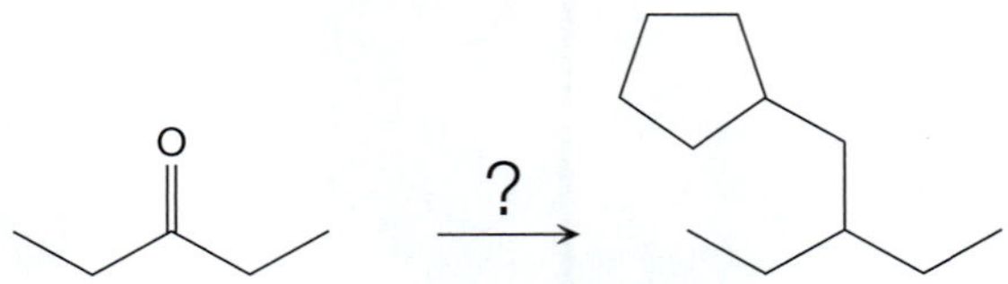

19.5b Catalytic Hydrogenation of Alkynes: Poisoned Catalysts

The C≡C triple bond of an alkyne can undergo catalytic hydrogenation because it, too, contains carbon–carbon π bonds. Similar to alkenes, the addition of H_2 removes a π bond and causes the total number of bonds between the alkyne carbons to decrease by one. Thus, an alkyne is reduced to an alkene, as shown in Equation 19-27.

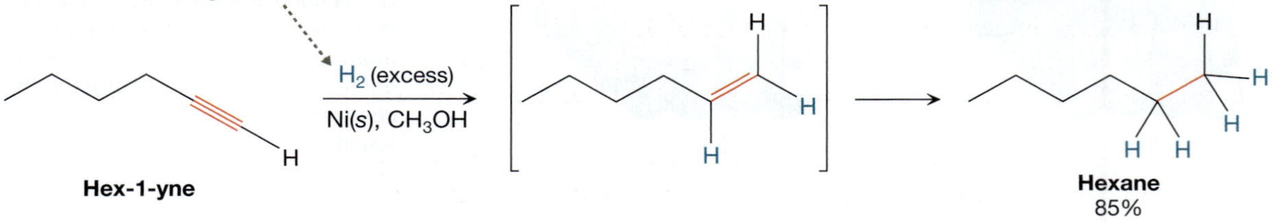

One addition of H₂ reduces an alkyne to an alkene.

A second addition of H₂ reduces the alkene to an alkane.

(19-27)

Hex-1-yne

Hexane
85%

The alkene that is produced is susceptible to catalytic hydrogenation, as indicated in Equation 19-27 and as we saw in Section 19.5a. Therefore:

- If equimolar amounts of H_2 and the alkyne react under normal conditions, we run the risk of producing a mixture of the alkene and alkane products—a generally undesirable result.
- If excess H_2 is present, then an alkyne will be reduced completely to the alkane.

If the alkene is the desired product, however, then:

Reduction of an alkyne can be stopped at the alkene stage by using a stoichiometric amount of H_2 and a *poisoned catalyst*.

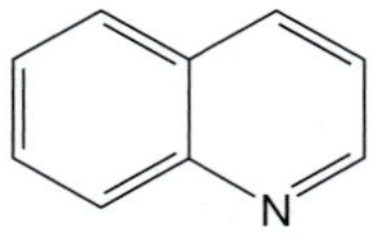

Quinoline

A **poisoned catalyst** is simply a metal catalyst that has been specially treated to decrease its catalytic ability, thus making possible a slower and more controlled reaction. One example is **Lindlar catalyst**—palladium adsorbed on calcium carbonate ($CaCO_3$) and treated with a small amount of quinoline and a lead salt. Barium sulfate ($BaSO_4$) can also be used instead of $CaCO_3$. Representative hydrogenation reactions with these kinds of catalysts are shown in Equations 19-28 and 19-29.

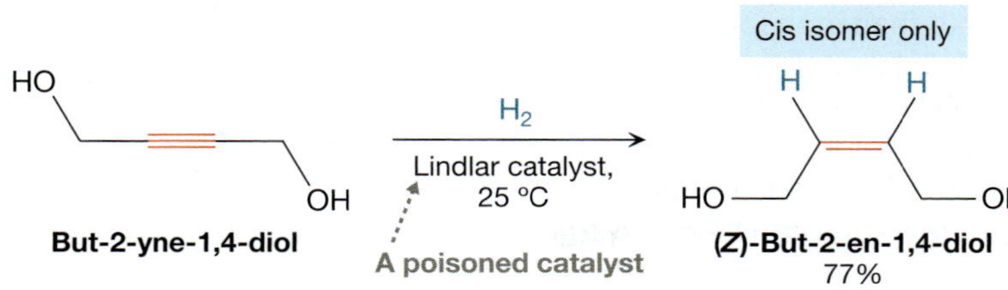

But-2-yne-1,4-diol
A poisoned catalyst
(Z)-But-2-en-1,4-diol
77%
Cis isomer only
(19-28)

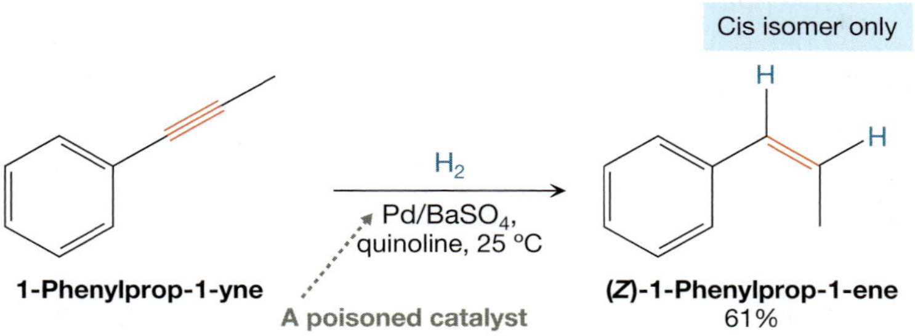

1-Phenylprop-1-yne
A poisoned catalyst
(Z)-1-Phenylprop-1-ene
61%
Cis isomer only
(19-29)

Equations 19-28 and 19-29 show that catalytic hydrogenation takes place *stereoselectively*.

The reduction of an alkyne to an alkene via catalytic hydrogenation favors production of the cis isomer over the trans.

This stereoselectivity can be understood from the simplified picture of the mechanism we saw previously in Figure 19-2 (p. 971). After the first H atom adds to one C of the multiple bond (Fig. 19-2d), the second H atom adds to the other C (Fig. 19-2e) before any significant changes occur in the orientation of the molecule. Thus, the individual H atoms end up on the same side of the newly formed C=C double bond; the H atoms are said to add in a *syn fashion*. (In Chapter 25, we will study a hydrogenation reaction that adds the H atoms *anti*.)

YOUR TURN 19.6

Which of molecules **A–C** could be produced from an alkyne and would require a poisoned catalyst?

PROBLEM 19.26 In Equation 19-29, why are the double bonds in the starting material unaffected by hydrogenation?

SOLVED PROBLEM 19.27

Outline a synthesis of (Z)-1-phenylhept-2-ene from (bromomethyl)benzene.

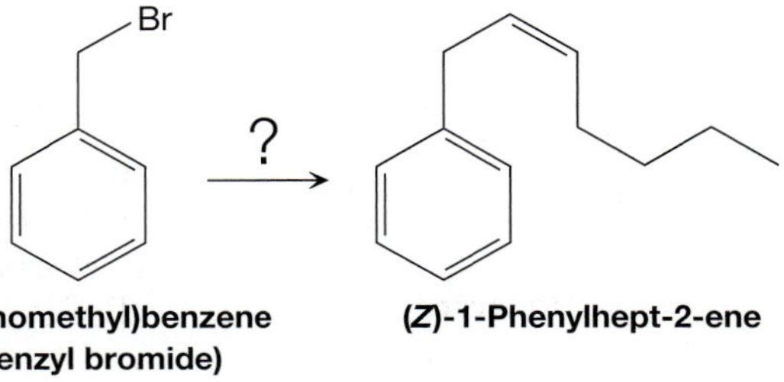

Think Does this synthesis require a carbon–carbon bond-forming reaction? If so, what species possessing an electron-rich carbon could be used? What species containing an electron-poor carbon could be used? Are there any stereochemical issues that must be considered? Can the product be formed in a single reaction, or should we consider making it from a precursor?

(Bromomethyl)benzene
(Benzyl bromide)
(Z)-1-Phenylhept-2-ene

Solve The carbon skeleton is different in the products than in the reactants, so a carbon–carbon bond-forming reaction is necessary. Unfortunately, we do not have at our disposal a way to form a carbon–carbon bond between an alkene carbon and the C atom of a C—Br bond in a single reaction. Therefore, let's perform a *retrosynthetic analysis*, seeking a precursor from which we can readily make the target. As we have seen previously in this section, a cis alkene can be made by hydrogenating an alkyne, suggesting the possibility of an alkyne precursor. That alkyne, in turn, can be the result of an S_N2 reaction that forms the necessary carbon–carbon bond. Overall, our retrosynthetic analysis appears as follows:

A cis double bond

Undo a catalytic hydrogenation.

Undo an S_N2 reaction.

The synthesis would then appear as follows:

Formation of a C—C bond.

1. NaNH₂
2.

Catalytic hydrogenation forms only the cis C=C bond.

H₂
Pd, BaSO₄, quinoline

A strong base deprotonates the terminal alkyne, thus converting it into a strong nucleophile. Addition of (bromomethyl)benzene results in an S_N2 reaction to generate the new C—C bond. Subsequent treatment with H₂ gas in the presence of a poisoned catalyst allows the reduction of the triple bond to stop at the cis double bond.

PROBLEM 19.28 Show how you would carry out this synthesis using an alkyne.

?

19.5c Addition of H₂ to Other Functional Groups

Catalytic hydrogenation is not limited to carbon–carbon multiple bonds; the addition of hydrogen to the π bonds of other systems is feasible as well. As Table 19-3 shows, a variety of functional group conversions can be performed using catalytic hydrogenation, though not with equal ease. Pay particular attention to the typical experimental conditions required.

Notice, in particular, that, using Pt as the catalyst:

The reduction of an aldehyde is essentially as easy as the reduction of an alkene or alkyne.

TABLE 19-3 Reduction of Various Functional Groups via Catalytic Hydrogenation

Reactant	Reaction Conditions	Product
Aldehyde	H_2 / Pt, 20 °C, 1 atm	1° Alcohol
Ketone	H_2 / Rh, 50 °C, 3 atm	2° Alcohol
Nitrile	H_2 / Raney Ni, 80 °C, 75 atm	1° Amine
1° Amide	1 equiv H_2 / Pt, 250 °C, 200 atm (<50% yield)	1° Amine

Other functional groups are more difficult to reduce and their reactions generally proceed with poorer yields. This is particularly true for the reduction of amides to amines, because the product amine poisons (i.e., deactivates) the metal catalyst. If you try to force the reaction by increasing the temperature or pressure of H_2, then the risk of unwanted side reactions increases. Fortunately, $LiAlH_4$ reduces amides to amines in good yield, as discussed in Chapter 20.

19.5d Selectivity in Catalytic Hydrogenation

Catalytic hydrogenation can reduce a variety of different functional groups, including those of alkenes, alkynes, aldehydes, ketones, nitriles, and amides. It can also take place selectively, so if two or more of these functional groups appear in a particular molecule, it is often possible to reduce just one of them. Equation 19-30 shows, for example, that treating limonene with one equivalent of H_2 reduces only the terminal C=C double bond.

The less sterically hindered C=C double bond is selectively reduced.

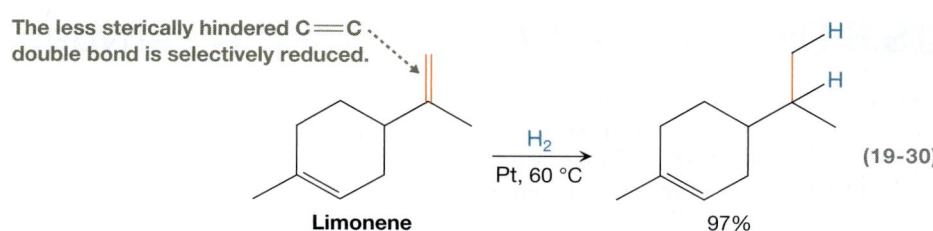

Limonene — H_2 / Pt, 60 °C → 97%   (19-30)

CONNECTIONS Limonene is a monoterpene that occurs naturally in the rinds of lemons and other citrus fruits, contributing to their distinctive aromas.

The terminal C=C bond is less highly substituted, so it is less sterically hindered. In general:

> Catalytic hydrogenation is more favored at a less sterically hindered multiple bond than at a more sterically hindered one.

With more steric bulk surrounding the double bond, it is more difficult for the alkene to adsorb to the surface of the metal catalyst—a critical step in catalytic hydrogenation (review Fig. 19-2c, p. 971).

PROBLEM 19.29 There are two possible syn-addition products in the catalytic hydrogenation of α-pinene, **A** and **B**, but one of them is formed exclusively. Which one? Why?

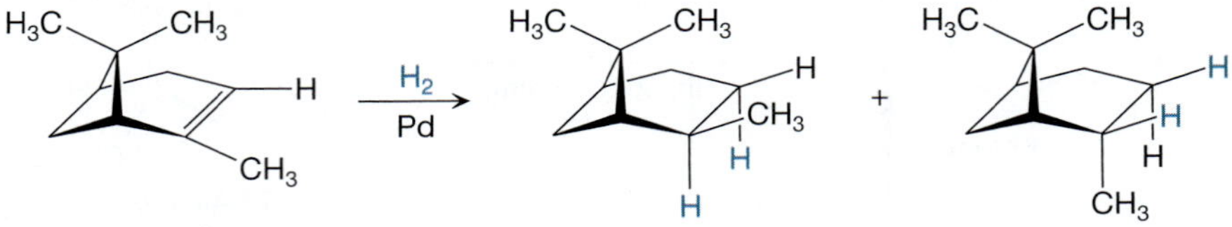

α-Pinene → A + B

Of all the functional groups in Table 19-3, only the C=O of an aldehyde is reduced under mild conditions, similar to those under which alkenes and alkynes are reduced. Therefore:

> The functional groups in alkenes, alkynes, and aldehydes can be selectively reduced over those in ketones, nitriles, and amides.

An example is shown in Equation 19-31.

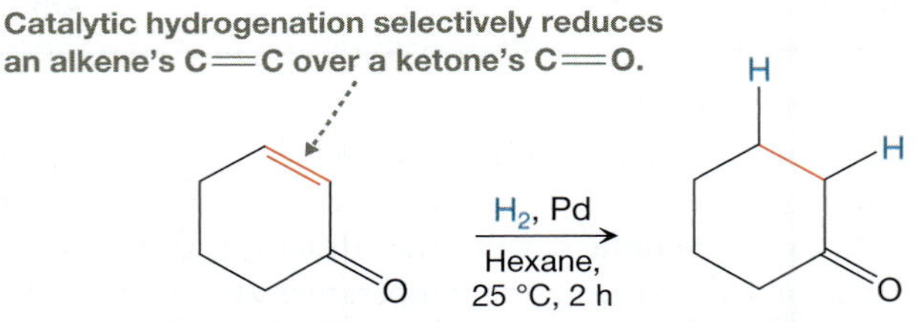

Catalytic hydrogenation selectively reduces an alkene's C=C over a ketone's C=O.

Cyclohex-2-enone — H₂, Pd, Hexane, 25 °C, 2 h → Cyclohexanone 99%

(19-31)

PROBLEM 19.30 Show how you would carry out this synthesis.

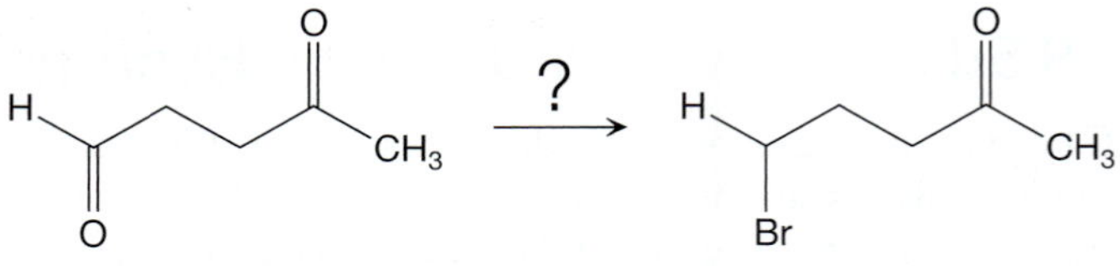

19.6 Oxidations of Alcohols and Aldehydes

In Chapter 17 and here in Chapter 19, we have discussed a variety of reduction reactions, including hydride reductions involving NaBH₄ and LiAlH₄, as well as catalytic hydrogenation. As shown in Figure 19-3, these reduction reactions increase the number of C—H bonds or decrease the number of C—O bonds, representing a progression toward the left in Figure 19-3.

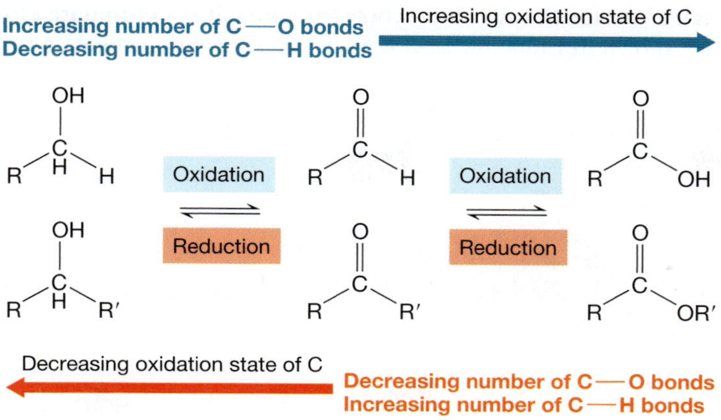

Increasing number of C—O bonds
Decreasing number of C—H bonds

Increasing oxidation state of C

$$
\begin{array}{ccc}
\underset{RH}{\overset{OH}{\underset{|}{\overset{|}{C}}}} & \xrightarrow[\text{Reduction}]{\text{Oxidation}} & \underset{RH}{\overset{O}{\overset{\|}{C}}} & \xrightarrow[\text{Reduction}]{\text{Oxidation}} & \underset{ROH}{\overset{O}{\overset{\|}{C}}}
\end{array}
$$

Decreasing oxidation state of C

Decreasing number of C—O bonds
Increasing number of C—H bonds

FIGURE 19-3 Oxidations and reductions Oxidation is represented by the conversion of a molecule into one toward the right, and is accompanied by an increase in the number of C—O bonds or a decrease in the number of C—H bonds. Reduction is represented by the conversion of a molecule into one toward the left, and is accompanied by a decrease in the number of C—O bonds or an increase in the number of C—H bonds.

To have greater flexibility in designing a synthesis, it is also important to be able to carry out oxidation reactions, which represent a progression toward the right in Figure 19-3. This can be done by carrying out reactions that increase the number of C—O bonds or decrease the number of C—H bonds. We present two types of oxidizing agents that can be used for these kinds of reactions: *chromic acid* and *potassium permanganate*.

19.6a Chromic Acid Oxidations

Chromic acid (H_2CrO_4) is prepared by dissolving either chromium trioxide (CrO_3) or sodium dichromate ($Na_2Cr_2O_7$) in an acidic aqueous solution (Fig. 19-4). A molecule of water adds to the starting material in both reactions. Furthermore, the oxidation state of chromium is +6 in all three forms — CrO_3, $Na_2Cr_2O_7$, and H_2CrO_4.

An example of a chromic acid oxidation is shown in Equation 19-32, in which cyclooctanol is oxidized to cyclooctanone.

$$CrO_3 \xrightarrow[H_2SO_4]{H_2O} H_2CrO_4$$

Chromium trioxide   **Chromic acid**

The oxidation state of Cr is +6.

$$Na_2Cr_2O_7 \xrightarrow[H_2SO_4]{H_2O} 2\ H_2CrO_4$$

Sodium dichromate   **Chromic acid**

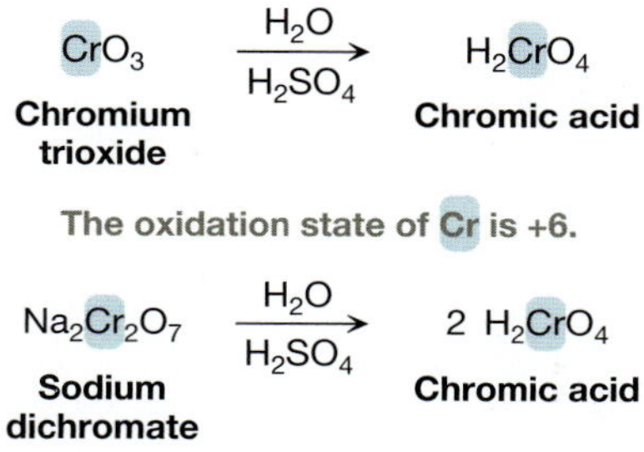

FIGURE 19-4 Chromium oxidizing agents CrO_3, $Na_2Cr_2O_7$, and H_2CrO_4 all contain Cr^{+6}. H_2CrO_4 is produced from either CrO_3 or $Na_2Cr_2O_7$ by treatment with aqueous acid.

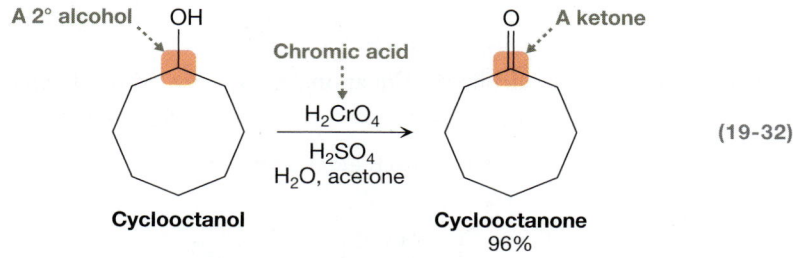

(19-32)

A 2° alcohol   OH   Chromic acid   O   A ketone

$$\xrightarrow[\substack{H_2SO_4 \\ H_2O,\ \text{acetone}}]{H_2CrO_4}$$

Cyclooctanol   **Cyclooctanone**
96%

In general:

Chromic acid oxidizes a secondary alcohol to a ketone.

YOUR TURN 19.7

Verify that the reaction in Equation 19-32 is an oxidation by calculating the oxidation state of the carbon atom screened in red in the reactant _____ and in the product _____. Did that carbon atom gain or lose electrons?

The mechanism for this reaction is believed to involve a **chromate ester** as a key intermediate, shown in Equation 19-33. An E2 step on the chromate ester then produces the ketone.

Partial mechanism for Equation 19-32

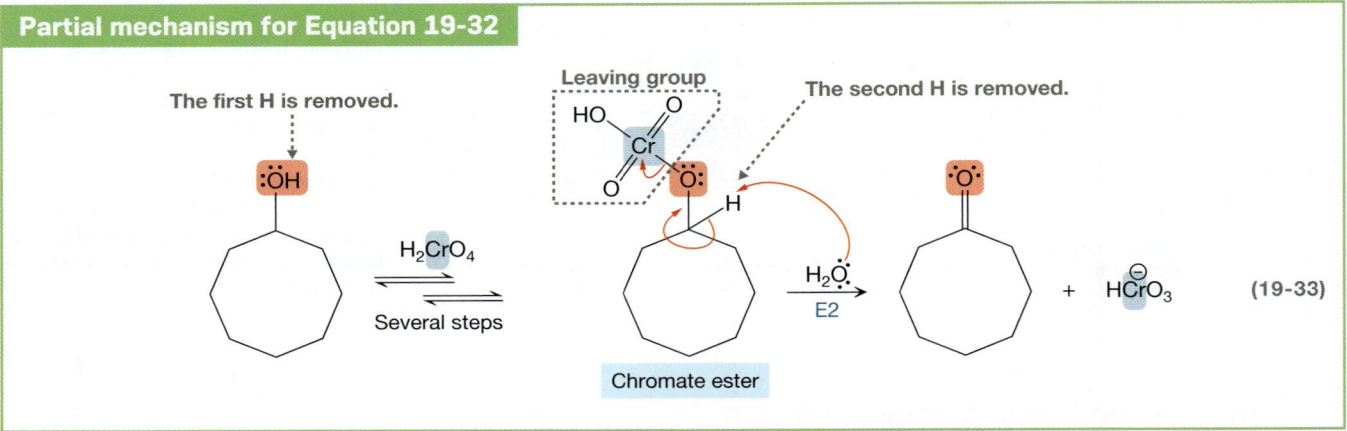

The final E2 step might appear a little peculiar. The leaving group in an E2 reaction is usually attached to a carbon atom, but in this case it is attached to O. Nevertheless, the leaving group and a proton on adjacent atoms are eliminated, as usual, leaving behind an additional π bond. In this case, the result is a new C=O double bond rather than a new C=C double bond.

Notice in Equation 19-33 that there are two H atoms that must be removed: one from the OH group and one from the adjacent C atom. Therefore:

Oxidation by H_2CrO_4 requires an OH group that is attached to a C atom bonded to at least one H atom:

This is why the tertiary alcohol in Equation 19-34 is *not* oxidized when treated with chromic acid.

A primary alcohol, on the other hand, *is* oxidizable because the alcohol carbon is bonded to at least one hydrogen. An example is shown in Equation 19-35.

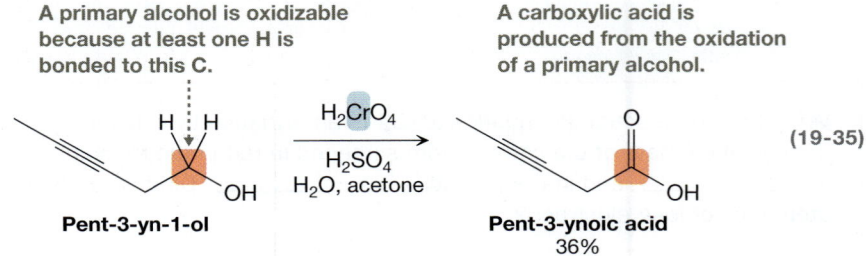

The oxidation product, a carboxylic acid, is different from the ketone obtained in the oxidation of a secondary alcohol.

Chromic acid oxidizes a primary alcohol to a carboxylic acid.

As shown in Equation 19-36, an aldehyde is initially produced when the H atoms on adjacent C and O atoms are removed. Water is present in solution, however, so the aldehyde equilibrates with its *hydrate* (Section 17.2). In that hydrate, an OH is attached to CH, which allows a second oxidation to take place.

H atoms are removed from adjacent OH and CH groups in a chromic acid oxidation.

No OH is present.

H atoms are removed from adjacent OH and CH groups in a chromic acid oxidation.

An aldehyde hydrate

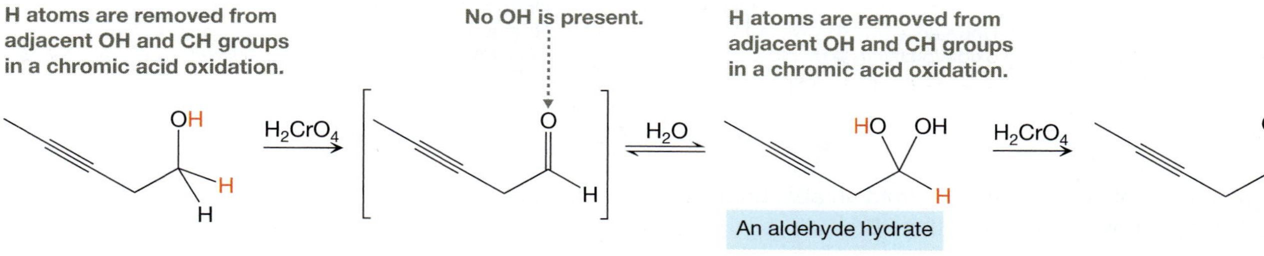

(19-36)

Given the role that water plays in the oxidation of a primary alcohol to a carboxylic acid, the oxidation would have to stop at the aldehyde stage in the absence of water. Chromic acid cannot be used in such an oxidation, however, because water is required to produce chromic acid in the first place (Fig. 19-4). Instead, Equation 19-37 shows that **pyridinium chlorochromate (PCC)** can be used as the oxidizing agent.

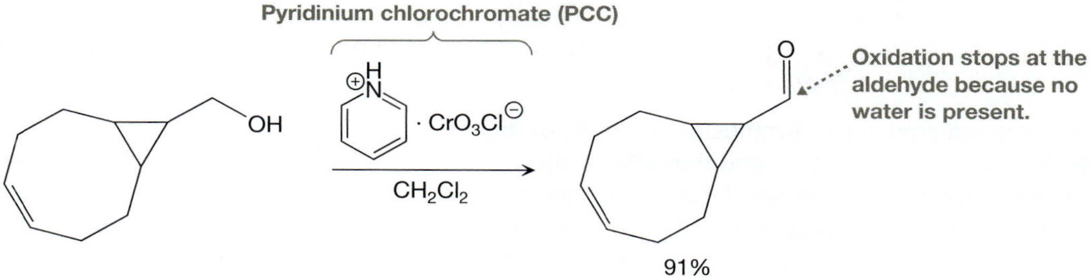

Pyridinium chlorochromate (PCC)

Oxidation stops at the aldehyde because no water is present.

(19-37)

91%

Pyridinium chlorochromate (PCC) oxidizes a primary alcohol to an aldehyde and a secondary alcohol to a ketone.

PCC contains the same Cr^{+6} species as chromic acid. Unlike chromic acid, however, PCC is soluble in nonaqueous solvents like dichloromethane (CH_2Cl_2), which allows oxidation to take place in the *absence of water*.

YOUR TURN 19.8

Which of oxidations **A** and **B** could be carried out using chromic acid? Using PCC?

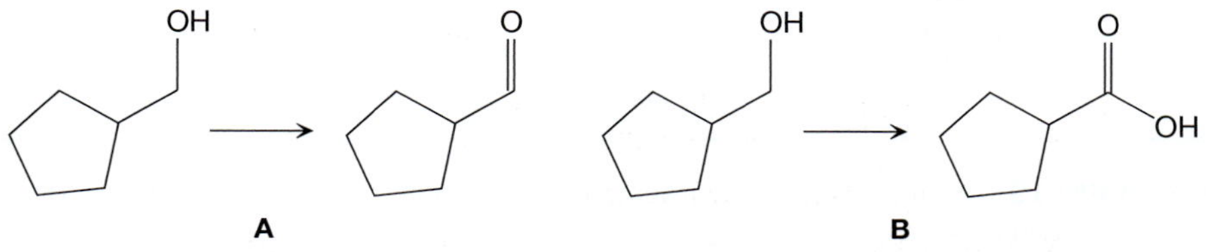

A

B

SOLVED PROBLEM 19.31

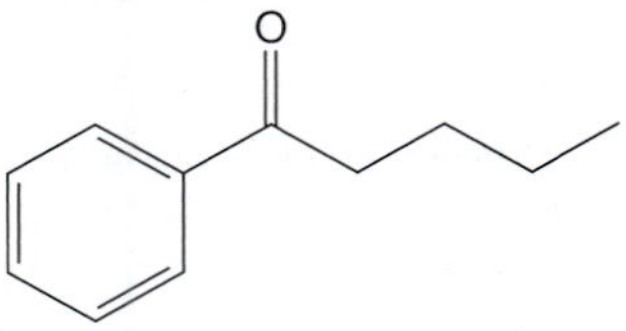

Show how to synthesize this compound using alcohols of seven or fewer carbons as your only carbon source.

Think Which C—C bond can we "disconnect" in a retrosynthetic analysis? By undoing what reaction? Are there any oxidations of alcohols that must be stopped at the aldehyde stage?

Solve A Grignard reaction forms a carbon–carbon bond, yielding an alcohol. Our target is a ketone that could be the result of oxidizing a Grignard product. Our retrosynthetic analysis might therefore begin with the following transforms.

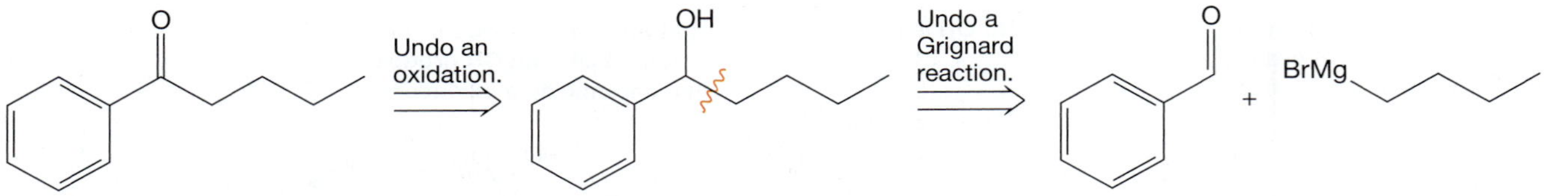

The Grignard reagent can be prepared from an alkyl bromide, which can be produced, in turn, from an alcohol using PBr_3.

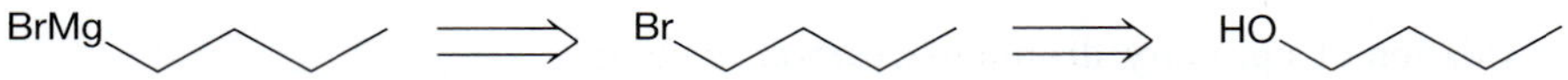

Benzaldehyde, on the other hand, can be made from benzyl alcohol via an oxidation.

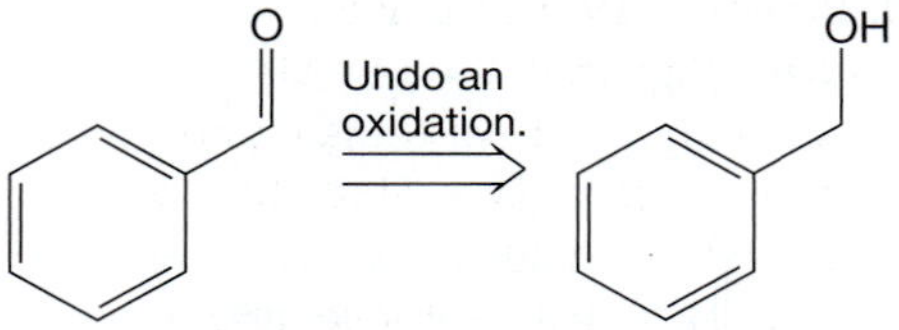

There are two oxidations of alcohols required in this synthesis. One is an oxidation to an aldehyde, which requires an aqueous-free solvent and thus PCC as the oxidizing agent. The other is an oxidation to a ketone, which can take place in water, thus allowing us to use H_2CrO_4 as the oxidizing agent. The overall synthesis can be written as follows:

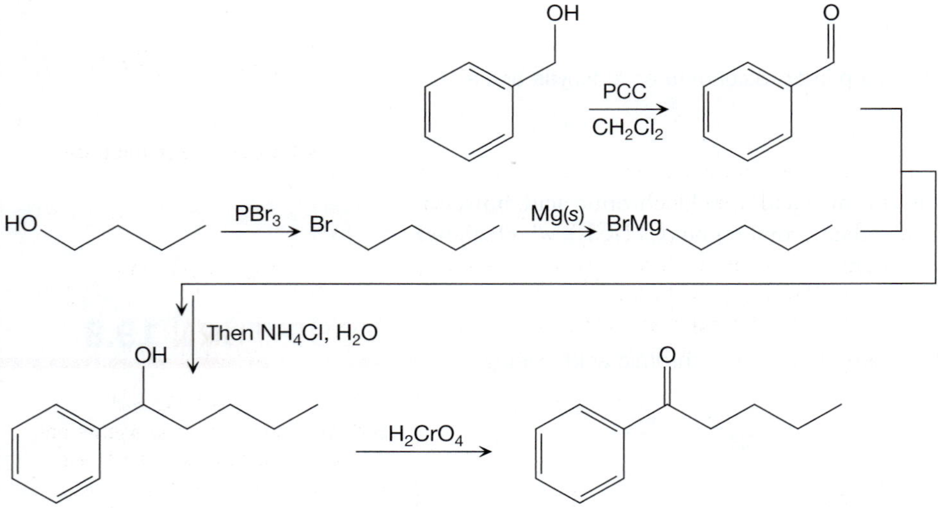

PROBLEM 19.32 Show how to synthesize the compound in Solved Problem 19.31 beginning with C_6H_5MgBr and any alcohol.

Chromic Acid Oxidation and the Breathalyzer Test

According to the National Highway Traffic Safety Administration, more than 10,000 people died in drunk-driving crashes in the United States in 2015. It is no wonder, then, that there are strict laws against driving under the influence of alcohol, which each state defines as having a blood alcohol content (BAC) above 0.08%. Law enforcement officers can measure the BAC of a suspected drunk driver using a Breathalyzer test, the basis of which is the same chromic acid oxidation reaction we studied here in Chapter 19.

To carry out a Breathalyzer test, a person blows into a mouthpiece and a fixed volume of breath is collected. The glass neck of a test vial is broken and the breath sample is forced into that vial. The test vial contains a mixture of $K_2Cr_2O_7$ and H_2SO_4 (i.e., H_2CrO_4), so any ethanol from the breath is quickly oxidized to acetic acid. Cr^{+6}, which is bright orange, is simultaneously reduced to Cr^{+3}, which is blue-green. Light is shined through the test vial and through a second vial containing the unreacted mixture, and the amount of transmitted light is detected by photocells. A difference in the amount of transmitted light through the two vials results in an electric current, which is used to determine BAC.

In more modern instruments, ethanol is oxidized by a fuel cell rather than chromic acid. Oxidation at the anode produces acetic acid, protons, and electrons, and the protons travel to the cathode, where atmospheric oxygen is reduced to water. The electric current that is produced is converted to a BAC reading.

19.6b Permanganate Oxidations of Alcohols and Aldehydes

Potassium permanganate ($KMnO_4$) is another common oxidizing agent; it is an ionic compound consisting of the potassium cation, K^+, and the permanganate anion, MnO_4^-. Structurally, MnO_4^- resembles H_2CrO_4 (Fig. 19-5). In both compounds, a central metal atom is bonded to four O atoms, thus giving the metal atom a high, positive oxidation state—namely, +7 for MnO_4^- and +6 for H_2CrO_4.

Like chromic acid, potassium permanganate can be used to oxidize alcohols, as shown in Equations 19-38 and 19-39.

Potassium permanganate

Chromic acid

FIGURE 19-5 $KMnO_4$ and H_2CrO_4 These oxidizing agents are similar structurally, and each has a metal atom in a high, positive oxidation state.

A secondary alcohol → → A ketone

1. $KMnO_4$, KOH, H_2O, 2.5 h, 0–5 °C
2. HCl, H_2O

(19-38)

60%

A primary alcohol

A carboxylic acid

$$\text{Phenylmethanol (Benzyl alcohol)} \xrightarrow[\text{2. HCl, } H_2O]{\text{1. } KMnO_4, KOH, H_2O} \text{Benzoic acid 75\%}$$ (19-39)

More generally:

> When treated with a basic solution of potassium permanganate ($KMnO_4$), followed by acid:
>
> - Primary alcohols and aldehydes are oxidized to carboxylic acids.
> - Secondary alcohols are oxidized to ketones.

These are the same outcomes we see with H_2CrO_4 as the oxidizing agent. It might therefore seem that $KMnO_4$ and H_2CrO_4 are completely interchangeable, but they are not.

> H_2CrO_4 is more selective than $KMnO_4$.

As discussed in Chapters 22 and 24, $KMnO_4$ can oxidize a variety of compounds in addition to alcohols and aldehydes, including alkenes, alkynes, and alkylbenzenes. As a result, chromic acid's greater selectivity makes it the oxidizing agent of choice when these other groups are present.

GREEN CHEMISTRY
Although H_2CrO_4 is the more selective oxidizing agent, it is also much more toxic than $KMnO_4$ and is a serious environmental hazard. Therefore, $KMnO_4$ is sometimes chosen as the oxidizing agent unless the selectivity of H_2CrO_4 is required in the synthesis.

PROBLEM 19.33 What are the structures of **A** and **B** in the following sequence of reactions?

$$\xrightarrow[\text{2. } H_2O, HCl]{\text{1. } KMnO_4, KOH, \Delta} \textbf{A} \xrightarrow[\text{2. } H_2O, H_2SO_4]{\text{1.}} \textbf{B}$$

19.7 Useful Reactions That Form Carbon–Carbon Bonds: Coupling and Alkene Metathesis Reactions

In Section 13.2, we learned about the importance of reactions that join carbon atoms together. Such carbon–carbon bond-forming reactions are vital in a synthetic chemist's ability to construct elaborate carbon frameworks like the ones we often see in biologically active compounds. To this point in the book, we have studied a handful of reactions that form new C—C bonds and, indeed, we have already seen how useful they are, having incorporated them in numerous organic syntheses.

Here in this section, we introduce some new carbon–carbon bond-forming reactions—namely, *coupling reactions* and *alkene metathesis*. Having these new reactions at our disposal will give us even greater flexibility in designing syntheses.

19.7a Coupling Reactions Involving Organocuprates

As we saw in Section 17.10, lithium dialkylcuprates (R_2CuLi), also called Gilman reagents, act as relatively weak R^- nucleophiles in the conjugate addition to α,β-unsaturated aldehydes, ketones, and other polar π bonds. As shown in Equation 19-40, lithium dialkylcuprates also react with alkyl halides.

In this reaction, two alkyl groups are joined together, one from the alkyl halide and the other from the dialkylcuprate, so these types of reactions are called **coupling reactions**.

Coupling reactions like this can involve a wide variety of alkyl halides and organocuprates:

In the coupling reaction R—X + R'_2CuLi ⟶ R—R′,

- R from the alkyl halide can be a methyl, primary, or secondary alkyl group; a vinylic group; or an aryl group.
- R′ from the dialkylcuprate can be an alkyl, vinylic, or aryl group.
- The halogen X can be Cl, Br, or I.

Examples illustrating some of these variations are shown in Equations 19-41 and 19-42.

Notice in Equation 19-42 that when R—X is a vinylic halide, the R′ group from R'_2CuLi assumes the position originally occupied by the halogen atom. Therefore:

Coupling reactions between R—X and R'_2CuLi are *stereospecific*, taking place with *retention of configuration*.

PROBLEM 19.34 Draw the major organic product for each of the following reactions.

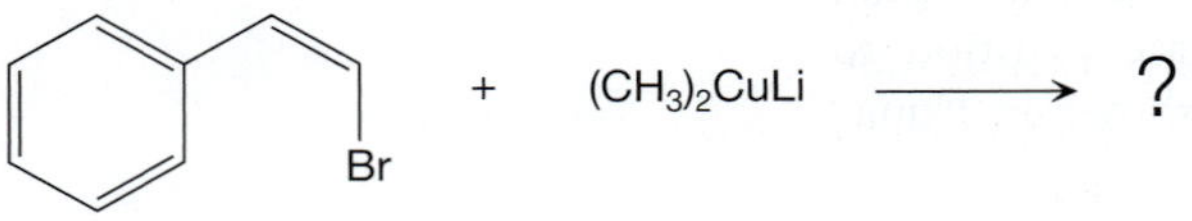

(a)

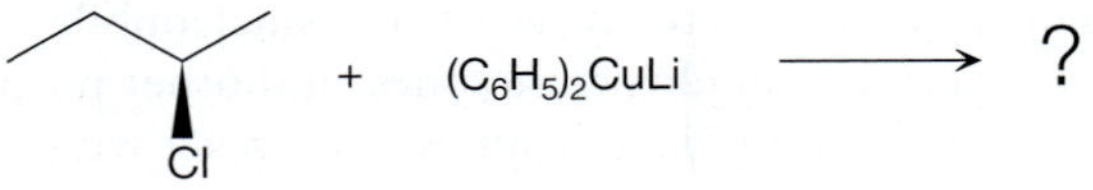

(b)

19.7b Palladium-Catalyzed Coupling Reactions

The organocuprate coupling reactions we examined in Section 19.7a are not the only type of coupling reactions used in organic synthesis. Numerous others have been developed, involving a range of metal atoms. We study two other coupling reactions here in Section 19.7b, both of which involve palladium: the *Suzuki reaction* and the *Heck reaction*. These reactions have found such widespread utility in organic synthesis that Akira Suzuki (b. 1930) and Richard F. Heck (1931–2015), the pioneers of the reactions, shared the 2010 Nobel Prize in Chemistry along with Ei-ichi Negishi (b. 1935).

There are several variations of the **Suzuki reaction**, but in the general reaction, a vinylic or aryl halide is treated with an organoboron compound and a palladium catalyst (PdL_n) under basic conditions:

Suzuki Coupling Reaction

$$R—X + R'—B(OR'')_2 \xrightarrow[\text{Base}]{PdL_n} R—R'$$

- R from RX can be a vinylic or aryl group.
- R' from the boron-containing compound can be an alkyl, vinylic, or aryl group.
- X can be Cl, Br, or I.

In the palladium catalyst, PdL_n, L represents any of a variety of ligands coordinated to the Pd metal center.

An example of the Suzuki reaction is shown in Equation 19-43.

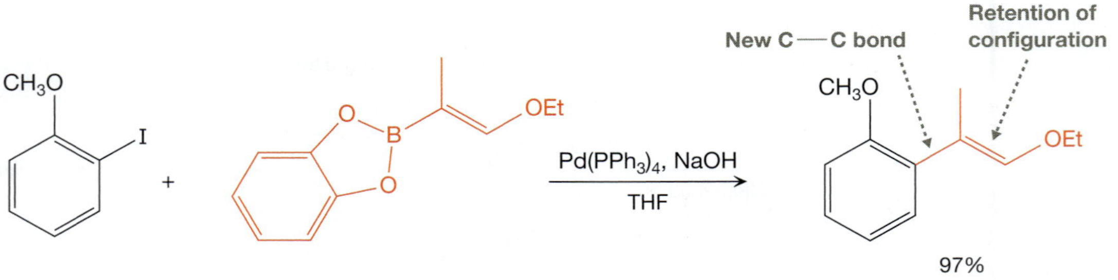

(19-43)

Notice that the Suzuki reaction is stereospecific, much like the coupling reactions involving dialkylcuprates.

Suzuki reactions involving possible *E/Z* isomerism usually proceed with *retention of configuration*.

An abbreviated mechanism for the general Suzuki reaction is shown in Equation 19-44, highlighting four major steps. Step 1 is oxidative addition, where the Pd metal atom essentially inserts between the R and X groups of the vinylic or aryl halide. In Step 2, the hydroxide anion displaces the halide leaving group. Meanwhile, another hydroxide anion coordinates with the $R'B(OR'')_2$ compound, which increases the nucleophilicity of R' and facilitates the transmetallation in Step 3. Finally, reductive elimination occurs in Step 4 to create the new R—R' bond and regenerate the PdL_n catalyst for another cycle.

Abbreviated mechanism for the Suzuki reaction

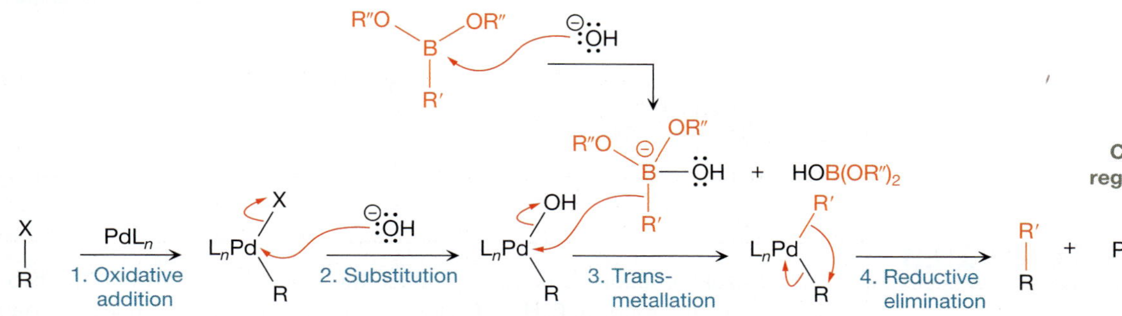

$$(19\text{-}44)$$

Similar to the Suzuki reaction, the **Heck reaction** can couple one vinylic or aryl group to another. An example is shown in Equation 19-45.

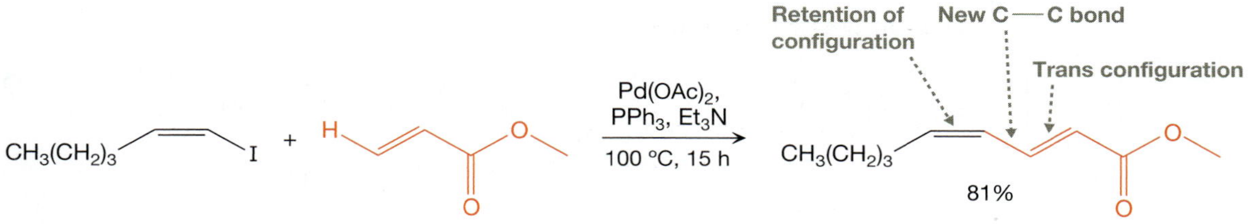

$$(19\text{-}45)$$

The general characteristics of a Heck reaction are as follows:

Heck Coupling Reaction

$$R—X + H—R' \xrightarrow[\text{Base}]{PdL_2} R—R' + H—X$$

- R from RX can be a vinylic or aryl group.
- R' from H—R' can be a vinylic or aryl group.
- X can be Cl, Br, or I.

In essence, R from R—X replaces H from H—R'.

The stereochemistry of the Heck reaction can be summarized as follows:

- If the R—X halide in a Heck reaction is vinylic, the configuration about the double bond is retained.
- If H—R' is vinylic and has an attached substituent (Sub) at the opposite end of the C=C, then R and the substituent will be trans to each other in the product.

Both of these aspects of stereochemistry are illustrated in Equation 19-45.

The abbreviated mechanism for the Heck reaction is shown in Equation 19-46, which highlights four main steps. Step 1 is oxidative addition of the PdL$_2$ catalyst to R—X. In Step 2, the double bond inserts between the Pd and R groups, temporarily converting the double bond to a single bond. Then, in Step 3, the double bond is reformed in an elimination and the organic product is produced. Step 4 regenerates the PdL$_2$ catalyst in a reductive elimination involving the amine base.

Abbreviated mechanism for the Heck reaction

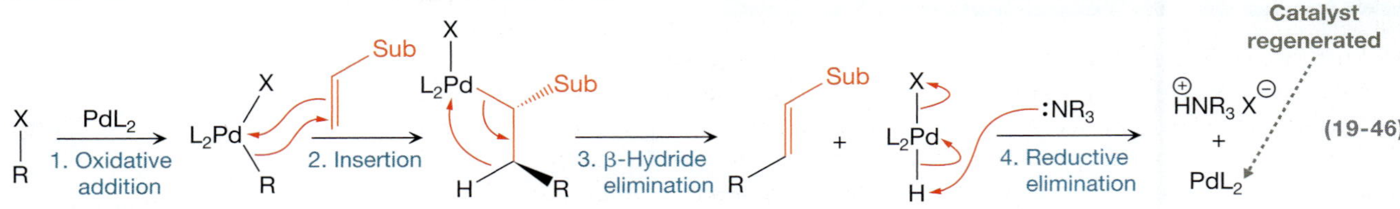

(19-46)

YOUR TURN 19.9

Can a Suzuki reaction be used to produce the single bond between the two phenyl rings in biphenyl, C$_6$H$_5$—C$_6$H$_5$? Can a Heck reaction be used? In each case, what would the precursors be?

PROBLEM 19.35 Draw the major organic product for each of the following reactions.

(a)

(b)

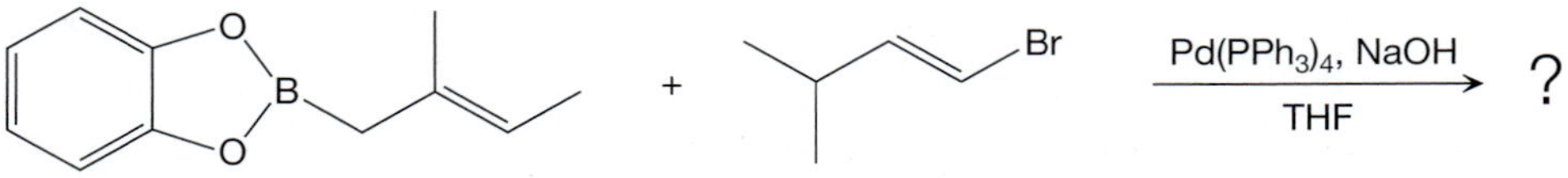

(c)

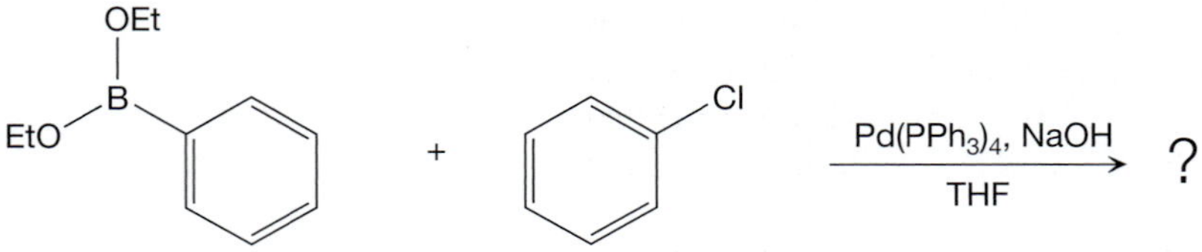

(d)

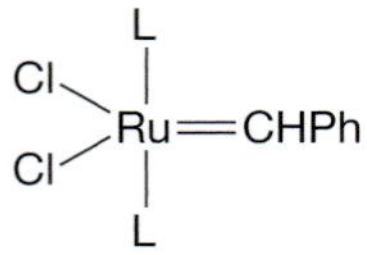

FIGURE 19-6 The general structure of a Grubbs catalyst Grubbs catalysts can vary in their ligands (L), but generally have in common a ruthenium–carbene bond (Ru=C) and two chloride ligands.

19.7c Alkene Metathesis

When dec-4-ene is treated with a **Grubbs catalyst** (Fig. 19-6), a mixture of dec-4-ene (starting material), oct-4-ene, and dodec-6-ene is produced, as shown in Equation 19-47. This is an example of an **alkene metathesis** or **olefin metathesis** reaction, in which the portions of the molecules joined by the C=C bond in the products were not initially joined in the reactant.

$$CH_3(CH_2)_2CH \!\!=\!\! CH(CH_2)_4CH_3 \;\underset{4\,h}{\overset{\text{Grubbs catalyst}}{\rightleftharpoons}}\; CH_3(CH_2)_2CH \!\!=\!\! CH(CH_2)_2CH_3 \;+\; CH_3(CH_2)_4CH \!\!=\!\! CH(CH_2)_4CH_3 \quad \text{(19-47)}$$

Dec-4-ene **Oct-4-ene** **Dodec-6-ene**

Mixture of *E* and *Z* isomers
39%

Alkene metathesis reactions generally take place under equilibrium conditions, as indicated in Equation 19-47. Therefore, with the product alkenes having similar thermodynamic stabilities to the reactant alkene, the yield is relatively low.

As is the case with any chemical equilibrium, Le Châtelier's principle can be exploited by removing products as they form. For a reaction like the one in Equation 19-47, where the products form as a mixture of liquids, this can be cumbersome and quite challenging. If the reactants are terminal alkenes, however, as in Equation 19-48, then one of the products is ethene, which is a gas that escapes the system as it forms. The reaction is then effectively irreversible.

Both C=C bonds are terminal.   **Ethene is a gas that leaves the system.**

$$+ \; H_2C \!\!=\!\! CH_2 \quad \text{(19-48)}$$

82%   **Ethene**

This aspect of alkene metathesis can be exploited to form a new ring from a compound containing two terminal C=C bonds. An example of this **ring-closing metathesis** is shown in Equation 19-49.

$$+ \; H_2C \!\!=\!\! CH_2 \quad \text{(19-49)}$$

86%

The general mechanism for alkene metathesis is shown in Equation 19-50 (next page), where R—CH=CH$_2$ represents a terminal alkene and M=CHPh represents the Grubbs catalyst (M is the RuCl$_2$L$_2$ portion of it). In Step 1, the Grubbs catalyst adds to the original alkene to produce a four-membered ring. In Step 2, the ring opens to produce a species in which M is doubly bonded to one of the fragments from the original alkene, represented as M=CH—R. The same two steps repeat in Steps 3 and 4, this time involving M=CH—R instead of the Grubbs catalyst.

The mechanism in Equation 19-50 shows explicitly how R—CH=CH—R is produced. The other alkene product of this reaction, H$_2$C=CH$_2$, is produced from the same mechanism, the difference being the relative orientations of the original alkene and the Grubbs catalyst in Step 1. To produce H$_2$C=CH$_2$, the CH$_2$ group must be oriented on the same end as M and the CH—R group must be oriented on the same end as CHPh.

YOUR TURN 19.10

Modify the mechanism in Equation 19-50 to show how H$_2$C=CH$_2$ is produced.

Mechanism for alkene metathesis

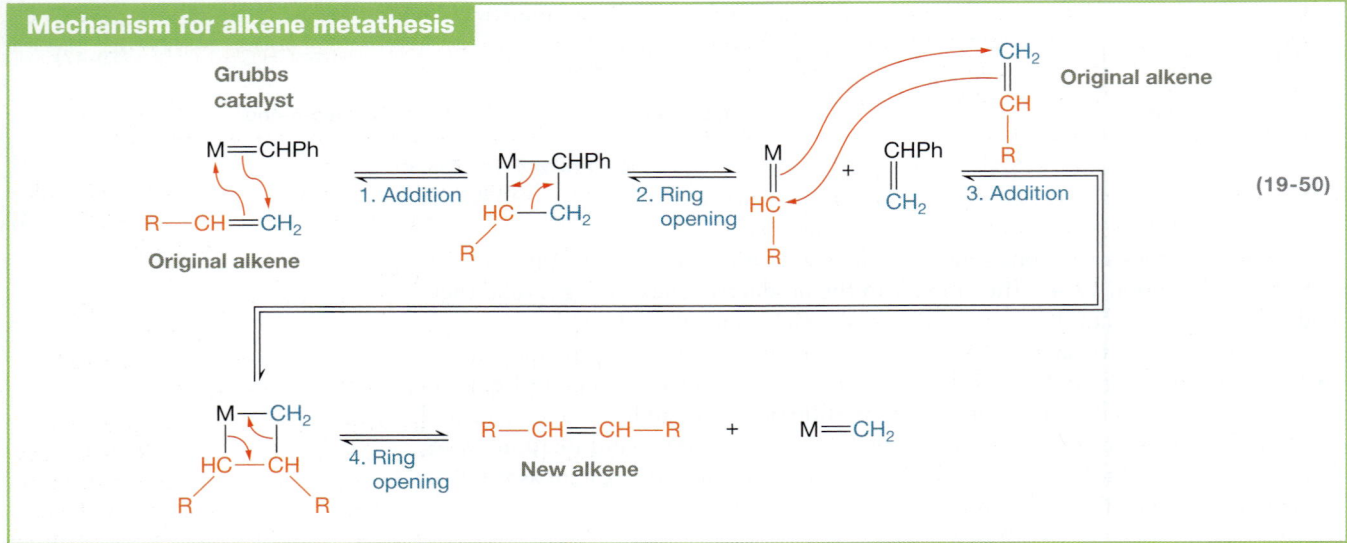

PROBLEM 19.36 Draw the major organic products that would be formed when each of the following is treated with Grubbs catalyst.

(a)   (b)   (c)

Chapter Summary and Key Terms

- The formation of a bond between two carbon atoms is a problem when the atoms are of like charge. To circumvent such a problem, we can first carry out a reaction that reverses the charge (polarity) of one atom via a process called **umpolung**. (Section 19.1; Objective 1)

- Umpolung takes place at a haloalkane carbon on treatment with a metal such as Mg or Li, producing a Grignard or alkyllithium reagent, respectively. (Section 19.1; Objective 1)

- Many carbon–carbon bond-forming reactions leave heteroatoms with a very specific relative positioning. Thus, target molecules that exhibit one such relative positioning provide clues as to which of those reactions should be considered in a synthesis. (Section 19.2; Objective 2)

- If a synthesis requires the formation of a new carbon–carbon bond, but the target does not exhibit a specific relative positioning of heteroatoms, then the synthesis may require a reaction that removes a functional group entirely from a molecule. (Section 19.3; Objective 3)

  - Reactions that remove the C=O functional group of a ketone or aldehyde include the *Wolff–Kishner reduction*, the

Clemmensen reduction, and the **Raney-nickel reduction**. These reactions take place under basic, acidic, and catalytic hydrogenation conditions, respectively, so care must be taken to avoid unwanted side reactions at other functional groups in the molecule. (Section 19.3)

- The C=C functional group of an alkene or the C≡C functional group of an alkyne can be removed via catalytic hydrogenation. (Section 19.5)

- If two or more reactive functional groups appear in a molecule, we can often carry out a reaction at just one of those groups by using either a *selective reagent* or a *protecting group*. (Section 19.4; Objectives 4, 5)

  - A **selective reagent** favors reaction with one functional group over another. (Section 19.4a)

  - A **protecting group** temporarily converts a functional group into one that is unreactive under the conditions necessary for a step in the synthesis. Ketones, aldehydes, and alcohols can often be protected in the form of an acetal or ether. (Section 19.4b)

- **Catalytic hydrogenation** is a reaction in which H_2 adds to a molecule in the presence of a metal catalyst such as Pt, Pd, or Ni. Catalytic hydrogenation readily reduces an alkene or an alkyne to an alkane. **(Sections 19.5a and 19.5b; Objective 6)**

- Using a **poisoned catalyst**, catalytic hydrogenation reduces an alkyne to a cis alkene. **(Section 19.5b; Objective 6)**

- Catalytic hydrogenation will readily reduce an aldehyde to an alcohol. Under moderate conditions, a ketone can also be reduced to an alcohol, and under more extreme conditions, other compounds such as nitriles and amides can be reduced. **(Section 19.5c; Objective 6)**

- Catalytic hydrogenation is selective toward alkenes, alkynes, and aldehydes. **(Section 19.5d; Objectives 4, 6)**

- **Chromic acid** (H_2CrO_4) will oxidize a primary alcohol to a carboxylic acid or a secondary alcohol to a ketone. **(Section 19.6a; Objective 7)**

- Oxidation of a primary alcohol by **pyridinium chlorochromate (PCC)** stops at the aldehyde because the reaction takes place in a nonaqueous medium. **(Section 19.6a; Objective 7)**

- Oxidation of an alcohol by **potassium permanganate** ($KMnO_4$) produces the same ketone and carboxylic acid products that would be produced by chromic acid. **(Section 19.6b; Objective 7)**

- A **coupling reaction** joins one alkyl, vinylic, or aryl group to another, forming a new C—C bond. **(Section 19.7; Objective 8)**

  - An alkyl, vinylic, or aryl halide (R—X) will react with a lithium dialkylcuprate (R'_2CuLi) to produce R—R'. **(Section 19.7a)**

  - In a **Suzuki reaction**, a vinylic or aryl halide (R—X) will react with an organoboron compound [R'—B(OR'')_2] under basic conditions in the presence of a palladium catalyst to produce R—R'. **(Section 19.7b)**

  - In a **Heck reaction**, the R group from a vinylic or aryl halide (R—X) will replace a vinylic or aryl H from H—R' to produce R—R' when treated with a palladium catalyst under basic conditions. **(Section 19.7b)**

- An **alkene metathesis** occurs when one or more alkenes is treated with **Grubbs catalyst**, which contains a Ru=C bond. In the product of an alkene metathesis, the fragments that are joined by a double bond were initially part of a double bond to other fragments in the reactant. These reactions are particularly useful when the reactants are terminal alkenes. When the reaction produces a new ring, it is called a **ring-closing metathesis**. **(Section 19.7c; Objective 8)**

Reaction Tables

TABLE 19-4 Functional Group Conversions[a]

	Starting Compound Class	Typical Reagents and Reaction Conditions	Compound Class Formed	Comments	Discussed in Section
(1)	R—X Haloalkane	$\xrightarrow[\text{Ether}]{\text{Mg}(s)}$	R—MgX Grignard reagent	Dissolving metal reduction	19.1
(2)	R—X Haloalkane	$\xrightarrow[\text{Ether}]{\text{Li}(s)}$	R—Li Alkyllithium reagent	Dissolving metal reduction	19.1
(3)	R—Li Alkyllithium reagent	$\xrightarrow[\text{Ether}]{\text{CuI}}$	R—Cu(Li)—R Lithium dialkylcuprate	—	19.1

[a]X = Cl, Br, I.

(continued)

TABLE 19-4 Functional Group Conversions[a] (continued)

	Starting Compound Class	Typical Reagents and Reaction Conditions	Compound Class Formed	Comments	Discussed in Section(s)
(4)	Ketone or aldehyde	Zn/Hg, HCl, H₂O, reflux	Alkane	Clemmensen reduction	19.3b
(5)	Ketone or aldehyde	1. HSCH₂CH₂SH, H⊕ 2. Raney Ni (H₂)	Alkane	Raney-nickel reduction	19.3c
(6)	Alkene	H₂, Pt, Pd, or Ni	Alkane	Catalytic hydrogenation	19.5a
(7)	Alkyne	H₂, Lindlar catalyst	cis-Alkene	Catalytic hydrogenation; poisoned catalyst	19.5b
(8)	Ketone or aldehyde	H₂, Pt, Pd, or Ni	Alcohol	Catalytic hydrogenation	19.5c
(9)	Primary alcohol	H₂CrO₄ or 1. KMnO₄, KOH 2. H₂O, HCl	Carboxylic acid	Oxidation	19.6a, 19.6b
(10)	Secondary alcohol	H₂CrO₄ or KMnO₄, KOH	Ketone	Oxidation	19.6a, 19.6b
(11)	Primary or secondary alcohol	PCC	Aldehyde or ketone	Oxidation	19.6a, 19.6b

[a]X = Cl, Br, I.

TABLE 19-5 Reactions That Alter the Carbon Skeleton[a]

	Starting Compound Class	Typical Reagents and Reaction Conditions	Compound Class Formed	Comments	Discussed in Section
(1)	R—X Alkyl, vinylic, or aryl halide	R'_2CuLi	R—R'	Coupling reaction	19.7a
(2)	R—X Vinylic or aryl halide	R'—B(OR")$_2$ PdL$_n$, base	R—R'	Suzuki reaction	19.7b
(3)	R—X Vinylic or aryl halide	H—R' PdL$_2$, base	R—R'	Heck reaction	19.7b
(4)	R—CH=CH$_2$ Terminal alkene	Grubbs catalyst	R—CH=CH—R Alkene	Alkene metathesis	19.7c

[a]X = Cl, Br, I.

Problems

Problems that are related to synthesis are denoted (SYN).

19.1 Umpolung in Organic Synthesis

19.37 For each pair of molecules, determine whether umpolung (polarity reversal) should be considered to join the highlighted carbons together.

(a)

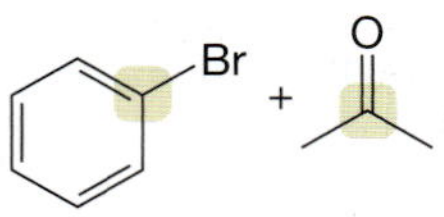

(b)

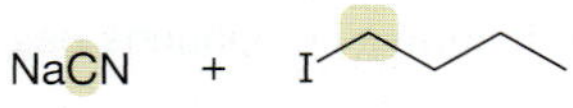

NaCN +

(c)

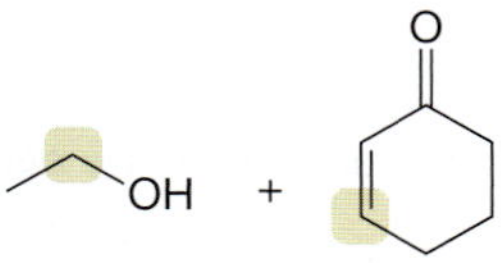

19.38 (SYN) Show how each of the following Grignard reagents can be synthesized from an alkyl, alkenyl, alkynyl, or aryl halide.

(a) MgBr

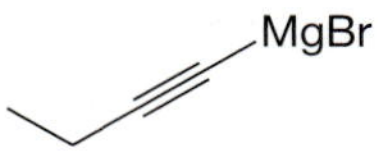

(b) MgI

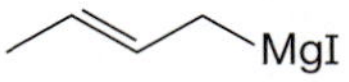

(c) MgBr

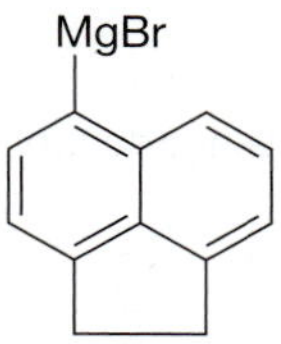

(d) MgBr

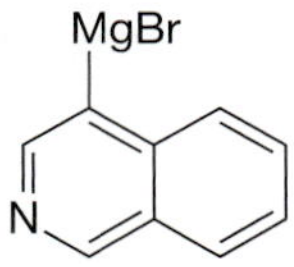

(e) MgI

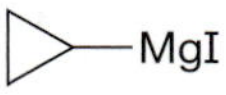

(f) ClMg

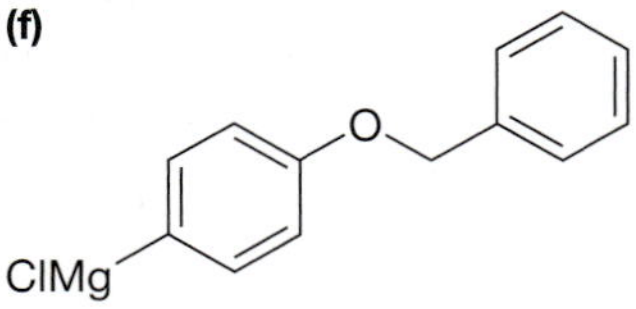

19.39 (SYN) Show how each of the following alkyllithium reagents can be synthesized from an alkyl, alkenyl, alkynyl, or aryl halide.

(a)

(b)

(c)

(d)

19.40 (SYN) Show how each of the following lithium dialkylcuprate reagents can be synthesized from an alkyl, alkenyl, alkynyl, or aryl halide.

(a)

(b)

(c)

(d)

19.41 Draw the organometallic compound that would be produced by each of the following reactions.

(a)

Br $\xrightarrow[\text{Ether}]{\text{Li(s)}}$?

(b)

$H_2C=C=CH$ $\xrightarrow[\text{THF}]{\text{Mg(s)}}$?
Br

(c)

$\xrightarrow[\text{THF}]{\text{Mg(s)}}$?

(d)

Cl $\xrightarrow[\text{Ether}]{\text{Mg(s)}}$?

(e)

Br $\xrightarrow[\text{2. CuI}]{\text{1. Li(s), THF}}$?

(f)

Br $\xrightarrow[\text{THF}]{\text{Li(s)}}$?

(g)

Cl $\xrightarrow[\text{2. CuI}]{\text{1. Li(s), THF}}$?

19.2 and 19.3 Relative Positioning of Heteroatoms in Carbon–Carbon Bond-Forming Reactions; Reduction of C=O to CH$_2$

19.42 (SYN) Suppose that each of the compounds below must be synthesized using compounds with six or fewer carbons. Which carbon–carbon bond-forming reaction(s) from Table 19-1 should you consider incorporating?

(a) Br Br

(b) Br ... Br

(c) Br Br

(d) Br ... Br

19.43 (SYN) Suggest how you would synthesize each of the following compounds beginning with pentan-2-one.

(a) OH OH

(b) OH ... OH

(c) O ... OH

19.44 Predict the product of each of the following reactions.

(a)

Br ... O $\xrightarrow[\text{HCl}]{\text{Zn/Hg}}$?

(b)

O $\xrightarrow[\text{NaOH, H}_2\text{O, }\Delta]{\text{H}_2\text{NNH}_2}$?

(c)

O ... OH ... OH $\xrightarrow[\text{2. Raney Ni}]{\text{1. HS} \frown \text{SH}, \text{H}^{\oplus}}$?

19.45 (SYN) Show how to carry out each of the following transformations.

(a)

(b)

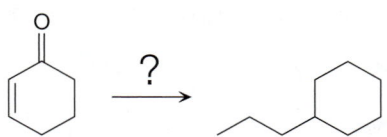

(c)

19.46 (SYN) Show how to carry out this transformation using (CH₃CH₂CH₂)₂CuLi as your only other carbon source.

19.4 Selective Reagents and Protecting Groups

19.47 Which carbonyl groups in the following reactions would require protection?

(a)

1. NaNH₂
2. CH₃I

(b)

Mg(s)

(c)

NaSCH₃

(d)

NaBH₄
Ethanol

19.48 Which hydroxyl groups in the following reactions would require protection?

(a)

Mg(s)

(b)

NaCN

(c)

1. NaH
2. C₆H₅CH₂I

(d)

NaBH₄
Ethanol

19.49 Here in Chapter 19, we learned that converting a ketone or aldehyde to an acetal is a good way to protect the carbonyl group, because an acetal is composed of ether linkages. In Problem 17.57, we saw that a ketone or aldehyde can be converted into an epoxide, which is a cyclic ether. Why would an epoxide be a poor choice as a protecting group?

19.50 (SYN) A student wants to carry out the reaction shown here. Explain the problem(s) associated with this synthesis scheme and suggest a way to carry out the transformation efficiently.

1. LDA
2. CH₃I

19.51 Suppose we want to carry out the following functional group conversions in which A is converted to B. In each case, however, the hydroxyl group is reactive under the reaction conditions that carry out the transformation, so protecting the hydroxyl group would be necessary. Here in Chapter 19, we learned more than one way to protect hydroxyl groups. Which ways do you think will be effective? Which ones will not? Why?

(a) A → B

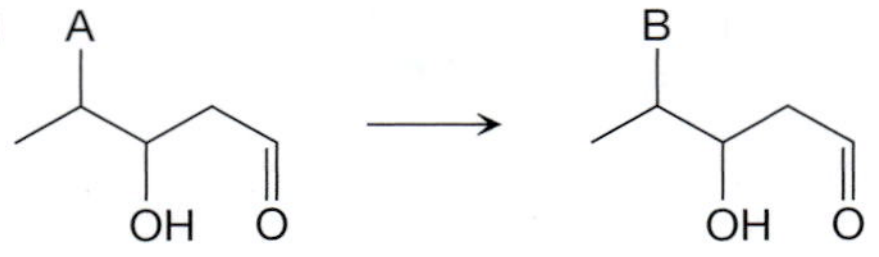

(b) A → B

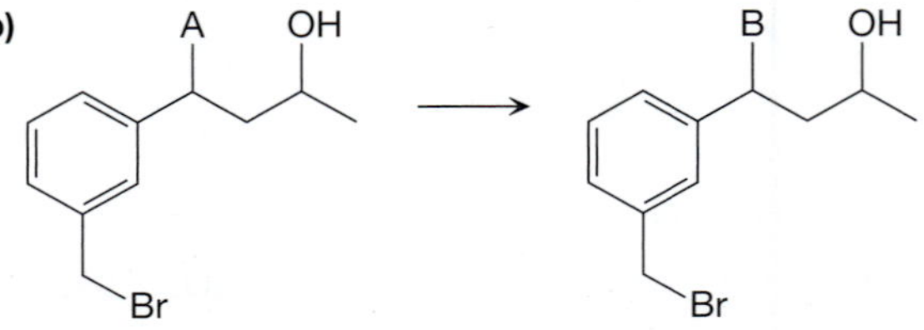

19.52 (SYN) Propose how you would carry out each of the following syntheses.

(a)

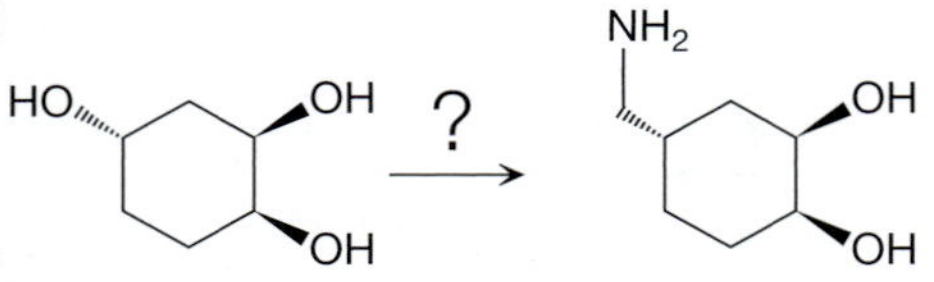

(b)

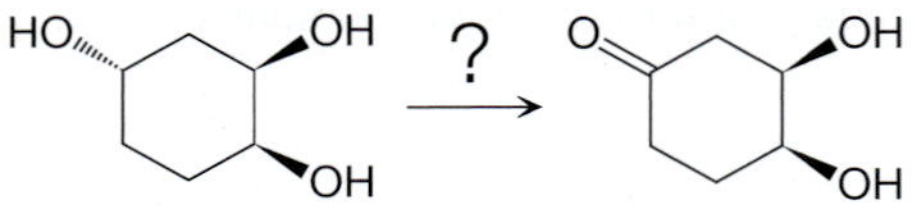

19.53 (SYN) Show how you would carry out this synthesis. *Hint:* What rearrangement occurs when an enol is formed?

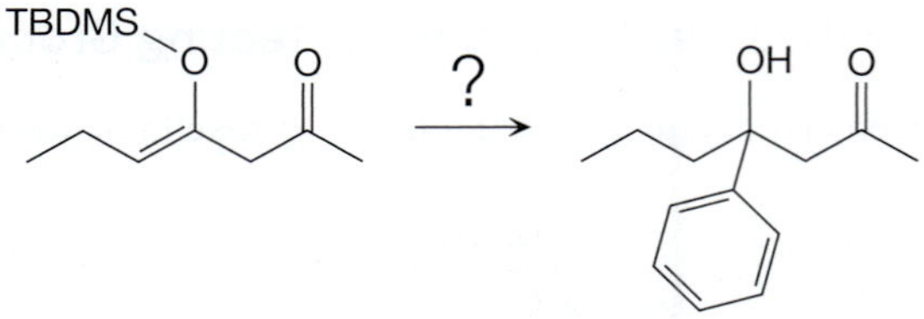

19.5 and 19.6 Catalytic Hydrogenation; Oxidation of Alcohols and Aldehydes

19.54 Predict the product for each of the following reactions. Unless otherwise indicated, you may assume that one molar equivalent of H_2 reacts.

(a) $\xrightarrow[\text{Pd}]{H_2}$?

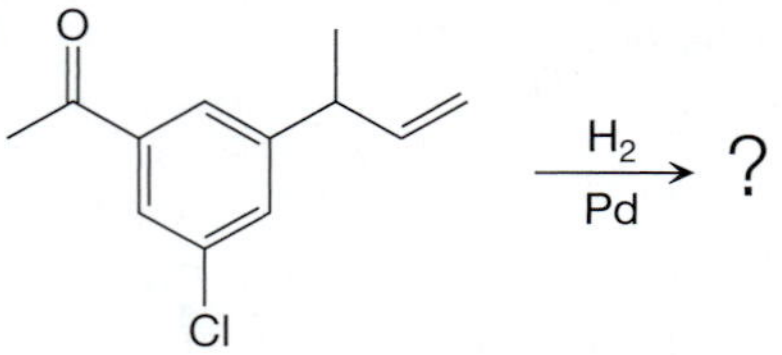

(b) $\xrightarrow[\text{Pt}]{H_2}$?

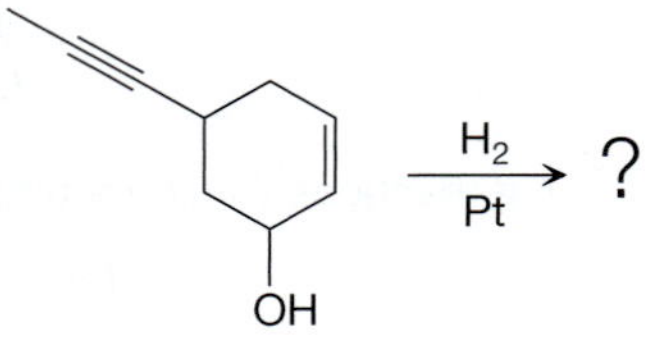

(c) $\xrightarrow[\text{Pt}]{H_2}$?

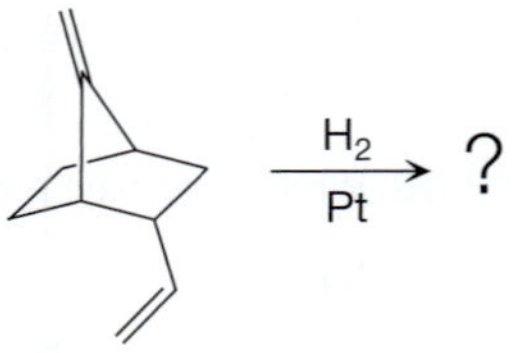

(d) $\xrightarrow[\substack{\text{Lindlar} \\ \text{catalyst}}]{H_2}$?

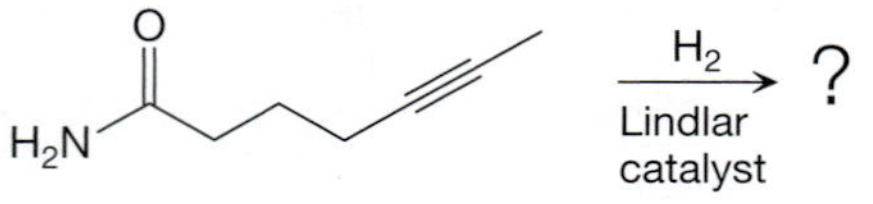

(e) $\xrightarrow[\text{Pt}]{H_2}$?

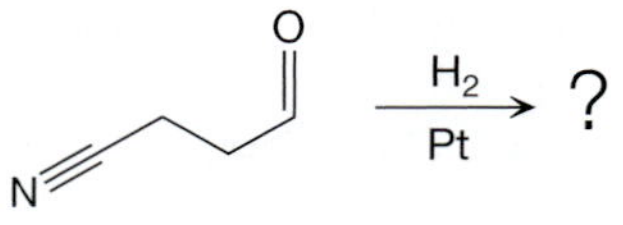

(f) $\xrightarrow[\text{Ni}]{H_2}$?

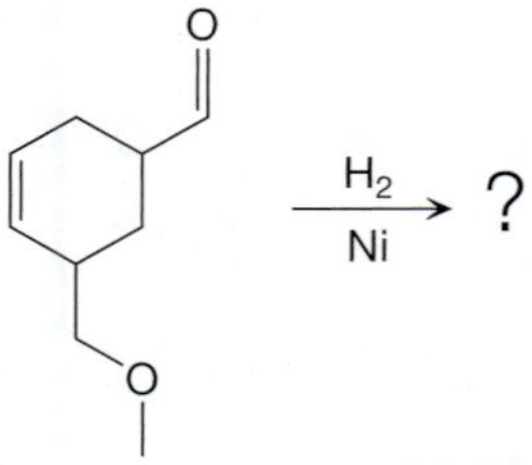

(g) $\xrightarrow[\text{Pd}]{H_2 \text{ (excess)}}$?

19.55 (SYN) Show how to carry out each of the following transformations.

(a)

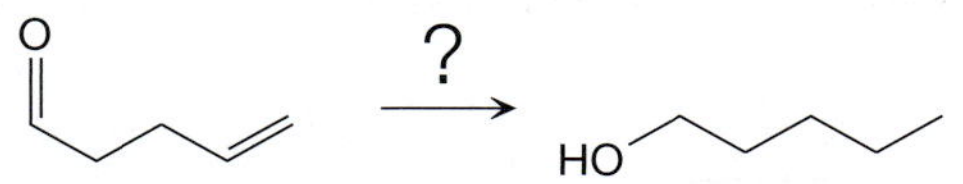

(b)

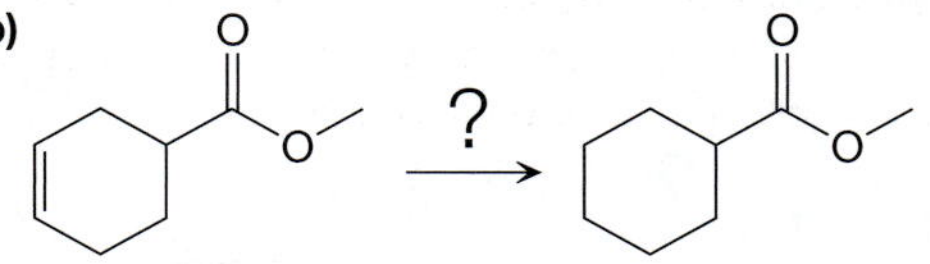

(c)

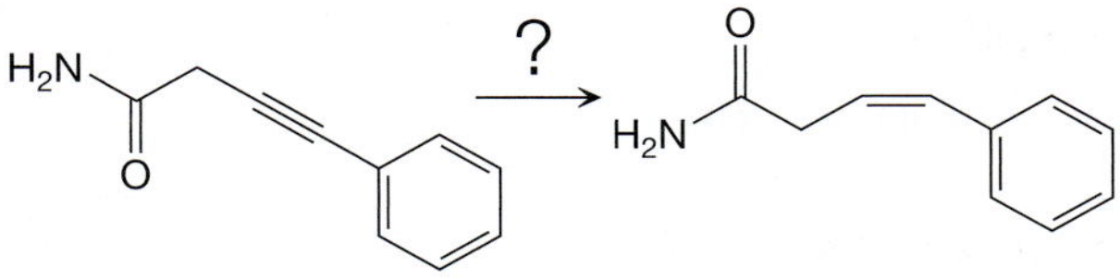

(d)

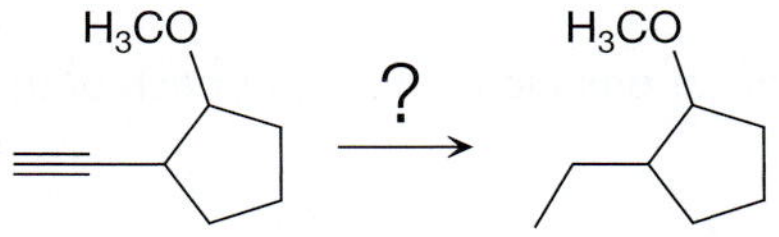

19.56 Predict the product of each of the following reactions.

(a)

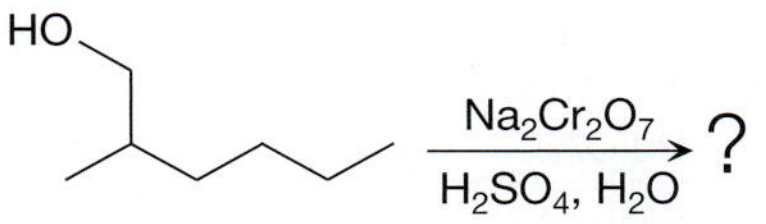

$$\xrightarrow[\text{H}_2\text{SO}_4, \text{H}_2\text{O}]{\text{Na}_2\text{Cr}_2\text{O}_7} \text{?}$$

(b)

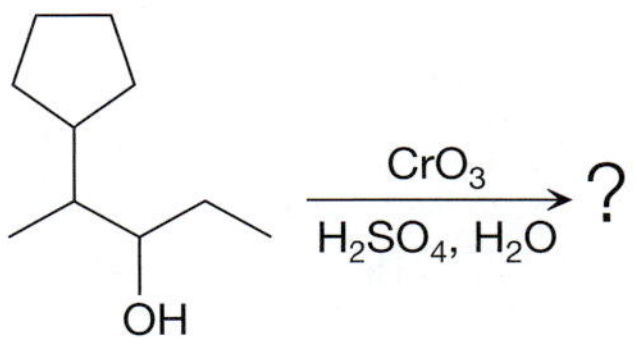

$$\xrightarrow[\text{H}_2\text{SO}_4, \text{H}_2\text{O}]{\text{CrO}_3} \text{?}$$

(c)

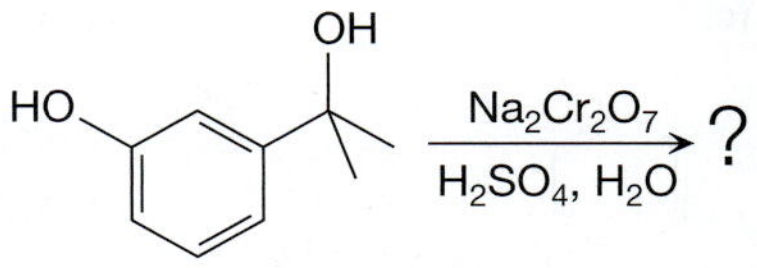

$$\xrightarrow[\text{H}_2\text{SO}_4, \text{H}_2\text{O}]{\text{Na}_2\text{Cr}_2\text{O}_7} \text{?}$$

(d)

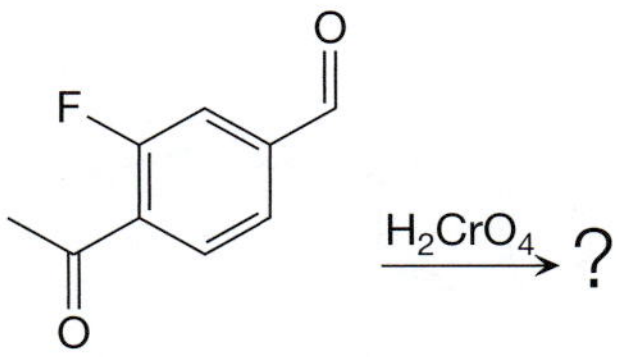

$$\xrightarrow{\text{H}_2\text{CrO}_4} \text{?}$$

(e)

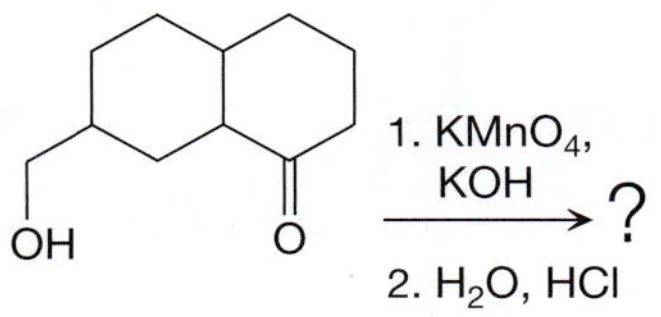

1. KMnO$_4$, KOH
2. H$_2$O, HCl

(f)

$$\xrightarrow[\text{CH}_2\text{Cl}_2]{\text{PCC}} \text{?}$$

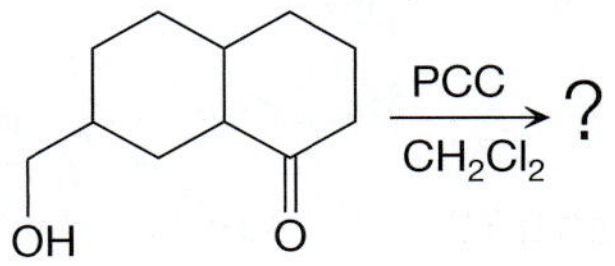

19.57 (SYN) Show how to carry out each of the following transformations.

(a)

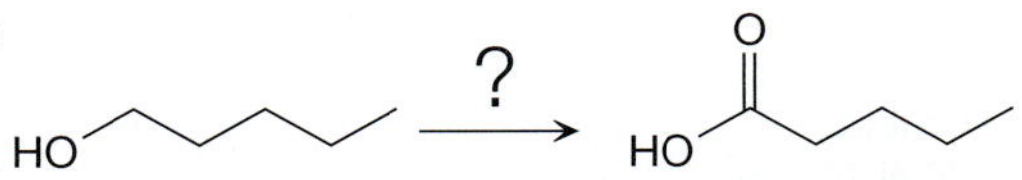

(b)

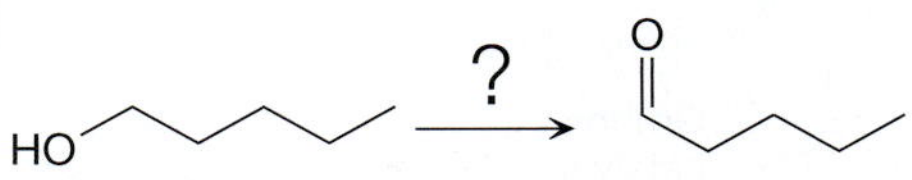

(c)

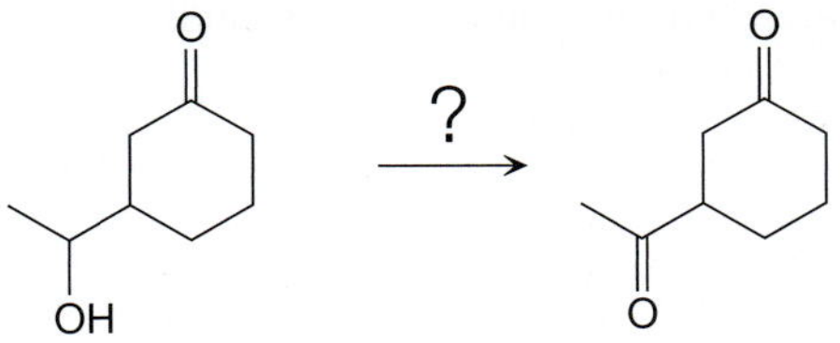

(d)

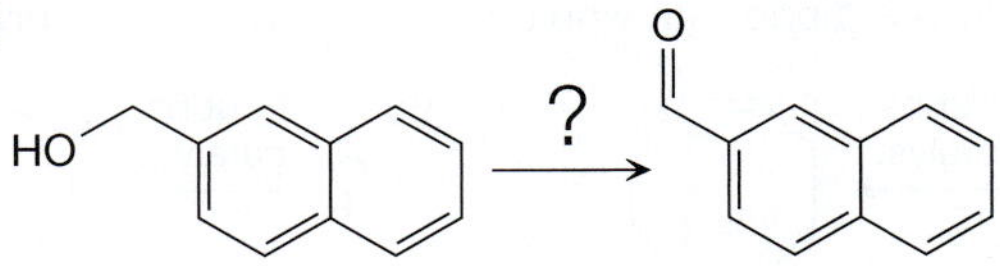

19.7 Coupling and Alkene Metathesis Reactions

19.58 Draw the major organic products for each of the following reactions. If no reaction occurs, state so.

(a)

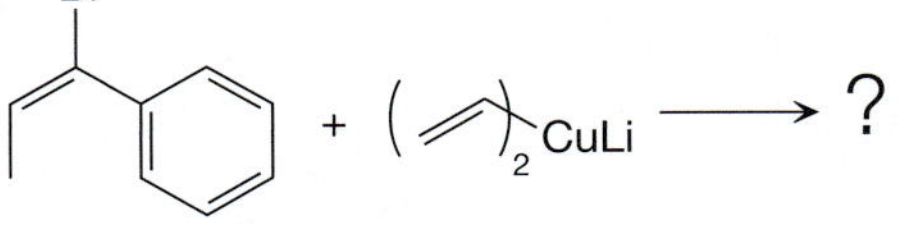

(b)

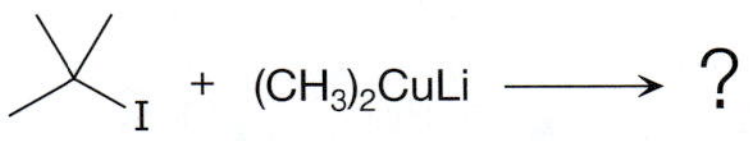

(c)

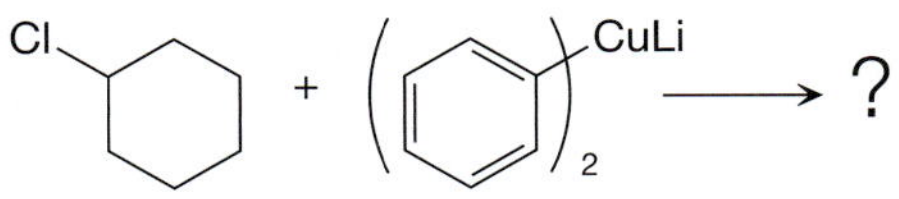

(d)

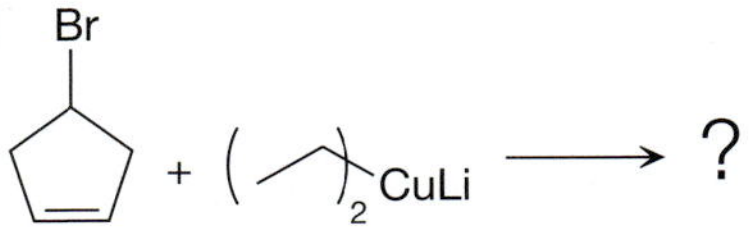

19.59 (SYN) Draw the missing lithium dialkylcuprate that would be necessary to carry out each of the following transformations.

(a)

Br

?

(b)

?

19.60 Draw the major organic products for each of the following reactions.

(a)

+ OCH₃

Pd(PPh₃)₄, NaOH / THF → ?

(b)

OEt

EtO—B

+

Pd(PPh₃)₄, NaOH / THF → ?

(c)

I

+ O—

Pd(OAc)₂ / PPh₃, Et₃N → ?

(d)

Br

+

Pd(OAc)₂ / PPh₃, Et₃N → ?

19.61 (SYN) Draw the missing reactant that would be necessary to carry out each of the following transformations.

(a)

OEt

EtO—B

+ ?

Pd(PPh₃)₄, NaOH / THF →

(b)

I

+ ?

Pd(OAc)₂ / PPh₃, Et₃N →

19.62 Draw the major organic products for each of the following reactions.

(a)

Grubbs catalyst → ?

(b)

Grubbs catalyst → ?

19.63 (SYN) Draw a diene that would react with Grubbs catalyst to produce each of the following compounds.

(a)

? Grubbs catalyst →

(b)

? Grubbs catalyst →

Integrated Problems

19.64 Supply the missing intermediates and reagents in the following synthesis.

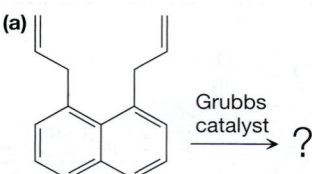

1. LiAlH₄
2. NH₄Cl, H₂O

A

1. PBr₃
2. Mg(s), ether

B

1. **C**
2. NH₄Cl, H₂O

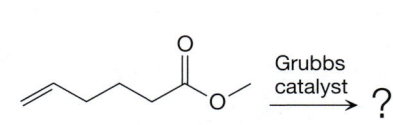

OH

1. NaH
2. **D**

19.65 Supply the missing intermediates and reagents in the following synthesis.

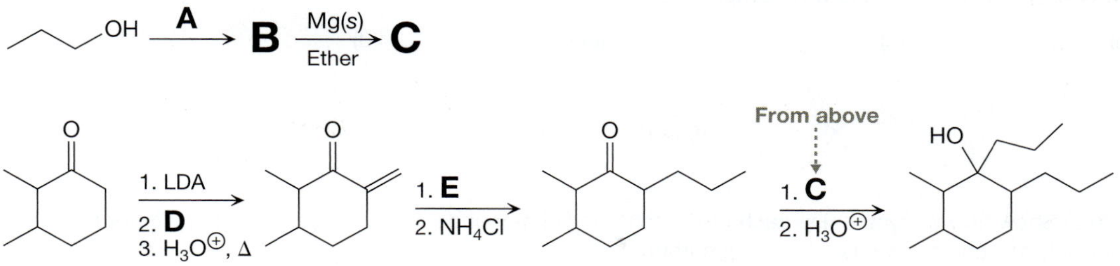

19.66 Provide the missing intermediates and reagents in the synthesis shown here.

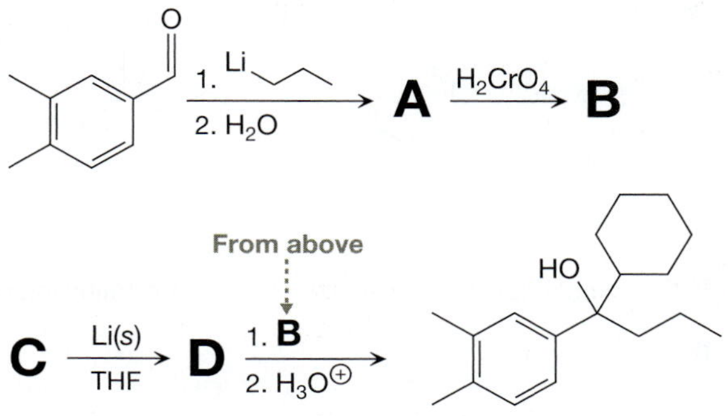

19.67 Supply the missing reagents, intermediates, and final product in the following synthesis.

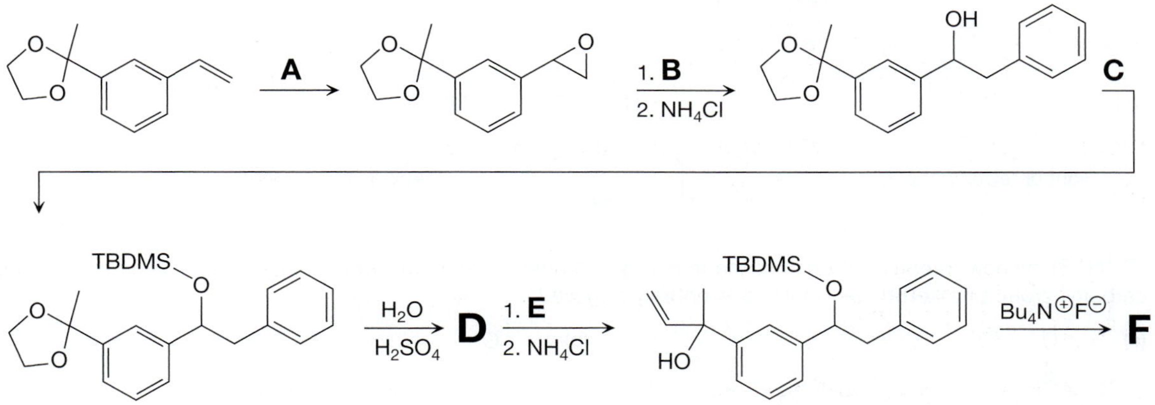

19.68 Provide the missing intermediates and final product in the following synthesis.

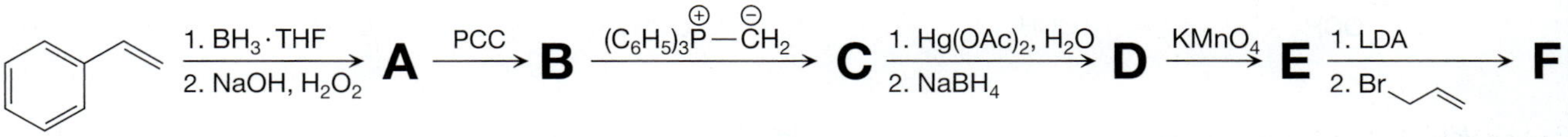

19.69 Provide the missing intermediates and final product in the following synthesis.

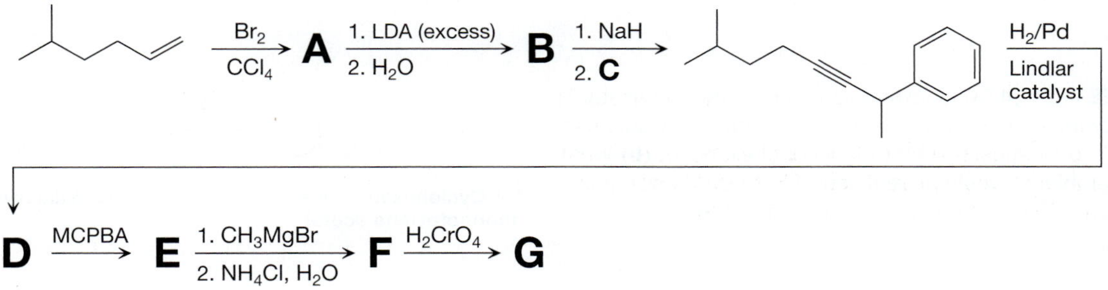

19.70 (SYN) Using acetone, any alcohol with six or fewer carbons, and any inorganic reagents necessary, show how to synthesize each of the following compounds.

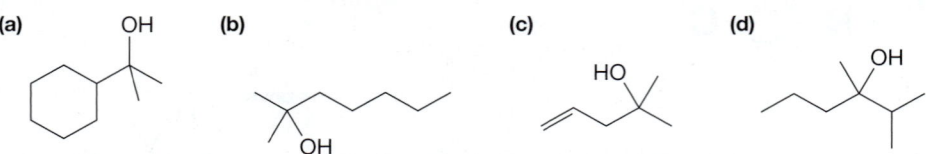

(a)   (b)   (c)   (d)

19.71 (SYN) Show how to synthesize each of the following compounds, using propanal and any other ketone or aldehyde as your only starting materials containing carbon.

(a)   (b)   (c)   (d)

19.72 (SYN) Show how to synthesize each of the following compounds, using the given restrictions on the starting materials.

(a)

Three or fewer carbons $\xrightarrow{?}$

(b)

Six or fewer carbons $\xrightarrow{?}$

(c)

Eight or fewer carbons $\xrightarrow{?}$

(d)

Acyclic compounds $\xrightarrow{?}$

19.73 (SYN) Show how to carry out each of the following syntheses, using any reagents necessary. *Hint:* In each case, the carbonyl group of a ketone or aldehyde is entirely removed.

(a) $\xrightarrow{?}$

(b) $\xrightarrow{?}$

(c) $\xrightarrow{?}$

(d) $\xrightarrow{?}$

19.74 (SYN) 1,4-Cyclohexanedione monoethylene acetal is commercially available. **(a)** Show how you would use it to synthesize 4-ethylidenecyclohexanone. **(b)** What problems would arise if you tried to synthesize the same target from 1,4-cyclohexanedione?

$\xrightarrow{?}$

1,4-Cyclohexanedione monoethylene acetal

4-Ethylidenecyclohexanone

19.75 (SYN) Show how to synthesize each of the following molecules beginning with phenylmethanol (benzyl alcohol).

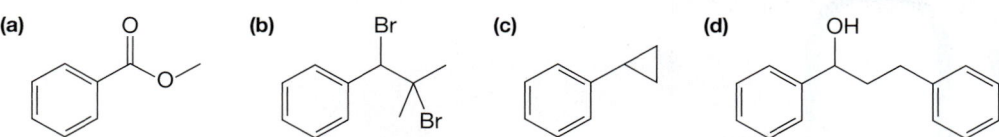

(a)   **(b)**   **(c)**   **(d)**

19.76 (SYN) Show how you would synthesize hexane-3,4-diol using propanal as your only source of carbon atoms.

19.77 (SYN) Show how you would carry out each of the following syntheses.

(a)

?

(b)

?

19.78 (SYN) Show how you would synthesize this compound using propanal as your only carbon source.

19.79 (SYN) Show how to carry out each of the following syntheses. *Hint:* You may need to use selective reagents or protecting groups.

(a)

?

(b)

?

(c)

?

(d)

?

(e)

?

The process for making soap involves one of the oldest organic reactions known, called saponification. Here in Chapter 20, we study saponification and other nucleophilic addition–elimination reactions.

Nucleophilic Addition– Elimination Reactions 1

The General Mechanism Involving Strong Nucleophiles

FIGURE 20-1 Some compound classes that undergo nucleophilic addition–elimination These compounds contain a leaving group (red type) attached to the electron-poor C atom of the polar π bond. The compound classes appearing in the box are collectively known as *carboxylic acid derivatives*.

I n Chapters 17 and 18, we discussed reactions in which nucleophiles add to compounds containing polar π bonds, such as ketones, aldehydes, imines, and nitriles. Those nucleophilic addition reactions are driven, in large part, by the flow of electrons from the electron-rich nucleophile to the electron-poor (i.e., electrophilic) atom of the polar π bond.

A variety of other compound classes contain polar π bonds, too, and are therefore susceptible to nucleophilic attack. For example, carboxylic acids have the polar carbonyl group (highlighted in Fig. 20-1), as do esters, amides, acid anhydrides, and acid halides (collectively known as carboxylic acid derivatives).

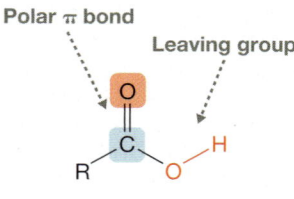

Polar π bond

Leaving group

Carboxylic acid

Carboxylic acid derivatives

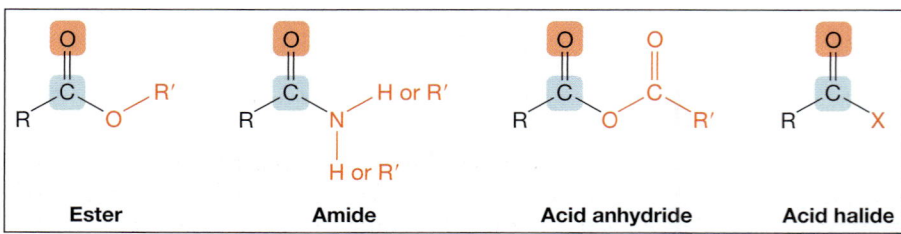

Ester · Amide · Acid anhydride · Acid halide

Chapter Objectives

On completing Chapter 20 you should be able to:

1. Identify the types of compounds capable of undergoing nucleophilic addition–elimination reactions and draw the general mechanism involving such compounds.

2. Draw an energy diagram for a general nucleophilic addition–elimination mechanism and identify the rate-determining step.

3. Predict the products of an acyl substitution reaction and determine whether such a reaction is energetically favorable.

4. Draw the mechanism for saponification and explain why it is irreversible.

5. Describe how an amide can be converted to a carboxylic acid under basic conditions, followed by acid workup, and show how to use this reaction to synthesize a primary amine in a Gabriel synthesis.

6. Explain why methyl ketones can undergo acyl substitution in a haloform reaction to produce a carboxylic acid.

7. Predict the outcomes of treating various acid derivatives with reducing agents, including $NaBH_4$, $LiAlH_4$, diisobutylaluminum hydride (DIBAH), and lithium tri-*tert*-butoxyaluminum hydride.

8. Predict the outcomes of treating carboxylic acids and various acid derivatives with organometallic reagents such as alkyllithium reagents (RLi), Grignard reagents (RMgX), and lithium dialkylcuprates (R_2CuLi).

Although carboxylic acids and carboxylic acid derivatives are susceptible to nucleophilic attack, they participate in different *overall* reactions from what we saw in Chapters 17 and 18. Whereas the compound classes in Chapters 17 and 18 primarily undergo nucleophilic addition, carboxylic acids and their derivatives tend to undergo what is called a *nucleophilic addition–elimination* mechanism. This is because they each contain a *leaving group* (red type, Fig. 20-1) bonded to the electron-deficient atom of the polar π bond.

> A compound (see margin) can undergo a nucleophilic addition–elimination reaction if it has a leaving group attached to the electron-poor atom of a polar π bond.

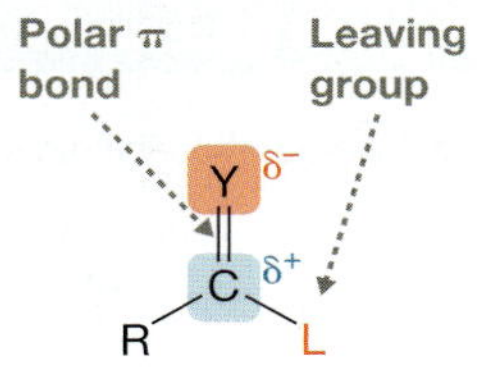

By contrast, ketones, aldehydes, imines, and nitriles do *not* possess a suitable leaving group; the electron-deficient carbon is bonded to H or alkyl (R) groups only (Fig. 20-2), so any leaving group would have to depart as H^- or R^-, which are too unstable to leave. Consequently, these compounds tend to undergo nucleophilic addition only.

As we will see throughout this chapter and Chapter 21, there are a wide variety of reactions that proceed via a nucleophilic addition–elimination mechanism. Here in Chapter 20, we focus just on nucleophilic addition–elimination reactions that involve strong nucleophiles bearing a negative charge. These include alkoxide anions (RO^-), hydride anions (H^-) from various reducing agents, and alkyl anions (R^-) from various organometallic reagents. In Chapter 21, we shift our focus to reactions involving weak nucleophiles and ones that require acid or base catalysis.

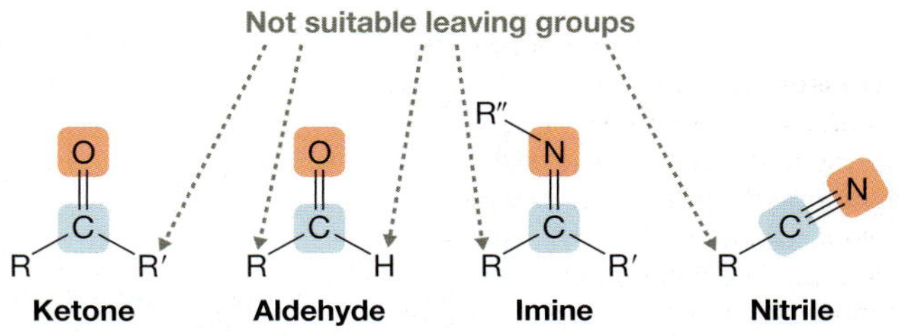

Polar π bond   Leaving group

Not suitable leaving groups

Ketone   **Aldehyde**   **Imine**   **Nitrile**

FIGURE 20-2 Compound classes that undergo nucleophilic addition only These compounds do not have a leaving group attached to the polar π bond. The H and R groups that are attached to the electron-poor C atom would have to depart as H^- or R^-, which are highly unstable.

20.1 An Introduction to Nucleophilic Addition–Elimination Reactions: Transesterification

We begin our discussion of **nucleophilic addition–elimination reactions** with *transesterification*, in which an ester is treated with an alkoxide anion (RO^-). Transesterifications are among the simplest nucleophilic addition–elimination reactions, but they

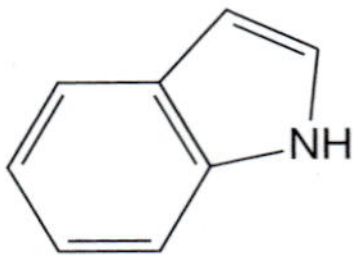

Indole

provide insight into many of the principles that underlie more complex reactions. Thus, transesterification acts as a model for the other nucleophilic addition–elimination reactions discussed throughout this chapter and Chapter 21.

20.1a The General Nucleophilic Addition–Elimination Mechanism

When an ester such as ethyl indole-2-carboxylate is treated with sodium methoxide, a different ester is produced (Equation 20-1).

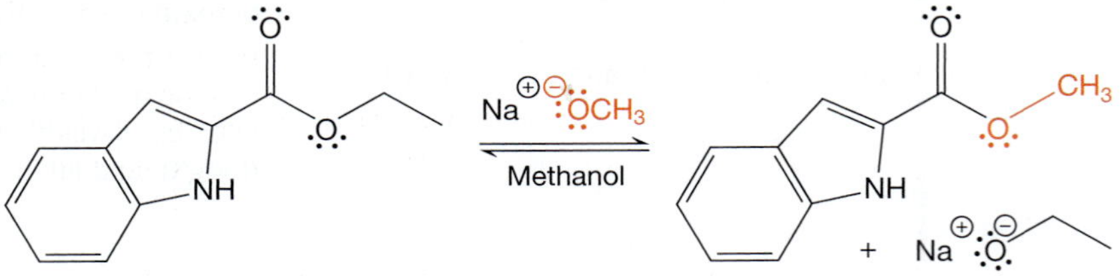

(20-1)

Ethyl indole-2-carboxylate   89%

This is an example of a **transesterification reaction**, so called because one ester is converted into another.

The mechanism for this transesterification, shown in Equation 20-2, is typical of reactions involving a nucleophile and a carboxylic acid derivative. It is called a *nucleophilic addition–elimination* mechanism, which describes the two steps that take place.

Mechanism for Equation 20-1

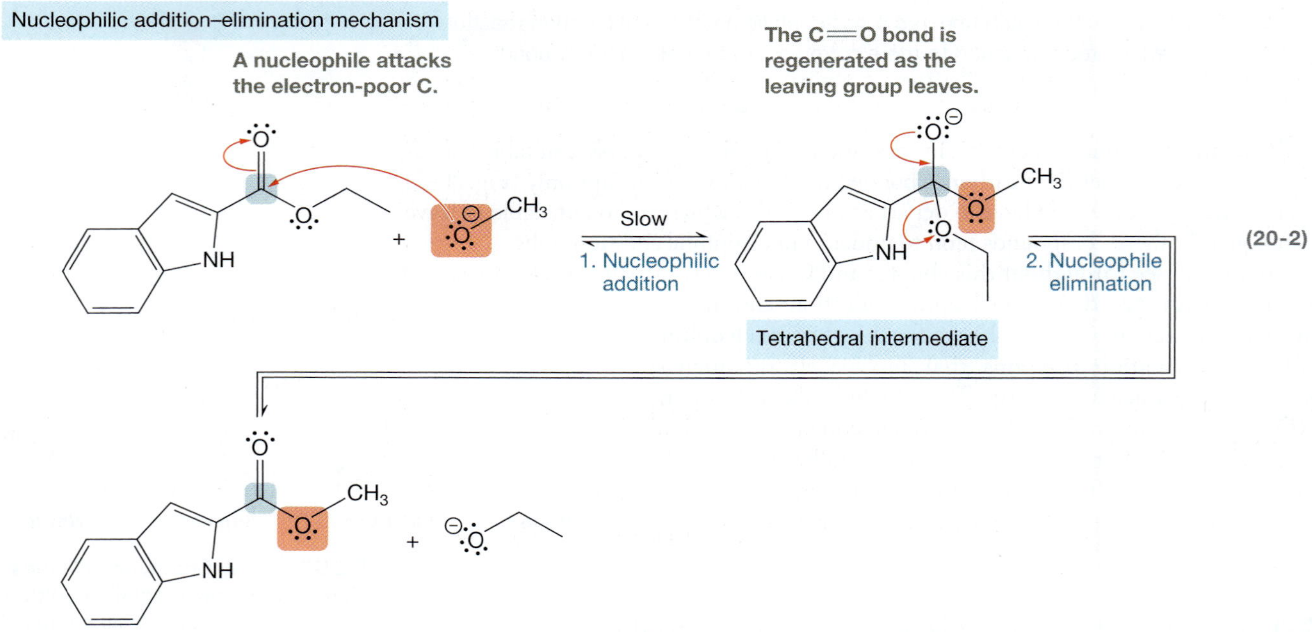

Nucleophilic addition–elimination mechanism

A nucleophile attacks the electron-poor C.

The C=O bond is regenerated as the leaving group leaves.

Slow

1. Nucleophilic addition

2. Nucleophile elimination

Tetrahedral intermediate

(20-2)

Step 1 is *nucleophilic addition* (Section 7.4), in which a nucleophile attacks the electron-poor carbonyl carbon. This forces the pair of electrons from the initial C=O π bond onto the O atom, which generates a negative charge on O. The product of that step is a **tetrahedral intermediate**, which, in Step 2, undergoes *nucleophile elimination* (Section 7.4). A lone pair of electrons on the negatively charged O atom is used to regenerate the C=O double bond, and the leaving group departs as $CH_3CH_2O^-$.

The following is the mechanism for another transesterification reaction, but the curved arrows have been omitted. Complete the mechanism by adding the necessary curved arrows and identify the tetrahedral intermediate. Below each reaction arrow, write the name of the elementary step that takes place.

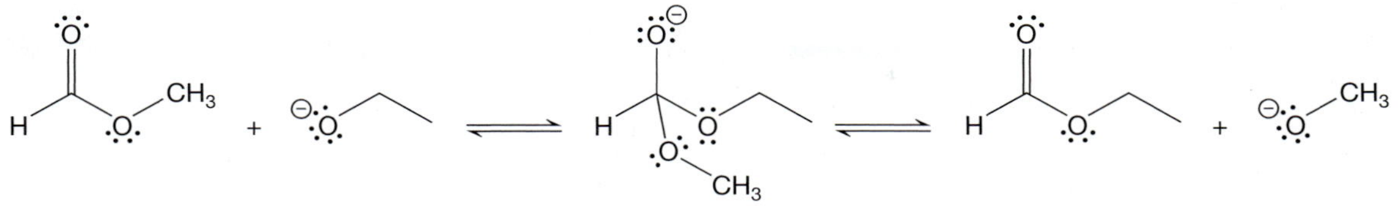

Answers to Your Turns are in the back of the book.

PROBLEM 20.1 Draw the complete, detailed mechanism and the major product(s) for each of the following reactions.

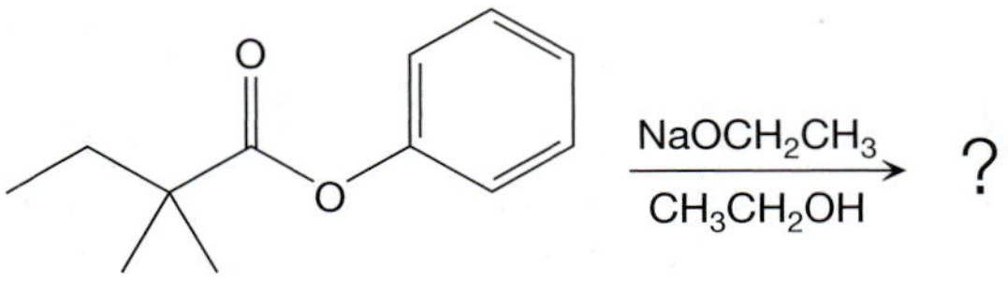

Overall, nucleophilic addition–elimination results in a *substitution* at the carbonyl carbon that is part of an acyl group (RC=O). In Equation 20-1 specifically, the –OCH$_2$CH$_3$ group attached to the carbonyl carbon is replaced by an –OCH$_3$ group. Thus, these kinds of reactions are often referred to as **nucleophilic acyl substitution**. Keep in mind, however, that this mechanism is quite different from the S$_N$2 and S$_N$1 mechanisms we encountered in Chapters 7 and 8. The leaving group in this case is attached to an *sp^2*-hybridized C atom and, as we learned in Chapter 9, these kinds of substrates do not undergo S$_N$2 or S$_N$1 reactions.

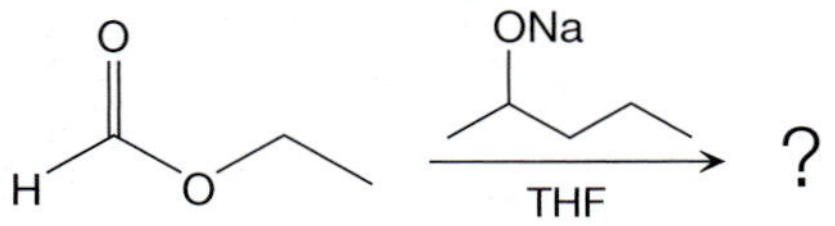

20.1b Kinetics of Nucleophilic Addition–Elimination: The Reaction Free Energy Diagram

Notice from Equation 20-2 that the first step of this nucleophilic addition–elimination mechanism is slow. In other words:

> In a nucleophilic addition–elimination mechanism, the first step (i.e., nucleophilic addition) is generally the rate-determining step.

The first step is slow because its transition state energy is so high, as shown in Figure 20-3 (next page).

Part of the reason for such a high transition state energy is the loss of the *resonance stabilization* from the ester reactant. The resonance in the ester involves the π bond of the carbonyl group and a lone pair of electrons from the leaving group, as shown in Figure 20-3. No such resonance stabilization exists in the tetrahedral intermediate, however, because the π bond is absent. Similarly, the transition state, which closely

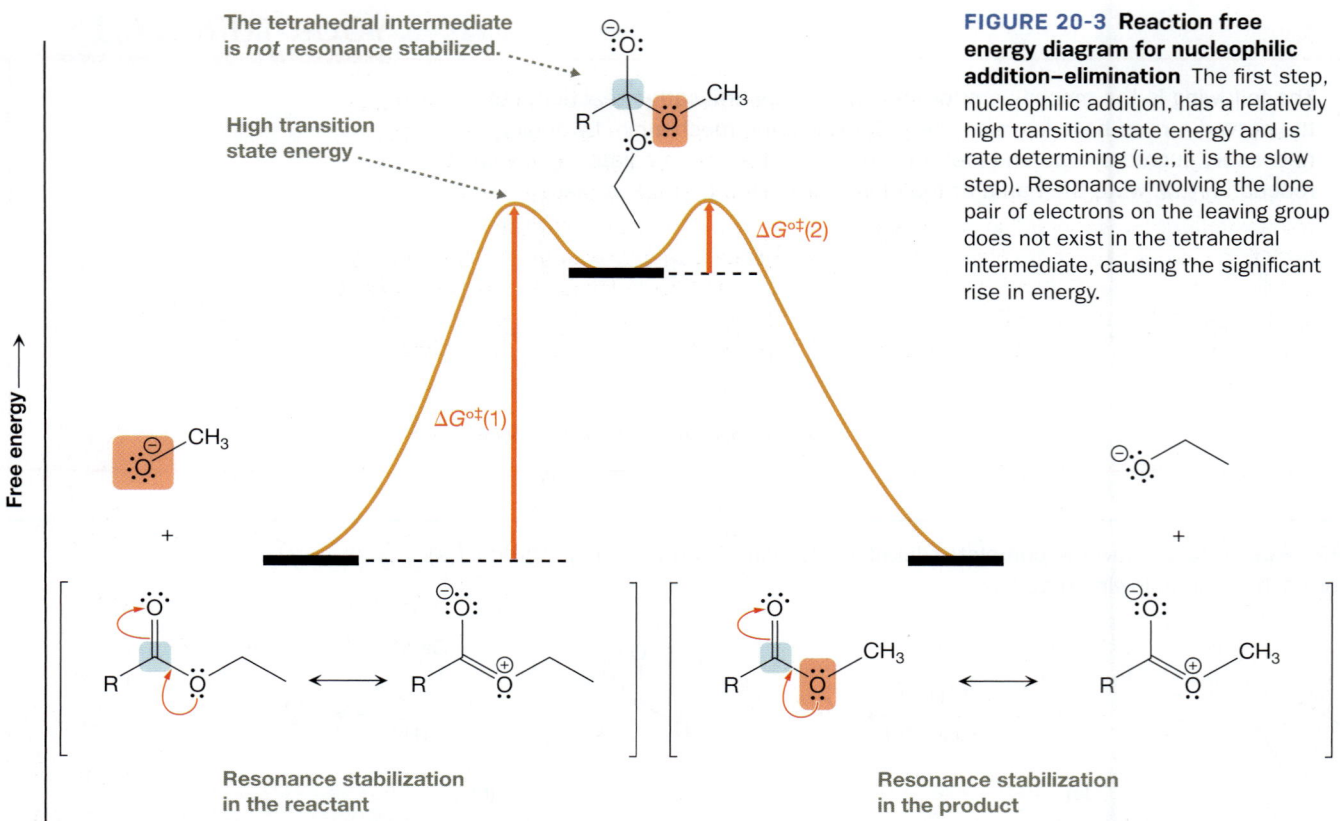

The tetrahedral intermediate is *not* resonance stabilized.

High transition state energy

ΔG°‡(1)

ΔG°‡(2)

Free energy ⟶

Reaction coordinate ⟶

Resonance stabilization in the reactant

Resonance stabilization in the product

FIGURE 20-3 **Reaction free energy diagram for nucleophilic addition–elimination** The first step, nucleophilic addition, has a relatively high transition state energy and is rate determining (i.e., it is the slow step). Resonance involving the lone pair of electrons on the leaving group does not exist in the tetrahedral intermediate, causing the significant rise in energy.

resembles the tetrahedral intermediate, lacks that resonance stabilization. When the products are formed, notice that the resonance stabilization is reestablished.

YOUR TURN 20.2

Construct a free energy diagram for the reaction in Your Turn 20.1. Include and label the overall reactants, overall products, and the tetrahedral intermediate, and draw a vertical arrow to represent each step's energy barrier.

20.1c Thermodynamics and Reversibility

Looking back at Equation 20-1, notice that the reactants and products are connected by an equilibrium reaction arrow (⇌). In other words:

Transesterification is a *reversible* reaction.

Thus, if the methyl ester on the product side of Equation 20-1 is treated with $CH_3CH_2O^-$, the ethyl ester will be produced via nucleophilic addition–elimination.

YOUR TURN 20.3

Draw the detailed mechanism for the reverse of the reaction in Equation 20-1.

Biodiesel and Transesterification

Biodiesel is an attractive alternative to petroleum-based fuels for a variety of reasons. Some studies indicate that its use results in lower net production of CO_2, a greenhouse gas, than petroleum diesel. Derived from plants, it can be produced domestically, and it is also *renewable*.

Biodiesel is produced from oils that are made by plants, using essentially the same transesterification process discussed here in Chapter 20. Plant oil is a triglyceride, which is a fatty acid triester of glycerol. Biodiesel, on the other hand, is a fatty acid monoester. To carry out the transesterification that produces the biodiesel, the triglyceride is treated with an alcohol under basic conditions. Methanol is the most popular alcohol used for this process, resulting in a *fatty acid methyl ester*, though other alcohols can be used, depending on cost and the specific properties desired of the biodiesel.

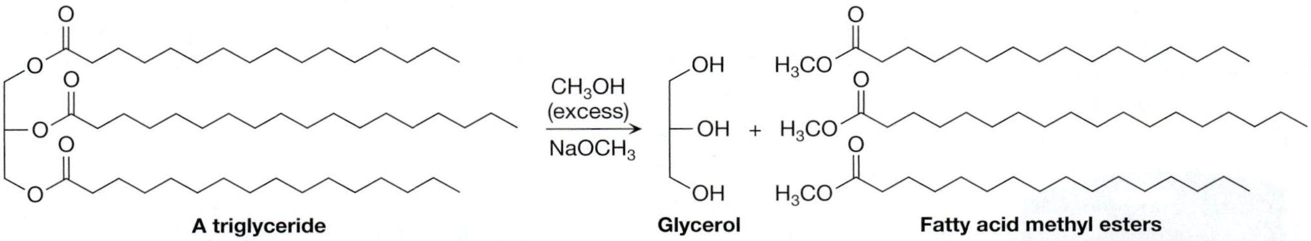

A triglyceride   **Glycerol**   **Fatty acid methyl esters**

Some issues surrounding biodiesel production still remain. One is that methanol is currently produced primarily from petroleum-based sources. However, ways of making renewable methanol are being developed and optimized. Another issue is that large amounts of glycerol are produced as a by-product, but researchers continue to find ways to put the compound to good use. Glycerol can potentially be used, for example, as the starting material in the large-scale production of other valuable compounds, such as ethanol, propylene glycol, acrolein, and hydrogen gas.

The reversibility of base-promoted transesterification is consistent with the reactants and products having similar energies, as shown previously in Figure 20-3. Therefore, the overall energy barrier that must be traversed in the reverse direction is roughly the same as that in the forward direction. The reactants and products are similar in energy because the negative charge that appears on either side of the reaction is stabilized comparably in the alkoxide anions. This is reflected by the fact that the pK_a values of their respective conjugate acids, which are 15.5 and 16, are not dramatically different (Fig. 20-4).

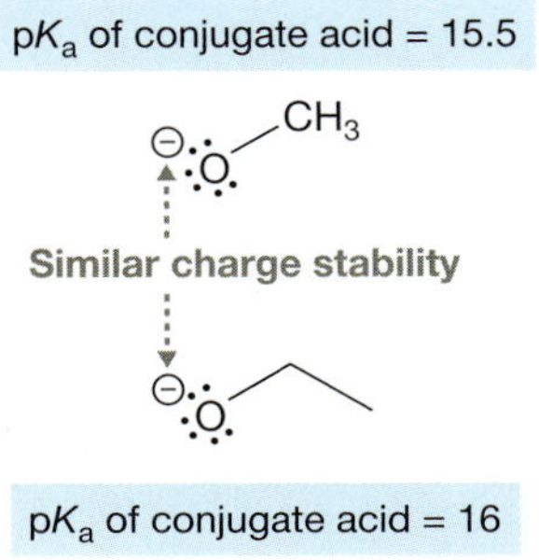

pK_a of conjugate acid = 15.5

Similar charge stability

pK_a of conjugate acid = 16

FIGURE 20-4 Charge stability and reversibility in nucleophilic addition–elimination
These alkoxide anions appear on the reactant and product sides in Equation 20-1. They exhibit similar charge stability, reflected by the similar pK_a values of their respective conjugate acids, which is largely why the transesterification in Equation 20-1 is reversible.

Although these transesterification reactions are reversible, this is *not* true of all nucleophilic addition–elimination reactions. As we will see, the relative stabilities of the nucleophile and the leaving group greatly affect the reversibility of the reaction.

20.2 Acyl Substitution Involving Other Carboxylic Acid Derivatives: The Thermodynamics of Acyl Substitution

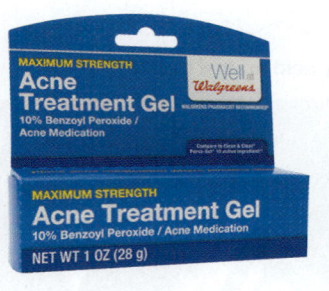

The transesterification reactions we examined in Section 20.1 are rather limited in scope, involving an ester as the carboxylic acid derivative and RO^- as the nucleophile. Numerous other acyl substitutions can be carried out, however, simply by using different combinations of carboxylic acid derivatives and nucleophiles. Equation 20-3 shows, for example, that CH_3O^- can displace Cl^- from an acid chloride such as benzoyl chloride to produce an ester.

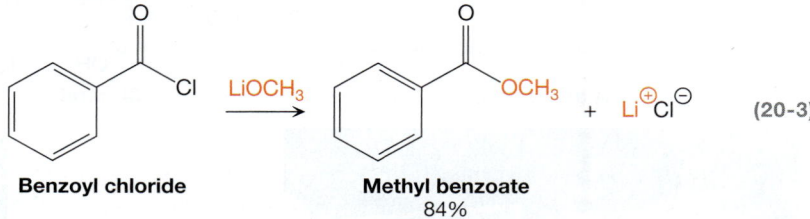

Benzoyl chloride   **Methyl benzoate**
                                    84%   (20-3)

The mechanism for this reaction is shown in Equation 20-4. It consists of the usual nucleophilic addition and elimination steps, and proceeds through a tetrahedral intermediate.

Mechanism for Equation 20-3

Tetrahedral intermediate

1. Nucleophilic addition
2. Nucleophile elimination

(20-4)

Not every combination of acid derivative and nucleophile leads to an effective acyl substitution reaction. Equation 20-5 shows, for example, that essentially no acyl substitution takes place when an ester is treated with chloride ion.

NaCl ⟶ No reaction   (20-5)

Methyl benzoate

An ester such as methyl acetate can react with $(CH_3)_2N^-$ to produce an amide. The nucleophilic addition–elimination mechanism that describes this reaction is as follows:

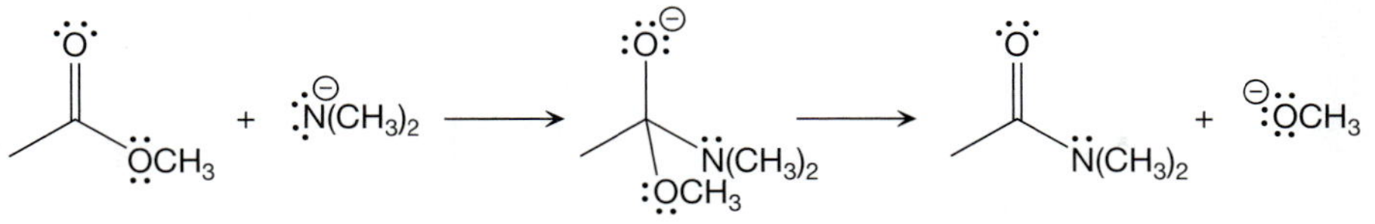

Complete the mechanism by adding the curved arrows to show the movement of the electrons. Under each reaction arrow, write the name of the elementary step taking place. Label the tetrahedral intermediate.

We can understand why there is no reaction in Equation 20-5 by examining the generic acyl substitution reaction in Equation 20-6. Notice that the nucleophile (Nu^-) bears a negative charge on the reactant side, and the leaving group (L^-) bears the negative charge in the products.

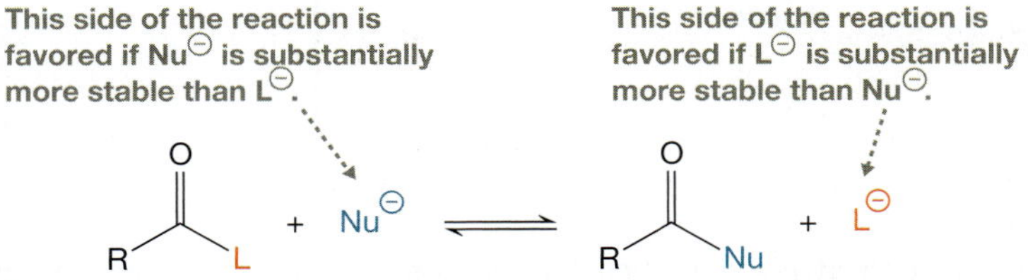

This side of the reaction is favored if $Nu^{\ominus}$ is substantially more stable than $L^{\ominus}$.

This side of the reaction is favored if $L^{\ominus}$ is substantially more stable than $Nu^{\ominus}$.

(20-6)

If L^- is more stable than Nu^- (reflected by the weaker basicity of L^-), then the reaction is energetically favorable and generally occurs readily. This is the case with the reaction in Equation 20-3, because the leaving group (Cl^-) is more stable than the nucleophile (CH_3O^-). If Nu^- is more stable than L^-, on the other hand, then the reaction is energetically *unfavorable*, and it generally does *not* proceed readily. This is the case with the reaction in Equation 20-5, in which the nucleophile is Cl^- and the leaving group is CH_3O^-.

Whether an acyl substitution reaction is energetically favorable can be summarized by the "stability ladder" shown in Figure 20-5 (next page). Descending a real ladder is generally easier than ascending one, so:

- An acyl substitution that converts an acid derivative from a higher rung on the stability ladder to one on a lower rung of the ladder is energetically *favorable*.
- An acyl substitution that converts an acid derivative from a lower rung on the stability ladder to one on a higher rung of the ladder is energetically *unfavorable*.

For example, the conversion of an acid chloride ($RCOCl$) to an ester (RCO_2R') represents descending the stability ladder (in this case, two rungs), so it is energetically favorable. This is consistent with the reaction we previously saw in Equation 20-3 taking place readily. By the same token, the conversion of an ester to an acid chloride is *unfavorable*, consistent with the lack of reaction in Equation 20-5.

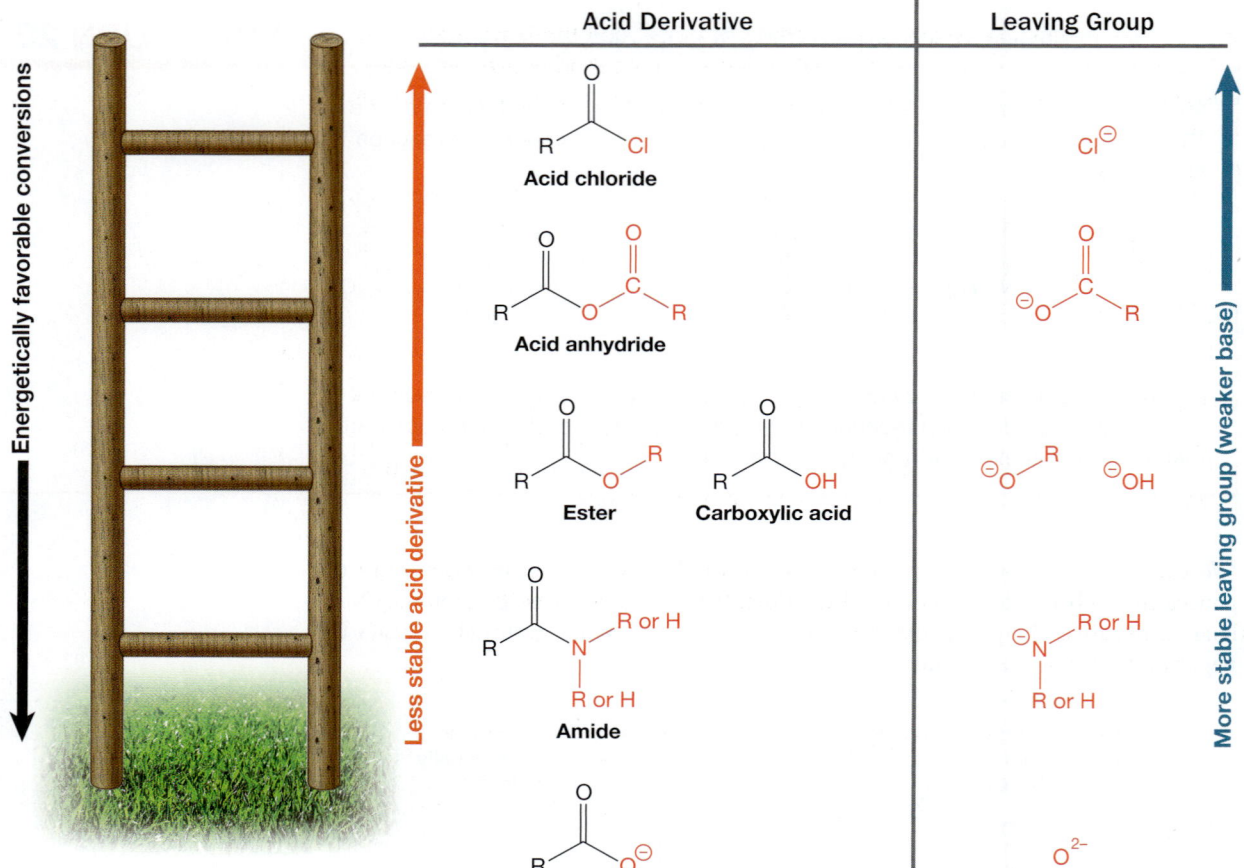

FIGURE 20-5 The "stability ladder" for carboxylic acid derivatives A reaction that converts an acid derivative from a higher rung to a lower rung is energetically favorable and generally takes place readily. Going from a lower rung to a higher rung is unfavorable, and is generally quite difficult to carry out.

YOUR TURN 20.5

Indicate whether each of the following conversions would be energetically favorable or unfavorable. Which reactions will likely occur readily?

(a)

(b)

(c)

(d)

SOLVED PROBLEM 20.2

Will the reaction shown here take place? If it will, draw the complete, detailed mechanism and predict the major product(s).

Think What is the nucleophile? What is the leaving group? Is the potential acyl substitution reaction energetically favorable?

Solve The nucleophile is the acetate anion, $CH_3CO_2^-$, and the leaving group is Cl^-, part of the acid chloride. In an acyl substitution reaction between these species, $CH_3CO_2^-$ replaces Cl^- in a nucleophilic addition–elimination mechanism, as follows:

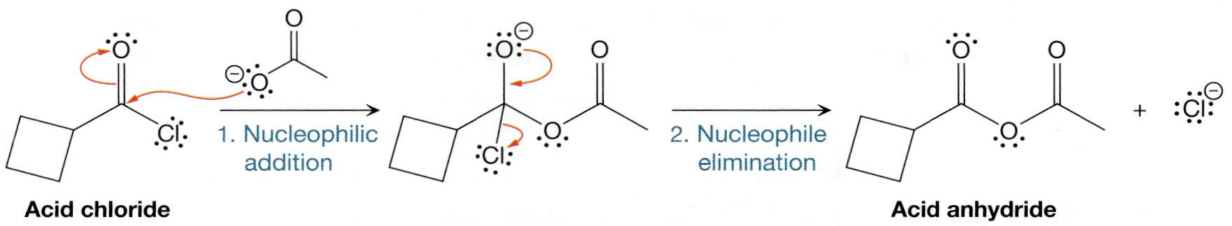

Acid chloride **Acid anhydride**

The product, an acid anhydride, is below the reactant, an acid chloride, on the stability ladder (Figure 20-5). This is consistent with Cl^- being a weaker base than $CH_3CO_2^-$ (the pK_a of HCl is -7; the pK_a of CH_3CO_2H is 4.75). As a result, this reaction is energetically favorable and will take place readily.

PROBLEM 20.3 Draw the complete, detailed mechanism and predict the major product(s) for each of the following reactions. If no reaction takes place, write "no reaction."

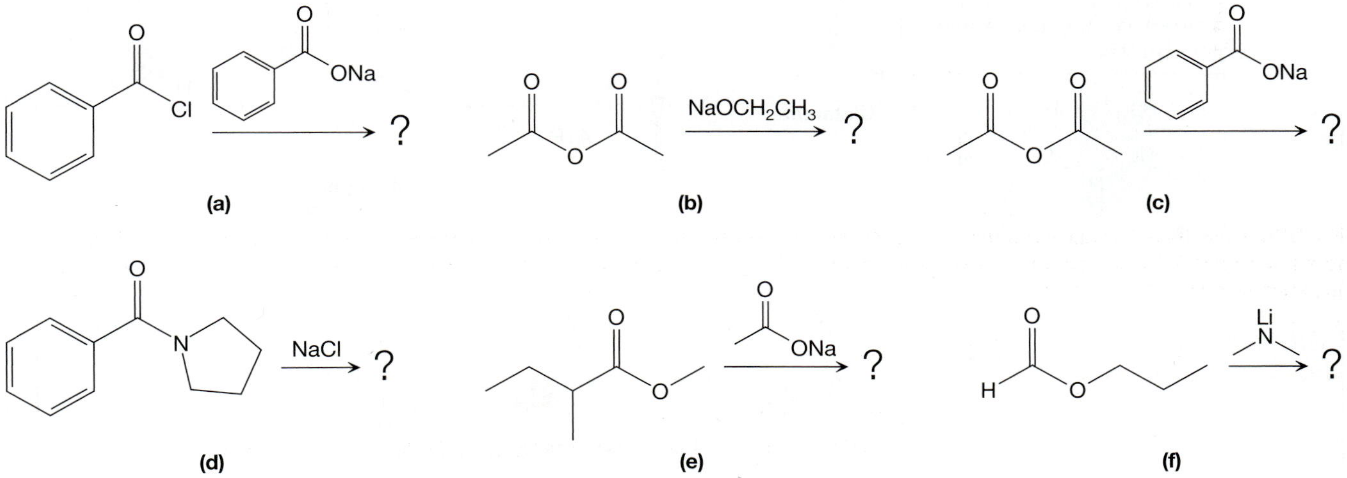

(a) (b) (c)

(d) (e) (f)

20.3 Reaction of an Ester with Hydroxide (Saponification) and the Reverse Reaction

Hydroxide (HO^-) and alkoxide (RO^-) ions generally have similar nucleophile strengths and leaving group abilities, so you might expect that a carboxylic acid ($R'CO-OH$) would react with an alkoxide anion to produce an ester ($R'CO-OR$). If such a reaction were to take place, RO^- would replace HO^-, much like we saw for transesterifications (Section 20.1), where one alkoxide ion replaces another. As shown in Equation 20-7, however, no such reaction occurs.

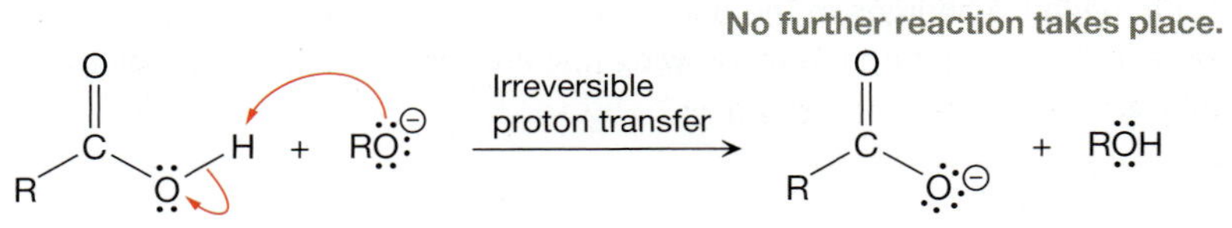

No further reaction takes place.

(20-7)

Instead, the alkoxide anion—a strong base—deprotonates the carboxylic acid *rapidly* and *irreversibly* to produce a carboxylate anion, after which no further reaction takes place.

YOUR TURN 20.6

To convince yourself that the proton transfer reaction in Equation 20-7 is irreversible, use the appropriate pK_a values from Table 6-1 to determine which side of that proton transfer step is favored. To what extent is it favored?

The Stability Ladder in Biochemical Systems

Functional groups characteristic of carboxylic acids and carboxylic acid derivatives play major roles in a wide variety of biochemical processes. Depending on its particular role, the functional group might need to be relatively stable. This is the case with proteins, for example, which are long chains of amino acids connected by $O{=}C{-}N$ functional groups that are characteristic of amides—so-called peptide linkages (shown below at left). As we saw in Figure 20-5, amides appear near the bottom of the stability ladder, so these peptide linkages are rather unreactive—something that helps a protein maintain its integrity, allowing it to carry out its specific function.

The functional group characteristic of an amide serves as a peptide linkage in a protein.

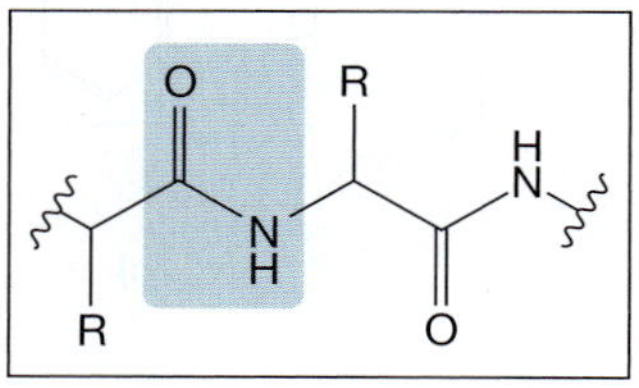

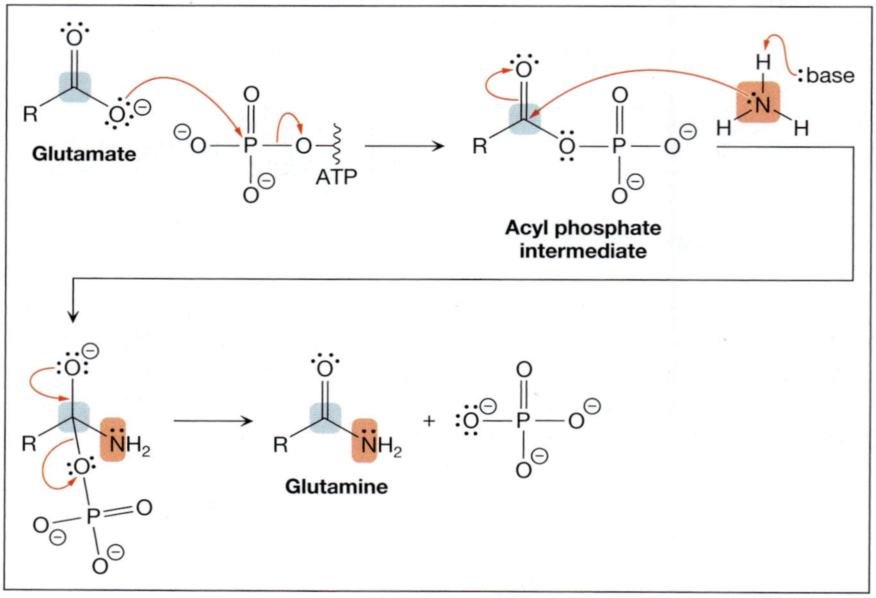

In other cases, a particular functional group might be too stable. Consider, for example, the scheme shown above on the right, which depicts the biosynthesis of glutamine—one of the 20 naturally occurring amino acids. The carboxylate (CO_2^-) portion of glutamate is converted into $O{=}C{-}N$, characteristic of amides, in a reaction catalyzed by the enzyme glutamine synthetase. Carboxylate anions are at the very bottom of the stability ladder (Fig. 20-5), however, so the conversion of a carboxylate anion to an amide is energetically *unfavorable* and thus would be quite difficult to carry out directly. To solve this problem, the carboxylate group is first phosphorylated by adenosine triphosphate (ATP) in a nucleophilic substitution reaction (believed to be S_N2). The product of that step is an acyl phosphate intermediate, which has a significantly better leaving group and is considerably less stable. This sets up nucleophilic addition–elimination involving NH_3 in the next two steps, yielding glutamine as the overall product.

It might seem that the S_N2 step between glutamate and ATP should be difficult because of the electrostatic repulsion by the negative charges in the phosphate group. However, a Mg^{2+} ion (not shown) is bound to O atoms in the phosphate group, drawing negative charge away and making the P atom more electrophilic.

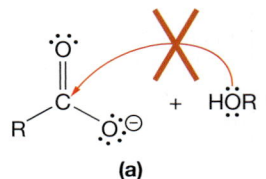

ROH is not a strong enough nucleophile to overcome the resonance stabilization in $RCO_2^{\ominus}$.

(a)

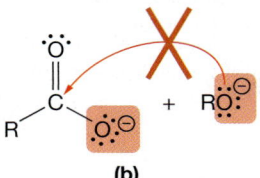

Charge repulsion prevents nucleophilic attack.

(b)

FIGURE 20-6 The unreactive nature of carboxylate anions Nucleophiles tend not to add to the carbonyl C of a carboxylate anion. (a) Uncharged nucleophiles like alcohols are not strong enough to overcome the resonance delocalizaton of the π electrons in the carboxylate anion. (b) Stronger nucleophiles, such as alkoxide anions, are repelled from the carboxylate anion due to the like charges.

No further reaction takes place because carboxylate ions are at the very bottom of the stability ladder. Thus, any acyl substitution reaction that converts a carboxylate ion to another carboxylic acid derivative in the stability ladder would be highly unfavorable. Carboxylate ions are unreactive in large part because they have equivalent resonance structures, and the resulting electron delocalization heavily stabilizes the C=O group's π electrons (Fig. 20-6a). Moreover, carboxylate anions are negatively charged, so they repel incoming nucleophiles (Fig. 20-6b).

YOUR TURN 20.7

To illustrate the stabilization of a carboxylate anion, draw the resonance structure and resonance hybrid in the spaces provided.

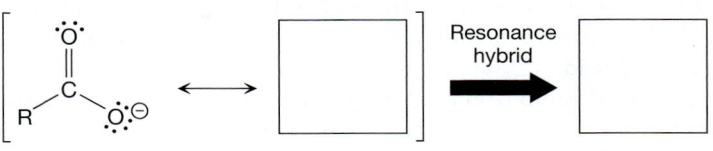

Although the deprotonation of a carboxylic acid poses a problem when trying to convert a carboxylic acid to an ester under basic conditions, this kind of deprotonation is advantageous when converting an ester to a carboxylic acid. Equation 20-8 shows, for example, that when ethyl acetate is treated with KOH, followed by acid workup, ethanoic acid (acetic acid) is produced in relatively high yield.

Saponification

Ethyl ethanoate
(Ethyl acetate)

1. KOH, CH₃OH, 60 min, 35 °C
2. HCl, H₂O

Ethanoic acid
(Acetic acid)
87%

(20-8)

CONNECTIONS Ethyl acetate is a common organic solvent used in liquid chromatography. It is also used as a decaffeinating agent for coffee because it can efficiently extract caffeine from the bean without excessively removing other compounds that would substantially alter the flavor of the coffee.

The first reaction in this sequence is known as **saponification**, which literally means "soap making" (*sapo* is the Latin root for "soap"). Ancient civilizations made crude soap by boiling animal fat (which is chiefly long-chain esters) together with wood ash, a source of HO⁻ ions. (See Problem 20.54 at the end of the chapter.)

Thus, generally speaking:

A carboxylic acid is produced when an ester undergoes saponification followed by acid workup.

The mechanism for this sequence of reactions is shown in Equation 20-9.

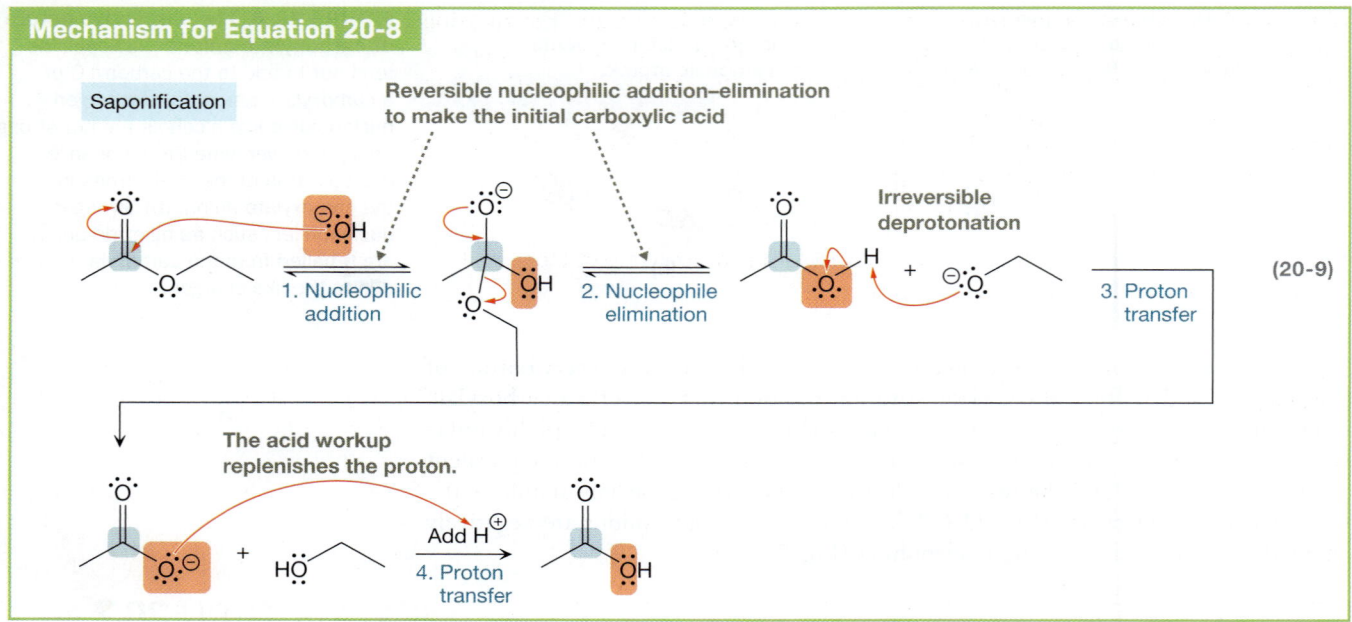

Mechanism for Equation 20-8

Saponification

Reversible nucleophilic addition–elimination
to make the initial carboxylic acid

1. Nucleophilic addition

2. Nucleophile elimination

Irreversible deprotonation

3. Proton transfer

(20-9)

The acid workup replenishes the proton.

Add H⊕

4. Proton transfer

FIGURE 20-7 Free energy diagram of a saponification reaction The direct products of nucleophilic addition–elimination, RCO_2H and RO^-, have roughly the same energy as the overall reactants. The products of the proton transfer in the third step, however, are substantially lower in energy, due largely to the resonance delocalization of the negative charge in the carboxylate anion.

Steps 1 and 2 are identical to the nucleophilic addition and elimination steps in Equation 20-2. Step 3 is the rapid, irreversible deprotonation of the newly formed carboxylic acid, similar to Equation 20-7 (p. 1009). Finally, the proton is replenished by adding a strong acid.

The irreversible deprotonation in Step 3 makes the *overall* saponification reaction irreversible. The initial carboxylic acid is produced reversibly, but it is continually removed via Step 3, which helps drive the reaction to completion.

We can better understand why saponification is irreversible by examining its reaction free energy diagram, shown in Figure 20-7. Notice that the immediate products

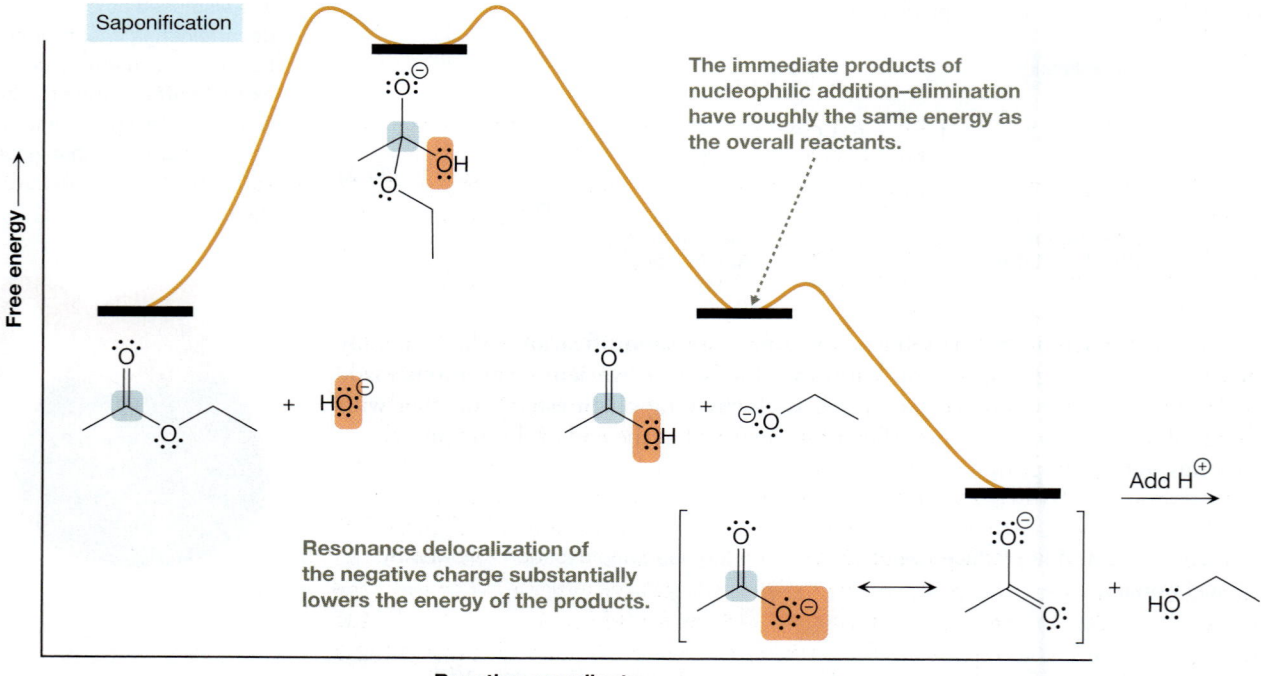

Saponification

The immediate products of nucleophilic addition–elimination have roughly the same energy as the overall reactants.

Resonance delocalization of the negative charge substantially lowers the energy of the products.

Add H⊕

Reaction coordinate ⟶

Free energy ⟶

of nucleophilic addition–elimination—namely, the carboxylic acid and the alkoxide anion—appear at roughly the same energy as the overall reactants, just as we saw in Figure 20-3. By contrast, the products of the proton transfer in Step 3—a carboxylate anion and an alcohol—are much lower in energy, due to the resonance stabilization in the carboxylate anion. That additional stability gained in Step 3 makes the overall reaction much more favorable energetically.

PROBLEM 20.4 Draw the complete, detailed mechanism and predict the overall products for each of the following reactions.

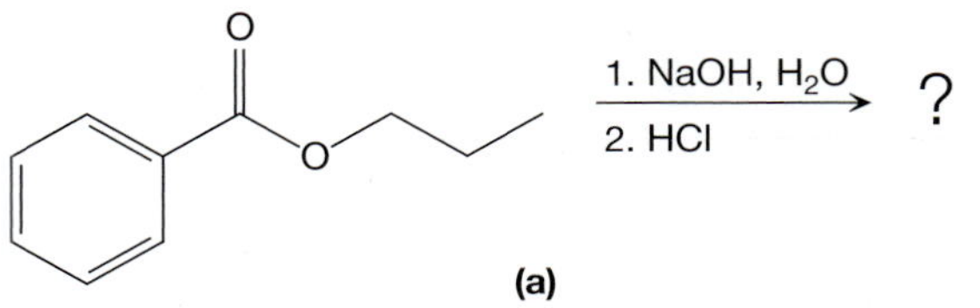

(a)

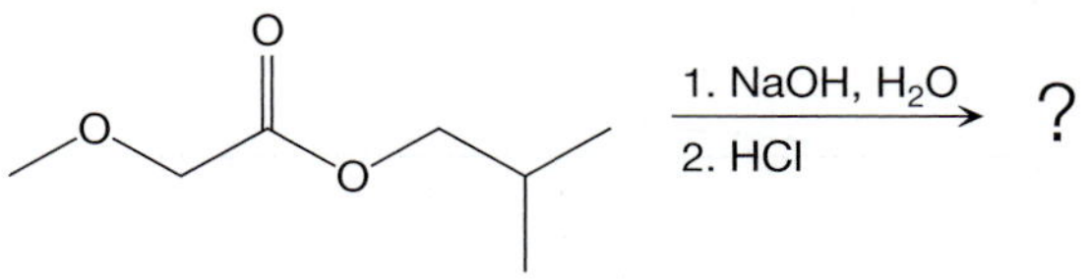

1. NaOH, H$_2$O
2. HCl

?

(b)

1. NaOH, H$_2$O
2. HCl

?

20.4 Carboxylic Acids from Amides; the Gabriel Synthesis of Primary Amines

In Section 20.2, we saw that an amide (RCONR$_2$) is on a lower rung than a carboxylic acid (RCO$_2$H) in the stability ladder (Fig. 20-5), so the conversion of an amide to a carboxylic acid is energetically *unfavorable*. Equation 20-10 shows, however, that we can carry out this transformation—an amide hydrolysis—by treating an amide with HO$^-$, followed by acid workup.

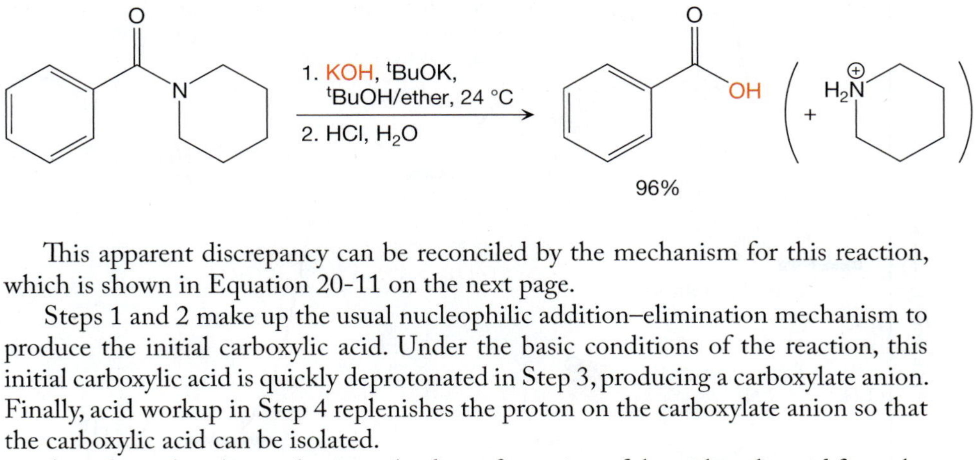

1. KOH, ^{t}BuOK,
   ^{t}BuOH/ether, 24 °C
2. HCl, H$_2$O

96%

(20-10)

This apparent discrepancy can be reconciled by the mechanism for this reaction, which is shown in Equation 20-11 on the next page.

Steps 1 and 2 make up the usual nucleophilic addition–elimination mechanism to produce the initial carboxylic acid. Under the basic conditions of the reaction, this initial carboxylic acid is quickly deprotonated in Step 3, producing a carboxylate anion. Finally, acid workup in Step 4 replenishes the proton on the carboxylate anion so that the carboxylic acid can be isolated.

As indicated in the mechanism, the direct formation of the carboxylic acid from the amide is energetically unfavorable, due to the negative charge being so poorly stabilized on N in the leaving group. Therefore, the reaction is reversible through the second step. The proton transfer in Step 3 is irreversible, however, just as we saw in the saponification mechanism in Equation 20-9, so it is the critical step that drives the reaction toward products: *Without that step, no appreciable amount of product would form.*

This can be better understood from the reaction free energy diagram in Figure 20-8. Notice that the immediate products from nucleophilic addition and elimination are higher in energy than the overall reactants. Thus, direct acyl substitution to form the carboxylic acid is energetically unfavorable. The irreversible proton transfer in the third step, however, is responsible for lowering the energy dramatically, which helps favor products.

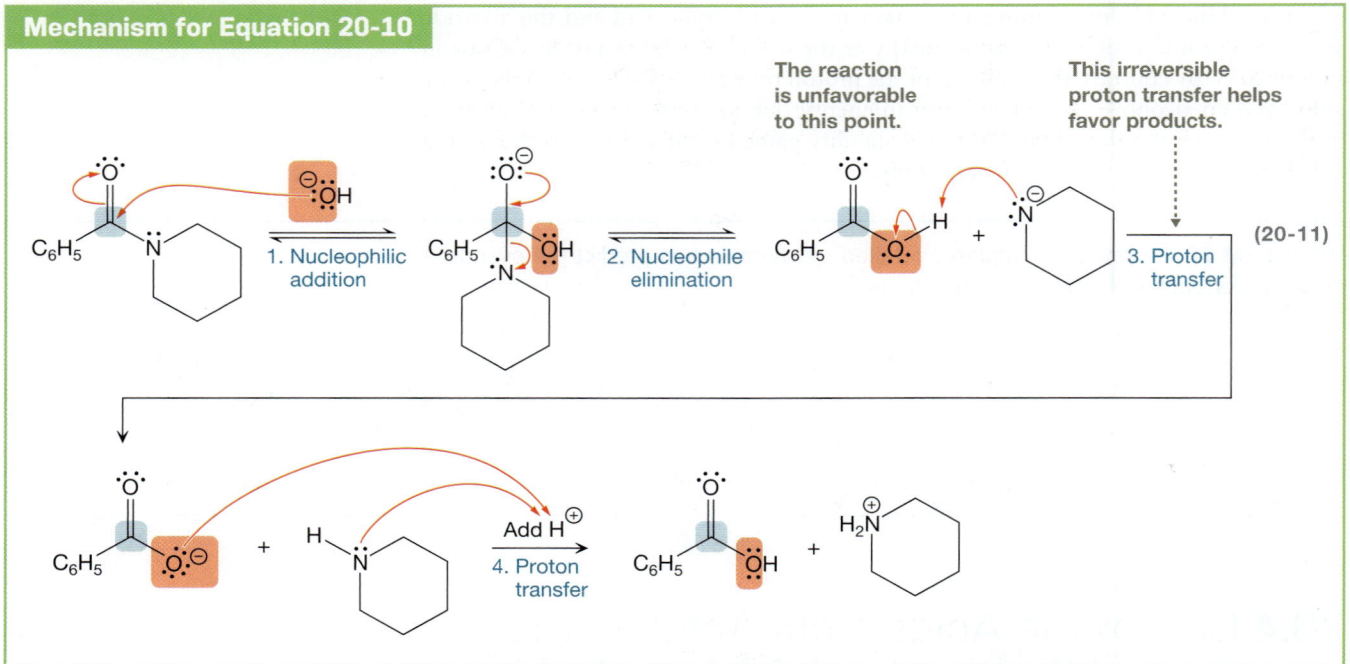

The reaction is unfavorable to this point.

This irreversible proton transfer helps favor products.

1. Nucleophilic addition

2. Nucleophile elimination

3. Proton transfer

(20-11)

Add H⊕
4. Proton transfer

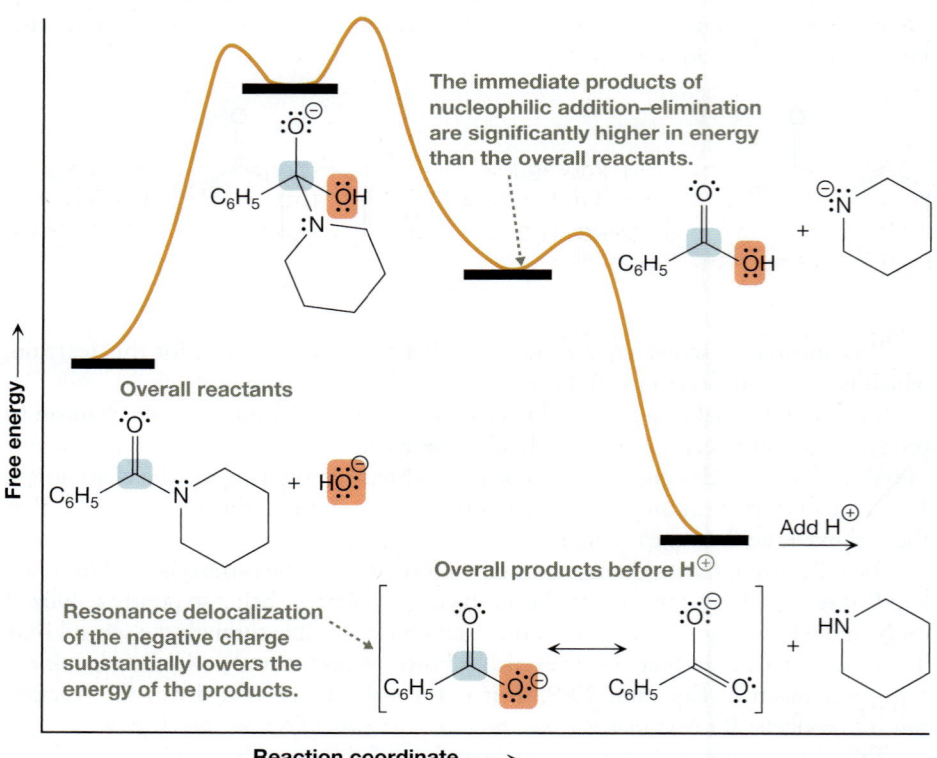

Overall reactants

The immediate products of nucleophilic addition–elimination are significantly higher in energy than the overall reactants.

Add H⊕

Overall products before H⊕

Resonance delocalization of the negative charge substantially lowers the energy of the products.

Reaction coordinate ⟶

FIGURE 20-8 Free energy diagram corresponding to the mechanism in Equation 20-11
The direct products of nucleophilic addition–elimination, RCO_2H and R_2N^-, are higher in energy than the overall reactants. Thus, the reaction is energetically unfavorable after the first two steps of the mechanism. The third step is energetically very favorable, however, which drives the reaction toward products. This is followed by an acid workup.

The following mechanism, which has the curved arrows omitted, is for another reaction that converts an amide to a carboxylic acid. Complete the mechanism by drawing the missing curved arrows, and identify the tetrahedral intermediate. Below each reaction arrow, write the name of the elementary step that takes place and indicate whether the step is reversible or irreversible.

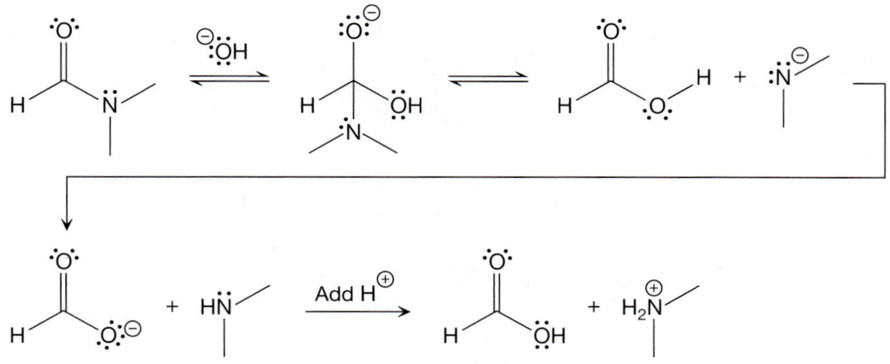

PROBLEM 20.5 Draw the complete, detailed mechanism and predict the major product(s) for each of the following reactions.

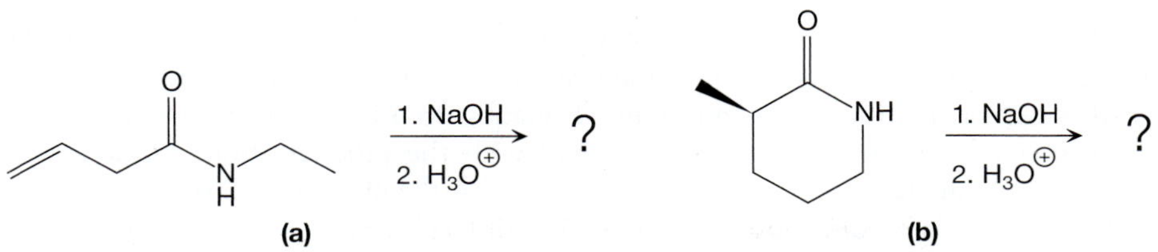

(a)   1. NaOH   2. H_3O^{+} ?

(b)   1. NaOH   2. H_3O^{+} ?

Although the carboxylic acid is often the desired product when an amide is treated with hydroxide, such as that in Equation 20-10, there are times when isolating the nitrogen-containing leaving group is desirable. This is the case for the **Gabriel synthesis**, shown in Equation 20-12.

The *Gabriel synthesis* is an effective way to synthesize a primary amine (RNH_2) from an alkyl halide.

The Gabriel synthesis

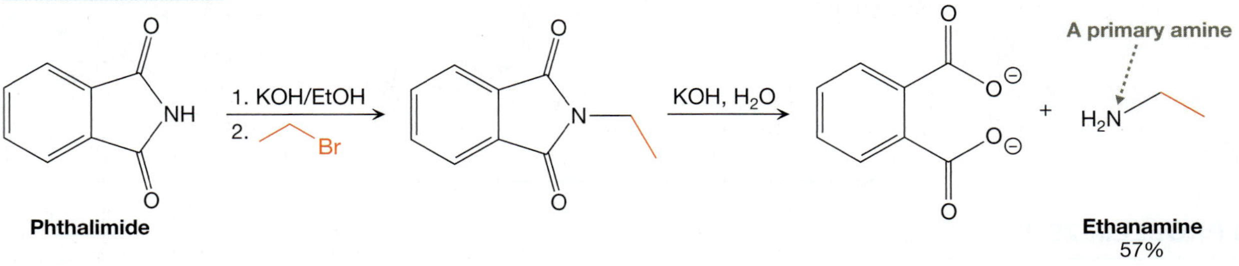

Phthalimide   1. KOH/EtOH 2. Br   KOH, H_2O + A primary amine H_2N (20-12)   **Ethanamine** 57%

In the Gabriel synthesis, phthalimide is first treated with a strong base such as potassium hydroxide, followed by an alkyl halide. Subsequent hydrolysis using hydroxide then liberates the primary amine.

In the mechanism shown in Equation 20-13, Step 1 is simply a proton transfer in which HO^- deprotonates phthalimide at the N atom. The resulting anion is strongly nucleophilic at N, so an S_N2 reaction occurs when the alkyl halide is added in Step 2. Steps 3 through 8 make up two successive acyl substitutions, each one similar to the first three steps of the mechanism shown in Equation 20-11.

Partial mechanism for Equation 20-12

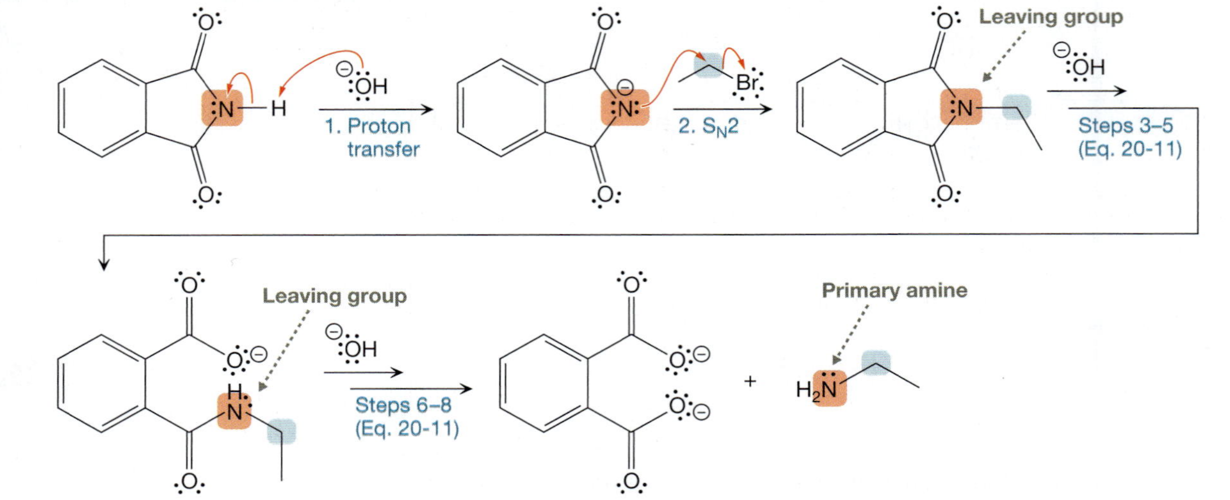

(20-13)

The benefit of synthesizing a primary amine via the Gabriel synthesis can be appreciated when we consider the complications that arise when a primary amine is synthesized via the alkylation of ammonia (Equation 20-14). As we saw in Section 10.2, the primary amine product is contaminated with the products from *polyalkylation* because each alkylated amine can react further with the alkyl halide. By contrast, the Gabriel synthesis produces only the intended primary amine.

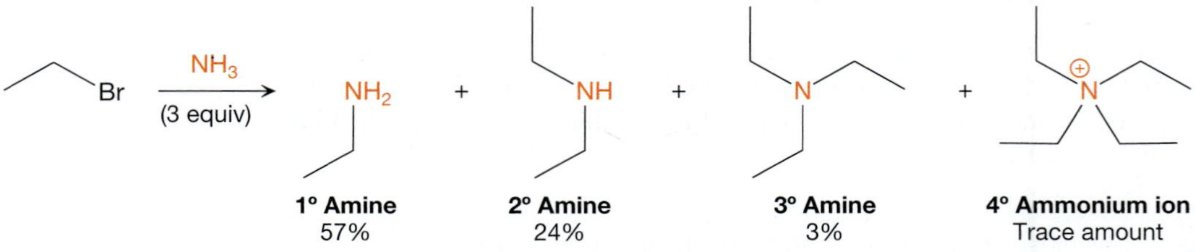

(20-14)

1° Amine	2° Amine	3° Amine	4° Ammonium ion
57%	24%	3%	Trace amount

PROBLEM 20.6 Predict the product of the sequence of reactions shown here.

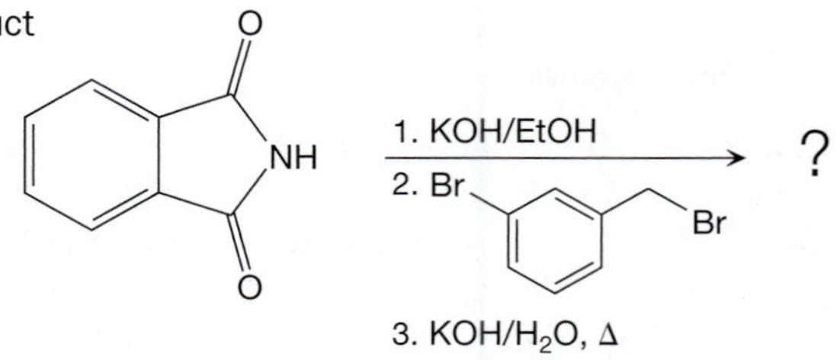

1. KOH/EtOH
2. Br〜〜Br
?
3. KOH/H_2O, Δ

SOLVED PROBLEM 20.7

Suggest how to carry out the synthesis shown here using a Gabriel synthesis.

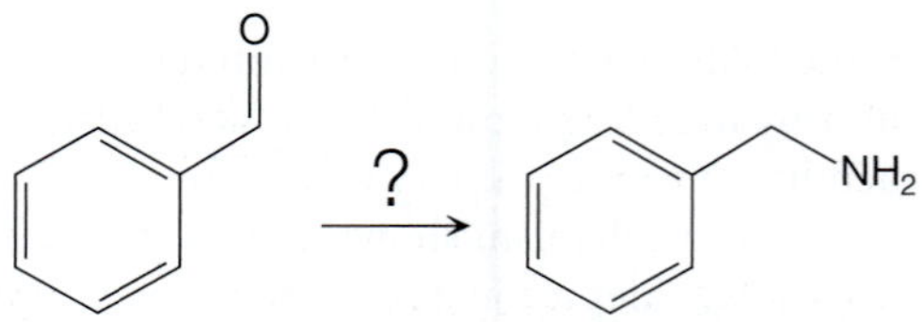

Think Is the target a primary amine that can be the product of a Gabriel synthesis? What alkyl halide can be used as a precursor in a Gabriel synthesis of the target amine? How can that alkyl halide be produced from the starting material?

Solve The target is a primary amine. To produce it in a Gabriel synthesis, a benzyl halide such as benzyl bromide is required, as shown in the following retrosynthetic analysis.

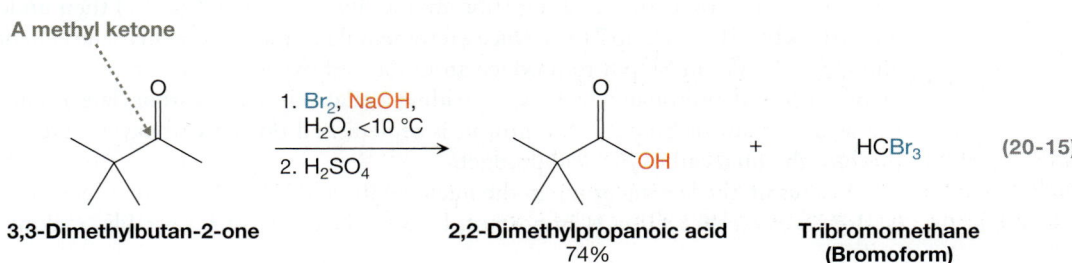

That halide can be synthesized from the corresponding primary alcohol, which can be produced, in turn, by reducing the starting aldehyde. To write the synthesis, we reverse the arrows and add the appropriate reagents. First we show how to synthesize benzyl bromide. Subsequently, benzyl bromide is incorporated along with phthalimide in a Gabriel synthesis.

PROBLEM 20.8 How would you carry out the transformation shown here using a Gabriel synthesis?

20.5 Haloform Reactions

In all of the acyl substitution reactions we have discussed so far, one reactant is an acid derivative in which the carbonyl carbon is bonded to a leaving group. Equation 20-15 shows, however, that a methyl ketone ($RCOCH_3$) can also undergo acyl substitution when treated with Br_2 in aqueous sodium hydroxide, followed by acid workup.

A haloform reaction

A methyl ketone

1. Br_2, NaOH, H_2O, <10 °C
2. H_2SO_4

+ $HCBr_3$ (20-15)

3,3-Dimethylbutan-2-one   **2,2-Dimethylpropanoic acid**   **Tribromomethane**
74%   **(Bromoform)**

One of the products, bromoform ($HCBr_3$), is an example of a **haloform**, which has the general formula HCX_3 (where X is a halogen atom). Thus, the reaction is called a **haloform reaction**. Other haloform reactions can take place when a methyl ketone is treated with Cl_2 or I_2, producing chloroform ($HCCl_3$) or iodoform (HCI_3), respectively. Thus:

> In general, a methyl ketone ($RCOCH_3$) can undergo a haloform reaction when treated with a molecular halogen (X_2) under basic conditions, followed by acid workup, producing a carboxylic acid (RCO_2H) and a haloform (HCX_3) by-product.

YOUR TURN 20.9

The following reactions are the same as the one in Equation 20-15; only the halogen molecule is different. Draw the products in the boxes provided.

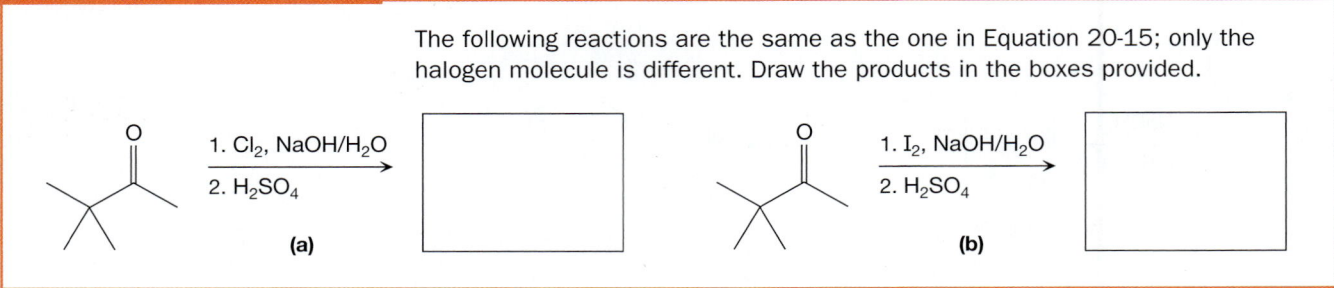

(a)

(b)

It appears in this acyl substitution reaction that HO^- replaces H_3C^-, a very poor leaving group. While this should seem peculiar, all is explained by the mechanism, shown in Equation 20-16.

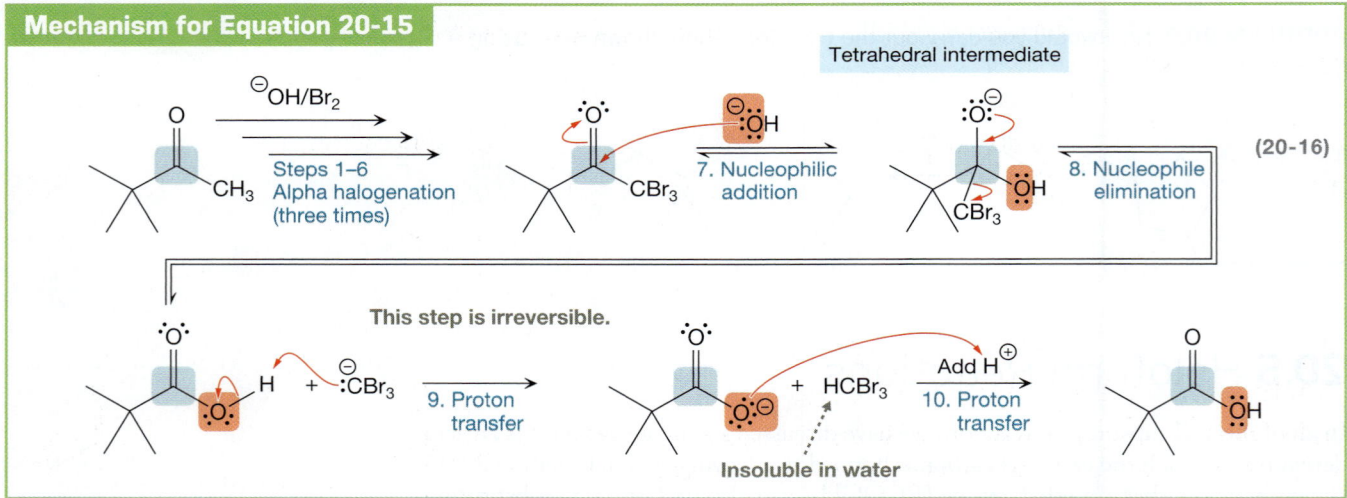

Mechanism for Equation 20-15

(20-16)

In Steps 1–6 of this mechanism, the three α hydrogens on the methyl group are replaced by bromine atoms via the alpha halogenation mechanism discussed previously in Section 10.4. The resulting tribromomethyl ketone ($RCOCBr_3$) then undergoes attack by HO^- in Step 7 to produce a tetrahedral intermediate, which subsequently eliminates Br_3C^- in Step 8 to produce an initial carboxylic acid. In Step 9, that carboxylic acid is deprotonated by Br_3C^-, yielding a relatively stable carboxylate anion. In the acid workup in Step 10, the proton is replenished on the carboxylate anion to produce the final carboxylic acid product.

As a result, the leaving group in the nucleophilic addition–elimination mechanism is Br_3C^-, *not* H_3C^-. This is important, because H_3C^- is very unstable, making it unsuitable as a leaving group. Br_3C^- is a suitable leaving group, however, because the three Br atoms significantly stabilize the negative charge on C.

All three α halogen atoms are necessary to make the carbanion a suitable leaving group. As shown in Equation 20-17, an ethyl ketone cannot undergo acyl substitution because it gains only two α halogens.

With only two halogen atoms on the α carbon, the leaving group is insufficiently stable for acyl substitution.

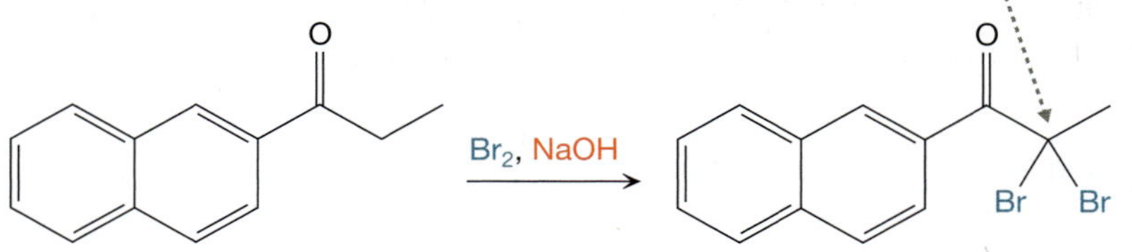

Br₂, NaOH   (20-17)

YOUR TURN **20.10**

Which of ketones **A–D** can undergo a haloform reaction?

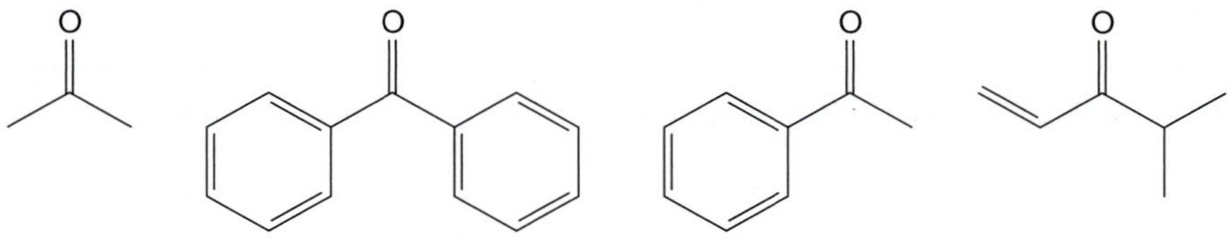

A   **B**   **C**   **D**

PROBLEM 20.9 Draw the complete, detailed mechanism and the major product(s) for each of the following reactions.

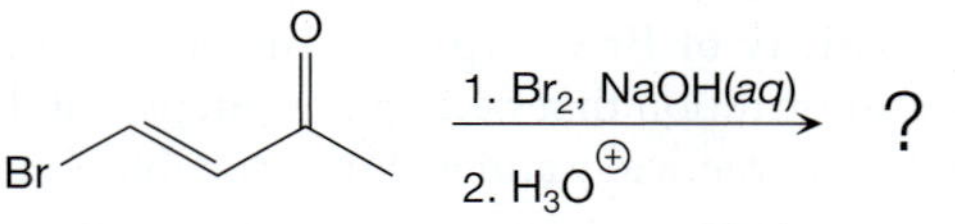

1. Br₂, NaOH(aq)
2. H₃O⊕
?

(a)

1. I₂, NaOH(aq)
2. H₃O⊕
?

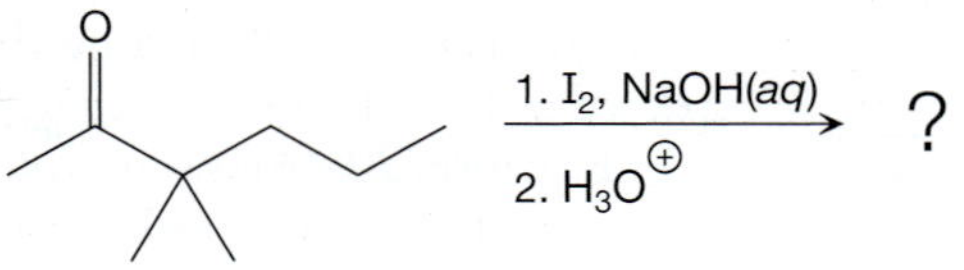

(b)

SOLVED PROBLEM 20.10

Show how to carry out the synthesis shown at right.

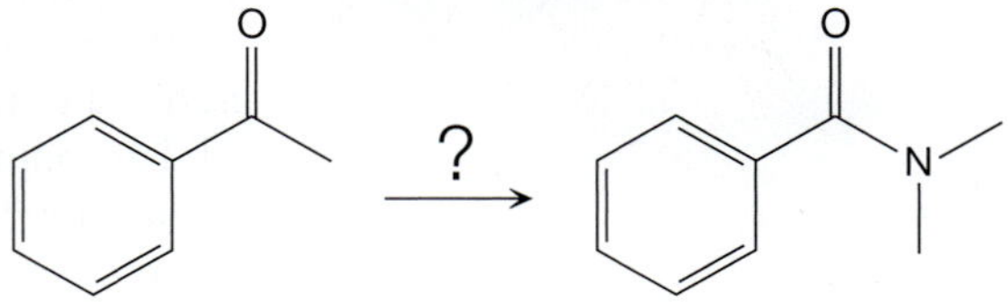

Think What reaction can convert a methyl ketone into a compound with a leaving group on the carbonyl carbon? Can the product of that reaction be converted directly into the amide target or must another precursor be used?

?

Solve The starting compound is a methyl ketone, which does not have a suitable leaving group. However, a haloform reaction can be used to convert that methyl ketone into a carboxylic acid.

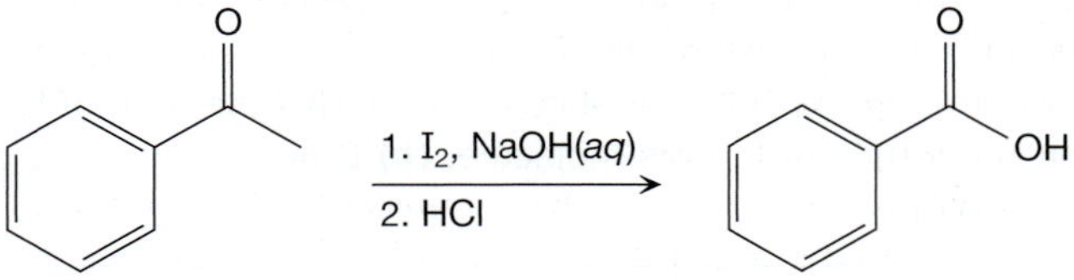

1. I₂, NaOH(aq)
2. HCl

We must now determine how to convert the carboxylic acid into the amide target. Thinking retrosynthetically, the amide can be produced from an ester via an acyl substitution. The ester, in turn, can be produced using diazomethane (Section 10.5).

Undo an acyl substitution. → Undo an ester formation (diazomethane). →

To complete the synthesis, reverse the direction and supply the appropriate reagents.

1. I_2, NaOH(aq)
2. HCl

CH_2N_2

Li–N

PROBLEM 20.11 Show how to carry out this synthesis using any reagents necessary.

?

(a) A negative iodoform test

(b) A positive iodoform test

FIGURE 20-9 The iodoform test
(a) When this sample was treated with a basic solution of I_2, no yellow precipitate of iodoform formed.
(b) When a basic solution of I_2 was added to this sample, however, iodoform precipitated, indicating that the sample contained a methyl ketone.

Despite the three Br atoms, the negative charge on Br_3C^- is less stabilized than that on HO^-, as reflected by the stronger basicity of Br_3C^- (pK_a of $Br_3CH \approx 20$; pK_a of $H_2O = 15.7$). Consequently, the acyl substitution that takes place (Steps 7 and 8 in Equation 20-16) is somewhat unfavorable energetically and is thus reversible. The subsequent proton transfer step (Step 9 in Equation 20-16) is energetically quite favorable, however, which helps drive the reaction toward products, just as we saw in saponification (Fig. 20-7, p. 1012). Moreover, the haloform that is produced in that step is insoluble in water, so it is effectively removed from the reaction mixture. According to Le Châtelier's principle, this drives the acyl substitution reaction even further toward products.

Prior to the advent of spectroscopy, the iodoform (HCI_3) reaction was commonly used as a test for methyl ketones. As shown in Figure 20-9, iodoform is a bright yellow solid at room temperature and is insoluble in water. Therefore, if a bright yellow precipitate appears on treating an organic compound with I_2 in basic solution, it is likely that the compound is a methyl ketone.

PROBLEM 20.12 A compound whose molecular formula is $C_6H_{12}O$ produces a bright yellow solid when it is treated with iodine and a basic solution of water. The IR spectrum of the compound shows a distinct peak at 1708 cm^{-1}. Its ^{1}H NMR spectrum contains only two singlets: one at 1.2 ppm and one at 2.2 ppm. Integration shows that the upfield signal has three times the area of the downfield signal. **(a)** Draw the structure for this molecule. **(b)** Draw the products of the reaction that is described.

20.6 Hydride Reducing Agents: Sodium Borohydride (NaBH$_4$) and Lithium Aluminum Hydride (LiAlH$_4$)

Not all nucleophiles that can attack a carbonyl group are capable of behaving as leaving groups. Hydride ions (H$^-$) from sources such as NaBH$_4$ and LiAlH$_4$, for example, are excellent nucleophiles but are not practical leaving groups. These kinds of nucleophiles can still react with acid derivatives in nucleophilic addition–elimination reactions, but some key differences exist.

Consider Equations 20-18 and 20-19, which show that NaBH$_4$ and LiAlH$_4$ can readily reduce an acid chloride.

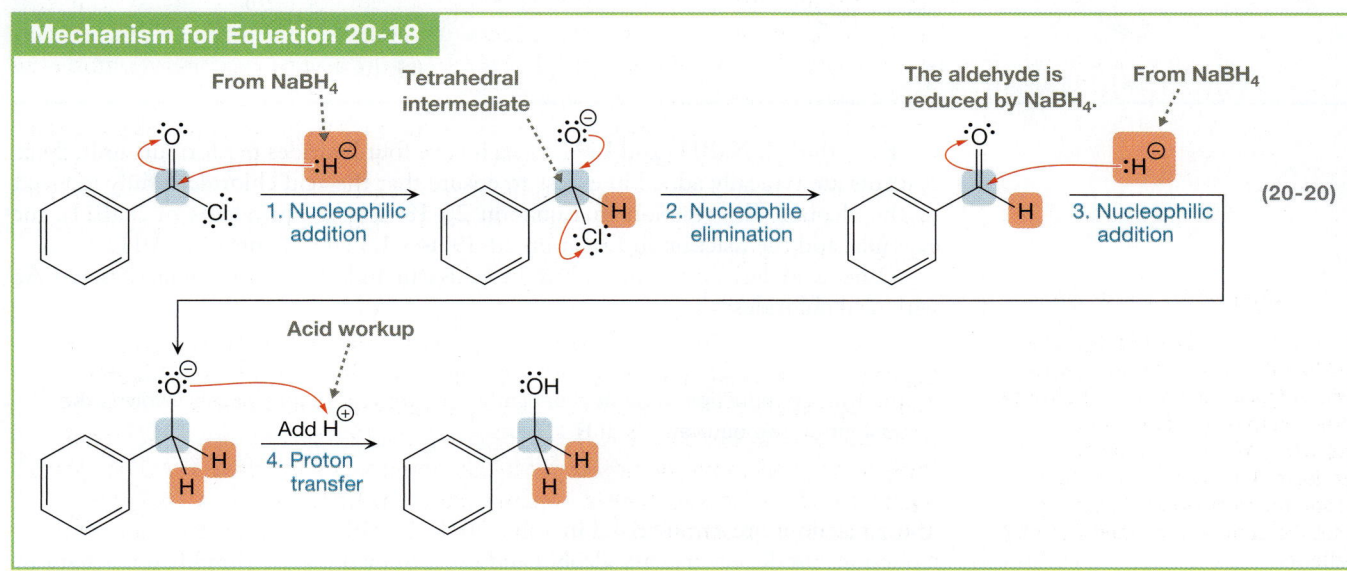

Notice that no carbonyl group is present in the product. Instead, a primary alcohol is produced. The simplified mechanism in Equation 20-20 shows how this occurs for the reduction using NaBH$_4$, and, as shown in Your Turn 20.11, the mechanism for the LiAlH$_4$ reduction is essentially the same. In each of these cases, we have represented NaBH$_4$ and LiAlH$_4$ as an H$^-$ nucleophile (review Section 7.1b), just as we did in Section 17.3 for the reductions of aldehydes, ketones, imines, and nitriles. In a more complete, detailed mechanism (review Section 17.3a), however, H$^-$ is transferred directly from the metal—H bond to the carbon atom.

Steps 1 and 2 make up the usual nucleophilic addition–elimination mechanism, producing an aldehyde as an intermediate. Under these reduction conditions, however, the aldehyde reacts rapidly with another equivalent of hydride (Section 17.3a) to produce an alkoxide anion. Subsequent acid workup yields the alcohol.

YOUR TURN 20.11

The following scheme outlines the mechanism for the LiAlH$_4$ reduction of the acid chloride in Equation 20-19. Supply the appropriate curved arrows, write the name of the elementary step underneath each reaction arrow, and identify the tetrahedral intermediate.

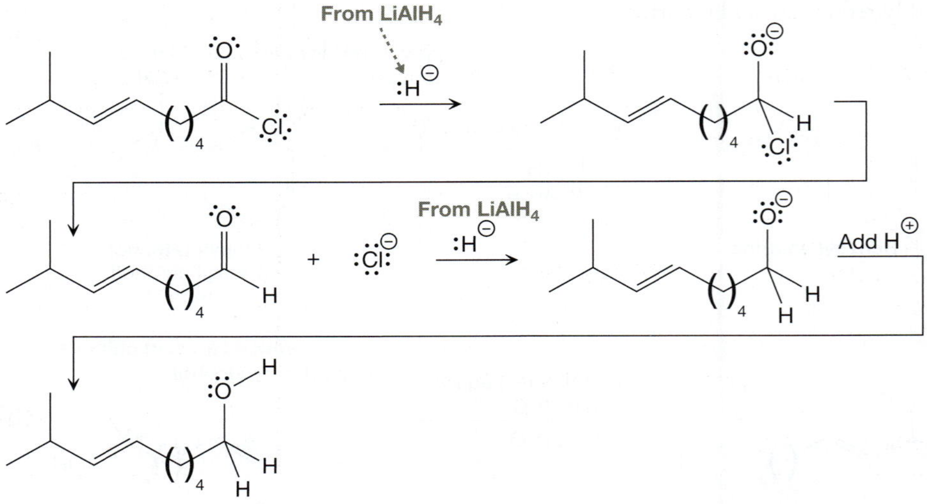

PROBLEM 20.13 Predict the major product and draw the complete, detailed mechanism for each of the following reactions.

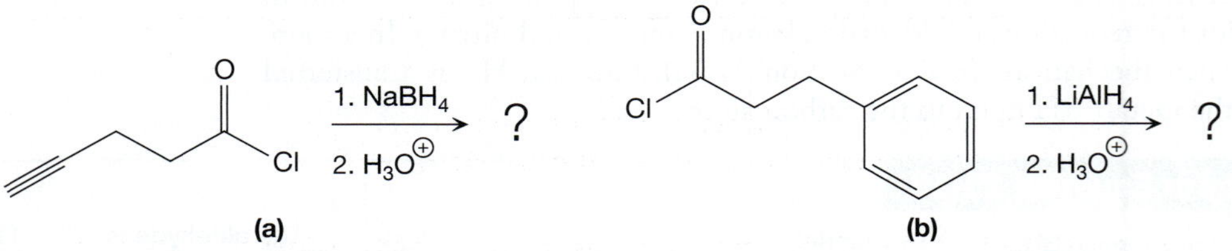

(a)

(b)

Even though NaBH$_4$ and LiAlH$_4$ each have four hydrides per formula unit, these reagents are typically added in excess to ensure that the acid chloride is fully reduced to the alcohol. The reaction in Equation 20-18 uses 3 equivalents of NaBH$_4$, for example, and the reaction in Equation 20-19 uses 1.3 equivalents of LiAlH$_4$.

Other acid derivatives and carboxylic acids can undergo hydride reduction, too. As with acid chlorides:

> The hydride reduction of an acid derivative or carboxylic acid typically involves the addition of two equivalents of H$^-$.

These reactions are summarized in Table 20-1. As with acid chlorides, many of these reductions produce a primary alcohol and proceed by the same simplified mechanism shown previously in Equation 20-20 (see Problem 20.14).

Reduction with NaBH$_4$	Reduction with LiAlH$_4$

Acid chloride → **Primary alcohol** (1. NaBH$_4$, 2. H$_3$O$^{\oplus}$)

Acid chloride → **Primary alcohol** (1. LiAlH$_4$, 2. H$_3$O$^{\oplus}$)

Acid anhydride → **Primary alcohol** (+ carboxylic acid) (1. NaBH$_4$, 2. H$_3$O$^{\oplus}$)

Acid anhydride → **Primary alcohol** (+ primary alcohol) (1. LiAlH$_4$, 2. H$_3$O$^{\oplus}$)

This reduction is very slow.

Ester → **Primary alcohol** (+ HOR′) (NaBH$_4$, EtOH)

Ester → **Primary alcohol** (+ HOR′) (1. LiAlH$_4$, 2. H$_3$O$^{\oplus}$)

Carboxylic acid → No reduction (NaBH$_4$)

Carboxylic acid → **Primary alcohol** (1. LiAlH$_4$, 2. H$_3$O$^{\oplus}$)

Amide → No reduction (NaBH$_4$)

Amide → **Amine** (1. LiAlH$_4$, 2. H$_2$O)

PROBLEM 20.14 Draw the complete, detailed mechanism for each generic reaction in Table 20-1 that converts an acid derivative to a primary alcohol.

The reactants in Table 20-1 appear in the same order as in the stability ladder (review Fig. 20-5, p. 1008), with the less stable species toward the top and the more stable species toward the bottom. Whereas LiAlH$_4$ reduces all of those species readily, the ability of NaBH$_4$ to carry out a reduction decreases substantially as the carboxylic acid derivative becomes more stable. We explore this distinction between NaBH$_4$ and LiAlH$_4$ in the following sections.

20.6a The Selectivity of NaBH$_4$

Notice in Table 20-1 that NaBH$_4$ reduces esters very slowly and does not reduce amides at all. This is a reflection of the resonance stabilization involving the carbonyl group and the leaving group, shown previously in Figure 20-3, which is more pronounced in esters and amides than in acid halides or acid anhydrides. (This is discussed further in Chapter 21.)

We can take advantage of the low reactivity of esters and amides to *selectively* reduce a more reactive acid derivative, such as an acid chloride or acid anhydride. Moreover, recall from Section 17.3 that NaBH$_4$ readily reduces aldehydes and ketones, so NaBH$_4$ can be used to selectively reduce aldehydes and ketones over esters and amides. In the β-keto ester in Equation 20-21, for example, the carbonyl group that has two attached carbons, characteristic of a ketone, is readily reduced to the alcohol, but the ester group is unaffected.

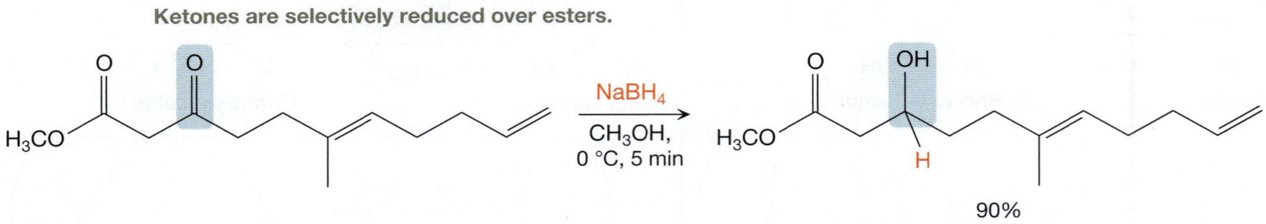

Ketones are selectively reduced over esters.

(20-21)

90%

SOLVED PROBLEM 20.15

Predict the major product of the reaction shown here.

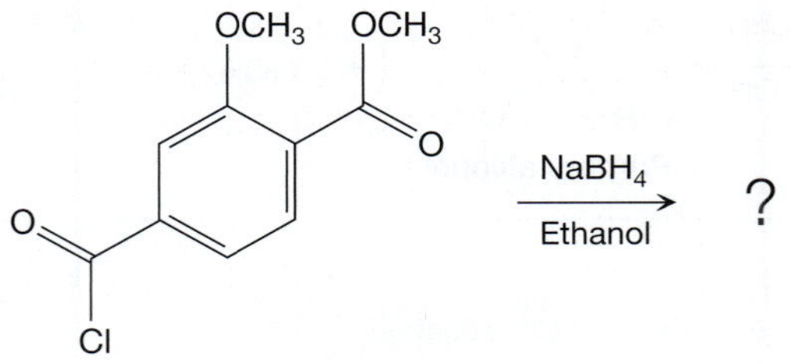

Think Which functional groups in the reactant can be reduced by NaBH$_4$? Does the reduction of one of those groups take place more readily than the other(s)?

Solve The COCl and CO$_2$CH$_3$ groups can be reduced by NaBH$_4$ to produce C—OH, but the ether group cannot be reduced by NaBH$_4$. Whereas NaBH$_4$ reduces the COCl group readily, the reduction of the CO$_2$CH$_3$ group is sluggish, so the COCl group will be reduced selectively as follows:

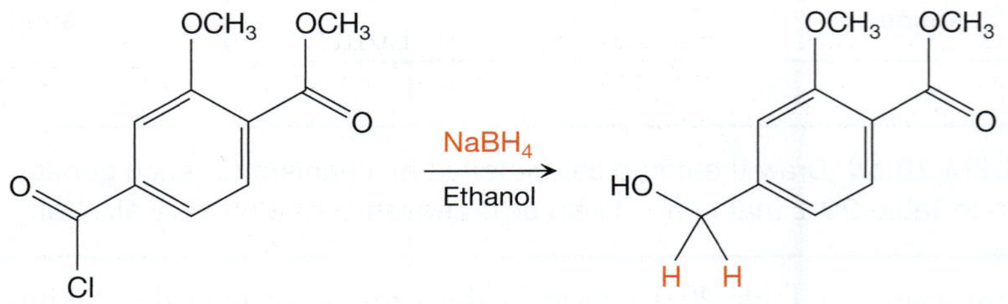

PROBLEM 20.16 Draw the complete, detailed mechanism and predict the major product for each of the following reactions.

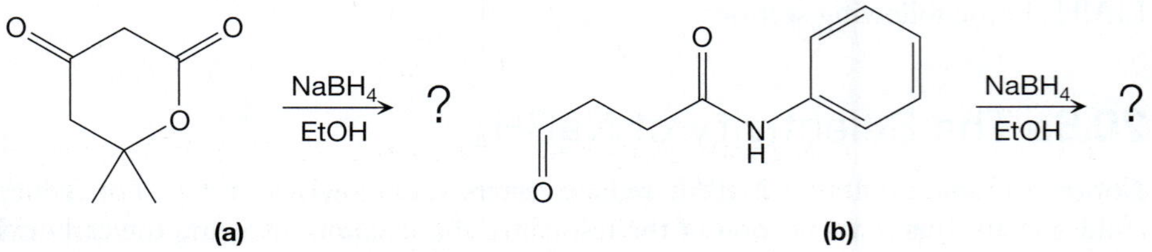

(a)

(b)

Notice, too, in Table 20-1 that NaBH$_4$ doesn't reduce a carboxylic acid even though it reduces esters slowly. This might seem peculiar because carboxylic acids are on the same rung of the stability ladder as esters (Fig. 20-5). $NaBH_4$ is somewhat basic, however, and a carboxylic acid is moderately acidic. Therefore, as shown in Equation 20-22, a rapid proton transfer converts the carboxylic acid into its carboxylate anion, which is heavily stabilized by resonance. Carboxylate anions, moreover, are at the very bottom of the stability ladder, so once the carboxylate is formed under these conditions, there is no further reaction.

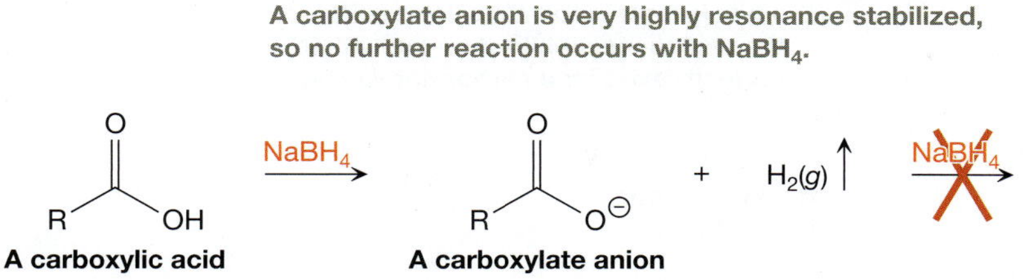

A carboxylate anion is very highly resonance stabilized, so no further reaction occurs with $NaBH_4$.

(20-22)

A carboxylic acid   **A carboxylate anion**

PROBLEM 20.17 Show how this carboxylic acid can be converted into the corresponding alcohol using $NaBH_4$ as the reducing agent. *Hint:* Can you convert the carboxylic acid into a different acid derivative first?

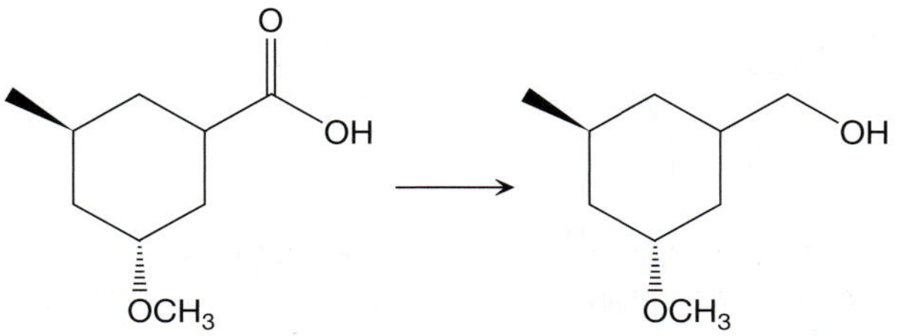

20.6b The Greater Reactivity of LiAlH$_4$

Whereas $NaBH_4$ cannot reduce carboxylic acids, $LiAlH_4$ is a much stronger reducing agent (Section 17.3a) and will reduce a carboxylic acid to a primary alcohol. An example is shown in Equation 20-23.

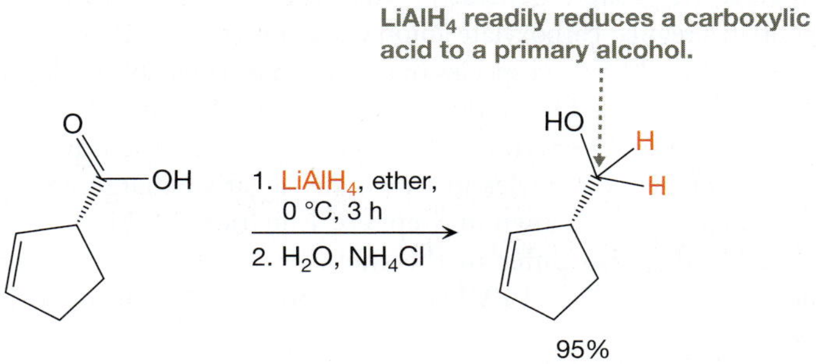

$LiAlH_4$ readily reduces a carboxylic acid to a primary alcohol.

1. $LiAlH_4$, ether, 0 °C, 3 h
2. H_2O, NH_4Cl

(20-23)

95%

The mechanism for this reaction is shown in Equation 20-24. Step 1 is a rapid proton transfer that produces a carboxylate anion, just as we saw in Equation 20-22 when a carboxylic acid is treated with $NaBH_4$. At the same time, the negatively charged O atom of the carboxylate anion forms a relatively strong bond to the Al atom. (This is analogous to the hydride transfer step we saw previously in Section 17.3a.) Subsequently, Steps 2–5 make up the back-to-back hydride reductions we saw previously for more reactive acid derivatives (Equation 20-20). For the nucleophilic addition steps in Steps 2 and 4, notice that we have simplified $LiAlH_4$ to a H^- nucleophile.

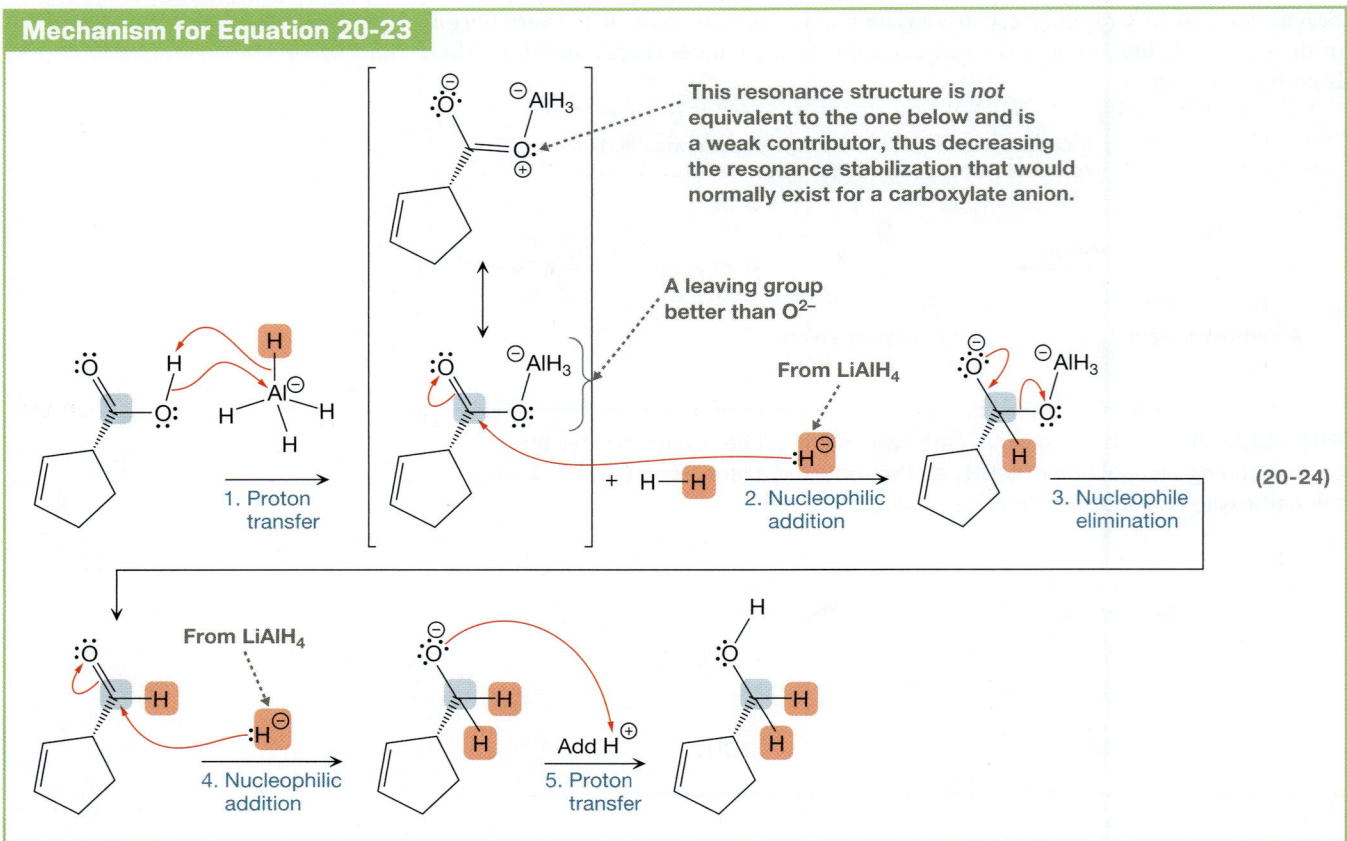

Mechanism for Equation 20-23

This resonance structure is *not* equivalent to the one below and is a weak contributor, thus decreasing the resonance stabilization that would normally exist for a carboxylate anion.

A leaving group better than O^{2-}

From $LiAlH_4$

1. Proton transfer

2. Nucleophilic addition

3. Nucleophile elimination

(20-24)

From $LiAlH_4$

4. Nucleophilic addition

Add $H^{\oplus}$

5. Proton transfer

The formation of the O—Al bond in Step 1 of Equation 20-24 facilitates the $LiAlH_4$ reduction of a carboxylic acid in two ways. First, as indicated in Equation 20-24, this bond removes the equivalence that is normally observed for the resonance structures of a carboxylate anion (RCO_2^-). Thus, the C=O is *not* as highly resonance stabilized as in a regular carboxylate anion and is therefore more susceptible to nucleophilic attack by H^-. Second, a better leaving group is generated; without the formation of the O—Al bond, the leaving group in a regular carboxylate anion would be an isolated oxygen atom bearing a -2 charge (review Fig. 20-5)—a species that is much too unstable to depart on its own.

Although not shown, the $O—AlH_3^-$ group produced in Step 1 can be the source of H^- in Step 2, becoming $O—AlH_2$ before departing. Thus, the leaving group would depart as $^-O—AlH_2$, which carries just a single negative charge and is even more stable than the leaving group shown in Step 3 of Equation 20-24.

$LiAlH_4$ and $NaBH_4$ also differ in the way they react with an amide. Whereas $NaBH_4$ cannot reduce an amide, $LiAlH_4$ can, as shown in Equation 20-25.

$LiAlH_4$ reduces an amide to an amine.

$\xrightarrow[\text{Ether}]{LiAlH_4}$

(20-25)

98%

More specifically:

> LiAlH$_4$ reduces an amide (RCONR$_2$) to an amine (RCH$_2$NR$_2$), preserving the overall carbon backbone of the original compound.

It is particularly striking that the product is an amine, because LiAlH$_4$ reduces all of the other acid derivatives to primary alcohols. This difference stems, once again, from the relatively strong O—Al bond that forms, as shown in the mechanism in Equation 20-26.

Mechanism for Equation 20-25

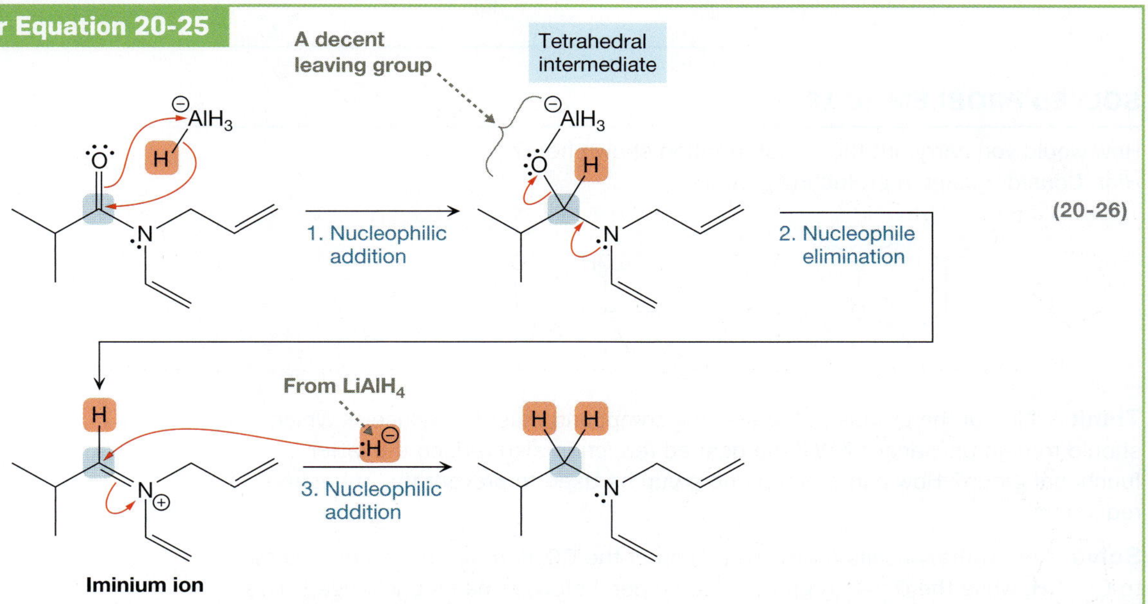

A decent leaving group

Tetrahedral intermediate

1. Nucleophilic addition

2. Nucleophile elimination

From LiAlH$_4$

3. Nucleophilic addition

Iminium ion

(20-26)

In Step 1, H$^-$ adds to the carbonyl group of the amide, and the O—Al bond forms at the same time. The tetrahedral intermediate that is produced contains the same O—AlH$_3^-$ leaving group we saw in the mechanism describing the reduction of carboxylic acids (Equation 20-24). In Step 2, the O—AlH$_3^-$ leaving group departs, yielding an iminium ion, which is subsequently attacked by H$^-$ in Step 3 to yield the final product.

Similar to the lithium aluminum hydride reduction of a carboxylic acid, the O—AlH$_3^-$ group in the product of Step 1 can donate a H$^-$ to become O—AlH$_2$. The leaving group in the subsequent nucleophilic addition–elimination, therefore, would be even more stable.

YOUR TURN 20.12

The following mechanism is for the LiAlH$_4$ reduction of another amide, but the curved arrows have been omitted. Supply the missing curved arrows, and identify the tetrahedral intermediate. Below each reaction arrow, write the name of the elementary step that is taking place.

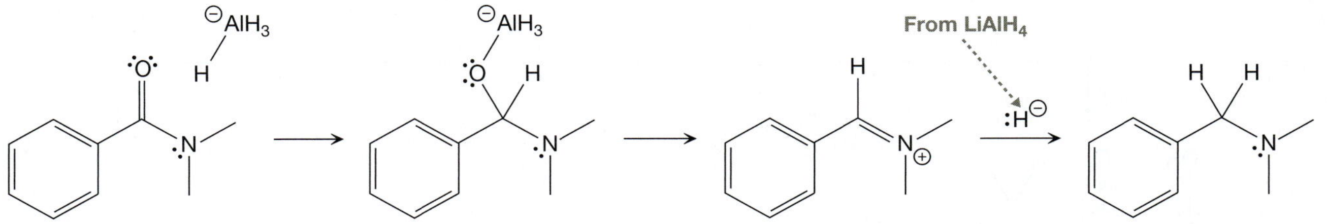

From LiAlH$_4$

Draw the complete, detailed mechanism and predict the major product for each of the following reactions.

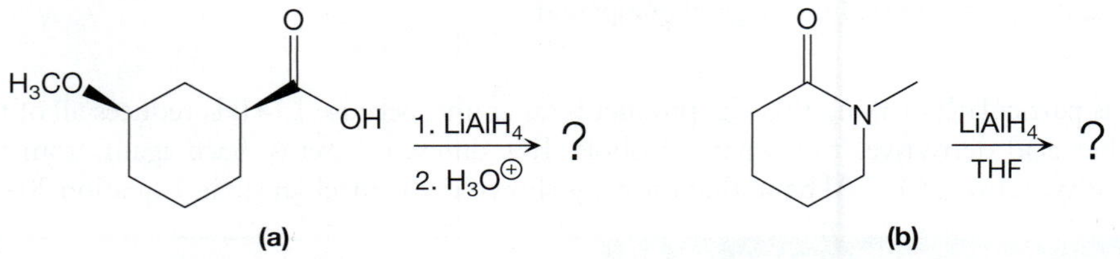

(a)

(b)

SOLVED PROBLEM 20.19

How would you carry out the transformation shown here?
Hint: Consider using a protecting group.

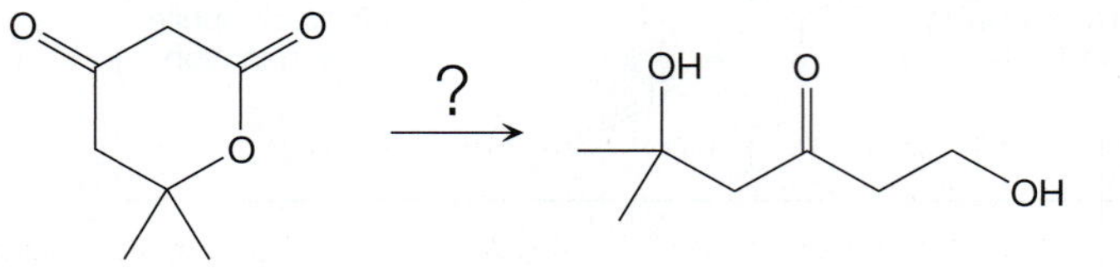

Think Which of the groups in the starting compound must be reduced? Which should remain unchanged? Will the desired reduction also reduce the latter functional group? How can a protecting group be used to prevent the undesired reduction?

Solve The synthesis calls for the reduction of the CO_2R group (red screen below) to C—OH, while the O=CC_2 group (blue screen below) remains unchanged. This kind of a reaction would proceed via nucleophilic addition–elimination at the CO_2R carbon and will open the ring. However, as shown at the right, reduction using $LiAlH_4$ will reduce the O=CC_2 as well (the same is true if we were to use $NaBH_4$).

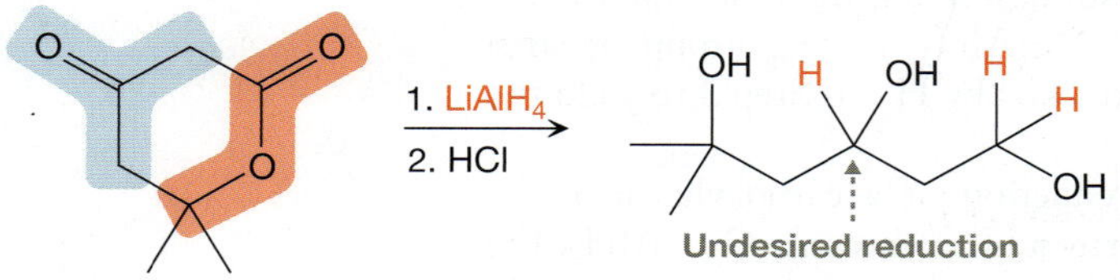

Undesired reduction

This problem can be circumvented by protecting the O=CC_2 group as an acetal, as follows:

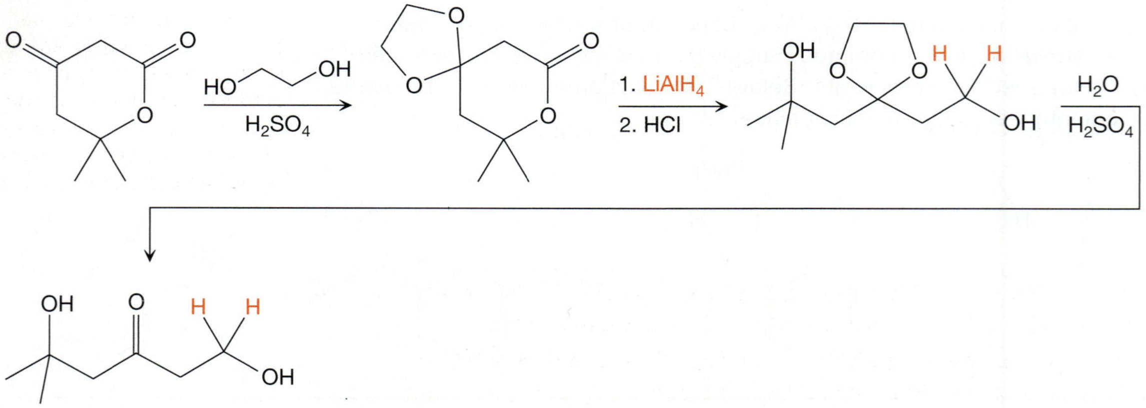

PROBLEM 20.20 Show how to carry out this synthesis. *Hint:* Consider using a protecting group.

$$\text{(structures)} \quad \xrightarrow{\ ?\ } \quad \text{(structures)}$$

20.7 Specialized Reducing Agents: Diisobutylaluminum Hydride (DIBAH) and Lithium Tri-*tert*-butoxyaluminum Hydride (LTBA)

The mechanism in Equation 20-20 (p. 1021), which describes the hydride reduction of an acid derivative to a primary alcohol, proceeds through an aldehyde intermediate. That is, once the aldehyde is formed, it quickly reacts with H⁻ in a second reduction. In some situations, however, it might be advantageous to stop at the aldehyde. Two specialized reducing agents are commonly used for such purposes: **lithium tri-*tert*-butoxyaluminum hydride (LTBA)** and **diisobutylaluminum hydride (DIBAH, or DIBAL-H)** (Fig. 20-10). Both of these compounds possess an Al—H bond and thus are hydride anion (H⁻) sources, much like LiAlH$_4$.

Examples of how these specialized reducing agents are used are shown in Equations 20-27 and 20-28.

LiAlH(O-*t*-Bu)$_3$ reduces an acid chloride to an aldehyde.

$$\text{Benzoyl chloride} \xrightarrow[\substack{(CH_3OCH_2CH_2)_2O, \\ -75\ °C,\ 1\ h, \\ 20\ °C,\ 1\ h}]{LiAlH(O\text{-}t\text{-Bu})_3} \text{Benzaldehyde}$$ (20-27)

Benzoyl chloride

Benzaldehyde
73%

DIBAH reduces an ester to an aldehyde.

$$\text{Ethyl 4-methylpent-4-enoate} \xrightarrow[\substack{2.\ HCl,\ H_2O/CH_3OH}]{\substack{1.\ DIBAH,\ CH_2Cl_2, \\ -78\ °C,\ 2.5\ h}} \text{4-Methylpent-4-enal}$$ (20-28)

Ethyl 4-methylpent-4-enoate

4-Methylpent-4-enal
95%

Specifically:

- Lithium tri-*tert*-butoxyaluminum hydride (LTBA) is commonly used to reduce an acid chloride (RCOCl) to an aldehyde (RCH=O) at low temperature.
- Diisobutylaluminum hydride (DIBAH) is commonly used to reduce an ester (RCO$_2$R) to an aldehyde (RCH=O) at low temperature.

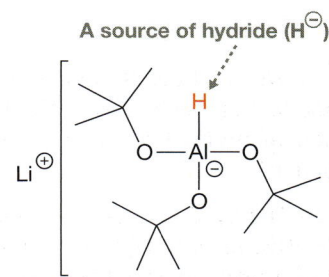

A source of hydride (H⁻)

(a) Lithium tri-*tert*-butoxyaluminum hydride [LiAlH(O-*t*-Bu)$_3$]

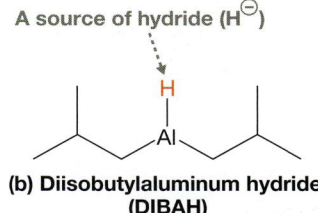

A source of hydride (H⁻)

(b) Diisobutylaluminum hydride (DIBAH)

FIGURE 20-10 Two specialized hydride reducing agents These hydride reducing agents can be used to reduce a carboxylic acid derivative to an aldehyde. (a) Lithium tri-*tert*-butoxyaluminum hydride can reduce an acid chloride to an aldehyde. (b) Diisobutylaluminum hydride can reduce an ester to an aldehyde.

The mechanism by which lithium tri-*tert*-butoxyaluminum hydride (simplified to H^-) reduces an acid chloride to an aldehyde is shown in Equation 20-29.

The reaction stops at the aldehyde because a second addition of $H^\ominus$ from LiAlH(O-*t*-Bu)₃ is much slower than the first.

From LiAlH(O-*t*-Bu)₃

1. Nucleophilic addition

2. Nucleophile elimination

Slow

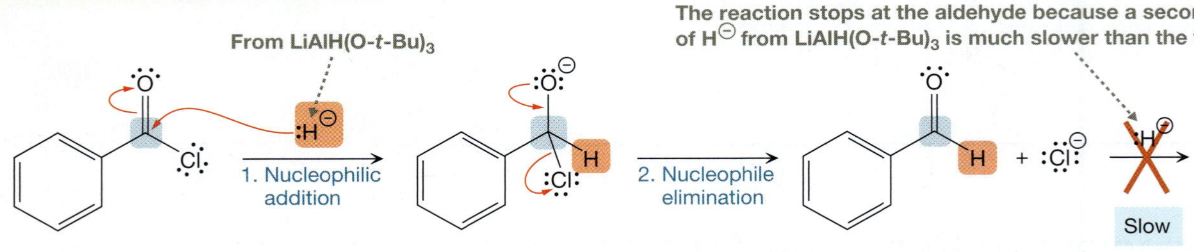

(20-29)

This is essentially the same nucleophilic addition–elimination mechanism as the one by which LiAlH₄ operates, the only difference being the rate.

Reduction by lithium tri-*tert*-butoxyaluminum hydride occurs much more slowly than by LiAlH₄, allowing the reaction to take place in a more controlled fashion.

Thus, an aldehyde is less reactive than an acid chloride, so the second reduction is slower than the first. Once the first reduction has come to completion, the reaction can be stopped before the second reduction can proceed.

There are two reasons why reduction with lithium tri-*tert*-butoxyaluminum hydride is significantly slower than reduction with LiAlH₄. One is the bulkiness of the *tert*-butoxy groups, which introduces significant steric hindrance in the nucleophilic addition step. The second reason is that the reaction is carried out at very cold temperatures, around $-75\ °C$. At room temperature, lithium tri-*tert*-butoxyaluminum hydride can rapidly reduce an acid chloride all the way to the alcohol, adding two equivalents of hydride (Equation 20-30).

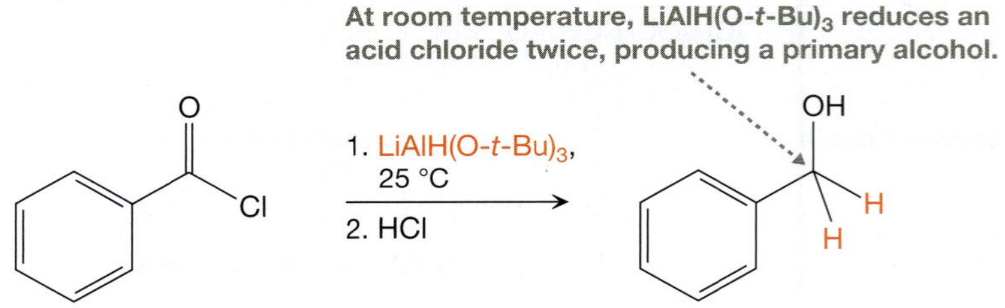

At room temperature, LiAlH(O-*t*-Bu)₃ reduces an acid chloride twice, producing a primary alcohol.

1. LiAlH(O-*t*-Bu)₃, 25 °C
2. HCl

(20-30)

The reduction of an ester by DIBAH stops at the aldehyde stage *not* because the second reduction is slower than the first, but because the tetrahedral intermediate that is formed (Equation 20-31) is relatively stable at $-78\ °C$. In that tetrahedral intermediate, the O atom from the initial carbonyl group has formed a bond to Al, much as we have seen previously in other mechanisms. This species persists until H_3O^+ is added, at which point the dialkylaluminum group bonded to O is replaced by a proton. A hemiacetal is produced, which equilibrates to the aldehyde. The H_3O^+ that is added in the acid workup also neutralizes any excess DIBAH. Consequently:

In the DIBAH reduction of an ester, the aldehyde product and the reducing agent are *not* present in the reaction mixture at the same time, so the aldehyde cannot be further reduced to the alcohol.

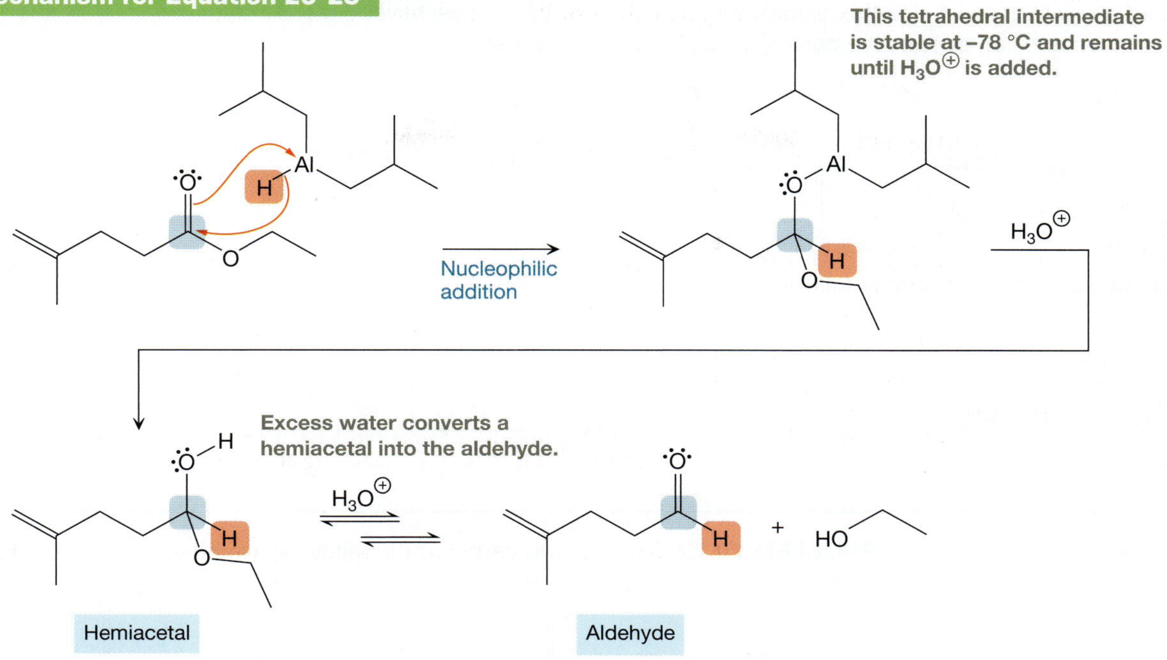

This tetrahedral intermediate is stable at –78 °C and remains until $H_3O^{\oplus}$ is added.

Nucleophilic addition

$H_3O^{\oplus}$

(20-31)

Excess water converts a hemiacetal into the aldehyde.

$H_3O^{\oplus}$

Hemiacetal

Aldehyde

+ HO

The relative stability of the tetrahedral intermediate in the DIBAH reduction of an ester can be rationalized in part by charge stability. As shown in Equation 20-32, elimination of the alkoxy leaving group from the tetrahedral intermediate would generate two additional charges: a negative charge on the O atom of the leaving group (RO⁻) and a positive charge on the O atom bonded to Al. The small size of O does little to stabilize those charges, so the uncharged tetrahedral intermediate persists.

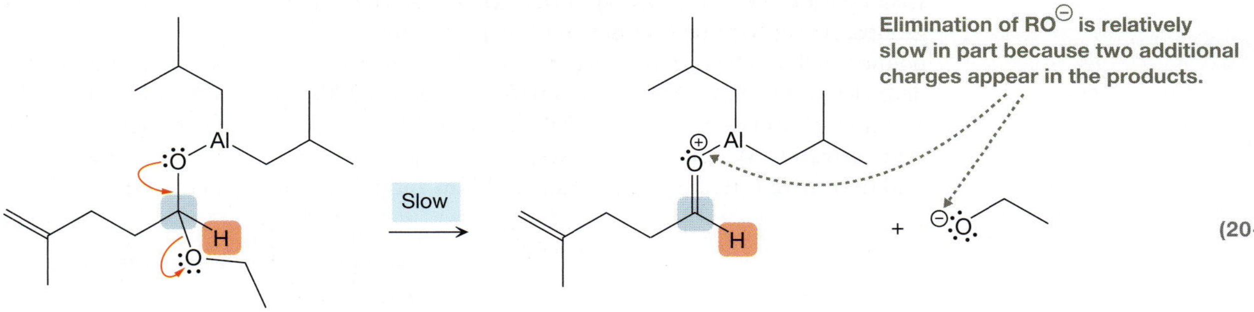

Elimination of $RO^{\ominus}$ is relatively slow in part because two additional charges appear in the products.

Slow

(20-32)

SOLVED PROBLEM 20.21

Show how you would convert this carboxylic acid into the corresponding aldehyde.

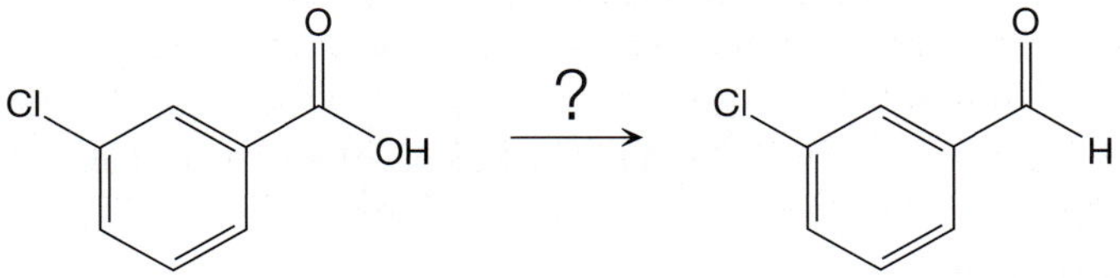

Think What reactions do we know that can convert an acid derivative into an aldehyde? Can this be accomplished directly from a carboxylic acid?

Solve We just learned that an aldehyde can be produced from an acid chloride using lithium tri-*tert*-butoxyaluminum hydride, or from an ester using DIBAH. Thus, we must convert the initial carboxylic acid into an acid chloride or an ester. We have not yet learned a way to convert a carboxylic acid into an acid chloride, but we have seen that diazomethane can convert the carboxylic acid into a methyl ester.

The synthesis would then be written as follows:

PROBLEM 20.22 Show how to carry out the following synthesis.

20.8 Organometallic Reagents

Like hydride (H⁻) anions, alkyl anions (R⁻) can act as nucleophiles but not as leaving groups. As we have seen in previous chapters, sources of R⁻ are organometallic compounds such as alkyllithium reagents (RLi), Grignard reagents (RMgX), and lithium dialkylcuprates (R₂CuLi). Thus, as with hydride reagents, the product of nucleophilic addition–elimination reaction involving one of these organometallic reagents is *not* an acid derivative. We can see this explicitly in Equations 20-33 and 20-34, in which an acid derivative is treated with an alkyllithium and a Grignard reagent, respectively.

In general:

> Acid chlorides (RCOCl), acid anhydrides (RCO$_2$COR), and esters (RCO$_2$R) can be treated with an alkyllithium (R′Li) or a Grignard (R′MgX) reagent to produce a tertiary alcohol (R$_2'$RCOH).

In each case, two equivalents of the R′$^-$ nucleophile must add to the acid derivative to produce a tertiary alcohol. This is explained by Equation 20-35, the simplified mechanism of the reaction in Equation 20-34.

Mechanism for Equation 20-34

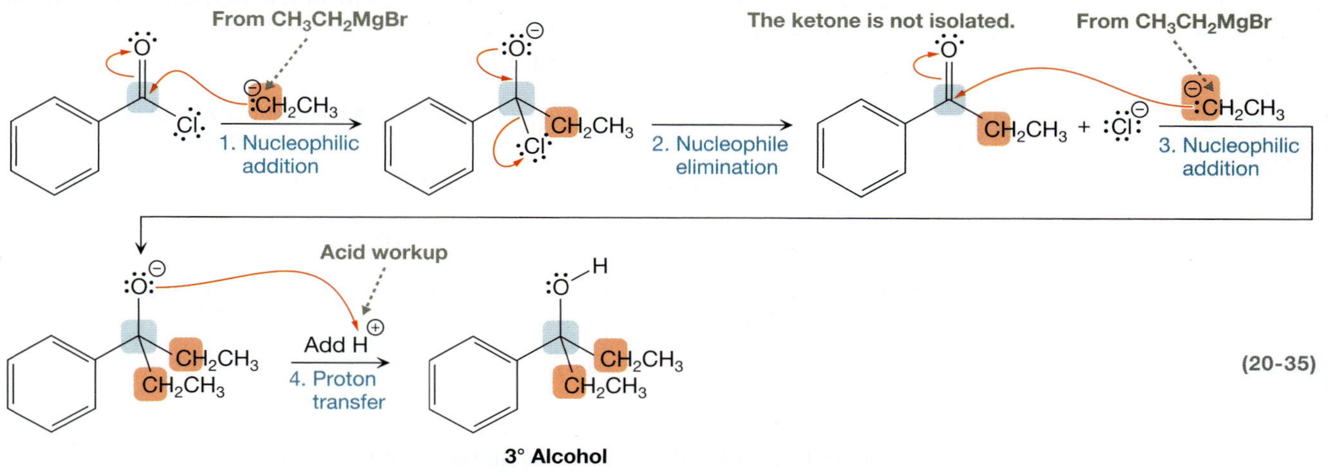

From CH$_3$CH$_2$MgBr

1. Nucleophilic addition

The ketone is not isolated.

2. Nucleophile elimination

From CH$_3$CH$_2$MgBr

3. Nucleophilic addition

Acid workup

Add H$^+$

4. Proton transfer

3° Alcohol

(20-35)

This mechanism is essentially the same as the one in Equation 20-20 (p. 1021), which describes the hydride reduction of an acid chloride to a primary alcohol. The first two steps make up the usual nucleophilic addition–elimination mechanism, which produces a ketone in this case. Once that ketone is produced, it reacts with a second equivalent of R$^-$ in Step 3 to produce a tertiary alkoxide anion, R$_3$CO$^-$. Acid workup in Step 4 yields the tertiary alcohol as the final product. Notice that Steps 3 and 4 compose the mechanism for a typical alkyllithium or Grignard reaction involving a ketone, which we previously discussed in Section 17.5.

YOUR TURN 20.13

Draw the mechanism for the reaction in Equation 20-33.

Although similar mechanisms can be drawn involving carboxylic acids or amides, these reactions are generally avoided because amides are significantly less reactive than esters, and carboxylic acids are acidic, so they protonate the R$^-$ nucleophile.

PROBLEM 20.23 Draw the complete, detailed mechanism and predict the major products for each of the following reactions.

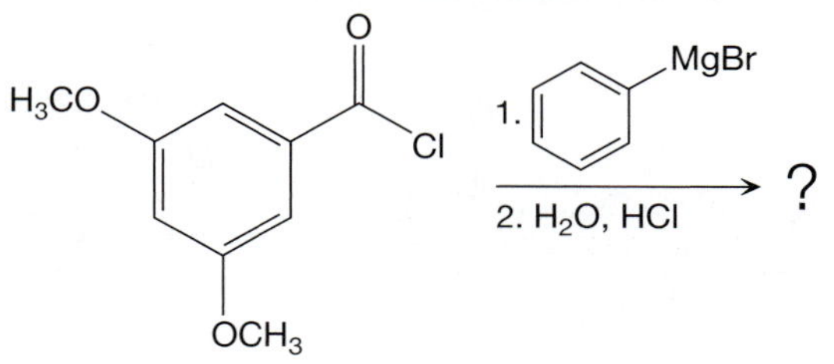

(a)

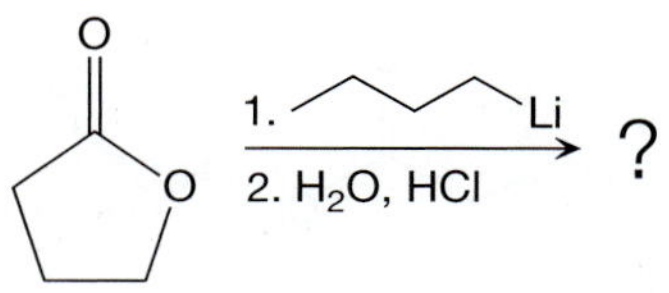

(b)

PROBLEM 20.24 Propose a synthesis for the following transformation.

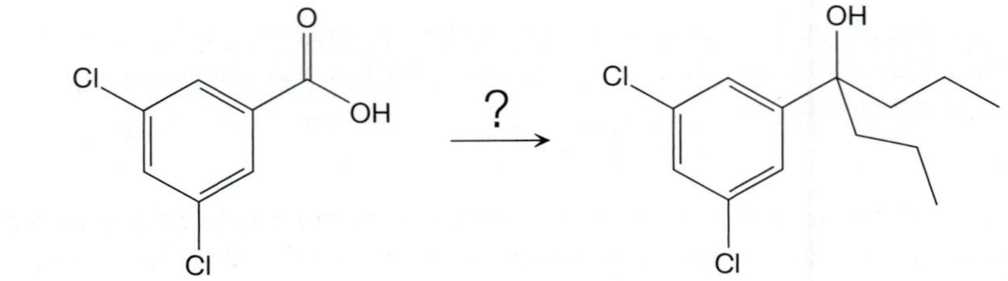

As with the hydride reductions in Section 20.7, it can be advantageous to carry out reactions in which only one equivalent of R^- adds to an acid derivative, thus producing a ketone that can be isolated. One way to carry out this kind of a reaction involves lithium dialkylcuprates, as shown in Equation 20-36.

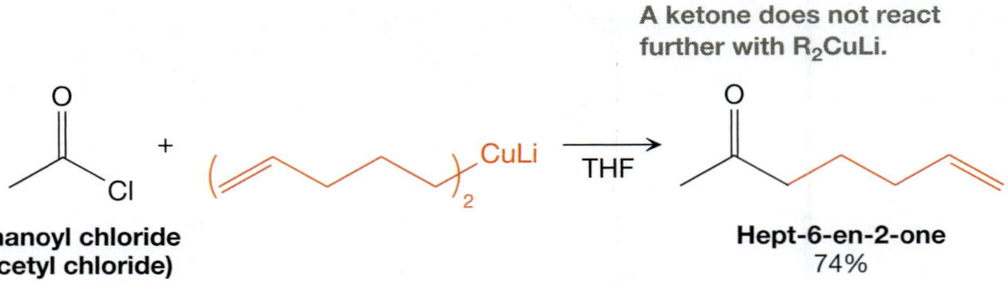

A ketone does not react further with R_2CuLi.

Ethanoyl chloride (Acetyl chloride)

Hept-6-en-2-one
74%

(20-36)

Thus:

An acid chloride (RCOCl) can generally be treated with a lithium dialkylcuprate (R_2CuLi) to produce a ketone ($R_2C=O$).

As we learned in Chapter 17, the mechanism involving lithium dialkylcuprates is different from the mechanisms involving Grignard or alkyllithium reagents, owing to the nucleophilic character at the Cu atom. Nevertheless, we can think of R_2CuLi as a weak source of R^- that tends not to react with ketones or aldehydes via direct addition. The R^- from R_2CuLi will add, however, to the $C=O$ group of an acid chloride, which is much more reactive than a ketone or aldehyde. If we think of R_2CuLi as a source of R^-, then nucleophilic addition–elimination produces a ketone and, under these conditions, that ketone does not react further.

Lithium dialkylcuprates do not react with esters or amides, which are significantly more stable than acid chlorides. Thus, to convert an ester or amide into a ketone, we must first convert it into an acid chloride. We discuss how to do so in Chapter 21.

PROBLEM 20.25 Draw the major organic product for each of the following reactions. If no reaction takes place, write "no reaction." For each reaction that does take place, draw its simplified mechanism, treating R_2CuLi as a weak source of R^-.

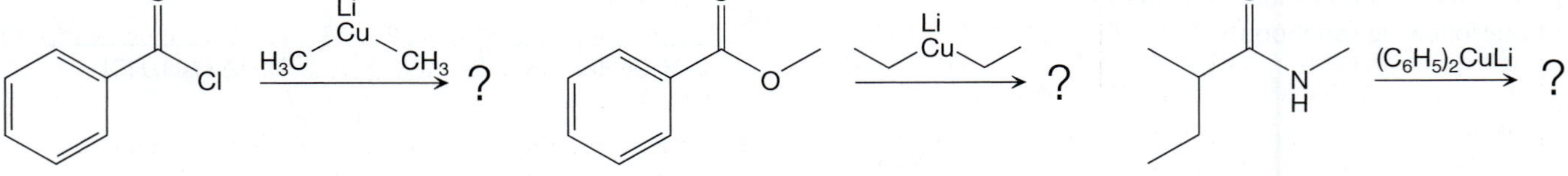

(a)   (b)   (c)

SOLVED PROBLEM 20.26

Show how to carry out this synthesis.

Think What precursor can be used to produce the α,β-unsaturated ketone? To make that precursor from the starting compound, which carbonyl-containing functional group must gain an additional carbon–carbon bond? What reagent can be used to selectively react with that functional group?

Solve The α,β-unsaturated ketone can be produced from an aldol condensation (Section 18.5), as shown in the following retrosynthetic analysis.

To make the precursor from the starting material, the Cl of the acid chloride must be replaced by a CH$_3$ group, which can be accomplished using (CH$_3$)$_2$CuLi. In the forward direction, the synthesis might be written as follows:

PROBLEM 20.27 Show how to carry out this synthesis.

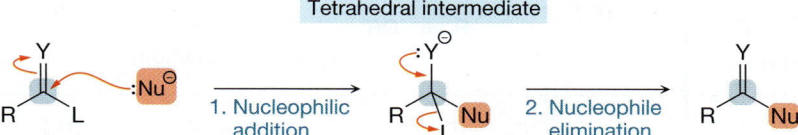

Chapter Summary and Key Terms

- The general mechanism of a **nucleophilic addition–elimination reaction** consists of two steps:

Tetrahedral intermediate

1. Nucleophilic addition
2. Nucleophile elimination

First a nucleophile attacks the electron-poor atom of a polar π bond (Y is an electronegative atom) in a nucleophilic addition step, producing a high-energy **tetrahedral intermediate**. Second, the tetrahedral intermediate eliminates a leaving group (L$^-$) initially bonded to the electron-poor atom of the polar π bond. **(Sections 20.1 and 20.1a; Objective 1)**

- When the leaving group is bonded to a carbonyl group, nucleophilic addition–elimination is called a **nucleophilic acyl substitution**. **(Section 20.1a; Objective 1)**

- In a **transesterification reaction**, an ester reacts with an alkoxide anion (RO$^-$) to produce a new ester. **(Section 20.1a; Objectives 1, 3)**

- The first step of a nucleophilic addition–elimination reaction is usually the rate-determining step. **(Section 20.1b; Objective 2)**

- A transesterification reaction is reversible because the reactants and products have roughly equal stabilities. **(Section 20.1c; Objective 3)**

- The stability of a carboxylic acid derivative decreases in the order: carboxylate anion > amide > ester ≈ carboxylic acid > acid anhydride > acid chloride. **(Section 20.2; Objective 3)**
- The conversion of one carboxylic acid derivative into another can be carried out with relative ease if the acid derivative on the product side is more stable (i.e., on a lower rung of the stability ladder) than the one on the reactant side. By contrast, the conversion is difficult if the acid derivative on the product side is higher in energy than the one on the reactant side. **(Section 20.2; Objective 3)**
- In a **saponification** reaction, an ester reacts with HO^- to produce an initial carboxylic acid that is rapidly and irreversibly deprotonated under the basic conditions of the reaction. **(Section 20.3; Objective 4)**
- The conversion of an amide ($RCONR_2$) to a carboxylic acid (RCO_2H) via nucleophilic addition–elimination is energetically unfavorable, but it can be carried out with relative ease by treating the amide with HO^-, followed by acid workup. The initial carboxylic acid that is formed is rapidly and irreversibly deprotonated under the basic conditions of the reaction. **(Section 20.4; Objective 5)**
- The **Gabriel synthesis** produces a primary amine from a corresponding alkyl halide. **(Section 20.4; Objective 5)**
- A **haloform reaction** converts a methyl ketone ($RCOCH_3$) into a carboxylic acid (RCO_2H). The CH_3 group is an unsuitable leaving group, but under the conditions of the reaction, it is first converted into CX_3 (X = Cl, Br, I), which is a suitable leaving group. Subsequent acyl substitution involving HO^- as the nucleophile produces an initial carboxylic acid

- that is rapidly and irreversibly deprotonated under the basic conditions of the reaction. **(Section 20.5; Objective 6)**
- $NaBH_4$ readily reduces high-energy acid derivatives, such as acid chlorides (RCOCl) and acid anhydrides (RCO_2COR), to primary alcohols. Esters, which are significantly more stable, are reduced to primary alcohols slowly. All of these reactions proceed through an aldehyde intermediate, which is reduced in a second reduction step to the alcohol. **(Sections 20.6 and 20.6a; Objective 7)**
- $LiAlH_4$ is a more powerful reducing agent than $NaBH_4$ and can thus reduce acid chlorides, acid anhydrides, and esters to primary alcohols, too. **(Section 20.6b; Objective 7)**
- $LiAlH_4$ reduces carboxylic acids (RCO_2H) to primary alcohols, and reduces amides ($RCONR_2$) to amines (RCH_2NR_2). These reactions are facilitated by the strong O—Al bond that forms in the mechanism. **(Section 20.6b; Objective 7)**
- **Diisobutylaluminum hydride** (**DIBAH, or DIBAL-H**) and **lithium tri-*tert*-butoxyaluminum hydride (LTBA)** are two specialized reducing agents that can be used to reduce an acid derivative to an aldehyde. DIBAH reduces an ester to an aldehyde, whereas lithium tri-*tert*-butoxyaluminum hydride reduces an acid chloride to an aldehyde. **(Section 20.7; Objective 7)**
- When an acid derivative such as an acid chloride, acid anhydride, or ester is treated with an alkyllithium reagent (RLi) or a Grignard reagent (RMgX), R^- adds twice to produce a tertiary alcohol. **(Section 20.8; Objective 8)**
- When an acid chloride (RCOCl) is treated with a lithium dialkylcuprate ($R_2'CuLi$), R'^- adds once to produce a ketone. **(Section 20.8; Objective 8)**

Reaction Tables

TABLE 20-2 Functional Group Conversions

	Starting Compound Class	Typical Reagents and Reaction Conditions	Compound Class Formed	Key Electron-Rich Species	Key Electron-Poor Species	Comments	Discussed in Section
(1)	Ester	NaOR″ / Ether	Ester	$^{\ominus}OR''$		Nucleophilic addition–elimination (transesterification)	20.1a
(2)	Acid chloride	LiOR′	Ester	$^{\ominus}OR'$		Nucleophilic addition–elimination	20.2
(3)	Acid anhydride	NaOR′	Ester	$^{\ominus}OR'$		Nucleophilic addition–elimination	20.2

TABLE 20-2 Functional Group Conversions (continued)

	Starting Compound Class	Typical Reagents and Reaction Conditions	Compound Class Formed	Key Electron-Rich Species	Key Electron-Poor Species	Comments	Discussed in Section
(4)	Ester	$LiNR_2''$	Amide	$^\ominus NR_2''$	Ester ($R\!-\!\delta^+\!-\!OR'$)	Nucleophilic addition–elimination	20.2
(5)	Ester	1. NaOH 2. HCl	Carboxylic acid	$^\ominus OH$	Ester ($R\!-\!\delta^+\!-\!OR'$)	Nucleophilic addition–elimination (saponification)	20.3
(6)	Amide	1. NaOH 2. HCl	Carboxylic acid	$^\ominus OH$	Amide ($R\!-\!\delta^+\!-\!NR_2'$)	Nucleophilic addition–elimination	20.4
(7)	Phthalimide	1. KOH/EtOH 2. RBr 3. KOH/H_2O	Primary amine $H_2N\!-\!R$	Phthalimide anion ($N^\ominus$)	$\delta^+ R\!-\!Br$	S_N2, then nucleophilic addition–elimination (Gabriel synthesis)	20.4
(8)	Acid chloride	1. $NaBH_4$ or $LiAlH_4$ 2. HCl	Primary alcohol	$^\ominus H$	Acid chloride ($R\!-\!\delta^+\!-\!Cl$)	Nucleophilic addition–elimination, then addition (reduction)	20.6
(9)	Acid anhydride	1. $NaBH_4$ or $LiAlH_4$ 2. HCl	Primary alcohol	$^\ominus H$	Acid anhydride ($R\!-\!\delta^+\!-\!O\!-\!R$)	Nucleophilic addition–elimination, then addition (reduction)	20.6
(10)	Ester	1. $NaBH_4$ or $LiAlH_4$ 2. HCl	Primary alcohol	$^\ominus H$	Ester ($R\!-\!\delta^+\!-\!OR'$)	Nucleophilic addition–elimination, then addition (reduction); very slow with $NaBH_4$	20.6
(11)	Carboxylic acid	1. $LiAlH_4$ 2. HCl	Primary alcohol	$^\ominus H$	Carboxylic acid ($R\!-\!\delta^+\!-\!O\!-\!{}^\ominus AlH_3$)	Nucleophilic addition–elimination, then addition (reduction)	20.6
(12)	Amide	$LiAlH_4$ Ether	Amine	$^\ominus H$	iminium ($R\!-\!CH\!=\!N^\oplus(R')(R')$)	Nucleophilic addition–elimination, then addition (reduction)	20.6

(continued)

TABLE 20-2 Functional Group Conversions (continued)

Starting Compound Class	Typical Reagents and Reaction Conditions	Compound Class Formed	Key Electron-Rich Species	Key Electron-Poor Species	Comments	Discussed in Section
(13) R—C(=O)Cl Acid chloride	LiAlH(O-t-Bu)₃ −78 °C	R—C(=O)H Aldehyde	⊖H	R—C(=O)δ⁺Cl	Nucleophilic addition–elimination (reduction); note cold *T*	20.7
(14) R—C(=O)OR′ Ester	1. DIBAH −78 °C 2. HCl	R—C(=O)H Aldehyde	⊖H	R—C(=O)δ⁺OR′	Nucleophilic addition–elimination (reduction); note cold *T*	20.7

TABLE 20-3 Reactions That Alter the Carbon Skeleton[a]

Starting Compound Class	Typical Reagents and Reaction Conditions	Compound Class Formed	Key Electron-Rich Species	Key Electron-Poor Species	Comments	Discussed in Section
(1) R—C(=O)CH₃ Methyl ketone	1. X₂, NaOH 2. HCl	R—C(=O)OH Carboxylic acid	⊖OH	R—C(=O)δ⁺CX₃	Halogenation, then nucleophilic addition–elimination (haloform reaction)	20.5
(2) R—C(=O)Cl Acid chloride	1. R′—Li or R′—MgX 2. HCl	R—C(OH)(R′)(R′) Tertiary alcohol	⊖R′	R—C(=O)δ⁺Cl	Nucleophilic addition–elimination, then addition (alkyllithium or Grignard reaction)	20.8
(3) R—C(=O)—O—C(=O)—R Acid anhydride	1. R′—Li or R′—MgX 2. HCl	R—C(OH)(R′)(R′) Tertiary alcohol	⊖R′	R—C(=O)δ⁺—O—C(=O)—R	Nucleophilic addition–elimination, then addition (alkyllithium or Grignard reaction)	20.8
(4) R—C(=O)OR Ester	1. R′—Li or R′—MgX 2. HCl	R—C(OH)(R′)(R′) Tertiary alcohol	⊖R′	R—C(=O)δ⁺OR	Nucleophilic addition–elimination, then addition (alkyllithium or Grignard reaction)	20.8
(5) R—C(=O)Cl Acid chloride	R′₂CuLi	R—C(=O)R′ Ketone	⊖R′	R—C(=O)δ⁺Cl	Nucleophilic addition–elimination	20.8

[a]X = Cl, Br, I.

Problems

20.1 and 20.3 Transesterification and Saponification

20.28 Predict the product for the reaction between methyl cyclohexylmethanoate and each of the following. If no reaction is expected to occur, write "no reaction." For each reaction that does occur, draw the complete, detailed mechanism. **(a)** NaOH, then H_3O^+; **(b)** $CH_3CH_2CH_2ONa$, $CH_3CH_2CH_2OH$; **(c)** C_6H_5OK, C_6H_5OH

20.29 Which products in Problem 20.28 will produce methyl cyclohexylmethanoate when treated with $NaOCH_3$?

20.30 Draw an energy diagram for each reaction in Problem 20.28, paying attention to the relative energies of the overall reactants, overall products, and any intermediates.

20.31 Draw the mechanism for each of the following reactions and predict the major organic product in each.

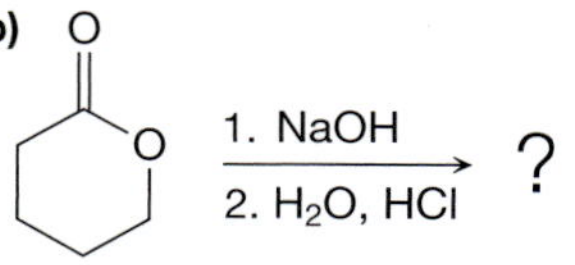

20.32 (SYN) Show how to carry out each of the following transformations.

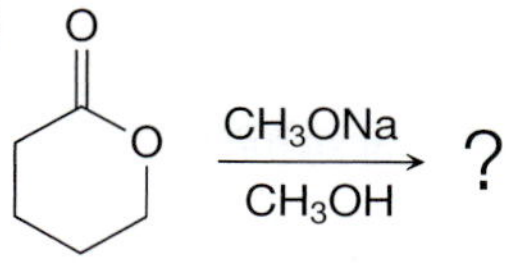

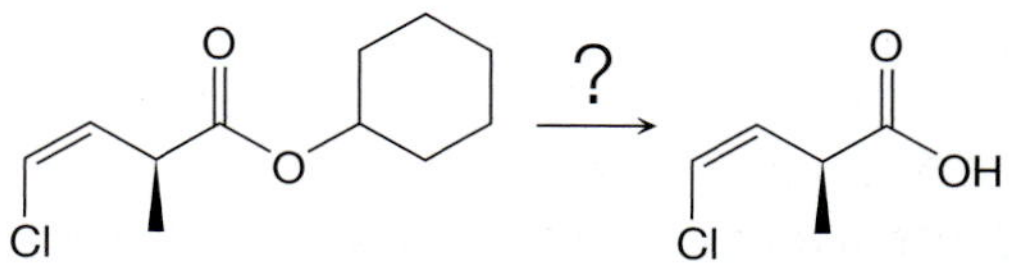

20.33 Which of the following esters, **A** or **B**, will undergo saponification faster? Why?

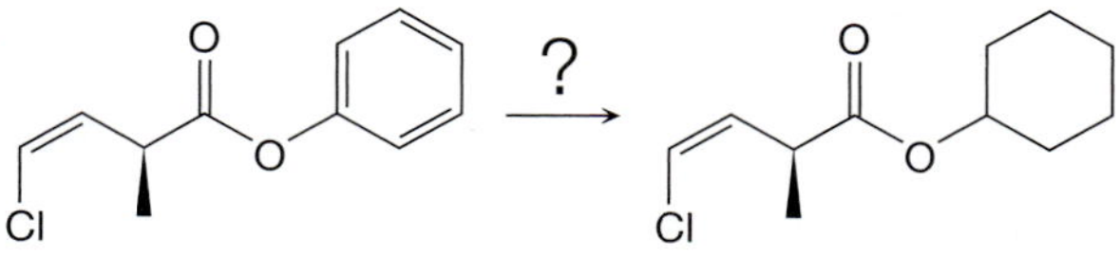

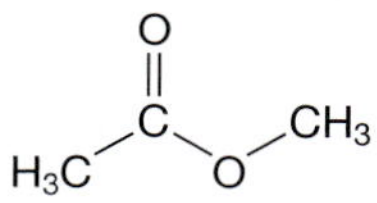

20.34 Draw a complete, detailed mechanism to account for the incorporation of ^{18}O twice into the acetate anion in the reaction shown here.

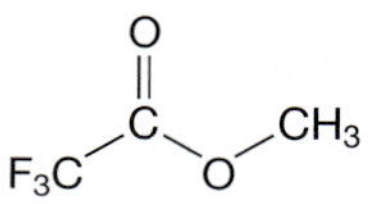

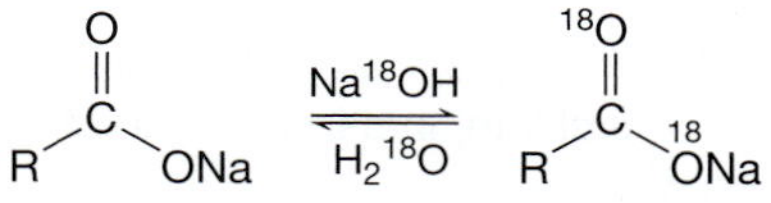

20.2, 20.4, and 20.5 Interconverting Carboxylic Acid Derivatives and the Stability Ladder; The Gabriel Synthesis of Primary Amines; Haloform Reactions

20.35 Predict the product for the reaction between *m*-ethylbenzoyl chloride and each of the following. Draw the complete, detailed mechanism for each reaction. If no reaction is expected to occur, write "no reaction." **(a)** NaOH, then H_3O^+; **(b)** CH_3NHLi; **(c)** CH_3CH_2OK; **(d)** $C_6H_5CO_2K$; **(e)** CH_3Cl; **(f)** CH_3OCH_3

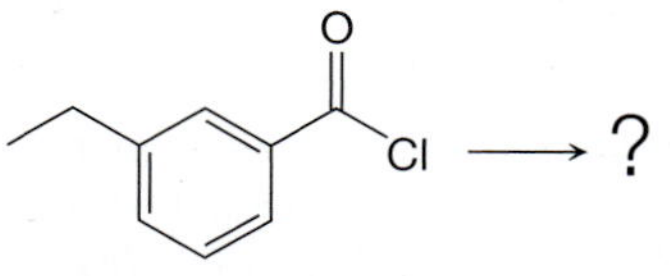

m-Ethylbenzoyl chloride

20.36 Predict the product for the reaction between acetic anhydride and each of the following. If no reaction is expected to occur, write "no reaction." For those reactions that do occur, draw the complete, detailed mechanism. **(a)** NaOH, then H_3O^+; **(b)** CH_3NHLi; **(c)** CH_3CH_2OK; **(d)** $C_6H_5CO_2K$; **(e)** NaBr; **(f)** $CH_3CH_2OCH_2CH_3$; **(g)** 3-chloropentane; **(h)** hexanal

20.37 Predict the product for the reaction between methyl benzoate and each of the following. If no reaction is expected to occur, write "no reaction." For those reactions that do occur, draw the complete, detailed mechanism. **(a)** NaBr; **(b)** $NaN(CH_3)_2$; **(c)** 3-chloropentane; **(d)** hexanal; **(e)** CH_3Cl

20.38 (SYN) How would you carry out the following transformation?

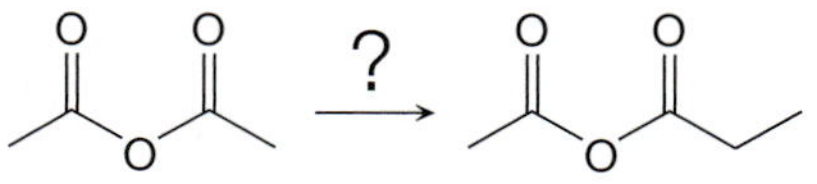

20.39 (SYN) Show how to carry out each of the following transformations.

(a)

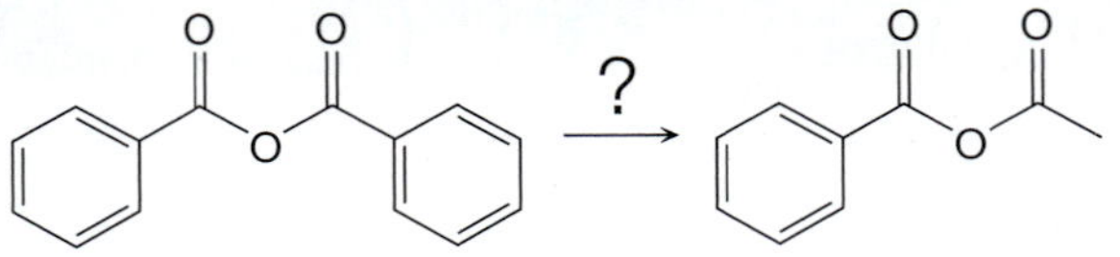

(b)

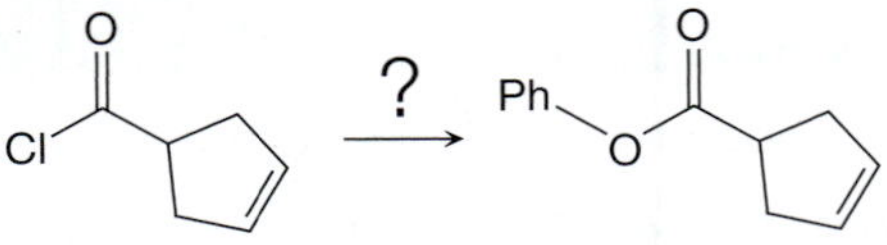

(c)

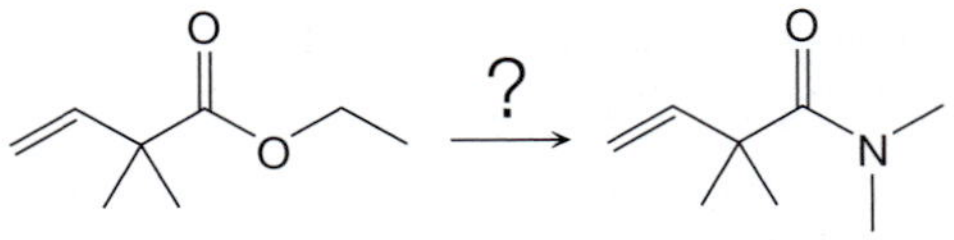

(d)

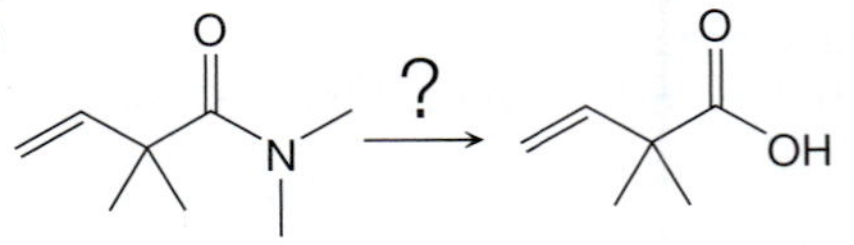

20.40 N,N-Diacylamides can be prepared by treating an acyl chloride with lithium nitride in a 3-to-1 ratio. Draw a complete, detailed mechanism for this reaction.

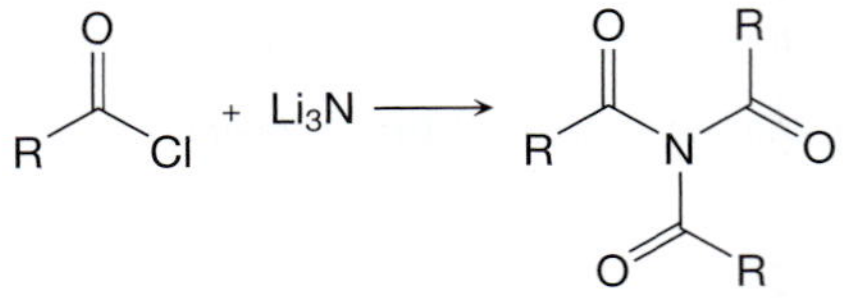

An N,N-diacylamide

20.41 Barbituric acid can be prepared from malonic ester and urea as follows. Provide a complete, detailed mechanism for this reaction.

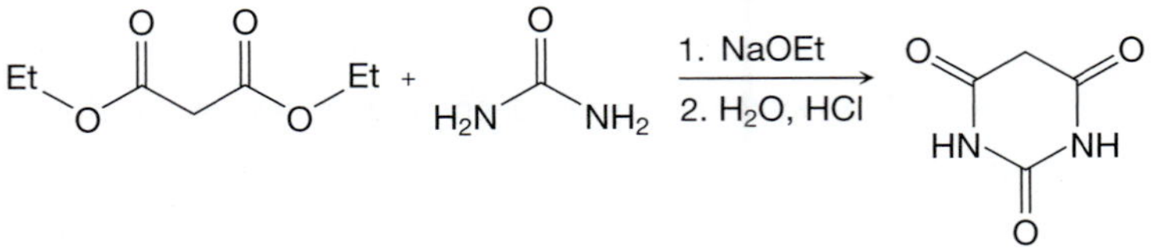

Barbituric acid

20.42 Which of the following compounds will form a yellow solid when dissolved in a basic, aqueous solution of I_2?
(a) butanoic acid; **(b)** pentan-2-one; **(c)** pentan-3-one; **(d)** cyclohexanone; **(e)** pentanal

20.43 (SYN) Show how to synthesize each of the following amines from an alkyl halide via a Gabriel synthesis.

(a)

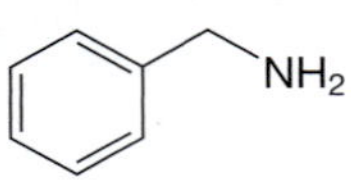

(b)

(c)

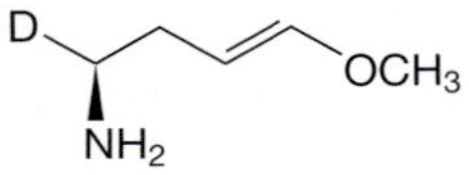

20.44 In a Gabriel synthesis, the yield of the amine can be compromised when hydrolysis is carried out under conditions that are too basic. The problem, in particular, is with the second hydrolysis (Steps 6–8 in Equation 20-13), not the first (Steps 3–5). Explain why.

20.6–20.8 Reactions with Hydride Reducing Agents and Organometallic Reagents

20.45 Predict the product for the reaction between m-ethylbenzoyl chloride (see Problem 20.35) and each of the following. Draw the complete, detailed mechanism for each reaction. If no reaction is expected to occur, write "no reaction."
(a) $(CH_3CH_2)_2CuLi$; **(b)** $LiAlH(O$-t-$Bu)_3$, $-78\ °C$; **(c)** $NaBH_4$, EtOH; **(d)** C_6H_5MgBr (excess), then H^+

20.46 Predict the product for the reaction between methyl benzoate and each of the following. If no reaction is expected to occur, write "no reaction." For those reactions that do occur, draw the complete, detailed mechanism. **(a)** $LiAlH_4$, then H_3O^+; **(b)** $CH_3CH_2CH_2Li$ (excess), then H_3O^+; **(c)** $(CH_3CH_2)_2CuLi$; **(d)** C_6H_5MgBr (excess), then H^+; **(e)** DIBAH, then H_3O^+

20.47 Predict the product for each of the following reactions. If no reaction is expected to occur, write "no reaction." For those reactions that do occur, draw the complete, detailed mechanism.

(a)

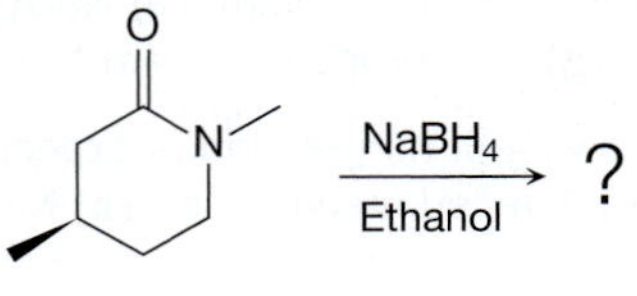

(b)

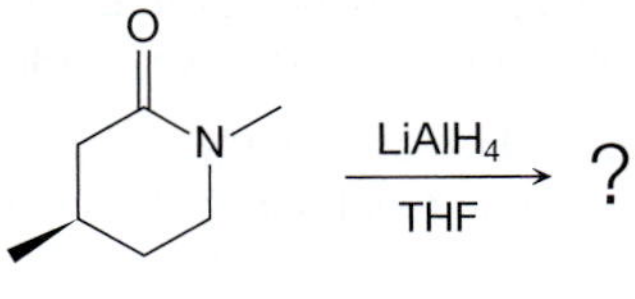

(c)

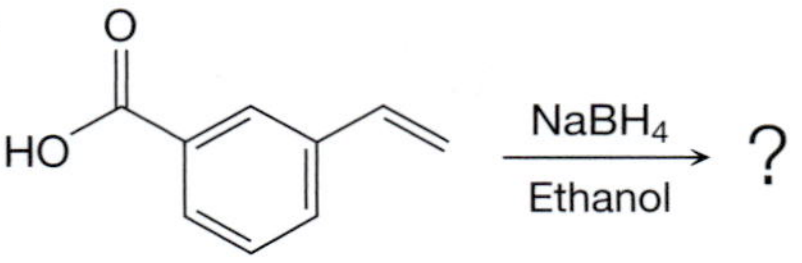

(d)

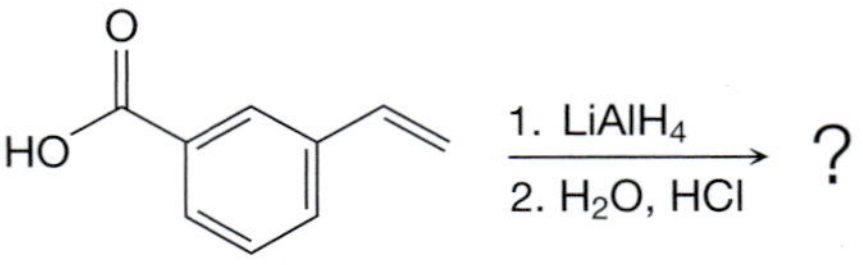

20.48 (SYN) Show how to carry out each of the following transformations.

(a)

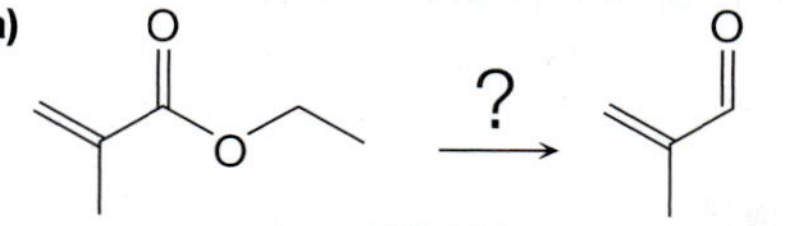

(b)

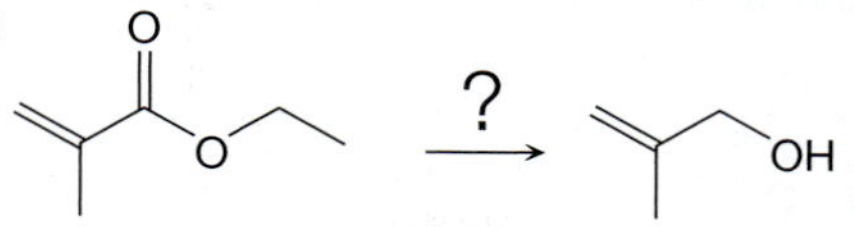

(c)

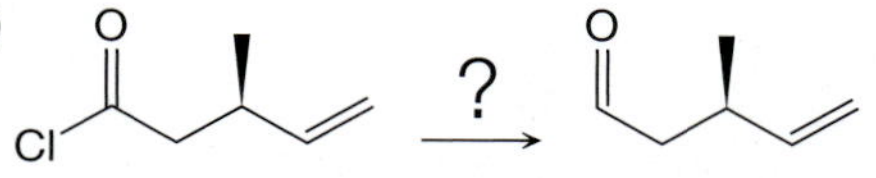

(d)

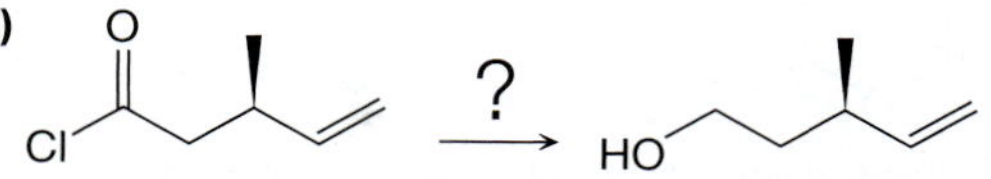

20.49 (SYN) Show how to carry out each of the following transformations.

(a)

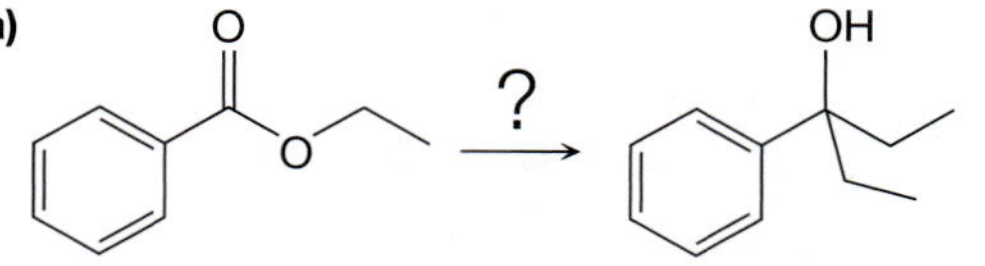

(b)

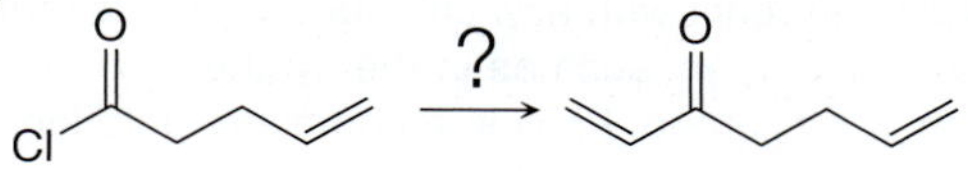

(c)

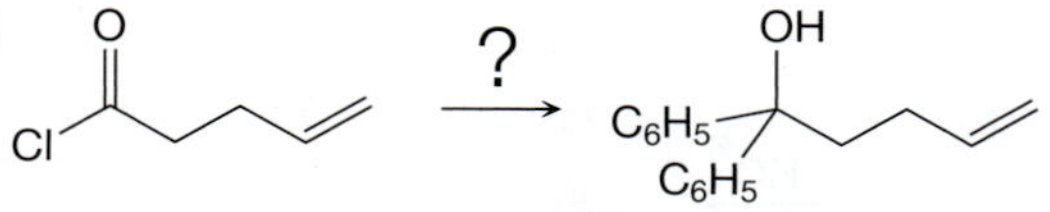

(d)

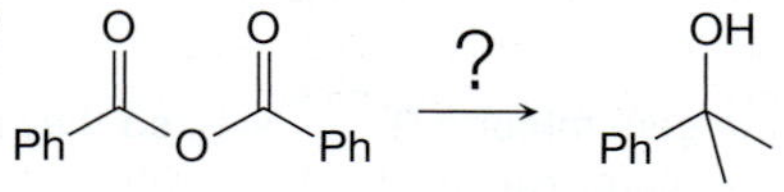

20.50 (SYN) How would you carry out the following transformation?

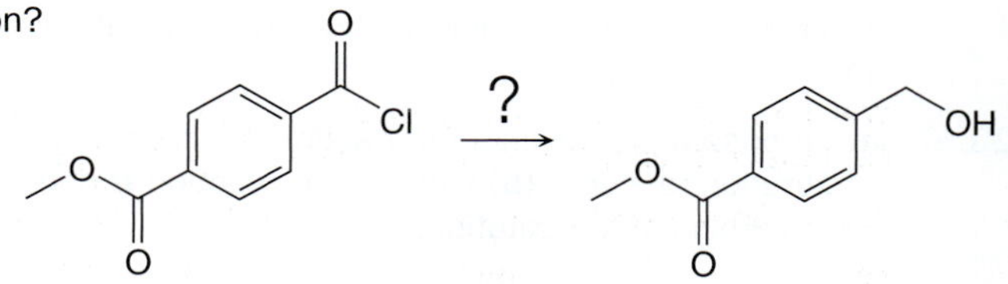

20.51 (SYN) How would you carry out each of the following transformations?

(a)

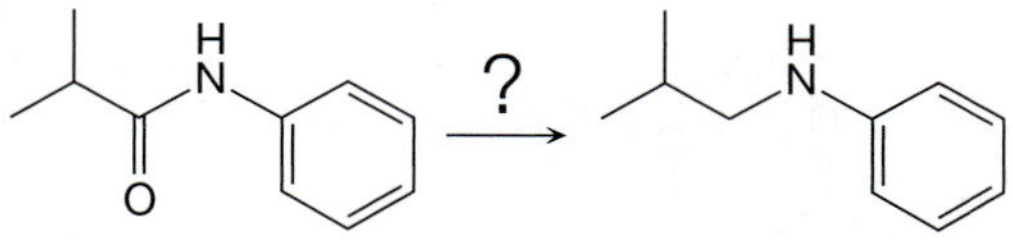

(b)

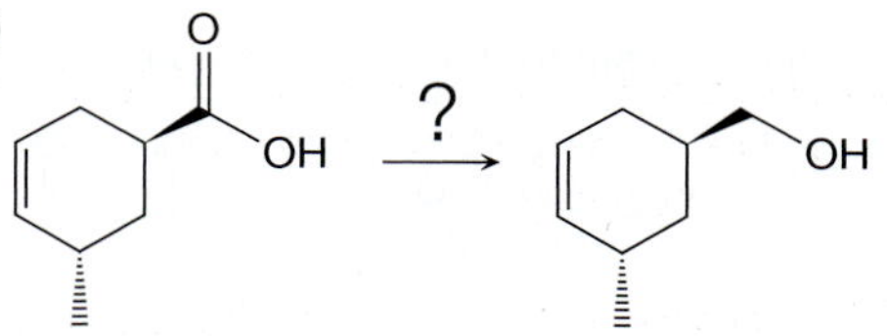

Integrated Problems

20.52 Predict the product of the following sequence of reactions.

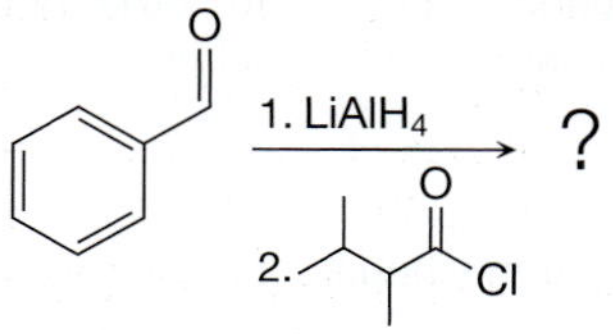

1. LiAlH$_4$

2.

20.53 Shown here is a proposed synthesis of a thioester from an ester. **(a)** Draw the mechanism for this reaction. **(b)** Would this reaction be energetically favorable? Why or why not?

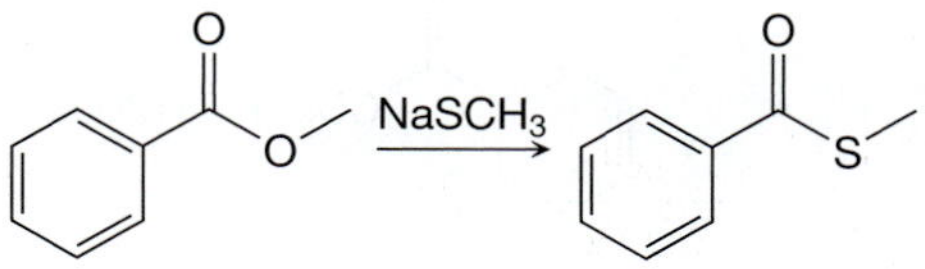

NaSCH$_3$

20.54 For thousands of years, civilizations have been synthesizing soap by heating animal fat with wood ash, a source of HO^-. Animal fat consists of triesters, known as triglycerides, an example of which is shown here on the left.

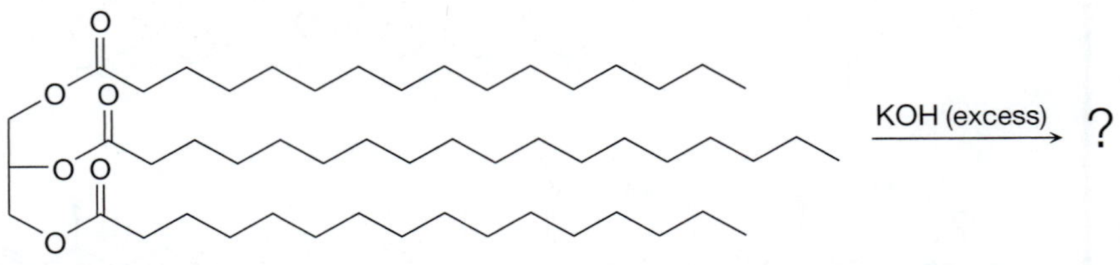

KOH (excess) → ?

(a) Assuming that the triglyceride reacts completely with HO^-, what are the products of this reaction?
(b) Draw the complete, detailed mechanism that leads to those products.
(c) Explain how those products can serve as soap. *Hint:* Review Section 2.10.

20.55 Enamines can react with acyl chlorides via nucleophilic addition–elimination, such as in the synthesis of the following 1,3-diketone. Provide a complete, detailed mechanism for this transformation.

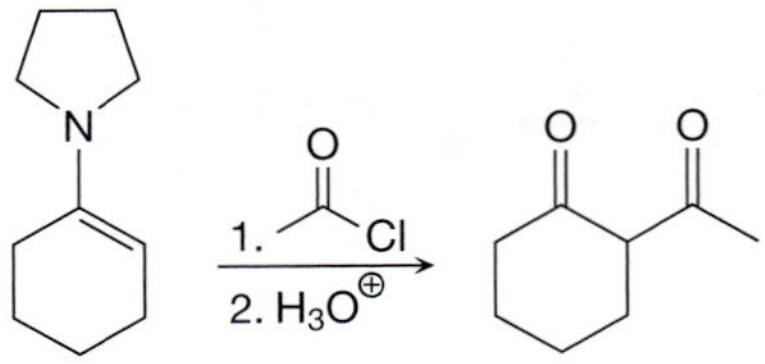

20.56 Amidines have the general form R $\overset{NH}{\underset{}{\|}}$ NH_2, so they are nitrogen analogs of carboxylic acids. As shown here, they can be hydrolyzed under basic conditions to form amides. Propose a mechanism showing the conversion of an amidine to an amide.

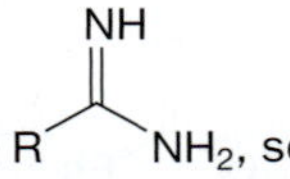

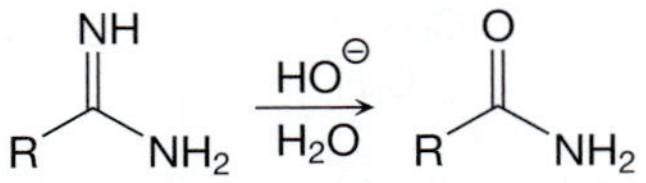

20.57 (a) Propose a mechanism for reaction **A**, which is a substitution reaction. **(b)** Explain why reaction **B** does not lead to a similar substitution.

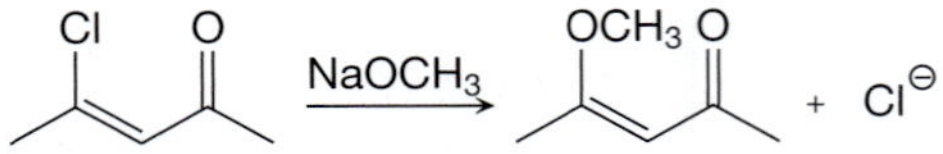

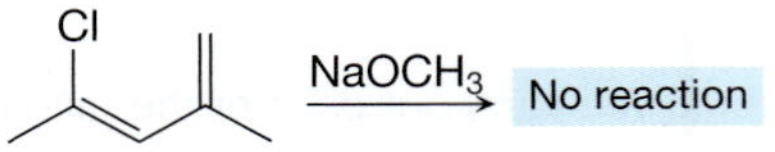

20.58 Propose a mechanism for the following reaction.

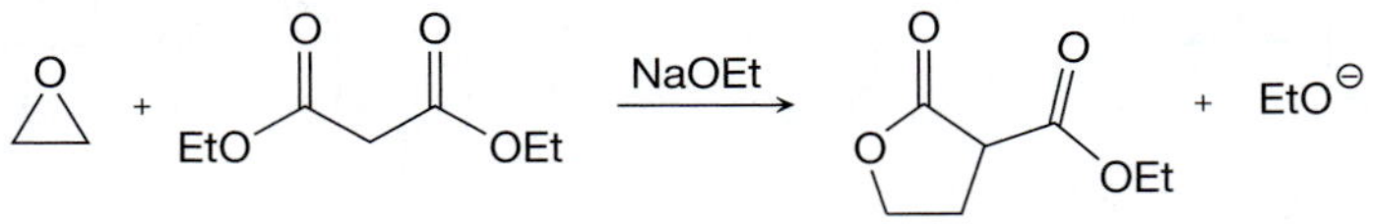

20.59 If a lactam (cyclic amide) contains an alkyl group with an amino group, treatment with lithium diisopropyl amide (LDA) results in a ring-expanded lactam, as shown here. Provide a detailed mechanism for this reaction.

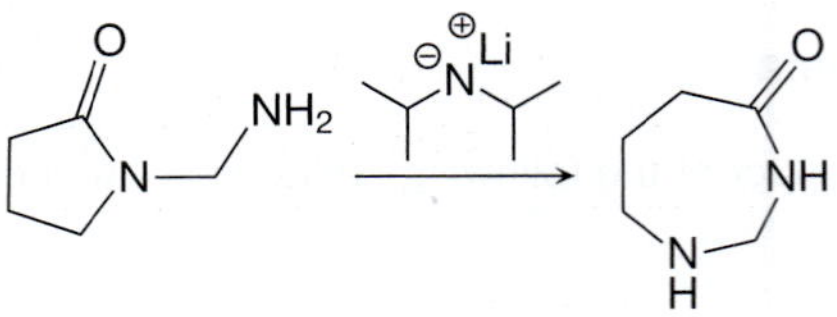

20.60 Draw the complete, detailed mechanism for the following reaction.

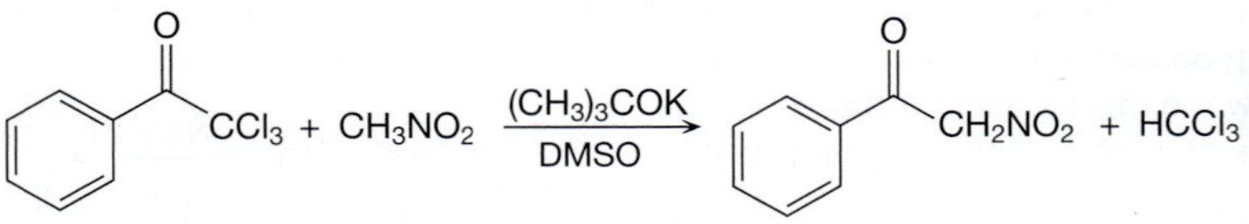

20.61 Supply the missing compounds **A** through **D**.

20.62 Supply the missing compounds **A** through **G**.

20.63 Supply the missing compounds **A** through **D**.

20.64 Supply the missing compounds **A** through **G**.

20.65 (SYN) Starting with acetyl chloride (CH_3COCl) and any other reagents necessary, how would you synthesize each of the following compounds?

20.66 (SYN) Using acetyl chloride as your only source of carbon, propose a synthesis for each of the following compounds. You may use any inorganic reagents necessary. **(a)** ethyl acetate; **(b)** butan-2-ol; **(c)** 3-methylpentan-3-ol; **(d)** butan-2-one; **(e)** ethanamine; **(f)** acetic anhydride

20.67 (SYN) Show how to synthesize the following molecule from any compounds containing two carbons. Draw the complete, detailed mechanism for the reaction.

20.68 (SYN) Show how to synthesize 2,4-diphenylbut-2-ene using phenylethanoic acid (phenylacetic acid) as your only source of carbon atoms.

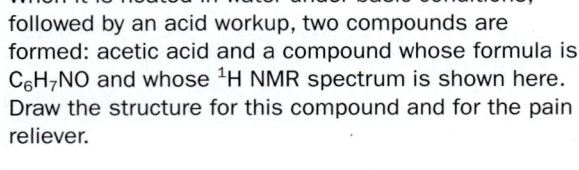

Phenylethanoic acid
(Phenylacetic acid)

2,4-Diphenylbut-2-ene

20.69 A pain reliever has the formula $C_8H_9NO_2$. Its IR spectrum is as follows:

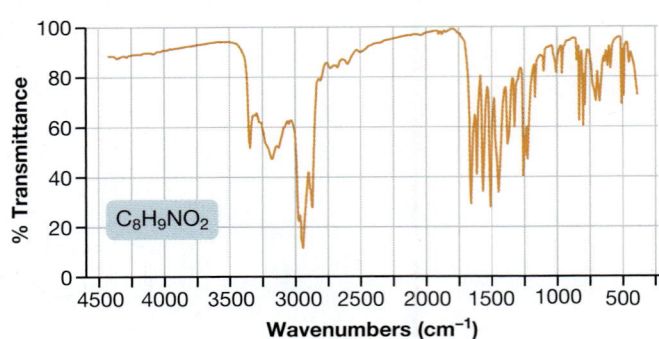

When it is heated in water under basic conditions, followed by an acid workup, two compounds are formed: acetic acid and a compound whose formula is C_6H_7NO and whose ^1H NMR spectrum is shown here. Draw the structure for this compound and for the pain reliever.

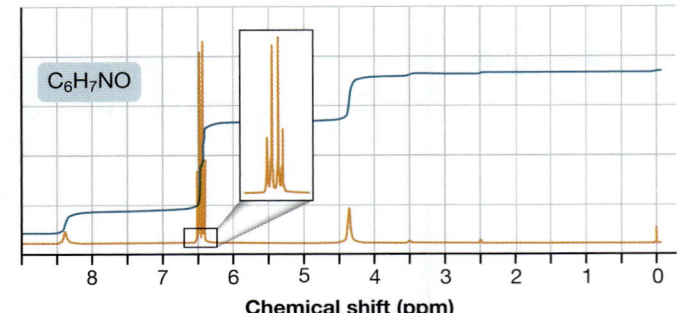

20.70 When methyl 5-oxopentanoate is treated with vinyl Grignard, a compound is produced whose formula is $C_7H_{10}O_2$. In the IR spectrum of $C_7H_{10}O_2$, an intense absorption appears at 1740 cm^{-1} and a weaker absorption appears at 1650 cm^{-1}. Seven signals appear in the ^{13}C NMR spectrum of $C_7H_{10}O_2$. The DEPT spectrum shows that there is one carbon that is bonded to no hydrogens, two CH carbons, and four CH$_2$ carbons. Draw the structure of $C_7H_{10}O_2$.

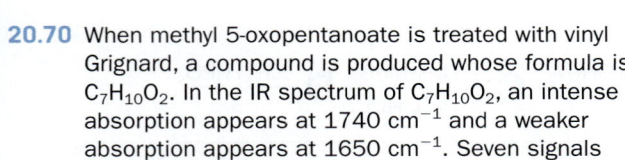

Methyl 5-oxopentanoate

1. ⌇MgBr, THF
2. NH₄Cl, H₂O

$C_7H_{10}O_2$

20.71 When dimethyl phthalate is treated with lithium aluminum hydride, a compound is produced whose ^1H NMR spectrum is shown below. Determine the product, and draw the complete, detailed mechanism of the reaction that produces it.

Dimethyl phthalate

LiAlH₄
Et₂O, THF,
−30 °C

?

?

Chemical shift (ppm)

Kevlar is a strong lightweight flexible material commonly used to make body armor. Kevlar is produced from an aminolysis reaction, a type of nucleophilic addition–elimination reaction we discuss here in Chapter 21.

Nucleophilic Addition–Elimination Reactions 2

Weak Nucleophiles

Chapter 20 discussed reactions that proceed by the nucleophilic addition–elimination mechanism:

Nucleophilic addition–elimination mechanism

A nucleophile attacks the electron-poor atom.

The leaving group is expelled, which regenerates the π bond.

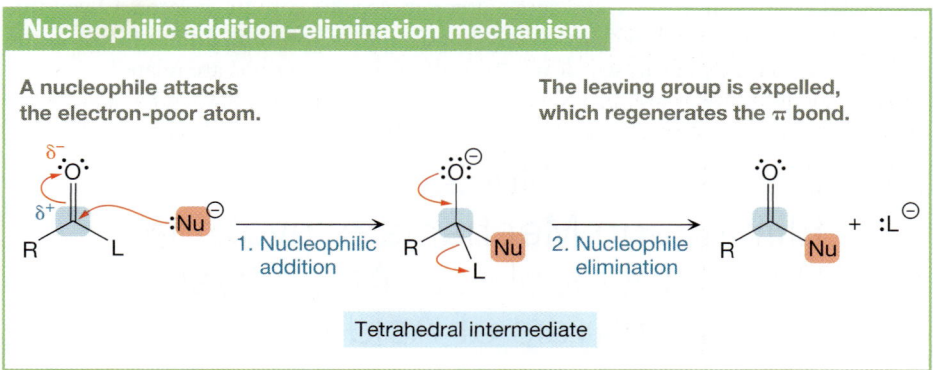

1. Nucleophilic addition

2. Nucleophile elimination

Tetrahedral intermediate

In Step 1, a nucleophile attacks the electrophilic atom of a polar π bond to generate an unstable tetrahedral intermediate. In Step 2, a leaving group is expelled from the tetrahedral intermediate, which regenerates the π bond.

Chapter Objectives

On completing Chapter 21 you should be able to:

1. Draw the general mechanism for the nucleophilic addition–elimination reaction of a weak nucleophile with a highly reactive polar π bond.

2. Rank the relative reactivities of carboxylic acid derivatives in nucleophilic addition–elimination reactions and explain what gives rise to that order.

3. Explain why the aminolysis of an acid chloride requires two molar equivalents of an amine unless one or more molar equivalents of a base such as pyridine or triethylamine are present.

4. Show how to convert a carboxylic acid to an acid halide and explain the significance of this reaction in the context of producing other carboxylic acid derivatives and halogenating the α carbon of a carboxylic acid.

5. Draw the mechanism for the production of a sulfonate ester from the corresponding sulfonyl chloride and an alcohol, and explain the significance of this reaction in the context of nucleophilic substitution and elimination reactions.

6. Determine whether a nucleophilic addition–elimination reaction involving an amide, ester, or carboxylic acid can be acid or base catalyzed, and draw the mechanisms for these reactions.

7. Predict the products of a Baeyer–Villiger oxidation and explain the driving force for that reaction in the context of an acid-catalyzed nucleophilic addition–elimination reaction.

8. Draw the mechanism for and predict the products of a Claisen condensation reaction.

9. Incorporate Claisen and crossed Claisen reactions effectively in a synthesis.

10. Determine whether an intramolecular Claisen condensation reaction will take place.

11. Design a synthesis of an α-substituted acetone and an α-substituted acetic acid using an acetoacetic ester synthesis or a malonic ester synthesis, respectively.

12. Design a synthesis that requires the protection of a carboxylic acid or amine.

The specific reactions we examined in Chapter 20 were limited to nucleophilic acyl substitutions in which *strong nucleophiles* are added as reagents. In a transesterification reaction, for example, an alkoxide anion (RO^-) serves as the nucleophile, and in a Grignard reaction involving an ester, an alkyl or aryl anion (R^-) serves as the nucleophile.

Here in Chapter 21, we expand our discussion of nucleophilic addition–elimination reactions to include ones that involve *weak nucleophiles* added as reagents. Fundamentally, nucleophilic addition–elimination reactions involving weak nucleophiles are quite similar to ones involving strong nucleophiles, but there are some key differences. Nucleophilic addition–elimination reactions involving weak nucleophiles added as reagents tend to proceed much slower than ones involving strong nucleophiles and, as a result, often require acid or base catalysis. Therefore, mechanisms that describe such reactions tend to consist of more steps.

Most of our discussion concerns reactions in which a nucleophile attacks the carbonyl carbon of a carboxylic acid derivative or a carboxylic acid, leading to acyl substitution. However, we also examine reactions that involve nucleophilic attack at other polar bonds that have π character, such as $S{=}O$. Some reactions we examine are functional group conversions, and some alter a molecule's carbon skeleton. Given this variety of reactions, it will be particularly important for you to maintain focus on the mechanism, so that you can see clearly how all of these reactions are related.

21.1 The General Nucleophilic Addition–Elimination Mechanism Involving Weak Nucleophiles: Alcoholysis and Hydrolysis of Acid Chlorides

In Section 20.2, we learned that an ester such as methyl benzoate can be produced from benzoyl chloride using methoxide anion (CH_3O^-), a relatively *strong* nucleophile (Equation 21-1). The mechanism for this reaction is simply the two-step nucleophilic addition–elimination mechanism just presented in this chapter's introduction.

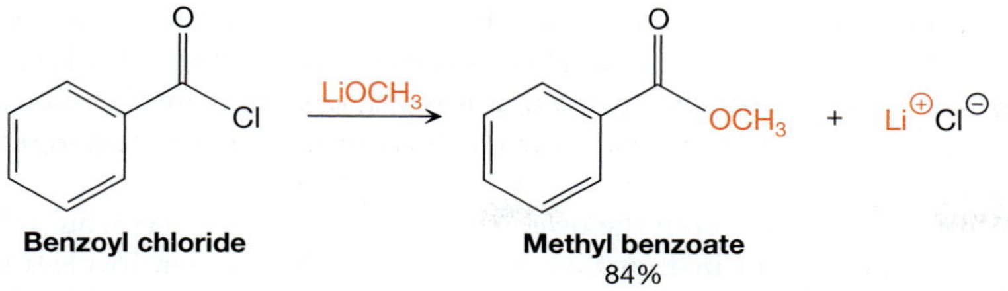

Benzoyl chloride → **Methyl benzoate** 84% + $Li^{\oplus}Cl^{\ominus}$ (21-1)

An ester can also be produced from an acid chloride by treating the acid chloride with an alcohol, a relatively *weak* nucleophile. An example is shown in Equation 21-2. Similarly, treating an acid chloride with water (another relatively weak nucleophile) produces a carboxylic acid, as shown in Equation 21-3.

An alcoholysis reaction

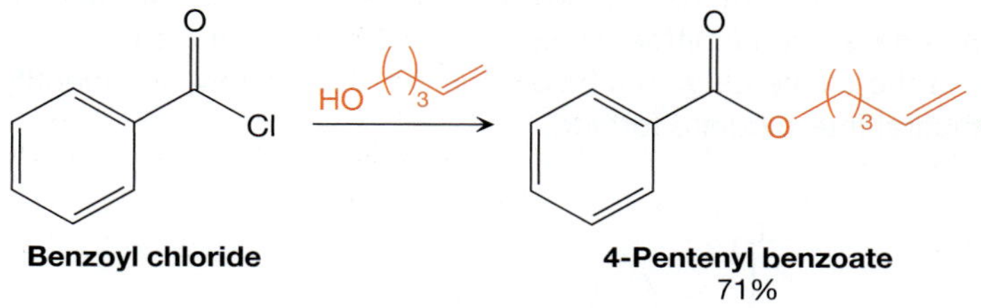

Benzoyl chloride → **4-Pentenyl benzoate** 71% + HCl (21-2)

A hydrolysis reaction

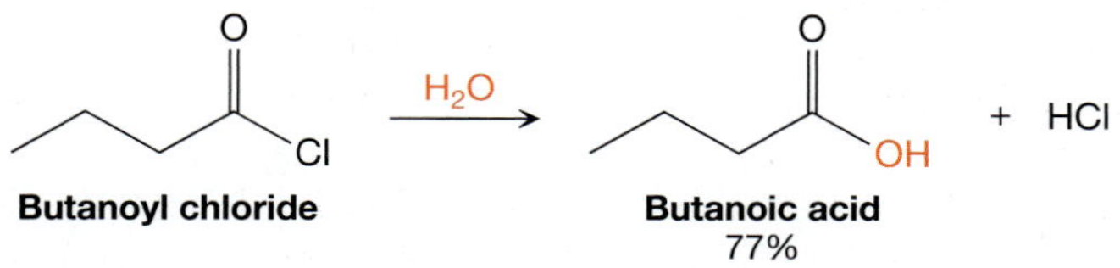

Butanoyl chloride → **Butanoic acid** 77% + HCl (21-3)

Equation 21-2 is called **alcoholysis** because the addition of the alcohol results in the breaking of a bond (*lysis* in Greek means "breaking")—in this case, the C—Cl bond of the acid chloride. Equation 21-3 is called **hydrolysis** because the bond in the acid chloride is broken as a result of the addition of water.

The mechanism for the hydrolysis of an acid chloride is shown in Equation 21-4. The mechanism for alcoholysis is essentially the same, and is presented in Your Turn 21-1.

Mechanism for Equation 21-3

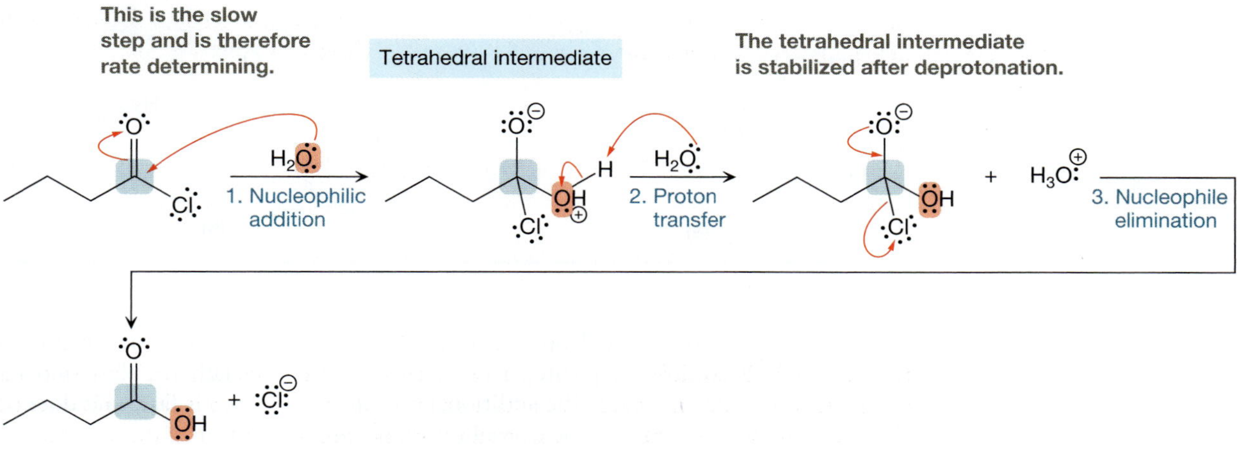

This is the slow step and is therefore rate determining.

Tetrahedral intermediate

The tetrahedral intermediate is stabilized after deprotonation.

1. Nucleophilic addition
2. Proton transfer
3. Nucleophile elimination

(21-4)

In Step 1 of Equation 21-4, H_2O attacks the carbonyl carbon to produce a tetrahedral intermediate. Unlike the general mechanism from Chapter 20, this tetrahedral intermediate contains an acidic proton that is rapidly deprotonated by another water molecule in Step 2. Finally, in Step 3, the Cl^- leaving group is expelled to reform the carbonyl bond.

HCl appears as a product in the overall reaction in Equation 21-3, but it does not appear in the mechanism in Equation 21-4 because the reaction involves water, in which HCl, a strong acid, exists almost entirely as hydronium ion (H_3O^+) and chloride ion (Cl^-). This is consistent with the production of H_3O^+ in Step 2 of the mechanism and Cl^- in Step 3.

YOUR TURN 21.1

The mechanism for the alcoholysis reaction in Equation 21-2 is shown here, but the curved arrows have been omitted. Draw in the appropriate curved arrows, write the name of the elementary step below each reaction arrow, and identify the initial tetrahedral intermediate formed.

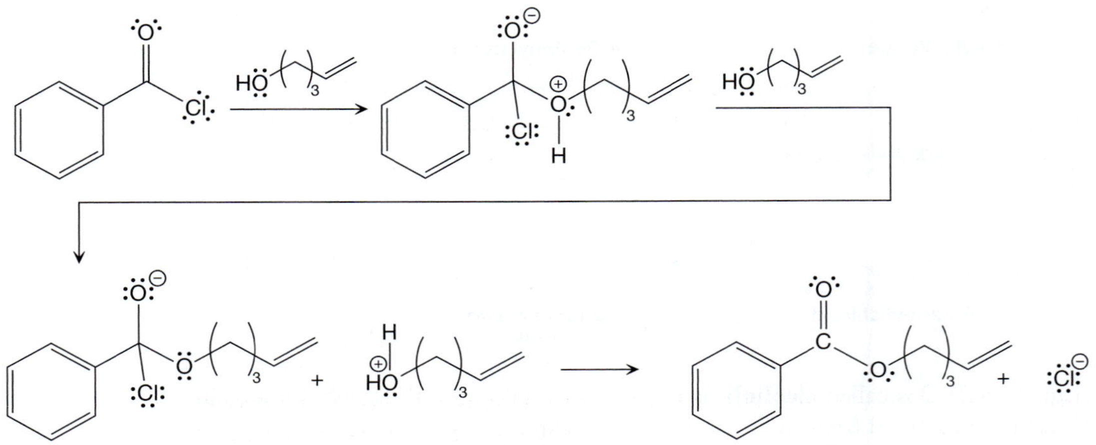

Answers to Your Turns are in the back of the book.

PROBLEM 21.1 Draw the complete, detailed mechanism and predict the major organic product for each of the following reactions.

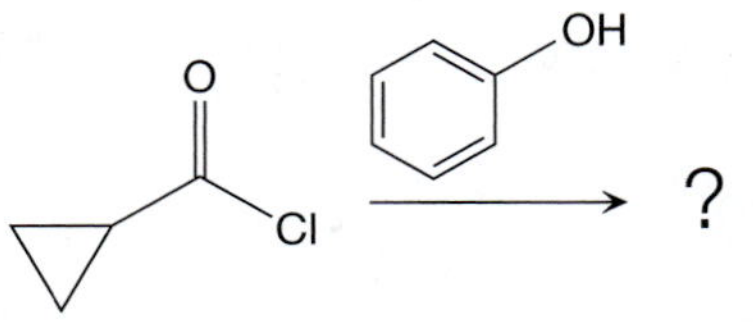

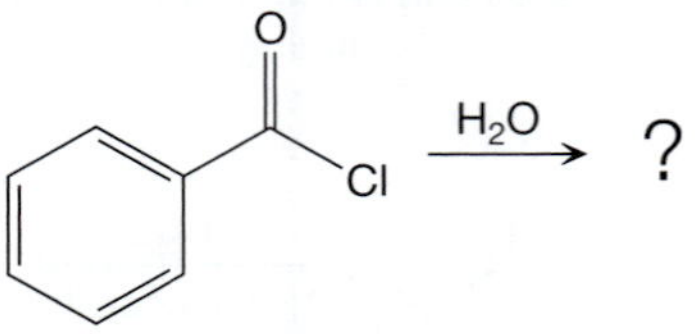

(a)   (b)

Just as we saw for the nucleophilic addition–elimination reactions in Chapter 20, the nucleophilic addition step (Step 1 in Equation 21-4) is usually the slow step and, therefore, is rate determining. The additional proton transfer step is fast, so it does not significantly affect the rate of the overall reaction. (See Your Turn 21.2.)

The free energy diagram for the hydrolysis reaction in Equations 21-3 and 21-4 is shown here.

(a) Draw the species that appear in the mechanism (Equation 21-4) in the appropriate boxes provided, and label the initial tetrahedral intermediate. *Hint:* Review Figure 20-3 (p. 1004).

(b) Explain how this diagram is consistent with the nucleophilic addition step being the slow step.

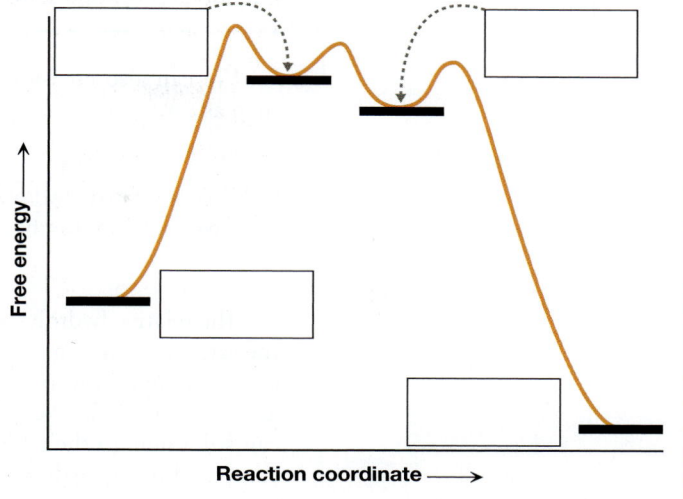

PROBLEM 21.2 Draw the free energy diagram for the alcoholysis reaction in Equation 21-2. Include the species that appear in its mechanism (see Your Turn 21.1) and label the initial tetrahedral intermediate.

21.2 Relative Reactivities of Acid Derivatives: Rates of Hydrolysis

Section 21.1 highlighted alcoholysis and hydrolysis reactions involving acid chlorides only, but these reactions are not limited to just acid chlorides. As shown in Equations 21-5 and 21-6, acid anhydrides can readily undergo alcoholysis and hydrolysis, producing esters and carboxylic acids, respectively. The mechanisms for these reactions are essentially the same as the one in Equation 21-4 that describes the hydrolysis of an acid chloride.

Alcoholysis of an acid anhydride

80%   (21-5)

Hydrolysis of an acid anhydride

(21-6)

Phthalic anhydride   **Phthalic acid**

> **CONNECTIONS** The monopotassium salt of phthalic acid is potassium hydrogen phthalate (KHP), a weak acid that is often used in analytical chemistry to standardize NaOH solutions.

We can also envision similar reactions taking place when other acid derivatives, such as esters or amides, are treated with water. As indicated in Table 21-1, however:

Ester and amide hydrolysis reactions are much too slow under neutral conditions to be useful for synthesis.

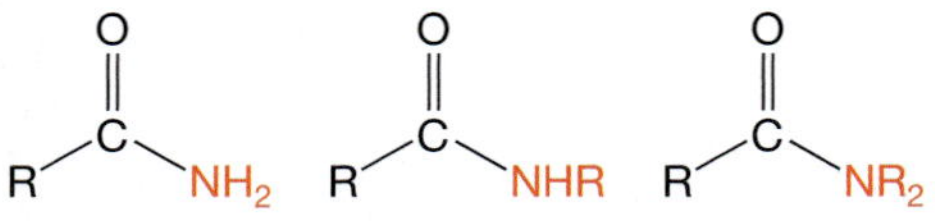

The relative hydrolysis rates of the species in Table 21-1 are essentially the rates of the rate-determining step — typically the nucleophilic addition of H_2O to the C=O group. Therefore, even though ketones and aldehydes don't undergo hydrolysis like the acid derivatives do (ketones and aldehydes don't have a suitable leaving group), they can still undergo the addition of H_2O to form a hydrate and that rate of hydration is compared to the hydrolysis rates of the acid derivatives:

Ketones and aldehydes are less susceptible to nucleophilic addition than acid chlorides or acid anhydrides, but are more susceptible than carboxylic acids, esters, amides, and carboxylates.

The hydrolysis rates in Table 21-1 tell us that the energy barrier for nucleophilic addition increases in the order: acid chlorides < acid anhydrides < carboxylic acids and esters < amides < carboxylate anions. This is explained by the increasing stability

TABLE 21-1 Relative Reactivities of Various Carbonyl-Containing Species

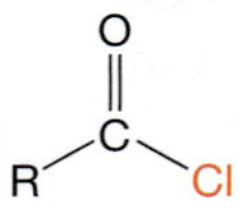

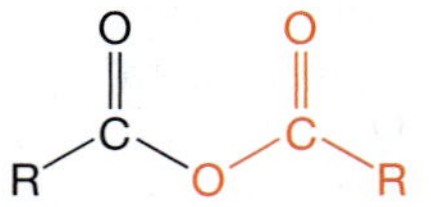

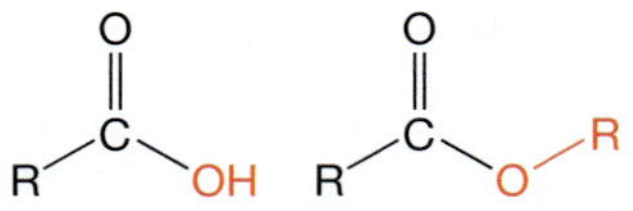

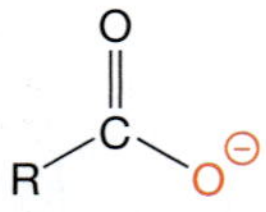

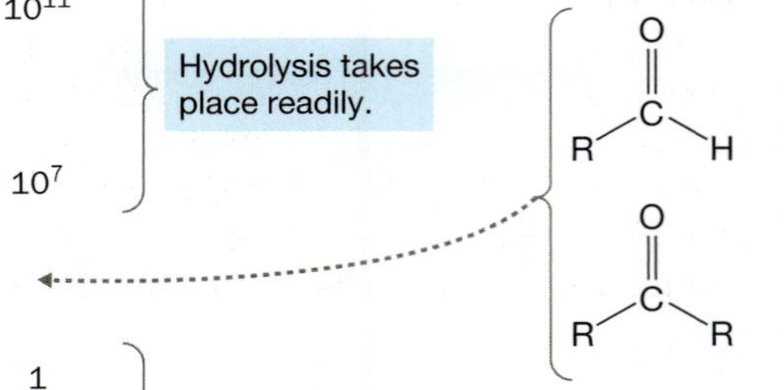

^aThe leaving groups are indicated in red.

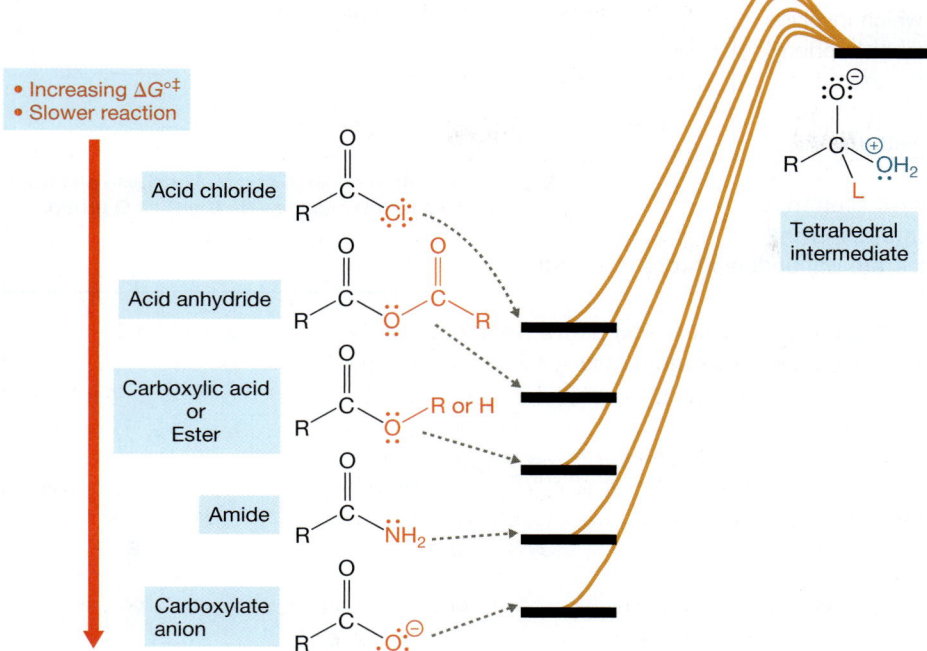

- Increasing $\Delta G^{\circ\ddagger}$
- Slower reaction

Acid chloride

Acid anhydride

Carboxylic acid or Ester

Amide

Carboxylate anion

Tetrahedral intermediate

of the acid derivatives, as shown in Figure 21-1, which is the same order we saw in the stability ladder in Figure 20.5 (p. 1008).

These relative stabilities of the acid derivatives are an outcome of the extent to which the π electrons in the C=O group participate in resonance with a lone pair of electrons on the leaving group. As shown in Figure 21-2a, those π electrons are delocalized the most in a carboxylate anion because it has two equivalent resonance structures. For the uncharged carboxylic acid derivatives, the second resonance structure (exhibiting a C=L bond) contributes less because it has both a positive and a negative formal charge, as shown in Figure 21-2b. Moreover, the contribution by the second resonance structure decreases in the order: amides > carboxylic acids ≈ esters > acid anhydrides > acid chlorides.

The second resonance structure contributes less for an ester than an amide because O, being more electronegative than N, does not accommodate a positive charge as well as N. In an acid anhydride, that contribution is even less because the central O atom's lone pair is tied up in resonance with *the other* carbonyl group simultaneously. Finally, the contribution is the least in an acid chloride because the lone pair on Cl belongs to a different shell ($n = 3$) than the valence electrons of the C and O atoms making up the C=O group ($n = 2$). This leads to a relatively poor overlap among the valence orbitals, and thus diminishes any resulting stabilization.

SOLVED PROBLEM 21.4

Is the hydrolysis of phenyl acetate faster or slower than the hydrolysis of ethyl acetate? Explain.

Phenyl acetate   **Ethyl acetate**

Think How does resonance play a role in the reactivity of carboxylic acid derivatives? Are the carbonyl groups stabilized differently by resonance? How so?

Carboxylate anions have equivalent resonance structures, so they are the most stable.

(a)

Contribution by this resonance structure decreases in the order:

$$L = NH_2 > OH \approx OR > OCR > Cl$$

(b)

FIGURE 21-2 **Resonance stabilization in the carboxylic acid derivatives** (a) Carboxylate anions (RCO_2^-) are stabilized the most because they have equivalent resonance structures. (b) For uncharged acid derivatives, the contribution by the resonance structure—and therefore the stability of the carboxylic acid derivative—decreases in the order: amides > carboxylic acids ≈ esters > acid anhydrides > acid chlorides.

Solve Both molecules have a resonance structure analogous to the one in Figure 21-2b, which involves the lone pair on the O of the leaving group. In phenyl acetate, that O atom's lone pair is partly tied up in resonance with the phenyl ring, too, making those electrons less available for resonance with the carbonyl group.

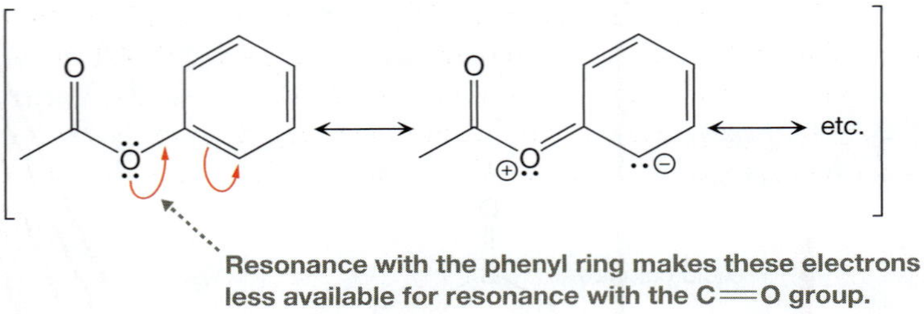

Therefore, the carbonyl group in phenyl acetate is not as stabilized as the one in ethyl acetate, making phenyl acetate more reactive; phenyl acetate will undergo hydrolysis faster.

Resonance with the phenyl ring makes these electrons less available for resonance with the C=O group.

PROBLEM 21.5 Which acid derivative in each of these pairs do you think undergoes hydrolysis more quickly? Explain.

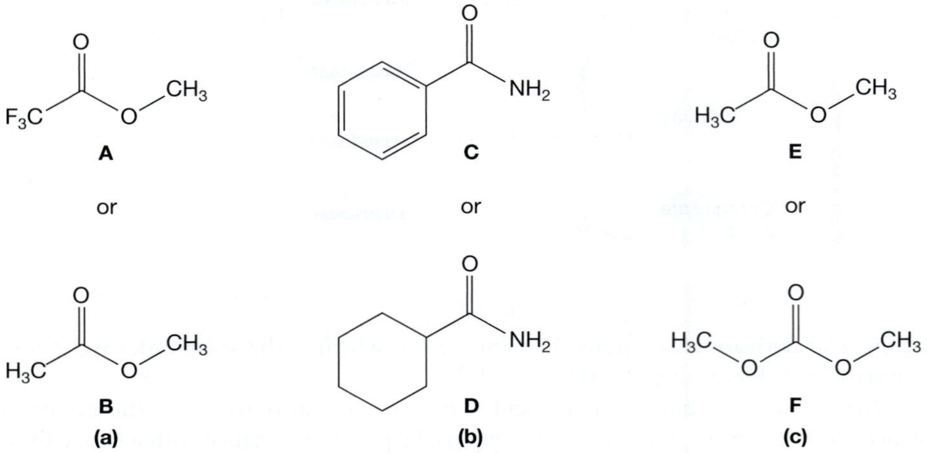

A

or

B

(a)

C

or

D

(b)

E

or

F

(c)

21.3 Aminolysis of Acid Derivatives

Amines are even more nucleophilic than water and alcohols, so they, too, will react with acid chlorides and acid anhydrides. As shown in Equations 21-7 and 21-8, these **aminolysis** reactions produce amides.

Aminolysis

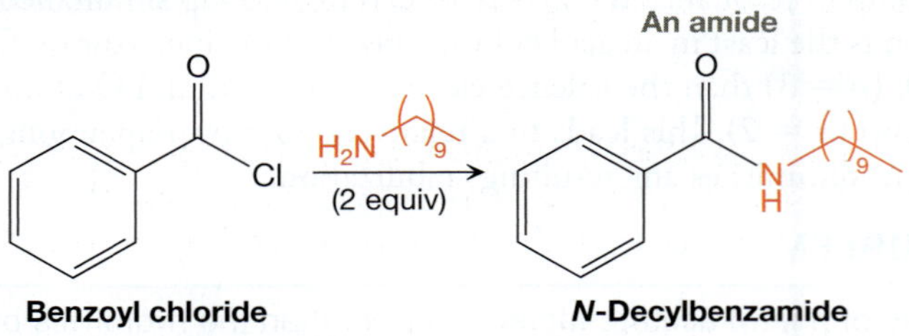

An amide

H_2N (2 equiv)

(21-7)

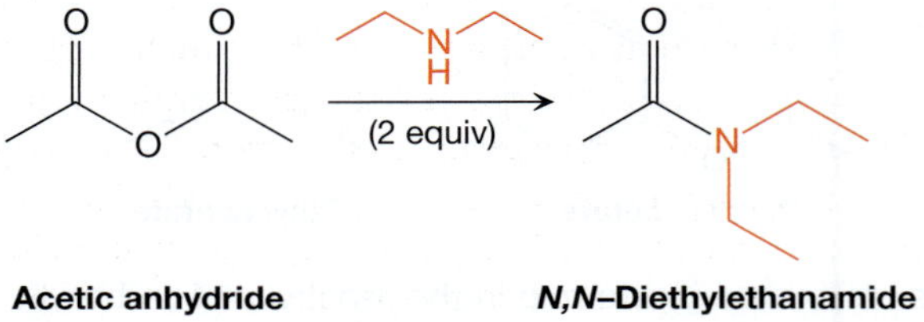

Benzoyl chloride

N-Decylbenzamide

(2 equiv)

(21-8)

Acetic anhydride

N,N–Diethylethanamide

The mechanisms for Equations 21-7 and 21-8 are similar to the one for the hydrolysis of an acid chloride, shown previously in Equation 21-4. Equation 21-9 shows the mechanism for the aminolysis of an acid chloride. Step 1 is nucleophilic addition of the amine to produce the tetrahedral intermediate, followed by deprotonation of the positively charged N. Finally, Cl^- is eliminated in Step 3.

Mechanism for Equation 21-7

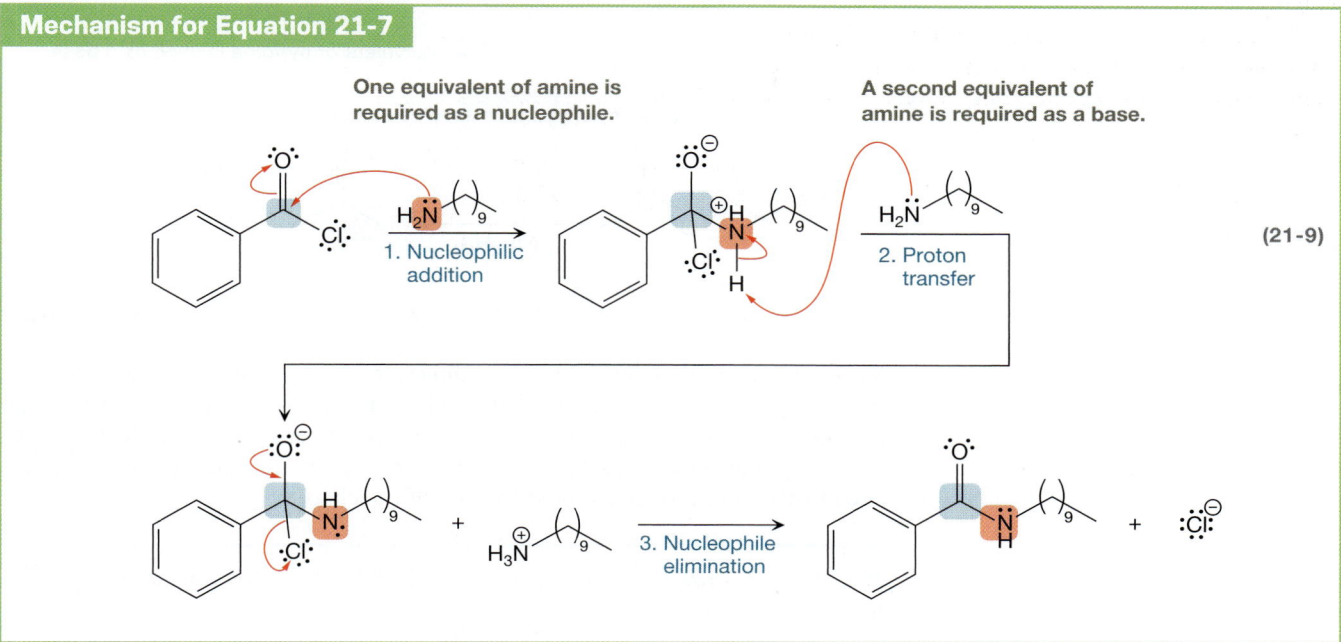

(21-9)

PROBLEM 21.6 Draw the complete, detailed mechanism for the reaction in Equation 21-8.

An important consideration for aminolysis reactions is the number of equivalents of the nucleophile required.

Aminolysis reactions require two equivalents of an amine.

According to the mechanism, the first equivalent is used in Step 1, in which the amine, acting as a nucleophile, attacks the carbonyl carbon of the acid chloride. The second equivalent is used in Step 2, where it deprotonates the positively charged N from the original nucleophile.

If the amine is readily available and inexpensive, it may be acceptable to carry out an aminolysis reaction that consumes two equivalents of the amine. If the amine is difficult to obtain or expensive, however, then it is advantageous to use only one equivalent. In these cases, another amine such as pyridine or triethylamine can be added to the reaction mixture to act as a base, freeing up the desired amine to act as the nucleophile. For example, the aminolysis reaction in Equation 21-10 requires only one equivalent of the amine when one equivalent of pyridine is also present. Similarly, the aminolysis reaction in Equation 21-11 requires only one equivalent of the desired amine in the presence of one equivalent of triethylamine.

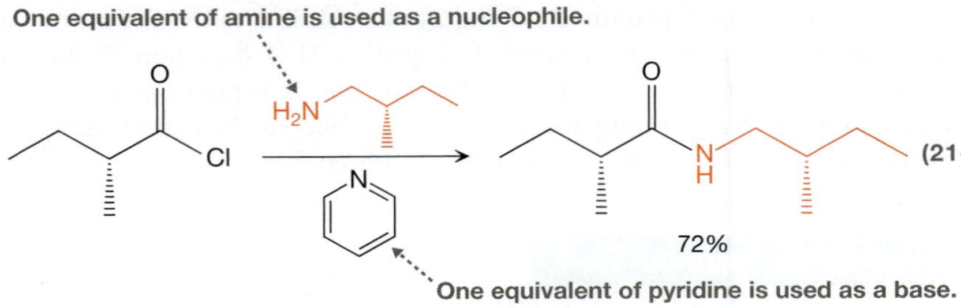
One equivalent of amine is used as a nucleophile.
O
Cl
H2N
N
O
N
H
72%
One equivalent of pyridine is used as a base.

(21-10)

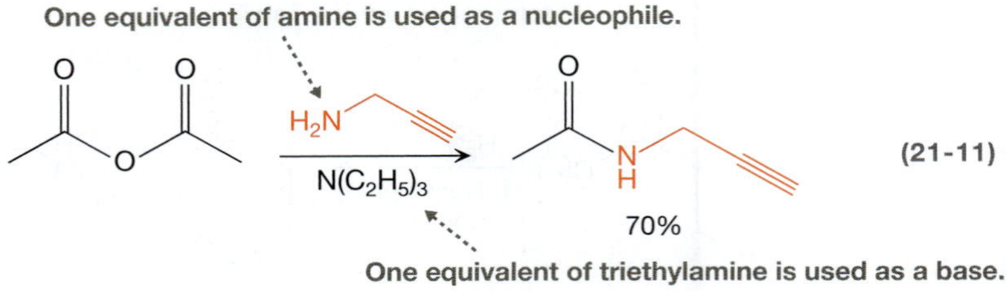
One equivalent of amine is used as a nucleophile.
O
O
O
H2N
N(C2H5)3
O
N
H
70%
One equivalent of triethylamine is used as a base.

(21-11)

Pyridine and triethylamine are good choices as bases in these reactions because neither compound's N atom is bonded to H. Thus, even though they can undergo nucleophilic addition–elimination with an acid chloride or acid anhydride, no subsequent proton transfer can take place to produce a stable uncharged amide. (This is explored further in Problem 21.44 at the end of the chapter.)

PROBLEM 21.7 Draw the mechanism for each of the following reactions and predict the major product.

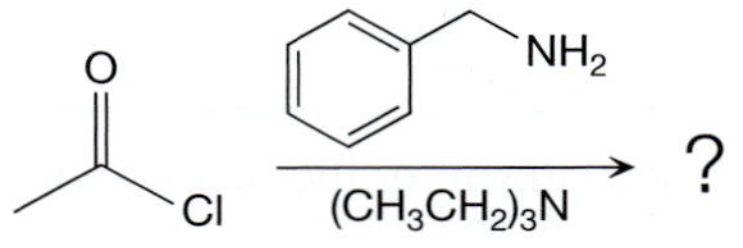
O
Cl
NH2
(CH3CH2)3N
?

(a)

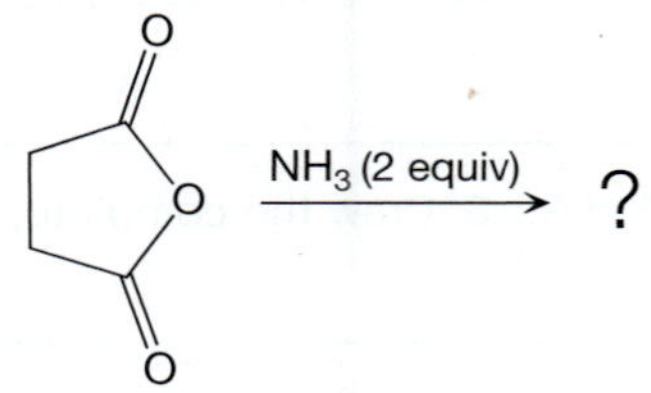
O
O
O
NH3 (2 equiv)
?

(b)

21.4 Synthesis of Acid Halides: Getting to the Top of the Stability Ladder

Recall from Section 20.2 that an acid chloride appears at the top of the stability ladder and can thus be converted into any of the other acid derivatives with relative ease. In light of this, it can be extremely useful to produce acid chlorides from compounds that are more readily available.

As shown in Equations 21-12 and 21-13, **thionyl chloride** ($SOCl_2$) and phosphorus trichloride (PCl_3) can be used to convert carboxylic acids into acid chlorides. Of these, thionyl chloride is the more widely used reagent, due largely to the fact that the two by-products, SO_2 and HCl, are gases that can irreversibly bubble out of an organic solvent as they are produced.

The reaction in Equation 21-12 is explained by the mechanism in Equation 21-14. It consists of back-to-back nucleophilic addition–elimination sequences. The first addition–elimination involves the carboxylic acid and $SOCl_2$. In Step 1, the carbonyl O of the carboxylic acid attacks the polar S═O bond of $SOCl_2$, producing a species in which the positive charge is resonance delocalized. In Step 2, a Cl^- leaving group is eliminated from S, thereby regenerating the S═O bond.

Gaseous SO₂ and HCl can bubble out of solution *irreversibly*.

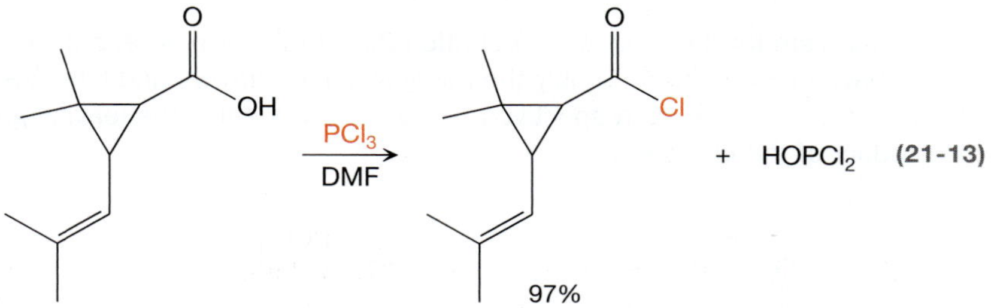

3,5-Dinitrobenzoic acid → **3,5-Dinitrobenzoyl chloride** 90%

$$\text{3,5-Dinitrobenzoic acid} \xrightarrow[\text{60 h, 90 °C}]{\text{SOCl}_2} \text{3,5-Dinitrobenzoyl chloride} + SO_2(g) + HCl(g) \quad \text{(21-12)}$$

$$\xrightarrow[\text{DMF}]{\text{PCl}_3} \quad + \quad HOPCl_2 \quad \text{(21-13)}$$

97%

Mechanism for Equation 21-12

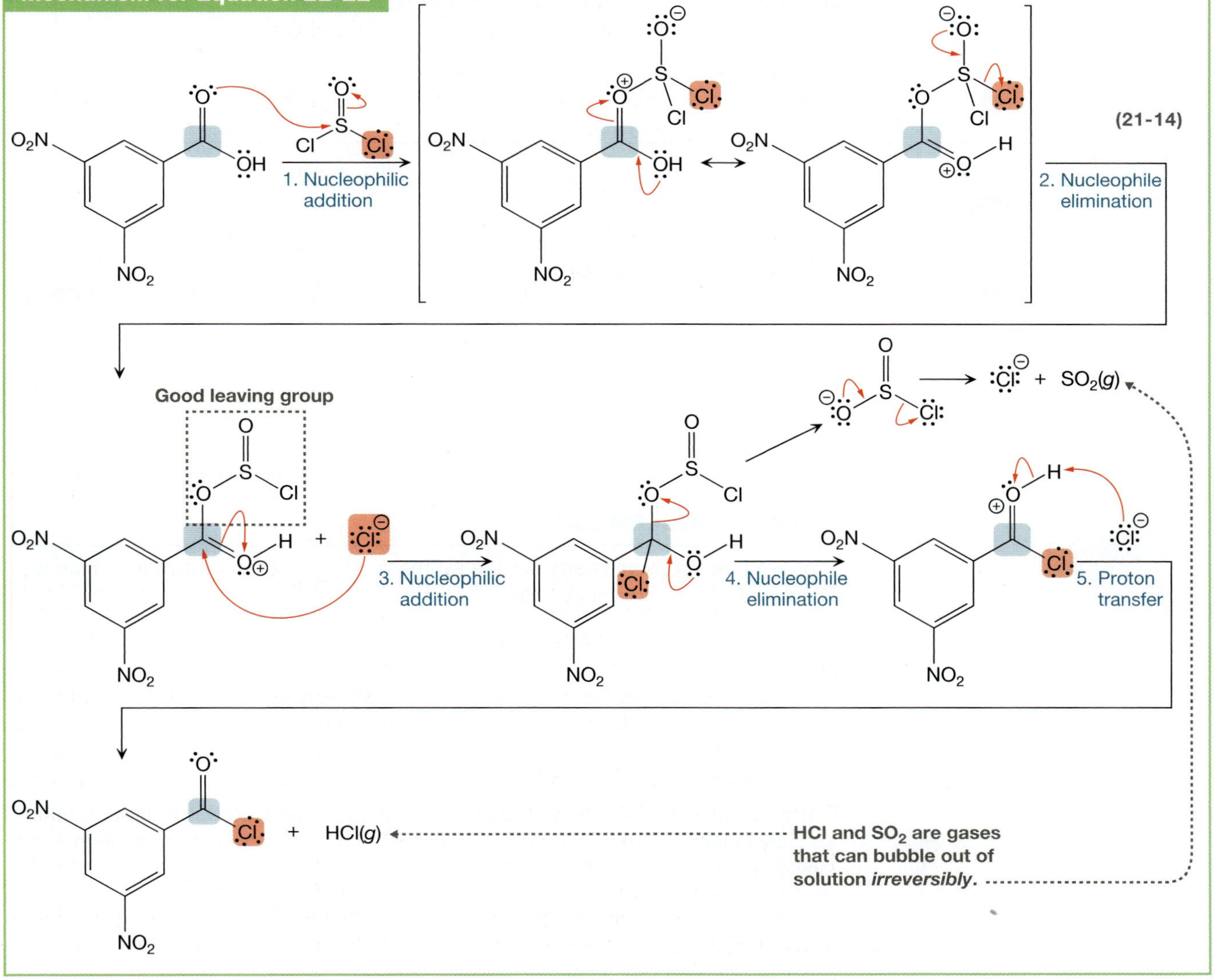

1. Nucleophilic addition

2. Nucleophile elimination

(21-14)

Good leaving group

3. Nucleophilic addition

4. Nucleophile elimination

5. Proton transfer

HCl and SO₂ are gases that can bubble out of solution *irreversibly*.

The product of Step 2 contains a very good leaving group bonded to the carbonyl C, as indicated in the mechanism. This sets the stage for the second nucleophilic addition–elimination sequence (Steps 3 and 4), in which the Cl$^-$ anion generated in Step 2 replaces ClSO$_2^-$. Finally, in Step 5, the carbonyl O is deprotonated.

The mechanism for the reaction involving PCl$_3$ is shown in Your Turn 21.3, and is similar to the one involving SOCl$_2$ (Equation 21-14). The difference is that PCl$_3$ does not possess a polar π bond, so Cl$^-$ is generated from an S$_N$2 step instead of an addition–elimination sequence. This is essentially the same way that Cl$^-$ is generated when PCl$_3$ is used to convert an alcohol to an alkyl chloride (review Section 10.1).

YOUR TURN 21.3

The mechanism for the reaction in Equation 21-13 is shown here, but the curved arrows have been omitted. Supply the missing curved arrows and label each elementary step below its respective reaction arrow. Identify the leaving group in the product of the first step.

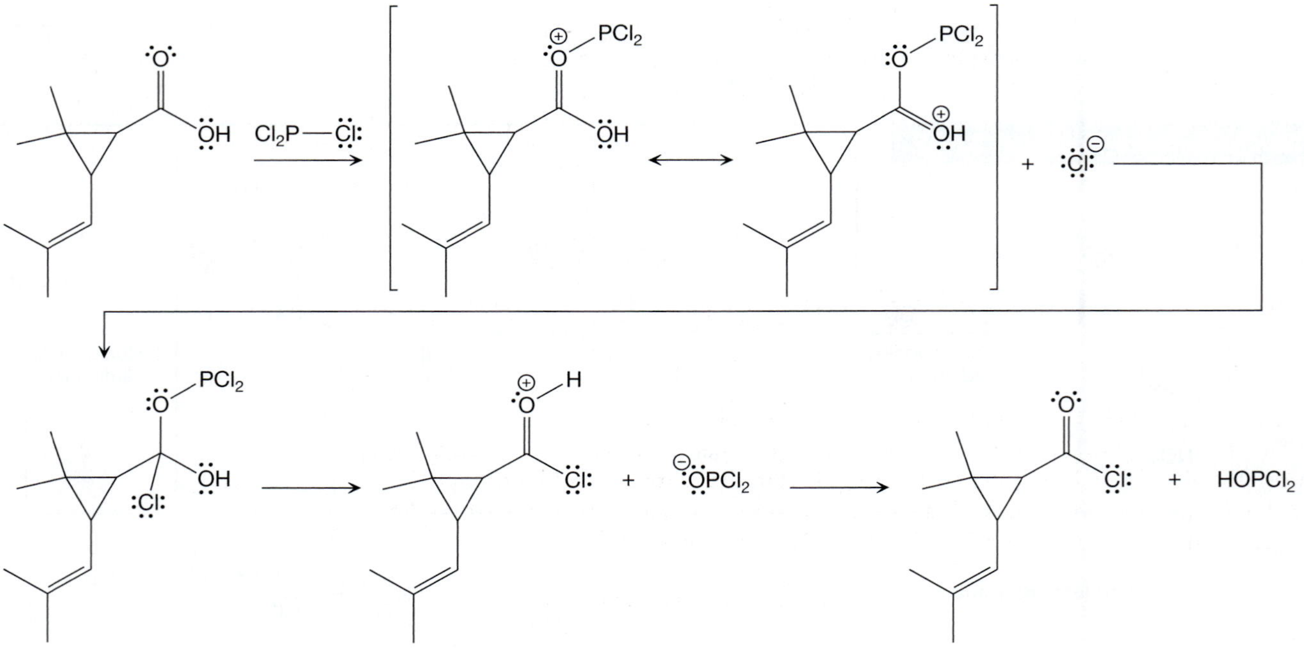

SOLVED PROBLEM 21.8

Propose how you would carry out the synthesis shown here, in which an amide is converted into an acid anhydride.

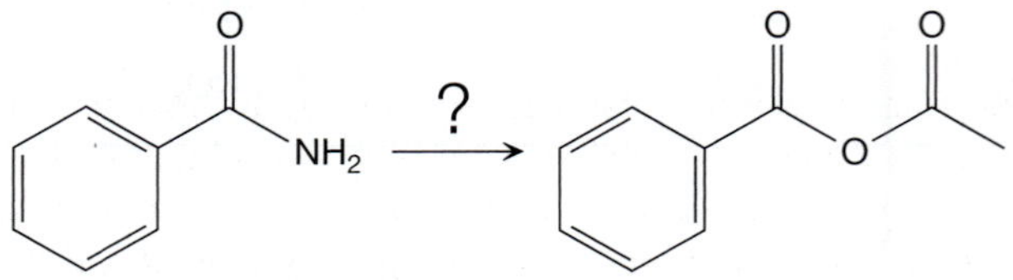

Think What reactions do we know that transform an acid derivative lower on the stability ladder to one higher on the ladder? Can this transformation be done in a single step?

Solve We know that SOCl$_2$ will transform a carboxylic acid into an acid chloride, from which any other acid derivative can be formed. In this case, we could treat benzoyl chloride with sodium acetate to arrive at the final anhydride, via an acyl substitution. Benzoyl chloride could be formed from benzoic acid using $SOCl_2$. Benzoic acid, in turn, could be formed from benzamide via hydrolysis.

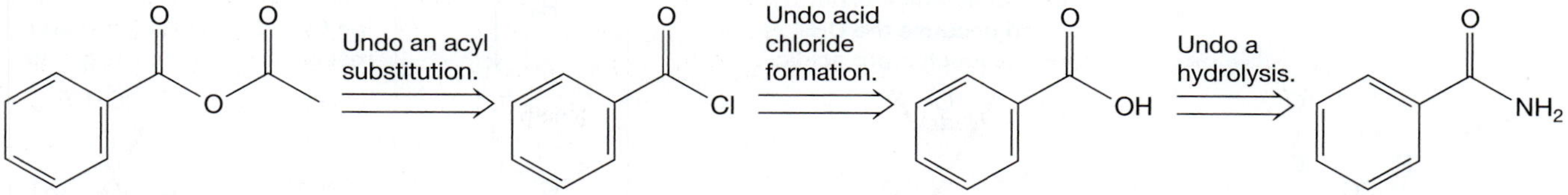

The overall synthesis would then be written as follows:

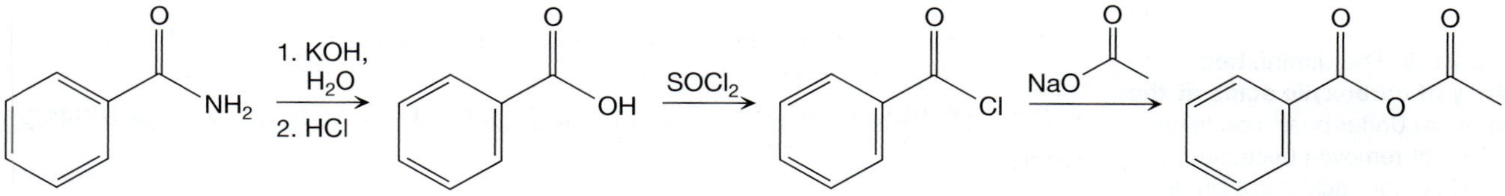

PROBLEM 21.9 Propose a synthesis that would carry out the transformation in Solved Problem 21.8, but does *not* involve benzoyl chloride as a synthetic intermediate.

21.5 The Hell–Volhard–Zelinsky Reaction: Synthesizing α-Bromo Carboxylic Acids

In Section 10.4, we learned that ketones and aldehydes undergo bromination at the α carbon by treating the ketone or aldehyde with molecular bromine (Br_2) in the presence of acid or base. In light of this, it might seem that similar conditions could be used to brominate carboxylic acids at the α carbon to produce an α-bromo carboxylic acid. As indicated in Equation 21-15, however, this does *not* occur.

Bromination of the α carbon does *not* take place under these conditions.

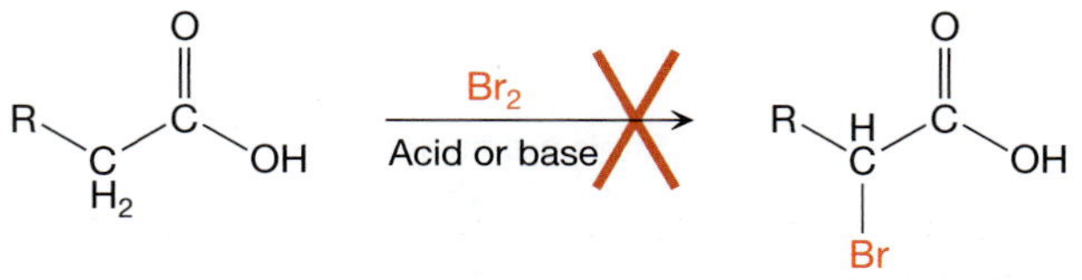

(21-15)

We can understand why if we recall from Section 10.4 that α-bromination requires a key intermediate in which the α carbon is nucleophilic. Under basic conditions, that intermediate is an enolate anion, whereas it is an uncharged enol under acidic conditions. To produce an enolate or enol, an α hydrogen must be removed, but this is not feasible for a carboxylic acid. As shown in Figure 21-3a (next page), the carboxyl hydrogen, being more acidic, is deprotonated under basic conditions instead of the α hydrogen. And, under acidic conditions (Fig. 21-3b), the enol is unfavorable due to the resonance stabilization of the carbonyl group involving the hydroxyl group.

A clever solution to this problem is to carry out the bromination in the presence of some phosphorus trihalide or elemental phosphorus, as shown in Equations 21-16

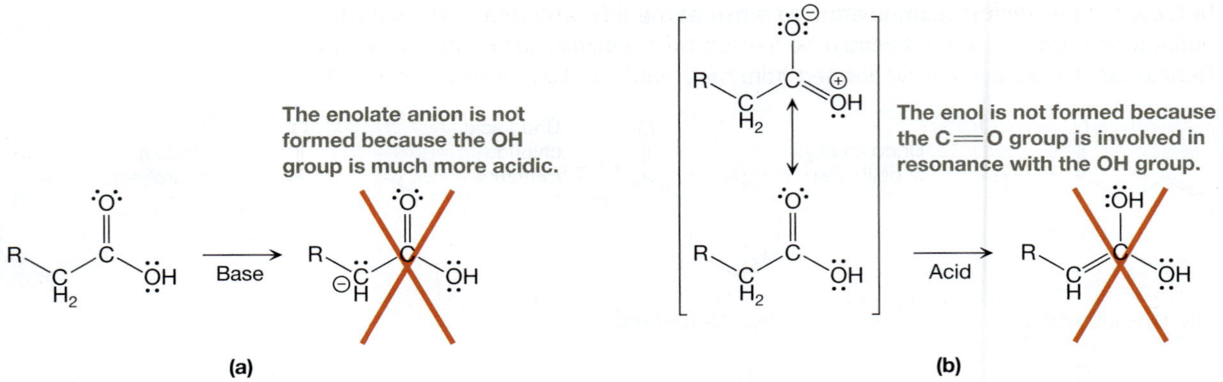

The enolate anion is not formed because the OH group is much more acidic.

The enol is not formed because the C=O group is involved in resonance with the OH group.

(a)

(b)

FIGURE 21-3 The diminished reactivity of carboxylic acids at the α carbon (a) Under basic conditions, the α H is not removed because the carboxyl H is more acidic. (b) Under acidic conditions, resonance disfavors the enol form.

and 21-17. This kind of bromination is called a **Hell–Volhard–Zelinsky (HVZ) reaction**.

The Hell–Volhard–Zelinksky reaction

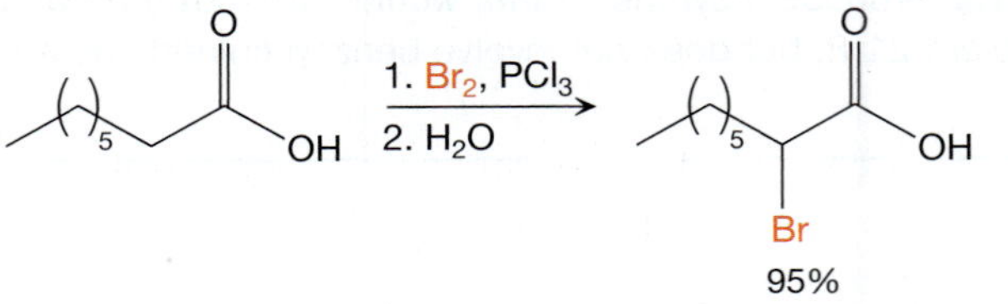

$$\text{1. Br}_2\text{, PCl}_3 \quad \text{2. H}_2\text{O}$$

(21-16)

95%

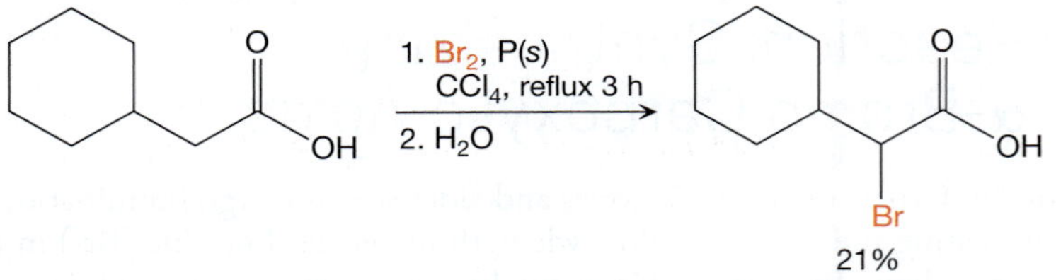

$$\text{1. Br}_2\text{, P}(s)\ \text{CCl}_4\text{, reflux 3 h} \quad \text{2. H}_2\text{O}$$

(21-17)

21%

In both cases, PBr$_3$ is produced [Br$_2$ + PCl$_3$ in Equation 21-16, or Br$_2$ + P(s) in Equation 21-17]. As shown in the partial mechanism in Equation 21-18, PBr$_3$ converts the carboxylic acid to an acid bromide (review Section 21.4).

Partial mechanism for Equation 21-16

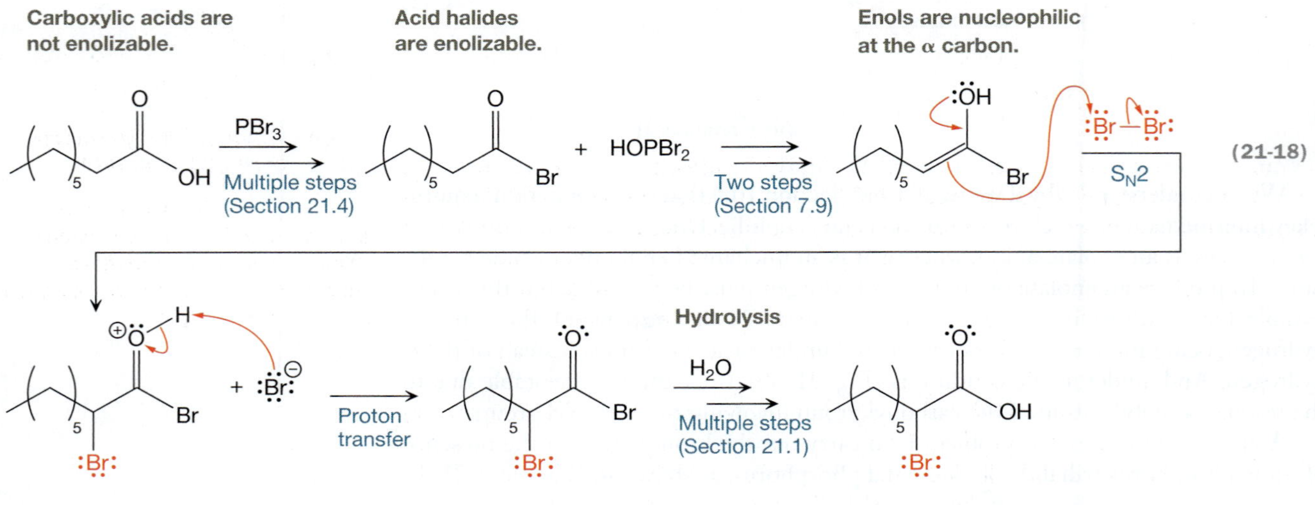

Carboxylic acids are not enolizable.

Acid halides are enolizable.

Enols are nucleophilic at the α carbon.

$$\text{PBr}_3 \quad \text{Multiple steps (Section 21.4)}$$

+ HOPBr$_2$

Two steps (Section 7.9)

S$_N$2

(21-18)

Proton transfer

Hydrolysis

H$_2$O

Multiple steps (Section 21.1)

In the acid bromide form, the enol becomes more favorable, which facilitates the substitution reaction in the subsequent steps to produce the α-bromoacid bromide. Then, treatment with water hydrolyzes the acid bromide to reform the carboxylic acid (Section 21.1).

YOUR TURN **21.4**

Draw the missing steps of the partial mechanism shown in Equation 21-18.

YOUR TURN **21.5**

Draw the partial mechanism for the reaction in Equation 21-17, similar to Equation 21-18.

The excellent ability of Br⁻ as a leaving group makes α-bromoacids useful synthetic intermediates. For example, Equation 21-19 shows that they can be used to synthesize α-amino acids, such as phenylalanine, via a nucleophilic substitution reaction with aqueous ammonia.

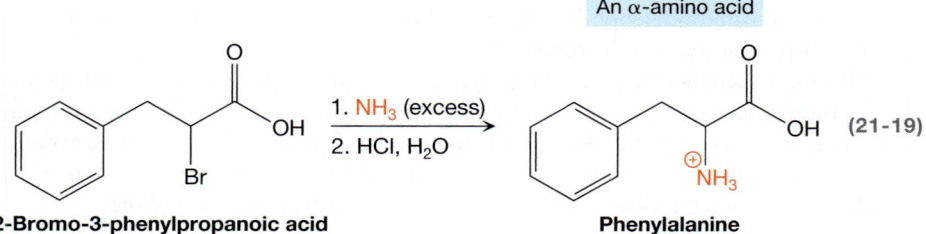

$$\text{An } \alpha\text{-amino acid}$$

1. NH₃ (excess)
2. HCl, H₂O

(21-19)

2-Bromo-3-phenylpropanoic acid   **Phenylalanine**
60%

PROBLEM 21.10 How would you synthesize each of the following compounds from butanoic acid?

(a)   (b)   (c)

21.6 Sulfonyl Chlorides: Synthesis of Mesylates, Tosylates, and Triflates

The three compounds in Figure 21-4 are **sulfonyl chlorides,** having the general form R—SO₂Cl. When one of these sulfonyl chlorides is treated with an alcohol, a **sulfonate ester** is produced via a sulfonation reaction, as shown in Equation 21-20.

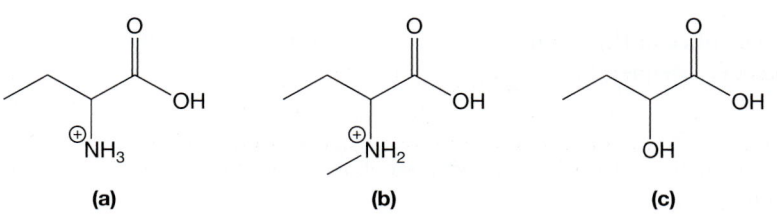

Sulfonate ester

R′—OH   (MsCl, TfCl, or TsCl)   →   R′—O—S—R   (21-20)

(R′—OMs, R′—OTf, or R′—OTs)

(a) Methanesulfonyl chloride (Mesyl chloride, MsCl)

(b) Trifluoromethanesulfonyl chloride (Triflyl chloride, TfCl)

(c) p-Toluenesulfonyl chloride (Tosyl chloride, TsCl)

FIGURE 21-4 Some common sulfonyl chlorides These sulfonyl chlorides can be used to make sulfonate esters, which have excellent leaving groups.

The mechanism in Equation 21-21 accounts for these sulfonation reactions, where the alcohol attacks the sulfonyl chloride in a nucleophilic addition–elimination sequence.

Mechanism for Equation 21-20

$$R'-\overset{..}{\underset{..}{O}}H \xrightarrow[\text{addition}]{\text{1. Nucleophilic}} R'-\overset{+}{\underset{..}{O}}-\overset{..}{S}-R \xrightarrow[\text{transfer}]{\text{2. Proton}} R'-\overset{..}{\underset{..}{O}}-\overset{..}{S}-R \xrightarrow[\text{elimination}]{\text{3. Nucleophile}} R'-\overset{..}{\underset{..}{O}}-S-R + :\overset{..}{\underset{..}{Cl}}:^{\ominus} \quad (21\text{-}21)$$

This mechanism is essentially identical to the one shown previously in Equation 21-4 for the hydrolysis of acid chlorides, but in this case the nucleophilic OH group attacks the S=O group rather than a C=O group. Notice, too, that pyridine is added to the reaction to act as a base in Step 3. As we saw in Section 21.3, pyridine will not compromise the desired nucleophilic addition–elimination reaction because pyridine has no N—H proton that can be removed.

Although there is experimental evidence to support the mechanism in Equation 21-21, there is also evidence suggesting that the addition and elimination steps occurs in a single S_N2 step instead. Neither mechanism has been established conclusively.

Recall from Chapter 9 that MsO^- (mesylate), TfO^- (triflate), and TsO^- (tosylate) are very good leaving groups for S_N2, S_N1, E2, and E1 reactions. Therefore:

> Sulfonation of an alcohol converts a poor HO^- leaving group into a very good RSO_3^- leaving group.

Moreover, notice in Equation 21-21 that no bonds to the C—OH carbon are broken or formed. Consequently:

> Sulfonation of an alcohol occurs with retention of configuration at the C—OH carbon.

A specific example is shown in Equation 21-22.

Preserves the stereochemistry

$$\xrightarrow[\text{pyridine}]{\text{TfCl}} \quad (21\text{-}22)$$

100%

Sulfonation of an alcohol, therefore, complements the PBr_3 bromination of an alcohol (Section 10.1), which converts a poor HO^- leaving group into a good Br^- leaving group via *inversion* of configuration at the C—OH carbon.

What is the major product of the sequence of reactions shown here?

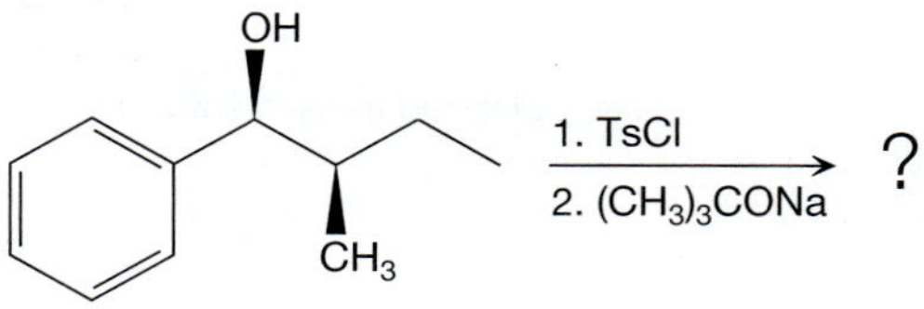

Think What is the product of sulfonation of an alcohol, and what is the stereochemistry that is associated with that reaction? What type of reaction is promoted by a strong, bulky base in the second step of this synthesis? What is the stereochemistry associated with that reaction?

Solve The first reaction is a sulfonation in which tosyl chloride (TsCl) converts the alcohol to an alkyl tosylate. Because the tosylate anion is an excellent leaving group, the presence of the strong, bulky base, $(CH_3)_3CONa$, promotes an E2 reaction. Recall that the alkene that is favored in an E2 reaction is produced from the substrate conformation in which the H and the leaving group on adjacent carbons are anti to each other (i.e., anticoplanar). That conformation is the one shown below, so the major product is the *E* alkene.

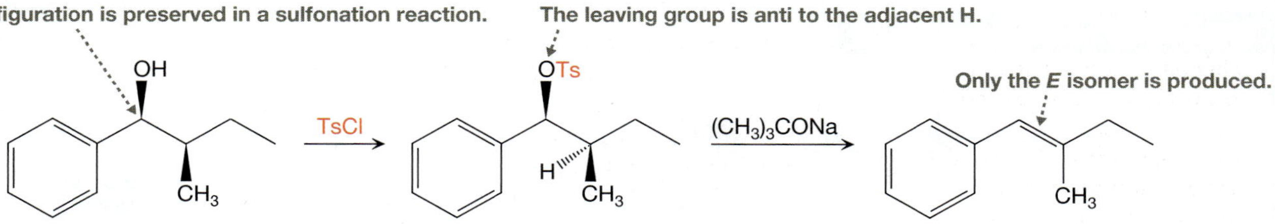

The configuration is preserved in a sulfonation reaction.

The leaving group is anti to the adjacent H.

Only the *E* isomer is produced.

PROBLEM 21.12 Show how to carry out the following synthesis.

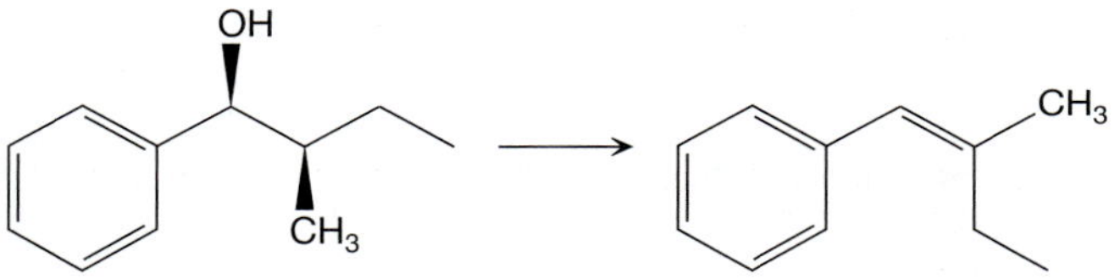

21.7 Base and Acid Catalysis in Nucleophilic Addition–Elimination Reactions

Thus far, we have seen that weak nucleophiles, such as water, alcohols, and amines, react readily with acid halides, acid anhydrides, thionyl chloride, and sulfonyl chlorides. Even though the nucleophiles are weak, the polar π bonds (C=O and S=O) in these compounds are reactive enough to compensate. By contrast, we saw in Section 21.2 that esters are significantly less reactive at the carbonyl carbon due to resonance involving the leaving group. Thus, Equation 21-23 shows that esters do not undergo nucleophilic addition–elimination under normal conditions with weak nucleophiles like alcohols.

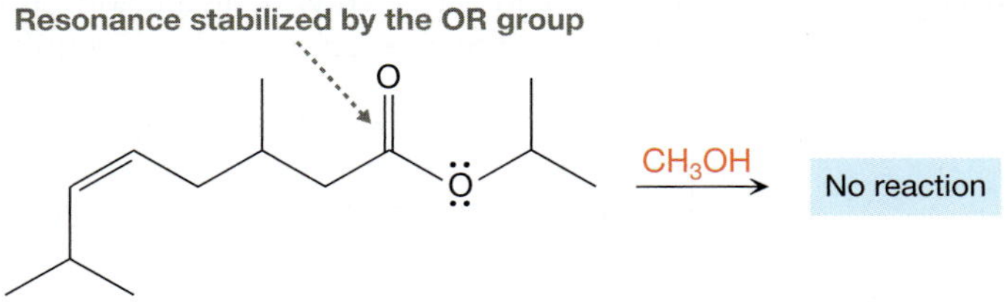

Resonance stabilized by the OR group

No reaction   (21-23)

Transesterification takes place readily, however, when a small amount (i.e., a catalytic amount) of a base (sodium methoxide, CH_3ONa, in Equation 21-24) is added.

Moreover, the base is not consumed overall in the reaction, so the reaction is *base catalyzed*.

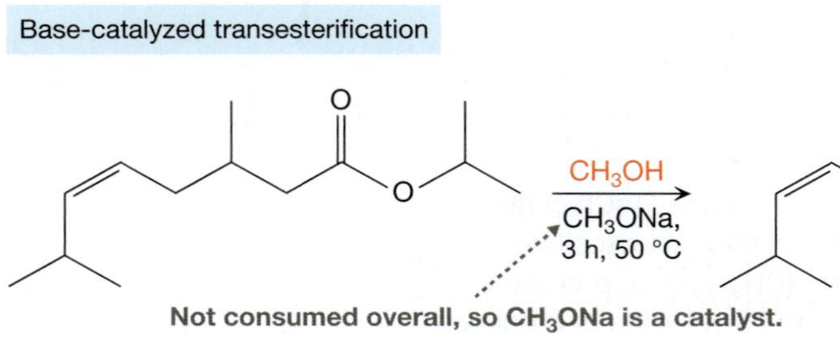

Base-catalyzed transesterification

(21-24)

Not consumed overall, so CH₃ONa is a catalyst.   69%

The mechanism for this reaction, presented in Equation 21-25, shows why the reaction is catalyzed by the methoxide anion base.

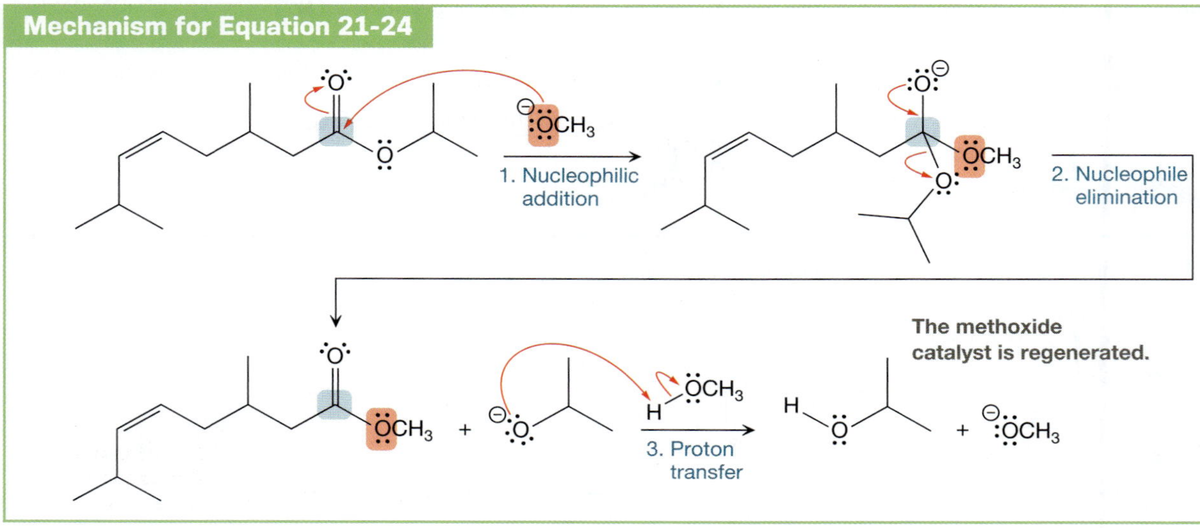

Mechanism for Equation 21-24

(21-25)

1. Nucleophilic addition

2. Nucleophile elimination

The methoxide catalyst is regenerated.

3. Proton transfer

The reaction rate is increased because the nucleophile that attacks the carbonyl carbon in the nucleophilic addition step (Step 1, the rate-determining step) is CH_3O^-, a strong nucleophile. In the absence of CH_3O^- (e.g., in Equation 21-23), the nucleophile would have to be methanol, a weak nucleophile. Moreover, there is no net consumption of CH_3O^- because it is regenerated in the proton transfer step (Step 3).

Because water and alcohols have similar behaviors, it might also seem that a base could be used to catalyze the interconversion between esters and carboxylic acids. But it can't. Although the presence of hydroxide facilitates the nucleophilic addition–elimination reaction, as shown in Equation 21-26, the carboxylic acid that is initially produced is acidic and thus reacts further with hydroxide. Consequently, the base is consumed in the overall reaction, making it a *base-promoted* reaction instead of a base-catalyzed reaction. In fact, this is an example of a *saponification reaction*, discussed previously in Section 20.3. As shown in Equation 21-27, a similar proton transfer takes place when a carboxylic acid is treated with an alkoxide anion, making it unfeasible to convert a carboxylic acid into an ester under basic conditions.

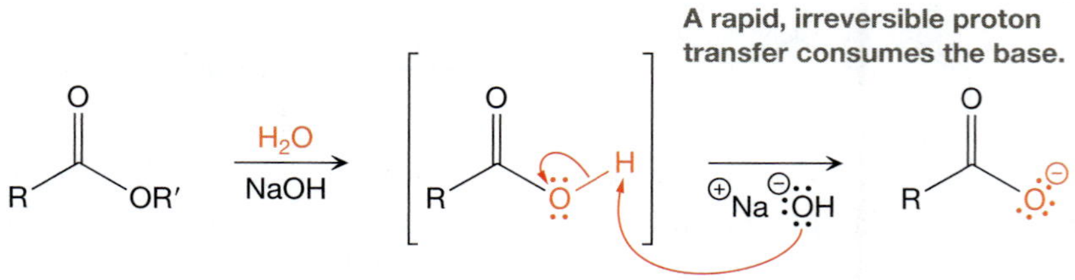

A rapid, irreversible proton transfer consumes the base.

(21-26)

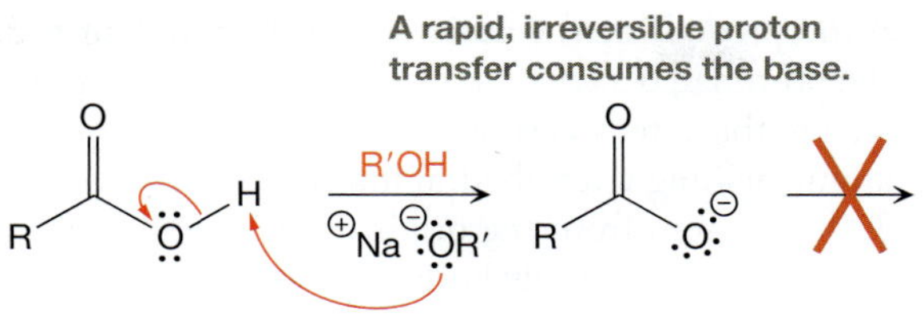

A rapid, irreversible proton transfer consumes the base.

No further reaction   (21-27)

Transesterification can also be *acid* catalyzed. Equation 21-28 shows, for example, that a catalytic amount of a strong acid, such as sulfuric acid, facilitates the production of hexyl acetate from methyl acetate and hexan-1-ol.

CONNECTIONS Hexyl acetate has a bittersweet, fruity taste and is a common flavoring agent, especially in hard candies.

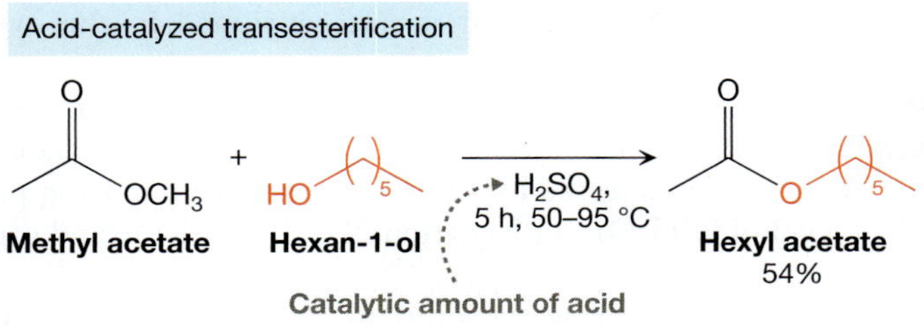

Acid-catalyzed transesterification

Methyl acetate   Hexan-1-ol H_2SO_4,   Hexyl acetate   + $HOCH_3$ (21-28)
                                        5 h, 50–95 °C   54%

Catalytic amount of acid

The mechanism for this reaction is shown in Equation 21-29.

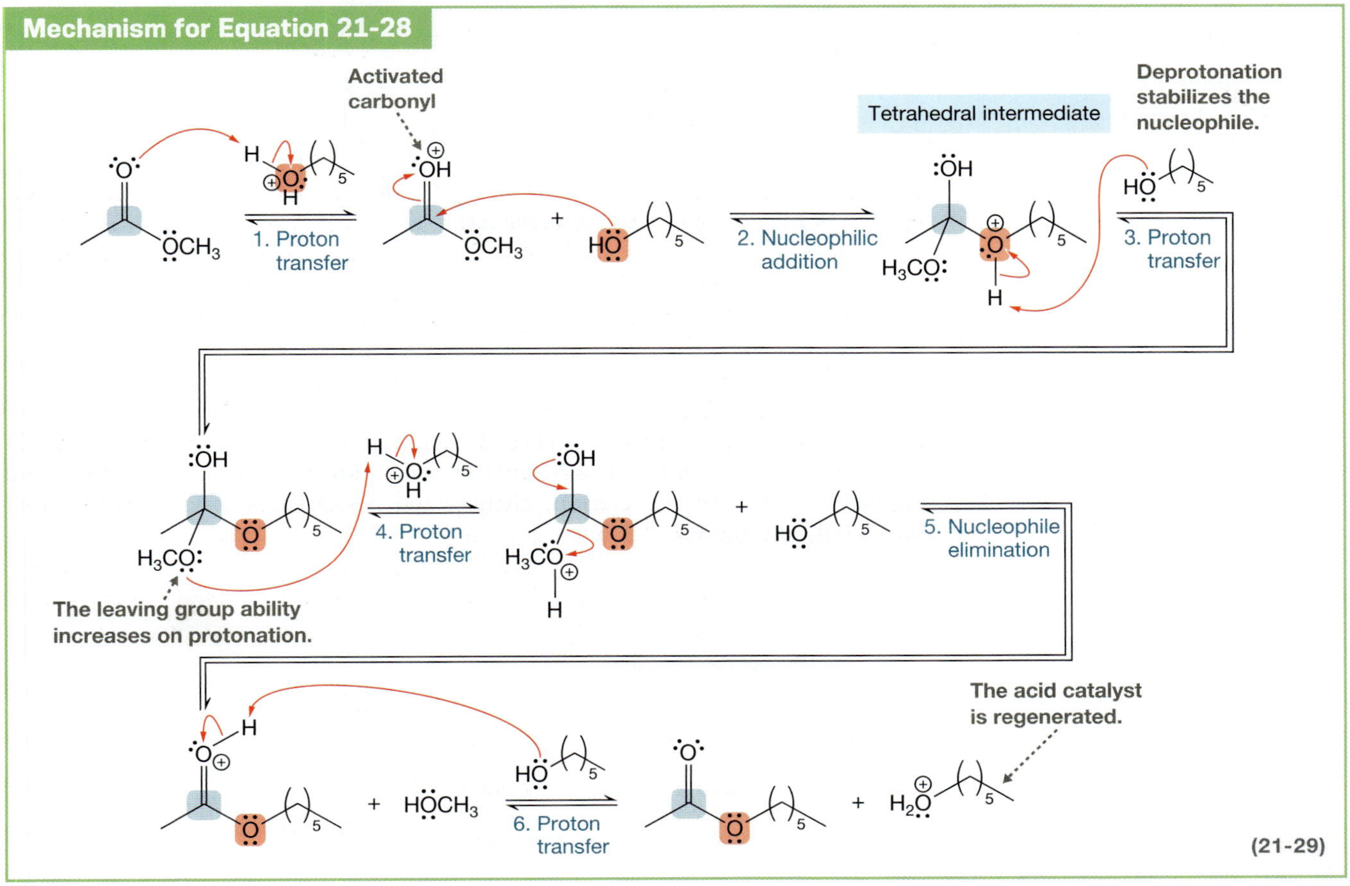

Mechanism for Equation 21-28

Activated carbonyl

Tetrahedral intermediate

Deprotonation stabilizes the nucleophile.

1. Proton transfer

2. Nucleophilic addition

3. Proton transfer

The leaving group ability increases on protonation.

4. Proton transfer

5. Nucleophile elimination

The acid catalyst is regenerated.

6. Proton transfer

(21-29)

Sulfuric acid is a strong acid, so the predominant acid in solution, according to the leveling effect (Section 6.2c), is the protonated alcohol, $CH_3(CH_2)_5OH_2^+$. In Step 1 of the mechanism, that acid protonates the ester's carbonyl group, and in Step 2, the alcohol attacks the carbonyl C atom, yielding a tetrahedral intermediate. Steps 3 and 4 are proton transfers. In Step 3, the O atom from the original nucleophile becomes uncharged, and thus is stabilized. In Step 4, the singly bonded O atom of the original ester gains a positive charge, which increases its leaving group ability. In Step 5, the leaving group departs and the C=O bond is reformed, and in Step 6, the carbonyl O is deprotonated, yielding the overall uncharged product.

Why does the presence of a strong acid catalyze this reaction? Notice in Step 2 (the rate-determining step) that the nucleophile is relatively weak, an alcohol, but the rate of nucleophilic addition is enhanced because the carbonyl group has become *activated* (review Section 18.1) by the proton transfer in Step 1—that is, the carbonyl C has become more electrophilic. Notice, too, that the strong acid used in Step 1 is regenerated in Step 6, so it is not consumed in the overall reaction.

Each step of the acid-catalyzed transesterification reaction in Equation 21-29 is reversible, so the overall reaction is reversible. Therefore, according to Le Châtelier's principle, the product yield can be increased by adding reactants or removing products. Often the product alcohol (CH_3OH in Equation 21-28) is volatile and can be removed by distillation.

Unlike what we saw previously under basic conditions, the conversion of an ester to a carboxylic acid can also be acid catalyzed. Equation 21-30 shows an example of this *ester hydrolysis*.

Acid-catalyzed ester hydrolysis

The ester is hydrolyzed to a carboxylic acid.

$$\xrightarrow[\text{H}_2\text{SO}_4, \text{ acetic acid,} \atop \text{reflux}]{\text{H}_2\text{O}}$$ (21-30)

94%

The *esterification* of a carboxylic acid can be acid catalyzed, too, as shown in Equation 21-31. This is an example of a **Fischer esterification reaction**, named after Emil Fischer (1852–1919), the German chemist and Nobel laureate who first reported these kinds of reactions.

Fischer esterification

$$\xrightarrow[\text{H}_2\text{SO}_4, \text{H}_2\text{O,} \atop \text{2 h, reflux}]{\text{CH}_3\text{CH}_2\text{OH}}$$ (21-31)

A carboxylic acid is converted to an ester.

86%

YOUR TURN 21.6

The mechanism for the Fischer esterification reaction in Equation 21-31 is shown here, but the curved arrows have been omitted. Supply the curved arrows for each step and write the name of each elementary step below its reaction arrow.

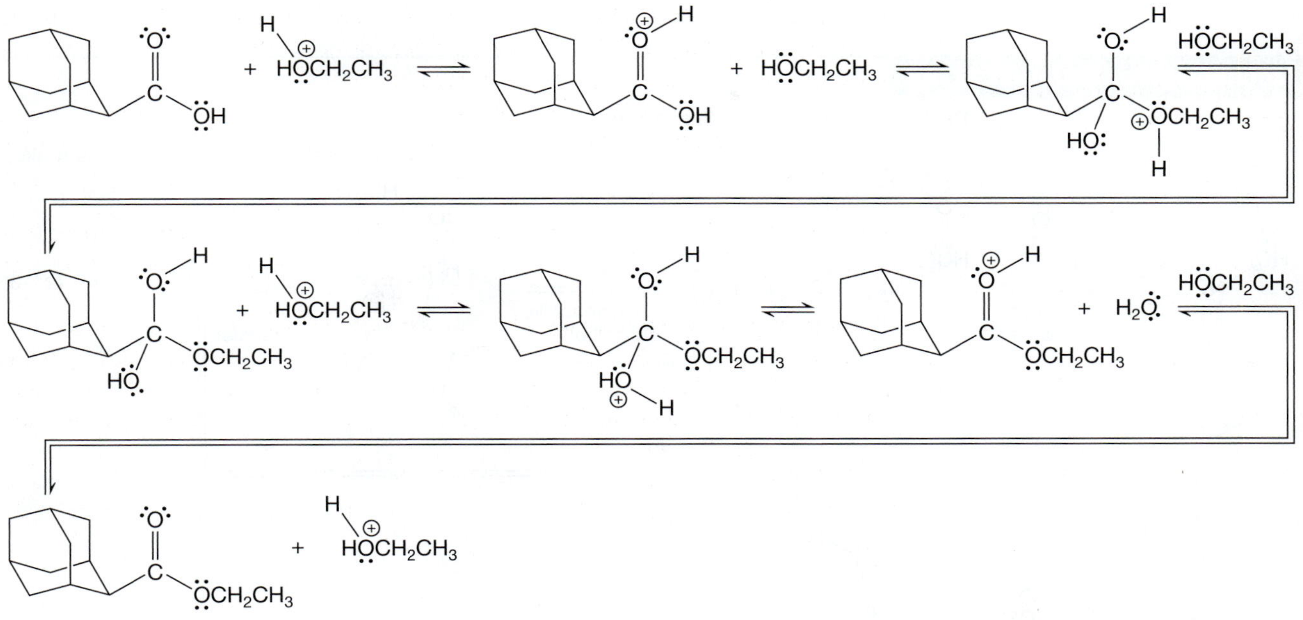

PROBLEM 21.13 Draw the complete, detailed mechanism for the acid-catalyzed ester hydrolysis in Equation 21-30.

Acyl substitutions under acidic conditions are not limited to esters and carboxylic acids. Amides, which are even less reactive than esters or carboxylic acids, do not react with weak nucleophiles under normal conditions either. Under acidic conditions, however, amides can undergo acyl substitution via an addition–elimination mechanism. This is exemplified by the amide hydrolysis shown in Equation 21-32.

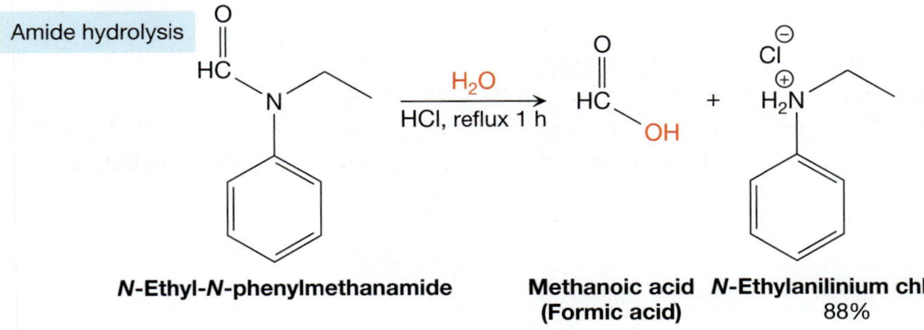
Amide hydrolysis
H2O
HCl, reflux 1 h
N-Ethyl-N-phenylmethanamide
Methanoic acid
(Formic acid)
N-Ethylanilinium chloride
88%

(21-32)

The mechanism for this reaction, presented in Equation 21-33, is essentially the same as the one that describes acid-catalyzed transesterification, shown previously in Equation 21-29.

The carbonyl group is activated in Step 1, which better facilitates nucleophilic addition of the weak nucleophile in Step 2. In Step 3, the charge on the O atom from the nucleophile goes from +1 to 0, so the O is stabilized. The reverse is true for the N atom of the leaving group in Step 4. In Step 5, the amine leaving group departs and the C=O bond is simultaneously regenerated. Finally, in Step 6, the newly formed amine, which is weakly basic, irreversibly deprotonates the carbonyl O. This irreversible step consumes the acid, so the reaction is *acid promoted* rather than acid catalyzed.

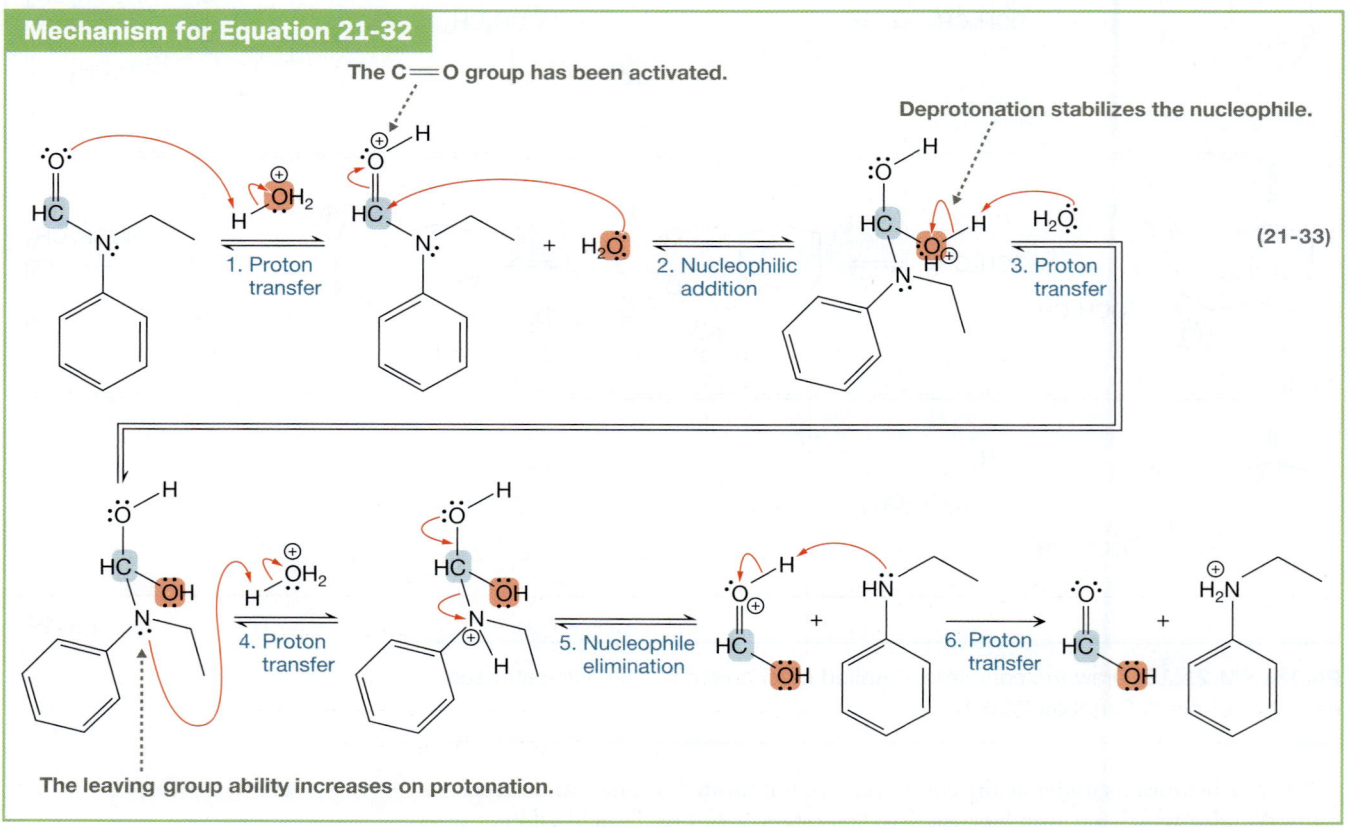

Mechanism for Equation 21-32

The C=O group has been activated.

Deprotonation stabilizes the nucleophile.

1. Proton transfer
2. Nucleophilic addition
3. Proton transfer

(21-33)

4. Proton transfer
5. Nucleophile elimination
6. Proton transfer

The leaving group ability increases on protonation.

PROBLEM 21.14 Draw the complete detailed mechanism and the overall products for each of the following reactions.

(a) $\xrightarrow{H_2O}{H_2SO_4}$?

(b) $\xrightarrow{H_2SO_4}$?

(c) $\xrightarrow{H_2O}{H_2SO_4}$?

(d) $\xrightarrow{H_2O}{H_2SO_4}$?

(e) $\xrightarrow{CH_3CH_2OH}{HCl}$?

(f) $\xrightarrow{H_2O}{HCl}$?

21.8 Baeyer–Villiger Oxidations

When diphenylmethanone (benzophenone) is treated with a *peroxyacid* (RCO_3H) such as *meta*-chloroperbenzoic acid (MCPBA), phenyl benzoate is produced, as shown in Equation 21-34. This is an example of a **Baeyer–Villiger oxidation**, named after the German chemist Adolf von Baeyer (1835–1917) and the Swiss chemist Victor Villiger (1868–1934).

Baeyer–Villiger oxidation

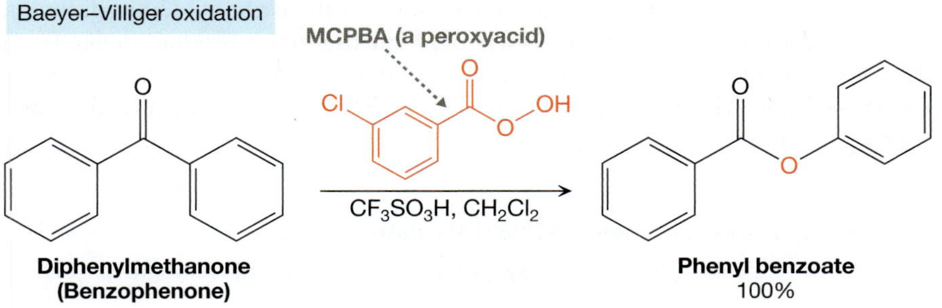

(21-34)

**Diphenylmethanone
(Benzophenone)**

Phenyl benzoate
100%

The mechanism for this reaction is shown in Equation 21-35 and is very similar to the mechanism for an acid-catalyzed transesterification (Equation 21-29).

Mechanism for Equation 21-34

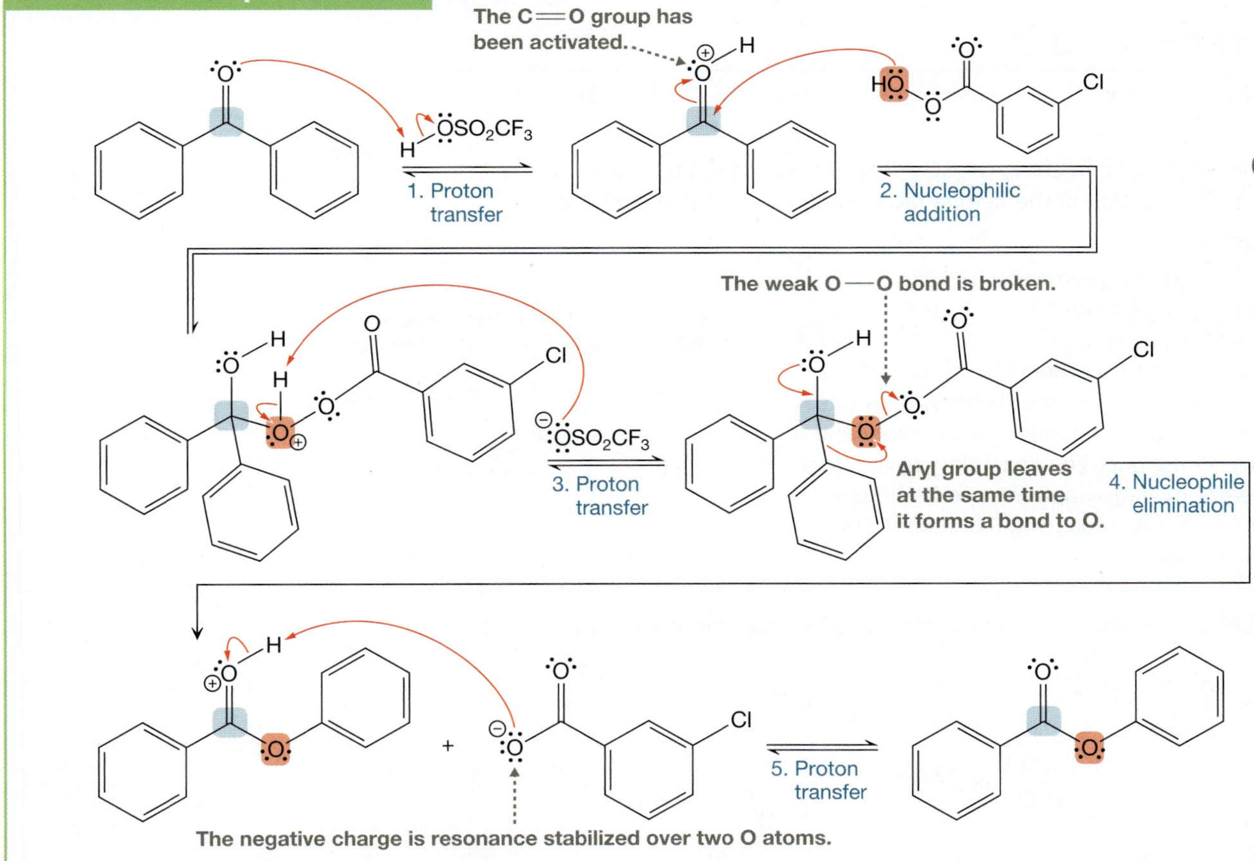

(21-35)

After the carbonyl group of the ketone has been activated by protonation in Step 1, the relatively weak peroxyacid nucleophile attacks the carbonyl C in Step 2. A proton is then removed from the peroxyacid's O atom in Step 3, and an aryl group departs from the carbonyl C in Step 4. Simultaneously, the aryl group forms a bond to an O atom from the peroxyacid, breaking the peroxyacid's O—O bond in the process. Finally, in Step 5, the ester's carbonyl group is deprotonated, yielding the overall uncharged ester product.

Even though ketones don't possess a good leaving group attached to the carbonyl carbon, the aryl group departs in Step 4 for two reasons. First, the O—O bond of the peroxyacid is weak and is replaced by a much stronger C—O bond. Second, the negative charge that is formed in that step appears on a carboxylate anion (RCO_2^-), so it is resonance stabilized over two O atoms.

The ketone oxidized in Equation 21-34 is *symmetric*—that is, the carbonyl group is bonded to identical groups. If the groups bonded to the carbonyl C are different, such as in an *unsymmetric* ketone or aldehyde, then the major product depends on which group leaves from the carbonyl C. The ability of a group to do so is called its **migratory aptitude**. Migratory aptitudes have been determined empirically for a variety of groups, as follows:

Migratory Aptitude in a Baeyer–Villiger Oxidation

Methyl group < 1° Alkyl group < 2° Alkyl group ≈ Aryl group < 3° Alkyl group < H

This roughly follows the order for cation stability. Indeed, the migrating group has been shown to acquire some cationic character in the transition state in the step in which it departs. This information is useful in predicting the major Baeyer–Villiger product, as shown in Solved Problem 21.15.

GREEN CHEMISTRY

Traditionally, Baeyer–Villiger oxidations are carried out under conditions that are not very green. The CH_2Cl_2 solvent may be carcinogenic, and some of the common peroxyacids are toxic. One green alternative is shown below, which is solvent free, incorporates hydrogen peroxide as the oxidizing agent, and is catalyzed by Fe_3O_4 nanoparticles that can be recycled.

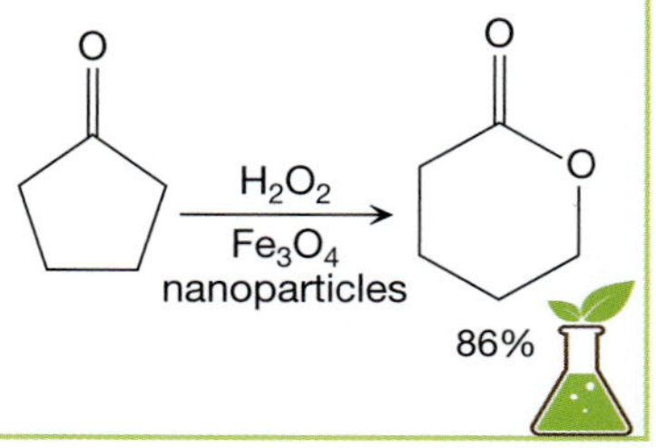

SOLVED PROBLEM 21.15

Predict the major product of the reaction shown here. (CF_3CO_3H is relatively acidic.)

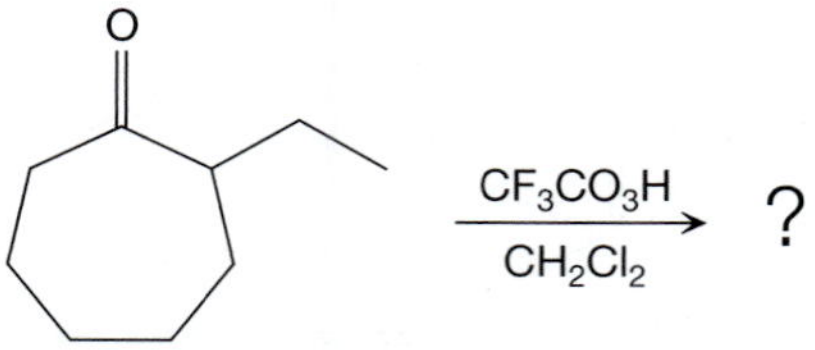

Think What type of reagent is CF_3CO_3H? How does it tend to react with a ketone? Does one side of the ketone favor reaction over the other? If so, which side?

Solve CF_3CO_3H is a peroxyacid and will react with a ketone in a Baeyer–Villiger oxidation. In such a reaction, an O atom from the peroxyacid is inserted between the carbonyl C and one of the groups initially bonded to the carbonyl C. In this case, the carbonyl C is bonded to a primary and a secondary alkyl group. The secondary alkyl group has a greater migratory aptitude, so its bond will preferentially break, producing the lactone (cyclic ester) shown here.

A secondary alkyl group has a greater *migratory aptitude* than a primary alkyl group.

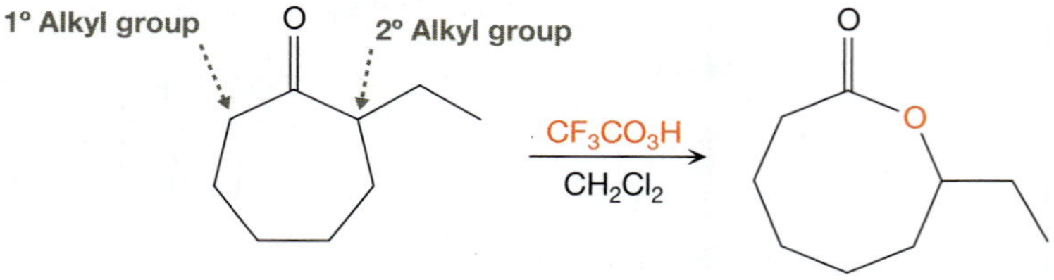

PROBLEM 21.16 Predict the major product of each of the following reactions.

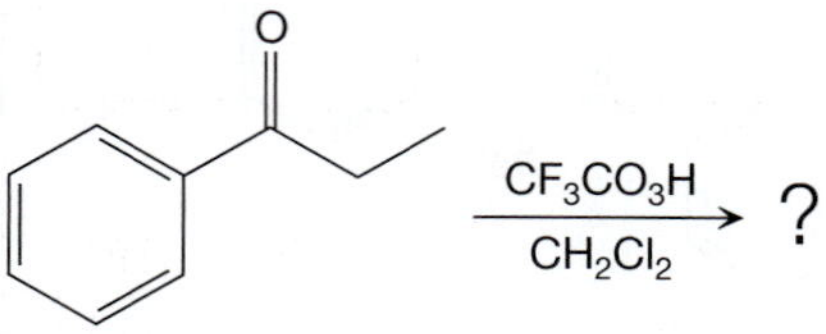

(a)

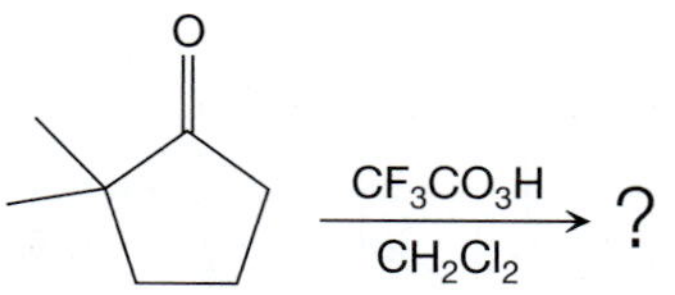

(b)

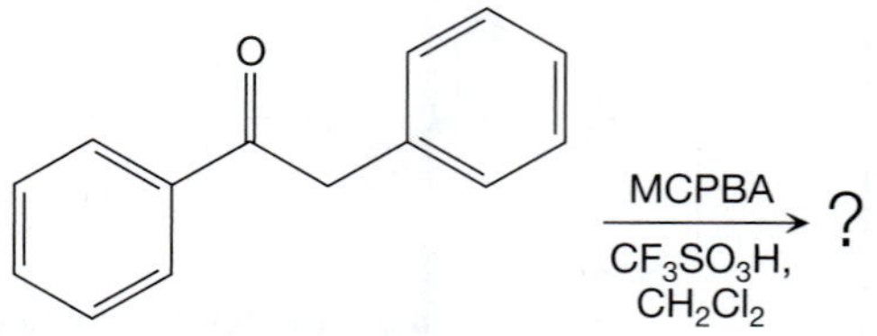

(c)

PROBLEM 21.17 In the mechanism in Equation 21-35, the hydroxyl (OH) oxygen of the peroxyacid is shown as the nucleophile. Why is that oxygen more nucleophilic than the adjacent oxygen?

21.9 Claisen Condensations

When ethyl ethanoate (ethyl acetate) is treated with sodium ethoxide, followed by acid workup, ethyl 3-oxobutanoate, a **β-keto ester**, is produced (Equation 21-36).

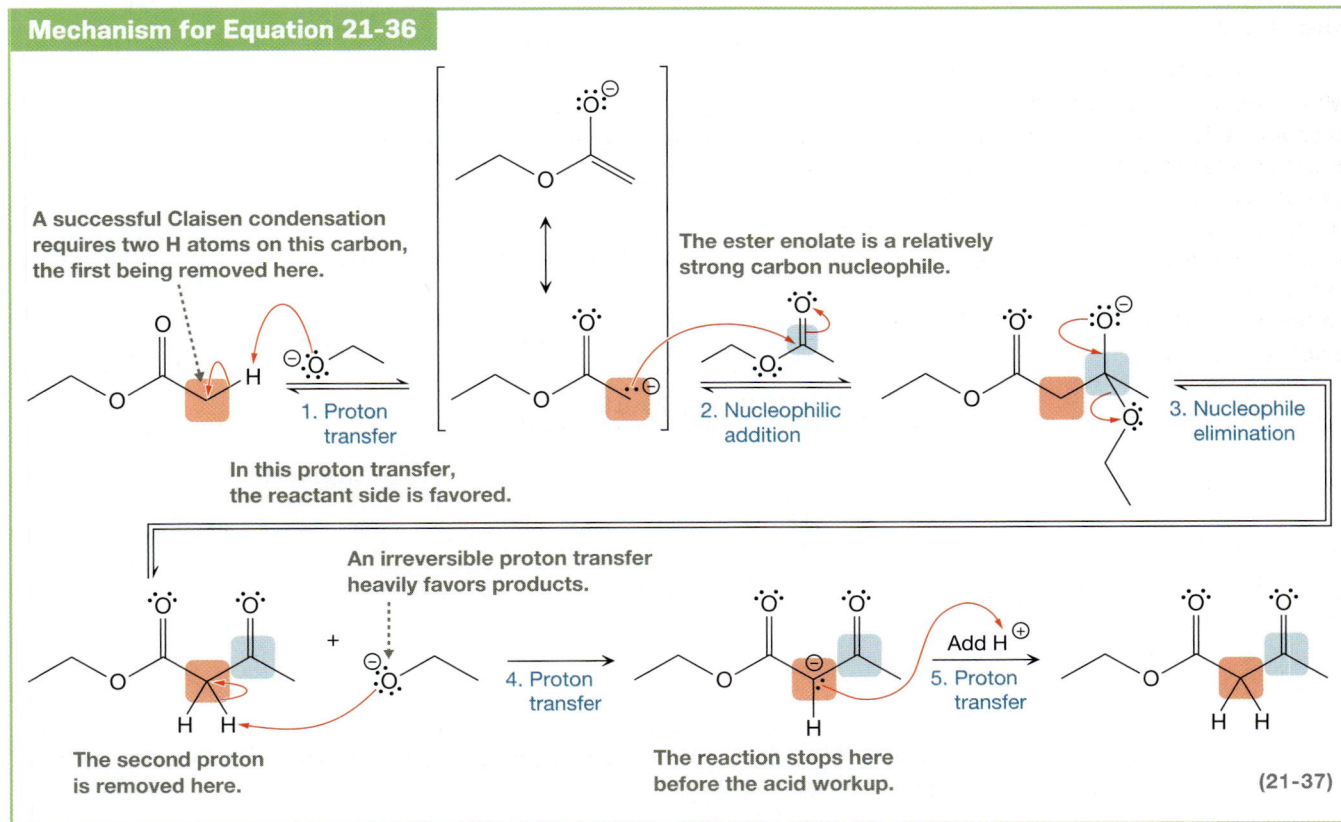

This is an example of a **Claisen condensation reaction**, named after Rainer Ludwig Claisen (1851–1930), who developed it. It is a *condensation* reaction because, overall, two ester molecules are fused together and a smaller molecule—in this case, a molecule of ethanol—is eliminated.

The mechanism for the Claisen condensation reaction in Equation 21-36, shown in Equation 21-37, is very similar to the mechanism for base-catalyzed transesterification (review Equation 21-25). Normally an ester is not strongly nucleophilic, but due to the basic conditions of the reaction, $CH_3CH_2O^-$ deprotonates the ester at its α carbon in Step 1 to produce a strongly nucleophilic enolate anion. In Step 2, that enolate anion attacks the carbonyl carbon of a second molecule of the same ester, forming a new C—C bond. The CH_3CH_2O group is then eliminated in Step 3, producing an initial β-keto ester, which is quickly deprotonated in Step 4. Acid workup in Step 5 replenishes the β-keto ester.

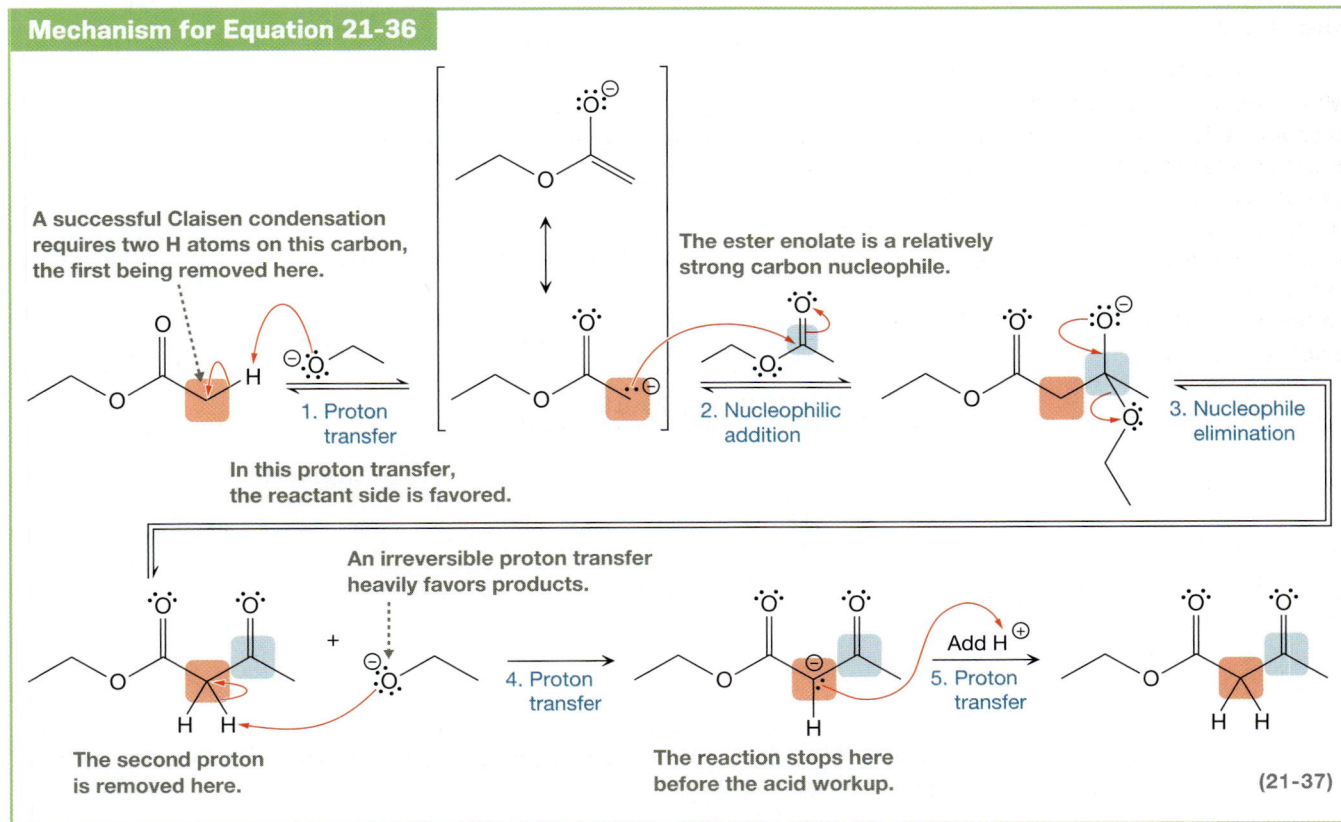

The first two proton transfer steps (i.e., Steps 1 and 4) play important roles in this mechanism. The first (Step 1) is reversible and favors the reactant side, much like we saw for aldol reactions (Section 18.4). Thus, only a small fraction of the initial ester exists in the form of the enolate anion at any given time, leaving a substantial amount of the uncharged ester available to react with the enolate anion. The second proton transfer (Step 4) is *irreversible*, heavily favoring the product side. That step is irreversible because the β-keto ester is much more acidic ($pK_a = 11$) than a normal ester, due to the substantial resonance delocalization of the negative charge in the enolate anion that is produced. This is important because the reaction is reversible through the first three steps, favoring the overall reactants. The deprotonation in Step 4 effectively removes the β-keto ester product and, according to Le Châtelier's principle, the first three steps shift continuously toward products.

The necessity for these proton transfer steps places a significant restriction on the esters that can participate in Claisen condensations.

> A successful Claisen condensation generally requires at least two α protons on the initial ester.

Without the first α proton, the enolate anion could not be produced. Without the second α proton, the equilibrium would favor the overall reactants and, as shown in Equation 21-38, no significant amount of the β-keto ester would be isolated.

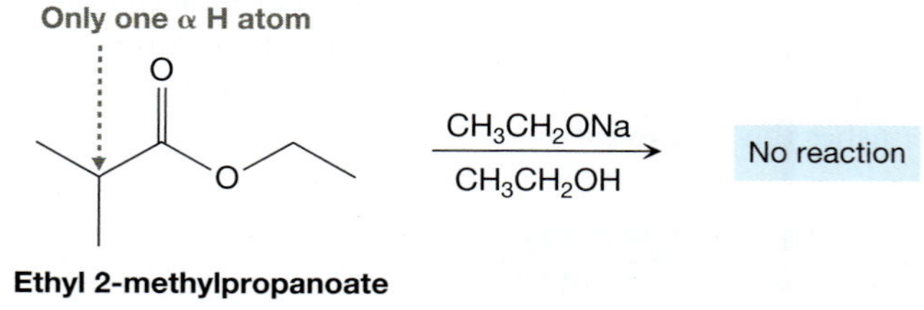

Only one α H atom

Ethyl 2-methylpropanoate → CH_3CH_2ONa / CH_3CH_2OH → No reaction   (21-38)

PROBLEM 21.18 Draw the complete, detailed mechanism for the reaction shown here and predict the major product.

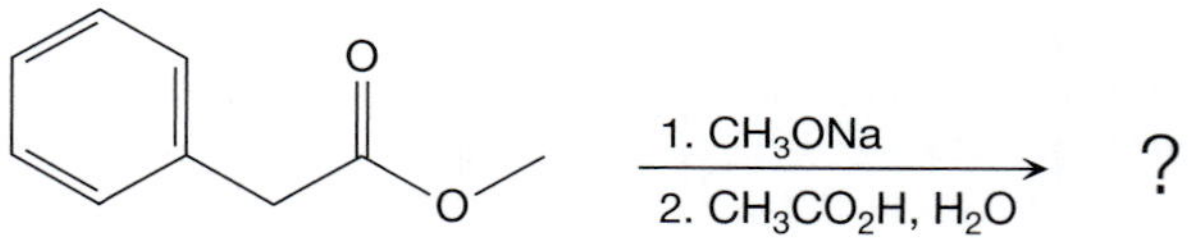

1. CH_3ONa
2. CH_3CO_2H, H_2O ?

SOLVED PROBLEM 21.19

Draw the ester that can be used to synthesize the compound shown here using a Claisen condensation reaction.

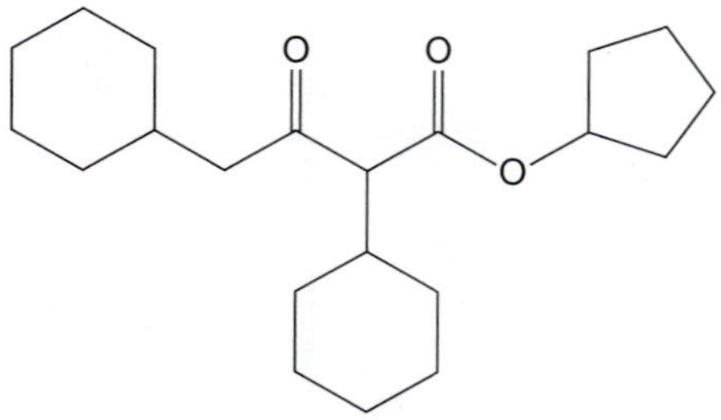

Think Which C—C bond in the β-keto ester product would have been formed in a Claisen condensation? Which C atom would have been bonded to the alkoxy leaving group?

Solve The new C—C bond is the one between the α and β carbons, as shown below on the left. Moreover, the leaving group would have been attached to the β C atom. Therefore, we can apply a transform that undoes a Claisen condensation by disconnecting that C—C bond and reattaching the leaving group. Doing so yields two ester precursors that are identical.

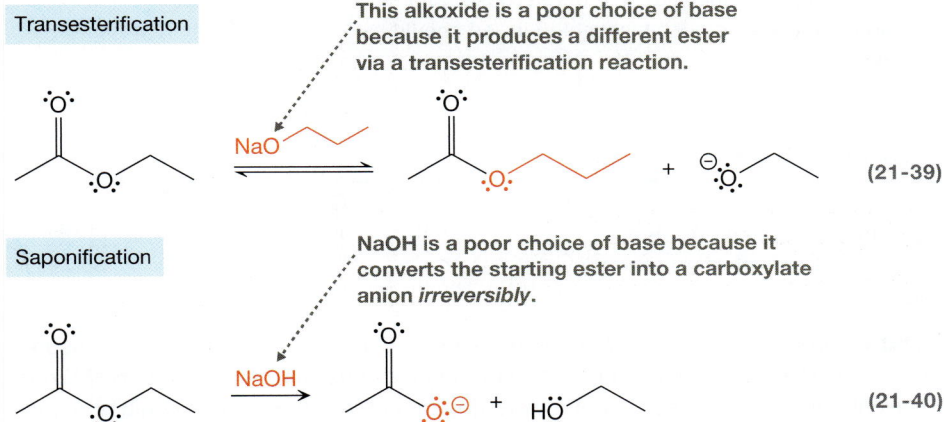

PROBLEM 21.20 Show how to synthesize this compound using a Claisen condensation reaction.

21.9a The Importance of the Solvent and Base in Claisen Condensations

In the Claisen condensation reaction in Equation 21-36, the base that is added (i.e., the ethoxide anion, $CH_3CH_2O^-$) is the same as the leaving group that departs in Step 3 of the mechanism in Equation 21-37. *Other choices of base could lead to undesired products.* If the base were an alkoxide other than ethoxide, for example, as in Equation 21-39, then transesterification would produce an ester that is different from the starting material. Alternatively, if hydroxide were used as the base, as in Equation 21-40, then *saponification* (Section 20.3) would irreversibly convert the ester into a carboxylate anion.

Transesterification

This alkoxide is a poor choice of base because it produces a different ester via a transesterification reaction.

(21-39)

Saponification

NaOH is a poor choice of base because it converts the starting ester into a carboxylate anion *irreversibly*.

(21-40)

YOUR TURN 21.7

Draw complete, detailed mechanisms for the reactions in Equations 21-39 and 21-40.

These problems are avoided when ethoxide is the base. Transesterification still occurs, but the transesterification product is identical to the initial ester.

YOUR TURN **21.8**

Draw the complete, detailed mechanism for the transesterification reaction that takes place among the species in Equation 21-36. What do you notice about the reactant and product esters?

Notice, too, that the solvent in Equation 21-36 (i.e., ethanol, CH_3CH_2OH) is the conjugate acid of the base that is added. *Other choices of solvent can also lead to an unwanted transesterification or saponification reaction.* If, for example, the solvent for the reaction in Equation 21-36 were another alcohol, such as propan-1-ol ($CH_3CH_2CH_2OH$), then the added ethoxide anion would deprotonate propan-1-ol to generate the 1-propoxide anion ($CH_3CH_2CH_2O^-$), thereby causing the transesterification reaction in Equation 21-39 to occur. Alternatively, if water were used as the solvent, then HO^- would be produced, and the saponification reaction in Equation 21-40 would take place. With ethanol as the solvent, this kind of proton transfer still takes place, but that proton transfer is not a problem because the base that is produced from it is the same as the base that is added as an overall reactant.

YOUR TURN **21.9**

Write the proton-transfer equilibrium that would take place if propan-1-ol were used as the solvent for the reaction in Equation 21-36. Do the same if water were used as the solvent and if ethanol were used as the solvent. What do you notice about the reactants and products of these proton transfer reactions?

In light of the issues that can arise with the wrong choice of base or solvent, Claisen condensation reactions are often carried out according to the following guideline:

For a Claisen condensation reaction, unwanted transesterification and saponification reactions can be avoided if the base that is added is the same as the alkoxide leaving group on the ester, and if the alcohol solvent is the conjugate acid of that base. In other words, the alkyl groups highlighted in red should be identical.

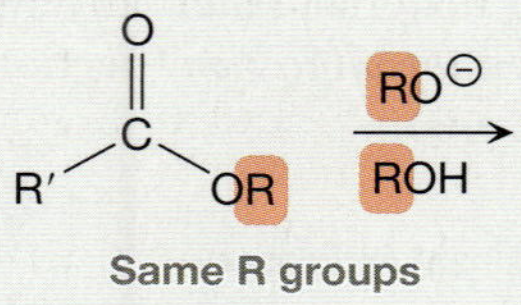

Same R groups

SOLVED PROBLEM 21.21

Identify the appropriate base–solvent pair to accomplish a Claisen condensation with the ester shown here.

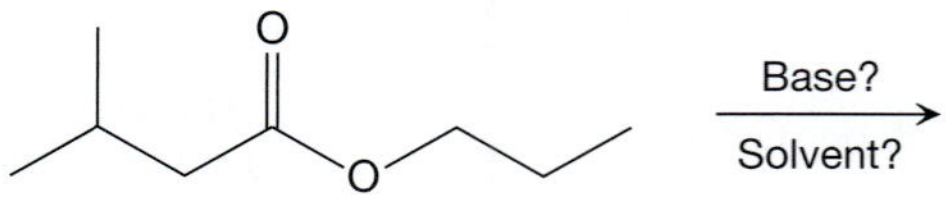

Base?
Solvent?

Think What is the leaving group on the ester? What choice of base would ensure that a nucleophilic addition–elimination would leave us with the same ester? What solvent could be used so that, if it were deprotonated, the resulting conjugate base would be the same as the base that is added?

Solve The leaving group is $CH_3CH_2CH_2O^-$. Therefore, if we were to add $NaOCH_2CH_2CH_3$ as the base, then transesterification would leave us with the same ester. $CH_3CH_2CH_2OH$ should therefore be the solvent, because if it were to be deprotonated, it would yield $CH_3CH_2CH_2O^-$, the same as the added base. Overall, the reaction would be as shown here:

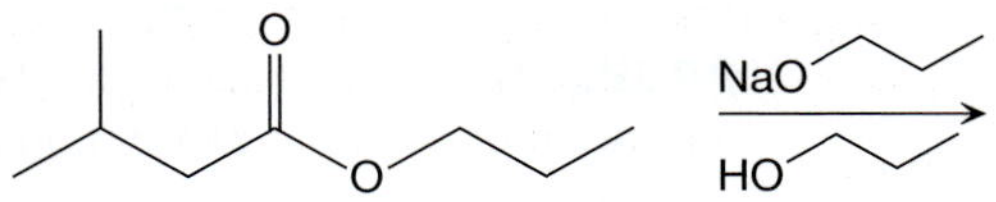

NaO
HO

PROBLEM 21.22 Identify the appropriate base–solvent pair to accomplish a Claisen condensation with each of the following esters.

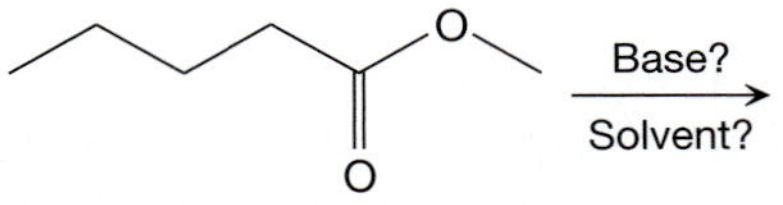

Base?
Solvent?

(a)

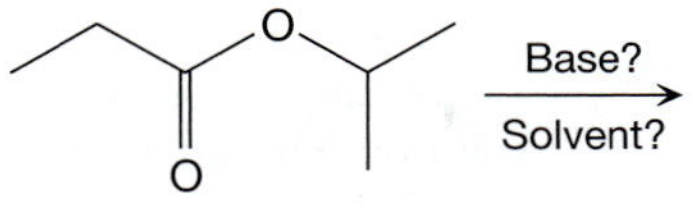

Base?
Solvent?

(b)

PROBLEM 21.23 The following Claisen condensation proceeds without an unwanted transesterification, even though the choice of base is not the same as the leaving group. Explain why.

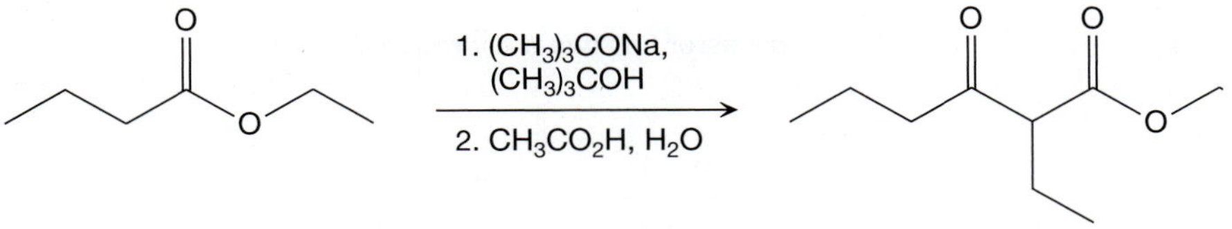

1. (CH3)3CONa, (CH3)3COH
2. CH3CO2H, H2O

21.9b Crossed Claisen Condensation Reactions

The Claisen condensation reactions discussed in Section 21.9a involve two molecules of the same ester—that is, they are *self-condensations*. A Claisen condensation can also involve two *different* esters—a so-called **crossed Claisen condensation reaction**. These reactions require special considerations, however, because crossed Claisen reactions can potentially lead to a mixture of condensation products (just as we saw with crossed *aldol* reactions in Section 18.7).

The reaction in Equation 21-41, for example, would produce a mixture of four β-keto esters, because two different ester enolates can be generated as nucleophiles and two different, uncharged esters can be attacked by those nucleophiles.

Four different β-keto esters can be produced from this crossed Claisen condensation reaction.

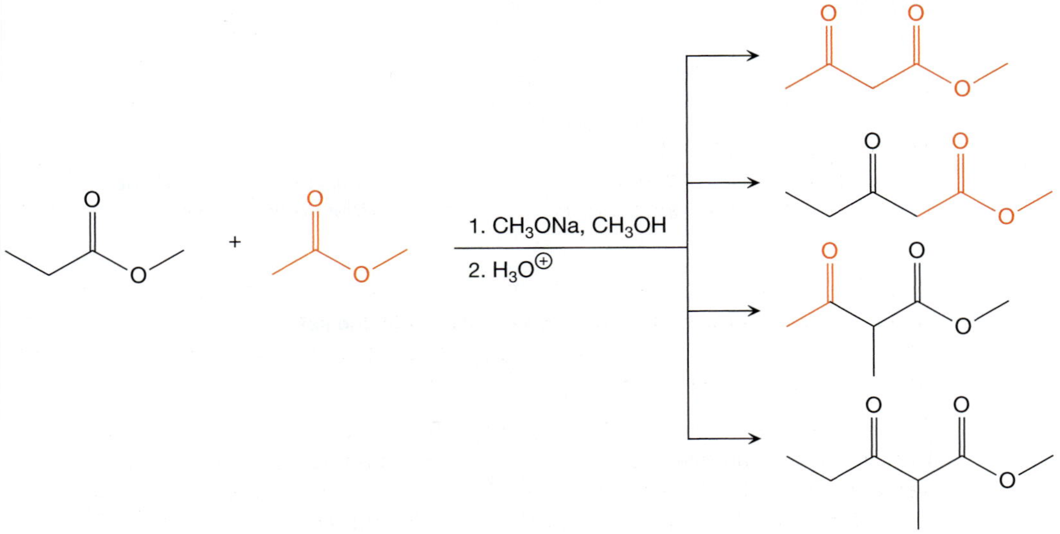

1. CH3ONa, CH3OH
2. H3O⊕

(21-41a)

(21-41b)

(21-41c)

(21-41d)

Diethyl carbonate

Ethyl formate

Ethyl benzoate

Diethyl oxalate

FIGURE 21-5 Esters with no α hydrogens
Without α hydrogens, these esters will not produce enolate anions that would act as nucleophiles in Claisen condensation reactions.

CONNECTIONS

1,3-Diphenylpropane-1,3-dione (Equation 21-43), commonly called dibenzoylmethane, is found in small amounts in licorice roots and exhibits a variety of anticancer properties.

PROBLEM 21.24 Draw a complete, detailed mechanism for the formation of *each* product in Equation 21-41.

One way to circumvent this problem is to ensure that one of the esters has no α hydrogens. (We previously used this strategy in Section 18.7 with crossed aldol reactions.) Some common examples of these kinds of esters are shown in Figure 21-5.

Because these esters have no α hydrogens, they can be dissolved in solution with an alkoxide base and no Claisen condensation takes place. A second ester that has α hydrogens can then be added slowly to this solution. As the ester with α hydrogens comes in contact with the base, its enolate anion is produced and the crossed Claisen condensation is initiated. An example is shown in Equation 21-42 using ethyl formate as the ester with no α hydrogens.

This ester has no α hydrogens.   **This ester has two α hydrogens.**

(21-42)

45%

Aldehyde and ketone enolate anions can also undergo nucleophilic addition–elimination with an ester, as shown in Equations 21-43 and 21-44, producing *1,3-dicarbonyl compounds*. The reaction in Equation 21-43 is synthetically useful because the ester has no α hydrogens. In Equation 21-44, the ester has α hydrogens, but the reaction is synthetically useful because an ester is significantly less acidic than a ketone. Therefore, when the ester is added to the ketone enolate that was first produced quantitatively by H_2N^-, the α protons largely remain on the ester. (See Your Turn 21.10.)

The ester has no α hydrogens.

1.

50 °C, 0.45 h

2. H_2SO_4, H_2O

(21-43)

1,3-Diphenylpropane-1,3,-dione (Dibenzoylmethane)
80%

The ester's α protons are less acidic than the ketone's.

1.

reflux, 2 h

2. H_2O, HCl

(21-44)

57%

Once the ketone enolate anion is produced in the reaction in Equation 21-44, the following proton transfer reaction is possible when the ester is added, and could result, therefore, in undesired condensation products.

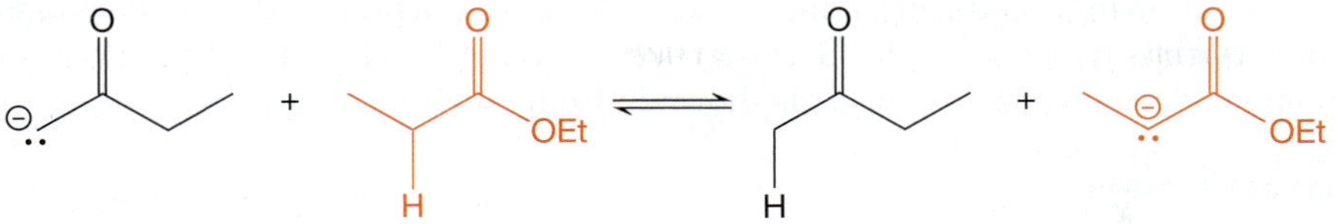

By looking up or estimating the appropriate pK_a values, determine which side of the reaction is favored and to what extent, and explain why this proton transfer reaction is not a major concern.

SOLVED PROBLEM 21.25

Show how you would synthesize this compound via a Claisen condensation reaction.

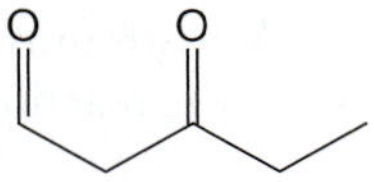

Think Which C—C bond can be "disconnected" in a transform that undoes a Claisen condensation? Does this transform suggest a self-Claisen or a crossed Claisen condensation reaction? Can we use a base that reversibly deprotonates an α hydrogen or should we use one that brings about an irreversible deprotonation?

Solve A transform that undoes a Claisen condensation disconnects a carbonyl's α C atom from the carbonyl group at the β position, as shown below; the carbonyl at the β position is part of an ester in the reactants. Given that the two carbonyl-containing precursors are not identical esters, a crossed Claisen condensation reaction is in order. One of the precursors is an ester that has no α hydrogens, which helps make a crossed Claisen reaction feasible.

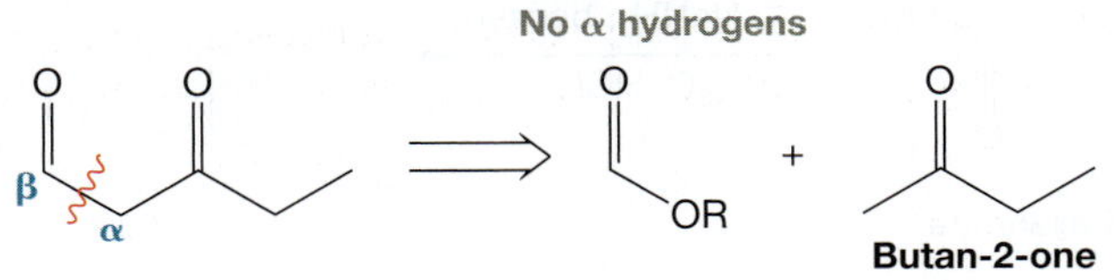

No α hydrogens

Butan-2-one

In the forward direction, butan-2-one would be converted into an enolate nucleophile via deprotonation at the α carbon. Because butan-2-one is unsymmetric, two isomeric enolate anions can be produced. This synthesis calls for the kinetic enolate, which can be produced by using LDA as the base. Overall, the synthesis would be written as follows:

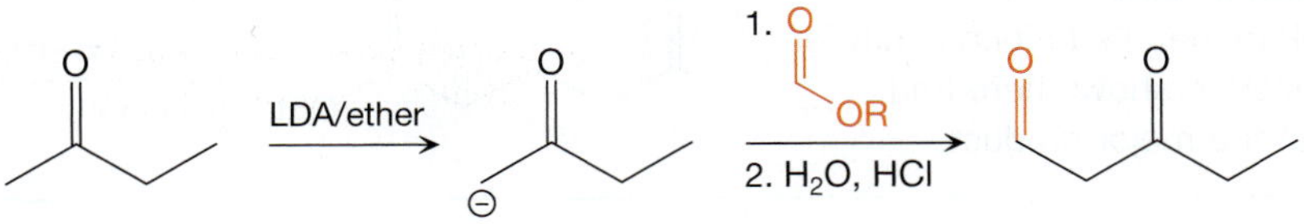

PROBLEM 21.26 Devise another synthesis of the target in Solved Problem 21.25 without using an ester of the form:

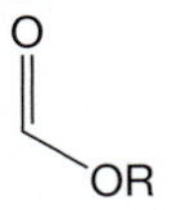

21.9c Intramolecular Claisen Condensation Reactions: The Dieckmann Condensation

The Claisen condensation, like the aldol reaction, has the potential for an *intramolecular* reaction, ultimately producing a ring. When it does, the reaction is called a **Dieckmann condensation**, named after the German chemist Walter Dieckmann (1869–1925). An example of one involving a diester is shown in Equation 21-45.

Dieckmann condensation

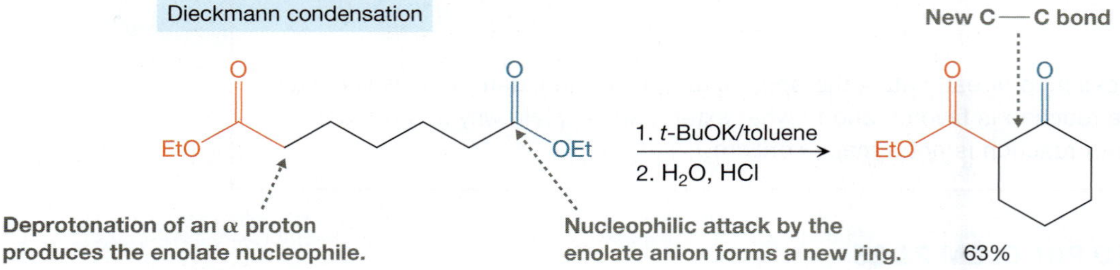

New C—C bond

1. *t*-BuOK/toluene
2. H_2O, HCl

(21-45)

63%

Deprotonation of an α proton produces the enolate nucleophile.

Nucleophilic attack by the enolate anion forms a new ring.

As with most cyclization reactions, Dieckmann condensations occur most readily when the reaction leads to the formation of a five- or six-membered ring.

PROBLEM 21.27 Draw the complete, detailed mechanism for the reaction in Equation 21-45.

An intramolecular nucleophilic addition–elimination reaction on an ester can also occur if the enolate anion derives from the portion of a molecule characteristic of a ketone or aldehyde. For example, methyl 6-oxo-6-phenylhexanoate cyclizes when treated with sodium amide to yield the five-membered ring product shown in Equation 21-46.

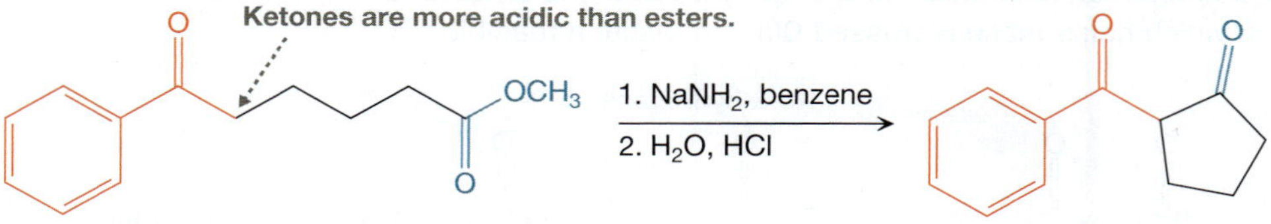

Ketones are more acidic than esters.

1. $NaNH_2$, benzene
2. H_2O, HCl

(21-46)

Methyl 6-oxo-6-phenylhexanoate

90%

Notice in the reactant in Equation 21-46 that there are H atoms on the carbons that are α to each carbonyl group. Deprotonation takes place almost exclusively at the α carbon on the left, however, because *ketones (and aldehydes) are substantially more acidic than esters.* (Compare their pK_a values from Appendix A.)

PROBLEM 21.28 Draw the complete, detailed mechanism for the reaction shown here and predict the major product.

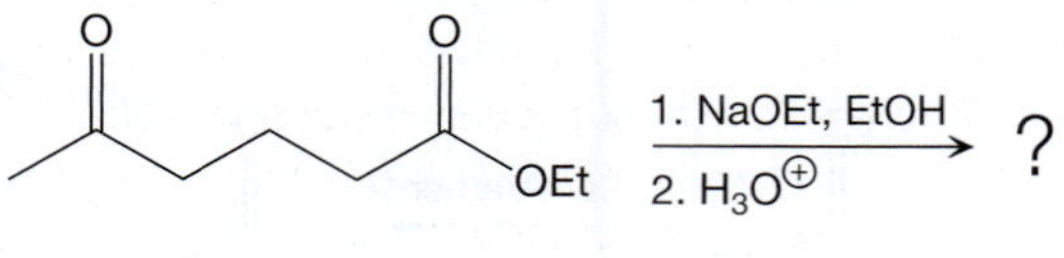

1. NaOEt, EtOH
2. $H_3O^{\oplus}$

?

PROBLEM 21.29 Draw the acyclic dicarbonyl compound that can be used to synthesize the cyclic β-keto aldehyde shown here.

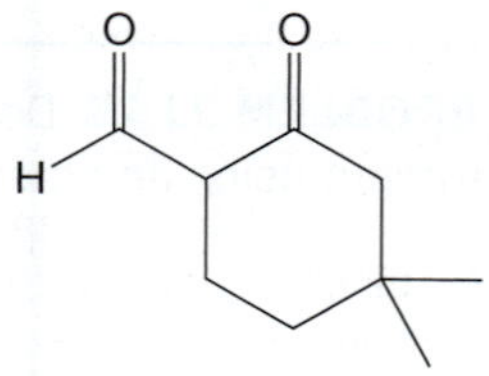

Biological Claisen Condensations

When people consume more calories in their diet than they burn, the excess fuel (typically from carbohydrates) is stored as fat. A fat is a triglyceride (Section 2.11a), which is a triester formed between glycerol (a triol) and three fatty acids. Each fatty acid is a carboxylic acid whose carbon chain has an even number of carbons, most commonly 14–20 carbons long. How do carbohydrates become converted to fat, and why do the carbon chains of fatty acids appear in increments of two carbons? The answer has to do with biological Claisen condensation reactions. An example between malonyl acyl carrier protein (ACP) and acetyl synthase is shown in Steps 1 and 2 of the following scheme. The acyl group indicated in malonyl ACP is produced from glycolysis, the metabolic pathway that breaks down glucose.

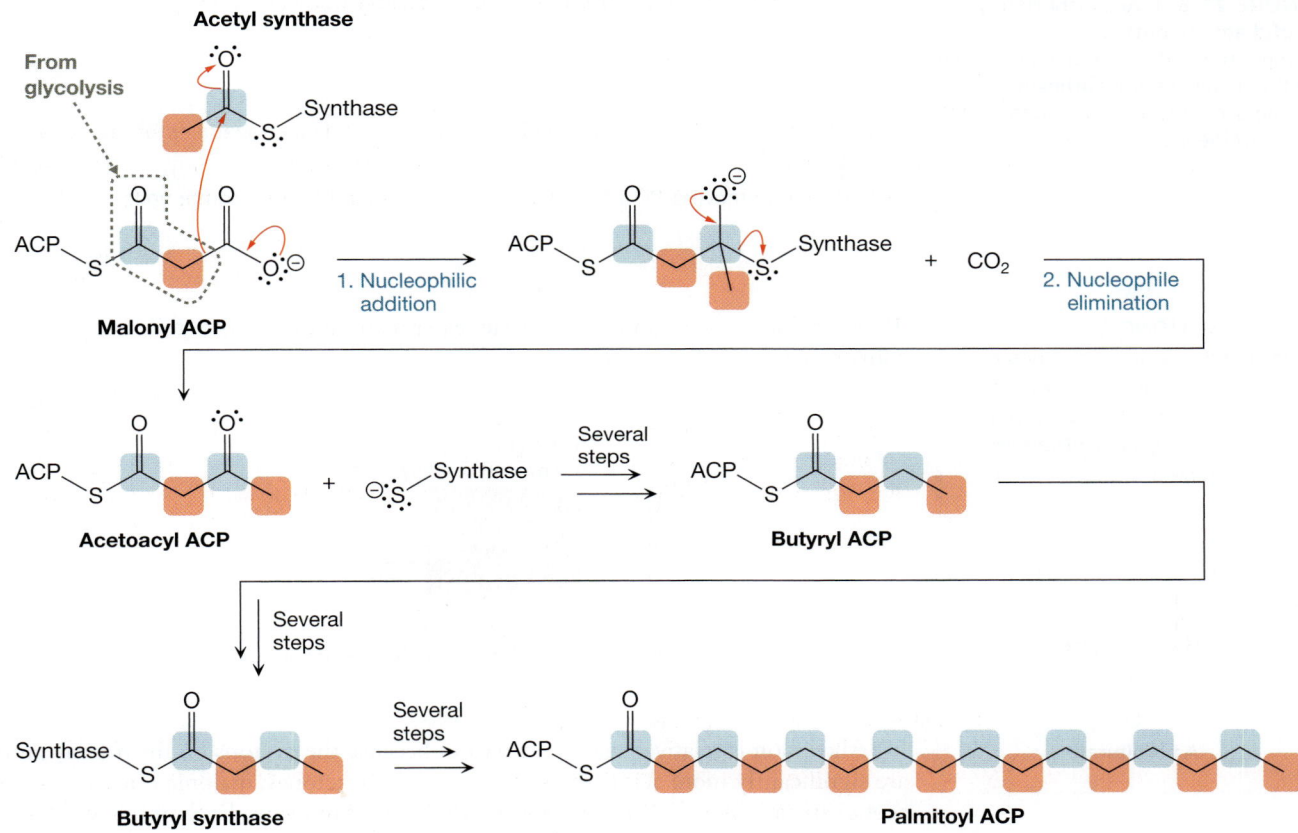

In Step 1, the α C of malonyl ACP undergoes nucleophilic addition to the carbonyl C of acetyl synthase to produce a tetrahedral intermediate. In Step 2, the S—Synthase leaving group is eliminated and the C=O bond is regenerated, producing acetoacyl ACP, a β-keto thioester. Through several subsequent steps, acetoacyl ACP is converted to butyryl ACP, which is the sulfur analog to the fatty acid butyric acid. Butyryl ACP is converted to butyryl synthase, which can participate in another Claisen condensation with malonyl ACP. Each time this process is repeated, the carbon chain grows by two, which explains the even number of carbons.

This scheme is validated in part by an experiment in which organisms are fed ^{14}C-labeled acetic acid (CH_3CO_2H). A cell uses acetic acid to synthesize acetyl-coenzyme A, from which both malonyl ACP and acetyl synthase are derived, so the ^{14}C label appears in the synthesized fatty acid. When the carbonyl C of acetic acid was labeled, the ^{14}C label appeared at the locations highlighted in blue above. When the label was at the α C, the label appeared at the locations highlighted in red.

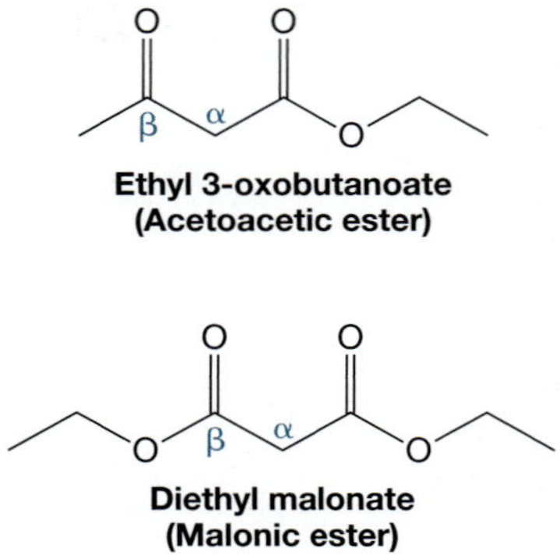

**Ethyl 3-oxobutanoate
(Acetoacetic ester)**

**Diethyl malonate
(Malonic ester)**

FIGURE 21-6 Two synthetically useful active methylene compounds Ethyl 3-oxobutanoate and diethyl malonate are starting materials for the acetoacetic ester and malonic ester syntheses, respectively.

CONNECTIONS Diethyl malonate is used to synthesize barbiturates, depressant drugs that act on the central nervous system. The generic structure is shown here.

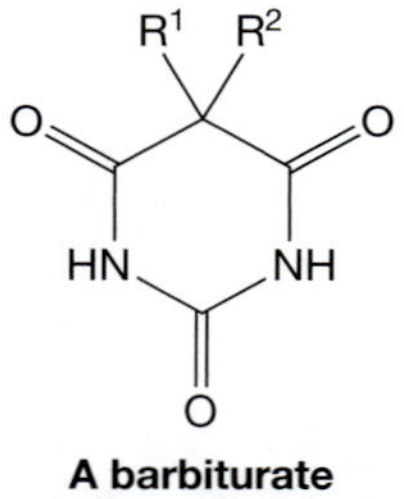

A barbiturate

21.10 Organic Synthesis: Decarboxylation, the Malonic Ester Synthesis, and the Acetoacetic Ester Synthesis

The main focus of Section 21.9 was producing β-keto esters via Claisen condensation reactions. Ethyl 3-oxobutanoate (acetoacetic ester), shown in Figure 21-6, is a β-keto ester that is particularly useful as a starting material in organic syntheses. Diethyl malonate (malonic ester), a β-diester, is structurally similar to acetoacetic ester and is also quite useful synthetically.

Their usefulness stems from our ability to carry out the following reactions:

- We can alkylate the α carbon of malonic ester and acetoacetic ester relatively easily.
- We can remove a CO_2Et group entirely from each of these compounds.

Thus, malonic ester and acetoacetic ester can undergo the following general transformation:

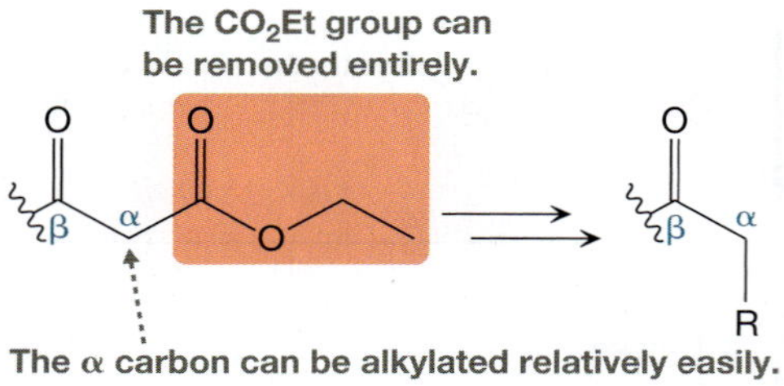

The CO_2Et group can be removed entirely.

The α carbon can be alkylated relatively easily.

Alkylation is relatively easy to carry out because the protons on the α CH_2 group are significantly more acidic than normal esters or ketones; malonic ester and acetoacetic ester are therefore called **active methylene compounds**. Each of the two C=O groups flanking the α CH_2 group significantly enhances the acidity via resonance and inductive stabilization in the enolate conjugate base. The α protons are sufficiently acidic that even a moderately strong base such as $CH_3CH_2O^-$ will produce the enolate anion *quantitatively* (Equation 21-47). That negatively charged enolate anion is highly nucleophilic and will react with an alkyl halide in an S_N2 reaction to produce the alkylated product.

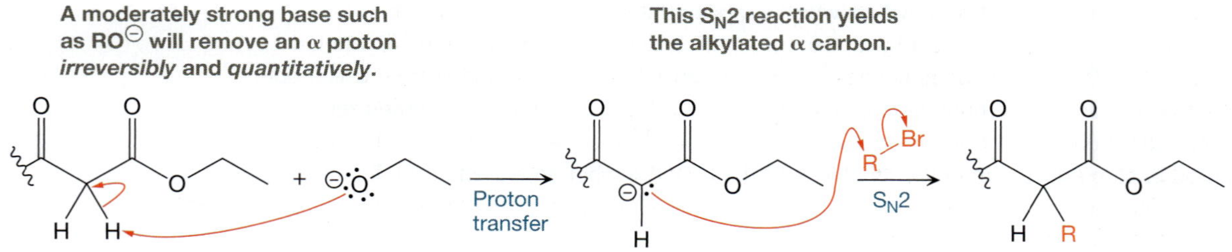

A moderately strong base such as $RO^{\ominus}$ will remove an α proton *irreversibly* and *quantitatively*.

This S_N2 reaction yields the alkylated α carbon.

Proton transfer

S_N2

(21-47)

Verify the preceding statement that an active methylene compound can be quantitatively deprotonated by using the appropriate pK_a values to determine the side of the following reaction that is favored. To what extent is that side favored?

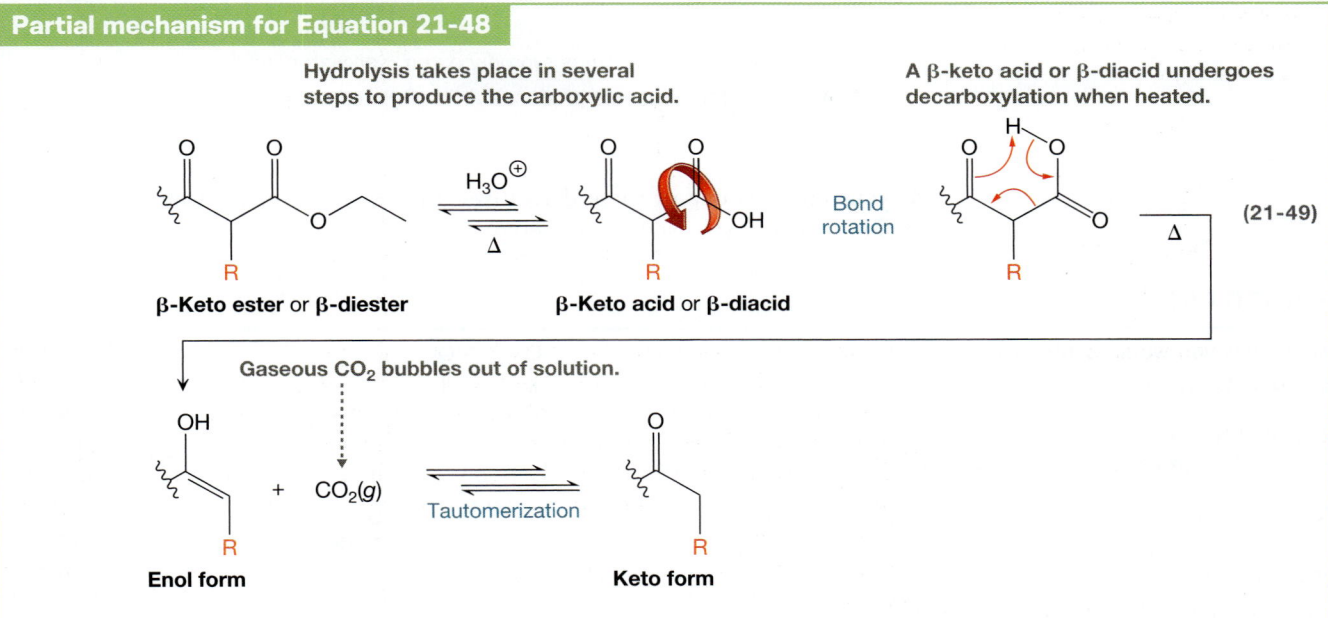

pK_a = [ ]

pK_a = [ ]

To remove the ester group entirely from a β-keto ester or a β-diester, we can simply heat it under aqueous, acidic conditions, according to Equation 21-48.

The CO$_2$Et group is removed entirely.

$$\xrightarrow[\Delta]{H_3O^\oplus}$$

(21-48)

The partial mechanism for this reaction is shown in Equation 21-49.

Partial mechanism for Equation 21-48

Hydrolysis takes place in several steps to produce the carboxylic acid.

A β-keto acid or β-diacid undergoes decarboxylation when heated.

$$\underset{\beta\text{-Keto ester or }\beta\text{-diester}}{} \xrightarrow[\Delta]{H_3O^\oplus} \underset{\beta\text{-Keto acid or }\beta\text{-diacid}}{} \xrightarrow{\text{Bond rotation}} \xrightarrow{\Delta}$$

(21-49)

Gaseous CO$_2$ bubbles out of solution.

$$\underset{\textbf{Enol form}}{} + CO_2(g) \underset{\textit{Tautomerization}}{\rightleftharpoons} \underset{\textbf{Keto form}}{}$$

The acidic, aqueous conditions hydrolyze esters to carboxylic acids (review Section 21.7)—more specifically, a β-keto ester is converted to a **β-keto acid** and a β-diester is converted to a **β-diacid**. Heating the β-keto acid or β-diacid drives a **decarboxylation** in which gaseous CO$_2$ is lost and an enol is formed. The enol rapidly tautomerizes to the keto form.

Beginning with acetoacetic ester, we can put all of these steps together in what is called an **acetoacetic ester synthesis**, as shown in Equation 21-50. Similarly, a **malonic ester synthesis** is shown in Equation 21-51.

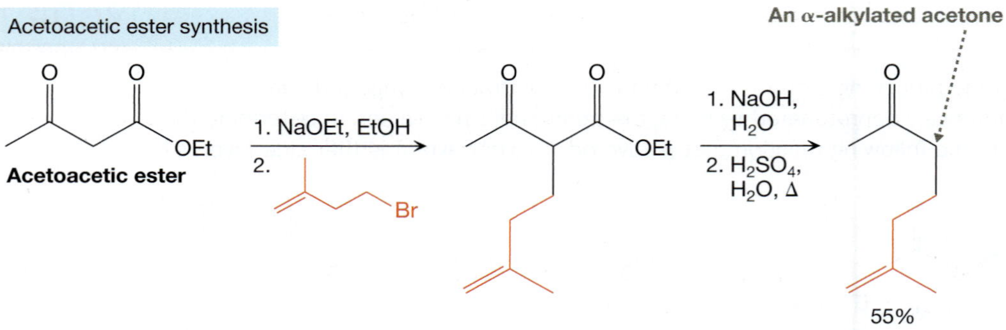

Acetoacetic ester synthesis

Acetoacetic ester

1. NaOEt, EtOH
2. [3-bromo-2-methylpropene structure]

An α-alkylated acetone

1. NaOH, H₂O
2. H₂SO₄, H₂O, Δ

(21-50)

55%

Malonic ester synthesis

EtO ... OEt
Malonic ester

1. NaOEt, EtOH
2. Br—[cyclohexylmethyl]

An α-alkylated acetic acid

1. KOH, H₂O
2. H₂SO₄, H₂O, Δ

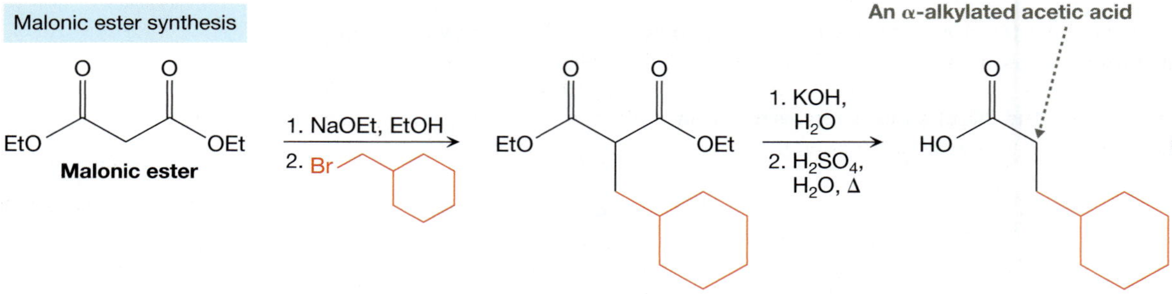

(21-51)

The acetoacetic ester and malonic ester syntheses have characteristic products.

- An acetoacetic ester synthesis produces an α-alkylated acetone.
- A malonic ester synthesis produces an α-alkylated acetic acid.

If your target is one of these derivatives, then consider using acetoacetic ester or malonic ester as your starting material.

SOLVED PROBLEM 21.30

Show how you would synthesize hexan-2-one beginning with acetoacetic ester.

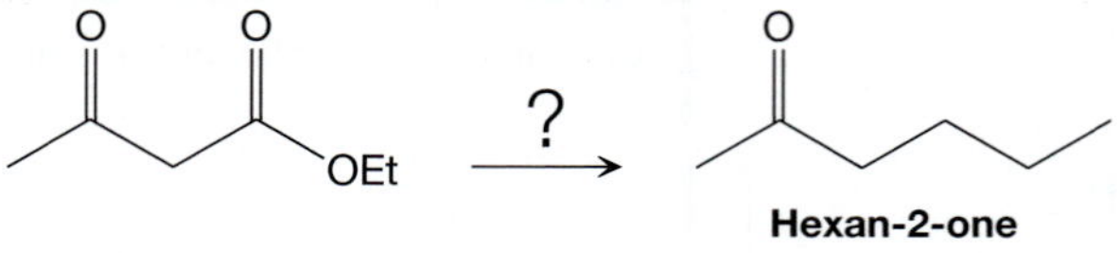

? **Hexan-2-one**

Think What would the precursor look like before decarboxylation? How could that precursor be produced from acetoacetic ester?

Solve Hexan-2-one is an α-alkylated acetone, so we can first perform a transform that undoes a decarboxylation, given that decarboxylation is the last step in an acetoacetic ester synthesis. This is shown at the right.

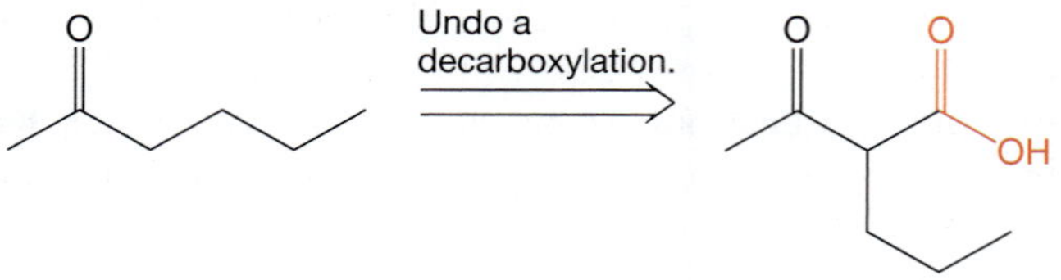

Continuing the retrosynthesis, let's undo a hydrolysis and an α-alkylation to bring us back to the starting acetoacetic ester.

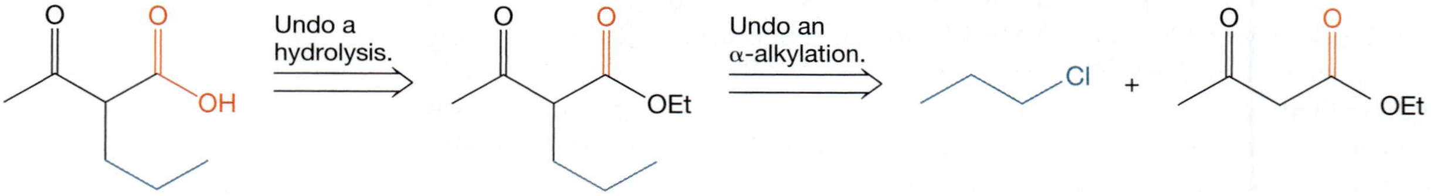

Undo a decarboxylation.

Undo a hydrolysis.

Undo an α-alkylation.

In the forward direction, the synthesis would be written as follows:

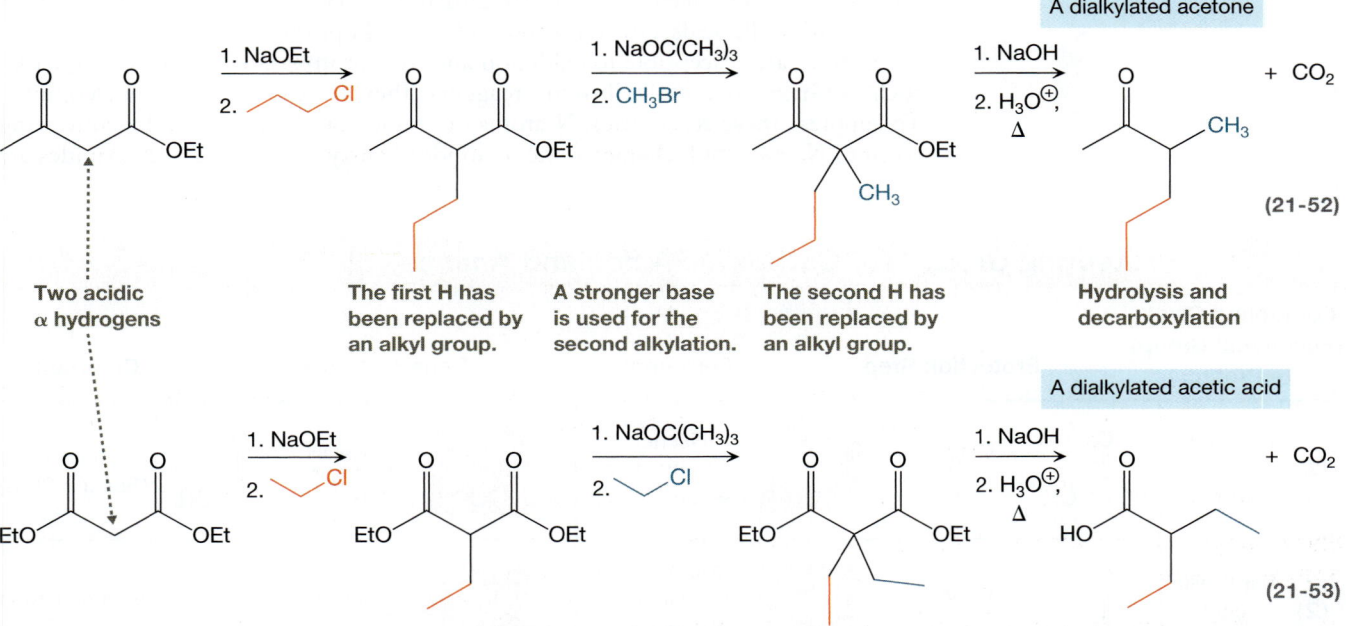

PROBLEM 21.31 Outline a synthesis of pent-4-enoic acid from either acetoacetic ester or malonic ester.

Pent-4-enoic acid

Both the acetoacetic ester synthesis and the malonic ester synthesis also lend themselves to *dialkylation* of the active methylene group, given that there are two acidic hydrogens. Examples are shown in Equations 21-52 and 21-53. Notice that the two alkyl groups can be the same or different.

A dialkylated acetone

$+ CO_2$

(21-52)

| Two acidic α hydrogens | The first H has been replaced by an alkyl group. | A stronger base is used for the second alkylation. | The second H has been replaced by an alkyl group. | Hydrolysis and decarboxylation |

A dialkylated acetic acid

$+ CO_2$

(21-53)

In both cases, alkylation of the methylene C is carried out twice prior to hydrolysis and decarboxylation. The base in the second alkylation is the *tert*-butoxide anion, $(CH_3)_3CO^-$, however, because the second deprotonation is more difficult than the first, so a stronger base compensates. The second deprotonation is more difficult because the alkyl group added in the first alkylation is electron donating, and thus destabilizes the resulting enolate anion.

YOUR TURN **21.12**

In the figure shown here, draw an arrow representing the inductive electron donation by the alkyl group. Do the alkyl group's effects increase or decrease the concentration of negative charge?

PROBLEM 21.32 Show how you would synthesize 2-ethylhexanoic acid from malonic ester.

21.11 Organic Synthesis: Protecting Carboxylic Acids and Amines

Chapter 19 introduced the concept of the *protecting group*. The general approach is as follows: In the anticipation of an undesired reaction involving a particular functional group, (1) protect that functional group by temporarily converting it into another one that is unreactive under the conditions for the desired reaction, (2) perform the desired reaction, and (3) deprotect the functional group by converting it back to its original form. In Chapter 19, we specifically examined how to protect C=O groups in ketones and aldehydes and OH groups in alcohols. Table 21-2 provides examples of how to protect and deprotect CO_2H groups in carboxylic acids and N atoms of amines.

The acidic nature of a carboxylic acid can interfere with a synthesis, so CO_2H groups are often protected by converting them to esters (Entries 1–3 in Table 21-2). Unlike carboxylic acids, esters have no acidic O—H protons.

Amines are susceptible to oxidation and to deprotonation by very strong bases, such as Grignard and alkyllithium reagents. They are also moderate nucleophiles. To suppress these reactivities, N atoms of amines can be protected by converting them to N—C=O, characteristic of amides (Entry 4 in Table 21-2). Amides are

TABLE 21-2 Protecting Groups for Carboxylic Acids and Amines

Compound Class (Functional Group) to Protect	Protection Step	Comments	Deprotection Step	Comments
(1) Carboxylic acid (Carboxy)	$R-CO-OH \xrightarrow[\text{H}^\oplus]{CH_3OH} R-CO-OCH_3$	Fischer esterification (Section 21.7)	$R-CO-OCH_3 \xrightarrow[\text{2. H}^\oplus]{\text{1. NaOH}} R-CO-OH$	Saponification (Section 20.3)
(2) Carboxylic acid (Carboxy)	$R-CO-OH \xrightarrow{CH_2N_2} R-CO-OCH_3$	Diazomethane is toxic and explosive. (Section 10.5)	$R-CO-OCH_3 \xrightarrow[\text{2. H}^\oplus]{\text{1. NaOH}} R-CO-OH$	Saponification (Section 20.3)
(3) Carboxylic acid (Carboxy)	$R-CO-OH \xrightarrow[\text{H}^\oplus]{HO-CH_2-Ph} R-CO-OBz$	Fischer esterification (Section 21.7)	$R-CO-OBz \xrightarrow[\text{Pd}]{H_2} R-CO-OH + \text{toluene}$	Catalytic hydrogenation (Section 19.5)
(4) Amine (Amino)	$NH \xrightarrow[\text{Pyridine}]{\text{Cl-CO-CH}_3} N-CO-CH_3$	Aminolysis (Section 21.3)	$N-CO-CH_3 \xrightarrow[\Delta]{H_3O^\oplus} NH$	Acid-catalyzed hydrolysis (Section 21.7)
(5) Amine (Amino)	$NH \xrightarrow[\text{NaHCO}_3]{\text{Boc}_2\text{O}} NBoc$	Carbamate formation (Section 21.11)	$NBoc \xrightarrow{H_3O^\oplus} NH$	Acid-catalyzed hydrolysis (Section 21.7)

less susceptible to oxidation and are less nucleophilic because of the resonance stabilization involving the carbonyl group and the lone pair on nitrogen (review Fig. 21-2, p. 1051). Furthermore, secondary amines result in tertiary amides, which have no N—H bonds and are therefore *not* susceptible at all to deprotonation of the nitrogen atom. Similarly, amines can be protected in the form of a **carbamate** (RO—CO—NR₂). Entry 5 of Table 21-2 shows, for example, how an amine N can be protected by a *tert*-butoxycarbonyl [—CO₂C(CH₃)₃, Boc] group using di-*tert*-butyl dicarbonate (Boc anhydride, Boc₂O).

You must be aware of the conditions necessary to protect and deprotect to ensure that no other undesired reactions take place. Notice, for example, that acidic conditions are required for protection via a Fischer esterification and deprotection via acid-catalyzed hydrolysis, which could result in undesired reactions with alcohols and alkenes. Protection via diazomethane could result in undesired reactions with alkenes (Section 12.2), and deprotection via catalytic hydrogenation could result in undesired reactions with alkenes, alkynes, and aldehydes (Section 19.5).

Even though an amine can be protected in the form of an amide or carbamate, the carbamate form is generally preferred because it is easier to deprotect than an amide. This is one of the reasons that the Boc group was chosen for the protection of amino groups in the automated synthesis of peptides (see Section 21.13).

SOLVED PROBLEM 21.33

Show how to carry out the following synthesis.

Think Without considering undesired side reactions, how might you produce a ketone in a way that also forms a C—C bond to the carbonyl C? When making the compound necessary to carry out that reaction, are there any undesired side reactions to be concerned about? How could a protecting group be incorporated to prevent such an undesired reaction from occurring?

Solve As shown in the retrosynthetic analysis below, we can produce a ketone from an acid chloride. The acid chloride can be produced from the corresponding carboxylic acid, which can be produced from a primary alcohol.

However, it would be difficult to react the dicarboxylic acid in a way that produces just the desired monoacid chloride.

To circumvent this problem, we can first convert the carboxylic acid into an ester, as shown in the first step below:

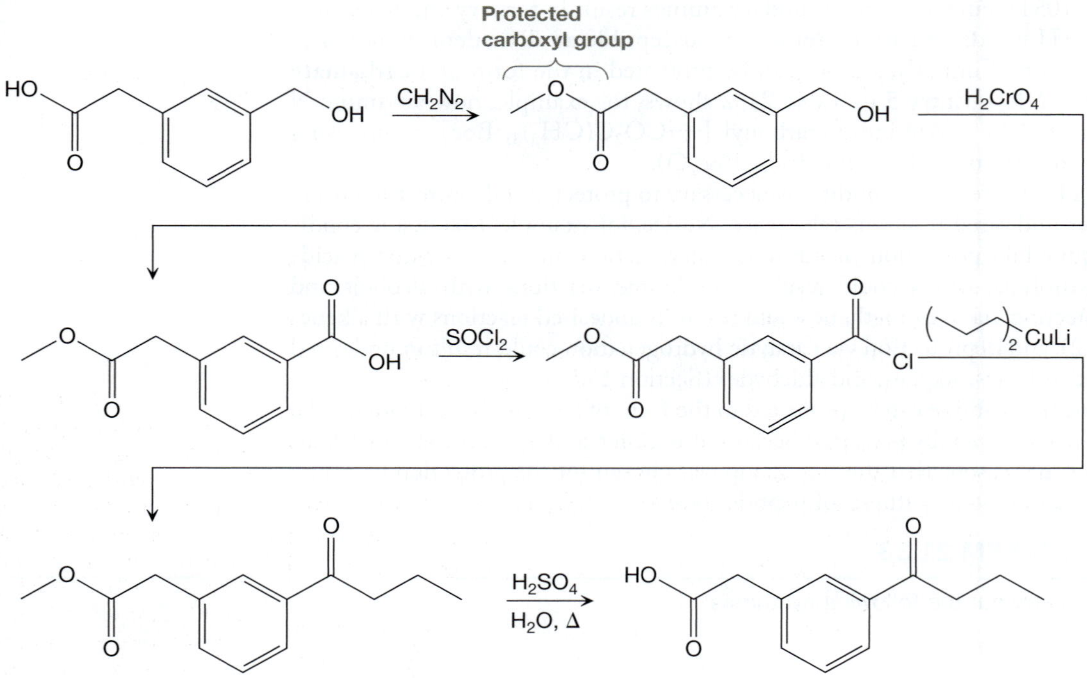

Then, oxidation of the primary alcohol produces just the monocarboxylic acid, not the diacid. The monoacid is then converted to the acid chloride using $SOCl_2$, after which $(CH_3CH_2CH_2)_2CuLi$ produces the desired ketone. Finally, hydrolysis of the ester restores the carboxylic acid.

PROBLEM 21.34 Show how to carry out the following synthesis.

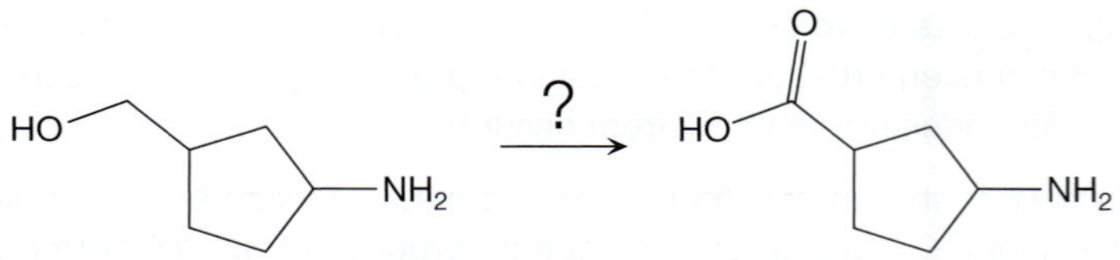

THE ORGANIC CHEMISTRY OF BIOMOLECULES

21.12 Determining a Protein's Primary Structure via Amino Acid Sequencing: Edman Degradation

Recall from Section 1.14a that proteins are large molecules constructed from α-amino acids, and each of the 20 possible amino acids is distinguished by the identity of its side group, R. As shown in Equation 21-54, each pair of amino acids in a protein is joined via an O═C—N functional group (characteristic of amides), also called a **peptide linkage** or a **peptide bond**.

The fact that amino acids are joined by groups characteristic of amides is critical to the stability of proteins. Recall from Figure 21-1 (p. 1051) that amides are among the least reactive acid derivatives, so they require relatively extreme conditions to be hydrolyzed. As shown in Equation 21-54, for example, the *complete* hydrolysis of a protein requires strongly acidic conditions and elevated temperatures.

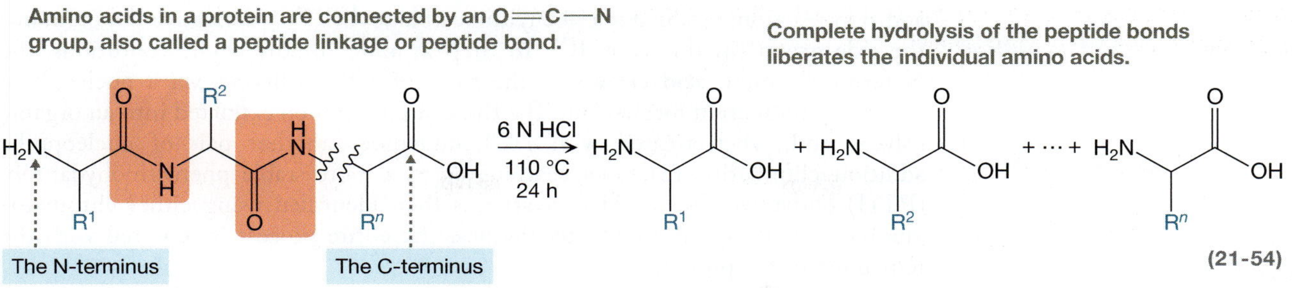

Amino acids in a protein are connected by an O=C—N group, also called a peptide linkage or peptide bond.

Complete hydrolysis of the peptide bonds liberates the individual amino acids.

The N-terminus

The C-terminus

$$\xrightarrow[\substack{110\ °C \\ 24\ h}]{6\ N\ HCl}$$

(21-54)

A complete hydrolysis of a protein can be used to determine the relative amounts of each type of amino acid in a protein. To do so, the product mixture is injected into an *amino acid analyzer*. However, complete hydrolysis does not reveal the **primary structure** of the protein—namely, the specific sequence of amino acids from the **N-terminus** (the end with the free amino group) to the **C-terminus** (the end with the carboxyl group). This sequence is important, because it is what gives a protein its key function; frequently, proteins that differ by the identity of just one amino acid at a key site, or **residue**, have very different biological properties.

One way to determine the amino acid sequence of a protein is with an **Edman degradation**. The basis of an Edman degradation is to remove and detect one amino acid at a time, from the N-terminus. The removal of each N-terminal amino acid requires the three steps shown in Figure 21-7.

In Step 1, the protein is treated with phenyl isothiocyanate. The free NH$_2$ group of the N-terminal amino acid adds to the polar C=N bond in a nucleophilic

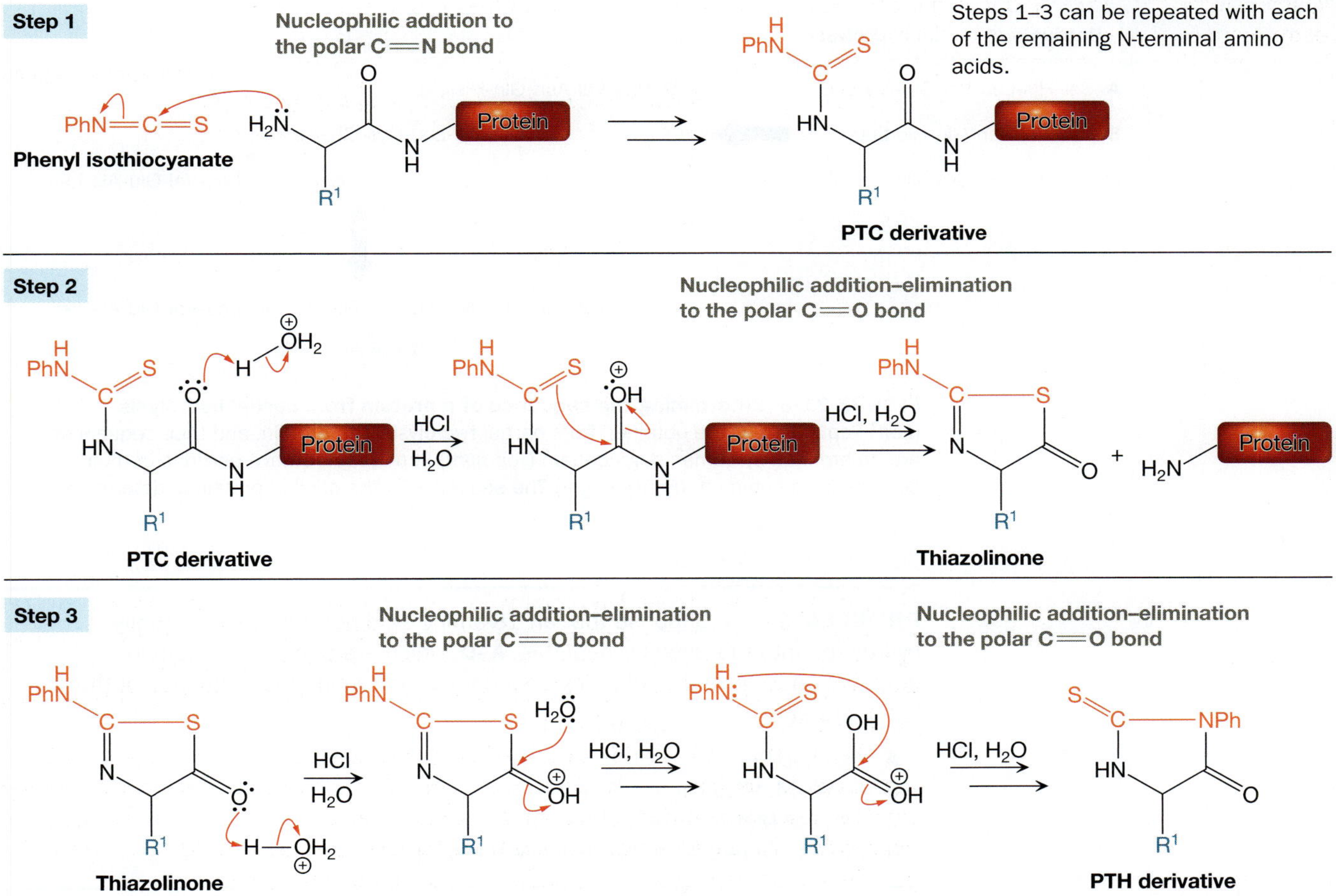

FIGURE 21-7 Partial mechanism for the Edman degradation (*Step 1*) The amino group of the N-terminal amino acid undergoes nucleophilic addition to phenyl isothiocyanate to yield the corresponding PTC derivative. (*Step 2*) The PTC derivative cyclizes when treated with HCl to yield the corresponding thiazolinone. (*Step 3*) Treatment of the thiazolinone with HCl produces the corresponding PTH derivative, which can be detected using chromatography or spectroscopy. Steps 1–3 can be repeated with each of the remaining N-terminal amino acids.

Step 1

Nucleophilic addition to the polar C=N bond

Phenyl isothiocyanate

PTC derivative

Step 2

Nucleophilic addition–elimination to the polar C=O bond

PTC derivative

Thiazolinone

Step 3

Nucleophilic addition–elimination to the polar C=O bond

Nucleophilic addition–elimination to the polar C=O bond

Thiazolinone

PTH derivative

addition reaction, producing a phenylthiocarbamoyl (PTC) derivative that contains the side group (in this case, R¹). In Step 2, under mildly acidic conditions, the N-terminal amino acid cleaves in the form of a thiazolinone, via a nucleophilic addition–elimination mechanism. The thiazolinone is then extracted into an organic solvent and, when treated with HCl, undergoes another pair of nucleophilic addition–elimination reactions, rearranging to a more stable phenylthiohydantoin (PTH) derivative. The PTH derivative is then identified using either chromatographic or mass spectrometric techniques. The entire process is repeated with the remainder of the protein.

Theoretically, the entire sequence of a protein can be determined in this manner. However, sequencing proteins via Edman degradation becomes impractical when the protein exceeds roughly 30 amino acids, because side products accumulate and interfere with the results. In these cases, the protein can be *partially* hydrolyzed using dilute acid, in which case hydrolysis takes place at essentially random locations in the protein. The smaller polypeptides that are produced then can be sequenced using an Edman degradation, and the sequence of the original protein subsequently can be pieced together from the results.

Suppose, for example, that the partial hydrolysis of a 15-amino acid polypeptide yielded three smaller polypeptides, **A–C**, whose sequences were determined to be the ones shown on the left of Figure 21-8. As shown on the top right of Figure 21-8, certain portions of the peptides have the same sequence—specifically, the Ser-His sequence in peptides **A** and **C** (red), and the Gln-His-Leu sequence in peptides **B** and **C** (blue). Assuming those sequences come from the same portions of the original protein, we can piece together the overall sequence, as shown at the bottom right of Figure 21-8.

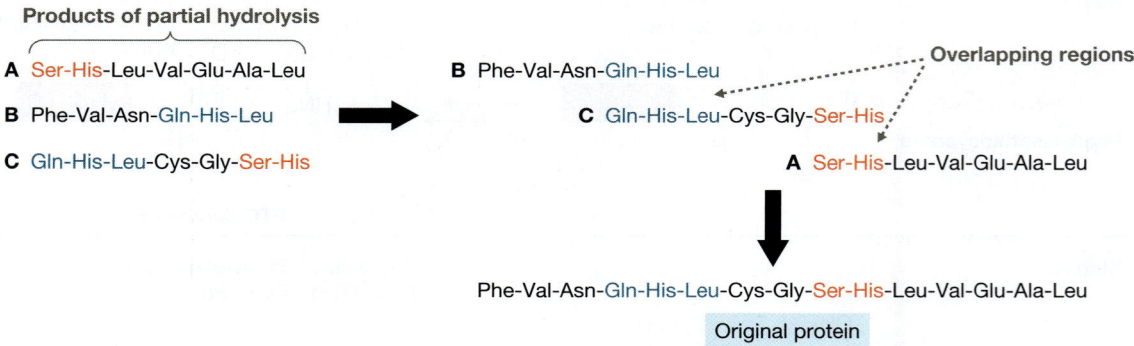

FIGURE 21-8 Determining the sequence of a protein from partial hydrolysis
(*Left*) Peptides **A–C** are obtained from partial hydrolysis of a protein, and their sequences are determined by Edman degradation. (*Top right*) Peptides **A–C** have portions of their sequences in common. (*Bottom right*) The sequence of the original protein is determined.

PROBLEM 21.35 Suppose that an 18-amino acid polypeptide is partially hydrolyzed into four smaller peptides, **A–D**, which are analyzed by Edman degradation to have the following sequences. Determine the sequence of the initial protein.

A Asp-Asp-Ser
B Ile-Val-Met-Pro-Val
C Asp-Ser-Met-Trp-Pro-Cys-Pro-Asn
D Pro-Asn-Gln-Asp-Cys-Phe-Ile-Val-Met-Pro-Val

21.13 Synthesis of Peptides

Historically, people with type 1 diabetes required routine injections of a protein called insulin that was isolated from pigs. In 1978, however, human insulin was synthesized, and soon after became less expensive and more readily available than pig insulin.

To synthesize a protein, amino acids must be joined together by peptide bonds, produced from the NH_2 group of one amino acid and the CO_2H group of another. Under normal conditions, these reactions are slow (see Section 21.2), so to facilitate them, **dicyclohexylcarbodiimide (DCC)** is typically used, as shown in Equation 21-55.

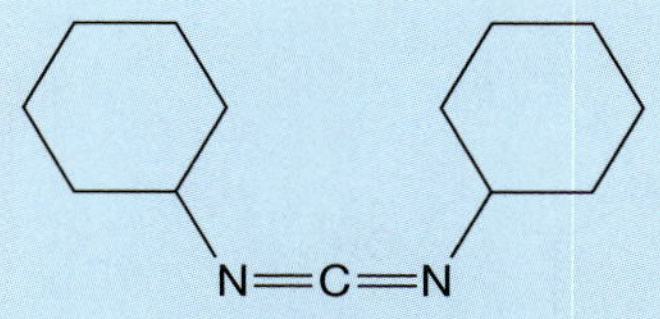

DCC

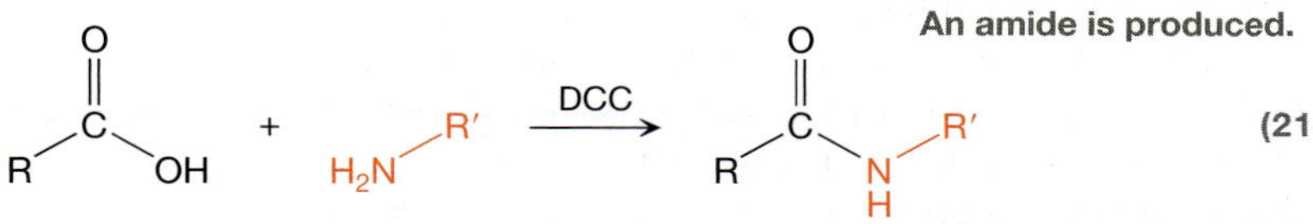

An amide is produced.

(21-55)

The mechanism for this reaction is shown in Equation 21-56.

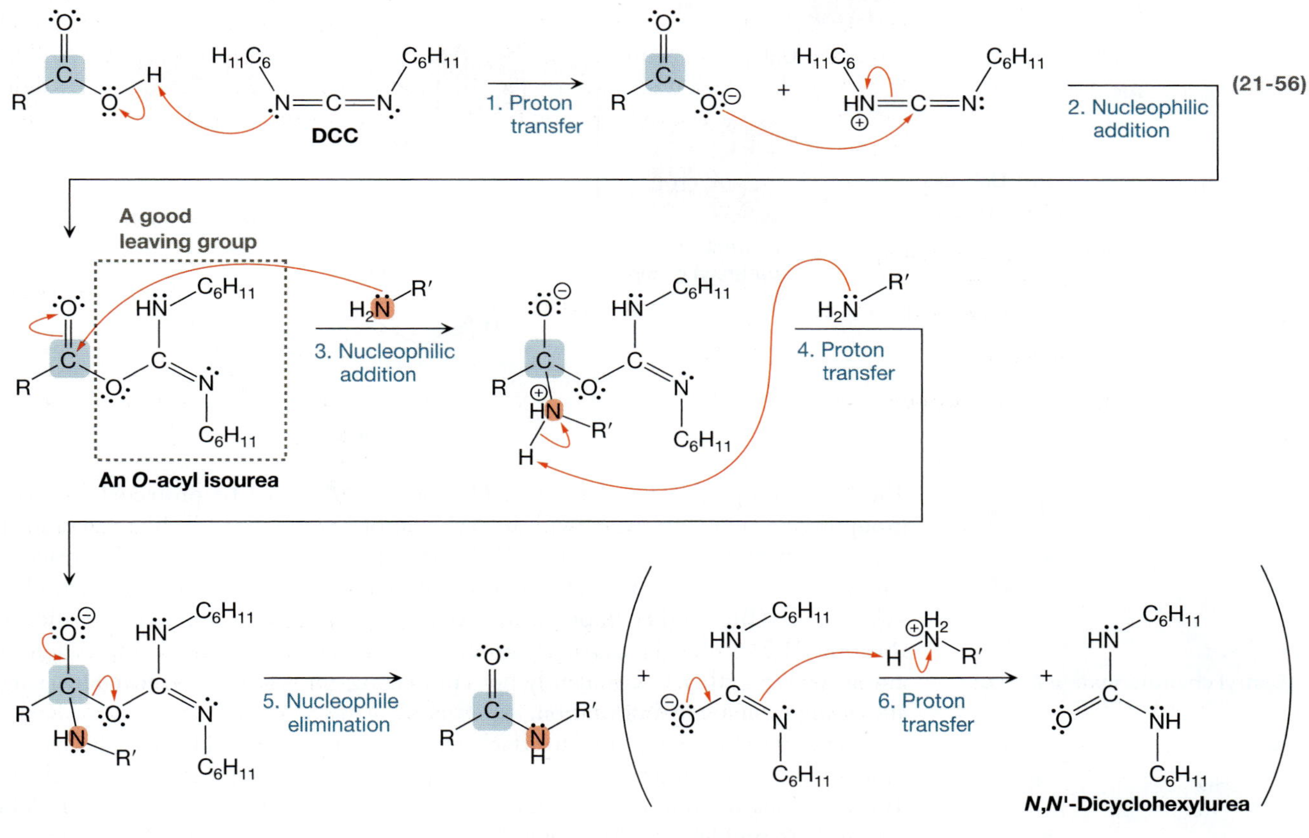

In Steps 1 and 2, DCC undergoes a nucleophilic addition reaction with the CO_2H group to produce an *O*-acyl isourea derivative, which contains a good leaving group on the carbonyl carbon. Subsequently, the *O*-acyl isourea undergoes a nucleophilic addition–elimination reaction with the NH_2 group, producing the desired amide along with *N,N'*-dicyclohexylurea.

Thus, DCC can be used to couple together amino acids, but each amino acid has both an NH_2 and a CO_2H group, so multiple different couplings can take place. For example, if a mixture of glycine and alanine is treated with DCC, four different couplings will take place to produce a mixture of four dipeptides, as shown in Equation 21-57.

A mixture of dipeptides is produced.

Alanine (Ala) + **Glycine (Gly)** $\xrightarrow{\text{DCC}}$ **Ala-Gly** + **Gly-Ala** + **Ala-Ala** + **Gly-Gly**

(21-57)

To carry out just one of these couplings, the groups that we don't want to react must be *protected*. Suppose, for example, that we want to couple together alanine (Ala) and glycine (Gly) to produce Ala-Gly (Equation 21-58).

(21-58)

Benzyl chloroformate = Z—Cl

Benzyl alcohol = Bn—OH

$(H_3C)_3C$ ... = **Boc**

The NH_2 group of Ala and the CO_2H group of Gly must be protected. The NH_2 group is commonly protected with benzyl chloroformate (Z—Cl) in a reaction that is essentially an aminolysis (Section 21.3) to produce a carbamate, which has reactivity similar to an amide. The CO_2H group is protected as an ester by treatment with benzyl alcohol (Bn—OH) under acidic conditions; this is a Fischer esterification (Section 21.7). Once the necessary groups are protected, DCC is used to couple the amino acids together. Conveniently, both protecting groups can be removed in a single step using catalytic hydrogenation.

These steps can be repeated to attach additional amino acids. However, because the process takes place entirely in solution, the products must be purified after each step. This is tedious and it leads to significant loss of product. R. B. Merrifield (1921–2006) solved these problems by designing a *solid-phase* synthesis, for which he was awarded the 1984 Nobel Prize in Chemistry.

The basis of the **Merrifield synthesis**, outlined in Figure 21-9, begins with polymer beads that terminate with benzyl chloride groups. The C-terminal amino acid, whose NH_2 group is protected with a *tert*-butoxycarbonyl (Boc) group (review Table 21-2), is then added to the polymer beads (Step 1). An S_N2 reaction ensues, joining the amino acid to the polymer bead via an ester linkage. In Step 2, the NH_2 group of the newly joined amino acid is deprotected under acidic conditions, and in Step 3, a second Boc-protected amino acid is added in the presence of DCC. A new peptide bond is produced, according to the mechanism in Equation 21-56, resulting

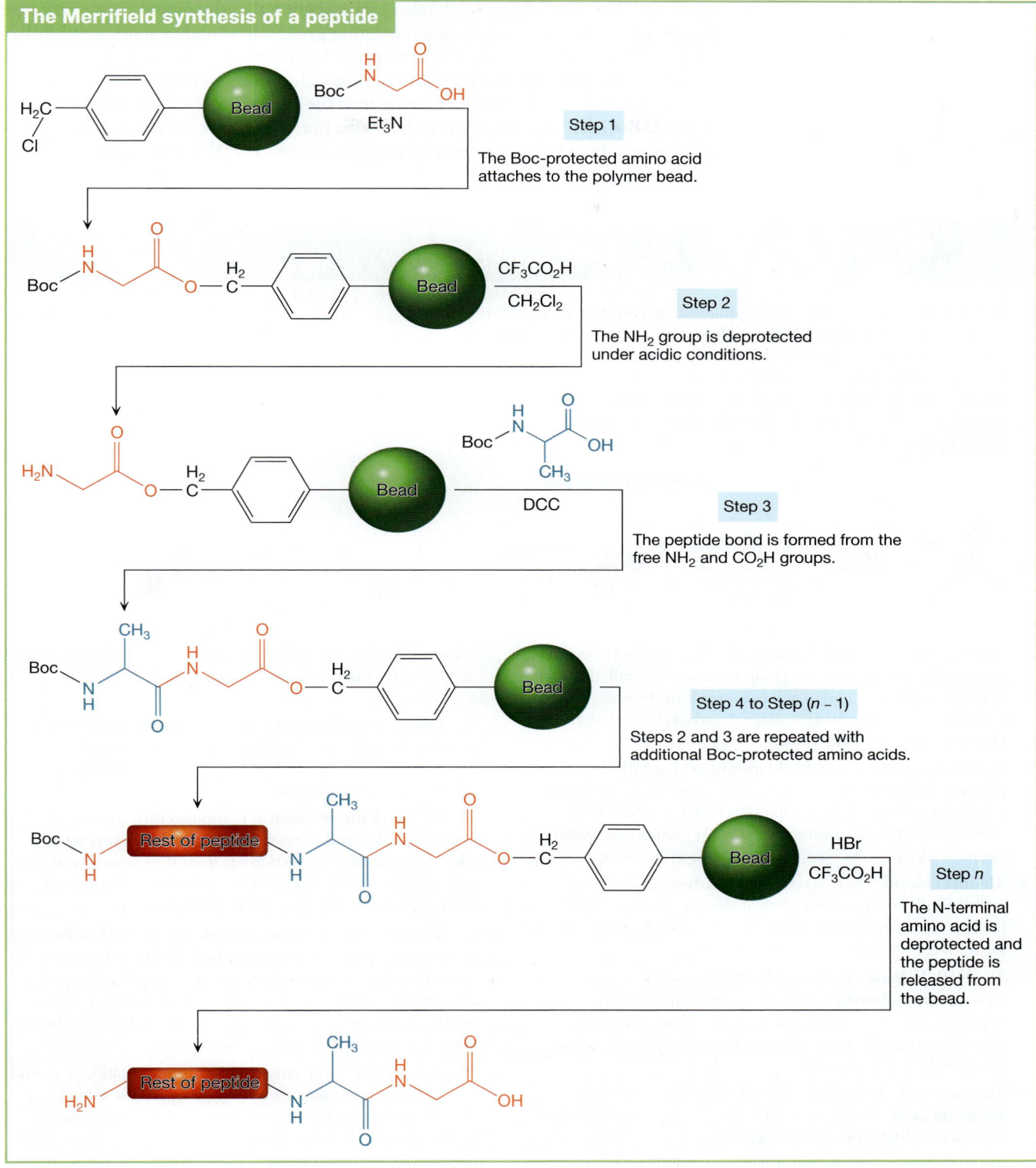

The Merrifield synthesis of a peptide

Step 1
The Boc-protected amino acid attaches to the polymer bead.

Step 2
The NH₂ group is deprotected under acidic conditions.

Step 3
The peptide bond is formed from the free NH₂ and CO₂H groups.

Step 4 to Step (n – 1)
Steps 2 and 3 are repeated with additional Boc-protected amino acids.

Step n
The N-terminal amino acid is deprotected and the peptide is released from the bead.

FIGURE 21-9 The Merrifield synthesis of a peptide The solid-phase synthesis protocol developed by R. B. Merrifield. The individual steps are described in the text.

in a Boc-protected dipeptide attached to the polymer bead. Steps 2 and 3 are then repeated for each amino acid that is to be added to the peptide, working from the C-terminus to the N-terminus. Once the protected N-terminal amino acid has joined, HBr is added, which simultaneously releases the peptide from the bead and deprotects the N-terminal amino acid.

Solid-phase synthesis is so simple in its execution that the process has been auto-mated. With improvements to the protecting groups and solid supports used in the original Merrifield synthesis, it is now possible to use a *peptide synthesizer* to make peptides of more than 100 amino acid residues. The addition of each residue takes up to 1 hour, which is about 50 times slower than the natural process that makes a protein in a cell. The Merrifield synthesis is quite effective, however, at making peptides when it is not possible to clone the protein using recombinant DNA techniques.

Chapter Summary and Key Terms

- The general mechanism for nucleophilic addition–elimination involving a weak nucleophile (H—Nu) is similar to the one involving a strong nucleophile, but includes a proton transfer step to produce an uncharged product. This mechanism describes the **hydrolysis**, **alcoholysis**, and **aminolysis** of both an acid chloride and an acid anhydride. **(Sections 21.1 and 21.3; Objective 1)**

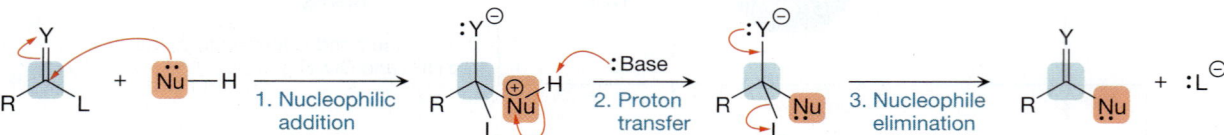

- The reactivity of an acid derivative decreases as the resonance stabilization of the carbonyl group increases. Overall, the relative reactivities of acid derivatives decrease in the order: acid chloride > acid anhydride > ester ≈ carboxylic acid > amide. **(Section 21.2; Objective 2)**

- In the aminolysis of an acid chloride or acid anhydride, which produces an amide, two equivalents of an amine are required: one to act as a nucleophile and other to act as a base. Only one equivalent of the amine is required if pyridine or triethyl-amine is added as the base. **(Section 21.3; Objective 3)**

- **Thionyl chloride** (SOCl₂) converts a carboxylic acid to an acid chloride (the acid derivative at the top of the stability ladder), from which any other carboxylic acid derivative can be produced. **(Section 21.4; Objective 4)**

- The **Hell–Volhard–Zelinsky (HVZ) reaction** produces an α-bromo carboxylic acid from a carboxylic acid. The key is to convert the carboxylic acid into an acid bromide, which is enolizable, and is thus nucleophilic at the α carbon. **(Section 21.5; Objective 4)**

- Treating an alcohol with a **sulfonyl chloride** produces a **sulfonate ester**. Thus, a poor HO⁻ leaving group is converted into an excellent alkylsulfonate leaving group for nucleophilic substitution and elimination reactions. Examples include mesylate (MsO⁻), tosylate (TsO⁻), and triflate (TfO⁻) leaving groups. **(Section 21.6; Objective 5)**

- Carboxylic acids, esters, and amides are relatively stable and do not react readily with weak nucleophiles under normal conditions. However, nucleophilic acyl substitution can be base catalyzed or acid catalyzed. Base catalysis involves deprotonating the nucleophile to convert it into a strong nucleophile. Acid catalysis involves protonating the carbonyl oxygen to activate the carbonyl carbon. **(Section 21.7; Objective 6)**

- The **Fischer esterification reaction** produces an ester from a carboxylic acid under highly acidic conditions, and is described by the acid-catalyzed acyl substitution mechanism. **(Section 21.7; Objective 6)**

- The **Baeyer–Villiger oxidation** produces carboxylic acids from aldehydes, and esters from ketones. In these reactions, a hydrogen or alkyl group departs from the carbonyl carbon, facilitated by the breaking of the weak O—O bond from a *peroxyacid* (RCO₃H). **(Section 21.8; Objective 7)**

- In a **Claisen condensation reaction**, an ester with at least two α protons is treated with a strong base, followed by acid work-up, to produce a **β-keto ester**. In these reactions, the base deprotonates the α carbon of one ester to produce an ester enolate anion, which acts as a nucleophile and attacks the carbonyl carbon of a second ester. **(Section 21.9; Objective 8)**

- If the base in a Claisen condensation is nucleophilic, it should be identical to the alkoxide leaving group on the ester. If the solvent can be deprotonated to become strongly nucleophilic, it should be the conjugate acid of the base that is used. **(Section 21.9a; Objective 9)**

- In a **crossed Claisen condensation reaction**, the enolate anion that acts as the nucleophile derives from an ester that is different from the ester it attacks. Crossed Claisen condensations are synthetically useful if only one ester enolate anion is present, and only one uncharged ester can be attacked. **(Section 21.9b; Objective 9)**

- A **Dieckmann condensation** is an intramolecular Claisen condensation, and is favored when a five- or six-membered ring can be formed. (Section 21.9c; Objective 10)
- A **β-keto acid** or **β-diacid** can undergo **decarboxylation** on heating under acidic conditions. The result is loss of CO_2, leaving behind a ketone or carboxylic acid. Decarboxylation is used in the **acetoacetic ester synthesis** to produce an alkyl-substituted ketone, as well as the **malonic ester synthesis** to produce an alkyl-substituted carboxylic acid. (Section 21.10; Objective 11)
- Carboxyl groups of carboxylic acids are frequently protected as a methyl or benzyl ester. Deprotection takes place on treatment of either of these esters with hydroxide in a saponification reaction. Alternatively, deprotection of a benzyl ester can take place by catalytic hydrogenation. (Section 21.11; Objective 12)
- Amino groups of amines are frequently protected as amides or **carbamates**. Deprotection typically requires acid-catalyzed hydrolysis. (Section 21.11; Objective 12)

Reaction Tables

TABLE 21-3 Functional Group Conversions

	Starting Compound Class	Typical Reagents and Reaction Conditions	Compound Class Formed	Key Electron-Rich Species	Key Electron-Poor Species	Comments	Discussed in Section
(1)	Acid chloride	H_2O	Carboxylic acid	H_2O	Acid chloride	Nucleophilic addition–elimination (hydrolysis)	21.1
(2)	Acid chloride	R'OH	Ester	R'OH	Acid chloride	Nucleophilic addition–elimination (alcoholysis)	21.1
(3)	Acid anhydride	H_2O	Carboxylic acid	H_2O	Acid anhydride	Nucleophilic addition–elimination (hydrolysis)	21.2
(4)	Acid anhydride	R'OH	Ester	R'OH	Acid anhydride	Nucleophilic addition–elimination (alcoholysis)	21.2
(5)	Acid chloride	$R_2'NH$, Et_3N or pyridine	Amide	$R_2'NH$	Acid chloride	Nucleophilic addition–elimination (aminolysis)	21.3
(6)	Acid anhydride	$R_2'NH$, Et_3N or pyridine	Amide	$R_2'NH$	Acid anhydride	Nucleophilic addition–elimination (aminolysis)	21.3

(continued)

TABLE 21-3 **Functional Group Conversions (continued)**

	Starting Compound Class	Typical Reagents and Reaction Conditions	Compound Class Formed	Key Electron-Rich Species	Key Electron-Poor Species	Comments	Discussed in Section
(7)	Carboxylic acid	SOCl₂	Acid chloride			Back-to-back nucleophilic addition–elimination	21.4
(8)	Carboxylic acid	Br₂, P	α-Bromo acid			Hell–Volhard–Zelinsky reaction	21.5
(9)	Alcohol		Sulfonate ester			Nucleophilic addition–elimination	21.6
(10)	Ester	R″OH / R″ONa	Ester			Base-catalyzed nucleophilic addition–elimination (transesterification)	21.7
(11)	Ester	R″OH / H₂SO₄	Ester			Acid-catalyzed nucleophilic addition–elimination (transesterification)	21.7
(12)	Ester	H₂O / H₂SO₄	Carboxylic acid			Acid-catalyzed nucleophilic addition–elimination (hydrolysis)	21.7
(13)	Carboxylic acid	R′OH / H₂SO₄	Ester			Acid-catalyzed nucleophilic addition–elimination (Fischer esterification)	21.7
(14)	Aldehyde		Carboxylic acid			Acid-catalyzed nucleophilic addition–elimination (Baeyer–Villiger oxidation)	21.8

TABLE 21-4 Reactions That Alter the Carbon Skeleton[a]

Starting Compound Class	Typical Reagents and Reaction Conditions	Compound Class Formed	Key Electron-Rich Species	Key Electron-Poor Species	Comments	Discussed in Section
(1) Ketone		Ester			Acid-catalyzed nucleophilic addition–elimination (Baeyer–Villiger oxidation)	21.8
(2) Ester	1. NaOR' 2. CH_3CO_2H	β-Keto ester			Base-promoted nucleophilic addition–elimination (Claisen condensation)	21.9
(3) Acetoacetic ester	1. NaOEt 2. R–X 3. $H_3O^{\oplus}$, Δ	Alkyl-substituted acetone		$\overset{\delta+}{R}$—X	Acetoacetic ester synthesis	21.10
(4) Malonic ester	1. NaOEt 2. R–X 3. $H_3O^{\oplus}$, Δ	Alkyl-substituted acetic acid		$\overset{\delta+}{R}$—X	Malonic ester synthesis	21.10

[a]X = Cl, Br, I.

Problems

Problems that are related to synthesis are denoted (SYN).

21.1–21.3 Alcoholysis, Aminolysis, and Relative Reactivities of Acid Derivatives

21.36 Predict the product of the reaction between *m*-ethylbenzoyl chloride and each of the following compounds. Draw the complete, detailed mechanism for each reaction. If no reaction is expected to occur, write "no reaction."
(a) H_2O; **(b)** CH_3NH_2; **(c)** (S)-butan-2-ol; **(d)** diethyl ether

m-Ethylbenzoyl chloride

21.37 Predict the product of the reaction between acetic anhydride and each of the following compounds. If no reaction is expected to occur, write "no reaction." Draw the complete, detailed mechanism for each reaction that does occur.
(a) H_2O; **(b)** CH_3NH_2, pyridine; **(c)** (S)-butan-2-ol; **(d)** diethyl ether

21.38 Aspirin (acetylsalicylic acid) is made by treating salicylic acid with acetic anhydride. Draw the complete, detailed mechanism for this reaction and draw the product.

Salicylic acid   **Acetylsalicylic acid**

Acetic anhydride

21.39 Draw the complete, detailed mechanism and the products for each of the following reactions.

(a)

(b)

(2 equiv)

(c)

(d)

21.40 (SYN) Show how each of the following compounds can be synthesized from an acid chloride and either water, an alcohol, or an amine. For each reaction, provide the complete, detailed mechanism.

(a)   **(b)**   **(c)**   **(d)**

21.41 A *thioester* is the sulfur analog of an ester. Do you think a thioester would undergo hydrolysis faster or slower than an ester under normal conditions? Explain your reasoning.

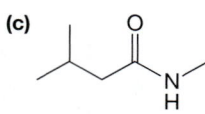

A thioester

21.42 Treating a δ-lactone (reaction **A**) with ammonia yields a hydroxyamide. If a β-lactone is treated with ammonia (reaction **B**), however, then a β-amino acid is formed. Provide the detailed mechanism for each of these reactions and explain these observations.

A   NH₃

B   NH₃

21.43 The reaction shown here is an example of the *Favorskii reaction*, which involves an R⁻ leaving group in a nucleophilic addition–elimination reaction.
(a) Draw the complete, detailed mechanism for this reaction and explain why R⁻ can act as a leaving group.
(b) Suggest how you can synthesize an ester from cyclopropanone using only this reaction.

NaOH
H₂O

21.44 In an aminolysis of an acid chloride, pyridine is often used to minimize the amount of the amine that is required, as shown here. Pyridine is a moderately strong nucleophile, however, so it can react with an

H₂N—R'

Pyridine

acid chloride via a nucleophilic addition–elimination mechanism. Nevertheless, this reaction involving pyridine does not interfere with the production of the desired amide.

(a) Draw the mechanism and the product for the nucleophilic addition–elimination reaction of pyridine with an acid chloride.

(b) Explain why this reaction involving pyridine does not interfere with the production of the desired amide product.

Hint: What reaction would ensue between the amine shown and the product of the reaction from part (a)?

21.4–21.6 Synthesis of Acid Halides, the Hell–Volhard–Zelinsky Reaction, and Sulfonyl Chlorides

21.45 Draw the complete mechanism and the major organic product for each of the following reactions.

(a)

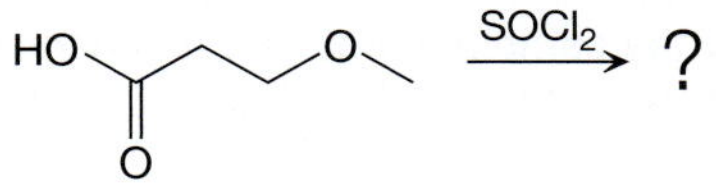

(b)

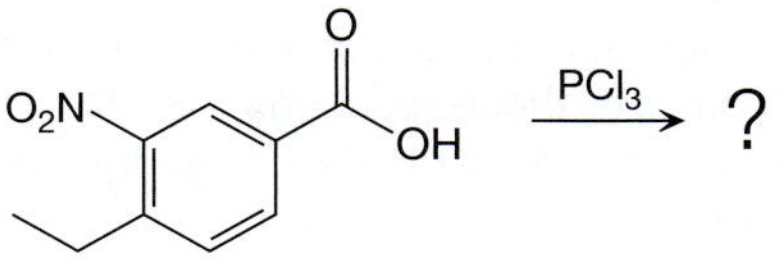

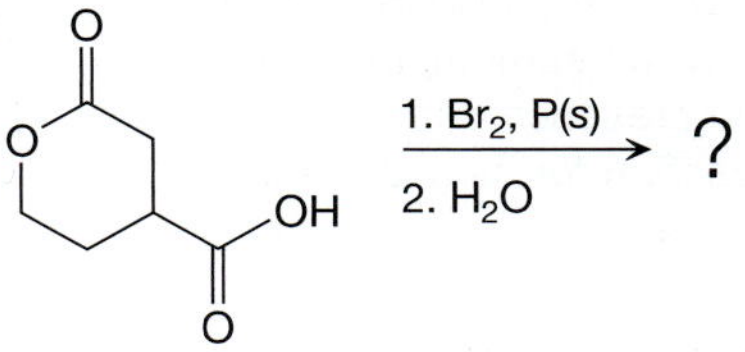

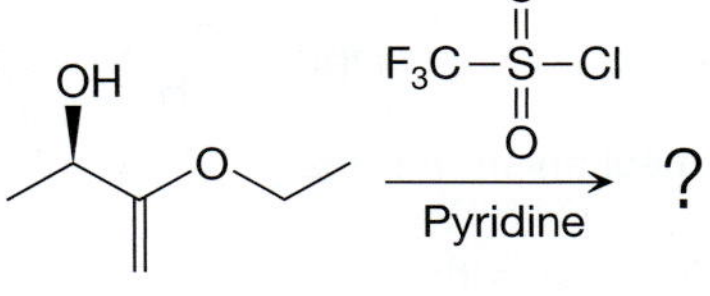

21.46 (SYN) For each acid chloride, draw the carboxylic acid that would produce it when treated with $SOCl_2$.

(a)

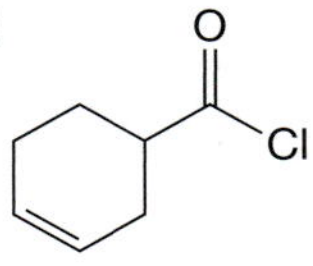

(b)

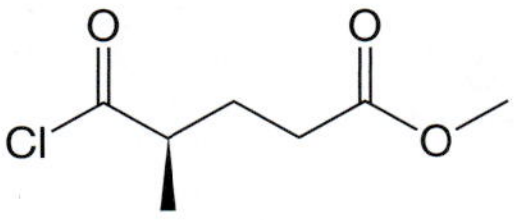

(c)

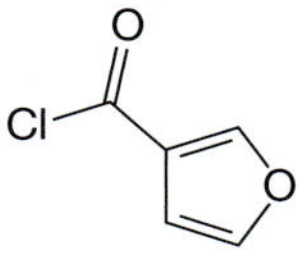

21.47 Draw the complete mechanism and the major organic product for each of the following reactions.

(a)

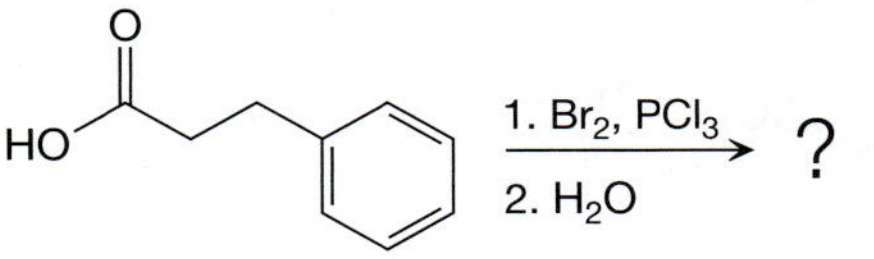

1. Br_2, PCl_3
2. H_2O

(b)

1. Br_2, P(s)
2. H_2O

21.48 (SYN) Show how to synthesize each of the following compounds from 3-methylpentanoic acid.

(a)

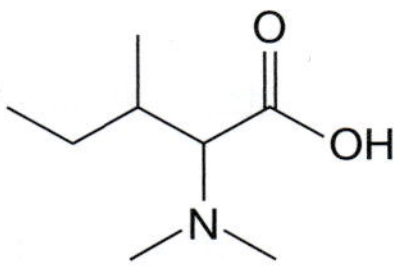

(b)

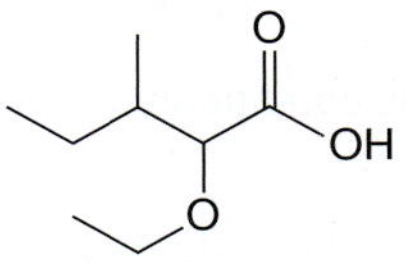

(c)

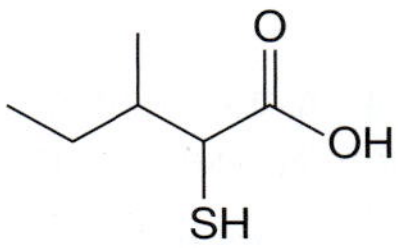

21.49 Draw the complete mechanism and the major organic product for each of the following reactions.

(a)

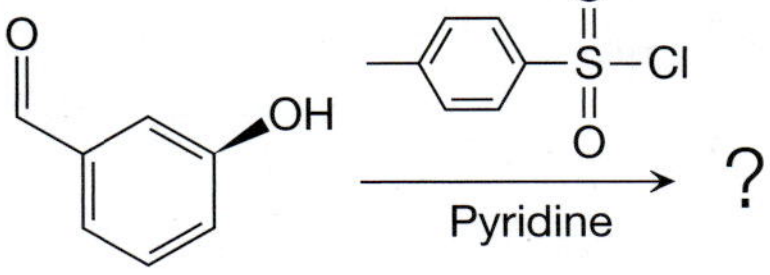

Pyridine

(b)

Pyridine

(c)

Pyridine

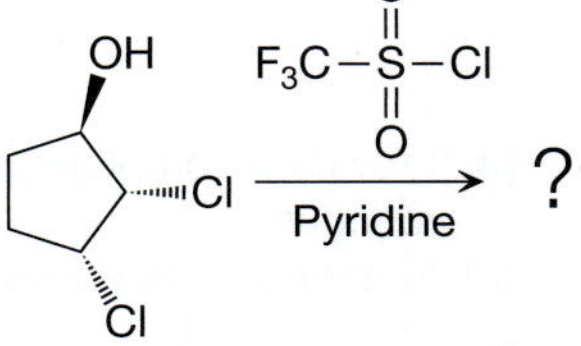

21.50 (SYN) Show how to carry out each of the following syntheses by first converting the alcohol into a sulfonyl chloride.

(a)

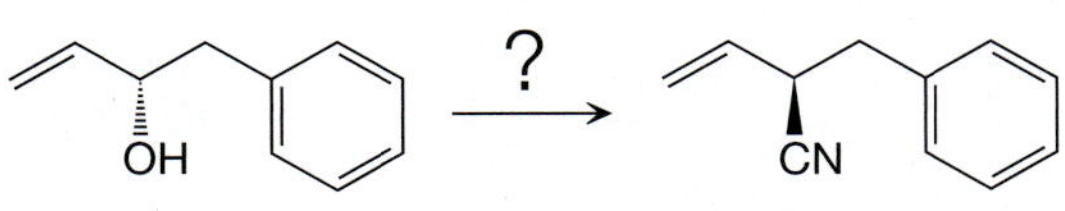

(b)

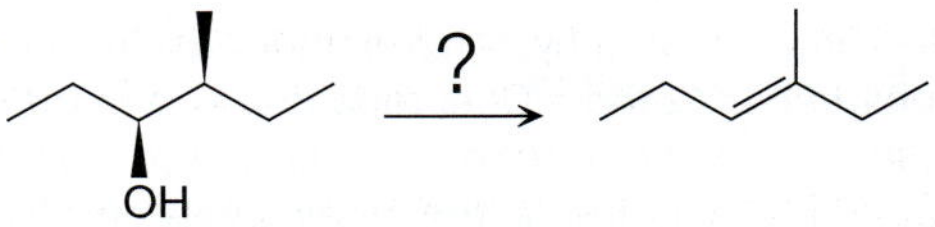

21.51 An imino chloride can be prepared from an amide according to the reaction shown here. Propose a mechanism for this reaction.

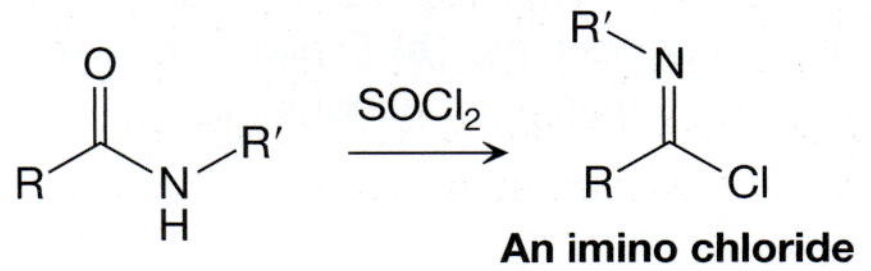

$SOCl_2$

An imino chloride

21.52 A *sulfonamide* is produced when a sulfonyl chloride is treated with an amine, as shown here. This reaction is a key step in the synthesis of *sulfa drugs*, which constitute an important class of antibiotics. Draw a complete, detailed mechanism for this reaction.

$$R-\underset{\underset{O}{\|}}{\overset{\overset{O}{\|}}{S}}-Cl \xrightarrow{\;H_2N{-}R'\;} R-\underset{\underset{O}{\|}}{\overset{\overset{O}{\|}}{S}}-\underset{R'}{N}H$$

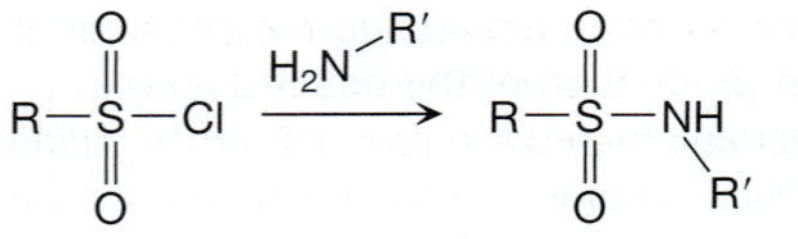

A sulfonamide

21.7 and 21.8 Base and Acid Catalysis in Nucleophilic Addition–Elimination Reactions; Baeyer–Villiger Oxidations

21.53 Predict the product of the reaction between methyl cyclohexylmethanoate and each of the following. If no reaction is expected to occur, write "no reaction." For those reactions that do occur, draw the complete, detailed mechanism. *Hint:* Pay attention to the reaction conditions. **(a)** H_2O, H^+; **(b)** H_2O, OH^-, then H_3O^+; **(c)** $CH_3CH_2CH_2OH$, $CH_3CH_2CH_2ONa$; **(d)** propan-2-ol, H^+; **(e)** propan-2-amine (excess), H^+; **(f)** propan-1-ol

21.54 Draw the mechanism and the major organic product for each of the following reactions.

(a)

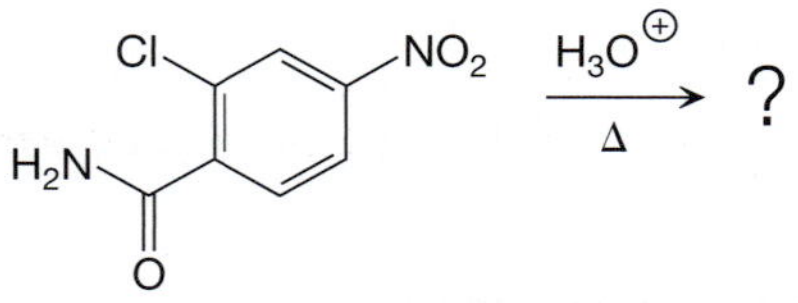

$$\xrightarrow[\Delta]{H_3O^{\oplus}} \;?$$

(b)

$$\xrightarrow[\Delta]{H_3O^{\oplus}} \;?$$

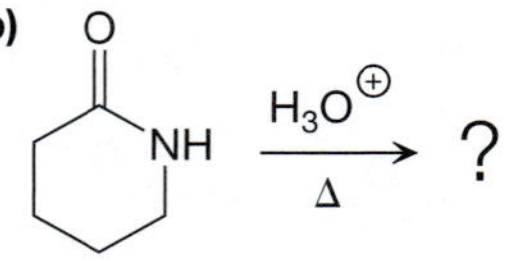

21.55 In each of the reactants shown here, a CO_2H group is separated from a OH group by the same number of carbons. When heated in the presence of acid, only the compound in the first reaction forms a lactone. Explain why.

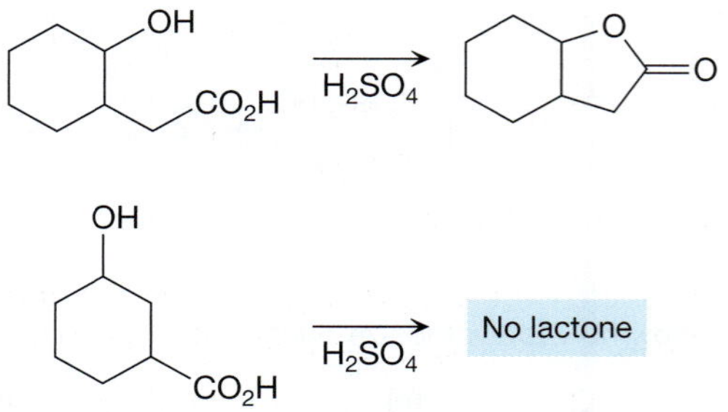

$$\xrightarrow{H_2SO_4}$$

$$\xrightarrow{H_2SO_4}$$ No lactone

21.56 The hydrolysis of an ester can be sped up by both acidic and basic conditions. Aminolysis of an ester can be sped up by acidic conditions, but not by basic conditions. Explain why.

21.57 An acid anhydride can be formed under equilibrium conditions by reacting an ester with a carboxylic acid, as shown below. Reasonable yield is achieved if the equilibrium can be shifted by exploiting Le Châtelier's principle.

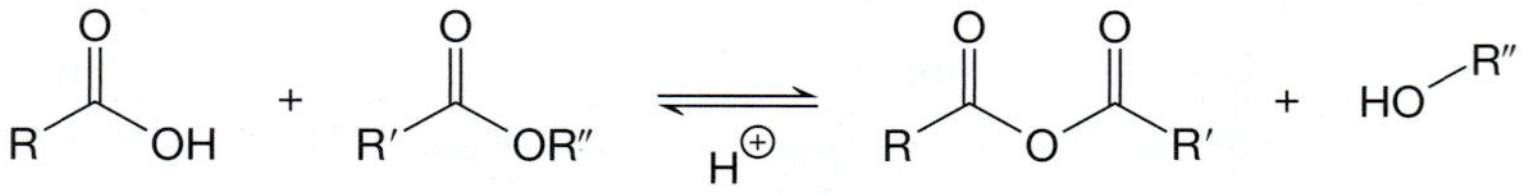

(a) Provide a complete, detailed mechanism for this reaction.
(b) In general, the above equilibrium favors the reactants. However, if the ester that is used is the one shown here, then the equilibrium favors the products. Explain why.

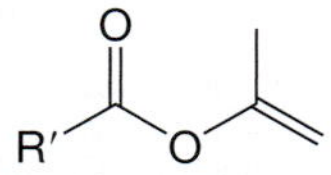

21.58 When a methyl ester is hydrolyzed under acidic conditions in $H_2^{18}O$, the ^{18}O isotope ends up in the carboxylic acid. When a *tert*-butyl ester is hydrolyzed under the same conditions, the labeled oxygen ends up in the alcohol product. **(a)** Propose mechanisms to account for these observations. **(b)** Explain why each ester undergoes the respective mechanism.

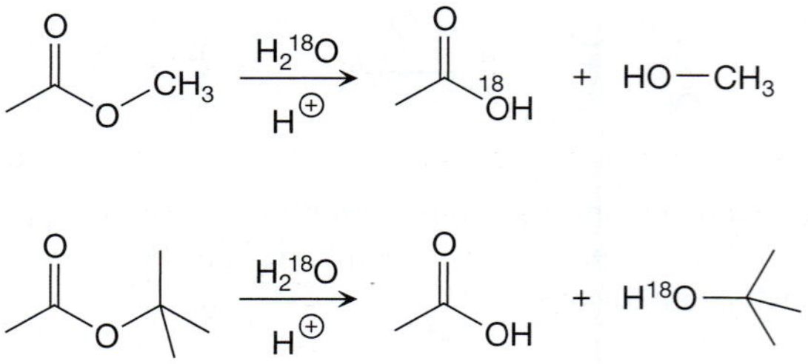

21.59 The Gabriel synthesis of primary amines discussed in Chapter 20 involves hydrolysis under basic conditions to release the amine. As shown in the reactions below, the amine can also be released by **(a)** hydrolysis under acidic conditions and **(b)** treatment with hydrazine. Draw the complete, detailed mechanisms for these reactions.

(a)

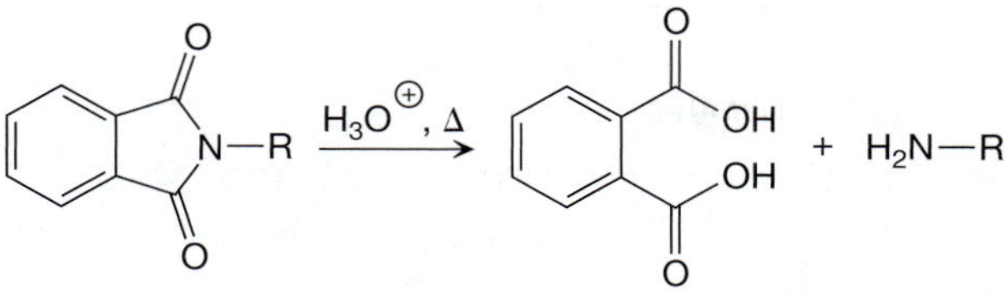

(b)

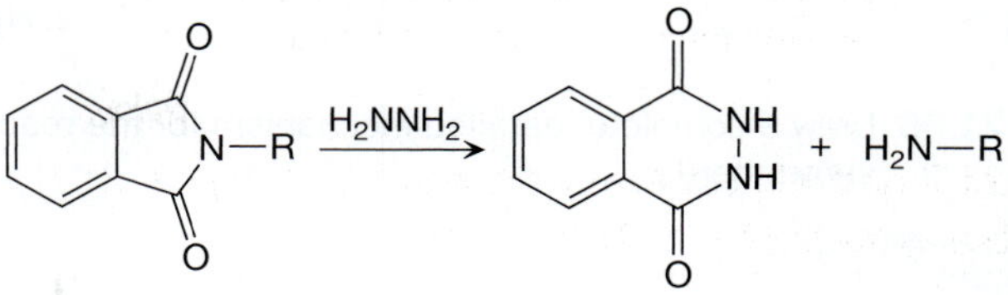

21.60 Draw the mechanism and the major organic product for each of the following reactions.

(a)

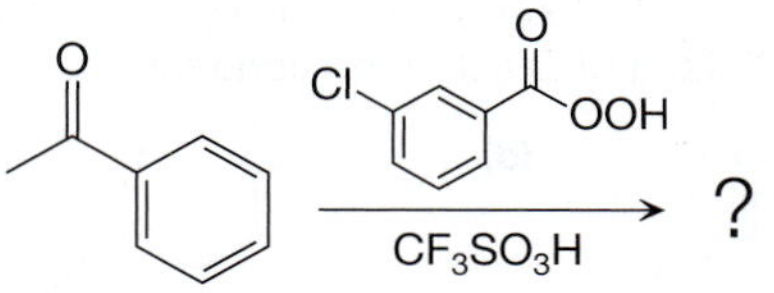

(b)

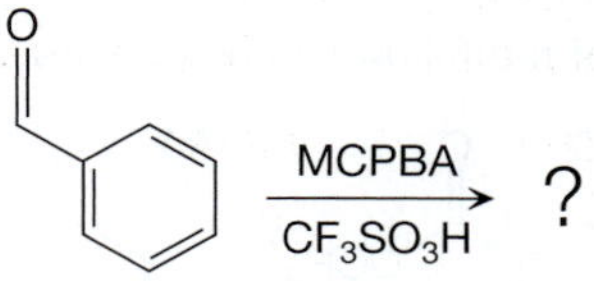

(c)

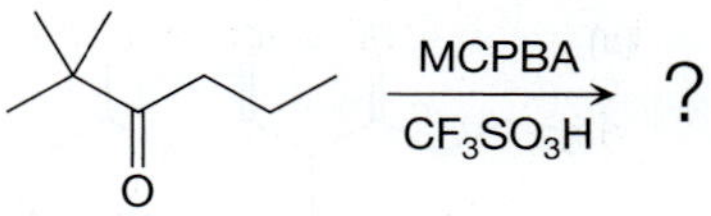

21.61 (SYN) For each compound below, draw the ketone or aldehyde that can be used to produce it when treated with *m*-chloroperbenzoic acid under acidic conditions.

(a)

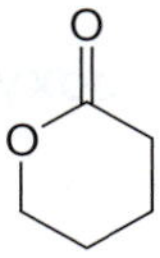

(b)

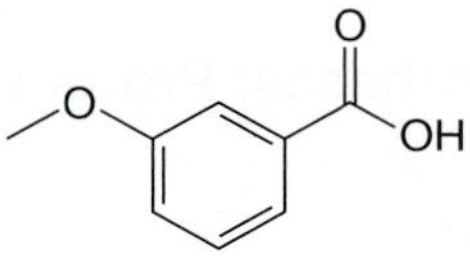

(c)

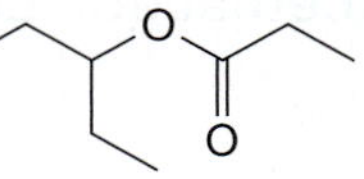

(d)

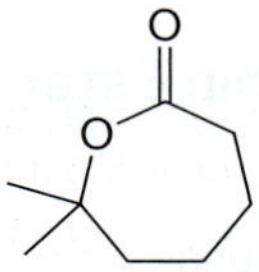

21.9 Claisen Condensations

21.62 Draw the complete, detailed mechanism for each of the following reactions and predict the major product.

(a) Ethyl 3-methylbutanoate $\xrightarrow[\text{EtOH}]{\text{NaOEt}}$?

(b) Ethyl propanoate + Ethyl benzoate $\xrightarrow[\text{EtOH}]{\text{NaOEt}}$?

(c) Ethyl butanoate + Diethyl carbonate $\xrightarrow[\text{EtOH}]{\text{NaOEt}}$?

21.63 A crossed Claisen reaction between methylpropanoate and methyl acetate is not synthetically useful because it produces a mixture of products. The reaction between methyl 2-methylpropanoate and methyl acetate, however, is synthetically useful. Explain why.

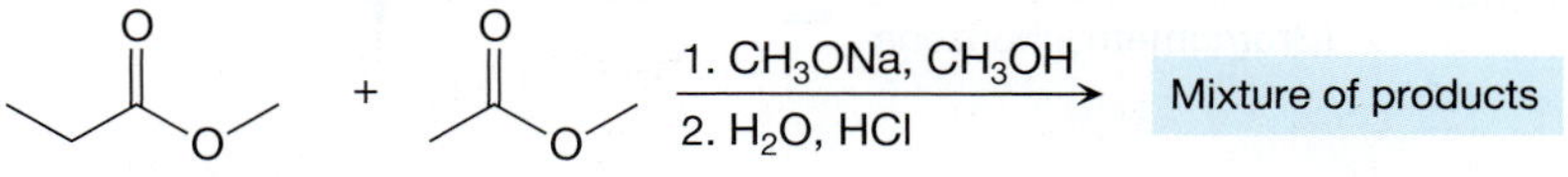

Methyl propanoate

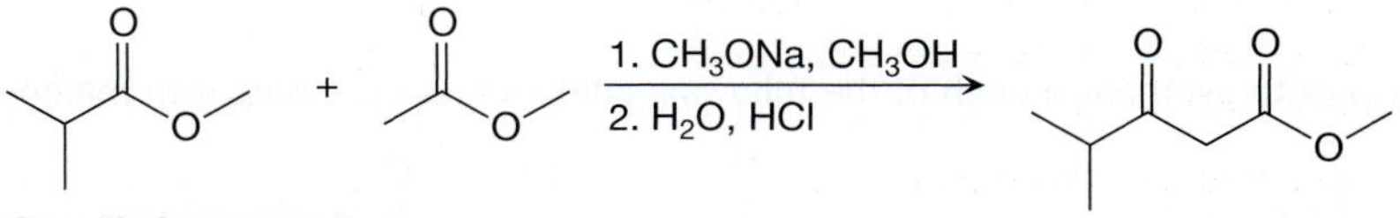

1. CH$_3$ONa, CH$_3$OH
2. H$_2$O, HCl

Mixture of products

Methyl 2-methylpropanoate

1. CH$_3$ONa, CH$_3$OH
2. H$_2$O, HCl

78%

21.64 Draw the complete, detailed mechanism for each of the following reactions and predict the major products.

(a)

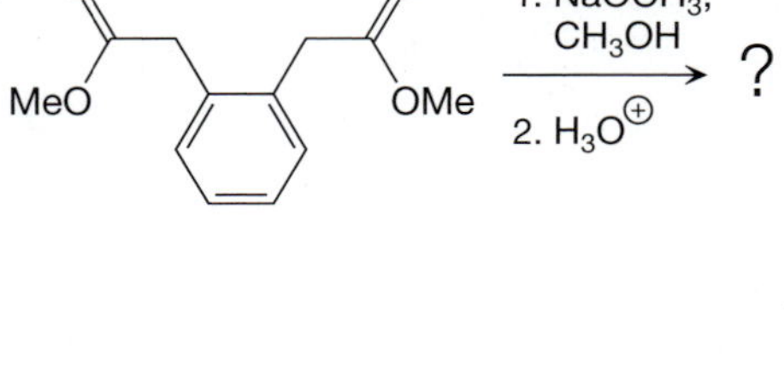

1. NaOCH$_3$, CH$_3$OH
2. H$_3$O$^{\oplus}$

?

(b)

1. LDA
2. Ethyl acetate
3. H$_3$O$^{\oplus}$

?

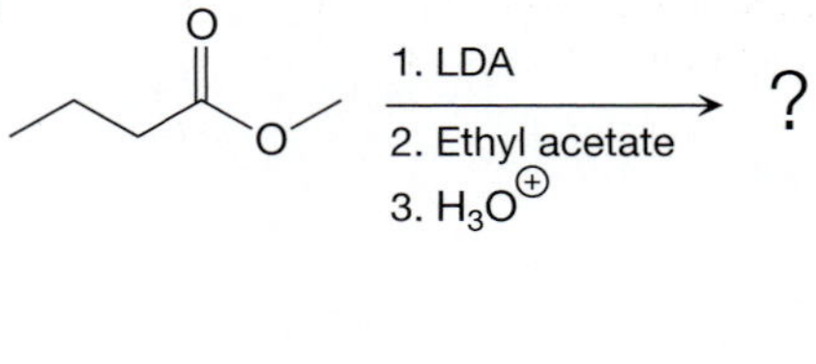

21.65 Explain why the following Claisen condensation does *not* work.

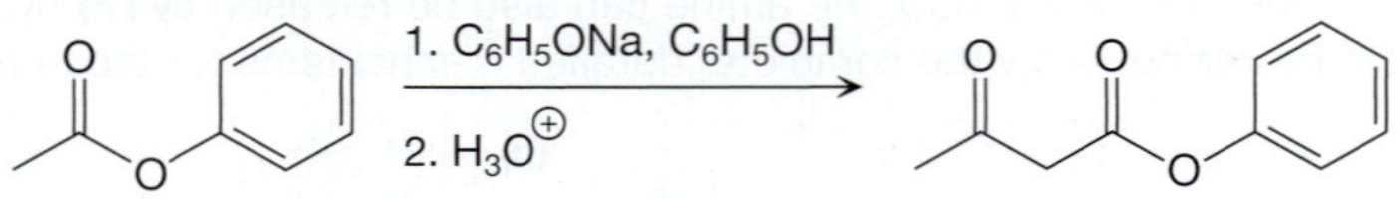

21.66 Draw a complete, detailed mechanism for the reaction shown here.

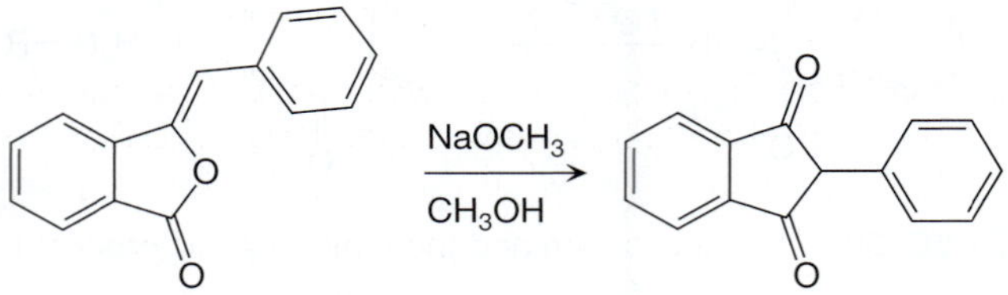

21.67 **(SYN)** Show how to synthesize each of the following compounds using a Claisen or Dieckmann condensation.

(a) (b) 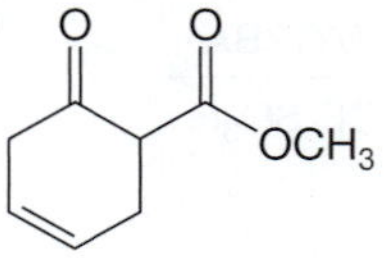 (c) 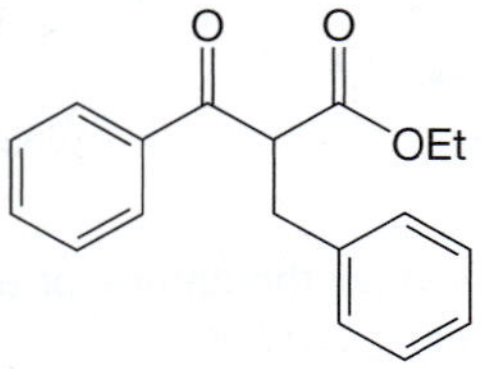 (d)

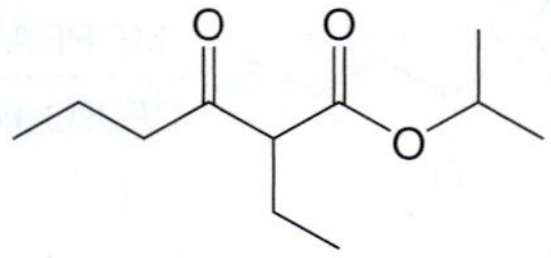

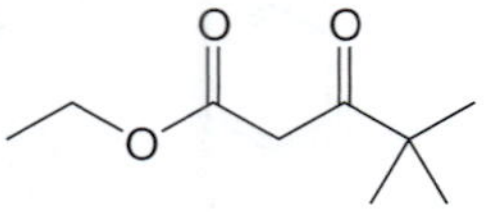

21.10 and 21.11 The Malonic Ester and Acetoacetic Ester Syntheses; Protecting Carboxylic Acids and Amines

21.68 Draw structures for compounds **A** through **D**.

(a) Ethyl acetoacetate $\xrightarrow[\text{2. 1-Chloro-4-methylpentane}]{\text{1. NaOEt, EtOH}}$ **A** $\xrightarrow[\text{2. } H_3O^{\oplus}, \Delta]{\text{1. NaOH, } H_2O}$ **B**

Same as above

(b) **A** $\xrightarrow[\text{2. 1-Chloropropane}]{\text{1. NaOC(CH}_3)_3, \text{(CH}_3)_3\text{COH}}$ **C** $\xrightarrow[\text{2. } H_3O^{\oplus}, \Delta]{\text{1. NaOH, } H_2O}$ **D**

21.69 Draw structures for compounds **E** through **H**.

(a) Diethyl malonate $\xrightarrow[\text{2. (Bromomethyl)benzene}]{\text{1. NaOEt, EtOH}}$ **E** $\xrightarrow[\text{2. } H_3O^{\oplus}, \Delta]{\text{1. NaOH, } H_2O}$ **F**

Same as above

(b) **E** $\xrightarrow[\text{2. 1-Iodopentane}]{\text{1. NaOC(CH}_3)_3, \text{(CH}_3)_3\text{COH}}$ **G** $\xrightarrow[\text{2. } H_3O^{\oplus}, \Delta]{\text{1. NaOH, } H_2O}$ **H**

21.70 **(SYN)** Show how to synthesize each of the following compounds beginning with malonic ester.

(a) (b) 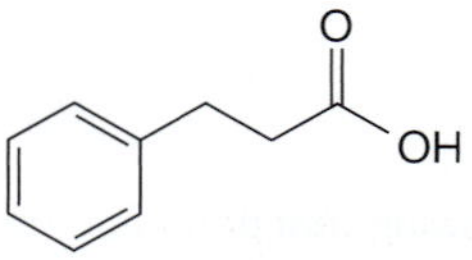 (c)

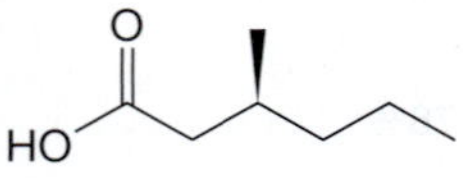

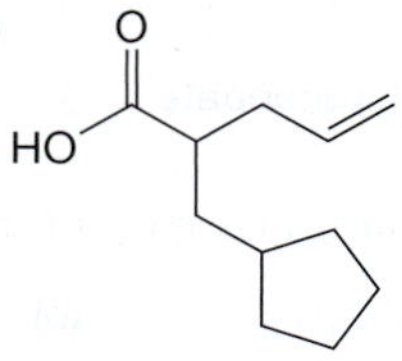

21.71 (SYN) Show how to synthesize each of the following compounds beginning with acetoacetic ester.

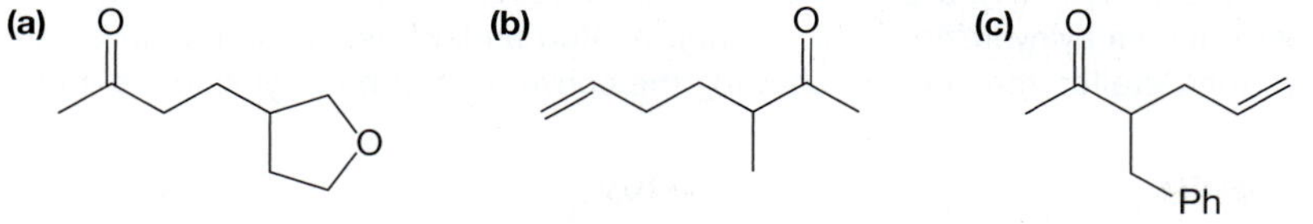

(a)

(b)

(c)

21.72 The Gabriel–malonic ester synthesis, shown here, is used to make α-amino acids. Draw complete, detailed mechanisms for this set of reactions and draw the structure of the intermediate **A**.

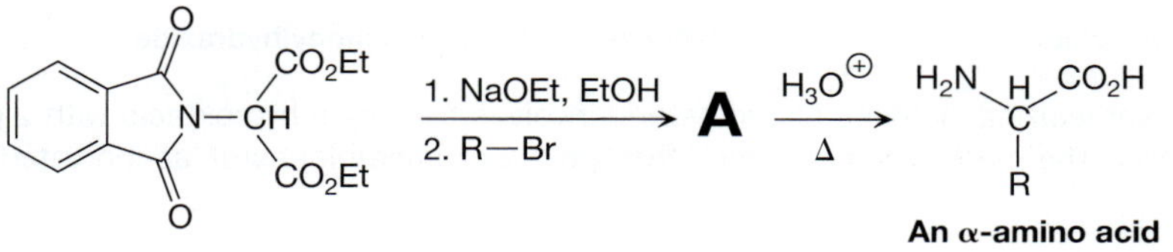

21.73 (SYN) Show how to carry out the following conversion.
Hint: Consider using a protecting group.

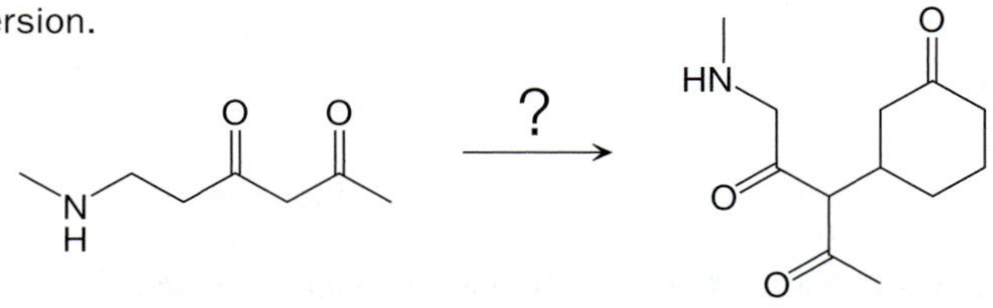

21.74 (SYN) Show how to carry out the following conversion. *Hint:* Consider using a protecting group.

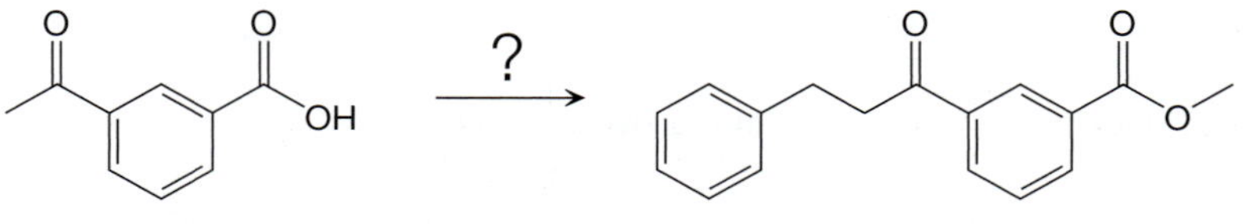

21.12 and 21.13 The Organic Chemistry of Biomolecules

21.75 A tripeptide undergoes complete hydrolysis and the resulting mixture contains only phenylalanine and glycine. Draw all possible sequences for the original tripeptide.

21.76 (SYN) Show how to synthesize **(a)** Ser-Phe and **(b)** Leu-Ala-Met, using the scheme similar to the one in Equation 21-58 (p. 1088).

21.77 A peptide containing 18 amino acid residues in its sequence was partially hydrolyzed and peptides **A–D** were detected in the resulting mixture. Draw the sequence of the original peptide.
A Phe-Gly-Ala
B Ser-Ser-Ser-Trp-Phe-Gly-Ala
C Phe-Phe-Met-Ala-Ala-Pro-Trp-Cys
D Met-Ala-Ala-Pro-Trp-Cys-Leu-Ile-Leu-Ser-Ser

21.78 During the hydrolysis of proteins, some amino acids, such as tryptophan, do not survive the reaction conditions. Other amino acids, such as asparagine and glutamine, are modified. Referring to Table 1-7 (p. 39), which shows the structures of the 20 common amino acids, write the structures of the two amino acids that are formed when asparagine and glutamine decompose in hot, concentrated HCl.

Integrated Problems

21.79 Predict the product of the sequence of reactions shown here.

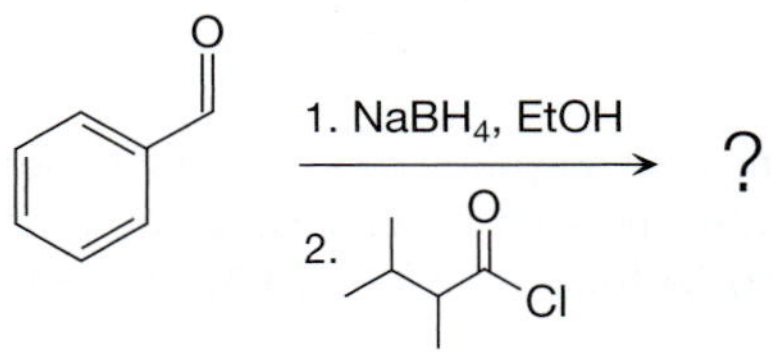

21.80 An example of the McFayden–Stevens reaction is shown below, in which an acyl chloride is converted to an aldehyde. First, benzoyl chloride is reacted with hydrazine, H_2NNH_2, the product of which is reacted with benzenesulfonyl chloride. The result is a 1-benzoyl-2-benzenesulfonylhydrazide, which, when heated under basic conditions, decomposes into the aldehyde. Provide the detailed mechanism showing the conversion of benzoyl chloride into 1-benzoyl-2-benzenesulfonylhydrazide.

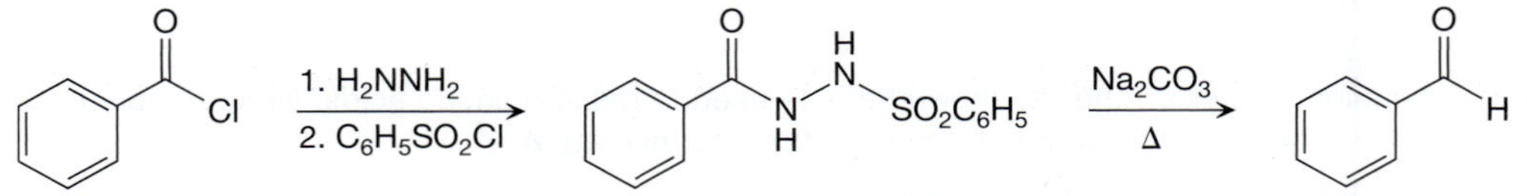

Benzoyl chloride   **1-Benzoyl-2-benzenesulfonylhydrazide**   **Benzaldehyde**

21.81 One method for synthesizing lactones (cyclic esters) involves treating a hydroxyacid with 2-pyridinethiol, followed by heating under reflux. The mechanism proceeds through a 2-pyridinethiol ester, as indicated.

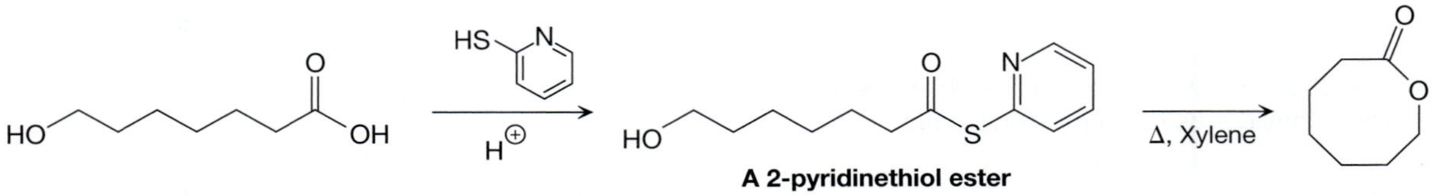

A 2-pyridinethiol ester

(a) Provide a complete, detailed mechanism for this reaction.
(b) If benzenethiol (C_6H_5SH) is used instead of 2-pyridinethiol, the conversion is much less effective. This suggests that the N atom is instrumental in the mechanism. Explain the role of the N atom.

21.82 Propose a mechanism for the reaction shown here.

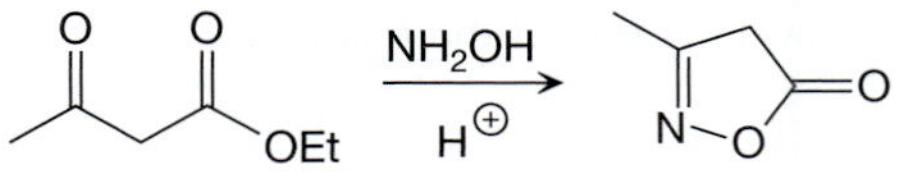

21.83 Draw the complete, detailed mechanism for the following reaction.

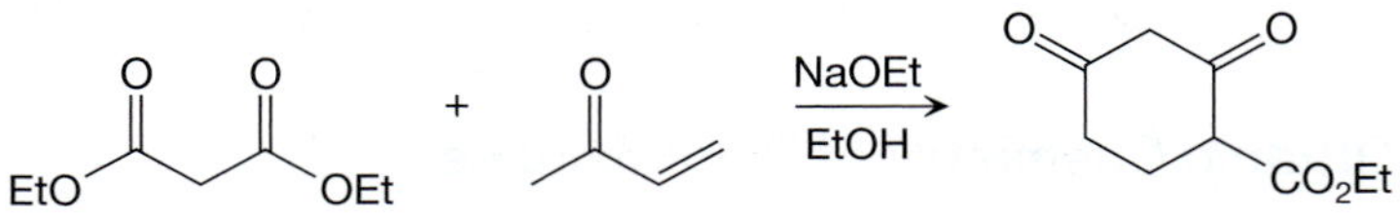

21.84 Diazomethane can be used to bring about a *ring expansion* of a cyclic ketone, as shown here. **(a)** Propose a mechanism for this reaction. **(b)** Suggest why this reaction is capable of converting a more stable six-membered ring to a less stable seven-membered ring.

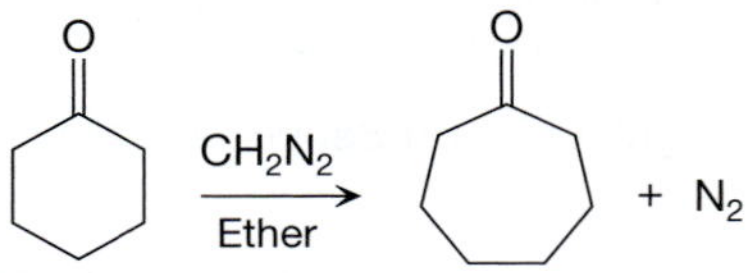

21.85 (SYN) Starting with acetic acid and using any other reagents necessary, show how you would synthesize each of the following compounds.

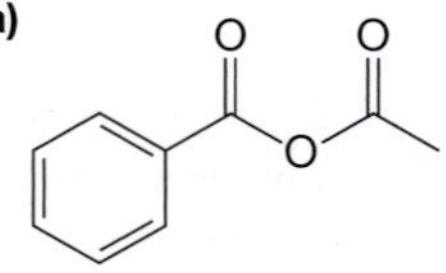

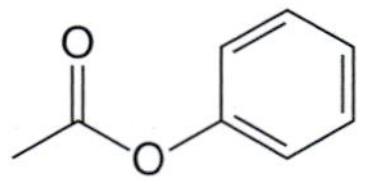

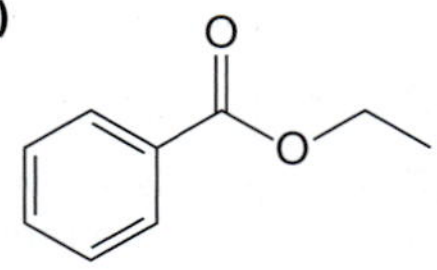

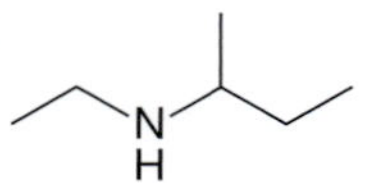

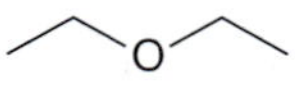

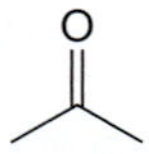

21.86 (SYN) Using acetic acid as your only source of carbon, propose a synthesis of each of the following compounds. You may use any inorganic reagent necessary. **(a)** ethyl acetate; **(b)** butan-2-ol; **(c)** 3-methylpentan-3-ol; **(d)** butan-2-one; **(e)** ethanamine; **(f)** *N*-ethylacetamide; **(g)** *N,N*-diethylacetamide

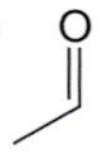

21.87 Draw the structures of **A**, **B**, and **C** in the following sequence of reactions.

1. LiAlH4
2. NH4Cl, H2O
A
TsCl
Pyridine
B
NaCN
DMSO
C

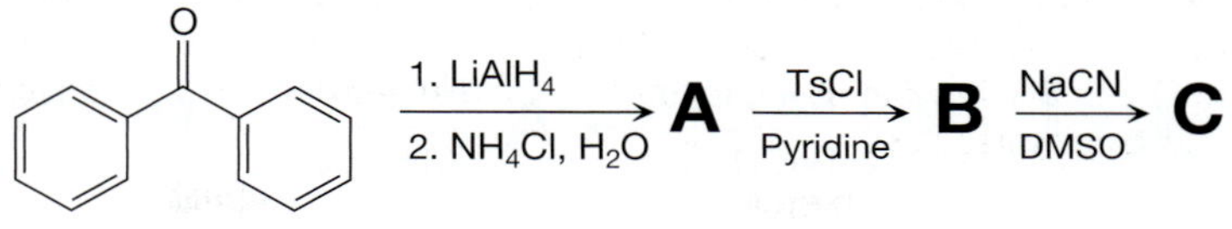

21.88 Draw the major product of each of the following sequences of reactions performed on methyl 3-cyclohexylpropanoate.

(a)

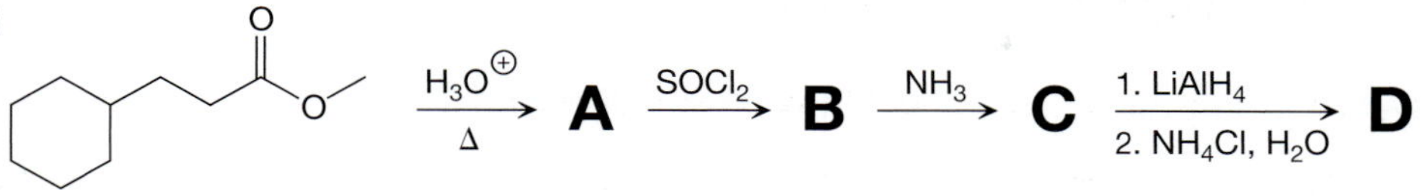

H3O⊕
Δ
A
SOCl2
B
NH3
C
1. LiAlH4
2. NH4Cl, H2O
D

(b)

Same as above
A
1. PBr3, Br2
2. H2O
E
NaCN
F
H3O⊕
Δ
G

(c)

1. DIBAH, –78 °C
2. H3O⊕
H
MgBr
1.
2. NH4Cl, H2O
I
H2CrO4
J

(d)

1. LiAlH4
2. NH4Cl, H2O
K
TsCl
Pyridine
L
CH3CH2SH
DMSO
M

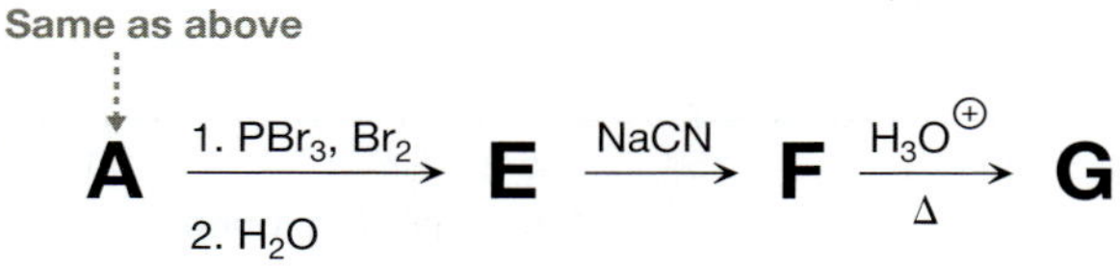

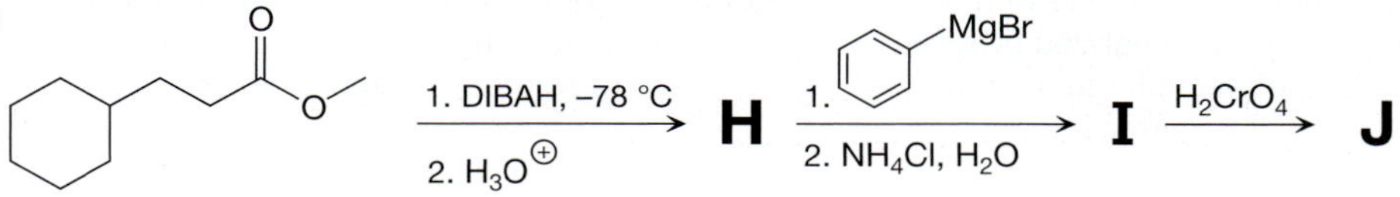

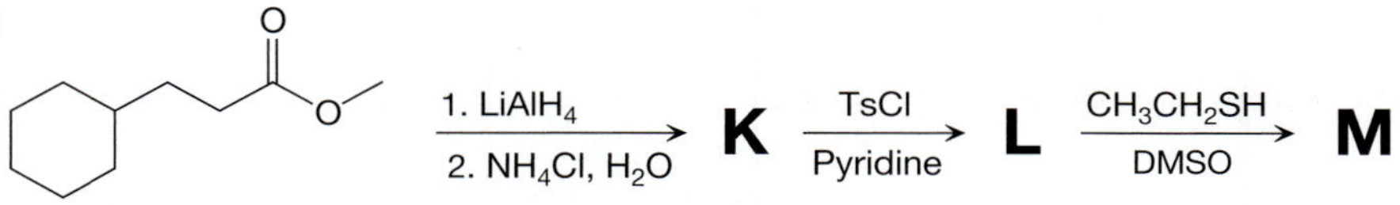

21.89 (SYN) Suggest how you should carry out each of the following syntheses.

(a)

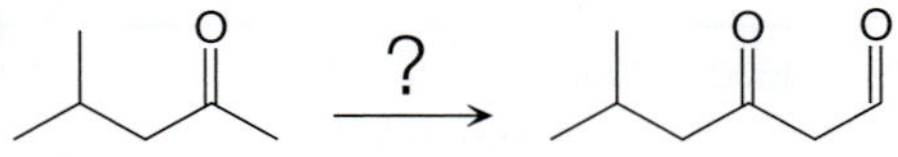

(b)

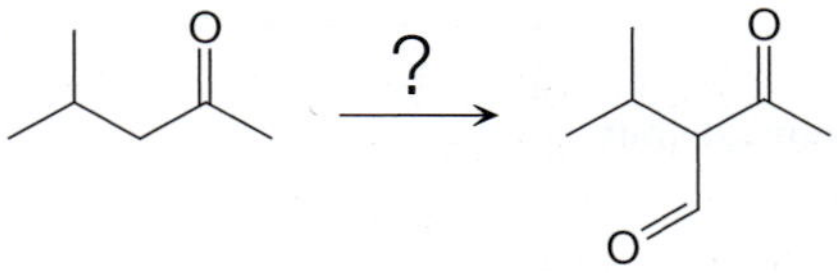

21.90 (SYN) Devise a synthesis of 2-methylpentane-1,3-diol. You may use any inorganic reagents, but your only carbon source must be alcohols containing three or fewer carbons.

21.91 (SYN) Show how you would synthesize 3-oxo-2-methylpentanal. You may use any inorganic reagents, but your only carbon source must be alcohols containing three or fewer carbons.

21.92 (SYN) Show how you would synthesize pentanoic acid from 1,3-propanedioic acid, using any reagents necessary.

21.93 (SYN) Show how you would carry out this synthesis, using any reagents necessary.

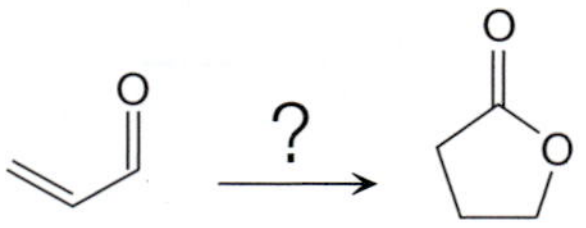

21.94 When phosgene is treated with excess methanol, a product is formed whose ^{1}H NMR spectrum shows one peak—a singlet at 3.8 ppm. Provide a complete, detailed mechanism for this reaction.

CH3OH
?

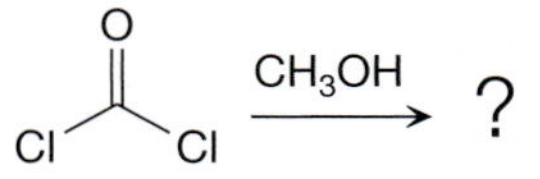

Phosgene

21.95 A student carries out the following sequence of reactions. The IR and ^{13}C NMR spectra are shown for the product **C**. Draw structures for **A–C**. (Remember that the ^{13}C NMR signal at 77 ppm is due to the $CDCl_3$ solvent.)

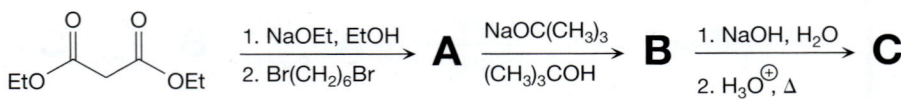

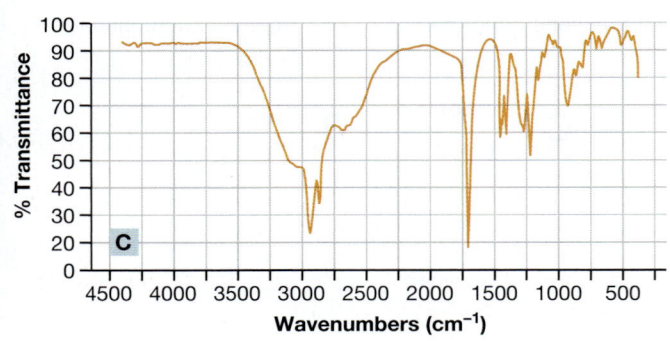
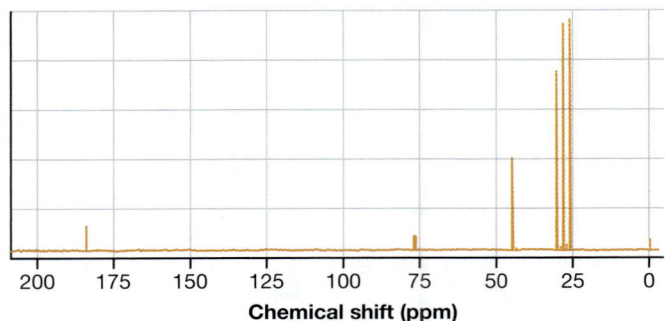

21.96 A student treated methyl 2-methylpropanoate with sodium methoxide dissolved in methanol. After the solution was refluxed for 2 hours, the mixture was analyzed by NMR, IR, and mass spectrometry. Those spectra are shown below.
(a) Draw the mechanism of the reaction the student was expecting and draw the expected product.
(b) What do the spectra suggest occurred?
(c) Explain these results.

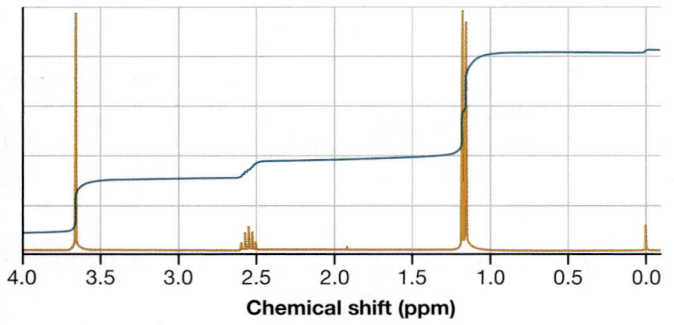
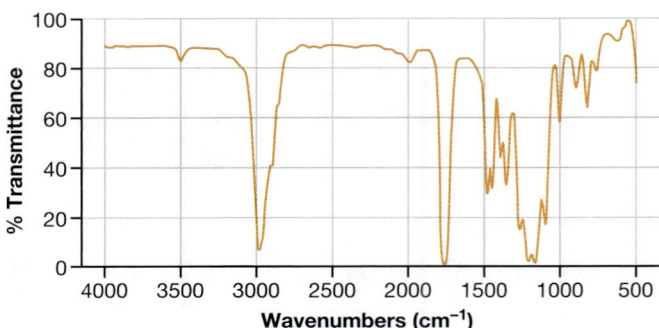

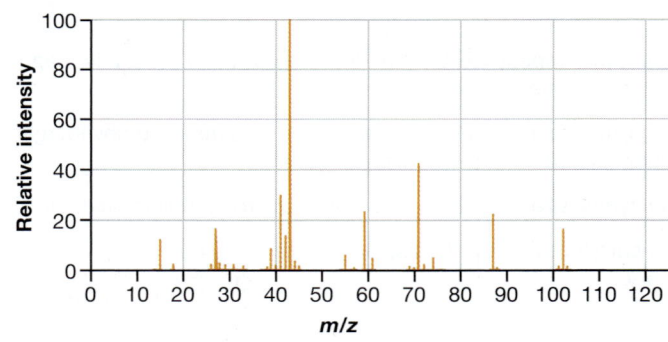

21.97 A liquid, which is insoluble in water, reacts in acidic water to form an insoluble solid product. That product is soluble in water under basic conditions. The IR and ^{1}H NMR spectra of the reactant are shown below. Provide a structure for both the reactant and the product.

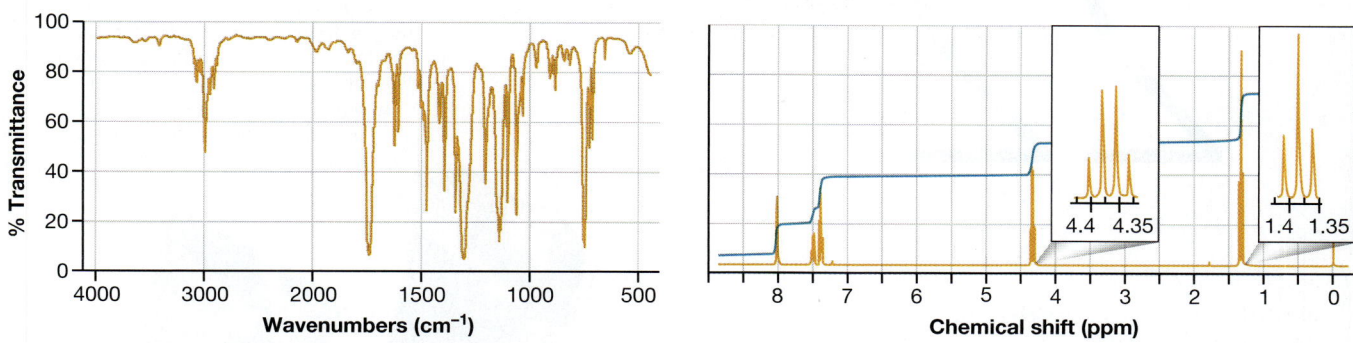

21.98 A compound is treated with excess ethanol. The product is a compound whose formula is $C_6H_{10}O_4$. The IR and ^{1}H NMR spectra of the product are shown below. Draw structures for both the product and the reactant.

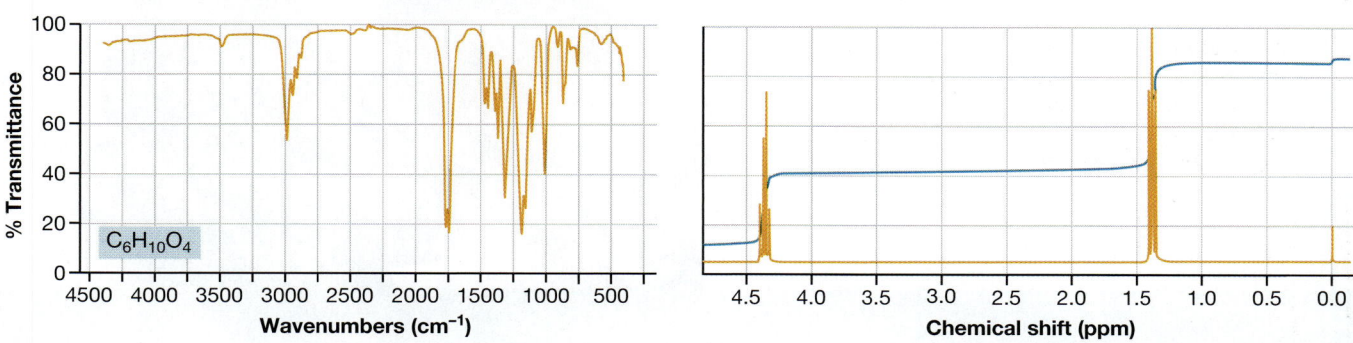

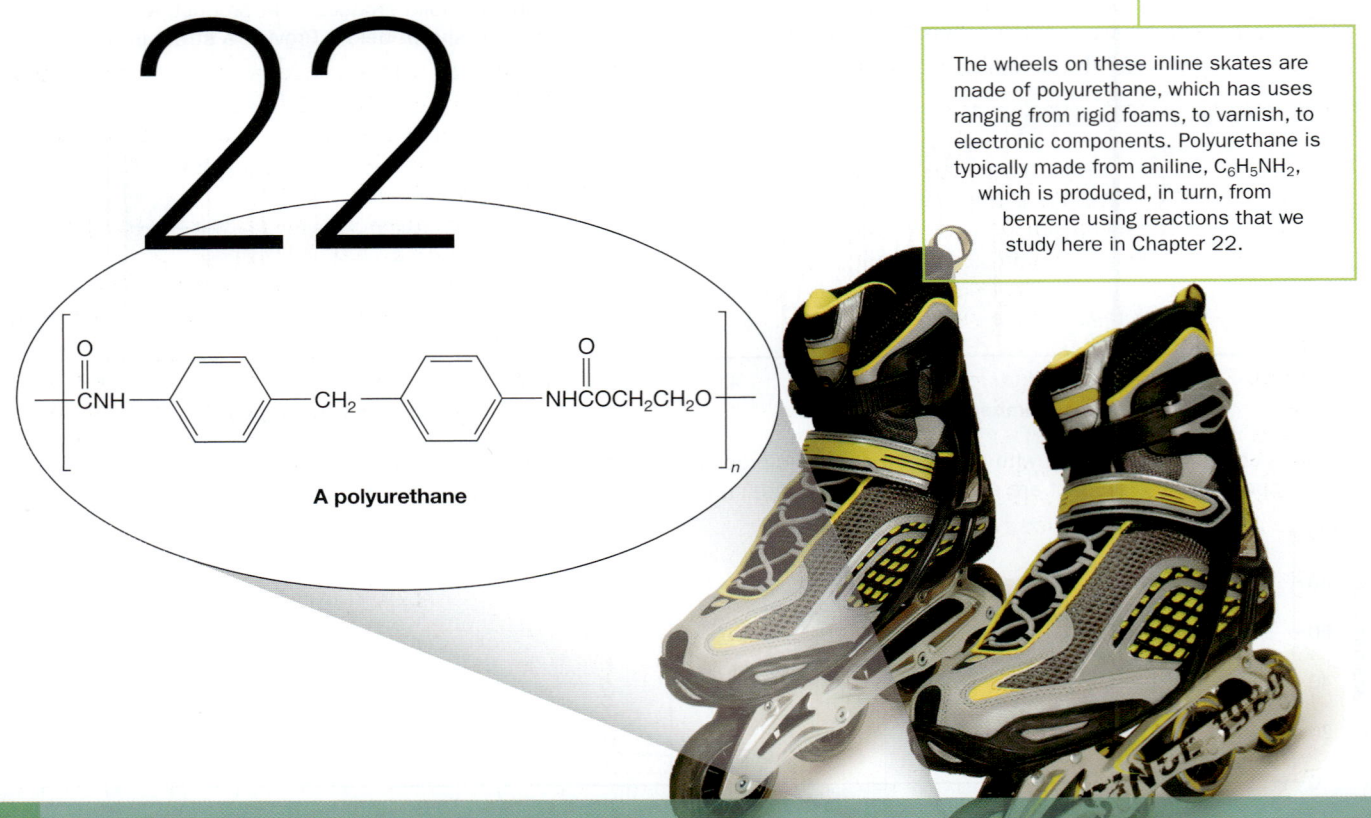

22

A polyurethane

Aromatic Substitution 1

Electrophilic Aromatic Substitution on Benzene; Useful Accompanying Reactions

The Lewis structure of benzene (also called its Kekulé structure) contains three conjugated double bonds arranged in a ring. We might expect, therefore, that benzene would undergo electrophilic addition like the various alkenes and alkynes in Chapters 11 and 12. But it doesn't. No reaction occurs, for example, when benzene is treated with HCl (Equation 22-1) or with Cl_2 (Equation 22-2) under normal conditions.

$$\text{benzene} \xrightarrow{\text{HCl}} \boxed{\text{No reaction}} \qquad (22\text{-}1)$$

$$\text{benzene} \xrightarrow{Cl_2} \boxed{\text{No reaction}} \qquad (22\text{-}2)$$

This lack of reactivity is attributed to the stability of benzene's aromatic π system (see Chapter 14); the addition of HCl or Cl_2 to benzene destroys that aromaticity by converting benzene to a *nonaromatic* compound. (See Your Turn 22.1.)

Chapter Objectives

On completing Chapter 22 you should be able to:

1. Explain why aromatic species resist electrophilic addition reactions.

2. Draw the general mechanism for electrophilic aromatic substitution. Describe the characteristics and the importance of the arenium ion intermediate.

3. Explain why electrophilic aromatic substitution tends to take place under kinetic control and specify the order of the reaction with respect to the electrophile and the aromatic ring.

4. Draw the detailed mechanism for the halogenation of benzene. Provide examples of Lewis acid catalysts and describe their role in the mechanism.

5. Draw the detailed mechanism for a Friedel–Crafts alkylation and a Friedel–Crafts acylation reaction. Specify the limitations of each kind of reaction.

6. Draw the detailed mechanism for a nitration reaction, and describe the role of sulfuric acid.

7. Draw the detailed mechanism for a sulfonation reaction. Describe the role of SO_3 in fuming sulfuric acid, and explain how sulfonation is reversed.

8. Show how to synthesize a primary alkylarene using a Friedel–Crafts acylation reaction.

9. Show how to synthesize aromatic carboxylic acids by using $KMnO_4$ oxidations of carbon side chains on aromatic rings.

10. Show how to synthesize aromatic amines via nitration.

11. Devise syntheses that incorporate substitution reactions on aryldiazonium ions. Identify which of these reactions are classified as Sandmeyer reactions.

YOUR TURN 22.1

Draw the electrophilic addition products for the reactions in Equations 22-1 and 22-2 and show that they are, indeed, nonaromatic.

Answers to Your Turns are in the back of the book.

Electrophilic addition reactions that *permanently* destroy benzene's aromaticity—such as those in Equations 22-1 and 22-2—are heavily disfavored.

When iron metal or iron(III) chloride ($FeCl_3$) is added to the reaction mixture, however, benzene does react with Cl_2, as shown in Equation 22-3.

$$(22\text{-}3)$$

Benzene   **Chlorobenzene**

The Cl in the product appears where there was initially a H, so this is a *substitution* reaction, *not* an addition. Because an electrophilic Cl and an aromatic species are involved, the reaction is called **electrophilic aromatic substitution**.

Here in Chapter 22, we study the general mechanism for the reaction in Equation 22-3 and other electrophilic aromatic substitution reactions involving benzene. Although the reactions can include a variety of different reagents, they follow the same general mechanism.

Aromaticity is not limited to benzene. Other aromatic compounds we examined in Chapter 14 can also undergo electrophilic aromatic substitution. The mechanisms for the reactions involving these other aromatic compounds are no different from the ones involving benzene, but they require additional consideration of reaction rates and regiochemistry, so we discuss them in Chapter 23. Here in Chapter 22, we focus on variations in reactions described by the electrophilic aromatic substitution mechanism. Then, toward the end of the chapter, we turn our attention toward organic synthesis, introducing reactions that are often used in conjunction with electrophilic aromatic substitution reactions.

22.1 The General Mechanism of Electrophilic Aromatic Substitutions

Just like alkenes and alkynes, the π system of benzene is relatively electron rich. In contrast, an electrophile is relatively electron poor, and may generically be represented as E^+. (The positive charge emphasizes the electron-poor nature of the electrophile.) When benzene encounters an electrophile, therefore, the two species can form a new bond, using a pair of π electrons from benzene. This is an electrophilic addition step, shown in Step 1 of Equation 22-4, the general mechanism for electrophilic aromatic substitution. It is the same first step we have seen previously in the addition of a Brønsted acid across an alkene (Chapter 11), in which the electrophile is a proton (H^+).

General mechanism for electrophilic aromatic substitution

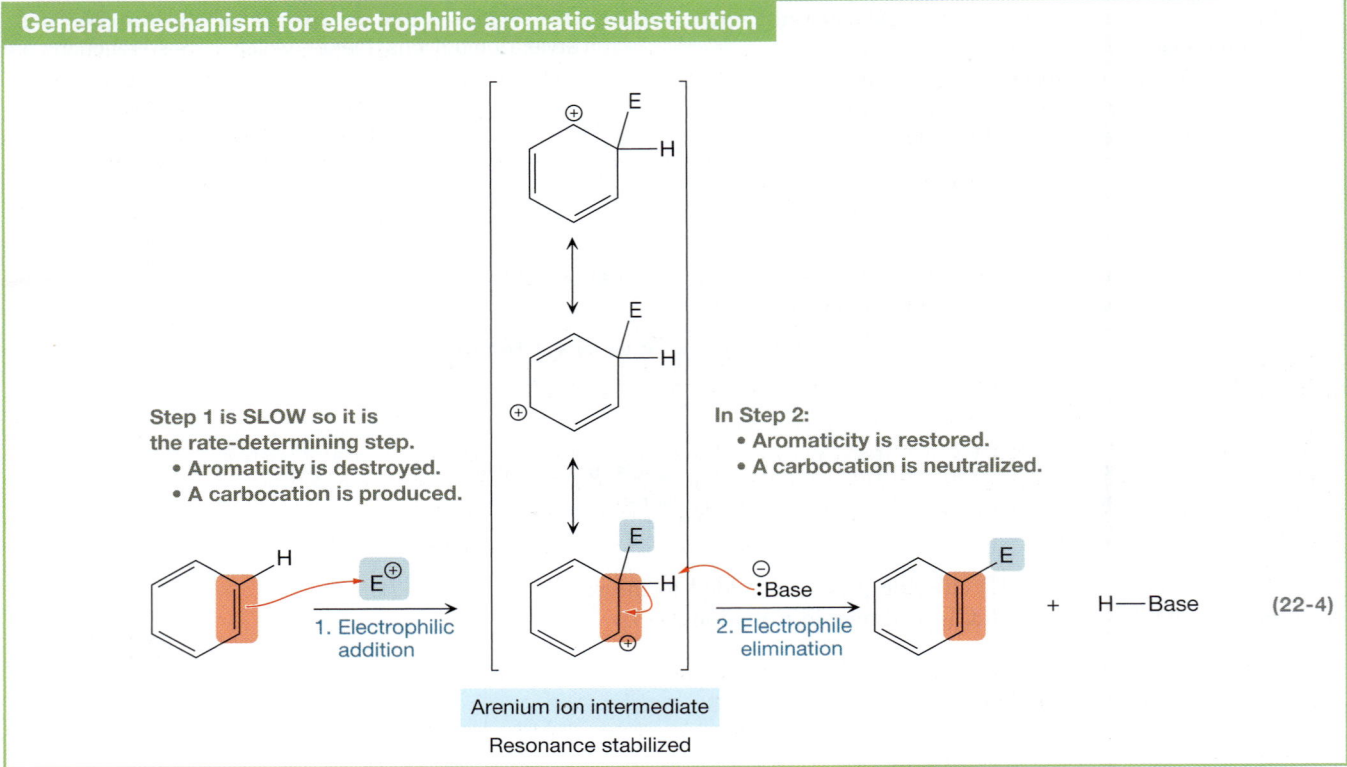

Step 1 is SLOW so it is the rate-determining step.
• Aromaticity is destroyed.
• A carbocation is produced.

In Step 2:
• Aromaticity is restored.
• A carbocation is neutralized.

1. Electrophilic addition

2. Electrophile elimination

Arenium ion intermediate
Resonance stabilized

(22-4)

The product of Step 1 is called an **arenium ion intermediate** or a **Wheland intermediate**. It is a carbocation intermediate consisting of five sp^2-hybridized C atoms and one sp^3-hybridized C atom. This intermediate participates in an electrophile elimination step in Step 2 of the mechanism. A proton (H^+) is eliminated, and is assisted by the formation of a bond to a base that is present. Overall, then, E^+ replaces H^+, converting benzene into a substituted benzene.

YOUR TURN 22.2

Label the hybridization on each carbon atom in this arenium ion intermediate.

The arenium ion intermediate is not aromatic, and it possesses a positively charged carbon that lacks an octet. As a result, this intermediate is much less stable

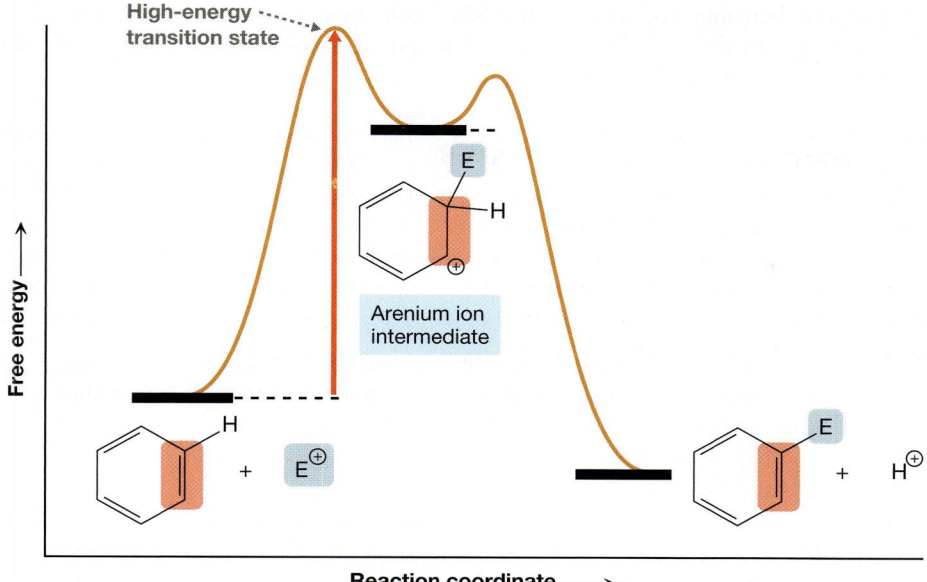

High-energy transition state

Free energy →

Arenium ion intermediate

Reaction coordinate →

FIGURE 22-1 Reaction energy diagram for an electrophilic aromatic substitution The arenium ion intermediate is significantly less stable than either the reactants or products, due to the loss of aromaticity and the formation of a positively charged C that lacks an octet. As a result, Step 1 has a very-high-energy transition state and is the slow step.

than the starting compound, benzene, as illustrated in the reaction energy diagram in Figure 22-1. Consequently, the first step is highly unfavorable. By contrast, elimination of H^+ from the sp^3-hybridized C atom in Step 2 is highly favorable, because aromaticity is restored and the carbocation is neutralized.

Given the loss of aromaticity and the formation of a carbocation in Step 1, you might wonder why electrophilic aromatic substitution takes place at all. The arenium ion intermediate is stabilized by the resonance delocalization of the positive charge (Equation 22-4), so it is significantly more stable than it otherwise would be.

YOUR TURN 22.3

Draw all of the resonance structures and the resonance hybrid of the arenium ion intermediate from Equation 22-4. Include the curved arrows that show how each resonance structure is converted to the next one. In the resonance hybrid, how many C atoms share the positive charge?

Hybrid

Notice in Figure 22-1 that the transition state for Step 1 is very high in energy and the energy barrier is quite large, which makes Step 1 the slow step of the mechanism. That is:

In electrophilic aromatic substitution, the electrophilic addition step (Step 1 in Equation 22-4) is the rate-determining step.

Because benzene and the electrophile both appear as reactants in the rate-determining step, the rate of the overall reaction depends on the concentration of both species, as indicated in Equation 22-5.

$$\text{Rate} = k \left[\underset{}{\bigcirc} \right] \left[E^{\oplus} \right] \tag{22-5}$$

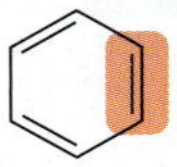

Specifically:

The general electrophilic aromatic substitution reaction is first order with respect to the concentration of benzene and first order with respect to the concentration of the electrophile.

PROBLEM 22.1 Suppose that the concentration of the electrophile is doubled and the concentration of benzene is tripled in an electrophilic aromatic substitution reaction. What happens to the rate of the overall reaction?

In situations where electrophilic aromatic substitution is involved in a competition, it is important to know whether that competition takes place under *kinetic control* or *thermodynamic control* (review Section 9.12). As indicated in Figure 22-1, the overall products appear significantly lower in energy than the reactants, so the overall reaction tends to be *irreversible*. Consequently:

Electrophilic aromatic substitution reactions generally take place under *kinetic control*.

These reactions tend to be irreversible because electrophiles (E^+) that are suitable for aromatic substitution tend to be significantly less stable than the H^+ that is removed, making the reaction highly favorable. Moreover, the bond that forms to H^+ in Step 2 further favors products.

So far, we have dealt only with the two-step electrophilic aromatic substitution mechanism in a general sense: An electrophile first adds to the benzene ring and a proton is then eliminated. In the sections that follow, we focus on specific types of substitution reactions. Each of the reactions we encounter involves the same two-step electrophilic aromatic substitution mechanism, but the identity of E^+ differs from reaction to reaction.

The high reactivity required of each of these electrophiles generally makes it difficult or impossible to add them directly as reactants. Instead:

Electrophiles in electrophilic aromatic substitution reactions typically must be generated in situ from more stable precursors that can be added as starting materials.

Therefore, as you encounter the mechanisms for the various kinds of electrophilic aromatic substitution reactions, pay particular attention to what the actual electrophile is and how it is generated.

22.2 Halogenation

Recall from the Introduction that benzene does not react with molecular chlorine, Cl_2 (Equation 22-2), under normal conditions. In the presence of Fe or $FeCl_3$, however, **chlorination** occurs to yield chlorobenzene (Equation 22-3); this is a type of **aromatic halogenation** reaction. Similarly, molecular bromine, Br_2, does not react with benzene under normal conditions (Equation 22-6), but **bromination** produces bromobenzene (Equation 22-7) in the presence of Fe or $FeBr_3$.

$$\text{benzene} \xrightarrow{\text{Br}_2} \text{No reaction} \qquad (22\text{-}6)$$

$$\text{benzene} \xrightarrow[\substack{\text{Fe or FeBr}_3, \\ 4 \text{ h, } 60\,°\text{C}}]{\text{Br}_2} \text{Bromobenzene} + \text{HBr} \qquad (22\text{-}7)$$

Bromobenzene
97%

The bromination of benzene behaves as if Br^+ is the electrophile that is involved, as shown in Equation 22-8.

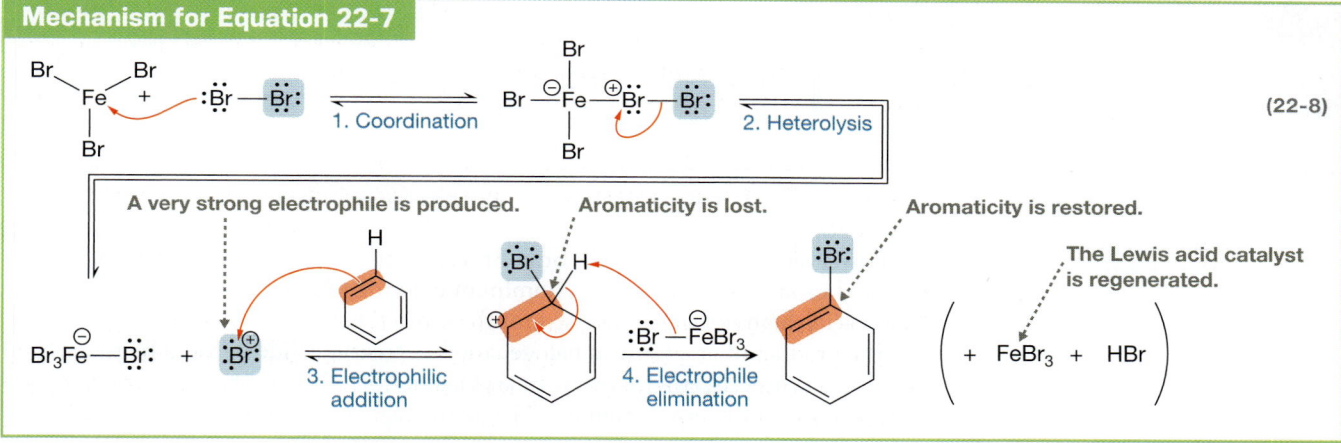

Mechanism for Equation 22-7

1. Coordination
2. Heterolysis

$(22\text{-}8)$

A very strong electrophile is produced. Aromaticity is lost. Aromaticity is restored.

The Lewis acid catalyst is regenerated.

3. Electrophilic addition
4. Electrophile elimination

$(+ \text{FeBr}_3 + \text{HBr})$

$FeBr_3$ acts as a Lewis acid in Step 1 when it complexes Br_2 in a coordination step. $FeBr_3$ can be added directly, or it can be produced in situ by combining Fe and Br_2. In Step 2, heterolysis of the Br—Br bond takes place *slowly*, producing Br^+ and $FeBr_4^-$. Finally, Steps 3 and 4 make up the electrophilic aromatic substitution (i.e., the mechanism presented previously in Equation 22-4), with Br^+ as the electrophile. Notice that $FeBr_3$ is regenerated in Step 4, so it is not consumed overall—it is a *catalyst*. Thus, $FeBr_3$ is an example of a **Lewis acid catalyst**.

Although the bromination of benzene *behaves* as if Br^+ is the electrophile, the arenium ion is more likely produced by the direct reaction between benzene and the Lewis acid–base adduct, $Br_3Fe^- - Br^+ - Br$, as shown below.

$$\cdots + {}^- FeBr_4$$

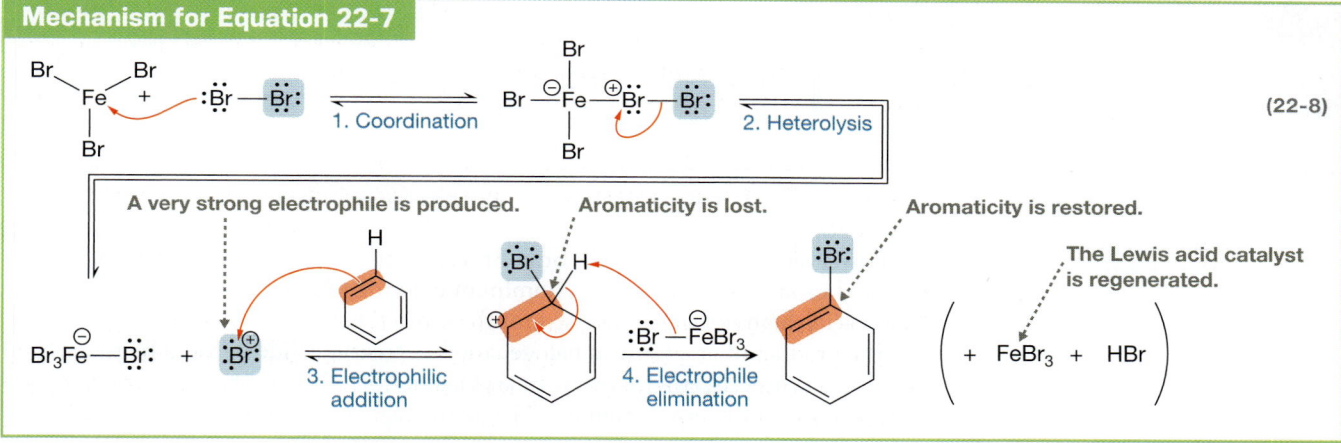

Effectively, the heterolysis and electrophilic addition steps (Steps 2 and 3) in Equation 22-8 take place in a single step, which avoids the production of the high-energy Br^+ species.

YOUR TURN 22.4

The mechanism of the chlorination of benzene is as follows, but the curved arrows have been omitted. Supply the missing curved arrows, and below each reaction arrow, write the name of the elementary step that is taking place. Which two steps make up the electrophilic aromatic substitution mechanism presented earlier in Equation 22-4?

Lewis acids other than $FeCl_3$ or $FeBr_3$ can be used to carry out aromatic chlorination and bromination reactions. Aluminum chloride ($AlCl_3$), for example, is another Lewis acid catalyst that can be used for these reactions.

The product of an aromatic halogenation—an aryl halide—could be the desired target in a synthesis. Alternatively, an aryl halide could be a useful synthetic intermediate, such as in the production of a Grignard reagent (see Solved Problem 22.2).

SOLVED PROBLEM 22.2

Show how to carry out this synthesis using an aromatic halogenation reaction.

Think What reaction can be used to form the necessary C—C bond? Does the product of that reaction contain the appropriate functional groups? What precursors are necessary to carry out the C—C bond-forming reaction? How must halogenation be incorporated to produce the appropriate precursor?

Solve The new C—C bond can be formed using a Grignard reaction, the product of which is an alcohol that can be oxidized to produce the ketone target. To carry out the Grignard reaction, a phenyl Grignard reagent must be used, which can be produced from bromobenzene. Bromobenzene can be produced by brominating benzene in an electrophilic aromatic substitution reaction.

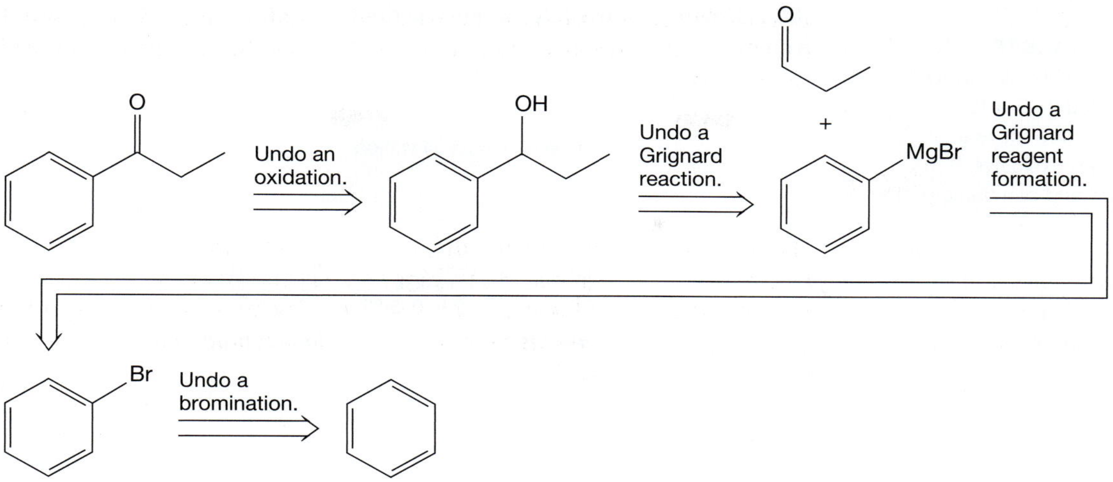

In the forward direction, the synthesis would be written as follows:

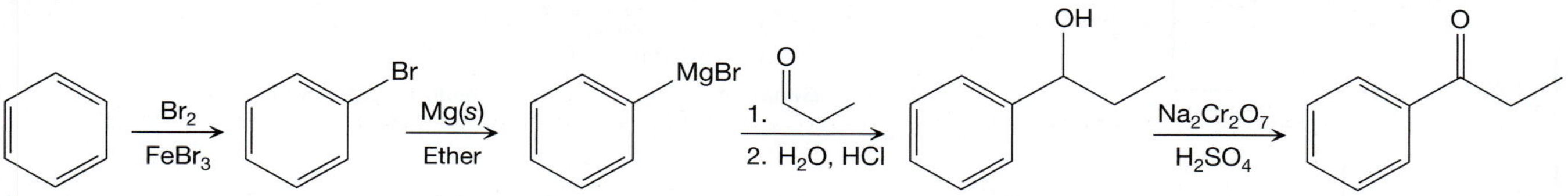

PROBLEM 22.3 Show how to carry out this synthesis using an aromatic halogenation reaction as one step.

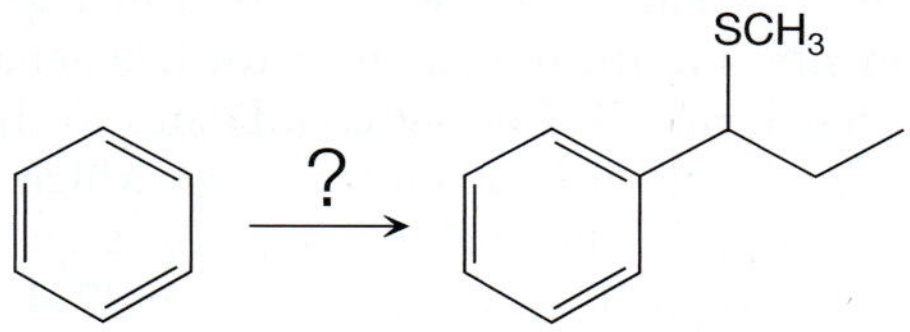

We can also envision the fluorination and iodination of benzene through a similar mechanism. However, just as we saw with the addition of molecular halogens across double bonds (Chapter 12), fluorine is very highly reactive, so other reagents are generally used to carry out aromatic fluorination. Iodination using one of the above Lewis acids, on the other hand, is too slow, but it can be carried out in the presence of an oxidizing agent like Cu^{2+} or hydrogen peroxide (Equation 22-9). Oxidation of I_2 generates I^+, which can then enter into the electrophilic aromatic substitution mechanism.

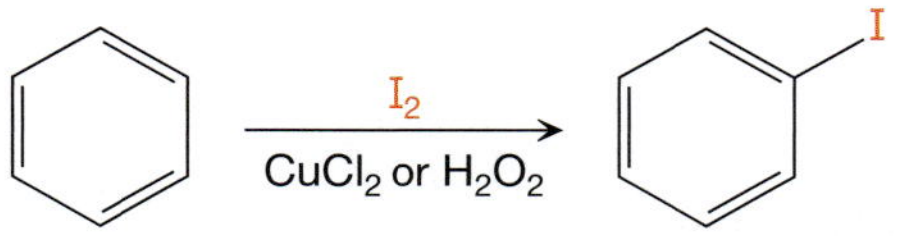

(22-9)

PROBLEM 22.4 With the understanding that the reaction conditions in Equation 22-9 generate a small amount of I^+, draw the two-step electrophilic aromatic substitution mechanism that leads to the formation of iodobenzene.

22.3 Friedel–Crafts Alkylation

In 1877, Charles Friedel (1832–1899) and James Crafts (1839–1917) reported that *alkylation* can take place when benzene is treated with an alkyl chloride in the presence of AlCl₃, which is a strong Lewis acid catalyst (Equations 22-10 and 22-11). In these **Friedel–Crafts alkylation** reactions, the electrophile is a carbocation, R^+. In Equation 22-10, the electrophile is a *tert*-butyl cation, $(CH_3)_3C^+$; in Equation 22-11, it is the cyclohexyl cation, $C_6H_{11}^+$.

Friedel–Crafts alkylation

Benzene

$\xrightarrow[\text{2 h, 0–5 °C}]{AlCl_3,}$

tert-Butylbenzene
72%

(22-10)

Benzene

$\xrightarrow[\text{3 h, 30 °C}]{AlCl_3,}$

Cyclohexylbenzene
93%

(22-11)

The production of these carbocation electrophiles is illustrated in Equation 22-12, the mechanism of the alkylation in Equation 22-10. This mechanism is essentially identical to the one for aromatic halogenation shown previously in Equation 22-8. In Step 1, the alkyl halide coordinates to the Al atom of $AlCl_3$. In Step 2, the Cl—C bond breaks in a heterolysis step, which takes place *slowly*, yielding the carbocation electrophile. The final two steps, once again, are the electrophilic aromatic substitution steps originally shown in Equation 22-4.

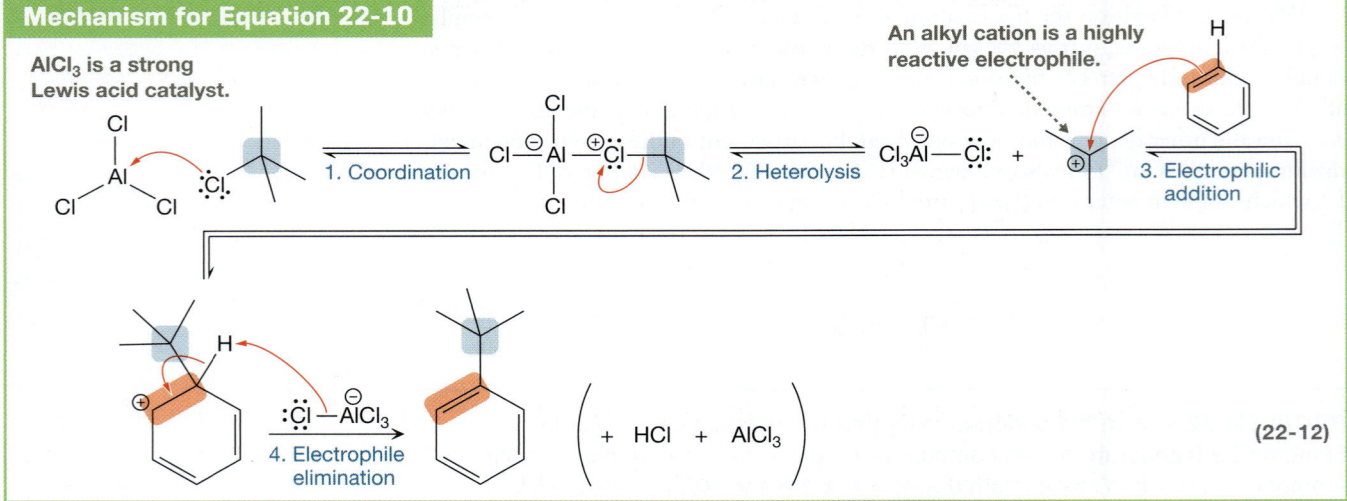

Mechanism for Equation 22-10

$AlCl_3$ is a strong Lewis acid catalyst.

1. Coordination

2. Heterolysis

An alkyl cation is a highly reactive electrophile.

3. Electrophilic addition

4. Electrophile elimination

(+ HCl + $AlCl_3$)

(22-12)

Friedel–Crafts alkylation reactions can take place in the presence of other Lewis acid catalysts, too, including $FeCl_3$ and $FeBr_3$. They can also take place under other conditions that generate carbocation intermediates. Examples are shown in Equations 22-13 and 22-14. In Equation 22-13, the acidic conditions convert the OH group into an excellent leaving group, which departs as H_2O. The carbocation is produced on loss of the water leaving group. In Equation 22-14, on the other hand, protonation of the double bond yields a carbocation intermediate directly. These are the same steps that make up the beginnings of the $S_N1/E1$ and electrophilic addition reactions, respectively.

$$\text{(benzene)} \xrightarrow[\text{H}_3\text{PO}_4]{\text{HO}} \text{(tert-butylbenzene)} \quad \text{(22-13)}$$

$$\text{(benzene)} \xrightarrow[\substack{\text{H}_2\text{SO}_4, \\ <10\ ^\circ\text{C}}]{\text{(cyclohexene)}} \text{(cyclohexylbenzene)} \quad \text{(22-14)}$$

68%

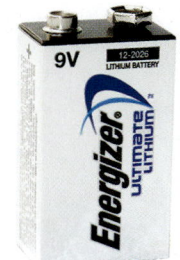
YOUR TURN 22.5

The mechanism of the reaction in Equation 22-11 is as follows, but the curved arrows have been omitted. Supply the missing curved arrows, and below each reaction arrow, write the name of the elementary step taking place. Identify the two steps that make up the general electrophilic aromatic substitution mechanism shown previously in Equation 22-4.

$$\text{AlCl}_3 + \text{(chlorocyclohexane)} \rightleftharpoons \text{Cl}-\overset{\ominus}{\text{Al}}-\overset{\oplus}{\text{Cl}}-\text{(cyclohexyl)} \rightleftharpoons \text{Cl}_3\overset{\ominus}{\text{Al}}-\ddot{\text{Cl}}: + \overset{\oplus}{\text{(cyclohexyl cation + benzene)}}$$

$$\xrightarrow{\text{Cl}-\overset{\ominus}{\text{AlCl}_3}} \text{(cyclohexylbenzene)} \left(+ \text{ HCl } + \text{ AlCl}_3 \right)$$

PROBLEM 22.5 Propose a mechanism for the reaction in Equation 22-13.

PROBLEM 22.6 Propose a mechanism for the reaction in Equation 22-14.

SOLVED PROBLEM 22.7

Show how to synthesize the following compound in a single reaction, using benzene as one reagent, and **(a)** an alkyl halide, **(b)** an alkene, or **(c)** an alcohol as the other reagent.

$$\text{(benzene)} \xrightarrow{?} \text{(diphenylmethyl compound)}$$

Think What is the electrophile that must substitute for H^+? How can that electrophile be produced from an alkyl halide, an alkene, or an alcohol?

Solve The electrophile that must substitute for H^+ is a carbocation, as follows:

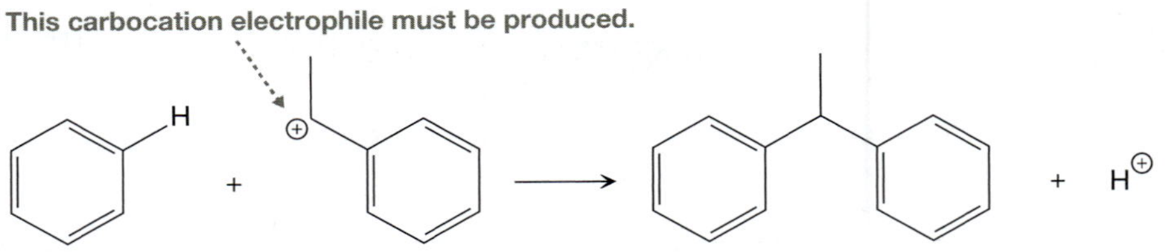

This carbocation electrophile must be produced.

The carbocation can be produced from an alkyl halide using $AlCl_3$, a strong Lewis acid catalyst, or it can be produced from an alkene or an alcohol under acidic conditions.

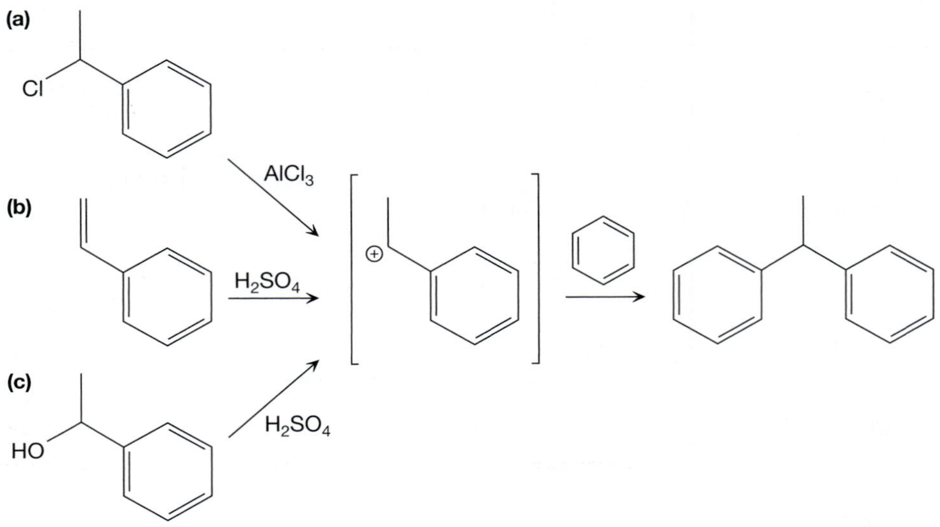

(a)

(b)

(c)

PROBLEM 22.8 Show how you can synthesize this alkylbenzene in a single reaction from benzene and **(a)** an alkyl halide, **(b)** an alkene, or **(c)** an alcohol.

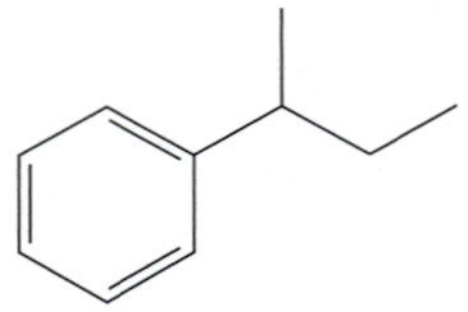

22.4 Limitations of Friedel–Crafts Alkylations

In the Friedel–Crafts alkylation on the next page, you might expect the product in Equation 22-15a to be produced, in which the new bond has replaced the C—Cl bond. However, the product in Equation 22-15b is produced instead.

The product in Equation 22-15a is not produced substantially because, as with any reaction in which a carbocation is produced:

Friedel–Crafts alkylations are susceptible to carbocation rearrangements.

This product would be expected if no carbocation rearrangement takes place.

Minor (22-15a)

(1-Methyl-2-phenylethyl)benzene

Carbocation rearrangement leads to this product instead.

Major (22-15b)

(1-Phenylpropyl)benzene

As shown in the partial mechanism in Equation 22-16, the carbocation that is initially formed in Equation 22-15 is a secondary carbocation in which the positive charge is localized. A 1,2-hydride shift rapidly produces a more stable carbocation that has the positive charge resonance delocalized onto the benzene ring. The product of that carbocation rearrangement is the actual electrophile that enters the electrophilic aromatic substitution mechanism.

Partial mechanism for Equation 22-15

3. 1,2-Hydride shift

4. Electrophilic addition

(22-16)

$Cl-\overset{\ominus}{AlCl_3}$

5. Electrophile elimination

Equation 22-17 shows another example of a rearrangement in a Friedel–Crafts alkylation.

(22-17)

(1-Methylethyl)benzene

CONNECTIONS

(1-Methylethyl)benzene, more commonly called cumene, is used industrially to produce phenol, C_6H_5OH, a starting material for polycarbonate plastics.

The product of this reaction is what we would expect if the C—Cl bond undergoes heterolysis to form a primary carbocation that then rapidly rearranges to a more stable secondary carbocation. However, as we learned in Section 9.6b, a primary carbocation with a localized positive charge is not a reasonable intermediate because it is too unstable. Instead, the hydride shift and heterolysis can occur together in the same step, as shown in the partial mechanism in Equation 22-18.

Partial mechanism for Equation 22-17

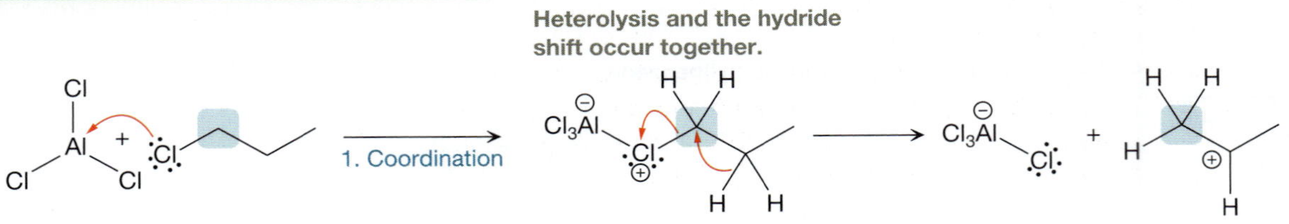

Heterolysis and the hydride shift occur together.

(22-18)

In some cases, no Friedel–Crafts reaction takes place at all, as shown in Equations 22-19 and 22-20.

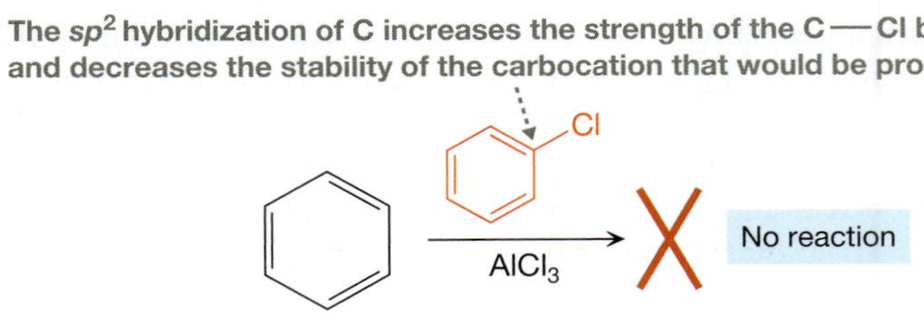

The sp^2 hybridization of C increases the strength of the C—Cl bond and decreases the stability of the carbocation that would be produced.

No reaction (22-19)

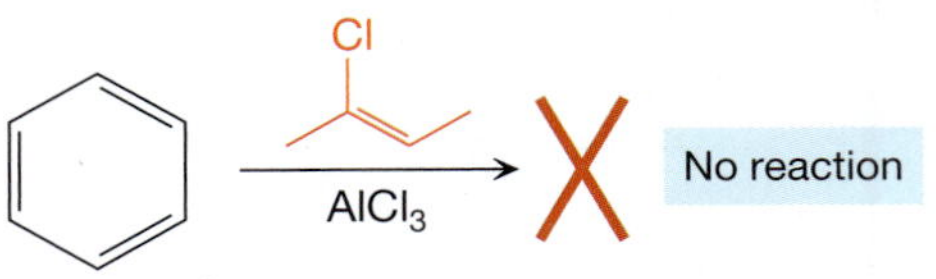

No reaction (22-20)

In general:

> A Friedel–Crafts alkylation reaction does not occur readily unless the halogen atom of the alkyl halide is bonded to an sp^3-hybridized carbon.

In Equations 22-19 and 22-20, which involve an aryl and a vinylic halide, respectively, the halogen atom is bonded to an sp^2-hybridized C. With the increased s character of an sp^2-hybridized C relative to an sp^3-hybridized C, the strength of the C—Cl bond is greater and so is the C atom's effective electronegativity (Section 3.11). Thus, the carbocation that would be produced would be excessively unstable. Recall from Section 9.6a that aryl and vinylic halides are resistant to nucleophilic substitution and elimination reactions for very similar reasons.

YOUR TURN 22.6

Verify the dependence of bond strength on atom hybridization by looking in Chapter 3 (p. 145) to find the C—H bond strength of an sp^3-hybridized carbon and comparing it to the bond strength of an sp^2-hybridized carbon.

sp^3 _____ sp^2 _____

SOLVED PROBLEM 22.9

Show how you can synthesize 2-phenylbut-2-ene from benzene.

Think Can the target be produced directly from a simple Friedel–Crafts alkylation? Is there another reaction that can be used to form the appropriate C—C bond?

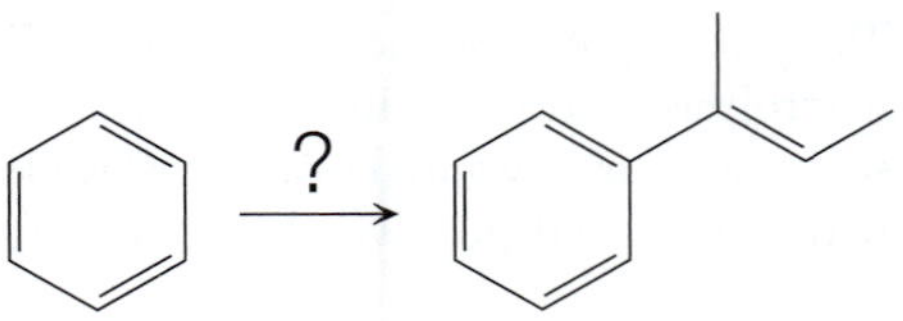

Solve For the target to be produced directly from a Friedel–Crafts alkylation, the precursors would be those in Equation 22-20. For reasons discussed previously, this kind of a reaction would not work. Another route must be sought.

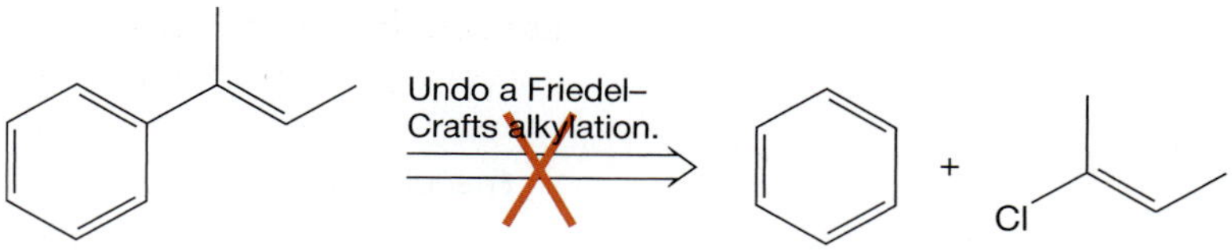

Another C—C bond-forming reaction we have used before is the Grignard reaction. The product of a Grignard reaction is an alcohol, which can undergo dehydration to form the final product. Here we can envision a Grignard reaction between phenylmagnesium bromide and butan-2-one. The Grignard reagent can be produced from bromobenzene, which can be synthesized, in turn, using a bromination reaction.

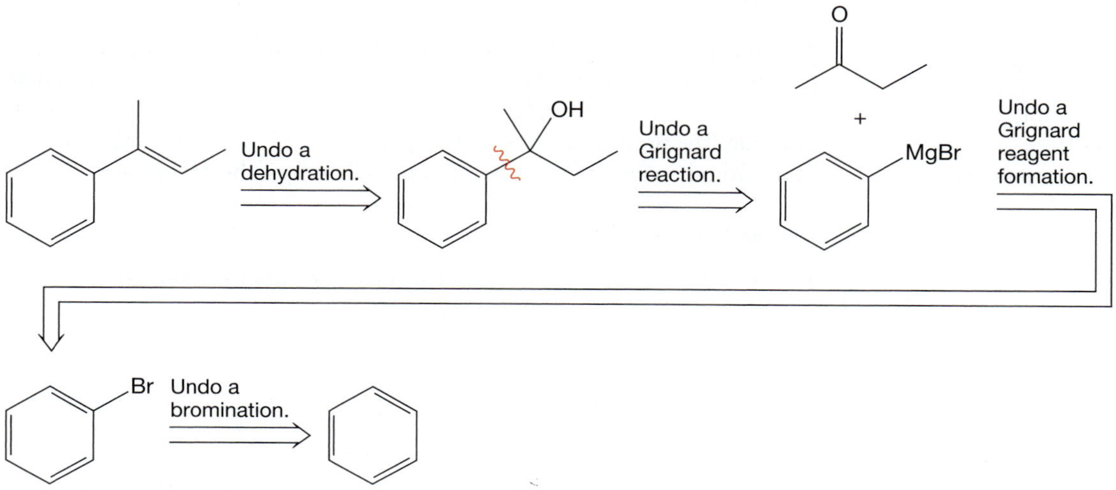

The synthesis might be written as follows.

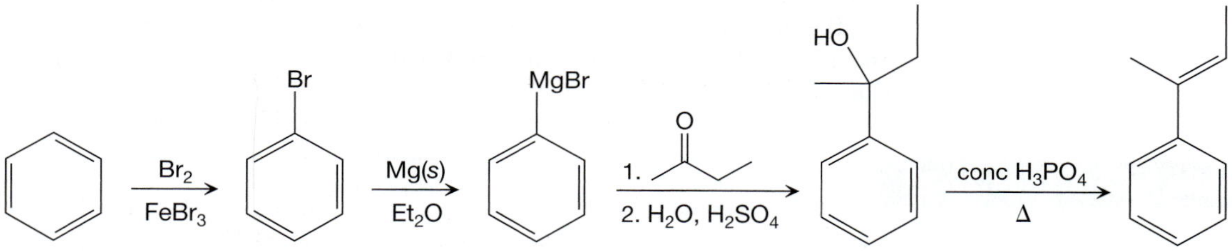

PROBLEM 22.10 Butylbenzene cannot be synthesized in good yield directly from benzene using a Friedel–Crafts alkylation. Why not? Propose an alternate synthesis of butylbenzene that does not use a Friedel–Crafts reaction.

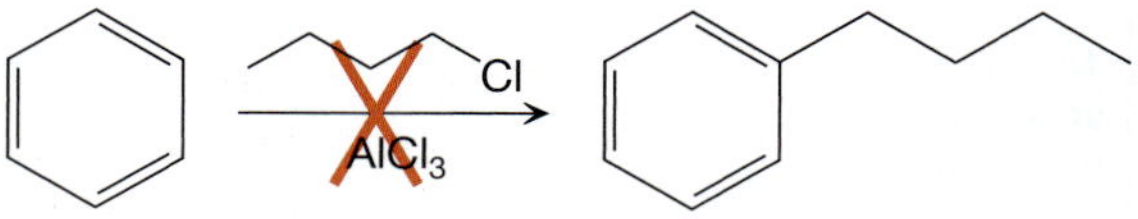

One additional limitation of Friedel–Crafts alkylations stems from the fact that each alkyl group attached to the aromatic ring makes the ring even *more* reactive toward electrophilic aromatic substitution. As shown below, therefore, this could result in a mixture of products.

Friedel–Crafts alkylation can result in polyalkylation.

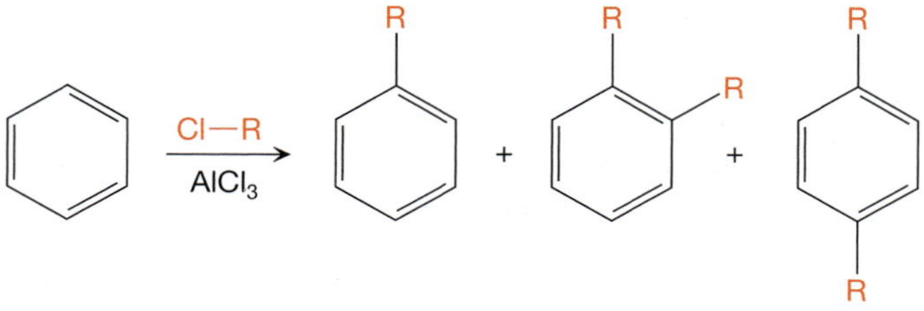

In general:

> Friedel–Crafts alkylations are susceptible to *polyalkylation*.

The reasons why will be discussed more fully in Chapter 23.

22.5 Friedel–Crafts Acylation

Friedel–Crafts acylation of benzene is carried out in much the same way as Friedel–Crafts alkylation. The difference is that the aromatic species is treated with an *acid chloride*, also called an *acyl chloride* (an acyl group is R—C=O), instead of an alkyl chloride. The product, as shown in Equation 22-21, is an *aromatic ketone*, in which a C atom from benzene forms a bond to the C atom of the acyl group.

An acyl group

Friedel–Crafts acylation

An acid chloride, also called an acyl chloride

An aromatic ketone

$$\text{(22-21)}$$

$\xrightarrow[\text{3 h, 0 °C}]{\text{AlCl}_3,}$

1-Phenylpropan-1-one
78%

The mechanism of Friedel–Crafts acylation, presented in Equation 22-22, is essentially identical to the one for Friedel–Crafts alkylation, shown previously in Equation 22-12.

Mechanism for Equation 22-21

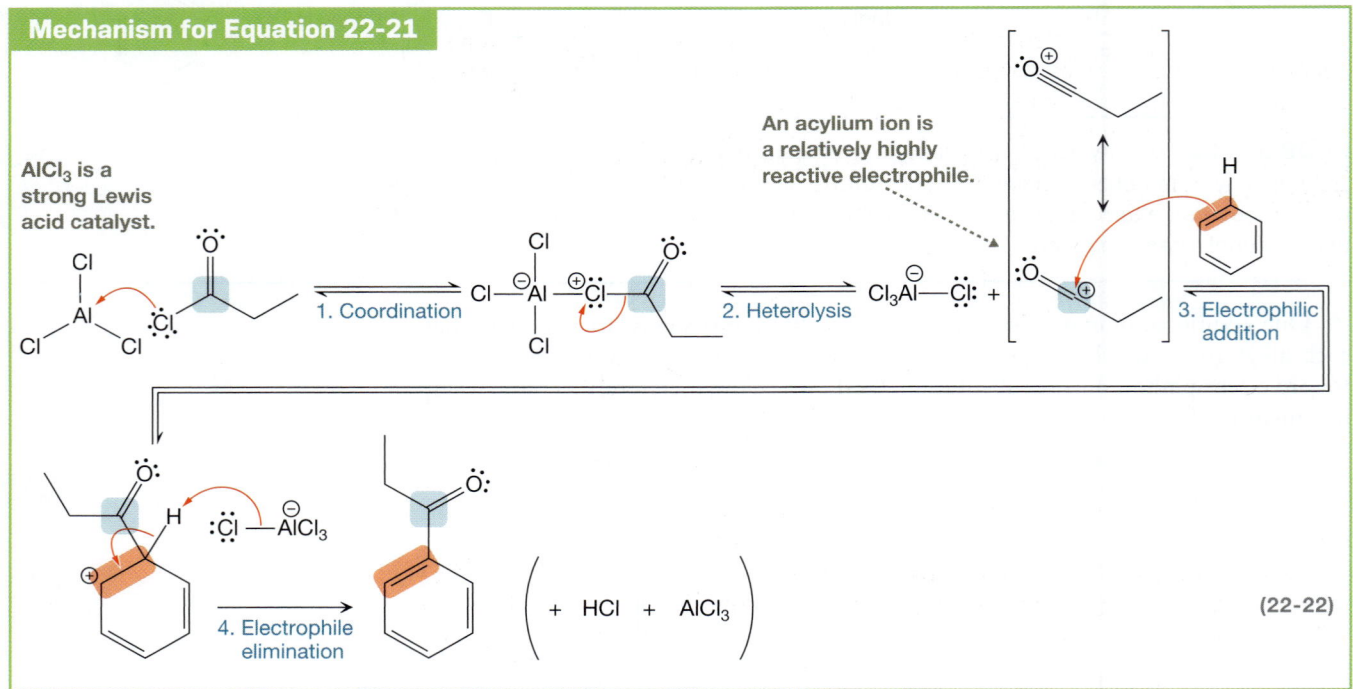

AlCl$_3$ is a strong Lewis acid catalyst.

1. Coordination

2. Heterolysis

An acylium ion is a relatively highly reactive electrophile.

3. Electrophilic addition

4. Electrophile elimination

$$\left(+ \text{ HCl } + \text{ AlCl}_3 \right)$$

$$\text{(22-22)}$$

The strong Lewis acid catalyst, AlCl$_3$, is responsible for generating the cationic electrophile, called an **acylium ion**, in the first two steps. In Step 1, the Cl atom of the acyl chloride coordinates to the electron-deficient Al atom. In Step 2, the C—Cl bond undergoes heterolysis. Step 3 is electrophilic addition of the acylium ion to the benzene ring, producing the arenium ion intermediate. Finally, in Step 4, removing the proton from the arenium ion yields the aromatic ketone product.

YOUR TURN **22.7**

The mechanism for the Friedel–Crafts acylation involving acetyl chloride is as follows, but most of the curved arrows have been omitted, as have the overall products. Supply the missing curved arrows and draw the overall products. Also, write the name of each elementary step below the appropriate reaction arrow and label the acylium ion.

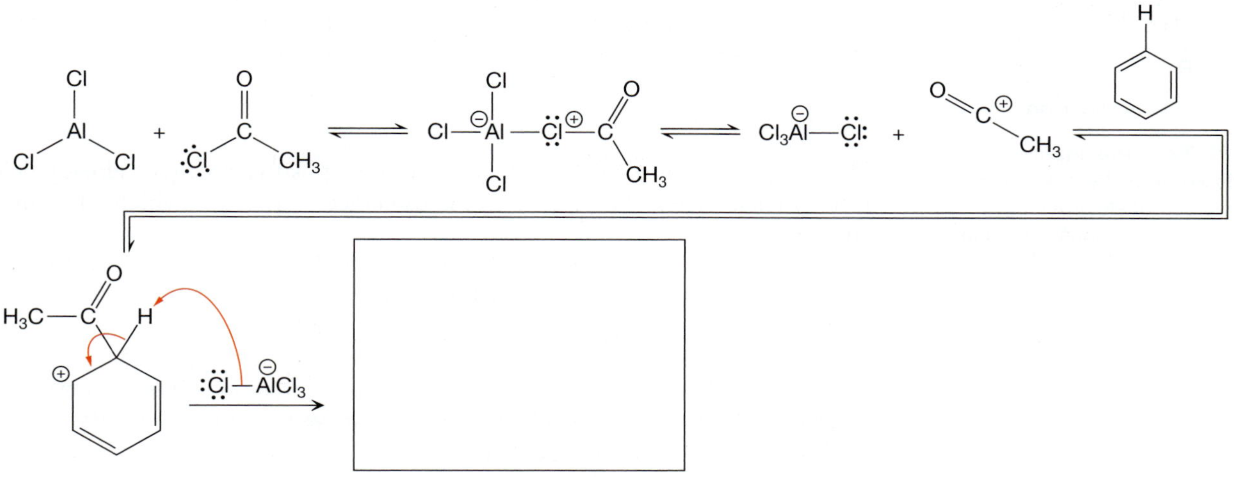

PROBLEM 22.11 As we learned in Chapters 20 and 21, carboxylic acid chlorides behave quite similarly to carboxylic acid anhydrides. Not surprisingly, then, aromatic acylation can be carried out using an acid anhydride in the presence of AlCl$_3$, analogous to the Friedel–Crafts acylation reaction in Equation 22-21. The example here shows this using acetic anhydride to produce phenylethanone. Propose a mechanism for this reaction.

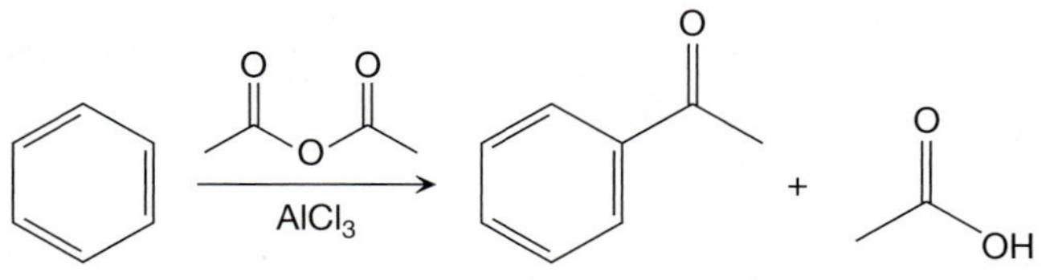

**Phenylethanone
(Acetophenone)**

PROBLEM 22.12 The compound shown here can be synthesized from benzene and a carboxylic acid anhydride in a single Friedel–Crafts reaction. What acid anhydride must be used? *Hint:* The two phenyl rings come from two different compounds.

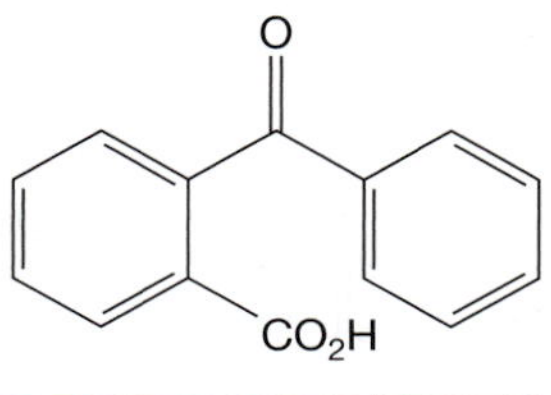

It might seem peculiar that Friedel–Crafts acylation can take place at all, given that the Cl atom in the acid chloride reactant is bonded to an sp^2-hybridized C. (Recall from Section 22.4 that this, in fact, prevents a Friedel–Crafts *alkylation* from taking place.) The acylium ion that is produced in Friedel–Crafts acylation, however, is *resonance stabilized* by a lone pair of electrons on the adjacent oxygen (Fig. 22-2). With this additional stability, the cationic electrophile can be produced and enter into the electrophilic aromatic substitution mechanism.

YOUR TURN 22.8

Draw the resonance hybrid of a generic acylium ion.

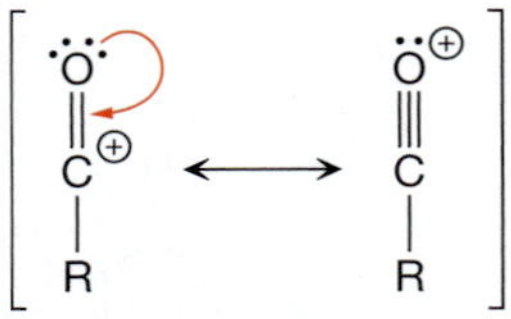

An acylium ion

FIGURE 22-2 Resonance stabilization of the acylium ion
The lone pair of electrons on O participates in resonance to delocalize the positive charge over O and C.

Unlike Friedel–Crafts alkylations:

> Friedel–Crafts acylation reactions are *not* susceptible to carbocation rearrangements.

The acylium ion does not rearrange because its positive charge is already delocalized by resonance and because it has a resonance structure in which all atoms have an octet.

Friedel–Crafts acylation reactions can be used with a variety of acyl halides to produce different aromatic ketones. However:

> Friedel–Crafts acylation as a means of *formylation* (replacing H by a formyl group, HC=O) requires in situ formation of the very unstable Cl—CH=O.

See Problem 22.31 at the end of the chapter for an example.

SOLVED PROBLEM 22.13

Show how to synthesize 1-phenylbutan-1-one from benzene and an alcohol, using a Friedel–Crafts acylation.

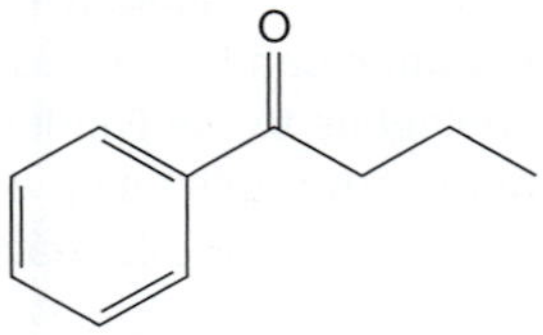

Think The product is an aromatic ketone, suggesting a Friedel–Crafts acylation. What precursors are necessary to produce the target from a Friedel–Crafts acylation reaction? How can those precursors be produced from the starting material?

Solve In a Friedel–Crafts acylation, an aromatic carbon forms a bond to an acyl carbon. We begin our retrosynthetic analysis by disconnecting that bond to arrive at the appropriate precursors—an aromatic ring and an acyl chloride. The acyl chloride can be produced by an acyl substitution on the corresponding carboxylic acid, which can be produced, in turn, by oxidizing a primary alcohol.

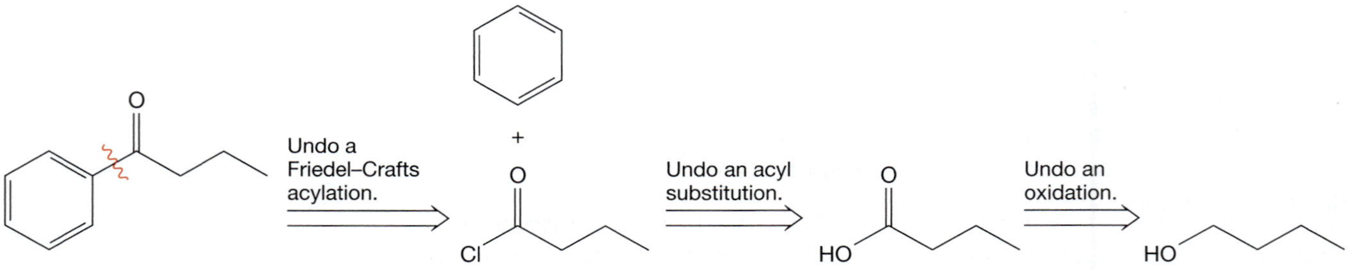

1-Phenylbutan-1-one

In the forward direction, the synthesis would be written as follows:

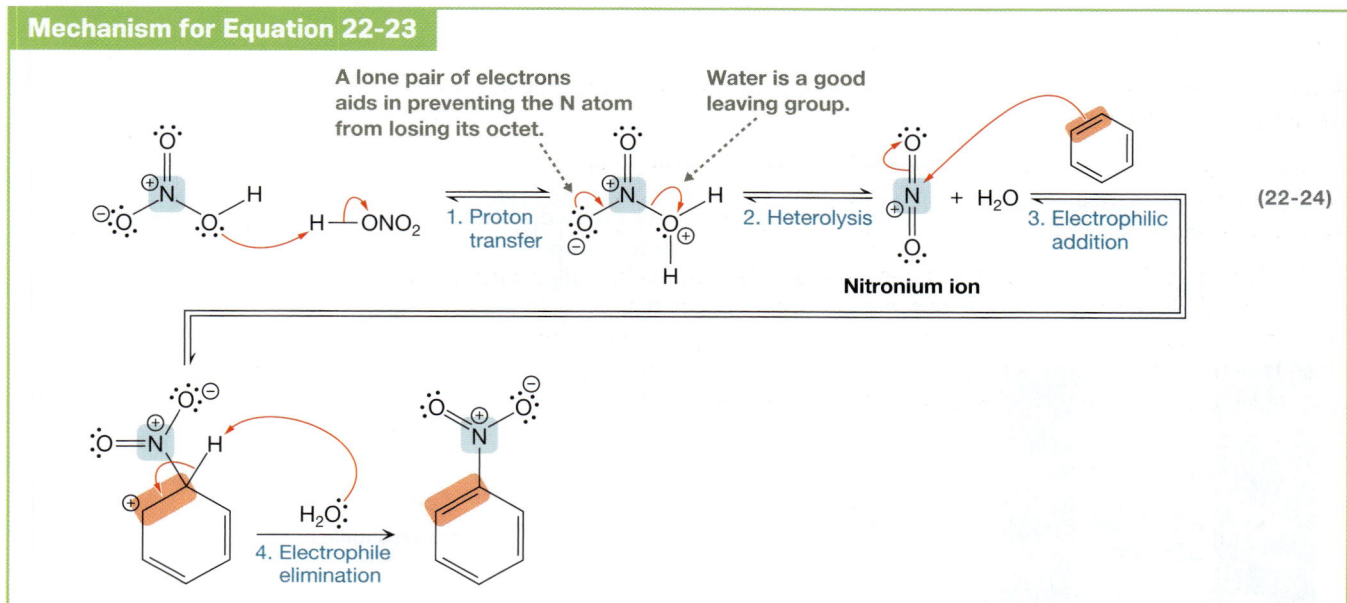

PROBLEM 22.14 Show how to synthesize benzophenone from benzene and any alcohol.

Benzophenone

CONNECTIONS Haloperidol, a commonly used antipsychotic drug, is a derivative of 1-phenylbutan-1-one (Solved Problem 22.13).

22.6 Nitration

When benzene is treated with concentrated nitric acid (HNO_3), nitrobenzene is produced (Equation 22-23). Benzene is said to undergo **nitration**.

Nitration

conc HNO_3
15 °C

(22-23)

Nitrobenzene
83%

The mechanism of this reaction is presented in Equation 22-24.

Mechanism for Equation 22-23

A lone pair of electrons aids in preventing the N atom from losing its octet.

Water is a good leaving group.

H—ONO_2

1. Proton transfer

2. Heterolysis

+ H_2O

3. Electrophilic addition

Nitronium ion

(22-24)

H_2O:

4. Electrophile elimination

Steps 1 and 2 of the mechanism are responsible for generating the powerful NO_2^+ electrophile, called the **nitronium ion**. First, HNO_3 is protonated to create a good

22.6 Nitration **1121**

H_2O leaving group. Next, H_2O departs via heterolysis. During the heterolysis step, an additional $N{=}O$ double bond is formed, using a lone pair of electrons from an O atom. This aids the elimination of the water leaving group by preventing the N atom from losing its octet. Once the NO_2^+ electrophile is created, it enters into the two-step electrophilic aromatic substitution mechanism (i.e., Steps 3 and 4).

Although this nitration reaction is effective:

> With concentrated nitric acid alone, the nitration of benzene tends to proceed relatively slowly.

The rate is slow because the amount of NO_2^+ present in concentrated HNO_3 is small—about 4%—and, as we saw in Equation 22-5, the rate of electrophilic aromatic substitution is directly proportional to the concentration of the electrophile.

To increase the rate of nitration, the reaction can be run at high temperatures. Alternatively:

> The rate of nitration increases by adding concentrated sulfuric acid.

Sulfuric acid (H_2SO_4) is much more acidic than HNO_3, so H_2SO_4 generates a substantially higher concentration of NO_2^+, the active electrophile.

YOUR TURN 22.9

Draw the complete mechanism for the nitration of benzene in which concentrated sulfuric acid has been added.

$$\text{benzene} \xrightarrow[\text{conc } H_2SO_4]{HNO_3} \text{nitrobenzene (} NO_2 \text{)}$$

CONNECTIONS

Benzenesulfonic acid is strongly acidic and produces a salt, called a besylate salt, when treated with a weakly basic compound. Some pharmaceutical drugs are sold as besylate salts, including amlodipine besylate (trade name Norvasc), which is used to treat angina.

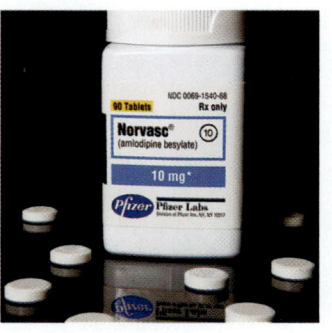

22.7 Sulfonation

Treating benzene with concentrated sulfuric acid yields benzenesulfonic acid (Equation 22-25), whereby a **sulfo group**, SO_3H, has been incorporated onto the ring. This kind of substitution reaction is called **sulfonation**.

Sulfonation

$$\text{benzene}{-}H \xrightarrow[\text{48 h, 20 °C}]{\text{conc } H_2SO_4} \text{benzenesulfonic acid (}SO_3H\text{)} + H_2O \qquad (22\text{-}25)$$

Benzenesulfonic acid
66%

Under these reaction conditions, SO_3H^+ is believed to be the electrophile involved in the electrophilic addition step of electrophilic aromatic substitution. This is illustrated in the mechanism presented in Equation 22-26.

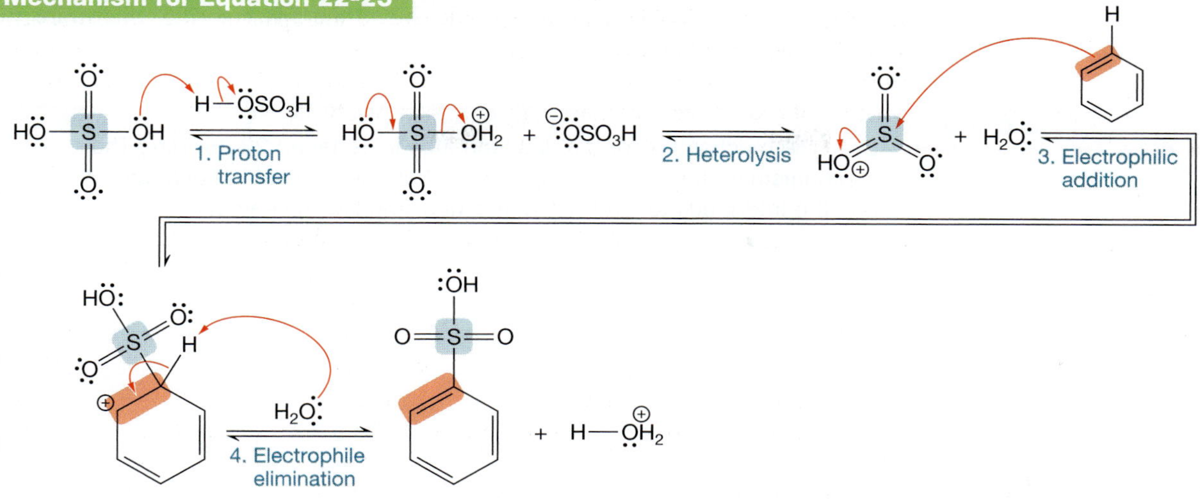

(22-26)

Concentrated H_2SO_4 contains a small amount of SO_3H^+ at equilibrium. It is produced in much the same way as NO_2^+ is generated in concentrated HNO_3 (Equation 22-24): an OH group is protonated in Step 1 to generate a very good H_2O leaving group, then H_2O leaves in Step 2. The resulting HSO_3^+ electrophile enters into the two-step electrophilic aromatic substitution mechanism (i.e., Steps 3 and 4).

As we saw with nitration, sulfonation can be slow because the concentration of the HSO_3^+ electrophile in concentrated H_2SO_4 is rather small. Sulfonation can be sped up, however, using **fuming sulfuric acid**, which is concentrated sulfuric acid infused with SO_3. Under these acidic conditions, SO_3 is protonated to produce HSO_3^+, according to the equilibrium in Equation 22-27.

The concentration of the $SO_3H^\oplus$ electrophile increases in fuming sulfuric acid.

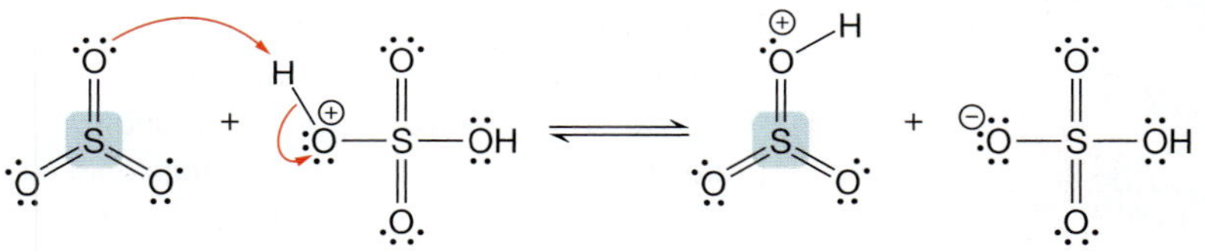

(22-27)

Although electrophilic aromatic substitution reactions are generally irreversible, sulfonations are not.

The sulfonation of an aromatic ring is *reversible*.

Notice that water is a product in the sulfonation reaction in Equation 22-25. Therefore, the reverse of a sulfonation reaction, called **desulfonation** (Equation 22-28), can be carried out by treating an aryl sulfonic acid with large amounts of water under acidic conditions.

Desulfonation

The sulfo group is removed.

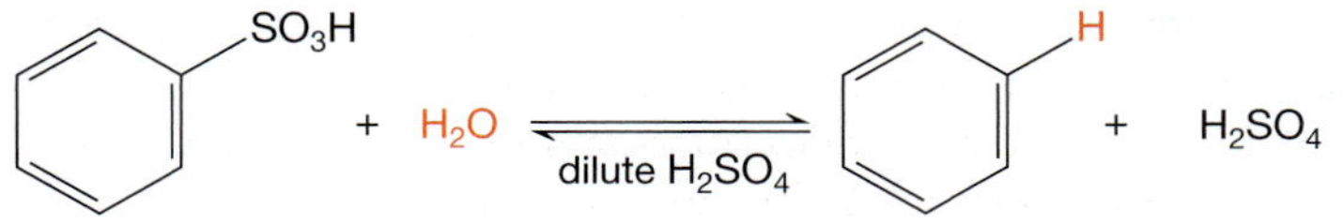

(22-28)

Because a SO_3H group can be added and removed from an aromatic ring in this way, sulfonation can be used to *protect* a specific site on the benzene ring. This strategy is discussed more fully in Chapter 23.

YOUR TURN 22.10

Draw the mechanism for the desulfonation in Equation 22-28. *Hint:* The mechanism is the reverse of the sulfonation mechanism in Equation 22-26, and the intermediates in the two mechanisms are the same.

Aromatic Sulfonation: Antibiotics and Detergents

Sulfa drugs, discovered in 1932, were among the first clinically useful antibiotics—and ultimately sparked a revolution in medicine. The general structure of these antibiotics, shown below on the left, exhibits sulfonamide ($-SO_2NHR'$) and amino groups that are para to each other on the phenyl ring. To synthesize the sulfonamide portion, a sulfo group ($-SO_3H$) is first attached to the aromatic ring via sulfonation, similar to the reaction we studied here in Chapter 22. Then, the sulfo group is converted to a sulfonyl chloride ($-SO_2Cl$), which is condensed with an amine (H_2NR') to produce the final sulfonamide.

Sulfa drugs resemble *para*-aminobenzoic acid (PABA, $H_2N-C_6H_4-CO_2H$), and it is this resemblance that is believed to be responsible for their antimicrobial properties. Bacteria, like humans, need folic acid to synthesize nucleic acids and proteins. Bacteria synthesize their folic acid from PABA in an enzymatic reaction. Sulfa drugs compete with PABA for the active site of that enzyme, and therefore inhibit the synthesis of folic acid. We humans, on the other hand, do not synthesize folic acid. Rather, it is a vitamin that is an essential part of our diets, so we can tolerate sulfa drugs.

Several detergents, too, are synthesized using aromatic sulfonation reactions. As we learned in Section 2.10, detergents have the general structure $RSO_3^- Na^+$, where R is a relatively large, nonpolar group. Some of the most common detergents have a benzene ring separating the SO_3^- portion from an alkyl portion, as shown above on the right. The alkyl portion can be linear or branched, but typically consists of 12 carbons in total, and is thus called a sodium dodecylbenzenesulfonate.

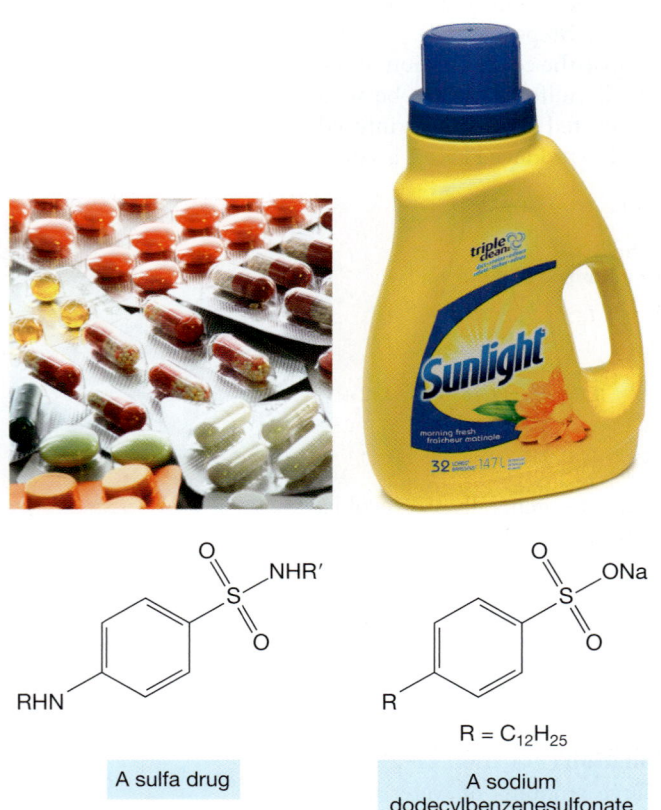

RHN

A sulfa drug

R

R = $C_{12}H_{25}$

A sodium
dodecylbenzenesulfonate

22.8 Organic Synthesis: Considerations of Carbocation Rearrangements and the Synthesis of Primary Alkylbenzenes

Recall from Section 22.4 that one of the limitations of Friedel–Crafts alkylations is the potential for carbocation rearrangements. As we saw in Equation 22-17 (p. 1115), for example, propylbenzene cannot be synthesized directly in good yield from a Friedel–Crafts alkylation because rearrangement of the cationic intermediate leads to (1-methylethyl)benzene instead.

To circumvent this problem, we can use Friedel–Crafts *acylation* instead, which forms a C—C bond to the aromatic ring without competing carbocation rearrangements (Section 22.5). Because the immediate product of an acylation reaction is an aromatic ketone, the carbonyl group must be reduced to produce an alkylbenzene. An example using the Clemmensen reduction (Section 19.3b) is shown in Equation 22-29.

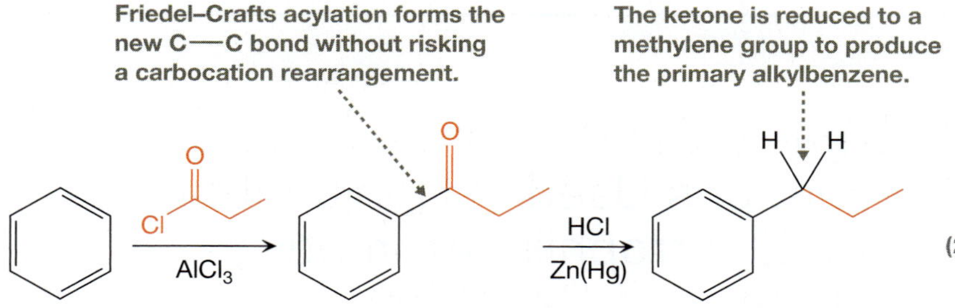

Friedel–Crafts acylation forms the new C—C bond without risking a carbocation rearrangement.

The ketone is reduced to a methylene group to produce the primary alkylbenzene.

(22-29)

SOLVED PROBLEM 22.15

Show how you would synthesize this compound, using benzene as the only aromatic starting compound.

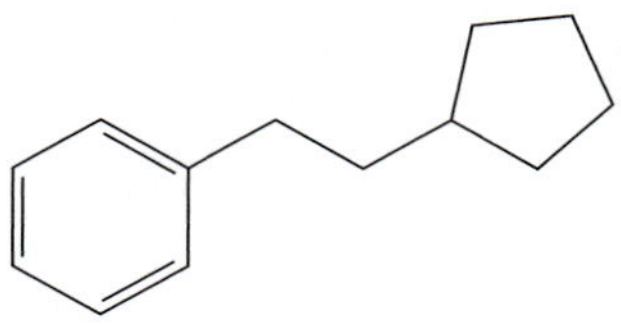

Think The product is an alkylbenzene. Can a Friedel–Crafts alkylation be used to form the C—C bond to benzene? If not, why not? How can a Friedel–Crafts acylation reaction be incorporated into this synthesis?

Solve A Friedel–Crafts alkylation reaction would require the equivalent of a primary carbocation intermediate, as shown below. This is unfeasible, however, because a primary carbocation intermediate, if produced, would rearrange to a more stable secondary carbocation.

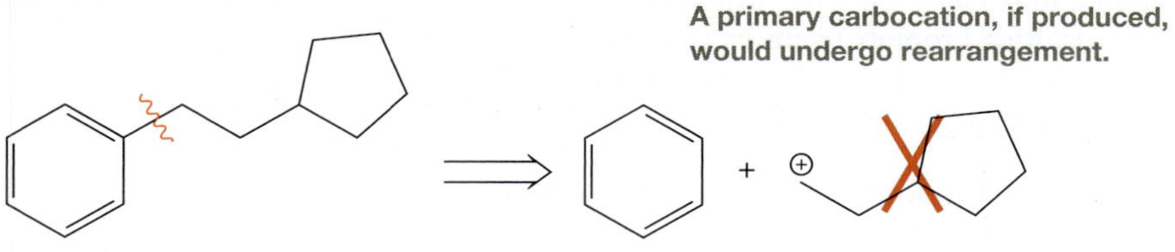

A primary carbocation, if produced, would undergo rearrangement.

Instead, we could incorporate a Friedel–Crafts acylation reaction to produce the new C—C bond. The C=O of the aromatic ketone product could then be reduced using a Clemmensen reduction.

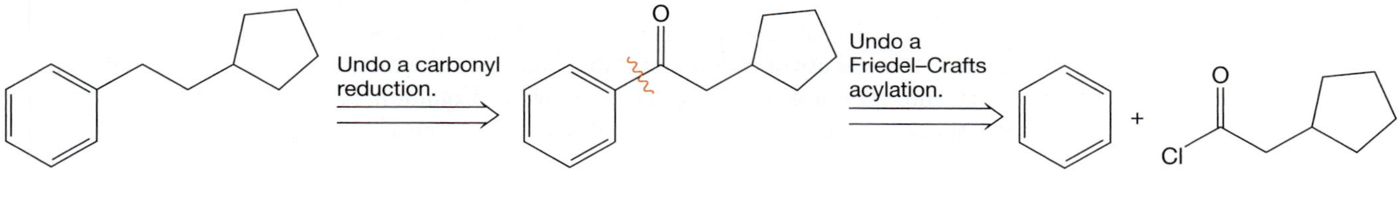

Undo a carbonyl reduction.

Undo a Friedel–Crafts acylation.

The final synthesis would be written as follows:

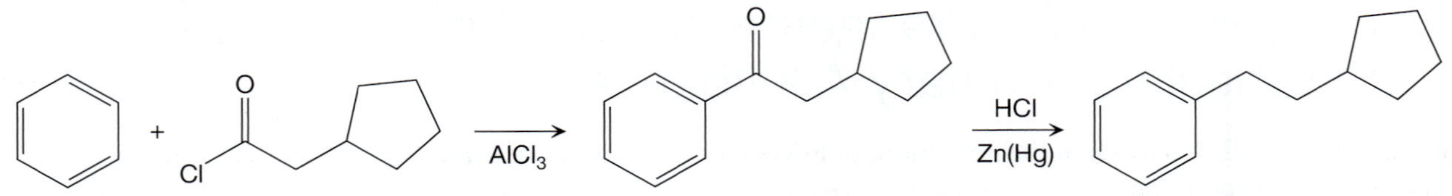

PROBLEM 22.16 Show how to carry out the synthesis in Solved Problem 22.15 without using the Clemmensen reduction. *Hint:* Are there other methods to reduce a ketone?

PROBLEM 22.17 Show how you would synthesize this compound, using benzene as the only aromatic starting compound.

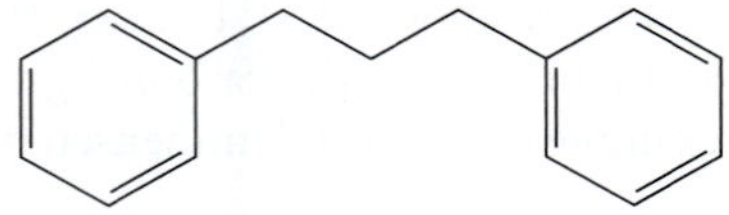

22.9 Organic Synthesis: Common Reactions Used in Conjunction with Electrophilic Aromatic Substitution Reactions

Electrophilic aromatic substitution reactions allow us to incorporate a variety of substituents onto an aromatic ring, but several substituents cannot be put in place directly via an electrophilic aromatic substitution reaction. Here we consider methods to incorporate these kinds of substituents by first using an electrophilic aromatic substitution reaction to incorporate a *different* substituent that is subsequently transformed into the desired substituent.

22.9a Oxidation of Carbon Side Chains: Synthesis of Benzoic Acids

Using the reactions we have learned thus far, how would you incorporate a carboxylic acid group into a benzene ring? Equation 22-30 shows that benzene can be brominated, and the resulting bromobenzene can be treated with Mg to produce a Grignard reagent. Bubbling carbon dioxide through the mixture leads to a Grignard reaction that produces the carboxylate, which is protonated via an acid workup.

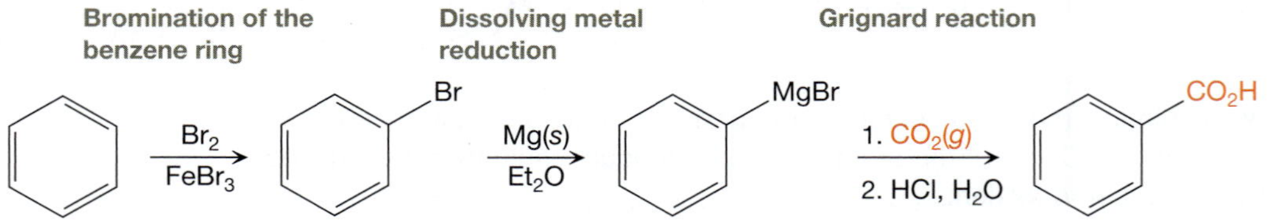

An alternative way to incorporate a carboxylic acid group into the ring begins with any of a variety of monosubstituted benzenes in which a C atom from the substituent is directly bonded to the ring, as shown in Equation 22-31a through 22-31e. Oxidation of these kinds of substituted benzenes using $KMnO_4$ heated in a basic solution produces benzoic acid after acid workup. Several of these precursors could be produced using a Friedel–Crafts reaction.

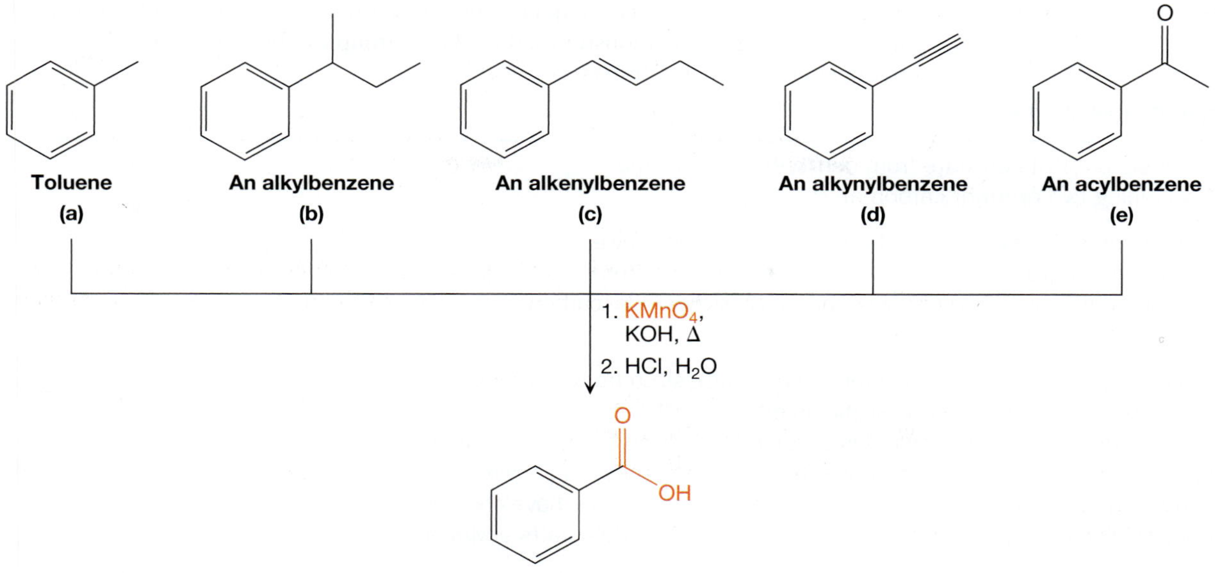
Toluene
(a)
An alkylbenzene
(b)
An alkenylbenzene
(c)
An alkynylbenzene
(d)
An acylbenzene
(e)
1. KMnO4,
KOH, Δ
2. HCl, H2O
O
OH

(22-31)

Benzoic acid

These oxidation reactions are believed to proceed through *free-radical intermediates* (species with an unpaired valence electron), which are discussed in detail in Chapter 25. The C atom bonded directly to the phenyl ring—the **benzylic carbon**—is particularly reactive under these conditions. This is why the product of each of the reactions in Equation 22-31 is the same, essentially independent of the groups that are bonded to the benzylic carbon.

This permanganate oxidation fails, however, when the benzylic C is *quaternary* (i.e., is bonded to four other C atoms). 2-Methyl-2-phenylpropane, for example, is *not* oxidized to benzoic acid (Equation 22-32).

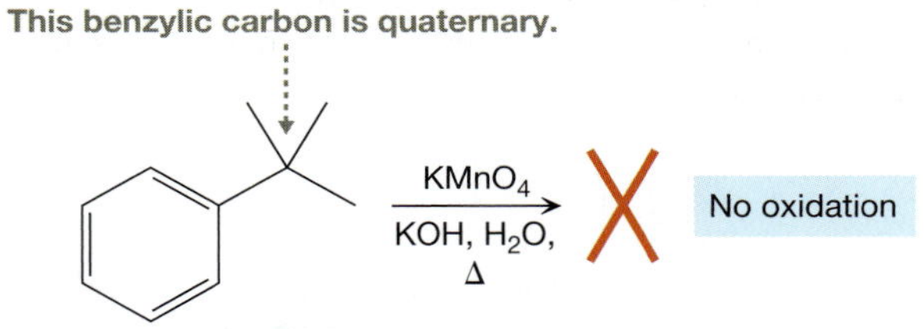
This benzylic carbon is quaternary.
KMnO4
KOH, H2O,
Δ
No oxidation

(22-32)

(1,1-Dimethylethyl)benzene

Thus:

> When a substituent on a phenyl ring is attached by a carbon atom, $KMnO_4$ oxidation requires the benzylic C to be bonded to at least one H atom or be part of a π bond.

YOUR TURN 22.11

A student proposed the following synthesis. Why does it not work?

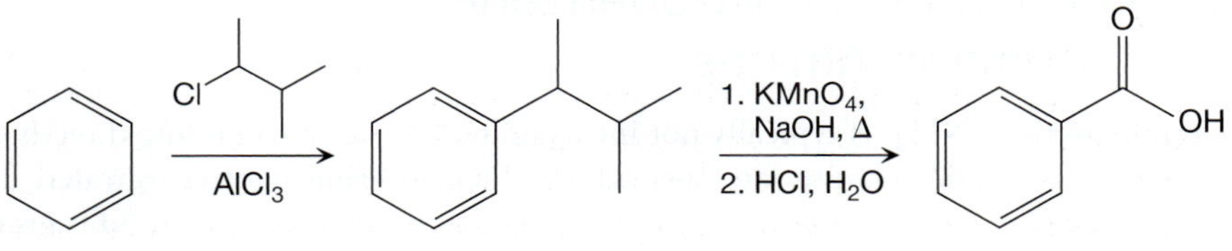
Cl
AlCl3
1. KMnO4,
NaOH, Δ
2. HCl, H2O
O
OH

In a synthesis, the CO_2H group need not appear in the target molecule. Instead, as Solved Problem 22.18 demonstrates, the CO_2H group can be transformed into other functional groups, such as esters.

SOLVED PROBLEM 22.18

Propose a synthesis of ethyl benzoate from benzene and any other compounds containing two or fewer carbon atoms.

Think How can an ester be synthesized from a precursor with fewer carbons? How can the appropriate precursor(s) be synthesized from benzene using compounds with two or fewer carbons? Is a carbon–carbon bond-forming reaction necessary?

Solve In Chapter 21 we learned that an ester can be synthesized from a carboxylic acid and an alcohol. Here, the necessary alcohol is ethanol, which contains two C atoms. The carboxylic acid that is required is benzoic acid, which can be generated from the $KMnO_4$ oxidation of any of a variety of substituted benzenes, as shown previously in Equation 22-31. In the following retrosynthetic analysis, we have chosen phenylethanone, which can be produced from benzene using a Friedel–Crafts acylation.

In the forward direction, the synthesis would be written as follows:

PROBLEM 22.19 Show how you could synthesize benzoic anhydride using benzene as the only aromatic starting compound.

22.9b Reduction of Nitrobenzenes to Aromatic Amines

An amino group ($-NH_2$) is typically not incorporated into a benzene ring directly via electrophilic aromatic substitution. Instead, the benzene ring is often nitrated first (Equation 22-23, p. 1121), followed by reduction of the NO_2 group to an NH_2 group. This reduction can be carried out either by catalytic hydrogenation (Equation 22-33a) or by treatment with a metal (e.g., Fe or Sn) under acidic conditions, followed by treatment with NaOH to neutralize the acid (Equation 22-33b). Reduction with a metal under acidic conditions resembles the Clemmensen reduction we first encountered in Chapter 19, in which the C=O group of a ketone or aldehyde is reduced to a methylene (CH_2) group by treatment with a Zn(Hg) amalgam in HCl.

The reduction of a nitro substituent to an amino group should be considered if the target is an aromatic amine. Moreover, the aromatic amine could be used as a synthetic intermediate toward the synthesis of a different target. In Section 22.9c, we see that aromatic amines are precursors to arenediazonium ions ($Ar—N_2^+$), which can be transformed into a variety of functionalized aromatic species. And in Chapter 23, we will see that amino groups can be important in influencing the incorporation of additional groups onto the aromatic ring.

SOLVED PROBLEM 22.20

Show how to synthesize *N*-phenylbenzamide, using benzene as the only aromatic starting compound.

Think What acid derivative and amine precursors will produce the amide target? Does each of those precursors have the same carbon backbone as the benzene ring, or will you need to produce a new carbon–carbon bond? What reactions add a carbon substituent to an aromatic ring?

***N*-Phenylbenzamide**

Solve An *N*-substituted amide can be produced from an amine and an acid chloride—in this case, aniline and benzoyl chloride. Aniline can be produced by the nitration of benzene, followed by reduction of the nitrobenzene product. Benzoyl chloride can be produced by carrying out an acyl substitution on benzoic acid, which can be produced, in turn, by an acylation of benzene, followed by $KMnO_4$ oxidation.

The synthesis in the forward direction would then be written as follows:

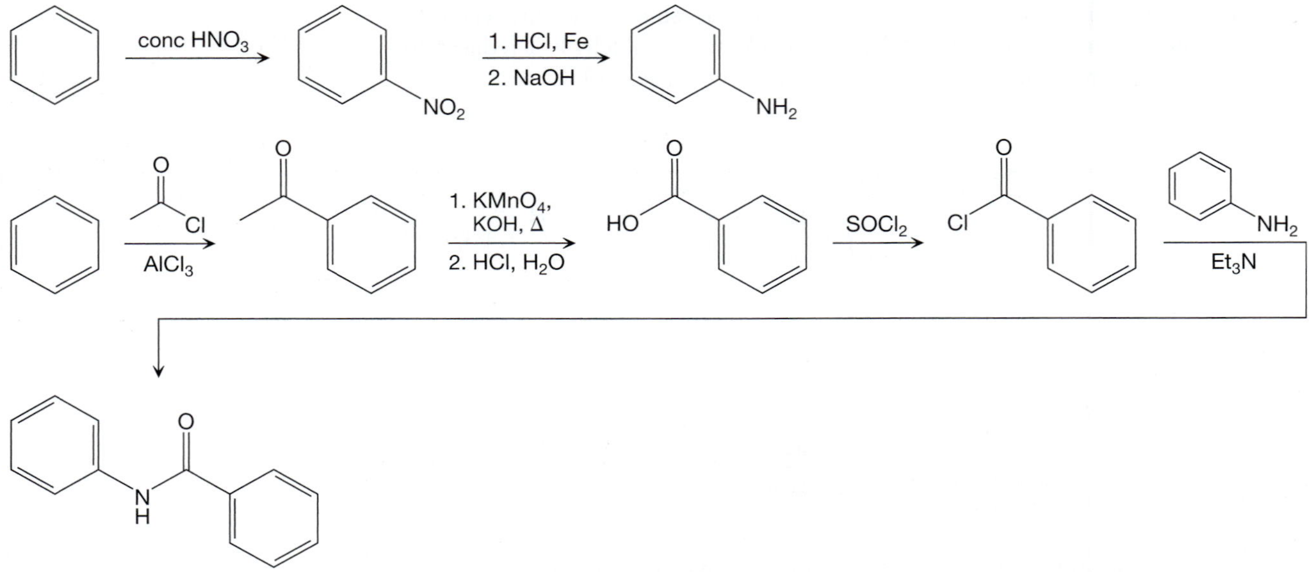

PROBLEM 22.21 Show how to synthesize each of these compounds, using benzene as the only aromatic starting compound.

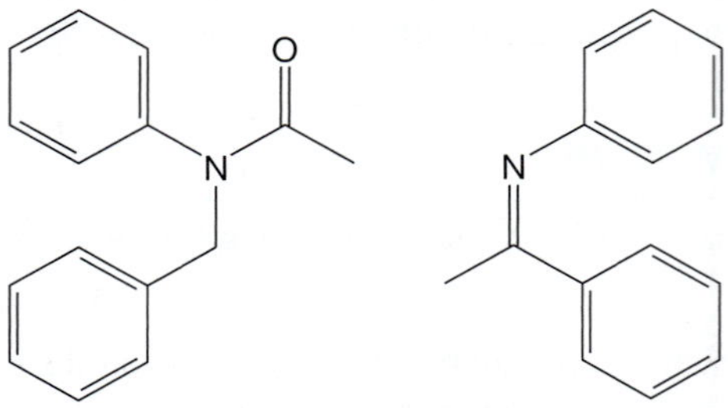

(a)   (b)

22.9c The Benzenediazonium Ion and the Sandmeyer Reactions

Bromobenzene is produced when aniline is treated with sodium nitrite ($NaNO_2$) under acidic conditions, followed by copper(I) bromide (CuBr) (Equation 22-34).

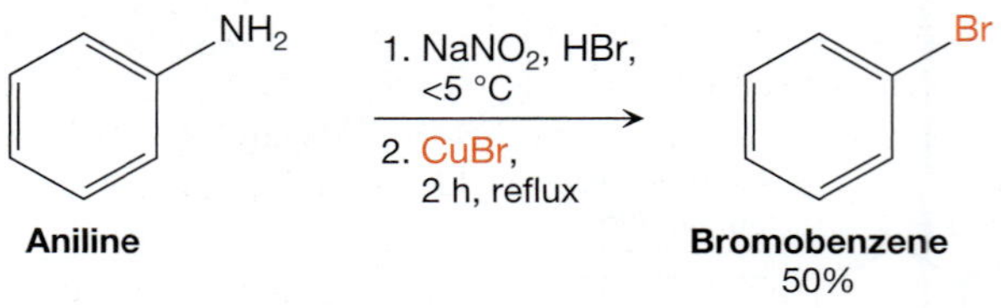

Aniline

1. $NaNO_2$, HBr, <5 °C
2. CuBr, 2 h, reflux

Bromobenzene
50%

(22-34)

Overall, Br has replaced NH_2, but the mechanism is not a simple electrophilic aromatic substitution. As shown in Equation 22-35, the first of the two reactions, called **diazotization**, is a multistep mechanism, which ultimately produces the **benzenediazonium ion**.

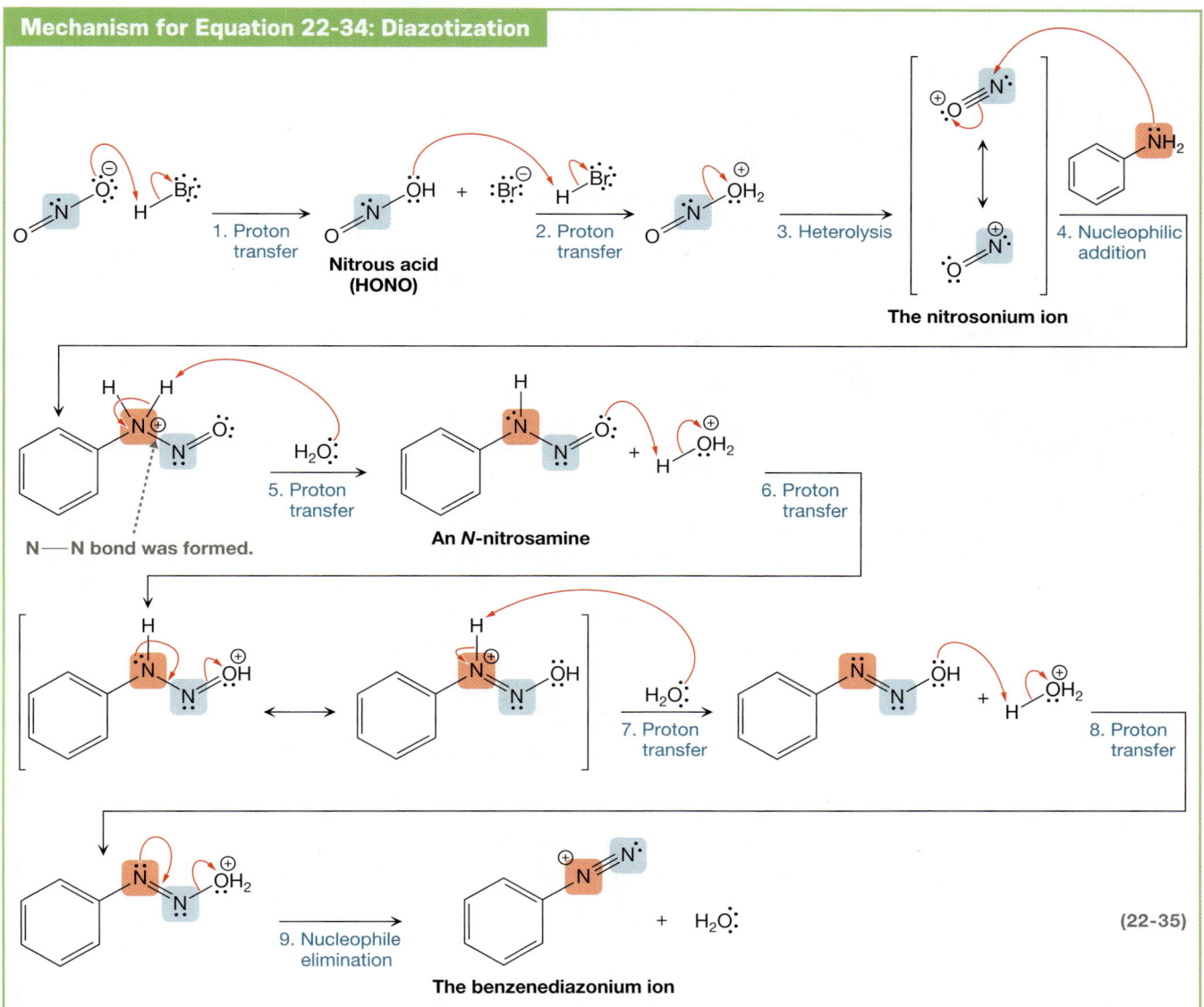

The nitrosonium ion

1. Proton transfer — **Nitrous acid (HONO)**

2. Proton transfer

3. Heterolysis

4. Nucleophilic addition

N——N bond was formed.

5. Proton transfer — **An *N*-nitrosamine**

6. Proton transfer

7. Proton transfer

8. Proton transfer

9. Nucleophile elimination

The benzenediazonium ion

(22-35)

In Step 1, the nitrite anion is protonated, which produces nitrous acid, HONO. Protonation in Step 2 converts a poor hydroxide leaving group into an excellent H_2O leaving group, which departs in Step 3 to produce the nitrosonium ion, NO^+. The aromatic amine attacks the nitrosonium ion in Step 4, in which a N——N bond is formed. Steps 5–8 are proton transfers that serve to remove protons from the initial amino N and to produce an excellent H_2O leaving group on the adjacent N, analogous to how protons are shuttled in a keto–enol tautomerization. Finally, in Step 9, the H_2O leaving group departs, producing the benzenediazonium ion.

The benzenediazonium ion possesses the N_2^+ group, which can behave as an excellent leaving group and depart as $N_2(g)$. Thus, the ion is highly reactive, and must be kept below ~10 °C to prevent decomposition or explosion.

At the same time, the high reactivity of the benzenediazonium ion makes it possible to carry out a subsequent substitution reaction simply by choosing the appropriate nucleophile. Examples are shown in Equation 22-36a through 22-36g. Some of these reactions require the presence of a Cu^+ catalyst; collectively, they are known as the **Sandmeyer reactions**, after the Swiss chemist Traugott Sandmeyer (1854–1922).

Sodium Nitrite and Foods: Preventing Botulism but Causing Cancer?

In Section 22.9c, we see how sodium nitrite ($NaNO_2$) is used to convert an aromatic amine to its corresponding diazonium ion. A very similar reaction is responsible for making most cured meats, such as ham and bacon, carcinogenic.

Sodium nitrite is added to a variety of cured meats to inhibit the growth of *Clostridium botulinum*, the bacterium whose toxin is responsible for botulism poisoning. It works by interfering with the metabolic processes of the bacteria. Just as we saw in Equation 22-35, however, sodium nitrite will also react with amino groups present in the meat, which are supplied by proteins. In Equation 22-35, a primary amine (RNH_2) is shown to proceed through an *N*-nitrosamine intermediate on its way to the diazonium ion. A secondary amine (R_2NH), on the other hand, stops at the *N*-nitrosamine, the actual compound believed to be carcinogenic. Since the 1950s, in fact, hundreds of these *N*-nitrosamines have been studied, and about 90% of them have been found to cause cancer in a variety of animals.

Notice in Equation 22-35 that the reaction requires an acidic environment, so it might seem that adding sodium nitrite to cured meats (which aren't highly acidic) shouldn't promote the formation of *N*-nitrosamines. The stomach, however, provides a highly acidic environment. Furthermore, the production of *N*-nitrosamines is facilitated by high temperatures, such as those typically used to fry bacon.

The benzenediazonium ion

Sandmeyer reactions

These reactions use a $Cu^{\oplus}$ catalyst.

(22-36)

CuCl	CuBr	CuCN	KI	1. HBF_4 2. Δ	H_2O, Cu_2O, $Cu(NO_3)_2$	H_3PO_2
Cl	Br	CN	I	F	OH	H
(a)	(b)	(c)	(d)	(e)	(f)	(g)

Even though these reactions involve nucleophiles and an excellent leaving group, they do not universally proceed by ionic nucleophilic substitution mechanisms. Some of these reactions do indeed resemble S_N1 mechanisms, but others are believed to involve *radical intermediates*, which contain unpaired electrons (Chapter 25). Because of the complex nature of these reactions, their mechanisms are not discussed here.

YOUR TURN 22.12

Diazonium ions can be produced from primary alkylamines, R—NH_2, when treated with $NaNO_2$ under acidic condition, but the resulting alkyldiazonium ions are generally much too reactive to be useful intermediates in Sandmeyer reactions. Draw the alkyldiazonium ion that would be produced from $(CH_3)_2CHNH_2$. Why do you think it is less stable than the benzenediazonium ion?

PROBLEM 22.22 How would you synthesize deuterobenzene from aniline?

Some of the products in Equation 22-36 can be produced using electrophilic aromatic substitution reactions we have seen previously here in Chapter 22. For example, bromobenzene and chlorobenzene can be produced using halogenation reactions (Section 22.2). However, the benzenediazonium ion intermediate provides much more flexibility in synthesis, allowing us to synthesize compounds that cannot be synthesized directly via electrophilic aromatic substitution. Recall, for example, that difficulties arise with the iodination and fluorination of benzene via electrophilic aromatic substitution. Beginning with the arenediazonium ion, however, iodobenzene (Equation 22-36d) and fluorobenzene (Equation 22-36e) can be synthesized rather straightforwardly. Similarly, cyanobenzene and phenol cannot be produced directly from benzene via electrophilic aromatic substitution, but they can be produced by incorporating the benzenediazonium ion as a synthetic intermediate. This is exemplified by Solved Problem 22.23.

SOLVED PROBLEM 22.23

Show how to synthesize cyanobenzene from benzene.

Think Can a cyano group replace H on a benzene ring directly? If not, what precursor is required? Can that precursor be synthesized from benzene using electrophilic aromatic substitution?

Solve We have not encountered a reaction in which a cyano group directly replaces a H^+ on benzene. However, cyanobenzene can be made from a benzenediazonium ion precursor, C_6H_5—N_2^+. The precursor to the benzenediazonium ion is an aromatic amine, which can be made, in turn, by reducing nitrobenzene. Finally, nitrobenzene can be made directly from benzene via nitration.

In the forward direction, the synthesis would be written as follows:

PROBLEM 22.24 Show how to synthesize phenol from benzene.

Chapter Summary and Key Terms

- Aromatic species tend to react with electrophiles (E^+) in an **electrophilic aromatic substitution** reaction instead of an addition reaction, because electrophilic aromatic substitution preserves aromaticity. **(Section 22.1; Objective 1)**

- The general mechanism for an electrophilic aromatic substitution reaction on benzene consists of two steps. First, E^+ undergoes electrophilic addition to the aromatic ring to produce an **arenium ion intermediate** or **Wheland intermediate**, which temporarily destroys the ring's aromaticity. Second, a proton (H^+) undergoes electrophile elimination, which restores aromaticity and results in a substituted aromatic ring. **(Section 22.1; Objective 2)**

- To temporarily destroy the aromaticity of the ring, the electrophiles that participate in electrophilic aromatic substitution tend to be very strong (in which case they must be generated in situ), so the reactions are generally irreversible. **(Section 22.1; Objective 3)**

- The first step of electrophilic aromatic substitution—addition of the electrophile—is slow, so it is the rate-determining step of the reaction. Thus, the reaction is first order with respect to both the aromatic ring and the electrophile. **(Section 22.1; Objective 3)**

- **Aromatic halogenation** typically requires a molecular halogen, Br_2 or Cl_2, in the presence of a strong **Lewis acid catalyst**, such as $FeBr_3$ or $FeCl_3$. These reactions behave as if Br^+ or Cl^+ is the electrophile. **(Section 22.2; Objective 4)**

- **Friedel–Crafts alkylation** involves a carbocation (R^+) electrophile, which can be produced from an alkyl halide in the presence of a strong Lewis acid catalyst, such as $AlCl_3$. **(Section 22.3; Objective 5)**

- Friedel–Crafts alkylation reactions are susceptible to carbocation rearrangements, so they cannot be used to produce primary alkylbenzenes other than methylbenzene and ethylbenzene. **(Section 22.4; Objective 5)**

- In a Friedel–Crafts alkylation, the halogen atom of the alkyl halide must be bonded to an sp^3-hybridized carbon atom. **(Section 22.4; Objective 5)**

- A **Friedel–Crafts acylation** reaction requires an acid halide or acid anhydride in the presence of a strong Lewis acid catalyst. The catalyst is responsible for the production of the electrophile, an **acylium ion**, $R—C\equiv O^+$. These reactions are not susceptible to carbocation rearrangements. **(Section 22.5; Objective 5)**

- **Nitration** of an aromatic ring requires concentrated nitric acid (HNO_3), which produces the **nitronium ion** (NO_2^+) as the electrophile. The addition of concentrated sulfuric acid (H_2SO_4) increases the concentration of the nitronium ion, and therefore increases the rate of the electrophilic aromatic substitution reaction. **(Section 22.6; Objective 6)**

- **Sulfonation** of an aromatic ring requires SO_3H^+ as the electrophile, which is present in small amounts in concentrated H_2SO_4. The rate of sulfonation is greater with **fuming sulfuric acid**, which is H_2SO_4 enriched with SO_3, because it increases the concentration of SO_3H^+. **(Section 22.7; Objective 7)**

- Sulfonation is *reversible*. A SO_3H group on an aromatic ring can be replaced by a hydrogen atom on treatment with H_3O^+. **(Section 22.7; Objective 7)**

- Although most primary alkylbenzenes cannot be produced directly from a Friedel–Crafts alkylation, they can be produced using a Friedel–Crafts acylation, followed by a reduction of the ketone's carbonyl group to a methylene (CH_2) group. **(Section 22.8; Objective 8)**

- A variety of carbon side chains on an aromatic ring can be oxidized to a CO_2H group by heating the aromatic compound in the presence of basic $KMnO_4$, followed by acid workup. **(Section 22.9a; Objective 9)**

- An aromatic amine can be produced by reducing a nitro-substituted aromatic ring, using either catalytic hydrogenation [e.g., $H_2(g)$, Pd] or dissolving metal reduction (e.g., Fe, HCl). **(Section 22.9b; Objective 10)**

- A **benzenediazonium ion** ($C_6H_5—N_2^+$) is produced when $C_6H_5—NH_2$ is treated with sodium nitrite ($NaNO_2$) under acidic conditions. The benzenediazonium ion can undergo a variety of substitution reactions readily, due to the excellent $N_2(g)$ leaving group. Reactions involving a Cu^+ catalyst are called **Sandmeyer reactions**. **(Section 22.9c; Objective 11)**

Reaction Tables

TABLE 22-1 Functional Group Conversions

	Starting Compound Class	Typical Reagents and Reaction Conditions	Compound Class Formed	Key Electron-Rich Species	Key Electron-Poor Species	Comments	Discussed in Section
(1)	Arene	Br$_2$, FeBr$_3$	Aryl bromide	Arene	$\cdot\ddot{Br}\cdot^{\oplus}$	Electrophilic aromatic substitution	22.2
(2)	Aryl amine (NH$_2$)	1. NaNO$_2$, H$_2$SO$_4$ 2. CuBr	Aryl bromide (Br)	CuBr	N$_2^{\oplus}$	Proceeds through a diazonium ion; Sandmeyer reaction	22.9c
(3)	Arene	Cl$_2$, FeCl$_3$	Aryl chloride (Cl)	Arene	$\cdot\ddot{Cl}\cdot^{\oplus}$	Electrophilic aromatic substitution	22.2
(4)	Aryl amine (NH$_2$)	1. NaNO$_2$, H$_2$SO$_4$ 2. CuCl	Aryl chloride (Cl)	CuCl	N$_2^{\oplus}$	Proceeds through a diazonium ion; Sandmeyer reaction	22.9c
(5)	Aryl amine (NH$_2$)	1. NaNO$_2$, H$_2$SO$_4$ 2. KI	Aryl iodide (I)	KI	N$_2^{\oplus}$	Proceeds through a diazonium ion	22.9c
(6)	Aryl amine (NH$_2$)	1. NaNO$_2$, H$_2$SO$_4$ 2. HBF$_4$, then Δ	Aryl fluoride (F)	HBF$_4$	N$_2^{\oplus}$	Proceeds through a diazonium ion	22.9c
(7)	Aryl amine (NH$_2$)	1. NaNO$_2$, H$_2$SO$_4$ 2. H$_2$O, Cu$_2$O, Cu(NO$_3$)$_2$	Aryl alcohol (OH)	H$_2$O	N$_2^{\oplus}$	Proceeds through a diazonium ion; Sandmeyer reaction	22.9c

(continued)

TABLE 22-1 Functional Group Conversions (continued)

Starting Compound Class	Typical Reagents and Reaction Conditions	Compound Class Formed	Key Electron-Rich Species	Key Electron-Poor Species	Comments	Discussed in Section
(8) Aryl amine (NH₂)	1. NaNO₂, H₂SO₄ 2. H₃PO₂	Arene (H)	–	N₂⁺ (aryl)	Proceeds through a diazonium ion	22.9c
(9) Nitroarene (NO₂)	1. HCl, Fe 2. NaOH	Aryl amine (NH₂)	–	–	Reduction	22.9b
(10) Arene	HNO₃ / H₂SO₄	Nitroarene (NO₂)	(arene)	⁺NO₂	Electrophilic aromatic substitution; nitration	22.6
(11) Arene	SO₃ / H₂SO₄	Arenesulfonic acid (SO₃H)	(arene)	SO₃H⁺	Electrophilic aromatic substitution; sulfonation	22.7
(12) Arenesulfonic acid (SO₃H)	H₃O⁺	Arene (H)	(aryl SO₃H)	H—OH₂⁺	Electrophilic aromatic substitution; desulfonation	22.7

TABLE 22-2 Reactions That Alter the Carbon Skeleton

Starting Compound Class	Typical Reagents and Reaction Conditions	Compound Class Formed	Key Electron-Rich Species	Key Electron-Poor Species	Comments	Discussed in Section
(1) Arene	R—Cl / AlCl₃	Alkylarene (R)	(arene)	R⁺	Electrophilic aromatic substitution; Friedel–Crafts alkylation	22.3

TABLE 22-2 Reactions That Alter the Carbon Skeleton (continued)

Starting Compound Class	Typical Reagents and Reaction Conditions	Compound Class Formed	Key Electron-Rich Species	Key Electron-Poor Species	Comments	Discussed in Section
(2) Arene	AlCl$_3$	Aromatic ketone		Acylium ion	Electrophilic aromatic substitution; Friedel–Crafts acylation	22.5
(3) Alkylarene	1. KMnO$_4$, KOH, Δ 2. HCl, H$_2$O	Aromatic carboxylic acid	—	—	Oxidation	22.9a
(4) Aromatic ketone/aldeyde	1. KMnO$_4$, KOH, Δ 2. HCl, H$_2$O	Aromatic carboxylic acid	—	—	Oxidation	22.9a
(5) Aryl amine	1. NaNO$_2$, H$_2$SO$_4$ 2. CuCN	Aryl nitrile	CuCN		Proceeds through a diazonium ion; Sandmeyer reaction	22.9c

Problems

Problems that are related to synthesis are denoted (SYN).

22.1–22.5 The General Mechanism, Halogenation, and Friedel–Crafts Reactions

22.25 Which of these isomers of trimethylbenzene will produce exclusively one monobrominated product when treated with Br$_2$ and FeBr$_3$? Explain.

A   B   C

22.26 In each of the following reactions, the aromatic ring has just one chemically distinct, aromatic H, so a single electrophilic aromatic substitution will lead to just a single product. With this in mind, predict the product of each of these reactions.

(a) $\xrightarrow[\text{AlCl}_3]{\text{Cl}}$?

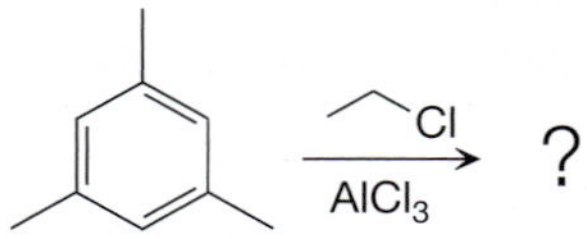

(b) $\xrightarrow[\text{FeCl}_3]{\text{Cl}_2}$?

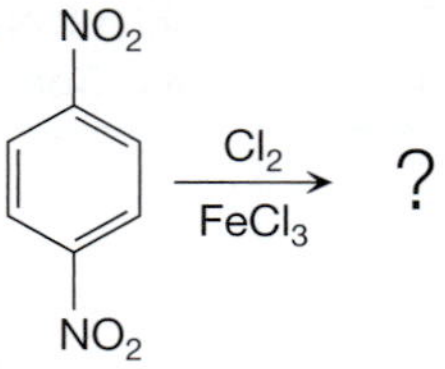

(c) $\xrightarrow[\text{AlCl}_3]{}$?

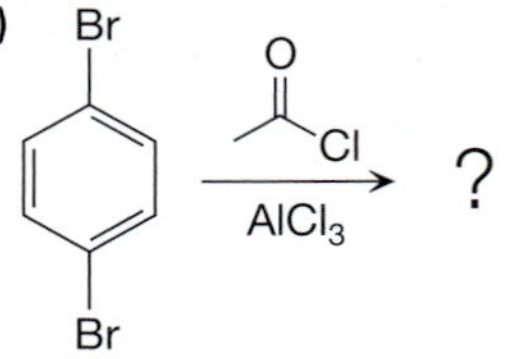

22.27 (SYN) In each of the following reactions, the aromatic ring has just one chemically distinct, aromatic H, so a single electrophilic aromatic substitution will lead to just a single product. With this in mind, supply the missing reagents needed to carry out each transformation.

(a) $\xrightarrow{?}$

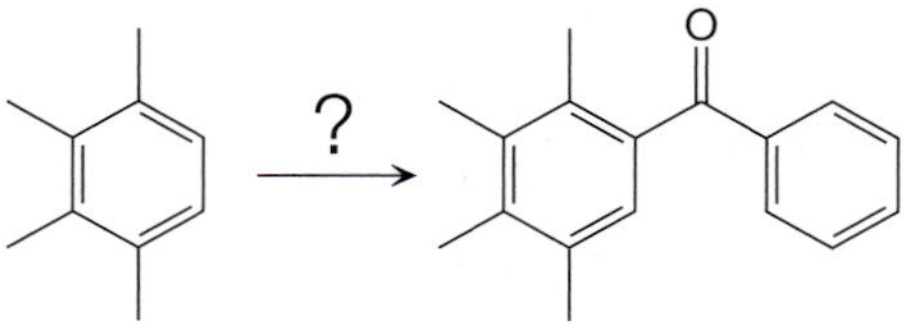

(b) $\xrightarrow{?}$

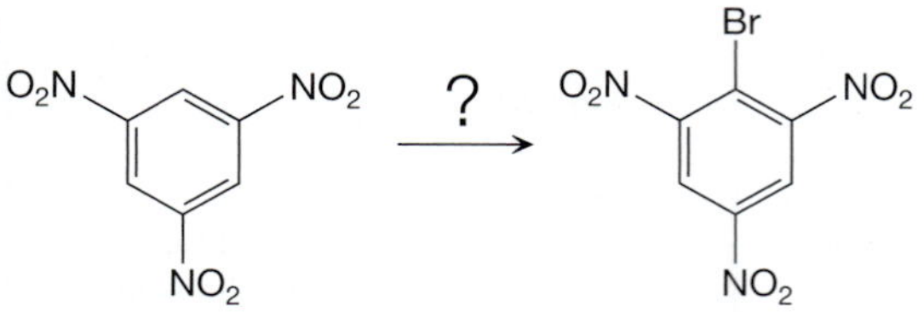

22.28 Propose a mechanism for the isotopic exchange reaction shown here.

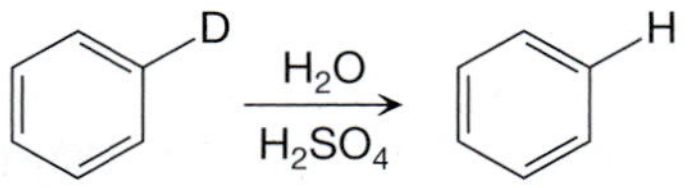

22.29 Halogenation using the mixed halogen ICl can feasibly lead to two different products, as shown here. Draw the complete, detailed mechanism for the formation of each product. Which do you think is the major product? Why?

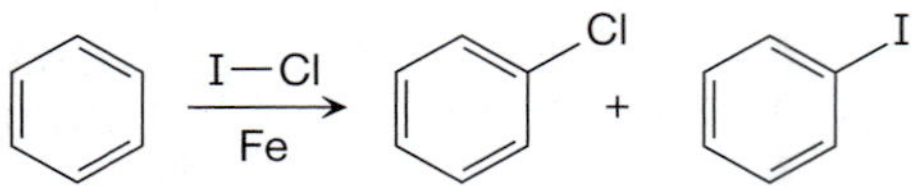

22.30 Draw the complete, detailed mechanism for the reaction shown here. Will the product be optically active? Explain.

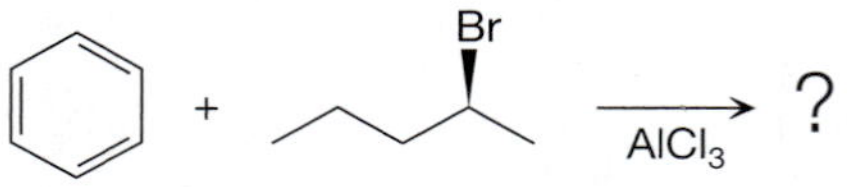

22.31 Previously (p. 1120) we mentioned that the formylation of benzene (i.e., the replacement of H by HC=O) cannot be carried out through a standard Friedel–Crafts acylation because methanoyl chloride (formyl chloride) cannot be added directly. The *Gattermann–Koch synthesis* circumvents this problem by making methanoyl chloride in situ (shown at the right), using a gaseous mixture of carbon monoxide and hydrochloric acid at high pressures. With this in mind, draw the detailed mechanism for the electrophilic aromatic substitution that takes place in the following reaction.

Formyl chloride is unstable, so it is produced only temporarily.

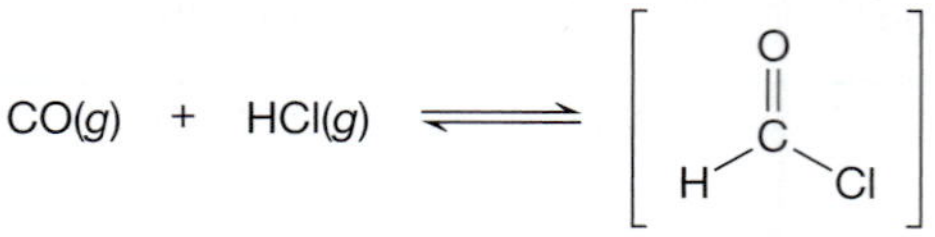

$$CO(g) \; + \; HCl(g) \; \rightleftharpoons \; \left[\text{} \right]$$

Formyl chloride

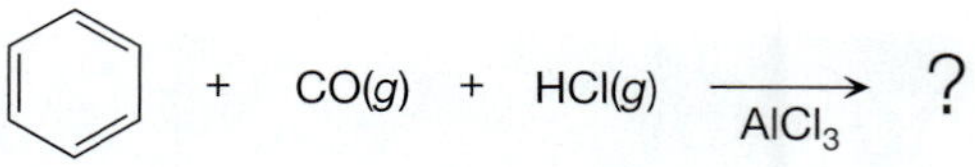

22.32 The reactant in the reaction shown here is aromatic and will undergo electrophilic aromatic substitution in much the same way that benzene does. This reaction leads to a single product only. Explain why. Predict the product of the reaction and draw the complete, detailed mechanism for its formation.

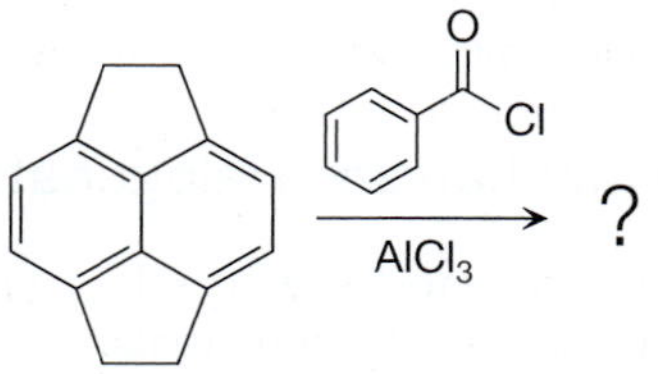

22.33 In the acid-catalyzed aromatic alkylation involving 1-methylcyclohexene and benzene, two isomeric products are possible, but only one is formed, as shown here. Draw the complete mechanism that leads to each product, and explain why only one isomer is formed.

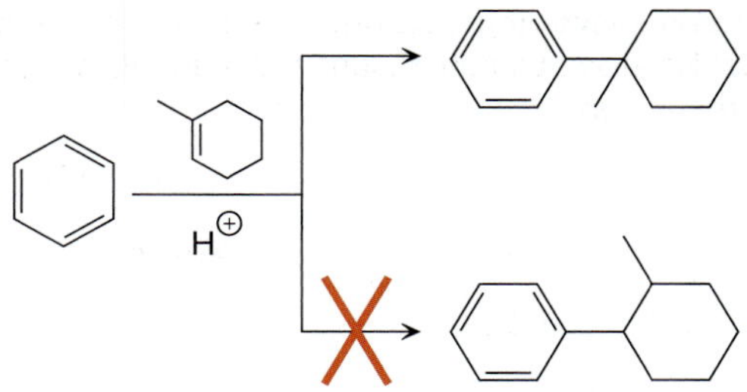

22.34 Two isomeric products can be produced from the reaction shown here. **(a)** Draw the complete, detailed mechanism showing the formation of each of those products. **(b)** Which of those products will be formed in greater abundance?

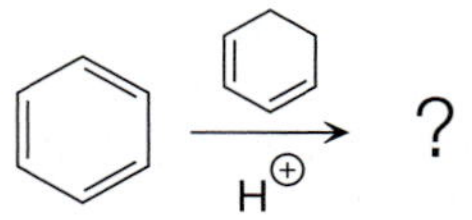

22.35 When benzene is treated with sulfur dichloride (SCl$_2$) in the presence of a Friedel–Crafts catalyst like AlCl$_3$, diphenyl sulfide is produced, as shown here. Propose a mechanism for this reaction.

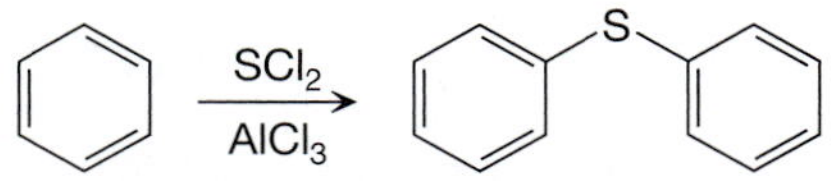

Diphenyl sulfide

22.6 and 22.7 Nitration and Sulfonation

22.36 An isomer of tetramethylbenzene undergoes nitration to yield a single product. Based on this information, which isomer(s) of tetramethylbenzene could the starting material have been?

22.37 Compounds **A**, **B**, and **C** are isomers of xylene (dimethylbenzene). When each of these isomers undergoes a single nitration, compound **A** produces just one product, **B** produces a mixture of two products, and **C** produces a mixture of three products. Identify which of compounds **A**, **B**, and **C** is the ortho isomer, which is the meta isomer, and which is the para isomer.

22.38 In each of the following reactions, the aromatic ring has just one chemically distinct, aromatic H, so a single electrophilic aromatic substitution will lead to just a single product. With this in mind, predict the product of each of these reactions.

(a)

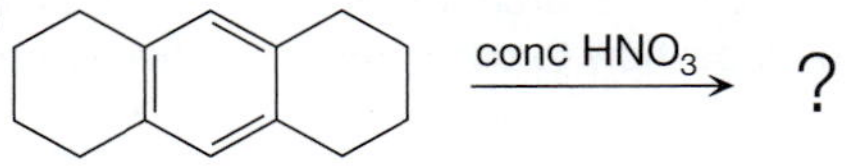

conc HNO$_3$?

(b) O$_2$N⎯ ⎯NO$_2$ / O$_2$N⎯ ⎯NO$_2$

conc HNO$_3$, H$_2$SO$_4$, Δ ?

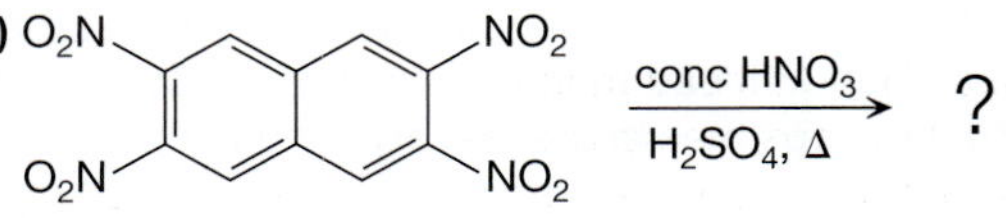

22.39 Draw the mechanism and the major organic product for each of the following reactions.

(a)

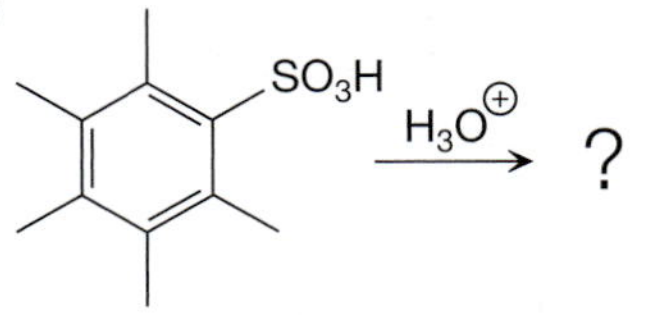

SO$_3$H

H$_3$O$^\oplus$?

(b) SO$_3$H ⎯ ⎯NO$_2$

H$_3$O$^\oplus$?

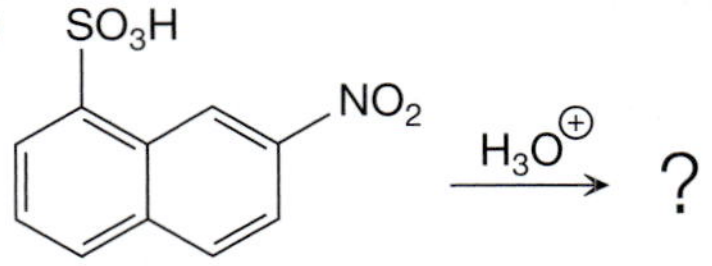

22.40 (SYN) In each of the following reactions, the aromatic ring has just one chemically distinct, aromatic H, so a single electrophilic aromatic substitution will lead to just a single product. With this in mind, supply the missing reagents needed to carry out each transformation.

(a)

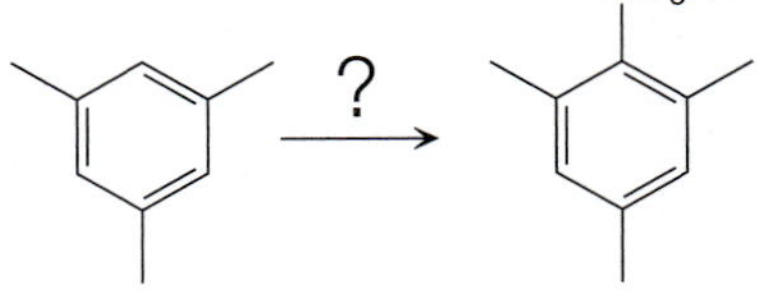

? ⟶ SO$_3$H

(b)

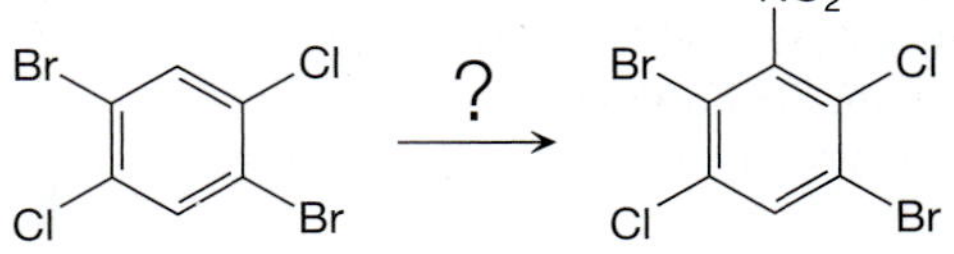

Br⎯ ⎯Cl / Cl⎯ ⎯Br ? ⟶ NO$_2$ / Br⎯ ⎯Cl / Cl⎯ ⎯Br

22.41 The reaction shown here is a halosulfonation, which is a useful variation of the sulfonation reaction. Draw the complete mechanism for this reaction.

H
HO–S–Cl
4 h, 20–25 °C
Cl

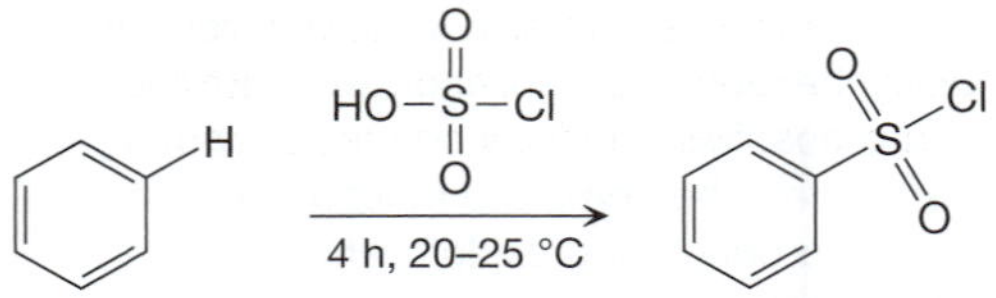

Benzenesulfonyl chloride
77%

22.8 and 22.9 Avoiding Carbocation Rearrangements; Common Reactions Used in Conjunction with Electrophilic Aromatic Substitution

22.42 Draw the major organic product of each of the following reactions.

(a)

NO2
1. HCl, Fe
2. NaOH
?

(b)

O2N
NO2
1. HCl, Fe
2. NaOH
?

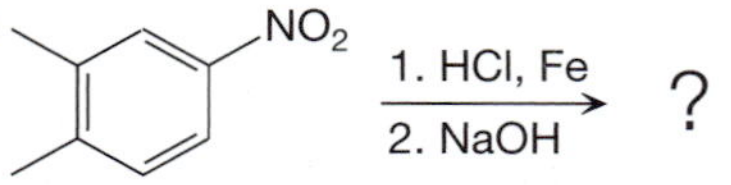

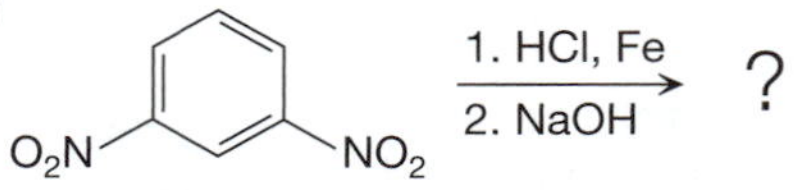

22.43 Draw the major organic product of each of the following reactions.

(a)

1. KMnO4, NaOH, Δ
2. HCl, H2O
?

(b)

O
1. KMnO4, NaOH, Δ
2. HCl, H2O
?

22.44 A dialkyl-substituted benzene, $C_{14}H_{22}$, is treated with basic potassium permanganate, followed by acid workup. The same dialkyl-substituted benzene was recovered afterward from the reaction mixture. Draw the structure of the compound.

22.45 **(SYN)** In the reaction shown here, the aromatic ring has just one chemically distinct, aromatic H, so a single electrophilic aromatic substitution will lead to just a single product. With this in mind, supply the missing reagents needed to carry out the transformation.

?

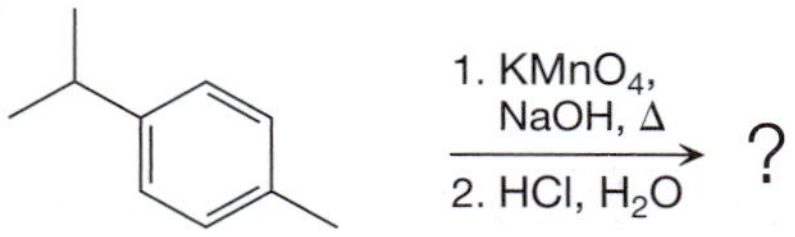

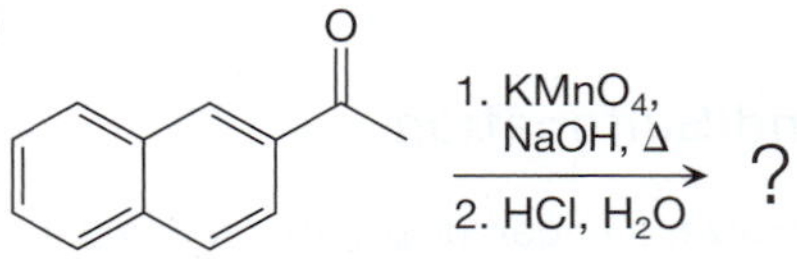

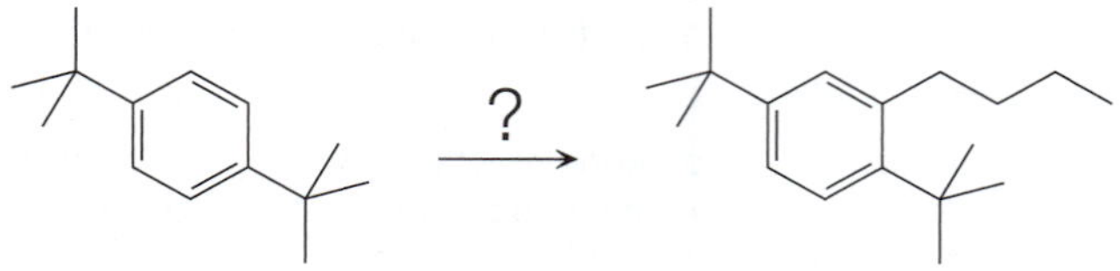

22.46 **(SYN)** In the following reactions, the aromatic ring has just one chemically distinct, aromatic H, so a single electrophilic aromatic substitution will lead to just a single product. With this in mind, show how to carry out these transformations. (Multiple synthetic steps may be necessary for each.)

(a)

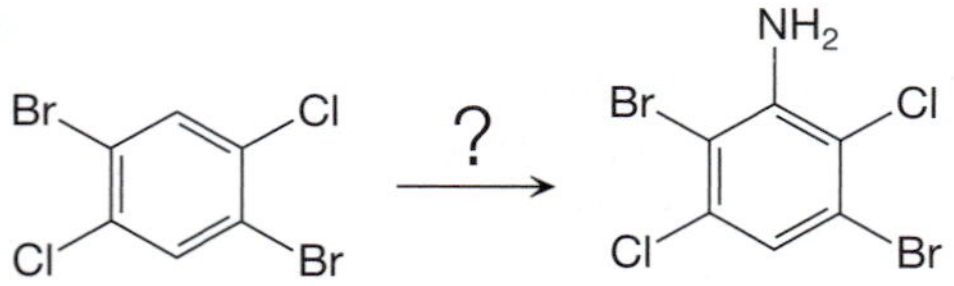

Br
Cl
Cl
Br
?
NH2

(b)

Br
Cl
Cl
Br
?
CO2H

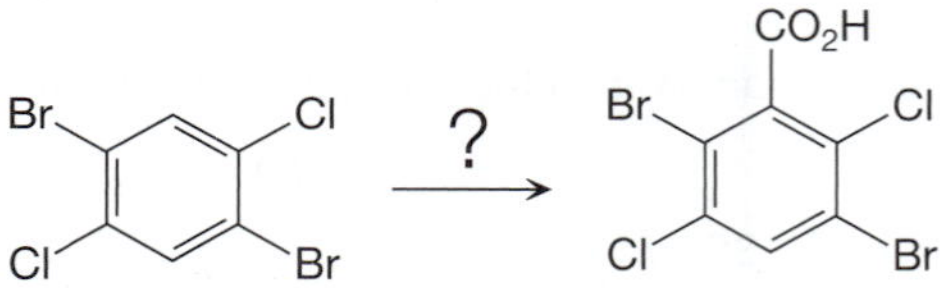

22.47 **(a)** Predict the product of the set of reactions shown here. **(b)** Draw the complete, detailed mechanism for the formation of the synthetic intermediate that is not shown.

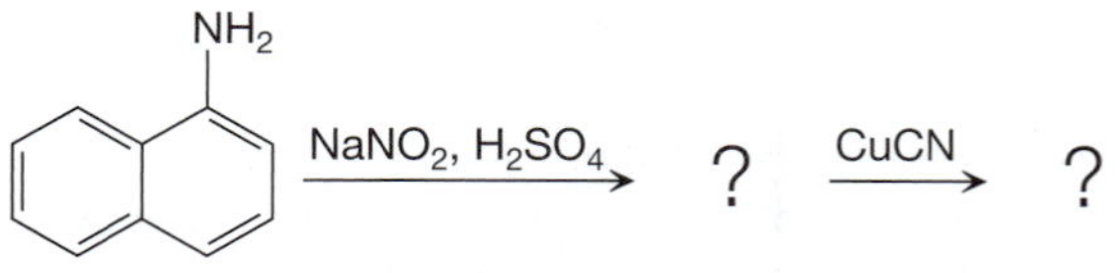

NH2
NaNO2, H2SO4
?
CuCN
?

Integrated Problems

22.48 No reaction occurs when benzene is treated with Br_2 in CCl_4. When anthracene is treated with Br_2 in CCl_4, however, then addition of Br_2 occurs, as shown here. Explain why.

Br2
CCl4
Br
Br

Anthracene

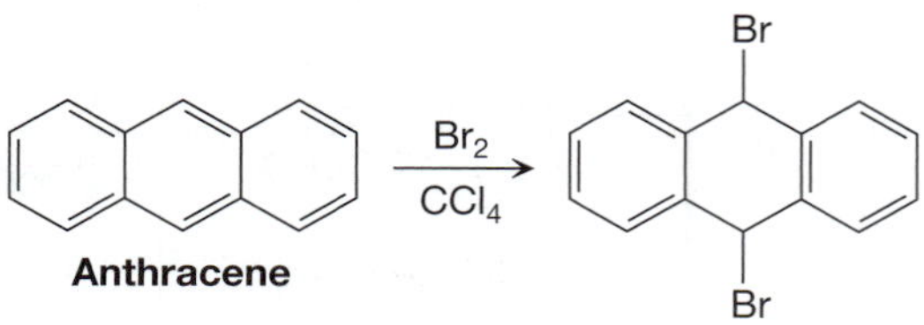

22.49 The reaction shown here is an example of a *deiodination*. Without AlCl₃ present, no reaction occurs. Draw a complete, detailed mechanism for this reaction.

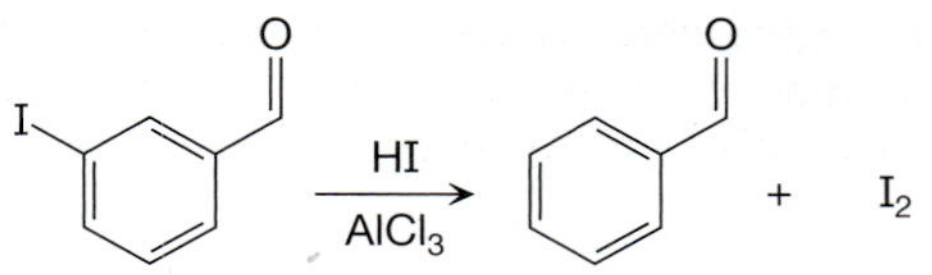

22.50 An example of a deiodination reaction is provided in Problem 22.49. Halogens other than iodine may be replaced on a benzene ring, but the reactions rates are slower; rates of substitution are observed to be: Ar—F < Ar—Cl < Ar—Br < Ar—I. Explain.

22.51 Propose a mechanism for the reaction shown here. Note that the reaction does *not* take place without the presence of AlCl₃.

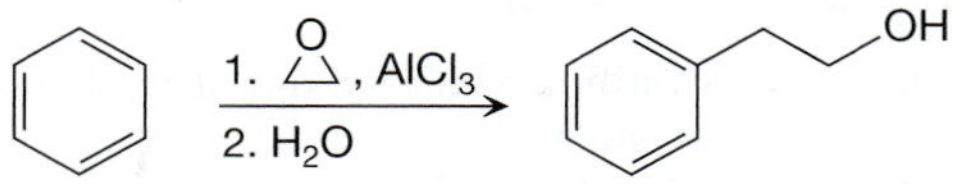

22.52 Propose a mechanism for the following reaction.

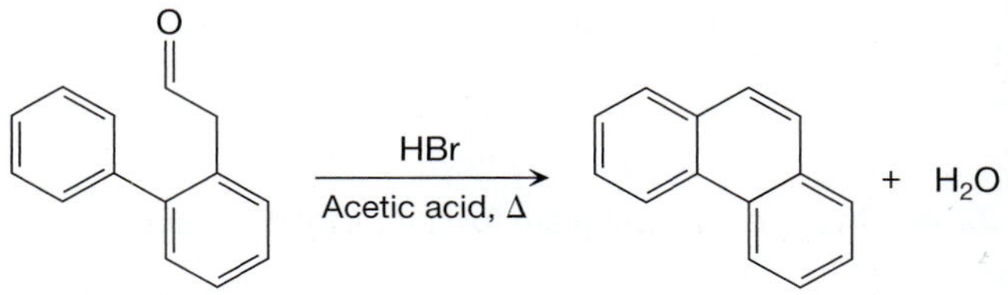

22.53 Identify compounds **A–G** in the following synthesis scheme.

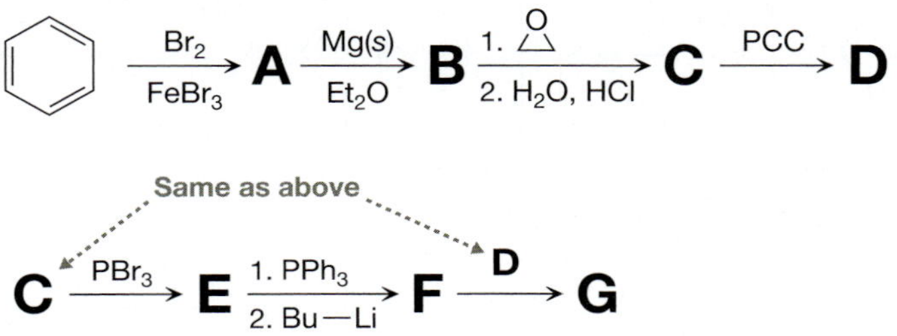

22.54 Identify compounds **A–G** in the following synthesis scheme.

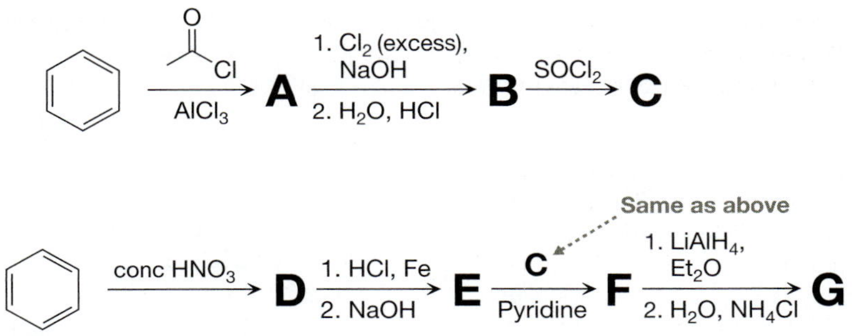

22.55 Identify compounds **A–I** in the following synthesis scheme.

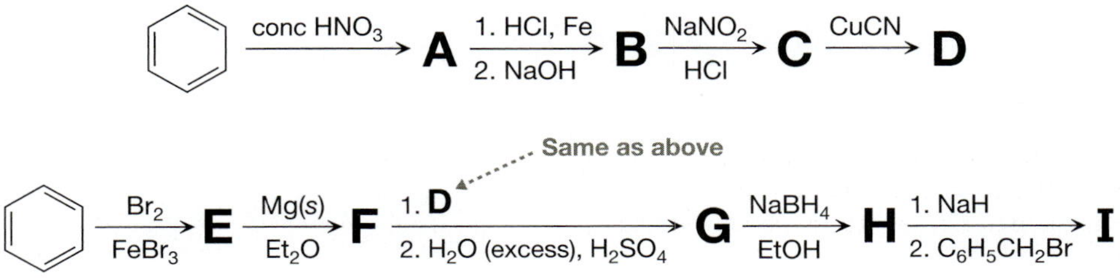

22.56 Predict the product of the following reaction.

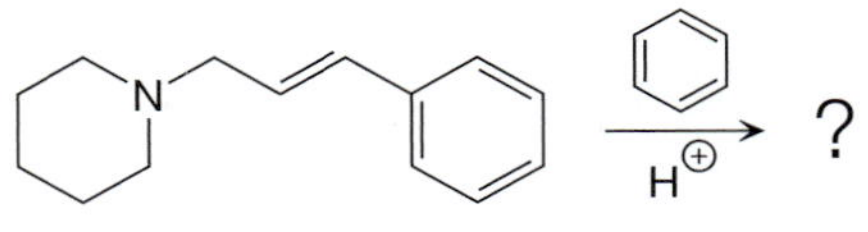

22.57 Draw the complete, detailed mechanism for the following reaction.

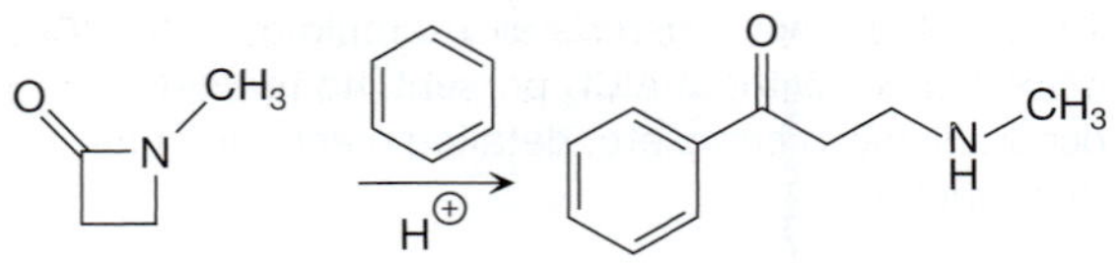

22.58 Benzene can be *hydroxylated* by treating it with hydrogen peroxide and a strong acid such as trifluoromethanesulfonic acid (TfOH). Propose a mechanism for this reaction.

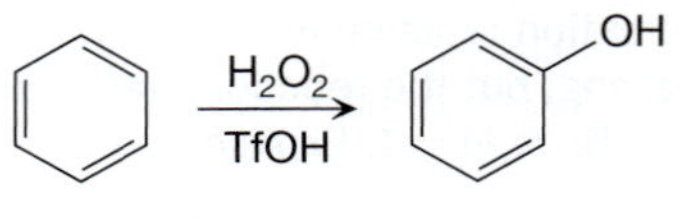

22.59 An aromatic imine is formed when 3,5-dihydroxyphenol is treated with ethanenitrile (CH_3CN) in HCl in the presence of a $ZnCl_2$ catalyst, as shown here. Propose a mechanism for this reaction.

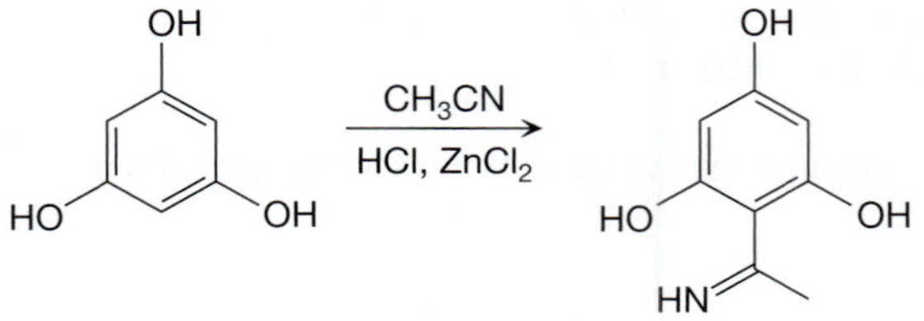

22.60 Propose a mechanism for the *Pictet–Spengler reaction*, an example of which is shown here. Note that a key intermediate is provided.

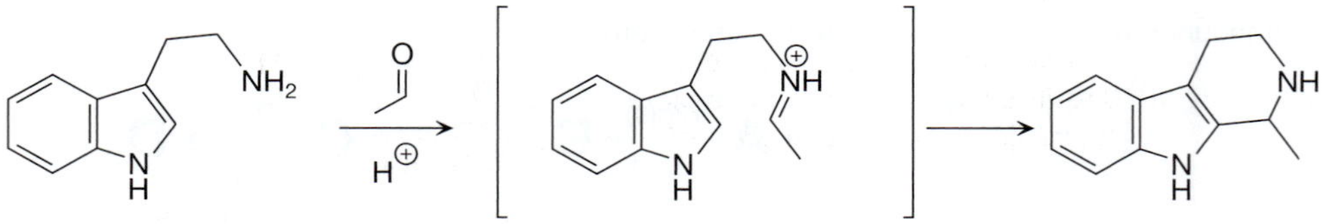

22.61 **(SYN)** Show how to carry out this synthesis using benzene and any alcohol as your only source of carbon.

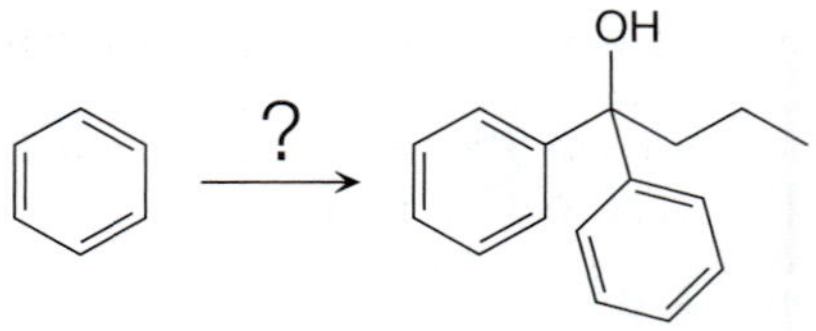

22.62 **(SYN)** Dibenzyl ether is used as a flavor and fragrance agent. Show how to synthesize dibenzyl ether using benzene and any ketone or aldehyde as your only source of carbon.

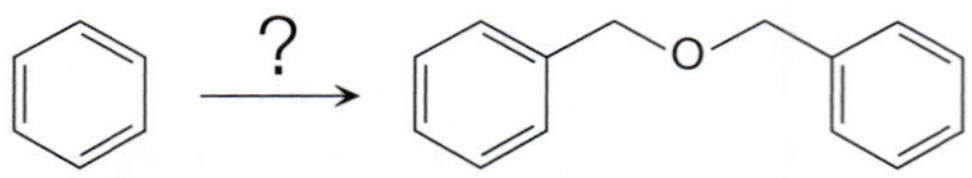

Dibenzyl ether

22.63 **(SYN)** Stilbene is used in the manufacture of dyes and also has estrogenic activity. Show how to synthesize stilbene using benzene and any carboxylic acid as your only sources of carbon.

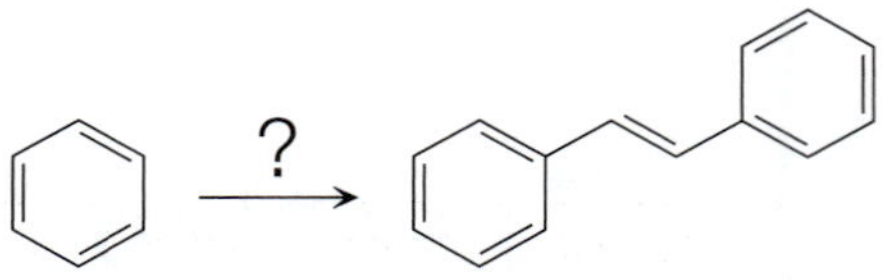

Stilbene

22.64 When diphenyl ether is reacted under the same conditions as in Problem 22.35, a compound is produced whose ^{13}C NMR spectrum shows six signals. Draw that product.

22.65 When the following acid chloride is treated with AlCl₃, followed by HCl and Zn(Hg), a product is formed whose ^{1}H NMR and IR spectra are shown here. Draw the product and propose a complete, detailed mechanism for the first of these two reactions.

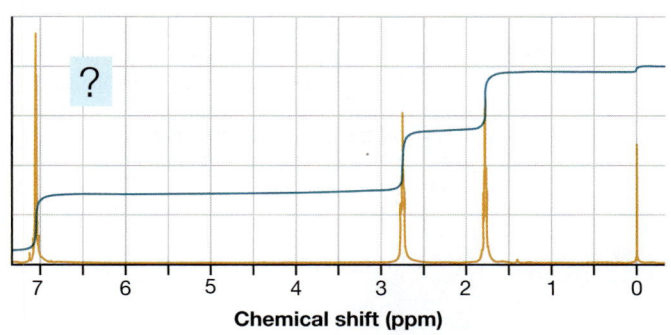

1. AlCl₃
2. HCl, Zn(Hg)

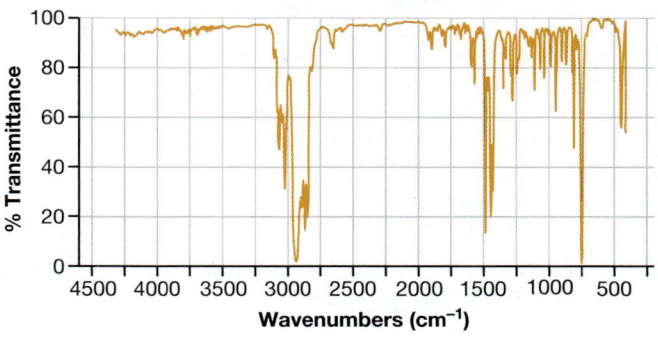

22.66 The product of the set of reactions shown here exhibits eight signals in its ^{13}C NMR spectrum. Draw a complete, detailed mechanism for the formation of that product.

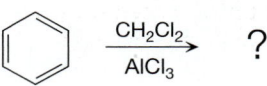

1. SOCl₂
2. AlCl₃

22.67 When benzene is treated with dichloromethane in the presence of aluminum trichloride, as shown here, a product is formed whose IR, ^{13}C NMR, and ^{1}H NMR spectra are shown below. Draw the product and propose a mechanism for this reaction.

CH₂Cl₂
AlCl₃

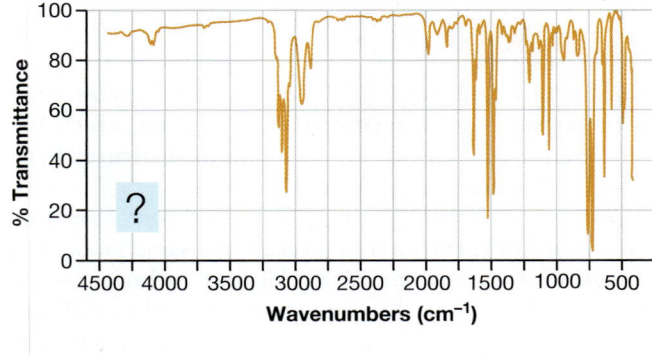

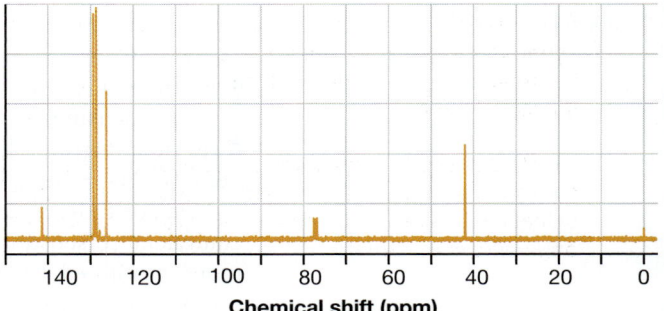

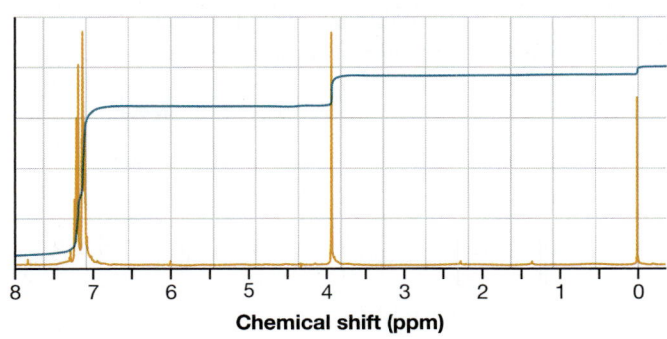

23

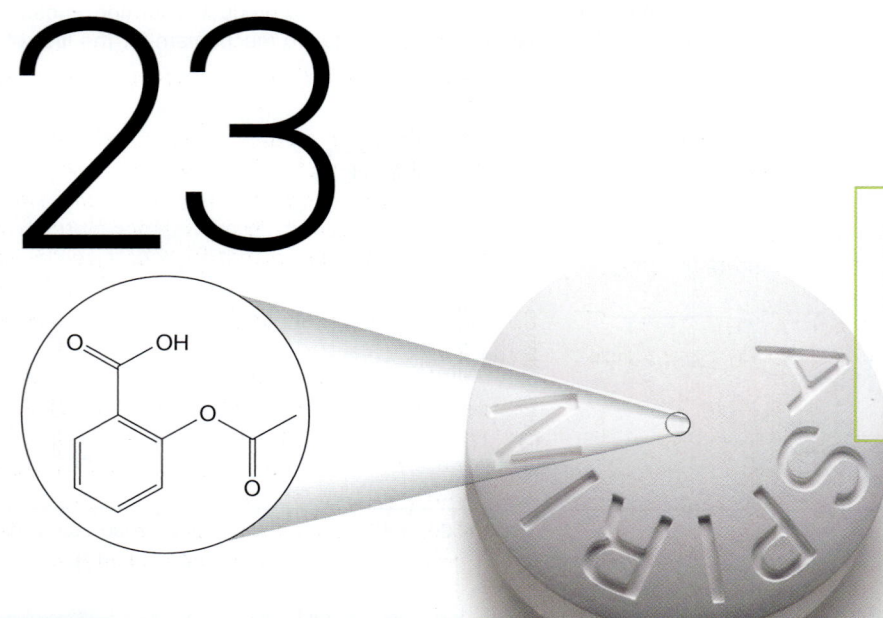

Aspirin, or acetylsalicylic acid, is one of the most widely produced and consumed medicines. A key step in its industrial synthesis is a regioselective electrophilic aromatic substitution reaction, one of the main concepts of Chapter 23.

Aromatic Substitution 2

Reactions of Substituted Benzenes and Other Rings

n Chapter 22, we examined a variety of *electrophilic aromatic substitution* reactions, focusing primarily on substitution involving benzene (C_6H_6), which has no substituents. Electrophilic aromatic substitution on a **monosubstituted benzene** (C_6H_5—Sub), however, involves additional issues.

One of those issues is *regiochemistry*. In benzene itself, all six hydrogen atoms are equivalent, so substitution of any one of them leads to precisely the same product. A monosubstituted benzene, however, has three chemically distinct hydrogens—the ortho, meta, and para hydrogens. Substitution, therefore, can lead to three different possible products—namely, the ortho-, meta-, and para-disubstituted benzenes (Equation 23-1), depending on which hydrogen is replaced. *(Notice the convention that is used to denote a generic disubstituted benzene without designating the specific isomer.)*

Ortho   Meta   Para

Sub → Sub + Sub + Sub   Same as   Sub   (23-1)

A generic monosubstituted benzene

A generic disubstituted benzene

Chapter Objectives

On completing Chapter 23 you should be able to:

1. Identify a substituent as an ortho/para- or meta-directing group based on the electrophilic aromatic substitution product ratio when that kind of a substituent is attached to a benzene ring.

2. Explain why a substituent tends to be an ortho/para director if the atom that attaches it to an aromatic ring has a lone pair of electrons.

3. Explain why electron-donating groups tend to be ortho/para directors, whereas most electron-withdrawing groups tend to be meta directors.

4. Characterize an aromatic ring as being activated or deactivated, based on the rate of electrophilic aromatic substitution relative to benzene, and explain why ortho/para directors tend to be activating, whereas meta directors tend to be deactivating.

5. Account for the effects that reaction conditions can have on a substituent's ortho/meta/para-directing

capabilities, as well as its activating/deactivating capabilities.

6. Predict the major product of electrophilic aromatic substitution on disubstituted benzene rings.

7. Draw the mechanisms for electrophilic aromatic substitution reactions involving aromatic rings other than benzene, and account for their regiochemistry and relative kinetics.

8. Identify azo dyes and propose how to synthesize them via azo coupling.

9. Draw the mechanisms of nucleophilic aromatic substitution reactions, and recognize the conditions that favor these reactions.

10. Devise effective syntheses that incorporate electrophilic aromatic substitution reactions.

11. Use protecting and blocking groups in a synthesis involving electrophilic aromatic substitution.

As we see here in Chapter 23, the substituent already in place on the benzene ring prior to substitution dictates the regiochemistry of the reaction. Some substituents are designated as ortho/para directors because they lead to product mixtures consisting primarily of the ortho- and para-disubstituted products. Other substituents are designated as meta directors because they favor the formation of the meta-disubstituted product.

A second issue with monosubstituted benzenes is the effect the substituent has on the *rate* of electrophilic aromatic substitution. Some substituents slow the reaction and therefore are called *deactivating groups*, whereas other substituents, called *activating groups*, cause the reaction to speed up.

The substituents attached to an aromatic ring can even alter the reaction mechanism involving the ring, making the ring susceptible to reaction with nucleophiles rather than electrophiles. Here in Chapter 23, we study two such *nucleophilic aromatic substitution reactions*.

Finally, we consider how the characteristics of the substituents on an aromatic ring factor into organic synthesis. Ultimately, you should be able to design efficient syntheses of aromatic rings with multiple substituents.

23.1 Regiochemistry of Electrophilic Aromatic Substitution: Defining Ortho/Para and Meta Directors

The nitration of phenol, a monosubstituted benzene, can lead to the ortho-, meta-, and para-disubstituted products. As shown in Equation 23-2 (next page), however, the ortho- and para-disubstituted products dominate the product mixture. The hydroxy (–OH) group, therefore, is an **ortho/para director**.

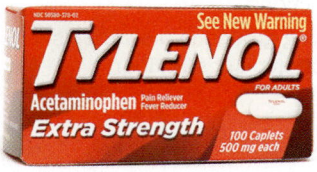
The ortho and para products dominate the product mixture, so OH is designated as an ortho/para director.

$$OH \xrightarrow[\text{CH}_3\text{CO}_2\text{H, 60 °C}]{\text{HNO}_3} \quad \text{ortho} + \text{meta} + \text{para} \qquad (23\text{-}2)$$

	Ortho	Meta	Para
Phenol Product ratio:	50%	~0%	50%

When nitrobenzene undergoes nitration, on the other hand, *m*-dinitrobenzene is the major product (Equation 23-3). Thus, the NO$_2$ group initially attached to the ring is a **meta director**.

The meta product dominates the product mixture, so NO$_2$ is designated as a meta director.

$$NO_2 \xrightarrow[\substack{\text{H}_2\text{SO}_4,\\ \text{30 min, 0 °C,}\\ \text{then 24 h, 25 °C}}]{\text{HNO}_3} \quad \text{ortho} + \text{meta} + \text{para} \qquad (23\text{-}3)$$

	Ortho	Meta	Para
Nitrobenzene Product ratio:	7%	91%	2%

The results from these and other nitration reactions are summarized in Table 23-1. The OH group and other ortho/para-directing substituents appear on the left side of the table; the NO$_2$ group and other meta-directing substituents appear on the right. To help illustrate which type of director each substituent is, the sum of the relative percentages of the ortho and para products ($o+p$) is also listed in the table.

In the next section, we examine why these substituents have the effects on regiochemistry that they do.

YOUR TURN 23.1

The iodo (–I) and formyl (–CHO) groups are not listed in Table 23-1. The relative amounts of ortho and meta nitration products obtained for each substituent are shown here.

Substituent	*o*	*m*	*p*	*o+p*	Type of Director
—I	45	1			
—CHO	19	72			

(a) Supply the missing information pertaining to the relative amounts of the para isomers that are produced, as well as the sum of the amounts of ortho and para products.

(b) Based on that information, determine whether each substituent is an ortho/para director or a meta director.

Answers to Your Turns are in the back of the book.

TABLE 23-1 **Product Distribution of the Nitration of Various Monosubstituted Benzenes**

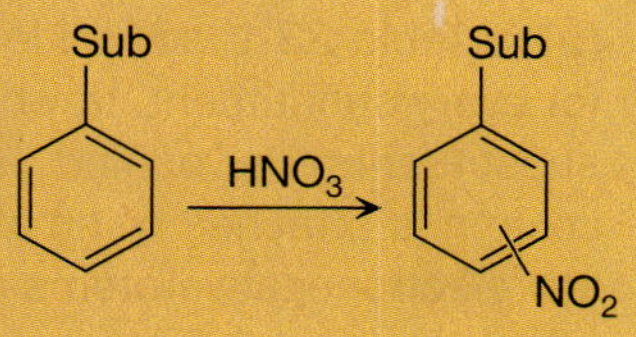

ORTHO/PARA-DIRECTING SUBSTITUENTS					META-DIRECTING SUBSTITUENTS				
Substituent	o	m	p	o+p	Substituent	o	m	p	o+p
—OH	50	0	50	100	—NO₂	7	91	2	9
—NHCOCH₃	19	2	79	98	—N⁺(CH₃)₃	2	87	11	13
—CH₃	63	3	34	97	—CO₂H	22	76	2	24
—F	13	1	86	99	—CN	17	81	2	19
—Cl	35	1	64	99	—CO₂Et	28	66	6	34
—Br	43	1	56	99	—COCH₃	26	72	2	28

PROBLEM 23.1 Based on the information in Table 23-1, predict the major product(s) for each of the following reactions. Draw the complete, detailed mechanism that leads to the formation of each major product.

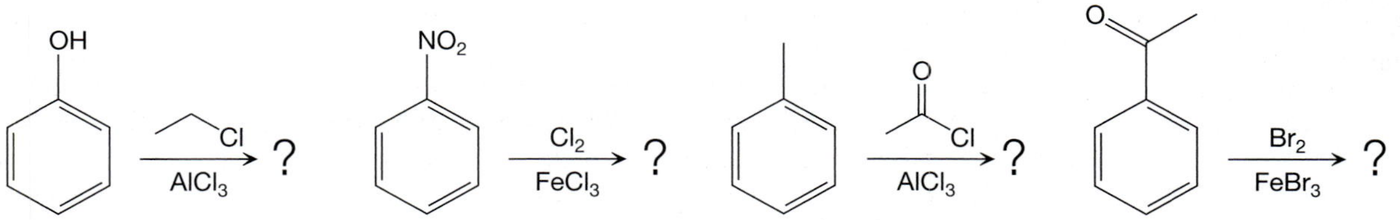

(a)   (b)   (c)   (d)

23.2 What Characterizes Ortho/Para and Meta Directors and Why?

As you look back at Table 23-1, notice the following characteristics of ortho/para and meta directors:

- Substituents attached by an atom having at least one lone pair of electrons are ortho/para directors.
- Substituents attached by an atom having no lone pairs of electrons are:
  - Ortho/para directors if they are alkyl groups.
  - Meta directors if the substituent contains an electronegative atom at or near the point of attachment.

(a) In Table 23-1, add all lone pairs to the atom at the point of attachment for every substituent. In which column (ortho/para- or meta-directing substituents) do you find those substituents? **(b)** Which substituent in that column is attached by an atom that has no lone pairs of electrons? Are there any electronegative atoms at or near the point of attachment? **(c)** In the other column, what electronegative atoms do you find at or near the point of attachment?

Why should these characteristics dictate the type of director each group is? To begin to answer this question, recall from Section 22.1 that electrophilic aromatic substitution reactions usually run under *kinetic control.* Therefore:

The relative amounts of ortho-, meta-, and para-disubstituted products that are produced in an electrophilic aromatic substitution reaction are proportional to the *rates* at which they are produced.

Recall also that the first step in an electrophilic aromatic substitution reaction—the formation of the arenium ion intermediate—is rate determining (Section 22.1). Moreover, because that step is highly endothermic, its energy barrier lowers and its rate increases as the stability of the arenium ion that is produced increases (Section 9.3a, the Hammond postulate). We therefore arrive at an important concept to help us understand the outcome of an electrophilic aromatic substitution:

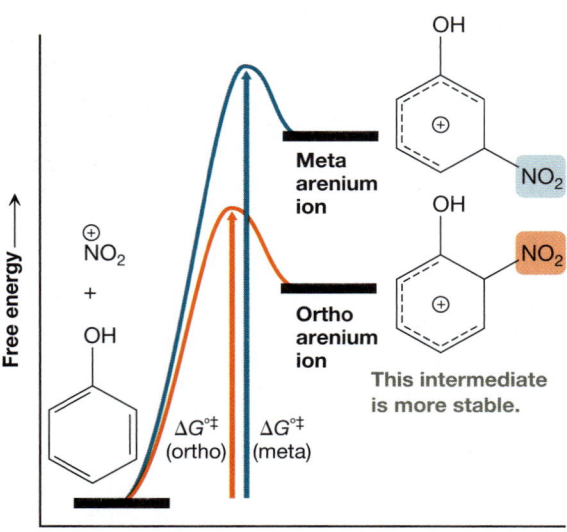

FIGURE 23-1 Energy diagrams for the rate-determining step in the nitration of phenol Nitration at the ortho position is represented by the red curve, and nitration at the meta position is represented by the blue curve. Because the ortho product is produced in greater abundance than the meta product (Table 23-1), we can say that the $\Delta G°^‡$ leading to the ortho intermediate is smaller. The *Hammond postulate* thus allows us to say that the ortho arenium ion is more stable than the meta arenium ion.

In general, the major product of electrophilic aromatic substitution is the one that is produced from the most stable arenium ion intermediate.

In the nitration of phenol (C_6H_5—OH), for example, the distribution of products is 50% ortho, ~0% meta, and 50% para. Because there is much more of the ortho isomer produced than the meta isomer, we know that the ortho arenium ion intermediate is lower in energy (more stable) than the meta intermediate, as shown in Figure 23-1. What can we say about the relative energies of the meta and para intermediates? (See Your Turn 23.3.)

The conclusions we have just drawn about the OH group, which is an ortho/para director, can be generalized to other ortho/para directors as well.

When an ortho/para director is attached to benzene, the ortho and para arenium ion intermediates are lower in energy (more stable) and are formed faster than the meta intermediate.

The opposite is true for meta directors, which lead predominantly to the meta-disubstituted product.

> When a meta director is attached to benzene, the meta arenium ion intermediate is lower in energy (more stable) and is formed faster than the ortho or para intermediate.

PROBLEM 23.2 Consider these arenium ion intermediates that are formed in the nitration of fluorobenzene. Using Table 23-1, determine which one is formed the fastest. Which is formed the slowest? Which is the most stable? Which is the least stable? Repeat this problem for the nitration of nitrobenzene, in which F has been replaced by NO_2 in each arenium ion intermediate.

To this point, we have learned how to identify ortho/para and meta directors *empirically*, and to use that information to determine the relative energies of the ortho, meta, and para arenium ion intermediates. However, we have yet to explain *why* ortho/para and meta directors influence the stability of the arenium ion intermediates in the ways that they do, which is the focus of the following sections.

23.2a Why Are Ortho and Para Arenium Ion Intermediates Lower in Energy When an Ortho/Para Director Is Attached?

We have encountered two kinds of ortho/para directors: (1) groups attached by an atom having at least one lone pair of electrons and (2) alkyl groups. When either of these groups is attached to benzene, we know that the ortho and para arenium ion intermediates are lower in energy than the meta intermediate. To see why this is true for the first type of ortho/para director, let's carefully examine the ortho and meta intermediates that are produced in the nitration of phenol, C_6H_5—OH.

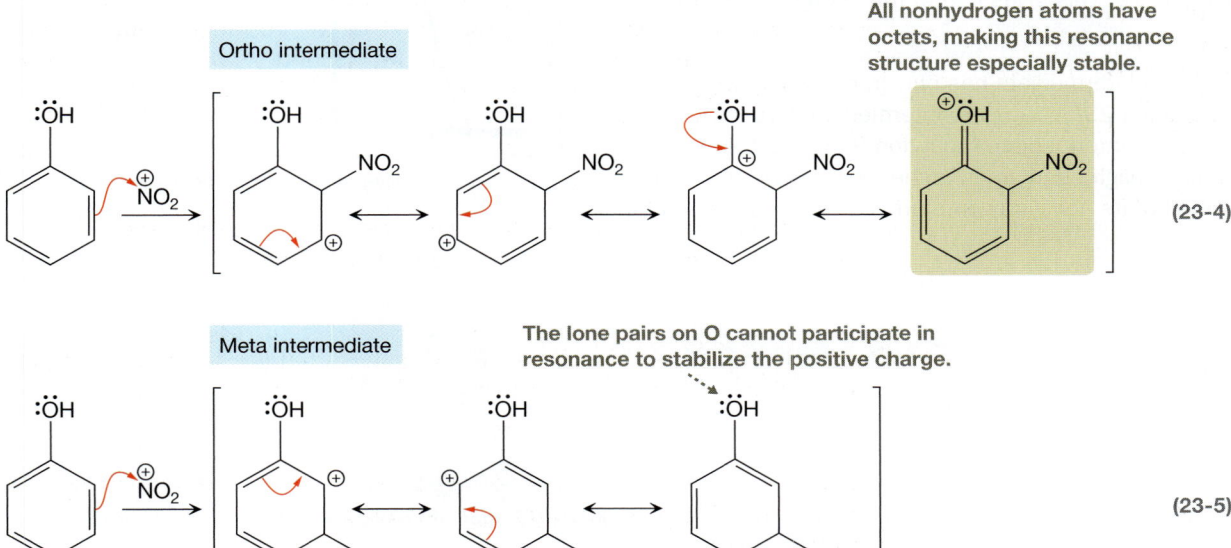

(23-4)

(23-5)

When nitration occurs at the ortho position (Equation 23-4), the arenium ion that is produced has four total resonance structures, allowing the positive charge to be delocalized over four separate atoms. By contrast, when nitration occurs at the meta position (Equation 23-5), the arenium ion intermediate has just three total resonance structures. The ortho arenium ion intermediate, with its additional resonance structure, is therefore lower in energy (more stable) than the meta intermediate.

The additional resonance structure for the ortho intermediate (highlighted in green in Equation 23-4) is due to the involvement of the lone pair of electrons from the OH group. The involvement of that lone pair is not possible for the meta intermediate (Equation 23-5) because the meta intermediate has no resonance structure in which C^+ is directly attached to OH.

All nonhydrogen atoms have an octet in the highlighted resonance structure in Equation 23-4, which is not true of the other three resonance structures. Therefore, the additional stability provided by that single resonance structure is extremely important.

YOUR TURN 23.4

Identify all of the atoms in Equations 23-4 and 23-5 that do *not* have an octet.

Similar reasoning allows us to rationalize why the para arenium intermediate in the nitration of phenol is lower in energy than the meta intermediate (see Solved Problem 23.3). The para intermediate has four resonance structures, including one in which all nonhydrogen atoms have an octet.

SOLVED PROBLEM 23.3

Draw a diagram similar to Figure 23-1 that illustrates the formation of the para and meta intermediates during the nitration of phenol. Which of those two arenium ions is more stable and why?

Think Which of the isomeric products—meta or para—is formed faster? What does that say about the stability of the respective intermediates? What role is played by resonance delocalization of the charge?

Solve According to Table 23-1, the para arenium ion is formed faster than the meta, so the para intermediate must be more stable. This gives rise to the energy diagram shown here.

Whereas the meta intermediate has only three resonance structures (Equation 23-5), the para intermediate has four, similar to the ortho intermediate (Equation 23-4). Furthermore, the one that is highlighted in green is the most important of the four, because all of its nonhydrogen atoms have complete octets.

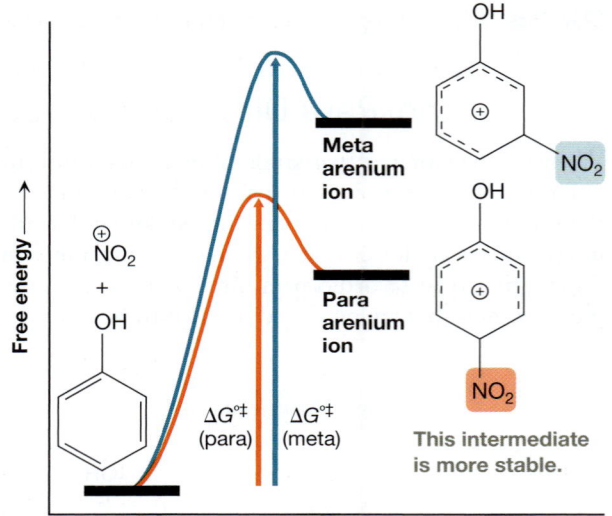

All nonhydrogen atoms have octets.

PROBLEM 23.4 Draw a plot similar to Figure 23-1 that shows the formation of the ortho and meta intermediates during the nitration of fluorobenzene. Use resonance arguments to explain the relative stabilities of the two isomeric intermediates. *Hint:* See Problem 23.2.

PROBLEM 23.5 Draw the mechanism that leads to the ortho-, meta-, and para-disubstituted products for each of the following reactions, and identify the major products.

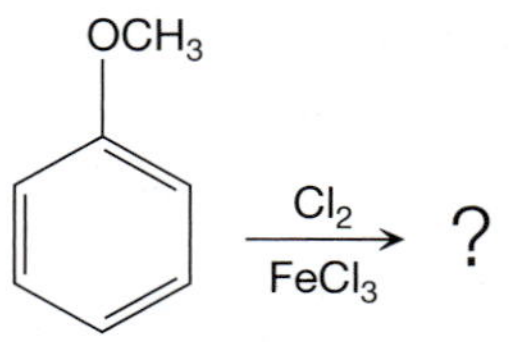

(a) 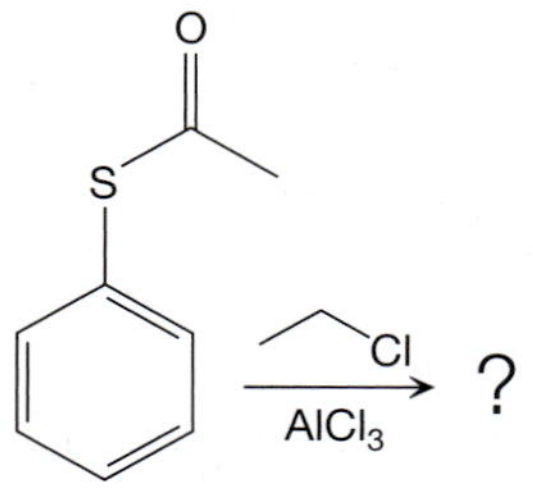   **(b)**

CONNECTIONS The reduction of *o*-nitrotoluene (a product of the nitration of toluene, Equation 23-6) produces *o*-toluidine, a precursor in the manufacture of the herbicides metolachlor and acetochlor.

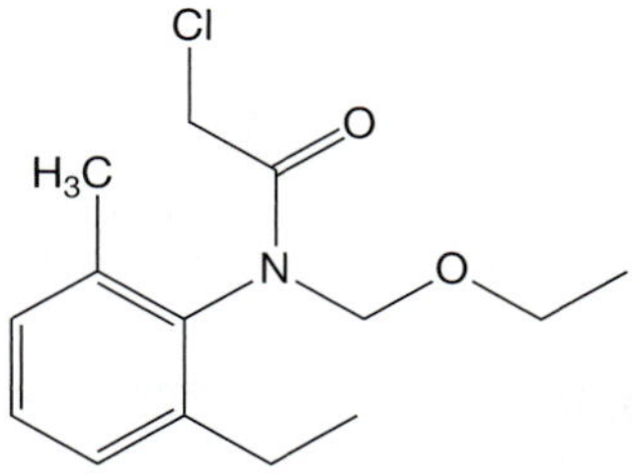

Acetochlor

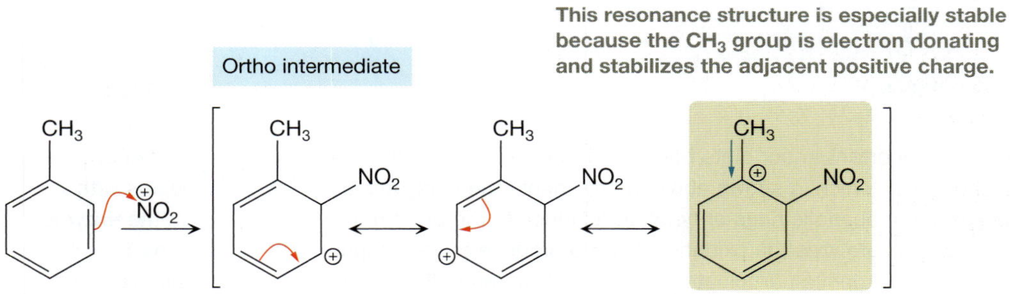

Let's now turn to the second kind of ortho/para directors: alkyl groups. An alkyl group is attached by a C atom that has no lone pairs of electrons, so, unlike what we saw in the case of an attached OH group, no *additional* resonance structure of the arenium ion intermediate can be drawn to give all nonhydrogen atoms their octet. Why, then, should alkyl groups cause the ortho and para arenium ion intermediates to be lower in energy than the meta intermediate? To answer this question, let's carefully examine the resonance structures of the ortho and meta intermediates in the nitration of toluene, C_6H_5—CH_3.

This resonance structure is especially stable because the CH_3 group is electron donating and stabilizes the adjacent positive charge.

Ortho intermediate

(23-6)

None of the resonance structures has the CH_3 group adjacent to the positive charge, so there is not as much stabilization.

Meta intermediate

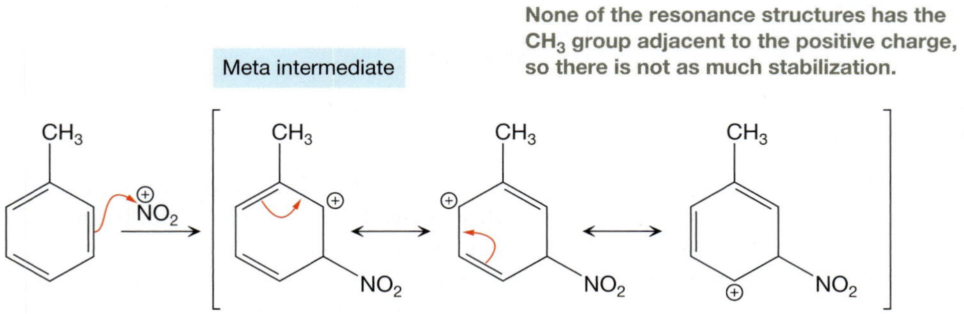

(23-7)

There are three resonance structures in each of the isomeric intermediates. If the NO_2^+ electrophile attaches ortho to the methyl group (Equation 23-6), then two of the resonance structures have the positive charge on an unsubstituted C atom of the ring, and the third (highlighted in green) has the positive charge on a C atom

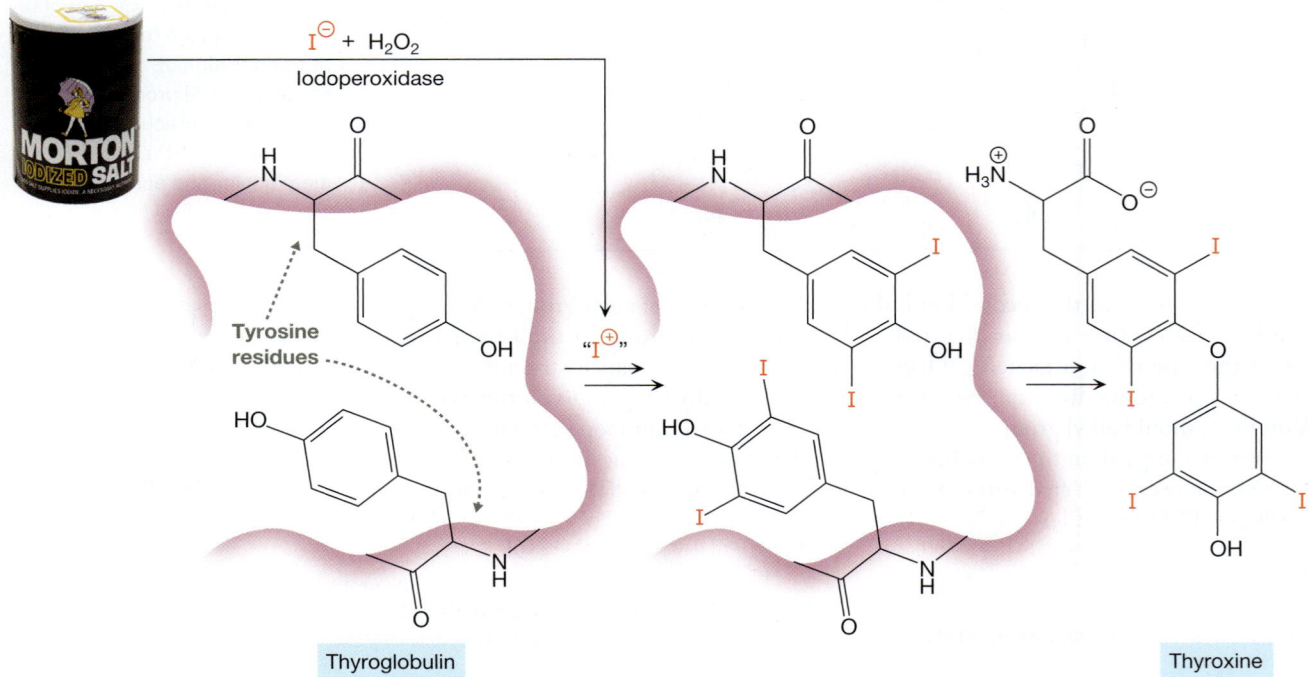
attached to a CH_3 group. The positive charge in the third resonance structure is stabilized by the electron-donating ability of the attached CH_3 group. If the NO_2^+ electrophile instead attaches meta to the methyl group (Equation 23-7), then all three of the intermediate's resonance structures have the positive charge on an unsubstituted C atom. Without contribution from a resonance structure with the positive charge adjacent to the CH_3 group, the meta arenium ion is not as stable as the ortho arenium ion.

As with the nitration of phenol, the arguments that hold for the ortho intermediate also hold for the para intermediate (see Your Turn 23.5). Thus, in the nitration of toluene, the para intermediate is also more stable than the meta intermediate.

The following resonance structures are for the para arenium ion intermediate produced during the nitration of toluene. Identify the one that is responsible for making this intermediate more stable than the meta arenium ion intermediate.

Para intermediate

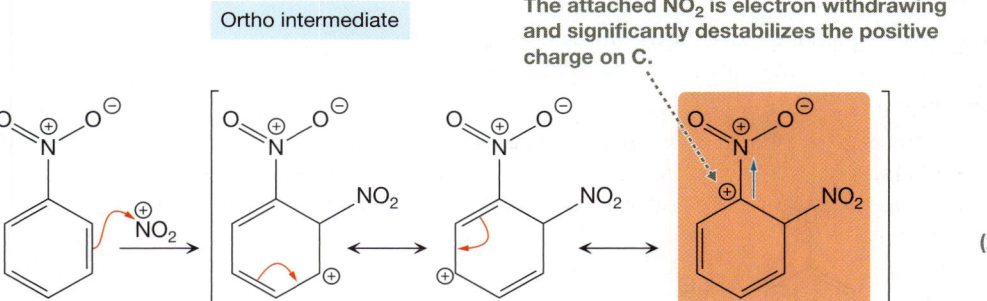

PROBLEM 23.6 Draw all possible products that can be produced in the nitration of ethylbenzene. Which are the major products? Explain.

23.2b Why Are Meta Arenium Ion Intermediates Lower in Energy When a Meta Director Is Attached?

Recall that, when a meta director is attached to benzene, electrophilic aromatic substitution forms predominantly the meta-disubstituted product. In such cases, the meta arenium ion intermediate must be lower in energy than the ortho or para intermediates. Why?

We can gain insight into the answer by examining the ortho and meta intermediates in the nitration of nitrobenzene, $C_6H_5-NO_2$.

Ortho intermediate

The attached NO_2 is electron withdrawing and significantly destabilizes the positive charge on C.

(23-8)

In no resonance structure is the positive charge on C destabilized by an attached NO_2.

Meta intermediate

(23-9)

Although each arenium ion has three resonance structures, the ortho intermediate has a resonance structure (highlighted in red in Equation 23-8) in which the positive charge on the ring is adjacent to the NO$_2$ group that was initially present. An NO$_2$ group is highly electron withdrawing, in large part due to the positive charge that appears on N, which destabilizes the adjacent positive charge on C. That resonance structure, therefore, is significantly higher in energy than the other two, which diminishes its contribution to the resonance hybrid. By contrast, none of the three resonance structures of the meta intermediate has the positive charge of the ring adjacent to the NO$_2$ group. Consequently, there is greater delocalization of the charge in the resonance hybrid of the meta intermediate, resulting in a lower energy.

Similar arguments explain why a meta director causes the para arenium ion intermediate to be higher in energy than the meta intermediate. (See Your Turn 23.6.)

YOUR TURN 23.6

Nitration of nitrobenzene at the para C atom proceeds through an arenium ion intermediate that has the following three resonance structures. Identify the *least* stable resonance structure.

Para intermediate

SOLVED PROBLEM 23.7

The trifluoromethyl group (CF$_3$) is not listed in Table 23-1. Predict whether CF$_3$ is an ortho/para director or a meta director.

Think Which of the arenium ions exhibit resonance structures in which the positive charge on the ring is directly adjacent to the CF$_3$ substituent? What effect does the CF$_3$ group have on an adjacent positive charge—is it stabilizing or destabilizing?

Solve The ortho and meta arenium ions, with all of their resonance structures, are shown here.

The ortho isomer has a resonance structure (highlighted in red) in which the positive charge is adjacent to the CF$_3$ group. Because CF$_3$ is electron withdrawing, it destabilizes the adjacent positive charge. Overall, then, the ortho intermediate is less stable than the meta intermediate. Analysis of the para intermediate in the same way shows that the para intermediate is also less stable than the meta intermediate. Hence the CF$_3$ group is a meta director.

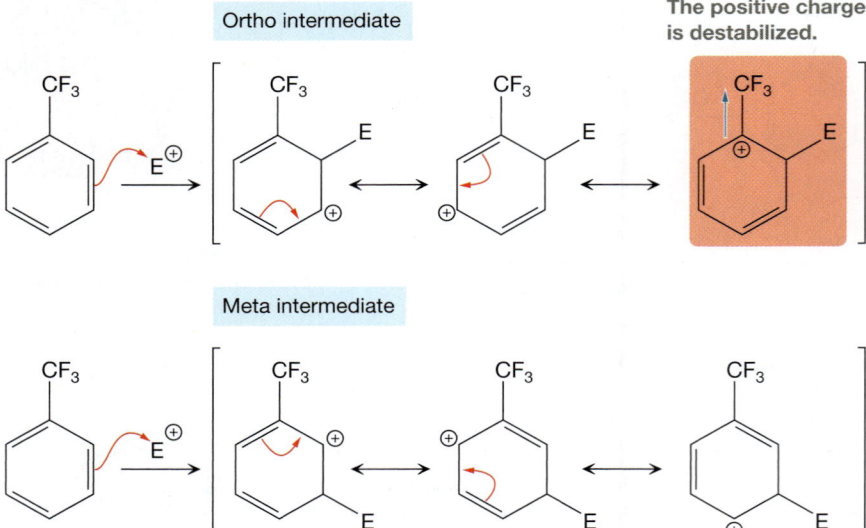

Ortho intermediate

The positive charge is destabilized.

Meta intermediate

PROBLEM 23.8 Verify the statement in Solved Problem 23.7 that the para intermediate is less stable than the meta intermediate. Begin by drawing all resonance structures of the para intermediate, and note the relative stability of each structure.

23.3 The Activation and Deactivation of Benzene toward Electrophilic Aromatic Substitution

The discussion to this point in the chapter has focused on how a substituent affects the rates of formation of the ortho, meta, and para products, *relative to one another*, for a single electrophilic aromatic substitution reaction. We have yet to discuss how substituents affect the *overall* rate of an electrophilic aromatic substitution reaction—that is, the rate of disappearance of the aromatic reactant relative to unsubstituted benzene. Let's begin by examining Table 23-2, which lists the relative overall rates of nitration for a variety of monosubstituted benzenes.

According to Table 23-2, phenol (C_6H_5—OH) undergoes nitration about 1000 times faster than benzene itself. Consequently, the OH group is classified as an **activating group**—we say that the group *activates the benzene ring* toward electrophilic aromatic substitution. The nitro (NO_2) group is a **deactivating group**, on the other hand, because nitrobenzene (C_6H_5—NO_2) undergoes nitration over

TABLE 23-2	Relative Rates of Nitration of Monosubstituted Benzenes	

Substituent	Relative Rate	Type of Group
—NH_2	—[a]	Strongly activating
—OH	1000	Strongly activating
—CH_3	25	Weakly activating
—H (benzene)	1 (reference)	—
—I	0.18	Weakly deactivating
—Cl	0.033	Weakly deactivating
—CO_2Et	0.0037	Moderately deactivating
—NO_2	6×10^{-8}	Strongly deactivating
—$\overset{\oplus}{N}(CH_3)_3$	1.2×10^{-8}	Strongly deactivating

[a]Aromatic amines are susceptible to protonation and oxidation under nitration conditions. The NH_2 group is determined to be a strongly activating group using other electrophilic aromatic substitution reactions.

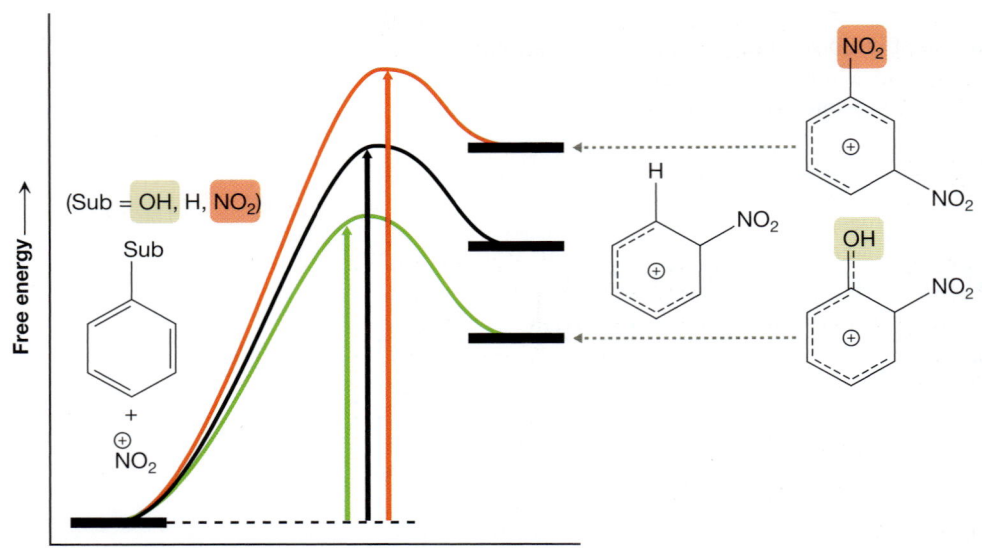

FIGURE 23-2 **Activation and deactivation of an aromatic ring**
Free energy diagrams for the addition of NO_2^+ to benzene (black curve), phenol (green curve), and nitrobenzene (red curve). The OH group in phenol stabilizes the arenium ion intermediate by involving the lone pair of electrons on the O atom in resonance. Hence, the OH group increases the rate of formation of the arenium ion intermediate, making it an *activating group*. A nitro group, on the other hand, destabilizes the arenium ion intermediate. Hence, the nitro group slows the rate of formation of the arenium ion, making it a *deactivating group*.

10 million times *slower* than benzene itself. Relative to H, the NO_2 group *deactivates the benzene ring* toward electrophilic aromatic substitution.

Why is the OH group so strongly activating? When the electrophile attaches to the ortho or para position on phenol's ring, a lone pair of electrons from the O atom of the OH group participates in resonance in the arenium ion intermediate. This lowers the energy barrier leading to the formation of the arenium ion, as shown in Figure 23-2 (green curve), which increases the rate of formation of the arenium ion.

Conversely, the NO_2 group of nitrobenzene destabilizes the arenium ion intermediate that is formed. This increases the energy barrier to the formation of the arenium ion (Fig. 23-2, red curve), which slows the overall reaction rate. As a result, the NO_2 group is a *deactivating group*.

PROBLEM 23.9 Use Table 23-2 to estimate the relative rate of nitration of bromobenzene under the same conditions as the other entries. Is it activating or deactivating?

SOLVED PROBLEM 23.10

Why does toluene undergo electrophilic aromatic substitution faster than benzene, thereby making CH_3 an *activating group*?

Think Are toluene's arenium ion intermediates more stable or less stable than those of benzene? How does that affect the energy barriers for the formation of toluene's arenium ions and, hence, the rates?

Solve The CH_3 group is electron donating, so it stabilizes the arenium ion intermediate. This lowers the activation energy for formation of the intermediates, which increases the rate at which they form. Hence, a CH_3 group increases the overall substitution rates.

The arenium ion is stabilized by the electron donation of CH_3.

PROBLEM 23.11 Determine whether the aromatic ring in trifluoromethylbenzene (C_6H_5—CF_3) is activated or deactivated.

Table 23-3 organizes substituents according to their ortho/para- or meta-directing ability as well as their activating or deactivating ability. The characteristics of the various groups suggest the following trends:

- Activating groups are generally ortho/para directors.
- Deactivating groups are generally meta directors.

These trends break down this way because both the regiochemistry and the extent of activation of the aromatic ring derive from essentially the same factors. Substituents that are electron donating, via either resonance or inductive effects, stabilize the arenium ion intermediate and increase the overall reaction rate, so they are activating. These groups also stabilize the ortho and para arenium ion intermediates better than they do the meta intermediate, so they are ortho/para directors. By the same token, substituents that destabilize the arenium ion intermediate by withdrawing electron density are deactivating groups. These groups destabilize the ortho and para intermediates more than they do the meta, making them meta directors.

The lone exceptions in Table 23-3 are the halogen atoms.

Halogen substituents are weakly deactivating, but they are ortho/para directors.

TABLE 23-3 Comparison of the Activating/Deactivating and the Ortho/Meta/Para-Directing Nature of Substituents

Substituent	Activating/Deactivating Nature	Ortho/Para- or Meta-Directing	Substituent	Activating/Deactivating Nature	Ortho/Para- or Meta-Directing
—O⁻	Strongly activating	Ortho/para	—C(=O)OR (or OH)	Moderately deactivating	Meta
—NH₂ / —NR₂	Strongly activating	Ortho/para	—C(=O)R (or H)	Moderately deactivating	Meta
—OH / —OR	Strongly activating	Ortho/para	—C≡N	Strongly deactivating	Meta
—NH—C(=O)—R (or H)	Moderately activating	Ortho/para	—S(=O)(=O)—OH	Strongly deactivating	Meta
—O—C(=O)—R (or H)	Moderately activating	Ortho/para	—N⁺(=O)O⁻	Strongly deactivating	Meta
—R	Weakly activating	Ortho/para	—N⁺(R or H)(R or H)(R or H)	Strongly deactivating	Meta
—H (Benzene)	—	—			
—Cl / —Br / —I	Weakly deactivating	Ortho/para			

Halogens are ortho/para directors because, like the OH group, halogens possess a lone pair of electrons and are electron donating via resonance. Consequently, a halogen atom substituent stabilizes the arenium ion intermediate substantially when an electrophile attaches to either the ortho or para position. Conversely, halogens are deactivators because they are inductively electron-withdrawing groups.

YOUR TURN 23.7

The ortho/para-directing and deactivating characteristics of a Cl substituent can be understood by considering the resonance structures and the resonance hybrid of the ortho arenium ion intermediate below. Supply the curved arrow notation that shows how the first resonance structure is converted into the second. How many atoms in the second resonance structure lack an octet? Draw an arrow along the C—Cl bond in the resonance hybrid to represent the inductive effect by Cl. How does that inductive effect impact the stability of positive charge in the intermediate?

We can gain more insight into the activating or deactivating character of a substituent by examining the effect the substituent has on the electron density of the aromatic ring prior to the reaction. As examples, the electrostatic potential maps of benzene, aniline, and nitrobenzene are shown in Figure 23-3.

The aromatic ring of aniline, which has an NH_2 activating group, has *greater* electron density (i.e., is more red) than the aromatic ring of benzene. The NH_2 group donates electron density to the ring via resonance, making it more electron rich than benzene itself and giving it greater affinity for an incoming electrophile

FIGURE 23-3 Effect of activating and deactivating groups on electron density in the benzene ring Electrostatic potential maps of (a) benzene, (b) aniline, and (c) nitrobenzene. The electron-donating ability of NH_2 via resonance increases the electron density of the aromatic ring (more red) and, hence, activates the ring toward electrophilic aromatic substitution. Conversely, the electron-withdrawing ability of the NO_2 group decreases electron density in the ring (more blue/green) and, hence, deactivates the ring toward electrophilic aromatic substitution.

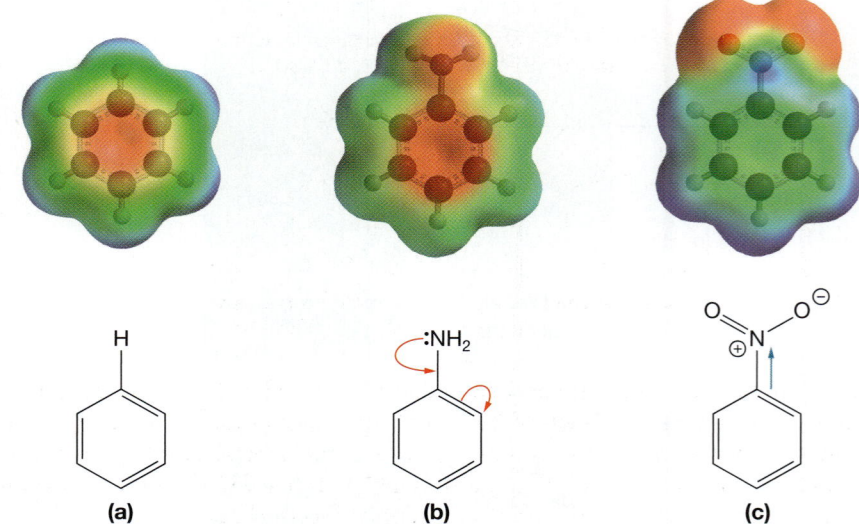

(E^+). This serves to enhance the rate of the aromatic ring's attack on the electrophile, thus increasing the overall rate of substitution (activating) compared to benzene.

Conversely, the aromatic ring of nitrobenzene, which has a NO_2 deactivating group, is less electron rich (i.e., is more blue/green) than benzene itself. The NO_2 group withdraws electron density from the ring, so the driving force for the aromatic ring to attack the electrophile is diminished relative to benzene, causing nitrobenzene to react more slowly.

YOUR TURN **23.8**

The electrostatic potential maps of benzene and a monosubstituted benzene are shown here. Is the aromatic ring of the monosubstituted compound activated or deactivated relative to benzene?

PROBLEM 23.12 Does the electrostatic potential map of the monosubstituted benzene in Your Turn 23.8 represent toluene or trifluoromethylbenzene? Explain.

23.4 The Impacts of Substituent Effects on the Outcomes of Electrophilic Aromatic Substitution Reactions

Compare the nitrations of benzene (Equation 23-10) and nitrobenzene (Equation 23-11).

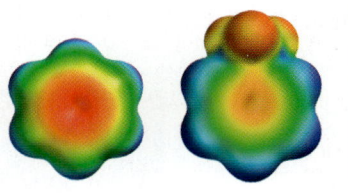

Nitrobenzene
83%

(23-10)

The ring is deactivated, so a stronger acid is necessary to carry out a second nitration.

NO_2 ... $\xrightarrow[\substack{H_2SO_4, \\ 25\,°C,\ 1\ h}]{HNO_3}$ ...

Nitrobenzene **m-Dinitrobenzene**
58%

(23-11)

The first nitration requires only concentrated HNO_3, but H_2SO_4 is added to carry out the second reaction. H_2SO_4 is added because the presence of the NO_2 group in nitrobenzene deactivates the ring substantially, making the second nitration slower and more difficult than the first. H_2SO_4 is a stronger acid than HNO_3 and increases the equilibrium concentration of the NO_2^+ electrophile (Section 22.6).

2,4,6-Trinitrotoluene (TNT)

2,4,6-Trinitrotoluene (TNT) is one of the most widely known explosives. Originally synthesized in the 1860s, TNT has had a long and rich history in military and industrial applications. It is a very useful explosive, in part because it is not very shock sensitive, which makes it relatively safe to handle. Moreover, it is a solid that melts well below the temperature at which it will spontaneously detonate, and, as a liquid, it can be cast into molds and mixed with other explosives.

In the industrial synthesis of TNT, toluene undergoes three successive nitration reactions, as we learned in Chapter 22. The first and second nitrations can be carried out with a nitric acid/sulfuric acid mixture. The third nitration is significantly more difficult, however, because the two NO_2 groups strongly deactivate the ring (Section 23.3), so fuming sulfuric acid is used to compensate.

TNT is explosive because its combustion reaction produces hot, gaseous products (e.g., N_2 and CO_2) *very* rapidly. The fast expansion of these gases is what produces the destructive shock wave, as can be seen over the water in the above photo. The exothermicity of the reaction stems from the fact that the CO_2 and N_2 products are highly stable. The fast reaction rate is facilitated by the NO_2 groups. In the combustion of a hydrocarbon—which is also highly exothermic—the reaction rate is limited to a large extent by the availability of gaseous O_2, which must come from the surroundings. TNT, on the other hand, does not rely as much on O_2 from the surroundings because the molecule itself has a rather high density of oxygen.

PROBLEM 23.13 Explain why a single sulfonation of benzene can be carried out with just concentrated H_2SO_4, but fuming sulfuric acid is needed to produce the disulfonated product. *Hint:* Review Section 22.7.

Benzene $\xrightarrow[\text{48 h, 20 °C}]{\text{conc } H_2SO_4}$ Benzenesulfonic acid (SO_3H)
66%

Benzene $\xleftarrow[\text{SO}_3, \text{4 h, <90 °C}]{\text{conc } H_2SO_4}$ 1,3-Benzenedisulfonic acid (HO_3S, SO_3H)
90%

Many electrophilic aromatic substitution reactions can still be carried out despite the presence of a moderately or strongly deactivating group on the ring, but this is not true for Friedel–Crafts reactions.

> Friedel–Crafts reactions do not readily take place on moderately or strongly deactivated aromatic rings (review Table 23-3, p. 1157).

So, for example, no alkylation takes place in Equation 23-12 because the C=O group is moderately deactivating.

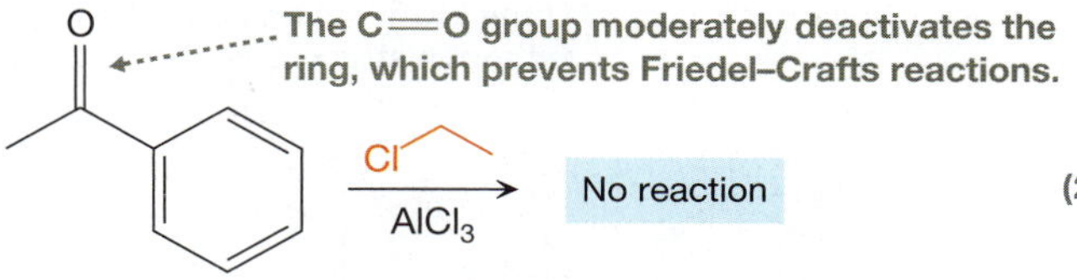

The C=O group moderately deactivates the ring, which prevents Friedel–Crafts reactions.

No reaction   (23-12)

The opposite problem exists when we want to carry out a Friedel–Crafts alkylation on benzene itself (Equation 23-13).

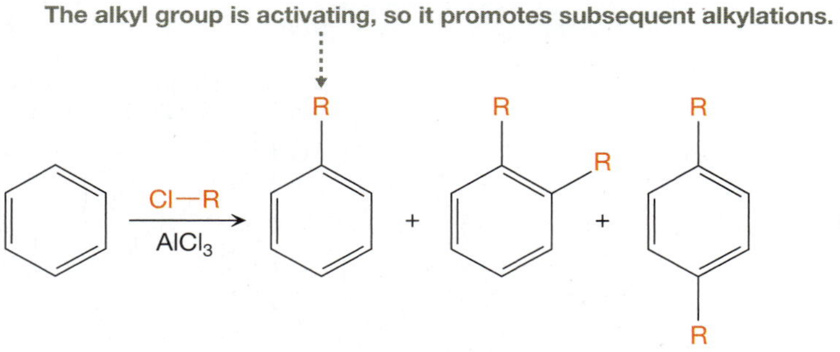

The alkyl group is activating, so it promotes subsequent alkylations.

(23-13)

Thus, as we first mentioned in Section 22.4:

> Friedel–Crafts alkylations are subject to polyalkylation.

Each alkyl group activates the aromatic ring toward electrophilic aromatic substitution, so with an alkyl group having been added in the first alkylation, subsequent alkylations become faster.

In light of this problem, it is often better to add an alkyl group by first acylating the ring and then reducing the C=O group to a CH$_2$ group, as shown in Equation 23-14, and as we saw previously in Section 22.8.

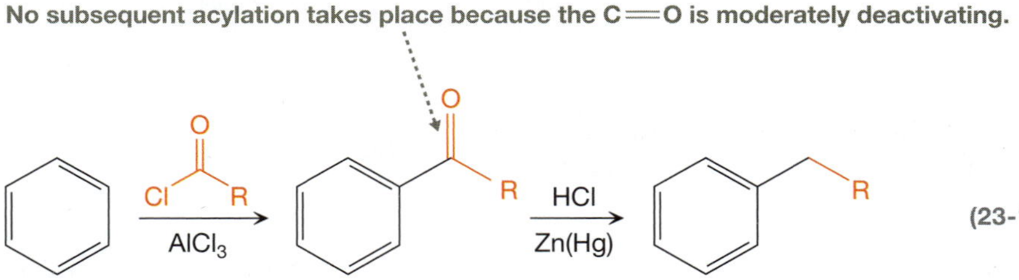

No subsequent acylation takes place because the C=O is moderately deactivating.

(23-14)

We are not concerned about multiple acylations because the aromatic ring in the product of the first acylation is deactivated.

PROBLEM 23.14 Show how to synthesize butylbenzene from benzene.

23.5 The Impact of Reaction Conditions on Substituent Effects

A substituent's effect on the regiochemistry and reaction rate of electrophilic aromatic substitution is not necessarily absolute. For some substituents, the reaction conditions play a significant role. For example, when phenol is treated with bromine in water, it undergoes multiple brominations, even in the absence of a Lewis acid catalyst (Equation 23-15). When acetic acid is added to the reaction mixture, however, only a single bromination occurs (Equation 23-16).

CONNECTIONS

2,4,6-Tribromophenol (the product of Equation 23-15) is used as a wood preservative and a fungicide, and is also used as a precursor in the manufacture of some flame retardants, such as brominated epoxy resins.

Three brominations occur, even without a strong Lewis acid catalyst.

(23-15)

97%

Only a single bromination occurs under mildly acidic conditions.

(23-16)

The nature of the substituent is affected by the pH of the solution. Phenol is a weak acid ($pK_a = 10.0$), so in the absence of acetic acid, phenol is in equilibrium with a small amount of its conjugate base, the phenoxide anion ($C_6H_5-O^-$), as shown in Equation 23-17. Even though very little of the phenoxide anion is present, the O^- substituent is a *very* powerful activating group—far better than the OH group—so it dramatically increases the rate of electrophilic aromatic substitution. As a result, the phenoxide anion is the dominant reactant in the reaction.

The $O^{\ominus}$ substituent is a *very* powerful activator.

(23-17)

The presence of acetic acid decreases the pH of the solution. The resulting increased concentration of H^+ substantially decreases the concentration of the phenoxide anion (Le Châtelier's principle), effectively cutting off the route by which the phenoxide anion participates in electrophilic aromatic substitution. The only route available is the one in which the electrophile reacts with the uncharged phenol itself—a much slower reaction that can be easily stopped after just a single bromination.

The pH can also dramatically affect the nitration of *N,N*-dimethylaniline. If *N,N*-dimethylaniline is treated with HNO_3 in acetic acid (Equation 23-18), then the ortho and para nitro products are predominantly formed. This is as expected, since the lone pair on the substituent's N atom makes it ortho/para directing and activating. However, if nitration is carried out in sulfuric acid (i.e., at a much lower pH), then the major product is the meta isomer (Equation 23-19).

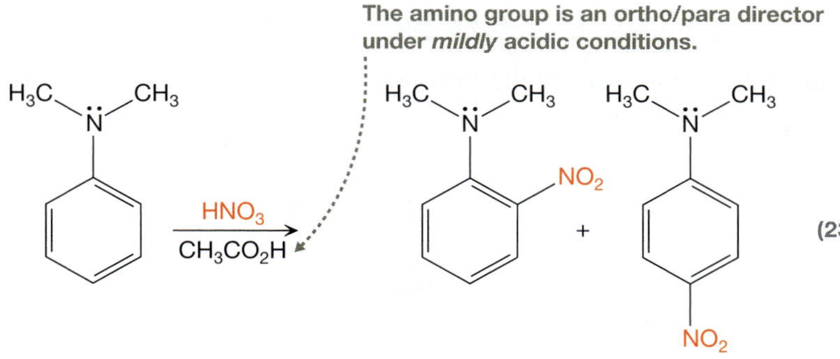

The amino group is an ortho/para director under *mildly* acidic conditions.

(23-18)

The amino group becomes a meta director under *strongly* acidic conditions.

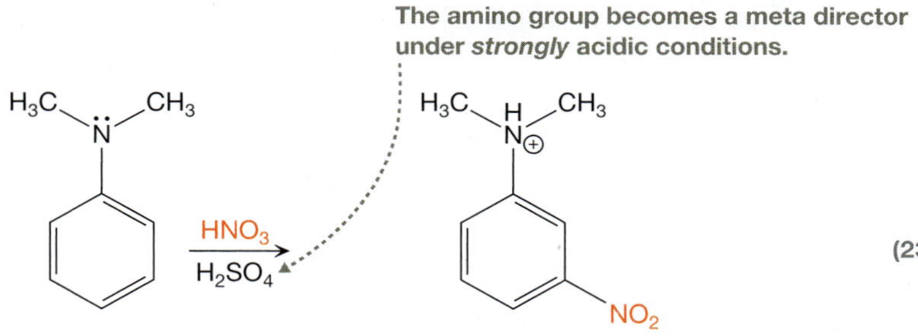

(23-19)

Sulfuric acid is much stronger than acetic acid, so it *quantitatively* protonates the amino group, generating an anilinium ion in which a $+1$ formal charge exists on N (Equation 23-20). In this form, the substituent is highly electron withdrawing, much like an NO_2 group, so it is a meta director and a deactivator.

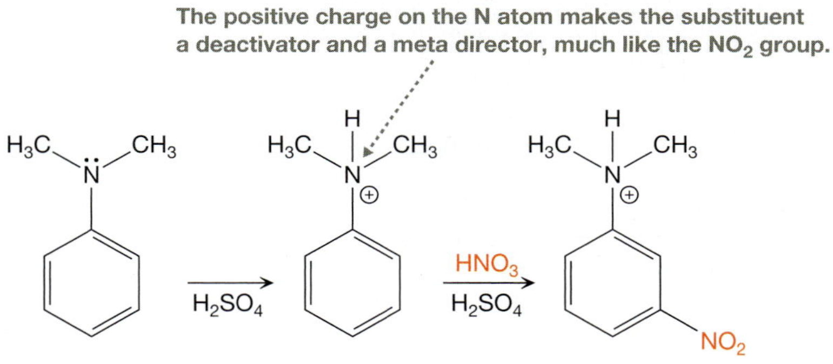

The positive charge on the N atom makes the substituent a deactivator and a meta director, much like the NO_2 group.

(23-20)

A similar phenomenon can occur when a strong Lewis acid is present. For example, even though aniline possesses a highly activating NH_2 group, it is incompatible with Friedel–Crafts reactions, as shown in Equation 23-21 (next page). The amino group is a relatively strong Lewis base, so it readily coordinates to the $AlCl_3$ Lewis acid. In that complexed form, the N atom possesses a $+1$ formal charge, making it a highly *deactivating* group that precludes Friedel–Crafts reactions altogether (Section 23.4).

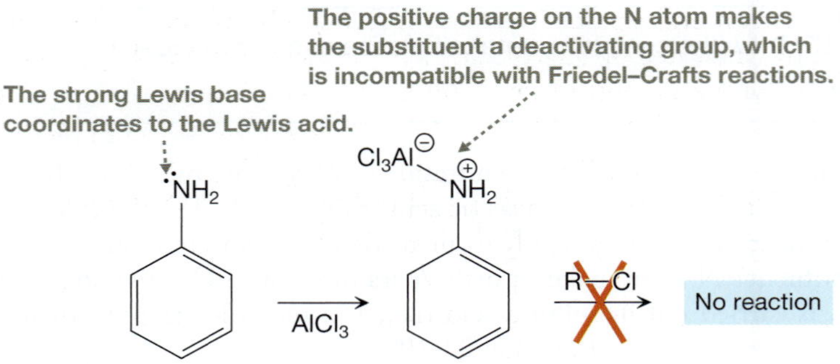

The positive charge on the N atom makes the substituent a deactivating group, which is incompatible with Friedel–Crafts reactions.

The strong Lewis base coordinates to the Lewis acid.

(23-21)

YOUR TURN 23.9

Determine whether each of the following reactions will proceed as indicated. Explain.

(a)

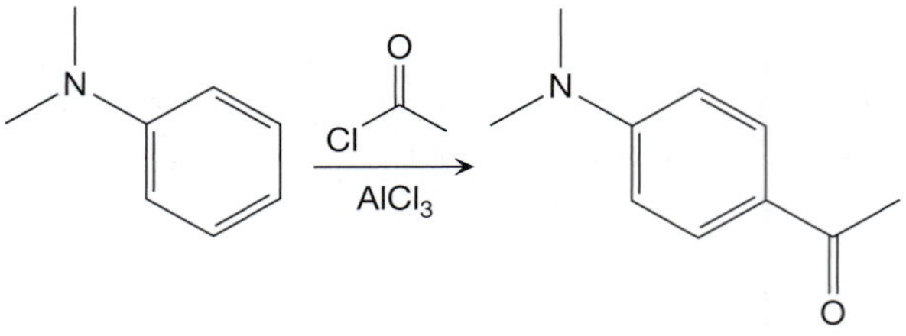

(b)

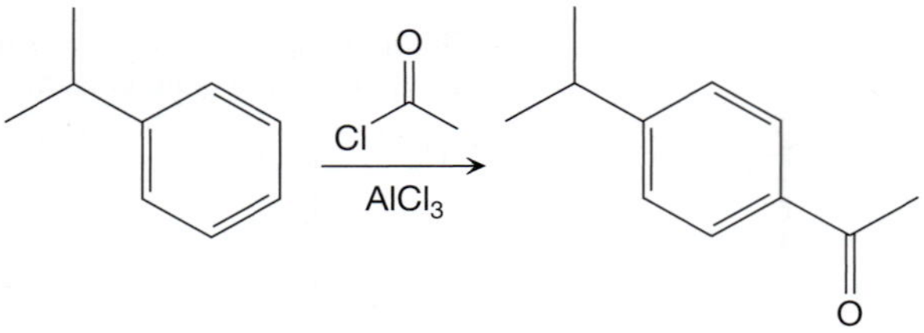

23.6 Electrophilic Aromatic Substitution on Disubstituted Benzenes

When electrophilic aromatic substitution takes place on a disubstituted benzene, effects from both of the substituents on the ring must be considered—both their activating/deactivating abilities and their ortho/meta/para-directing capabilities. Fortunately, these effects tend to be essentially additive.

The regiochemistry of these reactions is straightforward if the two substituents are "in agreement." For example, the nitration of *p*-nitrotoluene produces almost exclusively 2,4-dinitrotoluene (Equation 23-22).

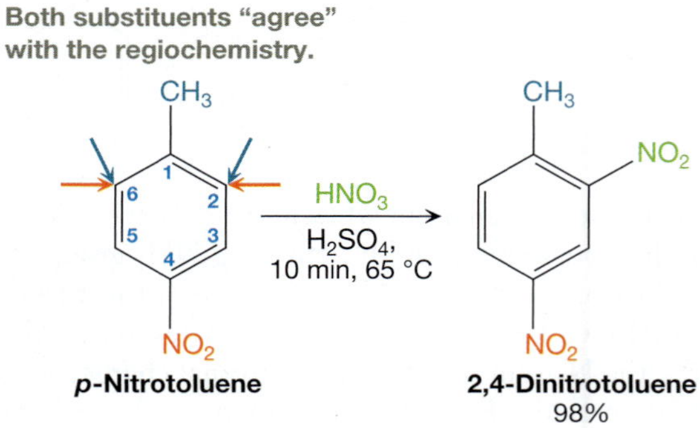

Both substituents "agree" with the regiochemistry.

(23-22)

p-Nitrotoluene

2,4-Dinitrotoluene
98%

The CH$_3$ group is an ortho/para director, so it favors substitution at the 2 and 6 positions on the ring (indicated by the blue arrows). Furthermore, the NO$_2$ group is a meta director, so it also favors substitution at the 2 and 6 positions on the ring (indicated by the red arrows). With agreement like this, we observe substitution primarily at the C atom ortho to the CH$_3$ group.

The situation is not quite so straightforward without this agreement among the substituents. In those situations, we can apply the following rule:

When two substituents attached to an aromatic ring "disagree" about where to direct the incoming electrophile, regiochemistry is usually dictated by the substituent that is the stronger activating group—that is, the group that appears nearer the beginning of Table 23-3 (p. 1157).

For example, when *m*-nitrotoluene undergoes nitration (Equation 23-23), the CH$_3$ group directs substitution to the 2, 4, and 6 positions (indicated by the blue arrows), whereas the NO$_2$ group directs substitution to the 5 position (indicated by the red arrow).

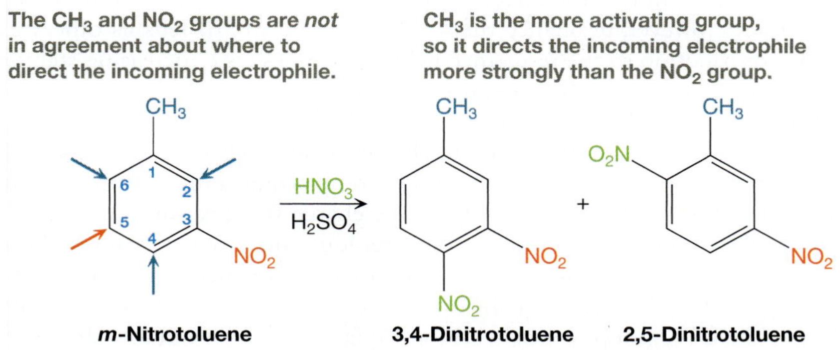

The CH$_3$ and NO$_2$ groups are *not* in agreement about where to direct the incoming electrophile.

CH$_3$ is the more activating group, so it directs the incoming electrophile more strongly than the NO$_2$ group.

(23-23)

m-Nitrotoluene   3,4-Dinitrotoluene   2,5-Dinitrotoluene

Experimentally, we observe that substitution occurs mainly at the 4 and 6 positions on the ring, suggesting that the CH$_3$ group has greater influence on the regiochemistry than the NO$_2$ group. This is consistent with the fact that CH$_3$ is a more strongly activating substituent than NO$_2$.

We can understand why the more activating group dictates the regiochemistry by examining the resonance contributors and resonance hybrids of the various arenium ion intermediates that are possible. When the electrophile attaches meta to the methyl group (Equation 23-24), then all resonance contributors of the resulting arenium ion have about the same stability—in each case, the positive charge appears on an unsubstituted carbon.

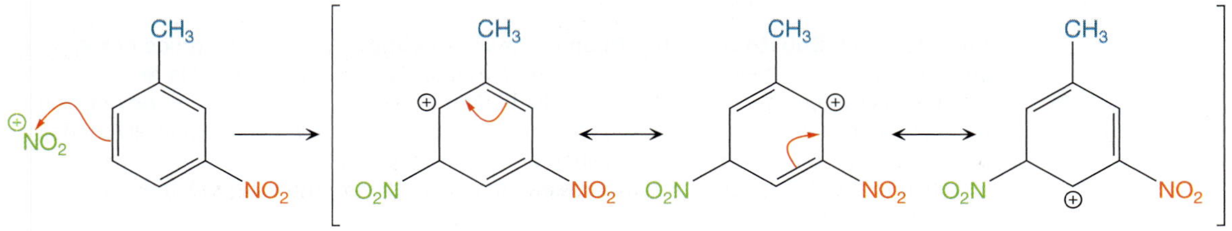

(23-24)

When the electrophile attaches para to the CH_3 group (Equation 23-25), the resulting arenium ion has one resonance structure that is especially stable (highlighted in green) because the positive charge is adjacent to the electron-donating CH_3 group. There is also one resonance structure that is particularly unstable (highlighted in red) because the positive charge is adjacent to the electron-withdrawing NO_2 group.

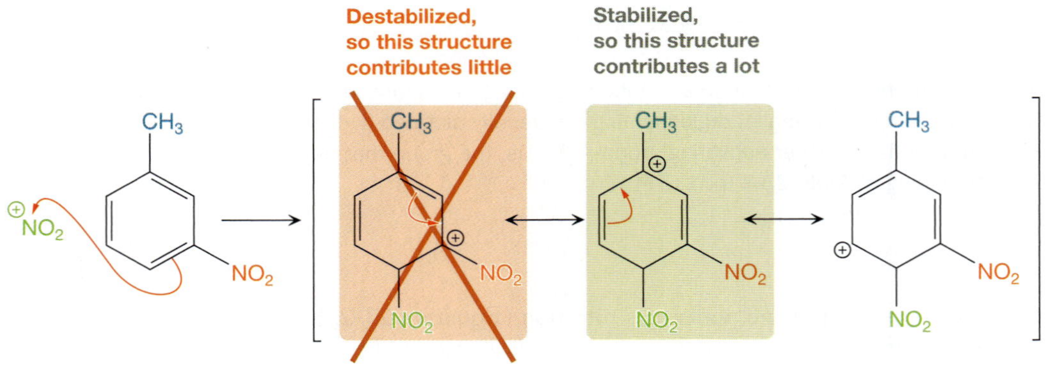

(23-25)

It is tempting to presume that the destabilization in the first resonance structure in Equation 23-25 counters the effects from the stabilization in the second resonance structure. However, resonance theory (Chapters 1 and 6) tells us otherwise, because the contribution to the resonance hybrid increases with increasing stability of the resonance structure. In this case, the resonance structure highlighted in green contributes substantially more to the resonance hybrid than does the resonance structure highlighted in red. Therefore, in the resonance hybrid (Fig. 23-4b), the C atom attached to the electron-donating CH_3 group acquires more of the positive charge than the other C atoms of the ring, so the charge stabilization is maximized. By contrast, the resonance hybrid of the arenium ion from Equation 23-24 (Fig. 23-4a) has similar contributions by each of its resonance structures, so the positive charge is shared roughly equally over three C atoms. None of those C atoms is attached to the CH_3 group, so the hybrid in Equation 23-24 is not as stable as the one in Equation 23-25.

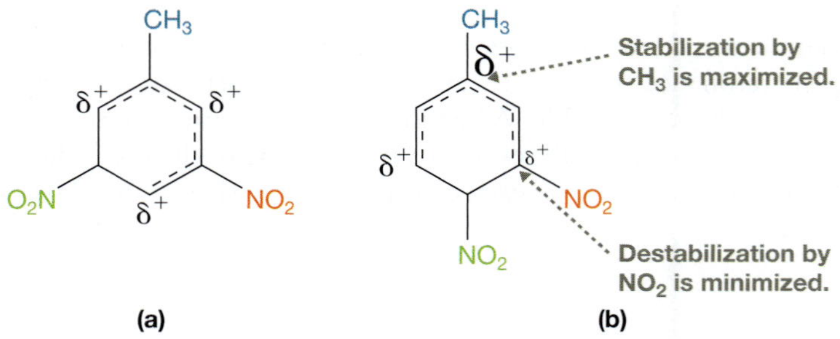

FIGURE 23-4 **Substituent effects on resonance hybrids** Resonance hybrids of the arenium ions from (a) Equation 23-24 and (b) Equation 23-25. When NO_2 adds meta to CH_3 (a), the positive charge is roughly equally distributed over the three C atoms of the ring and there is no special stabilization or destabilization by the attached groups. When NO_2 adds para to CH_3 (b), the greatest accumulation of the partial positive charge is on the C atom attached to the CH_3 group, which maximizes the stabilization of the charge. Hybrid (b) is therefore more stable than hybrid (a).

The following resonance structures are for the arenium ion produced when NO_2^+ attaches ortho to the CH_3 group in *m*-nitrotoluene. Identify the most stable and least stable resonance structures. Overall, is this arenium ion more stable or less stable than the one in Equation 23-24?

Although the methyl group in *m*-nitrotoluene directs the incoming electrophile to the 2, 4, and 6 positions, substitution is favored at just the 4 and 6 positions in Equation 23-23. This is due to *steric hindrance*. The 2 position is flanked by both the CH_3 group and the NO_2 group, making that site the least accessible to the incoming electrophile. The 4 and 6 positions are more accessible because each is flanked by only one substituent—either the CH_3 group or the NO_2 group.

The situation is not so clear if the two substituents on benzene disagree with the placement of the incoming electrophile *and* have roughly the same activating/deactivating ability—that is, if they are in roughly the same location in Table 23-3. This is the case when *p*-ethyltoluene undergoes electrophilic aromatic substitution, because the two alkyl groups have about the same activating capabilities. As indicated in Equation 23-26, a mixture of isomers is produced because the two alkyl groups are nearly equally good at directing the incoming electrophile to the position on the ring ortho to themselves.

The CH_3 and CH_2CH_3 groups are similar in their activating abilities, but direct the incoming electrophile to different carbons.

p-Ethyltoluene   4-Ethyl-2-nitrotoluene   4-Ethyl-3-nitrotoluene

Relative percentages: 56%   44%

(23-26)

FIGURE 23-5 Additivity of substituent effects on reaction rate With each additional alkyl group, the ring becomes increasingly activated toward electrophilic aromatic substitution, which causes the reaction rate to increase.

Substituent effects on the *overall* rate of electrophilic aromatic substitution are also additive. We can see this explicitly in Figure 23-5, which shows the rates of chlorination

Increasing reaction rate

Increasingly activated ring

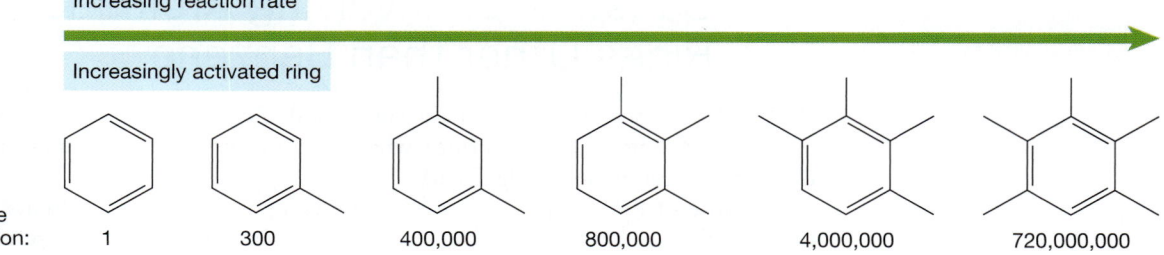

Relative rate of chlorination: 1   300   400,000   800,000   4,000,000   720,000,000

for several methylated benzenes. In general, the rate of chlorination increases with each additional methyl group, given that each methyl group is *activating*. (As explored further in Problem 23.45, the reaction rates are also sensitive to the particular locations of the substituents relative to one another about the ring.)

SOLVED PROBLEM 23.16

Rank the rate of electrophilic aromatic substitution from slowest to fastest for the following aromatic rings.

A B C D E

Think Are the substituents on the ring activating or deactivating? How many are there of each? Are all the activating and deactivating substituents of equal strength?

Solve According to Table 23-3 (p. 1157), the NO_2 group is strongly deactivating, the $CH_3C{=}O$ group is moderately deactivating, the CH_3 group is weakly activating, and the OCH_3 group is strongly activating. **A** will react the slowest because it is the only one with two deactivating groups. **D** and **E** will both react faster because each has one deactivating group and one activating group. **E** will react faster than **D**, however, because the deactivating groups are the same but the activating group in **E** is stronger. Finally, **B** and **C** will react faster still because they each have two activating groups, but **C** will react faster than **B** because both substituents in **C** are strongly activating. Therefore, the order of increasing reaction rate is: $A < D < E < B < C$.

PROBLEM 23.17 Rank the rate of electrophilic aromatic substitution from slowest to fastest for the following aromatic rings.

A B C D E F

23.7 Electrophilic Aromatic Substitution Involving Aromatic Rings Other than Benzene

Benzene is but one of many aromatic compounds. Naphthalene is an aromatic hydrocarbon consisting of two fused rings, whereas furan, pyrrole, and pyridine are *heterocyclic* aromatic compounds (Fig. 23-6).

Because of the aromatic nature of these compounds, their π systems, similar to benzene's, are resistant to electrophilic addition. Instead, treating these compounds with electrophiles tends to lead to substitution to preserve their aromaticity.

These compounds undergo many of the same electrophilic aromatic substitution reactions as benzene. Unlike benzene, however, each of these compounds has at least two chemically distinct C atoms, so more than one isomeric product can be produced even without a substituent initially attached to the ring. Moreover, the reaction rates for these compounds can be significantly faster or slower than that of benzene, characterizing their rings as either *activated* or *deactivated*.

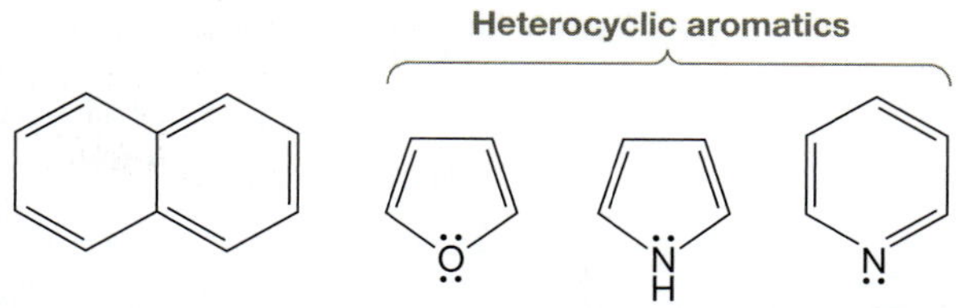

For example, Equation 23-27 shows that naphthalene has two chemically distinct H atoms—one designated α and the other β—and that the substitution of each yields a different product.

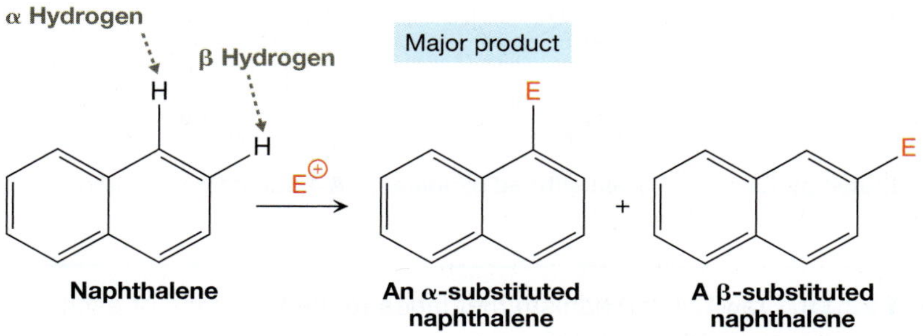

Heterocyclic aromatics

Naphthalene   **Furan**   **Pyrrole**   **Pyridine**

FIGURE 23-6 Some aromatic rings other than benzene These compounds have aromatic rings, so they, too, can undergo electrophilic aromatic substitution.

α Hydrogen
β Hydrogen
Major product

Naphthalene   **An α-substituted naphthalene**   **A β-substituted naphthalene**   (23-27)

As indicated in Equation 23-27:

> Electrophilic aromatic substitution on naphthalene generally favors the α product over the β product.

This means that the arenium ion produced when an electrophile attaches to the α position is more stable than when it attaches to the β position.

Moreover, electrophilic aromatic substitution reactions of naphthalene are generally faster than those of benzene, so:

> Naphthalene is considered to be activated toward electrophilic aromatic substitution.

Thus, naphthalene's arenium ion intermediate is more stable than benzene's.

The reasons for these outcomes are explored further in the problems at the end of the chapter (see Problems 23.59 and 23.60).

PROBLEM 23.18 Draw the mechanism that leads to the formation of each product in Equation 23-27.

Similarly, pyrrole has two chemically distinct H atoms and can undergo electrophilic aromatic substitution to produce two isomeric products, as shown in Equation 23-28. And pyridine has three chemically distinct H atoms, so it can undergo electrophilic aromatic substitution to produce three isomeric products, as shown in Equation 23-29.

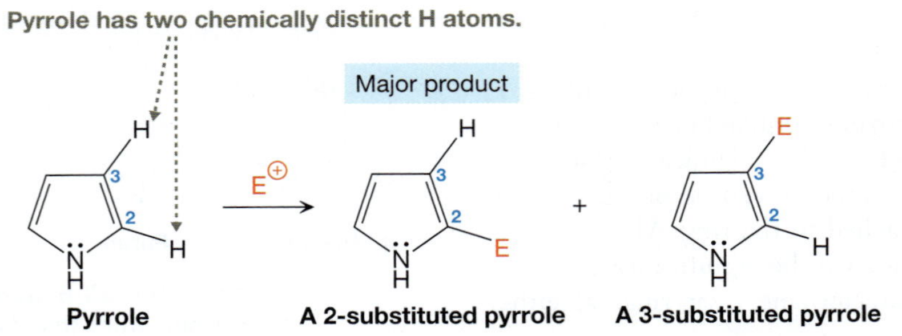

Pyrrole has two chemically distinct H atoms.

A 2-substituted pyrrole   A 3-substituted pyrrole

(23-28)

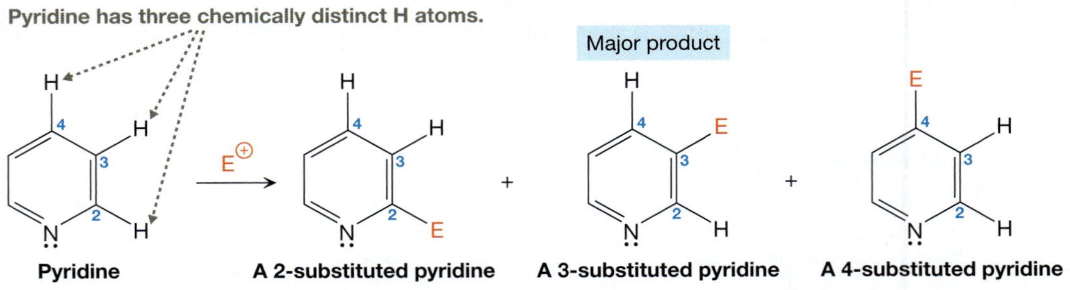

Pyridine has three chemically distinct H atoms.

A 2-substituted pyridine   A 3-substituted pyridine   A 4-substituted pyridine

(23-29)

PROBLEM 23.19 Draw the mechanism that leads to the formation of each product in Equations 23-28 and 23-29.

As indicated in Equations 23-28 and 23-29:

- Electrophilic aromatic substitution of pyrrole takes place primarily at the 2 position.
- Electrophilic aromatic substitution of pyridine occurs primarily at the 3 position.

Moreover, electrophilic aromatic substitution of pyrrole is typically faster than benzene, whereas electrophilic aromatic substitution of pyridine is typically slower than benzene. That is:

- Pyrrole is *activated* toward electrophilic aromatic substitution.
- Pyridine is *deactivated* toward electrophilic aromatic substitution.

Why should the regiochemistry and reaction rates differ so much as a result of different ring sizes? Just as we saw with the reactions involving substituted benzenes, the answer lies with the relative stabilities of the arenium ion intermediates.

Equation 23-30 shows the arenium ion intermediates produced when an electrophile attaches to pyrrole at positions 2 or 3. The arenium ion intermediate produced when the electrophile attaches to the 2 position (Equation 23-30a) has three total resonance structures, whereas the intermediate produced when the electrophile attaches to the 3 position has just two resonance structures (Equation 23-30b). Consequently, the arenium ion intermediate is more stable when substitution occurs at the 2 position, which is why the 2-substituted pyrrole is the major product.

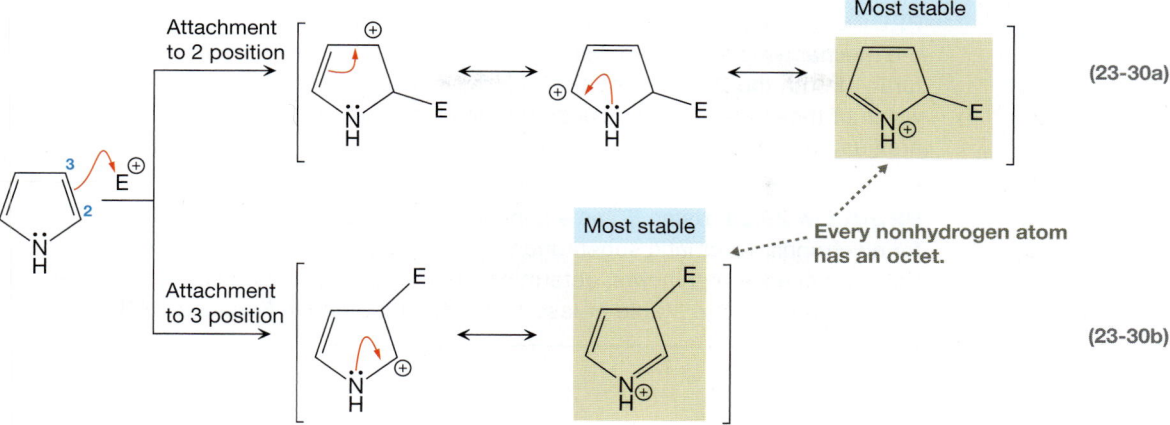

(23-30a)

(23-30b)

Notice, too, that each of pyrrole's arenium ion intermediates has a resonance structure in which *all nonhydrogen atoms have their octet*: the resonance structures highlighted in green in Equation 23-30. Thus, pyrrole's arenium ion intermediate is more stable than that of unsubstituted benzene, for which no such stabilized resonance structure exists. With a more stable arenium ion intermediate, pyrrole reacts faster than benzene in electrophilic aromatic substitution reactions.

Contrast this with the arenium ion intermediates of pyridine produced when an electrophile attaches to the 2 or 3 position, as shown in Equation 23-31.

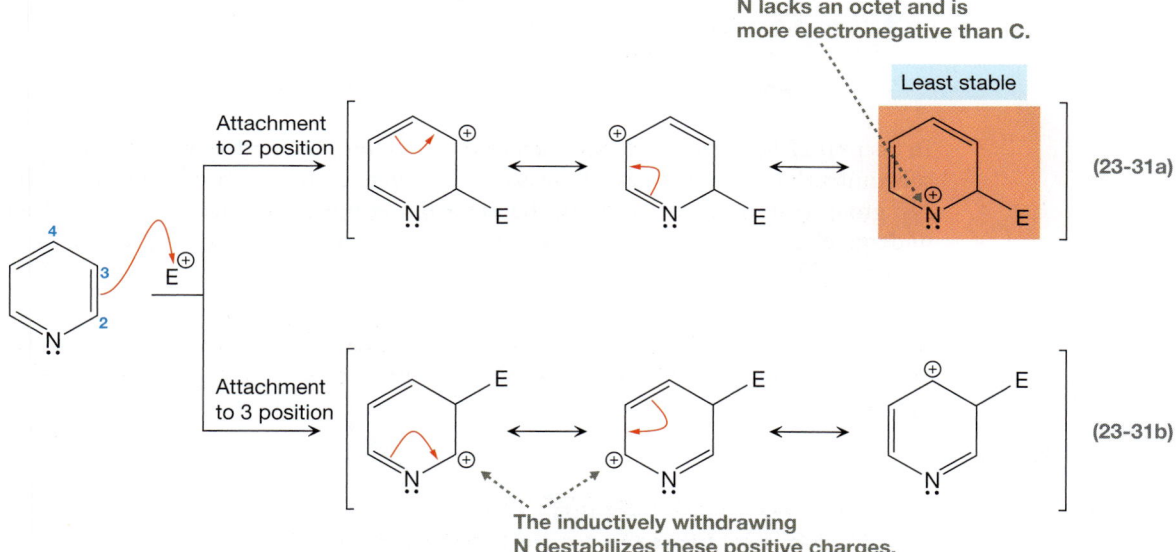

(23-31a)

(23-31b)

Each arenium ion intermediate has three total resonance structures, and in every resonance structure there is one positively charged atom that lacks an octet. The resonance structure highlighted in red in Equation 23-31a, however, is *much* less stable than the others because its atom lacking an octet is N. In the other resonance structures, the atom lacking an octet is C, which is less electronegative and can therefore better accommodate a positive charge.

Pyridine's most-stable arenium ion intermediate (Equation 23-31b) is still less stable than benzene's. In pyridine's arenium ion intermediate, the N atom is electron withdrawing and destabilizes the adjacent positive charge in two of the resonance structures. With a less stable arenium ion intermediate, pyridine undergoes electrophilic aromatic substitution slower than benzene.

23.8 Azo Coupling and Azo Dyes

Recall from Section 22.9c that an arenediazonium ion can be produced from the corresponding aromatic amine, as shown in Equation 23-32 for the conversion of aniline to the benzenediazonium ion:

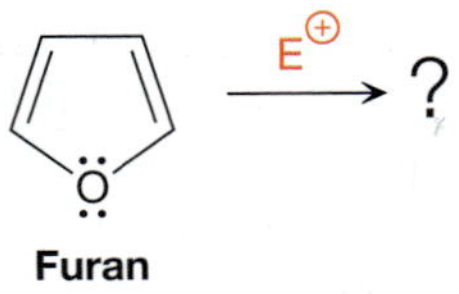

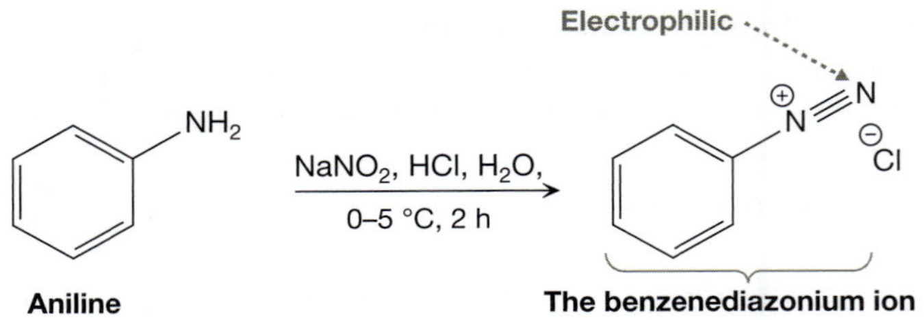

In Section 22.9c, we saw that such arenediazonium ions can undergo a wide variety of substitution reactions because of the excellent leaving group ability of $N_2(g)$. With the N_2^+ group still attached, however, the arenediazonium ion is electrophilic and can undergo electrophilic aromatic substitution with another aromatic ring, as shown in Equation 23-33.

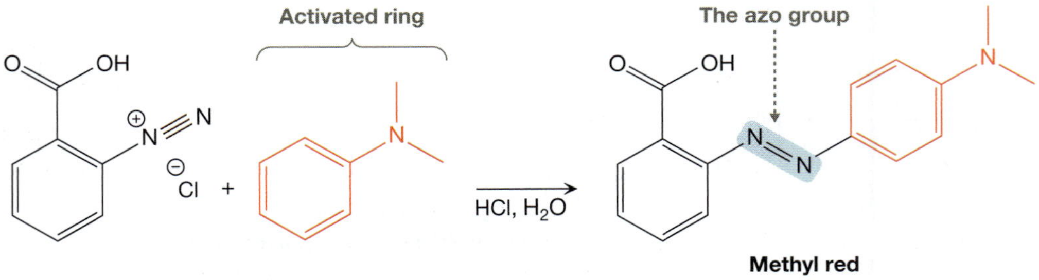

In the product, the —N=N— group, called the **azo group**, connects the two aromatic rings together, so this type of reaction is called **azo coupling**.

Arenediazonium ions are not very stable, so they are usually kept at relatively low temperatures, which slows the electrophilic aromatic substitution reaction. To compensate, the aromatic ring that reacts with the arenediazonium ion is generally activated, as indicated in Equation 23-33.

Notice that the azo group allows one aromatic ring to be conjugated to the other. As we learned in Section 15.3c, such extended conjugation can enable molecules to absorb visible light, which gives the compound a color that is detectable by the human eye. The product of Equation 23-33, for example, is methyl red, which is a pH indicator that is red when the solution's pH is less than about 4.4, and is yellow when the pH is above about 6.2.

Methyl red is an example of an **azo dye**: a compound with a distinct color that results from the azo coupling of aromatic rings. The particular color of the dye depends on the properties and the size of the aromatic rings as well as the substituents that are attached.

Many of the dyes used in the clothing industry are azo dyes, often having one or more sulfonate groups, as in the following examples.

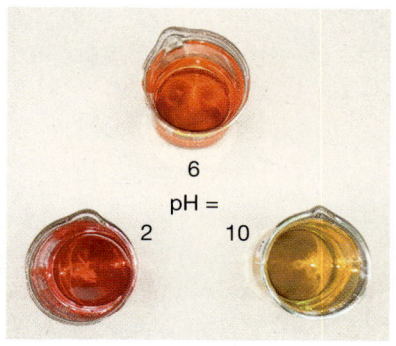

Methyl red

Methyl orange

Acid red 37

A sulfonate group serves two purposes. First, it enables the dye to be soluble in water. Second, the ionic character of the sulfonate group allows the dye to bind to the polymer molecules that make up the fabric. Thus, the dye is *colorfast*, meaning that it does not bleed substantially when the dyed item is washed.

PROBLEM 23.23 Show how methyl orange and acid red 37 can be synthesized beginning with an arenediazonium ion.

CONNECTIONS Prontosil, the azo dye shown below, was the first sulfa drug discovered. Sulfa drugs are antibiotics because they resemble *p*-aminobenzoic acid, PABA (see the box on p. 1124). Prontosil itself doesn't resemble PABA, but when it is metabolized, sulfanilamide is produced, which does resemble PABA.

Prontosil

Sulfanilamide

Metabolism

23.9 Nucleophilic Aromatic Substitution Mechanisms

The aromatic substitution reactions we have examined in Chapter 22 and thus far here in Chapter 23 proceed by mechanisms involving an aromatic ring attacking an electrophile—so-called *electrophilic aromatic substitution* reactions. Here in this section, we examine two other types of aromatic substitution reactions, in which the aromatic ring is attacked by a nucleophile instead; these are examples of **nucleophilic aromatic substitution** reactions. One of these reactions proceeds by a *nucleophilic addition–elimination* mechanism and the other by an *elimination–nucleophilic addition* mechanism.

These reactions are fundamentally different from the ones classified as electrophilic aromatic substitution. However, nucleophilic aromatic substitution reactions are presented here because they can be used in syntheses in conjunction with, or as an alternative to, electrophilic aromatic substitution reactions. Having had a thorough discussion of electrophilic aromatic substitution, we are now poised to have a better appreciation for what nucleophilic aromatic substitution has to offer.

23.9a Nucleophilic Aromatic Substitution via the Addition–Elimination Mechanism

When 1-chloro-2-nitrobenzene is treated with excess ethanamine, $CH_3CH_2NH_2$, *N*-ethyl-2-nitroaniline is produced (Equation 23-34). Overall, Cl is replaced by an ethylamino group ($NHCH_2CH_3$).

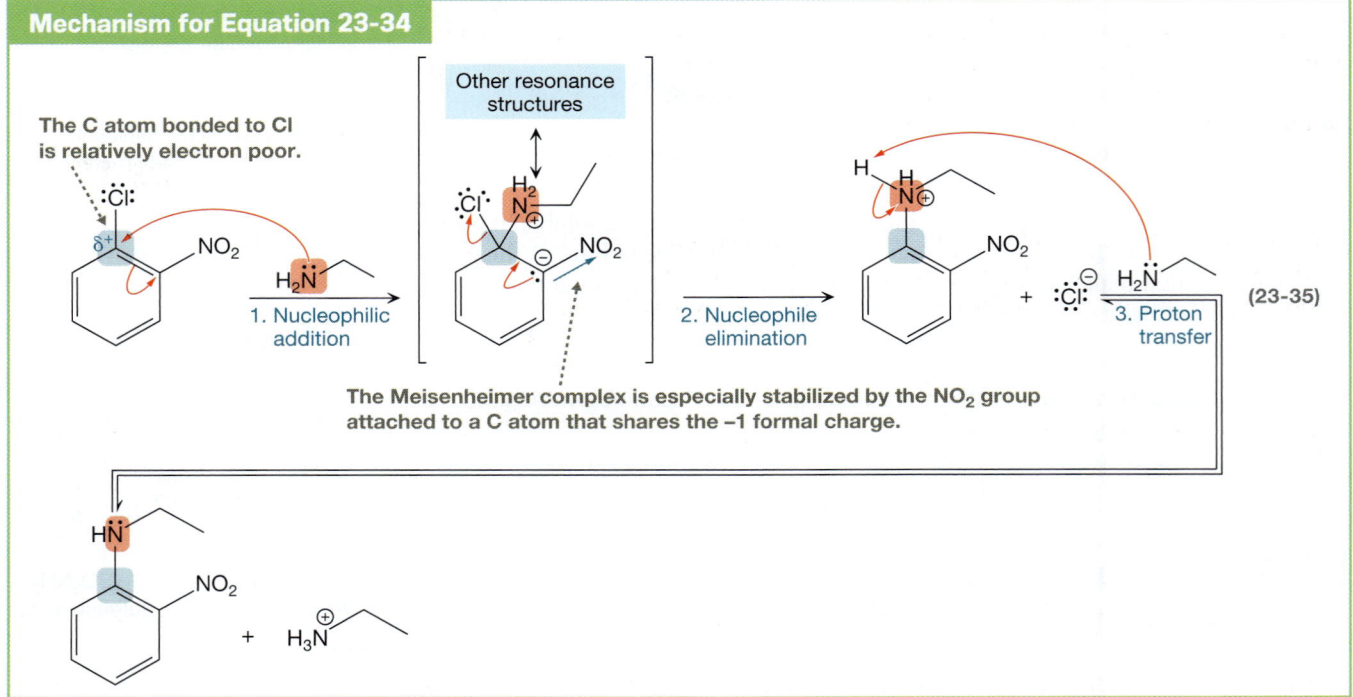

The mechanism for this aromatic substitution reaction is shown in Equation 23-35.

Mechanism for Equation 23-34

In Step 1, the amine acts as a nucleophile, attacking the C atom bonded to Cl. This nucleophilic addition step is the slow step and is therefore rate determining. In Step 2, Cl⁻ departs as a leaving group. Step 3, a proton transfer, is necessary to produce the overall uncharged product.

Step 1 is assisted partly by the resonance delocalization of the negative charge in the intermediate (see Your Turn 23.11), called a **Meisenheimer complex**, and is also assisted by the electron-poor nature of the aromatic ring. Not only does the leaving group impart a partial positive charge on the carbon atom to which it is attached, thereby attracting the electron-rich nucleophile, but the Meisenheimer complex is stabilized by the powerfully electron-withdrawing NO_2 group. This contrasts with *electrophilic* aromatic substitution, which is favored by electron-rich aromatic rings.

Draw the additional resonance structures of the Meisenheimer complex in Equation 23-35, indicating the negative charge on C shared over other atoms.

Not every nucleophilic aromatic substitution reaction is feasible:

> Nucleophilic aromatic substitution involving a substituted benzene generally requires at least one moderately or strongly electron-withdrawing group ortho or para to the leaving group.

This prevents the energy of the Meisenheimer complex from being excessively high by ensuring that the negative charge that develops is adjacent to the electron-withdrawing group in at least one of the resonance structures. Notice that this is, indeed, the case with the Meisenheimer complex shown in Equation 23-35. (See Your Turn 23.12, too.)

The nucleophilic addition of $CH_3CH_2NH_2$ to 1-chloro-4-nitrobenzene produces the following Meisenheimer complex. Draw the additional resonance structures of the intermediate. Which resonance structure is the most stable? Why?

The importance of the number and locations of the strongly electron-withdrawing groups relative to the leaving group is demonstrated by the reaction in Equation 23-36, which is a substitution similar to the one in Equation 23-34. The reaction requires less heat because the leaving group is ortho to one NO_2 group and para to the other.

The additional NO_2 group further stabilizes the negatively charged intermediate.

Less extreme conditions are required.

(23-36)

1-Chloro-2,4-dinitrobenzene

N-Methyl-2,4-dinitroaniline
96%

GREEN CHEMISTRY

I. Gallardo and G. Guirado of Barcelona have shown that nucleophilic aromatic substitution reactions can be carried out electrochemically, using a room temperature ionic liquid as the solvent and conducting electrolyte. These alternative reactions significantly reduce waste and decrease risks associated with high-temperature reactions.

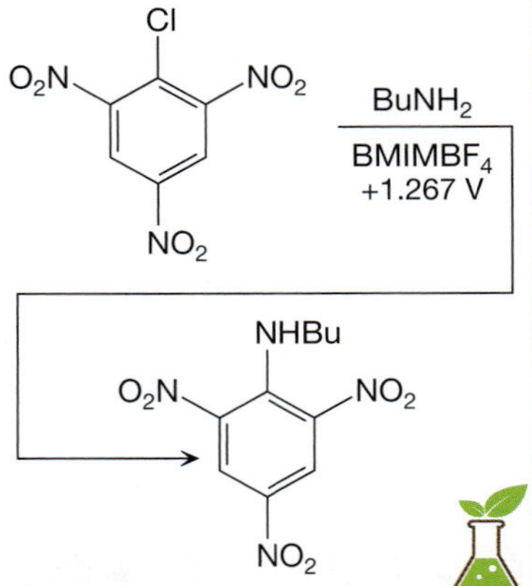

PROBLEM 23.24 Draw the detailed mechanism for this reaction, and predict the major product. Will this reaction be faster or slower than the one in Equation 23-36?

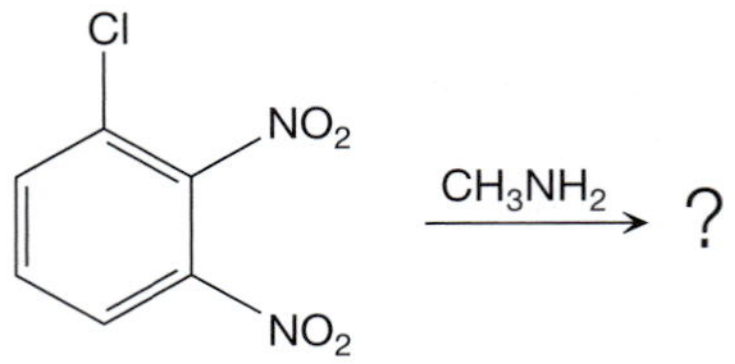

Nucleophilic aromatic substitution can be carried out with a variety of leaving groups—even F^- (Equation 23-37). The major restriction is that the leaving group needs to be more stable (i.e., less basic) than the incoming nucleophile, because the more stable of the two will depart in the elimination step.

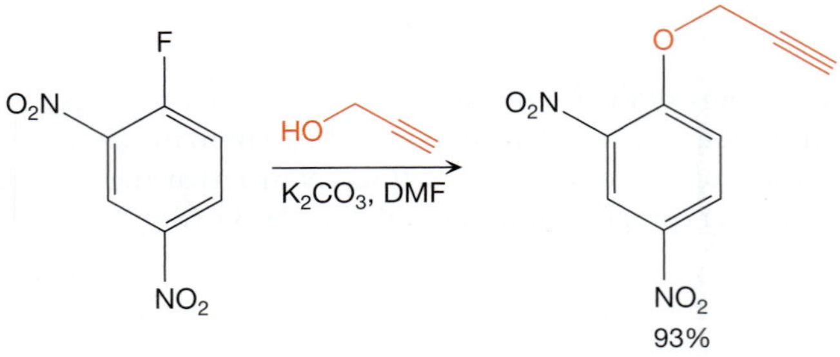

(23-37)

PROBLEM 23.25 Draw the mechanism and the major product of each of the following reactions.

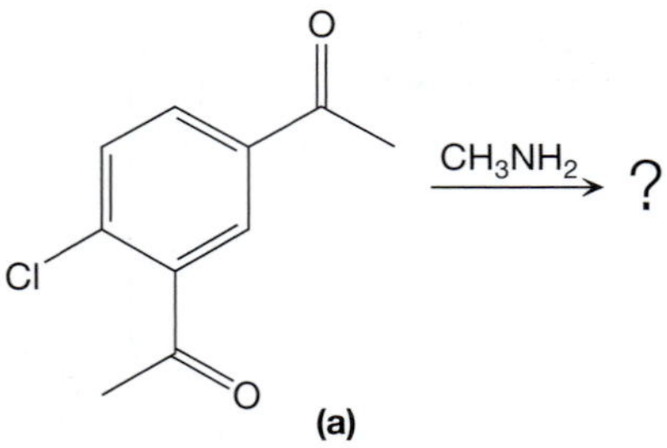

(a)

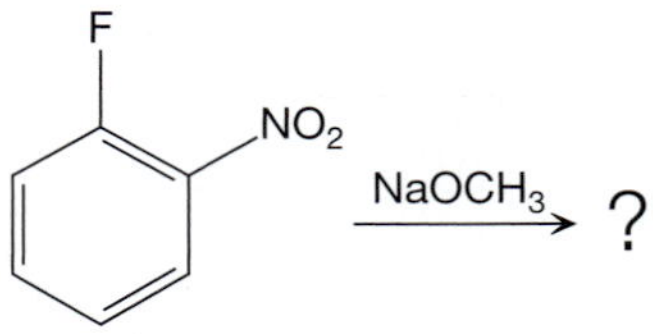

(b)

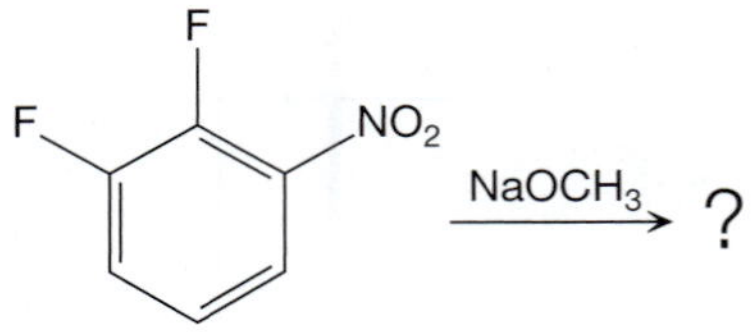

(c)

23.9b Nucleophilic Aromatic Substitution via the Elimination–Addition Mechanism: The Benzyne Intermediate

Even without a strongly electron-withdrawing group attached to the ring, a halobenzene can still undergo nucleophilic substitution. For example, chlorobenzene can react with sodium hydroxide to produce phenol (Equation 23-38), or it can react with potassium amide to produce aniline (Equation 23-39). These reactions typically require extreme conditions, however, such as the presence of a very strong base, high temperatures, or both.

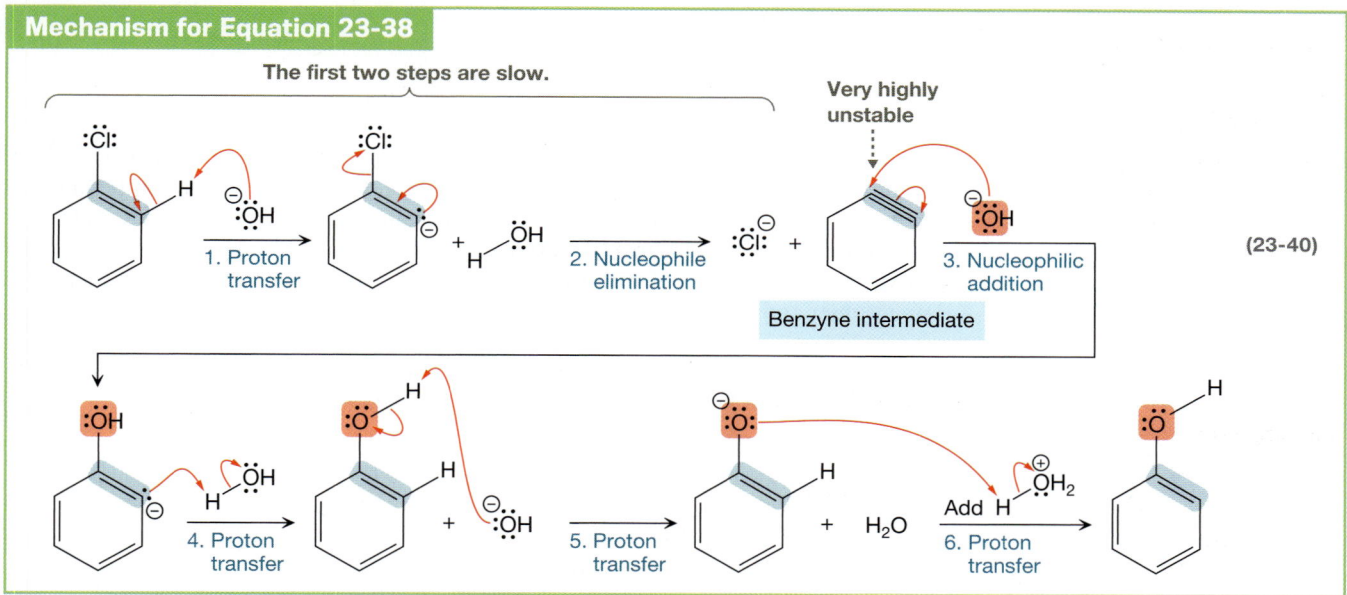

Chlorobenzene → Phenol 94%

1. NaOH (15%), 300 °C, 20 h
2. H₂O, HCl

(23-38)

Chlorobenzene → Aniline 60%

1. KNH₂, NH₃(ℓ), −33 °C
2. H₂O

(23-39)

The requirement for these extreme conditions is explained by Equation 23-40, the mechanism for the reaction in Equation 23-38.

Mechanism for Equation 23-38

The first two steps are slow.

Very highly unstable

1. Proton transfer
2. Nucleophile elimination

Benzyne intermediate

3. Nucleophilic addition

(23-40)

4. Proton transfer
5. Proton transfer

Add H

6. Proton transfer

In Step 1, the base removes a proton from the aromatic ring to produce a carbanion, and in Step 2, the leaving group departs. These two steps make up the same E1cb mechanism we saw in Section 18.5. The product of Step 2 is called a **benzyne intermediate**, because it has a C≡C triple bond in its Lewis structure. In Step 3, the benzyne intermediate undergoes nucleophilic addition to produce a deprotonated form of phenol. In Step 4, a proton from H₂O adds to produce phenol, whose OH group is irreversibly deprotonated in Step 5 due to the basic conditions. Acid workup in Step 6 replenishes that proton to produce the overall uncharged product.

Both Steps 1 and 2 are relatively slow. Step 1 is slow because a very unstable carbanion is produced, which is why a strong base is required. Step 2 is slow because the benzyne product is highly unstable due to the large amount of angle strain caused by the triple bond. The ideal bond angle for each triply bonded C atom is 180°, but the constraints of the ring require it to be ~120°. This has a major effect on the strength of the π bond that is in the plane of the ring, as seen in Figure 23-7. Adjacent p orbitals are parallel in a normal π bond, but they are ~60° apart in this case. As a result, the overlap of those adjacent p orbitals is significantly decreased, making the π bond abnormally weak and benzyne highly reactive.

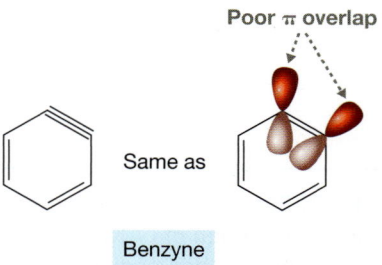

Poor π overlap

Same as

Benzyne

FIGURE 23-7 Instability of the benzyne intermediate The p orbitals in the plane of the ring intersect at ~60°. Thus, their overlap is less than ideal, making the resulting π bond rather weak and benzyne highly reactive.

YOUR TURN 23.13

The mechanism for the reaction in Equation 23-39 is as follows, but the curved arrows have been omitted. Supply the missing curved arrows, and write the name of the elementary step below each reaction arrow.

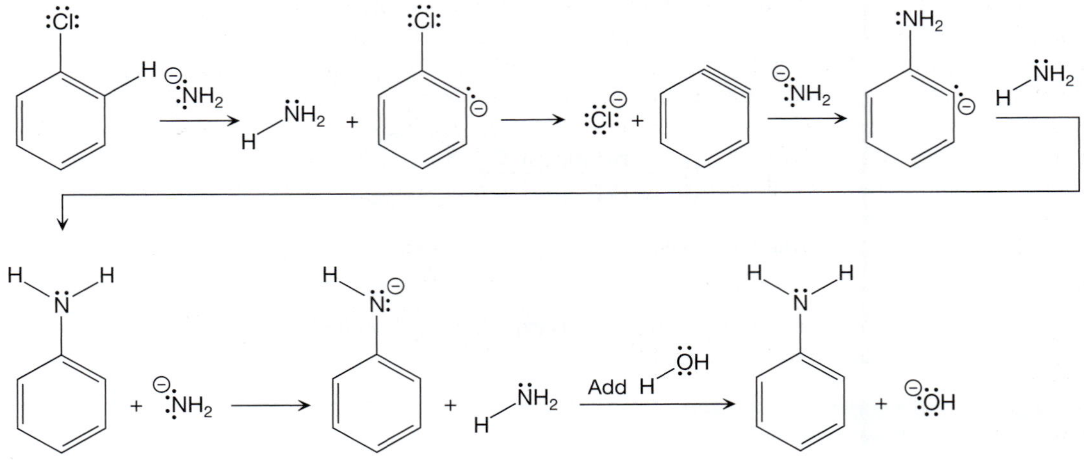

Because benzyne is highly reactive, it has not been isolated and purified, though some experiments indirectly support its existence. One of the most compelling is a carbon-14 labeling experiment, shown in Equation 23-41. When the Cl atom in chlorobenzene is bonded to a ^{14}C atom, reaction with KNH_2 produces a roughly equal mixture of two isomers of aniline, where the NH_2 group is bonded to the ^{14}C atom in one isomer and to a ^{12}C atom in the other.

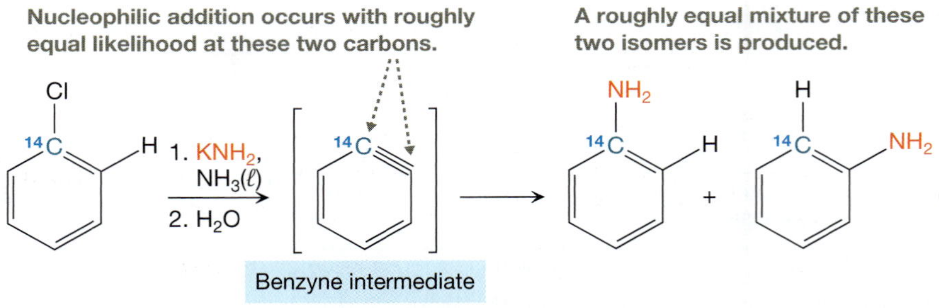

Nucleophilic addition occurs with roughly equal likelihood at these two carbons.

A roughly equal mixture of these two isomers is produced.

Benzyne intermediate

(23-41)

These results are consistent with benzyne's symmetry about the triple bond. Thus, nucleophilic addition to benzyne occurs with essentially equal likelihood at both alkyne C atoms. Addition to the alkyne ^{14}C atom produces one isomer, and addition to the alkyne ^{12}C atom produces the other.

PROBLEM 23.26 Draw the mechanism for the formation of phenol from bromobenzene and sodium hydroxide shown in the first reaction. Suppose, instead, that the deuterium-labeled bromobenzene shown in the second reaction were used. What percentage of the product would you expect to contain the deuterium atom? Explain.

23.10 Organic Synthesis: Considerations of Regiochemistry; Attaching Groups in the Correct Order

As we learned in Section 23.1, a substituent that is already on the aromatic ring greatly influences the site of reaction in a subsequent electrophilic aromatic substitution. In a synthesis that requires successive substitutions, then, we must choose wisely the order in which the reactions are carried out. For example, how would you synthesize *m*-chloronitrobenzene from benzene? Benzene can be nitrated by treating it with concentrated HNO_3 in H_2SO_4, and it can be chlorinated with Cl_2 in the presence of $FeCl_3$. The synthesis therefore must involve both a nitration and a chlorination, but in which order?

If chlorination takes place first, then we encounter a problem with regiochemistry, as shown in Equation 23-42. The Cl substituent in chlorobenzene is an ortho/para director (Table 23-1), so a subsequent nitration would yield as the major products *o*-chloronitrobenzene and *p*-chloronitrobenzene. Our target, however, is *m*-chloronitrobenzene.

(23-42)

Benzene   Chlorobenzene   o-Chloronitrobenzene   p-Chloronitrobenzene

If the nitration of benzene is carried out first (Equation 23-43), then the correct isomer is, indeed, produced as the major product, because the nitro group in nitrobenzene is a meta director.

(23-43)

Benzene   Nitrobenzene   m-Chloronitrobenzene

SOLVED PROBLEM 23.27

How would you synthesize *p*-isopropylnitrobenzene from benzene?

Think Is the NO_2 group an ortho/para director or a meta director? Is the isopropyl group an ortho/para director or a meta director? Which group should be on the ring prior to the second substitution?

Solution To obtain the target, benzene must be both alkylated and nitrated. Because an alkyl group is ortho/para directing and the NO_2 group is meta directing (Table 23-1), the order in which these substitutions are carried out is important. Given that we want to synthesize the para isomer, an ortho/para-directing substituent should be on the ring prior to the second substitution reaction. In other words, the second reaction must be a nitration of the alkylbenzene. That leaves alkylation as the first reaction. The overall synthesis would be written as follows:

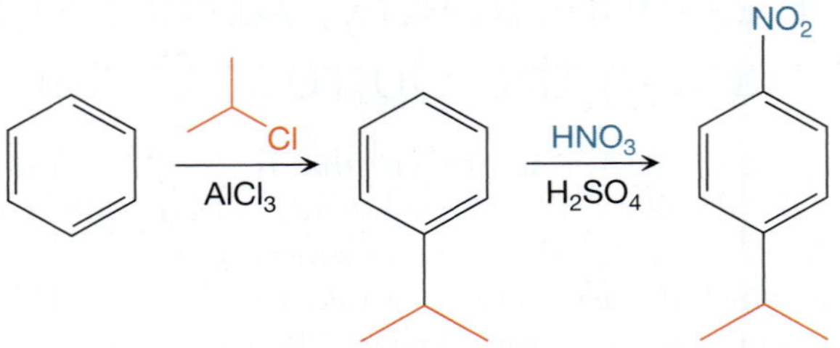

PROBLEM 23.28 Show how to synthesize this compound from benzene.

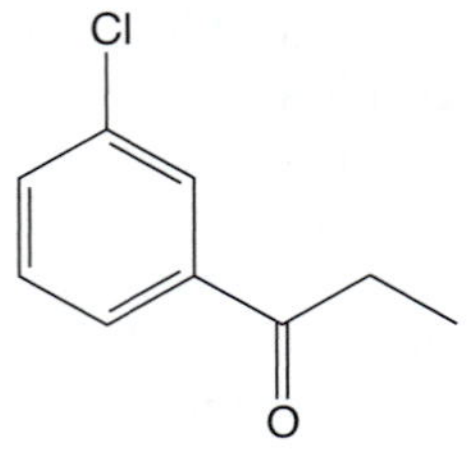

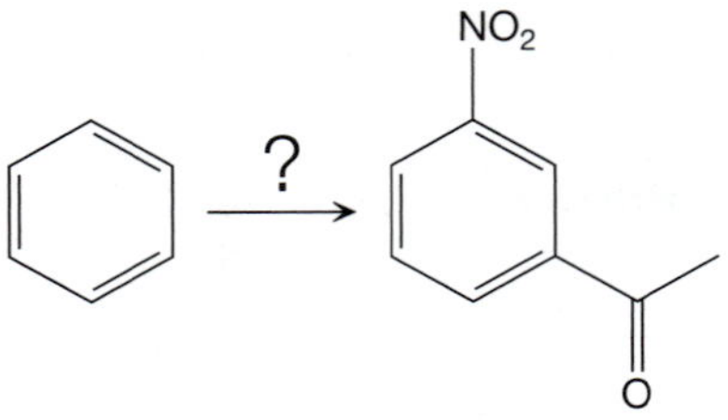

The order in which substitutions are carried out can be an issue, too, because some substituents on the ring make certain electrophilic aromatic substitution reactions unfeasible. Friedel–Crafts reactions, for example, do not occur with moderately or highly deactivated rings (Section 23.4). Therefore, we must avoid placing these kinds of deactivating groups on the ring prior to carrying out a Friedel–Crafts reaction.

How, then, would you synthesize m-nitroacetophenone from benzene?

This synthesis requires two substitutions—a nitration and a Friedel–Crafts acylation. Because both the NO_2 group and the acetyl group are meta directors (Table 23-1), regiochemistry will not depend on the order in which these substitutions are carried out. We do need to avoid attempting a Friedel–Crafts reaction on nitrobenzene, however, because the benzene ring would be highly deactivated (Equation 23-44).

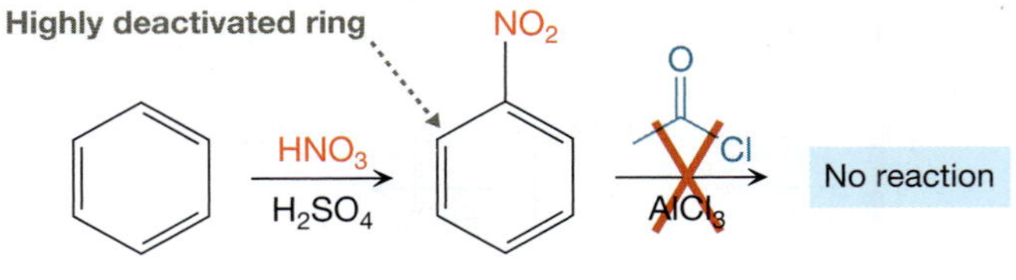

Benzene m-**Nitroacetophenone**

(23-44)

Instead, nitration should be the final step, leaving the Friedel–Crafts acylation as the first step. Equation 23-45 shows the best way to carry out this synthesis.

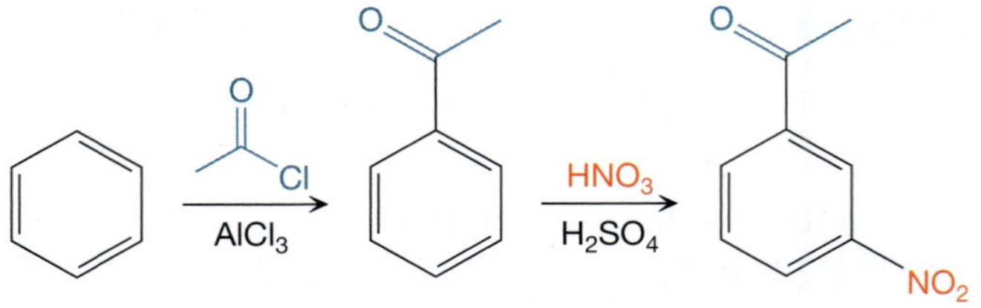

(23-45)

23.11 Organic Synthesis: Interconverting Ortho/Para and Meta Directors

In Section 23.10, we learned that, when devising a synthesis that requires carrying out successive electrophilic aromatic substitution reactions, the order in which ortho/para- and meta-directing groups are added to the ring can impact whether the synthesis is successful. With this in mind, we revisit some reactions we first encountered in Chapter 22, which can be used to interconvert ortho/para- and meta-directing substituents. Thus, the regiochemistry of a substitution that takes place later in a synthesis is not automatically predetermined by the ortho/meta/para-directing ability of a substituent added earlier.

Recall from Section 22.9a, for example, that treating an alkylbenzene with a hot, basic solution of $KMnO_4$ yields benzoic acid on acid workup (see Equation 23-46). The substituent in the reactant is an alkyl group, which is an ortho/para director, but the substituent in the product is a CO_2H group, which is a meta director.

Ortho/para director   Meta director

$$\text{An alkylbenzene} \xrightarrow[\text{2. HCl, H}_2\text{O}]{\text{1. KMnO}_4,\ \text{KOH, }\Delta} \text{Benzoic acid} \qquad (23\text{-}46)$$

How can we use this reaction in the synthesis of *p*-nitrobenzoic acid from benzene?

$$\text{benzene} \xrightarrow{?} \text{product}$$

These substituents are para to each other but are both meta directors.

The CO_2H substituent and the NO_2 substituent are both meta directors (Table 23-1, p. 1147), but the desired product is the para isomer. We can circumvent this problem by making sure that the second electrophilic aromatic substitution takes place with an ortho/para director already on the ring, such as an alkyl group. After the para-disubstituted benzene is produced, the alkyl group can then be oxidized to the carboxyl group. The retrosynthetic analysis is shown in Equation 23-47:

Ortho/para director

$$\text{(23-47)}$$

Undo an oxidation.   Undo a nitration.   Undo an alkylation.

The final synthesis would then be written as in Equation 23-48:

$$\xrightarrow[\text{AlCl}_3]{\text{R—Cl}} \xrightarrow[\text{H}_2\text{SO}_4]{\text{HNO}_3} \xrightarrow[\text{2. HCl, H}_2\text{O}]{\text{1. KMnO}_4,\ \text{KOH, }\Delta} \qquad (23\text{-}48)$$

CONNECTIONS

p-Nitrobenzoic acid is a precursor in the synthesis of procaine, the generic name of Novocain, a local anesthetic that was popular in dentistry in much of the 20th century.

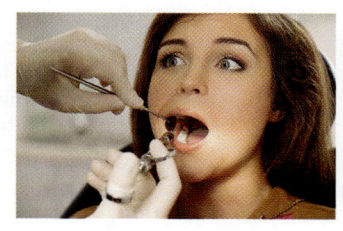

Procaine

Can you reorder the steps in the synthesis in Equation 23-48 to produce *m*-nitrobenzoic acid instead?

PROBLEM 23.29 Show how to synthesize benzene-1,4-dicarboxylic acid from benzene, using ethanol as your only other source of carbon.

Benzene $\xrightarrow{?}$ Benzene-1,4-dicarboxylic acid

We can achieve even greater flexibility in synthesis with the ability to convert a meta director into an ortho/para director. We have already encountered two types of reductions that accomplish this. One, which was discussed in Section 22.9b, is the reduction of a nitro (NO_2) group to an amino (NH_2) group. The second is the reduction of a carbonyl (C=O) group on a ketone or aldehyde to a methylene (CH_2) group. Recall from Section 19.3 that this carbonyl reduction can be done in either acidic conditions (Clemmensen reduction), basic conditions (Wolff–Kishner reduction), or neutral conditions (Raney-nickel reduction).

The utility of these reactions is demonstrated when *m*-bromoethylbenzene is synthesized from benzene. The substituents in the target are meta to each other, so the second of two electrophilic aromatic substitution reactions must be carried out with a meta director already on the ring. Both the Br and the CH_2CH_3 substituents are ortho/para directors (Table 23-1), however, so a meta director must be converted into an ortho/para director after the second substitution reaction is carried out.

Thinking retrosynthetically, the meta director could be an acetyl group that is reduced to the ethyl group, as in Equation 23-49.

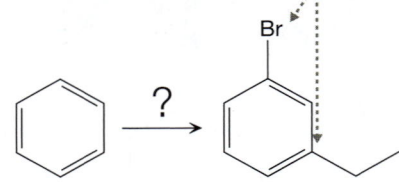

These substituents are meta to each other but are both ortho/para directors.

Benzene → *m*-Bromoethylbenzene

Br (compound) →[Undo a reduction.] Br (acetyl compound, *A meta director*) →[Undo a bromination.] acetophenone →[Undo an acylation.] Benzene (23-49)

In the forward direction, the synthesis would be written as in Equation 23-50:

Benzene $\xrightarrow[\text{AlCl}_3]{\text{Cl—C(=O)CH}_3}$ acetophenone $\xrightarrow[\text{FeBr}_3]{\text{Br}_2}$ bromoacetophenone $\xrightarrow[\text{Zn(Hg)}]{\text{HCl}}$ *m*-bromoethylbenzene (23-50)

SOLVED PROBLEM 23.30

Show how to synthesize *m*-(2-methylpropyl)aniline from benzene.

Think What is the relative positioning of the substituents on the ring in the target—ortho, meta, or para? Are the substituents ortho/para directors or meta directors? Will you need to incorporate a reaction that converts an ortho/para

Benzene $\xrightarrow{?}$ (NH$_2$-substituted product)

director to a meta director or vice versa? Does it matter which substituent is attached to the ring first?

Solve The substituents on the ring in the target are meta to each other, but both are ortho/para directors. Therefore, a meta director must be attached to the ring when the second substituent is added. Then, after adding the second substituent, we must convert the initial meta director to an ortho/para director. We know how to convert an NO_2 group to an NH_2 group, so we could envision an NO_2 group on the ring when we carry out a Friedel–Crafts acylation, as shown in Scheme **A** below. We also know how to convert an acyl group into an alkyl group, so we could envision an acyl group on the ring when we carry out a nitration reaction, as shown in Scheme **B**.

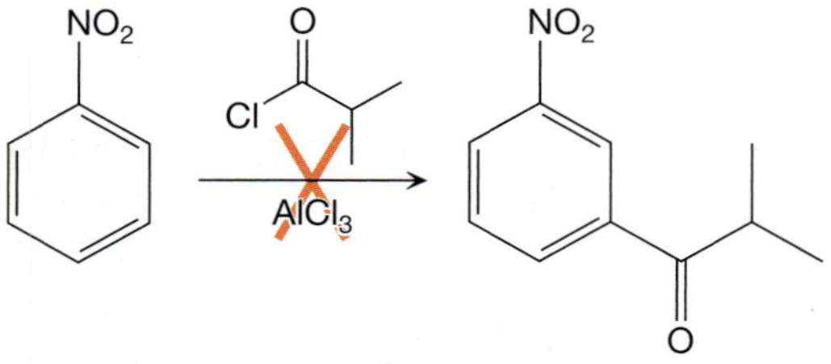

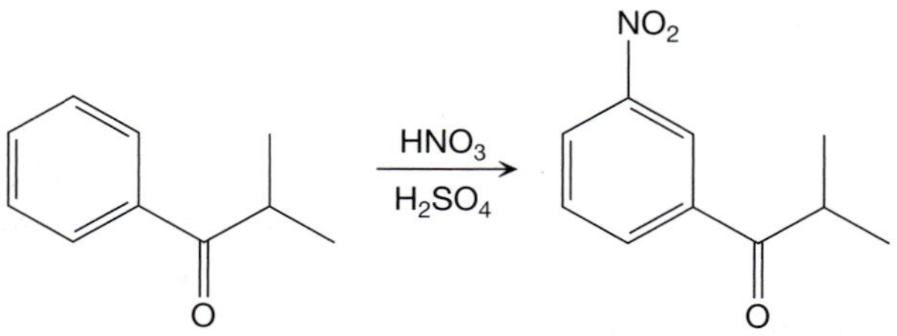

Scheme **A** will not work because, as we learned in Section 23.4, the strongly deactivated ring does not undergo a Friedel–Crafts reaction. Scheme **B**, on the other hand, will work, so the synthesis could be written as follows.

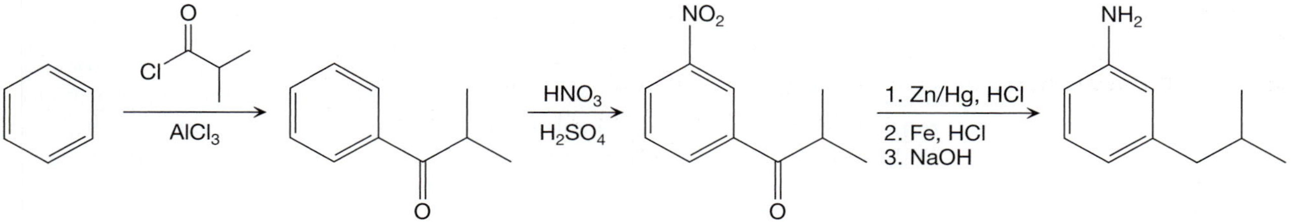

The Clemmensen reduction of the carbonyl group and the reduction of the nitro group are written as separate reactions, but we could also carry out both reductions together.

PROBLEM 23.31 Show how to synthesize *m*-chloroaniline from benzene.

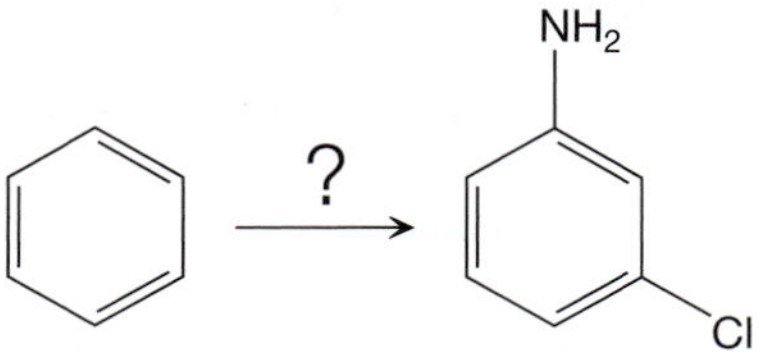

23.12 Organic Synthesis: Considerations of Protecting Groups

In the realm of aromatic substitution, there are two ways in which to incorporate a protecting group. One is to temporarily suppress the reactivity of a substituent attached to the ring, by reversibly converting it to a different functional group. The second is to block a particular site on the ring where substitution is not desired.

To illustrate the first scenario, how would you carry out the transformation at the left, where aniline is acylated at the para position? A simple Friedel–Crafts acylation would seem to be required, given that the NH_2 group on aniline is an ortho/para director. As we learned in Section 23.5, however, the reaction conditions that are required would convert the activating amino substituent into a strongly deactivating group (Equation 23-51), thereby making a direct Friedel–Crafts acylation unfeasible.

The problem here is that the NH_2 group is a good Lewis base; its lone pair of electrons is available to form a bond to $AlCl_3$, a strong Lewis acid. To circumvent this problem, the amino group can be converted *temporarily* into a different functional group (e.g., an O=C—N group, characteristic of an amide) in which the lone pair of electrons is less available. As we learned in Chapters 20 and 21, amides are relatively unreactive, due in large part to the resonance of the lone pair on the N atom with the adjacent carbonyl group.

Treating aniline with an acid chloride or an acid anhydride in the *absence* of a strong Lewis acid catalyst converts it to the amide (Equation 23-52, review Section 21.3). Even in the protected form, the substituent on the ring is an ortho/para director and a moderate activator (Table 23-3, p. 1157), due to the presence of the lone pair of electrons on N. (It would be undesirable for a protecting group to alter the regiochemistry of the reaction or to strongly deactivate the ring.) In the protected form, the acylation can then be carried out. Afterward, the amino group is deprotected by hydrolysis.

PROBLEM 23.32 Show how to carry out this synthesis.

In some instances, the reaction we want to carry out leads to possible substitution at two or more sites on the ring. This demonstrates the second use of protecting groups in electrophilic aromatic substitution—namely, we can guide the regiochemistry by protecting one of those sites. For example, how would you convert phenol into *o*-bromophenol, if the para product is undesired?

The OH group is an ortho/para director, so simple bromination will yield a substantial amount of both the ortho and para isomers, as shown in Equation 23-53.

One method that can be used to protect the para position is to place an SO_3H group there temporarily (Equation 23-54).

The SO₃H group "blocks" the para position.

As we learned in Section 22.7, sulfonation is *reversible*. Furthermore, it will take place almost exclusively at the para position due to the steric bulk of the SO_3H group, so subsequent bromination will take place ortho to the OH group (an ortho/para director). The OH group dictates regiochemistry because it is the more activating group on the ring. It can't direct Br to the para position because the SO_3H group is already there, blocking the para position. After bromination, the SO_3H group is removed with aqueous acid.

An NH_2 group can also be used as a blocking group, as shown in the synthesis of *m*-chlorotoluene from *p*-methylaniline (Equation 23-55).

(23-55)

p-Methylaniline

Prevents overhalogenation

***m*-Chlorotoluene**

In the first step, the NH_2 group (a strongly activating group) is converted to the $NHCOCH_3$ group (a moderately activating group) to prevent overhalogenation. In the second step, halogenation takes place ortho to the $NHCOCH_3$ group (an ortho/ para director), given that $NHCOCH_3$ is more activating than the CH_3 group. Next, the NH_2 group is deprotected by hydrolysis. Diazotization followed by treatment with H_3PO_2 converts the NH_2 group into an H (Section 22.9c). Therefore, even though the original NH_2 group is not in the final product, it was instrumental in directing the regiochemistry of chlorination.

PROBLEM 23.33 Show how you can carry out this synthesis without generating any of the para isomer.

Chapter Summary and Key Terms

- In electrophilic aromatic substitution reactions, certain substituents attached to a phenyl ring favor reaction at the ortho and para positions, and are characterized as **ortho/para directors**. Other substituents favor reaction at the meta position and are called **meta directors**. (Section 23.1; Objective 1)

- A substituent attached to a phenyl ring by an atom possessing a lone pair of electrons tends to be an ortho/para director. When the incoming electrophile attaches to a carbon at the ortho or para position of the ring, the arenium ion intermediate that is produced is stabilized by these substituents via resonance. (Section 23.2; Objective 2)

- Alkyl groups are ortho/para directors, because they stabilize ortho and para arenium ion intermediates via electron-donating effects. (Section 23.2; Objective 3)

- Meta directors such as the nitro group are electron withdrawing and are attached to the aromatic ring by an atom that has no lone pair of electrons. They destabilize the arenium ion intermediate when the incoming electrophile attaches to the ortho or para position. (Section 23.2; Objective 3)

- In general, ortho/para directors increase the rate of electrophilic aromatic substitution, and are called **activating groups**. Conversely, meta directors tend to slow the reaction rate, and are called **deactivating groups**. Halogen substituents are exceptions — they are ortho/para directors, but deactivating. (Section 23.3; Objective 4)

- Activating groups speed up an electrophilic aromatic substitution reaction because they *stabilize* the arenium ion intermediate that is produced. Deactivating groups *destabilize* the arenium ion intermediate. (Section 23.3; Objective 4)

- A deactivating group on an aromatic ring requires more extreme conditions to carry out an electrophilic aromatic substitution reaction such as nitration and sulfonation, and precludes Friedel–Crafts reactions altogether. The electron-donating property of an alkyl group makes polyalkylation a potential problem in Friedel–Crafts alkylations. (Section 23.4; Objective 5)

- The conditions under which an electrophilic aromatic substitution reaction takes place can alter a substituent's ortho/meta/para-directing capabilities, as well as its activating/deactivating capabilities. For example, the aromatic ring of phenol is much more activated under neutral conditions than it is under acidic conditions. More strikingly, an amino group is an ortho/para director under mildly acidic conditions, but it is a meta director under strongly acidic conditions. (Section 23.5; Objective 5)

- When multiple substituents are attached to a phenyl ring, their activating/deactivating qualities are additive, and the regiochemistry tends to be governed by the most activating of those substituents. (Section 23.6; Objective 6)

- Aromatic rings other than benzene can also undergo electrophilic aromatic substitution. These include naphthalene, furan, pyrrole, and pyridine. Just as in substituted benzenes, the regiochemistry and relative reaction rates are governed by the stabilities of the arenium ion intermediates that are produced. (Section 23.7; Objective 7)

- **Azo coupling** occurs when an aromatic ring undergoes electrophilic aromatic substitution with an arenediazonium ion,

$Ar—N_2^+$, producing a compound in which the aromatic rings are joined by an azo group, —N$=$N—. An **azo dye** is such a compound that exhibits extended conjugation involving the azo group and absorbs visible light, giving it a characteristic color. (Section 23.8; Objective 8)

- Aromatic rings with halogen substituents can undergo **nucleophilic aromatic substitution**, whereby the halogen atom is replaced by a nucleophile. (Section 23.9; Objective 9)

  - Rings that are strongly deactivated toward reaction with an electrophile (and thus electron poor) react via a nucleophilic addition–elimination mechanism under relatively mild conditions. (Section 23.9a)

  - Rings that are not very electron poor typically require strong bases, high temperatures, or both, and react via an elimination–nucleophilic addition mechanism, proceeding through a **benzyne intermediate**. (Section 23.9b)

- When incorporating successive electrophilic aromatic substitution reactions in a synthesis, it is important to consider the order in which those reactions are carried out. The incorrect order could lead to undesired regiochemistry, or could preclude some reactions entirely. (Section 23.10; Objective 10)

- Reactions that transform an ortho/para director into a meta director, and vice versa, can be instrumental in synthesis. An example of the former is the $KMnO_4$ oxidation of an alkyl side chain to a CO_2H group. An example of the latter is the reduction of a NO_2 group to an NH_2 group. (Section 23.11; Objective 10)

- The reactivity of an aromatic amine can be temporarily decreased by converting the NH_2 group into an O$=$C—N group. Additionally, the para position of a phenyl ring can be temporarily blocked by sulfonating that position. (Section 23.12; Objective 11)

Reaction Table

TABLE 23-4 Functional Group Conversions[a,b]

Starting Compound Class	Typical Reagents and Reaction Conditions	Compound Class Formed	Key Electron-Rich Species	Key Electron-Poor Species	Comments	Discussed in Section(s)
(1) Electron-poor aromatic halide	H_2NR	Substituted aromatic amine			Nucleophilic addition–elimination mechanism; EWG ortho or para to X	23.9a
(2) Electron-poor aromatic halide	NaOR	Substituted aromatic ether			Nucleophilic addition–elimination mechanism; EWG ortho or para to X	23.9a

[a]X = F, Cl, Br, I.
[b]EWG = electron-withdrawing group.

(continued)

TABLE 23-4 Functional Group Conversions[a,b] (continued)

	Starting Compound Class	Typical Reagents and Reaction Conditions	Compound Class Formed	Key Electron-Rich Species	Key Electron-Poor Species	Comments	Discussed in Section
(3)	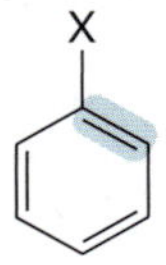 Aromatic halide	1. NaOH, Δ 2. H₂O, HCl	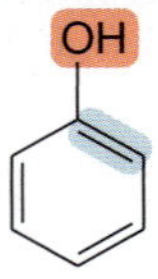 Phenol	:ÖH⁻		Proceeds through benzyne intermediate	23.9b
(4)	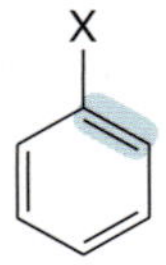 Aromatic halide	1. KNH₂, NH₃(ℓ) 2. H₂O	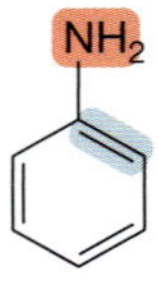 Aniline	:NH₂⁻	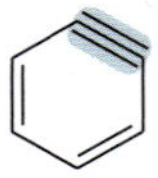	Proceeds through benzyne intermediate	23.9b

[a]X = F, Cl, Br, I.
[b]EWG = electron-withdrawing group.

Problems

Problems that are related to synthesis are denoted (SYN).

23.1 and 23.2 Ortho/Para and Meta Directors in Electrophilic Aromatic Substitution on Monosubstituted Benzenes

23.34 Predict the major product(s) of each of the following reactions. Draw the complete, detailed mechanism that leads to the formation of each of those products.

(a)

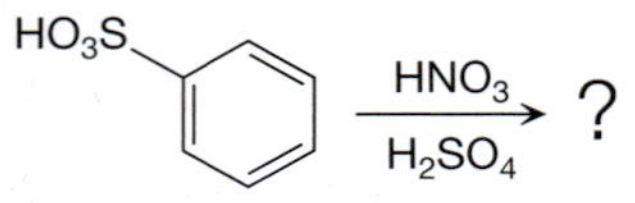

HO_3S — benzene, $\xrightarrow[\text{H}_2\text{SO}_4]{\text{HNO}_3}$?

(b)

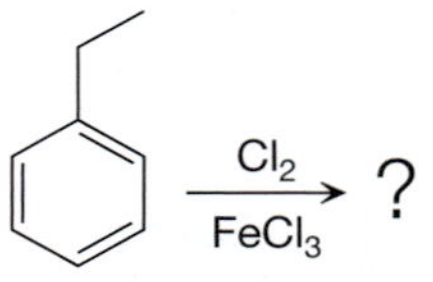

ethylbenzene, $\xrightarrow[\text{FeCl}_3]{\text{Cl}_2}$?

(c)

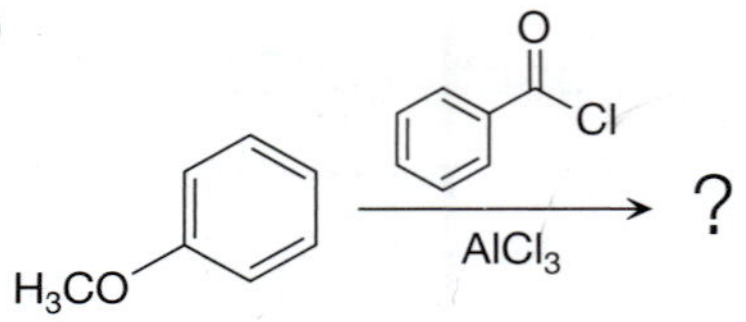

benzophenone chloride + H_3CO-benzene, $\xrightarrow{\text{AlCl}_3}$?

(d)

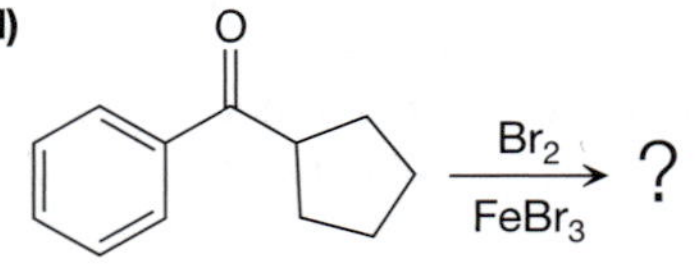

phenyl cyclopentyl ketone, $\xrightarrow[\text{FeBr}_3]{\text{Br}_2}$?

(e)

isopropylbenzene, $\xrightarrow{\text{conc HNO}_3}$?

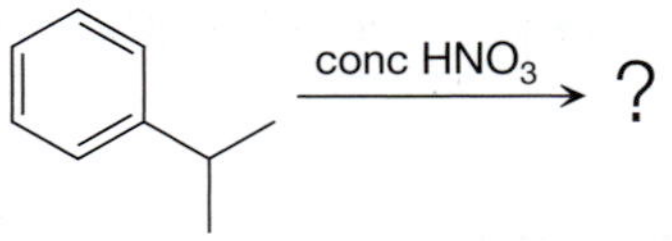

23.35 In Section 23.2, we learned that all alkyl groups are ortho/para directors. However, the relative amounts of ortho and para products depend on the specific identity of the alkyl group, as shown here for the nitration of various alkylbenzenes. What trend do you observe? What factor accounts for that trend?

	—CH₃	—CH₂CH₃	—CH(CH₃)CH₃	—C(CH₃)₂CH₃
% Ortho	63	45	30	16
% Meta	3	6	8	11
% Para	34	49	62	73

23.36 According to Table 23-1, the nitration of phenol results in a product mixture that is 50% ortho and 50% para. What would the major products be if one of the ortho positions of phenol were labeled with deuterium (D)? What would you expect the relative amounts of each of those products to be?

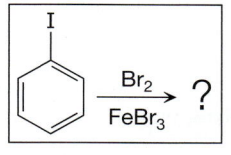

23.37 A Br substituent is an ortho/para director, so the halogenation of bromobenzene predominantly yields the ortho and para products, as shown in the following bromination and chlorination reactions:

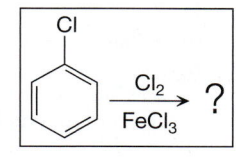

Explain why bromination yields more of the para product than chlorination.

23.38 Which of the reactions shown here do you think will produce the para product in the greater amount? Explain. *Hint:* See Problem 23.37.

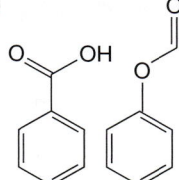

23.39 Predict the most likely site of electrophilic aromatic substitution in the compound shown here. *Hint:* How do you determine whether the substituent is an ortho/para-directing group or a meta-directing group?

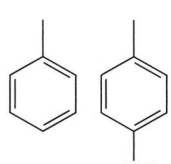

23.40 Predict whether the nitroso group (—N=O) is an ortho/para or meta director.

23.41 The phenyl group, C_6H_5, is known to be an ortho/para-directing group. **(a)** With that in mind, predict the product of the reaction shown here. **(b)** Justify why it is an ortho/para director by examining the ortho, meta, and para arenium ion intermediates that would be formed during the course of the reaction.

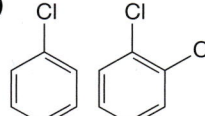

23.3 The Activation and Deactivation of Benzene toward Electrophilic Aromatic Substitution

23.42 The –NHCOR group of an amide is an activating group, but it is not as strongly activating as NH_2. **(a)** Explain why it is an activating group. **(b)** Explain why it is less activating than NH_2.

23.43 For each pair of aromatic compounds, determine which will undergo electrophilic aromatic substitution faster.

(a)
(b)
(c)
(d)
(e)
(f)
(g)
(h)

23.44 The OH group on phenol is an activating group, but the ring in phenylmethanol is deactivated. Explain.

23.45 Alkyl groups are activating, so both of the trisubstituted benzenes shown here undergo chlorination much faster than benzene. The 1,2,3-trisubstituted compound, however, undergoes chlorination faster than the 1,2,4-trisubstituted compound. Explain.

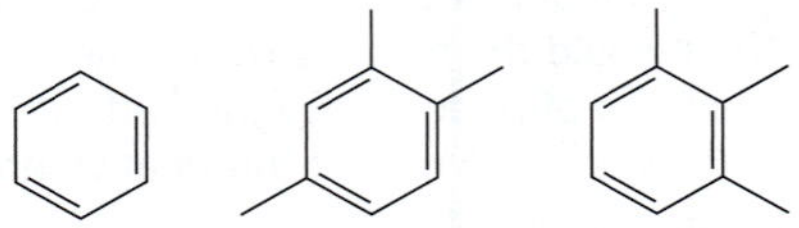

Relative rate of chlorination: 1   680,000   800,000

23.46 For each pair of isomers, determine the one that will undergo electrophilic aromatic substitution faster. Explain. *Hint:* See Problem 23.45. **(a)** *o*-Dimethylbenzene or *m*-dimethylbenzene; **(b)** *m*-dimethylbenzene or *p*-dimethylbenzene; **(c)** 1,2,3,4-tetramethylbenzene or 1,2,3,5-tetramethylbenzene

23.47 We learned that halogen atoms are one of the few substituents that are ortho/para directing but deactivating. The nitroso group (—N=O) is ortho/para directing (see Problem 23.40) and deactivating, too. Explain why it is deactivating.

23.48 As shown below, electrophilic aromatic substitution on *N,N*-dimethylaniline is faster than on the unsubstituted aniline. In other words, the aromatic ring in *N,N*-dimethylaniline is more activated than the ring is in aniline itself. If the ring is methylated at the 2 and 6 positions, however, then the *N,N*-dimethyl-substituted compound reacts more slowly in electrophilic aromatic substitution than the unsubstituted compound. Explain both of these results. *Hint:* It does *not* have to do with the number of H atoms on the ring.

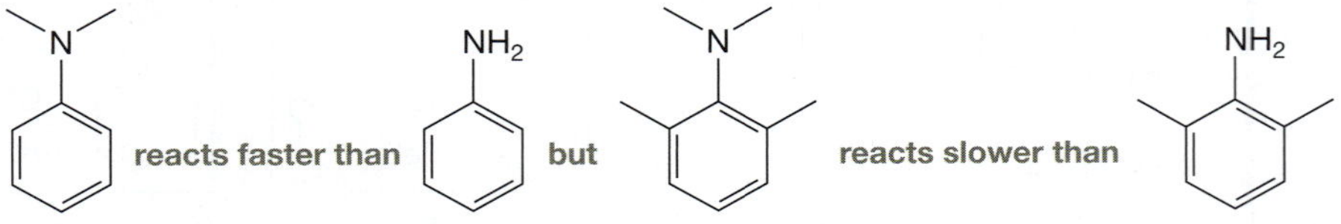

reacts faster than   but   reacts slower than

23.4–23.6 Reaction Conditions and Disubstituted Benzenes in Electrophilic Aromatic Substitution Reactions

23.49 (SYN) For each of the following substituted benzenes, determine whether sulfuric acid should be added to concentrated nitric acid to carry out a nitration.

(a)   **(b)** 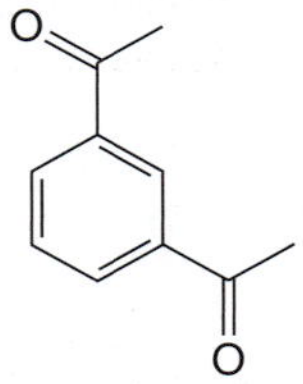   **(c)** 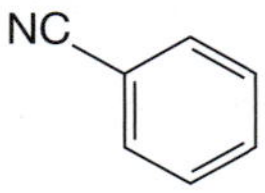   **(d)** SO_3H **(e)** SO_3H **(f)**

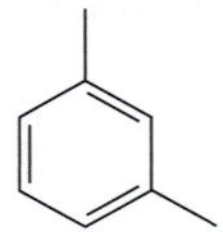

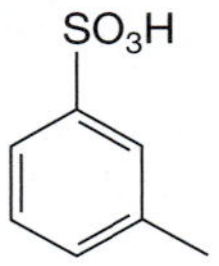

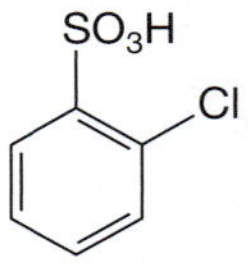

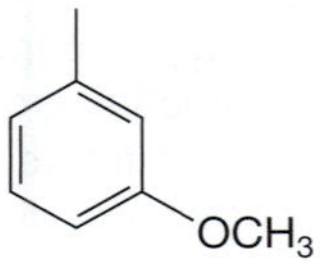

23.50 (SYN) For each compound in Problem 23.49, determine whether fuming sulfuric acid should be added to concentrated sulfuric acid to carry out a sulfonation.

23.51 (SYN) Which compounds in Problem 23.49 can undergo a Friedel–Crafts reaction? Explain.

23.52 In Section 23.5, we saw that phenol (C_6H_5—OH) undergoes three rapid, successive brominations, even without a strong Lewis acid catalyst. Under similar conditions, anisole (C_6H_5—OCH_3) undergoes just a single bromination. Explain why.

23.53 Predict the most likely site(s) of electrophilic aromatic substitution.

(a) OCH_3 OCH_3 **(b)** OH Cl **(c)** O Br **(d)** O Br

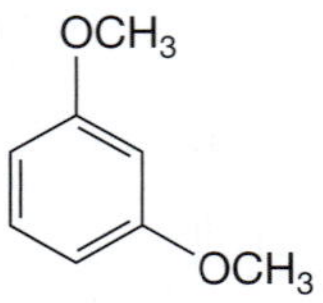

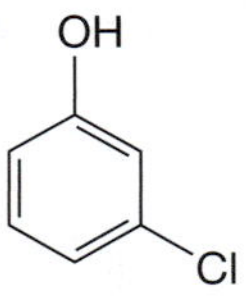

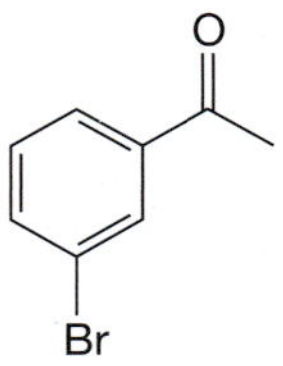

(e) NO_2 **(f)** 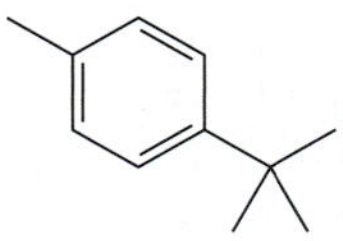   **(g)** NH_2

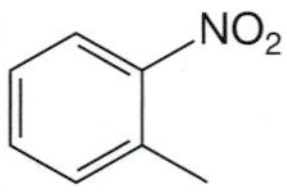

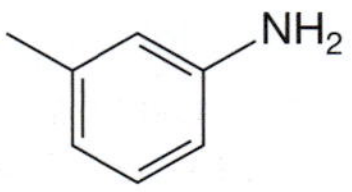

23.54 For each of the following reactions, draw the complete mechanism and the major organic product(s).

(a) 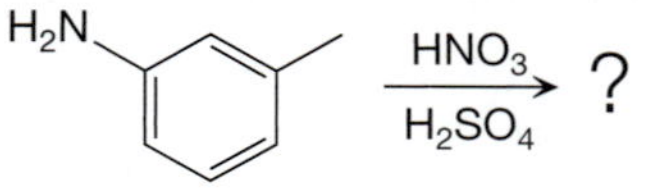 H$_2$N— (3-methylaniline) $\xrightarrow[\text{H}_2\text{SO}_4]{\text{HNO}_3}$ **?**

(b) 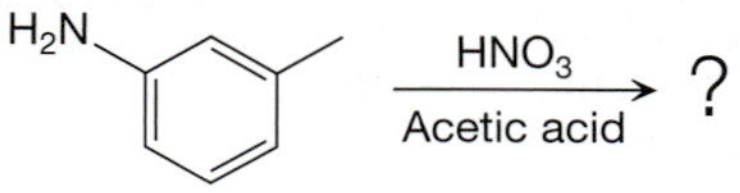 H$_2$N— (3-methylaniline) $\xrightarrow[\text{Acetic acid}]{\text{HNO}_3}$ **?**

23.55 (SYN) Show how to synthesize each of the following trisubstituted benzenes from a disubstituted benzene.

(a) 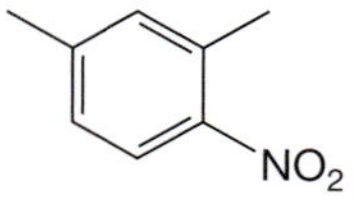 (2,4-dimethyl-1-nitrobenzene, NO$_2$)

(b) 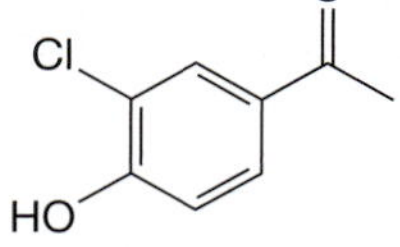 (3-chloro-4-hydroxyacetophenone, Cl, O, HO)

(c) 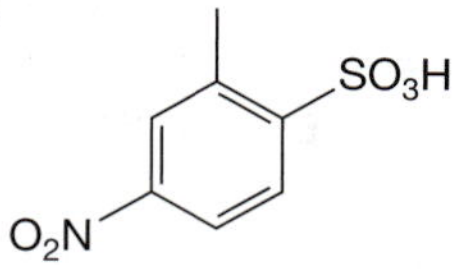 (4-methyl-2-nitrobenzenesulfonic acid, SO$_3$H, O$_2$N)

(d) 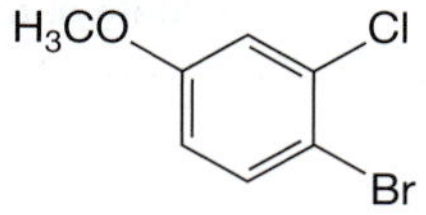 (H$_3$CO, Cl, Br)

23.7 Electrophilic Aromatic Substitution Involving Aromatic Rings Other than Benzene

23.56 The electrostatic potential maps of benzene and pyridine are shown here. Is the electrostatic potential map of pyridine consistent with the ring being activated or deactivated relative to benzene? Explain.

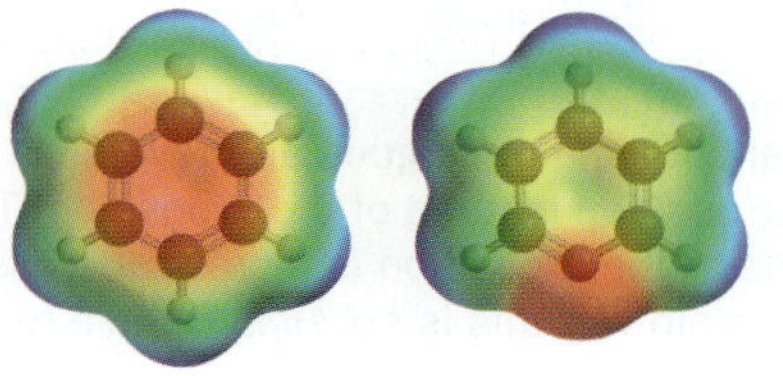

23.57 The electrostatic potential maps of benzene and pyrrole are shown here. Is the electrostatic potential map of pyrrole consistent with the ring being activated or deactivated relative to benzene? Explain.

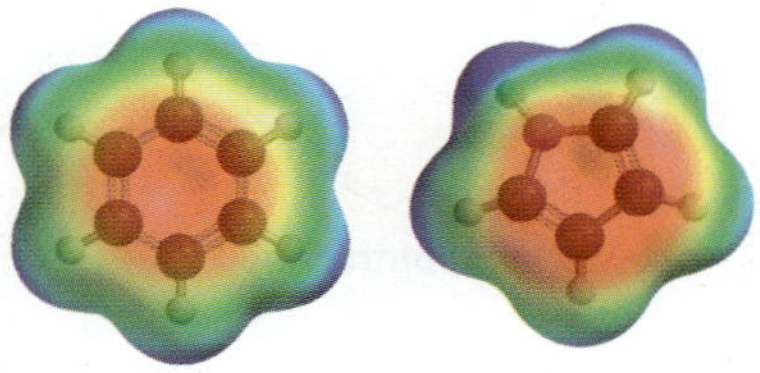

23.58 Here in Chapter 23, we learned that aniline becomes highly deactivated in the presence of a strong Lewis acid, due to coordination of the N atom to the Lewis acid. Thus, as shown below at the left, Friedel–Crafts reactions involving aniline are unfeasible. As shown in the reaction below at the right, though, this does not appear to be a problem with the N atom in pyrrole. Explain why.

(aniline, $\ddot{N}H_2$) + (acetyl chloride, O, Cl) $\xrightarrow[\text{catalyst}]{\text{Lewis acid}}$ No reaction

(pyrrole, H–N) + (acetyl chloride, O, Cl) $\xrightarrow[\text{catalyst}]{\text{Lewis acid}}$ (2-acetylpyrrole, H–N, O)

23.59 When naphthalene undergoes an irreversible electrophilic aromatic substitution, such as a Friedel–Crafts acylation, the major product is the kinetic product, which proceeds through the most stable arenium ion intermediate. In Section 23.7, we mentioned that substitution is generally favored at the α position over the β position, which means that the arenium ion is more stable when the electrophile attaches to the α position. Explain this difference in arenium ion stabilities. *Hint:* Draw out all resonance structures for each arenium ion intermediate. Does each one have the same number of resonance structures? How many resonance structures of each intermediate preserve the aromaticity?

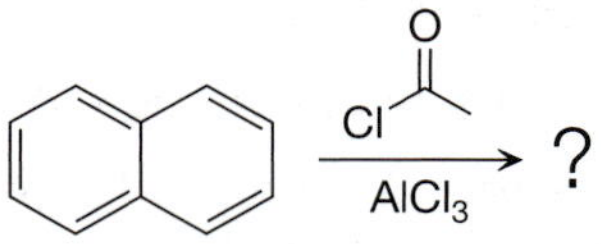

(naphthalene) + (Cl, O) $\xrightarrow{\text{AlCl}_3}$ **?**

23.60 The bromination of benzene requires a Lewis acid catalyst such as FeBr$_3$, but the bromination of naphthalene does not. Explain why.

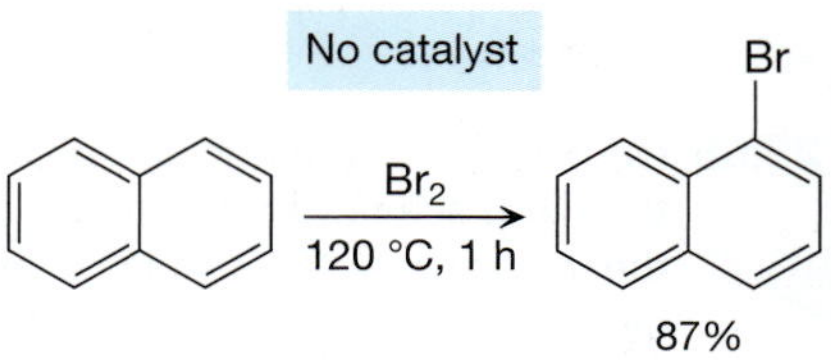

(naphthalene) $\xrightarrow[\text{120 °C, 1 h}]{\text{Br}_2}$ No catalyst (1-bromonaphthalene, Br) 87%

23.61 When an electrophilic aromatic substitution reaction on naphthalene is reversible, such as in a sulfonation reaction, the major product is the one that is most stable. With this in mind, predict the major product of the reaction shown here.

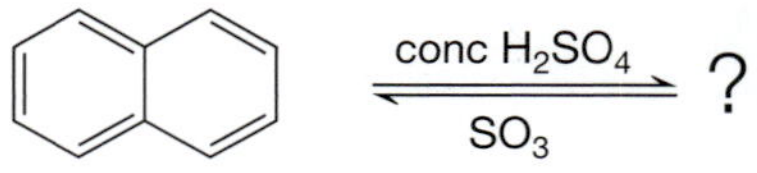

(naphthalene) $\xrightleftharpoons[\text{SO}_3]{\text{conc H}_2\text{SO}_4}$ **?**

23.62 Electrophilic aromatic substitution on a monosubstituted naphthalene tends to take place on the same ring as the substituent when the substituent is an activator like CH_3, and tends to take place on the unsubstituted ring when the substituent is a deactivator like NO_2. Explain.

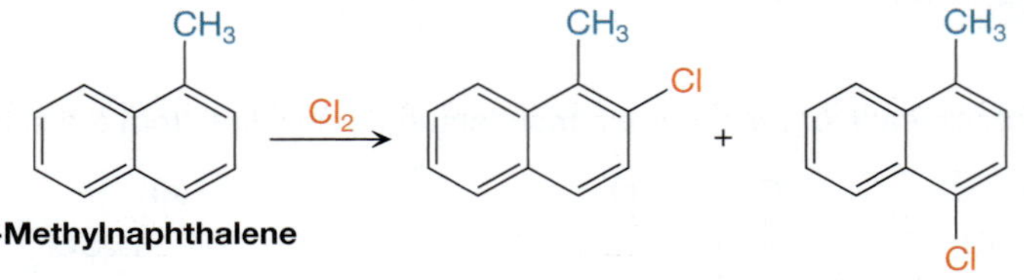

1-Methylnaphthalene

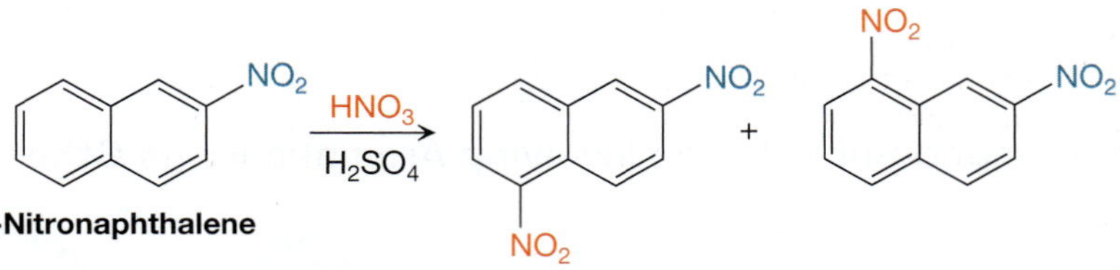

2-Nitronaphthalene

23.63 When 2-methylnaphthalene undergoes an irreversible electrophilic aromatic substitution, the electrophile predominantly attaches to the 1 position instead of the 3 position. This suggests that the arenium ion that is formed from attachment of the electrophile to the 1 position is more stable than the arenium ion formed from attachment of the electrophile to the 3 position. Explain why this is so. *Hint:* Simple resonance theory can explain why.

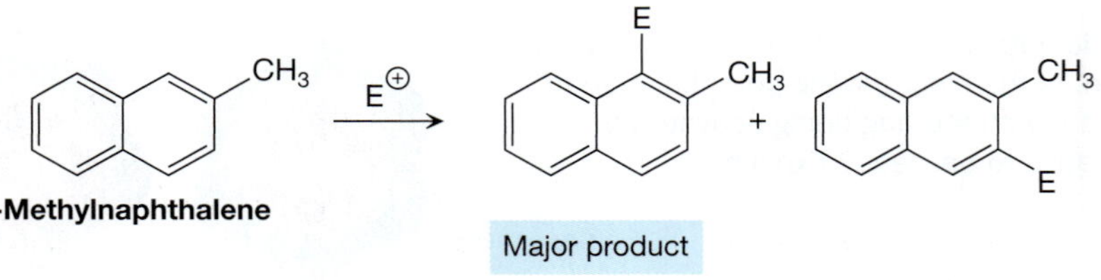

2-Methylnaphthalene

Major product

23.64 Draw the complete, detailed mechanism for the reaction shown here and predict the major product. *Hint:* See Problems 23.62 and 23.63.

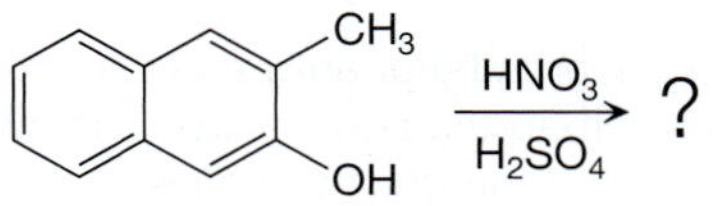

23.65 Draw the complete, detailed mechanism and predict the major product for each of the following reactions. *Hint:* See Problems 23.62 and 23.63.

(a)

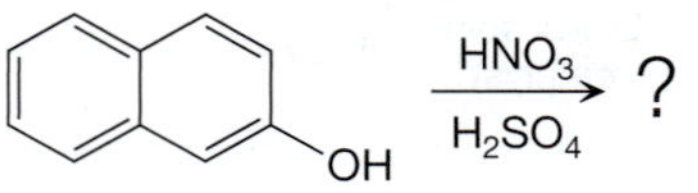

(b)

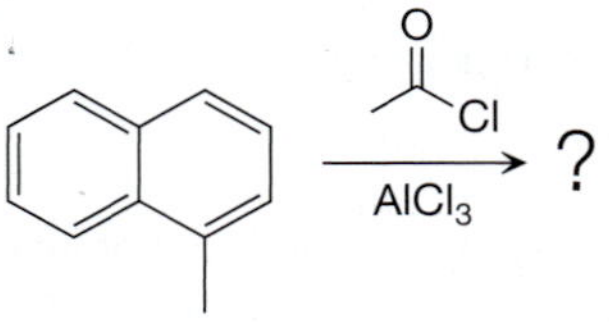

(c)

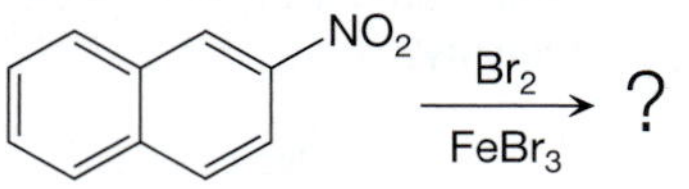

(d)

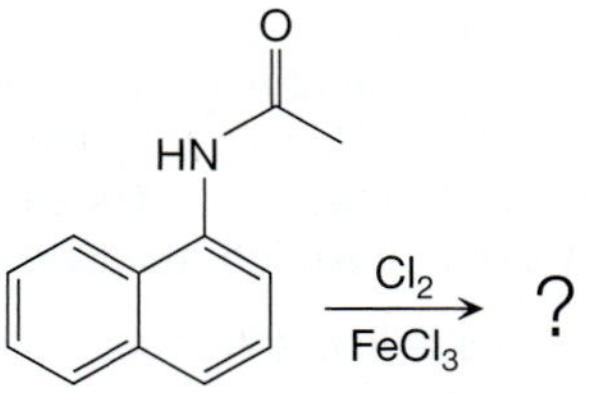

23.66 (a) For which aromatic compound do you expect nitration to take place faster: furan or thiophene? **(b)** For each of these compounds, at which C atom do you expect electrophilic aromatic substitution to predominantly take place? Explain your reasoning.

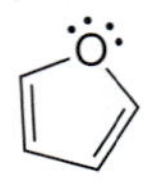

Furan 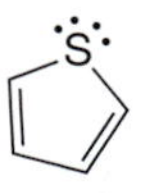   **Thiophene**

23.67 A thiophene ring is sufficiently activated that bromination may take place without the presence of a Lewis acid catalyst. With this in mind, draw the complete mechanism for the reaction shown here and predict the major product.

23.68 Draw the complete, detailed mechanism and predict the major product for each of the following reactions.

(a) Br₂ ? (b) Br₂ / FeBr₃ ? (c) HNO₃ / Acetic anhydride ? (d) HNO₃ / Acetic acid, Δ ?

23.69 Predict the site on each molecule that is most likely to undergo electrophilic aromatic substitution.

(a)   (b)

23.8 Azo Coupling and Azo Dyes

23.70 Butter yellow is an azo dye produced from the following reaction. Draw the complete mechanism for this reaction. Would you expect a significant amount of the meta product? Why or why not?

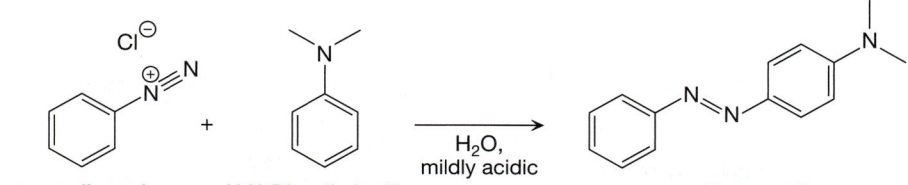

Benzenediazonium chloride   **N,N-Dimethylaniline**   H₂O, mildly acidic   **Butter yellow (An azo dye)**

23.71 (SYN) Tartrazine is an azo dye that is primarily used as a lemon-yellow food coloring. Draw the arenediazonium ion and the separate aromatic compound that would react to form the dye, and draw the mechanism for that reaction.

Tartrazine

23.72 (SYN) Sunset yellow FCF is an azo dye that, when used as a food coloring, is called yellow 6. Draw the arenediazonium ion and the separate aromatic compound that would react to form the dye, and draw the mechanism for that reaction.

Sunset yellow FCF

23.73 The reaction shown here is used to synthesize an azo dye called azo violet. Draw the mechanism for this reaction and the structure of azo violet.

NaNO₂ / p-MeC₆H₄SO₃H ?

23.9 Nucleophilic Aromatic Substitution Mechanisms

23.74 Draw the mechanism and the major organic product for each of the following reactions.

(a) NaO⟍ ? (b) HO⟍ / K₂CO₃, DMF ? (c) H₂N⟍⟍ ?

23.75 In nucleophilic aromatic substitution reactions that proceed by the nucleophilic addition–elimination mechanism, the reaction rate increases as the electronegativity of the halogen leaving group increases: Ar—I < Ar—Br < Ar—Cl < Ar—F. Which step does this suggest is the rate-determining step of the mechanism—the addition step or the elimination step? Explain.

23.76 Draw the detailed mechanism and predict the major product for each of the following reactions. *Hint:* See Problem 23.75.

(a)

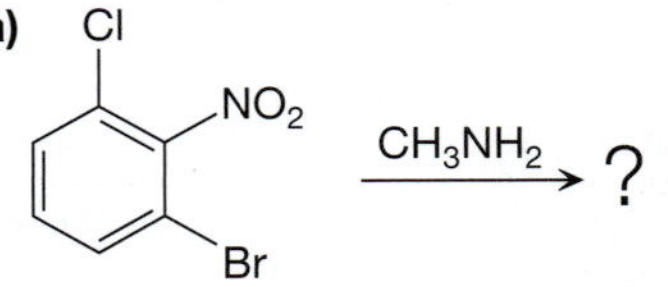

$$\xrightarrow{\text{CH}_3\text{NH}_2} ?$$

(b)

$$\xrightarrow[\text{2. H}_2\text{O, HCl}]{\text{1. NaOH}} ?$$

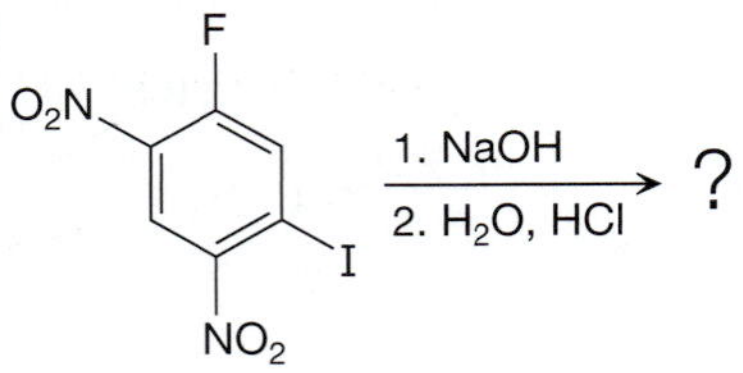

23.77 Draw the mechanism leading to the major organic product(s) for each of the following reactions.

(a)

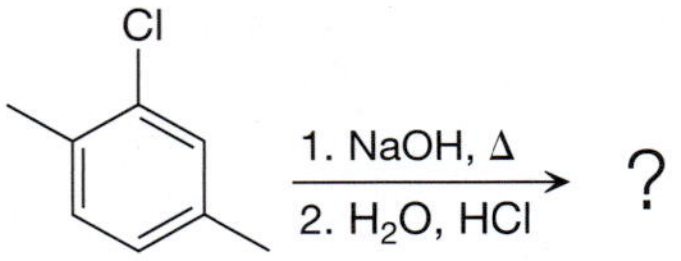

$$\xrightarrow[\text{2. H}_2\text{O, HCl}]{\text{1. NaOH, }\Delta} ?$$

(b)

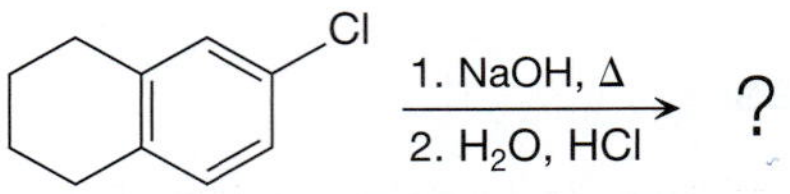

$$\xrightarrow[\text{2. H}_2\text{O, HCl}]{\text{1. NaOH, }\Delta} ?$$

(c)

$$\xrightarrow[\text{2. H}_2\text{O}]{\text{1. KNH}_2, \text{ NH}_2(\ell)} ?$$

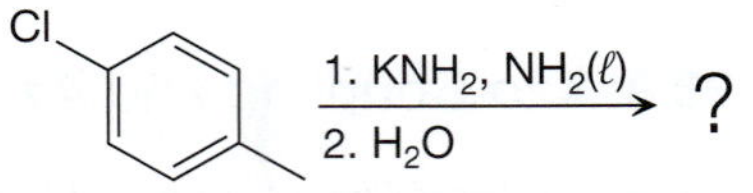

23.78 **(SYN)** Show how to synthesize each of the following compounds using only a single nucleophilic aromatic substitution reaction.

(a)

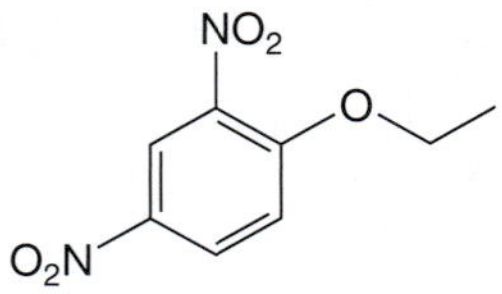

(b)

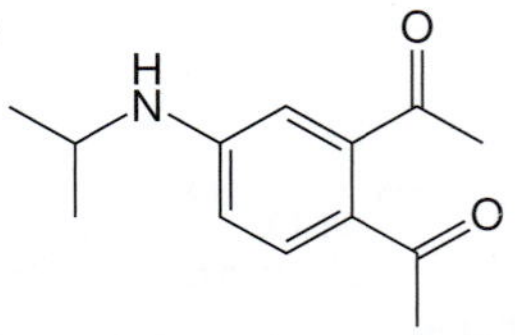

(c)

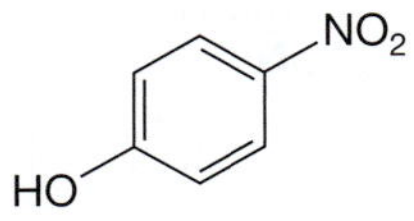

23.79 **(SYN)** Show how to synthesize each of the following compounds using only a single nucleophilic aromatic substitution reaction.

(a)

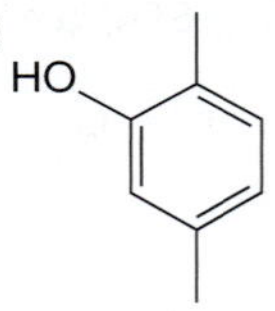

(b)

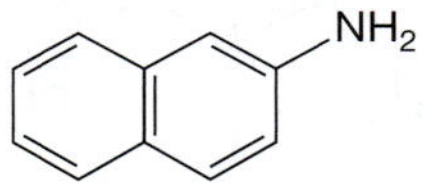

(c)

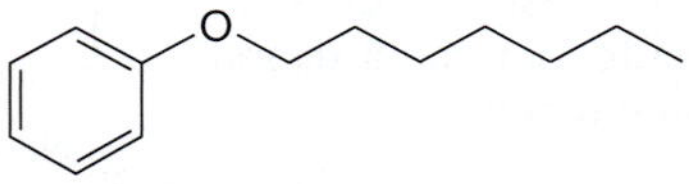

23.10–23.12 Organic Synthesis: Considerations of Regiochemistry; Attaching Groups in the Correct Order; Interconverting Ortho/Para and Meta Directors; Protecting Groups

23.80 **(SYN)** Show how you would synthesize each of the following compounds from benzene.

(a)

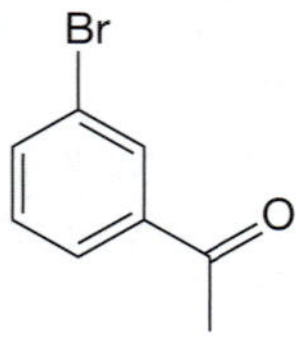

(b)

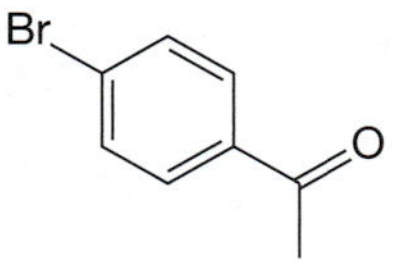

(c)

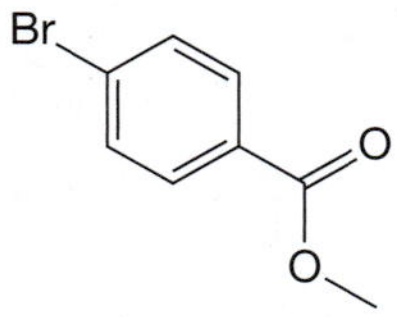

(d)

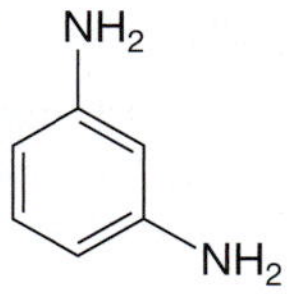

(e)

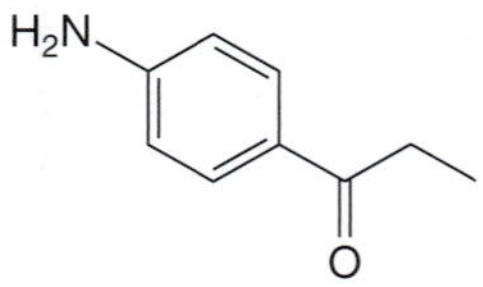

23.81 **(SYN)** Show how you would synthesize each of the following compounds from benzene.

(a)

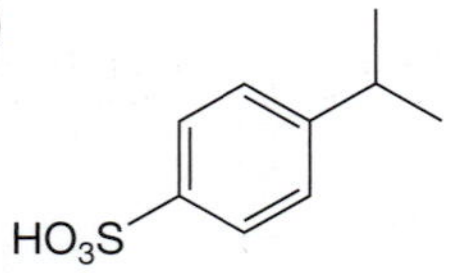

(b)

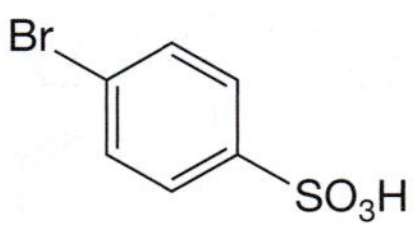

(c)

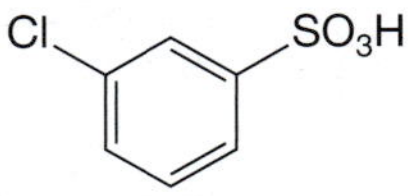

23.82 (SYN) Show how you would carry out each of the following transformations.

(a)

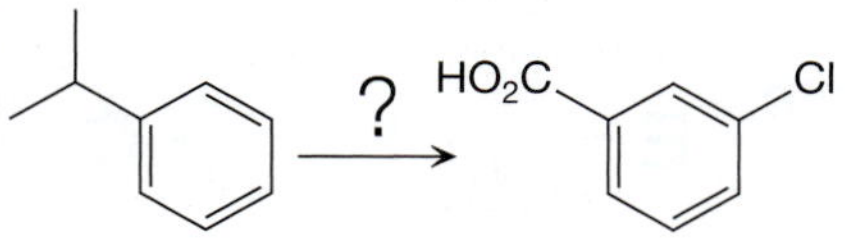

(b)

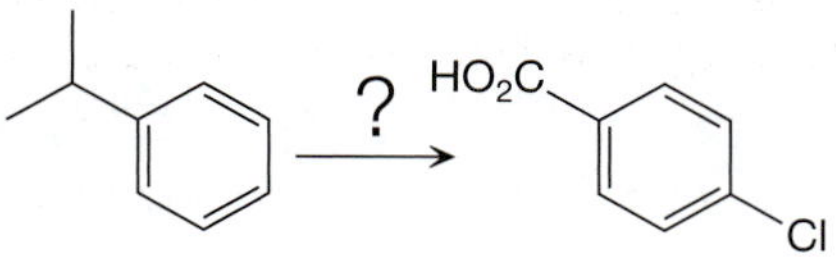

23.83 (SYN) Show how you would carry out each of the following transformations.

(a)

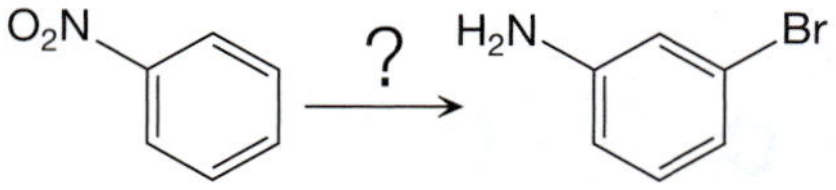

(b)

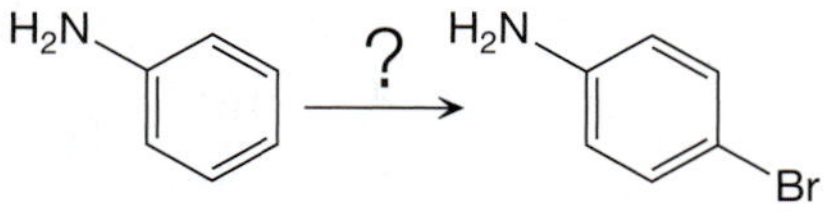

23.84 (SYN) Show how you would carry out this synthesis.

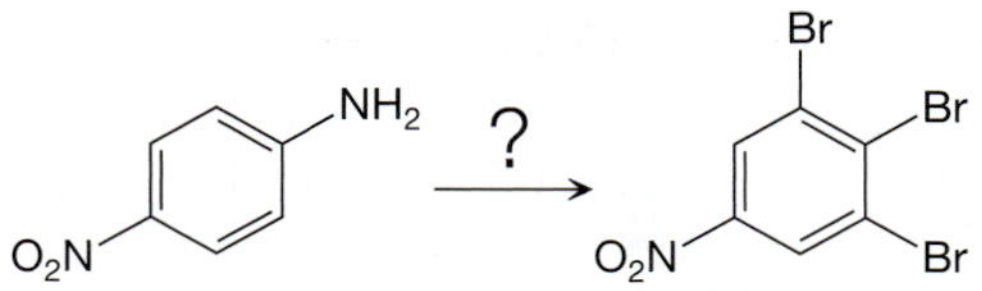

23.85 (SYN) Show how you would synthesize each of these compounds from benzene.

(a)

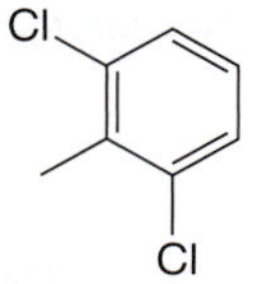

(b)

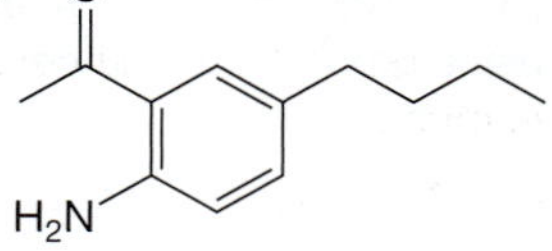

Integrated Problems

23.86 Predict the most likely sites of electrophilic aromatic substitution in each of the following molecules.

(a)

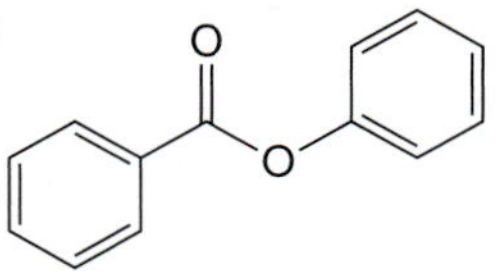

(b)

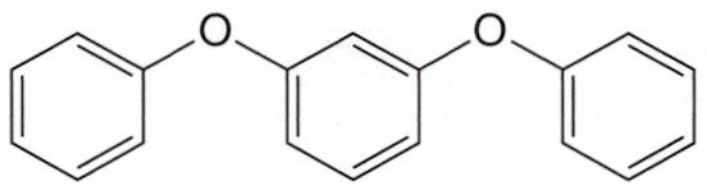

(c)

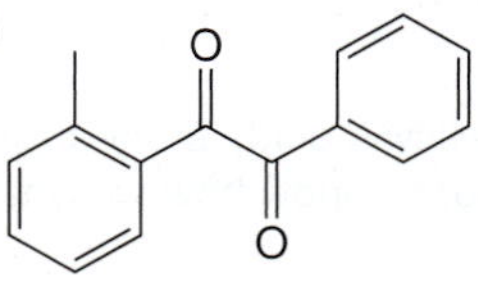

(d)

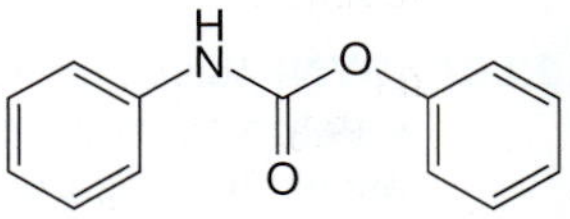

23.87 Draw a complete, detailed mechanism for this reaction.

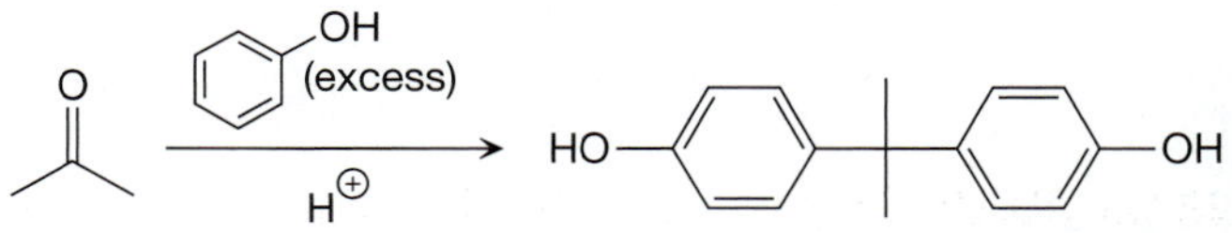

23.88 Draw a complete, detailed mechanism for the following reaction. A key intermediate is provided.

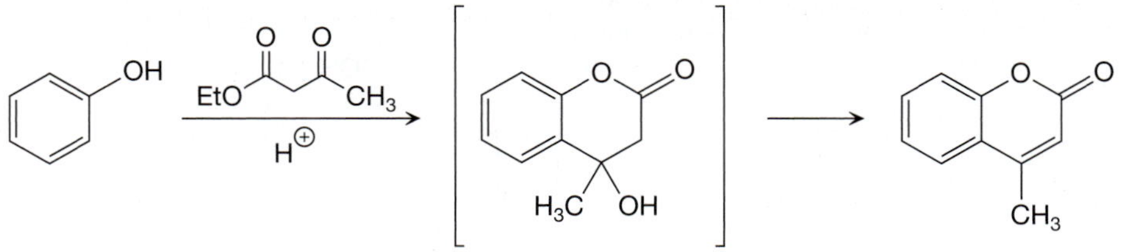

23.89 Draw the structures of compounds **A–F** in the following synthesis scheme.

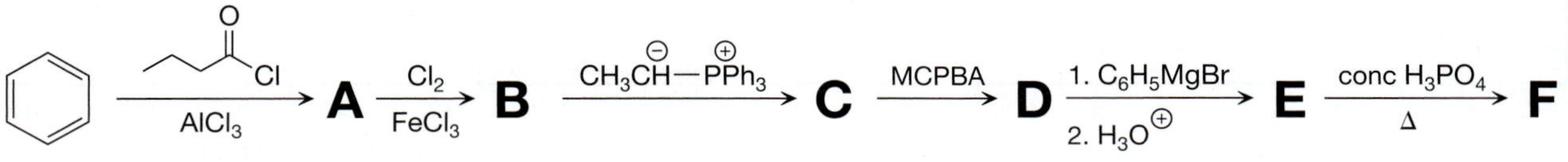

23.90 Draw the structures of compounds **A–H** in the following synthesis scheme.

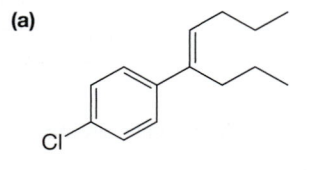

$$\text{benzene} \xrightarrow[\text{FeBr}_3]{\text{Br}_2} \textbf{A} \xrightarrow[\text{AlCl}_3]{\text{CH}_3\text{Cl}} \textbf{B} \xrightarrow[\text{2. HCl, H}_2\text{O}]{\text{1. KMnO}_4, \text{KOH, }\Delta} \textbf{C} \xrightarrow[\text{2. H}_3\text{O}^\oplus]{\text{1. LiAlH}_4} \textbf{D} \xrightarrow{\text{PCC}} \textbf{E} \xrightarrow[\text{NaOH, }\Delta]{\text{acetone}} \textbf{F} \xrightarrow[\text{H}_2\text{O}]{\text{KCN}} \textbf{G} \xrightarrow[\Delta]{\text{H}_3\text{O}^\oplus} \textbf{H}$$

23.91 Draw the structures of compounds **A–H** in the following synthesis scheme.

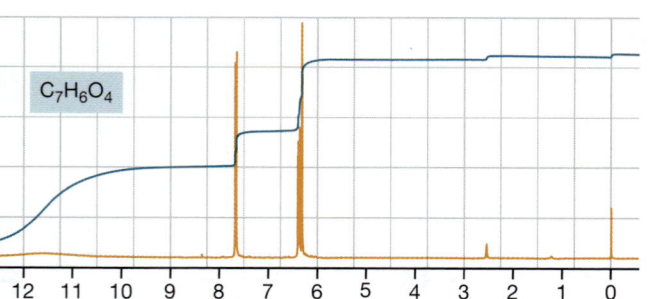

23.92 **(SYN)** Show how you would synthesize this compound, using propan-1-ol and benzene as your only sources of carbon.

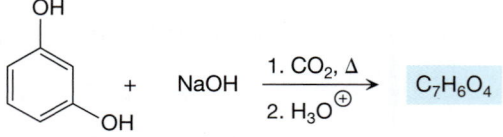

23.93 **(SYN)** Show how you would synthesize each of these compounds, using butan-1-ol and benzene as your only sources of carbon.

(a) **(b)**

23.94 **(SYN)** Aspirin, or acetylsalicylic acid, is one of the most widely produced medications. Show how to synthesize aspirin from benzene.

Acetylsalicylic acid
(Aspirin)

23.95 The reaction shown here yields a compound whose molecular formula is $C_7H_6O_4$. The ^{1}H NMR and ^{13}C NMR spectra of $C_7H_6O_4$ are shown below. Draw the product and the complete, detailed mechanism for this reaction.

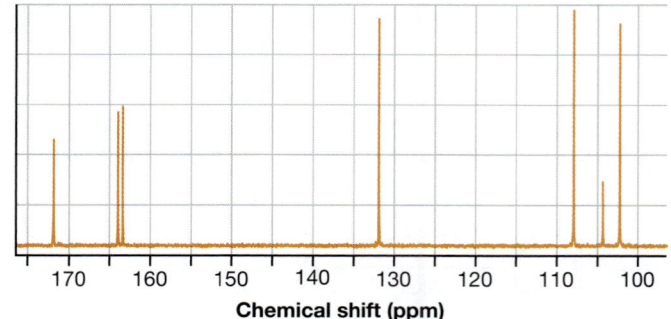

$C_7H_6O_4$

Chemical shift (ppm)

Chemical shift (ppm)

23.96 Compound **A**, whose formula is C_9H_{10}, dimerizes in the presence of acid to produce the compound shown here. The ^1H NMR and ^{13}C NMR spectra of **A** are shown below. Determine the structure of **A** and draw the complete mechanism for the reaction. (In the ^{13}C NMR spectrum, there are two signals >130 ppm.)

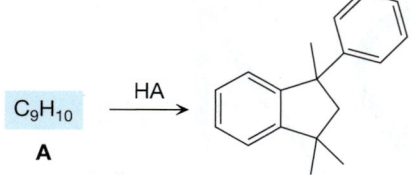

C_9H_{10} $\xrightarrow{HA}$

A

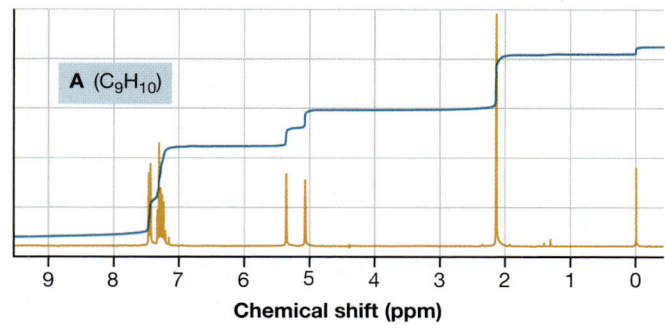

A (C_9H_{10})

Chemical shift (ppm)

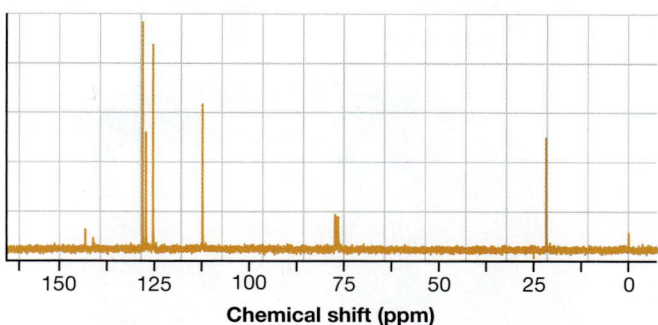

Chemical shift (ppm)

23.97 Heating compound **B** under acidic conditions produces compound **C**, whose formula is $C_{15}H_{12}$. The ^{13}C NMR spectrum of **C** exhibits one signal near 20 ppm and 14 signals between 120 and 140 ppm. The ^1H NMR spectrum exhibits one signal at 2.7 ppm and several overlapping signals between 7 and 9 ppm. The integrations of the two sets of signals are in a 1:3 ratio. Determine the structure of **C** and draw the complete mechanism for the reaction.

B $\xrightarrow[\Delta]{HA}$ $C_{15}H_{12}$

C

24

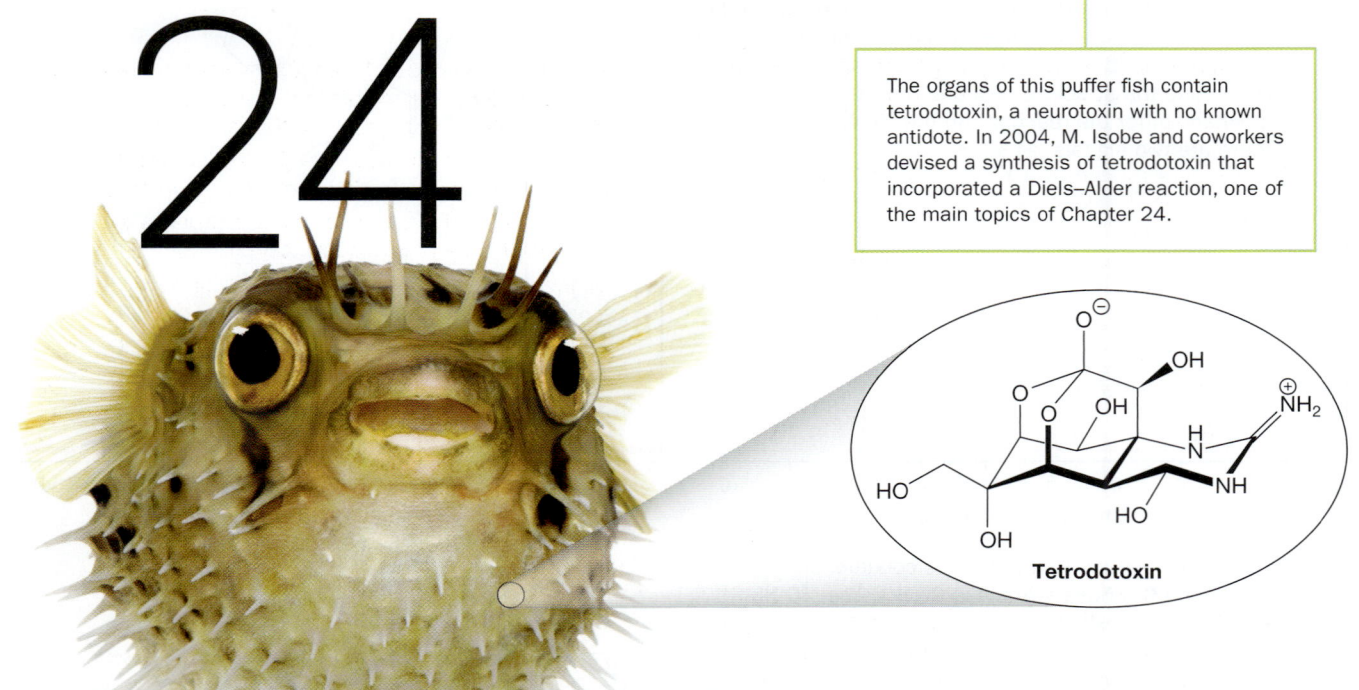

The organs of this puffer fish contain tetrodotoxin, a neurotoxin with no known antidote. In 2004, M. Isobe and coworkers devised a synthesis of tetrodotoxin that incorporated a Diels–Alder reaction, one of the main topics of Chapter 24.

Tetrodotoxin

The Diels–Alder Reaction and Other Pericyclic Reactions

thene reacts with buta-1,3-diene, a conjugated diene, to form cyclohexene, as illustrated in Equation 24-1. This is the simplest example of what is called the **Diels–Alder reaction**. This reaction is so important in organic chemistry that Otto Diels (1876–1954) and Kurt Alder (1902–1958), the two German chemists who first described the reaction, shared the 1950 Nobel Prize in Chemistry.

The Diels–Alder reaction

$$ \text{Buta-1,3-diene} + \text{Ethene} \xrightarrow[\text{high pressure}]{200\,°C,\ 17\ h,} \text{Cyclohexene} \quad 18\% $$

New C—C bonds

(24-1)

The Diels–Alder reaction is especially useful because it forms two new carbon–carbon bonds. (As we stressed in Chapter 13, these are *extremely* important types of reactions in organic synthesis.) Notice, too, that the product of the Diels–Alder

Chapter Objectives

On completing Chapter 24 you should be able to:

1. Recognize a Diels–Alder reaction as a pericyclic reaction and, more specifically, as a [4+2] cycloaddition.

2. Identify potential dienes and dienophiles for use in a Diels–Alder reaction.

3. Determine whether a cycloaddition reaction is thermally allowed or forbidden, based on whether the reaction proceeds through an aromatic transition state or an antiaromatic transition state.

4. Predict the viability of a Diels–Alder reaction based on the diene's ability to attain an s-cis conformation.

5. Predict relative rates of standard Diels–Alder reactions based on the presence of electron-donating and electron-withdrawing groups in the reactants.

6. Predict the major product of a Diels–Alder reaction in terms of its *stereochemistry* and *regiochemistry*.

7. Describe the roles of enthalpy and entropy in the reversibility of a Diels–Alder reaction.

8. Specify the conditions for the syn dihydroxylation of an alkene or alkyne, and draw the mechanism.

9. Identify various oxidative cleavage reactions, and explain how and why the products of these reactions can differ.

reaction is a six-membered ring, which is widely abundant in natural products. The reaction is also quite robust; it can be carried out with a variety of other functional groups present, and many Diels–Alder reactions can be carried out under mild conditions. Finally, the reaction is *stereospecific* and often *regioselective*, which can provide chemists substantial control over the products that are formed.

The mechanism of the Diels–Alder reaction cannot be described by the elementary steps we learned in Chapters 6 and 7. Therefore, we introduce a new elementary step and we explore it with the same systematic approach we took with the introduction of the previous elementary steps. We begin with the curved arrow notation and examples. We then investigate factors that affect the reaction's rate, and we examine the reaction's stereochemistry and regiochemistry.

Once we have examined the Diels–Alder reaction, we turn our attention to syn dihydroxylation and oxidative cleavage reactions—important reactions that involve elementary steps closely related to those of the Diels–Alder mechanism. Much of your knowledge gained from studying the Diels–Alder reaction is applicable to these reactions as well.

24.1 Curved Arrow Notation and Examples

The mechanism for the simplest possible Diels–Alder reaction (shown previously in Equation 24-1) is provided in Equation 24-2. This serves as the prototype for all other Diels–Alder reactions.

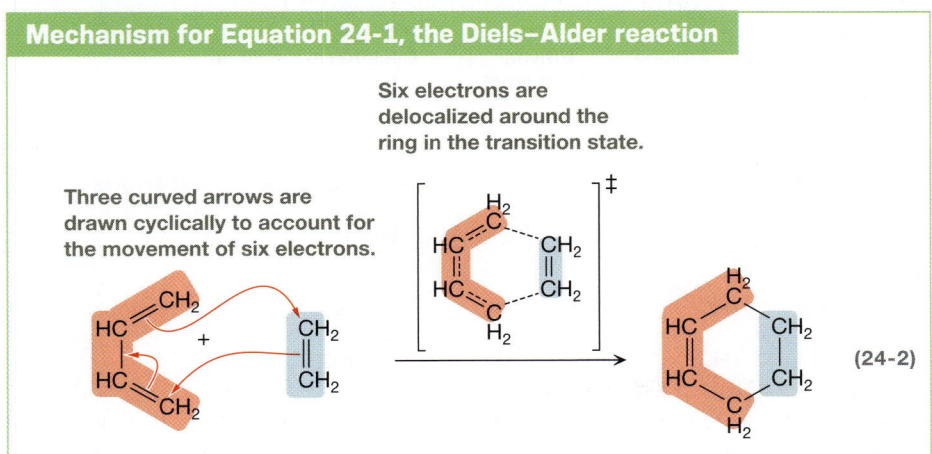

Mechanism for Equation 24-1, the Diels–Alder reaction

Three curved arrows are drawn cyclically to account for the movement of six electrons.

Six electrons are delocalized around the ring in the transition state.

(24-2)

The entire reaction takes place in a *single step*, so:

> The Diels–Alder reaction is concerted, whereby all of the bonds are formed and broken simultaneously.

For this to occur, six electrons must flow in a cyclic fashion. Thus, the reaction proceeds through a *cyclic transition state*, which makes it a **pericyclic reaction**. More specifically, because this cyclic movement of electrons joins together two separate species to form a ring, it is also called a **cycloaddition reaction**. (Other examples of pericyclic reactions are explored in Problem 24.38 at the end of the chapter.)

In Equation 24-2, the curved arrows depict a clockwise movement of the electrons. It would be equally correct, however, if the curved arrows were drawn to depict a counterclockwise movement of those electrons.

YOUR TURN 24.1

> Redraw the elementary step in Equation 24-2 to depict a counterclockwise movement of the electrons. Does this alternate way of drawing the curved arrows lead to the same product shown in Equation 24-2?
>
> *Answers to Your Turns are in the back of the book.*

A Diels–Alder reaction involves a *diene* and a *dienophile*:

- A **diene** contributes four π electrons from a pair of *conjugated* π bonds.
- A **dienophile** (meaning "diene loving") contributes two π electrons from a single π bond.

Consequently, a Diels–Alder reaction can be described as a **[4+2] cycloaddition**, where the numbers indicate the number of π electrons contributed by each species involved.

The Diels–Alder reaction can take place with an alkyne as the dienophile instead, as shown in Equation 24-3. Two electrons are contributed from one π bond of the triple bond, leaving the other π bond intact in the product.

Buta-1,3-diene + Ethyne (Acetylene) → Cyclohexa-1,4-diene (24-3)

YOUR TURN 24.2

> The reaction in Equation 24-3 is repeated here. Supply the curved arrows and label the diene and dienophile.

What is consistent among all Diels–Alder reactions is the cyclic flow of six π electrons. The number of electrons is important, because those electrons are delocalized over a complete ring in the transition state. Recall from Chapter 14 that six electrons is a *Hückel number* of electrons (i.e., an odd number of pairs), and the species is *aromatic* when the electrons are delocalized over an entire ring. Therefore:

Six electrons delocalized

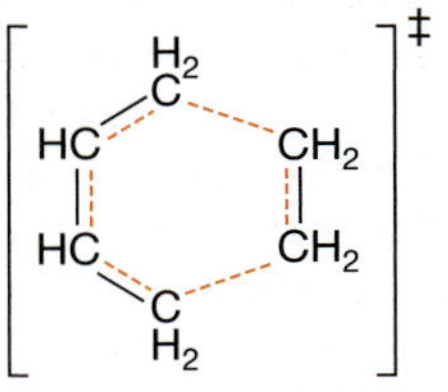

"Allowed"

> The Diels–Alder reaction proceeds through an *aromatic transition state*.

This substantially lowers the reaction's energy barrier, which helps make the reaction feasible. We say that the reaction is "allowed" under normal conditions.

For comparison, let's examine two other possible cycloaddition reactions. In Equation 24-4, two molecules of ethene combine in a **[2+2] cycloaddition**, and in Equation 24-5, a molecule of hexa-1,3,5-triene and a molecule of ethene combine in a **[6+2] cycloaddition**. Neither of these reactions occurs readily under normal conditions, because they would have to proceed through an *antiaromatic transition state*. In both cases, an even number of pairs of electrons is delocalized over the entire ring—two pairs for the [2+2] cycloaddition and four pairs for the [6+2] cycloaddition. Because these reactions do not occur readily, we say that they are "forbidden."

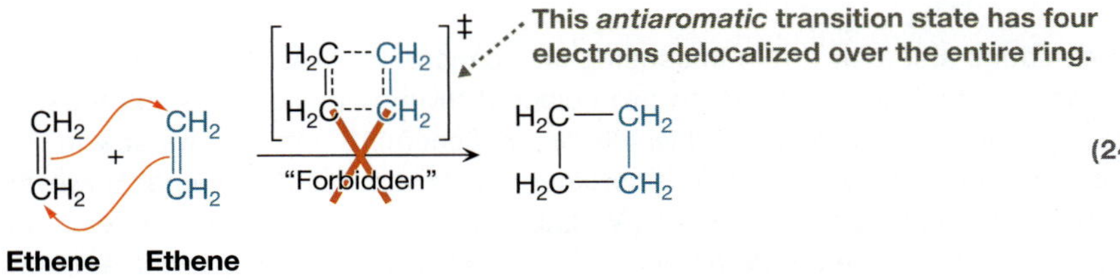

This *antiaromatic* transition state has four electrons delocalized over the entire ring.

(24-4)

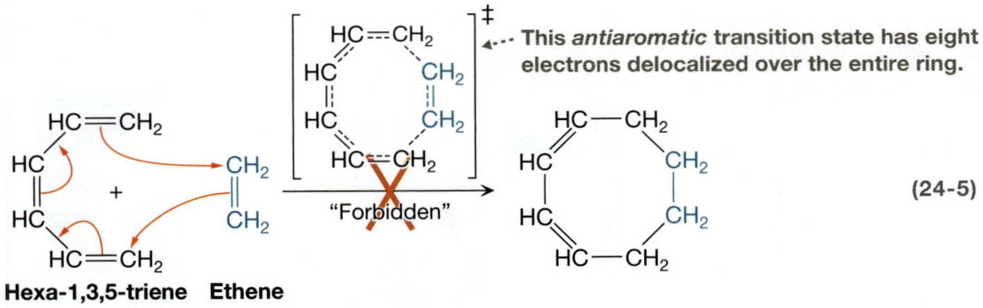

Ethene   Ethene

This *antiaromatic* transition state has eight electrons delocalized over the entire ring.

(24-5)

Hexa-1,3,5-triene   Ethene

The cycloaddition reaction involving two molecules of cyclopentene is as follows:

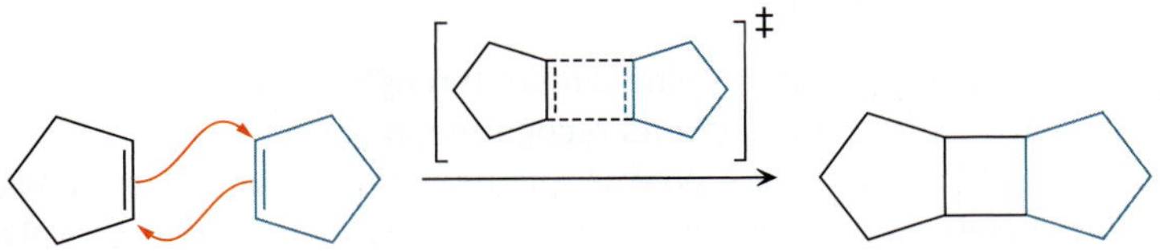

Is the transition state *aromatic* or *antiaromatic*?

Biological Cycloaddition Reactions

The Diels–Alder reaction was first documented in 1928, but nature has been using the reaction since long before then. One example is in the biosynthesis of lovastatin, a cholesterol-lowering natural product produced by the fungus *Aspergillus terreus* and marketed under the trade name Mevacor. The Diels–Alder step is shown in Scheme A below, which produces a bicyclic compound that goes on to form lovastatin. Cholesterol is synthesized in our bodies through a key step in which 3-hydroxy-3-methylglutaryl coenzyme A (HMG-CoA) is converted to mevalonate. Lovastatin binds to the active site of HMG-CoA reductase, which inhibits its enzymatic activity.

Scheme A

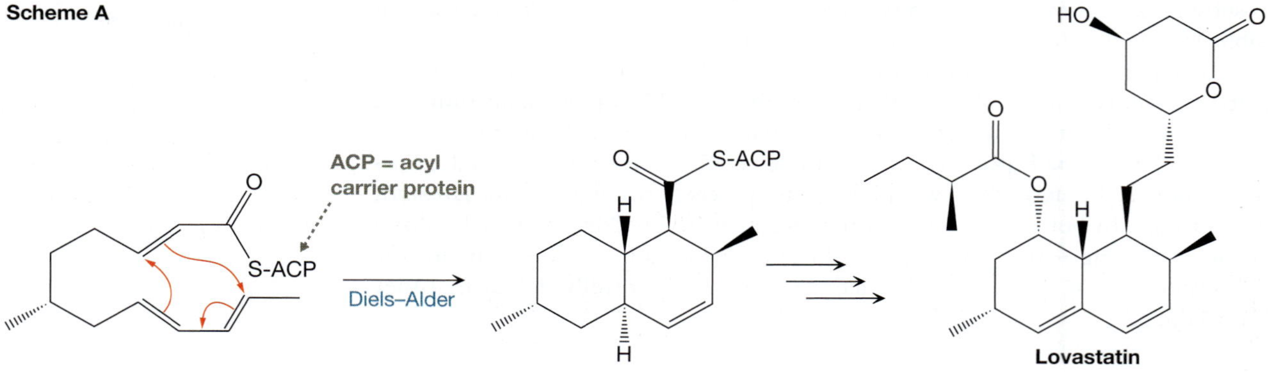

Cycloaddition reactions also play a role in causing skin cancer. In DNA, the thymine nitrogenous base has a C=C double bond, so two adjacent thymine nucleotides can feasibly undergo a [2+2] cycloaddition to produce a *thymine dimer* (Scheme B, left). As we saw in Section 24.1, this kind of [2+2] cycloaddition is *forbidden* under normal conditions. However, when one of the molecules absorbs a photon of UV light (such as from the sun), the reaction becomes *allowed*. (This is discussed in detail in Section 24.10.) The resulting thymine dimer causes a kink in the DNA strand (Scheme B, right), which interferes with the normal function of DNA.

Scheme B

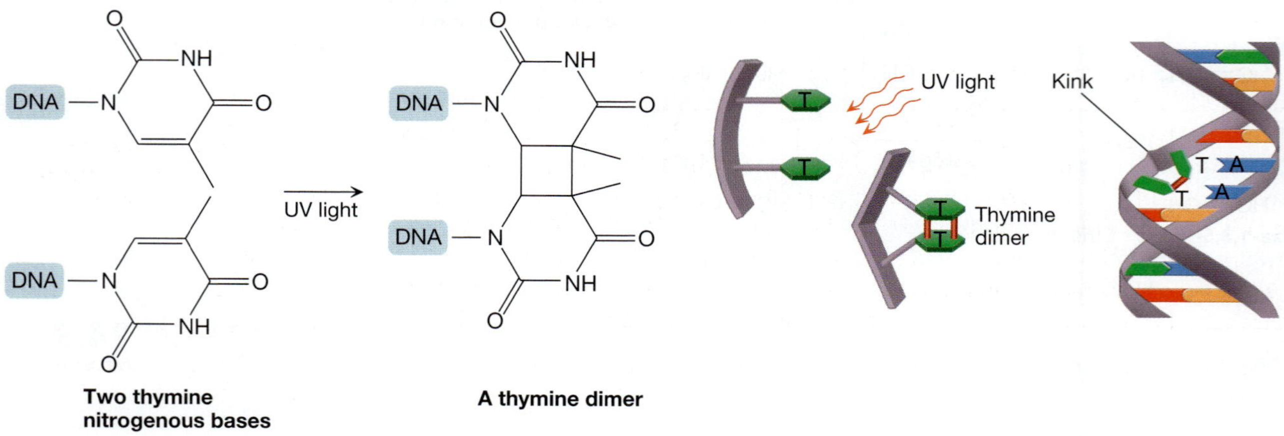

Two thymine nitrogenous bases

A thymine dimer

Fortunately, our bodies can remove these thymine dimers through a process called nucleotide excision repair. Once the thymine dimer is recognized, a short segment of damaged single-stranded DNA is removed. DNA polymerase uses the remaining undamaged single-stranded DNA as a template to produce the complementary sequence that is replaced using an enzyme called DNA ligase.

SOLVED PROBLEM 24.1

Is this cycloaddition reaction allowed or forbidden?

Think In the transition state, are there electrons delocalized over an entire ring? If so, how many electrons? Is that a Hückel number or an anti-Hückel number? How does that correspond to whether the reaction is allowed or forbidden?

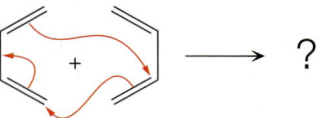

Solve The following transition state is somewhat of an average of the reactants and products.

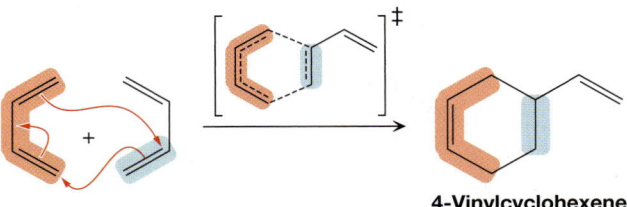

4-Vinylcyclohexene

CONNECTIONS The product of this reaction, 4-vinylcyclohexene, is used in the industrial manufacture of dodecanoic acid, a medium-sized fatty acid that is primarily used to produce soaps and cosmetics.

Six electrons are delocalized over an entire ring in the transition state (two electrons are represented by each of the three curved arrows). Since six is a Hückel number of electrons, the transition state is aromatic, making the reaction allowed.

PROBLEM 24.2 Draw the transition state for the cycloaddition reaction shown here. Does this reaction take place readily? Why or why not?

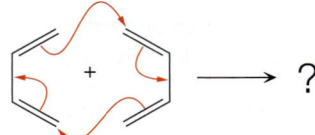

With the diene in the s-cis conformation, the ends of the diene are at the appropriate distance from the dienophile to produce two new C—C bonds.

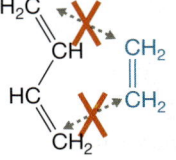

(a) s-Cis conformation

24.2 Conformation of the Diene

The concerted nature of the Diels–Alder reaction means that both of the new carbon–carbon σ bonds form simultaneously. Thus, both ends of the diene must be relatively close to the dienophile in the transition state. The appropriate distances are achieved only if the diene attains a geometry in which the two ends of the π system are pointing in nearly the same direction—the so-called **s-cis conformation** (Fig. 24-1a). (The s indicates that the cis designation describes the orientation of groups about a single bond, rather than a double bond or the plane of a ring.) In the **s-trans conformation** (Fig. 24-1b), the ends of the two reacting species are farther apart, which raises the energy of the transition state and makes the Diels–Alder reaction unfeasible. Thus:

With the diene in the s-trans conformation, the ends of the diene are too far from the dienophile.

> For a Diels–Alder reaction to take place, the diene must be in the s-cis conformation.

For buta-1,3-diene, the s-cis and s-trans conformations rapidly interchange via rotation about the C—C bond. The s-cis conformation is about 10 kJ/mol (2.4 kcal/mol) higher in energy than the s-trans conformation, so 98% of the molecules are in the s-trans conformation at equilibrium (Equation 24-6a), and the remaining 2% are

(b) s-Trans conformation

FIGURE 24-1 The diene conformation in Diels–Alder reactions A diene can exist in (a) the s-cis conformation and (b) the s-trans conformation. Diels–Alder reactions are feasible only when the diene is s-cis.

in the *s*-cis conformation (Equation 24-6b). As the *s*-cis conformation undergoes the Diels–Alder reaction, it is continually replenished via its equilibrium with the *s*-trans conformation (Le Châtelier's principle).

s-Trans

98%

No reaction   (24-6a)

Rotation about
C——C single bond

2%

(24-6b)

s-Cis

YOUR TURN 24.4

Which of the following conformations is s-cis and which is s-trans? Which can undergo a Diels–Alder reaction?

A   B

The requirement for the *s*-cis conformation precludes the Diels–Alder reactions in Equations 24-7 and 24-8 altogether.

The diene is locked in the
s-trans conformation.

No reaction   (24-7)

H_3C

CH_3

No reaction   (24-8)

The diene in Equation 24-7 is locked in the *s*-trans conformation by the fused ring system. The diene in Equation 24-8 is *not* locked in the *s*-trans conformation, but it is unable to attain the *s*-cis conformation due to severe steric strain, as illustrated in Equation 24-9.

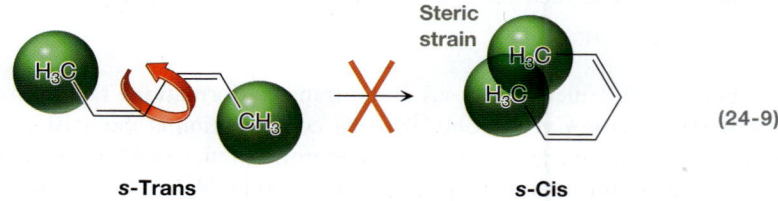

Steric
strain

H_3C

H_3C

H_3C

CH_3

(24-9)

s-Trans   *s*-Cis

Draw this diene as it would appear in its *s*-cis conformation.

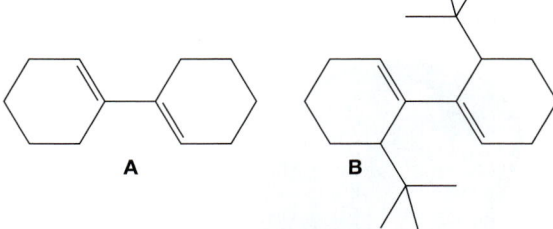

SOLVED PROBLEM 24.3

Which molecule, **A** or **B**, will react faster with ethene in a Diels–Alder reaction? Explain.

Think Are the dienes in an *s*-cis conformation? If not, can they undergo bond rotation to attain an *s*-cis conformation? How does steric strain in the *s*-cis conformation factor in?

Solve Neither diene is in the *s*-cis conformation as written, but the single bond in each molecule that connects the two double bonds can rotate.

A

B

Rotation about the single bond

s-Trans   **A**   *s*-Cis

s-Trans   **B**   *s*-Cis

Too much steric strain

No substantial strain exists in the *s*-cis conformation of molecule **A**. By contrast, severe steric strain from the overlapping *tert*-butyl groups prevents molecule **B** from attaining the *s*-cis conformation. Thus, diene **A** will undergo a Diels–Alder reaction with ethene, whereas diene **B** will not.

PROBLEM 24.4 Which molecule, **C** or **D**, will react faster with ethene in a Diels–Alder reaction? Explain.

C   **D**

PROBLEM 24.5 Although benzene's Lewis structure exhibits a pair of conjugated double bonds locked in an *s*-cis conformation, benzene does *not* react with ethene. Draw the product of this hypothetical reaction and explain why it does not occur.

Some dienes are conveniently locked in the *s*-cis conformation to allow them to undergo a Diels–Alder reaction. This is the case for cyclopentadiene in Equation 24-10 and cyclohexa-1,3-diene in Equation 24-11.

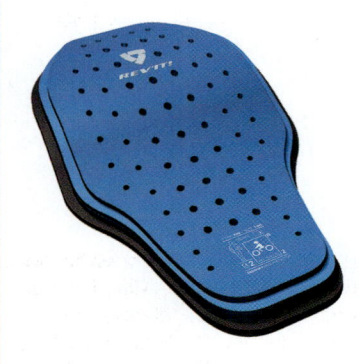

The diene is locked in the s-cis conformation.

Bridgehead carbons

Cyclopenta-1,3-diene + $CH_2 = CH_2$ → **Bicyclo[2.2.1]hept-2-ene** (24-10)

The diene is locked in the s-cis conformation.

Bridgehead carbons

Cyclohexa-1,3-diene + $CH = CH$ → **Bicyclo[2.2.2]octadiene** (24-11)

Both of these reactions produce a **bicyclic compound**, in which two **bridgehead carbons** are part of multiple rings. Notice that the bridgehead carbons are the C atoms from the reactant diene that form new bonds to the dienophile.

PROBLEM 24.6 Draw the mechanism and the product for the Diels–Alder reaction shown here. Do you think the reaction will take place as readily as the one in Equation 24-10? Why or why not?

⬡ + $CH_2 = CH_2$ ⟶ ?

24.3 Substituent Effects on the Reaction Rate

The prototypical Diels–Alder reaction (Equation 24-1) between ethene and buta-1,3-diene is quite sluggish, requiring temperatures around 200 °C to proceed at a reasonable rate. Even then, the yield is only 18%. Much of the difficulty with that reaction occurs because there is no well-established flow of electrons from an electron-rich site to an electron-poor site. In fact, as we saw in Chapter 11, both buta-1,3-diene and ethene often react as electron-rich species. One of those species can be made more electron rich, however, by attaching electron-donating groups, whereas the other species can be made electron poor by attaching electron-withdrawing groups. Thus, the flow of electrons is more clearly established, which increases the reaction rate.

In a *standard* Diels–Alder reaction, the diene is electron rich and the dienophile is electron poor:

- Electron-donating substituents bonded directly to the diene carbons (Fig. 24-2) facilitate *standard* Diels–Alder reactions. These are the same substituents that were classified as *activating* groups in electrophilic aromatic substitution (Table 23-3, p. 1157).
- Electron-withdrawing substituents bonded directly to the dienophile carbons (Fig. 24-3) facilitate *standard* Diels–Alder reactions. These are the same substituents that were classified as *deactivating* groups in electrophilic aromatic substitution.

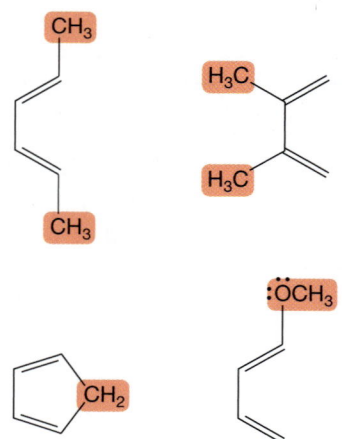

FIGURE 24-2 Electron-donating groups on the diene Electron-donating groups (highlighted in red) make the diene more electron rich, which facilitates *standard* Diels–Alder reactions.

In the reaction in Equation 24-12, for example, two electron-donating CH_3 groups are attached directly to the diene. The reaction takes place at a slightly lower temperature than the one in Equation 24-1, and the yield is much higher.

Electron-donating groups on the diene facilitate a *standard* Diels–Alder reaction.

$$ \text{(24-12)} $$

cis-3,6-Dimethylcyclohexene
60%

A similar outcome is observed with the reaction in Equation 24-13, in which a highly electron-withdrawing NO_2 group is attached to the dienophile.

An electron-withdrawing group on the dienophile facilitates a *standard* Diels–Alder reaction.

$$ \text{(24-13)} $$

4-Nitrocyclohexene
84%

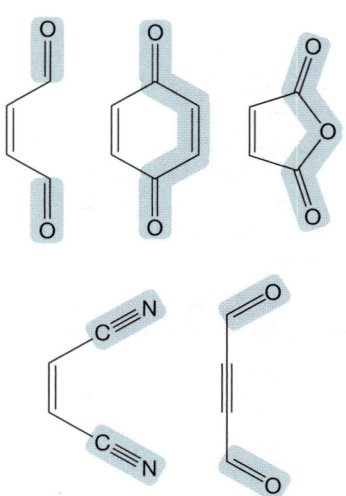

FIGURE 24-3 Electron-withdrawing groups on the dienophile Electron-withdrawing groups (highlighted in blue) make the dienophile electron poor, which facilitates *standard* Diels–Alder reactions.

SOLVED PROBLEM 24.7

Does diene **A** or diene **B** react faster in a standard Diels–Alder reaction? Explain.

Think Is the diene relatively electron rich or electron poor in a standard Diels–Alder reaction? Will electron-donating groups or electron-withdrawing groups on the diene speed up the reaction? What is the relative electron-donating or electron-withdrawing capability of each substituent?

Solve The diene is relatively electron rich in a standard Diels–Alder reaction, so the more electron donating a substituent is, the faster the reaction. In both dienes **A** and **B**, the substituent is electron donating, because each is attached to the diene by an atom with a lone pair of electrons. According to Table 23-3 (p. 1157), the alkoxy group in **B** is more strongly activating than the acylamino group (–NHCOR) in **A** (due to the presence of the carbonyl group). Therefore, diene **B** is more electron rich than diene **A**, and **B** will react faster.

PROBLEM 24.8 Does dienophile **C** or dienophile **D** react faster in a standard Diels–Alder reaction? Explain.

In addition to standard Diels–Alder reactions, the reaction rate can be increased by attaching electron-withdrawing groups to the diene and electron-donating groups to the dienophile. In these *inverse electron demand* Diels–Alder reactions, the diene is electron poor and the dienophile is electron rich, opposite to what we saw with standard Diels–Alder reactions. Although inverse electron demand Diels–Alder reactions are quite useful synthetically, we will not discuss them further.

24.4 Stereochemistry of Diels–Alder Reactions

In the prototypical Diels–Alder reaction between buta-1,3-diene and ethene, four carbon atoms rehybridize from sp^2 to sp^3—two from the diene and two from the dienophile (see Your Turn 24.6). These atoms, therefore, have the potential to become chiral centers in the product.

YOUR TURN 24.6

In the prototypical Diels–Alder reaction shown here, identify the C atoms in the product that have rehybridized from sp^2 to sp^3 making them *potential* chiral centers in other Diels–Alder reactions.

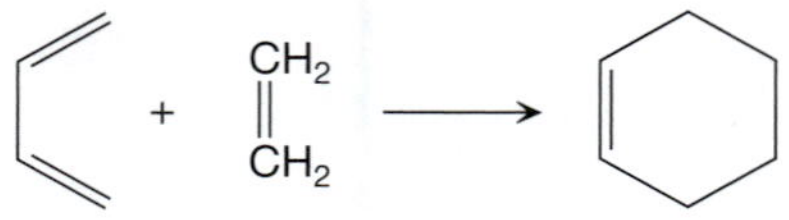

Two chiral centers are produced, for example, in each of the reactions in Equations 24-14 and 24-15.

A cis configuration in the dienophile → A cis configuration in the ring

Meso

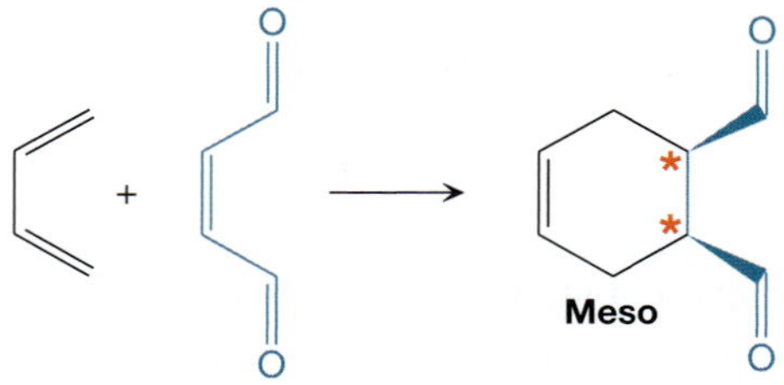

(24-14)

Racemic mixture of enantiomers

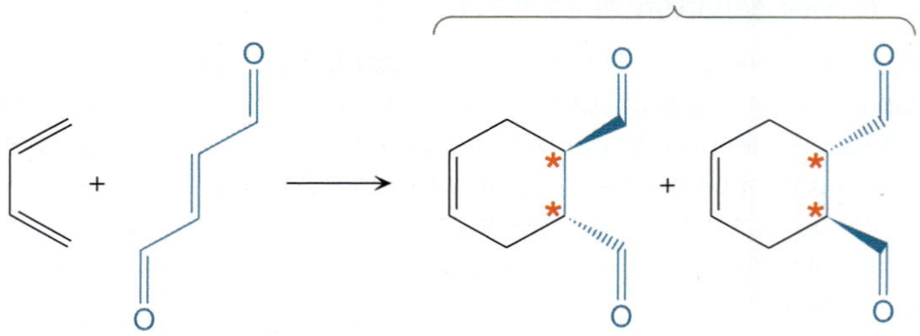

A trans configuration in the dienophile → A trans configuration in the ring

(24-15)

Only the cis configuration (meso) is produced in Equation 24-14, however, whereas only the trans configuration (a racemic mixture of enantiomers) is produced in Equation 24-15. These outcomes are due to the configurations of the C=C double bonds in the dienophile: The configuration is cis in Equation 24-14, but it is trans in Equation 24-15. In other words:

> The stereochemical configuration in the dienophile is conserved throughout the course of a Diels–Alder reaction.

In this regard, Diels–Alder reactions are *stereospecific*: the stereochemistry of the product is dictated by the stereochemistry of the reactants.

PROBLEM 24.9 Predict the product of the Diels–Alder reaction shown here, paying particular attention to the configuration of the C=C double bond from the dienophile.

PROBLEM 24.10 Draw the dienophile that would be required to generate each of the molecules shown here from buta-1,3-diene in a Diels–Alder reaction.

(a)   (b)

Similarly, the stereochemistry in the diene leads to specific stereochemical configurations in the products, as shown in Equations 24-16 through 24-18.

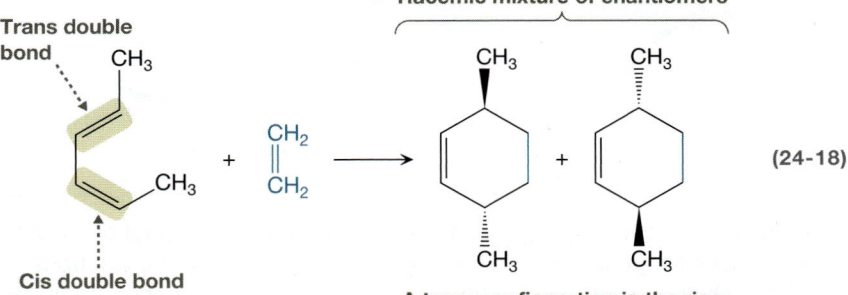

Both double bonds are trans.

$$CH_3 \quad + \quad CH_2{=}CH_2 \quad \longrightarrow \quad CH_3 \qquad (24\text{-}16)$$

Meso
A cis configuration in the ring

Both double bonds are cis.

$$+ \quad CH_2{=}CH_2 \quad \longrightarrow \qquad (24\text{-}17)$$

Meso
A cis configuration in the ring

Racemic mixture of enantiomers

Trans double bond

Cis double bond

$$CH_3 \quad + \quad CH_2{=}CH_2 \quad \longrightarrow \quad CH_3 \quad + \quad CH_3 \qquad (24\text{-}18)$$

A trans configuration in the ring

Namely:

- If the double bonds in the diene are either both trans (Equation 24-16) or both cis (Equation 24-17), then the substituents at the ends of the diene become cis to each other with respect to the plane of the ring in the product.
- Conversely, if one of the diene's double bonds is cis and the other is trans (Equation 24-18), then the two substituents become trans to each other in the product.

We obtain the stereochemical outcomes in Equations 24-14 through 24-18 when the diene and dienophile approach each other as shown in Equation 24-19 (alternatively, the dienophile can approach from above the diene). The Diels–Alder reaction is concerted—the two σ bonds that form at each end of the diene and dienophile are formed *simultaneously*—so the substituents about the dienophile (A, B, C, and D in Equation 24-19) cannot rotate relative to each other to scramble the stereochemistry. This explains our observations in Equations 24-14 and 24-15.

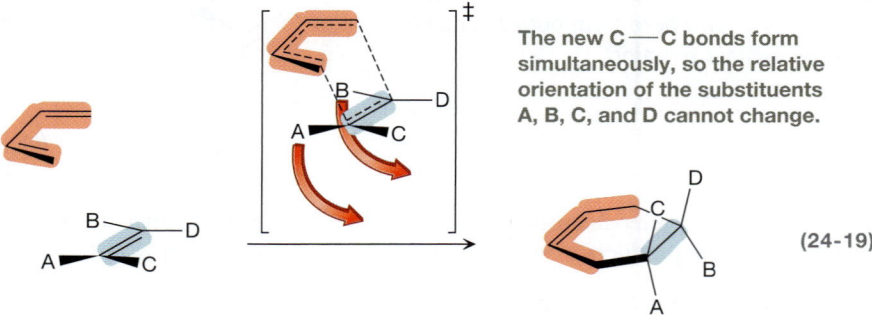

The new C—C bonds form simultaneously, so the relative orientation of the substituents A, B, C, and D cannot change.

(24-19)

The concerted nature of the Diels–Alder reaction also explains the stereochemistry observed in Equations 24-16 through 24-18, as shown in Equation 24-20 for the generic diene with substituents W, X, Y, and Z attached to the terminal carbon atoms. Those terminal carbons must rotate slightly as the new bonds are formed, allowing the six-membered ring to relax into its equilibrium geometry. More specifically, the carbon atoms rotate in *opposite directions* during the course of the reaction (i.e., if viewed from one end of the diene, one carbon rotates clockwise and the other counterclockwise). If substituents W and Z are both CH₃ groups and X and Y are both H atoms, for example, then both CH₃ groups will be cis to each other in the product, just as we saw previously in Equation 24-16.

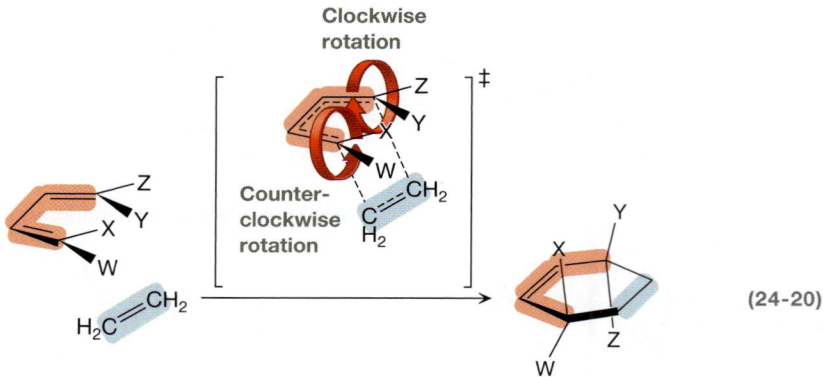

(24-20)

YOUR TURN 24.7

Match the generic groups (W, X, Y, and Z) on the diene in Equation 24-20 to the specific groups on the diene in Equation 24-18. Do the cis/trans relationships among the substituents in the product in Equation 24-20 agree with those in Equation 24-18?

SOLVED PROBLEM 24.11

Predict the product(s) of the Diels–Alder reaction shown here, paying particular attention to stereochemistry.

Think Do the carbons at the ends of the diene become chiral centers in the products? What are the configurations about each double bond in the diene? Do the alkene carbons of the dienophile become chiral centers? What is the configuration about the C═C double bond in the dienophile?

Solve If we draw the product without worrying about stereochemistry, notice that each end carbon in the diene becomes a chiral center, and so does each C=C carbon in the dienophile.

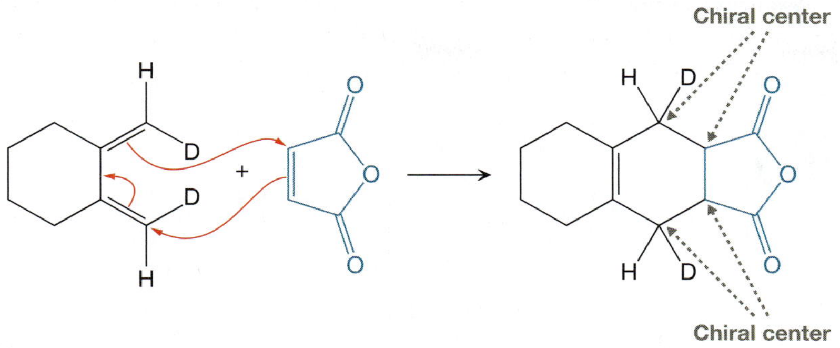

Each D—C=C—C configuration is cis along the diene skeleton, and each H—C=C—C configuration is trans. Therefore, the two D atoms and the two H atoms will both be cis to each other in the product. The C=C bond in the dienophile is in the cis configuration, so the carbonyl groups will be cis to each other in the product, too. Overall, then, there are four results to consider (**A–D**), which, due to the plane of symmetry along the horizontal axis of the molecule, reduce to two distinct stereoisomers (**A** and **B**).

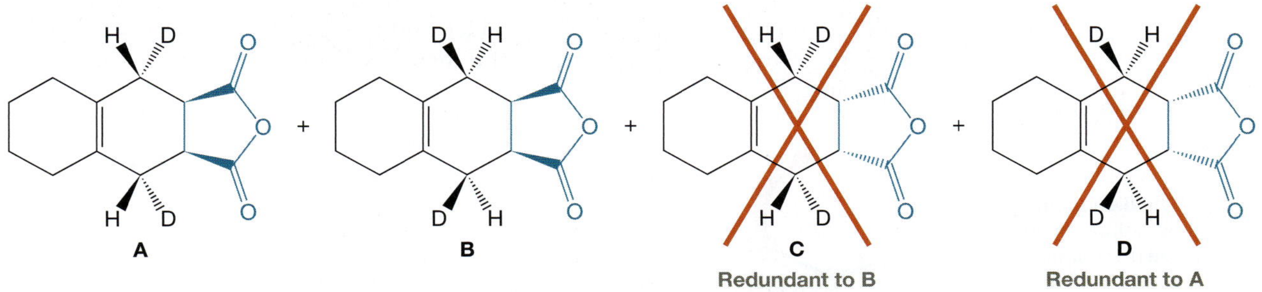

PROBLEM 24.12 Predict the Diels–Alder products for each of the following reactions in which the dienophile is isotopically labeled. Pay particular attention to stereochemistry.

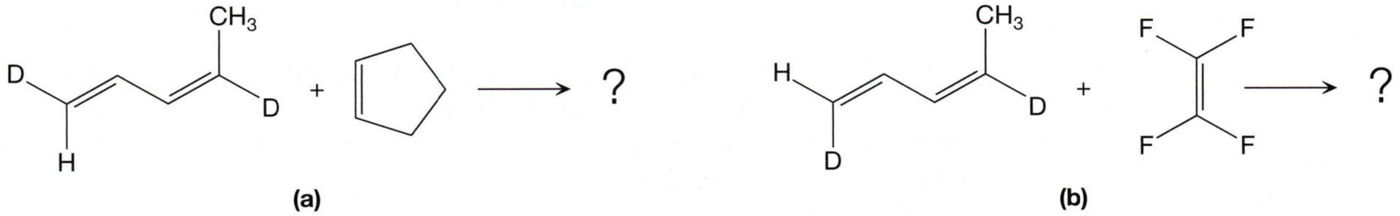

(a)

(b)

PROBLEM 24.13 If the compound shown here is one of two enantiomers produced from a Diels–Alder reaction with ethene, then what is the structure of the diene that would have been required?

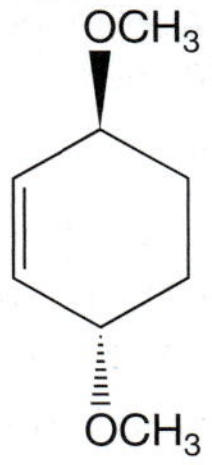

So far, the stereochemical aspects we have considered for the Diels–Alder reaction deal with the diene and dienophile independently. Another aspect of stereochemistry to consider pertains to the configurations of the carbon atoms from the diene *relative* to those from the dienophile. This is shown in Equation 24-21, in which a mixture of *diastereomers* is produced.

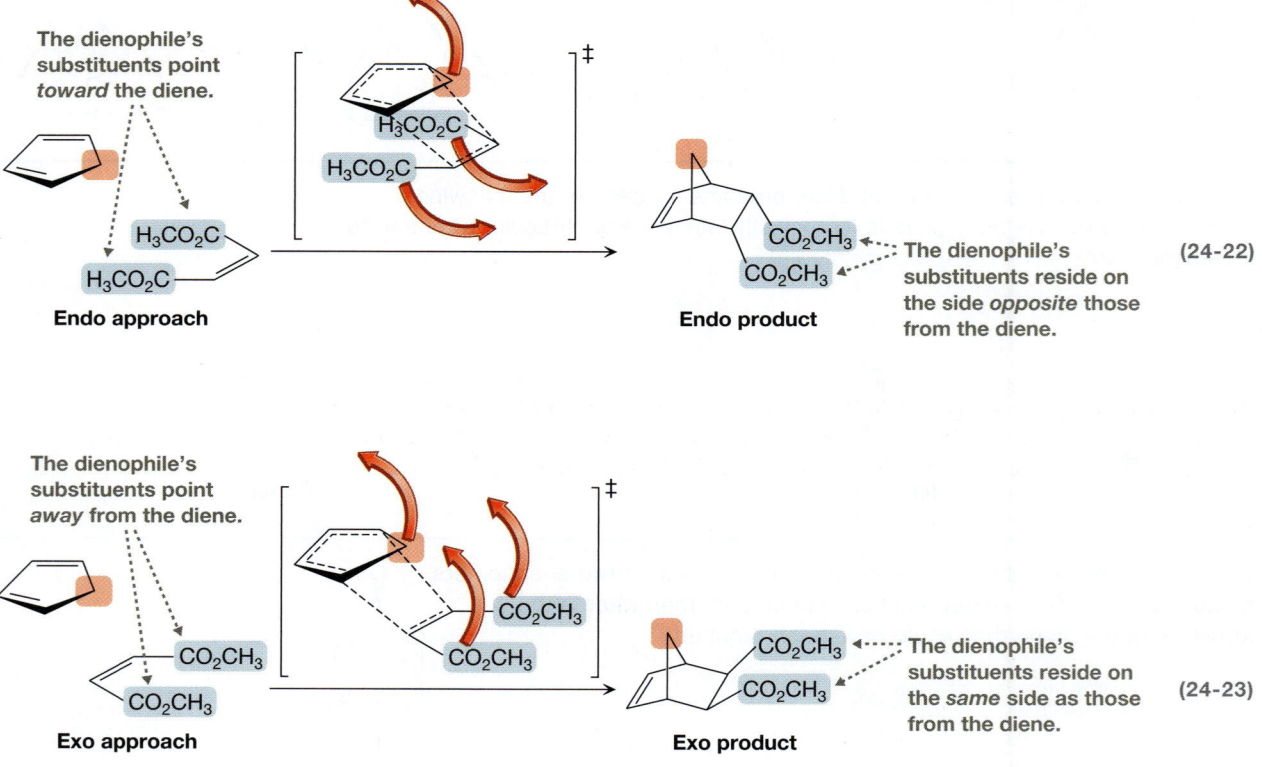

(24-21)

The endo product dominates.

Exo product
3%

Endo product
97%

Equations 24-22 and 24-23 show that these two products result from the two different approaches possible for the dienophile—one with the substituents pointed toward the diene and the other with it pointed away. When the dienophile's substituents point toward the diene (Equation 24-22), then it is called an **endo approach**, and the resulting product is called the **endo product**. Conversely, when the dienophile's substituents point away from the diene (Equation 24-23), then it is called an **exo approach**, and the resulting product is the **exo product**. In essence, if the dienophile's substituents (highlighted in blue) reside opposite the diene's substituents (highlighted in red) on the six-membered ring that is produced, then you have the endo product, and if the dienophile's substituents reside on the same side of the ring as the diene's substituents, then you have the exo product.

The dienophile's substituents point *toward* the diene.

Endo approach

Endo product

The dienophile's substituents reside on the side *opposite* those from the diene.

(24-22)

The dienophile's substituents point *away* from the diene.

Exo approach

Exo product

The dienophile's substituents reside on the *same* side as those from the diene.

(24-23)

Although both endo and exo approaches are feasible, the product mixture is usually dominated by one product, according to the **endo rule** (also called the **Alder rule**):

In a Diels–Alder reaction, the endo approach is generally favored over the exo approach, so the major product is usually endo.

This outcome can depend on temperature, however, as illustrated in Problem 24.47 at the end of the chapter.

Understanding the typical preference for the endo approach requires molecular orbitals, so it is discussed in Section 24.10.

PROBLEM 24.14 Draw the major product of the Diels–Alder reaction shown here, paying particular attention to stereochemistry.

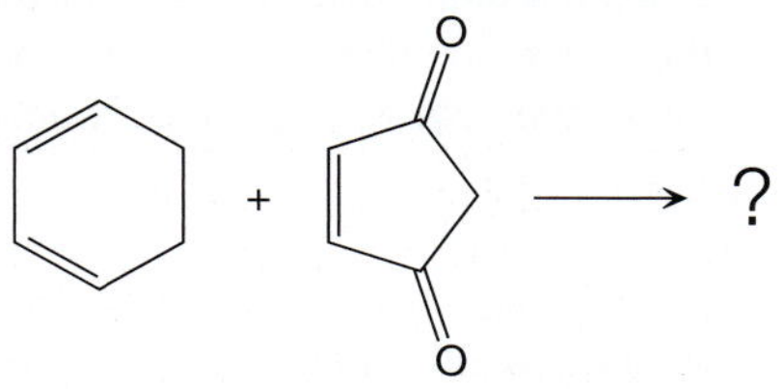

24.5 Regiochemistry of Diels–Alder Reactions

If both the diene and the dienophile of a Diels–Alder reaction are *unsymmetric,* then two constitutional isomers can be produced. An example is shown in Equation 24-24.

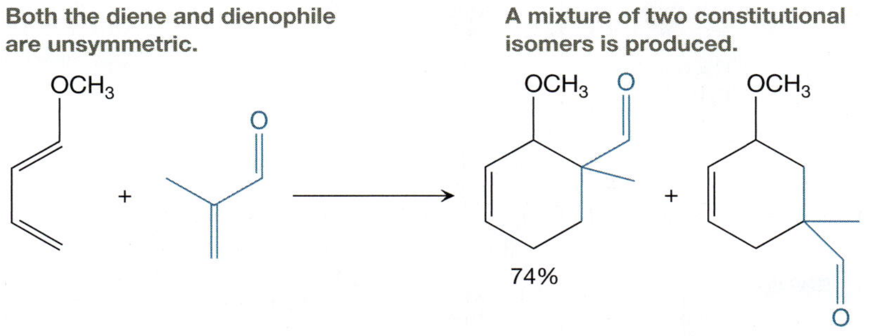

Both the diene and dienophile are unsymmetric.

A mixture of two constitutional isomers is produced.

74%

(24-24)

The isomers are produced from different relative orientations of the two reactants when they approach each other, as shown in Equations 24-25 and 24-26. The reaction is *regioselective,* and the major product is produced via the approach in Equation 24-25.

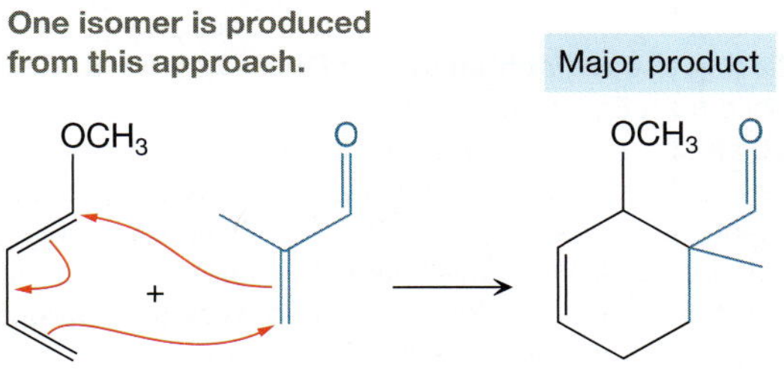

One isomer is produced from this approach.

Major product

(24-25)

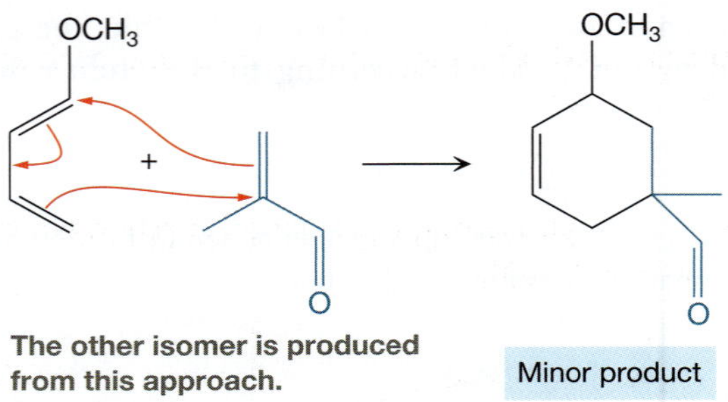

$$(24\text{-}26)$$

The other isomer is produced
from this approach.

Minor product

One way to understand this regiochemistry is to draw the resonance hybrids of the diene and dienophile and apply the following rule:

> When the diene and the dienophile of a Diels–Alder reaction are both unsymmetric, the major isomeric product is the one produced by the approach that exhibits the more favorable electrostatic attraction among atoms undergoing bond formation.

The resonance structures and the resonance hybrids are shown in Figure 24-4a and their two possible approaches are shown in Figure 24-4b. With the approach shown at the top of Figure 24-4b, there is a favorable electrostatic interaction between the atoms undergoing bond formation. The approach shown at the bottom of Figure 24-4b, how–

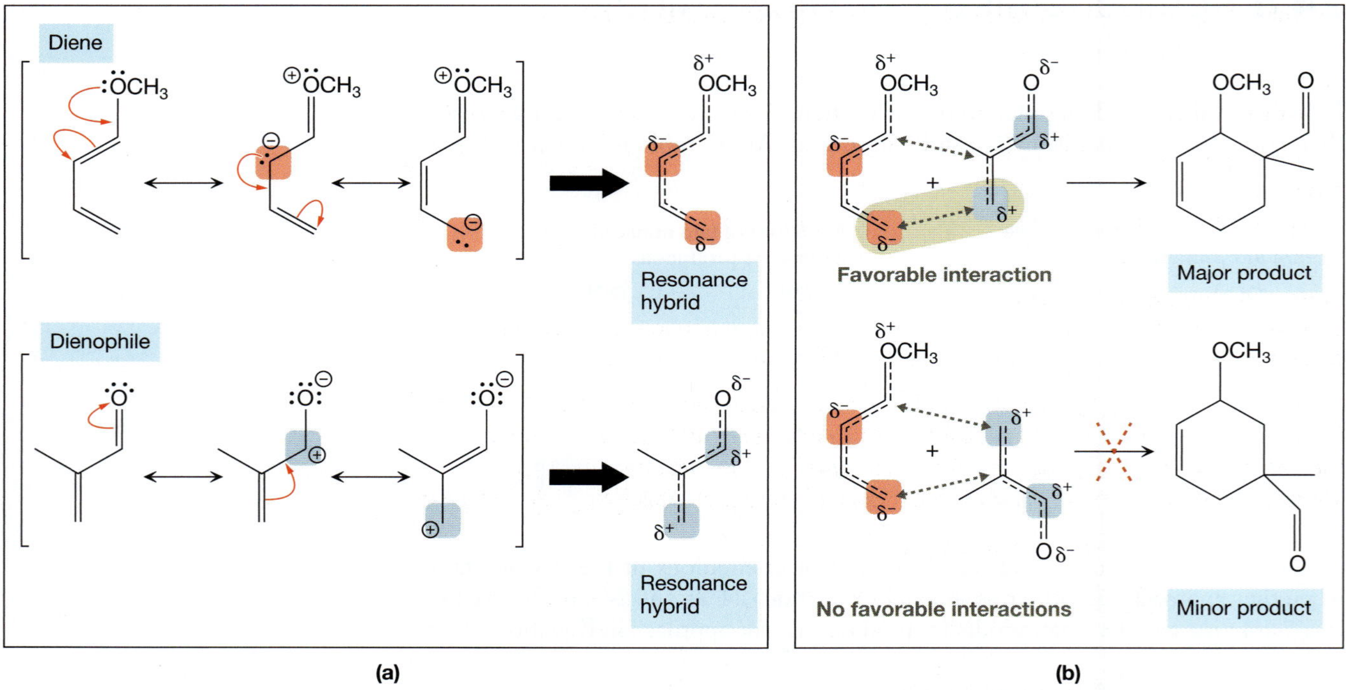

(a)

(b)

FIGURE 24-4 **Predicting the regiochemistry in Diels–Alder reactions** (a) Resonance in the diene and dienophile from Equation 24-24. The resonance structures and the resonance hybrid of the diene are shown at the top, with regions of excess negative charge highlighted in red. The resonance structures and the resonance hybrid of the dienophile are shown at the bottom, with regions of excess positive charge highlighted in blue. (b) Interactions between the diene and dienophile. With the approach shown at the top, there is a favorable electrostatic interaction between atoms forming a bond. With the approach shown at the bottom, there are no such favorable electrostatic interactions, so the approach at the top leads to the major product.

ever, does *not* exhibit such favorable interactions, so the approach at the top of the figure leads to the major product.

The following resonance structures and resonance hybrid are of a diene that is an isomer of the one in Figure 24-4.

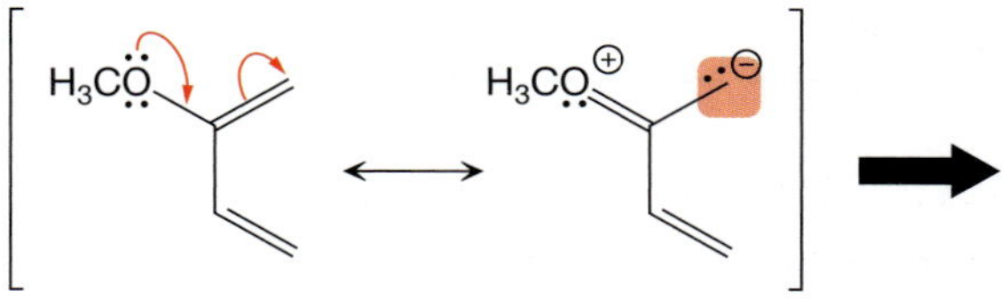

A and **B** represent the two ways this diene can approach the dienophile from Figure 24-4.

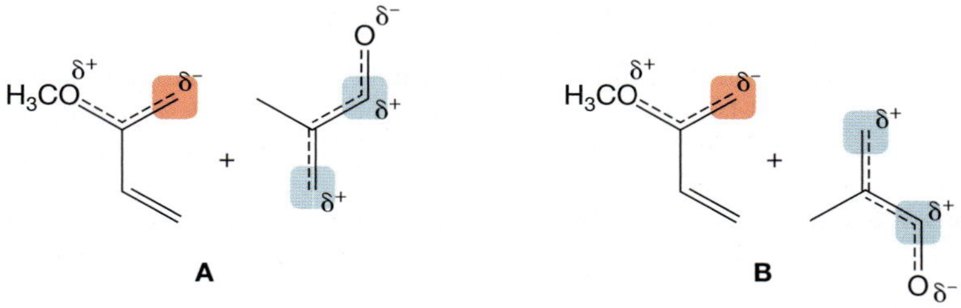

A   B

Which of these interactions will lead to the major Diels–Alder product?

SOLVED PROBLEM 24.15

Draw the mechanism and predict the major product for the Diels–Alder reaction shown here.

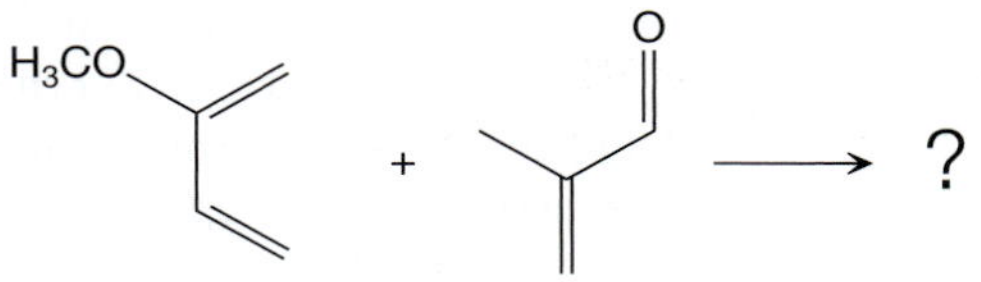

Think Where are the regions of excess negative charge and excess positive charge in these molecules? Which approach leads to favorable electrostatic interactions among the atoms undergoing bond formation?

Solve The resonance hybrids of these two species are shown in Your Turn 24.8 above. Interaction **B** in Your Turn 24.8 has a favorable electrostatic interaction among the carbon atoms undergoing bond formation, whereas interaction **A** does not. Therefore, the major product is from interaction **B**, yielding the following major product.

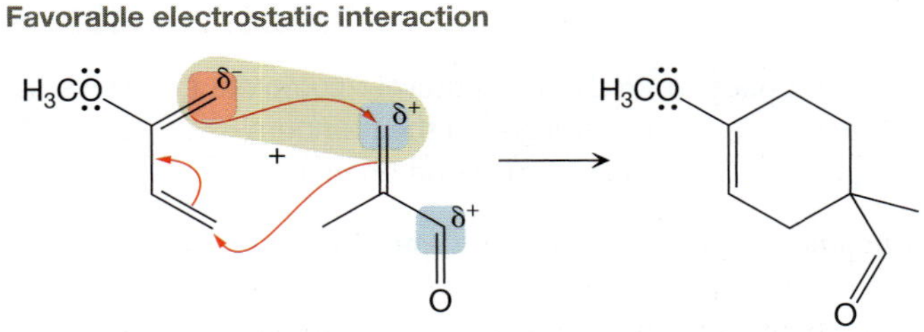

Favorable electrostatic interaction

Draw the mechanism that leads to the minor product of the Diels–Alder reaction in Solved Problem 24.15.

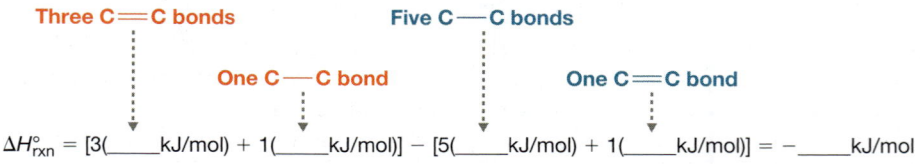

(a)   (b)

24.6 The Reversibility of Diels–Alder Reactions; the Retro Diels–Alder Reaction

Under normal conditions, the prototypical Diels–Alder reaction in Equation 24-1 (shown again in Equation 24-27) has a substantially negative change in standard free energy ($\Delta G_{rxn}^{\circ} = -113$ kJ/mol $= -27$ kcal/mol), giving it a very large equilibrium constant ($K_{eq} = 3.5 \times 10^{19}$). As a result, this reaction is *irreversible* (review Section 9.12).

Loss of 3 π bonds   **Gain of 1 π bond and 2 σ bonds**

$$\Delta G_{rxn}^{\circ} = -113 \text{ kJ/mol}$$
$$= -27 \text{ kcal/mol}$$
$$K_{eq} = 3.5 \times 10^{19}$$

(24-27)

As with any reaction, ΔG_{rxn}° can be broken down into a ΔH_{rxn}° term and a $T\Delta S_{rxn}^{\circ}$ term (i.e., $\Delta G_{rxn}^{\circ} = \Delta H_{rxn}^{\circ} - T\Delta S_{rxn}^{\circ}$). For the Diels–Alder reaction in Equation 24-27, $\Delta H_{rxn}^{\circ} = -168$ kJ/mol (-40 kcal/mol) and $T\Delta S_{rxn}^{\circ} = -55$ kJ/mol (-13 kcal/mol) at 25 °C. Thus, the large, negative ΔH_{rxn}° term dominates to make ΔG_{rxn}° substantially negative.

The large negative ΔH° for this reaction stems primarily from how the bond energies in the product differ from those in the reactants. Notice in Equation 24-27 that there are three C=C double bonds and one C—C single bond in the reactants, whereas there are five C—C single bonds and one C=C double bond in the product. In essence, then, three π bonds are converted into two σ bonds and one π bond, and, as we learned in Chapter 3, σ bonds are stronger than their corresponding π bonds.

YOUR TURN 24.10

To estimate ΔH_{rxn}° in Equation 24-27 quantitatively, complete and solve the following equation, which accounts for the changes in the numbers of C—C and C=C bonds. Use the average C—C and C=C bond energies in Table 1-3 (p. 10).

Three C=C bonds   **Five C—C bonds**

One C—C bond   **One C=C bond**

$\Delta H_{rxn}^{\circ} = [3(\underline{\quad}\text{kJ/mol}) + 1(\underline{\quad}\text{kJ/mol})] - [5(\underline{\quad}\text{kJ/mol}) + 1(\underline{\quad}\text{kJ/mol})] = -\underline{\quad}\text{kJ/mol}$

How does this estimate of ΔH_{rxn}° compare to the experimentally measured value of -168 kJ/mol (-40 kcal/mol)?

The entropy term in a Diels–Alder reaction is negative because the reaction decreases the number of independent molecules (i.e., the reaction leads to less "disorder"). This means that increasing the temperature makes ΔG°_{rxn} less negative (because $\Delta G^{\circ}_{rxn} = \Delta H^{\circ}_{rxn} - T\Delta S^{\circ}_{rxn}$). Even at 200 °C, however, the prototypical reaction in Equation 24-27 is not easily reversible due to its large, negative ΔH°_{rxn}. The Diels–Alder reaction in Equation 24-28, on the other hand, has a significantly smaller value of ΔH°_{rxn} (-77 kJ/mol $= -18$ kcal/mol), so it is easily reversible at moderately high temperatures.

This Diels–Alder reaction is *reversible*.

$$\Delta G^{\circ}_{rxn} = -17 \text{ kJ/mol}$$
$$= -4 \text{ kcal/mol}$$
(24-28)
$$K_{eq} = 120$$

(150 °C)

Cyclopentadiene   **Dicyclopentadiene**

YOUR TURN **24.11**

Supply the curved arrows missing from Equation 24-28.

YOUR TURN **24.12**

For the Diels–Alder reaction shown here, $\Delta H^{\circ}_{rxn} = -121$ kJ/mol (-29 kcal/mol). Do you think this reaction would be easily reversible at 150 °C? *Hint:* Compare ΔH°_{rxn} of this reaction to the one in Equation 24-27.

PROBLEM 24.17 What temperature would be required to achieve $K_{eq} = 120$ for the prototypical Diels–Alder reaction in Equation 24-27? (The values of ΔH° and ΔS° can be obtained from the text.)

Chemists take advantage of the reversibility of the Diels–Alder reaction in Equation 24-28 to prepare a fresh sample of cyclopentadiene. At room temperature, K_{eq} heavily favors the dicyclopentadiene product, and over about one week's time, cyclopentadiene *dimerizes* entirely into dicyclopentadiene. When a sample of dicyclopentadiene is heated to 150 °C or higher, the reaction is driven in the reverse direction, which allows the low-boiling cyclopentadiene (bp = 42 °C) to be removed and collected via distillation, whereas the higher-boiling dicyclopentadiene (bp = 170 °C) remains behind. Le Châtelier's principle continues to drive the reaction toward the side that has cyclopentadiene—that is, in the reverse direction.

Because this process "undoes" the Diels–Alder reaction in Equation 24-28, it is called a **retro Diels–Alder reaction** or a **[4+2] cycloelimination**. The mechanism for this reaction is simply the reverse of a Diels–Alder mechanism, so the entire reaction consists of a single (concerted) step in which six electrons move in a cyclic fashion, as shown in Equation 24-29. Notice that the three curved arrows are all

CONNECTIONS
Cyclopentadiene's conjugate base, the cyclopentadienyl anion (Cp^-), is a valuable ligand for organometallic complexes. In many cases, these complexes can be used as catalysts in reactions to produce pharmaceuticals or other organic compounds. Zirconocene dichloride (below), for example, can catalyze the polymerization of ethylene.

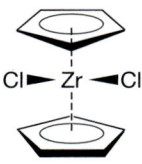

oriented in the same direction about the six-membered ring, and one curved arrow originates from the double bond. These curved arrows were drawn clockwise, but as we saw with cycloaddition reactions, it would be equally correct if they were drawn counterclockwise.

$$\text{Dicyclopentadiene} \xrightarrow[\text{200 °C}]{\text{Retro Diels–Alder}} \text{Cyclopentadiene} + \qquad (24\text{-}29)$$

Dicyclopentadiene **Cyclopentadiene**
70%

YOUR TURN 24.13

CONNECTIONS Dicyclopentadiene (Equation 24-29) can be polymerized to produce polydicyclopentadiene, which is used as an alternative to fiberglass/polyester composites in the manufacture of heavy-vehicle exterior components, such as this one.

CONNECTIONS (2R,3S)-2,3-Dihydroxybutanoic acid is a naturally occurring metabolite in humans.

When cyclohexene is heated to *very* high temperatures, the reverse of the reaction in Equation 24-27 takes place, producing ethene and buta-1,3-diene. Draw the curved arrows for this reaction.

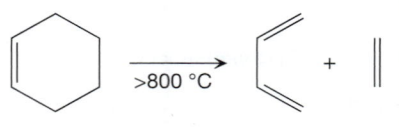

PROBLEM 24.18 When the compound shown here is heated, a single product with the formula C_6H_8 is produced. Draw the product and the curved arrows necessary to produce it.

$$\xrightarrow{\Delta} \quad C_6H_8$$

24.7 Syn Dihydroxylation of Alkenes and Alkynes Using OsO_4 or $KMnO_4$

What distinguishes the Diels–Alder reaction from other reactions we have studied previously is the cyclic movement of six electrons in a concerted fashion. Another reaction that incorporates a similar concerted elementary step is the **syn dihydroxylation** of alkenes and alkynes. An example is shown in Equation 24-30, in which two OH groups appear to have added across the C=C bond in a syn fashion.

Two OH groups add across the C=C bond in a syn fashion.

$$\xrightarrow[\substack{H_2O_2, (CH_3)_3COH, \\ 0\ °C, 16\ h}]{OsO_4} \qquad (24\text{-}30)$$

(E)-But-2-enoic acid   **(2S,3R)- and (2R,3S)-2,3-Dihydroxybutanoic acid**
53%

A partial mechanism for this reaction is shown in Equation 24-31.

Step 1 is the cycloaddition of OsO_4 across the C=C double bond in a syn fashion, producing an **osmate ester.** Then, over the course of several steps, the O—Os bonds are replaced by O—H bonds, thus producing a cis-1,2-diol and a reduced form of the osmium. Hydrogen peroxide, H_2O_2, oxidizes the reduced form of osmium back to OsO_4, making OsO_4 a *catalyst,* because it is not consumed overall.

Step 1 is key in this mechanism, because it attaches the two O atoms in a syn fashion to the initial alkene C atoms. This step involves the concerted movement of six

Both C—O bonds form on the same face of the initial alkene.

1. [4+2] Cycloaddition

Several steps

OsO$_4$ is regenerated.

$$\left(+ H_2OsO_4 \xrightarrow[\text{Oxidation}]{H_2O_2} OsO_4 \right)$$

(+ Enantiomer)

Osmate ester

(+ Enantiomer)

(24-31)

electrons in a cyclic fashion, so all bond formation and all bond breaking occur simultaneously; it is, therefore, a type of *pericyclic reaction*. In particular, the step involves the flow of four electrons from OsO$_4$ and two from the alkene, resulting in a new ring, so it is a [4+2] cycloaddition, much like the Diels–Alder reaction. Moreover, the concerted movement of the electrons requires the new bonds to the alkene to form on the same side (also like the Diels–Alder reaction).

Unlike in a Diels–Alder reaction, the electron movement in this [4+2] cycloaddition involves just five atoms—three atoms from one of the reacting species (in this case, OsO$_4$) and two atoms from the other (in this case, the alkene). As a result, two of the six electrons moving cyclically end up as a lone pair on the middle of the three atoms contributed from one reacting species (in this case, the Os atom from OsO$_4$).

An older method of producing cis-1,2-diols involves treating an alkene with a cold, basic solution of potassium permanganate (KMnO$_4$), as shown in Equation 24-32. It is not very synthetically useful, however, because the yields are typically low and the diol product can be further oxidized by KMnO$_4$ (see Section 24.8).

OH groups add across the C=C double bond in a syn fashion.

(24-32)

Cyclopentene

cis-**Cyclopentane-1,2-diol**
31%

GREEN CHEMISTRY OsO$_4$ is toxic and, when used as a heterogeneous catalyst in syn dihydroxylation reactions, can contaminate the products. One green alternative is to use elemental osmium confined inside a zeolite (a porous inorganic mineral) as the catalyst, with hydrogen peroxide as a co-oxidant. The zeolite-confined catalyst is easily removed and reused.

Zeolite–Os^0, H$_2$O$_2$

Acetone, H$_2$O

87%

YOUR TURN 24.14

The mechanism for the reaction in Equation 24-32 is believed to be similar to the one for OsO$_4$. The key step is the [4+2] cycloaddition of MnO$_4^-$ across the C=C double bond, making a *manganate ester*. Supply the curved arrows that are necessary to depict the formation of the manganate ester from the starting materials.

Hydrolysis

A manganate ester

GREEN CHEMISTRY Even though KMnO$_4$ is not as synthetically useful as OsO$_4$, KMnO$_4$ is much less toxic and less expensive, too. Therefore, when yield is not a high priority, KMnO$_4$ might be considered to carry out syn dihydroxylation rather than OsO$_4$.

The C≡C triple bond of an alkyne reacts with $KMnO_4$, too, generating a 1,2-dicarbonyl compound (not a 1,2-diol), as shown in Equation 24-33.

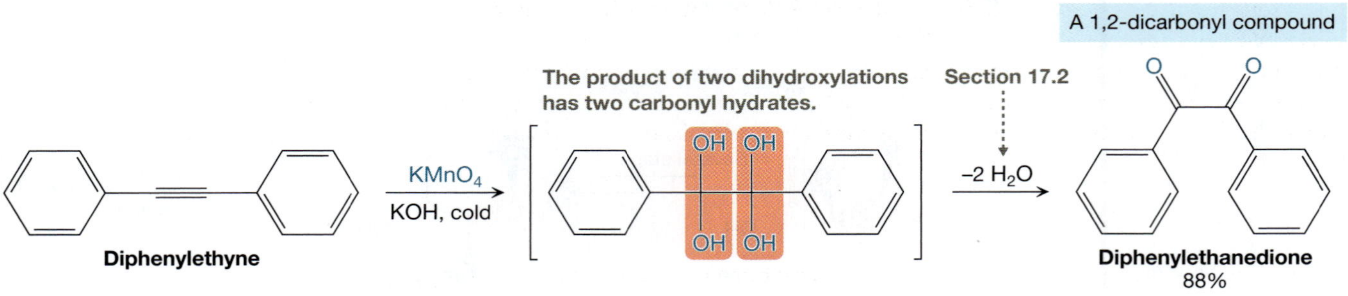

A 1,2-dicarbonyl compound

The product of two dihydroxylations has two carbonyl hydrates.   Section 17.2

Diphenylethyne $KMnO_4$ / KOH, cold   −2 H_2O **Diphenylethanedione** 88%

(24-33)

We can envision the alkyne as having undergone two syn dihydroxylations to yield a 1,1,2,2-tetraol intermediate. This intermediate has two adjacent hydrates (each characterized by two OH groups attached to the same carbon atom). Recall from Section 17.2 that hydrates like this equilibrate with their C=O forms through the loss of H_2O, resulting in the dicarbonyl product.

PROBLEM 24.19 Which of the following diols can be produced from an alkene using either OsO_4 or $KMnO_4$? For each one that can, draw the alkene that can be used to produce it.

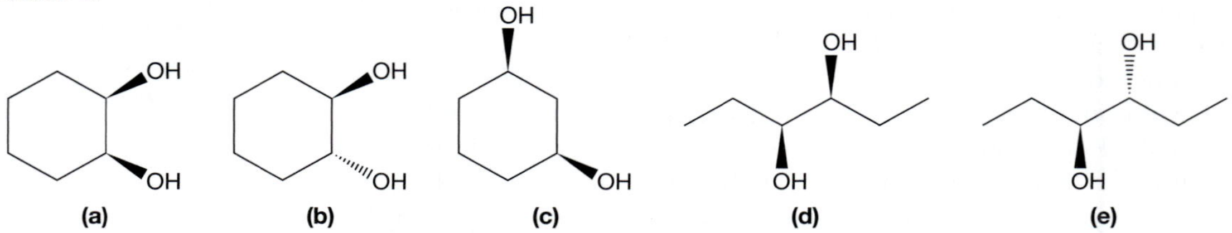

(a)   (b)   (c)   (d)   (e)

24.8 Oxidative Cleavage of Alkenes and Alkynes

Here we introduce a variety of **oxidative cleavage** reactions, whereby the carbon–carbon bond of an alkene or alkyne is broken and the initial alkene or alkyne carbons are oxidized in the process. The general process is illustrated in Equations 24-34 and 24-35, but, as we will see, oxidative cleavage reactions can generate ketones, aldehydes, carboxylic acids, or carbon dioxide, depending on the reactants, the reagents used, and the exact conditions of the reaction.

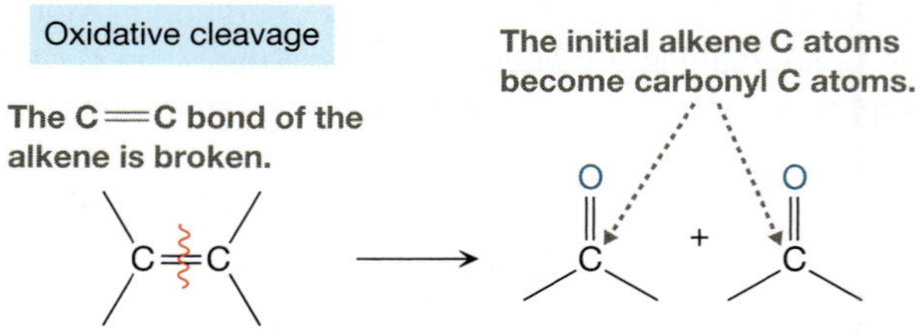

Oxidative cleavage

The C=C bond of the alkene is broken.

The initial alkene C atoms become carbonyl C atoms.

(24-34)

The C≡C bond of the alkyne is broken.

The initial alkyne C atoms become carboxyl C atoms.

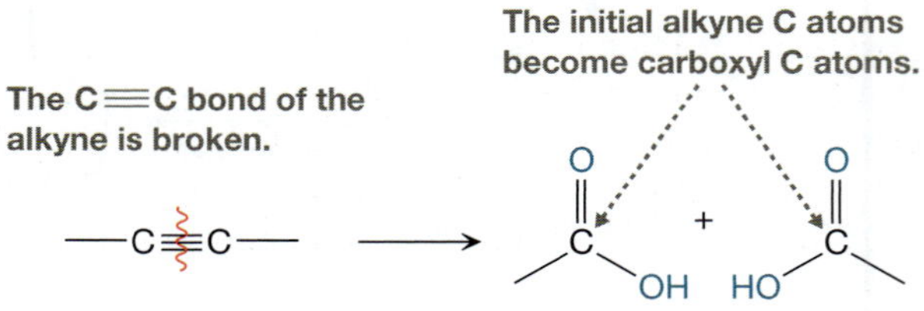

(24-35)

We examine oxidative cleavage reactions involving three different reagents: potassium permanganate ($KMnO_4$; Section 24.8a), the periodate anion (IO_4^-; Section 24.8b), and ozone (O_3; Section 24.8c). The mechanisms for these reactions involve at least one pericyclic step similar to the ones we have encountered thus far. Therefore, whereas the structural changes that take place in oxidative cleavage reactions might seem to differ significantly from what we have seen in the Diels–Alder and syn dihydroxylation reactions, all of these reactions are related through their mechanisms.

24.8a Oxidative Cleavage Involving $KMnO_4$

In Section 24.7, we saw that the C=C double bond of an alkene undergoes syn dihydroxylation when treated with cold (often <0 °C), basic $KMnO_4$. When the $KMnO_4$ solution is concentrated and the reaction mixture is heated significantly above room temperature, however, then *oxidative cleavage* of the C=C bond takes place, as shown in Equation 24-36.

The C=C bond is cleaved.

1. conc $KMnO_4$, KOH, H_2O, 50 °C
2. HCl

(24-36)

1-Methylcyclohexene

6-Oxoheptanoic acid
61%

The mechanism for this reaction is shown in Equation 24-37.

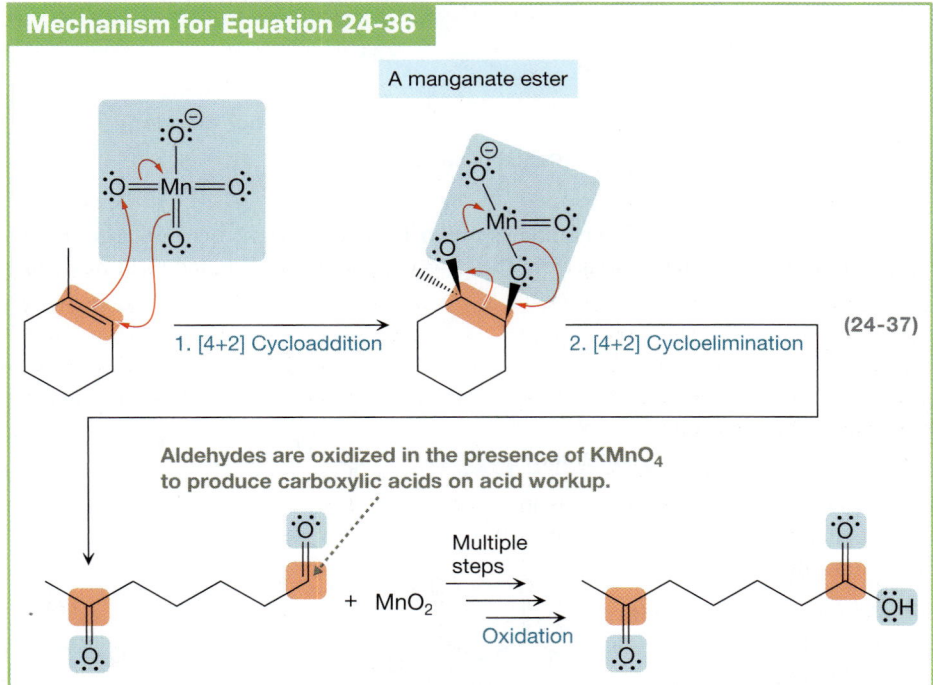

Mechanism for Equation 24-36

A manganate ester

1. [4+2] Cycloaddition
2. [4+2] Cycloelimination

(24-37)

Aldehydes are oxidized in the presence of $KMnO_4$ to produce carboxylic acids on acid workup.

+ MnO_2

Multiple steps

Oxidation

Just like the dihydroxylation reaction (Your Turn 24.14), MnO_4^- undergoes a [4+2] cycloaddition to the double bond in Step 1, yielding a manganate ester. The added heat, however, provides enough energy to break the C—C bond in Step 2 via another cyclic rearrangement of six electrons, thus eliminating MnO_2. Step 2, more specifically, is a [4+2] cycloelimination, similar to the retro Diels–Alder reaction we saw in Section 24.6.

At this stage of the reaction, the initial alkene C atoms have already been incorporated into carbonyl groups—one characteristic of a ketone and the other characteristic of an aldehyde. Under these oxidation conditions, however, an aldehyde reacts further to produce a carboxylic acid that is rapidly deprotonated under the basic conditions to produce a carboxylate anion, RCO_2^- (Section 19.6b). Subsequent acid workup produces the final carboxylic acid.

When one of the alkene carbons is part of a CH_2 group (i.e., when the alkene is terminal), then cleavage of the C=C double bond initially produces formaldehyde, $H_2C=O$, as shown in Equation 24-38. Further oxidation in the presence of basic $KMnO_4$, followed by acid workup, produces carbonic acid, which degrades into H_2O and CO_2.

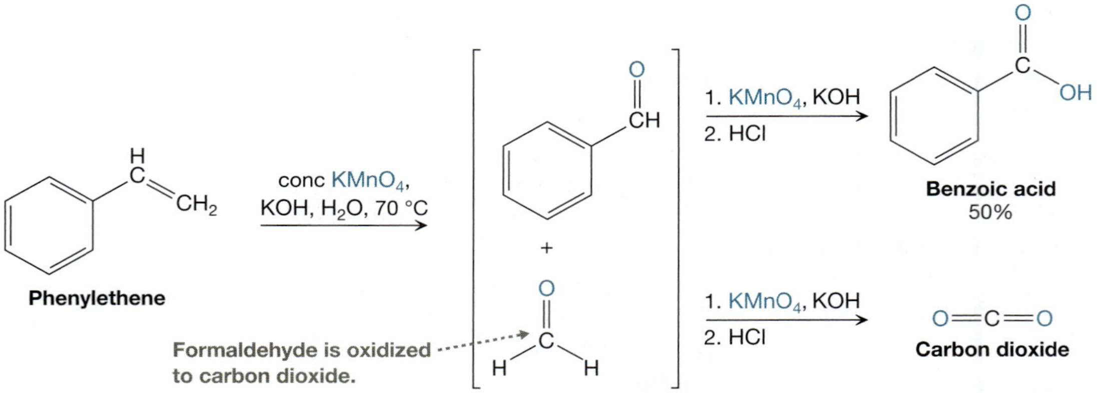

Phenylethene

conc $KMnO_4$, KOH, H_2O, 70 °C

Formaldehyde is oxidized to carbon dioxide.

1. $KMnO_4$, KOH
2. HCl

Benzoic acid
50%

1. $KMnO_4$, KOH
2. HCl

O=C=O
Carbon dioxide

(24-38)

Alkynes are also susceptible to oxidative cleavage with $KMnO_4$. In this case, though, oxidative cleavage followed by acid workup produces only carboxylic acids and (in the case of a terminal alkyne) CO_2, as shown in Equation 24-39.

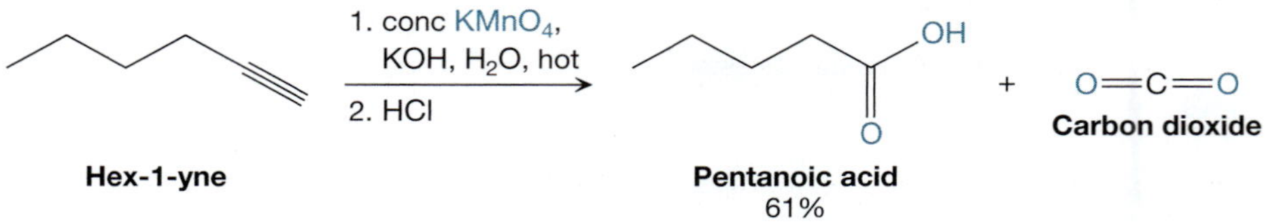

Hex-1-yne

1. conc $KMnO_4$, KOH, H_2O, hot
2. HCl

Pentanoic acid
61%

+ O=C=O
Carbon dioxide

(24-39)

Ketones cannot be formed from alkynes under these conditions because, after cleavage, each of the original C atoms from the C≡C bond can be bonded to, at most, one alkyl group.

SOLVED PROBLEM 24.20

A hydrocarbon having the formula C_8H_{14} is treated with hot, basic, concentrated $KMnO_4$, followed by acid workup. Bubbling is observed and the compounds shown at the right are recovered. What is a possible structure for the original hydrocarbon?

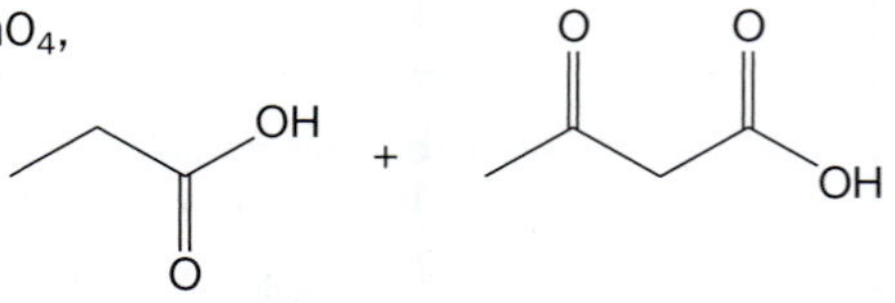

Think Is the number of C atoms in the product compounds the same as in the reactant? What could give rise to the bubbling? What functional group is the precursor to a CO_2H group under oxidative cleavage conditions?

Solve There are a total of seven C atoms in the two compounds that were recovered, whereas there are eight C atoms in the original hydrocarbon. The other C atom is lost in the form of CO_2, which bubbles out of solution.

Thinking backward, the two CO_2H functional groups must have come from the initial formation of HC=O groups, and CO_2 must have come from formaldehyde ($H_2C=O$).

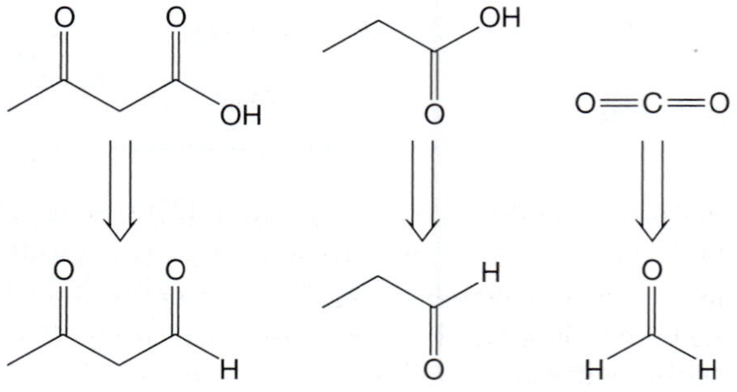

Thinking backward again, each pair of C=O bonds must have come from the cleavage of a C=C double bond. Therefore, a possible structure for the original hydrocarbon is as shown below.

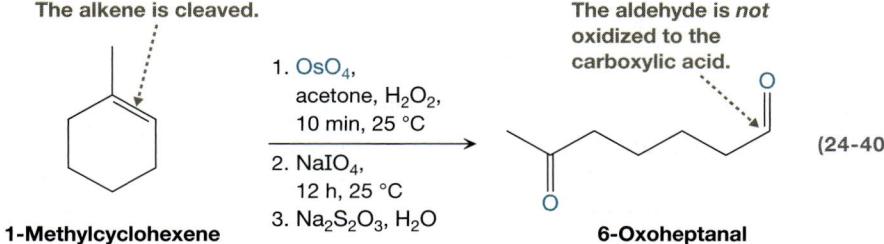

Undo oxidative cleavage.

PROBLEM 24.21 What is another possible structure for the original hydrocarbon in Solved Problem 24.20?

24.8b Oxidative Cleavage Involving IO_4^-

An alkene can also be cleaved oxidatively by treatment with OsO_4, followed by the **periodate anion**, IO_4^-. The periodate anion is most commonly introduced as $NaIO_4$ (Equation 24-40), but it can also be introduced in the form of **periodic acid**, HIO_4 (pronounced per-eye-OH-dik), which produces an equilibrium amount of IO_4^-.

The alkene is cleaved.

1. OsO_4,
   acetone, H_2O_2,
   10 min, 25 °C

2. $NaIO_4$,
   12 h, 25 °C

3. $Na_2S_2O_3$, H_2O

The aldehyde is *not* oxidized to the carboxylic acid.

(24-40)

1-Methylcyclohexene   **6-Oxoheptanal**

Unlike oxidative cleavage involving $KMnO_4$, however:

> Aldehydes that are produced on oxidative cleavage involving IO_4^- are *not* oxidized further.

A partial mechanism for this reaction is shown in Equation 24-41.

Partial mechanism for Equation 24-40

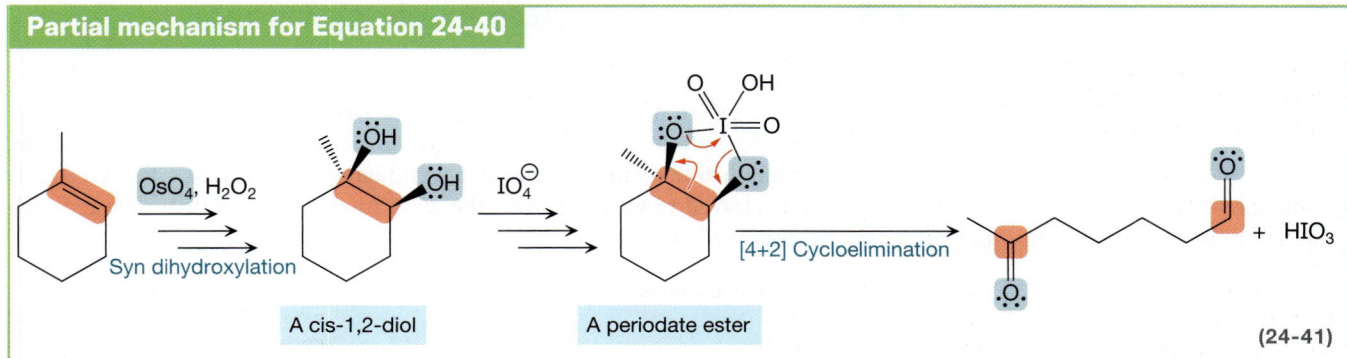

OsO_4, H_2O_2

Syn dihydroxylation

A cis-1,2-diol

IO_4^-

[4+2] Cycloelimination

A periodate ester

+ HIO_3

(24-41)

The first several steps make up the syn dihydroxylation mechanism from Equation 24-31 (Section 24.7), thus producing a cis-1,2-diol. Then, the addition of IO_4^- produces a cyclic **periodate ester**. The final step in the mechanism is key, because it cleaves the C—C bond and produces the two carbonyl groups. This step is a [4+2] cyclo-elimination, much like Step 2 in Equation 24-37 (p. 1221).

Notice in Equation 24-41 that the mechanism proceeds through a 1,2-diol. Therefore:

> Periodate oxidative cleavage can be carried out directly on a 1,2-diol.

Examples are shown in Equations 24-42 and 24-43.

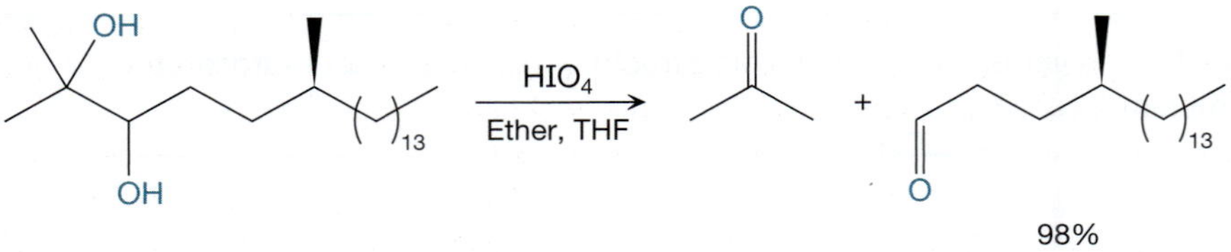

OH
OH
13
HIO_4
Ether, THF
O
+
O
13
98%

(24-42)

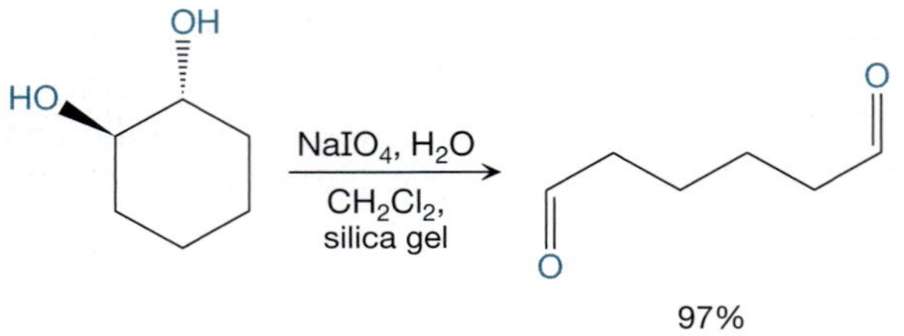

OH
HO
$NaIO_4$, H_2O
CH_2Cl_2,
silica gel
O
O
97%

(24-43)

A variety of 1,2-diols can undergo this reaction, the major requirement being the ability of the 1,2-diol to produce a cyclic periodate ester as in Equation 24-41. (See Problem 24.22.)

PROBLEM 24.22 Explain why **A** will undergo periodate oxidative cleavage but **B** will not. Draw the product of the reaction between **A** and HIO_4, including stereochemistry. *Hint:* It is not due to steric hindrance involving HIO_4.

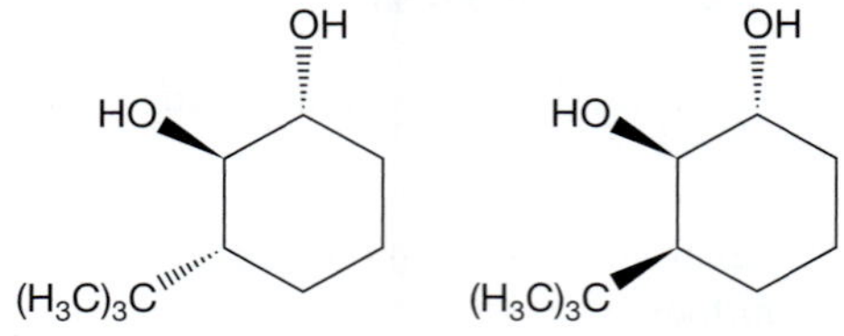

OH
HO
$(H_3C)_3C$
OH
HO
$(H_3C)_3C$

A **B**

24.8c Oxidative Cleavage Involving Ozone: Ozonolysis

Oxidative cleavage with **ozone**, O_3, a process called **ozonolysis**, is shown in Equation 24-44 for 1-methylcyclohexene. Treatment with ozone is followed by dimethylsulfide, CH_3SCH_3, though zinc metal in acetic acid could be used instead of CH_3SCH_3. The double bond is cleaved to yield precisely the same product as oxidative cleavage with IO_4^-, shown previously in Equation 24-40.

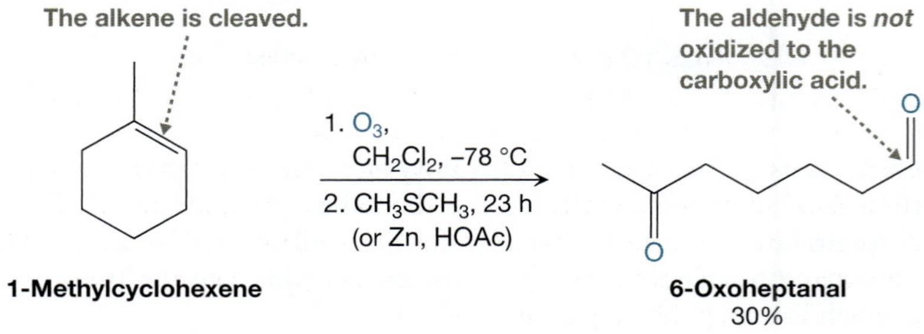

The alkene is cleaved.
The aldehyde is *not* oxidized to the carboxylic acid.
1. O_3,
CH_2Cl_2, –78 °C
2. CH_3SCH_3, 23 h
(or Zn, HOAc)
O
O
1-Methylcyclohexene
6-Oxoheptanal
30%

(24-44)

Ethene, KMnO₄, and Fruit Ripening

One of the marvels of today's supermarkets is their ability to offer wide selections of fruits from all over the world, regardless of the season. How is this possible? One of the keys is the reaction that potassium permanganate ($KMnO_4$) undergoes with alkenes, just as we have seen here in Chapter 24.

Ethene ($H_2C=CH_2$) is the particular alkene of interest. It is a gas, and although it is very simple in structure, it is a ripening hormone and has a profound effect on the development of fruit. When ethene is detected by an ethene receptor in the fruit, certain biochemical pathways are turned on, which change the fruit's color, odor, and texture. The fruit industry takes advantage of this by picking fruit when it is still green, shipping it to a warehouse, and then treating it with ethene a few days before its delivery to nearby supermarkets.

Because of its effect on ripening, ethene can pose a serious problem during the shipping process. Some fruits, such as tomatoes, bananas, and apples, are climactic fruit, meaning that when they detect ethene, they quickly make more. (If you place a green banana in a paper bag with a ripe apple, the banana will ripen in about a day!) One way to deal with this is to ship the fruit in containers in which $KMnO_4$ has been added. The $KMnO_4$ will react with ethene and effectively remove it from the surroundings. Another way to deal with the problem of ethene is to expose the fruit to 1-methylcyclopropene, which competes with ethene for the receptor. Treating the fruit with 1-methylcyclopropene for several hours can block the receptor for more than 10 days.

The mechanism for this reaction is shown in Equation 24-45.

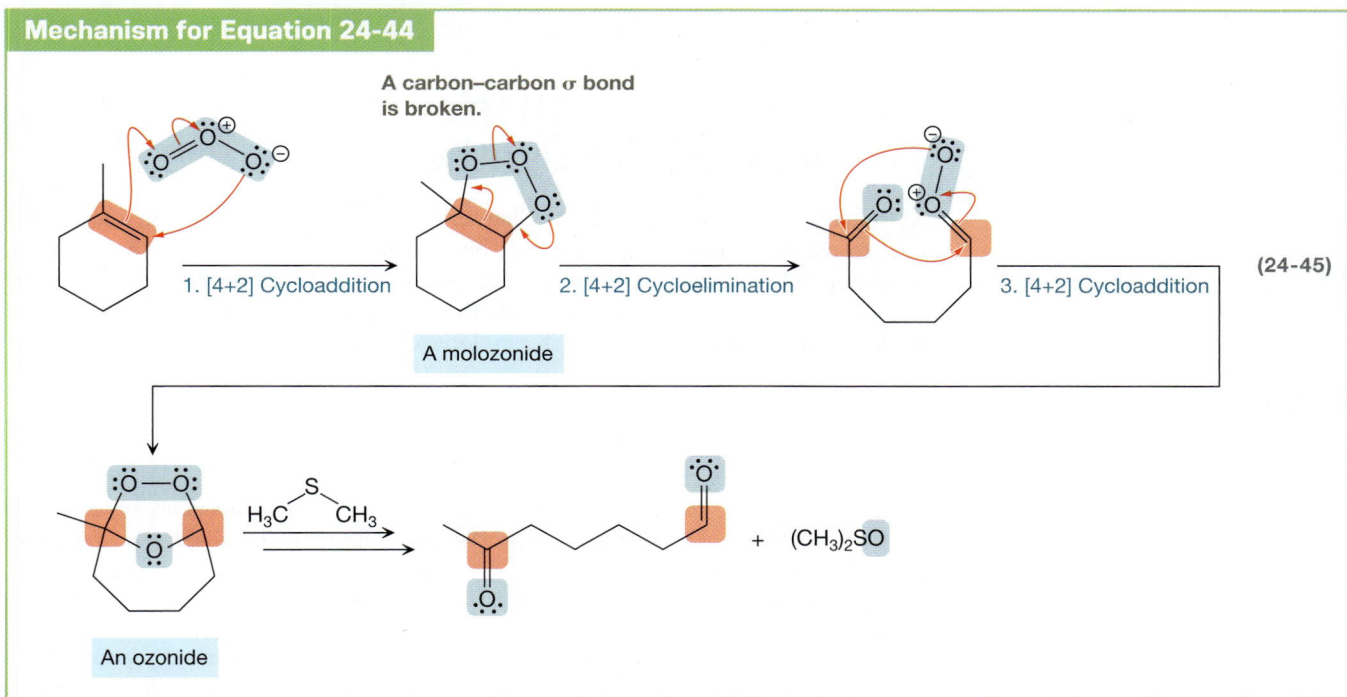

In Step 1, ozone undergoes a [4+2] cycloaddition to the alkene to produce a **molozon-ide**, much like OsO₄ adds to an alkene to produce an osmate ester (Equation 24-31, p. 1219) or MnO₄⁻ adds to an alkene to produce a manganate ester (Equation 24-37, p. 1221). Step 1, more specifically, is called a **1,3-dipolar cycloaddition** because ozone is a 1,3-dipolar compound, characterized by the resonance delocalization of charge over atoms that have a 1,3-positioning. Step 2 is a [4+2] cycloelimination, which breaks the carbon–carbon single bond. Step 3 is another [4+2] cycloaddition, also a 1,3-dipolar cycloaddition, producing an **ozonide**.

Ozonides are unstable and can be explosive, so they are generally not isolated. Rather, when the ozonide is treated with dimethyl sulfide, the ozonide decomposes into the final product. Dimethyl sulfide is a reducing agent, moreover, so it prevents aldehydes in the product from being oxidized to carboxylic acids.

If oxidizing the aldehyde product to a carboxylic acid is desirable, then hydrogen peroxide—an oxidizing agent—can be used in the ozonolysis reaction instead of dimethyl sulfide, as indicated in Equation 24-46.

The alkene is cleaved.

H_2O_2 oxidized the initial aldehyde to the carboxylic acid.

1. O_3, CH_2Cl_2, –78 °C
2. H_2O_2

(24-46)

1-Methylcyclohexene

6-Oxoheptanoic acid

PROBLEM 24.23 Draw the molozonide, the ozonide, and the products that are formed in each of these reactions.

1. O_3
2. H_2O_2

?

1. O_3
2. Zn, HOAc

?

(a)

(b)

24.9 Organic Synthesis: The Diels–Alder Reaction in Synthesis

As we have emphasized throughout Chapter 24, the Diels–Alder reaction is a valuable asset to organic synthesis, in large part because it makes it possible to synthesize a six-membered carbon ring from acyclic precursors. The prototypical Diels–Alder reaction—between buta-1,3-diene and ethene—is repeated in Equation 24-47; the fundamental motif in the product is a cyclohexene ring.

A C=C functional group

CH_2
HC
HC
CH_2

+

CH_2
CH_2

H_2
C
HC CH_2
HC CH_2
C
H_2

A six-membered ring of C atoms

(24-47)

The *alkene* functionality that is left in the product is an important feature of the Diels–Alder reaction. This allows further reaction to take place specifically at that site. For example, the C=C double bond can be hydrogenated (Section 19.5) to produce a cyclohexane ring. Or, it can undergo electrophilic addition to produce a new functional group that can be used further (Chapters 11 and 12); an example is provided in Equation 24-48, in which the Diels–Alder product is converted into a Grignard reagent.

1. HBr
2. Mg(s), ether

MgBr

(24-48)

Another option is to alter the size of the ring, as shown in Equation 24-49.

$$(24\text{-}49)$$

Oxidative cleavage (Section 24.8) opens the six-membered ring at the C=C double bond. An intramolecular aldol condensation (Section 18.8) follows, producing a five-membered ring.

The Diels–Alder reaction is not the first reaction we have encountered that can produce a six-membered carbon ring. In Section 18.10 we used the Robinson annulation reaction, which is shown again in Equation 24-50. The immediate product is an unsaturated cyclohexenone, but the carbonyl group can be selectively reduced to produce the cyclohexene ring.

$$(24\text{-}50)$$

Although a cyclohexene ring can be produced by the Robinson annulation, the Diels–Alder reaction generally offers numerous advantages. One advantage is the relatively mild conditions under which many Diels–Alder reactions can be run; by contrast, the Robinson annulation requires significantly basic conditions that may interfere with other functional groups present. Furthermore, reduction of a carbonyl group to a methylene (CH_2) group in the second step in Equation 24-50 can require harsh conditions, such as the presence of strong acid or strong base (see again Section 19.3).

Other advantages include the *stereospecificity* and the *regioselectivity* of the Diels–Alder reaction. In 1952, for example, R. B. Woodward and coworkers pioneered the use of the Diels–Alder reaction as a key step in the total synthesis of cholesterol, as indicated in Equation 24-51.

$$(24\text{-}51)$$

The Diels–Alder reaction was used to generate the six-membered ring that later becomes the D ring in the target. It was also used to establish the initial relative stereochemistry of the H and CH_3 groups indicated in red: Immediately after the Diels–Alder reaction they are cis to each other, and through a later step in the synthesis they become trans to each other.

Notice, too, the regiochemistry of the Diels–Alder reaction in Equation 24-51. There are two portions that can act as the dienophile in the starting material—namely, the two C=C double bonds on either side of the six-membered ring. Nevertheless, buta-1,3-diene selectively reacts with only one of them (the one to which the CH_3 group is attached), leaving the other one (to which the OCH_3 group is attached) available for further reaction.

Equation 24-51 provides a glimpse of the utility of the Diels–Alder reaction in total synthesis. If you are interested in learning more about its use in total synthesis, see the review by Nicolaou and coworkers (*Angew. Chem. Int. Ed.* **2002**, *41*, 1668–1698).

24.10 A Molecular Orbital Picture of the Diels–Alder Reaction

Lewis structures and resonance theory can describe the Diels–Alder and other pericyclic reactions quite well, but these theories alone cannot account for some aspects of the reaction, such as the *endo rule* for stereochemistry. Moreover, the presence of UV light changes the outcome of the Diels–Alder reaction and this, too, cannot be explained by simple resonance theory.

These results can be accounted for, however, by *frontier molecular orbital (FMO) theory*. As we saw in Interchapter D, FMO theory focuses on the interaction between the HOMO and LUMO of the reactants. A reaction is *allowed* if the HOMO and LUMO have significant interaction in the transition state and is *forbidden* if they do not.

Figure 24-5 shows the π MO energy diagrams for both buta-1,3-diene and ethene, and highlights one frontier orbital interaction between the HOMO of buta-1,3-diene and the LUMO of ethene.

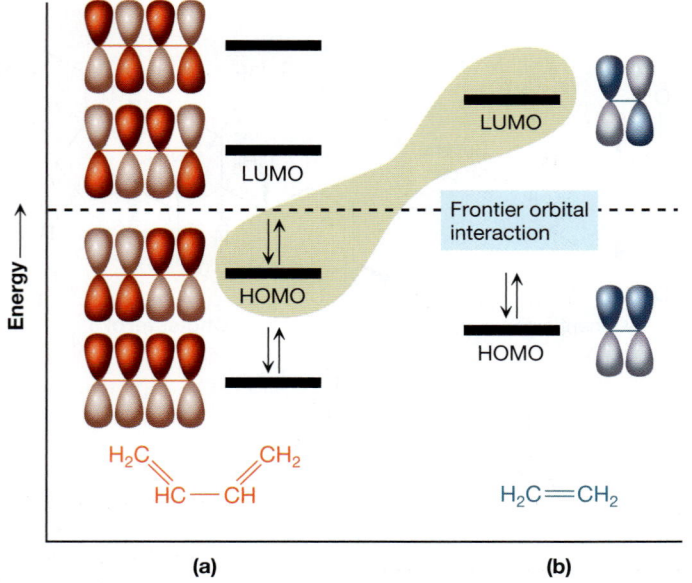

FIGURE 24-5 Frontier orbitals involved in a Diels–Alder reaction (a) MO energy diagram of buta-1,3-diene. (b) MO energy diagram of ethene. The green screen highlights the HOMO–LUMO interaction of interest.

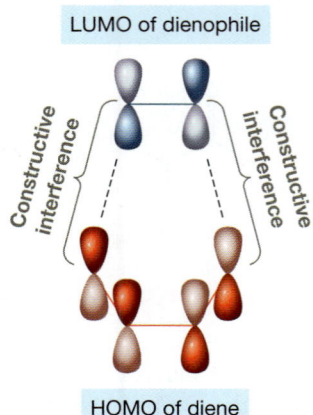

FIGURE 24-6 Frontier orbital interaction in a Diels–Alder reaction The HOMO of buta-1,3-diene (red) and the LUMO of ethene (blue) have the appropriate symmetries to result in significant net overlap. In the example shown, both regions of overlap result in constructive interference.

As the diene and dienophile approach each other, Figure 24-6 shows that the ends of the HOMO and LUMO π systems begin to overlap. Each region of overlap results in constructive interference. Thus:

> The frontier orbitals involved in a Diels–Alder reaction have the appropriate symmetries to interact, so the reaction is *allowed*.

This is in agreement with our analysis in Section 24.1, where we mentioned that the Diels–Alder reaction proceeds through an aromatic transition state.

YOUR TURN **24.15**

Figure 24-6 highlights one of two possible frontier orbital interactions, involving the HOMO of buta-1,3-diene and the LUMO of ethene. The drawing shown here highlights another, involving the HOMO of ethene and the LUMO of buta-1,3-diene.

Determine each type of interference resulting from the indicated regions of overlap. Are they both the same or are they different? Do the orbitals have the appropriate symmetries to interact?

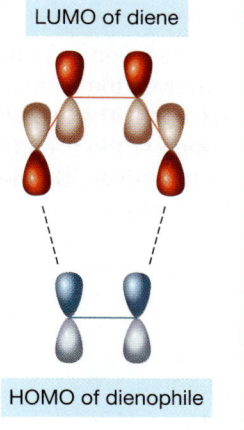

The story is quite different for the [2+2] cycloaddition reaction between two molecules of ethene. Figure 24-7 shows the MOs of π symmetry for each reactant, and highlights one possible frontier orbital interaction.

The overlap that takes place with these orbitals is illustrated in Figure 24-8. Unlike the Diels–Alder reaction (a [4+2] cycloaddition; Fig. 24-6), symmetry

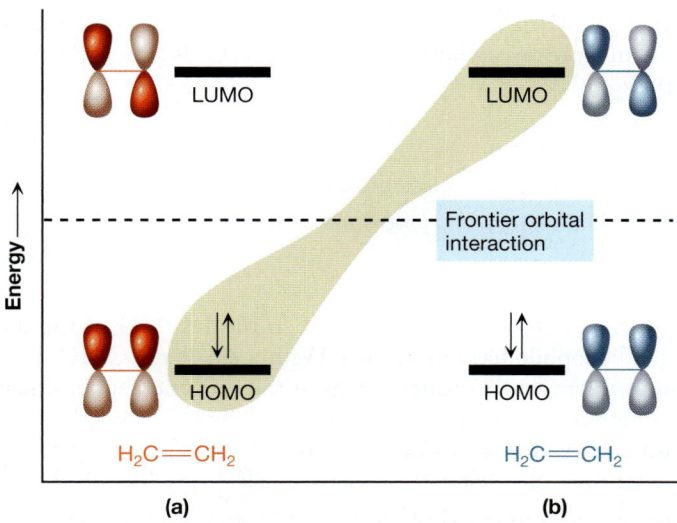

FIGURE 24-7 Frontier orbitals involved in a [2+2] cycloaddition (a) MO energy diagram of one molecule of ethene. (b) MO energy diagram of the second molecule of ethene. The green screen indicates a HOMO–LUMO interaction.

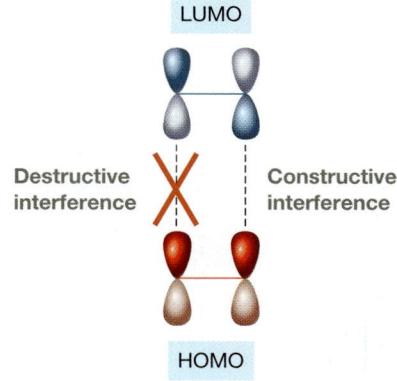

FIGURE 24-8 Frontier orbital interaction in a [2+2] cycloaddition reaction The net interaction for the HOMO–LUMO interaction is zero. Whereas overlap on the right leads to constructive interference, overlap on the left leads to destructive interference. The two effects cancel each other.

prevents the orbitals of the [2+2] cycloaddition from having a net interaction. Overlap on one end leads to constructive interference, but overlap on the other end leads to destructive interference.

> Because the frontier orbitals in a [2+2] cycloaddition do not have the appropriate symmetries to interact, the reaction is *forbidden*.

This outcome is consistent with the fact that a [2+2] cycloaddition proceeds through an antiaromatic transition state, as we saw in Section 24.1.

YOUR TURN 24.16

The second possible HOMO–LUMO interaction in Figure 24-7 involves the HOMO of the blue molecule and the LUMO of the red one. The overlap of those orbitals is shown here. Determine whether the overlap will lead to constructive or destructive interference. Do the orbitals have the appropriate symmetries to interact?

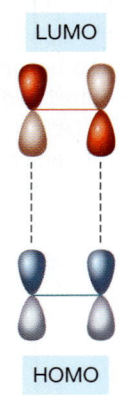

LUMO

HOMO

24.10a Substituent Effects on the Reaction Rate

Recall from Section 24.3 that standard Diels–Alder reactions are facilitated by electron-donating groups attached to the diene and electron-withdrawing groups attached to the dienophile. How does FMO theory account for these results?

The diagram in Figure 24-9a shows the energies of the MOs of buta-1,3-diene and ethene (reproduced from Fig. 24-5). The diagram in Figure 24-9b shows how the picture changes when an electron-donating group (EDG) is attached to buta-1,3-diene, and an electron-withdrawing group (EWG) is attached to ethene. Notice specifically that:

- An electron-donating group *raises* the orbital energies.
- An electron-withdrawing group *lowers* the orbital energies.

Consequently, the difference in energy between the HOMO of the diene and the LUMO of the dienophile has diminished. With a smaller HOMO–LUMO energy gap, the orbitals interact to a greater extent in the transition state, which lowers the energy barrier between reactants and products.

These MO energy effects by electron-donating and -withdrawing groups can be understood in terms of charge repulsion between electrons. Increased electron density from a nearby electron-donating group results in increased electron–electron repulsion, which increases the energy. Conversely, a nearby electron-withdrawing group reduces electron–electron repulsion by removing electron density from the orbital of interest, which decreases the energy.

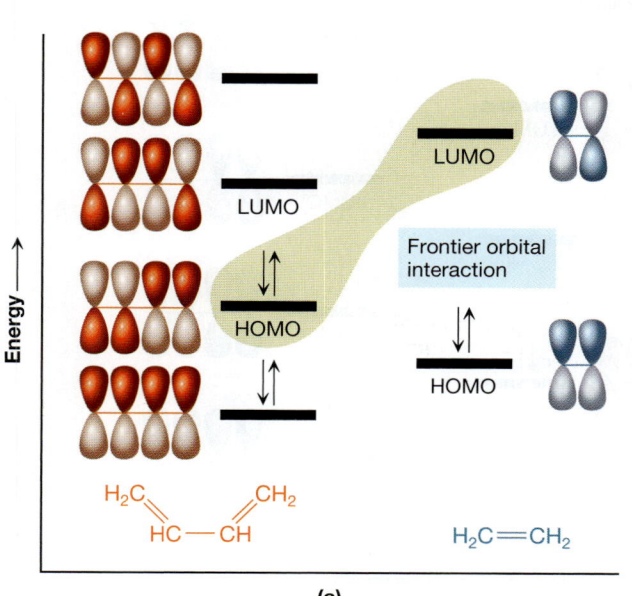

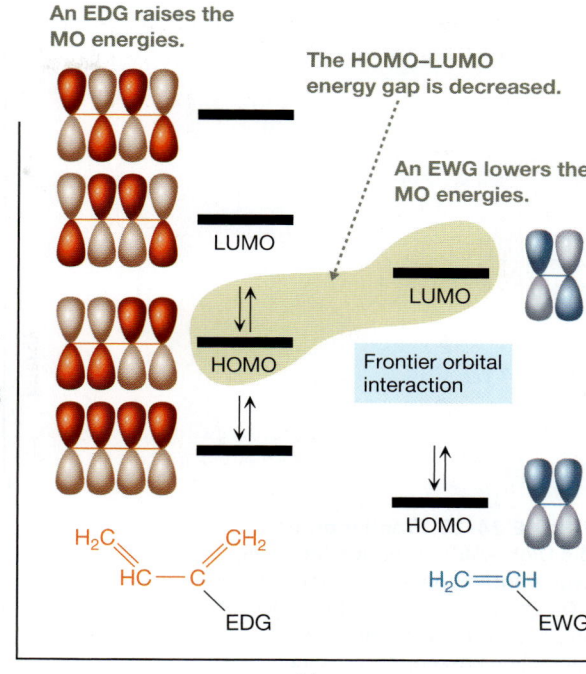

An EDG raises the MO energies.

The HOMO–LUMO energy gap is decreased.

An EWG lowers the MO energies.

LUMO

Frontier orbital interaction

HOMO

HOMO

H_2C
HC——CH
CH_2

H_2C==CH_2

(a)

LUMO

HOMO

Frontier orbital interaction

LUMO

HOMO

H_2C
HC——C
CH_2
EDG

H_2C==CH
EWG

(b)

FIGURE 24-9 Substituent effects on a frontier orbital interaction
(a) MO energy diagrams of unsubstituted buta-1,3-diene and ethene. (b) MO energy diagrams of buta-1,3-diene with an attached electron-donating group (EDG) and ethene with an attached electron-withdrawing group (EWG). The EDG raises the MO energies of the diene and the EWG lowers the MO energies of the dienophile, which decreases the HOMO–LUMO energy gap.

PROBLEM 24.25 Draw a diagram similar to that in Figure 24-9b, but assume instead that an EWG is placed on the diene and an EDG is placed on the dienophile (thus establishing an inverse electron demand Diels–Alder reaction). How does the HOMO(diene)–LUMO(dienophile) energy gap compare to the one in Figure 24-9a?

24.10b The Endo Rule for Stereochemistry

When both endo and exo products can be made in a Diels–Alder reaction, the endo products tend to dominate (Section 24.4). Simple resonance theory cannot account for these results, but FMO theory can.

The Diels–Alder reaction between cyclopentadiene and propenal can produce both endo and exo products (Equation 24-52).

Endo product   **Exo product**   (24-52)

The π MOs of the two molecules are shown in Figure 24-10. Notice that the HOMO and LUMO of cyclopentadiene (Fig. 24-10a) are essentially the same as those of buta-1,3-diene. (The presence of the extra CH_2 group modifies the energies only slightly, so we will ignore its effect.)

The HOMO and LUMO of propenal (Fig. 24-10b), on the other hand, are somewhat different from those of ethene, because the C==O bond in propenal is *conjugated* to the C==C bond. There are, therefore, four π orbitals in propenal—two bonding and two antibonding—similar to buta-1,3-diene. Propenal's MOs are lower in energy due to the presence of the electron-withdrawing C==O group.

Figure 24-11 explicitly shows the interaction between the HOMO of cyclopentadiene and the LUMO of propenal, in both the endo approach (Fig. 24-11a) and the

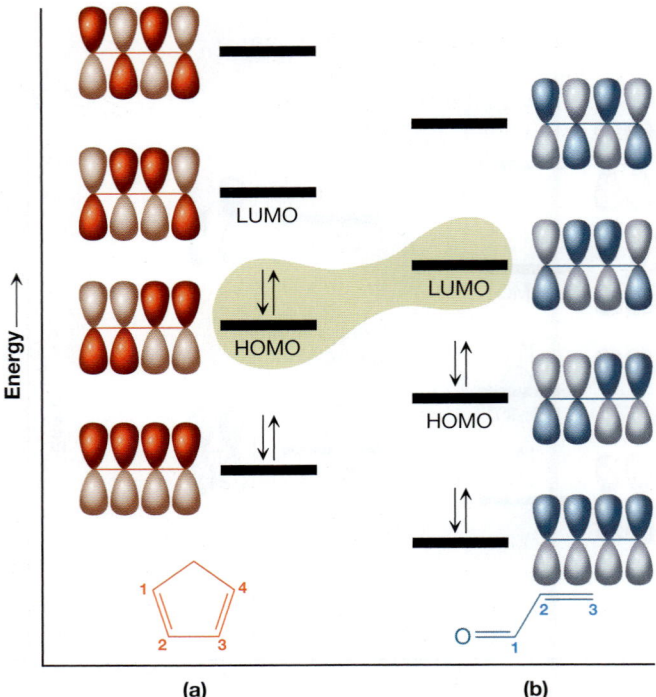

FIGURE 24-10 Frontier orbitals in the Diels–Alder reaction involving cyclopentadiene and propenal (a) π MOs of cyclopentadiene. (b) π MOs of propenal. The important HOMO–LUMO interaction between them is indicated by the green screen.

exo approach (Fig. 24-11b). Even though propenal has contributions from four p AOs in each of its π MOs, only the two p orbitals that make up the C=C double bond (i.e., those on C2 and C3) are directly involved in the reaction. (C2 and C3 are the ones that gain a σ bond during the course of the reaction.) Both the endo and exo approaches show the overlap of these two p orbitals on the dienophile with the p orbitals on C1 and C4 of the diene—the so-called **primary orbital overlap** for the reaction.

In the endo approach, the C=O group of the dienophile is oriented toward the π system of the diene, allowing the p orbital on the carbonyl carbon to interact with an unused p orbital on the diene (indicated by the dashed line in Fig. 24-11a). This **secondary orbital overlap** is not possible in the exo approach. Even though secondary overlap does not specifically lead to bond breaking or bond making in the reaction, it does provide stabilization.

FIGURE 24-11 Primary and secondary orbital overlap in a Diels–Alder reaction Interaction between the HOMO of cyclopentadiene with the LUMO of propenal when propenal approaches in (a) an endo fashion and (b) an exo fashion. Both approaches exhibit primary overlap of the orbitals used to form the two new σ bonds in the reaction, but only the endo approach exhibits secondary orbital overlap (dashed line), using a p orbital on the dienophile that is not directly involved in bond formation or bond breaking. This lowers the energy for the transition state, thereby favoring the endo approach over the exo approach.

> In a Diels–Alder reaction, the transition state produced via the endo approach tends to be lower in energy than the one produced via the exo approach. Thus, the endo approach tends to be favored, making the endo product the major product.

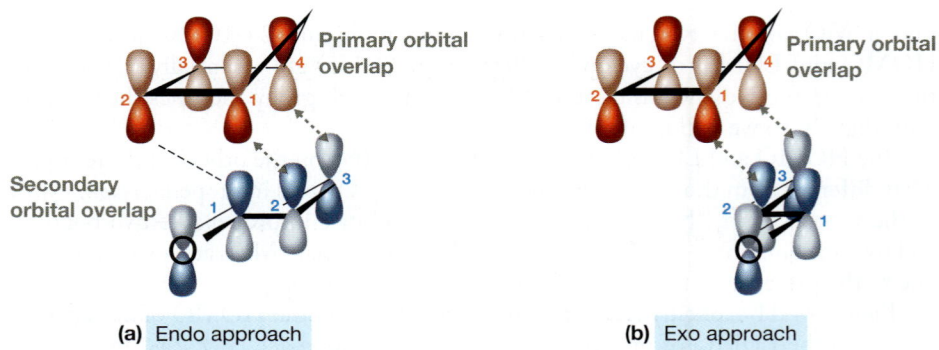

(a) Endo approach   (b) Exo approach

Figure 24-11b is reproduced here. Identify the two *p* orbitals that, in Figure 24-11a, are involved in secondary orbital overlap. Are those *p* orbitals closer together in Figure 24-11a or Figure 24-11b?

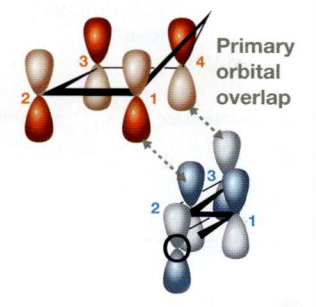

Primary orbital overlap

24.10c Thermal versus Photochemical Reactions

In all of the cycloaddition reactions we have considered thus far, the reactants have been in their ground state (i.e., lowest energy) electron configuration, where all electrons reside in the lowest-energy orbitals and are paired. When all reactants are in their ground state, the reaction is said to be **thermal**. Thus, the Diels–Alder reaction — a [4+2] cycloaddition reaction — is **thermally allowed**, due to the favorable frontier orbital interactions. By contrast, the [2+2] cycloaddition is **thermally forbidden**, because the frontier orbital interactions do not lead to net stabilization (see again Figs. 24-6 and 24-8).

It turns out, though, that [2+2] cycloaddition reactions, such as the one between two molecules of ethene, take place readily when irradiated with UV light (Equation 24-53).

$$H_2C = CH_2 \xrightarrow{\text{UV light}} \begin{array}{c} H_2C - CH_2 \\ | \quad\quad | \\ H_2C - CH_2 \end{array} \tag{24-53}$$

When ethene absorbs a photon with sufficient energy, an electron is promoted from the HOMO to the LUMO (Section 15.3a), thereby generating an **excited electronic state** (Fig. 24-12).

The new HOMO in that excited state is the π^* MO. Therefore, when an excited molecule of ethene encounters another ethene molecule still in the ground electronic state, the frontier orbital interaction involves the π^* orbital from each species (Fig. 24-13).

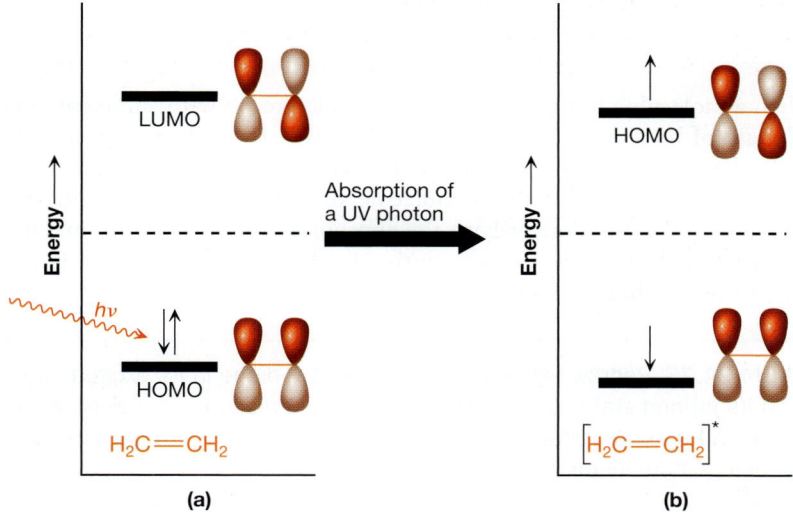

FIGURE 24-12 Impact of UV photon absorption on the identities of the HOMO and LUMO (a) Ground state electron configuration of ethene. (b) Excited electronic state of ethene, denoted by *. A UV photon with sufficient energy ($h\nu$) is capable of promoting an electron from ethene's HOMO to its LUMO to generate the excited state. The excited state's HOMO is the ground state's LUMO.

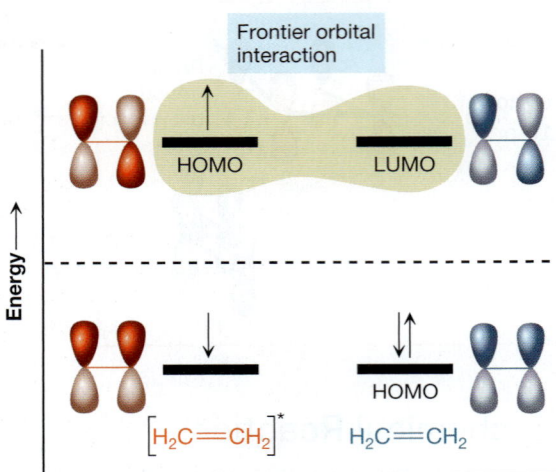

FIGURE 24-13 Frontier orbitals in a photochemical [2+2] cycloaddition reaction The π MOs of ethene in an excited state are shown in red, and the π MOs of ethene in its ground state are shown in blue. The resulting frontier orbital interaction is between the π* MO in each species.

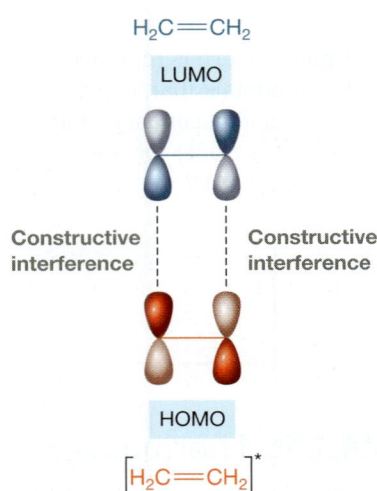

FIGURE 24-14 Frontier orbital interaction in a photochemical [2+2] cycloaddition reaction Both ends of the π systems exhibit the same type of interference—in this case, constructive interference. Therefore, the orbitals have the appropriate symmetries to interact, thus making this reaction *photochemically allowed.*

These HOMO and LUMO orbitals do, indeed, have the appropriate symmetries to interact, as illustrated in Figure 24-14. Consequently, the [2+2] cycloaddition reaction is *thermally forbidden*, but **photochemically allowed**.

Unlike the [2+2] cycloaddition reaction, the Diels–Alder reaction is impeded by UV light, as indicated in Equation 24-54.

(24-54)

When a molecule of buta-1,3-diene absorbs a photon with sufficient energy, an excited state of buta-1,3-diene is generated, in which an electron from what was originally the HOMO now occupies what was originally the LUMO. When this excited state species encounters a molecule of ethene in the ground state, the HOMO and LUMO no longer have the appropriate symmetries to interact, making the reaction forbidden. Consequently, the Diels–Alder reaction is *thermally allowed*, but **photochemically forbidden**.

PROBLEM 24.26 Redraw Figure 24-13 for buta-1,3-diene in its excited state and ethene in its ground state. Do the same for buta-1,3-diene in its ground state and ethene in its excited state. Are these results consistent with a Diels–Alder reaction being photochemically forbidden?

Chapter Summary and Key Terms

- The **Diels–Alder reaction** joins a conjugated **diene** and a **dienophile** (either an alkene or an alkyne) via the formation of two new σ bonds. The product is a six-membered ring of carbon atoms. (**Introduction and Section 24.1; Objectives 1 and 2**)

- The Diels–Alder reaction is concerted; all bonds that are formed and broken do so simultaneously. It requires the cyclic movement of electrons, so it is classified as a **pericyclic reaction**. (**Section 24.1; Objective 1**)

- The Diels–Alder reaction is thermally *allowed*, because it is a **[4+2] cycloaddition**, proceeding through an *aromatic transition state*. Reactions that proceed through an *antiaromatic transition state*, such as the **[2+2] cycloaddition** and the **[6+2] cycloaddition**, are thermally *forbidden*. (**Section 24.1; Objective 3**)

- For a Diels–Alder reaction to take place, the diene must be able to attain the **s-cis conformation**. Diels–Alder reactions cannot take place with the diene in the **s-trans conformation**. (**Section 24.2; Objective 4**)

- *Standard* Diels–Alder reactions are facilitated by electron-donating groups attached to the diene and electron-withdrawing groups attached to the dienophile. (**Section 24.3; Objective 5**)

- Diels–Alder reactions are stereospecific with respect to the stereochemistry of the diene and dienophile. (**Section 24.4; Objective 6**)

  - Substituents that are cis to each other about the C=C double bond of the dienophile end up cis to each other in the new ring that is produced. Otherwise, the substituents end up trans to each other in the ring.
  - Substituents attached to the terminal carbons of the diene end up cis to each other in the new ring that is produced if the double bonds of the diene are both cis or both trans. Otherwise, the substituents end up trans to each other in the ring.

- Diels–Alder reactions tend to favor an **endo product** over an **exo product**. (**Section 24.4; Objective 6**)

- When the diene and dienophile are both unsymmetric, two isomeric products can be produced. The major product is the one produced by the approach that exhibits the most favorable electrostatic attraction among atoms undergoing bond formation. (**Section 24.5; Objective 6**)

- Enthalpy tends to heavily favor the products of a Diels–Alder reaction, whereas entropy favors reactants. Therefore, under normal conditions, Diels–Alder reactions tend to be irreversible and proceed under kinetic control. At high temperatures, though, some can undergo the reverse reaction—a **retro Diels–Alder reaction** or **[4+2] cycloelimination**. (**Section 24.6; Objective 7**)

- An alkene can undergo **syn dihydroxylation** with OsO_4 or $KMnO_4$ (cold), resulting in a cis-1,2-diol. An alkyne that undergoes such a reaction results in a 1,2-dicarbonyl compound. (**Section 24.7; Objective 8**)

- **Oxidative cleavage** of an alkene or alkyne completely breaks the C=C double bond or C≡C triple bond, with the initial alkene or alkyne carbons becoming part of C=O groups in the final products. (**Section 24.8; Objective 9**)

  - Aldehydes that are initially produced from oxidative cleavage are further oxidized to carboxylic acids when cleavage is initiated with a hot, concentrated solution of $KMnO_4$ followed by workup with acid (**Section 24.8a**) or with **ozonolysis** followed by workup with H_2O_2 (**Section 24.8c**).
  - Aldehydes that are produced from oxidative cleavage are *not* oxidized further when cleavage involves the treatment of a 1,2-diol with $NaIO_4$ or HIO_4 (**Section 24.8b**) or is initiated with ozonolysis followed by workup with CH_3SCH_3 or $Zn/HOAc$ (**Section 24.8c**).

Reaction Tables

TABLE 24-1 Functional Group Conversions

	Starting Compound Class	Typical Reagents and Reaction Conditions	Compound Class Formed	Key Electron-Rich Species	Key Electron-Poor Species	Comments	Discussed in Section
(1)	C=C Alkene	OsO_4 / H_2O_2	HO—OH C–C Syn 1,2-diol	C=C	O=Os=O	An alternate method uses a cold, basic solution of $KMnO_4$	24.7
(2)	—C≡C— Alkyne	OsO_4 / H_2O_2	C–C 1,2-Dione	—C≡C—	O=Os=O	An alternate method uses a cold, basic solution of $KMnO_4$	24.7

TABLE 24-2 Reactions That Alter the Carbon Skeleton

	Starting Compound Class	Typical Reagents and Reaction Conditions	Compound Class Formed	Key Electron-Rich Species	Key Electron-Poor Species	Comments	Discussed in Section(s)
(1)	Conjugated diene	Dienophile	Substituted cyclohexene	N/A	N/A	Standard Diels–Alder reactions are facilitated by an electron-donating group on the diene and an electron-withdrawing group on the dienophile	24.1–24.6
(2)	Substituted cyclohexene	Δ	Conjugated diene + Alkene	N/A	N/A	Retro Diels–Alder reaction	24.6
(3)	Alkene	1. conc $KMnO_4$, KOH, H_2O, Δ 2. HCl	Ketone and/or carboxylic acid	(alkene)	$O=Mn=O$	Oxidative cleavage; formic acid product (R″ = H) oxidizes to CO_2	24.8a
(4)	Alkyne	1. conc $KMnO_4$, KOH, H_2O, Δ 2. HCl	Carboxylic acid	(alkyne)	$O=Mn=O$	Oxidative cleavage; formic acid product (R′ = H) oxidizes to CO_2	24.8a
(5)	Alkene	1. O_3 2. CH_3SCH_3 or Zn, HOAc	Ketone and/or aldehyde	(alkene)	(ozone)	Ozonolysis	24.8c
(6)	Alkene	1. O_3 2. H_2O_2	Ketone and/or carboxylic acid	(alkene)	(ozone)	Ozonolysis; oxidative cleavage; formic acid product (R″ = H) oxidizes to CO_2	24.8c
(7)	1,2-Diol	$NaIO_4$ or HIO_4	Ketone and/or aldehyde	—	—	Oxidative cleavage	24.8b

Problems that are related to synthesis are denoted (SYN).

24.1 and 24.2 The Diels–Alder Reaction and the s-Cis Conformation of the Diene

24.27 We mentioned in the chapter opener (p. 1198) that tetrodotoxin has been synthesized using a Diels–Alder reaction. The one incorporated in the synthesis by M. Isobe involved the following diene and dienophile. Draw the curved arrow notation for this reaction.

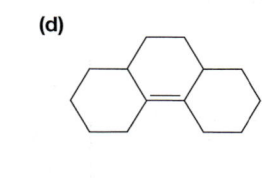

24.28 Draw the mechanism and product for each of the following Diels–Alder reactions.

(a)

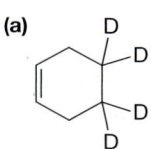

(b)

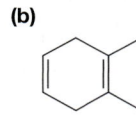

24.29 (SYN) Draw the diene that would react with ethene to produce each of the following compounds.

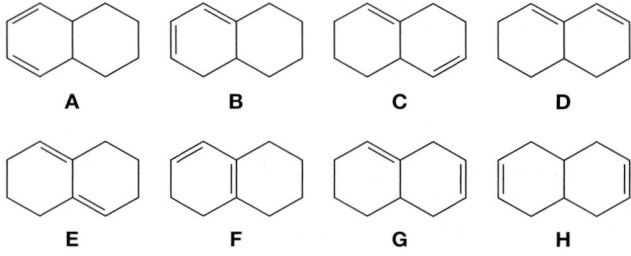

(a) (b) (c) D D (d)

24.30 (SYN) Draw the dienophile that would react with buta-1,3-diene to produce each of the following compounds.

(a) D D D (b) (c)

24.31 The following are several isomers of $C_{10}H_{14}$ with two fused six-membered rings.

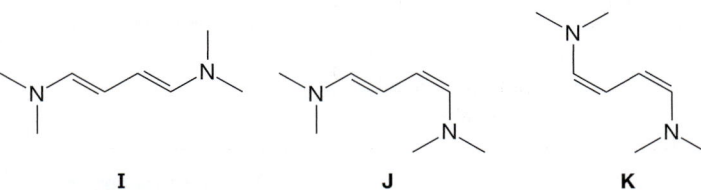

A   B   C   D

E   F   G   H

(a) Identify which will react with ethene in a Diels–Alder reaction and which will not.
(b) Draw one more isomer with two fused six-membered rings that *will* react with ethene in a Diels–Alder reaction.

24.32 Rank the following compounds in order from slowest to fastest rate of reaction in a Diels–Alder reaction with ethene.

I   J   K

24.33 Which of the molecules shown here will react faster as a diene in a Diels–Alder reaction with ethene? Explain.

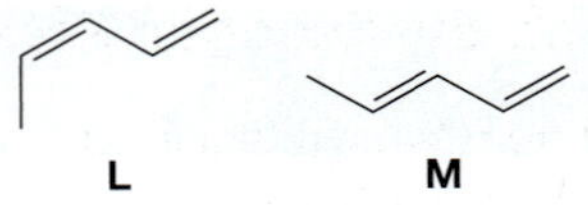

L   **M**

24.34 Anthracene readily undergoes a Diels–Alder reaction with tetracyanoethene, even though anthracene is aromatic.

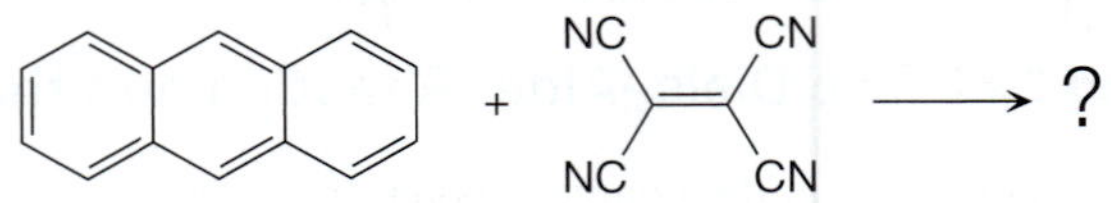

(a) Draw two possible products that can form from this reaction.

(b) Explain why anthracene can readily undergo a Diels–Alder reaction, whereas benzene does not.

24.35 The following is an example of a hetero Diels–Alder reaction, because a noncarbon atom (in this case, an N atom) is involved in bond formation and bond breaking. Draw the curved arrows necessary to account for this transformation.

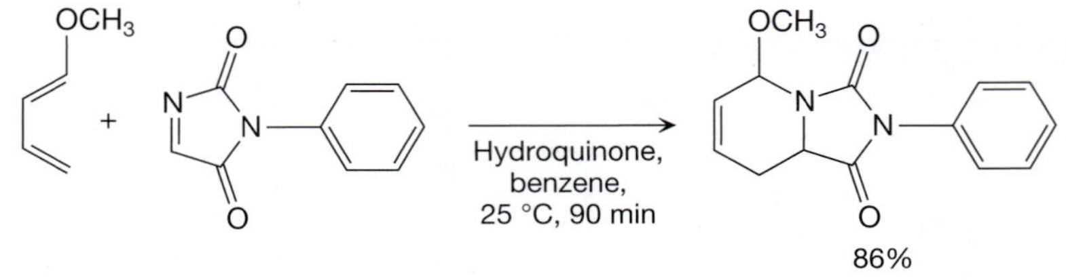

Hydroquinone, benzene, 25 °C, 90 min

86%

24.36 Draw the product of the following reaction, assuming that it takes place via a [6+4] cycloaddition.

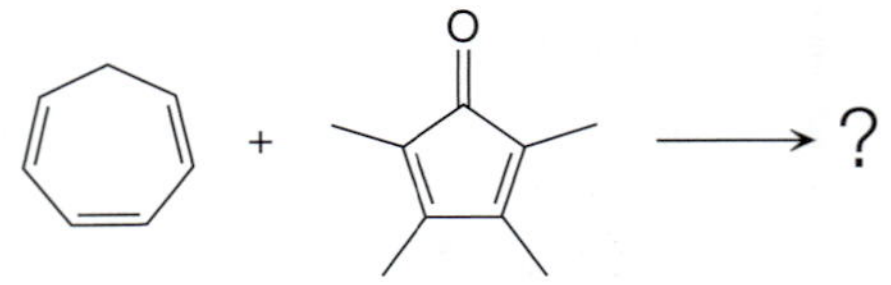

24.37 For each of the following reactions, (a) draw the curved arrows necessary for a concerted mechanism, (b) draw the transition state, (c) determine whether each transition state is *aromatic* or *antiaromatic*, and (d) based on your answer to part (c), determine whether the reaction is allowed or forbidden under normal (thermal) conditions.

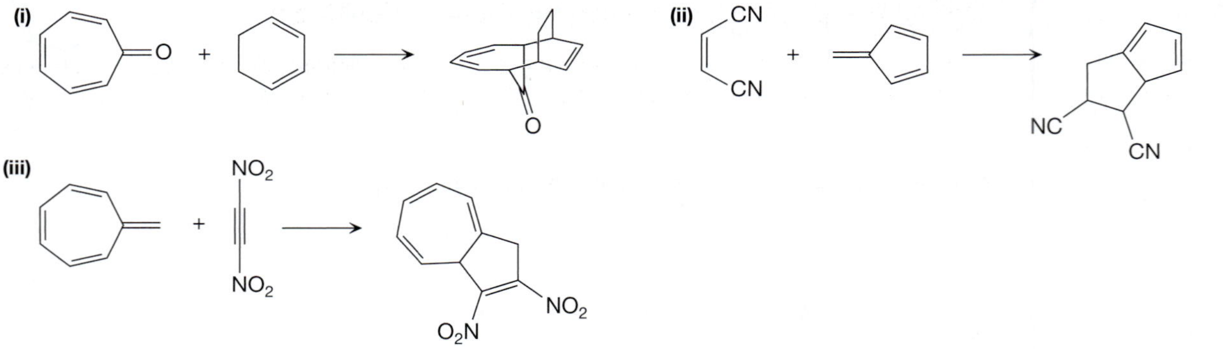

(i)

(ii)

(iii)

24.38 The Diels–Alder reaction is not the only pericyclic reaction involving the cyclic flow of six electrons over six atoms. Other examples include *electrocyclic reactions*, *Cope rearrangements*, and *Claisen rearrangements*, the curved arrow notations for which are shown here. Complete each of these reactions by drawing the product.

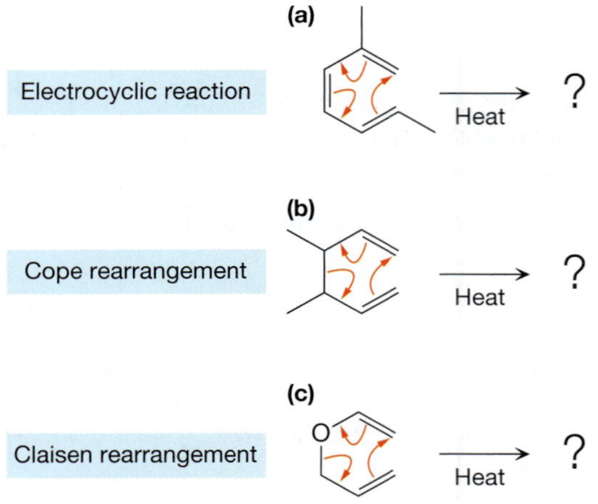

(a) Electrocyclic reaction

Heat

(b) Cope rearrangement

Heat

(c) Claisen rearrangement

Heat

24.3 Substituent Effects on the Reaction Rate

24.39 Rank the following in order from slowest to fastest rate of reaction in a Diels–Alder reaction with buta-1,3-diene.

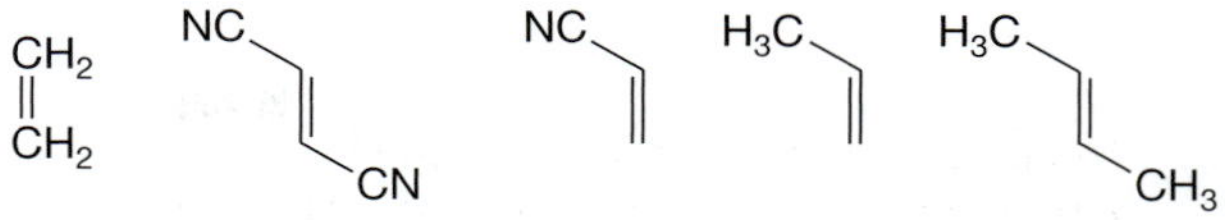

24.40 Consider the following dienes in a Diels–Alder reaction with ethene. Which will react the fastest? Which will react the slowest? Explain.

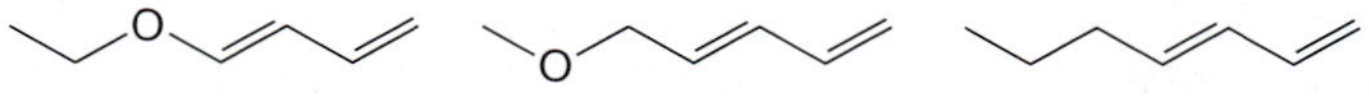

24.41 Which of the molecules shown here will react faster as a dienophile in a Diels–Alder reaction with buta-1,3-diene? Explain.

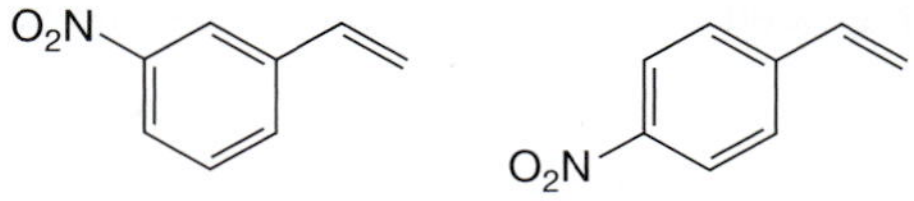

24.42 Which of the molecules shown here will react faster as a dienophile in a Diels–Alder reaction with buta-1,3-diene? Explain.

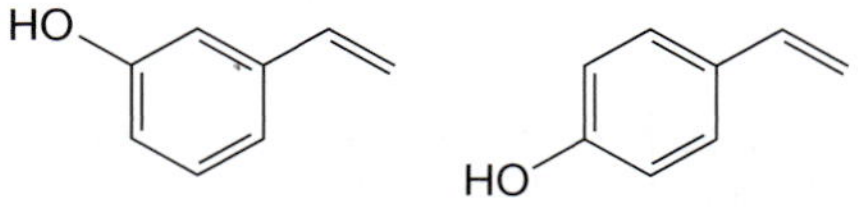

24.43 (SYN) The compound shown here can be produced from two *different* Diels–Alder reactions.

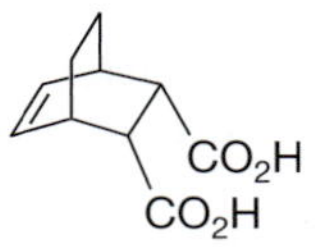

  (a) Draw the reactants that can be used for each Diels–Alder reaction.

  (b) Which would be the better route? Why?

24.4 and 24.5 Stereochemistry and Regiochemistry of Diels–Alder Reactions

24.44 Draw the mechanism and major product for each of the following Diels–Alder reactions. Pay attention to stereochemistry. *Hint:* You may need to consider conformations and orientations other than the ones shown.

(a) (b)

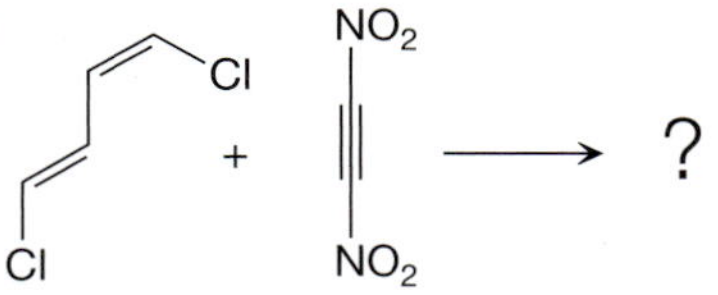

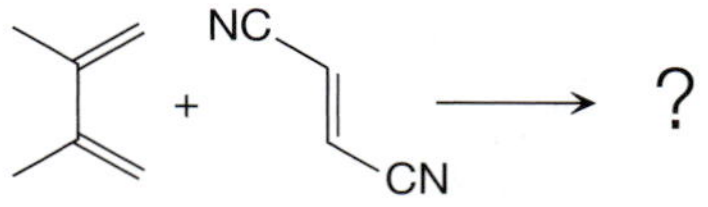

(c) 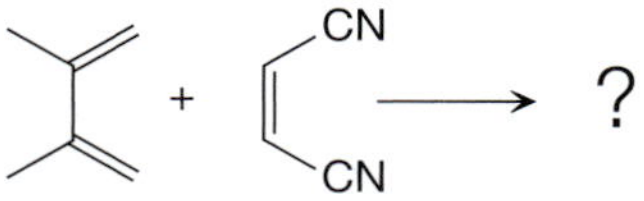 (d)

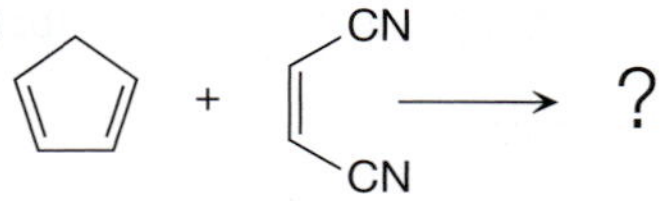

24.45 The following are examples of hetero Diels–Alder reactions (see Problem 24.35). Draw the mechanism and predict the product for each reaction, paying particular attention to stereochemistry.

(a) (b)

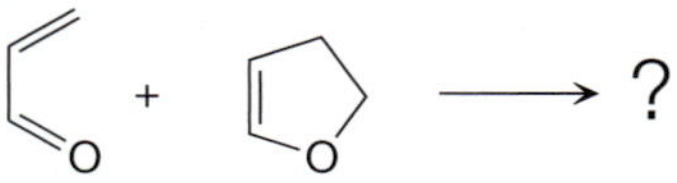

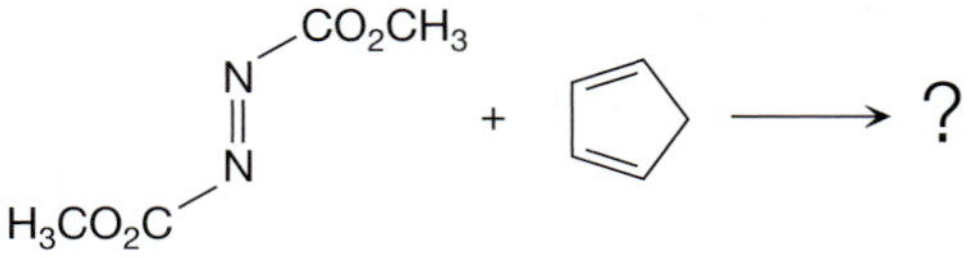

24.46 (SYN) The compound shown here can be produced from two different Diels–Alder reactions.

  (a) Draw the reactants that would be required for each reaction.

  (b) Which set of reactants would be the better choice? Why?

24.47 When the following Diels–Alder reaction takes place at 0 °C, compound **A** is produced as the major product. When the product mixture is heated to 60 °C, compound **B** is produced as the major product.

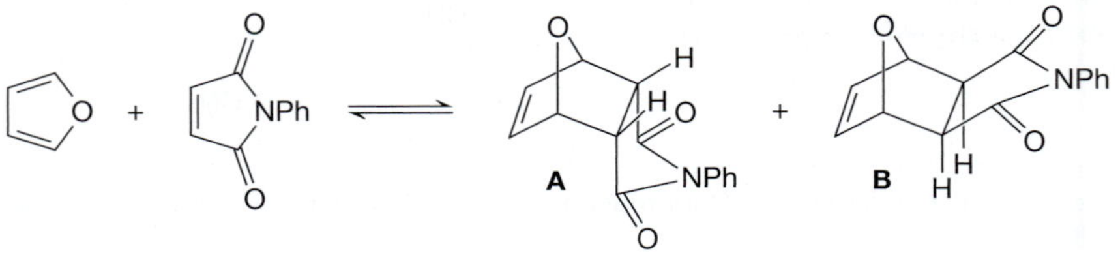

(a) Which product is produced faster in this Diels–Alder reaction?
(b) Which product is more stable? Why? *Hint:* It might help to build a model.

24.48 Draw the mechanism and major product for each of the following Diels–Alder reactions. Pay attention to regiochemistry. *Hint:* You may need to consider conformations and orientations other than the ones shown.

(a)

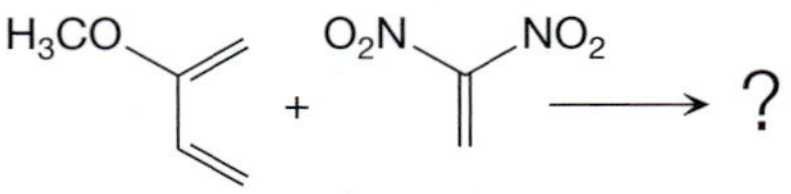

(b)

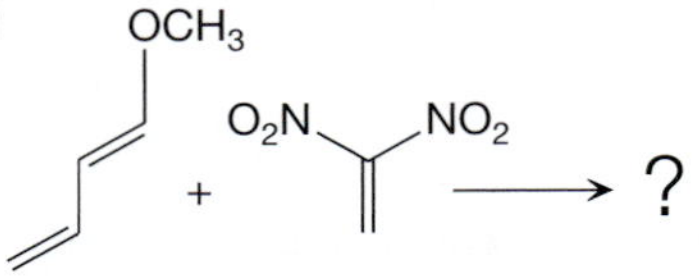

24.49 Two constitutional isomers can be produced from this Diels–Alder reaction.

(a) Draw the mechanism for the formation of each isomer.
(b) Determine which isomer is the major product.
  Hint: Is a phenyl ring electron rich or electron poor?

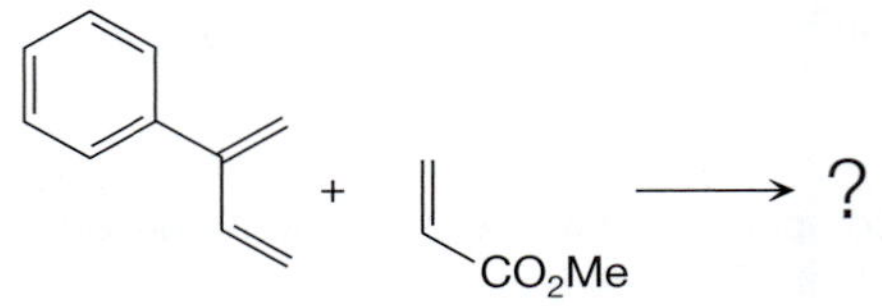

24.6 The Reversibility of Diels–Alder Reactions; the Retro Diels–Alder Reaction

24.50 Consider the Diels–Alder reaction below, for which $\Delta H^{\circ}_{rxn} = -54$ kJ/mol (-13 kcal/mol) and $\Delta S^{\circ}_{rxn} = -151$ J/mol·K (-36 cal/mol·K). (a) Do you think this reaction will be reversible at room temperature? (b) Estimate the temperature at which the equilibrium constant equals 1.

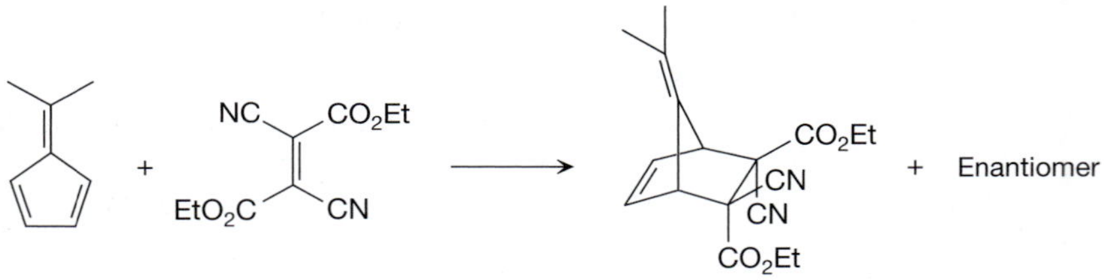

24.51 Draw the retro Diels–Alder mechanism for the product shown in Problem 24.50.

24.52 Draw the retro Diels–Alder mechanism and product for each of the compounds shown here. Which retro Diels–Alder reaction do you think would require the lower temperature? Why?

(a)

(b)

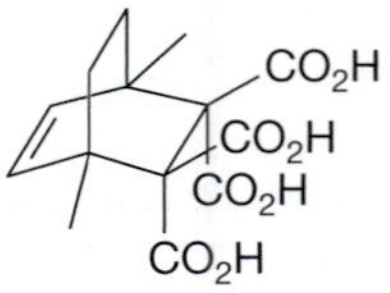

24.53 Draw a mechanism to account for the reaction shown here, which scrambles the isotopic labeling.

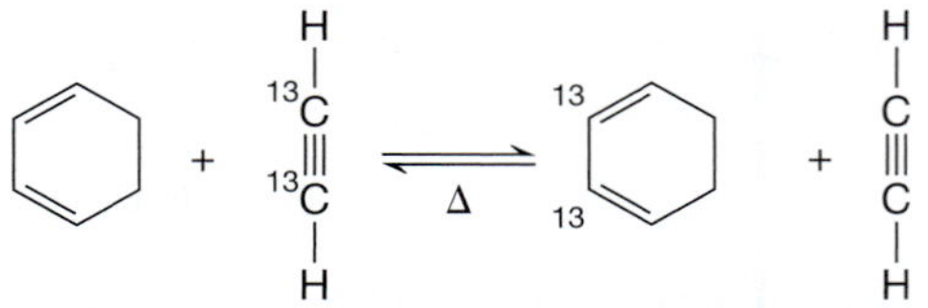

24.54 Using the reaction shown in Problem 24.47, draw the mechanism that converts **A** to **B** when the product is heated.

24.7 and 24.8 Syn Dihydroxylation and Oxidative Cleavage of Alkenes and Alkynes Using OsO_4 or $KMnO_4$

24.55 Draw the organic products of each of the following reactions.

(a)

HO
O
OH
O
OsO4
H2O2, (CH3)3COH
?

(b)

KMnO4
KOH, 0 °C
?

(c)

O
OsO4
H2O2, (CH3)3COH
?

(d)

KMnO4
KOH, 0 °C
?

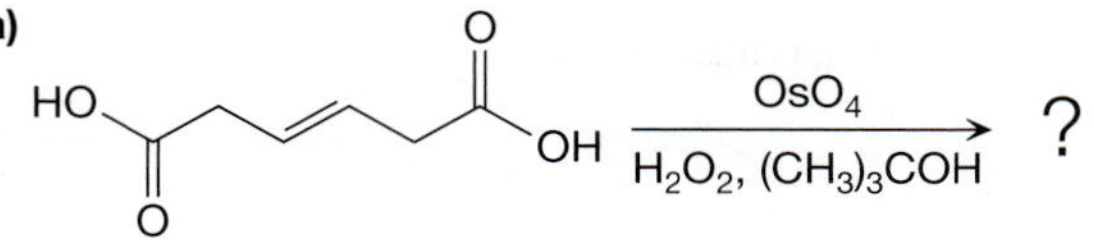

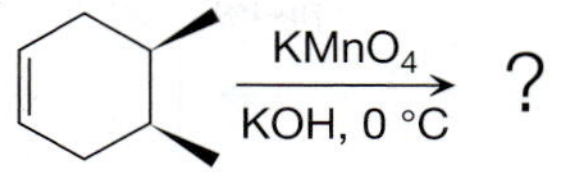

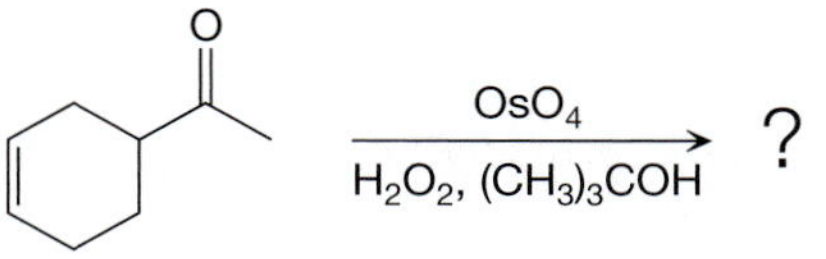

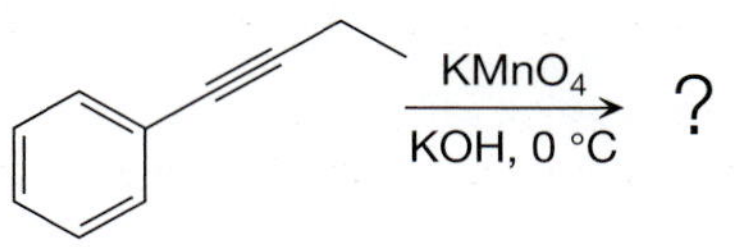

24.56 (SYN) Show how each of the following compounds can be produced from an alkene or alkyne.

(a)

OH
OH

(b)

HO
HO

(c)

O
O

(d)

OH
OH

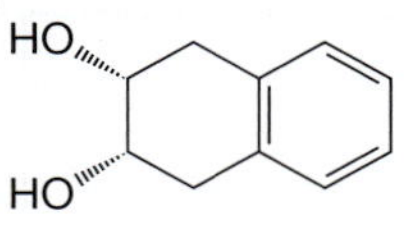

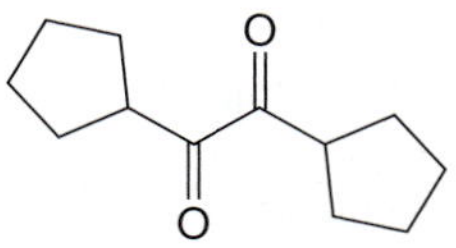

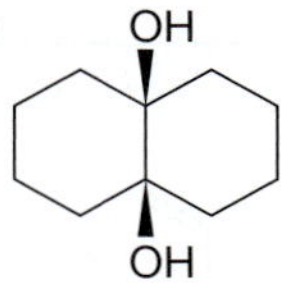

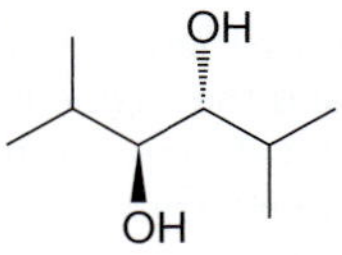

24.57 Draw the organic products of each of the following reactions.

(a)

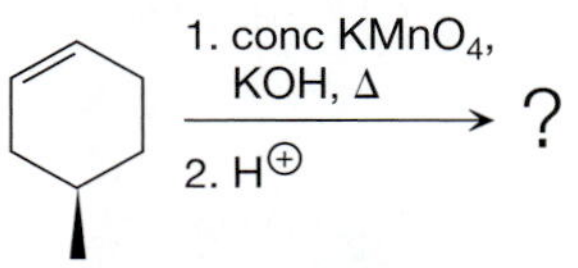

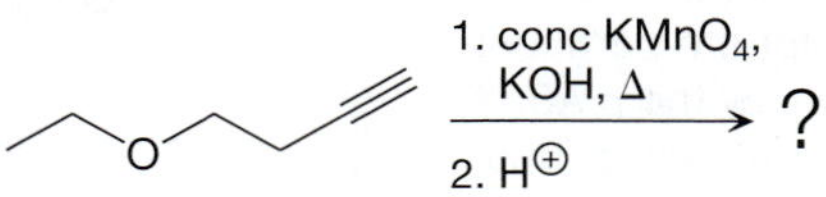

1. conc KMnO4,
KOH, Δ
2. H⊕
?

(b)

O
1. conc KMnO4,
KOH, Δ
2. H⊕
?

(c)

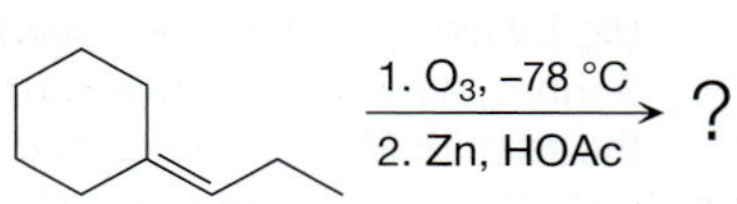

1. O3, −78 °C
2. Zn, HOAc
?

(d)

1. O3, −78 °C
2. H2O2
?

(e)

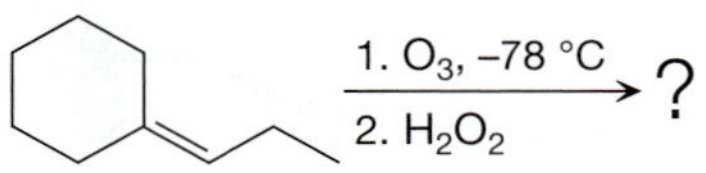

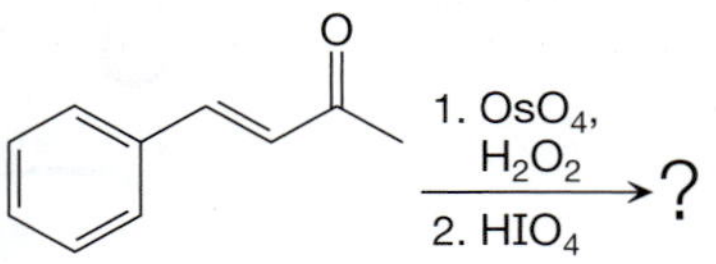

O
1. OsO4,
H2O2
2. HIO4
?

(f)

HO
conc KMnO4
KOH, Δ
?

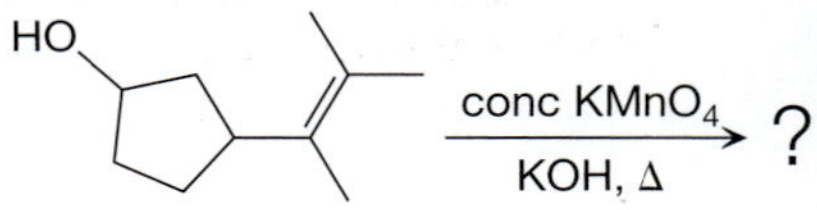

24.58 Draw the structure of the reactant that produces the molecules shown for each of the following reactions.

(a)

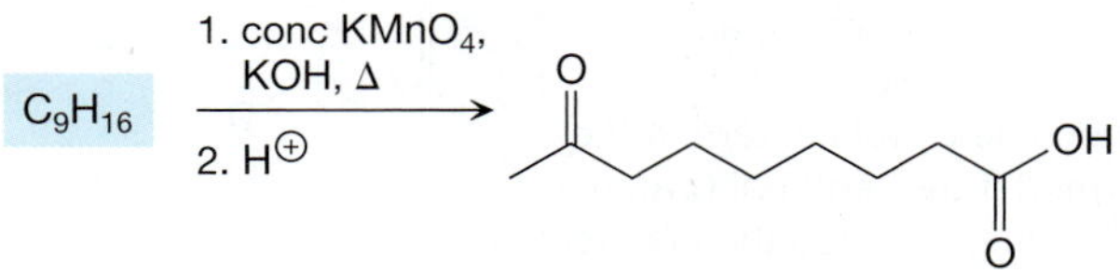

C9H16
1. conc KMnO4,
KOH, Δ
2. H⊕
O
OH
O

(b)

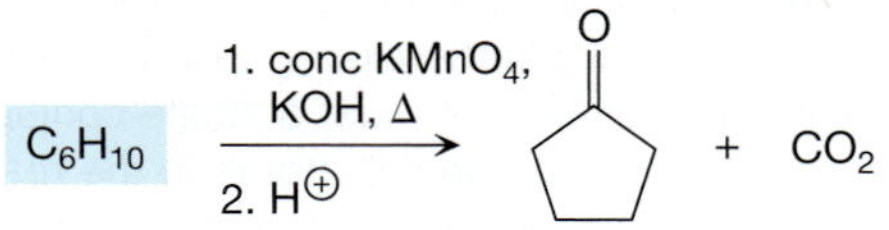

C6H10
1. conc KMnO4,
KOH, Δ
2. H⊕
O
+ CO2

(c)

C10H18
1. O3, −78 °C
2. (CH3)2S
O
+
O

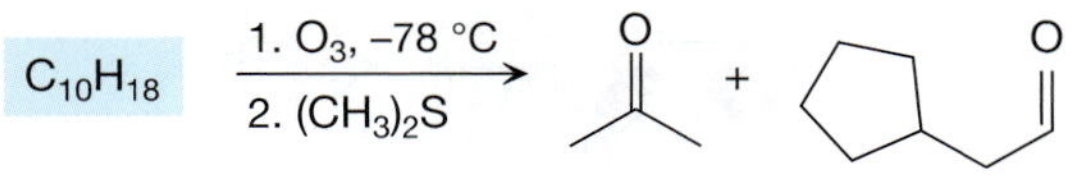

(d)

C7H12
1. O3, −78 °C
2. H2O2
O
O
HO
OH
+ CO2

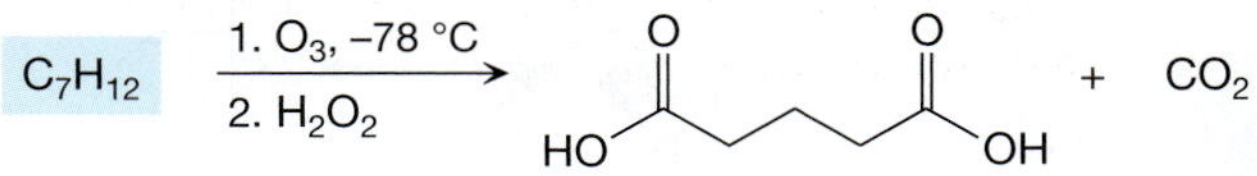

24.59 Draw the major product of the following reaction.

OH
OH
HO
HIO4
?

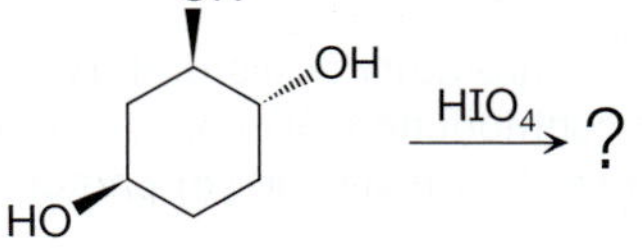

24.60 Rubber degrades when it is exposed to ozone for an extended time—a phenomenon called ozone cracking. Rubber is a natural polymer (a long-chain molecule with a regular repeating structural unit), a portion of which is shown here. Explain why ozone cracking occurs.

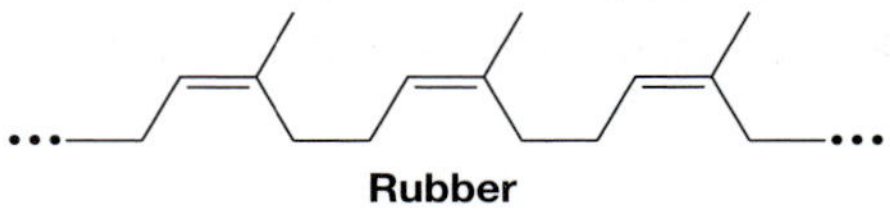

Rubber

24.10 A Molecular Orbital Picture of the Diels–Alder Reaction

24.61 For each of the following reactions, **(a)** draw the HOMO of one reactant and the LUMO of the other, assuming that both reactants are in their ground state, and **(b)** illustrate the HOMO–LUMO interaction and determine if such an interaction leads to a thermally allowed or forbidden reaction.

(i)

H_2C=CH_2 + [diene] ⟶ [cyclooctadiene]

(ii)

[diene] + [alkene] ⟶ [cyclooctadiene]

(iii)

[diene] + [diene] ⟶ [cyclooctadiene]

24.62 Repeat Problem 24.61, assuming that *one* of the reactants is in its lowest excited state. Based on your answers, which of the reactions are photochemically allowed?

24.63 In Section 24.5, we learned how to use resonance hybrids to understand the regiochemistry of Diels–Alder reactions. Frontier MO theory, on the other hand, explains the regiochemistry by taking into account the extent of orbital overlap between the frontier MOs. Consider the diene and dienophile shown here having the two different approaches indicated. In both cases, the *p* orbital contributions are shown for the pertinent frontier MOs. The relative contributions by the *p* AOs to the frontier MOs are indicated by the *p* AO sizes. Use this information to explain which approach is favored.

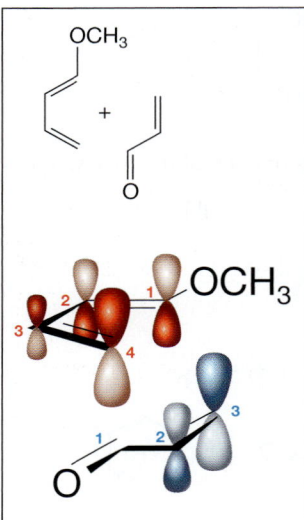

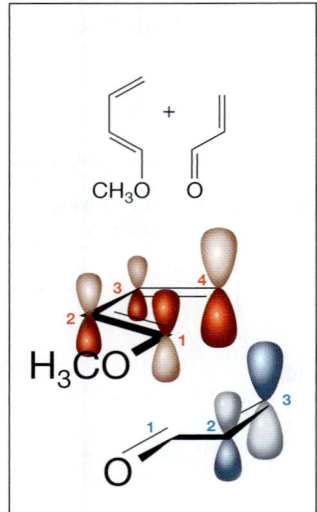

24.64 Similar to a Diels–Alder reaction, an electrocyclic reaction involves a cyclic flow of electrons, as shown in the example here. Unlike a Diels–Alder reaction, an electrocyclic reaction involves just a single conjugated π system. Notice in this example that the two terminal C atoms in the reactant become chiral centers in the product. As shown below, the trans product would be formed if the terminal C atoms rotate in the same direction (i.e., conrotatory) to close the ring, whereas the cis product would be formed if those C atoms rotate in opposite directions (i.e., disrotatory).

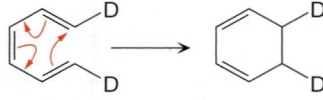

Conrotatory   **Disrotatory**

The favored mechanism (conrotatory or disrotatory) is the one for which the HOMO, during rotation, exhibits constructive interference among the contributing *p* AOs on the terminal C atoms. With this information, determine whether the cis or trans product is favored **(a)** thermally and **(b)** photochemically.

24.65 Repeat Problem 24.64 for the electrocyclic reaction shown here.

Integrated Problems

24.66 A reaction takes place when *cis*-1,3,9-decatriene is heated. The product of that reaction, when treated with excess Br_2 in CCl_4, yields a compound whose formula is $C_{10}H_{16}Br_2$. What is the product of the first reaction?

24.67 T. R. Hoye and coworkers reported the synthetic utility of the alkyne analog of a Diels–Alder reaction. An example is shown below.

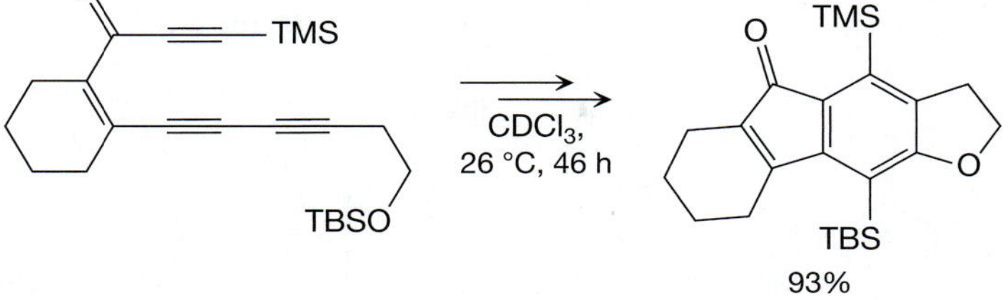

A [4+2] cycloaddition is believed to produce a benzyne intermediate that is quickly "trapped." Show the benzyne intermediate that would be produced in the above reaction.

24.68 Draw the product of the following reaction, assuming that it takes place via an [8+2] cycloaddition.

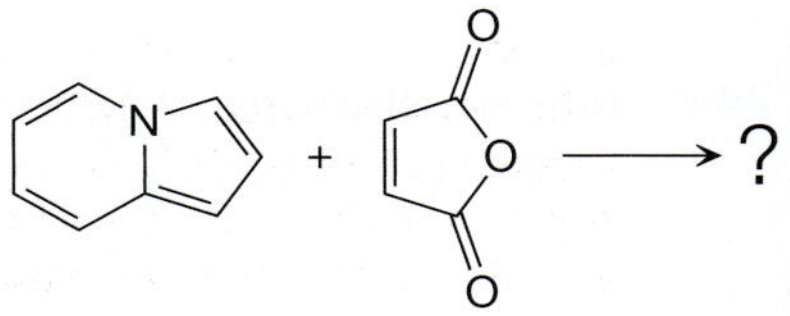

24.69 Strong support for the mechanism of the nucleophilic aromatic substitution reaction that proceeds through a benzyne intermediate comes from the reaction shown here, in which bromobenzene is treated with KNH_2 in the presence of cyclopentadiene. A product that is isolated has the formula $C_{11}H_{10}$. Draw the structure of that product and explain how it validates the production of a benzyne intermediate.

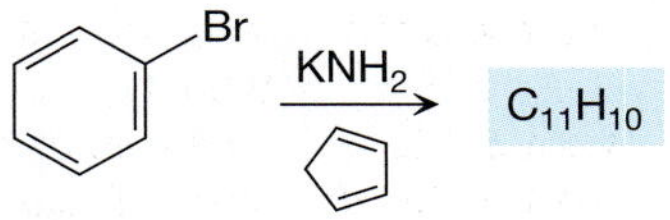

24.70 (SYN) Show how you would synthesize this compound beginning with benzene. *Hint:* See Problem 24.69.

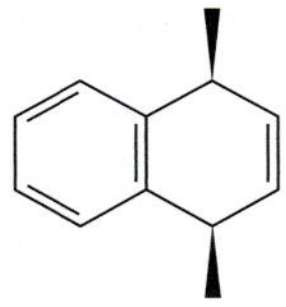

24.71 (SYN) The compound shown here cannot be synthesized *directly* from a Diels–Alder reaction.

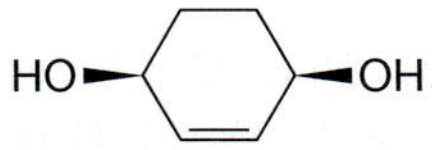

(a) Why not? *Hint:* Examine the reactants that would be required.
(b) What change(s) can be made to the synthesis to get around this problem?

24.72 (SYN) Show how to carry out this transformation using any reagents necessary.

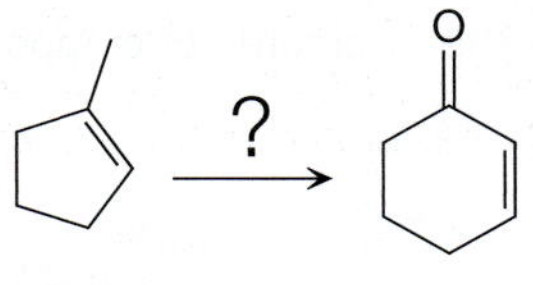

24.73 Gibberellic acid is a plant hormone that controls the development of plants. At various points in their synthesis of gibberellic acid, E. J. Corey and coworkers incorporated the Diels–Alder reactions below. Draw the mechanism and the product of each of these reactions. *Hint:* In the first reaction, which alkene is more electron poor?

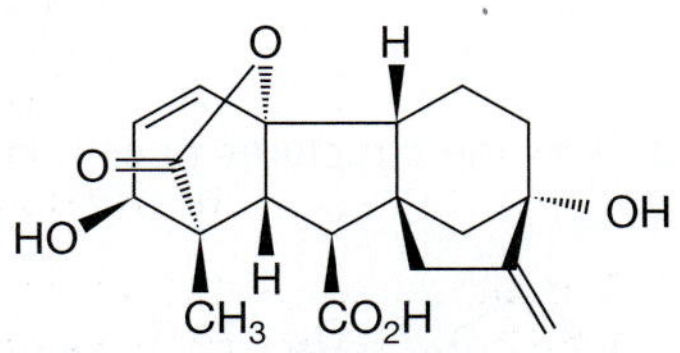

Gibberellic acid

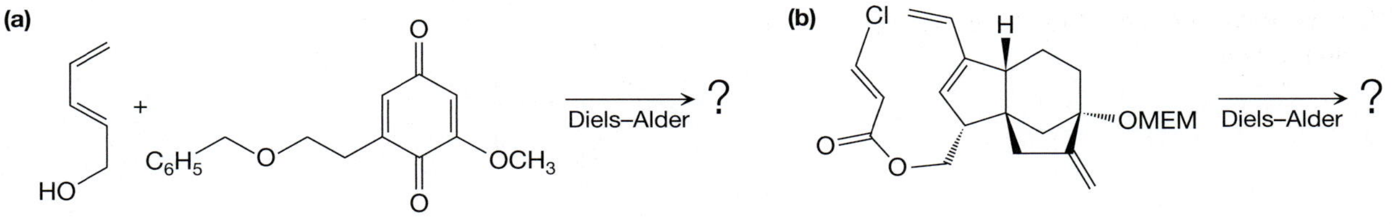

(a)

(b)

24.74 Myrocin C is an antitumor antibiotic. In their synthesis of myrocin C, S. J. Danishefsky and coworkers incorporated the Diels–Alder reactions below. Draw the mechanism and the product of each of these reactions.

Myrocin C

(a)

TBSO +   Diels–Alder   **?**

(b)

                          Diels–Alder   **?**

24.75 Draw the structures of compounds **A–I** in the following synthesis scheme.

$$\xrightarrow[40\ °C]{HBr} \textbf{A} \xrightarrow{NaCN} \textbf{B} \longrightarrow \textbf{C} \xrightarrow[2.\ H_2O,\ HCl]{1.\ NaOH,\ H_2O,\ \Delta} \textbf{D} \xrightarrow[2.\ DIBAH\ -78\ °C]{1.\ CH_2N_2}$$

$$\textbf{E} \xrightarrow[H^{\oplus}]{HO\diagdown OH} \textbf{F} \longrightarrow \textbf{G} \xrightarrow[Pd]{H_2} \textbf{H} \xrightarrow[\Delta]{H_3O^{\oplus}} \textbf{I}$$

24.76 Draw the structures of compounds **A–I** in the following synthesis scheme.

$$\xrightarrow[FeBr_3]{Br_2} \textbf{A} \xrightarrow[ether]{Mg(s)} \textbf{B} \xrightarrow[2.\ H_3O^{\oplus}]{1.\ \triangle} \textbf{C} \xrightarrow[H_2SO_4]{Na_2Cr_2O_7} \textbf{D} \xrightarrow{SOCl_2} \textbf{E} \xrightarrow{AlCl_3}$$

$$\textbf{F} \xrightarrow[EtOH]{NaBH_4} \textbf{G} \xrightarrow[\Delta]{conc\ H_3PO_4} \textbf{H} \xrightarrow{CH_3O\diagdown\diagup OCH_3} \textbf{I}$$

24.77 Draw the structures of compounds **A–F** in the following synthesis scheme.

$$\xrightarrow[\Delta]{NaOH} \textbf{A} \longrightarrow \textbf{B} \xrightarrow[KOH]{KMnO_4} \textbf{C} \xrightarrow[2.\ H_3O^{\oplus}]{1.\ C_6H_5MgBr} \textbf{D} \xrightarrow[\Delta]{conc\ H_2SO_4} \textbf{E} \xrightarrow[H_2O_2,\ H_2O]{OsO_4} \textbf{F}$$

24.78 Draw the structures of compounds **A–H** in the following synthesis scheme.

$$\xrightarrow[2.\ H_2O_2]{1.\ O_3,\ -78\ °C} \textbf{A} \xrightarrow{SOCl_2} \textbf{B} \xrightarrow[excess]{(CH_3)_2NH,} \textbf{C} \xrightarrow{NaBH_4} \textbf{D} \xrightarrow{PBr_3} \textbf{E} \xrightarrow[\Delta]{NaOH} \textbf{F} \xrightarrow{\Delta} \textbf{G} \xrightarrow[H_2O_2,\ H_2O]{OsO_4} \textbf{H}$$

24.79 (SYN) Show how you would synthesize each of the following compounds using ethene and buta-1,3-diene as your only sources of carbon.

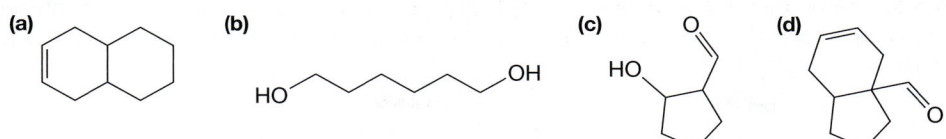

(a)　　　(b)　　　(c)　　　(d)

24.80 When the reactant at the right is warmed to 40 °C, it decomposes into a compound whose ^1H NMR spectrum exhibits one signal—a singlet at 7.3 ppm.

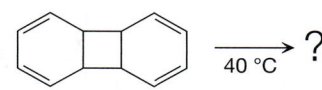

(a) Draw the mechanism for this reaction.
(b) Draw the transition state for this reaction, showing that it proceeds through an antiaromatic transition state.
(c) Explain why this reaction proceeds at mild temperatures, even though it proceeds through an antiaromatic transition state.

24.81 When the compound shown here is heated to 700 °C, a product with the formula C_4H_6 is collected. The ^1H NMR and ^{13}C NMR spectra of C_4H_6 are shown below. Draw the structure of C_4H_6, and draw the mechanism of the reaction that accounts for its formation.

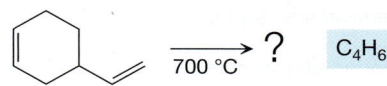

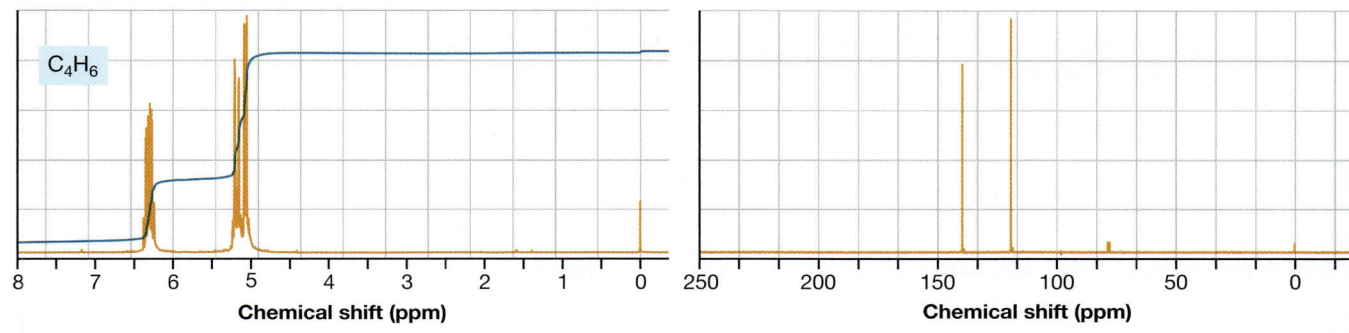

Chemical shift (ppm)　　　Chemical shift (ppm)

24.82 When the compound shown here is heated, ethene gas is evolved and a product with the formula $C_{14}H_8O_2$ is formed. The ^1H NMR and ^{13}C NMR spectra of $C_{14}H_8O_2$ are shown below. (There are two signals >150 ppm in the ^{13}C NMR spectrum. Recall that the ^{13}C NMR signal at 77 ppm is from the $CDCl_3$ solvent.)

(a) Draw the structure of $C_{14}H_8O_2$.
(b) Draw the mechanism that accounts for its formation.
(c) What is the main driving force that favors the products of this reaction?

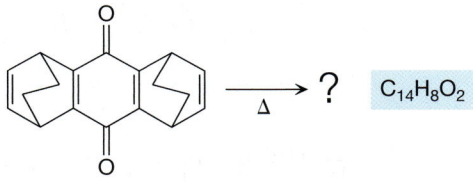

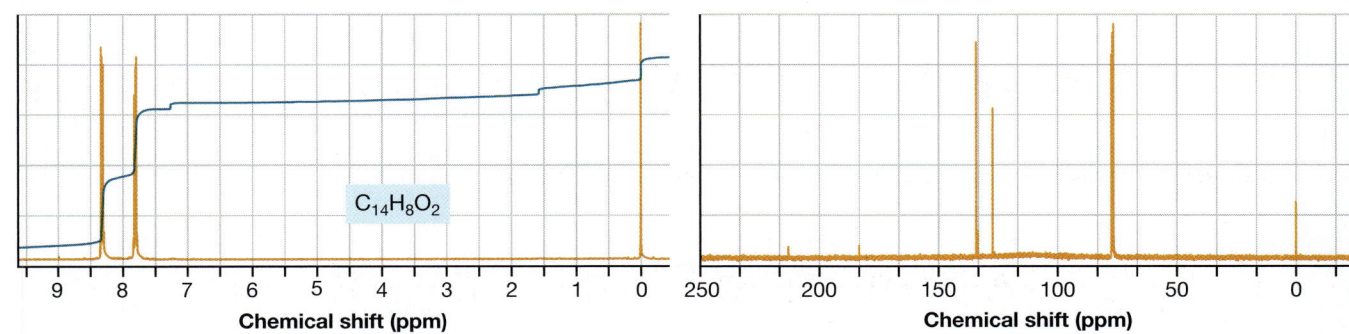

Chemical shift (ppm)　　　Chemical shift (ppm)

24.83 A compound with the formula C_6H_{10} is known to react with one molar equivalent of Br_2 in carbon tetrachloride. When C_6H_{10} was treated with a hot, basic solution of potassium permanganate, followed by acid workup, a product was formed whose IR spectrum exhibits a broad absorption of medium intensity from 2500 to 3300 cm^{-1} and a sharp, intense absorption near 1700 cm^{-1}. The ^1H and ^{13}C NMR spectra of the product are shown. (The multiplet at 39 ppm is from the solvent, DMSO-d_6.) Draw the structure of C_6H_{10}.

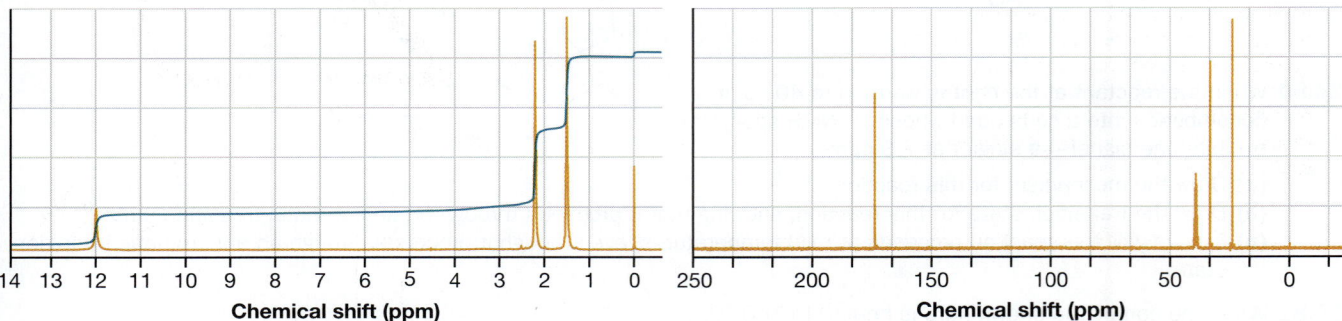
Chemical shift (ppm)   Chemical shift (ppm)

24.84 Heating the compound shown here at 225–235 °C for 8 h produced a new compound with the formula $C_{23}H_{16}N_2O$. The mechanism for this reaction is believed to consist of a [4+2] cycloaddition followed by a [4+2] cycloelimination. The ^1H NMR spectrum of the product contained the following signals: δ 8.33–8.25 and 7.58–6.97 (m, 14 H), 5.46 (s, 2 H). **(a)** Draw the mechanism for this reaction. **(b)** Draw the product.

C_6H_5 ... C_6H_5 ... $\xrightarrow{\Delta}$? $\boxed{C_{23}H_{16}N_2O}$

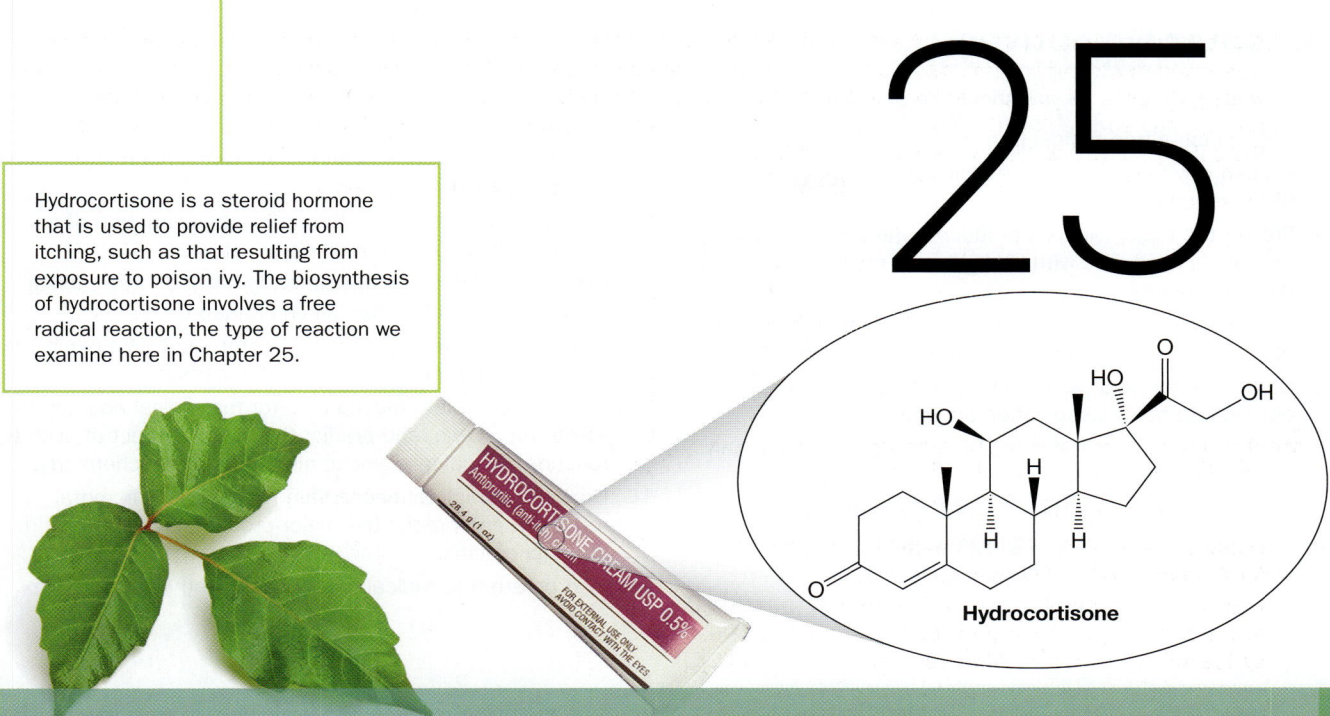

Hydrocortisone

25

Reactions Involving Free Radicals

Throughout this book, our focus has been on reaction mechanisms whose individual steps are driven by the flow of electrons from a site that is electron rich to one that is electron poor. Alkanes and alkyl (i.e., R) groups lack functional groups—they consist of only carbon and hydrogen atoms connected by relatively nonpolar single bonds—so they do not have sites that are especially electron rich or electron poor.

Alkanes, however, are not inert under all conditions. In fact, alkanes react *violently* via combustion. Methane, for example, is used as natural gas to heat our homes, propane is used in gas grills and camp stoves, and a mixture of relatively small hydrocarbons makes up the gasoline used in car engines. These reactions proceed by mechanisms that differ from what we have encountered thus far.

Although we will not explore combustion in depth, we examine other, related reactions that are important to synthetic organic chemistry. These include the halogenation of alkanes, alkene additions, and dissolving metal reductions. All of these reactions involve very highly reactive intermediates called *free radicals*.

A **free radical** is a species that possesses at least one unpaired electron.

Examples include the bromine atom and the methyl radical (Fig. 25-1).

Free radicals such as Br• and H_3C• behave somewhat differently from other species in which all of the electrons are paired—so-called **closed-shell species**.

Unpaired electron

Unpaired electron

Bromine atom

Methyl radical

FIGURE 25-1 Examples of free radicals Each of these species is a free radical because it has one unpaired electron.

1247

Chapter Objectives

On completing Chapter 25 you should be able to:

1. Distinguish a free radical from a closed-shell species.
2. Explain why free radicals are typically very unstable and highly reactive.
3. Predict the major radicals produced when a closed-shell species is irradiated with UV light, and recognize common radical initiators.
4. Use single-barbed curved arrows to describe the flow of electrons in elementary steps involving free radicals.
5. Determine the relative stabilities of free radicals, given appropriate bond dissociation energies.
6. Predict the relative stabilities of alkyl radicals, given only their Lewis structures.
7. Draw all resonance structures of a given free radical.
8. Describe the geometry and hybridization of a carbon atom that has an unpaired electron.
9. Outline the major characteristics of a free radical chain reaction.
10. Identify initiation, propagation, and termination steps of a radical chain-reaction mechanism.
11. Draw the mechanism for the radical halogenation of an alkane and predict the major product.
12. Explain the difference in selectivity between free radical bromination and free radical chlorination.
13. Explain the advantage of using *N*-bromosuccinimide in the free radical bromination of an allylic carbon.
14. Draw the complete mechanism for the radical addition of HBr to an alkene, and predict the major product of such a reaction, including regiochemistry and stereochemistry.
15. Draw the complete mechanism for a dissolving metal reduction and predict the major product, including stereochemistry.
16. Incorporate free radical reactions in synthesis.

Here in Chapter 25, we present a picture of how free radicals behave and develop the picture in a systematic way. We first discuss free radical structure and stability. Then, we take a close look at the most common elementary steps that free radicals undergo. Finally, we delve into the mechanisms of the useful free radical reactions just mentioned.

25.1 Homolysis: Curved Arrow Notation and Radical Initiators

Because free radicals contain an unpaired electron, it is impossible for all of their atoms to have a complete octet. Therefore:

> Free radicals are generally very unstable compared to analogous closed-shell species and are highly reactive, so they usually cannot be isolated for any substantial length of time.

Once a radical is produced, it will typically react very quickly with other species present, including the solvent.

By and large, radicals are produced from uncharged, closed-shell precursors via *homolytic bond dissociation*, or *homolysis*.

> **Homolytic bond dissociation**, or **homolysis**, is the breaking of a covalent bond, whereby the electrons making up that bond are distributed equally to the atoms that are disconnected.

Two bromine radicals (Br•), for example, can be produced from molecular bromine, Br_2, by homolysis of the Br—Br bond (Equation 25-1). Homolysis of a C—H bond in methane (Equation 25-2), on the other hand, produces a methyl radical ($H_3C•$) and a hydrogen radical (H•).

Homolysis of the Br——Br bond   **Two identical radicals are produced.**

$$Br\overset{\curvearrowright}{-}Br \xrightarrow{\Delta} Br\cdot + \cdot Br \qquad (25\text{-}1)$$

Homolysis of the C——H bond   **Two different radicals are produced.**

$$H-\overset{\overset{\displaystyle H}{|}}{\underset{\underset{\displaystyle H}{|}}{C}}\overset{\curvearrowright}{-}H \xrightarrow{h\nu} \overset{H}{\underset{H}{C}}{\cdot} + \cdot H \qquad (25\text{-}2)$$

The curved arrow notation used to describe homolysis differs from that used in previous chapters to describe elementary steps. Specifically, *single-barbed arrows* are used for homolysis.

> A **single-barbed arrow** (⌒) represents the movement of a *single electron*.

The curved arrow notation in Equations 25-1 and 25-2 explicitly shows that one electron from the covalent bond moves to one atom and the second electron moves to the other atom. In contrast, we have used *double-barbed arrows* (⌒) in previous elementary steps to show the movement of *pairs* of electrons in closed-shell species.

CONNECTIONS Cl_2 (see Your Turn 25.1) is often used to protect drinking water supplies from harmful bacteria because Cl_2 reacts with water to produce HOCl, a bactericide. In 1905 in Lincoln, England, this type of chlorination process helped stop a typhoid fever epidemic that was traced back to a contaminated water supply.

YOUR TURN 25.1

The homolysis of Cl_2 produces two chlorine radicals:

$$Cl-Cl \longrightarrow Cl\cdot + \cdot Cl$$

Add the appropriate curved arrows to depict this reaction.

Answers to Your Turns are in the back of the book.

PROBLEM 25.1 Draw the appropriate curved arrows and draw the products for the homolysis of each of the following bonds. **(a)** The C——C bond in ethane; **(b)** A C——H bond in ethane; **(c)** The C——Br bond in 2-bromopropane; **(d)** A C——H bond in benzene

The diagram in Figure 25-2 shows that it takes 439 kJ/mol (105 kcal/mol) of energy for a C——H bond in CH_4 to undergo homolysis and thereby produce $H\cdot$ and $H_3C\cdot$. This corresponds to a vanishingly small equilibrium constant of $<10^{-77}$, and should give you some idea of just how unstable these free radicals are.

Many other single bonds have similar homolytic bond dissociation energies, as shown in Tables 25-1 and 25-2 (next page). Table 25-1 lists bond dissociation energies of bonds involving hydrogen, whereas Table 25-2 lists bond dissociation energies of bonds that involve atoms other than hydrogen.

Because of the large amount of energy required for homolysis, the concentration of free radicals in solution is essentially zero under normal conditions for most species.

FIGURE 25-2 Reaction energy diagram for the homolysis of a C——H bond in CH_4 The homolytic bond dissociation energy of 439 kJ/mol (105 kcal/mol) is the difference in energy between the energy minimum and the energy of the separated free radicals at the right of the figure.

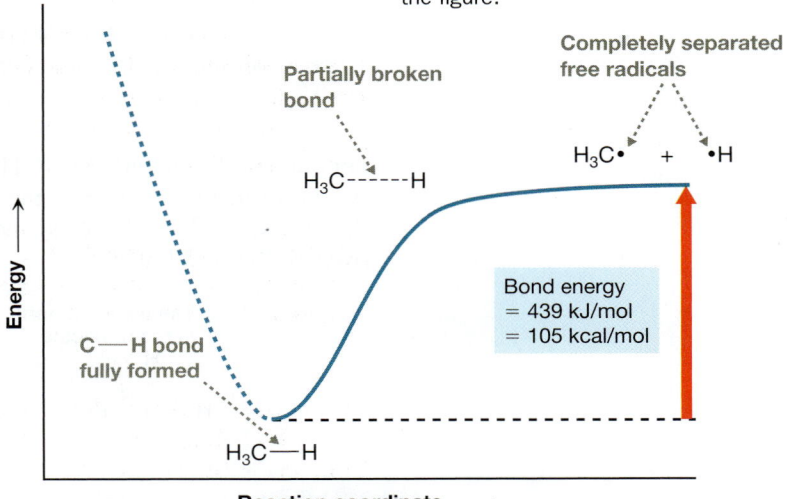

TABLE 25-1 Homolytic Bond Dissociation Energies Involving Bonds to Hydrogen[a]

Bond	HOMOLYTIC BOND DISSOCIATION ENERGY	
	kJ/mol	kcal/mol
H—H	436	104
H—F	569	136
H—Cl	431	103
H—Br	368	88
H—I	297	71
H—OCH_3	440	105
H—CH_3	439	105
H—CH_2CH_3	421	101
H—$CH_2CH_2CH_3$	422	101
H—$CH(CH_3)_2$	410	98
H—$C(CH_3)_3$	400	96
H—$CH=CH_2$	464	111
H—CH_2—$CH=CH_2$	369	88
H—CH_2—C_6H_5	376	90
H—CH_2Br	427	102
H—CH_2OH	402	96

[a]Adapted from Luo, Y.-R. *Comprehensive Handbook of Chemical Bond Energies*; CRC Press: New York, 2007.

TABLE 25-2 Homolytic Bond Dissociation Energies of Bonds Involving Atoms Other than Hydrogen[a]

Bond	HOMOLYTIC BOND DISSOCIATION ENERGY	
	kJ/mol	kcal/mol
H_3C—CH_3	377	90
Br—CH_3	294	70
HO—CH_3	385	92
HO—OH	211	50
F—F	159	38
Cl—Cl	243	58
Br—Br	192	46
I—I	151	36

[a]Adapted from Luo, Y.-R. *Comprehensive Handbook of Chemical Bond Energies*; CRC Press: New York, 2007.

> To produce any reasonable concentration of free radicals from a closed-shell species, energy can be supplied in the form of heat (Δ) or UV light ($h\nu$).

See how the notations Δ and $h\nu$ are used in Equations 25-1 and 25-2.

For a molecule that has only one type of covalent bond, such as Br_2 or CH_4, only one set of homolysis products is possible. For molecules that have more than one type of covalent bond, however, more than one set of homolysis products is possible, but one homolysis reaction generally dominates.

> The major products of homolysis derive from breaking the weakest bond in the molecule—that is, the bond with the smallest bond dissociation energy.

For example, bromomethane (CH_3Br) has three C—H bonds and one C—Br bond. As shown in Equation 25-3, the C—Br bond (294 kJ/mol, 70 kcal/mol) is weaker than the C—H bond (427 kJ/mol, 102 kcal/mol), so homolysis leads almost exclusively to Br• and H_3C•.

The weakest bond in the molecule

Homolysis of a C—H bond

$$H• + •CH_2Br \quad (25\text{-}3a)$$

Homolysis of the C—Br bond

$$H_3C• + •Br \quad (25\text{-}3b)$$

SOLVED PROBLEM 25.2

Predict the major homolysis products of ethane.

$$CH_3CH_3 \xrightarrow{h\nu} ?$$

Think What are the distinct types of bonds in the molecule? Using the tables of bond dissociation energies, can you determine which bond is the weakest? What are the products after homolysis of that bond?

Solve There are two distinct types of bonds. One is H—CH_2CH_3, and the other is H_3C—CH_3. The bond dissociation energy of H—CH_2CH_3 is 421 kJ/mol (101 kcal/mol) (Table 25-1), and that of H_3C—CH_3 is 377 kJ/mol (90 kcal/mol) (Table 25-2). Therefore, H_3C—CH_3 is weaker, and its homolysis leads to the major products.

$$H_3C{-}CH_3 \xrightarrow{h\nu} H_3C\bullet \ + \ \bullet CH_3$$

PROBLEM 25.3 Predict the major homolysis products of methanol.

$$CH_3OH \xrightarrow{h\nu} ?$$

Free radicals can conceivably be produced from the homolysis of a covalent bond in virtually any molecule, but in practice, some precursors are particularly well suited to producing free radicals. These **radical initiators** include molecular halogens (Cl_2, Br_2, and I_2), peroxides (RO—OR), **N-bromosuccinimide (NBS)** (Equation 25-4), and **2,2′-azobisisobutyronitrile (AIBN)** (Equation 25-5). They are often added to a reaction mixture to initiate a reaction that proceeds by a free radical mechanism. (See Section 25.4.)

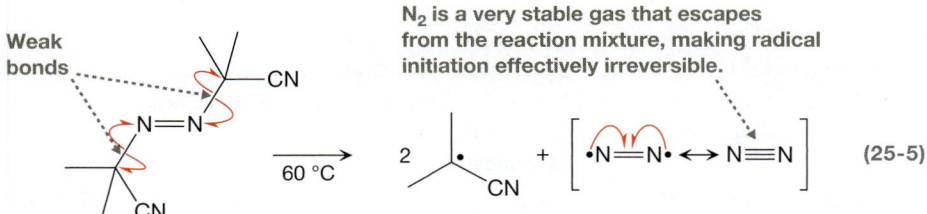

N-Bromosuccinimide (NBS)

(25-4)

Weak bonds —— CN

N_2 is a very stable gas that escapes from the reaction mixture, making radical initiation effectively irreversible.

2,2′-Azobisisobutyronitrile (AIBN)

(25-5)

An ether   **An ether hydroperoxide**

Radical initiators generally have a rather weak bond, making the bond relatively easy to break.

The bond dissociation energy of the N—Br bond in NBS, for example, is 276 kJ/mol (66 kcal/mol), which is a little over half the strength of a C—H bond in CH_4. AIBN is advantageous, too, because homolysis of the two C—N single bonds produces gaseous N_2, which permanently leaves the reaction mixture and makes homolysis effectively irreversible.

Verify that halogens and peroxides have weak bonds by comparing the bond energies of the following bonds to that of a typical C—H bond.

Cl—Cl _____   Br—Br _____   I—I _____

HO—OH _____   Compare to H_3C—H _____

25.2 Structure and Stability of Alkyl Radicals

Although free radicals tend to be very unstable, we can assign different *relative* stabilities to various radicals, based on the energy required to produce them via homolysis (Tables 25-1 and 25-2). For example, Equation 25-6 shows the C—H homolysis reactions of a variety of alkanes, yielding the H• atom and an alkyl radical (R•) as the products.

$H_3C•$ + •H **Methyl radical**	439 kJ/mol (105 kcal/mol)	(25-6a)
H_3C—$•CH_2$ + •H **1° radical**	421 kJ/mol (101 kcal/mol)	(25-6b)
H_3C—$•CH$ | CH_3 + •H **2° radical**	410 kJ/mol (98 kcal/mol)	(25-6c)
H_3C—$C•$ | CH_3 (CH_3) + •H **3° radical**	400 kJ/mol (96 kcal/mol)	(25-6d)

The free energy diagrams for these homolysis steps are shown in Figure 25-3. (For convenience, we have placed the various alkanes at the same energy. This is not actually true, but it allows us to better focus on the reactive species—the radicals.)

A H• radical is produced in all four reactions, so any difference in product energy must come from the different stabilities of the alkyl radicals that are produced. For example, it takes 18 kJ/mol (4 kcal/mol) of additional energy to produce a methyl radical than it does to produce an ethyl radical (439 vs. 421 kJ/mol; 105 vs. 101 kcal/mol), so •CH_3 is 18 kJ/mol (4 kcal/mol) higher in energy than •CH_2CH_3 in Figure 25-3. In other words, •CH_3 is 18 kJ/mol (4 kcal/mol) less stable than •CH_2CH_3. In turn, •CH_2CH_3 is 11 kJ/mol (3 kcal/mol) less stable than •$CH(CH_3)_2$, which is 10 kJ/mol (2 kcal/mol) less stable than •$C(CH_3)_3$.

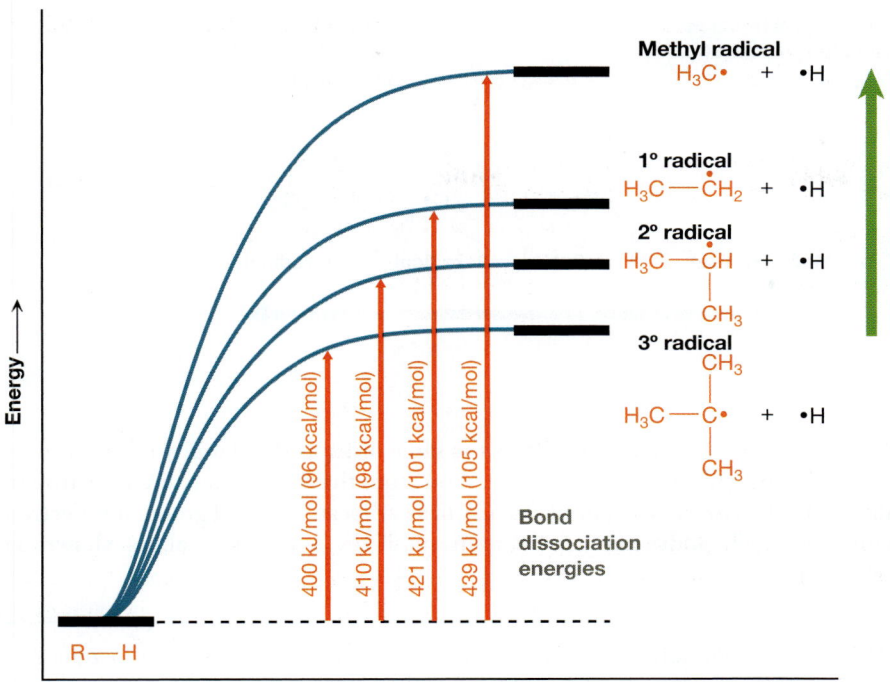

FIGURE 25-3 **Relative stabilities of alkyl radicals** Reaction energy diagrams for C—H homolysis in a variety of alkanes. Alkyl radical stability increases in the order $CH_3\bullet$ $< CH_3CH_2\bullet < (CH_3)_2CH\bullet < (CH_3)_3C\bullet$, where the unpaired electron is on a carbon atom in each case.

YOUR TURN 25.3

The following figure represents homolysis for the H—Cl and H—Br bonds, but it is incomplete. Write the homolytic bond dissociation energies in the boxes provided under the curves. Also, write the Br• and Cl• products in the appropriate boxes at the top of the figure. Which radical is more stable, Br• or Cl•?

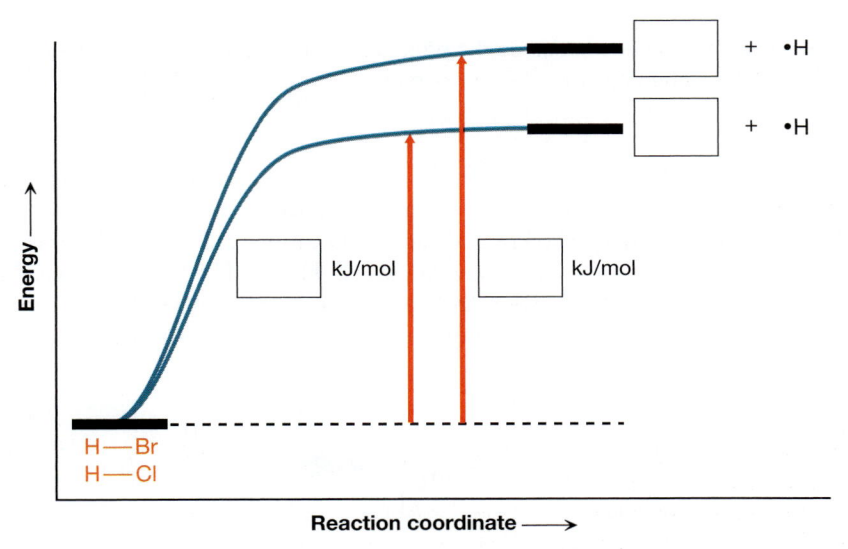

The relative stabilities shown in Figure 25-3 establish the following trend:

Alkyl radical stability increases in the order: Methyl < 1° < 2° < 3°.

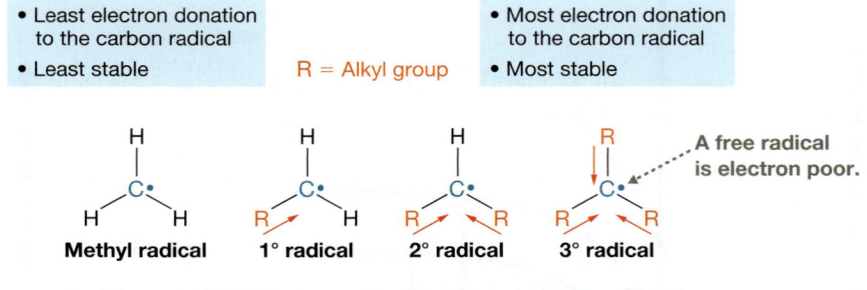

FIGURE 25-4 **Relative stabilities of alkyl radicals** A radical carbon (in blue) is stabilized by the electron-donating effect of attached alkyl groups (in red), so alkyl radical stability increases in the order: methyl radical < 1° radical < 2° radical < 3° radical.

The stability of alkyl radicals follows the same order as the stability of *carbocations* (R^+) (Section 6.6e). This is because the carbon atom that has the unpaired electron in an alkyl radical is *electron poor*, similar to C^+ in a carbocation. Alkyl groups are electron donating, so each additional alkyl group stabilizes a free radical, as shown in Figure 25-4.

PROBLEM 25.4 Homolysis of a C—H bond can occur at three distinct locations on 1,3,5-trimethylcyclohexane to yield three different free radicals. Draw each of the radicals and rank them in order from least stable to most stable.

Effective electronegativity (Section 3.11) also impacts the stability of an alkyl radical. Notice in Table 25-1 that the C—H bond dissociation energies for CH_3CH_3 and $H_2C=CH_2$ are 421 and 464 kJ/mol (i.e., 101 and 111 kcal/mol), respectively. According to these data, the ethyl radical ($\cdot CH_2CH_3$) is more stable than the vinyl radical ($\cdot CH=CH_2$). Thus:

> Alkyl radical stability decreases as the effective electronegativity of the carbon atom that gains the unpaired electron increases.

Just as the atom's ability to accommodate a positive charge decreases (Section 6.6c), so, too, does its ability to accommodate an unpaired electron.

The similarities between alkyl radicals and carbocations are further demonstrated with the allyl ($CH_2=CH-CH_2\cdot$) and benzyl ($C_6H_5-CH_2\cdot$) radicals. Recall that the allyl cation and the benzyl cation are more stable than ordinary primary carbocations, due to resonance delocalization of the positive charge, as shown in Equations 25-7 and 25-8.

Allyl cation (25-7)

Benzyl cation (25-8)

Complete Equation 25-8 by drawing the three missing resonance structures, including curved arrow notation.

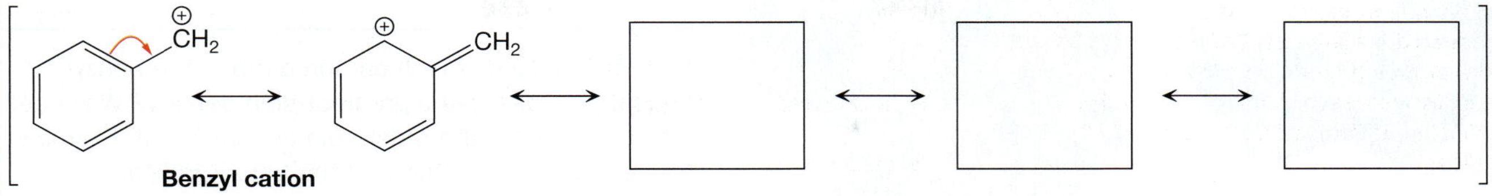

Benzyl cation

The allyl radical and the benzyl radical are more stable than ordinary primary alkyl radicals, too. This is evidenced by the fact that the methyl C—H bonds in CH_2=CH—CH_2—H and C_6H_5—CH_2—H are weaker than that in ethane by 52 and 45 kJ/mol (i.e., by 13 and 11 kcal/mol), respectively.

PROBLEM 25.5 Draw a plot similar to that in Figure 25-3 and Your Turn 25.3, illustrating the relative stabilities of the ethyl radical (CH_3—CH_2•) and the allyl radical.

Equations 25-9 and 25-10 show that the additional stabilization of the allyl and benzyl radicals is also due to resonance. Even though the curved arrow notation is different from that used to indicate resonance in the cations, the unpaired electron in each radical is shared among the same carbon atoms as the positive charge is in the analogous carbocations.

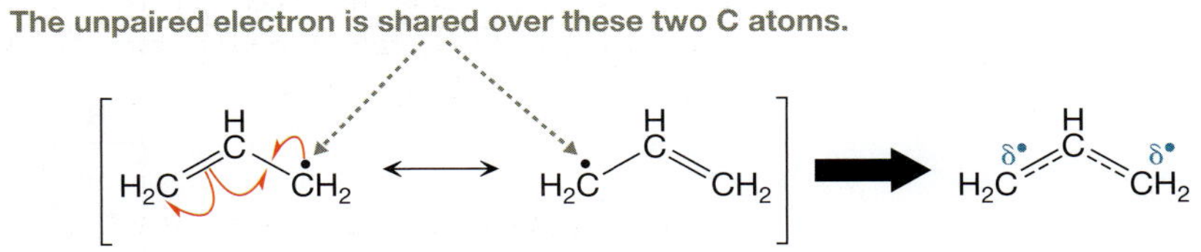

The unpaired electron is shared over these two C atoms.

(25-9)

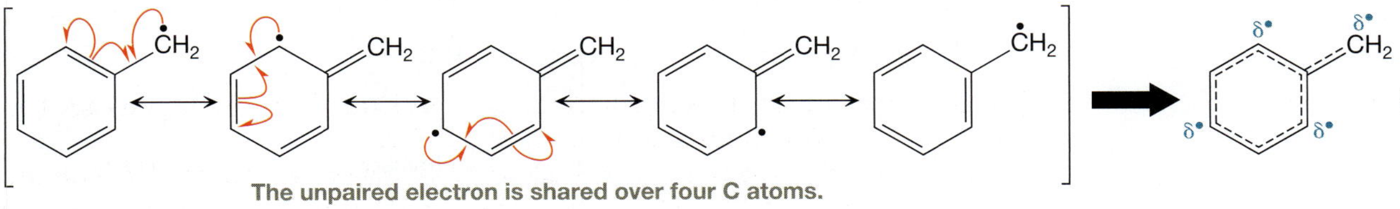

(25-10)

The unpaired electron is shared over four C atoms.

Equation 25-10 is missing the curved arrow notation that shows how the fourth resonance structure is converted to the fifth. Supply that curved arrow notation in the structure on the left that follows.

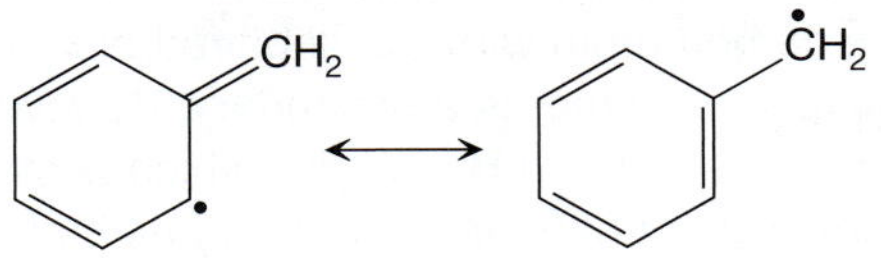

The *resonance hybrids* of the allyl and benzyl radicals (Equations 25-9 and 25-10) illustrate the delocalization of the unpaired electron about the entire species. Another way to depict this is through **electron spin density plots** like those in

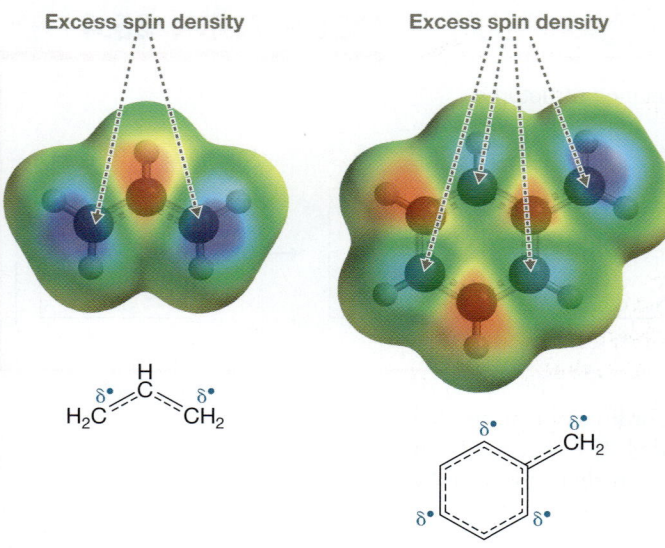

Excess spin density   **Excess spin density**

FIGURE 25-5. In these plots, *electron spin density* (i.e., unpaired electron character) is represented by the blue area—the deeper the blue color, the greater the spin density. Notice how well the electron spin density plots match the resonance hybrids.

PROBLEM 25.6 Which carbon atom in the benzyl radical possesses the greatest spin density? Why does that carbon bear the most spin density? *Hint:* Examine the resonance structures of the benzyl radical.

SOLVED PROBLEM 25.7

Which is the weakest C—C single bond in the molecule shown here?

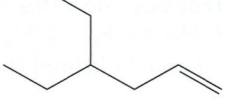

Think What are the distinct types of C—C bonds? What are the

FIGURE 25-5 Electron spin density plots Electron spin density plots of the allyl radical (*top left*) and the benzyl radical (*top right*). The deeper the blue, the greater the electron spin density. Spin density in the allyl radical is delocalized over the two terminal carbon atoms. Spin density in the benzyl radical is delocalized over four carbon atoms.

products on homolysis of each of those bonds? Can the stabilities of the product radicals be differentiated based on resonance and/or alkyl substitution?

Solve There are four distinct C—C single bonds, labeled **A–D** below on the left. The corresponding homolysis products are shown on the right.

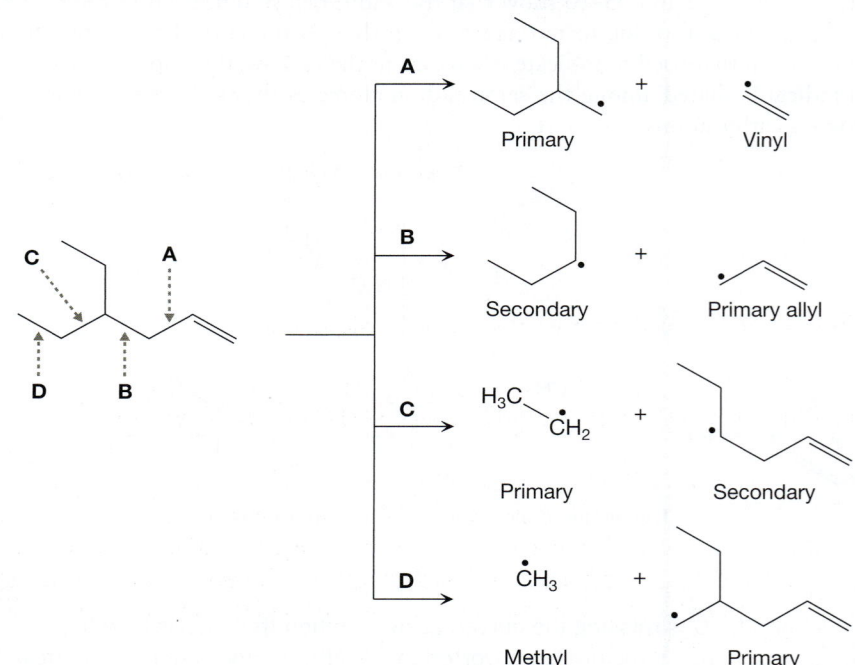

The alkyl radicals produced on homolysis of bonds **B** and **C** have the greatest alkyl substitution on the carbon atom with the unpaired electron—in each case, one is a primary radical, and the other is a secondary. However, the primary alkyl radical produced on homolysis of bond **B** is allylic, which is stabilized substantially by resonance. Thus, homolysis of bond **B** gives the most stable products, making bond **B** the weakest of the C—C bonds.

PROBLEM 25.8 Which is the weakest C—H bond in the molecule shown here?

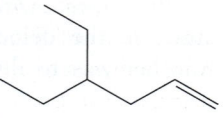

PROBLEM 25.9 For each of the molecules shown here, indicate which C—H bond is the weakest. Which do you think is the weaker of those two C—H bonds? Explain.

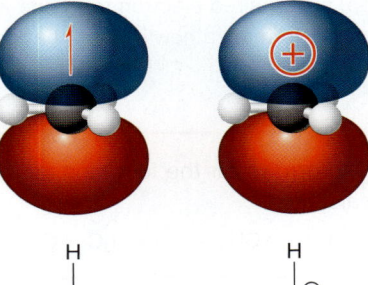

PROBLEM 25.10 Predict the most likely homolysis product when a *mixture* of cyclohexene, HBr, and hydrogen peroxide (H_2O_2) is heated.

Alkyl radicals and carbocations are structurally similar, too. The methyl cation and the methyl radical (Fig. 25-6), for instance, are both *entirely planar*. Because each sp^2 hybrid orbital is used to make a bond to hydrogen, the unpaired electron must reside in the carbon atom's unhybridized p orbital. By comparison, that p orbital is empty in H_3C^+.

$H_3C\cdot$ and $H_3C^\oplus$ are both planar, so each one's C atom is sp^2 hybridized, possessing a single unhybridized p orbital.

The unhybridized p orbital in $H_3C\cdot$ contains the unpaired electron (red arrow).

In $H_3C^\oplus$, the unhybridized p orbital is empty.

FIGURE 25-6 Structural similarities between the methyl radical and the methyl cation The three-dimensional structure of $H_3C\cdot$ is shown on the left, and that of H_3C^+ is shown on the right. Both structures are entirely planar, indicating that the C atom is sp^2 hybridized. The single, unpaired electron in $H_3C\cdot$ resides, therefore, in an unhybridized p orbital. The corresponding p orbital in H_3C^+ is empty.

This planarity is common to other carbon radicals, too.

> A carbon atom that has an unpaired electron tends to be planar and sp^2 hybridized.

Many carbon radicals do not achieve complete planarity, but rather are slightly pyramidal. Those radicals often behave as if they were planar, however, due to rapid inversion of the pyramid, which is important in the stereochemistry of reactions involving radicals (Section 25.6).

25.3 Common Elementary Steps That Free Radicals Undergo

Just as there are a small number of common elementary steps involving closed-shell species (Chapters 6 and 7), there are relatively few elementary steps that are common for free radicals, too. We examine those steps briefly here in Section 25.3, in particular looking at each step's curved arrow notation and aspects of its driving force. We will then be better equipped to tackle the mechanisms of the synthetically useful reactions beginning in Section 25.4, which are nothing more than specific sequences of these elementary steps.

25.3a Radical Coupling

Perhaps the simplest of steps that a free radical can undergo is *radical coupling*, also called *radical recombination*, an example of which is shown in Equation 25-11.

Radical coupling

A σ bond forms.

$$H_3C \cdot \quad \cdot H \longrightarrow H_3C-H \qquad (25\text{-}11)$$

Radical coupling involves two free radicals. In this case, the unpaired electron on C joins the unpaired electron on H to form a new C—H σ bond. This is indicated by the two single-barbed arrows pointing to the bond-forming region. (Just as with double-barbed arrows, the atoms are assumed to follow their own electrons.) In general:

> In a **radical coupling** step, an unpaired electron from one atom joins an unpaired electron from a second atom, forming a new σ bond that connects the two atoms.

YOUR TURN 25.6

> Supply the missing curved arrows for the following radical coupling step involving two $H_3C \cdot$ radicals.
>
> $$H_3C \cdot + \cdot CH_3 \longrightarrow H_3C-CH_3$$

Radical coupling is essentially the reverse of homolysis: Whereas homolysis produces two new free radicals by *breaking* a σ bond, radical coupling removes two free radicals by *forming* a new σ bond. Thus, in contrast to homolysis being very unfavorable:

> Radical coupling is usually very favorable and, therefore, *irreversible*.

FIGURE 25-7 Reaction energy diagram for radical coupling The completely separated $H_3C \cdot$ and $H \cdot$ radicals appear on the left, and the fully formed C—H bond appears on the right. This energy diagram is essentially the mirror image of the one for homolysis, shown previously in Figure 25-2. Unlike other reaction steps we have seen, radical coupling typically proceeds with no energy barrier between reactants and products.

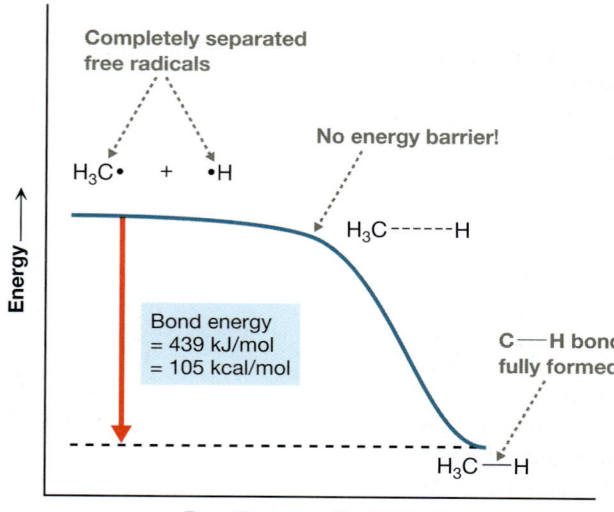

Completely separated free radicals

$H_3C \cdot$ + $\cdot H$

No energy barrier!

H_3C------H

Bond energy = 439 kJ/mol = 105 kcal/mol

C—H bond fully formed

H_3C—H

Energy →

Reaction coordinate →

PROBLEM 25.11 Two diphenylamino radicals undergo radical coupling *reversibly* to produce tetraphenylhydrazine. Explain why this reaction is an exception to the general rule just given.

$$\underset{Ph}{\overset{Ph}{>}}N \cdot \;+\; \cdot N \underset{Ph}{\overset{Ph}{<}} \;\rightleftharpoons\; \underset{Ph}{\overset{Ph}{>}}N-N\underset{Ph}{\overset{Ph}{<}}$$

Because radical coupling is the reverse of homolysis, the free energy diagram for a radical coupling step is precisely the reverse of one for homolysis. For example, the free energy diagram for the radical coupling of a methyl radical and a hydrogen radical, shown in Figure 25-7, is the reverse of CH_4 homolysis, shown previously in Figure 25-2.

Notice at every distance that the two radicals are separated, shortening that distance (i.e., moving to the right in the figure) results in a lower energy. In other words:

Radical coupling steps generally have no energy barrier!

In this regard, free radical coupling is distinct from every other elementary step we have seen. This is further evidence of just how reactive and unstable free radicals are.

25.3b Bimolecular Homolytic Substitution (S_H2)

As we saw in Section 25.3a, a radical coupling step involves the reaction of one free radical with another. It is also possible, however, for a free radical to react with a *closed-shell species*. One example is the **bimolecular homolytic substitution (S_H2)** step shown in Equation 25-12, which involves Br• and a molecule of methane, CH_4.

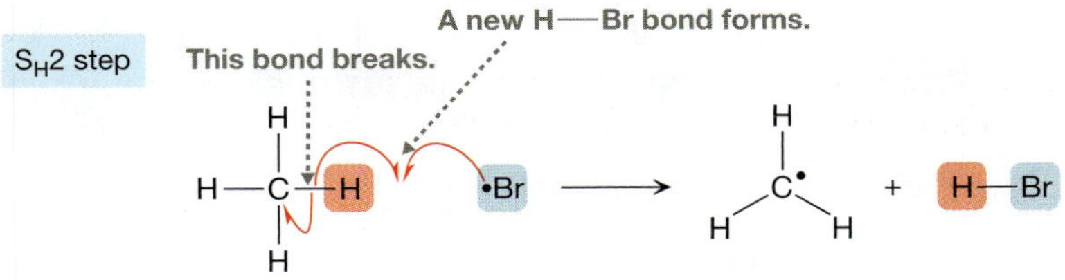

A new H—Br bond forms.

S_H2 step   **This bond breaks.**

$$H-C(H)(H)-H \quad \bullet Br \longrightarrow C(H)(H) \quad + \quad H-Br \qquad (25\text{-}12)$$

The curved arrows show explicitly that the C—H bond is broken, and that one of the two electrons from that bond ends up on the C atom. The other electron joins the unpaired electron on Br• to form a new H—Br bond. Thus, a H atom is transferred from the C atom to the Br atom. For this reason, the S_H2 step in Equation 25-12 is more specifically called a **hydrogen atom abstraction** (Br has *abstracted* a H atom).

YOUR TURN 25.7

Supply the missing curved arrows for the following hydrogen atom abstraction step.

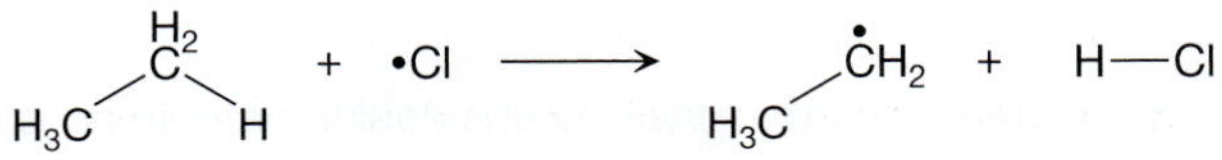

$$H_3C-\overset{H_2}{C}-H \quad + \quad \bullet Cl \longrightarrow H_3C-\overset{\bullet}{C}H_2 \quad + \quad H-Cl$$

An example of an S_H2 step that is *not* a hydrogen atom abstraction is shown in Equation 25-13. In this case, a chlorine radical, Cl•, is abstracted from a molecule of Cl_2.

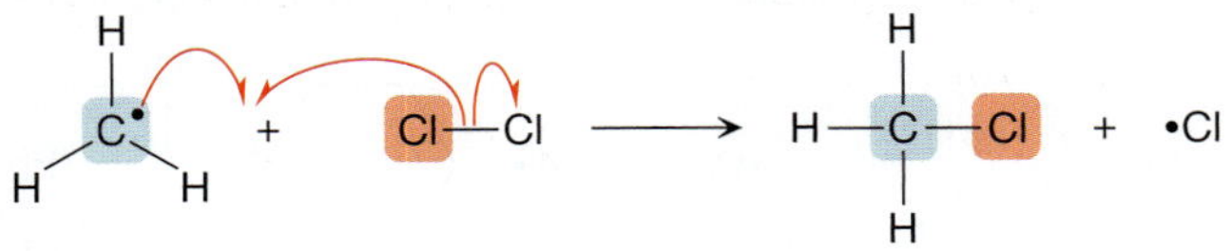

$$\overset{H}{\underset{H}{C}}\bullet \quad + \quad Cl-Cl \longrightarrow H-\overset{H}{\underset{H}{C}}-Cl \quad + \quad \bullet Cl \qquad (25\text{-}13)$$

Although the curved arrow notation for an S_H2 step differs from that of an S_N2 step, the dynamics of the two steps are essentially the same. In an S_N2 step, a nucleophile forms a new bond to an atom attached to a leaving group, thus breaking the bond to the leaving group. The leaving group typically leaves in the form of an anion. Analogously:

In an S_H2 step, a free radical forms a new bond to an atom, causing another bond to that atom to break. In the process, a new free radical is displaced from that atom.

25.3c Radical Addition to an Alkene or Alkyne

Recall from Chapter 11 that the C═C double bond of an alkene is relatively *electron rich*. When an alkene encounters an *electron-poor* free radical, therefore, a *radical addition* step can take place, such as the one shown in Equation 25-14. Likewise, a free radical can add to the C≡C triple bond of an alkyne, as shown in Equation 25-15.

Radical addition

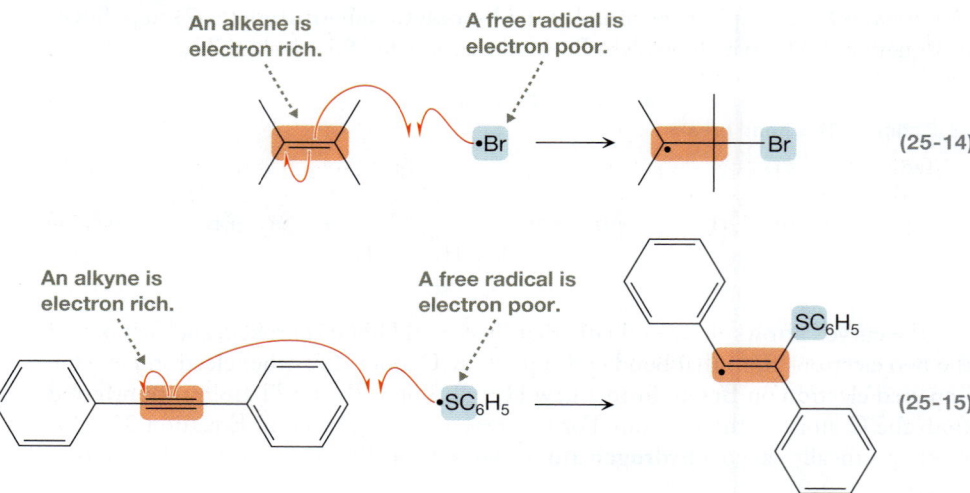

An alkene is electron rich.   A free radical is electron poor.   (25-14)

An alkyne is electron rich.   A free radical is electron poor. SC_6H_5 (25-15)

$\bullet SC_6H_5$

One electron from the multiple bond of an alkene or alkyne joins the unpaired electron from the free radical, and a second electron from the multiple bond ends up as an unpaired electron on carbon. Thus:

> In a **radical addition step**, one atom involved in a double or triple bond forms a new σ bond to a free radical, and the other atom of the double or triple bond gains an unpaired electron.

YOUR TURN 25.8

Supply the curved arrows necessary for the following radical addition step.

25.4 Radical Halogenation of Alkanes: Synthesis of Alkyl Halides

When an alkane such as cyclohexane is treated with a molecular halogen such as Cl_2, no reaction occurs (Equation 25-16). However, if the same mixture is irradiated with UV light, then *halogenation* takes place, producing chlorocyclohexane and HCl (Equation 25-17).

Cyclohexane + Cl$_2$ $\longrightarrow$ No reaction   (25-16)

Cyclohexane + Cl$_2$ $\xrightarrow[h\nu]{\text{40–75 °C}}$ Chlorocyclohexane (93%) + HCl   (25-17)

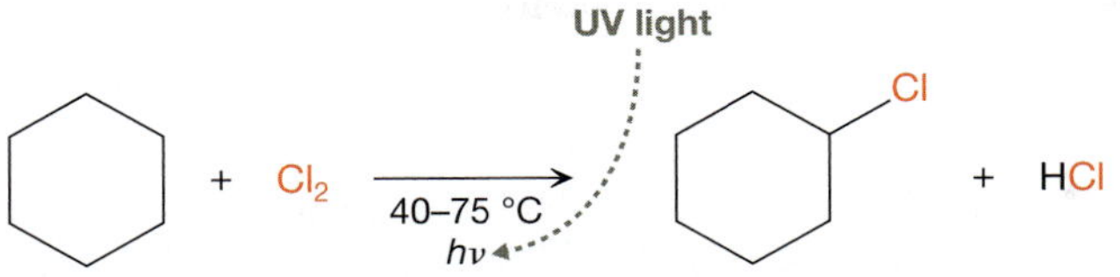

(UV light)

As we see in Section 25.4a, the mechanism involves free radical intermediates, so the reaction in Equation 25-17 is more specifically called a *radical halogenation*.

Radical halogenation can take place with a variety of alkanes, as well as other molecular halogens. In general:

> **Radical halogenation** is a substitution reaction in which a hydrogen atom from an alkane is replaced by a halogen atom from a molecular halogen, X$_2$, producing a hydrogen halide as a by-product.

Here, we discuss some important aspects of radical halogenations, beginning with a close look at the mechanism. Once we understand the mechanism, we then examine issues pertaining to the choice of halogen, including kinetics, thermodynamics, and selectivity.

25.4a The Mechanism of Radical Halogenation: An Introduction to Chain-Reaction Mechanisms

The mechanism for radical halogenation involves free radicals, but no radicals appear in the overall reaction in Equation 25-17. Recall from Section 25.1, however, that Cl$_2$ is a radical initiator and UV light can induce homolysis to produce free radicals. This is called the **initiation step** of the free radical mechanism and, in this case, produces two Cl• radicals (Equation 25-18).

Initiation step for Equation 25-17

UV light brings about the homolysis of the Cl——Cl bond.

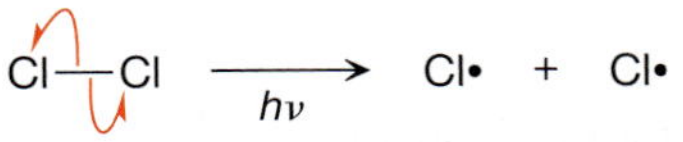

$$\text{Cl}-\text{Cl} \xrightarrow{h\nu} \text{Cl}\bullet \; + \; \text{Cl}\bullet \qquad (25\text{-}18)$$

YOUR TURN 25.9

Cyclohexane can be halogenated with Br$_2$ instead of Cl$_2$:

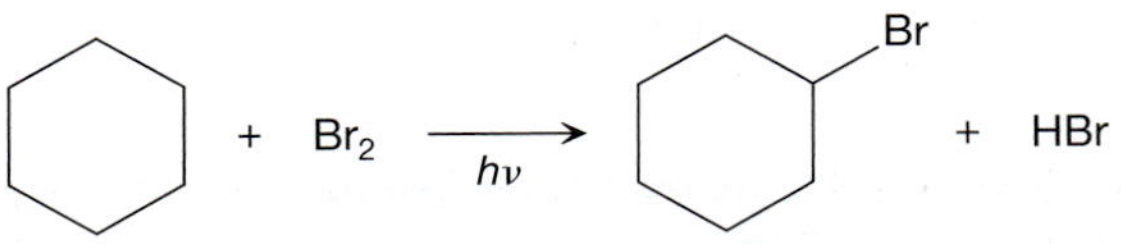

+ Br$_2$ $\xrightarrow{h\nu}$ (bromocyclohexane) + HBr

Draw the initiation step, including the necessary curved arrows.

The next two steps of the mechanism, shown in Equation 25-19a and 25-19b, are responsible for producing the overall products.

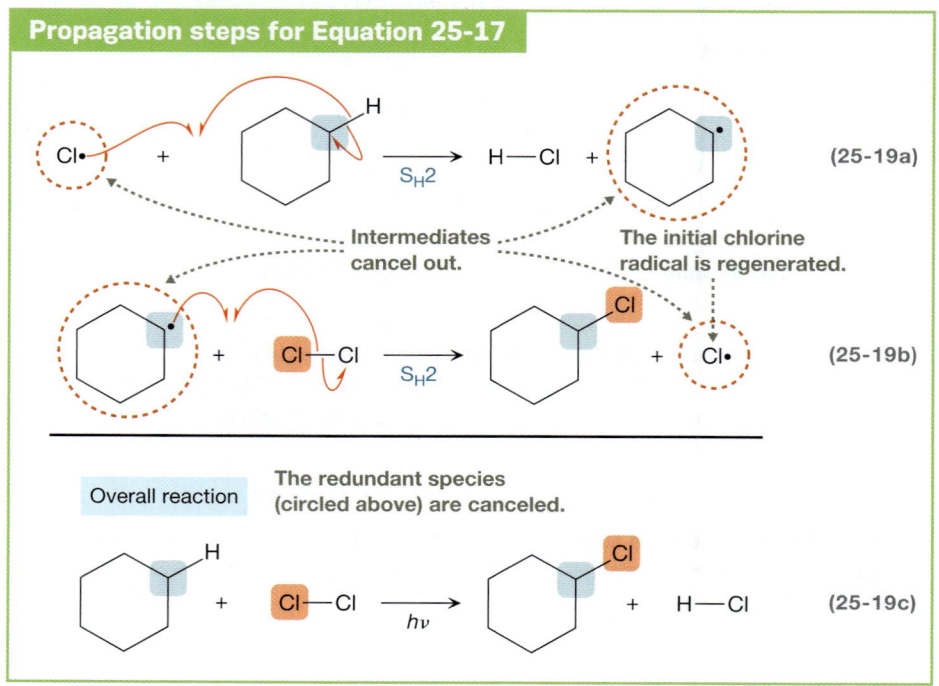

Propagation steps for Equation 25-17

(25-19a)

Intermediates cancel out.

The initial chlorine radical is regenerated.

(25-19b)

Overall reaction

The redundant species (circled above) are canceled.

(25-19c)

Equation 25-19a is an S_H2 step in which Cl• (formed by the homolysis in Equation 25-18) abstracts a H atom from cyclohexane (an overall reactant), producing a molecule of HCl (an overall product) and a cyclohexyl radical. Then, in Equation 25-19b, the cyclohexyl radical abstracts a Cl• from Cl_2 (an overall reactant) in another S_H2 step, producing a molecule of chlorocyclohexane (an overall product) and leaving behind another Cl•.

The Cl• that is produced in Equation 25-19b is available to react with another molecule of cyclohexane, so Equation 25-19a and 25-19b can be repeated many times. Each time these two elementary steps take place, one molecule of cyclohexane and one molecule of Cl_2 are converted to one molecule of chlorocyclohexane and one molecule of HCl. We can see this more clearly in Equation 25-19c, which is the sum of Equation 25-19a and 25-19b (the redundant free radical intermediates have been canceled), and is also the same conversion that is represented by the balanced *overall* reaction shown previously in Equation 25-17. In other words, the steps in Equation 25-19a and 25-19b are responsible for *propagating* the overall reaction, and are therefore called **propagation steps**; together they make up a cyclic process called a **propagation cycle**.

YOUR TURN 25.10

Draw the propagation steps for the reaction in Your Turn 25.9. *Hint:* They are very similar to the ones in Equation 25-19.

There are two ways for a propagation cycle to end. One is for the overall reactants to be completely consumed, as is the case for any chemical reaction. The second is for the radicals involved in the propagation steps to be destroyed.

Any reaction that is responsible for destroying free radicals that are involved in a propagation cycle is called a **termination step**.

Equation 25-20a and 25-20b are two possible termination steps for the radical chlorination of cyclohexane. Both of these are *radical coupling* steps that convert free radicals from the propagation cycle into closed-shell molecules.

Termination steps for Equation 25-17

$$Cl\cdot + \cdot Cl \xrightarrow[\text{Radical coupling}]{} Cl-Cl \qquad (25\text{-}20a)$$

(25-20b)

YOUR TURN **25.11**

Draw two termination steps for the reaction in Your Turn 25.9. *Hint:* They are similar to the ones in Equation 25-20.

PROBLEM 25.13 Draw a termination step for the reaction in Equation 25-17 that is different from the ones shown in Equation 25-20.

SOLVED PROBLEM 25.14

Methane can undergo radical halogenation:

$$CH_4 + Br_2 \xrightarrow{h\nu} H_3C-Br + H-Br$$

Draw the mechanism for this reaction, including the initiation step, the propagation steps, and three different termination steps.

Think For the initiation step, what is the weakest bond in the reactants that can undergo homolysis? For the propagation steps, what S_H2 step can the initial radical undergo to produce the HBr product? What S_H2 step can the resulting radical undergo to produce the CH_3Br product and regenerate the initial radical? For the termination steps, what possible radical coupling steps can take place?

Solve The weakest bond present is the Br—Br bond, which undergoes homolysis in an initiation step to produce two Br• radicals.

$$Br-Br \xrightarrow{h\nu} Br\cdot + Br\cdot$$

In the first of two propagation steps, a Br• abstracts a hydrogen from CH_4 to produce one overall product, HBr, and a methyl radical, $H_3C\cdot$. In the second propagation step, the $H_3C\cdot$ abstracts a Br atom from Br_2 to produce the second overall product, CH_3Br, and regenerate another Br•.

$$Br\cdot + H-CH_3 \longrightarrow Br-H + \cdot CH_3$$

$$H_3C\cdot + Br-Br \longrightarrow H_3C-Br + Br\cdot$$

Termination steps can be a radical coupling involving any radical appearing in a propagation step, such as the following three:

$$H_3C\cdot + \cdot Br \longrightarrow H_3C-Br$$

$$H_3C\cdot + \cdot CH_3 \longrightarrow H_3C-CH_3$$

$$Br\cdot + \cdot Br \longrightarrow Br-Br$$

PROBLEM 25.15 Ethane can undergo radical halogenation. Draw the mechanism for this reaction, including the initiation step, the propagation steps, and three different termination steps.

$$CH_3CH_3 + Cl_2 \xrightarrow[h\nu]{} CH_3CH_2Cl + HCl$$

The termination steps in a radical halogenation mechanism are often quite favorable energetically, so the free radical intermediates that are produced generally exist for only short times. Radical halogenation reactions can have high yields, however, because a single free radical that enters a propagation cycle can lead to many thousands of conversions of reactants to products. This is possible because there is no net consumption of free radicals after a complete propagation cycle—an idea that is captured in Figure 25-8, a schematic representation of the mechanism for the radical chlorination reaction in Equation 25-17.

YOUR TURN 25.12

Study the diagram in Figure 25-8 and match the initiation and propagation steps to either Equation 25-18, 25-19a, 25-19b, 25-20a, or 25-20b.

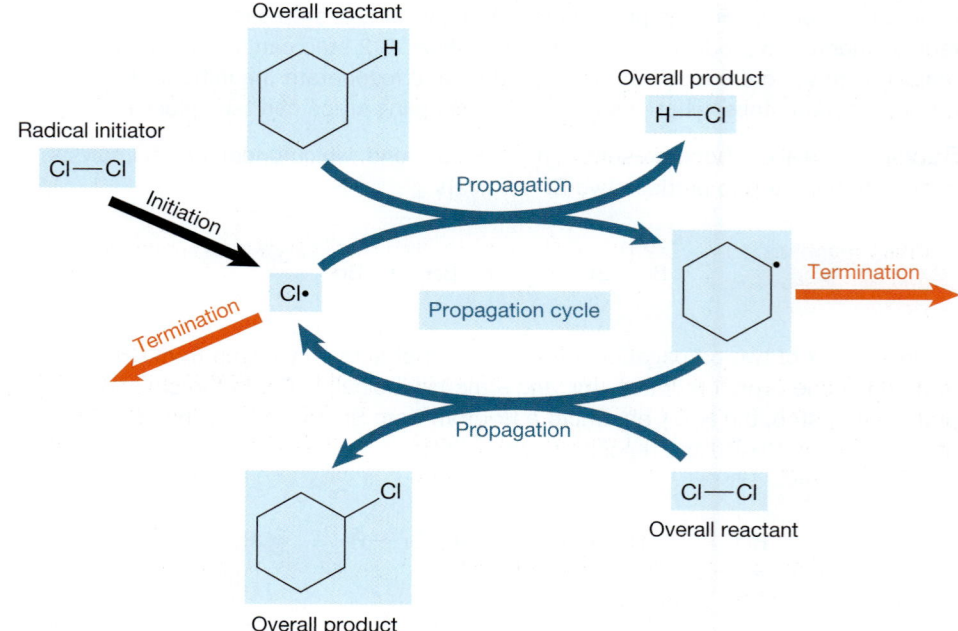

FIGURE 25-8 Schematic representation of the radical chlorination mechanism The initiation step (black arrow) represents the introduction of free radicals into the propagation cycle (blue arrows). Each time a full propagation cycle is completed, one molecule of cyclohexane (*top left*) and one molecule of Cl₂ (*bottom right*) are converted to a molecule of HCl (*top right*) and a molecule of chlorocyclohexane (*bottom left*). Termination steps (red arrows) represent the destruction of the radicals that participate in the propagation cycle.

Halogenated Alkanes and the Ozone Layer

Earth is constantly bombarded by harmful UV radiation from the sun, but a very small amount (<10 ppm) of gaseous ozone (O_3), located roughly 10–20 miles above sea level (in the region of the atmosphere called the stratosphere), protects us from much of that radiation. Ozone is believed to convert UV photons into heat via the mechanism depicted in Scheme A.

In the 1970s, however, the total amount of stratospheric ozone began to decrease an average of about 5% per decade, and that trend continued until the year 2000. Moreover, each year from mid-August to late November (i.e., winter/spring in the southern hemisphere), there is a dramatic decrease in the amount of ozone over Antarctica. This "ozone hole" is depicted in blue in the figure at the right.

Chlorofluorocarbons (CFCs) receive much of the blame. In 1928, certain CFCs, such as CCl_2F_2 and CCl_3F, were patented as types of Freon, and became the dominant coolants used in refrigerators and air conditioners throughout the 20th century. The problem with CFCs is that they are *very* stable. Therefore, as M. Molina and S. Rowland showed in their 1974 *Nature* article, when CFCs are released in the atmosphere, they eventually migrate to the stratosphere, where UV light causes the homolysis of a C—Cl bond to produce Cl•. Cl• can then enter the propagation cycle shown in Scheme B, the sum of which (shown on the third line) is the catalytic breakdown of O_3 to O_2.

The Montreal Protocol was established in 1987 to combat the problem by phasing out the production of CFCs and other ozone-depleting substances. Thanks to these measures, the ozone hole has shrunk more than 5% since it peaked in size in the year 2000. If the Montreal Protocol can be adhered to, then stratospheric ozone is projected to reach pre-1980 levels by 2050, which would be a great success story about the cooperative efforts of science and politics.

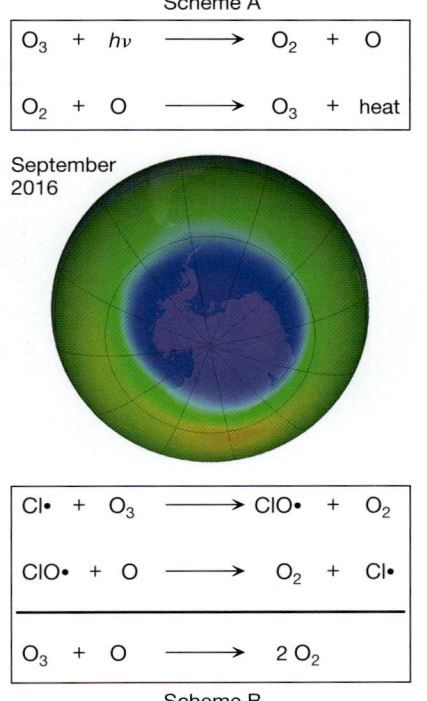

Scheme A

$$O_3 + h\nu \longrightarrow O_2 + O$$

$$O_2 + O \longrightarrow O_3 + heat$$

September 2016

$$Cl\bullet + O_3 \longrightarrow ClO\bullet + O_2$$

$$ClO\bullet + O \longrightarrow O_2 + Cl\bullet$$

$$O_3 + O \longrightarrow 2\,O_2$$

Scheme B

YOUR TURN 25.13

Construct a schematic representation of the free radical bromination of methane in Solved Problem 25.14. Use Figure 25-8 as a guide.

Free radical halogenation is not the only type of reaction whose mechanism consists of initiation, propagation, and termination steps. Rather, it belongs to a broader class of reactions that are called *chain reactions*.

Chain reactions generally have mechanisms that consist of *initiation*, *propagation*, and *termination* steps.

As we saw for free radical halogenation, the free radical intermediates that participate in any chain-reaction mechanism are produced by the initiation steps and are destroyed by the termination steps. Furthermore, the propagation cycle is responsible for converting the overall reactants into the overall products. That is:

The sum of the propagation steps of a chain-reaction mechanism yields the balanced overall reaction.

In Section 25.5, we see how these ideas are applied to the radical addition of HBr to alkenes, which also proceeds by a chain-reaction mechanism. And we will apply these ideas toward a number of other chain-reaction mechanisms in the problems at the end of the chapter.

25.4b Kinetics and Thermodynamics of Radical Halogenation

Although the mechanism of halogenation is the same regardless of the identity of the molecular halogen used, some halogenations are more feasible to carry out than others. One important aspect to consider is the relative reaction rates, which exhibit a periodic table trend, as shown in Equation 25-21a through 25-21d.

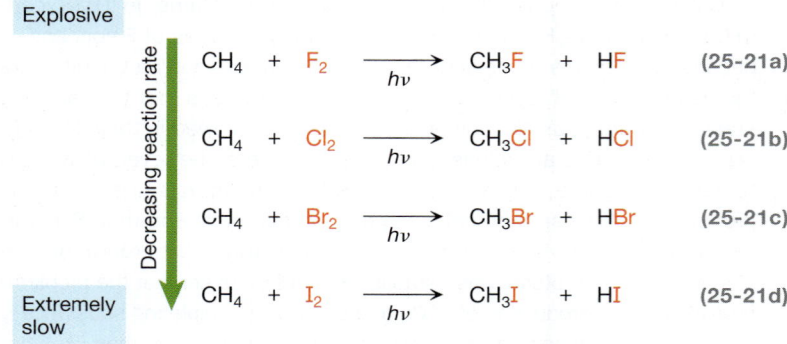

Namely:

> The rate of free radical halogenation decreases in the order: $F_2 > Cl_2 > Br_2 > I_2$.

Fluorination is explosive, even with dilute concentrations. Chlorination is slower than fluorination, but is still potentially explosive. Bromination is slower still and therefore quite controllable. Iodination is so slow that the reaction has to be heated for it to proceed at a reasonable rate.

To begin to understand this trend, recall from the discussion of kinetics in Chapter 8 that a reaction's rate depends on the size of the energy barrier between the reactants and products. To construct the energy diagrams for each of these halogenation reactions, we focus on the two propagation steps in each reaction, because those are the ones that are directly responsible for converting the overall reactants into overall products (review Equation 25-19). The value of ΔH°_{rxn} for each of these propagation steps and for each net reaction is listed in Table 25-3.

TABLE 25-3 Reaction Enthalpies for the Propagation Steps of $CH_4 + X_2 \rightarrow CH_3X + HX$

	ΔH°_{rxn}							
	$X\bullet = F\bullet$		$X\bullet = Cl\bullet$		$X\bullet = Br\bullet$		$X\bullet = I\bullet$	
Reaction	kJ/mol	kcal/mol	kJ/mol	kcal/mol	kJ/mol	kcal/mol	kJ/mol	kcal/mol
$X\bullet + CH_4 \rightarrow HX + CH_3\bullet$	−134	−32	+4	+1	+71	+17	+138	+33
$CH_3\bullet + X_2 \rightarrow CH_3X + X\bullet$	−301	−72	−107	−26	−102	−24	−88	−21
$CH_4 + X_2 \rightarrow CH_3X + HX$	−435	−104	−103	−25	−31	−7	+50	+12

The ΔH°_{rxn} for each propagation step in Table 25-3 can be obtained by subtracting the energy of the bond formed from the energy of the bond broken. To see how this is done, complete the following table for the propagation steps involved in the radical bromination of CH_4 and compare the values you obtain to the ones in Table 25-3. Bond energies can be found in Tables 25-1 and 25-2.

Reaction	Energy of Bond Broken	Energy of Bond Formed	(Energy of Bond Broken) − (Energy of Bond Formed)
$Br\bullet + CH_4 \rightarrow HBr + CH_3\bullet$	$H\text{---}CH_3 =$	$H\text{---}Br =$	
$CH_3\bullet + Br_2 \rightarrow CH_3Br + Br\bullet$	$Br\text{---}Br =$	$Br\text{---}CH_3 =$	

From these values, we can construct the reaction energy diagrams in Figure 25-9. The size of the energy barrier for these reactions decreases in the order: $I_2 > Br_2 > Cl_2 > F_2$. This is primarily because the first of the two propagation steps becomes increasingly exothermic in the same order, which, according to the Hammond postulate (Section 9.3a), should result in a smaller barrier. This is consistent with fluorination being the fastest of the halogenation reactions and iodination being the slowest.

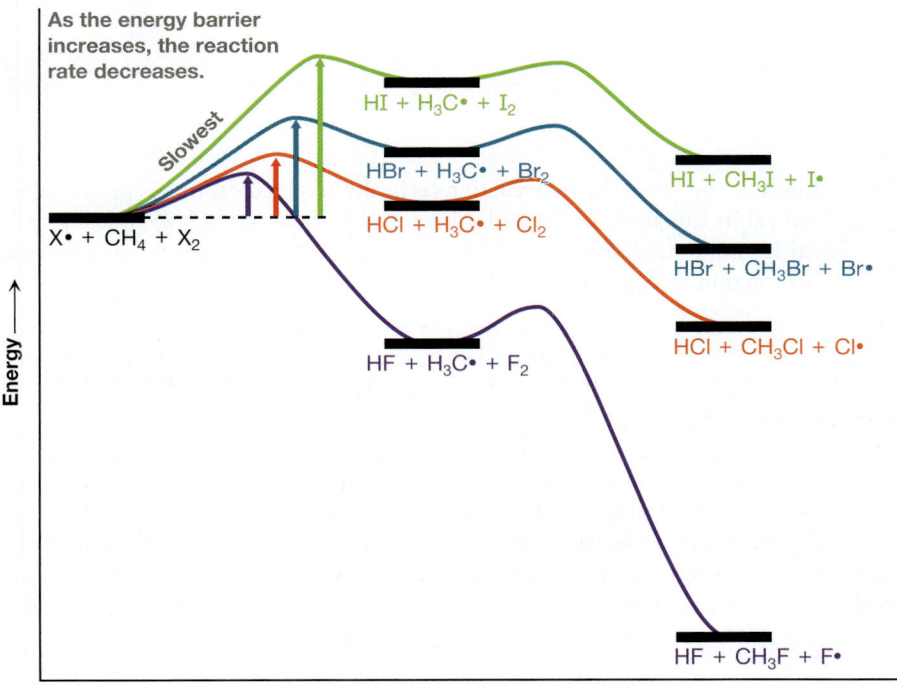

FIGURE 25-9 Energy diagrams for the halogenation of CH_4 The purple curve represents fluorination, the red curve represents chlorination, the blue curve represents bromination, and the green curve represents iodination. The overall energy barriers between reactants and products are indicated by the respective arrows. Fluorination has the smallest energy barrier, followed by chlorination, bromination, and iodination.

Another major factor contributing to the differences in halogenation rates is the *overall* ΔH°_{rxn}. The fluorination reaction is very exothermic *overall*—by about 435 kJ/mol (104 kcal/mol). Having this much heat generated raises the temperature of the reaction mixture, which further increases the reaction rate and contributes to making fluorination explosive. Iodination, on the other hand, is endothermic overall, which is why it must be heated to proceed.

25.4c Selectivity of Chlorination and Bromination

The chlorination and bromination of propane are shown in Equations 25-22 and 25-23 on the next page.

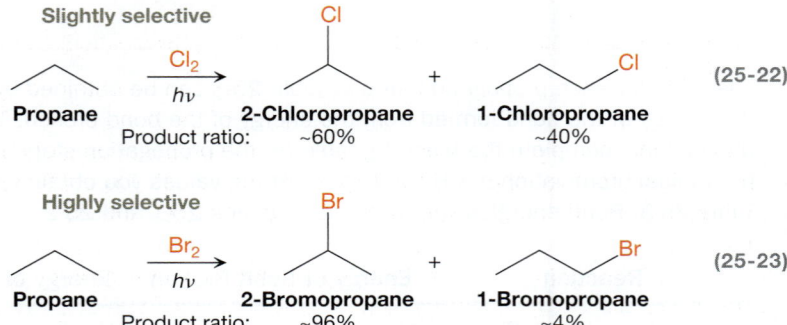

Slightly selective

Propane →(Cl₂, hv)→ 2-Chloropropane + 1-Chloropropane   (25-22)

Product ratio: ~60%   ~40%

Highly selective

Propane →(Br₂, hv)→ 2-Bromopropane + 1-Bromopropane   (25-23)

Product ratio: ~96%   ~4%

In both cases, the 2° halide is favored over the 1° halide. Why does halogenation exhibit this selectivity?

We must consider the first of the two propagation steps—the step in which the H atom is abstracted by the halogen radical. In Equation 25-24, for example, Cl• abstracts a H at the 2° C, and in Equation 25-25, Cl• abstracts a H at the 1° C.

A 2° radical is more stable than a 1° radical.

(25-24)

(25-25)

The only difference between the two reactions is in the alkyl radical that is produced—a 2° alkyl radical in Equation 25-24 and a 1° radical in Equation 25-25. Because a 2° alkyl radical is more stable (Section 25.2), there is a greater driving force for the hydrogen abstraction in Equation 25-24. Generally speaking:

> The rate of halogenation increases with increasing stability of the alkyl radical that is produced on hydrogen abstraction.

Thus, not only does radical halogenation tend to be faster at 2° carbons than 1° carbons, but it also tends to be faster at 3° carbons than 2° carbons. Halogenation is particularly favored when hydrogen abstraction produces a resonance-stabilized alkyl radical (see Your Turn 25.15).

YOUR TURN 25.15

Which of the following hydrogen atom abstraction steps will proceed faster?

A

B

Looking back at Equations 25-22 and 25-23, notice that the *extent* to which the 2° alkyl halide product is favored is much greater for bromination than it is for chlorination. In other words:

Free radical bromination of an alkane is much more selective than free radical chlorination.

Bromination is so much more selective than chlorination primarily because Br• is more stable and less reactive than Cl• (review Your Turn 25.3, p. 1253), reflected by the fact that the H—Br bond is substantially *weaker* than the H—Cl bond. This has a significant impact on the hydrogen abstraction step of the mechanism. As shown in Figure 25-10, the hydrogen abstraction by Cl• is significantly exothermic, whereas the hydrogen abstraction by Br• is significantly endothermic. According to the Hammond postulate (Section 9.3a), therefore, when Br• abstracts a hydrogen atom, the transition states leading to the primary and secondary products should differ significantly in energy (Fig. 25-10b), and the respective rates should differ significantly, too. When Cl• abstracts those H atoms, the transition states are similar in energy (Fig. 25-10a), as are the reaction rates.

YOUR TURN 25.16

Use Table 25-1 to calculate ΔH°_{rxn} for each hydrogen abstraction step depicted in Figure 25-10, which are written below. How do the steps involving Cl• differ from the ones involving Br•?

Reaction	Hydrogen Abstraction Step ΔH°_{rxn} (Energy of Bond Broken) − (Energy of Bond Formed)
Cl• + H—CH$_2$CH$_2$CH$_3$ → Cl—H + •CH$_2$CH$_2$CH$_3$	_____
Cl• + H—CH(CH$_3$)$_2$ → Cl—H + •CH(CH$_3$)$_2$	_____
Br• + H—CH$_2$CH$_2$CH$_3$ → Br—H + •CH$_2$CH$_2$CH$_3$	_____
Br• + H—CH(CH$_3$)$_2$ → Br—H + •CH(CH$_3$)$_2$	_____

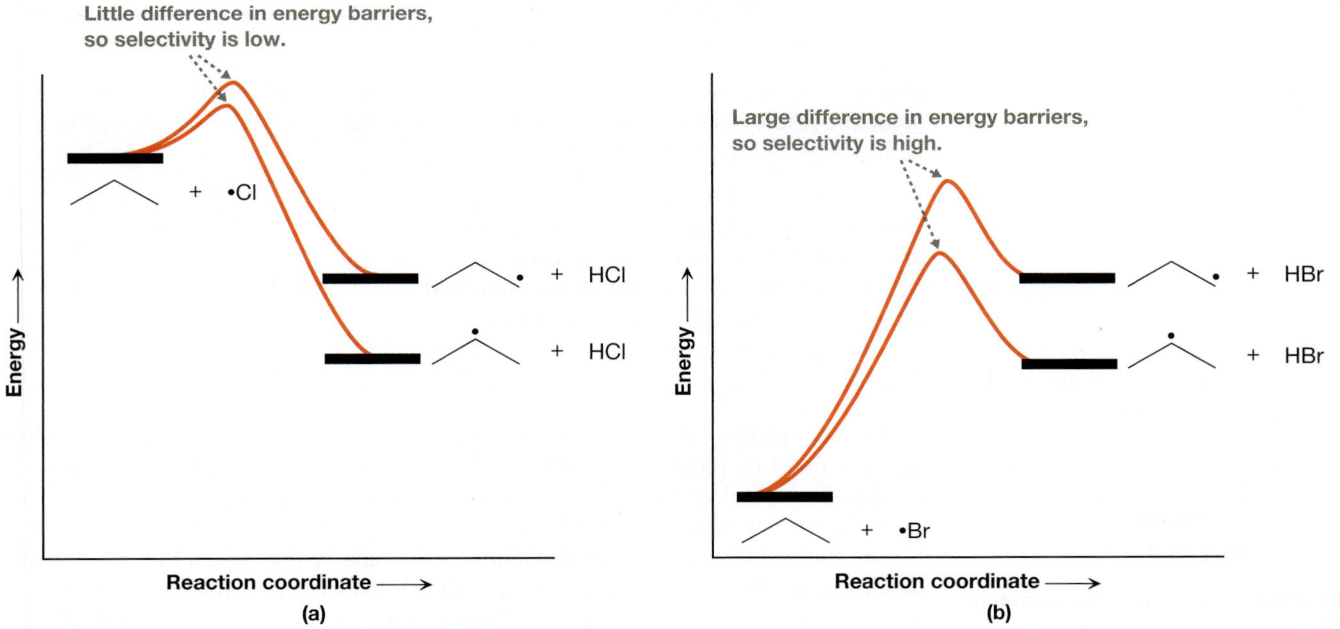

Little difference in energy barriers, so selectivity is low.

+ •Cl

+ HCl

+ HCl

Large difference in energy barriers, so selectivity is high.

+ HBr

+ HBr

+ •Br

Reaction coordinate ⟶

(a)

Reaction coordinate ⟶

(b)

FIGURE 25-10 Selectivity in radical halogenation reactions (a) Hydrogen abstraction by Cl• at a 2° and 1° C atom. (b) Hydrogen abstraction by Br• at a 2° and 1° C. The energy barriers are similar in size in (a), so there is little selectivity in chlorination. The energy barriers are significantly different in size in (b), so bromination occurs with high selectivity.

YOUR TURN 25.17

Construct energy diagrams similar to the ones in Figure 25-10, this time using 2-methylpropane instead of propane. Consider the fact that the abstraction of a hydrogen from 2-methylpropane can take place at either a 1° or 3° carbon. How do the energy diagrams you drew differ from the ones in Figure 25-10?

SOLVED PROBLEM 25.16

Predict the major product of the reaction shown here and draw the complete, detailed mechanism. Include all initiation and propagation steps.

$\xrightarrow[h\nu]{Br_2}$?

Think What is the initial radical that is formed? By what process is it formed? How will that radical interact with the uncharged molecules present? Is regiochemistry a concern?

Solve Br_2 will undergo homolysis (shown here), as suggested by the presence of light, so the initial radical is Br•. With the alkane present, radical halogenation will occur via two propagation steps. First the Br• abstracts a H atom to yield an alkyl radical, R•, then the alkyl radical abstracts a Br atom from Br_2 to yield R—Br and Br•.

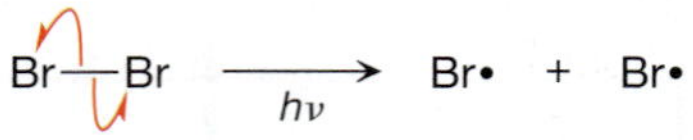

$$Br\!-\!Br \xrightarrow{h\nu} Br\bullet \;+\; Br\bullet$$

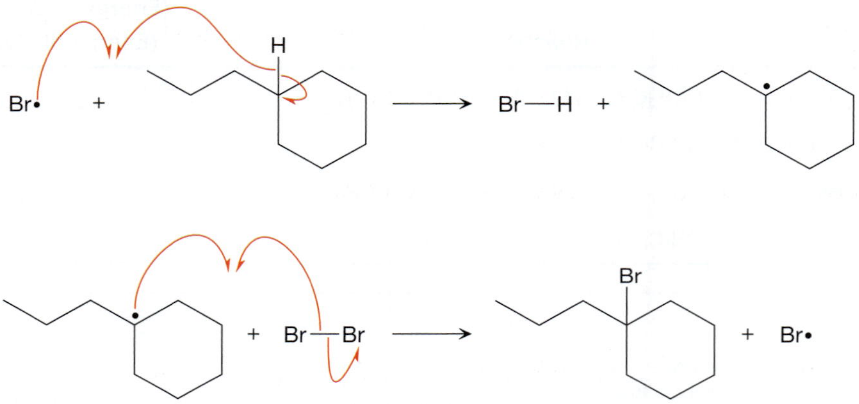

Regiochemistry is a concern because there are multiple, distinct H atoms that can be abstracted by Br•. Recall, however, that Br• is highly selective toward hydrogen atoms attached to 3° carbons, as shown.

PROBLEM 25.17 Predict the major product of the reaction shown here and draw the complete, detailed mechanism. Include all initiation and propagation steps. Draw two plausible termination steps as well.

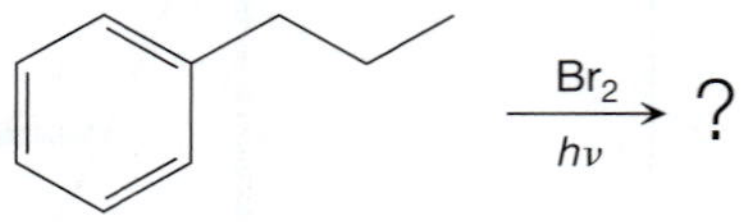

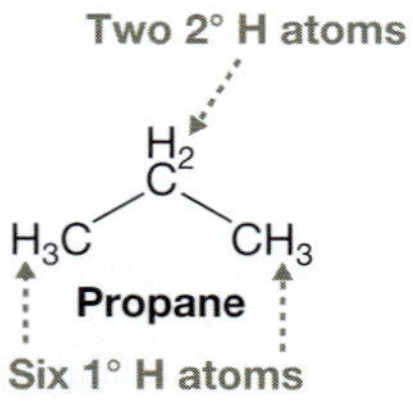

$\xrightarrow[h\nu]{Br_2}$?

We can quantify the selectivity of halogenation by comparing the actual product distribution of each reaction to what the distributions would be if the reactions were completely nonselective (i.e., if all positions were favored equally). Propane, for example, has two equivalent 2° H atoms and six equivalent 1° H atoms, as shown in Figure 25-11. If there were equal likelihood for each of the eight H atoms to react, then there would be a 25% chance (two of eight) that the reaction would take place at one of the 2° H atoms to produce 2-chloropropane, and there would be a 75% chance (six of eight) that the reaction would take place at one of the 1° H atoms to produce 1-chloropropane. In other words, a completely nonselective reaction would produce

Two 2° H atoms

Six 1° H atoms

FIGURE 25-11 Equivalent protons in propane The six 1° H atoms are equivalent and the two 2° H atoms are equivalent.

25% 2-chloropropane and 75% 1-chloropropane, giving a ratio of 1:3, not the observed ratio of 60:40 (i.e., 3:2).

To compute the selectivity of chlorination at the 2° carbon relative to the 1° carbon, we divide the projected nonselective ratio (i.e., 1:3) into the one that is actually observed (i.e., 60:40).

$$\text{Chlorination selectivity } (2°:1°) = \frac{60:40}{1:3} = \frac{60}{1} : \frac{40}{3} = 60:13\tfrac{1}{3} = 4.5:1$$

This selectivity of 4.5:1 means that if a molecule had the same number of 1° and 2° hydrogens, then the product mixture would have 4.5 times more of the secondary alkyl chloride than the 1°.

Repeating this for the bromination in Equation 25-23, we obtain:

$$\text{Bromination selectivity } (2°:1°) = \frac{96:4}{1:3} = \frac{96}{1} : \frac{4}{3} = 96:\tfrac{4}{3} = 72:1$$

Based on the two ratios for halogenation at a 2° versus a 1° carbon, bromination is about 72/4.5 = 16 times more selective than chlorination.

Product distributions from other halogenation reactions have also enabled us to determine the relative selectivities of halogenating 2° versus 3° carbons. These results can be combined with the ones from above to give the following relative selectivities for 3°, 2°, and 1° carbons.

Relative selectivities of halogenation at 3°, 2°, and 1° carbon atoms:

- Chlorination = 6:4.5:1
- Bromination = 1600:72:1

Notice how selective bromination is toward a 3° carbon!

25.4d Radical Bromination Using *N*-Bromosuccinimide

If an alkene undergoes radical bromination, then substitution will most likely take place at the allylic position, due to the resonance stabilization in the allylic radical that would be produced on hydrogen abstraction. Thus, radical bromination of cyclohexene would produce 3-bromocyclohexene, as shown in Equation 25-26a.

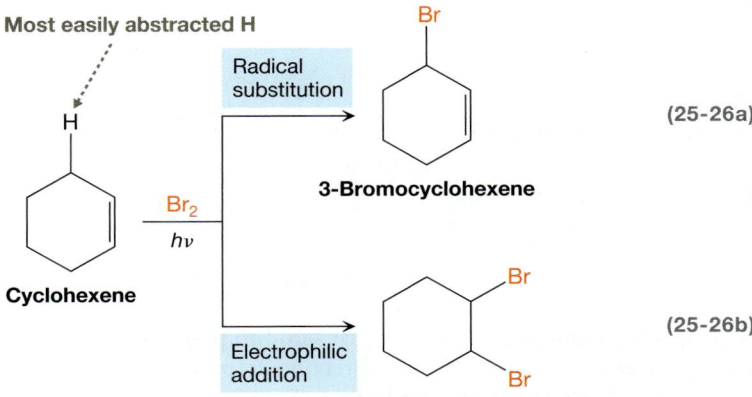

A problem arises, however, if we carry out a bromination by simply treating the alkene with molecular bromine and UV light. As we learned in Chapter 12, Br_2 will also add to the C=C double bond in an *electrophilic addition* reaction (Equation 25-26b).

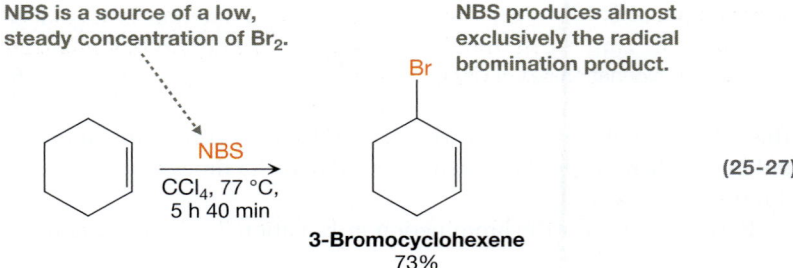

NBS = Br—N

FIGURE 25-12 *N*-Bromosuccinimide and bromination NBS is often used to carry out radical bromination at allylic positions without producing unwanted electrophilic addition products.

To avoid this problem, chemists carry out free radical brominations using *N*-bromosuccinimide (NBS) (Fig. 25-12). As shown in Equation 25-27, NBS reacts with cyclohexene to give only the allylic bromide.

NBS is a source of a low, steady concentration of Br₂.

NBS produces almost exclusively the radical bromination product.

$$
\text{cyclohexene} \xrightarrow[\substack{\text{CCl}_4,\ 77\ °\text{C}, \\ 5\ \text{h}\ 40\ \text{min}}]{\text{NBS}} \text{3-Bromocyclohexene}
$$

(25-27)

3-Bromocyclohexene
73%

N-Bromosuccinimide does not alter the mechanism of radical bromination, but it controls the rates of the competing reactions in Equation 25-26 by controlling the concentration of Br_2 (and, thus, the concentration of $Br\cdot$).

> In radical halogenation reactions, *N*-bromosuccinimide (NBS) is the source of a low, steady concentration of Br_2.

The way in which Br_2 is produced from NBS is shown in the mechanism in Equation 25-28.

Mechanism for the production of Br₂ from NBS

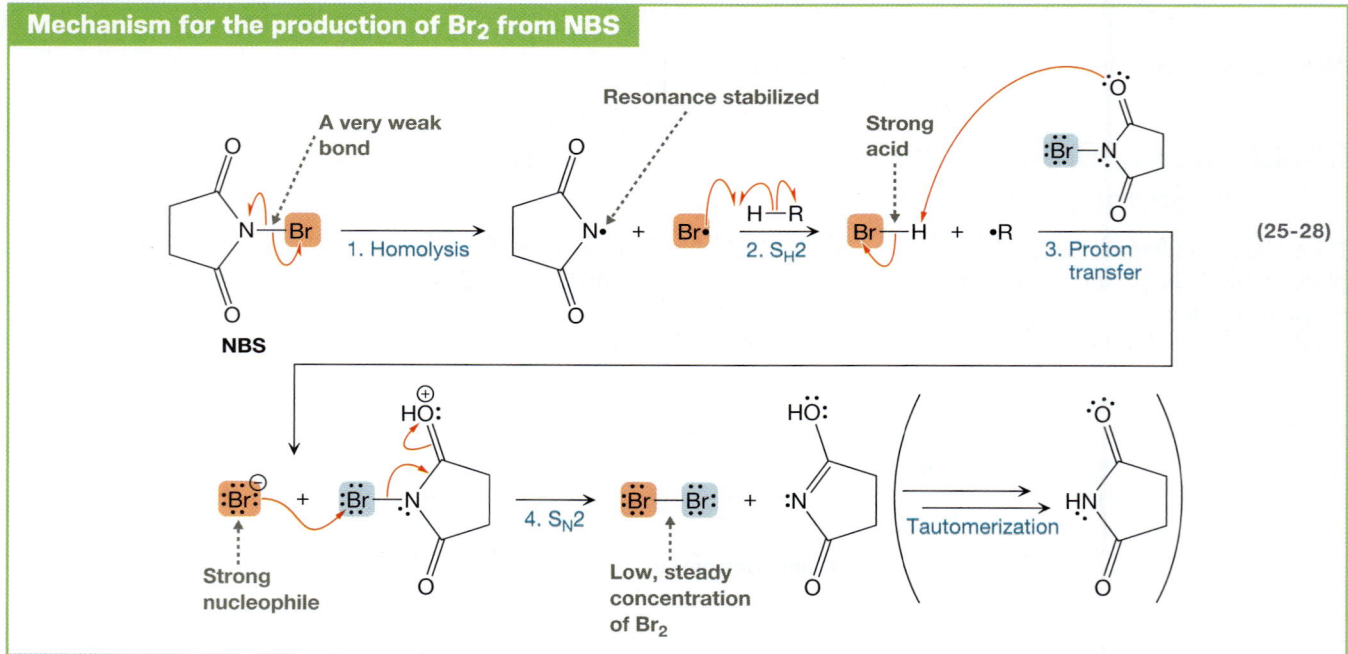

(25-28)

Step 1 is the homolysis of the N—Br bond to produce $Br\cdot$. This step requires only modest heating because the N—Br bond is particularly weak, but could also be accomplished by irradiation with light. In Step 2, the newly formed $Br\cdot$ abstracts a H atom from H—R, producing HBr, a strong acid. In Step 3, therefore, a carbonyl O on NBS is protonated, which also produces Br^-, a strong nucleophile. Finally, Step 4 is an S_N2 step, in which the Br^- nucleophile attacks a Br atom from a second molecule of NBS, yielding Br_2.

Step 1 in Equation 25-28, the homolysis of NBS, produces a resonance-stabilized free radical. In the space provided here, draw the additional resonance structures involving the unpaired electron, and include the appropriate curved arrows. Also, complete the mechanism by explicitly drawing the two proton transfer steps that make up the tautomerization in Equation 25-28. (Are the conditions acidic or basic?)

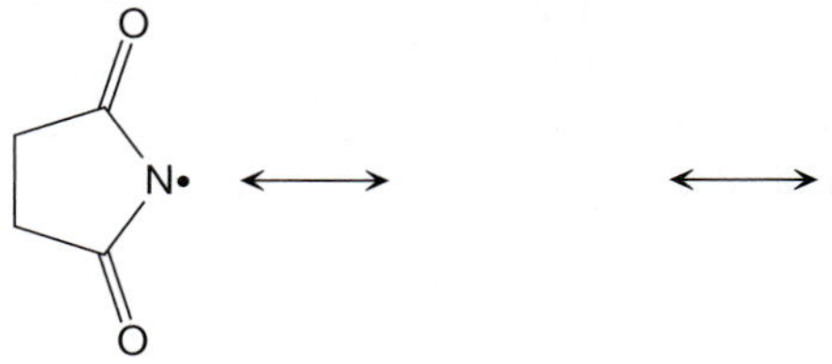

The low concentration of Br_2 maintained by NBS favors radical bromination over alkene addition because the rate of alkene addition is slowed more dramatically than radical bromination. To see why, recall from Section 12.3 that the electrophilic addition of Br_2 to an alkene proceeds through a bromonium ion intermediate (Equation 25-29):

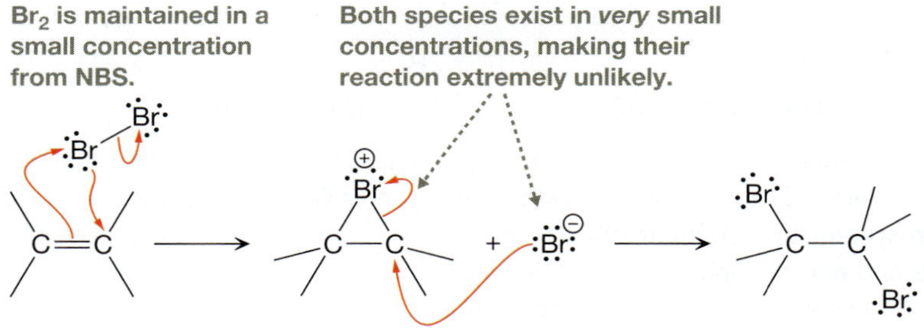

Br_2 is maintained in a small concentration from NBS.

Both species exist in *very* small concentrations, making their reaction extremely unlikely.

(25-29)

Because of the low abundance of Br_2, the bromonium ion and Br^- intermediates that derive from Br_2 are both present in *very* low concentrations, making it *extremely* unlikely that they will encounter each other to complete the reaction.

The story is somewhat different for radical bromination using NBS, whose initiation and propagation steps are shown in Equation 25-30.

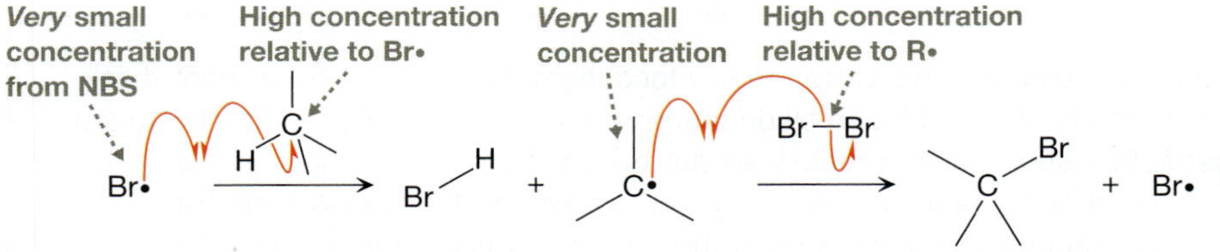

Very small concentration from NBS

High concentration relative to Br•

Very small concentration

High concentration relative to R•

(25-30)

None of these steps requires the reaction between two species produced from Br_2, so there are no steps that require the reaction between two species whose concentrations are excessively low.

PROBLEM 25.18 Predict the major product of the reaction shown here.

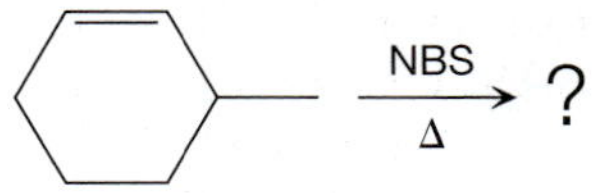

NBS, Δ, ?

Free Radicals in the Body: Lipid Peroxidation and Vitamin E

In the body, free radicals play important roles in cell signaling and protecting against invading bacteria. Because they are highly reactive, however, these free radicals can lead to unwanted side reactions that damage or kill cells. It is no wonder that free radicals are implicated in the aging process and are linked to diseases such as diabetes and cancer.

The hydroxyl radical (HO•) is particularly damaging via a process called lipid peroxidation. In Section 2.11b, we saw that the lipid bilayer of a cell membrane is made up of phospholipids, each containing two nonpolar hydrocarbon tails. As shown below on the left, cis C=C double bonds are common features of the hydrocarbon tails, and, as we learned here in Chapter 25, the allylic H atoms are susceptible to hydrogen abstraction.

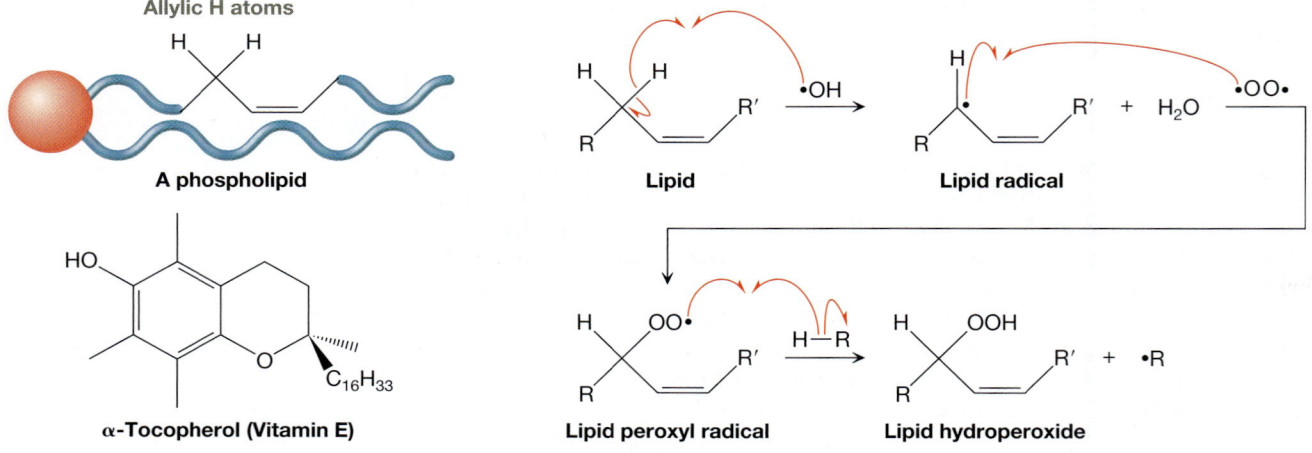

Allylic H atoms

A phospholipid

α-Tocopherol (Vitamin E)

Lipid

Lipid radical

Lipid peroxyl radical

Lipid hydroperoxide

As shown in the scheme above on the right, hydrogen abstraction produces an allylic radical, which subsequently undergoes radical coupling with a molecule of O_2 to produce a lipid peroxyl radical. The lipid peroxyl radical then abstracts a H atom from another phospholipid, resulting in a lipid hydroperoxide and another lipid radical that can go on to continue the propagation cycle.

Lipid peroxides and the compounds they produce in subsequent reactions are very toxic. Fortunately, vitamin E (one form of which is α-tocopherol, shown above), found abundantly in foods such as almonds, spinach, and avocados, is a natural antioxidant that is one of the body's most important defenses against lipid peroxidation. A lipid radical can readily abstract the phenolic H from vitamin E, leaving behind a relatively unreactive phenoxy radical (see Your Turn 25.19). In so doing, the lipid radical is converted to a closed-shell species and the propagation cycle is terminated.

YOUR TURN 25.19

In the preceding box, we stated that α-tocopherol is a natural antioxidant that terminates lipid peroxidation by donating H to a lipid radical (R•). This hydrogen abstraction step is shown below, as are two resonance structures of the resulting phenoxy radical. Add the curved arrows for the hydrogen abstraction step and also for the conversion of the first resonance structure into the second. Why do you think α-tocopherol is such an effective antioxidant?

R• +

α-Tocopherol (Vitamin E)

RH +

PROBLEM 25.19 Your Turn 25.19 shows two resonance structures of the phenoxy radical produced after abstraction of a hydrogen from α-tocopherol. Draw three more resonance structures of the phenoxy radical, including curved arrows.

25.5 Radical Addition of HBr: Anti-Markovnikov Addition

Recall from Section 11.3 that a hydrogen halide adds across the double bond of an alkene in a Markovnikov fashion under normal conditions. Thus, HBr will add to propene to produce 2-bromopropane, as shown in Equation 25-31.

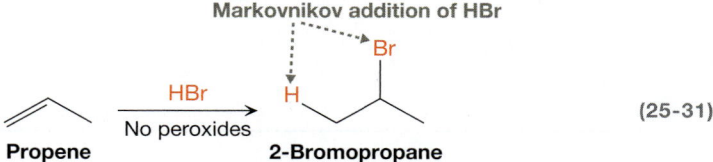

$$(25\text{-}31)$$

Overall, H^+ adds to the less alkyl-substituted C atom, whereas Br^- adds to the more alkyl-substituted C atom. The reason, as discussed in Chapter 11, is that the first step of the mechanism is protonation of the C=C double bond, which proceeds to give the *more stable carbocation intermediate*.

If a small amount of peroxide (RO—OR) is present, however, then 1-bromopropane is produced as the major product instead, as indicated in Equation 25-32.

$$(25\text{-}32)$$

A peroxide such as $(CH_3)_3COOC(CH_3)_3$ could be added directly, or peroxides could already be present as contaminants, which is often the case with ether solvents.

In Equation 25-32, HBr still adds across the C=C double bond, but with a regiochemistry opposite to that in Equation 25-31—that is, the H atom adds to the more substituted C, and the Br atom adds to the less substituted one. This is an example of an *anti-Markovnikov addition*.

Peroxides are *radical initiators* because they contain a weak O—O bond. Hence, the reaction in Equation 25-32 proceeds by a radical mechanism, which is outlined in Equation 25-33.

Mechanism for Equation 25-32

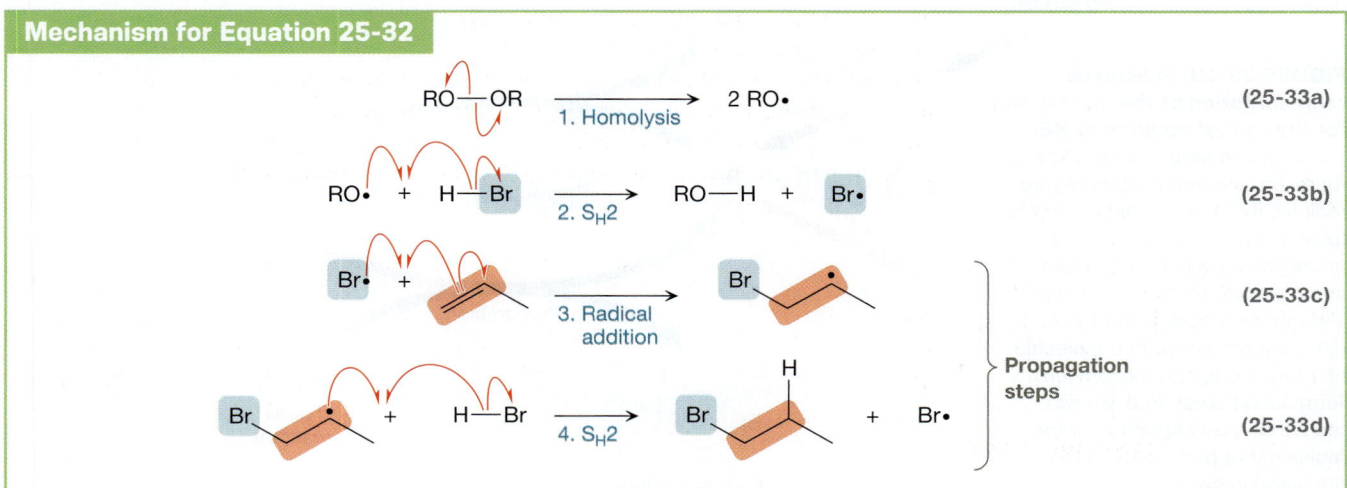

Step 1 of Equation 25-33 is homolysis of the peroxide to produce two alkoxy radicals (RO•). In Step 2, an RO• abstracts a H atom from HBr to produce a Br•. That Br• adds to the alkene in Step 3, and the resulting alkyl radical abstracts a H atom from HBr in Step 4, yielding the overall product.

This is a *chain reaction*, because Steps 3 and 4 make up a propagation cycle—that is, the Br• that reacts in Step 3 is regenerated as a product in Step 4. Summing Steps 3 and 4 yields the net reaction.

YOUR TURN 25.20

The propagation steps from Equation 25-33 are as follows. Cross out the redundant species, and sum the two steps to arrive at the net reaction.

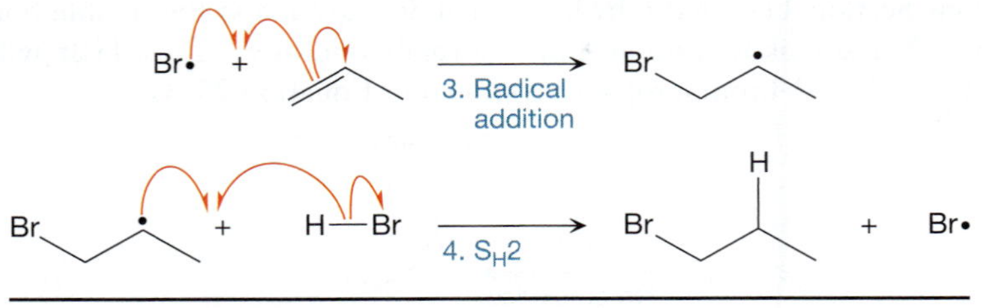

The initial Br• that is used in the propagation steps in Equation 25-33 is produced by the sequence of Steps 1 and 2. Hence, Steps 1 and 2 can be viewed as a *set* of initiation steps. As usual, a termination step is any step that removes a free radical from the propagation cycle (see Your Turn 25.21).

YOUR TURN 25.21

Draw three plausible termination steps for the reaction in Equation 25-32.

Just as we saw with radical halogenation, the mechanism for the radical addition of HBr can be represented schematically, as shown in Figure 25-13. Two steps (black arrows) are required to introduce Br• into the propagation cycle (blue arrows). With each completion of the propagation cycle, one molecule of propene and one molecule of HBr are converted to one molecule of 1-bromopropane. Termination

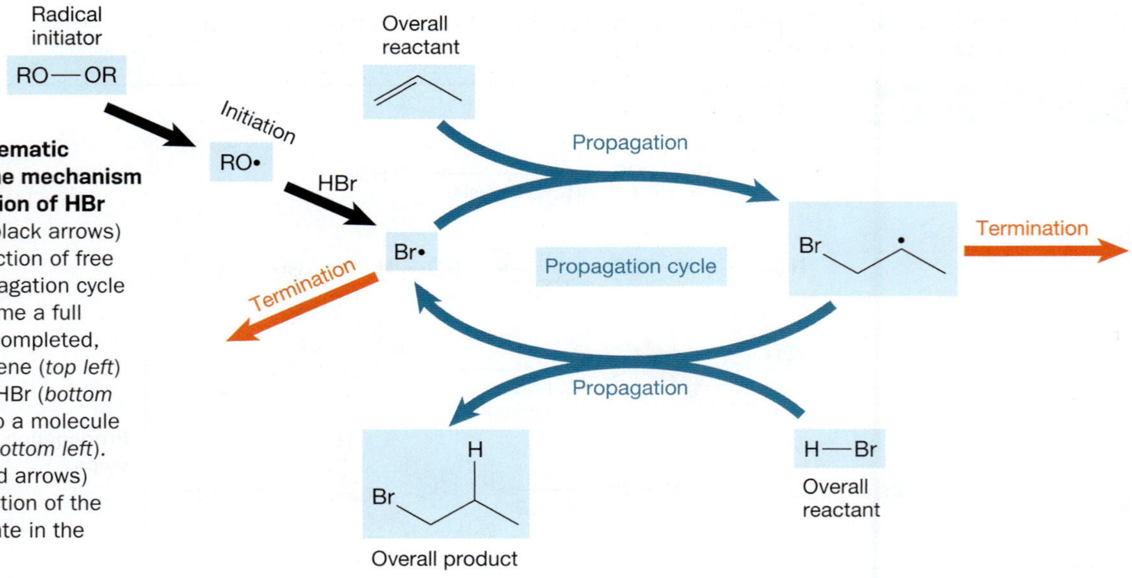

FIGURE 25-13 Schematic representation of the mechanism for the radical addition of HBr The initiation steps (black arrows) represent the introduction of free radicals into the propagation cycle (blue arrows). Each time a full propagation cycle is completed, one molecule of propene (*top left*) and one molecule of HBr (*bottom right*) are converted to a molecule of 1-bromopropane (*bottom left*). Termination steps (red arrows) represent the destruction of the radicals that participate in the propagation cycle.

steps (red arrows) include any steps that destroy the radicals that participate in the propagation cycle.

YOUR TURN 25.22

Study the diagram in Figure 25-13 and match the initiation and propagation steps to either Equation 25-33a, 25-33b, 25-33c, or 25-33d.

The reason that the HBr addition in Equation 25-32 proceeds in an anti-Markovnikov fashion has everything to do with Step 3 of its mechanism (Equation 25-33c), which is the first of the two propagation steps. Step 3 is a radical addition to the C=C double bond and, as shown in Equation 25-34a and 25-34b, Br• can add to the terminal C or to the central C.

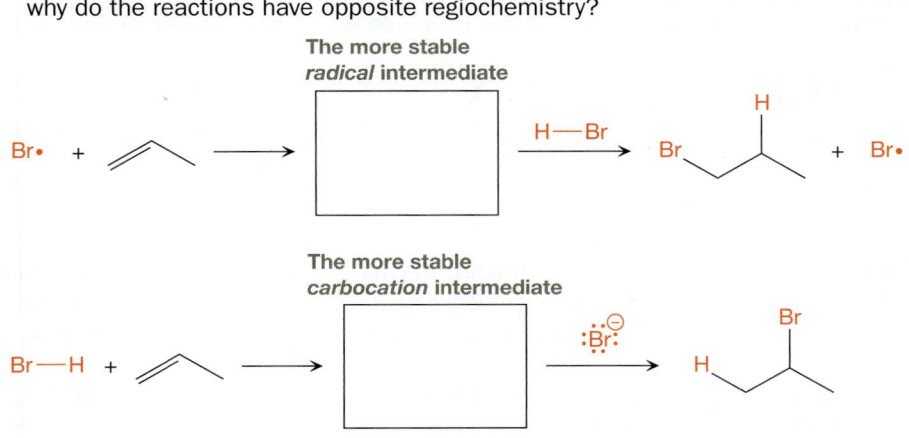

This alkyl radical is more stable.

(25-34a)

A 2° alkyl radical

3. Radical addition

Br

(25-34b)

A 1° alkyl radical

When Br• adds to a terminal C (Equation 25-34a), a 2° alkyl radical is produced, and when it adds to the central C (Equation 25-34b), a 1° alkyl radical is produced. The greater stability of the 2° radical provides more driving force.

Radical addition to an alkene generally takes place so as to produce the more stable alkyl radical intermediate.

This is the same idea behind why HBr addition to an alkene involving only closed-shell species (Section 11.3) takes place with Markovnikov regiochemistry (see Your Turn 25.23).

YOUR TURN 25.23

The radical and closed-shell mechanisms for the addition of HBr to propene are shown below, but the curved arrow notation and intermediate are omitted in both cases. Complete each mechanism by drawing the more stable intermediate in the box provided and adding the curved arrow notation. Realizing that both reactions proceed through the more stable intermediate, why do the reactions have opposite regiochemistry?

The more stable *radical* intermediate

Br• +

H—Br

Br

H

+ Br•

The more stable *carbocation* intermediate

Br—H +

:Br:⊖

H

Br

PROBLEM 25.20 Predict the major product of each of the following reactions.

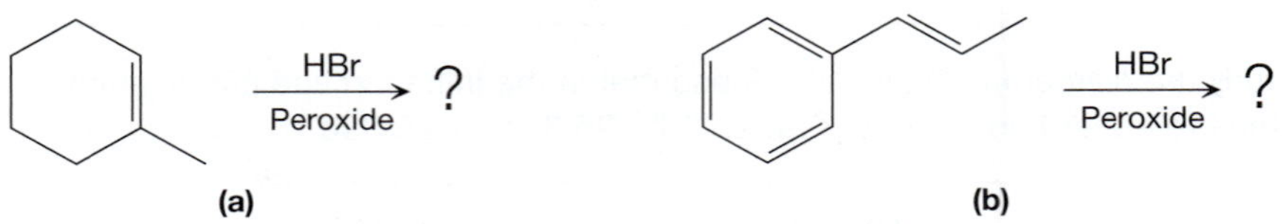

(a)   (b)

25.6 Stereochemistry of Free Radical Halogenation and HBr Addition

In the free radical reactions we have seen thus far, stereochemistry can be an issue if the reaction takes place at a carbon atom that is a chiral center in the reactants or becomes one in the products. In the radical halogenation in Equation 25-35, for example, a new chiral center appears in the products and both configurations are produced, resulting in a mixture of enantiomers.

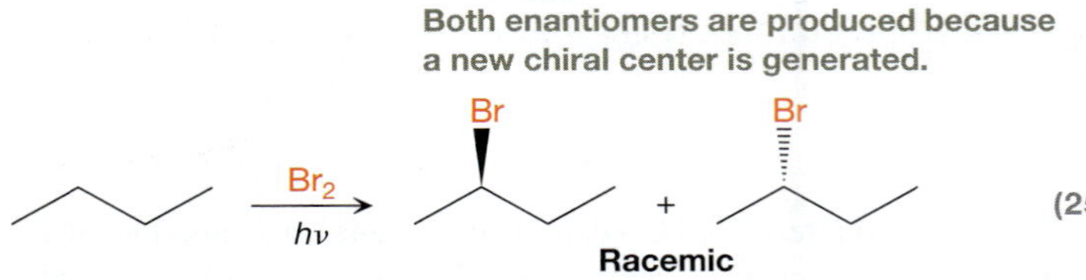

Both enantiomers are produced because a new chiral center is generated.

Racemic   (25-35)

The reaction results in a mixture of enantiomers because the chiral center is produced directly from the S_H2 step shown in Equation 25-36 (review Equation 25-19, p. 1262).

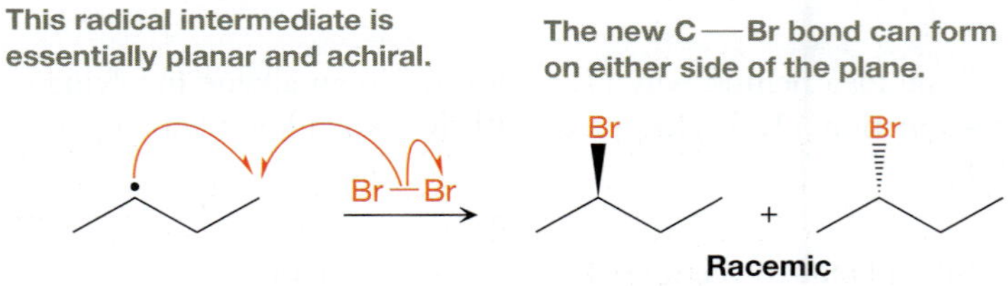

This radical intermediate is essentially planar and achiral.

The new C—Br bond can form on either side of the plane.

Racemic   (25-36)

Equation 25-37 shows an example in which halogenation takes place at a C atom that is initially a chiral center.

The chiral center becomes a planar radical intermediate after H is abstracted.

Both enatiomers are produced because a new chiral center is generated from the intermediate.

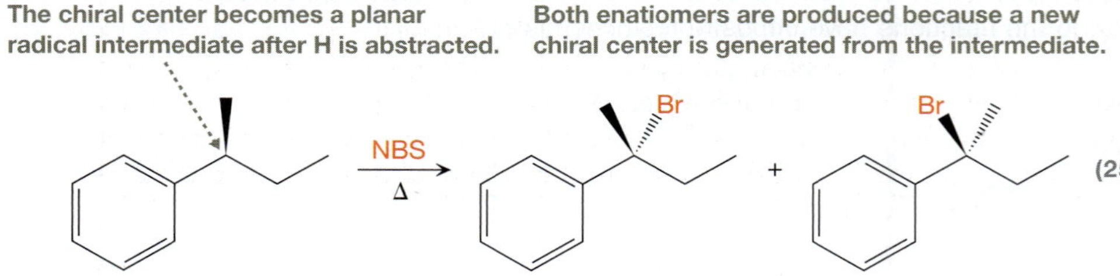

(25-37)

Once again, the reaction produces a mixture of enantiomers, because the C atom that undergoes halogenation becomes a chiral center. Even though the C atom is initially a chiral center, it becomes a planar free radical intermediate after the H atom is abstracted.

Similarly, if the free radical addition of HBr to an alkene results in a new chiral center, then both enantiomers will be produced. An example is shown in Equation 25-38.

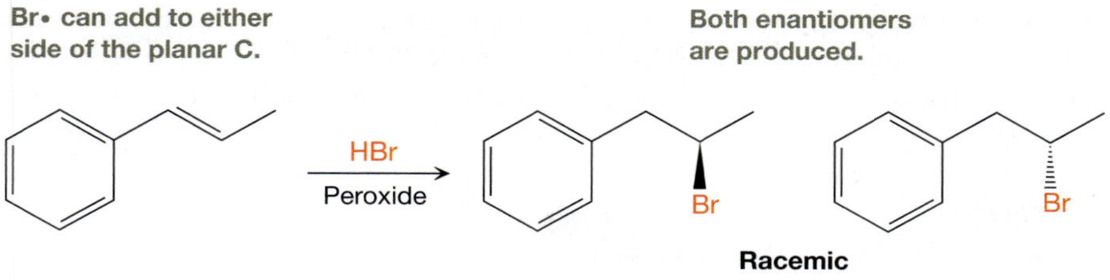

Br• can add to either side of the planar C.

Both enantiomers are produced.

Racemic

(25-38)

In this case, the chiral center is produced when Br• adds to the alkene C (review Equation 25-33). The alkene C is planar, so Br• can add to either side of the plane.

The lessons we learn from the stereochemistry of radical halogenation and HBr addition reactions can be extended to other radical reactions as well:

> In general, reactions that proceed by free radical chain-reaction mechanisms tend not to be stereoselective.

PROBLEM 25.21 Draw the major product of each of the following reactions, paying attention to stereochemistry. *Hint:* How many chiral centers does each reaction produce?

(a)   (b)

25.7 Dissolving Metal Reductions: Hydrogenation of Alkenes and Alkynes

To this point in Chapter 25, we have examined only free radical reactions that proceed by chain-reaction mechanisms. Here in Section 25.7, we study two types of hydrogenation reactions that are categorized as *dissolving metal reductions*, and neither reaction involves a chain-reaction mechanism.

Hydrogenation via a dissolving metal reduction takes place under conditions that differ from the catalytic hydrogenation discussed in Section 19.5. Catalytic hydrogenation uses a solid metal catalyst such as $Pt(s)$, $Pd(s)$, or $Ni(s)$. Thus, the catalyst is in a different phase from the reactants dissolved in solution, making catalytic hydrogenation with a solid metal catalyst an example of *heterogeneous catalysis*.

In a *dissolving metal reduction*, on the other hand, a metal such as $Na(s)$ or $Li(s)$ is *dissolved* in solution along with the reactants. When this takes place in a solvent such as liquid ammonia, $NH_3(\ell)$, for example, the metal loses an electron, and the ammonia solvent stabilizes the electron via extensive solvation. The result, called a **solvated electron**, is depicted in Equation 25-39.

$$ Na• \xrightarrow{NH_3(\ell)} Na^{\oplus} + e^{\ominus} \text{ (solvated)} \qquad (25\text{-}39) $$

In light of this:

> The active radical species in a dissolving metal reduction involving $NH_3(\ell)/Na(s)$ is simply an electron that does not formally belong to any atom.

CONNECTIONS A solvated electron can absorb visible light. Therefore, whereas liquid ammonia is clear and colorless, the solution turns blue when $Na(s)$ is dissolved.

In the discussion that follows, we see examples of how this solvated electron is involved in the reduction of a multiple bond. Section 25.7a discusses the reduction of alkynes to trans alkenes, and Section 25.7b discusses the Birch reduction of benzene rings.

25.7a Anti-Hydrogenation: Synthesis of Trans Alkenes

When oct-4-yne is treated with sodium metal dissolved in liquid ammonia (-78 °C), the triple bond is reduced to a double bond (Equation 25-40).

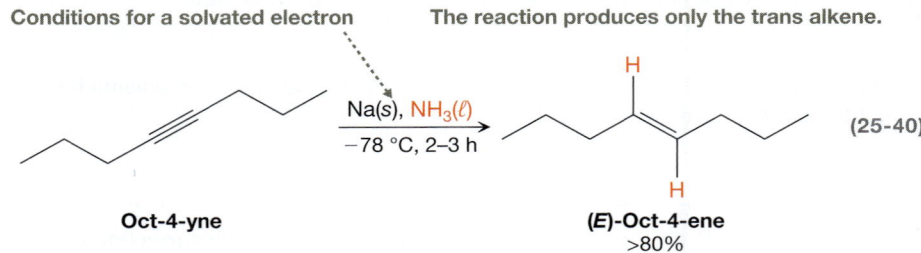

Conditions for a solvated electron

The reaction produces only the trans alkene.

Na(s), NH$_3$(ℓ)
-78 °C, 2–3 h

Oct-4-yne

(E)-Oct-4-ene
>80%

(25-40)

The reaction is highly stereoselective.

> A dissolving metal reduction of an alkyne produces the E alkene almost exclusively.

This is an **anti-hydrogenation** reaction in which two H atoms, overall, add to the triple bond in a trans fashion. It complements catalytic hydrogenation using a poisoned catalyst (Section 19.5b), which exclusively forms the cis product.

YOUR TURN 25.24

Draw the product of this reaction, in which oct-4-yne undergoes catalytic hydrogenation. (Review Section 19.5b.)

H$_2$(g)
Pd, BaSO$_4$,
Pb(OAc)$_2$

?

The mechanism for the dissolving metal reduction is shown in Equation 25-41. It begins with a solvated electron, which was produced on dissolving solid sodium in liquid ammonia. In Step 1, a solvated electron adds to an alkyne C, thus converting the triple bond into a double bond. One of the initial alkyne C atoms gains an unpaired electron and the other gains a negative charge. The carbanion is a very strong base and deprotonates NH$_3$ in Step 2 (notice the double-barbed curved arrows). This produces an uncharged vinylic radical, which, in Step 3, undergoes radical coupling with a second solvated electron to produce a carbanion. Finally, in Step 4, the carbanion deprotonates a second molecule of NH$_3$, yielding the overall product.

The stereochemistry of this reaction is established in the first step—the addition of a solvated electron to the C≡C triple bond. Although both the cis and trans forms of the vinylic radical can be produced, the two configurations are in rapid equilibrium (Fig. 25-14). The trans form has less steric strain, though, so it is more stable, and hence leads to the major product.

FIGURE 25-14 Stereoselectivity of anti-hydrogenation This radical anion, which is produced in Step 1 of Equation 25-41, exists in rapid equilibrium between the cis and trans forms. The trans form is more stable, which is why the reaction produces the trans alkene almost exclusively.

Steric strain makes this configuration less stable.

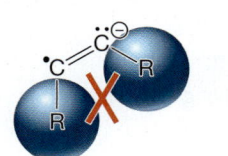

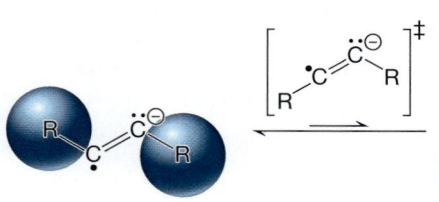

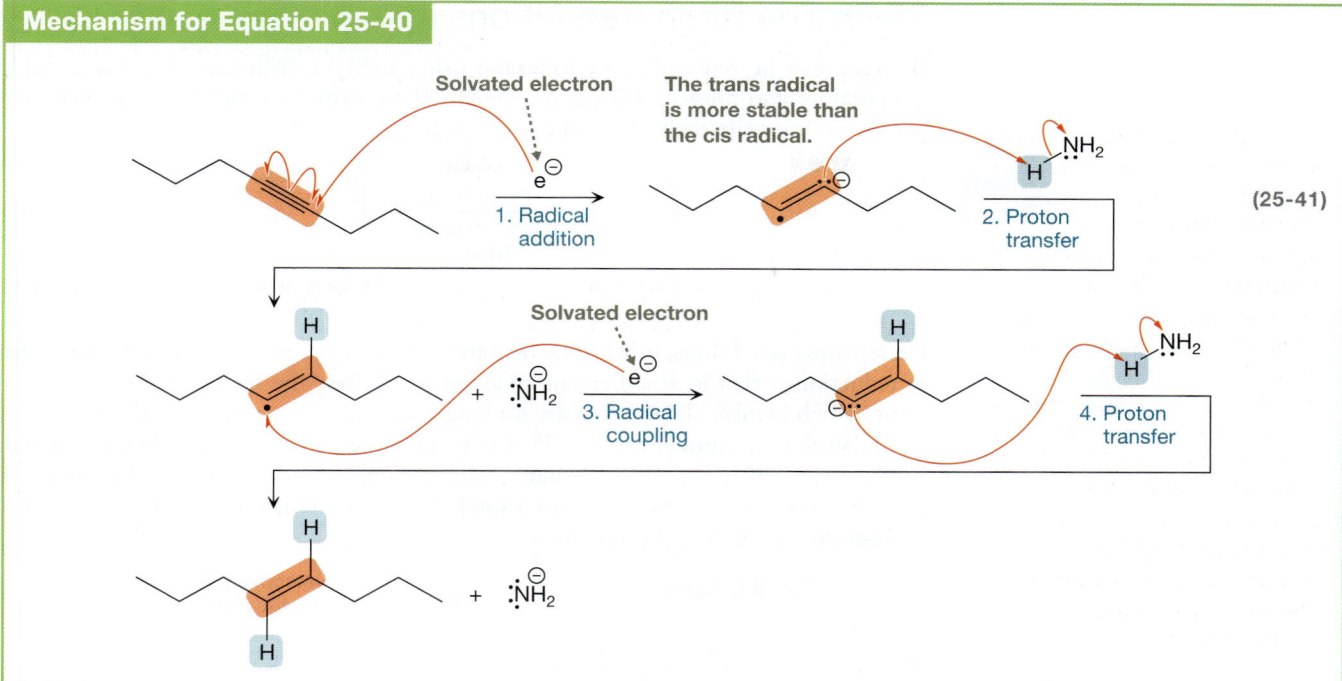

Solvated electron

The trans radical is more stable than the cis radical.

1. Radical addition

2. Proton transfer

(25-41)

Solvated electron

3. Radical coupling

4. Proton transfer

Dissolving metal reductions are much more sluggish with alkenes than with alkynes, in part because of charge stability. In the reduction of an alkyne, the C atom that gains the negative charge has a higher effective electronegativity (Section 3.11) than the C atom that gains the negative charge in the reduction of an alkene. As a result:

A dissolving metal reduction will selectively reduce an alkyne over an alkene.

An example is shown in Equation 25-42. This kind of selectivity can be very useful in synthesis.

The C≡C bond is selectively reduced over the C=C bond.

Na(s)
NH₃(ℓ), −78 °C

(25-42)

PROBLEM 25.22 Draw the alkyne from which the compound shown here can be produced via a dissolving metal reduction.

?
Li(s)
NH₃(ℓ), −78 °C

PROBLEM 25.23 A dissolving metal reduction can be used to reduce a ketone to an alcohol, as shown in the reduction here. Draw a complete, detailed mechanism for this reaction.

1. Na(s), THF, dilute ROH
2. H₂O, HCl

25.7b The Birch Reduction

Benzene can be reduced to cyclohexane using catalytic hydrogenation under high temperature and pressure (Equation 25-43). These extreme conditions are necessary because aromaticity makes benzene's π system quite stable.

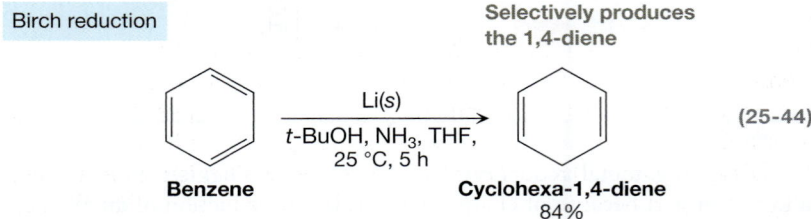

$$\text{(25-43)}$$

Benzene   **Cyclohexane**
                        100%

Under these conditions, it is impractical to stop the reaction at an intermediate stage of reduction—that is, at a diene or an alkene—because the second and third reductions, which involve alkenes that are no longer aromatic, are faster than the first.

A **Birch reduction** (Equation 25-44), named after the Australian chemist Arthur Birch (1915–1995), on the other hand, reduces benzene to cyclohexa-1,4-diene. Not only does the reaction stop after just a single hydrogenation, but reduction takes place *regioselectively*, yielding the 1,4-diene.

Birch reduction

Selectively produces the 1,4-diene

Li(s)
t-BuOH, NH₃, THF,
25 °C, 5 h

$$\text{(25-44)}$$

Benzene   **Cyclohexa-1,4-diene**
                         84%

The conditions for a Birch reduction are similar to those for the dissolving metal reductions discussed in Section 25.7a. One difference is that a small amount of an alcohol [typically *tert*-butyl alcohol, $(CH_3)_3COH$] is added to act as the proton source. As a result, the mechanism for the Birch reduction, shown in Equation 25-45, is similar to that for the dissolving metal reduction of an alkyne, shown previously in Equation 25-41.

Mechanism for the Birch reduction in Equation 25-44

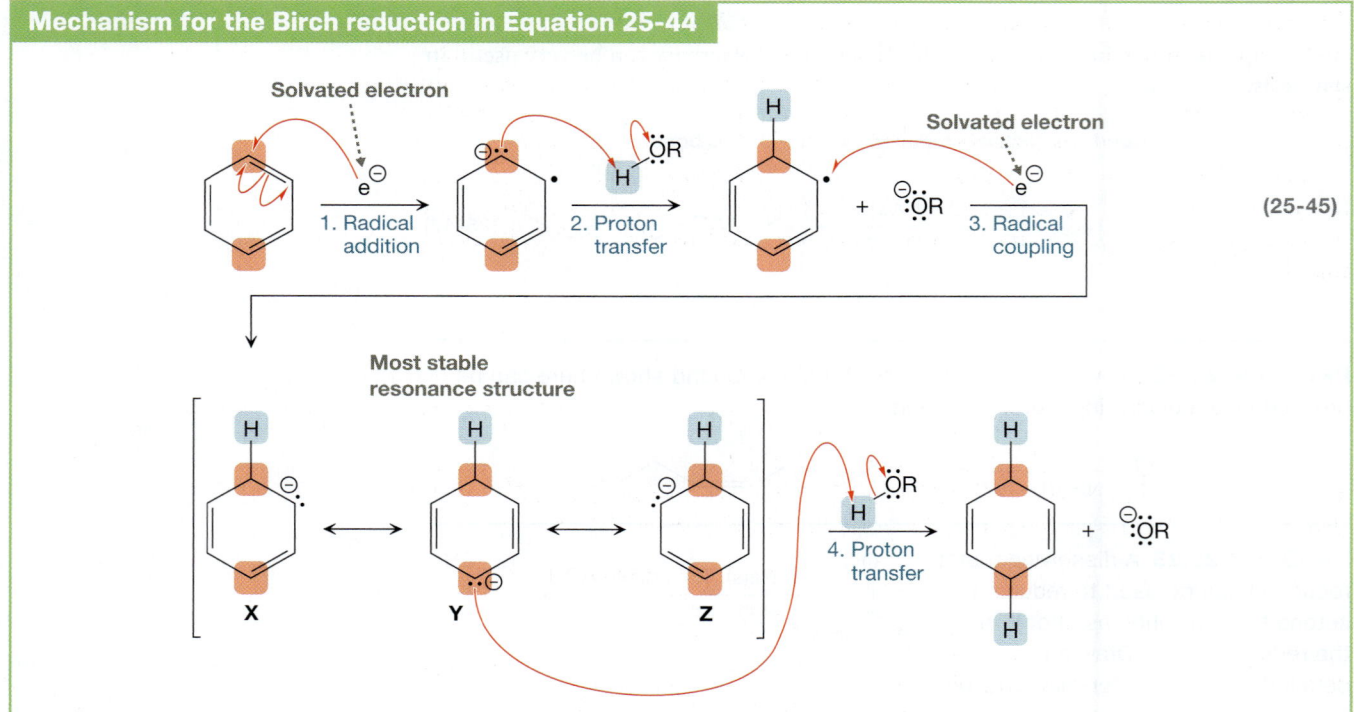

$$\text{(25-45)}$$

Step 1 is radical addition of a solvated electron to a C=C double bond of the benzene ring. The resulting radical anion intermediate is strongly basic, so it deprotonates the alcohol in Step 2 to produce an uncharged alkyl radical. Radical coupling takes place in Step 3, this time producing a closed-shell carbanion species in which the negative charge is delocalized over three C atoms. The carbanion then deprotonates a second molecule of the alcohol to produce the overall product.

Notice in the final step of the mechanism that protonation can occur at any of the three C atoms sharing the negative charge, but it takes place almost exclusively at the one indicated in resonance structure **Y**. Resonance structure **Y** is more stable than **X** or **Z**, giving **Y** the greatest contribution to the resonance hybrid. In **Y**, the negative charge is stabilized by two adjacent sp^2-hybridized C atoms, each of which has a relatively high effective electronegativity (Section 3.11). Thus, the C atom at the bottom of the ring has the greatest concentration of negative charge and will most strongly attract the proton from HOR. Moreover, notice that the major product resembles resonance structure **Y** more than it does **X** or **Z**. Therefore, the final step of the mechanism causes the least change in geometry when the bottom carbon is protonated—a result that is favored by the *principle of least nuclear motion*.

PROBLEM 25.24 The Birch reduction of a monosubstituted benzene can produce two isomeric products, as shown. Which one is the major product depends on whether the substituent is electron donating or electron withdrawing (see Problems 25.60 and 25.61 at the end of the chapter). Draw the mechanism that leads to each product.

25.8 Organic Synthesis: Radical Reactions in Synthesis

The free radical reactions we have examined in Chapter 25 are valuable synthetically for two main reasons. First, they allow us to carry out transformations that are impractical with reactions whose mechanisms involve only closed-shell species. Second, they allow us to carry out transformations we have seen in previous chapters, but with different regiochemistry or stereochemistry.

Free radical halogenation, for example, is one of the few practical reactions that can *functionalize* (i.e., add a functional group to) the 3 position of cyclohexene, as indicated in Equation 25-46, and it is the only reaction we have discussed in this book that does so.

(25-46)

This transformation can be achieved, in particular, with bromination using NBS (Equation 25-47), similar to what we previously saw in Equation 25-27.

(25-47)

Allylic bromination such as this was used in the synthesis of the natural product (+)-koninginin D, an antibiotic. A key step in the synthesis is shown in Equation 25-48.[1]

(25-48)

(+)-Koninginin D

Under the relatively mild conditions that are used, the other functional groups present are unaffected.

Radical bromination with NBS is also a popular method to functionalize the benzylic position of aromatic compounds in synthesis. Equation 25-49[2] shows how it was used in a key step in the synthesis of flavones, which are compounds that can contribute to plant pigments.

1. NBS (excess), benzoyl peroxide
2. Et₃N, reflux

(25-49)

A flavone precursor

PROBLEM 25.25 (a) Draw the product that is immediately formed in Equation 25-49 after reaction with excess NBS. **(b)** What kind of reaction takes place in the second reaction in Equation 25-49?

Another synthetically useful aspect of radical reactions is *regiochemistry* that can differ from that in other reactions we have encountered previously. An example is the hydrohalogenation of an alkene. Recall from Chapter 11 that HBr adds across a C=C double bond in a Markovnikov fashion, proceeding through the more stable carbocation intermediate. The addition of HBr in the presence of a radical initiator (such as peroxides), however, takes place in an *anti*-Markovnikov fashion, proceeding through the more stable *radical* intermediate. The importance of this difference is demonstrated in Solved Problem 25.26.

SOLVED PROBLEM 25.26

Show how you would carry out this transformation using at least one radical chain reaction.

Think What functional group can be used to halogenate a carbon that is two carbons away from a phenyl ring? How can that functional group be introduced on a carbon that initially is not functionalized?

[1]Liu, G.; Wang, Z. *Chem. Commun.* **1999**, *12*, 1129–1130.
[2]Silva, A. M. S.; Silva, A. M. G.; Tomé, A. C.; Cavaleiro, J. A. S. *Eur. J. Org. Chem.* **1999**, *1*, 135–139.

Solve As shown in the following retrosynthetic analysis, the alkyl bromide product can be made from ethenylbenzene via the anti-Markovnikov radical addition of HBr.

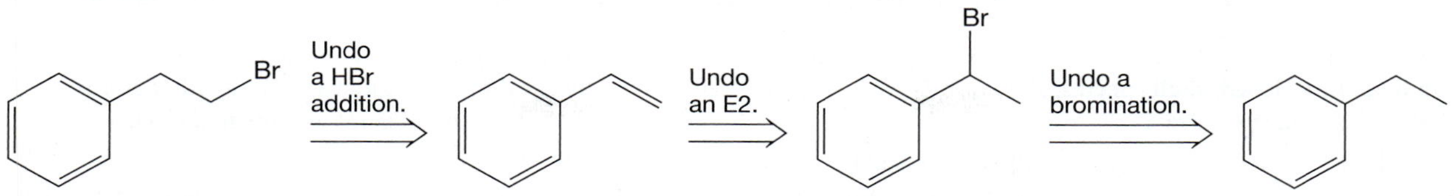

The C=C double bond can be made from (1-bromoethyl)benzene in an E2 reaction, and the alkyl bromide can be made, in turn, from a radical halogenation of phenylethane. Reversing the reactions and adding the appropriate reagents, the synthesis can be written as follows:

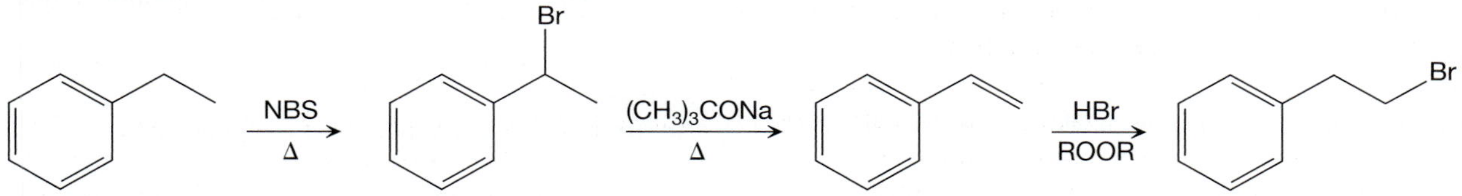

Finally, free radical reactions can offer *stereochemistry* that differs from other reactions we have seen previously. An important example of this is the dissolving metal reduction, which can reduce an alkyne to a trans alkene (Equation 25-50).

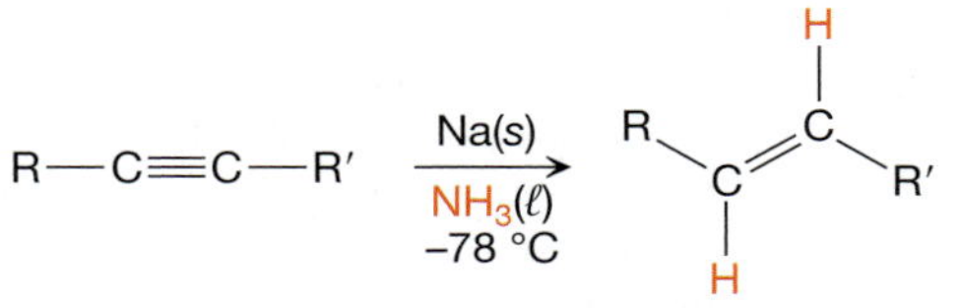

$$R—C\equiv C—R' \xrightarrow[\substack{NH_3(\ell) \\ -78\ °C}]{Na(s)}$$ (25-50)

A trans alkene

This contrasts with catalytic hydrogenation using a poisoned catalyst (Section 19.5b), which reduces an alkyne to a cis alkene (Equation 25-51).

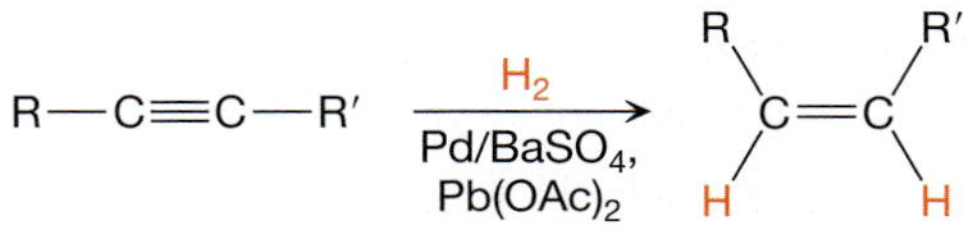

$$R—C\equiv C—R' \xrightarrow[\substack{Pd/BaSO_4, \\ Pb(OAc)_2}]{H_2}$$ (25-51)

A cis alkene

An application of the dissolving metal reduction of an alkyne is shown in Equation 25-52.[3] This is a step in the total synthesis of sphingosine, which undergoes phosphorylation in vivo to produce a potent signaling lipid. (Notice in Equation 25-52 that the reaction conditions also remove a protecting group.)

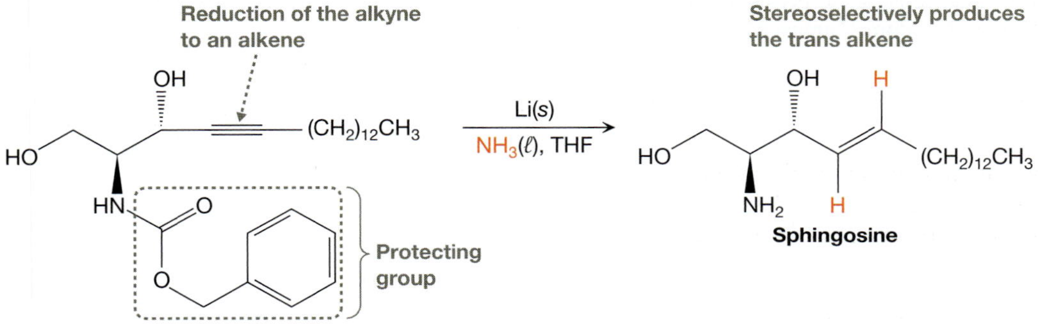

Reduction of the alkyne to an alkene

Stereoselectively produces the trans alkene

(25-52)

Sphingosine

Protecting group

---

[3]Boutin, R. H.; Rapoport, H. *J. Org. Chem.* **1986**, *51*, 5320–5327.

Chapter Summary and Key Terms

- A species that possesses at least one unpaired electron is called a **free radical**. In a **closed-shell species**, all electrons are paired. (Introduction; Objective 1)

- Free radicals lack an octet, so they are usually very unstable and highly reactive. They are generally introduced in a sample by **homolysis** of a covalent bond in a closed-shell precursor. (Section 25.1; Objective 2)

- Homolysis generally breaks the weakest bond in the compound. (Section 25.1; Objective 3)

- A **single-barbed arrow** (⌒) describes the movement of a single electron in an elementary step. (Section 25.1; Objective 4)

- A **radical initiator** is a precursor from which a free radical is produced and generally possesses a relatively weak covalent bond. Common initiators are molecular halogens, **N-bromosuccinimide (NBS)**, **2,2-azobisisobutyronitrile (AIBN)**, and peroxides (RO—OR). (Section 25.1; Objective 3)

- Radical stability can be determined from homolytic bond dissociation energies. In general, the greater the energy required to produce the radical via homolysis, the more unstable the radical is. (Section 25.2; Objective 5)

- The stability of a radical increases with:
  - Decreasing effective electronegativity of the atom that gains the unpaired electron.
  - Increasing resonance delocalization of the unpaired electron.
  - Additional alkyl groups attached to the atom with the unpaired electron. (Section 25.2; Objective 6)

- Resonance structures can be drawn for a radical when the atom having the unpaired electron is attached to a multiple bond or an atom with a lone pair. (Section 25.2; Objective 7)

- A carbon atom that has an unpaired electron tends to be sp^2 hybridized and has a planar geometry. (Section 25.2; Objective 8)

- In a **radical coupling** step, an unpaired electron from a radical joins an unpaired electron from a second radical to produce a new covalent bond. (Section 25.3a; Objective 4)
  - Radical coupling is typically irreversible, and proceeds to products with no energy barrier.

- In a **bimolecular homolytic substitution (S_H2)** step, a free radical forms a bond to an atom of a closed-shell species and displaces another free radical from that atom. (Section 25.3b; Objective 4)

- In a **radical addition step**, one atom of a double or triple bond forms a bond to a free radical, and the other atom of the double or triple bond gains an unpaired electron. (Section 25.3c; Objective 4)

- A free radical **chain reaction** involves free radicals in a cyclic sequence of steps, called **propagation steps**, which are responsible for converting overall reactants into overall products. (Section 25.4a; Objectives 9, 10)
  - Throughout the sequence of propagation steps, also called the **propagation cycle**, there is no net consumption of free radicals.
  - The overall reaction is obtained by summing just the propagation steps.
  - An **initiation step** is responsible for producing the free radical that enters the propagation cycle.
  - A **termination step** removes free radicals from the propagation cycle and is responsible for slowing the overall reaction.

- **Radical halogenation** is a chain reaction that replaces the H of a C—H bond with a halogen atom. It takes place when free radicals are generated in the presence of a molecular halogen and a compound containing a reactive C—H bond. (Section 25.4; Objectives 11, 16)
  - The identity of the molecular halogen plays a major role in the kinetics and thermodynamics of radical halogenation. Both the rate and exothermicity of halogenation decrease in the order: $F_2 > Cl_2 > Br_2 > I_2$. (Objective 12)
  - Bromination is highly regioselective, whereas chlorination is only slightly selective. (Objective 12)
  - N-Bromosuccinimide is commonly used to brominate an allylic carbon to avoid the addition of Br_2 to the alkene. (Objective 13)

- HBr adds to an alkene via *anti-Markovnikov addition* when free radicals are present. (Section 25.5; Objectives 14, 16)

- Free radical halogenation and the radical addition of HBr to an alkene produce a mixture of stereoisomers if new chiral centers are produced. (Section 25.6; Objective 14)

- In a *dissolving metal reduction*, a **solvated electron** behaves as the free radical species. (Section 25.7; Objectives 15, 16)
  - The dissolving metal reduction of an alkyne produces a trans alkene. (Section 25.7a)
  - The dissolving metal reduction of a benzene ring, called a **Birch reduction**, produces a cyclohexa-1,4-diene. (Section 25.7b)

Reaction Table

TABLE 25-4 Functional Group Conversions

	Starting Compound Class	Typical Reagents and Reaction Conditions	Compound Class Formed	Key Electron-Rich Species	Key Electron-Poor Species	Comments	Discussed in Section
(1)	R—H Alkane	$\xrightarrow[h\nu]{X_2}$	R—X Alkyl halide	X—X	R•	X = Cl, Br	25.4
(2)	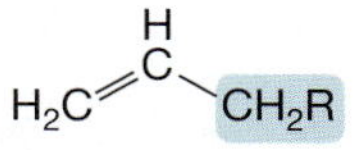 $H_2C=C(H)—CH_2R$ Alkene with allylic H	$\xrightarrow[\Delta \text{ or } h\nu]{NBS}$	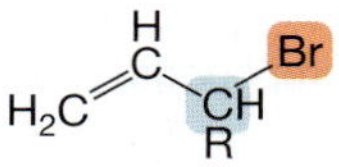 $H_2C=C(H)—CH(Br)R$ Allylic bromide	Br—Br	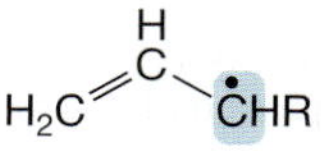 $H_2C=C(H)—\overset{\bullet}{C}HR$	Favors allylic substitution over bromination of C=C	25.4d
(3)	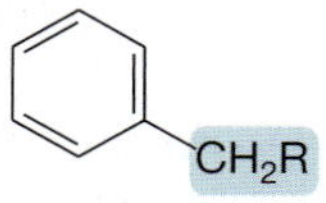 Alkylbenzene with benzylic H (Ph—CH₂R)	$\xrightarrow[\Delta \text{ or } h\nu]{NBS}$	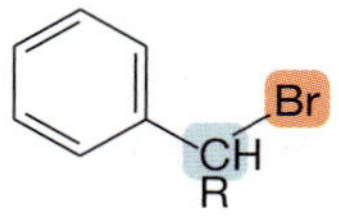 Benzylic bromide (Ph—CH(Br)R)	Br—Br	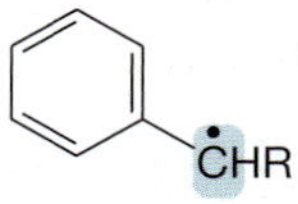 Ph—$\overset{\bullet}{C}$HR	Functionalizes a relatively unreactive carbon	25.4d
(4)	 Alkene	$\xrightarrow[\text{Peroxide}]{HBr}$	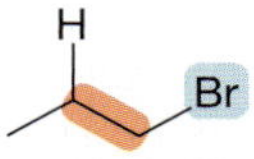 Alkyl bromide	 Alkene	Br•	Anti-Markovnikov regiochemistry	25.5
(5)	R—C≡C—R Alkyne	$\xrightarrow[NH_3(\ell),\ -78\ °C]{Na(s) \text{ or } Li(s)}$	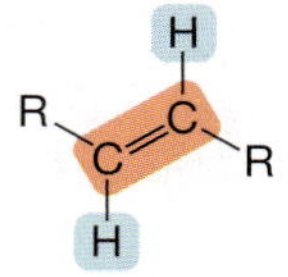 Trans alkene	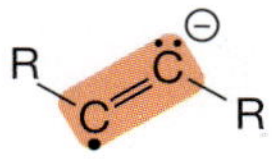 $R—C(\overset{\bullet}{})=\overset{\ominus}{C}—R$	$H—NH_2$	Dissolving metal reduction	25.7a
(6)	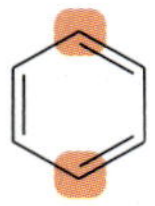 Benzene	$\xrightarrow[NH_3(\ell)]{Li(s),\ ROH}$	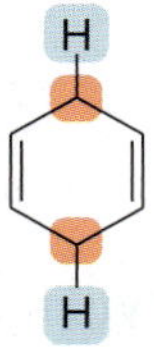 Cyclohexa-1,4-diene	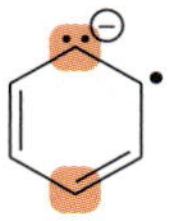 (cyclohexadienyl radical anion)	$H—OR$	Birch reduction; 1,4-positioning of the double bonds	25.7b

Problems

Problems that are related to synthesis are denoted (SYN).

25.1–25.3 Free Radicals: Curved Arrow Notation and Radical Initiators; Structure and Stability; Common Elementary Steps

25.27 The reaction shown here is an example of a *McLafferty rearrangement* (Interchapter G), which takes place in the gas phase in a single step. It is an important fragmentation process in mass spectrometry. Provide the curved arrow notation for this step.

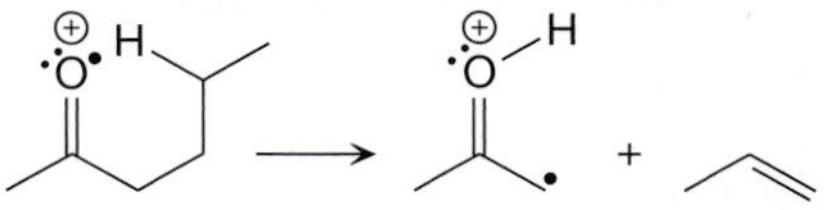

25.28 The radical cation product in Problem 25.27 has another resonance structure. Draw that resonance structure, using the proper curved arrow notation.

25.29 Acyl peroxides can be used to initiate alkyl radicals. Propose a mechanism for this reaction.

$$ R-C(=O)-O-O-C(=O)-R \longrightarrow 2\ R\bullet\ +\ 2\ CO_2 $$

25.30 The energy barrier for rotation of a CH_2 group in an allyl radical ($H_2C{=}CH{-}CH_2\bullet$) is about 66 kJ/mol (16 kcal/mol). The energy barrier for rotation of a CH_3 group in propane is ~14 kJ/mol (3.3 kcal/mol). Explain this difference.

25.31 Add the curved arrow notation for this elementary step. Classify the step as either homolysis, radical coupling, S_H2, or radical addition.

$$ Cl\bullet\ +\ \text{(cyclopropane structure)} \longrightarrow \text{(product structure)} $$

25.32 Determine the weakest C—H bond in each of the following compounds.

(a)   (b)   (c)

25.33 Determine the weakest C—C bond in each of the compounds in Problem 25.32.

25.34 Draw the major homolysis products when the compound shown here is irradiated with UV light.

25.35 Draw the curved arrow notation and product for each of the following when they undergo an S_H2 step.

(a)   + •Br ⟶ ?   (b)   + I—I ⟶ ?

25.36 For each of the following, there are two possible radical addition steps. In each case, draw the curved arrow notation and product for both radical additions and predict the more stable product.

(a)   + •Br ⟶ ?   (b)   + •Cl ⟶ ?

25.4 Radical Halogenation of Alkanes: Synthesis of Alkyl Halides

25.37 In each of the following compounds, which H would most likely be abstracted by a bromine radical, Br•?

(a)   (b)   (c)   (d)

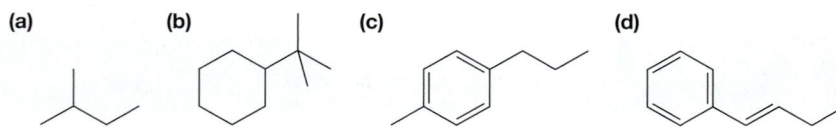

25.38 On treatment with $Cl_2/h\nu$, a compound with the formula C_9H_{12} yields only a single monochloride. What is a possible structure of the compound?

25.39 At which carbon in the molecule shown here will radical substitution predominantly take place? Explain.

25.40 Predict the major product of each of the following reactions and provide the complete, detailed mechanism.

(a)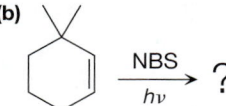

(b)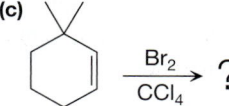

(c)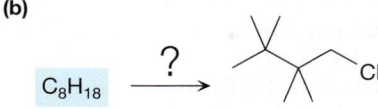

25.41 (SYN) Show how to produce each of the following compounds from a hydrocarbon that has the formula indicated.

(a)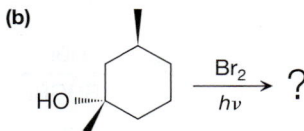

C₉H₁₂

(b) C₈H₁₈

25.42 A hydrogen radical, H•, is known to abstract a hydrogen atom from an alkane in the same manner that halogen atoms do. Like halogen atoms, H• abstracts a hydrogen atom from a secondary carbon over a hydrogen atom from a primary carbon, in this case with a selectivity of 5:1.
(a) Draw the two isomeric propyl radicals that are formed on hydrogen abstraction from propane by H•.
(b) Draw the curved arrow notation for the formation of each of those isomeric radicals.
(c) Compute the percentage of each propyl radical that is formed, taking into account the different numbers of each type of hydrogen on propane.

25.43 Draw the complete mechanisms leading to the two major products and determine whether the product mixture will be optically active for each of the following reactions.

(a)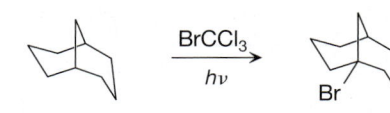

(b)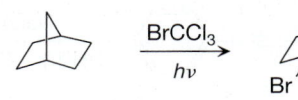

25.44 Similar to alkanes, hydrogen gas can undergo radical halogenation according to the following reaction, where X = F, Cl, Br, or I. Propose a chain-reaction mechanism for this reaction, including an initiation step, propagation steps, and two plausible termination steps.

$$\text{H—H} + \text{X—X} \xrightarrow[h\nu]{} 2\,\text{H—X}$$

25.45 In the halogenation of H₂, described in Problem 25.44, which halogen will react fastest? Defend your answer using the appropriate bond energies from Tables 25-1 and 25-2.

25.46 The selectivity of chlorination at 1°, 2°, and 3° carbons is about 1:4.5:6, a ratio that is relatively independent of the type of most solvents. If the reaction takes place in benzene, however, it is believed that Cl₂ forms a weak complex to the aromatic ring, thereby stabilizing the chlorine molecule. Does this lead to an increased selectivity or a decreased selectivity in chlorination? Explain.

25.47 Both bicyclo[3.3.1]nonane and bicyclo[2.2.1]heptane are saturated hydrocarbons composed of only 2° and 3° carbons. Recall that radical substitution generally takes place preferentially at a 3° carbon. Indeed, when bicyclo[3.3.1]nonane is treated with bromotrichloromethane under irradiation, substitution takes place 100% at the 3° carbon. Under the same conditions, however, no substitution is observed at the 3° carbon in bicyclo[2.2.1]heptane. Explain. *Hint:* Build molecular models of each of these compounds.

25.48 Propose a mechanism for the chain reaction in Problem 25.47, including initiation and propagation steps. Provide two plausible termination steps.

25.5 Radical Addition of HBr: Anti-Markovnikov Addition

25.49 Predict the major product of each of the reactions shown here and provide the complete, detailed mechanism.

25.50 (SYN) Show how to produce each of the following from a hydrocarbon that has the formula indicated.

(a)

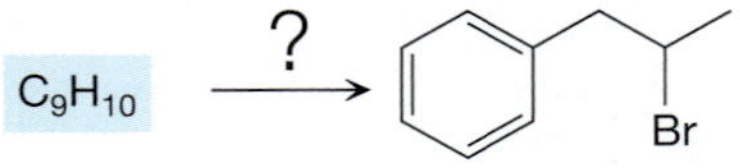

C_9H_{10} → (1-bromo-2-phenylpropane structure with Br)

(b)

$C_{10}H_{16}$ → (decalin with Br substituent)

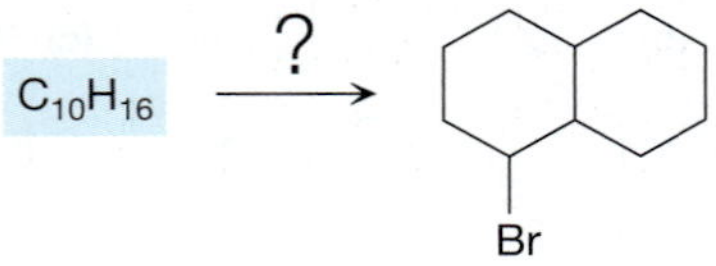

25.51 This addition of HBr to (Z)-2-bromobut-2-ene takes place regioselectively, with the Br preferentially adding to the alkene C that does *not* already possess a Br atom. **(a)** Provide a detailed mechanism for this reaction, including initiation and propagation steps. **(b)** Explain why this regiochemistry is observed.

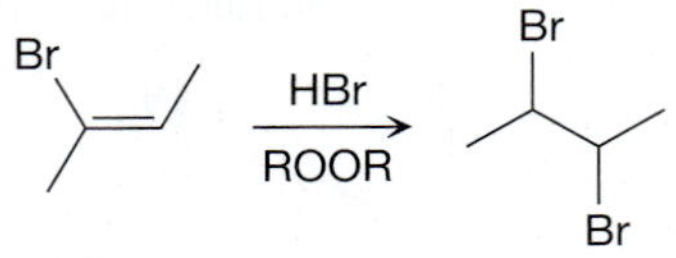

(2-bromobut-2-ene + HBr/ROOR → 2,3-dibromobutane)

25.52 The radical addition of HBr can be used to cyclize a carbon chain. Provide a detailed mechanism for this reaction. Include an initiation step, propagation steps, and two plausible termination steps.

(diene + HBr/ROOR → substituted cyclopentane with bromomethyl group)

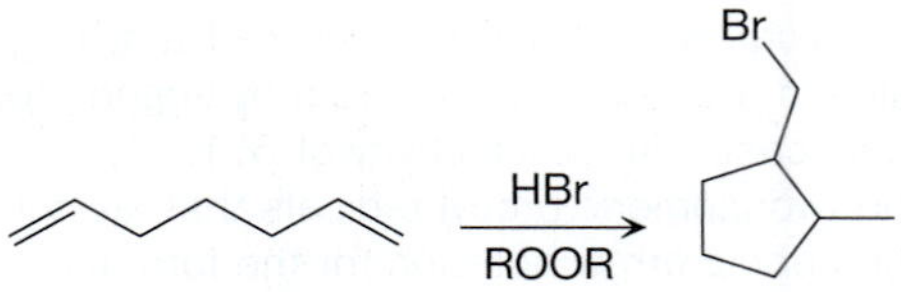

25.53 Provide a detailed mechanism to account for the following reaction. Include an initiation step, propagation steps, and two plausible termination steps. *Hint:* Consult Problem 25.52.

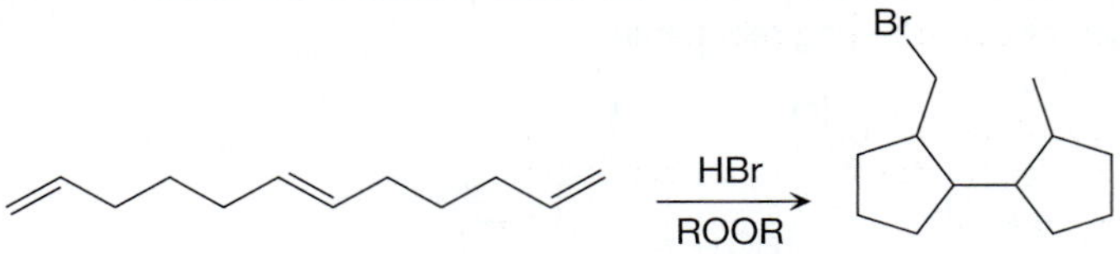

(triene + HBr/ROOR → bicyclopentane with bromomethyl group)

25.54 Propose a mechanism for the reaction shown here, which proceeds by a radical chain reaction. Include an initiation step, propagation steps, and two plausible termination steps.

$$H_2C{=}CH_2 + CCl_4 \xrightarrow{h\nu} Cl{-}\underset{H_2}{C}{-}\underset{H_2}{C}{-}CCl_3$$

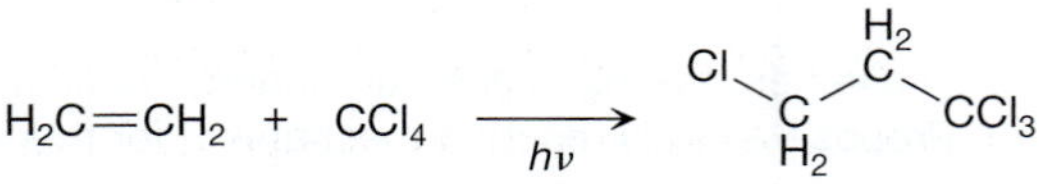

25.55 Propose a chain-reaction mechanism for the elimination of HI to form an alkene, showing reasonable initiation, propagation, and termination steps. *Hint:* Consider the mechanism for the reverse reaction.

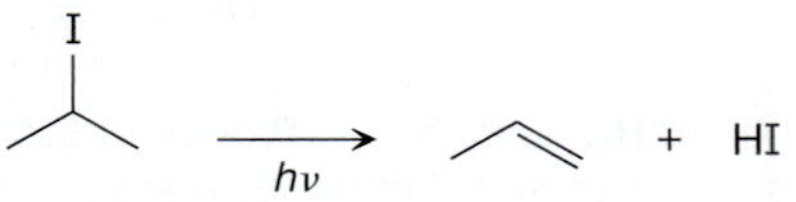

(2-iodopropane → propene + HI, hν)

25.56 The conversion of *cis*-1,2-diphenylethene to *trans*-1,2-diphenylethene is catalyzed when I_2 is added and the reaction mixture is irradiated with UV light. Provide a detailed mechanism for this reaction.

(cis-stilbene → trans-stilbene, I2/hν)

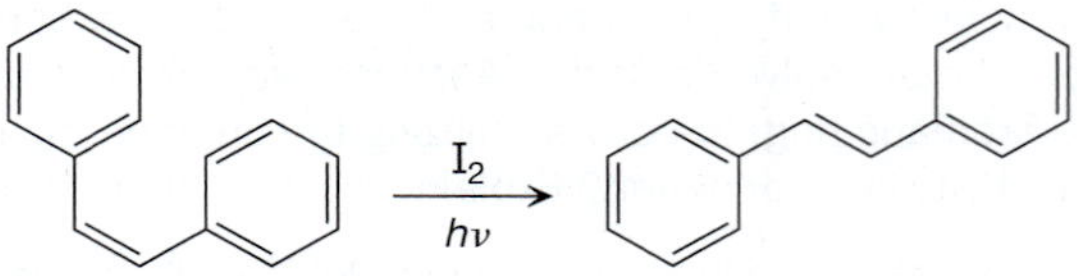

25.7 Dissolving Metal Reductions: Hydrogenation of Alkenes and Alkynes

25.57 Predict the major product of each of the following reactions and provide the complete, detailed mechanisms.

(a)

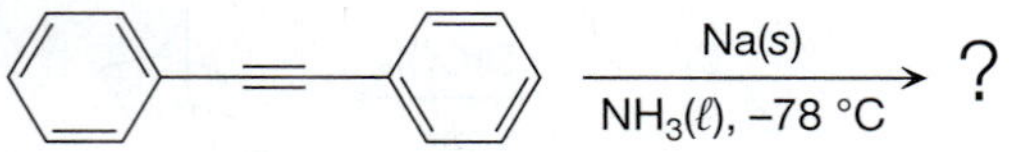

(diphenylacetylene) $\xrightarrow[NH_3(\ell),\,-78\,°C]{Na(s)}$?

(b)

(cyclopentyl-substituted alkyne) $\xrightarrow[NH_3(\ell),\,-78\,°C]{Na(s)}$?

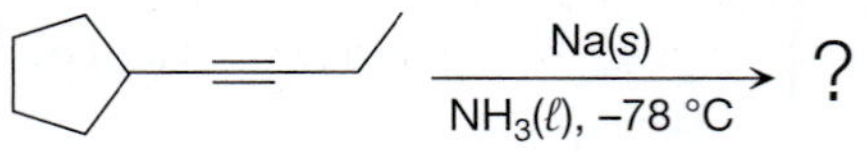

25.58 (SYN) Show how to produce each of the compounds shown here from 1-phenylprop-1-yne.

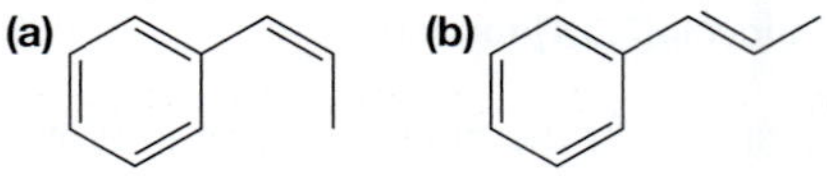

(a) (cis-1-phenylpropene) (b) (trans-1-phenylpropene)

25.59 As we saw in this chapter, dissolving metal reductions reduce alkynes to trans alkenes. Why, then, do these reactions fail with a terminal alkyne? *Hint:* What species appear in the mechanism?

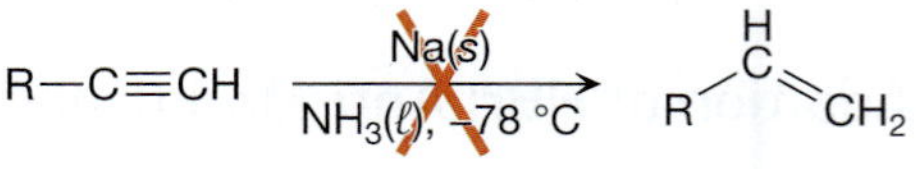

$$R{-}C{\equiv}CH \xrightarrow[NH_3(\ell),\,-78\,°C]{\cancel{Na(s)}} \underset{R}{\overset{H}{\diagdown}}C{=}CH_2$$

25.60 There are two isomeric cyclohexa-1,4-diene products when benzoic acid undergoes the Birch reduction (see Problem 25.24). **(a)** Draw the mechanism that leads to the formation of the major product. **(b)** Will the Birch reduction of benzoic acid occur faster or slower than the Birch reduction of benzene itself? *Hint:* Is –CO$_2$H an electron-donating or an electron-withdrawing group?

CO$_2$H

$\xrightarrow[\text{NH}_3(\ell)]{\text{Li}(s),\ \text{ROH}}$?

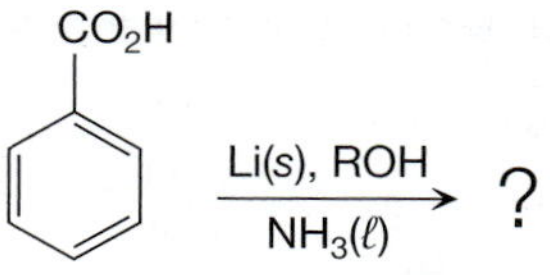

25.61 There are two isomeric cyclohexa-1,4-diene products when toluene undergoes the Birch reduction (see Problem 25.24). **(a)** Draw the mechanism that leads to the formation of the major product. **(b)** Will the Birch reduction of toluene occur faster or slower than the Birch reduction of benzene itself? *Hint:* Is –CH$_3$ an electron-donating or an electron-withdrawing group?

CH$_3$

$\xrightarrow[\text{NH}_3(\ell)]{\text{Li}(s),\ \text{ROH}}$?

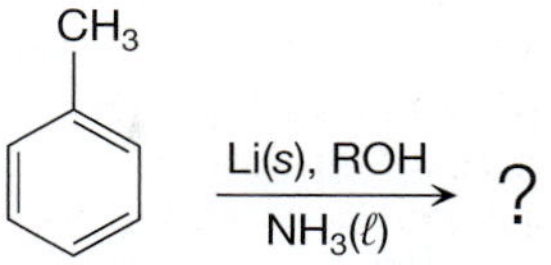

25.62 Why does a Birch reduction require ROH as the proton source, whereas a dissolving metal reduction of an alkyne uses NH$_3$ as the proton source? *Hint:* Which acid is stronger?

Integrated Problems

25.63 When an ether is exposed to air for a prolonged time, it undergoes a radical chain reaction with oxygen gas to form explosive *hydroperoxides*. The overall reaction between diethyl ether and O$_2$ is shown here. Provide a complete, detailed mechanism for this reaction, including an initiation step, propagation steps, and two plausible termination steps. *Hint:* In its ground state, O$_2$ is a diradical.

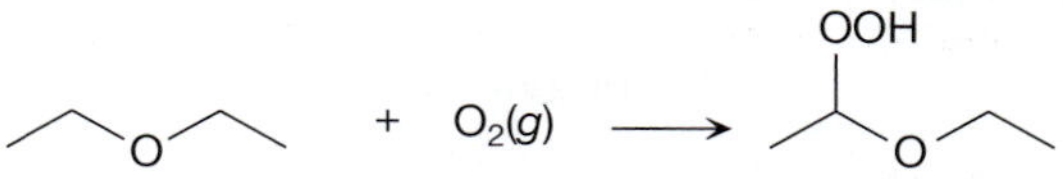

+ O$_2$(g) $\longrightarrow$

A hydroperoxide of diethyl ether

25.64 Notice in Problem 25.63 that the hydroperoxide forms at the C atom that is α to the O atom and not at the one that is β to it. Explain why. *Hint:* Write out the respective radicals that are formed on hydrogen abstraction.

25.65 The following aromatic substitution reaction has been proposed to proceed via a radical chain reaction.

$\xrightarrow{\text{H}_2\text{N}^{\ominus}}$ +

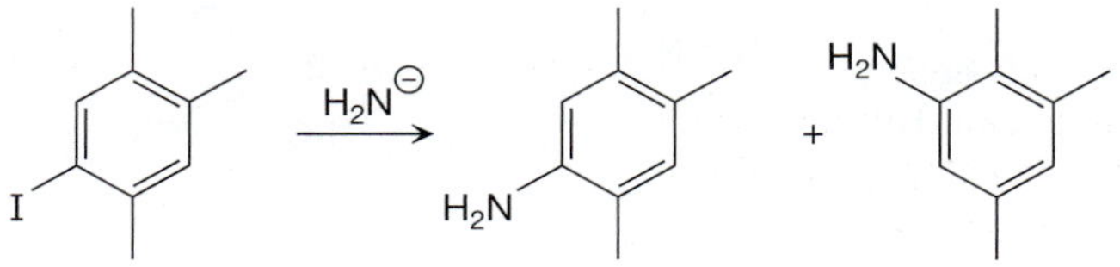

The mechanism shown here has been proposed, where Ar is the aromatic ring.

(a) Identify each step as either initiation, propagation, or termination.

(b) Sum the propagation steps to verify that the net equation matches that given in the problem statement.

(c) Propose a possible termination step.

ArI + e$^{\ominus}$ $\longrightarrow$ ArI$\overset{\ominus}{\cdot}$

ArI$\overset{\ominus}{\cdot}$ $\longrightarrow$ Ar• + I$^{\ominus}$

Ar• + H$_2$N$^{\ominus}$ $\longrightarrow$ ArNH$_2\overset{\ominus}{\cdot}$

ArNH$_2\overset{\ominus}{\cdot}$ + ArI $\longrightarrow$ ArNH$_2$ + ArI$\overset{\ominus}{\cdot}$

25.66 The steps shown here, which are in no particular order, have been proposed for a radical chain-reaction mechanism.

(a) Draw in the appropriate curved arrows for each step.

(b) Label each step as either initiation, propagation, or termination.

(c) Using this mechanism, write the balanced net reaction.

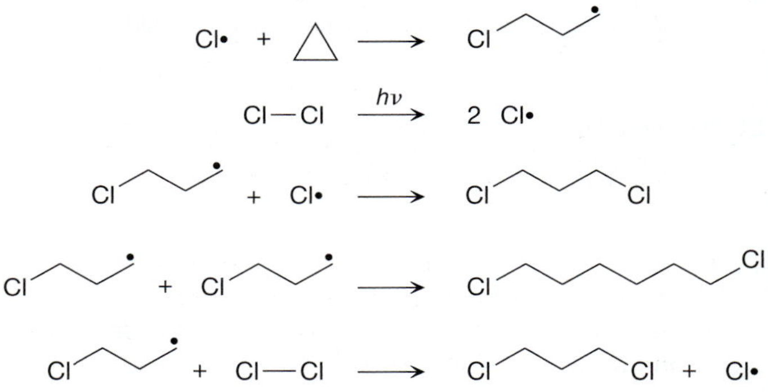

Cl• + $\triangle$ $\longrightarrow$ Cl⁀

Cl—Cl $\xrightarrow{h\nu}$ 2 Cl•

Cl⁀• + Cl• $\longrightarrow$ Cl⁀Cl

Cl⁀• + Cl⁀• $\longrightarrow$ Cl⁀⁀Cl

Cl⁀• + Cl—Cl $\longrightarrow$ Cl⁀Cl + Cl•

25.67 The steps shown here, which are in no particular order, have been proposed for the initiation and propagation of a radical chain-reaction mechanism.

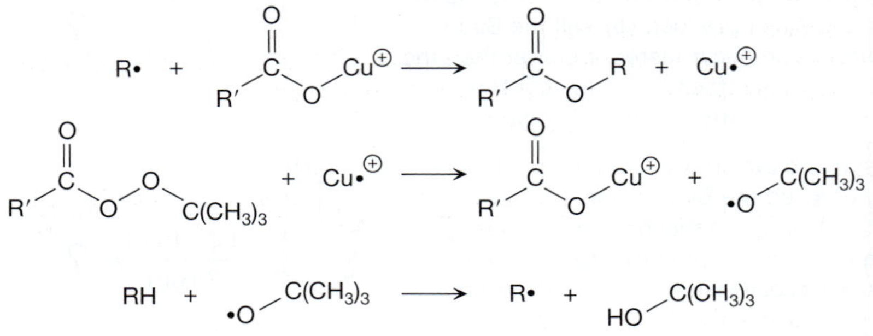

(a) Draw in the appropriate curved arrows for each step.
(b) Label each step as either initiation or propagation.
(c) Write the balanced net reaction.
(d) Provide two plausible termination steps, including curved arrows.

25.68 The following steps, which are in no particular order, have been proposed for the initiation and propagation of a radical chain-reaction mechanism.

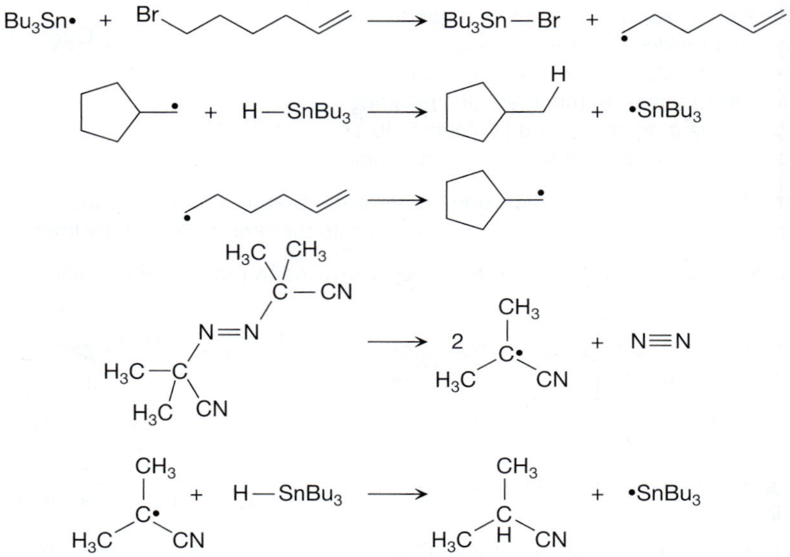

(a) Draw in the appropriate curved arrows for each step.
(b) Label each step as either initiation or propagation.
(c) Write the balanced net reaction.
(d) Provide two plausible termination steps, including curved arrows.

25.69 Propose a chain-reaction mechanism for the decomposition of dimethyl ether to form methane and formaldehyde.

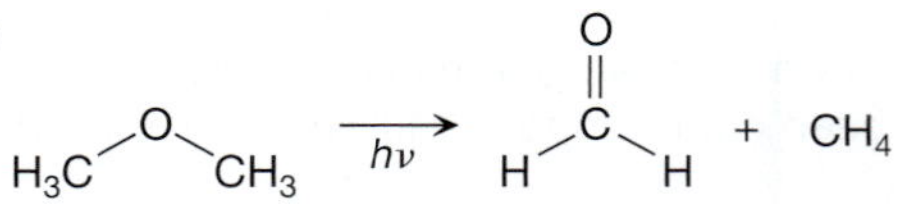

25.70 The following reaction proceeds by a chain-reaction mechanism. Propose an initiation step, propagation steps, and two plausible termination steps.

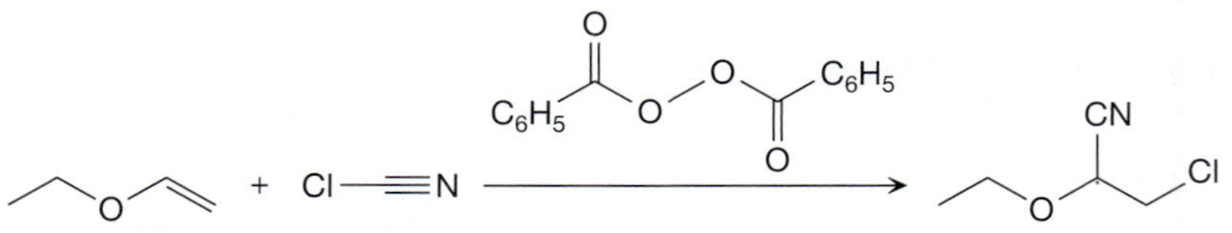

25.71 Radical halogenation generally yields a mixture of stereoisomers. When a Br atom is adjacent to the site of hydrogen abstraction, however, the reaction is stereospecific, as shown here. Provide a detailed mechanism for this reaction and propose a key intermediate that accounts for the reaction's stereochemistry. With that key intermediate, explain why this stereochemistry is observed.

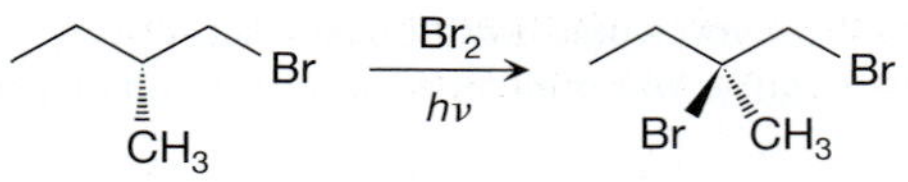

25.72 Supply the structures of compounds **A–I** in the following synthesis scheme.

$$\text{cyclohexane} \xrightarrow[hv]{Br_2} \textbf{A} \xrightarrow{(CH_3)_3COK} \textbf{B} \xrightarrow[CCl_4, \Delta]{NBS} \textbf{C} \xrightarrow{NaOH} \textbf{D} \xrightarrow{H_2CrO_4} \textbf{E} \xrightarrow[2.\ NH_4Cl]{1.\ (CH_3)_2CuLi} \textbf{F}$$

Same as above

$$\textbf{C} \xrightarrow{Mg(s)} \textbf{G} \xrightarrow[2.\ H_3O^{\oplus}]{1.\ \textbf{F}} \textbf{H} \xrightarrow[CCl_4]{Br_2} \textbf{I}$$

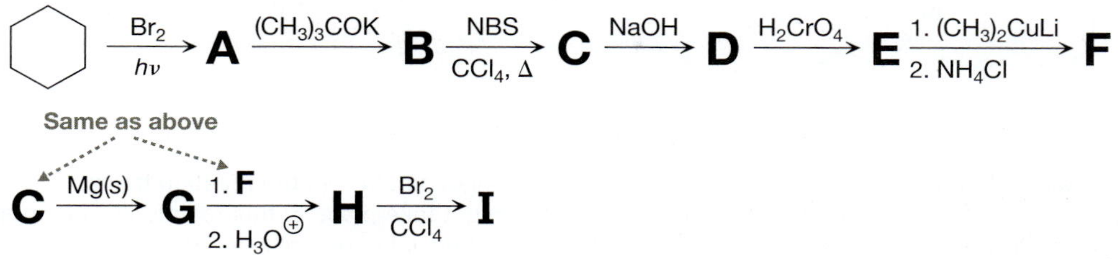

25.73 Supply the structures of compounds **J–U** in the following synthesis scheme.

$$H_3C\underset{CH_3}{\overset{H_2\ C}{\diagdown}}CH_3 \xrightarrow[hv]{Br_2} \textbf{J} \xrightarrow{NaOH}_{\Delta} \textbf{K} \xrightarrow[CCl_4]{Br_2} \textbf{L} \xrightarrow[2.\ H_2O]{1.\ LDA\ (excess)} \textbf{M} \xrightarrow[2.\ CH_3Br]{1.\ NaH}$$

$$\textbf{N} \xrightarrow[NH_3(\ell),\ -78\ °C]{Na(s)} \textbf{O} \xrightarrow{H_3O^{\oplus}} \textbf{P} \xrightarrow{KMnO_4} \textbf{Q}$$

Same as above

$$\textbf{K} \xrightarrow[Peroxide]{HBr} \textbf{R} \xrightarrow[2.\ Bu-Li]{1.\ PPh_3} \textbf{S} \xrightarrow{\textbf{Q}} \textbf{T} \longrightarrow \textbf{U}$$

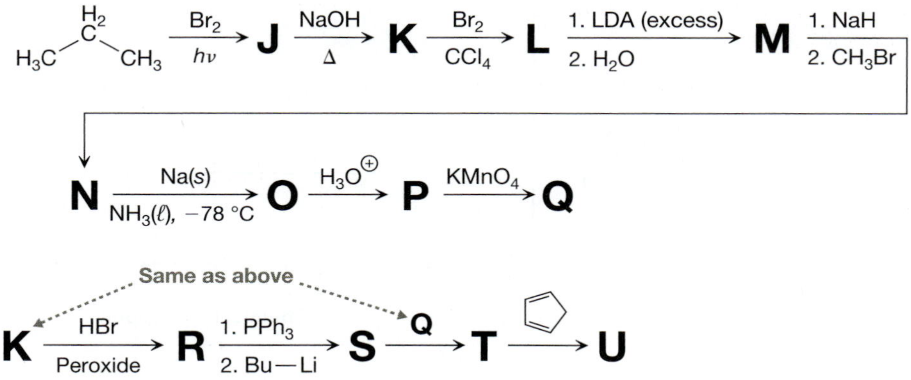

25.74 **(SYN)** Suggest how you would synthesize each of the compounds shown here beginning with propylbenzene. You may use any other reagents necessary.

(a)   (b) 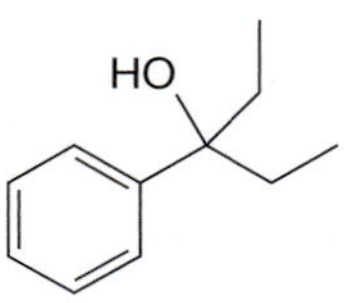   (c) 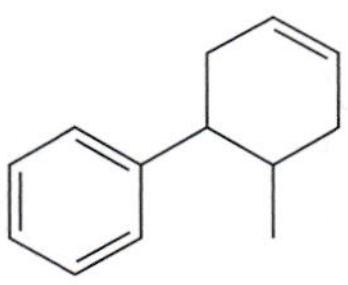   (d) 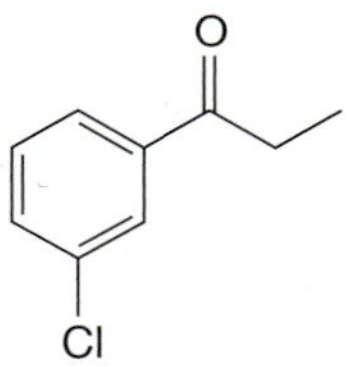   (e)

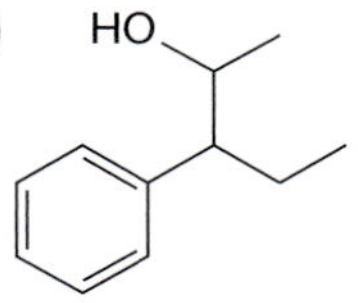

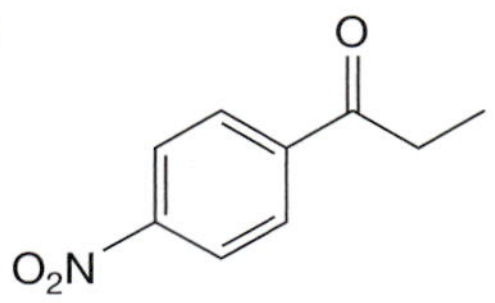

25.75 **(SYN)** How would you synthesize each of the following compounds beginning with 2-methylpropane? You may use any other reagents necessary.

(a)   (b) 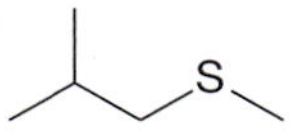   (c) 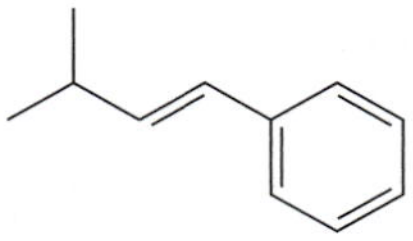   (d)

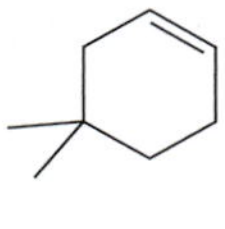

(e) 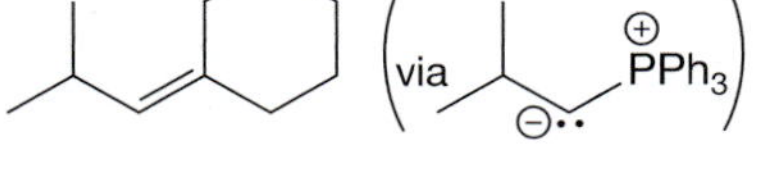 $\left(via\ \underset{\ominus}{\overset{\oplus}{\cdot\cdot}}PPh_3 \right)$ (f)

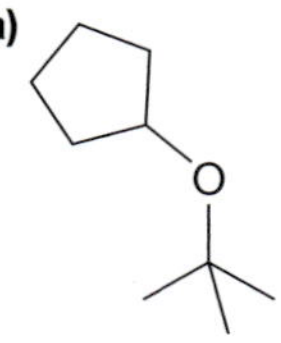

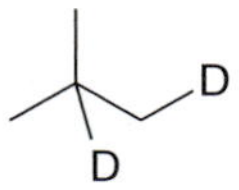

25.76 (SYN) Show how you would synthesize each of the following from 1-cyclopentylprop-1-yne. You may use any other reagents necessary. You may assume that each chiral target is produced as a racemic mixture.

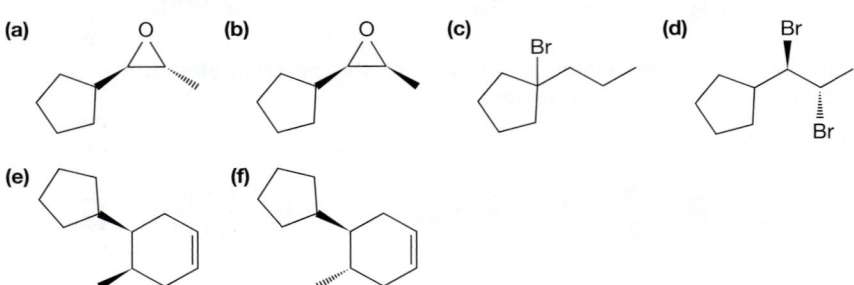

(a) (b) (c) (d)

(e) (f)

25.77 A student wanted to determine whether alkyl radicals rearrange, so she ran the reaction below.
  (a) Why is it plausible to think that radical rearrangement might take place in this reaction? To help answer this question, draw the mechanism of the radical reaction that would potentially take place.
  (b) The student acquired a ^{1}H NMR spectrum of the product, which is shown below. What should she conclude regarding alkyl rearrangements? Explain.

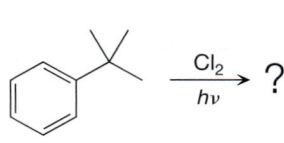

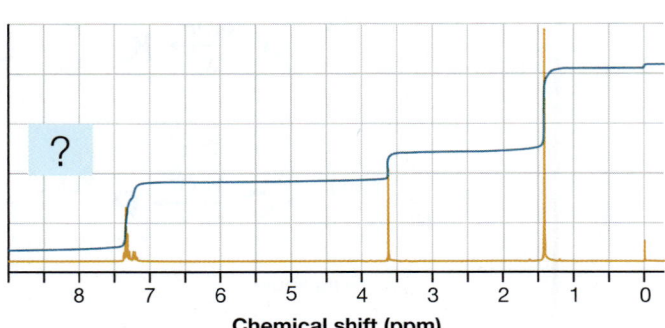

Chemical shift (ppm)

Fragmentation Pathways in Mass Spectrometry

Chapter 16 provided an introduction to mass spectrometry. There, we learned that once a sample is injected into a mass spectrometer, the acquisition of a mass spectrum typically consists of the four main stages shown in Figure G-1.

First, the sample, $M(\ell)$ or $M(s)$, is heated to produce gaseous molecules, $M(g)$. Second, the vaporized sample passes through an electron beam, where ionization takes place. When a high-energy electron collides with $M(g)$, an electron is ejected from the molecule, producing a molecular ion, $M^+(g)$. Third, ions are guided to the detector. Finally, the electronic signal generated by the collisions of the ions with the detector is converted into the ions' relative abundance, and the spectrum is produced.

In the ionization stage (depending on the type and parameters of the instrument), collisions between the electrons and the uncharged molecules can carry enough energy to break covalent bonds as well, resulting in fragmentation of the molecular ion. The fragment ions thus produced are also detected, which is why numerous mass peaks often appear in a mass spectrum.

In Chapter 16, we interpreted mass spectra primarily to determine the molecular mass of the compound that produces the spectrum. Doing so largely involves identifying the M^+ peak, as well as various $M + 1$ and $M + 2$ isotope peaks. Now we learn how to interpret additional structural information from the specific fragment peaks in a mass spectrum.

Interpreting fragment peaks requires an understanding of the chemical processes responsible for producing the corresponding fragment ions. Notably, these **fragmentation pathways** consist of elementary steps that involve *free radicals*. The ionization process causes the loss of a single electron from a closed-shell, uncharged molecule, so the molecular ion must have an unpaired electron, making it a free radical. More specifically, it also carries a positive charge, so the molecular ion is a **radical cation**, and M^+ is more accurately depicted as $M^{+\bullet}$.

Because our focus in this interchapter is on gathering structural information about a compound, our discussion is organized according to the compound classes involved. We begin by discussing the fragmentation of alkanes, which contain no functional groups at all. We then discuss alkenes and aromatic compounds, whose Lewis structures contain $C{=}C$ double bonds. Following that, we discuss alkyl halides, amines, ethers, and alcohols, all of which have functional groups containing a singly bonded heteroatom. Finally, we cover compounds that contain carbonyl groups, including ketones, aldehydes, and carboxylic acids.

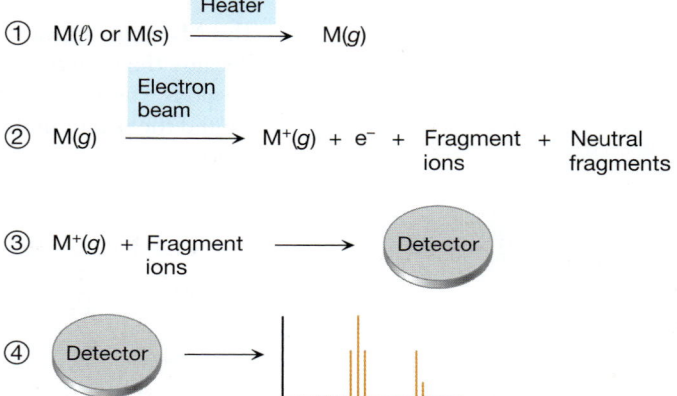

FIGURE G-1 How an electron ionization mass spectrum is produced (1) A sample, $M(\ell)$ or $M(s)$, is vaporized to $M(g)$. (2) $M(g)$ is ionized by an electron beam, producing $M^+(g)$. With enough energy from the electron beam, $M^+(g)$ will fragment into smaller ions. (3) The various ions are guided to the detector. (4) The signal generated by the detector is converted into a mass spectrum.

G.1 Alkanes

Let's begin by examining the mass spectrum of hexane (Fig. G-2), which we discussed previously in Section 16.17.

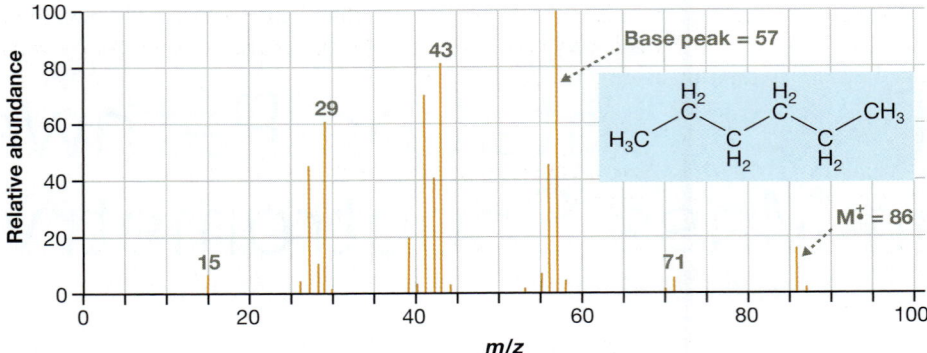

FIGURE G-2 Mass spectrum of hexane The M⁺̇ peak and the base peak are indicated. The values of m/z for selected peaks are indicated above the peaks.

The M⁺̇ peak appears at $m/z = 86$, which corresponds to hexane's molecular mass. Because the molecular ion is produced by the loss of an electron from one of the many σ bonds, the molecular ion is ambiguously represented by placing square brackets around the parent molecule, along with a dot and a positive charge to indicate a radical cation:

$$\text{(G-1)}$$

Several key fragment peaks in the spectrum can also be identified—at m/z values of 71, 57, 43, 29, and 15—which, as we saw in Section 16.17, correspond to the breaking of the various C—C bonds in hexane. For example, the base peak (i.e., the most intense mass peak) at $m/z = 57$ corresponds to the breaking of the C2—C3 bond to produce both a butyl cation and an ethyl radical:

The uncharged radical is not detected.

The fragment ion is detected.

$$\text{(G-2)}$$

Ethyl radical

Butyl cation
$m/z = 57$

The peak at $m/z = 15$ corresponds to the breaking of the C1—C2 bond to produce a methyl cation and a pentyl radical:

The fragment ion is detected.

The uncharged radical is not detected.

$$\text{(G-3)}$$

Methyl cation
$m/z = 15$

Pentyl radical

Because the peak at $m/z = 57$ is substantially more intense than the one at $m/z = 15$, we know that the fragmentation pathway in Equation G-2 is much more likely than the one in Equation G-3. Why is this?

The answer has to do with the stability of the fragmentation products. In each case, both a cation and an alkyl radical are produced. Cation stability favors Equation G-2, because the positively charged C is more highly alkyl substituted (Section 6.6e). Radical stability favors Equation G-3, however, because the C atom bearing the unpaired electron is more highly alkyl substituted (Section 25.2). Yet the cation stability wins out, because cations are more electron-deficient.

> In general, a fragmentation pathway becomes more likely as the stability of the fragment ion produced increases.

YOUR TURN **G.1**

Will the fragmentation shown here more likely produce a mass peak at $m/z = 43$ or $m/z = 57$? Explain.

This general rule can be very useful, particularly when distinguishing isomers by mass spectrometry. Consider 2-methylpentane, whose mass spectrum is shown in Figure G-3.

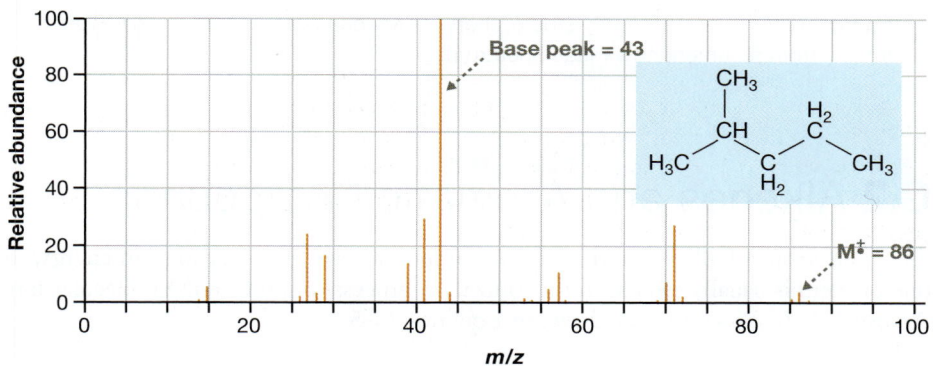

FIGURE G-3 Mass spectrum of 2-methylpentane The values of m/z for the molecular ion and base peaks are indicated above the peaks.

The $M^{+\cdot}$ peak appears at $m/z = 86$, the same as in the mass spectrum of hexane. The base peak for 2-methylpentane is at $m/z = 43$, however, whereas the one for hexane is at $m/z = 57$. The fragmentation that accounts for the base peak at $m/z = 43$ is shown in Equation G-4. It is favored because the cation that is produced is secondary, whereas the other potential cation would be primary.

The secondary carbocation is more stable than a primary or methyl carbocation.

Isopropyl cation $m/z = 43$ Propyl radical   (G-4)

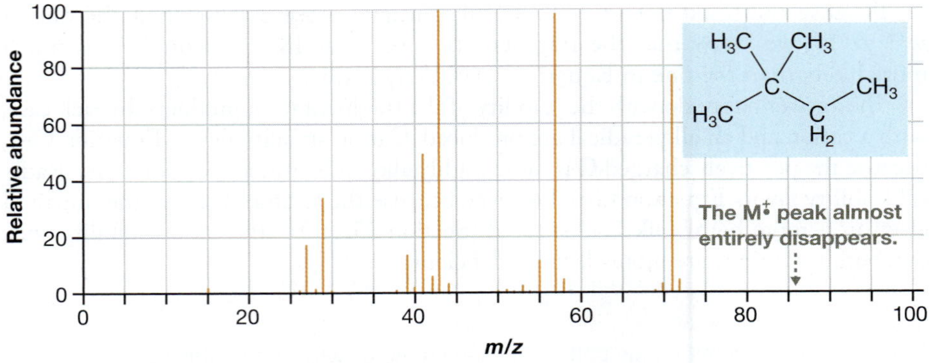

FIGURE G-4 **Mass spectrum of 2,2-dimethylbutane** The $M^{+\cdot}$ peak is almost entirely absent from the spectrum, because the significant branching of the alkyl chain means that fragmentation gives rise to relatively highly stable fragment ions.

The $M^{+\cdot}$ peak almost entirely disappears.

YOUR TURN G.2

Consider the mass peaks at $m/z = 71$ and $m/z = 29$ in Figure G-3. Which peak represents the more stable fragment ion? Explain why, accounting for the fragment ions that correspond to those peaks.

Another difference between the spectra in Figures G-2 (hexane) and G-3 (2-methylpentane) is the relative size of the $M^{+\cdot}$ peaks. The one for 2-methylpentane is much smaller than the one for hexane. Thus, fragmentation is more likely for the molecular ion of 2-methylpentane because, with additional branching, fragment ions that are more stable can be produced. With even more branching, as in 2,2-dimethyl-butane (Fig. G-4), the $M^{+\cdot}$ peak essentially disappears entirely.

> If a fragment ion is sufficiently stable, then the $M^{+\cdot}$ peak can essentially disappear entirely from a compound's mass spectrum.

G.2 Alkenes and Aromatic Compounds

The π electrons of alkenes and aromatic compounds are relatively high in energy, so one of them is usually ejected in the ionization process to produce the molecular ion. An example is shown for hex-2-ene in Equation G-5.

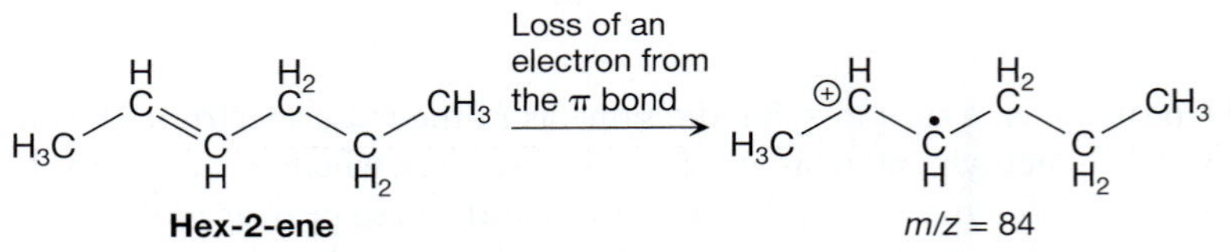

Loss of an electron from the π bond

Hex-2-ene

$m/z = 84$ (G-5)

Because of the relatively well-defined nature of an alkene's molecular ion, fragmentation pathways tend to be clear-cut. In fact:

> An alkene's molecular ion tends to expel an alkyl radical from an allylic carbon, producing an allylic cation.

Equation G-6 shows that the molecular ion of hex-2-ene can expel an ethyl radical to produce a resonance-stabilized 2-butenyl cation.

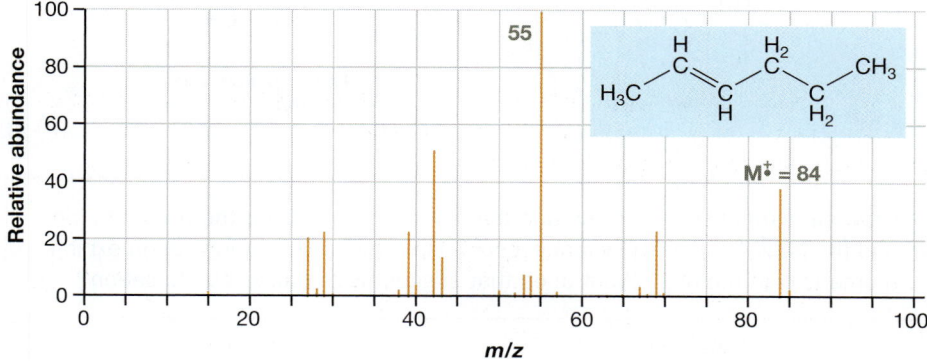

The allylic cation that is produced is resonance stabilized.

Allylic C from the original alkene

$m/z = 84$

2-Butenyl cation
$m/z = 55$

(G-6)

YOUR TURN **G.3**

Hex-1-ene undergoes ionization and fragmentation to expel a propyl radical, as shown below. Add the curved arrows for the fragmentation step.

Loss of an electron from the π bond

Fragmentation

The driving force for this process is the stability that is gained by the resonance delocalization of the positive charge in the resulting allylic cation. This fragmentation pathway is the most likely one for the molecular ion of hex-2-ene because the base peak in the mass spectrum of hex-2-ene (Fig. G-5) is found at $m/z = 55$, which corresponds to the 2-butenyl cation in Equation G-6.

Relative abundance

55

$M^{+\bullet} = 84$

m/z

FIGURE G-5 Mass spectrum of hex-2-ene The $M^{+\bullet}$ peak appears at $m/z = 84$. Loss of an ethyl radical produces a resonance-stabilized allylic cation that gives rise to the base peak at $m/z = 55$.

Similar fragmentations are observed for aromatic compounds.

The molecular ion of an alkylbenzene tends to expel an alkyl radical from a benzylic carbon, producing a benzylic cation.

The mass spectrum of butylbenzene is shown in Figure G-6. The molecular ion, produced by the loss of a π electron, gives rise to the peak at $m/z = 134$. Loss of a propyl

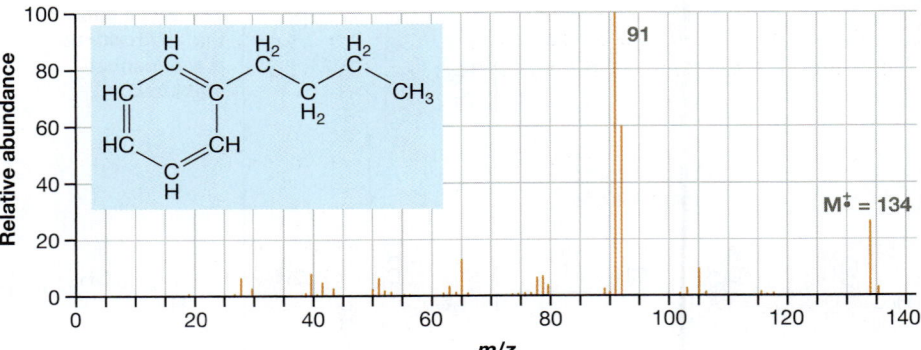

FIGURE G-6 Mass spectrum of butylbenzene The $M^{\ddagger}$ peak appears at $m/z = 134$. Loss of a propyl radical produces a resonance-stabilized benzyl cation that gives rise to the base peak at $m/z = 91$.

radical from the benzylic carbon (Equation G-7) produces the benzyl cation, whose mass peak appears at $m/z = 91$.

$$\text{(G-7)}$$

Benzyl cation
$m/z = 91$

In the gas phase, the benzyl cation in Equation G-7 is believed to rearrange to the **tropylium ion** (Equation G-8), which has an aromatic seven-membered ring.

Rearrangement $\longrightarrow$ $$\text{(G-8)}$$

$m/z = 91$

The tropylium ion
$m/z = 91$

YOUR TURN G.4

Draw all resonance structures and the resonance hybrid for the tropylium ion and the benzyl cation. How many resonance structures of each cation exhibit aromaticity in the ring? What does that suggest is the more stable cation?

G.3 Alkyl Halides, Amines, Ethers, and Alcohols

Many functional groups contain *heteroatoms* (i.e., atoms other than carbon or hydrogen), which can play major roles in fragmentation pathways, and, therefore, help govern the fragmentation patterns observed in a mass spectrum. In mass spectrometry, one of the most important features of a heteroatom is its lone pair of electrons. As nonbonding electrons, lone pairs are typically the least tightly bound electrons in the molecule.

Loss of a nonbonding electron from the Cl atom in 2-chloro-2-methylbutane produces the molecular ion, with $m/z = 106$, as shown in Equation G-9.

Loss of a nonbonding electron produces M⁺. **Heterolysis of the C—Cl bond**

$$\text{(G-9)}$$

$m/z = 106$ $m/z = 71$

The peak corresponding to M⁺ does not appear in the spectrum in Figure G-7, however, because fragmentation takes place too readily. One fragmentation pathway, shown in Equation G-9, is heterolysis of the C—Cl bond, producing an alkyl cation fragment with $m/z = 71$.

Another common fragmentation pathway is elimination of an alkyl group from the carbon atom bonded to the heteroatom—a process known as **α-cleavage**. There are two α-cleavage pathways for the molecular ion of 2-chloro-2-methylbutane, as shown in Equations G-10 and G-11.

$$\text{(G-10)}$$

$m/z = 106$ $m/z = 91$

$$\text{(G-11)}$$

$m/z = 106$ $m/z = 77$

These account for the mass peaks at $m/z = 91$ and $m/z = 77$. Notice that in the resonance structure shown for each fragment ion, *all nonhydrogen atoms have an octet*, which provides substantial driving force for the process.

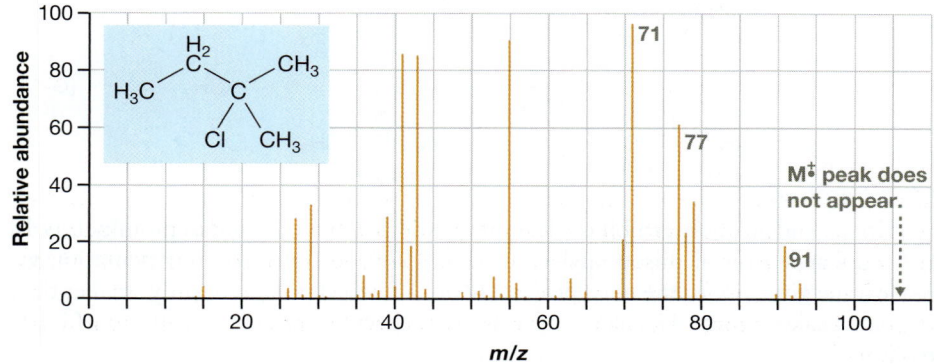

FIGURE G-7 Mass spectrum of 2-chloro-2-methylbutane The M⁺ peak, which would appear at $m/z = 106$, is absent. The masses of other significant fragment ions are labeled above their corresponding peaks.

Equation G-11 is repeated below. Draw the curved arrows that describe this α-cleavage step.

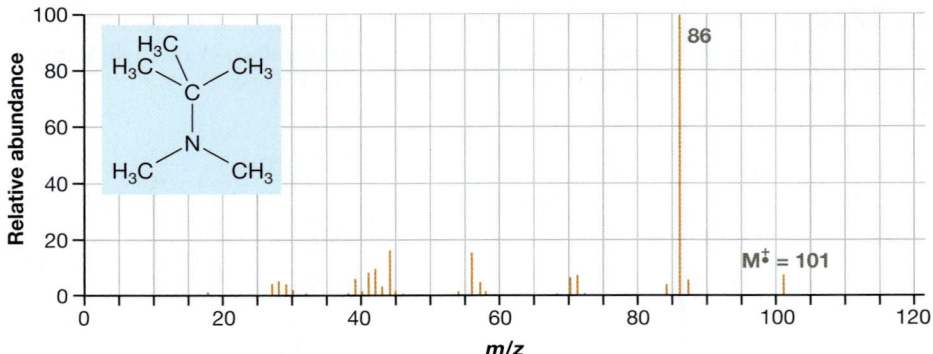

These two fragmentation pathways—heterolysis and α-cleavage—are also common to other compound classes that have functional groups with heteroatoms, such as amines, ethers, and alcohols. Depending on the identity of the functional group, as well as the specific structure of the compound, one of these fragmentation pathways can be highly favored over the other.

α-Cleavage is very common for amines, typically leading to the fragment ion that corresponds to the base peak. This is the case for *N,N,2*-trimethylpropan-2-amine, whose mass spectrum is shown in Figure G-8. The molecular ion appears at $m/z = 101$.

FIGURE G-8 **Mass spectrum of *N,N,2*-trimethylpropan-2-amine** The M⁺ peak appears at $m/z = 101$ and is relatively small. The base peak at $m/z = 86$ corresponds to a fragment ion produced by α-cleavage of the molecular ion.

Loss of a CH_3 group via α-cleavage (Equation G-12) produces the fragment ion giving rise to the base peak at $m/z = 86$.

This CH_3 group is eliminated.

An iminium ion is stable compared to M⁺.

(G-12)

$m/z = 101$ $m/z = 86$

Ethers and alcohols exhibit α-cleavage as well, but typically not as prominently. In the mass spectrum of diisopropyl ether (Fig. G-9), for example, fragmentation via α-cleavage gives rise to the mass peak at $m/z = 87$, which is significantly smaller than the base peak at $m/z = 45$. Likewise, in the mass spectrum of pentan-1-ol (Fig. G-10), α-cleavage gives rise to the relatively small peak at $m/z = 31$.

Draw the α-cleavage pathway of diisopropyl ether's molecular ion that accounts for the mass peak at $m/z = 87$. Do the same for pentan-1-ol's molecular ion to account for the mass peak at $m/z = 31$.

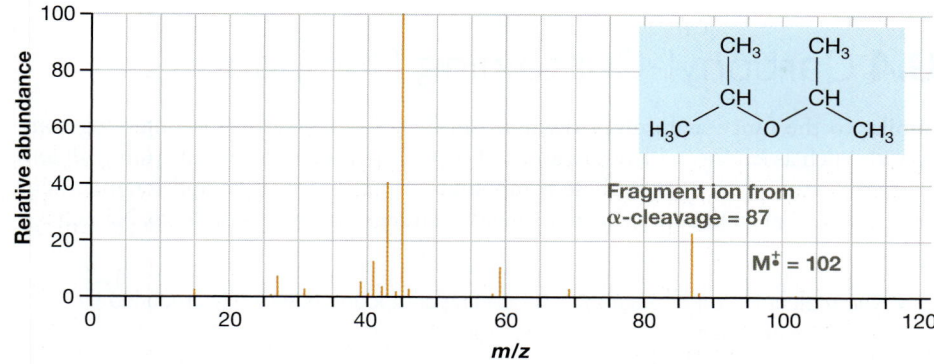

FIGURE G-9 Mass spectrum of diisopropyl ether The $M^{\ddagger}$ peak appears at $m/z = 102$. The peak at $m/z = 87$ corresponds to a fragment ion produced by α-cleavage of the molecular ion.

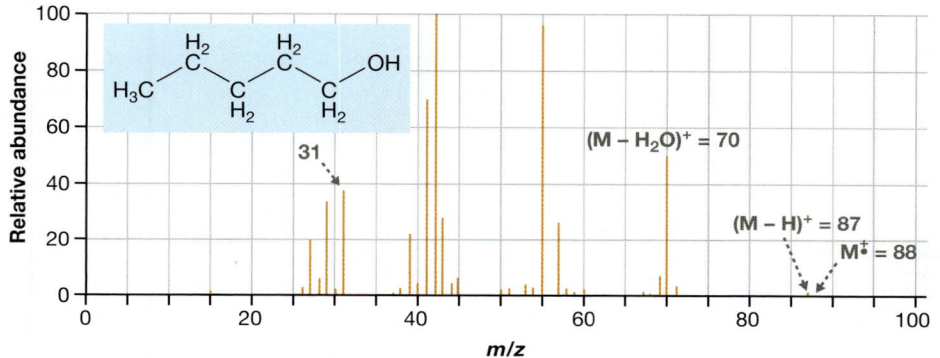

FIGURE G-10 Mass spectrum of pentan-1-ol The $M^{\ddagger}$ peak would appear at $m/z = 88$, but is absent. The peak at $m/z = 31$ corresponds to loss of an alkyl radical from $M^{\ddagger}$ via α-cleavage. The peak at $m/z = 70$ corresponds to loss of water from the molecular ion. The peak at $m/z = 87$ corresponds to a fragment ion produced by loss of a hydrogen atom.

There are two other fragmentation pathways characteristic of an alcohol.

An alcohol's molecular ion will typically undergo

- Loss of a hydrogen atom to give rise to an $(M - H)^+$ peak that is 1 u lighter than the molecular ion.
- Loss of a water molecule to give rise to an $(M - H_2O)^+$ peak that is 18 u lighter than the molecular ion.

For example, the molecular ion of pentan-1-ol is 88 u, but there is a small $(M - H)^+$ peak one mass unit lighter at $m/z = 87$ in Figure G-10. This is accounted for by the α-cleavage mechanism in Equation G-13, in which H• is expelled. There is a prominent peak at $m/z = 70$, too, which is 18 mass units less than the $M^{\ddagger}$ peak, and corresponds to the $(M - H_2O)^+$ ion. The curved arrow notation for this step is shown in Equation G-14. That step proceeds through a five-membered-ring transition state, which does not suffer from excessive strain or loss of entropy.

$$H_3C-CH_2-CH(-H)-CH_2-CH_2-\overset{\oplus}{O}H \xrightarrow{\text{Loss of } H_2O} H_3C-CH_2-CH\overset{\bullet}{-}CH_2-\overset{\oplus}{C}H_2 + H_2\overset{\bullet\bullet}{O} \quad (G\text{-}14)$$

$$m/z = 88 \qquad\qquad\qquad m/z = 70$$

G.4 Carbonyl-Containing Compounds

Similar to the functional groups discussed in Section G.3, carbonyl-containing compounds, such as ketones, aldehydes, and carboxylic acids, have an oxygen atom with lone pairs of electrons. As a result, these compounds undergo ionization and fragmentation processes similar to the ones that alkyl halides, amines, and ethers undergo. Namely:

- The molecular ion of a carbonyl-containing compound is typically produced by loss of a lone-pair (nonbonding) electron from the carbonyl oxygen.
- A common fragmentation pathway of this molecular ion is α-cleavage.

Equation G-15, for example, shows that hexan-2-one undergoes ionization to produce the molecular ion with $m/z = 100$ (Fig. G-11).

$$\text{Hexan-2-one} \xrightarrow[\text{nonbonding } e^-]{\text{Loss of}} [\,m/z = 100\,] \xrightarrow{\alpha\text{-Cleavage}} \;\text{An acylium ion} \quad (G\text{-}15)$$

$$m/z = 43$$

Subsequent α-cleavage can expel a butyl radical from the carbonyl carbon to produce a relatively stable *acylium ion*, giving rise to the base peak at $m/z = 43$. (This is the same type of ion that appears as an intermediate in a Friedel–Crafts acylation reaction; Section 22-5.) Alternatively, α-cleavage of a methyl radical produces the ion corresponding to the peak at $m/z = 85$.

YOUR TURN G.7

Draw the fragmentation pathway that gives rise to the peak at $m/z = 85$ in the mass spectrum of hexan-2-one (Fig. G-11). Use Equation G-15 as a guide.

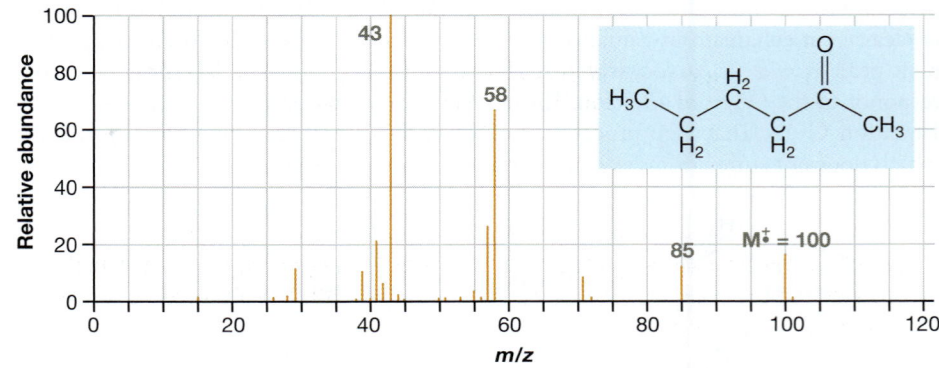

FIGURE G-11 Mass spectrum of hexan-2-one The M‡ peak appears at $m/z = 100$. The peaks at $m/z = 43$ and $m/z = 85$ correspond to fragment ions produced on α-cleavage of the molecular ion. The mass peak at $m/z = 58$ corresponds to a fragment ion produced by a McLafferty rearrangement.

In addition to α-cleavage, there is another fragmentation pathway characteristic of carbonyl-containing compounds.

A carbonyl-containing compound can undergo a **McLafferty rearrangement** if an alkyl group attached to the carbonyl carbon possesses a γ-carbon with at least one hydrogen atom.

This fragmentation pathway is named after Professor Fred W. McLafferty (b. 1923). As shown in Equation G-16, a H atom from the γ-carbon shifts to the carbonyl O atom, and the bond joining the α and β carbons is broken. The result is the ejection of an uncharged alkene molecule. In the case of hexan-2-one's molecular ion, the McLafferty rearrangement produces an enol radical cation of acetone, giving rise to the mass peak at $m/z = 58$.

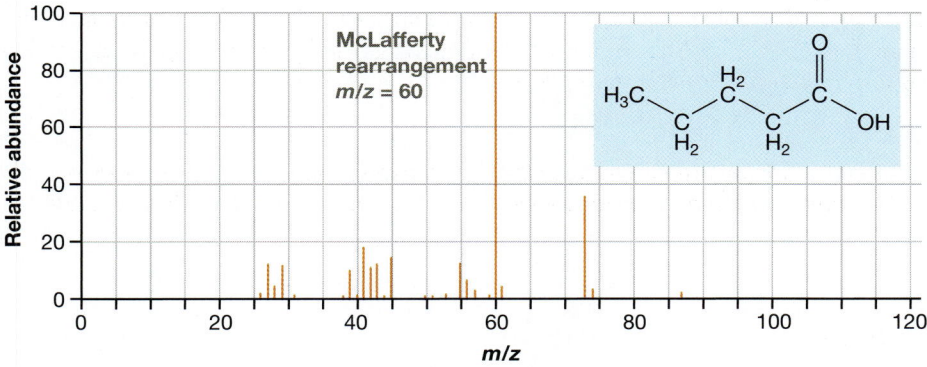

$$(G\text{-}16)$$

The McLafferty rearrangement is not limited to just ketones; it is characteristic of several compound classes containing the carbonyl group, including aldehydes, carboxylic acids, esters, and amides. For example, the molecular ion of pentanoic acid undergoes a McLafferty rearrangement, giving rise to the mass peak at $m/z = 60$ in Figure G-12.

FIGURE G-12 Mass spectrum of pentanoic acid The mass peak at $m/z = 60$ corresponds to a fragment ion produced by a McLafferty rearrangement.

YOUR TURN **G.8**

Draw the McLafferty rearrangement that the molecular ion of pentanoic acid undergoes to account for the mass peak at $m/z = 60$. Use Equation G-16 as a guide.

Problems

G.1 Mass spectra of butylcyclopentane and *tert*-butylcyclopentane were acquired. Spectrum **A** exhibited significant mass peaks at m/z values of 126, 97, 83, 69, 55, and 41. Spectrum **B** exhibited significant peaks at m/z values of 111, 69, 57, and 41. Match each spectrum with its compound.

G.2 The base peak in the mass spectrum of an alkane, C_7H_{16}, appears at $m/z = 57$. Draw a molecule that is consistent with these data.

G.3 A peak appears at $m/z = 83$ in the mass spectrum of hept-3-ene. Show how the fragmentation of hept-3-ene's molecular ion produces the ion that gives rise to this peak. Why do you think the peak at $m/z = 83$ is smaller than the one at $m/z = 69$?

G.4 The mass spectrum of an alkene, C_8H_{16}, exhibits a peak at $m/z = 41$. Draw two isomers that are consistent with these data.

G.5 Mass spectra were acquired for 1,4-diethylbenzene and 1-methyl-4-propylbenzene. The base peak of spectrum **A** is at $m/z = 105$ and that of spectrum **B** is at $m/z = 119$. Match each spectrum to its compound.

G.6 The mass spectrum of an amine ($C_6H_{15}N$) exhibits a base peak at $m/z = 30$. Which of the following amines could give rise to that mass spectrum?

G.7 The mass spectrum of an alcohol ($C_5H_{12}O$) exhibits a base peak at $m/z = 59$. Is the alcohol most likely pentan-1-ol, pentan-2-ol, or pentan-3-ol? Explain.

G.8 The mass spectrum of a ketone ($C_7H_{14}O$) exhibits a base peak at $m/z = 57$. Is the ketone most likely heptan-2-one, heptan-3-one, or heptan-4-one? Explain.

G.9 The base peak in the mass spectrum of hexanamide appears at $m/z = 59$. Draw the ion that corresponds to this mass peak, and show how it is produced from the molecular ion.

G.10 A mass peak at $m/z = 59$ appears in the mass spectrum of an amide, $C_5H_{11}NO$. Draw the structure of a molecule that is consistent with this result.

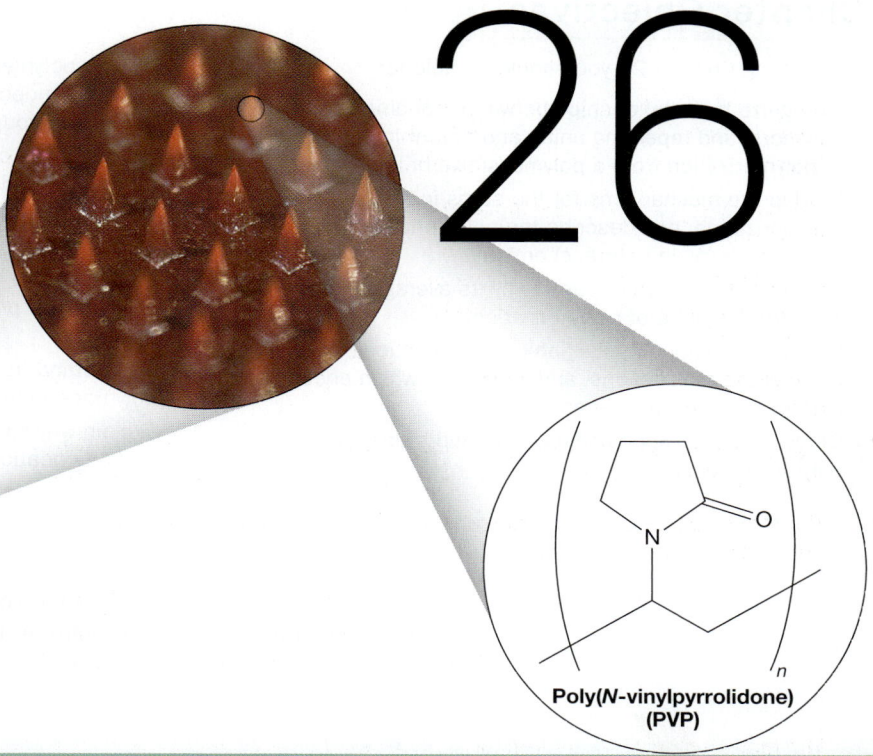

Microneedles of poly(*N*-vinylpyrrolidone) (PVP) are used to encapsulate vaccines and offer an alternative way to deliver vaccines and medications. PVP is water soluble and biocompatible; the polymer dissolves and releases the vaccine after the needles pierce the skin. Applications for influenza, measles, and other vaccines are under development. Here in Chapter 26, we describe how polymers such as PVP are made and how their properties and uses are related.

Poly(*N*-vinylpyrrolidone)
(PVP)

26

Polymers

In Chapter 25, we studied free radicals, with particular emphasis on the mechanism of chain reactions. Free radical chain reactions allow alkanes, which are unreactive under normal laboratory conditions, to undergo halogenation. We also saw that the free radical chain reaction of HBr with alkenes increases a chemist's control of regioselectivity, because the reaction provides the product of the anti-Markovnikov addition of HBr.

Here in Chapter 26, we apply the free radical chain reaction to the synthesis of **polymers**, which are large molecules formed from small molecules. Trademarked products such as Plexiglas, Kevlar, and Teflon are all polymers with special properties that make them desirable for specific uses, as shown in Figure 26-1.

The uses of polymers are determined principally by their physical properties, many of which depend on the intermolecular interactions present (Chapter 2) and the range of motion allowed by rotation about their single bonds (Chapter 4). We will see a variety of reactions from different chapters (particularly Chapters 21 and 25) that allow chemists to synthesize polymers and subsequently alter polymer molecules to introduce desirable properties.

We begin Chapter 26 with an examination of polystyrene (PS), a model for free radical polymerization and a polymer you are probably familiar with, and apply what we learn toward a variety of similar types of polymers. Then we study polymers made

(a)

(b)   (c)

FIGURE 26-1 Examples of commercially available polymers (a) The Dubai Mall aquarium walls are made of Plexiglas sheets. (b) Body armor made from Kevlar fiber. (c) A nonstick skillet with a Teflon lining. Teflon is the only known substance that geckos cannot climb (see p. 89).

Chapter Objectives

On completing Chapter 26 you should be able to:

1. Recognize the relationships between monomers, polymers, and repeating units, and determine the degree of polymerization from a polymer's size or molar mass.

2. Provide the mechanisms for the steps in free radical polymerization, and describe the polymerization's regiochemistry and stereochemistry.

3. Characterize a polymer sample by its average molar mass or average degree of polymerization.

4. Distinguish among radical, anionic, and cationic polymerization reactions, and determine which ones are ring-opening polymerizations.

5. Differentiate between chain-growth and step-growth polymerization.

6. Identify polymers by type (e.g., linear, network, cross-linked, homopolymers, and copolymers) and by the functional groups present.

7. Explain why it could be desirable to chemically modify a polymer's side groups (or pendant groups) after polymerization is complete.

8. Name simple polymers.

9. Recognize the relationships between polymer structure and physical properties, such as thermal transitions and solubility.

10. Understand the relationship between the structure of a polymer and its uses.

11. Explain why high temperatures lead to depolymerization and degradation.

from other types of reactions, and also learn some chemical modifications carried out after polymerization. Finally, we examine the relationship between the structure of polymer molecules and their properties.

26.1 Free Radical Polymerization: Polystyrene as a Model

Styrofoam (Fig. 26-2), the material used to make "packing peanuts" and insulated, disposable coffee cups, is a foam made from the synthetic polymer polystyrene.

Polystyrene has been available since the 1930s, but it was discovered in the 19th century when several chemists found that styrene, a liquid at room temperature, can produce polystyrene, which was originally obtained as a jelly-like substance (Equation 26-1). Because a polymer is produced, this type of reaction is called a **polymerization**.

FIGURE 26-2 Uses of Styrofoam (a) Styrofoam cups for hot beverages and (b) Styrofoam pellets for preventing damage to fragile objects during shipping.

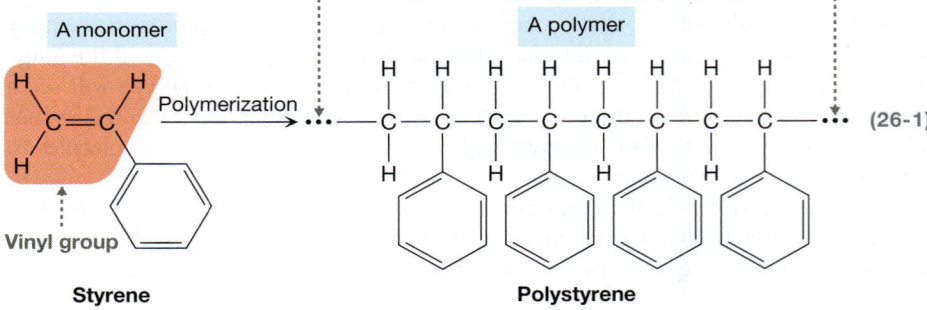

Only a partial structure of polystyrene is shown; the dotted lines (· · ·) indicate that the structure continues in both directions. A typical molecule of polystyrene is so large, in fact, that it would require the width of about 200 pages to depict the entire structure using the size scale shown!

26.1a Vinyl Polymers: Monomers, Repeating Units, and Degree of Polymerization

Equation 26-1 shows that polystyrene, the polymer, consists of many units of styrene, the **monomer**, held together by covalent bonds. Thus, styrene is the single part (*mono + mer*) that produces the large molecule with many parts (*poly + mer*). More specifically, polystyrene is a type of **vinyl polymer** because the atoms that characterize the vinyl group (i.e., the C=C bond) of the monomer are linked together on polymerization.

Polystyrene has a long, continuous chain of atoms that runs from one end of the molecule to the other, which is called the **polymer chain**, **main chain**, or **backbone** (Fig. 26-3). For polystyrene and other vinyl polymers, the main chain consists of only carbon atoms. For other polymers, heteroatoms can be included in the main chain, as we see in some examples later in this chapter.

When a polymer has a main chain, non-hydrogen atoms or groups attached to it are called **side groups** or **pendant groups**. Pendant means "hanging," so it appropriately describes these groups that are hanging off the polymer chain. In polystyrene, for example, the phenyl groups (blue in Fig. 26-3) are the polymer's pendant groups.

The *polymer chain*, or *polymer backbone*

The phenyl groups are *pendant groups* attached to the polymer chain.

Polystyrene

FIGURE 26-3 Polystyrene Polystyrene's main chain is screened in red and the phenyl pendant groups are drawn in blue.

YOUR TURN **26.1**

Identify the pendant groups in the partial structure of polypropylene shown here. What term, similar to *phenyl*, can you use to describe the pendant groups?

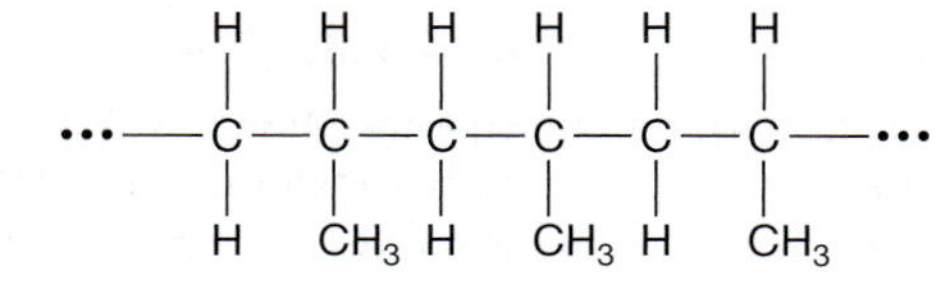

Answers to Your Turns are in the back of the book.

You can see the relationship between styrene and polystyrene more clearly by mentally cutting the main chain of the polymer into two-carbon units, as shown on the left in Figure 26-4. This gives you the *repeating unit* of the polymer, as shown in the polymer's *condensed formula* on the right of the figure.

FIGURE 26-4 The repeating unit of polystyrene Mentally cutting the backbone of polystyrene, as indicated on the left, produces the repeating unit shown on the right. The value of n is called the *degree of polymerization*.

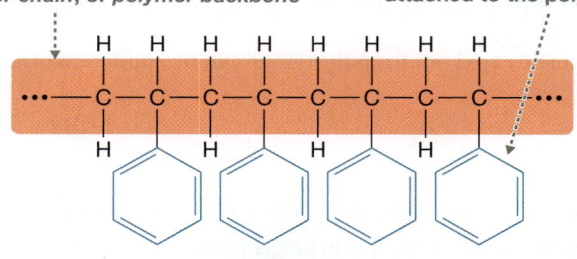

The *repeating unit* of polystyrene

Polystyrene

Same as

The *degree of polymerization*

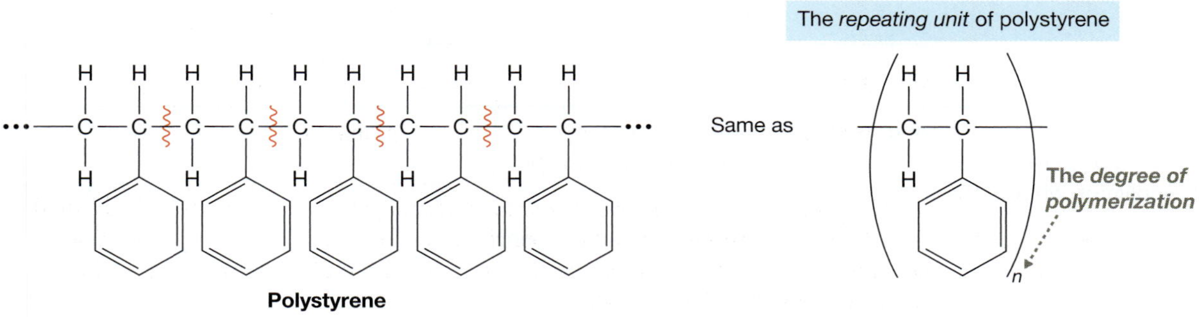

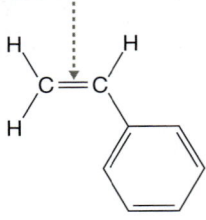

Monomer

Double bond

(a) Styrene

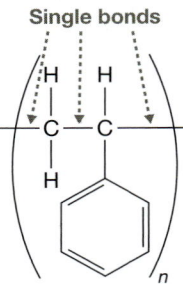

Repeating unit

Single bonds

(b) Polystyrene

FIGURE 26-5 Monomers versus repeating units The double bond that makes up the vinyl group in (a) styrene, the monomer, does not appear in the repeating unit of (b) polystyrene, the polymer. Each C atom along the polymer backbone gains a single bond.

- The **repeating unit** describes the connectivity that occurs over and over (repeatedly) in a polymer.
- The number of repeating units determines the **degree of polymerization (DP)**.

Using the notation on the right of Figure 26-4, DP is the value of the subscript n. If you know the molar mass of a polymer, you can calculate the approximate DP:

$$DP \approx \frac{\text{Molar mass of the polymer}}{\text{Molar mass of the repeating unit}}$$

For example, the atoms shown in the graphic on the left in Figure 26-4 can be represented as $C_{40}H_{40}$, which has a molar mass of 520 g/mol. The repeating unit's formula is C_8H_8, whose molar mass is 104 g/mol. Dividing 520 g/mol by 104 g/mol, we arrive at a DP of 5, the same as the number of repeating units depicted on the left of the figure. (The number is exact in this example, but in general a calculated DP will be an approximation because of variables we have yet to account for.)

Polystyrene's monomer and repeating unit are similar but not identical. Figure 26-5 shows, in particular, that the double bond in the monomer has disappeared in the repeating unit, and an additional single bond to each C atom has appeared. As we explain later, this is because the π electrons of the $C{=}C$ double bond are used to join the monomers together.

The exercises we just carried out for polystyrene can be applied to other vinyl polymers, too.

For vinyl polymers:
- The repeating unit includes two adjacent carbon atoms of the main chain.
- Those carbon atoms are involved only in single bonds in the polymer, but are connected to each other by a double bond in the vinyl monomer.

YOUR TURN 26.2

Given the partial structure of polypropylene shown here, draw its repeating unit and provide the appropriate value of n based on the atoms shown.

PROBLEM 26.1 Given the repeating units shown here, draw the structure of each polymer having $n = 3$, similar to the representation in Your Turn 26.2.

(a)   (b)

SOLVED PROBLEM 26.2

Poly(vinyl chloride) (PVC) is a vinyl polymer used for plumbing materials (e.g., pipe) and has the partial structure shown here. Based on the partial structure of the polymer, what is the structure of vinyl chloride?

Think What is the repeating unit of PVC? How is the repeating unit of a vinyl polymer related to the monomer?

Solve The C atoms in the structure make up the main chain of the polymer. The Cl atoms are pendant groups, or side groups, analogous to the phenyl rings in polystyrene. The repeating unit is shown at right.

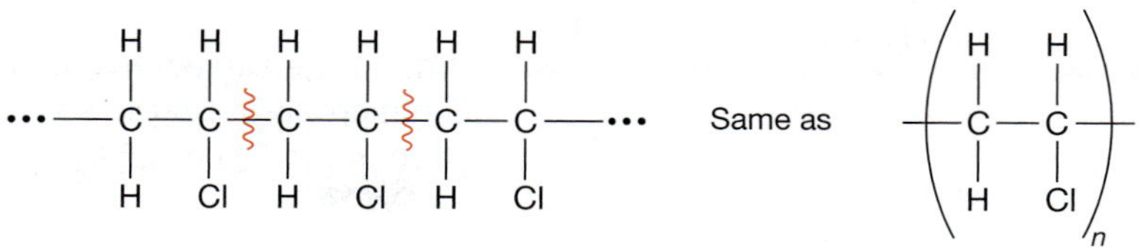

To determine the monomer from the repeating unit, remove the single bonds on the outside of the repeating unit and change the C—C single bond to a C=C double bond, as shown here.

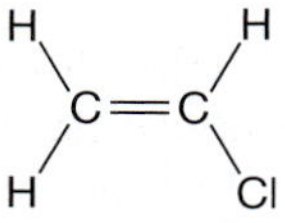

Vinyl chloride

PROBLEM 26.3 Poly(methyl methacrylate), or Plexiglas, has the partial structure shown here. What is the structure of the vinyl monomer from which Plexiglas can be made?

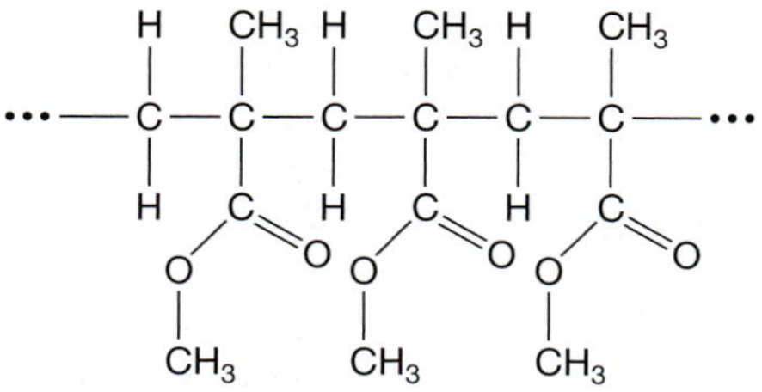

26.1b Free Radical Polymerization of Styrene and Other Vinyl Monomers

Manufacturers create polystyrene from styrene by means of **free radical polymerization**, which is a type of free radical chain reaction. An example is shown in Equation 26-2, in which styrene monomer is warmed in the presence of benzoyl peroxide, $C_6H_5CO_2$—$O_2CC_6H_5$.

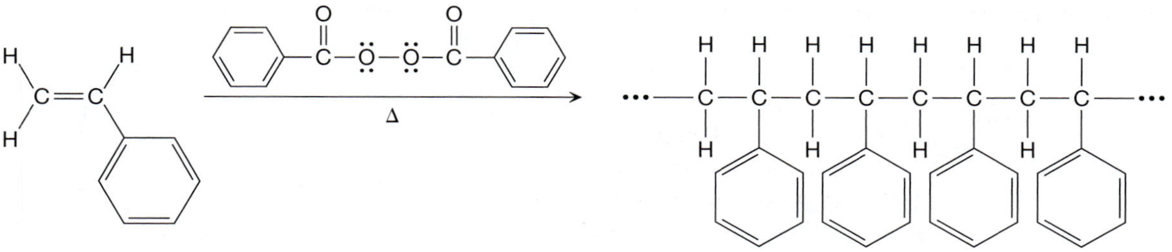

(26-2)

The mechanistic steps of this reaction are identical to the initiation, propagation, and termination steps you studied in Section 25.4a. The *initiation* step is needed to create the free radicals. Monomers such as styrene are usually incapable of producing free radicals readily, so we use benzoyl peroxide, a radical initiator (review Section 25.1). Because oxygen–oxygen single bonds are particularly weak (see Table 25-2, p. 1250), heating benzoyl peroxide results in the homolysis of the bond between the two O atoms (Equation 26-3):

Initiation

(26-3)

Benzoyl peroxide   **The benzoyloxyl radical**

The homolysis of benzoyl peroxide is further promoted by the additional resonance stabilization in the resulting radical, as shown in Figure 26-6.

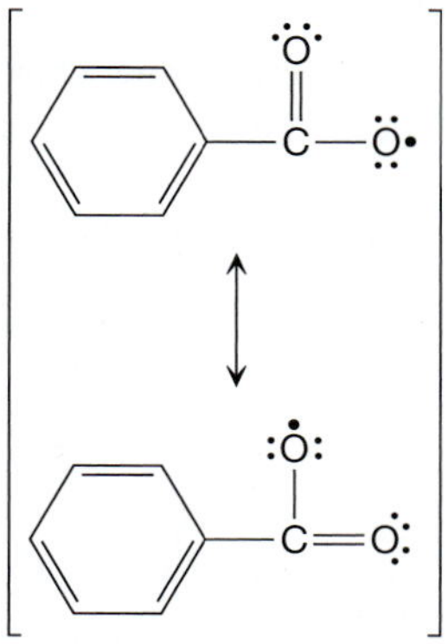

FIGURE 26-6 The benzoyloxyl radical Resonance delocalization of the unpaired electron in the benzoyloxyl radical facilitates the homolysis of benzoyl peroxide.

Draw in the curved arrows that show the movement of electrons in Equation 26-3. Do the same for the conversion of the top resonance structure in Figure 26-6 into the bottom structure.

After initiation, *propagation* occurs via radical addition (Section 25.4a). A radical, such as the benzoyloxyl radical produced directly from the initiator, adds to a styrene monomer, as shown in Equation 26-4. (The regiochemistry of this step will be discussed in Section 26.1c.)

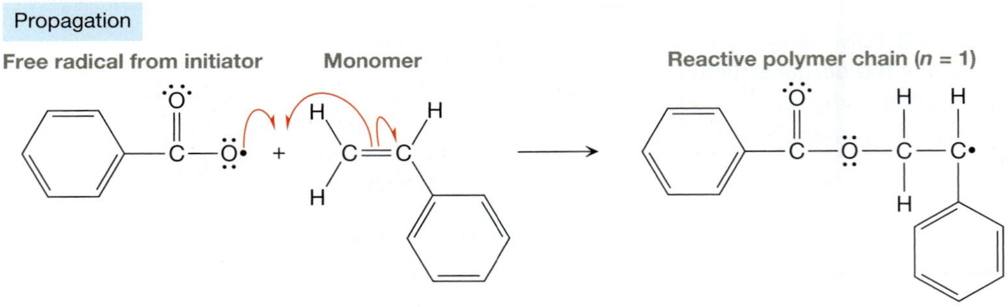

Propagation

(26-4)

What occurs next explains how the polymer chain grows — namely, the radical on the product side of Equation 26-4 adds to another monomer of styrene, increasing the length of the carbon chain, as shown in Equation 26-5.

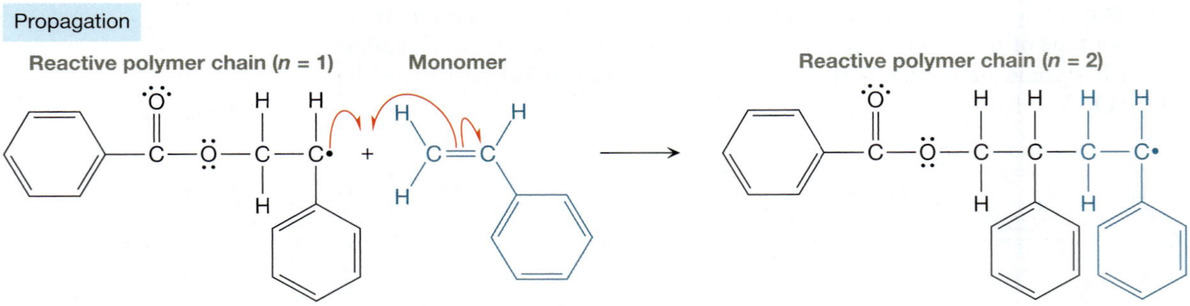

Propagation

(26-5)

This radical can continue to propagate, adding to a third styrene monomer (Equation 26-6).

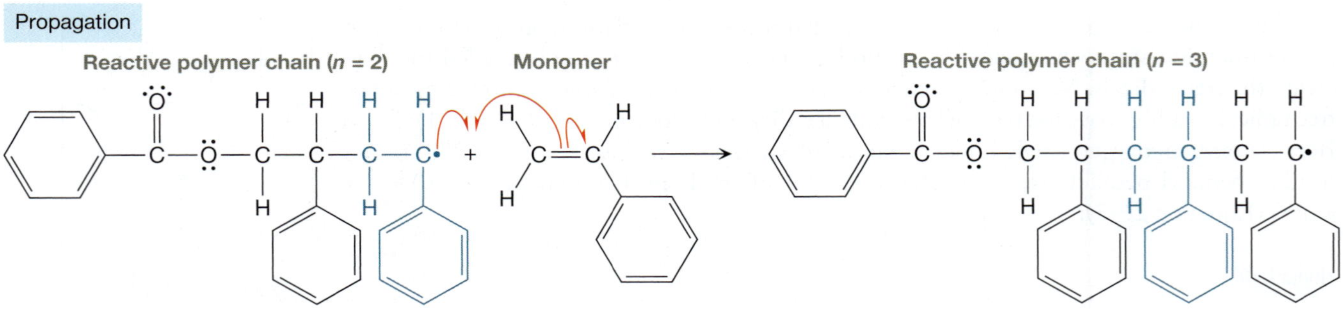

Propagation

(26-6)

This mechanism that produces polystyrene is called *chain-growth polymerization*.

In **chain-growth polymerization**, also called **addition polymerization**, one monomer adds at a time to the reactive site in the growing polymer chain.

Add the curved arrows, and draw the product of the next propagation step in the polymerization reaction, following Equations 26-4, 26-5, and 26-6:

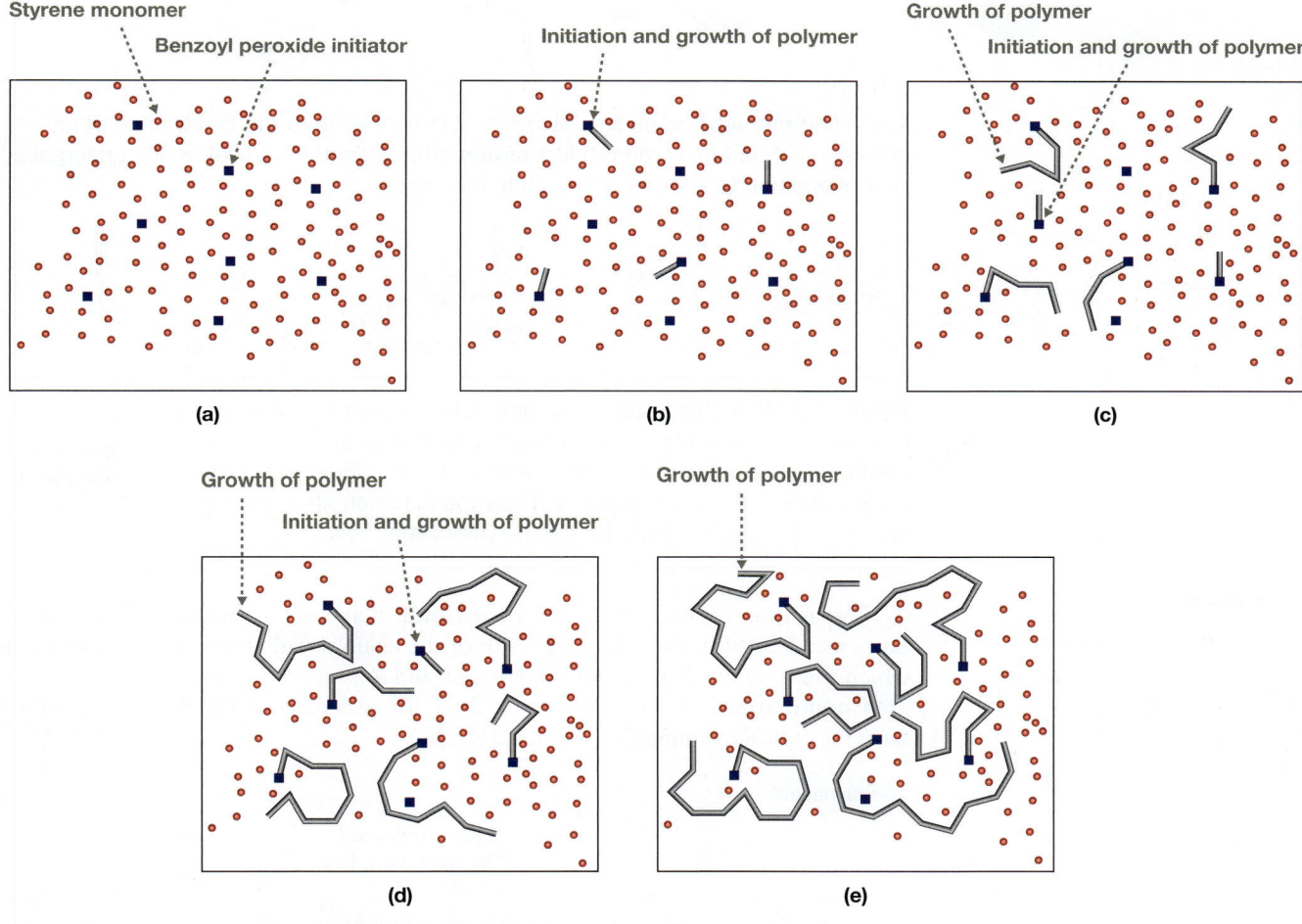

Chain-growth polymerization is depicted schematically in Figure 26-7, where the series of diagrams shows how propagation creates chains over time. The diagram begins with unreacted starting material in (a), proceeds to initiation and growth in (b), and then shows how propagation and initiation continue in (c) through (e).

Styrene monomer

Benzoyl peroxide initiator

Initiation and growth of polymer

Growth of polymer

Initiation and growth of polymer

(a) (b) (c)

Growth of polymer

Initiation and growth of polymer

Growth of polymer

(d) (e)

FIGURE 26-7 Chain-growth polymerization of styrene over time (a) A solution of styrene (red dots) and benzoyl peroxide (blue squares) prior to heating. (b) During heating, the benzoyl peroxide molecules undergo homolysis (initiation) and propagation. The gray lines represent bonds formed between an initiated radical and a styrene molecule. (c) Propagation continues for each growing chain. Initiation continues, but at a slower rate than propagation. (d) Propagation continues and each chain increases in length. Benzoyl peroxide molecules that remain also initiate chain growth. (e) Later in the reaction, initiation is minimal. Propagation continues and polymer chains grow longer.

In Figure 26-7e, draw a line indicating the initiation of a new polymer chain. Also draw a line indicating the propagation of two existing chains.

SOLVED PROBLEM 26.4

CONNECTIONS

Polyacrylonitrile is used to make ultrafiltration membranes for wastewater treatment and is also used to make hollow fiber membranes for reverse osmosis — a technique that converts seawater into fresh drinking water.

Acrylonitrile, shown here, is a vinyl monomer that produces the polymer polyacrylonitrile when treated with a small amount of benzoyl peroxide. Draw the initiation and first two propagation steps for this polymerization reaction.

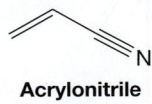
Acrylonitrile

Think How does benzoyl peroxide produce the initial free radicals? What elementary step occurs between a free radical and a vinyl monomer? Is the product of that step still a free radical?

Solve Benzoyl peroxide is a radical initiator that undergoes homolysis of the O—O single bond to produce benzoyloxyl radicals, as we saw previously in Equation 26-3.

In the first propagation step, a benzoyloxyl radical adds to the C=C bond of an acrylonitrile monomer. The product of that step is another radical, which undergoes radical addition to another acrylonitrile monomer.

PROBLEM 26.5 But-1-ene, commonly called butylene, is a vinyl monomer that, when treated with benzoyl peroxide, produces polybutylene, a polymer used to manufacture the pipes that carry modern municipal water supplies. Draw the initiation step and the first two propagation steps for this polymerization reaction.

But-1-ene (Butylene)

Propagation could continue, in theory, until there are no remaining molecules of monomer. In reality, though, the growth of the chain usually ends via *termination*, of which there are two basic types: *combination* and *disproportionation*.

Combination, shown in Equation 26-7 for two reactive PS chains, is another term for radical coupling (see Section 25.3a).

Combination

Chain terminated (No radical produced)

$$(26\text{-}7)$$

Disproportionation, shown in Equation 26-8, describes any reaction in which two identical species react to produce two different products.

Disproportionation

The radical in **black** abstracts a H atom from the radical in **red**.

A π bond was formed.

(26-8)

In the case of polystyrene, disproportionation occurs when one of the radicals abstracts a hydrogen atom from another radical. One electron from the C—H bond that is broken is used to form a new C—H bond, whereas the other electron joins with the unpaired electron on the adjacent carbon to form a C=C double bond.

Both combination and disproportionation produce closed-shell species from free radicals, so these steps are typically quite rapid. Which step dominates can depend on the nature of the monomer and the reaction conditions.

Figure 26-8 shows how a particular polystyrene molecule might appear after a combination step. Two growing polymer chains—one with n repeating units and the

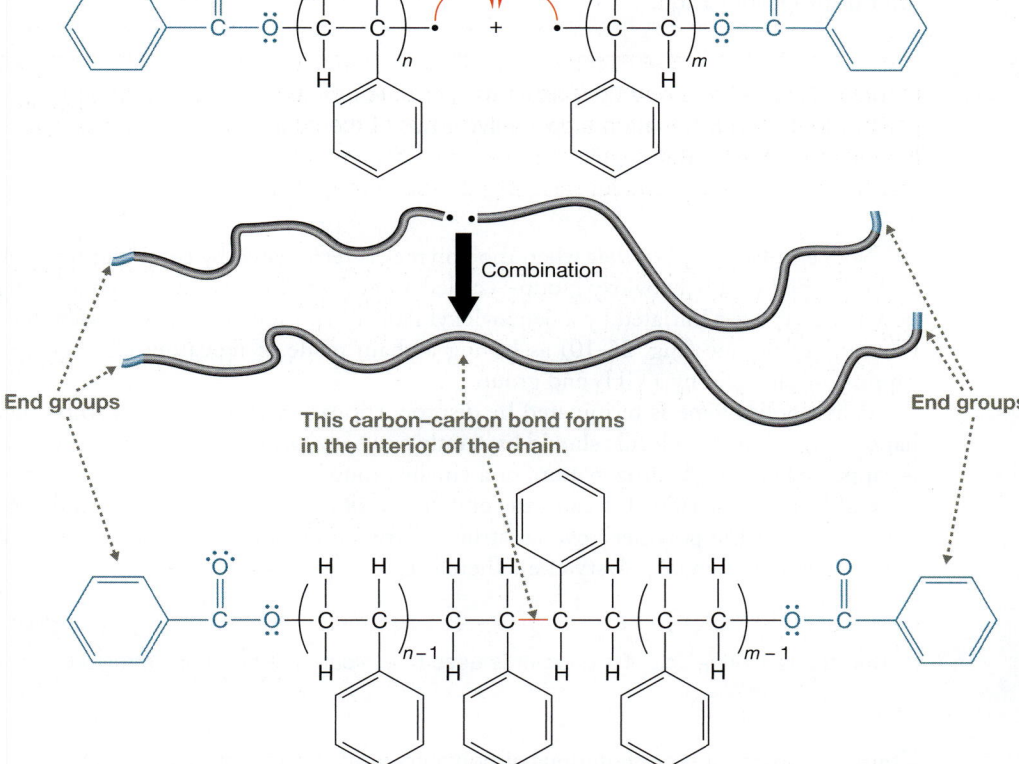

Two growing radicals combine.

+

Combination

End groups

This carbon–carbon bond forms in the interior of the chain.

End groups

FIGURE 26-8 Termination via combination (*Top*) Two growing polystyrene chains meet and the unpaired electrons join to produce a new carbon–carbon bond. One growing chain has n repeating units, and the other has m. (*Bottom*) The terminated polystyrene molecule has a total of $n + m$ repeating units, and has two benzoyloxy end groups (blue).

26.1 Free Radical Polymerization: Polystyrene as a Model **1315**

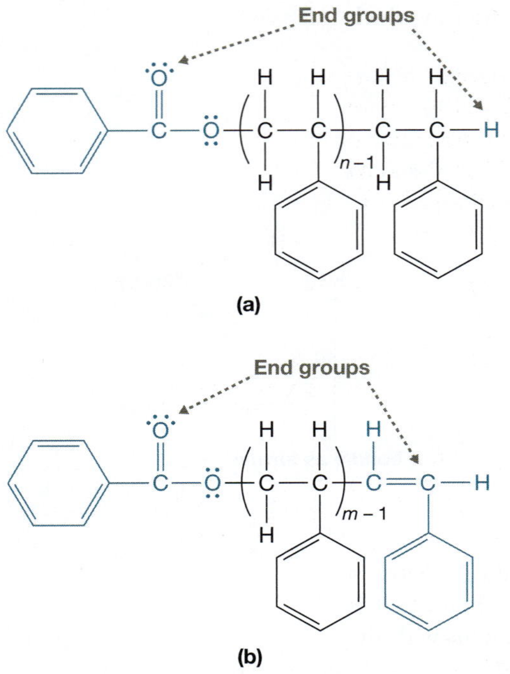

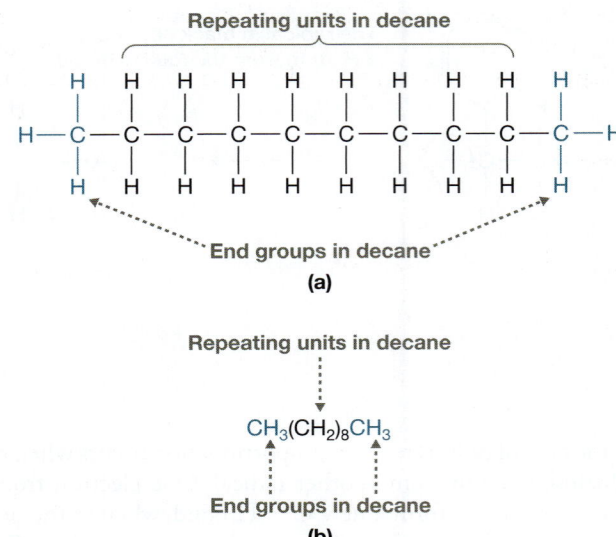

FIGURE 26-9 Termination via disproportionation A
disproportionation involving two growing polystyrene chains might
produce these two polystyrene molecules. Both have a benzoyloxy
end group on one end. On the other end, one molecule (a) is
capped by H and the other (b) is capped by a vinylic group.

(a) End groups

(b) End groups

Repeating units in decane

(a) End groups in decane

Repeating units in decane

$$CH_3(CH_2)_8CH_3$$

End groups in decane **(b)**

FIGURE 26-10 End groups in decane Decane represented
as its (a) Lewis structure and (b) condensed formula. Decane
can be viewed as having a chain of eight repeating CH_2 units,
capped by two CH_3 end groups.

other with m repeating units—join to produce a longer chain with $n + m$ total repeating units. Figure 26-9 shows the two polystyrene molecules that might be produced after disproportionation.

YOUR TURN 26.6

Using Figures 26-8 and 26-9 as guides, draw **(a)** the combination step and **(b)** the disproportionation step involving two of the final radicals shown in Solved Problem 26.4.

Notice in Figure 26-8 that when the polymer is terminated by combination, each end is capped by a benzoyloxy group—called an **end group**—because each growing polymer chain was initiated by a benzoyloxyl radical. This idea is similar to viewing a molecule of decane (Fig. 26-10) as having a chain made of repeating CH_2 groups, capped on each end by a CH_3 end group.

When polystyrene is terminated by disproportionation (Fig. 26-9), on the other hand, each polymer molecule should have only one benzoate end group. The other end is capped by either a hydrogen atom or a vinylic group.

Unlike in a molecule of decane, the end groups of a polymer are variable and contribute little to the polymer's overall structure (two units out of >1900 in a typical molecule in commercial polystyrene). Therefore:

Only the repeating unit of a polymer is used to evaluate the polymer's structure.

Consequently, most representations of polymers omit the end groups.

26.1c Regiochemistry of Free Radical Polymerization

Styrene is an unsymmetric vinyl monomer, so we can differentiate between the two carbons of the vinyl group. The less substituted carbon is called the *tail* and the more substituted carbon is called the *head*. As a result, *regiochemistry* comes into play during polymerization because there are two possible ways the radical can add to the vinyl group of styrene: **head-to-tail** (Equation 26-9) and **head-to-head** (Equation 26-10).

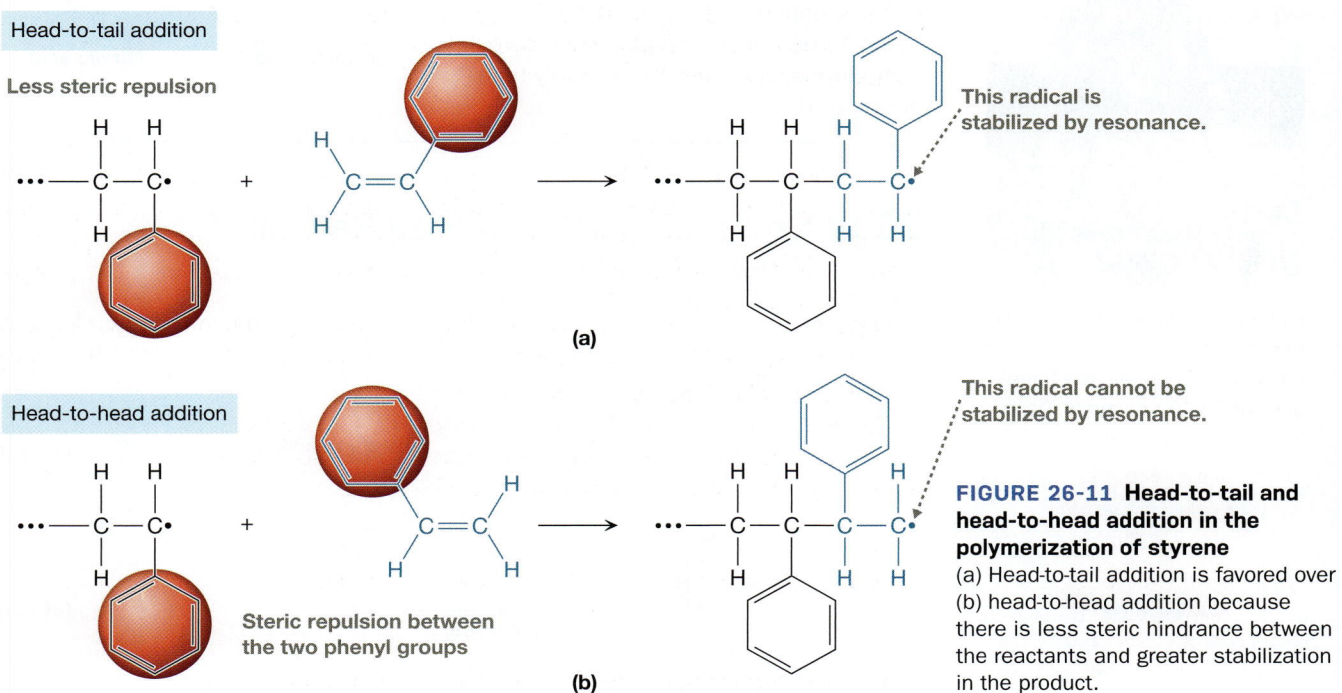

The head-to-tail addition (i.e., Equation 26-9) is favored in the polymerization of polystyrene, for the two reasons shown in Figure 26-11. First, steric repulsion between the two phenyl groups (one at the end of the propagating radical and one in the monomer) makes it less likely that a head-to-head collision will bring the

Head-to-tail addition

Less steric repulsion

This radical is stabilized by resonance.

(a)

Head-to-head addition

Steric repulsion between the two phenyl groups

This radical cannot be stabilized by resonance.

(b)

FIGURE 26-11 Head-to-tail and head-to-head addition in the polymerization of styrene (a) Head-to-tail addition is favored over (b) head-to-head addition because there is less steric hindrance between the reactants and greater stabilization in the product.

bond-forming carbons within a bond length. Second, the radical formed in the head-to-tail addition is more stable. The radical produced from head-to-tail addition is resonance stabilized, while the radical produced from head-to-head addition is not.

We often see similar contributions to the polymerization steps that form vinyl monomers other than styrene. Consequently:

The head-to-tail addition of vinyl monomers is usually favored in a free radical polymerization.

YOUR TURN 26.7

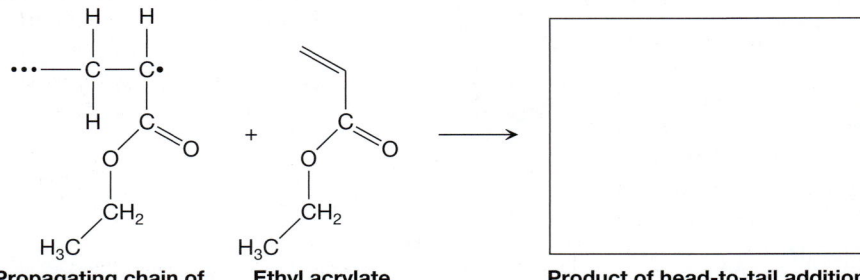

Propagating chain of poly(ethyl acrylate)   **Ethyl acrylate**   **Product of head-to-tail addition**

Draw all resonance structures for the product of head-to-tail addition shown in Figure 26-11. Then use curved-arrow notation to show the movement of electrons in the head-to-tail addition of a growing chain of poly(ethyl acrylate) to a molecule of ethyl acrylate. Draw resonance structures to explain why head-to-tail addition occurs instead of head-to-head addition.

PROBLEM 26.6 Using the generic symbol R• as the initial radical, draw the first two propagation steps in the polymerization of crotonic acid, paying close attention to regiochemistry. Do the same for acrylic acid. Explain why crotonic acid polymerizes much more slowly than acrylic acid.

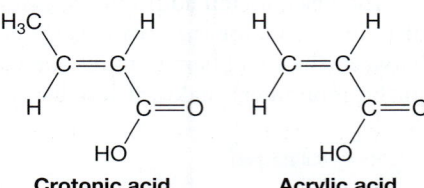

Crotonic acid   **Acrylic acid**

26.1d Stereochemistry of Free Radical Polymerization

When styrene polymerizes, the sp^2-hybridized C atoms of the vinyl group become sp^3-hybridized and the C atom attached to the phenyl ring becomes a new chiral center (Equation 26-11).

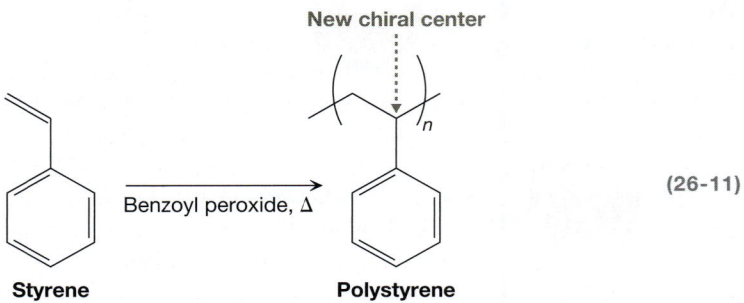

Styrene   Benzoyl peroxide, Δ   **Polystyrene**

(26-11)

There are two configurations possible for each monomer that adds, and the pattern of configurations—or lack thereof—that emerges along the polymer chain defines the polymer's **tacticity**. As shown in Figure 26-12, the polymer is **isotactic** if all configurations are the same, **syndiotactic** if the configurations alternate from one monomer to the next, and **atactic** if the configurations have no regular pattern.

The atactic form can be produced in the absence of a catalyst, simply by treating the monomer with a radical initiator such as benzoyl peroxide, as we saw previously in Equation 26-2. To produce the isotactic or the syndiotactic form, on the other hand, a **Ziegler–Natta catalyst** is used. Ziegler–Natta catalysts, originally pioneered by the German chemist Karl Ziegler (1898–1973) and the Italian chemist Giulio Natta (1903–1979), are typically mixtures of transition metal halides (such as those of Ti, V, Zr, and Cr) with organoaluminum compounds.

The three forms of polystyrene have different properties and, consequently, different uses. Atactic polystyrene is maleable and easily molded into solid objects, such as the yogurt container on the right in Figure 26-12. The syndiotactic form is rigid and can withstand relatively high temperatures. It is used to manufacture some of the components for the rice cooker shown in the middle of Figure 26-12. Isotactic polystyrene is rather crystalline and does not have commercial use, but is studied in thin films, such as the one shown on the left in Figure 26-12. We discuss the reasons for these different properties in Section 26.8.

Polystyrene is not the only polymer whose tacticity and properties can be controlled by Ziegler–Natta catalysts. In 1954, polypropylene was the first polymer synthesized in its isotactic form using a Ziegler–Natta catalyst (see Problem 26.7), and several others have been synthesized since then. Because of the impact that these catalysts have had on polymer chemistry and the polymer industry, Ziegler and Natta shared the 1963 Nobel Prize in Chemistry.

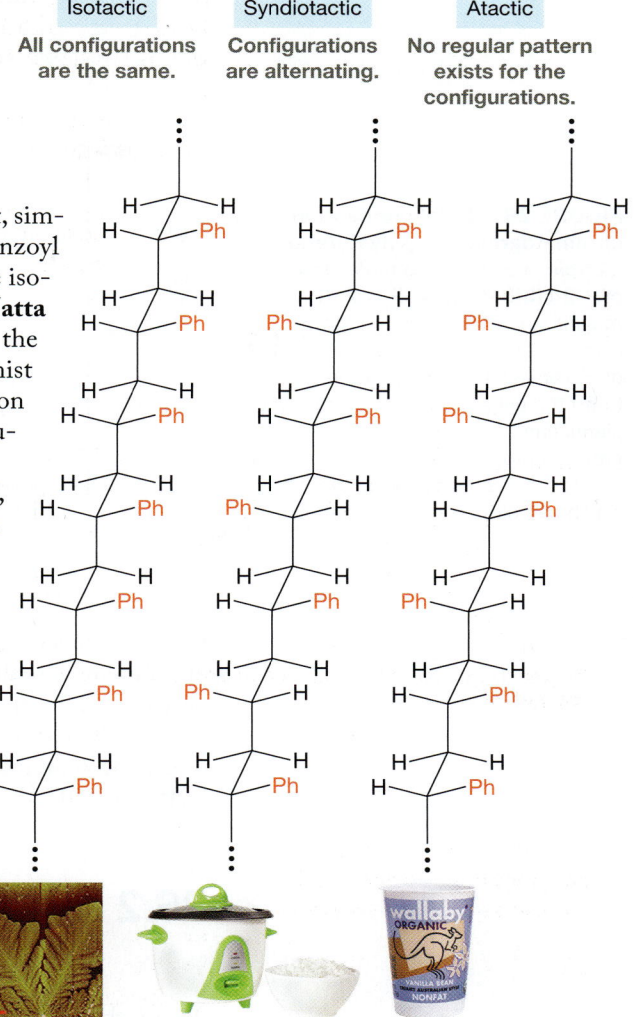

FIGURE 26-12 Tacticity in polystyrene (*Left*) In isotactic polystyrene, all stereochemical configurations are the same. (*Middle*) In syndiotactic polystyrene, the stereochemical configurations alternate. (*Right*) In atactic polystyrene, the stereochemical configurations do not establish a regular pattern.

PROBLEM 26.7 Polypropylene, whose repeating unit is shown here, has been synthesized in atactic and isotactic forms. Draw each polymer, using the ones shown in Figure 26-12 as a guide.

Polypropylene

26.1e Size Distribution of Polymers

The goal of most organic syntheses is to make a single product in high yield. In the synthesis of a polymer like polystyrene, however, this is difficult or impossible due to the variations that exist in the chain-growth mechanism. As we learned in Section 26.1b, for example, the lengths of the chains that are produced cannot be precisely controlled. Instead, a polymer is a mixture (a distribution) of chains of different lengths and can be described with an *average* molar mass or, alternatively, an average degree of polymerization. Commercial grade polystyrene, for example, has an average molar mass of >200,000 g/mol (i.e., an average degree of polymerization of >1920).

One way to characterize the distribution of a polymer sample uses **gel permeation chromatography (GPC)**, which works by size exclusion. Polymer samples in solution pass through a column with a porous gel. Smaller polymer molecules spend more time interacting with the gel, so they elute more slowly and are detected at later

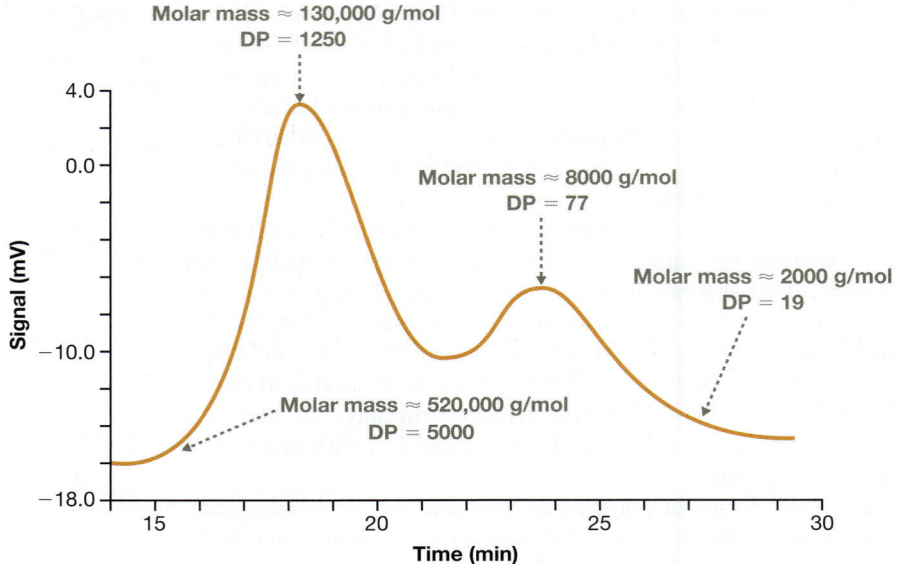

FIGURE 26-13 Gel permeation chromatogram of a polystyrene sample with average molar mass of ~24,000 g/mol (DP = 230) The magnitude of the signal on the *y* axis is proportional to the number of molecules that elute (i.e., exit from) the gel at a particular time. The signal can be used to determine the proportion of polymer molecules with a particular elution time on the *x* axis, which can be correlated with molar mass.

times. Consider Figure 26-13, for example, which shows the GPC results of a sample of polystyrene with an average molar mass of 24,000 g/mol, or an average degree of polymerization of 230. Molecular size varies widely, from ~2000 g/mol (DP = 19) to ~520,000 g/mol (DP = 5000), and the sample consists of two separate distributions, one centered at ~8000 g/mol (DP = 77) and the other at ~130,000 g/mol (DP = 1250).

26.2 Anionic and Cationic Polymerization Reactions

All of the chain-growth polymerization reactions we have studied so far have involved vinyl monomers and have proceeded through a free-radical mechanism. Chain-growth polymerizations can also be initiated by anions or cations, resulting in **anionic polymerization** or **cationic polymerization**, respectively. For example, styrene can undergo anionic polymerization using butyllithium as the initiator, which forms a negative charge on the growing chain instead of an unpaired electron (Equation 26-12).

The anion from butyllithium initiates the reaction by donating a pair of electrons to the vinyl group in styrene.

A new, more stable anion is formed.

$$\text{CH}_3\text{CH}_2\text{CH}_2\text{CH}_2^{\ominus} \quad + \quad \overset{\text{Initiation}}{\longrightarrow} \quad \text{CH}_3\text{CH}_2\text{CH}_2\text{CH}_2 - \text{C} - \text{C}\!:^{\ominus} \quad (26\text{-}12)$$

YOUR TURN **26.8**

Why is the anion on the product side of the reaction in Equation 26-12 more stable than the anion on the reactant side?

In the propagation steps, the reactive intermediates are anions, too, not free radicals, as shown in Equation 26-13.

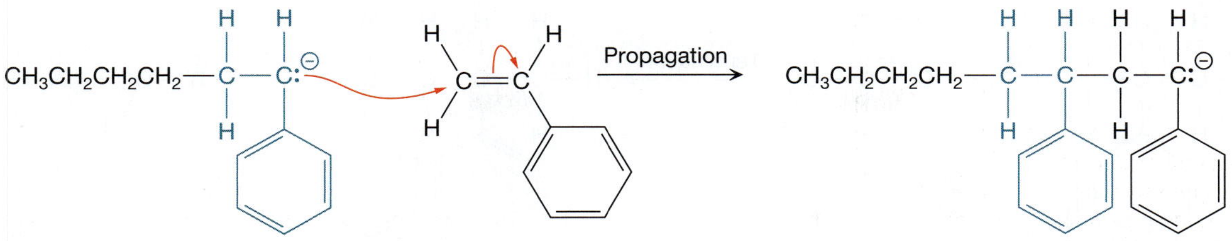

(26-13)

Termination occurs in anionic polymerization when an acid is added to the reaction mixture. The anion is neutralized in the resulting proton transfer, which ends the propagation cycle, as shown in Equation 26-14.

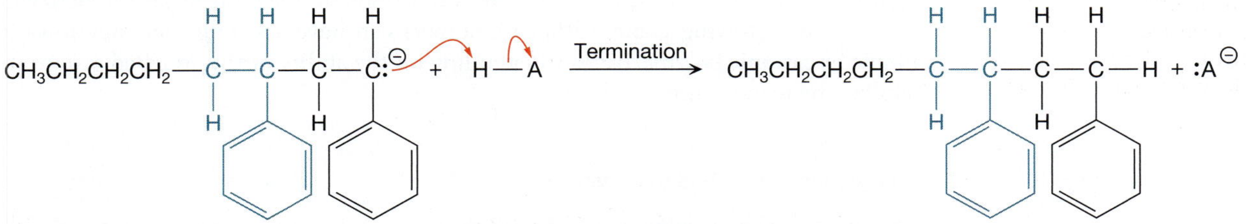

(26-14)

YOUR TURN 26.9

> Combination and disproportionation are not feasible termination steps for the anionic polymerization of styrene. Why not?

Cationic polymerization is similar to anionic polymerization except that the reactive intermediates are positively charged. For example, the polymerization of styrene can be initiated by treatment with $BCl_3 \cdot H_2O$ in a non-nucleophilic solvent like CH_2Cl_2. The $BCl_3 \cdot H_2O$ complex acts as a Brønsted acid to protonate the C=C bond of styrene, which produces the first carbocation intermediate (Equation 26-15).

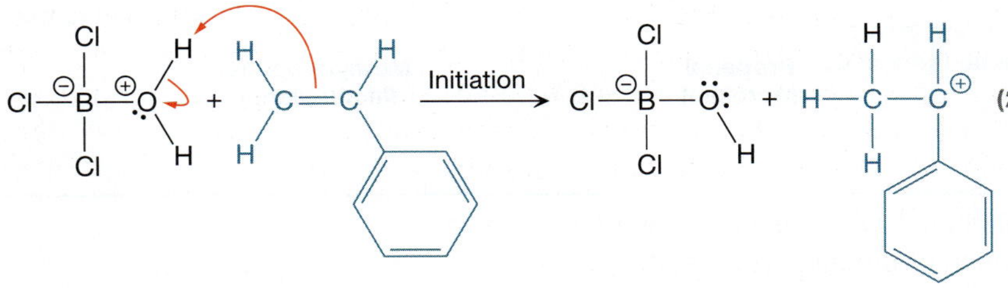

(26-15)

Propagation occurs when the carbocation of the growing chain undergoes electrophilic addition to another monomer (Equation 26-16).

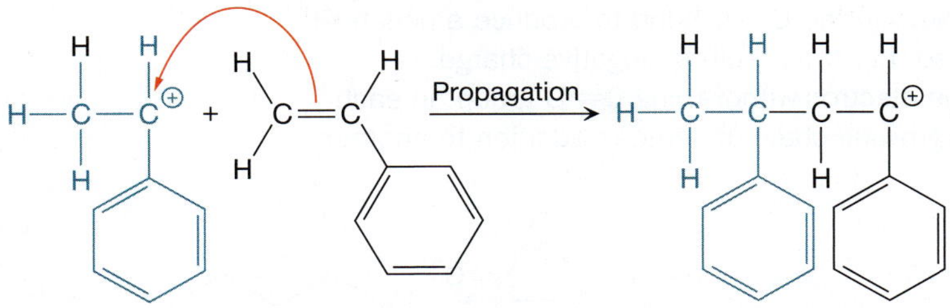

(26-16)

Polymerization of the chain terminates when all of its atoms have a complete octet. In this case, as shown in Equation 26-17, Cl^- can be donated by the conjugate base that was produced in the initiation step, probably through an addition–elimination sequence.

(26-17)

As we have seen, polystyrene can be produced from a chain-growth polymerization that proceeds through a radical, anionic, or cationic mechanism. All three types of mechanisms are available for this polymerization because an unpaired electron, a negative charge, and a positive charge are each resonance stabilized on the benzylic carbon of the growing chain. Other monomers can have a strong tendency toward anionic or cationic polymerization, depending on the ability of the attached groups to stabilize the reactive intermediate.

Vinyl monomers tend to undergo:

- Anionic polymerization if electron-withdrawing groups are attached to the C=C bond.
- Cationic polymerization if electron-donating groups are attached to the C=C bond.

The electron-withdrawing C=O group in propenal (acrolein), for example, will facilitate anionic polymerization, whereas the electron-donating methyl groups in methylpropene (isobutylene) will facilitate cationic polymerization.

Anionic polymerization

An electron-withdrawing group is attached to this C.

Propenal (Acrolein)

Cationic polymerization

Two electron-donating groups are attached to this C.

Methylpropene (Isobutylene)

SOLVED PROBLEM 26.8

Using a generic nucleophile, Nu:⁻, as the initiator, draw the initiation and the first two propagation steps for the anionic polymerization of propenal (acrolein).

Think What type of elementary step will take place between Nu:⁻ and acrolein? What kind of charge will be generated on one of the original vinylic carbons? Which vinylic carbon can better handle that charge? Can that step be repeated with another molecule of acrolein?

Solve The initiation step is the addition of Nu:⁻ to the C=C bond to produce a negative charge on one of the carbons. Nu:⁻ will add so that the resulting negative charge appears on the C atom that is attached to the electron-withdrawing C=O group. In each propagation step, the nucleophilic C⁻ on the growing chain undergoes addition to another molecule of acrolein.

PROBLEM 26.9 Using a generic strong Brønsted acid, HA, as the initiator, draw the initiation step and the first two propagation steps for the cationic polymerization of methylpropene (isobutylene).

26.3 Ring-Opening Polymerization Reactions

The chain-growth polymerization reactions we have examined so far have all involved vinyl monomers, but they can involve other types of monomers, too. Equation 26-18, for example, shows that oxirane undergoes polymerization when it is treated with calcium oxide.

$$(26\text{-}18)$$

Oxirane (Ethylene oxide)   **Poly(ethylene oxide)**

The monomer is cyclic, whereas the polymer is acyclic, so this is a type of **ring-opening polymerization**.

Equation 26-19 shows that the oxide anion in calcium oxide initiates polymerization, acting as a nucleophile. Recall from Section 10.7a that these S_N2 steps can occur under basic conditions because the relief of the ring strain compensates for the poor leaving group ability of RO^-.

$$(26\text{-}19)$$

The O atom from oxirane gains a negative charge, so it becomes strongly nucleophilic and can be the site of chain growth, as shown in Equation 26-20. The reaction, therefore, is a type of anionic polymerization.

$$(26\text{-}20)$$

Termination occurs when an acid is added to the reaction mixture and protonates the alkoxide anion, similar to the termination step we saw previously in Equation 26-14.

The polymer chain that is produced in Equation 26-20 will undergo two steps to complete its termination. Write the mechanism for these two steps.

Like oxirane, oxetane (a four-membered-ring ether) is strained and can undergo anionic ring-opening polymerization (see Problem 26.44 at the end of the chapter), but anionic polymerization is generally unfeasible for larger cyclic ethers. Cationic mechanisms are feasible for those monomers, however, because the leaving group is dramatically better. For example, as shown in Equation 26-21, tetrahydrofuran undergoes ring-opening polymerization when treated with trimethylsilyl trifluoromethanesulfonate, $CF_3SO_3Si(CH_3)_3$.

Tetrahydrofuran   **Poly(tetrahydrofuran)**   (26-21)

Initiation occurs when the nucleophilic O atom of tetrahydrofuran attacks the Si atom of $CF_3SO_3Si(CH_3)_3$ in an S_N2 step (Equation 26-22). Thus, the O atom becomes positively charged.

(26-22)

Propagation occurs when an uncharged molecule of tetrahydrofuran attacks the positively charged ring (Equation 26-23).

(26-23)

Each time a propagation step happens, a new positively charged O appears in the ring at the end of the growing chain until, finally, methanol is added to the reaction mixture to terminate the polymerization. (See Your Turn 26.11.)

Draw the curved arrows and product for the second propagation step in Equation 26-23. Also, draw the mechanism for the termination that would occur if methanol were to react with the product you drew.

The cationic ring-opening polymerization in Equation 26-21 is initiated by an S_N2 step, but it could also have been initiated using a strong Brønsted or Lewis acid. (See Problem 26.10.)

PROBLEM 26.10 Draw the initiation step and the first two propagation steps for the cationic ring-opening polymerization reaction that takes place when oxetane is treated with a Lewis acid like BF_3. Also, draw the repeating unit of the resulting polymer.

Oxetane

Ring-opening polymerization reactions are not limited to cyclic ethers for monomers. We explore these kinds of reactions involving other functional groups in Problems 26.45, 26.46, and 26.47 at the end of the chapter.

26.4 Step-Growth Polymerization

All of the polymerization reactions we have studied so far have proceeded through *chain-growth* mechanisms, in which one monomer adds at a time to the reactive site at the end of the growing polymer chain. Polymerization can also occur through a *step-growth* mechanism.

> **Step-growth polymerization** occurs when molecules throughout a mixture join with any other molecule that has an appropriate functional group available to react, whether it is a monomer unit or another polymer molecule.

Consider the synthesis of nylon-6,6 shown in Equation 26-24.

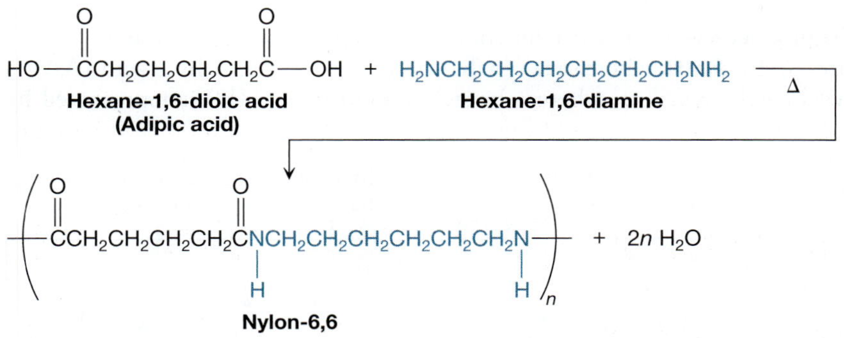

Hexane-1,6-dioic acid
(Adipic acid)   **Hexane-1,6-diamine**   (26-24)

Nylon-6,6

The reaction mixture contains two different compounds, adipic acid and hexane-1,6-diamine, each of which has two reactive functional groups. A CO_2H group from adipic acid reacts with an NH_2 group from hexane-1,6-diamine to form an $O{=}C{-}N$ group (characteristic of an amide) and a molecule of water, as shown in Equation 26-25.

The first steps in the growth of the polymer chain

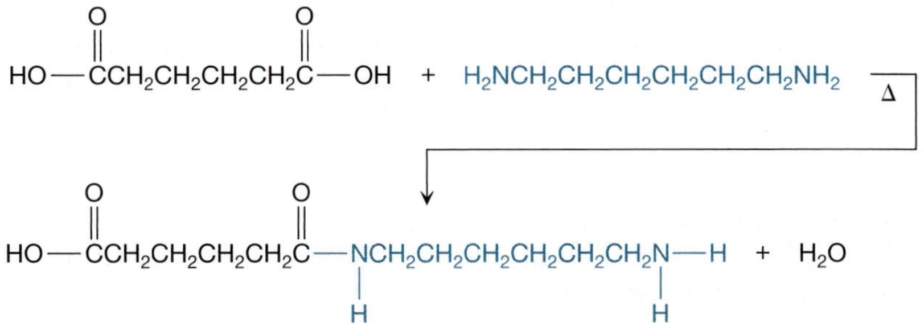

(26-25)

This reaction proceeds by a nucleophilic addition–elimination mechanism, similar to the ones we saw in Section 21.7, and forms a new $C{-}N$ bond that links the monomers together.

YOUR TURN **26.12**

The reaction in Equation 26-25 can be represented by the generic reaction shown here. Review Section 21.7 and then draw the complete mechanism for this reaction. You may assume that it is acid catalyzed.

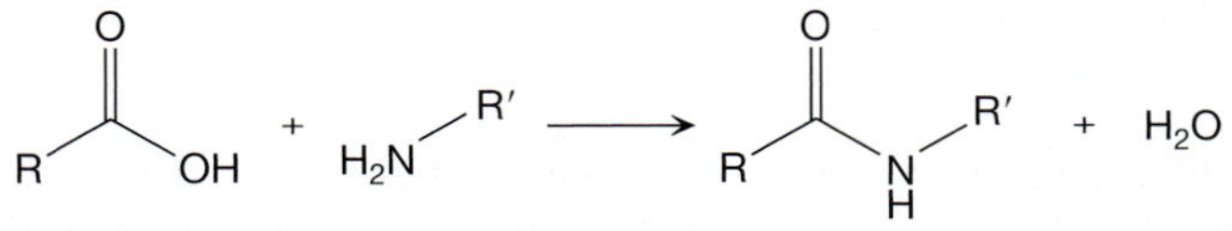

The reaction in Equation 26-25 continues and the chain length increases because the product also has two reactive functional groups. Each of those functional groups can therefore produce yet another O=C—N group. One possible way for this to occur is shown in Equation 26-26.

$$HO\text{—}\overset{\overset{O}{\|}}{C}CH_2CH_2CH_2CH_2\overset{\overset{O}{\|}}{C}\underset{\underset{H}{|}}{N}CH_2CH_2CH_2CH_2CH_2CH_2NH_2 \quad + \quad HO\text{—}\overset{\overset{O}{\|}}{C}CH_2CH_2CH_2CH_2\overset{\overset{O}{\|}}{C}\underset{\underset{H}{|}}{N}CH_2CH_2CH_2CH_2CH_2CH_2NH_2$$

$$\downarrow \Delta$$

(26-26)

$$HO\text{—}\overset{\overset{O}{\|}}{C}CH_2CH_2CH_2CH_2\overset{\overset{O}{\|}}{C}\underset{\underset{H}{|}}{N}CH_2CH_2CH_2CH_2CH_2CH_2\underset{\underset{H}{|}}{N}\text{—}\overset{\overset{O}{\|}}{C}CH_2CH_2CH_2CH_2\overset{\overset{O}{\|}}{C}\underset{\underset{H}{|}}{N}CH_2CH_2CH_2CH_2CH_2CH_2NH_2 \quad + \quad H_2O$$

Again the product has two reactive functional groups and can react further.

Figure 26-14 shows schematically what occurs over time. Adipic acid (AA) is represented by the red dots, whereas hexane-1,6-diamine (HD) is represented by

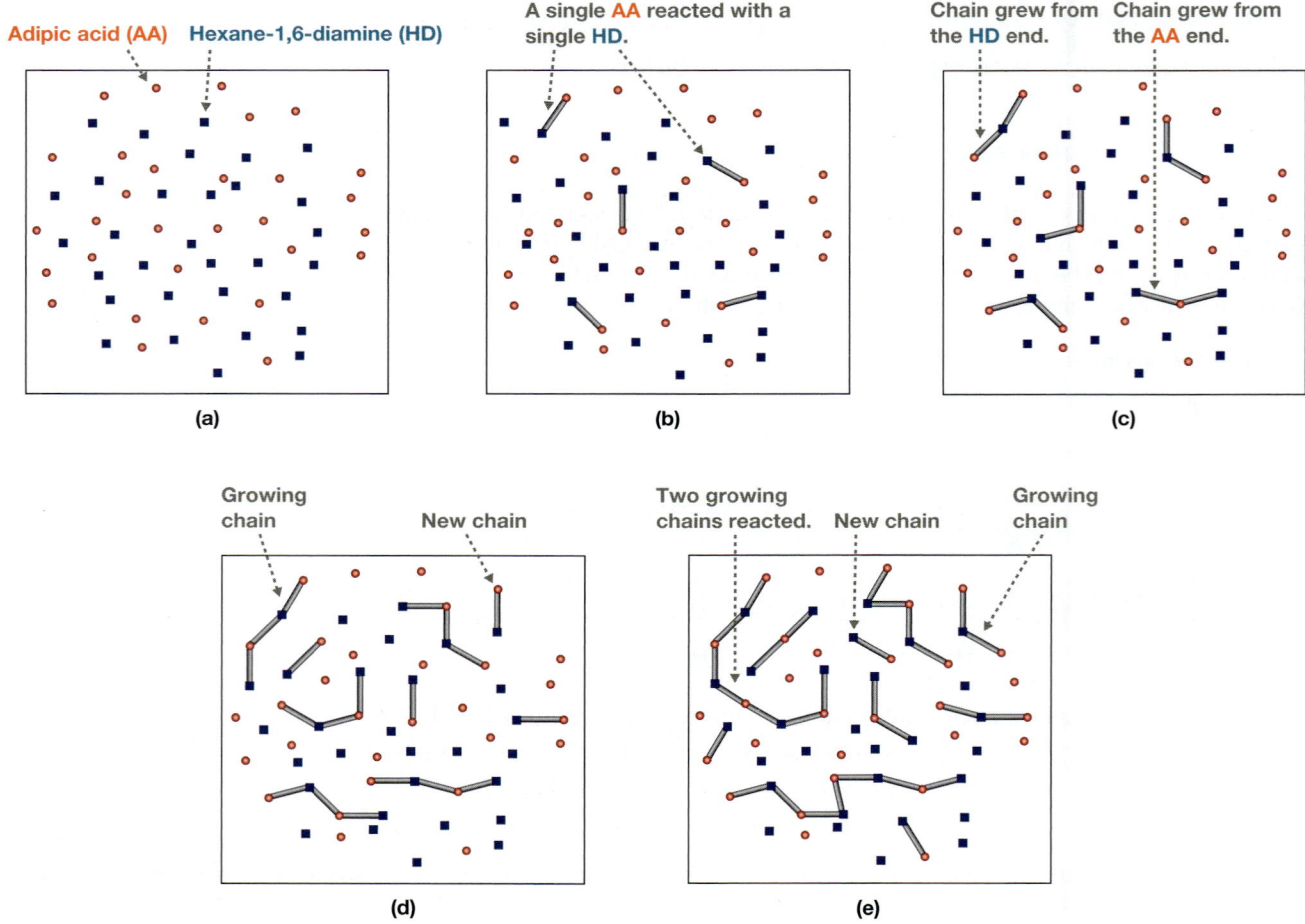

FIGURE 26-14 Step-growth polymerization in the synthesis of nylon-6,6 (a) A mixture of adipic acid (AA, shown as red dots) and hexane-1,6-diamine (HD, shown as blue squares). (b) When the reaction begins, chains begin to grow when a single AA reacts with a single HD. (c) The AA–HD units can grow at either end, reacting with either another AA or another HD. (d) The polymer chains grow stepwise, adding an AA or HD to the end of the chain. At the same time, single units of AA and HD continue to react to start new chains. (e) While the chains continue to grow through the addition of a single AA or HD, existing chains also react with each other.

the blue squares. Early in the reaction (Fig. 26-14a through 26-14d), chains are initiated and the size of the chain may increase by the length of a single AA molecule or a single HD molecule. As the reaction progresses, however, long chains may join (Fig. 26-14e), greatly increasing the size of the chain in a single step. This step is similar to combination in chain-growth polymerization (Equation 26-7, p. 1314), but the chains in step-growth polymerization are capable of further reactions.

YOUR TURN **26.13**

In Figure 26-14e, draw **(a)** a new line indicating two monomers linking together, **(b)** a new line indicating an existing polymer chain linking to a monomer, and **(c)** a new line indicating two existing chains linking together.

The distinctions between step-growth polymerization and chain-growth polymerization are important. Chain-growth polymerizations are typically fast, irreversible reactions, having products that are thermodynamically very favored. Step-growth polymerizations are typically slower and, because the reactions are often reversible, removal of water or other products can be necessary for the polymer to reach a high molecular weight. In the synthesis of nylon-6,6, for example, the reaction is reversible and so slow that heat is required to drive the reaction to completion.

Historically, these step-growth polymerizations (e.g., Equation 26-24) were referred to as **condensation polymerization**, because a small molecule (in this case, water) is formed as a product.

SOLVED PROBLEM 26.11

Kevlar is a high-strength polymer used in body armor and bicycle tires. It is a condensation polymer synthesized from benzene-1,4-diamine (*para*-phenylenediamine) and benzene-1,4-dicarbonyl chloride (terephthaloyl chloride). Draw the repeating unit for Kevlar.

Benzene-1,4-diamine
(*para*-Phenylenediamine)

Benzene-1,4-dicarbonyl chloride
(Terephthaloyl chloride)

Think What functional group does benzene-1,4-diamine contain? What functional group does benzene-1,4-dicarbonyl chloride contain? When those functional groups react, what functional group is produced? Does that product contain other reactive sites for polymerization to continue?

Solve Benzene-1,4-diamine has an NH_2 functional group and benzene-1,4-dicarbonyl chloride has a COCl functional group. We learned in Section 21.3 that molecules with these functional groups react to produce an amide that connects the two molecules together, so these two monomers will produce a dimer.

Each end of the dimer has an unreacted functional group that can react further. It can react with another monomer, or, as shown below, two dimers can react to increase the chain length. In the resulting tetramer, the repeating pattern becomes evident, which includes two phenyl rings: one that has two attached N atoms and the other that has two attached carbonyl groups.

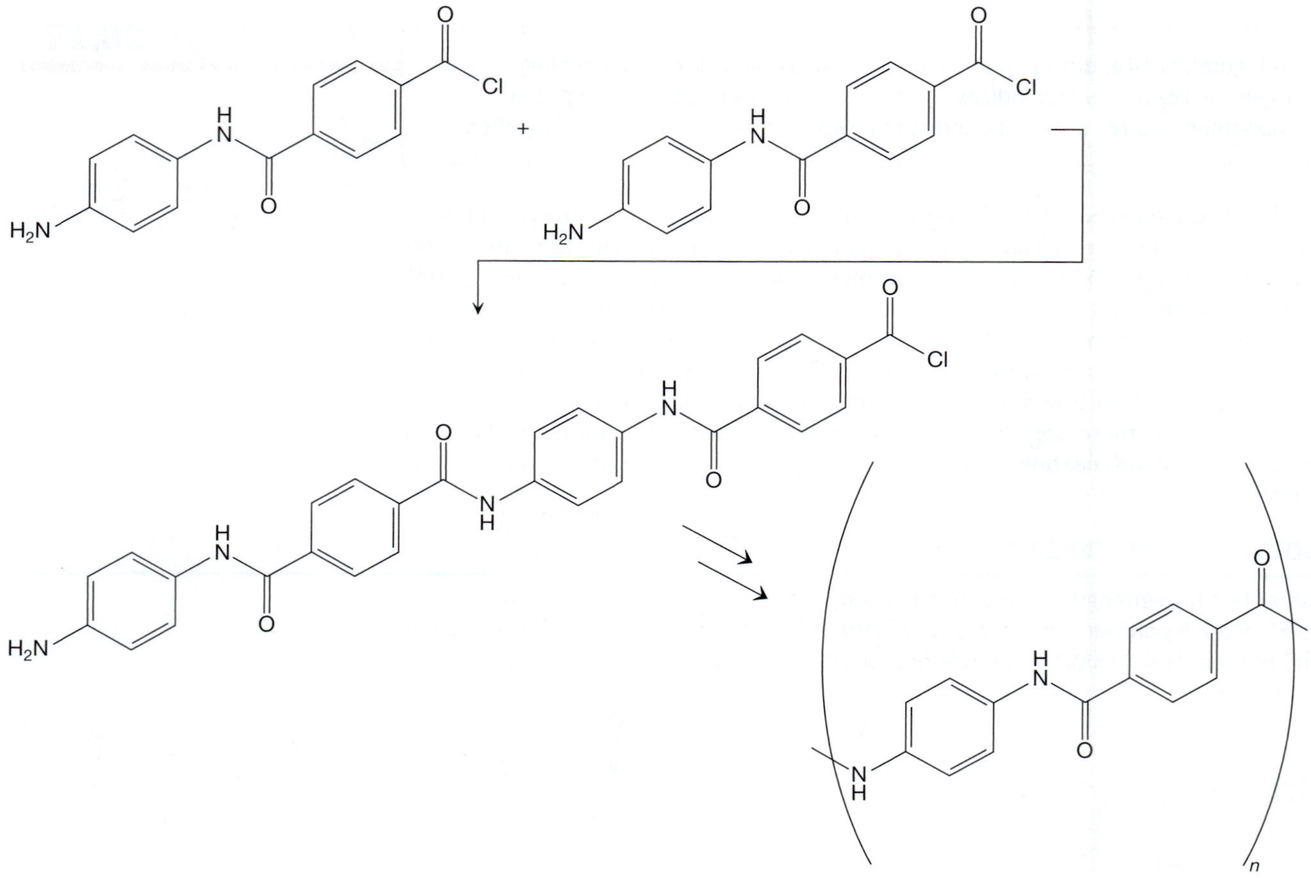

PROBLEM 26.12 Polyphthalamides are a class of polymers used in several components of automobile engines, and are produced using benzene-1,4-dioic acid and an aliphatic diamine. Draw the repeating unit for the polyphthalamide that uses hexane-1,6-diamine.

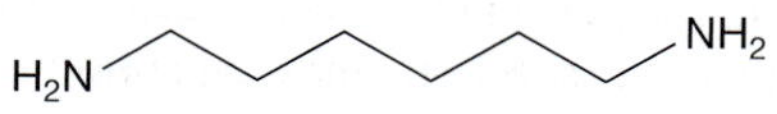

Hexane-1,6-diamine

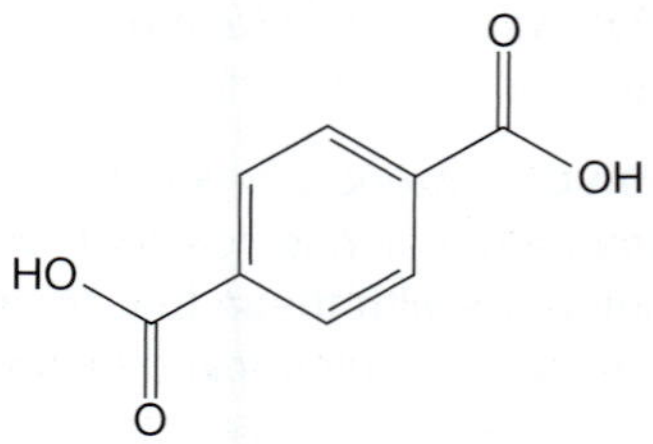

Benzene-1,4-dioic acid

Many nucleophilic addition–elimination reactions can be used to make step-growth polymers. For example, the Fischer esterification (Section 21.7) of terephthalic acid and ethylene glycol may be used to make polyesters, such as poly(ethylene tere-phthalate) (PET), as shown in Equation 26-27.

Terephthalic acid   **Ethylene glycol**

Acid catalyst
Δ

(26-27)

$$\cdots -CH_2CH_2-O-\overset{O}{\overset{\|}{C}}-\text{(ring)}-\overset{O}{\overset{\|}{C}}-O-CH_2CH_2-O-\overset{O}{\overset{\|}{C}}-\text{(ring)}-\overset{O}{\overset{\|}{C}}-O-CH_2CH_2-O-\overset{O}{\overset{\|}{C}}-\text{(ring)}-\overset{O}{\overset{\|}{C}}-O-\cdots$$

or $+\ 2n\ H_2O$

$$\left(-CH_2CH_2-O-\overset{O}{\overset{\|}{C}}-\text{(ring)}-\overset{O}{\overset{\|}{C}}-O-\right)_n$$

Poly(ethylene terephthalate) or **PET**
An ester of terephthalic acid and ethylene glycol

Alternatively, transesterification can be used to make PET from the dimethyl ester of terephthalic acid, as shown in Equation 26-28.

$$H_3CO-\overset{O}{\overset{\|}{C}}-\text{(ring)}-\overset{O}{\overset{\|}{C}}-OCH_3\ +\ HO-CH_2CH_2-OH \xrightarrow[\Delta]{\text{Acid catalyst}} \left(-CH_2CH_2-O-\overset{O}{\overset{\|}{C}}-\text{(ring)}-\overset{O}{\overset{\|}{C}}-O-\right)_n$$

$+\ 2n\ CH_3OH$ (26-28)

PROBLEM 26.13 In a third reaction for the synthesis of PET, the reactants are a diacid chloride and ethylene glycol. Provide the structures of the reactants and products for this reaction.

In the three step-growth polymerizations we have examined, water (Equations 26-24 and 26-27) or methanol (Equation 26-28) was a product of the reaction and each must be removed from the product mixture to isolate the polymer. Some step-growth polymerizations, however, do *not* produce a second product, as shown for the polyurethanes in Equation 26-29 (next page).

PROBLEM 26.14 The steps in the reaction in Equation 26-29 are: addition of a nucleophile to a polar double bond and two proton transfer steps. Draw the mechanism for this reaction.

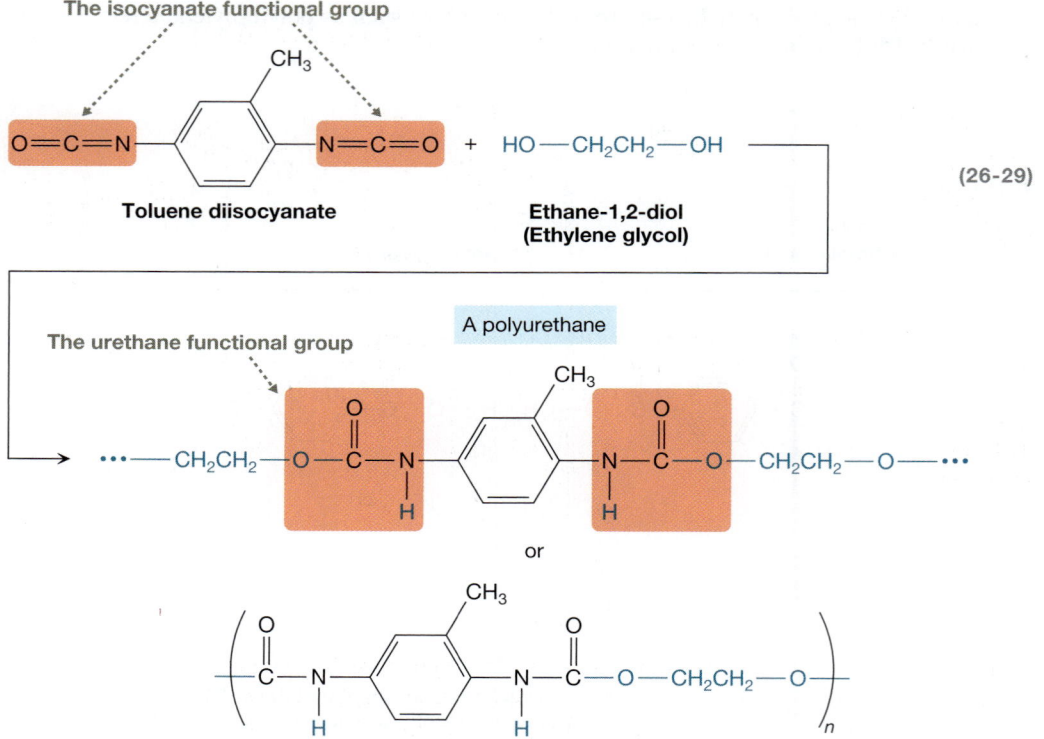

The isocyanate functional group

Toluene diisocyanate + Ethane-1,2-diol (Ethylene glycol)

(26-29)

The urethane functional group

A polyurethane

or

26.5 Linear, Branched, and Network Polymers

Polystyrene (Equation 26-1, p. 1308) and nylon-6,6 (Equation 26-24, p. 1325) are **linear polymers** because the monomers are linked end to end along the entire length of the polymer chain. **Branched polymers** can also be produced, in which smaller polymer chains split off from the main chain. Figure 26-15 shows schematically how linear and branched polymers are distinguished.

Branching can occur during chain-growth polymerization if the reactive site becomes located somewhere in the middle of a chain. Figure 26-16 shows schematically how such a **chain transfer** might occur during a free radical polymerization.

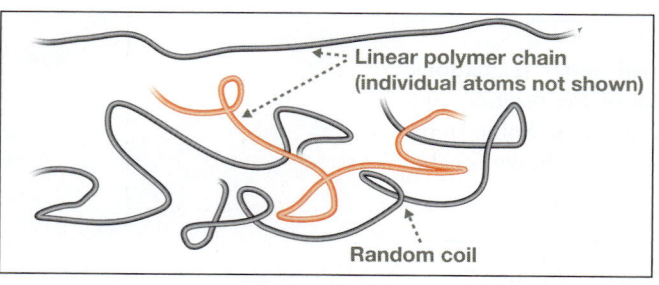

Linear polymer chain (individual atoms not shown)

Random coil

(a) A linear polymer

A branch coming off the main polymer chain

(b) A branched polymer

FIGURE 26-15 Branching in polymers (a) In a linear polymer, there is one continuous chain that contains the pendant groups. Linear polymer chains usually exist as random coils. (b) In a branched polymer, multiple smaller chains split off from the main chain.

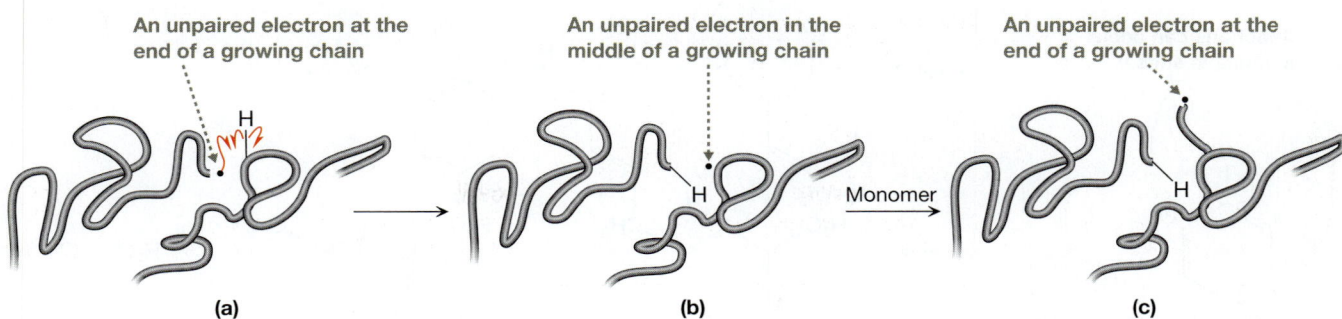

An unpaired electron at the end of a growing chain

An unpaired electron in the middle of a growing chain

An unpaired electron at the end of a growing chain

(a)   (b)   Monomer   (c)

FIGURE 26-16 **Formation of branches during free radical polymerization**
(a) The free radical at one end of one growing chain encounters a hydrogen atom in the middle of another chain. (b) After hydrogen atom abstraction, the original free radical has become a terminated closed-shell species and the reactive radical appears somewhere in the middle of another chain. (c) Further polymerization at the new radical creates a branch in the chain.

YOUR TURN **26.14**

The following two equations show polyethylene (PE) undergoing chain transfer, and so is a specific example of what is shown schematically in Figure 26-16. Use curved arrows to show the movement of electrons in both steps, and provide the product of the second step.

(a)

(b)

Branching can also occur during polymerization when a monomer has multiple reactive sites, such as in the reaction of phenol with formaldehyde to produce the hard material known as Bakelite (Equation 26-30, next page).

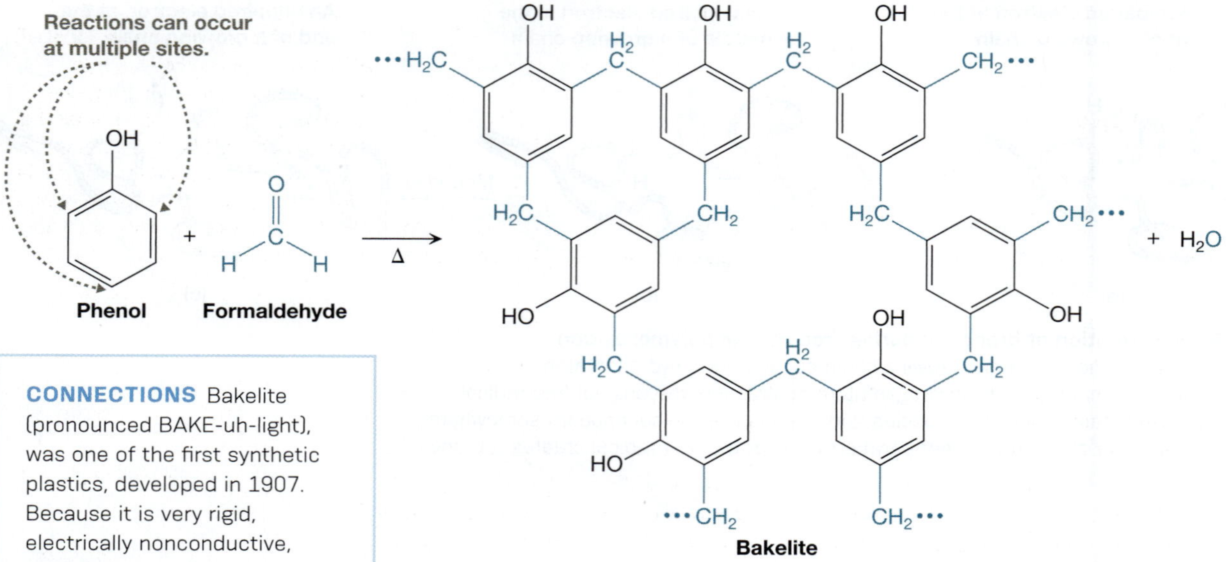
Reactions can occur at multiple sites.
OH
Phenol
+
O
C
H H
Formaldehyde
Δ
Bakelite
+ H_2O

(26-30)

CONNECTIONS Bakelite (pronounced BAKE-uh-light), was one of the first synthetic plastics, developed in 1907. Because it is very rigid, electrically nonconductive, and quite heat resistant, Bakelite found a wide variety of uses, including in electrical components, kitchenware, and even billiard balls.

The extent of branching in Bakelite is so high, in fact, that the polymer has no recognizable linearity. Bakelite, therefore, is a **network polymer** because a sample of it can be described as a single, *very* large molecule.

YOUR TURN 26.15

In Bakelite, what do you notice about the substitution pattern on the phenyl rings? What is the reason for that pattern?

26.6 Chemical Reactions after Polymerization

We have seen how to synthesize a variety of polymers, but they can undergo further reactions even after the polymerization reaction is complete. These reactions are designed to impart special properties to the polymers that make polymers more suitable for their purpose. Such reactions usually include modifications of the pendant groups in the polymer or *cross-linking* polymer chains.

26.6a Modification of Pendant Groups

In many cases, the monomer for a particular polymer is unattainable or it forms a polymer with inferior properties. Changes to the polymer after it has been made, particularly in its pendant groups, can address these problems. Here we highlight two such chemical modifications. One is part of the synthesis of poly(vinyl alcohol), whereas the other is part of the synthesis of poly(ether ether ketone).

26.6a.1 Transesterification of Poly(vinyl acetate) to Form Poly(vinyl alcohol)

Elmer's glue, eye drops, and other consumer products contain **poly(vinyl alcohol)**, or **PVA** (Fig. 26-17a). Because of the large number of hydroxyl groups off the main chain, PVA is soluble in water and is used in a large number of water-based products.

FIGURE 26-17 Poly(vinyl alcohol)
(PVA) (a) A portion of the PVA
polymer and (b) its repeating unit.
(c) Vinyl alcohol is the monomer that
would be derived from the repeating
unit. However, as described in the text,
PVA is not made from vinyl alcohol, an
enol that exists primarily in its keto
form, acetaldehyde.

(a) Poly(vinyl alcohol) (PVA)

(b) Repeating unit of PVA

(c) Vinyl alcohol

Supramolecular Polymers: Polymers That Can Heal Themselves?

All of the polymers we have discussed in this chapter are produced by the formation of *covalent* bonds between monomers. But similarly long arrangements of atoms can be built through *noncovalent* interactions among monomers, producing a *supramolecular* ("beyond the molecular") polymer. Chemists in the Netherlands produced such a polymer in which the monomer was capped at each end with a 2-ureidopyrimidone (UPy) group. Each UPy group has two hydrogen-bond donors and two hydrogen-bond acceptors, which are complementary to two donors and two acceptors from another UPy group. Therefore, a pair of these groups can bind to each other relatively strongly via four hydrogen bonds, as shown below.

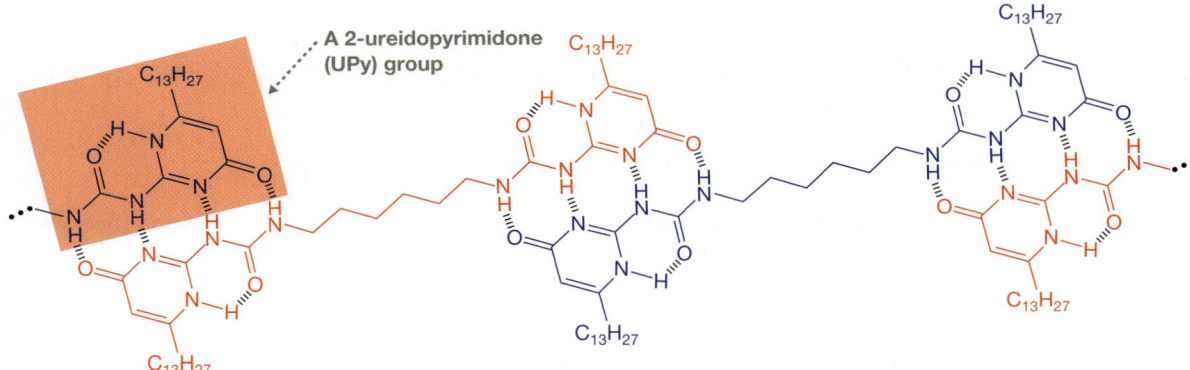

A 2-ureidopyrimidone (UPy) group

Stephen L. Craig of Duke University took this a step further, producing supramolecular polymers from monomers in which spacer chains were capped by relatively small, complementary segments of DNA. Thus, the properties of the resulting polymers could be tuned by varying the length and sequences of the DNA segments, and also by varying the length of the spacer units.

Remarkably, the hydrogen bonding in these supramolecular polymers is strong enough to give the polymers properties that are comparable to those of covalent polymers, but with the advantage that the supramolecular polymers can easily be broken down and reformed by heating or adding a solvent that disrupts the hydrogen bonding. Such chemical approaches are paving the way for polymers and plastics that are easily recycled, require less energy to process and mold, and are even *self-healing*! For example, in France, L. Leibler and coworkers have already synthesized a supramolecular polymeric rubber that can undergo many cycles of breaking and self-healing at room temperature, and you can imagine how this strategy could be applied to new types of adhesives and other materials.

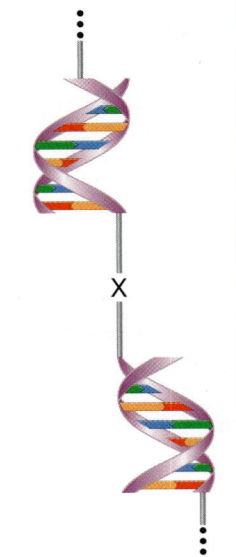

You cannot make PVA directly from vinyl alcohol (Fig. 26-17b), however, because vinyl alcohol does not exist in significant quantities. Instead, vinyl alcohol is an enol that readily tautomerizes to acetaldehyde, its keto form (Equation 26-31):

$$
\text{Enol form} \qquad\qquad \text{Keto form}
$$

Vinyl alcohol — Tautomerization → Ethanal (Acetaldehyde) (26-31)

YOUR TURN 26.16

Explain why vinyl alcohol is disfavored thermodynamically and isomerizes to a different product.

To prepare PVA, chemical plants begin with the synthesis of poly(vinyl acetate) from vinyl acetate, as shown in Equation 26-32.

Vinyl acetate — Benzoyl peroxide, Δ → Poly(vinyl acetate) or (26-32)

Poly(vinyl acetate) can then undergo base-catalyzed transesterification (Section 21.7) to remove the acetyl groups and produce PVA (Equation 26-33).

Poly(vinyl acetate) — NaOCH$_3$, CH$_3$OH, Δ → Poly(vinyl alcohol) (26-33)

The extent of deacetylation can vary from 80% to 99.3%, depending on the desired properties of the polymer.

YOUR TURN 26.17

Provide the mechanism for the reaction that occurs in Equation 26-33. What is the second product of the reaction?

26.6a.2 De-tert-butylation of Poly(ether ether ketone)

Polymers that can withstand high temperatures without degrading are highly desirable, so chemists often try to introduce thermal stability into polymers. An example is **poly(ether ether ketone)**, or **PEEK** (Fig. 26-18), which is stable to temperatures above

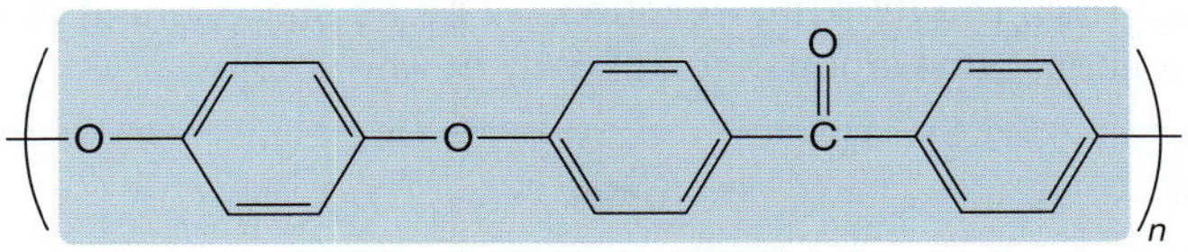

The continuous conjugated system imparts rigidity to the polymer chain.

Poly(ether ether ketone) or **PEEK**
Thermally stable >500 °C

FIGURE 26-18 The rigidity and thermal stability of PEEK The entire repeating unit of PEEK is conjugated (indicated by the blue screen), making the polymer rigid and thermally stable above 500 °C.

500 °C. PEEK is so thermally stable because its main chain is very rigid, and that rigidity is an outcome of the extended conjugation throughout the entire repeating unit.

YOUR TURN 26.18

All of the bond angles in PEEK are approximately 120°. Explain why the C—O—C bonds between the phenyl rings have a bond angle closer to 120° than 109.5°.

The advantages of PEEK become a disadvantage during synthesis. When chemists began to synthesize PEEK, they found that the polymer chains would crystallize out of solution before reaching a substantial length. These crystals form so readily because the polymer chains, being so rigid, allow the polymer molecules to easily align and maximize their intermolecular forces.

To address this problem, chemists incorporated a monomer with a *tert*-butyl group attached to the aryl ring, as shown in Figure 26-19, which decreases the strength of the intermolecular forces between chains. The *tert*-butyl groups prevent the polymer from crystallizing as it forms, so the reaction is able to proceed to a higher degree of polymerization.

To isolate PEEK, the *tert*-butyl groups must be removed. This is done by adding excess toluene under acidic conditions, as shown in Equation 26-34. The *tert*-butylated PEEK essentially undergoes the reverse of a Friedel–Crafts alkylation (Section 22.3), whereas toluene undergoes a Friedel–Crafts alkylation. The overall reaction is then a trans *tert*-butylation, in which the *tert*-butyl group moves from the polymer to toluene.

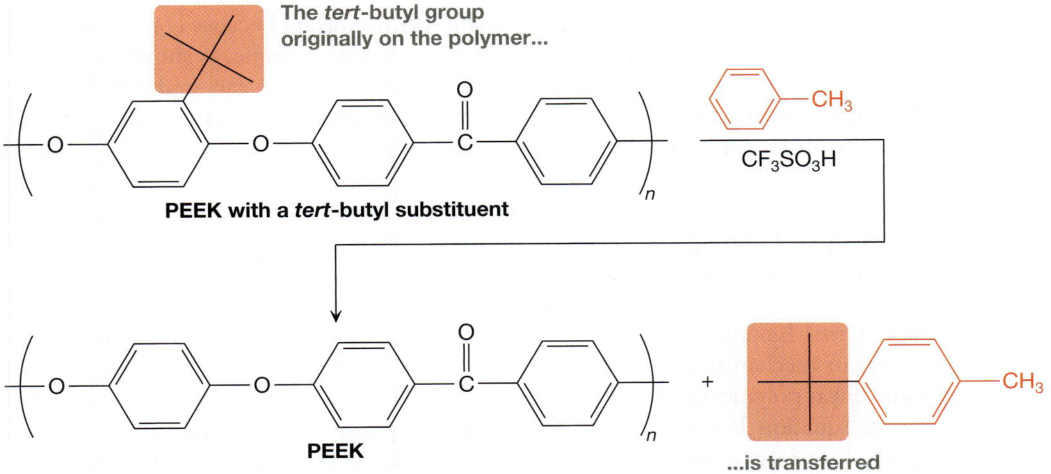

This *tert*-butyl group helps prevent crystallization.

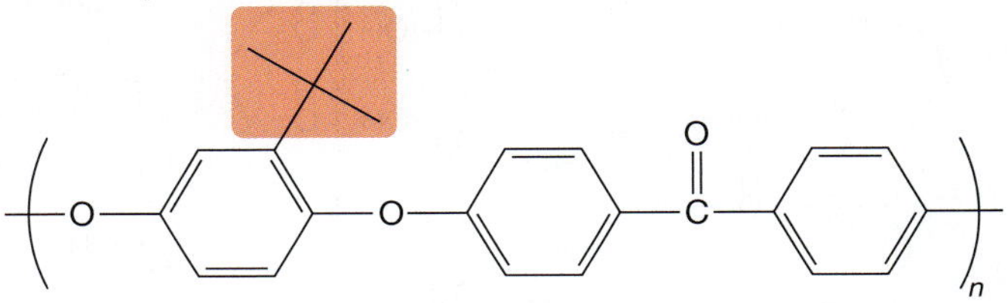

FIGURE 26-19 PEEK with a *tert*-butyl substituent PEEK is synthesized with a *tert*-butyl substituent attached to the ring to help prevent crystallization. This allows the polymer to be produced with a higher degree of polymerization. Subsequently, the *tert*-butyl group can be removed.

The *tert*-butyl group originally on the polymer...

PEEK with a *tert*-butyl substituent

(26-34)

PEEK

...is transferred to touene.

26.6 Chemical Reactions after Polymerization **1335**

PROBLEM 26.15 Propose a mechanism for the reaction that occurs in Equation 26-34. Do you think the reaction will be as successful if the *tert*-butyl group were a linear butyl group instead? Why or why not?

26.6b Cross-linking

In Section 26.6a, we discussed chemical reactions that change the structure and properties of linear polymers, but the polymer chains remained linear. Many polymers, however, undergo reactions that connect, or **cross-link**, linear polymer chains.

Borax ($Na_2B_4O_7$), for example, can react with hydroxyl groups from different chains of PVA to create *cross-linked* polymers, as shown in Equation 26-35.

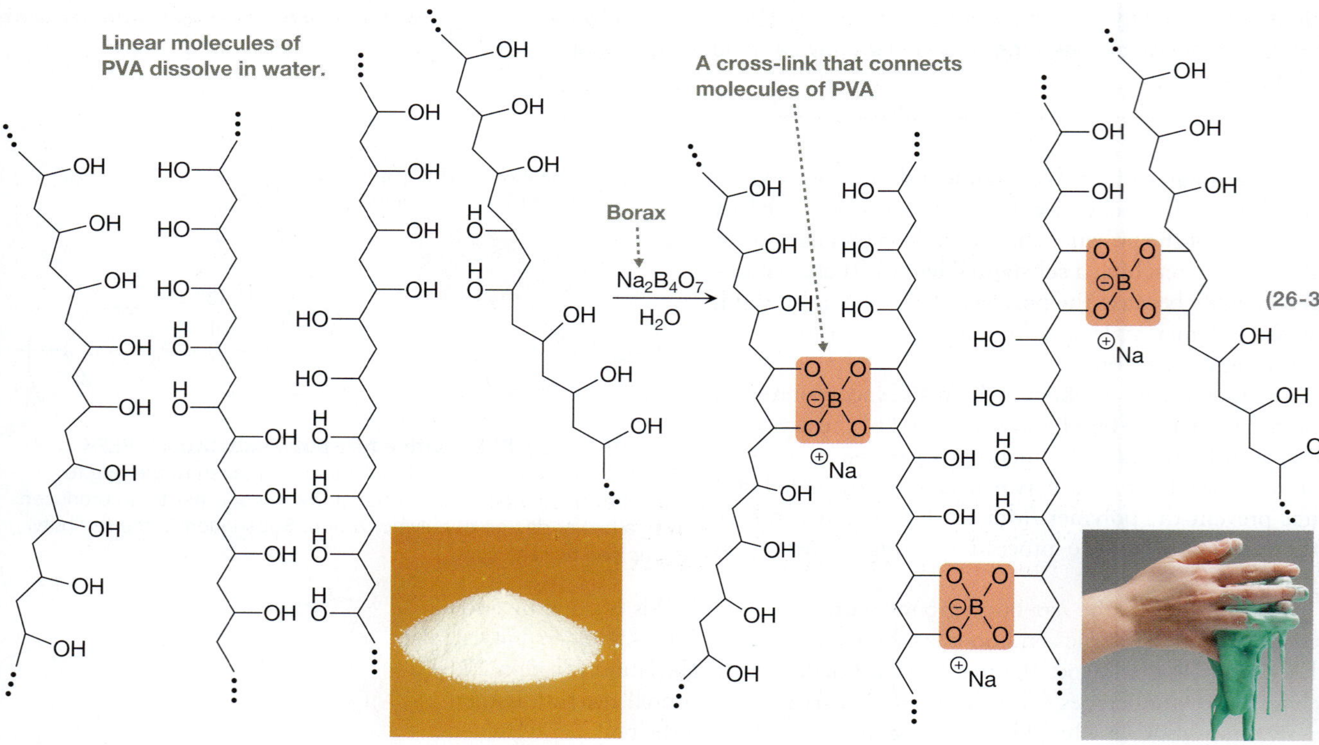

The ability to cross-link polymers can be very important because cross-linking can dramatically change the properties of the polymer. For example, as shown on the left of Equation 26-35, linear PVA (i.e., prior to cross-linking) can undergo extensive hydrogen bonding, making it a solid in its pure form and very water soluble. The cross-linked form, by contrast, interacts with water to form "slime," as shown on the right of Equation 26-35. Cross-linking holds the polymer chains together, so instead of the PVA molecules dispersing to dissolve in water, the water molecules now act as a solute in the cross-linked polymer. The result is a gel because liquid water is dispersed throughout the solid, cross-linked PVA.

The reaction between borax and alcohol groups is not the only way to create cross-links. Cross-linking reactions can involve a wide variety of functional groups and a wide variety of mechanisms—many of which we have studied in this book. Moreover, the same linear polymer can often be cross-linked by more than one choice of cross-linking agent. Equation 26-36 shows, for example, a second way to cross-link PVA, using formaldehyde to form acetal groups (review Section 18.2a) that act as cross-links.

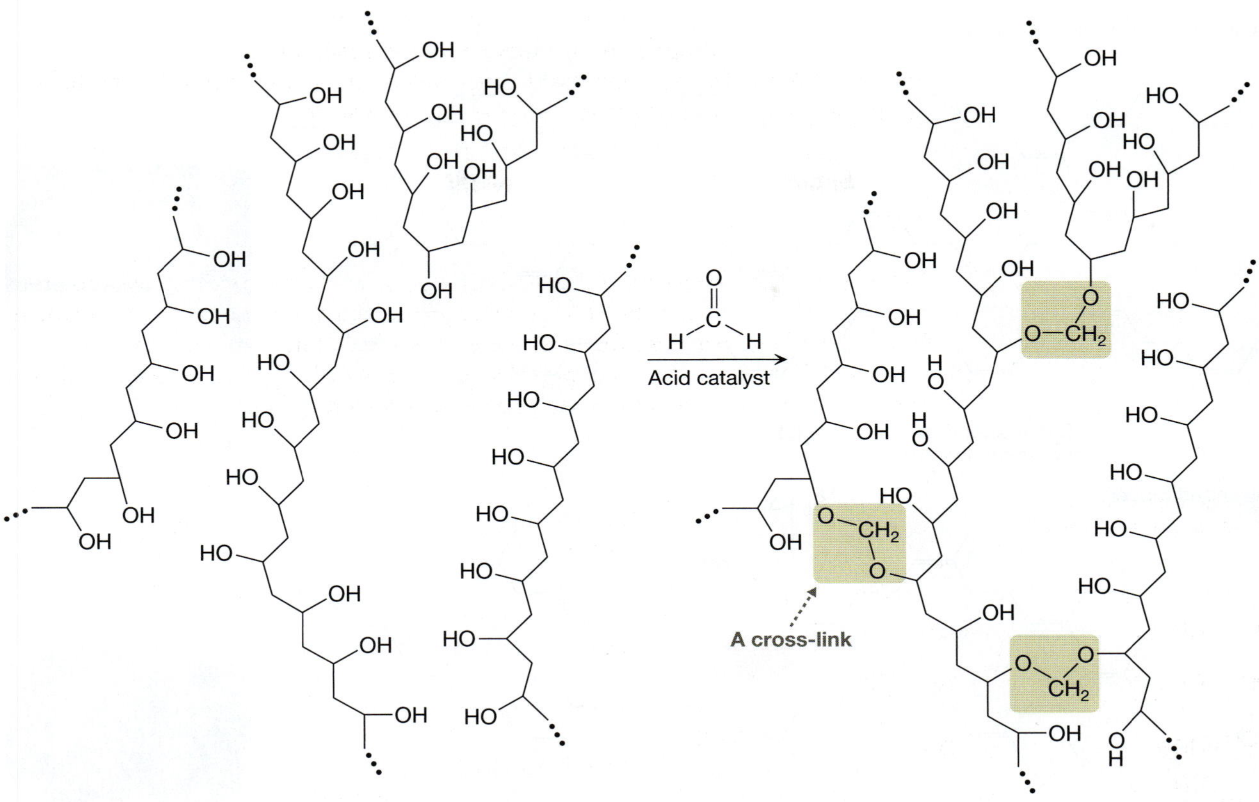

(26-36)

PROBLEM 26.16 Draw the mechanism for the cross-linking reaction between two PVA molecules and formaldehyde. Assume some acid, HA, is present to catalyze the reaction. (Only the reactive portions of the PVA molecules are shown in detail below.)

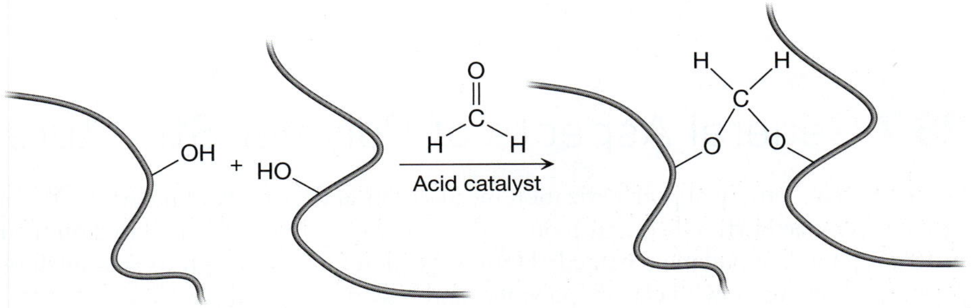

Rubber provides one particularly notable application of cross-linking. In the 19th century, natural rubber had limited uses because it became brittle at low temperatures and shapeless at high temperatures. When Charles Goodyear (1800–1860) accidentally dropped some natural rubber mixed with sulfur on a hot stove, however, he discovered a product that had the desired elasticity and did not lose its properties at extreme temperatures. The key to this transformation of rubber—a process called **vulcanization**—is the introduction of *disulfide bonds* that act as cross-links between the rubber chains, as shown in Figure 26-20 (next page).

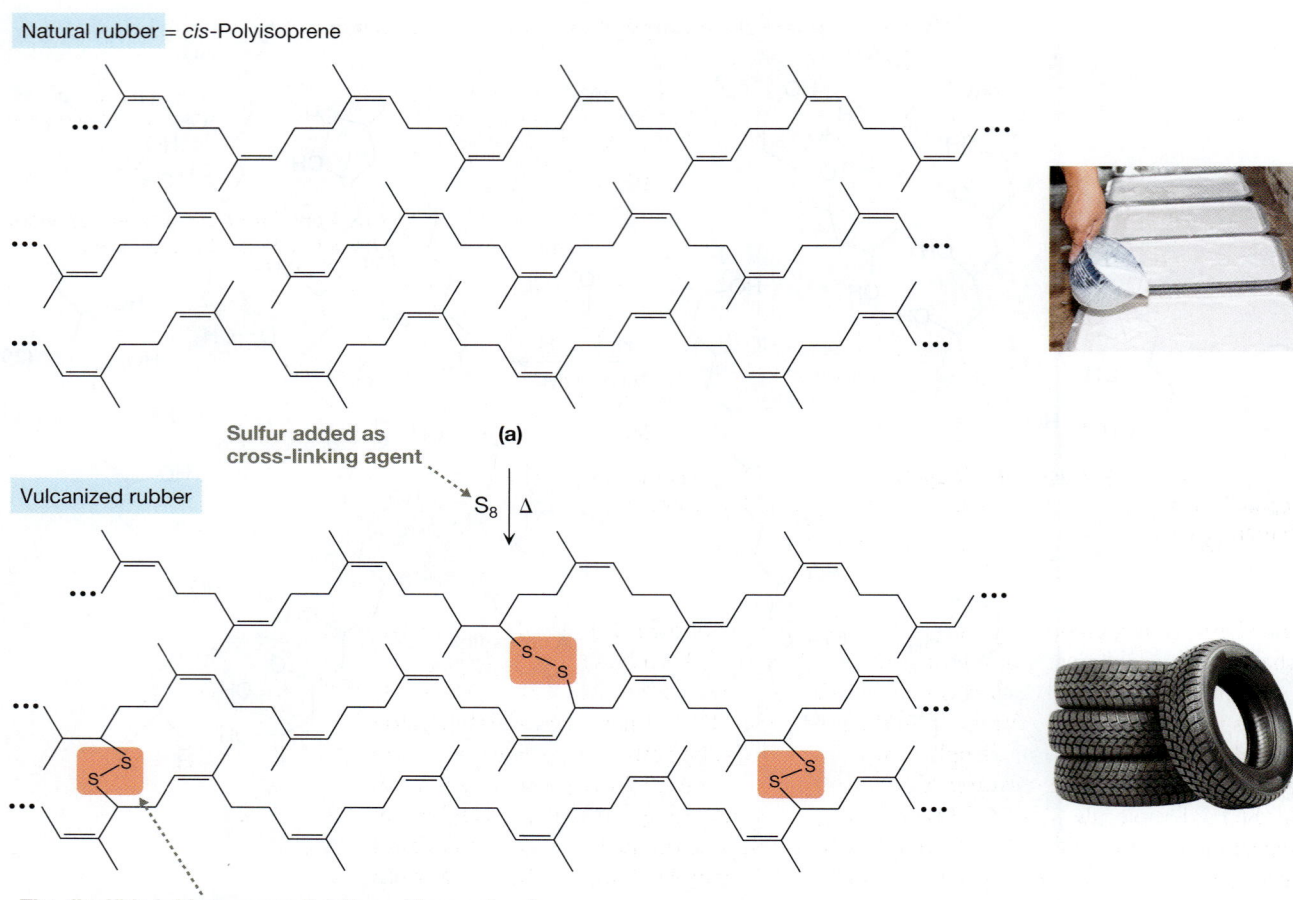

Natural rubber = *cis*-Polyisoprene

(a)

Sulfur added as cross-linking agent

S_8 | Δ

Vulcanized rubber

The disulfide bridges cross-link the rubber molecules.

(b)

FIGURE 26-20 **Vulcanization of rubber** (a) Natural rubber consists of independent polymer molecules. The addition of sulfur (S_8) results in cross-linking among the main chains to produce (b) vulcanized rubber. Natural rubber has high elasticity but minimal durability. Tires made of vulcanized rubber, on the other hand, are less elastic but much more durable.

26.7 General Aspects of Polymer Structure

We have seen a variety of polymerization mechanisms and a variety of polymers. What structural characteristics distinguish one type of polymer from another? Based on their structures, how are polymers named? Here in Section 26.7, we begin to answer these questions, learning how to classify polymers and how to name simple ones.

26.7a Classes of Polymers

Polymers can be classified according to the following criteria:

- Type of chain
- Functional group(s) in the chain
- Monomer(s) used to make the polymer
- Number of different monomers used

When polymers are classified by type of chain, we distinguish polymers with only carbons in the main chain (**carbon-chain polymers**) from those with heteroatoms in

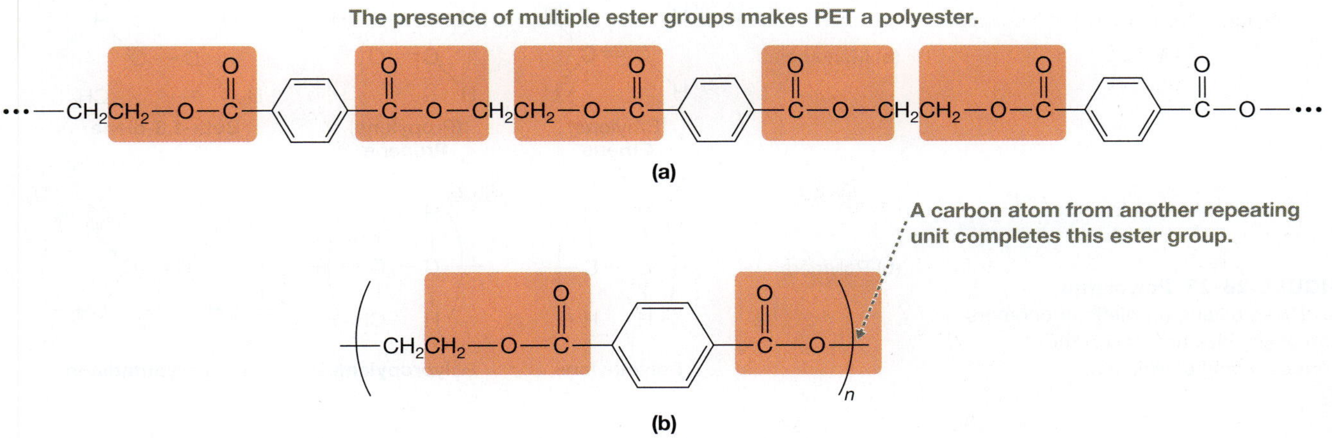

The presence of multiple ester groups makes PET a polyester.

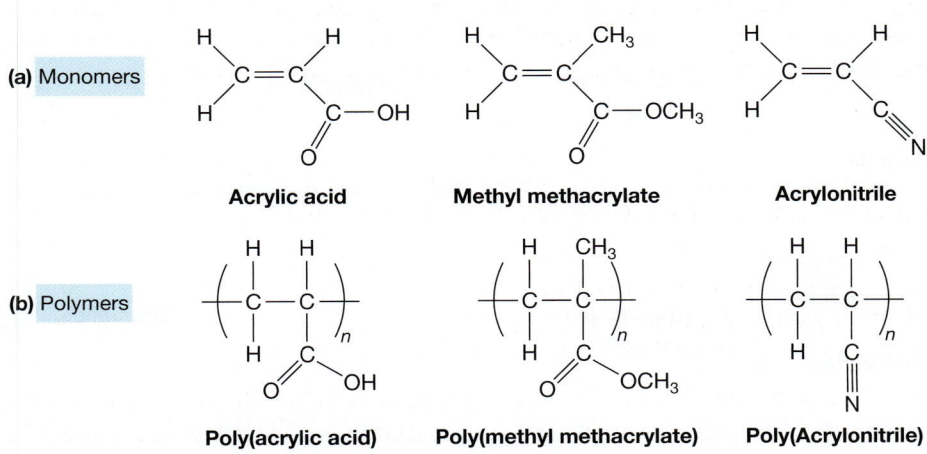

(a)

A carbon atom from another repeating unit completes this ester group.

(b)

FIGURE 26-21 Poly(ethylene terephthalate) or PET (a) A portion of PET and (b) its repeating unit. PET is a polyester because of the repeating ester groups in the main chain, screened in red.

the chain (**heterochain polymers**). Polystyrene (Equation 26-1, p. 1308) is a carbon-chain polymer, whereas nylon (Equation 26-24, p. 1325) is a heterochain polymer.

A more specific way to classify heterochain polymers is by the functional groups that are responsible for linking the monomers together to make the main chain. **Poly(ethylene terephthalate) (PET)**, for example, which is used to make bottles for carbonated beverages, is a **polyester**. The polyester structure is shown in Figure 26-21, both for a portion of the PET molecule (Fig. 26-21a) and its repeating unit (Fig. 26-21b). Notice that the repeating unit for PET contains two ester groups; the one on the left is a complete ester group, and the one on the right is incomplete but is completed by a carbon atom from an adjacent repeating unit.

YOUR TURN 26.19

Based on its functional groups, to what class of polymers does nylon-6,6 (Equation 26-24, p. 1325) belong? How many times does the functional group appear in the repeating unit?

Carbon-chain polymers are usually classified according to the monomer used to make them. For example, the polymers in Figure 26-22 are classified as **polyacrylates** or **acrylics** because the monomers are derivatives of acrylic acid.

(a) Monomers

Acrylic acid   Methyl methacrylate   Acrylonitrile

(b) Polymers

Poly(acrylic acid)   Poly(methyl methacrylate)   Poly(Acrylonitrile)

FIGURE 26-22 Polyacrylates (a) Acrylic acid and its derivatives can polymerize to form (b) the corresponding polyacrylate (or acrylic) polymers.

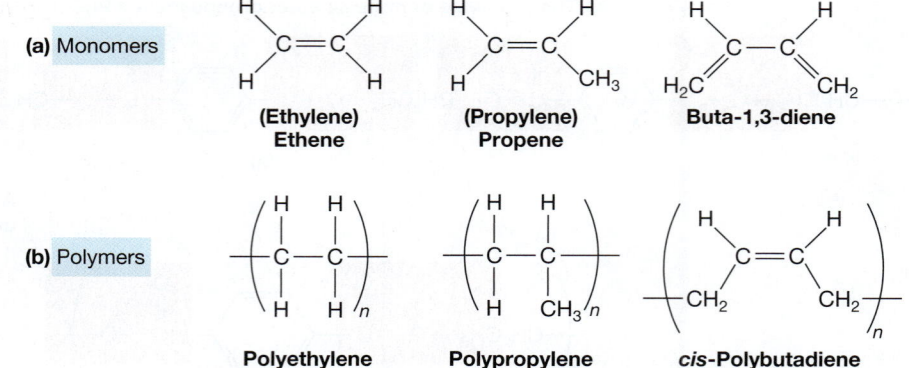

FIGURE 26-23 Polyolefins
(a) These alkene (or olefin) monomers can polymerize to form (b) the corresponding polyolefins.

Alkenes are commonly referred to as **olefins**, so polyethylene (PE), polypropylene (PP), and other polymers made from alkenes or dienes are referred to as **polyolefins**. Some examples are shown in Figure 26-23.

Polymers such as polystyrene and polyethylene are said to be **homopolymers** because each is synthesized from one monomer (i.e., styrene or ethylene, respectively). Other homopolymers are shown in Table 26-1 for comparison.

The prefix *poly* is not solely used to describe polymers; it can also be used to describe collections of similar molecules with several of a particular functional group. Figure 26-24, for example, shows a collection of polycarboxylic acids, which are not

TABLE 26-1 Comparison of Homopolymers and Their Descriptions

Monomer(s)	Polymer	Carbon–Chain or Heterochain Polymer	Description Based on Monomer	Description Based on Functional Group in Polymer
(Methyl vinyl ether) Methoxyethene	Poly(methyl vinyl ether)	Carbon–chain polymer	A poly(vinyl ether)	—
(Isobutylene) 2-Methylpropene	Polyisobutylene	Carbon–chain polymer	A polyolefin	—
1,4-Dichlorobenzene + Na₂S Sodium sulfide	Poly(phenylene sulfide)	Heterochain polymer	—	A polysulfide

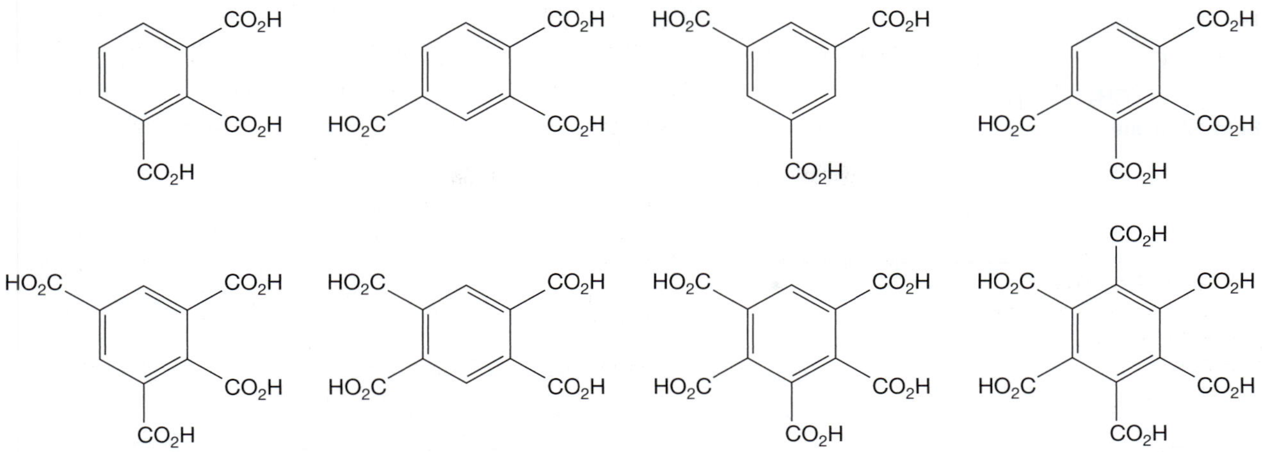

FIGURE 26-24 **A series of benzene polycarboxylic acids** Each molecule has the same core or backbone (benzene) but has multiple carboxyl groups attached. Since the number varies from one molecule to another, the term *poly* is used to indicate that each member of the series has more than one carboxyl group. None of these molecules is a polymer because none is formed by combining monomers to create a repeating unit.

polymers. In this case, the prefix *poly* refers to pendant groups, and not the monomer used to make a polymer or to the functional group in a polymer chain.

SOLVED PROBLEM 26.17

The repeating units of polymers **A** and **B** are shown here. Determine each one's polymeric classification according to Table 26-1.

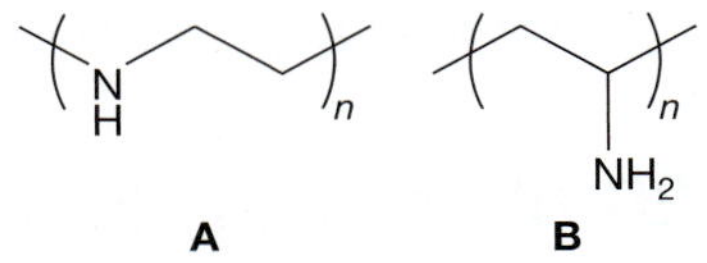

A   **B**

Think How are heterochain polymers distinguished from carbon-chain polymers?
What functional group is present in each repeating unit? Is that functional group responsible for connecting the monomers, or is it part of a pendant group?

Solve Polymer **A** has a nitrogen atom incorporated into the main chain, so it is a heterochain polymer. Furthermore, the N atom in polymer **A**, which is part of a secondary amine, is responsible for connecting monomeric units together, so polymer **A** is classified as a polyamine, too. Polymer **B**, on the other hand, is a carbon-chain polymer because its main chain consists of only carbon atoms. Polymer **B** has an NH_2 group characterizing an amine in its repeating unit, but the NH_2 group is a pendant group that repeats every two carbons of the main chain. Polymer **B**, therefore, is a poly(vinyl amine).

PROBLEM 26.18 Polyserine, the polymer made from the amino acid serine, can be described by more than one term that begins with the prefix *poly*. What are those terms? Which one refers to the polymeric nature of polyserine?

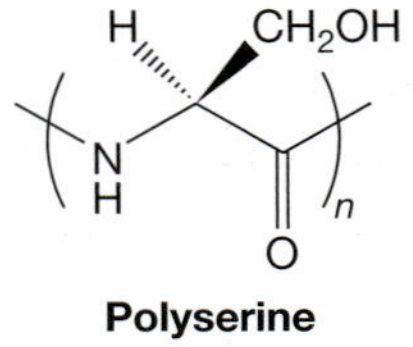

Polyserine

If two different monomers are polymerized together, and each monomer is capable of undergoing self-polymerization, then the resulting polymer is called a **copolymer**. A copolymer of styrene and butadiene, called styrene–butadiene rubber (SBR), was used during World War II as a suitable replacement for natural rubber, which was in limited supply. The formation of SBR is shown in Equation 26-37.

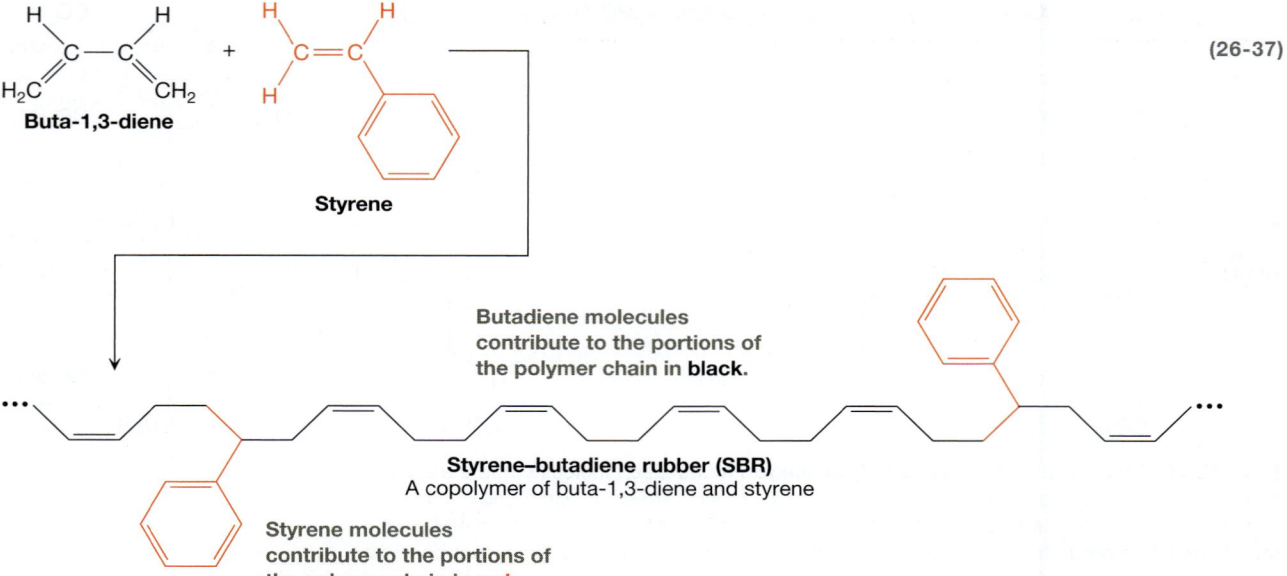

Buta-1,3-diene

Styrene

Butadiene molecules contribute to the portions of the polymer chain in **black**.

Styrene–butadiene rubber (SBR)
A copolymer of buta-1,3-diene and styrene

Styrene molecules contribute to the portions of the polymer chain in **red**.

As mentioned on the previous page, copolymers are possible when each monomer is capable of forming a homopolymer. For example, buta-1,3-diene is capable of forming a homopolymer (polybutadiene), as is styrene (polystyrene). The reaction of the two different monomers creates a copolymer.

Nylon-6,6 (see Section 26.4), on the other hand, is not a copolymer. Although two reactants are used to synthesize nylon-6,6, neither adipic acid nor hexane-1,6-diamine is capable of forming a polymer in the absence of the other reactant.

26.7b Polymer Nomenclature

Just as with other forms of chemical nomenclature, an IUPAC name of a polymer can differ from its trivial or common name.

> The IUPAC name of a polymer is based on its repeating unit, but its trivial name could be based on the monomer from which the polymer is made.

For example, if the repeating unit of a polymer is $-CH_2-$ (a methylene group), then the IUPAC name is polymethylene, but the trivial (and more common) name is polyethylene, because it is made by polymerizing ethylene (Equation 26-38).

$$\text{Ethylene} \xrightarrow[\Delta]{\text{Catalyst}} \text{Polymethylene} \quad \text{or} \quad \text{Polyethylene} \qquad (26\text{-}38)$$

Ethylene

Polymethylene
The IUPAC name is based on the simplest repeating unit.

Polyethylene
The common (trivial) name is based on the name of the monomer.

IUPAC recognizes the trivial names for approximately 20 common polymers. We use these trivial names in this text because you are more likely to encounter them in everyday life and because they help establish the relationship between the polymer and the monomer from which it is made.

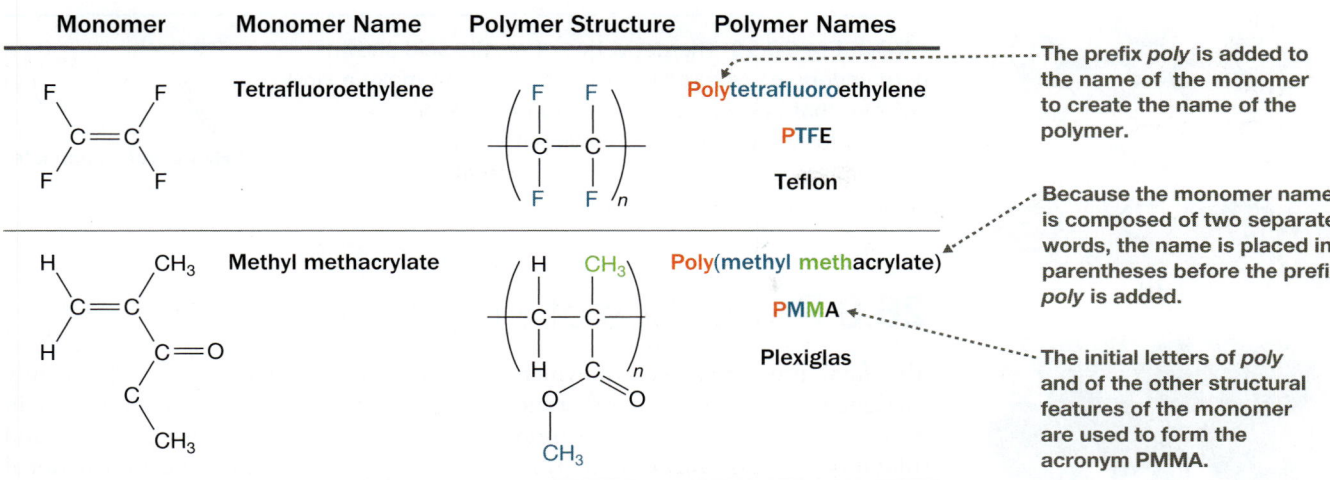

Monomer	Monomer Name	Polymer Structure	Polymer Names

Tetrafluoroethylene → Polytetrafluoroethylene / PTFE / Teflon

> The prefix *poly* is added to the name of the monomer to create the name of the polymer.

Methyl methacrylate → Poly(methyl methacrylate) / PMMA / Plexiglas

> Because the monomer name is composed of two separate words, the name is placed in parentheses before the prefix *poly* is added.

> The initial letters of *poly* and of the other structural features of the monomer are used to form the acronym PMMA.

FIGURE 26-25 Nomenclature of homopolymers In these examples, the name of a polymer is related to the name of the monomer.

In general:

- The names for homopolymers follow a *polymonomer* format (Fig. 26-25).
- If the name of the monomer consists of two or more words, then the name of the monomer appears in parentheses in the name of the polymer.

Thus, the name *polystyrene* does not contain parentheses, but *poly(methyl methacrylate)* does.

In addition to IUPAC and trivial names, there are also trademarked names for polymers. As shown in Figure 26-25, Teflon is the trademarked name for tetrafluoroethylene and Plexiglas is the trademarked name for poly(methyl methacrylate). These trademarked names, however, provide no information about the structure of the repeating unit.

Some polymers are referred to by their acronyms:

A polymer's acronym is usually derived from the prefix *poly* and the initial letters of the different portions of the monomer.

Teflon, for example, is also known as PTFE (*polytetrafluoroethylene*).

Some polymers use the name of the repeating unit as the basis for the name of the polymer. For example, the most common polyester is poly(ethylene terephthalate), or PET (Section 26.4). In PET, the repeating unit is named as an ester of terephthalic acid and the ethylene group (Equation 26-39).

Terephthalic acid + Ethylene glycol $\xrightarrow[\Delta]{\text{Acid catalyst}}$ Poly(ethylene terephthalate) PET

An ester of terephthalic acid and ethylene glycol

(26-39)

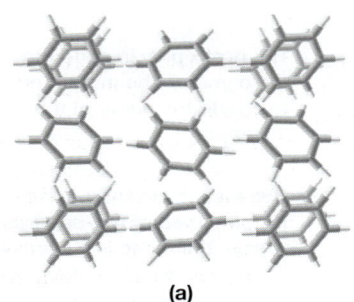

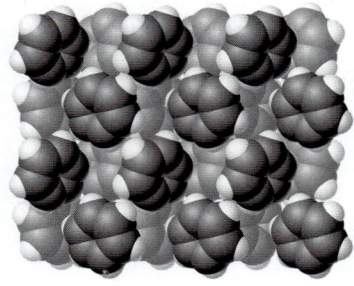

FIGURE 26-26 A crystal lattice of solid benzene Solid benzene is depicted with (a) stick models and (b) space-filling models.

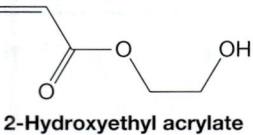

Glass

FIGURE 26-27 The structure of silicon dioxide, an amorphous solid There are four oxygen atoms attached to each silicon atom; each of those oxygen atoms is attached to two different silicon atoms. Note the lack of a regular or repeating arrangement of atoms.

PROBLEM 26.19 The derivative of acrylic acid shown here undergoes free radical polymerization to make a vinyl polymer that is water soluble. Draw the condensed formula for the polymer and provide its name and acronym.

2-Hydroxyethyl acrylate

26.8 Properties of Polymers

The final properties of a polymer are important, so polymers are frequently "designed," as are drugs and other molecules, to perform a specific function. Here in Section 26.8, we consider two properties of polymers—thermal transitions and solubility—as they relate to intermolecular forces (Chapter 2) and conformational isomerism (Chapter 4).

26.8a Thermal Transitions in Polymers

There are two fundamental types of solids: crystalline and amorphous. **Crystalline solids** are ones in which the molecules are arranged in a regular pattern that repeats throughout the solid. This pattern is referred to as the *crystal lattice*, a term that may be familiar to you from general chemistry. An example of the crystal lattice of solid benzene is shown in Figure 26-26, with two different representations of the benzene molecule.

Because of their high order:

> Crystalline solids have a well-defined **melting point (T_m)**, above which they become liquids.

In contrast to crystalline solids, the species that make up an **amorphous solid** have no long-range order to their arrangement. The most common example of an amorphous solid is glass, which is composed primarily of silicon dioxide (SiO_2). Figure 26-27 shows a representation of the arrangement of atoms in SiO_2.

Amorphous solids do not have a well-defined melting point. Instead:

> Amorphous solids have a **glass transition temperature (T_g)**, above which they become rubbery or viscous (i.e., they flow slowly).

Above the glass transition temperature, the atoms in the substance have sufficient kinetic energy that they can move past each other, which is why the solid exhibits a loss in mechanical strength.

Polymers range from being completely amorphous to being highly crystalline. They are rarely entirely crystalline, however, because polymer chains are generally very long, making it difficult for a single molecule to arrange itself entirely in an orderly fashion. Moreover, polymers are mixtures that vary in length, so it is difficult for them to create a regular, repeating crystal lattice.

Polymers can exhibit *regions* that are crystalline (depicted in Figure 26-28a), separate from ones that are amorphous (depicted in Figure 26-28b). These types of polymers can therefore have both a T_m that characterizes the crystalline regions and a T_g that characterizes the amorphous regions.

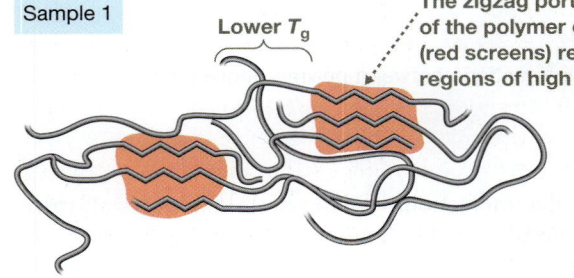

Crystalline region of a polymer

• Stronger intermolecular forces
• Lower entropy

Amorphous region of a polymer

• Weaker intermolecular forces
• Higher entropy

(a)   (b)

FIGURE 26-28 A comparison of a polymer's crystalline and amorphous regions (a) Polymer molecules in a crystalline region require significant kinetic energy to overcome intermolecular forces and to increase the entropy of the system as it enters the liquid phase. (b) Polymer molecules in an amorphous region are already at higher entropy and the intermolecular forces are not maximized, as they would be in a crystalline region. As a result, less energy is required for the molecules in an amorphous region to flow past each other.

As Figure 26-28 indicates, a crystalline region of a polymer generally has stronger intermolecular forces and lower entropy than an amorphous region, so the energy required to melt a crystalline region is greater than the energy required for an amorphous region to undergo a glass transition. Therefore:

A polymer's T_g is lower than its T_m.

The value of T_g increases, moreover, as the *crystallinity* of the polymer increases. T_g increases with the polymer's crystallinity because, as shown in Figure 26-29, the crystalline regions "anchor" the amorphous regions and impede the movement of the chains in the amorphous regions.

The T_m and the T_g of polymers are important data to know because polymers are generally heated before they are formed into a particular shape. In addition, these temperatures determine the range of temperatures over which the polymer can perform without losing its shape or function.

At room temperature, for example, the rubber in a rubber band or a racquetball is an **elastomer** because it can experience large deformations yet return to its original shape. A racquetball is deformed when it impacts a racquet or a wall, yet it resumes its shape and, in the process, forces itself away from the racquet or the wall (Fig. 26-30a, next page). Rubber can do this because room temperature is well above its T_g.

If you cool a racquetball in liquid nitrogen (77 K) to a temperature below its T_g (Fig. 26-30b), then the rubber molecules become locked in place. If you throw a frozen racquetball against the wall, it shatters like glass (Fig. 26-30c), because the molecules no longer have the mobility they had at room temperature. If you warm the broken pieces to room temperature, however, then the rubber passes through its glass transition temperature and regains its elasticity.

FIGURE 26-29 The effect of crystallinity on T_g Sample 2 has a greater degree of crystallinity than Sample 1. The noncrystalline regions in Sample 2 have a higher T_g than the noncrystalline regions in Sample 1, because the crystalline regions in Sample 2 constrain the noncrystalline regions more, which increases the amount of energy needed for those segments to flow.

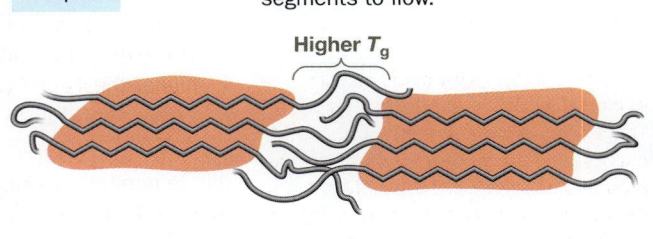

Sample 1

Lower T_g

The zigzag portions of the polymer chain (red screens) represent regions of high crystallinity.

Sample 2

Higher T_g

(a)   (b)   (c)

FIGURE 26-30 **A comparison of a racquetball's performance as an elastomer at high and low temperatures** (a) At room temperature, the racquetball has its normal elastic quality. (b) A ball immersed in liquid nitrogen is cooled below its T_g. (c) The polymer molecules in the chilled ball no longer bend or deform as easily; the ball shatters when thrown against a wall.

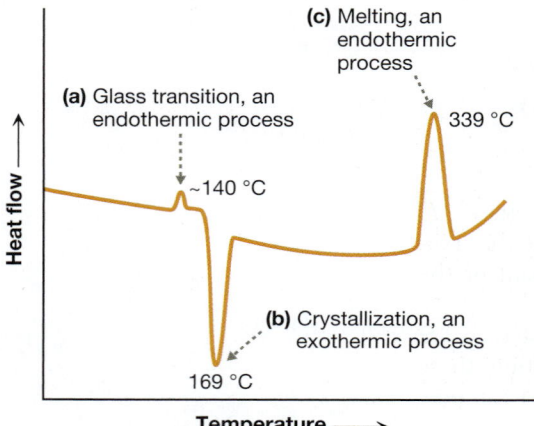

FIGURE 26-31 A differential scanning calorimetry (DSC) scan of a polymer (PEEK) (a) Above T_g (~140 °C), the polymer molecules have sufficient mobility to organize into a crystalline solid. (b) As the polymer molecules organize into a crystalline solid, which is more ordered and at lower energy, they give off energy in the form of heat. (c) At the melting point (339 °C), the polymer molecules absorb sufficient energy that they are no longer constrained in a crystal lattice, but change into a liquid and flow past each other.

In polymers, both melting and undergoing a glass transition are endothermic processes, whereas crystallization is an exothermic process. That is, when polymer molecules arrange themselves in an organized crystal lattice, they release energy as they go from a higher energy state to a lower energy state. This lower energy is achieved by maximizing intermolecular forces and minimizing steric repulsions.

We can see these points illustrated more clearly with differential scanning calorimetry (DSC), a technique that measures the heat that flows into or out of a sample as the temperature of the sample is changed. The curve shown in Figure 26-31 is the DSC result for a sample of PEEK. Around 140 °C, the polymer absorbs heat and goes through its glass transition. This gives the polymer chains enough mobility to become more ordered and, therefore, crystallize at 169 °C. When they do, heat is given off because they go from a higher energy state to a lower energy, more stable crystalline form. At about 339 °C, the crystalline form melts and the polymer molecules are able to flow.

26.8b Factors Affecting Thermal Transitions in Polymers

Synthetic polymers can be molded into useful objects. **Thermoplastic** polymers can be heated above their T_g or T_m and then forced into molds or pushed through slots to form films or threads (see Fig. 26-32). This is the origin of the term "plastic" to describe polymers.

Recall from Chapter 2 that the stronger the intermolecular interactions in a substance, the higher the temperature required for a phase transition, either from solid to liquid or from liquid to gas. This is also true for polymers:

> The stronger the intermolecular interactions between polymer molecules, the higher the T_g or T_m.

In general, stronger intermolecular interactions lead to a more stable crystal lattice, one with an arrangement of molecules that maximizes attractions between the

(a)

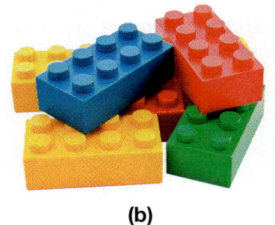

(b)

molecules and minimizes steric repulsions and other unfavorable arrangements. For the molecules in a solid to move past each other and flow, the solid must absorb sufficient heat to overcome the attractions between molecules and provide the molecules with enough energy to move away from the crystal lattice. Seven major factors affect the amount of heat required for this change.

1. An increase in the chain length of linear molecules raises T_g and T_m.

As we saw in Section 2.6d, larger molecules experience greater dispersion forces, so a higher temperature is required for melting.

2. Polymers that have stronger *types* of intermolecular interactions generally have a higher T_g and T_m.

Polar molecules can participate in dipole–dipole interactions, so they generally have higher melting points than nonpolar molecules with similar molar masses. Similarly, poly(vinyl chloride), a polar polymer, has a significantly higher T_g than polypropylene, a nonpolar polymer, even though both polymers have similar molar masses (Fig. 26-33).

The attraction between cations and anions is extremely strong, so polymers with ionic side groups generally have very high values of T_g and T_m. In fact, the energy required to overcome these ion–ion interactions can be greater than the energy required to break covalent bonds, in which case the polymer would decompose before it melts. Figure 26-34 (next page) shows, for example, that poly(acrylic acid) has a T_g of 106 °C, whereas poly(sodium acrylate) does not soften when heated and decomposes at temperatures above 260 °C.

YOUR TURN 26.20

Why do you think poly(acrylic acid) has a higher T_g than poly(vinyl chloride)?

3. Periodic or random branching of the polymer chain decreases the T_g and T_m.

Branched alkanes have lower melting points and boiling points than straight-chain alkanes, and polymers follow the same trend. In addition to reducing surface area for induced dipole–induced dipole attractions, branching interferes with the molecules' ability to organize into a regular crystal lattice; there is no longer a simple,

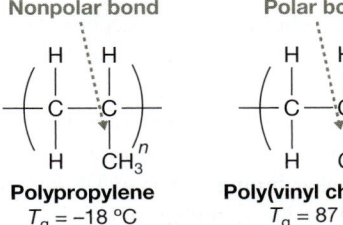

Nonpolar bond   Polar bond

Polypropylene
$T_g = -18$ °C

Poly(vinyl chloride)
$T_g = 87$ °C

FIGURE 26-33 **Polarity and thermal transitions** Because the C—Cl bond in poly(vinyl chloride) is polar, poly(vinyl chloride) has a higher T_g than polypropylene.

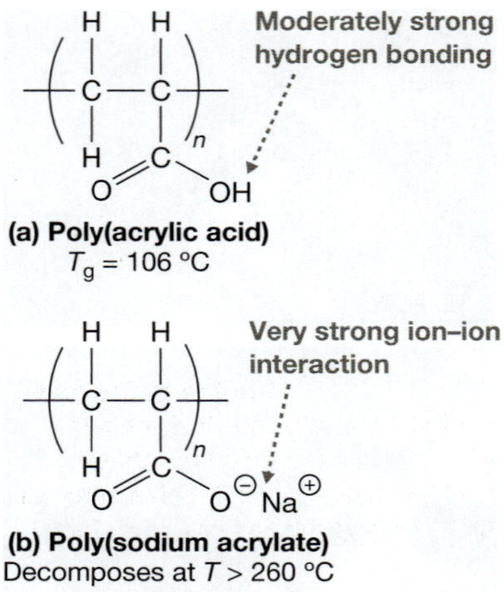

H   H
Moderately strong
hydrogen bonding

(a) Poly(acrylic acid)
$T_g = 106\ °C$

H   H
Very strong ion–ion
interaction

(b) Poly(sodium acrylate)
Decomposes at $T > 260\ °C$

FIGURE 26-34 Electrostatic attraction and thermal transitions (a) Hydrogen bonds between molecules of poly(acrylic acid) give it a relatively high T_g. (b) The much stronger electrostatic forces between the ions in poly(sodium acrylate), however, prevent the polymer from softening before it reaches a temperature at which it decomposes.

regular pattern for organizing the molecules into the solid. Moreover, in the solid form, branches push the neighboring molecules away from the main chain, decreasing the dispersion forces between them. Branched polyethylene, for example, has a melting point of 112 °C, whereas linear polyethylene has a melting point of 135 °C.

> **4.** The regular occurrence of pendant groups along a polymer chain increases T_g and T_m.

If the polymer has pendant groups that occur *regularly*, then the T_m will increase because these groups increase the barrier to rotation about single bonds. With less rotation about the main chain, the polymer can more easily arrange itself into a crystal lattice and achieve more extensive intermolecular interactions. For example, polypropylene has a higher melting point than polyethylene (Fig. 26-35).

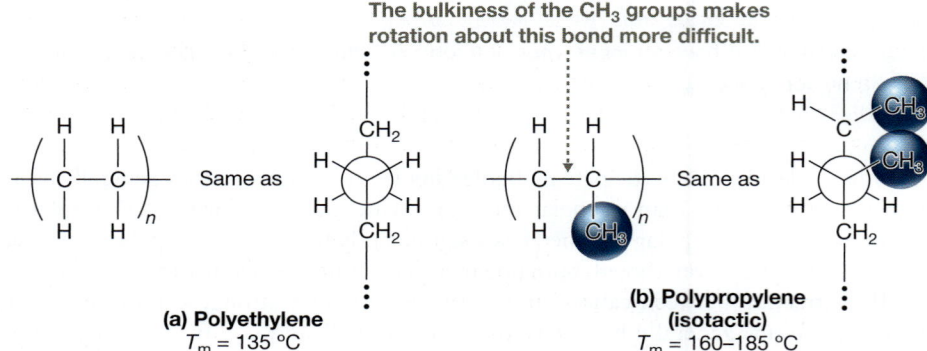

(a) Polyethylene
$T_m = 135\ °C$

(b) Polypropylene (isotactic)
$T_m = 160–185\ °C$

The bulkiness of the CH_3 groups makes rotation about this bond more difficult.

FIGURE 26-35 Regularity of pendant groups and thermal transitions (a) Polyethylene. (b) Polypropylene (isotactic). In each case, the repeating unit is shown on the left and a portion of the polymer chain is shown as a Newman projection on the right. The bulkiness of the CH_3 group in polypropylene hinders rotation about the single bonds in the polymer's main chain, and the cumulative effect along the polypropylene chain results in a significantly higher melting point.

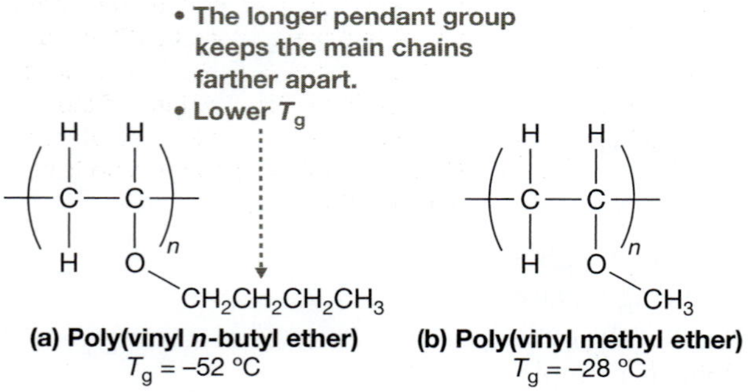

• The longer pendant group keeps the main chains farther apart.
• Lower T_g

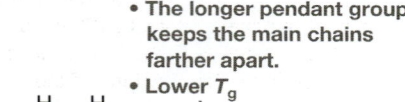

(a) Poly(vinyl *n*-butyl ether)
$T_g = -52\ °C$

(b) Poly(vinyl methyl ether)
$T_g = -28\ °C$

FIGURE 26-36 Pendant group length and thermal transitions Because the butyl group is longer than the methyl group, the chains of (a) poly(vinyl *n*-butyl ether) are kept farther apart than those of (b) poly(vinyl methyl ether). This diminishes the intermolecular attractions and lowers the glass transition temperature.

> **5.** Pendant groups that extend farther away from the main chain lower T_g and T_m.

Longer pendant groups push the polymer molecules apart, thus decreasing the intermolecular attractions and decreasing the T_m or T_g. Notice in Figure 26-36 that poly(vinyl *n*-butyl ether) has a longer pendant group, and thus a lower T_g, than poly(vinyl methyl ether).

> **6.** Rigid polymer chains have higher T_g and T_m than flexible chains.

If rotation about single bonds is not possible, then the polymer chains in a sample tend to adopt a more well-defined shape and the intermolecular interactions become stronger. This is why PEEK has a higher T_m than PET, as shown in Figure 26-37. Conjugation exists throughout the entire repeating unit in PEEK, whereas the repeating unit in PET has a flexible portion.

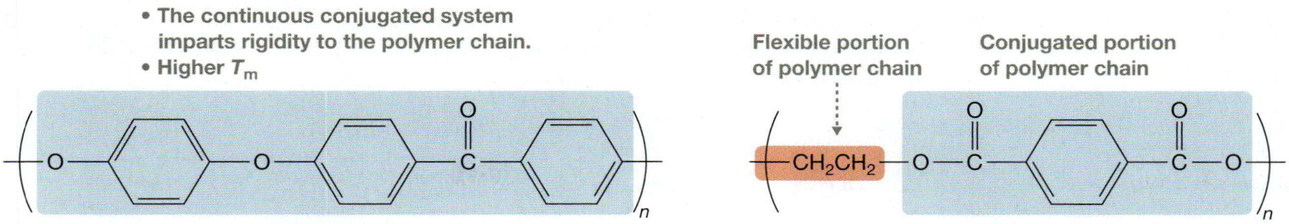

- The continuous conjugated system imparts rigidity to the polymer chain.
- Higher T_m

Flexible portion of polymer chain

Conjugated portion of polymer chain

(a) PEEK
$T_m = 365\ °C$

(b) PET
$T_m = 270\ °C$

FIGURE 26-37 Chain rigidity and thermal transitions The chain of (a) PEEK is more rigid than that of (b) PET (indicated by the blue screens), so PEEK has the higher melting point.

7. Greater regularity in structure along the polymer chain will usually increase T_g and T_m.

When polymer molecules have regularity in their structures, they organize more readily into crystal structures. A measure of regularity is *tacticity*, which describes the configurations of the chiral centers in the polymer chains (review Section 26.1d). For example, isotactic polypropylene (PP) (Fig. 26-38a) has a very regular structure—all configurations are the same—and has a relatively well-defined melting point that is well above room temperature. Atactic polypropylene (Fig. 26-38b), on the other hand, has no regular pattern of the configurations and neither does it have a well-defined melting point; it is tacky at room temperature and is used in adhesives.

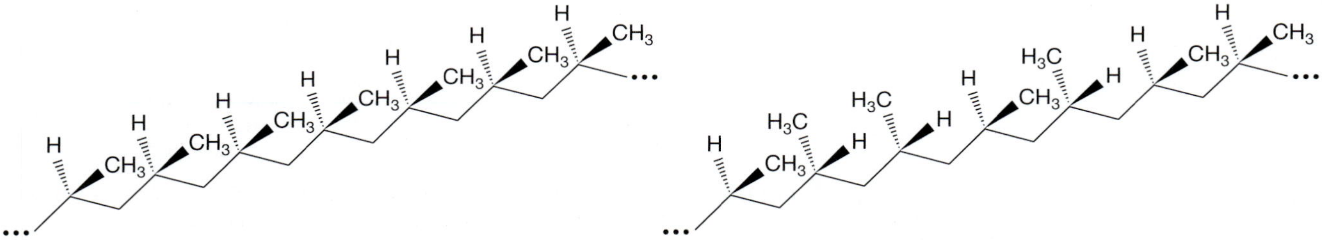

(a) Isotactic polypropylene
Melting point = 179 °C

(b) Atactic polypropylene
No true melting point

FIGURE 26-38 A comparison of isotactic and atactic polypropylene (PP) (a) In isotactic polypropylene, the methyl groups are all on the same side of the chain. Isotactic PP has a melting point of 179 °C. (b) In atactic polypropylene, the positions of the methyl groups are random, so atactic PP is not crystalline and has no true melting point.

SOLVED PROBLEM 26.20

Predict which polymer, **A** or **B**, has the higher T_g. Justify your choice.

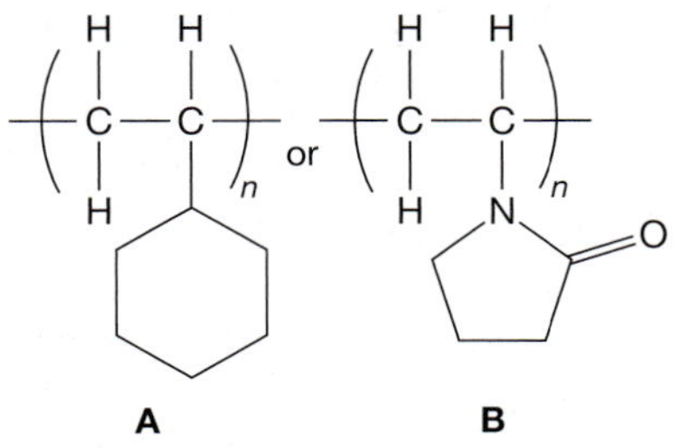

Think Does one polymer have a more rigid main chain than the other? Do the main chains participate in the same or different intermolecular forces? Do the pendant groups have functional groups that participate in the same or different intermolecular forces? If the intermolecular forces are different, for which polymer are the intermolecular forces stronger? Do the pendant groups extend away from the polymer chain similarly or significantly differently?

A   **B**

Solve Both **A** and **B** are carbon-chain polymers, consisting of only single bonds in their main chains, so both chains have similar rigidities and will participate in similar dispersion forces. The pendant groups are similar in size but differ in polarity. The pendant group in **A** is nonpolar, so it can participate in dispersion forces only, whereas the pendant group in **B** has a moderately polar O=C—N group, characteristic of an amide, which can participate in dipole–dipole interactions. This gives **B** stronger intermolecular interactions, which increases T_g. Therefore, we expect **B** to have the higher T_g.

PROBLEM 26.21 For each pair of polymers, predict which one has the higher T_g. Justify your choice.

(a)   (b)

PROBLEM 26.22 For each pair of polymers, predict which one has the higher T_g. Justify your choice.

G   or   H   I   or   J

(a)   (b)

26.8c Solubility

The solubility of polymers generally follows the same patterns that we discussed in Chapter 2. When the intermolecular interactions involving the solvent molecules are similar to those involving the polymer molecules, the polymer tends to dissolve. For example, polystyrene (Equation 26-1, p. 1308), a nonpolar aromatic polymer, dissolves in toluene ($C_6H_5CH_3$), a nonpolar aromatic solvent. Poly(vinyl alcohol) (Fig. 26-17, p. 1333) dissolves in water because its hydroxyl groups can hydrogen bond with water. Polymers with ionic side groups, such as poly(sodium acrylate) (Fig. 26-34b, p. 1348), dissolve even more readily in water due to ion–dipole interactions.

Any factors that limit the interactions between the solvent molecules and the polymer will decrease solubility. Network polymers (Section 26.5) are rarely soluble, for example, because the cross-links prevent sufficient interactions between the solvent and polymer. Linear poly(vinyl alcohol) is soluble in water and is used in water-based glues, but it can be used as a sponge when it is cross-linked. The cross-linked PVA (Section 26.6b) retains its ability to hydrogen bond with water, but is no longer soluble in water.

Highly crystalline polymers also have lower solubilities, because solvent–polymer interactions may not be strong enough to overcome polymer–polymer interactions. We saw this in the discussion of PEEK in Section 26.6a.2.

PROBLEM 26.23 Among the polymers shown in Solved Problem 26.20 and Problems 26.21 and 26.22, there are three that are water soluble and are used in pharmaceutical applications. Which polymers are they? Justify your choices.

26.9 Uses of Polymers: The Relationship between Structure and Function in Materials for Food Storage

Historically, food products were shipped in containers made of glass, metal, or ceramic. In an effort to improve shelf life, lower costs, and expand the range of foods that can be packaged, industry has turned to polymers for food packaging. One important property of materials for food storage is a lack of chemical reactivity—that is, you don't want to store food in a material that will react with it and affect its taste. The materials used for food storage need to be insoluble in water, too, and should show limited permeability toward water and air. You don't want the material to dissolve on the surface of the food, and you want to prevent spoilage. Consequently, polyolefins such as polyethylene and polypropylene, which are essentially really long alkanes, are logical choices for food storage.

26.9a Polyethylene

Polyethylene, the simplest polyolefin, comes in three types: low-density polyethylene (LDPE), linear low-density polyethylene (LLDPE), and high-density polyethylene (HDPE). LDPE is a flexible polymer that is used for packaging foods. As shown in Figure 26-39, it has a low T_g ($-120\,°C$) and a fairly low T_m ($\sim112\,°C$), properties that are attributed to its substantial branching (factor 3 in Section 26.8b).

In contrast to LDPE, HDPE is synthesized using a Ziegler–Natta catalyst (review Section 26.1d) that prevents branching, so only linear polymer is produced. Due to its increased crystallinity, HDPE has a higher T_m ($135\,°C$) and higher tensile strength than LDPE. This makes HDPE suitable for applications in which LDPE would fail, such as milk jugs and other containers for liquids (Fig. 26-40, next page).

Linear low-density polyethylene (LLDPE) combines properties of both LDPE and HDPE (Fig. 26-41, next page). LLDPE is made by copolymerizing ethylene and a small percentage of 1-alkenes such as but-1-ene. The copolymerization takes place with a catalyst that produces a linear polymer. The linearity of the polymer increases intermolecular forces, leading to a higher T_m ($T_m = 125\,°C$) and greater strength than that of LDPE. The alkyl pendant groups, however, keep the polymer chains apart, giving LLDPE a lower T_m than HDPE (factor 5 in Section 26.8b). LLDPE is used extensively in the manufacture of garbage bags, which need to be flexible and also need to have high tensile strength to prevent ripping.

26.9b Polystyrene

Whereas polyethylene (PE) is extremely flexible, polystyrene (PS) is not, due largely to the phenyl rings that increase the barrier to rotation around the main chain. Consequently (factor 6 in Section 26.8b), the T_g of PS ($100\,°C$) is much higher than that of PE ($-120\,°C$).

The mass of the phenyl groups makes PS denser ($1.05\ \text{g/cm}^3$) than HDPE ($0.94\ \text{g/cm}^3$). The rigidity and low cost of PS make it a good choice for a variety of applications, such as disposable razors (Fig. 26-42, next page). As we discussed earlier in the chapter, PS is also used in foams (Styrofoam) for insulation and for insulated food packaging such as coffee cups, "to-go" containers, and egg cartons.

PROBLEM 26.24 Why would HDPE be an inappropriate material for a disposable razor? Why is PS an inappropriate material for a milk jug?

Branching prevents crystallization and close packing of the polymer.

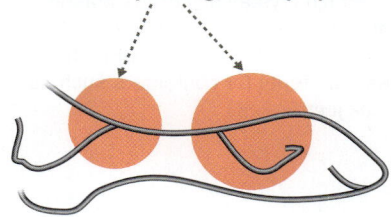

Low-density polyethylene (LDPE)

Density (g/cm³)	T_g (°C)	T_m (°C)
0.92	−120	112

FIGURE 26-39 Structure, properties, and use of LDPE The branching of the polymer prevents crystallization, which results in a relatively low density, low T_g, and low T_m. It also makes the plastic flexible, making it ideal for use as plastic wrap.

Highly linear polymer chains result in maximum crystallinity.

High-density polyethylene (HDPE)

Density (g/cm³)	T_g (°C)	T_m (°C)
0.94	~ −120	135

FIGURE 26-40 Structure, properties, and use of HDPE The polymer is relatively highly crystalline, so HDPE is more rigid than LDPE. The high crystallinity of the plastic also gives it a higher density and T_m than LDPE. This makes HDPE ideal as a material for containers such as milk jugs.

Regular branches from 1-alkenes that are copolymerized with ethylene

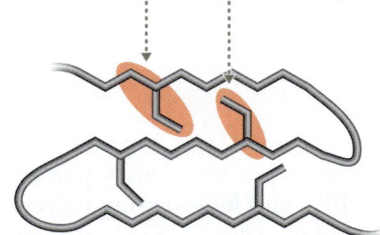

Linear low-density polyethylene (LLDPE)

Density (g/cm³)	T_g (°C)	T_m (°C)
0.92	~ −120	125

FIGURE 26-41 Structure, properties, and use of LLDPE The incorporation of 1-alkenes pushes the polymer chains apart, which leads to lower density than in HDPE. The absence of random branches of different lengths, however, leads to higher crystallinity, which is reflected in a melting point that is higher than that of LDPE. The regular branching of the polymer allows for more crystallinity in LLDPE and, consequently, a stronger plastic, making it ideal for use as garbage bags.

The phenyl groups hinder rotation of the polymer chain, which raises both T_g and T_m and creates a more rigid material.

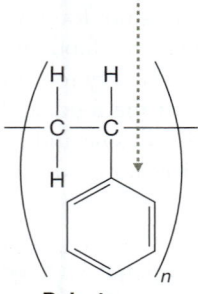

Polystyrene

Density (g/cm³)	T_g (°C)	T_m (°C)
1.05	100	—

FIGURE 26-42 Structure, properties, and uses of polystyrene The large pendant groups limit the flexibility of the polymer and make it rigid. Polystyrene is used to make disposable razors and is also used to make Styrofoam.

26.9c Poly(ethylene terephthalate)

Poly(ethylene terephthalate), PET, is the material used in bottles for carbonated beverages. Faced with the shipping costs of relatively heavy glass bottles and loss from breakage, manufacturers of carbonated beverages have found PET to be an effective replacement. However, an important requirement for a pop bottle is impermeability—that is, the carbon dioxide in carbonated beverages must be contained in the bottle at moderate pressures and not seep through the plastic bottle over time.

Plastics manufacturers have maximized the impermeability and strength of PET by maximizing its crystallinity (Fig. 26-43). When crystallinity is imposed on PET mechanically by stretching the material, the density of the plastic increases from 1.33 to 1.39 g/cm³. This compares to 0.94 g/cm³ for HDPE and 1.05 g/cm³ for PS. The high T_m of PET (260–265 °C) reflects its strength. PET has four times the tensile strength of HDPE and more than twice the tensile strength of PS.

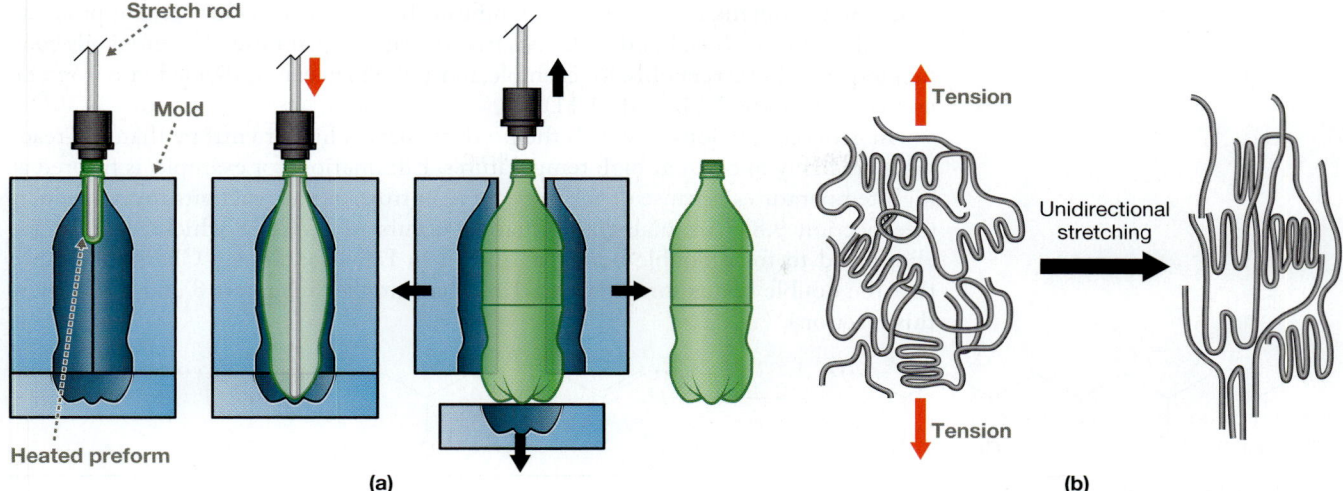

FIGURE 26-43 Molding of poly(ethylene terephthalate) bottles (a) Process for the manufacture of PET bottles. In the second stage, pressure is used to expand the PET, which stretches the polymer chains and facilitates crystallization. (b) Depiction of the crystallization that occurs when the PET is stretched.

26.10 Degradation and Depolymerization

Up to now, we have discussed ways of producing polymers from monomers. Under different circumstances, however, the ability to break a polymer into smaller units is also useful or, under certain conditions, unavoidable. A reaction that breaks a large molecule into smaller molecules is generally referred to as a **degradation reaction**. A reaction that causes polymer molecules to revert to monomer molecules is called a **depolymerization**.

The retro Diels–Alder reaction of dicyclopentadiene (see Section 24.6) is analogous to depolymerization, because the Diels–Alder adduct is a dimer that reverts to its constituent monomers. Recall that the retro Diels–Alder reaction is entropy driven and occurs at high temperatures. Consequently, dicyclopentadiene, which forms readily at room temperature, reverts to cyclopentadiene when heated to 200 °C (Equation 26-40).

A dimer, a molecule of two parts, made from two cyclopentadiene molecules — [structures] — 200 °C → — On heating, the dimer reverts to the monomers from which it was made. (26-40)

In 1860, Greville Williams (1829–1910) carried out a depolymerization of natural rubber by heating it in the absence of oxygen. The volatile product that he isolated turned out to be isoprene (Equation 26-41).

[structure] $\xrightarrow{\Delta}$ n [structure] (26-41)

Natural rubber   **Isoprene**

Although *cis*-polyisoprene (i.e., natural rubber) has been synthesized from isoprene, the rubber plant does not use isoprene to biosynthesize rubber. Nevertheless, the isolation of isoprene from rubber is significant because many other natural products are built from the same building blocks that the rubber plant uses. Essential oils contain terpenes and terpenoids, for example, most of which can be divided into isoprene units (see Sections 2.11c and 11.11).

In general, reactions in which the products have a higher entropy than the reactants are likely to occur at high temperatures. Elimination, for example, is favored at high temperatures because of the increase in entropy as the reactants form products (see Section 9.8). An analogous situation occurs in PVC, in which HCl can be eliminated to form double bonds, as shown in Equation 26-42. Chemists believe that the double bonds form via an E1 mechanism (see Chapters 8 and 9 to review this reaction).

$$\text{(26-42)}$$

Once double bonds begin to form, subsequent eliminations of HCl from the polymer chain occur more readily (see Problem 26.25), and the molecule "unzips" to form a polyene. PVC, therefore, poses a safety hazard when fires occur, because the HCl that

Plastic Made from Corn?

Tremendous excitement surrounded plastics throughout much of the 20th century because of their versatility, durability, and low cost of production. Plastics were being used for electrical and plumbing components, furniture, food storage, toys, automobile parts, and even clothing.

The benefits of plastics have come with severe downsides, too. Their durability means that they tend not to be biodegradable, so they pose serious environmental problems. Recycling helps, but cannot be the long-term solution because it is not very efficient and much of the energy it requires comes from burning fossil fuels, which adds large amounts of CO_2, a greenhouse gas, into the atmosphere. The low expense of producing plastics, moreover, is due to the fact that they are polymers whose monomers derive from petroleum sources, a resource that was cheap and seemingly endless in the early half of the 20th century. But petroleum is a nonrenewable resource, so making the same plastics from petroleum resources is unsustainable.

Fortunately, much progress has been made in recent decades developing polymers that are biodegradable and are synthesized from renewable resources. One such polymer is poly(lactic acid), PLA, which is made from corn! The most common synthesis is to begin with lactide, a dimer of lactic acid that is found in corn starch. Lactide is treated with a metal catalyst to carry out a ring-opening polymerization that yields PLA.

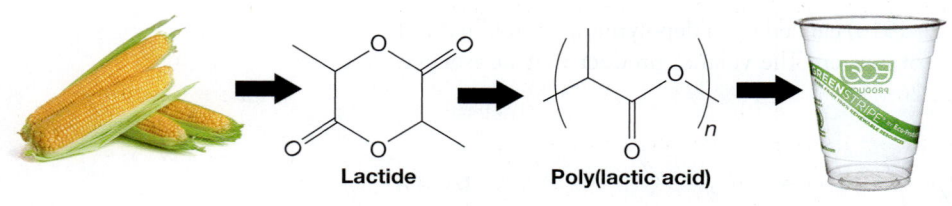

Lactide   **Poly(lactic acid)**

Even though PLA is biodegradable, it doesn't break down very quickly in a landfill, so PLA consumer products should not be thrown away as garbage. PLA does degrade rapidly, however, in the presence of certain bacteria or enzymes. And normal composting can completely degrade PLA in a matter of months.

is formed can be breathed in, leading to lung damage. This danger is offset by the addition of metal salts that react with HCl.

Polymer degradation is a major issue in polymer recycling. The heat and physical processing required to reform plastic also promote degradation reactions that reduce the desirable properties of plastics.

YOUR TURN **26.21**

Provide the mechanism for the reaction that forms the first double bond in the product in Equation 26-42.

PROBLEM 26.25 In the series of reactions shown in Equation 26-42, the second and subsequent double bonds form more quickly than the first double bond. Explain why this occurs.

THE ORGANIC CHEMISTRY OF BIOMOLECULES

26.11 Biological Macromolecules

Recall from Section 1.14 that proteins, carbohydrates (saccharides), and nucleic acids are constructed from relatively small molecular building blocks. Proteins are constructed from amino acids, carbohydrates from monosaccharides, and nucleic acids from nucleotides. Thus, proteins, carbohydrates, and nucleic acids can be thought of as **biopolymers** made up of amino acid, monosaccharide, and nucleotide monomers. More specifically, a protein is a **polyamino acid** or **polypeptide** (due to the peptide bond that links together amino acids), a large carbohydrate is a **polysaccharide**, and a nucleic acid is a **polynucleotide**.

In Section 9.13, we saw how bonds between monosaccharides—glycosidic linkages—can be formed and broken, and in Sections 21.12 and 21.13 we discussed the formation and breaking of bonds that connect amino acids, so-called peptide linkages. Here in Section 26.11, we turn our attention to the macroscopic properties of these biopolymers. Just as we saw with synthetic polymers throughout Chapter 26, the properties of biopolymers are governed by the nature of their monomers. We discuss examples of how this is so with polypeptides in Section 26.11a and with polysaccharides in Section 26.11b.

26.11a Polypeptides: Primary, Secondary, Tertiary, and Quaternary Structures

The specific function of a protein is governed by the four levels of its structure, which are designated as *primary*, *secondary*, *tertiary*, and *quaternary*. We examine each of these levels of structure in order.

26.11a.1 Primary Structure

The *primary structure* of a protein is its specific sequence of amino acids.

A protein is *characterized* by its primary structure; different proteins have different primary structures.

Recall from Section 21.12 that the amide functional group is responsible for connecting adjacent amino acids along a protein's main chain, and each of those amide groups

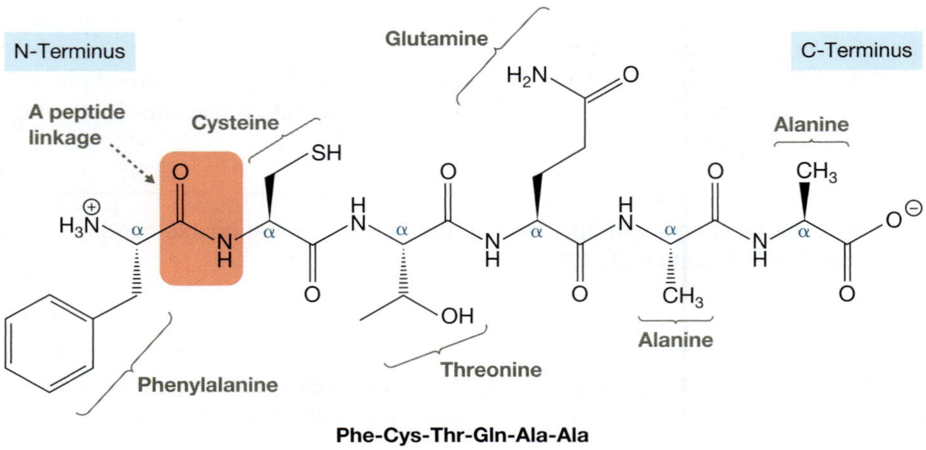

FIGURE 26-44 **Primary structure** In this hexapeptide, the peptide linkage that connects the first and second amino acids is highlighted in red. The side groups on the α carbons identify the amino acids, so the primary structure, which is read from the N-terminus to the C-terminus, is Phe-Cys-Thr-Gln-Ala-Ala.

Phe-Cys-Thr-Gln-Ala-Ala

is called a *peptide linkage* or *peptide bond* (highlighted in red in Fig. 26-44). When connected by a peptide linkage, the amino acids are referred to as *residues*. The side groups attached to the amino acids' α carbons typically remain intact in a residue, so the protein's primary sequence can be "read" by identifying the side groups in order from the N-terminus to the C-terminus. In the hexapeptide in Figure 26-44, for example, the primary sequence is Phe-Cys-Thr-Gln-Ala-Ala.

YOUR TURN 26.22

Identify all of the peptide linkages in the hexapeptide in Figure 26-44. How many are there in all?

Some amino acids can form *cross-links*, whereby one portion of the protein's backbone is covalently bonded to another portion of the backbone (review Section 26.6b). This occurs when the thiol (–SH) groups of two cysteine amino acids are oxidized, creating a **disulfide bond** (also called a **disulfide bridge**), as shown in Equation 26-43.

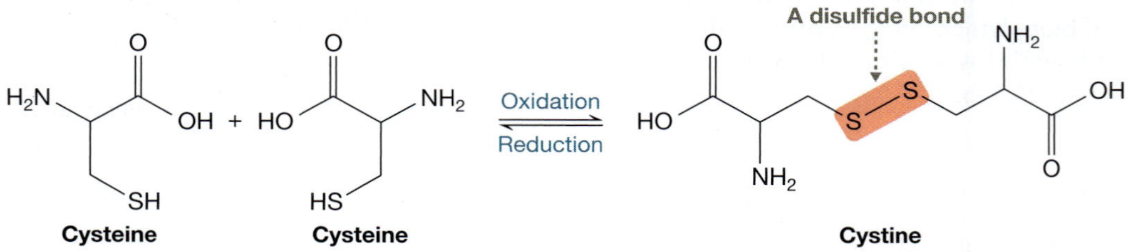

(26-43)

Vasopressin, a hormone protein that is responsible for the body's retention of water and is involved in the fight or flight response of mammals, has a disulfide bond. The sequence of amino acids that characterizes vasopressin is Cys-Tyr-Phe-Gln-Asn-Cys-Pro-Arg-Gly-NH₂. When oxidized, the two cysteine amino acids form a disulfide bond, giving rise to the structure in Figure 26-45.

SOLVED PROBLEM 26.26

Determine how many different tripeptides can be constructed from the naturally occurring amino acids.

Think What feature characterizes a polypeptide? How many possible amino acids could be located at the first position in the tripeptide? How many possibilities for the second position? The third? Does the identity of the amino acid at the first position cause any restrictions for the possible amino acids at the second or third positions?

Disulfide bond

Cys-Tyr-Phe-Gln-Asn-Cys-Pro-Arg-Gly-NH₂

(a)

Disulfide bond

(b)

Solve The primary structure characterizes a polypeptide. There are 20 naturally occurring amino acids that could exist at the first position of the tripeptide. The identity of the amino acid at that position causes no restriction on which amino acids could occupy the second position, so for each possible amino acid occupying the first position, there are 20 possible choices for the second position, giving $20 \times 20 = 20^2 = 400$ possible combinations in the first two positions. To take into account the third position, we multiply by 20 again, giving $20^3 = 8000$ possible combinations for a tripeptide.

PROBLEM 26.27 Determine how many different polypeptides 300 amino acids long (a modest number for a protein) can be constructed from the naturally occurring amino acids. How does that number compare to Avogadro's number?

PROBLEM 26.28 Glutathione is an antioxidant that helps prevent cell damage from free radicals. Under oxidizing conditions, glutathione forms a dimer via a disulfide bond. Draw the structure of that dimer.

Glutathione

26.11a.2 Secondary Structures

The secondary structure of a protein is the three-dimensional structural *pattern* of specific portions of the protein's backbone. The two most common secondary structures are the α-helix and the β-pleated sheet, shown in Figure 26-46a and 26-46b (next page), respectively.

An **α-helix** has the protein's backbone wound tightly into a spiral. This arrangement allows for efficient hydrogen bonding among amino acids that are four residues apart in the sequence: the NH group from one amino acid and the carbonyl O atom from the other. Moreover, the side groups of the amino acids in an α-helix project outward, maximizing the distances between them.

Whereas each α-helix involves a single portion of a protein's backbone, a **β-pleated sheet** (often called a *β-sheet*) is made from several portions of the backbone that align next to each other in the same region of space. These portions of the backbone can be many amino acids apart in the primary structure. As with the α-helix, a β-pleated sheet is held together by hydrogen bonds, each involving an NH group and a carbonyl O atom from different amino acids.

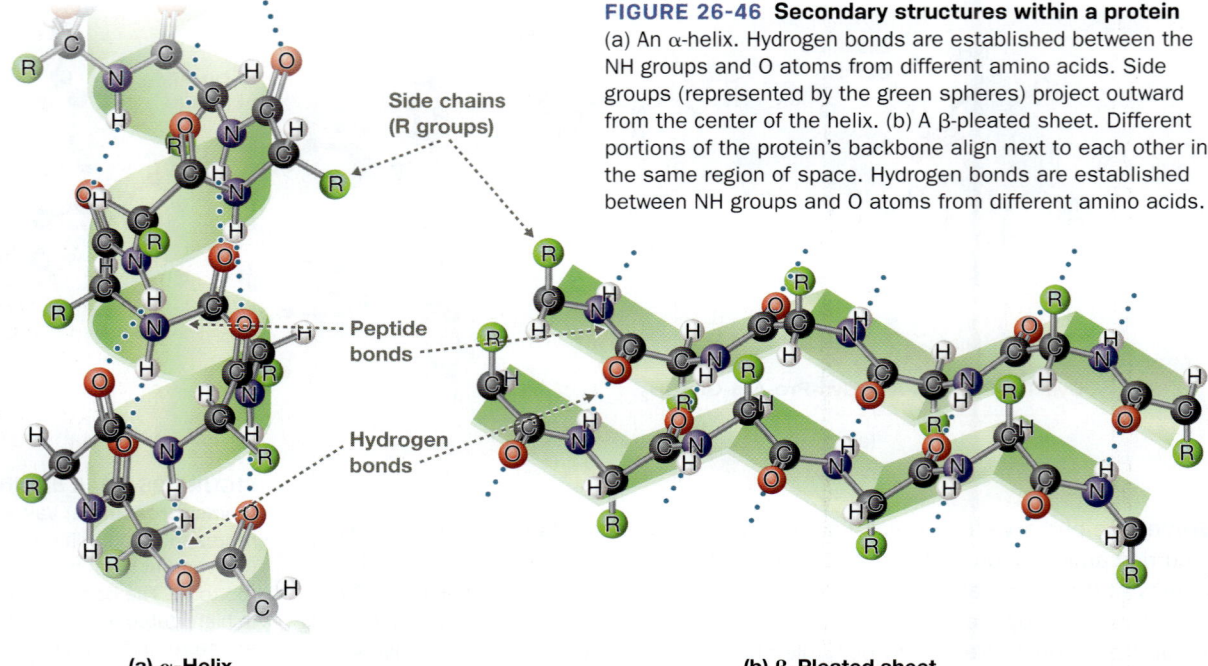

FIGURE 26-46 **Secondary structures within a protein**
(a) An α-helix. Hydrogen bonds are established between the NH groups and O atoms from different amino acids. Side groups (represented by the green spheres) project outward from the center of the helix. (b) A β-pleated sheet. Different portions of the protein's backbone align next to each other in the same region of space. Hydrogen bonds are established between NH groups and O atoms from different amino acids.

Side chains (R groups)

Peptide bonds

Hydrogen bonds

(a) α-Helix

(b) β-Pleated sheet

A kink between each peptide linkage establishes the "pleats."

Amide plane   Amide plane

Side chain

(a)

This contributor gives the resonance hybrid a planar structure.

(b)

FIGURE 26-47 Planarity of a peptide bond (a) In a β-pleated sheet, the planarity of each peptide linkage is maintained, and the rotation of one plane relative to the other establishes the kink that gives rise to the "pleats." (b) Resonance structures of an amide, showing the double bond character of the C—N bond.

The pleats—or folds—in a β-pleated sheet arise because each peptide linkage is planar, and adjacent planes are slightly rotated relative to each other, as shown in Figure 26-47a. The planarity of a peptide linkage results from the contribution of a resonance structure that gives the C—N bond significant double bond character, as shown in Figure 26-47b.

The nature of the amino acid side groups determines whether or not an α-helix or β-pleated sheet is favored.

- An α-helix tends to be disrupted when adjacent or nearby amino acids in a protein's sequence (within about three or four residues) have side groups that are either bulky or carry the same charge (remember, like charges repel).

- A β-pleated sheet tends to be disrupted if it requires bulky or like-charged side groups from different segments of the primary structure to occupy similar locations in space.

PROBLEM 26.29 Suggest whether alanine or isoleucine should be found more commonly in an α-helix. Explain why. (Consult Table 1-7 on p. 39 for structures.)

26.11a.3 Tertiary Structures

Even after a protein has been synthesized with a definite primary structure and segments of the protein have adopted specific secondary structures, the protein may still lack proper biological activity. That's because the function of a protein depends critically on its **tertiary structure**, which is the location of all of its atoms in space. Achieving the proper tertiary structure requires the protein to *fold* in a particular way.

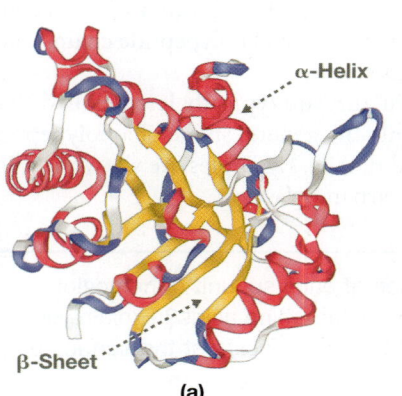

α-Helix

β-Sheet

(a)

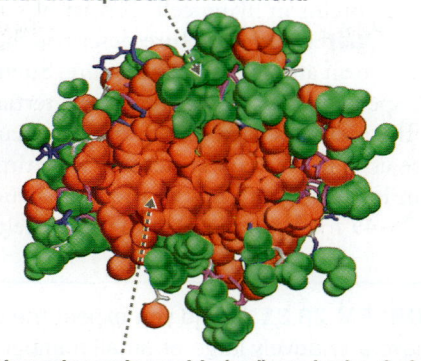

Polar amino acids (green) interact favorably with the aqueous environment.

Nonpolar amino acids (red) are hydrophobic and tend to be buried inside the protein.

(b)

FIGURE 26-48 Tertiary structure of phosphotriose isomerase
(a) Ribbon structure of phosphotriose isomerase. Magenta indicates an α-helix and gold indicates a β-sheet. (b) Cross section of a space-filling model of phosphotriose isomerase. Red indicates nonpolar amino acids and green indicates polar amino acids.

A **ribbon structure** is typically used to depict the tertiary structure of a protein, as shown in Figure 26-48a for phosphotriose isomerase, an enzyme used in glycolysis to catalyze the isomerization of three-carbon sugars with an attached phosphate group (see the box on p. 354).

In a ribbon structure, a trace of the backbone is shown explicitly, and the side groups are generally omitted. Regions of secondary structure are represented by helices (for α-helices) and by arrows (for β-sheets), where the arrows point in the direction from the N-terminal amino acid to the C-terminal amino acid. Secondary structures are further highlighted using a consistent color scheme. In Figure 26-48a, for example, the α-helices are indicated using magenta and the β-sheets are indicated using gold.

The advantage of using a ribbon structure can be seen by comparing Figure 26-48a to Figure 26-48b, which is a cross section of phosphotriose isomerase in which all atoms are shown explicitly as a space-filling model. The space-filling model makes it difficult to distinguish the secondary structures from the rest of the enzyme.

The tertiary structure of a protein tends to be the one that gives the lowest possible free energy, *G*. Factors such as intermolecular forces and steric crowding among amino acids contribute to a protein's overall free energy.

Another very important factor is the **hydrophobic effect**. Proteins typically reside in an aqueous environment, so amino acids that are polar—such as serine and aspartic acid (see Section 1.14a)—are *hydrophilic* and tend to reside on the exterior of a protein, whereas nonpolar amino acids—such as valine and phenylalanine—are *hydrophobic* and tend to reside in the protein's interior. This is shown in Figure 26-48b.

The hydrophobic effect is largely driven by entropy. Hydrophobic amino acids at the surface of a protein have very weak interactions with the surrounding water molecules, in which case the water molecules at the protein–water interface would favor strong hydrogen bonding among themselves. The water molecules would become highly ordered, causing an unfavorable decrease in entropy. Hydrophilic amino acids on the protein's surface, on the other hand, interact favorably with the water molecules, so the substantial decrease in water's entropy doesn't occur.

PROBLEM 26.30 For some analyses, proteins are first denatured (i.e., unfolded) by treatment with a detergent called sodium dodecyl sulfate (SDS). Explain why detergents like SDS have this effect. *Hint:* See Figure 2-29 (p. 105) for the structure of SDS.

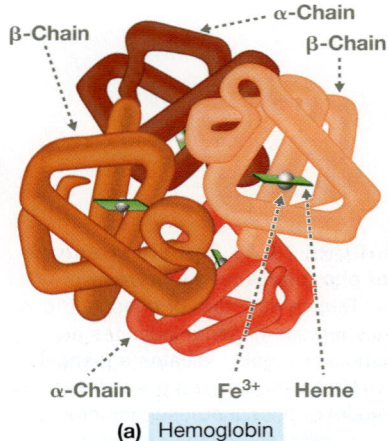

β-Chain

α-Chain

β-Chain

α-Chain   Fe³⁺   Heme

(a) Hemoglobin

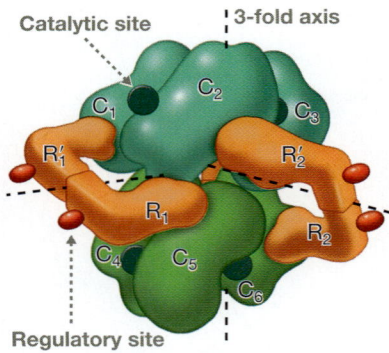

Catalytic site   3-fold axis

C_1 C_2 C_3

R_1' R_2'

R_1

R_2

C_4 C_5

C_6

Regulatory site

(b) Aspartate transcarbamoylase

FIGURE 26-49 Quaternary structures The quaternary structure of (a) hemoglobin and (b) aspartate transcarbamoylase (ATCase). Four distinct peptide chains are associated in hemoglobin (two α and two β, each chain with a Fe^{3+}–heme group), whereas ATCase contains ten peptide chains with catalytic (C_1 to C_6) or regulatory (R_1, R_1', R_2, and R_2') roles.

26.11a.4 Quaternary Structures

Many proteins in living cells consist of multiple, distinct peptide chains, not just one chain. **Quaternary structure** describes the way in which these protein *subunits* combine to form an even larger structure. Subunits are individual polypeptide chains, each with specific primary, secondary, and tertiary structures.

The methods used to depict quaternary structure are typically less detailed than those used for tertiary structure. Sometimes only the general shape of a polypeptide chain is drawn, with few of the finer structural details. Examples are shown in Figure 26-49 for hemoglobin and aspartate transcarbamoylase.

PROBLEM 26.31 Would you expect the surface of each subunit of hemoglobin to have a relatively large or small number of nonpolar amino acids? Explain your reasoning. *Hint:* Consider the interactions that the amino acids at the surface of the subunits have with water.

26.11b Polysaccharides

Some polysaccharides, such as *starch* and *glycogen*, are used to store energy, and are called **storage polysaccharides**. Others, such as cellulose, are responsible for maintaining the structure of organisms, and are called **structural polysaccharides**.

Starch, a main component of foods such as rice, potatoes, corn, and wheat, is actually a mixture of two different polysaccharides—namely, *amylose* (~20%) and *amylopectin* (~80%).

Amylose consists of unbranched chains of D-glucose, which, as shown in Figure 26-50, are connected by α-1,4′-glycosidic linkages (review Section 9.13). That is, the glycosidic linkage involves C1 of one glucose unit and C4 of the next, and the substituent on C1 is axial. These glycosidic linkages are hydrolyzed readily by α-glucosidase, an enzyme that is found in every known mammal. Hydrolysis of a glycosidic linkage releases a molecule of glucose, which is subsequently broken down further for energy.

The α configuration of the anomeric carbon also has implications for amylose's physical properties. Because of the α configuration, amylose favors a helical form, as shown in Figure 26-51a, which is reminiscent of the α-helix adopted by proteins. Based on the cross section in Figure 26-51b, there are numerous hydroxyl groups on the exterior of the helix, where they are available for extensive hydrogen bonding with water. Amylose, therefore, is highly soluble in water.

Amylopectin is very similar to amylose, having a main chain that consists of glucose units joined by α-1,4′-glycosidic linkages. The essential difference between the two is that amylopectin's chain is relatively highly branched. Branching takes place via α-1,6′-glycosidic linkages (Fig. 26-52), which occur roughly every 25–30 glucose units. As a result of this branching, there are more terminal glycosidic linkages, so α-glucosidase is capable of releasing glucose units much faster. This is beneficial when energy is needed moderately quickly.

In animals, excess glucose is stored in the form of *glycogen*, which is structurally very similar to amylopectin. The glucose units in glycogen are connected primarily by α-1,4′-glycosidic linkages to establish a main chain, and branching occurs via α-1,6′-glycosidic linkages. In glycogen, branching occurs even more frequently than in

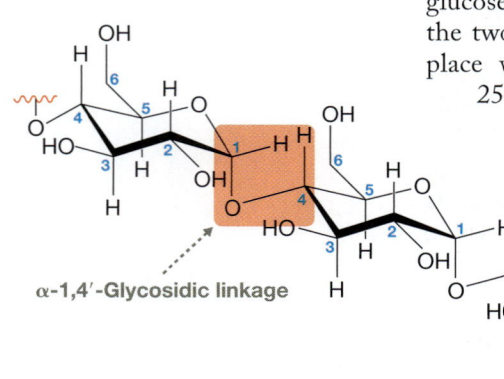

FIGURE 26-50 Partial structure of amylose Amylose consists of a large number of D-glucose units that are connected by α-1,4′-glycosidic linkages. One glycosidic linkage is highlighted in red, and the numbering systems for the rings are shown in blue.

α-1,4′-Glycosidic linkage

Extensive hydrogen bonding with water
(a)
(b)

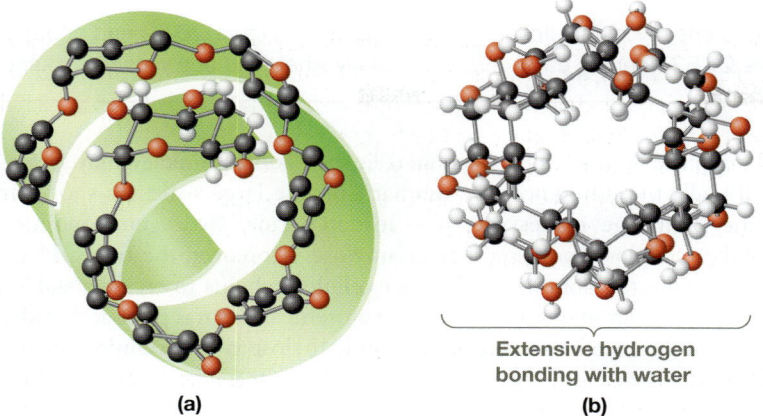

FIGURE 26-51 The helical structure of amylose (a) Representation of amylose as a helix. Most H atoms have been omitted. (b) Cross section of amylose, showing that the exterior of the helix has numerous OH groups available for hydrogen bonding with water.

amylopectin—roughly every 10 glucose units—allowing hydrolysis to take place even more quickly. This is important for organisms that need quick bursts of energy.

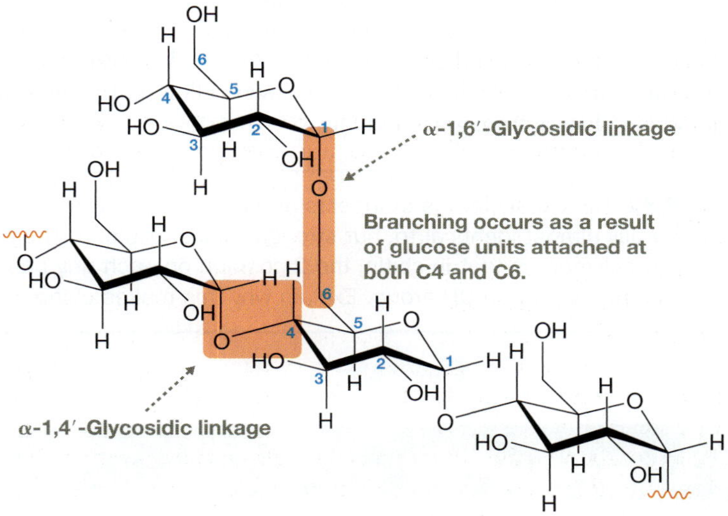

α-1,6′-Glycosidic linkage
Branching occurs as a result of glucose units attached at both C4 and C6.
α-1,4′-Glycosidic linkage

FIGURE 26-52 Partial structure of amylopectin Glucose units in the main chain are joined by α-1,4′-glycosidic linkages, whereas glucose units at the branch points are joined by α-1,6′-glycosidic linkages.

Cellulose, a structural polysaccharide, is the main component of plants' cell walls, making up about one-third of all plant matter. The specific cellulose content of a plant, however, depends on the specific nature of the plant. Wood, for example, is about 50% cellulose, whereas cotton fiber is about 90% cellulose.

Cellulose consists of long, unbranched chains of D-glucose. How, then, is cellulose different from amylose? The glucose units in amylose are connected by α-1,4′-glycosidic linkages, whereas the glucose units in cellulose are connected by β-1,4′-glycosidic linkages—that is, the substituent attached to C1 is equatorial in each unit, as shown in Figure 26-53.

This glucose unit is inverted relative to the adjacent ones.
β-1,4′-Glycosidic linkage
Internal hydrogen bonds add rigidity to the molecule.

FIGURE 26-53 Partial structure of cellulose Glucose units are connected by β-1,4′-glycosidic linkages.

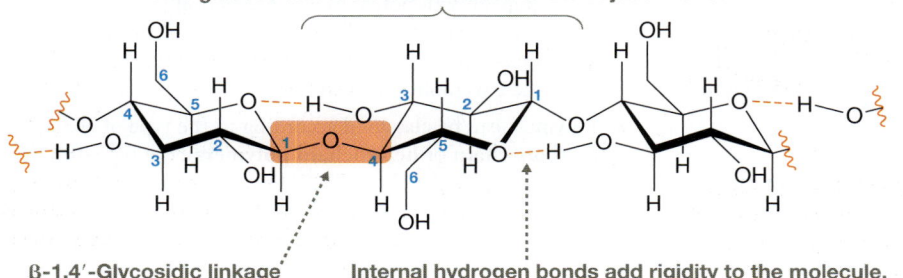

> There is one glycosidic linkage not highlighted in Figure 26-52 and another in Figure 26-53. Identify each one and classify what kind of glycosidic linkage it is.

Although the difference between an α linkage and a β linkage may seem slight on the small scale, the differences are dramatic on the large scale. As shown in Figure 26-53, the β linkage makes extensive intramolecular hydrogen bonding possible, which makes cellulose more rigid than amylose. Moreover, the free OH groups are situated for hydrogen bonding with other cellulose chains to create a stable polymer network. This is what makes cellulose such a useful structural polysaccharide.

Another difference between cellulose and amylose is that cellulose is insoluble in water. Whereas the helical shape of amylose allows its OH groups to remain free for hydrogen bonding with water, the OH groups in cellulose are tied up in hydrogen bonding with adjacent glucose units of the same chain, or with OH groups from a separate chain. Without substantial hydrogen bonding with water, there are not enough favorable interactions to make cellulose soluble.

Yet another difference between cellulose and amylose is that mammals cannot digest cellulose directly. Mammals lack β-glucosidase, the enzyme required for hydrolyzing the β-1,4′-glycosidic linkage of cellulose. In the digestive tracts of some animals, however, such as horses and cows, there exist certain bacteria that do have the enzyme and, therefore, are capable of breaking down cellulose. In this way, grazing animals can obtain glucose from grass and hay.

PROBLEM 26.32 The exoskeletons of insects and arthropods incorporate chitin, which is structurally almost identical to, but stronger than, cellulose. The difference between the two polymers is that in chitin, the 2-position on each ring has an $-NHCOCH_3$ group instead of an OH group. Explain why this makes chitin stronger.

Chapter Summary and Key Terms

- **Polymers** are large molecules made from smaller molecules called **monomers**; they contain a **main chain** and often have **pendant groups**. Polymers are usually depicted by showing their **repeating units** in a condensed formula, because the chains are so large that a full structure is impractical to draw. **(Section 26.1a; Objective 1)**

- The number of monomers making up a polymer's chain is the **degree of polymerization (DP)**. **(Section 26.1a; Objective 1)**

- Numerous polymers are made by **free radical polymerization**, involving *initiation*, *propagation*, and *termination* steps. Termination can occur by **combination** or **disproportionation**. Most monomers add in a **head-to-tail** fashion during propagation. **(Sections 26.1b and 26.1c; Objective 2)**

- **Tacticity** describes the regularity in a polymer's stereochemical configurations. The configurations are all the same in an **isotactic** polymer, and they alternate in a **syndiotactic** polymer. In an **atactic** polymer, there is no regular pattern. **(Section 26.1d; Objective 2)**

- Polymers are generally mixtures of different sizes but can be characterized by an average molar mass or average degree of polymerization. **(Section 26.1e; Objective 3)**

- **Anionic polymerization** and **cationic polymerization** proceed through mechanisms that involve negatively and positively charged intermediates, respectively. **(Section 26.2; Objective 4)**

- In a **ring-opening polymerization**, cyclic monomers produce an acyclic polymer. **(Section 26.3; Objective 4)**

- Free radical polymerization is an example of **chain-growth polymerization**, one general class of polymerization reactions. The other class of reactions is **step-growth polymerization**. **(Sections 26.1b and 26.4; Objective 5)**

  - In chain-growth polymerization, reaction occurs at the end of the growing chain, and each reaction lengthens the polymer chain by a single repeating unit.
  - In step-growth polymerization, reaction occurs between the ends of molecules—either monomers or polymer chains—and the increase in chain size depends on the size of the two molecules that react.

- Polymers can be classified according to the type of chain (**carbon-chain** or **heterochain polymers**), the type of functional group in the chain, the type of monomer used to make the polymer, and the number of different kinds of monomers used to make the polymer (**homopolymers** vs. **copolymers**).

Polymers can also be classified by the types of chain (**linear** vs. **branched polymers**) and the degree of cross-linking (**cross-linked** and **network polymers**). (**Sections 26.5, 26.6b, and 26.7a; Objective 6**)

- Polymers can undergo chemical reactions after polymerization is terminated. Most involve modifications to the pendant groups, an important one being **cross-linking**, in which polymer chains are joined together. (**Sections 26.6; Objective 7**)
- Names for polymers can be derived from the repeating unit or from the monomer used to create the polymer. (**Section 26.7b; Objective 8**)
- The **melting point (T_m)** and **glass transition temperature (T_g)** are thermal properties that quantify the crystalline and amorphous character of a polymer. (**Section 26.8a; Objective 9**)
- Thermal transitions and the solubility properties of a polymer depend on the length of the polymer molecules, the extent of branching, the rigidity of the polymer, the symmetry of the polymer, and the intermolecular forces. (**Sections 26.8b and 26.8c; Objective 9**)
- The structure of a polymer determines how it can function—that is, its properties dictate its uses. (**Objective 10**)
  - Polymers with low T_g, such as LDPE, are flexible and can be used as packaging films. (**Section 26.9a**)
  - Polymers with higher T_g are more rigid and are used for objects that need to retain their shapes. Polystyrene, for example, can be used for coffee cups and "to-go" boxes. (**Section 26.9b**)
  - Introducing crystallinity increases rigidity and density, often with retention of flexibility, so HDPE is suitable for milk jugs and PET is suitable for carbonated beverage bottles. (**Sections 26.9a and 26.9c**)
- At high temperatures, depolymerization (the reverse of polymerization) and degradation (the loss of atoms from the polymer) are thermodynamically favorable because of the higher entropy of the products. (**Section 26.10; Objective 11**)

Problems

26.1 Free Radical Polymerization: Polystyrene as a Model

26.33 Each of the monomers shown here can undergo free radical polymerization. For each polymerization:
  (i) Show the mechanism for the first two propagation steps, using benzoyl peroxide as the initiator.
  (ii) Draw the condensed formula for the polymer, showing the repeating unit.
  (iii) Provide a name for the polymer.

(a) 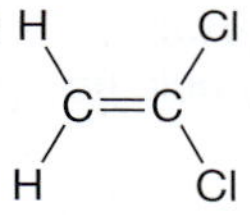 Vinylidene chloride (b) 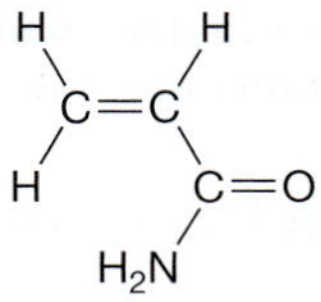 Acrylamide (c) 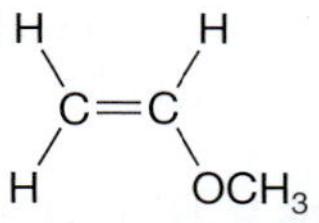 Methyl vinyl ether

26.34 Using each of the following polymers,

(a) 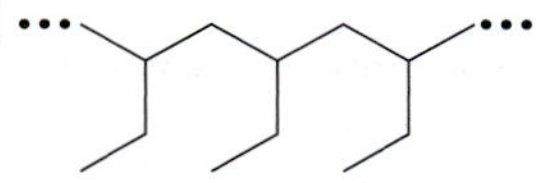   (b) 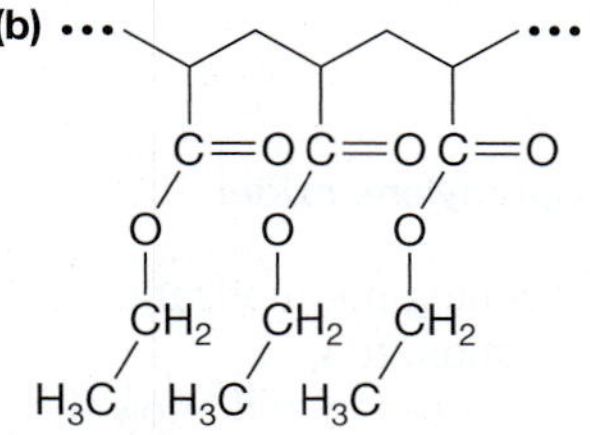   (c)

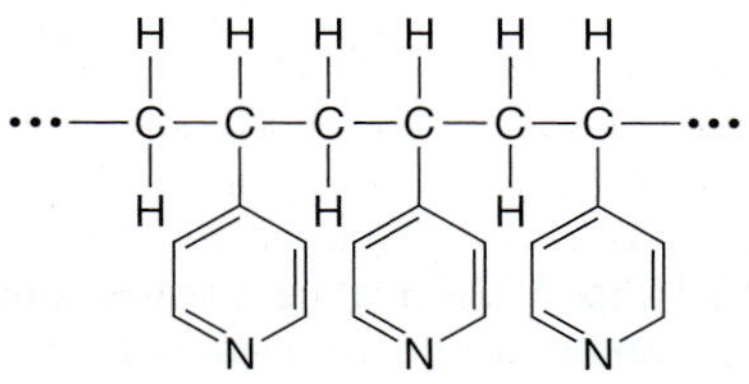

  (i) Draw the condensed formula for the polymer.
  (ii) Propose a structure for the monomer from which the polymer was synthesized.

26.35 Poly(N-vinylpyrrolidone), whose repeating unit is shown in the chapter opener on page 1307, can be synthesized by heating the monomer in the presence of a small amount of H_2O_2. **(a)** Draw the structure of N-vinylpyrrolidone. **(b)** Draw the initiation step and the first two propagation steps of this polymerization mechanism.

26.36 Teflon can be made from tetrafluoroethylene by means of a free radical polymerization. In the reaction, termination occurs only by combination and not by disproportionation. Explain this observation.

26.37 Propylene does not undergo free radical polymerization readily because there are two competing steps after initiation: propagation and hydrogen atom abstraction.
  (a) Using a generic radical R• as a reactant with propylene, draw the mechanism and products for the two competing steps.
  (b) Which step produces the more stable product?
  (c) How do your results explain propylene's poor reactivity in free radical polymerization?

26.2 Anionic and Cationic Polymerization Reactions

26.38 **(a)** Which monomer is most likely to undergo anionic polymerization? Justify your choice. **(b)** Which one is most likely to undergo cationic polymerization? Justify your choice.

Ethyl vinyl ether   **But-1-ene**   **Nitroethylene**

26.39 *N*-Vinylcarbazole undergoes cationic polymerization to produce poly(*N*-vinylcarbazole), or PVK, which is a photoconducting material used in laser printers (see the box on p. 572). Explain why the monomer readily undergoes cationic polymerization.

26.40 Do you think that 4-methoxystyrene will more easily undergo anionic or cationic polymerization? Explain your answer.

4-Methoxystyrene

26.41 For the polymerization you chose in Problem 26.40,
  (a) Propose a mechanism for the initiation step, using either a generic Brønsted acid (HA) or a generic nucleophile (Nu:$^-$) as the initiator.
  (b) Propose a mechanism for the first two propagation steps.
  (c) Draw the repeating unit for poly(4-methoxystyrene).

26.42 Cyanoacrylates like methyl 2-cyanoacrylate are used to make adhesives such as superglue. They readily undergo polymerization with an initiator such as hydroxide anion or an alkoxide anion.
  (a) Provide a mechanism for the initiation step.
  (b) Propose a mechanism for the first step of propagation.
  (c) Draw the structure of poly(methyl 2-cyanoacrylate).
  (d) Why is methyl 2-cyanoacrylate a better candidate for anionic polymerization than methyl acrylate?
  (e) Do you think methyl 2-cyanoacrylate would be a good candidate for cationic polymerization? Why or why not?

Methyl 2-cyanoacrylate

26.3 Ring-Opening Polymerization Reactions

26.43 **(a)** Show how poly(propylene oxide) can be synthesized via anionic ring-opening polymerization. **(b)** Draw the mechanism for the initiation and first two propagation steps.

Poly(propylene oxide)

26.44 Problem 26.10 (p. 1324) highlighted the cationic ring-opening polymerization of oxetane, producing polyoxetane. Polyoxetane can be synthesized via an anionic polymerization, too.
  (a) Propose a synthesis of polyoxetane via anionic polymerization and show the mechanism for the first two propagation steps.
  (b) Draw a condensed formula for polyoxetane.
  (c) Do you expect polyoxetane to be more or less soluble in water than poly(ethylene oxide)? Justify your answer.
  (d) Do you expect polyoxetane to have a higher or lower melting point than poly(ethylene oxide)? Justify your answer.

Oxetane

26.45 Lactones (cyclic esters) can undergo anionic ring-opening polymerization when treated with an alkoxide anion, as shown here. Draw the mechanism for the initiation and first two propagation steps for this polymerization, and draw the repeating unit of the resulting polymer.

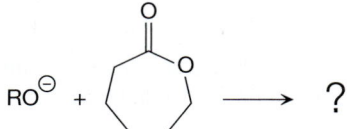

26.46 Lactams (cyclic amides) can undergo anionic ring-opening polymerization when treated with a small amount of a strong base such as NaH, as shown here. Draw the mechanism for the initiation and first two propagation steps for this polymerization, and draw the repeating unit of the resulting polymer.

26.47 A ketene acetal can undergo free radical ring-opening polymerization, as shown below. After the initial radical, R•, is produced, the catalytic cycle that adds each monomer to the growing chain consists of two steps. Draw the mechanism for those two steps and draw the repeating unit for the resulting polymer.

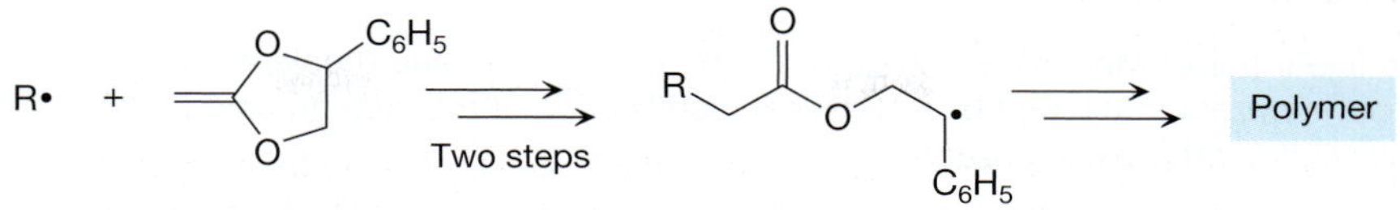

26.4 Step-Growth Polymerization

26.48 Propose a synthesis for each of the following polymers.

(a)

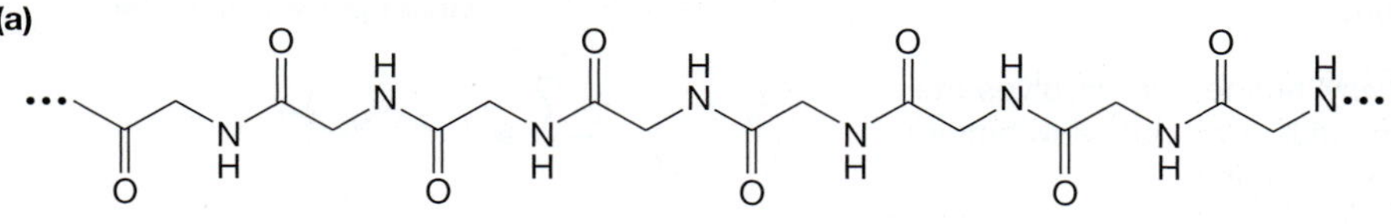

Polyglycine

(b)

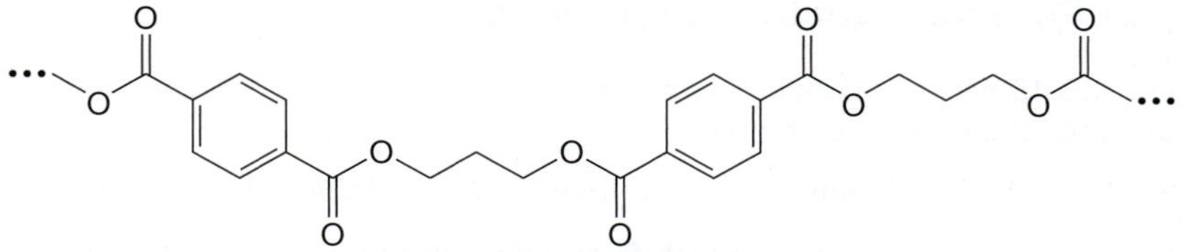

Poly(trimethylene terephthalate)

26.49 Nylon-6,6 (see Equation 26-24, p. 1325) can also be synthesized from adipoyl chloride and hexane-1,6-diamine, as shown below. What are the two products of this reaction?

$$Cl-\overset{\overset{\displaystyle O}{\|}}{C}CH_2CH_2CH_2CH_2\overset{\overset{\displaystyle O}{\|}}{C}-Cl \;+\; H_2NCH_2CH_2CH_2CH_2CH_2CH_2NH_2 \longrightarrow \;?$$

26.5 and 26.6 Linear, Branched, and Network Polymers; Chemical Reactions after Polymerization

26.50 When polyethylene is made by free radical polymerization, the resulting polymer has a significant amount of branching, resulting in low-density polyethylene (see Section 26.9a). When styrene undergoes free radical polymerization, on the other hand, very little branching occurs. Explain this phenomenon.

26.51 When treated with acid, aziridine will undergo ring-opening polymerization to produce branched polyethylenimine. Draw the mechanism that shows how a tetramer can be produced, in which three monomers form a chain and the fourth forms a branch.

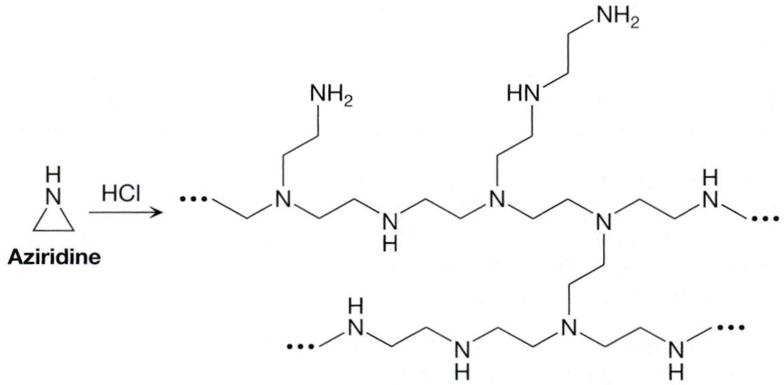

Branched polyethylenimine

26.52 Poly(ethylene terephthalate) is a linear polyester that can be synthesized from the condensation polymerization between terephthalic acid and ethylene glycol, as shown previously in Equation 26-27. What alcohol would you react with terephthalic acid to produce a *branched* polyester? Draw a portion of the resulting polymer that shows a main chain and one branching point.

26.53 To synthesize linear polyethylenimine instead of the branched form (see Problem 26.51), 2-ethyl-2-oxazoline can undergo ring-opening polymerization followed by a chemical modification of the side groups. Propose how to carry out the chemical modification after polymerization.

2-Ethyl-2-oxazoline   **Linear polyethylenimine**

26.54 Propose a synthesis of poly(vinyl amine) that involves the polymerization of *N*-vinyl formamide. Can poly(vinyl amine) be produced by polymerizing vinyl amine ($H_2C\!=\!CH\!-\!NH_2$)? Explain.

N-Vinyl formamide   **Poly(vinyl amine)**

26.7 and 26.8 General Aspects of Polymer Structure; Properties of Polymers

26.55 Classify the molecules in Problems 26.43 and 26.48 as either carbon-chain or heterochain polymers. Identify the repeating unit in each polymer and draw the condensed formula.

26.56 Identify the class of polymer, based on functional groups, for the polymers in Problems 26.43 and 26.48.

26.57 Consider the following series of poly(phenylene oxide) polymers **A–D**, and their glass transition temperatures. Explain how the differences in structure can account for the differences in T_g.

A	**B**	**C**	**D**
$T_g = 82$ °C	$T_g = 211$ °C	$T_g = 169$ °C	$T_g = 99$ °C

26.58 Consider polycarbonate polymers **E–G**. **(a)** Which one has the highest T_g? **(b)** Which one has the lowest T_g?

    E   **F**

G

26.59 Which polymer do you think has the higher T_g, Kevlar (see Solved Problem 26.11) or the polyphthalamide described in Problem 26.12 (p. 1328)? Explain your reasoning.

26.60 Poly(vinyl alcohol), PVA, is produced from poly(vinyl acetate) by converting the acetate groups into hydroxyl groups (Section 26.6a.1). As the number of hydroxyl groups in PVA increases, the solubility of the polymer in water decreases. How do you explain this behavior?

26.61 Which polymer in Problem 26.34 (p. 1363) is the most soluble in ethanol? Justify your choice.

26.62 Which polymer in Problem 26.34 is the most soluble in hexane? Justify your choice.

26.63 One method for the purification of polystyrene samples is to dissolve the polymer in benzene and then add methanol slowly to precipitate solid polystyrene. Explain why polystyrene precipitates under these conditions.

26.64 In the discussion of poly(ether ether ketone), or PEEK, in Section 26.6a.2, we observed that the addition of a *tert*-butyl group to the aromatic ring increases the solubility of a PEEK in organic solvents. However, when researchers added a second *tert*-butyl group, as shown here, they found that the solubility of the PEEK decreased. Explain this observation.

A PEEK with two *tert*-butyl groups per repeating unit

26.11 The Organic Chemistry of Biomolecules

26.65 The line structure of oxytocin, a hormone that regulates childbirth and breast feeding, is shown here. **(a)** Identify the disulfide bond, and **(b)** write the primary structure of oxytocin in abbreviated form, similar to Figure 26-45a (p. 1357).

Oxytocin

26.66 Some polypeptides can dimerize to form a coiled coil, a structural motif where one α-helix wraps around another (shown here). Typically the primary structure of each polypeptide has a seven–amino acid repeating pattern, in which the first and third amino acids are hydrophobic. As a result, a stripe of hydrophobic side chains extends down the length of each α-helix. Explain how this stabilizes the coiled coil in an aqueous environment.

26.67 Much of the strength of silk—a natural protein fiber—is attributed to β-pleated sheets stacking to form a crystalline structure. The sequence within those sheets exhibits a repeating pattern, Gly-Ser-Gly-Ala-Gly-Ala. Why do you think these amino acids are prevalent in those crystalline structures?

26.68 Arabinoxylan is a copolymer of arabinose and xylose—two five-carbon sugars. A portion of the polysaccharide is shown here. **(a)** Identify and classify each glycosidic linkage. **(b)** Based on the structure, do you think arabinoxylan functions as a storage polysaccharide or a structural polysaccharide? Explain your reasoning.

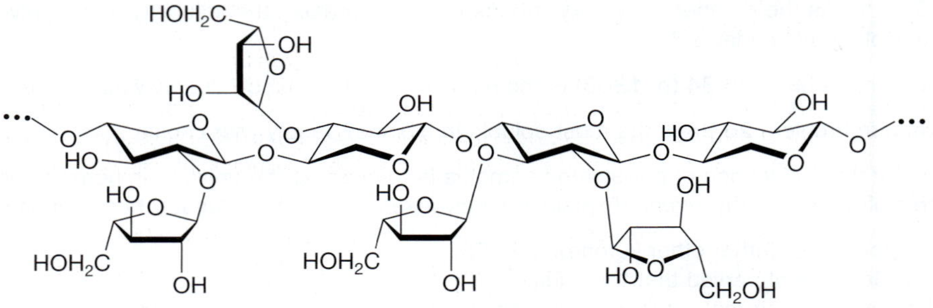

Integrated Problems

26.69 Methyl acrylate and methyl methacrylate react with radical initiators (R•) as shown here. The difference in their reactivities in free radical polymerizations is dramatic. For example, methyl acrylate is less reactive with radicals than methyl methacrylate, but free radicals formed from methyl acrylate are more reactive than those formed from methyl methacrylate. In other words, **C** is more reactive than **A**, and **B** is more reactive than **D**. Explain these observations. *Hint:* Draw an energy diagram for the two reactions.

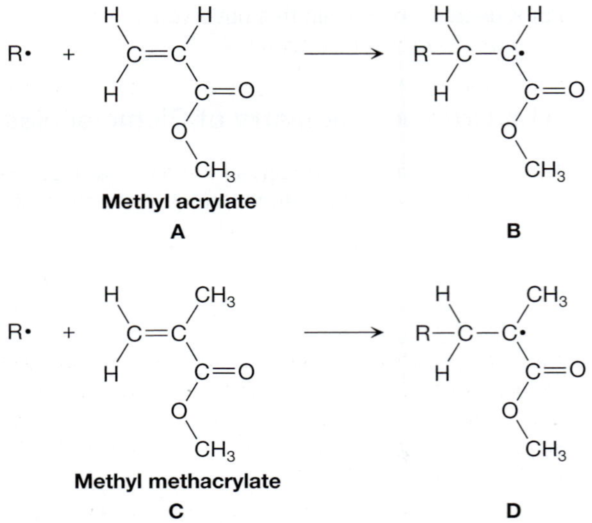

26.70 A polymer closely related to PET is PBT, which is made from terephthalic acid and butane-1,4-diol.
(a) Propose a structure for PBT and write the condensed formula for the structure.
(b) What does PBT stand for? That is, what do you expect the trivial name of PBT to be?
(c) Would you expect PBT to have a higher or lower melting point than PET? Justify your answer.
(d) PBT, rather than PET, is used in molding because it crystallizes faster. Explain why PBT crystallizes more quickly than PET.

26.71 Propose a mechanism for the following reaction, which is a synthesis of PEEK (see Section 26.6a.2).

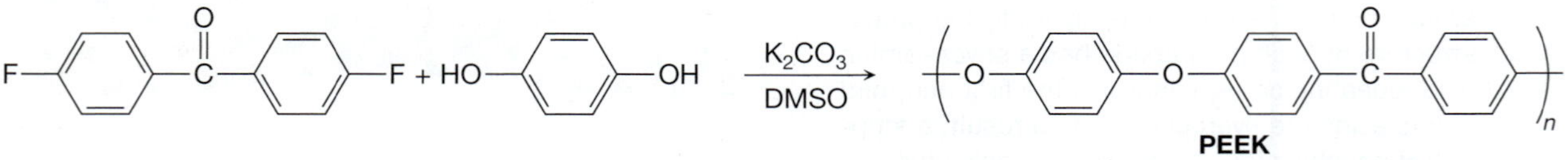

26.72 Polycarbophil is a dietary fiber supplement. It is the calcium salt of a copolymer of acrylic acid and divinyl glycol, a cross-linking agent. In the stomach, the calcium ions exchange for protons and, in the higher pH of the intestine, the polymer absorbs 70 times its mass in water.

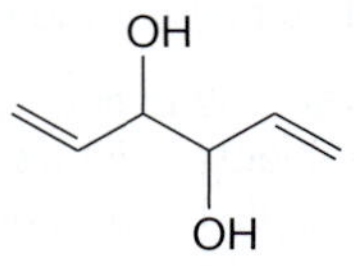

Divinyl glycol

(a) Propose a structure for polycarbophil.
(b) Why is cross-linking of the polymer necessary for this application?
(c) Why is polycarbophil effective at providing bulk in the intestine? That is, what occurs on a molecular level to polycarbophil?

26.73 Dry-erase boards, or whiteboards, in classrooms consist of a network polymer similar to Bakelite that is made from formaldehyde and melamine. **(a)** Propose a structure for the polymer formed. **(b)** Why is this type of polymer suitable for use in whiteboards?

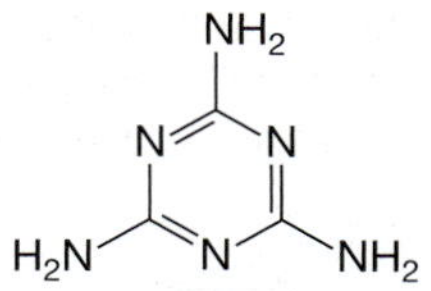

Melamine

26.74 Styrene can *autoinitiate* free radical polymerization and form polystyrene in the absence of an initiator. At temperatures >100 °C, radicals form due to homolysis of the π bond in the vinyl group. In the study of this reaction, researchers have found 1,2-diphenylcyclobutane in samples of heated styrene. The following mechanism for the formation of 1,2-diphenylcyclobutane has been proposed:

Step 1: Homolysis of the vinyl π bond in a styrene molecule
Step 2: Homolysis of the vinyl π bond in a second styrene molecule
Step 3: Tail-to-tail addition of the radicals from Steps 1 and 2
Step 4: Radical coupling of the diradical formed in Step 3

(a) Use curved arrow notation to show the steps in this mechanism.
(b) Are the steps in this mechanism consistent with those for free radical polymerization?
(c) Why is the use of an initiator such as benzoyl peroxide more efficient in the synthesis of polystyrene than autoinitiated polymerization?

26.75 Two of the compounds **X**, **Y**, and **Z** will react with phosgene to form a polycarbonate; the third one will not. Which compound will *not* form a polymer, and what will the product of its reaction with phosgene be? (See Problem 26.49.)

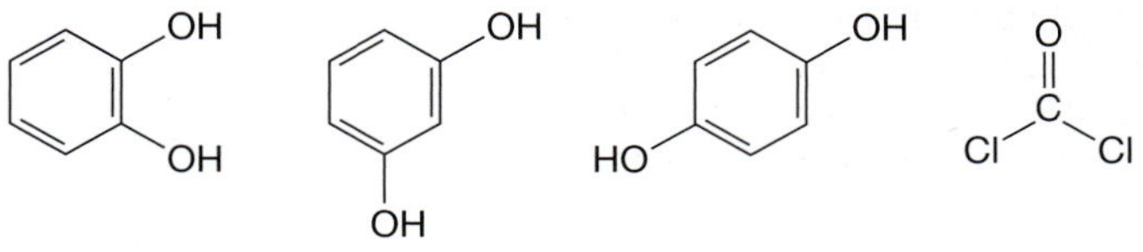

X   **Y**   **Z**   **Phosgene**

26.76 Urea has the structure shown here. Polyureas are used to make truck bed liners. Examine the synthesis of polyurethanes in Section 26.4 (Equation 26-29), then **(a)** propose a synthesis of a polyurea and **(b)** draw the condensed formula for the polymer.

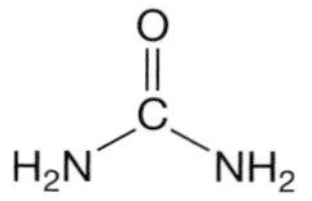

Urea

26.77 Research in the "unzipping" of poly(vinyl chloride) indicates that certain defects in the polymer contribute to the rate of unzipping. These defects are arrangements of atoms that are different from the arrangements in the repeating unit shown in the condensed formula. Researchers have found that defects involving allylic chlorides and tertiary chlorides, specifically, promote the reaction. Review the mechanism for the unzipping and explain this observation.

26.78 Explain how IR spectroscopy could be used to monitor the conversion of poly(vinyl acetate) to poly(vinyl alcohol). That is, how can a chemist use IR to tell if poly(vinyl acetate) has been converted to >99% poly(vinyl alcohol)?

APPENDIX A

Values of K_a and pK_a for Various Acids

Acid	Conjugate Base	K_a	pK_a
F$_3$C—S(=O)$_2$—OH Trifluoromethanesulfonic acid (TfOH)	F$_3$C—S(=O)$_2$—O$^{\ominus}$	1×10^{13}	-13
HI Hydroiodic acid	I$^{\ominus}$	1×10^{10}	-10
HO—S(=O)$_2$—OH Sulfuric acid	HO—S(=O)$_2$—O$^{\ominus}$	1×10^9	-9
HBr Hydrobromic acid	Br$^{\ominus}$	1×10^9	-9
HCl Hydrochloric acid	Cl$^{\ominus}$	1×10^7	-7
p-CH$_3$C$_6$H$_4$—S(=O)$_2$—OH p-Toluenesulfonic acid (TsOH)	p-CH$_3$C$_6$H$_4$—S(=O)$_2$—O$^{\ominus}$	1.6×10^{-3}	-2.8
H$_3$C—S(=O)$_2$—OH Methanesulfonic acid (MsOH)	H$_3$C—S(=O)$_2$—O$^{\ominus}$	1×10^2	-2
H$_3$O$^{\oplus}$ Hydronium ion	H$_2$O	55	-1.7
F$_3$C—C(=O)—OH Trifluoroethanoic acid (Trifluoroacetic acid)	F$_3$C—C(=O)—O$^{\ominus}$	1	0

Acid	Conjugate Base	K_a	pK_a
Cl$_3$C—C(=O)—OH Trichloroethanoic acid (Trichloroacetic acid)	Cl$_3$C—C(=O)—O$^{\ominus}$	0.17	0.77
H$_2$ClC—C(=O)—OH Chloroethanoic acid (Chloroacetic acid)	H$_2$ClC—C(=O)—O$^{\ominus}$	1.3×10^{-3}	2.87
HF Hydrofluoric acid	F$^{\ominus}$	6.3×10^{-4}	3.2
H—C(=O)—OH Methanoic acid (Formic acid)	H—C(=O)—O$^{\ominus}$	1.8×10^{-4}	3.75
C$_6$H$_5$—C(=O)—OH Benzoic acid	C$_6$H$_5$—C(=O)—O$^{\ominus}$	6.3×10^{-5}	4.2
CH$_3$—C(=O)—OH Ethanoic acid (Acetic acid)	CH$_3$—C(=O)—O$^{\ominus}$	1.8×10^{-5}	4.75
Pyridinium ion	Pyridine	6×10^{-6}	5.2
$^{\ominus}$O—C(=O)—OH Bicarbonate anion	$^{\ominus}$O—C(=O)—O$^{\ominus}$	2×10^{-6}	6.3

(continued)

Acid	Conjugate Base	K_a	pK_a
Thiophenol (C₆H₅SH)	C₆H₅S⁻	2.5×10^{-7}	6.6
4-Nitrophenol	conjugate base	6.3×10^{-8}	7.2
Hydrogen sulfide H_2S	HS^-	6.3×10^{-8}	7.2
2,4-Pentanedione	conjugate base	1.2×10^{-9}	8.9
Hydrocyanic acid $N \equiv C-H$	$N \equiv C^-$	6.3×10^{-10}	9.2
Ammonium ion H_4N^+	NH_3	4×10^{-10}	9.4
Trimethylammonium ion $(CH_3)_3\overset{+}{N}H$	$(CH_3)_3N$	1.6×10^{-10}	9.8
Phenol	conjugate base	1×10^{-10}	10.0
Nitromethane O_2N-CH_3	$O_2N-CH_2^-$	6.3×10^{-11}	10.2
Ethanethiol	conjugate base	2.5×10^{-11}	10.6
Methylammonium ion $H_3C-\overset{+}{N}H_3$	H_3C-NH_2	2.3×10^{-11}	10.63
Ethyl 3-oxobutanoate (Acetoacetic ester)	conjugate base	1×10^{-11}	11
2,2,2-Trifluoroethanol	conjugate base	4×10^{-13}	12.4
Diethyl propanedioate (Diethyl malonate)	conjugate base	3.2×10^{-14}	13.5
2-Chloroethanol	conjugate base	5.0×10^{-15}	14.3
Pyrrole	conjugate base	1×10^{-15}	15
Methanol CH_3OH	CH_3O^-	3.2×10^{-16}	15.5
Water H_2O	HO^-	2×10^{-16}	15.7
Ethanol	conjugate base	1×10^{-16}	16
Cyclopentadiene	conjugate base	1×10^{-16}	16
Propan-2-ol (Isopropyl alcohol)	conjugate base	3.2×10^{-17}	16.5
Ethanamide (Acetamide)	conjugate base	1×10^{-17}	17
Methylpropan-2-ol (tert-Butyl alcohol)	conjugate base	1×10^{-19}	19

Acid	Conjugate Base	K_a	pK_a
Propanone (Acetone)		1×10^{-20}	20
Ethyl ethanoate (Ethyl acetate)		1×10^{-25}	25
N≡C—CH₃ Ethanenitrile (Acetonitrile)	N≡C—$\overset{\ominus}{C}$H₂	1×10^{-25}	25
HC≡CH Ethyne (Acetylene)	HC≡C$^{\ominus}$	1×10^{-25}	25
Aniline (Phenylamine)		1×10^{-27}	27
H₂ Hydrogen gas	H$^{\ominus}$	1×10^{-35}	35
Dimethyl sulfoxide		1×10^{-35}	35
NH₃ Ammonia	H₂N$^{\ominus}$	1×10^{-36}	36

Acid	Conjugate Base	K_a	pK_a
N-Methylmethanamine (Dimethylamine)		1×10^{-38}	38
Toluene (Methylbenzene)		1×10^{-40}	40
Benzene		1×10^{-43}	43
H₂C=CH₂ Ethene (Ethylene)	H₂C=$\overset{\ominus}{C}$H	1×10^{-44}	44
Ethoxyethane (Diethyl ether)		$\sim 1 \times 10^{-45}$	~45
CH₃ Methane	H₃C$^{\ominus}$	1×10^{-48}	48
CH₃CH₃ Ethane	CH₃$\overset{\ominus}{C}$H₂	1×10^{-50}	50

APPENDIX B

Characteristic Reactivities of Particular Compound Classes

Alkanes

An alkane H is replaced by a halogen atom, X, when treated with $X_2/h\nu$. (25.4)

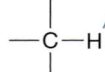

Alkenes

HBr, in the presence of radicals, will add across the double bond in an anti-Markovnikov fashion to produce an alkyl halide. (25.5)

OsO_4 or $KMnO_4$ adds to produce a cis 1,2-diol. (24.7)

A strong Brønsted acid, such as HCl, HBr, HI, or $H_3O^{\oplus}$, will add across the double bond to produce an alkyl halide or alcohol. (11.2)

Br_2 or Cl_2 will add to the double bond to produce a vicinal dihalide. (12.3)

Under conditions for catalytic hydrogenation, H_2 will add to the C=C bond to reduce an alkene to an alkane. (19.5)

Oxidized to a benzoic acid using $KMnO_4$ when R = C_6H_5. (22.9a)

An allylic H is replaced by a halogen atom, X, when treated with NBS/$h\nu$ or NBS/Δ. (25.4)

The C atoms of one C=C bond undergo alkene metathesis (Grubbs reaction) to form new double bonds with the C atoms of another C=C bond. (19.7)

The C of a vinylic C—H undergoes coupling with R—X (Heck reaction) to form a new C—R bond. (19.7)

BH_3 will add across the double bond to produce an alkylborane that can be oxidized to an alcohol. (12.6)

$KMnO_4$ or O_3 will cleave the C=C bond to produce separate carbonyl-containing compounds. (24.8)

RCO_3H will add an O atom to the C=C bond to produce an epoxide. (12.5)

Oxymercuration with $Hg(OAc)_2$ in water followed by reduction with $NaBH_4$ can produce an alcohol. (12.4)

A carbene will add to the C=C bond to produce a cyclopropane ring. (12.2)

Alkynes

Treatment with Li(s) or Na(s) in liquid NH_3 will reduce the alkyne to a trans alkene. (25.7a)

OsO_4 or $KMnO_4$ adds to produce an α-diketone. (24.7)

A strong Brønsted acid, such as HCl, HBr, HI, or $H_3O^{\oplus}$, will add across the triple bond to produce a vinylic halide or ketone. (11.7 and 11.8)

Br_2 or Cl_2 will add to the triple bond to produce a vinylic dihalide or a 1,1,2,2-tetrahalide. (12.3a)

Under conditions for catalytic hydrogenation, H_2 will add in a syn fashion to the C≡C to convert the alkyne into a cis alkene. A second addition produces an alkane. (19.5b)

$R-C≡C-H$

A strong base can deprotonate a terminal alkyne, generating $RC≡C^{\ominus}$, which has a highly nucleophilic C. (9.3b)

Oxidized to a benzoic acid using $KMnO_4$ when R = C_6H_5. (22.9a)

$KMnO_4$ or O_3 will cleave the C≡C bond to produce separate carboxylic acids. (24.8)

Oxymercuration with $Hg(OAc)_2$ in water followed by reduction with $NaBH_4$ can produce a ketone. (12.4)

BH_3 will add across the triple bond to produce a vinylborane that can be oxidized to an aldehyde. (12.7)

A carbene will add to produce a cyclopropene ring. (12.2)

Alkyl halides

Elimination of a halogen and a proton on an adjacent carbon results in an alkene. (7.5)

This C undergoes coupling with R_2CuLi, R—H (Heck reaction), or R—$B(OR')_2$ (Suzuki reaction) to form a new C—R bond. (19.7)

A metal such as Mg(s) or Li(s) produces an organometallic compound, making the C atom strongly nucleophilic and basic. (19.1)

H—C—C—X (Cl, Br, I)

A nucleophile will replace the halogen in a nucleophilic substitution reaction. (7.2)

Ethers

Elimination of H and ROH under acidic conditions produces an alkene. (8.6)

Under acidic conditions, a nucleophile will replace ROH. (9.5b)

H—C—C—OR

Alcohols

Elimination of H and H_2O under acidic conditions produces an alkene. (8.6)

The carbon–carbon bond of a cis 1,2-diol is cleaved by HIO_4, producing separate ketones and/or aldehydes. (24.8b)

H_2CrO_4 or $KMnO_4$ will oxidize a 1° alcohol to a carboxylic acid, and a 2° alcohol to a ketone. (19.6a and 19.6b)

PCC will oxidize a 1° alcohol to an aldehyde. (19.6a)

H—C—C—O—H

Deprotonation by a strong base converts ROH to $RO^{\ominus}$, a strong nucleophile. (8.6a)

Under acidic conditions, a nucleophile will replace H_2O. (9.5b)

OH is converted to OTs, OTf, or OMs, using the sulfonyl chlorides TsCl, TfCl, or MsCl. (21.6)

OH is converted to Br or Cl with PBr_3 or PCl_3. (10.1)

Aromatic rings

If the substituent is a halogen leaving group, it can be replaced by a nucleophile if the aromatic ring is highly deactivated, or if the nucleophile is also a strong base. (23.9)

If the subsitutent is a sulfo group, it is replaced by a proton under acidic aqueous conditions. (22.7)

This aromatic proton is replaced by a strong electrophile if the substituent is an ortho/para director. (23.2)

A Birch reduction (a type of dissolving metal reduction) will reduce the aromatic ring to a cyclohexa-1,4-diene. (25.7b)

The C of an aromatic C—H undergoes coupling with R—X (Heck reaction) to form a new C—R bond. (19.7)

A benzylic H is replaced by a halogen atom, X, when treated with NBS/$h\nu$ or NBS/Δ. (25.4)

This aromatic proton is replaced by a strong electrophile if the substituent is a meta director. (23.2)

A benzylic C is oxidized to a benzoic acid using $KMnO_4$. (22.9a)

This aromatic proton is replaced by a strong electrophile if the substituent is an ortho/para director. (23.2)

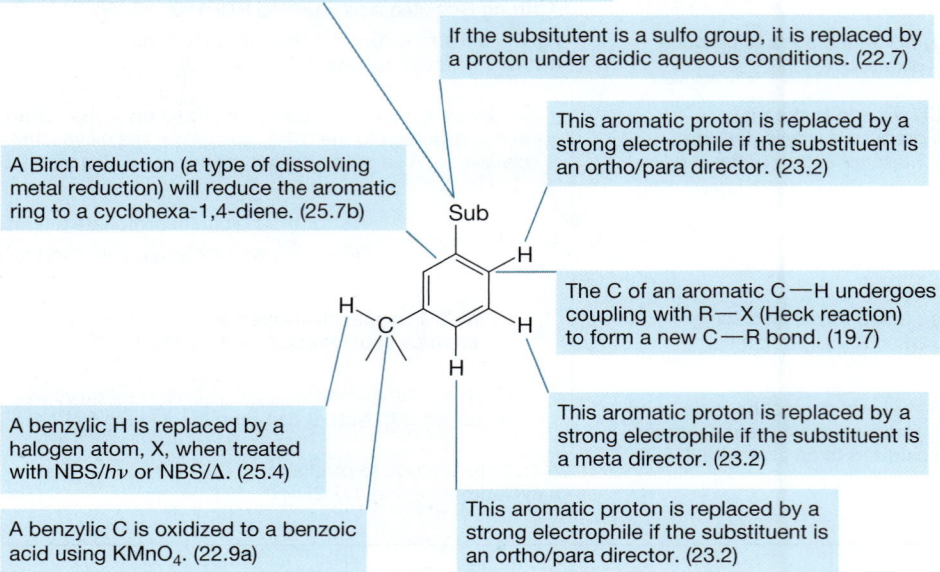

Ketones

The electron-rich O can be protonated by a strong acid, which activates the C=O carbon toward attack by a weak nucleophile. (18.1)

After nucleophilic addition at C, this O can be converted to a H_2O leaving group under acidic conditions, which can produce an acetal or imine via substitution or elimination, respectively. (18.2)

Under conditions for catalytic hydrogenation, H_2 adds to C=O under moderate conditions to produce an alcohol. (19.5)

Oxidized to a benzoic acid using $KMnO_4$ when R = C_6H_5. (22.9a)

A strong base can deprotonate an α C, generating a highly nucleophilic C. (18.4)

Oxidation with a peroxyacid can produce an ester. (21.8)

A methyl ketone is converted to a carboxylate anion by treatment with Cl_2, Br_2, or I_2 in basic, aqueous conditions. (20.5)

Nucleophiles add to the electron-poor C, converting C=O into C—O. Addition of a $H^{\ominus}$ or $R^{\ominus}$ nucleophile reduces a ketone to an alcohol. Addition of an $R^{\ominus}$ nucleophile results in the formation of a new C—C bond. (17.1, 17.3, 17.5)

Aldehydes

The electron-rich O can be protonated by a strong acid, which activates the C=O carbon toward attack by a weak nucleophile. (18.1)

After nucleophilic addition at C, this O can be converted to a H_2O leaving group under acidic conditions, which can produce an acetal or imine via substitution or elimination, respectively. (18.2)

Under conditions for catalytic hydrogenation, H_2 adds to C=O under moderate conditions to produce an alcohol. (19.5)

A strong base can deprotonate an α C, generating a highly nucleophilic C. (18.4)

In the presence of H_2O, the hydrate is oxidized to a carboxylic acid by H_2CrO_4 or $KMnO_4$. (19.6)

Ethanal (O=CH—CH$_3$) is converted to the methanoate anion ($HCO_2^{\ominus}$) by treatment with Cl_2, Br_2, or I_2 in basic, aqueous conditions. (20.5)

Oxidation with a peroxyacid can produce a carboxylic acid. (21.8)

Nucleophiles add to the electron-poor C, converting C=O into C—O. Addition of a $H^{\ominus}$ or $R^{\ominus}$ nucleophile reduces an aldehyde to an alcohol. Addition of an $R^{\ominus}$ nucleophile results in the formation of a new C—C bond. (17.1, 17.3, 17.5)

α,β-Unsaturated carbonyls

Direct addition is favored by nucleophiles that add irreversibly to the C=O. (17.9)

Conjugate addition is favored by nucleophiles that add reversibly to the C=O. (17.9)

Amines

Elimination of the H and the N-containing group can take place in the presence of a base when N is quaternary. (10.9)

Aromatic Ar—NH$_2$ produces an arenediazonium ion when treated with $NaNO_2/H^{\oplus}$. (22.9c)

N is moderately nucleophilic. (9.3)

N is weakly basic. (6.2)

Acid halides

Attack by $R^{\ominus}$ from an alkyllithium or Grignard reagent produces a 3° alcohol. (20.8)

Attack by $H^{\ominus}$ from $NaBH_4$ or $LiAlH_4$ produces a 1° alcohol. (20.6)

Attack by $R^{\ominus}$ from a lithium dialkylcuprate produces a ketone. (20.8)

Attack by $H^{\ominus}$ from $LiAl(O\text{-}t\text{-Bu})_3$ produces an aldehyde. (20.7)

Alpha halogenation takes place with the enol form. (21.5)

Acyl substitution by a strong or weak nucleophile produces an acid anhydride, ester, carboxylic acid, or amide. (21.1)

X (X = Cl, Br)

Acid anhydrides

Attack by R$^\ominus$ from an alkyllithium or Grignard reagent produces a 3° alcohol. (20.8)

Attack by H$^\ominus$ from NaBH$_4$ or LiAlH$_4$ produces a 1° alcohol. (20.6)

Acyl substitution by a strong or weak nucleophile produces an ester, carboxylic acid, or amide. (21.1)

Esters

Attack by R$^\ominus$ from an alkyllithium or Grignard reagent produces a 3° alcohol. (20.8)

The O can be protonated under acidic conditions to activate the carbonyl group toward attack by a weak nucleophile. (21.7)

Attack by H$^\ominus$ from NaBH$_4$ or LiAlH$_4$ produces a 1° alcohol. (20.6)

Attack by H$^\ominus$ from DIBAH produces an aldehyde. (20.7)

Under basic conditions, the α C of the enolate anion is strongly nucleophilic. (21.9)

Acyl substitution by a strong nucleophile under normal or basic conditions (20.1)

The α H is weakly acidic. (6.2)

Acyl substitution by a weak nucleophile under acidic conditions (21.7)

Carboxylic acids

The O can be protonated under acidic conditions to activate the carbonyl group toward attack by a weak nucleophile. (21.7)

Attack by H$^\ominus$ from LiAlH$_4$ produces a 1° alcohol. (20.6)

Attack by H$^\ominus$ from DIBAH produces an aldehyde. (20.7)

This H is acidic and can be deprotonated to produce a carboxylate anion, a moderately strong nucleophile. (9.3)

SOCl$_2$ converts the OH to a Cl. (21.4)

Acyl substitution by a weak nucleophile under acidic conditions (21.7)

Amides

The O can be protonated under acidic conditions to activate the carbonyl group toward attack by a weak nucleophile. (21.7)

Attack by H$^\ominus$ from LiAlH$_4$ produces an amine. (20.6)

Acyl substitution by a weak nucleophile under acidic conditions (21.7)

This H is mildly acidic and can be deprotonated by a strong base to produce a nucleophilic N. (6.2, 20.4)

Reactions That Alter the Carbon Skeleton

	Starting Compound Class	Typical Reagents and Reaction Conditions	Compound Class Formed	Key Electron-Rich Species	Key Electron-Poor Species	Comments	Discussed in Section(s)
(1)	Alkyne	1. NaH 2. R′—X (X = Cl, Br, I)	Alkyne	Alkynide anion	Alkyl halide	S$_N$2	7.2, 8.4, 8.5, 9,3b, 9.9
(2)	Alkyl halide (X = Cl, Br, I)	NaCN	Nitrile	Cyanide anion	Alkyl halide	S$_N$2	7.2, 8.4, 8.5, 9.9
(3)	Ketone or aldehyde	1. Base$^{\ominus}$ 2. R″—X (X = Cl, Br, I)	α-Alkylated ketone or aldehyde	Enolate anion	Alkyl halide	S$_N$2	10.3
(4)	Epoxide	1. R″—Li or R″—MgX 2. H$_3$O$^{\oplus}$ (X = Cl, Br)	Alcohol			S$_N$2	10.7a
(5)	Epoxide	NaCN H$_2$O	Nitrile (β-hydroxy)			S$_N$2	10.7a
(6)	R″C≡CH Alkyne (terminal)	1. NaH 2. 3. H$_3$O$^{\oplus}$	Alcohol (3-alkyn-1-ol)	R″C≡C:$^{\ominus}$		S$_N$2	10.7a
(7)	Oxetane	1. R—Li or R—MgX 2. H$_3$O$^{\oplus}$ (X = Cl, Br)	Alcohol	:R$^{\ominus}$		S$_N$2	10.7a

(continued)

	Starting Compound Class	Typical Reagents and Reaction Conditions	Compound Class Formed	Key Electron-Rich Species	Key Electron-Poor Species	Comments	Discussed in Section(s)
(8)	Alkene (C=C)	CH_2N_2, Δ or $h\nu$	Cyclopropane ring	C=C	:CH₂	Syn addition; retention of cis/trans configuration; not very useful in synthesis	12.2
(9)	Alkene (C=C)	$CHCl_3$, NaOH	Dichlorocyclopropane ring	C=C	:CCl₂	Syn addition; retention of cis/trans configuration	12.2
(10)	Ketone or aldehyde (R (or H) C=O R (or H))	1. R′—Li or R′—MgX 2. H_2O, H_2SO_4 (X = Cl, Br)	Alcohol (HO, R′)	:R′⁻	C=O, δ⁺	Nucleophilic addition	17.5
(11)	Nitrile (R—C≡N)	1. R′—MgX 2. CH_3OH (X = Cl, Br)	Imine (NH)	:R′⁻	—C≡N, δ⁺	Nucleophilic addition	17.5
(12)	Grignard reagent (R—MgX) (X = Cl, Mg)	1. $CO_2(s)$ 2. H_2O, H_2SO_4	Carboxylic acid (R—C(=O)—OH)	:R⁻	O=C=O, δ⁺	Nucleophilic addition	17.5
(13)	α,β-Unsaturated ketone or aldehyde	1. R′₂CuLi, THF 2. NH_4Cl, H_2O	Ketone or aldehyde	:R′⁻	HC=C, δ⁺	Conjugate nucleophilic addition	17.10
(14)	Ketone or aldehyde (R (or H) C=O R (or H))	Wittig reagent: ⊕P(C₆H₅)₃ ⊖C: R′ R″ (or H)	Alkene	⊕PPh₃ ⊖C:	C=O, δ⁺	Wittig reaction; cannot be used to produce tetraalkyl-substituted alkenes	17.7
(15)	Ketone or aldehyde (R (or H) C=O R (or H))	H—CN, NaOH or KCN	Cyanohydrin (HO, CN)	NC:⁻	C=O, δ⁺	Nucleophilic addition	18.1a

	Starting Compound Class	Typical Reagents and Reaction Conditions	Compound Class Formed	Key Electron-Rich Species	Key Electron-Poor Species	Comments	Discussed in Section(s)
(16)	Aldehyde	NaOH	β-Hydroxy aldehyde	Enolate anion	δ^+	Aldol addition; nucleophilic addition	18.4
(17)	Ketone	NaOH	β-Hydroxy ketone	Enolate anion	δ^+	Aldol addition; nucleophilic addition	18.6
(18)	Aldehyde (or H)	1. NaOH 2.	β-Hydroxy aldehyde	Enolate anion	δ^+	Crossed aldol addition; nucleophilic addition	18.7
(19)	Ketone	1. LDA 2.	β-Hydroxy ketone	Enolate anion	δ^+	Crossed aldol addition; nucleophilic addition	18.7
(20)	Nitrile or nitroalkane	NaOH		(or NO_2)	δ^+	Aldol-type addition; nucleophilic addition	18.9
(21)	Ketone	KOH		Enolate anion	δ^+	Robinson annulation	18.10
(22)	Alkyl, vinylic, or aryl halide (X = Cl, Br, I)	R'_2CuLi	R—R'	—	—	Coupling reaction	19.7a
(23)	Vinylic or aryl halide (X = Cl, Br, I)	$R'—B(OR'')_2$, PdL_n, base	R—R'	—	—	Suzuki reaction	19.7b

(continued)

	Starting Compound Class	Typical Reagents and Reaction Conditions	Compound Class Formed	Key Electron-Rich Species	Key Electron-Poor Species	Comments	Discussed in Section(s)
(24)	Vinylic or aryl halide (X = Cl, Br, I)	H—R' / PdL$_2$, base	R—R'	—	—	Heck reaction	19.7b
(25)	Terminal alkene R—CH=CH$_2$	Grubbs catalyst	Alkene R—CH=CH—R	—	—	Alkene metathesis	19.7c
(26)	Methyl ketone	1. X$_2$, NaOH 2. HCl (X = Cl, Br, I)	Carboxylic acid	⁻:ÖH	R—C(δ⁺)(=O)—CX$_3$	Nucleophilic addition–elimination (haloform reaction)	20.5
(27)	Acid chloride	1. R'—Li or R'—MgX 2. HCl (X = Cl, Br)	Tertiary alcohol	⁻:R'	R—C(δ⁺)(=O)—Cl	Nucleophilic addition–elimination (alkyllithium or Grignard reaction)	20.8
(28)	Acid anhydride	1. R'—Li or R'—MgX 2. HCl (X = Cl, Br)	Tertiary alcohol	⁻:R'	R—C(δ⁺)(=O)—O—C(=O)—R	Nucleophilic addition–elimination (alkyllithium or Grignard reaction)	20.8
(29)	Ester	1. R'—Li or R'—MgX 2. HCl (X = Cl, Br)	Tertiary alcohol	⁻:R'	R—C(δ⁺)(=O)—OR'	Nucleophilic addition–elimination (alkyllithium or Grignard reaction)	20.8
(30)	Acid chloride	R'$_2$CuLi	Ketone	⁻:R'	R—C(δ⁺)(=O)—Cl	Nucleophilic addition–elimination	20.8
(31)	Ketone	R''—C(=O)—O—OH	Ester	R''—C(=O)—O—Ö(δ⁻)H	R—C(=⁺OH)—R'	Acid-catalyzed nucleophilic addition–elimination (Baeyer–Villiger oxidation)	21.8
(32)	Ester	1. NaOR'' 2. CH$_3$CO$_2$H	β-Keto ester	R—C⁻H—C(=O)—OR'	R—CH$_2$—C(δ⁺)(=O)—OR'	Base-promoted nucleophilic addition–elimination (Claisen condensation)	21.9

	Starting Compound Class	Typical Reagents and Reaction Conditions	Compound Class Formed	Key Electron-Rich Species	Key Electron-Poor Species	Comments	Discussed in Section(s)
(33)	Acetoacetic ester	1. NaOEt 2. R—X 3. H₃O⁺, Δ (X = Cl, Br)	Alkyl-substituted acetone			Acetoacetic ester synthesis	21.10
(34)	Malonic ester	1. NaOEt 2. R—X 3. H₃O⁺, Δ (X = Cl, Br)	Alkyl-substituted acetic acid			Malonic ester synthesis	21.10
(35)	Arene	R—Cl / AlCl₃	Alkylarene		$R^{\oplus}$	Electrophilic aromatic substitution	22.3
(36)	Arene	/ AlCl₃	Aromatic ketone		Acylium ion	Electrophilic aromatic substitution; Friedel–Crafts acylation	22.5
(37)	Alkylarene	1. KMnO₄, KOH, Δ 2. HCl, H₂O	Aromatic carboxylic acid	—	—	Oxidation	22.9a
(38)	Aromatic ketone/aldehyde	1. KMnO₄, KOH, Δ 2. HCl, H₂O	Aromatic carboxylic acid	—	—	Oxidation	22.9a
(39)	Aryl amine	1. NaNO₂, H₂SO₄ 2. CuCN	Aryl nitrile	CuCN		Proceeds through a diazonium ion; Sandmeyer reaction	22.9c

(continued)

	Starting Compound Class	Typical Reagents and Reaction Conditions	Compound Class Formed	Key Electron-Rich Species	Key Electron-Poor Species	Comments	Discussed in Section(s)
(40)	Conjugated diene + Dienophile		Substituted cyclohexene	—	—	Standard Diels–Alder reactions are facilitated by an electron-donating group on the diene and an electron-withdrawing group on the dienophile.	24.1–24.6
(41)	Substituted cyclohexene	Δ	diene + dienophile	—	—	Retro Diels–Alder reaction	24.6
(42)	Alkene	1. conc KMnO$_4$, KOH, H$_2$O, Δ 2. HCl	Ketone and/or carboxylic acid	Alkene	O=Mn=O	Oxidative cleavage; formic acid product (R″ = H) oxidizes to CO$_2$	24.8a
(43)	Alkyne	1. conc KMnO$_4$, KOH, H$_2$O, Δ 2. HCl	Carboxylic acid	Alkyne	O=Mn=O	Oxidative cleavage; formic acid product (R = H) oxidizes to CO$_2$	24.8a
(44)	Alkene	1. O$_3$ 2. CH$_3$SCH$_3$ or Zn, HOAc	Ketone and/or aldehyde	Alkene	ozone	Ozonolysis	24.8c
(45)	Alkene	1. O$_3$ 2. H$_2$O$_2$	Ketone and/or carboxylic acid	Alkene	ozone	Ozonolysis; oxidative cleavage; formic acid product (R″ = H) oxidizes to CO$_2$	24.8c
(46)	1,2-Diol	NaIO$_4$ or HIO$_4$	Ketone and/or aldehyde	—	—	Oxidative cleavage	24.8b

APPENDIX D

Synthesizing Particular Compound Classes via Functional Group Transformations

	TABLE AppD-1 **Reactions That Produce or Alter Alkenes and Alkynes**						
	Starting Compound Class	Typical Reagents and Reaction Conditions	Compound Class Formed	Key Electron-Rich Species	Key Electron-Poor Species	Comments	Discussed in Section(s)
(1)	Alkene with allylic H	NBS, Δ or hν	Allylic bromide	Br—Br		Favors allylic substitution over bromination of C=C	25.4d
(2)	Conjugated diene	HX (1 equiv), cold, X = Cl, Br, I	Allylic halide (1,2-adduct)	:X⁻		Kinetic control; 1,2-addition	11.9, 11.10
(3)	Conjugated diene	HX (1 equiv), warm, X = Cl, Br, I	Allylic halide (1,4-adduct)	:X⁻		Thermodynamic control; 1,4-addition	11.9, 11.10
(4)	Benzene	Li(s), ROH / NH₃(ℓ)	Cyclohexa-1,4-diene		H—OR	Birch reduction; 1,4-positioning of the double bonds	25.7b
(5)	R—C≡C—R Alkyne	H₂ / Lindlar catalyst	Cis alkene	—	—	Catalytic hydrogenation; poisoned catalyst	19.5b
(6)	R—C≡C—R Alkyne	Na(s) or Li(s) / NH₃(ℓ), −78 °C	Trans alkene		H—NH₂	Forms trans alkene only	25.7a

(continued)

	Starting Compound Class	Typical Reagents and Reaction Conditions	Compound Class Formed	Key Electron-Rich Species	Key Electron-Poor Species	Comments	Discussed in Section(s)
(7)	Alkyne	HX (1 equiv) X = Cl, Br, I	Vinylic halide	:X:⁻		Markovnikov addition, predominantly trans; not useful for synthesis	11.7
(8)	Alkyl halide	(CH₃)₃CONa, Δ	Alkene	(CH₃)₃CO:⁻		E2	7.5, 8.4, 8.5, 9.3d, 9.9
(9)	Alcohol	H₃PO₄ or H₂SO₄, Δ	Alkene	H₂O δ⁻		E1	7.3, 7.6, 8.2, 8.4, 8.5, 9.5a, 9.9
(10)	Amine	1. CH₃I (excess) 2. Ag₂O 3. Δ	Alkene	⁻:OH		E2	10.9
(11)	Ketone or aldehyde	R₂NH (excess) H⁺ 2° amine	Enamine	R₂N̈H δ⁻		Nucleophilic addition, then E1	18.2b
(12)	β-Hydroxycarbonyl compound	H₂SO₄ or NaOH, Δ	α,β-Unsaturated carbonyl compound	—	—	E1 or E1cb	18.5
(13)	Vinylic halide	1. NaNH₂ or NaH 2. H₂O	Terminal alkyne R—C≡CH	⁻:NH₂ or ⁻:H		E2	10.8
(14)	Vinylic halide	NaNH₂ or NaH or NaOH, Δ	Internal alkyne R—C≡C—R′	⁻:NH₂ or ⁻:H or ⁻:ÖH		E2	10.8

	Starting Compound Class	Typical Reagents and Reaction Conditions	Compound Class Formed	Key Electron-Rich Species	Key Electron-Poor Species	Comments	Discussed in Section(s)
(1)	R—H Alkane	X_2, $h\nu$	R—X Alkyl halide	X—X	R•	X = Cl, Br	25.4
(2)	Alkylbenzene with benzylic H (CH₂R)	NBS, $h\nu$	Benzylic bromide (CH(Br)R)	Br—Br	•CHR	Functionalizes a relatively unreactive carbon	25.4d
(3)	C=C Alkene	HX, CCl_4 (X = Cl, Br, I)	H X on C—C Alkyl halide	:X:⊖	—C—C⊕ (with H)	Markovnikov addition	11.1
(4)	C=C Alkene	X_2 (X = Cl, Br), CCl_4	X / C—C / X Vicinal dihalide	:X:⊖	X⊕ bridged C—C	Anti addition	12.3a
(5)	H₂C=C(H)—CH₂R Alkene with allylic H	NBS, Δ or $h\nu$	H₂C=C(H)—CH(Br)R Allylic bromide	Br—Br	H₂C=C(H)—•CHR	Favors allylic substitution over bromination of C=C	25.4d
(6)	Arene	Cl_2, $FeCl_3$	Aryl chloride (Cl)	(arene)	:Cl:⊕	Electrophilic aromatic substitution	22.2
(7)	Arene	Br_2, $FeBr_3$	Aryl bromide (Br)	(arene)	:Br:⊕	Electrophilic aromatic substitution	22.2
(8)	Alkene	HBr, Peroxide	Alkyl bromide (H, Br)		Br•	Anti-Markovnikov regiochemistry	25.5

(continued)

APPENDIX D Synthesizing Particular Compound Classes via Functional Group Transformations   **APP-17**

	Starting Compound Class	Typical Reagents and Reaction Conditions	Compound Class Formed	Key Electron-Rich Species	Key Electron-Poor Species	Comments	Discussed in Section(s)
(9)	Conjugated diene	HX (1 equiv), cold (X = Cl, Br, I)	Allylic halide (1,2-adduct)	:X:⁻		Kinetic control; 1,2-addition	11.9, 11.10
(10)	Conjugated diene	HX (1 equiv), warm (X = Cl, Br, I)	Allylic halide (1,4-adduct)	:X:⁻		Thermodynamic control; 1,4-addition	11.9, 11.10
(11)	Alkyne	HX (1 equiv) (X = Cl, Br, I)	Vinylic halide	:X:⁻		Markovnikov addition, predominantly trans; not useful for synthesis	11.7
(12)	Alkyne	HX (2 equiv) (X = Cl, Br, I)	Geminal dihalide	:X:⁻		Markovnikov addition twice	11.7
(13)	Alkyne	X₂ (2 equiv) CCl₄ (X = Cl, Br, I)	1,1,2,2-Tetrahalide	:X:⁻		2 equivalents of halogen	12.3a
(14)	R—L L = Cl, Br, I, OTs, OMs, or OTf	NaX	R—X Alkyl halide	:X:⁻	R—CH₂—L or R⁺	S$_N$1 or S$_N$2	7.2, 7.3, 8.1, 8.4, 8.5, 9.9
(15)	R—OH Alcohol	HX	R—X Alkyl halide	:X:⁻	R—⁺OH₂ or R⁺	S$_N$1 or S$_N$2	7.2, 7.3, 8.1, 8.4, 8.5, 9.5a, 9.9

	Starting Compound Class	Typical Reagents and Reaction Conditions	Compound Class Formed	Key Electron-Rich Species	Key Electron-Poor Species	Comments	Discussed in Section(s)
(16)	1° or 2° alcohol	PBr₃	1° or 2° alkyl halide			Back-to-back S_N2 reactions; overall inversion of stereochemistry	10.1
(17)	Aryl amine	1. NaNO₂, H₂SO₄ 2. CuBr	Aryl bromide	CuBr		Proceeds through a diazonium ion; Sandmeyer reaction	22.9c
(18)	Aryl amine	1. NaNO₂, H₂SO₄ 2. CuCl	Aryl chloride	CuCl		Proceeds through a diazonium ion; Sandmeyer reaction	22.9c
(19)	Aryl amine	1. NaNO₂, H₂SO₄ 2. KI	Aryl iodide	KI		Proceeds through a diazonium ion	22.9c
(20)	Aryl amine	1. NaNO₂, H₂SO₄ 2. HBF₄, then Δ	Aryl fluoride	HBF₄		Proceeds through a diazonium ion	22.9c
(21)	Ketone or aldehyde	X₂ (X = Cl, Br, I) NaOH	α-Halogenated ketone or aldehyde	Enolate anion		Multiple S_N2 reactions	10.4
(22)	Ketone or aldehyde	X₂ (X = Cl, Br, I) H⁺	α-Halogenated ketone or aldehyde	Enol		Single S_N2 reaction	10.4
(23)	Carboxylic acid	Br₂, P		Enol		Hell–Volhard–Zelinsky reaction	21.5

Reactions That Produce Alcohols

	Starting Compound Class	Typical Reagents and Reaction Conditions	Compound Class Formed	Key Electron-Rich Species	Key Electron-Poor Species	Comments	Discussed in Section(s)
(1)	Alkene (C=C)	H_2O, $H^\oplus$	Alcohol (H, OH on C–C)	$H_2\ddot{O}:$	carbocation (H, C–C$^\oplus$)	Markovnikov addition, acid catalysis	11.6
(2)	Alkene (C=C)	X_2 (X = Cl, Br), H_2O	Halohydrin (X, OH on C–C)	$H_2\ddot{O}:$	halonium ion (X$^\oplus$ bridging C–C)	OH bonds to more highly substituted carbon; anti addition	12.3b
(3)	Alkene (C=C)	1. $Hg(OAc)_2$, H_2O 2. $NaBH_4$	Alcohol (H, OH on C–C)	$H_2\ddot{O}:$	mercurinium ion (OAc–Hg bridging C–C)	Markovnikov addition of water; no carbocation rearrangements	12.4
(4)	Alkene (C=C)	1. B_2H_6 or $BH_3 \cdot THF$ 2. H_2O_2, NaOH, H_2O	Alcohol (H, OH on C–C)	C=C	BH_3	Anti-Markovnikov addition of water	12.6
(5)	Alkene (C=C)	OsO_4, H_2O_2	Syn 1,2-diol (HO, OH on C–C)	C=C	$O=Os(=O)(=O)O$	An alternate method uses a cold, basic solution of $KMnO_4$.	24.7
(6)	Primary alkyl halide ($R–CH_2–X$); X = F, Cl, Br, I	NaOH	1° alcohol ($R–CH_2–OH$)	$H\ddot{O}:^\ominus$	$R–CH_2^{\delta+}–X$	S_N2	7.2, 8.4, 8.5, 9.9
(7)	Aromatic halide	NaOH, Δ	Aromatic alcohol	$H\ddot{O}:^\ominus$	benzyne	Proceeds through benzyne intermediate	23.9b
(8)	Tertiary alkyl halide ($R_3C–X$)	H_2O, $H^\oplus$	3° Alcohol ($R_3C–OH$)	$H_2\ddot{O}$	carbocation ($R_3C^\oplus$)	S_N1	7.3, 8.1, 8.4, 8.5, 9.5a, 9.9
(9)	Ether (R–O–R)	H_2O, $H^\oplus$	Alcohol (R–OH)	$H_2\ddot{O}$	$R^\oplus$ or $R–\overset{\oplus}{O}R$ (H)	S_N1 or S_N2	7.2, 7.3, 8.1, 8.4, 8.5, 9.5a, 9.9

TABLE AppD-3 Reactions That Produce Alcohols (continued)

	Starting Compound Class	Typical Reagents and Reaction Conditions	Compound Class Formed	Key Electron-Rich Species	Key Electron-Poor Species	Comments	Discussed in Section(s)
(10)	Epoxide	:Nu⁻ — Neutral or basic	Alcohol (2-substituted)	:Nu⁻	(epoxide δ+)	S_N2	10.7a
(11)	Epoxide	H—Nu — Acidic	Alcohol (2-substituted)	:Nu⁻	(protonated epoxide δ+)	S_N2	10.7b
(12)	Aryl amine	1. $NaNO_2$, H_2SO_4 2. H_2O, Cu_2O, $Cu(NO_3)_2$	Aryl alcohol	$H_2\ddot{O}$	(aryl N_2^+)	Proceeds through a diazonium ion; Sandmeyer reaction	22.9c
(13)	Ketone or aldehyde	1. $NaBH_4$, or $LiAlH_4$ 2. H_2O, H_2SO_4	Alcohol	:H⁻ Hydride anion	(carbonyl δ+)	Nucleophilic addition	17.3
(14)	α,β-Unsaturated ketone or aldehyde	1. $NaBH_4$, or $LiAlH_4$ 2. NH_4Cl, H_2O	Alcohol	:H⁻ Hydride anion	(carbonyl δ+)	Nucleophilic addition	17.9
(15)	Ketone or aldehyde	H_2 — Pt, Pd, or Ni	Alcohol	—	—	Catalytic hydrogenation	19.5c
(16)	Carboxylic acid	1. $LiAlH_4$ 2. HCl	Primary alcohol	:H⁻ Hydride anion	($R-C-O-AlH_3$ δ+)	Nucleophilic addition–elimination, then addition (reduction)	20.6
(17)	Ester	1. $NaBH_4$, or $LiAlH_4$ 2. HCl	Primary alcohol	:H⁻ Hydride anion	($R-C-OR'$ δ+)	Nucleophilic addition–elimination, then addition (reduction); very slow with $NaBH_4$	20.6

(continued)

	Starting Compound Class	Typical Reagents and Reaction Conditions	Compound Class Formed	Key Electron-Rich Species	Key Electron-Poor Species	Comments	Discussed in Section(s)
(18)	Acid chloride	1. $NaBH_4$, or $LiAlH_4$ 2. HCl	Primary alcohol	Hydride anion		Nucleophilic addition–elimination, then addition (reduction)	20.6
(19)	Acid anhydride	1. $NaBH_4$, or $LiAlH_4$ 2. HCl	Primary alcohol	Hydride anion		Nucleophilic addition–elimination (reduction)	20.6

	Starting Compound Class	Typical Reagents and Reaction Conditions	Compound Class Formed	Key Electron-Rich Species	Key Electron-Poor Species	Comments	Discussed in Section(s)
(1)	Alkene		Epoxide			Conservation of cis/trans configurations	12.5
(2)	Alkyl halide (X = Cl, Br, I)	$NaOR'$	Ether (symmetric or unsymmetric)			Williamson ether synthesis, S_N2	10.6
(3)	Halohydrin	NaOH	Epoxide			Intramolecular S_N2	10.6
(4)	Alcohol	$H^\oplus$ Δ	Ether (symmetric)			S_N1 or S_N2 (dehydration)	7.2, 7.3, 8.1, 8.4, 8.5, 9.5a, 9.9, 10.6
(5)	Ketone or aldehyde	$R'OH$ (excess) H_2SO_4	Acetal			Nucleophilic addition, then S_N1	18.2a

	Starting Compound Class	Typical Reagents and Reaction Conditions	Compound Class Formed	Key Electron-Rich Species	Key Electron-Poor Species	Comments	Discussed in Section(s)
(1)	Nitroarene (NO_2 on benzene)	HCl / Fe	Aryl amine (NH_2 on benzene)	—	—	Reduction	22.9b
(2)	Electron-poor aromatic halide (X, EWG)	H_2NR (or H)	Aromatic amine (HNR (or H), EWG)	$H_2\overset{\delta-}{\underset{..}{N}}R$ (or H)	$\overset{\delta+}{}$ aromatic ring with X and EWG	Nucleophilic addition–elimination mechanism; EWG ortho or para to X (X = F, Cl, Br, I)	23.9a
(3)	Aromatic halide (X)	LiNHR	Aromatic amine (NHR)	$\overset{\ominus}{\underset{..}{H}N}R$	benzyne	Proceeds through benzyne intermediate (X = F, Cl, Br, I)	23.9b
(4)	1° Alkyl halide (R—X)	NH_3 (excess)	R—NH_2 + other amines, 1° amine	$\overset{\delta-}{:}NH_3$	$\overset{\delta+}{R}$—X	S_N2 reaction; not synthetically useful	10.2
(5)	Ketone or aldehyde	R_2NH (excess) / $H^{\oplus}$	Enamine (NR_2)	$R_2\overset{\delta-}{\underset{..}{N}}H$	$\overset{\oplus}{O}H$ on C	Nucleophilic addition, then E1	18.2b
(6)	Imine	1. LiAlH$_4$ 2. H$_2$O	Amine (HN—R (or H))	$:H^{\ominus}$ Hydride anion	iminium ion	Nucleophilic addition	17.3
(7)	Amide (R—NR'_2)	LiAlH$_4$ / Ether	Amine (R—NR'_2)	$:H^{\ominus}$ Hydride anion	iminium ion	Nucleophilic addition–elimination, then addition (reduction)	20.6
(8)	Phthalimide	1. KOH/EtOH 2. RBr 3. KOH/H$_2$O	H_2N—R Primary amine	phthalimide anion $\overset{\ominus}{N:}$	$\overset{\delta+}{R}$—Br	S_N2, then nucleophilic addition–elimination (Gabriel synthesis)	20.4

(continued)

	Starting Compound Class	Typical Reagents and Reaction Conditions	Compound Class Formed	Key Electron-Rich Species	Key Electron-Poor Species	Comments	Discussed in Section(s)
(9)	R—C≡N Nitrile	1. LiAlH₄ 2. H₂O	H H / R—C—NH₂ 1° Amine	:H⊖ Hydride anion	δ+ —C≡N	Sequential nucleophilic additions	17.3
(10)	O ‖ C / R R (or H) (or H) Ketone or aldehyde	(H or) (H or) R R \ / N ‖ H NaBH₄	(H or) (H or) R R \ / N / C \ R R (or H) (or H) Amine	(H or) (H or) R R \ / N δ- H	O ‖ δ+ C / R R	Reductive amination; imine formation, then reduction	18.12

TABLE AppD-6 Reactions That Produce or Alter Ketones and Aldehydes

	Starting Compound Class	Typical Reagents and Reaction Conditions	Compound Class Formed	Key Electron-Rich Species	Key Electron-Poor Species	Comments	Discussed in Section(s)
(1)	—C≡C— Alkyne	H₂O TfOH, CF₃CH₂OH	H O \ ‖ H—C—C— Ketone	H₂O:	H \ / C=C⊕	Markovnikov addition of H₂O, keto–enol tautomerization	11.8
(2)	—C≡C— Alkyne	Hg(OAc)₂, H₂O	O ‖ C / \ C Ketone	H₂O:	OAc ⊕| Hg \ / C=C	Markovnikov addition of water	12.4
(3)	—C≡C— Alkyne	1. (C₅H₁₁)₂BH 2. H₂O₂, NaOH, H₂O	O ‖ H C \ / C Ketone or aldehyde	—C≡C—	(C₅H₁₁)₂BH	Anti-Markovnikov addition of water	12.7
(4)	—C≡C— Alkyne	OsO₄ H₂O₂	O O ‖ ‖ C—C 1,2-Dione	—C≡C—	O ‖ O=Os=O ‖ O	An alternate method uses a cold, basic solution of KMnO₄.	24.7
(5)	OH | CH / \ R R Secondary alcohol	H₂CrO₄ or KMnO₄, KOH	O ‖ C / \ R R Ketone	—	—	Oxidation	19.6a, 19.6b

Starting Compound Class	Typical Reagents and Reaction Conditions	Compound Class Formed	Key Electron-Rich Species	Key Electron-Poor Species	Comments	Discussed in Section(s)
(6) Primary or secondary alcohol	PCC	Aldehyde or ketone	—	—	Oxidation	19.6a
(7) Acetal	H₂O (excess) H₂SO₄	Ketone or aldehyde			S_N1, then E1	18.2a
(8) Ketone or aldehyde	X₂ (X = Cl, Br, I) NaOH	α-Halogenated ketone or aldehyde	Enolate anion		Multiple S_N2 reactions	10.4
(9) Ketone or aldehyde	X₂ (X = Cl, Br, I) H⁺	α-Halogenated ketone or aldehyde	Enol		Single S_N2 reaction	10.4
(10) α,β-Unsaturated ketone or aldehyde	H—Nu Nu = OR, SR, NR₂	β-Substituted ketone or aldehyde			Conjugate nucleophilic addition	18.1b
(11) Imine	H₂O (excess) HCl	Ketone or aldehyde			Nucleophilic addition, then E1	18.2b
(12) Acid chloride	LiAlH(O-t-Bu)₃ −78 °C	Aldehyde	Hydride anion		Nucleophilic addition–elimination (reduction); note cold T	20.7

(continued)

Starting Compound Class	Typical Reagents and Reaction Conditions	Compound Class Formed	Key Electron-Rich Species	Key Electron-Poor Species	Comments	Discussed in Section(s)
(13) Ester (R–CO–OR′)	1. DIBAH −78 °C 2. HCl	Aldehyde (R–CO–H)	Hydride anion (:H⁻)	R–$\overset{\delta+}{C}$(=O)–OR′	Nucleophilic addition–elimination (reduction); note cold T	20.7

TABLE AppD-7 Reactions That Produce or Alter Carboxylic Acids

Starting Compound Class	Typical Reagents and Reaction Conditions	Compound Class Formed	Key Electron-Rich Species	Key Electron-Poor Species	Comments	Discussed in Section(s)
(1) Primary alcohol (R–CH₂–OH)	H_2CrO_4 or 1. $KMnO_4$, KOH 2. H_2O, HCl	Carboxylic acid (R–CO–OH)	—	—	Oxidation	19.6a, 19.6b
(2) Aldehyde (R–CO–H)	R′–CO–O–OH	Carboxylic acid (R–CO–OH)	R′–CO–O–$\overset{\delta-}{\ddot{O}}$H	R–$\overset{\oplus}{C}$(OH)–H	Acid-catalyzed nucleophilic addition–elimination (Baeyer–Villiger oxidation)	21.8
(3) Carboxylic acid (R–CH₂–CO–OH)	Br_2, P	α-Bromo acid (R–CHBr–CO–OH)	R–$\overset{\delta-}{C}$(H)=C(OH)	$\overset{\delta+}{Br}$—$\overset{\delta-}{Br}$	Hell–Volhard–Zelinsky reaction	21.5
(4) Ester (R–CO–OR′)	1. NaOH 2. HCl	Carboxylic acid (R–CO–OH)	$\overset{\ominus}{:}\ddot{O}H$	R–$\overset{\delta+}{C}$(=O)–OR′	Nucleophilic addition–elimination (saponification)	20.3
(5) Ester (R–CO–OR′)	H_2O / H_2SO_4	Carboxylic acid (R–CO–OH)	$H_2\overset{\delta-}{\ddot{O}}:$	R–$\overset{\oplus}{C}$(OH)–OR′	Acid-catalyzed nucleophilic addition–elimination (hydrolysis)	21.7
(6) Amide (R–CO–NR′₂)	1. NaOH 2. HCl	Carboxylic acid (R–CO–OH)	$\overset{\ominus}{:}\ddot{O}H$	R–$\overset{\delta+}{C}$(=O)–NR′₂	Nucleophilic addition–elimination	20.4

TABLE AppD-7 Reactions That Produce or Alter Carboxylic Acids (continued)

	Starting Compound Class	Typical Reagents and Reaction Conditions	Compound Class Formed	Key Electron-Rich Species	Key Electron-Poor Species	Comments	Discussed in Section(s)
(7)	Acid chloride (R–C(=O)–Cl)	H_2O →	Carboxylic acid (R–C(=O)–OH)	$H_2\ddot{O}:^{\delta-}$	$R–C_{\delta+}(=O)–Cl$	Nucleophilic addition–elimination (hydrolysis)	21.1
(8)	Acid anhydride (R–C(=O)–O–C(=O)–R)	H_2O →	Carboxylic acid (R–C(=O)–OH)	$H_2\ddot{O}:^{\delta-}$	$R–C_{\delta+}(=O)–O–C(=O)–R$	Nucleophilic addition–elimination (hydrolysis)	21.2

TABLE AppD-8 Reactions That Produce or Alter Esters

	Starting Compound Class	Typical Reagents and Reaction Conditions	Compound Class Formed	Key Electron-Rich Species	Key Electron-Poor Species	Comments	Discussed in Section(s)
(1)	Carboxylic acid (R–C(=O)–OH)	$R'OH$ / H_2SO_4 →	Ester (R–C(=O)–OR')	$R'\ddot{O}H^{\delta-}$	protonated acid ($R–C(OH)(=OH^{+})$)	Acid-catalyzed nucleophilic addition–elimination (Fischer esterification)	21.7
(2)	Carboxylic acid (R–C(=O)–OH)	CH_2N_2 →	Methyl ester (R–C(=O)–O–CH$_3$)	Carboxylate anion (R–C(=O)–$\ddot{O}:^{-}$)	$H_3C–N_2^{+}$	S_N2	10.5
(3)	Ester (R–C(=O)–OR')	$R''OH$ / $R''ONa$ →	Ester (R–C(=O)–OR'')	$:\ddot{O}R''^{-}$	$R–C_{\delta+}(=O)–OR'$	Base-catalyzed nucleophilic addition–elimination (transesterification)	20.1a, 21.7
(4)	Ester (R–C(=O)–OR')	$R''OH$ / H_2SO_4 →	Ester (R–C(=O)–OR'')	$R''\ddot{O}H^{\delta-}$	protonated ester ($R–C(OH^{+})–OR'$)	Acid-catalyzed nucleophilic addition–elimination (transesterification)	21.7
(5)	Acid chloride (R–C(=O)–Cl)	$LiOR'$ →	Ester (R–C(=O)–OR')	$:\ddot{O}R'^{-}$	$R–C_{\delta+}(=O)–Cl$	Nucleophilic addition–elimination	20.2

(continued)

TABLE AppD-8 Reactions That Produce or Alter Esters (continued)

	Starting Compound Class	Typical Reagents and Reaction Conditions	Compound Class Formed	Key Electron-Rich Species	Key Electron-Poor Species	Comments	Discussed in Section(s)
(6)	Acid chloride	R′OH →	Ester	R′ÖH δ−	Acid chloride δ+	Nucleophilic addition–elimination (alcoholysis)	21.1
(7)	Acid anhydride	NaOR′ →	Ester	⊖:ÖR′	Acid anhydride δ+	Nucleophilic addition–elimination	20.2
(8)	Acid anhydride	R′OH →	Ester	R′ÖH δ−	Acid anhydride δ+	Nucleophilic addition–elimination (alcoholysis)	21.2

TABLE AppD-9 Reactions That Produce Amides

	Starting Compound Class	Typical Reagents and Reaction Conditions	Compound Class Formed	Key Electron-Rich Species	Key Electron-Poor Species	Comments	Discussed in Section(s)
(1)	Ester	$LiNR''_2$ →	Amide	⊖:NR''_2	Ester δ+	Nucleophilic addition–elimination	20.2
(2)	Nitrile R—C≡N	H_2O (1 equiv) / H_2SO_4 or NaOH	Amide	$H_2\ddot{O}:$ δ− or ⊖:ÖH	R—C≡N δ+ or R—C≡NH ⊕	Nucleophilic addition	18.2c
(3)	Acid chloride	R'_2NH / Et_3N or pyridine	Amide	$R'_2\ddot{N}H$ δ−	Acid chloride δ+	Nucleophilic addition–elimination (aminolysis)	21.3
(4)	Acid anhydride	R'_2NH / Et_3N or pyridine	Amide	$R'_2\ddot{N}H$ δ−	Acid anhydride δ+	Nucleophilic addition–elimination (aminolysis)	21.3

	Starting Compound Class	Typical Reagents and Reaction Conditions	Compound Class Formed	Key Electron-Rich Species	Key Electron-Poor Species	Comments	Discussed in Section(s)
(1)	Arene	Br_2 / $FeBr_3$	Aryl bromide		$\overset{..}{Br}\cdot{}^{\oplus}$	Electrophilic aromatic substitution	22.2
(2)	Aryl amine (NH_2)	1. $NaNO_2$, H_2SO_4 2. CuBr	Aryl bromide	CuBr	$N_2^{\oplus}$	Proceeds through a diazonium ion; Sandmeyer reaction	22.9c
(3)	Arene	Cl_2 / $FeCl_3$	Aryl chloride		$\overset{..}{\underset{..}{Cl}}\cdot{}^{\oplus}$	Electrophilic aromatic substitution	22.2
(4)	Aryl amine (NH_2)	1. $NaNO_2$, H_2SO_4 2. CuCl	Aryl chloride	CuCl	$N_2^{\oplus}$	Proceeds through a diazonium ion; Sandmeyer reaction	22.9c
(5)	Aryl amine (NH_2)	1. $NaNO_2$, H_2SO_4 2. KI	Aryl iodide	KI	$N_2^{\oplus}$	Proceeds through a diazonium ion	22.9c
(6)	Aryl amine (NH_2)	1. $NaNO_2$, H_2SO_4 2. HBF_4, then Δ	Aryl fluoride	HBF_4	$N_2^{\oplus}$	Proceeds through a diazonium ion	22.9c
(7)	Aryl amine (NH_2)	1. $NaNO_2$, H_2SO_4 2. H_2O, Cu_2O, $Cu(NO_3)_2$	Aryl alcohol	H_2O	$N_2^{\oplus}$	Proceeds through a diazonium ion; Sandmeyer reaction	22.9c
(8)	Arenesulfonic acid (SO_3H)	$H_3O^{\oplus}$	Arene	SO_3H	$H-\overset{\oplus}{O}H_2$	Electrophilic aromatic substitution; desulfonation	22.7
(9)	Aryl amine (NH_2)	1. $NaNO_2$, H_2SO_4 2. H_3PO_2	Arene	—	$N_2^{\oplus}$	Proceeds through a diazonium ion	22.9c

(continued)

	Starting Compound Class	Typical Reagents and Reaction Conditions	Compound Class Formed	Key Electron-Rich Species	Key Electron-Poor Species	Comments	Discussed in Section(s)
(10)	Nitroarene	HCl / Fe	Aryl amine	—	—	Reduction	22.9b
(11)	Arene	HNO_3 / H_2SO_4	Nitroarene		$\overset{\oplus}{N}O_2$	Electrophilic aromatic substitution; nitration	22.6
(12)	Arene	SO_3 / H_2SO_4	Arenesulfonic acid		$SO_3H^{\oplus}$	Electrophilic aromatic substitution	22.7
(13)	Electron-poor aromatic halide	H_2NR	NHR ... EWG	$H_2\overset{..}{N}R^{\delta-}$	X ... EWG	Nucleophilic addition–elimination mechanism; EWG ortho or para to X (X = F, Cl, Br, I)	23.9a
(14)	Electron-poor aromatic halide	NaOR	OR ... EWG	$:\overset{..}{\underset{..}{O}}R^{\ominus}$	X$^{\delta+}$ ... EWG	Nucleophilic addition–elimination mechanism; EWG ortho or para to X	23.9a
(15)	Aromatic halide	$:$Base$^{\ominus}$	Base	$:$Base$^{\ominus}$	(benzyne)	Proceeds through benzyne intermediate (X = F, Cl, Br, I)	23.9b
(16)	Alkylbenzene with allylic H (CH_2R)	NBS / Δ or $h\nu$	Benzylic bromide ($CHBr$/R)	Br—Br	$\overset{\bullet}{C}HR$	Functionalizes a relatively unreactive carbon	25.4d

TABLE AppD-11 Reactions That Produce Alkanes

	Starting Compound Class	Typical Reagents and Reaction Conditions	Compound Class Formed	Key Electron-Rich Species	Key Electron-Poor Species	Comments	Discussed in Section(s)
(1)	Alkene	H_2, Pt, Pd, or Ni	Alkane	—	—	Catalytic hydrogenation	19.5a
(2)	Ketone or aldehyde	1. H_2N—NH_2, $H^{\oplus}$ 2. KOH/diethylene glycol, Δ	Alkane	$H_2\overset{\delta-}{\ddot{N}}NH_2$	$\overset{\oplus OH}{C}$	Wolff–Kishner reduction: nucleophilic addition, then E1, then E2	18.3
(3)	Ketone or aldehyde	Zn/Hg, HCl, H_2O, reflux	Alkane	—	—	Clemmensen reduction	19.3b
(4)	Ketone or aldehyde	1. $HSCH_2CH_2SH$, $H^{\oplus}$ 2. Raney nickel (H_2)	Alkane	—	—	Raney-nickel reduction	19.3c

TABLE AppD-12 Reactions That Produce Other Species

	Starting Compound Class	Typical Reagents and Reaction Conditions	Compound Class Formed	Key Electron-Rich Species	Key Electron-Poor Species	Comments	Discussed in Section(s)
(1)	R—NH_2 Amine	excess R'X ($X = Cl, Br, I$)	Quaternary ammonium salt	R—$\overset{\delta-}{\ddot{N}}H_2$	$\overset{\delta+}{R'}$—X	Multiple S_N2 reactions	10.2
(2)	Alkyl halide	1. $P(C_6H_5)_3$ 2. R—Li	Wittig reagent	$:\overset{\delta-}{P}(C_6H_5)_3$ Triphenyl phosphine	$\overset{\delta+}{C}$ with H, Br	S_N2 followed by proton transfer	17.7

(continued)

	Starting Compound Class	Typical Reagents and Reaction Conditions	Compound Class Formed	Key Electron-Rich Species	Key Electron-Poor Species	Comments	Discussed in Section(s)
(3)	Haloalkane (X = Cl, Br)	Mg(s) / Ether	Grignard reagent	—	—	Dissolving metal reduction	19.1
(4)	Haloalkane (X = Cl, Br)	Li(s) / Ether	Alkyllithium reagent	—	—	Dissolving metal reduction	19.1
(5)	Alkyllithium reagent	CuI / Ether	Lithium dialkylcuprate	—	—	—	19.1
(6)	Carboxylic acid	$SOCl_2$	Acid chloride	(electron-rich species shown)	(electron-poor species shown)	Back-to-back nucleophilic addition–eliminations	21.4
(7)	Alcohol	(sulfonyl chloride reagent)	Sulfonate ester	$R\ddot{O}H$ δ^-	(electron-poor species shown)	Nucleophilic addition–elimination	21.6
(8)	Ketone or aldehyde	NH_3 or $R'NH_2$ (excess), $H^{\oplus}$	Imine	$R\text{—}\ddot{N}H_2$ δ^-	(electron-poor species shown)	Nucleophilic addition, then E1	18.2b

GLOSSARY

A

Absorbance (A) a measure of light absorbed by a sample for spectroscopy, ranging from 0 (no absorption) to infinity (complete absorption). $A = -\log(I_{\text{detected}}/I_{\text{source}}) = 2 - \log(\%T)$. (Ch. 15)

Absorption band (also known as *peak*) feature of a spectrum where absorbance is high. (Ch. 15)

Acetal a compound that has the bonding arrangement in which two alkoxy (—OR) groups are attached to the same carbon atom. (Ch. 1)

Acetoacetic ester synthesis a synthesis scheme that converts acetoacetic ester into an alkyl- or dialkyl-substituted acetone. (Ch. 21)

Achiral term used to describe a molecule that does not have an enantiomer; the molecule's mirror image is superimposable on itself. (Ch. 5)

Achiral environment environment that is superimposable on its mirror image. (Ch. 5)

Acid catalysis speeding up of a reaction in the presence of a small amount of acid, without the acid being consumed by the reaction. (Ch. 18)

Acid-catalyzed alkoxylation reaction a reaction in which an alcohol adds across a C≡C or C≡C bond, catalyzed by acid. (Ch. 11)

Acid-catalyzed hydration reaction a reaction in which water adds across a C≡C or C≡C bond, catalyzed by acid. (Ch. 11)

Acidity constant (K_a) an experimentally obtained constant that reflects the strength of an acid (i.e., an acid's propensity to donate a proton). (Ch. 6)

Acid workup a synthetic step in which acid is added to protonate a product or remove leftover basic components (e.g., extra Grignard or alkyllithium reagents) after the main reaction is completed. (Ch. 10)

Acrylic See *polyacrylate*. (Ch. 26)

Activated carbonyl group a C≡O group made more susceptible to nucleophilic attack, generally by protonation of the carbonyl O atom. (Ch. 18)

Activating group a substituent that increases the reactivity of a molecule in a particular reaction; often refers to substituents that make an aromatic ring more susceptible to electrophilic aromatic substitution. (Ch. 23)

Active methylene compound a CH_2 group with two adjacent C≡O groups; can be converted into a highly nucleophilic enolate anion relatively easily. (Ch. 21)

Acylium ion a cation of the form R—C≡O⁺. (Ch. 22)

1,2-Addition (also known as *direct addition*) a type of reaction in which two substituents add to two adjacent atoms. (Ch. 11)

1,4-Addition (also known as *conjugate addition*) a type of reaction in which two substituents add to two atoms that have a 1,4 relative positioning. (Ch. 11)

Addition polymerization See *chain-growth polymerization*. (Ch. 26)

Adduct the addition product of two molecules. (Ch. 11)

Alcohol a compound that contains the bonding arrangement C—OH. (Ch. 1)

Alcoholysis a reaction in which an alcohol reactant is responsible for breaking bonds; used to convert a carboxylic acid derivative into an ester. (Ch. 21)

Aldehyde a compound that contains the bonding arrangement HC≡O. (Ch. 1)

Alder rule See *endo rule*. (Ch. 24)

Aldohexose a six-carbon sugar in which the carbonyl group involves a terminal C, characteristic of an aldehyde. (Ch. 4)

Aldol a β-hydroxy aldehyde, characterized by an OH group that is β to a HC≡O group. (Ch. 18)

Aldol addition a reaction in which an enolate anion adds to the carbonyl group of a ketone or an aldehyde to produce a β-hydroxycarbonyl compound. (Ch. 18)

Aldol condensation a reaction consisting of an *aldol addition* followed by the elimination of water, resulting in the production of an α,β-unsaturated carbonyl compound. (Ch. 18)

Aldopentose a five-carbon sugar in which the carbonyl group involves a terminal C, characteristic of an aldehyde. (Ch. 4)

Aldose a sugar whose carbonyl group involves a terminal C, characteristic of an aldehyde. (Ch. 4)

Alkane an acyclic compound consisting of only C—C and C—H single bonds. (Ch. 1)

Alkanoate group an RCO_2 group containing a carbonyl (C≡O) that has a singly bound O atom. (Int. F)

Alkene a compound that contains the bonding arrangement C≡C. (Ch. 1)

Alkene metathesis reaction in which portions of molecules joined by C≡C bonds in the products were not initially joined in the reactant. (Ch. 19)

Alkoxy group the group —OR. (Int. A)

Alkoxymercuration–reduction the sequence of reactions in which an Hg^{+2} species adds to an alkene or alkyne, followed by reduction (typically with $NaBH_4$), resulting in the *Markovnikov addition* of an alcohol. (Ch. 12)

Alkylation a reaction in which a hydrogen atom is replaced by an alkyl group. (Ch. 10)

Alkylborane a compound R—BH_2, in which an alkyl group has replaced hydrogen on borane, BH_3. (Ch. 12)

Alkyl group (also called alkyl substituent) a portion of a molecule that consists only of C—C and C—H single bonds. (Ch. 1, Int. A)

Alkyl halide a compound that contains the bonding arrangement C—X, where X is a halogen atom. (Ch. 1)

Alkyllithium reagent (RLi) reagent with a C—Li bond wherein the C atom is strongly nucleophilic and strongly basic. (Ch. 17)

1,2-Alkyl shift a carbocation rearrangement wherein an alkyl group migrates to an adjacent atom. (Ch. 7)

Alkyl substitution the number of alkyl groups bonded to an isomer's alkene carbon atoms. (Ch. 5)

Alkyne a compound that contains the bonding arrangement C≡C. (Ch. 1)

Alkynide anion RC≡C⁻, a strong nucleophile. (Ch. 9)

All-anti conformation (also known as *zigzag conformation*) the lowest-energy conformation of an alkyl chain; conformation wherein an alkyl chain has the *anti conformation* at each C—C bond. (Ch. 4)

Allyl substrate compound of the form L—CH₂—CH=CH₂, where L is a leaving group. (Ch. 9)

α-Amino acid one of the relatively few types of small organic molecules from which proteins are constructed; an α-amino acid contains an amino group (NH₂), a carboxyl group (CO₂H), and a side chain (R) that are all attached to the α carbon of the amino acid. (Ch. 1)

α Anomer a diastereomeric form of a cyclic sugar in which the OH group attached to the anomeric carbon atom is on the opposite side of the ring as the substituent that is attached by C. (Ch. 18)

α,β-Unsaturated carbonyl compound a ketone or aldehyde that has a double or triple bond connecting the α and β carbons. (Ch. 17)

α Carbon a carbon atom that is attached to a carbonyl (C=O) carbon. (Ch. 1, Ch. 7)

α-Cleavage a fragmentation pathway in mass spectrometry in which a C—C bond adjacent to the radical is cleaved. (Int. G)

α-Elimination a type of reaction in which a H atom and a leaving group are eliminated from the same atom; mechanism by which dichlorocarbene is produced from trichloromethane. (Ch. 12)

α-Helix a type of secondary structure in a protein characterized by a tight winding of the protein's backbone around an axis. (Ch. 26)

α Spin state the state describing a nucleus that has a spin of $+\frac{1}{2}$ a.u. (Ch. 16)

Amide a compound that contains the bonding arrangement —C(O)N. (Ch. 1)

Amine a compound that contains the bonding arrangement C—N. (Ch. 1)

Aminolysis a reaction in which an amine reactant is responsible for breaking bonds; aminolysis of a carboxylic acid derivative produces an amide. (Ch. 21)

Amorphous solid a solid that does not have a well-defined crystal structure. (Ch. 26)

Analyte a sample that one wishes to analyze. (Ch. 15)

Angle strain the increase in energy that results from the deviation of a bond angle from its ideal valence shell electron pair repulsion (VSEPR) angle. (Ch. 2)

Anion a negatively charged ion. (Ch. 1)

Anionic polymerization a polymerization reaction in which the initiator is anionic and the propagation steps involve anionic species. (Ch. 26)

[*n*]Annulene one of a class of compounds that are monocyclic, fully conjugated, and contain only carbon and hydrogen. (Ch. 14)

Anode the positively charged pole of an electrochemical cell, such as the one used in electrophoresis. (Ch. 6)

Anomeric carbon the carbon atom of a cyclic sugar that is part of an acetal or a hemiacetal and is the carbonyl carbon of the sugar in its acyclic form. Depending on the stereochemical configuration of the anomeric carbon, the cyclic sugar can exist as one of two diastereomers called *anomers*. (Ch. 18)

Anomers diastereomers of a cyclic sugar that differ in the stereochemical configuration at the *anomeric carbon*. A cyclic sugar can exist as the α or β anomer. (Ch. 18)

Antiaromatic term used to describe compounds that have cyclic π systems that are unusually unstable; the cyclic π system contains an *anti-Hückel number* of electrons. (Ch. 14)

Antibonding contribution overlapping AOs with the opposite phase, which raises the energy of the MO they contribute to. (Ch. 14)

Antibonding MO a molecular orbital (MO) that is significantly higher in energy than its contributing atomic orbitals. (Ch. 3)

Anti conformation staggered conformation wherein bulky groups are 180° apart in a Newman projection. (Ch. 4)

Anticoplanar (also known as *antiperiplanar*) the conformation occurring when the H and leaving group on adjacent atoms are anti to each other and all four atoms reside in the same plane. (Ch. 8)

Anti-Hückel number a number in the set {4, 8, 12, 16, . . .}, which can be described as either an even number of pairs or a number that is consistent with 4*n*, where *n* is a positive integer. (Ch. 14)

Anti-hydrogenation reaction in which two H atoms, overall, add to a triple bond in a trans fashion. (Ch. 25)

Anti-Markovnikov addition a reaction in which the incoming halide in H—X, a Brønsted acid, adds to an alkene or alkyne at the less-substituted carbon (with more protons) and the proton adds to the other. See *Markovnikov's rule* and *Markovnikov addition*. (Ch. 11)

Anti-Markovnikov regiochemistry the regiochemistry obtained from *anti-Markovnikov addition*. (Ch. 12)

Antiperiplanar See *anticoplanar*. (Ch. 8)

Anti-Zaitsev product (also known as *Hofmann product*) the less-substituted alkene product of an elimination reaction, produced when the reaction does not follow Zaitsev's rule. (Ch. 9)

Aprotic solvent a solvent that does not possess a hydrogen-bond donor. (Ch. 2)

Arene a compound that contains an aromatic ring. (Ch. 1)

Arenium ion intermediate (also known as a *Wheland intermediate*) a cationic intermediate in an electrophilic aromatic substitution produced by the addition of an electrophile to an aromatic ring. (Ch. 22)

Aromatic term used to describe compounds that have cyclic π systems that are unusually stable; the cyclic π system contains a Hückel number of electrons. (Ch. 14)

Aromatic compound See *arene*. (Ch. 1)

Aromatic halogenation an electrophilic aromatic substitution reaction that replaces an aromatic hydrogen with a halogen. (Ch. 22)

Aromaticity the property of a molecule containing an unusually stable π system, and thus is consistent with Hückel's rules. (Ch. 14)

Arrow pushing See *curved arrow notation*. (Ch. 6)

Asymmetric carbon a carbon atom that is a chiral center, being bonded to four different groups. (Ch. 5)

Asymmetric stretch a vibrational mode involving two bonds within a molecule whereby one bond lengthens while the other shortens. (Ch. 15)

Asymmetric synthesis synthesis that favors the formation of a particular enantiomer or diastereomer. (Ch. 13)

Atactic term used to describe a polymer that has no regularity in the stereochemical configurations along its main chain. (Ch. 26)

Atomic number (Z) the number of protons in the nucleus. (Ch. 1)

Atomic orbital an *orbital* that is assigned to a single atom, such as *s*, *p*, or a hybrid orbital. (Ch. 1)

Attacking species any species that can act as a nucleophile or base to displace a leaving group from an atom. (Ch. 9)

Aufbau principle the principle that states that each successive electron must fill the lowest-energy orbital available. (Ch. 1)

Axial term used to describe a bond that is perpendicular to the plane that is roughly defined by the ring to which it is attached (i.e., perpendicular to the *equator* of the molecule). (Ch. 4)

2,2′-Azobisisobutyronitrile (AIBN) the molecule $(CH_3)_2C(CN)N=NC(CN)(CH_3)_2$, a common radical initiator. (Ch. 25)

Azo coupling reaction that connects two aromatic rings via the $—N=N—$ group. (Ch. 23)

Azo dye compound with a distinct color that contains the *azo group*. (Ch. 23)

Azo group the $—N=N—$ group. (Ch. 23)

B

Backbone the carbon chain or ring of a molecule to which substituents are attached. In polymers, also known as the *polymer chain* or *main chain*. (Ch. 26)

Backside attack occurs when a nucleophile approaches the substrate from the side opposite the leaving group, leading to an inversion of stereochemical configuration at the attacked carbon; occurs in all S_N2 reactions. (Ch. 8)

Baeyer–Villiger oxidation an oxidation reaction in which a peroxyacid (RCO_3H) converts a ketone or an aldehyde into an ester or a carboxylic acid. (Ch. 21)

Base catalysis the speeding up of a reaction in the presence of a small amount of base, without the base being consumed by the reaction. (Ch. 18)

Base peak the peak with the greatest intensity in a mass spectrum; it is assigned a relative abundance of 100% and used as a reference to assign the relative abundances of other peaks. (Ch. 16)

Beer–Lambert law law describing the direct relationship of absorbance (A) to sample concentration (C), molar absorptivity (ε), and light path length (l) in spectroscopy. (Ch. 15)

Bending one of two basic, independent types of vibrational motion. In bending, an angle—either a bond angle or a dihedral angle—becomes larger and smaller throughout one vibrational cycle. (Ch. 15)

Benzene C_6H_6, a nonpolar, aromatic ring; six-carbon annulene. (Int. B)

Benzene derivative relatively simple molecule with one or more substituents attached to benzene. (Int. B)

Benzenediazonium ion the reactive species $C_6H_5N_2^+$, used to produce a variety of substituted benzenes. (Ch. 22)

Benzyl (Bn) (also known as *phenylmethyl*) a $—CH_2—C_6H_5$ group. (Int. B)

Benzylic carbon a carbon atom attached directly to a phenyl ring. (Ch. 22)

Benzyl substrate a substrate of the form $L—CH_2—C_6H_5$, where L is a leaving group. (Ch. 9)

Benzyne intermediate a six-membered carbon ring whose Lewis structure exhibits two $C=C$ bonds and one $C≡C$ bond; an intermediate in a type of nucleophilic aromatic substitution reaction. (Ch. 23)

β Anomer a diastereomeric form of a cyclic sugar in which the OH group attached to the anomeric carbon atom is on the same side of the ring as the substituent that is attached by C. (Ch. 18)

β Carbon a carbon atom located two bonds away from an atom or group of interest—usually a $C=O$ group. (Ch. 17)

β-Diacid a species having two carboxyl ($—CO_2H$) groups separated by a single carbon atom: $HO_2C—C—CO_2H$. (Ch. 21)

β Elimination reaction in which substituents on adjacent atoms are eliminated from a molecule, resulting in a new π bond between those atoms. (Ch. 7)

β-Hydroxycarbonyl compound a species in which a hydroxyl ($—OH$) group is bonded to the carbon that is β to a carbonyl functional group. (Ch. 18)

Betaine a species in which a positive and a negative charge are separated by two uncharged atoms, and in which the positively charged atom has no attached hydrogens. (Ch. 17)

β-Keto acid a species in which the β carbon of a carboxylic acid is part of a carbonyl group that is characteristic of a ketone or aldehyde. (Ch. 21)

β-Keto ester a species in which the β carbon of an ester is part of a carbonyl group that is characteristic of a ketone or aldehyde. (Ch. 21)

β-Pleated sheet a type of secondary structure in a protein in which several portions of the backbone align next to each other in the same region of space. (Ch. 26)

β Spin state the state describing a nucleus that has a spin of $-\frac{1}{2}$ a.u. (Ch. 16)

B_{ext} abbreviation for *external magnetic field*. (Ch. 16)

Bicyclic compound a compound consisting of two rings. (Ch. 24)

Bimolecular elimination (E2) step an elementary step wherein a proton and a leaving group are eliminated from adjacent atoms, and a double bond or triple bond is formed. (Ch. 7)

Bimolecular homolytic substitution (S_H2) an elementary step in which a radical forms a bond to an atom in a closed-shell species and displaces another radical from that atom. (Ch. 25)

Bimolecular nucleophilic substitution (S_N2) step an elementary step wherein a nucleophile forms a bond to an atom that is attached to a leaving group and displaces the leaving group from that atom. (Ch. 7)

Biomolecule a particular organic molecule found almost exclusively in or produced by living organisms. (Ch. 1)

Biopolymer polymer (i.e., molecule consisting of monomers) that is produced biosynthetically; examples of biopolymers are polypeptides (proteins), polysaccharides (carbohydrates), and polynucleotides (nucleic acids) such as DNA and RNA. (Ch. 26)

Birch reduction a dissolving metal reduction that reduces a benzene ring to a cyclohexa-1,4-diene. (Ch. 25)

Bond dipole a separation of partial positive and negative charges along a covalent bond. (Ch. 1)

Bond energy See *bond strength*. (Ch. 1)

Bonding contribution overlapping AOs with the same phase, which lowers the energy of the MO they contribute to. (Ch. 14)

Bonding MO a molecular orbital (MO) that is significantly lower in energy than its contributing atomic orbitals. (Ch. 3)

Bonding pairs   pairs of electrons that make up covalent bonds. (Ch. 1)

Bond length   the internuclear distance at which energy is a minimum. (Ch. 1)

Bond strength   (also known as *bond energy*) the energy that would be required to increase the distance between two bonded atoms from the bond length to infinity. (Ch. 1)

9-Borabicyclo[3.3.1]nonane (9-BBN)   a reagent used in the hydroboration–oxidation of alkynes to produce aldehydes or ketones. (Ch. 12)

Borate ester   a class of compounds or functional groups, $B(OR)_3$. (Ch. 12)

Branched   describes a molecule in which carbon chains or alkyl groups are attached to the main chain. (Int. A)

Branched polymer   a polymer in which hydrogens or substituents on the main chain of the polymer are replaced by another chain of the polymer. (Ch. 26)

Bridgehead carbon   carbon atom that is simultaneously part of more than one ring. (Ch. 24)

Broadband decoupling   a technique used to "turn off" the coupling between ^{13}C and ^1H nuclei so that carbon signals are not split by attached protons in nuclear magnetic resonance spectroscopy; radio frequency radiation forces protons to rapidly switch between α and β spin states, producing a zero average magnetic field produced by each proton. (Ch. 16)

Bromination   a reaction that adds bromine atoms to a molecule; can be accomplished by substituting a bromine atom for a hydrogen atom or by adding molecular bromine (Br_2) to a C=C or C≡C bond. (Ch. 22)

Bromohydrin   a compound in which a bromine atom and a hydroxyl group (—OH) are attached to two adjacent carbons. (Ch. 12)

Bromonium ion intermediate   a positively charged intermediate containing a three-membered ring composed of two carbon atoms and a bromine atom. (Ch. 12)

***N*-Bromosuccinimide (NBS)**   a radical initiator with a particularly weak Br—N bond, which is the source of a small, steady concentration of Br_2. Commonly, NBS is used to brominate allylic and benzylic carbons. (Ch. 25)

Brønsted–Lowry acid   a species that donates a proton in a proton transfer reaction. (Ch. 6)

Brønsted–Lowry acid–base reaction   See *proton transfer reaction*. (Ch. 6)

Brønsted–Lowry base   a species that accepts a proton in a proton transfer reaction. (Ch. 6)

C

Cahn–Ingold–Prelog convention   a set of tie-breaking rules applied to substituents to determine their relative priorities in establishing stereochemical configurations. (Int. C)

Carbamate   a species with the general form RO—CO—NR$_2$. (Ch. 21)

Carbanion   an *anion* in which a negative formal charge appears on carbon. (Ch. 7)

Carbene   an uncharged species containing a carbon atom that possesses two bonds and nonbonded electrons. (Ch. 12)

Carbocation   a species that contains a positively charged carbon atom (C^+); carbocations are key reactive intermediates in a variety of chemical reactions. (Ch. 6)

Carbocation rearrangement   an elementary step in which the connectivity within a carbocation changes. Examples include a *1,2-hydride shift* and a *1,2-methyl shift*. (Ch. 7)

Carbohydrate   (also known as *saccharide*) a compound with the formula $C_xH_{2y}O_y$. (Ch. 1)

Carbon-chain polymer   a polymer in which the main chain consists only of carbon atoms. (Ch. 26)

^{13}C NMR spectroscopy   a spectroscopy method in which radio frequency electromagnetic radiation is used to cause spin flips in the nuclei of carbon-13 atoms. (Ch. 16)

Carbon skeleton   the bonding arrangement (connectivity) of the carbon atoms. (Ch. 13)

Carbonyl group   the C=O functional group. (Ch. 1, Int. E)

Carboxylation   a reaction in which a substrate gains a carboxyl group. (Ch. 17)

Carboxylic acid   a compound that contains the bonding arrangement —CO_2H. (Ch. 1)

Carboxylic acid derivative   a compound that can be produced relatively easily from a carboxylic acid; generally characterized by the presence of a leaving group attached to a carbonyl (C=O) carbon. (Int. F)

Catalytic hydrogenation   a reaction in which hydrogen atoms are added to a molecule in the presence of a catalyst. (Ch. 19)

Cathode   the negatively charged pole of an electrochemical cell, such as the one used in electrophoresis. (Ch. 6)

Cation   a positively charged ion. (Ch. 1)

Cationic polymerization   a polymerization reaction in which the initiator is cationic and the propagation steps involve cationic species. (Ch. 26)

Chain-growth polymerization   (also known as *addition polymerization*) a polymerization reaction in which the polymer grows one monomer at a time and the reaction takes place at specific sites on the growing polymer. (Ch. 26)

Chain reaction   a reaction whose mechanism consists of initiation, propagation, and termination steps and whose net reaction is the sum of its propagation steps. (Ch. 25)

Chain transfer   step of a chain-growth polymerization in which the site of reaction on a polymer chain changes; changing the site of reaction to the middle of a polymer chain can result in a branched polymer. (Ch. 26)

Chair conformation   a cyclohexane conformation in which all bond angles of the ring are about 111°, and all C—C bonds are staggered. (Ch. 4)

Chair flip   the process of single-bond rotations that converts one chair conformation of cyclohexane (or a substituted cyclohexane) into the other. (Ch. 4)

Chemical distinction test   a protocol used to determine whether atoms in a molecule are chemically distinct. (Ch. 16)

Chemical environment   the electron distribution surrounding a particular location in space, governing the chemical behavior of an atom. (Ch. 16)

Chemically distinct protons   See *heterotopic protons*. (Ch. 16)

Chemically equivalent protons   See *homotopic protons*. (Ch. 16)

Chemical reaction   the transformation of one substance (reactant) into another substance (product), typically through changes in chemical bonds. (Ch. 6)

Chemical shift (δ)   a measure of the extent to which a nuclear magnetic resonance (NMR) signal's frequency differs from that of a reference compound, usually tetramethylsilane (TMS). (Ch. 16)

Chiral   term used to describe a molecule that has an *enantiomer*. (Ch. 5)

Chiral center   a tetrahedral atom bonded to four different groups. (Ch. 5)

Chiral environment   an environment that is nonsuperimposable on its mirror image. (Ch. 5)

Chlorination   a reaction that adds chlorine atoms to a molecule; can be accomplished by substituting a chlorine atom for a hydrogen atom or by adding molecular chlorine (Cl_2) to a C$=$C or C$\equiv$C bond. (Ch. 22)

Chlorohydrin   a compound in which a chlorine and hydroxyl group (—OH) are attached to two adjacent carbons. (Ch. 12)

Chloronium ion intermediate   a positively charged species containing a three-membered ring composed of two carbon atoms and a chlorine atom. (Ch. 12)

***meta*-Chloroperbenzoic acid (MCPBA)**   a peracid (RCO_3H) reagent used in epoxidation reactions. (Ch. 12)

Chromate ester   a compound characterized by the arrangement O$=$Cr—OR; an intermediate in the chromium oxidation of alcohols and aldehydes. (Ch. 19)

Chromic acid   the compound H_2CrO_4, which is used to oxidize secondary alcohols to ketones and primary alcohols to carboxylic acids. (Ch. 19)

Cis   term that describes the configuration wherein two atoms are on the same side of a double bond or plane of a ring. (Ch. 3, 4)

Claisen condensation reaction   a reaction in which an enolate anion undergoes a nucleophilic addition–elimination with an ester, producing a β-keto ester. (Ch. 21)

Clemmensen reduction   a reaction that uses a zinc amalgam under acidic conditions to reduce the C$=$O group of a ketone or an aldehyde to a methylene (CH_2) group. (Ch. 19)

Closed-shell species   a molecule or ion in which all electrons are paired. (Ch. 25)

Combination   See *radical coupling*. (Ch. 26)

Common name   See *trivial name*. (Int. A)

Complementary nucleic acids   two DNA or RNA strands are complementary when their facing nitrogenous bases pair, forming strong hydrogen bonds and a characteristic double helix. In DNA, adenine pairs with thymine, and guanine with cytosine; in RNA, adenine pairs with uracil, and guanine pairs with cytosine. (Ch. 14)

Complex splitting   in NMR spectroscopy, splitting that arises when a nucleus is coupled to more than one distinct type of other nuclei. (Ch. 16)

Concerted   describes the breaking and forming of bonds in an elementary step that occur simultaneously. (Ch. 6)

Condensation polymerization   a type of *step-growth polymerization* in which a small molecule is eliminated when two monomers or growing polymer chains bond together. (Ch. 26)

Condensed formula   a way of representing molecules using a line of text with hydrogens written immediately to the right of the atom to which they are bonded. (Ch. 1)

Configurational isomers   isomers that have the same connectivity but differ in a way other than by rotations about single bonds. See the two types: *enantiomers* and *diastereomers*. (Ch. 5)

Conformational analysis   a plot of a molecule's energy as a function of one or more of its dihedral angles. (Ch. 4)

Conformers   nonsuperimposable molecules that have the same connectivity (see *stereoisomers*) and differ by rotations about single bonds. (Ch. 4)

Conjugate acid   the species that a base becomes after acquiring a proton. (Ch. 6)

Conjugate addition   See *1,4-addition*. (Ch. 11)

Conjugate base   the species that an acid becomes after losing a proton. (Ch. 6)

Conjugated   describes double or triple bonds that are separated by another bond; also describes *p* atomic orbitals or π molecular orbitals that are adjacent and overlap in a side-by-side fashion. (Ch. 11, 14)

Connectivity   (also called bonding scheme) information that conveys which atoms are bonded together and by what types of bonds (single, double, or triple). (Ch. 1)

Conservation of number of orbitals   the generalized concept that when *n* orbitals are mixed, *n* unique orbitals must be produced. See *molecular orbital theory*. (Ch. 3)

Constitutional isomers   (also known as *structural isomers*) compounds that share the same molecular formula but differ in their connectivity. (Ch. 4)

Constructive interference   interference that occurs when waves or orbitals overlap with the same phase; constructive interference produces a new wave or orbital that has been *built up* by the addition of the amplitudes from the contributing waves. (Ch. 3)

Contact surface area   the area over which two molecules interact. (Ch. 2)

Convergent synthesis   a method of synthesizing molecules, wherein portions of the target are synthesized separately and then assembled at a later stage, usually resulting in a higher yield than a corresponding *linear synthesis*. (Ch. 13)

Coordination step   an *elementary step* in which a single bond is formed using two electrons from the same atom and no bonds are broken. (Ch. 7)

Copolymer   a *polymer* that is produced from two different monomers that can each undergo self-polymerization. (Ch. 26)

Core electrons   inner, lower-energy shell electrons that are not used for bonding. (Ch. 1)

Coupled   term used to describe two or more atoms that exhibit *spin–spin coupling* in NMR spectroscopy. (Ch. 16)

Coupling constant (*J*)   the frequency difference between peaks of a split NMR signal, independent of the external magnetic field (B_{ext}). (Ch. 16)

Coupling reaction   reaction in which two groups are joined together and a new C—C bond is formed. (Ch. 19)

Covalent bond   one of two types of fundamental bonds in chemistry; it is characterized by the sharing of valence electrons between two or more atoms. (Ch. 1)

Crossed aldol reaction   an *aldol addition* reaction in which the carbonyl-containing reactants that join together are different. (Ch. 18)

Crossed Claisen condensation reaction   a *Claisen condensation reaction* in which the carbonyl-containing reactants that join together are different. (Ch. 21)

Cross-link   a portion of a polymer that is responsible for joining two separate main chains. (Ch. 26)

Crystal lattice   the regular array into which ions or molecules in a solid are arranged. (Ch. 1)

Crystalline solid   a solid that has a well-defined crystal structure (i.e., a structure in which the locations of the atoms form a regular pattern). (Ch. 26)

C-terminus   the end of a polypeptide (protein) chain that is part of a carboxyl group. (Ch. 21)

Curved arrow   the symbol that illustrates the electron movement necessary to change one resonance structure into another, or to form or break bonds in an elementary step of a mechanism. (Ch. 1)

Curved arrow notation (also known as *arrow pushing*) the notation used by organic chemists to describe the movement of electrons necessary to change one resonance structure into another or to form or break bonds in an elementary step of a mechanism. (Ch. 6)

Cyanohydrin a compound characterized by a hydroxyl (—OH) and a cyano (—C≡N) group attached to the same carbon. (Ch. 18)

Cycloaddition reaction a concerted reaction in which two separate species come together to produce a new ring. (Ch. 24)

[2+2] Cycloaddition reaction a concerted reaction in which two separate species come together to produce a new ring, with two electrons supplied by each species. (Ch. 24)

[4+2] Cycloaddition reaction a concerted reaction in which two separate species come together to produce a new ring; four electrons are supplied by one species, and two electrons are supplied by the other. (Ch. 24)

[6+2] Cycloaddition reaction a concerted reaction in which two separate species come together to produce a new ring; six electrons are supplied by one species, and two electrons are supplied by the other. (Ch. 24)

Cycloalkane a cyclic molecule consisting of only carbon and hydrogen atoms and only single bonds. (Int. A)

Cycloalkyl group a substituent consisting of a cyclic arrangement of only carbon and hydrogen atoms and only single bonds. (Int. A)

[4+2] Cycloelimination a concerted reaction in which a cyclic molecule is cleaved into two separate species; one species receives four electrons, and the other receives two. Also called a *retro Diels–Alder reaction*. (Ch. 24)

D

Dash–wedge notation a system of drawing to represent three-dimensional molecules using dashes (bonds point away from the viewer) and wedges (bonds point toward the viewer). (Ch. 2)

Deactivating group a substituent that decreases the reactivity of a molecule in a particular reaction; often refers to substituents that make an aromatic ring less susceptible to electrophilic aromatic substitution. (Ch. 23)

Debye (D) conventional unit of the dipole moment. (Ch. 2)

Decarboxylation a reaction that removes a carboxyl (—CO₂H) group from a molecule. (Ch. 21)

Degenerate orbital one of a set of orbitals that have identical energies. (Ch. 14)

Degradation reaction a reaction that breaks larger molecules into smaller molecules. (Ch. 26)

Degree of polymerization (DP) the number of *repeating units* in a polymer. (Ch. 26)

Degree of unsaturation See *index of hydrogen deficiency (IHD)*. (Ch. 4)

Dehydration the removal of water from a substance; often describes a reaction in which water is a product. (Ch. 9)

Delocalization the phenomenon of electrons or charges being less confined to a particular location within a species. Delocalization is generally stabilizing. (Ch. 1)

Delocalization energy See *resonance energy*. (Ch. 1)

Delocalized describes an electron or charge that is not confined to a particular location within a species. (Ch. 1)

Deoxyribonucleic acid (DNA) a double-helical pair of intertwined nucleic acids that stores genetic information. (Ch. 1)

Depolymerization a reaction that causes a polymer to lose its monomers; the reverse of a *polymerization*. (Ch. 26)

Deprotection step a reaction used in synthesis to selectively convert a protected functional group to the original functional group. (Ch. 19)

DEPT ¹³C NMR spectroscopy distortionless enhancement by polarization transfer, a technique used in ¹³C NMR spectroscopy to determine the type of carbon (CH₃, CH₂, CH, or C) responsible for producing a particular signal. (Ch. 16)

Deshielded describes protons whose shielding (see *shielded*) becomes diminished by nearby electronegative atoms or by magnetic anisotropy. Deshielding results in a higher signal frequency, and thus a larger *chemical shift*. (Ch. 16)

Destructive interference interference that occurs when waves or orbitals overlap with opposite phase; destructive interference between two orbitals produces a new orbital that is diminished in size. (Ch. 3)

Desulfonation a reaction that removes a sulfo (—SO₃H) group from a molecule. (Ch. 22)

Detergent one of a class of substances with long-chain molecules that are very hydrophilic on one end and very hydrophobic on the other. Detergents form micelles to emulsify dirt, grease, and oils; unlike soaps, detergents are effective cleansers in *hard water*. (Ch. 2)

Dextrorotatory (from Latin, meaning "rotating to the right") describes chiral compounds that rotate plane-polarized light clockwise (in the + direction). (Ch. 5)

Dialkylborane a compound R₂BH with two alkyl groups attached to boron. (Ch. 12)

Diastereomers *configurational isomers* that are *not* mirror images of each other. (Ch. 5)

Diastereotopic describes atoms such that Step 1 of the *chemical distinction test* yields *diastereomers*. (Ch. 16)

1,3-Diaxial interaction interaction between atoms or groups separately attached to a cyclohexane ring at relative positions numbered 1 and 3, and occupying axial positions. (Ch. 4)

Diazomethane the compound CH₂N₂; used as a reagent to produce a methyl ester from a carboxylic acid or to produce a cyclopropane ring from an alkene. (Ch. 10)

Diazotization a reaction that forms a compound with the diazo group (—N≡N—). (Ch. 22)

Diborane (B₂H₆) explosive, gaseous dimer of borane (BH₃) that consists of three-center, two-electron bonds. (Ch. 12)

Dichlorocarbene (also called dichloromethylene) the compound Cl₂C. (Ch. 12)

Dicyclohexylborane (C₆H₁₁)₂BH, a bulky dialkylborane used in the hydroboration of an alkyne to help ensure that the alkyne undergoes a single addition. (Ch. 12)

Dicyclohexylcarbodiimide (DCC) the molecule (C₆H₁₁)N≡C≡N(C₆H₁₁), which is used as a reagent to couple a carboxylic acid and an amine to produce an amide. (Ch. 21)

Dieckmann condensation an intramolecular *Claisen condensation reaction*. (Ch. 21)

Diels–Alder reaction a reaction between a conjugated diene and a dienophile that produces a cyclohexene ring. (Ch. 24)

Diene a compound with a pair of conjugated π bonds that contributes four π electrons in a *Diels–Alder reaction*. (Ch. 24)

Dienophile a compound that contributes a single π bond (two π electrons) in a *Diels–Alder reaction*. (Ch. 24)

Diglyceride a fatty acid diester of glycerol. (Ch. 2)

Dihedral angle (θ) the angle between a bond on the front atom of a Newman projection and a bond on the rear atom of the Newman projection; each angle of rotation defines a particular dihedral angle. (Ch. 4)

Diisobutylaluminum hydride (DIBAH or DIBAL-H) the compound (i-Bu)$_2$Al—H, which is used to reduce esters to aldehydes. (Ch. 20)

1,3-Dipolar cycloaddition a concerted reaction in which two separate species come together to produce a new ring and in which one species is characterized by the resonance delocalization of charge over atoms that have a 1,3-positioning. (Ch. 24)

Dipole arrow a labeling arrow depicting the direction from lower (the origin) to higher (the point) electronegativity. (Ch. 1)

Dipole–dipole interaction the intermolecular interaction that arises because the positive end of one molecule's *net dipole* is attracted to the negative end of another's. (Ch. 2)

Dipole moment a measure of the magnitude of a molecule's dipole. (Ch. 2)

Direct addition See *1,2-addition*. (Ch. 11)

Disiamylborane (C_5H_{11})$_2$BH, a bulky dialkylborane used in the hydroboration of an alkyne to help ensure that the alkyne undergoes a single addition. (Ch. 12)

Disproportionation a reaction in which two identical species react to form two different products. (Ch. 26)

Dissolving metal reduction a reduction reaction during which a metal dissolves in solution. Commonly used to reduce alkynes to trans alkenes. (Ch. 19)

Disubstituted having two substituents. (Ch. 4)

Disubstituted benzene benzene ring that has two substituents attached. (Int. B)

Disulfide bond (also known as *disulfide bridge*) a link characterized by two sulfur atoms singly bonded together (i.e., —S—S—); often found in proteins linking two cysteine residues. (Ch. 26)

Disulfide bridge See *disulfide bond*. (Ch. 26)

Diterpene a *terpene* that contains two pairs of isoprene units (i.e., four isoprene units), for a total of 20 carbons. (Ch. 2)

Doublet (d) the splitting pattern of an NMR signal into two peaks of essentially equal height. (Ch. 16)

Doublet of doublets the splitting pattern of an NMR signal when each of the peaks in a doublet brought about by a first coupling is split again into a doublet by a second coupling. (Ch. 16)

Downfield appearing in an NMR spectrum in a region of higher chemical shift. (Ch. 16)

Driving force a thermodynamic concept (not an actual force) describing the extent to which a reaction favors products over reactants under a particular set of conditions; it tends to increase with increasing stability (lower energy) of products relative to reactants; charge stability and total bond energy are important contributors. (Ch. 7)

Duet a set of two electrons in the $n = 1$ shell of an atom. (Ch. 1)

E

E1cb mechanism a type of β-elimination reaction in which the first step is deprotonation and the second step is departure of a leaving group. (Ch. 18)

Eclipsed conformation rotational conformation in which the bonds to the front atom in a Newman projection cover, or "eclipse," the bonds to the rear atom. (Ch. 4)

Edman degradation a reaction that removes the N-terminal amino acid from a polypeptide for analysis. (Ch. 21)

Effective electronegativity property describing the tendency of an atom to hold electrons more tightly as an outcome of the atom's hybridization. (Ch. 3)

Effective magnetic field (B_{eff}) the magnetic field that is "felt" by the nucleus; the sum of B_{ext} and B_{loc}. (Ch. 16)

Elastomer a polymer that has the ability to retain its original shape after large deformations; vulcanized rubber is a common example. (Ch. 26)

Electron configuration the way in which electrons are arranged in atomic or molecular orbitals. (Ch. 1)

Electron donating describes a substituent that adds electron density to an atom to which it is bonded; electron-donating ability is considered relative to hydrogen. (Ch. 6)

Electronegativity (EN) the ability of an atomic nucleus to attract electrons in a covalent bond. (Ch. 1)

Electron geometry the orientation of the electron groups about a particular atom. (Ch. 2)

Electron impact ionization the process of producing a molecular ion by colliding an electron beam with a gaseous molecule; used in mass spectrometry. (Ch. 16)

Electron spin density plot a plot of the electron cloud about a molecule, color coded to show regions where there is high probability of finding an unpaired electron. (Ch. 25)

Electron transition the change in the orbital occupied by an electron. (Ch. 15)

Electron withdrawing describes a substituent that removes electron density from an atom to which it is bonded; electron-withdrawing ability is considered relative to hydrogen. (Ch. 6)

Electrophile a strongly electron-deficient species. (Ch. 7)

Electrophile elimination step an elementary step in which an electrophile is eliminated from a carbocation, generating a new π bond; the reverse of an electrophilic addition step. (Ch. 7)

Electrophilic addition reaction an addition reaction whose mechanism involves an *electrophilic addition step* as a key step. (Ch. 11)

Electrophilic addition step an elementary step in which electrons from a nonpolar π bond (as part of a double or triple bond) are used to form a new σ bond to an electrophile. (Ch. 7)

Electrophilic aromatic substitution a reaction that consists of the addition of an electrophile to an aromatic ring followed by the elimination of another electrophile, resulting in a net substitution. (Ch. 22)

Electrophoresis a common way of separating a mixture of amino acids or proteins (often at a given pH) by using an electric field; it is often done in a gel-like medium (gel electrophoresis). (Ch. 6)

Electrostatic forces the forces by which opposite charges attract one another and like charges repel one another. (Ch. 1)

Electrostatic potential map the depiction of a molecule's electron cloud in colors that indicate its *relative* charge; it is one way to illustrate the distribution of charge along a covalent bond. (Ch. 1)

Elementary step a reaction that occurs in a single event and does not proceed through an intermediate; all bonds that break or form in an elementary step do so essentially simultaneously. (Ch. 6)

Empirical rate law an experimentally determined rate law, which describes how the rate of a reaction depends on reactant and product concentrations. (Ch. 8)

Emulsify to disperse a substance in a solvent in which it is normally insoluble. See *detergent*. (Ch. 2)

Enamine a compound characterized by the bonding arrangement C=C—N. (Ch. 18)

Enantiomeric excess (ee)   the fraction of a mixture of enantiomers that is not racemic; it is the fraction of the mixture that contributes to the rotation of plane-polarized light. (Ch. 5)

Enantiomers   *configurational isomers* that are mirror images of each other. (Ch. 5)

Enantioselective synthesis   a synthesis carried out in a way that would favor one enantiomer over another. (Ch. 13)

Enantiotopic   describes atoms in a species such that Step 1 of the *chemical distinction test* yields *enantiomers*. (Ch. 16)

Endergonic   term describing a reaction that gains free energy on producing products—i.e., $\Delta G^\circ_{rxn} > 0$—and is generally nonspontaneous. (Ch. 6)

End group   the collection of atoms that characterize the end of the main chain of a polymer. (Ch. 26)

Endo approach   in a Diels–Alder reaction, the orientation of the dienophile in which the substituents on the dienophile point toward the diene, resulting in the *endo product*. (Ch. 24)

Endo product   the diastereomeric product of a Diels–Alder reaction in which the substituents on the diene and the dienophile end up on opposite sides of the new cyclohexene ring. (Ch. 24)

Endo rule   (also known as *Alder rule*) the generality describing the tendency of a Diels–Alder reaction to favor the *endo product* over the *exo product*. (Ch. 24)

Endothermic   describes a reaction or process that absorbs heat; $\Delta H^\circ_{rxn} > 0$. (Ch. 6)

Enolate anion   a deprotonated *enol*, bearing a negative charge. (Ch. 7)

Enol form   a compound that has the bonding arrangement C$=$C—OH; the enol form of a molecule is the tautomer of its *keto form*. (Ch. 7)

Entropy   a thermodynamic quantity that increases with the number of equivalent ways the energy in a system can be arranged. Many people like to think of entropy as a measure of disorder; a system with greater entropy (i.e., one that is more disordered) tends to be more likely to occur than one with less entropy. (Ch. 2)

Envelope conformation   the lowest-energy conformation of cyclopentane where four of its five carbon atoms lie essentially in one plane, with the fifth carbon outside that plane—resembling an envelope. (Ch. 4)

Enzyme   protein that acts as a catalyst for biological reactions— that is, enzymes facilitate biological reactions, but are not consumed while doing so. (Ch. 1)

Epimers   compounds that differ in stereochemical configuration at only one chiral center. (Ch. 5)

Epoxidation reaction   a reaction wherein an epoxide is produced, often involving an alkene and a peroxy acid like *meta*-chloroperbenzoic acid (MCPBA). (Ch. 12)

Epoxide   a compound that has a three-membered ring made of two C atoms and one O atom. (Ch. 1, Ch. 10)

Equatorial   term used to describe bonds in a cyclic molecule that lie almost in the plane that is roughly defined by the ring (i.e., the *equator* of the molecule) and point outward from the center of the ring. (Ch. 4)

Equilibrium constant (K_{eq})   an experimentally obtained constant that describes a reaction's tendency to form products. (Ch. 6)

Essential fatty acid   one of the naturally occurring fatty acids that cannot be synthesized in the human body by any known chemical pathway; instead, it must be consumed. (Ch. 4)

Ester   a compound that contains the bonding arrangement C—CO$_2$C. (Ch. 1)

Ether   a compound that contains the bonding arrangement C—O—C. (Ch. 1)

Ethoxy group   the group —OCH$_2$CH$_3$; also written as —OEt. (Int. A)

Ethylene glycol   the compound HOCH$_2$CH$_2$OH, often used to protect a ketone or aldehyde. (Ch. 19)

Ethyl group   the group —CH$_2$CH$_3$; also written as —Et. (Int. A)

Excited electronic state   an electron configuration that is not the lowest-energy configuration. (Ch. 24)

Exergonic   term describing a reaction that releases free energy on producing products—i.e., $\Delta G^\circ_{rxn} < 0$—and is generally spontaneous. (Ch. 6)

Exo approach   in a Diels–Alder reaction, the orientation of the dienophile in which the substituents on the dienophile point away from the diene, resulting in the *exo product*. (Ch. 24)

Exo product   the diastereomeric product of a Diels–Alder reaction in which the substituents on the diene and the dienophile end up on the same side of the new cyclohexene ring. (Ch. 24)

Exothermic   describes a reaction or process that releases heat; $\Delta H^\circ_{rxn} < 0$. (Ch. 6)

Expanded octet   a group of more than eight valence electrons belonging to a single atom; only atoms in the third row of the periodic table and below can have an expanded octet. (Ch. 1)

External magnetic field (B_{ext})   a magnetic field applied to a sample by a magnet, such as a superconductor magnet in NMR spectroscopy. (Ch. 16)

F

Fat   one of a subclass of lipids whose most common biological function is to store energy. Fats contain three adjacent ester groups, each of which can be produced from a *fatty acid* and a hydroxyl group from *glycerol*. Thus, a fat or oil is often described as a *triacylglycerol* or a *triglyceride*. Animal fats, such as lard, are generally solids at room temperature. Fats from plants, however, are generally liquids at room temperature, and are thus more properly called *oils*. (Ch. 2)

Fatty acid   a long-chain carboxylic acid. (Ch. 2)

Fingerprint region   the region of a sample's infrared (IR) spectrum below ~1400 cm^{-1}. A fingerprint region is often difficult to analyze but is unique to each compound. (Ch. 15)

Fischer esterification reaction   an acid-catalyzed reaction between an alcohol and a carboxylic acid that produces an ester. (Ch. 21)

Fischer projection   a relatively convenient, two-dimensional representation of configurations about chiral centers in a given molecule; for each stereocenter represented in a Fischer projection, horizontal bonds point toward the viewer and vertical bonds point away from the viewer. (Ch. 5)

Flagpole interaction   the steric strain resulting from atoms or groups attached to C1 and C4 of cyclohexane in its boat conformation. (Ch. 4)

Formal charge   one of two methods used with Lewis structures to determine the charges of atoms involved in covalent bonds: each lone pair is assigned to the atom on which it appears in the Lewis structure, and in a given covalent bond, half the electrons are assigned to each atom involved in the bond. (Ch. 1)

Fourier transform   a mathematical algorithm used in NMR spectroscopy that decomposes a *free induction decay* into its individual frequencies, or *signals*. (Ch. 16)

Fragmentation the process of breaking molecules into smaller pieces, such as with an electron beam for mass spectrometry. (Ch. 16)

Fragmentation pathways in mass spectrometry, the reactions that account for the breaking apart of the molecular ion (M^+) into smaller pieces. (Int. G)

Free energy of activation ($\Delta G^{\circ\ddagger}$) the energy barrier that must be surmounted for reactants to form products; the difference in standard free energy between the reactants and the transition state. (Ch. 6)

Free induction decay (FID) a digitized record of amplitude versus time for radio frequency radiation re-emitted by a sample in NMR spectroscopy. (Ch. 16)

Free radical a species that contains at least one unpaired electron. (Ch. 25)

Free radical polymerization a polymerization reaction in which the initiator produces a *radical* and the propagation steps involve radical species. (Ch. 26)

Free rotation rotation about a bond that can occur relatively rapidly under a given set of conditions; free rotation can occur about single bonds at room temperature but not about double bonds. Rotation about a double bond can occur only if the π bond breaks. (Ch. 3)

Frequency (ν) the number of complete oscillations that occur in a given time. It is given in units of hertz (Hz), cycles/second, or s^{-1}, all of which are equivalent. (Ch. 15)

Friedel–Crafts acylation an electrophilic aromatic substitution reaction in which a proton on an aromatic ring is replaced by an acylium ion ($R—C\equiv O^+$). (Ch. 22)

Friedel–Crafts alkylation an electrophilic aromatic substitution reaction in which a proton on an aromatic ring is replaced by a carbocation (R^+). (Ch. 22)

Frontier molecular orbital a *HOMO* or *LUMO* orbital of a reacting species in an elementary step. (Int. D)

Frontier molecular orbital (FMO) theory a theory that invokes HOMO–LUMO interactions in the reacting species to establish whether an elementary step is allowed or forbidden. (Int. D)

Frontside attack describes a nucleophile in a nucleophilic substitution reaction attacking the substrate from the same side as the leaving group, leading to retention of stereochemical configuration. (Ch. 8)

Frost method a method to derive the relative energies of an [*n*]annulene's π MOs. (Ch. 14)

Fuming sulfuric acid concentrated H_2SO_4 infused with SO_3 liquid. (Ch. 22)

Functional group conversion or **transformation** a reaction that changes one functional group into another. (Ch. 13)

Functional group a common bonding arrangement of relatively few atoms. Functional groups dictate the behavior of entire molecules—that is, molecules with the same functional groups tend to behave similarly. (Ch. 1)

Furanose cyclic form of a monosaccharide characterized by a five-membered ring. (Ch. 18)

Fused rings rings that have bonds in common. (Ch. 14)

G

Gabriel synthesis a synthesis scheme that involves the reaction of phthalimide and an alkyl halide (RX) to produce a primary amine (RNH_2). (Ch. 20)

Gauche conformation a staggered conformation wherein bulky groups are 60° apart in a Newman projection. (Ch. 4)

Gel permeation chromatography (GPC) a technique used to analyze the distribution of molecular size within a sample; smaller molecules are detected later in the analysis because they interact more strongly with the gel. (Ch. 26)

Geminal dihalide description of a molecule in which two halide substituents are attached to the same carbon atom. (Ch. 11)

Gilman reagent See *lithium dialkylcuprate*. (Ch. 17)

Glass transition temperature (T_g) the temperature above which the species that make up an amorphous solid begin to flow past each other. (Ch. 26)

Glycolysis a metabolic pathway that breaks down simple carbohydrates for their energy. (Ch. 7)

Glycoside one of a class of compounds characterized by sugars with glycosidic linkages to another group. (Ch. 9)

Glycosidic linkage the C—O—C group that connects a sugar unit to another group, often another sugar. (Ch. 9)

Green alternative an innovative reaction, technique, or technology to avoid the production and accumulation of hazardous materials. (Ch. 13)

Green chemistry a set of guiding principles to prevent pollution and other health hazards associated with organic synthesis and the chemicals industry. (Ch. 13)

Grignard reaction a reaction in which a Grignard reagent adds in as a nucleophile to the electrophilic atom of a group containing a polar π bond. (Ch. 17)

Grignard reagent an alkylmagnesium halide (R—MgX, where X = Cl, Br, or I) nucleophile that favors direct addition to a polar π bond over conjugate addition. Grignard reagents are often used to make new carbon–carbon bonds. (Ch. 7)

Ground state the most stable (i.e., the lowest energy) electron configuration. (Ch. 1)

Grubbs catalyst a compound that has the bonding arrangement $Cl_2Ru\equiv C$, used in alkene metathesis reactions. (Ch. 19)

Gyromagnetic ratio (γ) a value that relates the energy separation between the spin states of a nucleus and an applied magnetic field. (Ch. 16)

H

^1H NMR spectroscopy See *proton NMR spectroscopy*. (Ch. 16)

Haloalkane an alkane with a halo substituent (—F, —Cl, —Br, or —I). (Int. A)

Haloform a compound having the general form HCX_3, where X = F, Cl, Br, or I. (Ch. 20)

Haloform reaction the reaction of a methyl ketone or an aldehyde with a molecular halogen (e.g., Cl_2, Br_2, or I_2) under basic conditions, producing a *haloform* and a carboxylate anion that can be protonated to a form a carboxylic acid. (Ch. 20)

Halogenation the process of adding halogens to a molecule. (Ch. 10)

Halohydrin one of a class of molecules in which a halogen atom and a hydroxyl group are on adjacent carbon atoms. (Ch. 10)

Hammond postulate postulate suggesting that if, in a reaction free energy diagram, two species lie near each other along the reaction coordinate (the *x* axis) and are similar in energy, then they will have very similar structures. (Ch. 9)

Hard water   water that contains a significant concentration of Mg^{2+} or Ca^{2+} ions. (Ch. 2)

Haworth projection   a two-dimensional representation wherein a ring is depicted as being planar, occupying the plane perpendicular to the page, and substituents are drawn perpendicular to that plane; they illustrate cis and trans relationships well, but they do not accurately convey three-dimensional relationships or steric strain. (Ch. 4)

Head-to-head   in a polymerization reaction, the orientation in which the heads of two monomers are facing each other. (Ch. 26)

Head-to-tail   in a polymerization reaction, the orientation in which the head of one monomer faces the tail of another monomer. (Ch. 26)

Heat of hydrogenation   the amount of heat released (ΔH°_{hyd}) when $H_2(g)$ adds to double or triple bonds. It is used to gain a sense of the stability of π systems. (Ch. 14)

Heat of combustion   the energy given off in the form of heat (ΔH°_{comb}) during a combustion reaction. (Ch. 4)

Heck reaction   coupling reaction similar to the *Suzuki reaction* in which one vinylic or aryl group joins with another in the presence of a palladium catalyst (PdL_2). (Ch. 19)

Heisenberg uncertainty principle   principle that states that the uncertainty in a measurement of an electron's position is inversely proportional to the uncertainty in a measurement of its momentum. In other words, the more precisely we know the electron's position, the less precisely we know where it is going, and vice versa. (Ch. 3)

Hell–Volhard–Zelinsky (HVZ) reaction   a reaction in which the α carbon of a carboxylic acid is brominated to produce an α-bromo acid. (Ch. 21)

Hemiacetal   a compound having the bonding arrangement in which one hydroxy (—OH) and one alkoxy (—OR) group are attached to the same carbon atom. (Ch. 1, Ch. 18)

Henderson–Hasselbalch equation   the equation that relates the pH of a solution to the ratio of the concentration of an acid's conjugate base to that of the acid. $pH = pK_a + \log([A^-]/[HA])$. (Ch. 6)

Heteroatom   a noncarbon or nonhydrogen atom. (Ch. 1)

Heterochain polymer   a polymer in which the main chain consists of carbon atoms and noncarbon atoms. (Ch. 26)

Heterocyclic aromatic compound   one of a class of aromatic compounds that have heteroatoms incorporated into their aromatic ring structures. (Ch. 14)

Heterogeneous catalyst   a catalyst that is in a different phase from the rest of the reaction mixture. (Ch. 19)

Heterolysis step   See *heterolytic bond dissociation step*. (Ch. 7)

Heterolytic bond dissociation step   an elementary step in which only a single bond is broken and both electrons from that bond end up on one of the atoms initially involved in the bond. (Ch. 7)

Heterotopic protons   (also known as *chemically distinct protons*) protons surrounded by different electron distributions—that is, ones that reside in different chemical environments and so absorb at different frequencies in NMR spectroscopy. (Ch. 16)

Hexose   a six-carbon sugar; nomenclature referring to the class of six-carbon sugars. (Ch. 4)

Highest occupied molecular orbital (HOMO)   the highest-energy MO that contains an electron. (Ch. 3)

Hofmann elimination reaction   a reaction creating a double bond and leading to an anti-Zaitsev elimination product: the less stable, less alkyl substituted alkene. (Ch. 10)

Hofmann product   (also known as *anti-Zaitsev product*) the product resulting from the Hofmann elimination reaction; the less substituted product. (Ch. 10)

Homolysis   (also known as *homolytic bond dissociation*) an elementary step in which only a single bond is broken and the two electrons from that bond are split evenly between the atoms that were initially bonded together. (Ch. 25)

Homolytic bond dissociation   See *homolysis*. (Ch. 25)

Homopolymer   a polymer made from a single monomer. (Ch. 26)

Homotopic protons   (also known as *chemically equivalent protons*) protons surrounded by identical electron distributions—that is, ones that reside in identical chemical environments and so absorb at the same frequency in NMR spectroscopy. (Ch. 16)

Hooke's law   description of the behavior of a spring, used to model molecular vibrations. Mathematically, $\nu_{spring} = \sqrt{(k/m)}$, where ν_{spring} is the vibrational frequency of the spring, k is a constant that describes the spring stiffness, and m is the mass undergoing vibration. (Ch. 15)

Hückel number   a number in the set $\{2, 6, 10, 14, \ldots\}$, which can be described as either an odd number of pairs or a number that is consistent with $4n + 2$, where n is a non-negative integer. (Ch. 14)

Hückel's rules   empirical rules for predicting the aromaticity of a species. (Ch. 14)

Hund's rule   principle stating that all orbitals at the same energy must contain a single electron before a second electron can be paired in the same orbital. (Ch. 1)

Hybrid atomic orbital   an orbital that results from mixing two or more pure atomic orbitals from the valence shell of a single atom. (Ch. 3)

Hydrazone   a compound characterized by the bonding arrangement C=N—N. (Ch. 18)

Hydride anion ($H:^-$)   an anion with a hydrogen nucleus and two electrons. (Ch. 17)

Hydride reagent   a reagent that contains or has the tendency to donate hydride. Common hydride reducing agents include lithium aluminum hydride ($LiAlH_4$) and sodium borohydride ($NaBH_4$). Sodium hydride (NaH) is a strong base but a poor nucleophile. (Ch. 7)

1,2-Hydride shift   a carbocation rearrangement wherein a hydride ion (H^-) migrates to an adjacent atom. (Ch. 7)

Hydroboration   the net addition of BH_3 across a double or triple bond; hydrogen adds to one atom of the multiple bond and BH_2 adds to the other. (Ch. 12)

Hydroboration–oxidation   a sequence of reactions (hydroboration followed by oxidation) that results in the *anti-Markovnikov addition* of water to an alkene or alkyne. (Ch. 12)

Hydrocarbon tail   the hydrophobic, nonpolar end of a species that is distinguished from its hydrophilic end; often used to describe the nonpolar end of a fatty acid carboxylate. (Ch. 2)

Hydrogen atom abstraction   an elementary step in which a hydrogen atom is removed from a molecule; a type of S_H2 step in which a radical forms a bond to a hydrogen atom in a molecule, and the initial bond to hydrogen is broken, generating a new radical. (Ch. 25)

Hydrogen bond   (H bond) intermolecular interaction that involves a *hydrogen-bond donor* (NH, FH, or OH) and a *hydrogen-bond acceptor* (typically N, O, or F). (Ch. 2)

Hydrogen-bond acceptor any atom with a large concentration of negative charge and a lone pair of electrons, such as N, O, or F. (Ch. 2)

Hydrogen-bond donor a hydrogen atom covalently bonded to either F, O, or N. (Ch. 2)

Hydrolysis a type of reaction in which water is a reactant that facilitates the breaking of bonds; water is generally a nucleophile in these reactions. (Ch. 21)

Hydroperoxide ion the HOO^- anion, a key nucleophile in the oxidation of a trialkylborane to produce an alcohol. (Ch. 12)

Hydrophilic "water loving"; tending to dissolve in water. (Ch. 2)

Hydrophobic "water fearing"; not tending to dissolve in water. (Ch. 2)

Hydrophobic effect an effect characterized by the tendency of hydrophobic side chains to reside in the interior of a folded protein to minimize interactions of those side chains with an aqueous environment. (Ch. 26)

Hydroxyl group the O—H functional group. (Ch. 1)

Hyperconjugation the delocalization of electrons in a σ bonding molecular orbital into an adjacent orbital of π symmetry. (Ch. 5)

Ideal bond angle the theoretical bond angle predicted by valence shell electron pair repulsion (VSEPR) theory. (Ch. 2)

Imine (also called a *Schiff base*) a compound characterized by the bonding arrangement —C≡N—R (or H). (Ch. 18)

Index of hydrogen deficiency (IHD) (also known as *degree of unsaturation*) *half* the number of hydrogen atoms missing from that molecule compared to an analogous, completely saturated molecule. (Ch. 4)

Induced dipole temporary dipole that arises within a molecule as a result of the electron distribution being altered by a nearby full or partial charge. (Ch. 2)

Induced dipole–induced dipole interactions (also known as *London dispersion forces*) the dominant intermolecular interactions between nonpolar molecules resulting from the temporary distortions of their electron distributions. (Ch. 2)

Induction the distortion of electron density along covalent bonds, brought about by the replacement of a hydrogen atom with another substituent. (Ch. 6)

Inductive effect the effect that *induction* has on stability or another property of a species. (Ch. 6)

Initiation step the step in a *chain reaction* mechanism that produces the type of intermediate found in the *propagation cycle*. (Ch. 25)

Inorganic compound a compound that is not organic; that is, a compound that is *not* composed primarily of carbon and hydrogen. (Ch. 1)

In-plane bending vibration vibrational motion of a molecule where bond angles change but the atoms involved in the vibration all remain in the same plane. (Ch. 15)

Instantaneous dipole temporary dipole that arises when there are more electrons on one side of a molecule than there are on the other at some instant. (Ch. 2)

Integral trace the line overlaid on an NMR spectrum that corresponds to the cumulative area under the peaks in the spectrum. The height of a stairstep in an integral trace is proportional to the number of nuclei that give rise to the corresponding peak. (Ch. 16)

Integration the operation of calculating the area under a curve, such as a peak in a spectrum; used to generate an integral trace in an NMR spectrum. (Ch. 16)

Intermediate a transitory species that appears in a reaction mechanism but does not appear in the overall reaction. (Ch. 8)

Intermolecular describes the involvement of two separate species. (Ch. 9)

Intermolecular forces See *intermolecular interactions*. (Ch. 2)

Intermolecular interactions (also known as *intermolecular forces*) interactions that arise between molecules due to their charge distributions. (Ch. 2)

International Union of Pure and Applied Chemistry (IUPAC) international chemistry authority that standardizes chemical nomenclature, constants, and processes. (Int. A)

Intramolecular term used to describe the involvement of separate portions of a single species. (Ch. 9)

Intramolecular proton transfer a reaction that results in the loss of a proton from one part of a species and the gain of a proton at another part of the same species. (Ch. 8)

Inverse centimeters (cm⁻¹) (also known as *wavenumbers* [$\bar{\nu}$] or *reciprocal centimeters*) a unit of frequency that is calculated by taking the reciprocal of the wavelength—that is, by dividing 1 by the wavelength (in cm). Physically, it corresponds to the number of waves that fit in 1 cm. (Ch. 15)

Ion a species that bears a net charge. (Ch. 1)

Ion–dipole interaction attraction between a positive or negative ion and a net molecular dipole; the intermolecular interaction whereby free ions can interact with the molecules of a polar solvent like water. (Ch. 2)

Ionic bonding one of two types of fundamental bonds in chemistry; in an ionic bond, the more electronegative atom acquires electrons given up by the less electronegative atom, forming oppositely charged ions. The electrostatic attraction between the positively charged cations and the negatively charged anions constitutes the ionic bond. (Ch. 1)

Ionic head group the charged (hydrophilic) end of a species that is distinct from its uncharged, nonpolar (hydrophobic) end; often used to describe the carboxylate portion of a fatty acid carboxylate. (Ch. 2)

Ion–ion interaction the attraction between two oppositely charged ions. (Ch. 2)

Ionizable describes a species that is able to gain or lose a proton to become charged. (Ch. 6)

Ion pair a pair of interacting, oppositely charged ions. (Ch. 8)

Irreversible describes a reaction that does not readily take place in the reverse direction. (Ch. 9)

Irreversible reaction arrow reaction arrow pointing in one direction (→), signifying that the reaction is irreversible. (Ch. 9)

Isoelectric focusing a type of electrophoresis experiment whereby amino acids or proteins are separated by charge using a pH gradient. (Ch. 6)

Isoelectric pH (also known as *isoelectric point [pI]*) the pH at which a substance (usually an amino acid or protein) has an average charge of zero. (Ch. 6)

Isoelectric point (pI) See *isoelectric pH*. (Ch. 6)

Isomer a molecular species that has the same formula as another molecular species but is different in some way; see *constitutional isomers, stereoisomers, conformers, configurational isomers, diastereomers,* and *enantiomers*. (Ch. 4)

Isomerism   the relationships among molecules that have the same formula but are not identical. (Ch. 4)

Isoprene unit   a five-carbon unit into which a terpene's carbon backbone can be divided; every carbon atom in a terpene can be assigned to separate isoprene units. (Ch. 2)

Isotactic   describes a polymer in which all chiral centers along the main chain have the same configuration. (Ch. 26)

K

Kekulé structure   a specific resonance structure of a species, usually referring to benzene. (Ch. 14)

Keto–enol tautomerization   the transformation of one keto or enol tautomer (isomeric form) into the other; usually the process is an equilibrium. (Ch. 7)

Keto form   a carbonyl compound that has an α hydrogen; the keto form of a molecule is the tautomer of its *enol form*. (Ch. 7)

Ketohexose   a six-carbon sugar in which the carbonyl group involves an internal C, characteristic of a ketone. (Ch. 4)

Ketone   a compound that contains the bonding arrangement $C_2C{=}O$. (Ch. 1)

Ketopentose   a five-carbon sugar in which the carbonyl group involves an internal C, characteristic of a ketone. (Ch. 4)

Ketose   a sugar whose carbonyl group involves an internal C, characteristic of a ketone. (Ch. 4)

Kinetic control   a reaction condition in which the major product is the one produced the fastest. (Ch. 9)

Kinetic enolate anion   the enolate ion that is produced most rapidly by deprotonation of the corresponding ketone. (Ch. 10)

L

Labile   describes a species that is reactive under a given set of conditions. (Ch. 19)

λ_{max}   the wavelength of light at which the *absorbance* of a particular peak is a maximum. (Ch. 15)

Law of conservation of energy   the law that states that energy cannot be created or destroyed. (Ch. 15)

Leaving group   the substituent displaced or eliminated from the substrate in a reaction. (Ch. 7)

Leaving group ability   the relative rate at which a leaving group will separate from its substrate in a reaction. (Ch. 9)

Le Châtelier's principle   principle that states that if a reaction at equilibrium experiences a change in reaction conditions (e.g., concentrations, temperature, pressure, or volume), then the equilibrium will shift to counteract that change. (Ch. 6)

Leveling effect   the phenomenon dictating the maximum strength of an acid or base that can exist in solution; the strongest acid is the protonated solvent and the strongest base is the deprotonated solvent. (Ch. 6)

Levorotatory   (from Latin, meaning "rotating to the left") describes chiral compounds that rotate plane-polarized light counterclockwise (i.e., in the − direction). (Ch. 5)

Lewis acid   an electron-pair acceptor. (Ch. 7)

Lewis acid–base reaction   See *coordination step*. (Ch. 7)

Lewis acid catalyst   a species that catalyzes a reaction by virtue of its ability to undergo coordination with a reactant; used in electrophilic aromatic substitution halogenation reactions and Friedel–Crafts reactions. (Ch. 22)

Lewis adduct   the product of a coordination step. (Ch. 7)

Lewis base   an electron-pair donor. (Ch. 7)

Lewis dot structure   representation of a molecule that conveys connectivity and depicts all valence electrons. (Ch. 1)

Lindlar catalyst   a catalyst prepared by treating palladium with calcium carbonate and a small amount of quinoline and a lead salt; it is a type of *poisoned catalyst* that is used to reduce an alkyne to a cis alkene. (Ch. 19)

Linear alkane   See *straight-chain alkane*. (Int. A)

Linear combination of atomic orbitals (LCAO)   the summing and subtraction of various atomic orbitals to generate molecular orbitals. (Ch. 3)

Linear polymer   a polymer that consists of a single main chain—i.e., linear polymers have no branching. (Ch. 26)

Linear synthesis   a synthesis composed of sequential synthetic steps. (Ch. 13)

Line structure   a standard chemical representation method wherein carbons are implied at the end of a line and at each intersection of two or more lines, hydrogens bonded to carbon are not drawn but any bonded to another atom are drawn, and heteroatoms are drawn. (Ch. 1)

Lipid   a biomolecule that is relatively insoluble in water. Most lipids are highly nonpolar, consisting primarily of carbon and hydrogen, with very little oxygen or nitrogen content. Consequently, they tend to be soluble in relatively nonpolar organic solvents, such as ether. (Ch. 2)

Lipid bilayer   the double-layer structure into which phospholipids organize in an aqueous environment; lipid bilayers are the basis of cell membranes. (Ch. 2)

Lithium aluminum hydride (LiAlH₄ or LAH)   a very strong hydride reducing agent that favors direct addition over conjugate addition. (Ch. 17)

Lithium dialkylcuprate   (also known as *Gilman reagent*) one of a class of reagents R_2CuLi, which are sources of weak R^- nucleophiles that tend to favor conjugate addition over direct addition. (Ch. 17)

Lithium tri-*tert*-butoxyaluminum hydride (LTBA)   the compound $LiAlH(O{-}t{-}Bu)_3$, commonly used to reduce a carboxylic acid chloride to an aldehyde. (Ch. 20)

Localized   describes an electron or charge that is confined to a particular location within a molecular species. (Ch. 1)

Local magnetic field (B_{loc})   a magnetic field that adds to or subtracts from an external magnetic field (B_{ext}); when subjected to B_{ext}, the electrons that surround an atom can undergo coherent motion to contribute to B_{loc}. (Ch. 16)

Local minimum   a location along a curve in a plot where the value on the *y* axis (often energy) rises on an increase or decrease of the value on the *x* axis. (Ch. 8)

Locator number   (also called locant) index to identify sequential carbons in a main chain or ring. (Int. A)

London dispersion forces   See *induced dipole–induced dipole interactions*. (Ch. 2)

Lone pairs   pairs of nonbonding electrons confined to a particular atom. (Ch. 1)

Long-range coupling   in NMR spectroscopy, the situation in which nuclei separated by more than three bonds exhibit weak coupling. (Ch. 16)

Lowest unoccupied molecular orbital (LUMO)   the lowest-energy MO that is empty. (Ch. 3)

M

M + 1 peak a peak in a mass spectrum representing a molecular ion containing a heavy isotope, the mass of which is 1 u heavier than the most abundant isotope. The relative intensity of this peak can be used to estimate the number of carbon atoms in a sample molecule. (Ch. 16)

M + 2 peak a peak in a mass spectrum representing a molecular ion containing a heavy isotope, the mass of which is 2 u heavier than the most abundant isotope. The relative intensity of this peak can be used to determine the presence of atoms such as Br, Cl, and S. (Ch. 16)

m/z See *mass-to-charge ratio (m/z)*. (Ch. 16)

Magnetic anisotropy in NMR spectroscopy, the phenomenon that contributes to the B_{loc} of nearby nuclei, resulting from ring current in double bonds and aromatic rings; it causes substantial deshielding of hydrogen and carbon nuclei associated with simple alkenes, aromatic rings, and aldehydes. (Ch. 16)

Main chain in a relatively small organic molecule, the longest continuous chain of carbon atoms that establishes the root of the IUPAC name. In a polymer, the chain of atoms forming a repeating pattern that runs from one end of the molecule to the other. (Int. A, Ch. 26)

Malonic ester synthesis a synthesis scheme that converts malonic ester (diethyl malonate) into an alkyl- or dialkyl-substituted carboxylic acid. (Ch. 21)

Markovnikov addition a reaction in which the incoming halide in H—X adds to an alkene or alkyne at the more substituted carbon and the proton adds to the adjacent alkene or alkyne carbon with more hydrogens. (Ch. 11)

Markovnikov's rule empirical rule describing the *regiochemistry* of addition of a Brønsted acid across a C=C bond: The addition of a hydrogen halide to an alkene favors the product in which the proton adds to the alkene carbon that is initially bonded to the greater number of hydrogen atoms. (Ch. 11)

Mass spectrometry an experimental process that provides insight into the mass of a molecule and the fragments that compose it; this information about a molecule is obtained by interpreting a *mass spectrum*. (Ch. 16)

Mass spectrum a plot of the relative abundances of gaseous ions produced from an analyte against the mass-to-charge ratio, *m/z*. (Ch. 16)

Mass-to-charge ratio (*m/z*) the ratio of a particle's mass to its charge; it is the property that a mass spectrometer distinguishes. (Ch. 16)

McLafferty rearrangement in mass spectrometry, a type of fragmentation pathway exhibited by molecular ions of compounds that have a carbonyl-containing functional group. (Int. G)

Measured angle of rotation (α) the angle by which plane-polarized light is rotated on passing through a sample of a chiral compound. (Ch. 5)

Mechanism the precise sequence of elementary steps that results in the conversion of the original reactants to the final products in a chemical reaction. (Ch. 6)

Meisenheimer complex a reactive intermediate in nucleophilic aromatic substitution reactions that proceed by the nucleophilic addition–elimination mechanism. (Ch. 23)

Melting point (T_m) the temperature above which the species that form a crystal lattice begin to flow past each other and behave as a liquid. (Ch. 26)

Mercurinium ion intermediate a positively charged, three-membered ring intermediate in which the ring consists of two carbon atoms and a mercury atom. (Ch. 12)

Merrifield synthesis the solid-phase synthesis of a peptide or protein; named after R. B. Merrifield. (Ch. 21)

Meso describes a molecule that contains at least two chiral centers but has a plane of symmetry that makes it achiral overall. (Ch. 5)

***meta*-Chloroperbenzoic acid (MCPBA)** a peracid reagent used in epoxidation reactions. (Ch. 12)

Meta director a substituent attached to a benzene ring that directs an incoming electrophile to the meta carbon of the ring. (Ch. 23)

Methoxy group the group —OCH_3; also written as —OMe. (Int. A)

Methyl cation the carbocation H_3C^+ that has only hydrogen atoms attached to the positively charged carbon. (Ch. 6)

Methyl group the group —CH_3; also written as —Me. (Int. A)

1,2-Methyl shift a carbocation rearrangement wherein a methyl group migrates to an adjacent atom. (Ch. 7)

Micelle a spherical aggregate formed among surfactant species (such as soaps or detergents) in solution to maximize the intermolecular interactions involving the surfactant and solvent. (Ch. 2)

Michael reaction (also called Michael addition) a reaction in which a carbon nucleophile adds via conjugate addition to form a carbon–carbon bond; general description of conjugate additions involving any nucleophile. See *conjugate addition*. (Ch. 17)

Migratory aptitude the tendency of a bond between a H or alkyl group and a carbonyl carbon to be broken in a Baeyer–Villiger oxidation. (Ch. 21)

Molar absorptivity (ε) an experimentally derived quantity that is characteristic of a given species' probability of absorbing a photon at a given wavelength of radiation. (Ch. 15)

Molecular geometry the arrangement of atoms or groups of bonding electrons about a particular atom. (Ch. 2)

Molecular ion, $M^+(g)$ an ion formed when an electron is ejected from a gaseous molecule, often via *electron impact ionization* for *mass spectrometry*. (Ch. 16)

Molecularity the number of separate reacting species in an elementary step. (Ch. 7)

Molecular modeling kit a kit with which you can construct real models of molecules and manipulate them in your hands. (Ch. 2)

Molecular orbital (MO) an orbital resulting from a *linear combination of atomic orbitals* (LCAO), delocalized over the entire molecule; each MO accommodates up to two electrons. (Ch. 3)

Molecular orbital theory the theory whose central concept is that all electrons in a molecule can be thought of as occupying orbitals called molecular orbitals (MOs), or *linear combinations of atomic orbitals* (LCAOs), which are delocalized over the entire molecule. (Ch. 3)

Molozonide the product of the cycloaddition of one molar equivalent of ozone (O_3) to an alkene or alkyne in a [4+2] cycloaddition, characterized by a five-membered ring consisting of three adjacent O atoms and two adjacent C atoms. In an ozonolysis reaction, it is reacted further to produce ketones, aldehydes, and carboxylic acids from an initial alkene. (Ch. 24)

Monomer one of the small reactant molecules that make up polymers. (Ch. 26)

Monosaccharide   (also known as *simple sugar*) a carbohydrate whose number of oxygen atoms is the same as the number of carbon atoms, giving it the general formula $C_xH_{2x}O_x$; polysaccharides are constructed from monosaccharides. (Ch. 1)

Monosubstituted benzene   a benzene ring that is bonded to one substituent, having the form C_6H_5—Sub. (Ch. 23)

Monosubstituted cyclohexane   the result when one of the hydrogen atoms in cyclohexane is replaced by another substituent. (Ch. 4)

Monoterpene   a terpene that contains one pair of isoprene units (i.e., two isoprene units), for a total of 10 carbons. (Ch. 2)

Monounsaturated   term that describes a fatty acid or other molecule that has just one C=C double bond. (Ch. 4)

Multiple-center molecular orbital   a molecular orbital (MO) that spans multiple atoms and involves the simultaneous interaction of multiple *atomic orbitals (AOs)*. (Ch. 14)

Multiplet (m)   the NMR splitting pattern of a signal when there is a large number (typically more than five) of coupled protons. (Ch. 16)

Multistep mechanism   a mechanism composed of two or more *elementary steps*. (Ch. 8)

Mutarotation   a change in the rotation of plane-polarized light due to the equilibration between epimers (stereoisomers that differ by the configuration at one stereocenter); observable when one anomer of a particular sugar equilibrates with its other anomer when dissolved in water. (Ch. 18)

N

N + 1 Rule   a rule stating that if a proton is coupled to N protons that are distinct from itself but equivalent to each other, then the signal produced by that proton will be split into $N + 1$ separate peaks. (Ch. 16)

n → π* Transition   term describing an electron from a nonbonding orbital being promoted to a π antibonding (π*) orbital. (Ch. 15)

Natural product   one of a class of compounds synthesized in biological systems having a biological function or role; often terpenes or terpenoids. (Ch. 13)

Net molecular dipole   (also known as *permanent dipole*) a separation of partial positive charge and partial negative charge that sum to zero; the dipole is said to point from the center of positive charge to the center of negative charge. (Ch. 2)

Net reaction   See *overall reaction*. (Ch. 8)

Network polymer   a polymer that is so highly branched it has no recognizable linearity and can be viewed as a single, very large molecule. (Ch. 26)

Newman projection   a two-dimensional representation of a molecule viewed down the bond of interest; one convenient way to illustrate rotational conformations. (Ch. 4)

Nitration   a type of electrophilic aromatic substitution reaction in which a proton is replaced by a nitro (—NO_2) group. (Ch. 22)

Nitrile   a compound that contains the bonding arrangement C—C≡N. (Ch. 1)

Nitroalkane   an alkane with a nitro substituent (—NO_2). (Int. A)

Nitrogen inversion   the process (an umbrella flip) by which the configurations about nitrogen interconvert in the species NRR′R″. (Ch. 5)

Nitrogen rule   principle that states that a compound tends to have an odd molecular mass if it contains an odd number of nitrogen atoms, and an even molecular mass if it contains an even number of nitrogen atoms or none at all. (Ch. 16)

Nitronium ion   the ion $^+NO_2$, which is the strong electrophile that replaces a proton in an electrophilic aromatic nitration reaction. (Ch. 22)

Noble gas   an element in group 8A, characterized by completely filled valence shells that provide substantial stability. (Ch. 1)

Nodal plane   a plane in which the electron density is zero—that is, a plane in which an electron has zero probability of being found. (Ch. 3)

Node   a location where electron density is zero. (Ch. 3)

Nomenclature   the naming of atoms and molecules. (Ch. 1, Int. A)

Nonaromatic   a catchall term used to describe compounds that are neither aromatic nor antiaromatic and, thus, do not have unusually stable or unusually unstable π systems. (Ch. 14)

Nonbonding MO   a molecular orbital (MO) that is similar in energy to its contributing atomic orbitals. (Ch. 3)

Nonpolar   describes a molecule having no net dipole moment. (Ch. 2)

Nonpolar covalent bond   a covalent bond between atoms that have the same electronegativity, such that the electrons that make up the bond are shared equally. (Ch. 1)

Nonsuperimposable   describes two molecules for which there is no orientation such that *all* atoms of both molecules can be superimposed (i.e., lined up perfectly). (Ch. 5)

N-terminus   the end of a polypeptide (protein) chain that is part of an amino group. (Ch. 21)

Nuclear magnetic resonance (NMR) spectroscopy   experimental process used for structure determination, in which radio frequency radiation is used to cause spin flips in the nuclei of atoms that have an odd number of protons, an odd number of neutrons, or both. (Ch. 16)

Nuclear spin   property of some atomic nuclei, which generates a small magnetic field, an outcome expected of a charge in motion. (Ch. 16)

Nucleic acid   a large molecular chain that is primarily associated with the storage and transfer of genetic information; it is constructed from relatively small molecular units called nucleotides. (Ch. 1)

Nucleophile   a "nucleus-loving" species (usually with a lone pair of electrons) that seeks a substrate with a full or partial positive charge to stabilize its full or partial negative charge. (Ch. 7)

Nucleophile elimination step   an elementary step wherein a new π bond is produced and the eliminated leaving group, in turn, has the characteristics of a nucleophile; the reverse of a *nucleophilic addition step*. (Ch. 7)

Nucleophilic acyl substitution   a nucleophilic addition–elimination mechanism in which a nucleophile replaces the leaving group attached to a C=O carbon. (Ch. 20)

Nucleophilic addition–elimination reaction   a reaction whose mechanism contains subsequent nucleophilic addition and nucleophile elimination steps, resulting in the net substitution of a nucleophile for a leaving group. (Ch. 20)

Nucleophilic addition reaction   an addition reaction in which a key step in the mechanism is a nucleophilic addition step. (Ch. 17)

Nucleophilic addition step   an elementary step in which a nucleophile forms a bond to the electron-deficient atom of a polar π bond. (Ch. 7)

Nucleophilic aromatic substitution a reaction in which a nucleophile replaces a leaving group attached to an aromatic ring; the reaction could have a nucleophilic addition–elimination mechanism or could proceed through a benzyne intermediate. (Ch. 23)

Nucleophilicity (also called nucleophile strength) the tendency of a nucleophile to promote an S_N2 reaction; typically determined by the rate of an S_N2 reaction between the nucleophile and a methyl halide. (Ch. 9)

Nucleotide a relatively small molecular unit from which nucleic acids are constructed. All nucleotides have three distinct components: an inorganic phosphate (PO_4) group, a cyclic monosaccharide (or sugar), and a nitrogenous base. (Ch. 1)

O

Octet a group of eight valence electrons; a complete valence shell of atoms in the second row of the periodic table. (Ch. 1)

Oil a triglyceride that is liquid at room temperature; oils are typically produced from plants. (Ch. 2)

Olefin See *alkene*. (Ch. 26)

Olefin metathesis See *alkene metathesis*. (Ch. 19)

Operating frequency the frequency at which a bare proton absorbs electromagnetic radiation when subjected to an NMR spectrometer's magnetic field. (Ch. 16)

Optically active describes a sample that rotates plane-polarized light. Optically active species are chiral. (Ch. 5)

Optically inactive describes a sample that does not rotate plane-polarized light. Achiral species are optically inactive, as are racemic mixtures of enantiomers. (Ch. 5)

Orbital a description of the particular state of an electron in an atom or molecule; an electron that belongs to a particular orbital has a high probability of being located in a region of space whose size and shape are characteristic of that orbital. (Ch. 3)

Order the exponent to which a reactant (or product) concentration is raised in the rate law of a reaction. (Ch. 8)

Organic chemistry the branch of chemistry involving *organic compounds* (compounds containing carbon and hydrogen atoms). (Ch. 1)

Organic compound a compound composed primarily of carbon and hydrogen. (Ch. 1)

Organometallic describes compounds that contain a metal atom bonded directly to a carbon atom. (Ch. 7)

Ortho/para director a substituent attached to a benzene ring that directs an incoming electrophile to the ortho and meta carbons of the ring. (Ch. 23)

Osmate ester a compound having the bonding arrangement RO—Os=O. A cyclic osmate ester is an intermediate in the oxidation of an alkene using osmium tetroxide (OsO_4). (Ch. 24)

Out-of-plane bending vibration a type of vibration in which the set of atoms undergoing vibrational motion occupy different planes throughout an oscillation. (Ch. 15)

Overall product a species that appears as a product in the overall (i.e., net) reaction; it is generally a product that can be isolated on the completion of the reaction. (Ch. 8)

Overall reactant a species that appears as a reactant in the overall (i.e., net) reaction; it is generally a starting material that is added to a reaction. (Ch. 8)

Overall reaction (also known as *net reaction*) the reaction that includes only the reactants that are added as starting materials and the products that are present when the reaction is complete; the sum of the elementary steps of the mechanism. (Ch. 8)

Oxaphosphetane an intermediate in a Wittig reaction characterized by a four-membered ring consisting of two carbon atoms, one phosphorus atom, and one oxygen atom. (Ch. 17)

Oxetane a four-membered ring ether. (Ch. 10)

Oxidation part of a redox reaction in which a species loses one or more electrons and becomes oxidized; accompanied by the reduction of another species. (Ch. 17)

Oxidation state one of two methods used to assign charge to atoms involved in covalent bonds. Lone pairs are assigned to the atom on which they appear in the Lewis structure; in a given covalent bond, all electrons are assigned to the more electronegative atom; if the two atoms are identical, the electrons are split evenly. (Ch. 17)

Oxidative cleavage an oxidation reaction that results in the severing of C=C or C≡C bonds into separate carbonyl-containing functional groups. (Ch. 24)

Oxidized describes a species that has lost one or more electrons in a redox reaction. (Ch. 17)

Oxymercuration–demercuration See *oxymercuration–reduction*. (Ch. 12)

Oxymercuration–reduction (also known as *oxymercuration–demercuration*) the sequence of reactions in which a Hg^{+2} species adds to an alkene or alkyne followed by reduction (typically with $NaBH_4$), resulting in the *Markovnikov addition* of water. (Ch. 12)

Ozone the compound O_3, which is a key reagent in ozonolysis reactions. (Ch. 24)

Ozonide a key intermediate in an ozonolysis reaction, characterized by a five-membered ring consisting of three O atoms and two C atoms, and in which only two of the O atoms are adjacent. An ozonide is reacted further to produce ketones, aldehydes, and carboxylic acids from an initial alkene. (Ch. 24)

Ozonolysis a reaction in which an alkene is reacted with ozone to cleave the C=C bond, producing two separate carbonyl-containing functional groups. (Ch. 24)

P

Parent chain See *main chain*. (Int. A, Ch. 26)

Pascal's triangle a pattern that describes the ratios of the relative peak heights in splitting patterns in accordance with the $N + 1$ rule. (Ch. 16)

Pauli's exclusion principle the principle that states that no more than two electrons (i.e., zero, one, or two electrons) can occupy a single orbital; two electrons in the same orbital must have opposite spins. (Ch. 1)

***p* Character** the resemblance of an orbital's characteristics (e.g., shape and energy) to those of a *p* orbital; the fraction of a hybrid atomic orbital that comes from a *p* orbital. (Ch. 3)

Peak See *absorption band*. (Ch. 15)

Pendant group (also known as *side group*) a nonhydrogen atom or group that is bonded to an atom of a polymer main chain. (Ch. 26)

Pentose a five-carbon sugar; nomenclature referring to the class of five-carbon sugars. (Ch. 4)

Peptide bond (also known as *peptide linkage*) the O=C—N group, characteristic of amides, that serves to connect two amino acids together in a peptide or protein. (Ch. 21)

Peptide linkage See *peptide bond*. (Ch. 21)

Percent atom economy   measure of the inherent efficiency of a reaction; in general, as the percent atom economy increases, the amount of accumulated waste decreases. It is calculated as follows: % Atom economy = [(Mass of atoms in desired product)/(Mass of atoms in all reagents)] × 100%. (Ch. 13)

Percent dissociation   the extent to which an acid or base dissociates into its ions in a solution (usually aqueous); related to the [A⁻]/[HA] fraction in the *Henderson–Hasselbalch equation*. (Ch. 6)

Pericyclic reaction   a concerted reaction in which all of the electrons involved in bond breaking and bond formation are delocalized in a ring in the transition state. (Ch. 24)

Periodate anion   the species IO_4^-, used as a reagent in the oxidative cleavage of 1,2-diols to produce aldehydes and ketones. (Ch. 24)

Periodate ester   a compound characterized by the bonding arrangement RO—I=O. A cyclic periodate ester is a key intermediate in the reaction between a 1,2-diol and the periodate ion (IO_4^-) to produce ketones and aldehydes. (Ch. 24)

Periodic acid   the reagent HIO_4, used to introduce periodate anion in the oxidative cleavage of 1,2-diols. (Ch. 24)

Permanent dipole   See *net molecular dipole*. (Ch. 2)

Peroxyacid   A molecule of the form RCO_3H. (Ch. 12)

Phase   (of an electron) an immeasurable quantity that plays an important role in how orbitals interact with one another via *constructive interference* or *destructive interference*. (Ch. 3)

Phenol   a compound that is characterized by an O—H group attached to a phenyl ring. (Ch. 1)

Phenyl (Ph)   the substituent —C_6H_5. (Int. B)

Phenylmethyl   See *benzyl (Bn)*. (Int. B)

Phospholipid   one of a subclass of lipids; it can efficiently store energy and is integral in the formation of cell membranes. Whereas a fat or an oil has three fatty acids, making it a *triglyceride*, a phospholipid has only two fatty acids, making it a *diglyceride*. (Ch. 2)

Phosphonium ylide   See *Wittig reagent*. (Ch. 17)

Photochemically allowed   describes a reaction that proceeds at a reasonable rate when irradiated with ultraviolet light; one or more of the reactants have an excited electron configuration. (Ch. 24)

Photochemically forbidden   describes a reaction that does not take place readily when one or more reactants has an excited electron configuration. (Ch. 24)

Photon   a particle that makes up light, which carries a specific quantity of energy that can be associated with its frequency and wavelength. (Ch. 5)

π Bond   a bond wherein a pair of electrons occupies a bonding molecular orbital (MO) of π symmetry—that is, an MO in which the overlap of atomic orbitals takes place on either side of a bonding axis. A double bond in a Lewis structure is generally a σ bond plus a π bond, and a triple bond is one σ and two π bonds. (Ch. 3)

π → π* Transition   occurs when an electron from a π molecular orbital (MO) is promoted to a π* MO; the HOMO–LUMO transition in an alkene or alkyne. (Ch. 15)

π Stacking   the interaction (believed to be stabilizing) that takes place between separate π systems when the face of one π system is in close proximity to the face of the other. (Ch. 14)

π Symmetry   the symmetry of a molecular orbital that is generated by AO overlap on *either side of* the bonding axis; the bonding axis itself contains a nodal plane. (Ch. 3)

pK_a   the negative logarithm of the *acidity constant*; pK_a = −log K_a. (Ch. 6)

Planck's constant (*h*)   the constant of proportionality that appears in the dependence of a photon's energy on its frequency: 6.626×10^{-34} J·s. (Ch. 15)

Plane of symmetry   a plane that bisects a molecule in such a way that one-half of the molecule is the mirror image of the other half. (Ch. 5)

Plane-polarized light   light whose photons have their electric fields oscillating in the same plane. (Ch. 5)

Poisoned catalyst   a specially treated catalyst that prevents a reaction from occurring at the rate at which it would occur with an untreated catalyst. A poisoned Pd catalyst is often used to reduce an alkyne to a cis alkene. (Ch. 19)

Polar   describes a molecule having a net dipole moment. (Ch. 2)

Polar covalent bond   a covalent bond between atoms of different electronegativity, such that the electrons that make up the bond are shared unequally. (Ch. 1)

Polarizability   the ease with which a species' electron cloud is distorted; it tends to increase with the total number of electrons a species has. (Ch. 2)

Polarizable   describes the ability of a species' electron cloud to be distorted. (Ch. 2)

Polarizer   a device that generates plane-polarized light by allowing through only photons whose electric field is oscillating in a specified plane, effectively filtering out light whose electric field oscillates in any other plane. (Ch. 5)

Polyacrylate   (also known as *acrylic*) a polymer derived from monomers that are derivatives of acrylic acid. (Ch. 26)

Polyamino acid   a biological macromolecule made of amino acids joined together by peptide linkages. See also *protein*. (Ch. 1, 26)

Polyatomic ion   an ion that contains more than one atom. Polyatomic ions typically consist only of nonmetals, and their atoms are held together by covalent bonds. (Ch. 1)

Polycyclic aromatic hydrocarbon (PAH)   one of a class of molecules characterized by fused, unsubstituted, aromatic rings and often having a single π system; PAHs are known atmospheric pollutants. (Ch. 14)

Polyester   a heterochain polymer with a repeating unit containing ester linkages. (Ch. 26)

Poly(ether ether ketone) (PEEK)   a polymer in which two C—O—C groups (characteristic of ethers) and a C_2C=O group (characteristic of ketones) make up the main chain in the repeating unit. (Ch. 26)

Poly(ethylene terephthalate) (PET)   a polyester with the repeating unit (—$CH_2CH_2O_2C$—C_6H_4—CO_2—)$_n$; commonly used in 2-liter bottles for carbonated beverages. (Ch. 26)

Polyhalogenation   the process of replacing multiple hydrogens on a single molecule with halogen atoms. (Ch. 10)

Polymer   a large molecule that is made from relatively small monomers and that typically has a recognizable repeating unit. (Ch. 26)

Polymer chain   (also known as *main chain* or *backbone*) the chain of atoms bonded together in a polymer, to which pendant groups are attached. (Ch. 26)

Polymerization   a reaction that produces a polymer from monomers. (Ch. 26)

Polynucleotide   a biological macromolecule made up of nucleotide monomers. See also *nucleic acid*. (Ch. 1, 26)

Polyolefin   a carbon-chain polymer made from the polymerization of alkene (i.e., olefin) monomers. (Ch. 26)

Polypeptide a large biomolecule with peptide linkages joining amino acid monomers. See also *protein*. (Ch. 1, 26)

Polysaccharide (type of *carbohydrate*) a very large molecule constructed from just a few types of smaller molecules, called *monosaccharides* or *simple sugars*. Starch and cellulose are examples of polysaccharides. (Ch. 1, 26)

Polyunsaturated describes a molecule that has more than one C=C double bond; often used to describe fatty acids. (Ch. 4)

Poly(vinyl alcohol) (PVA) a polymer with the repeating unit (—CH[OH]—CH$_2$—)$_n$. (Ch. 26)

1,2-Positioning describes substituents that are connected to adjacent atoms. (Ch. 19)

Potassium permanganate the compound KMnO$_4$, used principally as an oxidizing agent. KMnO$_4$ can be used in the oxidation of primary alcohols and aldehydes to carboxylic acids and secondary alcohols to ketones, and can cleave alkenes, alkynes, and C—C bonds to a benzylic carbon. (Ch. 19)

Precursor the preceding species from which a synthetic target can be synthesized. (Ch. 13)

Primary (1°) carbocation a carbocation that has one alkyl group directly attached to C$^+$. (Ch. 6)

Primary orbital overlap an overlap among orbitals that are directly involved in bond formation in a Diels–Alder reaction. (Ch. 24)

Primary structure the sequence of amino acids in a protein; it determines a protein's identity. (Ch. 21)

Principal quantum number (*n*) the number that defines a shell in an atom, given that *n* can assume any integer value from 1 to infinity. (Ch. 1)

Principle of microscopic reversibility the notion that a reaction has the same free energy diagrams in the forward direction and the reverse direction. (Int. D)

Propagation cycle in a chain reaction, the collection of *propagation steps* that sum to the net reaction. (Ch. 25)

Propagation steps in a chain reaction, the elementary steps that are principally responsible for converting the overall reactants into the overall products. (Ch. 25)

Protecting group a modified functional group that has been made unreactive under the conditions for a desired chemical reaction. The original functional group is restored by removing the protecting group. (Ch. 19)

Protection step a reaction used in synthesis to selectively change a labile group to another functional group that will not react under certain conditions. (Ch. 19)

Protein a biomolecule constructed from amino acids; proteins can be hundreds or thousands of amino acids long and play diverse roles in biological systems. (Ch. 1)

Protic solvent a solvent that possesses a hydrogen-bond donor. (Ch. 2)

Proton NMR spectroscopy (also known as *^{1}H NMR spectroscopy*) a spectroscopy method in which radio frequency electromagnetic radiation is used to cause spin flips in the nuclei of hydrogen atoms. (Ch. 16)

Proton transfer reaction (also known as *Brønsted–Lowry acid–base reaction*) a reaction in which a proton is transferred from a *Brønsted–Lowry acid* to a *Brønsted–Lowry base* in a single elementary step—that is, one bond is broken and another is formed simultaneously. (Ch. 6)

Pseudorotation partial rotations about the single bonds that compose a ring structure; envelope conformations of cyclopentane rapidly interconvert via pseudorotation. (Ch. 4)

Pyranose a cyclic form of a monosaccharide characterized by a six-membered ring. (Ch. 18)

Pyridinium chlorochromate (PCC) a specialized oxidizing agent that can be used in organic solvents and, therefore, can oxidize a primary alcohol to an aldehyde or a secondary alcohol to a ketone. (Ch. 19)

Q

Quantum mechanics a branch of chemistry and physics whose central idea is that very small particles, like electrons, have traits that are characteristic of both particles and waves. (Ch. 3)

Quartet (q) the splitting pattern of an NMR signal into four peaks with a 1:3:3:1 height ratio; occurs when $N = 3$ in the $N + 1$ rule. (Ch. 16)

Quartet of triplets the NMR splitting pattern in which each peak of a quartet is split into a triplet, resulting in a total of 12 peaks. (Ch. 16)

Quaternary ammonium salt An ionic compound in which the cation is of the form R$_4$N$^+$. (Ch. 10)

Quaternary structure description of the arrangement of multiple folded protein subunits that form a complex. (Ch. 26)

Quintet (qn) the splitting pattern of an NMR signal into five peaks; occurs when $N = 4$ in the $N + 1$ rule. (Ch. 16)

R

R group See *side chain*. (Ch. 1)

Racemic mixture a mixture that contains equal amounts of the (+) and (−) enantiomers of a chiral molecule. (Ch. 5)

Radical addition step an elementary step in which a radical adds to an atom of a π bond. (Ch. 25)

Radical cation a positively charged ion that has an unpaired electron. (Int. G)

Radical coupling (also known as *combination*) an elementary step in which unpaired electrons from two radicals join to make a new covalent bond. (Ch. 25)

Radical halogenation a reaction that results in the addition of a halogen atom to a molecule and proceeds by a mechanism that involves radicals. (Ch. 25)

Radical initiator a compound that has especially weak covalent bonds and can be used to produce radicals via heating or irradiation with ultraviolet light. (Ch. 25)

Radio frequency (RF) the region of the electromagnetic spectrum having wavelengths between $\sim 10^{-3}$ m and 10^5 m. RF radiation is used in NMR spectroscopy to cause spin flips among nuclei. (Ch. 16)

Raney nickel a specialized catalyst used in the Raney-nickel reduction of ketones and aldehydes to convert the carbonyl group to a methylene. It is produced by treating a 50:50 alloy of aluminum and nickel with hot sodium hydroxide, followed by treatment with hydrogen gas. (Ch. 19)

Raney-nickel reduction a reaction that reduces the carbonyl group of a ketone or aldehyde to a methylene group using *Raney nickel*. (Ch. 19)

Rate constant the constant of proportionality that relates the concentrations of reactants or products (raised to certain exponents) to the rate of a given reaction. (Ch. 8)

Rate-determining step the elementary step that establishes the rate of an *overall reaction*; usually identified as the slow step of a mechanism. (Ch. 8)

Reaction coordinate a variable that corresponds to geometric changes, on a molecular level, of the species involved in a reaction as reactants are transformed into products; as the reaction coordinate increases, the geometries of the species involved in the reaction increasingly resemble those of the products. (Ch. 6)

Reaction free energy diagram a plot of Gibbs free energy as a function of the *reaction coordinate*. (Ch. 6)

Reaction kinetics the study of the rates of chemical reactions. (Ch. 8)

Reciprocal centimeters (also known as *wavenumbers [$\overline{\nu}$]*) See *inverse centimeters (cm^{-1})*. (Ch. 15)

Rearrangement a reaction in which the reactant and product species are isomers of each other. (Ch. 7)

Reduced describes a species that has gained one or more electrons in a redox reaction. (Ch. 17)

Reducing agent a reagent used to donate electrons, thereby causing the oxidation state of an atom in another species to decrease. (Ch. 17)

Reduction a part of a redox reaction in which a species gains one or more electrons and becomes reduced, accompanied by the oxidation of another species. (Ch. 17)

Reductive amination the formation of an amine from a ketone or aldehyde by reducing an imine intermediate. (Ch. 18)

Regioselectivity the tendency of a reaction to take place at one site within a molecule over another. (Ch. 9)

Relative abundance the amount of a substance that is present relative to that of a standard; in mass spectrometry, the quantity that is plotted on the *y* axis. (Ch 16)

Repeating unit the bonding scenario that occurs over and over within a polymer. (Ch. 26)

Residue an amino acid in a polypeptide or protein. (Ch. 21)

Resolved describes items that are separated; enantiomers that are separated are said to be resolved, and peaks in a spectrum that are separated are said to be resolved, too. (Ch. 16)

Resonance contributor See *resonance structure*. (Ch. 1)

Resonance effect the impact that delocalization via resonance has on a property of a molecule or ion, such as stability or acid/base strength. (Ch. 6)

Resonance energy (also known as *delocalization energy*) the extent by which a species is stabilized due to the delocalization of electrons or charge. (Ch. 1)

Resonance hybrid a weighted average of all resonance structures that can be drawn for a species. (Ch. 1)

Resonance structure (also known as *resonance contributor*) one of two or more valid Lewis structures that represent a species. Resonance structures are imaginary and are related by the movement of valence electrons, not atoms; the one, true species is represented by the *resonance hybrid*. (Ch. 1)

Resonance theory the theory that reconciles the differences between a species' multiple Lewis structures and its observed characteristics. *Resonance* exists when two or more valid Lewis structures can be drawn for a given molecular species. (Ch. 1)

Retro Diels–Alder reaction a [4+2] cycloelimination reaction that is the reverse of a Diels–Alder reaction and produces a diene and a dienophile. (Ch. 24)

Retrosynthetic analysis the strategy for designing a synthesis in which the chemist begins by focusing on the target and asks questions pertaining to feasible precursors that can be used to produce the target. Each precursor becomes the new target until the chemist arrives at precursors that can be used as starting materials. (Ch. 13)

Retrosynthetic arrow an open arrow ($\Longrightarrow$) used to indicate a *transform* in retrosynthetic analysis; it is drawn from the target to the precursor and can be interpreted to mean "can be made from." (Ch. 13)

Reversible describes reactions that proceed in the reverse direction at a rate that is comparable to or faster than that in the forward direction. (Ch. 9)

Reversible reaction arrows reaction arrows pointing in both directions ($\rightleftharpoons$), signifying that the reaction is reversible. Distinct from an *irreversible reaction arrow*. (Ch. 9)

Ribbon structure a trace of a three-dimensional protein's backbone that highlights secondary structures such as α-helices and β-pleated sheets. (Ch. 26)

Ribonucleic acid (RNA) a pair of intertwined nucleic acids that participate in protein synthesis. (Ch. 1)

Ring-closing metathesis a type of *alkene metathesis* reaction that forms a new ring. (Ch. 19)

Ring current the motion of π electrons along a circular path parallel to the plane of the molecular ring to which they belong, giving rise to *magnetic anisotropy*; the motion is driven by forces from an external magnetic field. (Ch. 16)

Ring flip See *chair flip*. (Ch. 4)

Ring-opening polymerization a polymerization reaction in which the propagation steps involve the breaking of a bond that is part of a ring in the monomer. (Ch. 26)

Ring strain the increase in energy due to geometric constraints of the bonds and substituents involved in a ring; it is partly responsible for the instability of rings other than five- and six-membered rings; it can be quantified using *heats of combustion*. (Ch. 4)

Robinson annulation a reaction that is made up of a Michael addition followed by aldol condensation and that results in the formation of a new cyclohexenone ring. (Ch. 18)

Root (also called parent compound) the name of the alkane containing the same number of carbon atoms as the main chain for nomenclature. (Int. A)

Rotational conformation the distinct angle of rotation about a single bond. (Ch. 4)

Rotational energy barrier the amount of energy needed to convert one conformer into another via rotation about a specified bond. (Ch. 4)

Rotaxane one of a class of compounds in which a dumbbell-shaped molecule is threaded through a large cyclic molecule called a *macrocycle*. (Ch. 9)

Royal purple (also known as *Tyrian purple*) a clothing dye obtained by ancient Phoenicians from a type of aquatic snail called *Bolinus brandaris*. (Ch. 1)

S

Saccharide (also known as *carbohydrate*) a molecule characterized by the molecular formula $C_xH_{2y}O_y$. (Ch. 1)

Sandmeyer reaction reaction in which a nucleophile replaces the N_2 leaving group in an arenediazonium ion and in which a Cu^+ catalyst is involved. (Ch. 22)

Saponification   a reaction between hydroxide (HO^-) and an ester, resulting in the formation of a carboxylate anion. This is the same type of reaction used by ancient civilizations to make soap from animal fat. (Ch. 20)

Saturated   describes a molecule that has the maximum number of hydrogen atoms possible, consistent with the number and type of each nonhydrogen atom in the molecule, the octet rule, and the duet rule. A saturated molecule has no double bonds, triple bonds, or rings. (Ch. 4)

***s* Character**   the resemblance of a hybrid orbital's characteristics (e.g., shape and energy) to those of an *s* orbital; the fraction of a hybrid atomic orbital that comes from an *s* orbital. (Ch. 3)

Schiff base   See *imine*. (Ch. 18)

***s*-Cis conformation**   a rotational conformation in which substituents are on the same side of a single bond. (Ch. 24)

Secondary (2°) carbocation   a carbocation that has two alkyl groups attached directly to C^+. (Ch. 6)

Secondary orbital overlap   in a Diels–Alder reaction, overlap among orbitals that are *not* directly involved in bond formation. (Ch. 24)

Selective reaction   a reaction in which multiple products can be produced, but one product is favored over the others. (Ch. 19)

Selective reagent   a reagent that is used to favor the production of one product over another. (Ch. 19)

Self-aldol addition   an *aldol addition* reaction in which the reactants that join together are the same carbonyl compound. (Ch. 18)

Sesquiterpene   a terpene that contains one and one-half pairs of isoprene units (i.e., three isoprene units), for a total of 15 carbons. (Ch. 2)

Shell   the principal quantum number (*n*) associated with an atomic orbital; also referred to as the energy level of the orbital. (Ch. 1)

Shielded   term used in NMR spectroscopy to describe nuclei in a molecule that have a local magnetic field that opposes an external magnetic field; shielded nuclei have lower chemical shifts. (Ch. 16)

Side chain   (often referred to as *R group*) a group connected to an α carbon(s) in an amino acid or protein, which gives the amino acid its identity. (Ch. 1)

Side group   (also known as *pendant group*) a nonhydrogen atom or group attached to the main chain of a polymer. (Ch. 26)

σ Bond   a bond wherein a pair of electrons occupies a bonding molecular orbital (MO) of σ symmetry—that is, an MO in which the overlap of atomic orbitals takes place along a bonding axis. A single bond in a Lewis structure is generally a σ bond. (Ch. 3)

σ Symmetry   the symmetry of a molecular orbital (MO) that is generated by the interaction of atomic orbitals along a bonding axis. A σ bond occurs when two electrons occupy a σ MO. (Ch. 3)

Signals   in nuclear magnetic resonance spectroscopy, the rf radiation emitted by NMR-active nuclei, which are converted into peaks in the NMR spectrum. (Ch. 16)

Signal splitting   the splitting of an NMR signal into multiple peaks due to the *spin–spin coupling* between the nucleus giving rise to the signal and nuclei of nearby atoms. (Ch. 16)

Simple sugar   See *monosaccharide*. (Ch. 1)

Single-barbed arrow   curved arrow (⤵) used to show the movement of a single electron. (Ch. 25)

Singlet (s)   the splitting pattern of an NMR signal when there are no coupled protons; it appears as just a single peak. (Ch. 16)

Soap   a compound, typically a salt of a fatty acid (RCO_2^- Na^+ or RCO_2^- K^+), used in cleaning up dirt and oil. See also *surfactant*. (Ch. 2)

Soap scum   the precipitate of long-chain fatty acid carboxylates with cations from *hard water*. (Ch. 2)

Sodium borohydride ($NaBH_4$)   a moderately strong hydride reducing agent $NaBH_4$ that favors direct addition over conjugate addition. (Ch. 17)

Sodium hydride (NaH)   an ionic hydride reagent NaH that is a strong base but a poor nucleophile. (Ch. 17)

Solvated electron   an electron that is stabilized by solvent molecules and, therefore, does not formally belong to any one atom. Solvated electrons are used in dissolving metal reductions. (Ch. 25)

Solvated (ion)   an ion that is involved in multiple ion–dipole interactions with its solvent. (Ch. 2)

Solvation   the phenomenon wherein an individual ion participates in multiple ion–dipole interactions with its solvent. When this occurs, the ions are said to be *solvated*. (Ch. 2)

Solvent-mediated proton transfer   the transfer of a proton from one site to another assisted by a sufficiently acidic or basic solvent. (Ch. 8)

Solvolysis   a reaction in which the solvent participates as a reactant in breaking a bond (lysis); an S_N1 reaction in which the solvent is a nucleophile is an example of solvolysis. (Ch. 9)

Species   a particular collection of protons, neutrons, and electrons. (Ch. 1)

Specific rotation $[\alpha]_\lambda^T$ a constant term that is unique for a given chiral compound and describes the compound's propensity for rotating plane-polarized light; it is the angle of rotation of plane-polarized light of a given wavelength in nanometers, λ, if it passes through a sample whose concentration is 1 g/mL, whose length is 1 dm, and whose temperature is *T*°C. (Ch. 5)

Spectator ion   an ion that is relatively inert in solution and tends not to react. (Ch. 7)

Spectroscopy   the study of the interaction of electromagnetic radiation with matter. (Ch. 15)

***sp* Hybrid atomic orbital**   a hybrid atomic orbital that results from *sp* hybridization. (Ch. 3)

***sp* Hybridization**   the mixing of the single *s* orbital and one of the three *p* orbitals—$p_x, p_y,$ or p_z—from the valence shell to form what are called *sp* hybrid atomic orbitals. (Ch. 3)

***sp²* Hybrid atomic orbital**   a hybrid atomic orbital that results from sp^2 hybridization. (Ch. 3)

***sp²* Hybridization**   the mixing of the single *s* orbital and two of the three *p* orbitals—$p_x, p_y,$ or p_z—from the valence shell to form sp^2 hybrid orbitals. (Ch. 3)

***sp³* Hybrid atomic orbital**   a hybrid atomic orbital that results from sp^3 hybridization. (Ch. 3)

***sp³* Hybridization**   the mixing of the single *s* orbital and all three *p* orbitals—$p_x, p_y,$ and p_z—from the valence shell to form sp^3 hybrid orbitals. (Ch. 3)

Spin flip   a change in spin from one spin state to another (e.g., from the α state to the β state, or vice versa). (Ch. 16)

Spin–spin coupling   the interaction between nearby nuclei that have spin; the cause of signal splitting in NMR spectroscopy. (Ch. 16)

Splitting diagram   a diagram that illustrates the complex splitting of an NMR signal in a stepwise fashion. (Ch. 16)

Staggered conformation rotational conformation in which each single bond on the front carbon atom in a Newman projection bisects a pair of single bonds on the rear carbon. (Ch. 4)

Standard enthalpy difference $\Delta H^{\circ}_{\text{rxn}}$ the enthalpy difference between reactants and products under standard conditions; associated with a reaction's tendency to absorb or release heat. (Ch. 6)

Standard entropy difference $\Delta S^{\circ}_{\text{rxn}}$ the entropy difference between reactants and products under standard conditions; related to the gain or loss of the number of different states available to a system. (Ch. 6)

Standard Gibbs free energy difference $\Delta G^{\circ}_{\text{rxn}}$ the Gibbs free energy difference between reactants and products under standard conditions; associated with a reaction's tendency to favor products at equilibrium (i.e., to be spontaneous). (Ch. 6)

Starting material one of the reactants that is available to use in a synthesis. (Ch. 13)

Step-growth polymerization a polymerization reaction in which molecules throughout a mixture join with any other molecules that have appropriate functional groups available to react, whether they are monomer units or other polymer molecules. (Ch. 26)

Stereocenter an atom with the property that interchanging any two of its attached groups produces a different stereoisomer. See *chiral center*. (Ch. 5)

Stereochemical configuration the classification of the three-dimensional arrangement of substituents attached to a chiral center or a double bond. The configuration of a chiral center is designated as *R* or *S*, and that of a double bond is designated as *E* or *Z*. (Ch. 5)

Stereochemistry the aspects of chemistry associated with the arrangement of atoms in space. (Ch. 8)

Stereoisomers isomers that have the same connectivity. (Ch. 5)

Stereospecific describes reactions that produce one stereochemical configuration or arrangement of atoms (stereochemistry) exclusively, which is dictated by the stereochemistry of the reactant. (Ch. 8)

Steric hindrance the spatial restriction of movement (in rotation or of an incoming reacting species) from bulky substituents on a molecule. (Ch. 2)

Steric strain an increase in energy that results from electron repulsion between atoms or groups of atoms that are not directly bonded together but occupy the same space. (Ch. 4)

Steroid a type of lipid with four fused carbon rings—three six-membered rings and one five-membered ring—and having a biological function (e.g., cholesterol regulates cell membrane permeability in mammals). (Ch. 2)

Storage polysaccharide a polysaccharide, such as starch, whose main function is to store energy. (Ch. 26)

Straight-chain alkane (also known as *linear alkane*) an alkane (containing only C—C and C—H single bonds with no functional groups) that has no branching along the chain. (Int. A)

s-Trans conformation a rotational conformation in which substituents are on opposite sides of a single bond. (Ch. 24)

Stretching one of two basic, independent types of vibrational motion. In stretching, the distance between two atoms in a chemical bond grows longer and shorter throughout one period of vibration. (Ch. 15)

Strong absorption an absorption whose percent transmittance is near zero (near the bottom of the spectrum if percent transmittance is plotted on the *y* axis). (Ch. 15)

Structural isomers See *constitutional isomers*. (Ch. 4)

Structural polysaccharide a polysaccharide, such as cellulose, whose main function is to provide structural integrity. (Ch. 26)

Substituent an atom or group of atoms attached to a molecule in place of hydrogen. (Int. A)

Substrate a molecular species that undergoes a particular reaction with an attacking species; in nucleophilic substitution or elimination, the species that contains the leaving group. (Ch. 7)

Sulfo group a group with the bonding arrangement —SO$_3$H. (Ch. 22)

Sulfonate ester a functional group having the bonding arrangement R—SO$_3$R'. Sulfonate esters typically possess very good leaving groups for nucleophilic substitution and elimination reactions. (Ch. 21)

Sulfonation a reaction in which the group —SO$_3$H or the group —SO$_3$R is added to a molecule; one type of electrophilic aromatic substitution reaction. (Ch. 22)

Sulfonyl chloride a compound with the general formula R—SO$_2$Cl. (Ch. 21)

Surfactant one of a class of compounds that includes soaps and detergents. (Ch. 2)

Suzuki reaction a reaction that couples the carbon-containing group of a vinylic or aryl halide with that of an organoboron compound when treated with palladium catalyst (PdL$_n$) under basic conditions. (Ch. 19)

Symmetric stretch a type of vibration in which two or more bonds stretch simultaneously and compress simultaneously. (Ch. 15)

Syn dihydroxylation a reaction in which two OH groups appear to add across a C═C bond in syn fashion. (Ch. 24)

Syndiotactic describes a polymer in which stereochemical configurations alternate from one monomer to the next along its main chain. (Ch. 26)

Synthesis a specific sequence of chemical reactions that converts starting materials into a desired compound, called the target of the synthesis (or the *synthetic target*). (Ch. 13)

Synthetic step an overall (or net) reaction that is carried out in a synthesis. (Ch. 13)

Synthetic target the desired end product of a *synthesis*. (Ch. 13)

Synthetic trap a difficulty preventing the execution of a *synthesis* in the forward direction as planned. (Ch. 13)

T

Tacticity the pattern of stereochemical configurations, or lack thereof, along a polymer chain. (Ch. 26)

Target See *synthetic target*. (Ch. 13)

Tautomer one form of a pair of isomers that exist together in equilibrium. (Ch. 7)

Termination step an elementary step in a chain reaction that results in a decrease in the number of reactive intermediates involved in the propagation cycle. (Ch. 25)

Termolecular describes an elementary step that involves three reactant species simultaneously. (Ch. 8)

Terpene a naturally produced hydrocarbon whose carbon backbone can be divided into separate and distinct five-carbon units called *isoprene units*. (Ch. 2)

Terpenoid a natural product that is produced from chemical modifications to a terpene, in which the carbon backbone is altered or atoms other than just carbon or hydrogen are introduced. (Ch. 2)

Tertiary (3°) carbocation   a carbocation that has three alkyl groups attached directly to C^+. (Ch. 6)

Tertiary structure   describes the three-dimensional location of the amino acid residues that make up a particular protein. (Ch. 26)

Tesla (T)   the SI unit for the strength of a magnetic field. (Ch. 16)

Tetrahedral intermediate   a transient species formed in a nucleophilic addition–elimination mechanism; the product of a nucleophilic addition step and the reactant in the subsequent nucleophile elimination step. (Ch. 20)

Tetramethylsilane (TMS) $(CH_3)_4Si$, usually added to samples to provide a reference signal in NMR spectroscopy. (Ch. 16)

Theoretical rate law   the equation that relates the rate of a reaction to the concentrations of the species involved in the reaction, which is derived from a proposed mechanism. (Ch. 8)

Thermal   describes a molecule that is in its ground-state electron configuration. (Ch. 24)

Thermal energy   the average energy available through molecular collisions, which increases as temperature increases. (Ch. 4)

Thermally allowed   describes a reaction that can take place readily when the reactants are in their ground-state electron configurations. (Ch. 24)

Thermally forbidden   describes a reaction that does not take place readily when the reactants are in their ground-state electron configurations. (Ch. 24)

Thermodynamic control   the execution of a chemical reaction under equilibrium conditions, such that the major product is the one that is most stable. (Ch. 9)

Thermodynamic enolate anion   the more stable of the enolate ions that can be produced by deprotonation of the corresponding ketone. (Ch. 10)

Thermoplastic   describes a polymer that, when heated above its glass transition or melting temperature, can be forced into molds or pushed through slots to form films or threads. (Ch. 26)

Thiol   a compound that contains the bonding arrangement C—SH. (Ch. 1)

Thionyl chloride   the compound $SOCl_2$, used to convert alcohols to alkyl chlorides and carboxylic acids to carboxylic acid chlorides. (Ch. 21)

Three-center, two-electron bond   a bond formed by two electrons and shared by three atoms; diborane is an example. (Ch. 12)

Torsional stain   an increase in energy (i.e., decrease in stability) that appears in an eclipsed conformation. (Ch. 4)

Trans   describes the configuration wherein two atoms are on opposite sides of a double bond or plane of a ring. (Ch. 3)

Transesterification reaction   a reaction that converts one ester into another. (Ch. 20)

Transform   in a retrosynthetic analysis, the proposed undoing of a single reaction or set of reactions to arrive at a potential precursor. (Ch. 13)

Transition state   a high-energy species that corresponds to an energy maximum along the reaction coordinate on a reaction free energy diagram. (Ch. 6)

Transition state theory   the theory that relates the *free energy of activation* of a reaction to its rate constant. (Ch. 8)

Transmittance (%T)   in spectroscopy, the portion of light sent through an analyte that reaches a detector: %$T = (I_{detected}/I_{source}) \times 100$. (Ch. 15)

Triacylglycerol   (also known as *triglyceride*) a description of a fat or oil that contains three adjacent ester groups, each of which can be produced from a *fatty acid* and a hydroxyl group from *glycerol*. (Ch. 2)

Trialkylborane   a compound R_3B with three alkyl groups attached to boron. (Ch. 12)

Trialkylborate ester   a compound $(RO)_3B$ with three alkoxy groups attached to boron. (Ch. 12)

Triglyceride   See *triacylglycerol*. (Ch. 2)

Triplet (t)   the NMR splitting pattern of a signal into three peaks in a 1:2:1 height ratio. (Ch. 16)

Triplet of doublets   the NMR splitting pattern in which each peak of a triplet is split into a doublet, resulting in a total of six peaks. (Ch. 16)

Triplet of quartets   the NMR splitting pattern in which each peak of a triplet is split into a quartet, resulting in a total of 12 peaks. (Ch. 16)

Triterpene   a terpene that contains three pairs of isoprene units (i.e., six isoprene units), for a total of 30 carbons. (Ch. 2)

Trivial name   (also known as *common name*) a nonsystematic name given to a molecule based on its properties or origin. (Int. A)

Tropylium ion   the aromatic cation $C_7H_7^+$, which has a seven-membered ring. The benzylic cation $(C_6H_5CH_2^+)$ rearranges into $C_7H_7^+$ as part of a fragmentation pathway in mass spectrometry. (Int. G)

Tyrian purple   See *royal purple*. (Ch. 1)

U

Ultraviolet–visible (UV–vis) spectroscopy   spectroscopy that uses light from the ultraviolet (UV) and visible regions of the electromagnetic spectrum; typically causes electron transitions between orbitals. (Ch. 15)

Umpolung   a German word that means "polarity reversal"; it describes the conversion of an electron-poor carbon atom into an electron-rich one, and vice versa. (Ch. 19)

Unimolecular   having a molecularity of 1; it describes an elementary step in which there is one reactant species. (Ch. 8)

Unimolecular elimination (E1) reaction   a two-step reaction consisting of heterolysis followed by electrophile elimination, resulting in the net loss of a proton and a leaving group, and the production of a new π bond. (Ch. 8)

Unimolecular nucleophilic substitution (S$_N$1) reaction   a two-step reaction consisting of heterolysis followed by coordination, in which a leaving group is replaced by a nucleophile. (Ch. 8)

Unpaired electron   an electron occupying an orbital without the presence of a second electron in that orbital; unpaired electrons are found on species called free radicals. (Ch. 1)

Unpolarized   describes light whose photons travel in the same direction but have their electric fields oscillating in different planes. (Ch. 5)

Unsaturated   describes a molecule that has fewer than the maximum number of hydrogen atoms possible; such molecules have double bonds, triple bonds, or rings. (Ch. 4)

Upfield   appearing in an NMR spectrum in a region of lower chemical shift. (Ch. 16)

UV–vis spectrum   a plot of absorbance against wavelength of ultraviolet (UV) and visible light, showing *absorption bands* or *peaks*. (Ch. 15)

V

Valence bond (VB) theory a theory that describes orbitals in molecules as being produced by mixing just two atomic orbitals (AOs) at a time, one from each of two adjacent atoms; often one or both of the AOs involved in each mixing is a hybrid AO. (Ch. 3)

Valence electrons the electrons available to participate in bonding interactions, occupying the highest (outermost) energy shell. (Ch. 1)

Valence shell electron pair repulsion (VSEPR) theory a theory used to predict the geometry about an atom, based on the idea that the lowest-energy geometry about an atom is achieved by minimizing the electrostatic repulsions among the electron groups to which valence electrons can be assigned. (Ch. 2)

Vector a geometric entity that has both magnitude and direction. (Ch. 2)

Vinyl polymer a large molecule made from linking monomers containing vinyl groups (C=C) together. (Ch. 26)

Vitalism the outdated belief that organic compounds could *not* be made in the laboratory; instead, only living systems could summon up a mysterious "vital force" needed to synthesize them. (Ch. 1)

Vulcanization the reaction of natural rubber with sulfur that results in the cross-linking of the polymer chains. (Ch. 26)

W

Walden inversion the phenomenon that occurs on nucleophilic attack in an S_N2 reaction that causes inversion of stereochemistry. (Ch. 8)

Wavelength (λ) the distance between two successive maxima or minima in a wave's oscillating amplitude. (Ch. 15)

Wavenumbers ($\bar{\nu}$) (also known as *reciprocal centimeters*) See *inverse centimeters* (cm^{-1}). (Ch. 15)

Wax a secretion from plants or animals that is solid at room temperature but melts at relatively low temperatures; waxes are often long-chain hydrocarbons or esters. (Ch. 2)

Weak absorption an absorption whose percent transmittance is near 100% (near the top of the spectrum if percent transmittance is plotted on the *y* axis). (Ch. 15)

Wheland intermediate See *arenium ion intermediate*. (Ch. 22)

Williamson ether synthesis a reaction used to synthesize either symmetrical or unsymmetrical ethers via an S_N2 reaction between an alkoxide anion (RO^-) and an alkyl halide ($R—X$) under basic conditions. (Ch. 10)

Wittig reaction a reaction in which the negatively charged carbon of a Wittig reagent forms a C=C bond with the carbonyl carbon of a ketone or aldehyde to produce an alkene. (Ch. 17)

Wittig reagent (also known as a *phosphonium ylide*) one of a class of compounds having a $^+P—C^-$ bond, used to synthesize alkenes; characteristically, it is highly nucleophilic at the negatively charged carbon. (Ch. 17)

Wolff–Kishner reduction a reaction that reduces the carbonyl group of a ketone or an aldehyde to a methylene (CH_2) group through treatment with hydrazine followed by heating under basic conditions. (Ch. 18)

Z

Zaitsev product the most highly substituted alkene product that can be produced in an elimination reaction. (Ch. 9)

Zaitsev's rule an empirical rule for elimination reaction regioselectivity, producing the *Zaitsev product*: The major elimination product is the one produced by deprotonating the carbon atom initially attached to the fewest hydrogen atoms. In other words, elimination usually takes place so as to produce the most highly alkyl-substituted alkene. (Ch. 9)

Zeigler–Natta catalyst a catalyst used to produce isotactic or syndiotactic forms of polymers. Zeigler–Natta catalysts are typically mixtures of Ti, V, Zr, or Cr halides with organoaluminum compounds. (Ch. 26)

Zigzag conformation (also known as *all-anti conformation*) the lowest-energy conformation of an alkyl chain; it is the conformation wherein an alkyl chain has the *anti conformation* at each C—C bond. (Ch. 4)

Zinc amalgam an alloy of zinc and mercury, which is a key reactant in the Clemmensen reduction. (Ch. 19)

Zwitterion a species that has both a positive and a negative formal charge, but a net charge of zero. (Ch. 6)

ANSWERS TO YOUR TURNS

Chapter 1

Your Turn 1.1
The $2s$ and $2p$ orbitals are in the second shell. Electrons (3)–(10) in Figure 1-7 are in the second shell.

Your Turn 1.2
The valence electrons are in the second shell, $2s^2$ and $2p^2$, and the core electrons are in the first shell, $1s^2$.

Your Turn 1.3
Bond energy of $H_2 \approx 450$ kJ/mol $- 0$ kJ/mol $= 450$ kJ/mol.

Your Turn 1.4
(a) 586 kJ/mol (140 kcal/mol). (b) The Si—F bond. (c) 138 kJ/mol (33 kcal/mol). (d) The O—O bond. (e) 1072 kJ/mol (256 kcal/mol). (f) The C≡O triple bond.

Your Turn 1.5
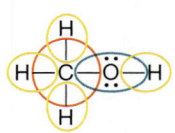

Your Turn 1.6
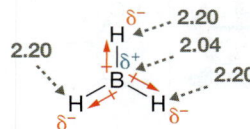

Your Turn 1.7
For NaCl, which is ionic, the difference is large: $3.16 - 0.93 = 2.23$. For CH_4, which is covalent, the difference is small: $2.55 - 2.20 = 0.35$.

Your Turn 1.8
Number of valence electrons (total number of electrons, number of protons, charge): 3 (5, 6, +1), 4 (6, 6, 0), 5 (7, 6, −1)

Your Turn 1.9
Sum of formal charges: $0 + 0 + 0 + (-1) = -1$. Sum equals overall charge on HCO_2^- ion, −1.

Your Turn 1.10
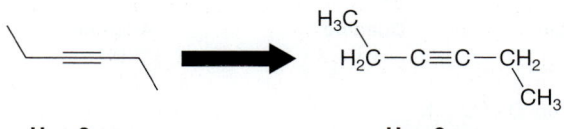

Your Turn 1.11

$$\left[HC\equiv C - CH_2 \longleftrightarrow HC = C = CH_2 \right]$$
(a)

$$\left[H_3C - \ddot{N} = BH \longleftrightarrow H_3C - N \equiv BH \right]$$
(b)

Your Turn 1.12
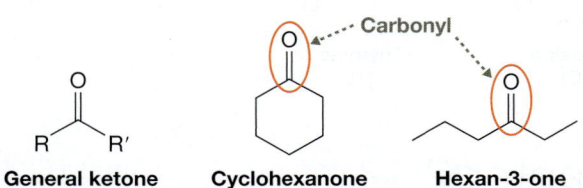

Your Turn 1.13
The compound has six carbon atoms, not four carbon atoms, and is hex-3-yne, $CH_3CH_2C\equiv CCH_2CH_3$.

Hex-3-yne → **Hex-3-yne**

Your Turn 1.14
Each compound has a carbonyl (C=O) functional group and is classified as a ketone.

General ketone **Cyclohexanone** **Hexan-3-one**

Your Turn 1.15

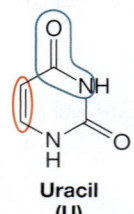

(a) Serine

(b) Aspartic acid

(c) Lysine

(d) Phenylalanine

Your Turn 1.16

The C=O group is characteristic of an aldehyde (RCH=O). The R—OH groups, characteristic of an alcohol, are circled.

Glucose, $C_6H_{12}O_6$

Your Turn 1.17

Uracil (U)

Guanine (G)

Adenine (A)

Alkene (red) C=C
Amide (blue) O=C—N

Cytosine (C)

Thymine (T)

Chapter 2

Your Turn 2.1

(a) Two electron groups surround the triply bonded C in Figure 2-1a, which has one single bond and one triple bond. **(b)** The central C in Figure 2-1b has two single bonds and one double bond, thus three

groups of electrons. **(c)** Each C atom in Figure 2-1c has four single bonds, thus four groups of electrons.

Your Turn 2.2

CH_3^+ has three single bonds and no lone pairs, thus three groups. CH_3^- has three single bonds and one lone pair, thus four groups.

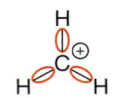 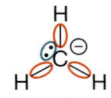

C = 3 groups   **C = 4 groups**

Your Turn 2.3

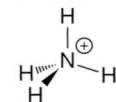 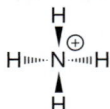

First structure   **Second structure**

Your Turn 2.4

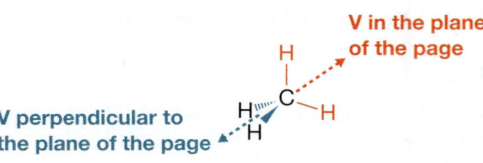

V in the plane of the page

V perpendicular to the plane of the page

Your Turn 2.5

Your Turn 2.6

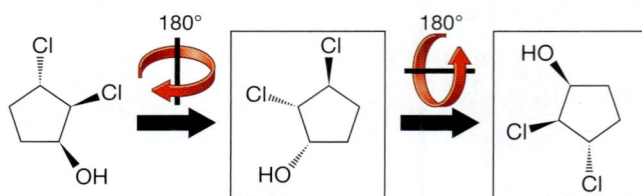

Your Turn 2.7

H—Be—H

Your Turn 2.8

The thin, red arrows indicate the bond dipoles. The thick, red arrows indicate the vector sum of each pair of bond dipoles. The thick, red arrows are equal in magnitude and point in opposite directions, so the molecule has no net dipole moment and, therefore, is nonpolar.

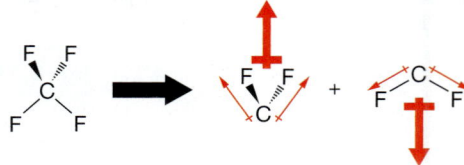

Your Turn 2.9

Methanoic acid has a CO_2H group and is a carboxylic acid; ethanol has an OH group and is an alcohol; ethanal has a C═O group and is an aldehyde; dimethyl ether has a C—O—C group and is an ether; propene has a C═C group and is an alkene; propane and ethane have no functional groups.

Your Turn 2.10

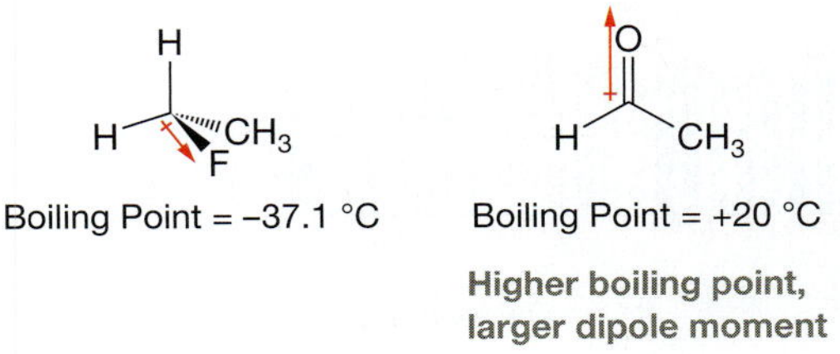
H
H
CH3
F
Boiling Point = −37.1 °C
O
H
CH3
Boiling Point = +20 °C
Higher boiling point,
larger dipole moment

Your Turn 2.11

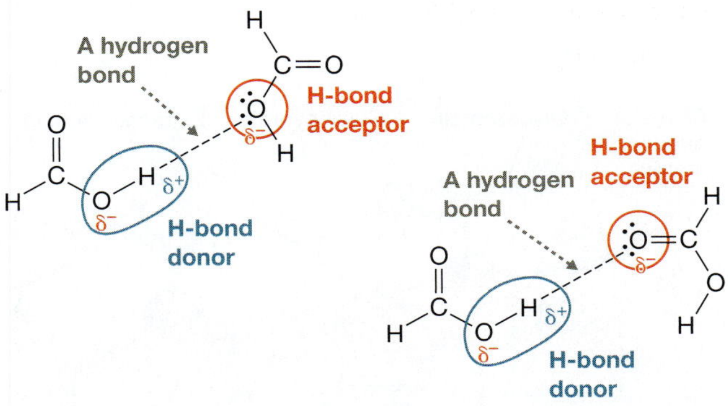
A hydrogen
bond
H-bond
acceptor
H-bond
donor
A hydrogen
bond
H-bond
acceptor
H-bond
donor
δ+
δ−

Your Turn 2.12

Cholesterol has one potential H-bond acceptor (the O atom) and one potential donor (the OH bond). Octanoic acid has two potential H-bond acceptors (each O atom) and one potential donor (the OH bond).

Your Turn 2.13

Boiling point increases as the number of electrons increases.

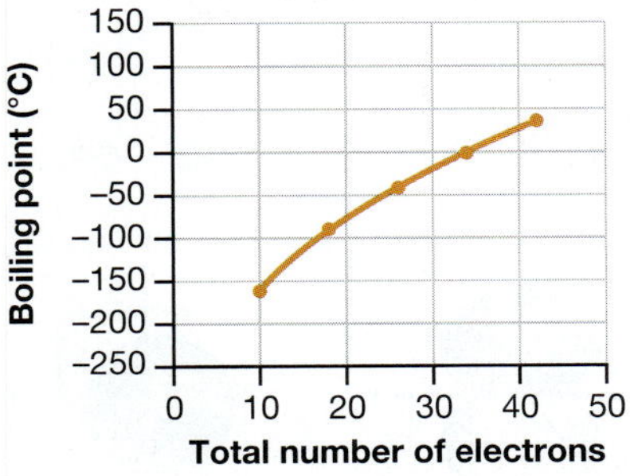
Boiling point (°C)
150
100
50
0
−50
−100
−150
−200
−250
0
10
20
30
40
50
Total number of electrons

Your Turn 2.14

As shown below, pentane has a greater intermolecular contact surface area than dimethylpropane.

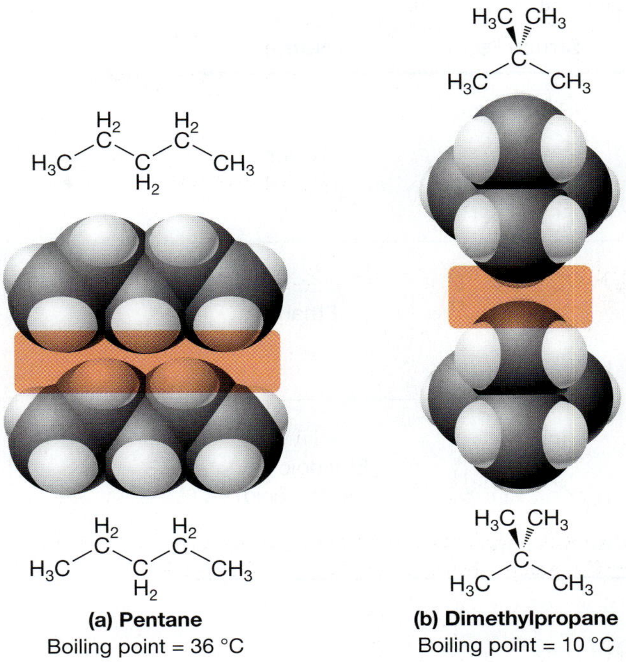
(a) Pentane
Boiling point = 36 °C
(b) Dimethylpropane
Boiling point = 10 °C

Your Turn 2.15

The amine **B** is more soluble in H_2O than the alkane **A** because RNH_2 is capable of hydrogen bonding with H_2O, whereas RCH_3 is not.

Your Turn 2.16

Each ion is shown with 6 ion–dipole interactions, giving 12 total for the two ions.

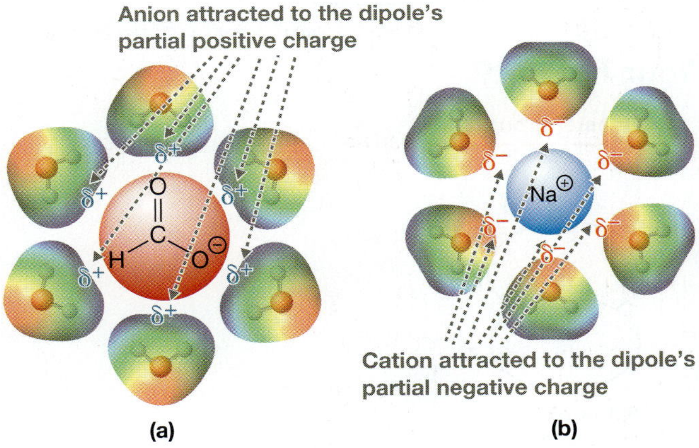
Anion attracted to the dipole's
partial positive charge
Cation attracted to the dipole's
partial negative charge
Na
(a)
(b)

Your Turn 2.17

Hydrogen-bond donors can be H—O, H—N, or H—F bonds. The only solvents in Table 2-7 that have a hydrogen-bond donor are water, ethanol, and ethanoic acid.

Structure	Name
Water	

Water

H-bond donor

Ethanol

H-bond donor

Ethanoic acid
(Acetic acid)

H-bond donor

Your Turn 2.18

The CO_2^- (red/orange region) is hydrophilic and the hydrocarbon tail (blue/green region) is hydrophobic.

Hydrophilic   **Hydrophobic**

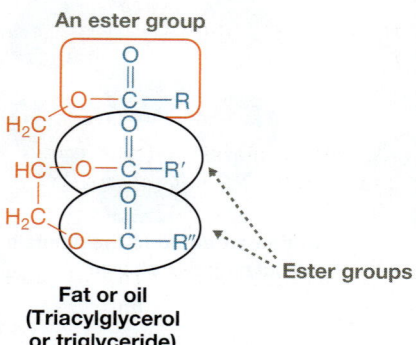

Your Turn 2.19

An ester group

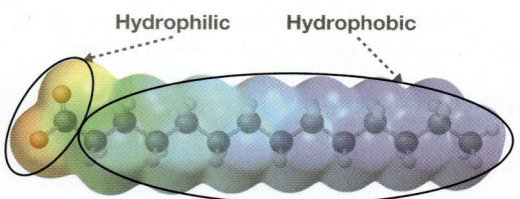

Ester groups

**Fat or oil
(Triacylglycerol
or triglyceride)**

Your Turn 2.20

The hydrophobic regions in Figure 2-31a and 2-31b are circled:

A lipid bilayer

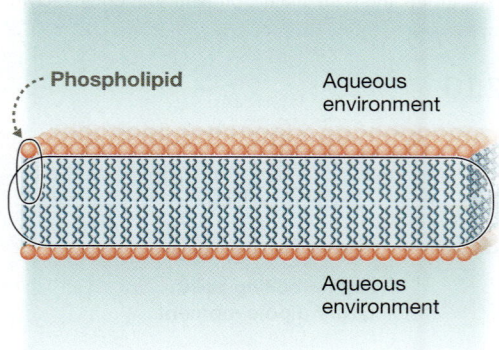

Phospholipid

Aqueous environment

Aqueous environment

(a)

Cell membrane

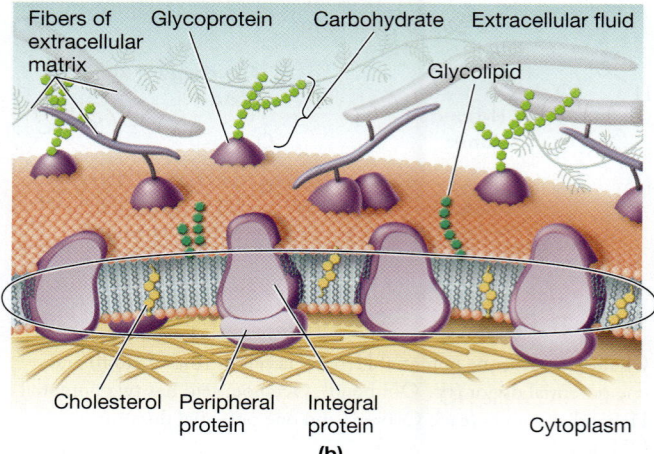

Fibers of extracellular matrix   Glycoprotein   Carbohydrate   Extracellular fluid

Glycolipid

Cholesterol   Peripheral protein   Integral protein   Cytoplasm

(b)

Chapter 3

Your Turn 3.1

p_x Orbital p_y Orbital p_z Orbital

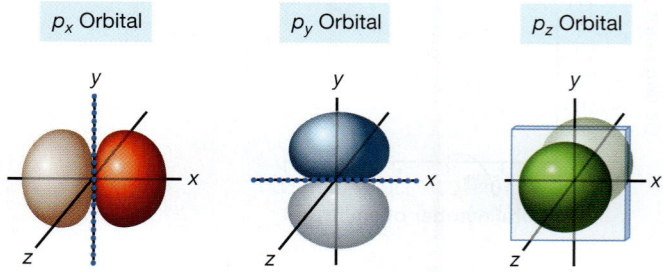

The nodal plane of the p_z orbital is parallel to the plane of the paper.

Your Turn 3.2

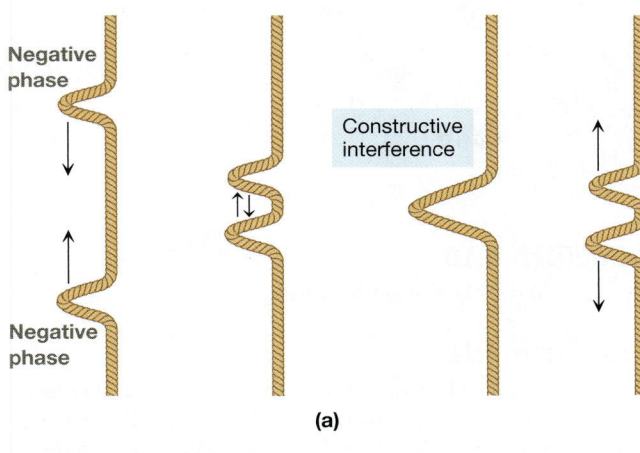

(a)

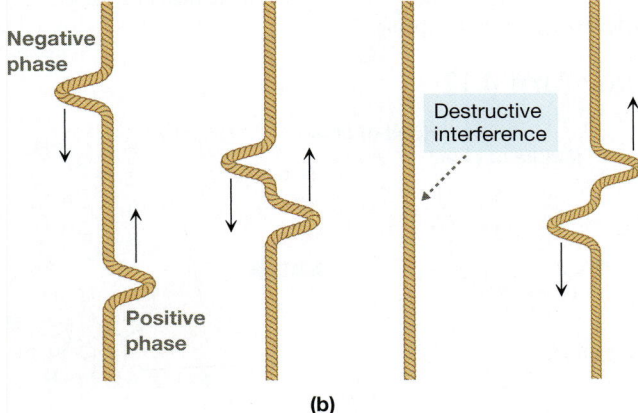

(b)

Your Turn 3.3

The σ* MO is the HOMO.

Your Turn 3.4

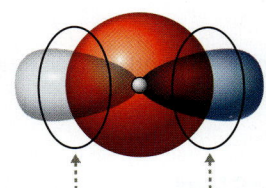

Destructive interference   Constructive interference

Your Turn 3.5

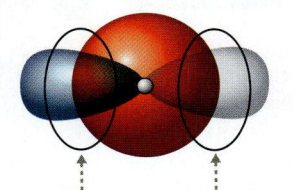

Constructive interference   Destructive interference

Your Turn 3.6

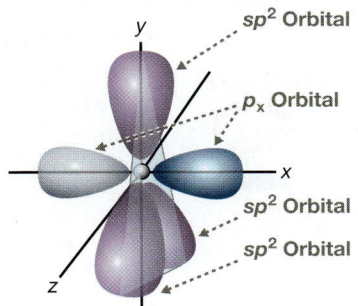

Your Turn 3.7

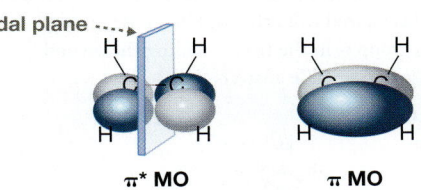

π* MO   π MO

Your Turn 3.8

The HOMO is the π MO, and the LUMO is the π* MO.

Your Turn 3.9

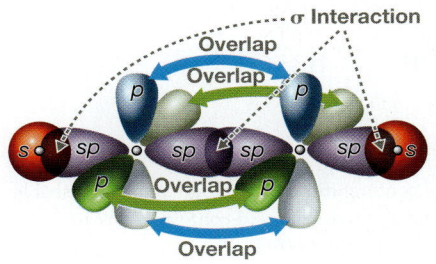

Your Turn 3.10

The HOMO is one of the π MOs, and the LUMO is one of the π* MOs.

Your Turn 3.11

If the molecule on the left is flipped vertically 180°, it appears to be identical to the one on the right, so the two molecules are identical.

Your Turn 3.12

The first molecule is planar. The second molecule is not; one CH_2 plane is perpendicular to the other.

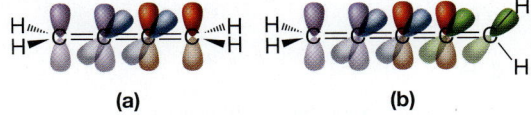

(a)   **(b)**

Your Turn 3.13

From left to right, the values are: $sp^3 = 25\%$, $sp^2 = 33.3\%$, and $sp = 50\%$ s character.

Chapter 4

Your Turn 4.1

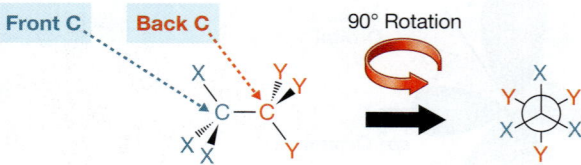

Your Turn 4.2

If the molecule in Figure 4-2 is rotated 90° clockwise (as viewed from the top), the CBr_3 group is in the front and the CH_3 group is in the back, yielding the first molecule below. If the molecule is rotated 90° counterclockwise (as viewed from the top), the CBr_3 group is in the back and the CH_3 group is in the front, yielding the second molecule below. Both are acceptable answers.

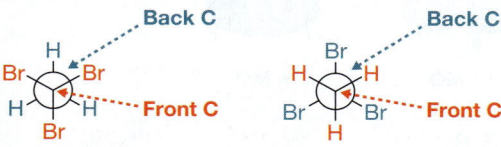

Your Turn 4.3

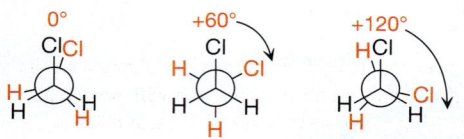

Your Turn 4.4

In Figure 4-4, all eclipsed conformations of C_2H_6 are indistinguishable (−120°, 0°, and +120°) and all three staggered conformations are indistinguishable (±180°, −60°, and +60°).

Your Turn 4.5

Your model should resemble the structures of C_2H_6 in Figure 4-5. The front and back C—H bonds are closer to each other in an eclipsed conformation, and, thus, are *higher* in energy, than in the staggered conformation.

Your Turn 4.6

Torsional strain ≈ ΔE ≈ 12 kJ/mol (2.9 kcal/mol).

Your Turn 4.7

Eclipsed conformations (−120°, 0°, and +120°), staggered gauche conformations (−60° and +60°), and staggered anti conformations (±180°) are shown in Figure 4-7.

Your Turn 4.8

In Figure 4-7, eclipsed conformation energy difference is: 40 − 27 = 13 kJ/mol (3.1 kcal/mol). Staggered conformation energy difference is: 12 − 0 = 12 kJ/mol (2.9 kcal/mol).

Your Turn 4.9

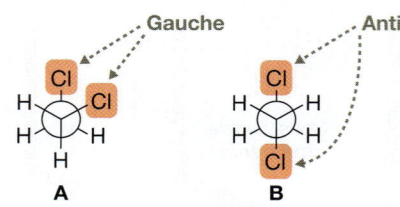

Your Turn 4.10

ΔE = 27 − 0 = 27 kJ/mol (6.5 kcal/mol).

Your Turn 4.11

Cyclooctane has 8 CH_2 groups. Heat per CH_2 group = 620.3 kJ/mol. Ring strain per CH_2 group = 620.3 − 615.1 = 5.2 kJ/mol. Total ring strain in C_8H_{16} = 8 × 5.2 = 41.6 kJ/mol. This value is less than ring strain in cyclopropane and cyclobutane, but more than in cyclopentane, cyclohexane, and cycloheptane.

Your Turn 4.12

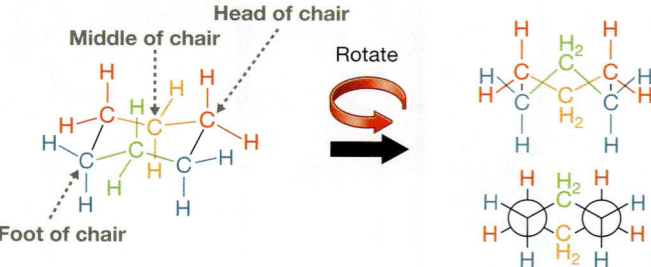

Your Turn 4.13

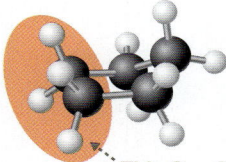

This C—C bond is eclipsed, highest torsional strain.

Your Turn 4.14

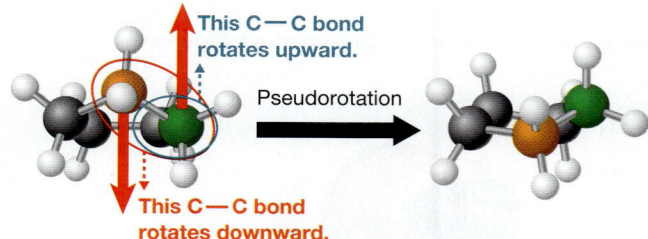

Your Turn 4.15

All of the C—C bonds undergo rotation as axial and equatorial groups change their positions. The "head" of the chair C′ rotates to become the foot of the chair and the foot of the chair C″ rotates to become the head of the chair. Clockwise from C′, the C—C bonds rotate up, down, up, down, up, and down during the chair flip illustrated going left to right.

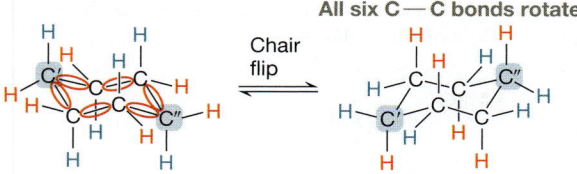

All six C—C bonds rotate.

Chair flip

Your Turn 4.16

Angle strain arises from three sp^3-hybridized C atoms having bond angles of ~120°. Torsional strain arises from eclipsed C—C bonds. The combined strains explain why the half-chair conformation of C_6H_{12} is so high in energy.

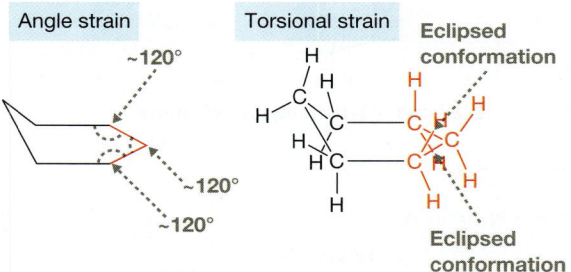

Angle strain

~120°
~120°
~120°

Torsional strain

Eclipsed conformation

Eclipsed conformation

Your Turn 4.17

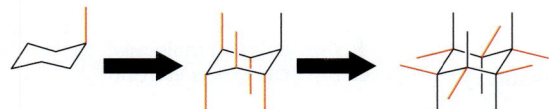

Your Turn 4.18

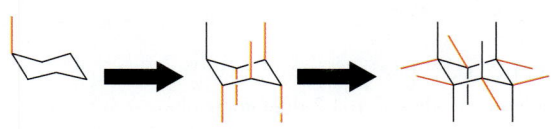

Your Turn 4.19

The CH_3 group is circled for clarity. Axial H atoms are labeled. All other H atoms bonded directly to the ring are equatorial.

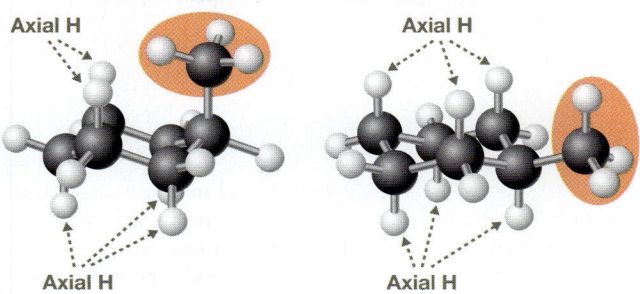

Axial H

Axial H

Axial H

Axial H

Your Turn 4.20

These H atoms are involved in 1,3-diaxial interactions with the CH_3 group.

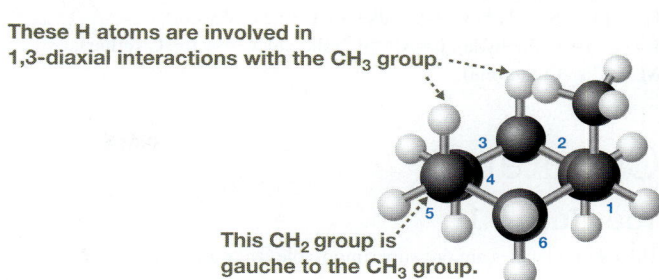

This CH_2 group is gauche to the CH_3 group.

Your Turn 4.21

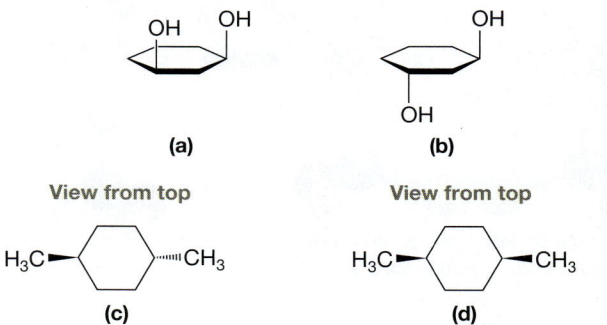

(a)

(b)

View from top

View from top

H_3C ———— CH_3

H_3C ———— CH_3

(c)

(d)

Your Turn 4.22

To translate the conformation of *trans*-1,3-dimethylcyclohexane on the left into the "chair-flipped" conformation on the right, without actually flipping the chair, carry out the following rotations of the molecule.

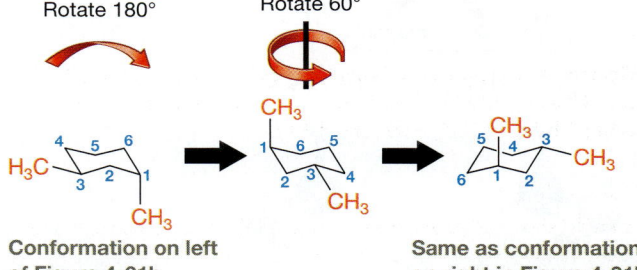

Rotate 180°

Rotate 60°

Conformation on left of Figure 4-31b

Same as conformation on right in Figure 4-31b

Your Turn 4.23

The longest continuous chain of carbons has six carbons. The first substituent is a CH_3 group located on C2. The second substituent is a CH_3 located on C4. This is the same molecule as the previous two molecules. Recall that there is free rotation about C—C single bonds. These three molecules are *conformational* isomers.

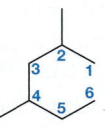

Your Turn 4.24

$H_3C—CH_3$ $H_3C—CH_3$ $H_3C—CH_2$ $H_3C—CH_2$
                                              $|$ $|$
                                              OH OH

(a)   (b)   (c)   (d)

Your Turn 4.25

G = **H** = **K** = **L** = C=C (alkene), O—H (alcohol).
I = C=O (aldehyde). **J** = C=C (alkene), C—O—C (ether).
M = C=O (ketone).

Chapter 5

Your Turn 5.1

The two molecules are not superimposable. They are enantiomers.

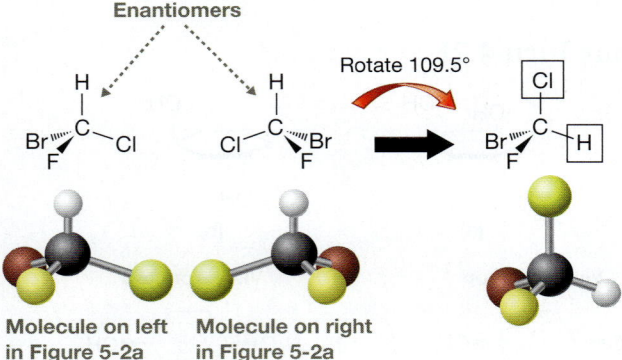

Your Turn 5.2

The mirror image is superimposable; therefore, the mirror image is the same as the original molecule. *Note:* H_A and H_B are labels assigned to the mirror images to aid in your visualization of the transformations performed.

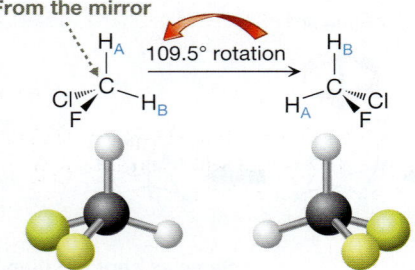

Your Turn 5.3

Use model kits for verification. There are no other rotations that would allow the two molecules to line up.

Your Turn 5.4

Chiral. The red and black molecules in Solved Problem 5.6 are mirror images of each other. From the solved problem, we have already verified that the two molecules are not the same and are, therefore, enantiomers.

Your Turn 5.5

trans-1,2-Dichlorocyclopropane does not possess a plane of symmetry because it is chiral. The C—Cl_A dash bond is pointing in and the C—Cl_B wedge bond is pointing out, and therefore are not reflections through a mirror.

Your Turn 5.6

Both C—Cl bonds in *cis*-1,2-dichlorocyclopropane are pointing out of the plane of the paper. Thus there is a plane of symmetry (as shown by the dotted line). The molecule is therefore achiral.

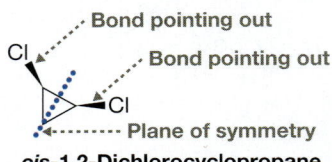

cis-**1,2-Dichlorocyclopropane**

Your Turn 5.7

The tetrahedral stereocenters (chiral centers) are marked below with an asterisk (*).

Butan-2-ol   **2-Bromo-1,1-dimethylcyclobutane**

Your Turn 5.8

Use model kits for verification.

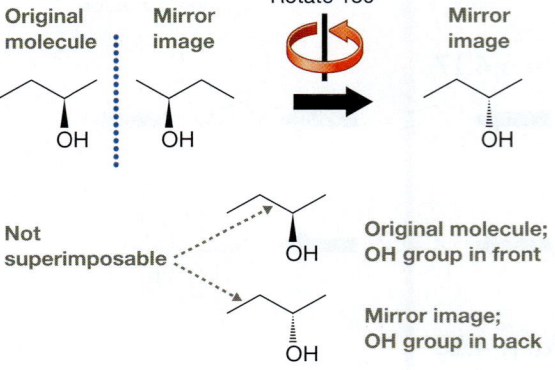

Your Turn 5.9

Use model kits for verification.

Your Turn 5.10

Construct molecular models of *cis*-1,2-difluorocyclohexane and its mirror image and carry out the rotations/chair flips shown in Figure 5-14.

Your Turn 5.11

Use model kits to verify that *trans*-1,2-difluorocyclohexane is chiral (Fig. 5-15b).

Your Turn 5.12

The first molecule does not line up perfectly with the second, regardless of rotations about single bonds or the orientations of the molecules.

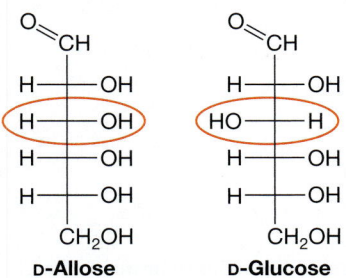

Your Turn 5.13

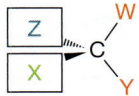

Your Turn 5.14

Z ∙∙∙ C — W
X — Y

Your Turn 5.15

The configurations differ at the circled C atoms.

D-Allose   D-Glucose

Your Turn 5.16

The pattern is that substituents on adjacent carbon atoms on the same side of the plane containing the carbon chain in a zigzag structure (substituents on C2 and C3) are on opposite sides in a Fischer projection and for C atoms that are separated by an additional bond (C2 and C4), substituents on the opposite sides of the plane in a zigzag structure are on opposite sides in a Fischer projection.

Your Turn 5.17

A solution of 30% A and 70% B is 60% racemic and its enantiomeric excess is %ee = 40% B.

Chapter 6

Your Turn 6.1

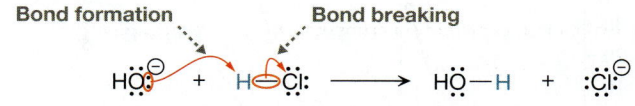

Your Turn 6.2

The curved arrows are shown below. The second reaction tends to form more products at equilibrium than the first (K_{eq} is larger).

$$K_{eq} = 7.1 \times 10^{-8}$$

$$K_{eq} = 4.0 \times 10^{-3}$$

Larger K_{eq}

Your Turn 6.3

HCl ($pK_a = -7$) is a stronger acid than H_3O^+ ($pK_a = -1.7$). The difference in pK_a values is $-1.7 - (-7) = 5.3$, which corresponds to a difference in acid strength of $>10^5$. Thus, HCl is $>100{,}000$ times stronger an acid than H_3O^+.

Your Turn 6.4

H_2O ($pK_a = 15.7$) is a stronger acid than $(CH_3)_2NH$ ($pK_a = 38$). The difference in pK_a values is $38 - 15.7 = 22.3$, which corresponds to a difference in acid strength of $10^{22.3}$. Thus H_2O is 2.0×10^{22} times stronger an acid than $(CH_3)_2NH$.

Your Turn 6.5

$(CH_3)_2NH$ ($pK_a = 38$) is a stronger acid than diethyl ether $(CH_3CH_2)_2O$ ($pK_a \approx 45$). The reactant side of the reaction is favored because the stronger acid is on the product side. This is not the same side that is favored in Equation 6-11. Therefore, diethyl ether is a suitable solvent for $(CH_3)_2N^-$ because the equilibrium lies to the left, indicating that diethyl ether is relatively inert in the presence of $(CH_3)_2N^-$.

Your Turn 6.6

The acid in Figure 6-1 is nearly 100% dissociated at ~2 pH units above the pK_a, or pH = 7. It is nearly 100% associated ~2 pH units below the pK_a, or pH = 3.

Your Turn 6.7

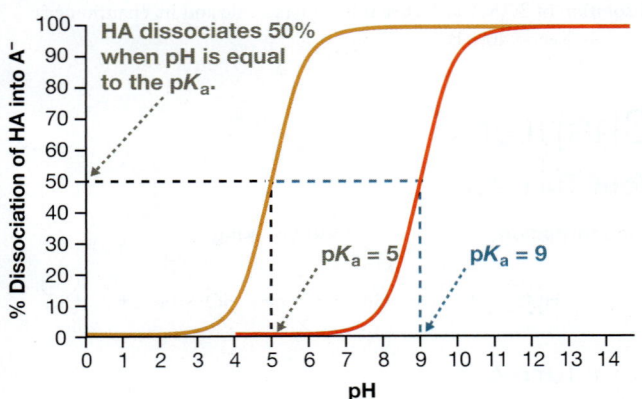

HA dissociates 50% when pH is equal to the pK_a.

$pK_a = 5$ $pK_a = 9$

Your Turn 6.8

The distance between Cl and H **decreases** and the distance between the O and H **increases**.

Your Turn 6.9

The free energy quantities are indicated in the diagram shown here. The free energy of activation is larger in Figure 6-2b compared to the reaction in Figure 6-2a.

Transition state

Free energy of activation ($\Delta G^{o\ddagger}$)

Energy barrier

$HCl + HO^{\ominus}$
Products

Overall free energy change (ΔG^o_{rxn})

Free energy

$Cl^{\ominus} + H_2O$
Reactants

Reaction coordinate ⟶

Your Turn 6.10

The pK_a of HCl is −7, and that of H_2S is 7.2. HCl is a stronger acid.

Your Turn 6.11

The pK_a of H_3O^+ is −1.7, and that of H_4N^+ is 9.4. H_3O^+ is a stronger acid.

Your Turn 6.12

The pK_a of H_3C—CH_3 is ~50, that of H_2C=CH_2 is ~44, and that of HC≡CH is ~25.

Your Turn 6.13

The pK_a of ethanoic acid (acetic acid) is 4.75. That of ethanol is 16.

Your Turn 6.14

Resonance hybrid

Your Turn 6.15

Resonance hybrid

Your Turn 6.16

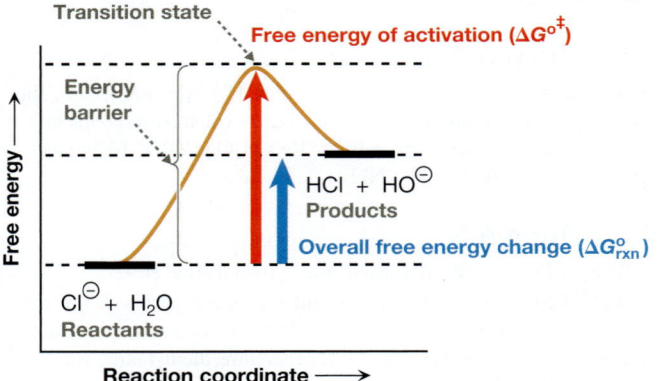

Your Turn 6.17

H_3C—$\overset{+}{N}H_3$

Your Turn 6.18

Protonated amines and alcohols are stronger acids compared to the uncharged species. R—NH_3^+ (10.63 estimated from $H_3CNH_3^+$); R—NH_2 [38 estimated from 2° amine $(CH_3)_2NH$]; R—OH_2^+ ($\approx$−2, similar to H_3O^+); R—OH (16 estimated from CH_3CH_2OH).

Your Turn 6.19

R—OH (16 estimated from CH_3CH_2OH); R—NH_2 [38 estimated from 2° amine $(CH_3)_2NH$]; R—CH_3 (50 estimated from CH_3CH_3), RC≡CH (25 estimated from HC≡CH).

Your Turn 6.20

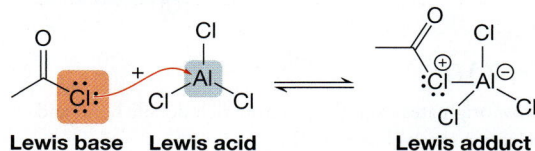

Chapter 7

Your Turn 7.1

Electron rich to electron poor

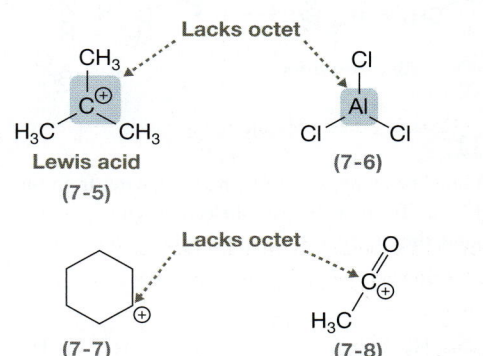

Electron-rich site   Electron-poor site

Your Turn 7.2

This is faulty because H cannot form two covalent bonds. A second arrow needs to be drawn to show the H—Cl bond breaking.

Your Turn 7.3

Negatively charged nucleophiles

Uncharged nucleophiles

Your Turn 7.4

Electron rich   Electron poor

Your Turn 7.5

Coordination step

Lewis base   Lewis acid   Lewis adduct

Your Turn 7.6

Lacks octet

Lewis acid
(7-5)

Cl
(7-6)

Lacks octet

(7-7)

(7-8)

Your Turn 7.7

Electron rich to electron poor

Electron poor

Electron rich

Your Turn 7.8

Electron rich

Electron rich to electron poor

Electron poor

Your Turn 7.9

E2 requires three arrows. The curved arrow originates from the electron-rich N of H_2N^- and points to the H adjacent to the C with the leaving group. The movement of electrons from the electron-rich site to the electron-poor site is depicted using two curved arrows.

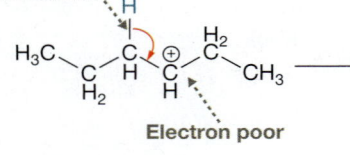

Your Turn 7.10

The curved arrow originates from the electron-rich double bond and points to the electron-poor H of H—Br.

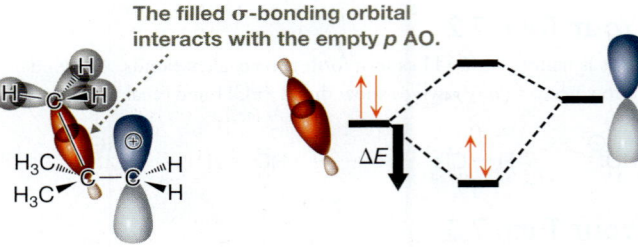

Your Turn 7.11

To show the C—H bond breaking, a curved arrow originates from the center of the C—H bond. To show the pair of electrons ending up in the C=C double bond, the curved arrow points to the center of the C—C bond.

Your Turn 7.12

The labels are provided in the answers to Your Turns 7.10 and 7.11.

Your Turn 7.13

The single curved arrow shows a C—H bond breaking from the tertiary carbon and simultaneously forming to the secondary carbon.

Your Turn 7.14

HCl $pK_a = -7$; NH_3 $pK_a = 36$. HCl is a stronger acid than NH_3 (it has a more negative pK_a) and the reaction favors the product side in Equation 7-29 (away from the stronger acid). This is confirmed by the large, positive equilibrium constant ($K_{eq} = 10^{43}$).

Your Turn 7.15

ΔBond energy (C=O) − (C=C) = 720 − 619 kJ/mol = 101 kJ/mol; ΔBond energy (C—C) − (C—O) = 339 − 351 kJ/mol = −12 kJ/mol; ΔBond energy (C—H) − (O—H) = 418 − 460 kJ/mol = −42 kJ/mol. From this it appears that, because the difference is greatest between the C=O and C=C bond energies, the C=O bond (being the stronger of the two) is the one that has the most influence on the outcome of the reaction.

Interchapter D

Your Turn D.1

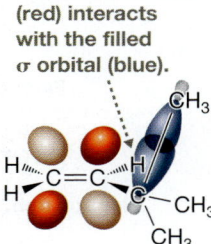

The filled σ-bonding orbital interacts with the empty p AO.

Your Turn D.2

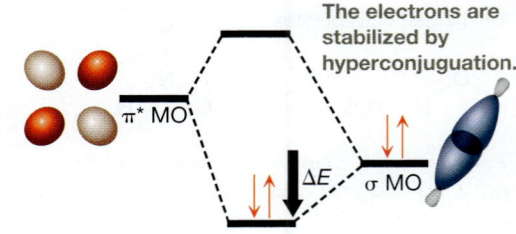

The empty π* MO (red) interacts with the filled σ orbital (blue).

The electrons are stabilized by hyperconjuguation.

Your Turn D.3

H is not hybridized and is s; O and Cl are both sp^3 hybridized. End-on overlap occurs in both H—O⁻ and H—Cl bonding and, therefore, results in a sigma bond. When the two AOs overlap, two MOs are produced: sigma bonding (σ) and sigma antibonding (σ*). The three sets of lone pairs on Cl and O are in sp^3 hybrid, nonbonding orbitals. The HOMO from HO⁻ is, therefore, a nonbonding orbital, and the LUMO from HCl is a σ* MO, in agreement with Figure D-5.

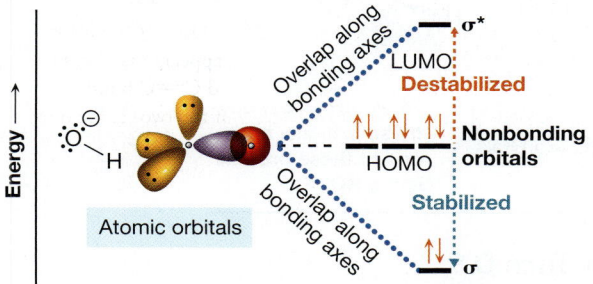

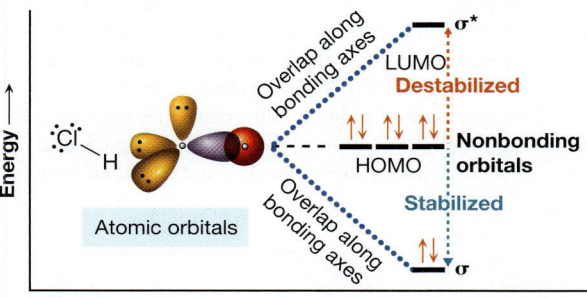

The nonbonding orbital of HO⁻ comes from a hybrid AO. The σ* MO of HCl is the result of destructive interference, leaving a node in the internuclear region:

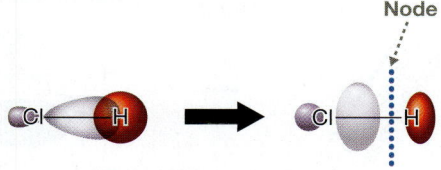

Your Turn D.4

Constructive interference occurs when two orbitals overlap with the same phases. Destructive interference occurs when two orbitals overlap with opposite phases.

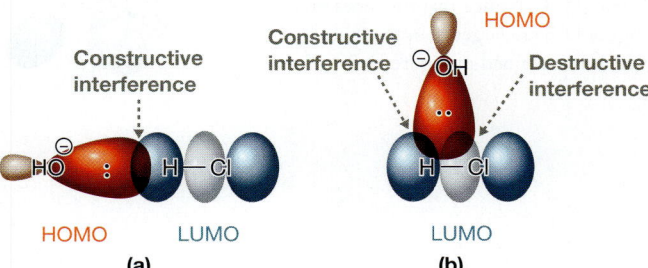

Your Turn D.5

C and Cl are both sp^3 hybridized and each H atom contributes a $1s$ orbital. The 11 AOs produce 11 new MOs. End-on overlap occurs in the C—Cl and C—H bonding and each overlap produces two MOs: σ bonding and σ* antibonding, which accounts for eight MOs. The remaining three orbitals are from the noninteracting sp^3 AOs from Cl, producing nonbonding orbitals. The LUMO is, therefore, a σ* MO, in agreement with Figure D-6.

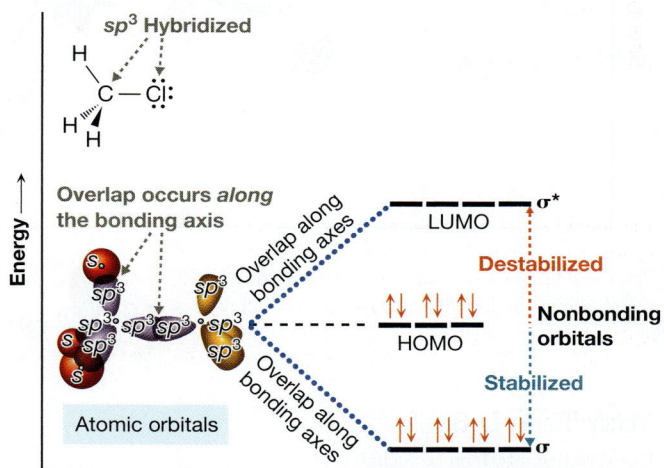

The C—Cl σ* can be thought of as resulting from the destructive interference between two sp^3 AOs, producing a node in the internuclear region:

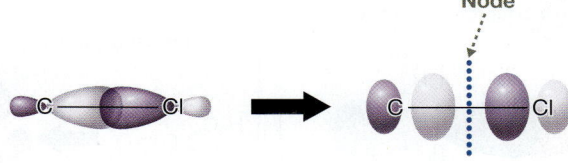

Your Turn D.6

Constructive interference occurs when two orbitals overlap with the same phases. Destructive interference occurs when two orbitals overlap with opposite phases.

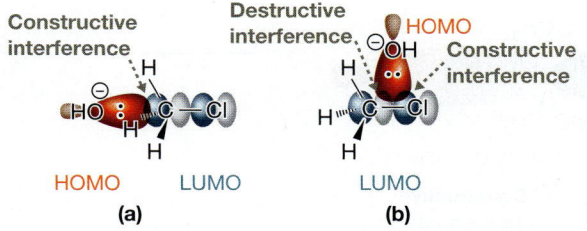

Your Turn D.7

The central C is sp^2 hybridized and has an unhybridized p orbital. The CH_3 carbons are all sp^3 hybridized and the H atoms each contribute a $1s$ orbital. End-on overlap occurs in the C—H and C—C bonding and, therefore, results in 12 σ bonds. When the 24 AOs overlap, 24 MOs are produced: 12 σ bonding and 12 σ* antibonding.

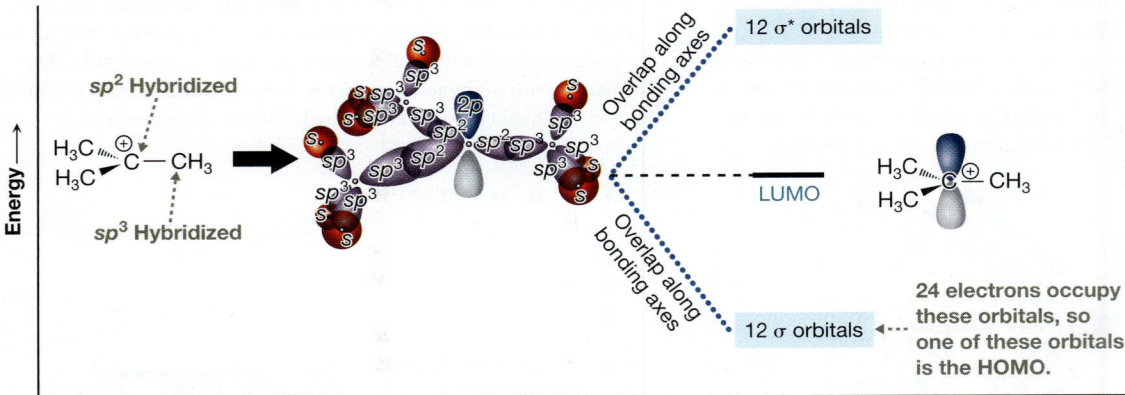

Yes, these figures are in agreement with Figure D-7, which shows the LUMO of $(H_3C)_3C^+$ as follows:

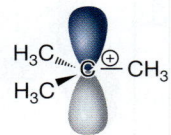

Your Turn D.8

Constructive interference occurs when two orbitals overlap with the same phases. Destructive interference occurs when two orbitals overlap with opposite phases. Both Figure D-7a and Figure D-7b exhibit constructive interference.

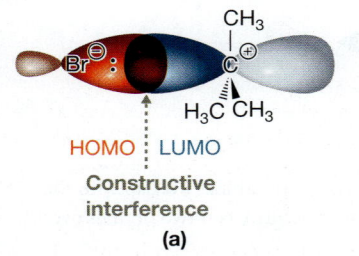

(a)

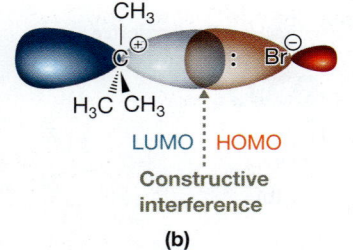

(b)

Your Turn D.9

The central C atom and the the O atom are both sp^2 hybridized and each has an unhybridized p orbital. The CH$_3$ carbon and the Cl are both sp^3 hybridized and the H atoms each contribute a 1s orbital. End-on overlap occurs in the C—H, C—C, C—Cl, and C—O bonding and, therefore, results in 6 σ bonds. Side-on overlap occurs between the two p orbitals of C and O. This results in π bonding and π* antibonding. When the 14 AOs overlap, 14 MOs are produced: six σ bonding, six σ* antibonding, one π bonding, and one π* antibonding.

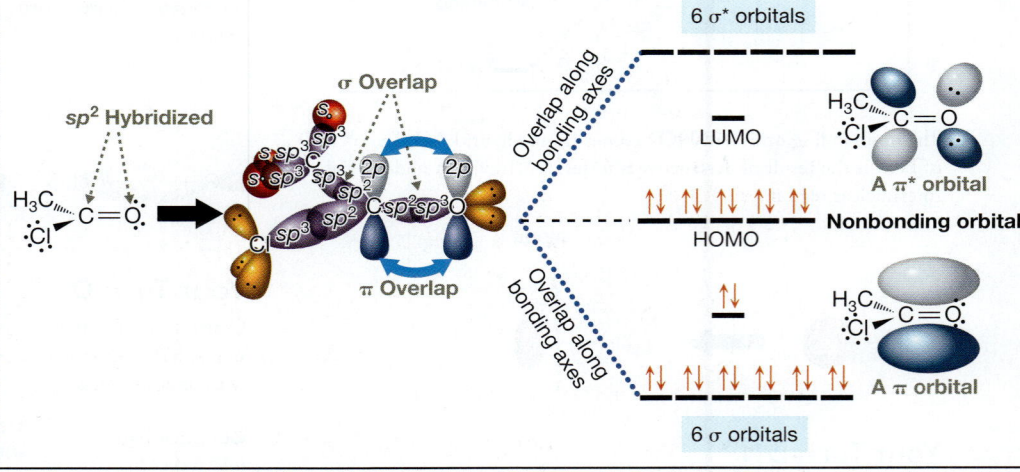

The LUMO is the π* MO, in qualitative agreement with the LUMO shown in Figure D-8a. Notice that the lobes on C in Figure D-8 are larger than the lobes on O, which is explained in Figure D-8.

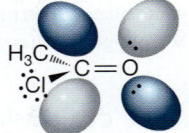

Your Turn D.10

Constructive interference occurs when two orbitals overlap with the same phases. Destructive interference occurs when two orbitals overlap with opposite phases. Constructive interference predominantly takes place in both Figure D-8a and Figure D-8b.

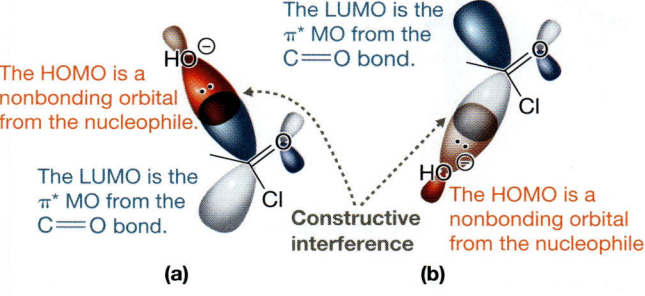

(a)　　　　　　　(b)

Your Turn D.11

The two C atoms and the Br atom are all sp^3 hybridized, and the H atoms each contribute a 1s orbital. End-on overlap occurs in the C—H, C—C, and C—Br bonding and, therefore, results in 7 σ bonds. When the 14 AOs overlap, 14 MOs are produced: 7 σ bonding and 7 σ* antibonding. This leaves three nonbonding orbitals.

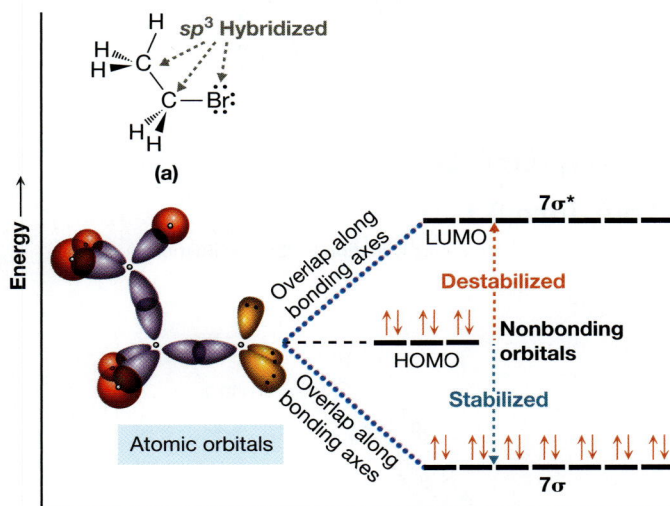

The LUMO of the molecule is a σ* MO, in agreement with Figure D-10a.

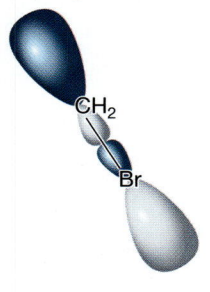

Although the HOMO of the entire molecule is a nonbonding orbital, the highest-occupied orbital involving the CH₃ group is a σ MO, in agreement with Figure D-10a.

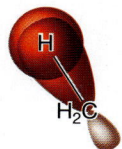

Your Turn D.12

Constructive interference occurs when two orbitals overlap with the same phases. Destructive interference occurs when two orbitals overlap with opposite phases. Predominantly constructive interference takes place in both Figure D-10a and Figure D-10b.

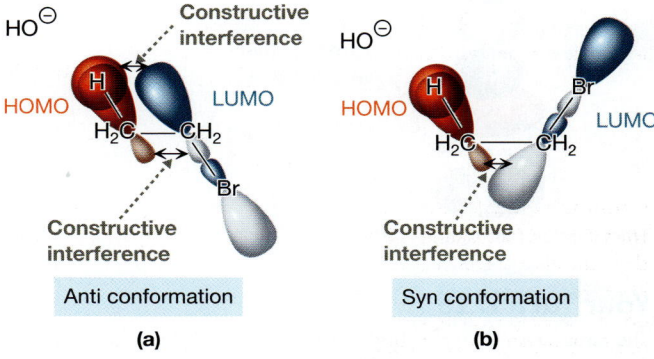

Anti conformation (a)　　　　Syn conformation (b)

Your Turn D.13

The two central C atoms are both sp^2 hybridized and each has an unhybridized p orbital. The CH₃ carbons are sp^3 hybridized and the H atoms each contribute a 1s orbital. End-on overlap occurs in the C—H and C—C bonding and, therefore, results in 11 σ bonds. Side-on overlap occurs between the two p orbitals of the two sp^2-hybridized C atoms. This results in π bonding and π* antibonding. When the 24 AOs overlap, 24 MOs are produced: 11 σ bonding, 11 σ* antibonding, one π bonding, and one π* antibonding. When the 24 total valence electrons fill the MOs, the π MO is the HOMO and the π* MO is the LUMO.

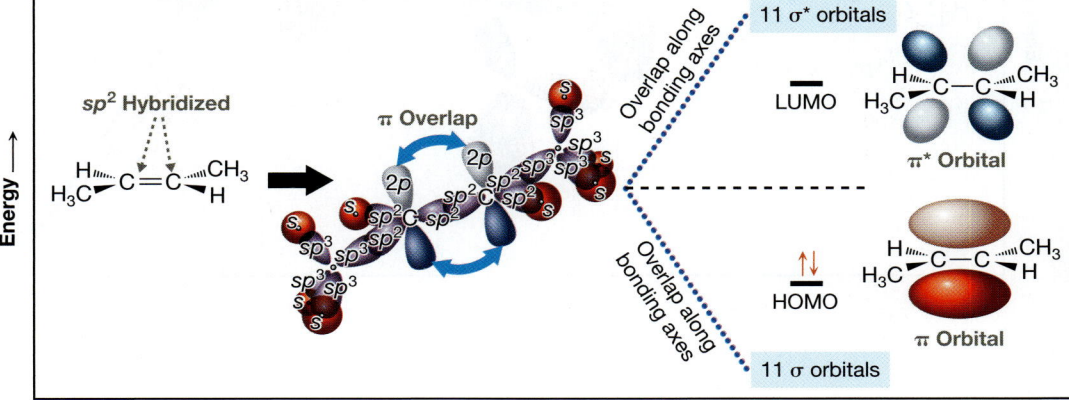

Yes, these figures are in agreement with Figure D-11, which shows the following as the HOMO of $CH_3CH{=}CHCH_3$.

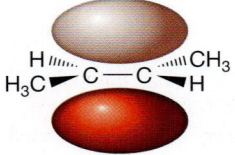

Your Turn D.14

Constructive interference occurs when two orbitals overlap with the same phases. Destructive interference occurs when two orbitals overlap with opposite phases. Constructive interference predominantly takes place in Figure D-11.

The LUMO of HCl is a σ* MO.

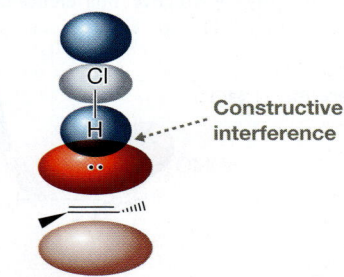

Constructive interference

The HOMO of an alkene is a π MO.

Your Turn D.15

The carbocation C is sp^2 hybridized and has an unhybridized p orbital. All of the other carbons are sp^3 hybridized and the H atoms each contribute a $1s$ orbital. End-on overlap occurs in the C—H and C—C bonding and, therefore, results in 15 σ bonds. The unhybridized p AO remains as a nonbonding orbital. The 30 valence electrons fill the 15 σ bonding MOs, so the HOMO is a σ bonding MO and the p AO is the LUMO.

Yes, these figures are in agreement with Figure D-12, which shows the HOMO and LUMO of $(CH_3)_2CHC^+HCH_3$ as follows:

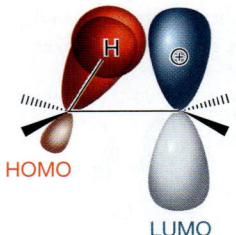

HOMO

LUMO

Your Turn D.16

Constructive interference occurs when two orbitals overlap with the same phases. Destructive interference occurs when two orbitals overlap with opposite phases. Predominantly constructive interference takes place in Figure D-12.

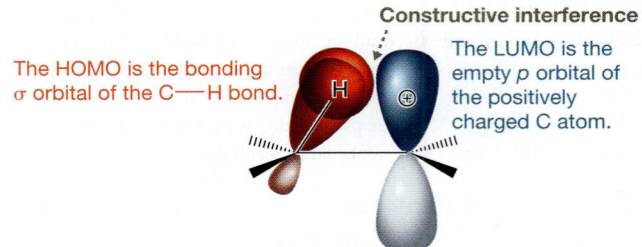

Constructive interference

The HOMO is the bonding σ orbital of the C—H bond.

The LUMO is the empty p orbital of the positively charged C atom.

Chapter 8

Your Turn 8.1

Equation 8-2a: heterolysis; Equation 8-2b: coordination

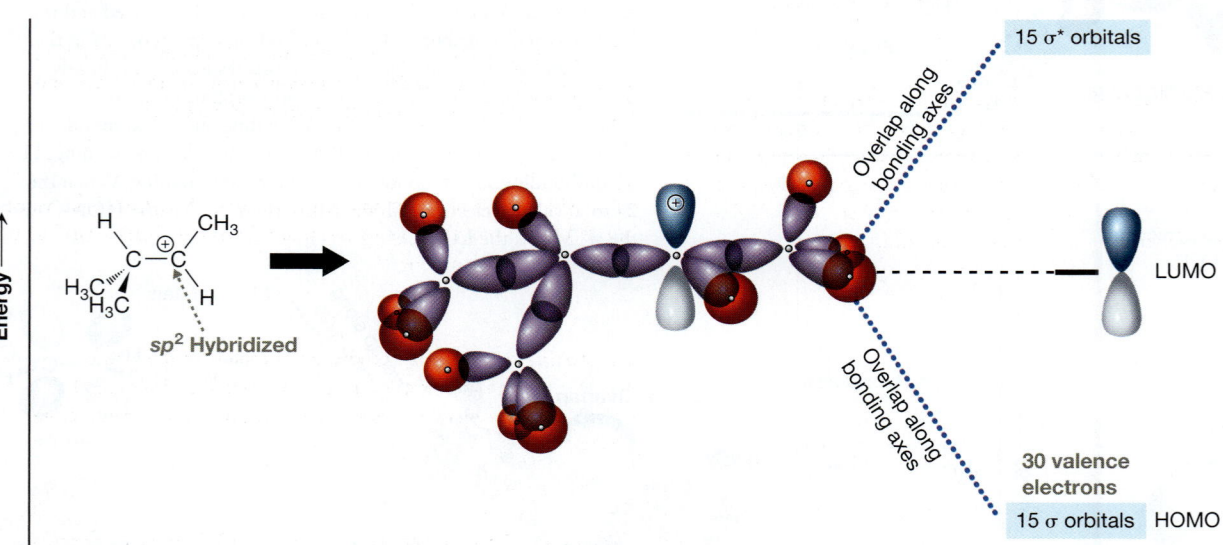

Your Turn 8.2

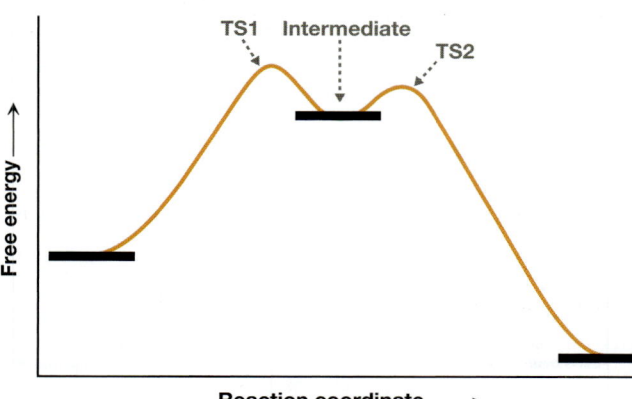

(a)

(b)

The allylic carbocation appears only in the reaction mechanism, not in the overall reaction, and is, therefore, an intermediate.

Your Turn 8.3

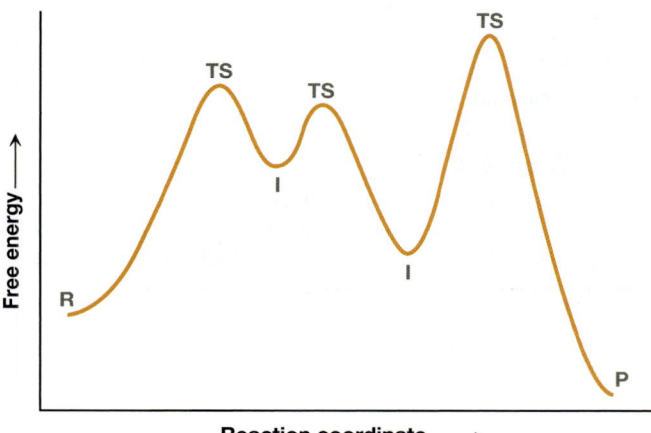

Your Turn 8.4

For a mechanism that contains n number of steps, there must be n number of transition states and $n - 1$ stages where intermediates appear. Transition states occur at energy maxima, and there are three transition states for this mechanism. Intermediates occur at energy minima, and there are two such minima on the curve for this mechanism. Therefore, there are three elementary steps in this mechanism.

Your Turn 8.5

Equation 8-5a: heterolysis; Equation 8-5b: electrophile elimination

Your Turn 8.6

The carbocation appears only in the reaction mechanism, not in the overall reaction, and is, therefore, an intermediate. B:⁻ and the R—L compound are overall reactants. B—H, the alkene, and :L⁻ are overall products.

Your Turn 8.7

Transition states appear at energy maxima and show the bond breaking and bond making in progress. Intermediates appear at energy minima. There are two transition states and one intermediate for a two-step mechanism.

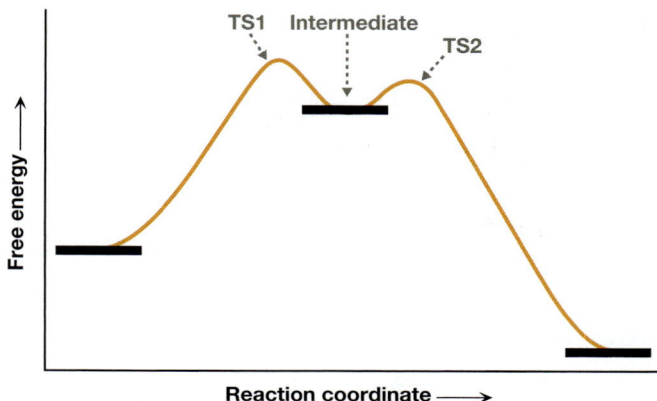

Your Turn 8.8

Doubling the concentration of $NaSCH_3$ doubles the rate of the reaction. The rate of an S_N2 reaction is directly proportional to the concentration of the nucleophile.

Your Turn 8.9

(a) The red curve (100 °C) has a value of ~2% of molecules able to surmount the energy barrier at an energy barrier of 15 kJ/mol, and **(b)** a value of <0.1% at an energy barrier of 25 kJ/mol.

Your Turn 8.10

(a) The blue curve (0 °C) at an energy barrier of 20 kJ/mol has a value of ~0.1% of molecules able to surmount the energy barrier. **(b)** The red curve (100 °C) has a value of ~0.4%.

Your Turn 8.11

The starting material and both products each have one chiral center (the C bonded to the halogen) and are, therefore, chiral. The carbocation intermediate has a plane of symmetry and is, therefore, achiral.

Your Turn 8.12

Carbocation **A** will undergo coordination with I⁻ to produce an *equal* mixture of enantiomers. Coordination of I⁻ with carbocation **B** results in an *unequal* mixture of diastereomers. Coordination with I⁻ and carbocation **C** results in a single, achiral product.

Your turn 8.13

The correct product of an S_N2 reaction is the one where the configuration is inverted, **B**, because the HO^- nucleophile must attack from opposite the Br^- leaving group.

Your Turn 8.14

With Equation 8-24's substrate in the anticoplanar conformation, the phenyl groups are on opposite sides of the H—C—C—L plane. That is consistent with the product alkene in Equation 8-24, which shows the phenyl groups on opposite sides of the C=C bond.

Your Turn 8.15

The first step yields a strong acid, R_2OH^+. This strong acid is incompatible in the basic, HO^- conditions.

Your Turn 8.16

The first step yields a strong base, RO^-. This strong base is incompatible in the acidic, H_3O^+ conditions.

Your Turn 8.17

The second step is an intramolecular proton transfer mechanism. This is an unreasonable step because there are typically solvent molecules that reside between the acidic and basic sites at any given time. This makes *direct* transfer of protons within a molecule unlikely to occur.

Your Turn 8.18

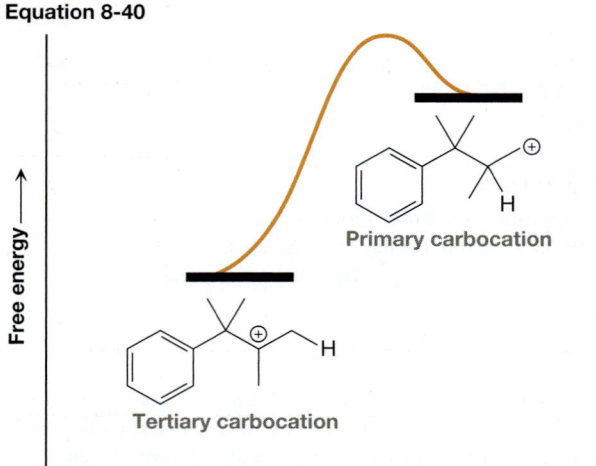

Benzylic C⁺

Your Turn 8.19

Equation 8-40

Equation 8-41

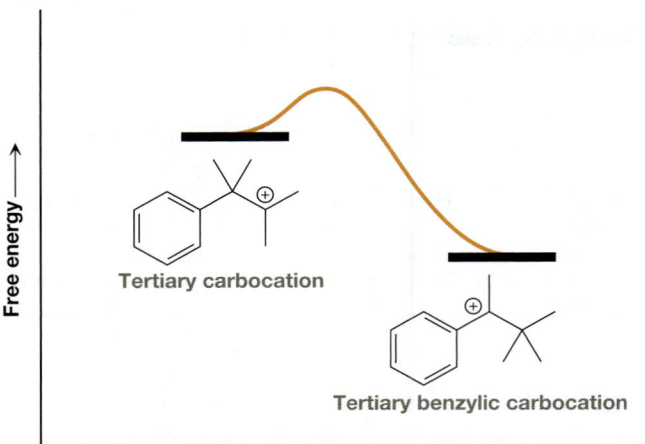

Chapter 9

Your Turn 9.1

NH_3 acts as a base (reaction **A**) when it bonds to the H and a nucleophile (reaction **B**) when it bonds to the electron-poor C.

Your Turn 9.2

The exergonic reaction in Figure 9-1a has a smaller energy-barrier compared to the endergonic reaction in Figure 9-1b, so the reaction is faster.

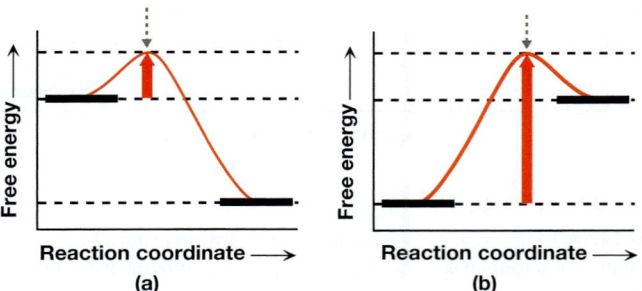

Your Turn 9.3

Your Turn 9.4

The weakest conjugate acid is that of hydroxide, H—OH, with a pK_a value of 15.7. This indicates that the hydroxide ion, HO^-, is the strongest base in Table 9-2 and has the fastest E2 rate. This suggests that the stronger base leads to a faster E2 reaction rate.

Your Turn 9.5

Strong, bulky bases | Groups contributing to steric hindrance

The *tert*-butoxide anion   The neopentoxide anion   Lithium diisopropylamide (LDA)

Your Turn 9.6

In Table 9-4, the strongest conjugate acid is that of TfO^- with a pK_a value of -13. This indicates that TfO^- is the weakest base and best leaving group, and therefore has the fastest S_N1 rate of reaction.

Your Turn 9.7

Methyl sulfonate (Mesylate) ($MsO^{\ominus}$)

Your Turn 9.8

$^{\ominus}\ddot{O}H$ is a *poor* leaving group. $H_2\ddot{O}$: is a *good* leaving group.

Your Turn 9.9

sp^3 C—H = 421 kJ/mol; sp^2 C—H = 464 kJ/mol; and sp C—H = 558 kJ/mol.

Your Turn 9.10

Hybrid

Your Turn 9.11

Br^- and N_3^- are reversed in ethanol compared to DMF. In DMF, N_3^- is a stronger nucleophile; but in ethanol, Br^- is the stronger nucleophile. This suggests that N_3^- is more strongly solvated by ethanol than Br^- is. Similarly, $CH_3CO_2^-$ is a stronger nucleophile in DMF than either Br^- or Cl^-; but in ethanol, the opposite is true, suggesting that the ethanol solvates $CH_3CO_2^-$ more strongly than it solvates Br^- or Cl^-.

Your Turn 9.12

The top reaction [products: $PhCH(OH)CH_3 + TsO^-$] results from substitution. The bottom reaction (products: $PhCH=CH_2 + TsO^- + H_2O$) results from elimination and has greater entropy (three products compared to two).

Your Turn 9.13

TABLE 9-13 Summary Table for the Reaction in Equation 9-40			
Factor	S_N1	S_N2	E1	E2
Strength				✓
Concentration				✓
Leaving group	✓		✓	
Solvent		✓		✓
Total	1	1	1	3

Your Turn 9.14

Equation 9-42

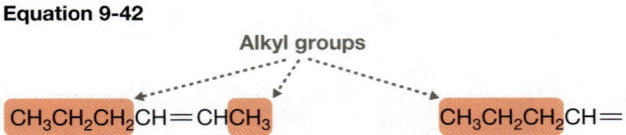

Hex-2-ene
(disubstituted)

Hex-1-ene
(monosubstituted)

Equation 9-43

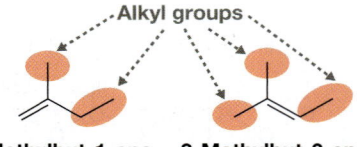

2-Methylbut-1-ene
(disubstituted)

2-Methylbut-2-ene
(trisubstituted)

Your Turn 9.15

An acetal group is an sp^3-hybridized carbon that is attached to two OR groups and either an R group or H at the other two bonds. In cellulose, amylose, and amylopectin, every C1 is bonded to two OR groups and, therefore, is part of an acetal.

Chapter 10

Your Turn 10.1

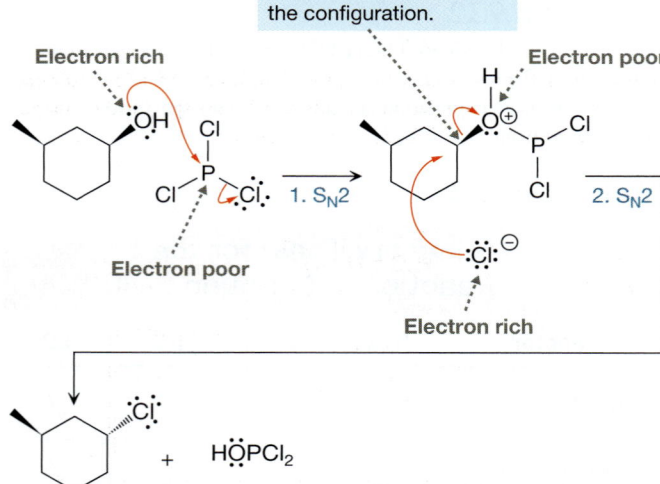

Your Turn 10.2

The nucleophilicities for NH_3 and Cl^- are 320,000 and 23,000, respectively. So NH_3 is, in fact, a stronger nucleophile in water than Cl^- is, making NH_3 a moderate nucleophile, too.

Your Turn 10.3

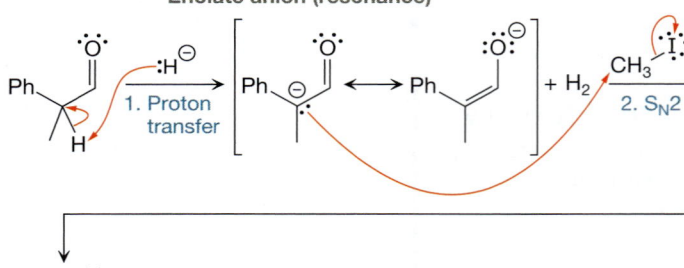

Your Turn 10.4

Alkyl groups are electron donating and, therefore, stabilize a nearby positive charge. Having two versus one electron-donating groups stabilizes the N^+ more in $Et_2NH_2^+$ compared to $EtNH_3^+$ (see the solution to Your Turn 10.3), so $Et_2NH_2^+$ should be drawn in the box on the bottom and $EtNH_3^+$ should be drawn in the box on the top. Electron donation by each Et group is represented by an arrow along the C—N^+ bond toward N^+. One such arrow should be drawn for $EtNH_3^+$ and two such arrows should be drawn for $Et_2NH_2^+$.

Your Turn 10.5

Your Turn 10.6

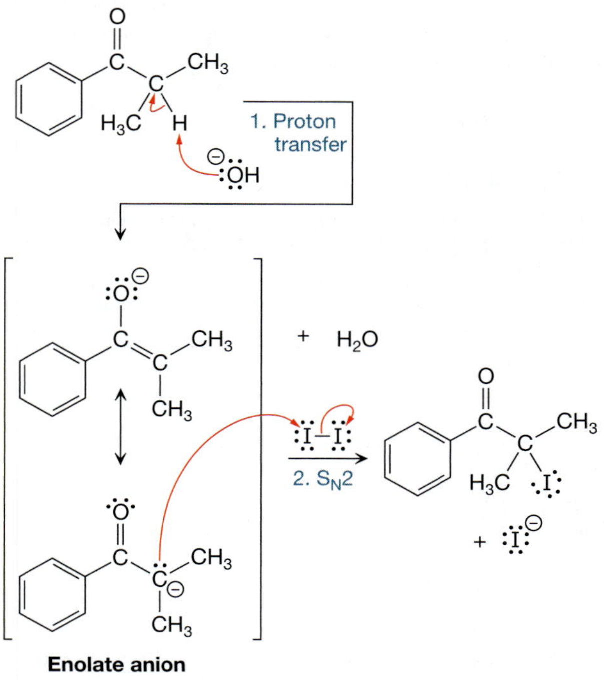

Enolate anion

Your Turn 10.7

Step 1 in Equation 10-29 is proton transfer; Step 2 is S_N2; Step 3 is proton transfer; Step 4 is S_N2.

Your Turn 10.8

Ketone **B** undergoes chlorination under basic conditions faster than **A** due to the electron-withdrawing Cl. Cl stabilizes the negative charge on the enolate anion that is produced.

Your Turn 10.9

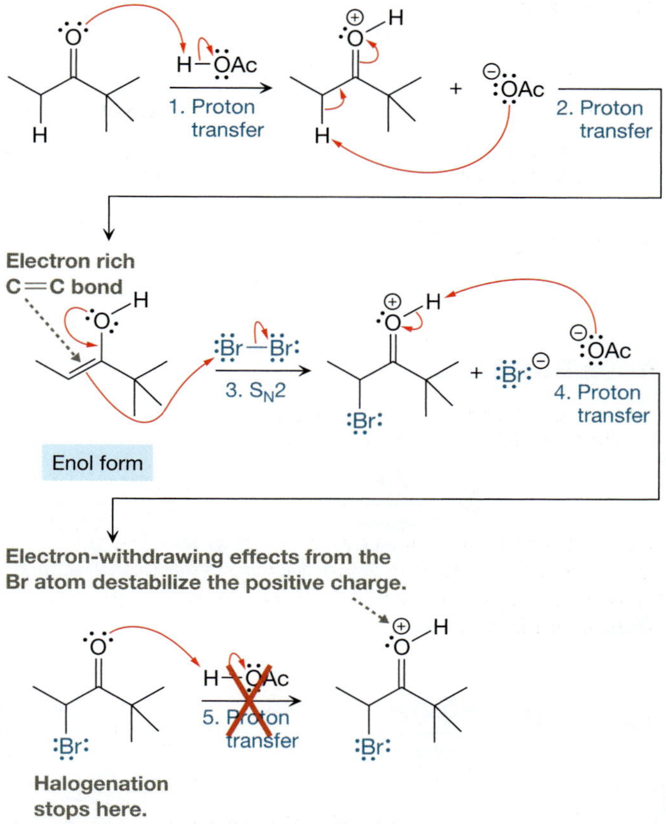

Enol form

Electron-withdrawing effects from the Br atom destabilize the positive charge.

Halogenation stops here.

Your Turn 10.10

Ketone **A** undergoes bromination faster than **B** under acidic conditions due to the electron-withdrawing Br. Br destabilizes the positive charge that develops when the carbonyl O is protonated.

Your Turn 10.11

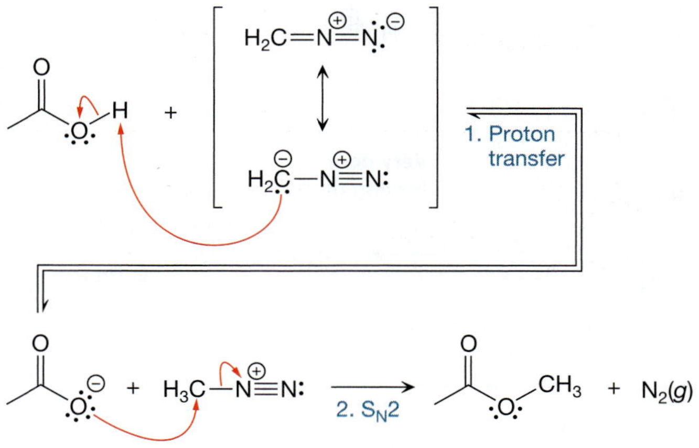

Your Turn 10.12

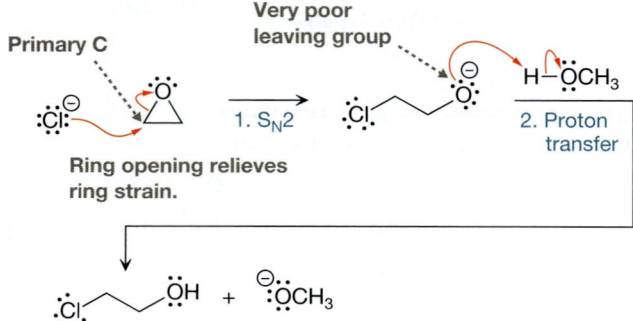

Your Turn 10.16

Your Turn 10.17

The pK_a of H_2 is 35, and the pK_a of RC≡CH is about the same as that of HC≡CH, which is 25. A lower pK_a indicates a stronger acid; thus, RC≡CH is a stronger acid compared to H_2 by a factor of 10^{10}. Therefore, the side opposite the terminal alkyne—the product side—is favored by a factor of 10^{10}. This makes the reverse reaction difficult, so the forward reaction is irreversible.

Your Turn 10.13

Your Turn 10.14

Your Turn 10.15

Your Turn 10.18

Mechanism for Equation 10-63

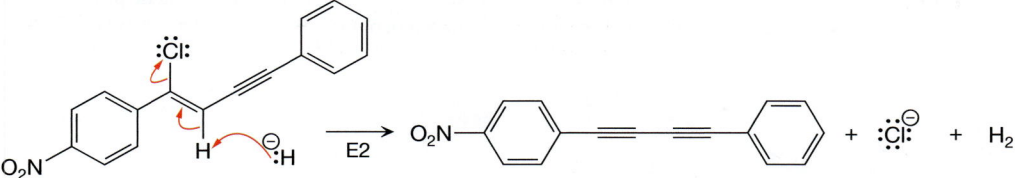

Mechanism for Equation 10-64

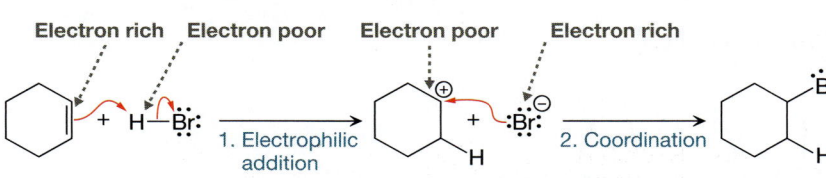

Your Turn 10.19

There is less steric strain in conformation **B** than in **A**.

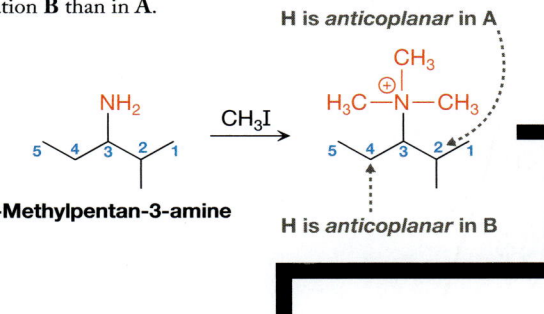

2-Methylpentan-3-amine

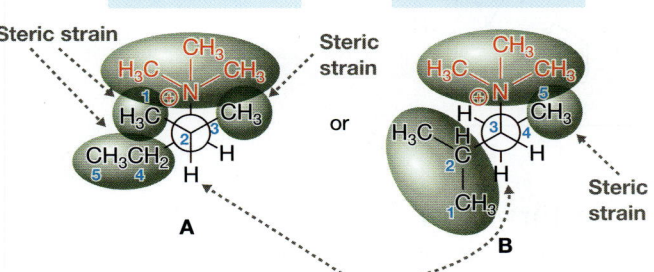

A   or   B

Anticoplanar with leaving group

Chapter 11

Your Turn 11.1

HBr adds across the C=C of (*E*)-but-2-ene to produce 2-bromobutane.

Your Turn 11.2

Electron rich   **Electron poor**   **Electron poor**   **Electron rich**

1. Electrophilic addition   2. Coordination

Your Turn 11.3

$\Delta H_{rxn}^{\circ} = (619 \text{ kJ/mol} + 431 \text{ kJ/mol}) - (339 \text{ kJ/mol} + 418 \text{ kJ/mol} + 331 \text{ kJ/mol}) = -38 \text{ kJ/mol}$; exothermic ($\Delta H_{rxn}^{\circ} < 0$), which favors products.

Your Turn 11.4

The primary carbocation is less stable (higher energy) and goes in the box with the red line:

The secondary carbocation is more stable (lower energy) and goes in the box with the blue line:

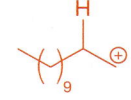

Your Turn 11.5

B

Your Turn 11.6

OSO₃H

Your Turn 11.7

The hypothetical mechanism is shown below. It is unfeasible because HO⁻ is too unstable to be produced along with a carbocation.

Unstable

Your Turn 11.8

Your Turn 11.9

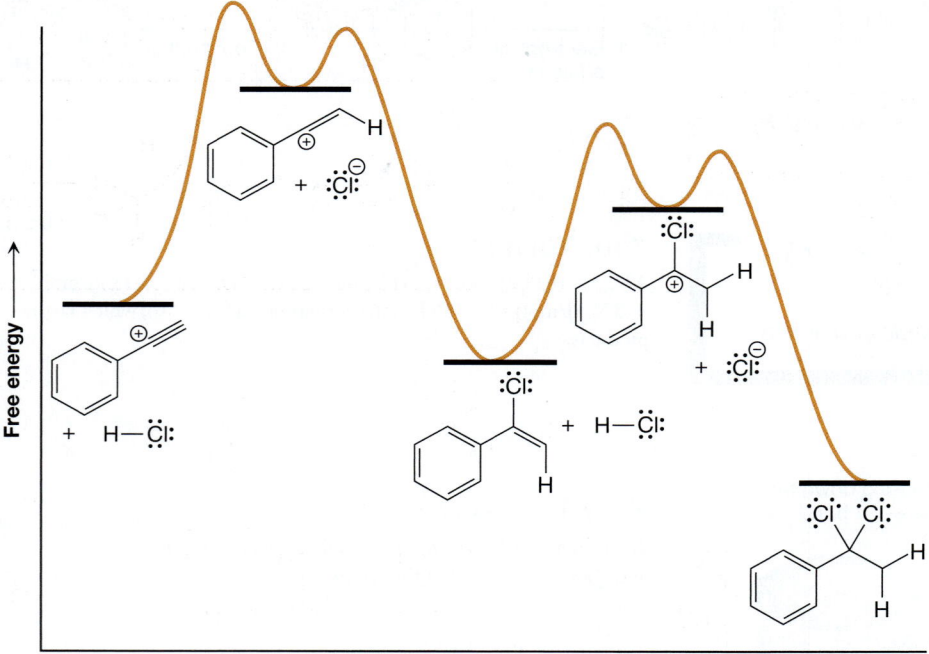

Reaction coordinate ⟶

Free energy ⟶

Your Turn 11.10

A less stable, vinylic carbocation would have to be generated, as shown below. This has the formal charge on a 1° C rather than a 2° C as in the more stable vinylic carbocation in Equation 11-21.

Electrophilic addition

Less stable vinylic C⁺

Your Turn 11.11

The strongest acid that can exist in solution is the protonated solvent, $CF_3CH_2OH_2^+$. This acid is stronger than H_3O^+ because the CF_3 group, which is electron withdrawing, destabilizes the O^+.

Your Turn 11.12

The first is from 1,2-addition and the second is from 1,4-addition.

Your Turn 11.13

Equation 11-27: The second product is the thermodynamic product (major product at high temperature, i.e., under thermodynamic control). Equation 11-28: The first product is the kinetic product (major product at low temperature, i.e., under kinetic control).

Your Turn 11.14

First product: thermodynamic, more stable alkene (tetrasubstituted is more stable than trisubstituted); second product: kinetic, 1,2-addition.

Chapter 12

Your Turn 12.1

Each C atom has four bonds, in both the reactant and the product. Therefore, each C atom maintains its octet.

Your Turn 12.2

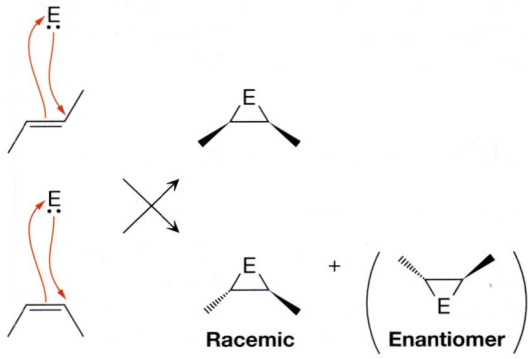

Your Turn 12.3

The carbene C is assigned one electron from each of the two bonds, and two electrons from the lone pair. The carbene C is assigned four valence electrons in all, the same as in an isolated C atom, so the formal charge is 0.

Your Turn 12.4

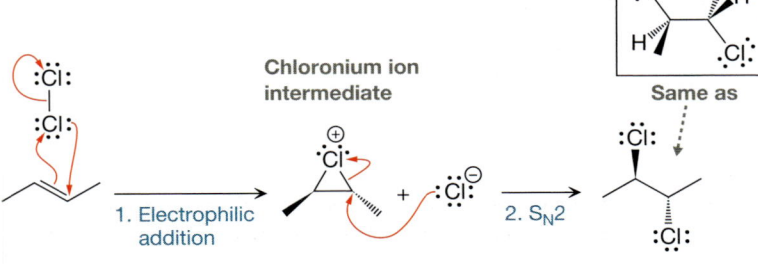

Your Turn 12.5

Your Turn 12.6

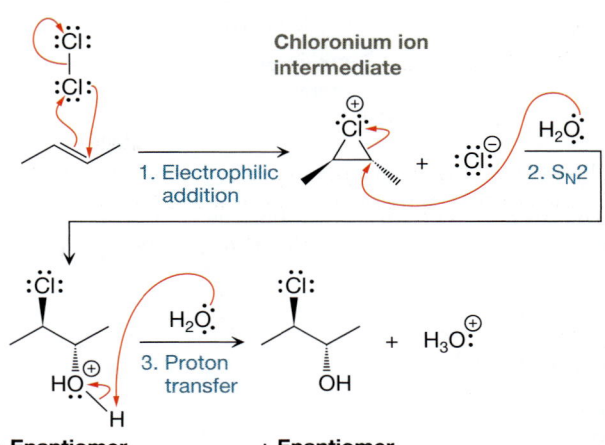

Your Turn 12.7

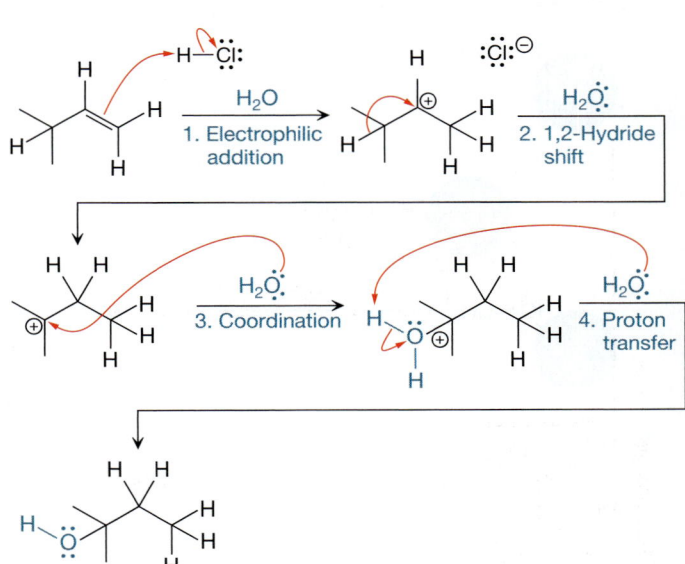

Your Turn 12.8

OH   OH

Equation 12-27   **Equation 12-28**

Your Turn 12.9

O—O = 138 kJ/mol; C—C = 339 kJ/mol

Your Turn 12.10

The electrophilic atoms are B in BH_3 and H in HCl. In both cases, the electrophilic atom adds to the less substituted C atom.

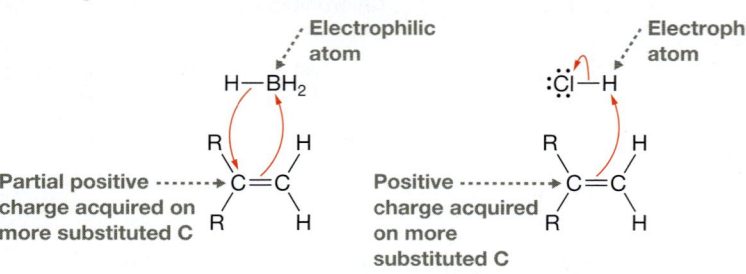

Your Turn 12.11

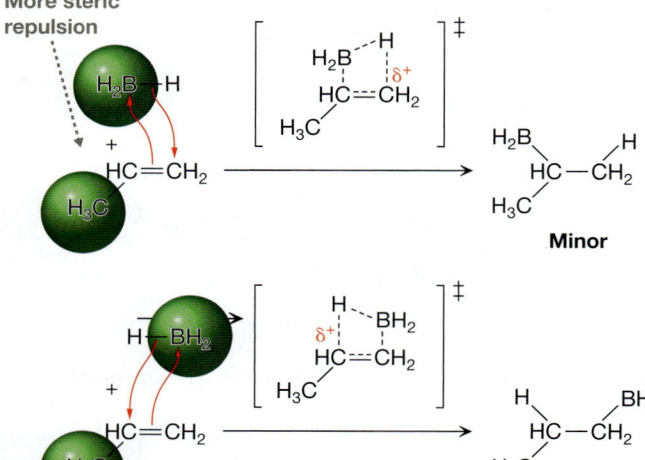

Your Turn 12.12

Each C—B bond is replaced by a C—OH bond, and the stereochemistry is retained.

Chapter 13

Your Turn 13.1

(a)

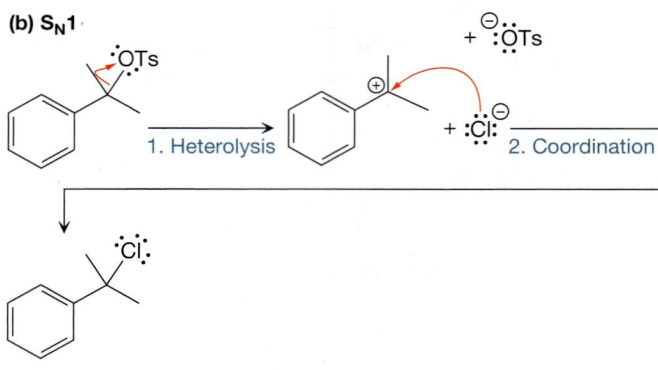

2-Phenyl-2-tosyloxypropane + NaCl ⟶ 2-Chloro-2-phenylpropane

(b) S_N1

1. Heterolysis 2. Coordination

Your Turn 13.2

H_2O is the solvent and the reaction condition is 70 °C. Therefore, both H_2O and 70 °C are written below the arrow: H_2O, 70 °C.

Your Turn 13.3

First, butanoic acid is treated with sodium hydroxide. Once that reaction has come to completion, bromoethane is added, which produces ethyl butanoate.

Your Turn 13.4

1. $Hg(OAc)_2$, H_2O
2. $NaBH_4$

(Z)-Hex-3-ene ⟶ **Hexan-3-ol**

Your Turn 13.5

Review the reaction tables in Chapters 9–12 and note any difficulties you have.

Your Turn 13.6

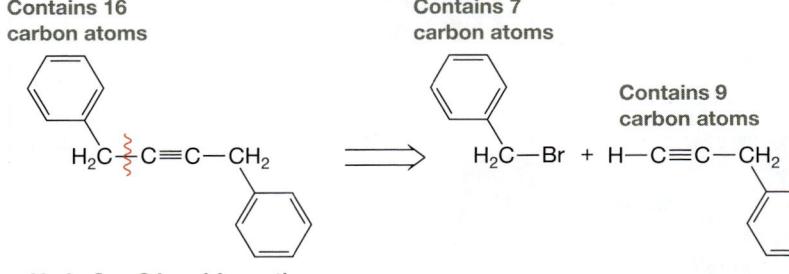

Epoxide → Functional group conversion → Alcohol and ether

Functional group conversion

Alkyl halide (Ether) → Alters carbon skeleton → Addition of C≡CH (Ether)

Functional group conversion

Aldehyde (Ether) → Alters carbon skeleton → Addition of CH₂CH₃ (Ether)

Your Turn 13.7

Synthetic step **B**: Appendix C Reactions That Alter the Carbon Skeleton, entry 4; synthetic step **C**: Appendix D, Table AppD-2 Reactions That Produce Alkyl and Aryl Halides, entry 16.

Your Turn 13.8

Contains 16 carbon atoms

$H_2C \overset{\xi}{\underset{}{\cdots}} C\equiv C-CH_2$ ⟹ H_2C-Br + $H-C\equiv C-CH_2$

Contains 7 carbon atoms

Contains 9 carbon atoms

Undo C—C bond formation reaction, entry 1 of Table 13-1

Your Turn 13.9

$\equiv$ —— O— ⟹ $\equiv$ —— ONa + CH_3—Br

CH_3—Br $\xrightarrow{\equiv \text{——ONa}}$ $\equiv$ —— O— $\xrightarrow[\text{2. } CH_3CH_2Br]{\text{1. NaH}}$ ———— O—

Your Turn 13.10

Ethanol, CH_3CH_2OH, would be an appropriate solvent. That way, if the nucleophile deprotonates the solvent, the product species would be identical to the reactant species, as shown below.

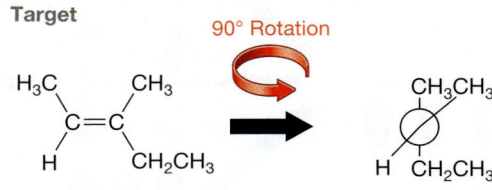

$CH_3CH_2\ddot{O}\!-\!H + {:}\ddot{O}CH_2CH_3 \longrightarrow CH_3CH_2\ddot{O}{:}^{\ominus} + H\!-\!\ddot{O}CH_2CH_3$

Your Turn 13.11

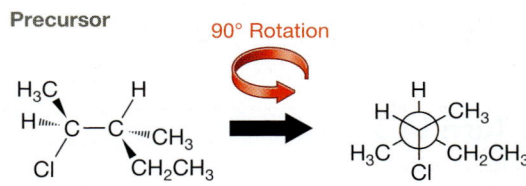

Target

90° Rotation

(*E*)-3-Methylpent-2-ene

Precursor

90° Rotation

(*2R,3R*)-3-Chloro-3-methylpentane

Your Turn 13.12

If each step's yield is 80%, the yield of a six-step synthesis would be $(0.80)^6 = 0.26$, or 26%.

Chapter 14

Your Turn 14.1

$H_2\overset{\oplus}{C}\!-\!CH$
$\overset{\|}{CH_2}$

Your Turn 14.2

The nodes are indicated by blue, vertical, dotted lines and planes in the MOs below. Note that the π_1 bonding MO has no vertical nodal plane.

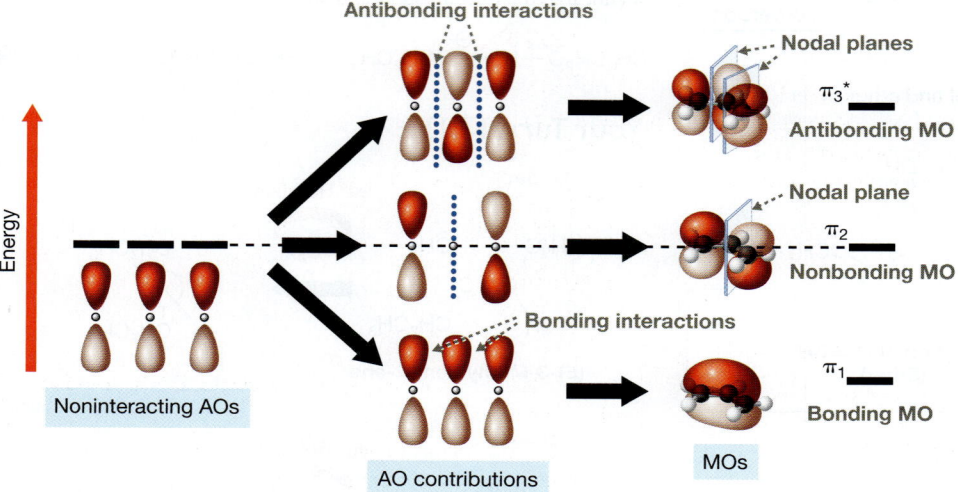

Your Turn 14.3

In this case, all of the antibonding interactions among atomic orbitals are indicated by the presence of a nodal plane (dotted blue line). The bonding interactions are indicated by green dashed ovals below.

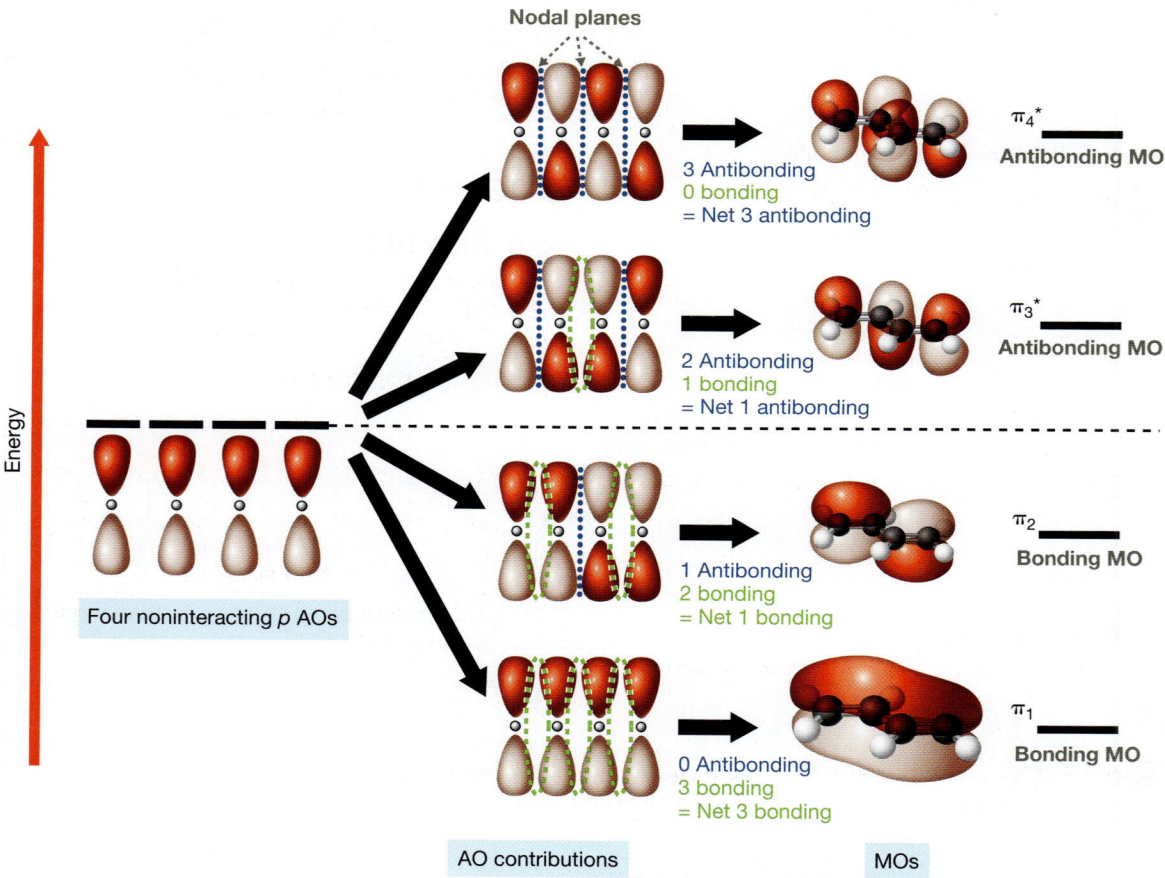

Your Turn 14.4

The resonance energy of 152 kJ/mol is 0.448, or 44.8%, as strong as the average C—C single bond energy of 339 kJ/mol (from Chapter 1).

Your Turn 14.5

These numbers are 132 pm for the C=C bond in ethene and 154 pm for the C—C, compared to 132 pm and 160 pm, respectively, for cyclobutadiene. Overall, the numbers are in very good agreement, suggesting that the π electrons are NOT significantly resonance delocalized.

Your Turn 14.6

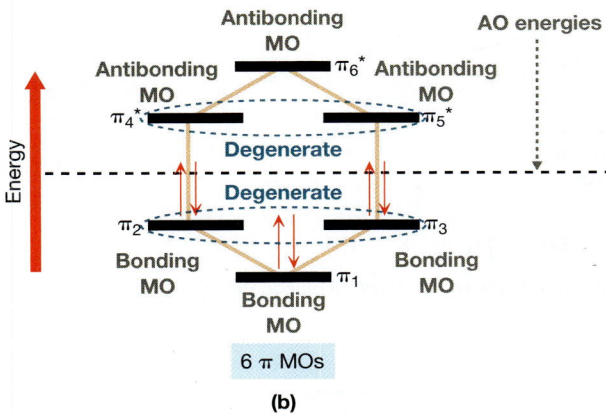

(b)

Your Turn 14.7

The number of nodal planes perpendicular to the bonding axis in the π_6 MO in Figure 14-20 is 3.

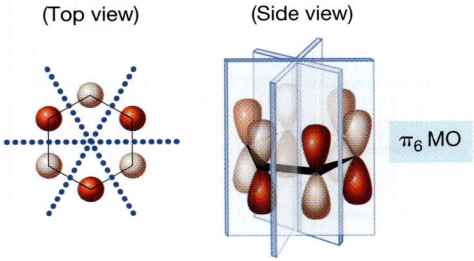

Your Turn 14.8

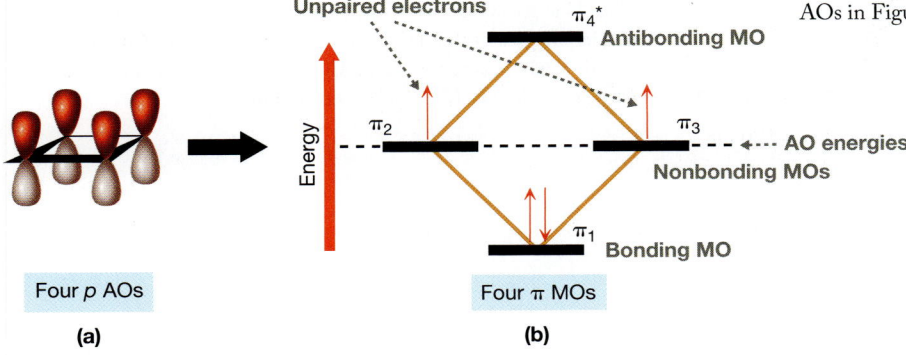

Your Turn 14.9

The H atoms are included in this figure. Notice that they do not crash into each other in [18]annulene, as they do in the other molecules in Figure 14-24.

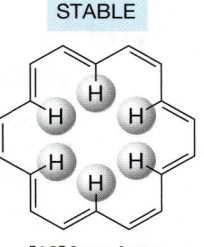

[18]Annulene

Your Turn 14.10

The dotted lines indicate a closed loop of orbitals. This is viewed as a single aromatic π system.

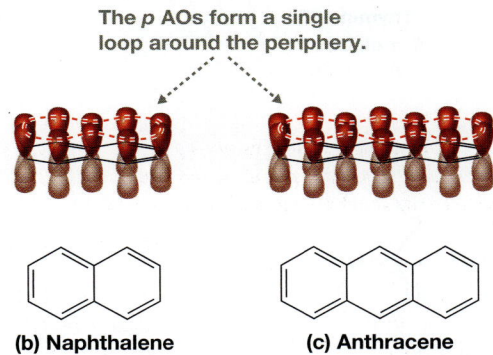

The p AOs form a single loop around the periphery.

(b) Naphthalene   (c) Anthracene

Your Turn 14.11

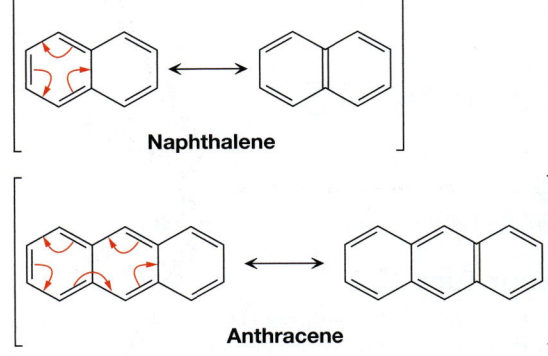

Naphthalene

Anthracene

The resonance structures do not affect the locations of the p AOs in Figure 14-25.

Your Turn 14.12

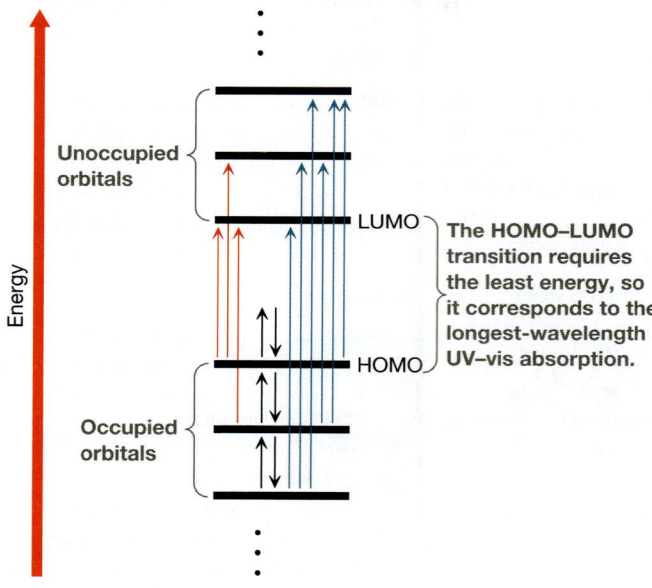

Guanine (G)
10 π electrons
Aromatic

Adenine (A)
10 π electrons
Aromatic

Thymine (T)
6 π electrons
Aromatic

Cytosine (C)
6 π electrons
Aromatic

Your Turn 14.13

Enol form
Guanine

Enol form
Cytosine

Hydrogen bonding is maximized when both G and C are in the keto form.

Chapter 15

Your Turn 15.1

(a) $\%T = 20\%$ represents more light being absorbed than $\%T = 40\%$.
(b) $A = 0.750$ represents more light being absorbed than $A = 0.500$.

Your Turn 15.2

If molar absorptivity (ε) increases by a factor of 3, absorbance also increases by a factor of 3 (Equation 15-3, $A = \varepsilon lC$, tells us they are directly related).

Your Turn 15.3

The wavelength of light is represented by λ_{photon}, which is inversely proportional to E_{photon} (Equation 15-5). So E_{photon} increases as λ_{photon} decreases. Therefore, a 375 nm photon has more energy than a 530 nm photon.

Your Turn 15.4

Additional transitions are shown in blue in the figure below.

Unoccupied orbitals

LUMO

The HOMO–LUMO transition requires the least energy, so it corresponds to the longest-wavelength UV–vis absorption.

HOMO

Occupied orbitals

Energy

Your Turn 15.5

All six blue arrows in Your Turn 15.4 are longer than (i.e., transition energies are greater than) the one representing the HOMO–LUMO transition, which is the shortest. All those transitions require more energy than the HOMO–LUMO transition.

Your Turn 15.6

The electron is circled below, and the transition is indicated by the curved blue arrow. The lowest two π MOs are bonding because they are below the energy of the isolated *p* orbitals (the horizontal dashed line), whereas the highest two π MOs are antibonding, because they are above that line.

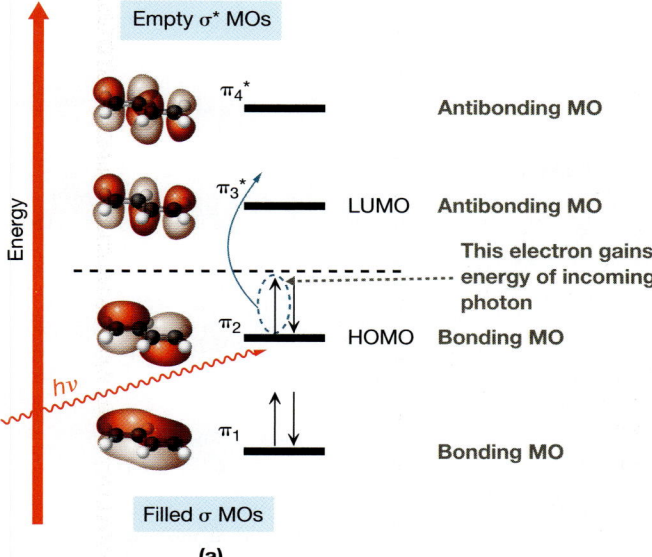

(a)

Your Turn 15.7

Refer to Table 15-1. One C═C π bond (no conjugation) ~180 nm, two conjugated C═C π bonds ~225 nm, three conjugated C═C π bonds ~275 nm, four conjugated C═C π bonds ~290, eleven conjugated π bonds >450 nm. As the number of conjugated π bonds increases, the λ_{max} value increases. The HOMO–LUMO energy gap, therefore, decreases.

Your Turn 15.8

The reaction proceeds by the E1 mechanism, not E2. For E1, the [Base] term is not part of the rate law and the rate of appearance of the product is directly proportional to the concentration of the organic substrate and not the base.

Your Turn 15.9

The peak at ~1320 cm^{-1} is labeled below. However, several other peaks could have been selected. The frequency value of the absorbed photon is the value of the peak at the *x* axis and the units are wavenumbers, cm^{-1}. These are the same frequencies for the mode of vibration responsible for photon absorption.

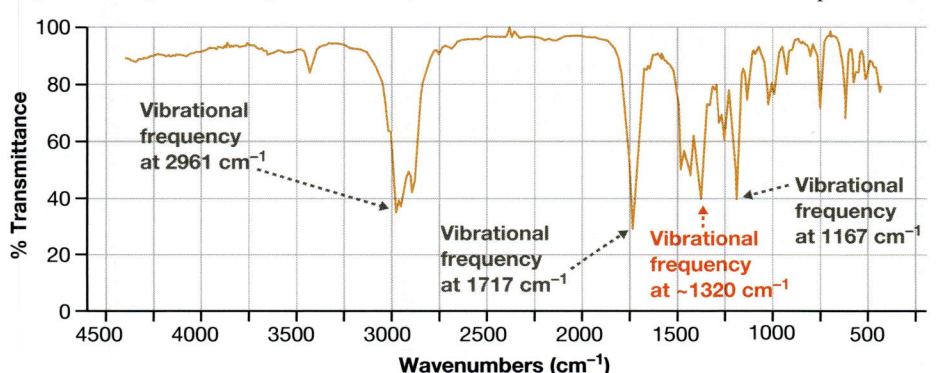

Your Turn 15.10

The C═O peak generally appears at 1720 cm^{-1} as a strong, sharp peak. The benzaldehyde C═O peak specifically occurs around 1700 cm^{-1}.

Your Turn 15.11

The fingerprint region is the region below ~1400 cm^{-1}. In Figure 15-14, the two regions that are in this range are the single-bond stretches and the bending modes (the light blue and light purple regions).

Your Turn 15.12

The general region for a C═C stretch is 1620–1680 cm^{-1}, so a C═C stretch for hept-3-ene would be expected near 1650 cm^{-1}.

Your Turn 15.13

The H—O stretch bands are intense, broad, and centered around 3300 cm^{-1} for RO—H (alcohols, Fig. 15-17) and 3000 cm^{-1} ROO—H (carboxylic acids, Fig. 15-18). In a carboxylic acid, the OH stretch is shifted to lower frequency and is much broader than in an alcohol; therefore, it overlaps the alkane C—H stretches (2800–3000 cm^{-1}).

Your Turn 15.14

The small bump around 3300 cm^{-1} in Figure 15-16b is due to a water or alcohol impurity in the hept-3-ene sample when the IR spectrum was taken. The OH stretch would appear much more intense if the OH were part of the molecule itself, instead of an impurity.

Your Turn 15.15

The N—H stretching modes are the moderately intense, moderately broad peaks near 3300 cm^{-1}. There are two in the first spectrum (1° amide, Fig. 15-20a), one in the second (2° amide, Fig. 15-20b), and none in the third (Fig. 15-20c).

Your Turn 15.16

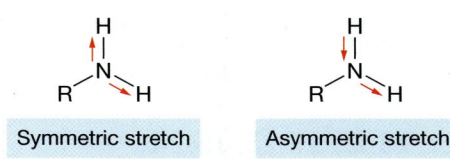

Symmetric stretch   Asymmetric stretch

Your Turn 15.17

Triple-bond stretching modes appear between 2000 and 2500 cm^{-1}. In Spectrum 3, that triple-bond absorption at ~2250 cm^{-1} is strong, whereas in Spectrum 4, it has weak intensity at ~2100 cm^{-1}.

Your Turn 15.18

The C—H stretches near 2720 and 2820 cm^{-1} in the spectrum of heptanal (Fig. 15-22) are indicative of an aldehyde C—H. In this case, the peak at ~2820 cm^{-1} appears at the edge of the alkane C—H band that appears between 2800 and 3000 cm^{-1}.

Your Turn 15.19

From Table 15-2, alkane, sp^3 C—H stretches occur between 2800 and 3000 cm^{-1} and vary in intensity. Alkene, sp^2 C—H stretches occur between 3000 and 3100 cm^{-1} and are generally weak. In Figure 15-23a, alkane C—H stretches are intense and occur at 2800–3000 cm^{-1}. In Figure 15-23b, alkane C—H stretches are moderately intense and occur at 2800–3000 cm^{-1}. In Figure 15-23c, alkene C—H stretches are weak and occur around 3100 cm^{-1}; no alkane C—H stretch is present.

Your Turn 15.20

(a) C—H bending at ~759 and 703 cm^{-1} [Table 15-3, Aromatic (monosubstituted)]

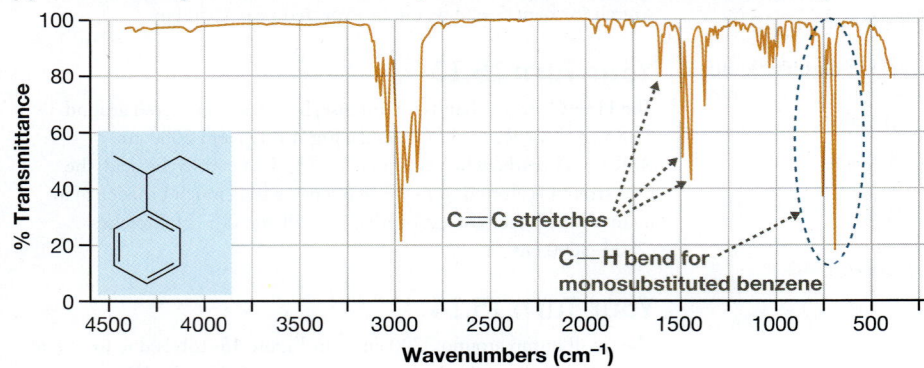

(b) C—H bending at 833 cm^{-1} [Table 15-3, Aromatic (para)]

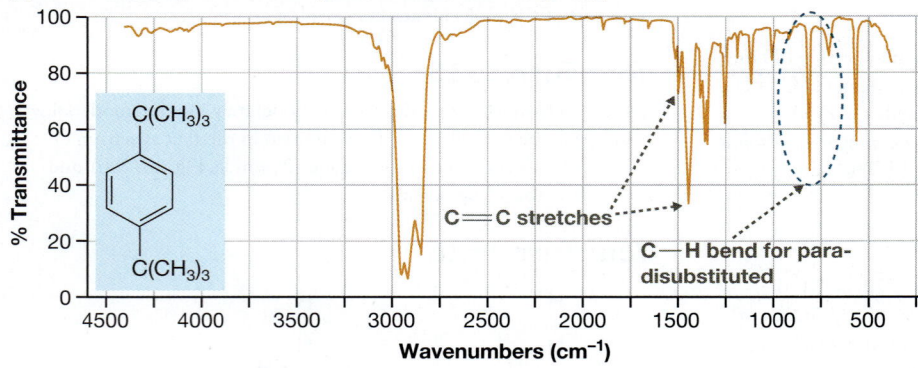

Your Turn 15.21, 15.22, 15.23, 15.24

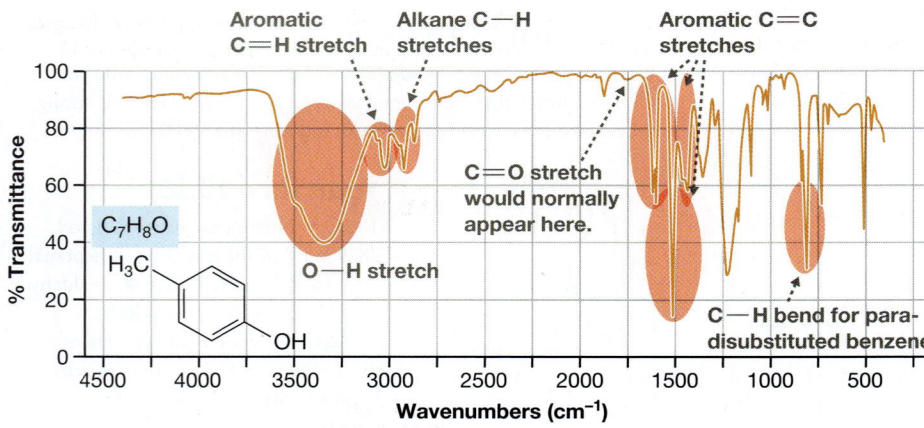

Your Turn 15.25, 15.26, 15.27, 15.28

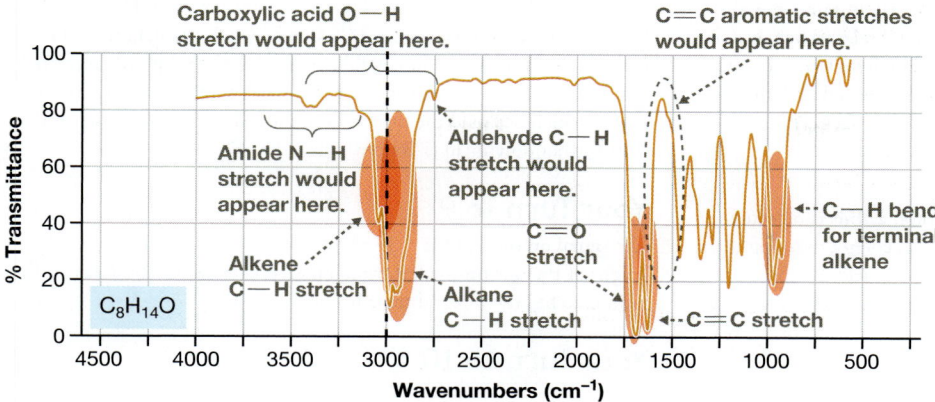

Your Turn 15.29

A saturated molecule with 6 C and 1 N is shown here. It has 15 H atoms, which is 8 more than in the molecular formula of Unknown 3.
Yes, this is consistent with Unknown 3 having an IHD of 4.

Your Turn 15.30

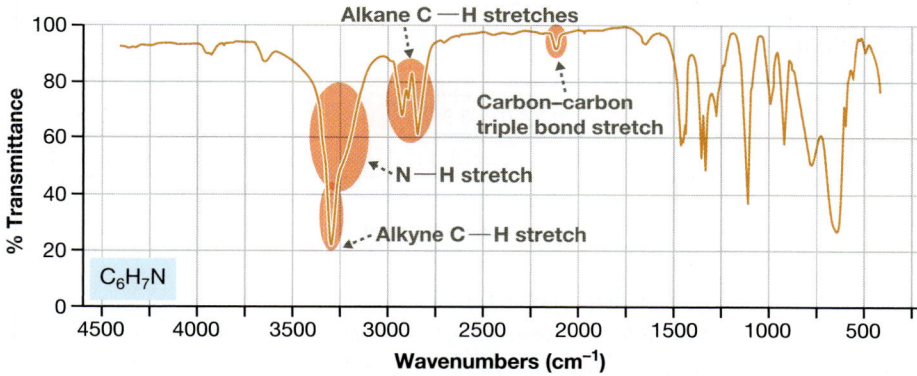

Chapter 16

Your Turn 16.1

A stronger magnetic field appears to the right of the states already shown in the figure. As shown below, this creates a larger energy difference between the α and β spin states.

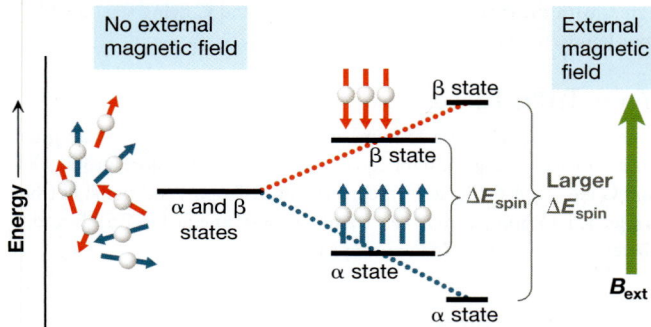

Your Turn 16.2

(a) One plane of symmetry that benzene has is indicated by the dashed line below. As shown, with respect to that plane of symmetry, H^2 and H^6 are mirror images, as are H^3 and H^5. (b) After benzene is rotated 60°, as indicated below on the right, the molecule appears unchanged, but all six H atoms end up occupying a location that a different H atom occupied before rotation.

Your Turn 16.3

When comparing signals in Table 16-1 to the CH_3Cl proton signal, the chemical shifts with larger values are downfield relative to CH_3Cl and the chemical shifts with smaller values are upfield relative to CH_3Cl. Entries 11–16, 19, and 20, therefore, are downfield relative to CH_3Cl, whereas entries 1–9 are upfield.

Your Turn 16.4

Proton 1 is shielded to a greater extent than proton 2. This is reflected by the fact that its spin states are closer together in energy, meaning that the nucleus experiences less of the external magnetic field. Proton 2 has a larger ΔE_{spin} than proton 1 and will require higher frequency radio waves to cause a spin flip.

Your Turn 16.5

FCH_3 (δ = 4.1 ppm, EN = 3.98), $ClCH_3$ (δ = 3.1 ppm, EN = 3.16), $BrCH_3$ (δ = 2.7 ppm, EN = 2.96). The trends are the same: Both chemical shifts and electronegativities increase as you go up the group, Br < Cl < F. Increased electronegativity leads to increased deshielding of a nearby CH proton. The electronegative atom is electron withdrawing and removes electron density from the nearby proton. Increased deshielding causes the signal to shift downfield.

Your Turn 16.6

The local magnetic field lines are the blue lines in Figure 16-10. The ones that are in the same direction as the external magnetic field are circled below. Notice that these local magnetic field lines appear where benzene's H atoms are located. This leads to additional deshielding of the protons on benzene, and the signal appears downfield.

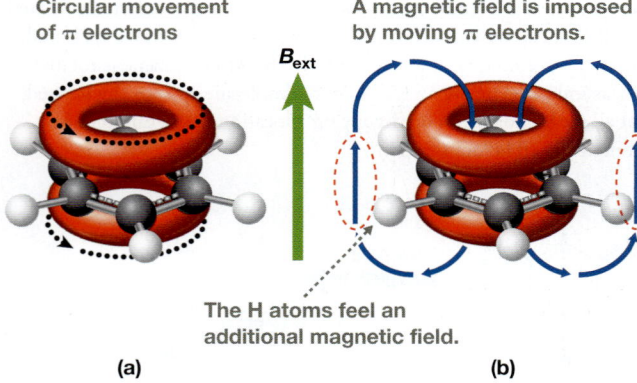

Circular movement of π electrons

A magnetic field is imposed by moving π electrons.

B_{ext}

The H atoms feel an additional magnetic field.

(a)   (b)

Your Turn 16.7

On each alkyne H atom, draw an arrow pointing directly downward, which opposes B_{ext}. This results in a shielding, not deshielding, effect.

Your Turn 16.8

The proton chemical shift in $CHCl_3$ is greater than in CH_2Cl_2. CH_2Cl_2 has two electronegative Cl atoms, which deshield the CH proton. This results in a downfield chemical shift (5.3 ppm) compared to CH_3Cl (Table 16-1; 3.1 ppm). $CHCl_3$ has three electronegative Cl atoms, which deshield the proton even more, resulting in a chemical shift further downfield (7.3 ppm).

Your Turn 16.9

The signal on the right is ~39 mm (~1.5 in.), roughly 1.5 times the height of the one on the left, which measures ~28 mm (~1.1 in.). This ratio matches the integral trace in Figure 16-14.

Your Turn 16.10

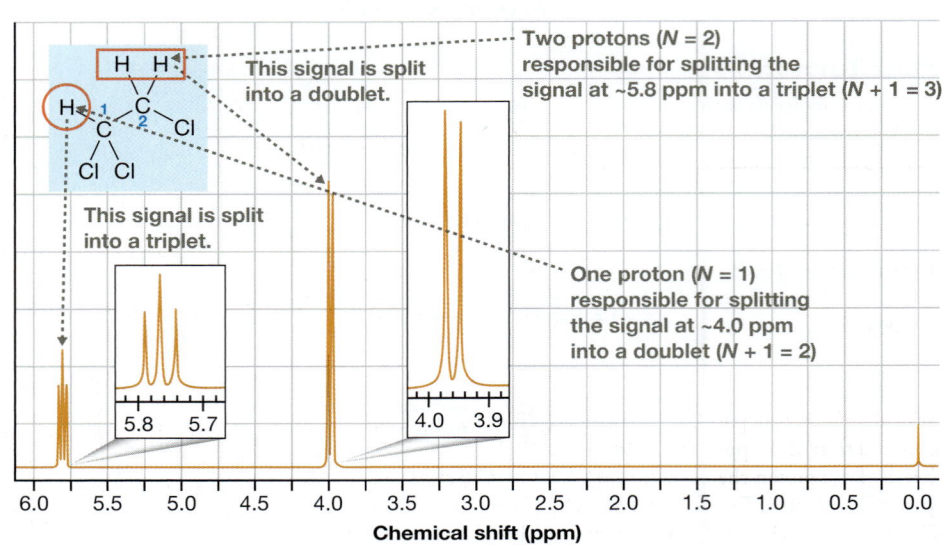

Two protons (N = 2) responsible for splitting the signal at ~5.8 ppm into a triplet (N + 1 = 3)

This signal is split into a doublet.

This signal is split into a triplet.

One proton (N = 1) responsible for splitting the signal at ~4.0 ppm into a doublet (N + 1 = 2)

Chemical shift (ppm)

Your Turn 16.11

When a proton is coupled to N protons that are distinct from itself, the resulting splitting pattern has N + 1 peaks. **(a)** Doublet (two peaks) = 2 = N + 1, N = 1 (i.e., one proton is coupled to the one giving rise to the signal); **(b)** quintet (five peaks) = 5 = N + 1, N = 4 (four coupled protons); **(c)** singlet (one peak) = 1 = N + 1, N = 0 (no coupled protons).

Your Turn 16.12

The coupling constants suggest that protons **B** and **C** are *not* coupled because the coupling constants, 8.5 Hz and 7.6 Hz, are different. This agrees with the structure showing protons **B** and **C** being separated by four bonds, too far for coupling to occur.

Your Turn 16.13

Signal **E** (the quartet at 4.15 ppm) and signal **G** (the triplet at 1.25 ppm) both have a coupling constant equal to 7.1 Hz and, therefore, protons **E** and **G** are coupled. Signal **F** (the quartet at 2.33 ppm) and signal **H** (the triplet at 1.15 ppm) both have a coupling constant equal to 7.5 Hz and, therefore, protons **F** and **H** are coupled.

Your Turn 16.14

There are a total of two H atoms coupled to the H^C proton. If both of those protons split the H^C signal equally, then, according the $N + 1$ rule, the H^C signal should be a triplet. No signals in the spectrum appear to consist of three signals in a $1:2:1$ ratio. So the $N + 1$ rule does not appear to apply in this case.

Your Turn 16.15

The CH would split the CH_2 signal in propanal into a doublet with a coupling constant of ~2 Hz. The CH_3 would split the CH_2 signal into a quartet with a coupling constant of ~6 Hz. Therefore, the signal is a quartet of doublets.

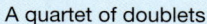
A quartet of doublets

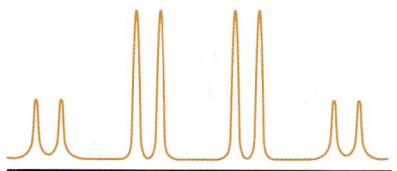

Your Turn 16.16

There are six signals in the ^{13}C NMR spectrum (not including the solvent or TMS) and, thus, six chemically distinct carbon atoms in the compound.

Your Turn 16.17

Several examples are possible. For example, CH_3R has a proton chemical shift of 0.9 ppm and a carbon chemical shift of 10–25 ppm. CH_2R_2 has a proton chemical shift of 1.3 ppm and a carbon chemical shift of 20–45 ppm. Thus CH_3R and CH_2R_2 have the same order for H and C chemical shifts (different scales). Other examples include $BrCH_3$ and $ClCH_3$; and alkenes and alkynes.

Your Turn 16.18

The signal at 30 ppm corresponds to the CH_2 and the signal at 18 ppm corresponds to CH_3.

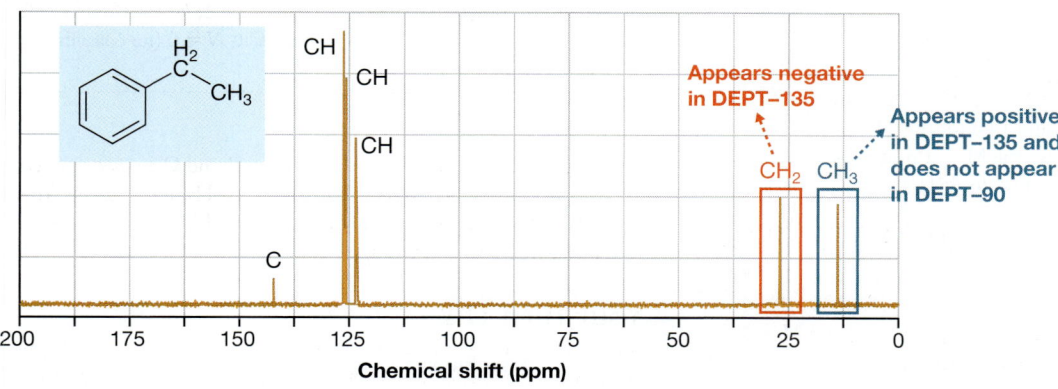

Your Turn 16.19

The first entry in Table 16-1 that has a proton part of a double bond is entry 13, $R_2C\!=\!CH\!-\!H$, $\delta = 4.7$ ppm. The first entry that has a proton on a carbon attached to an electronegative atom is entry 9, $BrCH_2\!-\!H$, $\delta = 2.7$ ppm.

Your Turn 16.20

The fragment $H\!-\!C\!-\!O$ is found in entry 11 in Table 16-1 and has a chemical shift $\delta = 3.3$ ppm, which is upfield compared to entry 3 of Table 16-5 with the same fragment $H\!-\!C\!-\!O$ that has a chemical shift $\delta = 4.2$ ppm. Proximity of protons to additional deshielding groups can cause a downfield shift. In this unknown the cause could possibly be a benzene ring or an additional oxygen.

Your Turn 16.21

The normal chemical shift of a proton on a benzene ring (Table 16-1, entry 11) is 7.3 ppm, which is upfield compared to entry 4 of Table 16-5 that has a range of chemical shifts, 7.4–8.1 ppm. The downfield shift is due to the oxygen atoms in the compound. Electronegative atoms deshield protons and cause downfield shifts.

Your Turn 16.22

Protons **A** correspond to the signals near 7.4–8.1 ppm in Figure 16-32; protons **B** correspond to the signal at ~4.3 ppm; protons **C** correspond to the signal at ~1.8 ppm; and protons **D** correspond to the signal at ~1.0 ppm.

Your Turn 16.23

The first entry in Table 16-4 that has a carbon as part of a double bond is entry 11, $C\!=\!C$, $\delta = 105$–150 ppm. The first entry that has a C attached to an electronegative atom is entry 6, $Br\!-\!CH_2R$, $\delta = 25$–35 ppm.

Your Turn 16.24

The chemical shift range for an alkyne carbon, $RC\!\equiv\!CR$ (Table 16-4, entry 10), is 65–85 ppm. The chemical shift range for a carbon singly bonded to an oxygen, $R_3C\!-\!OH$ (entry 9), is 55–70 ppm. Both of these ranges are reasonable to suggest that the signal of the unknown at ~73 ppm could be due to a C part of an alkyne or a C singly bonded to an O. The chemical shift range for a C part of a double bond (105–150 ppm) is much farther downfield than 73 ppm and therefore inconsistent with that signal.

Your Turn 16.25

In general, the H—O signal occurs between 3200 and 3600 cm^{-1} and gives rise to a broad and strong signal. *In general*, the $C\!\equiv\!C$ signal occurs between 2100 and 2260 cm^{-1} and gives rise to narrow signals varying in intensity. *In general*, the alkyne C—H signal occurs at ~3300 cm^{-1} and gives rise to a strong signal that is relatively narrow. In this example, the H—O signal is centered around 3400 cm^{-1} and no signals are present for a C≡C or an alkyne C—H (although it is possible that an alkyne C—H signal could be buried under a H—O signal).

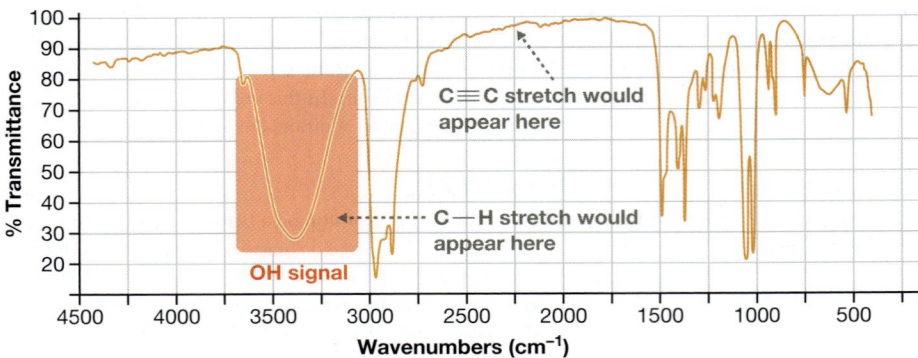

Your Turn 16.26

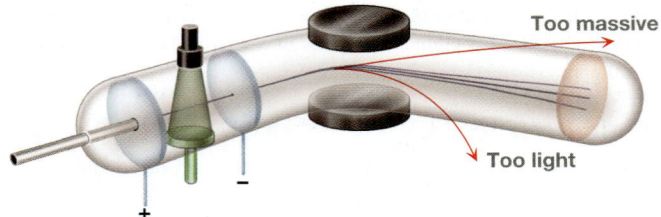

Your Turn 16.27

CH_3—CO—$N(CH_3)_2$ = C_4H_9NO = 12(4) + 9(1) + 14 + 16 = 87 u (odd number). There is one N atom, which is an odd number. This is consistent with the nitrogen rule.

Your Turn 16.28

$CH_3CH_2^+$ = m/z 29; CH_3^+ = m/z 15

Your Turn 16.29

The M + 1 peak in Figure 16-35 is the small peak at m/z 87. The M + 1 peak for ethylbenzene in Problem 16.33 is calculated to be at m/z 107: Molar mass of C_8H_{10} = (8 × 12) + (10 × 1) = M = 106 u. M + 1 = 107 u.

Your Turn 16.30

The mass of ^{12}C is 12 u and that of ^{13}C is 13 u. There are five ^{12}C (= 60 u) and one ^{13}C (= 13 u), and 14 u for the 14 H atoms present. This adds up to 87 u. This matches the M + 1 peak at m/z = 87 in the mass spectrum of hexane (C_6H_{14}, Fig. 16-35).

Your Turn 16.31

Hexane: M^+ = 86 (relative peak intensity 15.5%) and M + 1 = 87 (~1%); from Equation 16-7, [(1.0%)/(15.5%)] × (100%)/(1.1%) = 5.87, which rounds to 6. Dodecane: M^+ = 160 (5.9%) and M + 1 = 0.8%; [(0.8%)/(5.9%)] × (100%)/(1.1%) = 12.33, which rounds to 12.

Your Turn 16.32

The mass of ^{79}Br is 79 u and the mass of ^{81}Br is 81 u. The mass of the C_6H_5 portion is 77 u. The mass of $C_6H_5{}^{79}Br$ is 156 u and does correspond to the M^+ peak (Fig. 16-38). The mass of $C_6H_5{}^{81}Br$ is 158 u and corresponds to the M + 2 peak. The small peaks at m/z = 157 and m/z = 159 could be accounted for by substitution of one ^{12}C ~~ with ^{13}C (i.e., $^{12}C_5{}^{13}CH_5{}^{79}Br$ and $^{12}C_5{}^{13}CH_5{}^{81}Br$, respectively).

Your Turn 16.33

Plugging relative intensity values for M^+ (30%) and M + 1 (2.4%) peaks (Fig. 16-40) into Equation 16-7 gives an estimate of 7 carbons:

$$C \text{ atoms} \approx \frac{\text{Intensity of M + 1}}{\text{Intensity of M}^+} \times \frac{100\%}{1.1\%}$$

$$\approx \frac{2.4\%}{30.0\%} \times \frac{100\%}{1.1\%} = 7.27$$

Your Turn 16.34

A saturated compound with 7 C atoms and 1 N atom would have 17 H atoms: $CH_3CH_2CH_2CH_2CH_2CH_2CH_2NH_2$. Therefore, 7 C atoms and 1 N atom cannot accommodate 31 H atoms.

Chapter 17

Your Turn 17.1

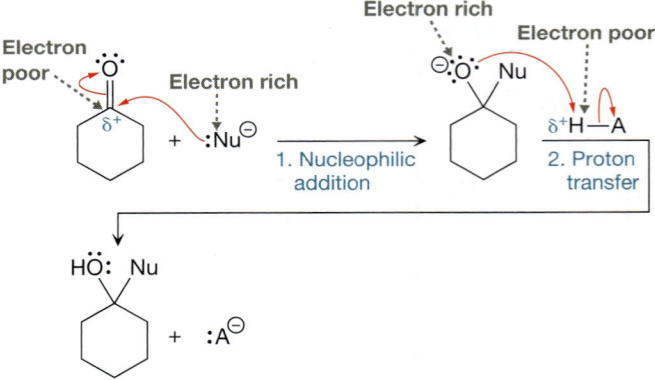

Your Turn 17.2

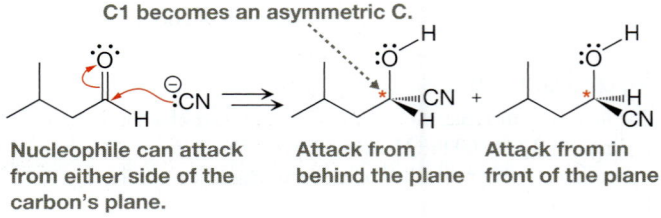

C1 becomes an asymmetric C.

Nucleophile can attack from either side of the carbon's plane.

Attack from behind the plane

Attack from in front of the plane

Your Turn 17.3

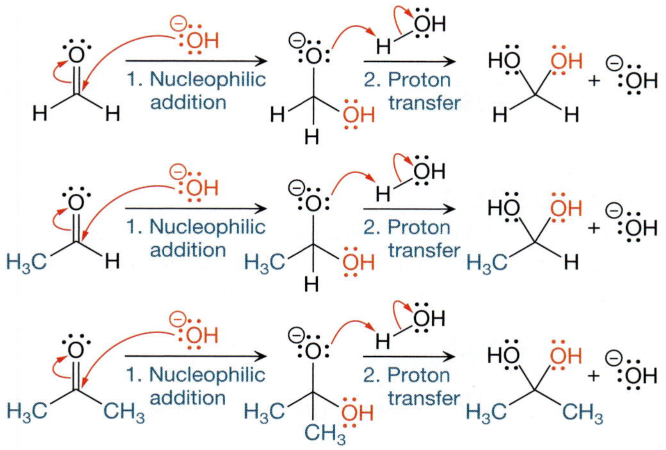

Your Turn 17.4

The most reactive imine is **B**; imine **C** is the least reactive.

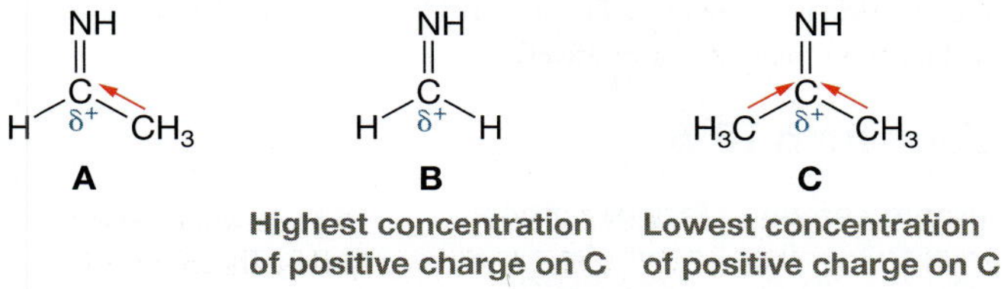

A

B
Highest concentration of positive charge on C

C
Lowest concentration of positive charge on C

Your Turn 17.5

Equation 17-4

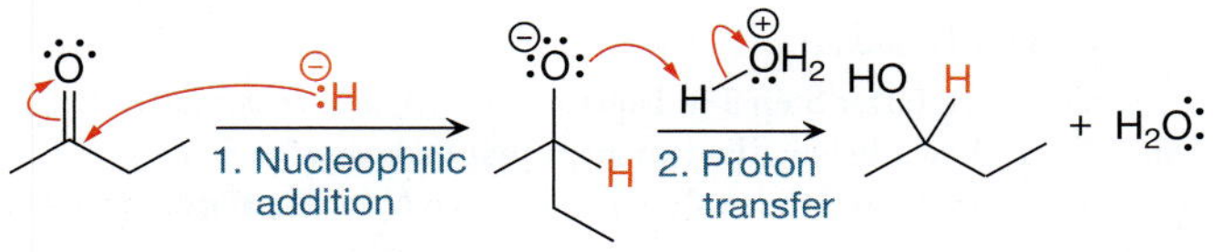

Equation 17-5

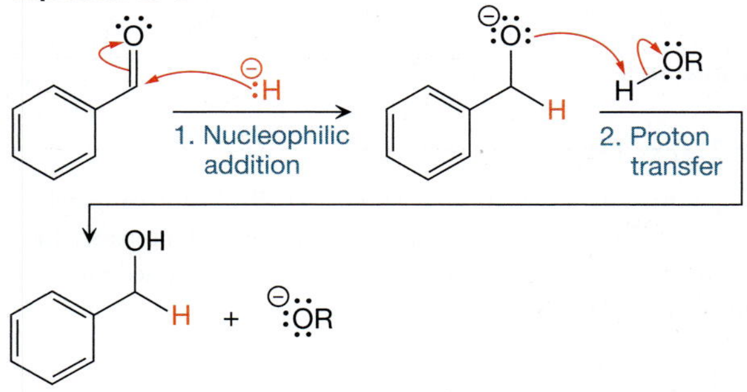

Your Turn 17.6

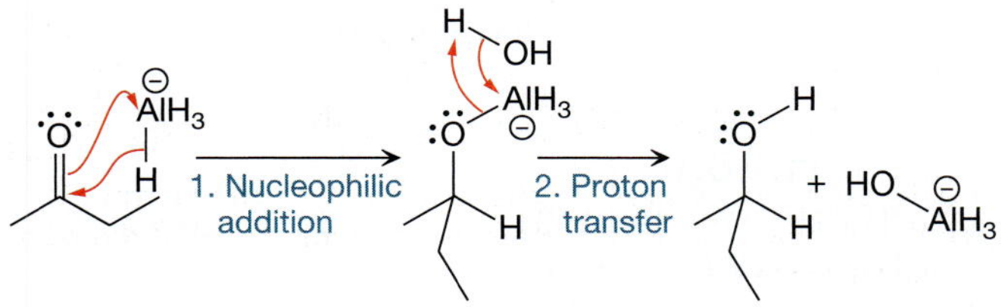

Your Turn 17.7

Butan-2-one (oxidation state of C2 = +2) gains 2 electrons when it is reduced to butan-2-ol (oxidation state of C2 = 0).

Your Turn 17.8

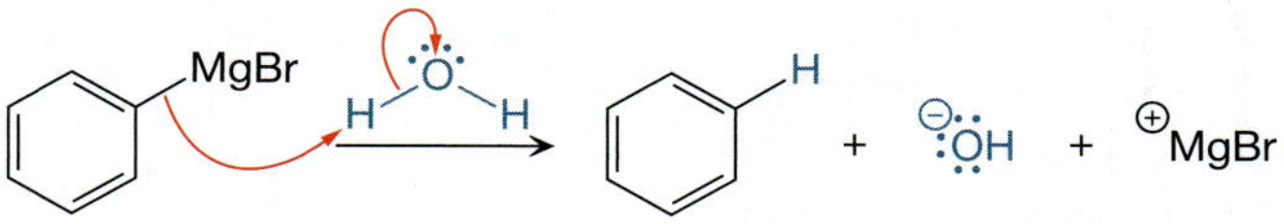

Your Turn 17.9

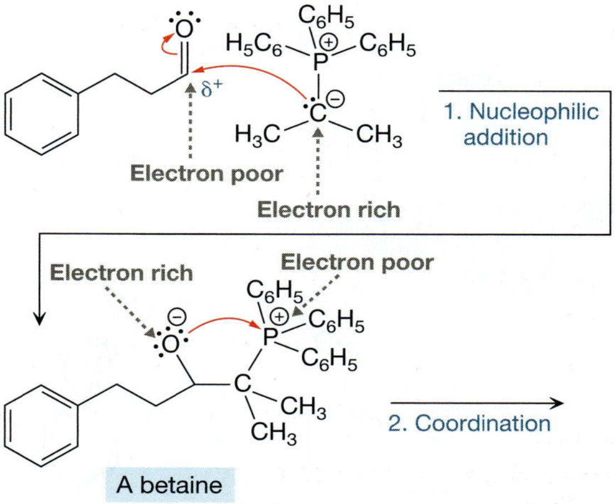

Electron poor

Electron rich

1. Nucleophilic addition

Electron rich

Electron poor

A betaine

2. Coordination

Your Turn 17.10

Table 9-10 lists the nucleophilicity (i.e., relative rate in S_N2 reaction) of Ph_3P as 10,000,000, and that of Br^- as 620,000. We know that Br^- is a very good nucleophile, so Ph_3P must be, too.

Your Turn 17.11

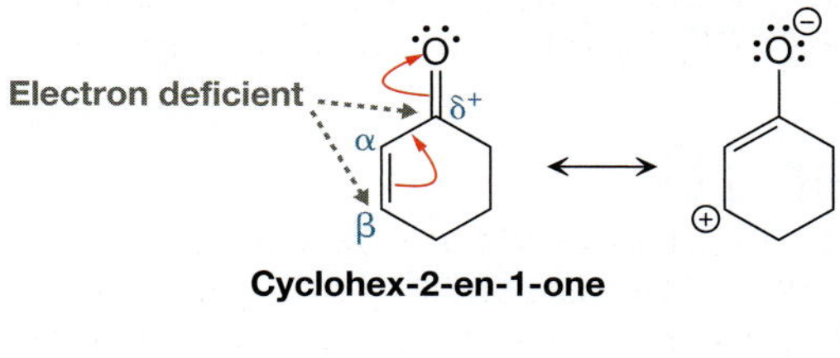

Electron deficient

Cyclohex-2-en-1-one

Your Turn 17.12

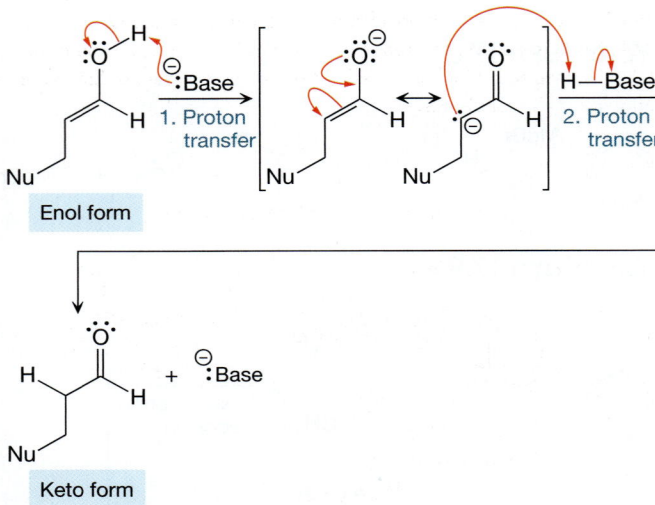

Enol form

Keto form

Your Turn 17.13

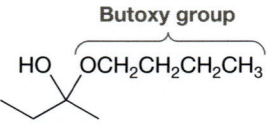

Conjugate addition
Thermodynamic product

Direct addition
Kinetic product

Your Turn 17.14

(a) The product has a Ph group beta to the C=O bond. The given compound can be generated by *conjugate* addition in which Ph⁻ attacks at the β carbon. **(b)** A C=C bond is adjacent to a C—OH group. The given compound can be generated by *direct* addition in which CH_3^- attacks the carbon of the carbonyl.

Chapter 18

Your Turn 18.1

Butoxy group

$HO \quad OCH_2CH_2CH_2CH_3$

Hemiacetal of butan-2-ol

Your Turn 18.2

Equation 18-3: 0, 2, 2, and 2 charges ($CH_3CH_2O^-$ produced in Step 2 is not shown but remains in solution); Equation 18-4: 1, 1, 1, and 1 charge; Equation 18-5: 1, 1, 1, and 1 charge.

Your Turn 18.3

The positively charged species is activated toward nucleophilic attack because nucleophiles form bonds to electron-poor species. It has a resonance structure that puts a positive charge on the C atom, making that C atom more susceptible to nucleophilic attack.

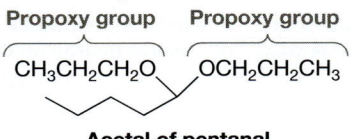

Your Turn 18.4

HCN has $pK_a = 9.2$ and H_3O^+ has $pK_a = -1.7$. The stronger acid (much lower pK_a) is on the product side, so the reactants in Equation 18-7 are favored by a factor of $10^{10.9} = 7.9 \times 10^{10}$.

Your Turn 18.5

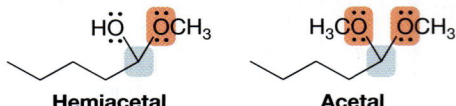

$pK_a = 9.2$ $pK_a = 15.7$

The stronger acid (lower pK_a) is on the reactant side, so the products are favored. The difference in pK_a values is $15.7 - 9.2 = 6.5$, so the product side is favored by a factor of $10^{6.5} = 3.2 \times 10^6$. Thus, the acid will be deprotonated quantitatively.

Your Turn 18.6

Propoxy group   **Propoxy group**

$CH_3CH_2CH_2O \qquad OCH_2CH_2CH_3$

Acetal of pentanal

Your Turn 18.7

The hemiacetal (after Step 3 in Equation 18-13) and acetal (after Step 7) are shown below. The two steps that make up the S_N1 mechanism are Step 5 (heterolysis) and Step 6 (coordination).

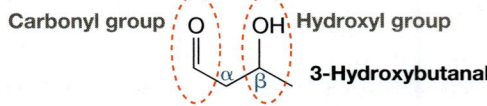

Hemiacetal   **Acetal**

Your Turn 18.8

In the reactant cyclopropylethanone (Equation 18-27), the carbonyl C has an oxidation state of +2; in the product ethylcyclopropane, the C has an oxidation state of −2. The C atom effectively gained four electrons, so this is a reduction.

Your Turn 18.9

Carbonyl group   O   OH **Hydroxyl group**

3-Hydroxybutanal

Your Turn 18.10

pK_a ($CH_3CH=O$) = 19; pK_a (H_2O) = 15.7. The stronger acid, H_2O, is on the product side, so the reactant side is favored. The difference in pK_a values is $19 - 15.7 = 3.3$, so the reactant side is favored by $10^{3.3} = 2 \times 10^3$.

Your Turn 18.11

Reactants: C = 6, H = 12, O = 2; products: C = 6, H = 10, O = 1. The difference is H_2O, which is eliminated during condensation (Equation 18-32).

Your Turn 18.12

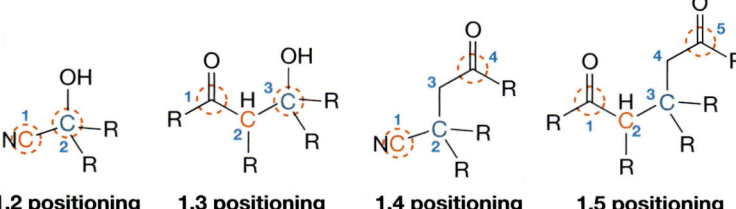

A β-hydroxycarbonyl compound is produced in Eq. 18-37.

4-Hydroxy-4-methylpentan-2-one

Your Turn 18.13

Ketone **A** (it is the only choice that has no α protons).

Your Turn 18.14

NaOH is a proper choice of base in the first reaction, where the more highly substituted α C is deprotonated. LDA is a proper choice of base in the second reaction, where the less substituted α C is deprotonated.

Your Turn 18.15

The product of an aldol condensation is an α,β-unsaturated carbonyl compound.

Alkene C=C   **Aldehyde C=O**

Chapter 19

Your Turn 19.1

(a)

(b)

Your Turn 19.2

1,2 positioning   **1,3 positioning**   **1,4 positioning**   **1,5 positioning**

Your Turn 19.3

A nitrile (R—C≡N) can be hydrolyzed to an amide under either acidic or basic conditions; therefore, the Wolff–Kishner reduction should not be carried out on a ketone or aldehyde containing a C≡N group due to the strongly basic conditions. Neither should the Clemmensen reduction be carried out on such a compound due to the strongly acidic conditions.

Your Turn 19.4

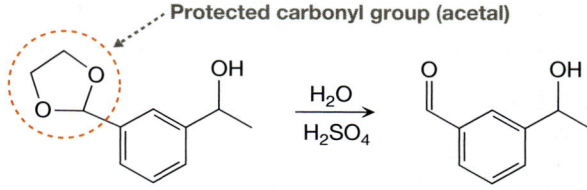

Protected carbonyl group (acetal)

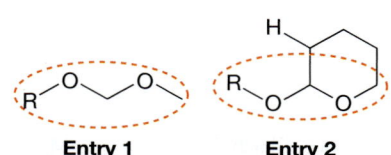

Your Turn 19.5

Entry 1   **Entry 2**

Your Turn 19.6

The cis alkene **B** could be produced from an alkyne using a poisoned catalyst. Alkene **A** does not have the correct stereochemistry (it is trans rather than cis) and **C** is a full hydrogenation to the alkane.

Your Turn 19.7

The secondary alcohol C is assigned four valence electrons and has an oxidation state of 0. The ketone C is assigned two valence electrons and has an oxidation state of +2. The oxidation state of C increased by 2, verifying that C lost electrons and was oxidized (remember that oxidation is loss of electrons; reduction is gain of electrons).

Your Turn 19.8

Oxidation **A** of the alcohol yields an aldehyde, which requires the absence of water; PCC will accomplish this. Oxidation **B** produces a carboxylic acid; this is accomplished using chromic acid.

Your Turn 19.9

The carbon–carbon bond of C_6H_5—C_6H_5 is formed from two aryl carbon atoms. In Suzuki coupling the R in R—X can be vinylic or aryl, whereas R′ can be alkyl, vinylic, or aryl. In Heck coupling the R in R—X can be vinylic or aryl and R′ can be vinylic or aryl. Therefore, C_6H_5—C_6H_5 can be formed by either the Suzuki or Heck coupling reactions.

Suzuki coupling

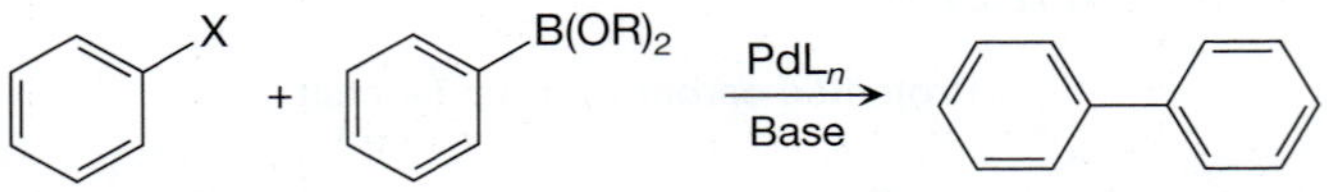

Heck coupling

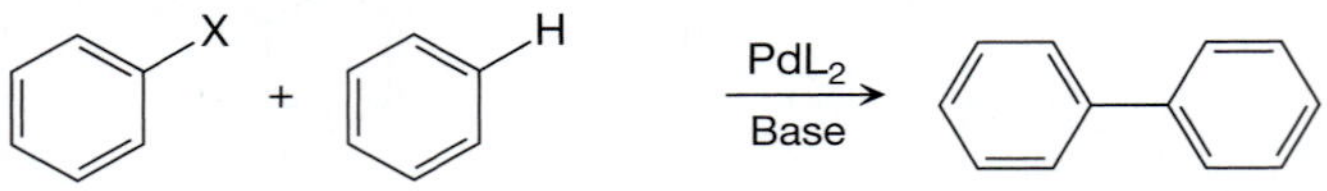

Your Turn 19.10

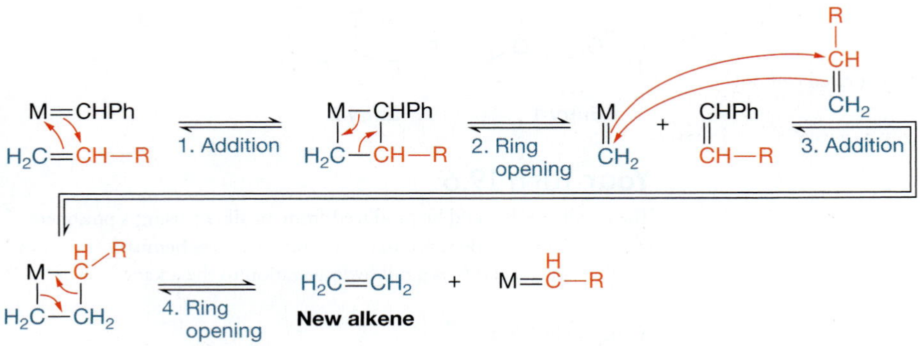

Chapter 20

Your Turn 20.1

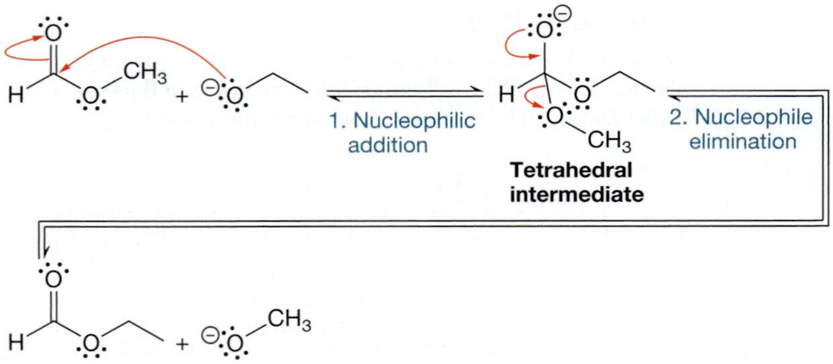

Your Turn 20.2

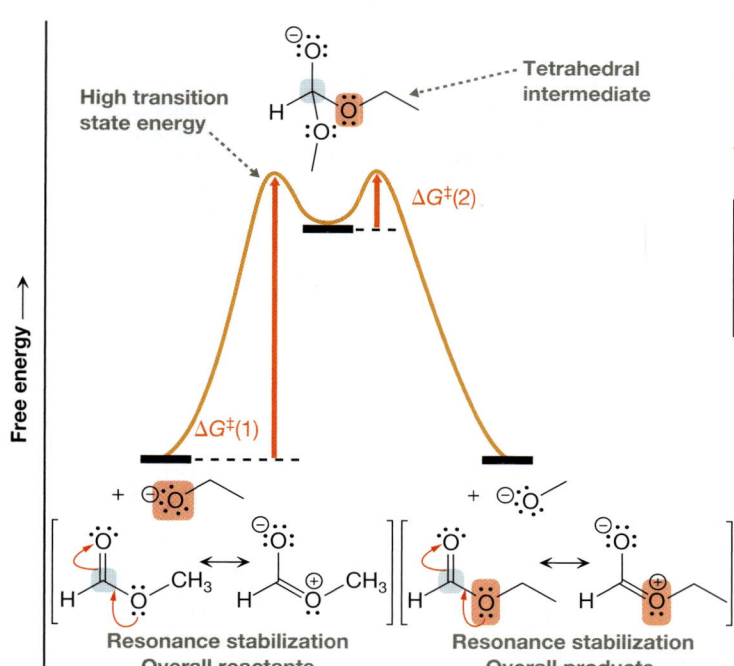

Your Turn 20.6

RCO$_2$H has p$K_a \approx$ 4.75 and ROH has pK_a = 16. The stronger acid (lower pK_a) is on the reactant side in Equation 20-7, so the products are very heavily favored. The difference in pK_a values is 16 − 4.75 = 11.25, so the product side is favored by $10^{11.25} = 1.8 \times 10^{11}$.

Your Turn 20.7

Your Turn 20.3

Your Turn 20.4

Your Turn 20.5

(a) Acid anhydride to acid chloride = up = unfavorable
(b) Amide to ester = up = unfavorable
(c) Acid chloride to ester = down = favorable
(d) Acid anhydride to carboxylic acid = down = favorable
Reactions (c) and (d) are likely to occur readily because they involve going down rungs on the stability ladder.

Your Turn 20.8

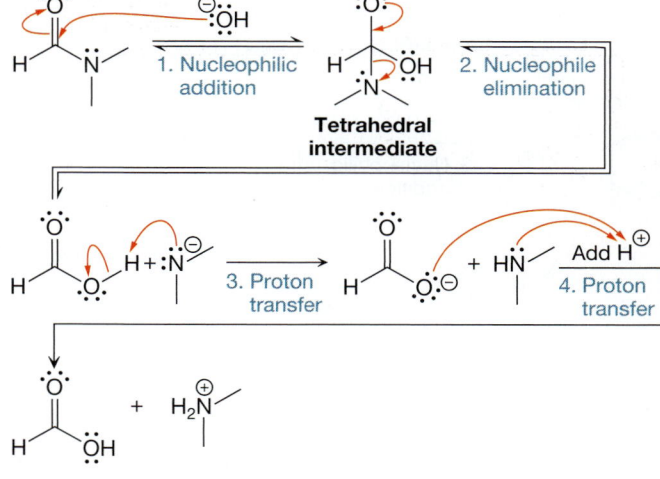

Your Turn 20.9

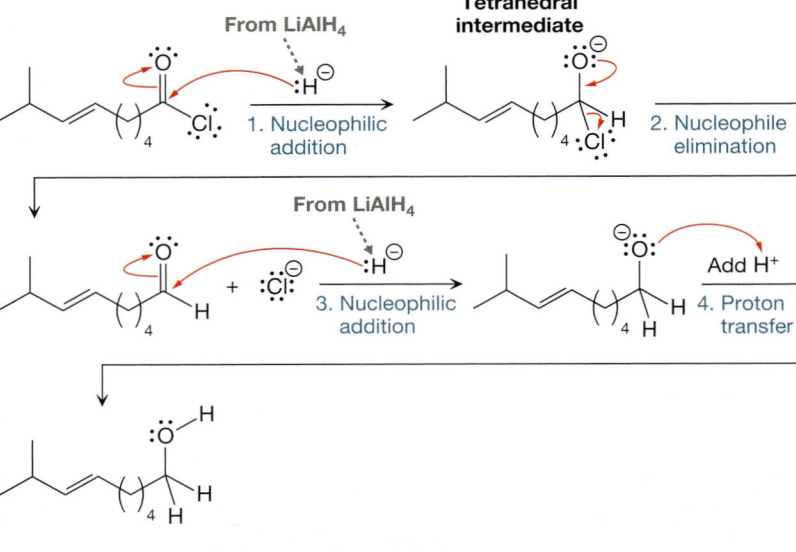

(a)   (b)

Your Turn 20.10

Only methyl ketones or aldehydes can undergo a haloform reaction.
Compounds **A** and **C** are methyl ketones and can undergo the reaction.

Your Turn 20.11

Your Turn 20.12

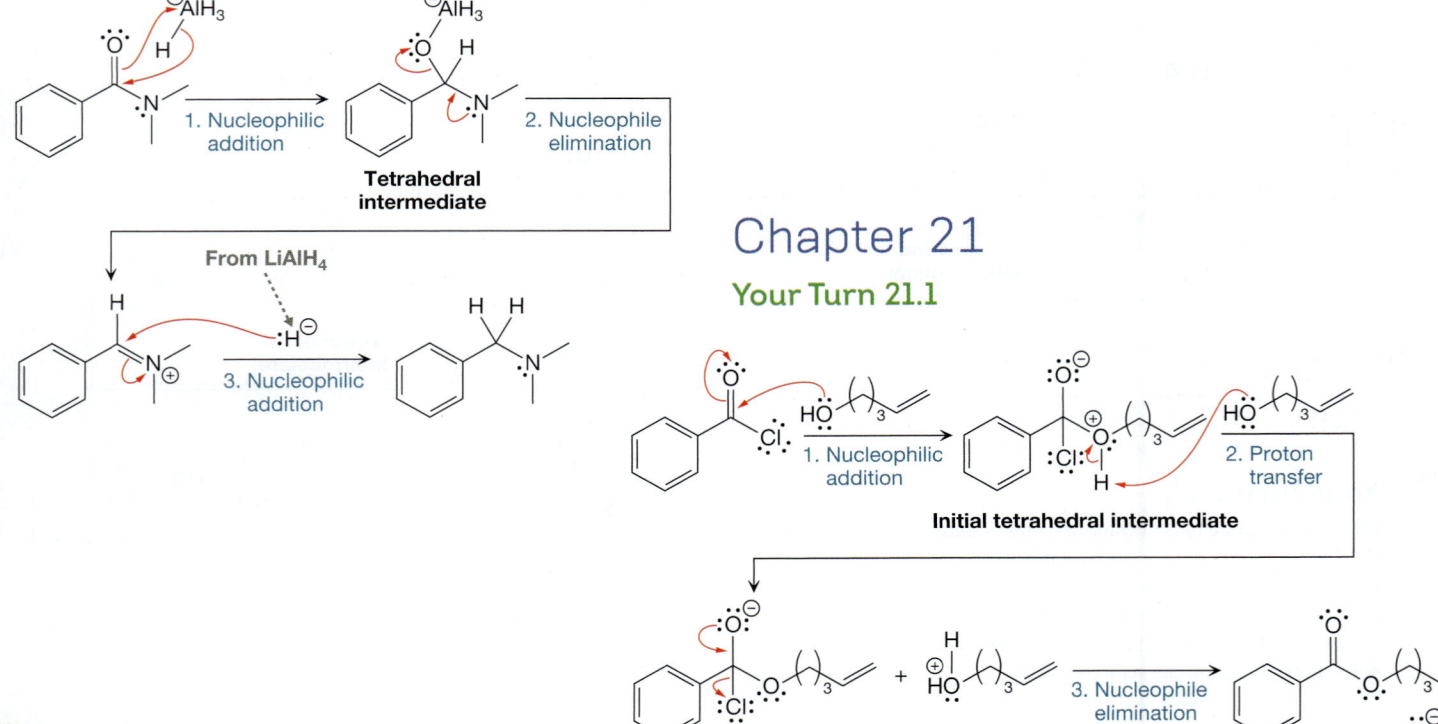

Your Turn 20.13

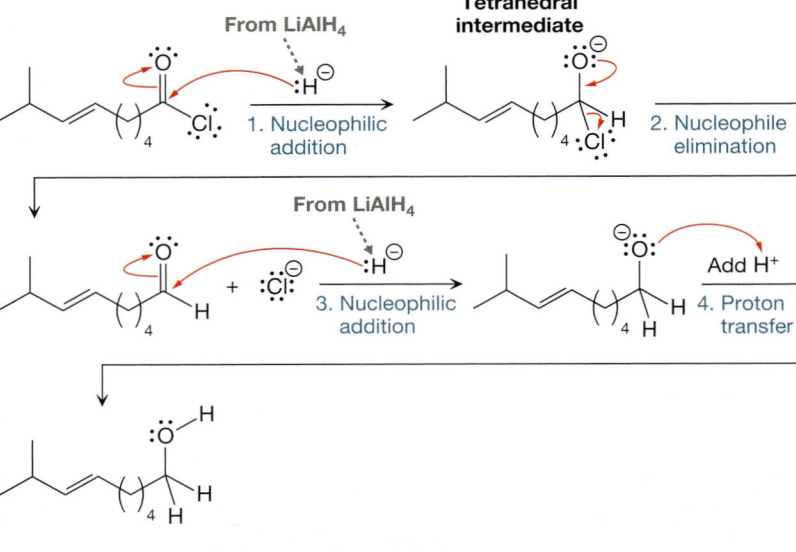

Methyl hept-6-enoate 1. Nucleophilic addition

Tetrahedral intermediate 2. Nucleophile elimination

3. Nucleophilic addition

4. Proton transfer

Two equivalents of CH₃⊖ have added in.

2-Methyloct-7-en-2-ol
99%

Chapter 21

Your Turn 21.1

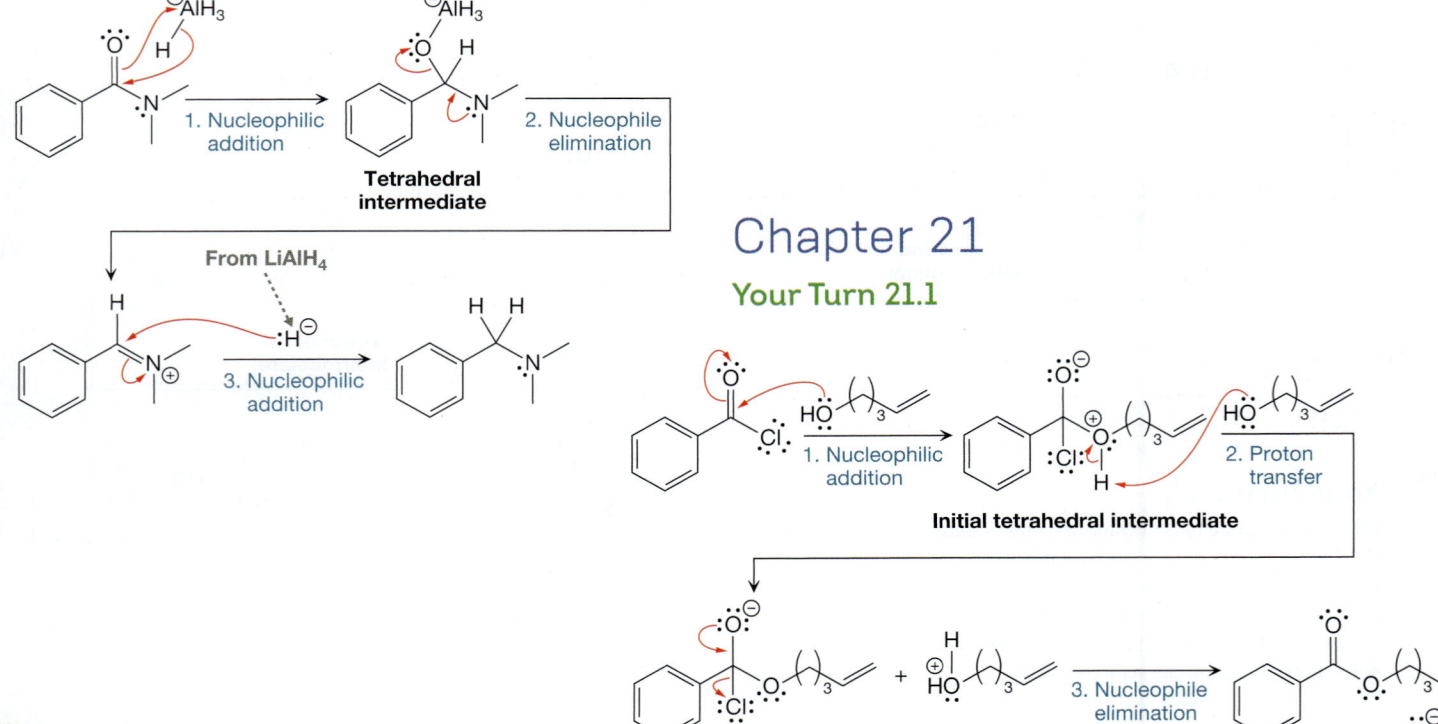

1. Nucleophilic addition

Initial tetrahedral intermediate

2. Proton transfer

3. Nucleophile elimination

Your Turn 21.2

(a) The species are added below. The initial tetrahedral intermediate is the product of nucleophilic addition, and has a relatively high energy.
(b) The first step (nucleophilic addition) has the highest energy transition state, which is consistent with it being the slow step.

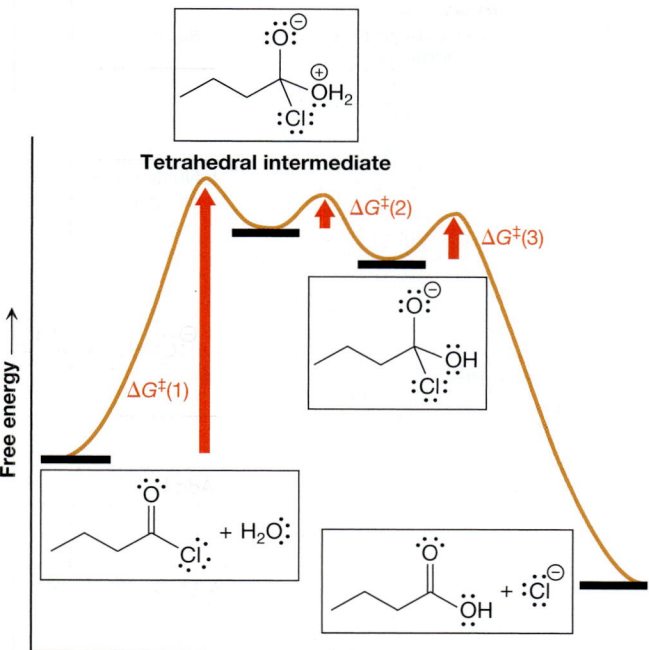

Your Turn 21.3

Your Turn 21.4

Your Turn 21.5

Your Turn 21.6

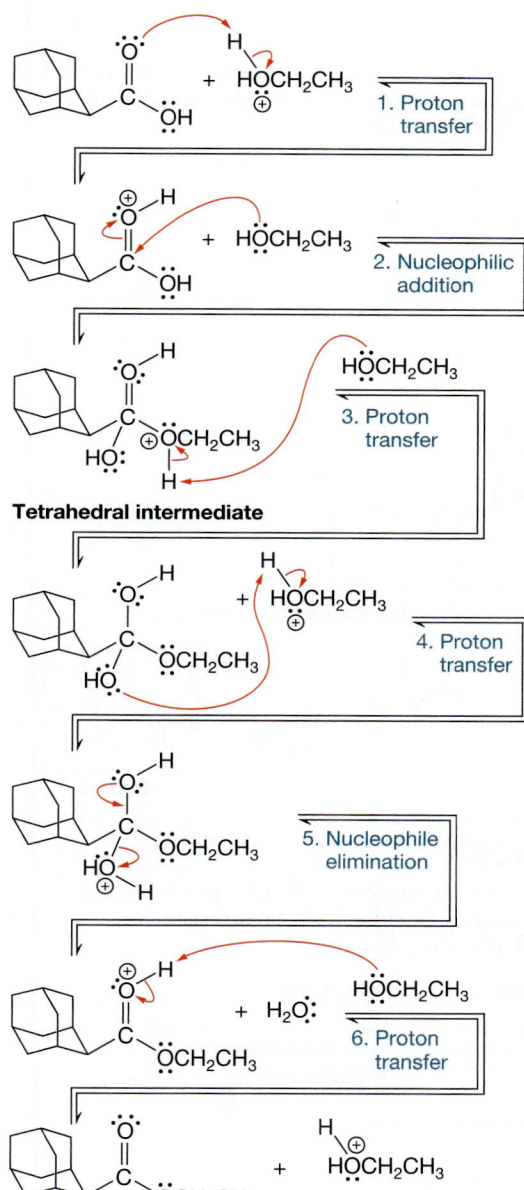

Your Turn 21.7

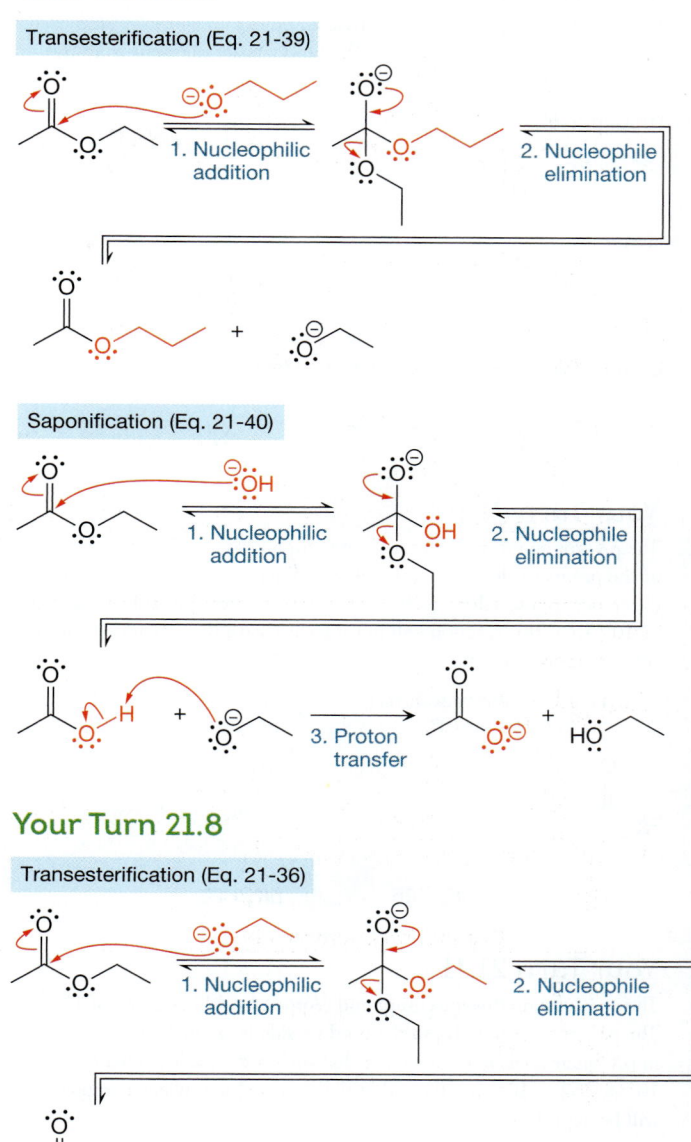

The ester that is produced is the same as the initial ester.

Your Turn 21.9

In each case, $CH_3CH_2O^-$ is the base and picks up a proton from each of the uncharged acidic solvents. The products are different from the reactants unless ethanol is the solvent.

Propan-1-ol

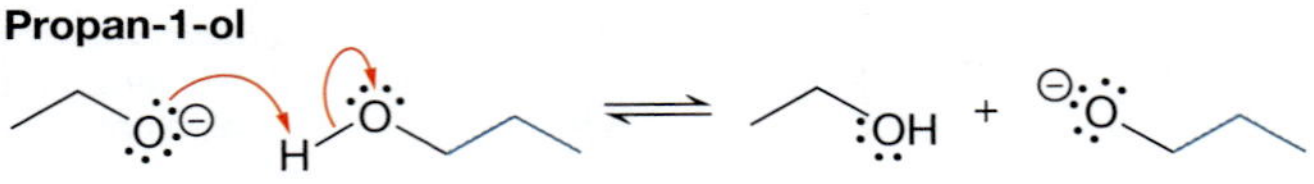

Water

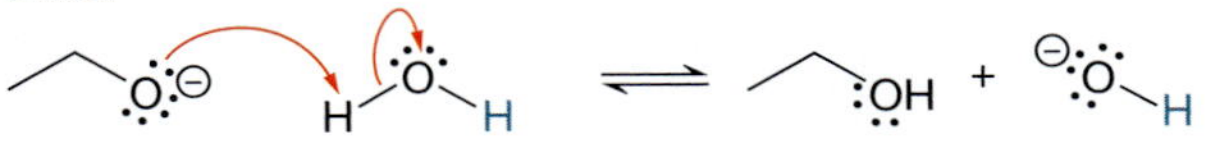

Ethanol (Notice that the products are identical to the reactants.)

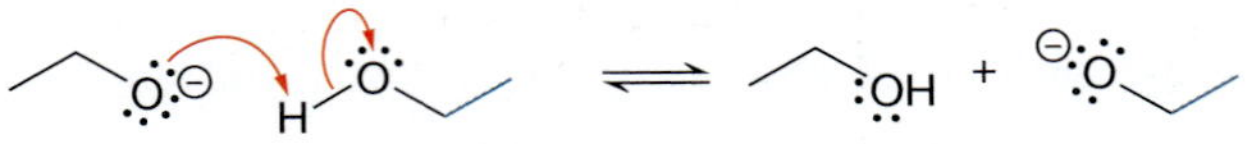

Your Turn 21.10

The pK_a for an ester is 25 and the pK_a for a ketone is 20. The pK_a of the product is lower and, therefore, the reactants are favored. The difference in pK_a values is $25 - 20 = 5$, so the reactant side is favored by 10^5. Thus, this reaction will not interfere with the intended Claisen condensation reaction.

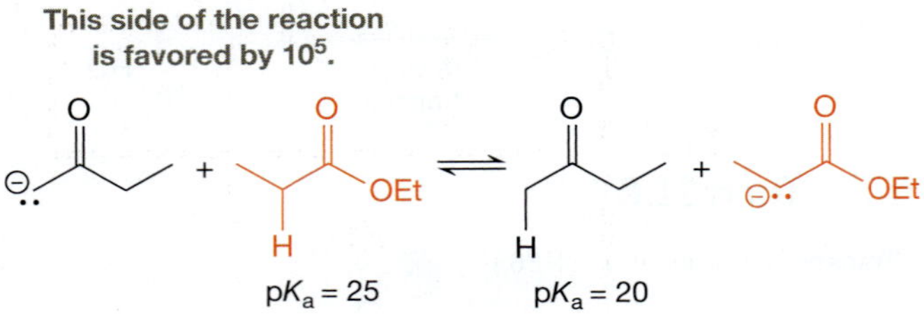

This side of the reaction is favored by 10^5.

$pK_a = 25$ $pK_a = 20$

Your Turn 21.11

The activated methylene compound (Appendix A) has a pK_a of 13.5. The pK_a of ethanol is 16, so the product side is favored. The difference in pK_a values is 2.5, so the extent that the product side is favored is $10^{2.5} = 3.2 \times 10^2$, signifying that the activated methylene compound will be deprotonated >99%.

Your Turn 21.12

The alkyl group is electron donating, as shown below, so it increases the concentration of negative charge on the nearby carbon, which destabilizes that anion.

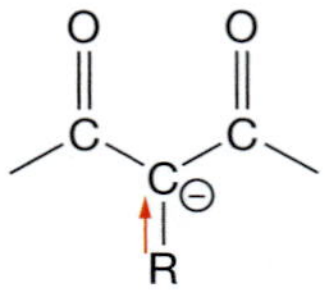

Chapter 22

Your Turn 22.1

In the products of HCl and Cl_2 addition, shown below, aromaticity has been lost because the π system is no longer fully conjugated around the ring. Thus the product species are *nonaromatic*.

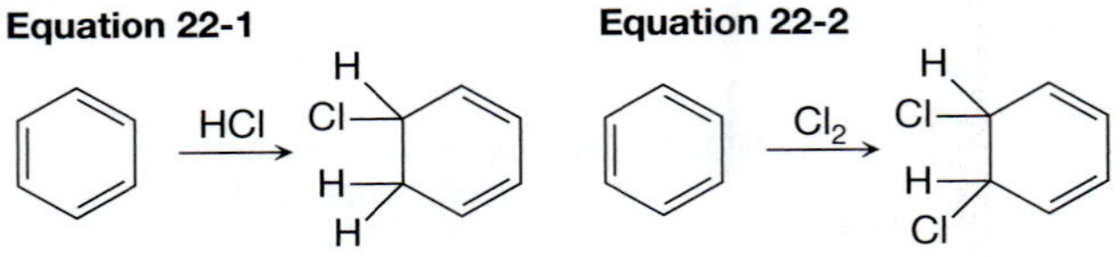

Equation 22-1   **Equation 22-2**

Your Turn 22.2

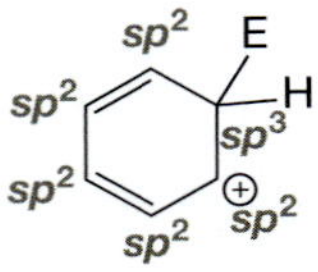

Your Turn 22.3

The positive charge is shared over three C atoms (δ^+).

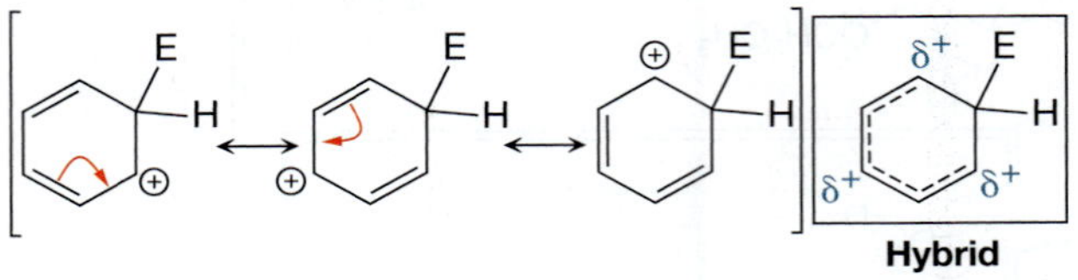

Hybrid

Your Turn 22.4

Steps 1 and 2 involve the formation of the Cl^+ electrophile. Steps 3 and 4 formally make up the electrophilic aromatic substitution mechanism with Cl^+ as the electrophile.

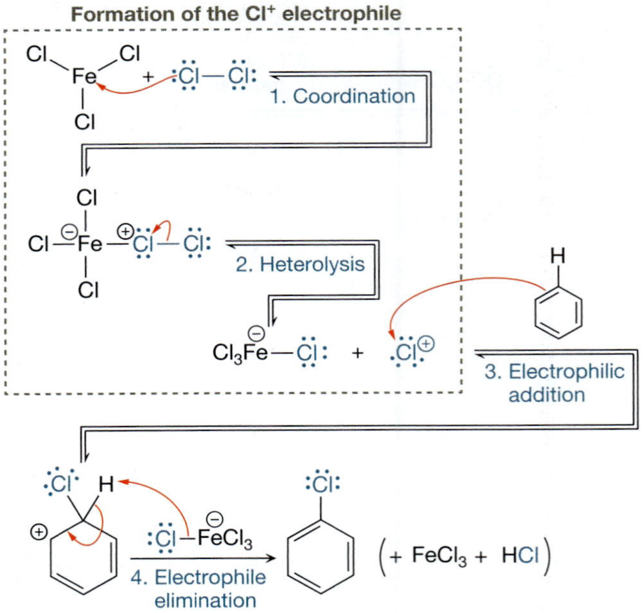

Formation of the Cl^+ electrophile

1. Coordination

2. Heterolysis

$Cl_3Fe—\ddot{Cl}: + \cdot\overset{\oplus}{\ddot{Cl}}$

3. Electrophilic addition

4. Electrophile elimination

$\left(+ \ FeCl_3 + \ HCl \right)$

Your Turn 22.5

Steps 1 and 2 involve the formation of the cyclohexyl C$^+$ electrophile. Steps 3 and 4 formally make up the electrophilic aromatic substitution mechanism with the cyclohexyl C$^+$ as the electrophile.

Formation of R$^+$ electrophile

1. Coordination

2. Heterolysis

3. Electrophilic addition

4. Electrophile elimination

$\left(+ \text{ HCl } + \text{ AlCl}_3 \right)$

Your Turn 22.6

The C—H bond energy for an sp^3-hybridized C is 410 kJ/mol, and that for an sp^2-hybridized C is 431 kJ/mol.

Your Turn 22.7

Formation of RCO$^+$ electrophile

1. Coordination

2. Heterolysis

Acylium ion

3. Electrophilic addition

4. Electrophile elimination

$\left(+ \text{ HCl } + \text{ AlCl}_3 \right)$

Your Turn 22.8

A resonance hybrid is an average of all resonance structures. The carbon–oxygen bond is between a double and triple bond. The positive charge is delocalized over both C and O.

Hybrid

Your Turn 22.9

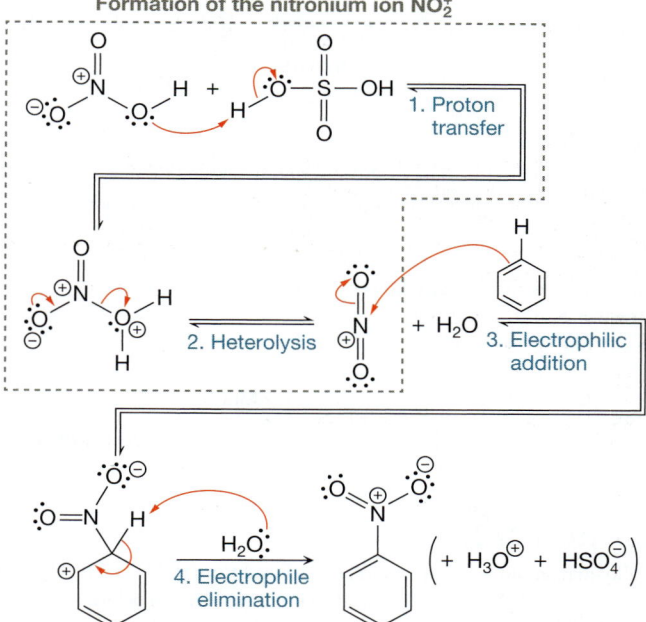

Formation of the nitronium ion NO$_2^+$

1. Proton transfer

2. Heterolysis   + H$_2$O   3. Electrophilic addition

4. Electrophile elimination

$\left(+ \text{ H}_3\text{O}^{\oplus} + \text{ HSO}_4^{\ominus} \right)$

Your Turn 22.10

Desulfonation is the reverse of Steps 1–4 in Equation 22-26.

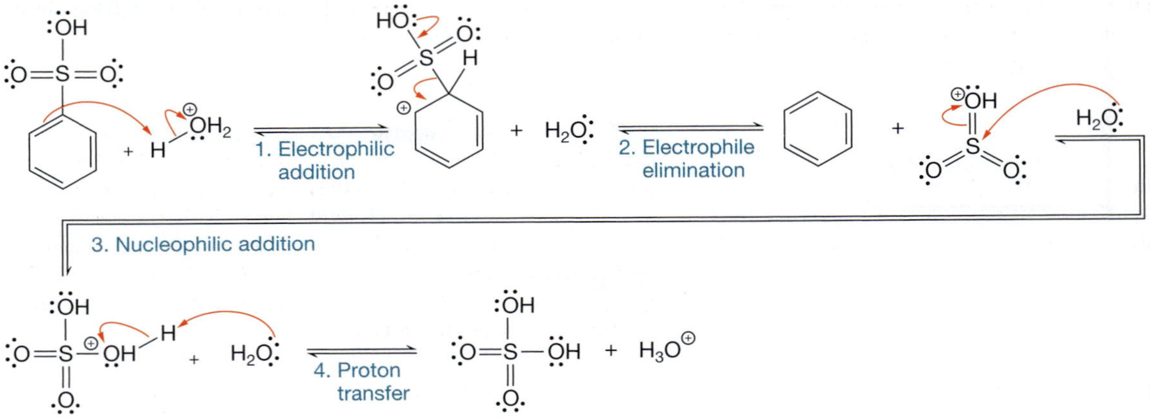

1. Electrophilic addition
2. Electrophile elimination
3. Nucleophilic addition
4. Proton transfer

Your Turn 22.11

The carbocation intermediate in this reaction is susceptible to a 1,2-hydride shift to convert the secondary carbocation into a more stable tertiary carbocation. The actual product of the Friedel–Crafts alkylation, shown below, yields a quaternary benzylic carbon. The permanganate oxidation fails, however, when the benzylic C is *quaternary*.

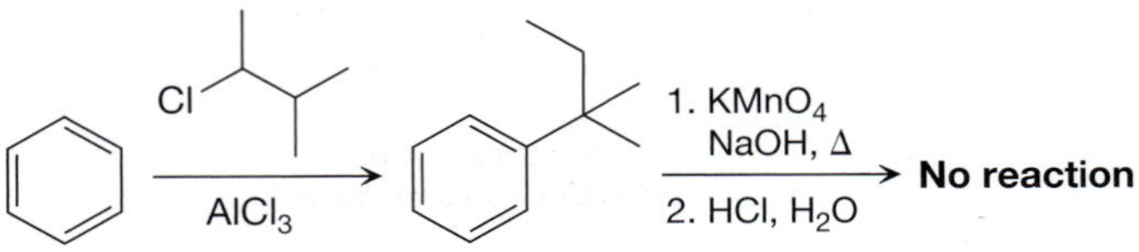

1. KMnO$_4$
   NaOH, Δ
2. HCl, H$_2$O
No reaction

Your Turn 22.12

The alkyldiazonium ion that would be produced from $(CH_3)_2CHNH_2$ when treated with NaNO$_2$ under acidic conditions, shown below, is less stable than the benzenediazonium ion because the positive charge on the nitrogen is localized. The positive charge on the nitrogen in the benzenediazonium ion is delocalized and is therefore more stable.

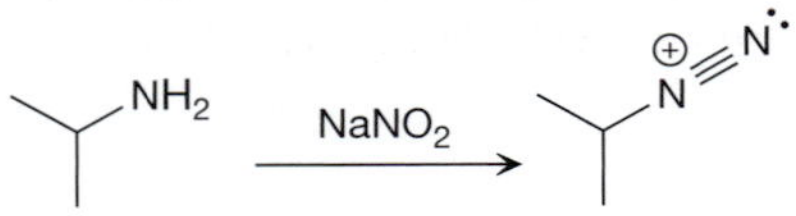

Chapter 23

Your Turn 23.1

Substituent	o	m	p	o+p	Type of Director
—I		45	1	54	99	Ortho/para
—CHO		19	72	9	28	Meta

Your Turn 23.2

(a) The lone pairs, which have been added to the atom at the point of attachment for every group in Table 23-1 in the list below, appear on substituents in the ortho/para-directing column. (b) The –CH$_3$ group, an ortho/para-directing substituent, is attached by an atom that has no lone pairs of electrons. There are no electronegative atoms at or near the point of attachment in the CH$_3$ group. (c) In the meta-directing column, the electronegative atoms at or near the point of attachment are N and O.

TABLE 23-2

ORTHO/PARA-DIRECTING SUBSTITUENTS				META-DIRECTING SUBSTITUENTS				
Substituent	o	m	p	o+p	Substituent	o	m	p	o+p
—ÖH	50	0	50	100	—NO$_2$	7	91	2	9
—N̈HCOCH$_3$	19	2	79	98	—N(CH$_3$)$_3$ (+)	2	87	11	13
—CH$_3$	63	3	34	97	—CO$_2$H	22	76	2	24
—F̈:	13	1	86	99	—CN	17	81	2	19
—C̈l:	35	1	64	99	—CO$_2$Et	28	66	6	34
—B̈r:	43	1	56	99	—COCH$_3$	26	72	2	28

Your Turn 23.3

From Table 23-1 the nitration of phenol (C_6H_5—OH) is 50% para and 0% meta. Because there is much more of the para isomer produced than the meta isomer, we know that the para arenium ion intermediate is formed faster and is lower in energy (more stable) than the meta intermediate.

Your Turn 23.4

Other than the H atoms, the positively charged C atoms in each resonance structure (Equations 23-4 and 23-5) do not possess an octet. The C^+ has three bonds and no lone pairs, and thus has a share of only six electrons.

Your Turn 23.5

The second resonance structure is especially stable due to the electron-donating CH_3 group attached directly to the C^+, which decreases the concentration of positive charge there.

Your Turn 23.6

When the positive charge on the ring is adjacent to the positive charge on the N, the arenium ion is destabilized (second structure). The second resonance structure, as a result, is higher in energy than the other two and is the least stable, and it contributes the least to the resonance hybrid.

Your Turn 23.7

The inductive effect leads to a larger concentration of positive charge on C^+ in the intermediate, and decreases stability, because Cl is an electron-withdrawing group.

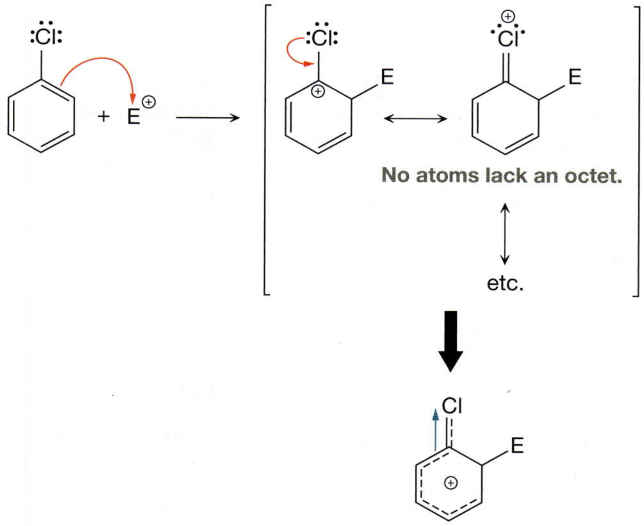

Your Turn 23.8

The aromatic ring of the monosubstituted species is deactivated relative to benzene, because the ring shows less negative charge (i.e., less red color) than benzene.

Your Turn 23.9

(a) The $-N(CH_3)_2$ amino group, although highly activating, is incompatible with Friedel–Crafts reactions, as shown below. The amino group is a relatively strong Lewis base, so it readily coordinates to the $AlCl_3$ Lewis acid. In that complexed form, the N atom has a +1 formal charge, making it a highly deactivating group that precludes Friedel–Crafts reactions altogether. **(b)** The $-CH(CH_3)_2$ isopropyl alkyl group is inert to the $AlCl_3$ Lewis acid and therefore will undergo the indicated Friedel–Crafts reaction.

Actual product formed

Product from Friedel–Crafts reaction

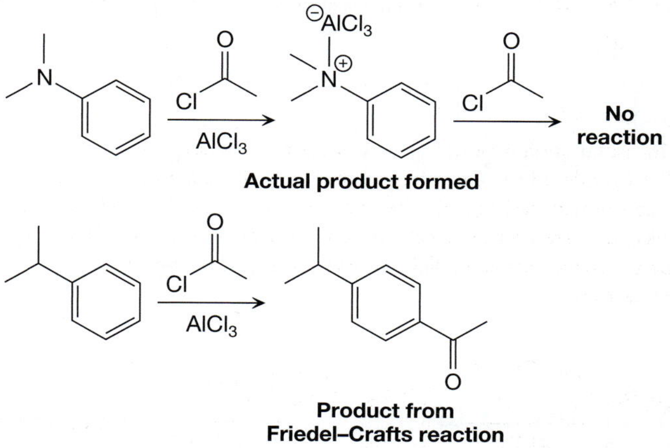

Your Turn 23.10

NO_2 groups are electron withdrawing, and the structure where the C^+ is adjacent to the N (second resonance structure) is the least stable of the three. CH_3 groups are electron donating and the structure that has C^+ adjacent to the CH_3 (third resonance structure) is the most stable. Overall, this intermediate, due to the stabilization provided by the third structure and the minimization of the contribution by the second structure, is more stable than the arenium ion in Equation 23-24.

Your Turn 23.11

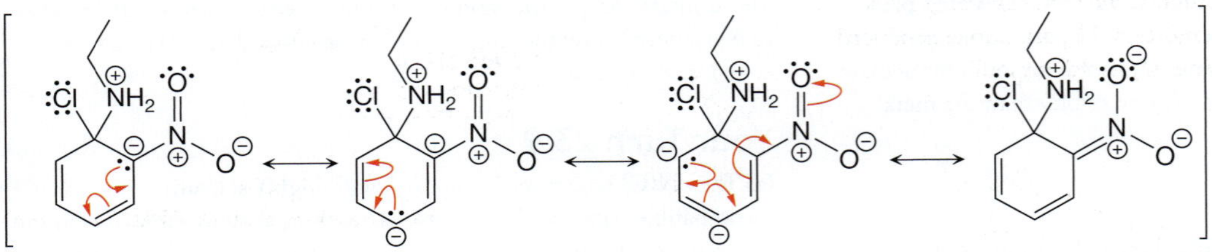

Your Turn 23.12

The fourth resonance structure is the most stable of the four, because the negative charge is on O, a more electronegative atom than C.

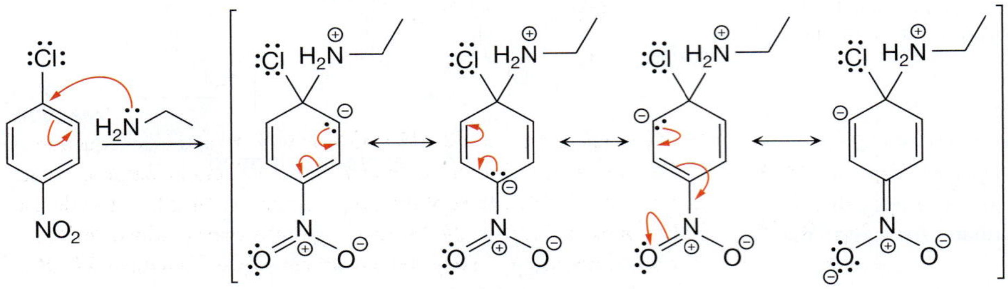

Your Turn 23.13

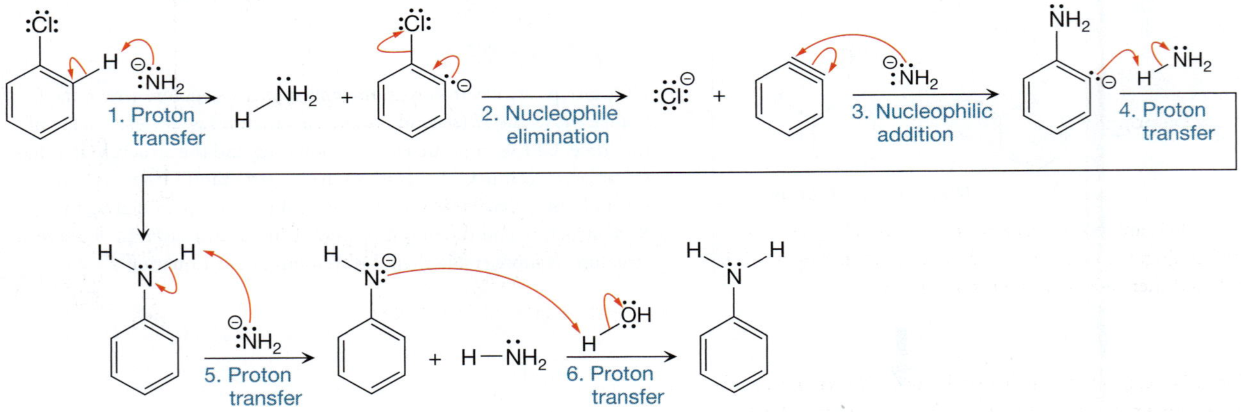

1. Proton transfer
2. Nucleophile elimination
3. Nucleophilic addition
4. Proton transfer
5. Proton transfer
6. Proton transfer

Your Turn 23.14

In an attempt to produce the meta product, the meta-directing NO_2 group should be on the ring before alkylation. However, the alkylation will not work because the nitrobenzene ring is deactivated.

Deactivated

Chapter 24

Your Turn 24.1

Yes, there are still three curved arrows, the same transition state, and overall the same product as in Equation 24-2.

Your Turn 24.2

Diene   **Dienophile**

Your Turn 24.3

The transition state is antiaromatic, because there are four electrons involved in the conjugated ring, and 4 is an anti-Hückel number (it is $4n$ with $n = 1$, and there are an even number of pairs).

Your Turn 24.4

Conformation **B** is *s*-cis, as the double bonds are on the same side of the single bond connecting them. Therefore, **B** can undergo a Diels–Alder reaction. Conformation **A** is *s*-trans and cannot undergo a Diels–Alder reaction.

Your Turn 24.5

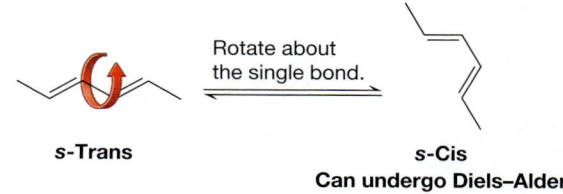

Rotate about the single bond.

s-Trans   **s-Cis**
Can undergo Diels–Alder

Your Turn 24.6

The two terminal C atoms of the diene and both C atoms of the dienophile are rehybridized from sp^2 to sp^3.

$$\text{diene} + \begin{array}{c} CH_2 \\ \| \\ CH_2 \end{array} \longrightarrow$$

Your Turn 24.7

The H atoms in Equation 24-18 match up with W and Y in Equation 24-20, whereas the CH_3 groups match up with X and Z. Because X and Z end up on opposite sides of the ring in Equation 24-20, so, too, do the CH_3 groups in Equation 24-18, resulting in the trans product. Yes, the cis/trans relationships among the substituents agree (Equations 24-18 and 24-20).

C—H bonds drawn in

Diene tilted into the paper

Your Turn 24.8

Approach **B** has a favorable interaction where opposite partial charges δ^+ and δ^- are located on atoms undergoing bond formation, whereas **A** does not. Therefore, **B** will lead to the major Diels–Alder product.

Your Turn 24.9

No favorable electrostatic interaction

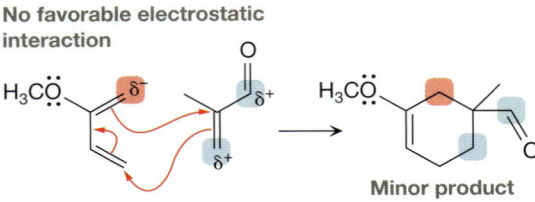

Minor product

Your Turn 24.10

From Table 1-3, [3(619 kJ/mol) + 1(339 kJ/mol)] − [5(339 kJ/mol) + 1(619 kJ/mol)] = −118 kJ/mol. This estimate is 50 kJ/mol more positive than the measured value of −168 kJ/mol.

Your Turn 24.11

Note: Arrows are drawn clockwise, but counterclockwise arrows would be equally correct.

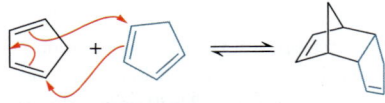

Diene   Dienophile

Your Turn 24.12

This Diels–Alder reaction takes place at the same temperature as the one in Equation 24-28, but ΔH°_{rxn} is much more negative (−121 kJ/mol vs. −17 kJ/mol). With ΔH°_{rxn} much more negative, ΔG°_{rxn} is much more negative. A significantly negative ΔG°_{rxn} is associated with an irreversible reaction, so whereas the reaction in Equation 24-28 is reversible, this one is irreversible.

Your Turn 24.13

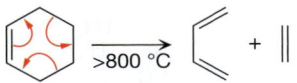

Your Turn 24.14

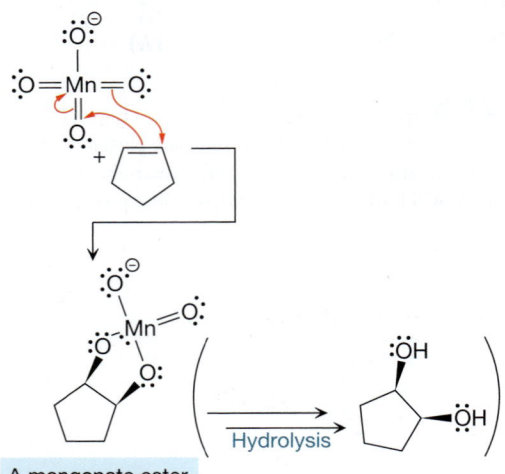

A manganate ester

Your Turn 24.15

In both regions of overlap, the orbitals have the same phase, which leads to constructive interference. Because both regions exhibit the same kind of interference, the orbitals have the appropriate symmetries to interact.

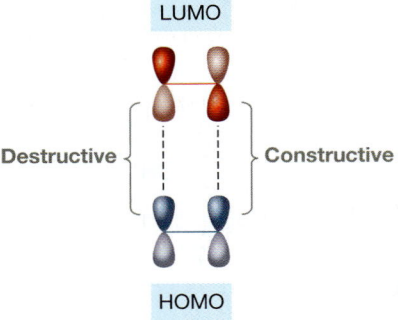

LUMO of diene

Constructive   Constructive

HOMO of dienophile

Your Turn 24.16

On the left, the orbitals have opposite phases, resulting in destructive interference. On the right, the orbitals have the same phase, resulting in constructive interference. Therefore, there is no net overlap and the orbitals do not have the appropriate symmetries to interact.

LUMO

Destructive   Constructive

HOMO

Your Turn 24.17

The *p* orbitals on C2 and C3 of the dienophile are involved in primary orbital overlap in both Figure 24-11a and 24-11b. In Figure 24-11a, the *p* orbital on C1 of the dienophile (blue) is involved in secondary orbital overlap with the *p* orbital on C2 of the diene (red), but that is not the case in Figure 24-11b (below, circled). Those orbitals are closer together in Figure 24-11a than they are in Figure 24-11b.

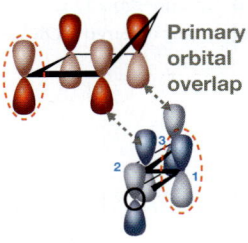

Primary orbital overlap

Chapter 25

Your Turn 25.1

Homolysis of the Cl—Cl bond
Two identical radicals are produced.

$$Cl \overset{\frown}{-} Cl \longrightarrow Cl\cdot \; + \; \cdot Cl$$

Your Turn 25.2

From Table 25-2, Cl—Cl = 243 kJ/mol, Br—Br = 192 kJ/mol, I—I = 151 kJ/mol, HO—OH = 211 kJ/mol; from Table 25-1, H_3C—H = 439 kJ/mol. A typical C—H bond (Table 1-2) is 418 kJ/mol. The larger the bond energy, the stronger the bond. This shows that bonds in molecular halogens and peroxides are much weaker than C—H bonds.

Your Turn 25.3

The H—Br and H—Cl bond energies are 368 and 431 kJ/mol, respectively. The H—Cl bond is stronger and requires more energy to cleave homolytically. The product chlorine radical ($\cdot$Cl) is therefore higher in energy and less stable than the bromine radical ($\cdot$Br).

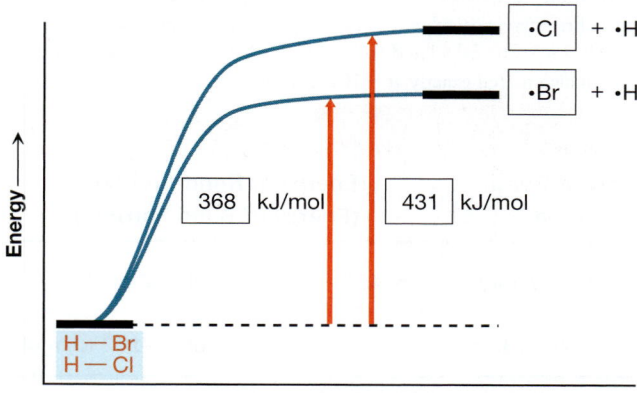

Your Turn 25.4

Benzyl cation

Your Turn 25.5

Your Turn 25.6

$$H_3C\cdot \; + \; \cdot CH_3 \longrightarrow H_3C-CH_3$$

Your Turn 25.7

Your Turn 25.8

Your Turn 25.9

The Br—Br bond is the weakest bond present and therefore will be broken in the initiation step.

Initiation $$Br \overset{\frown}{-} Br \xrightarrow{h\nu} Br\cdot \; + \; Br\cdot$$

Your Turn 25.10

Your Turn 25.11

Your Turn 25.12

The initiation step matches Equation 25-18, and the propagation steps match either Equation 25-19a or 25-19b (shown below). Equation 25-20a and Equation 25-20b are radical coupling or termination steps.

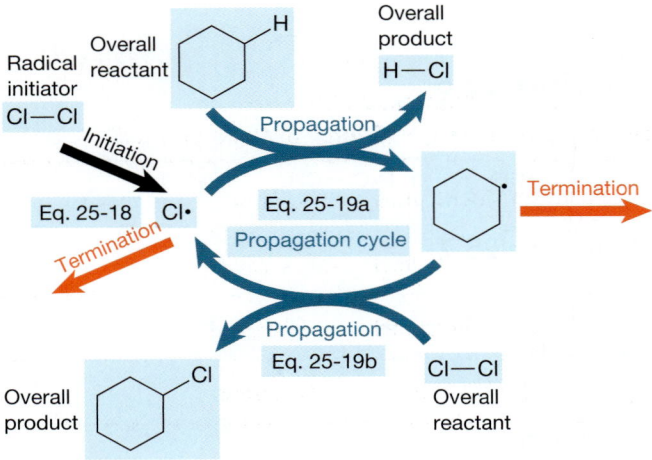

Your Turn 25.13

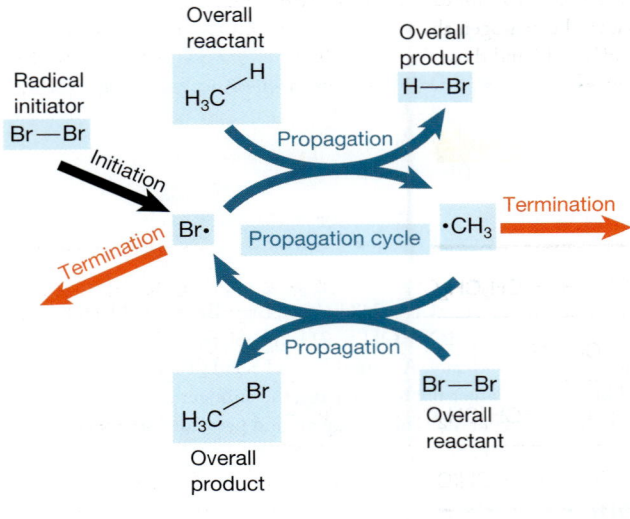

Your Turn 25.14

The values for the energy of the bonds broken and formed in the propagation steps of the bromination of CH_4 are given below. ΔH°_{rxn} for the first step is calculated to be 71 kJ/mol, the same as the value in Table 25-3. The calculated and Table 25-3 values for the second reaction also match exactly at -102 kJ/mol.

TABLE 25-14			
Reaction	Energy of Bond Broken	Energy of Bond Formed	(Energy of Bond Broken) – (Energy of Bond Formed)
$Br\cdot + CH_4 \longrightarrow HBr + CH_3\cdot$	H—CH_3 = **439 kJ/mol**	H—Br = **368 kJ/mol**	= 439 – 368 kJ/mol = **71 kJ/mol**
$CH_3\cdot + Br_2 \longrightarrow CH_3Br + Br\cdot$	Br—Br = **192 kJ/mol**	Br—CH_3 = **294 kJ/mol**	= 192 – 294 kJ/mol = **–102 kJ/mol**

Your Turn 25.15

The hydrogen atom abstraction in Step **B** will proceed faster because the allylic radical produced in Step **B** is more stable than the propyl radical produced in Step **A**, due to resonance delocalization of the unpaired electron. The rate of free radical halogenation increases with increasing stability of the radical that is formed.

Your Turn 25.16

The ΔH°_{rxn} values for each hydrogen abstraction step are given below. The hydrogen abstraction steps for $\bullet Cl$ are both exothermic ($\Delta H^\circ_{rxn} < 0$) and those for $\bullet Br$ are both endothermic ($\Delta H^\circ_{rxn} > 0$). This agrees with the free energy diagrams for these reactions (Fig. 25-10).

TABLE 25-16

Reaction	Hydrogen Abstraction Step ΔH°_{rxn} (Energy of Bond Broken) − (Energy of Bond Formed)
$Cl\bullet + H-CH_2CH_2CH_3 \longrightarrow Cl-H + \bullet CH_2CH_2CH_3$	$= 422 - 431 \text{ kJ/mol} = -9 \text{ kJ/mol}$
$Cl\bullet + H-CH(CH_3)_2 \longrightarrow Cl-H + \bullet CH(CH_3)_2$	$= 410 - 431 \text{ kJ/mol} = -21 \text{ kJ/mol}$
$Br\bullet + H-CH_2CH_2CH_3 \longrightarrow Br-H + \bullet CH_2CH_2CH_3$	$= 422 - 368 \text{ kJ/mol} = +54 \text{ kJ/mol}$
$Br\bullet + H-CH(CH_3)_2 \longrightarrow Br-H + \bullet CH(CH_3)_2$	$= 410 - 368 \text{ kJ/mol} = +42 \text{ kJ/mol}$

Your Turn 25.17

The 3° radical is more stable than a 2° or 1° radical. This is reflected in a larger energy difference in the two radicals compared to the 2° and 1° radicals formed from the reactions in Figure 25-10.

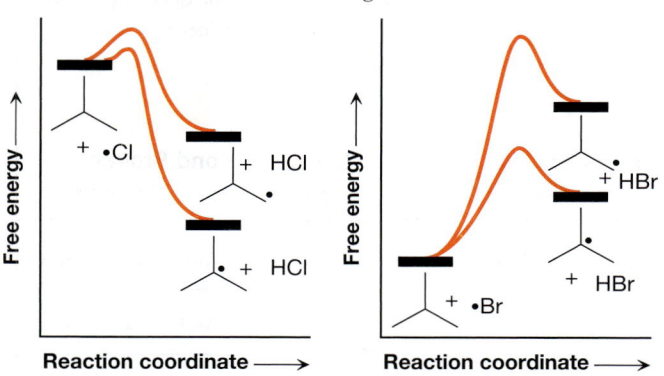

Your Turn 25.18

Resonance stabilization of radical from NBS

Tautomerization mechanism

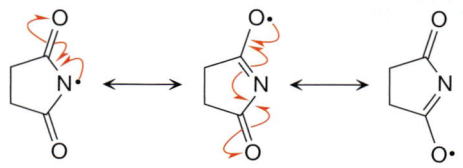

Your Turn 25.19

α-Tocopherol (Vitamin E)

α-Tocopherol is such an effective antioxidant because the radical it produces when a hydrogen atom is abstracted is heavily resonance stabilized, making it much less reactive than a lipid radical.

Your Turn 25.20

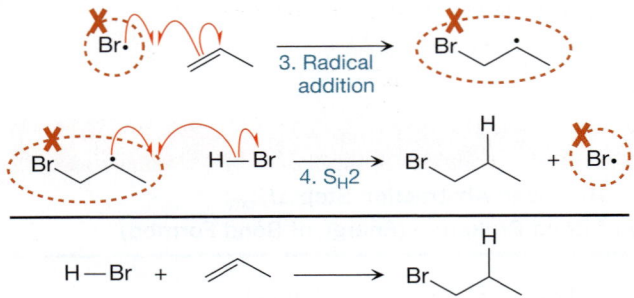

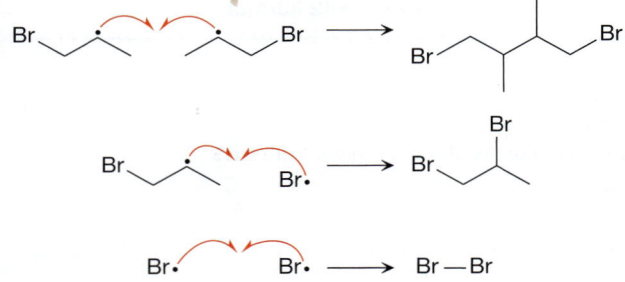

Your Turn 25.21

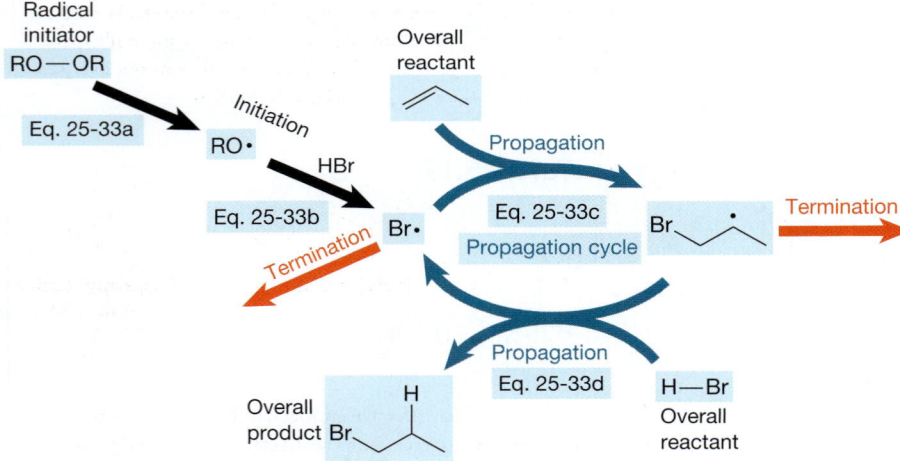

Your Turn 25.22

Your Turn 25.23

In the radical mechanism, the Br• radical adds to the alkene first at the 1° C to form the new σ Br—C bond and yield the more stable 2° C• intermediate. The new σ C—H bond is formed in the second step of the mechanism. In the closed-shell mechanism, H^+ of HBr adds to the alkene first at the 1° C to form the new σ H—C bond and yield the more stable 2° C^+ intermediate. The new σ C—Br bond is formed in the second step of the mechanism. The two reactions have different regiochemistry because the species that add to the primary carbon in the first step are different: Br• for the radical mechanism and H^+ for the closed-shell mechanism.

The more stable *radical* intermediate

The more stable *carbocation* intermediate

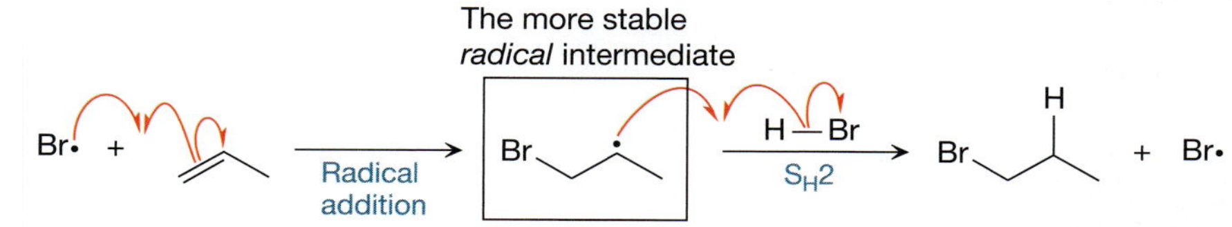

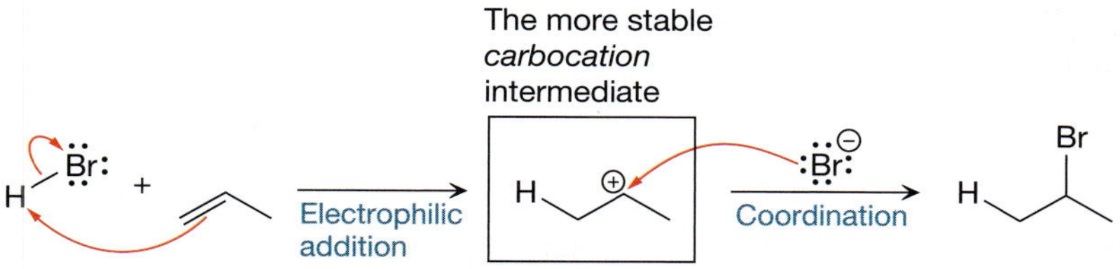

Your Turn 25.24

Catalytic hydrogenation produces the cis alkene.

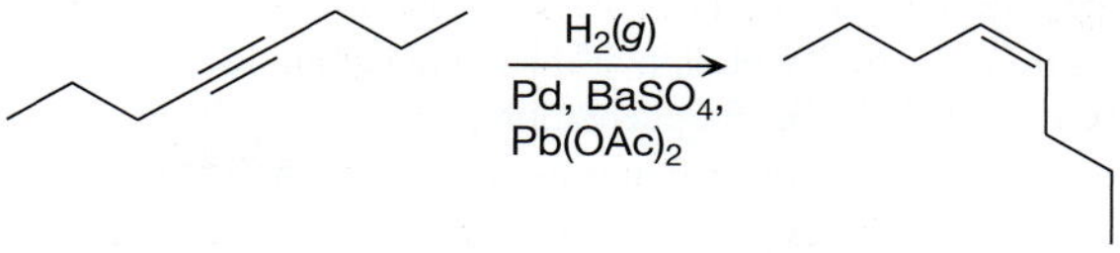

Interchapter G

Your Turn G.1

The mass peak at $m/z = 43$ corresponds to a tertiary isopropyl cation $(CH_3)_2CH^+$ and that at $m/z = 57$ corresponds to a primary butyl cation $CH_3CH_2CH_2CH_2^+$. The two possible fragmentation pathways are shown below. The tertiary isopropyl cation is more stable than the primary butyl cation; therefore, the fragmentation is more likely to produce a mass peak at $m/z = 43$. Cation stability wins out over radical stability, because cations are more electron deficient.

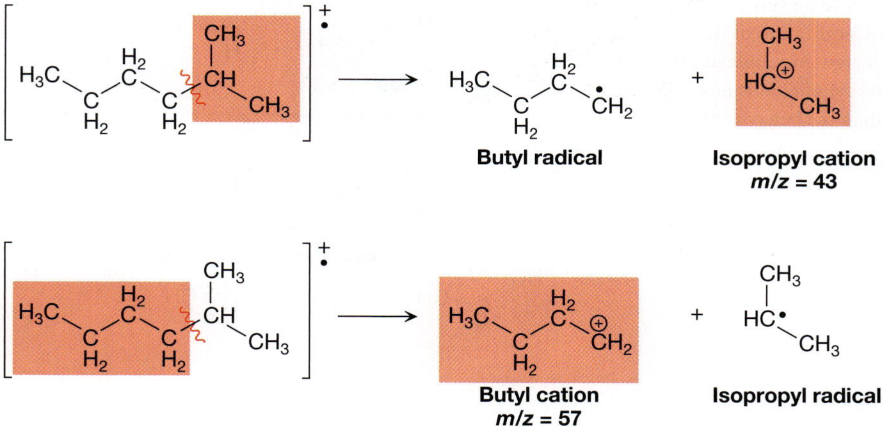

Butyl radical

Isopropyl cation
m/z = 43

Butyl cation
m/z = 57

Isopropyl radical

Your Turn G.2

Figure G-3 is the mass spectrum of 2-methylpentane. The peak at $m/z = 29$ is from an ethyl cation $CH_3CH_2^+$ and the peak at $m/z = 71$ is from the fragmentation of the methyl group to give $CH_3CH^+CH_2CH_2CH_3$. The fragmentation pathways are shown below. The secondary 2-pentyl cation is more stable than the primary ethyl cation; therefore, its peak is more intense.

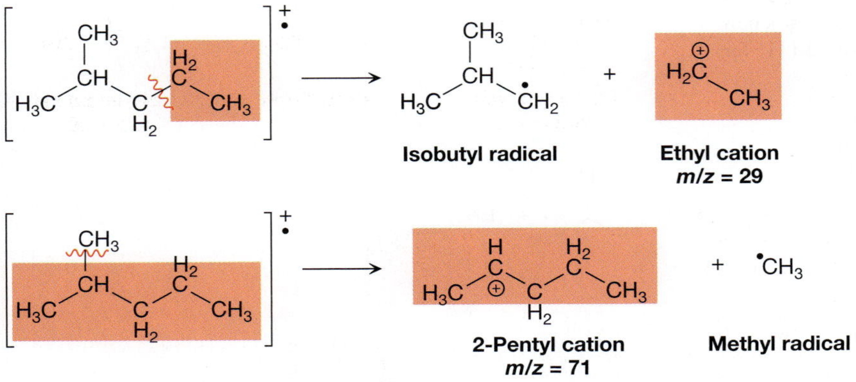

Isobutyl radical

Ethyl cation
$m/z = 29$

2-Pentyl cation
$m/z = 71$

Methyl radical

Your Turn G.3

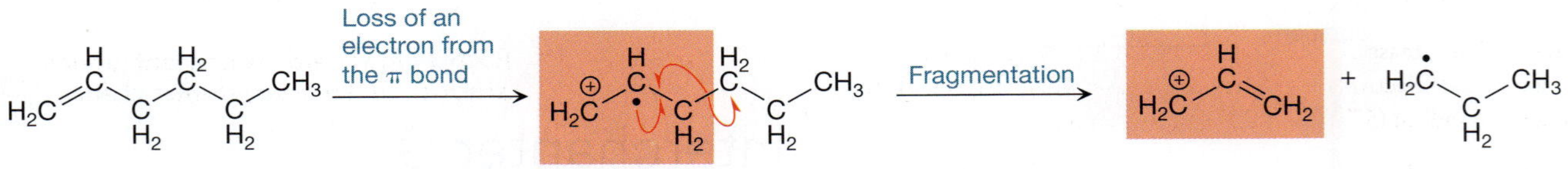

Loss of an electron from the π bond

Fragmentation

Your Turn G.4

For the tropylium ion there are seven equivalent resonance structures, all of which are aromatic (6 π electrons in a completely conjugated ring). For the benzylic cation there are four equivalent resonance structures, of which one shows aromaticity in the ring. Owing to the contribution of the especially stable aromatic resonance structures and the delocalization of the positive charge over more atoms, the tropylium ion is more stable than the benzylic cation.

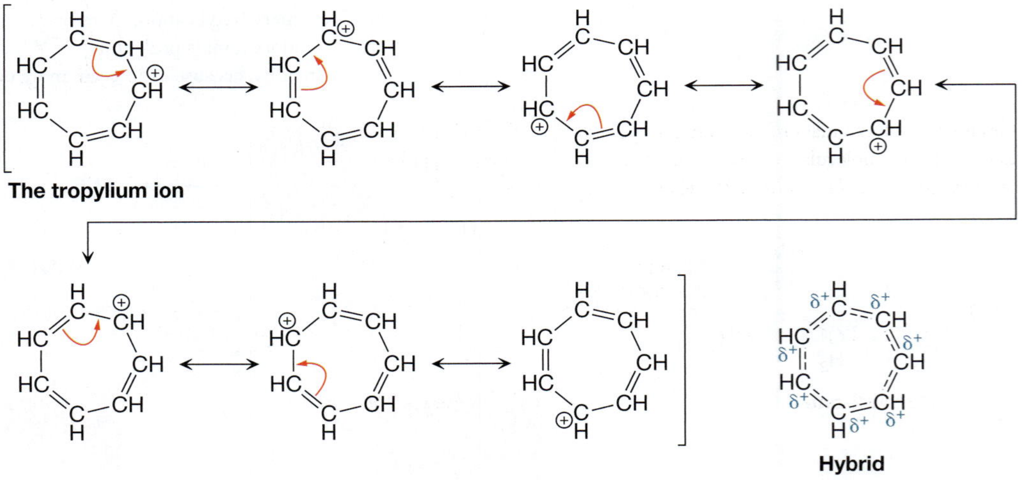

The tropylium ion

Hybrid

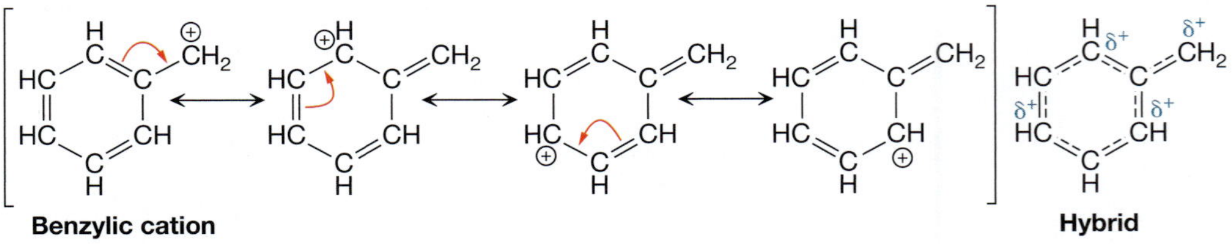

Benzylic cation

Hybrid

Your Turn G.5

The ethyl radical, bonded to the carbon atom bonded to the Cl, is eliminated. This is α-cleavage and in this example requires three single-barbed arrows as shown below.

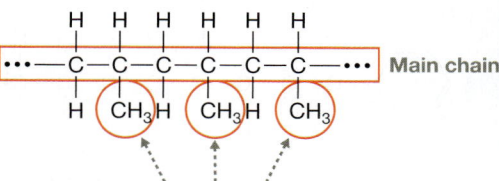

Your Turn G.6

Diisopropyl ether has a mass of 102 u. The parent ion is produced from the loss of a lone pair electron on O. The peak at $m/z = 87$ corresponds to loss of a methyl radical (15 u) from M^+ via α-cleavage.

$m/z = 102$ $m/z = 87$

Pentan-1-ol has a mass of 88 u. The parent ion is produced from the loss of a lone pair electron on O. The peak at $m/z = 31$ corresponds to loss of a butyl radical (57 u) from M^+ via α-cleavage.

$m/z = 88$ $m/z = 31$

Your Turn G.7

Hexan-2-one has a mass of 100 u. A lone pair electron from O is lost to produce the molecular ion. The peak at $m/z = 85$ corresponds to loss of a methyl radical (15 u) from M^+ via α-cleavage.

Hexan-2-one $m/z = 100$ An acylium ion $m/z = 85$

Your Turn G.8

The M^+ peak of pentanoic acid appears at $m/z = 102$. A McLafferty rearrangement results in the loss of prop-1-ene, $CH_3CH{=}CH_2$, and the formation of the enol radical cation $CH_2{-}C(^+OH)OH$, whose value of m/z is 60.

Pentanoic acid $m/z = 102$ Prop-1-ene   An enol radical cation $m/z = 60$

Chapter 26

Your Turn 26.1

Main chain

Methyl groups are the pendant groups attached to the main polymer chain.

Your Turn 26.2

The repeating unit is shown here. For the structure given, $n = 3$.

Your Turn 26.3

Your Turn 26.4

Your Turn 26.5

In Figure 26-7a the dots (red) are molecules of styrene and the squares (blue) are molecules of benzoyl peroxide, the initiator. Using Figure 26-7a–d as a reference, the initiation of a new polymer chain and the propagation of two existing chains are added to Figure 26-7e in the figure below.

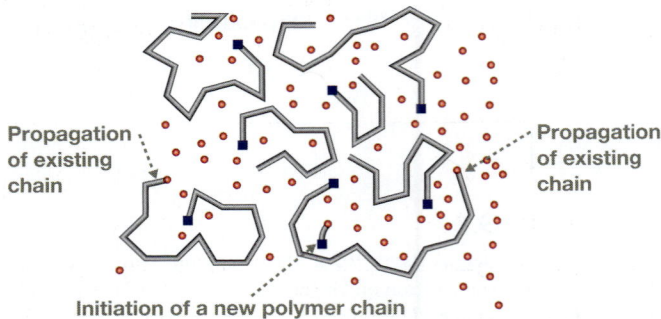

Propagation of existing chain

Propagation of existing chain

Initiation of a new polymer chain

Your Turn 26.6

(a) Combination involving two of the final radicals from Solved Problem 26.4 is shown below.

Combination

New C—C bond

(b) Disproportionation involving two of the final radicals from Solved Problem 26.4 is shown below.

Disproportionation

Your Turn 26.7

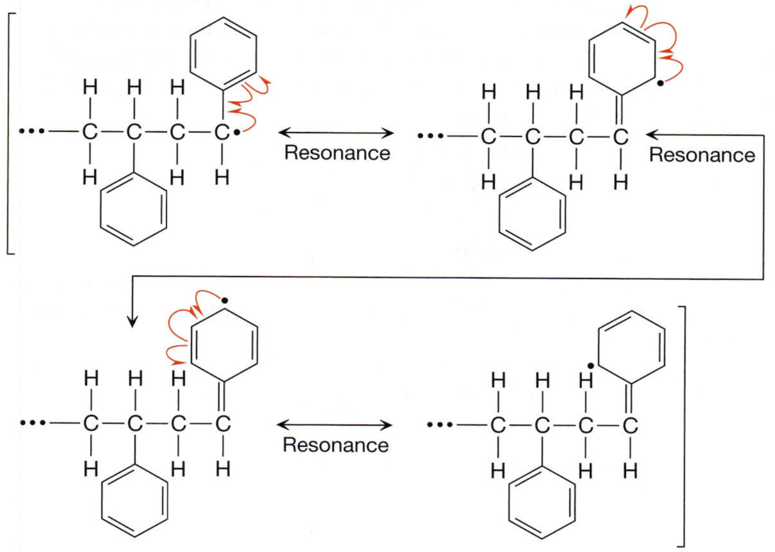

The head-to-tail addition is shown below, requiring three single-barbed curved arrows.

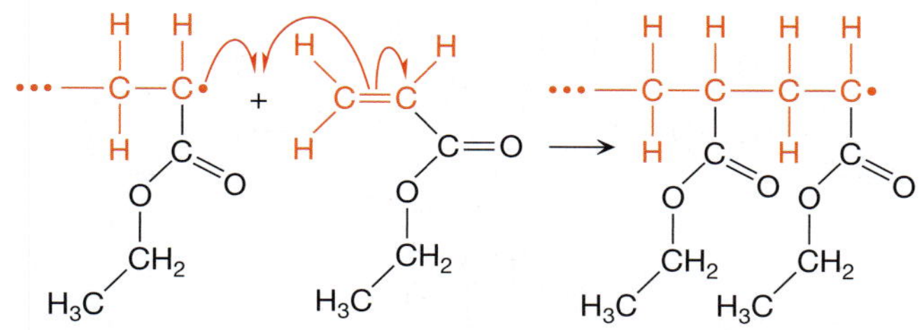

Propagating chain of poly(ethyl acrylate) **Ethyl acrylate** **Product of head-to-tail addition**

The product radical has the following two resonance structures:

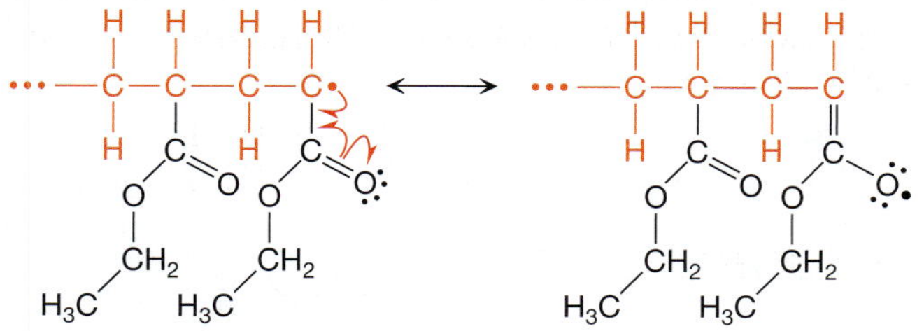

Product of head-to-tail addition

Your Turn 26.8

The product anion in Equation 26-12 is resonance stabilized due to the presence of π bonds next to the C^- electron pair. Resonance is possible around the entire phenyl ring. No such resonance exists in the anion on the reactant side.

Your Turn 26.9

In a radical mechanism, combination and disproportionation each require two radicals reacting. The equivalent species in an anionic mechanism are anions. Neither of these steps is likely because the like charges of the two anions would repel each other.

Your Turn 26.10

Each growing polymer will be protonated twice.

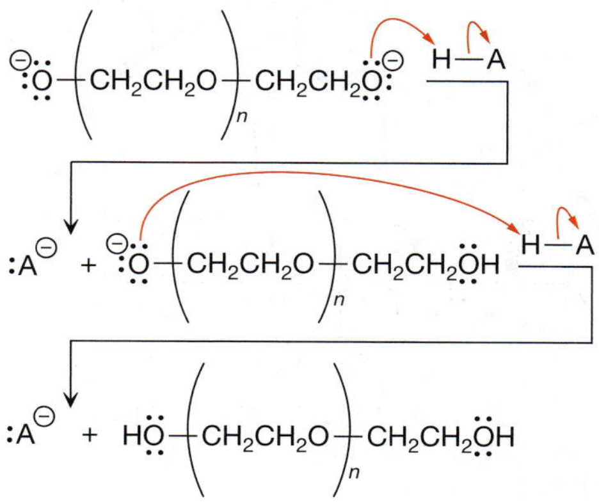

Your Turn 26.11

Propagation occurs when an uncharged molecule of tetrahydrofuran attacks the positively charged ring. The second propagation step occurs via the same mechanism as the first propagation step as shown below. A new positively charged ring appears at the end of the growing chain.

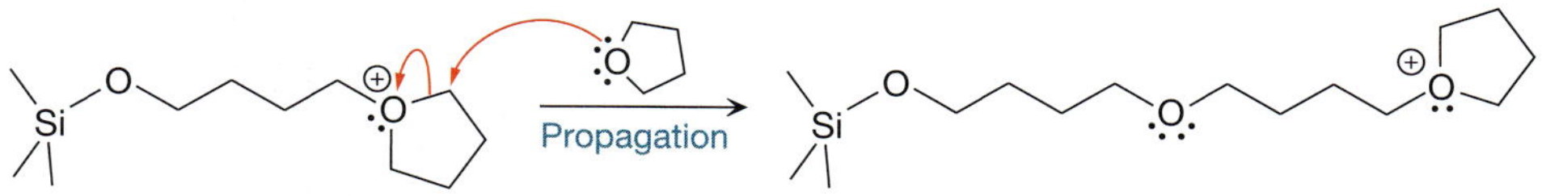

If methanol were added to the product shown above, the propagation is terminated resulting in the product shown below.

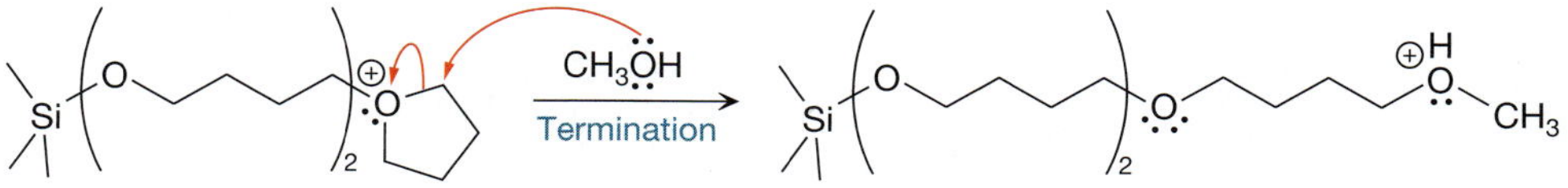

Your Turn 26.12

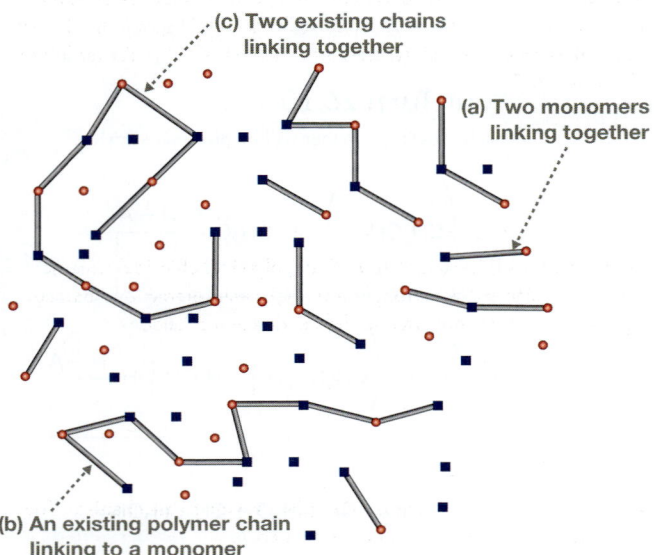

Your Turn 26.13

(c) Two existing chains linking together

(a) Two monomers linking together

(b) An existing polymer chain linking to a monomer

Your Turn 26.14

The first step is a hydrogen atom abstraction, whereby one C—H bond is formed and a second one is broken simultaneously. The second step is a radical addition, whereby a C—C single bond is formed and a C=C π bond is broken. Both steps require three single-barbed curved arrows.

(a)

(b)

Your Turn 26.15

In Bakelite, the starting monosubstituted benzene ring is a phenol and substitution takes place at the ortho or para sites because phenol is an ortho/para director (Section 23.1).

Your Turn 26.16

The total bond strength in an enol such as vinyl alcohol is less than in the analogous keto form. This is largely due to the strong C=O bond.

Your Turn 26.17

The reaction is a transesterification under basic conditions, which takes place in a nucleophilic addition–elimination mechanism. The second product is methyl acetate.

Tetrahedral intermediate

Methyl acetate or methyl ethanoate

Your Turn 26.18

According to VSEPR theory, the O atom of C—O—C is surrounded by four electron groups: two lone pairs and two single bonds. The electron geometry would therefore be tetrahedral and the bond angles would be 109.5°. This however is not the case and the bonds are actually closer to 120°. One lone pair of electrons on the oxygen in C—O—C is involved in resonance with the phenyl rings, giving each bond to O some double bond character. This double bond character is reflective of sp^2 hybridization, which gives rise to a bond angle of approximately 120°.

Your Turn 26.19

Nylon is a polyamide. The amide group appears twice in the repeating unit. One complete amide appears in the center of the repeating unit. The second amide is split between two repeating units: the N at the right of one repeating unit is connected to a C=O group that is part of the next repeating unit (represented by the C=O at the far left in Equation 26-24).

Your Turn 26.20

Poly(acrylic acid) has a higher T_g than poly(vinyl chloride) because it is more polar and has stronger intermolecular interactions, namely, hydrogen bonding compared to dipole–dipole interactions.

Your Turn 26.21

It is believed that this reaction takes place via an E1 mechanism. The Cl⁻ leaves first, then a proton is eliminated.

Your Turn 26.22

An O=C—N functional group, characteristic of an amide, is responsible for connecting adjacent amino acids along a protein's main chain, and each of those amide groups is called a peptide linkage. In the hexapeptide in Figure 26-44 there are 5 peptide linkages as shown by the red screens in the figure below.

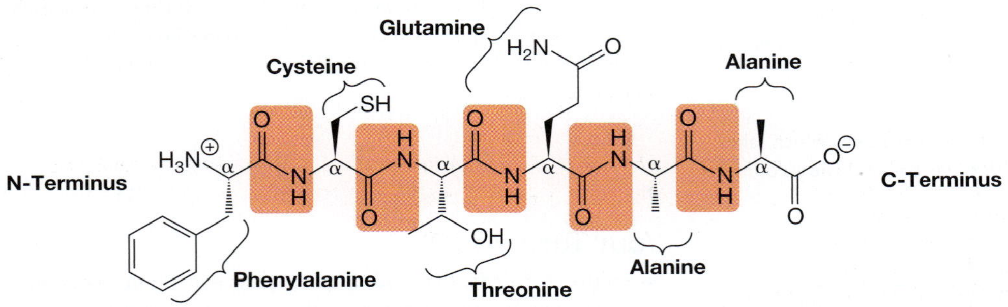

Phe-Cys-Thr-Gln-Ala-Ala

Your Turn 26.23

The missing labeled glycosidic linkage in Figure 26-52 is α-1,4′ and that in Figure 26-53 is β-1,4′ (shown below, red screens). These linkages involve C1 of one glucose unit and C4 of the next, and the substituent on C1 is axial in amylopectin and equatorial in cellulose.

Figure 26-52

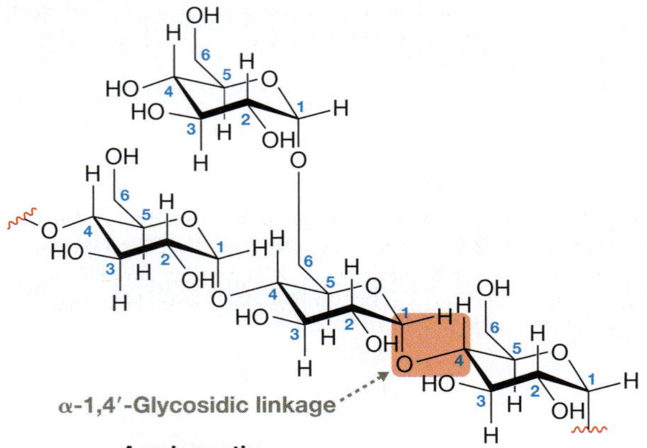

α-1,4′-Glycosidic linkage

Amylopectin

Figure 26-53

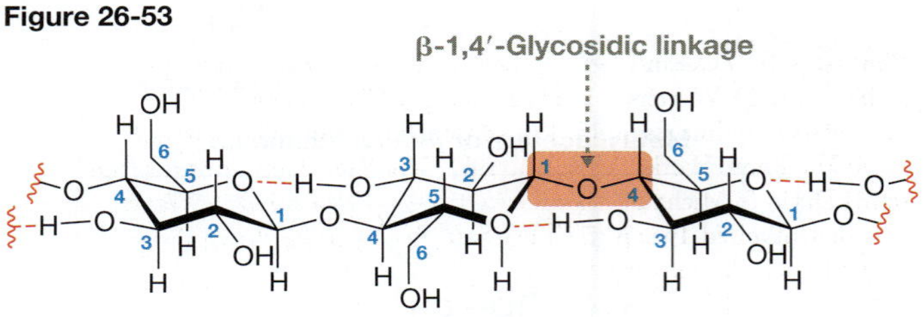

β-1,4′-Glycosidic linkage

Cellulose

CREDITS

Front Matter

Photos: p. i: iStock/Getty Images Plus; **p. iii:** Photo by Joshua Karty and Jacob Karty; **p. vii:** © Ocean/Corbis; **p. ix:** Skyhobo/Getty Images; **p. xi:** Spencer Platt/Getty Images; **p. xiv:** SPL/Science Source; **p. xvi:** Mopic/Shutterstock; **p. xvii:** James Steidl/Shutterstock; **p. xix:** iStock/Getty Images Plus; **p. xxi:** iStock/Getty Images Plus; **p. xxxix:** anopdesignstock/Getty Images; **p. xlvii:** Science VU/Frederick Mertz/Visuals Unlimited, Inc.; **p. xlviii:** (left) Pavel Ilyukhin/Shutterstock; (second from left) Phanie/Science Source; (third from left) Martin Brayley/Alamy Stock Photo; (right) one-image photography/Alamy Stock Photo; (top) ThomasDeco/Shutterstock; (bottom) © Columbia University, From Groundbreaking Research on Molecular Electronics Reported in Journal Science, By Alex Lyda; **p. xlix:** (left) age fotostock/Alamy Stock Photo; (top) NASA; (bottom) Sunpix Travel/Alamy Stock Photo.

Chapter 1

Photos: p. 1: © Ocean/Corbis; **p. 3:** © Philippe Clement/naturepl.com; **p. 4:** Charles D. Winters/Science Source; **p. 8:** Sam Shere/Hulton Archives/Getty Images; **p. 13:** Leo Mason/Popperfoto/Getty Images; **p. 14:** The Photo Works; **p. 23:** (top left) Greg Elms/Getty Images; (top right) David & Micha Sheldon/Getty Images; (bottom) Dave Watts/Visuals Unlimited, Inc.; **p. 25:** The Photo Works; **p. 36:** Alexander Chelmodeev/Shutterstock.

Interchapter A

Photos: p. 53: (left) Daniel Hurst Photography/Getty Images; (second from left) Ken Welsh/Design Pics/Getty Images; (third from left) Bill Hatcher/National Geographic/Getty Images; (right) Media Bakery.

Chapter 2

Photos: p. 70: (left) Michel Gunther/Science Source; (middle) Perennou Nuridsany/Science Source; (right) Eye of Science/Science Source; **p. 73:** Dusan Jankovic/Shutterstock; **p. 78:** Courtesy of Joel Karty; **p. 80:** tam_odin/Shutterstock; **p. 81:** Triduza Studio /Shutterstock; **p. 89:** Defense Advanced Research Projects Agency; **p. 92:** CreVis2/Getty Images; **p. 111:** (left) Valentina Proskurina/Shutterstock; (second from left): © Lena Trindade/BrazilPhotos; (third from left): Digital Zoo/Getty Images; (right): Qrt/Alamy Stock Photo.

Chapter 3

Photos: p. 119: Hermann Eisenbeiss/Science Source; **p. 123:** Nicole Fuller/NSF (National Science Foundation-Public Domain); **p. 125:** Tramino/Getty Images; **p. 136:** Editorial Image/Science Source; **p. 141:** Photo by Mark Ridgway; **p. 143:** Karl Martin/Alamy Stock Photo; **p. 144:** (top) Africa Studio/Shutterstock; (bottom) Courtesy of Joel Karty; **p. 145:** Courtesy of Joel Karty.

Interchapter B

Photos: p. 158: Justin Sullivan/Getty Images; **p. 159:** Gaby Kooijman/Shutterstock; **p. 161:** winnond/Shutterstock.

Chapter 4

Photos: p. 165: Skyhobo/Getty Images; **p. 172:** Robert Pickett/Visuals Unlimited, Inc./Getty Images; **p. 181:** Hank Morgan/Science Source; **p. 185:** Andrey_Kuzmin/Shutterstock; **p. 190:** Courtesy of Joel Karty; **p. 191:** Mike Jaquish, Realty Arts; **p. 194:** shironosov/Getty Images.

Chapter 5

Photos: p. 208: vesilvio/iStockphoto; **p. 214:** (left) Courtesy of Joel Karty; (middle) Wikimedia Commons; (right) Lawrence Lawry/Science Source; **p. 220:** The Photo Works; **p. 222:** Gregory James Van Raalte/Shutterstock; **p. 235:** Mtsaride/Shutterstock; **p. 236:** John Kelly Green Photography; **p. 245:** NASA/JPL-Caltech.

Chapter 6

Photos: p. 274: Spencer Platt/Getty Images; **p. 277:** The Photo Works; **p. 279:** anopdesignstock/Getty Images; **p. 286:** (left) The Photo Works; (right) Sherry Yates Young/Shutterstock; **p. 291:** Marilyn Howell/Stockimo/Alamy Stock Photo; **p. 307:** © Rene Fluger/AgeFotostock; **p. 310:** Elnur/Shutterstock.

Chapter 7

Photos: p. 328: 123RF.com; **p. 331:** hartphotography/Shutterstock; **p. 344:** Aksenova Natalya/Shutterstock; **p. 347:** Aubrie Pick/SLAC National Accelerator Laboratory.

Interchapter E

Photos: p. 378: evemilla/Getty Images; **p. 381:** Gary Bell/OceanwideImages.com; **p. 387:** Camera Press Ltd/Alamy Stock Photo; **p. 388:** The Photo Works; **p. 390:** Billion Photos/Shutterstock.

Chapter 8

Photos: p. 393: Zerbor/Shutterstock; **p. 401:** The Photo Works; **p. 425:** Mariusz Szczygiel/Shutterstock; **p. 426:** Image Point Fr/Shutterstock; **p. 429:** Everett Collection/Shutterstock; **p. 432:** D. Kucharski K. Kucharska/Shutterstock.

Chapter 9

Photos: p. 442: Whitebox Media/Alamy Stock Photo; **p. 452:** Milos Luzanin/Shutterstock; **p. 454:** Prasad, V.; Semwogerere, D.; Weeks, E. R. Confocal Microscopy of Colloids. *J. Phys.: Condens. Matter* **2007**, *19*, 113102 (25pp). ©IOP Publishing, LTD.; **p. 469:** SHSPhotography/Getty Images; **p. 477:** winnond/Shutterstock; **p. 483:** Kongsky/Shutterstock; **p. 492:** (left) xiaorui/Shutterstock; (right) Nataly Studio/Shutterstock.

Interchapter F

Photos: p. 504: WIN-Initiative/Getty Images; **p. 506:** blickwinkel/Alamy Stock Photo; **p. 507:** eye35.pix/Alamy Stock Photo; **p. 508:** Grafner/Getty Images; **p. 510:** The Photo Works.

Chapter 10

Photos: p. 515: Adam Berry/Getty Images; **p. 517:** vz maze/Shutterstock; **p. 534:** Dusan Jankovic/Shutterstock; **p. 541:** Jaromir Chalabala/Shutterstock; **p. 542:** (top) kolesniks/Shutterstock; (bottom) Vladimir Godnik/Getty Images.

Chapter 11

Photos: p. 563: (left) BSIP SA/Alamy Stock Photo; (right) SPL/Science Source; **p. 571:** Fidel/Shutterstock; **p. 575:** Mike Flippo/Shutterstock; **p. 580:** NASA/MSFC; **p. 584:** Westend61/Getty Images; **p. 589:** (left) Carol and Mike Werner/Science Source; (right) C. E. V./Science Source.

Chapter 12

Photos: p. 601: flowerphotos/Alamy Stock Photo; **p. 615:** (top) Andrew J. Martinez/Science Source; (bottom) Dr. D.P. Wilson/Science Source; **p. 619:** Igor Normann/Shutterstock; **p. 622:** Jody Ann/Shutterstock; **p. 623:** (left) Peter Dazeley/Getty Images; (right) Wikimedia Commons; **p. 624:** Chaikom/Shutterstock; **p. 625:** Bret Hartman/AP Images for Hyundai; **p. 631:** Robyn Mackenzie/Shutterstock.

Chapter 13

Photos: p. 641: Sally Scott/Shutterstock; **p. 646:** digidreamgrafix/FeaturePics; **p. 656:** Marevision/AgeFotostock; **p. 661:** (top) AP Photo; (bottom): Schuler, B.; Fatayer, S.; Mohn, F.; Moll, N.; Pavli, N.; Meyer, G.; Peña, D.; Gross, L. Reversible Bergman Cyclization by Atomic Manipulation. *Nature Chemistry* **2016**, *8*, 220–224. doi:10.1038/nchem.2438. © 2016 Nature Publishing Group; **p. 664:** The Photo Works.

Chapter 14

Photos: p. 682: LONG WEI/FEATURECHINA/Newscom; **p. 685:** zhudifeng/Getty Images; **p. 697:** James King-Holmes/Science Source; **p. 706:** Grafner/Getty Images; **p. 707:** Dan Kitwood/Getty Images; **p. 708:** karandaev/Getty Images; **p. 709:** (left) Crevis/Shutterstock; (right) Raimundo79/Shutterstock.

Chapter 15

Photos: p. 723: Mopic/Shutterstock; **p. 732:** The Photo Works; **p. 733:** PlanilAstro/Shutterstock; **p. 734:** (top) iStock/Getty Images Plus; (bottom) Steve Cavalier/Alamy Stock Photo; **p. 749:** Courtesy of Heinrich Pniok; **p. 750:** Chris Gallagher/Getty Images; **p. 758:** iStockphoto.
Drawn art: Fig. 15.12: © Sigma-Aldrich Co. LLC. Used with permission of Sigma-Aldrich Co. LLC; **Fig. 15.13:** © Sigma-Aldrich Co. LLC. Used with permission of Sigma-Aldrich Co. LLC; **YT15.10:** © Sigma-Aldrich Co. LLC. Used with permission of Sigma-Aldrich Co. LLC; **Fig. 15.16:** © Sigma-Aldrich Co. LLC. Used with permission of Sigma-Aldrich Co. LLC; **P15.20:** © Sigma-Aldrich Co. LLC. Used with permission of Sigma-Aldrich Co. LLC; **Fig. 15.17:** © Sigma-Aldrich Co. LLC. Used with permission of Sigma-Aldrich Co. LLC; **Fig. 15.18:** © Sigma-Aldrich Co. LLC. Used with permission of Sigma-Aldrich Co. LLC; **P15.21:** © Sigma-Aldrich Co. LLC. Used with permission of Sigma-Aldrich Co. LLC; **Fig. 15.19:** © Sigma-Aldrich Co. LLC. Used with permission of Sigma-Aldrich Co. LLC; **Fig. 15.20:** © Sigma-Aldrich Co. LLC. Used with permission of Sigma-Aldrich Co. LLC; **Fig. 15.21:** © Sigma-Aldrich Co. LLC. Used with permission of Sigma-Aldrich Co. LLC; **P15.23:** © Sigma-Aldrich Co. LLC. Used with permission of Sigma-Aldrich Co. LLC; **YT15.17:** © Sigma-Aldrich Co. LLC. Used with permission of Sigma-Aldrich Co. LLC; **Fig. 15.22:** © Sigma-Aldrich Co. LLC. Used with permission of Sigma-Aldrich Co. LLC; **P15.25:** © Sigma-Aldrich Co. LLC. Used with permission of Sigma-Aldrich Co. LLC; **Fig. 15.23:** © Sigma-Aldrich Co. LLC. Used with permission of Sigma-Aldrich Co. LLC; **P15.26:** © Sigma-Aldrich Co. LLC. Used with permission of Sigma-Aldrich Co. LLC; **Fig. 15.24:** © Sigma-Aldrich Co. LLC. Used with permission of Sigma-Aldrich Co. LLC; **Fig. 15.25:** © Sigma-Aldrich Co. LLC. Used with permission of Sigma-Aldrich Co. LLC; **Fig. 15.26:** © Sigma-Aldrich Co. LLC. Used with permission of Sigma-Aldrich Co. LLC; **P15.53:** © Sigma-Aldrich Co. LLC. Used with permission of Sigma-Aldrich Co. LLC; **P15.55:** © Sigma-Aldrich Co. LLC. Used with permission of Sigma-Aldrich Co. LLC; **P15.59:** © Sigma-Aldrich Co. LLC. Used with permission of Sigma-Aldrich Co. LLC; **P15.60:** © Sigma-Aldrich Co. LLC. Used with permission of Sigma-Aldrich Co. LLC; **P15.61:** © Sigma-Aldrich Co. LLC. Used with permission of Sigma-Aldrich Co. LLC; **P15.62:** © Sigma-Aldrich Co. LLC. Used with permission of Sigma-Aldrich Co. LLC; **P15.63:** © Sigma-Aldrich Co. LLC. Used with permission of Sigma-Aldrich Co. LLC; **P15.64:** © Sigma-Aldrich Co. LLC. Used with permission of Sigma-Aldrich Co. LLC; **P15.65:** © Sigma-Aldrich Co. LLC. Used with permission of Sigma-Aldrich Co. LLC; **P15.66:** © Sigma-Aldrich Co. LLC. Used with permission of Sigma-Aldrich Co. LLC; **P15.67:** © Sigma-Aldrich Co. LLC. Used with permission of Sigma-Aldrich Co. LLC.

Chapter 16

Photos: p. 771: James Steidl/Shutterstock; **p. 778:** Rasmus Loeth Petersen/Alamy Stock Photo; **p. 780:** TFoxFoto/Shutterstock; **p. 791:** iStock/Getty Images; **p. 803:** (top) Donna Beeler/Shutterstock; (bottom) The Photo Works; **p. 823:** Mitch Fuqua/USAF/Getty Images; **p. 828:** iStock/Getty Images.
Drawn art: P16.64: © Sigma-Aldrich Co. LLC. Used with permission of Sigma-Aldrich Co. LLC; **P16.65:** © Sigma-Aldrich Co. LLC. Used with permission of Sigma-Aldrich Co. LLC.

Chapter 17

Photos: p. 839: Photo by Ted Kinsman; **p. 841:** Petr Goskov/Alamy Stock Photo; **p. 863:** Jack Jelly/Shutterstock; **p. 869:** David Cobb/Alamy Stock Photo.

Chapter 18

Photos: p. 888: Sergiy Bykhunenko/Shutterstock; **p. 897:** Pressmaster/Shutterstock; **p. 913:** Courtesy of Business Wire; **p. 916:** govindji/Shutterstock.

Chapter 19

Photos: p. 946: John Kelly Green Photography; **p. 948:** FUN FUN PHOTO/Shutterstock; **p. 955:** John Jeddore/Flickr; **p. 975:** ntstudio/Shutterstock; **p. 981:** TRL ltd./Science Source.

Chapter 20

Photos: p. 1000: iStock/Getty Images Plus; **p. 1005:** (left) Dave Reede/Getty Images; (right) Sean Gallup/Getty Images; **p. 1006:** The Photo Works; **p. 1011:** Courtesy of Star International Holdings, Inc.; **p. 1020:** Courtesy of Joel Karty.

Chapter 21

Photos: p. 1045: Fotokosic/Shutterstock; **p. 1060:** The Photo Works; **p. 1074:** O.Bellini/Shutterstock.
Drawn art: P21.95: © Sigma-Aldrich Co. LLC. Used with permission of Sigma-Aldrich Co. LLC; **P21.96:** © Sigma-Aldrich Co. LLC. Used with permission of Sigma-Aldrich Co. LLC; **P21.97:** © Sigma-Aldrich Co. LLC. Used with permission of Sigma-Aldrich Co. LLC; **P21.98:** © Sigma-Aldrich Co. LLC. Used with permission of Sigma-Aldrich Co. LLC.

Chapter 22

Photos: p. 1104: graficart.net/Alamy Stock Photo; **p. 1113:** The Photo Works; **p. 1115:** CSP_drohn/AgeFotostock; **p. 1121:** Robert Brook/Science Source; **p. 1122:** Bloomberg/Getty Images; **p. 1124:** (left) © Andrey Kiselev/Fotolia RF; (right) Helen Sessions/Alamy Stock Photo; **p. 1132:** (left) Tetra Images/Alamy Stock Photo; (right) © the food passionates/Corbis.
Drawn art: P22.65: © Sigma-Aldrich Co. LLC. Used with permission of Sigma-Aldrich Co. LLC; **P22.67:** © Sigma-Aldrich Co. LLC. Used with permission of Sigma-Aldrich Co. LLC.

Chapter 23

Photos: p. 1144: Last Resort/Getty Images; **p. 1146:** The Photo Works; **p. 1152:** GIPhotoStock/Getty Images; **p. 1160:** Wikimedia Commons; **p. 1173:** Photo by Dan Wright and Ashley Moreno; **p. 1181:** Nejron Photo/Shutterstock.

Chapter 24

Photos: p. 1198: iStock/Getty Images Plus; **p. 1206:** Courtesy of REV'IT! Sport International B.V.; **p. 1218:** Courtesy of Kenworth/PACCAR, Inc.; **p. 1225:** Lighttraveler/Shutterstock.
Drawn art: P24.81b: © Sigma-Aldrich Co. LLC. Used with permission of Sigma-Aldrich Co. LLC; **P24.82b:** © Sigma-Aldrich Co. LLC. Used with permission of Sigma-Aldrich Co. LLC; **P24.83:** © Sigma-Aldrich Co. LLC. Used with permission of Sigma-Aldrich Co. LLC.

Chapter 25

Photos: p. 1247: (left) Shutterstock RF; (right) Leonard Lessin/Getty Images; **p. 1249:** imageBROKER/Alamy Stock Photo; **p. 1265:** NASA.
Drawn art: P25.77: © Sigma-Aldrich Co. LLC. Used with permission of Sigma-Aldrich Co. LLC.

Chapter 26

Photos: p. 1307: (opener) Jeong-Woo Lee, Georgia Institute of Technology; (top) Jan Wlodarczyk/Alamy Stock Photo; (left) risteski goce/Shutterstock; (right) Oksana Shufrych/Shutterstock; **p. 1308:** (left) iStockphoto; (right): iStockphoto; **p. 1318:** (top) Dmitry Kalinovsky/Shutterstock; (bottom) Duplass/Shutterstock; **p. 1319:** (left) Gránásy, L.; Pusztai, T.; Börzsönyi, T.; Warren, J. A.; Douglas, J. F. A General Mechanism of Polycrystalline Growth. *Nature Materials* **2004**, *3*, 645–650. doi:10.1038/nmat1190. © 2004 Nature Publishing Group; (middle) oneclearvision/iStock/Getty Images; (right) The Photo Works; **p. 1336:** (left) Courtesy of Joel Karty; (right) oliver leedham/Alamy Stock Photo; **p. 1338:** (top) PhilipYb Studio/Shutterstock; (bottom) Dmitry Kalinovsky/Shutterstock; **p. 1344:** Madlen/Shutterstock; **p. 1346:** (left) iStockphoto; (middle) D. Dockstader/M. Noble/S. Pruett; (right) D. Dockstader/M. Noble/S. Pruett; **p. 1347:** (top) Arne Hückelheim/Wikimedia Commons; (bottom) jirasaki/Shutterstock; **p. 1351:** iStockphoto; **p. 1352:** (left) Michael Flippo/Alamy Stock Photo; (second from left) Feng Yu/Alamy Stock Photo; (second from right) The Photo Works; (right) C Squared Studios/Getty Images; **p. 1354:** (left) Nataly Studio/Shutterstock; (right) © Eco-Products, Inc.
Drawn art: p. 1333: Reprinted with permission from Xu, J.; Fogleman, E. A.; Craig, S. L. Structure and Properties of DNA-Based Reversible Polymers. *Macromolecules* **2004**, *37* (5), 1863–1870. Copyright © 2004 American Chemical Society.

INDEX

Note: Material in figures or tables is indicated by *italic* page numbers.

acylium ions, *1118*, 1119–20

acyl substitution

nucleophilic acyl substitution, reaction mechanism, 1002–3

thermodynamics of, 1006–9

See also nucleophilic addition–elimination reactions

addition polymerization, 1312–13

adducts, defined, 565

adenine, 44, 715, 716, 717

adenosine diphosphate (ADP), 420

adenosine triphosphate (ATP), 420, 852, 1010

adipic acid (hexane-1,6-dioic acid), 1325, *1326*, 1327, 1342

aerogel, *682*

alanine, *39*

alcohol dehydrogenase, 194

alcohols

acetals as protecting groups, 965–68

acid-catalyzed dehydration, 463, 538–40

from acid-catalyzed hydration of alkenes, 576–77, 614, 630–31

charge stability and pK_a of, 309

chemical shift in ^{13}C NMR, 808

chromic acid oxidations, 977–79, 980–81, 982

conversion of hydroxide leaving group to water, 461–62, 516

conversion to alkyl halides, acidic conditions, 461–62, 516

conversion to alkyl halides by PBr_3 or PCl_3, 517–20, 653, 656, 665, 670, 873, 950

converting bad leaving groups to good leaving groups, 461–62

dehydration to form ethers, 538–40

from epoxides under acidic conditions, 544–46

from epoxides under basic conditions, 541–44

from Grignard reactions, 872, 964

hemiacetal formation by reaction with ketones or aldehydes, 889–92, 897, 899, 929

from hydride reduction of aldehydes and ketones, 845–48, 1023–24

from hydroboration reactions, 628–31, 656

hydroxyl group as poor leaving group, 459–60, 516

lack of coupling in NMR spectra, 796–97

as leaving groups in S_N2 reactions, 335

mass spectrometry, 1302–4

nomenclature, 378, *379*, 380, 381, 386–87, *504*

O—H stretch absorption bands in IR spectroscopy, *740*, 741, 746

permanganate oxidation of, 981–82

primary (1°) alcohols, 387

primary alcohols from reduction of carboxylic acid derivatives, 1021–25

protecting groups and, 655, 965–68

resonance effects on pK_a, 309

retrosynthetic analysis, 873–74, 876

secondary (2°) alcohols, 387

sulfonation reactions, 1059–61

synthesis by oxymercuration, 616–18, 630, *631*

tertiary (3°) alcohols, 387

tertiary alcohols from organometallic reagents and carboxylic acid derivatives, 1032–33

trivial names, 386–87

water solubility, 94–95

in Williamson ether synthesis, 538, 872, 932, 961, 965–66

See also specific types

alcoholysis reactions, 1047, 1048, 1049

aldehyde hydrates, 979

aldehydes

acetals as protecting groups, 899, 963–65, 968

aldol reactions, 908–10, 914–17

alkylation of α carbon, 523–28, 960–61

alkylation of enolates, 524–25

carbonyl (C=O) stretch absorption bands in IR spectroscopy, 738–39, 740, 749–50, 752

carbonyl hydration by nucleophilic addition, 843–44, 892–93

catalytic hydrogenation, 974, *975*

chromic acid oxidations, 979

C—H stretch absorption bands in IR spectroscopy, *740*, 752

Clemmensen reduction, 957–58, 1129, 1182

conversion to alcohols by hydride reagents, 845–48, 849–50, 1023–24

conversion to alkenes in Wittig reaction, 859–61

crossed aldol reactions, by aldehydes with no α hydrogen, 915–16

cyanohydrin formation, 842, 893–94, 952–53

from diisobutylaluminum hydride reduction of esters, 1029–32

enamine formation by secondary amine addition, 902

halogenation of α carbon, in acid, 532–33, 1057

halogenation of α carbon, in base, 528–32, 1057

hemiacetal formation from reaction with alcohols, 889–92, 897, 899, 929

from hydroboration of alkynes, 631–32

imine production from NH_3 or primary amine addition, 900–902, 903, 928

lack of leaving groups for nucleophilic addition–elimination reactions, 1001

from lithium tri-*tert*-butoxyaluminum hydride reduction of acid chlorides, 1029–30

magnetic anisotropy, 788

nomenclature, *379*, 384–86, 389, *504*

polar π bond in, *840*

polyhalogenation of α carbon, in base, 530–31

from pyridinium chlorochromate oxidation of primary alcohols, 979–80

Raney-nickel reduction, 957, 958–59, 1182

reductive amination, 927–28

regioselectivity of α alkylations, 525

relative reactivity of ketones and aldehydes in nucleophilic addition, 842–44

selective reduction by sodium borohydride, 1023–24

thioacetal formation from reaction with thiols, 899, 958

trivial names, 389

Wolff–Kishner reduction, 906–7, 955–57, 958–59, 1182

See also aldol reactions

Alder, Kurt, 1198

aldohexoses, 199, 248, *249*

aldol (β-hydroxycarbonyl), defined, 908

aldol reactions

aldol addition

aldehydes as reactants, 908–13

defined, 908

β-hydroxycarbonyl product, 908, 910–11, 926

ketones as reactants, 913–14, 918–19

nitriles and nitroalkanes, 922–23

overview, 908–10

reaction mechanisms, 909–10

self-aldol addition, 914, 915, 916

aldol condensation

dehydration, 911–12, 916–17, 923, 924, 925, 927

E1cb mechanism, 911, 912, 923, *924*

E1 reaction under acidic conditions, 912–13

β-hydroxycarbonyl as reactant, 911, 917

overview, 911–13

α,β-unsaturated carbonyl as product, 911–13, 917, 924–25, 926

atom efficiency, 908

crossed aldol reactions, 914–19, 1073–74

five- and six-membered rings as products, 919, 921–22, 924–25

green chemistry and, 908, 914

intramolecular aldol reactions, 919–22, 924, 929–30

organic synthesis guidelines and overview, 925–27

1,3-positioning of functional groups in products, 954

retrosynthetic analysis, 910, 917, 926–27

Robinson annulation, 924–25, 1227

in synthesis, 910, 917, 925–28

See also nucleophilic addition to polar π bonds

aldopentoses, 199, 248, *249*

aldoses, 199, 248–50

See also carbohydrates; monosaccharides

aldosterone, 108, *109*

aldotetroses, 248, *249*

aldotriose, 247, 248, *249*

alkanes

all-gauche conformation, 176

branched alkanes (alkyl-substituted alkanes), 58–60

chemical shift in ^{13}C NMR, 808

C—H stretch absorption bands in IR spectroscopy, *740*, 742, 753–54, *755*

combustion, 1247

isomers and numbering systems, 55–58, 191

mass spectrometry, 1296–98

nomenclature, 53–62, 64–65

overview, 34

radical halogenation, 1260–74

straight-chain alkanes (linear alkanes), 53–54

substituted alkanes, nomenclature, 54–58

trivial names, 64–65

zigzag conformation, 174, 176

See also specific types

alkanoate group, 508

alkene metathesis, 986–88

allyl cation *(cont.)*
  highest occupied molecular orbital (HOMO), 689
  lowest unoccupied molecular orbital (LUMO), 689
  molecular orbital energy diagrams, *688, 689*
  molecular orbitals, 686–91
  nodal planes, 687–88
  nonbonding molecular orbital, 688
  partial positive charges on terminal carbons, 684, 691
  planarity, 684, 690
  resonance-stabilized charge, 469, 1254
  resonance structures, 683–84
  shortcomings of valence bond theory, 683–84, 690–91
  σ molecular orbitals, 689
allyl chloride, 469
allyl radical, 1254–55, *1256*
allyl substituents (2-propenyl substituents), *160*
α (alpha) carbon, defined, 350
α,β-unsaturated carbonyl. *See under* carbonyl compounds
α-bromo carboxylic acids, 1057–59
α-cleavage (mass spectrometry), 1301–3, 1304–5
α-elimination, 606
α-helix, 1357–58, 1359
aluminum chloride (AlCl₃)
  electron deficiency of Al, 338
  in Friedel–Crafts acylation, 1118–21, *1125–26, 1128*
  in Friedel–Crafts alkylation, 1111–16, 1161, 1163–64
  green alternatives to, 1112
  as Lewis acid catalyst, 1110, 1114, 1118–19
amides
  acid-catalyzed amide hydrolysis, 1065–66
  from aminolysis reactions, 1052–54, 1082–83
  carbonyl (C=O) stretch absorption bands in IR spectroscopy, 740, 749–50
  carboxamides, 506
  catalytic hydrogenation of 1° amides, 975
  conversion to carboxylic acids, 1013–15
  formation by acyl substitution of esters, 1007
  hydrogen bonding, 742
  leaving groups, *840, 1000*
  N—H stretch absorption bands in IR spectroscopy, 740, 747–49
  from nitrile hydrolysis, 904–5
  nomenclature, *379, 504*, 506, *512*
  in peptide linkages, 1010, 1084, 1087, 1355–56
  polar π bond in, *840*
  primary (1°) amides, *505*
  reduction by hydride reagents, *1023*
  reduction by LiAlH₄, *1023*, 1026–27
  resonance stabilization, 1051
  secondary (2°) amides, *505*
  stability, *1007*, 1010, 1084, 1184
  tertiary (3°) amides, *505*
amines
  alkylation reactions, 520–23, 927
  charge stability and pK_a of, 309
  chemical shift in ¹³C NMR, 808
  equivalents used in aminolysis reactions, 1053–54

Gabriel synthesis of primary amines, 1015–17
  hydrogen bonding, 742
  lack of coupling in NMR spectra, 796–97
  mass spectrometry, 1302
  N—H stretch absorption bands in IR spectroscopy, 740, 747–49
  nomenclature, 379–81, 383, 388–89, *504*
  nucleophilic substitution or elimination reaction (resistance to), 463
  primary (1°) amines, 388, 520–21, 748, 749, 762, 1015–17
    protecting groups for, 1082–83
    from reduction of imines, 928
    reduction of nitrobenzenes to aromatic amines, 1128–30, 1182
    resonance effects on pK_a, 309
  secondary (2°) amines, 388, 521, 748, 749, 762, *1016*
  synthesis by reductive amination, 927–28
  synthesis from alkyl halides and ammonia, 520–23, 927, 1016
  tertiary (3°) amines, 388, 521, 748, 749, *1016*
  trivial names, 388–89
amino acids
  α-amino acids, defined, 38, 511
  alpha carbon, 38
  chirality, 246–47
  constitutional isomers, 198
  D/L system for classifying enantiomers, 247
  electrophoresis, 317–20
  hydrophilic amino acids, 1359
  hydrophobic amino acids, 1359
  ionizable side chains, 315, 317, 318
  isoelectric pH, 318
  isoelectric point (pI), 318
  naturally occurring amino acids, *39*
  peptide linkages, 1010, 1084–85, 1088–89, 1355–56, 1358
  pK_a values, 315, *316*, 318
  protecting groups for, 1088–89
  residues in proteins, 1085, 1356
  side chains (R groups), 38–40, *316*
  structure, *38, 39, 314*
  structure, as function of pH, 314–17
  zwitterions, 315, 317, 318–19
  See also proteins
2-aminoethanol (ethanolamine), 73, *74*, 425–26
(2*S*,3*E*,5*R*)-5-aminohex-3-en-2-ol, *384*
aminolysis reactions, *1045*, 1052–54, 1082–83
3-amino-2-(2-methylpropyl)-hex-5-ene-1,4-diol, *382*
5-aminopentane-2,4-diol, *382*
5-aminopentan-1-ol, 381
ammonia
  alkylation reactions, 520–23, 1016
  borane–ammonia complex, 625
  Brønsted–Lowry base, 277
  deprotonation, 292–94
  nucleophilicity, 474, 520
  pK_a, 292, 293
  poor leaving group, 463
  reaction with oxirane, mechanisms, 425–26
  solvated electrons in, 1279–80
  uses, 277
ammonium cyanate, 2

ammonium ion
  deprotonation, 292–93, 295
  pK_a, 281, *292*
  proton transfer equilibrium in water, 295
  quaternary ammonium ion formation, 521, 522–23, 551, 554, *1016*
  relative stabilities of H₃O⁺ and H₄N⁺, 295
amorphous solids, 1344–45
amphetamines, 750, 808
amylopectin, 492, 1360–61
amylose, 491, *492*, 1360–62
analytes, defined, 725
androsterone, 174
angle strain, 74–75, 178–79, *180*, 705
aniline (phenylamine or aminobenzene)
  benzenediazonium ion production, 1130–31
  conversion to amide, 1184
  conversion to benzenediazonium ion, 1172
  electrostatic potential map, *1158*
  incompatibility with Friedel–Crafts reactions, 1163–64, 1184
  NH₂ activating group, 1158–59
  nomenclature, 388
  pK_a, 280, 309
  polyurethane from, *1104*
  production from benzene, *1104*, 1129, *1130*
  production from chlorobenzene, 1176–77, *1178*
  production from nitrobenzene, 344, 1129, *1130*
  reaction with benzoyl chloride, 1129, *1130*
  resonance in conjugate base, 309
  structure, *309*, 388
  uses, 310, 344
anilinium ion, 1163
anionic polymerization, 1320–21, 1322, 1323–24
anions, defined, 5
anisole (methoxybenzene), 160, 387, 535
annulenes ([*n*]annulenes), 705
anodes, defined, 317
anomeric carbons, 931
anomers, defined, 931
anthracene, 706, 707
anti addition, 607–12, *617*
antiaromatic compounds, defined, 699
antiaromatic ions, 710
antiaromatic transition states, 1201, 1230
antibonding molecular orbitals, *125, 126, 127*, 686, 688
anti conformation, *171*
anticoplanar (antiperiplanar) conformation, 417–20
anti-Hückel numbers, 699–700, 705, 710, 713
anti-Markovnikov addition, 573, 623–31, 656, 1275–77, 1284–85
  See also Markovnikov addition
anti-Zaitsev (or Hofmann) products, 484, 551, 659–60
Apheloria corrugate, *839*
aprotic solvents
  anions not strongly solvated, 471
  common aprotic solvents, *101*, 471, *472*
  competition between S_N2, S_N1, E2, and E1 reactions, 471–73, 479
  defined, 100
  relative nucleophilicity of common species in, 473–74

benzenesulfonic acid, 1122, *1160*

benzo[*a*]pyrene, 624

(+)-benzo[*a*]pyrene-7,8-dihydrodiol-9,10-epoxide, 624

(+)-benzo[*a*]pyrene-7,8-epoxide, 624

2-benzofuran-1,3-dione (phthalic anhydride), *512*

benzoguanamine, 855

benzoic acid (phenylmethanoic acid)
    from benzamide, 1057
    benzoyl chloride formation, 1057
    from benzyl alcohol oxidation, *982*
    as meta director, 1181
    nomenclature, *510*
    from oxidative cleavage of phenylethene, *1222*
    pK_a, 290
    structure, *290*, *510*, *1126–27*
    synthesis of benzoic acids by oxidation, 1126–28, 1181

benzoic ethanoic anhydride, *510*

benzonitrile, *512*, 855

benzophenone (diphenylmethanone, diphenyl ketone), 389, 390, 754, 1067, *1121*

benzoyl chloride (benzenecarbonyl chloride)
    alcoholysis reactions, 1047, 1048
    amide production by reaction with aniline, 1129, *1130*
    aminolysis reactions, 1052–53
    ester production by acyl substitution, 1006, 1046–47, 1129, *1130*
    formation from benzoic acid, 1057, 1129, *1130*
    nomenclature, *512*
    reaction with Grignard reagents, 1032–33
    reduction by hydride reagents, 1021–22, 1029–30
    See also acid chlorides

benzoyloxyl radical, *1311*, 1312, 1314, 1316

benzoyl peroxide, 1006, 1311, *1313*, 1314, 1319, 1334

benzyl alcohol (phenylmethanol)
    from benzaldehyde reduction, 845, 849
    from benzoyl chloride reduction, *1021*
    conversion to methoxyphenylmethane under basic conditions, 422–23
    Fischer esterification reaction, 1088
    oxidation to benzaldehyde, 980
    oxidation to benzoic acid, 982
    structure, 387
    uses, 387

benzyl cation, resonance-stabilized charge, 469, 1254

benzyl chloride, 402, 808

benzyl chloroformate, 1088

benzylic carbon, reactivity, 1127

2-benzyl-3-phenylpropanal, 654, *654*

benzyl radical, 1254–55

benzyl substituents (Bn; phenylmethyl group), 159

benzyne intermediates, 1176–78

β (beta) carbon, defined, 863–64

β elimination, 342

betaines, 860

β-pleated sheets (β-sheets), 1357–58, 1359

bicyclo[2.2.1]hept-2-ene (norbornene), 1206

bicyclo[2.2.2]octadiene, *1206*

bimolecular elimination steps. *See* E2 reactions

bimolecular homolytic substitution (S_H2), 1259, 1262, 1263, *1272*, *1275*, 1278

bimolecular nucleophilic substitution. *See* S_N2 reactions

biodiesel, 1005

biomolecules
    biological macromolecules, 1355–62
    biopolymers, 1355–62
    chirality, 214, 236, 245–47
    defined, 37
    polynucleotides, 1355
    polypeptides or polyamino acids, 1355–60
    polysaccharides, 40, *41*, 105, 1355
    See also carbohydrates; lipids; nucleic acids; proteins

biphenyl, 706

Birch reduction, 1282–83

bisulfate anion as leaving group, 459

boat conformation, 182

Body Worlds, *515*

boiling points
    dipole–dipole interactions and, 83–84
    intermolecular interactions and, 83–84, 96–98
    nonpolar compounds, *89*
    ranking boiling points of compounds, 96–98
    of representative compounds, *81*

Bolinus brandaris, 3

bond dipoles
    alkyl groups, 305
    defined, 16
    dipole moment, 78
    electronegativity, 16–17, 78–79
    electrostatic potential map, defined, 17
    and net molecular dipole, 78–80
    vector addition, 79

bond energy (bond strength), defined, 8

bond length, defined, 8

9-borabicyclo[3.3.1]nonane (9-BBN), 632

borane
    alkylborane formation, 627
    bond dipoles, *17*
    borane–ammonia complex, 625
    borane–dimethyl sulfide complex, 625
    borane–tetrahydrofuran complex, 623, 625
    dialkylborane formation, 627
    diborane, 623, 625
    disiamylborane, 631–32
    electrostatic potential map, *17*
    octet rule exception, 14
    trialkylborane formation, 627, *628*
    trialkylborane oxidation, 628–31
    uses, 14
    See also hydroboration

borate esters, 629

borax, 1336

Bordwell, F. G., 454

botulism, 1132

branched polymers, 1330–32, 1347–48, 1351, *1352*, 1360, *1361*

Breathalyzer test, 981

bridgehead carbons, 1206

broadband decoupling, 806–7, 808, 809, *810*, *811*

bromine
    aromatic bromination, 1109–10
    Br$_2$ electrophilic addition to alkenes, 607–9, 613–14

Br$_2$ electrophilic addition to alkynes, 610–11

Br$_2$ nucleophilic substitution, 528–30, 533, 1057

Br$^-$ as leaving group, 335, 410, 413, 415, 459, 516, 1059

Br$^-$ as nucleophile, 448–49, 473–74

bromination of benzene, 1109–10, 1126

bromination of phenol, 1162, 1185

bromination reaction mechanism, 608

bromonium ions, 608, 609, 611–12, 613, 615, 617
    induced dipole, 529, 608
    mass spectrometry, 825
    radical bromination using *N*-bromosuccinimide, 1251, 1271–74, 1283–84
    solvent effects in bromination reactions, 611, 662

bromine radical
    bimolecular homolytic substitution (S_H2), 1259, 1263, *1272*, *1275*, 1278
    Lewis structure, *1247*
    from *N*-bromosuccinimide (NBS), *1251*, 1271–74, 1283–84
    production by homolysis, 1248, *1249*, 1250, 1263
    radical halogenation of methane, 1263–64
    selectivity of chlorination and bromination, 1267–71

bromobenzene
    from aniline, 1130
    from bromination of benzene, 1109, 1117, 1126, 1133
    dissolving metal reduction, 1126
    Grignard reagent production, 948, 1110, 1117, 1126
    mass spectrum, 825
    structure, *948*, *951*, *1130*
    uses, 948

1-bromobutane, 781

2-bromobutane (*sec*-butyl bromide), 66

4-bromobutan-1-ol, 537

1-bromo-2-chlorobenzene (*ortho*-bromochlorobenzene), *158*

trans-1-bromo-2-chlorocyclohexane, 613

3-bromo-4-chloro-1,1-dimethylhexane, 61–62

(*S*)-1-bromo-1-chloroethane, 260–61

bromochlorofluoromethane (CHFClBr)
    chirality, 214, 215, *216*, 217, 219
    enantiomers, 210, *213*
    (*R*)-1-bromo-1-chloro-(*S*)-4-fluoro-4-nitrobutane, 261–62

bromocyclohexane, 454

trans-2-bromocyclohexanol, *611*

3-bromocyclohexene, 1271, *1272*

1-bromo-1-cyclopentylpentane, 877

2-bromo-1-cyclopentylpentane, 877, *878*

(*R*)-6-bromo-1,3-dichlorohexane, 263–64

2-bromo-1,1-dimethylcyclobutane, 219

(1*S*,2*R*)-1-bromo-1,2-diphenylpropane, *414*, *416*

bromoethane, 520, *650*, 651

bromoethene, 801–2

1-bromo-4-ethylbenzene, 799

m-bromoethylbenzene, 1182

3-bromo-3-ethylpentane, 479–80

1-bromo-3-fluoropropane, *58*

bromoform (tribromomethane, methyl tribromide), *66, 1017,* 1018

3-bromoheptane, *648*

1-bromohexane, *646*

bromohydrins, 611–12, 613–14, 662

bromomethane, 56, 1250

bromomethoxymethane, *652*

(bromomethyl)benzene (benzyl bromide, bromophenylmethane), 654, 973–74

(*R*)-(bromomethyl-*d*)-benzene, 482

1-bromo-4-(1-methylethyl)benzene, 825

2-bromo-4-methylhexane, 237

2-bromo-2-methylpropane, 536

bromonium ion intermediate, 608, 609, 611–12, 613, 615, 617

(*R*)-2-bromopentane, 479

o-bromophenol, *1185*

p-bromophenol, *1185*

2-bromo-1-phenylethan-1-ol, *613, 614*

2-bromo-2-phenylethan-1-ol, *613*

bromophenylmethane (benzyl bromide, (bromomethyl)benzene), 654, 973–74

2-bromo-2-phenylpropane, 643–44

2-bromo-3-phenylpropanoic acid, *1059*

1-bromopropane, 222, 670, 971, *1268,* 1275, 1276

2-bromopropane, 476, 488, 861, *948, 1268,* 1275

3-bromopropan-1-ol, *654*

3-bromoprop-1-ene (allyl bromide), *160*

N-bromosuccinimide (NBS), 1251, 1271–74, 1283–84

Brønsted–Lowry acid–base reactions. *See* acid–base chemistry; proton transfer reactions

Brønsted–Lowry acids

  addition to alkenes, 565–67, 569–73, 601

  addition to alkynes, 578–81, 601

  defined, 275

  See also acid–base chemistry; proton transfer reactions

Brønsted–Lowry bases, defined, 275

  See also acid–base chemistry

buckminsterfullerene, 5

buta-1,3-diene

  1,2-addition and 1,4-addition, 583–84, 586–87, 660, *661*

  antibonding π molecular orbitals, *692,* 693

  bonding π molecular orbitals, *692,* 693

  bond lengths, 685, 694–95

  Brønsted acid addition, 583–84, 586–87

  conjugated *p* orbitals, 692

  Diels–Alder reactions, 1198–99, 1200–1201, 1203–4, 1206, 1208, 1226

  electrophilic addition reactions, 583–84, 586–87

  excited electronic state, 1234

  highest occupied molecular orbital (HOMO), 693

  HOMO–LUMO transitions and energies, 729, 731–32

  λ_{max} in UV–vis spectrum, 726, *727,* 728, 730, *731*

  lowest unoccupied molecular orbital (LUMO), 693

  molecular orbital energy diagrams, *692, 693, 729, 1228, 1231*

  molecular orbitals, 692–95, 1228–29, *1231*

monomer in polybutadiene, *1340,* 1342

monomer in styrene–butadiene rubber (SBR), 1341–42

nodal planes, 692–93

photon absorption, MO energy diagram, *729*

planarity, 685, 692, 694

reaction with ethene, 1198–99, 1203–4, 1208, 1226

reaction with ethyne, 1200–1201

resonance structures, 685

s-cis and *s*-trans conformations, 1203–4

shortcomings of valence bond theory, 685, *686*

σ molecular orbitals, 693

structure, *584*

thermodynamic and kinetic control of addition reactions, 586–87

uses, 584, 685

UV–vis absorption by conjugated double bonds, 730

See also conjugated dienes

butanal (butyraldehyde)

  aldol reaction, 909–10

  chlorination, 531

  proton nuclear magnetic resonance spectroscopy, 802

  structure, *389*

  in Wittig reaction, 862–63

butan-1-amine (*n*-butylamine), *388*

butan-2-amine (*sec*-butylamine), 551

butane (*n*-butane), 64, 213

butanedioic acid (succinic acid), *511*

(2*S*,3*S*)-butane-2,3-diol, *237*

meso-butane-2,3-diol, *237*

butanediol fermentation, 219

butanoic acid (butyric acid), 200, *510,* 646, *1047*

butan-1-ol, 92–93, 461–62, *542*

butan-2-ol, 76, 219, 235, 845, 850

butanone (methyl ethyl ketone, 2-butanone, butan-2-one), 76, 389, 390, 845, 849, 889, 1075

butanoyl chloride, *1047*

(*Z*)-but-2-en-1,4-diol, *973*

but-1-ene

  acid-catalyzed hydration, 576–77

  from Hofmann elimination reaction, 551–52

  structure, 190–91, *551, 1314*

but-2-ene

  trans-but-2-ene epoxidation, *621*

  diastereomers from E2 reaction, 417–18, 419–20

  (*E*)-but-2-ene reaction with Cl₂, *611*

  electrophilic addition of acid, 584

(*E*)-butenedioic acid (fumaric acid), *511*

(*Z*)-butenedioic acid (maleic acid), *511*

trans-but-2-ene oxide, *621*

(*E*)-but-2-enoic acid, *1218*

but-3-en-1-ol, 235

butenone (methyl vinyl ketone), *389,* 390

Butler, A., 615

tert-butoxide anion

  anti-Zaitsev product, 484

  E2 reactions favored over S_N2 reactions, 455–56, 480–81, 960

  nucleophilicity, 455–56

  regioselective α alkylations, *525, 526, 527,* 658, 961

steric hindrance, 456, 465

as strong base, 455, 658

tert-butoxycarbonyl (Boc) group, 1083, 1088–89

butylbenzene, 1117, 1299–1300

sec-butylbenzene, *750,* 755

tert-butylbenzene, *1112*

butyl cation, 1296

tert-butyldimethylsilyl chloride (TBDMSCl), *966,* 967

butyl (*n*-butyl) group, *58, 64*

butyllithium, 2, 854, 861, 1320

tert-butylbenzene, *1112*

but-1-yne, 652

but-2-yne (2-butyne, dimethylacetylene), 33

but-2-yne-1,4-diol, *973*

butyric acid (butanoic acid), 200, *510,* 646, *1047*

butyryl acyl carrier protein (ACP), 1077

butyryl synthase, 1077

C

Cahn–Ingold–Prelog system, 258–73

  See also R/S designation; *Z/E* designation

cancer

  benzo[*a*]pyrene as procarcinogen, 624

  detection with mass spectrometry, 828

  DNA alkylation, 547

  epoxide binding to DNA, 624

  eriolangin reaction with DNA polymerase, 869

  imaging with gold nanoparticles (GNP), 11

  mechlorethamine chemotherapy, 547, 869

  Michael reactions, 869

  thalidomide for multiple myeloma, 209

carbamates, *1082,* 1083, 1088

3-carbamoylbutanedioic acid, *505*

carbanions

  defined, 333

  organometallic compounds as carbanions, 333, 451, 854

  as strong nucleophiles, 451–52

carbenes, 604–6, 607, 644, 666

  See also cyclopropane

carbocation rearrangements

  1,2-alkyl shifts, overview, 346

  charge stability and total bond energy as driving forces, 349, 429

  curved arrow notation, 346

  E1 reactions, 346, 431–32

  electron-rich and electron-poor species, 346

  electrophilic addition reactions, 346, 573–74

  as fast reactions, 430, 573

  in Friedel–Crafts acylation, lack of, 1120, 1125

  Friedel–Crafts alkylation, 1114–16, 1120, 1125

  frontier molecular orbital theory, 375

  general rule for reasonableness of mechanisms, 429–32

  HOMO–LUMO interactions, 375

  1,2-hydride shifts, 346, 375, 430–31, 573, 1115–16

  mechanisms, 345–47, 429–32

  1,2-methyl shifts, 346, 430–31, 573–74

  overview, 345–47

  in primary alkylbenzene synthesis, 1125–26

  S_N1 reactions, 346, 429–31, 517

carbocations

  tert-butyl carbocation (CH₃)₃C⁺, 371, 400,
      1112
  comparison to free radicals, 1254, 1257
  coordination reactions, 371, 410, 429
  defined, 305
  electrophile elimination steps, 344, 375
  electrophilic addition steps, 343
  electrophilic addition to alkenes, *565, 569–71,
      573–74, 576–77, 584–85, 589–94, 601*
  electrophilic addition to alkynes, 579–80, 582,
      601
  in Friedel–Crafts alkylation, 1112–14
  hyperconjugation and stability, 305, 364–65
  inductive effects and stability, 305, 364
  instability of methyl and primary
      carbocations, 469, 478, 539, 1115–16, 1125
  intermediates in E1 mechanisms, 399, 415,
      429
  intermediates in S_N1 reactions, 394–95, 396,
      407–8, 410, 429, 432–33, 517
  as Lewis acids, 337, 345
  methyl cation, 74, 305, 364, *365*, 1257, 1296
  octet rule violation, 30, 345, 429
  primary (1°) carbocations, 305, 364–65
  resonance-stabilized intermediates, 432–33,
      469–70
  secondary (2°) carbocations, 305, 364
  stability in superacids, 307
  stability order, 305, 364
  tertiary (3°) carbocations, 305, 364
  See also carbocation rearrangements;
      Markovnikov addition; *specific reaction
      mechanisms*

carbohydrates

  acetal groups in carbohydrates, 490, 491
  anomeric carbons, 931
  α and β designations, 490
  cyclic and acyclic forms, 42, 929
  diastereomers, 931–32
  D/L system for classifying, 247–48
  glycolysis, 40, 354
  glycosidic linkages in complex carbohydrates,
      491–92, 493, 1355, 1361–62
  overview, 37, 40–42
  polysaccharides, 40, *41*, 105, 1355, 1360–62
  simple sugars, 40
  size hierarchy, *41*
  stability of five- and six-membered rings,
      929–30
  See also glycosides; monosaccharides; *specific
      types*

carbon, 3–4, 5

carbon-13 (¹³C) isotope, 804–8, 823, 1178
  See also carbon nuclear magnetic resonance
      spectroscopy; mass spectrometry

carbon dioxide

  boiling point, 70
  bond dipoles, *17, 78, 79*
  carboxylation reaction with Grignard
      reagents, 1126
  electrostatic potential map, *17*
  greenhouse gas, 9, 135, 1005
  infrared radiation absorption, 9
  net molecular dipole, *78*
  nonpolar molecule, *78, 79*
  organic/inorganic classification, 2

solubility, 70
structure, *70*

carbon nanotubes, 5, 147

carbon nuclear magnetic resonance spectroscopy
    (¹³C NMR)
  broadband decoupling, 806–7, 808, 809, *810,
      811*
  ¹³C NMR signal, 804–6
  chemically distinct carbon atoms, defined,
      804–6
  chemical shifts (δ), 807–8
  1-chloropropane, 804, *805,* 806
  DEPT (distortionless enhancement by
      polarization transfer) ¹³C NMR, 809–10,
      811
  deshielding, 808
  effects of electronegative atoms, 808, 811
  ethylbenzene, 809, *810, 811*
  gyromagnetic ratio (γ), 804
  integration of signals, 808, *809*
  2-methylbutanal, 808, *809*
  nuclear Overhauser effect, 808
  overview, 776, 804
  shielding, 808
  signal averaging, 804, *805*
  signal splitting, 806–7
  structure elucidation, 811, 815–18
  tetramethylsilane (TMS) standard, 804, 807,
      816
  See also nuclear magnetic resonance
      spectroscopy

carbon nucleophiles, 451–52, 542, 853, 865

carbon tetrachloride. *See* tetrachloromethane

carbonyl compounds
  β-hydroxycarbonyls
      as aldol addition product, 908, 910–11,
          926
      as aldol condensation reactant, 911, 917
      1,3-positioning of heteroatoms, *953*
  mass spectrometry, 1304–5
  reactivities of carbonyl-containing species,
      1050
  reduction of C=O groups to CH₂, 955–59
  α,β-unsaturated carbonyls
      addition by lithium dialkylcuprates
          (Gilman reagents), 869–71, 876, 956–57,
          961, 983
      as aldol condensation product, 911–13,
          917, 924–25, 926
      conjugate addition (1,4-addition),
          mechanisms, *864, 895*
      direct addition (1,2-addition), mechanism,
          864
      direct *vs.* conjugate addition, 863–68,
          869–72, 874–76, 895–96, 961
      electrostatic potential map, *864*
      Michael reactions, 865, 869, 924
      regioselectivity involving 1,2- *vs.*
          1,4-addition, 865–67, *868,* 870–72, 876,
          961
      resonance structures, 863, *864*
      reversibility in nucleophilic addition,
          865–67
      as Robinson annulation reactant, 925
  β,γ-unsaturated carbonyls, 912–13
  See also acid anhydrides; aldehydes; amides;
      carbohydrates; carboxylic acids; esters;

ketones; nucleophilic addition to polar π
    bonds

carbonyl group, defined, 34

carboxylate anions
  as nucleophile in S_N2 reactions, 534
  in nucleophilic addition–elimination
      reactions, 1007, 1010, 1018
  resistance to nucleophilic attack, 1007, 1010–
      11, 1025, 1051
  resonance delocalization of negative charge,
      1007, 1010–11, 1025, 1051, 1068
  in soaps and detergents, 101–3
  stability and low reactivity, *1008,* 1010–11
  as weak bases, 455

carboxylation reactions, 857, 1126

carboxylesterases, 1220

carboxylic acid derivatives
  α-bromo carboxylic acids, 1057–59
  aminolysis reactions, 1052–54
  free energy diagrams, nucleophilic addition,
      1051
  leaving groups in nucleophilic addition–
      elimination reactions, *1000, 1001, 1008*
  leaving groups in nucleophilic addition
      reactions, *840*
  nomenclature, 503–14
  organometallic reagents and, 1032–35
  rates of hydrolysis, 1049–52
  reduction by hydride reagents, 1021–28
  reduction by specialized reducing agents,
      1029–32
  relative reactivities, 1049–52
  resonance stabilization, 1003–4, 1051
  stability ladder, 1007, *1008,* 1010, 1011, 1013,
      1023, 1025
  trivial names, 511–12
  See also acid anhydrides; acid chlorides; acid
      halides; amides; esters; nitriles; nucleophilic
      addition–elimination reactions

carboxylic acids
  from amides, 1013–15, 1057
  carbonyl (C=O) stretch absorption bands in
      IR spectroscopy, 740, 749–50
  from chromic acid oxidation of primary
      alcohols, 978–79
  conversion to acid chlorides, 1054–57
  diazomethane formation of methyl esters,
      533–35, 675, 1020
  dimer formation, 746
  from ester hydrolysis, 1064
  Fischer esterification reaction, 535, 1064–65,
      1083, 1088, 1328–29
  from haloform reactions, 1017–20
  from hydrolysis of acid anhydrides, 1049
  β-keto acids, *511,* 1079
  nomenclature, *379,* 503–5, *504,* 506–7,
      510–11
  O—H stretch absorption bands in IR
      spectroscopy, *740,* 746
  from permanganate oxidations of primary
      alcohols, 982
  polar π bond in, *840*
  protecting groups for, 1082–84
  reduction by LiAlH₄, *1023*
  resistance to reduction by sodium
      borohydride, *1023,* 1024–25
  resonance effects on pK_a, 309

Markovnikov addition, 584
  regiochemistry, 584–86, 660, *661*
  UV–vis absorption by conjugated double
    bonds, 730
  See also buta-1,3-diene; Diels–Alder reactions
conjugated double bonds, defined, 583, 710
conjugated linoleic acids (CLAs), 697
conjugated orbitals, defined, 686
connectivity, 12, 190, 771
constitutional isomers
  amino acids, 198
  for C₄H₈F₂, 196–97
  for C₃H₆O, 197–98
  double bond location in alkenes, 191
  drawing all for a given formula, 195–98
  index of hydrogen deficiency (IHD), 193–95
  monosaccharides, 198–99
  overview, 190–92, 198, 779
  physical and chemical properties, 234–35
constructive interference
  atomic orbitals (AOs), 122–26, 128–29, *130*,
    137, 682
  frontier molecular orbital (FMO) theory,
    369–70
  hybrid atomic orbitals, 128–29, *130*, 137
  hyperconjugation, 365
  multiple-center molecular orbital theory, 687
  two-center molecular orbital theory, 122–24,
    682, 686
contact surface area, 70, 89, 90, *91*
convergent synthesis, 672–73
coordination (bond forming) steps
  acid-catalyzed dehydration of alcohols,
    538–40
  carbocations, 371, 410, 429
  charge stability and total bond energy as
    driving forces, 348
  curved arrow notation, *337*, 338
  electron-rich and electron-poor species, 338,
    371, 947
  frontier molecular orbital theory, 371
  HOMO–LUMO interactions, 371
  Lewis acids and bases, 337, 371
  overview, 337–38
  reaction mechanisms, 337–38
  S_N1 reactions, 394, 399, 410, *424*, 429
  See also reactions that alter the carbon
    skeleton
copolymers, 1341–42
copper, electronegativity, 332
core electrons, 7–8
Corey, Elias J., 650
corn, polymers made from, 1354
cortisone, 108, *109*
coumarone-indene resin, 571
coupling constants (*J*), 798–804, 814
  See also spin–spin coupling
coupling reactions
  defined, 983
  Heck reaction, 985–86
  organocuprates and, 983
  palladium-catalyzed reactions, 984–86
  stereochemistry, 983, 984, 985
  Suzuki reaction, 984–85
covalent bonds
  bond energy (bond strength), 4, 8–11
  bond length, 8–9, *10*, 11

defined, 8
formal charge method of assigning electrons,
  19–21
formal charges on atoms, various bonding
  scenarios, *30*
nonpolar covalent bonds, 16
number of bonds and lone pairs for selected
  atoms, *15*
polar covalent bonds, 16–18, 78–79
spring model of, 9
valence shell atomic orbitals, 125
See also double bonds; Lewis structures; triple
  bonds
Crafts, James, 1111
Craig, Stephen L., 1333
Creutzfeldt–Jakob disease (CJD), 589
Crick, Francis, 717
Crosby, Alfred, 89
crossed aldol reactions, 914–19, 1073–74
crossed Claisen condensation reactions,
  1073–75
cross-linking and cross-links, 553, *888*, 1332,
  1336–37, *1338*, 1350, 1356–57
crotonaldehyde, 31
crotonic acid, *1318*
crystal lattice, benzene, 1344
crystal lattice, sodium chloride, 18
crystalline solids, defined, 1344
C-terminus of peptide, 1085, 1089
cubane, 183
curved arrow notation
  1,2-alkyl shifts, 346
  carbocation rearrangements, 346
  coordination (bond forming) steps, *337*, 338
  Diels–Alder reactions, 1199–1203, 1217–18
  double-barbed arrows, 276, 1249, 1258
  E2 reactions (bimolecular elimination steps),
    341, 342
  electron-rich and electron-poor species, 330–
    31, 336
  electrophile elimination steps, *344*, 345
  electrophilic addition steps, *343*, 345
  free radicals, 1249, 1255, 1259
  halogenation of α carbon of aldehydes and
    ketones, 530
  homolysis, 1249
  1,2-hydride shifts, 346
  nucleophile elimination steps, *339*, 340–41
  nucleophilic addition steps, *339*, 340
  proton transfer reactions, 276–77, 330–34,
    336, 368
  resonance structures, 25–27
  rules for using, 276
  single-barbed arrows, 1249, 1258
  S_N2 reactions, 336–37, *369*, 370
  See also reaction mechanisms
cyanide anion, 447, *448*, 542, 842, 875, 894, 947
cyanobenzene, 1133
cyanohydrins
  alpha-hydroxy acid formation by hydrolysis,
    893
  carbon–carbon bond formation during
    synthesis, 893
  defined, 842, 893
  formation from aldehydes and ketones, 842,
    893–94, 951–53

HCN addition catalysis by KCN or bases, 894
  and millipedes, *839*
  1,2-positioning of heteroatoms, 951–52, *953*
  as synthetic intermediates, 951–52
3-cyanopropanoyl chloride, 505
cyclic alkanes. *See* cycloalkanes
cycloaddition reactions
  aromatic and antiaromatic transition states,
    1201, 1229–30
  [2+2] cycloaddition, 1201, 1202, 1229–30,
    1233–34
  [4+2] cycloaddition in Diels–Alder reactions,
    1200, 1217, 1229, 1233
  [4+2] cycloaddition in dihydroxylation
    reactions, 1219
  [4+2] cycloaddition in oxidative cleavage
    reactions, 1221, 1225
  [6+2] cycloaddition, 1201
  defined, 1200
  [1,3]-dipolar cycloaddition, 1225
  frontier molecular orbital theory, 1228–34
  oxaphosphetane formation, 860
  See also Diels–Alder reactions
cycloalkanes
  chair conformations of alkylcyclohexanes, 187
  cycloalkyl groups, 60–61
  disubstituted cyclohexanes, cis and trans
    isomers, 188–89
  heats of combustion (Δ*H*°), 175–77
  monosubstituted cyclohexanes, conformers,
    184–87
  nomenclature, 60–62
  ring strain, 174–78, 182
  strain calculation from heat of combustion
    (Δ*H*°), 175–76
  substituted cycloalkanes, nomenclature,
    61–62
cycloalkenes
  nomenclature, 152–55, 271
  ring strain, 271
  structures, *271*
  See also alkenes
cycloalkyl groups, 60–61
cycloalkynes, 152, 154
cycloalkynes, nomenclature, 152–55
cyclobutadiene
  angle strain, *74*, 75
  antiaromatic compound, 698, 703–4
  anti-Hückel number of π electron pairs, 699,
    700
  bond lengths, *698*, 699
  conjugated *p* orbitals, 699, *700*
  free radicals, 703
  instability, 698–99, 703
  molecular orbital energy diagrams, *703*, *704*
  molecular orbitals, 702–4
  nodal planes, 703, *704*
  structure, *698*, 699
  unpaired electrons, 703
cyclobutadienyl dication, 703–4
cyclobutane
  angle strain, 179
  but-1-ene as constitutional isomer, 190–91
  Newman projection, *179*
  ring strain, *175*, 177, 179
  structure, *174*, 179, 190–91

ethanenitrile (acetonitrile), 73, *512*

ethanethiol, 895

ethanoic acid (acetic acid)
　condensed formula, *31*, 32
　diazomethane formation of methyl acetate, 534
　NMR spectroscopy, 782
　pK_a, 290, 301
　resonance effects on acid strength, 298–99
　resonance structures, 25
　structure, *31*, *290*, *510*
　uses, 25
　in vinegar, 25

ethanoic anhydride (acetic anhydride), *512*, *1052*, 1119

ethanol (ethyl alcohol)
　C—H stretch absorption bands in IR spectroscopy, 753–54
　hydrogen bonding, 85–86, 87, 92
　inductive effects from nearby atoms, 302–3, 310
　metabolism of, 194
　nomenclature, *378*, 387
　oxidation to acetic acid, 981
　physical properties, *81*, *87*
　pK_a, 290, 302, 310
　saturation, 193
　S_N2 reaction rates in, 473
　solubility, *81*, 92, 94
　structure, *83*, *193*, *290*, *387*
　uses, 84
　as weak nucleophile, 889

ethanoyl chloride (acetyl chloride), *512*

ethene (ethylene)
　from alkene metathesis, 987
　bonding, 136–38
　[2+2] cycloaddition reaction, 1233–34
　Diels–Alder reactions, 1198–99, 1203–4, 1206, 1208, 1226
　electrostatic potential map, *146*
　energy diagram for MO formation, *137*
　excited electronic state, 1233–34
　and fruit ripening, 1225
　HOMO–LUMO transitions and energies, 731–32, 733
　index of hydrogen deficiency (IHD), *193*, 194
　λ_{max} in UV–vis spectrum, 730, *731*, 733
　Lewis structure, *136*, *138*, *159*
　molecular orbitals, 136–38, 1228–30, 1233–34
　monomer in polyethylene, *1340*
　p atomic orbital overlap, *136*, *137*, 142
　π (pi) symmetry, 137–38
　planar molecular geometry, 119, *136*, 141
　potassium permanganate and, 1225
　reaction with buta-1,3-diene, 1198–99, 1203–4, 1206, 1208, 1226
　relative acidities of ethane, ethene, and ethyne, 297
　restricted rotation about double bond, 141, *142*, 144
　sp^2 hybrid atomic orbitals, *136*, *146*
　uses, 136
　See also alkenes

ethenyl substituents (vinyl substituents), *160*

ether hydroperoxides, 1251

ethers
　from acid-catalyzed alkoxylation of alkenes, 577
　from acid-catalyzed dehydration of alcohols, 538–40
　alkoxy groups, 62–63
　cyclic ethers, nucleophilic substitution, 540–48
　cyclic ethers, synthesis, 537–38
　mass spectrometry, 1302–3
　nomenclature, 62–63, 66–67
　oxetanes, 540–41, 546, 1324
　protonation, 462–63
　stability of five- and six-membered rings, 537, 548
　trivial names, 66–67
　Williamson ether synthesis, 535–38, 539, 651–52, 872, 952, 961, 965–66
　See also epoxides

ethoxide ion, 299, *303*

ethoxyethane (diethyl ether)
　anesthetic use, 66
　nomenclature, *62*, 63, 67
　solubility, 92–93, 94
　as solvent, 66
　from Williamson ether synthesis, 535

ethoxy group, 62

2-ethoxypropane, 488

ethyl acetate (ethyl ethanoate), *103*, *512*, 1011, 1051–52, 1069

N-ethylanilinium chloride, *1065*

ethylbenzene
　NMR spectroscopy, 789–90, 809, *810*, *811*
　structure, *790*
　synthesis by Clemmensen reduction, 957–58
　uses, 789

ethyl benzoate (ethyl benzenecarboxylate), *512*, *1074*, 1128

ethyl butanoate, 508, *646*

N-ethylcyclohexanamine (cyclohexylethylamine), 388

ethylcyclohexane, *34*

3-ethylcyclopentanone, 876

ethylcyclopropane, 906

1-ethyl-2,2-dimethylbutyl group, 59

ethylene. *See* ethene

ethylene glycol (1,2-ethanediol), 87, 899, 963, 965, *1329*, *1330*

ethylene oxide (oxirane), *193*, 194, 425–26, 540–41, 543, 1323

ethyl ethanoate (ethyl acetate), *103*, *512*, 1011, 1051–52, 1069

ethyl formate, 1074

ethyl group (Et), *58*

2-ethylhexanal, 738, *739*

2-ethylhexanenitrile, *648*

2-ethylhexan-1-ol, 753, *754*

ethyl indole-2-carboxylate, 1002

ethylmagnesium bromide, 950

N-ethyl-5-methoxy-N,3-dimethylpentan-2-amine, *383*

N-ethyl-N-methylhept-3-ynamide, *505*

ethyl 4-methylpent-4-enoate, *1029*

ethyl 2-methylpropanoate, *1070*

N-ethyl-2-nitroaniline, 1174

2-ethyl-6-nitro-3-propylhepta-1,3,5-triene, *156*

4-ethyl-2-nitrotoluene, *1167*

4-ethyl-3-nitrotoluene, *1167*

ethyl 2-oxobutanoate (acetoacetic ester), 1078–81

ethyl 3-oxobutanoate (ethyl acetoacetate), 1069

(S)-2-ethylpentanenitrile, 665

3-ethylpent-2-ene, *480*

2-ethyl-4-pentyltoluene, *161*

N-ethyl-N-phenylmethanamide, *1065*

ethyl radical, 1296

ethyl *tertiary*-butyl ether (ETBE), 159

p-ethyltoluene, *1167*

ethyne (acetylene)
　bonding, 140–41
　electrostatic potential map, *146*
　energy diagram for MO formation, *140*
　index of hydrogen deficiency (IHD), *193*, 194
　Lewis structure, *140*, *141*, *159*
　molecular orbitals, 140–41
　p atomic orbital overlap, 140
　reaction with buta-1,3-diene, 1200
　relative acidities of ethane, ethene, and ethyne, 297
　structure, *145*, *159*, *193*
　triple bonds, 140–41
　uses, 140
　See also alkynes

ethynylbenzene, 578–79, 581–83

eucalyptus leaf beetle larvae, *141*

eugenol, 641

excited electronic state, 1233–34

exergonic reactions, 287

exo approach in Diels–Alder reactions, 1212–13, 1232

exo product in Diels–Alder reactions, 1212, 1231

exothermic reactions, 288

expanded octets, 14

extinction coefficient (molar absorptivity, *ε*), 726, 734

extraterrestrial life, search for, 758

F

farnesol, 592

farnesyl pyrophosphate, 592

fats (triacyl glycerols or triglycerides), 105–6, 1005, 1077

fats, synthetic, 946

fat-soluble vitamins, 105

fatty acid methyl esters, 1005

fatty acids
　biological functions and effects, 106
　carboxylate ions, 101–3
　cis conformation of double bonds, 200
　common fatty acids, *107*
　conjugated linoleic acids (CLAs), 697
　essential fatty acids, 200
　in fats, 106
　hydrocarbon tail, 103
　ionic head group, 103
　monounsaturated fatty acids, 200
　names and structures, *107*
　polyunsaturated fatty acids (PUFAs), 200, 697
　salts of fatty acids in soaps, 101
　saturated fats, 199–200

electron-withdrawing groups, 302–3, 305–7, 309

electrophilic addition reactions, *595, 633–34*

electrophilic aromatic substitution, *1135–36, 1187–88*

free radicals, *1287*

hydrocarbon groups, effect on solubility, 94–96

hydrophilic groups, 94

hydrophobic groups, 95

labile functional groups, defined, 960, 962

leaving groups in nucleophilic addition–elimination reactions, *1000*

most common groups, 35

nomenclature for compounds, 377–92

nucleophilic addition–elimination reactions, 1000–1001, *1036–38, 1091–92*

nucleophilic addition to polar π bonds, *879, 934–35*

nucleophilic substitution and elimination reactions, *494–95, 555–56, 652*

overview, 34–37

polar π (pi) bonds in, 840

relative positioning of heteroatoms in C—C bond forming reactions, 951–54

substituents in compounds whose IUPAC names have suffixes, 378, 381, 383, 385–86

suffixes for naming compounds, 377–84

See also reaction tables; *specific types*

functional group transformations (tables)

cataloging reactions and planning synthesis, 647

Diels–Alder reactions, *1235*

electrophilic addition reactions, *595, 633–34*

electrophilic aromatic substitution, *1135–36, 1187–88*

free radicals, *1287*

nucleophilic addition–elimination reactions, *1036–38, 1091–92*

nucleophilic addition to polar π bonds, *879, 934–35*

nucleophilic substitution and elimination reactions, *494–95, 555–56, 652*

in organic synthesis, *989–90*

See also reaction tables

furan, 708, 714, *1168, 1169, 1172*

G

Gabriel synthesis of primary amines, 1015–17

Gallardo, I., 1176

gamma rays, 724

gauche conformation, *171,* 172–73, 185–86

geckos, *70,* 89, *1307*

Geckskin, 89

gel creation from poly(vinyl alcohol), 1336

gel permeation chromatography (GPC), 1319–20

geminal dihalides, 578–80

geraniol, 52, *53,* 111, 591

geranyl pyrophosphate, 590, 591, 592

Gibbs free energy ($G°$)

autoionization of water, 291

Hammond postulate and $\Delta G°_{rxn}$, 447–50

K_{eq} relationship to Gibbs free energy differences, 287, 526

reaction free energy diagram, 288–89

relationship to standard enthalpy difference ($\Delta H°$), 288

relationship to standard entropy difference ($\Delta S°$), 288, 477

reversible and irreversible reactions and $\Delta G°_{rxn}$, 488, *489*

standard free energy of activation ($\Delta G°‡$), 289, 366, 405–6, 488

standard Gibbs free energy difference ($\Delta G°_{rxn}$), 287

See also activation energy

Gilman, Henry, 869

Gilman reagents (lithium dialkylcuprates)

addition to α,β-unsaturated carbonyls, 869–71, 876, 956–57, 961

conjugate addition by, 870–71, 876

coupling reactions with alkyl halides, 983

nucleophilic addition–elimination reactions, 1034–35

simplifying assumptions regarding, 332–33, 870

synthesis, 949

glass transition temperatures (T_g), 1344–45, 1346–47, 1348–49

D-glucose

chiral centers, 246

constitutional isomers, 198–99

conversion to glycoside, 490–91

Fischer projection, *231, 929*

α-D-glucopyranose, *490, 492,* 931–33

β-D-glucopyranose, 490, *492,* 931–33

glycolysis, 354

mechanism of cyclization to five- and six-membered rings, *930*

mutarotation, 932

pyranose and furanose rings, 929–33

specific rotation, 932

structure, *3, 41, 231, 246, 250*

L-glucose, 231, *250*

glucose-6-phosphate, 354

glutamic acid, *39*

glutamine, *39*

glutamine biosynthesis, 1010

glutamine synthetase, 1010

glyceraldehyde, 247–48, *249*

(R)-glyceraldehyde-3-phosphate (G3P), 102

glycerol (glycerin) backbone in lipids, 106, *107*

glycerol (glycerin) from biofuel production, 1005

glycine, *39,* 318, 319

glycogen, 40, 420, 1360

glycogen phosphorylase, 420

glycolic acid, 893

glycolysis, 40, 354

glycosides, 490–93

See also carbohydrates

glycosidic linkages in complex carbohydrates, 491–92, 493, 1355, 1361–62

goiter, 1152

gold nanoparticles (GNP), 11

gold–sulfur bonding, 11

Goodyear, Charles, 1337

graphene, 5, 147, *682*

green chemistry

aldol reactions and, 908, 914

alternatives to AlCl₃ catalyst, 1112

alternatives to Baeyer–Villiger oxidations, 1068

alternatives to dissolving metal reduction, 1280

alternatives to Grignard reactions, 855

chromic acid *vs.* permanganate oxidations, 982

dihydroxylation reactions, 1219

green alternatives, 674, 855, 1068

hydride reagents, reactions with ketones or aldehydes, 847

less toxic reagents and solvents, 674–75

minimizing by-products and waste, 962

nucleophilic aromatic substitution, 1176

overview, 673–74

percent atom economy, 676

reduction of C=O groups to CH₂, 958

safer synthetic routes, 675

selective reactions and protecting groups, 962

Wittig reactions *vs.* E1 and E2 reactions, 877

Grignard reactions

acid workup afterward, 856, 857

carbon–carbon bond formation, 332, 542, 855, 872

carboxylation reaction with carbon dioxide, 857, 1126

conjugate addition (1,4-addition) to α,β-unsaturated carbonyls, 871

direct addition (1,2-addition) to α,β-unsaturated carbonyls, 865, *866,* 871, 876, 961

epoxide ring opening, 541–42

formation of alcohols, 872, 964

green alternatives to, 855

irreversibility in nucleophilic addition, 86, 865

nucleophilic addition to polar π bonds, overview, 854–56

overview, 855

reaction mechanism, 855–56

in synthesis, 872–74

Grignard reagents

aprotic solvents, 542

carbon–carbon bond formation by, 332, 542, 855, 872

carboxylation reaction with carbon dioxide, 857, 1126

conjugate addition (1,4-addition) to α,β-unsaturated carbonyls, 871

direct addition (1,2-addition) to α,β-unsaturated carbonyls, 865, *866,* 871, 876, 961

epoxide ring opening, 541–42

feasible and unfeasible Grignard reagents, 857–58

formation from alkyl halides, 948–49, 964–65

formation from bromobenzene, 948, 1110, 1117, 1126

formation from Diels–Alder product, 1226

incompatible groups, 858, 964, 967

limitations of, 857–58

nucleophilic addition–elimination reactions, 1032–33

nucleophilic addition to polar π bonds, overview, 854–56

oxetane ring opening, 546

protected Grignard reagent, synthesis, 964–65

Grignard reagents (*cont.*)
proton transfer reactions, 332, 333
reaction with ketones, 855–56
reaction with nitriles, 855–56
reaction with water, 854
simplifying assumptions regarding, 332–33, 854
as strong bases, 455, 542, 854–55
as strong nucleophiles, 451, 542, 546, 854–56
use in synthesis, 872–74
Gross, Leo, 661
ground state, 6
Grubbs catalyst, 986–88
guanine, *43*, 44, *547*, 715, 716, 717
Guirado, G., 1176
gyromagnetic ratio (γ), 775, 783, 784, 804

H

Haber–Bosch process, 277
haloacid dehalogenase, 475
haloalkanes, nomenclature, 54–58, 66
See also alkyl halides
haloform reactions, 1017–20
halogenation
aldehydes and ketones, in acid, 532–33, 1057
aldehydes and ketones, in base, 528–32, 1057
α carbon halogenation by molecular halogens, 528–33, 1018, 1057
aromatic bromination, 1109–10
aromatic chlorination, 1109, 1110
aromatic halogenation, 1109–11
bromination of phenol, 1162, 1185
bromination reaction mechanism, 608
electrophilic addition by molecular halogens, 607–14
electrophilic aromatic substitution, 1109–11
energy diagrams, halogenation of CH_4, *1267*
geminal dihalides, 578–80
halogenated metabolites of red algae, 615
halohydrin synthesis, 611–14, 667
polyhalogenation of the α carbon of aldehydes and ketones, 530–31
of propane, 1267–71
radical halogenation of alkanes, 1260–74
selectivity of chlorination and bromination, 1267–71
solvent effects, 611, 662
stability of alkyl radicals and halogenation rates, 1268
vicinal dihalides (1,2-dihalides), 607–11, 613, 662, 667
See also alkenes, addition reactions; radical halogenation
halogens
as deactivating ortho/para directors, 1157–58
induced dipole, 529–30, 608
See also halogenation
halohydrins
bromohydrins, 611–12, 613–14, 662
chlorohydrins, 611
epoxide formation, 538, 620
regiochemistry, 613–14
stereochemistry, 612, 613–14, 667
structure, 538
synthesis, 538, 611–14, 662, 667

haloperidol, 1121
Hammond, George S., 447
Hammond postulate, 447–51, 454, 477, 483, 569, 1148, 1269
hard water, 104–5
Haworth projections, overview, 188
heat, shorthand notation, 477
heat of combustion ($\Delta H°$), 175–77, 694
See also enthalpy change
heat of hydrogenation, 696, *697*, 698
Heck, Richard F., 984
Heck reaction, 985–86
Heisenberg uncertainty principle, 120–21, 123
Hell–Volhard–Zelinsky (HVZ) reaction, 1057–59
heme group, 709
hemiacetals
acetal formation, 897–99
in cyclic monosaccharides, 929
defined, 889
from ester reduction by hydride reagent, 1030
formation by acid catalysis, 889, 890–92, 897, 899
formation by base catalysis, 889, 890, 891–92, 897, 899
formation from ketones or aldehydes, 897, 899, 929
mechanisms of formation reactions, 890–91
reaction rates of formation under acidic or basic conditions, 891–92
hemoglobin, 709, 1360
Henderson–Hasselbalch equation, 285
hepta-4,6-dien-2-ol, *380*
heptanal, 752, *753*
heptanedial, 920
heptan-1-ol, 623
heptan-3-ol, *648*
heptan-2-one, 619
heptan-3-one, *619*
hept-1-ene, *623*, 743–44
hept-3-ene, 743–44, 751
hept-2-yne, *619*
hept-1-yn-4-ol, *380*
heteroatoms, defined, 32, 1300
heterochain polymers, 1338–39
heterocyclic aromatic compounds, 707–9, 1168–72
heterogeneous catalysts, 970, 1219, 1279
heterolytic bond dissociation steps (heterolysis steps), 338, 350, 371–72
heterotopic protons, 777
hexa-1,3-diene, 694
hexa-1,4-diene, 694
hexa-2,4-diene, 711
hexanal (capronaldehyde), *389*, 631, *632*
hexane, 91–92, 98–99, 174, 820, 821–22, *972*, 1296–97
See also cyclohexane
hexanedial, 919–20
hexane-1,6-diamine, 1325, *1326*, 1327, *1328*, 1342
hexane-1,6-dioic acid (adipic acid), *1326*, 1327, 1342
hexane-2,4-diol, 968
hexanoic acid (caproic acid), *510*
hexan-1-ol, 1063–64
hexan-2-one, *583*, *619*, 738, *739*, *1080*, 1304–5

hexan-3-one, *36*, 797–98
hexa-1,3,5-triene, 731–32, 733, *1201*
hexenes
heats of combustion of isomers, *238*, 239
hex-1-ene, *238*, 239, 483, 484, *648*
hex-2-ene, 483, 484, 1298–99
relative stability of isomers, 238–39
See also cyclohexene
hex-1-en-4-yne, *156*
hex-4-en-1-yne, *156*
hexoses, defined, 199
hexyl acetate, 1063–64
hexyl group, *58*
hex-1-yne, *583*, *619*, *632*, *972*, *1222*
hex-2-yne, 646
hex-4-yn-1-ol, *654*
highest occupied molecular orbital (HOMO)
allyl cation, 689
buta-1,3-diene, 693
carbocation rearrangements, 375
coordination (bond forming) steps, 371
cyclopentadiene, 1231, *1232*
defined, *126*, 127
E2 reactions, 373–74
electrophilic addition steps, 374–75
frontier molecular orbital theory, 366–68
heterolysis (bond breaking) steps, 371–72
HOMO–LUMO interactions in elementary steps, 368–75
HOMO–LUMO transitions and energies, 728–29, 731–32, 733
nucleophile elimination steps, 373
nucleophilic addition steps, 372–73
proton transfer reactions, 368–69
S_N2 reactions, 370
See also lowest unoccupied molecular orbital; molecular orbitals
Hindenburg, *8*
histidine, *39*
Hofmann elimination reaction, 523, 551–52, 554, 659–60
Hofmann products (anti-Zaitsev products), 484, 551, 659–60
Holton, Robert A., 642
HOMO. *See* highest occupied molecular orbital
homolysis (homolytic bond dissociation)
curved arrow notation, 1249
defined, 1248, 1258
homolytic bond dissociation energies, 1249, *1250*, 1252
overview, 1248–52
reaction energy diagrams, *1249*, *1253*, 1258
weakest bond and, 1250–51, 1256, 1263
See also free radicals
homopolymers, 1340, 1342, 1343
homotopic protons, 777, *779*
Hooke's law, 742
Hückel, Erich, 699
Hückel numbers, 699–700, 705, 706–7, 708–9, 715, 1201, 1203
Hückel's rules, 699–700, 705, 707, 710, 715
See also aromaticity; Frost method
Hund's rule, 7, 703
hybrid atomic orbitals
bond length and, 145–46
bond polarity and, 146
bond strength and, 145, 146

constructive interference, 128–29, *130*, 137
defined, 128
destructive interference, 128–29, 137
effective electronegativity, *146*, 147, 297–98, 309, 579
orientations and geometries of, 130, 131, 132, *133*
overview, 119, 128
p character of, 128, 131, 132
relationship to acid strength, 297–98
relationship to VSEPR theory, 132–33
s character of, 128, 131, 132, 145–46
sp hybrid atomic orbitals, 128–31
sp² hybrid atomic orbitals, 131–32
sp³ hybrid atomic orbitals, 132
valence bond theory, 134–36
See also atomic orbitals; electron configuration
hydration
acid-catalyzed hydration of alkenes, 576–77, 614, 623, 630–31
acid-catalyzed hydration of alkynes, 581–83, 614
comparison of acid-catalyzed, hydroboration, and oxymercuration methods, 623, 630–31
oxymercuration–reduction, 616–20, 623, 630, *631*
hydrazine, 906
hydrazones, 906–7
hydride anion, 844–45, 1001
hydride reagents
diisobutylaluminum hydride (DIBAH or DIBAL-H), 1029–32
epoxide ring opening, 543
leaving groups and, 1021, 1026–27, 1031
lithium tri-*tert*-butoxyaluminum hydride (LTBA), 1029–30
NADH as biological hydride reducing agent, 852
nucleophilic addition–elimination reactions, 1021–28
nucleophilic addition to aldehydes and ketones, 845–48, 849–50
as reducing agents, 849–50
reduction of carboxylic acid derivatives, 1021–28
simplifying assumptions regarding, 333
See also lithium aluminum hydride; sodium borohydride; sodium hydride
1,2-hydride shifts, 346, 375, 430–31, 573, 1115–16
hydride transfers in nucleophilic addition, 846
hydroboration
alcohol synthesis, 628–31, 656
alkenes, 623–31
alkylborane formation, 627
alkynes, 631–32
concerted reaction, 625, 630
defined, 623
dialkylborane formation, 627
hydroboration–oxidation of alkenes, 623–28, 656, *667*
regiochemistry, 623, 625, 626, 630, 656
stereochemistry, 623, 625, 667–68
steric hindrance and steric repulsion, 625–26
trialkylborane formation, 627, *628*

trialkylborane oxidation, 628–31
See also alkenes, addition reactions; borane
hydrobromic acid, pK_a, 296
hydrocarbon tails of soaps and detergents, 103, 105
hydrochloric acid
addition to cyclohexene, 565, *575*
addition to propene, *569*
as Brønsted acid, 343, 565
pK_a, 296
as strong acid, 296
hydrocortisone, *1247*
hydrocyanic acid (HCN), 332, 452, 893–94
hydrofluoric acid, pK_a, 293, 296
hydrogen
covalent bonding in H_2, 8–10, 125
electronegativity, 78
in the *Hindenburg*, 8
Lewis structure, *127*
molecular orbital energy diagrams, *126*, 127
molecular orbitals of H_2, 125–27
quantized magnetic dipoles of nucleus, 774
uses, 8, 125
hydrogenation
benzene, heat of hydrogenation, 696, *697*
benzene conversion to cyclohexane, 696, 1282
catalytic hydrogenation
of alkenes, 970–72
of alkynes, 972–74
of benzene, 696, 1282
cis products of, 973, 1280, 1285
of cyclohexene, 696
fats and oils, *946*
mechanism, 970, 971
of nitrobenzene, 1128–29
overview, 969–70
selectivity in, 975–76
of various functional groups, 974–75
cyclohexene, 696
heats of hydrogenation, 696, *697*, 698
See also alkenes, addition reactions; alkynes, addition reactions
hydrogen atom abstraction, 1259, 1268–70, 1271, 1274
hydrogen bonding
alcohols, 85–86
amines, 747
defined, 84
deoxyribonucleic acid (DNA), 716–17, 735
electronegativity and, 85, 87
enzyme active sites, 475
ethanamine, 87
ethanol, 85–86, 87, 91
hydrogen bond acceptors, *84*, 85–87
hydrogen bond donors, *84*, 85–87, 97–98, 100
intermolecular interactions, overview, 84–88
methanoic acid, 85–86
and NMR chemical shifts, *785*, 796
nucleic acids, 716–17
O—H stretch absorption bands in IR spectroscopy, 746
protic solvents, 100
water, *91*, 92, 94
hydrogen cyanide (HCN), 141
hydrogen fluoride, *17*, 78
hydrogen peroxide, 615, 623, 628, 631–32, 1068, 1111

hydrogen radical
hydrogen atom abstraction, 1259, 1268–70, 1271, 1274
production by homolysis, 1248, *1249*
hydrolysis reactions
acid anhydrides, 1049
acid-catalyzed amide hydrolysis, 1065–66
acid-catalyzed ester hydrolysis, 1064
acid chlorides, 1047–49
ester hydrolysis, 1064
free energy diagrams, nucleophilic addition for acid derivatives, *1051*
glycosides, acid-catalyzed hydrolysis, 492
imines, hydrolysis to aldehydes or ketones, 855, 901, 903
nitriles, 904–5
proteins, 1084–85, 1086
rates of hydrolysis of carbonyl-containing species, *1050*
reaction mechanisms, 905, 1047–48, 1065–66
reactivities of carboxylic acid derivatives, 1049–52
hydronium ion (H_3O^+)
acid strength, 283, 295, 576–77
autoionization of water, 291
leveling effect of water, 283, 576
pH calculations, 285
pK_a, 291
relative stabilities of H_3O^+ and H_4N^+, 295
See also water
hydroperoxide ion, 629
hydrophilic functional groups, 94
hydrophobic effect, 1359
hydrophobic functional groups, 95
hydroxide anion
as base in E2 reactions, *452*, 453–54
base strength, *452*, 453–54, 459–60
covalent bonding in, 18, *19*
leaving group in aldol condensation, 912
poor leaving group, 459–60, 516, 899
hydroxyacetaldehyde, 87
hydroxybenzene (phenol). *See* phenol
3-hydroxybutanal, 908–9, 912, *915*
β-hydroxycarbonyls
as aldol addition product, 908, 910–11, 926
as aldol condensation reactant, 911, 917
1,3-positioning of heteroatoms, *953*, 954
See also aldol reactions
4-hydroxy-3,3-dimethylbutanone, *918*
2-hydroxyethyl acrylate, *1344*
hydroxy group
as activating group, 1155
leaving group of alcohols, conversion to water, 461–62, 516
in multiple functional groups, 34
O—H stretch absorption bands in IR spectroscopy, *740*, 745–47
as ortho/para director, 1145–46, 1148, 1149–51
as poor leaving group, 459–60, 516
protecting groups and, 655, 965–68
See also alcohols; carbohydrates
hydroxyl radical, 1274
3-hydroxy-2-methylbutanal, 914–15
3-hydroxy-3-methylglutaryl coenzyme A (HMG-CoA), 1202
5-hydroxy-2-methylheptan-3-one, *918*

M + 1 peaks, 823–24, 826–27
M + 2 peaks, 825–27
mass spectrometer schematic, *819*
mass spectrum features, 820–22
mass-to-charge ratio (*m/z*), 819
2-methylpentane, 1297–98
molecular formula determination, 826–28
molecular ion, M$^+$(*g*), 818–19, 820
molecular mass determination, 820, 826, 1295
molecules containing heteroatoms,
  1300–1302
nitrogen rule, 820–21, 827
overview, 723–24, 771–72, 818–19, 1295
pentanoic acid (valeric acid), 1305
pentan-1-ol, 1302–4
radical cation formation, 822, 1295, 1305
relative abundance, 819, 820
structure elucidation, 723
N,*N*,2-trimethylpropan-2-amine, 1302
use in criminal investigations, 724, 828
use in medicine, 828
use in synthesis, 724
McLafferty, Fred W., 1305
McLafferty rearrangement, 1305
measured angle of rotation (plane-polarized
  light), *242*, 243
mechanically generated acid, 553
mechanisms of reactions. *See* reaction
  mechanisms
mechlorethamine, 547, 869
Meisenheimer complex, 1174–75
melting points (*T*$_m$), *81*, 83–84, *89*, 1344–49
menthol, 111, 174, 641
mercurinium ion intermediate, *616*, 617–18,
  619, 623
mercury(II) acetate [Hg(OAc)$_2$], 616, 618, 674
mercury(II) catalyst, 583, 616, 618–19
Merrifield, R. B., 1088
Merrifield synthesis of peptides, 1088–91
meso molecules, 218
mesylate (methyl sulfonate) anion, 459, 1060
mesyl chloride (methanesulfonyl chloride,
  MsCl), *1060*
meta-chloroperbenzoic acid (MCPBA), 620,
  621, 622, 624, *648*, 1067–68
meta substitution
  comparison of activating/deactivating and
    ortho/meta/para-directing substituents,
    1157
  electron-withdrawing groups as meta
    directors, 1153–54
  interconverting ortho/para and meta
    directors, 1180–83
  meta directors, defined, 1146
  meta directors, identifying, 1146–48
  meta (*m*) substitution, defined, 158, 1144
  NO$_2$ group as meta director, 1146, 1153–54
  trifluoromethyl group as meta director, 1154
  See also electrophilic aromatic substitution;
    ortho/para substitution; substituted
    benzenes
methamphetamine, 750
methanal (formaldehyde)
  Bakelite synthesis, *1332*
  crossed aldol reactions, 915, 916
  electrostatic potential map, *844*
  λ$_{max}$ in UV–vis spectrum, 730, *731*, 733, 734

molecular orbitals, 139, 733
nomenclature, 389
nonbonding orbitals, 139, 733
propan-1-ol synthesis from, 950
sp^2 hybrid atomic orbitals, 139
types of vibration, 736–37
methanamide (formamide), *512*
methanamine (methylamine), 282, *388*, 733
methane
  bond dipoles, *17*
  deprotonation, 293–94, 296–97
  electrostatic potential map, *17*
  energy diagrams, halogenation of CH$_4$, *1267*
  greenhouse gas, 9, 135
  infrared radiation absorption, 9
  melting and boiling points, *89*, 90
  molecular orbitals, 135–36
  in natural gas, 135, 1247
  p*K*$_a$, 293
  radical halogenation, 1263–64, 1266–67
  reaction enthalpies, halogenation of CH$_4$,
    1266
  saturation, 193
  sp^3 hybrid atomic orbitals, 135
  structure, *34*, 75, *193*
  use as fuel, 135, 1247
  valence bond theory, 135–36
methanesulfonyl chloride (mesyl chloride,
  MsCl), *1060*
methanethiol, 427
methanoic acid (formic acid), 70, 81, 85–86,
  510, 1065
methanol (methyl alcohol)
  acid-catalyzed alkoxylation of alkenes, 577
  biodiesel production and, 1005
  electron-poor species, 331
  glycoside formation and, 490–91
  p*K*$_a$, 280, 305
  as protic solvent, 480, 850
  solubility, 94, *95*
  solvolysis reactions, 479–80, 482, 645
  structure, *331*, *387*
  uses, 13
methionine, 39
methohexital (Brevital), 517
methoxide anion (CH$_3$O$^-$)
  in base-catalyzed transesterification, 1002,
    1061–62
  conjugate addition (1,4-addition) to
    α,β-unsaturated carbonyls, 866
  covalent bonding in, 18, *19*
  as electron-rich species, 332
  as leaving poor group, 540
  as nucleophile, 540, 870, 1046
methoxybenzene (anisole), 160, 387, 535
2-methoxycyclohexa-1,3-diene, *155*
2-methoxyethanol, 541
methoxyethene (methyl vinyl ether), *1340*
3-methoxy-3-ethylpentane, 480
methoxy group, 62
methoxymethane (dimethyl ether), *62*, 63, *81*,
  83, 84
1-methoxy-2-methylpropane (isobutyl methyl
  ether), *66*
1-methoxypent-2-yne synthesis, 650–53
2-methoxyphenol (guaiacol), 462, *463*
methoxyphenylmethane, 422

1-methoxypropane (methyl propyl ether), *63*, 67
2-methoxypropane (isopropyl methyl ether), *63*
3-methoxyprop-1-yne, *650*, 651
2-methoxy-1,1,3,3-tetramethylcyclopentane, *63*
methyl acetate (methyl ethanoate), 534, 1007,
  1063–64
methylammonium ion (CH$_3$NH$_3$$^+$), 18, *19*
p-methylaniline, 1185, *1186*
methyl anion, 74
methyl benzoate, *1006*, 1046, *1047*
2-methyl-2-bromopropane, *423*
2-methylbuta-1,3-diene (isoprene), 109–10,
  589–90, 1353–54
  See also isoprene units
2-methylbutanal, 808, *809*
3-methylbutanal (isovaleraldehyde), 841–42
methylbutane (2-methylbutane, isopentane),
  31–32, 64
2-methylbutan-2-ol, 429, 483, *484*, 614
3-methylbutan-2-ol, *616*
3-methylbutanone (methyl isopropyl ketone),
  389, 390, 918
2-methylbut-1-ene, 483, *484*
2-methylbut-2-ene, 483, *484*
3-methylbut-1-ene, 573, *614*
2-methylbut-3-en-2-ol, 432
3-methylbutyl (isopentyl) group, *64*
methyl carbon, 465
methyl cation, 74, 305, 364, *365*, 1257, 1296
  See also carbocations
methylcyclohexane, 185–86
2-methylcyclohexanone, 525–27
N-methylcyclohex-3-en-1-amine, *383*
1-methylcyclohexene, *744*, *1221*, *1223*, 1224,
  1226
trans-2-methylcyclopentanol, *623*
1-methylcyclopentene, *623*
1-methylcyclopropene, 1225
N-methyl-2,2-dimethylpropanamide, *748*
N-methyl-2,4-dinitroaniline, *1175*
methylene chloride (dichloromethane, methyl
  dichloride), *66*, 79, 80
methylenecyclohexane, *744*
methylenecyclopentane, 656, *657*
N-methylethanamine, 223, *224*
methyl ethanoate (methyl acetate), 534, 1007,
  1063–64
(1-methylethyl)benzene (isopropylbenzene,
  cumene), *160*, 1115
1-methylethyl group, 59
1-methylethyl (isopropyl) group, *64*
3-(1-methylethyl)-2-nitrohexane, 60
methyl β-D-glucopyranoside, *492*
methyl group (Me), *58*
2-methylhept-2-ene, *142*
methyl hept-6-enoate, *1032*
3-methylhexane, *58*, 217
(*S*)-3-methylhex-1-ene, 266
(1-methylhexyl)cyclooctane, *61*
methyl ketones, haloform reactions, 1017–20
methyl methacrylate (methyl formate), *512*,
  1339, *1343*
N-methylmethanamine (dimethylamine), 387,
  388
methyl methanoate (methyl formate), *512*
methyl (*E*)-2-methylbut-2-enoate, 617
2-methyloctane, *34*

2-methyloct-7-en-2-ol, *1032*

methyl orange, *1173*

methyl 6-oxo-6-phenylhexanoate, 1076

2-methylpentan-3-amine, 554

2-methylpentane (isohexane), *90*, 1297–98

3-methylpentane, 59

2-methylpentane-1,3-diol synthesis, 954

3-methylpentan-3-ol, *951*

4-methylpentan-2-one, 749

2-methylpent-2-enal, *977*

4-methylpent-4-enal, *1029*

(*E*)-3-methylpent-2-ene, 666

4-methylpent-3-en-2-one, 749

4-methylphenol, 279, *281*

(1-methyl-2-phenylethyl)benzene, *1115*

2-methyl-2-phenylpropane, 1127

2-methylpropanamide, *748*

methylpropane (isobutane), 64

methylpropan-2-ol (*t*-butyl alcohol), 282, *387*, *423*

methylpropene (isobutylene), 159, 536, 1322, *1340*

m-(2-methylpropyl)aniline, 1182–83

1-methylpropyl (*sec*-butyl) group, *64*, 65

2-methylpropyl (isobutyl) group, *64*

1-methylpropyl 3-oxobutanoate, *508*

methyl radical
    halogenation, 1263–64
    Lewis structure, *1247*
    production by homolysis, 1248, *1249*, 1250
    radical coupling, 1258, 1263–64
    structure and stability, *1253*, 1254, 1257

methyl red, *1172*, 1173

1,2-methyl shifts, 346, 430–31, 573–74

methyl *tertiary*-butyl ether (MTBE), 159

3-methylthiopropanoic acid, *427*

mevalonic acid, *590*

micelles, 103, 105

Michael, Arthur, 865

Michael additions (Michael reactions), 865, 869, 924

microcin J25, 470

microscopic reversibility, 368, 372, 373, 375

migratory aptitude, 1068

molar absorptivity (ε), 726, 734

molecular dipoles, 78–80

molecular geometry, 72–74, 144–45
    See also valence shell electron pair repulsion (VSEPR) theory

molecular ion, M$^+$(*g*), 818–19, 820

molecular modeling kits, 77–78, 144–45, 189–90

molecular orbitals (MOs)
    allyl anion (allylic anion), 689–90, 691
    allyl cation, 686–91
    antibonding contribution from atomic orbitals, 686
    antibonding molecular orbitals, *125*, *126*, 127, 686, 688
    benzene, 700–702
    bonding contribution from atomic orbitals, 686
    bonding molecular orbitals, *125*, *126*, 127, 686, 688
    buta-1,3-diene, 692–95
    conjugated orbitals, defined, 686
    conservation of numbers of orbitals, 126

cyclobutadiene, 698–99, *700*

degenerate orbitals, 701

empty orbitals, 367

ethane, 134, *135*

ethene, 136–38

ethyne, 140–41

frontier molecular orbital (FMO) theory, 366–75

frontier molecular orbitals, defined, 367

HOMO–LUMO transitions and energies, 728–29, 731–32, 733

hydrogen molecule (H$_2$), 125–27

hyperconjugation, 364–66

linear combination of atomic orbitals (LCAOs), 124

methane, 135–36

molecular orbital theory, overview, 119, 124–27, 682–83

multiple-center MOs, 682–714

nodal planes, 125, *126*, 137, 687–88, 692–93, 701–2, 703–4

nodes, 126, 137

nonbonding molecular orbitals, 127, 139, 686, 688, 733

n → π* transitions, 733

π (pi) bonds, introduction, 136–38

π (pi) MOs, allyl cation, 687–91

π (pi) symmetry, 126, 137–38

π → π* transitions, 729, 732, 733

rotation around double and single bonds, 141–44

σ (sigma) bonds, defined, 135

σ (sigma) symmetry, 126

two-center molecular orbitals, 124–27, 133–41, 369, 682, 686

    See also atomic orbitals; highest occupied molecular orbital; lowest unoccupied molecular orbital; valence bond theory

molozonides, 1225

monomers, 1309
    See also polymers

monosaccharides
    aldohexoses, 199, 248, *249*
    aldopentoses, 199, 248, *249*
    aldoses, 199, 248–50
    aldotetroses, 248, *249*
    anomeric carbons, 931
    α and β designations, 490, 931–33
    chirality, 246–47
    constitutional isomers, 198–99
    cyclic and acyclic forms, 42, 929
    definition and overview, 40–42, 490
    diastereomers, 931, 932
    D/L system for classifying, 247–48
    furanose ring structure, 931
    glyceraldehyde, 247–48, *249*
    hemiacetals in cyclic monosaccharides, 929
    hexoses, 199
    ketohexoses, 199
    ketopentoses, 199
    ketoses, 199
    mechanism of cyclization to five- and six-membered rings, *930*
    mutarotation, 932
    nomenclature for ring forms, 930
    pentoses, 199
    pyranose and furanose rings, 929–33

pyranose ring structure, 931

ring opening and closing, 929–33

specific rotation, 932–33

stability of five- and six-membered rings, 929–30

    See also carbohydrates; *specific types*

monoterpenes, *110*

monounsaturated fatty acids, 200

Montreal Protocol, 1265

Moore, Jeffrey, 553

morphine, 52, *53*, 286, 865

Mrozack, S. R., 454

multiple-center molecular orbitals, 682–714
    See also molecular orbitals

multiple myeloma, 209

multiplet (m) splitting pattern, 793

multistep mechanisms, introduction to, 393

mutarotation, 932

Mycobacterium tuberculosis, *563*

N

NADH (nicotinamide adenine dinucleotide), 852

naming conventions. *See* nomenclature

nanocars, 225

nanotubes, 5, 147

naphthalene, 29, 194, 698, 706, 1168–69

2-naphthol, *95*, 96

naproxen, 664

natural products, overview, 641

N-bromosuccinimide (NBS), 1251, 1271–74, 1283–84

Negishi, Ei-ichi, 984

neopentane. *See* dimethylpropane

neopentoxide anion, 456

(*E*)-(+)-nerolidol, 615

neryl pyrophosphate, 590–91

net molecular dipole (permanent dipole), 77–80

net reaction, defined, 395

network polymers, 1332, 1350, 1362

neutrons, properties and location, 4, 5

Newman projections, 166–68

nicotinamide adenine dinucleotide (NADH), 852

nitration
    of benzene, 1121–22, 1159
    of *N,N*-dimethylaniline, 1163
    p-ethyltoluene, 1167
    of monosubstituted benzenes, product distribution, 1146, *1147*, 1148
    of monosubstituted benzenes, relative rates, *1155*
    of nitrobenzene, 1146, 1153–54, 1155–56, 1159
    m-nitrotoluene, 1165–67
    p-nitrotoluene, 1164–65
    of phenol, 1145–46, 1148–51, 1152, 1155–56
    of toluene, 1151–53, 1156, 1160

nitric acid, 1121–22

nitriles
    aldol addition reactions, 922–23
    carbonitriles, 506
    catalytic hydrogenation, *975*

nucleic acids
    base pairs, 715–16
    bases, 42–44, 715–17
    DNA double helix, 42, 715, 716
    hydrogen bonding, 716–17
    nucleotides, composition, 42, *43*, 44, 714
    overview, 37, 42–44, 105
    π stacking in DNA, 716
    as polynucleotides, 1355
    ribonucleic acid (RNA), 42, *43*, 44, 107, 174
    size hierarchy, *43*
nucleophile elimination steps
    curved arrow notation, *339*, 340–41
    electron-rich and electron-poor species,
        340–41
    frontier molecular orbital theory, 373
    HOMO–LUMO interactions, 373
    leaving groups, 339–40
    microscopic reversibility, 373
    nucleophilic addition–elimination reactions,
        1002
    overview, 339–41
    as part of multistep reactions, 340
nucleophiles
    attraction to atoms with positive charge, 335
    attributes of nucleophiles, 335
    carbon nucleophiles, 451–52, 542, 853, 865
    common uncharged nucleophiles, 335
    and competition between S_N2 and S_N1
        reactions, 450–51
    enzyme-regulated reactivity, 475
    formal charge and nucleophile strength, 451
    intrinsic strength of nucleophiles, 450–51,
        840
    nucleophile strength in acid or basic
        conditions, 892
    nucleophile strength in S_N2 and S_N1
        reactions, 447–51
    in nucleophilic addition steps, 339–41
    relative nucleophilicity in protic and aprotic
        solvents, 473–74
    reversibility in nucleophilic addition, *865*
    in S_N1 reaction mechanisms, 394–95,
        408–10
    and S_N1 reaction rates, 401, 403, 446, 450–51
    in S_N2 reaction mechanisms, 334–37, 394,
        411–13, 422
    and S_N2 reaction rates, 446, 447–51
    weakening by protic solvents, 471–72, 473–74
    See also Lewis bases
nucleophilic acyl substitution, reaction
    mechanism, 1002–3
    See also nucleophilic addition–elimination
        reactions
nucleophilic addition–elimination reactions,
    1000–1044, 1045–1103
    acetoacetic ester synthesis, 1078–81
    acid-catalyzed amide hydrolysis, 1065–66
    acid-catalyzed transesterification, 1063–66
    alcoholysis, 1047, 1048
    alkyllithium reagents, 1032–33
    aminolysis reactions, *1045*, 1052–54
    Baeyer–Villiger oxidations, 1067–68
    base-catalyzed transesterification, 1001–6
    carboxylic acids from amides, 1013–15
    decarboxylation, 1078–81

Dieckmann condensation, 1076–77
ester hydrolysis, 1064
Fischer esterification reaction, 535, 1064–65,
    1083, 1088, 1328–29
free energy diagrams, nucleophilic addition
    for acid derivatives, *1051*
functional group transformations, *879*,
    1036–38, *1091–92*
Gabriel synthesis of primary amines,
    1015–17
Grignard reagents, 1032–33
haloform reactions, 1017–20
Hell–Volhard–Zelinsky (HVZ) reaction,
    1057–59
hydride reductions of carboxylic acids and
    derivatives, 1021–28
hydrolysis reactions, 1047–48, 1049–52
kinetics, 1003–4, 1048–52
leaving groups, *1000*, 1001
malonic ester synthesis, 1078–81
nucleophile elimination step, 1002
nucleophilic aromatic substitution, 1174–76
organometallic reagents, 1032–35
rate-determining nucleophilic addition step,
    1003, 1048, 1050
reaction free-energy diagrams, *1004*, *1012*,
    1014
reaction mechanism with strong nucleophile,
    1002–3, *1006*, 1045
reaction mechanism with weak nucleophile,
    general, 1047–48
reactions that alter the carbon skeleton, *1038*,
    1093
reactivities of carbonyl-containing species,
    1050
saponification, *1000*, 1009–13, 1062, 1072
specialized reducing agent reductions of acid
    chlorides and esters, 1029–32
stability ladder of carboxylic acid derivatives,
    1007, *1008*, 1010, 1011, 1013, 1023, 1025
step-growth polymerization, 1325, 1328–29
strong nucleophiles and, 1000–1044
sulfonation reactions of alcohols, 1059–61
tetrahedral intermediates, 1002–4, 1006–7,
    1018, 1027, 1030–31, 1045, 1048–49
thermodynamics and reversibility, 1004–6,
    1006–9
weak nucleophiles and, 1045–1103
See also Claisen condensations
nucleophilic addition steps
    curved arrow notation, *339*, 340
    electron-rich and electron-poor species, 340,
        372
    frontier molecular orbital theory, 372–73
    HOMO–LUMO interactions, 372–73
    nucleophiles in, 339–41
    overview, 339–41
    as part of multistep reactions, 340
    polar π (pi) bonds, 339, 372
nucleophilic addition to polar π bonds, 839–87,
    888–945
    acetal formation, 897–99, 900–901
    acid-catalyzed addition, overview, 889–92
    acid workup afterward, 841, 845, 846, 851,
        856–57
    base-catalyzed addition, overview, 889–92

carbonyl hydration in ketones and aldehydes,
    843–44, 892–93
charge stability and total bond energy as
    driving forces, 867
conjugate addition by weak nucleophiles,
    895–96
cyanohydrin formation by HCN addition,
    893–94
direct *vs.* conjugate addition to α,β-
    unsaturated carbonyls, 863–68, 869–72,
    874–76, 895–96, 961
enamine formation, 902
functional group transformations, *879*,
    934–35
general reaction mechanism for strong
    nucleophiles, 841–42
hemiacetal formation from ketones or
    aldehydes, 889–92, 897, 929
hydride transfers, 846
imine formation, 855, 900–904
inductive effects in, 844
Michael reactions, 865, 869, 924
nucleophile and electrophile strength in
    acidic or basic conditions, 892
racemic mixtures in products, 842, 845
reactions that alter the carbon skeleton, *880*,
    936
regioselectivity involving 1,2- *vs.*
    1,4-addition, 865–67, *868*, 870–72, 876,
    961
relative reactivity of ketones and aldehydes,
    842–44
reversibility in nucleophilic addition, 865–67
stereochemistry, 842
steric repulsion in ketones, 843
steric strain, 843, 913
strong nucleophile addition, 839–87
thermodynamic and kinetic control of
    addition reactions, 866–67, *868*, 872
weak nucleophile addition with acid and base
    catalysis, 888–945
Wittig reaction, 858–61
Wolff–Kishner reduction, 906–7, 955–57,
    958–59, 1182
See also aldol reactions; alkyllithium reagents;
    Grignard reactions
nucleophilic aromatic substitution
    benzyne intermediates, 1176–78
    elimination–nucleophilic addition
        mechanism, 1176–78
    green alternatives, 1176
    leaving groups, 1174–75, 1176–77
    Meisenheimer complex, 1174–75
    nucleophilic addition–elimination
        mechanism, 1174–76
    nucleophilic addition–elimination reactions,
        1174–76
    overview, 1173
nucleophilicity
    alkyllithium reagents, 451, 542, 546
    ammonia, 520
    decrease in strong, bulky bases, 455–56
    Grignard reagents, 451, 542, 546
    organometallic compounds, 451
    relative nucleophile strength of common
        species, 473–74, 840
    S_N2 reaction rate and, 447, *448*

nucleophilic substitution and elimination
   reactions
     competition among S_N2, S_N1, E2, and E1
       reactions, 442–502
     elimination reactions that are useful for
       synthesis, 548–54
     functional group transformations, *494–95,*
       555–56, 652
     reactions that alter the carbon skeleton, *495,*
       557
     substitution reactions that are useful for
       synthesis, 515–48
     thermodynamic control, 445, 487, 526–27
     See also specific topics
nucleotides, 42, *43,* 44, 246, 714
numbering systems. *See under* nomenclature
nylon, synthesis from cyclohexanone, 36
nylon-6,6, 1325–27, 1330, 1339, 1342

O

octa-3,5-dien-1-yne, 712
octanedial, 920
octanoic acid (caprylic acid), 86
octa-1,3,6-triene, 711
oct-4-ene, 862–63, 986, *987, 1280*
octet rule, 8, 12, *14,* 193, 330
octets (electron configuration), 6
oct-1-yne, 610
oct-4-yne, 1280, *1281*
oils, 105–6
Olah, George A., 400, 430
olefin metathesis, 986–88
olefins, 1340
   See also alkenes
oleic acid, 200, 201
operating frequency of NMR spectrometer,
   783–84, 797–800
opsin, 903
optical activity, overview, 241–45
orbital interactions, 119–51
   See also hybrid atomic orbitals; molecular
     orbitals
organic chemistry, definitions and origins, 1–3
organic compounds, defined, 1, 2, 37
organic compounds, physical properties, 80–82
organic photoconductors (OPCs), 572
organocuprates in coupling reactions, 983
organometallic compounds
   acid workup after treatment of acid chlorides,
     1033
   base strength, 455
   as carbanion (R^-) donors, 332–33, 451, 854
   carbon–carbon bond formation by, 332
   carboxylic acid derivatives and, 1032–35
   formation from alkyl halides, 948–49
   metal-containing portion as spectator ion, 332
   nucleophilic addition–elimination reactions,
     1032–35
   nucleophilicity, 451
   overview, 332
   regioselectivity of, 870–72
   relative reactivities of, 870
   simplifying assumptions regarding, 332–33
   tertiary alcohol production from carboxylic
     acid derivatives, 1032–33

   See also alkyllithium reagents; Grignard
     reagents
ortho/para substitution
   comparison of activating/deactivating and
     ortho/meta/para-directing substituents,
     1157
   hydroxyl group as ortho/para director, 1145–
     46, 1148, 1149–51
   interconverting ortho/para and meta
     directors, 1180–83
   lone electron pairs on substituents, 1147–48,
     1149–51, 1158
   methyl group as ortho/para director, 1151–53
   nomenclature, 158
   ortho (*o*) substitution, defined, 158, 1144
   ortho/para directors, defined, 1145–46
   ortho/para directors, identifying, 1146–48
   para substitution, defined, 158, 1144
   resonance structures and charge stability,
     1149–53
   See also electrophilic aromatic substitution;
     meta substitution; substituted benzenes
oseltamivir (Tamiflu), 507
osmate esters, 1218, *1219,* 1225
osmium(VIII) oxide (OsO_4), 1218–20, 1223,
   1225
otter, *23*
overall products, defined, 396
overall reactants, defined, 396
overall reaction, defined, 395
oxaphosphetanes, 860
oxetanes, 540–41, 546, 1324
   See also ethers
oxidation–reduction reactions, overview, 848–50
oxidation state, 848–50
oxidative cleavage of alkenes and alkynes
   in organic synthesis, 1227
   overview, 1220–21
   with ozone, 1224–26
   with periodate anion (IO_4^-), 1223–24
   potassium permanganate ($KMnO_4$),
     1221–23
oximes, 21
oxiranes
   anionic polymerization, 1323
   (2*S*,3*R*)-2-ethyl-2,3-dimethyloxirane, 543
   (2*S*,3*S*)-2-ethyl-2,3-dimethyloxirane, *544*
   index of hydrogen deficiency, *193,* 194
   (*R*)-2-methyloxirane, 544–45
   (*S*)-2-methyloxirane, 545–46
   oxirane (ethylene oxide), *193, 194,* 425–26,
     540–41, *543,* 1323
   reaction with ammonia, mechanisms, 425–26
   ring opening, acidic conditions, 544–46
   ring opening, neutral or basic conditions,
     541–44, 545–46
   ring opening, reaction mechanisms, 425–26,
     541, 543, 545–46
   ring opening, steric hindrance, 543, 544, 546
   ring strain, *541*
   structure, *194, 425, 541*
   uses, 194
   See also epoxides; ethers
(*E*)-9-oxodec-2-enoic acid, 504
4-oxo-2,2-diphenylpentanal, *618*
6-oxo-heptanal, *921, 1223, 1224*
6-oxoheptanoic acid, *1221, 1226*

oxymercuration–reduction (oxymercuration–
   demercuration), 616–20, 623
ozone, 758, 1224–26, 1265
ozone hole, 12, 1265
ozonides, 1225–26
ozonolysis, 1224–26

P

Pacific yew tree (*Taxus brevifolia*), *641,* 642
palladium-catalyzed coupling reactions, 984–86
palmitoyl acyl carrier protein (ACP), 1077
para-aminobenzoic acid (PABA), 1124, 1173
para substitution, defined, 158, 1144
   See also ortho/para substitution
Pasteur, Louis, 240–41
Pauling, Linus, 14
Pauli's exclusion principle, 7
p character of hybrid atomic orbitals, 128, 131,
   132
PEEK (poly[ether ether ketone]), 1334–35,
   1346, 1348, *1349*
(*E*)-penta-1,3-diene (*trans*-penta-1,3-diene),
   271, 731, 732
penta-1,4-diene, *731,* 732
cis-penta-1,3-diene, 730, *731*
trans-penta-1,3-diene, *271, 731,* 732
pentanal (valeraldehyde), *389,* 897
pentan-3-amine, *379*
pentane (*n*-pentane), 64, 90, *91*
pentanedial, 920
pentane-2,4-dione, 301–2
pentanoic acid (valeric acid), 36, *510, 1222,*
   1305
pentan-1-ol, 1302–4
pentan-3-one (diethyl ketone), *389,* 852–53,
   954
penta-1,2,3-triene, 712
pent-1-ene (1-pentene), 234–35
pent-2-ene (2-pentene), 234–35, 269
pent-4-enoic acid, *1081*
pent-4-en-2-ol, 912
4-pentenyl benzoate, *1047*
pentoses, defined, 199
(*R*)-2-pentylbutane-1,4-dioic acid, *504*
pentyl (*n*-pentyl) group, *58,* 64
pentyl radical, 1296
pent-1-yne, *550,* 646
pent-3-ynoic acid, *978*
pent-3-yn-1-ol, *978*
peppermint oil, 641
peptide linkages (peptide bonds), 1010, 1084–
   85, 1088–89, 1355–56, 1358
peptides
   cross-links, *888,* 1356, *1357*
   C-terminus, 1085, 1089
   disulfide bonds (disulfide bridges), 1356,
     1357
   lasso peptides, 470
   Merrifield synthesis, 1088–91
   N-terminus, 1085, 1089
   peptide linkages, 1010, 1084–85, 1087–88,
     1089, 1355–56, 1358
   primary structure, 1085, 1355–57
   synthesis, 1087–90
   See also amino acids; proteins

peptide synthesizers, 1090
peracids (peroxyacids), 620–22, 1067–68
percent atom economy, 676
percent dissociation of acids, 285–86
percent yield, 671–73, 675–76
pericyclic reactions
  defined, 1200
  oxidative cleavage of alkenes and alkynes,
    1221
  syn dihydroxylation reactions, 1218–20
  See also Diels–Alder reactions
periodate anion, 1221, 1223–24
periodate esters, 1223, 1224
periodic acid, 1223
permanent dipole (net molecular dipole),
  77–80
peroxides as radical initiators, 1275–76, 1284
peroxyacids (peracids), 620–22, 1067–68
pH, 285–86
phases of electrons, 121, 123, *124*
phases of matter, at molecular level, *82*, 83
phenol (hydroxybenzene)
  Bakelite synthesis, *1332*
  bromination, 1162, 1185
  nitration, 1145–46, 1148–51, 1152, 1155–56
  nomenclature, 387
  pK_a, 290, 1162
  production from chlorobenzene, 1176–77
  structure, *281, 290, 387*
  sulfonation, 1185
  synthesis, 1133, 1176–77
  uses, 281
phenoxide anion, 455, 1162
phenyl acetate, 1051–52
phenylalanine, *39, 40*, 1059
phenylalanine hydroxylase, 1059
N-phenylbenzamide (benzanilide), 1129, *1130*
phenyl benzoate, 1067
(2*R*,3*R*)-3-phenylbutan-2-ol, 667–68
4-phenylbutan-1-ol, 951
1-phenylbutan-1-one, 1120–21
2-phenylbut-2-ene, 667, *860*, 1116–17
(3*S*,4*S*)-phenyl 3,4-dihydroxypentanoate, *508*
phenylethanal, *917*
2-phenylethanol, 542
phenylethanone (methyl phenyl ketone,
  acetophenone), *389, 390*, 581–83, *619, 957,
  1119*, 1128
phenylethene, *613, 1222*, 1285
phenylethyne (phenylacetylene), *550, 619*
(*Z*)-1-phenylhept-2-ene, 973–74
3-phenylhept-2-ene, *159*
phenyl isothiocyanate, 1085
phenylketonuria, 1059
phenylmethanamine (benzylamine), 388
phenylmethanoic acid (benzoic acid). *See*
  benzoic acid
phenylmethanol (benzyl alcohol)
  from benzaldehyde reduction, 845, 849
  from benzoyl chloride reduction, *1021*
  conversion to methoxyphenylmethane under
    basic conditions, 422–23
  Fischer esterification reaction, 1088
  oxidation to benzaldehyde, 980
  oxidation to benzoic acid, 982
  structure, 387
  uses, 387

phenylmethyl substituents (benzyl group; Bn),
  159
3-phenylpentan-3-ol, *1032*
1-phenylpropan-1-one, *907, 955, 1118*
1-phenylpropan-2-one (phenylacetone), 750
(*Z*)-1-phenylprop-1-ene, *973*
(1-phenylpropyl)benzene, *1115*
1-phenylprop-1-yne, *973*
phenyl substituents (phenyl group; Ph), 159
phenylthiocarbamoyl (PTC) derivatives, *1085,*
  1086
phenylthiohydantoin (PTH) derivatives, *1085,*
  1086
2-phenyl-2-tosylpropane, 643–44
pH gradients, 319–20
phosphatidylcholines, *107*
phosphohexose isomerase, 354
phospholipids, 106–8, 1274
phosphonium ylides
  generation from alkyl halides, 861–63, 873,
    948, 971
  resonance structures, 859
  in Wittig reactions, 859–61
  See also Wittig reagents
phosphorus tribromide (PBr₃)
  alcohol conversion to alkyl halide, 517–20,
    653, 656, 665, 670, 873, 950
  carboxylic acid conversion to acid bromide,
    1058
  inversion of configuration, *517*, 518–20,
    1060
  methohexital synthesis, 517
phosphorus trichloride, 517, 520, 665, 1054–56,
  1058
phosphorylation and enzyme regulation, 420
phosphotriose isomerase, 354, 1359
photochemically allowed reactions, 1234
photochemically forbidden reactions, 1234
photons, 241, 727
photosynthesis, *723*
phthalic acid, 1049
phthalic anhydride, *1049*
phthalimide, 1015–16
physical properties of compounds, 80–82
physical properties of phases of matter, *82*, 83
α-pinene, 110, *976*
π (pi) bonds
  ethene, 136–38
  introduction, 136–38
  polar, overview, 339, 839–40
  reactivity of, 563–64
π (pi) stacking, 716
pK_a
  and absorption and secretion of drugs, 286
  and acid strength, 279, 309–12
  alcohols, 290, 309
  amines, 309
  amino acids, 315, *316*, 318
  and base strength, 281
  carboxylic acids, 290, 309
  defined, 279
  effect of charge, 309
  effect of type of atom attached to proton,
    308, 309
  functional groups and acidity, 289–91
  inductive effects, 308, 310

K_{eq} relationship to product and reactant pK_a
  values, 281–82
pH and, 285–86
predicting outcome of proton transfer
  reactions, 281–82
relative importance of charge-stability factors,
  309–11
resonance effects, 308, 309
values for various acids, *280*
See also acid–base chemistry
Planck's constant (*h*), 727, 775
plane of symmetry test for chirality, 215–16,
  218–19, *222*
plane-polarized light, overview, 241–42
plastination, *515*
Plexiglas [poly(methyl methacrylate)], 1307,
  1311, 1343
poisoned catalysts, 972–74, 975, 1280, 1285
poison ivy, *1247*
polar covalent bonds, 16–18, 78–79
polarimetry, 243–44, 245
polarity reversal (umpolung) in organic
  synthesis, 947–51
polarizability, 89
polarizers, 241–42
polar π (pi) bonds, overview, 339, 839–40
  See also nucleophilic addition to polar π
    bonds
pollen, *328*
polyacrylates, 1339
poly(acrylic acid), *1339*, 1347, *1348*
polyacrylonitrile or poly(acrylonitrile), 194, 542,
  1314, *1339*
polyatomic anions, 18
polyatomic cations, 18
polyatomic ions, *19*, 20
polybutadiene, 584, 854
polybutylene, 191, 1314
polycarboxylic acids, 1340
polycyclic aromatic hydrocarbons (PAHs),
  706–7
polydicyclopentadiene, 1218
polydimethylsiloxane, 89
polyester, defined, 1339
poly(ether ether ketone) (PEEK), 1334–35,
  1346, 1348, *1349*
poly(ethyl acrylate), 1318
polyethylene (polymethylene)
  branching in low-density polyethylene, 1351
  catalytic synthesis, 1351
  chain transfer, 1331
  flexibility, 1351
  hex-1-ene comonomer in, 483
  high-density polyethylene (HDPE), 483
  linear low-density polyethylene (LLDPE),
    483, 1351, *1352*
  low-density polyethylene (LDPE), 1351,
    1352
  in materials for food storage, 1351
  melting points, 1351, *1352*
  monomer in, *1340*
  as polyolefin, 1340, 1351
poly(ethylene oxide), 1323
poly(ethylene terephthalate) (PET)
  via Fischer esterification, 1328–29
  flexible portion of repeating unit, 1348

linear polymer chain, *1339*
in materials for food storage, 1352, *1353*
melting point, 1348, *1349*
as polyester, 1339, 1343
step-growth polymerization, 1328–29
synthesis, 1328–29
polyisobutylene (butyl rubber), 159, *1340*
cis-polyisoprene, 1354
See also rubber
poly(lactic acid) (PLA), 1354
polymerization, defined, 1308
polymers, 1307–69
acrylics, 1339
addition polymerization, 1312–13
amorphous solids, 1344–45
anionic polymerization, 1320–21, 1322, 1323–24
atactic stereochemistry, *1318*, 1319
biological macromolecules, 1355–62
biopolymers, 1355–62
branched polymers, 1330–32, 1347–48, 1351, *1352*, 1360, *1361*
carbon-chain polymers, 1338, 1339
cationic polymerization, 1321–22, 1324
chain-growth polymerization, 1312–13, 1320–22, 1323, 1327, 1330
chain transfer, 1330–31
classes of, 1338–42
condensation polymerization, 1327
condensed formulas, 1309, *1316*
copolymers, 1341–42
from corn, 1354
cross-linking and cross-links, 553, *888*, 1332, 1336–37, *1338*, 1350, 1356–57
crystalline solids, 1344–45, 1346, 1350
defined, 1307
degradation reactions, 1353–55
degree of polymerization, 1309–10, 1319–20, 1335
depolymerization, 1353–55
elastomers, 1345, *1346*
end groups, *1315*, 1316
free radical polymerization, 1311–16
gel permeation chromatography (GPC), 1319–20
glass transition temperatures (T_g), 1344–45, 1346–47, 1348–49
head-to-head addition, 1317–18
head-to-tail addition, 1317–18
heterochain polymers, 1338–39
homopolymers, 1340, 1342, 1343
isotactic stereochemistry, *1318*, 1319
linear polymers, 1330, 1336, 1351, *1352*
melting points (T_m), 1344–45, 1346–49
modification of side groups, 1332–35
monomers, 1309
network polymers, 1332, 1350, 1362
nomenclature, 1342–43
polyacrylates, 1339
polyester, defined, 1339
polymer chain (main chain, backbone), 1309
properties, 1344–50
recycling plastic, 1333, 1354, 1355
regiochemistry of free radical polymerization, 1317–18
repeating units, 1309–10
ring-opening polymerization, 1323–24

self-healing polymers, 553
side groups (substituents, side chains, pendant groups), 1309, 1332–35
size distribution of, 1319–20
solubility, 1350
step-growth polymerization, 1325–30
stereochemistry of free radical polymerization, 1318–19
syndiotactic stereochemistry, 1319
termination via combination, 1314, 1315–16
termination via disproportionation, 1315, 1316
thermal stability, 1334–35
thermal transitions, 1344–46
thermal transitions, factors affecting, 1346–50
thermoplastic polymers, 1346, *1347*
uses in food storage, 1351–52
vinyl polymers, 1309–11
poly(methyl acrylate), 553
poly(methyl methacrylate) (Plexiglas), 1307, 1311, *1339*, 1343
poly(methyl vinyl ether), *1340*
polynorbornene, 1206
polynucleotides, 1355
See also nucleic acids
poly(*N*-vinylcarbazole) (PVK), 572
polyolefins, 1340, 1351
polypeptides or polyamino acids, 1355–60
See also proteins
poly(phenylene sulfide), *1340*
polyphthalamides, 1328
polypropylene, 1309, 1310, 1318, 1319, 1340, 1347–49
polysaccharides, 40, *41*, 105, 1355, 1360–62
See also carbohydrates
polyserine, 1341
poly(sodium acrylate), 1347, *1348*, 1350
polystyrene
addition polymerization, 1312–13
anionic polymerization of styrene, 1320–21
atactic polystyrene, *1318*, 1319
cationic polymerization of styrene, 1321–22
chain-growth or addition polymerization, 1312–13
free radical polymerization of styrene, 1311–16
gel permeation chromatography (GPC), 1319–20
head-to-head addition, 1317–18
head-to-tail addition, 1317–18
in materials for food storage, 1351, *1352*
physical properties, 1351, *1352*
regiochemistry of free radical polymerization, 1317–18
repeating units, 1309–10
size distribution of, 1319–20
solubility, 1350
stereochemistry of free radical polymerization, 1318–19
structure, *1308*, *1309*, 1339
styrene monomer, 789, 1308, *1310*
styrofoam, 1308, 1351, *1352*
syndiotactic polystyrene, 1319
termination via combination, 1314, 1315–16
termination via disproportionation, 1315, 1316

uses, 161
polytetrafluoroethylene (PTFE, Teflon), 1307, 1343
poly(tetrahydrofuran), *1324*
poly(tetramethylene ether) glycol, 235
polyunsaturated fatty acids (PUFAs), 200, 697
polyurethane, *1104*, 1329, *1330*
poly(vinyl acetate), 1334
poly(vinyl alcohol) (PVA), 1332–34, 1336, 1337, 1350
poly(vinyl chloride) (PVC), 780, 1310–11, 1347, 1354
polyvinyl fluoride (Tedlar), 143
poly(vinyl methyl ether), 1348
poly(vinyl *n*-butyl ether), 1348
poly(*N*-vinylpyrrolidone) (PVP), *1307*
porphyrin, 709
potassium hydrogen phthalate (KHP), 1049
potassium oleate, *101*
potassium permanganate
oxidation of alcohols, 981–82
oxidation of alkenes, 1221–23
oxidation of substituted benzenes, 1126–27, 1128, 1129, *1130*, 1181
reaction with ethene, 1225
syn dihydroxylation of alkenes and alkynes, 1218–20, 1221
precursors in retrosynthetic analysis, 650
primary (1°) alcohols, 387
primary (1°) amides, *505*
primary (1°) amines, 388, 521, 748, 749, 762, 1015–17
primary (1°) carbon, 65, 465
primary structure of proteins, 1085, 1355–57
principal quantum number (*n*), 6
principle of microscopic reversibility, 368, 372, 373, 375
prion diseases, 589
proline, *39*
prontosil, 1173
propa-1,2-diene (allene), 145
propanal (propionaldehyde), 94, *389*, 803, 911, 914–16, 952–53
propane, *81*, *88*, *168*, 1247, 1267–71
propanedioic acid (malonic acid), *511*
propane-1,2-diol, 383
propane-1,3-dithiol, *899*
propanoic acid (propionic acid), *510*
propanoic anhydride, *510*
propan-1-ol (*n*-propanol, propyl alcohol), 94, *378*, 387, 538–39, 950
propan-2-ol (isopropyl alcohol, isopropanol), 290, *378*, 387, 538, 539, 663
propanone (dimethyl ketone, acetone)
aldol reactions, 913
electron and molecular geometries, *73*
electrostatic potential map, *844*
nomenclature, 385, 389, 390
pK_a, *280*, 282
as polar aprotic solvent, *101*, 471
in protection of 1,2-diols and 1,3-diols, 968
synthesis, 670
propargyl substituents (2-propynyl substituents), *160*
propenal (acrolein), 730, *731*, 863–64, *868*, 869, 1230–31, 1322

propene (propylene), *81, 159,* 365–66, 569, 1275–77, *1340*

propenoic acid (acrylic acid), 510, *1318,* 1339, 1344

2-propenyl substituents (allyl substituents), *160*

β-propiolactone, 426, 427

4-propoxyanisole (*p*-propoxyanisole), *161*

propoxybenzene, 814–15

propoxycyclohexane (cyclohexylpropyl ether), *62,* 63

propoxypropane (dipropyl ether), *539*

propylbenzene, 803, *907,* 955–56, 1125

propyl benzoate, 815

propylene (propene), *81, 159,* 365–66, 569, 1275–77, *1340*

propyl ethanoate (propyl acetate), 508

propyl (*n*-propyl) group, *58,* 64

2-propylpenta-1,4-diene, *155*

3-propylpentane-2,4-dione, *385*

propyne (methylacetylene), 580, *654*

propynoic acid, NMR spectroscopy, 777–78, 779–80

prop-2-yn-1-ol (propargyl alcohol), structure, *74*

2-propynyl substituents (propargyl substituents), *160*

prostaglandin D$_2$ receptor antagonists, 1002

protecting groups
  acetals as protecting groups for aldehydes or ketone, 899, 963–65, 968, 1028
  acetone for protecting 1,2-diols and 1,3-diols, 968
  for alcohols, 655, 965–68
  for amino acids, 1088–89
  for carboxylic acids and amines, 1082–84
  deprotection steps, 962, 963–64, 965–66, 967–68, 969, 1184
  in DNA synthesis, 969
  effect on yield and waste from synthesis, 962
  electrophilic aromatic substitution, 1183–86
  general strategy for using, 962–63, 1082
  limitations, 965–67
  protected Grignard reagent, synthesis, 964–65
  protection step, defined, 962
  selective reactions and protecting groups, 962
  sulfo group, 1123–24

proteins
  α-helix, 1357–58, 1359
  β-pleated sheets (β-sheets), 1357–58, 1359
  cross-links, *888,* 1356, *1357*
  C-terminus, 1085, 1089
  disulfide bonds (disulfide bridges), 1356, *1357*
  Edman degradation and, 1085–86
  hydrolysis, 1084–85, 1086
  hydrophobic effect, 1359
  Merrifield synthesis of peptides, 1088–91
  N-terminus, 1085, 1089
  overview, 37, 38–40, 105
  peptide linkages, 1010, 1084–85, 1087–88, *1089,* 1355–56, 1358
  as polypeptides or polyamino acids, 1355–60
  primary structure, 1085, 1355–57
  prion diseases and protein folding, 589
  proton/deuteron exchange, in drug research, 427

quaternary structures, 1360

ribbon structures, 1359

secondary structures, 1357–58

sequencing via Edman degradation, 1085–86

size hierarchy, *38*

subunits, 1360

synthesis of peptides, 1087–90

tertiary structures, 1358–59

See also amino acids

protic solvents
  common protic solvents, 101, 470–71, *472*
  competition between S$_N$2, S$_N$1, E2, and E1 reactions, 471–73
  defined, 100
  hydrogen-bond donors, 100, 471
  ion–dipole interactions, 99, 103, 471
  nucleophile weakening, 471–72, 473–74
  relative nucleophilicity of common species, 473–74
  solvation of anions, 471, 473
  solvation of leaving groups, 472–73
  solvation of nucleophiles, 471–72

proton nuclear magnetic resonance spectroscopy (^{1}H NMR)
  acetic acid, 782
  alkynes, 788
  benzene, 777, 779, 787, 794
  1-bromobutane, 781
  bromoethene, 801–2
  1-bromo-4-ethylbenzene, 799
  butanal, 802
  chemical distinction, 778–81
  chemical distinction test for hydrogen atoms, 779
  chemically distinct protons, defined, 777
  chemically equivalent protons, defined, 777
  chemical shifts for different types of protons, 784, *785*
  common splitting patterns and relative peak heights, 793–96
  complex signal splitting, 801–4
  coupling constants (*J*), 798–804, 814
  cyclohexane, 181, 781–82
  defined, 776
  deshielding, 784, 786–88, 789–90, 815
  deuterium, 776, 784, 797
  diastereotopic hydrogen atoms, 779, 780
  dichloroacetic acid, 778
  1,4-dimethylbenzene (*p*-xylene), 791
  (1*R*,2*R*)-1,2-dimethylcyclobutane, 781
  doublet (d) splitting pattern, 793, 794
  doublet of doublets, 801–2
  effective magnetic field (*B*$_{eff}$), 785–86, 787–88, 792, 794–95
  effects of electronegative atoms, 786–87, 811, 814
  enantiotopic hydrogen atoms, 779
  energy difference between spin states (ΔE_{spin}), 774–75, 798
  ethanamine, 796
  ethylbenzene, 789–90
  gyromagnetic ratio (γ), 775, 783, 784
  heterotopic protons, 777
  hexan-3-one, 797–98
  homotopic protons, 777, *779*
  inductive effects, 786, 787, 789–90
  integration of signals, 790–92

lack of coupling of OH and NH protons, 796–97

long-range coupling, 799, 800

magnetic anisotropy, 787–88

magnetic field strength and signal resolution, 797–800

multiplet (m) splitting pattern, 793

N + 1 rule, 792–96, 803–4

number of protons and area under peaks, 790–92

number of signals, 777–78

overview, 772–73

propylbenzene, 803

propynoic acid, 777–78, 779–80

quartet (q) splitting pattern, 793, 795

quintet (qn) splitting pattern, 793

shielding, 784, 785–86, 788

signal splitting by spin–spin coupling, 792–97, 801–4

singlet (s) splitting pattern, 793

spin flips, 775–76, 803

splitting diagrams, 801–2

symmetry molecules, 780–81, 788

tetramethylsilane (TMS) standard, 777, 783–84, *785*

time scale of NMR spectroscopy, 781–82

1,1,2-trichloroethane, 792

triplet of doublets, 802

triplet (t) splitting pattern, 793, 794–95

vinylic hydrogens, 788

See also chemical shifts; nuclear magnetic resonance spectroscopy; spin–spin coupling

protons, properties and location, 4, 5–6

proton transfer reactions, 274–327, 329–34
  acidic and basic conditions, general rules of compatibility, 421–25
  acidity constant (*K*$_a$), 278–79, *280*
  acid strength and periodic table, 293–95
  acid strength and p*K*$_a$, 279, 309–12
  acid strengths of charged and uncharged acids, 291–93, 309, 310
  charge stability and total bond energy as driving forces, 348
  concerted reaction, 276
  curved arrow notation, 276–77, 330–34, 336, 368
  discovery of new drugs, 427
  electron-donating groups, effects of, 304–7, 309, 314
  electron-rich and electron-poor species, 330–34, 336, 368
  electron-withdrawing groups, effects of, 302–3, 305–7, 309
  as elementary steps, 275, 276, 289, 329–34
  equilibrium constant (*K*$_{eq}$), overview, 277–78
  as fast reactions, 421–25, 427, 430, 853
  free energy diagrams, *311*
  frontier molecular orbital theory, 368–69
  functional groups and acidity, 289–91
  hybridization, effect on attached protons, 297–98
  hydride reagents, 333–34
  inductive effects from nearby atoms, 302–7, 310
  intramolecular reactions as unreasonable mechanisms, 425–26

substituted benzenes *(cont.)*
  phenyl substituents (phenyl group; Ph), 159
  regiochemistry and attaching groups in
    correct order, 1179–80
  substituent effects, impact of reaction
    conditions, 1162–64
  substituent effects on outcomes of reactions,
    1159–61
  trifluoromethyl group as meta director, 1154
  trivial names, 160–61
  See also aromatic compounds; benzene
substrates in S$_N$2 reactions, 334, 335
sucrose, 95
Sudan-1 dye, *96*
sugars. *See* carbohydrates
sulfa drugs, 1124, 1173
sulfanilamide, 1173
sulfonamide groups, 1124
sulfonates, 459
sulfonation reactions
  of alcohols, 1059–61, 1185
  of aromatic rings, 1122–24
  desulfonation, 1123–24
  detergents, 1124
  sulfa drugs, 1124
sulfonyl chlorides, 1059–61, 1124
sulfonyl esters, 1059–61
sulfo group, 1122–24
sulfuric acid, *274*, 301, 307, 565, 1122–23, 1159,
  1163
sulfuric acid, fuming, 1123, 1160
superacids, 307, 400
supramolecular polymers, 1333
Suzuki, Akira, 984
Suzuki reaction, 984–85
syn addition, 608, *617*, 623, 627–28, *631*
syn dihydroxylation reactions, 1218–20, 1221
syndiotactic polystyrene, 1319
synthesis, 641–81, 946–99
  acid halides, 1054–57
  aldol reactions in, 910, 917, 925–28
  amines via reductive amination, 927–28
  aromatic amines from nitrobenzene
    reduction, 1128–30, 1182
  asymmetric syntheses, 664–68
  benzoic acids via oxidation of side chains,
    1126–28
  carbocation rearrangements in primary
    alkylbenzene synthesis, 1125–26
  cataloging reactions and planning synthesis,
    647–49
  catalytic hydrogenation, 969–76
  choosing the best synthesis scheme, 671–76
  chromic acid oxidations of alcohols, 977–79,
    980, 982
  Claisen condensation reaction, 1075
  conventions for writing synthesis schemes,
    642–46
  convergent synthesis, 672–73
  decarboxylation, 1078–81
  Diels–Alder reactions, 1226–28
  direct *vs.* conjugate addition to
    α,β-unsaturated carbonyls, 874–76, 961
  enantioselective synthesis, 668
  functional group conversions (table), *989–90*
  Grignard and alkyllithium reactions, 872–74

interconverting ortho/para and meta
  directors, 1180–83
linear synthesis, 671–73
malonic ester and acetoacetic ester synthesis,
  1078–81
nucleophilic addition–elimination reactions,
  1054–57, 1059–61, 1075, 1078–81
overview, 642, 946–47
oxidations of alcohols and aldehydes, 976–82
percent yield, 671–73, 675–76
permanganate oxidations of alcohols and
  aldehydes, 981–82
protecting carboxylic acids and amines,
  1082–84
pyridinium chlorochromate oxidations,
  979–80
radical reactions in, 1283–85
reactants with more than one reactive
  functional group, 654–55
reactions that alter the carbon skeleton
  (table), *991*
reduction of C=O groups to CH$_2$, 955–59
regiochemistry and attaching groups in
  correct order, 1179–80
regiochemistry and synthetic traps, 656–60
relative positioning of heteroatoms in C—C
  bond formation reactions, 951–54
Robinson annulation, 924, 1227
Sandmeyer reactions, 1131–33
selective reactions and selective reagents,
  960–61, 962, 970
solvent effects, 611, 662–64
spectroscopy and mass spectrometry use, 724
stereochemistry considerations, 664–68
strategies for solving multistep syntheses,
  668–70
sulfonation reactions, 1059–61
synthetic traps, 654–61, 946, 960
transforms, use in retrosynthetic analysis,
  overview, 650–53, 873–74
umpolung (polarity reversal), 947–51
Wittig reaction and alkene synthesis, 859–61,
  877–78, 971
See also protecting groups; reaction tables;
  retrosynthetic analysis; *specific substances*
synthetic target, 642
synthetic traps, 654–61, 946, 960

T

tacticity, 1318–19, 1349
target of synthesis, 642
tautomerization
  nitrile hydrolysis, 905
  Wolff–Kishner reduction, 906
  See also keto–enol tautomerization
tautomers, defined, 350
taxol, 540, 641–42, 672
taxonomy, 52
Teflon (polytetrafluoroethylene, PTFE), 1307,
  1343
terephthalic acid, 791, 1328–29, 1343
terpenes, 109–11, 589–94
  See also isoprene
terpenoids, 108–11
α-terpineol, *591*

terpin hydrate, 591
tertiary (3°) alcohols, 387
tertiary (3°) amides, *505*
tertiary (3°) amines, 388, 521, 748, 749, *1016*
tertiary (3°) carbon, 65, 465
tertiary structures of proteins, 1358–59
tert- prefix, 64–66
tesla (T), 775
testosterone, *3*, 108, *109*, 246
1,1,2,2-tetrabromooctane, *610*
tetrachloromethane (methyl tetrachloride,
  carbon tetrachloride)
  bond dipoles, 79
  electrostatic potential map, 79
  environmental effects, 674–75
  nomenclature, 2, *66*
  nonpolar molecule, 79
  solvent effects in bromination reactions, 611,
    662–64
  as a solvent in 1,2-dihalide synthesis, 607,
    674
  uses, 79
tetracosyl hexadecanoate, 111
tetradecane, 176
tetraethylammonium bromide, 522
tetrahydrofuran (THF)
  borane–tetrahydrofuran complex, 623, 625
  melting and boiling points, *235*
  resistance to nucleophilic substitution, 548
  ring-opening polymerization, 1324
  solvent in oxymercuration–reduction
    reactions, 616
  structure, *235*, 537, *1324*
  synthesis, 537
  uses, 235, 948
tetrahydropyran (THP), 485, 537, 548
tetramethylsilane (TMS), 777, 783–84, *785*,
  804, 807, 816
tetraphenylhydrazine, 1258
tetrodotoxin, *1198*
thalidomide, 209, 245–46
thermal Diels–Alder reactions, 1233–34
thermal energy, 170–71, 173
thermally allowed Diels–Alder reactions, 1233,
  1234
thermally forbidden Diels–Alder reactions,
  1233, 1234
thermodynamic control
  electrophilic addition to conjugated dienes,
    586–89, 660, *661*
  nucleophilic addition to polar π bonds,
    866–67, *868*, 872
  nucleophilic substitution and elimination
    reactions, 445, 487, 526–27
  protein folding, 589
  reversible reactions, 487, 488, *489*, 526
thermodynamic enolate anion, 526–28, 919
thermodynamics
  endergonic reactions, 287
  endothermic reactions, 288
  entropy change (Δ$S°$), 91–92, 288, 476–77,
    486
  exergonic reactions, 287
  exothermic reactions, 288
  K_{eq} relationship to Gibbs free energy
    differences, 287
  See also enthalpy change; Gibbs free energy

valence shell electron pair repulsion (VSEPR) theory *(cont.)*
  electron groups, 71–72
  ideal bond angle, 74–75
  molecular geometry, 72–74
  relationship to hybridization, 132–33
  shortcomings of theory, 119
  value for studying molecules, 119
δ-valerolactone, 36
valine, *39*
vanadium bromoperoxidase (V–BrPO), 615
vanillin, 52, *53*, 968
vasopressin, 1356, *1357*
vicinal dihalides, 607–11, 662, 667
Villiger, Victor, 1068
vinyl acetate, 1334
vinyl alcohol, *1333*, 1334
vinylbenzene (styrene)
  anionic polymerization, 1320–21
  free radical polymerization, 1311–16
  monomer in polystyrene, 789, 1308, *1310*, 1340
  monomer in styrene–butadiene rubber (SBR), 1341–42
  structure, *160*, *1308*, *1310*
  uses, 161
N-vinylcarbazole, 572
vinyl chloride (chloroethene), 780, 1311
4-vinylcyclohexene, 1203
vinylic position in alkenes, 548
vinyl polymers, 1309–11
vinyl substituents (ethenyl substituents), *160*
vision, chemistry of, 903
vitalism, 2
vitamin E (α-tocopherol), 1274
von Hagens, Gunther, *515*
VSEPR. *See* valence shell electron pair repulsion (VSEPR) theory
vulcanization, 897
vulcanized rubber, 897, 1337

W

Walden inversion, 412, *543*, 544
Washington, George (statue), *274*
water
  as an acid, 277, 847
  anti-Markovnikov syn addition to alkenes, 623–31

autoionization, 291
as a base, 278
bond dipoles, *78*, 79
hard water, 104–5
hexane solubility in, 91–92
hydrogen bonding, *91*, 92, 94
infrared (IR) spectroscopy, 747
as leaving group, 335, 459, 462, 538, 1112
leveling effect of, 282–84, 576, 1064
micelles in, 103, 105
net molecular dipole, *78*, 99
as nucleophile, 577, 582, 612, 614
nucleophilic addition–elimination reactions, 1047–49, 1062
pK_a, 281, 291, 293
and primary alcohol oxidation, 979
as protic solvent, 99–100, *471*, 474
in solvent-mediated proton transfer, 426–27
  See also acid–base chemistry; hydration; hydronium ion; solubility
water softeners, 105
Watson, James, 717
wavelength (λ), 726–27
wavenumbers, 737
wave–particle duality, *119*, 120
waxes, 111
Wheland intermediates, 1106
  See also arenium ion intermediates
Williams, Greville, 1353
Williams, Stewart O., 347
Williamson ether synthesis, 535–38, 539, 651–52, 872, 952, 961, 965–66
Wittig, Georg, 858
Wittig reactions
  alkene synthesis from aldehydes and ketones, 859–61, 877–78, 971
  betaine production, 860
  C═C bond formation, 859
  E1 and E2 reactions as greener alternative, 877
  E/Z isomerism in products, 860–61
  oxaphosphetane production, 860
  reaction mechanisms, 860
Wittig reagents (phosphonium ylides)
  generation from alkyl halides, 861–63, 873, 948, 971

overview, 858–59
  See also phosphonium ylides
Wöhler, Friedrich, 2
Wolff–Kishner reduction, 906–7, 955–57, 958–59, 1182
Woodward, R. B., 1227

X

xylene (dimethylbenzene, methyltoluene), 161, 791

Y

Yakobson, Boris, 147
ylides
  generation of phosphonium ylides from alkyl halides, 861–63, 873, 948, 971
  phosphonium ylides in Wittig reactions, 859–61
  resonance structures, 859
  See also Wittig reagents

Z

Zaitsev, Alexander M., 483
Zaitsev products, 483, 551, *552*, 590, 659–60
Zaitsev's rule, 483–84, 590, 877
Z/E designation
  alkenes as part of rings, 271
  Cahn–Ingold–Prelog system, 268–71
  functional groups that require a suffix, 384
  limitations of cis and trans designations, 268
  molecules with more than one alkene group, 270–71
  overview, 268–70
Ziegler–Natta catalysts, 1319, 1351
zigzag conformation, 174, 176, 231–34
zinc amalgam, 957–58
zingiberene, 110
zirconocene dichloride, 1217
zusammen (Z) isomers, 269
  See also cis isomers; *Z/E* designation
zwitterions, 315, 317, 318–19